OXFORD MEDICAL PUBLICATIONS

Cunningham's
Textbook of Anatomy

Cunningham's
Textbook of Anatomy

Twelfth edition

Edited by

G. J. ROMANES, C.B.E.

B.A., Ph.D., M.B., Ch.B., F.R.C.S.Ed., F.R.S.E.

Professor of Anatomy in the University of Edinburgh

OXFORD UNIVERSITY PRESS

Oxford New York Toronto

1981

Oxford University Press, Walton Street, Oxford OX2 6DP

OXFORD   LONDON   GLASGOW
NEW YORK   TORONTO   MELBOURNE   WELLINGTON
IBADAN   NAIROBI   DAR ES SALAAM   LUSAKA   CAPE TOWN
KUALA LUMPUR   SINGAPORE   JAKARTA   HONG KONG   TOKYO
DELHI   BOMBAY   CALCUTTA   MADRAS   KARACHI

© *Oxford University Press 1964, 1972, 1981*

First edition 1902
Twelfth edition 1981

British Library Cataloguing in Publication Data
Cunningham, Daniel John
Cunningham's textbook of anatomy.—12th ed.
1. Anatomy, Human
I. Romanes, George John
II. Textbook of anatomy
611   QM 23.2   80–41214
ISBN 0–19–263134.9

Typeset by CCC in Great Britain by
William Clowes (Beccles) Limited
Beccles and London
Printed in Hong Kong

Preface

This edition of *Cunningham's Textbook of Anatomy* follows the same general principles as the last in order to retain its suitability for undergraduate medical students and postgraduates studying for higher qualifications. Foreshortened medical courses increase the need for this type of text, and though the modern trend is to denigrate factual information in education, the corpus of information essential for a medical practitioner is not easily obtained unless some details are studied.

The entire text has been revised to remove unnecessary information and to keep it as short as is consonant with clarity. Three of the sections, Digestive, Respiratory, and Radiographic, have been entirely rewritten, while two others, Joints and Central Nervous System, have been extensively altered. One hundred and sixteen new illustrations have been introduced. The Radiographic sections appear as short parts at appropriate positions in the text of each section, separated from the text by double rules. The radiographic illustrations, previously gathered together in plates, have been repositioned in association with the corresponding anatomical illustrations and text so that an immediate comparison of the two images can be achieved and the tedious process of searching for the plates overcome. Illustrations of some new radiographic techniques have been included, and because of the increasing importance of computerized axial tomography in diagnosis and assessment of treatment, an atlas of twenty-eight photographs of transverse sections of the head, neck, and trunk has been introduced. This series supplements the illustrations of such sections which already appear in the text, and permits the student to visualize anatomy as it is displayed in axial tomography and to follow structures through the body. It also shows the relative positions of the major structures without the displacement which dissection inevitably produces. Twelve of these pictures have corresponding, though inevitably not identical, tomographic scans placed beside them for comparison, and the sections are viewed from below to fit with the convention used in axial tomography. This series also supplies the information necessary to interpret the pictures obtained with ultrasound techniques.

I am indebted to the Contributors, new and old, for their forbearance with editorial foibles and their unstinted help, and to the Staff of the Medical Department of the Oxford University Press for their continuing assistance. I am also grateful to Professor E. Samuel and Dr R. A. McKail for obtaining many of the new radiographs and tomographic scans, and to Dr G. T. Vaughan for supplying several special radiographs and scans.

Edinburgh   GJR
September 1979

Contents

Contents

Contributors

R. G. HARRISON, M.A., D.M.
Derby Professor of Anatomy in the University of Liverpool.

R. J. HARRISON, M.A., M.D., D.Sc., F.R.S.
Professor of Anatomy in the University of Cambridge.

F. R. JOHNSON, M.D.
Professor of Anatomy, University of London at the London Hospital Medical College.

R. A. McKAIL, M.B., Ch.B., D.M.R.
Lecturer in Radiographic Anatomy in the University of Edinburgh.

G. J. ROMANES, C.B.E., B.A., Ph.D., M.B., Ch.B., F.R.C.S.Ed., F.R.S.E.
Professor of Anatomy in the University of Edinburgh.

ERIC SAMUEL, B.Sc., M.D., F.R.C.S.Eng., F.R.C.S.Ed., F.R.C.P.Ed., F.F.R., D.M.R.E.
Emeritus Professor of Medical Radiology in the University of Edinburgh.

R. J. SCOTHORNE, M.D., B.Sc., F.R.S.E.., F.R.C.S. Glasg.
Regius Professor of Anatomy in the University of Glasgow.

D. C. SINCLAIR, M.A., M.D., D.Sc., F.R.C.S.Ed.
Director of Post-Graduate Medical Education, Sir Charles Gairdner Hospital, Nedlands, Western Australia.

R. A. STOCKWELL, B.Sc., M.B., B.S., Ph.D.
Reader in Anatomy in the University of Edinburgh.

E. W. WALLS, M.D., B.Sc., F.R.S.E., F.R.C.S.Eng.
Emeritus Professor of Anatomy, University of London.

1 Introduction

G. J. ROMANES

ANATOMY deals with all those branches of knowledge which are concerned with the study of bodily structure. Originally it meant the cutting up of the body for the purpose of determining the character and arrangement of its parts. It is concerned with knowledge of the architecture and interrelation of the parts of the body whether obtained by dissection with scalpel and forceps (gross or macroscopic anatomy) or uncovered by magnification with the hand lens or microscopes of all types (microscopic anatomy).

In the process of gross dissection, the body is studied region by region producing a sort of geography of the body—regional or topographical anatomy. This demonstrates that each region consists of the same kinds of tissues (bones, muscles, nerves, blood vessels, etc.) arranged in different ways for particular purposes. Thus the whole body is seen to be composed of a limited number of different tissues each of which comprises a system, the study of which constitutes systematic anatomy. Within each system, similarity of structure in some of its different parts indicates a similarity of function of these parts, while differences in the arrangement of other parts argue for differences in function. Similarly, the varying associations of the different systems in the regions of the body give clues to their functional importance. Thus, though the study of structure is a science (morphology) in itself, the functional inferences which can be drawn from it, and which are susceptible to experimental investigation, form one of its most important features (functional anatomy).

Microscopic anatomy differs from gross anatomy only in the size of structure which may be studied. It deals with the architecture of tissues, histology, and of the basic elements or cells of which they are constructed—cytology. The great increase in magnification which is obtained with the electron microscope has uncovered structural details within individual cells which are at once of great functional importance and structural beauty, and has made it possible even to visualize some of the larger molecules within cells, thus bringing anatomy within the realms of biochemistry and allowing of the elucidation of structural details invisible with the light microscope. It should be stressed that such magnifications enlarge not only the true structures but also the artefacts which arise from the processes necessary for the preparation of tissues, but it should also be clear that at this level, as in histology and in gross anatomy, no functional theory can be accepted until it has been shown to be compatible with the structure of the organ or system concerned.

No living body is a static structure, but is subject to change from the moment of conception till the time of death, and these changes, which are common knowledge, represent the alterations in structure which differentiate the fertilized egg from the embryo, the child, the adult, and the aged. These processes, which constitute the development and growth of the individual (ontogeny), proceed at very different rates and involve different tissues at the various stages. They can be divided into prenatal or intra-uterine development or embryology, and postnatal changes. Some postnatal changes are immediate and allow the foetus to adapt itself to extra-uterine life, while others persist during growth and involve every tissue for the purpose of increasing the size of the individual. Yet others continue throughout life for replacing those tissues which are subject to wear and tear, and for altering the strength or volume of tissues in relation to the demands placed upon them. The last two continue throughout life, and are exemplified by the continuous replacement of the skin and the increased size of the muscles and strength of the bones in manual labourers or their decrease in the sedentary.

In the growing individual the rate of growth is not uniform in all tissues at the same time; thus the postnatal growth of the limbs is proportionately much greater than that of the head, and the rate of growth is not equal in all parts of the limbs at the same time, either in the same or different individuals. Rates of growth are different in the sexes, females reaching their final height and puberty earlier than males, and showing periods of rapid growth at different ages from those in the male. Some structures such as the sex organs lie dormant only to make a rapid spurt of growth at puberty when the secondary sex characters appear, while the placenta, which is derived from the fertilized egg and is useful only for the period of intra-uterine life, shows many age changes by the end of that period (10 lunar months) which the tissues of the body, developed from the same egg, will show only scores of years later.

The structural changes which involve the individual (ontogeny) are superimposed on another developmental process which involves every member of the animal group collectively, and through which gradual changes in the structure of animals are believed to take place. This evolution constitutes the ancestral history or phylogeny of the individual, and just as Man is part of the Animal Kingdom, so Human Anatomy is part of a larger subject, the study of which is known as comparative anatomy and comparative embryology. These subjects have produced much of the evidence on which the theory of evolution is based and, drawing information from a wide spectrum of animals, have tended to formulate laws governing the relation of form and structure which, for the vertebrates, is known as Vertebrate Morphology.

The student of medicine is not in a position to become conversant with every aspect of anatomy, but he should understand that a knowledge of anatomy is essential to him in the practice of medicine, since he will be unable properly to examine a patient and recognize abnormality without a sound knowledge of the normal. He must be able to transfer information obtained on the cadaver to the living body so that he can visualize the arrangement of structures within the body from the surface—surface anatomy. In this he can derive considerable help from the use of X-rays and slices of the body, but he should lose no opportunity of confirming on himself and his friends the various structures which can be felt or seen and of trying to analyse the movements which the body can perform. The student will inevitably fail to gain from his studies any real interest and enlightenment if he allows the learning of detailed facts from a book to obscure the primary purpose of attempting to discover the function of the body by the examination of its structure.

TOPOGRAPHICAL ANATOMY

Anatomy is a descriptive science which uses a series of clearly defined and unambiguous terms to indicate the positions of structures relative to each other and to the body as a whole. In human anatomy the body is always described erect with the palms of the hands facing forwards, that is, in the standard or anatomical position, shown on the left of FIGURE 1.1. Three artificial sets of planes are described which are the co-ordinates forming the basis of any description; of these, the sagittal and coronal planes run parallel to the long axis of the body but at right angles to each other, while the transverse or horizontal plane cuts across the body at right angles to the other two.

1. **Sagittal** planes run from the front to the back of the body, and one of these divides the body into two apparently equal halves, the middle or **median plane** [FIG. 1.1]. Any structure which lies in a sagittal plane nearer to the median plane than another is said to be **medial** to it, while the other is said to be **lateral**; structures lying in the median plane are said to be **median**.

2. **Coronal** planes also run parallel to the long axis of the body and pass through it from side to side. Those nearer the front of the body are **anterior** or **ventral**, while those which lie nearer the back are **posterior** or **dorsal**. The terms anterior and posterior are used in human anatomy while dorsal and ventral are applicable also to the quadrupedal position and are used particularly in comparative anatomy. Anterior and posterior are used for the most part as relative terms but are also absolute when they refer to the surfaces of organs or of the body. The surfaces of the body are cut by the median plane at the anterior and posterior median lines respectively.

3. **Transverse** or horizontal planes need no description. Of any two, that lying nearer the head is said to be **superior** or **cephalic** (headward), while the other is **inferior** or **caudal** (tailward).

Thus the relative position of any structure in the body is described as medial, lateral, anterior, posterior, superior or inferior to another, or any combination may be used, e.g. superomedial, posterolateral, etc. It should be appreciated that in quadrupeds the terms anterior and posterior are synonymous with cephalic or headward and caudal or tailward. Certain advantages are therefore to be gained in using the latter terms together with dorsal and ventral since they are applicable to all circumstances.

Certain other terms are also used:

1. In any two adjacent structures, that nearer to the surface of the body is **superficial** or **external** while the other is **deep** or **internal**, irrespective of the plane in which they may lie; thus the teeth are deep to the lips and the scalp is superficial to the skull.

2. The relation of a third structure lying between two others is described as intermediate (middle) in whatever plane they may be arranged; thus they may be medial, intermediate and lateral, or anterior, intermediate and posterior, etc.

3. In the limbs certain additional terms are commonly used to overcome the difficulties of description arising out of the mobility of these structures and of using the term caudal especially in the lower limb. Though all the other terms are applicable in the anatomical position, they become difficult to apply when, for example, the arm is raised above the head.

i. The parts of the limbs further from the trunk are **distal** while those nearer are **proximal**. The hand is distal to the forearm and the leg is proximal to the foot.

ii. In the hand and foot the surfaces corresponding to the palm and sole are known as the **palmar** and **plantar** surfaces, while the

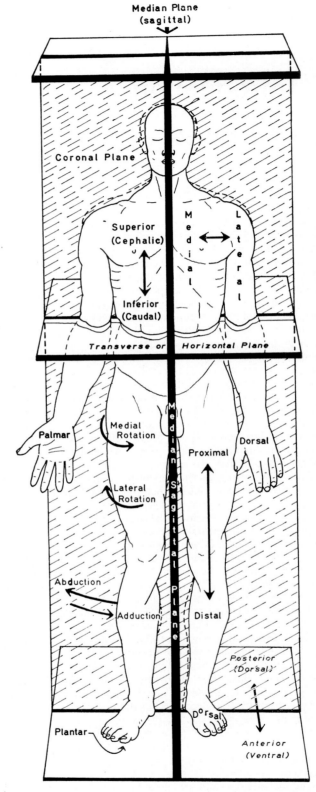

FIG. 1.1. A diagram illustrating the use of some anatomical terms referring to position and movement.

opposite surfaces, the dorsum of the hand and the dorsum of the foot, are the **dorsal** surfaces, even though the dorsal surface of the foot is superior in the anatomical position; a situation which arises from the rotation of the lower limbs which brings the plantar surfaces into contact with the ground.

iii. The lateral or thumb side of the forearm and hand may be designated the radial side from the radius, a bone which occupies this position in the forearm, while the medial or little finger side similarly may be called the ulnar side. Similarly in the leg and foot the terms fibular and tibial may replace lateral and medial respectively.

MOVEMENTS

The terms used to describe movements are also related to the above planes. Thus *movements of the trunk* or neck in the sagittal plane are known as **flexion**, forward bending, and **extension**, backward bending, and in a coronal plane as **lateral flexion**. Turning movements around the long axis are known as **rotation**.

The *movements of the limbs* are more complicated, but again those in the sagittal plane are known as flexion and extension. Flexion carries the limbs anteriorly and folds them. Extension is the reverse movement. On the other hand movement of the limbs in a coronal plane is almost limited to the proximal joint and carries the limb either further away from the median plane, **abduction**, or closer to it, **adduction**. Similar movements occur at the wrist when the hand is slewed on the forearm in the plane of the palm, and these are known as abduction or **radial deviation** and adduction or **ulnar deviation**. Abduction and adduction in the fingers and toes are named with reference to movement away from or towards the median plane of the hand, which passes through the middle finger, and the median plane of the foot, which passes through the second toe. Rotation in the limbs, as in the trunk, is represented by a movement of the whole limb or a part of it around its long axis. Medial or lateral rotation referring to the direction of movement of the anterior surface of the limb.

A special example of rotation occurs in the thumb and to a lesser extent in the little finger. Medial rotation of the thumb brings its palmar surface into opposition with the palmar surfaces of the fingers. This movement of **opposition** is characteristic of the hand, is an essential part of grasping and holding, and the basis of many skilled movements.

A very special form of rotation is the movement of one forearm bone (the radius, which carries the hand with it) around the other (the ulna). This is not simple rotation of the forearm on its long axis since the ulna remains stationary during the movement. At the limits of this movement the palm of the hand faces either anteriorly, the **supine position**, or posteriorly, the **prone position**. Thus movement to the prone position is known as **pronation** while **supination** describes the reverse movement. At the ankle joint the movements are in the sagittal plane and in either direction are known as flexion; **plantar flexion** if the movement is towards the sole, **dorsiflexion** if it is towards the dorsum. Distal to the ankle joint, the foot may be moved so that its sole is turned to face inwards or, to a lesser degree, outwards; these movements are named **inversion** and **eversion** respectively.

SYSTEMATIC ANATOMY

The description of the several systems of organs separately and in logical order, constitutes systematic anatomy, and is the plan upon which this book is based. The several parts of each system not only present a certain similarity of structure but are also associated in specialized functions. As already pointed out, functional anatomy merges insensibly into physiology. It begins with simple ideas, such as that the skeleton has the primary function of a supporting framework of the body, and the muscles have the primary function of moving the parts of the framework; it advances by deductions about the function of parts from their anatomical arrangement (such as Harvey's famous discovery of the circulation of the blood (1628), from observations and simple experiments on the valves of the veins and of the heart): it is also concerned with the wider field of the interrelations of parts belonging to different systems—for example, the anatomical localization in the nervous system of the origin of nerve fibres concerned with the regulation and control of the functions of different organs. Anatomy and physiology are indeed but two different aspects of one subject, separated by their methods of investigation. Structure and function are in reality indissolubly associated; and that is the basis of systematic anatomy. Thus there are:

1. THE LOCOMOTOR SYSTEM, which includes:

(i) The bones and certain cartilaginous and membranous parts associated with them.
(ii) The joints or articulations.
(iii) The muscular system. With the muscles are usually included fasciae, synovial sheaths of tendons, and bursae.

2. THE DIGESTIVE SYSTEM, which consists of the alimentary canal and its associated glands from the mouth to the anus.

3. THE RESPIRATORY SYSTEM, the nasal passages, larynx, windpipe, and lungs.

4. THE UROGENITAL SYSTEM, composed of the urinary organs and the genital organs—the latter differing in the two sexes.

5. THE DUCTLESS GLANDS, which, though heterogeneous in their origin, structure, and particular functions, are conveniently grouped together as a 'system', for they share the common feature of releasing into the blood stream secretions which are distributed throughout the body and have a profound influence on its functions. They include the thyroid and parathyroid glands, the thymus, the hypophysis cerebri and pineal body (these two are attached to the brain), and the suprarenal glands.

The term splanchnology denotes the knowledge of the organs included in the digestive, respiratory, and urogenital systems, and the ductless glands.

6. THE NERVOUS SYSTEM, which is divided into:

(i) The central nervous system—the brain and the spinal medulla (*cord*).
(ii) The peripheral nervous system—the nerves and their ganglia.
(iii) The autonomic nervous system, comprising the sympathetic and parasympathetic systems of nerves and ganglia.
With the nervous system may be included:
(iv) The organs of the special senses (sight, hearing, smell, taste).
(v) The skin and its appendages (nails, hair, etc.), which is the largest organ in the body and an important sensory apparatus.

7. THE BLOOD VASCULAR SYSTEM, including the heart and blood vessels (arteries, veins, and capillaries).

8. THE LYMPHATIC SYSTEM of lymph vessels, lymph nodes, and spleen.

MICROSCOPIC ANATOMY (HISTOLOGY)

The organs which constitute the various systems of the body are constructed of tissues which can only be studied satisfactorily under the microscope. This reveals them to be formed of two distinct but closely interrelated elements: the cells, and the intercellular materials which are the products of the cells themselves. Each **cell** consists of

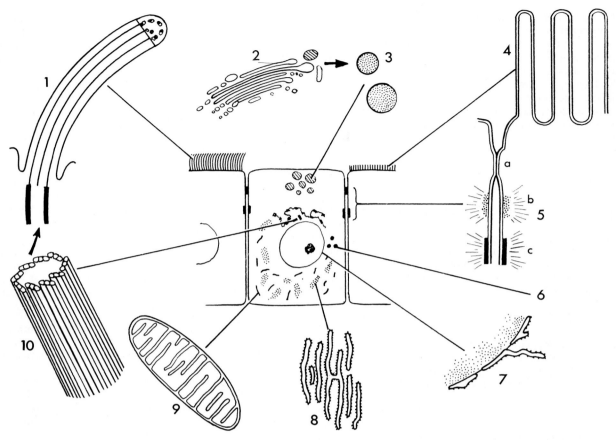

FIG. 1.2. Some of the surface and cytoplasmic modifications of cells. Not all the structures shown are present in every cell, and the diagrammatic cells shown are unlikely to be found adjacent to each other.

1. Cilia. These long, motile processes are found in many epithelia, (e.g. respiratory epithelium). Each contains nine pairs of tubules around its periphery (5 of the 9 are shown in the longitudinally sectioned cilium) and a central pair. The basal body from which each arises appears to be a centriole. Thus in ciliated epithelium the usual two centrioles which make up the centrosome (10) seem to have multiplied considerably.

2. The Golgi complex lies close to the nucleus, towards the free surface in secretory cells, but may surround the nucleus, e.g. in nerve cells. It consists of layers of flattened, membranous sacs, and gives rise to the secretion granules of the cell. It is believed that materials produced by the rough endoplasmic reticulum (8) are further elaborated in the Golgi complex to produce the secretory material.

3. Secretory granules are enclosed in a membrane derived from the Golgi complex.

4. Where absorption is taking place the surface of the cells frequently shows an array of fine, non-motile, finger-like processes, **microvilli,** which greatly increase the surface area of the cell. These are found in the epithelial cells of the intestine and the proximal convoluted tubules of the kidney.

5. Three types of adhesions between cells. a. A tight junction where the outer layer of the trilaminar cell membranes fuse. b. A zone of adherence forms a continuous strip where intestinal epithelial cells come together. The cell cytoplasm is granular close to the membrane, and fibrils in the cytoplasm pass into the granular region. c. A desmosome or point of adherence between cells. The inner layer of the cell membrane is thickened and fibrils in the cytoplasm pass into it. These are widely distributed in the

tissues of the body and may appear as half desmosomes where an epithelial cell is attached to the underlying connective tissue, e.g. in skin.

6. Lipid droplets in the cytoplasm.

7. Part of the nuclear membrane to show the nuclear pores.

8. Rough endoplasmic reticulum. These membrane sacs are studded with ribosomes (particles of ribonucleoprotein). They are mainly responsible for the basophilia of the cytoplasm, e.g. in secretory and nerve cells, and are concerned with protein synthesis, including enzyme production. Scattered ribosomes are also found free in the cytoplasm, and there is endoplasmic reticulum which does not carry ribosomes; smooth endoplasmic reticulum.

9. Mitochondria. These structures, found in all cells, are principally concerned with oxidative mechanisms, but they may also play a part in protein synthesis.

10. Centrioles. All cells capable of division contain a pair of cylindrical centrioles which form the centrosome and usually lie close to the nucleus at right angles to each other. Early in mitosis they are duplicated, one pair passing to each end of the cell and forming the basis of the spindle. Each consists of nine triplets of tubules, and where it forms the basal body of a cilium, each triplet appears to be continuous with a twin tubule of the cilium. There is no central structure in the centriole corresponding to the central pair of tubules in the cilium.

The cytoplasm also contains fibrils, glycogen granules, lysosomes, etc. The latter contain proteolytic enzymes in a membrane, but may cause autolysis of the cell if they escape into the cytoplasm. They are probably concerned, amongst other activities, with the digestion of phagocytosed foreign material.

a minute mass of colourless, watery, living substance or protoplasm whose external surface is a delicate, lipoid-containing, cell-membrane about 1/1 000 000 mm thick which separates the cell from its neighbours and from the intercellular substances. The cell contains one (occasionally several) smaller, often centrally placed, body, the nucleus, which is surrounded by a membrane and embedded in the remainder of the protoplasm, cytoplasm. Cells vary in size from 8 to 200μm in diameter (a μm = a micrometre, which is 1/1000 of a millimetre); the majority lie between 10 and 30μm and are invisible to the naked eye. They also show many different shapes and some have processes which extend over considerable distances from the cell body (up to 1 m in the case of some nerve cells) while striated muscle cells may be several centimetres long and have many nuclei. It is a mistake to assume that cells have a static shape just because they always appear similar after the processes of preparing them for microscopy. In fact, all cells undergo changes in shape and constitution and many of them are capable of movement. The nucleus differs from the cytoplasm in containing all the deoxyribo-nucleic acid of the cell, and has within it one or more small bodies (the nucleoli) which contain ribose nucleic acid. The nuclear membrane has numerous pores through which nuclear material is believed to enter the cytoplasm [FIG. 1.2], but these are not simple apertures for they do not allow free ionic exchange between nucleus and cytoplasm. The cytoplasm contains a number of structures within it which are similar in all cells but of different degrees of development in each.

1. Mitochondria [FIG. 1.2(9)] are minute rod-like structures which appear at high magnification to consist of a sac of membranes the inner layer of which is folded into the interior (cristae) to a greater or lesser degree. These structures contain the energy-producing oxidative enzymes present in the cytoplasm.

2. Golgi complex [FIG. 1.2(2)] has the appearance of a collection of flattened vesicles within the cell, usually aggregated near the nucleus. It is intimately associated with the formation of secretory granules in many cells.

3. Endoplasmic reticulum [FIG. 1.2(8)] consists of membrane-like structures arranged in tubes or sheets, often carrying minute granules of ribonucleic acid (ribosomes). It seems to be concerned with protein synthesis in the cells and is present in large amounts in secretory cells such as those of the pancreas, liver, and nervous system. Such a high content of nucleic acid makes the cytoplasm take up basic dyes and thus appear blue or purple when stained with basic dyes. Endoplasmic reticulum without ribosomes is present in many cells, and is concerned with lipid metabolism. Ribosomes are also found free in the cytoplasm, as are granules of glycogen.

4. Fibrils [FIG. 5.2] of many kinds are found in several different types of cells, but particularly in muscle cells where each myofibril is formed of a number of even finer filaments, myofilaments (in this case composed of actin and myosin). In nerve cells there are fine neurofibrillae and also neurotubules. Tubules are also found arranged in pairs in cilia (the motile processes of some cells) and in sperm tails, and in triplets in the basal bodies of these structures as in the centrioles from which they appear to be formed.

5. The centrosome is a minute structure which contains two small, dense, cylindrical bodies, the centrioles [FIG. 1.2(10)], which lie at right angles to each other. It lies close to the nucleus and forms the achromatic spindle in cell division. It is absent from those cells which have become highly differentiated and have lost the power of division, e.g. most nerve cells.

6. Vesicles in scavenging or phagocytic cells may be ingestion vacuoles which are produced where a part of the cell membrane is engulfed by the surrounding cell surface and nipped off, thus containing material which lay outside the cell. This process is very active in some cells and is a method whereby particulate and other matter may enter the cell and even be transferred across it. In the latter case the vacuole is discharged at another surface of the cell; a process which is believed to occur in many situations where transport of large molecules or particles though a cellular membrane is involved, notably in the endothelium of capillaries and in the lining cells of the intestine. Vesicles may also contain substances produced by the cells, and their contents can likewise be discharged at the cell surface, the lining of the vesicle being incorporated in the cell membrane. One special type of membrane-enclosed vesicle, the lysosome, is very variable in appearance under the electron microscope. It contains numerous enzymes which destroy proteins (acid hydrolases) and are known to discharge their contents into phagocyctic vesicles, thus digesting their contents. Such action is important in the destruction of phagocytosed bacteria, etc. Normally the enzymes are isolated from the cell cytoplasm, but may escape when the cell is damaged, thus leading to its destruction.

7. The cell-membrane is a lipoid-containing structure which undergoes many alterations in shape and may be thrown into complicated folds or even into numerous thin finger-like processes or microvilli [FIG. 1.2(4)] which markedly increase the surface area of the cell. These are therefore found in situations where cells are concerned with absorption, as in the gastro-intestinal tract and in parts of the urinary apparatus. Here they constitute what is known as a brush border because of the appearance of the microvilli under the light microscope. Other thinner, longer, and motile processes of the cell called cilia [FIG. 1.2(1)] are present in many situations, including the surface cells of the respiratory tract and parts of the genital tract. In the former they are concerned with the movement of a surface film of moisture so that particles of dust which are breathed in may be carried away, and in the latter with the transport of ova along the oviduct. The cell membranes of adjacent cells may be separated by variable amounts of intercellular substance or they may be applied and even adherent to each other, e.g. in epithelia and endothelia [FIG. 1.2], and where the former are attached to connective tissue. Apart from well-defined structures such as the cilia, the cell membrane during life is in a state of constant movement.

Growth of the body as a whole is achieved either by the division of cells and their consequent increase in number, or by the growth of the individual cells (e.g. muscle), or by the increase in the amount of intercellular substances of all kinds. The normal division of cells is known as mitosis and is a complicated process whereby the material in a cell is equally shared between the two daughter cells. The stages of this are shown diagrammatically in FIGURE 1.3A. The first sign of division, the prophase, is seen in the duplication and movement of the centrosomes to opposite ends of the cell. Condensation of the nuclear material into a series of long, nodular, intertwined, double filaments occurs, and these pairs of identical filaments, the chromosomes, steadily shrink in length and increase in thickness and density. The nuclear membrane disappears, a series of fibrils radiates from each centrosome, and apparently passing through the condensed chromosomes, forms the achromatic spindle on the equator of which the chromosomes are arranged in a disc.

The two filaments of each chromosome now begin to separate, apparently drawn apart (the metaphase) by the fibrils of the achromatic spindle. Thus the single disc of chromosomes splits into two discs, one moving towards each centrosome. This is the anaphase. Towards the end of the anaphase the cell becomes constricted in the region of the original equatorial disc of chromosomes and this constriction gradually deepens until the cell finally divides into two daughter cells, each containing one disc of chromosomes. These gradually elongate to form a tangled skein of fine filaments as in the prophase, except that each filament is now single. The nuclear membrane re-forms and all evidence of the chromosomes as separate bodies disappears, leaving two daughter cells which are small replicas of the original cell. These grow to the parent size and synthesize the materials necessary for a further division, including duplication of the chromosomal filaments.

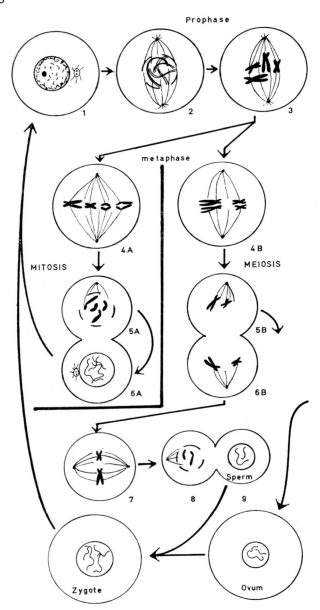

Prophase

metaphase

MITOSIS

MEIOSIS

Sperm

Zygote   Ovum

FIG. 1.3A. A diagram to indicate the processes involved in the division of a cell. On the left, the common division of mitosis which gives rise to two daughter cells with the same number of chromosomes as the parent cell. On the right, reduction division or meiosis, the process used in the production of ova and spermatozoa whereby four daughter cells are formed, each with half the number of chromosomes present in the parent cell. The figure shows the formation of spermatozoa, four cells of equal size, the only difference in the production of ova being the formation of one large ovum and three small polar bodies. To avoid confusion, a hypothetical cell containing only four chromosomes is shown.

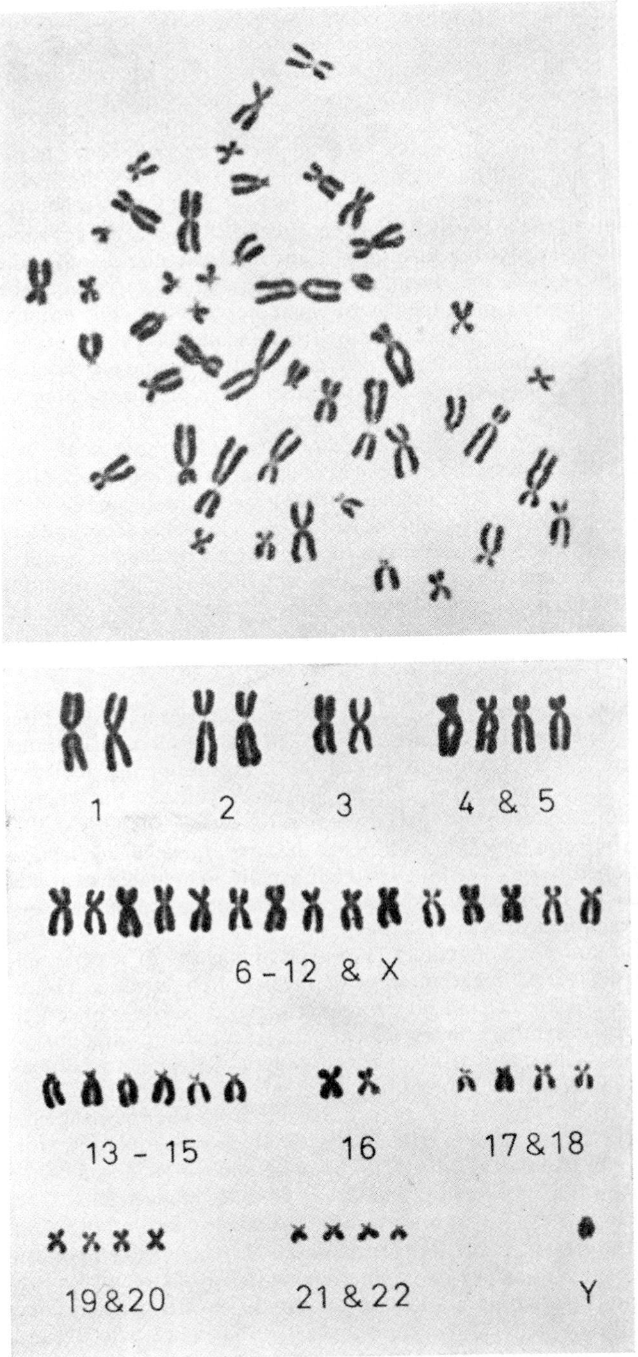

1   2   3   4 & 5

6 – 12 & X

13 – 15   16   17 & 18

19 & 20   21 & 22   Y

FIG. 1.3B. *Above.* A photograph of human male chromosomes arrested in mitosis. *Below.* The same chromosomes arranged in corresponding pairs. By courtesy of Dr W. M. Court Brown.

Mitotic division is found in all growing cells with the exception of the germ cells in the ovary and testis. These undergo a very similar process [FIG. 1.3A] except that in the prophase the chromosomes, each consisting of two filaments, arrange themselves in corresponding pairs [FIG. 1.3B], each pair representing four filaments of the mitotic prophase. In the subsequent separation of metaphase and anaphase, these pairs separate from each other, and half of the original chromosomes passes into each daughter cell. This process constitutes meiosis or reduction division and is a preparation of

germ cells with half the normal complement of chromosomes ready to have this number made up by a similar reduced complement from the germ cell of the opposite sex at fertilization.

In each species of animal the number, shape, and arrangement of the chromosomes is constant and in the normal cell is always an even number. In human cells there are forty-six chromosomes, composed in the male of twenty-two similar pairs and one dissimilar pair known as the X (large) and Y (small) chromosomes [FIG. 1.3B], and in the female of twenty-three similar pairs, there being a pair of

X chromosomes in that sex. In mitotic division each of the forty-six chromosomes is split longitudinally into two equal parts so that a replica passes to each daughter cell; thus their chromosome content and therefore the material which they inherit from the parent cell are identical and consist of an equal share of the chromosome material derived originally from the ovum and sperm at fertilization. In meiosis, on the other hand, the chromosomes are not split but arrange themselves in pairs which are similar in appearance (with the exception of the X and Y chromosomes in the male cell) but different in origin, one of each pair having been derived from the ovum, the other from the sperm. When these pairs of chromosomes separate as the cell divides, each daughter cell receives half the number that the parent cell had, and since some of the chromosomes which each daughter cell receives were originally derived from the ovum and some from the sperm, the chromosomal content of the two cells is not the same and the inheritance (genes) which each carries is different.

In addition to this relatively simple process the chromosomes lying in pairs prior to the metaphase of meiosis fuse at a point along their length and may separate again in such a fashion that they exchange a corresponding part. Thus some of the chromosomes passing into each daughter cell may be a mixture of the chromosomes derived originally from the ovum and sperm and therefore different from both. The separation of homologous chromosomes in meiosis is brought out most clearly by the pairing of the X and Y chromosomes and their passage into opposite halves of the dividing testicular germ cells from which the spermatozoa are derived, so that half the sperms contain X and half Y chromosomes, and the chromosomal determination of sex depends on the type of spermatozoon, X- or Y-containing, which fertilizes the X-containing ovum. Complete or partial failure of separation of the pairs of chromosomes (non-disjunction) can occur. Thus a complete pair may pass to one daughter cell so that this chromosome is missing from the other. Zygotes arising from fertilization with such gametes would either have three such chromosomes (trisomy) or only one. Such non-disjunctions are usually lethal to the zygote except in the case of the sex chromosomes and in some of the autosomes trisomy of which usually produces profound abnormalities. Thus trisomy of chromosome 21 leads to mongolism (Down's syndrome) with 47 chromosomes in those that survive. A similar result may arise from translocation of a part of one chromosome to a member of another pair, e.g. part of 21 to 14. At meiosis, the normal 21 may pass into the same daughter cell as the abnormal 14. In this case, fertilization will lead to trisomy 21 with a normal number (46) of chromosomes, though there is one unduly large chromosome $(14 + 21)$ and a single normal 14.

It will be appreciated that all cells derived from the fertilized ovum by mitotic division must of necessity have an equal complement of chromosomes and yet a wide variety of different tissues are formed from these apparently uniform cells. This is an interesting but unexplained phenomenon since all the daughter cells of the developing ovum are capable of forming any tissue up to a certain phase in development, but after this stage is reached they are capable of forming only the tissue which is appropriate to their position in the developing embryo—a probable source of identical twins each with its own amnion and chorion [p. 33]. Thus separation of the first two daughter cells derived from the amphibian ovum leads to the formation of two complete animals. At a later stage, transplantation of cells, to a new situation, is followed by their development into a tissue appropriate to their new position, provided that the transplantation is made at a sufficiently early phase. Each cell is, however, a complex of structures, and the differences between cells represent a difference in degree of development of these different characters (e.g. mitochondria, endoplasmic reticulum, fibrils, etc.) rather than a basic difference in the constituents of the cells. Once a cell has developed the characteristics of a particular tissue, it and its daughter cells retain these unless some profound change occurs in its basic constitution, as can arise in cancer.

The apparent similarity of the chromosomal pattern in a single species no doubt determines the similarity of its members, but does not prevent the cells which make up any individual member from developing the marked variations which occur in different tissues. It should be appreciated, however, that just as individual members of a species are different, so all the cells of each individual have certain common chemical characteristics which are unique to that individual and which represent, it is believed, minor differences in the molecular arrangements of the complex molecules which make up the chromosomes and which are faithfully reproduced from cell to cell. Only two individuals derived from a single fertilized ovum, identical twins, can have exactly the same chromosomal pattern and it is only in these cases that tissues can be transferred from one to another and survive. In all other cases, including non-identical twins, brothers and sisters, the chemical specificity of the individual tissues is recognized by the host to which it is transplanted and it is unable to survive because of the host reaction to this foreign, though similar, tissue.

TISSUES OF THE BODY

The animal body consists essentially of a thick-walled cylinder lined internally and covered externally with a continuous layer of cells, or epithelium, called entoderm and ectoderm respectively. The tissue which they enclose is the mesoderm and the two epithelia are continuous at the mouth and anus, the entoderm forming the lining of the alimentary and respiratory passages and the ectoderm the skin. Both epithelia send prolongations into the mesoderm which are thereby divorced from the protective or absorptive functions of the surface layer and undertake different activities, including the formation of glands which discharge their secretions to the surface of the epithelium from which they are derived along the epithelial stalk, or duct, which connects them. Examples of these are the liver, pancreas, sweat and mammary glands, and hair follicles. Some downgrowths, however, become separated from the surface and discharge their secretions into the blood stream, endocrine glands, or take on quite other functions, e.g. the nervous system, which is derived in this fashion from the ectoderm.

The mesoderm forms all the cells which originate neither in ectoderm nor in entoderm, thus including all the epithelia lining the spaces which develop in mesoderm, the blood and lymph vessels, the serous cavities which surround the heart, lungs, alimentary canal and central nervous system, bursae, tendon sheaths and joint cavities. It also forms muscle, blood, and connective tissue cells, the gonads, and the lining epithelia of most of the urogenital tract.

Connective tissue cells are the source of the intercellular substances of the body including connective tissue fibres, the numerous inelastic, white, collagen fibres and their slender counterparts, the reticular fibres, as well as the yellow elastic fibres.

It should be appreciated that the intercellular substance fills the entire extracellular space, permeates every organ and is continuous with the surface of all the cells which are exposed to it. It is responsible for holding the various organs in place and yet for allowing the requisite amount of movement between them. Thus in some situations where free movement is essential the extracellular material is in the form of a fluid, as in the blood and lymph vessels, in the cavities of joints, tendon sheaths, and bursae and in the spaces surrounding the heart, lungs, abdominal viscera, and central nervous system. Even these fluids are variable in consistence, ranging from the thin watery lymph to the thick glutinous material in some joints. Elsewhere the fluid contains a fine meshwork of connnective tissue fibres (loose **areolar tissue**) which is sufficiently delicate to allow a considerable amount of movement between adjacent tissues. This is found between the bundles of fibres in a muscle, and surrounds many other tissues such as nerves and ligaments which have to slide on adjacent structures during movement, and organs which have to be capable of considerable distension such as the gullet and the urinary bladder. This loose areolar tissue forms most of the planes of cleavage in the body and allows the rapid spread of infection through it.

In other situations the extracellular material is stronger. This can be either by an increase in the number of fibres which it contains, or by the deposition of different substances produced by the connective tissue cells in the interstices of the meshwork it forms. Thus in many situations the connective tissue cells become loaded with fat, or the fibres are surrounded with mucopolysaccharides to form the gelatinous basement membranes which surround some cells, or with the more solid sulphated mucopolysaccharides to form cartilage, or with calcium salts to form bone.

Where both strength and flexibility are required, only the fibrous elements are increased to form:

1. Compact layers of interlacing collagen fibres. This is the deep **fascia** which invests the entire body and sends sheets inwards between the various organs to surround them with a sheath of greater or lesser density which, because it is felted, is equally strong in all directions. It therefore produces structures such as the epimysium and peritendineum (within which each muscle and tendon respectively slides) the epineurium which confers on peripheral nerves their great strength, and the tough fibrous capsules of such organs as the kidneys and testes.

2. Muscles may be attached to some of the sheets described above, and where they are thus subjected to a pull in one direction, large numbers of collagenous fibres are laid down in this direction, conferring on the sheet a glistening appearance and forming a flat tendon or **aponeurosis**. In an exactly similar fashion the fibrous capsule which surrounds each type of movable joint is subjected to forces unequally in its several parts; where these are greatest there is a similar addition of fibres to form the **ligaments** which give stability to the joints and which are usually therefore a part of the capsule. Most of the ligaments so formed are inelastic or collagenous but a few contain a preponderance of elastic fibres and are yellowish in colour; to this group belong some of the ligaments of the vertebral column. The thickness of such ligaments and aponeuroses is directly related to the forces which they are called upon to resist. The thick rounded **tendons** which arise from muscles of fusiform shape are formed in the same way and have the same structure as ligaments and aponeuroses.

Since the ligaments, aponeuroses, and tendons are but part of the general extracellular material, they are continuous with the same tissue which forms the fibrous basis of cartilage and bone and are thereby directly attached to these tissues by fibres which penetrate them as part of their structure. In a similar fashion, extracellular materials are continuous with the surfaces of the cells which they surround but do not penetrate. Thus the collagen fibres of a tendon are attached to the outer membrane of muscle cells (sarcolemma), and the collagen fibres of the dermis are attached to the deepest cells of the epithelium (epidermis) which covers them, binding the outer protective tissue to the underlying structures.

Reticular tissue is a special kind of areolar tissue formed of very fine collagen (reticular) fibres in a loose network supporting the proper tissue of various organs. It is especially seen in lymph nodes, tonsils, the spleen, the liver, and bone marrow. The connective tissue cells associated with reticular fibres are frequently phagocytic (**macrophages**), that is, they are capable of ingesting particles injected into the living animal. These cells and others of similar properties in the blood and elsewhere are collectively known as the reticulo-endothelial system because of their frequent association with reticular fibres and the endothelium of lymph and blood vessels.

The cells of the connective tissue in some situations may take up and store large quantities of fat, thus forming **adipose tissue**. The individual cells are swollen with the accumulation of fat which forms a single globule within the cytoplasm. Groups of them form small nests enclosed in the collagen network which may be either very delicate, as in perirenal fat, or dense as in the palm of the hand, the back of the neck, and the scalp, where they increase the resilience of the superficial fascia. Everywhere they help to control heat loss. Another type of fat storage, brown fat, is found especially in hibernating animals in which the fat forms a large number of fine globules in the cytoplasm giving the cells a foamy appearance.

In the embryo the connective tissue forms a loose cellular network or mesenchyme from which are developed the connective tissues, cartilage, bone, fat, smooth muscle, lymph nodes, endothelium, blood cells, and macrophages of the adult. Then these cells are very different from each other and from the mesenchyme cells, except in the case of fibrous tissue cells. As a general rule the greater the degree of differentiation in the development of a cell, the more it loses its primitive potentiality to form cells of various kinds and in a few cases it may even lose its ability to undergo mitosis (e.g. nerve cells). The similarity of fibrous tissue cells in the adult to the mesenchyme cells of the embryo, has led to the assumption that some of them at least retain the ability to form other kinds of cells, but it is doubtful if in fact they have more than a very limited ability to do so. Certainly if the blood-forming or lymph cells are destroyed they are not replaced, and injuries involving smooth muscle and other specialized tissues are repaired with fibrous tissue. However, it does seem that the connective tissue cells which are most closely allied to each other can fulfil different roles in some circumstances. Thus fibrous tissue cells, cartilage cells, and bone cells, which share the ability to form collagen and probably also elastic fibres, seem to be partly interchangeable, since cartilage masses can be formed in a broken bone, even when this is developed originally in membrane and no cartilage cells would normally be present (Girgis and Pritchard 1955), and ossification can occur in connective tissue in abnormal situations. However, ossification is difficult to produce experimentally in abnormal situations, and cannot be produced in many sites where there is an adequate supply of connective tissue cells no different in appearance from those in areas, such as the kidney, where abnormal ossification is easily induced.

Whether they are interchangeable or not, the cells derived from mesenchyme retain to a very full measure the ability to react to changing circumstances. Thus, in the rabbit's ear, Clark and Clark (1934, 1940) have shown the remarkable changes which can occur in the vascular pattern; growth of new capillaries, the development in their wall of muscle fibres, and a complete change in the pattern of blood vessels, even to the temporary formation of arteriovenous anastomoses, as a result of changes in the flow of blood through the ear. Again the thickness of tendons and bones can be modified in relation to the stresses placed upon them, the blood-forming bone marrow can react to blood loss by the rapid production of new cells, and under an adequate stimulus the fixed tissue macrophages appear in large numbers, and becoming motile, move towards and ingest foreign material.

These changes and many others are going on constantly in the body, and the study of the cadaver and of histological preparations alone gives very little idea of the vital processes which no student of anatomy should ignore if he is to understand the significance of the structures he studies.

One of the most formidable tasks which face the medical student at the beginning of his studies is the need to acquire the vocabulary of medicine and to use it with precision. For this purpose it is essential to obtain a clear understanding of anatomical terms and this the student can obtain only by repeated reference to a good medical dictionary. The following short glossary defines some of the terms in general use but cannot replace the need for a medical dictionary.

A SHORT GLOSSARY OF SOME GENERAL ANATOMICAL TERMS

In some cases the translation of the term from Latin or Greek only is given, in others this is followed by its modern meaning where this is different, and in a third group where there is no clear derivation or the original meaning has been distorted, only the modern usage is given.

Acinus. A berry growing in clusters. Hence the smallest unit of a compound gland; more commonly called an alveolus.

Aditus. The entrance into a cavity.

Afferent. Used to indicate that a structure leads towards the organ it supplies. Of nerves, corresponds to sensory nerves, cf. efferent.

Ala. A wing-like projection.

Alveolus. A small pail. Hence applied to a tooth socket, the smallest air spaces of the lungs; synonymous with acinus.

Ampulla. A flask. Hence the dilated portion of a tube.

Ansa. A loop.

Antrum. A cave. Hence a cavity or hollow filled with air and lined with mucous membrane in the interior of a bone.

Anulus. A ring.

Aponeurosis. A tendinous sheet covering a muscle or extending from it to the attachment of the muscle.

~blast. A builder. Hence an ending indicating an immature or stem cell.

Brachium. An arm.

Bursa. A collapsed sac of fluid especially where a tendon or skin slides over bone. Such structures enclose the tendons where these lie in osteofascial tunnels and are then known as synovial sheaths. In general arrangement they are similar to the cavities of synovial joints.

Canaliculus. Diminutive of canalis, a canal. Also used to indicate a tunnel.

Commissure. A joining together. Used in the nervous system to indicate bundles of nerve fibres crossing the midline from side to side, but also for a ridge, etc., joining the two halves of an organ across the median plane.

Condyle. A knuckle. A smooth rounded eminence covered with articular cartilage.

Corona. A crown. Hence any encircling structure, e.g. coronary arteries in the coronal sulcus.

Cortex. A rind or outer covering, cf. Medulla.

Crista. A sharp upstanding ridge.

Crus. A leg. Used (as in 'brachium') to indicate any projecting process.

Cystic. Appertaining to a bladder (cyst).

Dens. A tooth. Odontoid—tooth-like.

Efferent. The reverse of afferent. Of nerves, corresponds to 'motor' nerves.

Epicondyle. The prominence or projection situated above a smooth articular eminence, though that eminence may not be called a condyle.

Facies. A surface.

Falciform. Sickle-shaped. Falx, a sickle.

Fascia. A bandage or swaddling-cloth. Hence the membrane of fibrous tissue which sheaths tissues.

Filum. A thread.

Folliculus. A small bag.

Foramen. A hole.

Fornix. An arch.

Fossa. A shallow depression.

Fovea. A pit. Also used in place of facet—usually a small articular surface.

Frenulum. A small bridle or ligament. Usually applied to a fold of skin limiting the separation of the structures to which it is attached.

Fundus. The base of a hollow organ, usually opposite to its outlet.

Ganglion. A swelling, usually small and round. In the nervous system any collection of nerve cells outside the central nervous system.

Genu. A knee. Hence any bend on a structure.

Glomus. A ball or tight meshwork, usually of vessels. Diminutive, glomerulus.

Hamulus. A hook.

Hiatus. A slit or gap.

Hilus. A depression where blood vessels and nerves enter an organ.

Incisura. A notch.

Infundibulum. A funnel.

Isthmus. A narrow strip of tissue joining two larger pieces. Hence a narrowing of a canal.

Labium or Lip. The raised margin of an orifice.

Labrum. A brim.

Labyrinth. A maze of communicating spaces or canals.

Lacuna. A pit or hole.

Lamina or Lamella. A thin plate or sheet.

Ligament. A band or tie joining two structures. Most commonly a fibrous band but may be composed of any tissue.

Limbus. A border or margin.

Lingula. A tongue-like projection.

Macula. A spot or stain.

Meatus. A passage. A short canal.

Medulla. Bone marrow or the pith of plants. Hence the central portion of any organ where its structure is different from the outer layer or cortex.

Mesentery. The fold of tissue which supports the bowel (enteron) in the belly. Has been expanded to include any such supporting structure, e.g. the mesentery of the ovary, more correctly called the mesovarium.

Nodus or Node. Diminutive, nodulus or nodule. A knot. A swelling or protuberance, a spherical aggregation of cells. A discontinuity in the myelin sheath of a nerve fibre.

Nucleus. The kernel of a nut. The internal body of a cell. In the central nervous system refers to a collection of nerve cells.

Ostium. An entrance. Hence the opening into a tube or space.

Papilla. The nipple. Any nipple-shaped elevation.

Parenchyma. The proper tissue of an organ as distinct from accessory structures such as its fibrous capsule.

Paries. A wall. The walls of the abdomen. The parietal bone—the side wall of the skull.

Pedicle. Diminutive of *pes*, a foot. **Pediculus.** The stalk of a fruit. A narrow rod or tube joining two structures.

Pelvis. A basin.

Plexus. A plaited or braided structure of vessels or nerves.

Plica. A fold.

Porus. A pore or opening, e.g. of a meatus.

Radix. A root or origin.

Ramus. A branch.

Raphe. A seam. The line of union of two soft tissues, e.g. the interlocking aponeuroses of two muscles.

Rete. A net. Often a labyrinth of communicating channels.

Retinaculum. A stay or tie. Usually a band of connective tissue (ligament) which holds a tendon in place.

Rima. A chink or cleft, e.g. the slit between the eyelids, lips, or vocal cords.

Scaphoid. Scapha, a boat. Any hollowed-out structure.

Septum. A hedge, fence, or dividing wall. A partition, usually thin.

Sinus. A hollow or creek are two of its many meanings. In anatomy it is applied to the air-filled cavities of the cranial bones, to large venous spaces within the skull and elsewhere, and to dilatations of blood vessels.

Spina or Spine. A sharp-pointed projection.

Somatic. Belonging to the body wall.

Splanchnic. Belonging to the gut tube.

Squama. A fish scale or armour plate. Squamous, scaly.

Stria. A stripe or line.

Styloid. Any structure resembling a stylus or stake.

Sulcus. A rut or furrow.

Synovia. The fluid in joint cavities, bursae, and tendon sheaths.

Taenia. A band or ribbon. Hence any ribbon-like structure.

Tegmen. A cover.

Tela. A web. Hence any thin mesh, usually of fibrous tissue.

Torus. A heap or protuberance.

Trabecula. A small beam. Used especially for the pieces which make up the lattice of cancellous bone.

Trochlea. A pulley or pulley-shaped surface.

Tuber. A bump or swelling. Also tubercle and tuberosity. Three terms used, without much distinction, for any kind of rounded swelling or eminence.

Vagina. A sheath.

Velum. A covering or curtain.

Vesica. A bladder, especially the urinary bladder, cf. cystic.

Villus. Shaggy hair. Hence applicable to fine processes projecting from a surface.

Visceral. Belonging to the gut tube.

Zona. A girdle or belt. A circular or ring-like structure.

REFERENCES AND FURTHER READING

ABBIE, A. A. (1946). *The principles of anatomy*, 2nd edn. Sydney.

ANDERSON, J. E. (1978). *Grant's atlas of anatomy*, 7th edn. Williams and Wilkins, Baltimore.

BLOOM, W. and FAWCETT, D. W. (1968). *A textbook of histology*, 9th edn. Saunders, Philadelphia.

CLARK, E. R. and CLARK, E. L. (1934). The new formation of arteriovenous anastomoses in the rabbit's ear. *Am. J. Anat.* **55**, 407.

—— —— (1940). Microscopic observations on the extra-endothelial cells of living mammalian blood vessels. *Am. J. Anat.* **66**, 1.

CLARK, W. E. LE GROS (1971). *The tissues of the body*, 6th edn. Oxford University Press.

DONALD, I. (1970). Prenatal development. In *Child life and health* (ed. R. G. Mitchell). Churchill Livingstone, London.

FIELD, E. J. and HARRISON, R. J. (1968). *Anatomical terms*, 3rd edn. Cambridge.

GARRISON, F. H. (1929). *An introduction to the history of medicine*, 4th edn. Philadelphia.

GIRGIS, F. G. and PRITCHARD, J. J. (1955) Cartilage in repair of the skull vault. *J. Anat. (Lond.)* **90**, 573.

GRANT, J. C. B. (1975). *A method of anatomy by regions, descriptive and deductive* (ed. J. V. Basmajian). Williams and Wilkins, Baltimore.

GUTHRIE, D. (1958). *A history of medicine*, 2nd edn. Edinburgh.

HAM, A. W. (1974). *Histology*, 7th edn. Lippincott, Philadelphia.

HARVEY, W. (1628). *Exercitatio Anatomica de Motu Cordis et Sanguinis in Animalibus*. Frankfurt.

HILTON, J. (1863). *Lectures on rest and pain*, 6th edn. (1950) (eds. E. W. Walls and E. E. Philipp). London.

JONES, F. W. (1941). *The principles of anatomy as seen in the hand*, 4th edn. London.

—— (1949). *Structure and function as seen in the foot*, 2nd edn. London.

KEITH, A. (1948). *Human embryology and morphology*, 6th edn. London.

SCAMMON, R. E. (1923). A summary of the anatomy of the infant and child. In *Pediatrics* (ed. J. A. Abt) Vol. 1, Chapter 3. Philadelphia.

SHELDON, W. H. (1940). *The varieties of human physique.* New York.

SINCLAIR, D. C. (1961). *A student's guide to anatomy.* Oxford.

—— (1969). *Human growth after birth.* London.

SINGER, C. (1925). *The evolution of anatomy. A short history of anatomical and physiological discovery to Harvey.* London.

—— and UNDERWOOD E. A. (1962). *A short history of medicine*, 2nd edn. Oxford.

SYMINGTON, J. (1887). *The topographical anatomy of the child.* Edinburgh.

—— (1917). *An atlas illustrating the topographical anatomy of the neck, thorax, abdomen and pelvis* (with supplement: *Head*). Belfast. Reprinted (1956) for the *Anat. Soc. Gt Brit. and Ireland.* Edinburgh.

TANNER, J. M. (1978). *Foetus into man; physical growth from conception to maturity.* Open Books, London.

TONER, P. G. and CARR, K. E. (1968). *Cell structure.* Edinburgh.

WHITNALL, S. E. (1939). *The study of anatomy.* 4th edn. London.

Radiographic anatomy and organ imaging

R. A. McKAIL AND E. SAMUEL

	Radiolucent		Intermediate Density	Radiopaque	
Very	Moderately			Moderately	Very
Gas Lung	Fat		Muscle Blood Viscera, e.g. liver, spleen, etc.	Bone Teeth	Heavy metals Barium

The study of anatomy in the living subject has lately entered a new and exciting phase. Using waves of energy as dissecting tools, it is now possible with comparative ease to inspect structures that a few years ago could not have been seen.

Most of these methods stem from the exploitation of well-known properties of X-rays and radioactive substances. Others, such as ultrasound, have no relation to X-rays at all.

The method of imaging by the direct use of X-rays is the oldest. The part to be examined is interposed between the source of the rays, the X-ray tube, and an X-ray film. The different structures of the body, because of their differing atomic constituents, absorb X-rays in varying degrees as they pass through them; the photographic film being blackened in proportion to the amount of radiation reaching it. For example, a bone like the femur, because of its thickness and high atomic number, absorbs most of the X-rays traversing it and allows few to pass through. Thus the film overlapped by it shows little or no blackening, unless X-rays or very high energy are used. On the other hand, the film left uncovered by the part examined appears completely black, if adequate exposure has been given.

The levels of X-ray absorption of various tissues and other substances commonly found in the body are set out in the following table, modified from Meschan (1975).

The table is accurate only if each substance or tissue is of the same thickness, is exposed to X-rays of the same amount and penetrating power and if there are no overlying opacities or translucencies. These conditions are virtually never fulfilled, but the table is useful as a general guide.

The following factors must be considered in using the table:

1. Variations of thickness of tissue and variations in the amount of exposure to X-rays are reciprocal factors and are considered together. FIGURE 1.4A is a radiograph of a specimen of a scapula which has been relatively 'overexposed'. The thin parts appear absent but in the thicker parts it is easy to see the detail of the arrangement of the trabeculae. FIGURE 1.4B is an 'underexposed' radiograph of the same specimen. In this the thin parts are satisfactorily shown (small cracks in the specimen may be apparent) but there is little detail in the thicker parts. Thus it is possible to view certain structures and obliterate others by varying the amount and penetration of the X-rays. The mammogram [FIG. 1.5] is another example of the same process. Here rays of very low penetrating power are used to produce 'underexposure', and consequently there is good detail in the radiograph. The breast is too readily penetrated by X-rays of

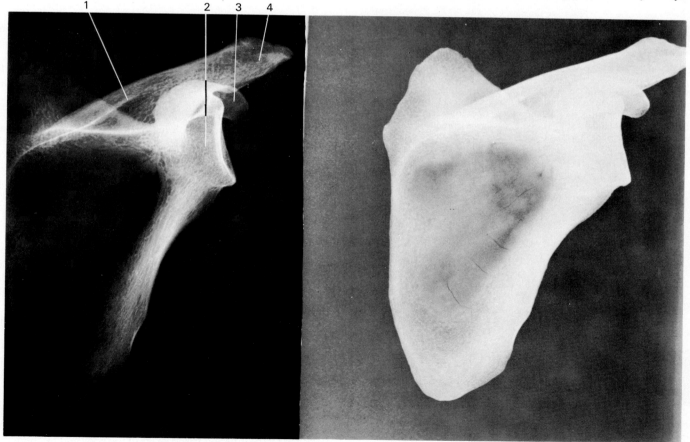

FIG. 1.4. Two radiographs of the same scapula at different exposures. The 'overexposed' radiograph (left) shows no detail over a large area where the bone is thin, but is adequate for bone detail in the thicker parts of the bone (spine, acromion, and neck), and is even underexposed where the coracoid process overlaps the spine. The 'underexposed' radiograph shows no detail in the bone except in the thinnest areas. Here three 'hair-line' cracks are visible, though they could not have been seen had there been any more exposure.

1. Spine.   2. Neck.   3. Coracoid process.   4. Acromion.

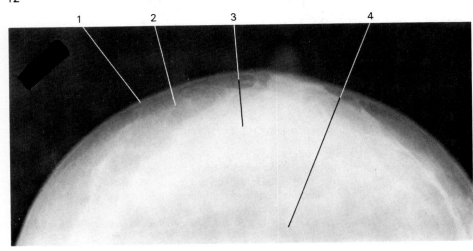

FIG. 1.5. Mammogram. A supero-inferior radiograph of the breast taken with low-energy X-rays in a middle-aged patient. This accentuates minor differences in the X-ray density of tissues which would not be apparent in conventional radiographs.
1. Skin.
2. Fatty subcutaneous tissue.
3. Remnants of glandular tissue mixed with fat.
4. Retromammary fatty connective tissue.

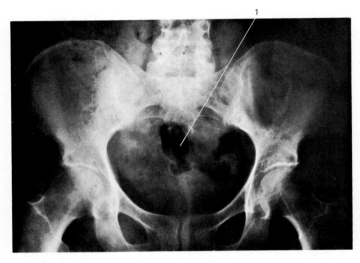

FIG 1.6. Anteroposterior radiograph of the pelvis to show the effects of intestinal contents on the visualization of bone detail. Gas in the rectum has obliterated part of the sacrum and coccyx (local overexposure) while semifluid material mixed with gas in the caecum has produced a mottled appearance in the right iliac fossa which could be interpreted as a bony change.
1. Gas in rectum.

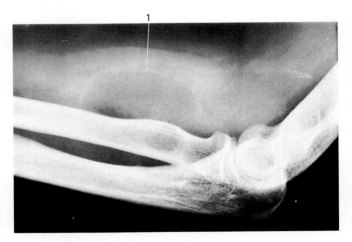

FIG. 1.7. Lateral radiograph of the forearm containing a fat tumour (lipoma). This shows as a dark area relative to the less radiolucent muscle and bone,
1. Edge of fat tumour.

average penetration, as may be seen in a radiograph of the chest [FIG. 7.42]. (See page 584 for further details about mammography.)
2. Effects of overlying translucencies and opacities.
In a radiograph of the pelvis [FIG. 1.6] the gas filled segment of rectum allows X-rays to pass freely, thus producing a translucent area superimposed on the sacrum and detailed structure of that bone is lost.
3. Effect of contrast.
When placed between muscle and bone, a collection of fat (lipoma) appears as a translucent area [FIG. 1.7] as indicated in the table. However a similar collection of fat at the apex of the heart [FIG. 1.8], contrasted with the air in the lungs, appears opaque.

Fat is also sometimes visible as a translucent line around viscera such as the kidney, if not overlain by colon containing gas.

Usually adjacent tissues of intermediate density cannot be distinguished unless one of them is made opaque by a contrast medium.

X-rays emitted from an X-ray tube diverge from a point source and therefore structures which lie at a distance from the film are enlarged relative to those lying adjacent to it. Also the closer the tube is to the object the greater is the degree of magnification. Structures which are nearer to the recording film not only show less magnification but are also sharper because they are less affected by X-ray scatter. Hence the part to be examined is placed as close to the film as possible.

The conventional radiograph has the disadvantage that a three dimensional object is reduced to two dimensions. This can be overcome to some extent by taking lateral and/or oblique radiographs in addition to the routine postero-anterior views. Sometimes neither of these views provides the required information because of confusion from overlying tissues. Then *tomography* is often useful. This allows optical sections of part of the body to be made. The X-ray source (tube) and the film are attached at opposite ends of a moving lever whose fulcrum is at the level of the tissue to be investigated. The tube and film move in opposite directions for the duration of the exposure so that the structures in the selected plane form a sharp image whereas those on either side of it are blurred [FIG. 7.21].

More recently (instead of relying on the direct production of an image by X-ray) other machines (computerized tomographic scanners) have been developed which measure the amount of X-rays absorbed by the different tissue layers of the body and by computing these findings, are able to reconstruct a transverse section (scan) of the body at any desired level [ATLAS FIG. 20A]. The head scans are extensively used in the investigation of neurological disease, because they readily give information previously obtainable only by means of costly, time consuming and sometimes hazardous procedures. The whole body scanner is becoming more widely used. These scans are very similar in appearance to anatomical sections [ATLAS].

Methods of imaging employing X-rays or radioactive substances are all potentially hazardous to some degree. Ill effects depend on the amount absorbed and the frequency of exposure. With due precaution, most investigations can be carried out without risk as long as examinations requiring prolonged exposure are not repeated often, and exposure of vulnerable tissues such as the gonads is avoided, especially in the young. Above all, the foetus in the early weeks of development must not be irradiated, because of the risk of foetal abnormality.

ULTRASOUND

The use of ultrasonic waves is subject to limitation because of their failure to penetrate gas and their almost complete inability to penetrate bone, but the method has one outstanding virtue; as far as is known it has no harmful effects on adult or embryonic tissues when used diagnostically. Its harmlessness is of obvious importance in abdominal examination of patients who may be pregnant.

These pulsed sound waves can penetrate tissues and where one tissue meets another, for example where the liver meets the kidney (an interface), a reflection of the sound wave occurs producing an echo. This is recorded and translated into an electronic pattern, which is displayed on a cathode ray tube. Ultrasound sections of the body can be made in both transverse and longitudinal planes. Ultrasound has the disadvantage that its waves are immediately reflected by large collections of air so that it cannot be used in the lung. Abdominal investigation also is sometimes difficult because of gas in the intestinal tract. Fluid collections, on the other hand, offer little resistance to ultrasonic waves, thus it is advantageous to have the bladder distended when examinations of the pelvis are undertaken—the bladder acting as a 'window' through which the pelvic organs may be seen clearly.

The records of ultrasonic scans are more difficult to interpret than those of computerized tomography.

RADIOISOTOPE SCANNING

This method depends on the uptake of certain short-lived radioactive substances by some tissues and their detection by reason of the X-rays which they give off. When these substances are injected into the bloodstream, their absorption by certain tissues can be detected by a recording instrument passed over the tissue. More recently a larger fixed type of detector has been devised for the same purpose.

A record of such a scan is shown [FIG. 1.9]. Though anatomical structure is not faithfully reproduced, the value of such examinations may sometimes be considerable in pathological states. For example, there can be no uptake in tissues which do not receive a blood supply because their arteries are blocked, the area lacking a blood supply showing as an area of no radiation. The scans are usually simple to perform and not hazardous.

SUGGESTIONS FOR FURTHER READING

MESCHAN, I. (1975). *An atlas of anatomy basic to radiology.* Philadelphia.
SQUIRE, L. F. (1975). *Fundamentals of radiology.* Cambridge, Mass.
WEIR, J. AND ABRAHAMS P. (1978). *An atlas of radiographic anatomy.* London.

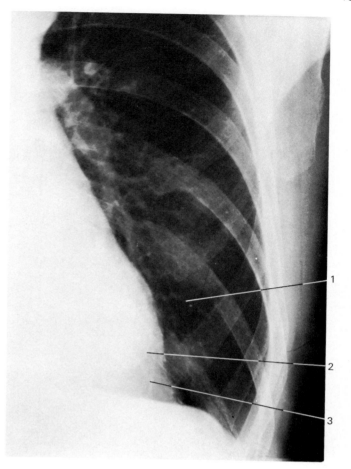

FIG. 1.8. Radiograph of fat pad at apex of heart. This shows as a light area by comparison with the highly radiolucent lung tissue (dark). The fat is more radiolucent than the adjoining heart muscle, the edge of which is visible against the fat.

   1. Lung tissue.   2. Edge of heart muscle.   3. Fat pad.

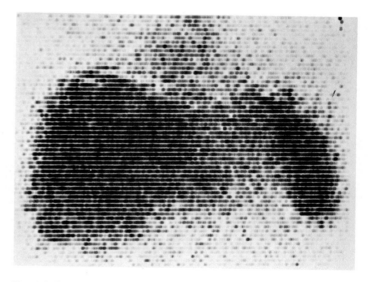

FIG. 1.9. Radioactive isotope scan of the upper abdomen. Injected isotope is taken up in different amounts by different tissues. These can then be visualized by measuring the X-ray emission of the isotope over a sequential series of narrow strips and recording the result photographically.

2 Introduction to human embryology

R. G. HARRISON

HUMAN embryology—the study of the human embryo and its development—is concerned with the phenomena of the intra-uterine period of life, from the fertilization of the ovum to the birth of the child. But there is a cycle of events in succeeding generations; and so it is necessary first to consider the life-history of the germ cells which link one generation to the next.

All the cells of the body are the descendants of a single cell which is the starting-point of each individual. The innumerable cells derived from the fertilized ovum are separable into two main groups, a larger group—the somatic cells—from which the body of the individual is developed, and a much smaller group—the germ cells—upon which the continuation of the species depends. The germ cells are lodged in the body of the individual in the sex glands (gonads)—ovaries in the female, testes in the male.

The germ cells undergo a process of ripening, known as maturation; and when mature they are called gametes. A new individual is initiated by the union of two gametes—a male gamete, or spermatozoon, and a female gamete, or ovum. The new cell thus formed is a fertilized ovum and is known as a zygote—a term conveniently indicating that it is formed by the union of two individuals and that it contains therefore the potentialities of both.

The newly formed human zygote is about $150\mu m$ in diameter and just visible to the naked eye. From it are produced not only a new individual and the germ cells of the next generation, but also a series of membranes and appendages necessary for growth and development during intra-uterine life.

The ontogenetic or developmental history of the individual is usually considered to extend from the zygote to the adult form. The period of development thus defined is divided into an intra-uterine or prenatal period and an extra-uterine or postnatal period.

The intra-uterine life of Man lasts for about 9 months, and is itself divided into three secondary periods:

1 The pre-embryonic period, during which the growing zygote is implanted in the mucous membrane of the uterus (endometrium), and is differentiated into embryonic and non-embryonic portions; this period lasts rather less than 2 weeks.
2. The embryonic period, in which the rudiments of all the main organs of the adult are developed, although the embryo has not yet assumed a definitely human form. This period extends to the end of the second month.
3. The foetal period, from the end of the second month, when the embryo begins to assume a definitely human appearance and is called, thenceforth, a foetus. The foetal period ends at birth, when the foetus becomes a child and passes into the stage of postnatal development.

During the foetal period growth proceeds rapidly. It is especially rapid during the fourth month; and at birth the child usually weighs about 3 to 3.5 kg.

During the first stages of the postnatal period—infancy, childhood, adolescence—growth and development still proceed until the adult condition is attained; then follows a period of maturity, which passes insensibly into the last stage of all—senescence—which ends in natural death. In the following pages it is the intra-uterine period of life which will be considered and more especially the phenomena of the first 2 months—the pre-embryonic and embryonic periods.

The aim of this Section is to provide the student with a general account of the formation of the embryo—embryogenesis—so that he may be in a position to understand the more detailed paragraphs on the development of organs—organogenesis—which he will find throughout the textbook. Some account will be given also of those important extraembryonic organs peculiar to intra-uterine life—the foetal membranes—which are responsible for the protection of the embryo and foetus and provide for its physiological needs by the connection established between one of them and the uterus (Barcroft 1946, 1952; Huggett and Hammond 1952).

For further information on human development the student should consult the special works on the subject included in the list of References at the end of this Section.

Growth and differentiation

In the development of the embryo two main processes go hand in hand, but growth—cell division leading to augmentation of size—must be carefully distinguished from differentiation—the specialization of cells in the formation of tissues and the rudiments of organs. Growth depends upon a proper supply of nutriment, and the growth of some parts of the body depends further upon special chemical factors, either derived from food-material or secreted into the blood by organs of the body itself. The differentiation of groups of cells to form the rudiments of organs is a progressive diversification, and may be due to potentialities located in each particular group—such rudiments of organs are said to be 'self-differentiating'; or it may depend on the situation of the group of cells and the influence of neighbouring cells and rudiments upon them. Experiments have been made by the simple excision, or the transplantation into other regions, of portions of embryos, and by the cultivation of excised portions, either as grafts on other growing tissues or inserted in suitable nutritive media (tissue-culture). It has thus been shown that the type of differentiation which embryonic cells undergo in a particular situation depends on the influence of adjacent structures which have already differentiated or are differentiating.

The formation of the axial structures in the embryo depends upon the influence exerted by chemical compounds (organizers) residing in the tissues of the primitive streak: these are believed to reside in the dorsal lip of the blastopore, the corresponding structure in lower forms. The formation of the lens vesicle of the eye from the ectoderm depends, in some species, on the presence of the underlying optic cup—a self-differentiating structure which grows out from the forebrain; if the optic cup is removed the lens does not appear, but a lens vesicle may be induced to form from another part of the ectoderm by transplanting the cup (Lewis 1904).

Progressive differentiation of the cells of a region is illustrated by experiments on limb buds in Amphibia. Part of a limb bud which, if left *in situ,* would form only a portion of the limb, may form a complete limb if transplanted early enough to another site. After a time, however, the cells of the limb bud differentiate, chemically if not morphologically, so that any part has its destiny fixed and is incapable of forming a whole limb; yet, for a time, it may still be capable of producing the whole of the segment of the limb of which it would normally form only a part. Thus, a portion of the thigh region of a chick embryo, transplanted (Murray and Huxley 1925) or grown by tissue-culture methods (Fell and Robison 1929), will produce not a complete limb but only a femur.

For further information on the control of development see the works of Spemann (1938), Weiss (1939), and Waddington (1956).

There are also many general chemical problems in the development of the embryo, and for information on this aspect of the subject and on the history of embryology the student should consult Needham (1931, 1934, 1950).

Germ cells

The majority of the cells derived from a zygote are somatic cells, which form the tissues of the body; but a minority, which retain all the inherited potentialities of their parents, remain as primordial germ cells or stem cells. These primordial germ cells segregate from the somatic cells at a quite early period, and migrate from the wall of the yolk sac to the genital ridge of the embryo (Witschi 1948). Though the functions and the life-histories of germ cells are quite different from those of somatic cells, the structural characters of the two groups of cells are very similar; both contain cytoplasmic inclusions, the Golgi complex and mitochondria, and both contain the same number (forty-six) of chromosomes in their nuclei. The forty-six chromosomes in human cells consist of twenty-two pairs of chromosomes, the **autosomes**, which are common to both males and females, and one pair of sex chromosomes.

Each primordial germ cell produces many descendants which are lodged in the gonads—female sex cells or oogonia in the ovaries, male sex cells or spermatogonia in the testes—where they undergo further multiplication to form **primary oocytes** or **primary sper-matocytes** respectively. Each of these cells is capable of producing four descendants by two final divisions—the maturation divisions. The four descendants of an oocyte are a mature ovum and three small structures called **polar bodies**; the four descendants of a spermatocyte are four equal **spermatids** which undergo metamorphosis into spermatozoa.

MEIOSIS

The number of chromosomes in the dividing cells varies greatly in different species, but is characteristically constant in each. The chromosomes differ in size and shape, but are arranged in pairs of **homologous chromosomes**, one member of each pair being derived from the male parent and the other from the female parent. Human chromosomes have been carefully classified according to their length and other characteristics (Böök *et al.* 1960; Lennox 1960).

The first maturation division is heterotypical because each chromosome does not split in the process, as in normal cell division

or mitosis, but each pair of chromosomes separates, one passing to each resultant cell. This is termed meiotic division because the outcome is a reduction of the number of chromosomes in each resultant cell to half that in the dividing cell. In Man and other mammals meiosis occurs in the first of the two maturation divisions. The result of this division is two secondary spermatocytes or one secondary oocyte and a polar body.

During the early prophase of meiosis which is short in the primary spermatocyte, but may be very prolonged in the primary oocyte, the chromosomes appear and lie side by side—a stage known as **conjugation of the chromosomes**. At the end of the prophase of meiotic division the paired chromosomes are assembled at the equator of the spindle, where they are attached to the achromatic fibrils.

There is not true metaphase, since the chromosomes do not split longitudinally as in ordinary mitosis, and the anaphase begins when the homologous chromosomes separate from each other, a process of **disjunction**. They thus pass undivided to the opposite ends of the achromatic spindle; therefore, in the telophase, when the whole cell divides, one resultant cell contains a group of chromosomes which are the homologues of the chromosomes in the other resultant cell, and each of the two resultant cells has only half (the haploid number) of the chromosomes which were present in the parent cell (the diploid number).

The result of this reduction division is that the dividing cell's complement of chromosomes is equally divided between the two resultant cells (secondary spermatocytes, or secondary oocyte and polar body). The second maturation division is not meiotic: each haploid secondary spermatocyte divides mitotically to form two haploid spermatids, and each haploid secondary oocyte undergoes mitosis to form one haploid mature ovum and a second polar body. The first polar body also divides, so that three polar bodies are present at the end of the second maturation division [FIG. 1.3A].

MECHANISM OF INHERITANCE AND DETERMINATION OF SEX

In human primary oocytes and spermatocytes there are twenty-two pairs of autosomes, in addition to a special pair in each (oocyte, XX: spermatocyte, XY) which has a specific relation to the inheritance of sex. The statistical study of the mode of inheritance of pairs of characters, of which one may be dominant and the other recessive, is in accord with the behaviour of the chromosomes. As it appears to be a matter of chance which resultant cell will receive which of any pair of chromosomes, it may be calculated that in the maturation and union in fertilization of any two human germ cells there are over 281 billions of different combinations of the chromosomes alone.

Since interchange of genes or groups of genes—crossing-over—is known to occur between pairs of chromosomes during their conjugation in the prophase of the reduction division, it is abundantly clear that the chromosome theory of inheritance is consistent with the infinite variety of individual characteristics.

Sex is a heritable character, and is determined at the time of the union of the gametes. It is dependent upon special chromosomes, which are known as **sex chromosomes**, although they bear other genes than those that determine sex. In Man, the male sex is characterized by the presence of sex chromosomes of unequal size, termed X (large) and Y (small), whereas each female cell contains two X chromosomes.

The human oocyte has twenty-two pairs of autosomes and one pair of X chromosomes—forty-six in all; its chromosome-constitution may thus be written 44XX. The secondary oocyte and mature ovum therefore each possess an X chromosome; the chromosome formula of the mature ovum is 22X. The formula of the primary spermatocyte, on the other hand, is 44XY; when the reduction division occurs the X chromosome passes into one secondary spermatocyte, the Y chromosome to the other. Therefore, of the four spermatids two will have an X chromosome and two a Y chromosome. There are thus two kinds of spermatozoa, of 22X and 22Y constitution respectively. If an X-bearing spermatozoon unites with an ovum a female results, but if a Y-bearing spermatozoon fertilizes an ovum a male results.

All these changes depend on normal disjunction of the sex chromosomes during the first maturation division. If this fails to occur, abnormal spermatozoa with the constitution 22XY and 22 may occur, and non-disjunction in the meiosis of primary oocytes can result, theoretically, in ova having the constitution 22XX or 22. Such non-disjunction is said to be responsible for the production of certain clinical syndromes. Thus, if a 22Y sperm fertilizes an ovum with a 22XX chromosome constitution, or a normal 22X ovum is fertilized by a non-disjunctive 22XY sperm, the resultant zygote will have 44XXY chromosomes, a condition found in Klinefelter's syndrome. This syndrome consists of an apparent male having gynaecomastia (enlarged mammary glands), small testes, poor facial hair growth and a high-pitched voice. In like manner, Turner's syndrome, with a constitution 44X (22X sperm fertilizing a 22 ovum or a 22 sperm fertilizing a 22X ovum), and the 'super female' with 44XXX chromosomes, may result, the last condition being caused by fertilization of a 22XX ovum with a 22X sperm. Chromosomal anomalies may also occur in autosomes, and mongolism is a condition in which the body cells each contain forty-seven chromosomes; the sex chromosomes are normal but forty-five autosomes are present. It is possible for an individual to be mosaic—the body cells being of two or more stem lines with different chromosome numbers.

The cytological and other evidence upon which the chromosome theory of sex is based is confirmed by the observation that uniovular twins, i.e. twins developed from a single ovum, are always of the same sex, whilst binovular twins, resulting from fertilization of two ova, may be of the same or of opposite sex. It is also strikingly confirmed by the phenomena of 'sex linkage' in inheritance. The sex chromosomes are not solely concerned with sex; they bear genes which control other heritable characters, and these characters are linked to sex in their transmission from generation to generation. Sex-linked characters are of some importance in human inheritance, the best-known example being the condition called haemophilia, in which there is an abnormal tendency to spontaneous or excessive bleeding. It is believed that the determining factor of this condition is recessive and is carried on the X chromosome, which explains its characteristic mode of inheritance; it is not transmitted from an affected father to his son, but only to a grandson through a 'carrier' daughter who remains herself unaffected. The gene for red–green colour-blindness is a sex-linked recessive on the X chromosome, and many cases of Turner's syndrome show such colour-blindness.

GAMETOGENESIS

Gametes are the final descendants of a line of germ cells which pass through successive stages of specialization and are finally transformed into mature ova and spermatozoa by means of the two maturation divisions; the whole process is known as gametogenesis.

Both ovum and spermatozoon are specialized cells, inasmuch as each possesses only half the number of chromosomes present in the oogonium or spermatogonium; moreover, each is incapable, under ordinary circumstances, of undergoing cell division, and must either die or unite with the other to produce a new, rejuvenated cell—the zygote.

The ovum and the spermatozoon have their own special characteristics. The phenomena of oogenesis and spermatogenesis must, therefore, be considered separately.

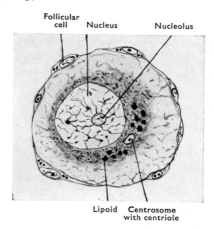

FIG. 2.1. Human oocyte at beginning of growth period. Diameters, 38µm × 33µm.

OOGENESIS AND THE OVUM

The structure of the ovary and the relation of the primary oocytes to the 'germinal epithelium' are described with the Urogenital System.

In the cortical part of the ovary, each primary oocyte lies surrounded by a single layer of follicular cells. Together they constitute a primary ovarian follicle [FIG. 2.1].

Gradually the oocyte increases in size until it attains a diameter of 100 to 200µm, and so becomes a relatively large cell. During the growth period changes occur both in the surroundings and in the contents of the oocyte.

At first the follicular cells are flattened. Then they become cubical and increase in number by mitosis until they form several layers (the stratum granulosum; FIG. 2.2). Thereafter, a cavity filled with a fluid called the liquor folliculi appears amidst the cells. It increases rapidly in size and separates the cells into two multicellular parts. One part forms the boundary of the cavity and is still called the stratum granulosum. The other, with the oocyte embedded in it, projects into the cavity and is the cumulus oophorus. This results in a vesicular ovarian (or Graafian) follicle [FIG. 2.3].

When the cells of the stratum granulosum have formed two layers, a clear envelope, the zona pellucida, begins to appear around the oocyte, separating it from the cells of the cumulus. It increases in thickness until the maturation of the oocyte begins, when it is an elastic, pellucid membrane varying from 7 to 10µm in thickness [FIGS. 2.5 and 2.6].

The changes which take place in the oocyte itself during the growth period concern the contents of the cytoplasm, and the position and relative size of the nucleus.

At the beginning of the growth period the oocyte (called a primary oocyte to avoid confusion with its two immediate descendants) is an almost spherical cell; but it soon becomes ovoid and retains this form in all subsequent periods. The cell body contains a relatively large, eccentrically placed nucleus and a centrosome. The centrosome lies near the central pole of the nucleus and contains one or two centrioles [FIG. 2.1].

The nucleus possesses one or two nucleoli. The Golgi complex, mitochondria, and globules of lipoid are to be found in the cytoplasm around the nucleus, and are most abundant around the centrosome [FIG. 2.1]. As growth proceeds they become more diffused throughout the cytoplasm.

When growth is completed the full-grown primary oocyte, enclosed in the zona pellucida, lies in the cumulus of a vesicular follicle in the cortical part of the ovary [FIG. 2.6]. It consists of a cell body which contains a nucleus, mitochondria, and numerous

Stratum granulosum   Nucleus of primary oocyte   Nucleolus

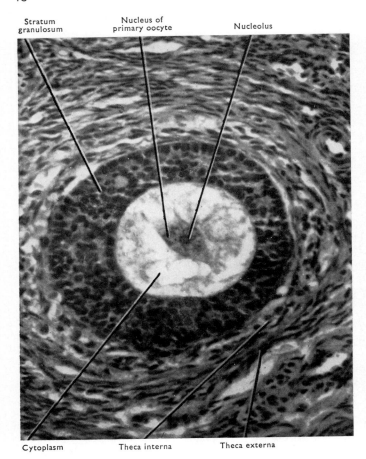

Cytoplasm   Theca interna   Theca externa

FIG. 2.2. Human ovarian follicle from the ovary of a woman aged 38. The follicular cells of the stratum granulosum have multiplied to form a layer of granulosa cells 5–6 cells thick (× 390).

Theca externa   Theca interna   Oocyte

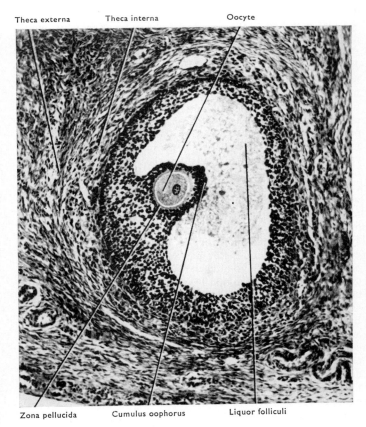

Zona pellucida   Cumulus oophorus   Liquor folliculi

FIG. 2.3. Human vesicular ovarian follicle (× 180).

globules of yolk; but during maturation a centrosome may not be obvious.

The diameter of the cell body, measured along its major axis, varies from 100 to 200μm, for full-grown oocytes are not all of the same size; the average diameter is 130–140μm (Hartman 1929). The diameter of the nucleus varies from 25 to 50μm.

The amount of **yolk** in human and most mammalian ova is relatively small compared with those of fishes, amphibians, reptiles, and birds. The amount of yolk determines whether the segmentation of the fertilized ovum is complete or incomplete, and special terms are employed in comparative embryology to denote differences in the amount and distribution of the yolk. It is sufficient here to note that the amount of yolk in the human ovum is so small that it does not in the least interfere with its complete division in segmentation.

The yolk serves as a store of nutritive material which is utilized during the early stages of the growth of the zygote in mammals, and until the time of hatching in oviparous vertebrates.

Apart from the question whether any new ova are produced in the human ovary after birth—as is apparently the case in some animals—there is evidence that a large number of developing follicles atrophy and disappear so that 'radical selective elimination of ova occurs in human ovaries' (Allen, Pratt, Newell, and Bland 1930).

MATURATION OF THE OVUM

Maturation is the term applied to the phenomena of the two cell divisions which take place after the primary oocyte has attained its full growth.

The first of the two divisions is heterotypical and meiotic. During it the nuclear membrane and the nucleolus or nucleoli disappear, and an achromatic spindle appears at one pole of the oocyte, in the situation previously occupied by the nucleus.

The chromosomes aggregate together in pairs as twin chromosomes [FIGS. 2.7 and 2.8]. There appear therefore to be only half the number of the chromosomes originally present in the oocyte; at the end of the prophase they lie at the periphery of the equator of the spindle. When this condition is attained the spindle rotates on its transverse axis [FIGS. 2.9 and 2.10].

During the metaphase the two halves of each twin chromosome separate from each other, the two chromosomes thus formed being each equivalent to a whole chromosome of an ordinary cell.

In the anaphase the halves of each twin chromosome travel to the opposite poles of the achromatic spindle [FIG. 2.11].

Following the telophase the first maturation division is completed, the primary oocyte being divided into a large segment, the **secondary oocyte**, and a small segment called the **first polar body**, both of which lie inside the zona pellucida and each of which contains half the number of chromosomes originally present in the primary oocyte [FIGS. 2.12 and 2.15]. Human ova which have attained this stage of maturation have been seen (Rock and Hertig 1942).

After the first maturation division is completed a new achromatic spindle is formed in the secondary oocyte; this lies at the periphery of the secondary oocyte near the first polar body, the chromosomes become grouped at its equator [FIG. 2.12], and the second maturation division is commenced.

The ovarian follicle then ruptures, and the liquor folliculi carrying with it the secondary oocyte undergoing its second maturation

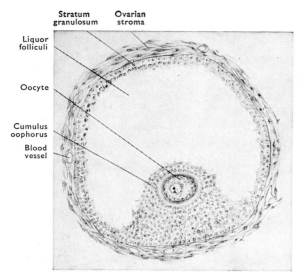

Stratum granulosum
Ovarian stroma
Liquor folliculi
Oocyte
Cumulus oophorus
Blood vessel

FIG. 2.4. Human vesicular ovarian follicle. Diameters of follicle, 620μm × 465μm × 465μm.

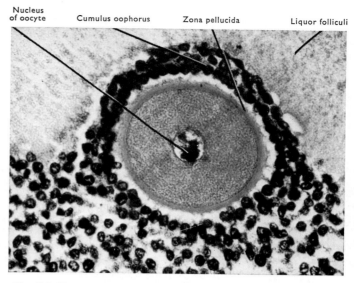

Nucleus of oocyte
Cumulus oophorus
Zona pellucida
Liquor folliculi

FIG. 2.5. Human primary oocyte from the ovary of a woman aged 25 (×450).

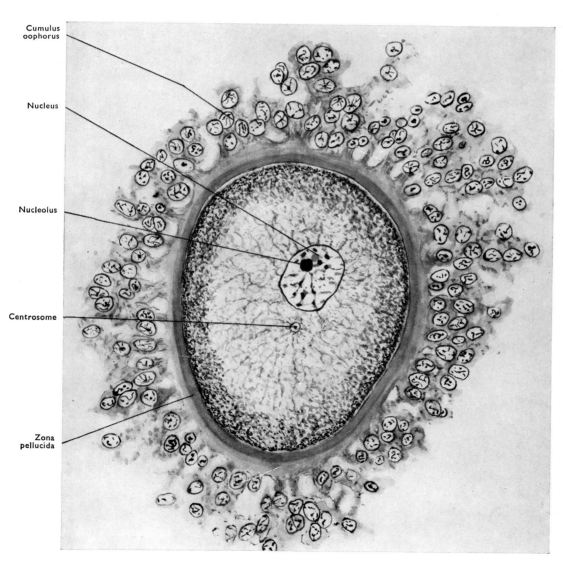

Cumulus oophorus
Nucleus
Nucleolus
Centrosome
Zona pellucida

FIG. 2.6. Human oocyte near the end of the growth period. Diameters of oocyte exclusive of zona pellucida, 140μm × 110μm × 84μm. Average thickness of zona pellucida, 5μm.

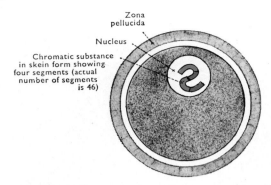

FIG. 2.7. Schema of maturation of ovum. Early part of prophase of first division. Only four chromosomes are detailed, whereas the actual number is 46.

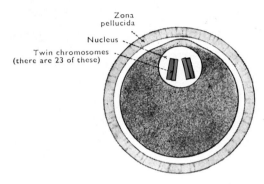

FIG. 2.8. Later prophase of first division. The chromatic thread has divided into twin chromosomes.

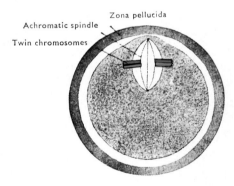

FIG. 2.9. End of prophase of first division. The twin chromosomes lie at the equator of the achromatic spindle.

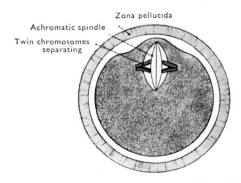

FIG. 2.10. Metaphase of first division. One pole of the spindle lies in the first polar projection, and the paired chromosomes are separating from one another.

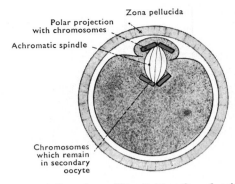

FIG. 2.11. End of anaphase of first division. One of each pair of chromosomes now lies in the polar projection. The other remains in the larger part of the cell body which becomes the secondary oocyte.

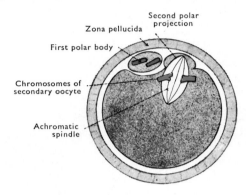

FIG. 2.12. Beginning of metaphase of second division. The chromosomes of the secondary oocyte are dividing.

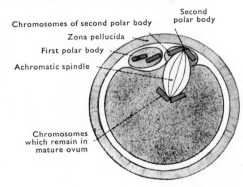

FIG. 2.13. End of anaphase of second division. The chromosomes of the secondary oocyte have separated into equal parts which have passed to the opposite poles of the spindle.

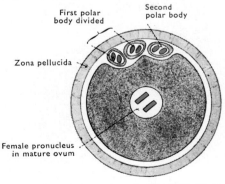

FIG. 2.14. Schema of maturation of ovum. End of telophase of second division. The four descendants of the primary oocyte are the mature ovum, with half the original number of chromosomes, and three polar bodies.

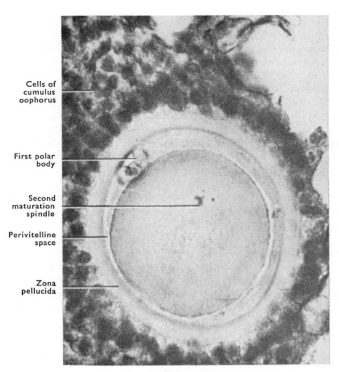

Cells of
cumulus
oophorus

First polar
body

Second
maturation
spindle

Perivitelline
space

Zona
pellucida

FIG. 2.15. Section of a human ovum after the first maturation division. (Dixon 1927.)

division and surrounded by its **corona radiata**, composed of attached cumulus cells, is forced through the breach in the surface of the ovary and is swept into the uterine tube. This process, known as **ovulation**, has a definite relation to the menstrual cycle.

If the oocyte is penetrated by a spermatozoon it then completes the second maturation division. The second maturation division is a homotypical mitosis. The achromatic spindle, already present in the secondary oocyte, carries chromosomes (which, it must be remembered, are only half as numerous as those originally present in the primary oocyte), each of which splits longitudinally into equal parts. The opposite halves of the divided chromosomes then travel to the ends of the achromatic spindle; consequently, the secondary oocyte is divided into a mature ovum and the **second polar body** [FIG. 2.13], each of which contains half the number of chromosomes (the haploid number) present in the primary oocyte, which is diploid.

As soon as the second polar body is separated off, the maturation is completed [FIG. 2.14], and a mature ovum is formed.

THE OVUM

The mature ovum, like the full-grown oocyte, differs from a typical animal cell on account of its large size, and also because it is surrounded by a special protective envelope—the zona pellucida. An additional and important difference is the fact that the mature ovum under ordinary circumstances is incapable of further cell division because it possesses no centrosome—a deficiency which is supplied by the spermatozoon if fertilization occurs.

The fine surface layer of the cytoplasm of the ovum is a membrane which becomes apparent when the polar bodies are given off [FIG. 2.20] and has been termed the **vitelline membrane**. The space between the body of the oocyte and the zona pellucida, into which the polar bodies are discharged, is called the **perivitelline space**; in fixed specimens it may appear greatly increased by retraction of the cell body.

As the second polar body is separated off, the first polar body not uncommonly divides into two parts. When that occurs, the result of the two maturation divisions of the oocyte is the formation of one large cell (the mature ovum) and three polar bodies, all of which are enclosed within the zona pellucida [FIG. 2.14].

If the secondary oocyte does not meet with a spermatozoon it degenerates or passes through the genital passages and is cast off and lost; but if it unites with a spermatozoon a zygote is formed, from which a new individual may arise. In that case the polar bodies persist until the zygote has undergone one or two divisions; but sooner or later they disappear, probably breaking down into fragments which are absorbed by the cells of the zygote.

The polar bodies were originally so named in the belief that they indicate the line along which the first segmentation of the fertilized ovum took place; but their exact position does not appear to be of any significance in the mammalian ovum. The production of a polar body ensures conservation of the cytoplasm in a single ovum which has ejected half its chromosomes.

SPERMATOGENESIS AND THE SPERMATOZOON

The **spermatogonia** lie on the membrana propria of the seminiferous tubules of the testes. Their descendants become converted into spermatozoa [FIG. 2.16]. Primary spermatocytes are first formed and by two maturation divisions these form secondary spermatocytes and then spermatids (Bishop and Walton 1960).

Spermatocytes differ from oocytes in four important respects: (1) they have no protective membrane corresponding to the zona pellucida of the oocyte; (2) they are not enclosed in follicles; (3) they are not surrounded by definite encircling layers of cells similar to the cells of the stratum granulosum; (4) the spermatozoa to which they give origin are smaller and motile and the centrosome is concerned in the formation of the tail of the spermatozoon.

In the tubules of the testes the spermatocytes are intermingled with **sustentacular** (supporting) **cells (cells of Sertoli)**, amidst which they undergo their maturation divisions. The spermatids become embedded in the supporting cells, where they mature into spermatozoa, and probably receive nutrition from material in the cytoplasm of these cells.

After it has reached its full growth each primary spermatocyte, like each primary oocyte, can produce only four descendants; the descendants, as in the case of the primary oocyte, are formed by two successive divisions, of which the first is meiotic and produces reduction of the chromosomes, and the second is mitotic.

These divisions differ from the corresponding divisions of the oocyte in the important respect that the four spermatids are of equal size and value, so that they are all eventually capable of uniting with an ovum to form a zygote.

In the prophase of the first or heterotype division the nucleus and nucleolus disappear in the ordinary way. The centrosome divides, and an achromatic spindle appears, which has the daughter centrosomes at its poles and apparently half the typical number of chromosomes at its equator. But the chromosomes are arranged in pairs, and during the metaphase the two segments of each twin chromosome separate from each other. In the anaphase they travel to the opposite poles of the achromatic spindle, and consequently, when the cell divides in the telophase, each daughter cell or secondary spermatocyte contains a centrosome and half the typical number of chromosomes. It is therefore a haploid cell.

The second maturation division, which takes place without the intervention of a resting stage, is homotypical. The centrosome divides, a new achromatic spindle appears, and the chromosomes gather at its equator. In the metaphase the chromosomes divide longitudinally into equal parts, which travel to the opposite poles of the spindle during the anaphase, and when the telophase is completed each granddaughter cell, now called a spermatid, possesses a centrosome and half the typical number of chromosomes.

THE SPERMATOZOON

Ova become ready for conjugation with spermatozoa directly after the first maturation division is completed, but spermatids which result from the second maturation division have still to undergo a complicated process of transformation before they become converted to spermatozoa. The process of transformation takes place in association with the supporting cells in which the developing spermatozoa become embedded. Before these details are considered it is necessary that the student should be acquainted with the anatomy of an adult spermatozoon [FIG. 2.17].

A spermatozoon is a minute organism possessing a head, a middle piece, and a flagellum or tail. Its total length is about 50μm, that is, its length is about the same as the diameter of the nucleus of the ovum.

The head has the form of a compressed ovoid. It contains the nucleus of the spermatid and is completely covered by a head-cap or galea capitis (Schultz-Larsen 1958). The mean length of the head is 4.4μm (Duijn 1958).

The middle piece is about 5μm in length, and its constituent parts are: (1) the centriole, (2) a portion of the axial fibre, and (3) the mitochondrial sheath.

The axial fibre, which extends from the centriole through the middle piece into the flagellum, consists of one pair of central filaments surrounded by nine pairs of peripheral filaments. The mitochondrial sheath surrounds the axial fibre and is formed by numerous fused mitochondria; it is composed of two bands twisted spirally five to seven times round the axial fibre.

The flagellum or tail is long. It consists of prolongations of the axial fibre surrounded by a tail sheath, and terminates in a short thin end-piece, in which the axial fibre appears free from the sheath.

SPERMIOGENESIS

The phase of spermatogenesis involving successive cell divisions up to the production of spermatids is termed **spermatocytogenesis**. The maturation of a spermatid into a spermatozoon involves no further cell division, but a modification of the spermatid cell into a highly organized cell, lacking cytoplasm and modified for motility

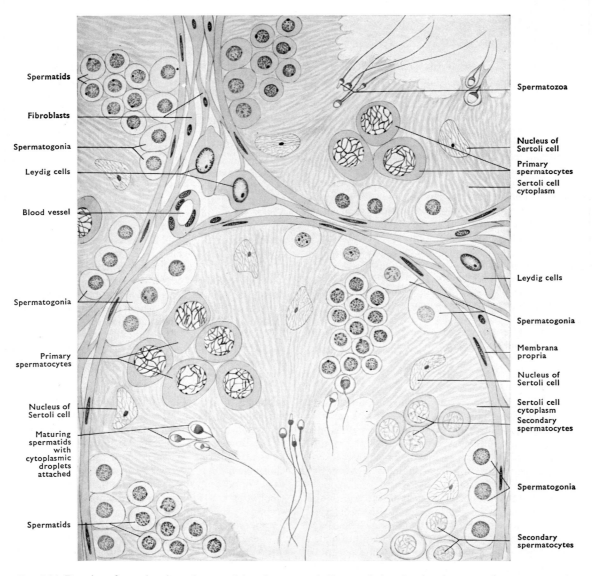

FIG. 2.16. Drawing of a section through parts of three human seminiferous tubules, showing the stages of spermatogenesis.

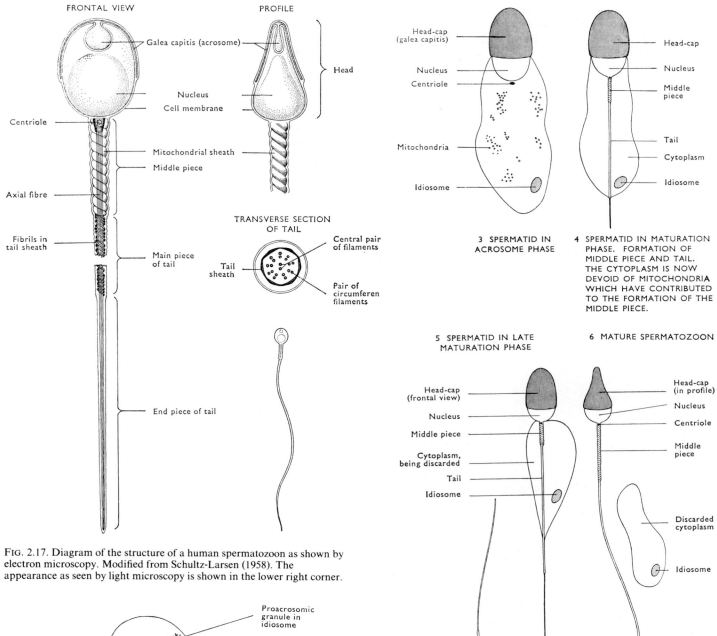

FRONTAL VIEW

PROFILE

Galea capitis (acrosome)

Head

Nucleus
Cell membrane

Centriole

Mitochondrial sheath

Middle piece

Axial fibre

Fibrils in tail sheath

Main piece of tail

TRANSVERSE SECTION OF TAIL

Tail sheath

Central pair of filaments

Pair of circumferen filaments

End piece of tail

FIG. 2.17. Diagram of the structure of a human spermatozoon as shown by electron microscopy. Modified from Schultz-Larsen (1958). The appearance as seen by light microscopy is shown in the lower right corner.

Proacrosomic granule in idiosome

Idiosome

Mitochondria

Nucleolus in nucleus

Centriole

I SPERMATID IN THE GOLGI PHASE

Idiosome

Acrosomic granule

Nucleolus in nucleus

Mitochondria

Centriole

2 SPERMATID IN EARLY CAP PHASE

Head-cap (galea capitis)

Head-cap

Nucleus

Nucleus

Centriole

Middle piece

Mitochondria

Tail

Cytoplasm

Idiosome

Idiosome

3 SPERMATID IN ACROSOME PHASE

4 SPERMATID IN MATURATION PHASE. FORMATION OF MIDDLE PIECE AND TAIL. THE CYTOPLASM IS NOW DEVOID OF MITOCHONDRIA WHICH HAVE CONTRIBUTED TO THE FORMATION OF THE MIDDLE PIECE.

5 SPERMATID IN LATE MATURATION PHASE

6 MATURE SPERMATOZOON

Head-cap (frontal view)

Head-cap (in profile)

Nucleus

Nucleus

Middle piece

Centriole

Cytoplasm, being discarded

Middle piece

Tail

Idiosome

Discarded cytoplasm

Idiosome

Tail

FIG. 2.18. Diagram of the process of spermiogenesis.

and eventually fertilization. This process of maturation is termed **spermiogenesis** or **spermateliosis**. The head of the spermatozoon is formed from the nucleus of the spermatid, which becomes encased in a cap at the future anterior end of the spermatozoon. The Golgi complex assembles on the surface of the nucleus as a bead which is deposited on the nucleus to form the **acrosome** from which the head-cap is derived [FIG. 2.18].

After the second maturation division the centriole of the spermatid passes towards the side of the cell opposite the attachment of the acrosome and from the centriole the axial fibre grows out through the surface of the cell.

The middle piece has still to be formed and the flagellum completed. The mitochondria crowd around the anterior 5μm of the axial fibre to form the mitochondrial sheath. The remains of the cytoplasm of the spermatid, carrying other mitochondria and a remnant of the Golgi body, are then stripped off, and the spermatozoon is complete (Clermont and Leblond 1955).

FERTILIZATION

The process of union of male and female gametes to form a zygote is known as fertilization of the ovum. It begins [FIG. 2.19] when a spermatozoon enters an ovum (either mature or in the last stage of maturation), and it is completed when the nuclear elements of the two have combined.

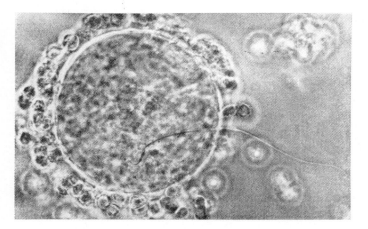

FIG. 2.19. Oocyte of a rat at the moment of fertilization. The oocyte is surrounded by cells of the corona radiata, and the head of the spermatozoon has penetrated into the perivitelline space; most of its tail still lies outside the oocyte (× 700). By permission of Dr C. R. Austin.

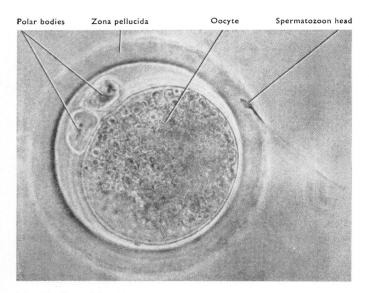

Polar bodies   Zona pellucida   Oocyte   Spermatozoon head

FIG. 2.20. Guinea-pig ovum, showing polar bodies, and spermatozoon head lacking acrosome, embedded in zona pellucida (× 900). By permission of Dr C.R. Austin.

It is believed that fertilization takes place in the ampulla of the uterine tube, and that the entry of one spermatozoon causes some change in the surface of the ovum that prevents the entry of others [FIG. 2.20].

It has already been pointed out that the maturation of the oocyte is usually not completed until after the spermatozoon has entered; thereafter, the second polar body is extruded [FIG. 2.20] and the nucleus of the mature ovum, now called the **female pronucleus**, takes up a central position. It contains half the number of chromosomes present in the nucleus of the oocyte before the maturation began. As the female pronucleus forms, the tail of the spermatozoon disappears, the head is transformed into the **male pronucleus**, and two centrosomes arise from the middle piece. The male pronucleus is smaller than the female pronucleus, and it also contains only half the number of chromosomes present in the primary spermatocyte from which it is derived.

As soon as the female and male pronuclei are established they approach each other, meet, and lose their nuclear membranes. The chromosomes of each pronucleus appear, become arranged on the equator of a spindle, and split along their length; the half chromosomes then pass towards their corresponding centrosomes, and the first cleavage division of the zygote ensues.

When fertilization is completed, therefore, a new structure—the **zygote**—is formed. It lies, together with the polar bodies, inside the zona pellucida. There are several phenomena occurring at fertilization. The first is the reconstitution of the diploid number of chromosomes that is typical for the cells of the species to which the zygote belongs, half of them derived from the female parent and half from the male. The second factor is the stimulus to division (**segmentation** or **cleavage**) which the spermatozoon provides. Thirdly, the sex of the resultant embryo is determined by the chromosome constitution of the sperm effecting fertilization—whether it contains an X or a Y chromosome. Finally, various changes occur in the structure and chemistry of the fertilized ovum; these have been collectively termed **activation** (see Rothschild 1956; Austin and Bishop 1957; Austin and Walton 1960).

The pre-embryonic period

SEGMENTATION

At the time of fertilization, segmentation of the zygote is initiated. There follows a series of consecutive mitotic divisions to result in the formation of twelve to sixteen cells called **blastomeres**; these are grouped together in the form of a spherical mass, called a morula [FIGS. 2.21 and 2.23] on account of the mulberry-like appearance of its surface. The segmentation of the human ovum has not yet been followed in detail, but a perfect two-cell stage [FIG. 2.22], an eight-cell stage (Rock and Hertig 1948), and a morula containing twelve cells (Hertig, Rock, Adams, and Mulligan 1954) have been described.

The **morula** is formed inside the zona pellucida so that its total size does not exceed that of the mature ovum; consequently, the size of the individual cells is progressively reduced by each division and the proportion of cytoplasm to nucleus becomes steadily smaller during segmentation. The proportion is very high in the ovum, and its reduction within the normal range of the somatic cells appears to be an essential preliminary to the next stage of development. This proportionate change involves a transfer of cell proteins, cytoplasmic ribonucleic acid being transferred into the nucleus. At the end of segmentation the zona pellucida disintegrates.

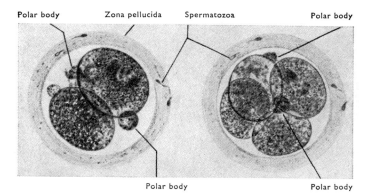

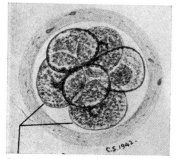

Cells about to divide

FIG. 2.21. Two-cell, four cell, and six-cell stages of segmentation of living ovum of macaque monkey. (After Lewis and Hartman 1933.) The two larger cells in the six-cell stage were about to divide to produce the eight-cell stage. For the two-cell human stage, see Fig. 2.22.

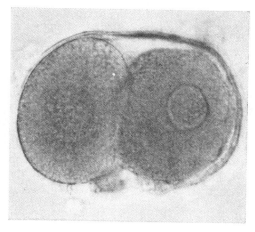

FIG. 2.22. Segmentation of intact human ovum at two-cell stage. The smaller of the two polar bodies is visible between the two blastomeres (× 500). (Carnegie No. 8698; by special permission of Dr Arthur T. Hertig.)

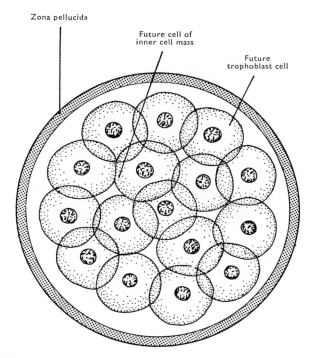

FIG. 2.23. Diagram of a morula.

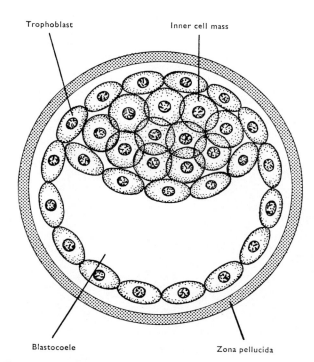

FIG. 2.24. Diagram of a blastula.

TROPHOBLAST AND INNER CELL MASS

The first structural change that occurs in the morula is the differentiation of its cells into an outer layer, the trophoblast, and an inner cell mass [FIGS. 2.24 and 2.25] by the appearance of a cavity

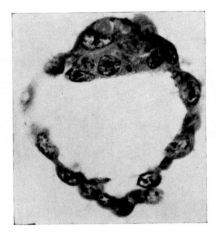

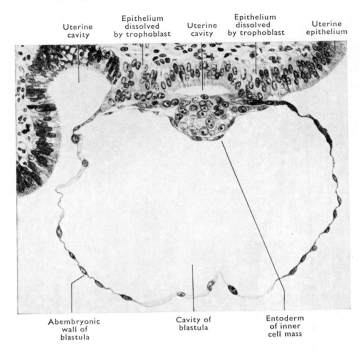

FIG. 2.25. Section of free human blastula, 5–6 days old, from uterine cavity (× 600). Note the arrangement of the inner cell mass in relation to the expanding trophoblastic wall and the segmentation cavity. (Carnegie No. 8663; by special permission of Dr Arthur T. Hertig.)

filled with fluid (the **segmentation cavity** or **blastocoele**) between the cells of the morula. As soon as this segmentation cavity appears, the zygote becomes a **blastula** or **blastocyst**. The fluid which fills the blastocoele also acts as a source of nutrition, and is derived from the uterine secretions. By this time the zona pellucida has disintegrated and the blastula is free to expand with the further intake of fluid so that the trophoblastic cells are flattened out into a thin membrane, except where the inner cell mass is attached [FIG. 2.26]. At this stage—the zygote having entered the uterus as a morula, probably on the third or fourth day from ovulation (see Hertig, Rock, and Adams 1956)—the blastula becomes attached to the uterine epithelium [FIG. 2.27]. It is probable that this occurs on the sixth or seventh day, as the earliest implanting human zygote yet observed [FIG. 2.28] was recovered seven and a half days after the presumed date of fertilization; it is still in the blastula stage.

FIG. 2.27. Early stage of attachment of macaque blastula (about 9 days) to the uterine epithelium. Note the dissolution of the endometrial cells in contact with the trophoblast and the spread of the entoderm from the inner cell mass (× 185). (After Heuser and Streeter 1941.)

The trophoblast or trophoblastic ectoderm, by its active proliferation [FIGS. 2.28–2.33], plays a most important part in enabling the zygote to burrow into and embed itself in the **decidua** (the term applied to the endometrium after it has become hypertrophied in preparation for implantation of the zygote), and in the nutrition of the embryo and foetus. It enters into the formation of the chorion or outermost envelope of the growing zygote. The chorion serves, in the first instance, as both a protective and a nutritive covering [p. 33], but later a part of it forms the **placenta**, the organ through which gaseous and nutritional exchange between mother and foetus takes place.

The inner cell mass contains the cells from which the embryo itself will be developed; but it is concerned also with the formation of the other extra-embryonic organs—amnion, yolk sac, and allantois—which, with the placenta and chorion, constitute the foetal membranes.

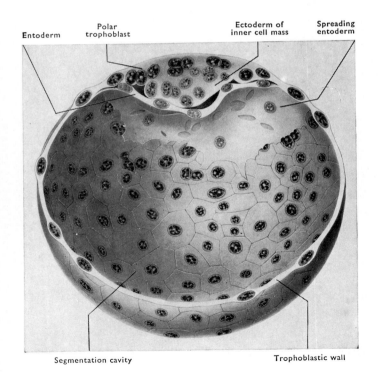

FIG. 2.26. Reconstruction of 9-day-old blastula of macaque monkey. The entodermal part of the inner cell mass is flattened and its cells are spreading over the inner surface of the trophoblast (× 280). (After Heuser and Streeter 1941.)

DIFFERENTIATION OF INNER CELL MASS

As the blastula begins to attach itself to the decidua and to be implanted, the fate of the cells of the inner cell mass is already indicated by the relative positions they occupy [FIGS. 2.28 and 2.29]. The group that adheres for a time to the trophoblast, itself of ectodermal origin, produces the **ectoderm**, destined to take part in the formation of both the embryo and its amnion; the remainder, projecting at first into the blastocoele but rapidly flattening out into a plate of cells one layer thick, constitutes the **entoderm**, which takes part in the formation of the embryo, and of its yolk sac.

The delimitation of the ectodermal and entodermal elements of the inner cell mass is followed by the formation of two vesicles, one from each group of cells. The cavity in the ectodermal group is

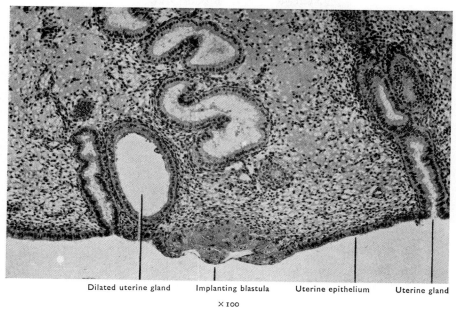

Dilated uterine gland   Implanting blastula   Uterine epithelium   Uterine gland

× 100

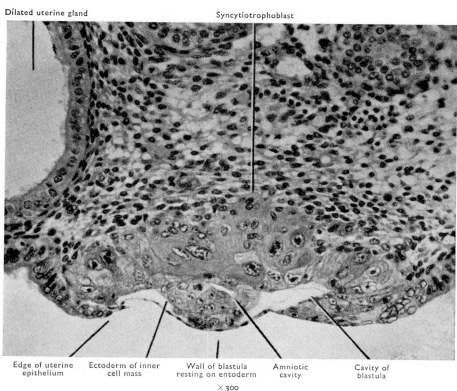

Dilated uterine gland   Syncytiotrophoblast

Edge of uterine   Ectoderm of inner   Wall of blastula   Amniotic   Cavity of
epithelium   cell mass   resting on entoderm   cavity   blastula

× 300

FIG. 2.28. Section of human zygote, 7½ days old, in early superficial stage of implantation. The trophoblast is proliferating wherever it is in contact with the endometrium, and the thin abembryonic wall of the blastula, still exposed to the uterine cavity, is collapsed on the inner cell mass. (Hertig and Rock 1945.)

formed by a loosening of the cells in the middle of the group and by the partial separation of a layer of 'amniogenic' cells from the overlying trophoblast; the cells are reoriented around a space in which fluid accumulates. This amniotic cavity is already present in the 7½-day human blastula [FIG. 2.28].

The entodermal vesicle or yolk sac is at first very much larger than the ectodermal vesicle and the question of its origin is complicated by the simultaneous appearance of the loose, semifluid tissue known as the primary (or extra-embryonic) mesoderm.

The entodermal portion of the inner cell mass appears in the 7½-day human blastula as a simple layer of cells [FIG. 2.28]. In a 9½-day specimen [FIG. 2.29] there is little change in this layer beyond a slight extension at its edges, although the ectodermal plate has grown thicker, the amniotic cavity is better defined, and the implantation of the whole zygote has advanced considerably. Yet in quite a number of slightly older specimens a complete 'entodermal' vesicle is present, larger than the ectodermal vehicle with its amniotic cavity, and occupying a varying, but considerable, part of

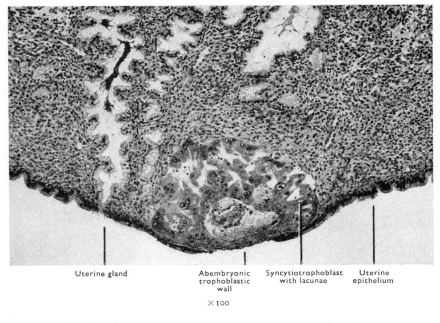

Uterine gland   Abembryonic   Syncytiotrophoblast   Uterine
                       trophoblastic   with lacunae   epithelium
                          wall

× 100

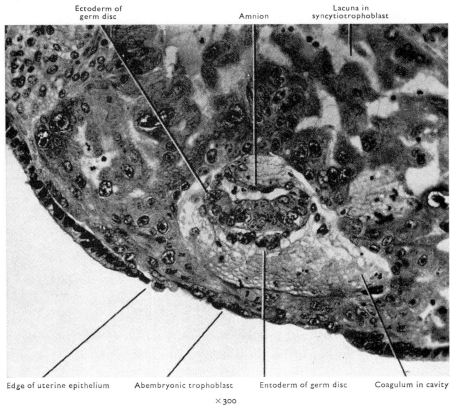

Ectoderm of   Amnion   Lacuna in
germ disc   syncytiotrophoblast

Edge of uterine epithelium   Abembryonic trophoblast   Entoderm of germ disc   Coagulum in cavity

× 300

FIG. 2.29. Section of human zygote, $9\frac{1}{2}$ days old, more deeply implanted than the $7\frac{1}{2}$-day specimen shown in Fig. 2.28. Only a small portion of the trophoblastic wall is still exposed and the uterine epithelium is beginning to regenerate over it. There are numerous anastomosing lacunae in the syncytiotrophoblast. (Hertig and Rock 1945.)

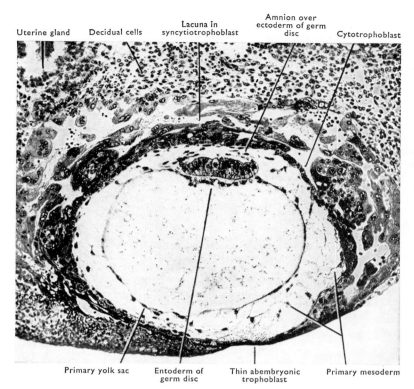

Uterine gland   Decidual cells   Lacuna in syncytiotrophoblast   Amnion over ectoderm of germ disc   Cytotrophoblast

Primary yolk sac   Entoderm of germ disc   Thin abembryonic trophoblast   Primary mesoderm

FIG. 2.30. Section of human zygote, 12½ days old. The general features are the same as in the 11½-day specimen, but the primary yolk sac is much larger and the 'germ disc' more advanced. The implantation, however, is more superficial (×125). (Hertig and Rock 1941.)

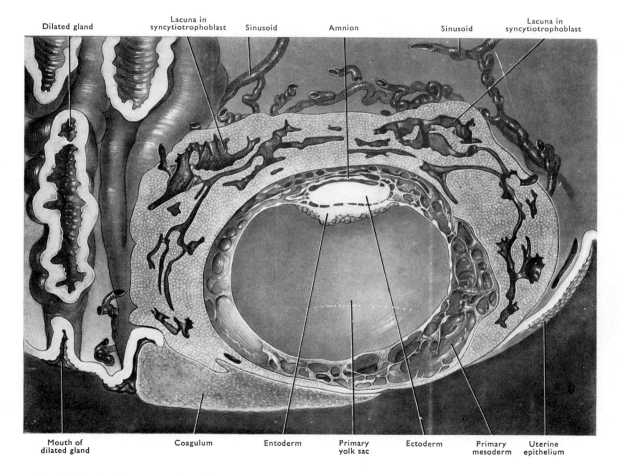

Dilated gland   Lacuna in syncytiotrophoblast   Sinusoid   Amnion   Sinusoid   Lacuna in syncytiotrophoblast

Mouth of dilated gland   Coagulum   Entoderm   Primary yolk sac   Ectoderm   Primary mesoderm   Uterine epithelium

FIG. 2.31. Sectional reconstruction of the 12½-day human zygote in Fig. 2.30. The reconstruction shows the main features of the embryonic rudiment and of the trophoblast, in particular its relation to the endometrium (×125). (After Hertig and Rock 1941, from a drawing by James F. Didusch.)

the original cavity of the blastula, which is now filled with 'primary mesoderm' [FIGS. 2.31 and 2.36]. In mammals, in general, the entoderm grows round the inside of the trophoblastic shell before any mesoderm—which arises later from the primitive streak of the embryonic disc—is present, but in Man the development of primary mesoderm precedes the formation of the yolk sac.

There is, however, an obvious difference between the cuboidal cells that form the roof of the vesicle (primary yolk sac) in contact with the ectodermal plate and the extremely tenuous remainder of the wall of the vesicle. The thin, drawn-out cells that form it are of mesothelial character and are indeed indistinguishable from the cells of the primary mesoderm that now line the trophoblastic shell and occupy the rest of the cavity in wisps of mesenchymatous tissue.

FOETAL MEMBRANES AND PLACENTA

Nutrition and protection of embryo

While the zygote is passing along the uterine tube, before the zona pellucida disintegrates and the blastula stage is reached, and for a brief period after it enters the uterus, it depends for its nutrition upon the store of lipoid and other food material in the cytoplasm of the ovum.

As the human ovum is small, and as it contains but little food store, there is urgent necessity for an external source of nutritive supply, of oxygen for tissue respiration, and a provision for the removal of waste products of its metabolism. The morula obtains some food material from the uterine secretions as well as the fluid essential for its transformation into a blastula; but development cannot proceed beyond that stage until an intimate connection between the zygote and the uterus is established by the critical

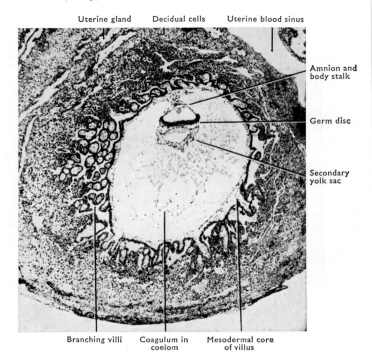

FIG. 2.33. Section of human zygote, 16½ days old. The secondary yolk sac is fully formed with an outer covering of primary mesoderm. The 'germ disc' is more advanced than in the 13½-day specimen (Fig. 2.32) with a primitive streak, early formation of intra-embryonic mesoderm, and a head-process. There is an indication of an 'amnion stalk' in the condensation of mesoderm that attaches amnion to chorion, and an allantoic diverticulum was present. The chorionic villi with their mesodermal cores are branching freely (× 30). (Heuser, Rock, and Hertig 1945.)

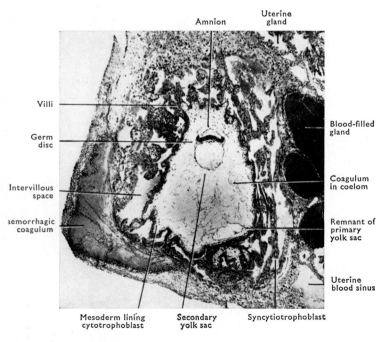

FIG. 2.32. Section of human zygote 13½ days old. The embryonic rudiment is clearly defined with amnion and secondary yolk sac, and a remnant of the primary yolk sac is seen. There is an unusually large 'closing coagulum', and chorionic villi have begun to form (× 30). (Heuser, Rock, and Hertig 1945.)

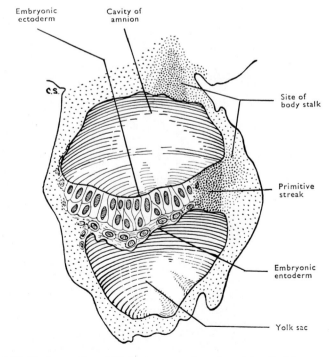

FIG. 2.34. Median section of 15-day human embryo (reconstruction) to show the earliest stage of the primitive streak at the caudal end of the embryonic area. (After Brewer 1938.)

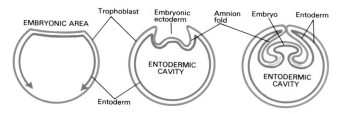

FIG. 2.35. Diagrams of stages in the evolution of the foetal membranes. For detailed explanation, see text. (Adapted from Graham Kerr's *Text-book of embryology: Vertebrata.*) Trophoblast, amnion, and embryonic ectoderm, blue; entoderm, red; mesoderm, yellow.

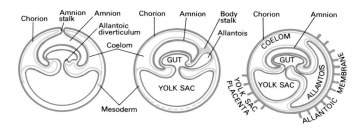

operation of implantation or embedding in the decidua. This connection is provided by the placenta, which is one of the **foetal membranes**.

Evolution of foetal membranes

The foetal membranes are parts of the original body wall (amnion and chorion) and parts of the original gut wall (yolk sac and allantois) which are set aside to provide organs that adapt the embryo and foetus to its temporary environment. These organs are extra-embryonic structures concerned with protection, nutrition, respiration, and excretion; and the manner in which they perform these functions, as well as the details of their arrangement, differ in different groups of animals. The four membranes are found only in the higher classes of vertebrates—the reptiles, the birds, and the mammals—whose embryos develop either within an egg-shell or within the body of the mother. Since they all possess the characteristic protective membrane called the amnion, the reptiles, birds, and mammals are grouped together as the Amniota.

The **amnion** provides a cavity full of fluid in which the embryo may develop under conditions which resemble the watery environment of the embryos of lower vertebrates. The embryos of fishes and amphibia do not require an amnion, since for the most part they hatch from the egg and develop in water; and the only 'foetal membrane' they possess is a yolk sac which contains enough food material for the initial stages of development and, becoming highly vascularized, also forms a 'respiratory foetal membrane'. All the Amniota also possess a **yolk sac**, and in the embryos of all those that are oviparous—including the lowest order of mammals (the monotremes)—it is filled with food material. Although the yolk sac ceases to have any importance as a food store in the viviparous mammals, including Man, it is retained in them because of its

intimate relation to the development of the gut and also because the earliest blood vessels and blood cells are formed in its wall. In some mammals it is retained, at least temporarily, as a yolk sac placenta.

The third membrane—the **allantois**—is as characteristic of the Amniota as the amnion itself; and it has a very remarkable history. It grows out as a diverticulum from the ventral wall of the hindgut [FIGS. 2.35 and 2.38], and in reptiles and birds it serves primarily as a urinary bladder for the embryo; its relation to the gut is exactly the same as that of the permanent urinary bladder of the amphibia, with which it is probably homologous. But the allantois in these animals, as it increases in size, comes into contact with the fourth foetal membrane, the **chorion** [FIG. 2.35], and fuses with it to form a combined membrane which lines the inside of the egg-shell and surrounds the embryo. This membrane, by virtue of the blood vessels it contains, provides for the respiration of the embryo by gaseous exchange through the porous shell between the air and the blood. Thus the allantois of oviparous animals comes secondarily, through fusion of its wall with the chorion, to act functionally as an embryonic 'lung'. In most viviparous mammals the chorion is vascularized in the same manner by blood vessels carried to it by the allantois and thus is enabled to act likewise as a respiratory membrane; but in this case the exchange of oxygen and carbon dioxide is not with the air but with the blood of the mother; and, in addition, food materials also can be obtained by the embryo from the same source by the formation of a **placenta**.

AMNION

The human amnion is formed from that portion of the wall of the ectodermal or amnio-embryonic vesicle which does not take part in the formation of the embryo. It consists of ectoderm cells covered externally by a layer of extra-embryonic mesoderm, and it is

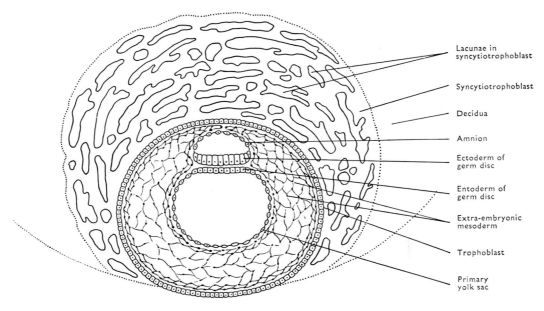

Lacunae in
syncytiotrophoblast

Syncytiotrophoblast

Decidua

Amnion

Ectoderm of
germ disc

Entoderm of
germ disc

Extra-embryonic
mesoderm

Trophoblast

Primary
yolk sac

FIG. 2.36. Diagram of early implanted human zygote.

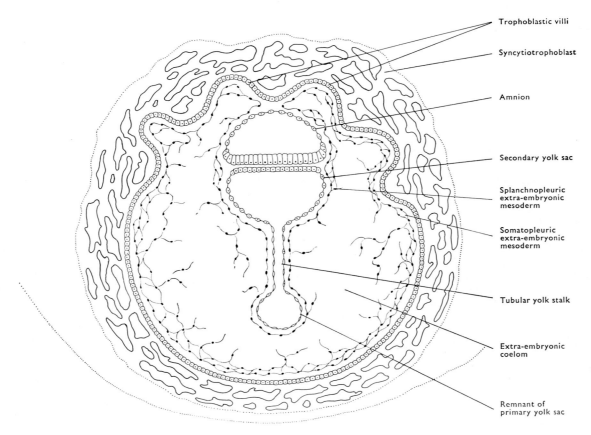

Trophoblastic villi

Syncytiotrophoblast

Amnion

Secondary yolk sac

Splanchnopleuric
extra-embryonic
mesoderm

Somatopleuric
extra-embryonic
mesoderm

Tubular yolk stalk

Extra-embryonic
coelom

Remnant of
primary yolk sac

FIG. 2.37. Diagram of implanted human zygote showing formation of extra-embryonic coelom.

continuous with the margin of the embryonic area or germ disc [FIGS. 2.36–2.38].

The cavity of the ectodermal vesicle, enclosed by the amnion and the embryonic area, is the amniotic cavity; it is filled with fluid which raises the amnion in the form of a cupola over the embryonic area [FIG. 2.35]. The amnion progressively enlarges as the amount of fluid increases (Harrison and Malpas 1953) until it fuses with the chorion; by that time its attachment to the embryo is around the umbilical orifice and it forms a sheath for the umbilical cord [FIG. 2.42].

YOLK SAC

The wall of the primary yolk sac is not entirely of entodermal origin, since its greater part—the so-called exocoelomic membrane—is composed of cells which appear to be derived from those of the primary mesoderm. But the whole of this vesicle does not participate in the formation of the 'yolk sac' of later stages. The primary yolk sac reaches its final state of development about the twelfth day, when it is much larger than the ectodermal vesicle [FIGS. 2.30 and 2.31]; thereafter, it is quite rapidly reduced in size so that in a 13½-day embryo [FIG. 2.32] it is not much larger than the ectodermal vesicle, and soon, by the growth of that vesicle, the yolk sac appears about the same size or even, for a time, relatively smaller [FIGS. 2.33 and 2.34]. It may now be called the secondary yolk sac. The term definitive yolk sac is retained for the yolk sac after separation from the alimentary canal [p. 48].

The reduction in size of the primary yolk sac to a smaller secondary yolk sac is occasioned by a considerable portion of the primary yolk sac being cut off and subsequently degenerating. In several young embryos small detached vesicles, sometimes connected by strands, can be recognized as degenerating portions of the primary yolk sac [FIG. 2.32]. The earliest record of the stage of constriction of the primary yolk sac, with a suggestion of its evolutionary significance, is in the description of the 'Teacher-Bryce II' embryo of Bryce (1924). Stieve (1931) figured hypothetical stages in the reduction in size of the yolk sac; and his interpretation is borne out by the 13½–14-day specimens described by Lewis and Harrison (1966) and Morton (1949). The primary yolk sac appears to correspond to the yolk sac of the lower mammals, and its replacement by the secondary yolk sac is a specialized developmental process in Man and other Primates.

When the embryonic area is folded into the form of the embryo [p. 48: see FIG. 2.63] the entodermal vesicle (secondary yolk sac) is differentiated into three parts: (1) a part enclosed in the embryo, where it forms the primitive entodermal alimentary canal; (2) a part which lies external to the embryo in the extra-embryonic coelom [p. 34]—this is the yolk sac proper or definitive yolk sac; (3) the third portion is the vitello-intestinal duct, which connects the yolk sac with the primitive alimentary canal [FIGS. 2.59 and 2.63].

The cavity of the yolk sac is therefore in free communication with that of the alimentary canal, and their walls are continuous with each other and identical in structure, each consisting of an internal layer of entodermal cells and an external layer of mesoderm.

Free communication between the yolk sac and the primitive alimentary canal appears to exist in the human embryo until it is 4 weeks old and about 2.5 mm long [FIG. 2.63]. During the fifth week the vitello-intestinal duct is elongated into a relatively long, narrow tube, lodged in the umbilical cord, and the yolk sac, which has become a relatively small vesicle, is placed between the outer surface of the amnion and the inner surface of the chorion, in the region of the placenta. By the end of the fifth week, when the embryo has attained a length of about 5 mm, the vitello-intestinal duct begins to undergo atrophy, and it separates from the intestine when the

embryo is about 11 mm long; but remnants of it may be found in the umbilical cord up to the third month.

The definitive yolk sac itself persists until birth, when it may be found as a minute object lying between the amnion and the placenta near the attachment of the umbilical cord.

The human yolk sac, since it contains no yolk, is in a sense a vestige from the evolutionary point of view; yet, in the early stages of human development, it is relatively quite large [FIG. 2.89], and it retains its special importance in the development of the blood and blood vessels [p. 62]. The vessels which develop in the wall of the yolk sac are soon connected with others in the body of the embryo to form a vitelline circulation. This circulation is mainly responsible for the transfer of nutriment from the yolk to the embryos of birds and reptiles; in mammals the vessels are formed in the same way since they are the basis of the vascular arrangements of the gut. At a very early period, a number of primitive vitelline arteries are distributed to the yolk sac from the primitive aortae, and the blood is returned from the yolk sac to the embryo by a pair of vitelline veins [FIG. 2.75].

After a time the arteries are reduced to a single pair, which then become converted into a single trunk; this passes through the umbilical orifice along the vitello-intestinal duct to the definitive yolk sac [FIG. 2.77]. After the umbilical cord is formed the extra-embryonic parts of the vitelline veins disappear and can no longer be traced in the cord. The same fate overtakes the vitelline artery—all except a portion of its intra-embryonic part, which persists as the superior mesenteric artery.

ALLANTOIS

The human allantois, like the yolk sac, is a greatly reduced structure, and may therefore be termed the allantoic diverticulum. In its relation to the chorion and the formation of the placenta, it too may be considered an evolutionary vestige; but it does not disappear entirely, for it is related to the development of the urinary bladder [p. 61]. The allantoic diverticulum grows into the body stalk but does not reach the chorion [FIG. 2.49]; it arises from the primary entodermal vesicle and so appears at first to come from the yolk sac; but its site of origin is included in the tail fold of the embryo and so is transferred to the ventral wall of the cloacal part of the hindgut [FIGS. 2.58 and 2.59]. The allantoic diverticulum is included in the umbilical cord [FIG. 2.58], where it shrivels and disappears; its intra-abdominal portion becomes the urachus, which atrophies to form the median umbilical ligament.

CHORION

When the human blastocyst is first implanted, its wall consists solely of trophoblast [FIGS. 2.28 and 2.29]; but the primary mesoderm soon appears and, after the formation of the secondary yolk sac [p. 33], the primary mesoderm splits into two layers—one lining the inner surface of the trophoblast and the other covering the outer surfaces of the inner cell mass [FIGS. 2.37, 2.38, and 2.49]; the space between them is the extra-embryonic coelom. The extra-embryonic coelom, however, never completely separates the inner cell mass from the inner surface of the trophoblast; continuity is maintained at one point to provide a pathway for blood vessels between the chorion and the embryo. The basis of that pathway is provided by a part of the primary mesoderm which is not split by the extra-embryonic coelom and continues to connect the inner surface of the chorion with the inner cell mass in the region that will become the caudal end of the embryonic area. The mesodermal connecting link is called the body stalk [FIGS. 2.38, 2.49, and 2.58].

The trophoblast and its lining of mesoderm together constitute the chorion, and the essential function of that membrane, in addition to providing an outer protective sac within which the embryo may develop in its amnion, is the formation of the placenta. The development and vascularization of the chorion are described with the account of the development of the placenta itself.

BODY STALK

In the body stalk lies the allantoic diverticulum of the entodermal vesicle [FIG. 2.58], on either side of which pass the umbilical arteries and veins (one of which atrophies very early in development), by means of which blood is conveyed between the embryo and the chorion. It is important to note that the blood vessels which thus pass through the body stalk enter or leave the body of the embryo at the margin of the embryonic disc. As the embryo enlarges into the amniotic cavity, this margin becomes the edge of the umbilical orifice which is, at first, a relatively large aperture [FIG. 2.56] through which these vessels therefore pass.

As the embryonic area is folded into the form of the embryo, the amnion increases in extent, filling more and more of the extra-embryonic coelom, and the embryo rises into its cavity. Thus the walls of the amnion bulge ventrally round the margin of the umbilical orifice [FIGS. 2.42, 2.58, and 2.59], tending to cover the body stalk and its enclosed structures and force the yolk sac further and further away from the embryo. The vitello-intestinal duct is elongated and surrounded by the external surface of the amnion and its covering layer of extra-embryonic mesoderm, which form a hollow tube around the duct. The cavity of this tube—the proximal part of it becomes the **umbilical sac** (see below)—is an elongated part of the extra-embryonic coelom, surrounded by the amnion [FIGS. 2.72–2.74]. The caudal wall of the tube necessarily consists of the elongated body stalk.

UMBILICAL CORD

As the distension of the amnion continues, the mesoderm lining of the amniotic tube which surrounds the vitello-intestinal duct and

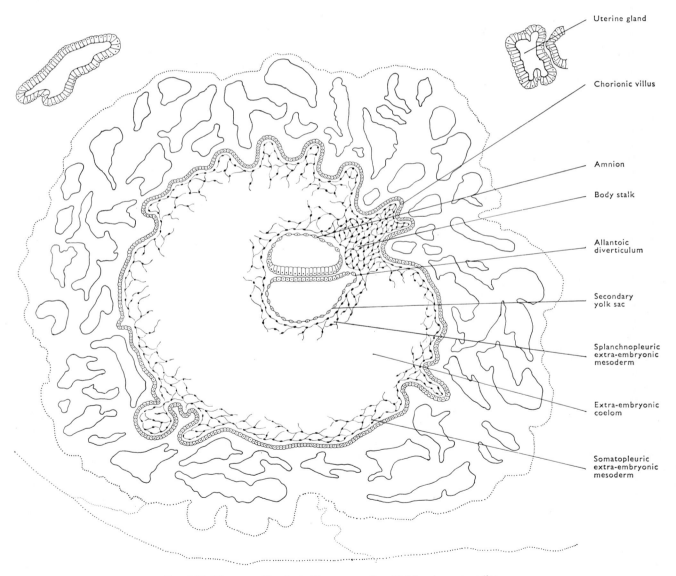

Uterine gland

Chorionic villus

Amnion

Body stalk

Allantoic diverticulum

Secondary yolk sac

Splanchnopleuric extra-embryonic mesoderm

Extra-embryonic coelom

Somatopleuric extra-embryonic mesoderm

FIG. 2.38. Diagram of implanted human zygote with bilaminar germ disc.

body stalk fuses with the mesoderm surrounding that duct and with the body stalk. When the fusion is complete, a solid cord—the umbilical cord—is formed [FIG. 2.42]; it consists of an external covering of amniotic ectoderm, and a core of mesoderm in which lie two umbilical arteries, a single umbilical vein, and the remains of the vitello-intestinal duct and the vitelline vessels. One of the arteries is absent in about 1 per cent of umbilical cords (Armitage, Boyd, Hamilton, and Rowe 1967), and there is considerable evidence to suggest that this is associated with congenital malformations of the foetus. One end of the umbilical cord is connected with the embryo; the other end is attached to the chorion. Near the latter attachment, in the mesoderm of the chorion, lies the yolk sac, now a relatively small vesicle. The allantoic diverticulum, continuous with the urachus of the urogenital chamber, projects into the umbilical cord from the embryo [FIGS. 2.42, 2.73, and 2.74]. For some time the part of the extra-embryonic coelom persisting in the embryonic end of the cord is distended to form the umbilical sac. This is continuous with the intra-embryonic coelom and the intestines develop in it for a time [p. 60].

As the amnion enlarges [FIG. 2.94] the part which does not enclose the umbilical cord is pressed against the inner surface of the chorion, and fusing with it, obliterates the cavity of the extra-embryonic coelom [FIG. 2.42].

The outer wall of the complete vesicle now consists of the fused chorion and amnion, and contains in its interior the amniotic cavity and the embryo, which is attached to the chorion by the umbilical cord.

When the umbilical cord is first formed, it is comparatively short, but as the amniotic cavity increases the cord elongates, until it attains a length of from 30 to 129 cm (average 61 cm (Malpas 1964)). This allows the embryo to float freely in the fluid in the amniotic cavity, and to obtain nutrition from the placenta by the flow and return of blood, through the vessels within the umbilical cord.

The Placenta

The placenta is an organ developed to provide the foetus with food and oxygen and to remove the waste products of metabolism. It is formed partly from the chorion and partly from the mucous membrane (endometrium) of the uterus, altered to form the uterine 'decidua'.

In the placenta the blood vessels of the foetus and the blood of the mother are brought into intimate relation, so that free interchanges may readily take place between the two blood streams; but *the blood of the foetus and the blood of the mother are always separated from each other by two or more layers of foetal cells*.

ENDOMETRIAL PREPARATION

The endometrium undergoes changes in preparation for the reception and retention of the zygote. If implantation occurs the modified endometrium is known as the uterine decidua, a term based on the belief that it is to a great extent 'cast off' when the placenta is ultimately detached.

The changes which take place are, for the most part, hypertrophic in character. The vascularity of the endometrium is increased, mainly by the dilatation of its capillaries. The tubular glands are elongated; they become tortuous and dilated, and the interglandular tissue increases in amount. As a result of these changes the decidua is thicker, softer, more spongy, and more vascular than secretory endometrium.

Partly on account of the dilatation of the deep parts of the glands and partly on account of differences in texture, the decidua is usually described as consisting of three layers: (1) an internal layer, next the

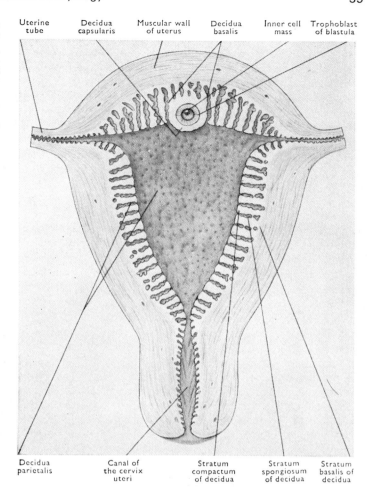

FIG. 2.39. Diagrammatic coronal section of uterus, showing the various parts of the decidua and a blastula embedded in it.

cavity, containing the necks of the uterine glands and called the compact layer; (2) a middle layer—the spongy layer—formed largely by the dilated parts of the glands; (3) an external unchanged layer, in which lie the comparatively unaltered bases of the glands—the basal layer.

These changes are an intensification of the periodic hypertrophy and secretory changes in the endometrium which precede menstruation or implantation. During menstruation the superficial part of the mucous membrane disintegrates, and the whole of the compact layer with much of the spongy layer is cast off together with some blood, constituting the menstrual flow which lasts a variable period, but usually about 3 to 4 days. Repair ensues and the endometrium is built up again to the premenstrual state, being regenerated from the basal layer under the influence first of the oestrogenic hormones and later of progesterone—produced respectively by the growing ovarian follicle and by the corpus luteum, the structure formed in the ovary from the ruptured follicle after ovulation. The evidence that ovulation occurs about 14 days before the expected onset of the next menstruation (Rock and Hertig 1944, 1948) is reviewed by Davies (1948). This relation is of great importance in estimating the age of young embryos. If the ovum is not fertilized and implantation does not occur, the corpus luteum begins to degenerate about twelve days after ovulation, menstruation occurs, and thereafter the corpus luteum becomes infiltrated with fibrous tissue to form a corpus albicans. If fertilization and implantation take place, the corpus luteum continues to grow and maintains its secretion of progesterone; this induces the final changes in the endometrium, constituting

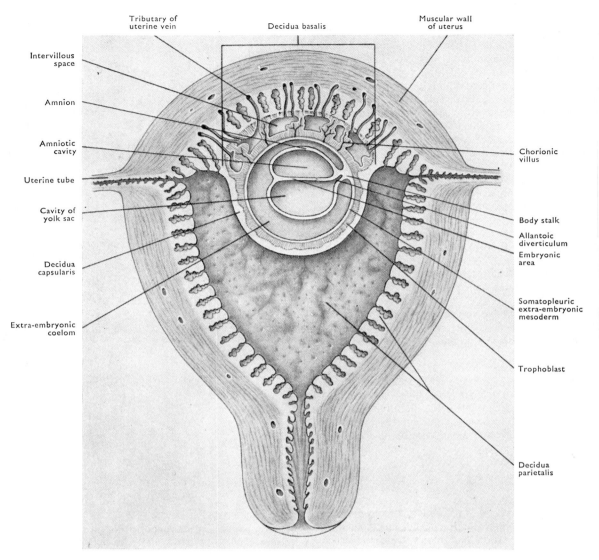

Tributary of
uterine vein

Decidua basalis

Muscular wall
of uterus

Intervillous
space

Amnion

Amniotic
cavity

Uterine tube

Cavity of
yolk sac

Decidua
capsularis

Extra-embryonic
coelom

Chorionic
villus

Body stalk

Allantoic
diverticulum

Embryonic
area

Somatopleuric
extra-embryonic
mesoderm

Trophoblast

Decidua
parietalis

FIG. 2.40. Diagrammatic coronal section of a pregnant uterus after the formation of the intervillous spaces.

the **decidual reaction**, which characteristically accompany the implantation of the blastula.

IMPLANTATION AND FORMATION OF PLACENTA

In the uterus the trophoblast of the blastula first sticks to the surface of the decidua, then proliferates and produces enzymes which erode the decidua, so as to implant itself in the compact layer [FIGS. 2.28 and 2.29]. It may penetrate the decidua at any point on the wall of the uterine cavity; but it usually enters on the fundus or on an adjoining part of the intestinal or vesical surface. The aperture through which it passes is closed by proliferation of a portion of the trophoblast which acts as a sealing plug—the **operculum** (Teacher 1924); and there is sometimes a **closing coagulum** [FIG. 2.31], which may be partly blood clot, superficial to the operculum and projecting from the aperture. Abnormally, implantation may occur lower down on the uterine wall, so that the placenta overlies the cervical canal: this is termed **placenta praevia**. Implantation may also occur in a site outside the uterus, such as the uterine tube; this is known as ectopic gestation.

The portion of the decidua in which the blastula is embedded soon becomes thicker than the other parts of the decidua; it is separated by the blastula into an internal part, between the zygote and the uterine lumen, called the **decidua capsularis**, and an external part between the blastula and the myometrium—the **decidua basalis**. The remainder of the decidua, lining the uterus—by far the larger portion—is the **decidua parietalis** [FIG. 2.39].

As soon as the blastula becomes embedded in the decidua its trophoblast undergoes rapid proliferation and becomes several cells thick wherever it is in direct contact with maternal tissue. Presently the trophoblast differentiates into two layers—an inner cellular layer or **cytotrophoblast**, and an outer syncytial layer or **syncytiotrophoblast** (sometimes incorrectly called plasmoditrophoblast). In the syncytial layer, remnants of cell boundaries are visible by electron microscopy—an indication of the origin of the multinucleated syncytioplasm by fusion of cells which are derived from the division of cytotrophoblast cells (Boyd and Hamilton 1970).

As development proceeds, the trophoblast, now the **chorionic epithelium**, increases in thickness and continues to invade and destroy the decidua, probably by enzyme action. At this stage the

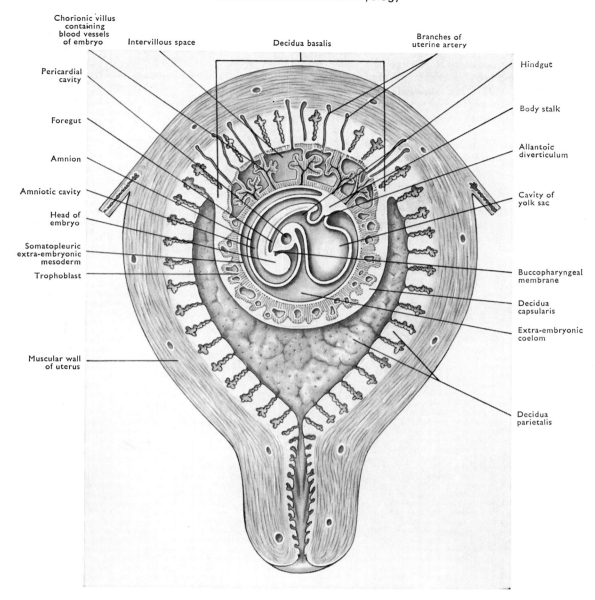

FIG. 2.41. Diagrammatic coronal section of a pregnant uterus at period of formation of embryo. The amnion has begun to enlarge.

decidual reaction becomes apparent. Its characteristic feature is the multiplication of the connective tissue cells of the endometrium which become swollen and have the appearance of epithelial rather than connective tissue cells. These decidual cells [FIGS. 2.30 and 2.33] contain glycogen and lipoids, and they congregate in close proximity to the advancing trophoblast. If the balance between trophoblast and decidua is upset, then invasion may proceed without check—a rare occurrence that may result in a chorionepithelioma, one of the most malignant tumours known—or, on the other hand, the zygote may fail to implant properly and be cast off as an early abortion.

As the invasion of the decidua extends, the syncytiotrophoblast erodes the walls of the maternal blood vessels and itself becomes permeated with lacunae or spaces into which the maternal blood flows. The maternal blood does not clot in the trophoblastic lacunae and thus begins to circulate slowly through them. The type of placenta in different mammals depends on the extent to which the maternal tissues are destroyed by the action of the trophoblast. In the human placenta, the uterine blood vessels are broken into, so

that chorion comes into direct contact with the maternal blood, and the placenta is of the haemochorial type (Mossman 1937; Boyd and Hamilton 1952; Amoroso 1952).

The spaces in the syncytium enlarge rapidly after the maternal blood begins to circulate within them, and the trophoblast becomes divided into three parts. (1) The primary chorionic villi, which lie between adjacent blood spaces and are formed by the proliferation and extension of cytotrophoblastic cells into the syncytiotrophoblastic trabeculae separating the lacunae. (2) The parts which lie in contact with the mesoderm of the chorion and with it form the chorionic plate. (3) The cytotrophoblastic shell (Hamilton and Boyd 1960) which covers the maternal tissues and forms the outer boundaries of the blood spaces. The cytotrophoblastic shell is produced by further proliferation of the cytotrophoblastic cells within the primary villi towards the junctional zone between the syncytiotrophoblast of these villi and the decidua basalis. Here, the cytotrophoblastic cells mushroom outwards, and fusion between these lateral proliferations from adjacent villi constitutes the cytotrophoblastic shell. The lacunar blood spaces in the trophoblastic syncytium then become confluent to form the **intervillous space** of the placenta [FIGS. 2.42 and 2.44].

After a time each primary villus differentiates into a cellular core and syncytial periphery, and thereafter the villi are invaded by the mesoderm of the chorion and are thus converted into **secondary chorionic villi** [FIGS. 2.40–2.42]. The secondary villi consist, therefore, of a mesodermal core continuous with the mesoderm of the chorion, and covered by a layer of cytotrophoblast and a layer of syncytiotrophoblast.

The umbilical arteries extend through the body stalk into the mesoderm of the chorion, and branches from them enter the mesodermal cores of the villi, which thus become vascularized [FIG. 2.43].

The proximal end of each villus is continuous with the chorionic plate, the inner boundary of the intervillous space; the distal end is attached to the cytotrophoblastic shell, which forms the outer boundary of the intervillous space and is fused with the maternal decidual tissue.

In the region of the decidua basalis, the villi become longer and more complex, each villus constituting a **truncus chorii** or main stem villus of the later, definitive placenta. With increasing length of the trunci there is concomitant longitudinal division of their maternal extremities to produce **rami chorii** of second or third order, which in turn subdivide to form **ramuli chorii** or minor villous stems. After a time the ramuli send out numerous branches into the intervillous space, and thus increase greatly in complexity and surface area [FIGS. 2.40–2.42]. In this way two sets of secondary villi are differentiated: (1) the **anchoring villi** which cross from the chorion to the trophoblastic shell and are attached to it by cell columns which are the outer parts of the primary villi not yet invaded by the primary mesoderm; (2) **free** or **absorbing villi** which extend from the sides of the original ramuli into the blood in the intervillous space [FIGS. 2.42–2.44].

The development of the chorionic villi does not proceed equally in all parts of the membrane. The trophoblast thickens only where it is in direct contact with the uterine tissues. The part of the

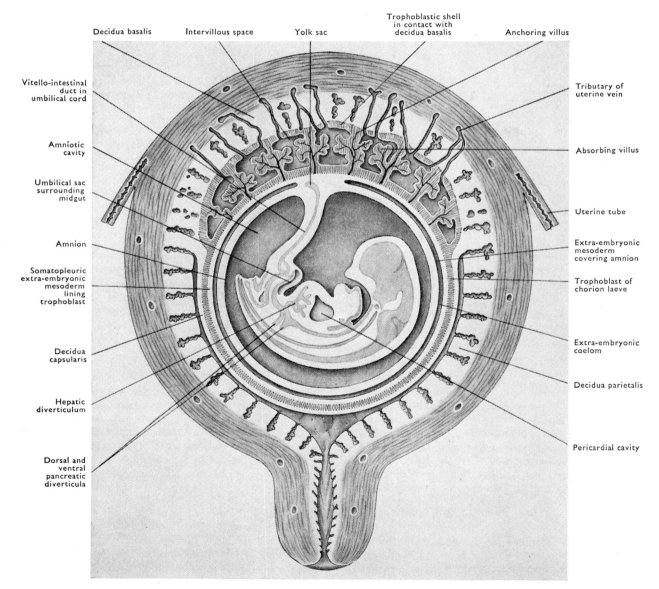

FIG. 2.42. Diagrammatic coronal section of a pregnant uterus after formation of the umbilical cord. The expanding amnion has almost obliterated the part of the extra-embryonic coelom which lies between it and the chorion. The expansion of the decidua capsularis has brought it in contact with the decidua parietalis, so that the uterine cavity is almost obliterated.

blastocyst containing the inner cell mass (the embryonic pole of the blastocyst) advances first into the decidua. Thus this trophoblast which has the body stalk of the embryo—through which vascular connections are established—attached to it, is the first part to be affected, and maintains the lead thus acquired [FIGS. 2.28–2.31]. As implantation proceeds and the decidua closes over the abembryonic surface of the zygote, the trophoblast proliferates on that surface also. For a time, therefore, the whole surface of the chorion is covered with villi. But the capsular portion of the decidua has a relatively poor maternal blood supply which is progressively diminished as the chorionic vesicle continues to expand; thus the villi in that region undergo atrophy and disappear. When these degenerative changes have occurred, the portion of the chorion in association with the thinned decidua capsularis presents a relatively smooth surface and is known as the **chorion laeve**. In the meantime the decidua basalis increases in thickness, and the villi associated with it increase in size and in the complexity of their branches. The portion of the chorion from which these large villi spring is termed the **chorion frondosum**. It is this portion of the chorion which takes part in the formation of the **foetal portion of the placenta**, the **maternal (uterine) part** of that organ being formed by the decidua basalis.

The placenta, therefore, is formed mainly by the chorion frondosum and to a small extent only by maternal tissues. The interchanges between the foetal and the maternal blood take place in the substance of the foetal part of the placenta through the trophoblast which covers the surfaces of the villi. Food and oxygen pass from the maternal to the foetal blood, whilst the waste products of foetal metabolism pass from the foetal to the maternal blood.

At this period the cavity of the amnion is bounded by a wall composed of the fused amnion, chorion, and decidua. As the

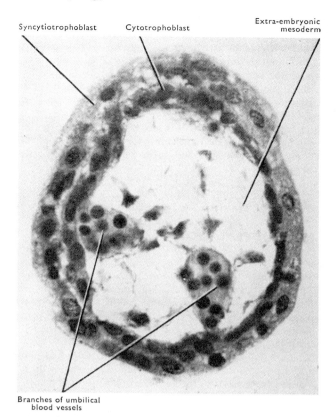

FIG. 2.43. Chorionic villus of a 5-mm human embryo (× 760).

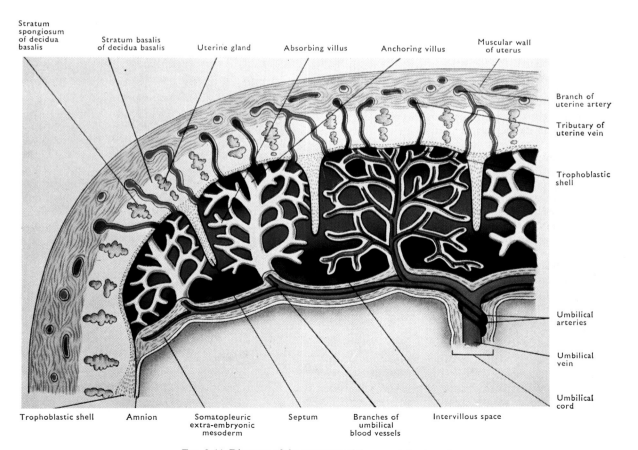

FIG. 2.44. Diagram of the structure of the complete placenta.

distension of the amnion continues, the decidua capsularis bulges more and more into the cavity of the uterus, until it is forced against the surrounding wall of the uterine cavity, and fuses with the decidua parietalis, obliterating the cavity of the uterus. This fusion takes place towards the end of the second month, and as soon as it has occurred the discoid mass of placental tissue is continuous at its margin with the fused amnion, chorion, and decidua parietalis [FIG. 2.42].

After the second month the foetus lies in the **amniotic cavity**, which is bounded by the fused amnion, chorion, and uterine wall except where the cavity of the body of the uterus communicates with the canal of the cervix; there the amniotic cavity is bounded by a membrane formed by the fused amnion, chorion laeve, and decidua capsularis only.

COMPLETION OF PLACENTA

Each absorbing villus consists of a vascular, mesodermal core covered by cytotrophoblast and syncytiotrophoblast, the latter lying next to the maternal blood in the intervillous space. As development proceeds and the intervillous space becomes larger, the villi become longer and more complicated, and at the same time the cellular layer of the trophoblast largely disappears, until in the majority of the villi the syncytiotrophoblast alone covers the mesodermal core.

In still later stages, degeneration occurs not only in the villi, but also in the chorionic plate and in the cytotrophoblastic shell. One of the results of the degenerative process is the deposition of fibrinoid material in the place originally occupied by the trophoblast; the fibrinous layers on the surfaces of adjacent villi adhere together, and the villi thus connected fuse into masses of intermingled fibrinous and vascular tissue.

When the **foetal (chorionic) part** of the placenta is completed it consists of: (1) the chorionic plate closing the intervillous space internally; (2) the villi; (3) the intervillous space; and (4) the cytotrophoblastic shell, which closes the intervillous space externally and is perforated by the maternal vessels passing to and from the space. The cytotrophoblastic shell, together with the decidua basalis, constitutes the **basal plate** of the placenta, the site of junction between maternal and foetal tissues. Since the embryo and trophoblast constitute a graft with a genetic endowment which is different from that of the mother, some immunological reaction might be expected. Such a reaction would damage the embryo or placenta, and perhaps even cause their death and expulsion from the uterus. There are many possible reasons why this does not normally occur, but the most likely explanation is the absence of antigenicity in the trophoblast.

The **maternal (uterine) part** of the complete placenta consists of the decidua basalis of earlier stages. It is fused internally with the cytotrophoblastic shell to constitute the basal plate.

Projections of the basal plate into the intervillous space form incomplete septa which serve to separate the placenta into lobes or **cotyledons**—from ten to thirty-eight in number. The lobes are spoken of as placental units, and each contains several trunci chorii from which the free villi branch in the manner of a tree [FIG. 2.44]. Most of the free villi take a recurved course back towards the chorion (Spanner 1935).

The maternal blood vessels pass from the muscular wall of the uterus directly into the placenta, where they traverse the basal plate and open into the intervillous space. The arteries usually open on or near the septa and the veins begin in the areas between them. The veins of the pregnant uterus are greatly enlarged.

At the end of the pregnancy the membranes, under the increased pressure of the amniotic fluid produced by muscular contraction of the uterus, dilate the cervix, and when they rupture the fluid is expelled through the vagina—the 'breaking of the waters'. Next the

child is born, but it remains attached to the placenta by the umbilical cord, which is usually ligatured and then divided. Later the ruptured membranes and placenta (the after-birth) are expelled from the uterus.

Detachment of the placenta is caused by contraction of the myometrium and rupture of the stratum spongiosum of the decidua. As the detached placenta is expelled, the decidua parietalis is torn through along the plane of the stratum spongiosum. The fused amnion and chorion and the inner part of the decidua parietalis, which are attached to the margin of the placenta, constitute 'the membranes', and are expelled with it.

PLACENTA AT BIRTH

At birth the placenta weighs about 500 g, has a diameter of 10 to 24 cm, and is about 3 cm thick at its centre. It may be fairly uniform in thickness throughout, but usually thins off rapidly at the periphery where it is continuous with the 'membranes'. Its inner surface is covered with the amnion, which fused with the chorion towards the end of the second month of pregnancy. Its rough, outer surface is formed by the remains of the stratum spongiosum of the decidua, and is divided into areas by fissures which correspond in position with the septa by which the organ is divided into lobes. One or more accessory (succenturiate) lobes of the placenta may occur occasionally.

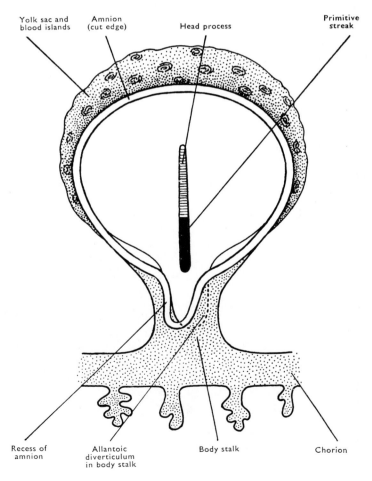

FIG. 2.45. Reconstructed surface view of embryonic area of young human embryo of 18 days to show position of head process and its relation to primitive streak.

EMBRYONIC AREA

When the extra-embryonic coelom is developed, the zygote then consists of three vesicles, one large and two small. The large **chorionic vesicle** is bounded by the chorion, composed of the trophoblast lined with primary mesoderm. It contains the two small vesicles projecting into the coelom and connected by the body stalk to the part of the chorion furthest from the uterine cavity. The **ectodermal** or **amniotic vesicle** and the **entodermal** or **yolk sac vesicle** remain in contact where the primary mesoderm has not penetrated between the original ectodermal and entodermal elements of the inner cell mass. Here they form a bilaminar plate known as the **embryonic area** or **germ disc**—the rudiment of the embryo itself. The ectodermal lamina is **embryonic ectoderm**—it will take part in the formation of the embryo; the **amniotic ectoderm** completes the ectodermal vesicle. Similarly, the entodermal vesicle consists of a lamina of **embryonic entoderm** and extra-embryonic or **yolk sac entoderm** [FIGS. 2.33 and 2.38]. The embryonic area is indicated as soon as the amniotic cavity is formed; it is well defined in the classical Peters embryo of 13 days, and is clearly demarcated in all young embryos in which the secondary yolk sac is formed, for example, the 13½-day embryo [FIG. 2.32] and the embryos of 15 days [FIG. 2.34] and 16½ days [FIG. 2.33] in which the primitive streak [see below] is already in evidence.

With the formation and spread of the intra-embryonic mesoderm from the primitive streak [p. 44], the embryonic area becomes trilaminar.

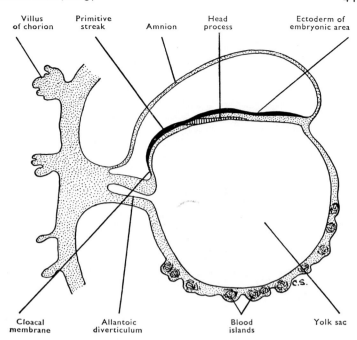

FIG. 2.46. Reconstructed median section of same young human embryo as in Fig. 2.45 (× 50). (Thompson and Brash 1923).

Differentiation of the embryonic area and formation of primary axial structures

The embryonic area is the area of contact between the ectodermal and the entodermal vesicles so, at first, it is circular in outline. As growth continues the area becomes oval, and a linear, median thickening of the ectoderm, the **primitive streak**, appears in that part of the oval which becomes the caudal end of the area [FIGS. 2.45, 2.48, 2.51, and 2.83]. Bilateral symmetry is thus impressed upon the embryonic area and the line of the axial structures of the developing embryo is indicated. From this point in development we can speak of cephalic and caudal ends of the embryonic area, which elongates and for a time assumes a pear-shaped outline [FIGS. 2.45 and 2.83].

The deeper cells of the primitive streak adhere to the entoderm and migrate laterally into the interval between the ectoderm and entoderm of the embryonic area to form the **intra-embryonic** or **secondary mesoderm**. They also give rise, after a series of remarkable developmental events, to the **notochord**—an axial structure characteristic of all vertebrates—the formation of which is presaged by a proliferation of the cephalic end of the primitive streak, the **primitive node** (Hensen's node). A small dimple, the **primitive pit**, soon appears in the dorsal aspect of this node.

Immediately after the formation of the primitive streak an ectodermal groove, called the **neural groove**, appears in the cephalic part of the embryonic area [FIGS. 2.50, 2.57, and 2.58]. It is formed by the longitudinal folding of a thickened plate of ectoderm—the **neural plate**—which is the rudiment of almost the whole nervous system, the only exceptions being the olfactory nerves, some parts of the ganglia of the cranial nerves, and the end-organs of the sensory nerves. From it also are derived the neurolemmal cells of peripheral nerve fibres, the medullary cells of the suprarenal glands, and other structures [p. 51]. The lateral parts of the neural plate form the **neural folds**. Almost from the first, the cephalic extremities of the neural folds are united a short distance behind the cephalic end of the embryonic area. Their caudal ends, which remain separate

for a time, embrace the cephalic part of the primitive streak, and later converge and fuse together.

FORMATION OF NOTOCHORD AND INTRA-EMBRYONIC MESODERM

The notochord and the intra-embryonic mesoderm are formed from the primitive streak—the notochord from its cephalic end, and the mesoderm from its lateral margins and later its caudal end. The primitive streak probably represents the fused and elongated lips of the blastopore in lower vertebrates, and as a continued line of contact between ectoderm and endoderm remains, in part, as the cloacal membrane [p. 61].

As soon as the primitive streak is established its cephalic end becomes a node or centre of growth—the **primitive (Hensen's) node**—by means of which mesoderm is produced. The portion of the mesoderm formed by the activity of the primitive node in the midline is the dorsomedian portion, from the posterior part of the roof of the nose to the caudal end of the trunk. The mesoderm of the perineum and the ventral body wall from the perineum to the umbilicus are formed from the caudal part of the primitive streak. Nevertheless, the primitive streak undergoes little or no increase in length; indeed, as growth continues, it becomes relatively shorter compared with the total length of the embryonic region; for the new material, formed by its lateral and cephalic borders, is transformed into the tissues of the embryo as rapidly as it is created.

NOTOCHORD

The notochord, or primitive skeletal axis, is formed indirectly from the primitive node. It appears first as a narrow, median rod of cells, called the **head process**, which projects headwards from the primitive node between the ectoderm and the entoderm [FIGS. 2.45, 2.46, and 2.50], and continued proliferation from the primitive node [FIG. 2.48] adds to its caudal end.

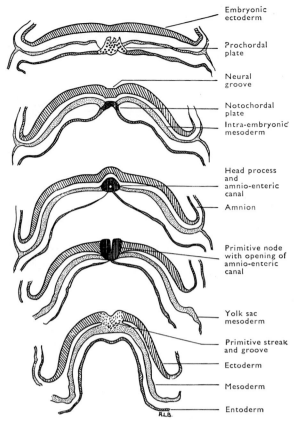

FIG. 2.47. Transverse sections of 18-day presomite embryo to show relation of head process to primitive streak and prochordal plate (× 70). (After Heuser 1932a.)

The head process is at first solid, but presently a cavity appears in the rod of cells and tunnels it from the ectodermal surface of the primitive pit headwards, though not quite to its cephalic end [FIG. 2.47], to form the notochordal canal. The floor of the tunnel, i.e. the cells next the cavity of the entodermal vesicle, together with the roof of the entodermal vesicle in the midline, then break down, the roof flattens out, and the head process is then represented by a plate of cells intercalated in the dorsal wall of the entodermal vesicle—the notochordal plate [FIGS. 2.47 and 2.52]; for a time, therefore, there is a passage from the cavity of the amniotic vesicle through the primitive pit to the cavity of the entodermal vesicle, the amnio-enteric canal, later to become the transitory neurenteric canal when the neural folds [p. 50] close over it.

At a later period the notochordal cells become separated from the entoderm, and they again form a cylindrical rod of cells in the median plane which has been more or less moulded off from the dorsal part of the entodermal sac [FIG. 2.54B], and lies between the floor of the ectodermal neural groove and the reformed entodermal roof of the primitive alimentary canal. This is the true notochord. The cephalic end of the notochord is continuous with the ectoderm and entoderm of a small portion of the embryonic area which lies immediately beyond the cephalic end of the neural groove. This part of the embryonic area, indicated at an earlier date by enlargement of its cells, has been named the prochordal plate [FIG. 2.47]. The ectoderm and entoderm of the prochordal plate are in direct contact without mesoderm between its two layers. This bilaminar region, because it afterwards forms the boundary membrane between the cephalic end of the primitive entodermal canal and the primitive oro-nasal cavity or stomodaeum, is later called the buccopharyngeal membrane [FIG. 2.59]. It disappears in the fourth week of embryonic life. The caudal end of the notochord remains continuous with the primitive streak until the formation of the neural tube is completed.

After a time the cylindrical notochordal rod is surrounded by secondary mesoderm which becomes converted into the vertebral column. As the vertebral column is formed the notochord is enlarged in the regions of the intervertebral discs and for a time assumes a nodulated appearance.

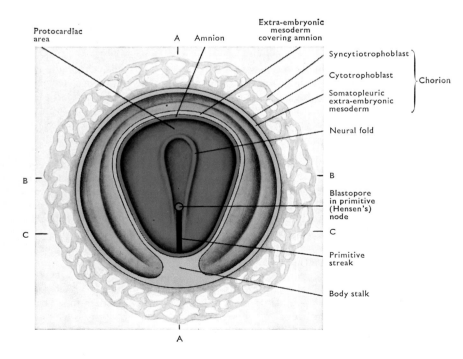

FIG. 2.48. Diagram of dorsal surface of embryonic area of early zygote after removal of part of the chorion and amnion.

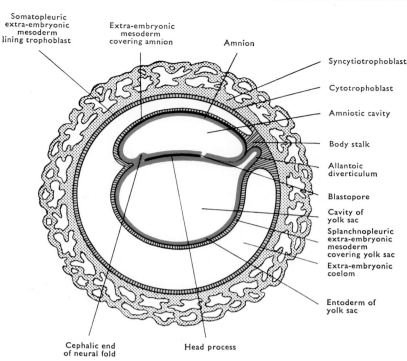

Somatopleuric
extra-embryonic
mesoderm
lining trophoblast

Extra-embryonic
mesoderm
covering amnion

Amnion

Syncytiotrophoblast

Cytotrophoblast

Amniotic cavity

Body stalk

Allantoic
diverticulum

Blastopore

Cavity of
yolk sac

Splanchnopleuric
extra-embryonic
mesoderm
covering yolk sac

Extra-embryonic
coelom

Entoderm of
yolk sac

Cephalic end
of neural fold

Head process

FIG. 2.49. Longitudinal section of zygote along line A–A in Fig. 2.48.

Ultimately the notochord disappears as a distinct structure, but remnants of it are believed to persist as the pulpy centres of the intervertebral discs. The extension of the notochord into the region of the head is of interest from a morphological and possibly also from a practical point of view. It extends through the base of the cranium from the anterior border of the foramen magnum into the posterior part of the body of the sphenoid bone. Its presence in the posterior part of the skull-base suggests that that region was, primitively, of vertebral nature. The notochord lies within the occipital bone but emerges from it for a short distance to lie between it and the dorsal pharyngeal wall. Proliferation of remnants of its cranial portion may give rise to tumours known as chordomata.

Differentiation of the intra-embryonic mesoderm

It has already been noted that the **primary mesoderm**—derived either from the inner cell mass or directly from the trophoblast—is formed outside the embryonic area, and is therefore **extra-embryonic mesoderm**.

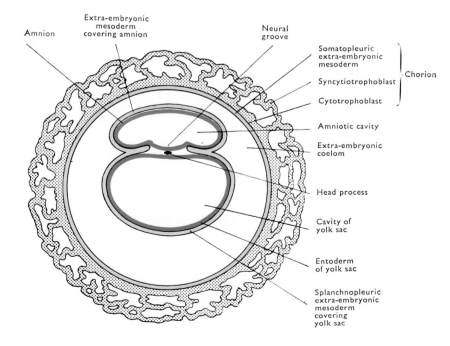

Amnion

Extra-embryonic
mesoderm
covering amnion

Neural
groove

Somatopleuric
extra-embryonic
mesoderm

Syncytiotrophoblast

Cytotrophoblast

Chorion

Amniotic cavity

Extra-embryonic
coelom

Head process

Cavity of
yolk sac

Entoderm
of yolk sac

Splanchnopleuric
extra-embryonic
mesoderm
covering
yolk sac

FIG. 2.50. Transverse section of zygote along line B–B in Fig. 2.48.

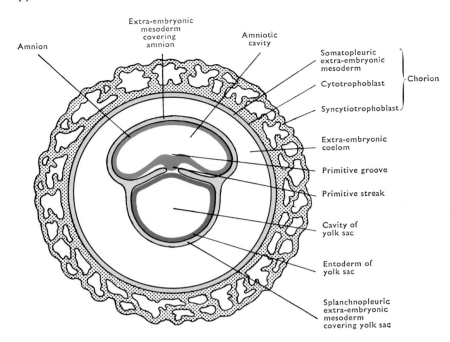

FIG. 2.51. Transverse section of zygote along line C–C in Fig. 2.48.

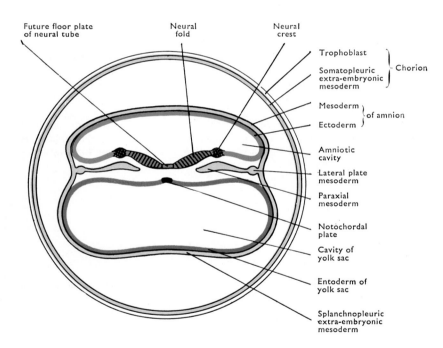

FIG. 2.52. Diagrammatic transverse section of zygote to show formation of neural folds and position of intra-embryonic mesoderm.

After the formation of the embryonic area, the ectodermal thickening which forms the primitive streak becomes the main source of the mesoderm. As this mesoderm appears after the primary mesoderm, and is formed in the embryonic area, it may be termed **secondary** or **intra-embryonic mesoderm**.

The deeper cells of the primitive streak [FIGS. 2.48 and 2.51], which are in contact with the entoderm, produce the intra-embryonic mesoderm, which passes into the interval between the ectoderm and the entoderm [FIG. 2.51]. The superficial cells of the primitive streak remain as part of the surface ectoderm of the embryo.

The intra-embryonic mesoderm soon forms a continuous sheet of cells which spreads laterally and headwards in the embryonic area on each side of the median plane. Each of these lateral sheets is

thickest where it abuts against the notochord and the wall of the neural groove, and thinnest at its peripheral margin, where it is continuous with the extra-embryonic mesoderm [FIG. 2.52].

At the cephalic end of the embryonic area the medial margins of the mesodermal sheets pass around the prochordal plate and fuse together across the median plane cephalic to it, forming a transverse bar called the **pericardial (cardiogenic) mesoderm** [FIG. 2.58A] because the pericardium is afterwards developed in it. The area in which this mesoderm lies is named the **pericardial (protocardiac; cardiogenic) region** of the embryonic area [FIG. 2.48].

Between the prochordal plate and the cephalic end of the primitive streak the medial margins of the mesodermal plates are separated from each other by the notochord and the neural groove

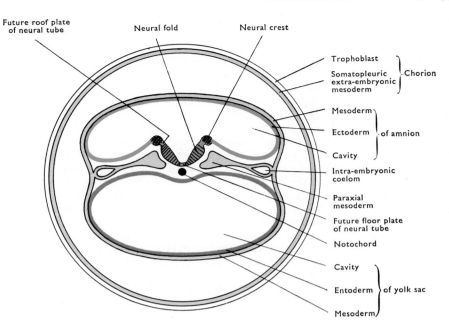

FIG. 2.53. Diagrammatic transverse section of zygote to show development of neural groove and intra-embryonic coelom.

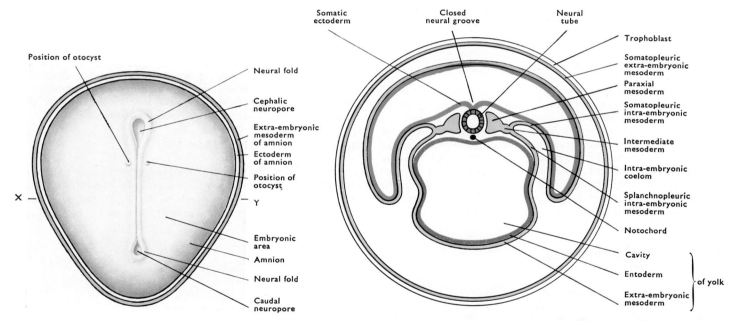

FIG. 2.54A. Diagram of embryonic area showing closure of neural groove with cephalic and caudal neuropores.

FIG. 2.54B. Corresponding transverse section through the plane X–Y in Fig. 2.54A. The intra-embryonic coelom has appeared and communicates with the extra-embryonic coelom.

[FIG. 2.53], and still more caudally they are continuous with the sides of the streak [FIG. 2.51].

INTRA-EMBRYONIC COELOM

After the intra-embryonic mesodermal sheets are definitely established, a series of cleft-like cavities appears in their peripheral parts. These cavities, on each side, soon run together, splitting the

mesoderm to form the rudiments of the intra-embryonic coelom [FIG. 2.53].

At the lateral border of the embryonic area on each side the mesoderm remains unsplit for a time, and separates the intra-embryonic from the extra-embryonic coelom. This septum soon disappears, and the two coelomic cavities become continuous [FIG. 2.54B].

The intra-embryonic coelom extends medially also, but the medial extension ceases at some distance from the median plane, except at the cephalic end of the embryonic area, where the two halves of the coelom become continuous with each other through the interior of the pericardial mesodermal bar [FIGS. 2.58A and 2.59]. In this region the septum of mesoderm separating the intra- and extra-embryonic parts of the coelom persists at the anterior margin of the embryonic disc as the **septum transversum**.

As the intra-embryonic coelom is forming and extending, a longitudinal constriction appears in each half of the mesoderm a short distance from its medial border. This constriction separates each plate into three parts: (1) a medial bar called the **paraxial mesoderm**, which lies at the side of the neural groove and the notochord; (2) the constricted portion, called the **intermediate cell mass** (intermediate mesoderm); (3) the part lateral to the constriction called the **lateral plate** [FIG. 2.54B].

The intra-embryonic coelom is confined, in the human embryo, to the lateral plate, which it divides into a superficial layer next the ectoderm, called the **somatic (somatopleuric) mesoderm**, and a deeper layer next the entoderm, called the **splanchnic (splanchnopleuric) mesoderm**. Ectoderm and somatic mesoderm together constitute the **somatopleure**; entoderm and splanchnic mesoderm together are known as the **splanchnopleure**.

The medial borders of the somatic and splanchnic mesoderm are continuous with each other round the medial border of the coelom. The lateral border of the somatic mesoderm is continuous, at the margin of the embryonic area, with the extra-embryonic mesoderm which covers the outer surface of the amnion; the lateral border of the splanchnic layer is continuous with the extra-embryonic mesoderm on the wall of the extra-embryonic portion of the entodermal vesicle or yolk sac.

PARAXIAL MESODERM

Each paraxial mesodermal bar soon assumes the shape of a prism lying lateral to the neural tube. The cephalic portion of each paraxial bar, as far as the middle of the hindbrain [p. 51], remains unsegmented, but the more caudal part is divided by a series of transverse clefts into a number of segments, called the **mesodermal somites** [FIGS. 2.55 and 2.61]. The first cleft appears in the region of the hindbrain, and the others are formed successively, each caudal to its predecessor. Only three or four somites lie in the region of the head. The segmentation of the paraxial bars begins before they have reached their full length, and somites continue to be separated off as

the paraxial bars are extended by addition to their caudal extremities from the cephalic end of the primitive streak. The total number of somites in the human embryo is thirty-eight or thirty-nine, not counting a few that develop in the tail and soon disappear.

When the somites are first defined they are solid masses of cells, but in a short time a cavity appears in each mass. This is called the **myocoele** because, in addition to other structures, muscles are derived from the somites.

The ventromedial portion of the hollow mesodermal somite is known as the **sclerotome** (Goodsir 1857), since it is responsible for the formation of 'hard' skeletal structures. The cells of the scleratogenous section of the somite undergo rapid proliferation and assume the character of embryonic connective tissue to which the special name of **mesenchyme** has been given. Some of these cells invade the myocoele while others migrate towards the notochord. Finally, the scleratogenous cells separate from the remainder of the somite, and, as they increase in number, they migrate around the sides of the notochord and neural tube (which has been formed in the meantime from the neural groove) and mingle with those of the opposite side and with those derived from adjoining cephalic and caudal somites [FIG. 2.56]. In this way a continuous sheath of mesoderm is formed around the neural tube and the notochord. It will differentiate into the vertebral column including its ligaments, and the dura mater of the spinal medulla.

The part of a mesodermal somite left after the separation of the sclerotome is called a **myotome**. Each myotome gives rise to a flat plate with incurved dorsal and ventral margins, known as a muscle plate because from these plates voluntary muscle fibres are derived [FIG. 2.56]. The outer portion of each of the myotomes develops into the cells of subcutaneous connective tissue and dermis; consequently it is spoken of as a cutis plate, or **dermatome**. The dermis of the skin is thought thus to arise segmentally, each segmental contribution derived from a dermatome corresponding to the area of skin supplied by one spinal nerve—also called a dermatome.

INTERMEDIATE CELL MASS

This continuous tract of cells, lateral to the paraxial mesoderm on each side, remains unsegmented; but as it gives rise to a series of excretory structures it corresponds to the segmented portions of mesoderm in lower forms known as **nephrotomes**. The intermediate

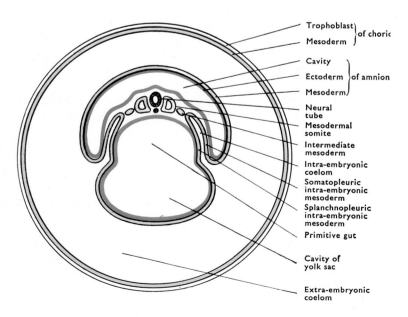

Trophoblast ⎫
Mesoderm ⎬ of chorion

Cavity ⎫
Ectoderm ⎬ of amnion
Mesoderm ⎭

Neural tube
Mesodermal somite
Intermediate mesoderm
Intra-embryonic coelom
Somatopleuric intra-embryonic mesoderm
Splanchnopleuric intra-embryonic mesoderm
Primitive gut
Cavity of yolk sac
Extra-embryonic coelom

FIG. 2.55. Diagrammatic transverse section of zygote shown in Fig. 2.61.

cell mass gives rise to the urogenital system [FIG. 8.66], with the exception of the germinal cells of the gonads and the lining of the urinary bladder, urethra, and prostate.

LATERAL PLATES

The cells of the lateral plates give origin to: (1) the lining mesothelial cells of the great serous cavities of the body—the pleurae, the pericardium, and the peritoneum; (2) the majority of the connective tissues (with the exception of those of the vertebral column and the head); (3) the greater part or all of the mesoderm of the limbs; and (4) the plain muscle fibres of the walls of the alimentary canal and the blood vessels. Most of these tissues are derived from mesenchymatous cells budded off from the lateral plates.

CEPHALIC MESODERM

It has already been noted that the mesoderm of the head becomes segmented only in the region of the caudal part of the hindbrain, where four cephalic mesodermal somites are formed on each side. From the scleratogenous portions of these somites the occipital part of the skull and the corresponding part of the dura mater of the brain are developed. Their muscle plates give rise to the intrinsic muscles of the tongue. Three myotomes develop in front of the otic capsule on each side and give rise to the extrinsic ocular muscles (musculi bulbi).

The unsegmented part of the cephalic mesoderm gives rise to the remaining muscles and connective tissues of the head region. It has been suggested that this mesoderm, which is concerned in the development of the pharyngeal arches, arises from the neural crest [p. 51].

MESENCHYME

In the early stages of development many cells, of irregular shape and wandering habits, appear as a loose mesh between the more definite layer of mesoderm and the adjacent ectoderm or entoderm. They are

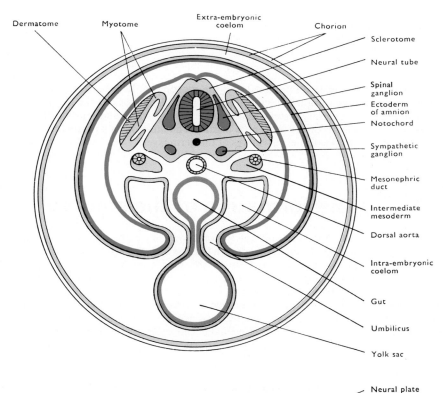

Dermatome   Myotome   Extra-embryonic coelom   Chorion
Sclerotome
Neural tube
Spinal ganglion
Ectoderm of amnion
Notochord
Sympathetic ganglion
Mesonephric duct
Intermediate mesoderm
Dorsal aorta
Intra-embryonic coelom
Gut
Umbilicus
Yolk sac

FIG. 2.56. Diagrammatic transverse section of a zygote showing formation of umbilicus and ventral extension of amnion.

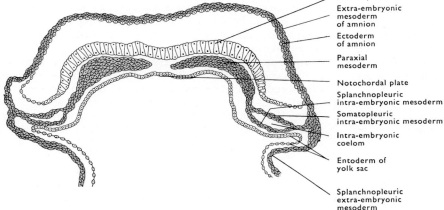

Neural plate
Extra-embryonic mesoderm of amnion
Ectoderm of amnion
Paraxial mesoderm
Notochordal plate
Splanchnopleuric intra-embryonic mesoderm
Somatopleuric intra-embryonic mesoderm
Intra-embryonic coelom
Entoderm of yolk sac
Splanchnopleuric extra-embryonic mesoderm

FIG. 2.57. Diagram of a transverse section through the M'Intyre embryo (Bryce 1924).

Derivatives of germ layers

Ectoderm	Mesoderm	Entoderm
Epidermis of skin; hair follicles and hairs; nails.	Connective tissues, including bone and cartilage.	Epithelium of alimentary tract (except ectodermal parts of mouth and anal canal); tongue and taste-buds.
Epithelium and myoepithelial cells of sebaceous, sweat, and mammary glands.	Synovial membranes; bursae and tendon sheaths.	Epithelium of glands of alimentary canal (except salivary): liver; gall-bladder; bile-passages; pancreas.
Epithelium of sense organs.	Serous membranes; pleura; pericardium; peritoneum; tunica vaginalis testis.	
Eye: retina (including ciliary and iridial parts); lens; conjunctiva; lacrimal gland and ducts.	Muscle—plain, striated, and cardiac (except plain muscle of iris and sweat glands).	Epithelium of pharyngeal organs: thyroid; parathyroids; thymus.
Ear: epithelium of membranous labyrinth and external acoustic meatus.	Endothelium of heart, blood vessels, and lymph vessels.	Epithelium of respiratory tract (except nasal cavity); nasal part of pharynx; auditory tube; tympanum; mastoid antrum; mastoid air cells.
Nose: epithelial lining and glands of nasal cavity and paranasal sinuses, including olfactory area and vomeronasal organ.	Blood and bone marrow.	
Nervous system (central and peripheral): nerve cells; nerve fibres; neuroglia (except microglia); ependyma; neurolemmal cells; leptomeninges (pia and arachnoid); pineal body; both lobes of hypophysis cerebri; medulla of suprarenal glands; chromaffin tissues.	Lymph tissue: tonsils; lymph nodes.	Epithelium of urinary bladder (except trigone); urethra (except terminal part in male); prostate; bulbo-urethral glands; vagina (part).
	Spleen.	
	Coats and contents of eye (except ectodermal parts specified).	
Epithelium of lips, cheeks, gums, and hard palate (including glands); salivary glands.	Dura mater and microglia.	
Enamel of teeth.	Teeth (except enamel).	
Epithelium of lower one-third of anal canal; terminal part of male urethra; vestibule of vagina and vestibular glands.	Cortex of suprarenal glands.	
Plain muscle of iris.	Urogenital organs (except epithelium of most of bladder, of prostate, urethra, and part of vagina).	

called mesenchyme cells. The scleratogenous cells that wander out from the somites are of this nature; the remainder of the mesenchyme is derived largely from the lateral plates of mesoderm, though it may have other sources, possibly even from ectoderm and entoderm.

The complete role of the mesenchyme in development has not yet been elucidated, but in addition to connective tissues in general, it forms the tissues of the vascular and lymphatic systems, the reticulo-endothelial system in general, and plain muscle fibres in the walls of the alimentary canal and elsewhere. It appears to be concerned with the development of striated muscles in the limbs, which have not been proved to be directly derived from myotomes.

GERM LAYERS AND THEIR DERIVATIVES

The formation and development of the embryonic area have now been traced to a point at which the general relation of the three main cellular sheets or layers to each other can be understood [FIGS. 2.50–2.56]. The ectoderm, the entoderm, and the mesoderm are distinct sheets of cells in the early stage of development, each with its own part in the formation of the embryo and its organs; they are therefore known as the primary germ layers. It must be emphasized, however, that the germ layers are neither independent elements in the structural organization of the embryo, nor even completely specific. On the contrary, the interaction and co-operation of the layers are essential for normal development, the destiny of any particular group of cells being partly due to its position in the embryo. Thus one group may deputize for another of the same or a different germ layer under abnormal or experimental conditions at an early stage in development.

Before the description of the formation of the embryo and of the initial stages in the development of the principal organs, it is useful to summarize the parts played by the three germ layers. In general, it may be said that there is a functional distinction between them: thus the ectoderm, the original external covering of the body, has a protective function, and, since it is in contact with the external environment, it is also the source of the essential elements of the sense organs and of the whole of the nervous system. The entoderm, the lining of the alimentary and respiratory tracts, has digestive and

absorptive functions; and the mesoderm, between the other two, gives origin to connective and muscular tissues, is the basis of the circulatory system, and provides for excretion and reproduction by means of the urogenital organs. The table on this page gives some details of the derivatives of the germ layers.

The formation of the embryo

The transformation of the relatively flat embryonic area into the form of the embryo is due, in the first instance, to the rapid extension of the area compared with the slow growth of the immediately adjacent parts; and the later modelling of the various parts of the embryo is due to different rates of growth in different regions.

By the rapid proliferation of cells from the cephalic end of the primitive streak, the surface length of the embryonic area is increased, whilst its cephalic and caudal ends remain relatively fixed; consequently the area bulges dorsally into the amnion. At the same time, the cephalic end of the neural groove becomes convex dorsally, so that it appears to rise out of the embryonic area and projects beyond its cephalic border. As a result of this movement the buccopharyngeal and the pericardial areas become reversed in position, and a head fold is formed. This fold is bounded dorsally by what is now the head of the embryo, and ventrally by the reversed pericardial region [FIG. 2.58B].

The same rapid growth process not only produces a head fold, but at the same time it forces the rest of the primitive streak over the caudal end of the embryonic area, thus forming a tail fold.

As the head and tail folds are produced by the longitudinal growth flexing the embryonic area ventrally, transverse growth of the area results in the formation of right and left lateral folds [FIGS. 2.54B and 2.55], the embryo rising like a mushroom into the cavity of the amnion.

The portion of the secondary yolk sac which is enclosed within the embryo by this folding of the embryonic area is the primitive alimentary canal. The part which remains outside the embryo is the definitive yolk sac; and the connecting passage between the two is

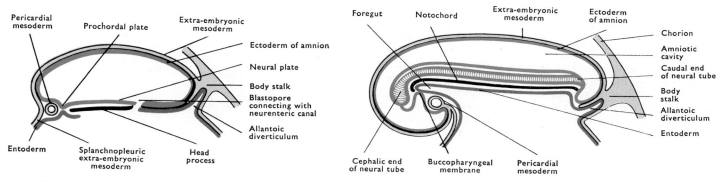

FIG. 2.58A. Diagram of longitudinal median section of embryonic area before folding.

FIG. 2.58B. Diagram of longitudinal median section of embryo after folding has commenced.

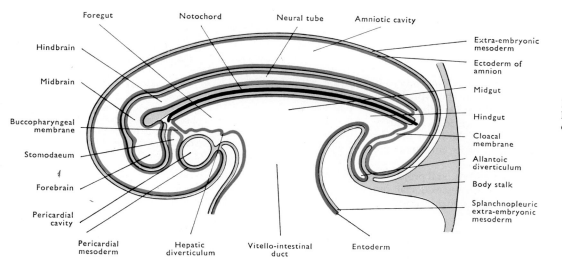

FIG. 2.59. Diagrammatic longitudinal section of embryo during folding.

the vitello-intestinal duct [FIG. 2.63].

That portion of the primitive alimentary canal which lies in the head fold is termed the **foregut**, the part in the tail fold is the **hindgut**, and the middle portion, which at first is in open communication with the yolk sac, is the **midgut**.

As the expansion of the embryonic area and its folding proceed, the margin of the area, which remains relatively stationary, becomes the edge of an orifice on the ventral aspect of the embryo (the **umbilical orifice**) through which the primitive alimentary canal of the embryo communicates with the yolk sac, the intra-embryonic coelom is continuous with the extra-embryonic coelom, and the component structures of the body stalk enter or leave the embryo.

At first the umbilical orifice is relatively large, but as the embryo rapidly extends in all directions the orifice soon becomes relatively small. Ultimately the margins of the orifice fuse together, closing the opening and forming the area of attachment of the umbilical cord which, after birth, becomes a scar on the ventral abdominal wall known as the **umbilicus** or navel.

The embryo

While the embryonic area is folding into the form of the embryo, the neural groove on its ectodermal surface is being converted into the neural tube. When the closure of the neural tube is completed (third week) the elongating embryo has a larger cephalic end and a smaller

caudal end, and is attached by the body stalk to the chorion. Its dorsal surface is continuous and unbroken, but its ventral surface is separated into cephalic and caudal portions by the umbilical orifice. The embryo contains three cavities: (1) the cavity of the neural tube; (2) the primitive alimentary canal [FIGS. 2.58B and 2.59; page 45]; (3) the intra-embryonic coelom.

The intra-embryonic coelom consists of a median pericardial portion, which lies ventral to the foregut between the bucco-pharyngeal membrane and the cranial margin of the umbilical orifice, and is continued dorsolaterally into right and left portions, which extend caudally at the sides of the foregut, the midgut, and the hindgut.

The right and left portions, in addition to communicating with each other through the pericardial coelom, ventral to the foregut, also communicate with the extra-embryonic coelom through an aperture between the midgut and the lateral and caudal margins of the umbilical orifice.

By this time the embryo has become easily distinguishable from the remainder of the zygote, and indications of its general plan of organization are discernible. As yet, it has no limbs, but the general contour of the head and body are defined. It possesses a primitive skeletal axis—the notochord. On the dorsal aspect of the notochord lies the neural tube, which is the rudiment of the brain and the spinal medulla (spinal cord).

At the sides of the neural tube and the notochord are the mesodermal somites and the ganglia of the cranial and spinal nerves [FIGS. 2.55 and 2.56].

Ventral to the notochord is the primitive alimentary canal closed at its cephalic and caudal ends by the **buccopharyngeal** and **cloacal** membranes, which separate the amniotic cavity from the foregut and the hindgut respectively [FIGS. 2.59 and 2.63]. The caudal part of the hindgut becomes the **entodermal cloaca** into which open the primary excretory (pronephric) ducts [p. 61].

At the sides of the primitive alimentary canal are the right and left parts of the intra-embryonic coelom, and between the dorsal angle of each half of the coelom and the mesodermal somites of the same side lies the intermediate cell mass mesoderm, which is the rudiment of the greater part of the urogenital system [FIGS. 2.55 and 2.56].

Ventral to the foregut is the mesoderm surrounding the pericardial cavity, the caudal part of which thickens to form the **septum transversum** [page 46 and FIG. 2.73]. This portion of mesoderm, attaches the caudal portion of the foregut to the ventral abdominal wall. Ventral to the hindgut is the cloacal membrane. Between the septum transversum cranially and the cloacal membrane caudally is the umbilical orifice [FIGS. 2.56 and 2.59].

THE LIMBS

For some time after its general form is well defined the embryo is entirely devoid of limbs [FIGS. 2.87–2.89]. During the fifth week a slight ridge appears along each side of the embryo. On this ridge the rudiments of the limbs—the **limb buds**— are formed as secondary elevations [FIG. 2.60]; the upper limb precedes the lower limb in time of appearance and in the development of its corresponding features [FIGS. 2.90–2.92].

Each limb bud early assumes a semilunar outline, and projecting at right angles from the surface of the body, possesses dorsal and ventral surfaces, and cephalic or preaxial, and caudal or postaxial borders.

As the limb rudiment increases in length the segments of the limb are differentiated; at the same time the limbs are folded ventrally, so that their original ventral surfaces become medial, and their original dorsal surfaces lateral, with the convexities of the elbow and knee directed laterally. At a later period, on account of a rotation which takes place in opposite directions in the two limbs, the convexity of the elbow is turned towards the caudal end of the body and that of the knee towards the head [FIG. 2.93].

The distal segment of each limb is, at first, a flat plate with a rounded margin, but it soon differentiates into a proximal or basal part and a more flattened marginal portion [FIG. 2.91]. It is along the line where these two parts are continuous that the rudiments of the **digits** appear as small elevations on the dorsal surface of the limb bud about the sixth week; the ridges extend peripherally, and by the seventh week the fingers project beyond the margins of the hand segment [FIG. 2.92], but the toes do not attain a corresponding stage of development until the early part of the eighth week [FIG. 2.93]. The nails are later developments.

Each limb bud is essentially an extension from a definite number of segments of the body. It consists, at first, of a core of mesenchyme covered with ectoderm. As it grows, the ventral rami of the spinal nerves of the corresponding segments are prolonged into it, together with branches of a number of intersegmental blood vessels. The nerves remain as the nerves of the fully developed limb, but the blood vessels are reduced in number and are modified to form the permanent main vascular trunks.

The mesenchymatous core of the primitive limb rudiment is derived mainly from somatic mesoderm of the lateral plate. As development proceeds it is differentiated into the cartilaginous and other connective tissue elements which are the rudiments of the skeletal framework and the fasciae of the fully formed limb.

The rudiments of the muscles of the limbs appear in the mesenchyme as pre-muscle masses which have no direct connection

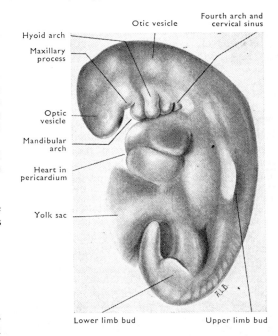

FIG. 2.60. Human embryo, 5.3 mm long, to show the development of the limb buds. The rudiment of the upper limb, more advanced than that of the lower limb, appears as a longitudinal ridge on the side of the embryo. (After G. L. Streeter, *Contrib. Embryol. Carneg. Instn*, 1945; Embryo no. 8066.)

with the myotomes of the segments to which the limb buds are related. Nevertheless, the limb muscles become innervated by the nerves of the segments to which they belong.

OUTLINE OF DEVELOPMENT OF NERVOUS SYSTEM

EARLY STAGES

No definite trace of the nervous system is present until the primitive streak has appeared and the embryonic area has passed from a circular to a pear-shaped form. An area of thickened ectoderm, called the **neural plate**, then appears in the longitudinal axis of the embryonic area dorsal to the notochord. It begins to differentiate just caudal to the buccopharyngeal membrane, and its caudal extremity embraces the nodal end of the primitive streak. Its margins fade into the surrounding ectoderm; but, while the plate lengthens with the elongation of the embryonic area, its margins are elevated by the thickening paraxial mesoderm beneath them, and so they become distinct.

As its margins are raised the plate is necessarily folded longitudinally, and the sulcus is called the **neural groove** [FIGS. 2.50 and 2.54A]. Each half of the neural plate, as it is raised to form a lateral wall for the groove, is then named a **neural fold**. After the neural groove is thus defined, the neural folds approach each other until they meet and their dorsal margins fuse in the median plane, so converting the neural plate into the **neural tube**. The ventral wall of the tube, formed by the central part of the original neural plate, is called the **floor plate**: the dorsal wall, formed where the neural folds unite, is called the **roof plate**, and it is soon separated from the surface ectoderm, which is closed over and completed dorsal to the neural tube by this fusion.

The fusion of the margins of the neural folds to form the roof plate begins in the cervical region, and from there it extends headwards and tailwards. The last parts of the roof plate to be formed are, therefore, its cephalic and its caudal extremities; consequently, for a time, the neural canal, which is the cavity of the tube, opens on the surface at the two ends; the openings are called the cephalic and the caudal neuropores [FIG. 2.54A]. Eventually, however, about the third week of embryonic life, both apertures are closed and, for a time, the neural canal becomes a completely closed cavity.

Failure of union of the neural folds to form a closed neural canal (myeloschisis or rachischisis) may occur at any point but is commonest at the two ends, and it is often associated with malformations of the skull (anencephaly) or vertebral column (spina bifida). The union of the neural folds is an example of the process of 'embryonic healing' by which growing edges come together and unite, either to form a continuous structure—as in the case of the face and palate [p. 52]—or to enclose a cavity—as in the case of the neural canal, or the lens of the eye [FIG. 12.33], or the otic vesicle [FIG. 12.62].

As the neural folds rise and approach each other their dorsal margins give origin on each side to a column of cells which comes to lie in the angle between the neural tube and the surface ectoderm. This column is called the neural crest [FIGS. 2.52 and 2.53].

The neural crest is the rudiment of the cranial and spinal nerve ganglia, the sympathetic ganglia, the cells of chromaffin tissue, the medulla of the suprarenal glands and the sheath (Schwann) cells of the peripheral nerves. Melanoblasts and even some of the mesoderm of pharyngeal arches (Horstadius 1950) are said to develop from the neural crest.

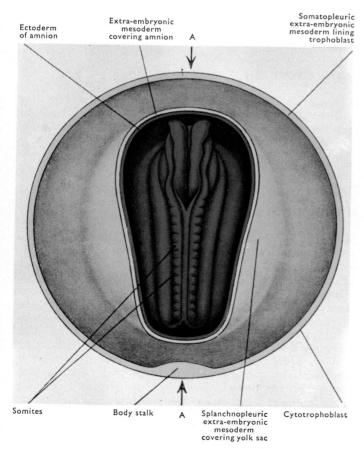

FIG. 2.61. Diagram of dorsal aspect of embryo, showing partial closure of the neural groove.

The neural tube becomes transformed into the brain and spinal medulla, and the brain portion gives origin also on each side to the essential part of the eye—the retina—and the optic nerve.

DIFFERENTIATION OF THE NEURAL TUBE

Before the neural plate is converted into a closed tube, an expansion of its cephalic part indicates the distinction between brain and spinal medulla.

While the cerebral portion is still unclosed, three secondary dilatations of its walls indicate its separation into three vesicles—the primitive forebrain, the midbrain, and the hindbrain [FIGS. 2.59 and 2.71], which are continuous with each other.

Shortly after the three sections of the brain are defined, and before it becomes a closed tube, a vesicular evagination forms at the cephalic end of each side-wall of the primitive forebrain. These evaginations are the primary optic vesicles, and they are the rudiments of the optic nerves, the retinae, and the posterior epithelium of the ciliary body and of the iris. As the optic vesicle approaches the surface of the head, the lens of the eye develops from the overlying ectoderm [FIG. 12.33], which thickens and forms a depression which is finally cut off from the surface by the union of its edges. The lens vesicle, thus produced, is received into a depression in the optic vesicle which deepens until the vesicle is transformed into the optic cup.

After the three brain vesicles are formed, diverticula grow out from the cephalic end of the primitive forebrain; these are the right and left cerebral vesicles, which are the rudiments of the cerebral hemispheres of the adult brain.

After their formation the cerebral vesicles expand rapidly in all directions. They soon overlap the primitive forebrain and the midbrain, and eventually the hindbrain also, and each gives off from the cephalic end of its ventral wall an olfactory diverticulum, which becomes converted, later, into the olfactory bulb and olfactory tract.

FLEXURES OF BRAIN AND HEAD   [FIGS. 2.60, 2.62, and 2.72]

In conformity with the general curvature of the embryo, but due mainly to the rapid, unequal growth of the three brain vesicles, the head region of the embryo bends ventrally as the brain develops. At first the curvature is continuous, but very soon it is accentuated at the midbrain to form the cephalic flexure. This makes the small midbrain the most prominent part of the head, and causes the forebrain and the hindbrain to approach each other until their ventral walls lie almost parallel.

At the same time a cervical flexure develops at the junction of the hindbrain with the spinal portion of the neural tube; it affects the whole head and keeps the facial region in close relation with the pericardium.

At a later stage the exuberant growth of the hindbrain causes that vesicle to bend in a direction opposite to the other flexures, so that its ventral wall forms a prominent convexity. This bend affects the brain only, and it is called the pontine flexure as it occurs in the region where the pons is formed.

The cervical flexure eventually disappears as the neck is formed; but the cephalic and pontine flexures have a permanent influence on the form of the brain.

CAVITIES AND MENINGES OF NEURAL TUBE

The cavity of the spinal portion of the neural tube becomes the central canal of the spinal medulla. The cavities of the primitive

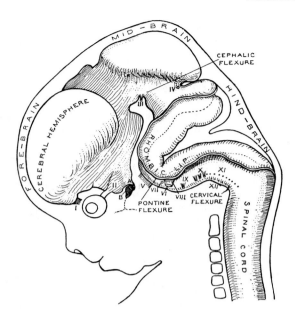

FIG. 2.62. Head of human embryo of 8 weeks to show the primary parts of the brain and the cephalic, cervical, and pontine flexures. (His.)

A and B, Cerebral and buccal rudiments of hypophysis cerebri. The cranial nerves are indicated by numerals.

brain vesicles are transformed into the ventricles, the interventricular foramina, and the **aqueduct of the cerebrum** in the midbrain. All these cavities are filled with **cerebrospinal fluid**, secreted by the **choroid plexuses** of blood vessels which invaginate the persistently thin parts of the walls of the ventricles at certain places.

The cavities of the developing cerebral hemispheres become the right and left **lateral ventricles** of the brain. The cavity of the central portion of the forebrain becomes the **third ventricle**, and the region where the third and lateral ventricles communicate is the **interventricular foramen**.

The cavity of the hindbrain vesicle becomes the **fourth ventricle**, and is connected to the third ventricle by the aqueduct of the cerebrum which traverses the midbrain.

After the neuropores [p. 51] are closed, the cavity of the neural tube is, for a time, a completely enclosed space. Subsequently three membranous sheaths or **meninges** are developed around the tube; they are the **pia mater**, the **arachnoid**, and the **dura mater**. The dura mater, which is the outermost, is derived from the sclerotomes; but the pia mater, which is the innermost, and the intervening arachnoid are possibly formed in part by ectodermal cells derived from the neural crest.

As the meninges are differentiated, narrow **subdural** and more extensive **subarachnoid** spaces are formed between them. After a time, a **median** and two **lateral apertures** appear in the dorsal wall of the fourth ventricle and in the overlying pia mater. These connect the fourth ventricle with the subarachnoid space and permit the passage of cerebrospinal fluid between them.

EARLY DEVELOPMENT OF ALIMENTARY CANAL AND FORMATION OF STOMODAEUM AND PROCTODAEUM

The greater part of the permanent alimentary canal is derived from the entodermal vesicle and is therefore lined with entodermal cells.

The foregut and hindgut parts are entirely enclosed in the embryo, while the midgut remains in continuity with the extra-embryonic part of the yolk sac [FIGS. 2.59 and 2.67].

The cephalic extremity of the alimentary canal is formed from a depression of the ectodermal surface, the **stomodaeum**, which lies, at first, between the ventrally bent head and the bulging pericardial region [FIGS. 2.59 and 2.64]. When the stomodaeum first appears it is separated from the entodermal alimentary canal by the buccopharyngeal membrane, but when this disappears, during the fourth week, the stomodaeum communicates with the foregut. Later, it is separated into nasal and oral portions by the formation of the palate, and the oral portion forms that part of the mouth in which the gums and the teeth are developed.

The terminal portion of the permanent alimentary canal is formed by a pit-like, ectodermal hollow, the **proctodaeum** or **anal pit** [FIG. 2.74], which is separated from the blind end of the hindgut until about the eighth week by the **anal membrane**. This is the dorsal portion of a more extensive cloacal membrane [p. 61].

DERIVATIVES OF THE STOMODAEUM

When the stomodaeum is definitely established, it is bounded cranially by the frontal end of the head, caudally by the conjoined ends of the **mandibular arches** [p. 55], and laterally by the **maxillary processes**, which grow ventrally from the dorsal parts of the mandibular arches. The space is closed dorsally by the buccopharyngeal membrane until that membrane disappears [FIG. 2.63].

In the roof of the stomodaeum, anterior to the attachment of the buccopharyngeal membrane, there is a depression which is deepened by the growth of mesoderm around it so that it appears as a diverticulum of the roof. It is lined with ectoderm and is known as Rathke's pouch.

RATHKE'S POUCH

The blind end of this pouch is in contact with the floor of the third ventricle of the brain where the hypophysial diverticulum arises and forms the posterior lobe of the hypophysis. Rathke's pouch forms the anterior lobe of the **hypophysis** or **pituitary gland** [FIGS. 2.72–2.74].

Development of face and separation of nose and mouth

The frontal end of the head lies in the cephalic boundary of the stomodaeum and is called the **frontonasal process**. The upper part of the frontonasal process remains undivided and represents the region of the future forehead; but a pair of shallow depressions—the **olfactory pits**—appear in the lower part. Each pit therefore becomes bounded by a **lateral nasal process** and a **medial nasal process** [FIG. 2.64A]. As the margins of these processes increase in height the olfactory pits deepen.

The lateral boundary of the stomodaeum is at first formed by the maxillary process which springs from the dorsal part of the mandibular arch and grows forwards to approach the lateral nasal process. Between these two processes lie the projecting eye and a groove leading downwards and forwards from it to the olfactory pit—the **nasolacrimal sulcus** [FIG. 2.64A].

Each maxillary process grows ventrally to fuse with the lateral nasal process along the line of the nasolacrimal sulcus, burying the surface epithelium of the sulcus which thus forms a cord of cells between the eye and the olfactory pit; this later canalizes to become the **nasolacrimal duct**. Continuing its growth medially, each maxillary process fuses with and overgrows the corresponding

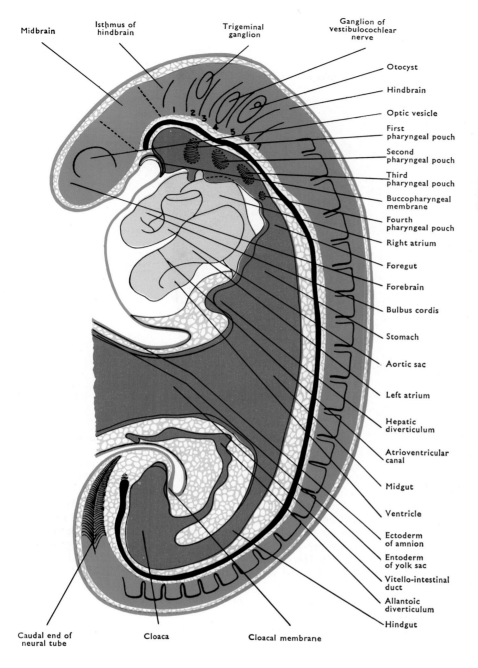

Midbrain

Isthmus of hindbrain

Trigeminal ganglion

Ganglion of vestibulocochlear nerve

Otocyst

Hindbrain

Optic vesicle

First pharyngeal pouch

Second pharyngeal pouch

Third pharyngeal pouch

Buccopharyngeal membrane

Fourth pharyngeal pouch

Right atrium

Foregut

Forebrain

Bulbus cordis

Stomach

Aortic sac

Left atrium

Hepatic diverticulum

Atrioventricular canal

Midgut

Ventricle

Ectoderm of amnion

Entoderm of yolk sac

Vitello-intestinal duct

Allantoic diverticulum

Hindgut

Caudal end of neural tube

Cloaca

Cloacal membrane

FIG. 2.63. Diagrammatic reconstruction of 2.5-mm human embryo (Thompson 1907) with twenty-three pairs of somites. 1–7: neuromeres.

medial nasal process to reach the opposite maxillary process in the midline. The olfactory pits are thus completely separated, for a time, from the stomodaeum, and they lie above the newly constituted ledge which now forms its superior boundary. This ledge consists on each side of the maxillary process fused with the frontonasal process near the midline, and excludes the lateral nasal process from the margin of the stomodaeum [FIG. 2.64B].

Thus, the upper parts of the cheeks are formed by the maxillary processes and the **upper lip** by their fusion over the lower part of the medial nasal processes. At a later stage, the **maxillae**—the bones of the upper jaw—are formed in the mesodermal core of these united processes. The **premaxillae** ossify separately and unite with the maxillae, but do not form separate bones on the facial surface [FIG. 3.49]. The lower parts of the cheeks and the lower lip are derived from the mandibular arch. The **aperture of the mouth**, which at

first is wide, is gradually reduced by the progressive fusion of the maxillary processes with the mandibular arch.

The **external nose** is developed from the nasal processes; the lateral nasal process forms the lateral part of the external nose, including the ala, and the prominence of the nose rises gradually from the root of the frontonasal process where it blends with the forehead. The external openings of the olfactory pits become the **nostrils**.

The olfactory pits at first are blind and are separated from the cavity of the stomodaeum by thin **bucconasal membranes**; but these soon disappear, and the olfactory pits then communicate directly with the cavity of the stomodaeum through openings which are called the **primitive posterior apertures (choanae) of the nose**. The olfactory pits now form the primitive nasal cavities; they extend upwards and backwards towards the roof of the main cavity of the

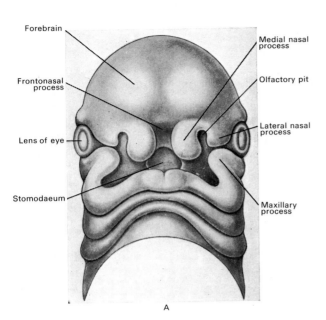

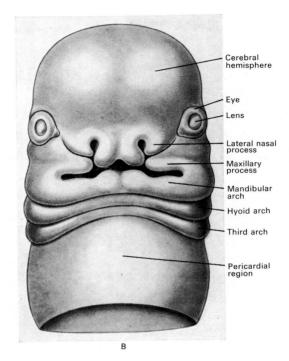

FIG. 2.64. Two stages in development of the face.

A, Boundaries of stomodaeum before completion of primitive upper lip.

B, Completion of upper lip. Note the union of maxillary process with lateral and medial nasal processes to form cheek and upper lip.

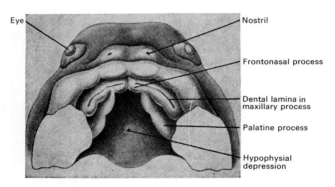

FIG. 2.65. Portion of the head of a human embryo about 8 weeks old (His) The lips are separated from the gums, and the line of the dental lamina is visible in the gums. The palatine processes are growing inwards from the maxillary processes.

stomodaeum on each side. Mesoderm growing dorsally in the dorsolateral walls of the stomodaeum (future nasal cavities) from each maxillary process forms the roof of this region after meeting its fellow; this roof mesoderm then grows ventrally in the midline to form the posterior part of the **nasal septum**.

After the formation of the primitive choanae of the nose, a horizontal ledge grows from the medial surface of each maxillary process towards the median plane. These ledges—the **palatine processes**—are separated for a while by the developing tongue, which bulges superiorly between them, but they meet and fuse with each other and with the nasal septum during the third month of intra-uterine life, the fusion beginning anteriorly and being completed posteriorly in the region of the uvula. In this way the palate is formed, and bone develops in the anterior portion of it to form the **hard palate**; the posterior portion becomes the **soft palate**. The original stomodaeum is thus separated into definitive nasal and oral cavities. The **nasal cavity** is divided into lateral halves by the

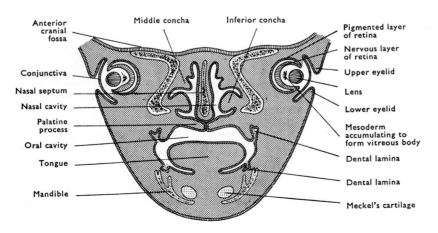

FIG. 2.66. Diagrammatic coronal section through face of human embryo about 9 weeks old.

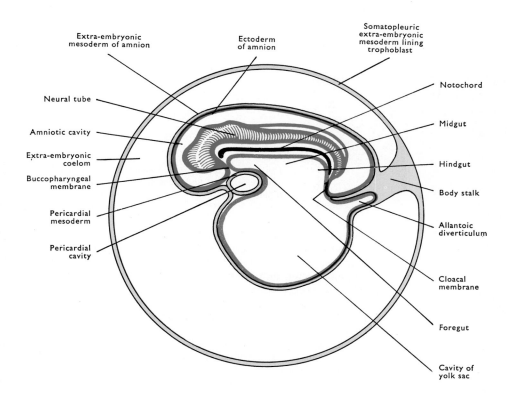

Extra-embryonic mesoderm of amnion

Ectoderm of amnion

Somatopleuric extra-embryonic mesoderm lining trophoblast

Neural tube

Amniotic cavity

Extra-embryonic coelom

Buccopharyngeal membrane

Pericardial mesoderm

Pericardial cavity

Notochord

Midgut

Hindgut

Body stalk

Allantoic diverticulum

Cloacal membrane

Foregut

Cavity of yolk sac

FIG. 2.67. Median longitudinal section through plane A–A of the embryo shown in Fig. 2.61.

nasal septum [FIG. 2.66]. The **oral cavity** is continuous with the ventral part of the primitive pharynx. It forms the cavity and the vestibule of the mouth and from its walls the gums and teeth are developed [FIG. 2.65]. These serial changes have been illustrated by Kraus, Kitamura, and Latham (1966).

Failure of union of the various elements from which the face and palate are formed causes a variety of malformations. The commonest are **hare lip** and **cleft palate**, which often occur together. Hare lip is due to failure of union of the maxillary and medial nasal processes. It may occur on one or both sides and rarely in the midline due to failure of fusion of the maxillary processes in the formation of the upper lip. When it is bilateral, the premaxillary part of the upper jaw is isolated.

Differentiation of foregut

DERIVATIVES OF THE SIDE-WALL

While still separated from the stomodaeum by the buccopharyngeal membrane the cephalic part of the foregut, situated dorsal to the pericardium [FIG. 2.63], dilates to form the primitive pharynx, the walls of which develop alternating ridges and depressions by proliferation of bars of mesoderm. In each lateral wall there are five such depressions, the **pharyngeal pouches**. In the ventral wall there are two: the first is the rudiment of the thyroid gland, and the other, situated more caudally, is the origin of the respiratory system and provides the epithelial lining of larynx, trachea, bronchi, and lungs.

Simultaneously with the formation of the pharyngeal pouches internally a series of grooves appears externally. They correspond in position with the first four pharyngeal pouches, and they are called the **external pharyngeal grooves** [FIGS. 2.60 and 2.89]. There is no external groove corresponding to the fifth pouch, which has a common opening into the pharynx with the fourth [FIGS. 2.68 and 2.70].

The pharyngeal pouches and the external grooves divide each side-wall of the primitive pharynx into a series of mesodermal bars called the **pharyngeal (branchial) arches**. The bars are six in number, but the fifth is rudimentary and the sixth can only be seen by virtue of the structures it contains.

The first pharyngeal bar is the rudiment of the maxillary and mandibular processes; it is called the **mandibular arch**. The second is the **hyoid arch**, and between them lie the first pharyngeal groove and pouch. The others are numbered as the third to sixth arches. Each pharyngeal arch consists at first of simple mesoderm—covered externally by ectoderm, internally by entoderm—but a series of similar structures soon develop in them. Each arch then contains a cartilaginous skeleton, a muscle rudiment, two nerves (main and subsidiary), and an artery. The cartilage of the first arch (Meckel's cartilage) forms a basis for the **mandible** which ossifies intramembranously on its external surface [FIG. 3.66]; those of the second and third give rise to the **hyoid bone** [FIG. 3.68]; the fourth to sixth arches belong to the larynx. The arteries are the **aortic arches** which connect the ventral and dorsal aortae [FIG. 2.78].

When the arches first appear, they extend from the level of the dorsal wall of the foregut to the pericardium. As the neck develops, the mandible is separated from the pericardium, and the ventral ends of the other arches are lifted away from it and exposed on the surface, though the mandibular arch is the only one that meets its fellow of the opposite side externally. The growth of the mandibular and the hyoid arches soon greatly exceeds that of the other arches, which gradually recede from the surface until, on each side, they lie at the bottom of a depression—the **cervical sinus**—overlapped by the caudal border of the hyoid arch [FIGS. 2.68 and 2.90]. The increasing growth of the hyoid arch reduces the opening of the cervical sinus to a narrow channel called the **cervical duct**, but this is soon obliterated, and then the sinus becomes the **cervical vesicle**. The cervical vesicle, lying at the side of the third pharyngeal groove, is associated with the second and fourth grooves by narrow canals called the 'branchial' ducts, which are the remains of the external

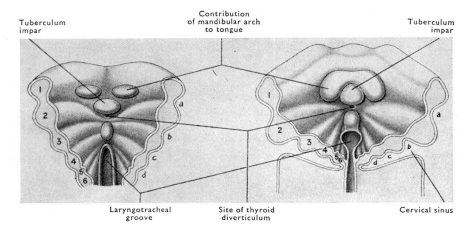

Tuberculum impar

Contribution of mandibular arch to tongue

Tuberculum impar

Laryngotracheal groove

Site of thyroid diverticulum

Cervical sinus

FIG. 2.68. Diagrams showing formation and closure of cervical sinus, and the development of the tongue. Numbers 1–6: first to sixth pharyngeal arches; a–d: first to fourth pharyngeal grooves.

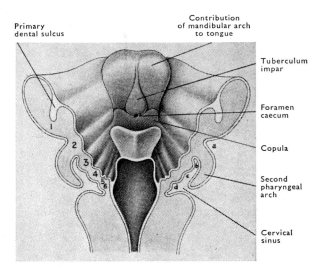

Primary dental sulcus

Contribution of mandibular arch to tongue

Tuberculum impar

Foramen caecum

Copula

Second pharyngeal arch

Cervical sinus

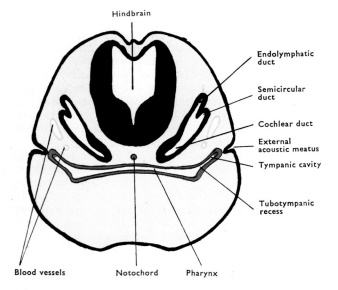

Hindbrain

Endolymphatic duct

Semicircular duct

Cochlear duct

External acoustic meatus

Tympanic cavity

Tubotympanic recess

Blood vessels   Notochord   Pharynx

FIG. 2.69. Diagram of transverse section through the head of an embryo. Showing the rudiments of the three parts of the ear and their relation to the tubotympanic recess and the first pharyngeal groove.

pharyngeal grooves or 'branchial clefts'. Ultimately this submerged space (lined with ectoderm) is obliterated; but it may on occasion give rise to a **branchial cyst**, or to a **branchial sinus** (if the communication with the exterior remains) or to a **branchial fistula** (if a separating membrane [see below] also breaks down).

The portion of the wall of the primitive pharynx which lies between any two adjacent arches and separates the external groove from the internal pouch is called the **separating membrane**. In the earliest stages it consists of ectoderm and entoderm separated by mesenchyme; then the mesenchyme disappears so that, for a time, the membranes consist of ectoderm and entoderm only. At a later period the ectoderm and entoderm are again separated by ingrowing mesoderm.

In vertebrates with gills the membranes break down so that **gill-clefts** are formed; but, except as a malformation, the separating membranes are not usually perforated in mammalian development, so that a complete 'cleft' rarely exists.

The first external pharyngeal groove is the only one that is not submerged in the cervical sinus [FIG. 2.68]; it is the site of the formation of the **external acoustic meatus**, and a series of tubercles which appear at its margins develop into the **auricle** of the external ear. In man, as in other mammals, the organ of hearing consists of: the **internal ear** or labyrinth; the **middle ear** or tympanum, which is connected to the pharynx by the **auditory tube**; and the external ear, which includes the external acoustic meatus and the auricle. The **tympanic cavity** and the **auditory tube** are developed as a lateral extension of the upper part of the cavity of the primitive pharynx, between the first and third arches. The anterolateral wall is formed by the first arch, the posterior wall and floor by the second arch, and the medial wall by the third arch. The dorsal parts of the first and second pouches form the cavity—the **tubotympanic recess** (Frazer 1914). A part of the cavity of the second pharyngeal pouch may be represented in the adult by the **intratonsillar cleft**, which passes into the upper part of the tonsil in the side-wall of the pharynx [FIG. 6.33] The **tympanic membrane**, which separates the tympanic cavity from the external acoustic meatus, is formed in the position of a 'separating membrane' between the tubotympanic recess and the first external groove [FIGS. 2.69 and 2.70]. But it is a secondary formation, since the deeper part of the meatus is first formed by a solid plug of ectodermal cells which later breaks down in its centre to form the meatus.

The development of the **auditory ossicles** is intimately associated with the development of the primitive pharynx, since they develop from the cartilages of the first and second pharyngeal arches. The organ of hearing and balance, the internal ear, has already developed dorsomedial to the tubotympanic recess from the ectoderm of the side of the head in close relation to the hindbrain and the vestibulocochlear (eighth cranial) nerve [FIGS. 2.63 and 2.71]. This **otic vesicle**, formed by thickening and invagination of the ectoderm,

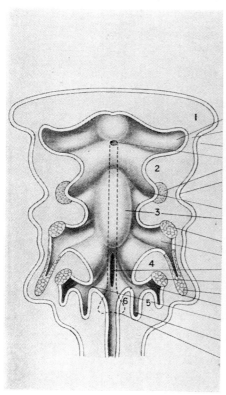

First pharyngeal pouch
First separating membrane

First pharyngeal groove
Site of thyroid diverticulum
Tonsil
Second pharyngeal groove

Second pharyngeal pouch
Hypobranchial eminence

Inferior parathyroid

Third pharyngeal pouch
Laryngotracheal groove
Superior parathyroid
Thymus
Fourth pharyngeal pouch

Duct of fourth and fifth pouches

Fifth pharyngeal pouch

Thyroid gland

FIG. 2.70. Diagram showing the floor and lateral walls of the embryonic pharynx and their derivatives.

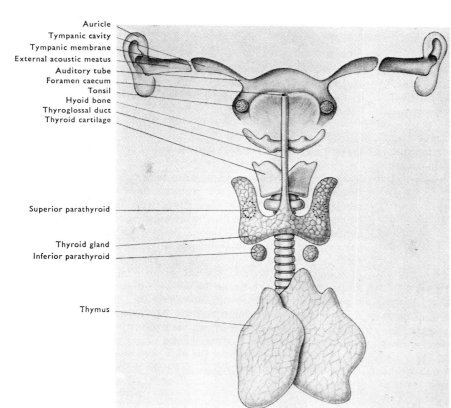

Auricle
Tympanic cavity
Tympanic membrane
External acoustic meatus
Auditory tube
Foramen caecum
Tonsil
Hyoid bone
Thyroglossal duct
Thyroid cartilage

Superior parathyroid

Thyroid gland
Inferior parathyroid

Thymus

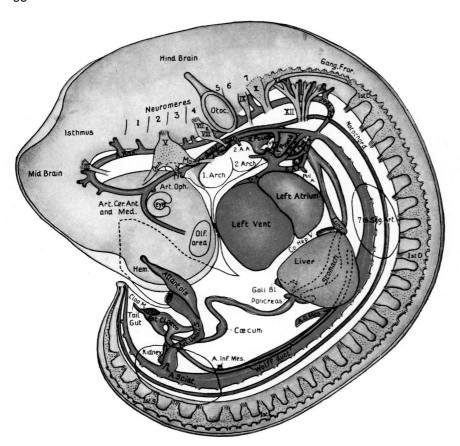

FIG. 2.71. Reconstruction of human embryo, 7 mm long.

1st D, First thoracic spinal ganglion: *Co. Hep. V.*, common hepatic vein; *Gang. Fror.*, Froriep's ganglion (XII); *Hem.*, cerebral hemisphere; *Otoc.*, otic vesicle; *Wolff. duct*, mesonephric duct. (Peter Thompson, *Studies in Anatomy*, University of Birmingham, 1915.)

is a small epithelial sac at the side of the hindbrain, which is converted into the membranous labyrinth. This includes the utricle, the saccule, and the three semicircular ducts, all concerned with balance, and the duct of the cochlea which contains the essential organ of hearing.

The third pharyngeal pouch is a direct extension of the cavity of the foregut, but as the pharyngeal wall thickens the communication is drawn out into a duct-like passage. The fourth and fifth pouches similarly communicate through a common recess with the cavity of the primitive pharynx [FIG. 2.70].

The cavities of the third, fourth, and fifth pouches ultimately disappear. Before they disappear a diverticulum, at first hollow but afterwards solid, grows from the ventrolateral wall of each, and a solid epithelial body buds from the dorsolateral wall of the third and fourth pouches [FIG. 2.70].

ORGANS DERIVED FROM PHARYNGEAL POUCHES

The diverticula and the epithelial bodies of the third and fourth pouches lose their attachment to the pharyngeal wall and give rise to a series of important organs [FIG. 2.70]. The diverticulum from the third pouch is concerned with the development of the thymus; and the epithelial bodies of the third and fourth pouches become the parathyroid glands. Changes in the situation of these organs occur as the head grows away from the pericardium and the neck is formed; thus the thymus, which maintains its relation to the pericardium, extends into the thorax, and the parathyroids alter their relative position, that derived from the third pouch being drawn down with the thymus so that it comes to lie inferior to the other and may occasionally become intrathoracic.

The diverticulum from the fifth pharyngeal pouch is transformed into the ultimobranchial body, which receives its name because it

is the last of the series of organs derived from the branchial region of the pharynx. It is associated with the corresponding lobe of the thyroid gland and contributes a specific type of cell to it [see thyroid gland].

DERIVATIVES OF VENTRAL WALL

The rudiment of the thyroid gland appears at first as a diverticulum from the ventral wall of the primitive foregut. It begins in the median plane and passes ventrally between the mandibular and hyoid arches. It grows caudally in the substance of the neck, passing ventral to the arch cartilages which develop into the hyoid bone and the cartilages of the larynx. When the caudal end of the diverticulum reaches the level of the origin of the trachea its epithelium proliferates and it becomes bilobed, thus expanding into the isthmus and the two lateral lobes of the permanent gland [FIG. 2.70]. The stalk of the diverticulum is the thyroglossal duct; the position of its original upper end is indicated by the foramen caecum of the tongue. The caudal end of the stalk sometimes persists and is transformed into the pyramidal lobe of the thyroid gland, and persistence of other parts of the duct may give rise, at any point between the gland and the tongue, to swellings known as thyroglossal cysts.

RESPIRATORY SYSTEM

A more caudally situated diverticulum from the ventral wall of the foregut is the rudiment of the respiratory system [FIGS. 2.68 and 2.72]. It first appears behind the developing tongue as a longitudinal groove bounded on each side by the sixth pharyngeal arches. The caudal end of this laryngotracheal groove soon dilates into a pouch, the **tracheobronchial diverticulum**. As the foregut elongates with the growth of the neck, the pouch and groove are separated from the more dorsal part of the foregut, which becomes the **oesophagus**. The tracheobronchial diverticulum elongates, but remains in communication with the pharynx at the permanent **inlet of the larynx**. As the groove elongates and is separated from the primitive foregut it is converted into a tube which is gradually differentiated into the **larynx** and the **trachea**. The pouch at the caudal end of the tube soon divides into right and left lobes, each of which is the rudiment of the epithelium lining the bronchi, bronchioles and alveoli of the lung of the corresponding side.

At the cranial margin of the laryngotracheal groove is a swelling, the **hypobranchial eminence** [FIG. 2.70], which develops into the epiglottis.

THE TONGUE

The mucous membrane of the tongue is formed by four separate rudiments which appear in the ventral wall of the primitive pharynx. Two of these are elevations formed on the ventral ends of the mandibular arches, the **lateral lingual swellings**. The third is a median elevation, called the **tuberculum impar**, which is situated on the caudal surface of the conjoined ventral ends of the mandibular arches. The fourth rudiment, the **copula of His**, formed by the conjoined ventral ends of the third pair of arches which overgrow the second arches, is separated from the tuberculum impar by the orifice of the thyroid rudiment [FIG. 2.68].

The lateral lingual swellings submerge the tuberculum impar and unite to form the greater part of the anterior two-thirds of the tongue—the part on which all the papillae are developed. The posterior third of the tongue, which lies in the ventral or anterior wall of the permanent pharynx, is formed by a V-shaped swelling which rises from the copula and, overgrowing the second arch, blends with the anterior two-thirds along the line of the **sulcus terminalis**. The foramen caecum is a depression left in the median plane at the junction of the two parts of the tongue, and it therefore indicates the site of origin of the thyroid gland.

ABDOMINAL PORTION OF THE FOREGUT

The portion of the foregut that has now been considered gives rise to the floor of the mouth (with the exception of the lips, teeth, and gums), the pharynx, the thyroid gland, the thymus, the parathyroid glands, the respiratory organs, and the oesophagus. The oesophagus extends through the part of the embryo that will become the thoracic region of the body, and the remainder of the foregut is situated in the upper part of the abdominal region. The abdominal portion of the foregut is differentiated into the **stomach**, the superior part and upper portion of the descending part of the **duodenum**, and from it the liver and pancreas take origin as outgrowths.

The gut tube is attached to the posterior wall by its dorsal mesentery (splanchnopleure), the part attached to the stomach being known as the **dorsal mesogastrium** and that supporting the duodenum as the **mesoduodenum**. It lies on the caudal surface of the **septum transversum**, the mesodermal partition which separates the abdominal foregut from the pericardium [FIG. 2.72] and from which a **ventral mesentery** for the abdominal foregut is produced as that structure shrinks away from the septum.

LIVER AND PANCREAS

When the embryo is about 4 weeks old and has attained a length of 2.5 mm a ventral diverticulum, which forms the **liver**, the **gall-bladder**, the system of bile-ducts and part of the **pancreas**, appears in the angle between the widely open vitello-intestinal duct and the ventral wall of the foregut [FIG. 2.63] and grows into the septum transversum between the pericardium and the foregut. The septum transversum is thus divided into three parts: a midzone containing this liver diverticulum; a cranial part separating the liver from the pericardium, which forms part of the diaphragm; and a caudal portion attaching the liver to the foregut, the ventral mesentery [FIG. 2.73]. When the length of the embryo has increased to about 4 mm another diverticulum is formed from the dorsal wall of the foregut a little nearer the stomach. This is the **dorsal pancreas** which forms the greater part of the adult pancreas [FIGS. 2.73 and 2.74].

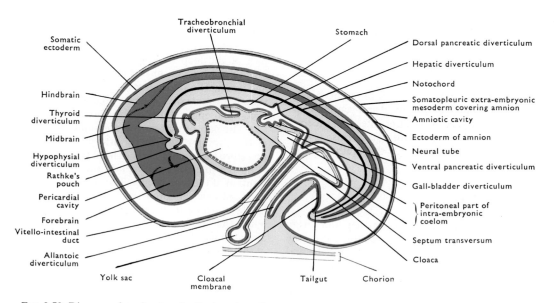

FIG. 2.72. Diagram of median longitudinal section of a 5-mm human embryo (about 5 weeks old). (After Mall.)

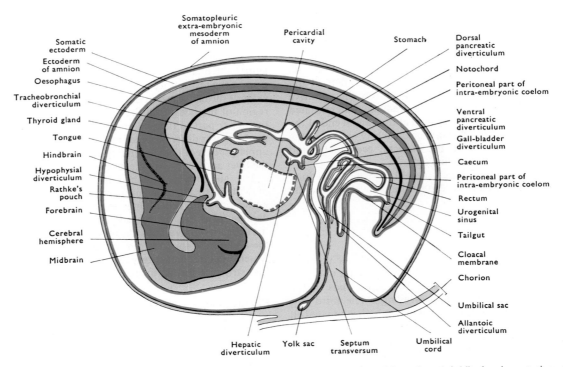

FIG. 2.73. Diagram of median longitudinal section through a 10-mm human embryo (about 6 weeks old), showing rotation of the midgut loop lying in umbilical sac (After Mall.)

Differentiation of midgut and hindgut

DERIVATIVES OF MIDGUT

The midgut is that part of the primitive alimentary tract which is in free communication with the yolk sac by the vitello-intestinal duct. It is transformed into the greater part of the small intestine (from the entrance of the liver diverticulum, the bile duct, into the duodenum to the end of the ileum) and part of the large intestine (caecum and appendix, ascending colon, and most of the transverse colon).

DERIVATIVES OF HINDGUT

The part of the alimentary tract found in the tail of the embryo is the hindgut. Like the foregut, it is divided into dorsal and ventral parts. The dorsal part forms the remainder of the large intestine, except a small portion of the anal canal. The ventral part forms: (1) the urachus and part of the urinary bladder; (2) the urethra in the female, and part of the urethra in the male.

DEVELOPMENT OF MIDGUT

As development proceeds the midgut elongates and separates from the posterior abdominal wall to which it remains attached by the dorsal mesentery. It then forms a U-shaped tube with cephalic and caudal limbs, and a ventral knuckle which points towards the umbilical orifice, through which it remains connected with the yolk sac by the narrowed and elongated vitello-intestinal duct [FIG. 2.72].

In embryos about 8 or 9 mm long (between 5 and 6 weeks old) there appears about the middle of the caudal limb of the loop an enlargement which is the rudiment of the caecum and vermiform appendix, thus demarcating the large intestine from the small.

But before this rudiment appears rotation of the loop has begun. The growth of other structures in the abdomen, especially of the liver, is relatively so great that there is insufficient room for the developing intestines. When the embryo is between 5 and 8 mm long, the loop begins to pass out through the umbilical orifice into an umbilical sac which has been formed in the umbilical cord by the enclosure of part of the extra-embryonic coelom [FIGS. 2.72–2.74].

As the loop enters this sac its limbs already lie side by side—the cephalic limb on the right, the caudal on the left—and there it remains for a considerable time while a further stage of its rotation and further development take place.

The chief change that now occurs is the rapid elongation of the part of the loop (mainly the cephalic limb) that becomes small intestine, which is thrown into the numerous coils of the jejunum and ileum. The vitello-intestinal duct has meanwhile separated from the loop and disappeared together with the vitelline artery accompanying it, thus leaving the coiling gut free in the umbilical sac. The part of that artery remaining in the mesentery of the loop is the superior mesenteric artery, which supplies the loop and therefore the greater parts of both the small and the large intestine.

As the intestines return into the abdomen the last stage of the rotation of the loop occurs. The small intestine, returning first, falls back into the abdominal cavity inferior and to the right of the hindgut, which is thereby pushed to the left and superiorly. Thus when the large intestine part of the midgut loop returns to the abdominal cavity it lies transversely, above and in front of the coils of small intestine, with the caecum immediately below the right lobe of the liver. The rotation as a whole is usually described as taking place round the superior mesenteric artery as an axis. It is completed by the passage of the caudal limb of the loop to the right, in front of the trunk of that artery and in front of the most cephalic part of the midgut loop which, because of the rotation around the artery, loops behind it to form the horizontal part of the duodenum.

Subsequently, certain parts of the large intestine are fixed in their permanent positions by peritoneal adhesions. The last 60 cm of the

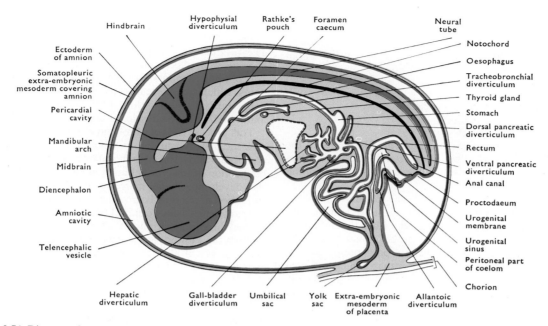

Hindbrain
Hypophysial diverticulum
Rathke's pouch
Foramen caecum
Neural tube

Ectoderm of amnion
Somatopleuric extra-embryonic mesoderm covering amnion
Pericardial cavity
Mandibular arch
Midbrain
Diencephalon
Amniotic cavity
Telencephalic vesicle

Notochord
Oesophagus
Tracheobronchial diverticulum
Thyroid gland
Stomach
Dorsal pancreatic diverticulum
Rectum
Ventral pancreatic diverticulum
Anal canal
Proctodaeum
Urogenital membrane
Urogenital sinus
Peritoneal part of coelom
Chorion

Hepatic diverticulum
Gall-bladder diverticulum
Umbilical sac
Yolk sac
Extra-embryonic mesoderm of placenta
Allantoic diverticulum

FIG. 2.74. Diagram of a 25-mm human embryo (about 8 weeks old), showing herniation of the midgut loop into the umbilical sac.

ileum, the **ascending colon** and the **transverse colon** are derived from the caudal limb of the loop, whilst the proximal part of the hindgut is transformed within the abdomen into the **descending colon**, and the **sigmoid colon**. The developing intestines remain in the umbilical sac until the embryo is about 40 mm long and about 10 weeks old, when they return rather suddenly to the abdomen, and the coelomic space in the umbilical cord is obliterated.

Occasionally the site of attachment of the vitello-intestinal duct is indicated by the persistence of a portion of it as a **diverticulum of the ileum** (Meckel's diverticulum) some 60 cm from the caecum. This diverticulum may rarely be attached to the back of the umbilicus by a fibrous thread; such a thread, from the distal small intestine to the umbilicus and continuous with the end of the superior mesenteric artery, is normally present in many new-born animals (e.g. cats).

The umbilical sac and its communication with the peritoneal cavity may fail to close and the midgut may remain in the umbilical coelom. Thus the child is born with the malformation of the abdominal wall known as congenital umbilical hernia.

DEVELOPMENT OF CAUDAL PART OF HINDGUT

When the hindgut is first enclosed, its blind end and its ventral wall are bounded by the caudal portion of the primitive streak—the **cloacal membrane**—which is bent ventrally during the folding-off of the embryo [FIG. 2.72].

The terminal part of this portion of the gut becomes expanded to form a chamber called the **entodermal cloaca**, because the **pronephric ducts** from the primitive kidneys open into it, one on each side, as in the permanent **cloaca** of the lower vertebrates.

The ventral part of the cephalic end of the cloaca is continuous with the allantoic diverticulum, and the dorsal part with that portion of the hindgut which forms the descending colon and the sigmoid colon.

As the temporary tail is formed and turned ventrally, a diverticulum of the caudal end of the dorsal part of the cloaca is prolonged into it, forming the **tailgut**. This soon becomes shut off from the cloaca and entirely disappears before the temporary tail is absorbed into the caudal end of the body [FIGS. 2.72–2.74].

At a later period the cloaca itself is divided by a septum (the urorectal septum) into a dorsal part which becomes the **rectum**, and part of the anal canal, and a ventral part called the **urogenital chamber** [FIG. 2.74]. The septum begins in the angle between the allantoic diverticulum and the ventral wall of the hindgut, and grows towards the surface till it reaches and fuses with the cloacal membrane; the membrane is thus separated into urogenital and anal portions, both of which disappear about the eighth week.

In both sexes the urogenital chamber is separable into three portions: (1) a cephalic portion called the **urachus**, which is continuous with the remnant of the allantoic diverticulum and becomes the **median umbilical ligament**; (2) a middle portion, the **vesico-urethral canal**, which becomes the greater part of the urinary bladder, the whole of the female urethra and the upper part of the prostatic urethra in the male; (3) a caudal portion, known as the **urogenital sinus**, which becomes the vestibule of the vagina in the female and the lower part of the prostatic urethra and a large part of the penile urethra in the male.

UROGENITAL SYSTEM

The greater part of the urogenital system is derived from the intermediate cell mass of the mesoderm [p. 46]. That includes the series of three excretory organs or kidneys known as the **pronephros**, the **mesonephros**, and the **metanephros**—the last forming the permanent kidney—and the two ducts, the mesonephric and paramesonephric ducts, which have a different fate in the two sexes [FIGS. 8.71–8.73].

DERIVATIVE OF PROCTODAEUM

The proctodaeum is a surface depression which owes its origin to the elevation of the surface around the margin of the anal portion of the cloacal membrane [FIG. 2.74]. It forms the lowest portion of the anal canal of the adult.

BLOOD VASCULAR SYSTEM

In the early stages of implantation, the trophoblast utilizes the dissolving decidual tissues as a food supply, and fluid is absorbed and transmitted into the interior of the zygote to fill its expanding cavities.

Nutritive materials in the fluid suffice for the requirements of the embryonic and non-embryonic parts of the zygote so long as both consist of comparatively thin layers of cells. But when the embryonic area increases in thickness and begins to be moulded into the embryo, and as the development of its various parts progresses, the supply of food and oxygen required is greater than can be provided by diffusion from the adjacent fluid media. A method of food supply adequate for the increasing requirements of continued development and growth becomes necessary and the blood vascular system is formed. In its earliest stages it consists of a series of simple vessels but, after a short time, parts of the vessels undergo changes which result in the formation of the heart, contractions of the myocardium then producing circulation of blood.

The tissues of the vascular system are derived from **mesenchyme cells** [p. 47]. The development of the blood vessels takes place first in the wall of the yolk sac and in the body stalk, but it occurs later in the mesenchyme of the embryo also.

On the wall of the yolk sac the mesenchyme cells which are the rudiments of the blood vascular system become spherical with enlarged nuclei, and aggregate into rounded clumps called **blood islands** [FIG. 2.46]. As soon as the cells have attained their distinctive appearance they are known as **angioblasts**, and from them are derived both the endothelial walls of blood vessels and blood corpuscles.

When the blood islands first become distinct in the walls of the yolk sac they are isolated from one another, but the endothelium which develops around the different islands soon joins together to form a network of endothelial tubes in which fluid appears. This establishes a plexus of anastomosing tubes filled with blood plasma in which the primitive nucleated blood corpuscles are suspended.

Most of the corpuscles of the blood eventually become non-nucleated and form the red corpuscles or **erythrocytes**. In the early stages they are being formed very rapidly and retain their nuclei while circulating. A small proportion (about 0.2 per cent) remain nucleated and form the white corpuscles or **leucocytes**, which are separable into several groups having distinctive characteristics. The blood vessels also change as development proceeds, for the original plexuses are transformed into distinct stems and branches of varying size and importance. Therefore two groups of events in the development of the vascular system have to be considered—the evolution of the different kinds of blood corpuscles and the development of the main embryonic blood vessels.

DEVELOPMENT OF BLOOD CORPUSCLES

The view that the angioblast is the parent cell not only of the endothelium of blood vessels but also of all the blood corpuscles both red and white was elaborated by Maximow (1924). It implies that blood formation in the growing embryo and foetus takes place by the development of angioblastic cells from the mesenchyme in the same way as on the yolk sac, and that the stem cells (haemocytoblasts) that give rise to all the varieties of blood corpuscles in the adult are carried from the yolk sac into the embryo to multiply there.

The red blood corpuscles pass through several stages to maturity. The original nucleated cells (**erythroblasts**) become laden with haemoglobin, pass through a **normoblast** stage (in which the nuclei are contracted), and after the loss of their nuclei become **reticulocytes**, which are anucleate cells containing a fine network stainable with vital dyes. With the loss of this reticulum these cells become the mature red blood corpuscles or **erythrocytes**.

This process goes on throughout life; it begins in the wall of the yolk sac, is very active in the liver from the third month onwards, and occurs in the spleen in the later stages of intrauterine life. When ossification begins, formation of blood cells is actively carried on in the developing bone marrow; and after birth the continuous formation of red blood corpuscles is entirely confined to that special haemopoietic (blood-forming) tissue.

The three main kinds of white blood corpuscles are **granular leucocytes**, **monocytes**, and **lymphocytes**. The bone marrow is the great factory of all these, though the monocytes and lymphocytes may also be formed in lymph tissue wherever it may be situated—in the spleen, in the mucous membrane of the alimentary canal, and in the lymph nodes. All these cells reach the blood stream by passing through the endothelial walls of blood capillaries, but lymphocytes also reach it in great numbers via the lymph vessels when the lymph nodes have developed, mainly after birth.

In very young embryos both immature and mature blood cells are present in the blood stream. After birth only mature erythrocytes pass into the blood stream; but under pathological conditions, in which there may be excessive formation of red corpuscles in the bone marrow, normoblasts may again appear in the blood.

FORMATION OF PRIMITIVE BLOOD VESSELS

It has been pointed out that a plexiform system of blood vessels is formed from angioblasts in the yolk sac area of the entodermal vesicle.

Similar but less easily demonstrated plexuses are formed in the mesenchyme of the body stalk, and of the chorion, and also later in the embryonic area.

It is probable that the various plexuses become connected with one another very soon after they appear, so that a continuous capillary network is formed. By enlargement in some places and diminution in others the general vascular network becomes more open, and from this **retiform** arrangement, by further enlargement of definite channels, there are evolved what may be called the primitive stem vessels of the embryo, through which the blood circulates from one area to another.

In the embryonic area two sets of vessels are differentiated—the **vitelline veins**, which emerge from the vascular plexus in the wall of the yolk sac, and the **primitive aortic vessels**, which extend from the pericardial area along the dorsal aspect of the entodermal vesicle as the **primitive dorsal aortae**. In the pericardial area the primitive aortic vessels form the paired primitive **endocardial heart tubes**.

As the aortae pass tailwards in the embryonic area, ventral to the paraxial bars of mesoderm, they give off a ventral series of paired branches, called **vitelline arteries**, which terminate in the vascular plexus in the yolk sac wall [p. 33]. From the caudal and dorsal part of that plexus a pair of vessels—the **umbilical arteries**—pass along the body stalk, one on each side of the allantoic diverticulum, to the vascular plexus in the chorion [FIG. 2.75]. Another stem vessel, called the **unpaired umbilical vein**, begins in the vascular plexus in the chorion and passes along the body stalk to the caudal margin of the embryonic area, where it divides into **right and left umbilical veins**. These two umbilical veins run along the corresponding margins of the embryonic area in the somatopleure and fuse with the corresponding vitelline veins in the septum transversum, cranial to the pericardial area, but caudal to it when reversal of the embryo occurs. Together they form the **vitello-umbilical trunks**, each of which, right and left, enters the caudal aspect of the developing heart. Therefore, when rhythmic contraction begins in the cardiogenic region, the blood is driven through the primitive aortic vessels and passes through their vitelline and umbilical branches to the yolk sac and the chorion, whence it is returned to the endocardial heart tubes by the vitelline and umbilical veins.

In the meantime lateral branches have been given off from the dorsal aortae, and in later stages, as the mesodermal somites are defined, they become the intersegmental arteries [FIG. 2.76]; but for a time no purely intra-embryonic veins are distinguishable.

As the head fold is formed and the pericardial region is reversed into the ventral wall of the foregut [p. 48], the cephalic parts of the primitive aortic vessels are carried ventrally in the fold, and so are bent into the shape of hooks. That condition is well seen in an embryo of 1.35 mm length which has six mesodermal somites [FIG. 2.75].

Several parts of the bilaterally symmetrical embryonic vascular system are now defined: the short ventral limb of the hook is the **primitive ventral aorta**; the long or dorsal limb is the **primitive dorsal aorta**. The bend, which connects the two limbs, is the **first**

aortic arch; it runs along the side of the buccopharyngeal membrane in the substance of the mandibular arch. The primitive ventral aorta emerges from the cephalic end of the endocardial heart tube.

Heart. Three further important changes take place in the vascular system during the time in which the embryo grows little more than another millimetre in length, the tail fold and lateral folds are formed and the number of its mesodermal somites increases to fourteen pairs.

1. The paired endocardial heart tubes fuse together to form a single, median tube, which comes to be surrounded by a loose reticulum (the **subendocardial tissue**) and by developing muscle (the future myocardium). The single heart tube now formed is divided by dilatations and constrictions into six parts, named, from the caudal towards the cephalic end, the **sinus venosus** in the septum transversum, the **atrium**, the **atrioventricular canal**, the **ventricle** and the **bulbus cordis** in the pericardial cavity, and the **truncus arteriosus** in the mesoderm cephalic to the pericardium. At the same time the heart tube increases in length more rapidly than the pericardial cavity; it therefore becomes bent both in the longitudinal and the transverse direction, and its caudal and cephalic ends come close together [FIGS. 2.76 and 2.77].

2. The origins of the umbilical arteries are transferred from the vitelline plexus to the primitive dorsal aortae from which they arise by three roots for each [FIG. 2.76].

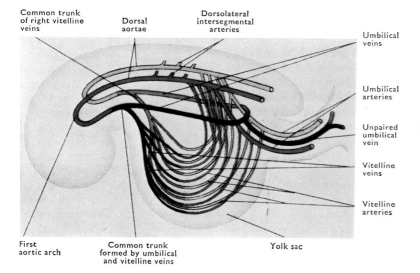

Common trunk of right vitelline veins

Dorsal aortae

Dorsolateral intersegmental arteries

Umbilical veins

Umbilical arteries

Unpaired umbilical vein

Vitelline veins

Vitelline arteries

First aortic arch

Common trunk formed by umbilical and vitelline veins

Yolk sac

FIG. 2.75. Diagram of the vascular system of a human embryo, 1.35 mm long. Although the umbilical arteries and veins are shown red and blue, they are carrying deoxygenated and oxygenated blood respectively. (After Felix 1910.)

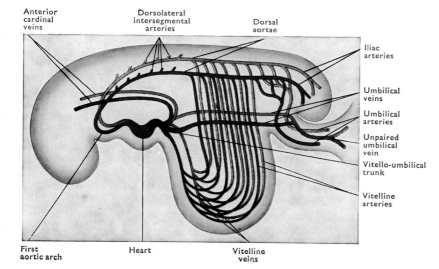

Anterior cardinal veins

Dorsolateral intersegmental arteries

Dorsal aortae

Iliac arteries

Umbilical veins

Umbilical arteries

Unpaired umbilical vein

Vitello-umbilical trunk

Vitelline arteries

First aortic arch

Heart

Vitelline veins

FIG. 2.76. Diagram of the vascular system of a human embryo, 2.6 mm long. (After Felix 1910.)

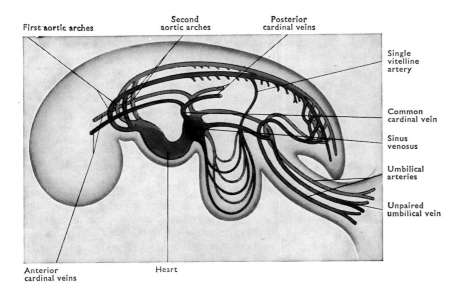

First aortic arches

Second
aortic arches

Posterior
cardinal veins

Single
vitelline
artery

Common
cardinal vein

Sinus
venosus

Umbilical
arteries

Unpaired
umbilical vein

Anterior
cardinal veins

Heart

FIG. 2.77. Diagram of the vascular system of a human embryo showing a single vitelline artery. (After Felix 1910.)

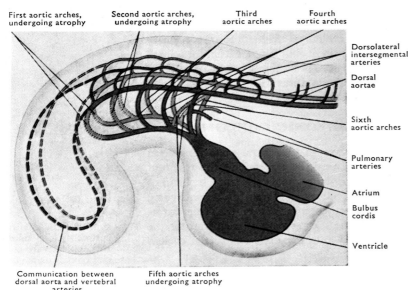

First aortic arches,
undergoing atrophy

Second aortic arches,
undergoing atrophy

Third
aortic arches

Fourth
aortic arches

Dorsolateral
intersegmental
arteries

Dorsal
aortae

Sixth
aortic arches

Pulmonary
arteries

Atrium

Bulbus
cordis

Ventricle

Communication between
dorsal aorta and vertebral
arteries

Fifth aortic arches
undergoing atrophy

FIG. 2.78. Diagram showing the aortic arches of a human embryo.

3. The **anterior cardinal veins** are defined [FIG. 2.76], and are the first pair of purely intra-embryonic veins to appear. They begin in the head and end in the sinus venosus, which they enter, in the septum transversum, close to the common vitello-umbilical veins. They bring back to the heart the blood which has been distributed to the head, neck, and headward part of the trunk of the embryo.

Thereafter, numerous changes take place in the cardiac, the arterial, and the venous parts of the vascular system, and the rudiments of all the main blood vessels of the adult are defined [for details see The Blood Vascular System].

ARTERIES

The primitive dorsal aortae fuse, from the region of the tenth (seventh cervical) somite tailwards, to form the permanent descending aorta [FIG. 2.77]. In this region also, as the gut tube becomes relatively smaller, the paired ventral branches of the aortae fuse to form single, median trunks, and become reduced in number. In the abdominal region they are reduced to three main vessels—the coeliac trunk, the superior mesenteric artery (which for a time passes

to the yolk sac as the vitelline artery) [FIG. 2.77], and the inferior mesenteric artery.

Branches are given off from the lateral aspects of the descending aorta to the rudiments of the urogenital system, developing from and in association with the intermediate cell mass.

Four additional pairs of **aortic arches** connecting the unfused portions of the primitive ventral and dorsal aortae appear in the following sequence, the second, the third, the fourth, and the sixth—the temporary fifth pair appearing later [FIG. 2.78]. These aortic arterial arches run in the **pharyngeal arches** of the embryo [FIG. 2.71], and they undergo a series of transformations as the arterial system of the thorax and neck develops. The first and second arch arteries disappear, the pulmonary arteries arise from the sixth [FIG. 2.78], and the truncus arteriosus and bulbus cordis divide so that a main pulmonary trunk communicates with the right side of the divided heart and the permanent aorta with the left side.

VEINS

Two additional pairs of intra-embryonic venous trunks are formed for the drainage of the hinder parts of the body.

The posterior cardinal veins [FIG. 2.77] appear in the thoracic and abdominal regions of the embryo dorsolateral to the intermediate cell mass.

They drain blood from the body walls and from the intermediate mesoderm, and their cephalic ends join the anterior cardinal veins in the thoracic region at the point where the latter turn ventrally to enter the septum transversum [FIG. 2.77].

As soon as the union of the anterior and posterior cardinal veins is completed, the parts of the anterior cardinals ventral to the points of union are called the common cardinal veins. The right common cardinal vein later becomes the terminal part of the superior vena cava. In the meantime the common vitello-umbilical venous trunks have been absorbed into the sinus venosus, and the vitelline and umbilical veins open independently into it; so that three pairs of veins now open into the sinus venosus of the heart.

The subcardinal veins appear in the abdominal region ventro-medial to the intermediate cell mass. They communicate both at their cephalic and caudal ends with the posterior cardinal veins, and drain the intermediate mesoderm.

In addition to the posterior cardinal and subcardinal veins, two further pairs of intra-embryonic veins are established at a later period in the abdominal and thoracic regions. They are called the supracardinal and azygos veins and they appear dorsolateral to the descending aorta. These veins are concerned in the complex manner of development of the inferior vena cava and azygos veins [FIG. 13.126].

COELOM AND DIAPHRAGM

Now that an outline of the formation of the body of the embryo and of the principal organs of the trunk—alimentary canal, respiratory system, excretory organs, heart and great vessels—has been given, the division of the coelomic cavity into peritoneal, pleural, and pericardial cavities, and the manner in which these are separated from each other with the formation of the diaphragm may be considered.

The extra-embryonic coelom and the intra-embryonic coelom are derived independently as clefts in the mesoderm. At first separate from each other [FIGS. 2.53 and 2.57], they become continuous, for a time, through the umbilical orifice [FIG. 2.54B], but are separated from each other again when the umbilical orifice closes and the extra-embryonic coelom disappears. The intra-embryonic coelom remains as the cavities of the pericardium, pleura, and peritoneum.

Extra-embryonic coelom

The extra-embryonic coelom is formed in the primary mesoderm [p. 33], but is entirely obliterated, except for a small rudiment in the root of the umbilical cord, when the outer surface of the expanding amnion fuses with the body stalk and the inner surface of the chorion [FIGS. 2.41 and 2.42].

Intra-embryonic coelom

When the cleft-like spaces in the lateral plate mesoderm become confluent at their cephalic extremities, the resultant ∩-shaped cavity is the intra-embryonic coelom. The bend of the ∩-shaped cavity is in the cephalic part of the embryonic region, and has no direct communication with the extra-embryonic coelom, but the greater part of each limb of the cavity soon opens laterally into the extra-embryonic coelom [FIGS. 2.54B and 2.79].

The transverse portion of the ∩ is the pericardial cavity. The adjacent parts of the limbs, the pericardio-peritoneal canals, are

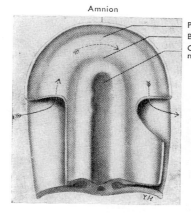

FIG. 2.79. Diagram of intra-embryonic coelom seen from dorsal surface before folding of embryonic area. Portions of the somatopleure are cut away to show the continuity of the intraembryonic coelom of the two sides through the pericardial tube.

invaded later by the growing lungs and are transformed into the pleural cavities. The remaining portions of the two limbs unite ventrally, before the umbilical orifice closes, to form the single peritoneal cavity.

As the head fold forms, the pericardial part of the cavity is carried ventrally and caudally into the ventral wall of the foregut [FIGS. 2.59 and 2.80]. The mesoderm which originally formed its peripheral boundary then lies in the cephalic margin of the umbilical orifice, and is thickened to form an important mass, the septum transversum [FIGS. 2.63 and 2.72].

On each side, the pericardial cavity is still continuous with the pericardioperitoneal canals which lie dorsal to the septum transversum, between the foregut and the body wall [FIG. 2.80].

SEPARATION OF PERICARDIAL, PLEURAL, AND PERITONEAL PARTS OF THE COELOM

Each pericardioperitoneal canal lies lateral to the oesophagus and trachea in the mesentery, ventral to the intermediate mesoderm, and dorsomedial to the curved dorsal edge of the septum transversum which contains the common cardinal vein and the phrenic nerve [FIG. 2.80C]. On each side a lung bud grows laterally from the caudal end of the trachea into the corresponding canal which becomes the primitive pleural cavity. The cavity and lung bud then extend laterally into the dorsomedial edge of the septum transversum, caudal to the common cardinal vein and the phrenic nerve. Thus they split this part of the septum transversum into two ridges, a cranial pleuropericardial membrane and a caudal pleuroperitoneal membrane, which meet ventrally in the unsplit part of the septum transversum. Having split this part of the septum transversum to form the definitive pleural cavity within it, that cavity and the contained lung bud expand into the body wall, passing cranially (lateral to the common cardinal vein and the phrenic nerve), caudally, and ventrally round the pericardium, till the two pleural cavities abut on each other ventral to the pericardium in the midline, separated only by mesoderm which becomes the anterior medias-tinum and anterior part of the superior mediastinum [FIGS. 2.81 and 2.82]. Thus the definitive pleural cavity is formed by an extension of the pericardioperitoneal canal into the body wall, the canal remaining as a narrow space around the lung root. As the growing lung passes upwards it lies to the lateral side of the fold containing the common cardinal vein and phrenic nerve, which is thus pressed against the pleuropericardial opening, compressing it towards the median plane until it is obliterated by fusion of the fold with the

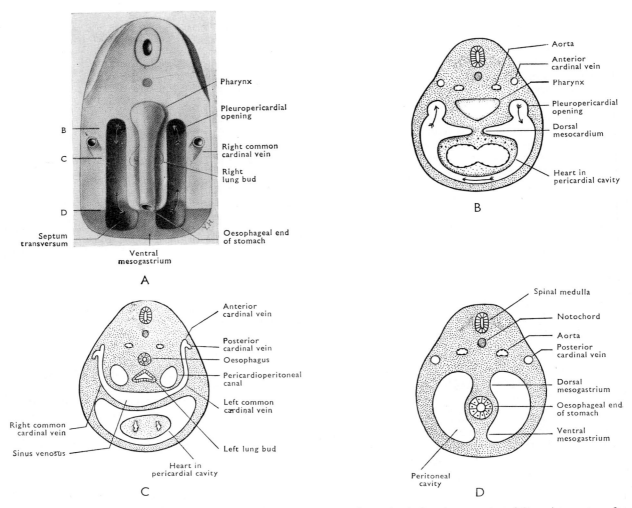

FIG. 2.80. Diagrammatic representation of embryonic coelom after folding of embryonic area but before the separation of the various parts. A. from dorsal surface; B, C, and D (after Patten 1953), at levels indicated in A.

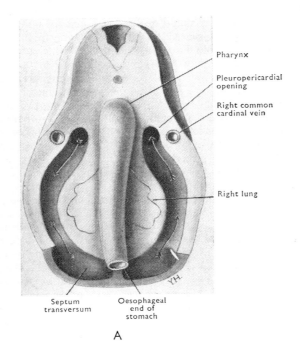

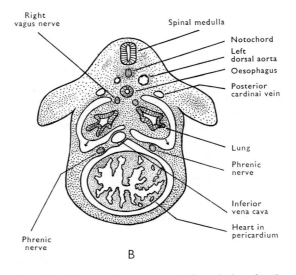

FIG. 2.81. Diagram of later stage of differentiation of coelom. A, from dorsal surface; B (after Patten 1953), transverse section cut at the level of lung bud in A, showing commencing ventral extension of pleural cavities.

mesoderm surrounding the trachea and lung root. When this occurs the pericardial cavity is entirely shut off from the remainder of the coelom and it becomes a completely closed space. Very rarely, as an abnormal condition, the pericardial cavity remains in communication on one or other side, more often the left, with the pleural cavity.

The pleuroperitoneal folds, which increase in size as the pleural cavities and lung buds expand caudally lateral to them, are rudiments of the lateral parts of the diaphragm connected ventromedially with the septum transversum. They extend medially until they fuse with the mesoderm of the side wall of the foregut and with the dorsal mesentery [FIGS. 2.81 and 2.82]. With this fusion the **pleuroperitoneal opening** is closed and the cavity of the portion of the coelom surrounding the lung root—the original pericardioperitoneal canal—is separated from the more caudal part of the coelom, which now becomes the peritoneal cavity.

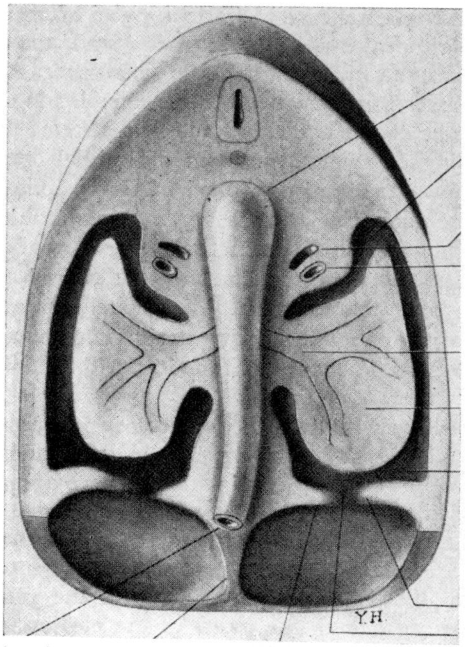
Pharynx
Growing apex of lung
Pleuropericardial opening
Right common cardinal vein
Right bronchus
Right lung
Pleural cavity
Pleuroperitoneal fold
Pleuroperitoneal opening
Oesophageal end of stomach
Ventral mesogastrium
Dorsal rudiment of diaphragm
YH
A

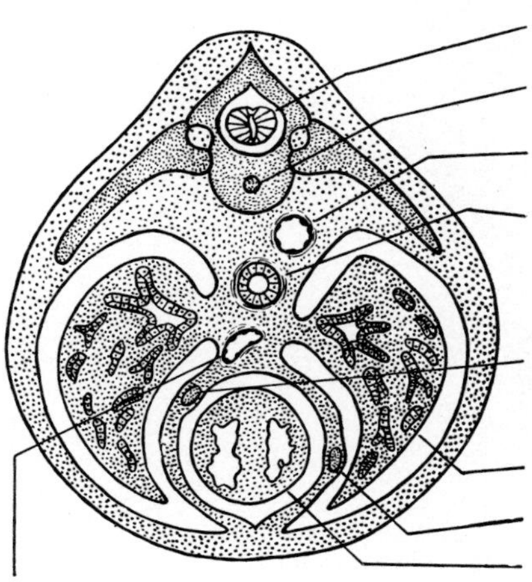
Spinal medulla
Notochord
Aorta
Oesophagus
Phrenic nerve
Lung in pleural cavity
Phrenic nerve
Heart in pericardium
Inferior vena cava
B

FIG. 2.82. Diagram of still later stage of differentiation of coelom. The pleurae are separated from the pericardium, but still communicate with the peritoneum.

A, from dorsal surface; B (after Patten 1953), transverse section at the level of lung roots in A, showing ventral extension of the pleural cavities.

FORMATION OF DIAPHRAGM

There are four main developmental parts of the diaphragm—ventral, dorsal, and right and left lateral.

The ventral part is formed from the septum transversum, which is gradually differentiated into caudal, middle, and cephalic parts. The caudal part is transformed into: (1) the mesodermal tissue of the liver; and (2) the ventral mesentery of the abdominal part of the foregut. The cephalic part becomes the caudal or diaphragmatic wall of the pericardium. The middle part is transformed into the ventral portion of the diaphragm.

The dorsal part of the diaphragm, including the part surrounding the oesophagus, is developed from the mesoderm of the dorsal mesentery of the foregut. The left gastric artery, which originally lay in the dorsal mesentery, thus comes to lie on the oesophageal part of the diaphragm. Each lateral part is derived from the pleuroperitoneal folds mentioned above. The two lateral portions grow towards the median plane till they fuse with the dorsal portion; but sometimes, especially on the left side, the fusion is not completed. The pleuroperitoneal opening then remains unclosed in the region of the vertebrocostal trigone [FIG. 5.102] close to the mesoderm forming the cranial pole of the kidney and suprarenal. A portion of the abdominal contents may pass through it into the pleural sac, constituting a congenital diaphragmatic hernia.

Summary of external features of human embryo and foetus at different periods of development

FIRST MONTH

During the first 14 days after fertilization the human ovum descends through the uterine tube, enters the uterus as a morula, and is transformed into a blastula which attaches itself to the uterine epithelium [FIG. 2.27] and penetrates the decidua compacta [FIG. 2.28]. About the **seventeenth day** the primitive node has appeared; secondary mesoderm is spreading in the embryonic area from the primitive streak (now well formed); and the allantoic diverticulum has grown into the body stalk [FIG. 2.83].

By the **eighteenth** or **nineteenth day** the area, though variable in both size and shape, may be from 1 to 1.5 mm long and 1 mm or less in breadth [FIGS. 2.45 and 2.84]. The primitive streak is now fully developed with a very distinct primitive groove on its surface. The head process is growing headwards from the primitive node and is progressively tunnelled by the notochordal canal. The tail fold is carrying the primitive streak over the caudal margin of the embryonic area, and the body stalk, containing the allantois, appears to be bent dorsally at right angles to the region of the cloacal membrane [FIG. 2.46]. There may be an indication of the formation of neural folds [FIG. 2.84].

By the **nineteenth** or **twentieth day** the embryonic area is still variable in size, but in length (1.5–2 mm) is then definitely greater than the breadth. The tail fold and neural folds are very evident [FIG. 2.85].

By the end of the **third week** the head fold and tail fold are distinctly formed, the neural folds are well developed especially in

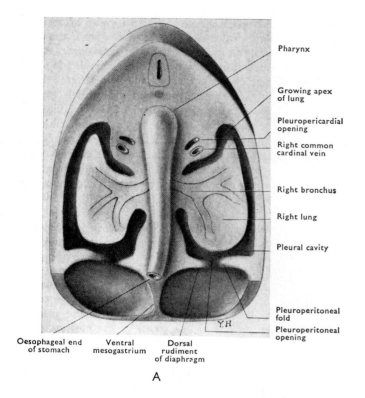

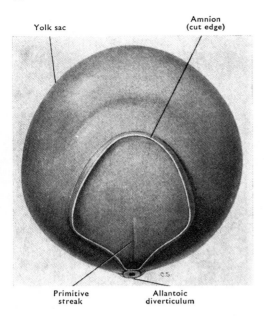

FIG. 2.83. Dorsal view of human embryo (Mateer) of 17 days (× 35). The amnion is cut away to show the embryonic area from above. The primitive streak is seen in the caudal part of the area and beyond that is a section of the body stalk with the allantoic diverticulum in it. (After G. L. Streeter, *Contrib. Embryol. Carneg. Instn*, 1920.)

the head region, the neural groove is still completely open or just beginning to close, and the formation of mesodermal somites has begun [FIG. 2.86]. The length of the embryo may now be about 2 mm, but there is some variation in the relation of length to state of development at this stage.

In the next few days the primary parts of the brain begin to be evident, the neural groove closes except in the cephalic and caudal regions where the neuropores [p. 51] are seen. The mesodermal somites increase in number, and the cephalic region begins to bend ventrally as the cephalic flexure forms [FIGS. 2.87 and 2.88].

By the end of the fourth week the length of the embryo is about 2.5–3 mm, the head is bent at right angles to the body, and shows the projection of the forebrain. The number of somites increases to over twenty. The rudiments of the otic vesicles have appeared as slight ectodermal depressions in the region of the hindbrain, and the cephalic and caudal neuropores are closed or closing. The yolk sac is still relatively large and in free communication with the midgut. The pharyngeal arches and grooves are appearing in the wall of the foregut dorsal to the bulging pericardium [FIGS. 2.63 and 2.89].

SECOND MONTH

During the fifth week the embryo attains a length of 5 to 6 mm. The somites increase to thirty-eight or thirty-nine; the rudiments of the limbs appear and become quite distinct, the fore limb in advance of the hind limb. The otic vesicles sink into the interior of the head. The tail becomes a very definite appendage, and the bulgings caused by the optic vesicles are quite obvious on the surface of the head. The cervical flexure remains acute, and the head bends at right angles upon itself in the region of the midbrain, forming the cephalic flexure, with the result that the frontal region is turned towards the tail [FIG. 2.90].

By the end of the sixth week the length of the embryo has increased to 11 or 12 mm CR (CR indicates the crown-rump measurement which corresponds with the sitting-height (Mall 1910)). During this week the lens of the eye appears as a thickening

of the surface ectoderm which becomes a vesicle and separates from the surface, sinking into the interior of the eyeball. The three segments of the upper limb become visible, and the rudiments of the fingers appear. The lower limb is less advanced. The third and fourth pharyngeal arches lie in the depths of the cervical sinus, overlapped by the caudal margin of the second arch [FIG. 2.68].

During this week also the olfactory pits appear between the medial and the lateral nasal processes, and grow dorsally into the roof of the stomodaeum; the maxillary processes of the mandibular arches, growing towards the median plane, fuse with the lateral and medial nasal processes, so completing the lateral parts of the primitive upper lip [FIGS. 2.64 and 2.65].

The nodular outgrowths which form the rudiments of the auricle appear on the margins of the first pharyngeal groove and begin to fuse together [FIG. 2.91].

By the seventh week the embryo has attained a length of 17–18 mm (CR). The cervical flexure has begun to straighten. The rudiments of the eyelids have appeared. The margins of the auricles are now well defined; the thighs and the toes have appeared [FIGS. 2.92 and 2.94].

At the end of the eighth week, when the embryo becomes a foetus, it has attained a length of about 25 to 30 mm (CR). The

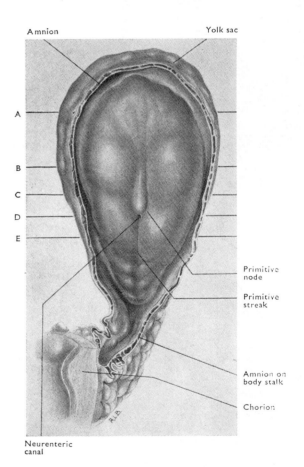

FIG. 2.84. Dorsal view of human embryo of 18 days (× 50).
(After C. H. Heuser, *Contrib. Embryol. Carneg. Instn*, 1932a.)
The amnion is cut away to show the embryonic area now elongated (1.53 × 0.75 mm) and with the caudal part slightly bent ventrally. The primitive node, with opening of neurenteric canal, and primitive streak are seen in the caudal half of the area and there is an indication of commencing neural folds in the cranial half. Cf. Fig. 2.45; and for sections of this embryo, see Fig. 2.47.

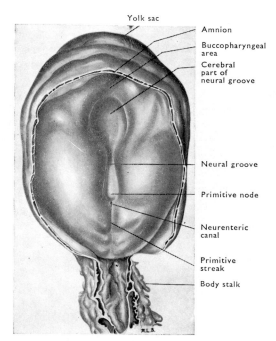

FIG. 2.85. Dorsal view of human embryo of 19 days (× 50). (After W. C. George, *Contrib. Embryol. Carneg. Instn*, 1942.)

The amnion is cut away to show the embryonic area (1.16 mm long) with primitive node and streak, neurenteric canal, and developing neural groove.

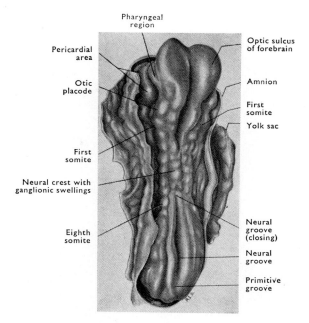

FIG. 2.87. Dorsal view of 2.2 mm human embryo of 22 days (× 40). (After F. Payne, *Contrib. Embryol. Carneg. Instn*, 1925.)

The neural groove is closed in the future cervical region, the brain vesicles though still open are beginning to take shape, the site of origin of the optic vesicle is indicated, and there are seven pairs of somites present.

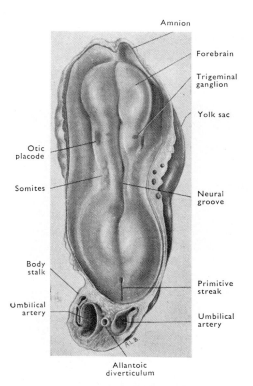

FIG. 2.86. Dorsal view of human embryo of 21 days (× 50). (After N. W. Ingalls, *Contrib. Embryol. Carneg. Instn*, 1920.)

The broad cephalic end of the neural plate is elevated with commencing formation of the brain. The neural groove is still open and its caudal end is enclosing the cephalic end of the primitive streak. The length of the embryonic area has increased to 1.4 mm and somite formation has begun.

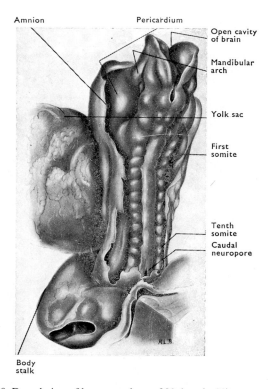

FIG. 2.88. Dorsal view of human embryo of 23 days (× 30). (After G. W. Corner, *Contrib. Embryol. Carneg. Instn*, 1929.)

The closure of the neural groove has advanced almost to the stage of cephalic and caudal neuropores. The three brain vesicles are forming, the pericardium and the mandibular arch are evident, and there are ten pairs of somites present.

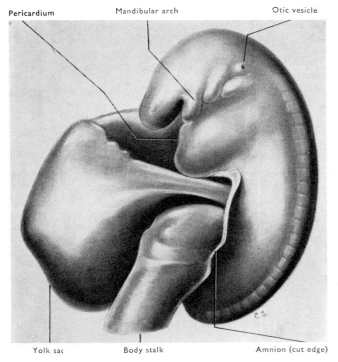

Pericardium   Mandibular arch   Otic vesicle

Yolk sac   Body stalk   Amnion (cut edge)

FIG. 2.89. Human embryo, 2.5 mm long, with twenty-three pairs of somites (× 34). (After Thompson 1907.)

The head is small and flattened; the optic vesicles project slightly from the forebrain, and the position of the otic vesicle is indicated. The yolk sac, still relatively large, is connected to the alimentary canal through the vitelline duct which, with the body stalk, emerges from the umbilical orifice. The position of the somites is indicated by slight elevations of the surface. There is no trace of the limbs.

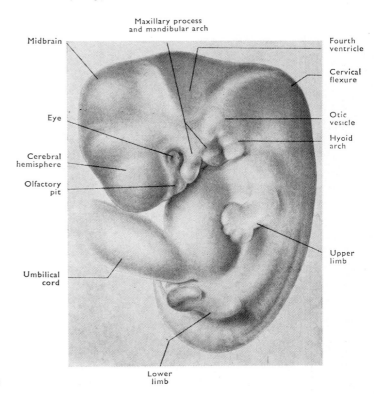

Maxillary process and mandibular arch

Midbrain   Fourth ventricle

                                            Cervical flexure

Eye   Otic vesicle

                                            Hyoid arch

Cerebral hemisphere

Olfactory pit

                                            Upper limb

Umbilical cord

Lower limb

FIG. 2.91. Human embryo, 13 mm CR length (After G. L. Streeter, *Contrib. Embryol. Carneg. Instn*, 1948; Embryo no. 8101.)

The limbs are bent ventrally and show differentiation into segments. The cervical sinus is closing and rudiments of the auricle of the external ear are present on the first and second arches. The pontine flexure of the hindbrain, the fourth ventricle, and the olfactory pit are evident.

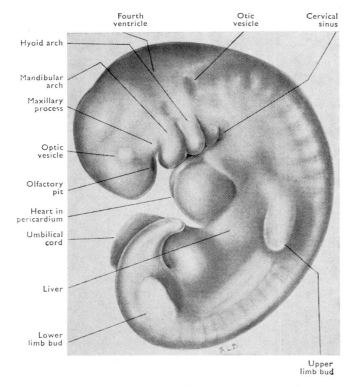

Fourth ventricle   Otic vesicle   Cervical sinus

Hyoid arch

Mandibular arch

Maxillary process

Optic vesicle

Olfactory pit

Heart in pericardium

Umbilical cord

Liver

Lower limb bud   Upper limb bud

FIG. 2.90. Human embryo, 5.5 mm CR length. (After G. L. Streeter, *Contrib. Embryol. Carneg. Instn*, 1945; Embryo no. 6830.)

The limb buds are now quite distinct [cf. Fig. 2.60] and the upper limb is beginning to elongate and curve forwards. The maxillary process has grown forwards below the eye, and the cervical sinus is evident.

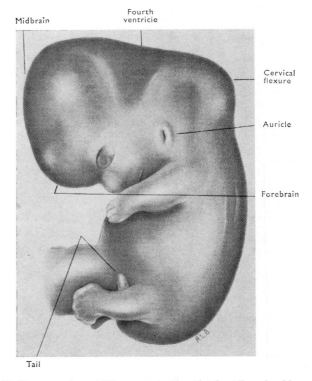

Midbrain   Fourth ventricle

                                            Cervical flexure

                                            Auricle

                                            Forebrain

Tail

FIG. 2.92. Human embryo, 18.5 mm greatest length, about 7 weeks old. The abdomen is very prominent on account of the rapid increase of the liver. The digits of the hand and foot are distinct but not separated from one another. The margin of the auricle is completed. The eyelids have begun to form. [Cf. Fig. 2.94.]

external genital organs are developing and a **genital tubercle** is evident [FIG. 8.74].

THIRD MONTH [FIG. 2.93].

The head grows less rapidly, and is relatively smaller in proportion to the whole body. The eyelids close, and their margins fuse together. The neck increases in length. Nails appear on the fingers and toes. The proctodaeum is formed and the external genital organs are sufficiently differentiated for the sex to be distinguished. By the end of the third month the CR length of the foetus is about 100 mm and it weighs about 50 g.

FOURTH MONTH

Fine, downy hairs (lanugo) are developed. The CR length of the foetus is about 145 mm and it weighs 200 g.

Between the **fifth** and **eighth months** the hairs become more developed, there is less disproportion between the upper and lower limbs and a sebaceous deposit (**vernix caseosa**) increases in amount to cover the skin. The CR length increases from 190 to 200 mm at the end of the fifth month, to 300 to 320 mm at the end of the eighth

month. The weight of the foetus increases from a mean of 460 g to 2 to 2.5 kg in the same period.

NINTH MONTH

The hair begins to disappear from the trunk, but it often remains long and abundant on the head. The skin becomes paler, the plumpness increases, and the umbilicus reaches the centre of the trunk. During this month the nails reach the ends of the fingers. At the end of the month, when the foetus is born, its CR length is from 340 to 360 mm and it weighs from 3 to 3.5 kg.

AGE, LENGTH, AND WEIGHT OF EMBRYO AND FOETUS

The lengths and weights of embryos and foetuses at increasing ages given in the foregoing summary are compiled from a number of sources. It must be understood, however, that they are only approximate, since embryos and foetuses of the same length, or of the same weight, or even at the same stage of development, are not necessarily of the same age (Streeter 1921, 1942). Moreover, it is usually impossible to calculate the **actual age** of an embryo (dating

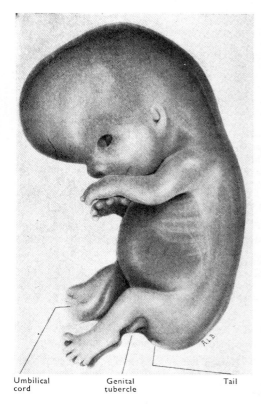

Umbilical   Genital   Tail
cord   tubercle

FIG. 2.93. Human foetus at the beginning of the third month. The cervical flexure is now slight. The upper limbs have assumed a characteristic position with the hands in front of the face, and the toes are free. The tail has almost disappeared, and the genital tubercle is evident.

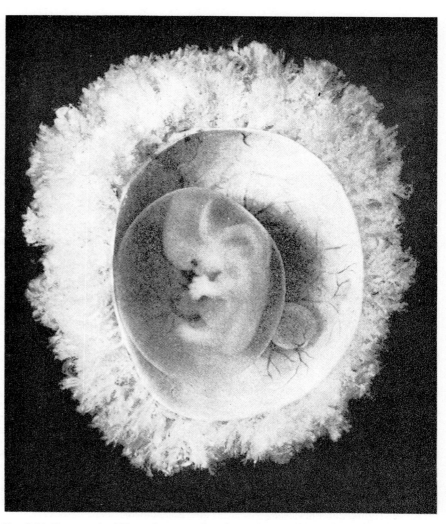

FIG. 2.94. Photograph of 21-mm human embryo in amnion and chorion (×2.5). For the form of the embryo, cf. Fig. 2.92; and note the position of the definitive yolk sac between the expanding amnion and the chorion. (Carnegie no. 8537A; by special permission of Dr George W. Corner.)

from the time of fertilization) and the age assigned (except in the case of some young embryos for which the data necessary for the calculation of fertilization age are available) is then the **menstrual age**, dating from the first day of the last menstruation. Owing to the average time of ovulation in relation to the menstrual cycle, the menstrual age of the majority of embryos is probably at least 14 days more than the actual age. The usual method of calculating the duration of pregnancy by counting 280 days from the first day of the last menstruation gives in any case only the approximate date of parturition.

It should also be noted that the 'greatest length' of an embryo (excluding the lower limbs) depends upon the cervical flexure. While the head of the embryo is fully flexed the crown-rump measurement (CR) is less than the greatest length, but later the two measurements are the same.

Formulae for the calculation of age from length have been devised, but it is more satisfactory to use tables and graphs (constructed from extensive data) from which the probable range of age can be read at a glance (Streeter 1921).

References

ALLEN, E., PRATT, J. P., NEWELL, Q. U., and BLAND, L. J. (1930). Human ova from large follicles: including a search for maturation divisions and observations on atresia. *Am. J. Anat.* **46**, 1.

AMOROSO, E. C. (1952). Placentation. In *Marshall's physiology of reproduction* (ed. A. S. Parkes) Vol. II, Chapter 15. Longmans, London.

ARMITAGE, P., BOYD, J. D., HAMILTON, W. J., and ROWE, B. C. (1967). A statistical analysis of birth-weights and placental weights. *Hum. Biol.* **39**, 430.

AUSTIN, C. R. and BISHOP, M. W. H. (1957). Fertilization in mammals. *Biol. Rev.* **32**, 296.

—— and WALTON, A. (1960). Fertilisation. In *Marshall's physiology of reproduction* (ed. A. S. Parkes) Vol. I, Part 2, Chapter 10. Longmans, London.

BARCROFT, J. (1946). *Researches on pre-natal life.* Oxford.

—— (1952). Foetal respiration and circulation. In *Marshall's physiology of reproduction* (ed. A. S. Parkes) Vol. II, Chapter 17. Longmans, London.

BISHOP, M. W. H. and WALTON, A. (1960). Spermatogenesis and the structure of mammalian spermatozoa. In *Marshall's physiology of reproduction* (ed. A. S. Parkes) Vol. I, Part 2, Chapter 7. Longmans, London.

BÖÖK, J. A., LEJEUNE, J., LEVAN, A., CHU, E. H. Y., FORD, C. E., FRACCARO, M., HARNDEN, D. G., HSU, T. C., HUNGERFORD, D. A., JACOBS, P. A., MAKINO, S., PUCK, T. T., ROBINSON, A., TJIO, J. H., CATCHESIDE, D. G., MULLER, H. J., and STERN, CURT (1960). A proposed standard system of nomenclature of human mitotic chromosomes. *Lancet* **i**, 1063.

BOYD, J. D. and HAMILTON, W. J. (1952). Cleavage, early development and implantation of the egg. In *Marshall's physiology of reproduction* (ed. A. S. Parkes) Vol. II, Chapter 14. Longmans, London.

—— —— (1970). *The human placenta.* Longmans, Heffer, Cambridge.

BREWER, J. I. (1938). A human embryo in the bilaminar blastodisc stage. (The Edwards–James–Brewer ovum). *Contrib. Embryol. Carneg. Instn* (No. 162) **27**, 85.

BRYCE, T. H. (1924). Observations on the early development of the human embryo. *Trans. R. Soc. Edinb.* **53**, 533.

CLERMONT, Y. and LEBLOND, C. P. (1955). Spermiogenesis of man, monkey, ram and other mammals as shown by the 'Periodic Acid–Schiff' technique. *Am. J. Anat.* **96**, 229.

DAVIES, F. (1948). Ovulation and the menstrual cycle. *Lancet* **ii**, 720.

DAWES, G. S. (1968). *Foetal and neonatal physiology.* Chicago.

DIXON, A. F. (1927). Normal oocyte showing first polar body and metaphase stage in formation of second polar body. *Irish J. Med. Sci.* 6th Ser., p. 149.

DUIJN, C. VAN, JR. (1958). *Fertilization.* London.

FELIX, W. (1910). Zur Entwicklungsgeschichte der Rumpfarterien des menschlichen Embryo. *Morphol. Jahrb.* **41**, 577.

FELL, H. B. and ROBISON, R. (1929). The growth, development and phosphatase activity of embryonic avian femora and limb-buds cultivated *in vitro. Biochem. J.* **23**, 767.

FRAZER, J. E. (1914). The second visceral arch and groove in the tubotympanic region. *J. Anat. Physiol.* **48**, 391.

GOODSIR, J. (1857). On the morphological constitution of the skeleton of the vertebrate head. In *Anatomical memoirs* (ed. W. Turner) (1868), Vol. 2, p. 88. Edinburgh.

GRAY, S. W. and SKANDALAKIS, J. E. (1972). *Embryology for surgeons.* London.

HAMILTON, W. J. and BOYD, J. D. (1960). Development of the human placenta in the first three months of gestation. *J. Anat. (Lond.)* **94**, 297.

—— —— and MOSSMAN, H. W. (1972). *Human embryology.* Heffer, Cambridge.

HARRISON, R. G. (1978). *Clinical embryology.* London.

—— and MALPAS, P. (1953). The volume of human amniotic fluid. *J. Obstet. Gynaecol. Br. Emp.* **60**, 632.

HARTMAN, C. G. (1929). How large is the mammalian egg? *Q. Rev. Biol.* **4**, 373.

HERTIG, A. T. and ROCK, J. (1941). Two human ova of the pre-villous stage, having an ovulation age of about eleven and twelve days respectively. *Contrib. Embryol. Carneg. Instn* (No. 184) **29**, 127.

—— —— (1945). Two human ova of the pre-villous stage, having a developmental age of about seven and nine days respectively. *Contrib. Embryol. Carneg. Instn* (No. 200) **31**, 65.

—— —— and ADAMS, E. C. (1956). A description of 34 human ova within the first 17 days of development. *Am. J. Anat.* **98**, 435.

—— —— —— and MULLIGAN, W. J. (1954). On the preimplantation stages of the human ovum. *Contrib. Embryol. Carneg. Instn* (No. 240) **35**, 199.

HEUSER, C. H. (1932a). A presomite human embryo with a definite chorda canal. *Contrib. Embryol. Carneg. Instn* (No. 138) **23**, 251.

—— (1932b). An intrachorionic mesothelial membrane in young stage of the monkey (*Macacus rhesus*). *Anat. Rec.* **52**, Suppl. 15.

—— and STREETER, G. L. (1941). Development of the macaque embryo. *Contrib. Embryol. Carneg. Instn* (No. 181) **29**, 15.

—— ROCK, J., and HERTIG, A. T. (1945). Two human embryos showing early stages of the definitive yolk-sac. *Contrib. Embryol. Carneg. Instn* (No. 201) **31**, 85.

HORSTADIUS, S. (1950). *The neural crest.* Oxford University Press, London.

HUGGETT, A. ST. G. and HAMMOND, J. (1952). Physiology of the placenta. In *Marshall's physiology of reproduction* (ed. A. S. Parkes) Vol. II, Chapter 16. Longmans, London.

KRAUS, B. S., KITAMURA, H., and LATHAM, R. A. (1966). *Atlas of developmental anatomy of the face.* New York.

LENNOX, B. (1960). Chromosomes for beginners. *Lancet* **i**, 1046.

LEWIS, B. V. and HARRISON, R. G. (1966). A presomite human embryo showing a yolk-sac duct. *J. Anat. (Lond.)* **100**, 389.

LEWIS, W. H. (1904). Experimental studies on the development of the eye in Amphibia. I. On the origin of the lens. *Rana palustris. Am. J. Anat.* **3**, 505.

—— and HARTMAN, C. G. (1933). Early cleavage stages of the egg of the monkey (*Macacus rhesus*). *Contrib. Embryol. Carneg. Instn* (No. 143) **24**, 187.

MALL, F. P. (1910) Determination of the age of human embryos and fetuses. In *Manual of human embryology* (eds. F. Keibel and F. P. Mall) Vol. I, p. 180.

MALPAS, P. (1964). Length of the human umbilical cord at term. *Br. med. J.* **i**, 673.

MAXIMOW, A. A. (1924). Relation of blood cells to connective tissues and endothelium. *Physiol. Rev.* **4**, 533.

METZ, C. B. and MONROY, A. (1967). *Fertilization.* London,

MORTON, W. R. M. (1949). Two early human embryos. *J. Anat. (Lond.)* **83**, 308.

MOSSMAN, H. W. (1937). Comparative morphogenesis of the fetal membranes and accessory uterine structures. *Contrib. Embryol. Carneg. Instn* (No. 158) **26**, 129.

MURRAY, P. D. F. and HUXLEY, J. S. (1925). Self-differentiation in the grafted limb-bud of the chick. *J. Anat. (Lond.)* **59**, 379.

NEEDHAM, J. (1931). *Chemical embryology.* Cambridge University Press.

—— (1934). *A history of embryology.* Cambridge University Press.

—— (1950). *Biochemistry and morphogenesis.* Cambridge University Press.

O'RAHILLY, R. (1973). *Developmental stages in human embryos,* Part A. Carnegie Institution, Washington.

PATTEN, B. M. (1953). *Human embryology.* Philadelphia.

ROCK, J. and HERTIG, A. T. (1942). Some aspects of early human development. *Am. J. Obstet. Gynecol.* **44**, 973.

—— —— (1944). Information regarding the time of human ovulation derived from a study of 3 unfertilized and 11 fertilized ova. *Am. J. Obstet. Gynecol.* **47**, 343.

—— —— (1948). The human conceptus during the first two weeks of gestation. *Am. J. Obstet. Gynecol.* **55**, 6.

ROTHSCHILD, LORD (1956). *Fertilization.* Methuen, London.

SCHULTZ-LARSEN, J. (1958). The morphology of the human sperm. *Acta pathol. microbiol. Scand.* Suppl. **128.**

SPANNER, R. (1935). Mütterlicher und kindlicher Kreislauf der menschlichen Placenta und seine Strombahnen. *Z. Anat. EntwGesch.* **105**, 163.

SPEMANN, H. (1938). *Embryonic development and induction.* New Haven.

STIEVE, H. (1931). Die Dottersackbildung beim Ei des Menschen *Anat. Anz.* **72**, Ergänzhft. 44.

STREETER, G. L. (1920). A human embryo (Mateer) of the presomite period. *Contrib. Embryol. Carneg. Instn* (No. 43) **9**, 389.

—— (1921). Weight, sitting height, head size, foot length, and menstrual age of the human embryo. *Contrib. Embryol. Carneg. Instn* (No. 55) **11**, 143.

—— (1942). Developmental horizons in human embryos. Description of age group XI, 13 to 20 somites, and age group XII, 21 to 29 somites. *Contrib. Embryol. Carneg. Instn* (No. 197) **30**, 211.

TEACHER, J. H. (1924). On the implantation of the human ovum and the early development of the trophoblast. *J. Obstet. Gynaecol. Br. Emp.* **31**, 166.

THOMPSON, P. (1907). Description of a human embryo of twenty-three paired somites. *J. Anat. Physiol.* **41**, 159.

—— and BRASH, J. C. (1923). A human embryo with head-process and commencing archenteric canal. *J. Anat. (Lond.)* **58**, 1.

WADDINGTON, C. H. (1956). *Principles of embryology.* Allen and Unwin, London.

WEISS, P. A. (1939). *Principles of development.* New York.

WILLIS, R. A. (1958). *The borderland of embryology and pathology.* Butterworths, London.

WITSCHI, E. (1948). Migration of the germ cells of human embryos from the yolk sac to the primitive gonadal folds. *Contrib. Embryol. Carneg. Instn* (No. 209) **32**, 67.

WOOLLAM, D. H. M. (1967–72). *Advances in teratology,* Vols 1–5. Logos Press, London and Paul Elek, New York.

3 Bones

R. J. HARRISON

INTRODUCTION

OSTEOLOGY is the study of bone and bones, the hard supporting tissues of the body. All the bones present in an animal are together called the skeleton. The endoskeleton is that part of the supporting structure within the animal; the exoskeleton is the outer superficial component, well developed in many invertebrates but represented in Man only by hair, nails, and teeth.

The word skeleton is derived from *skeletos* (Gk.) = dried, and regrettably it implies that all that is worth studying is the dead, dry bones left after crude preparation. In fact, living bones are plastic tissues, with organic and inorganic components, that change dramatically with age, that exhibit remarkable variations in shape and size, that can show sexual dimorphism and, although mainly supporting, have other important functions. Bones are affected by genetic, external, and internal environmental factors, and are thus subject not only to direct effects of disease but also indirectly to physiological and pathological processes involving other organs. When studying any bone, or a whole skeleton, a student must realize that information can be derived about the original owner of the bones on matters pertaining to age, sex, stature, body habitus, health, diet, race, genetic and endocrinological conditions, and even personal idiosyncrasies. Moreover, techniques have been developed to assess the antiquity of skeletal remains, thus providing information of anthropological interest.

The bony skeleton cannot be considered solely as a collection of variously shaped components. Each bone is linked to another, or to several others, by joints or articulations. This enables the skeleton, apart from providing support for soft tissues, to act as a system of levers to which muscles and other structures are attached. Where bones surround structures such as the brain, heart, lungs, or other organs they provide protection. The interior of bones is used to lodge and protect the bone marrow in which blood cells develop and the bone substance itself provides a store of calcium, and other salts that can be drawn upon when necessary.

Living bone consists of connective tissue fibres impregnated with mineral substances, making it almost equally resistant to compression and to tension (Thompson 1942), while giving it considerable elasticity. If a bone is decalcified by soaking it in dilute acid, it retains its shape completely but becomes flexible owing to the removal of mineral substances from the fibrous framework. If, on the other hand, the fibrous tissue is destroyed by burning, it still retains its shape but becomes brittle, inelastic and may crumble.

The connective tissue element consists of collagen fibrils in a ground substance, containing polysaccharides, which serves to bind the fibrils together. The mineral element resembles a phosphoric limestone or apatite, the principal ions being calcium and phosphate, though others are deposited in smaller amounts. The mineral salts are present as minute flattened crystallites which provide an enormous surface area. This, along with the vascularity of bone, facilitates the rapid ion exchange which is known to occur in bone. Radioactive isotopes of calcium and phosphorus and certain other elements (e.g. lead and strontium) may be included in this process.

ARTICULATIONS

To provide movement in an otherwise rigid framework, and to allow for growth, the bony skeleton is divided into separate parts joined by tissues which may be sufficiently flexible to allow movement to take place. The union between two or more adjacent bones is called a joint or articulation. Bones are said to articulate with each other at any joint. The surface of a bone which meets another at a joint is its articular surface and, where the bones can move freely on each other, a joint space is present. The articular surfaces are then covered by an adherent layer of smooth, well-lubricated, articular cartilage. The underlying bone is also smooth so that the articular surface of a dried bone is easily identified although the cartilage has gone. Where less movement occurs, the joint is made by some form of connective tissue such as fibrous tissue or fibrocartilage, and the articular surface is usually rough. The union may, however, be made complete by fusion of the bones so that movement or further growth between them is prevented.

Movements at joints are produced and controlled mainly by muscles which are attached to the bones, but the bones are held together by all the surrounding tissues. Each bone is enclosed in a dense layer of fibrous tissue (periosteum), and this sleeve runs from bone to bone, constituting the fibrous membrane of the capsules of joints. Here it may be thickened to form connecting bands or ligaments. These ligaments, together with muscles and other soft tissues, maintain the continuity of the skeleton and transmit tensile stresses. Compression stresses demand something more solid and the ends of bones are usually modified in shape and enlarged so as to provide a good bearing surface which distributes pressure.

ARCHITECTURE OF BONE

The form and structure of any bone are adapted to its functions of support and of resisting mechanical stresses. During its growth, and during adult life, each bone is continually being modified to maintain these functions as the stresses to which it is subjected change. Some bones, such as the mandible, alter dramatically in shape with advancing age, others exhibit little change once established. Alteration in shape and structure is brought about by changes in the rates of deposition and absorption of bone substance. The factors controlling these processes are complex and not only involve local activity by various cell types but are also affected by more general hormonal and metabolic changes.

The structural arrangements in any bone must not be considered without reference to its articulations with adjacent bones, and the attachment of muscles and ligaments to it. These related structures can cause or transmit either compression or tension forces which vary when a joint is moved to a different position. The internal structure of bones is arranged so as to resist all such forces.

Two fundamental factors lead to strength and economy of material. The first of these is an intimate combination of mineral salts and fibrous tissue. The other is the nature of the basic constructional units (osteones) of concentric microscopic tubular lamellae [FIG. 3.4] which resist bending equally in any direction and, being hollow, allow access of blood vessels to living tissue. Pritchard (1956) makes a further point that fibres in each lamella change in direction as it is traced longitudinally, the fibre bundles being plaited or woven, and that some regularly leave one lamella to pass into the next, binding them together so that bone is a continuum.

The femora shown in FIGURES 3.1 and 3.2 demonstrate certain essential architectural features. The bone substance forms trabeculae running in directions suited to their functions. In the **body** (*shaft*) of the femur the trabeculae are concentrated at the periphery and merge to form thick **compact bone** capable of resisting bending and other stresses, while the interior is free of bone substance and constitutes the **marrow cavity**. At the ends of long bones the thick tubular body is continued into an expanded meshwork of thin trabeculae, **cancellous** or spongy bone, within a thinned-out cortex. The trabeculae are strongly developed in regions subjected to compression or tensile stresses, as is shown in the upper end of the femur in FIGURE 3.2. The trabeculae spring from the compact bone, diverge, and arch over to reach the head of the bone and greater trochanter.

Lining the compact body there is a transitional zone of coarse cancellous bone, and when the bone is fresh the cavity is filled with soft, yellow marrow. The cylindrical structure of the body of a long bone provides almost as much strength as a heavier solid rod of the same diameter and material. Where the body is curved, as it usually is to a slight extent, there is a compensatory thickening of the compact bone in the concavity. This thickening may show as a ridge on the surface of the bone.

At the expanded ends of a long bone, the wide surface area distributes pressure over the articular cartilage and transmits shocks through a large number of suitably arranged trabeculae. The spaces between trabeculae accommodate red bone marrow.

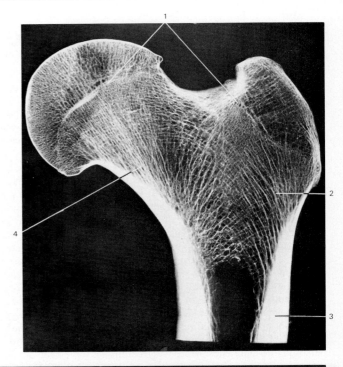

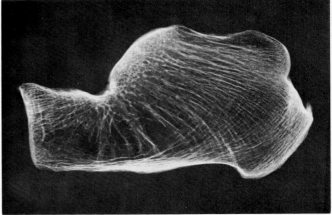

FIG. 3.2. Radiographs of sections of the upper end of femur and of calcaneus. The compact bone of the femoral cortex tapers to the thin cortex on the head, upper part of the neck, and greater trochanter. Note that the trabeculae of the cancellous bone are thickened in the lines of maximum stress in both.
1. Fusion lines for epiphyses of head and greater trochanter.
2. Tension trabeculae.
3. Thick cortex of body.
4. Compression trabeculae of neck.

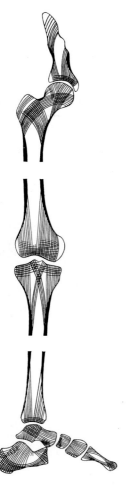

FIG. 3.1. Diagram of the architecture of the bones of the lower limb showing the direction of trabeculae. (After Meyer 1867.)

Bones without a body (e.g. the calcaneus, FIGURE 3.2) have a thin shell of compact bone, thickened where necessary, and are filled with cancellous bone and red marrow. Trabeculae in the calcaneus transmit the weight of the body from its upper articulations back towards the heel and forwards to the ball of the foot. Others curve forwards from the heel to the front of the bone. These knit together the other two columns and transmit the pull of the tendon of the calf muscles. In the space between these two systems there is less stress and consequently a more open texture of trabeculae.

CLASSIFICATION OF BONES

For descriptive purposes bones are classified according to their shape.

Long bones are found in the limbs and where they are small, as in the fingers and toes, are miniature long bones. Between the ends of each long bone is the body, a thick-walled tube of compact bone with a continuous, central medullary (marrow) cavity.

The remaining bones are classed as short, flat, or irregular, according to their shape, and consist of cancellous bone within a shell of compact bone. They have no body and no medullary cavity, but the interstices of the cancellous bone are filled with vascular red marrow.

Short bones include the carpal and tarsal bones of the wrist and foot respectively, and the sesamoid bones. The sesamoids are found within tendons and are named from their resemblance to sesame seeds. The patella is usually regarded as a sesamoid bone.

Flat bones are thin rather than flat, for all are curved to some extent. They include the ribs, the shoulder blade, and the bones of the skull vault. In the skull the cancellous substance is sandwiched between plates of compact bone, contains red bone marrow and many venous channels, and is referred to as diploë, the veins being called diploic veins.

Irregular bones include the vertebrae and many of the skull bones.

Pneumatic bones. In birds, in particular, many bones are invaded by air-containing sacs from the respiratory system to provide lightness with strength. In Man, a number of the skull bones are so invaded from the nasal cavity. The spongy part of the bone is absorbed and replaced by air-filled cavities with walls of compact bone lined by mucous membrane. This membrane is continuous through small openings with the lining of the nose and the cavities thus communicate permanently with the nose. These cavities are the paranasal sinuses. In some sinuses all the cancellous bone is not replaced, with the result that the air spaces form a series of smaller pockets or 'cells' as in the ethmoid bone. Of a similar nature are the mastoid cells which extend from the middle ear cavity into the mastoid process of the temporal bone.

FRESH BONE

Bone which has recently been removed from the body differs from the usual dried bones available for study.

The articular parts of most bones are covered by articular cartilage and, where the bones have been freely movable, the cartilage is smooth and slippery. But many bones have no articular cartilage and in the skull cap, for example, the rough articulating margins are united by fibrous tissue.

The whole of the bone, except the parts covered with articular cartilage, is invested by a membranous covering of periosteum. This consists of a layer of fibrous tissue containing connective tissue cells, with a second, deeper, osteogenic layer containing bone-forming cells, the osteoblasts. These cells are evident only when bone is actively being formed, during which process many of them are enclosed in the bone substance they produce, thus becoming osteocytes.

The periosteum is adherent to the bone because many of its collagenous bundles penetrate the bone and become incorporated in it. Vessels (and some nerves) pass from the periosteum into the bone to supply both the bone substance and the bone marrow. If periosteum is partially stripped off a living bone by accident or disease the loss of blood supply may be severe enough to cause death of the underlying bone.

Structures which are attached to bone (muscles, tendons, intermuscular septa, and ligaments) are in continuity with it. The constituent fibres blend with the periosteum and many pass into the bone to join its constituent collagen fibre bundles. Muscles with 'fleshy' attachments have many minute fibrous continuations into the periosteum and bone. Vessels which enter bone tend to follow ligaments or muscles to their point of entry so that tearing of these structures is likely to be associated with damage to the blood supply.

The medullary cavity is filled with yellow marrow and the spaces in the cancellous bone with red marrow. The cavities in cancellous bone and the medullary cavity are lined with a layer of osteogenic cells and fibrous tissue (endosteum) continuous with the lining of the canals which traverse compact bone [p. 80] and so indirectly with the periosteum.

BONE MARROW

Red and white blood cells are formed in the red bone marrow. After birth the marrow is the only normal source of red corpuscles and the chief source of granular leucocytes and of the stem cells of lymphocytes and monocytes.

In infants, red marrow pervades all the bones including the medullary cavities and even the larger canals. It is gradually replaced by yellow marrow, and at puberty red marrow is found only in the cancellous bone. As age advances even this is replaced by yellow marrow in the peripheral parts, such as the distal ends of long bones, and in the lower end of the vertebral column. Red marrow persists throughout life in most vertebrae, in the hip bones, proximal ends of the femur and humerus, ribs, skull, and sternum. Samples of this marrow may be obtained by sternal puncture.

Yellow marrow is simple in structure and is mainly composed of fatty tissue. A few blood-forming cells remain, however, and may increase in numbers as a result of prolonged blood loss or continued blood destruction.

Red marrow is an active blood-forming organ and its colour is due to red cells. It consists of large numbers of cells contained in a spongework of reticular tissue in which numerous blood vessels ramify and anastomose.

In addition to reticular cells and normal blood cells, the main groups of cells found in red marrow are haemocytoblasts which give rise to white and red blood cells by division, nucleated red blood cells at an intermediate stage of development not normally found in the blood stream, and megakaryocytes credited with the production of blood platelets. Large multinucleated osteoclasts are found on or near the bone surfaces and are believed to absorb bone.

Gelatinous marrow is the degenerated marrow found in the skull bones of aged people.

Nutrient blood vessels piercing the bodies of bones are few and relatively small. They supply the yellow marrow and the inner two-thirds of the bone substance, the remainder of the compact bone being supplied by periosteal vessels. In adult long bones, these nutrient vessels extend to the cancellous ends, but during growth they are separated by a plate of cartilage (epiphysial plate) from the separate, cancellous extremities, i.e. epiphyses [p. 82], which have an independent blood supply.

Vessels supplying cancellous bone pierce the thin shell of compact bone and ramify in the marrow. Where the marrow is red, these vessels are numerous, relatively large, and form a terminal mesh of wide, thin-walled sinusoids which slow the blood flow and discharge into large veins. Many of the sinusoids appear to be partially or completely collapsed, and are referred to as intersinusoidal capillaries. Small nodules of lymph tissue are present throughout red marrow but lymph vessels have not been found in either yellow or red marrow. Bones containing red marrow often have large foramina, but this is not invariably true, e.g. ribs, cranium, and sternum.

DRIED BONES

Bones are prepared for anatomical study by soaking in alkali solutions, degreasing, and drying (Edwards and Edwards 1959). Attached structures, including periosteum and cartilage, are removed in the processing; the endosteum and marrow shrivel, and fat is removed.

Where articular cartilage of a synovial joint [FIG. 4.3] has been present the underlying bone is smooth with a polished appearance so that articular areas can usually be distinguished. Where the cartilage on the end of a bone is adherent to a disc of fibrocartilage, as in the case of the upper and lower surfaces of a vertebra, the underlying bone is flat and has the appearance of unglazed porcelain. When bones have been joined together by fibrous tissue without intervening cartilage the articular surfaces are rough and frequently serrated.

Where fibrous tissue has been attached to bone, a rough line or ridge is likely to be raised, but a solid mass such as a tendon or ligament may leave a smooth circumscribed area, depressed or elevated, which can resemble an articular facet. Where fleshy muscle is directly attached, the bone is smooth unless the fibrous sheath forms a raised outline or a fibrous intramuscular septum raises a ridge.

For any one bone the markings and proportions vary in different individuals, but they are sufficiently alike for a general description to be applicable to all.

Cartilage

Cartilage (gristle) is, like bone, a supporting or strengthening tissue. In many places it is a temporary formation which is later replaced by bone, but in others it is permanent, persisting throughout life.

Cartilage is not hard like bone; nor is it as strong. It consists of living cells, chondrocytes, contained in spaces or lacunae and surrounded by intercellular substance containing collagen. The amorphous component stains with basic dyes due to the chondroitin sulphates present. These mucopolysaccharides and the collagen give firmness and some elasticity to cartilage but it can easily be cut or damaged by pressure over a small area.

Cartilage is relatively non-vascular and is nourished normally by tissue fluids. The chondrocytes divide and grow within the cartilage but this interstitial growth is limited to a certain size by the need for diffusion of nutrients through the matrix. Increase beyond this limiting size can only be achieved by penetration of blood vessels into the cartilage. Such a vascular invasion, however, is often associated with the deposition of calcium salts in the ground substance, and this calcification of cartilage usually results in death of its cells followed by removal of the calcified tissue and its replacement with bone, ossification of cartilage. Growth at the surface is provided for by the presence of chondrogenic cells underlying a more fibrous layer of its covering, the perichondrium. The articular cartilage of synovial joints is an exception in that it has no perichondrium on its articular surface, which is in effect a raw surface of cartilage.

CLASSIFICATION OF CARTILAGE

There are three main varieties of cartilage, namely, hyaline cartilage, and white and yellow fibrocartilage.

Hyaline cartilage appears homogenous, is bluish-white and translucent owing to the quality of its intercellular substance. It forms the temporary cartilage models from which many bones develop. Remains of these models are: (1) the articular cartilages of synovial joints; (2) the plates of cartilage between separately ossifying parts of a bone during growth [p. 82]; and (3) the xiphoid process of the sternum and the costal cartilages, all of which ossify late or not at all.

Hyaline cartilage is also found in the respiratory system; in the nasal septum, in most of the cartilages of the larynx and in the supporting rings of the trachea and bronchi. Calcification occurs frequently in old age in the costal cartilages and in those of the larynx, and may be followed by ossification. Such deposition of calcium salts can be detected radiologically.

White fibrocartilage contains strong bundles of white fibrous tissue with more or less widely distributed groups of cartilage cells, but little matrix, among the bundles. This structure gives the tissue greater tensile strength than found in hyaline cartilage and also sufficient elasticity to resist considerable pressure. It is found: (1) as sesamoid cartilages in certain tendons; (2) as articular discs in the wrist joint and the joints at each end of the clavicle: (3) as a labrum or rim deepening the sockets of the shoulder and hip joints; (4) as two semilunar cartilages in each knee joint; (5) as intervertebral discs joining the adjacent surfaces of the vertebral bodies; (6) in a plate or disc which unites the two hip bones at the pubic symphysis; and (7) as a covering to the articular surfaces of certain bones that ossify in membrane such as the clavicle and mandible. Almost any grade of cartilage between hyaline and virtually pure collagenous tissue can be encountered at these sites.

Yellow fibrocartilage contains bundles of yellow elastic fibres but little or no white fibrous tissue. It is found in the external ear, in the auditory tube associated with the middle ear, and in the epiglottis.

Bone as a tissue

Compact bone is hard but possesses some resilience. It consists of numerous cylindrical units, called Haversian systems after Clopton Havers who described them in 1691. Each system, also known as an osteone, has a central canal, about 0.05 mm in diameter, which is surrounded by up to ten, or more, concentric lamellae of bony tissue [FIG. 3.4]. The central canal contains a small artery and vein, areolar tissue, bone forming cells or osteoblasts, and fine nerve filaments. The larger canals contain lymph vessels. The Haversian systems are orientated in line with the long axis of a bone and may branch and communicate at irregular intervals.

Between the lamellae of a Haversian system are numerous minute spaces or lacunae, connected with each other and with the central canal by fine radiating canaliculi [FIG. 3.4]. Lying in the intervals between the Haversian systems are varying numbers of irregularly arranged interstitial lamellae [FIG. 3.4], which also possess lacunae and canaliculi and often have the appearance of partially eroded Haversian systems.

Bone has two principal components, blended together physically but each with its own functional importance. There is an organic matrix, largely of collagen, but with some mucopolysaccharide, which accounts for about 25 per cent of the weight of fully formed bone. There is also a mineral matrix consisting of calcium, phosphate, and variable amounts of magnesium, sodium, carbonate, citrate, and fluoride. The mineral matrix has a crystalline structure, and X-ray and electron microscope studies have shown it to be characteristic of crystals of hydroxyapatite. The mineral component is influenced by various factors, such as parathyroid hormone, and is able to exchange its ions with those of the body fluids; thus forming a ready store of calcium.

Development of bone

Bone develops in mesoderm [p. 46] as a sequel to the deposition of mineral salts in connective tissue or in a previously formed cartilage model.

When this process of calcification, followed by ossification, takes place initially in cartilage, the bone so formed is referred to as **cartilage bone**. When it takes place in fibrous tissue without the intervention of cartilage, the bone is referred to as **membrane bone**.

FORMATION OF CARTILAGE BONE

The commonest method of bone formation is for a preliminary foundation or model of cartilage to be laid down. This becomes calcified and then ossified to give rise to true bone.

The cartilage model

The first visible step is an aggregation of mesodermal cells which become visibly modified and rounded, **chondroblasts**, and lay down an amorphous intercellular substance between them.

Each cell enlarges and the surrounding amorphous substance forms a capsule round it. The cells divide and a capsule is laid down round each daughter cell, two new capsules being formed inside each older one. As the cells multiply and the formation of intercellular substance further separates them, the cartilage enlarges by interstitial growth, that is, growth of the cells and of the intercellular substance of the cartilage.

The surrounding mesenchyme forms a sheath of **perichondrium** with a superficial, more fibrous layer and a deeper cellular layer which forms chondroblasts. These provide for growth on the surface of the cartilage by apposition.

Bone formation

The cartilage model of a **long bone** grows largely at its ends. The oldest part is near the middle and it is here that the oldest cartilage cells lie. After a time the amount of intercellular substance has increased considerably. The cartilage cells swell, alkaline phosphatase appears in them (Greep 1948) and in the extracellular substance in which lime salts are then deposited as fine crystals less than 50 nm in length (Robinson and Watson 1952). The cartilage matrix thus becomes calcified and its cells, cut off from their nutriment, degenerate. The perichondrium becomes vascularized, and a clear ground substance is deposited between it and the cartilage. This substance differs from cartilage and is uncalcified bone or **osteoid tissue** which later becomes impregnated with calcium salts. The perichondrial chondroblasts survive as **osteoblasts** and some of them are buried in the osteoid tissue which they now produce and are then called **osteocytes**. In this way the body of the model comes to be encased in bone derived from the erstwhile perichondrium, now known as **periosteum**.

The calcified cartilage is short-lived. Cells from the periosteal region, including multinucleated **osteoclasts**, which can destroy bone, eat into the calcified cartilage. Blood vessels with bone-forming osteoblasts and blood-forming cells follow into the channels and spaces so formed, and the calcified cartilage is excavated and largely removed. The osteoblasts begin forming bone on the surface of the remaining spicules of calcified cartilage to form trabeculae. Some of the osteoblasts are trapped in the bone substance to become osteocytes. Each lies in a **lacuna** and sends many fine processes into

the bone to form **canaliculi**. Other cells multiply to fill the spaces between bone trabeculae with marrow.

Ultimately, the continued process of excavation by osteoclasts and rebuilding by osteoblasts leads to complete removal of all the calcified cartilage and, if the bone is long enough and wide enough, the excavated spaces will run together to form a **marrow cavity** supported externally by a sheath of periosteal bone.

The cartilage at the ends of the bone continues to grow by multiplication of its cells, particularly where it is continuous with the newly formed bone. Cartilage thus formed calcifies and is, in its turn, excavated by cells advancing from the new bone, invaded by blood vessels, and converted into cancellous bone. Multiplication of cartilage cells abutting on the calcified part keeps up a supply of cartilage, the cells of which are arranged in the characteristic columns seen in FIGURE 3.3. The calcified matrix lying between the columns forms the basis of the bony trabeculae which are later laid down around them.

The bone grows in length as a result of invasion of calcified cartilage produced by the active cartilage at its ends, and in width by deposition of periosteal bone on its surface. The cartilaginous ends increase in size partly by interstitial growth and partly by addition of new cells from the surrounding perichondrium.

Bone when first laid down is cancellous. It consists of a framework of delicate trabeculae surrounding a labyrinth of vascular spaces. It is known as woven, or woven fibred bone, and is seen in foetal bones, in certain situations in adult bones where cartilage is ossifying, and in early stages of repair in fractured bones.

New bone is laid down in two ways:

1. As successive thin concentric lamellae round blood vessels to form Haversian systems. These are associated with blood vessels which grow into the relatively wide channels between the bony trabeculae. The osteoblasts line these channels and lay down successive layers of bone substance so as to fill the space available.

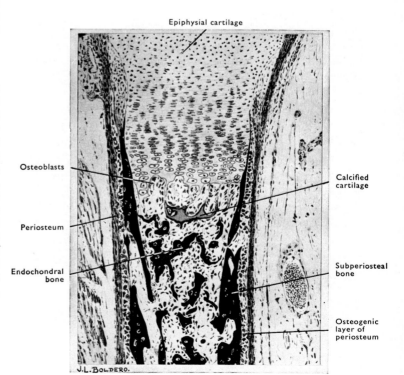

FIG. 3.3. Ossification in cartilage (Le Gros Clark 1971). Longitudinal section of metatarsal bone of foetal kitten. Note the arrangement of the zones described in the text.

Some of them become trapped between the lamellae so formed and, with their processes, produce **lacunae** and **canaliculi** in the bone substance. The innermost lamella is a tube surrounding the blood vessel and is lined with cells which through their processes link with others further out in the system. Completion of the

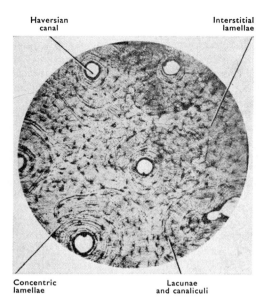

FIG. 3.4A. Photograph of ground transverse section of compact bone, showing Haversian systems (Haversian canal, concentric lamellae, lacunae, and canaliculi). × approx. 75.

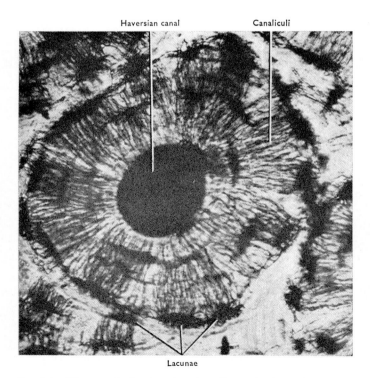

FIG. 3.4B. Photograph of a ground section of bone showing a single Haversian system. The lacunae, canaliculi, and the central tubular cavity all appear black owing to their content of air. Note how the canaliculi stream towards the central cavity from which the blood supply of the osteocytes is derived.

Haversian systems leads to formation of compact bone from cancellous bone. Small vessels which pass vertically in from the periosteal surface into the bone, are surrounded by the growing bone and produce spaces known as Volkmann's canals. Longitudinal periosteal vessels on the surface become submerged and give rise to superficial Haversian canals.

2. As successive thin layers or lamellae under the periosteum. The periosteal lamellae are also referred to as circumferential lamellae because they lie parallel to the surface and surround the bone [FIG. 3.5]. Bundles of collagen fibres from the surrounding tissue are incorporated in the bone at right angles to the periosteal lamellae and are called **perforating fibres** (Sharpey 1856).

Bone is a vascular tissue richly supplied with capillaries, and canals containing capillaries in bone must be close enough to adjacent cells to keep them alive. About 0.2 mm is the greatest distance that can be tolerated and in practice the thickness of bone substance between an osteocyte and its supply is not more than 0.1 mm.

Osteoblasts and osteoclasts are the cells which lay down and destroy bone respectively. The osteoblast was noted as long ago as 1845 by Goodsir and Goodsir and the osteoclasts in 1849 by Robin. Many details about the origin, differentiation, and activities of these cells remain obscure, and it may well be that one type can act like the other under certain conditions (Bourne 1956; Ham 1969).

Osteoblasts remain near the surface of bone and form a layer of cuboidal cells with large nuclei and basiphilic cytoplasm. They increase by mitosis and possibly by recruitment from neighbouring mesenchymal cells. They lay down the organic collagenous matrix and their ultrastructural characteristics are similar to those of other cells that produce protein. When active, alkaline phosphatase is present in osteoblasts but its function in mineralization is not clear. When inactive, osteoblasts take on an inconspicuous spindle-shaped appearance but are believed capable of renewed osteoblastic activity in response to certain stimuli, e.g. fractures.

Osteoclasts are usually large and multinucleated. The current view is that they arise by fusion of smaller cells (Tonna and Cronkite 1961). Their presence is indicated by eroded areas in nearby bone, especially at places where remodelling occurs, and they are found in small pits called resorption or Howship's lacunae. Their character-

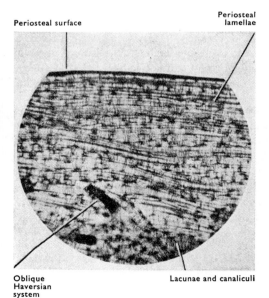

FIG. 3.5. Photograph of ground longitudinal section of compact bone, showing periosteal lamellae parallel to surface above and Haversian systems cut obliquely below. × approx. 75.

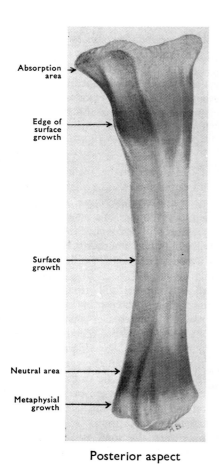
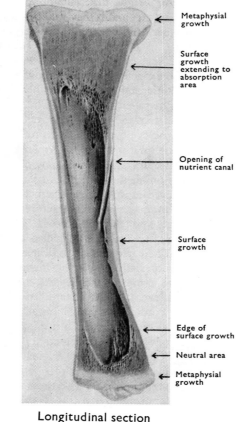

Absorption area

Edge of surface growth

Surface growth

Neutral area

Metaphysial growth

Posterior aspect

Metaphysial growth

Surface growth extending to absorption area

Opening of nutrient canal

Surface growth

Edge of surface growth

Neutral area

Metaphysial growth

Longitudinal section

FIG. 3.6. Diaphysis of tibia of madder-fed pig to illustrate the mode of growth of a long bone. (After Brash 1934.)
The animal was 40⅔ weeks old, and the madder had been omitted from the food for 30 days. The new bone added during that period is white, in contrast to the old, red bone.

istics include a conspicuous brush border, numerous mitochondria, and vacuolated cytoplasm. They increase in number in response to parathyroid hormone. After their activity has ceased, they seem to disappear or to revert to their original cell type. Reviews by Blackwood (1964), Frost (1964), and Richelle and Dallemagne (1965) are concerned with these cells and with various aspects of mineralization.

INTRAMEMBRANOUS OSSIFICATION

When bone is laid down in connective tissue other than cartilage the site is indicated by a condensation of cells and their associated collagen fibres; organic bone matrix is laid down, and later impregnated with bone salts. In this manner spicules or trabeculae are formed in association with cells which differentiate into osteoblasts. Some of these cells become incorporated in the ground substance laid down and so become osteocytes. Others continue to lay down what can now be called bone so as to thicken and increase the number of the trabeculae. Vascular connective tissue is incorporated in the spaces between the trabeculae, and as more and more bone is laid down concentrically in the spaces, Haversian systems are produced and the bone becomes compact where so induced; elsewhere it remains cancellous.

After development in membrane, the formation of new bone continues much as it does in bone developed in cartilage. The surrounding connective tissue condenses to form a limiting layer of periosteum. In the deeper part of this are osteogenic cells and blood vessels which permit the formation of Haversian systems and continued growth or reconstruction as required. Intramembranous ossification occurs in certain skull bones, including the vault and the mandible, and in part of the clavicle.

GROWTH OF BONE

The absorption of bony substance in some places, and its deposition in others, are processes which continue after the initial formation of the bone to provide for growth. These processes may be observed in bones taken from animals fed for suitable periods on substances, such as madder, which colour the bone, so that new bone laid down is white by contrast [FIGS. 3.6 and 3.67]. This continuation of controlled growth and modelling absorption permits alterations in the shape of bones, alterations which are necessary if proper proportions and the correct relative position of individual parts of the bone are to be preserved.

The principles involved have been stated by Brash (1934), Lacroix (1949), and Frost (1963). One essential feature is that bone is laid down on or absorbed from a surface, whether this is an outer surface, the surface of a cavity, or the surface of a lamella in cancellous bone. Thus, if bone is laid down on one side of a given part of a bone and absorbed on the opposite side, the effect will be to move that part bodily in one direction. This will hold good whether it is a projection or a hollow, the wall of a cavity such as the skull, or the whole thickness of part of a long bone.

There is no interstitial growth of bone, as distinguished from interstitial changes in density which are constantly taking place. The proof of this is threefold:

1. The insertion of two metallic marks in a long bone of an animal, as performed by John Hunter (1772) and others, shows that these remain the same distance apart as the bone grows in length. Metallic rings placed round the bone under the periosteum become embedded in the bone by surface accretion and may be found later in the marrow cavity which has enlarged by absorption.

2. Sometimes the long bones of a child show 'lines of arrested growth'. These are transverse zones of greater density caused by heavier calcification at the ends of a bone where it has grown at a slower rate during an illness. Repeated radiographic observations at increasing ages show that these markings remain the same distance apart as the bone grows in length (Harris 1933).

3. The indirect madder method of colouring the bones completely and then omitting the dye for a period, leads to the same conclusion. The increase in length is accounted for by the new white bone which appears at the ends, and the increase in girth by the subperiosteal deposit [Fig. 3.6].

All three methods, as well as modern techniques using radioactive isotopes and tetracycline as markers, show that growth in length of a long bone occurs at the ends only.

Measurements can be used to calculate the rates of growth (Payton 1932) and show that growth is greater at one end than the other. This is termed the 'growing end' of the bone. Both ends may be growing but the 'growing end', if damaged during growth, will lead to greater deformity than would the other if similarly damaged.

MODELLING OF BONE

Long bones grow in girth by subperiosteal deposition of bone, and the thickness of the cortex increases in relation to the stresses placed on the bone. This is achieved by the slower absorption of bone from the medullary aspect, so that the cavity enlarges less rapidly. Growth in length is by the continued ossification of the growing cartilage at the ends of the bone. As growth takes place at the junction between bone and cartilage, the bone which is laid down and not absorbed by the extension of the marrow cavity is left behind to become part of the body, as the process of ossification moves further away from the centre of the body. This bone has, however, been formed at a wide part of the bone, and modelling absorption must take place on the surface if the proper shape is to be maintained. This process is particularly important at the necks of long bones and in the ramus of the mandible and in similar situations where it is essential that the overall *shape* of the bone remains the same while the bone grows. Failure of such absorption may occur occasionally, even in bones that appear to be normal otherwise, and the result is a characteristic club-shaped appearance at the ends of the bones.

In bones such as those of the skull which are developed in membrane, growth takes place under the periosteum, mainly by deposition of new bone on the outer surface and edges and absorption from the inner surface.

Centres of ossification

Places where bone begins to be laid down are referred to as centres of ossification and the process spreads from such centres. The earliest, and usually principal centre in the body or shaft of a bone is referred to as a primary centre. This may result from fusion of several closely spaced smaller centres and, if they are further apart, there may be more than one primary centre appearing about the same time in one bone. Ossification centres can be detected by dissection, by staining with alizarin, and, when large enough, on X-ray flims.

Primary centres of ossification appear at different dates in different developing bones but most of them before the end of the fourth month (135 mm stage) of intra-uterine life; the majority appear between the seventh and twelfth weeks. Virtually all are present before birth. The date is fairly constant in different subjects for any one bone, and centres appear in an orderly sequence (Noback and Robertson 1951). In the following descriptions all dates of ossification stated in weeks or months refer to the antenatal period.

Secondary centres occur at a much later date than primary centres and are formed in parts of the cartilage model into which ossification from the primary centre has not spread. All long bones and many others acquire secondary centres. Nearly all of these secondary centres appear after birth, and the ossification which extends from them is the same as from primary centres, except that in this case the whole of the ossification takes place within the cartilage, which continues to enlarge by multiplication of its peripheral cells until growth ceases. The bone so formed is almost entirely cancellous.

The part of a bone ossified from a primary centre is called the diaphysis. The part ossified from a secondary centre is called an epiphysis, and a secondary centre is therefore often referred to as an epiphysial centre.

Ossification proceeds from the primary and secondary centres until only certain parts of the cartilage model are left.

1. A thin plate persists for a time between the diaphysis and each epiphysis. It is called the epiphysial cartilage, growth cartilage, or epiphysial plate [Fig. 4.3], and its edge at the surface of the bone forms the epiphysial line. So long as this plate of cartilage persists the diaphysis can continue to grow in length [Figs. 3.168 and 3.169], for it is the growth of this cartilage which produces the material into which the diaphysial ossification extends.

2. Where each epiphysis articulates with another bone, the articular layer of cartilage survives on that surface.

Epiphysial growth

Since the epiphysis develops within cartilage it is initially surrounded by this substance which can be said to have three parts, all of them proliferating but in different ways.

1. A plate of cartilage lying, as mentioned above, between the epiphysis and the diaphysis, the margin of which forms the epiphysial line at the surface of the bone. This cartilage produces new cells which allow for longitudinal growth of the diaphysis only.

2. If an epiphysis is articular, the cartilage from the margin of the epiphysial plate to the edge of the articular area is covered by perichondrium from which new cartilage cells may be formed to allow lateral expansion of the epiphysis.

3. The articular surface is devoid of perichondrium and here the cartilage cells grow in two directions:

   i. Towards the epiphysis. These cells calcify and ossify, allowing for expansion of the epiphysis towards the articular surface.

   ii. Towards the articular surface. These cells form the articular surface; they neither calcify nor ossify but act as replacements for the worn cartilage.

With the exception of the latter cells, which persist throughout life, but do not proliferate after growth of the epiphysis as a whole ceases, all the other parts of the epiphysial cartilage cease to grow, are invaded by the ossification process and are turned into bone. Extension of diaphysial ossification into the epiphysial plate finally obliterates it, leading to fusion of the epiphysis and the diaphysis, and thus preventing any further growth of the diaphysis in length. Extension of the epiphysial ossification to the perichondrium converts it to periosteum and prevents further endochondral growth of the epiphysis in this direction. With the cessation of growth, no further cartilage cells proliferate towards the epiphysis from its articular surface and the epiphysis therefore ceases to grow towards this surface. The cartilage which remains is entirely concerned with the continued production of new material for the bearing surface of the bone, though no new cells are formed by mitosis.

If an epiphysis is non-articular, as in the iliac crest, the cartilage is usually all replaced by bone.

Diaphysial growth

The diaphysis continues to grow by virtue of cartilage growth at each epiphysial plate, and this part of the diaphysis where ossification is spreading rapidly is sometimes called the **metaphysis**. Growth in width and thickness takes place under the periosteum and endosteum. Healthy periosteum and endosteum retain their ability to produce new bone when required, and absorption continues to take place where necessary. Bones are remodelled to meet increased exercise, or may atrophy from lack of it, and broken bones can be repaired by the renewed activity of osteoblasts beneath the periosteum and in the medullary cavity. During growth, the longitudinal trabeculae are consolidated before the transverse trabeculae are laid down, and conversely, in cases of atrophy, the transverse trabeculae are absorbed before changes occur in the longitudinal ones (Harris 1933). This keeps the length of the formed bone constant and so prevents interference with the function of joints and muscles.

OSSIFICATION AND FUSION OF EPIPHYSES

Every long bone has an epiphysis at one end, most have an epiphysis at each end, and some have more than one epiphysis at one or both ends, e.g. femur and humerus.

Most epiphyses are ossified from a single centre but if there is more than one centre they usually fuse before union with the diaphysis.

Secondary centres of ossification have been classified into: pressure, traction, and atavistic epiphyses (Parsons 1904, 1905, 1908). These are supposed to be associated respectively with pressure on articular surfaces, with the pull of muscles, and with bone elements which have lost their independence during evolution, but the evidence is lacking. Section of a tendon or muscle, before appearance of its 'traction epiphysis', does not arrest subsequent development of that epiphysis (Barnett and Lewis 1958). Members of the third group may remain as separate ossicles and cause doubt as to whether a fracture has occurred.

DATES OF OSSIFICATION AND FUSION OF EPIPHYSES

The dates at which certain epiphysial centres appear and those at which they fuse with each other or with the diaphysis, particularly the latter, may be of considerable use clinically.

The apposed surfaces of diaphyses and epiphyses have pronounced elevations and depressions which fit into each other so that the bones are to some extent locked together, and can withstand knocks and twists during the period of growth. However, an epiphysis may be damaged or broken off as a result of violence. When this occurs the epiphysial cartilage may be injured so that the bone stops growing prematurely at that part or grows unevenly. It is therefore important, when a bone is broken near one end in a young person, to recognize at the outset whether the condition is an ordinary fracture or a separation of an epiphysis. It is consequently useful to remember the approximate times at which epiphyses usually join.

Postnatal changes in epiphyses of long bones may be placed in three main periods.

1. Secondary centres of ossification appear from the time of birth to 5 years of age (Francis, Werle, and Behm 1939).
2. Ossification spreads from these centres until the age of about 12 years in girls or 15 years in boys (Francis 1940).
3. From 12 or 14 years to 25 years epiphyses fuse with the diaphyses, and growth ceases as fusion occurs.

The process is speeded up in girls at 5 years and again at 10 years (Todd 1933, 1937), so that after these ages the female is in advance of the male and epiphysial fusions are earlier.

Exceptions of practical use include the appearance of a centre in the distal end of the femur in the last month of foetal life. The proximal end of the tibia often develops a centre late in that month, as may the upper end of the humerus. A child is assumed to be 'full-term' if one of these centres is present. The clavicle has an epiphysis at its medial end the centre of which appears late (18–20 years). Fusion is delayed until between 20 and 30 years. A clavicle with an epiphysis present but unfused would be from a person between 18 and 30 years old.

As a general rule an epiphysis which appears early fuses late and vice versa. In the humerus, for example, a centre is present in the head shortly after birth but fusion does not take place until the age of 18 to 21 years. The centres at the lower end appear later (2–14 years) and fuse comparatively early (14–18 years).

The end of a bone with an epiphysis which fuses late grows for a longer time and more rapidly than the other end, and is the 'growing end' [p. 82].

Most of the epiphysial centres in the vertebrae and ribs, and in the shoulder and pelvic girdles do not appear until puberty, and fuse about the age of 25. The dates of ossification of the parts of the scapula and hip bone entering into the formation of the shoulder and hip joints conform more with the ends of the adjacent long bones than with the rest of the shoulder and hip girdles.

The ring-like epiphyses on the upper and lower surfaces of the vertebral bodies fuse at the fully adult stage, which may not be reached till the twenty-fifth year or later. The back, therefore, continues to grow after the limb bones have ceased, for these usually fuse before the twenty-first year.

Variations

For any one epiphysis, the date of appearance of its centre varies in different children by weeks or months, and sometimes by 1 or 2 years. The date of union varies even more, especially in the vertebral column. It is common knowledge that some people reach their full height in their teens, while others do not do so until after the age of 25.

The sequence of dates of union is remarkably constant and the intervals between them remain proportionately the same in different people (Stevenson 1924). If the first union takes place earlier than usual, ossification and growth will be completed at an early age. Dates given for individual bones in this text are approximations based largely on Schinz, Baensch, Friedl, and Uehlinger (1971), but also on Paterson (1929), Todd (1930), Flecker (1932), Greulich and Pyle (1959), and Krogman (1962).

FACTORS AFFECTING BONE GROWTH

As a result of the growth processes described above, the bones of an individual skeleton normally reach a particular length and shape by the time of epiphysial fusion. These features are usually maintained until affected by age or pathological changes. There are many factors, however, that influence bone growth while the skeleton is attaining its adult characteristics. They exert their actions chiefly on the regions of growth, and essentially affect the rate and duration of the proliferation of cartilage at these regions.

Genetic, hormonal, nutritional, mechanical, and neural factors influence bone growth. The general shape and over-all dimensions of a bone are genetically determined and genetic factors are primarily responsible for final adult height or stature. Experiments have shown that the development of the main anatomical features of a bone depends on intrinsic factors residing in the tissue itself (Murray 1936). The pattern and multiplicity of ossification centres in the skeleton reveal much about its phylogenetic history in that the centres are surprisingly persistent and may indicate where fusion has occurred between elements which are separate in other animals.

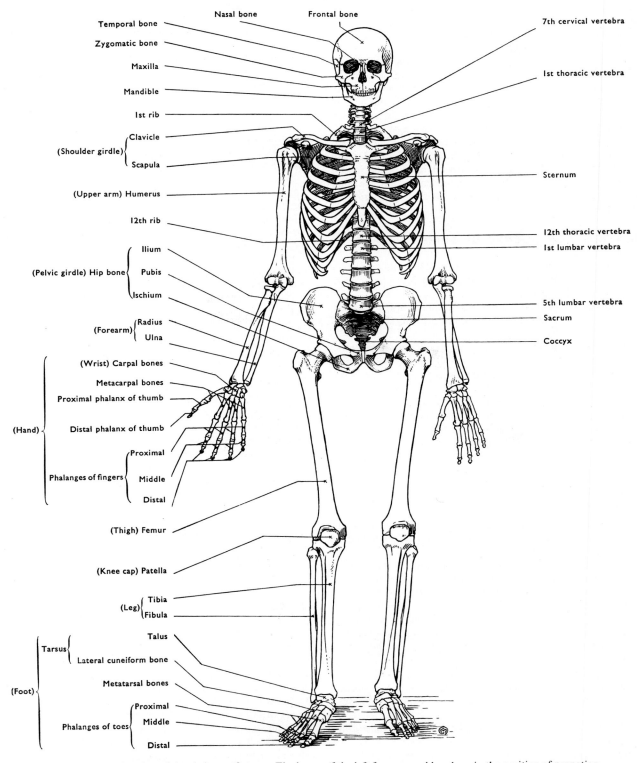

Temporal bone
Zygomatic bone
Maxilla
Mandible
1st rib

Nasal bone   Frontal bone

7th cervical vertebra

1st thoracic vertebra

(Shoulder girdle) {
Clavicle
Scapula

Sternum

(Upper arm) Humerus

12th rib

12th thoracic vertebra
1st lumbar vertebra

(Pelvic girdle) Hip bone {
Ilium
Pubis
Ischium

5th lumbar vertebra
Sacrum
Coccyx

(Forearm) {
Radius
Ulna

(Wrist) Carpal bones
Metacarpal bones
Proximal phalanx of thumb

(Hand)
Distal phalanx of thumb

Phalanges of fingers {
Proximal
Middle
Distal

(Thigh) Femur

(Knee cap) Patella

(Leg) {
Tibia
Fibula

Tarsus {
Talus
Lateral cuneiform bone

Metatarsal bones

(Foot)

Phalanges of toes {
Proximal
Middle
Distal

FIG. 3.7. Anterior view of the skeleton of a man. The bones of the left forearm and hand are in the position of pronation.

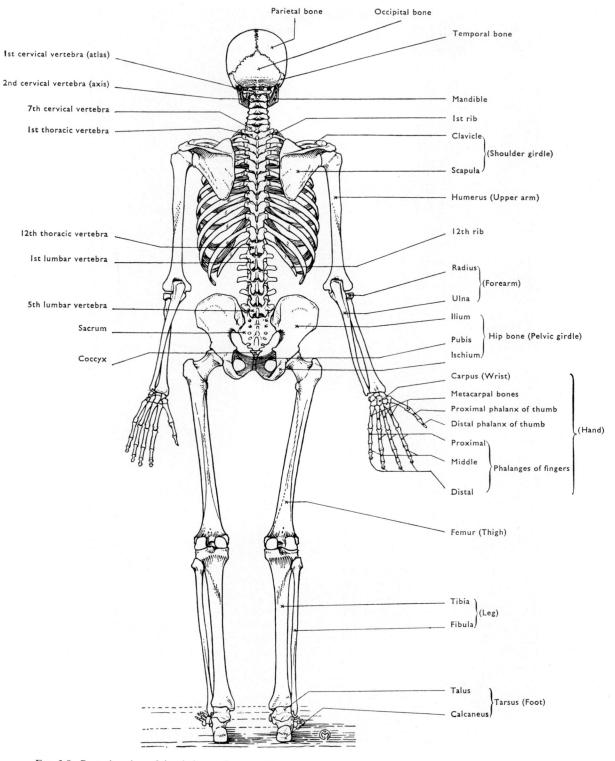

FIG. 3.8. Posterior view of the skeleton of a man. The bones of the left forearm and hand are in the position of pronation.

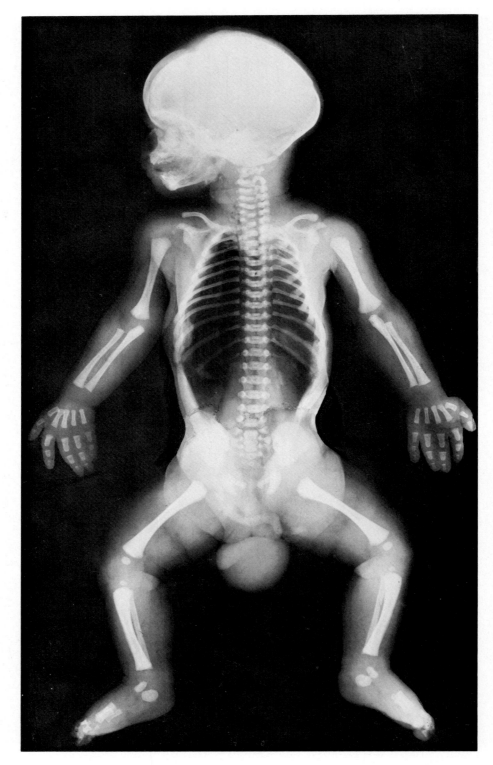

FIG. 3.9. Radiograph of full-term (nine month) foetus. Compare the proportions of the body at birth with those in the adult.

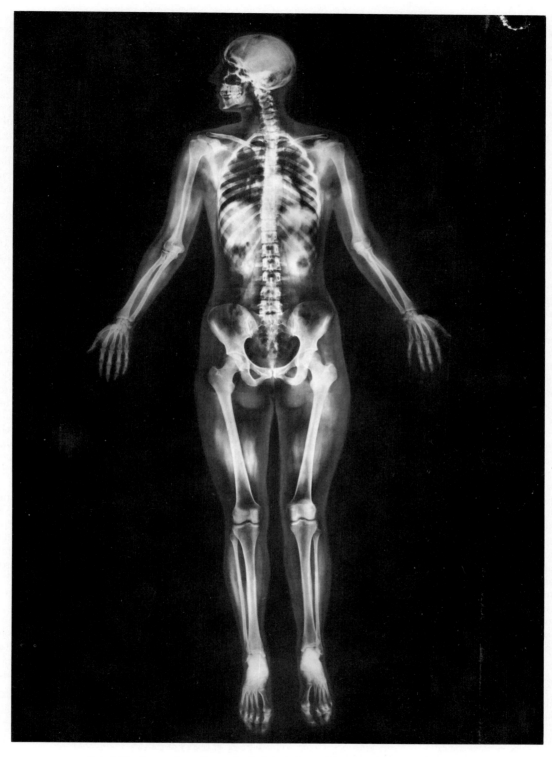

FIG. 3.10. Composite radiograph of adult skeleton. (*Courtesy of Messrs. Ilford Ltd.*)

The somatotrophic or growth hormone from the anterior pituitary has an effect on proliferation of cartilage and thus on growth of long bones. Hypersecretion of the hormone before the epiphysial cartilages are lost leads to gigantism, after they are lost to acromegaly. In the latter condition, growth continues in bones which grow by apposition, such as the mandible and the carpal and tarsal bones. Reduced secretion can produce a form of dwarfism associated with premature fusion of epiphyses. Reduction of testicular androgen secretion during the growing period, as after castration in the young, results in prolonged growth at the epiphysial cartilages of long bones, and thus accounts for the disproportionately long limb bones of some eunuchs.

Thyroid hormones affect cartilage cell proliferation, thus thyroidectomy in immature animals results in defective growth of long bones. Parathyroid hormone causes proliferation of osteoclasts, mobilizes bone calcium and phosphate (causing an increase of their concentrations in the blood), and later resorbs the collagen framework. Calcitonin, from the parafollicular or light cells of the thyroid, appears to inhibit the breakdown of bone, perhaps by an effect on the protein matrix, and to cause a fall in blood calcium.

Normal bone growth will obviously depend on an adequate supply of calcium, phosphorus, and vitamin D in appropriate form. Deficiency of these interrelated substances leads to failure of proper ossification, osteoporosis, and loss of mechanical strength; the cartilage, continuing to grow or even hypertrophy, causes enlarged and weakened ends to the soft bones (rickets). The region of the calcifying cartilage displays irregularity in the arrangement of cartilage cells, the bone formed there is imperfectly ossified, and an increased amount of vascular tissue is often present. This poor quality bone is unable to withstand normal stresses and may become grossly deformed. Vitamin C is required to stimulate normal production of the collagenous matrix. Healing of fractures is delayed even in mild scurvy. Lack of vitamin A has been shown experimentally to lead to abnormal skeletal growth with thickening of bones, and to cause interference in remodelling.

The influence of mechanical factors in certain situations indicates how plastic growing bone can be. The primary shape and main features of a bone are genetically determined but less conspicuous markings and final modelling are in the main conditioned by extrinsic factors (Le Gros Clark 1971; Ham 1969). Removal of the temporal muscles and some neck muscles in newborn rats results in bony markings such as the temporal line and nuchal crests failing to appear. Certain postural activities, such as squatting, may be the cause of extension of articular facets and changes in shape of a bone. Abnormal pressure, exerted by an enlarging tumour or an aneurysm, can cause extensive bone erosion. A completely denervated bone does not grow as much in thickness as it should, nor show such pronounced markings, though there is only slight loss in length: muscular paralysis plays a major part in this type of atrophy. Severe illness may leave its mark in lines of arrested growth discernible in radiographs as a thin region of dense bone deposition. Loss of normal stresses on bone leads to rapid demineralization, e.g. in prolonged weightlessness. Increased stress due to growth or muscularity causes periosteal and endosteal deposition of bone.

Radiology

The gross internal structure of bones is clearly seen on radiographs. The cortex, in virtue of its high calcium content, appears as a white line of varying thickness, in contrast to the cancellous bone which is of a more open structure and has a lace-like appearance. In some bones, the trabeculae of the cancellous bone have a linear arrangement, corresponding, it is thought, to the stress patterns to which the bone is subjected, e.g. in weight bearing. These are particularly obvious in the neck of the femur [FIG. 3.2].

Large muscle attachments can produce bony ridges on the cortex, e.g. the linea aspera on the femur, which appear as areas of greater opacity.

The Haversian systems are too small to be demonstrated on radiographs. The canal for the nutrient artery can often be seen, for example in the tibia and femur; the Y-shaped branching of the nutrient channel in the ilium is characteristic when present.

Bone also acts an an important reservoir of calcium and phosphorus in the body. As such it is influenced by processes, e.g. rickets or renal disease, which affect the metabolism of these salts.

The degree of mineralization tends to be proportional to stresses to which the bone is exposed (e.g. forces due to muscular pull or weight bearing). Thus immobility of a limb due to disease or fracture can result in a considerable loss of these elements.

The skeleton gradually loses calcium and phosphorus during ageing. Senile or post-menopausal *osteoporosis* represents such a loss and also a loss of cortical and trabecular structures. The cause of this bone loss is not fully understood but physiological decrease in muscular activity and involutional changes in the ovaries play a part.

During the growth period the epiphysial cartilages appear as translucent bands across a bone. Familiarity with common appearances will enable the student to distinguish most of these from fractures without difficulty, but in some areas it is necessary to be wary. In the upper part of the humerus, for example, the epiphysial line is an inverted cone and parts of it may appear at different levels where they lie parallel to the X-ray beam [FIG. 3.124]. If the observer is not familiar with this appearance and the individual concerned has suffered recent trauma, it may be interpreted as a fracture. In trauma to a child's elbow, it is often very difficult to decide if one of the numerous epiphyses has been displaced or not, especially if they show little evidence of ossification. Expert opinion is usually required before a definite opinion is reached, but as an interim measure it is often helpful to have radiographs of the other side for comparison.

Unfused epiphyses and accessory bones can sometimes simulate fractures on radiographs of adults. For example, in the foot the posterior process of the talus may appear as a separate bone (os trigonum), or there may be an apparently detached fragment of bone adjacent to the styloid process of the fifth metatarsal, representing an unfused epiphysis or a small accessory bone.

It is possible for quite large amounts of cancellous bone to be destroyed by disease, without this being apparent in the radiograph, because of the normal bone overlying the defect. Erosion of the thick cortical bone is more likely to be visible at an earlier stage. Also, areas not exposed to stress may contain so little cancellous bone that bone destruction is suspected if the normal is not known, e.g. the central region of the calcaneus. Radioisotope scanning provides, at present, the earliest evidence of defects in bone.

The skeleton

The main subdivision of the skeleton is into axial and appendicular parts. The axial skeleton includes vertebral column, sternum, twelve pairs of ribs, and skull. With the skull are associated the mandible, the hyoid bone, and three pairs of small ear bones or tympanic ossicles. The vertebral column is common to all vertebrate animals—fishes, amphibians, reptiles, birds, and mammals. The appendicular skeleton includes the bones of the upper and lower limbs.

The vertebral column

The vertebral column extends from the base of the skull through the whole length of the neck and trunk. It consists of a number of vertebrae, placed in series and connected together by ligaments and discs of fibrocartilage to form a flexible curved support for the trunk. In a child, the normal number of separate vertebrae is 33. In an adult, 5 have fused to form the sacrum and 4 to form the coccyx [FIGS. 3.29 and 3.31]. Of the 24 vertebrae remaining free, 7 are in the cervical region, 12 articulate with ribs and are thoracic, and 5 are in the lumbar region [FIG. 3.12]. They are numbered from the head end of each region.

Features common to most vertebrae

With the exception of the first two cervical vertebrae, all vertebrae consist of a large, anterior, weight-bearing body, and a posteriorly

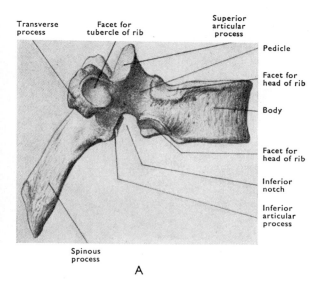

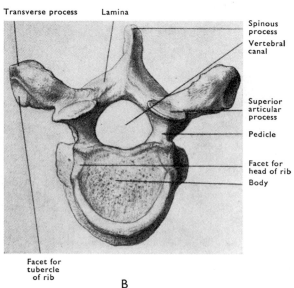

FIG. 3.11. Fifth thoracic vertebra. A, as viewed from the right side; B, as viewed from above.

placed vertebral arch. The arch springs from the posterolateral aspects of the body, and with it surrounds a large hole, the vertebral foramen. When the vertebrae are placed in series, these foramina, together with the ligamenta flava that unite the adjacent laminae (q.v.), form the vertebral canal which lodges the spinal medulla and its coverings.

The vertebral arch consists of two pedicles and two laminae [FIG. 3.11]. The sloping laminae meet in the midline posteriorly where they are continuous with the posteriorly directed spinous process, many of which can be felt in the back of the neck and trunk in the midline.

Each pedicle joins the anterolateral extremity of the corresponding lamina to the upper part of the posterolateral aspect of the vertebral body. Where each pedicle and lamina meet is a mass of bone from which three processes arise—the transverse process, directed laterally, and two articular processes, one directed superiorly, the other inferiorly. When adjacent vertebrae are fitted together [FIG. 3.21] the superior articular processes are in contact with the anterior surfaces of the inferior articular processes of the vertebra above, forming synovial joints with them. The apposed surfaces are, therefore, smooth articular foveae (facets) which are covered with articular cartilage during life.

Since the movements between vertebrae are about axes passing through the intervertebral discs, these movements are controlled by the arrangement of the joints between the articular processes.

There is a deep notch (incisura vertebralis inferior) on the inferior margin of each pedicle [FIG. 3.11A]. This forms the intervertebral foramen [FIG. 3.21] with the shallow notch (incisura vertebralis superior) on the upper aspect of the pedicle below when the vertebrae are articulated. The intervertebral foramen is closed antero-inferiorly by the intervertebral disc which unites the vertebral bodies. Thus the disc, the joint between the articular processes, and (in the thoracic region) the costovertebral joint all abut on the foramen, which transmits the spinal nerve and the blood vessels entering and leaving the vertebral canal.

THE BODY

This is approximately cylindrical, but its flat superior and inferior surfaces are slightly flared so that the anterior and lateral surfaces are concave from above downwards. These surfaces are perforated by many small vascular foramina, while the posterior surface is perforated by foramina for the large basivertebral veins from the red marrow in the body, and is flat or concave from side to side.

Each superior and inferior surface is rough for the attachment of the fibrocartilaginous intervertebral disc, and is not usually marked by its nucleus pulposus [FIG. 4.12]. At the periphery there is a smooth, slightly raised ring produced by the fused epiphysis.

THE SPINES (SPINOUS PROCESSES)

These posterior projections act as attachments for the interspinous and supraspinous ligaments (which strengthen the vertebral column posteriorly) and give attachments to many muscles. They are thickest and strongest where these attachments are maximal, particularly in the lumbar region.

TRANSVERSE PROCESSES

These act as attachments for muscles, helping to increase their leverage over the vertebral column. They form one articulation for most of the ribs, thus supporting them and controlling their movements. In regions other than the thoracic, the rib elements are fused with the transverse processes [FIG. 3.35].

STRUCTURE

The vertebrae are irregular bones consisting of cancellous bone enclosed in a thin shell of compact bone which is thickest at the margins of the intervertebral foramen, the inner surfaces of the laminae, and the superior edge of the spine.

The vertebral column as a whole

The vertebral column varies in length but is about 70 cm in a man and 60 cm in a woman.

It can diminish by nearly 2 cm during the day and measurements of height on a supine cadaver require a deduction of 2 cm to obtain the living height owing to straightening of the curves after death. Length diminishes in old age owing to reduction in thickness of the discs and to exaggeration of the curvatures, particularly in the thoracic region. The discs account for nearly a quarter of the total length.

The column can bear a weight of nearly 355 kg without crushing and a tearing strain of nearly 152 kg. Its weakest part is in the neck, which normally carries least weight, but the mode of application of a force is most important. Damage is particularly apt to occur at the junction of relatively fixed and freely movable parts such as in the

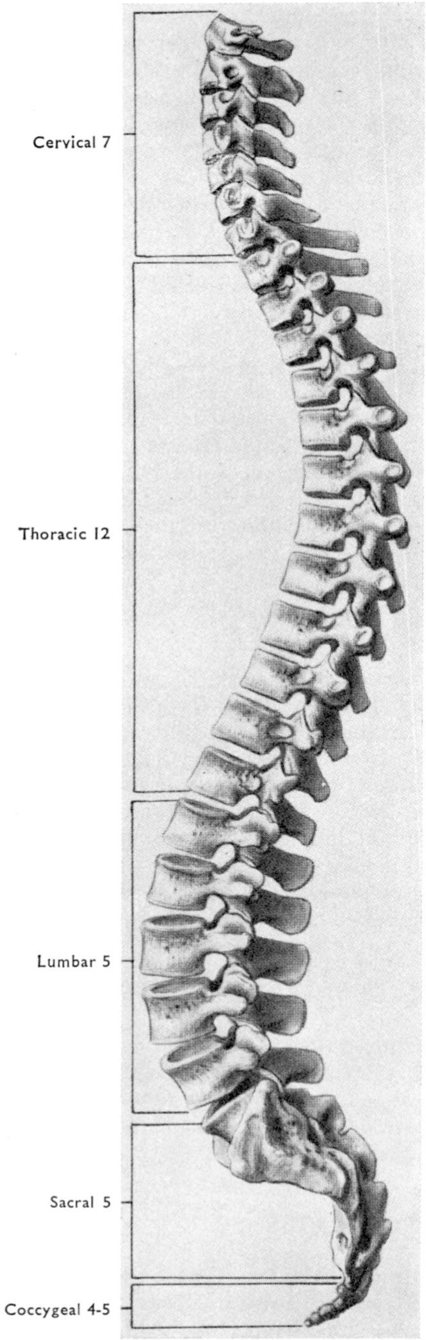

Cervical 7

Thoracic 12

Lumbar 5

Sacral 5

Coccygeal 4-5

FIG. 3.12. Left side of the vertebral column.

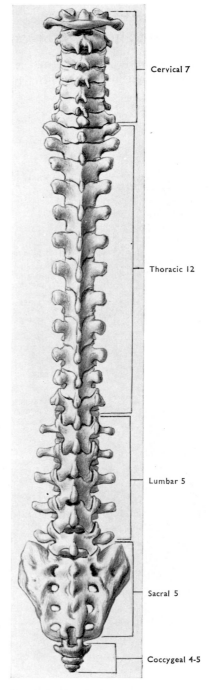

Cervical 7

Thoracic 12

Lumbar 5

Sacral 5

Coccygeal 4-5

FIG. 3.13. Vertebral column from behind.

thoracolumbar region or where a localized force can be applied by leverage (e.g. the dens or an intervertebral disc) or by direct contact (e.g. the coccyx).

Taken as a whole the vertebral column appears as in FIGURES 3.12 and 3.13. The vertebral bodies increase in size from the skull down to the point where the weight of the body is passed from the sacrum to the pelvic girdle, and then diminish rapidly [FIG. 3.7].

The transverse processes form a line of projections on each side which vary considerably in form, size, and direction. That of the first cervical is large and palpable below the mastoid process [FIG. 3.45]: those of the other cervical vertebrae are less easily felt at the side of the neck. The thoracic transverse processes decrease in size from the first to the twelfth; and the lumbar ones are again more prominent but are mainly equivalent to small fixed ribs.

Posteriorly, the spines project to a varying degree and are palpable in the midline from the seventh (vertebra prominens) or sixth cervical down to the sacrum and coccyx. The large second cervical spine is palpable through the ligamentum nuchae and muscles. The spines become increasingly oblique to the mid-thoracic region, and thus, like the laminae, overlap progressively, though both are widely separated in the lumbar region. In the cervical region gaps are opened up when the neck is flexed. The widest interlaminar gaps are between the atlas and axis, the fourth and fifth lumbar, and fifth lumbar and sacrum.

The third thoracic spine is level with the root of the spine of the scapula. The twelfth is midway between the levels of the lower angle of the scapula and the top of the iliac crest. The second sacral spine is level with the posterior superior iliac spines and the sacral promontory. The sacrum forms a bony resistance between the two iliac bones but its details are masked by strong connective tissue and ligaments.

The vertebral groove is the wide furrow alongside the spines. Its floor is formed by the laminae, the articular processes and the transverse processes.

Below the atlanto-axial joints the articular masses and the superior and inferior processes form a series of short rods articulated together and joined to the vertebral arch, but from the back they are scarcely distinguishable from the laminae. They lie deeply under the erector spinae.

CURVATURES

The vertebral column has four anteroposterior curvatures of which two are regarded as primary.

The fundamental curvature is concave ventrally from an early stage of development. At about the fifth month of intra-uterine life an additional curve, also concave ventrally, appears in the sacral and coccygeal regions and accommodates the pelvic viscera. As the head is raised from the thorax in the first few months after birth, a secondary curve, convex anteriorly, is formed in the neck and extends as far as the second thoracic vertebra. Later, towards the end of the first year, when a sitting posture, followed by an upright walking posture, is adopted, the cervical curvature is increased and the lumbar part of the column also becomes convex forwards to produce another secondary curve. This curve extends from the twelfth thoracic vertebra to the sacral promontory.

The cervical curve is the least marked and is undone as the neck is flexed. The lumbar curve is pronounced in the erect position and brings the bodies of the vertebrae very close to the anterior abdominal wall, its most anterior point being opposite the umbilicus, between the third and fourth lumbar vertebrae, or a little lower.

When upright, the alternating curves of the vertebral column provide for absorption of vertically directed shocks. These are not transmitted straight through the column but are absorbed by the

intervertebral discs and by slight bending of the curvatures which is resisted by muscles and ligaments.

Cervical vertebrae

The cervical vertebrae form a bony axis for the neck and are in a flexible part of the column. Their distinctive character is the foramen transversarium in each transverse process. They are small and do not carry much weight. The free ends of the transverse processes are little more than 2.5 cm from the median plane.

The lower border of the third cervical vertebra lies at the level of the upper border of the thyroid cartilage. The sixth is at the level of the cricoid cartilage (felt below the thyroid cartilage).

THIRD TO SIXTH CERVICAL VERTEBRAE

The small body of these cervical vertebrae is about one and a half times wider from side to side than from front to back. The upper surface is lipped up at the sides and the lower surface is correspondingly bevelled off. Between the lip and the bevel on

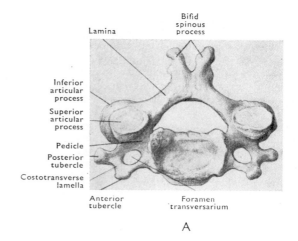

A

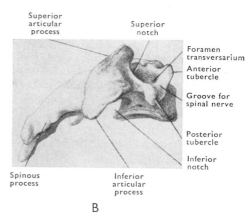

B

FIG. 3.14. Fourth cervical vertebra. A, from above; B, from the right side.

adjacent vertebrae there may be a small synovial joint at each side of the intervertebral disc.

The anterior surface is marked by the anterior longitudinal ligament near the midline and projects down to overlap the disc below. Laterally it is hollowed by the attachment of longus colli.

The posterior surface is almost flat and the short rounded pedicles project laterally as well as backwards so that the **vertebral foramen** is nearly triangular and is relatively large and roomy.

In the second to the fifth the **spine** is short and bifid at its free end where it gives attachment to the ligamentum nuchae and various muscles. The first has no spine.

The **transverse processes** are composite and project from the lateral side of the body and pedicle. The anterior portion is a rib element which has fused with the body and with the true transverse process. It passes laterally and slightly forwards to end in a prominent **anterior tubercle**, in a typical cervical vertebra. From this, a curved costotransverse lamella of bone, representing the tubercle of the rib, passes back to fuse with the true transverse process. The true transverse process element passes forwards and laterally from the end of the pedicle to join the rib element. It then terminates as a prominent **posterior tubercle** which projects laterally to give attachment to muscles.

The costotransverse lamella forms the floor of a groove which accommodates the ventral ramus of a spinal nerve as it emerges from behind the vertebral artery to pass between the tubercles.

The **foramen transversarium** is bounded by the pedicle posteriorly and by the two roots of the transverse process laterally and anteriorly. It transmits the vertebral artery and plexuses of veins and of sympathetic nerves from the inferior cervical ganglion, except in the seventh vertebra which transmits the accessory vertebral vein.

The **articular processes** are large. On each side they form the ends of a short pillar of bone, the articular mass, which carries the flat, oval facets lying in an oblique coronal plane on its extremities.

The **superior** and **inferior incisurae** are shallow and nearly equal in depth, the emergent nerves lie very close to the intervertebral articulations, especially their dorsal rami; those of the third and fourth nerves grooving the corresponding articular masses.

THE ATLAS

The atlas or first cervical vertebra is a ring of bone consisting of slender anterior and posterior arches united on each side by a lateral mass on which are situated the articular facets and the transverse process. Most of the body (centrum) of the atlas joins the axis to form the dens, and the anterior arch and lateral masses represent that part of the body normally formed by the vertebral arch ossification [FIGS. 3.33 and 3.35]. The facets articulate with the skull above and the axis below, and correspond to the small synovial joints at the lateral edges of the bodies of most of these vertebrae. Hence the spinal nerves emerge posterior to them. The normal articular facets are missing.

The **posterior arch** represents the pedicles and laminae of a typical vertebra. It is a curved bar of bone which joins the lateral mass and root of the transverse process anteriorly. Posteriorly, a **posterior tubercle** represents the spine and gives attachment to the nuchal ligament and a rectus capitis posterior minor muscle on each side.

Where it joins the lateral mass, the superior surface of the arch is grooved obliquely by the vertebral artery curving medially below the posterior edge of the articular facet to enter the skull. Here the first cervical nerve lies between the bone and the artery. The upper border gives attachment to the posterior atlanto-occipital membrane. The fibres bordering the gap which transmits the vertebral artery may be ossified from the arch to the edge of the articular facet, so as to convert the groove into a foramen.

The **anterior arch** is shorter, less curved, and flattened anteroposteriorly. A facet on the middle of its posterior surface is for articulation with the dens of the axis. An **anterior tubercle** on the middle of its anterior surface gives attachment to the longus cervicis muscles [FIG. 3.15]. The anterior atlanto-occipital membrane and anterior longitudinal ligament are attached to the upper and lower margins respectively.

The **lateral masses** project medially into the vertebral foramen, and have a rounded tubercle on the anterior part of the projection to which the transverse ligament of the atlas is attached. This ligament holds the dens against the facet on the anterior arch and separates it from the vertebral foramen.

The superior surface of each mass carries an elongated, kidney-shaped **articular facet** which is concave to receive the corresponding occipital condyle. The articular surfaces are closer together anteriorly than posteriorly, and face slightly medially as well as upwards. They may be notched or even divided into two facets. The lateral mass is slightly larger than the facets, leaving narrow rough areas laterally and medially and a tubercle posteriorly which overhangs the vertebral artery. The fibrous capsule of the joint is attached immediately beyond the margins of the facet.

The anterior surface of the lateral mass merges with the anterior arch and with the root of the transverse process. Near that root is a small swelling which corresponds to the anterior tubercle of a typical cervical transverse process, and has rectus capitis anterior attached to its anterior surface.

The **lateral aspect** is grooved below the superior articular facet by the vertebral artery, and immediately inferior to that the transverse process takes origin.

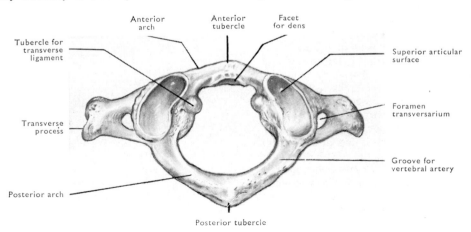

FIG. 3.15. The atlas from above.

The strong **transverse process** gives attachment to muscles which stabilize and help to rotate the head and atlas. Its free end may be tuberculated, or even bifid, but represents only the posterior tubercle of a typical cervical transverse process. The **foramen transversarium** lies close to the lateral mass and transmits the vertebral artery, its veins and sympathetic nerves.

The bar of bone limiting the foramen anteriorly represents the costotransverse lamella; the ventral ramus of the first cervical nerve emerges between the vertebral artery and the bone, and passes above the lamella.

The **inferior surface** of the lateral mass carries a nearly circular facet for articulation with the axis. It faces downwards and medially and is slightly concave.

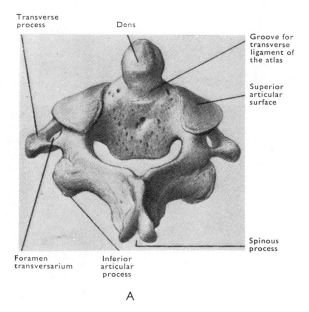

A

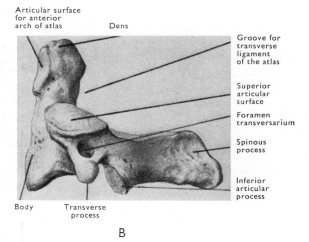

B

FIG. 3.16. Axis vertebra. A, from behind and above; B, from the left side.

THE AXIS

The second is the strongest of the cervical vertebrae. The separated centrum [p. 100] of the atlas is fused with the upper surface of the axis as a toothlike process, or **dens** [FIG. 3.16].

The **laminae** and **spine** are thick and strong, and the inferior surface of the body and the inferior articular process and facet resemble those of a typical cervical vertebra.

The **transverse process** is the smallest of the series, but resembles that of the atlas in that its rounded end represents the posterior tubercle, while the anterior boundary of the foramen transversarium is a bar of bone representing the costotransverse lamella. There is no anterior tubercle.

The **superior articular facets** are large, slightly convex, and face upwards and laterally. They lie at the junction of body and pedicle, with the second cervical nerves posterior to them, and they transmit the weight of the head to the body of the axis, leaving the dens free to act as a pivot. The vertebral arteries deviate laterally to pass round these large superior facets.

The **dens** projects up to lie within the 'ring' of the atlas. It is constricted where the transverse ligament crosses its posterior surface, and then widens slightly to form a head which tapers sharply to an apex.

The anterior aspect of the head carries a curved facet which is held against the facet on the anterior arch of the atlas by the transverse ligament, thus leaving the dens free to rotate. An apical ligament continues the line of the bodies from the apex of the dens to the anterior margin of the foramen magnum. A pair of strong alar ligaments run from the sloping sides of the apex to tubercles on the medial sides of the occipital condyles.

THE SEVENTH CERVICAL VERTEBRA

This has a long non-bifid spine which forms the upper of two, or the middle of three, prominences at the back of the root of the neck (vertebra prominens). The anterior root of the transverse process is slender and the anterior tubercle is small or absent. Occasionally it may be large (cervical rib) [FIG. 3.20] extending as a fibrous or bony strip which usually ends on the upper surface of the first rib, but may pass to the sternum inferior to the subclavian artery and the first thoracic ventral ramus passing to the brachial plexus. The posterior root is large and triangular. Its posterior tubercle extends as far laterally as the first thoracic transverse process.

The foramen transversarium is usually small and only transmits the accessory vertebral vein which is occasionally large.

Thoracic vertebrae

The twelve thoracic vertebrae lie in the posterior wall of the thorax, each with a pair of ribs attached to it.

The distinctive feature of these vertebrae is the presence of articular facets on the sides of the body for the heads of the ribs. Most of the ribs (2 to 9) articulate with the corresponding body, with an intervertebral disc *and* with the body superior to it. The first rib articulates with the body of the first thoracic vertebrae, and rarely just reaches the lower border of the seventh cervical. The last three ribs articulate only with the bodies to which they correspond numerically.

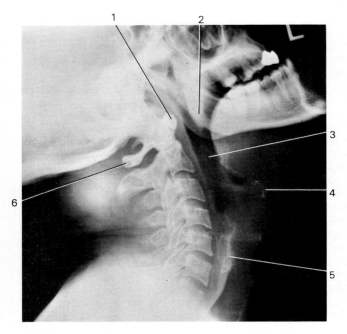

FIG. 3.17A. Lateral radiograph of the cervical vertebral column in extension.
1. Anterior arch of atlas.
2. Soft palate.
3. Pharynx.
4. Hyoid bone.
5. Cricoid cartilage, partly ossified.
6. Posterior arch of atlas.

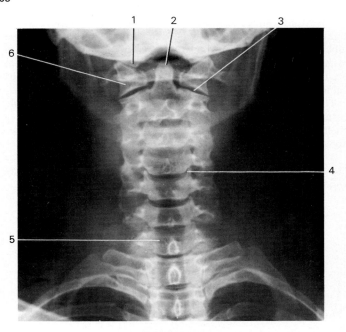

FIG. 3.18. Anteroposterior radiograph of cervical and upper thoracic vertebral column. Note that the image of the mandible has been reduced by moving it while the radiograph was taken.
1. Occipital condyle.
2. Dens of axis.
3. Atlanto-axial joint.
4. Synovial joint between vertebral bodies.
5. 7th cervical vertebral body.
6. Lateral mass of atlas.

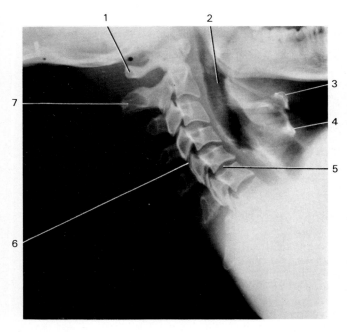

FIG. 3.17B. Lateral radiograph of the cervical vertebral column in flexion.
1. Posterior arch of atlas.
2. Pharynx.
3. Hyoid bone.
4. Thyroid cartilage, partly ossified.
5. Intervertebral disc between 5th and 6th cervical vertebrae.
6. Joint between articular processes.
7. Spine of axis.

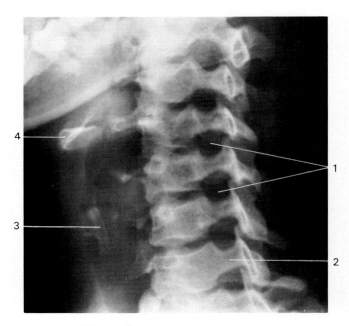

FIG. 3.19. Oblique radiograph of cervical vertebral column.
1. Intervertebral foramina.
2. Pedicle.
3. Thyroid cartilage.
4. Hyoid bone.

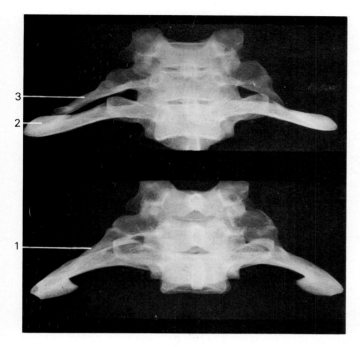

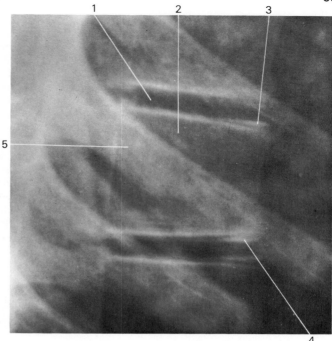

FIG. 3.20. Radiograph of specimen with a right and a left cervical rib. Both views are of the same specimen taken at slightly different angles. Note that the cervical ribs are almost invisible in the second picture because their images overlap those of the normal first ribs.
1 & 3. Cervical rib.
2. Normal 1st rib.

FIG. 3.22. Lateral radiograph of a mid-thoracic vertebral body. Age 16 years.
1. Intervertebral disc.
2. Vertebral body.
3 & 4. Epiphysial rings seen edge on.
5. 6th rib.

◀ FIG. 3.21. Left side of the first, ninth, tenth, eleventh, and twelfth thoracic vertebrae.

Superior ⎫   ⎧ mamillary
          ⎬ tubercles correspond to ⎨   processes of
Inferor ⎭   ⎩ acessory lumbar
                                              vertebrae
Lateral tubercle = posterior part of root of transverse ⎭

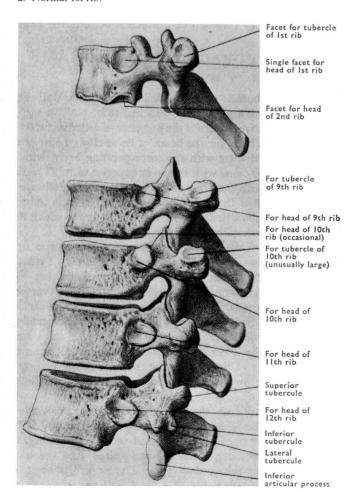

Facet for tubercle of 1st rib

Single facet for head of 1st rib

Facet for head of 2nd rib

For tubercle of 9th rib

For head of 9th rib
For head of 10th rib (occasional)
For tubercle of 10th rib (unusually large)

For head of 10th rib

For head of 11th rib

Superior tubercule

For head of 12th rib

Inferior tubercule
Lateral tubercule

Inferior articular process

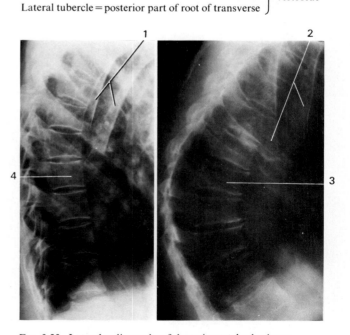

FIG. 3.23. Lateral radiographs of thoracic vertebral column.
(i) In mid-adult life.
(ii) In old age.
Note the increase in curvature (kyphosis), the 'wedging' of the vertebral bodies, and an anterior narrowing of the intervertebral discs in (ii).
1 & 2. Scapulae.
3 & 4. 7th thoracic vertebra.

The **bodies** of the thoracic vertebrae increase in size from the third downwards and bear increasing weight. Typically heart-shaped, they vary in size and shape from that of a cervical vertebra superiorly, to that of a lumbar vertebra inferiorly, the first having a lipped upper surface like a cervical vertebra. The body of the first and ninth thoracic vertebrae each have an articular facet near the upper border and a small part of one at the lower border, but the upper facet is on the body in the case of the first and at the base of the pedicle in the ninth. The single facet is practically on the pedicle in the twelfth [FIG. 3.21].

The **spines** are long, three-sided, and sloping, especially in the middle of the series where they are almost vertical. The upper and lower ones are -shorter and much less sloping, the upper ones resembling lower cervical spines, and the lower ones approaching the lumbar type. The twelfth is shaped like the first lumbar spine.

The **pedicles** share in the general increase in size and strength from above downwards. They pass almost directly back and are attached to the upper part of the body so that the inferior notch is deep.

The **laminae**, though narrow from side to side, are deep and overlap protectively like roof tiles.

The **vertebral foramen** is nearly circular and smaller than in the other regions.

The **transverse process** is long, thick, and rounded, and the free end is roughened for ligaments and muscle attachments. It is inclined backwards and a little upwards, and has the articular facet for the tubercle of the corresponding rib facing anterolaterally near its tip. The lower, flatter facets also face slightly upwards, but the eleventh and twelfth transverse processes have no facets. A typical twelfth transverse process is very short and has a trituberculated appearance owing to the presence of lumbar features, the mamillary and accessory processes.

The facets on the **superior articular processes** face mainly backwards but slightly upwards and laterally to permit rotation round the long axis of the vertebral column, as well as flexion and extension. These are overlapped by similar facets, facing the opposite direction, situated on the short **inferior articular processes** and extending on to the laminae. In the case of the twelfth thoracic vertebra the inferior articular processes are lumbar in type and fit between the first pair of lumbar processes, thus limiting rotation.

Lumbar vertebrae

The lumbar vertebrae lie in the 'small of the back' or loins, carry considerable weight, and are large. The rib element is completely incorporated in the transverse process, thus there are no articular facets for ribs and no foramina transversaria.

The fourth lumbar vertebral body is at the level of the highest part of the iliac crest [FIG. 3.7], whilst the umbilicus normally lies at the level of the disc above it.

The **bodies** are about 3 cm in depth and nearly 5 cm in width. The upper and lower surfaces are kidney-shaped in outline. They are flat

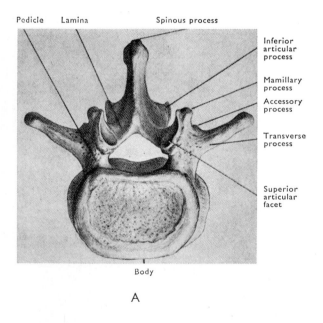

Pedicle   Lamina   Spinous process

Inferior articular process

Mamillary process

Accessory process

Transverse process

Superior articular facet

Body

A

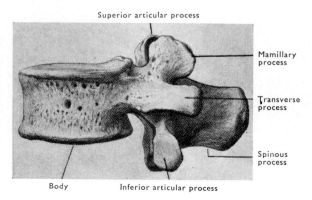

Superior articular process

Mamillary process

Transverse process

Spinous process

Body   Inferior articular process

B

FIG. 3.24. Third lumbar vertebra. A, From above; B, from the left side.

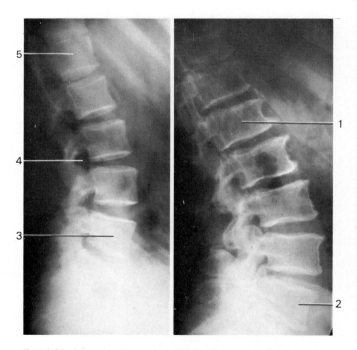

FIG. 3.25. Lateral radiographs of the lumbar vertebral column.
(i) In mid-adult life.
(ii) In old age.
Note the increase in curvature (lordosis), 'wedging' of the vertebral bodies, and posterior narrowing of the intervertebral discs in (ii).
1. 1st lumbar vertebra.
2 & 3. 5th lumber vertebra.
4. Intervertebral foramen.
5. 1st lumbar vertebra.

and almost parallel to each other, except in the case of the fifth, which is deeper in front than behind. Consequently, the normal lumbar curvature forwards is accounted for mainly by slight wedging of the intervertebral discs.

The short, stout **pedicles** project almost straight backwards. The laminae are relatively narrow. The vertebral foramen is triangular, but larger than in the thoracic region because of the wide lumbar body, and smaller than in the cervical region.

The **superior vertebral incisurae** are distinct but shallow; the inferior ones are deep and the structures emerging through the intervertebral foramen are related to two joints, the intervertebral disc anteriorly and the joint between inferior and superior articular processes posteriorly.

The **articular processes** are reciprocally curved in a horizontal plane but are almost straight in a vertical plane. The inferior facets face laterally and anteriorly, and fit between the superior ones, which are more widely set and face medially and posteriorly, thus allowing flexion and some extension, but limiting rotation severely. The inferior facets on the fifth are more widely set, flatter, and face in an anterolateral direction to fit the superior articular facets of the sacrum.

The **transverse processes** are flat and spatula-shaped. They project laterally and slightly backwards; the lower two appearing more oblique because their thick bases extend forwards towards the body. The third is the longest and the fourth and fifth are inclined upwards. The fifth is stout, short, and pyramidal, mainly for the attachment of the iliolumbar ligament and it may be fused with the lateral part of the sacrum.

Occasionally the rib element in a first lumbar vertebra may be separate as a **lumbar rib**. Posteriorly, at the root of each transverse process is a small tubercle called the **accessory process** (true transverse process) and a more rounded **mamillary process** lies on the posterior edge of the superior articular process. They give attachment to muscles.

The rectangular **spine** of a lumbar vertebra projects almost horizontally back, but owing to the slope of the laminae is level with the lower half of the body. It is wide from above downwards with a rounded posterior edge to give it a hatchet-shaped appearance.

The **laminae** in the lumbar region are widely separated, leaving diamond-shaped spaces between them filled in by the ligamenta flava, through which a lumbar puncture may be made.

In the lumbar region, psoas muscle lies in the angle between the bodies and the transverse processes; erector spinae fills the angle between the spines and transverse processes.

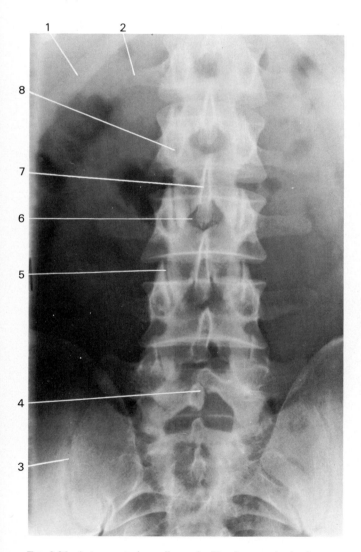

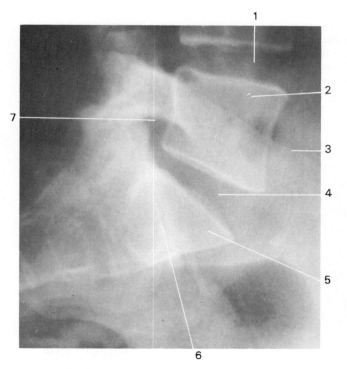

FIG. 3.26. Anteroposterior radiograph of lumbar vertebral column.
1. 12th rib.
2. Transverse process of 1st lumbar vertebra.
3. Sacro-iliac joint.
4. Spinous process of 5th lumbar vertebra. Note anomalous arrangement of laminae. This is common and not significant.
5. Joint between articular processes.
6. Interlaminar space.
7. Spinous process of 2nd lumbar vertebra.
8. Pedicle of 2nd lumbar vertebra.

FIG. 3.27. Lateral radiograph of the lumbosacral region. Note wedge shape of lumbosacral intervertebral disc.
1. Intervertebral disc between 4th and 5th lumbar vertebra.
2. Body of 5th lumbar vertebra.
3. Iliac crest.
4. Lumbosacral intervertebral disc.
5. Body of 1st sacral vertebra.
6. Surfaces of lateral parts of sacrum overlapping first sacral vertebral body.
7. Intervertebral foramen for 5th lumbar nerve.

RADIOLOGY

At birth the main ossification centres for the **vertebral bodies** are present with wide intervening spaces representing the surrounding cartilage and intervertebral discs. The **neural arches** of the vertebrae are seen to be incomplete. In the lateral view the bodies are convex superiorly and inferiorly and the anterior and posterior borders are often indented [FIG. 3.34].

At puberty the **epiphysial rings** appear as thin white lines on the superior and inferior surfaces of the vertebral body, most obvious at the edges because of the length of overlapping ring. They usually unite with the bodies of the vertebrae in the early twenties [FIG. 3.22].

In an anteroposterior radiograph of adult cervical vertebrae, the body is outlined as an elongated rectangle; synovial joints ('neuro-central') are seen at the lateral margins of the bodies [FIG. 3.18]. When the discs lose their elasticity, these joints commonly show pathological changes. In the thoracic and lumbar regions the bodies appear nearly square. The **pedicles** show as oval outlines of cortical bone nearer the upper than the lower surfaces of the bodies. The **spinous processes** are seen through the bodies. Those in the cervical region are mostly bifid (the second being the most obvious), in the thoracic region they are triangular, and in the lumbar region they are more massive and rectangular.

The **articular processes** and their synovial joints can sometimes be seen in anteroposterior views, especially in the thoracic region, but they are most obvious in oblique views [FIG. 3.28]. (For appearances in the lateral views see FIGURES 3.17B and 3.25.) However, oblique views are also of value in the cervical region for the demonstration of the intervertebral foramina, pedicles and 'neurocentral' joints [FIG. 3.19] and in the lumbar region for articular processes and related joints [FIG. 3.28].

Congenital anomalies are common in the vertebral column, fusion of one or more vertebral bodies (block vertebra) or persistence of a secondary epiphysial centre are amongst the commonest. Failure of development of one of the two primary centres of ossification in a vertebral body gives rise to a **hemivertebra** which has considerable

clinical significance because of the lateral curvature of the vertebral column (scoliosis) which results.

Failure of fusion of the neural arch (*spina bifida*) is a common anomaly which may be symptomless. Larger and more extensive examples of this defect are associated with imperfect development of the central nervous system.

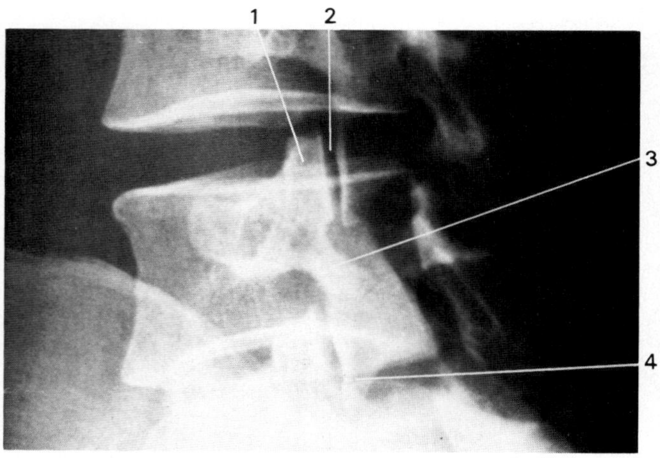

Sacrum and coccyx

SACRUM

The five fused sacral vertebrae [FIG. 3.29] provide strength and stability to the pelvis and transmit the weight of the body to the pelvic girdle through the sacro-iliac joints. The joint involves the upper three vertebrae and the bone tapers below that level to a blunt apex formed by the body of the fifth sacral vertebra about 2 cm above the level of the upper border of the symphysis pubis.

The sacrum is roughly wedge-shaped from base to apex and from pelvic to dorsal surface, and the base is directed more forwards than upwards. It is bent so that the pelvic surface of the anterior part of the bone makes an angle of about 10 degrees with the horizontal when the body is erect, while the posterior part lies at right angles to it.

The **pelvic surface** faces downwards and forwards in the upright position and the **dorsal surface** faces in the reverse direction [FIG. 8.27].

The **base of the sacrum** is the anterosuperior surface of the first sacral vertebra. The body of this vertebra forms an oval elevation occupying about one-third of the width or less in the female, more

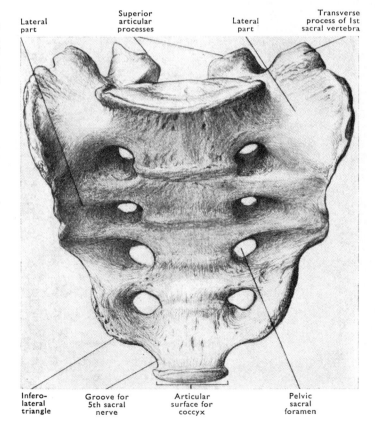

FIG. 3.28. Oblique radiograph of 4th lumbar vertebra. The narrow 'interarticular' part of the lamina is sometimes fractured or deficient, and may subsequently allow forward slipping on the vertebra below, commonly at this level (spondylolisthesis).
1. Superior articular process.
2. Joint between articular processes.
3. Narrow 'interarticular' part of lamina.
4. Inferior articular process.

Lateral part — Superior articular processes — Lateral part — Transverse process of 1st sacral vertebra

Infero-lateral triangle — Groove for 5th sacral nerve — Articular surface for coccyx — Pelvic sacral foramen

FIG. 3.29. The sacrum (pelvic surface).

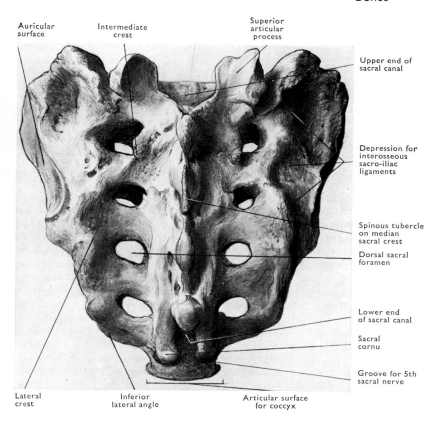

Auricular surface

Intermediate crest

Superior articular process

Upper end of sacral canal

Depression for interosseous sacro-iliac ligaments

Spinous tubercle on median sacral crest

Dorsal sacral foramen

Lower end of sacral canal

Sacral cornu

Groove for 5th sacral nerve

Lateral crest

Inferior lateral angle

Articular surface for coccyx

FIG. 3.30. The sacrum (dorsal surface).

than one-third in the male. Its anterior border projects into the superior aperture of the pelvis to form the **promontory** of the sacrum [FIG. 3.153]. On each side the costal elements are fused with the body, pedicle, and transverse processes, and are enlarged to form the **lateral part** of the sacrum which is wing-shaped on its abdominal aspect. Its rounded inferior edge forms part of the linea terminalis of the pelvis. Posteriorly, the pedicle and transverse process are incorporated in the lateral part but a pit or, occasionally, a foramen may be present between them and the enlarged rib element. From the region of the pedicles large flattish articular processes project to articulate with the fifth lumbar inferior processes, and beyond these the laminae converge to fuse in a short spine. The vertebral foramen so formed is triangular and is the upper end of the **sacral canal**.

The dorsal surface is rough, and has a series of longitudinal ridges formed from the fused posterior parts of the vertebrae [FIG. 3.30]. In the midline, the **median sacral crest** carries tubercles representing the spines, but is missing to a variable extent inferiorly where the laminae do not fuse, leaving the **sacral hiatus**. During life this is filled with fibrous tissue through which local anaesthetic may be injected into the sacral canal (epidural anaesthesia).

Lateral to the median sacral crest is a rough plate of bone formed by the fused laminae and ossified ligamenta flava. Immediately lateral to this is a row of small tubercles (**intermediate sacral crest**) representing the articular masses, and beyond this a row of four dorsal sacral foramina. These transmit the dorsal rami of the sacral nerves and small blood vessels, and are separated by bars of bone representing parts of the transverse processes. Lateral to this is a double row of small tubercles representing the tips of the transverse processes (**lateral sacral crest**). The most lateral part, representing the ribs, is well developed in the upper three sacral vertebrae. This part articulates with the ilium, and is roughened by the strong interosseous sacro-iliac ligaments immediately behind the articular, **auricular surface** [FIG. 3.30]. Below this region the sacrum narrows, and ends in a small, oval apex which is joined to the coccyx by a

narrow intervertebral disc. At the sides of the lower end of the sacral hiatus the **sacral cornua** project down to articulate with the corresponding processes of the coccyx [FIG. 3.30].

In lateral view the angulation is obvious between the massive, horizontal, anterior part and the thinner, postero-inferior, vertical part which descends so that the tip of the coccyx is level with the upper margin of the pubic symphysis [FIG. 3.154]. The L-shaped, auricular surface, covered by cartilage, forms a tight-fitting synovial joint between the first three (or two) sacral vertebrae and the corresponding surface on the ilium. It has ventral sacro-iliac ligaments attached to its anterior and inferior borders, and the interosseous sacro-iliac ligament attached to its posterosuperior border [FIG. 4.44]. The dorsal part of the sacrum is slightly narrower than the ventral part and the auricular surfaces are sloped accordingly. The interosseous ligaments hold it wedged between the ilia.

The remainder of the lateral part of the bone is an edge which gives attachment to ligaments. The sacrospinous and sacrotuberous ligaments, in particular, are attached to this edge and to the dorsal surface. They hold down the posterior part of the sacrum and resist the tendency of the body weight to rotate the base into the pelvis [FIG. 4.44]. The pelvic surface is concave and relatively smooth. It is limited anteriorly by the promontory and rounded margin of the lateral part of the sacrum on each side. In the midline is a column of bone formed by the fused bodies of the five sacral vertebrae and ossified intervertebral discs which form ridges between the bodies.

On each side, lateral to the intervertebral discs, lies a row of four **pelvic sacral foramina** which transmit the ventral rami of the sacral nerves, branches of the lateral sacral arteries, and communications between the pelvic veins and the internal vertebral venous plexus. These foramina are larger than the dorsal ones and lead from the sacral canal towards the greater sciatic notch, grooving the lateral part of the sacrum. The fifth sacral nerve sends its ventral ramus round the projecting apex of the bone.

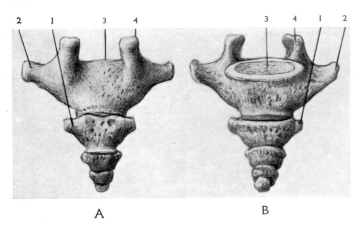

FIG. 3.31. The coccyx.
A, Dorsal surface. B, Pelvic surface.
1. Transverse process.
2. Transverse process.
3. For sacrum.
4. Cornu.

COCCYX

The coccyx can be felt as the lowest part of the backbone in the groove between the buttocks. It is roughly triangular in outline, tapering to an apex below, and consists of four more or less fused coccygeal vertebrae [FIG. 3.31]. The **base** is formed by the first vertebra and shows an oval body for articulation with the apex of the sacrum, two **cornua**, or superior articular processes, articulating with the sacral cornua and two transverse processes. These parts become progressively more indefinite in the succeeding vertebrae until the terminal one is only a nodule of bone. The number may be reduced to three or increased to five.

SEXUAL DIFFERENCES

In the female the sacrum is often wider in proportion to its length than in the male and the lateral parts form nearly two-thirds of the width of the base. The body of the first sacral vertebra is large in the male. In the female, the auricular surface tends to be limited to the first two vertebrae, and the joints, including the sacrococcygeal, are more mobile and less likely to be fused. The curvature of the sacrum in the male is fairly uniform; in the female it is flatter anteriorly and its angulation is more pronounced.

Ossification of vertebrae

A typical vertebra is ossified in cartilage from three primary centres and five epiphyses.

The **primary centres** appear in the body and in each half of the vertebral arch. That for the **centrum**, the larger, median part of the body, appears dorsal to the notochord first in the lower thoracic region (about 10th intra-uterine week), then spreads up and down the column so that virtually all centra have begun to ossify by the twentieth week. The last two pieces of the sacrum are delayed as late as the thirtieth week and the coccygeal centres appear after birth. The centra are usually formed from two ossific centres; they may fail to unite so that the centrum is formed in two separate halves, or one half only may ossify, producing a **hemivertebra**.

A primary centre appears in each half of the **vertebral arch** (at the junction of pedicle and lamina) in the upper cervical region (approximately seventh week) and sequentially in the lower

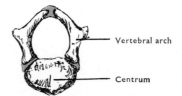

FIG. 3.32. Ossification of vertebrae.

vertebrae to reach the sacrum (approximately twentieth week). Ossification from each centre spreads into the lamina and pedicle, the latter extending on to the centrum to complete the body and articulate with the rib (thoracic region).

At birth the centrum and the two halves of the vertebral arch are ossified to form three pieces which are united by cartilage [FIG. 3.32]. The cartilage between the centrum and each vertebral arch forms a **neurocentral joint** (vertebral arch = neural arch) ossification of which begins in the cervical region (third year) and extends to the other regions by the seventh year. The cartilage between the laminae is continuous with the cartilaginous spine. The laminae begin to unite soon after birth in the lumbar region. The process spreads to the cervical region by the second year (except for the arch of the atlas) and is not completed in the sacrum till the seventh or tenth year. The degree of fusion in the sacrum is variable. After fusion of the laminae, ossification spreads into the spine.

Secondary centres, the vertebral epiphyses, appear about the age of puberty; those for the bodies at about 9 years, those for the tips of the spines and transverse processes at about 18 years. Those for the bodies are multiple and fuse to form flat rings on the circumferential parts of the superior and inferior surfaces.

The various epiphyses so formed begin to fuse with the rest of the bone at 18 years and fusion should be complete at 25 years of age. Primary centres for the coccygeal bodies appear between birth and puberty and ossification spreads without formation of secondary centres.

In the **atlas** a primary centre for each half of the vertebral arch and the corresponding lateral mass appears about the ninth week, and the two halves unite through the posterior arch in the third or fourth year. The anterior arch is cartilaginous at birth and a centre appears during the first or second year, to fuse with the lateral masses soon after the seventh year. It includes part of the superior articular surface [FIG. 3.33]. An epiphysis for the transverse process unites after puberty.

In the **axis** primary centres for the vertebral arch appear about the seventh week, and one or two for the lower part of the body early in the fifth month. Two more appear side by side for the dens and upper part of the body later in that month, and fuse together by the seventh month.

At birth the axis is therefore in four pieces which unite between 3 and 6 years of age. Secondary centres appear in the tip of the dens between 2 and 6 years and for the lower surface of the body about puberty. These epiphyses fuse before 12 years and between 18 and 25 years respectively.

The tip of a bifid cervical spine has a pair of epiphyses and the sixth and seventh cervical vertebrae may have separate primary centres for the costal elements. These normally join the rest of the bone early at the fifth year but the seventh in particular may remain separate to form a cervical rib.

Lumbar vertebrae have centres for the mamillary processes and the first lumbar may have separate primary centres for its transverse processes. These may remain separate to form a lumbar rib on either or both sides.

The fifth may have two primary centres in each half of the vertebral arch, the two parts of each half then being united by cartilage set obliquely between the superior and inferior articular

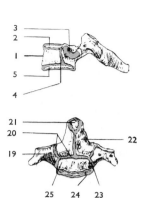

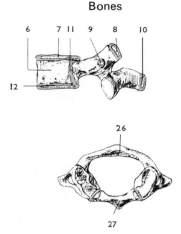

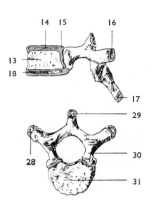

FIG. 3.33. Ossification of vertebrae.

Cervical vertebra
1. Centre for body.
2. Superior epiphysial ring.
3. Anterior bar of transverse process ossified by lateral extension from pedicle.
4. Neurocentral joint.
5. Inferior epiphysial ring.

Axis
19. Centre for transverse process and vertebral arch.
20. Joints close about 3rd year.
21. Centre for summit of dens.
22. Centre for lower part of dens.
23. Neurocentral joint.
24. Inferior epiphysial ring.
25. Single or double centre for body.

Lumbar vertebra
6. Centre for body.
7. Superior epiphysial ring.
8. Centre for mamillary process.
9. Centre for transverse process.
10. Centre for spine.
11. Neurocentral joint.
12. Inferior epiphysial ring.

Atlas
26. Posterior arch and lateral masses ossified from a single centre on each side. In this figure the posterior arch is represented complete by the union posteriorly of its posterior elements.
27. Anterior arch and portion of superior articular surface ossified from single or double centre.

Thoracic vertebra
13. Centre for body.
14. Superior epiphysial ring.
15. Neurocentral joint.
16. Centre for transverse process.
17. Centre for spine.
18. Inferior epiphysial ring.

Thoracic vertebra
28. Centre for transverse process.
29. Centre for spine.
30. Centre for vertebral arch on each side. The arch is here shown complete posteriorly.
31. Centre for body.

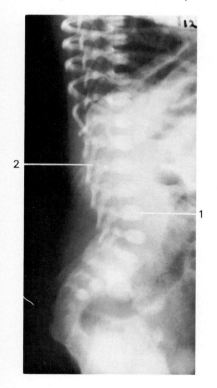

FIG. 3.34. Lateral radiograph of an infant to show vertebral column. The convexity of the upper and lower surfaces of the vertebral bodies is obvious in the lumbar region. The notching of the anterior and posterior surfaces shows in the lower thoracic region.
1. Centrum.
2. Neural arch.

processes. There is a risk of separation through this joint which is usually temporary.

The sacrum has numerous centres but they are of little importance, except for the varying degrees of fusion or consolidation which may occur. Primary centres appear between the third and eighth months, one for each of the centra and half vertebral arches and for each costal element in the upper three or four vertebrae. The costal parts fuse with the arches at about 5 years, the arches with the centra a little later, and the halves of the arches join between the seventh and tenth years.

The segments of the lateral mass fuse together at puberty and epiphysial centres for the bodies appear about the same time. The epiphyses and the bodies all fuse together between 18 years and 25 years. Numerous epiphysial centres appear on the ends of the costal and transverse processes and from them an epiphysis is formed which covers the auricular surface, while another completes the margin below that.

Variations in vertebrae

Variation in the total number of vertebrae is usually due to a reduction or, more rarely, an increase in number of the coccygeal vertebrae. The number of cervical vertebrae is exceedingly constant although a seventh cervical vertebra may carry a cervical rib which is usually incompletely formed.

The number of thoracic vertebrae may be increased by the formation of ribs on the first lumbar vertebra, and the number of the lumbar vertebrae may be reduced in this manner or by fusion of a fifth lumbar vertebra to the sacrum. This sacralization may be partial or complete and quite frequently is incomplete on one side.

Bones

The sacrum may have an additional piece, becoming longer than usual, or may lose a vertebra at one end while gaining one at the other, so retaining five pieces.

In addition, it is not uncommon to find suppression of half a vertebra or intercalation of an additional half, making a wedge-shaped body. Such hemivertebrae cause an abnormal lateral curvature (scoliosis) of the column. The laminae may fail to meet and fuse at any or all parts (spina bifida), but most commonly in the lumbosacral region. The dens of the axis may be a separate bone, an os odontoideum, connected to the body of the atlas by ligaments. The spine, laminae, and inferior articular processes of the fourth and fifth lumbar vertebrae may only be united to the rest of the vertebra by cartilage, thus predisposing to a separation which leaves the vertebral body (particularly the fifth) free to slide forwards (spondylolisthesis). This condition is thought to arise as a fracture in many cases.

The standard work on vertebral variations is by Le Double (1912); see also Brash (1915) and Brailsford (1948).

Serial homologies of vertebrae

Since each part of a vertebra is potentially represented in every other vertebra it is possible to work out the homologies in detail for each region. The essential features are shown in FIGURE 3.35.

The body is the only part represented throughout the series, and even there the centrum of the atlas has been detached and fused with that of the axis. The upper cervical vertebrae require special mention in that the articular processes proper are not represented above the inferior process of the axis. The missing joints are replaced by an enlargement of those found in the neurocentral region lower down. The anterior arch of the atlas is ossified from a persistent hypochordal bow [FIG. 3.43] which disappears in all the other vertebrae. The ribs are represented down to the lower pieces of the sacrum but are normally free only in the thoracic region.

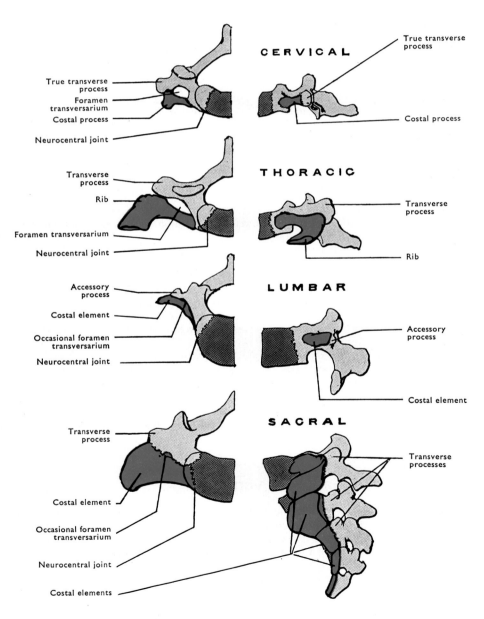

FIG. 3.35. Diagram to illustrate homologous parts of vertebrae.

From above   From the side

The sternum

The sternum is an elongated, nearly flat bone, 15–20 cm long. It lies in the midline of the anterior wall of the thorax and extends from the root of the neck into the abdominal wall. It can be felt through the skin in its whole length and articulates on each side with the clavicles at its upper end and with the upper seven costal cartilages at its lateral borders. Its three main parts are the manubrium, body, and xiphoid process. These are united by cartilage and in youth the body is in four pieces, also united by cartilage.

THE MANUBRIUM

This is the widest piece of the sternum, is roughly quadrilateral, lies inferior to the root of the neck, and forms the anterior wall of the superior mediastinum. It is usually a separate bone and only fuses with the body in advanced age.

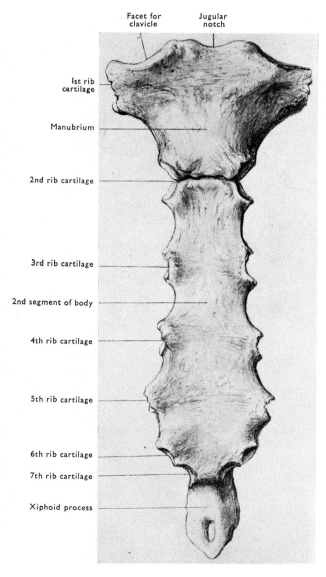

FIG. 3.36. Sternum (anterior view).

The middle part of the superior border forms the rounded jugular notch. On each side of this is the clavicular notch, the sternal part of the sternoclavicular joint. Immediately inferior to each clavicular notch is a rough projection which receives the first costal cartilage, and below this the manubrium narrows to its lower border, the lateral ends of which each carry a half facet for the second costal cartilage. The posterior surface, smooth and slightly concave, has sternohyoid attached at the level of the clavicular notch and sternothyroid at the level of the first costal cartilage.

The anterior surface is rougher and has a raised area below the medial part of the clavicular notch for the tendon of sternocleido-mastoid. Below that, a concave area on each side gives rise to the upper sternal fibres of pectoralis major.

The manubrium is joined to the superior border of the body by the fibrocartilage of the manubriosternal joint. The two parts are angulated to each other. This obtuse sternal angle is easily palpated even in the obese. It is a useful landmark, for it indicates the level of articulation of the second costal cartilage with the manubrium and body of the sternum, and the junction of the superior and inferior mediastina in the thorax.

THE BODY

This consists of four segments fused together and is approximately twice as long as the manubrium. Its shape is variable but it usually widens from about 2.5 cm at its upper end to about 4 cm at its fourth segment, which tapers sharply to its lower end. The posterior surface is slightly concave in its long axis and is smooth, the transversus thoracis muscle [FIG. 5.101] and the sternopericardial ligaments are attached to it. The rougher anterior surface has three transverse ridges indicating the lines of union between the segments or sternebrae. The attachment of pectoralis major is continued down from the manubrium to the sixth costal cartilage just short of the xiphoid process.

Articular facets for the synovial joints with the second to seventh costal cartilages are present on the lateral margins: at the manubriosternal joint (half of the second), between the segments of the body (third to fifth), on the fourth segment (sixth), and at the xiphosternal joint (seventh).

The lower end of the body is continuous with the xiphoid process at the xiphosternal joint. The process is a flat, pointed piece of cartilage which ossifies slowly from a central core until this joins the body of the sternum after middle age. Its shape is very variable and it is often perforated in the middle. It is thinner than the body of the sternum and is set flush with its posterior surface. The depression so formed is the epigastric fossa, or pit of the stomach, and the site of the joint can be felt as a ridge in the upper margin of the fossa. The xiphoid process lies in the anterior abdominal wall in front of the liver, and has the linea alba attached to its inferior end. It therefore has not only transversus thoracis but also all the adjacent abdominal muscles attached to its edge and anterior surface and slips of the diaphragm to its deep surface.

The sternum as a whole slopes antero-inferiorly, the manubrium slanting about 20 degrees more than the body so as to be almost in the same plane as the first ribs. The jugular notch is level with the lower border of the body of the second thoracic vertebra [FIG. 3.41], the sternal angle with the upper border of the fifth and with the interval between the third and fourth spines. The xiphosternal joint is level with the ninth thoracic vertebra and the tip of the eighth spine.

OSSIFICATION

Ossification of the sternum is variable (Ashley 1956) and any centre of ossification may be single or double and can affect the width of

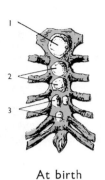

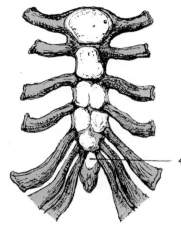

At birth

At 3 years

FIG. 3.37. Ossification of sternum.
1. Appears at 5th intra-uterine month.
2. At 6th and 7th month.
3. At 8th and 9th month.
4. At 3rd year.

the bone. Centres appear in the manubrium in the fifth intra-uterine month and then in the sternebrae in order from above downwards in the sixth, seventh, eighth, and ninth months [FIG. 3.37]. The xiphoid process does not ossify until the third year and does not join the body until middle age.

The remainder join in sequence from below in childhood, at puberty, and at 21 years. The manubrium should not fuse until old age is reached.

STRUCTURE AND VARIATIONS

The sternum is composed of cancellous bone enclosed in thin compact bone and is used to obtain samples of red marrow by sternal puncture. The vascular foramina are mostly on the deep surface of the manubrium and body where the bone is supplied by the internal thoracic artery.

The sternum arises from two separate mesodermal masses. Failure of fusion leads to centres of ossification being double, with a hole in the body, or more rarely a cleft sternum, the gap being wider above.

Some degree of asymmetry is common and may be associated with unequal physical development of the upper limbs. The sternal angle may be situated one rib lower down than usual. The seventh costal cartilage may fail to reach the sternum or the eighth may reach it.

The ribs

The ribs (costae) are curved and slightly twisted strips of bone which lie between the segments of the body. There are twelve of them on each side in both sexes and they are classified as 'flat' bones. They curve round from the thoracic vertebrae posteriorly, slant downwards and forwards at the sides, and anteriorly each becomes continuous with a costal cartilage. With the vertebral column and sternum they thus form a thoracic cage which supports the thoracic wall and encloses part of the upper abdomen.

Since the thorax widens from the root of the neck downwards, the upper seven ribs increase similarly in length and are referred to as

true ribs because their cartilages reach the sternum. The lower five ribs fail to reach the sternum, become progressively shorter from above downwards, and are referred to as false ribs. The cartilages of the eighth, ninth, and tenth each articulate with the cartilage next above, but the cartilages of the eleventh and twelfth lie free and the ribs are called floating ribs [FIG. 3.42].

The first, second, tenth, eleventh, and twelfth ribs have special features, but the third to the ninth have sufficient in common to be described together.

A typical rib has its posterior end slightly enlarged to form a head which articulates with the upper border of the body of its own vertebra and the lower border of the vertebral body next in series *above* it. It has therefore two sloping articular facets with a ridge between [FIG. 3.39]. The articular surfaces are surrounded by a joint capsule and the ridge gives attachment to an intra-articular ligament which connects it to the intervertebral disc.

A neck about 2.5 cm long immediately succeeds the head and is continuous with a body. The rough posterior surface is connected to the transverse process directly behind by the costotransverse ligament; and the crest on the upper border gives attachment to a superior costotransverse ligament passing to the vertebra above.

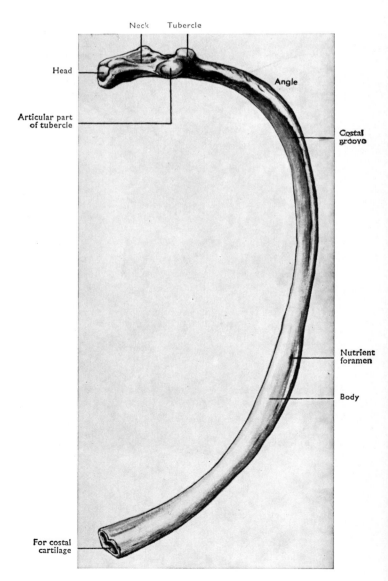

FIG. 3.38. Fifth right rib from below.

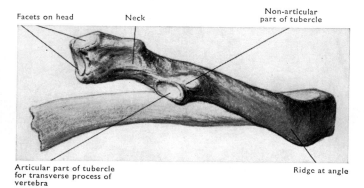

Facets on head   Neck   Non-articular
                                part of tubercle

Articular part of tubercle
for transverse process of   Ridge at angle
vertebra

FIG. 3.39. Fifth right rib from behind.

The smooth part of the anterior surface, covered by pleura, is limited above by a faint ridge produced by the internal intercostal membrane [FIG. 5.100].

The **tubercle**, on the posterior surface at the junction of neck and body, articulates with the corresponding transverse process, and has a rough, lateral part for the attachment of the **lateral costotransverse ligament** [FIG. 3.39].

The **body** continues laterally to reach a roughened vertical line on its external surface made by the thoracolumbar fascia covering erector spinae, and situated where the rib curvature is accentuated to form the **angle of the rib**. It coincides with the tubercle of the first rib but is placed increasingly further laterally as the ribs are followed down to the eighth or ninth. Below this, both the line and the angle become indistinct. This lateral displacement corresponds with the increasing bulk of erector spinae (attached between the tubercle and the angle) as the lumbar region is approached.

Beyond the angle of the rib the body continues in an even curve round the thorax, sloping down, until it reaches the costal cartilage anteriorly. It has a convex external surface and an internal surface which is concave lengthwise. Anteriorly the rib is somewhat twisted on its long axis so as to evert the lower border slightly.

The rounded superior border of the rib has the intercostal muscles and internal intercostal membrane attached to it [FIG. 5.100]. The sharp inferior border forms the external margin of the **costal groove** (which fades out anteriorly), the internal margin is higher on the deep surface of the rib. Both margins and the groove give attachment to the intercostal muscles. The rest of the deep surface of a rib is smooth and lined with pleura, except posteriorly, where it may be covered by subcostal muscles and marked by their attachments.

The body of a rib widens at its anterior end and terminates as a rough oval pit where it becomes continuous with its costal cartilage. The external surface of the body is covered with muscles and is smooth except near its anterior end where there may be an oblique ridge between the interdigitating origins of serratus anterior and obliquus externus abdominis [FIG. 5.56], and a sudden increase in curvature may form the **anterior angle**.

The internal surface of the thorax is concave supero-inferiorly, hence the pleural surfaces of the various ribs face in different directions: the first inferiorly, the twelfth anterosuperiorly, and the intermediate ribs with varying degrees of obliquity [FIG. 3.41].

The **first rib** is short, sharply curved, and flattened so as to present an anterosuperior surface and inner and outer borders on its body. It is felt as a bony resistance in the neck behind the middle of the clavicle. Its cartilage lies deep to pectoralis major in the angle between the sternum and the clavicle. Its **head** is small and normally articulates only with the first thoracic vertebra by means of a single facet. The relatively long, narrow and horizontal **neck** has the sympathetic trunk, the highest intercostal artery, and the first thoracic ventral ramus separating it from the pleura and the apex of the lung. The large tubercle lies on the outer border, and from this level the body slopes downwards and forwards forming the lateral margin of the superior aperture of the thorax.

The body has anterosuperior and postero-inferior surfaces, and passes beneath the medial part of the clavicle to reach the sternum. The postero-inferior (pleural) surface is smooth, but may be roughened by the internal intercostal muscle. The anterosuperior surface is grooved obliquely by the subclavian artery and vein [FIG. 3.40]. Between these grooves, the **tubercle for scalenus anterior** projects from the inner border, and a triangular, rough area, also for that muscle, extends laterally separating the artery and vein. The arterial groove also contains most of the ventral ramus of the first thoracic nerve which curves up over the first rib and separates the artery from scalenus medius which is attached to the first rib between the nerve and the tubercle of the rib. The first digitation of serratus anterior also arises from this surface close to its outer border, level with and posterior to the grooves. The costoclavicular ligament binds the clavicle down to the rib at the costochondral junction with subclavius muscle close to it.

The sharply curved **second rib** is about twice as long as the first rib but it is much narrower. It is not twisted and the outer surface of its body faces distinctly upwards and outwards. The outer surface has a rough tubercle for serratus anterior [FIG. 3.40]. The second rib is easily recognized in the living subject because its cartilage is level with the sternal angle. Both the angle and the cartilage may be visible as well as palpable 5 cm below the jugular notch.

The **tenth, eleventh**, and **twelfth ribs** articulate only with their own vertebrae and so each has only one facet on its head. The tenth may, however, resemble a typical rib. The eleventh has only a small tubercle without an articular facet and its angle is poorly defined. The twelfth rib is small and usually slender. It has no definite tubercle, angle, or groove, and the body tapers off at its free end in contrast to the widening of the others.

Tuberosity for
serratus anterior

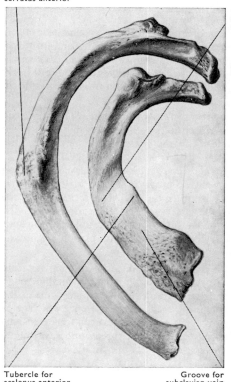

Groove for
subclavian
artery

Tubercle for   Groove for
scalenus anterior   subclavian vein

FIG. 3.40. First and second ribs as seen from above.

COSTAL CARTILAGES

The hyaline costal cartilages have their perichondrium continuous with the periosteum of the ribs. Rupture of the cartilage may occur within the perichondrium without displacement.

The cartilages continue the line of the ribs towards the sternum and the upper seven pairs reach the side of the sternum. The first is continuous with the sternum, the perichondrium being again continuous with the periosteum. The remaining six are joined by means of synovial joints to the edge of the sternum. The fifth to the tenth continue their direction for 2–3 cm and then turn up. The eighth to the tenth taper to a point and articulate edge to edge with the cartilage next above by a synovial joint in the case of the eighth and ninth, by fibrous tissue in the case of the tenth.

The cartilages of the eleventh and twelfth ribs do not articulate and form the ends of the ribs.

Thoracic cage

The wall of the thorax surrounds and protects many structures and its chief function is to provide support against the elastic pull of the lungs, which would otherwise collapse. It is also part of a suction apparatus which can counteract that elasticity actively and so allow air to enter the lungs.

The necessary rigidity for this purpose is provided by the thoracic cage, which is formed by the vertebral column behind, the sternum in front and, between them, the curving ribs and cartilages. The slope of the ribs from their evenly spaced posterior ends means that, when the anterior ends of the ribs are raised, the ribs are separated and the intercostal spaces become wider. At the same time the sternum is pushed forwards and so increases the anteroposterior diameter of the chest cavity. The cage also gives attachment, particularly along the rib margin, to the diaphragm and muscles of the abdominal wall, which are active in respiration. For movements of the ribs see page 228.

The thorax has the shape of a cone with the apex and base cut off obliquely: so the superior aperture slopes downwards and forwards, and the inferior aperture (base) slopes downwards and backwards. It is, therefore, much shorter in front than at the back and sides, and is also flattened anteroposteriorly, being widest from side to side. In children it is more circular in transverse section.

The **superior aperture of the thorax** lies at the root of the neck and measures about 5 cm anteroposteriorly and some 10 cm transversely. It is bounded by the manubrium sterni, the first thoracic vertebra, and the first ribs and costal cartilages. The projection of the vertebral body gives it a kidney-shaped outline and the slope of the ribs leaves the first two thoracic vertebral bodies above the level of the sternum [FIG. 3.41]. In the middle lie the trachea, oesophagus, and the vessels and nerves entering and leaving the thorax. Each more circular, lateral part is occupied by the apex of the lung and pleura, supported by the suprapleural membrane [FIG. 7.37].

The **inferior aperture** is large and is bounded by the xiphosternal junction and the lower six pairs of costal cartilages which form the **infrasternal angle**. The shafts of the twelfth ribs slope up to the twelfth thoracic vertebra to complete the circumference which gives attachment to the edge of the diaphragm (except for extensions down to the xiphoid process in front and three lumbar vertebrae behind) and to the muscles of the anterior abdominal wall.

The **posterior wall** is formed by the vertebral column and by the posterior parts of the ribs which sweep back to their angles before turning medially and then forwards, thus forming a deep, paravertebral, **pulmonary sulcus**. Posteriorly, the curve of the ribs is nearly

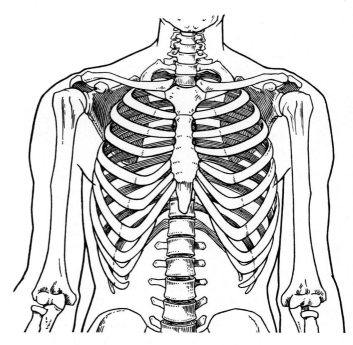

FIG. 3.41. Front of portion of skeleton showing thorax.

flush with the vertebral spines, and this allows Man to lie easily on his back.

The **side-walls** are formed by bodies of the ribs which are not quite equally spaced. The upper spaces are wider than the lower ones and the obliquity of the ribs increases from the first to the tenth, which makes the spaces slightly wider towards the front. The degree of obliquity varies with muscularity, age, and sex. The side walls of the thorax slope outwards to the level of the eighth or ninth rib and, as a rule, slightly inwards below that level.

The **anterior wall** is covered by muscles except near the midline. It is formed by the anterior ends of the ribs, costal cartilages, and sternum, and is shorter than the other walls. The third cartilage is almost horizontal; those above slope slightly down, those below are angulated and slope up towards the sternum, the spaces between them becoming narrower towards the lower end of the sternum [FIG. 3.41].

Apart from counting down from the second rib, other guides may be used to estimate the position of some of the ribs. Except in the adult female, the nipple usually lies opposite the fourth intercostal space, between the ends of the fourth and fifth ribs. The lower border of the pectoralis major muscle slopes medially to the sixth costal cartilage and the seventh is normally the lowest to reach the sternum. The tip of the ninth cartilage usually lies in the midclavicular line, where the lateral border of rectus abdominis crosses the costal margin. The tenth cartilage is the lowest seen from the front and lies 2–3 cm from the iliac crest. The costochondral junctions lie deep to a line between the medial end of the clavicle and a point 2–3 cm behind the bend of the tenth costal cartilage.

At the back [FIG. 3.42] the scapula overlies the second to the seventh ribs, or even the eighth, and the triangular area at the medial end of the root of the scapular spine overlies the fourth rib, level with the tip of the third thoracic spine. The twelfth rib is seldom palpable because of the mass of the overlying erector spinae muscle. It lies about half-way between the inferior angle of the scapula and the crest of the ilium, level with the twelfth thoracic or first lumbar spine.

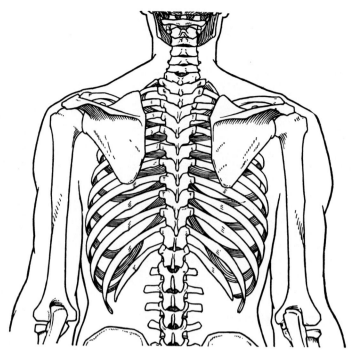

FIG. 3.42. Back of portion of skeleton showing thorax.

Ossification of ribs

In its ossification, a rib behaves like a long bone, the first ossification being subperichondrial and appearing almost throughout its length. At the same time, an endochondral centre appears near the angle, first in the sixth rib about the sixth week of intra-uterine life. Secondary centres appear about puberty and fuse with the rest of the bone by the twenty-fifth year. The first rib and the seventh to the tenth have one for the head and one for the tubercle. The second to the sixth have one for the head and one each for the articular and non-articular parts of the tubercle. The eleventh and twelfth have one for the head.

Structure and variations

A rib is composed of highly vascular cancellous bone in a flattened tube of compact bone which is thicker on its two surfaces, thinner at its borders. The compact bone is thickened at the angle where indirect bending stresses tend to cause fracture. The vascular foramina are numerous at the back of the neck, the vessels following the ligaments, and the nutrient canals are in the costal groove, directed towards the angle.

A cervical rib may reach the sternum or fuse with the first costal cartilage or with the first rib. Part of it may be represented by a fibrous band and it may be present only at its sternal end.

Occasionally an extra rib may be present in the lumbar region but a reduction in number is rare. The first rib may be rudimentary or fuse with the second to form a bicipital rib.

Ribs may be wide and partly fused together, particularly near their anterior ends, and a widened rib may have a hole in it, as may a costal cartilage. The posterior parts of two ribs may be united by projections which articulate with an intervening nodule of bone.

There may be some variation in the number of cartilages reaching the sternum. In old age, bone tends to be laid down around the cartilages especially the first where this normally happens. Some degree of asymmetry is normal.

DEVELOPMENT OF VERTEBRAE, RIBS, AND STERNUM

VERTEBRAE

During development the paraxial mesoderm becomes divided by transverse clefts into segments or mesodermal somites and the ventromedial, or scleratogenous, parts of these somites proliferate to form a sheath round the previously formed notochord and neural tube [p. 41]. This sheath is the membranous stage of the vertebral column and is soon followed by procartilaginous and cartilaginous stages which ultimately lead to ossification.

In the procartilaginous stage the vertebrae form as a series of horseshoe-shaped vertebral bows [FIG. 3.43] which spread round the notochord but are at first open dorsally. The tissue which will form a vertebral centrum surrounds the notochord opposite an intersegmental septum. The hypochordal bow is ventral to the cephalic part of the centrum to which it belongs. The limbs of the vertebral bow slant back to enclose the neural tube dorsal to the caudal part of the vertebral centrum and ultimately meet to enclose the neural tube and form a vertebral arch.

About the fourth week of intra-uterine life the scleratogenous tissue between the vertebral centra undergoes fibrous and chondrifying changes to form intervertebral discs. This tissue is opposite the middle of a somite in each case and the main part of each vertebral body is derived from halves of two adjacent somites. The beginning of chondrification in the centrum and vertebral bow is followed rapidly by fusion of centrum and bow together; chondrification of the bow is completed during the fourth month of intrauterine life. The limbs of the bow form the vertebral arch and its processes and also the adjacent part of the vertebral body. The hypochordal portion of the bow disappears except in the case of the atlas where it forms the anterior arch. The atlas is therefore a vertebral ring and its centrum becomes the dens. The transverse processes and spine project into the intersegmental septum between somites.

The parts of the notochord included in the bodies of the vertebrae become constricted and disappear after ossification begins, but the part enclosed in each disc takes part in the formation of a nucleus pulposus.

The coccygeal vertebrae develop from small rudiments in which the vertebral bows are deficient as they are in the sacral hiatus. The same deficiency at higher levels is known as spina bifida.

RIBS

The ribs develop in the intersegmental septa and pass through the usual stages of condensation, chondrification, and ossification. Each

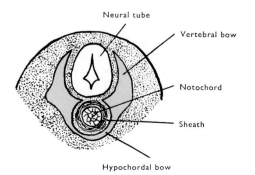

FIG. 3.43. Transverse section showing development of the membranous basis of a vertebra. (After Keith 1948.)

rib element is first formed opposite the corresponding vertebra, later the vertebral ends of most are shifted slightly headwards to reach their final articulations.

STERNUM

The sternum is formed from mesodermal tissue in front of the pericardium, the manubrium probably in association with the shoulder girdle, the remainder from a pair of sternal bars which grow from the manubrial rudiment and fuse in the median plane. After chondrification it becomes connected with the costal cartilages by synovial joints. See also Paterson (1904) and Ashley (1956).

The skull

The skull is made up of twenty-one bones which are closely fitted together. In addition, there are other bones, developmentally associated with the pharyngeal arches [p. 55], which remain separate but are grouped with the skull bones. The **mandible** forms the skeleton of the lower jaw. The **hyoid bone** lies in the root of the tongue where it gives attachment to lingual muscles and to those of the floor of the mouth. Three small ossicles are found in each middle ear in the temporal bone. They are described in the section on Sensory Organs.

Skull denotes the skeleton of the head, with or without the mandible. **Cranium** has been used in the same general sense, but it is better to restrict its meaning to skull without mandible. **Calvaria** usually means the top part of the skull or skull-cap.

The skull consists of the bony walls of a large cranial cavity which contains the brain, the bones of the face which complete the walls of the orbits and nasal cavity and form the roof of the mouth, and a separate mandible [FIG. 3.45].

Phylogenetically the skull evolved from cartilage enclosing the brain and the organs of special sense. This origin is reflected by the cartilage bone retained in the base of the skull. As a more recent development, membrane bone is added at the vault of the skull to cover the enlarged space occupied by the brain. The cartilage capsules of the ear and nose are incorporated in the base. In a young person the skull bones are separable with difficulty. Except in the case of the mandible and the ear ossicles, the bones are either united by fibrous tissue (**sutures**) or by cartilage (**synchondroses**) [FIG. 4.3]. With advancing years the sutures and synchondroses become ossified.

Most skulls used for study have had the upper part of the skull-cap removed by a circumferential saw cut and the compact outer and inner tables can be seen with the **diploë** (cancellous bone and marrow spaces) sandwiched in between. The thickness of the skull varies in different places in any one skull and the sites of thickening vary in different skulls. Areas covered with muscles are thinner than more exposed parts, e.g. the temporal fossa and the floor of the posterior cranial fossa. The thickness of the skull also varies with age, sex, and race, being thin in children and in the aged. In children the cortex is thin, in the aged the diploë is reduced. In the young a greater elasticity reduces the risk of fracture. The skull is usually thinner in the female than in the male.

THE SKULL AS A WHOLE

1. The **orbits** are the pair of pyramidal cavities in the front of the skull [FIG. 3.49], so named because the eyes rotate in them.

2. The anterior bony aperture of the nose or **piriform aperture** lies between and below the orbits.

3. The **external acoustic meatus** [FIG. 3.45] is the small round or oval passage of the ear seen in the lower part of the side of the skull about midway between front and back.

4. The **zygomatic arch** spans the interval between the acoustic meatus and the orbit.

5. The **foramen magnum** is the large round hole in the base of the skull a little behind its centre [FIG. 3.51].

The exterior of the skull is described as it is seen from five aspects, and each is spoken of as a **norma**. From the front it is norma frontalis; from the back, norma occipitalis; from above, norma verticalis; and from below, norma basalis; from the side, norma lateralis.

For purposes of description or measurement the skull is orientated in a standard position with the lower border of the orbit and the upper border of the external acoustic meatus in a horizontal plane. This is known as the Frankfurt plane, and the coronal and sagittal planes at right angles to it are used to orientate the skull in obtaining the normae.

Norma verticalis

The top of the skull [FIG. 3.44] is rounded or ovoid in most Europeans, the widest part being towards the back, but there is considerable variation in the proportions and often some asymmetry. Its convexity serves to distribute and so minimize the effects of a blow, but hard contact with a flat surface can lead to splits in the scalp resembling cuts.

The frontal bone forms the forehead and part of the top of the skull. Behind this, separated from it by the **coronal suture**, the two parietal bones extend up from the sides to meet in the median **sagittal suture** [FIG. 3.44]. This suture extends backwards from the point where it meets the coronal suture (**bregma**) to the apex of a

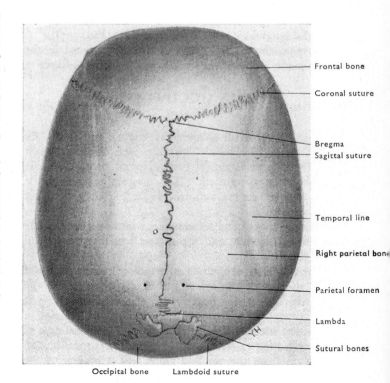

Frontal bone
Coronal suture
Bregma
Sagittal suture
Temporal line
Right parietal bone
Parietal foramen
Lambda
Sutural bones

Occipital bone   Lambdoid suture

FIG. 3.44. Superior aspect of the skull (norma verticalis).

protuberant, triangular part of the occipital bone. Though the occipital bone forms the back of the skull, this triangular part can be seen from above where it intrudes between the parietal bones posteriorly, being separated from them by the lambdoid suture. The meeting point of the lambdoid and sagittal sutures is termed lambda. The coronal suture extends antero-inferiorly from bregma to pterion, where the parietal, frontal, sphenoid, and temporal bones almost meet at one point [FIG. 3.45]. Near the coronal suture, the parietal bone is flattened or slightly concave, whilst it may form a low ridge along the sagittal suture. A parietal foramen, which transmits an emissary vein, may be found in each parietal bone close to the sagittal suture.

At bregma and lambda the skull remains membranous after birth, constituting respectively the anterior and posterior fonticuli. These ossify from the surrounding bones in the second and first years. Bregma lies just anterior to a vertical plane passing through the external acoustic meatus; lambda lies a little behind the site of the parietooccipital sulcus of the brain [FIG. 10.190].

The most convex parts of the frontal and parietal bones, frontal and parietal tubera [FIGS. 3.74 and 3.76], are easily identified by feeling with the palm of the hand and are useful as landmarks.

The temporal lines [FIG. 3.45] arch up over the side of the frontal bone and across the parietal to run down towards its postero-inferior angle (see norma lateralis).

The vertex, the highest point of the skull placed in the Frankfurt plane, lies near the middle of the sagittal suture.

Norma occipitalis [FIG. 3.8]

The back of the skull is markedly convex and is usually widest at the parietal tubera. The bones seen are the posterior parts of the two parietal and two temporal bones with the occipital bone between them. The mastoid portion of the temporal bone and its mastoid process are visible at the lower corner on each side.

The parietal tubera and foramina, the posterior part of the sagittal suture, lambda, and the entire lambdoid suture are visible in this view.

The lambdoid suture passes inferolaterally to the mastoid angle of the parietal bone, and is then continued as the occipitomastoid suture between the occipital and temporal bones [FIG. 3.51]. The mastoid foramen, which varies in size, lies in the latter suture or in the mastoid temporal near it, 5–6 cm behind the external acoustic meatus. It transmits an emissary vein from the sigmoid sinus and a branch of the occipital artery.

Half-way between lambda and the foramen magnum is the median external occipital protuberance, from which a curved superior nuchal line passes laterally to the mastoid process on each side. The lines, which give origin to trapezius and sternocleidomastoid muscles, mark the junction between the scalp region and that for attachment of the muscles of the back of the neck. Above each of these lines is a poorly marked highest nuchal line for the attachment of the epicranial aponeurosis of the scalp and the occipital belly of occipitofrontalis muscle. The centre of the external occipital protuberance is called inion.

Norma lateralis [FIGS. 3.45 and 3.58]

The side view of the skull shows clearly the division into a large, ovoid brain-case and a smaller, uneven but almost triangular facial skeleton attached to it antero-inferiorly by the zygomatic bone. This triradiate bone extends: (1) upwards (forming the lateral margin of the orbit and the anterior margin of the temporal fossa) to unite with the zygomatic process at the antero-inferior corner of the frontal bone; (2) backwards (forming the anterior part of the zygomatic arch and the lower margin of the temporal fossa) to unite by an oblique suture running downwards and backwards with the zygomatic process of the temporal bone, which completes the zygomatic arch; (3) antero-inferiorly, where it is buttressed to the maxilla (the main part of the facial skeleton) and forms a large part of the inferior margin of the orbit. It thus forms the main lateral strut attaching the facial skeleton to the brain-case and it is not uncommonly fractured in severe blows to the face.

The superior temporal line begins at the posterior edge of the zygomatic process of the frontal bone (there is a palpable notch at the suture with the zygomatic bone), arches backwards over the frontal and parietal bones, and curves forwards above the external acoustic meatus where it forms the supramastoid crest before becoming continuous with the upper margin of the zygomatic arch. The anterior end of this margin bends sharply upwards (jugal point) and continues as the posterior margin of the frontal process of the zygomatic bone. The temporal fascia is attached to this entire margin, and covers the temporalis muscle which is attached below the inferior temporal line [FIG. 5.34].

The region enclosed by the superior temporal line is the temporal fossa. The anterior wall of the fossa is formed by the zygomatic bone, by the zygomatic process of the frontal bone and by the greater wing of the sphenoid. A small foramen in the zygomatic bone here transmits the zygomaticotemporal nerve.

The medial wall of the fossa is formed by the parietal and frontal bones, and by the upturned parts of two bones in the base of the skull. One is the squamous part of the temporal bone, a scale-like plate of bone which lies above the acoustic meatus and overlaps the lower edge of the parietal bone to form the squamosal suture. The other is the greater wing of the sphenoid which turns upwards from the base of the skull, between the squamous temporal and frontal bones to meet the antero-inferior corner of the parietal bone (sphenoidal angle) at the sphenoparietal suture.

Four bones thus articulate within an area less than 3 cm in diameter. Usually the group of sutures is H-shaped as shown in FIGURE 3.45, but variations occur and epipteric (sutural) bones may be present. The area is named pterion. Its centre lies approximately 4 cm above the zygomatic arch and 3 cm behind the zygomatic process of the frontal bone. In this region the frontal branch of the middle meningeal artery lies in a groove or tunnel on the inside of the bone and it is liable to be torn in fractures.

The temporal surface of the greater wing of the sphenoid is limited below by an infratemporal crest, a horizontal ridge which separates the temporal surface from the pterygoid region on the inferior surface of the skull, and separates the temporal fossa from the infratemporal fossa.

The zygomatic process of the temporal bone arises from the squamous part anterosuperior to the external acoustic meatus. Posteriorly continuous with the supramastoid crest, its root forms part of the articular fossa for the head of the mandible inferiorly, together with a small postglenoid tubercle which projects downwards behind the fossa. Further anteriorly, the inferior surface of the root of the zygomatic process forms a prominent articular tubercle which provides the anterior part of the articular area for the head of the mandible. The tubercle is palpable laterally, medially it continues as a ridge to the squamous temporal bone.

The external acoustic meatus lies in the angle between the squamous part of the temporal and its mastoid process. The bony meatus is 1.5 cm long, but cartilage attached to its edge during life increases its length to 2.5 cm. The external opening is oval, 5–6 mm across its narrow anteroposterior diameter. The canal passes medially and slightly upwards, with a gentle superior convexity, to end at the tympanic membrane which separates it from the middle ear cavity during life [FIG. 12.34]. In a dried skull, the medial wall of the middle ear can be seen directly through the meatus.

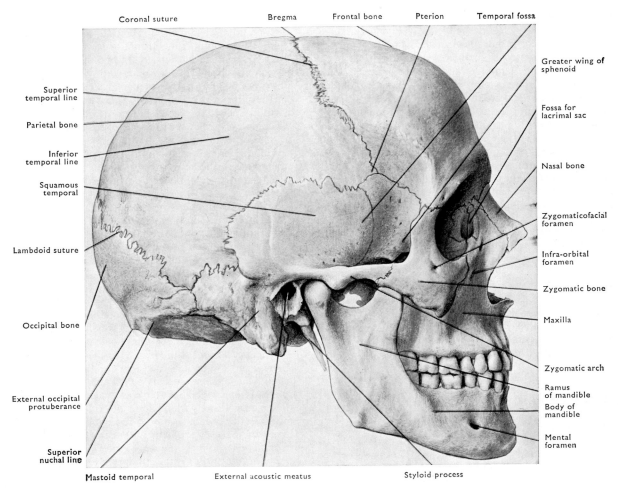

FIG. 3.45. Lateral aspect of the skull.

The bony meatus is formed superiorly and to a small extent posteriorly (postauditory process) by the squamous part of the temporal bone. The remaining walls are formed by the tympanic part of the temporal bone, a curved, triangular plate which fits between the mastoid and squamous parts of the same bone. Between it and the mastoid process is the **tympanomastoid fissure**. Anteriorly, the edge of the plate fits in behind the postglenoid tubercle and, more medially, is separated from the mandibular fossa of the squama by the **squamotympanic fissure** [p.116]. In the infant the tympanic plate is a mere ring, so that the meatus is short. The **suprameatal triangle** is a small depression lying at the meeting of tangents to the superior and posterior walls of the meatus. It lies inferior to the suprameatal crest and marks the position of the aditus to the mastoid antrum (an extension of the middle ear cavity to the air cells in the mastoid process) which is 12 mm deep to it [FIG. 3.47].

The **styloid process** is a tapering spicule of variable length projecting from behind the tympanic plate which partially surrounds its base. It is continuous antero-inferiorly with the stylohyoid ligament and it lies deeply in the interval between the mastoid process and the mandible [FIG. 3.45].

The **mastoid region** of the temporal bone is the lateral part of the petrous temporal bone where it appears on the surface of the skull between the external acoustic meatus in front, the squamous part of the temporal bone and the mastoid angle of the parietal bone above, and the occipital bone behind. It is fused with the squamous part along a line behind and below the supramastoid crest [FIG. 3.81] but is separated from the others by sutures. It carries the prominent **mastoid process** which projects antero-inferiorly behind the lower

part of the ear and is readily palpable. The process varies in size with age and muscularity, is absent at birth and small in a child. It is associated with an upright posture and the resultant balance of the head. After puberty the mastoid process becomes permeated by small **mastoid cells** which are air-filled spaces extending from the mastoid antrum, a process which has started at birth but does not extend into the mastoid until this time. They then increase rapidly and reach their full size in a year or two. The first appearance and extent of these air cells is very variable [FIG. 3.46].

The auricular branch of the vagus nerve pierces the **tympano-mastoid fissure**, anterior to the mastoid process, and the mastoid emissary foramen lies in or near the occipitomastoid suture.

The mastoid process has sternocleidomastoid, splenius capitis, and longissimus capitis muscles attached to it [FIG. 5.21]. Behind the mastoid, the superior nuchal line forms an almost horizontal line running to the external occipital protuberance.

The **infratemporal fossa** lies below the horizontal part of the greater wing of the sphenoid. The anterior wall of the fossa is the rounded posterior aspect of the maxilla with the **tuberosity of the maxilla** projecting from its lowest part above and behind the last molar tooth. About the middle of this surface are small **alveolar foramina** for the posterior superior alveolar vessels and nerves entering to supply the molar teeth.

Projecting antero-inferiorly from the sphenoid is the **lateral lamina** of its **pterygoid process** which forms the medial wall of the infratemporal fossa. It is an almost flat plate of bone 3 cm in length and over 1 cm wide. Anteriorly and medially, it is continuous with a **medial lamina**, concealed from the present viewpoint. The pterygoid process and maxilla are separated by a narrow V-shaped

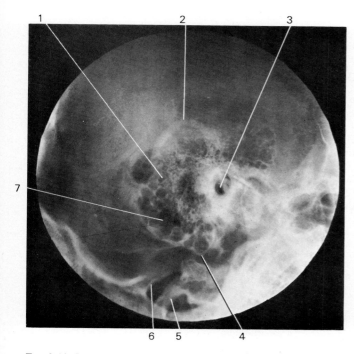

FIG. 3.46. Lateral radiograph of mastoid process of temporal bone.
1. Small mastoid air cell.
2. Helix of auricle.
3. External and internal acoustic meatus superimposed.
4. Tip of mastoid process.
5. Posterior arch of atlas.
6. Posterior margin of foramen magnum.
7. Large mastoid air cell.

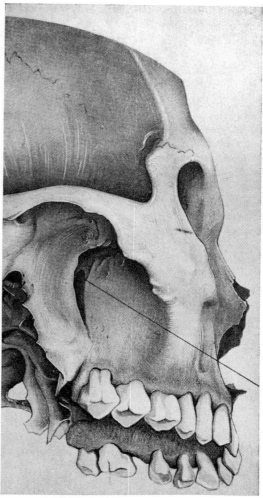

Sphenopalatine foramen

FIG. 3.48. Lateral view of the pterygoid region to show the pterygomaxillary fissure and sphenopalatine foramen.

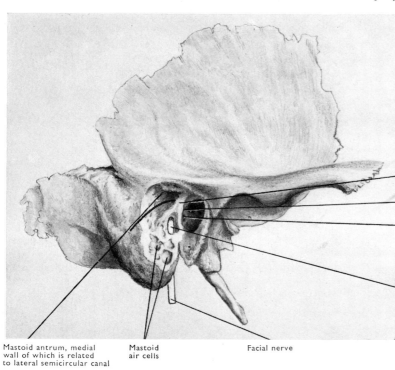

Medial part of posterior wall of external acoustic meatus left *in situ*

Points to epitympanic recess

Fenestra cochleae

Facial canal laid open displaying facial nerve within

Mastoid antrum, medial wall of which is related to lateral semicircular canal

Mastoid air cells

Facial nerve

FIG. 3.47. Right temporal bone prepared to show the position of the mastoid antrum.

The greater part of the posterior wall of the external acoustic meatus has been removed, but a bridge of bone has been left at its medial end; under cover of this a bristle (thick black) passes through the mastoid antrum to the tympanic cavity.

gap, the **pterygomaxillary fissure**. It leads into the narrow **pterygopalatine fossa**, so called because it lies between the pterygoid process behind and the vertical plate of the palatine extending laterally over the back of the maxilla to form the anterior wall of the fossa.

At the lower end of the pterygopalatine fossa, the **pyramidal process of the palatine bone** intervenes between the maxilla and the pterygoid process of the sphenoid, and all three are fused. At the upper end of the fossa, the greater wing of the sphenoid passes laterally, separated from the maxilla by a horizontal cleft which leads into the orbit, the **inferior orbital fissure**.

The vertical plate of the palatine bone has a notch at its upper end, converted into a foramen by contact with the sphenoid and therefore called the **sphenopalatine foramen**. A probe passed

through the pterygopalatine fossa enters the nasal cavity through this foramen.

Norma frontalis [FIG. 3.49]

The front of the skull comprises the anterior part of the brain-case above, and the skeleton of the face below. It varies greatly in shape and proportions with age, race, and sex.

The adult **mandible** carries a set of sixteen teeth and closes these against the same number in the two **maxillae** which together form the upper jaw. Anteriorly the upper parts of the maxillae diverge to enclose the **piriform aperture** of the nose, and then pass up to reach the frontal bone on each side of the nasal bones which form the bridge of the nose. The maxilla forms the greater part of the floor of

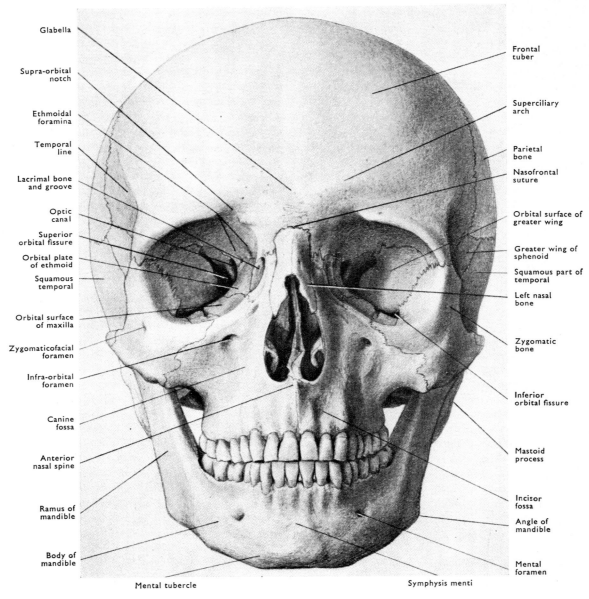

FIG. 3.49. Anterior aspect of the skull (norma frontalis).

an orbit, but only a small part of its inferior margin for it is overlapped laterally by the zygomatic bone.

THE MAXILLARY REGION

Below the piriform aperture the two maxillae meet in the median plane and a sharp median spur, the anterior nasal spine, projects forwards at the lower margin of the nasal aperture. It may be bifid, and it is embedded in the lower, mobile part of the septum of the nose.

The external margin of each maxilla projects down to form, with its counterpart, a U-shaped alveolar process which carries a set of eight upper teeth on each side [FIG. 6.5] embedded in a row of sockets on its inferior surface. If the walls of the sockets have not been absorbed, vertical ridges caused by the roots of the teeth extend upwards from the alveolar margin. The canine tooth causes the largest of these, the canine eminence, and above and lateral to it is a wide, shallow depression—the canine fossa—which reaches almost to the orbital margin. The shape of this and of the alveolar process markedly affect the facial appearance.

The infra-orbital foramen lies 6 mm below the orbital margin and 1.5 cm from the side of the nose. It transmits the infra-orbital vessels and nerve.

The margin of the piriform aperture is formed by the two maxillae and the two nasal bones. It is oval or pear-shaped and lies in a plane which slopes downwards and backwards before curving forwards to the region of the anterior nasal spine. The edges give attachment to the soft parts of the external nose [FIG. 7.2], including cartilage.

Above the piriform aperture, the two nasal bones form the bridge of the nose which varies considerably in size and prominence. They articulate: (1) with the frontal bone (nasion) superiorly; (2) with the frontal processes of the maxillae laterally; and (3) with each other in the midline. Their inferior ends are wider and irregular and form the upper margin of the piriform aperture. The external nasal nerve passes to the skin between this margin and the nasal cartilage [FIG. 11.17]. A minute foramen may transmit a vein from the skin through the nasal bone to reach the superior sagittal sinus inside the skull, thus providing a possible pathway for infection.

To the lateral side of the face the maxilla has a buttress-like zygomatic process, which supports the zygomatic bone.

The zygomatic bone forms the hard, prominent part of the cheek and is often called the cheek bone [FIG. 3.45]. It has three surfaces, lateral, orbital, and temporal. The lateral surface is rounded and subcutaneous. It presents a small zygomaticofacial foramen, for the nerve of that name. Anteriorly, this surface blends with that of the maxilla; anterosuperiorly it forms the lateral half of the lower orbital margin; posteriorly its temporal process enters the zygomatic arch; while superiorly, the frontal process extends to articulate with the frontal bone between the temporal fossa and the orbit. A flange passes deeply from the frontal process to the greater wing of the sphenoid and, with it, completes the separation of the orbit (orbital surface) from the temporal fossa (temporal surface). The flange also articulates inferiorly with the maxilla and extends medially into the floor of the orbit, closing off the lateral end of the inferior orbital fissure.

Thus the orbital surface of the zygomatic bone forms part of the floor and lateral walls of the orbit, and it transmits the zygomatic nerve through a small foramen of that name. The temporal surface is best seen from the side of the skull and is described on page 109.

The margins of the orbit are formed by the zygomatic bone below and laterally, by the maxilla below and medially. Above, the frontal bone forms both margin and roof of the orbit and sends down a maxillary and a zygomatic process to articulate on the medial and lateral sides with the corresponding processes from below.

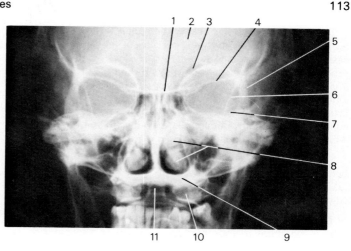

FIG. 3.50. Postero-anterior radiograph of skull with hard palate horizontal.
1. Central part of floor of anterior cranial fossa.
2. Margin of frontal sinus.
3. Extension of frontal sinus into orbital plate of frontal bone.
4. Lesser wing of sphenoid.
5. Frontozygomatic suture.
6. Medial wall of temporal fossa (greater wing of sphenoid).
7. Petrous temporal bone, superior margin.
8. Middle and inferior conchae.
9. Hard palate.
10. Lateral mass of atlas. ⎫
11. Dens of axis. ⎭ Obvious because of absence of maxillary teeth.

The whole of the frontal region is formed by the frontal bone. Superiorly it merges with the top of the skull and the frontal tubera are visible on each side [FIG. 3.49]. The narrowest part of the brain-case normally lies immediately posterior to the zygomatic processes. The upper margin of the orbit (supra-orbital margin) forms a curved border. Its outer two-thirds are sharp and definite. Its medial third is much less so and, at the junction of these two parts there is usually a supra-orbital notch which can be felt from below. The supra-orbital vessels and nerves lie in the notch as they turn up into the scalp. A small foramen in the floor of the notch transmits an emissary vein to connect the frontal diploic vein with the supra-orbital vein. The notch may be divided or converted into a foramen.

Glabella is the smooth rounded eminence above nasion [FIG. 3.73] and between the two supra-orbital margins and the two eyebrows in the living subject. From glabella a smooth elevation on each side arches laterally above the orbital margin to form the superciliary arch. This overlies the frontal sinus [p. 124]. The surface of glabella frequently shows the remains of the lower end of a suture between the two halves of the frontal bone. Fusion normally occurs at this frontal suture between 6 and 10 years. When persistent, it is referred to as a metopic suture.

Norma basalis [FIG. 3.51]

After removal of the mandible, the inferior surface of the skull shows the under surface of the upper jaw in front, with the palate and teeth, and the supporting zygomatic arches curving back on each side. Behind is the wide nuchal plane of the occipital bone, marked by the muscles of the neck. In between, the surface is exceedingly irregular because of foramina, articulations, and processes.

The posterior part of the base of the skull is formed mainly by the occipital bone, but the mastoid processes of the temporal bone are easily identified [FIGS. 3.45 and 3.51]. Level with the mastoid processes the **foramen magnum** pierces the base of the skull. It is about 3 cm wide by 3.5 cm anteroposteriorly. On each side its anterolateral margin is encroached on by an occipital condyle which articulates with the atlas. These obliquely set, oval condyles are convex longitudinally, but are relatively flat in the transverse plane which slopes inferomedially.

The anterior edge of the foramen is slightly thickened and lies between the anterior ends of the condyles. Its midpoint is used in taking measurements and is named **basion** since it is on the posterior edge of the basi-occipital.

The posterior half of the edge is thin and semicircular. The **atlanto-occipital membranes** are attached to the anterior and posterior margins of the foramen and joint capsules to the edges of the condyles. The **apical ligament** of the dens is attached to the anterior margin, but the upper fasciculus of the **cruciate ligament** and the **membrana tectoria** pass through the foramen before gaining attachment [FIG. 4.17].

Superior to the anterior half of each condyle, a **hypoglossal canal** transmits the hypoglossal nerve from the cranial cavity. Behind the posterior end of the condyle is a **condylar fossa**, usually with a **condylar canal** which carries a venous communication backwards from the sigmoid sinus in the skull to the vertebral veins in the neck. Below this the vertebral artery, veins, and sympathetic nerves curve round the lateral mass of the atlas to reach the foramen magnum and enter the skull. Here the spinal medulla is continuous with the medulla oblongata as are the spinal and cranial meninges. Spinal vessels run down through the foramen and the spinal parts of the accessory nerves pass up through it.

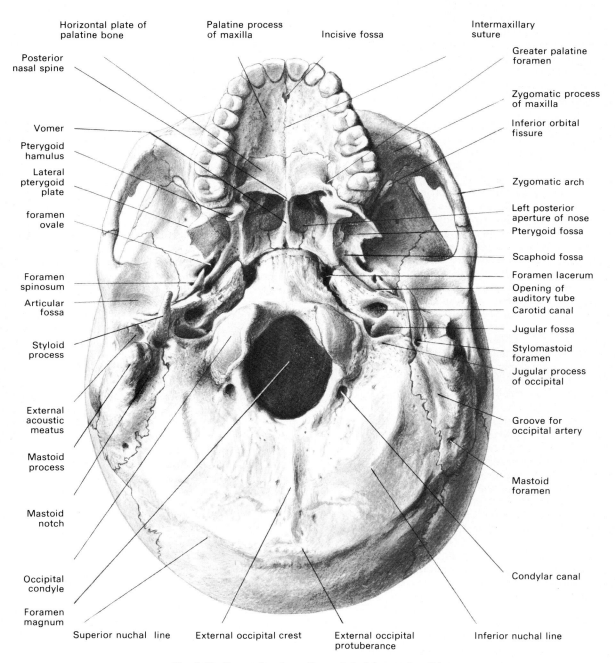

FIG. 3.51. External surface of base of skull (norma basalis).

The nuchal plane of the occipital bone lies behind the foramen magnum and is divided into right and left halves by an **external occipital crest** which runs from the external occipital protuberance to the edge of the foramen in the midline. Each half is separated into two areas by an **inferior nuchal line** which curves laterally from the crest, midway between the superior nuchal line and the foramen magnum. Each of these four areas gives attachment to two muscles and the external protuberance and the nuchal crest give attachment to the elastic nuchal ligament [FIGS. 3.51, 3.78, and 5.17]. The muscles and the nuchal ligament support and balance the skull on the vertebral column.

The anterior part of this nuchal area is continued forwards lateral to the condyle on to the condylar part of the bone and ends abruptly at a transverse edge as the jugular foramen is reached. The quadrilateral area behind this edge and lateral to the condyle is the **jugular process** [FIG. 3.51]. It gives attachment to rectus capitis lateralis which connects it to the transverse process of the atlas.

The part anterior to the foramen magnum, including the anterior ends of the articular facets, is the **basilar part** or basi-occiput, as distinguished from the **lateral** (condylar) and **squamous** (posterior) **parts** of the occipital bone. Anteriorly, the basilar part tapers to a flattened bar of bone which is joined to the body of the sphenoid by a growth cartilage until that cartilage ossifies at puberty. A small **pharyngeal tubercle** projects downwards from the basilar part in the midline. It is about 1 cm in front of the foramen magnum, and gives attachment to the pharyngeal raphe in the posterior wall of the pharynx. Longus capitis is attached anterolateral to it with rectus capitis anterior on each side, immediately anterior to the condyle [FIG. 5.17].

The occipital and sphenoid bones form the solid central part of the base of the skull, and the **petrous temporal bones** are wedged anteromedially between their lateral parts. The line of junction between the occipital bone and the petrous part of the temporal (mastoid portion) runs forward, a short distance medial to the mastoid process, as far as the anterior edge of the jugular process. Here the large **jugular foramen** faces downwards and forwards between the two bones; the edge of the occipital bone forms a **jugular notch**, and the petrous temporal is excavated to form a **jugular fossa**, which accommodates the superior bulb of the jugular vein. The fossa and the vein are close to the middle ear, as can be judged by looking through the external acoustic meatus, and infection may spread through the bone from the middle ear to the vein.

The jugular foramen lies medial to the external acoustic meatus, and almost level with its lower border. The foramen may be partly or completely divided into three compartments by small spicules of bone. The anteromedial compartment transmits the inferior petrosal sinus and a meningeal branch of the ascending pharyngeal artery. The middle compartment contains the ninth, tenth, and eleventh cranial nerves, from before backwards. The large, posterolateral compartment transmits the sigmoid sinus on its way to become the internal jugular vein, meningeal branches of the occipital artery and of the vagus nerve, and some lymph vessels.

On the lateral wall of the jugular foramen, a small pinpoint hole forms the medial end of the **mastoid canaliculus** for the auricular branch of the vagus which emerges through the **tympanomastoid fissure**.

On the edge of the ridge separating the jugular foramen from the carotid canal [FIG. 3.51] lie small foramina for vessels. One is the entrance to a **tympanic canaliculus** for the tympanic branch of the glossopharyngeal nerve passing to the tympanic cavity.

Anteromedial to the jugular foramen, the articulation between the petrous temporal and the occipital bones runs forwards and medially and leads to an irregular space enclosed by the apex of the petrous temporal, the basi-occipital, and the body of the sphenoid

bone. This **foramen lacerum** is closed with cartilage, which is pierced vertically by small blood and lymph vessels only.

All parts of the temporal bone are visible from below. The petrous part extends anteromedially from the mastoid process to the foramen lacerum. Anterior to the jugular foramen its inferior surface is pierced by the round opening of the **carotid canal** which turns at right angles within the bone to reach the side of the foramen lacerum. It transmits the internal carotid artery with its plexus of veins and sympathetic nerves.

Immediately above the lower aperture of the carotid canal, on its posterolateral wall, are the openings of two small **caroticotympanic canaliculi** which convey tympanic branches of the carotid artery and sympathetic nerves.

The tympanic plate lies under the petrous part of the bone and joins the anterior surface of the mastoid process at the **tympano-mastoid fissure**. This may be reduced to a small canaliculus for the auricular branch of the vagus nerve.

More medially, behind and partly sheathed in the tympanic plate, the tapering **styloid process** projects downwards and forwards lateral to the internal jugular vein [FIGS. 3.51 and 5.25] to be

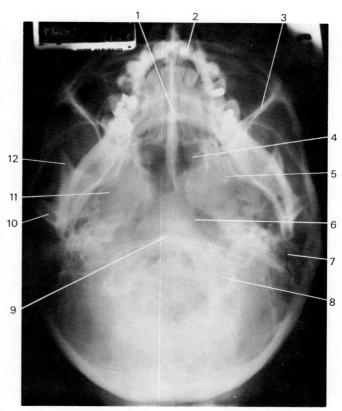

FIG. 3.52. Radiograph of the base of the skull taken with the neck fully extended.
1. Nasal septum.
2. Maxillary incisor tooth.
3. Superimposed lateral wall of orbit and posterolateral wall of maxillary sinus.
4. Sphenoidal sinus.
5. Lateral pterygoid lamina.
6. Lateral margin of clivus.
7. Mastoid air cells.
8. Occipital condyle.
9. Anterior arch of atlas.
10. Head of mandible.
11. Foramen ovale.
12. Coronoid process of mandible.

continued as the stylohyoid ligament. Between the styloid and the mastoid processes lies the **stylomastoid foramen**. This transmits the facial nerve, emerging from the facial canal in the petrous temporal. With the nerve are small branches of the posterior auricular vessels. Swelling and consequent pressure in the narrow rigid canal can produce a facial paralysis.

Behind the stylomastoid foramen and medial to the mastoid process, a distinct groove, the **mastoid notch**, gives attachment to the posterior belly of the digastric muscle. Medial to the notch a slight groove may be produced by the occipital artery which ascends to the skull deep to the posterior belly of the digastric.

Anterior to the tympanic plate lies the **squamotympanic fissure** and the fossa for the head of the mandible on the squamous part of the bone [FIGS. 3.51 and 3.84]. Between the two edges of the fissure a down-turned edge of the roof of the tympanic cavity (tegmen tympani) appears in the fissure [FIG. 3.81] and divides it into **petrotympanic** and **petrosquamous** parts. The chorda tympani branch of the facial nerve emerges from the middle-ear cavity, through the medial end of the petrotympanic fissure, and passes medial to the spine of the sphenoid.

The **mandibular fossa** is wide and smooth but short from front to back. The articular surface extends over the posterior and inferior surfaces of the **articular tubercle** [p. 109] which limits the fossa anteriorly. Laterally, the fossa extends on to the root of the zygomatic process.

In front of the articular tubercle, the bone slopes upwards into the temporal fossa, whilst anteromedial to it is the roof of the infratemporal fossa. The latter extends anteriorly to the inferior orbital and pterygomaxillary fissures, laterally to the infratemporal crest on the greater wing of the sphenoid, and medially to the lateral pterygoid lamina.

The posterior part of the greater wing on the sphenoid projects between the squamous and petrous parts of the temporal bone, articulating with the edge of the petrous temporal posteromedially. The **auditory tube** lies in a groove along this line or articulation and enters the bone in the apex of the angle between the squamous and petrous parts of the temporal bone, immediately behind the posterior tip of the greater wing of the sphenoid which projects downwards as the **spine of the sphenoid**. The bony semicanal for the auditory tube is partially separated by a thin, bony shelf from the semicanal for the tensor tympani muscle superior to it [FIG. 12.46].

Immediately anteromedial to the spine of the sphenoid is the **foramen spinosum** which transmits the middle meningeal artery and the meningeal branch of the mandibular nerve. Further anteromedially, in the greater wing of the sphenoid immediately lateral to the groove for the auditory tube, lies the **foramen ovale** which transmits the mandibular nerve, an accessory meningeal artery, and venous communications between the cavernous sinus and the pterygoid plexus. The foramen ovale lies 4 cm deep to the tubercle of the root of the zygoma.

On each side, a **pterygoid process** hangs down from the root of the greater wing of the sphenoid, lateral to the posterior aperture of the nose [choana; FIG. 3.51]. It consists of a medial and a lateral **pterygoid lamina** which are joined anteriorly in their upper half but separated by the **pyramidal process of the palatine bone** inferiorly. The hollow between the two plates is the **pterygoid fossa**.

The **lateral pterygoid lamina** gives attachment to the pterygoid muscles; the lateral muscle arising from its lateral surface, the medial muscle from its medial surface. It slopes laterally as well as backwards to a thin edge which often bears a prominent spine. A layer of fascia extends from this posterior edge to the spine of the sphenoid and may be ossified.

The **medial pterygoid lamina** is narrower and lies in an anteroposterior plane. Where its posterior edge meets the base of the skull, it divides to enclose a **scaphoid fossa** which extends along the posterior edge of the greater wing of the sphenoid, medial to the

foramen ovale and lateral to the groove for the auditory tube. The tensor palati muscle arises from it. The lower end of the posterior border of the medial lamina is continued as a slender, curved **pterygoid hamulus**. The tendon of tensor palati hooks around the hamulus into the palate. The hamulus and the edge of the lamina immediately above it give rise to the superior constrictor muscle of the pharynx, which interdigitates with the buccinator muscle to form the pterygomandibular raphe below the level of the hamulus. This raphe connects the hamulus to the mandible posterior to the last molar tooth. The medial lamina forms the lateral wall of the nasal cavity where the cavity becomes continuous with the nasal part of the pharynx, and has the opening of the auditory tube against the upper part of its posterior margin.

A **pterygoid tubercle** projects posteriorly from the base of the pterygoid process beneath the foramen lacerum and the pterygoid canal, which runs through the base of the process from the foramen lacerum to the pterygopalatine fossa. Medially, the process extends across the inferior surface of the body of the sphenoid and forms a **vaginal process** which articulates with the vomer. A vomerovaginal sulcus with a small artery may lie between them.

In the midline the **vomer** [p. 147] articulates with the inferior surface of the sphenoid by expanding laterally (**alae of the vomer**). From this articulation the vomer projects antero-inferiorly to form the posterior part of the nasal septum. Thus the vomer in the midline, the medial pterygoid plates laterally, the body of the sphenoid superiorly with the vaginal and alar processes applied to it, and the palate inferiorly, surround the posterior apertures of the nose or **choanae**. These oval openings (2.5 cm high and 1.2 cm wide) connect the nasal cavities with the nasal part of the pharynx.

The spheno-occipital synchondrosis lies on a line joining the two pterygoid tubercles.

A prominent feature in the anterior part of the norma basalis is the inferior surface of the upper jaw [FIG. 3.51]. The maxillae bear the alveolar processes which together project down as a thick, arched ridge. Filling in the concavity of the arch is the bony palate which forms the floor of the nose and the roof of the mouth. Each maxilla in a normal adult carries a set of eight teeth: three molars, two premolars, one canine, and two incisors [FIG. 6.5].

If teeth have been removed some time before death the walls of their sockets are absorbed so that part of the alveolar process becomes nearly level with the palate.

Behind the last molar tooth on the free end of the alveolar margin, is the **tuberosity of the maxilla**, well developed only if the third molar, or wisdom tooth, has erupted. Posterior to the tuberosity, the **pyramidal process of the palatine bone** is wedged between the medial and lateral pterygoid laminae and the maxilla. It is usually fused with the laminae and with the maxilla and, medially, is continuous with the horizontal plate of the palatine bone. Its inferior surface has, as a rule, two **lesser palatine foramina** which transmit lesser palatine nerves and vessels to the soft palate and adjacent structures from the greater palatine branches in the greater palatine canal.

The **bony palate** is formed from three sources; a premaxilla, two maxillary palatine processes, and, posteriorly, the **horizontal plates of the palatine bones** which form a horizontal strip about 6 mm in anteroposterior width. The common posterior margin is sharp and concave on each side with a **posterior nasal spine** projecting back from it in the midline.

The soft palate is attached to the posterior margin, and the **palatine crest** on the inferior surface just in front of it takes part of the insertion of tensor palati muscle. Laterally, opposite the last molar tooth, in front of the end of the crest, lies the **greater palatine foramen**, the lower end of the **greater palatine canal**. The canal leads from the pterygopalatine fossa, down between the maxilla and a groove on the perpendicular plate of the palatine bone. It conveys the greater palatine vessels and nerves which then run forwards on the bony palate medial to the alveolar process.

The **maxillary part of the palate** lies in front of the palatine bones, and its two halves join in the midline. It is slightly concave and the hollow appearance is increased by the prominence of the alveolar borders. The surface is rough and pitted by the many mucous glands in the mucoperiosteum. Scattered small foramina are more numerous in front.

Anteriorly, the median suture divides and in a young bone may be seen to demarcate a triangular area of bone carrying the incisor teeth. This area is the inferior surface of the **premaxilla** which is a separate bone in some animals. The median suture and its two anterior continuations to the spaces between the incisor teeth and the canines, indicate the lines along which a developmental failure to fuse may lead to a cleft palate.

Behind the incisor teeth in the midline is a pit of varying depth and width, called the **incisive fossa**, into which four foramina open. Two in the midline transmit nasopalatine nerves, the left being in front. The lateral foramina transmit terminal branches of the greater palatine arteries to anastomose with sphenopalatine vessels in the nose.

THE INTERIOR OF THE SKULL

The walls of the cranial cavity are formed by the frontal, parietal, occipital, temporal, and sphenoid bones, already seen on the exterior of the skull, with the ethmoid which also enters into the formation of orbit and nose.

The capacity of the cranial cavity has been much used in anthropological work. Over 1450 cm³ is a large capacity; below 1350 cm³ is small.

Measurements of the thickness of 448 male white American crania (Todd 1924) ranged from 3 mm to 6 mm at the posterior border of the foramen magnum and at the vertex. At the position of greatest thickness the range was from nearly 5 mm to 11 mm.

The cranial cavity is lined by two closely adherent layers of fibrous membrane sometimes referred to as outer and inner layers of the dura mater. The outer layer, in fact periosteum (or **endocranium**), lines the interior of the cranium and is continuous with the periosteum on the outside (or **pericranium**) through the foramina and fissures.

The inner layer is the **dura mater** proper and forms a continuous covering for the brain. It is folded between the major parts of the brain, but elsewhere is attached to the endocranium.

The roof of the cranial cavity (Skull cap, or calvaria)

It is formed by the frontal, parietal, and occipital bones, and a small part of the squamous temporal bone on each side.

The internal or cerebral surface is fairly smooth and highly concave, particularly from side to side. Except at the sutures, the endocranium can be stripped from it quite easily, particularly in a young skull.

The **sutures** may be indistinct or obliterated if the skull is from an old person, for fusion begins on the inside between the ages of 20 years and 30 years; 10 years earlier than on the outside. The sutures are the same as on the outside, but the squamous suture is at a lower level because of the considerable overlap of the squamous temporal on the outside of the parietal bone.

The **frontal crest** [FIG. 3.53] is a median ridge on the lower part of the internal surface of the frontal bone. It extends upwards to become continuous with the **sulcus for the superior sagittal sinus**;

a shallow, median furrow extending upwards and backwards on the frontal bone, then along the internal surface of the sagittal suture, and down on the occipital bone as far as the internal occipital protuberance, where it turns to one side [FIG. 3.79]. The groove becomes progressively wider as the sinus enlarges in its course towards the occiput [FIG. 10.182]. The dura mater leaves the endocranium at each edge of the furrow to form a fold (falx cerebri) which extends inferiorly between the cerebral hemispheres to a free edge below. The sinus lies between its layers and the endocranium [FIG. 13.99].

Arachnoid granulations [FIG. 10.186] project into the cerebral sinuses and particularly into venous spaces at the side of the superior sagittal sinus (lateral lacunae). They project sufficiently to indent the bone [FIG. 3.77]. These small **granular pits**, lie alongside the groove for the sagittal sinus. They increase in size and number with age.

The **parietal foramen**, if present, lies close to the groove for the sinus on one or both sides 2.5–4 cm above lambda. In some bones a groove for meningeal vessels leads up to the foramen.

Narrow, branching **vascular grooves** for the meningeal vessels extend from the lower edge of the vault up to the top. The largest are for the middle meningeal vessels and their direction is upwards and backwards [FIG. 3.57]. Small vascular foramina are numerous especially in or near the grooves.

The floor of the cranial cavity [FIG. 3.53]

This is formed by the base of the skull and is divided into three main **fossae**, separated by prominent ridges and set at different levels. The ridges are arranged as a flattened **X**, and the posterior fossa which lies behind and below its posterior limbs has the foramen magnum in its floor.

The middle fossa lies at a higher level. It has a median part hollowed out in the middle of the X and two lateral parts between its limbs, one on each side. The anterior fossa lies between the anterior limbs of the X and at a still higher level, above the nose and the two orbits.

THE POSTERIOR CRANIAL FOSSA

This is the largest and deepest of the three. The floor and posterior wall are formed chiefly by the concave surface of the occipital bone, whilst the posterior surface of the petrous part of the temporal bone forms the anterolateral wall, and the body of the sphenoid, continuous with the anterior surface of the basi-occipital, forms the anterior wall with it. The mastoid angle of the parietal bone enters the fossa at the side. Posteriorly, an **internal occipital protuberance** corresponds to the external protuberance and is joined to the edge of the foramen magnum by the **internal occipital crest** which divides the posterior part of the fossa into two. A small fold of dura mater, the falx cerebelli, projects from it between the two cerebellar hemispheres. Each hemisphere occupies a thin-walled cerebellar fossa lateral to the crest and to the posterior half of the foramen magnum, above the attachment of the muscles of the back of the neck. The **sulcus for the superior sagittal sinus** passes down to the internal occipital protuberance and then turns abruptly to one side, usually the right, as a transverse sinus. This forms a wide groove on the bone and a similar groove for the opposite transverse sinus is formed by the continuation of the straight sinus. The **groove for the transverse sinus** [FIG. 3.79] passes laterally above the cerebellar fossa, from the internal occipital protuberance to the superior margin of the base of the petrous temporal bone, and then turns down to continue as the S-shaped **groove for the sigmoid sinus** which leads to the jugular foramen.

The transverse sinus crosses and grooves the mastoid angle of the parietal bone. The sigmoid sinus grooves the petrous temporal (mastoid part) close to the mastoid air cells and to the mastoid antrum [FIG. 3.47]. It then grooves the occipital bone as it curves towards the jugular foramen. A foramen is usually present in the floor of the groove for the sigmoid sinus and transmits a vein which communicates with the mastoid veins.

The superior margin of the petrous temporal bone runs anteromedially from the groove for the transverse sinus to the apex of the petrous part of the temporal bone where it articulates with the body of the sphenoid. This margin has a groove for the superior petrosal sinus which runs between the cavernous sinus [FIG. 13.100] and the transverse sinus. A fold of dura mater forming the tentorium cerebelli extends inwards in a continuous line (attached margin) from the bone at the edges of the grooves for the transverse and superior petrosal sinuses and from the posterior clinoid process.

The groove for the superior petrosal sinus is absent near its medial end where the trigeminal nerve runs inferior to the edge of the tentorium and the sinus and grooves the bone inferior to them.

The inferior edge of the posterior surface (posterior margin of the petrous part) of the petrous temporal bone is irregular but mainly convex downwards. Medially it articulates with the sphenoid and basi-occipital and along the suture both bones are grooved by the inferior petrosal sinus. The groove leads into the anterior part of the jugular foramen, whilst the groove for the sigmoid sinus enters its posterior part. Just lateral to the end of the sulcus for the inferior petrosal sinus the margin is notched for the glossopharyngeal nerve, and from the bottom of a pit in the notch, a canaliculus leads up to the cochlea (canaliculus cochleae).

The jugular foramen lies between the edge of the petrous temporal bone and, on this aspect, the jugular notch in the occipital bone [FIG. 3.79]. Medial to the foramen the occipital bone forms a ridge, with a rounded jugular tubercle and a shallow groove behind it for the ninth, tenth, and eleventh cranial nerves passing to the intermediate part of the jugular foramen. Between this raised edge of bone and the lower margin of the foramen magnum is the internal opening of the hypoglossal canal for the twelfth cranial nerve.

In the posterior surface of the petrous part of the temporal bone, is the opening of the internal acoustic meatus, a cylindrical canal about 10 mm long and 3–5 mm wide. Its opening is oblique, but the canal runs directly laterally in the same line as its fellow of the opposite side. It is also opposite the external meatus at a depth of

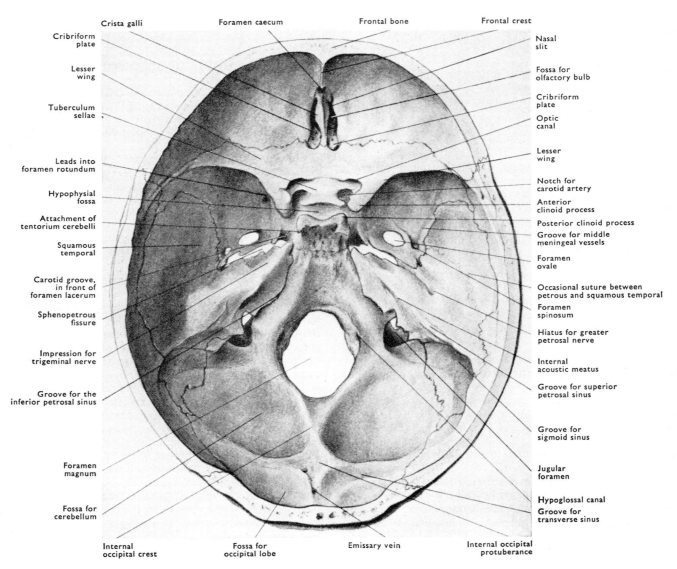

Crista galli · Foramen caecum · Frontal bone · Frontal crest

Cribriform plate · Lesser wing · Tuberculum sellae · Leads into foramen rotundum · Hypophysial fossa · Attachment of tentorium cerebelli · Squamous temporal · Carotid groove, in front of foramen lacerum · Sphenopetrous fissure · Impression for trigeminal nerve · Groove for the inferior petrosal sinus · Foramen magnum · Fossa for cerebellum

Nasal slit · Fossa for olfactory bulb · Cribriform plate · Optic canal · Lesser wing · Notch for carotid artery · Anterior clinoid process · Posterior clinoid process · Groove for middle meningeal vessels · Foramen ovale · Occasional suture between petrous and squamous temporal · Foramen spinosum · Hiatus for greater petrosal nerve · Internal acoustic meatus · Groove for superior petrosal sinus · Groove for sigmoid sinus · Jugular foramen · Hypoglossal canal · Groove for transverse sinus

Internal occipital crest · Fossa for occipital lobe · Emissary vein · Internal occipital protuberance

FIG. 3.53. Internal surface of base of skull.

Cochlea   Geniculate ganglion   Malleus   Incus

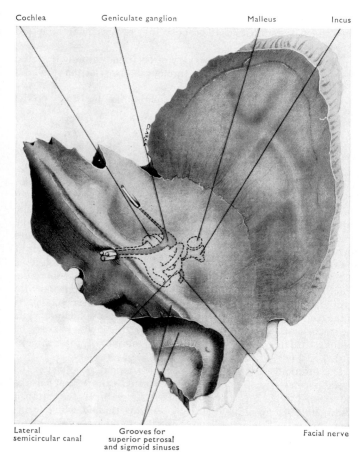

Lateral
semicircular canal   Grooves for
superior petrosal
and sigmoid sinuses   Facial nerve

FIG. 3.54. Internal aspect of a right temporal bone to show the position of structures within the bone.

50 mm. The **fundus** or bottom of the meatus is formed by the thin bony wall of the internal ear and this is pierced by the facial nerve, by branches of the vestibulocochlear nerve, and by small blood vessels [FIG. 12.51]. The facial nerve passes laterally in the bone, then turns backwards (at its genu) to a point above the stylomastoid foramen where it turns vertically downwards through the bone to emerge at that foramen [p. 116]. The notch for the glossopharyngeal nerve is inferior to the internal acoustic meatus.

Above and posterolateral to the opening of the meatus is an irregular depression with one or two small foramina for vessels opening into it. It is best seen in young bones [FIG. 3.86] and is the **subarcuate fossa**, which indicates the concavity of the anterior semicircular canal. Further laterally and at a lower level is a narrow fissure which conceals the external aperture of the **aqueduct of the vestibule**. This carries small vessels and the endolymphatic duct which ends in a dilated endolymphatic sac.

Anterior to the foramen magnum, the basilar part of the occipital bone and the body of the sphenoid slope upwards as the **clivus** to the **dorsum sellae**. This upstanding, square plate of bone, forms the posterior wall of the **sella turcica**, the 'Turkish saddle', the median part of the middle fossa. The upper corners of the dorsum sellae project as **posterior clinoid processes** which give attachment to the anterior ends of the fixed margin of the tentorium cerebelli.

The clivus is separated by meninges from the pons and the basilar artery above, and from the medulla oblongata and vertebral arteries below. The abducent nerve runs up on the clivus for more than 1 cm and **turns forward at the apex of the petrous temporal bone under a small petrosphenoid ligament which may be ossified.

THE MIDDLE CRANIAL FOSSA

This lies anterior and superior to the posterior fossa. It consists of a median and two lateral parts and is formed by the temporal and sphenoid bones.

The lateral parts contain the temporal lobes of the brain, and extend from the superior margin of the petrous part of the temporal bone to the lateral wall of the orbit formed by the greater wing of the sphenoid bone. Here its anterior part is overlapped by the sharp posterior edge of the anterior cranial fossa, the lesser wing of the sphenoid bone [FIG. 3.53].

Laterally, the greater wing of the sphenoid and the squamous part of the temporal bone turn up to form the lateral wall of the skull in the temporal region. Medially is the raised body of the sphenoid bone.

The floor of the lateral part of the fossa sinks to the level of the upper border of the zygomatic arch. Posteriorly it is formed by the sloping anterior surface of the **petrous part of the temporal bone**. Laterally it articulates with the squamous part at the **petrosquamous suture**, medially with the greater wing of the sphenoid to reach the **foramen lacerum**. Medial to the petrosquamous suture, the petrous part forms the roof of the middle ear cavity, the **tegmen tympani**. Further medially and near the superior border, a distinct prominence, the **eminentia arcuata**, overlies the anterior semicircular canal of the internal ear [FIG. 3.82]. Medial to this again, and close to the petrosphenoid suture, a groove runs towards the foramen lacerum. At the posterior end of the groove, midway between the foramen lacerum and the side of the skull, lies a **hiatus for the greater petrosal nerve**. The nerve comes from the geniculate ganglion of the facial nerve and runs in the groove to enter the cartilage filling in the foramen lacerum and join the deep petrosal nerve there.

Anterolateral to the hiatus is a second, smaller **hiatus for the lesser petrosal nerve**. The nerve runs to the petrosphenoid suture and passes through it, or through the foramen ovale, to reach the otic ganglion. The anterior surface of the petrous temporal bone is slightly hollowed near the foramen lacerum where the trigeminal ganglion lies against it—the **trigeminal impression**.

The lateral part of the middle fossa in front of the petrous part of the temporal bone is formed by the squamous temporal and the greater wing of the sphenoid bone and is marked by the under surface of the brain which produces the digital impressions. Anteriorly the **greater wing of the sphenoid** passes laterally and sweeps posterosuperiorly in the lateral wall of the skull. Above it the lesser wing of the sphenoid extends horizontally from the body to join the greater wing [FIG. 3.87]. The comma-shaped gap between them is the **superior orbital fissure**. It is wide below and medially, and tapers to a point laterally where the wings meet. The fissure opens directly into the back of the orbit and transmits numerous structures to and from the orbital cavity [FIG. 11.11].

Immediately below the wide medial end of the superior orbital fissure, the **foramen rotundum** passes forwards through the greater wing to open into the upper part of the pterygopalatine fossa. It transmits the maxillary nerve which crosses the fossa to reach the orbit. The larger **foramen ovale** lies further back, lateral to the foramen lacerum. The mandibular nerve, including its motor branch and accompanied by veins, passes vertically down through the foramen ovale to the pterygoid region, and the accessory meningeal artery ascends through it. Posterolateral to the foramen ovale, passing through the corner of the greater wing of the sphenoid, is the **foramen spinosum** for the middle meningeal vessels and a meningeal branch of the mandibular nerve.

The **middle meningeal vessels**, the largest nutrient vessels of the skull, lie in the periosteum and groove the interior of the skull deeply [FIG. 3.57]. A parietal branch passes almost horizontally back across the squamous temporal and parietal towards the occipital region. A

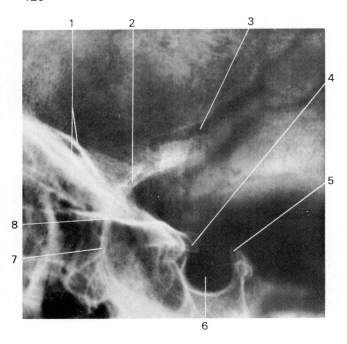

FIG. 3.55. Lateral radiograph of body of sphenoid bone and adjacent structures (natural size).
1. Orbital roofs.
2. Lesser wing of sphenoid.
3. Anterior temporal diploic vein.
4. Anterior clinoid process.
5. Posterior clinoid process.
6. Sella turcica.
7. Anterior wall of middle cranial fossa (greater wing of sphenoid).
8. Anterior cranial fossa, central part of floor.

frontal branch curves upwards and usually becomes enclosed in a bony tunnel in the vicinity of the end of the lesser wing of the sphenoid and of the pterion [p. 109]. Thereafter, the frontal branch runs posterosuperiorly (separated from the motor area of the brain by the meninges) towards the top of the skull midway between nasion and inion. Fractures of the skull may tear the vessels, particularly where the grooves are deep.

At the medial side of the foramen lacerum the internal carotid artery, with its plexus of veins and sympathetic nerves, emerges from the cartilage and turns forwards. It grooves the body of the sphenoid and the edge of the groove projects back over the foramen lacerum as the lingula. The artery subsequently turns upwards and backwards completing an S-bend and passing medial to the anterior clinoid process.

The median part of the middle cranial fossa is formed largely by the sella turcica of the sphenoid bone. Posteriorly it is limited by the dorsum sellae [p. 119] with its posterior clinoid processes. The wide smooth depression in the body of the sphenoid is the hypophysial fossa which lodges the hypophysis [FIG. 3.57]. Anteriorly an oval elevation, the tuberculum sellae forms the 'pommel of the saddle' [FIG. 3.55].

In front of the tuberculum a transverse sulcus runs between two round openings, the optic canals. The sulcus is the sulcus chiasmatis, but it does not lodge the optic chiasma.

Each optic canal runs forwards and laterally between the two roots of the lesser wing of the sphenoid, to the corresponding orbit. The roots run to the body of the bone, and the posterior root and edge of the wing continue back as a projection with a rounded point, the anterior clinoid process. This gives attachment to the free border of the tentorium cerebelli, and is grooved on its medial aspect by the internal carotid artery.

The term clinoid, bed-shaped, refers to the old four-poster bed, but a third, or middle clinoid process, may be present on each side [p. 145].

A craniopharyngeal canal is very rarely found running from the bottom of the hypophysial fossa through the body of the sphenoid [p. 135].

The cavernous sinus is a lateral relation of the body of the sphenoid [FIG. 13.100].

THE ANTERIOR CRANIAL FOSSA

This lodges the lower part of the frontal lobes of the brain and lies above and in front of the middle fossa, separated from it on each side by the concave, posterior edge of the lesser wing of the sphenoid, which is wedged into the stem of the lateral sulcus of the brain [FIG. 10.96]. The walls of the fossa and most of the floor are formed by the frontal bone. In the floor, the convex orbital parts of the frontal, roughened to correspond to the overlying convolutions of the brain, are separated in the midline by a narrow ethmoidal notch in which lie the cribriform plates of the ethmoid bone with its crista galli projecting between them. Posteriorly, the orbital parts of the frontal articulate with the lesser wings of the sphenoid, while the smooth jugum sphenoidale which unites these wings across the midline, articulates with the posterior margin of the ethmoid [FIG. 3.53].

The ethmoid bone is recessed in the ethmoidal notch but its lateral parts (ethmoidal cells) are overlapped by the orbital parts of the frontal bone [FIG. 3.75]. The crista galli is the upward continuation of the perpendicular plate of the ethmoid, which forms the upper part of the nasal septum inferior to the cribriform plates [FIG. 3.56]. The crista galli gives attachment to the falx cerebri [FIG. 10.183].

Immediately in front of the crista galli is the foramen caecum, a small pit which occasionally transmits a vein through an aperture leading to the nose.

Anterior to the foramen caecum, the frontal crest, the median ridge on the frontal bone [p. 117], extends upwards on the vault. The edges of the foramen caecum and the crest also give attachment to the anterior end of the falx cerebri.

Each narrow cribriform plate of the ethmoid forms part of the roof of the corresponding nasal cavity and transmits the olfactory nerve fibres through its numerous foramina from olfactory mucous membrane to the olfactory bulb on the superior surface of the plate. On each side of the crista galli is a nasal slit [FIG. 3.91] which transmits the anterior ethmoidal vessels and nerve to the nasal mucous membrane. These reach the cribriform plate and anterior cranial fossa by passing from the orbit between the ethmoid bone and the overlapping orbital part of the frontal bone, in the anterior ethmoidal foramen [FIG. 3.49]. The posterior ethmoidal foramina also transmit vessels and nerves from the orbit but these are distributed mainly to the ethmoidal cells. Either of the ethmoidal vessels may be of importance in nasal haemorrhage.

THE ORBIT

The orbit is an irregular pyramidal cavity in which the eyeball rotates [FIG. 3.49]. The pyramid is laid on its side; the apex is placed posteriorly and to the medial side and the base opens forwards on to the face. It has, in consequence, a roof, a floor, a medial wall, and a lateral wall, all of which are approximately triangular. The medial walls of the two orbits are nearly parallel, the lateral walls are at right angles to each other.

The concave **roof** is formed mainly by the orbital part of the frontal bone. Posteriorly, it is also formed by the lesser wing of the sphenoid which articulates with the orbital part of the frontal bone. At its lateral side the roof turns down to articulate with the greater wing of the sphenoid bone behind and with the zygomatic bone in front. These three bones form the **lateral wall** and separate the orbit from the temporal fossa. At the back the lesser wing of the sphenoid is separated from the greater wing by the **superior orbital fissure**, which leads into the middle cranial fossa. The orbital surface of the zygomatic bone is pierced by a foramen for the zygomatic nerve, which divides in the bone to emerge as zygomaticotemporal and zygomaticofacial branches.

The frontal process of the zygomatic bone with the zygomatic process of the frontal bone [FIG. 3.74] form the lateral margin of the orbit. Within the margin of the orbit, about the middle of the lateral wall (10 mm below the frontozygomatic suture), a slight eminence may be felt with a finger-tip. It is for attachment of the lateral palpebral ligament and the check and suspensory ligaments of the eye. At the lateral angle of the roof there is a shallow **fossa for the lacrimal gland**. The lateral two-thirds of the supra-orbital margin of the frontal bone are sharp and the medial one-third is smooth and ill defined. At their junction the supra-orbital notch or foramen transmits the nerves and vessels of that name. At the medial angle of the roof, on the frontal bone, is a small **trochlear fossa** or a **trochlear spine**, where the trochlea or pulley for the superior oblique muscle of the eyeball is attached.

The indefinite medial margin of the orbit is formed by the frontal process of the maxilla and the maxillary process of the frontal bone which meet at the frontomaxillary suture.

If the **medial wall** of the orbit is traced back from the margin, an edge (anterior lacrimal crest) is reached which gives attachment to orbicularis oculi and the medial palpebral ligament. It forms the anterior limit of the lacrimal groove which passes from the orbital surface of the maxilla on to its medial surface and so reaches the nose. In the orbit, the groove forms the anterior part of the **fossa for the lacrimal sac**. The fossa is completed by a groove in the lacrimal **bone** which lies posterior to the frontal process of the maxilla. The lacrimal bone is a thin flake of bone which fills in a rectangular area between the maxilla in front and below, the ethmoid behind and the frontal bone above. A vertical **posterior lacrimal crest** separates its anterior grooved part, which forms part of the lacrimal fossa, from a smooth posterior orbital surface. The lower part of the posterior lacrimal crest turns forwards to meet the anterior crest and so closes the lower border of the fossa for the lacrimal sac laterally [FIG. 3.94], leaving room for the **nasolacrimal canal** to pass down to the nose.

Posterior to the lacrimal bone the **ethmoid** forms the medial wall of the orbit with the adjacent parts of the frontal and maxillary bones with which it articulates; while posteriorly it meets the sphenoid where the two roots of the lesser wing enclose the optic canal.

The orbital surface, or **lamina orbitalis** of the ethmoid, separates the orbit from the ethmoidal cells and is so thin that it has been called the lamina papyracea. The suture between it and the frontal bone is interrupted by two foramina, the anterior and posterior **ethmoidal foramina**, which transmit vessels and nerves of the same name [FIG. 11.11].

The **optic canal**, about 4 mm in diameter, lies at a slightly lower level than the ethmoidal foramina as it enters the orbit through the most posterior part of the medial wall. It transmits the optic nerve and the ophthalmic artery.

Between the greater and lesser wings of the sphenoid the **superior orbital fissure** opens into the back of the orbit, and separates the posterior parts of its roof and lateral wall. The fissure is narrow above and laterally and curves downwards and medially to become wide and rounded at the apex of the orbit. It transmits many structures between the middle cranial fossa and the orbit [FIGS. 11.9 and 11.11].

The **inferior orbital fissure** passes forwards and laterally, from the upper part of the pterygopalatine fossa between the lateral wall and floor of the orbit. The maxillary nerve crosses the fossa, passes laterally along the medial half of the fissure, and turns anteriorly into the floor or the orbit as the infra-orbital nerve. Through the fissure, veins may pass from the orbit to the pterygoid plexus, the infra-orbital artery enters the infra-orbital sulcus, and small branches pass up to the orbit from the pterygopalatine ganglion.

The **floor of the orbit** is formed by the orbital surface of the maxilla, with the zygomatic bone anterolaterally, and the orbital process of the palatine bone at the apex. The floor also forms the roof of the maxillary sinus [FIG. 3.60]. The anterior parts of the zygomatic bone and maxilla are thick and together form the inferior margin of the orbit.

Posteriorly, the **infra-orbital sulcus** runs forwards and medially from the middle of the inferior orbital fissure and becomes continuous with the **infra-orbital canal** which emerges on the face 6 mm below the medial one-third of the lower margin of the orbit. The sulcus and canal transmit the infra-orbital nerve and vessels. In the canal an anterior superior alveolar branch of the nerve enters a canaliculus a few millimetres from the infra-orbital foramen and takes a sinuous course through the bone to the lower part of the nasal septum. It gives off branches which run through the bone to the teeth as far back as the first molar.

A shallow depression on the medial part of the floor, just lateral to the lacrimal fossa, gives attachment to the inferior oblique muscle. The remaining extrinsic ocular muscles take origin far back in the orbit surrounding the optic canal and the adjacent part of the superior orbital fissure [FIG 11.12].

THE CAVITY OF THE NOSE

In the dried skull the cavity of the nose extends backwards from the piriform aperture to the choanae or posterior nares. The floor is nearly horizontal [FIG. 3.56].

The cavity is divided by a median septum into right and left halves and the term nasal cavity is used to mean either the whole cavity or one or other of the halves. The interior of the nose has little space in it between the septum and the lateral wall because of projections from the wall and because, on each side, the roof of the cavity is narrow while the floor is only 10 mm wide.

The **external nose** consists largely of soft parts, supported by cartilage and attached to the edges of the piriform aperture and septum [FIG. 7.2]. The elongated **nasal bones** support the bridge of the nose and represent the external nose in the skull. They articulate with the frontal bone above, with the frontal process of the maxilla at each side, and with each other in the midline [FIG. 3.96]. Here their edges are turned down to increase their area of contact, to form a minute part of the nasal septum, and to articulate with the nasal spine of the frontal bone, the perpendicular plate of the ethmoid, and the septal cartilage [FIG. 7.5]. Laterally, the internal surface of the bone is grooved by the anterior ethmoidal nerve [p. 120].

The **septum of the nose** separates the cavity into approximately equal halves, but it is seldom quite in the midline and is often considerably distorted. It is cartilaginous in front [FIG. 7.5]. The bony part is formed mainly by the perpendicular plate of the ethmoid and the vomer. The **perpendicular plate of the ethmoid** is attached between the cribriform plates of the ethmoid above, and reaches the cartilaginous septum below and in front, the body of the sphenoid posteriorly, and the vomer below and behind.

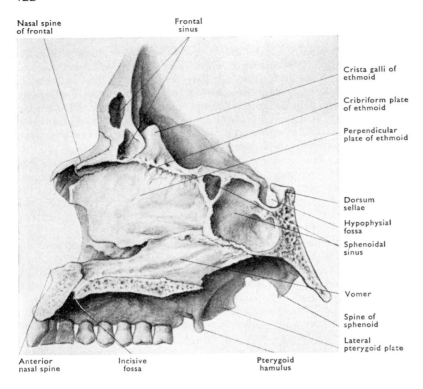

Nasal spine of frontal

Frontal sinus

Crista galli of ethmoid

Cribriform plate of ethmoid

Perpendicular plate of ethmoid

Dorsum sellae

Hypophysial fossa

Sphenoidal sinus

Vomer

Spine of sphenoid

Lateral pterygoid plate

Anterior nasal spine

Incisive fossa

Pterygoid hamulus

FIG. 3.56. Roof, floor, and septum of nose.

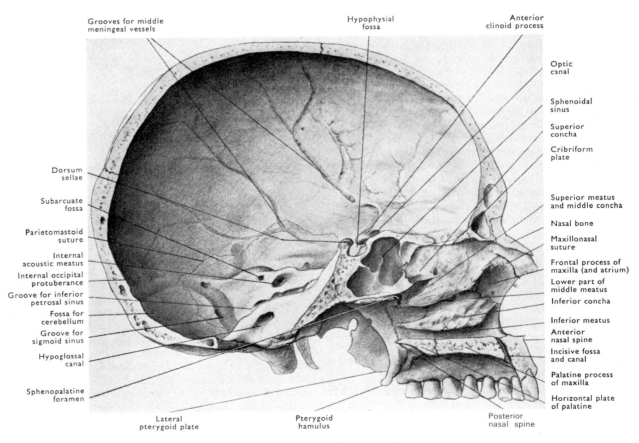

Grooves for middle meningeal vessels

Hypophysial fossa

Anterior clinoid process

Optic canal

Sphenoidal sinus

Superior concha

Cribriform plate

Superior meatus and middle concha

Nasal bone

Maxillonasal suture

Frontal process of maxilla (and atrium)

Lower part of middle meatus

Inferior concha

Inferior meatus

Anterior nasal spine

Incisive fossa and canal

Palatine process of maxilla

Horizontal plate of palatine

Dorsum sellae

Subarcuate fossa

Parietomastoid suture

Internal acoustic meatus

Internal occipital protuberance

Groove for inferior petrosal sinus

Fossa for cerebellum

Groove for sigmoid sinus

Hypoglossal canal

Sphenopalatine foramen

Lateral pterygoid plate

Pterygoid hamulus

Posterior nasal spine

FIG. 3.57. Medial aspect of left half of skull sagitally divided.

The **vomer** [p. 147] slopes downwards and forwards from the inferior surface of the body of the sphenoid to the superior surface of the bony palate in the midline [FIG. 3.56]. Its superior border is widened to form everted **alae** which articulate with the sphenoid, enclose its rostrum, and reach over the roof of the nose to be overlapped by the expanded bases of the pterygoid processes. The posterior border slopes downwards and forwards and forms a posterior edge for the nasal septum. Its inferior edge fits between the two parts of the **nasal crest** on the superior surface of the bony palate. The anterior edge slopes downwards and forwards to the anterior part of the palate and articulates with the perpendicular plate of the ethmoid and the septal cartilage. The surfaces of the vomer are grooved by the sphenopalatine vessels and by the nasopalatine nerve [FIG. 11.14].

The **roof of the nasal cavity** on each side of the septum is only 1 or 2 mm wide except posteriorly where it measures about 10 mm between each pterygoid process and the septum. It consists of a sloping, anterior part formed by the frontal and nasal bones, a middle, horizontal part formed by the cribriform plate of the ethmoid, and a steeply sloping, angulated, posterior part formed by the anterior and inferior surfaces of the body of the sphenoid [FIG. 3.56] separating it from the sphenoidal sinus. The anterior wall of the body of the sphenoid is perforated by an opening, 4 or 5 mm in diameter, through which the sinus is continuous with the nasal cavity [FIG. 3.57]. The original size of the opening is much reduced by a thin piece of bone, the sphenoidal concha [p. 144], which is originally separate but joins with the bones adjacent to it in early youth. The inferior surface of the body of the sphenoid is partly separated from the mucosa by the alae of the vomer, the vaginal processes of the pterygoid process, and part of the sphenoidal process of the palatine bone.

The middle, horizontal part of the roof is formed by the cribriform plate. Here the close proximity of the nasal mucosa to the membranes of the brain constitutes a danger and the plate is a weak spot mechanically.

The anterior part of the roof lies inferior to the frontal sinus, the frontal bone forming only a narrow strip leading down to the nasal bones.

The **floor of the nasal cavity** (10–12 mm wide) is formed by the palatine processes of the two maxillae and by the horizontal plates of the palatine bones [FIGS. 3.51 and 3.101] which form the posterior edge of the hard palate. Where the pairs of processes meet in the midline they project up a little to form a continuous **nasal crest** which lodges the vomer in a longitudinal groove. The anterior end of the crest is prominent and leads to the **anterior nasal spine**.

Near the anterior end of the septum an **incisive canal** passes down through the palate on each side and divides into two so that four foramina open into the **incisive fossa** on the inferior surface of the palate. The foramina transmit the terminal parts of the right and left nasopalatine nerves to the oral surface of the palate and the termination of the greater palatine arteries up into the nose.

The **lateral wall of the nose** is a fairly flat surface with a number of openings which lead to the paranasal air sinuses. It is usually described as uneven because of three folds which hang down from it. These are supported by thin, curved, bony laminae or **conchae** which project from the wall [FIG. 3.57].

Normally there are three well formed conchae. Each projects inwards from the lateral surface and turns down, thus partially enclosing a space below it which forms a **meatus**, or passage, through the nose [FIG. 3.57].

The **inferior concha**, with the **inferior meatus** beneath it, passes back, sloping slightly downwards, from near the piriform aperture almost to the posterior edge of the bony palate. The **middle concha** lies above, but separated from the inferior one by the **middle meatus**. It is shorter and the uncovered space in front of it is called the **atrium** of the middle meatus. The small **superior concha** is

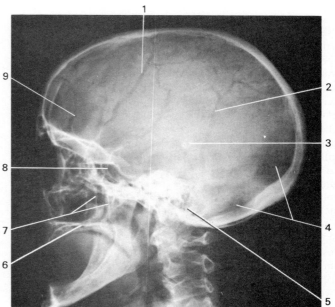

FIG. 3.58. Lateral radiograph (slightly oblique) of adult, edentulous head. Note that the sella turcica is poorly shown in such oblique views, cf. FIGURE 3.59.
1. Anterior temporal diploic vein.
2. Posterior temporal diploic vein.
3. Calcified pineal body.
4. Thin areas of occipital bone.
5. Mastoid air cells.
6. Hard palate.
7. Posterior walls of maxillary sinuses.
8. Sphenoidal sinuses.
9. Frontal diplotic vein in thin area of frontal bone.

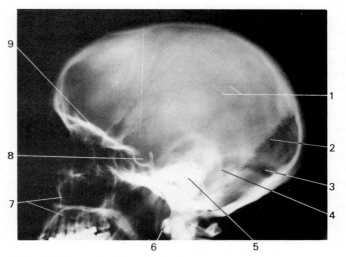

FIG. 3.59. Lateral radiograph of adult skull.
1. Diploic sinuses.
2. Lambdoid suture.
3. Transverse sinus.
4. Helix of auricle.
5. Petrous temporal bone.
6. Anterior arch of atlas.
7. Anterior wall of maxillary sinus.
8. Hypophysial fossa.
9. Roof of orbits,

situated far back in the nose, above and behind the middle concha, and separated from it by the **superior meatus**. The space above the superior concha is the **spheno-ethmoidal recess**.

The *lower half of the lateral wall* of the nose is formed in front by the nasal surface of the maxilla which separates the nose from the maxillary sinus [FIG. 3.60] but leaves a large opening into the sinus posterosuperiorly. The opening is much reduced in size by overlapping parts of the lacrimal, ethmoid, inferior concha, and palatine bones.

Behind and below the opening of the sinus is a rough area on the maxilla for articulation with the **perpendicular plate of the palatine bone**. This area is grooved reciprocally with the palatine bone by a **greater palatine sulcus**, in which the greater palatine nerve and vessels pass down into the palate [FIG. 3.98] sending posterior nasal branches forwards to pierce and supply the lateral wall of the nose. The perpendicular plate encroaches on the opening of the sinus from behind and forms part of the lateral wall of the nose.

Posteriorly, the palatine bone articulates with the medial pterygoid lamina which in turn forms the most posterior part of the lateral wall of the nose [FIG. 3.57].

Two horizontal crests project from the nasal surface of the palatine bone. The lower or **conchal crest** supports the inferior concha. The upper or **ethmoidal crest** is reached by the middle concha and it is just above and behind this that posterior nasal and nasopalatine vessels and nerves reach the nose through the **sphenopalatine foramen** [FIG. 3.57].

The **nasal surface of the maxilla** is smooth where it forms part of the inferior meatus below and in front of the large opening into the sinus. Above this area the root of the frontal process carries a **conchal crest** for the inferior concha. Posterior to the frontal process, and between it and the opening of the sinus, is the lacrimal groove leading down to the inferior meatus.

The **inferior concha** is a separate bone which articulates with the two conchal crests, maxillary and palatine, and bridges over the lacrimal groove so that the **nasolacrimal canal** opens under cover of the concha into the inferior meatus 30–35 mm from the edge of the nostril. The base of the concha [p. 146] closes the greater part of the opening in the medial surface of the maxilla, and forms a considerable part of the medial wall of the maxillary sinus [FIG. 3.60].

The *upper half of the lateral wall of the nose* contains the openings of most of the paranasal sinuses. It consists largely of ethmoid bone, the labyrinth of which lies between the orbit and the nose on each side [FIG. 3.60]. The labyrinth consists of small air sinuses (cells) enclosed by exceptionally thin bone so that access from the nose to the orbit can be perilously easy if the bone is damaged by injury or disease.

The *anterior part* of this region is formed by the nasal bone and the frontal process of the maxilla and has the frontal bone and frontal sinus immediately above it.

The *posterior part* is formed by the ethmoid bone which fills in the space between sphenoid bone behind, maxilla below and in front, and frontal bone above. The ethmoid covers the nasal aspect of the lacrimal bone and sends down a process to assist in the reduction of the upper part of the opening through the maxilla into its sinus. The residual opening, therefore, lies inferior to the ethmoid labyrinth and superior to the inferior concha.

The **middle concha** and the **superior concha** project from the ethmoid bone. If the middle concha is removed, the lateral wall of the **middle meatus** shows a swelling (bulla ethmoidalis) caused by the lower middle ethmoidal air cells which open on its surface. Immediately below and in front of the bulla is a curved groove, the **hiatus semilunaris**. This is limited inferiorly by a thin curved bone, the **uncinate process**, which extends postero-inferiorly from the ethmoid to reach the inferior concha [FIG. 3.92].

Under cover of the superior concha a posterior group of ethmoidal cells opens into the **superior meatus** through one or more small openings.

The frontal sinus opens into, or anterior to, the upper, anterior end of the hiatus via the **infundibulum**. This is an ethmoid air cell, or cells, which forms a funnel connecting the frontal sinus with the hiatus. Other anterior and some middle ethmoidal cells open into the wall of the hiatus.

THE PARANASAL SINUSES

The paranasal sinuses are air-filled cavities produced by evagination of mucous membrane from the nasal cavity into adjacent skull bones. They all communicate, therefore, with the nasal cavity and their mucous linings are continuous with the nasal mucosa. Infection may spread from the nose to the sinuses with the production of localized pain or headache. The relation of nerves to the walls is thus of importance.

The walls of the sinuses are of compact bone, the cancellous bone being displaced by the sinuses. At birth, the sinuses are small or absent and they grow only slowly until puberty, after which they grow rapidly to their adult size. In old age, absorption of diploë leads to further enlargement. They are normally larger in men, and serve to lighten what would otherwise be heavy bone. They also act as resonating chambers which affect the quality of the voice.

THE MAXILLARY SINUS [FIGS. 3.60 and 3.98]

This is a large, pyramidal cavity in the maxilla. It is set with its apex in the zygomatic process of the maxilla and its base or medial wall is the lower part of the lateral wall of the nose. Its roof is the floor of the orbit. Its narrow floor lies over the alveolar process of the maxilla and the roots of the molar and premolar teeth, with its deepest part overlying the second premolar and first molar teeth. Variations in size may limit its extent to the three molars or increase it to include all the molar, premolar, and canine teeth. The roots of the teeth, particularly the first two molars, may produce eminences in the floor or even penetrate the bone.

The lowest part of the floor lies as much as 10 mm below the level of the floor of the nose. The opening of the sinus passes through the upper part of its medial wall so that, in the absence of effective ciliary action, drainage cannot occur in the erect position.

The very large aperture seen in FIGURE 3.98 is much reduced by the overlap of adjacent bones and opens into the hiatus semilunaris or into the middle meatus posterior to it. The opening in the mucosa is normally single, though it may be double in the bone.

At birth the maxillary sinus is still no more than a groove between the conchae although it is detectable at the fourth month of intra-uterine life. Growth is slow until puberty. After puberty it grows rapidly to reach an approximate size of 35 mm in height, 30 mm anteroposteriorly, and 25 mm in width. It usually extends anterior to the nasolacrimal canal as a **lacrimal recess**, and otherwise may be partially subdivided [p. 149].

The greater palatine, posterior superior alveolar, infra-orbital, and anterior superior alveolar nerves and vessels lie in the posterior wall, roof, and anterior wall of the sinus respectively [FIGS. 11.17 and 11.18].

THE FRONTAL SINUSES

These are situated in the frontal bone near the midline, above the supra-orbital margin and the root of the nose. They are separated by

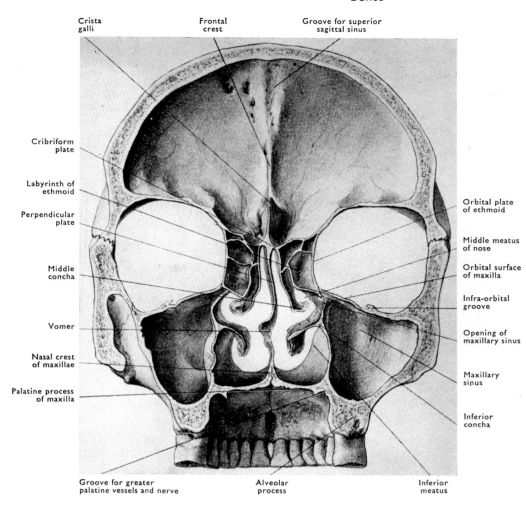

Crista galli

Frontal crest

Groove for superior sagittal sinus

Cribriform plate

Labyrinth of ethmoid

Perpendicular plate

Middle concha

Vomer

Nasal crest of maxillae

Palatine process of maxilla

Orbital plate of ethmoid

Middle meatus of nose

Orbital surface of maxilla

Infra-orbital groove

Opening of maxillary sinus

Maxillary sinus

Inferior concha

Groove for greater palatine vessels and nerve

Alveolar process

Inferior meatus

FIG. 3.60. Coronal section passing inferiorly between the first and second molar teeth.

a bony septum which usually deviates to one or other side. They vary considerably in size and are frequently lobulated. Their presence may be detectable at the second year but they are, as a rule, not clearly recognizable until the seventh year. They grow slowly until shortly before puberty and more rapidly to adult size thereafter. In the adult, however, they vary from the size of a pea to large spaces. They may extend close to the zygomatic process, or upwards to the frontal bone, or in the roof of the orbit to the lesser wing of the sphenoid. An average size would be about 25 mm by 25 mm as seen from the front, but the shape of the supra-orbital ridges and the forehead are no indication of their size.

The frontal sinuses normally drain through the ethmoid [p. 124], but may open directly at the side of the nasal septum.

THE ETHMOIDAL SINUSES [FIG. 3.50]

They constitute the ethmoidal labyrinth, between the nasal cavity and the orbit. They lie near the midline, inferior to the anterior cranial fossa, but are separated from it by the orbital plate of the frontal bone. This roofs in some of the superior cells which extend into the frontal bone, and the same extension occurs where the ethmoid articulates with other bones [p. 146].

For descriptive purposes the ethmoidal cells are divided into **posterior, middle**, and **anterior** groups which drain, respectively, on

to the surface of the superior meatus, the bulla ethmoidalis and the hiatus semilunaris, and the infundibulum [p. 124].

The ethmoidal cells begin to develop after the fifth month of intra-uterine life and are formed in the cartilaginous nasal capsule. Ossification spreads into the cell walls before birth. The cells are, therefore, already formed at birth.

THE SPHENOIDAL SINUSES

These are two cavities which lie side by side in the body of the sphenoid bone with a bony septum between them which is almost always bent to one side [FIG. 3.56]. The surrounding bony walls separate them from the nasopharynx and posterior part of the nose below, from the upper part of the nose and posterior ethmoidal cells in front, from the frontal lobes of the brain, olfactory tracts, and hypophysis above, and from the pons and basilar artery behind. Laterally the bony wall separates it from the optic nerve in the optic canal, the cavernous sinus and the structures which pass through it. The paired apertures [FIG. 3.88], about 4 mm in diameter, open on each side of the nasal septum directly into the **spheno-ethmoidal recesses** [p. 124].

The rudiments of the sphenoidal sinuses appear in the fifth month of intra-uterine life as recesses enclosed in the sphenoidal conchae [p. 144]. These extend into the body of the sphenoid in the seventh of eighth year, and undergo their main growth after puberty.

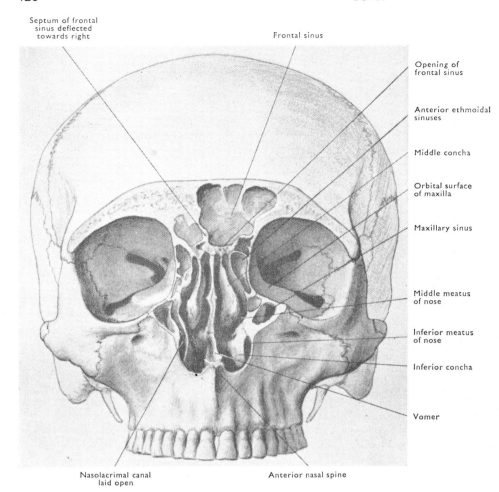

Septum of frontal sinus deflected towards right

Frontal sinus

Opening of frontal sinus

Anterior ethmoidal sinuses

Middle concha

Orbital surface of maxilla

Maxillary sinus

Middle meatus of nose

Inferior meatus of nose

Inferior concha

Vomer

Nasolacrimal canal laid open

Anterior nasal spine

FIG. 3.61. Part of the frontal, nasal, and maxillary bones removed in order to display the relation of the various cavities exposed.

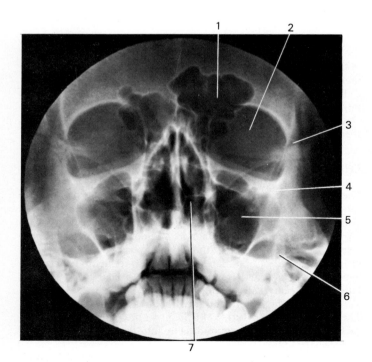

FIG. 3.62. Postero-anterior radiograph of skull taken with head tilted backwards.
1. Frontal sinus.
2. Orbit.
3. Zygomaticofrontal suture.
4. Zygomatic bone.
5. Maxillary sinus.
6. Coronoid process of mandible.
7. Nasal cavity and sphenoidal sinus superimposed.

THE PALATINE SINUSES

When present these are air spaces, one on each side, in the orbital processes of the palatine bones. They open into the sphenoidal sinus or into a posterior ethmoidal cell, and can be regarded as extensions of these into the palatine bone [FIG. 3.100].

THE MANDIBLE

The mandible, the skeleton of the lower part of the face, reaches from the chin to the mandibular fossa [FIG. 3.45]. It consists of a horseshoe-shaped body joined to a pair of flat rami which project up behind the posterior ends of the body. The upper end of each ramus carries an anterior, triangular coronoid process and a posterior condylar process, separated by the mandibular notch. The condylar process consists of a neck surmounted by an oval head with an articular surface.

The right and left halves of the mandible are joined at their anterior ends by fibrous tissue at birth, but bony fusion occurs during the second year at the symphysis menti. The posterior parts of the two halves are widely separated [FIG. 6.9].

The upper part of the body is a curved bar of bone, the alveolar arch, which carries the lower teeth. It normally contains sockets, or alveoli, for the roots of the teeth, unless the teeth have been lost during life [p. 116]. The sockets for the canine teeth produce a vertical ridge on the surface on each side, and the incisor teeth may do the same. In old age, when all the teeth have been lost, the alveolar arch becomes absorbed almost completely [FIG. 3.65].

The lower part of the body is strong and rounded. It extends from the ramus to the symphysis and forms a wider curve than the alveolar arch, thus projecting beyond it and producing the prominence of the chin.

THE BODY OF THE MANDIBLE

On the outer surface of the bone, along the line of the symphysis, a faint ridge runs down into a triangular mental protuberance. The lower angles of this form a mental tubercle on each side, the forward curvatures of which produce the chin. From the tubercle on each

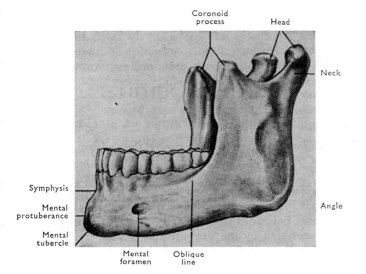

FIG. 3.63. Mandible as seen from the left side.

side an oblique line runs back and slightly upwards to become continuous with the anterior border of the ramus. Between the tubercle and the incisor teeth, is an incisor fossa. Further back, above the oblique line and below the second premolar tooth, or the space between the first and second, the mandibular canal opens on the surface as a mental foramen. This faces upwards and slightly backwards, and transmits a terminal branch of the inferior alveolar nerve, the mental nerve, and the corresponding vessels. It lies midway between the upper and lower borders of the bone, provided the alveolar process has not been absorbed. If it has been absorbed the foramen will be close to the upper border. The anterior part of the body, the oblique line and the lower margin give attachment to facial muscles [FIG. 5.27].

The lower border of the body is the base of the mandible. It is smooth and rounded except close to the symphysis where a small but distinct depression, the digastric fossa, lies on or immediately posterior to it. Posteriorly the base passes back to the angle of the mandible.

THE RAMUS OF THE MANDIBLE

On each side, this flat plate of bone forms the mandible behind the last molar tooth. It extends from the base to the coronoid and condylar processes, and its anterior limit at the base is marked by the slight groove produced by the facial artery crossing it.

Its sharp anterior border is continued down from the coronoid process to become continuous laterally with the oblique line on the body [FIG. 3.63], whilst a rounded ridge leads to the posterior end of the mylohyoid line on the medial surface [FIG. 3.64]. The triangular area between these lines behind the last molar tooth is often ridged where it gives attachment to the buccinator and superior constrictor muscles and the inferior end of the pterygomandibular raphe. The lingual nerve slips under the lower border of the superior constrictor muscle close to the medial side of the bone.

The posterior border of the ramus is rounded. It descends from the back of the neck, and turning through approximately a right angle, angle of the mandible, becomes continuous with the base of the mandible [FIG. 3.63]. The middle of the convexity of the angle of the mandible is used as a landmark.

The superior border of the ramus exhibits the coronoid and condylar processes. Between them is the sharp concave edge of the mandibular notch which runs from the coronoid process in front, to the lateral aspect of the condylar process behind.

The coronoid process is a flat, triangular plate which gives attachment to the temporalis muscle. Its base is continuous with the anterosuperior corner of the ramus, and the muscle is inserted into its medial surface and margins as well as to the anterior margin of the ramus and the blunt ridge on its medial side [FIG. 5.23].

The condylar process consists of an articular head supported on a neck which is flattened anteroposteriorly and widens as it passes up to the head. The posterior surface is smooth and is continuous with the posterior surface of the head as well as the posterior border of the ramus. The anterior surface is rough, concave, and overhung by the anterior margin of the head. This pterygoid fovea is the insertion of part of the tendon of the lateral pterygoid muscle. Lateral to the fovea a strengthening flange of bone sweeps downwards and forwards to become continuous with the ramus. The lateral ligament of the joint is attached to the lateral and posterior aspects of the neck [FIG. 4.18].

The head of the mandible is convex but considerably elongated from side to side and narrow from front to back. Its long axis is slightly oblique and passes medially and a little backwards. It is separated from the mandibular fossa by an articular disc [FIG. 4.19].

The flat lateral surface of the ramus [FIG. 3.63] is roughened by the insertion of the masseter muscle which is attached to the whole

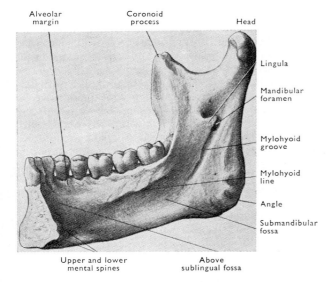

FIG. 3.64. Medial surface of right half of mandible.

surface, except the condylar process which is not covered by the muscle [FIG. 5.33]. The facial artery may groove the bone at the antero-inferior corner of the area for the muscle.

THE INNER SURFACE OF THE MANDIBLE

This includes the medial surface of the ramus and body and the posterior aspect of the region near the symphysis [FIG. 3.64].

The *medial surface of the ramus* is strongly marked near the angle of the jaw by the insertion of the medial pterygoid muscle. The stylomandibular ligament reaches the posterior border behind this area [FIG. 4.18] while anterior to it a mylohyoid groove slopes down towards the lower part of the ramus and is occupied by the mylohyoid vessels and nerve.

The groove runs down from the mandibular foramen which leads into the mandibular canal for the inferior alveolar vessels and nerve. The lingula is a spur of bone which projects upwards and backwards over the foramen. It gives attachment to the sphenomandibular ligament which also spreads out to the adjacent bone [FIG. 4.21] and bridges the mylohyoid groove. The smooth bone behind this is in contact with the parotid gland.

On the *inner surface of the body* the mylohyoid line extends as an oblique ridge which runs downwards and forwards from a point below the socket of the third molar tooth to the anterior part of the lower border. It divides this area into two triangular zones, one above the line and the other below it. The two zones are separated by the mylohyoid muscle which is attached to the whole length of the line. Above the mylohyoid line, near the symphysis, is the sublingual fossa, which accommodates part of the sublingual gland [FIG. 3.64]. The remainder of this area is covered by mucous membrane except at its posterior extremity where the lingual nerve intervenes. Below the line and extending back on the the ramus and down to the lower borders of the mandible is a shallow depression, the submandibular fossa for the submandibular gland.

On the inner surface of the symphysial region, above the mylohyoid line, there is a prominence which is sometimes divided into two pairs of mental spines (genial tubercles). Genioglossus and geniohyoid are attached to the upper and lower pairs, respectively. The small digastric fossae lie inferior to the mylohyoid line.

THE STRUCTURE OF THE MANDIBLE

The mandible consists of cancellous bone with an outer layer of compact bone which is exceptionally thick, especially at the lower border. The labial walls of the tooth sockets are considerably thinner than their lingual walls except in the case of the third molar, a point

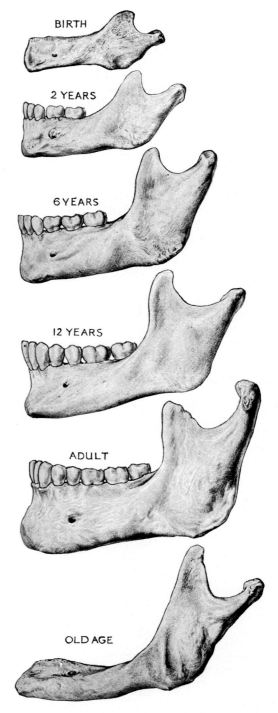

FIG. 3.65. Form of mandible at different ages. By the 6th year the first permanent molar has erupted behind the milk teeth; by the 12th year the milk teeth have been replaced by the corresponding permanent teeth, and the second permanent molar has erupted.

to be remembered in the extraction of teeth. The mandibular canal is in the spongy substance about the level of the mylohyoid line but does not have a definite wall. Branches of the canal carry vessels and nerves to the teeth and gums and to the mental foramen.

AGE CHANGES IN THE MANDIBLE

At birth the two halves of the mandible are separate and the body is a shell of bone enclosing the developing teeth, which are not completely separated from each other. The mandibular canal and mental foramen are near the lower border and the foramen is opposite the cavity for the first deciduous molar tooth. The ramus is short and the angle is very obtuse so that the coronoid process is almost in line with the body [FIG. 3.65].

The halves begin to unite in the first year and union is completed in the second year. As the teeth erupt, the body becomes stronger and deeper, the rami enlarge and the angle becomes reduced, to about 140 degrees by the fourth year.

Progressive increase in depth and elongation, especially behind the mental foramen, provides room for the permanent molars, the mental foramen assumes its adult position, and the angle becomes reduced to 110 degrees or less.

With the loss of teeth in old age, absorption of the sockets makes the chin appear prominent and the mental foramen is approached by the upper border. The angle opens out again by remodelling to 140 degrees and the condylar process is bent back so that the mandibular notch is widened.

DEVELOPMENT OF THE MANDIBLE

The mandible is laid down for the most part as membrane bone which lies lateral to the cartilage of the first pharyngeal arch (Meckel's cartilage). Some of the cartilage becomes incorporated in the bone, and small accessory cartilages may appear [FIG. 3.66]. Ossification starts during the sixth week of intra-uterine life, spreads

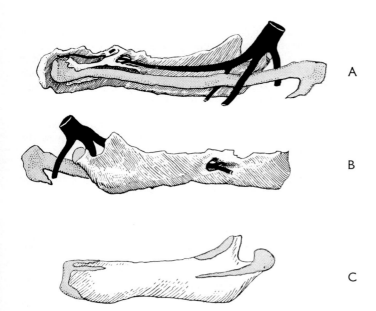

A

B

C

FIG. 3.66. Development of mandible.
A, As seen from the medial side; B from the lateral side; C, showing accessory (metaplastic) cartilage (blue).
(In A and B Meckel's cartilage is coloured blue and mandibular nerve is shown.)

through the membrane and envelops the nerves and vessels. Small mental ossicles may appear at the symphysis shortly before birth, but soon fuse with the mandible.

The upper and posterior end of Meckel's cartilage reaches the skull and forms the malleus, one of the ear ossicles. The original joint of the lower jaw lies between malleus and incus (Scott 1951). The sphenomandibular ligament between the spine of the sphenoid and the lingula of the mandible is usually regarded as representing a part of Meckel's cartilage.

GROWTH OF THE MANDIBLE

The mandible grows by surface accretion and absorption. It grows in width between its angles as well as in length, height, and thickness. Increase of thickness is brought about chiefly by addition to the outer surfaces. Increase in height of the body is due mainly to growth at the alveolar border. There is an associated continuous upward and forward movement of the teeth in the bone before, during, and after their eruption. The lengthening process takes place partly by addition to the posterior border, and, owing to the lateral slope of the body from before backwards, that accounts for increase in width also. The main increase depends upon the obliquity of the ramus and its processes in a young jaw. The growth of the condylar process is not only upwards but also backwards and sideways, and it thus contributes to the total length as well as to the height and width. As the coronoid process grows, its sloping, anterior margin is encroached upon by the rising alveolar border, which thus becomes longer. This, along with absorption from the anterior edge of the ramus, is the chief means of providing more room for the teeth as they rise with the alveolar border. Modelling maintains the shape of the condyle and the undulating curves of the anterior margin of the ramus and coronoid process. During growth the mandibular foramen maintains its relative position by a corresponding extension of its anterior lip, and the mandibular canal is thus lengthened. The mental foramen also changes its position during growth by moving backwards relative to the teeth [FIG. 3.65]; its sharp edge is below and behind in the young jaw, but in the adult it is below and in front, an indication of its direction of movement. (For further details and historical references see Brash 1934; Brash, McKeag, and Scott 1956.)

The teeth

There are normally two sets of teeth. The twenty deciduous or milk teeth appear between 6 months of age and the end of the second year. These are replaced between the sixth and twelfth year by most of the permanent set of thirty-two teeth. The four third molars appear later, usually between 17 and 21 years. The teeth, their arrangement, and their dates of eruption are described under Digestive System.

The hyoid bone

The hyoid bone is a U-shaped piece of bone situated below the tongue and above the larynx [FIG. 7.15]. It can be felt in the front of the neck 1.5–2 cm above the laryngeal prominence when the chin is held up. It is attached to the tongue, moves up and down with it during swallowing, and is held in position by muscles and by ligaments [FIG. 5.24].

It consists of a body, the curved piece of bone felt in the front of the neck, a pair of greater horns, and a pair of lesser horns

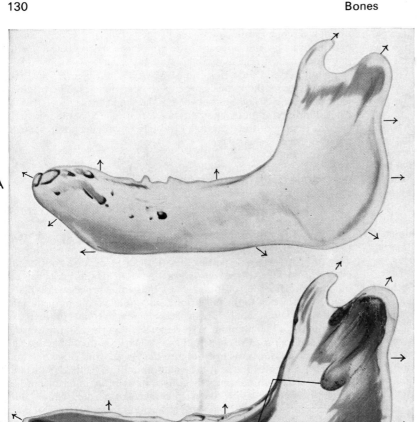

FIG. 3.67. Mandible of madder-fed pig to show the sites of growth. (See Brash 1934.)

The animal was 20½ weeks old and the madder had been omitted from the food for 19 days. The new bone added during that period is white in contrast to the old, red bone; and the main directions of growth are indicated by arrows.

A, Lateral surface of left half to show the distribution of growth, neutral, and absorption areas.

B, Medial surface of right half. Note particularly the indication of growth at the alveolar and posterior borders of the condyle and coronoid process, and of the upward and backward growth-movement of the mandibular foramen.

A

B

Growing edge of mandibular foramen

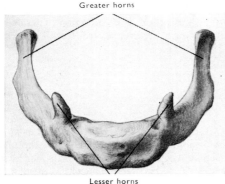

Greater horns

Lesser horns

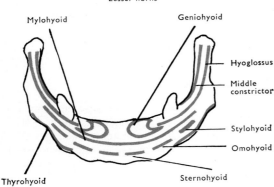

Mylohyoid

Geniohyoid

Hyoglossus

Middle constrictor

Stylohyoid

Omohyoid

Thyrohyoid

Sternohyoid

FIG. 3.68. Hyoid bone.

[FIG. 3.68]. The **greater horns** are continued backwards and upwards from the ends of the body and can be felt through the skin below the mandible. They lie just external to the mucous membrane and can be palpated by a finger on the internal pharyngeal wall. The tips of the greater horns are rounded and slightly swollen, but are not always easily felt because they are overlapped by the anterior edges of the sternocleidomastoid muscles. The **lesser horns** are small nodules of bone or cartilage which project upwards and backwards from the junction of the body and greater horn on each side.

The infrahyoid muscles, attached at its lower border, pull the hyoid down. The thyrohyoid membrane passes up, deep to the muscles and the bone, to be attached near its upper border. The stylohyoid ligament runs from the styloid process to the lesser horn and the middle constrictor of the pharynx sweeps back from the angle between the greater horn and the lesser horn and that ligament. The remaining muscles connect the hyoid to the tongue, the mandible, and the base of the skull and so tend to depress the tongue or raise the hyoid bone. The epiglottis is connected to the deep surface of the body by a hyo-epiglottic ligament in the midline [FIG. 7.26].

Development and ossification

The greater horn and most of the body of the hyoid bone are derived from the cartilage of the third arch. The lesser horn and probably the upper part of the body are from the cartilage of the second, or hyoid, arch which continues up as the stylohyoid ligament.

Ossification of the hyoid bone begins shortly before birth as a pair of centres for the body (which soon unite) and one for each greater horn. During the first year, or considerably later, centres appear for the lesser horn. The greater horn and body unite in middle age; the lesser horn is united with them by a synovial joint which only disappears in old age.

The skull at birth

At birth the cranium is relatively large but the face is small, about one-eighth of the whole cranium as compared with one-third in the adult. The teeth are not fully formed and the paranasal sinuses are rudimentary so that both the jaws and the nasal cavities are small.

The edges of the bones on the vault are not yet serrated and are separated by strips of fibrous tissue continuous with the periosteum covering the bones (pericranium and endocranium). This provides some mobility for the bones and allows overlap during birth of the child. The bones are thin but are not liable to fracture because of their mobility and flexibility.

In certain places, principally at the angles of the parietal bones, ossification does not spread into the connective tissue as rapidly as elsewhere, thus leaving large, unossified areas named fonticuli at birth.

The anterior fonticulus is large, diamond-shaped, and easily felt in the newborn child. It lies at bregma between the separate halves of the frontal bone and the two parietal bones. It is about 4 cm long by 2.5 cm wide. Ossification spreads from the edges of the surrounding bones until the gap is closed some time during the second year.

The posterior fonticulus is at the apex of the occipital, between the two parietal bones, and is closed 2 months after birth. A sphenoidal fonticulus lies at the sphenoid angle of each parietal bone and is closed 3 months after birth. Similarly a mastoid

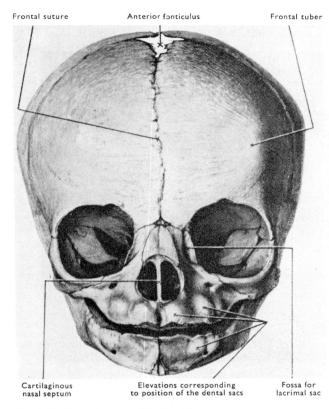

FIG. 3.69. Frontal aspect of skull at birth.

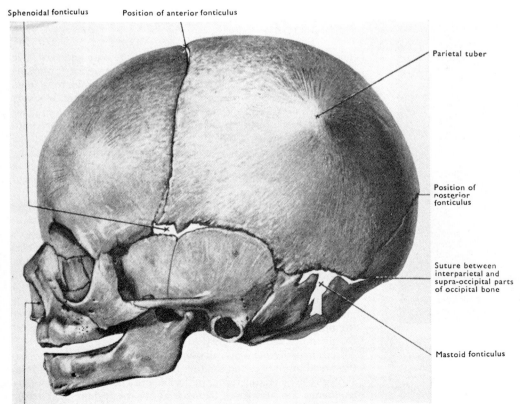

FIG. 3.70. Lateral aspect of skull at birth.

A

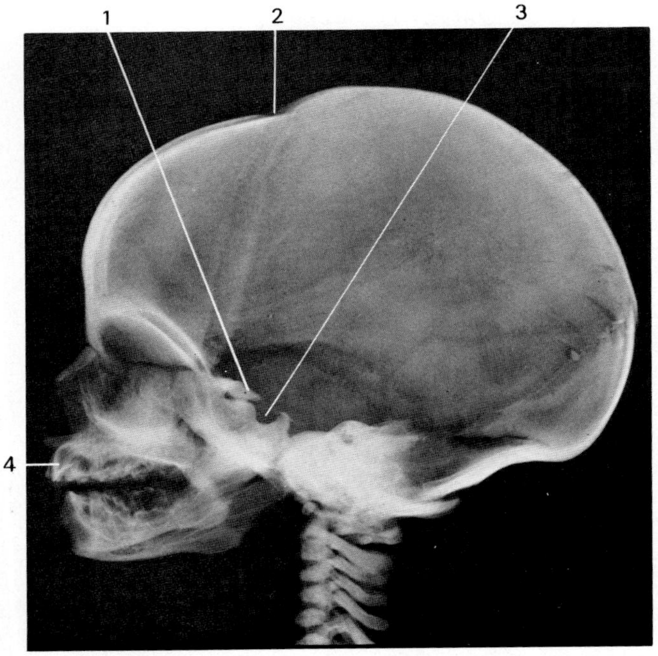

1. Anterior clinoid process.
2. Anterior fonticulus.
3. Sella turcica.
4. Maxillary tooth germs.

B

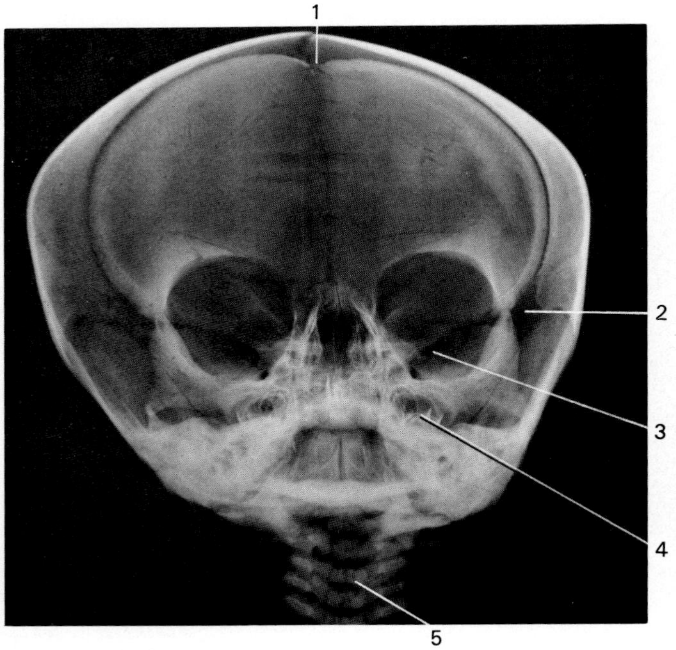

1. Anterior fonticulus.
2. Sphenoidal fonticulus.
3. Superior orbital fissure.
4. Maxillary tooth germs.
5. Centrum of 3rd cervical vertebra.

FIG. 3.71. Lateral (A) and anteroposterior (B) radiographs of the skull at birth. Note the large calvaria compared with the facial skeleton. The overlapping of skull bones is a post mortem feature.

fonticulus is found at the mastoid angle of each parietal and does not close until the end of the first year.

The parietal foramina are derived from, and indicate the position of, an unossified space which extends laterally into both parietal bones. It is often closed at birth but, if present, is called a **sagittal fonticulus**.

Sutural bones may result from separate ossification in various sutures and fonticuli. Those formed in the sphenoidal fonticulus, in the region of pterion, are called **epipteric bones**.

At birth the individual bones are separable, being joined by cartilage or membrane, and in many cases they are in several pieces in accordance with their ossification. On the outer surface of the vault, the parietal and frontal tubera are prominent, the frontal bone is in two parts and the occipital squama is partly separated into an upper (interparietal) part and a lower (supraoccipital) part by a fissure on each side [FIG 3.80].

The tympanic part of the temporal bone is still only an incomplete **tympanic ring** and the acoustic meatus is therefore short. Also, the tympanic membrane is set obliquely, facing more downwards than laterally, and is almost on the inferior surface of the skull. There is no **mastoid process** at this time. Thus the styloid process and stylomastoid foramen are near the side of the head, but gradually recede after the second year as the mastoid process and the tympanic plate begin to grow.

The maxilla is shallow because of the absence of its sinus. It consists largely of the alveolar process with its developing teeth. Of the other air sinuses only the ethmoidal are well formed.

The nasal bones are flat so that the child has little bridge to its nose and this, with the absence of superciliary arches, adds to the prominent appearance of the forehead. The orbits are relatively large and the nasal cavity lies almost entirely between them.

RADIOLOGY

The *newborn* skull vault is large relative to the base and face. The vault bones appear widely separated and the **anterior fonticulus** (between the frontal and parietal bones) and the **posterior fonticulus** between the developing occipital and parietal bones are relatively large with ill-defined margins [FIG. 3.71]. The posterior fonticulus closes soon after birth but the anterior fonticulus remains open until the age of two years.

The membrane bones of the **vault** consist at first of thin bony plates. In the first few years of life they may show irregularities (a beaten silver appearance). This must be recognized as a normal variation which gradually disappears before adult life.

All **sutures** of the skull remain visible until early middle age when fusion usually begins. Some sutures, e.g. parieto-occipital (lambdoid) may remain visible throughout life, even though fused internally.

As the vault develops its adult form, its inner and outer tables become better defined and the intervening **diploic space** is often more clearly seen in the lateral films. Considerable variation in the *thickness* of the vault occurs in different parts of the skull. The squamous portion of the temporal bone is commonly thin so that it may appear absent in radiographs of this area. Similar areas of thinning may be noted on either side of the external occipital protuberance and elsewhere [FIG. 3.58].

Certain *vessel markings* can sometimes be seen on the skull bones, the broad grooves of the sagittal, transverse, and sigmoid **venous sinuses** can be readily recognized when present and the sphenoparietal sinus can sometimes be made out in the region of the lesser wing of the sphenoid bone [FIG. 3.55]. Grooves caused by the branches of the **middle meningeal vessels** are not infrequently seen crossing the parietal and squamous temporal bones. The **diploic** veins can be recognized from their spider-like arrangement, they appear as channels often of varying calibre. The most constant group is in the parietal region [FIGS. 3.58 and 3.59] but similar systems can be found in the frontal and occipital areas.

Emissary veins connect the venous system of the scalp with that of the meninges and are especially important because they are a potential route for spread of infection into the cranial cavity. The mastoid emissary foramen appears as a curvilinear marking in the parieto-occipital region of the skull in lateral radiographs.

As the facial bones develop, the air spaces of the maxillary, frontal, and ethmoid sinuses grow in proportion to the facial bones. The developing maxillary sinuses are readily seen about the age of three years, and the frontal sinuses develop soon afterwards. The pneumatization of the mastoid occurs at the same time as the development of the frontal sinuses.

The paranasal sinuses are readily seen in radiographs because of their air content. The frontal sinuses have a thin well-defined bony cortex appearing as a thin outline, their scalloped margins representing ridges on the roof of the sinus from which project bony or membranous septa which may partially or completely divide the sinus [FIG. 3.62]. The frontal sinuses may be absent when the original suture (metopic) separating the developing frontal bones remains unfused. Asymmetry of the frontal sinuses is common, and occasionally they may extend backwards into the orbital plate of the frontal bone [FIG. 3.50] sometimes as far as the lesser wing of the sphenoid.

The maxillary sinuses show less variability in development and full growth does not occur until the eruption of the permanent teeth, which lie in their floor.

The ethmoid air cells are variable in extent, and cells may extend into the roof of the maxillary sinus, into the frontal bone, or posteriorly into the sphenoid bone.

The sphenoidal sinuses usually occupy the anterior two thirds of the body of the sphenoid bone. They may, however, develop to such a degree that the anterior and posterior clinoid processes and the body of the occipital bone are pneumatized.

The sella turcica is variable in size and shape, but is usually demarcated by a dense thin white line in lateral radiographs [FIG. 3.55]. It is sometimes more important to recognize this feature than to estimate the size of the fossa. Obvious enlargement or 'ballooning' of the fossa is easily recognized (not all enlargements are caused by abnormalities of the hypophysis), but minor changes are more difficult to see [FIG. 3.58] and impossible to assess if the head has not been accurately positioned to give a true lateral radiograph [FIG. 3.59].

In lateral views of the skull the mastoid processes overlap, but with angulation of the incident beam the cellular development of a mastoid process can be appreciated [FIG. 3.46]. The extent of the mastoid air cells is extremely variable. They may extend upwards into the squamous temporal bone, downwards to the mastoid tip, anteriorly towards the zygoma, into the petrous temporal bone around the labyrinth, or they may be restricted to the zone around the mastoid antrum.

Growth and age changes in the skull

The skull grows rapidly from birth until the seventh year, but the greatest increase in the size of the brain takes place during the first year. At the seventh year the orbits are almost as large as in the adult; the cribriform plates of the ethmoid, the body of the sphenoid, the petrous parts of the temporal bones, and the foramen magnum have reached their full size. The jaws have enlarged in preparation for the eruption of the permanent teeth and enlarge during their eruption, chiefly in the alveolar processes. For some years after the seventh, growth is slower, but at puberty a rapid increase in the rate of growth takes place in all directions, especially in the frontal and facial regions, related to the great increase in size of the paranasal sinuses.

In the base of the skull, growth takes place in the cartilage between centres of ossification and so, for a time, continues between bones such as the basisphenoid and basi-occipital [FIG. 3.72]. After the cartilage has ossified, growth can only take place on a surface of the bone covered with membrane—periosteum, endosteum, or endocranium, according to its position.

In the vault of the skull growth takes place between the edges of the bones, but the essential mode of growth in the vault is by deposition of bone on the external surface and absorption from the internal surface. Consequently, sutures which are perpendicular to the surface, such as the sagittal suture or, to a less extent, the coronal, maintain their position during growth. Sutures which are set obliquely to the surface, with a bevel to the edges of the bones, migrate as the skull grows. This is because growth of the overlapping superficial edge of one bone, such as the squamous temporal, carries it over the surface of its neighbour. The deep edge of the neighbouring bone, being part of its internal surface, is absorbed according to the rate of absorption on the inner surface of the skull. If the two processes keep pace the suture will retain its width.

Accretion outside and absorption inside produce enlargement of the cranial cavity with a simultaneous reduction in curvature. Differential growth and modelling produce and maintain the proportions and shape of the skull and the correct thickness of the bones. In the bones of the face, growth and modelling are also accomplished by surface accretion and absorption wherever the surface contours change. It has been known since the time of John Hunter that modelling absorption [p. 82] can occur in a marrow cavity or on the surface of a bone. Brash (1934) used bones from madder-fed animals to demonstrate the essential principles of bone growth in the skull. Enlow (1968) and Duterloo and Enlow (1970) should be consulted for further details.

Up to maturity, age can be ascertained approximately from the skull and the teeth. During the *first year* the separate parts unite in the temporal and sphenoid bones; the jugum sphenoidale is ossified by extension medially from the lesser wings, and the perpendicular plate of the ethmoid begins to ossify. During the *second year*, the halves of the mandible unite more firmly, the anterior fonticulus closes, the cribriform plate ossifies and unites with the other parts of the ethmoid, and the mastoid process appears. The lateral parts of the occipital bone unite with the occipital squama during the *third year*, and with the basi-occiput during the *fourth or fifth year*. The halves of the frontal bone have almost completely united at the *sixth year*, and during that year the first permanent molar teeth erupt [see page 421].

If the third molar has erupted and the cartilage between the body of the sphenoid and the occipital bone is present, the age is between 17 and 25. If the sphenoid and occipital are completely fused the age is likely to be over 25, for the cartilages that unite the body of the sphenoid to the ethmoid and to the occipital bone normally disappear before that.

FUSION OF SUTURES

After maturity, the wear of the teeth and the degree of obliteration of the sutures may give a rough indication of the age. The condition of the sutures in the interior of the skull vault is a more reliable indication of age than the condition on the exterior. Fusion of the outer tables is less regular, slower, and often incomplete. It is commonly stated that the inner table starts to fuse about 10 years earlier than the outer table (Todd and Lyon 1924–5).

In the interior of the skull, fusion of the sagittal suture normally begins at about 22 years of age and should be completed by 31 years. The coronal suture follows at 24 years and is for the most part fused by about 30 years, though not completely so till about 40 years. Its lower, pteric part is delayed about 2 years. The lambdoid suture does not start to fuse until about 26 years, but joins rapidly at first and then lags so that fusion is not complete until after 40 years. Other sutures close considerably later: there is much variation, even on the two sides of one skull.

Failure of one suture to conform to the usual pattern is likely to be associated with similar anomalies in the remaining sutures, i.e. there is likely to be a slowing up or an acceleration of closure in other sutures (Todd and Lyon 1924–5; Bolk 1915). However, the normal mode of growth by accretion and absorption can produce a normally shaped skull without the assistance of sutures.

In old age the skull vault may be thicker owing to deposit of bone on the inner surface; but in nearly all cases the skull bones are thinner, and the skull is lighter owing to absorption of the diploë and the associated extension and enlargement of the paranasal sinuses. Consequent on loss of teeth there is diminution of the size of the jaws owing to absorption of the walls of the sockets; the chin protrudes and the angle of the mandible becomes more obtuse. For details on age changes in the skull see Montagu (1938) and Krogman (1962).

Sex differences in the skull

There is little difference between the skulls of boys and girls till the age of puberty; but the skull of a woman is, on the whole, smaller than that of a man, and the paranasal sinuses are small relative to the size of the skull. The capacity of the cranial cavity averages 200 cm³ less than in a man of the same race, and that is more or less in conformity with the relative size of the whole body of women and men of average build. Very often it is not possible to say with certainty whether a given adult skull is that of a man or of a women.

Several of the criteria appear or become marked at puberty and many are affected as the individual ages. The sex differences given below are best seen, if present at all, in skulls of individuals 30–50 years of age.

The skull of a woman is lighter than that of a man and the architecture is smoother; the muscular ridges are less pronounced; the mastoid processes are relatively small. The glabella and superciliary arches are less prominent; the forehead is therefore more vertical, and the frontal eminences appear to bulge more; the upper margin of the more rounded orbital opening is sharper (a fact appreciated better by touch than by sight); the parietal eminences are more convex. The facial region is rounder, the zygomatic region is lighter and more compressed. The palate tends to be smaller and to be parabolic in shape. The mandible is smaller and the height at the symphysis is less.

The smoother the external surface of the squama of the occipital bone, the more likely the skull is female. The occipital condyles are smaller. The vertex is said to be more flattened and the relative height of the skull to be less.

Attempts to sex skulls by comparing selected dimensions have revealed considerable overlap in skulls of known sex. No single dimension is absolutely reliable but if all dimensions are either large or small then confidence increases in the assessment of sex of an unknown skull (Krogman 1962). Cranial traits and dimensions have been assessed by Keen (1950) and discriminant analysis has been used to sex Japanese skulls by Hanihara (1959).

Development of the skull

At an early stage of mammalian development the cerebral vesicles are enclosed in a membranous envelope, the **primordial membranous cranium**, derived from the mesoderm. In the elasmobranch fishes a complete cartilaginous capsule, the **primordial cartilaginous cranium**, is developed. In mammals, the cartilaginous part is confined to the base and the rapidly growing brain is covered by

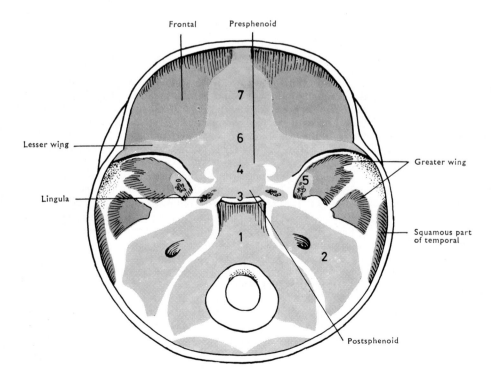

FIG. 3.72. Ossification on base and side-walls of skull of a 4½ months foetus (Schultze's method).
Cartilage, blue; membrane bone, red. Cartilage bone not shown. For explanation of numbers, see text.

membrane in which bone is laid down. The roof, most of the sides, and even part of the base of the skull are formed from this **membrane bone**.

The notochord [p. 41] is embedded in the basal part of the mesenchyme from which the skull develops, reaching cranially to the developing hypophysis [FIG. 10.31], i.e. extending from the foramen magnum to the root of the dorsum sellae. This divides the base into a chordal and a prechordal part.

In the human embryo chondrification begins early in the second month and is well developed by the end of the third month. In the following description the numbered paragraphs correspond to the numbered areas in FIGURE 3.72.

1. A basal or parachordal plate of cartilage forms behind the hypophysis cerebri and round the notochord. The notochord grooves the cerebral surface of the plate, pierces it to reach its pharyngeal surface and then turns up to terminate in the anterior end of the plate. Extensions from the sides of the basal plate form the condylar parts of the occipital bone and pass back to form the plate of cartilage in which ossification of the occipital squama takes place. This ossification in cartilage is all below the superior nuchal lines. The halves of this plate are slow to unite posteriorly and so the region behind the foramen magnum remains membranous till shortly before birth. This is the commonest site for a cerebral meningocele.

2. Cartilage appears in the mesoderm round the internal ear to produce the auditory capsule in which the petrous (including mastoid) part of the temporal bone ossifies.

3. A small strip of cartilage in front of the parachordal plate forms the dorsum sellae.

4. Cartilage around and in front of the end of the notochord forms the body of the sphenoid. It sends an extension forwards on each side to surround the craniopharyngeal canal (by which the stomodaeal part of the hypophysis enters the cranial cavity) and these extensions join to form the anterior part of the body of the sphenoid. They usually occlude the canal during the third month, but failure to do so may result in a patent craniopharyngeal canal [p. 120].

5. Chondrification occurs in the region of the root of the pterygoid process and the greater wing of the sphenoid including the lingula and the formen rotundum.

6. Chondrification occurs in the whole of the region of the lesser wings.

7. A cartilaginous nasal capsule forms during the third month. A paranasal cartilage appears in the lateral wall of the nasal cavity and gives rise to the ethmoidal labyrinth, all the conchae, and part of the maxilla. Part of it persists as the cartilage in the side of the nose but the septum is formed from the cartilaginous sphenoid and from independent paraseptal cartilages on each side of the anterior part of the septum. A pair of subvomerine cartilages are persistent parts of the paraseptal cartilages.

The cartilaginous roof of the nose is formed by fusion of the side-wall with the septum and by chondrification round the olfactory nerve filaments. Chondrification of the floor is very incomplete.

All the chondrified areas fuse to form a platform in the base of the skull from which the pharyngeal arches are suspended.

Ossification

In FIGURE 3.72 all the parts coloured red are ossified in membrane as are all the parts above this level. They include the occipital squama above the nuchal lines, which is designated interparietal, and the zygomatic bone.

Most of the cartilage ossifies from centres which are indicated in the description of individual bones but some in the nose are ossified from the membrane overlying the cartilage, namely, the vomer, lacrimal, and nasal bones and the upper part of the maxilla.

The lower part of the maxilla, the palatine bones, and the greater part of the pterygoid processes are ossified from the membrane external to the mucous lining of the oropharyngeal region.

The tympanic plate ossifies in membrane overlying the auditory capsule.

SEGMENTATION

Although the skull never shows conclusive evidence of segmentation, probably owing to the need for stability even in its early forms, it is assumed that the chordal part of the base has arisen by fusion of segments equivalent to vertebrae. The reasons for this assumption are the presence of myotomes in the head region, the connexion with the series of pharyngeal arches, and the segmental arrangement of nerves. The points of exit of nerves through the dura mater, which represents the primitive brain envelope, are the best indication of the disposition of segments. Evolution has altered some of the apertures in the bone, e.g. in the case of the oculomotor and abducent nerves.

The mammalian occipital bone is usually regarded as the equivalent of four fused vertebrae and some vertebrates may have acquired extra occipital elements from the vertebral column. The chordal part of the primordial cranium is related to the primitive cranial nerves, excluding the olfactory and optic nerves which are parts of the brain. The latter are enclosed in capsules developed in the prechordal part of the skull and protect the enlarging brain and the organs of smell and sight.

The outstanding features that distinguish the human skull are the large size of the brain-case, the small size of the face, and the way the skull is balanced on the vertebral column. This poising of the skull on a vertebral column is associated with alteration in its shape and, in particular, its muscular attachments. The occipitovertebral joints are so spaced that the fore and hind parts of the head nearly balance each other.

The small size of the face is associated with a reduction in the size of the teeth and a corresponding reduction in the bulk and length of the jaws. Diminution of the muscles of mastication also reduces the fossae, surfaces, and crests which provide attachment for them in animals with stronger jaws.

The human cranium also exhibits a number of minor morphological variations: some have already been mentioned in the text and others are noted in the description of individual skull bones which follows (highest nuchal line, metopic suture, sutural bones, parietal foramen, incomplete or accessory foramina). Although some variants are the result of pathological changes, most result from normal development processes and are genetically determined (Berry and Berry 1967).

Craniometry

The external size of the skull and cranial capacity vary considerably in different individuals. Varying thicknesses of bone and size of the paranasal sinuses reduce the degree of correlation between the two. Skull capacity is best measured in practice by filling the skull with beads or lead shot, after closing up the larger foramina. The volume required to fill the cavity is then measured. For comparison skulls are grouped according to their cranial capacity as follows.

Microcephalic skulls have a capacity below 1350 cm^3.

Mesocephalic skulls range from 1350 to 1450 cm^3.

Megacephalic skulls have a capacity over 1450 cm^3.

Other systems of grouping have been suggested which take

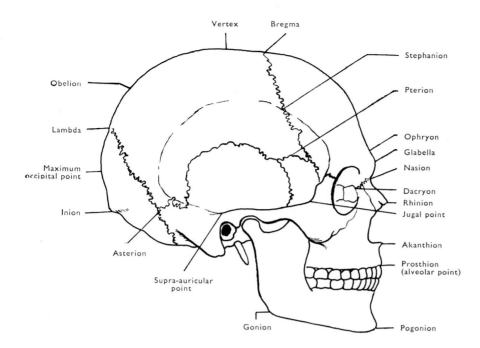

FIG. 3.73. Points on the skull from which the most important measurements are taken.

account of the difference in average cranial capacity between the sexes.

The shape, as well as the size, of the cranium is used as a means of comparing skulls. The differences are expressed by comparing measurements between given points: the more important are indicated in FIGURE 3.73.

The relation of the maximum breadth of a skull to its maximum length is a commonly used index and is expressed as the cephalic index.

$$\frac{\text{Maximum breadth}}{\text{Maximum length}} \times 100 = \text{cephalic index.}$$

The results are as follows:
Dolichocephalic skulls have an index below 75.
Mesaticephalic skulls range from 75 to 80.
Brachycephalic skulls have an index above 80.

The skulls of most Europeans fall within the mesaticephalic range. Many other measurements and indices from the skull have been established and tabulated for the various groups of mankind (Duckworth 1915; Hrdlicka 1952; Trevor 1955; Comas 1960). The height/breadth index of the orbit and the breadth/height index of the nasal aperture are of use together with the cephalic index. Discriminant analysis has been applied to Negro and White skulls with 90 per cent success in one sample (see Krogman 1962).

The individual bones of the cranium

It is more important that a student should understand the construction of the skull as a whole than that he should acquire detailed knowledge of individual bones disarticulated from the cranium. It is assumed that the reader is already acquainted with the skull as a whole and in the following brief descriptions of each bone reference should be made to the earlier section and especially to the illustrations.

The frontal bone

The main part of the frontal bone is the convex shell which forms the front of the skull above the orbital openings. It is deeply concave to accommodate the frontal lobes of the brain.

The **external surface** [FIG. 3.74] is rounded with an added convexity forming a frontal tuber on each side. Above and posteriorly, it is limited by the serrated edge which forms the coronal suture with both parietal bones. Inferolaterally the surface narrows to reach a prominent zygomatic process which projects downwards and laterally to articulate with the zygomatic bone.

Inferiorly, the surface is limited by the concave supraorbital margins, each with a supra-orbital notch, or a foramen, two thirds of the way from its lateral end. The medial end of each orbital margin leads down to a nasal part from which a **nasal spine** projects downwards. The anterior edge of this area and the nasal spine articulate with a nasal bone on each side of the midline and with the maxilla posterolaterally. Above this the smooth bulge of glabella is marked by the remains of a frontal suture and is continued laterally as the **superciliary arch**.

The **temporal surface** is nearly flat and passes down from the temporal line to the lower edge of the bone where it meets the sphenoid.

The **inferior aspect** [FIG. 3.75] shows the articular area for the zygomatic bone continued posteriorly to articulate with the greater wing of the sphenoid. The apex of this articulation reaches the end of the coronal suture; its base reaches the roof of the orbit. The **orbital surface** is on the inferior aspect of a horizontal orbital plate which projects back from the supra-orbital margin. It is smooth, concave, and triangular in outline and its blunt apex reaches posteriorly to articulate with the lesser wing of the sphenoid. Near the lateral end is the **fossa for the lacrimal gland**, and near the medial end is the small **trochlear fossa** or spine [p. 121].

Medially each orbital surface has a series of depressions where it articulates with the ethmoid bone and roofs in the ethmoidal cells. Anteriorly, the **nasal spine** projects downwards and forwards between two small, smooth, nasal areas which help to make up the roof of the nose. Posterior to this a deep **ethmoid notch** separates

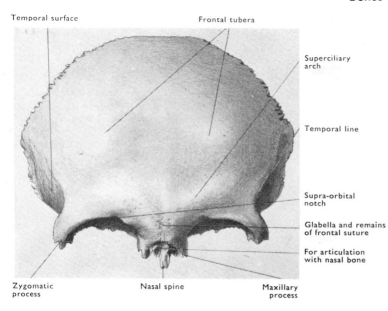

Temporal surface   Frontal tubera

Superciliary
arch

Temporal line

Supra-orbital
notch

Glabella and remains
of frontal suture

For articulation
with nasal bone

Zygomatic   Nasal spine   Maxillary
process   process

FIG. 3.74. Frontal bone (anterior view).

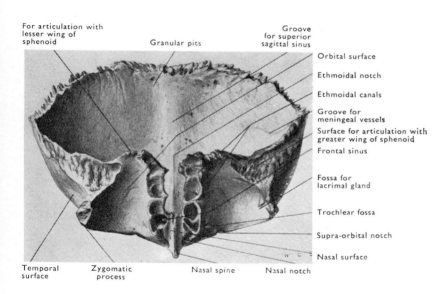

For articulation with   Groove
lesser wing of   for superior
sphenoid   Granular pits   sagittal sinus

Orbital surface

Ethmoidal notch

Ethmoidal canals

Groove for
meningeal vessels

Surface for articulation with
greater wing of sphenoid

Frontal sinus

Fossa for
lacrimal gland

Trochlear fossa

Supra-orbital notch

Nasal surface

Temporal   Zygomatic   Nasal spine   Nasal notch
surface   process

FIG. 3.75. Frontal bone (inferior aspect).

the two ethmoidal areas and is occupied by the cribriform plate of the ethmoid bone. On each side, two grooves run from the notch to the orbit between the ethmoidal cells. They form the **anterior** and **posterior ethmoidal foramina**.

The **nasal margin** is a small rough area, anterior to the ethmoidal cells and the nasal spine, for articulation with the nasal bones.

The **internal surface** of the frontal bone, above the orbital parts, shows structures already described [pp. 117 and 120]. The **foramen caecum** may be wholly in the frontal bone. The frontal sinuses are described on page 124.

OSSIFICATION

Ossification in membrane spreads from centres appearing above the zygomatic processes in the sixth or seventh week of foetal life and from additional centres for the zygomatic processes, the nasal spine, and the trochlear fossae. Fusion in each half takes place in the sixth or seventh foetal month. The **frontal suture** is present at birth but is normally closed by the fifth or sixth year.

VARIATIONS

The zygomatic process is dense throughout but varies in strength with that of the jaws. The frontal suture may persist as a **metopic suture** (up to 8 per cent in Europeans) and sutural bones may alter the line of the sutures.

The parietal bones

The parietal bones are the curved plates of bone, convex on their outer surface, which fill in the vault of the skull between the frontal, occipital, and temporal bones [FIG. 3.45]. Below and in front they also reach the sphenoid.

The **squamosal border** is inferior and has a sharp, concave, bevelled edge where it is overlapped by the squamous part of the temporal bone and, in its anterior part, by the greater wing of the sphenoid. The outer surface is radially grooved where it is covered by these bones. Where it is covered by the sphenoid the border runs down to join the frontal border at the **sphenoidal angle**. The frontal

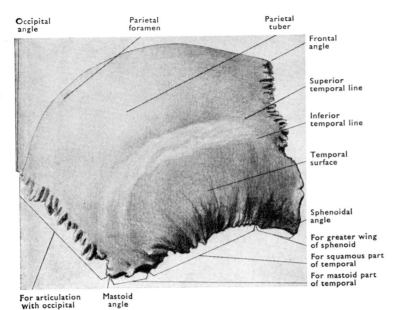

FIG. 3.76. Right parietal bone (external surface).

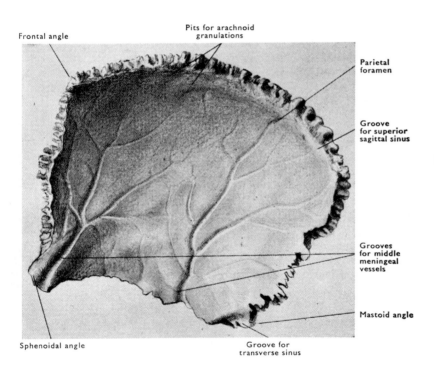

FIG. 3.77. Right parietal bone (internal surface).

border is bevelled in this region on its deep surface and overlaps the frontal bone. This corner of the bone projects downwards and forwards a little and is deeply grooved on its interior aspect by meningeal vessels.

The frontal border runs up from the sphenoidal angle to meet the sagittal border at a right angle. It articulates with the frontal bone to form half of the coronal suture and is slightly overlapped by the frontal bone in its upper part, and vice versa in its lower part. The right-angled anterosuperior angle is named the frontal angle.

The sagittal border articulates with its neighbour by a series of small pegs and sockets to form the sagittal suture [p. 108]. The occipital angle of the parietal lies between the sagittal and occipital borders.

The occipital border articulates with the occipital bone to form half of the lambdoid suture [p. 109]. The mastoid angle is at the posterior end of the squamosal border where the parietal articulates with the mastoid region of the temporal bone, and the transverse sinus grooves the deep surface of the bone close to the parietomastoid suture.

The outer surface is convex with an increased convexity in its posterior half, the parietal tuber or eminence. A parietal foramen for an emissary vein is usually found near the sagittal suture, about 3.5 cm from its posterior end. At a lower level are the superior and inferior temporal lines [p. 109].

The internal surface of the bone is concave. The frontal branch of the middle meningeal artery and its vein groove the bone near the frontal border, the parietal branches are further back and the transverse sinus grooves it near the mastoid angle. The superior sagittal sinus and arachnoid granulations mark the inner surface near the sagittal border [FIG. 3.77].

OSSIFICATION

Ossification spreads from two centres which appear in the eighth week and unite in the fourth foetal month. Its spread is delayed at the angles and this produces the fonticuli. Ossification round the parietal foramina is also slow and a sagittal fonticulus may result from this.

VARIATIONS

The outer table may be very thin above the temporal lines. The bone may be in two or even three pieces and ossification in the region of the parietal foramina may be poor or absent.

The occipital bone

The occipital forms the posterior part of the base of the skull with the basilar part anterior to the foramen magnum. The leaf-like squama, behind the foramen magnum, is bent so as to turn up into the back of the head. The two condylar parts lie between the squama and the basilar part, one on each side of the foramen magnum.

The superior part of the **squama**, or interparietal part, is wedged between the parietal bones and curves up to reach the posterior end of the sagittal suture at lambda. Inferiorly, the surface reaches the **superior nuchal lines** where it bends almost through a right angle into the base of the skull [FIG. 3.45]. A **highest nuchal line** for occipitofrontalis may be seen above the superior nuchal line. The main features of the inferior aspect are shown in FIGURE 3.78. It is almost horizontal and the squama narrows so that each lateral border is concave where the mastoid region of the petrous temporal articulates with the **mastoid border**. The surface posterior to the centre of the foramen magnum is marked by the curved superior and **inferior nuchal lines** [FIG. 5.17], the two halves being separated by

an **external occipital crest** which terminates posteriorly as the **external occipital protuberance**.

In the **lateral parts** the two **occipital condyles** encroach on the anterior half of the foramen magnum and the articular surfaces, elongated and convex in shape, converge anteriorly so that rotation is severely limited. Posterior to each condyle is a condylar fossa, to receive the lateral mass of the atlas during extension at the joint. Opening into the fossa there is usually a **condylar canal** for transmission of a vein. The bone lateral to the posterior half of the condyle is the **jugular process**. Its free anterior margin is the posterior margin of the jugular foramen. Its inferior surface gives attachment to rectus capitis lateralis. Laterally it articulates, by means of cartilage, with the petrous part of the temporal bone.

The **hypoglossal canal**, often double, for the twelfth cranial nerve comes through the bone superior to the anterior end of the condyle to open medial to the jugular notch.

The lateral parts of the occipital bone converge to join the **basilar part** which continues as a flattened bar of bone to the body of the sphenoid. The inferior surface of the basilar part presents, at its centre, a **pharyngeal tubercle**, which receives the raphe of the pharyngeal constrictors, and the bone may be marked by rectus capitis anterior and by longus capitis muscles [FIG. 5.17].

The **internal surface** of the occipital shows the features already described for the posterior cranial fossa [p. 117]. The groove for the superior sagittal sinus runs down to the **internal occipital protuberance** and swings to one or other side as a groove for a transverse sinus. A groove for the other transverse sinus runs laterally from the protuberance to the other side. These mark off two cerebral fossae in the upper part of the squama from two cerebellar fossae inferiorly [p. 117]. The falx cerebri, tentorium cerebelli, and a small falx cerebelli are attached to the edges of the grooves. The superior surface of the jugular process is grooved by the sigmoid sinus. This groove undercuts the bone in the jugular foramen (jugular notch) to form part of the jugular fossa. Medial to this a raised ridge carries the **jugular tubercle**. Its posterior part is often grooved by the ninth, tenth, and eleventh cranial nerves. In front of the tubercle the ridge

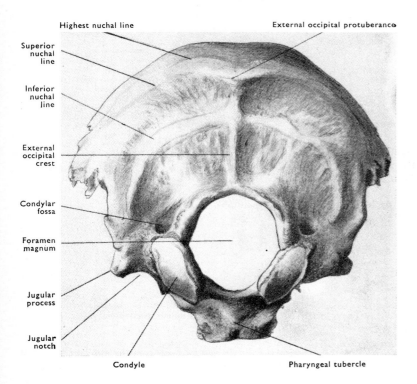

FIG. 3.78. Occipital bone (inferior aspect).

Highest nuchal line

External occipital protuberance

Superior nuchal line

Inferior nuchal line

External occipital crest

Condylar fossa

Foramen magnum

Jugular process

Jugular notch

Condyle

Pharyngeal tubercle

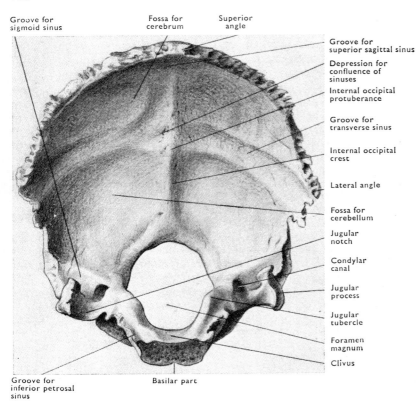

Groove for sigmoid sinus

Fossa for cerebrum

Superior angle

Groove for superior sagittal sinus

Depression for confluence of sinuses

Internal occipital protuberance

Groove for transverse sinus

Internal occipital crest

Lateral angle

Fossa for cerebellum

Jugular notch

Condylar canal

Jugular process

Jugular tubercle

Foramen magnum

Clivus

Groove for inferior petrosal sinus

Basilar part

FIG. 3.79. Occipital bone (internal surface).

is marked by a groove for the inferior petrosal sinus. The basilar part of the bone forms the floor of a wide groove which lodges the medulla with its coverings and the vertebral arteries. At the side, between the foramen magnum and the ridge, the hypoglossal canal enters the bone. A small bony tubercle on the anterior margin of the foramen magnum may be present to indicate the position of the apical ligament of the dens.

OSSIFICATION

The basilar part begins to ossify in cartilage in the sixth week of intra-uterine life, and includes a portion of each condyle [FIG. 3.80]. Centres for the lateral part appear in the eighth week and the squama below the highest nuchal line (supra-occipital) ossifies from centres near the protuberance in the seventh week. The interparietal part is ossified in membrane from centres appearing in the eighth week. Union of the two parts of the squama may be incomplete at birth. The four main components are united by cartilage until after birth. The lateral parts fuse with the squama during the third year and with the basilar part during the fourth or fifth year. The basilar part begins to fuse with the sphenoid between 18 and 20 years, or earlier in the female, and fusion should be complete by the twenty-fifth year.

VARIATIONS

The superior nuchal line and external protuberance may form a torus occipitalis transversus and an emissary vein may pierce the protuberance. The hypoglossal canal may be double or multiple. The upper part of the squama may remain incompletely united laterally, may be separated as an interparietal bone, or it may be in right and left halves or multiple. An epiphysis may occur between the basi-occipital and the sphenoid.

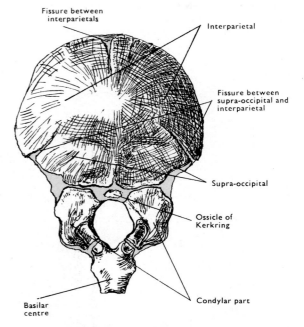

Fissure between interparietals

Interparietal

Fissure between supra-occipital and interparietal

Supra-occipital

Ossicle of Kerkring

Condylar part

Basilar centre

FIG. 3.80. Ossification of occipital bone.
The occasional ossicle of Kerkring fuses with the squama before birth. The basilar and the squamous centres may be bilateral or multiple.

A third occipital condyle may project from the anterior border of the foramen magnum to articulate with the dens and a **paramastoid process** may project from the jugular process and even be long enough to articulate with the transverse process of the atlas.

The temporal bones

Each temporal bone articulates with the lower lateral margin of the squama of the occipital bone, and with the side of its basilar part. It is wedged between these behind, and the sphenoid bone in front, and helps in the formation of the lateral wall of the skull by passing up to join the parietal bone.

It consists of squamous, tympanic, and petrous parts, the last including the mastoid region and mastoid process which are continuous with it.

The **squamous part** is a thin, vertical plate of bone applied to the lateral aspect of the petrous part. Its lateral aspect is smooth above with a thin semicircular **parietal border**. This merges anteriorly with a thicker **sphenoidal border** which articulates with the greater wing of the sphenoid. Posteriorly the **squamosomastoid suture** can usually be seen passing downwards and forwards [FIG. 3.81]. The external acoustic meatus forms a landmark below the squama which projects down behind it as a postauditory process and in front of it as a postglenoid process. Above this, the root of the zygomatic process is continued forwards from the supramastoid crest, to pass above the meatus and the mandibular fossa. The suprameatal triangle, above and behind the meatus, the **articular tubercle** in front of the fossa, and the root of the zygoma are described on page 109–10.

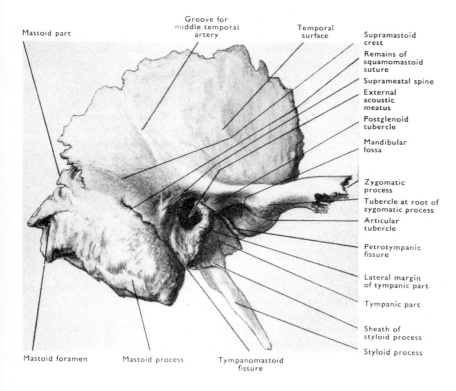

Mastoid part
Groove for middle temporal artery
Temporal surface
Supramastoid crest
Remains of squamomastoid suture
Suprameatal spine
External acoustic meatus
Postglenoid tubercle
Mandibular fossa
Zygomatic process
Tubercle at root of zygomatic process
Articular tubercle
Petrotympanic fissure
Lateral margin of tympanic part
Tympanic part
Sheath of styloid process
Styloid process
Mastoid foramen
Mastoid process
Tympanomastoid fissure

FIG. 3.81. Right temporal bone (lateral aspect).

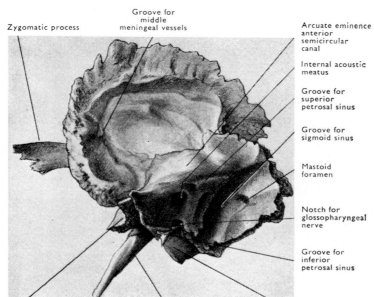

Zygomatic process
Groove for middle meningeal vessels
Arcuate eminence anterior semicircular canal
Internal acoustic meatus
Groove for superior petrosal sinus
Groove for sigmoid sinus
Mastoid foramen
Notch for glossopharyngeal nerve
Groove for inferior petrosal sinus
Carotid canal
Styloid process
Medial surface of mastoid process

FIG. 3.82. Right temporal bone (medial aspect).

The **zygomatic process** curves forwards to reach the zygomatic bone [FIG. 3.45]. The squama seen above the level of the zygomatic process is the temporal surface. In front and below, it runs medially to form a small infratemporal surface.

The **cerebral surface** of the squama is marked by gyri of the cerebrum and by grooves for middle meningeal vessels. Above is the bevelled articulation for the parietal bone and below a thin line indicates the **petrosquamous suture** [FIG. 3.82].

The **tympanic part** [FIGS. 3.81 and 3.84], referred to as the tympanic plate, is a bent, triangular plate of bone which has the petrous part behind and medial to it and the squamous part above and anteriorly. A lamina (the **tegmen tympani**) projecting from the petrous part intervenes between squama and tympanic part at the petrotympanic fissure [FIG. 3.81].

The upper surface of the tympanic plate is concave and forms most of three walls of the external acoustic meatus, the squama forming its roof and a little of the posterior wall. The curved lateral edge gives attachment to cartilage of the meatus [FIG. 12.38]. At the medial end of the meatus a **tympanic sulcus** grooves the plate and gives attachment to the tympanic membrane. Posteriorly the plate reaches the mastoid process at the **tympanomastoid fissure** which transmits the auricular branch of the vagus. More medially it is moulded over the styloid process as a sheath. The upper border of the anterior surface joins the postglenoid tubercle laterally but is separated from the squama more medially by the **squamotympanic fissure** [p. 116].

The **petrous part** of the temporal bone consists of a mastoid portion from which the mastoid process projects downwards and forwards [FIG. 3.81], and a three-sided pyramid of hard bone which contains the internal ear. It tapers a little as it passes forwards and medially towards the foramen lacerum at the side of the body of the sphenoid [FIG. 3.51]. The lateral part, anterior to the mastoid region, is excavated to form the **cavity of the middle ear** which is closed in laterally by the tegmen tympani which turns down between the cavity and the squamous part and by the tympanic membrane attached to the tympanic plate [FIGS. 12.38–12.42]. The cavity is continued forwards as the auditory tube, and backwards through the antrum to the air cells in the mastoid process [FIG. 3.83].

The mastoid region of the petrous part forms the rounded bony resistance behind the ear. A **mastoid foramen** is usually present in its upper part near the occipitomastoid suture. Sternocleidomastoid muscle is its principal attachment with splenius and longissimus capitis deep to it.

INFERIOR SURFACE [FIG. 3.84]

This is irregular and rough and most of the structures have already been noted [p. 115], including the grooves for the digastric muscle and the occipital artery. The **styloid process**, with the **stylomastoid foramen** behind it, emerges between the tympanic plate and digastric groove on the petrous part. Medial to the foramen a deep notch receives the corner of the jugular process of the occipital bone and must not be confused with the **jugular fossa** which lies still further medially and helps to form the jugular foramen. In the lateral part of the wall of the fossa is the medial end of the **mastoid canaliculus** for the auricular branch of the vagus which runs to the tympanomastoid fissure. The lower end of the **carotid canal** is immediately anterior to the jugular fossa, and between them, a **tympanic canaliculus** transmits a tympanic branch of the glossopharyngeal nerve to the tympanic cavity [FIG. 3.84]. **Caroticotympanic canaliculi** for sympathetic nerves from the carotid canal also lead to the tympanic cavity. Towards the apex of the petrous temporal the floor of the carotid canal is formed by a rough quadrilateral area, and at the tip the opening of the carotid canal can be seen; the canal having turned at right angles to run anteromedially in the bone. In the angle between the petrous and squamous parts, anterior to the carotid canal, lie the openings of the pair of **semi-canals** for the auditory tube and tensor tympani muscle.

Between the inferior surface and the posterior surface the articular surface for the occipital bone is divided into two by the jugular fossa.

ANTERIOR AND POSTERIOR SURFACES [FIG. 3.82; pp. 118 and 119]

On the medial part of the **anterior surface** in the middle cranial fossa there is a distinct impression for the trigeminal nerve and ganglion. About halfway between this and the base of the pyramid, the **arcuate eminence** indicates the position of the anterior semicircular canal. The bone between this and the squama is the tegmen tympani forming a roof for the middle ear [FIG. 3.83]. The bone anterior to the eminence covers the cochlea and the facial nerve. Passing anteromedially from two small openings in this region, two small grooves carry the greater and lesser petrosal nerves towards the foramen lacerum. The anterior surface is separated

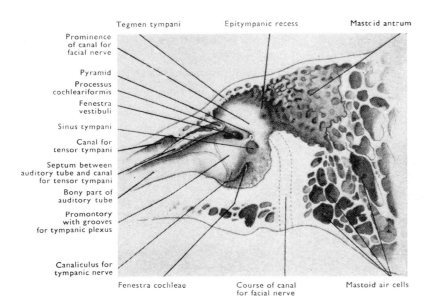

Tegmen tympani   Epitympanic recess   Mastoid antrum

Prominence
of canal for
facial nerve

Pyramid

Processus
cochleariformis

Fenestra
vestibuli

Sinus tympani

Canal for
tensor tympani

Septum between
auditory tube and canal
for tensor tympani

Bony part of
auditory tube

Promontory
with grooves
for tympanic plexus

Canaliculus for
tympanic nerve

Fenestra cochleae   Course of canal   Mastoid air cells
                           for facial nerve

FIG. 3.83. Section through left temporal bone, showing medial wall of tympanic cavity.

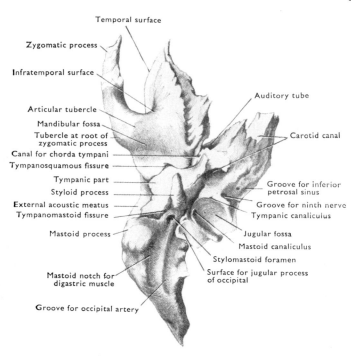

FIG. 3.84. Right temporal bone (inferior aspect).

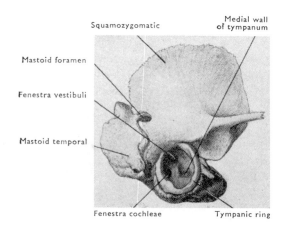

FIG. 3.85. Temporal surface of right temporal bone at birth.

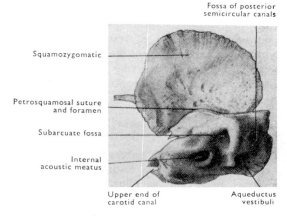

FIG. 3.86. Cerebral surface of right temporal bone at birth.

from the squama by a faint irregular **petrosquamous suture**, and from the posterior surface by a **superior border** grooved by the superior petrosal sinus, except where the trigeminal nerve intervenes near the medial end.

The **posterior surface** has as its most prominent feature the **internal acoustic meatus** which passes laterally so that its opening is oblique. It is about 10 mm deep and its lateral end or fundus has openings for the facial nerve and for vessels and nerves of the internal ear [FIG. 12.51]. Above and lateral to the meatus lies a **subarcuate fossa** [p. 119]. More laterally, between the meatus and the sigmoid sinus, is another fossa with a ridge above it which indicates the position of the upper part of the posterior semicircular canal. A small slit overhung by a scale of bone is the opening of the aqueductus vestibuli. Inferior to the meatus a notch in the lower border of the surface accommodates the glossopharyngeal nerve and has at its apex a canaliculus cochleae which leads up to the cochlea of the internal ear. Medial to this the edge of the bone is grooved by the inferior petrosal sinus. The **groove for the sigmoid sinus** lies deep to the posterior half of the mastoid process, and frequently has a foramen for a vein opening into it.

OSSIFICATION

The petromastoid part ossifies from four centres which appear in the cartilage during the fifth foetal month and are more or less fused by the end of the sixth month.

The **styloid process** ossifies from two centres which appear in the upper part of the second pharyngeal arch cartilage. The upper, or tympanohyal, centre appears before birth and fuses with the petrous part during the first year. The lower or stylohyal one appears shortly after birth but ossifies slowly. It fuses with the upper part after puberty if at all.

The **tympanic part** ossifies from a centre in the lateral wall of the tympanum in the third month and forms a curved tympanic ring which is, however, incomplete superiorly [FIG. 3.85]. After birth ossification spreads medially, laterally, and downwards to form the tympanic plate.

The **squamous part** is ossified in membrane from a centre at the root of the zygoma which appears about the end of the second month.

VARIATIONS

The markings vary considerably in different bones. A venous sinus may lie along the petrosquamous suture internally, leading through the bone to the transverse sinus posteriorly. Anteriorly, it may pass through a canal which traverses the root of the zygoma to appear above the squamotympanic fissure. Variations in the ossification of the squamous part include absence of the zygoma. Abnormalities of the petrous part are likely to be associated with deaf-mutism or idiocy. The squama may articulate with the frontal bone, as in apes.

The sphenoid bone

The sphenoid bone is wedged in among other bones of the skull and consists of a central body, two wings extending out on each side of the body, and a pterygoid or wing-like pair of processes extending down from the body, one on each side. Most of the bone has been described as part of the skull as a whole. The main features which become evident in a separate bone are sutures and openings which were difficult to see in the intact skull.

On the *posterior aspect* [FIG. 3.87] the body presents a sawnoff surface or an articular area for the basilar part of the occipital. Its

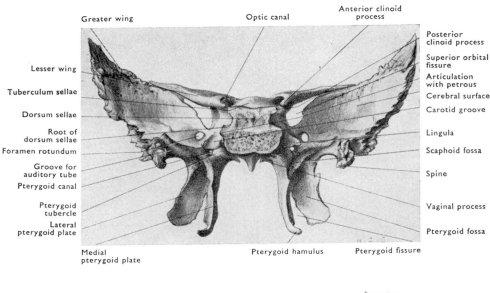

Greater wing   Optic canal   Anterior clinoid process

Lesser wing
Tuberculum sellae
Dorsum sellae
Root of dorsum sellae
Foramen rotundum
Groove for auditory tube
Pterygoid canal
Pterygoid tubercle
Lateral pterygoid plate
Medial pterygoid plate

Posterior clinoid process
Superior orbital fissure
Articulation with petrous
Cerebral surface
Carotid groove
Lingula
Scaphoid fossa
Spine
Vaginal process
Pterygoid fossa

Pterygoid hamulus   Pterygoid fissure

FIG. 3.87. Sphenoid bone (posterior aspect).

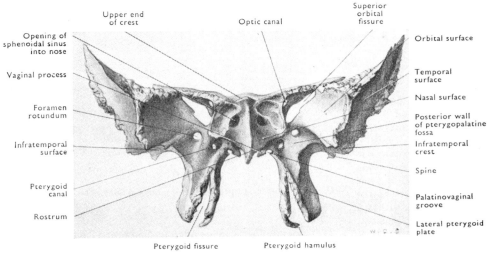

Upper end of crest   Optic canal   Superior orbital fissure

Opening of sphenoidal sinus into nose
Vaginal process
Foramen rotundum
Infratemporal surface
Pterygoid canal
Rostrum

Orbital surface
Temporal surface
Nasal surface
Posterior wall of pterygopalatine fossa
Infratemporal crest
Spine
Palatinovaginal groove
Lateral pterygoid plate

Pterygoid fissure   Pterygoid hamulus

FIG. 3.88. Sphenoid bone (anterior aspect).

upper surface is deeply excavated by the hypophysial fossa. The dorsum sellae slopes up to overhang the fossa and extend laterally as the posterior clinoid processes.

The lesser wings form tapering processes, each attached to the body by two roots with the optic canal between them. The anterior roots are joined by the jugum sphenoidale. The posterior roots join the body, one on each side, and the tuberculum sellae lies between them and behind the sulcus chiasmatis. The anterior edge of the wing articulates with the frontal and ethmoid bones. The posterior edge curves back to the prominent anterior clinoid process. The sharp tip of the wing reaches towards the greater wing and the region of pterion. The gap between the lesser and greater wings is the superior orbital fissure.

The greater wing springs from the lower part of the body and, expanding laterally, passes forwards and then upwards to reach the frontal bone and the corner of a parietal bone. It has a concave cerebral surface, a concave temporal surface, and a nearly flat orbital surface. The edge between the orbital and temporal surfaces articulates with the zygomatic bone to separate the orbit from the temporal fossa. The posterior margin of the greater wing terminates as a sharp angle which fits in between the squamous and petrous parts of the temporal bone. Its end projects down as the spine of the

sphenoid. The foramen spinosum and foramen ovale pierce the bone near the posterior edge of the wing. The most medial part of the edge is prolonged backwards as a lingula over the carotid groove on the side of the body. This almost horizontal part of the wing helps to form the floor of the middle cranial fossa and the roof for the infratemporal fossa. Close to the body and below the medial part of the superior orbital fissure, the foramen rotundum passes forwards through the wing to reach the pterygopalatine fossa. Below the lingula and the carotid groove a pterygoid canal passes horizontally through the root of the pterygoid process from the foramen lacerum to the pterygopalatine fossa.

The greater wing is grooved by the middle meningeal vessels. Its concave squamosal margin is crossed by their parietal branches, while the frontal branches groove or penetrate its anterosuperior corner.

The *anterior aspect* [FIG. 3.88] shows the posterior parts of the nose and orbit and the posterior wall of the pterygopalatine fossa.

The thin anterior wall of the body is perforated on each side by the opening of a sphenoidal sinus [p. 125] into the sphenoethmoidal recess of the nose. Each opening is normally covered by a thin, inverted triangle of bone. This sphenoidal concha is perforated by a smaller opening 4 or 5 mm in diameter. Laterally this wall articulates with the ethmoidal labyrinth and the orbital process of the palatine bone, and its inferolateral edge forms the upper margin of the sphenopalatine foramen. Between the conchae, the median crest of the sphenoid continues down to become the **sphenoidal rostrum**. The crest articulates with the perpendicular plate of the ethmoid. The rostrum fits into a recess between the alae of the vomer.

The roof and medial wall of the orbit articulate with the lesser wing and with the sphenoidal concha respectively. The smooth orbital surface of the greater wing passes posteromedially from the zygomatic articulation, and forms the curved inferior margin of the superior orbital fissure. Superiorly, it reaches the frontal articulation. The foramen rotundum opens into the upper part of the posterior wall of the pterygopalatine fossa, and the pterygoid canal opens a little inferomedial to it. This wall of the fossa is continued down on the **pterygoid process** [Fig. 3.88] the medial and lateral laminae of which are united above, but separated below by a V-shaped pterygoid fissure which is filled by the pyramidal process of the palatine bone. This is fused anteriorly to the posterior surface of the maxilla and thus the pterygopalatine fossa is closed off below.

The **lateral pterygoid lamina** is turned outwards and gives attachment to medial and lateral pterygoid muscles, from its corresponding surfaces. The **medial lamina** forms the most posterior part of the lateral wall of the nose and gives attachment, in the lower half of its posterior margin, to the superior constrictor of the pharynx. Anterior to it the wall of the nose is formed by the palatine bone. The other features of the medial lamina are described on page 116.

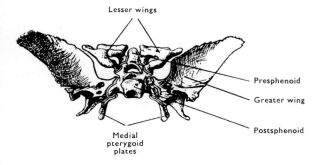

Fig. 3.89. Ossification of sphenoid.

OSSIFICATION

The sphenoid ossifies in cartilage except for the lateral part of the greater wing and part of the pterygoid processes. Centres appear from the eighth to the tenth foetal week for most of the bone. Those for the conchae are delayed until the fifth month or later. The hamulus chondrifies in the third month.

At birth the bone is in three pieces, the greater wing and pterygoid process on each side, the body and lesser wings between them. These have fused by the end of the first year.

VARIATIONS

Foramen spinosum and foramen ovale are frequently incomplete posteriorly and the superior orbital fissure may extend down into foramen rotundum. The optic canal may be divided into compartments for the optic nerve and ophthalmic artery. Accessory ossification in relation to the anterior clinoid process may convert the groove for the internal carotid artery on its medial side into a caroticoclinoid foramen.

Rarely, a craniopharyngeal canal [p. 135] may persist between the pharynx and the hypophysial fossa.

The ethmoid bone

The ethmoid bone consists mainly of two lateral masses, one on each side of the nasal cavity, excavated by air cells and called **ethmoidal labyrinths**. Each is joined by a cribriform plate to a median perpendicular plate [Fig. 3.90].

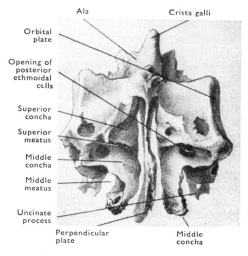

Fig. 3.90. Ethmoid (posterior aspect).

The **perpendicular plate** forms part of the nasal septum [Fig. 3.56] and has irregular edges for articulation with the crest of the sphenoid and the vomer behind and below, the cartilaginous septum in front and below, and the nasal spine of the frontal bone and the median crest of the nasal bones, above and in front. The plate is continued up as the crista galli in the anterior cranial fossa [p. 120].

On each side a **cribriform plate** forms part of the nasal roof through which the olfactory and anterior ethmoidal nerves pass by way of the perforations.

Each **labyrinth** consists of three groups of thin-walled air **cells**, anterior, middle, and posterior, between nasal cavity and orbit.

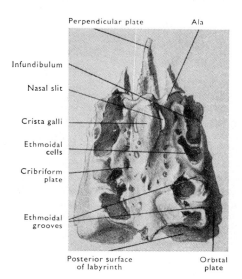

Fig. 3.91. Ethmoid (superior aspect).

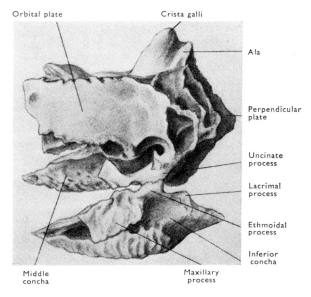

FIG. 3.92. Lateral surface of ethmoid and inferior concha.

Many of the cells are incomplete when the bones are disarticulated. Posteriorly it articulates with the body of the sphenoid and the orbital process of the palatine bone; anteriorly, with the lacrimal bone and the frontal process of the maxilla; superiorly, with the orbital plate of the frontal bone; and inferiorly, with the maxilla. On the upper aspect, two narrow grooves correspond to those seen on the frontal bone for the anterior and posterior ethmoidal vessels and nerves. The smooth, thin **orbital plate** is on the lateral surface and forms part of the medial wall of the orbit [FIG. 3.60]. The inferior surface of the ethmoidal labyrinth reaches the maxilla and the **uncinate process** passes down on its medial side to help in the closure of the opening of the maxillary sinus, and to form the inferior margin of the hiatus semilunaris.

The labyrinth forms part of the lateral wall of the nasal cavity and from it the **superior** and **middle conchae** project towards the septum and then turn down over the superior and middle meatuses [FIG. 3.57]. The middle concha extends far enough back to articulate with the palatine bone and forwards to reach the maxilla. The middle cells bulge into the middle meatus as the **ethmoidal bulla** and open into the meatus through it. The groove below the bulla and above the uncinate process is the **hiatus semilunaris** which receives the openings of the anterior cells, including the one which forms the ethmoidal infundibulum [p. 124]. The posterior cells open into the superior meatus.

OSSIFICATION

Centres of ossification appear in the cartilage of the nasal capsule in the fourth and fifth months for each labyrinth and the labyrinth is completely ossified at birth, including the walls of its cells which have already appeared. Bone appears in the perpendicular plate at the end of the first year and has spread through it by 5 or 6 years. Bone extends from the labyrinth to the cribriform plate during the second year but does not fuse with the sphenoid until after 25 years.

VARIATIONS

The number of conchae may vary from one to four. The orbital plate varies in size and shape: it may be narrowed from above downwards or may fail to articulate with the lacrimal, as in anthropoid apes.

The inferior nasal conchae

Each inferior concha is a thin, curled leaf of bone, pointed at both ends. It has a rough, convex, free, medial surface [FIG. 3.93] and a concave lateral surface attached by its upper edge to the lateral wall of the nasal cavity [FIG. 3.92]. Here it articulates with the conchal crests on the maxilla anteriorly, on the perpendicular plate of the palatine bone posteriorly. Between these the attachment expands into a flat plate extending upwards to the lacrimal and ethmoid as **lacrimal and ethmoidal processes** and down to the maxilla (**maxillary process**) where it becomes part of the medial wall of the maxillary sinus [FIG. 3.92]. The lacrimal process articulates with the lacrimal bone and the maxilla to complete the wall of the **nasolacrimal canal**. The ethmoidal process reaches the uncinate process of the ethmoid bone. The lateral surface is smooth below the lacrimal process where the nasolacrimal canal enters the inferior meatus.

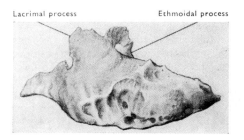

FIG. 3.93. Medial surface of right inferior concha.

OSSIFICATION

Ossification is in the cartilage of the nasal capsule from a centre which appears about the fifth foetal month.

The lacrimal bones

Each lacrimal bone is a thin scale of bone found in the medial wall of the orbit behind the frontal process of the maxilla. It is roughly quadrangular and its two surfaces are medial and lateral. It articulates with the frontal bone, the maxilla, the inferior concha, and the ethmoid. The medial surface is uneven where it articulates with and helps to close some ethmoidal cells. Anterior to this it is smooth where it enters into the lateral wall of the nose above the inferior concha. The lateral surface is shown in FIGURE 3.94. Its posterior part is the smooth orbital surface. Its anterior part, or **fossa for the lacrimal sac**, is deeply concave to form the posterior half of the fossa and is limited behind by the prominent **posterior lacrimal crest**. At the lower end, the orbital surface and the crest curve forwards as a **lacrimal hamulus** which reaches the maxilla

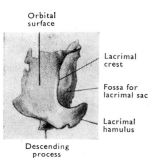

FIG. 3.94. Right lacrimal bone (lateral surface).

and forms the posterolateral boundary of the beginning of the nasolacrimal canal. A small descending process forms the medial boundary of the opening.

OSSIFICATION AND VARIATIONS

One centre, appearing in the third foetal month, spreads in the membrane covering the nasal capsule.

The lacrimal bone may be absent, small, or divided. The hamulus may reach the orbital margin, as in lemurs.

The vomer

The vomer is shaped, as its name indicates, like a ploughshare [FIG. 3.95]. It lies in the nasal septum and has one smooth, free edge which forms the posterior margin of the septum. The remaining edges articulate as shown in FIGURES 3.56 and 7.5. The superior border is split into a pair of everted **alae** which fit over the rostrum of the sphenoid and a small sphenovomerine canal begins at the posterior end of the groove between the alae. It runs horizontally through the bone and transmits nutrient vessels. The inferior border fits into the nasal crest on the upper surface of the palate and the truncated anterior end reaches the region of the incisive canals where it sends a pointed process between them.

Grooves for vessels mark the surfaces of the bone and one, usually more distinct, lodges the nasopalatine nerve.

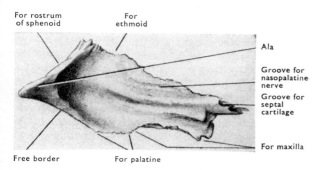

FIG, 3.95. Vomer, from the right side.

OSSIFICATION

The vomer begins to ossify in the eighth week from plate-like centres in the membrane on each side of the septal cartilage. Fusion, from below upwards, begins in the third foetal month and is almost complete by puberty. Ossification at the anterior end extends into the paraseptal cartilages.

VARIATIONS

Ossification may be imperfect with cartilage remaining between the two bony laminae and the groove along the anterior border may be deep. The sphenoidal air sinus may extend between the laminae.

The nasal bones

The two nasal bones are elongated pieces of compact bone, smooth and convex on their outer surfaces, concave and a little rough on

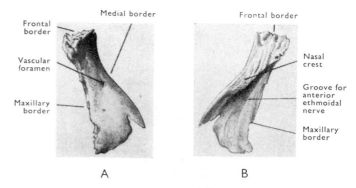

FIG. 3.96. Right nasal bone. A, outer aspect; B, inner aspect.

their inner surfaces. Each bone is wide and thin inferiorly where it joins the nasal cartilage, narrower and thicker above at the root of the nose and slightly waisted near its upper end.

The articulations are shown in FIGURES 3.49 and 3.96. The medial border which articulates with its fellow is deepened above to increase its area. This widened part projects a little at its posterior edge to form, with its neighbour, a median nasal crest which articulates with the nasal spine of the frontal bone, the perpendicular plate of the ethmoid, and the septal cartilage. The inner surface of the bone is triangular in outline and is grooved by branches of the anterior ethmoidal nerve.

OSSIFICATION

Each nasal bone ossifies from a single centre, appearing about the end of the second foetal month, in the membrane covering the nasal capsule.

VARIATIONS

Both racially and individually the nasal bones vary considerably. They tend to be prominent in white races, small and flat in Mongolians and Negroes. They may be absent or divided. Obliteration of the nasal suture is rare though it is the usual condition in apes.

The maxillae

The two maxillae join in the midline to form the upper jaw and each has a hollow pyramidal body with four processes projecting from it.

The surfaces of the body correspond to the walls of the pyramidal cavity of the maxillary sinus [p. 124]. The **nasal surface** is the base of the pyramid and has the wide opening of the sinus in its posterosuperior part [FIG. 3.98]. Its smooth antero-inferior part forms most of the lateral wall of the inferior meatus of the nose. Behind these two parts the surface articulates with the perpendicular plate of the palatine, which thereby forms the medial wall of the greater palatine canal [FIG. 3.99].

The anterosuperior corner of the nasal surface has a projecting frontal process and immediately behind that is a short, deep, vertical **lacrimal groove**. The angle between the frontal process and the body of the bone is filled in by the lacrimal bone so that the upper end of the groove is converted into a canal. The medial surface of the frontal process has on its lower part a **conchal crest** for the inferior concha and, in its upper part, an **ethmoidal crest** for the middle concha. Between them is the atrium of the middle meatus.

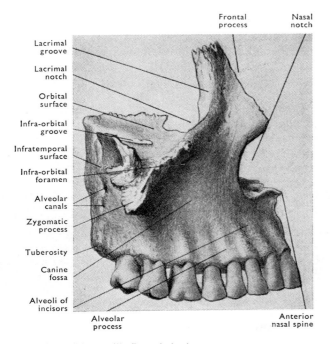

Frontal process   Nasal notch

Lacrimal groove
Lacrimal notch
Orbital surface
Infra-orbital groove
Infratemporal surface
Infra-orbital foramen
Alveolar canals
Zygomatic process
Tuberosity
Canine fossa
Alveoli of incisors

Alveolar process   Anterior nasal spine

FIG. 3.97. Right maxilla (lateral view).

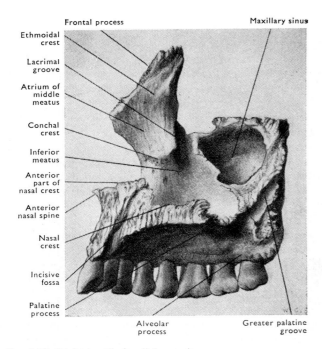

Frontal process   Maxillary sinus

Ethmoidal crest
Lacrimal groove
Atrium of middle meatus
Conchal crest
Inferior meatus
Anterior part of nasal crest
Anterior nasal spine
Nasal crest
Incisive fossa
Palatine process

Alveolar process   Greater palatine groove

FIG. 3.98. Right maxilla (medial aspect).

The nasolacrimal canal thus lies lateral to the lacrimal, ethmoid, the middle meatus, and the inferior concha, and opens into the inferior meatus. The opening of the maxillary sinus is reduced by encroaching bones.

Each **palatine process** extends medially to articulate with its neighbour in the midline, and with it forms a **nasal crest**, which projects forwards as the **anterior nasal spine**. The rough posterior edge of the process articulates with the horizontal plate of the palatine bone. The anterior part of its upper surface curves up a little in front to form a lower edge for the piriform aperture, but it thickens peripherally, the lower surface turning down into the

alveolar process. The articulating surfaces of the maxillae behind the alveolar process are grooved to form an inverted, funnel-shaped **incisive fossa** with a small **incisive canal** leading down into it on each side [p. 117; FIG. 11.15]. The inferior surface of the palatine process is pitted by palatine glands. Near its lateral margin a groove may be present for the greater palatine vessels and nerve.

The **alveolar process** projects down from the margin of the inferior surface of the pyramidal body [FIG. 3.60]. It meets its fellow in front to form a semi-elliptical **alveolar arch** with sockets for eight teeth on each side. Posteriorly, the process ends as the **tuber** (*tuberosity*) of the maxilla. Laterally, the alveolar process is ridged by the roots of the teeth, particularly the canine.

The **anterior surface** [FIG. 3.97] extends from the midline to the zygomatic process. The body has two other surfaces: a horizontal, **orbital surface**, and a posterior or **infratemporal surface**. These three surfaces meet at the hollow **zygomatic process**, and this has a rough articular surface for the zygomatic bone. This articular area extends back on to the edge between the orbital and infratemporal surfaces so that the anterior part of the orbit is separated from the infratemporal region by the junction of the two bones.

The **infratemporal surface**, on the posterior aspect, is convex and rather rough but becomes continuous laterally with the smooth, concave, posterior wall of its zygomatic process and forms the anterior wall of the infratemporal fossa and pterygomaxillary fissure. The orbital and pyramidal processes of the palatine bone articulate with the superomedial and inferomedial parts of this surface respectively; the pyramidal process separating it from the pterygoid laminae. Above the tuberosity, foramina in the maxilla transmit the posterior superior alveolar vessels and nerves to the molar teeth [FIG. 11.18].

The posterolateral margin of the smooth, slightly concave, **orbital surface** articulates with the zygomatic bone anterolaterally and the orbital process of the palatine bone posteromedially, and forms the anterior margin of the inferior orbital fissure between these. This margin is notched by the **infra-orbital groove**, which is continuous anteriorly with the **infra-orbital canal** and **foramen**. They transmit the infraorbital vessels and nerve, which send anterior superior alveolar branches into a small opening in the floor of the canal. The medial edge of the orbital surface articulates with the ethmoid and the lacrimal bones.

The sloping **anterior surface** is continued up from the alveolar process to the infra-orbital margin and merges laterally with the zygomatic process. Medially it is limited by the piriform aperture with the **anterior nasal spine**. The incisor fossa is below the edge of the piriform aperture and above the incisor teeth. The wider fossa between the ridge produced by the canine and the zygomatic process is called the **canine fossa**. The infra-orbital foramen [p. 113] lies above this fossa. Above and medially, the surface is continued on to the frontal process, which extends up to articulate with the frontal bone. The lower part of the anterior edge of the frontal process bounds part of the piriform aperture of the nose; its upper, posterior margin articulates with a lacrimal bone down the middle of the lacrimal fossa. The anterior edge of the fossa is formed by the **anterior lacrimal crest**, a ridge continuing the orbital margin upwards. The groove behind the ridge passes down medial to the body of the bone as the **lacrimal sulcus**. The medial surface of the process is smooth between the ethmoidal and conchal crests already noted and forms the side-wall of the atrium of the middle meatus. The anterior end of the ethmoidal crest produces the agger nasi [FIG. 7.8].

The **premaxillae** are normally fused with the maxillae in Man. They lie in front of the incisive fossa and carry the incisor teeth on each side. Up to middle age their independent origin may be indicated by a faint suture from the hinder part of the incisive fossa to the interval between the canine and incisor sockets.

OSSIFICATION

The maxilla ossifies in membrane from a centre which appears in the sixth week in the wall of the oral cavity above the germ of the canine tooth. The infra-orbital nerve is above the bone until the second foetal month when the groove and canal begin to form. For the growth of the sinus see page 124.

The premaxilla ossifies from at least two centres, the first appearing above the incisor tooth germs in the sixth week. It forms most of the premaxilla and a small part of the frontal process, and fusion with the maxilla proper begins almost at once. The second centre appears in the twelfth week, to form the wall of the incisive canal, and soon joins the rest.

VARIATIONS

The infra-orbital canal may remain open as a groove or a vertical suture may be present in the infra-orbital margin above it. A ridge on the oral surface of the median suture of the palate, or torus palatinus, is not uncommon. Cleft palate may occur along the suture joining the two halves of the maxilla or on one or both sides of the premaxilla, i.e. between canine and incisor teeth [p. 55]. The maxillary sinus may be loculated by septa to a varying extent and the anterior part may be constricted off by the lacrimal canal to produce a lacrimal recess. Absence of the premaxilla and incisor teeth has been recorded.

The palatine bones

Each palatine bone is L-shaped with a horizontal and a perpendicular plate. At their junction, posteriorly, there is a pyramidal process and at its upper end the perpendicular plate carries two processes, sphenoidal and orbital, separated by the sphenopalatine notch.

The **horizontal plate** projects medially to form the posterior third of the hard palate [FIG. 3.101]. It has a smooth upper **nasal surface**, a rougher lower **palatal surface**, and an anterior articular border for the palatine process of the maxilla. Medially, where it meets its neighbour, the articular part is raised up to form the posterior third of the **nasal crest**. The posterior border is a smooth concave edge which ends medially on the prominent **posterior nasal spine** of the

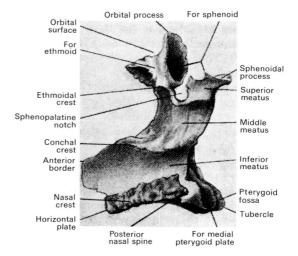

FIG. 3.100. Right palatine bone (medial surface).

palate. Seen from below, a sharp palatine crest runs laterally near the posterior border, and curves towards the pyramidal process, passing the inferior end of the greater palatine groove on the way [FIG. 3.51].

The thin **perpendicular plate** is continued up from the horizontal plate to form part of the lateral wall of the nose. It is applied to the surface of the maxilla and articulates with the pterygoid process of the sphenoid posteriorly [FIG. 3.57].

The medial or nasal surface has a smooth area which is part of the inferior meatus, a **conchal crest** for the inferior concha, another smooth area forming part of the middle meatus and above that, a smaller, **ethmoidal crest** for the middle concha [FIG. 3.100].

At the upper anterior corner of the plate, an irregular, more or less triangular, **orbital process** (sometimes pneumatic) is applied to the posterior end of the orbital surface of the maxilla and so enters the floor of the orbit. Medially it reaches the sphenoid. Its concave, posterior edge forms the anterior limit of the deep sphenopalatine notch.

The **sphenoidal process** projects from the posterosuperior corner of the perpendicular plate. It curves upwards and medially and its surface is applied to the sphenoid at the root of the medial pterygoid

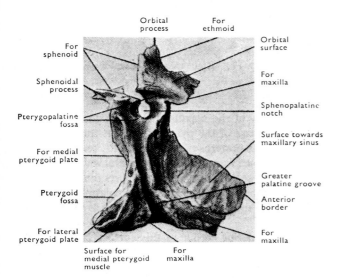

FIG. 3.99. Right palatine bone (lateral surface).

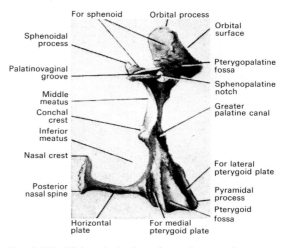

FIG. 3.101. Right palatine bone (posterior aspect).

plate. Its upper surface is grooved by the palatovaginal canal. The canal carries pharyngeal vessels and nerves from the pterygopalatine fossa to the roof of the pharynx. The medial edge of the process reaches the ala of the vomer.

The **sphenopalatine notch**, converted into a foramen by the edge of a sphenoidal concha, is the main communication between the pterygopalatine fossa and the nose.

The lateral or **maxillary surface** is applied to the maxilla [FIG. 3.99] and helps to reduce the opening of the maxillary sinus, but posterosuperiorly it forms the medial wall of the pterygopalatine fossa. From this free surface, which is smooth, a distinct **greater palatine groove** runs downwards and forwards. With the corresponding groove on the maxilla it forms the **greater palatine canal**.

Posterior to the palatine canal the **pyramidal process** projects backwards from the lower end of the perpendicular plate, and articulating with the lower ends of the pterygoid plates, fills the gap between them and unites them to the maxilla. From the greater palatine groove two **lesser palatine canals** descend through the process to open on to its inferior surface.

OSSIFICATION

Ossification spreads from a centre which appears in the eighth intra-uterine week in the membrane of the side-wall of the nasal cavity, but the orbital process may ossify separately. Until the maxilla enlarges in the third year the vertical height is small.

VARIATIONS

The maxillary sinus may extend between plates of cortical bone forming the horizontal plate. The orbital process varies in size, and may join the sphenoidal process, and bridge the sphenopalatine notch.

The zygomatic bones

Each zygomatic or cheek bone lies below and lateral to the orbit in the most prominent part of the cheek. It connects the maxilla with the temporal bone behind and the frontal bone above by means of a horizontal temporal process and an ascending frontal process.

Its **lateral surface** is slightly convex. Its anterior border is the edge of its rough triangular articulation with the maxilla, which meets the evenly curved orbital border at an acute angle. This forms

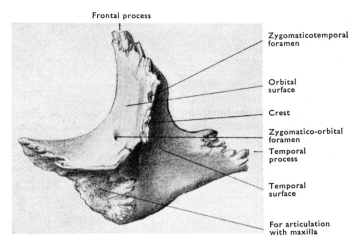

FIG. 3.103. Right zygomatic bone (medial surface).

most of the lateral and inferior margins of the orbit. The smooth border joining the two processes is more sharply curved than the orbital margin. The free inferior border continues back on to the temporal process.

The **temporal process** extends back to a very oblique, serrated articulation with the zygomatic process of the temporal bone. The **frontal process** narrows sharply as it approaches its articulation with the frontal bone and this usually produces a **marginal tubercle** on its posterior border [FIG. 3.102].

The deep or **medial aspect** of the bone is separated into two concave surfaces by the articular area for the maxilla and its extension upwards as a crest which articulates with the sphenoid. The more anterior smooth **orbital surface** can be identified by the orbital margin. The more irregular posterior area is the **temporal surface**. The named foramina are shown in FIGURES 3.102 and 3.103.

OSSIFICATION

The zygomatic bone ossifies in membrane from a centre which appears between the eighth and tenth weeks of intra-uterine life. The bone may be divided horizontally or vertically by a suture.

Sutural bones

Sutural bones are isolated pieces of bone sometimes found in the sutures or in the region of the fonticuli. They arise from independent centres, may occur in any suture, and usually include the whole thickness of the bone. They are most frequent at lambda and in the lambdoid suture, but are also common in the region of pterion as **epipteric bones**.

The bones of the upper limb

The shoulder region is supported by the two bones of the shoulder girdle, which are the **scapula** and **clavicle**. The clavicle acts as a prop for the scapula which is otherwise connected to the trunk only by muscles and is freely movable. The **humerus** is the bone of the

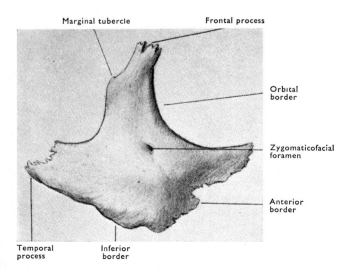

FIG. 3.102. Right zygomatic bone (lateral surface).

arm or brachium and the radius and ulna are the two bones in the forearm or antebrachium.

The hand or manus is divided into three parts. The first or proximal part is the wrist or carpus made up of eight small carpal bones arranged in two rows. It differs from the colloquial 'wrist' which refers to the distal part of the forearm bones. The second is the hand proper, including the palm and the back of the hand, which is supported by five metacarpal bones, largely concealed by soft tissues but clearly seen when they form the knuckles of the clenched fist. They are numbered, one to five, beginning on the lateral side with the thumb. The third or distal part is composed of five digits numbered, like the metacarpals, from the lateral side. The first digit is the thumb or pollex; the second is the forefinger or index; the third or middle finger is digitus medius; the fourth is the ring finger or anularis; the fifth or 'little finger' is digitus minimus. The bones of the digits are the fourteen phalanges of which two are in the thumb and three are in each 'finger'.

All the upper limb bones exhibit variations in size and degree of marking depending on sex and race but the differences are seldom reliable criteria for accurate diagnosis. They are often shorter, smoother, and more slender in women but the degree of muscular development in both sexes influences the bony characteristics.

THE SCAPULA

The scapula is a triangular bone which lies obliquely over the second to the seventh ribs on the back and in the posterior wall of the axilla. Its lower end is easily felt and may be used as a landmark. The main flattened part is the body and is concave on its costal surface, convex and divided obliquely into two parts by a triangular spine on its dorsal surface. The spine is continuous laterally with a shelf of bone, the acromion, which turns forwards and overhangs the shoulder joint. The lateral angle can be identified because it is truncated and thickened to form a head. This has a concave articular surface, the glenoid fossa for the humerus and is continuous with the body through an ill-defined neck [FIG. 3.104]. A thickened lateral border runs down to meet the medial border at the inferior angle which is the lower end of the scapula.

The medial border, a thin edge, ascends to the superior angle, and is slightly angulated at the medial end of the spine. The levator scapulae muscle is attached above the root of the spine, rhomboideus minor opposite the root of the spine, and rhomboideus major from there to the inferior angle. The superior angle [FIG. 3.106] is nearly a right angle and the thin superior border passes almost horizontally towards the lateral angle. From the upper part of the head and neck a thick coracoid process projects forwards, and medial to the root of this process, a distinct scapular notch interrupts the superior border and is converted into a foramen by the superior transverse scapular ligament. The suprascapular nerve lies below the ligament, and the artery lies above it, and the omohyoid muscle is attached to the ligament and the adjoining border.

The costal surface [FIG. 3.104 and 3.107] has a smooth narrow strip along the entire medial border to which the serratus anterior muscle is attached. The rest of the surface is marked by the fibrous septa in the subscapularis muscle, and gives attachment to it except for a smooth area on and near the neck. A bursa extending from the joint lies here between the muscle and the bone.

The dorsal surface [FIGS. 3.106 and 3.108] is divided by the spine into a supraspinous fossa and an infraspinous fossa for the

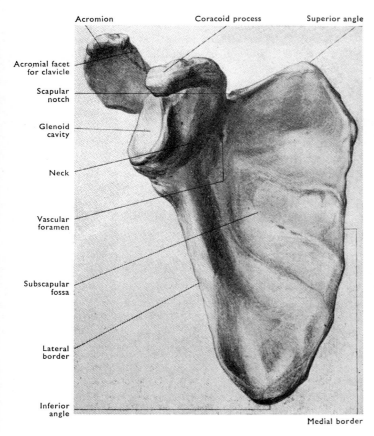

FIG. 3.104. Right scapula (costal surface).

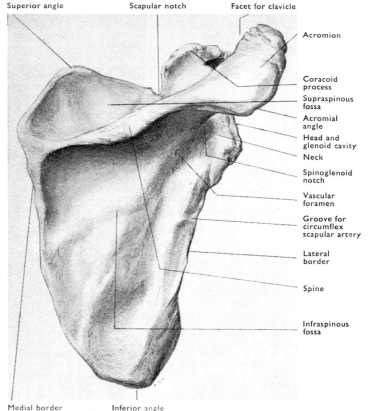

FIG. 3.105. Right scapula (dorsal surface).

attachment of supraspinatus and infraspinatus muscles respectively. These muscles extend on to the sides of the spine, but leave a smooth area on and near the neck. A bursa usually lies under infraspinatus at the neck.

Between infraspinatus and the lower end of the lateral border is an area for attachment of teres major. Inferior to this, latissimus dorsi gets a small attachment. Between teres major and the neck of the bone, are two areas for teres minor separated by a groove for the circumflex scapular vessels.

The **spine of the scapula** is continuous with the acromion where the rounded lateral border of the spine [FIG. 3.105] curves towards the neck of the bone and makes with it a **spinoglenoid notch** through which vessels and nerves pass to the infraspinous fossa.

The posterior border (or crest) of the spine is flattened, readily palpable, and makes edges for the supraspinous and infraspinous fossae. At its medial end, covered by fibres of trapezius, it forms a smooth triangular area which is easily felt and is usually at the level of the third thoracic vertebral spine. The upper edge of the posterior border becomes continuous with the medial border of the acromion and both give attachment to part of trapezius which marks the surface as well as the edge of the bone.

Near its medial end, the inferior edge projects as a tubercle and the lower fibres of trapezius run down from this area [FIG. 3.108]. The inferior edge becomes continuous with the lateral border of the acromion, and (lateral to the tubercle) both give attachment to the deltoid muscle. Where the edge of the spine becomes continuous with the lateral border of the acromion is an easily felt **acromial angle**.

The **acromion** slopes up to its tip from the acromial angle [FIG. 3.106]. From the tip, the coraco-acromial ligament fans out to reach the lateral border of the coracoid process and complete the coraco-acromial arch. A small, oval, articular facet for the clavicle lies on the anterior part of the medial border of the acromion. The facet faces slightly upwards and medially, and the clavicle projecting above it makes the joint easily felt through the skin. The acromion and the posterior border of the spine can be felt in their whole length between the attached muscles.

The **coracoid process** has a broad base which is directed upwards and forwards, and a narrow horizontal part which passes forwards, laterally, and slightly downwards from the upper edge of the base. Its tip can be felt inferior to the junction of middle and lateral thirds of the clavicle, deep to the edge of the deltoid muscle.

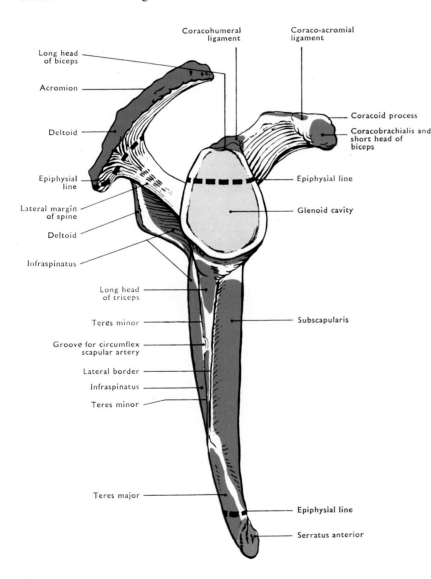

FIG. 3.106. Right scapula (lateral aspect).

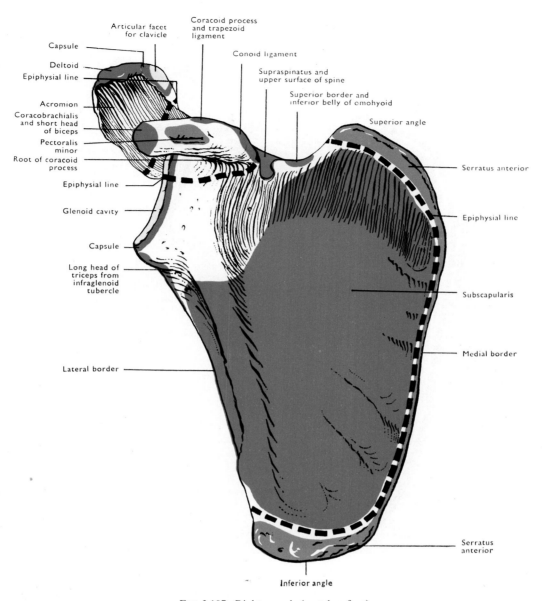

Articular facet
for clavicle

Capsule

Deltoid

Epiphysial line

Acromion

Coracobrachialis
and short head
of biceps

Pectoralis
minor

Root of coracoid
process

Epiphysial line

Glenoid cavity

Capsule

Long head of
triceps from
infraglenoid
tubercle

Lateral border

Coracoid process
and trapezoid
ligament

Conoid ligament

Supraspinatus and
upper surface of spine

Superior border and
inferior belly of omohyoid

Superior angle

Serratus anterior

Epiphysial line

Subscapularis

Medial border

Serratus
anterior

Inferior angle

FIG. 3.107. Right scapula (costal surface).

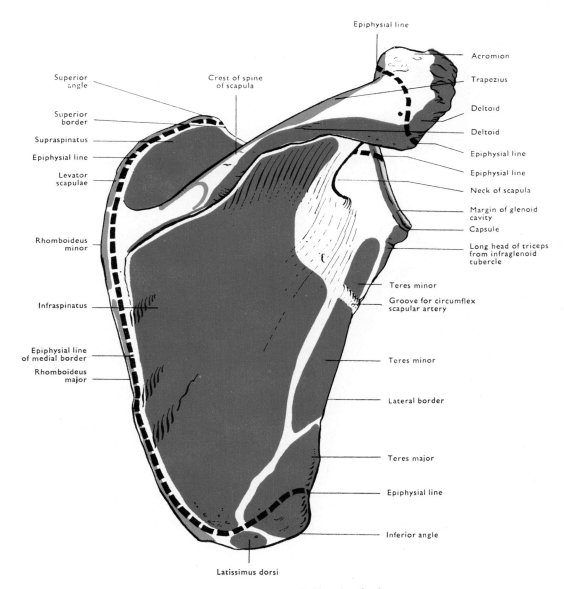

Superior angle

Crest of spine of scapula

Epiphysial line

Acromion

Superior border

Trapezius

Supraspinatus

Deltoid

Epiphysial line

Deltoid

Levator scapulae

Epiphysial line

Epiphysial line

Rhomboideus minor

Neck of scapula

Margin of glenoid cavity

Capsule

Long head of triceps from infraglenoid tubercle

Infraspinatus

Teres minor

Groove for circumflex scapular artery

Epiphysial line of medial border

Teres minor

Rhomboideus major

Lateral border

Teres major

Epiphysial line

Inferior angle

Latissimus dorsi

FIG. 3.108. Right scapula (dorsal surface).

The concave inferior surface of the coracoid process is smooth where a bursa intervenes between it and the subscapularis tendon—the horizontal part is flattened supero-inferiorly. The lateral margin of the upper surface has the coracohumeral and coraco-acromial ligaments attached to it—the coracohumeral to the intermediate part (opposite and continuous with the attachment of pectoralis minor to the medial margin [FIG. 3.107]) the coraco-acromial on either side of it. Also attached to the medial margin is the clavipectoral fascia and to the tip the coracobrachialis and short head of biceps. The posterior part of the upper surface of the coracoid is marked by strong coracoclavicular ligaments (conoid tubercle and trapezoid line).

The shallow articular glenoid cavity is pear-shaped with the narrow end uppermost. Its anterior border is slightly notched where developmentally separate parts of the bone join. The glenoid fossa has its depth slightly increased by a fibrous lip, the labrum glenoidale; the fibrous capsule of the shoulder joint is attached to the rim of the cavity and to the labrum. Above the cavity, a supraglenoid tubercle provides attachment for the long head of biceps which is also incorporated in the labrum. Below the cavity, the lateral edge of the bone is thickened to form an infraglenoid tubercle for attachment of the long head of triceps which is also attached to the labrum.

OSSIFICATION

A primary centre appears in cartilage about the end of the eighth week and at birth the acromion, coracoid process, glenoid cavity, inferior angle, and medial border are still cartilaginous. The times of appearance and fusion of the numerous secondary centres are given in FIGURE 3.109.

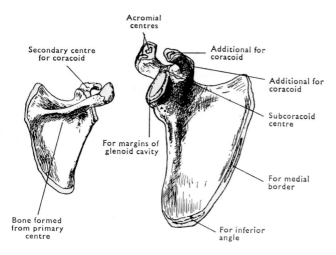

AT END OF FIRST YEAR   ABOUT AGE OF PUBERTY

FIG. 3.109. Ossification of scapula.
Secondary centres appear in the coracoid process in the first year (sometimes even before birth), and in the root of the coracoid and upper third of the glenoid cavity at 10 or 11 years. These fuse during or just after puberty, at which time other centres appear for the margins of the glenoid cavity, the inferior angle, the medial border, and two for the acromion; all these fuse between 20 and 25 years. Additional small scaly epipyses may appear on the surface of the coracoid after puberty, fusing at about 20 years.

VARIATIONS

The acromion and the coracoid process may remain separate throughout life. In old people the bone may be absorbed in parts, though the periosteum remains.

THE CLAVICLE

The clavicle extends from the sternum to the acromion and serves as a strut to prevent the shoulder falling forwards and downwards. The part forming its medial two-thirds is convex forwards and roughly triangular in section. Its lateral third is concave forwards and flattened from above downwards. The medial convexity is in conformity with the curvature of the superior thoracic aperture, the lateral concavity with the shape of the shoulder.

The inferior surface [FIG. 3.111] is irregular and strongly marked by ligaments at each end. The lateral part has a rough ridge, the trapezoid line, which runs forwards and laterally. At its posterior end is a rounded conoid tubercle at the posterior edge of the bone. These give attachment to the conoid and trapezoid parts of the coracoclavicular ligament which binds the clavicle and scapula together. The medial part of the inferior surface has a rough impression for the costoclavicular ligament. It holds down the medial end of the clavicle. The intermediate portion is marked by a groove for subclavius muscle and the clavipectoral fascia splits to enclose it before reaching the edges of the groove.

Looked at from above, the bone is comparatively smooth. The sinuous anterior border is roughened in its medial half, or more, by pectoralis major and in its lateral third by deltoid. The posterior border, between the posterior and inferior surfaces, is smooth except in its lateral third which is marked by the anterior fibres of trapezius.

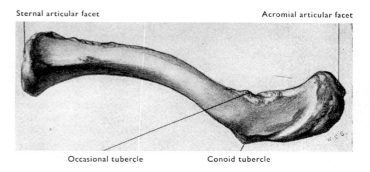

FIG. 3.110. Right clavicle (superior surface).

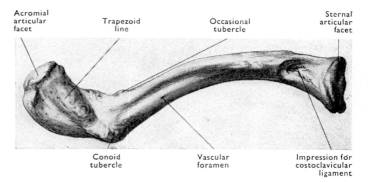

FIG. 3.111. Right clavicle (inferior surface).

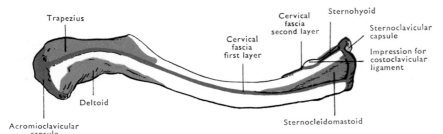

FIG. 3.112. Right clavicle (superior surface).

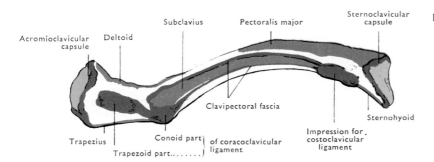

FIG. 3.113. Right clavicle (inferior surface).

Medially, the superior part of the bone is raised so as to form a superior border which gives attachment to the clavicular head of sternocleidomastoid. In the middle third of the bone the superior border becomes rounded and indefinite. Cervical deep fascia is attached to the superior and to the posterior borders. Sternohyoid gains attachment to the posterior surface near the medial end. The curvature of the clavicle allows the brachial plexus and subclavian vessels to pass freely between it and the first rib.

The lateral or **acromial end** has an oval articular facet for the acromion process and is set a little obliquely, facing downwards and laterally to fit the articular disc between it and the acromial facet. The capsule of the joint is attached to the ridge on the margin of the facet.

The medial or **sternal end** is enlarged. The inferior three quarters are bevelled off where it moves on an articular disc, separating it from the clavicular notch on the sternum and on the first costal cartilage. The superior quarter is rough for attachment of the disc, which acts as a ligament, and for the interclavicular ligaments. The articular capsule is attached to the edge of the facet. Since the notch on the sternum is shallow, the medial end of the clavicle stands up from it and is easily felt.

The double curvature of the body of the bone is a source of weakness when it is compressed lengthwise as may occur in a fall on the shoulder or outstretched arm.

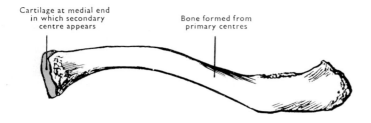

FIG. 3.114. Ossification of clavicle.

A secondary centre appears in cartilage at the medial end from about 14 to 18 years: it fuses as early as 18 to 20 years in females and about 23 to 25 in males. A small epiphysis may appear at the acromial end about puberty but it soon fuses.

OSSIFICATION

The clavicle is the first bone to begin ossification. Two primary centres appear during the fifth embryonic week in mesoderm. They unite and ossification then spreads into the formed cartilage nearer the ends. A secondary centre appears in the cartilage at the medial end [FIG. 3.114], and is often the last to unite.

VARIATIONS

Its outer end may be raised or depressed according to muscular development or position of the body. Occasionally one of the supraclavicular nerves pierces the clavicle. Macerated bones often have the medial epiphyses missing and have a flat, roughened appearance.

THE HUMERUS

The humerus extends from the scapula to the elbow joint and has a **body** and two ends. It is almost covered with muscles but can be felt through them in all its length. The upper end includes an almost hemispherical **head** and two distinct **tubercles** with the deep **intertubercular groove** between them [FIG. 3.115].

The smooth, rounded surface of the head articulates with the shallow and smaller glenoid cavity on the scapula and may be felt indistinctly if the fingers are pushed well up into a relaxed axilla. With the arm by the side, the articular surface faces medially and upwards and also slightly backwards to meet the glenoid surface which is turned slightly forwards [FIG. 3.116]. The head is joined to the upper end by the **anatomical neck**, a slightly constricted region which encircles the bone at the edge of the articular surface and separates it from the tubercles. The **surgical neck** is the region below the head and tubercles where they join the shaft, and where fractures often occur. The capsule of the shoulder joint is attached to the anatomical neck above but extends down 1.5 cm on to the surgical neck inferiorly [FIG. 3.119].

The **greater tubercle** is the prominence on the lateral side of the upper end of the bone. It merges with the body below, and can be

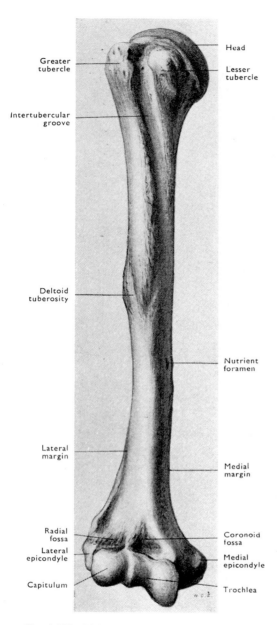

FIG. 3.115. Right humerus (anterior aspect).

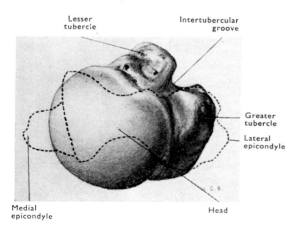

FIG. 3.116. Proximal aspect of right humerus (with the outline of lower end in relation to it in dotted line).

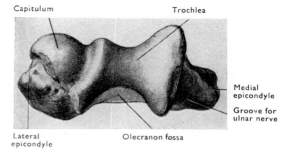

FIG. 3.117. Right humerus (distal aspect).

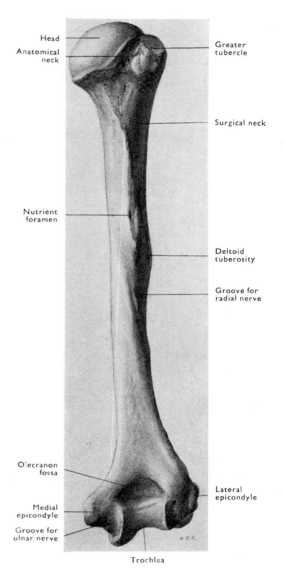

FIG. 3.118. Right humerus (posterior aspect).

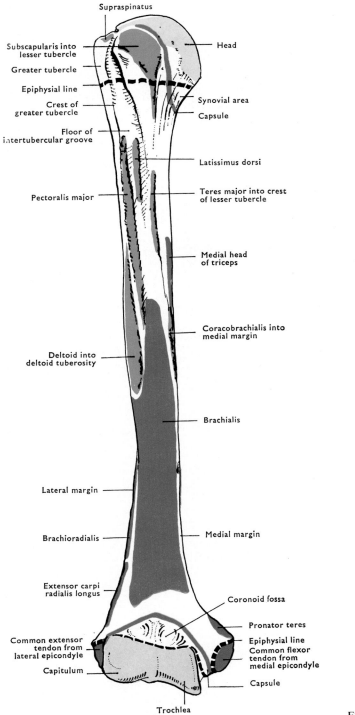

Supraspinatus

Subscapularis into
lesser tubercle

Greater tubercle

Epiphysial line

Crest of
greater tubercle

Floor of
intertubercular groove

Pectoralis major

Deltoid into
deltoid tuberosity

Lateral margin

Brachioradialis

Extensor carpi
radialis longus

Common extensor
tendon from
lateral epicondyle

Capitulum

Head

Synovial area

Capsule

Latissimus dorsi

Teres major into crest
of lesser tubercle

Medial head
of triceps

Coracobrachialis into
medial margin

Brachialis

Medial margin

Coronoid fossa

Pronator teres

Epiphysial line
Common flexor
tendon from
medial epicondyle

Capsule

Trochlea

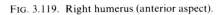

FIG. 3.119. Right humerus (anterior aspect).

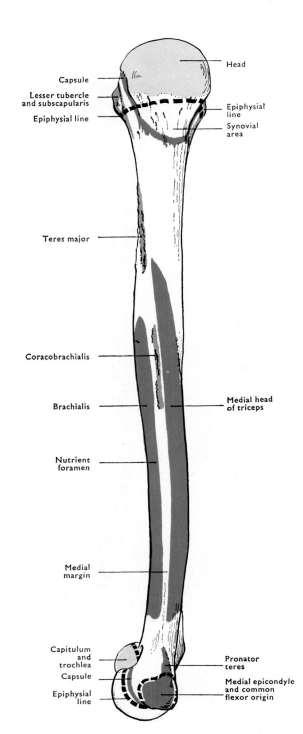

Capsule

Lesser tubercle
and subscapularis

Epiphysial line

Teres major

Coracobrachialis

Brachialis

Nutrient
foramen

Medial
margin

Capitulum
and
trochlea

Capsule

Epiphysial
line

Head

Epiphysial
line

Synovial
area

Medial head
of triceps

Pronator
teres

Medial epicondyle
and common
flexor origin

FIG. 3.120. Right humerus (medial aspect).

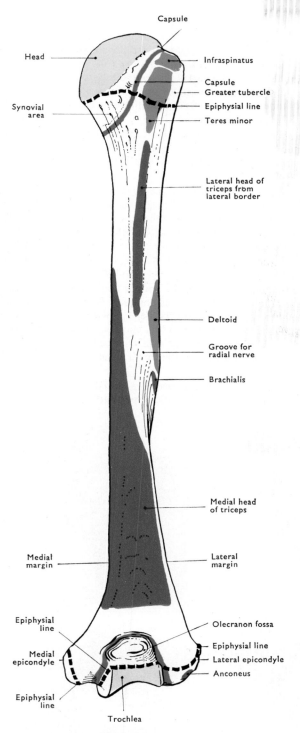

FIG. 3.121. Right humerus (posterior aspect).

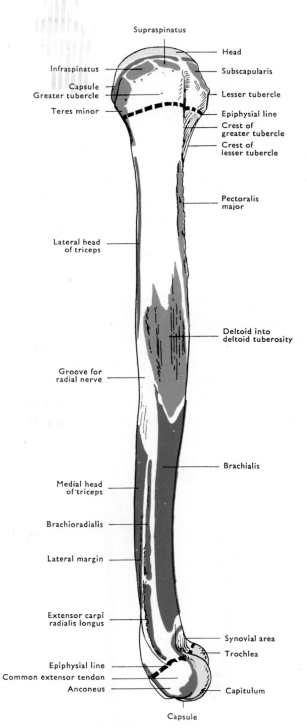

FIG. 3.122. Right humerus (lateral view).

felt through deltoid below the edge of the acromion process. The tendons of the supraspinatus and infraspinatus are inserted into impressions on its superior aspect, and teres minor posteriorly [FIG. 3.122]. Teres minor also extends down on to the surgical neck.

The **lesser tubercle** is the distinct prominence on the front of the upper end of the bone. It can be felt through the deltoid just lateral to the tip of the coracoid process [p. 152], moving further away from it when the arm is rotated laterally. The medial side of the tubercle has a well-marked impression on it for the tendon of subscapularis which is continued down on to the surgical neck.

The **intertubercular groove** passes on to the body between the **crests of the greater** and **lesser tubercles** which continue down from the anterior borders of the tubercles to form lateral and medial lips to the groove. The tendon of the long head of biceps brachii, in a synovial sheath, is held in the groove by a transverse ligament uniting the two tubercles.

Pectoralis major crosses over to reach the crest of the greater tubercle, teres major is inserted into the crest of the lesser tubercle, and latissimus dorsi curves round the latter to reach the floor of the groove [FIG. 3.119].

The **body** is triangular with distinct **medial** and **lateral borders** in its lower part. From the front, two **surfaces** are visible, anteromedial and anterolateral, separated by a smooth, rounded, median ridge which divides below and is continuous with the crest of the greater tubercle above. The intertubercular groove is thus in continuity with the anteromedial surface and the **medial border** begins as the crest of the lesser tubercle. It ends below by curving medially to the prominent medial epicondyle, and its lower third is referred to as the medial supracondylar ridge or line. Half-way down, the medial border is roughened by the insertion of coracobrachialis [FIG. 3.119].

The **lateral border** is prominent in its lower third where it runs down to the lateral epicondyle as the lateral supracondylar ridge or line. Above this it is interrupted by the smooth **groove for the radial nerve** which passes downwards and forwards. The groove is limited above and in front by a part of the **deltoid tuberosity**. The remainder of the lateral border continues up to the posterior border of the greater tubercle and gives attachment to the lateral head of triceps [FIG. 3.121].

The borders of the body give attachment to the corresponding intermuscular septa which separate the anterior and posterior compartments of the arm and give an increased area of attachment to adjacent muscles.

The anteromedial and anterolateral surfaces of the body give origin to fibres of brachialis in the lower half. The smooth anterolateral surface has a **deltoid tuberosity** [FIG. 3.115] for insertion of deltoid. This can be felt half-way up the lateral aspect of the arm. The anterior surface of the lateral supracondylar line gives attachment to brachioradialis in its upper two-thirds and to extensor carpi radialis longus in its lower third.

The **posterior surface** of the body is marked by a spiral **groove for the radial nerve** which runs obliquely across the upper half of the body to reach the lateral border below the deltoid tuberosity. The radial nerve and profunda brachii vessels lie in the groove. The body between the groove and the inferior end of the bone is smooth and gives attachment to the medial head of triceps brachii [FIG. 3.121].

The **nutrient foramina** are usually situated anterior to the middle of the medial border and in the upper part of the groove for the radial nerve.

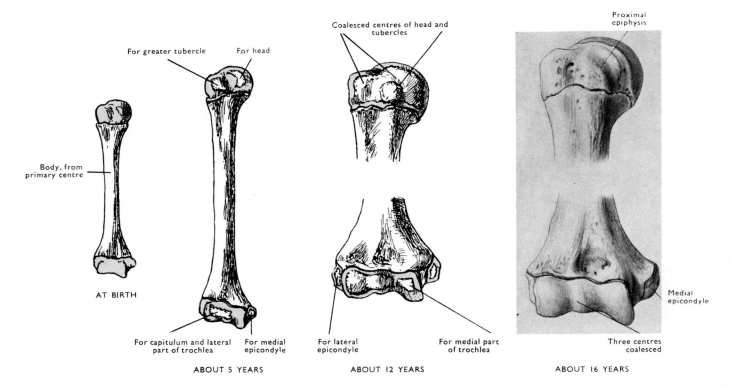

FIG. 3.123. Ossification of humerus.

For greater tubercle   For head

Coalesced centres of head and tubercles

Proximal epiphysis

Body, from primary centre

AT BIRTH

For capitulum and lateral part of trochlea   For medial epicondyle

For lateral epicondyle

For medial part of trochlea

Three centres coalesced

Medial epicondyle

ABOUT 5 YEARS   ABOUT 12 YEARS   ABOUT 16 YEARS

Secondary centres appear for the head (early in first year), the greater tubercle, (about 3 years), and lesser tubercle (about 5 years): these join into a single cap of bone at 6 to 8 years and fuse at 18 to 20 years (female) and 20 to 22 years (male).

In the lower end secondary centres appear for the capitulum (2nd year), trochlea (9 to 10 years) and lateral epicondyle (12 to 14 years): these join together at about 14 years and fuse with the shaft about 15 years (female) and 18 years (male). A separate centre forms for the medial epicondyle at 6 to 8 years and fuses at 15 to 18 years with a spicule of the shaft projecting down medial to the trochlea (see third figure above).

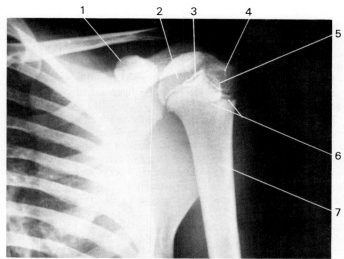

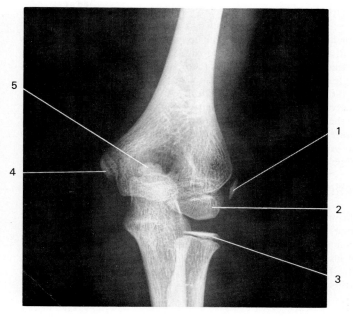

FIG. 3.124. Anteroposterior radiograph of shoulder. Age 10 years. Note the apparent duplication of the epiphysial line between the greater tubercule and the diaphysis. This is due to the anterior and posterior edges of the epiphysial cartilage lying in the same plane (lower dark line) while the intermediate part lies at a higher level (upper dark line) as the conical end of the diaphysis flattens out laterally. This appearance should not be mistaken for an undisplaced fracture.
1. Coracoid process.
2. Epiphysis.
3. Internal part of epiphysial line surmounting conical diaphysis.
4. Greater tubercule.
5. Lesser tubercule.
6. Surface position of epiphysial line.
7. Diaphysis of humerus.

FIG. 3.125. Anteroposterior radiograph of elbow joint. Age 10 years.
1. Epiphysis of lateral epicondyle.
2. Epiphysis for capitulum and lateral part of trochlea.
3. Epiphysis for head of radius.
4. Epiphysis for medial epicondyle.
5. Epiphysis for olecranon.

The **lower end** of the humerus is wide, flattened anteroposteriorly, and bent slightly forwards. Approximately the middle third of its distal edge forms the pulley-shaped **trochlea** which articulates with the ulna and has a prominent medial lip. Above the bone is thinned (or defective) by the presence of the **coronoid fossa** in front and the **olecranon fossa** behind, into which the corresponding parts of the ulna fit during extreme flexion or extension.

Lateral to the trochlea, a rounded **capitulum** for articulation with the radius is continuous with the anterior and distal aspects of the trochlea but is not continued on to the posterior surface of the bone. A small **radial fossa** lies above it anteriorly. Behind the capitulum, the bone is roughened and projects laterally as the **lateral epicondyle**. Its posterior surface is smooth, subcutaneous, and easily felt; its distal aspect is marked by the radial collateral ligament with anconeus attached posterior to it [FIG. 3.122]. The anterior surface is marked by the common extensor tendon of the forearm muscles and above this the extensor carpi radialis longus continues the line of attachments up on to the supracondylar part of the lateral border.

The **medial epicondyle** is very prominent, partly because the supracondylar part of the medial border is less prominent than the lateral one. Its posterior surface is smooth and has on it a shallow **sulcus for the ulnar nerve** and collateral vessels. Its distal margin gives attachment to the ulnar collateral ligament and, in front of this, the anterior surface has an impression for the common flexor tendon. Pronator teres arises from the flatter upper part and from the adjacent lower end of the medial border [FIG. 3.119].

The anterior part of the fibrous capsule is attached to the upper margins of the coronoid and radial fossae. The posterior part crosses the olecranon fossa. Between these it is attached to the epicondyles close to the trochlea and capitulum in front, and close to each side of the trochlea behind.

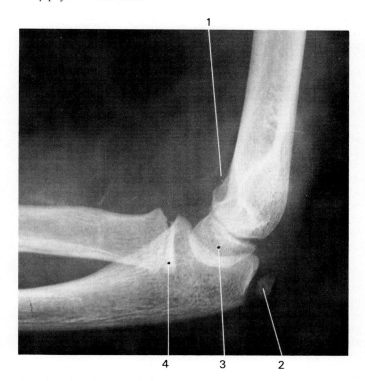

FIG. 3.126. Lateral radiograph of elbow joint. Age 10 years.
1. Fat protruding from coronoid fossa of humerus.
2. Epiphysis of olecranon.
3. Epiphysis for capitulum and lateral part of trochlea.
4. Epiphysis for head of radius.

OSSIFICATION

A primary centre appears in the body in the eighth week and spreads until, at birth, only the ends are cartilaginous [FIG. 3.123]. Secondary centres for the head, greater tubercle, and lesser tubercle form a single epiphysis by ossification spreading from the centre in the head of the bone (Paterson 1929). Details of appearance and fusion of secondary centres are given in FIGURE 3.123. Note that the centre for the medial epicondyle is entirely outside the joint capsule, and does not fuse with the adjacent epiphysis directly. Much variation can occur in times of fusion: most of the growth in length of the humerus occurs at its upper end.

VARIATIONS

The only distinct variation is the presence of a **supracondylar process**. This is found occasionally 3–4 cm above the medial epicondyle. It is a hooked process connected to the epicondyle by a fibrous band under which the median nerve and brachial vessels pass. The ligament may be ossified, a condition normally found in Carnivora.

THE ULNA

The ulna is the medial bone of the two in the forearm. It is a long bone with a body, an upper, and a lower end. It can be felt in its whole length down the back of the forearm.

The **upper end** is large with two projecting processes enclosing a concavity [the **trochlear notch**; FIG. 3.127] which fits on to the trochlea of the humerus.

The **olecranon**, the larger of the two processes, forms the proximal end of the bone. It is continuous with the body and makes the point of the elbow which is felt at the back of the elbow joint. Posteriorly it is smooth and subcutaneous and this area extends down on to the body. The proximal surface extends forwards to a point. Posteriorly it is marked by the strong tendon of triceps; anteriorly is the attachment of the capsule of the elbow joint, and between these is a smooth area for a bursa.

The anterior surface of the olecranon forms part of the articular surface of the trochlear notch and is divided by a rounded ridge which fits into the groove on the trochlea of the humerus. The borders are thick and rough, the lateral for the attachment of anconeus and the capsule of the elbow joint [FIG. 3.132]; the medial, for flexor carpi ulnaris, flexor digitorum profundus, and the ulnar collateral ligament.

The **coronoid process** projects from the front of the body and has an articular upper surface which completes the trochlear notch and is divided by a ridge into a larger medial part and a smaller lateral part. The lateral part is continuous over its lateral edge with the articular surface of the radial notch on the lateral surface of the process. The posterior part of the articular surface in the trochlear notch is usually separated from the articular surface on the olecranon by a rough strip.

Where the medial and anterior edges of the articular surface on the coronoid process meet, is a small tubercle to which the anterior part of the ulnar collateral ligament is attached, and from which the medial margin of the process extends distally, giving attachment to the flexor digitorum superficialis, pronator teres, and flexor pollicis longus muscles [FIG. 3.130].

The irregular, anterior surface of the coronoid process ends inferiorly in a rough **tuberosity of the ulna**. Both the surface and the tuberosity provide attachment for brachialis.

On the lateral side of the process the **radial notch** lodges the head of the radius and has the anular ligament attached to its anterior and posterior edges, while the quadrate ligament is inferior to it. Below this and extending on to the body, is a triangular fossa which gives origin to supinator and is bounded posteriorly by a distinct **supinator crest**.

The body of the ulna is triangular in section and tapers down to a slender cylindrical part which widens again a little to form the **head**.

The prominent lateral edge is the **interosseous border** where the interosseous membrane is attached. It runs down from the apex of the fossa for supinator and is prominent in its upper part. The **anterior border** runs down from the medial margin of the coronoid process but is indefinite. The **posterior border** is sinuous, subcutaneous, and prominent in its upper part. It is continuous above with the apex of the subcutaneous triangle on the olecranon and upper part of the shaft [FIG. 3.131], and the aponeurosis of flexor carpi ulnaris is attached to its upper three-quarters and to the medial

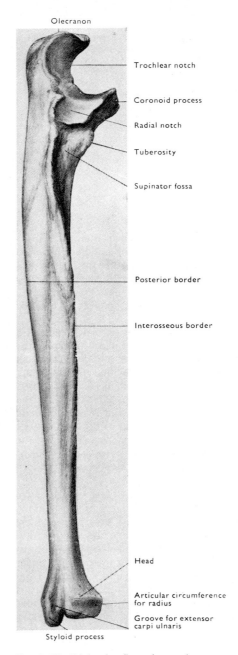

Olecranon

Trochlear notch

Coronoid process

Radial notch

Tuberosity

Supinator fossa

Posterior border

Interosseous border

Head

Articular circumference for radius

Groove for extensor carpi ulnaris

Styloid process

FIG. 3.127. Right ulna (lateral aspect).

margin of the triangle. Extensor carpi ulnaris is attached to the posterior border on the lateral side.

The anterior and medial surfaces are continuous at the rounded anterior border, and are covered by flexor digitorum profundus which arises from their upper two-thirds and also from the medial surfaces of both olecranon and coronoid processes [FIG. 3.130]. An oblique ridge extending downwards and medially marks the lower quarter of the anterior surface where pronator quadratus arises [FIG. 3.130].

The **posterior surface**, between the interosseous and posterior borders, has anconeus attached from the olecranon to an oblique ridge extending downwards and posteriorly from the back of the radial notch to the posterior border [FIG. 3.132]. The remainder of the surface is divided into a smooth medial area, covered by extensor carpi ulnaris, and a lateral area which has three faint ridges separating the origins of abductor pollicis longus, extensor pollicis longus, and extensor indicis [FIG. 3.132].

The **lower end** of the ulna consists of a small, rounded head and a conical styloid process which projects downwards.

The **head** is the widened lower end of the body, and is the knob felt on the back of the wrist when exposed by the movement of the radius into full pronation. It has an articular surface for the radius round its anterior and lateral aspects, and is smooth medially. Posteriorly, it is separated from the styloid process by a groove which lodges the tendon of extensor carpi ulnaris. The distal surface of the head is smooth and nearly flat. It articulates with an articular disc which intervenes between the head and the lunate. The disc is approximately triangular in shape and is attached by its apex to a pit between the head of the ulna and the base of the styloid process.

The **styloid process** projects downwards from the posteromedial part of the distal end of the bone, and its tip is fully 1 cm proximal to the corresponding part of the radius. It gives attachment to the ulnar collateral ligament of the carpus.

OSSIFICATION

The body, coronoid process, and nearly all of the olecranon are ossified from a primary centre. There are secondary centres for the head of the ulna and for the top of the olecranon. The head usually fuses last and is at the 'growing' end. [See FIG. 3.128 for details.]

VARIATIONS

The olecranon epiphysis varies considerably, and the epiphysial line may pass through the articular part of the olecranon [FIG. 3.126].

THE RADIUS

The radius lies lateral to the ulna. It extends from the humerus to the carpus but, since it does not overlap the humerus, it is shorter than the ulna. It has a body and two ends of which the inferior one is the larger. The superior end consists of a head which narrows abruptly to join the body by a cylindrical neck.

The **head** is a thick disc with a concave superior surface for articulation with the capitulum of the humerus. Its **articular circumference** is encircled by the radial notch of the ulna with the

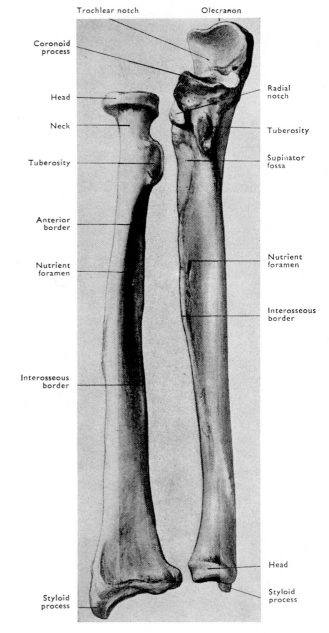

FIG. 3.129. Right radius and ulna (anterior surface).

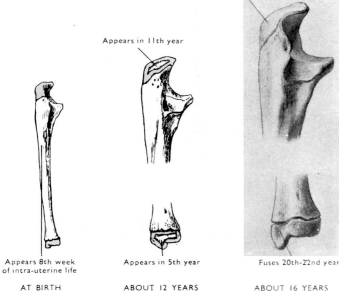

FIG. 3.128. Ossification of ulna.

anular ligament which holds it against the notch while still leaving it free to rotate. The anterior, medial, and posterior parts of the edge bear on the ulna and are distinctly articular. The remainder is narrower and smooth where it bears on the ligament.

The neck is smooth and slopes a little medially as it approaches the body. The anular and quadrate ligaments are attached loosely to the neck so as to permit free rotation.

The body is round where it joins the neck but becomes triangular in section lower down. Along with the neck it is curved with a slight medial convexity in its upper fifth and has a lateral convexity in its remaining lower part. The radial tuberosity lies on the medial aspect of the uppermost part of the body, at the maximum convexity of the medial curve. This provides leverage for the biceps tendon which is inserted into its posterior part and produces lateral rotation of the radius. The anterior part of the tuberosity is smooth where a bursa intervenes between the tendon and the bone [FIG. 4.32]. Pronator teres, which has the opposite action, is inserted into a rough mark on the maximum convexity of the lateral curvature, half-way down the lateral surface of the body [FIG. 3.132].

The interosseous border is the prominent edge which extends from just below the tuberosity to the medial side of the inferior

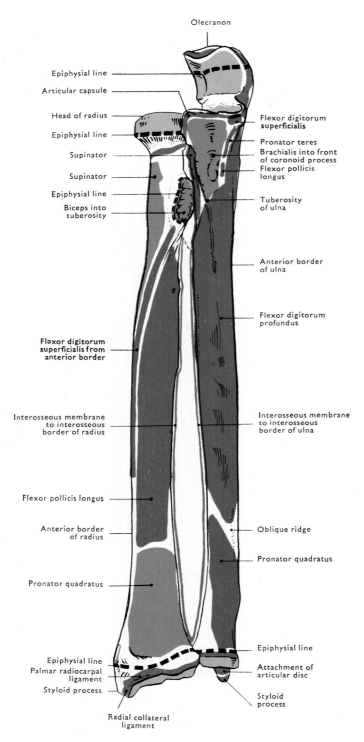

FIG. 3.130. Right radius and ulna (anterior aspect).

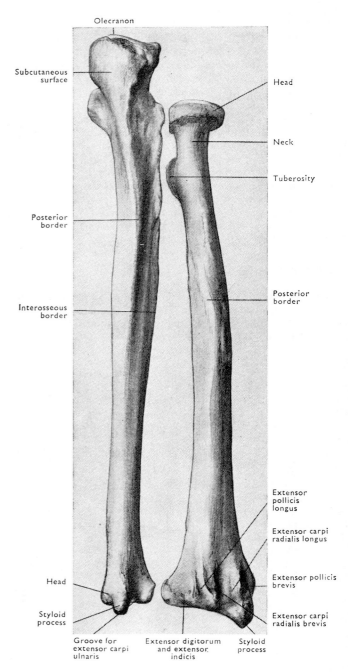

FIG. 3.131. Right radius and ulna (posterior surface).

extremity of the bone. In its lower part it splits into two diverging ridges which limit the anterior and posterior surfaces of the end of the bone and are continuous with anterior and posterior margins of the ulnar notch. The interosseous membrane is attached to the border and to the posterior ridge. The oblique cord passes superomedially from the bone anterior to the upper end of the membrane [FIG. 4.33].

The **anterior** and **posterior borders** pass in front of and behind the rough mark for the pronator teres on the lateral side of the body.

They become more distinct above this level as two oblique lines which ascend superomedially to the anterior and posterior parts of the tuberosity [FIGS. 3.129 and 3.131]. The rounded lateral surface thus encroaches on the anterior and posterior aspects of the body above the impression for pronator teres muscle. The supinator is attached here after passing behind the radius from the ulna [FIG. 3.130].

The **lateral surface** of the body has no attachments below the pronator teres and the posterior border is very indistinct here. The

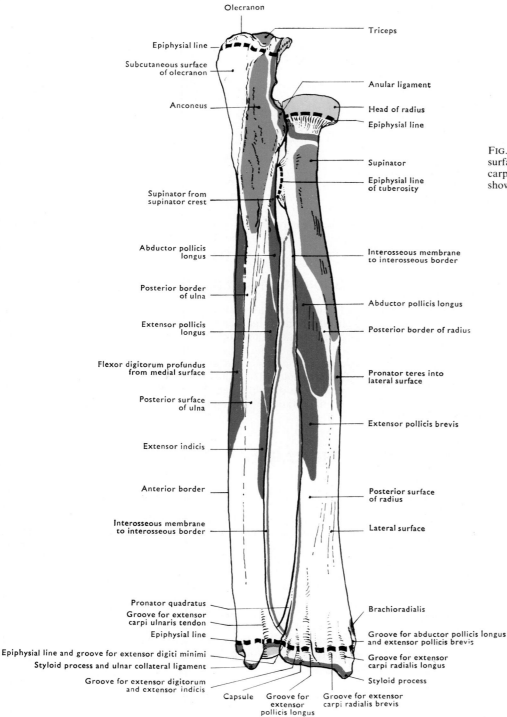

FIG. 3.132. Right radius and ulna (posterior surface). The attachment of flexor and extensor carpi ulnaris to the posterior border of ulna is not shown,

anterior border becomes quite distinct at the inferior end of the bone. It is accentuated in its upper half by the attachment of the radial head of flexor digitorum superficialis.

The anterior surface of the body lies between the anterior and interosseous borders and is continuous inferiorly with the anterior end of the bone. Its upper three-quarters provide attachment for flexor pollicis longus. Its lower part receives the insertion of pronator quadratus [FIG. 3.130].

The posterior surface is flatter and merges with the lateral surface in its inferior half where the overlying tendons give the bone a smooth, rounded appearance. In its middle third it gives part origin to abductor pollicis longus and extensor pollicis brevis [FIG. 3.132].

At the inferior end, the anterior surface curves forwards to a distinct anterior margin which can be felt about 2.5 cm above the ball of the thumb. The margin gives attachment to the palmar radiocarpal ligament and is a guide to the position of the wrist joint. The lateral surface, about 1 cm wide, is a shallow groove for the tendons of abductor pollicis longus and extensor pollicis brevis, with the tendon of brachioradialis inserted deep to them. The edge between this groove and the anterior surface is sharp and easily felt. It gives attachment to the end of the extensor retinaculum which binds down the tendons.

The lateral part of the inferior end of the radius is prolonged downwards as a thick styloid process which is palpable at the lateral side of the wrist in the anatomical snuff box. The apex of this styloid process is 1 cm lower than that of the ulna. Absence of this feature indicates shortening of the radius, e.g. in fractures. The styloid process gives attachment to the radial collateral ligament of the carpus.

The medial surface forms the concave ulnar notch for articulation with the head of the ulna. It is shaped to allow movement of the radius round the ulna and its edges give attachment to the capsule of the joint except inferiorly. The inferior edge has the base of the articular disc attached to it, between the radio-ulnar and radiocarpal surfaces of the radius.

The posterior surface of the lower end of the radius is convex but is much grooved by tendons. Its lateral half continues down on to the styloid process and has two shallow grooves for the tendons of extensor carpi radialis longus, laterally and brevis, medially. The prominent ridge in the middle of the posterior surface projects as a landmark, the dorsal radial tubercle, which can be felt above the back of the wrist nearly in line with the forefinger. The ridge and tubercle have the extensor retinaculum attached to them, and lie lateral to the tendon of extensor pollicis longus which changes direction here to reach the thumb, and forms a clear-cut, slightly oblique groove on the tubercle. Medial to this is a wide shallow groove for the extensor digitorum and indicis tendons. The extensor digiti minimi lies over the joint between the radius and ulna and extensor carpi ulnaris lies in the groove between the head and styloid process of the ulna [FIG. 3.132].

The carpal articular surface is concave to fit the carpal bones. It extends on to the styloid process and is divided by a ridge into a triangular lateral area for the scaphoid and a quadrilateral medial area for the lunate. In life the articular surface is continuous medially with that of the articular disc separating the lunate from the head of the ulna. The forward curve of the inferior end of the bone brings its posterior border to a more distal level than the anterior border.

OSSIFICATION

The radius is ossified from a primary centre and at birth only the head, the lower end, and the tuberosity are cartilaginous. Secondary centres appear first in the lower 'growing' end and later in the head.

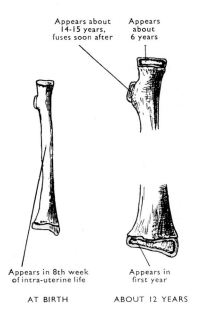

FIG. 3.133. Ossification of radius.

A centre may appear for the tuberosity but soon fuses with the body. [See FIG. 3.133 for dates.]

Supination and pronation [p. 237]

The normal anatomical position of the forearm is that of supination in which the radius and ulna are parallel. The back of the hand is posterior, the palm is anterior and the thumb is directed laterally. In full pronation the hand is turned so as to reverse its position by rotating the upper end of the radius and swinging its lower end round the ulna. The radius and ulna thus become crossed. The slight supination present when the limb hangs naturally is masked by a medial rotation of the humerus.

THE CARPUS

The eight small bones of the wrist are referred to collectively as the carpus and individually as the carpal bones. They articulate with each other and are bound together by ligaments so as to form a compact mass which is curved with a posterior convexity and a pronounced anterior concavity, the carpal sulcus, which is converted into a carpal canal or tunnel by the flexor retinaculum. The proximal surface of the carpus is rounded for articulation with the radius, and with the articular disc which extends from radius to ulna, to form the wrist joint proper; its distal surface is wide and irregular for articulation with the five metacarpal bones [FIG. 4.40]. The tunnel is completely filled during life by tendons and the median nerve. Because of the mobility of the hand frequent use is made of the terms dorsal and palmar in place of posterior and anterior.

The carpus as a whole

The student should know the carpus as a whole rather than have a precise knowledge of every carpal bone, but some knowledge of each of them is a necessity in learning the whole. Their clinical importance lies partly in the frequency of injury to some of them, particularly the lunate and scaphoid bones, and partly in the need for recognizable landmarks in the region of the wrist and hand.

The bones of the carpus are arranged round the capitate but they are commonly described as forming two rows. The proximal row contains the scaphoid, lunate, and triquetral bones which together form the distal surface of the wrist joint. The pisiform is also included in this row although it is often regarded as a sesamoid bone (in the tendon of flexor carpi ulnaris) which articulates loosely with the triquetrum. The distal row contains the trapezium, the trapezoid, the capitate and the hamate bones. The trapezium has a saddle-shaped articular surface for the first metacarpal and reaches the edge of the second. The trapezoid bone fits into the deep notch in the second metacarpal. The capitate articulates with the third and reaches the edges of the second and fourth metacarpals. The hamate bone articulates with both the fourth and fifth metacarpals.

The margins of the carpus are narrow and receive the collateral ligaments of the wrist [p. 239]. For muscle attachments, see FIGURE 3.136.

The concavity of the **carpal sulcus** is concealed by its contents enclosed by the **flexor retinaculum** which stretches between the pisiform bone and the hook of the hamate bone, medially, and the tubercles of the scaphoid and trapezium, laterally. The hook of the hamate is felt by deep pressure 15–25 mm along a line directed from the pisiform towards the centre of the palm. The superficial division of the ulnar nerve can be felt slipping from side to side on the hook. The deep division is against its base and cannot be felt. The ulnar

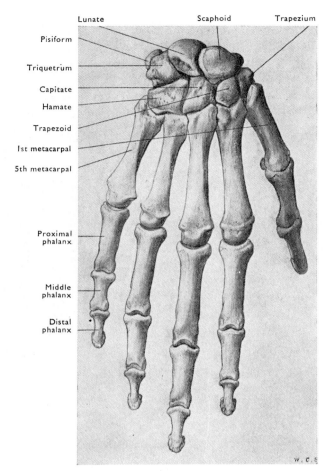

FIG. 3.135. Dorsal aspect of bones.

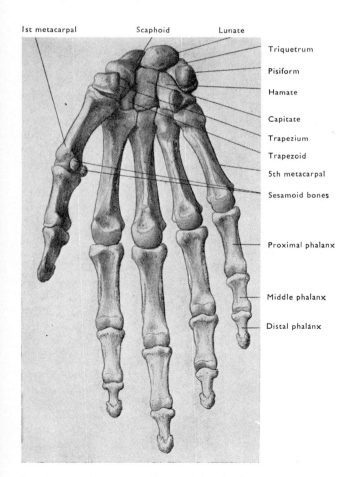

FIG. 3.134. Palmar aspect of bones of right hand.

vessels and nerve lie on the lateral side of the pisiform bone. The tuberosity of the scaphoid can be found by following the tendon of flexor carpi radialis which passes medial to it. Distal to this, the tubercle on the trapezium can be felt as a deeper bony resistance through the attachments of the thenar muscles.

The dorsal surface of the carpus is convex and uneven but is covered by ligaments and tendons so that individual parts are difficult to identify [FIG. 5.81]. The styloid process at the lateral side of the base of the third metacarpal can be identified if the body of the bone is traced up to the carpus. It makes a small prominence where it projects between the capitate and trapezoid bones and receives the tendon of extensor carpi radialis brevis. The radial styloid, the distal part of the scaphoid bone, and the trapezium, lie in the floor of the anatomical snuff box.

The blood supply to the carpus reaches the bones along the ligaments, and foramina can be seen on the non-articular surfaces between areas for their attachment and between such areas and the articular facets. Fracture or dislocation, particularly with displacement and tearing of ligaments, may interfere with the blood supply.

The capitate

The capitate is the largest carpal bone and is centrally placed with a rounded head set into the cavities on the lunate and scaphoid bones. Flatter articular surfaces for the hamate medially and the trapezoid laterally are present at the sides. The dorsal surface is

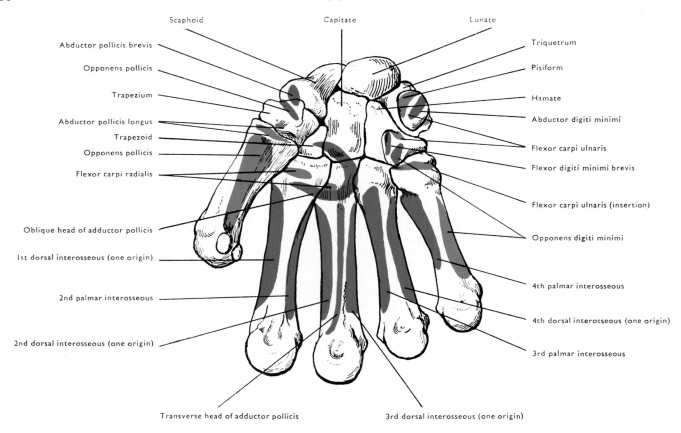

Scaphoid　Capitate　Lunate

Abductor pollicis brevis
Opponens pollicis
Trapezium
Abductor pollicis longus
Trapezoid
Opponens pollicis
Flexor carpi radialis

Oblique head of adductor pollicis
1st dorsal interosseous (one origin)

2nd palmar interosseous

2nd dorsal interosseous (one origin)

Triquetrum
Pisiform
Hamate
Abductor digiti minimi
Flexor carpi ulnaris
Flexor digiti minimi brevis
Flexor carpi ulnaris (insertion)
Opponens digiti minimi

4th palmar interosseous

4th dorsal interosseous (one origin)

3rd palmar interosseous

Transverse head of adductor pollicis　　3rd dorsal interosseous (one origin)

FIG. 3.136. Palmar surface of right carpus and metacarpus with attachments marked.

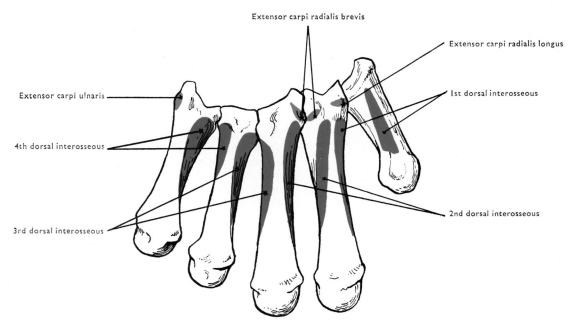

Extensor carpi radialis brevis
Extensor carpi radialis longus
1st dorsal interosseous
Extensor carpi ulnaris
4th dorsal interosseous
2nd dorsal interosseous
3rd dorsal interosseous

FIG. 3.137. Dorsal surface of the right metacarpus with attachments marked.

nearly flat but the palmar aspect is rough and protuberant for the attachment of ligaments [FIG. 4.39]. The distal surface, or base, is articular and is mainly for the base of the middle metacarpal but the capitate reaches and narrowly articulates with the second and fourth metacarpals at the edges of its distal surface. The styloid process of the third metacarpal overlaps its dorsal surface and may be felt on the back of the wrist.

The hamate

The hamate is distinguished by its wedge shape and by the curved **hamulus** (hook) which projects from its palmar surface near the base of the fifth metacarpal. The hamulus is concave on its lateral side and forms part of the carpal canal. Apart from muscles and the pisohamate ligament [FIG. 3.136], the flexor retinaculum is attached to its free end. The base of the wedge is articular and carries two fused articular surfaces for the bases of the fifth and most of the fourth metacarpals. The wedge lies between the capitate and the triquetrum and its edge reaches the lunate. The articulation with the capitate is flat, that with the triquetrum is sinuous. The dorsal surface is flat; the palmar surface curves on to the lateral aspect of the hamulus.

The triquetrum

The triquetrum lies in the angle between the hamate and the lunate and has, therefore, a sinuous articular surface for the side of the hamate. The articular surface for the lunate is almost square and is set at right angles to that for the hamate. The triquetrum is distinguished by the circular surface for the pisiform which is set on the distal part of the palmar surface close to the hook of the hamate. The remaining surfaces form a convex area which is marked by the ulnar collateral ligament. The proximal part enters the wrist joint and articulates, during ulnar deviation of the hand, with the articular disc which separates it from the ulna.

The pisiform

The pisiform is a small bone which can be felt as a slightly mobile prominence at the medial side of the front of the wrist. It articulates with the palmar surface of the triquetrum by a circular or oval facet with an articular capsule attached to its margin, often in a narrow groove. The remainder of the bone is rounded, but its anterior surface, having the flexor retinaculum attached to it, projects distally and laterally, forming a lateral concavity which is part of the carpal canal.

The flexor carpi ulnaris tendon surrounds the non-articular part of the bone, and is continued to the hook of the hamate and the base of the fifth metacarpal as the **pisohamate** and **pisometacarpal** **ligaments**.

The lunate

The lunate has a square articular surface for the triquetrum on its medial side, and a crescent-shaped surface for the scaphoid on its lateral side. Distally it has a deep concavity for the head of the capitate and proximally it is convex where it articulates with the radius and the articular disc. The dorsal and palmar surfaces have ligaments and nutrient vessels and are small compared with the articular area. The palmar surface is convex, smoother, and, unlike the other carpal bones, larger than the dorsal surface. An articular

surface for the edge of the hamate may be found as a narrow strip between the areas for triquetrum and the capitate.

The scaphoid

The scaphoid has a concave articular surface for the head of the capitate and at the edge of this is a crescentic surface for the corresponding area on the lunate. Proximally, it has a wide convex articular surface for the radius. Distal to these surfaces, the bone has a prominent **tubercle** which projects forwards. It can be felt on the anterior aspect of the wrist immediately proximal to the thenar eminence (especially with the wrist extended) at the junction of middle and lateral thirds of the most distal transverse crease in the wrist. The remaining articular surface is to the lateral side of the tubercle. It faces laterally and carries the trapezium and trapezoid.

The narrow strip between the radial and trapezial surfaces and the tubercle gives attachment to the radial collateral carpal ligament. The tubercle receives part of the flexor retinaculum. This small, non-articular surface is all that is available for the entry of blood vessels. It is a common site of fracture, the result of which may be to deprive the bone, or part of it, of a blood supply.

The trapezoid

The trapezoid is small and irregular and articulates with the second metacarpal. It lies in the space bounded by the metacarpal, the capitate, scaphoid, and trapezium [FIG. 3.135].

It has a relatively large, slightly convex, non-articular dorsal surface. The small, palmar surface is also non-articular. This surface

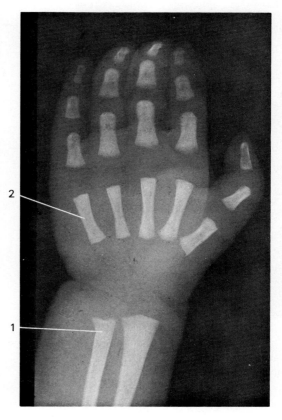

FIG. 3.138. Radiograph of hand at birth.
1. Distal end of ulnar diaphysis.
2. Diaphysis of 5th metacarpal.

is rough and extends on to the lateral aspect of the bone as a V-shaped area between the articular surfaces for the second metacarpal and the trapezium.

The distal articular surface is approximately kidney-shaped and fits into the notch on the base of the metacarpal. The remaining three sides are formed by articular surfaces for the adjacent carpal bones.

The trapezium

The trapezium is the most irregular of the carpal bones. Its main feature is a saddle-shaped articular surface for the metacarpal of the thumb. This faces distally, laterally, and slightly forwards. On the opposite side of the bone, are two articular surfaces set at an angle to each other—a larger one for the trapezoid, a smaller one for the scaphoid. Proximal to the surface for the metacarpal of the thumb, the anterior aspect of the bone carries a distinct ridge (tubercle of the trapezium) which runs longitudinally and has a groove on its medial side for the flexor carpi radialis tendon.

The tubercle can be felt indistinctly, distal to the tuberosity of the scaphoid, through the muscles attached to it [FIG. 3.136]. The flexor

retinaculum splits, to be attached to the tubercle and the opposite edge of the groove, thus enclosing the tendon of flexor carpi radialis.

Ossification and variations

The bones of the carpus are ossified from one centre each. These all appear after birth, so that the carpus is cartilaginous at birth. Centres appear in the first 6 months for the capitate and hamate and thereafter for the triquetrum (2–4 years), lunate (3–5 years), scaphoid, trapezium, and trapezoid (all 4–6 years). The pisiform is late in appearance (between 9 and 14 years). Ossification is not complete until between 20 and 25 years.

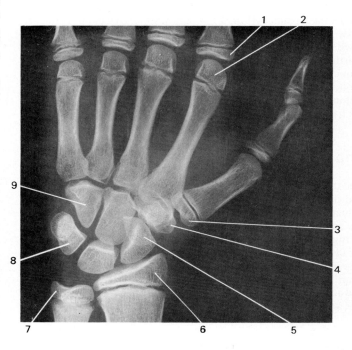

FIG. 3.140. Anteroposterior radiograph of hand. Age 12 years.
1. Epiphysis of proximal phalanx.
2. Epiphysis of metacarpal.
3. Epiphysis of metacarpal of thumb.
4. Trapezium.
5. Scaphoid.
6. Distal epiphysis of radius.
7. Distal epiphysis of ulna.
8. Triquetrum.
9. Hamate.

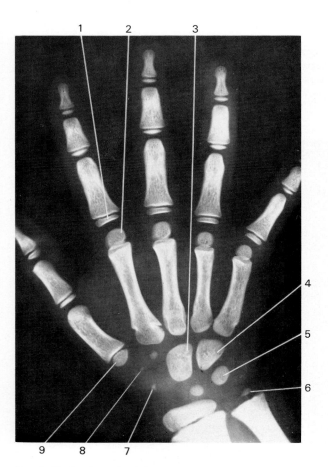

FIG. 3.139. Radiograph of the wrist and hand of a boy aged 7 years.
1. Phalangeal epiphysis.
2. Metacarpal epiphysis.
3. Capitate.
4. Hamate.
5. Triquetrum.
6. Distal epiphysis of ulna.
7. Scaphoid.
8. Trapezium.
9. Epiphysis of first metacarpal, normal position.

The hook of the hamate may be separate. An os centrale, normally incorporated in the scaphoid, may be present between the scaphoid, capitate, and trapezoid. Other small additional nodules may be present, and the styloid process of the third metacarpal may be separate or fused with the capitate or trapezium.

THE METACARPUS

The skeleton of the hand, or metacarpus, consists of five miniature long bones, the metacarpal bones. One corresponds to each digit and they are numbered in sequence from the lateral side, that of the thumb being the first. They extend from the carpus to the fingers and are concealed from the palmar aspect but, though covered by tendons [FIG. 5.88], they can be felt in their whole length on the dorsum of the hand. Their distal ends form the proximal row of the knuckles which become evident when the fist is clenched, and can also be felt in the palm as a series of bony resistances 2 cm proximal to the crease at the root of each finger.

The metacarpal of the thumb differs considerably from the rest (see below). The main features of the remaining metacarpals are that each has a base, approximately cubical in shape with various articular surfaces, a body with a smooth triangular area on the greater part of the posterior surface and a distal end with a large rounded head for articulation with a phalanx.

The **head** of the metacarpal is capped by a smooth, rounded articular surface which extends further on the palmar than the dorsal aspect to allow for the pronounced flexion occurring at the joint. The palmar articular margin is notched in the midline. The base of the flat, triangular area on the dorsal surface of the body extends on to the head, and its corners form **tubercles** which project at each side. The collateral ligaments are attached to these tubercles and to the hollow antero-inferior to each of them.

The **body** is curved with a longitudinal palmar concavity. The sides extend from the edges of the smooth posterior triangle forwards to a palmar border and provide areas for attachment of the interosseous muscles [FIGS. 5.89 and 5.91]. The lateral surface of the first and the medial surface of the fifth receive the opponens muscles, and the palmar border of the third gives origin to the transverse head of the adductor pollicis muscle.

The **bases** of the metacarpals differ from each other considerably. The metacarpals are arranged fanwise round the distal row of the carpal bones, and their bases articulate with the carpals and with each other [FIGS. 3.136 and 3.137]. The consequent variations in

shape provide an easy means of distinguishing them and of associating them with the proper carpal bones. Examine FIGURES 3.141–3.145 for these details.

The metacarpal bases are bound to the carpals by palmar and dorsal carpometacarpal ligaments and, except in the case of the thumb, by palmar, dorsal, and interosseous ligaments. For the attachment of muscles and tendons see FIGURES 3.136 and 3.137

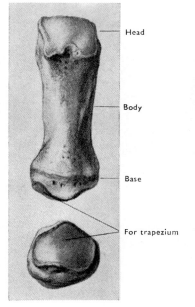

FIG. 3.141. First right metacarpal bone: its base has a *saddle-shaped* facet for articulation with the trapezium.

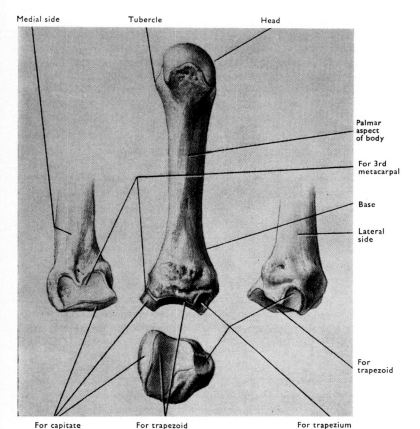

FIG. 3.142. Second right metacarpal bone: its base is *notched* to receive the trapezoid.

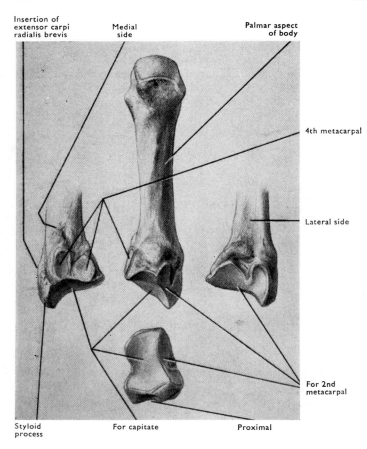

FIG. 3.143. Third right metacarpal bone: its base has a *styloid process* projecting from its posterolateral corner.

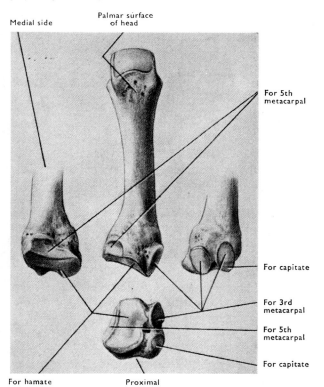

FIG. 3.144. Fourth right metacarpal bone: its base has a *flat facet* for the hamate, sometimes shared with the capitate, and facets on each side for the third and fifth metacarpals.

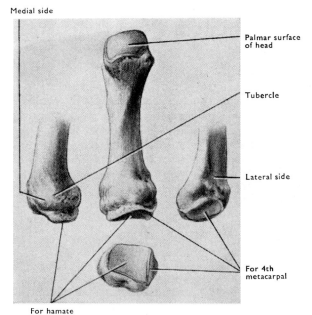

FIG. 3.145. Fifth right metacarpal bone: its base has a proximal facet *concave* in its medial part for articulation with the hamate a facet on its lateral side for the fourth metacarpal and a *tuberosity* on its medial side for attachment of the extensor carpi ulnaris and ligaments.

THE METACARPAL OF THE THUMB

The first metacarpal is shorter than the others and its body is broad and flat [FIG. 3.141]. It is more freely movable than the others, diverges from them and is set with its dorsal surface looking more in a lateral than a posterior direction. Its position facilitates the characteristically human movement of opposing the thumb to other digits.

The **head** is similar to the heads of the other metacarpals except that it is wider. It has the usual tubercles and markings for ligaments at the sides of the head. Two sesamoid bones, normally found in the short tendons which run over the flexor surface of the joint, articulate with the palmar part of the joint surface and may groove it.

The **body** is nearly as wide as the base and has a slightly rounded dorsal surface. Its palmar surface is divided by a blunt ridge into a larger lateral part, which gives attachment to opponens pollicis, and a smaller medial part for one head of the first dorsal interosseous muscle.

The **base** has a saddle-shaped articular surface which fits the corresponding surface on the trapezium and is surrounded by a loose articular capsule. Its dorsal surface is flush with the dorsal surface of the body but its palmar aspect projects considerably. Abductor pollicis longus is inserted into the lateral side of the base.

OSSIFICATION

The metacarpal bones ossify from primary centres which appear in the ninth week and are well ossified at birth. The secondary centres for the second to fifth metacarpals appear in the heads at the age of 2–3 years. The secondary centre for the first appears in the base only slightly later, and all of them fuse between 17 and 19 years [FIG. 3.139].

In rare cases the first metacarpal may have a secondary centre in its head. Even more rarely the second or one of the others may have a secondary centre for its base.

The **styloid process** of the third metacarpal may develop as a separate ossicle.

THE PHALANGES OF THE FINGERS

Each phalanx is a miniature long bone with a **body**, a larger proximal end or **base**, and a smaller distal end or **head** [FIG. 3.146]. Each finger has a skeleton of three phalanges, **proximal**, **middle**, and **distal**, jointed to each other and to the appropriate metacarpal head. The thumb has only two phalanges, proximal and distal.

The sizes of individual bones are so variable that it is not possible to estimate which finger they come from even if a complete set is available for comparisons to be made.

The phalanges of the thumb are shorter and broader than those of the fingers. The middle finger is normally the longest; the little finger is the shortest; the ring and index fingers are nearly equal, but the ring finger is usually a little longer than the index. The relative lengths of the phalanges are such that the tips of the fingers fit across the palm when all the interphalangeal and metacarpophalangeal joints are flexed.

PROXIMAL PHALANX

This is the largest (35–45 mm in length) and has a concave, oval, articular surface on its base for the head of a metacarpal, thus

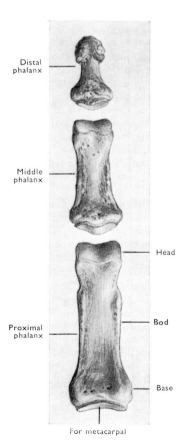

FIG. 3.146. Phalanges of a finger (palmar view).
Comparison with the phalanges of a foot will show that in the foot the bases of the proximal phalanges are relatively large and the bodies are much thinner. The middle and distal phalanges are so short that they should not be mistaken. The bones of the big toe have special features.

allowing abduction and adduction as well as flexion and extension. Except posteriorly, the **base** has an articular capsule attached round the articular surface and collateral ligaments to its sides. The palmar ligament is attached firmly to the edge of the base, only loosely to the metacarpal bone. The bases receive insertions from some of the interosseous muscles (see Muscle Section) with abductor digiti minimi to the medial side of the fifth in place of an interosseus. Flexor digiti minimi joins the abductor.

The **body** is slightly curved along its length and is convex dorsally. It is markedly convex from side to side on its dorsal surface, flat or concave on its palmar surface, and often has its edges raised by the attachment of the fibrous flexor sheath.

The **head** has a pulley-shaped articular surface, wide from side to side with a groove in the middle and markings for the collateral ligaments on each side.

MIDDLE PHALANX

This is similar to a proximal one except that it is shorter and has a ridge on the articular surface of the base. The ridge fits the groove on the head of a proximal phalanx. The edges of its palmar surface are raised and roughened near the middle by the insertion of flexor digitorum superficialis and the fibrous flexor sheath. The dorsal region of the base receives part of the insertion of an extensor expansion.

DISTAL PHALANX

This is small and may be referred to as a terminal or ungual phalanx because it is at the end of the finger and carries a finger nail.

Its **base** is relatively large and has a tendon of flexor digitorum profundus inserted into its palmar aspect, collateral ligaments to its sides, and part of an extensor tendon to its dorsal aspect.

The **body** tapers rapidly to reach a rough spatulate **tuberosity** at the end of the bone.

PHALANGES OF THE THUMB

These are similar to the proximal and distal phalanges of a finger but they are shorter and broader. They are connected to each other and to the metacarpal by similar ligaments.

The muscles attached differ from those going to a finger. The base of the proximal phalanx receives the short muscles of the thumb except the opponens [FIG. 5.89], but including a palmar interosseous muscle and also the tendon of extensor pollicis brevis. The terminal phalanx has the tendons of flexor and extensor pollicis longus attached to it. The fibrous flexor sheath is attached to the edges of the body of the proximal phalanx and to the base of the distal phalanx.

OSSIFICATION

Primary centres for all the phalanges appear between the eighth and twelfth weeks and the phalanges are well formed at birth. The distal phalanges ossify first. Secondary centres appear in the bases in the second or third year and fuse between 17 and 19 years. In rare cases a centre may appear for the head as well as the base.

Occasionally a distal phalanx is bifurcated or a phalanx may be small or absent, and the thumb may have three phalanges.

The sesamoid bones

As a rule the only two sesamoids found in the hand are small nodules in the tendons of insertion of flexor pollicis brevis and adductor pollicis. They blend with the palmar ligament of the metacarpophalangeal joint and move on the head of the metacarpal bone. Occasionally small sesamoids are found in the other metacarpophalangeal joints or in the interphalangeal joint of the thumb. The sesamoids are cartilaginous in childhood and begin to ossify after 13 years.

The bones of the lower limb

The skeleton of the lower limb is connected to the vertebral column by the pelvic girdle which is formed by an **os coxae** or hip bone on each side. They articulate posteriorly with the **sacrum** and meet below and in front at the symphysis pubis. With the sacrum, the two hip bones form the skeleton of the pelvis which surrounds the lowest part of the abdominal cavity.

The hip joint is on the lateral aspect of the hip bone and from this the **femur** or thigh bone extends to the knee. The bulge of the gluteal muscles running between the hip bone and the upper part of the femur form, with their covering tissues, a **buttock** or natis on each side, but the region is usually named **gluteal**, from the muscles.

The free lower limb is divided into **femur** or thigh between the hip and the knee; **crus** or leg, between knee and ankle, and **pes** or foot distal to the ankle joint.

The bones of the leg are the medial, **tibia** and the slender, lateral, **fibula**. They articulate with each other at both ends, but, of the two, only the tibia articulates with the lower end of the femur. Both tibia and fibula take part in the ankle joint. The term peroneal (to be distinguished from perineal) is synonymous with the adjective fibular.

The foot is distal to the ankle joint and is divided into the foot proper and the toes. It is normally set almost at right angles to the leg and so has a superior surface or **dorsum** and an inferior, **plantar surface** which is the sole of the foot. The bones of the foot include the seven **tarsal bones** or tarsus, corresponding to the carpus in the hand, and five **metatarsals** corresponding to the metacarpals.

The big toe or **hallux** is on the medial side and has two phalanges, like the thumb. The remaining toes each have three phalanges.

The toes and the metatarsals are numbered in sequence from the medial side, starting with the hallux.

Variations

The skeleton of the lower limb is adapted for bipedal gait and weight bearing. The bones vary in their characteristics in relation to muscular development and build of the body. The proportions of each bone, its robustness, degree of development of muscular markings, and protuberances vary accordingly. Many bones, but especially the hip bone and to a lesser extent the femur, display sexual differences [p. 183]. Variations in the dimensions of the female pelvis [p. 182] may be important to detect in the management of labour.

The length of the femur is an indication of stature: roughly, stature is ×4 the length of a femur. More accurate estimates use regression formulae derived from particular measurements of long bones from individuals whose stature was known before death. Trotter and Gleser (1952, 1958) and Krogman (1962) critically discuss the use of measurements of bones to assess stature, age, sex, race, and identity.

The tibia of prehistoric and primitive peoples may be compressed from side to side (platycnemia).

Additional ossicles, or divided navicular or cuneiform bones, may be found in the tarsus. An **os trigonum** may be present at the back of the talus.

THE HIP BONE

The hip bone [FIGS. 3.147 and 3.149] consists of three parts, the ilium, ischium, and pubis, fused together. The ilium is the broad, sinuously curved portion which joins the other two at the acetabulum, a cup-shaped cavity situated on the lateral aspect of the bone. The ilium is superior and is set at right angles to the other two parts.

The ischium and pubis are fused with each other and with the ilium in the acetabulum [FIG. 3.148]. The ischium is the heavier of the two parts; the pubis is lighter with a flattened body anteriorly. The two parts, pubis and ischium, surround the oval or triangular obturator foramen by fusing with each other.

The ilium

The ilium is the bone felt in the flank 4–6 cm inferior to the ribs. It is somewhat fan-shaped with the hinge of the fan in the acetabulum. The margins of the fan form anterior and posterior borders with a curved superior border or iliac crest between them.

The sinuous **iliac crest** is palpable in the flank and separates the abdominal wall from the gluteal region. Its anterior end is the **anterior superior iliac spine** and can be felt as a landmark at the lateral end of the groove of the groin [FIG. 5.104]. It gives attachment to sartorius and to the inguinal ligament which lies deep to that groove. The iliac crest has an external and an internal lip, with an intermediate line between them. The **external lip** is prominent 5–7 cm above and behind the anterior superior spine and this forms another palpable landmark, the **tubercle of the crest**. The **internal lip** becomes very indefinite posteriorly. The crest widens posteriorly (tuberosity of the ilium) and ends as a **posterior superior iliac spine** which is at the bottom of a dimple level with the second sacral spine, 4 cm from the median plane. The muscles of the lateral and posterior abdominal walls and the fascia lata of the thigh are attached to the iliac crest [FIGS. 3.148 and 3.150].

The anterior border of the ilium extends downwards and backwards from the anterior superior spine to the anterior margin of the acetabulum. Just above the acetabulum it has a considerable prominence called the **anterior inferior iliac spine** [FIG. 3.147]. Below and medial to this, a low iliopubic eminence on the outer wall of the acetabulum marks the junction of the ilium and pubis. Rectus femoris is attached to the upper half of the anterior inferior spine, and the strong iliofemoral ligament to the lower half [FIG. 3.148].

The posterior border of the ilium runs down from the posterior spine for a short distance to the **posterior inferior iliac spine**. Thence it arches forwards, and then downwards and backwards (the **greater sciatic notch**) to meet the posterior border of the ischium [FIG. 3.147]. Where the two bones join, the outer surface of the acetabulum again shows a rounded eminence. The principal attachments are ligamentous to the posterior spines, particularly the sacrotuberous ligament to the inferior spine, but piriformis gains some attachment to the edge of the notch.

The **gluteal surface** of the ilium is the undulating, external surface between the acetabulum and the crest. Three distinct **gluteal lines**, formed by the fascia between the gluteal muscles, curve upwards and forwards over it from the deepest part of the greater sciatic notch. The **inferior gluteal line** arches towards the anterior inferior iliac spine. The **anterior gluteal line** curves up and then forwards towards the anterior superior iliac spine. The **posterior gluteal line** sweeps backwards and then upwards and forwards close to the external lip of the crest for 5–7 cm [FIG. 3.147].

There is a rough mark above the acetabulum for the reflected head of rectus femoris. The areas giving attachment to the glutei and tensor fasciae latae are shown on FIGURE 3.148.

The medial aspect of the ilium [FIG. 3.149] consists of: (1) a smooth, concave **iliac fossa** occupied by iliacus, which arises from the upper, larger part of its surface; and (2) a **sacropelvic surface**, which lies postero-inferior to the iliac fossa and at an angle to it. The sacral portion has a rough L-shaped articular area (**auricular surface**) with well-defined edges, which articulates with the sacrum at the sacro-iliac joint. Posteriorly the joint reaches the posterior inferior iliac spine, and antero-inferior to the joint surface there may be a preauricular sulcus, especially in female bones.

Between the auricular surface and the posterior part of the crest, the rough bone is raised to form the **iliac tuberosity**, to which the sacro-iliac ligaments and erector spinae are attached. The iliolumbar ligament reaches the ilium at a triangular area between the iliac fossa, the crest, and the tuberosity [FIG. 3.150].

The remaining pelvic portion is smooth and forms the part of the bony wall of the lesser pelvis [p. 180] between the greater sciatic notch and the curved **arcuate line** [FIG. 3.150]. The arcuate line separates iliac fossa from lesser pelvis and is continuous with the pecten pubis to form the iliopubic part of the linea terminalis [p. 180].

The ischium

The ischium is the L-shaped part of the hip bone which passes down from the acetabulum and then turns forwards to join the pubis. The thick vertical part which enters into the acetabulum is the **body**. The lighter arm which joins the corresponding part of the pubis is the **ischial ramus**. Posteriorly, on the angle between these parts and extending up on the body, is a rough area, the **ischial tuber** (or tuberosity), which gives attachment to the hamstring tendons. A lateral impression is for semimembranosus, and a vertical line separates it from a more medial impression for semitendinosus and biceps (Martin 1968). An inferior impression for the hamstring part of adductor magnus is placed laterally on the inferior part of the tuberosity. The medial part is covered by fibrous tissue and a bursa. The medial border of the tuberosity gives attachment to the strong sacrotuberous ligament which is continued (falciform process) along the medial margin of the ischial ramus, raising a small ridge. Quadratus femoris arises from the lateral border of the tuberosity, which is deep to the edge of gluteus maximus when the thigh is extended.

Superiorly, the **body of the ischium** forms the postero-inferior part of the acetabulum and joins the ilium at the rounded dorsum acetabuli. Between the acetabulum and the ischial tuberosity, the medial margin of the bone projects backwards and medially as a triangular **ischial spine** [FIG. 3.147]. This separates the **greater sciatic notch** above from a **lesser sciatic notch** below (sciatic is derived from ischiadic). The lesser notch, between the ischial spine and the ischial tuberosity, is smooth where the obturator internus tendon turns over it.

The sacrospinous ligament runs to the tip of the ischial spine and, with the sacrotuberous ligament and sacrum, converts the sciatic notches into foramina. The **pelvic surface** of the ischium is smooth where it forms the wall of the pelvis and is partly covered by the origin of the obturator internus [FIG. 3.150]. Coccygeus and levator ani take origin from the pelvic surface of the spine and the two gemelli from the margins of the lesser notch. The ramus of the ischium passes forwards and

upwards to meet the inferior ramus of the pubis and form the **conjoined ramus**.

The pubis

Each pubis has a flattened **body**, and together with the disc between them, they form the antero-inferior wall of the lesser pelvis at the lowest part of the front of the abdomen. The **superior ramus** is a bar of bone passing upwards and laterally to form the antero-inferior part of the acetabulum. The **inferior ramus** passes backwards, laterally, and slightly downwards to join the ramus of the ischium.

The body of the pubis has a smooth pelvic surface, slightly convex anteroposteriorly, which faces upwards and only slightly backwards. It has the anterior end of the levator ani muscle and fascial ligaments for the bladder or prostate attached to it, but these do not mark it appreciably. Medially, it has a rough oval **symphysial surface** for the attachment of a fibrocartilaginous interpubic disc. This connects it to its neighbour and so forms the pubic symphysis. The inferior or femoral surface faces into the thigh and is rough for muscle attachments (see below). Postero-inferiorly, the body becomes continuous with the inferior ramus and the arcuate ligament arches between the two sides. The lateral edge of the body is thin and helps to bound the obturator foramen.

The anterior border of the body of the pubis is thickened to form a **pubic crest** which receives tendons and aponeuroses of the abdominal muscles [FIG. 5.105]. Its lateral end projects as the **pubic tubercle** 2.5–3 cm from the median plane. It provides the main pubic attachment for the inguinal ligament and is a useful landmark.

The superior ramus widens as it passes laterally to reach the iliopubic eminence and unite with the ilium above and the ischium behind at the acetabulum. The part between the body and the acetabulum is three-sided. Its pelvic surface is smooth and continuous with the rest of the pelvic surface of the bone. It is limited at the superior aperture of the pelvis by the **pecten pubis**, a sharp edge which runs laterally from the pubic tubercle to the iliopubic eminence. It provides a continuation of the line of attachment of the structures going to the pubic crest, including the conjoint tendon, the lacunar ligament, the reflected ligament [FIG. 5.108], the pectineal ligament [q.v.], and the fascia transversalis.

The triangular area between the pubic tubercle and the iliopubic eminence (**pectineal surface**) gives attachment to the fleshy fibres of pectineus [Fig. 3.148]. Anteriorly, it is limited by the **obturator crest**, a ridge which runs from the pubic tubercle to the lower part of the acetabulum and provides attachment for the pubofemoral ligament. Below this, the femoral surface of the body becomes continuous with the inferior aspect of the ramus at the **obturator groove** (see below).

The **conjoined ramus** of the pubis and the ischium forms the lower boundary of the obturator foramen, and its superior edge is thin where the obturator membrane is attached. Its *femoral surface* is continuous with that of the body of the pubis in front and the body of the ischium behind. The muscle attachments are continued on to these surfaces. The obturator externus is attached near the edge of the foramen and to the membrane. The adductor group of muscles have their origin inferior and anterior to obturator externus [FIG. 3.148] and the fascia lata is attached to the inferior edge of the bone.

The **medial surface** gives attachment to obturator internus superiorly [FIG. 3.150]. Inferiorly, the bone is everted and grooved by the attachment of the crus of the penis or clitoris. The superior edge of the groove has the **inferior fascia of the urogenital diaphragm** (perineal membrane) attached to it, and is continuous posteriorly with the ridge for the falciform part of the sacrotuberous ligament. Ischiocavernosus and transversus perinei superficialis and the fascia covering them are attached to the posterior end of the

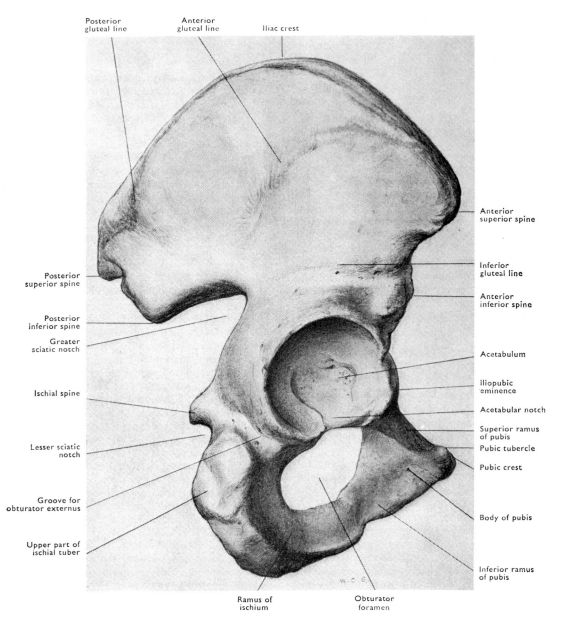

Posterior
gluteal line

Anterior
gluteal line

Iliac crest

Anterior
superior spine

Inferior
gluteal line

Anterior
inferior spine

Posterior
superior spine

Posterior
inferior spine

Greater
sciatic notch

Acetabulum

Iliopubic
eminence

Ischial spine

Acetabular notch

Superior ramus
of pubis

Lesser sciatic
notch

Pubic tubercle

Pubic crest

Groove for
obturator externus

Body of pubis

Upper part of
ischial tuber

Inferior ramus
of pubis

Ramus of
ischium

Obturator
foramen

FIG. 3.147. Right hip-bone (lateral aspect).

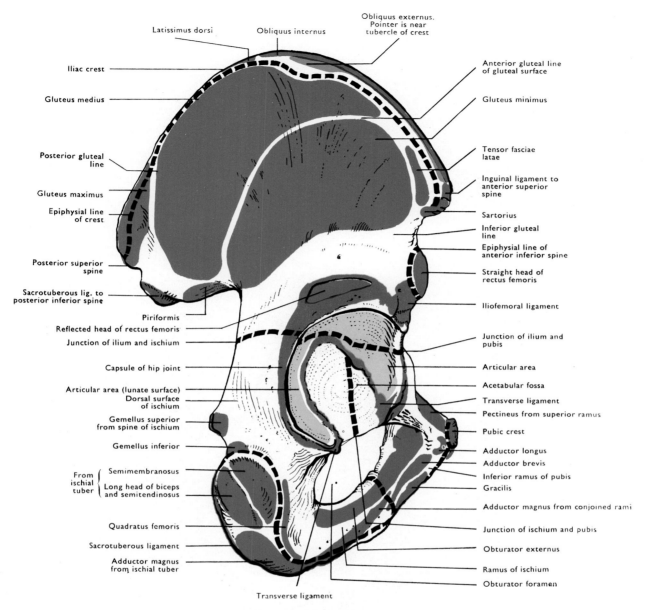

Latissimus dorsi

Obliquus internus

Obliquus externus. Pointer is near tubercle of crest

Iliac crest

Gluteus medius

Anterior gluteal line of gluteal surface

Gluteus minimus

Posterior gluteal line

Tensor fasciae latae

Gluteus maximus

Inguinal ligament to anterior superior spine

Epiphysial line of crest

Sartorius

Inferior gluteal line

Posterior superior spine

Epiphysial line of anterior inferior spine

Straight head of rectus femoris

Sacrotuberous lig. to posterior inferior spine

Iliofemoral ligament

Piriformis

Reflected head of rectus femoris

Junction of ilium and ischium

Junction of ilium and pubis

Capsule of hip joint

Articular area

Articular area (lunate surface)

Acetabular fossa

Dorsal surface of ischium

Transverse ligament

Pectineus from superior ramus

Gemellus superior from spine of ischium

Pubic crest

Gemellus inferior

Adductor longus

Adductor brevis

From ischial tuber { Semimembranosus

Inferior ramus of pubis

Long head of biceps and semitendinosus

Gracilis

Adductor magnus from conjoined rami

Quadratus femoris

Junction of ischium and pubis

Sacrotuberous ligament

Obturator externus

Adductor magnus from ischial tuber

Ramus of ischium

Obturator foramen

Transverse ligament

FIG. 3.148. Right hip-bone (lateral aspect).

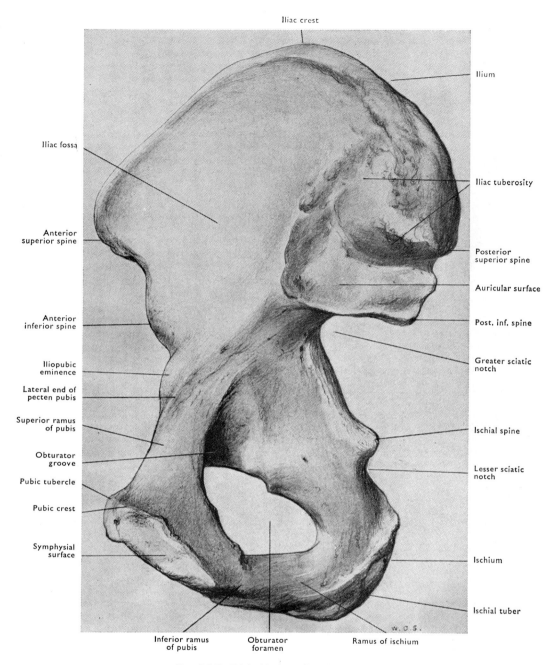

Iliac crest

Ilium

Iliac fossa

Iliac tuberosity

Anterior
superior spine

Posterior
superior spine

Auricular surface

Anterior
inferior spine

Post. inf. spine

Iliopubic
eminence

Greater sciatic
notch

Lateral end of
pecten pubis

Superior ramus
of pubis

Ischial spine

Obturator
groove

Lesser sciatic
notch

Pubic tubercle

Pubic crest

Symphysial
surface

Ischium

Ischial tuber

W.O.S.

Inferior ramus
of pubis

Obturator
foramen

Ramus of ischium

FIG. 3.149. Right hip-bone (medial aspect).

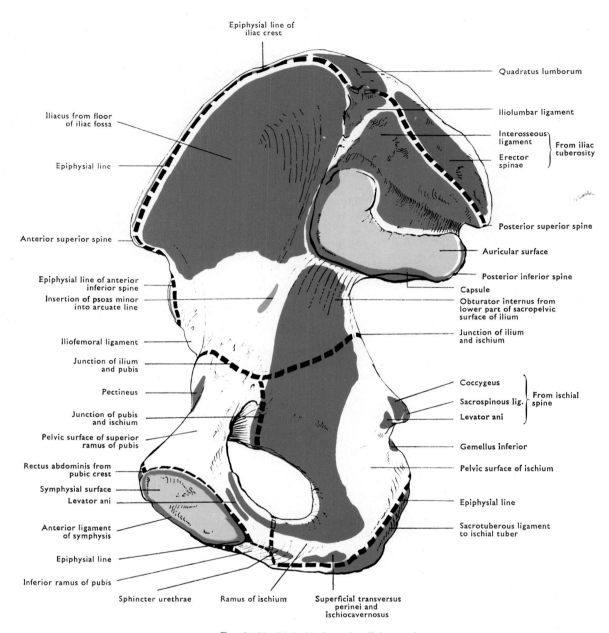

Epiphysial line of
iliac crest

Quadratus lumborum

Iliacus from floor
of iliac fossa

Iliolumbar ligament

Interosseous
ligament

Erector
spinae

From iliac
tuberosity

Epiphysial line

Posterior superior spine

Anterior superior spine

Auricular surface

Posterior inferior spine

Epiphysial line of anterior
inferior spine

Capsule

Insertion of psoas minor
into arcuate line

Obturator internus from
lower part of sacropelvic
surface of ilium

Junction of ilium
and ischium

Iliofemoral ligament

Junction of ilium
and pubis

Coccygeus

Sacrospinous lig.

From ischial
spine

Pectineus

Levator ani

Junction of pubis
and ischium

Gemellus inferior

Pelvic surface of superior
ramus of pubis

Pelvic surface of ischium

Rectus abdominis from
pubic crest

Symphysial surface

Epiphysial line

Levator ani

Sacrotuberous ligament
to ischial tuber

Anterior ligament
of symphysis

Epiphysial line

Inferior ramus of pubis

Sphincter urethrae   Ramus of ischium   Superficial transversus
perinei and
ischiocavernosus

Fig. 3.150. Right hip-bone (medial aspect).

inferior edge, whilst the muscles of the deep perineal space are attached above the inferior fascia of the urogenital diaphragm.

The obturator foramen is the oval or triangular space surrounded by the bodies and rami of the pubis and ischium. It lies inferior to the acetabulum and is nearly closed by the obturator membrane. The membrane gives origin to the muscles attached to the margins of the foramen, obturator externus, and obturator internus. The obturator groove under the superior ramus of the pubis leads to a gap between the anterior part of the membrane and the bone, through which the obturator nerve and vessels emerge. This free edge of the membrane may produce a small anterior obturator tubercle where it reaches the superior ramus of the pubis and, more frequently, a rough posterior obturator tubercle where it reaches the junction of pubis and ischium.

The acetabulum is the rounded pit situated on the lateral side of the hip bone over the Y-shaped junction of ilium, ischium, and pubis. Its upper part forms a projecting surface on the lower end of the ilium for transmission of weight to the femur. A fibrocartilaginous labrum acetabulare is attached to the bony rim within the capsule and deepens the cavity. The inferior wall of the pit is notched (acetabular notch) and this is bridged by a transverse acetabular ligament which completes the rim but leaves a gap between the ligament and the bone.

The floor of the acetabulum above the notch is rough and depressed to form the acetabular fossa. The remainder, between rim and fossa, is a smooth horseshoe-shaped, articular, lunate surface for the head of the femur. The synovial ligament of the head of the femur spreads out and is attached to the transverse ligament and the margins of the notch and fossa.

Ossification

The hip bone is ossified before birth from three primary centres: (1) above the greater sciatic notch of the ilium early in the third month; (2) in the ischium, below the acetabulum, in the fourth or fifth month; (3) in the superior ramus of the pubis, in the fifth or sixth month.

At birth ossification has spread to include part of the acetabulum [FIG. 3.151]. By the tenth year the cartilage in the acetabulum is

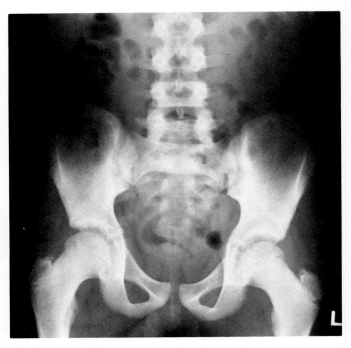

FIG. 3.152. Anteroposterior radiograph of male pelvis. Age 17.

reduced to a triradiate strip, growth of which separates the three components and increases the size of the acetabulum. Secondary centres appear in the acetabular cartilage, and the triradiate strip is usually consolidated by the age of puberty or shortly after. The bones of the ischial and pubic rami normally unite by the tenth or eleventh year, but may be delayed until about 16 years, or may fail to unite.

The remaining secondary centres are shown in FIGURE 3.151: all appear by the age of puberty and join between 20 and 25 years of age.

THE PELVIS

The pelvis is the lower division of the abdomen [Fig. 6.44] and its skeleton is referred to as the bony pelvis or simply the pelvis. The bony pelvis is made up of the two hip bones, the sacrum, and the coccyx, bound together by ligaments. The walls are padded with muscles and other structures so that the living pelvis is very different from a bony specimen. The term pelvis means a basin, but the pelvis, in the anatomical position, is tilted so that: (1) the anterior superior iliac spines and the pubic tubercles are in the same coronal plane; (2) the coccyx is level with the upper margin of the symphysis pubis; and (3) the anterior part of the pelvic surface of the sacrum is set at an angle of only 10 or 12 degrees from the horizontal.

The primary functions of the bony pelvis are to provide a base for attachment of the lower limb and trunk muscles; to transmit the weight of the body from the vertebral column to the femora; and to contain, support, and protect the pelvic viscera. Childbirth in a female demands a cavity of the pelvis large enough to allow passage of a full term foetus.

The pelvis consists of two parts, the greater pelvis, and the lesser pelvis. The plane separating these two parts passes through the arcuate line of the ilium and the pecten pubis at each side, the lateral parts and promontory of the sacrum behind, and the upper part of the pubic bodies in front. All these are grouped together as the linea terminalis, and surround the superior aperture of the lesser pelvis.

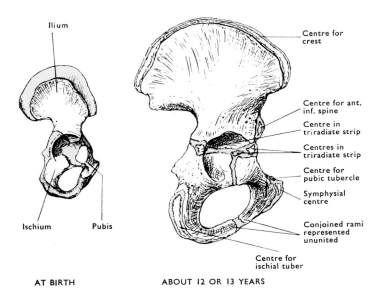

Ilium

Centre for crest

Centre for ant. inf. spine

Centre in triradiate strip

Centres in triradiate strip

Centre for pubic tubercle

Symphysial centre

Conjoined rami represented ununited

Centre for ischial tuber

Ischium   Pubis

AT BIRTH   ABOUT 12 OR 13 YEARS

FIG. 3.151. Ossification of hip-bone.
The more anterior of the secondary centres in the acetabular cartilage may remain separate as an os acetabuli.

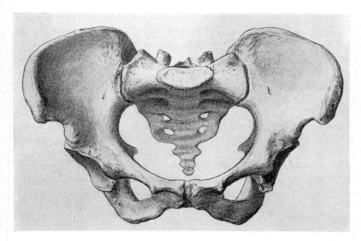

FIG. 3.153. Male pelvis seen from in front.

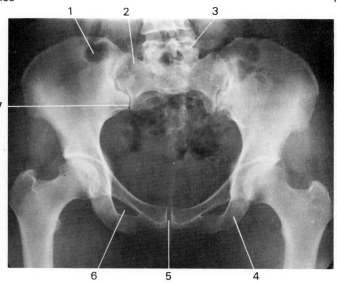

FIG. 3.155. Anteroposterior radiograph of adult pelvis, female.
1. Gas in caecum overlying ilium.
2. Lateral part of sacrum.
3. 5th lumbar vertebra.
4. Ischial tuberosity.
5. Pubic symphysis.
6. Obturator foramen.
7. Sacro-iliac joint.

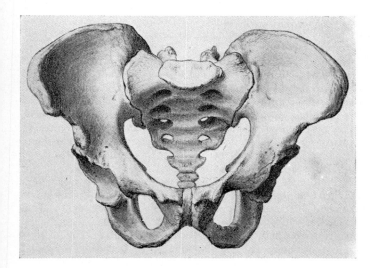

FIG. 3.154. Female pelvis seen from in front.

The greater pelvis is the upper part, formed mainly by the iliac fossae, and is the skeleton of the lower part of the abdomen proper. It is partly filled by the iliopsoas muscles.

The lesser pelvis is the cavity behind and below the linea terminalis, and its contents are usually referred to as the contents of the pelvis. Its posterosuperior wall is formed by the sacrum and coccyx and so measures 12–15 cm long and is concave. Its lateral wall is formed by the obturator membrane, the smooth surfaces of all three bones below the linea terminalis, and the gap between them and the sacrum and coccyx, partly filled by the sacrospinous and sacrotuberous ligaments [FIGS. 3.153 and 4.44], which complete the sciatic foramina.

The greater sciatic foramen is partly filled by the piriformis, the lesser sciatic foramen by obturator internus. In the lower part of the lateral wall, the obturator membrane is covered, as is most of the bony lateral wall, by obturator internus.

The antero-inferior wall is formed by the pubic bodies and the interpubic disc between them. It is convex and only 4–5 cm long and faces more upwards than backwards into the cavity. The conjoined rami pass backwards, laterally, and a little inferiorly to form the pubic arch, and where they diverge from the disc is the subpubic angle. The rami are united here by the arcuate ligament of the pubis, and after a slight gap, by the urogenital diaphragm [FIG. 5.119] which stretches between them.

The cavity of the lesser pelvis is a short, wide canal, curved in conformity with the sacrum and coccyx and is continuous with the abdominal cavity at the superior aperture of the pelvis. This is bounded by the linea terminalis and looks forwards and slightly upwards. The plane of the superior aperture makes an angle of about 60 degrees with the horizontal.

The inferior aperture of the pelvis is bounded by the coccyx behind; by the pubic symphysis in front; and on each side by the sacrotuberous ligament and the conjoined ramus. These are also the boundaries of the perineum. The plane of the inferior aperture of the pelvis makes an angle of only 10–15 degrees with the horizontal.

Because the cavity of the lesser pelvis is curved, numerous planes of the cavity can be described in addition to the planes of the apertures, including one through the middle of the third sacral vertebra and the middle of the pubic symphysis.

The spines of the ischia project a little into the pelvis and, when this is pronounced, the plane passing through the spines and the posterior margin of the pubic symphysis may be of importance as it lies in the narrowest part of the canal.

The axis of the superior aperture is at right angles to the centre of its plane and, if produced, would pass approximately through the umbilicus and the tip of the coccyx. The axis of the inferior aperture drawn similarly is directed towards the promontory of the sacrum. The axis of the pelvis is a line drawn through the centres of all the planes and is therefore curved and nearly parallel to the surface of the sacrum and coccyx.

Measurements of the pelvis

The intercristal diameter is the furthest distance between the outer lips of the iliac crests. The interspinous diameter is the distance between the anterior superior iliac spines. The external anteroposterior diameter is measured between the first sacral spine and the anterior (upper) margin of the pubic symphysis. These three measurements are usually slightly greater in a man than in a woman but the difference between intercristal and interspinous measurements is more important than their absolute size and should be more than 3–4 cm in a normal female.

The diameters of the lesser pelvis are of more importance, though not easily measured in the living subject. They are measured anatomically as follows.

DIAMETERS OF THE SUPERIOR APERTURE

Anteroposterior or conjugate: from the middle of the sacral promontory to the anterior (upper) margin of the symphysis pubis. Transverse: across the aperture where it is widest. Oblique: from one sacro-iliac joint to the iliopubic eminence of the opposite side.

In the living subject the anteroposterior or conjugate measurement may be increased by extension at the sacro-iliac joint.

DIAMETERS OF THE CAVITY

Anteroposterior: from the centre of the pelvic surface of the middle piece of the sacrum to the middle of the pelvic surface of the symphysis pubis. Transverse: across the cavity where it is widest. Oblique: from the posterior end of the sacro-iliac joint to the centre of the obturator membrane of the other side.

DIAMETERS OF THE INFERIOR APERTURE

Anteroposterior: from the tip of the coccyx to the posterior (lower) margin of the symphysis pubis (variable owing to mobility of the coccyx). Transverse: between the ischial tuberosities. Oblique: from the crossing of the sacrotuberous and sacrospinous ligaments to the junction of pubic and ischial rami on the other side.

The diameters of the lesser pelvis should be greater in a woman than in a man, but measurements vary considerably within one sex and there may be some overlap. The term conjugate applied to the anteroposterior diameters should only be applied to the superior aperture of the lesser pelvis.

In obstetrical work two other conjugates are used in addition to the anatomical or true conjugate measured to the anterior margin of the symphysis. The obstetric or available conjugate is measured from the promontory of the sacrum to the nearest part of the posterosuperior surface of the symphysis, and is almost 0.5 cm shorter. The diagonal conjugate is taken to the postero-inferior margin of the arcuate ligament of the pubis. It provides a rough estimate of the available conjugate on deduction of 0–3 cm, depending on the depth and angle of the pubic symphysis and the height of the sacral promontory (Chassar Moir 1964).

The figures in the following table are *average* measurements on skeletal material in centimetres (Bryce 1915) and serve to indicate the differences in proportion.

	Female			Male		
Intercristal	28			29		
Interspinous	25			24		
External anteroposterior	18			18		
Lesser Pelvis						
	Superior aperture	Cavity	Inferior aperture	Superior aperture	Cavity	Inferior aperture
Anteroposterior	11	13	11	10	11	9
Oblique	13	14	11	12	11	10
Transverse	13	13	12	13	12	9

The coccyx is normally movable so that it may increase (or reduce) the anteroposterior measurement of the inferior aperture by its movement by as much as 2.5 cm. Ligaments are not rigid structures, thus the oblique measurement of the inferior aperture is not a constant one.

Nicholson (1943, 1945) measured 640 living subjects, all primigravidae, by means of stereometric radiography. He measured the obstetrical conjugate and the transverse diameter of the pelvis, and found a decrease in the obstetrical conjugate during the First World War (when nutrition was poor) compared with the subsequent 7 years, though the transverse diameter was not similarly affected.

Available figures indicate that not only do pelvic measurements vary between selected groups, but (see Nicholson) may vary within homogeneous groups as a result of various conditions.

Sexual differences

The female pelvis differs from the male in many particulars and distinctive sex characters are already present in the foetal pelvis by the third or fourth month of intra-uterine life (Boucher 1957). They are poorly marked in childhood but become fully developed after puberty (Krogman 1962).

The essential differences are in the lesser pelvis. In the female it is absolutely wider in all its diameters, particularly those of the inferior aperture. The cavity in the female is roomier and more rounded, shorter and less funnel-shaped.

It is difficult to find samples which are female in all respects. Pelves are often intermediate in many characters, or are female in type in one part (e.g. anteriorly) but show male characters elsewhere. It is advisable to know the types of pelvis to expect and the significance of particular features in obstetrics.

In the normal adult female pelvis the pubic arch is wide with a rounded apex and can be represented fairly well by the angle between outstretched thumb and forefinger. The depth from pectineal line to ischial tuberosity averages 90 mm. In the male the angle is narrow and is generally less than 90 degrees. The depth is greater, averaging 101 mm. A wide space between the ischial tuberosities and a shallow pelvis increases the subpubic angle. A slender ramus set at the side of a pubic body widened by growth at the pubic symphysis causes a wide, rounded apex to the subpubic angle. The subpubic angle, then, indicates the form of the ischiopubic part of the pelvis and, in particular, a narrow angle indicates that the side-walls converge towards the inferior aperture. A wide angle indicates a female type of pelvis anteriorly but is not a reliable indication of the size of the pelvic cavity.

In the female the greater sciatic notch is wide (about 90 degrees); in a massive male pelvis its outline is U-shaped. Massive bone and an extensive sacro-iliac articulation tend to depress the posterior part of the upper margin of the notch and the posterior inferior iliac spine, and a sacrum with its lower end tilted in has a similar effect on the margin of the notch. A wide angle then indicates a roomy posterior part of the pelvis, a narrow angle indicates a long, narrow pelvis.

In the female the space between the spine of the ischium and the sacrum should accommodate three fingers with ease; in the male the distance is about two finger-breadths, indicating a short sacrospinous ligament and a spine which projects into the cavity. For further information on this subject and its clinical and other applications see Caldwell and Moloy (1933), Caldwell, Moloy, and D'Esopo (1934, 1935), and Krogman (1962).

Abnormal softening of bone (e.g. in rickets) can produce a flat pelvis due to compression by the bodyweight, with a reduced anteroposterior diameter while, at the same time, widening the measurement between the anterior superior iliac spines and reducing the difference between it and the intertubercular measurement.

There are numerous differences in detail but *in the female* the bones are lighter and thinner, and muscular markings are less evident. The ilium is less everted and the iliac fossa shallower. The superior aperture is larger and kidney-shaped rather than heart-shaped. The plane of the superior aperture forms a greater angle with the horizontal and the tilt of the pelvis is greater so that the anterior superior iliac spine may be further forward than the pubic tubercles. The sacral promontory projects less and the pubic tubercles are wider apart. The cavity is less funnel-shaped, shorter, and roomier. The sacrum is shorter and wider and its curve is less uniform, being flatter anteriorly; each lateral part nearly equals the first sacral body in its transverse measurement. The inferior aperture is much larger, the coccyx is more movable, and the ischial spine is less inturned. The greater sciatic notch is wider. The obturator foramen is smaller, and is often triangular rather than oval. The acetabula are further apart and are smaller, especially relatively, for the femoral head is small and the hip bone is large. The width of the acetabulum is 2–3 cm less than the distance between its anterior border and the symphysis. A pre-auricular sulcus is more commonly present than in the male.

Growth of the pelvis

At birth the pelvis is relatively small in both sexes. The narrow ilium is more upright, and the shallow iliac fossa looks more forwards than in the adult. The subpubic angle is more acute, the ischia are closer together and the cavity is consequently more funnel-shaped. The sacrum is narrow, more upright and less curved and its promontory projects less. The lower limbs are only a quarter of the length of the body but after birth they grow rapidly, especially after the first year when the child learns to walk. The whole pelvis keeps pace with the lower limbs and, since it gives origin to muscles moving the limbs, grows concurrently to provide adequate attachment. At birth, the lesser pelvis is small, and the urinary bladder and the uterus and ovaries are in the abdomen proper. By the sixth year the pelvis contains the greater part of them, but the pelvic organs are not wholly pelvic until puberty.

Radiographic studies of the pelvis in early infancy (Reynolds 1945) have shown that the pelvis of boys appears to be larger in measurements relating to the outer structures, whereas that of girls is large in measurements relating to the inner structures of the pelvis, including a relatively larger inlet. At puberty growth is modified to produce the distinctive sex characters of the adult pelvis. Such growth takes place at the cartilaginous surfaces of the bones in the acetabulum, at the sacro-iliac joint and the symphysis pubis and between the pubic and ischial rami, and commences in the prepubertal period (Reynolds 1947). Absorption and accretion on the surfaces produces modelling such as is required to maintain proper strength and shape (Payton 1935).

Changes occur when the child, learning to walk, straightens out the lower limbs and adopts the erect posture. The upper part of the pelvis tilts forwards and the anterior sacrovertebral angle becomes more obtuse. The base of the sacrum, taking the weight of the trunk obliquely, sinks more deeply between the hip bones to accentuate the sacrovertebral angle still further. The sacrum becomes more curved as the base is thrust downwards, while the sacrotuberous and sacrospinous ligaments hold down the apical part. At the same time the hip bone, where it bounds the lesser pelvis, becomes curved so as to widen the cavity. The sacrum lies between the upper ends of the V-shaped pelvis, the apex of the V being at the symphysis.

The sacrum is slung to the upper ends of the V by ligaments and the weight of the trunk acting through the sacrum, draws the ends medially, thus tending to lever out the lower part of the wall and so widen the pelvis. Growth between the bones and remodelling are probably more important factors.

The acetabulum is relatively shallower at birth than at the sixth month of intra-uterine life, but becomes deeper and makes the joint more stable, especially when the child begins to run and jump actively. The part of the ilium between the auricular surface and the acetabulum grows stronger and the arcuate line on the ilium becomes better defined in association with the transmission of weight to the femur.

RADIOLOGY OF THE PELVIS

The hip bones and sacrum are well ossified at birth, but are widely separated by cartilage at the sacro-iliac joints and pubic symphysis. The three parts of each hip-bone are widely separated for the first two years by a Y-shaped growth cartilage where they meet in the acetabulum. Such growth cartilages persisting between primary centres of ossification grow towards both, unlike the growth cartilages between primary and secondary (epiphyses) centres of ossification which only grow towards the primary centre. The Y-shaped cartilage gradually narrows, but only disappears after adolescence when the bones fuse as the adult size of the acetabulum is attained [FIGS. 3.166 and 3.167]. The synchondrosis between the ramus of the ischium and the inferior ramus of the pubis is often associated with an overgrowth of cortical bone in children about eight years of age. This may be mistaken for a reaction to injury or disease.

Epiphyses appear in the cartilage of the crest of the ilium and the ischial tuberosity about puberty, and fuse with the hip-bone between the late teens and twenty-five years, the epiphysis on the tuberosity before that on the crest.

Radiographs of the adult *male pelvis* show a relatively narrow ilium and a narrow subpubic angle. The *female pelvis* has wider ilia, a more rounded superior aperture of the lesser pelvis, and a wider subpubic angle representing a greater width to height ratio in the female pelvis than in the male [FIG. 3.155]. These differences are obvious in the majority of cases, but some pelves are intermediate in their characteristics.

Important obstetrical measurements of the anteroposterior diameter of the lesser pelvis are obtained from lateral radiographs. These are taken at the superior aperture, at the mid-plane, and at the inferior aperture (inferior aspect of pubis to coccyx), and help to determine any disproportion between foetal head and maternal pelvis. The magnification due to the divergent X-ray beam has to be taken into account.

THE FEMUR

The femur extends from a rounded head, which fits the acetabulum, down to the knee joint where the two large condyles on its inferior end have extensive articular surfaces for the tibia. Its body is long and strong enough to take the full weight of the trunk and much more. The body has a thick tubular wall and is slightly convex forwards with a ridge of bone, the linea aspera, which buttresses it posteriorly.

The upper end consists of the head, a neck which passes downwards and laterally to the body and the greater and lesser trochanters, both situated at the junction of the neck and body.

The head forms about two-thirds of a sphere, and is articular except for a small superolateral area adjacent to the neck and at a pit (fovea capitis femoris) on its posteromedial aspect, where the

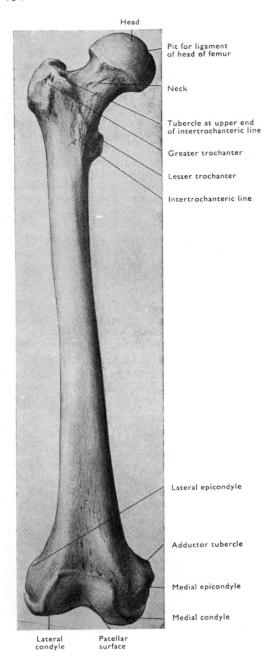

Head

Pit for ligament
of head of femur

Neck

Tubercle at upper end
of intertrochanteric line

Greater trochanter

Lesser trochanter

Intertrochanteric line

Lateral epicondyle

Adductor tubercle

Medial epicondyle

Medial condyle

Lateral   Patellar
condyle   surface

FIG. 3.156. Right femur (anterior aspect).

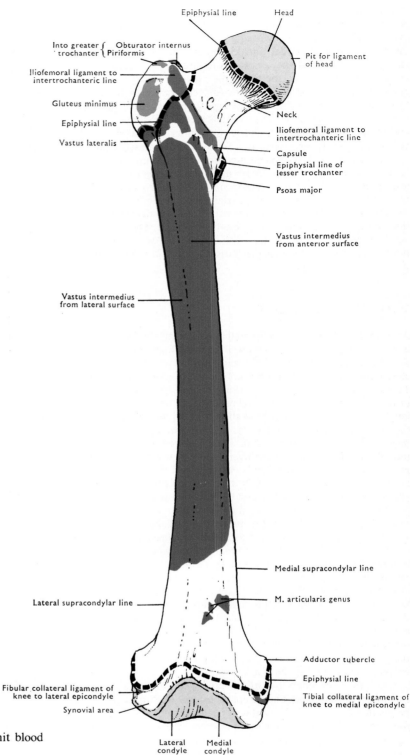

Epiphysial line   Head

Into greater ⌠ Obturator internus
trochanter ⌡ Piriformis

Iliofemoral ligament to
intertrochanteric line

Gluteus minimus

Epiphysial line

Vastus lateralis

Pit for ligament
of head

Neck

Iliofemoral ligament to
intertrochanteric line

Capsule

Epiphysial line of
lesser trochanter

Psoas major

Vastus intermedius
from anterior surface

Vastus intermedius
from lateral surface

Medial supracondylar line

M. articularis genus

Lateral supracondylar line

Adductor tubercle

Epiphysial line

Fibular collateral ligament of
knee to lateral epicondyle

Synovial area

Tibial collateral ligament of
knee to medial epicondyle

Lateral   Medial
condyle   condyle

FIG. 3.157. Right femur (anterior aspect).

ligament of the head is attached. Both areas may transmit blood vessels to the head.

The **neck** runs inferolaterally to meet the body at an angle of about 125 degrees. This angle varies with age, stature, and width of pelvis, being smaller in the adult, in short-limbed people, and in women. The angle is correlated with the width of the pelvis (i.e. the distance separating the two femoral heads) and with the obliquity of the shafts when the knees are together in the normal standing position.

The anterior part of the neck is grooved and perforated, particularly above and below, by many small vessels which run in the periosteum and enter the neck and head. The posterior surface

is smoother and may be grooved by the tendon of obturator externus passing to the trochanteric fossa [see below].

The neck is developmentally part of the body but is marked off from it in front by a wide, rough **intertrochanteric line** which runs downwards and medially to be continued as a spiral line below the lesser trochanter. Medially, the intertrochanteric line gives attachment to the inferior end of the iliofemoral ligament and, at its ends,

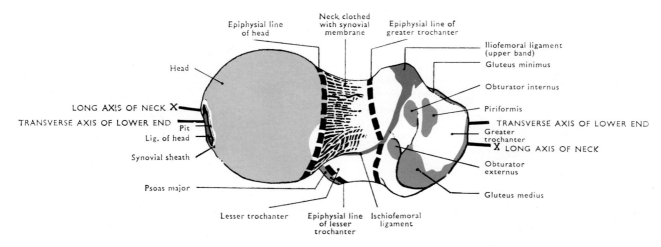

FIG. 3.158. Right femur (superior aspect).

there may be tubercles for parts of the iliofemoral and pubofemoral ligaments. The neck is covered by synovial membrane up to the intertrochanteric line.

The posterior part of the neck is marked off from the shaft by a prominent **intertrochanteric crest** and only the medial half of the neck is covered by synovial membrane and joint capsule.

The **greater trochanter** is the large prominence on the upper extremity of the body. It is felt at the side of the hip a hand's breadth below the crest of the ilium. Its concave medial surface meets its sloping lateral surface at a superior border, which is on the same level as the centre of the hip joint and the upper border of the symphysis pubis. The highest point forms the apex of the trochanter, the medial surface, just below the apex, is hollowed out into a **trochanteric fossa** which receives the obturator externus tendon [FIG. 3.158]. Obturator internus is inserted into the medial surface *in front* of this fossa.

The greater trochanter also has anterior, lateral, and posterior surfaces which are continuous with the body and are largely for attachment of the gluteal muscles. Piriformis and gluteus medius pass laterally to reach the superior border of the trochanter [FIG. 3.158]. The insertion of gluteus medius is continued down on the lateral surface to the antero-inferior angle. Gluteus minimus goes to the large impression on the anterior surface. At the lateral side the greater trochanter overhangs the body at a rough ridge where the origin of vastus lateralis is continued round to the back [FIG. 3.161].

The intertrochanteric crest passes inferomedially to the lesser trochanter and, near the upper end, has a rounded eminence, the quadrate tubercle, for quadratus femoris.

The **lesser trochanter** is a conical projection at the inferior end of the intertrochanteric crest, and is situated posteromedially in the angle between the body and the neck. It is lateral to the normal axis of rotation of the femur which passes through the head, although medial to the long axis of the body. Thus psoas major which passes backwards to the lesser trochanter, medial to the body, can produce medial rotation of the intact femur, though it is mainly a flexor. Iliacus is attached to the psoas tendon and to the bone distal to the trochanter.

The **body of the femur** as seen from the front is smooth and rounded. It is wider above and below.

Seen from the side it is convex forwards, and this variable curvature accounts for the fullness of the front of the thigh. The concavity at the back is filled in by a ridge of bone, the **linea aspera**, which gives a piriform outline to a transverse section and makes the posterior surface practically straight in its middle third.

The surface of the body gives attachment to the fleshy fibres of vastus intermedius on its anterior and lateral aspects in its upper two-thirds. The medial aspect is covered by vastus medialis. The articularis genus muscle arises from the lower quarter of the anterior aspect.

The posterior surface of the body narrows in its middle third to form the **linea aspera** (a rough linear strip with a medial and a lateral lip), but widens above as it approaches the trochanters, and below where the lips diverge as two supracondylar lines enclosing the **popliteal surface** [FIG. 3.160].

The **medial lip of the linea aspera** is continued up as a spiral line to reach the intertrochanteric line. Vastus medialis is attached to this lip of the linea aspera and the spiral line, as well as to the lower part of the intertrochanteric line. Part of iliacus is inserted below the lesser trochanter, and pectineus goes to a vertical, **pectineal line** leading down to the linea aspera [FIG. 3.159]. The **lateral lip of the linea aspera** is continued up as the **gluteal tuberosity**, which reaches the greater trochanter. It is prominent and may be sufficiently raised to form a **third trochanter**. The gluteal tuberosity is for part of gluteus maximus. Vastus lateralis arises from its lateral edge as well as from the lateral lip of the linea aspera, the root of greater trochanter and the upper part of the intertrochanteric line. The attachment of adductor magnus is from the lower border of the insertion of quadratus femoris down to the adductor tubercle [FIG. 3.160], lying medial to the gluteal tuberosity, on the middle of the linea aspera, and on the medial supracondylar line. The adductor brevis and longus are attached to the medial lip of the linea aspera medial to magnus [FIG. 3.160] and, in the case of brevis, behind the pectineal line. The short head of the biceps is attached to the lateral lip, lateral to adductor magnus, and medial to vastus lateralis and to vastus intermedius inferiorly. The bone may be grooved by the four perforating arteries.

Inferiorly, the linea aspera expands into the flat, triangular, **popliteal surface**, bounded at the sides by the supracondylar lines. The **lateral supracondylar line** is well defined and extends down to the lateral condyle. The **medial supracondylar line** is less well defined because it is smoothed in its upper part by the femoral vessels passing between the bone and adductor magnus to enter the popliteal space as the popliteal vessels. Inferiorly, this line reaches the **adductor tubercle**, a sharp prominence on the upper part of the medial condyle for the insertion of the tendinous, hamstring part of adductor magnus. The tubercle can be found by following down the tendon with the knee flexed.

Inferiorly, the popliteal surface curves back to a condyle on each side and to a prominent **intercondylar line** between them. This

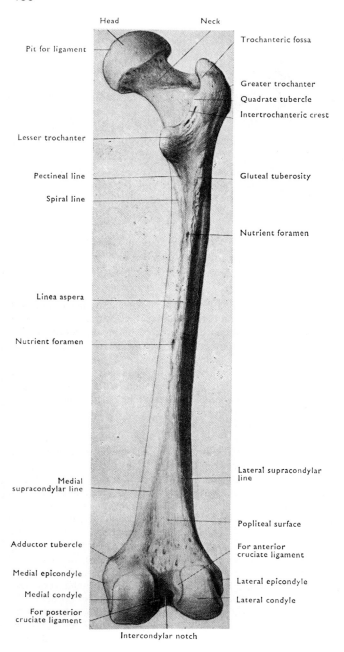

Fig. 3.159. Right femur (posterior aspect).

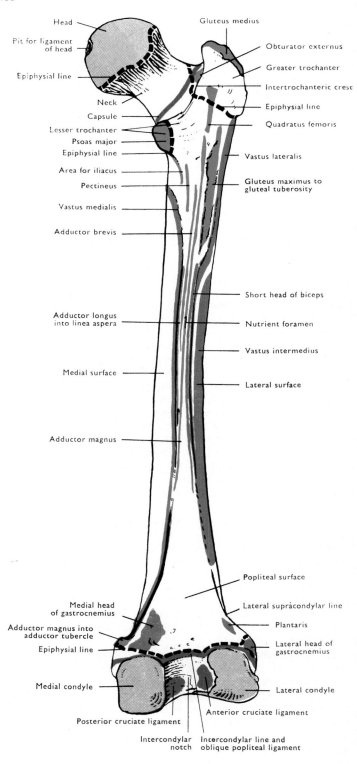

Fig. 3.160 Right femur (posterior aspect).

lower part of the surface is pitted by foramina and has at its medial side a rough mark for the medial head of gastrocnemius. Plantaris arises from the corresponding area at the lateral side.

The structures attached to the lower part of the linea aspera are continued down on the supracondylar lines; vastus medialis and adductor magnus on the medial line, biceps and vastus intermedius on the lateral line.

The medial and lateral intermuscular septa pass to the bone; the former between the anterior, extensor compartment and the medial adductor group of muscles, and the latter between the same extensor compartment and the hamstring group of muscles posteriorly. The septa reach both the linea aspera and the supracondylar lines and increase the area of muscular attachment.

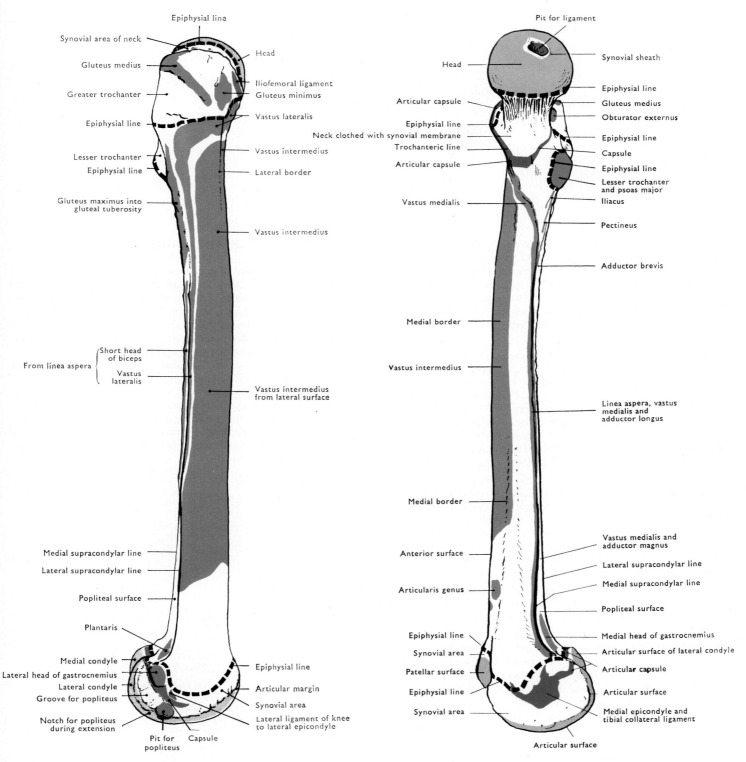

FIG. 3.161 Right femur (lateral aspect).

FIG. 3.162. Right femur (medial aspect).

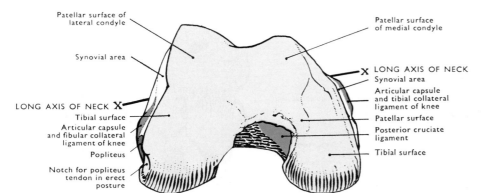

Patellar surface of
lateral condyle

Synovial area

LONG AXIS OF NECK X

Tibial surface

Articular capsule
and fibular collateral
ligament of knee

Popliteus

Notch for popliteus
tendon in erect
posture

Patellar surface
of medial condyle

X LONG AXIS OF NECK

Synovial area

Articular capsule
and tibial collateral
ligament of knee

Patellar surface

Posterior cruciate
ligament

Tibial surface

FIG. 3.163. Right femur (distal aspect).

The inferior end of the femur consists of two large condyles which blend with each other anteriorly and with the body superiorly. They project backwards and are separated posteriorly by a deep **intercondylar fossa**, and can be felt as large prominences at the sides of the knee. When the femur is in its normal position the distal surfaces of the two condyles should touch the same horizontal plane so that the body is directed obliquely outwards and upwards from the knee to the hip.

The **articular surfaces of the condyles** for the tibia and its menisci extend over the distal and posterior aspects so as to provide for full flexion. From below and behind, the surface of the lateral condyle is rounded although the parallel medial and lateral edges may give its outline an oblong appearance. The medial condyle is elongated anteroposteriorly, and turning laterally at its anterior end, is curved, in a horizontal plane, round a centre situated on the middle of the lateral condylar surface. The anteroposterior curvature of the articular surfaces is spiral with the inferior part flatter, the posterior part more convex.

Anteriorly, the two articular surfaces are confluent, but each is separated from this **patellar surface** by a slight groove. In flexion of the knee joint, the patella articulates with the distal aspect immediately anterior to the grooves, but in extension, is in contact with the anterior aspect. The midline of the patellar surface is deeply grooved and its lateral part is much more prominent and extensive than the medial part. The lateral and medial margins of the patellar surface can be felt when the knee is bent.

The **medial surface** of the lower end of the femur has a prominent **medial epicondyle** which gives attachment to the tibial collateral ligament. Above is the adductor tubercle.

The **lateral surface** has a prominent **lateral epicondyle** on its posterior half often showing a smooth hollow where the fibular collateral ligament is attached. Above this, an impression marks the attachment of the lateral head of gastrocnemius. Below and behind these markings, running parallel to the margin of the articular surface, is a **groove for popliteus**. The tendon of popliteus is attached to the antero-inferior end of the groove and lies in it only in flexion. A notch in the edge of the articular surface below the attachment of the tendon may show where it crosses the edge during extension of the joint.

Posteriorly, the anterior wall of the intercondylar fossa is rough and marked by vascular foramina. It is limited above by the **intercondylar line**; below, by the articular edge. At the sides its walls are the condyles. The posterior part of the lateral wall of the fossa (lateral condyle) is smooth for the upper attachment of the anterior cruciate ligament. The anterior part of the medial wall (medial condyle) similarly is smooth for the posterior cruciate ligament. The oblique popliteal ligament reaches the intercondylar line and is attached to it.

THE STRUCTURE OF THE FEMUR

The ends are composed of cancellous bone enclosed in a thin shell of compact bone. The body is a thick tube of compact bone, thickest above the middle of the bone and along the linea aspera. It encloses a medullary cavity which extends from the level of the lesser trochanter to approximately 8 cm above the distal articular surface.

The upper end of the femur is particularly important in walking and weight-bearing and the structure, architecture, and vascularization of the bone have been studied in great detail (Dixon 1910; Koch 1917; Murray 1936; Thompson 1942; Crock 1967).

The compact bone of the body is continuous with trabeculae which branch off, diverge, and intersect to form the cancellous bone of the neck, the trochanters, and the head. Many of these trabeculae are concentrated to form the walls of a curved tube which is a continuation of the tubular compact bone of the body through the

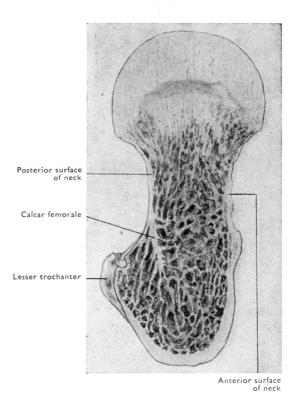

Posterior surface
of neck

Calcar femorale

Lesser trochanter

Anterior surface
of neck

FIG. 3.164. Section through head and neck of femur to show calcar femorale.

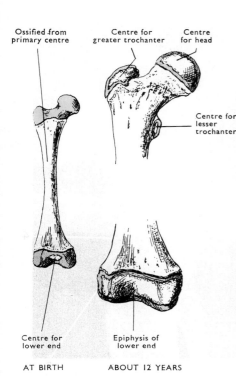

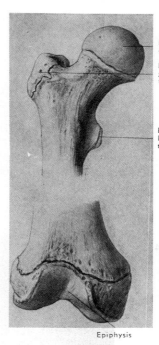

Ossified from primary centre

Centre for greater trochanter

Centre for head

Centre for lesser trochanter

Epiphysis of head

Epiphysis of greater trochanter

Epiphysis of lesser trochanter

Centre for lower end

Epiphysis of lower end

Epiphysis

AT BIRTH   ABOUT 12 YEARS   ABOUT 16 YEARS

FIG. 3.165. Ossification of femur.

Secondary centres appear for the head at 6–10 months, for the greater trochanter at 2–3 years, and for the lesser trochanter at 9–12 years: all these centres fuse with the neck or body from 16–20 years, usually in the reverse order of appearance. A centre for the lower end appears just before or just after birth and does not usually fuse until 20 years.

neck to the head. If the trochanters and intertrochanteric crest are removed carefully it is possible to demonstrate the continuity of this tube from the body up into the neck. The part which passes deep to the lesser trochanter to reach the posterior aspect of the neck is thickened to form a buttress or spur, the calcar femorale [FIG. 3.164], and is of considerable importance in surgical procedures on the neck. Where the tube lies deep to the greater trochanter it has been referred to as the internal femoral lamina. Trabeculae branch off from the deep and superficial surfaces of the tube to form the adjacent cancellous bone the arrangement of which has been likened to ossified lines of force. The strong trabecular systems are:

1. Trabeculae which pass superomedially from the compact bone of the medial aspect of the body and the inferior aspect of the neck mainly to the upper, weight-bearing part of the articular surface of the head. These are obviously subjected to compression from the weight of the body.

2. Trabeculae which arch across from the compact bone of the lateral aspect of the body, and passing deep to the greater trochanter and through the upper part of the neck, cross the first series approximately at right angles, and reach the articular surface of the head, mainly the lower part. These must be subject to tension when the body weight is applied to the head of the bone.

3. A series of trabeculae arch across from the medial wall of the body to the greater trochanter.

4. Trabeculae also sweep up from the lateral side into the greater trochanter and upper part of the neck of the bone.

The first two of these systems form the inferomedial and superolateral walls of the bent tubular continuation of the body. The area enclosed by the first three systems is relatively free of trabeculae and forms the internal triangle of the femur illustrated by Ward as early as 1838. Many of the trabeculae are arranged spirally as intersecting right- and left-handed series, an exceedingly efficient arrangement. Dixon (1910) found that adult femora could stand compression (applied to the lower end and head of the bone) amounting to between 810 and 1125 kg. Weights of this order produced shearing of the neck before fracture from bending stresses took place.

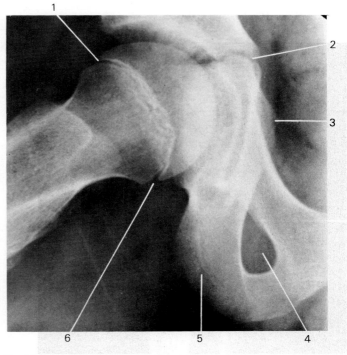

FIG 3.166. Oblique radiograph of hip joint. Age 17 years.
1 & 6. Margins of epiphysial line of head of femur.
2. Triradiate cartilage of hip bone (consolidating).
3. Ischial spine.
4. Obturator foramen.
5. Ischial tuberosity.

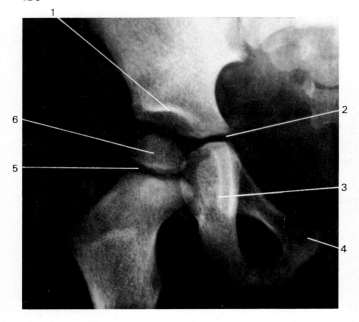

FIG. 3.167. Anteroposterior radiograph of hip joint. Age 4 years.
1. Margin of acetabulum.
2. Triradiate cartilage in acetabulum.
3. Acetabulum at acetabular notch.
4. Body of pubis.
5. Epiphysial cartilage of head of femur.
6. Epiphysis of head of femur.

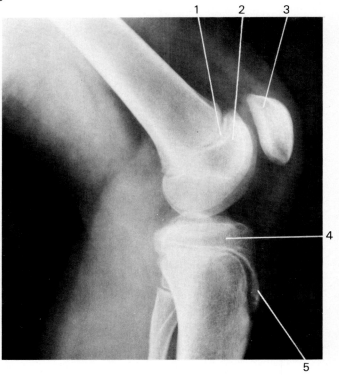

FIG. 3.169. Lateral radiograph of knee. Age 12 years.
1. Epiphysial cartilage at distal end of femur.
2. Distal epiphysis of femur.
3. Patella.
4. Proximal epiphysis of tibia.
5. Anterior and distal extension of proximal tibial epiphysis (tibial tuberosity).

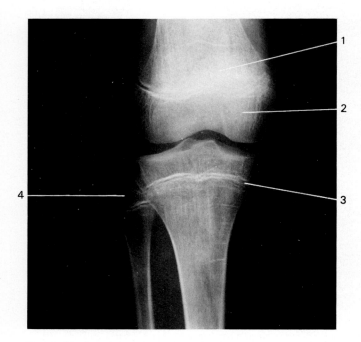

FIG. 3.168. Anteroposterior radiograph of knee. Age 12 years.
1. Patella superimposed on femur.
2. Epiphysis of distal end of femur.
3. Epiphysial cartilage at proximal end of tibia.
4. Epiphysis of proximal end of fibula.

OSSIFICATION

A primary centre for the body and neck appears during the seventh week. A secondary centre for the lower end appears about birth and its presence suggests that a foetus is mature. Details of secondary centres are given in FIGURE 3.165. The order in which the epiphyses fuse is lesser trochanter, greater trochanter, head, lower end. The epiphysial line at the lower end passes through the adductor tubercle, immediately above the patellar surface, and along the intercondylar line [FIG. 3.160].

THE PATELLA

The patella is a sesamoid bone about 5 cm in diameter, set in the tendon of quadriceps femoris. It is oval in outline but has a pointed **apex** projecting downwards, and so is commonly described as triangular. The **ligamentum patellae**, which is the continuation of the quadriceps tendon, is attached to the apex and the inferior part of the adjacent posterior surface. *The apex indicates the level of the knee joint when the ligament is taut.* The quadriceps tendon is attached to the upper border or base and some of its fibres are continued over the slightly convex **anterior surface** which is vertically striated by them as they pass to join the ligamentum patellae.

The lateral and medial edges are rounded and receive fibres of vastus lateralis and medialis respectively. These, with associated

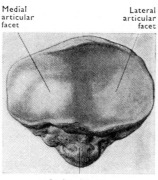

A

B

FIG. 3.170. Right patella.
A, Anterior surface.
B, Posterior surface.

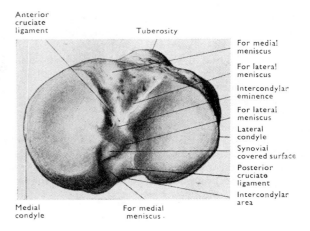

FIG. 3.171. Right tibia (proximal surface).

fibres of the fascia lata, form the **retinacula of the patella** [see Knee Joint].

The posterior surface of the patella is largely occupied by an oval **articular surface** for the femur. This is divided by a vertical ridge into a larger lateral part and a smaller medial part for the corresponding areas on the patellar surface of the femur.

These areas may be subdivided further to correspond to the parts in contact with the femur in different phases of flexion and extension.

The attachment of the ligamentum patellae below and of the quadriceps tendon above are each separated from the articular surface by a non-articular area related to synovial membrane and fat. These areas are frequently perforated by blood vessels.

OSSIFICATION

The patella is cartilaginous at birth but ossifies from one or more centres between the third year and puberty (Walmsley 1940).

VARIATIONS

The patella may be absent and this is said to be hereditary. It has also been found with a large upper part and a smaller lower part embedded in the ligamentum patellae. The articular facets can be modified by habitual postures such as squatting.

THE TIBIA

The tibia is the larger and medial of the two bones of the leg. Almost all of its anteromedial surface is subcutaneous.

The proximal end is expanded to form the medial and lateral condyles. Both condyles project backwards and the upper part of the anterior surface slopes upwards and backwards, thus giving the upper end a curved appearance. The lateral condyle projects further laterally. On its posterolateral part, a round articular surface faces downwards and laterally for the head of the fibula. The medial condyle is grooved posteromedially where semimembranosus tendon is attached.

Each condyle has a concave **superior articular surface**. Between these is an **intercondylar eminence** with uneven **intercondylar areas** anterior and posterior to it [FIG. 3.171].

The lateral superior articular surface is rounded and slightly smaller than the ovoid medial one. Each articulates centrally with

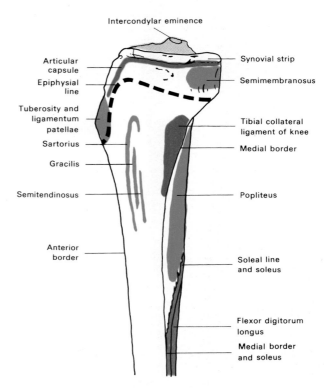

FIG. 3.172. Right tibia (medial aspect of upper half).

a condyle of the femur, while the peripheral part articulates with a meniscus and is flatter and crescentic [FIG. 4.56]. The two menisci are attached to the margins of the condylar articular surfaces via the articular capsule, the medial being reinforced by the tibial collateral ligament.

The **intercondylar eminence** projects up on each side as medial and lateral **intercondylar tubercles** with a ridge passing backwards and laterally from the medial one to give attachment to the posterior horn of the lateral meniscus. Behind this a smooth area gives attachment to the posterior cruciate ligament. A further smooth area in front of the eminence is for the attachment of the anterior cruciate ligament. The remaining attachments between the two articular surfaces [FIG. 4.56] are: the anterior horn of the medial meniscus in front; the two horns of the lateral meniscus and the posterior horn of the medial meniscus between the cruciate ligaments; the posterior cruciate ligament passing well down the sloping hinder part of the posterior intercondylar area [FIG. 4.55].

Anteriorly, the upper end slopes downwards and forwards to reach the **tuberosity of the tibia** [FIG. 3.172] which leads down to the anterior border of the body. The smooth upper part of the tuberosity gives attachment to the patellar ligament. The rough lower part is subcutaneous and takes most of the weight of the body in the kneeling position. Its superior edge is raised where it receives the superficial fibres of the patellar ligament.

The anterolateral aspect of the **lateral condyle** is marked by the attachment of the iliotibial tract of the fascia lata: posteriorly its inferior surface carries the **articular surface for the fibula**. Below this, the interosseous border runs down the body and a ridge for the fascia covering the tibialis anterior muscle connects its upper end to

the area for the iliotibial tract. Biceps femoris gains an attachment to the condyle between the fibular articular surface and the above-mentioned ridge.

The **medial condyle** is roughened medially for the attachment of the articular capsule, fascia lata, and fascia of the leg. It is continuous below with the medial surface of the body and this part is limited posteriorly by the **medial margin of the tibia** where the tibial collateral ligament roughens the bone [FIG. 3.172]. Behind this, the posterior surface of the medial condyle is grooved horizontally for the tendon of semimembranosus which also inserts into the rough area below the groove [FIG. 3.177].

Posteriorly, the space between the condyles is smooth and passes down as the popliteal surface of the tibia, limited below by the oblique **soleal line** on the body.

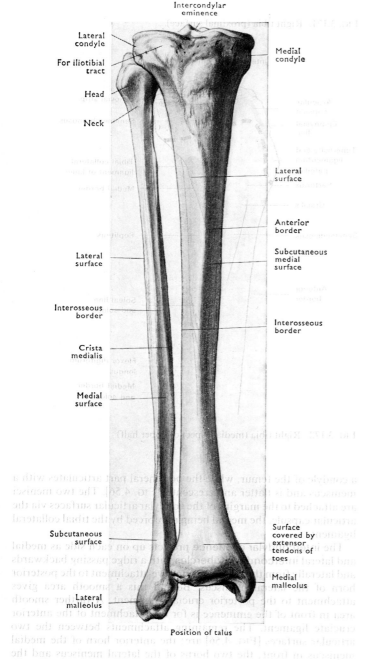

FIG. 3.173. Right tibia and fibula (anterior aspect).

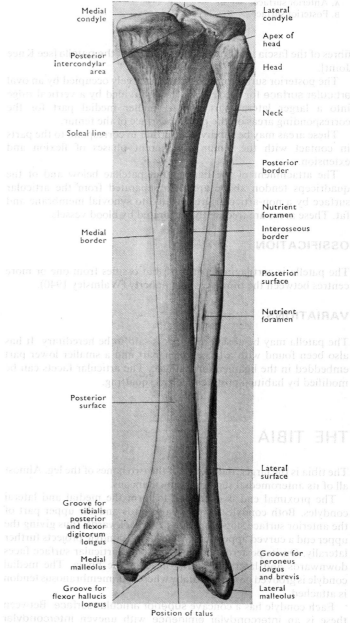

FIG. 3.174. Right tibia and fibula (posterior aspect).

The body of the tibia tapers slightly from the condyles downwards for two-thirds of its length and then widens again near the lower end.

It has a prominent subcutaneous anterior border which begins below the tuberosity and runs down to end by passing medially towards the medial malleolus [FIG. 3.173].

The smooth and almost wholly subcutaneous medial surface slopes posteromedially from the anterior border. At its upper end, the rough mark for the tibial collateral ligament lies at the posterior edge of the surface, and sartorius, gracilis, and semitendinosus are attached between this and the tibial tuberosity without marking the bone appreciably [FIG. 3.172].

The lateral surface slopes back from the anterior border. Its upper two-thirds are concave and give attachment to tibialis anterior. Inferiorly, it is rounded and becomes continuous with the anterior surface of the lower end of the bone. This lateral surface is limited posteriorly by the sharp interosseous border, to which the interosseous membrane is attached. This border begins inferior to the fibular articular surface, and finishes below as a triangular area for the interosseous ligament.

The posterior surface of the body lies between the interosseous border and a less defined medial margin which separates it from the medial surface and runs from the rough mark for the tibial collateral ligament to the posterior edge of the medial malleolus. An oblique line for soleus runs from the fibular articular surface on the lateral condyle to the medial margin a third of the way down. Soleus is attached to this line and to the middle third of the medial margin, whilst the fleshy fibres of popliteus arise from the triangular popliteal part of the posterior surface above the line, and the fascia covering popliteus is attached to the line.

A groove for the tendon of popliteus may be found above the fibular articular surface. Below the oblique line, the posterior surface is divided into a lateral area for tibialis posterior and a medial area for flexor digitorum longus by a vertical line to which the fascia covering tibialis posterior is attached. The nutrient foramen is usually found near the vertical or oblique lines.

The lower end of the tibia has a prominent medial malleolus projecting down from its medial side. The malleolus has a rounded medial surface which is continuous with the medial surface of the shaft and which forms the prominence at the medial side of the ankle. Its distal border has a pit or notch in its posterior part for attachment of the strong medial (deltoid) ligament of the ankle joint. The distal extremity of the malleolus is 1–2 cm higher than that of the lateral malleolus.

Posteriorly, the malleolar sulcus lodges the tendon of tibialis posterior as it descends on to the medial ligament. The lateral part of the posterior surface of the lower end may also be grooved by the tendon of flexor hallucis longus [FIG. 3.174]. The inferior edge of the posterior surface and the adjacent margin of the malleolus provide attachment for the articular capsule of the ankle joint and the transverse fibres of the posterior tibiofibular ligament.

Anteriorly, the inferior end of the bone and the malleolus are smooth and slightly convex. The tendons of the muscles passing down to the foot cover the bone but do not mark it. It may have a pressure facet near its lateral end caused by full dorsiflexion of the ankle joint.

The inferior articular surface has an anteroposterior ridge which fits the groove on the trochlea of the talus. Medially, the articular surface is continuous with the malleolar articular surface and laterally it usually turns up on to the lateral surface where the fibula articulates with it. The lateral surface is concave anteroposteriorly (fibular notch) where it receives the fibula and is continued up as a rough triangular area for the attachment of the interosseous ligament—the thick lower end of the interosseous membrane. A synovial fold usually extends up between the lower part of the notch and the fibula.

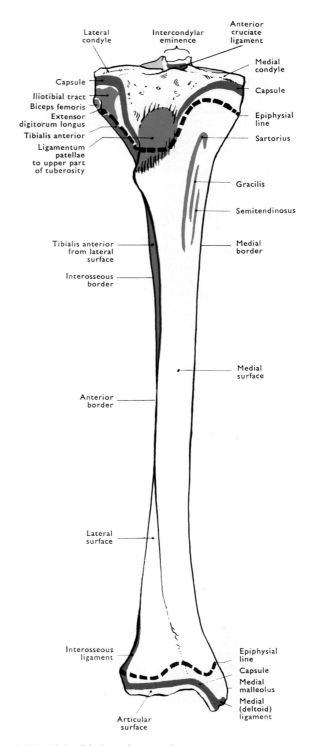

FIG. 3.175. Right tibia (anterior aspect).

OSSIFICATION

A primary centre for the tibia appears during the seventh week and spreads so that only the ends are cartilaginous at birth. Times of appearance and fusion of the upper and lower epiphyses are given in FIGURE 3.178.

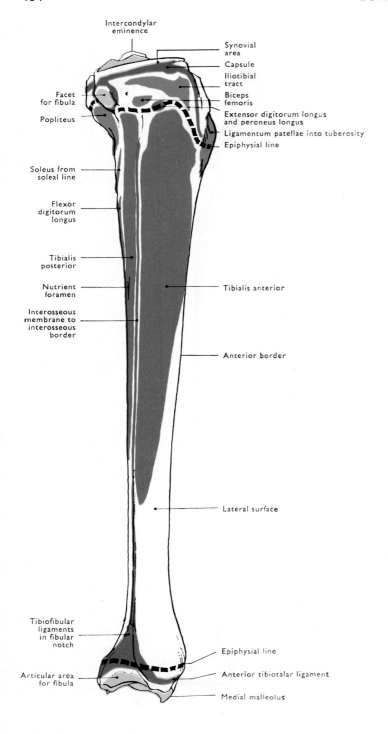

Intercondylar
eminence

Synovial
area

Capsule

Iliotibial
tract

Facet
for fibula

Biceps
femoris

Popliteus

Extensor digitorum longus
and peroneus longus

Ligamentum patellae into tuberosity

Epiphysial line

Soleus from
soleal line

Flexor
digitorum
longus

Tibialis
posterior

Tibialis anterior

Nutrient
foramen

Interosseous
membrane to
interosseous
border

Anterior border

Lateral surface

Tibiofibular
ligaments
in fibular
notch

Epiphysial line

Anterior tibiotalar ligament

Articular area
for fibula

Medial malleolus

FIG. 3.176. Right tibia (lateral aspect).

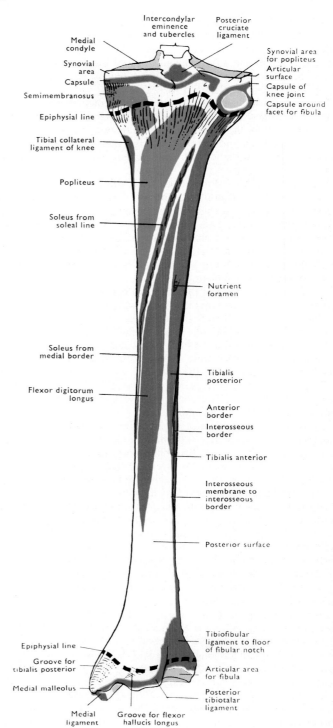

Intercondylar
eminence
and tubercles

Posterior
cruciate
ligament

Medial
condyle

Synovial
area

Synovial area
for popliteus

Articular
surface

Capsule

Capsule of
knee joint

Semimembranosus

Capsule around
facet for fibula

Epiphysial line

Tibial collateral
ligament of knee

Popliteus

Soleus from
soleal line

Nutrient
foramen

Soleus from
medial border

Tibialis
posterior

Flexor digitorum
longus

Anterior
border

Interosseous
border

Tibialis anterior

Interosseous
membrane to
interosseous
border

Posterior surface

Tibiofibular
ligament to floor
of fibular notch

Epiphysial line

Groove for
tibialis posterior

Articular area
for fibula

Medial malleolus

Posterior
tibiotalar
ligament

Medial
ligament

Groove for flexor
hallucis longus

FIG. 3.177. Right tibia (posterior aspect).

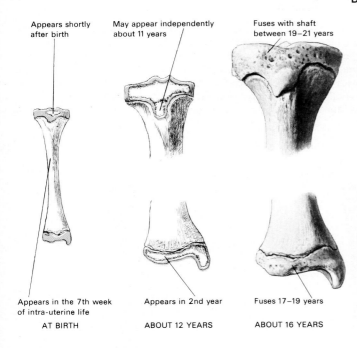

Appears shortly after birth

May appear independently about 11 years

Fuses with shaft between 19–21 years

Appears in the 7th week of intra-uterine life

AT BIRTH

Appears in 2nd year

ABOUT 12 YEARS

Fuses 17–19 years

ABOUT 16 YEARS

FIG. 3.178. Ossification of tibia.

The upper epiphysis includes the tuberosity of the tibia and ossification spreads down to the tuberosity after the tenth year. Fusion with the shaft occurs at 16 to 19 years. Several separate centres may appear for the tuberosity.

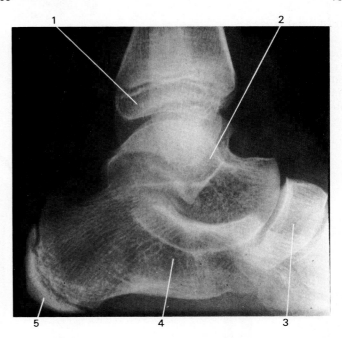

FIG. 3.180. Lateral radiograph of ankle. Age 12 years.
1. Distal epiphysis of tibia.
2. Talus.
3. Navicular.
4. Calcaneus.
5. Epiphysis of calcaneus.

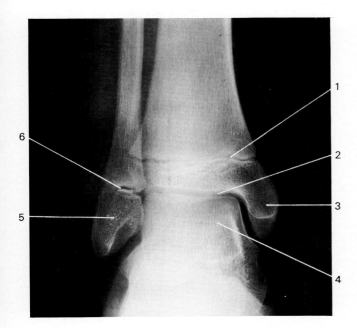

FIG. 3.179. Anteroposterior radiograph of ankle. Age 12 years.
1. Epiphysial cartilage of distal end of tibia.
2. Ankle joint 'space'.
3. Medial malleolus.
4. Talus.
5. Distal epiphysis of fibula.
6. Epiphysial cartilage of distal end of fibula.

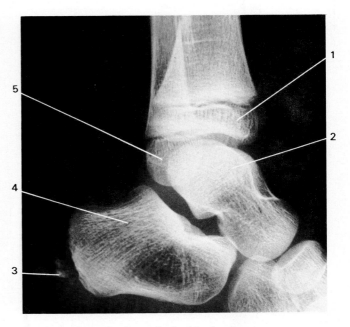

FIG. 3.181. Lateral radiograph of ankle. Age 8 years.
1. Distal epiphysis of tibia.
2. Talus.
3. Epiphysis of calcaneus.
4. Calcaneus.
5. Distal fibular epiphysis.

THE FIBULA

The fibula is slender but has slightly expanded ends and is twisted in appearance. The upper end, or head, can be felt as a knob behind and below the lateral side of the knee where the tendon of biceps femoris leads down to it. The lower end is the lateral malleolus, a subcutaneous prominence felt on the lateral side of the ankle joint.

The **head** is roughly cuboidal and tapers below through a neck to the body. It has a conical **apex of the head**, which projects up from the posterolateral part. Medial to this is a round **articular surface of the head** which faces upwards and slightly medially to articulate with the lateral condyle of the tibia. Anterior to the apex, the bone is hollowed where it receives the fibular collateral ligament and part of the insertion of biceps femoris tendon. The main part of the

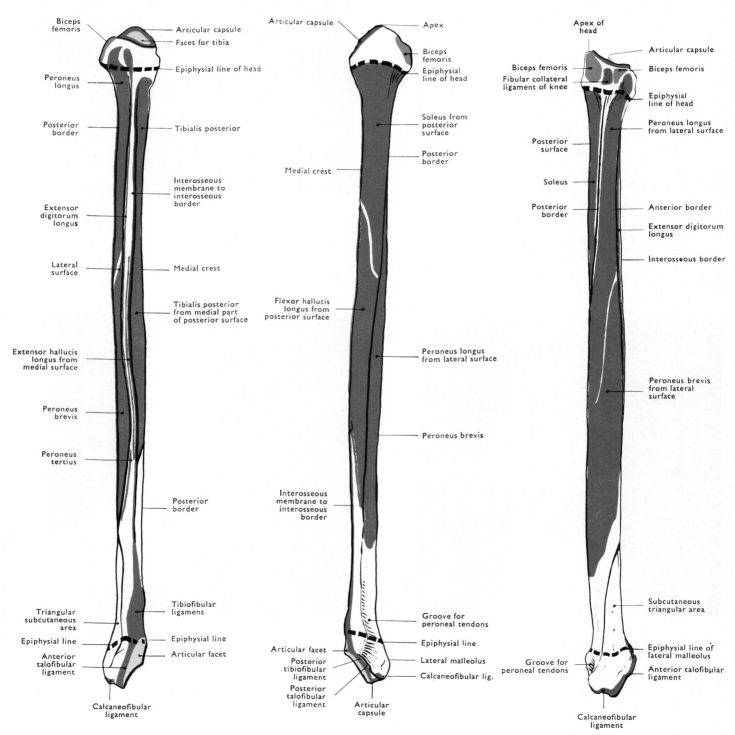

FIG. 3.182. Right fibula (anterior aspect).   FIG. 3.183. Right fibula (posterior aspect).   FIG. 3.184. Right fibula (lateral aspect).

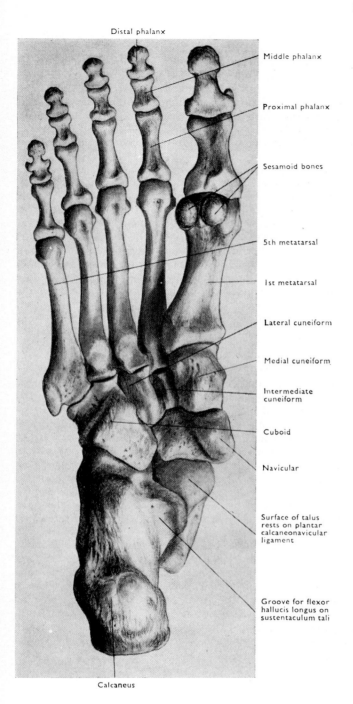

Distal phalanx

Middle phalanx

Proximal phalanx

Sesamoid bones

5th metatarsal

1st metatarsal

Lateral cuneiform

Medial cuneiform

Intermediate cuneiform

Cuboid

Navicular

Surface of talus rests on plantar calcaneonavicular ligament

Groove for flexor hallucis longus on sustentaculum tali

Calcaneus

FIG. 3.188. Bones of right foot (inferior or plantar surface).

border of the foot to touch the ground. The **medial**, highly arched part extends from the calcaneus through the body, neck, and head of the talus, the navicular, the three cuneiforms, and the medial three metatarsals. The most medial element of this part, calcaneus, talus, navicular, medial cuneiform, and the massive first metatarsal, is the highest part of the arch. It receives support from the tendons of tibialis posterior and anterior, and from flexor hallucis, all of which hold the bones together and help to maintain this part of the arch, the anterior pillar of which carries the thrust of the kick-off in walking and running.

The anterior part of the **transverse arch** is formed by the heads of the metatarsals and is maintained largely by the subjacent transverse structures. More posteriorly, it is formed by the sloping cuboid and the wedge-shaped cuneiforms. This part derives considerable support from all the tendons on the medial side of the sole of the foot, particularly tibialis posterior because of its extensions to most of the tarsal bones. The curvature is also maintained by the bowstring effect of peroneus longus tendon.

The weight of the body is transmitted downwards and backwards through the body of the talus to the calcaneus and the heel, downwards and forwards through the head and neck of the talus towards the ball of the foot and the ball of the great toe. Laterally, the weight passes through the cuboid to the metatarsals [see also FIG. 5.161]. In these bones as elsewhere, the supporting trabeculae are so arranged as to resist the forces to which they are subjected, and are particularly developed in the calcaneus where the load is applied to a relatively small area compared with the fore part of the foot.

THE TARSUS

The talus

The talus is characterized by a squarish body, a short neck, and a head with an oval articular surface [FIGS. 3.189–3.192].

The **body** lies between the malleoli of the tibia and fibula and its upper surface is grooved longitudinally to form a pulley-shaped **trochlea** which articulates with the inferior surface of the tibia and is continuous with the medial and lateral **malleolar surfaces** at the sides. The body is thus held between the two malleoli and its upper surface takes the weight transmitted from the tibia [FIGS. 3.189–3.192 for details]. The articular surface may extend posteriorly or anteriorly beyond the normal limit, thus indicating a wide range of movement at the ankle joint.

The **medial malleolar surface** [FIG. 3.192] is comma-shaped, with the head of the comma extending towards or on to the neck anteriorly, where it may be distinctly concave to receive the medial malleolus in full dorsiflexion.

Inferior to the concave edge of the comma, the medial surface of the bone is uneven and is continuous posteriorly with the prominent **medial tubercle of the posterior process** [FIGS. 3.189 and 3.192]. This gives attachment to the deep part of the medial ligament of the ankle joint and the attachment is continued forwards on to the neck.

The **lateral malleolar surface** [FIG. 3.191] is triangular and is continued from the superior surface down on to a projecting **lateral process**. The narrow area anterior to the surface provides attachment for the anterior talofibular ligament. The similar area posterior to the surface is for the posterior talofibular ligament and in each case the surface cartilage may be deeply notched by an extension into it. The posterior talofibular ligament is also continued on to the posterior aspect of the body, and the bone in this region projects back as a variable **lateral tubercle of the posterior process** [see p. 204].

a high arch, and eversion does the opposite, the muscles which produce these movements also play a part in the shape and arrangement of the arches.

Because the foot is a half dome, arches may be traced over it in any direction. However, the architecture of the bones is such that obvious **longitudinal and transverse arches** are present. The foot is narrow posteriorly (the heel) and broad anteriorly (the ball of the foot and of the great toe). Thus there are lateral and medial parts of the longitudinal arch which have a common posterior end (the calcaneus) but separate anteriorly. The **lateral part of the longitudinal arch** extends from the calcaneus through the cuboid and the lateral two metatarsals, and is sufficiently flat to allow the lateral

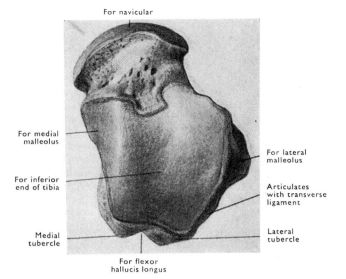

FIG. 3.189. Right talus (superior surface).

The superior articular surface is bevelled at its posterior part by contact with the transverse tibiofibular ligament so that the trochlea is considerably narrower at its posterior end.

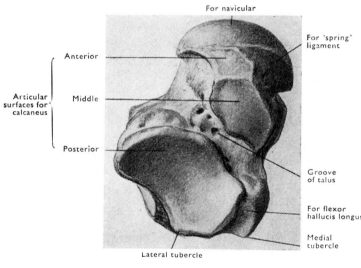

FIG. 3.190. Right talus (inferior surface).
The groove of talus is the sulcus tali.

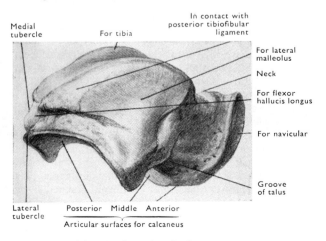

FIG. 3.191. Right talus (lateral surface).

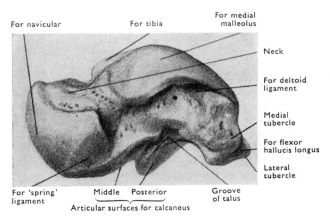

FIG. 3.192. Right talus (medial surface).

The **posterior process** extends inferiorly from the trochlea, and is grooved obliquely by the tendon of flexor hallucis longus descending between the lateral and medial tubercles, and covered posteriorly by ligamentous fibres [FIG. 3.189].

The **inferior surface** of the body is mainly occupied by an oval, deeply concave **posterior articular surface for the calcaneus**. This is set obliquely with its long, concave axis extending from the inferior aspect of the posterior process to that of the lateral process. The surface is frequently modified in shape by habitual movements of the foot.

Anterior to the posterior articular surface is a deep **sulcus tali**. The sulcus forms the **sinus tarsi** with the corresponding sulcus on the calcaneus, and the anterior and posterior walls of the sinus are completed by the capsules of the adjacent joints.

The **middle articular surface for the calcaneus** is medially placed in front of the sulcus tali, and articulates with the sustentaculum tali of the calcaneus. It is flat, nearly oval, and may meet the posterior articular surface. Anteriorly, it is separated from the anterior articular surface for the calcaneus by a ridge or, occasionally, by a non-articular gap. Laterally, the sulcus tali widens inferolateral to the neck of the talus, and contains the cervical ligament of the talus, [p. 257] and the stem of the inferior extensor retinaculum [FIG. 4.68].

The **anterior articular surface for the calcaneus** lies on the inferior surface of the head and articulates with the anterior part on the calcaneus. It is continuous with the main articular surface of the head and is usually continuous with the middle calcanean surface. Medial to the anterior calcanean surface, the articular surface of the head lies against the plantar calcaneonavicular (spring) ligament supported by the tendon of tibialis posterior.

The **neck of the talus** projects forwards, downwards, and medially, and the medial border is shorter and lower than the lateral border. The angle between the axes of the neck and body, as seen from above, is normally about 18 degrees. It varies, however, between 0 degrees in old age and 30 degrees in the new-born and may be as much as 50 degrees in cases of clubfoot (varus deformity).

The inferior aspect of the neck forms part of the sinus tarsi laterally; medially it is covered by the middle articular surface for the calcaneus.

The prominent lateral border gives attachment to ligaments, including the anterior talofibular and cervical ligaments. The short medial border is frequently encroached on by the medial malleolar surface. It carries a ridge which passes from the medial surface of the body to the superior surface of the neck where it provides attachment for the capsules of the ankle and talonavicular joints. The upper surface frequently shows medial or lateral extensions of the trochlear surface which indicate habitual strong dorsiflexion of the ankle joint [FIG. 3.189]. There may also be a lateral facet or smoothing on the neck, or, much more rarely, a pressure facet on the medial part of the ridge for the capsule of the ankle joint.

The **head of the talus** is oval in outline and evenly convex, except for the anterior calcanean surface. The articular surface for the navicular faces forwards and slightly medially and downwards, and its long axis slopes downwards and medially. It forms the navicular part of the talocalcaneonavicular joint.

The calcaneus

The calcaneus lies inferior to the talus and projects back to form the prominence of the heel. Anteriorly, it articulates with the cuboid bone and is strongly bound to it and to the remaining tarsal bones by ligaments.

The internal, supporting trabeculae of bone pass downwards and backwards to the heel, downwards and forwards to the anterior end, and horizontally from the heel to the anterior end, thus leaving between these an area comparatively transparent to X-rays which may suggest bone resorption by disease [FIG. 3.1].

The anterior end of the calcaneus is raised a little in its normal position to make an angle of a few degrees with the horizontal and so contributes to the longitudinal arch of the foot. The intermediate part of the upper surface projects medially as a thick shelf which supports the neck of the talus and so is called the **sustentaculum tali** [FIGS. 3.188 and 3.196].

The **posterior surface** is enlarged and presents three areas. The uppermost is smooth where a bursa lies between it and the tendo calcaneus. The middle, also smooth and convex, receives that tendon and usually has an irregular or jagged inferior edge. The lowest, **tuber calcanei**, is rough and covered by the strong fibrous tissue and fat of the heel pad. Its inferior part transmits the weight of the body from the heel to the ground, and curves forwards as a raised area on the inferior surface. This part of the tuber has a larger **medial process** and a smaller but distinct **lateral process** extending anteriorly from it. These give attachment to the strong fascia and short muscles of the sole of the foot. To the medial process are

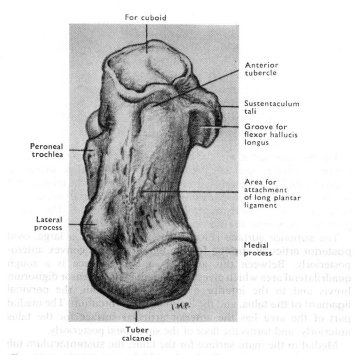

FIG. 3.194. **Right calcaneus (inferior surface).**

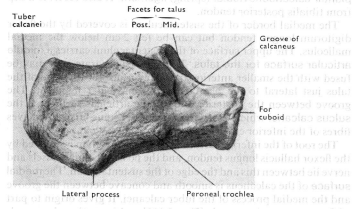

FIG. 3.195. **Right calcaneus (lateral surface).**

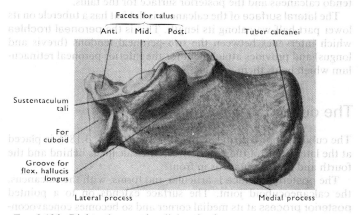

FIG. 3.196. **Right calcaneus (medial surface).**

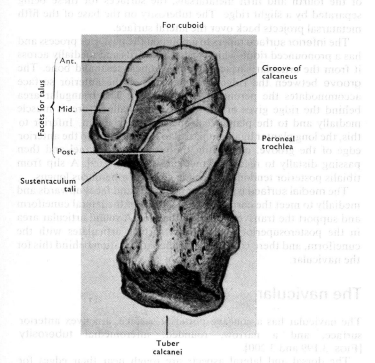

FIG. 3.193. **Right calcaneus (superior surface).**

attached parts of abductor hallucis and flexor digitorum brevis, abductor digiti minimi, the flexor retinaculum, and the plantar aponeurosis. The abductor digiti minimi and the aponeurosis also arise from the lateral process [FIG. 3.209].

The **inferior surface** is continued forwards as a rough area which is 2–3 cm wide and terminates as an **anterior tubercle**. The strong, **long plantar ligament** is attached to this rough area. The **plantar calcaneocuboid ligament** passes forwards from the anterior aspect of the tubercle and from the depression between it and the anterior end of the bone.

The **anterior surface** of the calcaneus is occupied by a concavo-convex articular surface for the cuboid, which is overhung superomedially by the anterior surface for the talus. The medial part of the articular surface extends on to the medial surface to accommodate a posterior projection on the cuboid. The shape prevents any simple sliding movement.

The **superior surface** [FIG. 3.193] is marked by a large, oval **posterior articular surface** for the talus, which is convex antero-posteriorly. Between this and the anterior surface is a rough quadrilateral area which gives attachment to the extensor digitorum brevis and to the **inferior extensor retinaculum**, the **cervical ligament** of the talus, and the **ligamentum bifurcatum**. The medial part of the area has the anterior articular surface for the talus anteriorly, and forms the floor of the **sinus tarsi** posteriorly.

Medial to the main surface for the talus, the **sustentaculum tali** projects with a rough medial border which gives attachment to the medial ligament of the ankle joint and, further forwards, to the **plantar calcaneonavicular (spring) ligament**. It also receives a slip from tibialis posterior tendon.

The medial border of the sustentaculum is covered by the flexor digitorum longus tendon but can be felt 2 cm below the medial malleolus. The upper surface of the sustentaculum carries a **middle articular surface** for the talus. This may be separate or it may be fused with the smaller anterior articular surface for the head of the talus just lateral to the plantar calcaneonavicular ligament. The groove between the posterior and middle articular surfaces is the **sulcus calcanei** which forms the floor of the sinus tarsi. It receives fibres of the inferior extensor retinaculum [FIG. 4.68].

The root of the inferior surface of the sustentaculum is grooved by the flexor hallucis longus tendon, and the posterior tibial vessels and nerve lie between this and the edge of the sustentaculum. The **medial surface** of the calcaneus is smooth and concave between the groove and the medial process of the tuber calcanei. It gives origin to part of flexor accessorius muscle [FIG. 5.156] and is continued upwards and backwards to a saddle-shaped area between the insertion of the tendo calcaneus and the posterior surface for the talus.

The **lateral surface** of the calcaneus is flat but has a tubercle on its lower part half-way along its length. This is the **peroneal trochlea** which intervenes between the two peroneal tendons (brevis and longus) and provides attachment for the inferior peroneal retinaculum which holds them in place.

The cuboid

The cuboid, approximately cubical in shape [FIG. 3.187], is placed at the lateral side of the foot between the calcaneus behind and the fourth and fifth metatarsals in front.

The **posterior surface** is articular and forms, with the calcaneus, the calcaneocuboid joint. The surface extends on to a pointed **posterior process** at its medial corner and so becomes concavoconvex. The process projects towards the medial surface of the calcaneus under the plantar calcaneonavicular ligament.

The **superior surface** is roughened by the attachments of ligaments passing to adjacent bones but is nearly flat and faces laterally as well as upwards.

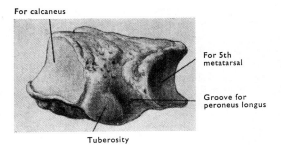

FIG. 3.197. Right cuboid bone (lateral surface).

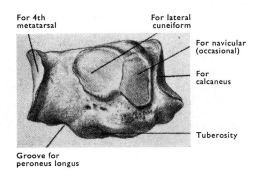

FIG. 3.198. Right cuboid bone (medial surface).

The **lateral surface** is short and narrow and lies in the lateral border of the foot between the base of the fifth metatarsal and the calcaneus. It is encroached on by a **groove for peroneus longus tendon** on the inferior surface. The **tuberosity** on the lateral end of the ridge forming the posterior wall of the groove carries an articular surface for a sesamoid bone in the tendon where it turns into the groove.

The **anterior surface** is nearly flat and articulates with the bases of the fourth and fifth metatarsals, the surfaces for these being separated by a slight ridge. The tuberosity on the base of the fifth metatarsal projects back over the lateral surface.

The **inferior surface** tapers to a point on the posterior process and has a pronounced **ridge** which passes forwards and medially across it from the articular surface for the peroneal sesamoid bone. The **groove** between the ridge and the edge of the anterior surface accommodates the peroneus longus tendon. The triangular area behind the ridge gives origin to the flexor hallucis brevis muscle medially and to the plantar calcaneocuboid ligament. Inferior to this, the long plantar ligament passes to the ridge and to the anterior edge of the groove, converting the groove to a tunnel and then passing distally to reach the metatarsals [FIG. 4.69]. A slip from tibialis posterior tendon reaches the groove for peroneus longus.

The **medial surface** is rough for ligaments and faces upwards and medially to meet the corresponding surface of the lateral cuneiform and support the transverse arch of the foot. A round articular area in the posterosuperior angle of the surface articulates with the cuneiform, and there is usually a smaller articulation behind this for the navicular.

The navicular

The navicular has a concave posterior surface, a convex anterior surface, and a narrow, rounded, inferomedial **tuberosity** [FIGS. 3.199 and 3.200].

The dorsal and lateral aspects are rough near their edges for ligaments but together they form a curved surface. The **tuberosity**

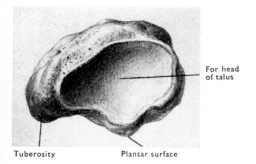

FIG. 3.199. Right navicular bone (posterior surface).

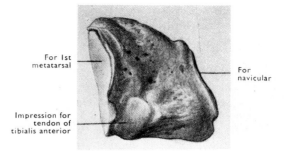

FIG. 3.202. Right medial cuneiform (medial side).

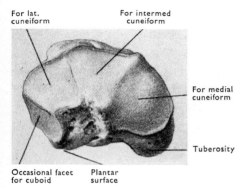

FIG. 3.200. Right navicular bone (anterior surface).

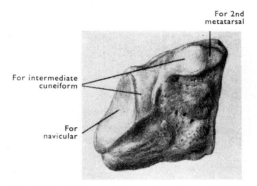

FIG. 3.203. Right medial cuneiform (lateral side).

can be felt, and is usually visible, on the medial side of the foot 3 cm in front of the sustentaculum tali, 3–4 cm below and in front of the medial malleolus. Tibialis posterior tendon has its main insertion into the tuberosity.

The inferior surface is rough and is prominent in its middle part for the attachment of ligaments, particularly the plantar calcaneo-navicular ligament. Anteriorly it slopes up to the anterior articular surface. The posterior surface is deeply concave to fit the convexity of the head of the talus.

The anterior surface is separated by ridges into three areas, arranged in a convex curve, to articulate with the three cuneiforms. Laterally there may be an additional articular surface for the cuboid.

The cuneiform bones

The three cuneiform bones are wedge-shaped. The **medial** one is the largest with a convex medial surface and a large kidney-shaped

articular surface on its anterior end. The **intermediate** one is distinctly wedge-shaped and has a square dorsal surface as a base to the wedge. The **lateral** one is also distinctly wedge-shaped but the dorsal surface is longer than it is broad and has an angulated lateral border.

The three cuneiforms articulate with the navicular bone posteriorly and with the bases of the first three metatarsals anteriorly. The line of the tarsometatarsal joints runs obliquely backwards and laterally, the anterior ends of the cuneiforms and the cuboid bone being nearly flush. However, the short intermediate cuneiform allows the base of the second metatarsal to extend proximally between the medial and lateral cuneiforms and the fourth metatarsal

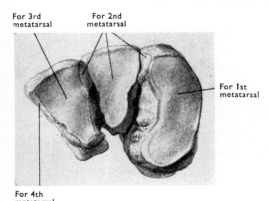

FIG. 3.201. Anterior aspect of cuneiform bones of right foot.

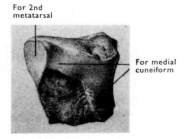

FIG. 3.204. Right intermediate cuneiform (medial side).

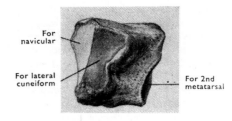

FIG. 3.205. Right intermediate cuneiform (lateral side).

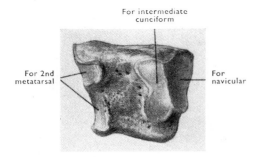

FIG. 3.206. Right lateral cuneiform (medial side).

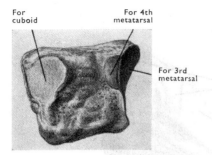

FIG. 3.207. Right lateral cuneiform (lateral side).

overlaps the lateral surface of the lateral cuneiform for a short distance [FIG. 3.187].

The **medial cuneiform** is thicker and more rounded on its plantar than on its dorsal aspect. Its superior, medial, and inferior aspects form a continuous curved surface which forms part of the medial side of the foot. It is roughened by ligaments and has, at the antero-inferior corner of its medial aspect, a smooth impression over which the tendon of tibialis anterior runs to its insertion on the plantar aspect of the cuneiform and the base of the first metatarsal. A bursa lies between the bone and the tendon near its insertion. Posteriorly, the inferior aspect receives much of the tibialis posterior tendon [FIG. 5.145].

The **anterior surface** has a large kidney-shaped articular surface, often divided by a groove, for the base of the first metatarsal. The posterior surface has a smaller pear-shaped concave facet for the medial impression on the navicular.

The **lateral surface** is rough for the interosseous ligaments and has an L-shaped articular surface for the intermediate cuneiform. This surface is continuous with that for the navicular, and is continued along the upper border of the lateral surface [FIG. 3.203] to a small articular surface for the base of the second metatarsal at the anterior end. The peroneus longus tendon is attached to the antero-inferior corner of this surface.

The **intermediate cuneiform** is wedge-shaped with its base uppermost, and is shorter than the other two. Its superior surface is roughened and almost square in outline. Posteriorly, its triangular articular surface is concave to fit the middle impression on the navicular. Anteriorly, a flat triangular surface articulates with the base of the second metatarsal.

The medial surface is rough for ligaments, but has a large L-shaped articular surface for the medial cuneiform. The lateral surface has a slightly concave, B-shaped articular surface at its posterior border for the lateral cuneiform. Both these articular surfaces are continuous with the articular surface for the navicular.

The inferior edge receives a slip from tibialis posterior tendon.

The **lateral cuneiform**, also wedge-shaped, has a rectangular base uppermost. It is roughened by ligaments and has the posterolateral corner cut off by its articulation with the cuboid.

The posterior surface is oval and slightly concave where it articulates with the navicular. The anterior articular surface is triangular for the third metatarsal. This articular surface is prolonged a little on to the medial, and sometimes the lateral, surface where it is overlapped by the second and fourth metatarsals. The medial and lateral surfaces are both rough for ligaments but in addition to the above small articular extensions, they have articular surfaces for the adjacent bones. The posterior part of the medial surface articulates (B-shaped surface) with the intermediate cuneiform, while the posterior part of the lateral surface carries a large round articular surface for the cuboid.

The inferior edge receives a slip from tibialis posterior tendon.

OSSIFICATION

The bones of the tarsus are each ossified from one primary endochondral centre, with a secondary centre in the case of the calcaneus.

The **primary centres** for the calcaneus and talus appear before birth, in the sixth and eighth months or a little earlier. The centre for the cuboid may be present at birth or soon after. These can therefore be used to estimate the degree of maturity of a new-born child. The centre for the lateral cuneiform does not appear until the end of the first year, to be followed by the navicular and the other two cuneiforms in the third and fourth years. Ossification is completed after puberty.

The **secondary centre** appears in the posterior end of the calcaneus about the ninth year, and extends to include the medial and lateral processes. Fusion occurs between 15 and 20 years. The lateral tubercle of the posterior process of the talus may ossify separately as an **os trigonum**.

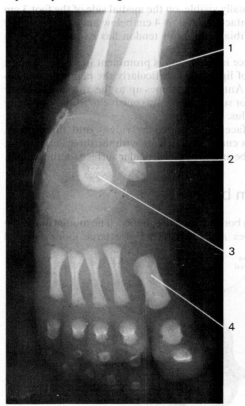

FIG. 3.208. Radiograph of foot at birth.
1. Distal end of tibial diaphysis.
2. Talus.
3. Calcaneus.
4. Diaphysis of first metatarsal.

THE METATARSUS

The metatarsus consists of five metatarsal bones numbered from the medial side. They are miniature long bones, each with a base proximally, a body, and a head distally. The flat triangle seen on the dorsum of a metacarpal body is absent [FIG. 3.187, cf. FIG. 3.135].

The first metatarsal is shorter than the rest but is much stronger. Its base has a kidney-shaped articular surface for the medial cuneiform, the convex border being medial. The bone close to the circumference of the articular surface is rough for ligaments which bind it to the cuneiform and to the second metatarsal. An articular surface may be present on the lateral side if the base touches the second metatarsal.

The inferior part of the base forms a downwards projection which finishes as a tuberosity on the lateral side. This projection provides attachment for part of tibialis anterior on the medial side, and the greater part of peroneus longus on the lateral side [FIG. 5.145], both tendons going also to the adjacent part of the first cuneiform.

The body is nearly straight on its dorsal and medial aspects and the smooth, rounded surface has no attachments. Its lateral surface is flat and triangular, owing to the downward projection of the base, and provides attachment for one head of the first dorsal interosseous muscle.

The head of the first metatarsal is large, wide, and helps to form the ball of the great toe with the base of the first phalanx and two sesamoid bones. Laterally and medially it is hollowed and tuberculated for the attachment of strong collateral ligaments. It is covered on its dorsal, distal, and plantar aspects by articular cartilage which extends furthest proximally on the plantar surface. This surface is grooved, on each side of a prominent, central ridge, by the sesamoid bones in the tendons of the short muscles which pass inferior to it.

The second, third, fourth, and fifth metatarsals are all similar in that they have long, tapering, slightly curved bodies and narrow heads. The second is the longest and provides a long axis for this part of the foot. The concavity of the curvature is plantar.

The bases articulate with each other as well as with the intermediate and lateral cuneiform bones and the cuboid bone. They are quite distinctive and can be identified by their shape and articular surfaces [FIGS. 3.211 and 3.212]. The proximal surfaces slope backwards and laterally, in keeping with the sloping tarsometatarsal line of joints. This indicates the medial and lateral sides of each bone, and so the foot from which it has been taken. The two double articular surfaces on the lateral aspect of the second are for the lateral cuneiform and the third metatarsal [FIG. 3.209].

The sides and plantar aspects of the bases are strongly marked by interosseous and plantar ligaments, and their shapes are in conformity with the curve of the transverse arch of the foot. The second and third are triangular in outline. The fourth is cuboidal but is set a little obliquely to articulate with the sloping cuboid bone. The fifth is even more obliquely set so as to fit the shape of the lateral border of the foot. It has a large tuberosity which projects back over the edge of the cuboid and provides attachment for the peroneus brevis tendon. It is also marked on the medial part of its dorsal surface by the smaller tendon of peroneus tertius which extends on to the body as well. The plantar aspect gives attachment laterally to the strong fascia (with some muscle fibres) which passes from the calcaneus and represents an abductor of the metatarsal, although normally fused with the abductor of the little toe [FIG. 5.153]. The flexor digiti minimi brevis and a slip from tibialis posterior are attached to the plantar surface medial to this.

The oblique head of adductor hallucis and slips from tibialis posterior are attached to the plantar surface of the second, third, and fourth bases [FIG. 3.209] and the long plantar ligament reaches them also.

The bodies of the metatarsals are sufficiently separated to accommodate the dorsal interosseous muscles, each of which arises from the adjacent surfaces of two metatarsals. The plantar interossei come from the medial side of the third, fourth, and fifth bodies [FIG. 5.157].

The heads, although narrow, are similar to those of the metacarpals with a convex articular surface covering their dorsal, distal, and plantar surfaces, the plantar surface being more extensive than the dorsal surfaces to facilitate flexion. The sides of the heads have, like the metacarpals, a depression for ligamentous attachment and tubercles placed posteriorly [FIG. 3.211]. They rest on plantar ligaments which have only a loose attachment to the metatarsal, proximal to the head.

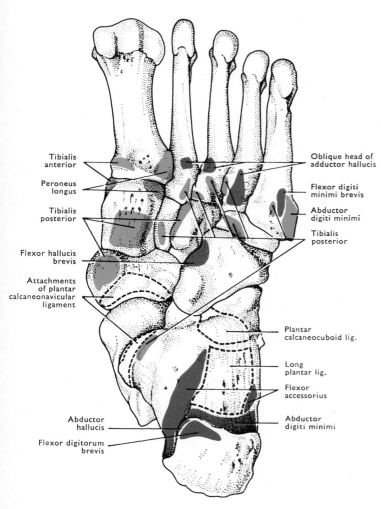

Tibialis anterior
Peroneus longus
Tibialis posterior
Flexor hallucis brevis
Attachments of plantar calcaneonavicular ligament
Abductor hallucis
Flexor digitorum brevis

Oblique head of adductor hallucis
Flexor digiti minimi brevis
Abductor digiti minimi
Tibialis posterior
Plantar calcaneocuboid lig.
Long plantar lig.
Flexor accessorius
Abductor digiti minimi

FIG. 3.209. Muscle attachments to left tarsus and metatarsus (plantar aspect).

OSSIFICATION

A primary centre appears for the body of each metatarsal in the ninth week and they are well ossified at birth. Secondary centres appear for the base of the first and for the heads of the others during the second and third years, and they all fuse with the bodies between 15 and 18 years. Epiphyses may, however, be found in the bases of lateral metatarsals instead of in the heads.

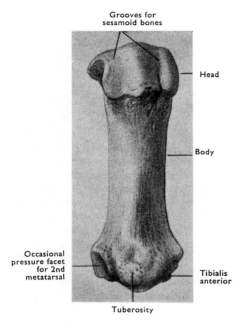

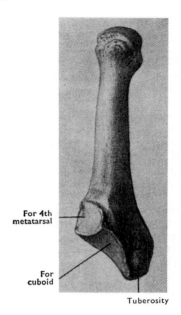

FIG. 3.212. Right fifth metatarsal bone (dorsal aspect).

FIG. 3.210. Right first metatarsal bone (plantar aspect).

2ND METATARSAL

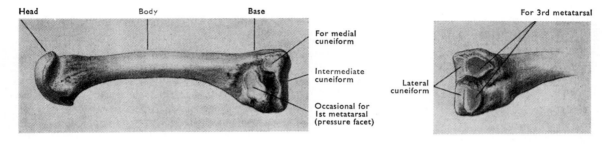

3RD METATARSAL

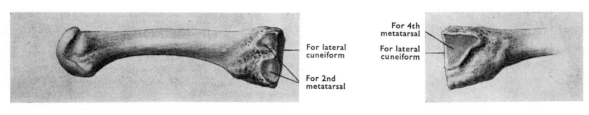

4TH METATARSAL

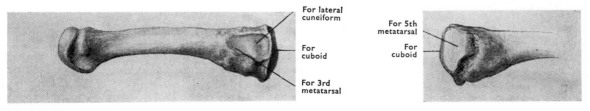

FIG. 3.211. View of bases and bodies of second, third, and fourth metatarsal bones of right foot. Note that the second is the longest.

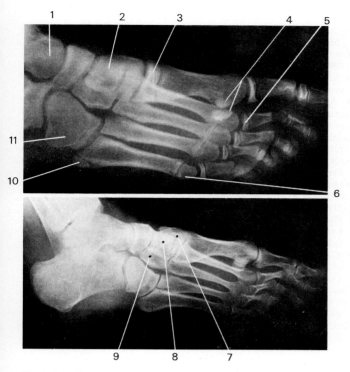

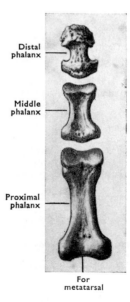

FIG. 3.213. Oblique radiographs of the foot at 12 years and in adult.
1. Head of talus.
2. Medial cuneiform superimposed on intermediate cuneiform.
3. Epiphysis of first metatarsal.
4. Sesamoid bones.
5. Epiphysis of proximal phalanx.
6. Epiphysis of 5th metatarsal.
7. Medial cuneiform.
8. Intermediate cuneiform overlapped by medial cuneiform.
9. Lateral cuneiform.
10. Epiphysis of tuberosity of 5th metatarsal.
11. Cuboid.

to the fibrous flexor sheath and part of the flexor digitorum brevis to each edge.

The **distal phalanx** only has a proximal articular surface which is slightly ridged to fit the pulley-shaped head of the middle phalanx. Its distal end is expanded and flattened with its plantar surface roughened by the attached fibrous tissue of the toe. The base with its articular surface is slightly expanded and gives attachment to flexor digitorum longus on its plantar aspect, the extensor expansion on its dorsal aspect, and a collateral ligament at each side.

The **phalanges of the great toe** are much larger than the others in association with the forces to which the great toe is subjected in walking and the consequent increase in muscle attachments. Because of the stresses there is commonly some degree of deformity, particularly hallux valgus (lateral deviation of the phalanges) even in persons who have never worn shoes (Lake 1956; Barnicott and Hardy 1955).

The **proximal phalanx** of the great toe is short and strong with a large base which assists in producing the ball of the great toe. The base has a large, oval, articular surface for the head of the metatarsal and is marked on each side of its plantar surface by the tendon of flexor hallucis brevis combined with abductor hallucis medially and adductor hallucis laterally. Dorsally it is marked by the insertion of extensor hallucis brevis. The plantar ligament is largely replaced by the tendons and sesamoid bones mentioned above.

The wide **body** is slightly concave on its plantar surface and provides attachment at its edges for a fibrous sheath for flexor hallucis longus tendon. It tapers slightly to a head with a pulley-shaped articular surface and markings for collateral ligaments on each side.

THE PHALANGES OF THE TOES

The phalanges are similar to those of the fingers but their proportions vary considerably. They are miniature long bones with a proximal base, a body, and a distal head. The great toe has two strong phalanges, proximal and distal. The remaining toes have three, a proximal, a middle, and a distal, or ungual, carrying the toe nail: they diminish from medial to lateral side of the foot. A proximal phalanx is about 2.5 cm long; the others are much shorter and, in the little toe, may be no more than nodules of bone.

The **proximal phalanx** has a relatively large base with a concave articular surface for the metatarsal. It provides attachment for interosseous muscles at the sides and for a plantar ligament inferiorly. Its dorsal aspect is covered by an extensor expansion. The body, except in the great toe, is slender, rounded, and convex dorsally with an ill-defined plantar surface which has the fibrous flexor sheaths attached to its edges. The head is relatively wide and has a pulley-shaped articular surface with rough areas at the sides for collateral ligaments.

The **middle phalanx** is ridged on the articular surface of its base and pulley-shaped on its head. The body is short and frequently ill-formed, especially at the lateral side of the foot, but gives attachment

FIG. 3.214. Phalanges of a toe (plantar aspect).

The **terminal phalanx** of the great toe is similar to those of the other toes except for its larger size. The plantar surface of its base is roughened by the attachment of flexor hallucis longus tendon and extensor hallucis longus is attached to the dorsal part of the base.

The **phalanges of the little toe** are similar to those of the other toes except that they are apt to be poorly developed and distorted. Flexor digiti minimi brevis and abductor digiti minimi are attached to the lateral side of the base of the proximal phalanx and the remaining attachments are similar to the other toes.

OSSIFICATION

Primary centres for the distal and proximal phalanges appear between the end of the third month and the end of the fourth month, the distal ones appearing first. The centre for the middle phalanx appears later, between the sixth month and the time of birth. Secondary centres for the bases appear during the second and third years and fuse with the bodies between 15 and 21 years.

References

ASHLEY, G. T. (1956). The relationship between the pattern of ossification and the definitive shape of the mesosternum in Man. *J. Anat.* (*Lond.*) **90**, 87.

BARNETT, C. H. and LEWIS, O. J. (1958). The evolution of some traction epiphyses in birds and mammals. *J. Anat.* (*Lond.*) **92**, 593.

BARNICOTT, N. A. and HARDY, R. H. (1955). The position of the hallux in West Africans. *J. Anat.* (*Lond.*) **89**, 355.

BERRY, A. C. and BERRY, R. J. (1967). Epigenetic variation in the human cranium. *J. Anat.* (*Lond.*) **101**, 361.

BLACKWOOD, H. J. J. (1964). *Bone and tooth*. Pergamon Press, Oxford.

BOLK, L. (1915). On the premature obliteration of the sutures in the human skull. *Am. J. Anat.* **17**, 495.

BOUCHER, B. J. (1957). Sex differences in the foetal pelvis. *Am. J. phys. Anthropol.* **15**, 581.

BOURNE, G. H. (1956). *The biochemistry and physiology of bone*. New York.

BRAILSFORD, J. F. (1948). *The radiology of bones and joints* (4th edn.). Churchill, London.

BRASH, J. C. (1915). Vertebral column with six and a half cervical and thirteen true thoracic vertebrae, with associated abnormalities of the cervical spinal cord and nerves. *J. Anat. Physiol.* **49**, 243.

—— (1934). Some problems in the growth and developmental mechanics of bone. *Edinb. Med. J.* **41**, 305, 363.

—— McKEAG, H. T. H., and SCOTT, J. H. (1956). *The aetiology of irregularity and malocclusion of the teeth*. The Dental Board of the United Kingdom.

BRYCE, T. H. (1915). Osteology and arthrology, in *Quain's elements of anatomy* (11th edn.), Vol. 4, Pt. I, p. 177. London.

CALDWELL, W. E. and MOLOY, H. C. (1933). Anatomical variations in the female pelvis and their effect on labour, with a suggested classification. *Am. J. Obstet. Gynecol.* **26**, 479.

—— —— and D'ESOPO, D. A. (1934). Further studies on the pelvic architecture. *Am. J. Obstet. Gynecol.* **28**, 482.

—— —— —— (1935). A roentgenologic study of the mechanism of engagement of the fetal head. *Am. J. Obstet. Gynecol.* **28**, 824.

CHASSAR MOIR, J. (1964). *Munro Kerr's operative obstetrics*. London.

CLARK, W. E. le GROS (1971). *The tissues of the body* (6th edn.). Oxford University Press.

COMAS, J. (1960). *Manual of physical anthropology*. C. C. Thomas, Springfield, Ill.

CROCK, H. V. (1967). *The blood supply of the lower limb bones in Man*. E. & S. Livingstone, Edinburgh.

DIXON, A. F. (1910). The architecture of the cancellous tissue forming the upper end of the femur. *J. Anat. Physiol.* **44**, 223.

DUCKWORTH, W. H. L. (1915). *Morphology and anthropology*. Cambridge University Press.

DUTERLOO, H. S. and ENLOW, D. H. (1970). A comparative study of cranial growth in *Homo* and *Macaca*. *Am. J. Anat.* **127**, 357.

EDWARDS, J. J. and EDWARDS, M. J. (1959). *Medical museum technology*. Oxford University Press, London.

ENLOW, D. H. (1968). *The human face: an account of the postnatal growth and development of the craniofacial skeleton*. Harper Row, New York.

FLECKER, H. (1932). Roentgenographic observations of the times of appearance of the epiphyses and their fusion with the diaphyses. *J. Anat.* (*Lond.*) **67**, 118.

FRANCIS, C. C. (1940). The appearance of centres of ossification from 6–15 years. *Am. J. phys. Anthropol.* **27**, 127.

—— WERLE, P. P., and BEHM, A. (1939). The appearance of centres of ossification from birth to five years. *Am. J. phys. Anthropol.* **24**, 273.

FROST, H. M. (1963). *Bone remodelling dynamics*. C. C. Thomas, Springfield, Ill.

—— (1964). *Bone biodynamics*. London.

GOODSIR, J. and GOODSIR, H. D. S. (1845). *Anatomical and pathological observations*. Edinburgh.

GREEP, R. O. (1948). Alkaline phosphatase in bone formation. *Anat. Rec.* **100**, 667.

GREULICH, W. W. and PYLE, S. I. (1959). *Radiographic atlas of skeletal development of the hand and wrist* (2nd edn.). Stanford.

HAM, A. W. (1969). *Histology* (6th edn.). Lippincott, Philadelphia.

HANCOX, N. M. (1956). The osteoclast, in *The biochemistry and physiology of bone* (ed. G. H. Bourne). New York.

HANIHARA, K. (1959). Sex diagnosis of Japanese skulls and scapulae by means of discriminant functions. *J. anthropol. Soc. Nippon* **67**, 21.

HARRIS, H. A. (1933). *Bone growth in health and disease*. Oxford University Press, London.

HAVERS, C. (1691). *Osteologia nova*. London.

HRDLICKA, A. (1952). *Anthropometry*. Lippincott, Philadelphia.

HUNTER, J. (1772). Experiments and observations on the growth of bones, in *Palmer's edition of 'Works of John Hunter'*, Vol. IV (1837), p. 315. London.

KEEN, J. A. (1950). Sex difference in skulls. *Am. J. phys. Anthropol.* **8**, 65.

KOCH, J. C. (1917). The laws of bone architecture. *Am. J. Anat.* **21**, 177.

KROGMAN, W. H. (1962). *The human skeleton in forensic medicine*. C. C. Thomas, Springfield, Ill.

LACROIX, P. (1949). *L'organisation des os*. Liège.

LAKE, N. C. (1956). The problem of hallux valgus. *Ann. r. Coll. Surg. Engl.* **19**, 23.

LE DOUBLE, A. F. (1912). *Traité des variations des os de la colonne vertébrale de l'homme et de leur signification au point de vue de l'anthropologie zoologique*. Paris.

LERICHE, R. and POLICARD, A. (1926). *Les problèmes de la physiologie normale et pathologique de l'os*. Paris.

MARTIN, B. F. (1968). The origins of the hamstring muscles. *J. Anat.* (*Lond.*) **102**, 345

MEYER, G. H. (1867). Die Architectur der Spongiosa. *Arch. Anat. Physiol.* (*Lpz.*) 615.

MONTAGU, M. F. A. (1938). Ageing of the skull. *Am. J. phys. Anthropol.* **23**, 355.

—— (1960). *An introduction to physical anthropology*. C. C. Thomas, Springfield, Ill.

MURRAY, P. D. F. (1936). *Bones. A study of the development and structure of the vertebrate skeleton*. Cambridge University Press.

NICHOLSON, C, (1943). Accurate pelvimetry. *J. Obstet. Gynaecol. Brit. Emp.* **50**, 37.

—— (1945). The two main diameters at the brim of the female pelvis. *J. Anat.* (*Lond.*) **79**, 131.

NOBACK, C. R. and ROBERTSON, G. G. (1951). Sequences of appearance of ossification centers in the human skeleton during the first five prenatal months. *Am. J. Anat.* **89**, 1.

PARSONS, F. G. (1904). Observations on traction epiphyses. *J. Anat. Physiol.* **38**, 248.

—— (1905). On pressure epiphyses. *J. Anat. Physiol.* **39**, 402.

—— (1908). Further remarks on traction epiphyses. *J. Anat. Physiol.* **42**, 388.

PATERSON, A. M. (1904). *The human sternum*. London.

PATERSON, R. S. (1929). A radiological investigation of the epiphyses of the long bones. *J. Anat.* (*Lond.*) **64**, 28.

PAYTON, C. G. (1932). The growth in length of the long bones in the madder-fed pig. *J. Anat.* (*Lond.*) **66**, 414.

—— (1933). The growth of the epiphyses of the long bones in the madder-fed pig. *J. Anat.* (*Lond.*) **67**, 371.

—— (1935). The growth of the pelvis in the madder-fed pig. *J. Anat.* (*Lond.*) **69**, 326.

PRITCHARD, J. J. (1956). General anatomy and histology of bone, in *The biochemistry and physiology of bone* (ed. G. H. BOURNE). New York.

REYNOLDS, E. L. (1945). The bony pelvic girdle in early infancy. A roentgenometric study. *Am. J. phys. Anthropol.* **3**, 321.

—— (1947). The bony pelvis in prepuberal childhood. *Am. J. phys. Anthropol.* **5**, 165.

RICHELLE, L. J. and DALLEMAGNE, M. J. (1965). *Calcified tissues*. University of Liège.

ROBIN, C. (1849). Sur l'existence de deux espèces nouvelles d'éléments anatomiques qui se trouve dans le canal médullaire des os. *C.R. Soc. Biol.* (*Paris*) **1**, 149.

ROBINSON, R. A. and WATSON, M. L. (1952). Collagen-crystal relationships in bone as seen in the electron microscope. *Anat. Rec.* **114**, 383.

SCOTT, J. H. (1951). The development of joints concerned with early jaw movements in sheep. *J. Anat.* (*Lond.*) **85**, 36.

SHARPEY, W. (1856). *Quain's elements of anatomy* (6th edn.), Vol. 1. London.

STEVENSON, P. H. (1924). Age order of epiphyseal union in Man. *Am. J. phys. Anthropol.* **7**, 53.

THOMPSON, D'ARCY W. (1942). *Growth and form* (2nd edn.). Cambridge. University Press.

TODD, T. W. (1924). Thickness of the white male cranium. *Anat. Rec.* **27**, 245.

—— (1930). The anatomical features of epiphysial union. *Child Dev.* **1**, 186.

—— (1933). Growth and development of the skeleton, in *Growth and development of the child*, *Pt. II. Anatomy and physiology*. New York.

—— (1937). *Atlas of skeletal maturation*, Introduction: Pt 1, Hand. St. Louis.

—— and LYON, D. W. (1924–5). Endocranial suture closure, its progress and age relationship. *Am. J. phys. Anthropol.* **7**, 325; **8**, 23, 149.

TONNA, E. A. and CRONKITE, E. P. (1961). Use of tritiated thymidine for study of the origin of the osteoclast. *Nature (Lond.)* **190**, 459.

TREVOR, J. C. (1955). Anthropometry, in *Chambers's encyclopaedia*, **1**, 458–62. London.

TROTTER, M. and GLESER, G. C. (1952). Estimation of stature from long bones of American Whites and Negroes. *Am. J. phys. Anthropol.* **10**, 463.

—— —— (1958). A re-evaluation of estimation of stature based on measurements of stature taken during life and of long bones after death. *Am. J. phys. Anthropol.* **16**, 79.

WALMSLEY, R. (1940). The development of the patella. *J. Anat. (Lond.)* **74**, 360.

WARD, F. O. (1838). *Outlines of human osteology*, p. 370. London.

4 Joints

R. A. STOCKWELL

THE bones of the body are joined to one another at joints or articulations and the study of these is called syndesmology or arthrology. Although syndesmology (*syndesmos* (Gk.) = a ligament) is the recognized anatomical term, arthrology more specifically indicates a discussion of 'joints'. The Greek word 'arthron' has been widely adopted in standard and accepted clinical terminology as in 'arthritis' and 'arthrodesis' and for this reason also the term 'arthrology' is to be preferred.

Joints vary considerably in structure and function but fundamentally provide for the stability and movement of the skeleton. The form of the articulating bone surfaces and the type and arrangement of the connecting tissues intervening between the bones at a joint are the principal factors that determine the amount of movement which may occur at it. At many joints in the adult skull the articulating bones are so interlocked and so united by a continuous band of fibrous tissue that little or no movement is permitted at them; this example is given in this introductory statement merely to emphasize that the anatomical designation 'joint' does not always imply movement. Other joints, and especially those in the limbs, are much more complex in their structure, and permit considerable movement. The structure of a joint is an expression of the function that it subserves, and the function and structure of joints should always be considered concurrently.

The variation that exists in the form and function of the various joints of the body allows them, for descriptive purposes, to be grouped into well-defined classes. This classification is, moreover, dependent on the manner in which bones are ossified and some knowledge of the principal features of bone formation is a necessary prelude to a consideration of the classes of joints [pp. 79–88].

Not all joints occur where bones meet. For example, the important joints of the larynx occur between hyaline cartilages [p. 501].

Classification of joints

Three distinct classes of joints are recognized and each class may have two or more types. The three classes are:

1. **Fibrous joints**, where the bones are joined by fibrous tissue.
2. **Cartilaginous joints**, where the bones are united by hyaline cartilage or fibrocartilage.
3. **Synovial joints**, in which the articular surfaces of the bones are covered with articular cartilage and are separated by a potential space enclosed by an articular capsule which holds the bones together. The space is filled with synovial fluid, hence the designation 'synovial joint'.

FIBROUS JOINTS

Fibrous joints are of three types, namely **sutures**, **gomphoses**, and **syndesmoses**.

Sutures

A suture is a type of fibrous joint which is found only in the skull and at which no active movement occurs. The fibrous layers of periosteum [p. 77] on the outer and inner surfaces of the articulating bones bridge the gap between the bones and constitute the main bond between them [FIG. 4.1]. Between the articular surfaces of the bones there is an intermediate layer of vascular fibrous tissue which also unites the bones and together with the two layers of periosteum constitutes the **sutural ligament**. In the vault of the skull, where the inner periosteal layer is fused with the dura mater, the veins of the

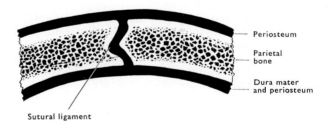

Periosteum

Parietal bone

Dura mater and periosteum

Sutural ligament

FIG. 4.1. Vertical section through a suture (sagittal).

sutural ligaments communicate with the dural venous sinuses, the diploic veins of the adjacent bones, and the extracranial veins. The histological structure of the sutures has been described by Pritchard, Scott, and Girgis (1956) in some detail and they have stressed that the sutures not only constitute a firm bond of union between neighbouring bones but that they are also one of the sites at which active bone growth occurs. The sutures are not usually permanent structures as obliteration of the simplest of them commences before the thirtieth year. The obliteration is effected by the growth of the articulating bones towards each other. This process of **synostosis** first occurs on the deep aspect of the suture and slowly extends to its superficial part. In a middle-aged skull some sutures may therefore be obliterated on their endocranial aspect but distinctly visible on their external (pericranial) surface. Complete obliteration of the sutures does not occur until an advanced age.

Different varieties of suture are named from the form of the articulating surfaces and the ways in which the opposed surfaces are fitted to each other. The opposed surfaces are seldom smooth and when the roughness is not too exaggerated and does not have a characteristic form it is termed a **plane** suture, e.g. the suture between the horizontal plates of the two palatine bones. Sutures are commonly strengthened by fairly regular projections of one bone fitting between the projections of the opposing bone and these are termed **serrate** sutures, e.g. the sagittal suture between the two parietal bones. In a **squamous** suture the margin of one bone overlaps that of the opposing bone to a greater or lesser extent; the joints of the squamous temporal bone represent examples of this variety of suture. A 'schindylesis' is a form of suture in which one bone edge fits into the grooved margin of the other, as in the joint between the upper border of the vomer and the sphenoidal rostrum.

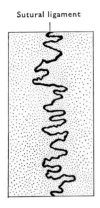

FIG. 4.2. Surface view of serrate suture (sagittal).

Gomphosis

This is a type of fibrous joint in which a peg fits into a socket, e.g. the roots of the teeth each suspended by a fibrous periodontal membrane in its socket in the maxilla or mandible.

Syndesmosis

A syndesmosis is a fibrous joint in which the uniting fibrous tissue, much greater in amount and more dense than in a suture, forms an interosseous membrane or ligament. The inferior tibiofibular joint is a syndesmosis where the bones are joined by an interosseous ligament; the attachment of the shaft of the radius to that of the ulna may likewise be regarded as an example of syndesmosis in which an interosseous membrane unites the bones. However it is difficult to define a syndesmosis precisely since there is no adequate criterion which precludes the extension of the term to almost any ligament joining two bones. Movement may be permitted to a small extent at a syndesmosis by the flexibility of an interosseous ligament.

CARTILAGINOUS JOINTS

In a cartilaginous joint the bones are united by a continuous plate of hyaline cartilage or disc containing fibrocartilage. Cartilaginous joints are of two types, namely **synchondrosis** and **symphysis**, which differ in development, structure, and function.

Synchondrosis

The cartilage of the synchondrosis is a part of the continuous cartilaginous mass in which the bones are formed and, as long as the joint persists, it retains its original hyaline nature. Hence this is regarded as a *primary cartilaginous* joint. The commonest synchondroses are the epiphysial (or growth) cartilages between the diaphysis and epiphyses of a long bone and the neurocentral joints in a vertebra; the joint between the occipital and sphenoid bones (spheno-occipital synchondrosis) has the same pattern and fate [FIG. 4.3]. These joints are concerned with the growth of the bones and are obliterated by the fusion of the related bones certainly before the age of 25 and often considerably earlier. While they persist these joints are virtually immobile because hyaline cartilage

is relatively rigid and the plates which they form are not very thick. Two synchondroses, however, may persist throughout life; these are the first sternocostal and the petrobasilar.

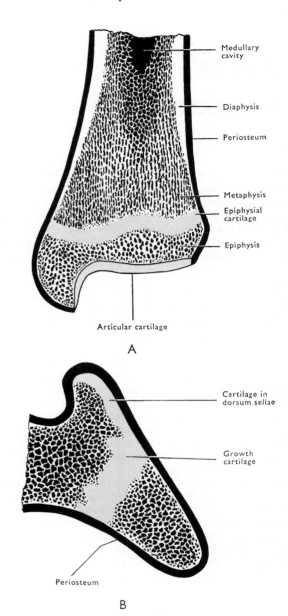

FIG. 4.3. Synchondroses.
A, diagrammatic section of joint between diaphysis and epiphysis of growing long bone.
B, median section of joint between occipital and sphenoid bones in child aged 1 year.

Symphysis

In a symphysis the joint is more specialized, occurs only in the median plane, and persists throughout life. In its early stages of development it is represented by a plate of cellular mesenchyme between separate cartilaginous rudiments [FIG. 4.4], but later this plate differentiates into fibrocartilage. It is therefore classified as a *secondary cartilaginous* joint. When the cartilaginous rudiments ossify, a thin lamina of the cartilage adjoining the fibrocartilage remains unossified and persists as a lamina of hyaline cartilage between the bone and the fibrocartilage. The joints between the bodies of the vertebrae and the joint between the bodies of the two pubic bones are the only typical examples of symphyses. Fibrocartilage, if sufficient in amount, allows a slight movement and it is for that reason that movement in the vertebral column is possible. Both the symphysis pubis and the intervertebral discs between the bodies of the vertebrae show further modifications and these are considered in the discussion of these joints.

There is one joint in the body which, although belonging to the class of cartilaginous joints, is not typical of either a synchondrosis or a symphysis. It is the manubriosternal joint. The sternum at the fourth month of intra-uterine life is formed of hyaline cartilage and thereafter ossific centres appear in the manubrium and the body. The manubriosternal joint is therefore a synchondrosis in the first instance, but as its hyaline cartilage becomes fibrocartilage at a later date, its definitive form is that of a symphysis.

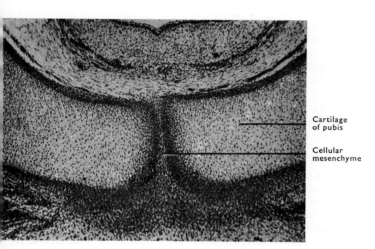

FIG. 4.4. Photomicrograph of symphysis pubis of 35-mm human foetus to illustrate development of a symphysis. × 60.

SYNOVIAL JOINTS

Synovial joints permit free movement. The great majority of the permanent joints of the limbs are therefore synovial whereas many of the joints in the rest of the body are either fibrous or cartilaginous. Movement of one bone on another, particularly in joints of the lower limb, involves problems of load-bearing, lubrication, and stability, more acute than in the fibrous and cartilaginous joints. The structural features of synovial joints are specialized to withstand the mechanical stresses involved, at least for a large fraction of the lifespan.

GENERAL FEATURES

The articular end of a bone is usually enlarged; this provides more surface area for muscle and ligamentous attachments. The large articular surface also results in a diminished load per unit area of bone. The articular surfaces of the bones are covered by a thin lamina of **articular cartilage**. This contains no lymphatics or nerves and has blood vessels only in its deepest layer adjacent to the bone. The absence of these vulnerable structures and the elastic deformability of cartilage make it an ideal cushioning material to distribute the load on the end of the bone. The bones are connected to each other by a tubular ligament, the **articular capsule**, which surrounds their articular surfaces and encloses the joint cavity. The outer layer of the articular capsule is formed of dense fibrous tissue, is richly innervated, and is called the **fibrous capsule**. This holds bone to bone at a joint and often exhibits thickenings which form named

anatomical ligaments. It is aided by separate **accessory ligaments** which may be either intra- or extracapsular. The inner layer of the articular capsule is formed by the highly vascular **synovial membrane** which secretes a small amount of **synovial fluid (or synovia)** into the joint cavity. Synovial fluid acts as a lubricant for the articulating surfaces and as a source of nourishment for the articular cartilage.

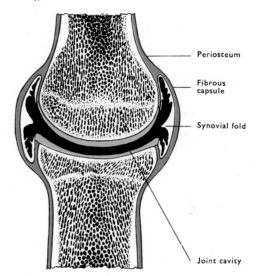

FIG. 4.5. Diagram of a synovial joint. Articular cartilage, blue; synovial membrane, white.

Classification of synovial joints

Synovial joints may be subdivided for further consideration in terms of the shape of the articulating surfaces and the movements performed at the joint. These two factors are intimately related and may be used together in a system of classification. From the point of view of form only, a distinction can be made between **homomorphic** and **heteromorphic** joints. The articular surfaces in a homomorphic joint are similar in conformation; those in a heteromorphic joint are dissimilar. Combining further consideration of shape with an analysis of movement, we distinguish **plane** and **saddle** joints of homomorphic pattern, and **hinge, pivot, ball-and-socket, condyloid**, and **ellipsoid** joints in the heteromorphic division.

Articulatio plana or plane joint

In a typical plane joint the opposed surfaces are flat, or relatively flat, and are approximately equal in extent. Movement is of a simple gliding type or the twisting of the one bone on the other within the narrow limits permitted by a slight laxity of the fibrous capsule. Such joints are found between some of the carpal and tarsal bones, and between the articular processes of some adjacent vertebrae.

Articulatio sellaris or saddle joint

Here the concavoconvex articular surface of the one bone is reciprocally homomorphic with the convexoconcave surface of the other. The only good example is the carpometacarpal joint of the thumb, and the multiple movements that occur at it are considered in the description of this joint [p. 240].

Ginglymus or hinge joint

In this, the articulating surfaces are arranged to permit movement around only one axis—and that a transverse one—as in the hinge of the lid of a box. The interphalangeal joints are good examples. Usually, the articulating bones are in line with each other in the normal anatomical position; movement around the transverse axis causing angulation at the joint is termed **flexion**, while movement in the reverse direction is called **extension**. In cases where the normal position is angular, as at the ankle joint, suitably modified terms may describe the movements with less ambiguity, as **dorsiflexion** and **plantar flexion**. Two features in the disposition of the ligaments are common to all hinge joints. In the first place, the fibrous capsule must be of sufficient length in front and behind to permit the full range of movement; in either full flexion or full extension part of the fibrous capsule will be taut and part will be lax. In the second place, strong ligaments are placed at the sides of the joint and these are called **collateral ligaments**. One end of each ligament is attached in the axis of the hinge and some parts of these ligaments are therefore taut in all positions of the joint without restricting its movement.

Articulatio trochoidea or pivot joint

Here, also, movement can take place around only one axis, but it is a vertical one, as in the hinge of a gate. A more or less cylindrical articular surface **rotates** within a ring formed partly by bone, partly by ligament, as in the proximal radio-ulnar joint and in the joint between the dens and the anterior arch and transverse ligament of the atlas.

Articulatio spheroidea or ball-and-socket joint.

Here, as in the hip and shoulder joints, a 'ball' formed by a spheroidal surface on the one bone rotates in a 'socket' provided by a concavity in the other. This variety of joint allows a greater range of movement than any other, as not only are there an almost infinite number of axes but the amount of movement in any one of them is considerable. It is this freedom of movement at a ball-and-socket joint that distinguishes it functionally from other types of joint. For descriptive convenience three principal axes at right angles to one another are specially considered. Around a transverse axis, movements of **flexion** and **extension** take place as in a hinge joint. Movements sideways around an anteroposterior axis are named **abduction** in the case of a limb raised laterally, away from the median plane of the body, and **adduction** when the limb is brought towards or across the median plane. Around a vertical axis through the centre of the ball, the limb **rotates, medially** or **laterally**. **Circumduction** is a composite movement around a number of axes whereby the limb describes the side of a cone, the apex of which corresponds to the centre of the 'ball'. The stability of such a joint depends on the depth of the 'socket', but the greater the depth of the socket, the more the movement is limited.

Articulatio condylaris or condyloid joint

In this variety the articular surfaces are of the ball-and-socket type; but, owing to the disposition of the ligaments and muscles, the movement of active rotation around a vertical axis is absent. **Flexion** and **extension**, **abduction** and **adduction**, and **circumduction** take place. The metacarpophalangeal joints of the fingers (but not of the thumb) are condyloid joints.

Articulatio ellipsoidea or ellipsoid joint

This is a further modification of the ball-and-socket joint; the articular surfaces, instead of being spheroidal, are ellipsoid, as in the radiocarpal joint. The curvature in the long axis of the joint surfaces has a greater radius than in the short axis. Movement takes place around two principal axes at right angles, corresponding to the greatest and least diameters of the joint, and to a slight extent around intermediate axes. Circumduction also is allowed, but the form of the ellipsoid surfaces prevents rotation.

Surfaces and movements

The method of classification of joints and their movements described above is valuable when considering the joint as a whole, either anatomically or clinically. However such a classification tends to neglect the variation in curvatures of the surfaces of a joint and can be misleading because in some cases it introduces erroneous ideas about the movements actually occurring between joint surfaces. A rather different approach assumes only two basic types of joint surface (MacConaill 1946). An ovoid surface is convex (the female surface concave) in any vertical section cut across it. A sellar surface has one direction where a section will show a convex curve and another direction, in general at right angles to the first, showing a concave curvature (the female surface is reciprocally concavoconvex but is the smaller of the two surfaces). The curvature of a joint is never constant. This is more apparent in some joints than others, for example the 'spiral' of the femoral condyle seen in anteroposterior section (Goodsir 1858). With respect to the path traced out by a point on one surface over the other surface, variation in curvature is probably true of all joints. It follows that in all joints, the degree of congruity of the two surfaces changes as one surface travels over the other. In many joints, the surfaces are most incongruent, and therefore the joint most mobile, in the middle of the range of movement. Conversely it is most stable at the extremes of the range when the surfaces become most congruent. Here the contact area is greatest (the 'close-packed' position of the joint) and the load is distributed most effectively; hence the weight-bearing position of the joint tends to be at one extreme of movement, for example full extension at the hip joint.

It is said that all joint movements can be resolved into two basic kinds of motion of one articular surface on another (Barnett, Davies, and MacConaill 1961)—spinning and sliding. In spinning the moving surface rotates about an axis normal to the fixed surface at the point of contact, while sliding comprises all movements other than spinning. Anatomical rotation often approximates to spinning but at the shoulder and hip joints, the heads of the bones spin in their sockets as the shafts of the bones follow the anatomical movements of flexion and extension. The other kind of displacement, sliding, can occur along a 'chord' or along an 'arc', i.e. the slide path on the stationary articular surface when viewed *en face* is straight (a chord) or curved (an arc). It follows that if the slide path is an arc then the moving bone undergoes a turning movement or spin. Such a spin occurring in conjunction with another movement is known as 'conjunct rotation', as distinct from 'adjunct rotation' which is not dependent on another movement for its existence. In an ovoid joint the amount of spin is dependent on the length of the arc; in sellar joints it is determined and constrained by the shape of the surfaces and the degree of congruity. If the slide is such that when viewed from the side any point on the moving bone (except on the axis of movement) traces out a curved path in space, then the slide may be termed a 'swing'; this implies that there is a change in angulation of the bones. If when viewed from the side there is no such curved path then the slide is a form of 'translation', i.e. change in position of the point of contact of the two articular surfaces with little or no change in angulation of the bones.

These types of motion underlie even the most complicated joint movements, usually occurring together. For example, the 'rolling' of the femoral on the tibial condyle during flexion/extension of the knee is really a combination of swing and translation, while spin around a vertical axis through the lateral femoral condyle occurs mainly during the last few degrees of extension. The value of these more generalized concepts of joint movement lies in emphasizing the real motions occurring at the articular surfaces, particularly in relation to patterns of wear and damage to the surfaces during life. In this context, movement at a small area of the articular surface is either perpendicular, i.e. approximation and separation of the two surfaces on application and removal of load, or tangential, i.e. sliding. Spinning in reality implies that all parts of the contact area, except the still centre at the axis, are sliding along an arc. This is especially true of joints under load where cartilage deformation occurs and the contact area enlarges as the surfaces flatten.

Components

ARTICULAR CARTILAGE

In any joint, all of the surface of a bone that comes into contact with another during movement is covered with articular cartilage. This is usually of the hyaline variety although the articular surfaces in joints such as those of the clavicle and in the temporomandibular joint are covered with fibrocartilage. Its thickness varies considerably. In large joints such as the knee and hip it lies in the range 1.5–4 mm while in small joints such as the interphalangeal and those between the auditory ossicles it is only a fraction of a millimetre.

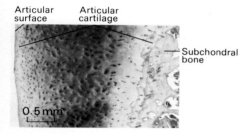

FIG. 4.6. Photomicrographs illustrating the structure of articular cartilage. A, Section of human femoral condylar cartilage. Note the thickness of the cartilage of the knee joint compared with that of the interphalangeal joint in B. × 15.

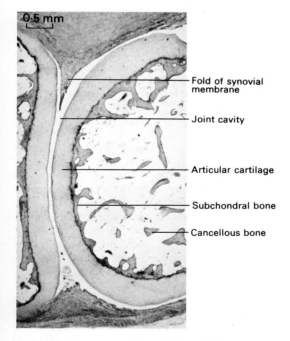

B, Section of human distal interphalangeal joint. The base of the distal phalanx is to the left. The articular cartilage lies on a thin plate of compact bone which is supported by cancellous bone. The 'space' between the cartilages, normally occupied by synovial fluid, is enlarged due to shrinkage during preparation of the specimen. × 15.

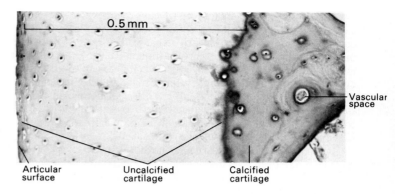

C, Section of articular cartilage of human distal interphalangeal joint (enlargement of part of B). The chondrocytes are flattened just beneath the articular surface but are more spheroidal in the deeper tissue. A basophilic line (the 'tidemark') demarcates the junction of calcified with uncalcified cartilage. The lamellar subchondral bone forms an irregular junction with the calcified zone of the cartilage. Note the vascular space in the subchondral bone. × 80.

Within a joint, the greatest thickness is usually at the centre of the convex surface and at the margins of the concave surface. Thicker cartilage is found in joints where the surfaces are highly incongruent, thus increasing the potential for deformation and hence the area of contact between the apposed articular cartilages (Simon, Friedenberg, and Richardson 1973).

Cartilage consists of a small number of cells (chondrocytes) embedded in matrix. This is composed of water (75 per cent), collagen, and ground substance—mostly mucopolysaccharide. Most of the cartilage is uncalcified, except for a thin lamina adjacent to the subchondral bone. The chondrocytes lie in lacunae and are most numerous near the joint surface. Here they are discoidal in shape, flattened in a plane parallel to the articular surface. In the deeper tissue they are spheroidal and occur in groups. Their function is to produce and maintain the matrix. Their environment is avascular and their oxygen requirement is low, nevertheless they are active in the synthesis and degradation of mucopolysaccharide molecules; collagen turnover is minimal or absent in adult articular cartilage so that collagen fibres may undergo ageing. The cellularity of articular cartilage varies almost inversely with its thickness (Stockwell and Meachim 1973). Hence cartilage of the hip is much less cellular than that of small joints with thin cartilage. The mechanical properties of articular cartilage are a consequence of the anatomical and physicochemical organization of the matrix macromolecules. Collagen fibres have a high tensile strength; in cartilage they are of narrow diameter (30–200 nm) and of a characteristic chemical variety (Type II). The mucopolysaccharides or glycosaminoglycans are mostly chondroitin sulphate and keratansulphate. These are linear carbohydrate polymers with a high negative charge resulting from their numerous sulphate and carboxyl groups. The glycosaminoglycans form part of larger molecules termed proteoglycans. Each of these consists of a protein core with glycosaminoglycans attached as side-chains at intervals along its length. In cartilage, proteoglycans themselves are attached as side-chains to long molecules of hyaluronic acid—also a glycosaminoglycan. The resulting proteoglycan–hyaluronic acid aggregates are very large indeed, with molecular weights of $30–50 \times 10^6$ and over a micrometre in diameter in their fully expanded form (Muir 1977).

The reversible compressibility of cartilage is largely due to the properties of the middle zone of the tissue, where proteoglycan concentration is highest and collagen fibres are randomly orientated. They form a mesh in which the large proteoglycan aggregates are entangled and restrained. Compression deformation of the cartilage under prolonged load depends on the movement of tissue water away from the stressed area. Proteoglycans retain water and endow cartilage with an internal 'swelling pressure' by virtue of their colloidal nature and fixed negative charge. Hence they resist the movement of water and are in turn immobilized by the collagen mesh. The amount of water lost and the extent of cartilage deformation is dependent on the relative magnitudes of the load and the internal swelling pressure (Maroudas 1973). On removal of load, the original equilibrium is regained. Thus loss of the integrity of the collagen mesh or depolymerization of the proteoglycans will result in reduction of cartilage resilience.

The precise nature of the articular surface is still in doubt. To the naked eye and even transmission electron microscopy, the cartilage surface can appear perfectly smooth, in harmony with the extremely low coefficients of friction found in biological joints. There is evidence, however, that the surface may have an undulating contour, resembling that of a dimpled golf-ball. The hollows and mounds may be caused by the underlying superficial cells since their frequency and diameter correspond. Minor irregularities and fissures of the surface, seen only at high magnification, occur in otherwise perfectly normal joints. These may be associated with normal wear and tear and it is uncertain whether these features necessarily progress to frank pathological change. Fine collagen fibres separated by a small amount of ground substance occupy the acellular zone immediately beneath the articular suface. Their orderly arrangement parallel to the articular surface may account for the brilliant transmission of light through this zone in cartilage sections, dubbed the 'lamina splendens'. The dense array of fibres resists the tangential tensile stresses produced by deformation of the surface under load and by shearing forces set up during joint motion. The interface with the synovial fluid may be an electron-dense lamina 8 nm thick, regarded as typical of an intact surface (Meachim and Stockwell 1973). Thicker layers of filamentous and electron-dense amorphous material are not uncommon, lying either on or within the cartilage.

The deep surface of the cartilage abuts on the subchondral bone. The articular end of the bone is made up of a thin shell of compact bone supported by an organized meshwork of cancellous bone. This gives the bone sufficient mechanical strength at little cost in weight of material. Furthermore, dynamic stresses transmitted through the joint are absorbed more effectively by cancellous bone which is more elastic than thick compact bone. The interface with the calcified layer of the cartilage is very irregular. Collagen fibres do not cross from the bone into the calcified cartilage. Hence the mechanical strength of the junction depends on the interlocking of calcified cartilage and bone and on an intermediate mineralized 'cement line' which itself shows undulations of the order of a micrometre (Sokoloff 1973). The calcified cartilage in most joints is about 0.1 mm thick and contains viable chondrocytes. The junction with the uncalcified cartilage is relatively smooth compared with the osseochondral boundary and in histological sections is indicated by a basophilic line, the 'tidemark'. Collagen fibres 'anchored' in the calcified lamina run radially through the tidemark into the uncalcified tissue. The tidemark and the adjacent uncalcified cartilage form a plane of weakness (relative to the osseochondral junction), presumably due to the abrupt change in elastic moduli there. Blood vessels in the subchondral marrow spaces occasionally penetrate the bone to enter the calcified cartilage. Such vascular contacts are of potential significance in nutrition of the cartilage and in the remodelling of the osseochondral boundary.

At the periphery of the articular area, the cartilage becomes continuous with periosteum and the deep fibrous layer of the synovial membrane through a fibrocartilaginous region known as the marginal transitional zone. The connective tissue in this region is thought to be less specialized than elsewhere. It is particularly well nourished because of a layer of vascular synovial membrane which overlaps the superficial aspect of this zone and contains capillary loops of the *circulus articuli vasculosus* of William Hunter (1743). Hence the marginal transitional zone possesses a marked proliferative capacity and this facilitates its overgrowth in osteoarthrosis, resulting in the 'osteophytes' which produce 'lipping' of the articular margin.

The central part of the articular area can obtain its nutrient only from two sources, the synovial fluid and the subchondral marrow vasculature. The relative importance of these two routes has long been debated but it now seems clear that the subchondral route is of much less significance in mature than in immature joints. Numerous observations show that the synovial fluid, provided that it is well-stirred by joint motion, is adequate to nourish the full depth of adult articular cartilage. Passage of nutrients through the cartilage occurs by diffusion which is not increased by intermittent compression of the cartilage. It is believed that joint immobilization is disadvantageous because synovial fluid stagnation results in poorer diffusion and not because movement produces pumping mechanisms which aid nutrition (Maroudas, Bullough, Swanson, and Freeman 1968). While the central part of the articular area may be less well nourished than the periphery, and the more severe and progressive degenerative cartilage changes occur in this area, degeneration is observed earlier at the articular margin.

The intrinsic repair capacity of articular cartilage is poor. Repair may be considered in relation to 'full thickness' defects where the underlying bone marrow is exposed and 'partial thickness' defects which are limited to the uncalcified cartilage. In both forms, chondrocytes in the zone adjoining the defect degenerate while the adjacent cells react by forming multicellular clusters. However this intrinsic response rarely leads to healing. In full thickness lesions, extrinsic repair tissue, usually from the subchondral bone but sometimes also from the peripheral synovial tissue, fills the defect and healing eventually occurs by fibrocartilage. In the case of partial thickness defects, it is believed that synovial membrane can be a source of repair tissue, although healing rarely occurs.

FIBROUS CAPSULE (MEMBRANA FIBROSA)

The fibrous capsule is the outer layer of the articular capsule. It is a barrel-shaped tube of dense, white fibrous (collagenous) tissue which by its flexibility permits movement at a joint but by its great tensile strength resists dislocation. The fibrous capsules of the joints of the ear ossicles are predominantly elastic tissue. The size and shape of the tube varies with that of the articulating bones. At each end the tube is firmly attached to the periosteum of the articulating bones at a variable distance from the edge of the articular cartilage. This line of attachment varies in its relation to the epiphysial (growth) cartilage [if present, p. 82], which may be intracapsular, or extracapsular, or partly the one and partly the other. Small nerves and vessels pierce the fibrous capsule, and there may be larger openings through which synovial pouches protrude. The fibrous capsule may be strengthened or even replaced by adjacent tendons or their expansions. The fibres of the capsule may form a dense feltwork; but where any recurring strain must be resisted, the fibres are increased and arranged in parallel bundles in the line of the tensile stress; such specialized parts are usually called ligaments and are given special names. In addition to these, separate ligaments may unite the bones. These **accessory ligaments** are usually **extracapsular** as at the costoclavicular and sternoclavicular joints, but may be **intracapsular**, like the cruciate ligaments of the knee joint. In some joints, such as the acromioclavicular, accessory ligaments constitute the principal ligamentous connections between the articulating bones. The fibres in ligaments usually follow an oblique or even spiral course. Although ligaments are usually regarded as inelastic they may extend by a small proportion of their length when subjected to a load, and return to their original length when the load is removed (Annovazzi 1928; Smith 1954); ligaments

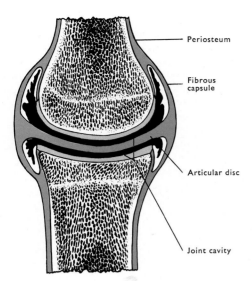

FIG. 4.7. Diagram of a synovial joint and articular disc dividing the joint cavity into two compartments. Articular cartilage, blue; synovial membrane, white.

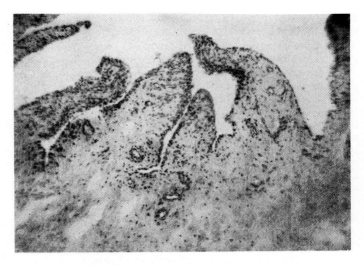

FIG. 4.8. Synovial folds from human knee joint.
The fibrous capsule (just appearing at lower end of photograph) is separated from the synovial membrane by loose areolar tissue. Note numerous blood vessels in the synovial membrane. × 100.

are, therefore, elastic in the true sense of the word. Ligaments and fibrous capsules have a poor blood supply and are slow to heal.

Inside some synovial joints, pads of fibrocartilage lie between the articular surfaces and blend peripherally with the fibrous capsule. Such a fibrocartilage may form a complete **articular disc** and so divide the joint cavity into two separate compartments [FIG. 4.7]; e.g. the mandibular and sternoclavicular joints. The intra-articular fibrocartilage may be wedge-shaped as in the acromioclavicular joint, or crescentic as in the knee joint, with a free edge projecting into the joint cavity. Articular discs, for example the menisci in the knee joint, add their resilience to that of the articular cartilages in cushioning the bone-ends, and assist in maintaining a layer of synovial fluid between articular surfaces (MacConaill 1932; see p. 218). In the sternoclavicular joint the articular disc is so attached to the bones and to the capsule that it also plays the part of an intracapsular ligament.

SYNOVIAL MEMBRANE

Synovial membrane is one of the characteristic features of synovial joints. It is a highly vascular connective tissue which lines the inner surface of the fibrous capsule and with it constitutes the articular capsule. At the attachments of the fibrous capsule the synovial membrane is reflected on to the bones to clothe them as far as the edge of the articular cartilage or of articular discs. Thus the synovial membrane covers those internal surfaces of the joint which are not formed by articular cartilage or discs. The surface of the synovial membrane is smooth, moist, and pink. It forms projections such as the small synovial villi and the larger synovial folds which help to increase its surface area, and fill small angular recesses between the edges of the articulating surfaces.

A varying amount of fatty tissue may separate the synovial membrane from the structures which it lines. When plentiful, such extrasynovial fatty tissue bulges the synovial membrane towards the joint cavity and thus may form soft and highly compressible fat pads or folds. At some joints the fat pads occupy an intra-articular depression in one bone, but may be displaced during movement to allow a projection on the moving bone to occupy the depression. Good examples of such movable fat pads are seen at the elbow joint. Articular fat pads tend to remain constant in size even in severe malnutrition. Rather than acting as a fat store, they probably have

a role in the mechanics of the joint by assisting the flow of synovial fluid during movement. The synovial membrane is also folded over intracapsular tendons and ligaments, which are thus excluded from the joint cavity. Where fibrocartilages or ligaments within a joint are exposed to fairly continuous pressure, the synovial membrane does not cover them. Thus, the menisci of the knee joint, under pressure between the articulating bones, have no synovial covering.

Microscopically, synovial membrane is composed of vascular connective tissue. Its thickness and cell content vary in different situations. Invariably it is more cellular near the joint surface and this 'intimal' lining rests on a 'sub-intimal' ('sub-synovial') layer which may consist of areolar, fibrous, or adipose tissue. Numerous mast cells are found in the subintimal layer and the more mobile parts of synovial membrane have many elastic fibres which prevent nipping of the membrane by the articular surfaces during joint movement (Davies 1945). The surface cells are very irregular in shape and do not form an unbroken continuous lamina, as in epithelia. Even where they form a layer several cells thick, electron microscopy demonstrates intervals 0.1–1.0 µm filled with amorphous material and some collagen fibres between the cells. Hence the blood plasma is separated from the joint cavity by endothelium and a variable thickness of the amorphous material. The nature of the blood–synovial barrier, and the large surface area and rich blood supply of the membrane [p. 219] accounts for the rapidity of absorption of low molecular weight substances from the joint cavity.

The surface cells appear to have a dual role. They secrete the mucin (hyaluronic acid) found in the synovial fluid and they 'scavenge' the joint cavity by phagocytosis of particles. Ultrastructural studies have shown that most (Type A) of the synovial lining cells contain prominent Golgi complexes and many vacuoles but little granular endoplasmic reticulum, while a minority (Type B) have profuse granular endoplasmic reticulum but scant Golgi membranes and few vacuoles. It is almost certain that the Type A and B appearances are dependent on functional activity and that they do not represent two distinct races of cells. Although the function of cells in the Type B state is uncertain the A cells both synthesize the synovial mucin and act as phagocytes. Macrophages in the subintimal layer are probably derived from intimal cells which have migrated into the deeper tissues following uptake of material from the joint fluid. Removal of colloidal particles thus occurs via the lining cells and thereafter by the lymphatic system. The rate of removal depends on particle size: large haemosiderin granules may remain in the subsynovial macrophages for long periods.

It will be clear that the joint or synovial cavity, bounded by synovial membrane and articular cartilage, is a closed cavity. Where an opening in the fibrous capsule allows the synovial layer to protrude, the imperforate continuity of the synovial membrane is always maintained. But the synovial cavity is not completely closed in a histological sense: the synovial membrane, unlike the pleura or the peritoneum of the coelomic cavities, has no complete epithelial lining as a barrier between the cavity of the joint and the tissue spaces around. The cavity is, in effect, an enormously enlarged tissue space from which rapid absorption can take place—an important feature in acutely infected joints (Ghadially and Roy 1969).

The **synovia** or **synovial fluid** is a clear, slightly yellow, glairy fluid, sufficient only to form a thin film over all surfaces within the joint cavity. Even in a large joint such as the knee joint less than 0.5 ml can be aspirated. The synovia is a dialysate of blood plasma with the addition of hyaluronic acid–protein complex (about 0.4 per cent) which endows the fluid with its characteristic viscosity. The fluid contains proteins but no fibrinogen since substances with a molecular weight in excess of 160 000 are excluded from it. Hence synovial fluid does not clot when removed from the joint. Synovial fluid is a source of nourishment for the articular cartilage. The few cells it contains (about 60 per mm³), with those of the synovial lining, act to remove micro-organisms and the debris of wear and tear from the joint (Coggeshall, Warren, and Bauer 1940). It also serves to lubricate the joint surfaces.

Effective lubrication depends on the maintenance of adequate lubricant between the articular surfaces during movement and under load. The viscosity of synovial fluid is non-Newtonian—it varies inversely with shear rate. However, under physiological loading conditions, shear rates are too high for the non-Newtonian property alone to be effective in maintaining fluid between the joint surfaces. In mechanical bearings, fluid-film lubrication results in lower coefficients of friction than in boundary lubrication where the thickness of the film is reduced to molecular proportions. A fluid film can be sustained between the surfaces in various ways, for example, by rapid movement of the articulating surfaces (hydrodynamic lubrication), or by entrapment ('squeeze-film' lubrication); alternatively a monomolecular layer (boundary lubrication) could be retained by adsorption on to the surface. It is probable that several modes of lubrication operate in synovial joints under different conditions of joint usage (Dowson *et al.* 1975) but the situation under load is of particular interest. Under high load either squeeze-film or boundary lubrication may occur. The formation of a squeeze-film is possible because hyaluronic acid solutions are viscous and because the articular surface of cartilage is permeable to water but impervious to high molecular weight hyaluronic acid. Hence a gel of hyaluronic acid forms on the articular surface as water is filtered away from the area under load, either through the cartilage or between the articular surfaces. On removal of load, the squeeze-film is immediately rehydrated (Maroudas 1973). On the other hand, hyaluronic acid itself may not be required for boundary lubrication under experimental loading conditions (McCutchen 1967). Although degradation of hyaluronic acid results in loss of viscosity, the coefficient of friction is not changed; proteolysis, however, causes an increase in coefficient but little change in viscosity. Hence the essential boundary lubricant may be a protein (Swann 1975).

Bursae

Closely akin to synovial membrane are the walls of certain closed sacs, known as **synovial bursae**, that are found in many situations in the body. The bursae contain fluid resembling synovia, and they serve to facilitate the play of one structure upon another.

Subcutaneous bursae are found between the skin and underlying bony prominences such as the olecranon and the patella. **Subfascial bursae** are similarly placed beneath the deep fascia. **Subtendinous bursae** are found where one tendon overlies another tendon or a projection of bone. The walls of such bursae in the neighbourhood of a joint may be continuous with the synovial membrane of the joint through an aperture in the fibrous capsule.

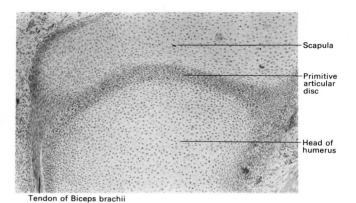
Scapula
Primitive articular disc
Head of humerus
Tendon of Biceps brachii

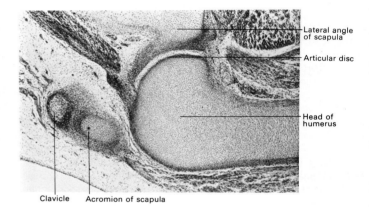
Lateral angle of scapula
Articular disc
Head of humerus
Clavicle   Acromion of scapula

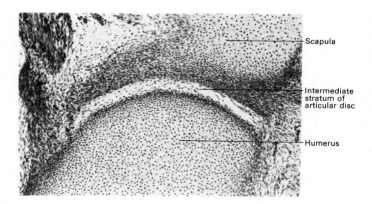
Scapula
Intermediate stratum of articular disc
Humerus

FIG. 4.9. Photomicrographs to illustrate the development of synovial joints.
A, Shoulder (glenohumeral) joint of a 20-mm human embryo. × 160. The primitive articular 'disc' or interzone, from which the joint cavity is formed, is very cellular at this stage and merges imperceptibly with the adjoining cartilages.
B, Shoulder joint of a 25-mm human embryo. × 64. The acromioclavicular joint is seen to the left of the glenohumeral joint in which the articular 'disc' is at a later stage of development.
C, Shoulder joint of a 25-mm human embryo. × 160. This shows the same specimen as in B. The nature of the articular 'disc' is clearly seen, and in places the intermediate stratum of loose tissue is becoming absorbed.

Of the same nature are the **synovial sheaths** that surround tendons running in osteofascial tunnels, e.g. the tendons passing behind the flexor retinaculum at the wrist, and the flexor tendons lying on the front of the fingers. The synovial layer lines the tunnel and covers the tendon that lies within it.

Development of a synovial joint

After the cartilaginous precursors of the bones have been formed, their ends are separated from each other by an equally avascular disc of very cellular mesenchyme that is called the **primitive articular disc** or interzone [FIG. 4.9]. There are no sharp lines of demarcation between the ends of the cartilages and the articular disc, and indeed the cells of a disc that are adjacent to cartilage form a chondrogenic zone that is associated with the growth in length of the cartilage (Haines 1947). As development proceeds, the disc laminates into two chondrogenetic layers separated from each other by an intermediate stratum of loose tissue in which the cells are relatively scanty [FIG. 4.9]. The intermediate stratum of loose tissue later undergoes liquefaction so that a joint cavity appears between the ends of the cartilages [FIG. 4.9]. The cartilages in turn lose the chondrogenic layers over the articular surfaces though a thin fibrillar layer may persist over them until birth. The vascular mesenchymatous tissue that surrounds the primitive articular disc differentiates into a dense outer layer that forms the **fibrous capsule**, and an inner loose layer that forms the **synovial membrane**. The synovial layer is invaded by extensions of the joint cavity so that the typical relationship of cavity to synovial membrane is established. **Intra-articular ligaments** and **fibrocartilages** arise from thickenings that project inwards from the wall of the primitive articular cavity. A complete **articular disc** arises from a transverse mesenchymatous septum between double cavities which appear in the primitive articular disc.

Blood and lymph vessels

A synovial joint is freely supplied by branches of the main arteries that are adjacent to it. These branches perforate the fibrous capsule and break up within the synovial membrane into capillaries which form a rich and intricate network. Many of the capillaries are extremely close to the free surface of the synovial membrane [FIG. 4.10] and this accounts for the frequency with which haemorrhage occurs into the articular cavity after even minor trauma (Davies 1945). Adjacent to the peripheral margin of the articular cartilage the larger vessels of the synovial membrane branch and anastomose to form the **circulus articuli vasculosus**. Arteriovenous anastomoses occur in the articular vessels as they do in so many other places in the body.

The **lymph vessels** form a plexus within the synovial membrane, and from it efferent vessels pass towards the flexor aspect of the joint.

Nerve supply

The fibrous capsule and, to a lesser extent, the synovial membrane are both supplied with nerves. The correlation that exists between the nerves of a joint and the nerves to the overlying tissues is expressed in Hilton's Law (Hilton 1863): 'The same trunks of nerves, whose branches supply the groups of muscles moving a joint, furnish also a distribution of nerves to the skin over the insertions of the same muscles; and the interior of the joint receives its nerves from the same source.' Some of the nerves in the fibrous capsule have encapsulated nerve endings and others have free nerve endings. There are no Pacinian corpuscles in joint capsules but many other encapsulated proprioceptive nerve endings occur. They are concerned not only with the reflex control of posture but also with the constant awareness of the precise position of a joint, as, for example, its degree of flexion or extension. Injury to joint ligaments

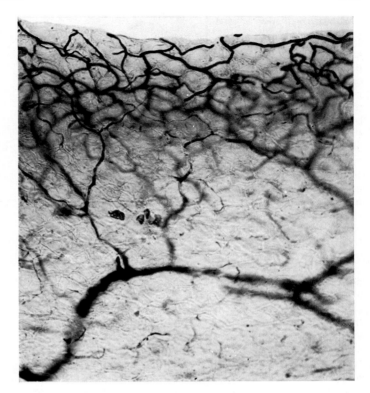

FIG. 4.10. Section of synovial membrane of rabbit (100 μm thick) after vessels had been injected with Indian ink. The close relationship of the vessels to the free (upper) surface of the synovial membrane is apparent. × 100.

is notoriously painful and the pain receptors are generally regarded as the free nerve endings. The synovial membrane is much less sensitive and most of the nerves in it are vasomotor.

Joint approximation forces

There are several factors which keep the articular surfaces in contact with each other in synovial joints. Two of these are minor factors and are common to all synovial joints, namely cohesion because the smooth joint surfaces are in contact with each other except for an intervening capillary layer of fluid, and atmospheric pressure because the atmosphere is excluded from the closed joint cavity. Another factor of very different value in different joints is the interlocking of the co-apted articulating surfaces; this is well marked in the hip joint, but is completely absent in plane joints.

The most important factors maintaining the apposition of joint surfaces are the strength of the joint ligaments and the tension of the muscles, although all tissues around a joint play a minor part. Ligaments and muscles act together in varying proportions; in the shoulder joint, with its relatively lax capsule, the ligamentous factor is small, whereas in the hip joint it is of greater importance. At the knee joint the fibrous capsule is largely formed of tendinous expansions from muscles acting on the joint, and this joint complex represents an excellent example of the close structural and functional co-operation of muscles and ligaments at a synovial joint. As the muscles increase in size in graduated exercise, so too, more and more collagen is simultaneously laid down in the ligaments, thereby increasing their strength. Even where the ligaments are not directly connected with muscles, they may increase in thickness if they are subject to the stress of graduated exercise. It must be emphasized, however, that the fibrous tissue of ligaments cannot indefinitely resist major stresses without the harmonious support of muscles—a feature of many flat-foot deformities.

Limitation of movement

Four factors are concerned in the limitation of movement at synovial joints: the apposition of soft parts, the locking of the articulating bones, the tightness of ligaments, and the tension of fully stretched or contracted muscles. **Apposition of soft parts** obviously limits movements in such joints as the elbow when, in flexion, the front of the forearm is pressed against the front of the arm. **Locking of the bones** probably occurs only exceptionally; thus, in the elbow joint, the occasional presence of pressure facets at the sides of the olecranon fossa of the humerus indicates that extension must have been limited by actual contact of the olecranon with the humerus. However, deformation of the cartilage and increasing congruence of the articular surfaces at the extremes of joint movement act as minor braking forces in all joints. The **tension of ligaments** is an important factor in the limitation of movement. The various parts of a fibrous capsule are tense only in certain positions of a joint and they are concerned not only in the final limitation of a movement but also in directing the movement of the articular surfaces on each other. In the hip and knee joints the major ligaments are lax in flexion and tense in extension. Furthermore, it is in the position of extension at these joints that the articular surfaces are in fullest contact with each other. Tense ligaments and maximal contact of the articular surfaces are the essential features of a 'locked' joint, and it is when a joint is 'locked' that it is most stable and best adapted for weight-bearing. At many joints **muscular tension** reinforces the restraint on movement imposed by the ligaments, if it does not actually check movement before the ligaments are fully stretched. This effect of the muscles is called their *passive insufficiency* or *ligamentous action*. A good example of limitation of movement by muscular tension is the check imposed on flexion of the hip by the tension of the hamstrings while the knee is fully extended; flexion of the knee lessens the tension of the hamstrings, and further flexion of the hip can then take place [FIG. 5.10]. The effect of muscles in limiting movement in the ordinary person can be appreciated when one considers the greater range of movement possessed by acrobats, who by constant practice conserve or even increase the full extensibility of muscle possessed by all in early life. In such people, however, the ligaments and articular surfaces also are modified to allow the extreme movements. The restraint that muscles normally exert on movement, apart from that exercised by ligaments, is shown by the extreme movements sometimes produced in the convulsions of tetanus or epileptic fits, when the safeguarding muscular restraint operating in ordinary voluntary action is withdrawn.

RADIOLOGY OF SYNOVIAL JOINTS

Cartilage has a radiographic density similar to that of other soft tissues, and so it cannot be demonstrated in joints except as an apparent 'space' between the articulating bones. The accuracy with which these 'joint spaces' can be examined in radiographs depends on the extent to which the cartilage is represented on the radiograph free from overlying bone and whether or not the subchondral bone forms a linear opacity. Ideally, the centre of the X-ray beam must pass exactly through the 'joint space' if an accurate assessment of the thickness of the articular cartilage is to be made [cf. FIGS. 4.27, 4.54, and 4.67 (subtalar joint)].

Introduction of contrast medium (air and/or iodine contrast medium) allows the articular cartilage to be seen more clearly. Whilst this technique (arthrography) has been used extensively in the knee joint [FIGS. 4.59 and 4.60] and less frequently in the shoulder, ankle, and hip joints, there are very many other joints to which it is not applicable.

In complicated articulations such as those of the **vertebral column**, such techniques are not possible and reliance on oblique tomographs and other projections must be made to visualize the joint spaces.

The intervertebral discs are a type of joint which can only be assessed on the plain films by alterations in the space between the vertebral bodies (the disc space). However, injection of contrast media into the disc (discography) will show changes in the disc caused by degenerative or other conditions. Further information concerning possible prolapse of the nucleus pulposus into the spinal canal can be obtained by myelography [p. 625].

The vertebral column shows changes with increasing age more commonly than any other skeletal structure and the intervertebral discs are often narrowed [FIGS. 3.23 and 3.25]. The stooping posture of the aged represents an increase in the normal curvature of the thoracic vertebral column with consequent compensatory increase in the cervical and lumbar curves. The narrowing of the intervertebral disc spaces is so common that it represents the normal appearance of old age and consists of posterior narrowing of the disc in the lower cervical and lumbar spine with anterior narrowing in the mid-thoracic region.

Age changes are common in many joints, more particularly those concerned with weight bearing. Often there is some diminution in cartilage thickness and an increase in radiographic density of the subchondral bone.

Tendons may be seen on radiographs if the adjacent tissue is of lesser density, e.g. the tendo calcaneus or ligamentum patellae.

Joints of the vertebral column and head

THE INTERVERTEBRAL JOINTS

The joints between the free vertebrae are arranged in a common plan with the exception of the specialized joints between the atlas and axis. The bodies of adjacent vertebrae are bound together mainly by the strong and extremely important intervertebral discs. The more posterior parts of the vertebral arches are also united, partly by the synovial joints between the articular processes and partly by ligaments. The intervertebral discs are separated from the joints between the vertebral arches by the intervertebral foramina through which the spinal nerves pass, and the upper and lower boundaries of these foramina are formed by the pedicles of the arches.

The joints between the bodies of the vertebrae and the connections between the vertebral arches are considered separately, but functionally they must be considered together, as they are both concerned in the structure of the flexible part of the vertebral column which is not only strong but also resilient.

Joints between vertebral bodies

These are secondary cartilaginous joints or symphyses as the intervertebral discs are composed of fibrocartilage.

INTERVERTEBRAL DISCS

The adjacent surfaces of the vertebral bodies are united by a fibrocartilaginous intervertebral disc [FIG. 4.11]. The peripheral part, the **anulus fibrosus**, is composed of fibrous tissue externally with an increasing amount of fibrocartilage internally. It is so densely fibrous that the cartilaginous element is appreciated only

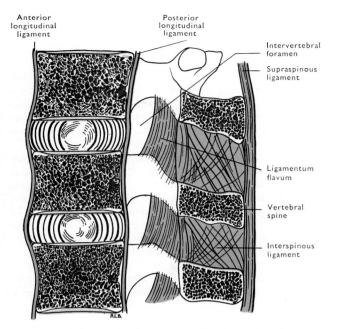

FIG. 4.11. Median section through part of lumbar region of vertebral column. Hyaline cartilage, blue.

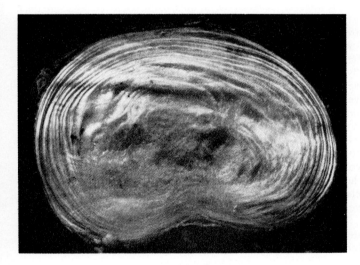

FIG. 4.12. Transverse section of the lumbosacral intervertebral disc of a young adult. The laminated form of the anulus fibrosus is apparent but the nucleus pulposus shows no such organization
(Reproduced by courtesy of the Editor, *Edinb. med. J.*)

with difficulty under the microscope. The term anulus fibrosus is therefore extremely appropriate. The fibres of the anulus run obliquely between the two vertebrae in concentric layers; the fibres in successive layers having different obliquities. The central part of the disc, the nucleus pulposus, is soft and gelatinous in young people but it is gradually replaced by fibrocartilage with increasing age. The disc is united to both vertebral bodies by a thin plate of hyaline cartilage, calcified near its ossified surface; the anular (ring) epiphysis of the vertebral body develops in its margin. The surface of the vertebra in contact with the cartilage plate is formed of modified cancellous bone perforated by numerous small foramina occupied by vascular tissue. These are most apparent in the zone adjacent to the anular epiphysis during development, but in the adult are much more numerous next to the nucleus pulposus. The blood vessels at the periphery of the anulus and those perforating the osseo-chondral interfaces are equally important in nutrition of

the disc, although the nucleus pulposus is almost wholly dependent on diffusion through the cartilage plates (Maroudas, Stockwell, Nachemson, and Urban 1975). In the cervical region the discs do not extend quite to the lateral edges of the vertebral bodies, where there is a small synovial joint between the lipped edge of the body below and the bevelled edge of the one above. The discs account for nearly a quarter of the length of the vertebral column; they are thinnest in the upper thoracic region and thickest in the lumbar region. The disc between the fifth lumbar vertebra and the sacrum is the largest avascular structure in the body and the degenerative changes which are so liable to occur in it may partly be due to its nutrition being dependent on a process of diffusion. The discs are slightly wedge-shaped, in conformity with the curvature of the vertebral column in their neighbourhood. This is particularly marked in the cervical and lumbar regions, where the anterior convexities are due chiefly to the greater thickness of the discs in front than behind. It is of considerable practical importance that each intervertebral disc forms one of the anterior boundaries of an intervertebral foramen [FIG. 4.11], and so lies immediately anterior to the corresponding spinal nerve as it passes through the foramen.

The nucleus pulposus is customarily described as a white, glistening body composed of fibrocartilaginous tissue embedded in a gelatinous matrix, but from the time of its first appearance in the foetus until late adult life, it undergoes a continuous structural change. It is stated to be a remnant of the notochord but this, as shown by Keyes and Compere (1932) and others, is not strictly correct. Early in foetal life the notochordal cells which atrophy in the vertebral regions, proliferate rapidly in each intervertebral region. Later they are surrounded by a plentiful mucoid matrix producing a gelatinous nucleus pulposus. At birth and during the first year of life, the nucleus pulposus forms a greater proportion of the intervertebral disc [FIG. 4.13] than subsequently. Thereafter, the anulus fibrosus grows more rapidly than the nucleus which becomes relatively smaller. As the notochordal cells degenerate the nucleus is gradually invaded by the surrounding fibrocartilaginous tissue. In young subjects the nucleus is still largely formed of gelatinous tissue and hence has a translucent appearance that contrasts sharply with the fibrocartilaginous anulus fibrosus. With increasing age the gelatinous tissue is gradually replaced by fibrocartilage which blends with the anulus fibrosus in such a way that there is no longer a sharp line of demarcation between the two parts of the disc in old age.

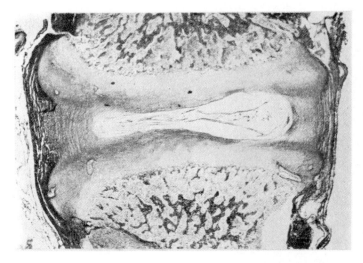

FIG. 4.13. Median microscopic section of lumbar intervertebral disc of full-term foetus. The nucleus pulposus forms the greater part of the disc, but will become relatively smaller after birth as the anulus fibrosus increases in thickness. × 15.
(Reproduced by courtesy of the Editor, *Edinb. med. J.*)

There is a progressive decrease in water content of the nucleus pulposus, from 88 per cent at birth to 70 per cent at 70 years (Püschel 1930). Changes also occur in the intercellular substance, notably increases in the collagen and keratansulphate content (Hallen 1962). These features are associated with reduced flexibility of the disc.

The nucleus pulposus forms a larger proportion of the intervertebral disc in the more mobile cervical and lumbar regions of the vertebral column. It is normally under considerable pressure and, when the disc of a young or middle-aged subject is cut across, it protrudes above the level of the anulus fibrosus. The nucleus pulposus may be regarded as a fluid ball around which movements between vertebrae take place and which transmits compressive forces to the vertebral bodies and the anulus fibrosus for absorption (Beadle 1931; Bradford and Spurling 1945; Walmsley 1953).

LONGITUDINAL LIGAMENTS

The anterior and posterior longitudinal ligaments also help to unite the bodies of the vertebrae. The anterior ligament [FIG. 4.23] is a broad thick band that runs over the anterior surfaces of the vertebral bodies. It is narrowest at its upper end, where it has a pointed attachment to the anterior tubercle of the atlas, but widens as it descends, to end below by spreading on to the pelvic surface of the sacrum. It is firmly attached to the intervertebral discs and adjacent parts of the vertebral bodies. The posterior longitudinal ligament [FIG. 4.14] is constructed on a similar plan. It is placed on the back of the vertebral bodies, in the anterior wall of the vertebral canal. It is broadest above where it continues as the membrana tectoria, behind the dens and its ligaments to be attached to the occipital bone. Opposite the bodies of the thoracic and lumbar vertebrae the posterior ligament narrows markedly, but widens at the intervertebral discs, so that each edge presents a series of pointed dentations. Its narrow, lower end passes on to the anterior wall of the sacral canal. The posterior ligament is attached to the discs and adjoining edges of the vertebral bodies but is separated from the rest of the bone of the bodies by an interval which transmits the thin-walled basivertebral veins [FIG. 4.14]. The posterior surface of the ligament is separated from the spinal dura mater by loose areolar tissue containing the internal vertebral plexus of veins; but the ligament is rather more intimately connected to the dura by fibrous bands in the lower half of the canal. In both ligaments the most superficial longitudinal fibres run over several vertebrae; the deepest pass only from one bone to the next.

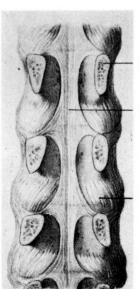

Pedicle
divided

Posterior
longitudinal
ligament

Intervertebral
disc

FIG. 4.14. Posterior longitudinal ligament of vertebral column.

Joints between vertebral arches

The vertebral arches have synovial joints between their articular processes.

The shape and orientation of the joint surfaces on the articular processes vary in the different regions of the column [p. 223]. Each joint is surrounded by a thin fibrous capsule lined with synovial membrane. The joint capsules are lax, particularly in the cervical region, so that slight gliding movements may occur between the bones.

The accessory ligaments are the supraspinous, the interspinous, the intertransverse, and the ligamenta flava. The interspinous ligaments are relatively weak bands that run between adjacent vertebral spines; they are longest and strongest in the lumbar region [FIG. 4.11]. The strong supraspinous ligament runs over the tips of the spines and is fused with the posterior edges of the interspinous ligaments; short fibres in the ligament join adjacent spines; longer ones connect spines several vertebrae apart.

In the neck, the supraspinous ligament merges into the triangular ligamentum nuchae. In Man, this is merely a thin fibro-elastic septum between the muscles of the two sides of the back of the neck. It is attached above to the external occipital crest and deeply to the cervical spines, while its posterior border runs between the external occipital protuberance and the spine of the last cervical vertebra. In quadrupeds the ligamentum nuchae is a well-developed elastic ligament which helps to support the head.

The intervals between the laminae of adjacent vertebrae are filled by the yellow elastic ligamenta flava. Each is attached above to the front of the lower border of a lamina [FIG. 4.15] and below to the superior margin of the next lamina. There are two—a right and a left—in each vertebral interval; their medial borders are separated by a narrow, median chink transmitting veins that connect the venous plexus within the vertebral canal to the posterior external vertebral plexus. Laterally, they extend to the capsules of the intervertebral synovial joints, but do not blend with them. The ligamenta flava are the only markedly elastic ligaments in Man. They permit separation of the laminae in flexion of the vertebral column and on account of their elasticity do not form folds when the column returns to the erect position; such folds might press upon the dura mater or be caught between the laminae they connect. The elastic tension in the ligamenta flava also assists the posterior vertebral muscles in maintaining the erect attitude.

The intertransverse ligaments are insignificant bands between the transverse processes. They are readily recognizable only in the lumbar region. In the upper part of the column they tend to be replaced by the intertransverse muscles.

Movements of vertebral column

The adult vertebral column in the erect posture has four anteroposterior curves—cervical, thoracic, lumbar, and sacrococcygeal [p. 91]. The compensatory or secondary curves (cervical and lumbar) are due mainly to the shape of the discs, whereas the thoracic curvature, which is a primary one, is caused mainly by the shape of the vertebrae. The sinuous form of the vertebral column, made up of alternate bony vertebrae and resilient intervertebral discs, is a structure that is well adapted to its supporting and shock-absorbing functions. The intervertebral discs undergo a diurnal variation in thickness, but in old age they narrow and the vertebral column tends to revert to its original primary curve. Thus it is that some people in advanced years have a 'bowed back' and appear to become smaller.

The movement between any two successive vertebrae is very slight; but the sum total of movement in the whole column is considerable [FIG. 3.17]. Movement is due to the slight flexibility of the intervertebral discs coupled with the laxity of the capsules of the

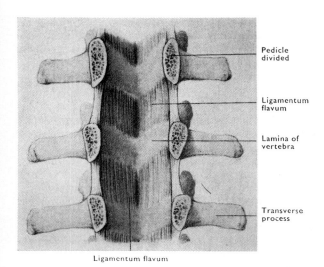

Pedicle divided

Ligamentum flavum

Lamina of vertebra

Transverse process

Ligamentum flavum

FIG. 4.15. Ligamenta flava seen from the front after removal of the bodies of the vertebrae by saw cuts through the pedicles.

synovial joints between the articular processes. Where the discs are thick relative to the horizontal diameter of the vertebral body, the range of movement is increased, though the arrangement of the fibres in the discs strictly limits movement between adjacent vertebrae in any direction. The semifluid nature of the nucleus pulposus tends to equalize pressure within each disc during movements between the vertebrae.

The movements of the vertebral column are customarily described as forward **flexion**, backward **extension**, **lateral flexion**, and **rotation**. Lateral flexion and rotation are always associated and neither can take place independently of the other. The direction and range of movements in the different regions are largely determined by the thickness of the intervertebral discs and the 'set' and shape of the articular processes. During flexion the anterior borders of the vertebrae are approximated, whereas the distance between the posterior borders is increased. As full flexion is reached the anterior parts of the intervertebral discs are tightly compressed, but the posterior parts of the discs, the posterior longitudinal ligament and the ligaments of the vertebral arches, all become taut, though the principal limiting factor to the movement is the tension of the posterior vertebral muscles. In extension, the vertebrae rock backwards on one another so that the distance between the anterior borders of the vertebrae is increased and the posterior borders are approximated [FIG. 3.17]. During this movement the anterior longitudinal ligament becomes increasingly taut whereas the other ligaments of the column are relaxed. The movement of extension of the column has a wider range than the movement of flexion, and is greatest in the cervical and lumbar regions. Much of the apparent movement of flexion of the vertebral column during the bending forwards of the body is in reality due to flexion at the hip and atlanto-occipital joints.

In the **cervical region**, where the discs are relatively thick, all movements are fairly extensive. The articular surfaces slope anterosuperiorly, and lie in the same oblique plane. They permit flexion and extension, but cause lateral flexion and rotation to occur together. Thus, when the neck is bent to one side, each lower articular process on that side glides downwards and backwards on the upper process of the vertebra below, while on the other side each lower process mounts upwards and forwards upon the superior process below it; hence bending to one side is accompanied by a slight rotation to that side.

In the **thoracic region** movement is limited by thin intervertebral discs, by the presence of ribs and sternum, and by the imbrication of the vertebral spines. Rotation is possible because the articular surfaces lie on an arc of a circle with its centre in the vertebral body; but flexion is extremely limited by the presence of the ribs and sternum. Lateral flexion is always associated with rotation although these movements are greatly diminished, if not altogether lost in complete extension of the vertebral column.

In the **lumbar region** the discs are thick but since the inferior articular processes lie between the superior processes of the vertebra below the movement of rotation in particular is prevented. Fairly free flexion and extension occur, especially at the lumbosacral joint where the intervertebral disc is thickest. When the vertebral column is flexed the lumbar curve is lost and lateral flexion, which is possible in the erect posture, is greatly diminished.

The atlanto-occipital joints

The skull is joined to the atlas by a pair of atlanto-occipital synovial joints of the ellipsoid types. Here the occipital condyles lying anterolateral to the foramen magnum articulate with the superior articular facets of the atlas. The anterior and posterior atlanto-occipital membranes connect the arches of the atlas with the margins of the foramen magnum [FIG. 4.16].

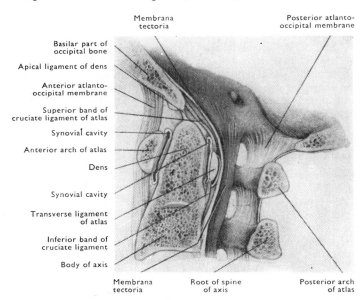

Membrana tectoria

Posterior atlanto-occipital membrane

Basilar part of occipital bone

Apical ligament of dens

Anterior atlanto-occipital membrane

Superior band of cruciate ligament of atlas

Synovial cavity

Anterior arch of atlas

Dens

Synovial cavity

Transverse ligament of atlas

Inferior band of cruciate ligament

Body of axis

Membrana tectoria

Root of spine of axis

Posterior arch of atlas

FIG. 4.16. Median section through atlanto-occipital and atlanto-axial region.

Atlanto-occipital membranes

The anterior atlanto-occipital membrane is attached above anterior to the margin of the foramen magnum, and below to the anterior arch of the atlas; its central portion is thickened by fibres prolonged upwards from the anterior longitudinal ligament; the right and left margins blend with the fibrous capsule of the atlanto-occipital joints. The posterior atlanto-occipital membrane likewise reaches the fibrous capsule at each side. Its upper border is attached to the posterior margin of the foramen magnum, its lower border to the posterior arch of the atlas. The thick inferolateral margins on each side, are sometimes ossified and arch over the corresponding vertebral artery and first cervical nerve as they cross the posterior arch of the atlas.

Articular surfaces

Each of the pair of synovial joints is enclosed by a fibrous capsule which is attached to the margins of the articular surfaces and is lined with **synovial membrane**. The atlantal articular surface is concave; the occipital surface is convex. Both surfaces have their shorter diameter in the coronal plane. But the surfaces of the joints of the two sides are really segments of one ellipsoid whose transverse diameter is the longer, and functionally the joints act together as a single ellipsoid.

Movements at the atlanto-occipital joints are described under Movements of the Head.

The atlanto-axial joints

The atlas and the axis have **bilateral synovial joints** between their opposed articular processes, and two **median synovial joints** formed by the articulation of the dens with the anterior arch of the atlas and with the transverse ligament of the atlas. In addition, various accessory ligaments connect the axis to the atlas, and the dens to the occipital bone.

Lateral atlanto-axial joint

Each of the paired joints is of the plane variety, enclosed by a loose fibrous capsule with a synovial lining. The articular surfaces have slight reciprocal curvatures, and are inclined obliquely laterally and downwards. Applied to the back of the capsule is an **accessory atlanto-axial ligament** which runs obliquely downwards and medially from the back of the lateral mass of the atlas to the back of the body of the axis [FIG. 4.17].

The anterior arch of the atlas is connected to the front of the body of the axis by a membranous expansion from each side of the pointed upper part of the anterior longitudinal ligament as it descends from the anterior tubercle of the atlas. The gap between the posterior arch of the atlas and the arch of the axis is bridged by a membrane, in series with the ligamenta flava, which is pierced at each side by the second cervical nerve. The position of this nerve behind the

atlanto-axial joint and of the first cervical nerve posterior to the atlanto-occipital joint indicates that these joints do not correspond morphologically to the joints between the articular processes of other vertebrae but may be homologous with the small synovial joint at each side of the cervical intervertebral disc.

Median atlanto-axial joint

The articulation of the dens with the atlas is a pivot joint provided with two little synovial cavities [FIG. 4.16], the smaller between the facet on the front of the dens and the facet on the back of the anterior arch of the atlas, and the larger between the posterior facet on the dens and the transverse ligament of the atlas. Each cavity is enclosed by a thin fibrous capsule lined with synovial membrane.

The **transverse ligament of the atlas** is a stout band that passes behind the dens between the tubercles on the medial sides of the lateral masses of the atlas [FIG. 4.17]. From the middle of the transverse ligament a small longitudinal band passes upwards to the anterior edge of the foramen magnum, and another passes downwards to the back of the body of the axis [FIG. 4.16]. These three bands constitute the **cruciform ligament of the atlas**.

The occipital-axial ligaments

The axis is connected with the occipital bone indirectly by the longitudinal bands of the cruciform ligament and, more directly, by the apical and alar ligaments of the dens and by the membrana tectoria.

Ligaments of the dens

Immediately in front of the upper band of the cruciform ligament, the slender **apical ligament of the dens** stretches from the apex of the dens to the anterior edge of the foramen magnum. The **alar ligaments** of the dens are short, stout bands that run, one from each side of the apex of the dens, laterally and slightly upwards to the medial side of each occipital condyle [FIG. 4.17].

Membrana tectoria

The dens and the cruciform and alar ligaments are covered posteriorly by the membrana tectoria [FIG. 4.16]. This broad sheet is continuous below with the upper end of the posterior longitudinal ligament and is attached above to the occipital bone medial to the hypoglossal canals. A small synovial bursa sometimes lies between the membrana and the median part of the transverse ligament of the atlas. The lateral margin of the membrana tectoria overlies and blends with the accessory atlanto-axial ligament. The upper end of the spinal dura is closely attached in front to the posterior surface of the membrana tectoria, and, behind, to the anterior surface of the posterior atlanto-occipital membrane.

Movements of the head

Movement of the head at the atlanto-occipital joints takes place in two principal directions—nodding or bending the head forwards and backwards and lateral flexion of the head on the neck. No rotation takes place between the skull and the atlas beyond a little slipping of the occipital condyles, the one forwards the other backwards, on the atlantal facets, giving the face a slight oblique tilt; the joint surfaces are then in better apposition, and this is the natural pose of greatest ease and stability.

The combined movements at the three **atlanto-axial joints** facilitates rotation of the head around a vertical axis, the skull and the atlas moving as one. While the ring formed by the anterior arch and transverse ligament of the atlas pivots round the dens, the

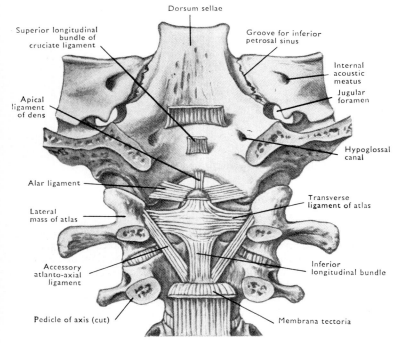

Dorsum sellae

Superior longitudinal bundle of cruciate ligament

Groove for inferior petrosal sinus

Internal acoustic meatus

Jugular foramen

Apical ligament of dens

Alar ligament

Lateral mass of atlas

Hypoglossal canal

Transverse ligament of atlas

Accessory atlanto-axial ligament

Inferior longitudinal bundle

Pedicle of axis (cut)

Membrana tectoria

FIG. 4.17. Dissection from behind to show the main ligaments that connect the occipital bone, the atlas, and the axis.

lateral masses of the atlas glide, the one forwards and the other backwards, on the upper articular facets of the axis. Owing to the oblique disposition of these lateral joint surfaces, rotation of the head to one side is accompanied by a slight vertical descent of the head, so that the median pivot joint around the dens resembles in action a rising-butt hinge. Excessive rotation is checked by the alar ligaments and to a lesser extent by the accessory atlanto-axial ligaments. But the descent of the head, by approximating the attachments of the alar ligaments, delays their check action, and this action can be counteracted by tilting the head backwards and to the opposite side during a further degree of extreme lateral rotation.

THE TEMPOROMANDIBULAR JOINT

The temporomandibular joint is a synovial joint of the condyloid variety. The head of the mandible articulates with the mandibular fossa and the articular tubercle of the temporal bone through an articular disc which divides the articular cavity into upper and lower compartments. Both bones are covered by fibrocartilage, i.e. the superficial layer of articular tissue consists of dense fibrous tissue containing isolated chondrocytes while hyaline cartilage is found in the deeper layers.

The temporal articular surface is concavoconvex from behind forwards, extending from the squamotympanic fissure over the articular fossa and eminence to the anterior border of the eminence [FIG. 4.19]. The head of the mandible is a narrow ellipsoid of bone directed medially and slightly backwards, markedly convex from before backwards but only slightly convex from side to side.

The articular disc separates the incongruent articular surfaces of the bones and moulds itself upon them during movements at the joint. It is an oval plate of avascular fibrous tissue. Thin centrally, it thickens towards its peripheral attachment to the fibrous capsule and especially at the back due to a thick layer of vascularized connective tissue which connects it to the capsule.

The fibrous capsule is attached above to the limits of the temporal articular surface and below to the neck of the mandible. Above the attachment to the disc the capsule is lax to permit free movement of the disc on the temporal bone. Below its attachment to the disc, the fibrous capsule is much more taut and the fibres of the medial and lateral parts of the capsule are so short that the disc appears to be attached to the medial and lateral ends of the condyle. These attachments permit the rotatory movement that occurs between the condyle and the under surface of the articular disc, and at the same time assure the forward and backward movement of the disc and the condyle as one unit on the articular surface of the temporal bone (Sarnat 1951). A part of the tendon of the lateral pterygoid muscle is inserted through the fibrous capsule into the anterior margins of the articular disc. In the embryo the lateral pterygoid tendon passes backwards between the head of the mandible and the squamous temporal bone to be attached to the malleus, and it is the part between the bones that becomes compressed and forms the disc (Harpman and Woollard 1938). The fibrous capsule is strengthened laterally by a well-defined lateral ligament, which passes downwards and backwards from the lower border and tubercle of the zygoma to the lateral and posterior aspect of the neck of the mandible [FIG. 4.18].

The synovial membrane lines the non-articular structures in each compartment of the joint; it fuses with the periphery of the disc but is absent from its upper and lower surfaces.

Accessory ligaments

There are two accessory ligaments which contribute little to the strength of the joint. The sphenomandibular ligament is a thin

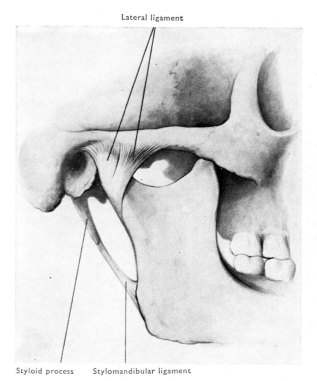

Lateral ligament

Styloid process   Stylomandibular ligament

FIG. 4.18. Temporomandibular joint.

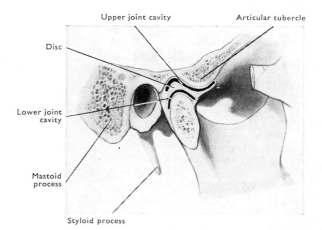

Upper joint cavity   Articular tubercle

Disc

Lower joint cavity

Mastoid process

Styloid process

FIG. 4.19. Anteroposterior section through the temporomandibular joint.

band that runs from the spine of the sphenoid bone downwards and forwards to the lingula and the adjoining part of the deep surface of the ramus of the mandible [FIG. 4.21]. The ligament is separated from the medial side of the joint capsule by the maxillary artery and the first parts of its middle meningeal and inferior alveolar branches, the auriculotemporal and inferior alveolar nerves, and a process of the parotid gland. The ligament is developed from the sheath of that part of the cartilage of the mandibular arch (Meckel's cartilage) that lies between the base of the skull and the mandibular foramen.

The stylomandibular ligament is a thickened band of deep cervical fascia between the styloid process and the lower part of the posterior border of the ramus of the mandible [FIG. 4.21].

Nerve supply

The mandibular joint is supplied from the mandibular nerve by twigs from its auriculotemporal and masseteric branches.

Movements

An understanding of movements at the temporomandibular joint is dependent, in some measure at least, on an appreciation of the relationship that the teeth of the upper and lower jaws have to each other when the jaws are closed. The teeth are then in contact with one another and this is called the **occlusal position**. In this position the incisor teeth of the upper jaw lie in front of the lower incisors

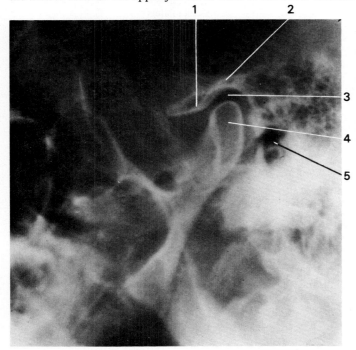

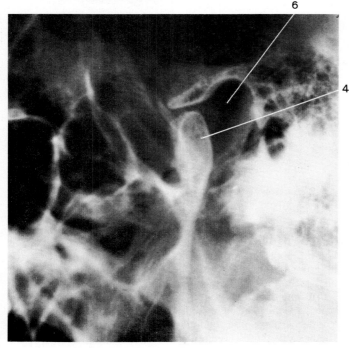

FIG. 4.20. Lateral radiographs of the temporomandibular joint with the mouth closed and open.
1. Articular tubercle.
2. Temporal bone.
3. Articular disc in mandibular fossa.
4. Head of mandible.
5. External acoustic meatus.
6. Region occupied by head of mandible when mouth closed.

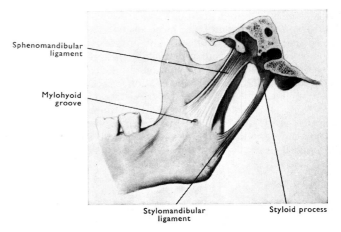

FIG. 4.21. Accessory ligaments of temporomandibular joint.

and there is usually some contact between their opposing surfaces. In opening the mouth from the occlusal position, the edges of the lower incisors pass downwards and forwards until the edges of the two sets of incisors are directed towards each other. This downward and forward movement of the mandible, in initiating opening of the mouth, is indicative of the two distinct movements that occur at the temporomandibular joint. These are a forward gliding motion when the head of the mandible and the articular disc move as a unit downwards and forwards, and a simultaneous hinge-like rotation between the head of the mandible and the articular disc. In small movements of the mandible in chewing and quiet speech, most of the movement is between the head of the mandible and the disc, with its axis through the condyle. In larger movements, both elements of the joint motion are brought into play and the mandible moves about a line passing through the mandibular foramina. In protrusion, the greater part of the movement occurs in the upper compartment, and the mandible is carried forward without opening the mouth. Retraction is the reverse of this movement. The side-to-side grinding movements of chewing are brought about by alternate and opposite movements in the upper compartments of the two joints, combined with minor hinge movements in the lower compartments.

The upper attachment of the anterior band of the lateral ligament is on the axis of the curved, downward-and-forward gliding movement in the upper compartment, and therefore, to some extent, the ligaments of the two sides resemble the collateral ligaments of a hinge joint. Their tension will not vary much throughout this upper compartment movement, and hence they do not completely prevent the possibility of dislocation of the mandibular head on to the front of the articular eminence. As this danger, however, arises only in excessively wide opening of the mouth, the associated rotation that then occurs in the inferior compartment causes some tension in the ligament which has a certain safeguarding action. The gross, microscopic and functional anatomy of the temporomandibular joint is fully described by Sarnat (1951).

Joints of the ribs and sternum

The ribs articulate posteriorly with the thoracic vertebrae. Each typical rib articulates by its head with the bodies of its own vertebra and of the vertebra above, and by its tubercle with the transverse process of its own vertebra [FIG. 4.22]. The first and last two or three ribs, which themselves deviate in their shape from a typical rib, do not conform to the typical rib in their articulations with the vertebrae. Anteriorly, the ribs end in costal cartilages, which, with the exception of the last two, make further connections with the sternum or with one another.

THE COSTOVERTEBRAL JOINTS

Joints of the heads of ribs

The head of each typical rib has two articular facets, superior and inferior, separated by the crest of the head. The crest of the head of the rib is bound to the intervertebral disc by a short, thick intra-articular ligament, while the facets on the head articulate with those on the vertebral bodies.

The whole articulation is surrounded by a fibrous capsule and is divided into separate upper and lower synovial cavities by the intra-articular ligament. The anterior part of the capsule is thickened to form the radiate ligament of the head of the rib. This fans out medially from the front of the head to the adjacent parts of the disc and vertebral bodies, under the edge of the anterior longitudinal ligament [FIG. 4.23]. Three radiating bands are often to be seen, especially in the middle joints of the series, a central one blending with the intervertebral disc, and upper and lower bands attached to the vertebral bodies. The back of the capsule is connected to the adjacent denticulation of the posterior longitudinal ligament.

Costotransverse joints

A small costotransverse synovial joint surrounded by a thin fibrous capsule is formed between the medial part of the costal tubercle and the circular facet on the anterior surface of the corresponding transverse process near the tip. The facets on the upper ribs are slightly convex while those on the lower ribs are flatter and directed slightly medially and downwards. Strengthening the joint, on its posterolateral aspect, there is the lateral costotransverse ligament, a stout band running between the rough lateral part of the costal tubercle and the tip of the transverse process. It is in contact with the fibrous capsule but is not fused with it.

The costotransverse ligament is composed of very short fibres that bridge the narrow interval between the back of the neck of the rib and the front of the corresponding transverse process.

The superior costotransverse ligament [FIG. 4.23] consists of an anterior and a posterior band that spring from the upper border (crest) of the neck of the rib and are attached to the lower border of the transverse process above. Laterally, the two bands are separated by the external intercostal muscle and the anterior band blends laterally with the internal intercostal membrane.

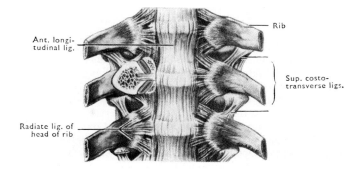

FIG. 4.23. The anterior longitudinal ligament and costovertebral joints from in front. One joint has been opened by an oblique slice through the head of the rib.

The vertebral connections of the first rib and of the last two or three ribs are atypical. The head of each of these ribs has a single facet for the side of the corresponding vertebra only. There is a single synovial cavity, the radiate ligament is poorly developed, and there is no intra-articular ligament. The superior costotransverse ligament of the first rib is represented by feeble bands attached to the seventh cervical transverse process. The tubercles of the lowest ribs do not form synovial joints with the transverse processes; their costotransverse ligaments are progressively less well defined and those of the last rib may be absent altogether.

THE STERNOCOSTAL AND INTERCOSTAL JOINTS

The anterior end of each rib is continuous with a bar of hyaline cartilage, called the costal cartilage, which is firmly united to a conical cavity in the end of the rib. The first costal cartilage is directly united to the sternum; the second to the seventh costal cartilages articulate with facets on the side of the sternum; and the lower cartilages, except for the eleventh and twelfth, articulate with one another.

Sternocostal joints

Little synovial cavities usually develop in the joints between the sternum and the costal cartilages, except in the case of the first, which remains a continuous cartilaginous joint [FIG. 4.24]. The cavity of each of the others is divided into two by an intra-articular ligament until the corresponding sternal segments become fused together. The second sternocostal joint cavity usually remains double throughout life and the cavities in the remaining joints tend to become obliterated in old age. The fibrous capsules of these little joints are strengthened in front and behind by the anterior and posterior radiate ligaments. The fibres of the anterior ligaments interlace with those of the opposite side to form a felted membrane which is fused with tendinous fibres of the pectoralis major muscles.

Interchondral joints

The tips of the eighth and ninth costal cartilages form little synovial joints, each with the lower border of the cartilage above. In addition, synovial joints are formed between slight bosses developed on the adjacent margins of the fifth to the eighth or ninth cartilages. All these joints between cartilages are enclosed by short fibrous capsules strengthened in front and behind by oblique ligamentous bands. The terminal part of the tenth cartilage is united to the ninth by a syndesmosis.

The last two costal cartilages are short conical structures ending freely amongst the muscles of the flank.

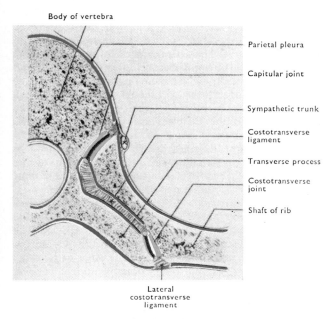

FIG. 4.22. Transverse section through a typical costovertebral joint.

THE STERNAL JOINTS

The manubriosternal joint lies between the manubrium and the body of the sternum and is an unusual type of joint. Prior to its ossification the sternum is a continuous bar of hyaline cartilage, but the manubriosternal joint, at the level of the second costal cartilage, is transformed early into fibrocartilage. When ossification in the sternum is complete a thin layer of the hyaline cartilage above and below the fibrocartilage remains unossified so that the joint has an appearance similar to that of a symphysis. Sometimes the central part of the fibrocartilage is absorbed and a cavity is formed within it, and less frequently the joint is obliterated by synostosis in old age. Special longitudinal fibres and the neighbouring sternocostal ligaments strengthen this joint in front and behind. The cartilaginous xiphosternal joint between the body of the sternum and the xiphoid process is usually ossified in early adult life.

Movements of the ribs and sternum

In respiration there are two factors concerned in increasing and diminishing the capacity of the thorax: these are the movement of the diaphragm and the movements of the ribs and sternum. It is the movements of the ribs and sternum that are considered in this section but the respiratory action of the diaphragm should be studied concurrently [FIG. 5.103]. A rib and its costal cartilage together form a costal arch and the upper seven of these articulate with the sternum anteriorly; the eighth, ninth, and tenth cartilages each articulate with the costal cartilage that lies immediately above; the eleventh and twelfth costal cartilage are free so that the corresponding ribs are called 'floating' ribs. The ribs that articulate with the sternum increase in length from above downwards and their anterior ends lie at a lower level than the posterior ends. Owing to the length of the eighth and ninth ribs and the manner in which each of their costal cartilages articulates with the one lying immediately above, the thoracic skeleton continues to increase in width certainly down to the level of the ninth rib.

The ribs are fixed posteriorly by their articulation with the vertebral column, and because they slope downwards and forwards from the articulation, their anterior ends move upwards and forwards when they are elevated, thus increasing the anteroposterior diameter of the thorax. However, the capitular and costotransverse joints of most of the ribs together form a hinge with an axis which passes backwards and laterally through the two joints. Hence, when the rib is elevated around this axis, the anterior end moves outwards as well as upwards and forwards, thus increasing the transverse as well as the anteroposterior diameter of the thorax.

In the upper ribs, where the axis is less oblique and the costotransverse facets permit rotation, the lateral movement is slight or absent and their shorter cartilages meet the sternum approximately at a right angle. The sternum therefore moves upwards and forwards in inspiration. As the first rib is so much shorter than those which succeed it, the upward and forward excursion of its anterior end is less in extent, and the movement of the upper border of the sternum is correspondingly reduced (Haines 1946). The greater travel—especially in the forward direction—imposed upon the rest of the sternum by the longer ribs causes a bending of the breast bone at the manubriosternal joint. The result is to increase the anteroposterior diameter of the thorax. The obliquity of the axis of movement posteriorly is most marked in the case of the middle ribs, which possess long, inclined cartilages, and where the plane of the costotransverse facets causes a greater lateral movement of the anterior ends of these ribs. This causes an increase in the transverse diameter of the thorax and a demonstrable widening of the infrasternal angle in deep inspiration.

The transverse diameter of the chest is also appreciably increased on inspiration due to the position of the ribs midway between full expiration and inspiration. Each costal arch then has its lateral part lying below a line joining its anterior and posterior ends. As the sternal and interchondral attachments of the costal arches hamper movement of their anterior ends, the raising of the ribs is in part translated into a twisting of the costal cartilages and a rotation at the sternocostal joints. This causes the intermediate part of the rib to be raised upwards and outwards as in lifting a low-lying bucket handle to a horizontal position. In inspiration, therefore, the increase in anteroposterior diameter of the thorax is accompanied by an increase in its transverse diameter. The lower ribs, held down by abdominal muscles, act to prevent elevation of the diaphragm.

Expiration is accompanied by reverse movements whereby the anteroposterior and transverse diameters of the thorax are diminished. The main force in quiet expiration is the elastic recoil of the lungs. Forced expiration requires positive muscular effort.

Apart from respiratory movements, the ribs move passively with changes in the thoracic part of the vertebral column. Flexion of the column causes a crowding together of the ribs; extension spreads them apart; bending of the column to one side crowds the ribs on that side and spreads them on the opposite side of the trunk.

Joints of the upper limb

THE SHOULDER GIRDLE

The shoulder girdle is formed by the clavicle and the scapula which articulate with each other at the acromioclavicular joint. The girdle in turn articulates with the skeleton of the trunk only at the sternoclavicular joint which has a degree of mobility that greatly increases the range of movement of the upper limb.

The sternoclavicular joint

At the sternoclavicular joint the medial end of the clavicle articulates with the clavicular notch of the manubrium sterni through an articular disc and with the upper surface of the first costal cartilage. The clavicular is larger than the sternochondral articular surface, hence the medial end of the clavicle projects above the upper margin of the manubrium sterni. The articular surfaces are reciprocally concavoconvex with that on the clavicle being convex vertically and slightly concave horizontally. On functional grounds and despite the form of the articular surfaces, the joint should be regarded as a ball and socket around which the clavicle can be moved in many directions. Movement of the clavicle in a vertical plane around the joint is free with a range of about 60 degrees; anteroposterior movement is more restricted, approximately 30 degrees; at least 20 degrees of rotation of the clavicle around its long axis occurs during rotation of the scapula. The articular cartilage, unlike that of most articular surfaces, is fibrocartilage and not hyaline cartilage.

A fibrous capsule surrounds the whole joint and is attached around the clavicular and the sternochondral articular surfaces; the epiphysis at the sternal end of the clavicle is intracapsular. The weak inferior part of the capsule passes between the clavicle and the superior surface of the first costal cartilage; the other parts of the capsule are strong, and are reinforced in front, behind, and above, by thickenings called the anterior and posterior sternoclavicular ligaments and the interclavicular ligament.

The anterior and posterior sternoclavicular ligaments, of which the anterior is the stronger, run downwards and medially from

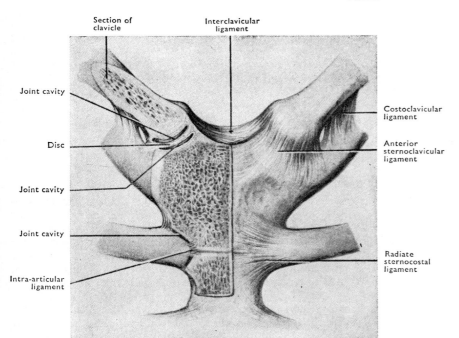

Section of clavicle

Interclavicular ligament

Joint cavity

Disc

Joint cavity

Joint cavity

Intra-articular ligament

Costoclavicular ligament

Anterior sternoclavicular ligament

Radiate sternocostal ligament

FIG. 4.24. Sternoclavicular and sternocostal joints.

clavicle to sternum [FIG. 4.24]. The origin of the sternohyoid muscle extends across the posterior ligament; this muscle and the sternothyroid separate the back of the left joint from the left brachiocephalic vein and the left common carotid artery, and the right joint from the brachiocephalic trunk. The interclavicular ligament inserts into and passes across the floor of the jugular notch forming thickenings of the upper parts of the fibrous capsules of both joints.

Within the joint there is a complete fibrocartilaginous articular disc, which blends with the fibrous capsule in front and behind, but has a firm attachment to the clavicle above and to the first costal cartilage below, near the sternal articular surface. Hence the disc, not only cushions the articular surfaces from forces transmitted along the clavicle from the shoulder, but acting as a ligament prevents the sternal end of the clavicle riding upwards and medially over the sternum. Also, when the lateral end of the clavicle is depressed, as in carrying a heavy weight, the tendency for the medial end of the bone to be levered out of its sternal socket is resisted by the disc and the costoclavicular ligament.

There are two separate synovial cavities within the sternoclavicular joint, though occasionally the thinner central part of the articular disc is perforated.

The accessory extracapsular costoclavicular ligament ascends from the upper surface of the first costal cartilage near its lateral end to a rough tubercle on the lower aspect of the medial part of the clavicle. The anterior fibres of the ligament pass upwards and laterally whereas the posterior fibres pass upwards and medially. This cruciate arrangement limits elevation and horizontal movement of the clavicle and compensates for the weakness of the inferior part of the sternoclavicular fibrous capsule.

The stability of the sternoclavicular joint is dependent on the strength and integrity of its ligaments. The form of the articular surfaces and the surrounding muscles gives little additional security to the joint, and when dislocation does take place the ligaments are strained so that it is liable to recur.

Nerve supply

The sternoclavicular joint is supplied by twigs from the medial supraclavicular nerve and the nerve to subclavius.

The acromioclavicular joint

The acromioclavicular joint is a small synovial joint, of the plane type, between oval facets on the lateral end of the clavicle and on the medial border of the acromion [FIG. 4.29]. The articular surfaces are covered with fibrocartilage and they both slope downwards and medially so that the clavicle tends to override the acromion and normally projects above it.

A weak fibrous capsule surrounds the joint; it is strongest above, where it is reinforced by fibres from the trapezius. A wedge-shaped fibrocartilaginous articular disc, attached to the upper part of the capsule, partially divides the articular cavity in most joints. Only rarely does the disc form a complete partition and it may be absent altogether.

Coracoclavicular ligament

This powerful ligament [FIG. 4.29] is the principal bond of union between clavicle and scapula. It anchors the lateral end of the clavicle to the coracoid process by its two parts, the conoid and trapezoid ligaments. They are continuous with each other posteriorly but are separated anteriorly by a synovial bursa.

The conoid ligament passes upwards and slightly backwards from an apical attachment on the coracoid process [FIG. 3.107] to a wider insertion on the conoid tubercle of the inferior surface of the clavicle. The trapezoid ligament lies anterolateral to the conoid ligament. Its lower attachment is to a short ridge on the posterior part of the superior surface of the coracoid process; its clavicular end is wider, and is attached to the trapezoid ridge on the inferior surface of the acromial end of the clavicle. The trapezoid ligament is more nearly horizontal than vertical.

Both ligaments, but more particularly the trapezoid, prevent displacement of the acromion medially below the lateral end of the clavicle when blows fall upon the lateral surface of the shoulder.

Nerve supply

The acromioclavicular joint is supplied by the lateral pectoral, suprascapular, and axillary nerves.

Movements of the shoulder girdle

Movements of the shoulder girdle accompany most movements at the shoulder joint and increase the range of movement of the upper limbs by altering the position and orientation of the scapular socket for the head of the humerus. Reference is made to shoulder girdle movement at this stage merely as a prelude to shoulder joint movement. Forward movement of the scapula round the chest causes the glenoid cavity to face more directly forwards; backward movement causes the cavity to face laterally; upward movement of the scapula, combined with rotation so that the inferior angle passes forwards and upwards, causes the glenoid cavity to be turned upwards. In all these changes of position the shoulder is kept boomed out from the trunk by the thrust of the clavicle upon the acromion, thus permitting greater freedom of shoulder movement. The lateral angle of the scapula, carrying the glenoid cavity, travels in an arc of a circle whose radius is the clavicle: but the medial part of the scapular blade follows the chest wall. As a result, small variations in the position of the scapula relative to the clavicle occur at the acromioclavicular joint,

The clavicle is pivoted at its sternal end while its acromial end moves with the scapula. During rotation of the scapula as in elevation of the upper limb, the clavicle rotates so that the anterior surface of the bone is increasingly directed upwards. Any impairment of clavicular rotation due to a lesion at either the sternoclavicular or acromioclavicular joints will interfere with the free movement of the scapula (Inman, Saunders, and Abbott 1944).

The scapular ligaments

The following ligaments are attached wholly to the scapula and therefore are not connected with any joint. The **superior transverse scapular ligament** bridges the scapular notch, and so continues the line of the superior border of the scapula laterally to the root of the coracoid process. It may be partly or completely ossified. Fibres of the inferior belly of the omohyoid are attached to its medial portion. The suprascapular nerve enters the supraspinous fossa through the foramen completed by the ligament, while the suprascapular vessels pass over the ligament to reach the fossa.

The **inferior transverse scapular ligament** is a weak band that stretches between the lateral border of the scapular spine and the back of the neck of the bone, over the suprascapular nerve and vessels. It may be regarded as formed from a fusion of the fasciae over the contiguous supraspinatus and infraspinatus muscles.

The **coraco-acromial ligament** is functionally related to the shoulder joint, and will be described with it.

THE SHOULDER JOINT

The shoulder joint is a synovial joint of the ball-and-socket variety in which the spheroidal surface of the head of the humerus articulates with the shallow glenoid cavity of the scapula, each surface being covered with articular cartilage. Freedom of movement is developed at some expense to stability. Thus the form of the surfaces contributes very little to the security of the joint since the area of the scapular 'socket' is little more than a third of that of the humeral 'ball'. The fibrous capsule helps little since it must be lax enough to permit the required range of movement. The strength and stability of the joint depends on the muscles which surround it. These muscles are attached close to the articular areas and are intimately related to the fibrous capsule; they are, therefore, well placed to keep the articular surfaces in firm contact in all positions of the joint.

Labrum glenoidale [Fig. 4.29]

The labrum glenoidale is a fibrocartilaginous rim, triangular in transverse-section, attached to the edge of the glenoid cavity, which is slightly deepened by it. The tendon of the long head of biceps arises within the joint from the supraglenoid tubercle, and is there fused to the labrum glenoidale, as is the long head of triceps at its attachment to the infraglenoid tubercle.

Fibrous capsule

The fibrous capsule is attached to the scapula external to the labrum glenoidale and partly to the labrum itself, especially above and behind. On the humerus the capsule is attached above to the

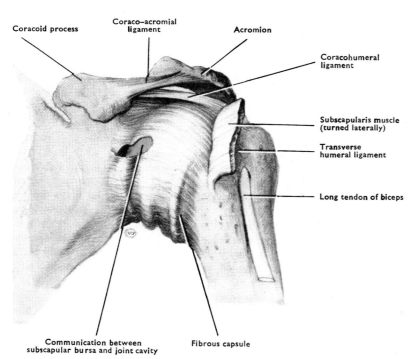

FIG. 4.25. Articular capsule of left shoulder joint from the front.

anatomical neck immediately medial to the tubercles, but below to the medial surface of the surgical neck at least 1 cm from the head of the bone. The upper epiphysial line of the humerus is therefore chiefly extracapsular, but comes within the joint on the medial side. When the arm is by the side the lower part of the capsule forms a redundant fold [FIG. 4.30], but when the arm is abducted, this part of the capsule becomes increasingly taut. Anteriorly the fibrous capsule is immediately in contact with the tendon of subscapularis, superiorly with the supraspinatus tendon, and posteriorly with the tendons of infraspinatus and teres minor. All these tendons blend with the capsule towards their insertion. Since they are intimately concerned with the maintenance of the head of the humerus in its correct relationship to the glenoid cavity whether at rest or during movement, they have been variously termed 'articular' muscles, 'rotator cuff', and 'musculotendinous cuff', and might be regarded as ligaments of variable length and tension. The inferior part of the fibrous capsule is the weakest and it is also relatively unsupported by muscles; however, as the arm is raised from the side, the long head of the triceps and teres major are increasingly applied to this surface of the joint.

There are two **openings** in the fibrous capsule [FIG. 4.25]. A gap in its attachment to the humerus at the upper end of the intertubercular sulcus transmits the tendon of the long head of biceps. In front of the fibrous capsule there is an opening beneath the subscapularis tendon, through which the **subscapular bursa** communicates with the synovial cavity of the joint.

The fibres of the fibrous capsule run for the most part from bone to bone, though a few fibres run transversely round the joint. The **transverse humeral ligament** is a special bundle of these transverse fibres that is attached to the greater and lesser tubercles and holds the tendon of long head of the biceps brachii in the intertubercular sulcus as it leaves the joint [FIG. 4.25].

Glenohumeral ligaments

These three thickenings may be seen on the internal surface of the anterior part of the capsule [FIG. 4.29]. They are seldom prominent, and may be absent. The **superior** glenohumeral ligament is attached to the upper part of the labrum glenoidale immediately anterior to the tendon of the biceps, and passes laterally, alongside the tendon, to reach the humerus near the upper surface of the lesser tubercle. The **middle** glenohumeral ligament is attached to the scapula close to the superior ligament and reaches the humerus at the front of the lesser tubercle. The opening in the fibrous capsule under the subscapularis tendon is between the superior and middle bands. The **inferior** glenohumeral ligament, although usually considered to be the best developed of the three ligaments, is frequently indistinct or even absent. It passes downwards and laterally from the middle of the anterior border of the labrum glenoidale to be attached to the lowest part of the front of the anatomical neck.

Accessory ligaments

The **coracohumeral ligament** [FIG. 4.25] passes from the lateral side of the coracoid process to the upper part of the anatomical neck of the humerus. Laterally it fuses with the tendon of the supraspinatus and they both blend with the capsule. It strengthens the upper part of the fibrous capsule which is under tension when the arm hangs by the side.

The **coraco-acromial ligament** is separate from the fibrous capsule of the shoulder joint, but forms a horizontal shelf above the joint, producing with the coracoid and acromion a kind of secondary socket for the humerus [FIGS. 4.25 and 4.29]. It is triangular and its base is attached to the lateral border of the horizontal part of the coracoid process and its apex to the tip of the acromion in front of the acromioclavicular joint. The anterior and posterior borders of the ligament are stronger than the middle portion. Sometimes a

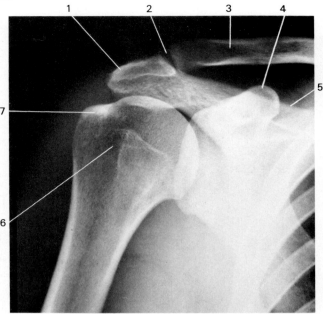

FIG. 4.26. Anteroposterior radiograph of adult shoulder.
1. Acromion.
2. Acromioclavicular joint, position of disc.
3. Clavicle.
4. Coracoid process.
5. Spine of scapula.
6. Intertubercular sulcus.
7. Greater tubercle, site of insertion of supraspinatus.

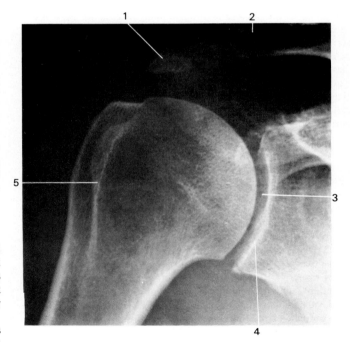

FIG. 4.27. Oblique radiograph of adult shoulder joint.
1. Acromion.
2. Clavicle.
3. Joint 'space' representing cartilage on head of humerus and glenoid fossa.
4. Margin of glenoid cavity.
5. Crest of lesser tubercle.

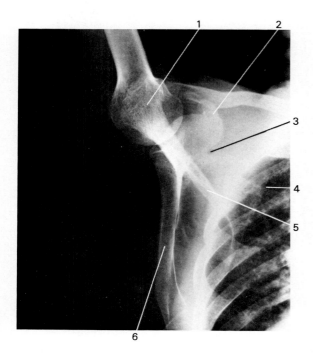

FIG. 4.28. Anteroposterior radiograph of adult shoulder in full abduction.
1. Acromion superimposed on head of humerus.
2. Coracoid process.
3. Scapular notch.
4. Medial border of scapula.
5. Spine of scapula.
6. Lateral border of scapula.

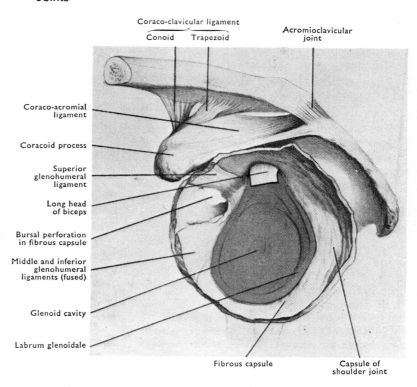

FIG. 4.29. Articular capsule of shoulder joint cut across and humerus removed.

prolongation of the pectoralis minor tendon pierces the base of the ligament and blends, below it, with the coracohumeral ligament. Superiorly, the coraco-acromial ligament is covered by the deltoid muscle; inferiorly, it is separated from the supraspinatus tendon on the upper surface of the shoulder joint by the subacromial bursa. The acromion and the coraco-acromial ligament form a protective arch against which the head of the humerus is pressed (the articular capsule and overlying muscle intervening) as when supporting the weight of the body on the down-stretched arms.

Synovial membrane

The synovial membrane of the shoulder joint lines the fibrous capsule and thus extends downwards as a pouch medial to the humerus when the arm is by the side. From the fibrous capsule the membrane is reflected on the humerus to the edge of the articular cartilage. The membrane protrudes through the opening in the front of the capsule to form the bursa posterior to subscapularis tendon. The size of this bursa varies; when large it wraps round the upper border of the subscapularis tendon underneath the coracoid process, but this upper, subcoracoid extension may be replaced by an independent bursa of that name. Occasionally the synovial membrane protrudes through an opening in the back of the fibrous capsule to form a small bursa beneath the infraspinatus tendon.

The intracapsular part of the tendon of the long head of biceps [FIG. 4.30] is covered with a sheath of synovial membrane which is continued along the tendon in the upper part of the intertubercular sulcus. The synovial sheath is then reflected upwards as a lining to the osteofascial tunnel in which the tendon runs, and so is continuous with the synovial lining of the fibrous capsule. As the tendon runs over the head of the humerus it has a steadying influence upon movements of the shoulder joint.

A large subacromial bursa separates the coraco-acromial arch and the deltoid muscle from the upper and lateral aspect of the shoulder joint and the tendons lying upon it; this bursa does not communicate with the interior of the joint.

Nerve supply

The shoulder joint is supplied by branches from the suprascapular, axillary, and lateral pectoral nerves (Gardner 1948c).

Movements at the shoulder joint

The form of the articular surfaces and the laxity of the fibrous capsule allow a wider range of movement at the shoulder joint than at any other joint in the body. Being a ball-and-socket joint, movement can take place around an infinite number of axes intersecting in the centre of the globular head of the humerus. Anatomically the principal axes are: (1) a transverse axis around which the arm moves forward in flexion and backwards in extension; (2) an anteroposterior axis around which occur the movements of abduction (away from the side of the trunk) and adduction (towards the side of the trunk); and (3) a vertical axis around which the arm rotates medially and laterally. Circumduction is a combination of anteroposterior and lateral movements around successive axes, when the arm swings round the side of a cone whose apex is the point of intersection of the various axes in the head of the humerus. It is more convenient for functional purposes to adopt another set of axes of movement (except for the axis of rotation) because in the anatomical position the glenoid cavity is directed forwards and laterally and the humeral head backwards and

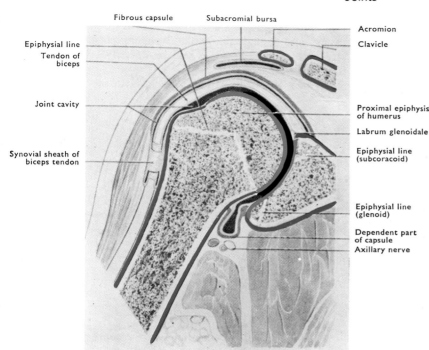

Fibrous capsule Subacromial bursa

Epiphysial line
Tendon of
biceps

Joint cavity

Synovial sheath of
biceps tendon

Acromion

Clavicle

Proximal epiphysis
of humerus

Labrum glenoidale

Epiphysial line
(subcoracoid)

Epiphysial line
(glenoid)

Dependent part
of capsule
Axillary nerve

FIG. 4.30. Coronal section through the
right shoulder joint. The parietal and
visceral layers of the synovial sheath of
the biceps tendon have been partially left
in place.

medially. The true 'plane of the joint' is therefore inclined backwards and laterally at an angle of approximately 45 degrees to the median plane, i.e. at right angles to the plane of the scapula. Thus in the movement of flexion the arm swings forwards and medially at an angle of 45 degrees to the median plane and in extension it is carried backwards and laterally. It may be noted that in this movement the humeral head in actuality 'spins' in the glenoid cavity [p. 214]. Similarly, in abduction the arm is carried forwards and laterally in the 'plane of the scapula'. Abduction in this plane does not involve torsion of the fibrous capsule and therefore provides the most restful position in the treatment of certain injuries at the shoulder (Johnston 1937). It is in this plane that the subscapularis, supraspinatus, infraspinatus, and teres minor muscles are least stretched.

The axis of rotation of the humerus passes through the centre of the head and the centre of the capitulum at the lower end of the bone. The range of rotation varies with the position of the humerus. When the arm is by the side it can be rotated about 170 degrees but when the arm is vertically upright the range of rotation is greatly reduced. The elbow must be flexed when the range of these movements is assessed [p. 238], otherwise the forearm movements of pronation and supination will summate with it.

The amount of movement that occurs at the shoulder joint during movement of the upper limb is usually difficult to estimate because almost all free movements at the shoulder joint are accompanied by associated movements of the shoulder girdle. In raising the arm into a vertically upright position, the arc of movement of the humerus is made up of its real movement upon the scapula at the shoulder joint (120 degrees) plus rotation of the scapula upon the chest wall (60 degrees) [p. 230], with associated movements at the clavicular joints. Thus the effective mobility of this ball-and-socket joint is increased by the mobility of the socket itself. The scapula begins to move at the same time as the shoulder joint (Cleland 1884; Lockhart 1930) and the two movements are associated throughout. However, shoulder joint movement is more concerned with the first half of the activity and scapular movement with the second half.

Nevertheless, it is not until the arm is in the vertically upright position that the humerus is brought approximately into line with the spine of the scapula—the fully abducted position of the shoulder blade (Lockhart 1930).

The combined shoulder girdle and shoulder joint activity also increases the power of this movement because of the involvement of powerful scapular muscles with considerable leverage. The importance of scapular movements is strikingly illustrated by the retention of a considerable degree of mobility of the arm upon the trunk when there is complete fixation of the shoulder joint as the result of disease or surgical treatment.

When the arm is raised to the vertically upright position, the medial epicondyle of the humerus is always directed forwards and slightly medially, and this direction of the epicondyle is constant irrespective of the plane (sagittal, coronal, or any intermediate) in which the limb has been raised. When the limb is raised in the coronal plane the humerus undergoes a lateral rotation of more than 90 degrees, whereas if it is raised to the upright in the sagittal plane it undergoes a slight medial rotation if the initial dependent position of the limb is the anatomical one with the palm of the hand directed forwards. The causal factor of the lateral rotation is most probably the action of the scapulohumeral muscles; the contact of the greater tubercle of the humerus with the sloping coraco-acromial arch when the humerus nears its limit of full abduction may also contribute (Lockhart 1933; Johnston 1937; Martin 1940; MacConaill 1946). In any case it is impossible to raise the arm to the vertical position if the humerus is held in full medial rotation, in which position the greater tubercle abuts on the acromion.

THE ELBOW JOINT

This large synovial joint appears to be of the ginglymus or hinge variety, the bones of the forearm articulating with the lower end of the humerus. The upper ends of the radius and ulna are bound together by the **anular ligament of the radius** in such a way as to permit movement between these two bones at what is described separately as the proximal radio-ulnar joint. The elbow and the proximal radio-ulnar joints have a common fibrous capsule and synovial cavity, and though the anular ligament plays a part in the structure of both joints, it is described with the proximal radio-ulnar joint.

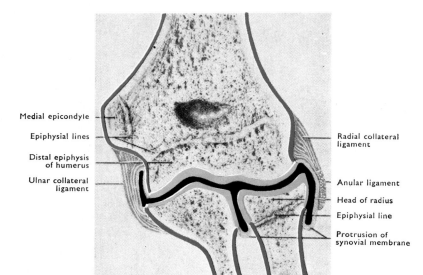

Medial epicondyle

Epiphysial lines

Distal epiphysis
of humerus

Ulnar collateral
ligament

Radial collateral
ligament

Anular ligament

Head of radius

Epiphysial line

Protrusion of
synovial membrane

FIG. 4.31. Coronal section through the
elbow and proximal radio-ulnar joints.

Articular surfaces

The humeral articular surface at the elbow comprises the grooved
trochlea, the spheroidal **capitulum**, and the sulcus between them.
This composite surface is covered with a continuous layer of
articular cartilage. The capitulum is confined to the anterior and
distal aspects of the bone, but the trochlea extends round the distal
end of the bone from the lower edge of the coronoid fossa on the
front of the humerus to the lower edge of the olecranon fossa on the
back [FIG. 4.34]. The ulnar surface of the elbow joint [FIG. 4.37] is
the **trochlear notch**, covered with articular cartilage which is
interrupted along a transverse line across the deepest part of the
notch. The trochlear notch articulates with the trochlea of the
humerus making a saddle-shaped joint with it. The radial surface is
the slightly concave proximal surface of the **head** of the radius
which articulates with the capitulum while its raised margin bears
on the capitulotrochlear groove. This surface of the head is covered
with articular cartilage which is continuous with that round the
sides in the radio-ulnar joint. The radial and ulnar surfaces are most
fully in contact with the corresponding humeral surfaces when the
forearm is in a position midway between full pronation and full
supination and the elbow is flexed to a right angle.

Fibrous capsule

A fibrous capsule completely invests the joint and is relatively weak
in front and behind, but strengthened at the sides to form the radial
and ulnar collateral ligaments.

The anterior part of the fibrous capsule [FIG. 4.32] is composed of
longitudinal, transverse, and oblique fibres: it is thicker in the
middle than at the sides. This part of the capsule is attached to the
front of the humerus immediately above the radial and coronoid
fossae, to the anterior border of the coronoid process of the ulna,
and to the anterior part of the anular ligament of the radius. The
brachialis muscle covers the greater part of the front of the capsule,
and some of its deep fibres, inserted into the capsule, draw it
upwards when the muscle contracts to flex the joint.

The posterior part of the capsule is very weak in its median part,
but the overlying tendon of the triceps attaches to it supporting and
drawing it upwards in extension. The fibres in the posterior part of
the capsule run mainly from the sides of the olecranon fossa to the
margin of the olecranon; some fibres stretch across the fossa above
the olecranon as a transverse band with a free upper border which
falls short of the upper margin of the fossa. Beneath these transverse

fibres a few longitudinal strands pass to the upper part of the fossa
and afford slight support to the synovial pouch within it. The lateral
fibres of the posterior part of the capsule pass from the back of the
lateral epicondyle to the ulna at the posterior border of the radial
notch and also to the anular ligament.

The **radial collateral ligament** [FIG. 4.32] is a strong, triangular
thickening of the capsule. Its apex is attached above to the antero-
inferior aspect of the lateral epicondyle of the humerus in close
relation to the overlying common origin of the extensor muscles.
Distally, the broad base of the ligament blends with the anular
ligament of the radius, and is attached in front and behind to the
margins of the radial notch on the ulna.

The **ulnar collateral ligament** [FIG. 4.33] is composed of three
fairly distinct thickenings of the capsule which are continuous with
one another. An **anterior band** passes from the front of the medial
epicondyle of the humerus to the medial edge of the coronoid process
of the ulna; it is closely associated with the common origin of the
superficial flexor muscles and gives rise to fibres of the flexor
digitorum superficialis. A **posterior band** is attached above to the
back of the medial epicondyle and below to the medial edge of the
olecranon. A **transverse band** stretches between the attachments of
the anterior and posterior bands on the coronoid process and the
olecranon. The lower edge of this transverse ligament is free, and
through the narrow gap between this edge and the bone the synovial
membrane protrudes slightly during movement at the joint.

The middle, thinner, triangular part of the ligament lies between
these bands, and its external surface is grooved by the ulnar nerve
as it passes from upper arm to forearm. The apex of the middle part
is attached to the under surface of the medial epicondyle, and its
base is fixed distally to the upper border of the transverse band.

Synovial membrane

The synovial membrane of the elbow joint lines the fibrous capsule
and is reflected on to the humerus to line the radial and coronoid
fossae in front and the olecranon fossa behind [FIG. 4.34]. Distally,
it is prolonged to the upper part of the deep surface of the anular
ligament; it covers the lower part of the deep surface of the anular
ligament and is then reflected on to the neck of the radius. This
reflection is supported by a few loose fibres which pass from the
lower border of the anular ligament to the neck of the radius. The
synovial membrane passing from the medial side of the radial neck
to the lower border of the radial notch on the ulna is supported by a

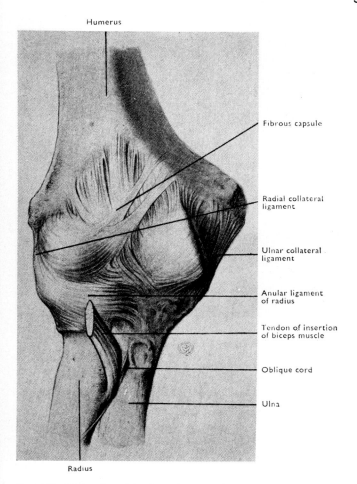

FIG. 4.32. Right elbow joint (anterior surface).

lax band of fibres called the **quadrate ligament**, which stretches between these two points. Well-marked extrasynovial fatty pads fill the radial and coronoid fossae in extension of the joint and the olecranon fossa in flexion. These pads are displaced when the upper ends of the forearm bones occupy the fossae [FIG. 4.34]. A synovial fold, which forms the greater part of a ring, also overlies the peripheral part of the head of the radius and is interposed between it and the corresponding part of the capitulum of the humerus. There are no large defects in the fibrous capsule of the elbow joint, but slight pouching of the synovial membrane may occur beneath the edge of the transverse band of the ulnar collateral ligament, and above the transverse fibres that bridge the upper part of the olecranon fossa.

Nerve supply

The elbow joint derives its supply anteriorly from the musculocutaneous, median, and radial nerves, and posteriorly from the ulnar nerve and from the branch of the radial nerve to the anconeus muscle.

Movements at the elbow joint

Movement occurs principally around a transverse axis—a movement of **flexion** when the forearm makes anteriorly a diminishing angle with the upper arm, and of **extension** when this angle is opened out again. The axis of movement passes through the humeral epicondyles and is not at right angles with either the humerus or the bones of the forearm. In full extension, with the forearm supinated, the arm and forearm form an angle of more than 180 degrees medially. The angle by which this is more than 180 degrees is called the 'carrying angle' which is said to be more pronounced in women than in men. The axis of movement bisects this angle; hence as the forearm is flexed it comes into line with the arm, the distal end of the ulna moves medially and the 'carrying angle' disappears. If the humerus is rotated slightly medially, as it is with the arm hanging naturally by the side, the hand approaches the mouth when the forearm is flexed;

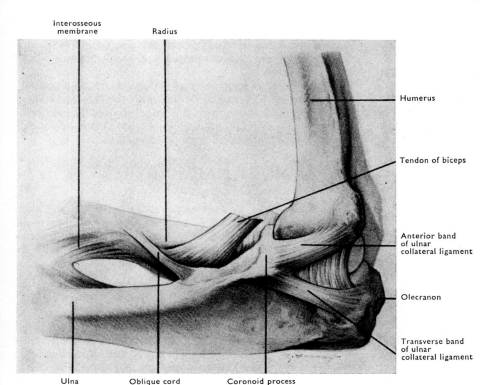

FIG. 4.33. Right elbow joint (medial aspect). The anterior and posterior parts of the articular capsule were removed.

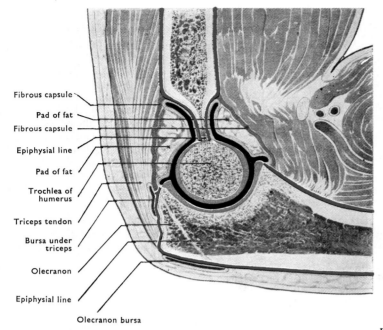

Fibrous capsule

Pad of fat

Fibrous capsule

Epiphysial line

Pad of fat

Trochlea of
humerus

Triceps tendon

Bursa under
triceps

Olecranon

Epiphysial line

Olecranon bursa

FIG. 4.34. Sagittal section through the elbow joint. The synovial
membrane is shown in grey, the articular cartilage in blue, and the
periosteum in green. Note that the pads of fat, though intracapsular, are
extrasynovial.

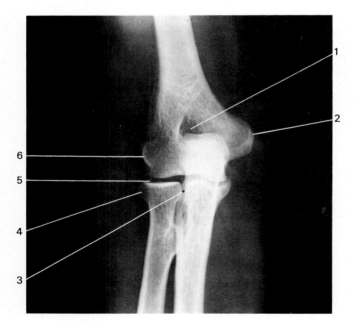

FIG. 4.35. Anteroposterior radiograph of adult elbow joint.
1. Olecranon and coronoid fossae superimposed.
2. Medial epicondyle of humerus.
3. Proximal radio-ulnar joint.
4. Head of radius.
5. Elbow joint 'space'.
6. Lateral epicondyle of humerus.

bringing the hand to the mouth is usually associated with some
abduction of the arm at the shoulder joint. The 'carrying angle' is
masked when the forearm is pronated [p. 237], and in that position
the wrist, elbow, and shoulder joints are all in line with one another
when the elbow joint is extended—this is the usual position of the
upper limb in which pushing and pulling movements are performed.

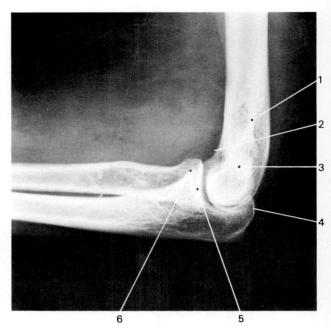

FIG. 4.36. Lateral radiograph of adult elbow joint.
1. Superimposed medial and lateral margins of humerus.
2. Fat protruding from coronoid fossa.
3. Medial epicondyle.
4. Olecranon.
5. Head of radius.
6. Coronoid process of ulna overlapped by head of radius.

Flexion proceeds until checked by apposition of arm and forearm
and by the tension of the posterior muscles and collateral ligaments.
Extension cannot occur beyond the straight position of the limb; it
is then limited by the tension of the anterior muscles and collateral
ligaments. Collateral ligaments are fairly tense in all positions their
anterior parts being specially tight in extension and their posterior
parts in flexion.

Limitation of movement by locking of the bones seldom occurs.
That it does take place in some persons is shown by the occasional
presence of small cartilage-covered facets at the bottom of the
coronoid fossa and at the sides of the olecranon fossa which
obviously must have made contact during life with the coronoid and
olecranon processes.

Unlike most other joints the joint surfaces are in closest contact
in a position of right-angled flexion, with the forearm midway
between pronation and supination. This is therefore the position of
greatest stability and is the position which is most naturally assumed
when the hands are engaged in fine manipulations.

THE RADIO-ULNAR JOINTS

The bones of the forearm are united at their proximal and distal
ends by synovial joints which act together to allow movement of the
radius on the ulna around a vertical axis; these joints are therefore
of the pivot type. In addition, the shafts of the two bones are
connected by a fibrous interosseous membrane.

The proximal radio-ulnar joint

At the proximal radio-ulnar joint the cylindrical head of the radius
rotates within the ring formed by the radial notch on the ulna and
the anular ligament of the radius. The notch on the ulna is lined with
articular cartilage which is continuous with that on the lower part of
the trochlear notch in the elbow joint; the surface of the radial notch

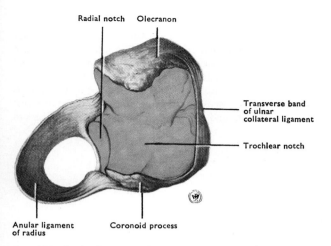

Radial notch   Olecranon

Transverse band
of ulnar
collateral ligament

Trochlear notch

Anular ligament
of radius

Coronoid process

FIG. 4.37. Anular ligament of radius and proximal articular surfaces of ulna.

is concave anteroposteriorly and is almost flat vertically [FIG. 4.37]. The head of the radius is covered with a continuous layer of cartilage above and around the sides. The upper surface is slightly concave and rotates against the capitulum of the humerus, while the sides of the head bear on the anular ligament and radial notch.

Anular ligament of radius

This is a strong, well-defined, curved band attached by its ends to the anterior and posterior margins of the radial notch on the ulna so as to form nearly four-fifths of a ring which is completed by the notch itself. This ring narrows slightly towards the neck and clasps the bevelled lower margin of the head of the radius. This arrangement tends to prevent the radius from being pulled downwards through the ring. The anular ligament is supported above by its fusion with the radial collateral ligament and the fibrous capsule of the elbow joint in front and behind [FIG. 4.32]. Below, it is only feebly attached to the neck of the radius, distal to the epiphysial line, by fibres which are too loose to interfere with movement. The upper part of the anular ligament is lined with fibrocartilage which is continuous above with the synovial membrane of the elbow joint. The lower part of the anular ligament is lined with synovial membrane which continues downwards on the upper part of the neck of the radius; the membrane lies on the upper surface of the quadrate ligament [p. 235] as it passes medially from the radius to the lower border of the radial notch on the ulna. The synovial cavities of the elbow and proximal radio-ulnar joints are freely continuous with each other.

The distal radio-ulnar joint

The distal radio-ulnar joint is formed between the head of the ulna and the ulnar notch on the radius. The bony surfaces are covered with articular cartilage; the lateral and distal ulnar surfaces are continuous with each other over a rounded border.

Articular disc

The chief uniting structure is the triangular articular disc of fibrocartilage. This is attached by its base to the sharp edge on the radius between the ulnar and carpal surfaces and by its apex to the lateral side of the root of the ulnar styloid process [FIG. 4.38]. The upper surface of the disc articulates with the distal aspect of the head of the ulna. Thus, the joint cavity is L-shaped in vertical section, with a horizontal limb between the ulna and the articular disc [FIG. 4.40] and a vertical limb between the radius and ulna. The

distal surface of the disc forms part of the proximal articular surface of the wrist joint only. The cavities of the two joints communicate if the disc is perforated.

Fibrous capsule

The fibrous capsule is represented only by transverse bands of no great strength which stretch from radius to ulna across the front [FIG. 4.39] and back of the joint. The distal edges of these bands blend with the margins of the articular disc, but proximally they are separated from each other by a pouch of the synovial lining of the joint, called the **recessus sacciformis**, which extends upwards a little way between the radius and ulna [FIG. 4.40].

Nerve supply

The distal radio-ulnar joint is supplied by twigs from the anterior and posterior interosseous nerves.

Connection between the shafts of the radius and ulna

The shafts of the radius and ulna are connected by the oblique cord and the interosseous membrane of the forearm.

Oblique cord

This is a slender fibrous band which passes from the lateral border of the tuberosity of the ulna downwards and laterally to the radius just distal to its tuberosity [FIG. 4.33]; it is sometimes considered to be a degenerated portion of the flexor pollicis longus but is more probably a degenerate part of the supinator muscle (Martin 1958).

Interosseous membrane

This is a strong fibrous sheet which stretches between the interosseous borders of the radius and ulna. Its fibres run medially and downwards from radius to ulna. It has a free oblique upper border attached to the radius about 2.5 cm below its tuberosity and passing to a slightly more distal part of the interosseous border of the ulna. The posterior interosseous vessels pass to the back of the forearm between the upper border of the interosseous membrane and the oblique cord. Distally, the interosseous membrane is continuous with the fascia on the dorsal surface of the pronator quadratus muscle where it is attached to the posterior of the two lines into which the lower part of the interosseous border of the radius divides. The distal part of the membrane is pierced by the anterior interosseous vessels.

The interosseous membrane provides extra surface area for origin of the deep muscles of the forearm and braces the radius and ulna together. Its oblique fibres transmit any force passing upwards from the hand along the radius to the ulna.

Movements at the radio-ulnar joints

In the supine position of the extended forearm, the radius and ulna lie parallel, and the palm is directed forwards with the thumb lateral. In the prone position the radius crosses in front of the ulna and the palm is directed backwards. Movement of the radius on the ulna takes place around an axis that passes through the centre of the head of the radius above and the apical attachment of the articular disc below. The movement is chiefly on the part of the radius; its upper end rotates within the ring formed by the anular ligament and the radial notch on the ulna, while its lower end, bearing the hand, travels round the lower end of the ulna, to which it is bound by the articular disc. In the movement of pronation, starting from the supine position, the lower end of the radius is carried forwards and medially round the lower end of the ulna until the palm is turned backwards and the shafts of the radius and ulna cross each other. Movement in the reverse direction is termed **supination**. These movements are normally associated with an accompanying rotation of the humerus at the shoulder joint—medially with pronation and laterally with supination—unless the elbow joint is flexed. However, the ulna may also change position during pronation and supination and this alters the axis of movement in space. Movement of the lower end of the ulna is the magnified result of a very slight 'slewing' movement of its upper end. When using a screwdriver in the right hand, the axis of movement passes through the thumb and forefinger and considerable ulnar 'slewing' occurs; by contrast in certain knitting movements the axis of pronation and supination passes through the little finger. The difference in axes may help to explain why pronation is said to be stronger than supination (Darcus 1951) even though the design of screws and screwing instruments obviously implies that supination (of the right forearm) is the more powerful movement.

When the upper limb is straight, the axis of humeral rotation is in the same line as the axis of radio-ulnar movement, and therefore pronation and supination can be supplemented by the full extent of rotation at the shoulder joint; it is then possible to turn the hand through almost 360 degrees.

THE RADIOCARPAL OR WRIST JOINT

The radiocarpal or wrist joint proper is a synovial joint of the ellipsoid variety. It is formed between the distal surface of the radius and the articular disc, and the proximal row of carpal bones, the scaphoid, lunate, and triquetrum.

Articular surfaces

The surface of the radius and the disc together form an oval concavity shallower in its long axis from side to side than in its short axis from before backwards [FIG. 4.38]. The distal convex surface is formed by the proximal articular areas of the scaphoid, lunate, and triquetrum, closely united by interosseous ligaments which are flush with the articular cartilage on the proximal surfaces of the bones [FIG. 4.40]. The articular cartilage on the radius is divided by a low ridge into a triangular lateral area and a quadrangular medial area. At rest the scaphoid bone is opposite the lateral radial area, the lunate is opposite the medial area and the disc, and the triquetrum is in contact with the medial portion of the articular capsule [FIG. 4.40]. When the hand is bent to the ulnar side, the carpus rotates so that the triquetrum lies opposite the disc [FIG. 4.41B].

Fibrous capsule

A fibrous capsule surrounds the joint attached close to the articular areas. Its proximal attachment is distal to the inferior epiphysial lines of the radius and ulna.

The fibrous capsule has a number of thickenings.

The **palmar radiocarpal ligament** is a broad band of fibres that spreads downwards and medially from the anterior edge of the distal end of the radius to the first carpal row—some longer fibres extending to the capitate bone in the second row [FIG. 4.39].

The **palmar ulnocarpal ligament** is a thickening of the fibrous capsule which extends downwards and laterally from the anterior edge of the articular disc and the base of the ulnar styloid process to the carpus.

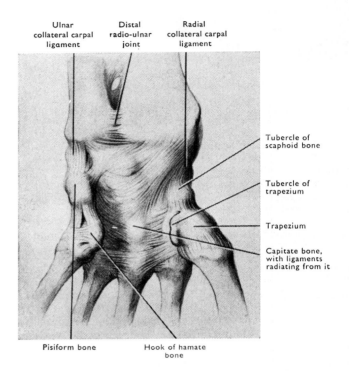

FIG. 4.39. Ligaments on the front of radiocarpal, carpal, and carpometacarpal joints.

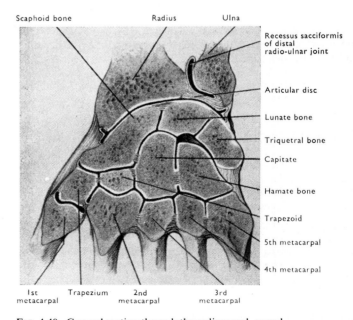

FIG. 4.40. Coronal section through the radiocarpal, carpal, carpometacarpal, and intermetacarpal joints, to show joint cavities and interosseous ligaments (diagrammatic).

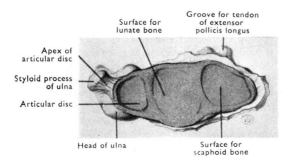

FIG. 4.38. Proximal articular surface of wrist joint.

The dorsal radiocarpal ligament lies on the back of the joint. Its fibres run mainly from the posterior edge of the lower end of the radius downwards and medially to the bones of the first carpal row and particularly to the triquetrum.

At the sides, the fibrous capsule is strengthened to form the radial and ulnar collateral carpal ligaments.

The radial collateral carpal ligament runs from the radial styloid process to the scaphoid immediately lateral to its proximal articular surface; the radial artery crosses the ligament deep to the long abductor and short extensor tendons of the thumb.

The ulnar collateral carpal ligament passes downwards from the ulnar styloid process to the medial, non-articular border of the triquetrum and to the pisiform bone.

Synovial membrane

Synovial membrane lines the fibrous capsule. Exceptionally the synovial cavity communicates with the distal radio-ulnar joint through a perforated articular disc, or with the intercarpal joint when one of the interosseous ligaments of the proximal carpal row forms an incomplete barrier.

Nerve supply

The radiocarpal joint (in common with the intercarpal and carpometacarpal joints) receives its supply from the median nerve, through its anterior interosseous branch; from the radial nerve, through its posterior interosseous branch; and from the ulnar nerve, through its dorsal and deep branches.

Movements at the radiocarpal joint

In movements of the hand at the wrist much of the apparent movement at the radiocarpal joint is the result of movement between the proximal and distal rows of carpal bones (midcarpal or transverse intercarpal joint). Hence movements at the radiocarpal and midcarpal joints are best considered together.

THE INTERCARPAL JOINTS

The carpal bones are articulated in proximal and distal rows that form between them the important midcarpal joint. The joints between the individual bones of the carpus are mostly of the plane variety, but the capitate rotates considerably during movements of the hand.

Joints of the proximal row

The scaphoid, lunate, and triquetrum are bound together by palmar, dorsal, and interosseous bands. The palmar and dorsal intercarpal ligaments pass between neighbouring parts on those aspects of the bones. The interosseous intercarpal ligaments are short bands which unite the proximal parts of the contiguous surfaces in their whole palmar-dorsal depth, but leave joint clefts between their distal parts, which are therefore coated with articular cartilage [Fig. 4.40].

Pisiform joint

The pisiform bone articulates with the palmar surface of the triquetrum by a separate little synovial joint surrounded by a thin but strong fibrous capsule. It is further anchored by the strong pisometacarpal and pisohamate ligaments to the base of the fifth metacarpal and the hook of the hamate bone. These ligaments resist the pull of the flexor carpi ulnaris muscle upon the pisiform bone, and, in effect, provide additional insertions for that muscle.

Joints of the distal row

All four bones of the distal carpal row are connected by palmar, dorsal, and interosseous intercarpal ligaments. The interosseous ligaments [Fig. 4.40] are, in general, not so extensive as those of the proximal row, but leave joint clefts between the bones which are continuous proximally with the midcarpal joint, and distally with the carpometacarpal joint. These two larger joints sometimes communicate with each other between the bones of the distal row when an interosseous ligament is absent (most often that between the trapezium and trapezoid) or when it does not extend from the dorsal to the palmar surfaces of the bones it unites.

Midcarpal joint [Fig. 4.40]

This is the joint between the two carpal rows. On the lateral side the line of the joint is convex distally where the trapezium and trapezoid are opposed to the rounded distal surface of the scaphoid. In the larger medial part, the joint is deeply concave distally: the triquetrum, lunate, and medial surface of the scaphoid form the concavity occupied by the rounded head of the capitate and the proximal angle of the hamate.

The joint is surrounded by a fibrous capsule made up, in front and behind, of irregular bands which run between the two rows of bones and constitute the palmar and dorsal intercarpal ligaments; the bands in the palmar ligament radiate chiefly from the head of the capitate bone [Fig. 4.39]. At the sides of the midcarpal joint the fibrous capsule is strengthened as it passes between the scaphoid and trapezium and between the triquetrum and hamate.

Intercarpal joint cavity

The intercarpal synovial cavity [Fig. 4.40] is large and complicated. The main part extends from side to side between the two rows of bones in the midcarpal joint; this may be partially interrupted by an interosseous ligament connecting contiguous parts of the scaphoid and capitate. As indicated above the cavity ramifies proximally between the three main bones of the proximal row as far as the interosseous ligaments and distally between the four bones of the distal row. It may also communicate with the radiocarpal and carpometacarpal joint cavities.

Nerve supply

The intercarpal joints are supplied by twigs from the anterior and posterior interosseous nerves and the dorsal and deep branches of the ulnar nerve.

Flexor retinaculum

The carpus forms a transverse arch with a palmar concavity. The concavity is bridged by the flexor retinaculum, which is attached medially to the pisiform and the hook of the hamate, and laterally to the scaphoid tubercle and the tubercle of the trapezium. The retinaculum is a major factor in maintaining the carpal bones in their normal position and so is considered an accessory ligament of the intercarpal joints. The median nerve and the flexor tendons enter the palm through the tunnel deep to the retinaculum, while the ulnar vessels and nerve pass superficial to it.

Movements at the radiocarpal and intercarpal joints

The principal movements between the carpal bones are those that occur at the midcarpal joint and this is usually associated with movement at the radiocarpal joint.

Movements at these joints occur around transverse and antero-posterior axes. Around the transverse axes the wrist is bent forward in flexion and backwards in extension. Around the anteroposterior

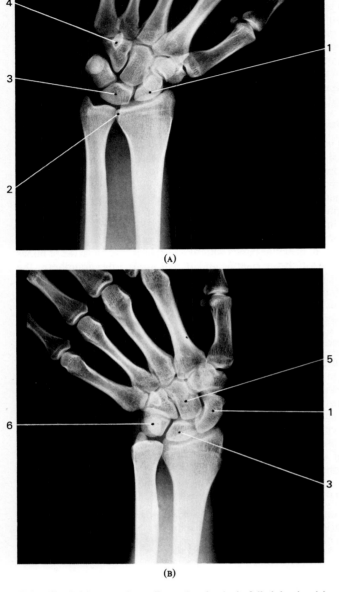

(A)

(B)

FIG. 4.41. Anteroposterior radiographs of wrist in full abduction (A) and adduction (B).

1. Scaphoid bone.
2. Distal radio-ulnar joint.
3. Lunate bone.
4. Hook of hamate bone.
5. Capitate bone.
6. Pisiform bone superimposed on triquetral bone.

axes the hand is deflected towards the ulnar border of the forearm in **adduction** (ulnar deviation) or towards the radial border in **abduction** (radial deviation). Oblique movements and circumduction are also possible around intermediate axes.

In **flexion**, movement occurs both at the radiocarpal and midcarpal joints, but in full flexion, which can only occur when the fingers are extended, the greater part of the movement occurs at the midcarpal joint. **Extension** is also a compound movement, but the range of movement is greater at the radiocarpal than at the midcarpal joint. In both flexion and extension the distal row of

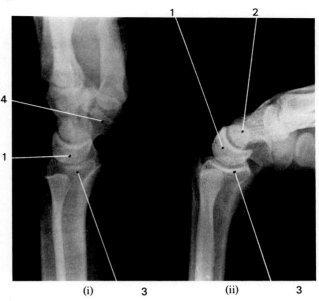

(i)   3   (ii)   3

FIG. 4.42. Lateral radiographs of straight (i) and fully flexed (ii) wrist joints. Note the range of movement of the lunate on the radius and of the capitate on the lunate.

1. Lunate bone.
2. Capitate bone.
3. Distal surface of radius.
4. Scaphoid bone.

carpal bones moves around a transverse axis that passes through the middle of the head of the capitate, and the scaphoid moves on the lunate to accommodate the head of the capitate. In adduction the greater part of the movement occurs at the radiocarpal joint, the proximal row of carpal bones gliding laterally so that the entire proximal surface of the lunate lies distal to the radius [FIG. 4.41B]. Abduction is less extensive than adduction and occurs almost entirely at the midcarpal joint. In adduction and abduction, movement at the midcarpal joint is round an anteroposterior axis that passes through the centre of the head of the capitate. (For a full consideration of these movements see Wright 1935; MacConaill 1941.)

It is important to note that when the hand is used for grasping purposes the wrist naturally assumes a position of slight extension. Therefore, if the wrist joint is likely to become fixed through disease, fixation should be secured in a position of slight extension in order to conserve grasping power.

THE CARPOMETACARPAL AND INTERMETACARPAL JOINTS

Carpometacarpal joint of the thumb [FIG. 4.40]

This is the only highly mobile carpometacarpal joint and is a separate, self-contained joint between the trapezium and the base of the first metacarpal bone. It is a synovial joint of the saddle variety, the articular surfaces being reciprocally concavoconvex. The joint is enclosed by a strong but rather loose fibrous capsule [FIG. 4.39], lined with synovial membrane. Deep to the fibrous capsule lie three discrete ligaments, termed the **anterior** and **posterior oblique** and the **radial carpometacarpal** ligaments (Haines 1944). The radial ligament is attached to the adjacent radial surfaces of the trapezium and the first metacarpal. The oblique ligaments are attached to the anterior and posterior surfaces of the trapezium and converge distally to be attached to the ulnar side of the base of the first metacarpal.

Movements

This is a saddle joint and movement can occur around two principal axes at right angles to each other; movements around intermediate axes when combined allow the movement of circumduction of the thumb. For the understanding of the movements of the thumb it is necessary to stress the inclination of its palmar and dorsal surfaces. When the thumb is in a position of rest, its dorsal surface is directed laterally and its palmar surface medially. In **flexion** the thumb is moved to the ulnar side in the plane of the palm, and in **extension** it is carried to the radial side in the same plane. In **adduction** the thumb is carried directly backwards and in **abduction** it is carried forwards. In the movement of full flexion the thumb is carried directly in front of the palm of the hand, the metacarpal bone undergoing a **medial rotation** of about 30 degrees so that the palmar surface of the thumb becomes opposed to the corresponding surfaces of the fingers. Conversely, the first metacarpal undergoes **lateral rotation** in extension. Flexion is of necessity accompanied by medial rotation of the metacarpal, since this bone travels along the curved groove on the articular surface of the trapezium (Kuczynski 1974). Thus the shape of the articular surfaces ensures precision of movement. The movement of medial rotation is accentuated by the posterior oblique carpometacarpal ligament which becomes increasingly taut during flexion of the metacarpal; similarly lateral rotation is also dependent on the anterior oblique carpometacarpal ligament becoming taut during extension (Haines 1944). During the movement of opposition a certain amount of medial rotation of the first phalanx occurs at the metacarpophalangeal joint and this supplements the rotation of the metacarpal (Bunnell 1938).

Common carpometacarpal joint [FIG. 4.40]

This is formed by the bases of the medial four metacarpal bones and the distal row of carpal bones. The line of the joint is highly irregular. Laterally, the base of the second metacarpal fits into a recess formed by the trapezium, trapezoid, and the capitate. The third metacarpal articulates with the capitate bone at a small transverse segment of the joint line. The fourth metacarpal causes the joint line to become angular again as it articulates to a small extent with the capitate and to a greater extent with the hamate bone. Most medially the fifth metacarpal articulates only with the medial curved facet on the hamate bone.

The joint is surrounded by a fibrous capsule in which various **palmar** and **dorsal carpometacarpal** thickenings pass from the distal carpal bones to the metacarpal bases. An **interosseous ligament** is usually present, stretching from contiguous parts of the capitate and hamate bones to the third or the fourth metacarpal base or to both; this ligament may divide the joint into separate medial and lateral compartments.

The synovial cavity of the common carpometacarpal joint extends proximally into the carpal joints. Distally, the cavity is continuous with the little joints between the bases of the medial four metacarpal bones [FIG. 4.40].

Intermetacarpal joints [FIGS. 4.39 and 4.40]

These three joints are formed between small articular facets on the contiguous sides of the bases of the medial four metacarpal bones. They are closed in front, behind, and distally by **palmar, dorsal,** and **interosseous** ligaments that pass transversely between adjacent bones.

Nerve supply

The carpometacarpal and intermetacarpal joints are supplied by twigs from the anterior and posterior interosseous nerves and the dorsal and deep branches of the ulnar nerve.

Movements

Very little movement is possible between the carpus and the medial four metacarpal bones. It is more appreciable in the case of the fifth metacarpal, whose articulation with the hamate bone is of a flattened saddle type. This metacarpal bone can be slightly flexed and rotated, moving the little finger across the palm towards the thumb—opposition of the little finger.

THE METACARPOPHALANGEAL JOINTS

These are synovial joints of the condyloid variety. Each is formed between the slightly cupped base of the proximal phalanx and the rounded metacarpal head, which is covered with articular cartilage distally and on the front but not on the back.

The **fibrous capsule** is strengthened on each side by a **collateral ligament**, an oblique band which radiates fanwise from the tubercle and adjacent depression on the side of the metacarpal head to the side of the base of the proximal phalanx and to the front of the capsule [FIG. 4.43].

On the front there is a dense fibrous plate, called the **palmar ligament**, which is firmly fixed distally to the base of the phalanx but is only weakly attached proximally to the neck of the metacarpal bone. At its margins the palmar plate in each of the medial four digits is attached to the deep transverse metacarpal ligaments, the fibrous flexor sheath, slips of the palmar aponeurosis, the collateral ligaments, and the transverse fibres of the extensor expansion. The plate is grooved in front by the long flexor tendons. In the thumb there are two sesamoid bones embedded in it; there is usually one such sesamoid in the radial side of the plate on the index finger and occasionally in the ulnar side in the little finger.

The fibrous capsule is deficient **dorsally**, where it is replaced by the extensor expansion which blends at the sides with the collateral ligaments.

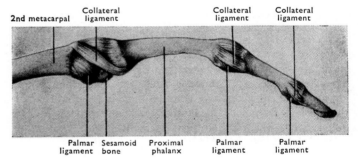

FIG. 4.43. Metacarpophalangeal and interphalangeal joints.

Deep transverse metacarpal ligaments

Three deep transverse metacarpal ligaments unite the palmar ligaments of the medial four metacarpophalangeal joints. Indirectly they bind the heads of the medial four metacarpals together. The interossei tendons descend behind these bands; the lumbrical tendons pass in front of them. It is noteworthy that there is no such binding ligament between the metacarpophalangeal joints of the thumb and the index finger. The thumb is therefore free to move independently of the other fingers.

Movements

Flexion, extension, abduction, and adduction can all take place at the metacarpophalangeal joints. The proximal phalanges can be **flexed** to at least a right angle with the metacarpal bones but can be **extended** little beyond the line of the metacarpals. In the extended position the fingers can be **adducted** or **abducted** towards or away from the centre line of the middle finger; considerable passive rotation is possible in extension when the collateral ligaments are relaxed. When the fingers are flexed, lateral and rotatory movements become impossible because of increased tension in the collateral ligaments; these are fixed nearer to the dorsal and distal surfaces than to the palmar surface of the metacarpal heads, and are more stretched in flexion as the phalanx moves onto the wider palmar aspect of the metacarpal articular surface. At the metacarpophalangeal joints, the phalanges converge during flexion and diverge during extension. This is independent of the action of the interossei and occurs because the metacarpals are arranged in an arc convex dorsally. The metacarpophalangeal joint of the thumb has much less extensive movement than the others—hardly any adduction and abduction being possible—though a slight medial rotation occurs at it during the movement of opposition of the thumb to the fingers [p. 241].

INTERPHALANGEAL JOINTS [Fig. 4.43]

These joints are constructed, as regards ligaments, in exactly the same fashion as the metacarpophalangeal joints. The articular surfaces have a bicondylar shape, a medial and a lateral ridge on the head of the more proximal phalanx fitting into grooves on the base of the more distal phalanx. Hence the interphalangeal joints are pure hinge joints, capable only of **flexion** and **extension**. Although the fingers are of unequal length, the phalanges are so proportioned that in flexion the finger tips come into line and meet the palm simultaneously.

Nerve supply

The metacarpophalangeal and interphalangeal joints of the thumb, index, and middle fingers are supplied mainly by the median nerve, but where the radial nerve supplies the skin over these joints, it too may send twigs to them. The corresponding joints of the ring and little fingers are supplied mainly by the ulnar nerve.

Joints of the lower limb

THE PELVIC GIRDLE

The pelvic or lower limb girdle is formed by the two hip bones which are firmly, but not quite immovably, joined to the sacrum behind and to each other in front. It is joined to the skeleton of the trunk so that it provides great strength for weight transference from trunk to girdle at the sacrifice of almost all mobility. The pelvic joints are the sacro-iliac, interpubic, sacrococcygeal, and lumbosacral articulations.

The lumbosacral and sacrococcygeal joints

Lumbosacral joint [Fig. 3.27]

At the lumbosacral joint the fifth lumbar vertebra articulates with the first piece of the sacrum. As with typical intervertebral joints they are united by a very thick wedge-shaped intervertebral disc, anterior and posterior longitudinal ligaments, synovial joints between articular processes, ligamenta flava, and interspinous and supraspinous ligaments. In addition there is a special ligament that springs from the fifth lumbar transverse process on each side and is called the iliolumbar ligament.

The strong **iliolumbar ligament** [Fig. 4.45] spreads laterally from the fifth lumbar transverse process to the posterior part of the inner lip of the iliac crest. It is really the thickened lower border of the anterior and middle layers of the thoracolumbar fascia which enclose the quadratus lumborum muscle. A less distinct fibrous band, the **lateral lumbosacral ligament**, continuous with the lower border of the iliolumbar ligament, passes downwards and laterally from the lower border of the fifth lumbar transverse process to the lateral part of the sacrum, intermingling with the ventral sacro-iliac ligament.

Sacrococcygeal joint

At the sacrococcygeal joint there is an intervertebral disc between the last sacral and first coccygeal segments. This is reinforced all round by longitudinal strands called the **sacrococcygeal ligaments** [Fig. 4.45]. These have been designated as distinct ventral, dorsal, and lateral sacrococcygeal ligaments. The lateral ligament forms the lateral boundary of the foramen for the fifth sacral ventral ramus. The sacrococcygeal joint frequently becomes obliterated or partially obliterated in old age.

The sacro-iliac joint

This synovial joint is formed between the **auricular surfaces** on the sacrum and the ilium. The joint has a complete fibrous capsule lined with synovial membrane, but movement is greatly restricted by the reciprocal irregularities of the articular surfaces and the thickness and disposition of the dorsal sacro-iliac ligaments. The shape and the irregularity of the auricular surfaces not only vary considerably in different individuals, but may do so on the two sides in the same individual. The auricular surface of the sacrum is covered with hyaline cartilage, but the cartilage on the corresponding facet of the ilium is usually of a fibrous type (Schunke 1938). In later life, particularly in males, it is usual to find fibrous or fibrocartilaginous adhesions between the articular surfaces, with partial or complete obliteration of the cavity of the joint (Brooke 1924).

The **ventral sacro-iliac ligament** is a broad band, of no great thickness, which closes the joint in front, both above and below the linea terminalis. It stretches from the anterior and inferior surfaces of the lateral part of the sacrum to the adjoining surfaces of the ilium.

The strong **sacro-iliac ligaments** lie **behind** and **above** the joint and are divisible into two distinct strata. The deeper layer, the **interosseous sacro-iliac ligament**, is short, thick, and very strong. It fills the narrow cleft between the rough areas on the bones immediately above and behind the articular surfaces [Fig. 4.44]. Occasionally one or two small accessory joint cavities are found in the substance of this ligament between facets near the posterior superior iliac spine and bosses on the sacrum in the position of transverse tubercles (Jazuta 1929).

Superficial to the interosseous fibres, longer bands run obliquely medially and downwards from the ilium constituting the **dorsal sacro-iliac ligament** [Figs. 4.44 and 4.45]. The longest and most superficial fibres of this ligament pass almost vertically downwards from the posterior superior iliac spine to the third and fourth segments of the sacrum. The lateral part of this ligament is indistinguishable from the upper part of the sacrotuberous ligament.

Nerve supply

The sacro-iliac joint is supplied: (1) by twigs directly from the sacral plexus and the dorsal rami of the first two sacral nerves; and (2) by branches from the superior gluteal and obturator nerves.

ACCESSORY LIGAMENTS

The gap left in the bony pelvis between the sacrum and the ischial part of the hip bone is bridged by two important accessory ligaments of the sacro-iliac joint called the sacrotuberous and sacrospinous ligaments [FIGS. 4.44 and 4.45].

The **sacrotuberous ligament** has an extensive attachment to the posterior superior and posterior inferior iliac spines, the back and side of the sacrum, and the upper part of the coccyx. Its fibres converge as they pass downwards and laterally towards the ischial tuberosity, but, twisting upon themselves, they diverge again to be attached to the medial margin of the tuberosity and the lower margin of the ramus of the ischium. The fibres to the ramus form a sickle-shaped extension of the ligament called the **falciform process**.

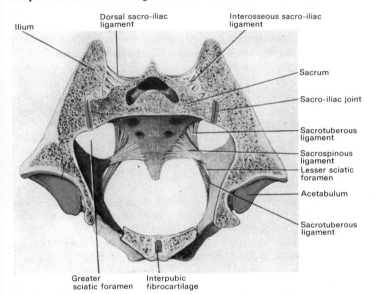

FIG. 4.44. Coronal section of pelvis.

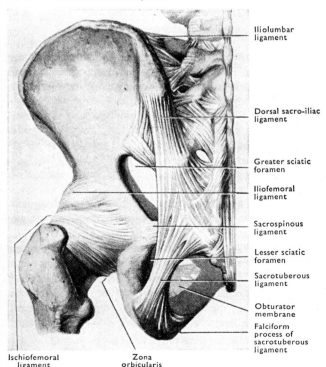

FIG. 4.45. Posterior view of pelvic ligaments and hip joint.

The most superficial of the fibres which are attached to the ischial tuberosity are intimately associated with the long head of the biceps femoris muscle.

The **sacrospinous ligament** lies on the pelvic surface of the sacrotuberous ligament. It is triangular in outline. Its base is attached to the lateral margin of the lower sacrum and upper coccyx, in front of the sacrotuberous ligament; its apex is attached to the ischial spine. This ligament is closely blended with the coccygeus muscle, and may be a fibrous derivative of the dorsal part of that muscle.

Sciatic foramina

The sacrotuberous and sacrospinous ligaments convert the greater and lesser sciatic notches of the hip bone into the greater and lesser sciatic foramina. The **greater sciatic foramen** is bounded above and in front by the greater sciatic notch, behind and medially by the sacrotuberous ligament, and below by the sacrospinous ligament and the ischial spine. The triangular **lesser sciatic foramen** is bounded by the lesser sciatic notch, the ischial spine and sacrospinous ligament above, and the sacrotuberous ligament below [FIG. 4.45]. The greater sciatic foramen leads from the lesser pelvis into the gluteal region while the lesser sciatic foramen leads from the gluteal region into the perineum. The piriformis muscle runs out through the **greater sciatic foramen** with the superior gluteal vessels and nerve above the muscle, and below it the sciatic nerve, the posterior cutaneous nerve of the thigh, the inferior gluteal vessels and nerve, the internal pudendal vessels and the pudendal nerve, and the nerves to the quadratus femoris and obturator internus muscles. The tendon of the obturator internus emerges from the **lesser sciatic foramen**; the nerve to that muscle and the internal pudendal vessels and pudendal nerve enter the perineum above the tendon, by hooking round the ischial spine at its junction with the sacrospinous ligament.

The pubic symphysis

The pubic symphysis is a median joint between the bodies of the pubic bones [FIG. 4.44]. Each pubic articular surface is covered with a thin layer of hyaline cartilage united to the cartilage of the opposite side by a thick fibrocartilaginous **interpubic disc**. In the posterosuperior part of this disc a slit-like cavity appears during early life and may become more extensive, especially in women. This cavity has no synovial lining.

Ligaments of pubic symphysis

The fibrocartilaginous disc of the joint is strengthened by the superior and arcuate ligaments.

The **superior ligament** is attached to the pubic crests and tubercles on each side and strengthens the anterior aspect of the symphysis. The **arcuate ligament** arches across between the inferior pubic rami, and so rounds off the subpubic angle. It is separated from the urogenital diaphragm by an interval through which the deep dorsal vein of the penis (or clitoris) enters the pelvis.

The obturator membrane

The interlacing fibres of this membrane [FIG. 4.46] close the obturator foramen and are attached to its pubic and ischial margins except anterosuperiorly where the **obturator canal** passes through the foramen. The canal is bounded anterosuperiorly by the obturator sulcus and completed below by the short, free edge of the membrane; it transmits the obturator vessels and nerve. The obturator muscles arise from the surfaces of the membrane and the adjacent bone.

Pelvic mechanics

The bony pelvis protects the pelvic viscera and gives attachment to the muscles of the trunk and lower limb, but its primary skeletal function is to provide for stable transference of body weight from the vertebral column to the thigh bones. The weight is transmitted from the upper part of the sacrum through the sacro-iliac joint and along a pelvic arch of thickened bone, corresponding to the posterior half of the linea terminalis, to bear on the head of the femur at the hip joint. The ventral pelvic bar is formed by the superior rami and bodies of the pubic bones and acts as a horizontal tie-beam connecting the bases of the pillars of the arches. The force applied to the acetabula by the femora tends to compress the pelvis inwards. Looked at in this way, the ventral pelvic bar and the bone of the linea terminalis are struts which prevent this compression, but which may deform if softened by conditions such as rickets. The weight of the body tends to displace the anterosuperior part of the sacrum downwards into the pelvis and so tilt the postero-inferior part upwards. This tendency is resisted by a number of factors. (1) As the sacrum is depressed, the interosseous and dorsal sacro-iliac ligaments tighten and draw the ilia together, thus compressing the irregular surfaces of the sacro-iliac joint against each other. (2) The sacrum is wedge-shaped where it articulates with the ilia, the base of the wedge being superior in the anterosuperior part [FIG. 4.44] and inferior in the postero-inferior part. Thus the tendency to depress the upper part and elevate the lower is resisted by the shape of the articulation alone, particularly when the joint surfaces are tightened together. (3) The strong sacrotuberous ligaments prevent any upward displacement of the postero-inferior part of the sacrum relative to the hip bones. The slight amount of gliding movement and the articular cartilage of the sacro-iliac joints have a mild cushioning action against jarring shocks, and the iliolumbar ligaments help to prevent the fifth lumbar vertebra from slipping forwards on the sloping upper surface of the first sacral vertebra. Thus the various joints of the pelvis give it a degree of resilience, but tend to make it progressively more rigid as greater loads are applied to it.

The female pelvis is modified in association with the requirements of child-bearing. Differences are seen in the joints as in the bones [p. 182]. At the sacro-iliac joint there is less interlocking by reciprocal irregularity of the bones, and more movement is permitted than in the male pelvis; the limitation of movement by fibrous ankylosis in later life is not nearly so marked in women as in men. The larger cavity in the interpubic disc is characteristic of the female joint and allows greater separation of the bones. The coccyx, which in men is usually fused to the sacrum by synostosis in later life, is more mobile and preserves its mobility longer in women.

During pregnancy, softening and relaxation of the pelvic ligaments occur which allow the separation of the pelvic bones and an increase in the movements at the pelvic joints. These joint changes were formerly regarded as of importance during child-birth, but radiographic observations indicate that they do not materially affect the diameters of the pelvis (Young 1940).

THE HIP JOINT [FIG. 3.155]

At the hip joint the globular head of the femur articulates with the cup-like acetabulum of the hip bone, and provides the most striking example in the body of a ball-and-socket joint.

The mechanical requirements at this joint are severe. It must be capable not merely of supporting the entire weight of the body—as in standing on one leg—but of stable transmission of forces amounting to several times the body weight during movement of the trunk upon the femur, such as occurs during walking and running

(Paul 1966). The joint must therefore possess great strength and stability even at the expense of limitation of range of movement. Accordingly, the deep socket securely holding the femoral head, the strong, tense fibrous capsule, the insertion of the controlling muscles at some distance from the centre of movement, are all in marked contrast to the conditions at the shoulder joint.

Articular surfaces

The femoral articular surface forms nearly two-thirds of a spheroid. The covering cartilage, thickest superiorly and interrupted at the pit on the head of the femur, ends at the commencement of the neck of the femur along a sinuous line which sometimes encroaches on the front of the neck. The acetabular articular surface [FIG. 3.147] does not occupy the entire acetabulum but is limited to a broad horseshoe-shaped belt which is covered with cartilage (the lunate surface); the cartilage and the acetabular wall are absent below and in front at the acetabular notch. The depressed non-articular part of the acetabulum within the horseshoe, the acetabular fossa, lodges a pad of fat covered with synovial membrane.

Transverse ligament of the acetabulum and labrum acetabulare

The acetabular notch is bridged by the transverse ligament of the acetabulum. The superficial edge of the ligament is flush with the acetabular rim; its deep edge does not reach the bottom of the notch but helps to bound an aperture which admits articular vessels and nerves. The acetabulum is deepened all round by a fibrocartilaginous lip called the labrum acetabulare, which is firmly attached to the bony rim and the transverse ligament. The thin, free edge of the labrum forms a slightly smaller circle than its attached base, and so is able to grasp the femoral head [FIG. 4.47].

Fibrous capsule [FIGS. 4.45 and 4.46]

The cylindrical sleeve of the fibrous capsule encloses the joint and the greater part of the femoral neck. It is very strong and tense in full extension—in contrast to the thin and lax capsule of the shoulder joint. The tendons of surrounding muscles are much less intimately connected with it, but it does receive expansions from the rectus femoris, the gluteus minimus, and the piriformis muscles.

Proximally, the fibrous capsule surrounds the acetabulum and is attached above and behind directly to the hip bone just beyond the labrum, while below and in front it is fixed to the bone, the outer surface of the labrum, and the transverse ligament of the acetabulum. Distally, the capsule is attached in front to the intertrochanteric line at the junction of the femoral neck and shaft; above and below it is attached to the neck close to its junction with the trochanters; but, behind, it covers only the medial two-thirds of the neck. Thus, the whole of the neck of the femur is intracapsular in front, but the lateral third is extracapsular behind. The epiphysial line of the head is entirely intracapsular; the trochanteric epiphysial lines are extracapsular.

The fibres of the fibrous capsule mostly run longitudinally from pelvis to femur. Some deeper fibres, however, pass circularly round the joint, constituting the zona orbicularis; this is marked only on the back of the capsule, where it appears on the surface, winding round behind the femoral neck [FIG. 4.45].

Some of the deepest longitudinal fibres, on reaching the neck of the femur, turn upwards upon the neck towards the articular margin; these reflected fibres form bundles, the retinacula. These transmit blood vessels to the head and neck and are best marked on the upper and lower surfaces of the neck [FIG. 4.47].

The main longitudinal parts of the capsule are thickened to resist the tensile stresses to which the capsule is subjected. They are named after the regions of the acetabulum to which they are attached, and are called the iliofemoral, pubofemoral, and ischiofemoral ligaments. These ligaments are thickened parts of the fibrous capsule and are not always readily identifiable.

The **iliofemoral ligament** is of great strength and considerable thickness; it is a triangular band attached proximally by its apex to the lower part of the anterior inferior iliac spine and adjoining part of the acetabular rim, and distally by its base to the intertrochanteric line. It occupies all the front of the capsule except at the medial side above. The sides of this triangular ligament, which are stronger than the middle part, diverge below from a common stem above. This gives the ligament the appearance of an inverted Y. The upper or lateral of the two diverging bands is attached distally to a special tubercle on the front of the greater trochanter at the upper end of the intertrochanteric line [FIG. 3.156].

In full extension of the hip joint the iliofemoral ligament becomes taut. It therefore resists the tensile stress put upon the anterior part of the fibrous capsule in the standing position, when the body weight tends to roll the pelvis backwards on the femoral heads.

The **pubofemoral ligament** arises from the pubic part of the acetabular rim and the superior pubic ramus [FIG. 4.46]. As it passes distally it blends with the inferior part of the fibrous capsule, although some of its fibres may be traced to the lower part of the femoral neck. This ligament, like the iliofemoral, becomes taut in extension of the joint and it also assists the adductor muscles in checking excessive abduction of the thigh.

The **ischiofemoral ligament**, less well-defined than the others, springs from the ischial wall of the acetabulum on the posterior and lower aspect of the joint [FIG. 4.45]. The upper fibres pass horizontally across the back of the joint, the lower ones ascend spirally; both sets of fibres converge on the upper and lateral part of the femoral neck where they are attached medial to the root of the greater trochanter, close behind the upper band of the iliofemoral ligament. The spiral course of the ischiofemoral ligament upwards and laterally across the back of the joint causes this ligament, like the other two, to be most tense in the position of extension.

Ligament of the head of the femur

This is a weak, flattened ligament within the hip joint [FIG. 4.47] consisting of delicate connective tissue surrounded by synovial membrane. Hence, although it is within the fibrous capsule of the joint, it is extrasynovial. It is attached to the margin of the acetabular fossa and the adjacent edge of the transverse ligament of the acetabulum. It narrows as it passes to its apical attachment in the pit on the head of the femur and lies in the acetabular fossa below the lower part of the femoral head.

The ligament of the head of the femur varies in size, and in rare cases it is absent. Its function is uncertain but may help to distribute synovial fluid. It is stretched when the flexed thigh is adducted or rotated laterally, but in many cases it is too weak to have any definite ligamentous action. No appreciable disability results from its rupture or absence.

A small artery is always found within its substance; this vessel takes little, if any, part in the blood supply of the femoral head in children, though in adults it may supplement the supply derived from the vessels within the retinacula (Tucker 1949). After fusion of the femoral head with the femoral diaphysis, there may be an anastomosis of the vessels of the ligament of the head of the femur with the metaphysial vessels (Trueta 1957).

Synovial membrane

The synovial membrane of the hip joint lines the fibrous capsule and covers the labrum acetabulare. At the acetabular notch it is attached to the medial margin of the transverse ligament, and covering the fatty tissue in the acetabular fossa, extends as a funnel-shaped sheath upon the ligament of the head of the femur to the rim of the pit on the femoral head. At the femoral attachment of the fibrous capsule, the synovial membrane is reflected upwards on the neck of

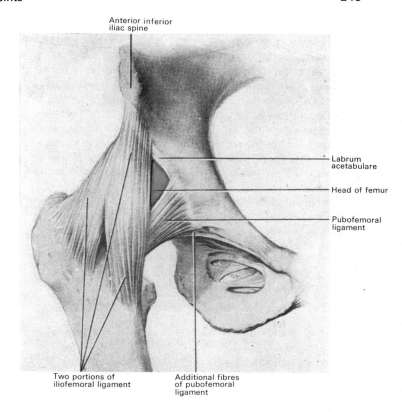

FIG. 4.46. Dissection of the right hip joint from the front.

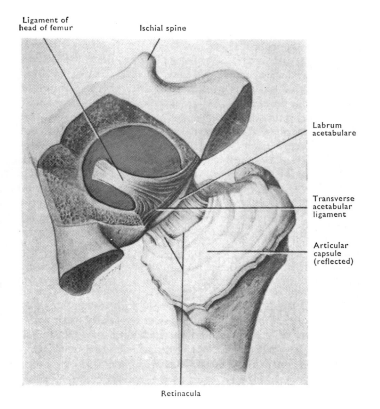

FIG. 4.47. Dissection of right hip joint from pelvic side. The floor of the acetabulum has been removed, and the articular capsule of the joint thrown laterally towards the trochanters.

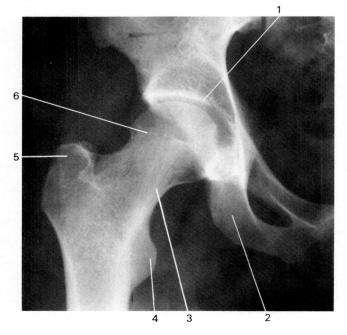

FIG. 4.48. Anteroposterior radiograph of adult hip joint.
1. Acetabulum.
2. Ischial tuberosity.
3. Neck of femur.
4. Lesser trochanter.
5. Greater trochanter.
6. Line of fusion of epiphysis for head of femur with neck.

the femur as far as the margin of the articular cartilage on the head. The retinacula raise this reflected part into prominent folds [FIG. 4.47], within which blood vessels run upwards to the head of the bone. These vessels enter the head of the femur through the non-articular part of the epiphysis [FIG. 3.158]. These vessels are of clinical importance as they constitute the principal arterial supply to the head; they have been extensively investigated by many workers (Trueta 1957).

At the back of the joint a fold of synovial membrane may protrude beneath the lateral border of the fibrous capsule, which is here composed mainly of fibres passing circularly round the femoral neck; the few longitudinal fibres that reach the back of the neck afford some slight support to this synovial protrusion. Where the ligaments are liable to be subjected to pressure from the femoral head, as on the front of the capsule, the synovial membrane is not usually recognizable as a definite layer. The presence of articular cartilage on the front of the neck of the femur may be a reaction to the pressure of the iliopsoas tendon as it crosses the iliofemoral ligament.

There may be an extra-articular extension of the synovial membrane: the psoas tendon passes over the thin portion of the capsule between the upper ends of the iliofemoral and pubofemoral ligaments, with a small bursa intervening. Sometimes the bursa communicates with the synovial cavity of the joint [FIG. 4.46].

Nerve supply

The hip joint is supplied: (1) from the lumbar plexus by twigs from the femoral and obturator nerves; and (2) from the sacral plexus by twigs from the superior gluteal nerve and the nerve to the quadratus femoris muscle (Gardner 1948b).

Movements at the hip joint

The movements are those typical of a ball-and-socket joint (1) flexion and extension around a transverse axis; (2) abduction and adduction around an anteroposterior axis; (3) medial and lateral rotation around a vertical axis; with other movements around intermediate axes. In the movement of circumduction, the limb swings round the side of a cone the apex of which is in the hip joint. All axes pass through the centre of the head of the femur.

Flexion is limited by the tension of the hamstring muscles when the knee joint is extended; but these muscles are relaxed when the knee is bent, and the thigh can then be brought against the anterior abdominal wall. All parts of the fibrous capsule are relaxed in flexion, and dislocation is then most liable to occur. Little extension takes place beyond the vertical position with the limb in line with the trunk. The limitation of extension is in part due to the tension of the ligaments and in part to the form of the articular surfaces. The head of the femur is not truly spherical but is an ellipsoid that is only fully congruent with the acetabulum in the position of complete extension. In bringing the limb into full extension from the position of flexion, the principal ligaments become increasingly tense and force the head of the femur tightly into the acetabulum. When the femur has passed about 15 degrees behind the vertical, the articular surfaces of femur and acetabulum are fully congruent, the major ligaments of the joint are taut and no further extension of the joint normally takes place—the joint is then said to be locked (Walmsley 1928). Because ligaments are slightly extensible and articular cartilage may be distorted when subjected to a compression stress, the precise position of locking will vary, to a slight extent at least, with the extending force (Smith 1956). A force corresponding to at least half the body weight may be necessary for full contact of the superior surface of the femoral head with the zenith of the acetabulum. In some subjects this is said to occur for as little as 5 per cent of the 24-hour period (Bullough, Goodfellow, and O'Connor 1973). This part of the articular area also experiences the highest contact pressures (Day, Swanson, and Freeman 1975). During the screwing home of the head of the femur, part of the pad of fat in the acetabular fossa is displaced from the fossa and passes outwards between the transverse ligament and the acetabular notch, but when the limb is again flexed the fat is drawn into the acetabulum to help to occupy the space that is created by the lateral movement of the head of the femur. In the erect position a perpendicular line through the centre of gravity of the trunk falls behind a line joining the centres of the femoral heads; hence, the pelvis, bearing the trunk, tends to roll backwards on the femora, but is prevented from so doing by the locking that occurs at the hip joints and by the iliofemoral ligaments. Abduction is limited by tension in the adductor muscles and the pubofemoral ligament; adduction, by tension in the abductors and the lateral part of the iliofemoral ligament. Rotation is termed medial or lateral according to the direction in which the toes are turned; it is rather more free when combined with flexion than in the extended position of the hip. Medial rotation is limited by tension in the lateral rotator muscles and the ischiofemoral ligament; lateral rotation, by the medial rotators and the iliofemoral ligament. A small amount of rotation at the hip joint occurs automatically in association with the termination of extension and the commencement of flexion at the knee joint [p. 253] when the foot is on the ground.

THE KNEE JOINT

The knee joint is the largest and most complicated joint in the body. Although it is generally considered to be a modified hinge joint which permits flexion and extension, the movements also involve gliding and rotation of the articular surfaces on each other; rotation of the leg may be demonstrated convincingly when the knee joint is flexed. Despite these extensive and complex movements the human knee joint possesses great stability, especially in extension.

Articular surfaces

Functionally, as well as phylogenetically, the knee joint comprises three articulations—an intermediate one between the patella and the patellar surface of the femur, and lateral and medial articulations between the femoral and tibial condyles.

The articular surface of the femur has three parts: the condylar areas, which are opposed to the tibia and are separated behind by the intercondylar fossa; and the patellar surface, which unites the condyles in front and is opposed to the patella. Each condylar surface is delimited from the patellar surface by a shallow groove on the articular cartilage. On the lateral condyle this groove is almost transverse and is more prominent at each end. On the medial condyle the medial end of the groove begins further forward and, passing obliquely backwards across the condyle, disappears before reaching the lateral edge, where a narrow crescentic facet is marked off which engages with the patella in full flexion [FIG. 4.49]. The lateral condyle viewed *en face* is approximately rectangular, narrowing posteriorly. The medial condylar surface has two parts, a posterior part parallel and equivalent to the lateral condylar surface and an anterior, oblique extension which swings laterally; this area is approximately triangular and has no corresponding part on the lateral condyle. Each condylar surface is convex from side to side and from before backwards and when viewed from either lateral or medial side is seen to have a profile which is spiral in form. The manner in which the condyles become increasingly flatter as they are followed from back to front is a factor of particular significance in the mechanism of the joint. The patellar surface is divided by a well-marked groove into a small medial part and a larger and more prominent lateral part [FIG. 4.49], the importance of which is considered in the next paragraph.

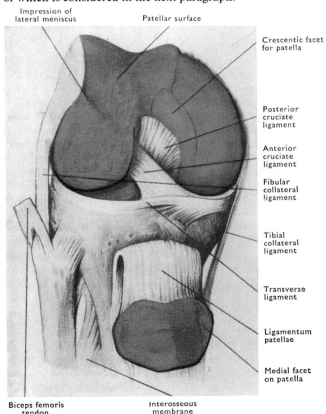

Impression of lateral meniscus

Patellar surface

Crescentic facet for patella

Posterior cruciate ligament

Anterior cruciate ligament

Fibular collateral ligament

Tibial collateral ligament

Transverse ligament

Ligamentum patellae

Medial facet on patella

Biceps femoris tendon

Interosseous membrane

FIG. 4.49. Dissection of right knee from the front: patella turned down.

The articular surface on the patella is broadly oval, and it is divided into a larger lateral area and a smaller medial area by a rounded vertical ridge apparent even on the macerated bone. The cartilaginous covering reveals a further subdivision of the surface; on each side of the vertical ridge there are two faint transverse ridges which separate three facets on each side; another faint vertical ridge delimits the medial perpendicular part of the medial area. This may be termed the 'medial facet' or 'odd facet' of orthopaedic parlance. In extreme flexion, this medial facet rests on the crescentic facet on the medial femoral condyle; the other facets engage in succession from above downwards with the patellar surface of the femur as the joint moves towards full extension [FIG. 4.51]. In the erect posture with the heels together, the tibiae are almost vertical but each femur is directed downwards and medially at an angle of approximately 10 degrees to the medial plane: this may be larger in women than in men. Thus femur and tibia meet at a definite angle at the knee joint. The quadriceps femoris muscle follows the alignment of the femur, but the ligamentum patellae, like the tibia, is vertical. Hence the patella tends to be displaced laterally during forced extension of the knee joint, as in rising from a chair. This is prevented partly by the buttress of the prominent lateral patellar surface of the femur, and partly by the action of the lower fibres of the vastus medialis muscle which are inserted into the medial border of the patella.

The articular surfaces of the tibia are the cartilage-covered areas on the upper surface of each tibial condyle. These surfaces are separated by the triangular intercondylar areas anterior and posterior to the intercondylar eminence. The medial articular surface is oval and concave. The lateral surface, smaller and more circular, is concave from side to side and concavoconvex from before backwards; posteriorly it is prolonged downwards over the back of the condyle where it is in contact with the popliteus tendon [FIG. 4.55]. A peripheral, flattened crescentic strip on each condyle underlies a meniscus [p. 250]. The tibial articular surfaces are far from congruent with the femoral condyles, which rest upon them; but the effect of this incongruence is lessened by the interposition of the menisci.

Fibrous capsule

There is no complete, independent fibrous capsule uniting the bones. Instead, the joint is surrounded by a thick ligamentous sheath constructed largely of tendons or expansions from them. The synovial lining of the joint is in places widely separated from the ligamentous stratum by fatty pads or special intra-articular structures.

At the front of the joint this ligamentous sheath is composed mainly of the fused tendons of insertion of the rectus femoris and vasti muscles. These descend to the patella from above and from each side to insert around the margins of the upper half of the bone. Superficial tendinous fibres are continued downwards over the front of the patella into the strong ligamentum patellae which in reality is the distal part of the quadriceps insertion. It is attached above around the apex of the patella and below to the upper part of the tuberosity of the tibia. Other thinner bands pass down from the sides of the patella diverging to be attached to the front of each tibial condyle; these constitute the medial and lateral patellar retinacula. Tendinous fibres of the vasti muscles pass obliquely downwards across the patella into the retinaculum of the other side. Deeper fibres from each side of the patella pass across to the anterior part of each femoral epicondyle. The manner in which the quadriceps muscle and its expansions, the patella, and the ligamentum patellae participate in replacing the fibrous capsule is a most important feature of the structure of the knee joint.

Strong expansions of the fascia lata lie more superficially. The iliotibial tract descends across the anterolateral aspect of the joint to

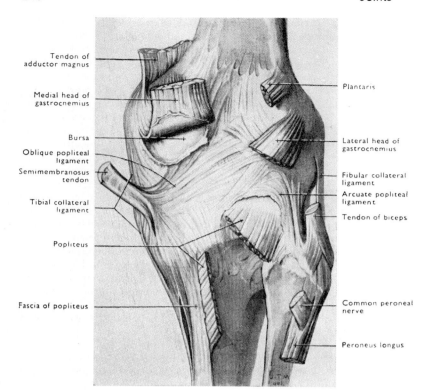

Tendon of
adductor magnus

Medial head of
gastrocnemius

Bursa

Oblique popliteal
ligament

Semimembranosus
tendon

Tibial collateral
ligament

Popliteus

Fascia of popliteus

Plantaris

Lateral head of
gastrocnemius

Fibular collateral
ligament

Arcuate popliteal
ligament

Tendon of biceps

Common peroneal
nerve

Peroneus longus

FIG. 4.50. Right knee joint from behind.
In the specimen the lateral part of the arcuate
popliteal ligament was present as a separate
band: and the continuity between the fibular
collateral ligament and the peroneus longus
muscle was specially marked.

the lateral tibial condyle and sends a band forwards to the upper part of the lateral edge of the patella; this is sometimes called the superior patellar retinaculum.

The whole of the anterior covering of the knee joint, including the patella, is kept tense by contractions of the extensor muscles, and is tightly braced up when these muscles contract in extension. In a typical hinge joint the extensor portion of the capsule must be loose in the extended position to permit the movement of flexion. At the knee, however, this part of the capsule is continuous with the muscle above, instead of being attached directly to the proximal bone; thus its tension is fully adjustable in all positions.

The **posterior** part of the fibrous capsule is also a composite structure. The true capsular fibres are attached to the femur immediately above the condyles and to the intercondylar line, and pass vertically downwards to be attached to the posterior border of the upper end of the tibia. Centrally it is strengthened by an expansion which passes upwards and laterally from the semimembranosus tendon and constitutes the **oblique popliteal ligament**. The lower lateral part of the back of the joint is strengthened by the **arcuate popliteal ligament**. This springs from the back of the head of the fibula, arches upwards and medially over the popliteus tendon as it emerges from the joint, and spreads out on the back of the capsule.

At the **sides of the joint** true capsular fibres unite the femoral condyles to the tibial condyles. The **collateral ligaments** characteristic of hinge joints lie superficial to these fibres.

The **tibial collateral ligament** is a broad, flat band. Its upper end has an extensive attachment to the medial epicondyle of the femur and some fibres may be traced upwards into the adductor magnus tendon; it may represent in part at least, an original tibial insertion of this muscle. The ligament passes downwards and slightly forwards to the medial surface of the tibia. The superficial fibres descend below the level of the tibial tubercle and are separated inferiorly from the tibia by the inferior medial genicular vessels and nerve. The deeper fibres pass directly from femur to tibia and are fused to the medial meniscus. A downward expansion from the semimem-

branosus tendon reaches the shaft of the tibia and lies partly under cover of the posterior border of the tibial collateral ligament. The expansions of the semimembranosus tendon add considerable strength to the medial and posterior aspects of the fibrous capsule.

The **fibular collateral ligament** [FIGS. 4.49 and 4.50], round and cord-like, stands clear of the thin, lateral part of the fibrous capsule and is enclosed by an expansion of the fascia lata. It passes from the lateral epicondyle of the femur to the head of the fibula in front of its highest point, and splits the tendon of the biceps femoris. The ligament is sometimes partly continued into the upper end of the peroneus longus muscle [FIG. 4.50] and may be regarded as a stranded femoral origin of this muscle.

On the lateral side of the joint the true capsular fibres are short and weak, and bridge the interval between the femoral and tibial condyles. The popliteus tendon intervenes between the lateral meniscus and the capsule. The inferior lateral genicular vessels and nerve run forwards between the capsule and the fibular collateral ligament.

The tibial and fibular collateral ligaments prevent disruption of the joint at the sides. They are most tightly stretched in extension, and then their direction—the fibular ligament downwards and backwards, the tibial downwards and forwards—prevents rotation of the tibia laterally or of the femur medially. Rotation may, however, be demonstrated easily in the flexed knee.

Intra-articular ligaments [FIGS. 4.49, 4.55, and 4.56]

In addition to the capsular ligamentous formation, there are two strong, cord-like, intra-articular ligaments that stretch between the tibia and femur. These are the cruciate ligaments—so-named because they cross each other between their attachments. They are termed anterior and posterior according to their relative positions on the tibia, from which they ascend to the sides of the intercondylar fossa of the femur. The **anterior cruciate ligament** extends obliquely upwards and backwards from the rough, anterior intercondylar area of the tibia to the medial side of the lateral femoral condyle. The **posterior cruciate ligament** passes upwards and forwards medial to

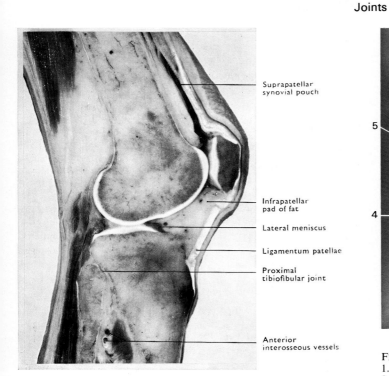

Suprapatellar
synovial pouch

Infrapatellar
pad of fat

Lateral meniscus

Ligamentum patellae

Proximal
tibiofibular joint

Anterior
interosseous vessels

FIG. 4.51. Sagittal section through the lateral condyle of femur, the tibia,
and fibula. The section was cut slightly obliquely so as to show the
proximal tibiofibular joint and also the extent of the suprapatellar synovial
pouch.

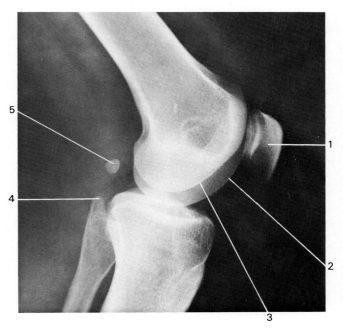

FIG. 4.53. Lateral radiograph of adult knee joint.
1. Patella.
2. Margin of medial condyle of femur.
3. Margin of lateral condyle of femur.
4. Apex of head of fibula.
5. Sesamoid (fabella) in lateral head of gastrocnemius.

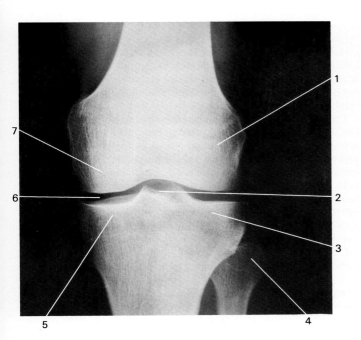

FIG. 4.52. Anteroposterior radiograph of adult knee joint.
1. Lateral condyle of femur.
2. Intercondylar area of tibia.
3. Lateral condyle of tibia.
4. Head of fibula.
5. Medial condyle of tibia.
6. Joint 'space'.
7. Medial condyle of femur.

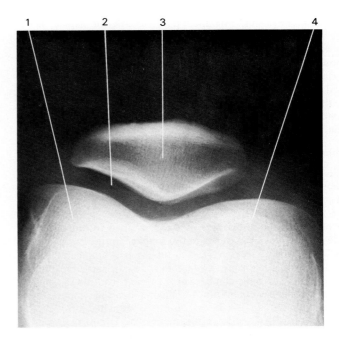

FIG. 4.54. Axial radiograph of the joint between the patella and the femur
to show the thickness of the cartilage (joint space).
1. Lateral condyle of femur.
2. Articular cartilage of patella and femur.
3. Patella.
4. Medial condyle of femur.

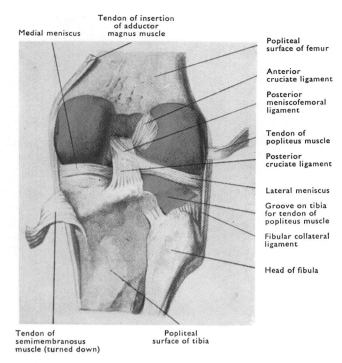

Medial meniscus

Tendon of insertion
of adductor
magnus muscle

Popliteal
surface of femur

Anterior
cruciate ligament

Posterior
meniscofemoral
ligament

Tendon of
popliteus muscle

Posterior
cruciate ligament

Lateral meniscus

Groove on tibia
for tendon of
popliteus muscle

Fibular collateral
ligament

Head of fibula

Tendon of
semimembranosus
muscle (turned down)

Popliteal
surface of tibia

FIG. 4.55. Right knee joint opened from behind by removal of the posterior part of the articular capsule.

the anterior cruciate ligament from the posterior intercondylar area of the tibia to the lateral side of the medial condyle of the femur. The anterior cruciate prevents posterior displacement of the femur on the tibia and the posterior ligament prevents anterior displacement.

The ligaments are more taut in extension than in flexion. They allow a small amount of passive anteroposterior movement of the tibia on the femur if the thigh is fixed and the leg is forcibly moved with the knee joint flexed. The ligaments tighten when the tibia is rotated medially, although they slacken when it is rotated laterally since the ligaments tend to be uncrossed (Haines 1941). Thus the cruciate ligaments tend to rotate the tibia laterally (or the femur medially) as they tighten in full extension.

Menisci [FIGS. 4.55 and 4.56]

Within the joint a medial and a lateral meniscus are interposed between the femoral and tibial condyles, helping to compensate for the incongruence of the bones. They are formed almost entirely of fibrous tissue with relatively few cartilage cells interspersed. Each is a flattened, crescentic pad and is wedge-shaped in cross-section, with the thin edge of the wedge at the concave border.

The thick, convex border of each meniscus is connected to the deep surface of the fibrous capsule, and the capsular fibres which attach the menisci to the condyles of the tibia constitute the **medial and lateral coronary ligaments**. The medial meniscus is much more firmly anchored than the lateral, for it is attached through the capsule to the tibial collateral ligament. The lateral meniscus is attached only to the weak capsular fibres of the lateral side of the joint which are separate from the fibular collateral ligament. Even this attachment is interrupted where the lateral meniscus is crossed by the popliteus tendon. The popliteus muscle arises not only from the femur but also has an attachment to the posterior edge of the lateral meniscus; this attachment shows great variation in size. The extremities or horns of each meniscus are firmly attached to the intercondylar areas on the upper surface of the tibia [FIG. 4.56]. The **anterior horn** of the medial meniscus is fixed to the anterior intercondylar area in front of the anterior cruciate ligament; its **posterior horn** is attached to the posterior intercondylar area,

between the posterior cruciate ligament and the posterior horn of the lateral meniscus. The **two horns** of the lateral meniscus are attached close together, in front of and behind the intercondylar eminence, so that this meniscus is more nearly a circle than the medial one.

The anterior parts of the two menisci are joined by a fibrous band of variable thickness called the **transverse ligament of the knee** [FIG. 4.56]. The posterior part of the lateral meniscus frequently gives a slip to the posterior cruciate ligament [FIG. 4.55]; this slip splits on the lateral side of the cruciate ligament, the anterior part forming the **anterior meniscofemoral ligament** and the posterior part the **posterior meniscofemoral ligament**. The posterior is the more constant of the two ligaments.

The connection between the lateral meniscus and the posterior cruciate ligament is in keeping with the typical lower mammalian condition, where the lateral meniscus is attached posteriorly, not to the tibia but to the medial femoral condyle behind the posterior cruciate ligament.

Synovial membrane

The synovial cavity of the knee is the largest in the body. It is best described in four parts which all communicate freely with one another—a central part and three extensions in the form of recesses.

The **central** part lies between the patella in front and the patellar surface of the femur and the cruciate ligaments behind. Laterally and medially it passes into the posterior recesses between the femoral and tibial condyles and also both above and below the menisci. The synovial membrane passes from the margins of the articular surface of the patella in all directions on the deep aspect of the anterior wall of the joint. Below the patella, the membrane covers a mass of fatty, fibrous tissue, called the **infrapatellar pad of fat**, which lies on the deep surface of the ligamentum patellae and, inferiorly, extends towards the intercondylar notch. In the middle of the joint, the synovial membrane on this pad extends backwards as a vertical, crescentic fold, the **infrapatellar fold**, which becomes continuous with the synovial membrane covering the cruciate ligaments in the intercondylar fossa. On each side of the infrapatellar pad of fat is a horizontal synovial fold, called the **alar fold**, which is usually readily identifiable [FIG. 4.58].

The **three recesses** of the synovial cavity open off its central part. The median **suprapatellar synovial pouch** [FIG. 4.51] extends upwards for about one hand's breadth above the level of the patella between the quadriceps muscle and the lower part of the shaft of the femur. It develops as a separate bursa, but it eventually communicates with the joint. The articularis genus muscle, composed of fleshy slips that arise from the femur above the pouch, is inserted into its wall and draws it upwards during extension of the joint. The other two recesses extend backwards behind the posterior part of each femoral condyle.

The posterior recesses are separated from each other by a thick, median septum which projects into the joint cavity from behind. This septum is formed by the cruciate ligaments which carry forward a broad, vertical fold of synovial membrane continuous antero-inferiorly with the infrapatellar fold. The posteromedian part of the fibrous capsule is therefore devoid of synovial membrane which passes forwards to cover the sides and fronts of the cruciate ligaments.

Bursal extension

As the popliteus tendon passes backwards from its femoral attachment deep to the lateral part of the fibrous capsule, the synovial membrane covers its medial side. This separates it from the adjacent part of the lateral meniscus except where it is attached to the meniscus. Where the tendon leaves the joint through the opening in the posterior part of the capsule, a bursa-like **synovial pouch** is

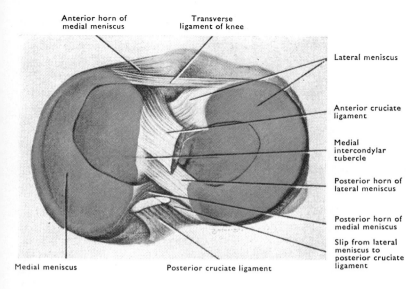

Anterior horn of medial meniscus

Transverse ligament of knee

Lateral meniscus

Anterior cruciate ligament

Medial intercondylar tubercle

Posterior horn of lateral meniscus

Posterior horn of medial meniscus

Slip from lateral meniscus to posterior cruciate ligament

Medial meniscus

Posterior cruciate ligament

Fig. 4.56. Upper end of tibia with menisci and attached portions of cruciate ligaments.

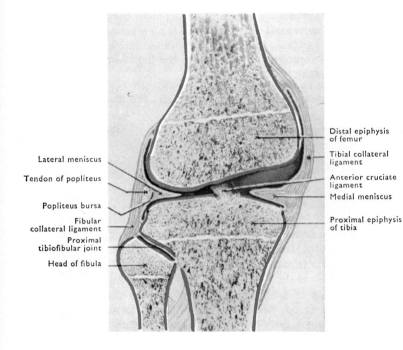

Lateral meniscus

Tendon of popliteus

Popliteus bursa

Fibular collateral ligament

Proximal tibiofibular joint

Head of fibula

Distal epiphysis of femur

Tibial collateral ligament

Anterior cruciate ligament

Medial meniscus

Proximal epiphysis of tibia

Fig. 4.57. Oblique section through the right knee joint.

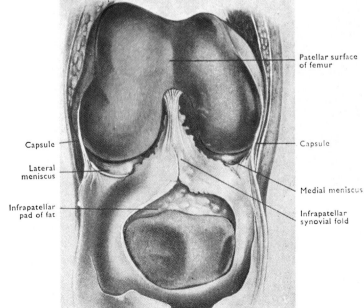

Capsule

Lateral meniscus

Infrapatellar pad of fat

Patellar surface of femur

Capsule

Medial meniscus

Infrapatellar synovial fold

Fig. 4.58. Right knee joint opened from the front to show the infrapatellar and alar synovial folds. The joint is flexed to a right angle and the patella turned down.
Articular cartilage, blue.

carried down on its deep surface across the back of the proximal tibiofibular joint. Sometimes the cavity of this joint communicates with the pouch, and therefore also with the cavity of the knee joint.

Additional bursae

A bursa usually separates the fibrous capsule from the overlying lateral and medial heads of the gastrocnemius muscle. A bursa is more common on the medial side [FIG. 4.50], where it sometimes communicates with the joint cavity through an opening in the capsule. Many other bursae are found in the vicinity of the knee joint. The more constant are: the **subcutaneous prepatellar bursa**, under the skin over the lower part of the patella; the **subfascial prepatellar bursa**, between the fascial and tendinous expansions in front of the patella; the **subtendinous prepatellar bursa**, between the superficial and deep layers of the tendinous fibres that cross the patella; the **deep infrapatellar bursa**, between the upper part of the tibia and the lower part of the ligamentum patellae, separated from the knee joint by the infrapatellar pad of fat. A bursa between the

tibial collateral ligament and the overlying tendons of the sartorius, gracilis, and semitendinosus forms part of the **anserine bursa**; and there are bursae between the semimembranosus tendon and the tibial collateral ligament, between that tendon and the medial head of the gastrocnemius, and between the biceps tendon and the fibular collateral ligament.

Nerve supply

The knee joint is supplied by several branches from the femoral nerve and from both divisions—common peroneal and tibial—of the sciatic nerve, and by a filament from the obturator nerve [see Gardner 1948a].

Movements at the knee joint

The movements that occur at the knee joint are flexion and extension and a limited amount of active rotation; the joint is justifiably regarded as a modified hinge joint. It differs from a typical hinge joint not only in that some rotation occurs throughout flexion and extension, but also because the axis of movement and the zone of contact between the articular surfaces move forwards during extension and backwards during flexion principally with respect to the femur; whereas in a typical hinge joint the axis of movement is constant. The axis changes its position during flexion-extension owing to a difference in curvature of different parts of the femoral condyles.

In voluntary flexion the leg may be moved through 140 degrees or more from the position of full extension, and this may be increased

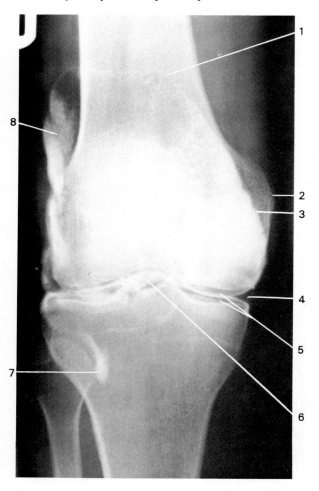

FIG. 4.59. Anteroposterior arthrogram of knee joint produced by the introduction of air and contrast material into the knee joint cavity. Knee joint extended.
1. Upper limit of suprapatellar bursa.
2. Medial femoral condyle.
3. Medial margin of knee joint cavity.
4. Medial meniscus.
5. Articular cartilage of femur and tibia with layer of contrast material between.
6. Contrast material in synovial cavity of knee joint.
7. Contrast material in superior tibiofibular joint which is continuous with knee joint in 25 per cent of cases.
8. Contrast material in posterolateral part of suprapatella bursa.

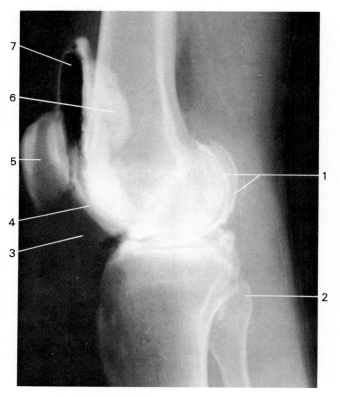

FIG. 4.60. Lateral arthrogram as same knee joint as FIGURE 4.59.
1. Contrast material in synovial cavities posterior to femoral condyles.
2. Head of fibula.
3. Infrapatellar pad of fat.
4. Contrast material in synovial cavity posterior to infrapatellar fold.
5. Patella.
6. Contrast material in posterolateral part of suprapatellar bursa.
7. Air in suprapatellar bursa which has distended it.

by forced movement or by the adoption of the squatting position. In full flexion the posterior surfaces of the femoral condyles articulate with the posterior parts of the articular surfaces of the tibia and the corresponding parts of the menisci.

As the joint passes from flexion to extension the femoral condyles roll forwards and slip backwards upon the tibia, if the tibia is fixed as in rising from a chair with the feet planted on the floor. Both the anteroposterior and transverse curvatures of the femoral condyles become progressively more flattened as they are traced forwards. Hence there is an increasingly greater contact with the tibia as the femur passes into full extension. Just before full extension is reached, the tibial surface of the lateral femoral condyle is almost exhausted, although the anterior, oblique part of the surface on the medial condyle has still to be brought into play. While the lateral condyle completes its forward roll, the medial condyle, continuing the movement, glides backwards until its oblique anterior surface is in contact with the tibia. The gliding backwards of the medial condyle, assisted by the tightening of the posterior cruciate ligament, occurs about a vertical axis through the lateral condyle (held forwards by the anterior cruciate ligament) so that there is a medial rotation of the femur. The increasing amount of rotatory movement during terminal extension tightens the collateral ligaments of the knee and the joint is then, as it were, screwed home. In the fully extended position all ligaments of the knee joint are taut and the articular surfaces of the femur and tibia are in maximal contact; the joint is then said to be locked. On account of the deformation that occurs in articular cartilage and the extensibility of the ligaments the precise position in which locking occurs varies slightly with the extending force. During flexion lateral rotation of the femur on the fixed tibia is accomplished by the action of the popliteus muscle. If both bones are free to move (e.g. in kicking), the tibia rotates laterally during extension and medially during flexion (Goodsir 1858).

When the body is erect, the centre of gravity of the body lies in front of a line between the knee joints; gravity therefore tends to produce extension at the knee joint. When standing in the natural upright position, however, the knee joint is not fully extended but is maintained in a position a little short of it. The ilio-tibial tract is then taut although the quadriceps muscle is relaxed and the patella can be moved freely from side to side on the femur. The hamstring and gastrocnemius muscles, on the other hand, contract sufficiently to maintain the joint in this position against the action of gravity. In reaching forwards, however, the knee joint often becomes fully extended and locked, and in this position the quadriceps remains relaxed.

In full extension of the knee, all rotation is prevented by the tension of the collateral and cruciate ligaments. In other positions of the joint some active rotation of the leg around a vertical axis can take place. About 50 degrees of rotation is possible (Fick 1911), when the knee is flexed to a right angle. Neither cruciate ligament is then very tight and the tension of the collateral ligaments is reduced both by descent of their femoral attachments and the conjunct rotation accompanying flexion.

The menisci compensate partly for the incongruity of the opposed femoral and tibial surfaces and adjust themselves to the varying shapes of the femoral condyles resting upon them in different positions of the joint. In extension the menisci move forwards slightly and open out their curve under the flatter, anterior parts of the condyles then applied to them. The lateral meniscus, in particular, moves forward, pivoting about its closely set attachments to the tibia, so that its anterior part glides down on the cambered surface of the lateral tibial condyle (Barnett 1953). In flexion the menisci shift backwards and curl in upon the more rounded posterior parts of the femoral condyles. The difference in the attachments of the menisci to the collateral ligaments is important in one form of the commonest derangement of the knee joint—torn medial meniscus. This may happen in a sudden unguarded movement, for

example if the body is turned to the opposite side while taking the weight on the semi-flexed knee. Due to its attachment to the tibial collateral ligament, the medial meniscus is drawn medially as the ligament tightens during extension. Thus held and before it can open out as in normal extension, the meniscus may be pinned between the bones as the femur rotates medially. This splits the meniscus in its long axis producing the familiar 'bucket handle' tear.

According to MacConaill (1932) the menisci play an important part in relation to efficient lubrication in the joint, the slight mobility of the menisci permitting them to assume the proper angulation to the joint surfaces necessary for the maintenance of wedge-shaped synovial films. The menisci also have an essential role in load bearing. The total load acting on the knee joint is transmitted through the articular cartilage and the menisci and the fraction carried by the menisci rises with increasing load. After meniscectomy the stress per unit area is said to be three times that in a knee with intact menisci (Seedhom 1975).

THE TIBIOFIBULAR JOINTS

The fibula is closely bound to the tibia by a joint at each end; and the shafts of the two bones are connected by an interosseous membrane.

The proximal tibiofibular joint [FIGS. 4.55 and 4.57]

This is a small synovial joint, of plane type, between a flat, circular or oval facet on the head of the fibula and a similar facet on the tibia placed posterolaterally on the inferior aspect of the overhanging lateral condyle. The fibrous capsule which surrounds the joint is strengthened in front and behind by the anterior and posterior ligaments of the head of the fibula which run upwards and medially from the fibula to the adjoining part of the tibia.

The tendon of the popliteus crosses the posterosuperior aspect of the joint; and the pouch of synovial membrane prolonged under the tendon from the knee joint sometimes communicates with the synovial cavity of the tibiofibular joint through an opening in the upper part of the capsule.

Nerve supply

The superior tibiofibular joint is supplied by twigs from the common peroneal nerve and from the branch of the tibial nerve to the popliteus.

The distal tibiofibular joint [FIGS. 4.61–4.65]

This is a fibrous joint. The rough, triangular, opposed surfaces of the bones are united by a strong interosseous ligament. In addition, the joint is strengthened in front and behind by the anterior and posterior tibiofibular ligaments. These bands run laterally and downwards from the lower border of the tibia, in front and behind, to the front and back of the distal end of the fibula. Under cover of the posterior ligament there is a longer band, called the transverse tibiofibular ligament, which is attached to the whole length of the posterior edge of the inferior surface of the tibia and to the upper end of the malleolar fossa on the fibula. This band closes the posterior angle between the tibia and fibula and articulates with the posterior part of the lateral border of the trochlea tali [FIG. 4.61].

A recess of the cavity of the ankle joint usually extends upwards between the tibia and fibula for about 0.5 cm. It is blocked above by the base of the interosseous ligament and is occupied by a fold of synovial membrane of the ankle joint. Sometimes the articular cartilage on the lower ends of the tibia and fibula extends upwards for a little way on the walls of this recess.

Nerve supply

The inferior tibiofibular joint is supplied by twigs from the deep peroneal and tibial nerves.

The crural interosseous membrane

The interosseous membrane of the leg is tightly stretched between the interosseous borders of the two bones. The general direction of its fibres is downwards and laterally from tibia to fibula. The membrane reaches up to the under aspect of the proximal joint and downwards to blend with the upper edge of the interosseous ligament of the distal joint. There is an oval aperture near the upper end of the membrane through which pass the anterior tibial vessels [FIG. 4.49], and a smaller one at the lower end for the perforating branch of the peroneal artery [FIG. 4.61].

Movement at the tibiofibular joints is slight and entirely passive. It is occasioned by movement at the ankle joint, with which it will be described [p. 256].

THE TALOCRURAL OR ANKLE JOINT

The ankle joint is a synovial joint of typical hinge pattern. The lower ends of the tibia and fibula provide a socket in which the upper part of the talus rocks around a transverse axis. The sides of the socket and the powerful ligaments and tendons around the joint provide great strength and stability.

Articular surfaces

The proximal articular surface formed by the cartilage-covered areas on the lower ends of the leg bones provides the socket [FIG. 4.62]. Its roof is formed by the distal surface of the tibia; this is wider in front than behind, is slightly convex from side to side, and is concave from before backwards. Posteriorly the articular surface projects downwards slightly and the socket is deepened

further by the posterior tibiofibular ligament. The medial wall is the lateral surface of the medial malleolus and its articular cartilage is continuous with that of the roof at a rounded angle. The lateral wall is the triangular facet on the anterior part of the medial side of the lateral malleolus. In the angle between this wall and the roof is the narrow cleft between the tibia and fibula distal to the interosseous ligament.

The distal articular surface is formed entirely by the upper part of the body of the talus—the trochlea tali—which has the approximate shape of the upper part of a short cylinder placed transversely. However, the radius of curvature decreases and the trochlea tali is slightly broader in front, the lateral surface often sloping medially at the back. The upper surface articulates with the roof of the socket (tibia), which also is broader in front. The upper trochlear surface is slightly concave transversely to fit the convexity of the tibial surface. The cartilage on top of the trochlea is continued on to its sides. The medial articular surface is vertically set, and is shaped like a comma laid on its side, tail backwards; it is opposed to the medial malleolar wall of the proximal articular surface. The lateral articular area of the trochlea is triangular and very much larger than the medial articular area [FIG. 4.63]. Inferiorly, it curves laterally to a projecting apex, while posteriorly it slopes medially. It articulates with the facet on the lateral malleolus and, posterosuperiorly, with the inferior part of the posterior tibiofibular ligament (transverse tibiofibular ligament) where the angle between the upper and lateral surfaces of the trochlea is bevelled [FIG. 4.61].

Fibrous capsule

This surrounds the joint and is attached superiorly to the margins of the articular surfaces on the tibia and fibula, passing below to the talus close to the articular areas on the upper, medial, and lateral surfaces of its body. As is customary in a hinge joint, the capsule is weak in front and behind, but medially and laterally it is reinforced by strong collateral ligaments.

The medial (deltoid) ligament [FIG. 4.64] is the capsular thickening on the medial side of the joint. This powerful, triangular ligament is attached by its apex to the pit on the lower border of the

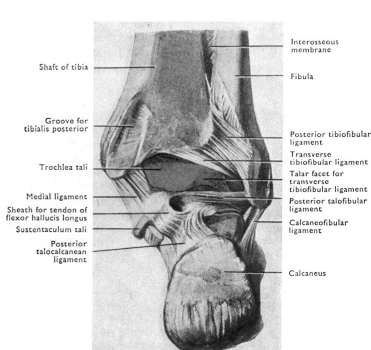

Shaft of tibia

Groove for tibialis posterior

Trochlea tali

Medial ligament

Sheath for tendon of flexor hallucis longus

Sustentaculum tali

Posterior talocalcanean ligament

Interosseous membrane

Fibula

Posterior tibiofibular ligament

Transverse tibiofibular ligament

Talar facet for transverse tibiofibular ligament

Posterior talofibular ligament

Calcaneofibular ligament

Calcaneus

FIG. 4.61. Ankle joint dissected from behind and part of articular capsule removed.

medial malleolus. Its broad base is attached from the navicular bone in front to the posterior part of the medial surface of the body of the talus behind. From front to back these basal attachments are to the rounded upper and medial part of the navicular bone (tibionavicular part), to the medial part of the neck of the talus (anterior tibiotalar part), to the plantar calcaneonavicular ('spring') ligament, to the medial border of the sustentaculum tali (tibiocalcaneal part), and to the medial side of the talus, under the 'tail' of the comma-shaped facet (posterior tibiotalar part). The tibiotalar parts are on a deeper plane than the rest.

The lateral ligaments [FIGS. 4.61 and 4.65] consist of three separate bands which are in the shape of a T. The stem of the T—the calcaneofibular ligament—runs from in front of the tip of the lateral malleolus downwards and slightly backwards to the middle of the lateral surface of the calcaneus a little above and behind the peroneal trochlea.

The anterior talofibular ligament runs from the anterior border of the lateral malleolus forwards and medially to the neck of the talus. The posterior talofibular ligament is attached to the fibula at the bottom of the malleolar fossa and passes medially and slightly backwards to the lateral tubercle of the posterior process of the talus. The calcaneofibular ligament is separate from the fibrous capsule of the joint but the two talofibular ligaments are fused with it.

Synovial membrane

The synovial membrane of the joint lines the fibrous capsule and covers fat pads lying anteriorly and posteriorly. A synovial fold [FIG. 4.62] occupies the cleft between the tibia and fibula below the base of the interosseous tibiofibular ligament; the sides of this cleft may be lined by an extension of the articular cartilage on the tibia and fibula.

Nerve supply

The ankle joint is supplied by twigs from the deep peroneal and tibial nerves.

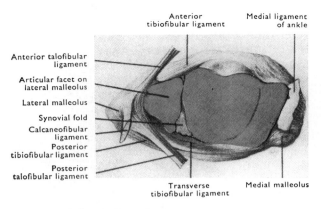

FIG. 4.62. Articular surfaces of the right tibia and fibula which are opposed to the talus at the ankle joint.

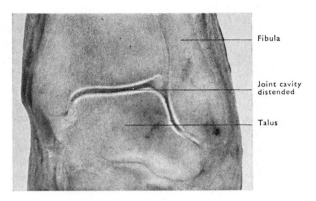

FIG. 4.63. Coronal section of ankle joint in which the joint has been distended. The lateral malleolus is seen to be much larger than the medial.

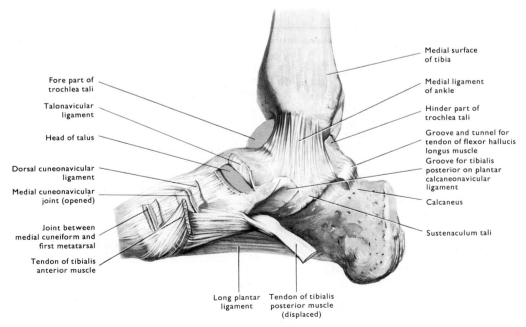

FIG. 4.64. Ankle joint and tarsal joints from the medial side.

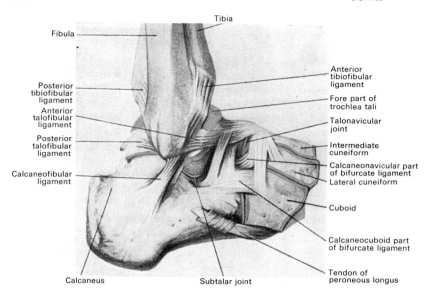

Fibula

Tibia

Posterior
tibiofibular
ligament

Anterior
talofibular
ligament

Posterior
talofibular
ligament

Calcaneofibular
ligament

Calcaneus

Subtalar joint

Anterior
tibiofibular
ligament

Fore part of
trochlea tali

Talonavicular
joint

Intermediate
cuneiform

Calcaneonavicular part
of bifurcate ligament
Lateral cuneiform

Cuboid

Calcaneocuboid part
of bifurcate ligament

Tendon of
peroneous longus

FIG. 4.65. Ligaments of lateral side of ankle joint
and dorsum of tarsus.

Movements at the ankle joint

Movement takes place around a transverse axis passing through the tip of the lateral malleolus and slightly below the level of the medial malleolus. In the normal, standing position the foot makes a right angle with the leg. To avoid confusion the terms plantar flexion and dorsiflexion are to be preferred to flexion and extension of the ankle joint. In **dorsiflexion** the foot is drawn upwards, and the trochlea tali turns backwards in its socket; movement in the opposite direction is **plantar flexion**. The range of plantar flexion (about 55 degrees) is greater than that of dorsiflexion (about 35 degrees) but there is considerable individual variation in the extent of the movements. The axis of movement is not quite horizontal and changes during movement; it is inclined slightly downwards and laterally in extreme dorsiflexion and downwards and medially in extreme plantar flexion. In many cases, slight medial rotation of the tibia occurs during plantar flexion (Barnett and Napier 1952).

In **dorsiflexion** the broader part of the trochlea tali is forced back into the narrower part of the tibiofibular socket; this separates the tibia and fibula slightly, causing increased tension of the interosseous and tibiofibular ligaments. The talus is then most securely held between the malleoli. In **plantar flexion**, as the narrower part of the trochlea turns forward into the broader part of the socket, the malleoli come together again thus adjusting to the width of the talus; in full plantar flexion the talus is only loosely held and a little side-play can be demonstrated in the joint. The wedge-shaped form of the joint surfaces, as viewed from above, helps to prevent backward displacement of the foot on the leg when coming to a sudden stop in running or jumping. Maintenance of the integrity of the tibiofibular socket depends on the strong ligaments of the inferior tibiofibular joint and the strength of the malleoli. The ligaments provide the fulcrum through which the resilience of the fibula superior to them gives the necessary spring to the lateral malleolus. The superior tibiofibular joint permits slight gliding and lateral rotational movement in association with the displacement of the lower ends of the bones of the leg.

The ankle joint is strongly supported by the tendons of the muscles of the leg which descend in close relation to it. In front are the tendons of tibialis anterior, extensor hallucis longus, extensor digitorum longus, and peroneus tertius, in that order mediolaterally. Posteriorly and medially are the tibialis posterior and flexor digitorum longus tendons. Posteriorly is the tendon of the flexor hallucis longus. Posteriorly and laterally are the tendons of the peroneus longus and peroneus brevis.

When the body is erect its centre of gravity falls in front of the ankle joints and muscular effort, mainly by soleus and gastrocnemius, is needed to prevent falling forwards. This effect of gravity is minimized in so far as the feet are usually turned a little outwards.

THE JOINTS OF THE FOOT

The joints between the tarsal bones, between the tarsus and metatarsus, and between the metatarsal bases are structurally intricate and complicated in their movements. They can only be understood in the light of the architecture of the foot as a whole. The account of the foot in this text is, of necessity, curtailed, and for further study the reader is advised to consult works which have been devoted entirely to it (Morton 1935; Jones 1949).

The arches of the foot

The human foot is highly specialized for the support of the body in the erect posture and for its propulsion during movement. It must therefore be able to adapt itself to the shape of the ground in many different positions and must be capable of absorbing mechanical shocks. This is achieved by the presence of a series of resilient arches which in each foot form half of a dome, convex above; this becomes slightly flattened by the body weight in standing or during progression, and resumes its original curvature when relieved of load. The dome is formed by the tarsal and metatarsal bones, and conventionally is described as consisting of both longitudinal and transverse arches.

The longitudinal arch is higher on the medial than on the lateral side, and is sometimes regarded as having two parts called the lateral and medial arches. The posterior part of the calcaneus forms a common posterior pillar for both arches [FIG. 3.213]. The low lateral arch passes from this through the anterior part of the calcaneus, the cuboid bone, and the lateral two metatarsal bones, which articulate with the cuboid. The medial arch is much higher and more important. The head of the talus forms the keystone of the arch at its summit between the sustentaculum tali and the navicular bone. The anterior pillar is continued through the navicular to the three cuneiforms and the medial three metatarsal bones.

The **transverse arch** is placed across the anterior part of the tarsus and the posterior part of the metatarsus, and is best marked along the line of the tarsometatarsal joints. Here the cuneiforms and

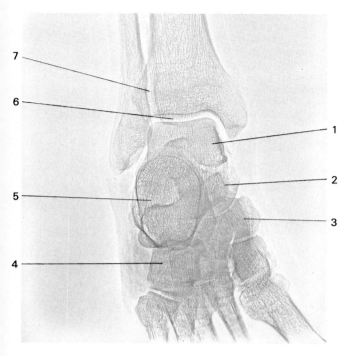

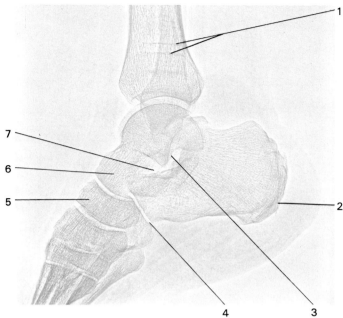

FIG. 4.66. Anteroposterior radiograph (xerogram) of planter flexed ankle joint and tarsus.
1. Body of talus.
2. Head of talus.
3. Navicular.
4. Cuboid.
5. Calcaneus.
6. Ankle joint.
7. Tibiofibular syndesmosis.

FIG. 4.67. Lateral radiograph (xerogram) of plantar flexed ankle joint and tarsus.
1. Lines of arrested growth in tibia.
2. Calcanean epiphysis (fused). Note direction of trabeculae in relation to pull of tendo calcaneus.
3. Subtalar joint.
4. Calcaneocuboid joint.
5. Navicular.
6. Head of talus.
7. Sinus tarsi.

metatarsal bases narrow inferiorly and their wedge shapes help to build the arch. The arch is gradually lost as the metatarsals are traced forwards towards their heads which are all in functional contact with the ground. These arches are preserved by the maintenance of the tarsal, tarsometatarsal, and intermetatarsal joints. The bones are held in their proper relationship as segments of the arches, and collapse of the arches prevented, by the joint ligaments. The most important and strongest of these lie on the plantar aspect of the joints, that is, on the concave side of the arches, the plantar calcaneonavicular ligament in particular supporting the head of the talus.

The body weight is transmitted from the bones of the leg to the foot through the talus. The talus forms two joints with the rest of the tarsus [FIG. 4.68], each concerned with the transference of weight to one of the longitudinal arches although together they constitute a single functional unit. Posteriorly the subtalar joint transmits weight to the lateral arch and to the common posterior pillar—the heel. Anteriorly, the talocalcaneonavicular joint transmits weight to the medial arch. The talus rests on the calcaneus at both joints.

The intertarsal joints

SUBTALAR JOINT [FIGS. 4.65 and 4.68]

This joint is formed between the cylindrical facet on the lower surface of the body of the talus and the posterior facet on the upper surface of the calcaneus. The facet on the talus is concave anteroposteriorly, the other is convex.

The fibrous capsule surrounding the joint is made up of short fibres attached close to the articular surfaces all round, strengthened by thickenings named the medial and lateral talocalcanean ligaments. Anteriorly the fibrous capsule is thin and is attached to

the roof and floor of the sinus tarsi. This is a narrow tunnel between the opposing surfaces of the grooves on the talus and the calcaneus [FIG. 4.70] which runs obliquely forwards and laterally in front of the subtalar joint. At its anterolateral end the sinus opens on to the dorsum of the foot. The anterior part of the fibrous capsule of the subtalar joint and the posterior part of the capsule of the talocalcaneonavicular joint are both attached to the sinus tarsi. The two capsules are thickened where they are adjacent to each other and there constitute the interosseous talocalcanean ligament. Between the two parts of this ligament lies the deep extension of the lateral limb of the inferior extensor retinaculum which is attached to the floor of the sinus tarsi [FIGS. 4.68 and 4.70]. At the lateral end of the sinus tarsi the strong ligamentum cervicis (Jones 1944) is attached above to the neck of the talus and below to the calcaneus. This ligament forms a strong bond between the two bones which tightens in the movement of inversion [FIG. 4.65].

The calcanean parts of the medial and lateral ligaments of the ankle joint act as accessory ligaments of the subtalar joint.

The fibrous capsule is lined with synovial membrane; the synovial cavity does not communicate with any other joint [FIG. 4.70].

TALOCALCANEONAVICULAR JOINT [FIG. 4.68]

At this joint a large facet on the head and lower surface of the neck of the talus articulates with a deep socket formed partly by bone and partly by ligaments. Anteriorly the socket is formed by the concave articular surface of the navicular bone, and posteriorly by the upper surface of the sustentaculum tali together with a facet that is usually present on the upper surface of the calcaneus anterolateral to the sustentaculum. The interval between the navicular and the calcaneal articular surfaces is occupied medially by the plantar calcaneonav-

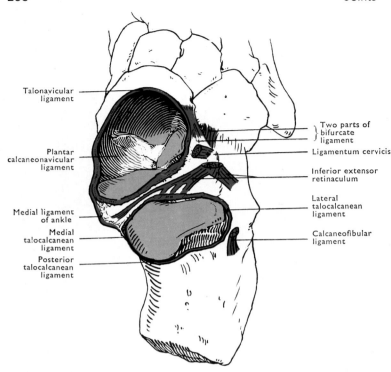

Talonavicular
ligament

Plantar
calcaneonavicular
ligament

Medial ligament
of ankle
Medial
talocalcanean
ligament
Posterior
talocalcanean
ligament

Two parts of
bifurcate
ligament
Ligamentum cervicis

Inferior extensor
retinaculum

Lateral
talocalcanean
ligament

Calcaneofibular
ligament

FIG. 4.68. Ligaments and inferior
articular sufaces of the right subtalar and
talocalcaneonavicular joints seen from
above after removal of the talus.

icular ligament and laterally by the dorsal calcaneonavicular fibres of the bifurcate ligament.

The plantar calcaneonavicular ligament is a fibrocartilaginous band of great strength [FIGS. 4.64, 4.68, and 4.69] often called the 'spring' ligament because of the resilience it imparts to the medial longitudinal arch. It runs forwards from the medial border of the sustentaculum tali to the plantar, the medial, and the adjacent parts of the upper surface of the navicular bone. Its medial part is joined by fibres of the medial ligament of the ankle joint. The upper surface of the ligament articulates with the head of the talus, while the lower surface is supported by the tendon of tibialis posterior (as it spreads into the sole) and by the flexor hallucis longus tendon [FIGS. 4.64 and 4.69]. The tension of the plantar calcaneonavicular ligament, aided by the sling of tibialis posterior, resists the tendency of the weight of the body to drive the head of the talus downwards between the bones of the medial longitudinal arch.

The dorsal calcaneonavicular part of the bifurcate ligament completes the socket on the lateral side. It is composed of short fibres that pass from the upper surface of the anterior end of the calcaneus to the adjacent lateral surface of the navicular bone [FIG. 4.65].

The rounded head of the talus has a covering of articular cartilage that extends on to the lower surface of the neck and the body. This articular surface is marked off by faint ridges into four facets for the navicular bone in front, for the sustentaculum tali postero-inferiorly, for the anterior part of the calcaneus anterolaterally, and for the plantar calcaneonavicular ligament inferomedially.

Fibrous capsule

On account of the extensive form of the bony and ligamentous socket a true fibrous capsule is only present posteriorly and dorsally.

The posterior part of the fibrous capsule is slightly thickened as the anterior component of the interosseous talocalcanean ligament [p. 257].

On the dorsal aspect of the joint the fibrous capsule is completed by a thin sheet, the talonavicular ligament [FIG. 4.64]. This passes from the neck of the talus to the dorsal surface of the navicular bone. It blends medially with the anterior fibres of the deltoid ligament that go to the navicular bone.

Synovial membrane

The synovial membrane of this joint lines all the non-articular surfaces and the enclosed synovial cavity does not communicate with neighbouring articulations [FIG. 4.70].

CALCANEOCUBOID JOINT [FIGS. 4.65 and 4.70]

This joint is the highest point in the lateral part of the longitudinal arch. The body weight falls on this part of the arch behind its summit, at the subtalar joint, and the strain is felt most at the joint immediately in front of the calcaneus—the calcaneocuboid joint. This joint is reinforced on its lower aspect, like the talocalcaneonavicular joint at the summit of the medial arch, by the strong long plantar and plantar calcaneocuboid ligaments. The tendon of the peroneus longus passes under the cuboid bone in front of the articulation and may act as a sling.

The articular facets on the opposed surfaces of the calcaneus and cuboid bone are quadrilateral and reciprocally concavoconvex.

Fibrous capsule

A fibrous capsule completely surrounds the joint. The calcaneocuboid part of the bifurcate ligament is applied to the dorsomedial aspect of the joint. This springs from the anterosuperior part of the calcaneus, in common with the dorsal calcaneonavicular part of the ligament, and is attached to the adjacent dorsomedial angle of the cuboid bone [FIG. 4.65].

The plantar calcaneocuboid ligament [FIG. 4.69], a broad band of short fibres, is immediately applied to the plantar aspect of the joint. It runs from the anterior part of the inferior surface of the calcaneus to the cuboid bone behind the ridge that bounds the peroneal groove. The long plantar ligament lies more superficially, separated from the plantar calcaneocuboid ligament by areolar tissue. It attaches to the whole length of the rounded, keel-like ridge on the plantar surface of the calcaneus and passes anteriorly to the ridge on the cuboid bone. The more superficial fibres pass on over the peroneus longus tendon to the bases of at least the lateral three metatarsal bones. This ligament therefore stretches under nearly the whole length of the lateral part of the arch and strengthens the

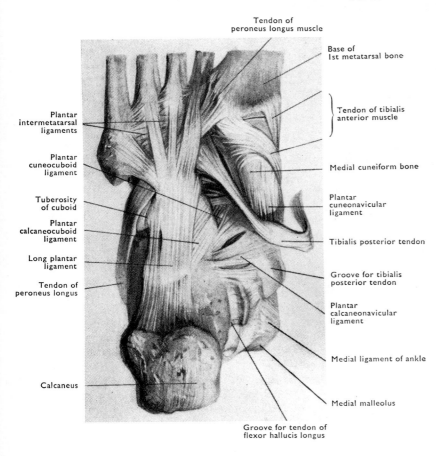

Tendon of
peroneus longus muscle

Base of
1st metatarsal bone

Tendon of tibialis
anterior muscle

Plantar
intermetatarsal
ligaments

Plantar
cuneocuboid
ligament

Tuberosity
of cuboid

Plantar
calcaneocuboid
ligament

Long plantar
ligament

Tendon of
peroneus longus

Calcaneus

Medial cuneiform bone

Plantar
cuneonavicular
ligament

Tibialis posterior tendon

Groove for tibialis
posterior tendon

Plantar
calcaneonavicular
ligament

Medial ligament of ankle

Medial malleolus

Groove for tendon of
flexor hallucis longus

FIG. 4.69. Plantar aspect of the right
tarsal and tarsometatarsal joints.

plantar aspect of all the joints in the arch. As this arch has little height, a long band of this nature stretching, like a tie-beam, from pillar to pillar is more effective in preserving the arch than short ties between adjacent segments of the arch.

Synovial membrane covers the inside of the fibrous capsule. The joint cavity is self-contained, making no communication with any other [FIG. 4.70].

TRANSVERSE TARSAL JOINT

The talocalcaneonavicular joint and the calcaneocuboid joint, although they are not in communication, together extend across the tarsus in an irregular transverse plane, between the talus and calcaneus behind and the navicular and cuboid bones in front [FIG. 4.70]. This articular plane is termed the transverse tarsal joint.

The anterior two bones involved—navicular and cuboid—are bound together by dorsal [FIG. 4.65] and plantar cuboideonavicular ligaments between adjacent parts of the corresponding surfaces, and by an interosseous ligament between their contiguous sides [FIG. 4.70]. A synovial cavity between cartilage-covered facets on the bones, when present, may open anteriorly into the cuneonavicular joint.

CUNEONAVICULAR JOINT

This is the synovial joint between the convex anterior surface of the navicular bone and the concave articular surface provided by the posterior ends of the three cuneiform bones.

The joint is surrounded by a fibrous capsule except laterally, towards the cuboid bone, where the joint may communicate with the cuneocuboid joint and always with a cuboideonavicular joint

when present. Distally, the joint cavity forms recesses between the cuneiform bones [FIG. 4.70]. In the upper and medial parts of the capsule short and relatively weak dorsal cuneonavicular ligaments [FIGS. 4.64 and 4.65] pass from the navicular to each cuneiform bone. Similar stronger bands on the lower aspect of the joint blending with slips of insertion of the tibialis posterior tendon, constitute the plantar cuneonavicular ligaments.

INTERCUNEIFORM JOINTS

At the intercuneiform joints the three cuneiform bones are bound together by weak, transverse dorsal ligaments and much stronger interosseous and plantar ligaments. The interosseous ligaments form the anterior boundaries of the small joint cavities that lie between the posterior parts of the cuneiform bones.

CUNEOCUBOID JOINT

This is a joint between the circular facets on the lateral cuneiform bone and the cuboid bone. It is surrounded by a weak dorsal cuneocuboid ligament between the dorsal surfaces of the bones, by a plantar cuneocuboid ligament [FIG. 4.69] between their adjacent plantar borders, and by a strong interosseous cuneocuboid ligament [FIG. 4.70] which closes the cavity anteriorly. Posteriorly, the joint may or may not open into the cuneonavicular joint.

The cuboid and cuneiform bones, placed side by side across the distal part of the tarsus, form the tarsal part of the transverse arch of the foot. The strong interosseous and plantar intercuneiform and cuneocuboid ligaments maintain the segments of the arch in position, and additional strength is obtained from the peroneus longus tendon as it passes transversely across the sole of the foot.

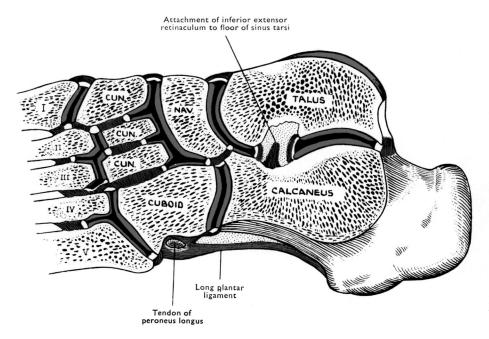

FIG. 4.70. Oblique section of left foot to show the synovial cavities of the tarsal and tarsometatarsal joints.

The foot was in strong inversion when sectioned and does not reveal the communication that exists betwen the cuneonavicular and intermediate tarsometatarsal joints. An unusually large cuboideonavicular joint cavity is present.

Articular cartilage, blue; synovial membrane, white; ligaments and tendon, green.

Nerve supply

The tarsal joints are supplied on the dorsal aspect by the deep peroneal nerve, and on the plantar aspect by the medial and lateral plantar nerves.

Movements in the tarsus

The movements of dorsiflexion and plantar flexion occur at the ankle joint and have been described on page 256. In these movements the talus rotates in the tibiofibular socket and the tarsus and metatarsus move as a unit. In the other movements of the foot the calcaneus and navicular, carrying with them the more distal tarsal bones (between which movements also occur), move on the talus.

The movements that occur at the tarsal joints summate to produce inversion and eversion.

In inversion, the foot is adducted and twisted so that the medial border is raised and the lateral border depressed until the sole is turned slightly medially. Inversion is usually accompanied by plantar flexion. In eversion, the foot is abducted, the lateral border is raised and the medial border lowered, so that the sole is turned slightly laterally. Eversion is usually accompanied by dorsiflexion. Inversion and eversion occur principally at the subtalar and talocalcaneonavicular joints; fixation of the calcaneus greatly limits these movements (Inkster 1938). The axis of movement of the subtalar and talocalcaneonavicular joints runs obliquely forwards, upwards, and medially through the sinus tarsi, and the complicated movements that occur at each separate joint have been described in detail by Manter (1941). Some movement occurs also at the other tarsal (especially the transverse tarsal) and at the tarsometatarsal joints during inversion and eversion, but this diminishes in amount in a proximodistal direction (MacConaill 1945), and is principally concerned with conferring resilience on the foot.

The tarsometatarsal and intermetatarsal joints

TARSOMETATARSAL JOINTS [FIG. 4.70]

The four anterior tarsal bones—the cuboid and three cuneiform bones—articulate with the bases of the metatarsal bones along an irregular line which 'presents the outline of an indented parapet both on its tarsal and its metatarsal aspects'. The indented form of the tarsometatarsal articulations strengthens the transverse arch of the foot which is most pronounced in this region.

The first metatarsal bone articulates only with the anterior surface of the medial cuneiform bone. The second metatarsal articulates in a socket with all three cuneiform bones—the lateral side of the medial bone, the anterior end of the intermediate, and the medial side of the lateral bone. The third metatarsal articulates with the lateral cuneiform bone. The fourth metatarsal articulates with the cuboid and to a small extent with the lateral cuneiform bone. The fifth metatarsal articulates with the cuboid bone only.

The articulations are all synovial joints of a plane variety, closed by dorsal and plantar ligaments. In addition interosseous ligaments delimit three separate synovial cavities.

The short dorsal tarsometatarsal ligaments pass between the adjoining dorsal surfaces of the tarsal and metatarsal bones, each metatarsal bone receiving a slip from each tarsal bone with which it articulates.

The plantar tarsometatarsal ligaments are similar bands on the inferior aspect of the joints, but are less regularly disposed. Those for the medial two metatarsal bones are the strongest, the lateral tarsometatarsal joints obtaining support from fibres of the long plantar ligament. The tarsometatarsal joint of the big toe is strengthened below by the insertion of the tibialis anterior medially and the peroneus longus laterally into the medial cuneiform and the base of the first metatarsal. Slips from the tibialis posterior tendon also support the lower aspect of the joints of the middle three metatarsal bones [FIG. 4.69].

Two strong interosseous tarsometatarsal ligaments are always present [FIG. 4.70]. One passes from the anterior and lateral part of the medial cuneiform bone to the contiguous medial side of the second metatarsal base. The other passes from the anterolateral angle of the lateral cuneiform to the adjacent medial surface of the fourth metatarsal base. These two ligaments separate **three tarsometatarsal synovial cavities.**

The **medial synovial** cavity is entirely confined to the joint between the first metatarsal and medial cuneiform bones. The **intermediate cavity** separates the second and third metatarsals from the intermediate and lateral cuneiform bones. This cavity is prolonged forwards for a little way between the second and third, and third and fourth metatarsal bases; posteriorly, it usually communicates, between the medial and intermediate cuneiform bones, with the cuneonavicular joint. The **lateral joint cavity** separates the cuboid from the fourth and fifth metatarsal bones, and is prolonged forwards between these two metatarsal bases.

INTERMETATARSAL JOINTS

The intermetatarsal joints are small synovial joints between cartilage-covered facets on the contiguous sides of the bases of the lateral four metatarsal bones. The joint cavities [FIG. 4.70] are prolonged from the tarsometatarsal cavities—that between the fourth and fifth bones from the lateral tarsometatarsal joint, those on each side of the third bone from the intermediate joint.

Above, below, and in front, these little joints are closed by **dorsal, plantar,** and **interosseous intermetatarsal ligaments,** composed of short, transverse fibres between corresponding surfaces of the bases of the metatarsal bones. The interosseous ligaments are very strong, and play an important part in holding together the metatarsal bases as segments of the metatarsal span of the transverse arch of the foot.

The first metatarsal bone is connected to the second by interosseous fibres only. Sometimes a bursal sac is formed amidst these fibres, between indistinct facets on the bones; its cavity may communicate with the medial tarsometatarsal joint.

Nerve supply

The tarsometatarsal and intermetatarsal joints are supplied by twigs from the **deep peroneal** and the **lateral** and **medial plantar nerves.**

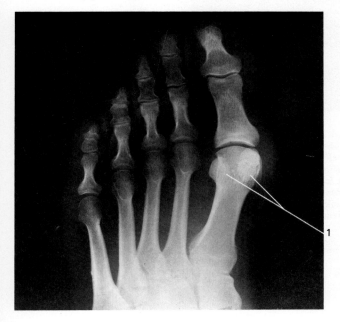

FIG. 4.71. Plantar view of adult metatarsus and phalanges. Note angulation of phalanges on metatarsal of hallux. 1. Sesamoid bones.

Metatarsal movements

The interlocking of the bones and the intermetatarsal interosseous ligaments permit only small gliding movements of the metatarsal bones. This contributes very slightly to inversion and eversion of the foot. At the first tarsometatarsal joint slight rotatory movement and a limited amount of vertical slide facilitate the adaptation of the foot to the ground in varying degrees of inversion and eversion.

The metatarsophalangeal joints

These joints closely resemble the metacarpophalangeal joints in the shape of the articular surfaces and the disposition of the ligaments. At rest the phalanges are slightly extended relative to the metatarsals and the concave base of each phalanx is applied to the convex dorsal portion of the metatarsal head to form a shallow ball and socket joint.

The **fibrous capsule,** lined with synovial membrane, is formed at the sides by strong, fan-shaped **collateral ligaments** attached to neighbouring parts of the bones, and, on the plantar aspect, by the thick fibrous plate of the plantar ligament; dorsally the capsule is replaced by the expansion of the extensor tendon.

Each **plantar ligament** is grooved by the long flexor tendon. In the joint of the big toe this ligament has two fairly large sesamoid bones embedded in it; they are covered with cartilage on their upper surfaces which articulate with grooves on the lower surface of the head of the first metatarsal bone. The plantar ligament is attached by its margins to the fibrous flexor sheath, the collateral ligaments, the margins of the extensor expansions, and to slips of the plantar aponeurosis. The plantar ligaments of all the joints are joined by the **deep transverse metatarsal ligaments** which are arranged like the deep transverse metacarpal ligaments of the palm, except that in the foot the joint of the big toe is linked with the others. As a result, the big toe is much less mobile than the thumb.

Movements

The movements of these condyloid joints are similar to the movements at the corresponding joints in the hand but are less extensive. In the foot, however, extension of the proximal phalanges can be carried beyond the line of the metatarsal bones; and abduction and adduction are centred on the second toe because of the slightly different arrangement of the interosseous muscles. During flexion the toes are drawn together while in extension they tend to spread apart and incline slightly in a lateral direction.

The interphalangeal joints

The interphalangeal joints also are similar to the corresponding joints in the hand. At each articulation the proximal bone presents a median groove separating two convexities which fit into reciprocal concavities parallel to a median ridge on the distal bone.

Collateral ligaments at the sides, a thick **plantar ligament** and the expansion of the extensor tendon dorsally complete the **fibrous capsule,** lined with synovial membrane. The distal interphalangeal joint of the little toe is often obliterated by synostosis.

Nerve supply

The metatarsophalangeal and interphalangeal joints are supplied by the dorsal and plantar digital nerves.

Movements

These are hinge joints. Movement is limited to **flexion** towards the sole of the foot and **extension** in the opposite direction, around a transverse axis.

MECHANISM OF THE FOOT

The foot is specialized to support the body in the erect position, both in standing and in progression. The skeletal arrangement in arches, longitudinal and transverse, has already been described. The presence of the joints in the arches introduces a spring mechanism analogous to the half-elliptical springs of a carriage. They yield slightly when weight is put upon them, and recoil when it is removed [FIG. 4.72]. The medial side of the foot has a greater resilience than the lateral side because the medial longitudinal arch is higher and has more joints than the lateral arch. This is in accordance with the propulsive function of the medial part of the foot in walking and running, whereas the lateral part of the foot is more in the nature of a stabilizing flap. The arched form of the foot is established in early foetal life, but the height of the arches shows considerable individual variation.

Attention has been drawn to the ligaments which support the arches of the foot [pp. 258–61]. These are supplemented by the plantar aponeurosis [FIG. 5.165] but the action of muscles in safeguarding the ligaments from being stretched under, for example, the heavy loading of propulsion, is of great importance. The tendon of the tibialis posterior muscle passes as a sling below the summit of the medial part of the longitudinal arch, and inserts on the plantar aspect of nearly all the tarsal and metatarsal bones; hence contraction of the muscle must tend to increase the general concavity of the sole. Similarly peroneus longus may act as a sling beneath the lateral part of the longitudinal arch and as a contractile tie-beam between the extremities of the transverse arch. The tibialis anterior and peroneus tertius muscles pull upon the medial and lateral sides of the longitudinal arch from above. The long flexors of the toes act as contractile ties to the longitudinal arch. The short muscles of the sole help to brace the arches, according to their disposition.

In walking, the weight is first applied to the common posterior pillar of the two longitudinal arches as the heel strikes the ground. As the body moves forward the weight falls on the lateral border of the foot, is rapidly transferred across the metatarsal heads to the medial part of the longitudinal arch as the heel leaves the ground, and a final propulsive thrust by the big toe is obtained by extension of the hip and knee joints, plantar flexion of the ankle, and flexion of the joints of the big toe. In running, the weight reaches the ground only at the distal end of the longitudinal arch, which then acts as a quarter-elliptical spring, and, by its recoil, reinforces the final thrust derived from flexion of the medial toes.

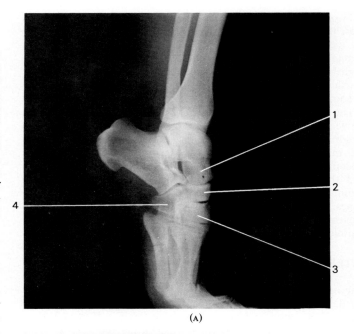

(A)

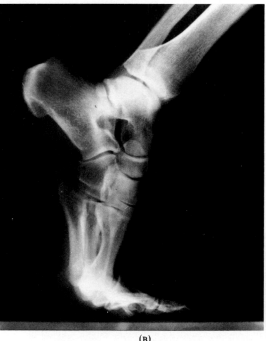

(B)

FIG. 4.73. Lateral radiographs of a foot in 'tip-toe' (A) and 'push-off' (B) positions. In A note extreme plantar flexion at ankle joint with flexion at the calcaneocuboid joint and forced extension of the phalanges. The last tightens the plantar fascia and flexes the calcaneus on the cuboid. In B there is further extension of the metatarsophalangeal joints which maintain the flexion of the calcaneus on the cuboid in spite of some dorsiflexion of the ankle.

1. Head of talus.
2. Navicular.
3. Medial cuneiform.
4. Cuboid.

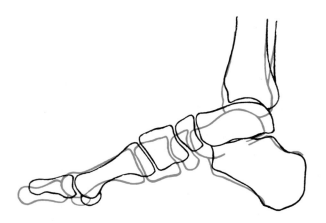

FIG. 4.72. Tracing of radiographs superimposed to show the position of the longitudinal arch of the foot during relaxation (red) and contraction (black) of the muscles. [See also FIG. 5.161, footprints 2 and 3.]

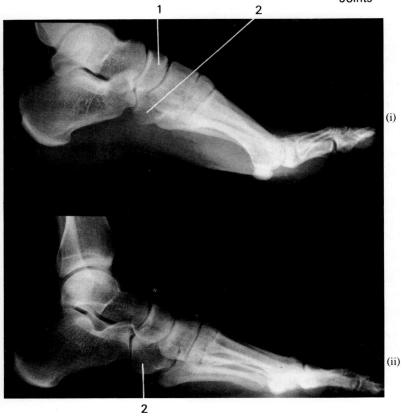

FIG. 4.74. Lateral radiographs of a foot non-weight bearing (i) and weight bearing (ii). Note the flattening of the arches.
1. Navicular.
2. Cuboid.

References

ANNOVAZZI, G. (1928). Osservazioni sulla elasticita dei legamenti. *Arch. Sci. biol.* (*Bologna*) **11**, 467.

BARNETT, C. H. (1953). Locking at the knee joint. *J. Anat.* (*Lond.*) **87**, 91.

—— and NAPIER, J. R. (1952). The axis of rotation at the ankle joint in Man. Its influence upon the form of the talus and the mobility of the fibula. *J. Anat.* (*Lond.*) **86**, 1.

—— DAVIES, D. V., and MACCONAILL, M. A. (1961). *Synovial joints.* Longmans, London.

BEADLE, O. A. (1931). The intervertebral discs. *Spec. Rep. Ser. med. Res. Coun.* (*Lond.*), No. 161. HMSO, London.

BRADFORD, F. K. and SPURLING, R. G. (1945). *The intervertebral disc* (2nd edn.). Thomas, Springfield.

BROOKE, R. (1924). The sacro-iliac joint. *J. Anat.* (*Lond.*) **58**, 299.

BRYCE, T. H. (1915). *Quain's elements of anatomy* (11th edn.). Vol. IV, Pt I. Longmans, London.

BULLOUGH, P., GOODFELLOW. J., and O'CONNOR, J. (1973). The relationship between degenerative changes and load-bearing in the human hip. *J. Bone Jt Surg.* **55B**, 746.

BUNNELL, S. (1938). Opposition of thumb. *J. Bone Jt Surg.* **20**, 1072.

CLELAND, J. (1884). Notes on raising the arm. *J. Anat. Physiol.* **18**, 275.

COGGESHALL, H. C., WARREN, C. F., and BAUER, W. (1940). The cytology of normal human synovial fluid. *Anat. Rec.* **77**, 129.

DARCUS, H. D. (1951). The maximum torques developed in pronation and supination of the right hand. *J. Anat.* (*Lond.*) **85**, 55.

DAVIES, D. V. (1945). Anatomy and physiology of diarthrodial joints. *Ann. rheum. Dis.* **5**, 29.

DAY, W. H., SWANSON, S. A. V., and FREEMAN, M. A. R. (1975). Contact pressures in the loaded human cadaver hip. *J. Bone Jt Surg.* **57B**, 302.

DOWSON, D., WRIGHT, V., UNSWORTH, A., and GVOZDANOVIC D. (1975). An overall view of synovial joint lubrication. *Ann. rheum. Dis.* **34**, Suppl 2, 94.

FICK, R. (1904–11). *Handbuch der Anatomie und Mechanik der Gelenke. Bardeleben's Handbuch der Anatomie des Menschen*, Bd. II. Jena.

GARDNER, E. (1948a). Innervation of knee joint. *Anat. Rec.* **101**, 109.

—— (1948b). Innervation of hip joint. *Anat. Rec.* **101**, 353.

—— (1948c). Innervation of shoulder joint. *Anat. Rec.* **102**, 1.

GHADIALLY, F. N. and ROY, S. (1969). *Ultrastructure of synovial joints in health and disease.* Butterworth, London.

GOODSIR, J. (1858). On the mechanism of the knee joint. *Proc. roy. Soc. Edinb.*; also in *Anatomical memoirs* (ed. W. Turner) (1868), Vol. II, p. 231. Black, Edinburgh.

HAINES, R. W. (1941). A note on the actions of the cruciate ligaments of the knee joint. *J. Anat.* (*Lond.*) **75**, 373.

—— (1944). The mechanism of rotation at the first carpometacarpal joint. *J. Anat.* (*Lond.*) **78**, 44.

—— (1946). Movements of the first rib. *J. Anat.* (*Lond.*) **80**, 94.

—— (1947). The development of joints. *J. Anat.* (*Lond.*) **81**, 33.

HALLEN, A. (1962). The collagen and ground substance of human intervertebral disc at different ages. *Acta Chem. Scand.* **16**, 705.

HARPMAN, J. A. and WOOLLARD, H. H. (1938). The tendon of the lateral pterygoid muscle. *J. Anat.* (*Lond.*) **73**, 112.

HILTON, J. (1863). *Lectures on rest and pain* (6th edn. 1950) (eds. E. E. Philipp and E. W. Walls) p. 166. London.

HUNTER, W. (1743). On the structure and diseases of articular cartilage. *Phil. Trans. B* **42**, 514.

INKSTER, R. G. (1938). Inversion and eversion of the foot and the transverse tarsal joint. *J. Anat.* (*Lond.*) **72**, 612.

INMAN, V. T., SAUNDERS, J. B. DE C. M., and ABBOTT, L. C. (1944). Observations on the function of the shoulder joint. *J. Bone Jt Surg.* **26**, 1.

JAZUTA, K. (1929). Die Nebengelenkflächen am Kreuz- und Hüftbein. *Anat. Anz.* **68**, 137.

JOHNSTON, T. B. (1937). The movements of the shoulder joint. *Br. J. Surg.* **25**, 252.

JONES, F. W. (1944). The talocalcanean articulation. *Lancet* **ii**, 241.

—— (1949). *Structure and function as seen in the foot* (2nd edn.). Baillière, London.

KEYES, D. C. and COMPERE, E. L. (1932). The normal and pathological physiology of the nucleus pulposus. *J. Bone Jt Surg.* **14**, 897.

KUCZYNSKI, K. (1974). Carpometacarpal joint of the human thumb. *J. Anat.* (*Lond.*) **118**, 119.

LOCKHART, R. D. (1930). Movements of the normal shoulder joint. *J. Anat.* (*Lond.*) **64**, 288.

—— (1933). A further note on movements of the shoulder joint. *J. Anat.* (*Lond.*) **68**, 135.

MacConaill, M. A. (1932). The function of intra-articular fibrocartilages. *J. Anat.* (*Lond.*) **66,** 210.

—— (1941). The mechanical anatomy of the carpus. *J. Anat.* (*Lond.*) **75,** 166.

—— (1945). The postural mechanism of the foot. *Proc. roy. Irish Acad. B* **50,** 14, 265.

—— (1946). Studies in the mechanics of synovial joints. *Irish J. med. Sci.* **246,** 190; **247,** 223; **249,** 620.

McCutchen, C. W. (1967). Physiological lubrication. *Proc. Instn Mech. Eng.* **181** (3J), 55.

Manter, J. T. (1941). Movements of the subtalar and transverse tarsal joints. *Anat. Rec.* **80,** 397.

Maroudas, A. (1973). Physicochemical properties of articular cartilage. In *Adult articular cartilage* (ed. M. A. R. Freeman) p. 131. Pitman, London.

—— Bullough P. G., Swanson, S. A. V., and Freeman, M. A. R. (1968). The permeability of articular cartilage. *J. Bone Jt Surg.* **50B,** 166.

—— Stockwell, R. A., Nachemson, A., and Urban, J. (1975). Factors involved in the nutrition of the human lumbar intervertebral disc: cellularity and diffusion of glucose *in vitro. J. Anat.* (*Lond.*) **120,** 113.

Martin, B. F. (1958). The oblique cord of the forearm. *J. Anat.* (*Lond.*) **92,** 609.

Martin, C. P. (1940). The movements of the shoulder joint. *Am. J. Anat.* **66,** 213.

Meachim G. and Stockwell, R. A. (1973). The matrix. In *Adult articular cartilage* (ed. M. A. R. Freeman) p. 1. Pitman, London.

Morton, D. J. (1935). *The human foot.* New York.

Muir, H. (1977). Molecular approach to the understanding of osteoarthrosis. *Ann. rheum. Dis.* **36,** 199.

Paul, J. P. (1966). The biomechanics of the hip joint and its clinical relevance. *Proc. roy. Soc. Med.* **59,** 943.

Pritchard, J. J., Scott, J. H., and Girgis, F. G. (1956). The structure and development of cranial and facial sutures. *J. Anat.* (*Lond.*) **90,** 73.

Püschel, J. (1930). Der Wassergehalt normaler und degenerierter Zwischenwirbelscheiben. *Beitr. path. Anat.* **84,** 123.

Sarnat, B. G. (1951). *The temporomandibular joint.* Thomas, Springfield.

Schunke, G. B. (1938). The anatomy and development of the sacro-iliac joint in Man. *Anat. Rec.* **72,** 313.

Seedhom, B. B. (1975). Load bearing function of the menisci of the knee joint. *Ann. rheum. Dis.* **34,** Suppl. 2, 118.

Simon, W. H., Friedenberg, S., and Richardson, S. (1973). Joint incongruence. *J. Bone Jt Surg.* **55A,** 1614.

Smith, J. W. (1954). The elastic properties of the anterior cruciate ligament of the rabbit. *J. Anat.* (*Lond.*) **88,** 369.

—— (1956). Observations on the postural mechanism of the human knee joint. *J. Anat.* (*Lond.*) **90,** 236.

Sokoloff, L. (1973). A note on the histology of cement lines. In *Perspectives in biomedical engineering* (ed. R. M. Kenedi) p. 135. Academic Press, London.

Steindler, A. (1955). *Kinesiology of the human body.* Thomas, Springfield.

Stockwell, R. A. and Meachim, G. (1973). The chondrocytes. In *Adult articular cartilage* (ed. M. A. R. Freeman) p. 51. Pitman, London.

Swann, D. A. (1975). Purification and properties of articular lubricant. *Ann. rheum. Dis.* **34,** Suppl. 2, 91.

Trueta, J. (1957). The normal vascular anatomy of the human femoral head during growth. *J. Bone Jt Surg.* **39 B,** 358.

Tucker, F. R. (1949). Arterial supply to the femoral head and its clinical importance. *J. Bone Jt Surg.* **31 B,** 82.

Walmsley, R. (1953). The development and growth of the intervertebral disc. *Edinb. med. J.* **60,** 341.

Walmsley, T. (1928). The articular mechanism of the diarthroses. *J. Bone Jt Surg.* **10,** 40.

Wright, R. D. (1935). A detailed study of movement of the wrist joint. *J. Anat.* (*Lond.*) **70,** 137.

Young, J. (1940). Relaxation of the pelvic joints in pregnancy. *J. Obstet. Gynaecol. Br. Emp.* **47,** 493.

5 Muscles and fasciae

D. C. SINCLAIR

General introduction

SOMATIC MUSCLE

THE somatic muscles are derived from the mesoderm of the embryo, and make up about 40 per cent of the total weight of the human body. Somatic muscle is sometimes called skeletal, voluntary, or striated muscle, but these alternative names are not always appropriate. In the first place, some somatic muscles, such as the orbicularis oris [p. 290] which surrounds the mouth, have no connection with the skeleton. Others, such as the muscles of the pharynx and the upper part of the oesophagus, are not wholly under the control of the will, and normally operate involuntarily. Finally, the word striated, or striped, is unspecific because there are also longitudinal striations in visceral muscle, and cardiac muscle, like somatic muscle, shows both longitudinal and transverse striations [FIG. 5.1].

STRUCTURE

Somatic muscle, like the other types of muscle, is made up of cells in which the ability of protoplasm to contract is both highly developed and organized directionally. When these cells are stimulated they contract along their long axis, and, since their volume remains unaltered, they expand and thicken at the same time.

The functional unit of somatic muscle is the muscle fibre, an elongated, cylindrical, multinucleated cell, varying in length from a few millimetres to more than 30 centimetres (Lockhart and Brandt 1938; Barret 1962), and in width from about 10 to 100 micrometres (μm). The fibres do not usually branch.

When examined by the light microscope, a muscle fibre exhibits both a longitudinal striation and a very characteristic transverse striping by alternate dark and light bands. A long fibre may contain several hundred nuclei, which lie towards the side of the fibre rather than centrally; there are no cell boundaries between the nuclei. The cytoplasm of the fibre is called the sarcoplasm, and is enclosed within a cell membrane called the sarcolemma. This has the typical three-layered structure characteristic of boundary membranes, and at intervals it sends into the fibre slender invaginations which make up the T-system of tubules seen in electron micrographs [FIG. 5.3].

Under the electron microscope, the contractile elements of the muscle fibres, which are called myofibrils, are revealed. They are long unbranched threads varying from 1 to 3 μm in diameter, and they run the whole length of the fibre; between them lie variable numbers of mitochondria [FIG. 5.3]. Each myofibril exhibits alternate dark and light banding, and the bands are 'in step' with each other, so producing the typical cross-striation of the muscle fibre. The light bands are called isotropic (I) bands, and the dark

ones anisotropic (A) bands. In the middle of the isotropic band is a thin dark line known as the Z disc, and the segment of the fibril lying between two adjacent Z discs is called a sarcomere. In the middle of the anisotropic band there is a relatively clear H disc, and in the centre of this lies a dark line known as the M line [FIG. 5.2].

Higher magnification shows that each myofibril is composed of a number of myofilaments [FIG. 5.2], which are of two kinds. The thicker filaments, made of the protein myosin, are about 10 nm in diameter, and 1.5 μm long, and lie in the A band. The thinner filaments, about 5 nm in diameter, are composed of the protein actin, and constitute the I band, extending into the adjacent A bands for some distance and lying in the spaces between the myosin filaments [FIG. 5.4].

Outside the sarcolemma of each muscle fibre is a very delicate connective tissue called endomysium, which serves to separate the fibre from its neighbours and at the same time to connect it to them. The fibres are arranged parallel to each other in bundles called fasciculi, and each fasciculus is bound together by a rather more dense connective tissue called the perimysium. Finally, the assembly of fasciculi is enclosed in a fibrous wrapping called the epimysium, which forms a sheath for the whole muscle. In some muscles this is thick and strong, and in others thin and more or less transparent.

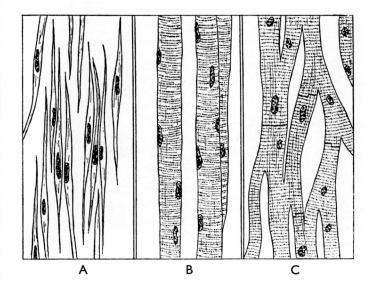

FIG. 5.1.   Three varieties of muscle (human; × 250).
A, Visceral. B, Somatic. C, Cardiac.

There are two kinds of somatic muscle fibres, red and white. The red fibres owe their colour to their content of myoglobin, a substance similar to haemoglobin, and their nuclei are sometimes centrally placed. They have more mitochondria than the white fibres, and thicker Z bands. In man, red and white fibres are always mixed together in any given muscle, but in some animals the muscles which perform rapid movements are much whiter than the muscles which are concerned with slow and prolonged contractions to maintain posture. The advantage red fibres have over white fibres when contracting in this way is that the myoglobin provides them with a store of oxygen. In general, therefore, red fibres are more numerous in postural muscles and white fibres in muscles which perform rapid movement.

Quite apart from the difference between red and white muscle, some fibres contract more rapidly than others. Three histochemical types are now recognized in mammalian muscle on the basis of the enzyme activity they exhibit, which is correlated with their mode of contraction. Type A are large, white, 'fast-twitch' fibres; type C are small, red, 'fast-twitch' fibres; and type B are low-speed fibres (Close 1972; Taylor and Calvey 1977). The proportions of the three types vary from muscle to muscle and from one region to another within a given muscle (Pullen 1977; Lobley, Wilson, and Bruce 1977).

The attachment of a muscle to bone or to other tissues is always indirect, through the medium of the connective tissue elements in the muscle. Sometimes the perimysium and epimysium unite directly with such elements as the periosteum of bone, the connective tissue underlying the skin, or the capsule of a joint. If this intermediate tissue is inconspicuous, the muscle is said to have a fleshy attachment. In other instances the connective tissue components of

the muscle fuse together to form an intervening **tendon**, composed of bundles of collagen fibres. Each collagen fibre unites at one end with the fibres of the structure to which the muscle gains attachment, and at the other with the thickened and corrugated end of the sarcolemma of a muscle fibre; there is no direct continuity between an individual muscle fibre and an individual tendon fibre, and the pull is transmitted from one to the other through the sarcolemma.

Tendons in general are immensely strong. They may be round cords, flat bands, or thin sheets of dense connective tissue (**aponeuroses**). The transition between muscle and tendon does not necessarily take place uniformly across the width of the muscle, so that there may be prolongations of the tendon among the muscle fibres. Sometimes flat sheets of dense connective tissue penetrate the muscle, forming **septa** to which the muscle fibres are attached. In other cases patches of tendon may develop in situations where the muscle is exposed to friction; for example, the deep surface of the trapezius [p. 311] is tendinous where it rubs against the spine of the scapula. If a tendon is subjected to friction it may develop in its substance a **sesamoid bone** [p. 77], e.g. the peroneus longus [p. 391] tendon in the sole of the foot. Other tendons, although equally exposed to friction, do not develop sesamoid bones, and such bones may appear where friction does not seem to be a primary exciting agent, though external pressure on the tendon may play a part.

BLOOD SUPPLY

In between the muscle fibres, and largely parallel to them, there runs a very fine plexus of capillaries. The capillaries are said to be more

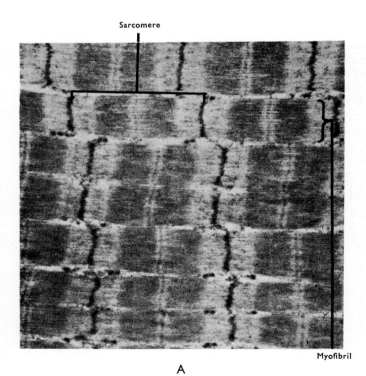

A

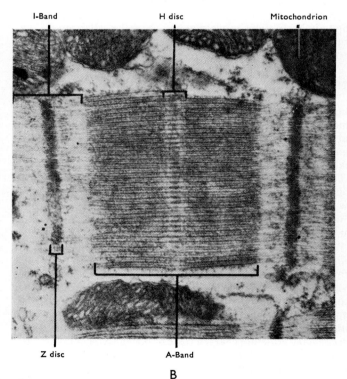

B

FIG. 5.2. Electron micrographs of part of a single muscle fibre in longitudinal section.
A, Parts of seven myofibrils in somatic muscle.
B, A single sarcomere of cardiac muscle. This is the unit of a myofibril between two Z discs. Centrally it contains thick myosin filaments (A band) interspersed with thin actin filaments which reach the Z discs and form the I bands with them (*By courtesy of Professor A. R. Muir.*)

numerous in red muscle than in white, and in red muscle they show dilatations which may form a reservoir of blood when the capillaries are compressed by sustained contraction of the muscle. During exercise the capillaries dilate, and the amount of blood they can contain may be increased up to about 800 times.

The vessels which supply the capillary plexus may enter the muscle either towards the ends or in the middle: because muscles have to slide on the surrounding connective tissue, it is not possible to have blood vessels entering them at several different points. Muscles supplied or drained by only a single vessel are liable to be wholly or partly destroyed if this vessel is damaged. Other muscles are supplied by a series of anastomosing arteries, and within the muscle itself there are numerous communications between the smaller branches (Blomfield 1945). However, such collateral circulations are not particularly good, and parts of a muscle can be functionally damaged by tying off one of its arteries of supply.

The tendon of a muscle, being composed of a relatively inactive tissue, has a much less profuse blood supply, but vessels may enter it at intervals from the surrounding connective tissue and run longitudinally between its fibres (Edwards 1946; Brockis 1953).

NERVE SUPPLY

The nerve supply to a muscle usually enters along with the arterial supply at a neurovascular hilus (Brash 1955), which is usually situated on the deep surface of the muscle, and so protected to some extent from injury. There may be several nerves supplying different parts of the one muscle, as in the segmentally innervated muscles of the abdominal wall [p. 357].

The 'motor' nerve entering the muscle contains both sensory and motor nerve fibres in approximately equal proportions, though some muscles may receive separate sensory branches—for example, brachialis [p. 323]. Muscles supplied by spinal nerves commonly receive motor and sensory fibres from more than one segment of the spinal cord, but some muscle groups, such as the small intrinsic muscles of the hand, are supplied almost entirely by a single segment.

The motor nerve fibres form a plexus in the muscle from which individual myelinated branches emerge to supply groups of muscle fibres. One nerve cell through its branching fibres supplies several muscle fibres, but each muscle fibre receives only one terminal branch of a nerve fibre. As this approaches the muscle fibre, it loses its myelin sheath, and divides into a number of terminal arborizations, which together form what is called a **motor end plate**. Each separate part of the nerve ending fits into a gutter on the surface of the sarcolemma, which is thrown into folds at right angles to the gutter, and between each part of the nerve ending there are many nuclei which lie in an accumulation of sarcoplasm called the **sole plate**. The whole mechanism is covered over by extensions from the last Schwann cell of the sheath of the nerve fibre.

FIG. 5.3. Longitudinal section of human skeletal muscle showing numerous mitochondria between the myofibrils. Representatives of the T-system of tubules (arrows) are also shown. Electron micrograph (× 48 000) by courtesy of Dr J. M. Papadimitriou.

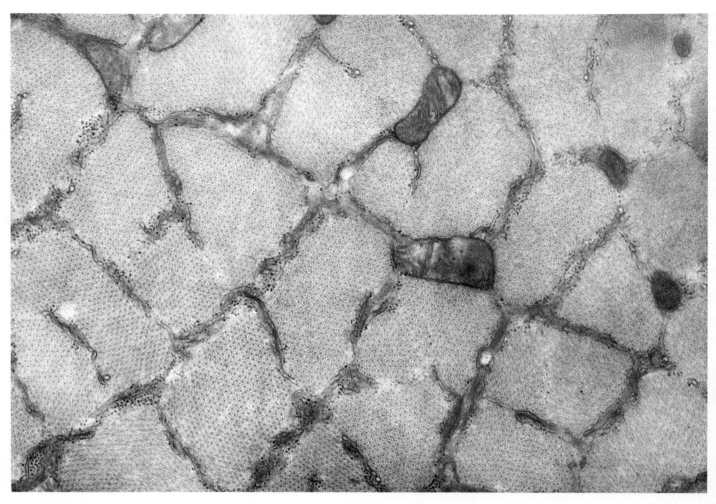

FIG. 5.4. Transverse section through a group of myofibrils in a human somatic muscle. The section passes through the A band, and shows the thin actin filaments surrounding the thicker myosin filaments. Mitochondria lie between the fibrils. Electron micrograph (×48 000) by courtesy of Dr J. M. Papadimirou.

The apparatus consisting of a single motor nerve cell and the muscle fibres which it innervates is called a **motor unit**. The size of the motor unit varies in different muscles (Feinstein, Lindegård, Nyman, and Wohlfart 1955), and determines the possibility of grading the contraction of the muscle coarsely or finely. Since individual muscle fibres probably obey the all-or-none law, and either contract fully or not at all, grading the strength of contraction of a muscle can only be achieved by throwing varying numbers of fibres into action, and this is done by varying the number of nerve cells stimulated. The strength of contraction is therefore increased in 'steps' dependent on the size of the motor units concerned. Where fine gradations of movement are essential, as in the muscles of the eyeball, the larynx, and the tongue, the number of muscle fibres supplied by a single nerve cell is small. Where relatively gross movements are needed, as in the muscles of the lower limb, every nerve cell may supply several hundred muscle fibres.

When a maximum effort is required, nearly all the motor units may be stimulated at the same time, and individual units may be used several times in sequence. But under normal conditions the motor units probably work in relays, so that in prolonged contractions some are always resting while others are contracting. The muscle fibres comprising a single unit do not necessarily lie side by side, and may be scattered in such a way that adjacent units interlock with each other. In the upper limb muscles each unit is confined to a cylinder of muscle 5–7 mm in diameter; in the lower limb the corresponding cylinders are 7–10 mm in diameter. Territories of this size can accommodate the fibres of 25 interlocking units each comprising 500–2000 fibres (Buchthal, Erminio, and Rosenfalck 1959).

The sensory fibres which enter a muscle also form a dense plexus, and there are simple branching terminals between the muscle fibres (Stacey 1969). In addition to these there are more complicated endings such as the **neuromuscular spindles** and the **tendon endorgans**. The number of nerve fibres destined for these specialized terminals indicates the importance of a feed-back system informing the nervous system of the state of contraction of the muscle. The sensory organs in a muscle respond either to an increase in tension of the muscle or its tendon, or to stretching of the muscle. The spindles consist basically of several specialized muscle fibres enclosed in a complex connective tissue sheath and profusely innervated [FIG. 12.10]. Raising the tension in these fibres increases the sensitivity of the apparatus and decreasing the tension lowers it.

Tendons receive their sensory supply from the nerve fibres in the muscle, and also from nearby sensory nerves (Stilwell 1957).

Muscle tone

A healthy muscle at rest is firm, with a characteristic elastic resistance to pressure. This is called muscle tone, and was formerly

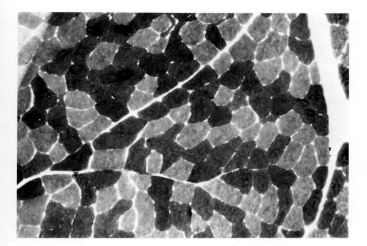

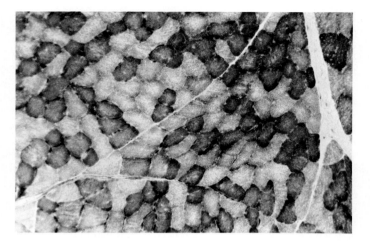

FIG. 5.5A. Transverse section through a human somatic muscle stained to show ATPase activity. (× 150.)

B. An adjacent section stained to show nicotinamide adenine dinucleotide dehydrogenase activity. (× 150.) Contrast the appearance of the same fibres in the two sections. Photomicrographs by courtesy of Dr J. M. Papadimitriou.

attributed to a resting activity of the muscle fibres. However, no electrical activity can be detected in a resting normal muscle, and what causes tone is now a matter for discussion (Joseph 1960; Basmajian 1961, 1967). If the nerve supply to the muscle is cut, the normal tone disappears and the muscle becomes flaccid and unresisting. Conversely, if there is destruction of the neurons leading from the brain to the motor neurons in the spinal medulla that innervate the muscle, there is often an increase in tone and the muscle is said to become spastic. In either case postural deformities may be produced; in flaccid paralysis the innervated antagonists pull the part out of its normal resting posture, and in spastic paralysis the paralysed muscles themselves do so.

SHAPE AND CONSTRUCTION

The fasciculi in a muscle can be arranged in various ways. The simplest arrangement, that of parallel fasciculi extending throughout the length of the muscle, is found in the so-called strap muscles, such as sternohyoid [p. 284]. Rectus abdominis [p. 361] is a strap muscle divided up into shorter segments by three or more fibrous intersections, and a similar arrangement exists in other strap muscles on a microscopic scale. Some of the fasciculi in the human gracilis and sartorius muscles contain units of from two to six short fibres arranged end to end in series and joined by bridges of fibrous tissue (Schwarzacher 1959; Barrett 1962).

In a fusiform muscle the fleshy belly of the muscle is spindle-shaped, and there is a tendon at one or both ends. In such a muscle not all the fibres extend throughout the whole length of the muscle belly. A modification of this type of muscle is found when a fleshy belly develops at either end, with a tendon in the middle. Such muscles are said to be digastric.

The fasciculi of some muscles are attached at an angle to the tendon, an arrangement which, because of its similarity to the structure of a feather, leads to their being called pennate muscles. If the tendon develops on one side of the muscle, the pattern is said to be unipennate; if the tendon is in the middle and fibres reach it at an angle from both sides, the muscle is bipennate; if there are several tendinous intrusions into the muscle with fibres reaching them from several directions, the muscle is multipennate [FIG. 5.6].

When a muscle fibre contracts, it shortens to about 55 per cent of its fully stretched length (Haines 1934). It follows that the longer the fibres in a given muscle, the greater the range of movement it can produce. The strength of a muscle, on the other hand, depends on the total number of fibres it contains, for a short fibre can pull as hard as a long one.

Now a given volume of muscle can contain either a small number of long fibres or a greater number of short fibres. The arrangement of the fasciculi within a muscle is therefore a compromise between power and range of movement. In the pennate muscles the increased power provided by the greater number of short fibres accommodated is to some extent offset by the fact that these fibres do not exert a straight pull on the tendon. Nevertheless, pennate muscles are generally strong, whereas strap muscles have the greatest possible range of movement for their size.

A long tendon permits a muscle to act at a distance without the movement being impeded by the swelling of the fleshy belly. For example, the long muscles which operate on the fingers have their bellies safely removed from the scene of action and located in the upper part of the forearm. However, for a given length of muscle, the longer the inextensible tendon, the shorter the contractile belly has to be, and this reduces the range of movement considerably. A tendon also permits an alteration of the direction of pull of a muscle, since it can be made to turn round corners. The tendon of tensor veli palatini [p. 301] converts the vertically delivered pull of the muscle

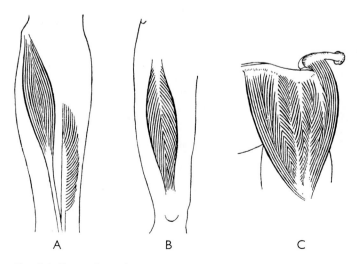

FIG. 5.6. Types of muscle structure.
A, Fusiform (flexor carpi radialis) and unipennate (flexor pollicis longus).
B, Bipennate (rectus femoris).
C, Multipennate (deltoid).

belly into a horizontal force by swinging at right angles round the pterygoid hamulus, and the posteriorly directed pull of the superior oblique muscle of the eye [p. 293] is converted into an anteriorly and medially directed force because the tendon doubles back on itself at the trochlea.

ATTACHMENTS

Many muscles have only two attachments, one at each end, but more complex muscles may be attached to several different structures. If these structures are separated from each other by an interval, the muscle is said to have two or more **heads**. Biceps brachii [FIG. 5.7] has two heads at its upper end, one from the coracoid process of the scapula, and the other from the supraglenoid tubercle. Triceps brachii [p. 323] has three heads, and quadriceps femoris [p. 385] has four. The internal architecture of such muscles is obviously somewhat more complicated than that of simple strap muscles, and patches of tendon may develop between their components to facilitate independent movement.

Most somatic muscles are attached to bone and pass across one or more joints; in this way they can produce or prevent movements of the bones. Some paired muscles may be attached to each other. An example is levator ani [p. 364], which acts as a diaphragm supporting the floor of the pelvis. Others again, such as the muscles of facial expression [p. 287], may have an attachment to skin. The lumbrical muscles of the fingers and toes [pp. 339, 398] run from the tendon of one muscle to the tendon of another.

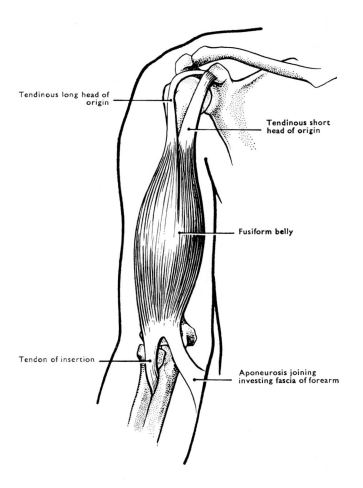

Tendinous long head of origin

Tendinous short head of origin

Fusiform belly

Tendon of insertion

Aponeurosis joining investing fascia of forearm

FIG. 5.7. The construction of the biceps brachii.

GROWTH AND REPAIR

There is some argument as to whether the number of fibres in a given muscle increases after birth. The fibres certainly grow longer, by the addition of sarcomeres, as the muscle increases in length, but it is thought that in most mammals muscles grow thicker simply by increasing the diameter of their individual fibres without increasing their numbers (Rowe and Goldspink 1969); this may not hold for marsupials (Bridge and Allbrook 1970). Regular exercise of a muscle readily increases the average diameter of its fibres, and in this way the whole muscle belly can be made to grow in thickness—it seems that the smaller fibres are brought to about the same diameter as the larger ones, which do not grow appreciably.

In contrast, if a muscle is deprived of its nerve supply the average diameter of its fibres decreases, and if the innervation is not restored they may degenerate and be replaced by a mass of fat or fibrous tissue (Sunderland 1968). This is not wholly explicable on the basis of disuse and consequent vascular deprivation. It is possible that the nerve fibres reaching the muscle may in some way influence its nutrition, perhaps by liberating chemicals within the muscle.

A muscle which has been injured, with destruction of some of its fibres, makes considerable efforts to repair itself, and new fibres are produced by 'satellite cells' lying between the fibres (Church 1969). Unfortunately, scar formation occurs more easily than regeneration of muscle fibres, and damaged muscle tissue thus tends to be replaced by scar tissue. If this happens, the muscle may exhibit **contractures**, which pull its attachments closer together, so producing deformities. Similar deformities can be produced if the muscle is immobilized for a long time in a position which allows its fibres to shorten, or if continued abnormal stimulation of the muscle occurs, as in cases of spastic paralysis.

ACTIONS AND FUNCTIONS

When a muscle is stimulated, naturally or artificially, its fibres attempt to contract, so as to bring its attachments closer to each other. If such a movement actually takes place, the length of the muscle alters but the tension generated inside it remains approximately constant; such contractions are called **isotonic**. If the muscle is, for one reason or another, unable to approximate its attachments, the tension generated in it increases, but the length remains constant; such contractions are called **isometric**. For example, in the upright position, the location of the centre of gravity of the body produces a tendency for the body to fall forward at the ankle. The muscles of the calf are in constant activity to prevent this (Joseph 1960), but they do not shorten appreciably. Again, the centre of gravity of the skull falls in front of the atlanto-occipital joint, and the muscles at the back of the neck contract isometrically to keep the gaze of the eyes level.

Isotonic contractions are of two kinds, **concentric** and **excentric**. In a concentric contraction the muscle shortens, but in an excentric one it actually lengthens. A good example of excentric contraction is the gradual relaxation of the flexors of the elbow during the movement of setting down a glass of water on the table. In such movements the muscle gradually 'pays out', and this active relaxation is of great importance in controlling the movement.

The bones, joints, and muscles form a system of levers in which the muscles apply the force, the joints act as the fulcra, and the bones bear the weight of the part of the body which is to be moved. The **mechanical advantage** of a muscle therefore depends on the distance of its attachment from the fulcrum relative to the distance of the weight which the muscle has to move. A muscle attached close to the fulcrum will be relatively weaker than an equivalent muscle which is attached further away [FIG. 5.8]. On the other hand, it will be able to produce a faster movement of greater range, since

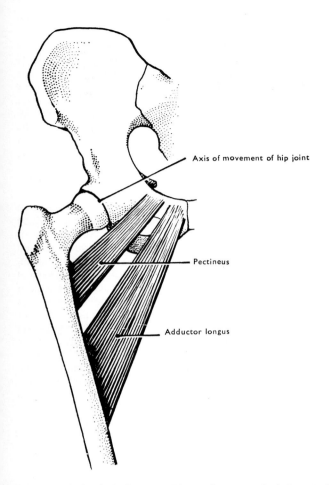

FIG. 5.8. Mechanical advantage. The pectineus, attached closer to the axis of movement, is a weaker adductor of the hip joint than the adductor longus, but produces a greater movement of the lower limb for every centimetre of contraction.

surfaces being pulled apart by centrifugal force. Such muscles are sometimes called 'shunt' muscles, and must be contrasted with the 'spurt' muscles, which act at right angles to the moving bone. For example, during rapid flexion of the elbow, brachioradialis [p. 331] is a shunt muscle, while biceps brachii [p. 322] and brachialis [p. 323] are spurt muscles (MacConaill 1946; Basmajian 1959). It has recently been argued (Stanier 1977) that such a distinction is of little value, but there is no doubt that for a given movement at a joint the muscles primarily engaged may differ according to whether the movement is rapid or slow.

In many cases the capacity of a muscle to contract is of less functional importance than its ability to maintain its tension throughout the movement of a joint, so acting as what is termed an **extensile ligament**. In contrast to ligaments composed of inextensible collagen, these muscular 'ligaments' can protect the joint during the whole movement and not merely at the end of its range. Perhaps the best example of this function is afforded by the short muscles which surround the very mobile shoulder joint and protect it during movement by holding the head of the humerus into the glenoid socket of the scapula [p. 318]. Indeed, the stability of any joint depends very largely on the tension in the muscles which surround it. If the muscles are paralysed, the ligaments of the joint are unable to resist the forces applied to it, and the joint becomes 'flail', allowing a much greater range of movement than normal, and becoming much more liable to dislocation.

As a joint approaches the extremes of its range of movement, the muscles which oppose the particular movement concerned come into action to slow the movement down to a halt, so preventing damage to the ligaments, which might otherwise give way under the

the distance travelled by its moving attachment will be magnified by the length of the lever. Every muscle, therefore, exhibits a mechanical compromise between the power and the speed of the movement it produces.

Again, the power exerted by a muscle necessarily varies according to the position of the moving bone. In the initial stages of flexion of the elbow, the pull of brachialis [FIG. 5.9] is more or less in line with the long axis of the bones of the forearm. Later in the movement, when the elbow has been bent to about a right angle, the power of the pull becomes maximal. Finally, when the elbow has been flexed past the right angle, the flexing force exerted by the brachialis diminishes again as its direction comes to lie parallel to the forearm bones once more.

A knowledge of the positions of the limbs in which maximum thrust with the foot or pull with the hand is attainable is clearly of importance in industrial design. For a seated subject, using an isometric foot pedal, maximum push is obtained with the hip flexed to 75 degrees and the knee extended to about 160 degrees. Using an isometric hand lever the maximum pull or push is possible when the elbow is extended to approximately 135 degrees, the lever grip is at elbow height, and the lever moves in a vertical plane passing through the shoulder joint (Hugh-Jones 1947).

In rapid movements of a joint the muscles which exert their pull more or less parallel to the moving bone help to prevent the bony

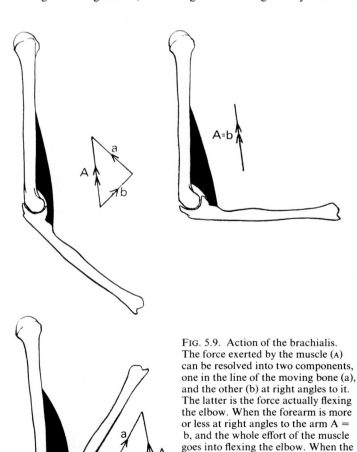

FIG. 5.9. Action of the brachialis. The force exerted by the muscle (A) can be resolved into two components, one in the line of the moving bone (a), and the other (b) at right angles to it. The latter is the force actually flexing the elbow. When the forearm is more or less at right angles to the arm A = b, and the whole effort of the muscle goes into flexing the elbow. When the forearm is extended or flexed beyond this point, b becomes less than A, and the muscle becomes a less efficient flexor.

sudden strain. In some instances the contraction of these muscles is a hindrance to the movement. For example, when the knee is straight, the untrained hamstring muscles [p. 378] do not permit full flexion of the hip joint; it is for this reason that it is difficult to perform the 'high kick' or to touch the toes without bending the knees [FIG. 5.10]. This situation, in which muscles prevent the full movement of the joint, is sometimes called **passive insufficiency**.

The opposite condition, **active insufficiency**, results from the inability of a muscle to contract by more than a fixed percentage of its length. For example, the length of the hamstrings is such that they can fully flex the knee when the hip is flexed, but cannot do so completely when the hip is extended. In the first case the muscles are stretched at the outset of the movement, and the whole of the contraction is used to produce movement. In the second, the attachments of the muscles are passively approximated at the outset, and some of the contraction has to be used to take up the slack.

In most movements one attachment of a muscle remains more or less stationary while the other attachment moves. The relatively stationary attachment is called the **origin** of the muscle, and the other is known as its **insertion**. A spring which closes a door could be said to have its origin on the door frame and its insertion into the door itself. In the body the arrangement is never so clear cut as this, and the terms origin and insertion are only relative. For example, the muscles which pass between the chest wall and the upper limb are customarily used to move the limb about relative to the trunk; they are therefore said to have their origins on the trunk and their insertions on the limb. But when the upper limb is used in climbing, these same muscles are used to pull the trunk up on the fixed limb. They are thus acting from their now relatively fixed 'insertions' on their relatively movable 'origins'. It is for this reason that many

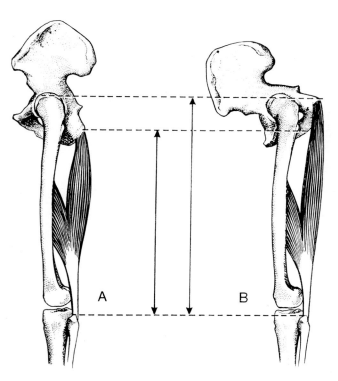

FIG. 5.10. Passive insufficiency.
A, Normal standing.
B, Bending forward, as in touching the toes.
The origin of biceps femoris is pulled further away from its insertion by the movement of the hip bone, so stretching the muscle. A continuation of the movement, if the muscle has not been trained to relax, causes the knee to flex.

people prefer simply to speak of the attachments of a muscle, not specifying either origin or insertion.

Nevertheless, the attachments of a muscle which lie more proximally, towards the base of a limb, or on the trunk, are usually referred to as the origin, and those which lie more distally as the insertion. An action in which the insertion remains fixed and the origin moves is sometimes called a **reversed action**, and in the lower limb it is often the reversed actions which are the more important. Thus, the ability to invert the foot while it is off the ground by the use of tibialis anterior [p. 388] acting from its origin on its insertion is in normal circumstances of little functional value. But when the foot is fixed to the ground by the body weight, tibialis anterior, acting from its insertion on its origin, plays a most important part in preventing overbalancing during walking.

The action of a muscle necessarily depends on its relationship to the joint which it crosses, and also on the types of movement which that joint permits. At a hinge joint, for example, all the muscles which cross on one aspect are flexors, and all those which cross on the other are extensors; there is no other possibility. But at a ball and socket joint, a muscle may be capable of exerting several different actions, depending on its position relative to the axis of movement. For example, the anterior fibres of deltoid, which cross the front of the shoulder joint obliquely [p. 318], act as flexors and medial rotators; the posterior fibres, which lie behind the joint, act as extensors and lateral rotators, while the central portion of the muscle, which lies superior and lateral to the joint, acts as a pure abductor. Furthermore, the action of a muscle may depend on the position of the joint at the time. Thus, obturator internus [p. 366] is a lateral rotator of the thigh when the hip joint is extended, but an abductor of the thigh when the hip joint is flexed.

Very seldom is a movement wholly entrusted to a single muscle; among the few exceptions is the movement of flexion of the distal interphalangeal joints of the fingers and toes. Conversely, it is unusual to find a muscle with but a single action. A common arrangement is that a muscle may be of major importance in relation to one movement, being helped by several other muscles, and at the same time contribute in a minor capacity towards other movements. It follows that if one muscle is paralysed, no single movement will be completely abolished, though several may be weakened. The remaining muscles may satisfactorily compensate for the missing one, so that the resulting disability is small or even unnoticeable.

The fact that several muscles normally collaborate in the production of a movement means that any investigation of weakness of movement must be based on a knowledge of the group concerned, and it is essential to think of muscles as members of functional associations rather than isolated individuals. It also means that if a tendon is to be surgically transplanted to compensate for a paralysed muscle the transplant must be selected from the group of muscles which normally act in association with the paralysed one (Dunn 1920).

If a muscle has several actions, it normally produces them all when it is activated; it cannot usually contract in such a way as to single out one of them. But the unwanted movements can be neutralized by other members of the group set in action. Thus, flexor carpi ulnaris [p. 328] both adducts and flexes the wrist, and when stimulated produces both results together. If pure adduction of the wrist is required, the flexion can be cancelled out by setting in action at the same time an extensor of the wrist. The obvious choice in this instance is of course extensor carpi ulnaris [p. 334], which not only extends the wrist but also contributes to the desired movement of adduction. Similarly, if pure flexion is required, the adduction can be neutralized by the collaboration of an abductor.

A muscle is said to be a **prime mover** when it contracts for the primary purpose of producing the specified movement. A muscle which opposes the specified movement is called an **antagonist**, and often the antagonists play a leading part in the control of the movement [p. 270]. Muscles which help to prevent unwanted movements inherent in the attachments of the prime movers are called **synergists**, and the term is also applied to muscles which contract in order to provide a stable base for the action of the prime movers. A good example is afforded by the muscles which stabilize the scapula during movements of the upper limb [p. 345]. This kind of action is also spoken of as 'fixation' and the muscles performing it are termed **fixators**.

In most movements the role of the prime movers, antagonists, synergists, and fixators does not vary throughout the movement. For example, when the fist is clenched, the extensors of the wrist contract to prevent the flexors of the fingers from also flexing the wrist. This is a vital part of the action, for flexion of the wrist added to flexion of the fingers stretches the extensors of the fingers until they can stretch no more; continued flexion at the wrist then causes the fingers to open out and the grip to relax. This is how an opponent can be made to drop a weapon by forcibly flexing his wrist. The flexors and extensors of the elbow must also contract in order to fix the origins of the flexors of the fingers and the extensors of the wrist, and the scapula must be fixed in order to stabilize in turn the origins of the muscles stabilizing the elbow. In such a manner even an apparently simple movement can have repercussions throughout the musculature of almost the whole body.

In other movements the functional significance of a given muscle may change in successive phases of the movement. For example, in the initial stages of abduction of the humerus the supraspinatus acts as a prime mover, but later it has a more important function as an extensile ligament.

In many instances gravity acts as a prime mover. Thus, in lowering the arm to the side from an abducted position, there is no contraction of the adductor muscles unless the movement is conducted against resistance. Instead, the antagonist to the movement, the deltoid muscle [p. 318] gradually pays out and the movement is effected as a result of the combination of this and the action of gravity. When the arm is adducted against resistance, the deltoid become flaccid, and the adductors are thrown into contraction [FIGS. 5.11 and 5.12].

The muscles in one part of the body often act in co-ordination with muscles in more distant parts, a phenomenon known as **associated action**. In walking, the balance of the trunk is maintained by the erector spinae muscles [p. 278] and the upper

FIG. 5.12. Flaccidity of deltoid when antagonists pull on the limb against resistance, shown by weight sinking into the muscle. Note the activity of pectoralis major in contrast to its passive state in FIGURE 5.11. (*By courtesy of Professor R. D. Lockhart.*)
1. Calvicular part of pectoralis major.
2. Sternocostal part of pectoralis major.

limb of one side swings in concert with the lower limb of the opposite side. Again, when we look round, the displacement of the head and eyes is accompanied by movements of the feet and trunk. Paralysis of a single muscle may thus upset movements which are not necessarily located in the vicinity of the muscle itself.

There is very little direct voluntary control over individual muscles without training; it is only the movement desired that is voluntary, and the pattern of activity of the executive muscles is automatically organized by the brain. A simple example emphasizes this. Stretch out the upper limb horizontally and then bend the forearm up towards the ceiling. In the phase before the forearm becomes vertical, the biceps can be felt contracting, and the triceps is relaxed. But as soon as the forearm reaches the vertical and starts descending towards the shoulder, the triceps is felt to harden while the biceps relaxes. We are completely unaware of the change-over.

Finally, it is important to distinguish between the actions of which a muscle is capable and the functions which it usually performs (MacConaill and Basmajian 1969). The interossei of the foot [p. 400] can abduct and adduct the toes, but their chief function is probably to support the anterior arch of the foot and prevent pressure on the vessels and nerves running to the toes. The mere presence of some muscles is of functional importance. For example, in patients who are lying in bed, gluteus maximus [p. 375] forms a soft pad intervening between the skin and the bony prominence of the buttock. If this muscle is thinned and atrophied by paralysis, the skin is exposed to pressure from the underlying bones, and this may lead to the production of bedsores.

Types of movement

In some movements, the antagonists are in action throughout, relaxing in a graduated manner to control the activity of the prime movers. But in certain circumstances, as when a sprinter starts a race, a **maximum effort**, unrestrained by the antagonists, is suddenly required. In the initial stages of such movements the antagonists may be almost completely relaxed, so as to reduce the opposing force to a minimum.

Ballistic movements are those in which the prime movers relax once the effort has been made. This type of movement is well seen

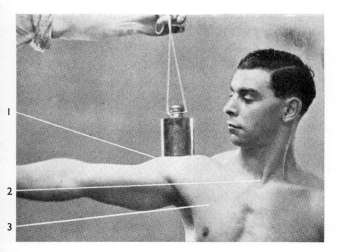

FIG. 5.11. Activity of deltoid in controlling gravitational descent of upper limb, shown by weight riding upon the muscle. Cf. FIGURE 5.12. (*By courtesy of Professor R. D. Lockhart.*)
1. Deltoid.
2. Sternal head of sternocleidomastoid.
3. Pectoralis major.

in serving at tennis, or in kicking a football. The 'follow-through' is due to the momentum produced in the limb at the time of impact.

Contrasting with these movements are the very **fine controlled movements** required by activities such as writing, sewing, painting, etc. Here the prime movers and the antagonists are set to oppose each other equally, and the actual movements are due to temporary slight alterations of the balance of power between the two groups. Movements can be very finely graduated by this means, and the net force exerted can be made very small.

Finally, the occurrence of **involuntary movements** and their occasional dissociation from voluntary movements must be noted. In certain cases of brain injury, muscles which are paralysed for voluntary movement may be found to contract during involuntary or emotional acts. For example, an upper limb which has become paralysed can be moved with its fellow on the other side during the stretching which accompanies a yawn, and facial muscles which cannot be voluntarily activated are found to contract when the patient is asked to smile. The medial rectus muscle of the eye [p. 292] may be unable to contract during convergence of the eyes, but yet be active when the eyes are turned to the opposite side (Beevor 1904). The explanation of such phenomena is not yet established.

Methods of investigating muscle action

It is often very difficult to determine the part played by a muscle in different movements, and there is still argument about the precise actions of many muscles.

The actions of a muscle may be deduced from its attachments, and it is sometimes helpful to imagine a sheet of elastic passing between them. The bellies and tendons of superficial muscles can be palpated during a given movement to find out whether they are contracting. The standard method of demonstrating a muscle is for the examiner to resist the movement concerned; then the muscles performing it will stand out as they contract forcibly. If the resistance is increased additional muscles may be recruited. Electromyography, in which the electrical activity taking place in a muscle during contraction is recorded, is only a more refined example of this method, and has the advantage that it can be applied to muscles lying deeper in the body.

In both cases it is difficult to be sure what the muscle is actually doing when it contracts; is it a prime mover, or a synergist? For example, the tensor fasciae latae [p. 376] contracts during abduction of the thigh, but does not help to abduct the thigh when it is stimulated electrically.

Direct stimulation, whether electrical or mechanical—as when a muscle is pulled upon during a surgical operation—is open to criticism because, in naturally produced movements, muscles are never singled out in this way. However, the method does at least reveal what the muscle is capable of doing, though it does not necessarily show what its functions are in the intact body. A similar objection applies to observations made on the results of paralysis of single muscles or groups of muscles, for the loss of one member of a group at once creates an artificial situation, in which other muscles may take over some or all of its functions.

Testing for paralysis

When a muscle is paralysed, the movements in which it normally takes part are weakened, some being more affected than others, according to the relative importance of its participation in them. The paralysed muscle may never contract again, or it may gradually recover its power, depending on the nature and extent of the damage inflicted upon it or upon its nerve supply. It is naturally important to be able to detect which muscles are paralysed following disease or injury, and also to be able to follow the progress of recovery in them.

The only infallible guide to the integrity of a muscle is to see and feel it contract (Sunderland 1944b), but this is not always possible. True contraction of a muscle must not be confused with displacement or distortion of a paralysed muscle caused by contraction of neighbouring muscles or by a pull imparted by fascial attachments. It is also very necessary to be on guard against '**trick' movements** (Jones 1919; Sunderland 1944b). These may be produced in several ways.

An intact muscle may be able to pull on the tendon of a paralysed muscle because it has an accessory slip of attachment to this tendon. Thus, the abductor pollicis brevis can be used to extend the terminal phalanx of the thumb when all the true extensors are paralysed because some of its fibres are inserted into the tendon of the extensor pollicis longus [p. 343].

Again, an intact muscle which does not normally take part in the affected movement may be recruited to help. The abductor pollicis longus, which does not normally flex the wrist, may do so when the flexors are paralysed because its position allows such an action.

A paralysed prime mover may be brought passively into action by its antagonist, providing the muscles concerned are 'two-joint' muscles. Perhaps the best example is seen when active extension of the metacarpophalangeal joints of the hand is paralysed [p. 345]. By flexing the wrist, the extensor tendons can be put on the stretch and the joints can be made to extend passively. Conversely, if the wrist is fully extended, the tendons of the flexors of the digits can be made to pull the fingers into mild flexion at the interphalangeal joints, even though their muscle bellies may be paralysed. Another important example is afforded in paralysis of flexor pollicis longus (Sunderland 1944b); by hyperextending the wrist and fully abducting the thumb the tendon of flexor pollicis longus is put on the stretch and the terminal joint of the thumb flexes passively.

If the patient is allowed any leverage he may be able to deceive the examiner. For example, when the muscles of the lower limb are being investigated, the heel must not be allowed frictional contact with the couch since movements which are in reality completely paralysed may yet be produced by any muscle capable of taking a leverage from this fixed point (Jones 1919).

Finally, a sudden relaxation of strongly contracted antagonists may allow a 'rebound' movement in the opposite direction which may simulate an active contraction. Thus, a degree of dorsiflexion of the toes may follow relaxation of the plantar flexors even though the dorsiflexors are paralysed.

When assessing the recovery of a muscle from paralysis, or when attempting to detect a degree of contraction in a grossly weakened muscle, it is obviously inappropriate to resist the movement concerned, as when trying to demonstrate muscle actions. Instead, the weakened muscle is put in a position where gravity does not act against it and the patient attempts to carry out the movement without any resistance. A weak muscle may be incapable of executing a voluntary movement but yet be capable of maintaining the part in a position into which it has been moved passively.

The recovery of muscles is often assessed by a grading system. Thus 0 represents no contraction; 1 represents a flicker or trace of contraction; 2 is active movement with gravity eliminated; 3 is active movement against gravity; 4 is active movement against both gravity and resistance, and 5 is normal power. Unfortunately such grading systems are by no means uniformly accepted (Salter 1955a).

It is a commonplace that muscle strength and size can be increased by use, but there is still controversy about the best method of increasing the power of normal muscles and assisting the recovery of paralysed ones. Isotonic and isometric training both have their proponents, but the results are often equivocal or conflicting (Salter 1955b; Petersen 1960).

NOMENCLATURE

The name given to a muscle usually conveys valuable information regarding its position, structure, functions, etc. A muscle may receive its name because of its attachments (sternothyroid), its position (subclavius, supraspinatus), its shape (deltoid, trapezius), its construction (semimembranosus), its action (extensor digitorum), or its size (gluteus maximus). Many names involve a combination of features (extensor carpi radialis longus). Before the eighteenth century few muscles had names, and Galen and Vesalius used numbers, and Leonardo da Vinci letters, to designate the muscles in their illustrations.

VARIATIONS

All muscles are subject to a certain amount of variation, but some are more often affected than others. For example, one such muscle, the palmaris longus [p. 327] is not present in every individual, and is thought to be disappearing in the course of evolution. Others are possibly becoming converted into ligaments; with the disappearance of the tail the coccygeus muscle [p. 365] has been partly replaced by fibrous tissue to form the sacrospinous ligament.

Conversely, certain evolutionarily discarded elements may appear in a given individual, either in their primitive form or in an apparently intermediate stage. The muscle belly of the coracobrachialis [p. 323], which normally ends by being attached to the humerus half-way down its length, may be prolonged downwards to the elbow as it is in lower animals. Again, the muscles of the fingers and toes may exhibit anomalous attachments which are characteristic of lower animals.

In a brief account it is impossible to mention more than a very few such variations, and for more details the works of Testut (1884) and Le Double (1897) should be consulted.

SUPERFICIAL FASCIA

All over the body the dermis of the skin merges into a layer of areolar tissue which in most regions allows the skin to move more or less freely over the underlying structures. Except in the eyelids, the nose and external ear, the penis, and the scrotum, this superficial fascia contains a variable quantity of fat, and, from the second year of life onwards, this is more plentiful in females than in males. At puberty the female body stores additional fat in the secondary sexual distribution, chiefly in the breasts and buttocks, but also in the lower abdominal wall, the outer sides of the thighs, and the backs of the arms, as well as in less clearly marked deposits elsewhere. The difference between male and female superficial fascia is thus exaggerated, and in the adult female the subcutaneous fat is almost twice as heavy as in a male of the same total weight. As age advances, more fat may be deposited in the superficial fascia of both sexes.

The deeper part of the superficial fascia is often more fibrous than the rest, and in the lower region of the abdominal wall this stratum condenses into a membranous layer containing elastic fibres, which may play a part in supporting the abdominal viscera against the pull of gravity [p. 371]. In places such as the sole of the foot and the palm of the hand, the deeper layers of the superficial fascia are tightly fused with the deep fascia, but elsewhere the connection is loose.

The superficial fascia conducts the blood vessels and nerves which supply the skin and is an important factor in preventing the loss of body heat. In the face and neck it contains numerous somatic muscles [p. 287], and in the scrotum the visceral dartos muscle [p. 372], which corrugates the scrotal skin and helps to support the testes, as well as altering the surface area of the scrotum to control heat loss from it.

DEEP FASCIA

Deep to the superficial fascia there is a stratum of much denser fibrous tissue known as the **investing layer** of the deep fascia. The thickness of this relatively inelastic material varies considerably from place to place; in the palm and sole it is very thick and strong. In contrast is the thin and delicate deep fascia on the medial side of the thigh. As the investing layer passes over superficial bony prominences it often becomes attached to them, and from its deep surface there arise sheets of similar material which penetrate between the muscles, forming **intermuscular septa** to which the muscles are often attached [FIG. 5.13]. These septa are themselves directly or indirectly attached to the underlying bone, and the whole arrangement serves to maintain the shape of the body and to exert a slight compression force on the contents of the **osteofascial compartments** so formed.

The existence of such compartments has an important effect on the return of blood to the heart. As the muscles contract their bellies widen, and because the walls of the compartments are inelastic the veins within them are compressed. The venous valves ensure that the blood can only be squeezed in one direction, towards the heart.

Similar sheets of deep fascia may enclose glands and other viscera, and they serve to conduct to them, and to the muscles, their innervation and vascular supply. Sometimes the fascia covering a muscle is virtually indistinguishable from the epimysium, but often there is a clear separation, so that the muscle is able to move freely against the partition formed by the fascia. In the living body the septa of fascia are very smooth and slippery, so facilitating movement of this kind.

Where the deep fascia gives attachment to muscles it becomes thickened and aponeurotic, and the distinction between such thickened tracts of fascia and true aponeuroses is often impossible to make. A good example is the iliotibial tract of the fascia lata [p. 405].

The deep fascia is not easily penetrated by fluid, and for this reason infection tends to spread from one part of the body to another along 'fascial planes'. Some of these planes are rather artificial

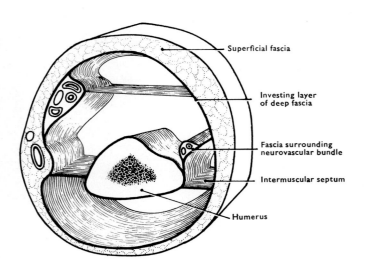

FIG. 5.13. Schematic cross-section of an arm from which the muscles have been removed, showing the arrangement of superficial and deep fascia.

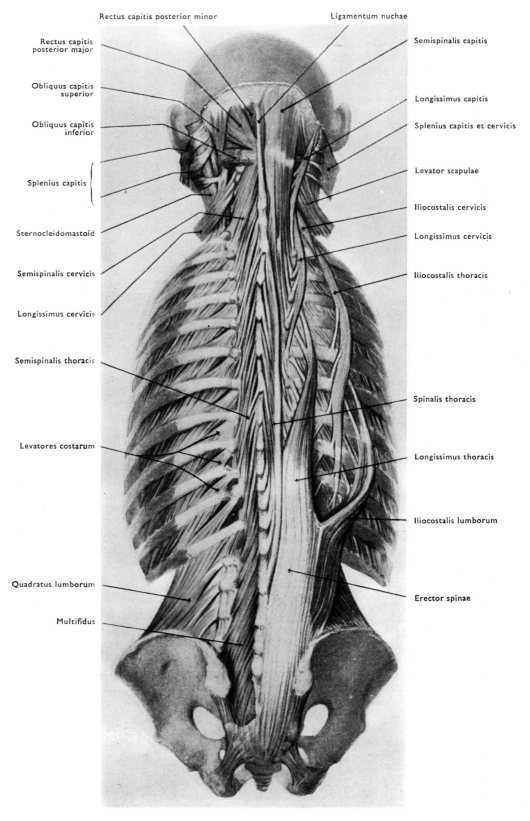

Rectus capitis posterior minor

Ligamentum nuchae

Rectus capitis posterior major

Obliquus capitis superior

Obliquus capitis inferior

Splenius capitis

Sternocleidomastoid

Semispinalis cervicis

Longissimus cervicis

Semispinalis thoracis

Levatores costarum

Quadratus lumborum

Multifidus

Semispinalis capitis

Longissimus capitis

Splenius capitis et cervicis

Levator scapulae

Iliocostalis cervicis

Longissimus cervicis

Iliocostalis thoracis

Spinalis thoracis

Longissimus thoracis

Iliocostalis lumborum

Erector spinae

FIG. 5.14. Deep muscles of the back.

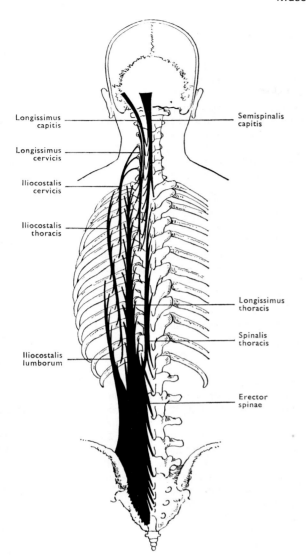

Longissimus
capitis

Longissimus
cervicis

Iliocostalis
cervicis

Iliocostalis
thoracis

Iliocostalis
lumborum

Semispinalis
capitis

Longissimus
thoracis

Spinalis
thoracis

Erector
spinae

FIG. 5.15. Schematic representation of the parts of the left erector spinae muscle.

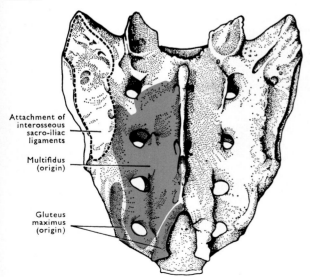

Attachment of
interosseous
sacro-iliac
ligaments

Multifidus
(origin)

Gluteus
maximus
(origin)

FIG. 5.16. Muscle attachments to the sacrum (dorsal surface).

products of the process of dissection in dead bodies, and the detailed description which was formerly given of the topography and attachments of the deep fascia can now be abbreviated without serious loss. Connective tissue pervades the whole body and there is no agreed definition of how dense it must be in a given site before it can be regarded as forming a fascial plane. For this reason descriptions of the arrangement of the fascial planes differ from book to book. Similarly the 'fascial spaces', in which fluid may accumulate, are merely regions of relatively loose connective tissue. Nevertheless, a knowledge of the main fascial 'planes and spaces' is of clinical importance (Gallaudet 1931; Singer 1935).

In certain places the deep fascia becomes strengthened to form retention bands or **retinacula**, which serve to strap down the underlying tendons to the bone and so to prevent them from 'bow-stringing' away from the joint on which they operate when their muscle bellies contract. Such retinacula are found in their most characteristic form at the wrist and the ankle [pp. 347, 406]. In the fingers and toes the flexor tendons are confined within a tunnel formed by the attachment to the bones of a dense **fibrous sheath** derived from the deep fascia [p. 348]. Elsewhere specialized **pulleys** may be formed, as for example the trochlea which changes the direction of pull of the superior oblique muscle of the eye [p. 293] and the tunnel of fascia which holds down the digastric muscle [p. 284] to the hyoid bone.

When a tendon crosses a bone it may be separated from it by a closed 'bag' of connective tissue containing a few drops of lubricant fluid and known as a **bursa**. The fluid which a bursa contains is similar to synovial fluid, and indeed many bursae communicate with the cavity of an adjacent joint. The walls of the bursa are extremely smooth, and ride on each other with a minimum of friction. Similar bursae may separate tendons which are attached very close to each other, and others may separate skin from underlying bone. Where a tendon rubs against the walls of a tight tunnel, as happens at the wrist and ankle, it may be surrounded by an elongated form of bursa called a **synovial sheath**. Such a sheath is wrapped round the tendon, and has a visceral layer adherent to it; the parietal layer of the sheath adheres to the walls of the tunnel. When the tendon moves, the visceral and parietal walls of the sheath slip over each other with very little friction, so facilitating the movement.

Muscles of the vertebral column, neck, and head

POSTVERTEBRAL MUSCLES

The muscles of the back can be divided into three groups. Most superficial are those which connect the upper limb with the trunk, comprising trapezius, latissimus dorsi, rhomboids, and levator scapulae: they belong to and are dealt with in the section on the upper limb [p. 310]. Deep to this group are two small muscles called serratus posterior superior and serratus posterior inferior, which are described along with the muscles of the thorax.

The third, and deepest, group consists of muscles which run longitudinally on the vertebral column, and vary considerably in thickness and importance in its different regions. They play a vital part in the maintenance of posture and in the movements of the vertebral column, but so complicated are their attachments that for most purposes only the broad outlines of their origins and insertions are necessary.

The superficial members of this group are arranged in more or less separate longitudinal columns, the fibres of which travel for a considerable distance between origin and insertion. In the deeper layers the fibres are shorter, and the deepest muscles stretch only between one vertebra and its immediate neighbours.

ERECTOR SPINAE

This large, complicated, and very powerful muscle takes origin by a strong tendon which is attached along a U-shaped line circumscribing the origin of the multifidus muscle [FIGS. 5.14 and 5.16]. The medial limb of the U springs from the lower two thoracic spines, all the lumbar spines, and the median sacral crest. The lateral limb extends upwards along the lateral sacral crest to the posterior superior iliac spine and the posterior part of the iliac crest. Deep to this lateral limb, the erector spinae has a fleshy attachment to the iliac tuberosity and the inner lip of the iliac crest. The tendinous fibres blend with the dorsal sacro-iliac, the sacrotuberous, and the sacrococcygeal ligaments, as well as with the origin of gluteus maximus.

The muscle has a relatively narrow sacral attachment, but the muscle fibres arising from the tendon swell out to form a broad belly which is prominent in the lumbar region and has a well marked lateral margin which can be detected in the living subject [FIG. 5.53]. As this vertically running mass approaches the twelfth rib, it divides into three parallel columns. The most lateral of these is iliocostalis, the intermediate one is longissimus, and the most medial column is spinalis. All three columns are in turn divided into three relays of fibres, each relay arising just as the lower fibres are inserted [FIG. 5.15].

Iliocostalis lumborum inserts by six slips into the lower six ribs near their angles. Medial to the insertion of each of these slips arises iliocostalis thoracis, which is inserted by similar slips into the upper six ribs. Medial to these in turn arises iliocostalis cervicis, a narrow band which inserts into the posterior tubercles of the transverse processes of the lower cervical vertebrae.

The intermediate column, as its name suggests, is the longest (and also the thickest) element of erector spinae. Longissimus thoracis is inserted by two sets of slips, lateral and medial, to the ribs, and the transverse processes of all the thoracic vertebrae and the accessory processes of the upper lumbar vertebrae. Longissimus cervicis originates from the transverse processes of the upper six thoracic vertebrae, medial to the insertions of longissimus thoracis, and inserts into the posterior tubercles of the transverse processes of all the cervical vertebrae except the first and the seventh. In the neck it lies deep to the iliocostalis cervicis. Longissimus capitis, the only portion of erector spinae to reach the skull, takes origin from the transverse processes of the upper thoracic vertebrae, in common with longissimus cervicis, and also from the articular processes of the lower four cervical vertebrae. Its narrow tendon runs between splenius capitis and semispinalis capitis muscles, and is inserted into the posterior aspect of the mastoid process, deep to splenius capitis.

The spinalis column is relatively insignificant. Spinalis thoracis, which is its only well demarcated component, runs from the vertebral spines in the lower thoracic and upper lumbar regions to the upper thoracic spines. Spinalis cervicis is not often present as a distinct entity, and spinalis capitis is usually fused with the most medial part of semispinalis capitis.

Nerve supply

The components of erector spinae are all supplied by the dorsal rami of the spinal nerves according to their situation.

Actions

The main bulk of erector spinae crosses the secondary lumbar curvature of the spine, and is responsible (particularly longissimus thoracis) for maintaining this curvature in the erect and sitting positions. It is concerned in extension and rotation of the vertebral column (Morris, Benner, and Lucas 1962), and comes into play in lateral flexion, but only if forward bending is allowed (Pauly 1966). It is a most important controlling agent during flexion of the column, which is largely produced by active relaxation of erector spinae and the other postvertebral muscles. During walking, erector spinae on both sides contracts to steady the vertebral column on the pelvis. This action is readily seen and felt.

Longissimus capitis can extend and laterally flex the skull on the neck, and rotates the face to the same side.

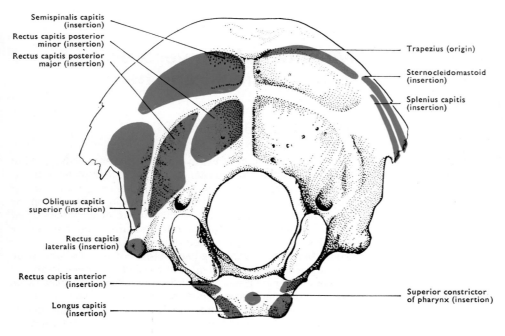

Semispinalis capitis (insertion)
Rectus capitis posterior minor (insertion)
Rectus capitis posterior major (insertion)
Trapezius (origin)
Sternocleidomastoid (insertion)
Splenius capitis (insertion)
Obliquus capitis superior (insertion)
Rectus capitis lateralis (insertion)
Rectus capitis anterior (insertion)
Longus capitis (insertion)
Superior constrictor of pharynx (insertion)

FIG. 5.17. Muscle attachments to the occipital bone.

SPLENIUS CAPITIS

Splenius capitis arises from the lower part of the ligamentum nuchae, and from the spines of the seventh cervical and the upper three or four thoracic vertebrae. It appears in the floor of the posterior triangle [p. 286] before inserting into the posterior aspect of the mastoid process and the lateral part of the superior nuchal line [FIG. 5.17], deep to the attachment of sternocleidomastoid. Splenius capitis is covered posteriorly by trapezius, rhomboideus major and minor, and serratus posterior superior. In turn, it lies superficial to the cervical portions of erector spinae, and posterior to semispinalis capitis.

Nerve supply

Lateral branches of the dorsal rami of the cervical nerves.

Actions

When both muscles act together they extend the head and the neck; when one acts individually, it extends the head and neck, laterally flexes the neck, and turns the face to the same side. It therefore assists the sternocleidomastoid of the opposite side [p. 282].

SPLENIUS CERVICIS

This muscle arises in common with splenius capitis, but rather lower down, from the spines of the third to the sixth thoracic vertebrae. It passes upwards to be inserted into the posterior tubercles of the transverse processes of the upper three or four cervical vertebrae.

Nerve supply

Lateral branches of the dorsal rami of the lower cervical nerves.

Actions

With its fellow it helps to extend the neck; acting on its own it laterally flexes and slightly rotates the cervical portion of the vertebral column to the same side.

TRANSVERSOSPINALIS

Deep to erector spinae lies a group of muscles loosely classed together as the transversospinalis system. This name is given them because their fibres in general pass upwards and medially from the transverse processes towards the vertebral spines (contrast the fibres of the erector spinae). Like erector spinae, transversospinalis consists of three components. These, however, do not lie side by side, but successively deeper from the surface. The most superficial is the semispinalis system, which extends from the loin to the skull, and is described as three separate muscles.

Semispinalis thoracis arises from the transverse processes of the lower thoracic vertebrae, and is inserted into the spines of the upper thoracic and the lower cervical vertebrae. Semispinalis cervicis, a larger muscle, arises from the transverse processes of the upper thoracic and the articular processes of the lower cervical vertebrae, and is inserted into the spines of the cervical vertebrae. The largest of the three is semispinalis capitis, which lies under cover of splenius capitis, taking origin from the tips of the transverse processes of the upper six thoracic and the articular processes of the lower four cervical vertebrae. The most medial part of the muscle is often separated from the rest and is sometimes referred to as a separate muscle under the name of spinalis capitis [p. 278]; it may be attached to some of the lower cervical spines. Semispinalis capitis broadens as it passes upwards towards the skull, and is inserted into the medial impression between the superior and inferior nuchal lines of the occipital bone [FIG. 5.17].

Semispinalis capitis forms the bulge on either side of the nuchal furrow in the midline of the neck, and it covers over semispinalis cervicis and most of the suboccipital triangle. Usually its vertically running fibres may be seen at the upper angle of the posterior triangle of the neck, appearing from under cover of splenius capitis [FIG. 5.21].

The second component of the transversospinalis group is multifidus, which lies in the furrow between the spines of the vertebrae and their transverse processes. Its deeper fasciculi extend only from one vertebra to the next, though the more superficial ones may cross several vertebrae. The muscle arises from the dorsal surface of the sacrum [FIG. 5.16], under cover of the tendon of erector spinae, from the posterior sacro-iliac ligaments, and from the mamillary processes of the lumbar vertebrae, the transverse processes of the thoracic vertebrae, and the articular processes of the lower cervical vertebrae. It is inserted into the spines of all the vertebrae from the fifth lumbar up to the axis. Throughout its extent the muscle lies deep to the semispinalis system and erector spinae.

The third component of the transversospinalis group, the rotatores, may be said to represent the deepest layer of multifidus, with which their fibres are often continuous. They are only developed to any extent in the thoracic region, but some small variable bundles with similar attachments may be found in the cervical and lumbar regions. Each rotator muscle arises from the transverse process of one vertebra and is inserted into the lamina of the vertebra directly above it.

Nerve supply

The transversospinalis system is supplied by the dorsal rami of the spinal nerves.

Actions

The thoracic and cervical portions of semispinalis extend the thoracic and cervical parts of the vertebral column. Semispinalis capitis is the most powerful extensor of the head on the neck.

Multifidus is concerned in extension, lateral flexion, and rotation of the vertebral column, but its main function is probably to act as a series of extensile ligaments guarding the movements of the vertebral column produced by the more powerful and superficial prime movers. As their name implies, the rotatores may be concerned in rotatory movements of the vertebral column, but functionally they too are probably extensile ligaments.

INTERSPINALES AND INTERTRANSVERSARII

The interspinales are short and insignificant bands extending from one spinous process to the next, lying on either side of the interspinous ligament. They are best developed in the cervical and lumbar regions, and may be absent in the thoracic part of the column. The intertransverse muscles are equally small slips extending between the transverse processes in the cervical and lumbar regions. In the cervical portion of the column they lie both anterior and posterior to the emerging ventral rami of the spinal nerves, but in the lumbar region they are wholly posterior to the ventral rami, divided into a medial and a lateral component.

Nerve supply

The interspinales are supplied by the dorsal rami of the spinal nerves, as are the medial portions of the lumbar intertransversarii. The lateral portions of the lumbar and all the cervical intertransversarii are supplied by the ventral rami of the spinal nerves.

Actions

They are probably extensile ligaments.

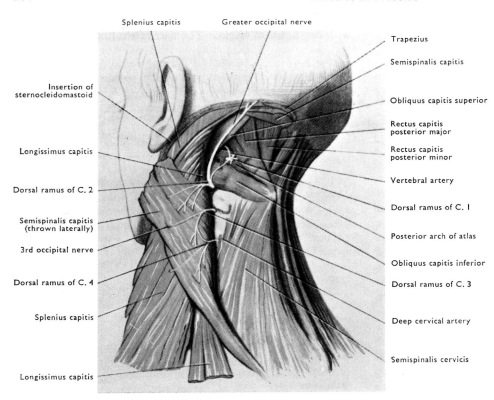

Splenius capitis
Greater occipital nerve
Trapezius
Semispinalis capitis
Insertion of sternocleidomastoid
Obliquus capitis superior
Rectus capitis posterior major
Longissimus capitis
Rectus capitis posterior minor
Dorsal ramus of C. 2
Vertebral artery
Semispinalis capitis (thrown laterally)
Dorsal ramus of C. I
3rd occipital nerve
Posterior arch of atlas
Dorsal ramus of C. 4
Obliquus capitis inferior
Splenius capitis
Dorsal ramus of C. 3
Deep cervical artery
Longissimus capitis
Semispinalis cervicis

Fig. 5.18. Suboccipital triangle of the left side.

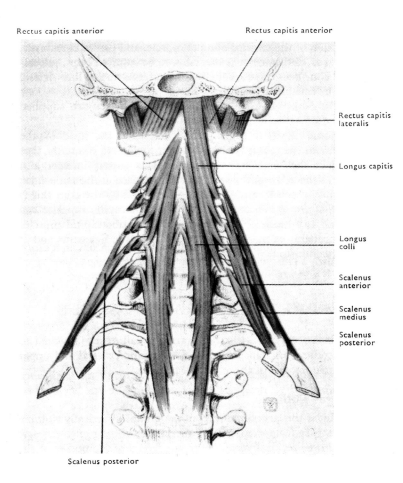

Rectus capitis anterior
Rectus capitis anterior
Rectus capitis lateralis
Longus capitis
Longus colli
Scalenus anterior
Scalenus medius
Scalenus posterior
Scalenus posterior

Fig. 5.19. Prevertebral and scalene muscles.

SUBOCCIPITAL GROUP

The last group of postvertebral muscles lies deep in the neck [FIGS. 5.17 and 5.18], anterior to the semispinalis capitis, longissimus capitis, and splenius capitis. They enclose a triangular space known as the suboccipital triangle, in which lie the posterior arch of the atlas, the posterior atlanto-occipital membrane, the third part of the vertebral artery, the dorsal ramus of the first cervical nerve, and the suboccipital plexus of veins.

Rectus capitis posterior major arises from the spine of the axis and is inserted into the occipital bone deep to the obliquus capitis superior and semispinalis capitis, below the inferior nuchal line. Rectus capitis posterior minor arises from the posterior tubercle of the atlas and is inserted into the occipital bone medial and deep to the insertion of the rectus capitis posterior major.

Obliquus capitis inferior, the largest muscle of the group, originates from the spine of the axis and passes upwards, laterally and forwards to be inserted into the posterior aspect of the transverse process of the atlas. Despite its name, none of its fibres reaches the skull. Obliquus capitis superior arises from the transverse process of the atlas at the site of insertion of the inferior oblique muscle, and passes backwards, upwards, and somewhat medially, to be inserted into the occipital bone between the superior and inferior nuchal lines, lateral to semispinalis capitis.

Nerve supply

All the suboccipital muscles are supplied by the dorsal ramus of the first cervical nerve.

Actions

Both rectus muscles can extend the head on the neck, and the rectus capitis posterior major can rotate the face towards the same side. Obliquus capitis inferior also rotates the atlas on the axis, turning the face to the same side; because of the length of the lever afforded by the atlas this is quite a strong movement. Obliquus capitis superior can extend the head. Nevertheless, all the suboccipital muscles are probably of more importance as extensile ligaments and in the maintenance of posture.

PREVERTEBRAL MUSCLES

The prevertebral muscles form a small group attached to the bodies and transverse processes of the cervical and upper thoracic regions of the vertebral column.

LONGUS COLLI

This is the largest member of the group. Vertically running fibres pass from the fronts of the bodies of the upper thoracic and lower cervical vertebrae to the fronts of the bodies of the upper cervical vertebrae. Some oblique fibres run from the upper three thoracic bodies to the transverse processes of the fifth and sixth cervical vertebrae, and others from the transverse processes of the third, fourth, and fifth cervical vertebrae to the anterior tubercle of the atlas [FIG. 5.19].

Nerve supply

From the ventral rami of the cervical nerves.

Actions

Longus colli bends the neck forwards; its inferior oblique fibres may help to produce lateral flexion and rotation to the opposite side.

LONGUS CAPITIS

Longus capitis travels from the transverse processes of the middle three or four cervical vertebrae to the basilar part of the occipital bone lateral to the pharyngeal tubercle [FIGS. 5.17 and 5.19]. It lies in front of the upper oblique fibres of longus colli.

Nerve supply

From the ventral rami of the first three or four cervical nerves.

Actions

Longus capitis flexes the head and the upper part of the cervical spine. This movement is commonly produced merely by relaxation of the extensor muscles [p. 287], and active flexion is usually needed only against resistance.

RECTUS CAPITIS ANTERIOR

This short strap muscle [FIGS. 5.17 and 5.19] passes from the anterior surface of the lateral mass of the atlas to the basilar part of the occipital bone between longus capitis and the occipital condyle. It corresponds to an anterior intertransverse muscle, but is considerably larger.

Nerve supply

From the loop between the ventral rami of the first and second cervical nerves.

Actions

It can flex the head on the neck, but probably serves mainly to hold the articular surfaces of the atlanto-occipital joint in close apposition during movements.

RECTUS CAPITIS LATERALIS

Another strap muscle, in series with the posterior intertransverse muscles, rectus capitis lateralis passes from the transverse process of the atlas to the inferior surface of the jugular process of the occipital bone [FIGS. 5.17 and 5.19].

Nerve supply

From the loop between the ventral rami of the first and second cervical nerves.

Actions

It can bend the head to the same side, but its stabilizing action on the atlanto-occipital joint is probably more important.

LATERAL MUSCLES OF NECK

SCALENE GROUP

Three muscles arise from the transverse processes of the cervical vertebrae and pass downwards to the ribs.

Scalenus anterior, an important landmark in the neck, arises from the anterior tubercles of the transverse processes of the third to the sixth cervical vertebrae and runs downwards and laterally behind the prevertebral fascia to be inserted into the scalene tubercle on the first rib [FIGS. 5.19, 5.20, and 5.98].

Scalenus medius [FIGS. 5.19 and 5.20] is the largest of the scalenes, and usually arises from the posterior tubercles of the transverse processes of all the cervical vertebrae. Its fibres descend in the floor of the posterior triangle [p. 286], behind the brachial plexus, to insert into a rough impression on the first rib behind the groove for the subclavian artery.

Scalenus posterior, a relatively small muscle, arises behind scalenus medius from the posterior tubercles of the fourth, fifth, and sixth cervical transverse processes, and is inserted into the outer side of the second rib posterior to the origin of serratus anterior [FIG. 5.20]. It is often difficult to separate scalenus posterior from scalenus medius, which completely hides it, except in the lowest part of the floor of the posterior triangle.

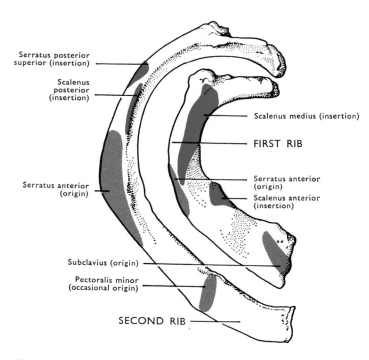

FIG. 5.20. Muscle attachments to upper surface of first rib and outer surface of second rib (right side).

Nerve supply

By branches from the ventral rami of the cervical nerves: scalenus anterior from the fourth to the sixth, scalenus medius from the lower five or six, and scalenus posterior from the last three.

Actions

Acting from their origins on their insertions, the scalene muscles steady the first two ribs during respiration [p. 367], and may assist in elevating them. Acting from their insertions, they help to produce lateral flexion of the cervical part of the vertebral column.

STERNOCLEIDOMASTOID

Sternocleidomastoid [FIG. 5.21] is a long strap muscle with two heads of origin. The narrow tendinous sternal head arises from the anterior surface of the manubrium sterni, and the broad clavicular head from the upper surface of the clavicle in its medial third: there may be an interval between the two. The clavicular head passes deep to the sternal head, and the united muscle is inserted by a short tendon into the outer surface of the mastoid process and into the lateral third of the superior nuchal line of the occipital bone [FIG. 5.17].

The muscle is the major landmark in the neck. The spinal part of the accessory nerve enters the deep aspect of its upper third, and runs within its substance to emerge about half way down its lateral border.

Nerve supply

By motor fibres from the spinal part of the accessory nerve, and by sensory fibres from the ventral rami of the second and third cervical nerves through the cervical plexus: these fibres may join the accessory.

Actions

It tilts the head towards the same side and rotates it so that the face is turned towards the opposite side and also upwards. The anterior fibres flex the head on the neck at the atlanto-occipital joint, but the posterior fibres may extend this joint (Last 1954). The two

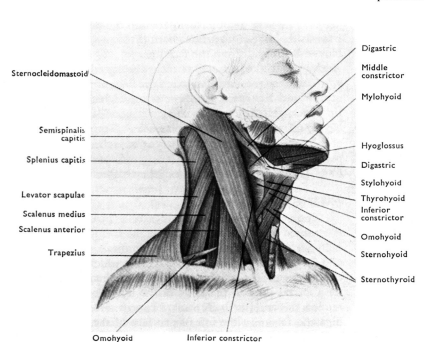

FIG. 5.21. Muscles of side of neck (anterior and posterior triangles).

sternocleidomastoids acting together flex the neck, as when the head is raised from the pillow. If the head is fixed on the neck by other muscles the sternocleidomastoids can act as accessory muscles of inspiration, since they raise the manubrium sterni and consequently the ribs [p. 367].

The muscle is sometimes injured at birth, and may be wholly or partly replaced by fibrous scar tissue, which contracts to produce the condition of torticollis (wry neck). Temporary wry neck may result from irritation of the muscle or its motor nerve, and spasms may occur which cause jerking movements of the head; these are often associated with a similar spasm of the clavicular portion of trapezius [p. 311].

HYOID MUSCLES

The hyoid bone is tethered by muscles [FIGS. 3.68 and 5.21] to the mandible and skull above and to the sternum, thyroid cartilage, and scapula below. These muscles are primarily concerned with steadying or moving the hyoid bone, and hence the larynx and tongue, both of which are attached to it. They are usually described in two groups. The suprahyoid group includes mylohyoid, geniohyoid, stylohyoid, and digastric, and the infrahyoid group comprises sternohyoid, sternothyroid, omohyoid, and thyrohyoid. Members of both groups of hyoid muscles co-operate to depress the mandible when the mouth has to be opened against resistance.

MYLOHYOID

Mylohyoid arises from the mylohyoid line on the inner surface of the body of the mandible [FIG. 5.23], and its fibres run downwards and

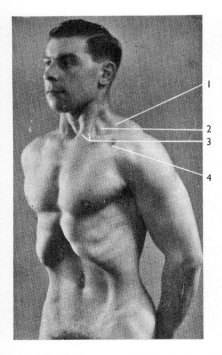

FIG. 5.22. Subject attempting inspiration with closed glottis. Note the effect of atmospheric pressure on the abdomen, intercostal spaces, the suprasternal and the supraclavicular and infraclavicular fossae. (*By courtesy of Professor R. D. Lockhart.*)
1. Clavicular fibres of trapexius.
2. Scalene muscle in greater supraclavicular fossa.
3. Heads of sternocleidomastoid.
4. Infraclavicular fossa.

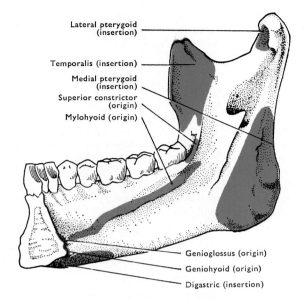

Lateral pterygoid (insertion)
Temporalis (insertion)
Medial pterygoid (insertion)
Superior constrictor (origin)
Mylohyoid (origin)
Genioglossus (origin)
Geniohyoid (origin)
Digastric (insertion)

FIG. 5.23. Muscle attachments to deep surface of mandible.

medially, the posterior ones to be inserted into the body of the hyoid bone and the remainder into a median raphe which extends from the hyoid bone to the chin. The two mylohyoid muscles therefore form a sling or diaphragm which supports the floor of the mouth [FIG. 5.37]. Geniohyoid lies superior to this sling, and the anterior belly of digastric lies inferior to it.

Nerve supply

From the mylohyoid branch of the inferior alveolar nerve.

Actions

The mylohyoids raise the floor of the mouth in swallowing; they elevate and fix the hyoid bone, and help to press the tongue upwards and backwards against the roof of the mouth (Whillis 1946). They can depress the mandible if the hyoid bone is fixed by the infrahyoid group of muscles, but this action is not called into play unless there is some resistance to opening the mouth.

GENIOHYOID

The geniohyoid muscles arise from the lower part of the mental spine of the mandible [FIG. 5.23] and run backwards between the mylohyoid and the genioglossus muscles to be inserted into the body of the hyoid bone [FIG. 3.68]. The muscles of the two sides are often fused.

Nerve supply

From fibres of the first cervical nerve conveyed by the hypoglossal nerve.

Actions

They pull the hyoid bone forwards and slightly upwards, so shortening the floor of the mouth, and widening the pharynx for the reception of food. If the hyoid bone is fixed, they can help to retract the mandible and possibly to depress it.

STYLOHYOID

This small muscle arises from the posterior border of the styloid process near its root. It runs antero-inferiorly on the posterior belly of digastric and splits into two slips enclosing the tendon of this muscle [FIGS. 5.21 and 5.24]. The slips reunite to insert into the body of the hyoid bone at the root of its greater horn [FIG. 3.68].

Nerve supply

From the facial nerve.

Actions

It elevates and retracts the hyoid bone, carrying with it the tongue, and elongating the floor of the mouth. The stylohyoid ligament is a fibrous cord which runs from the tip of the styloid process to the lesser horn of the hyoid bone, closely related to the stylohyoid muscle. It represents a portion of the cartilage of the second pharyngeal arch, and is sometimes partly ossified. It usually gives origin to some of the highest fibres of the middle constrictor of the pharynx [FIGS. 5.24 and 5.39].

DIGASTRIC

The posterior belly of digastric [FIGS. 5.21, 5.24, and 5.25] is attached to the mastoid notch of the temporal bone. This belly, closely related to stylohyoid, runs downwards and forwards to an intermediate tendon which passes through the insertion of stylohyoid

and at this point is strapped down by a fascial sling to the body of the hyoid bone. From this tendon the anterior belly runs forwards and upwards to the digastric fossa, on the inferior surface of the mandible, close to the symphysis.

Nerve supply

The posterior belly is supplied by the facial nerve, and the anterior belly by the mylohyoid branch of the inferior alveolar nerve.

Actions

It raises the hyoid bone, or, if the hyoid bone is fixed, can depress and retract the mandible (Last 1954).

STERNOHYOID

Sternohyoid [FIGS. 5.21 and 5.24] takes origin from the posterior surface of the manubrium sterni, the back of the sternoclavicular joint, and the medial end of the clavicle. It is a strap muscle, which runs upwards, inclining towards its fellow of the opposite side, to be inserted into the medial part of the lower border of the body of the hyoid bone [FIG. 3.68]. Most of the muscle is superficial, but inferiorly it is covered by the sternum, clavicle, and sternocleido-mastoid and elsewhere by the investing layer of the deep fascia.

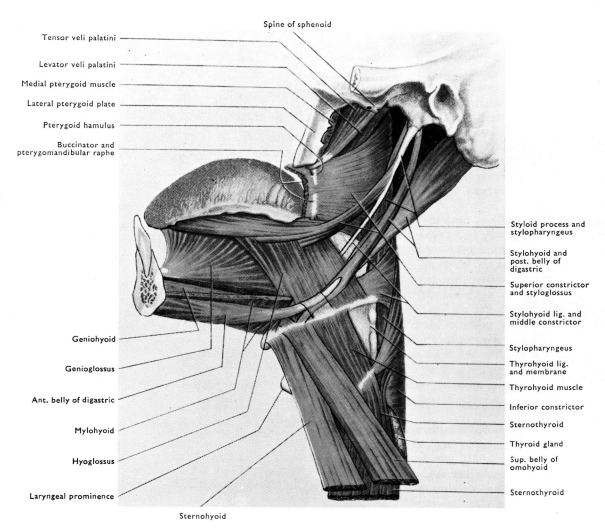

FIG. 5.24. Muscles of the tongue, hyoid bone, and pharynx.

Nerve supply

From the ventral rami of the first three cervical nerves, by fibres conveyed throught the ansa cervicalis.

Actions

Sternohyoid depresses the hyoid bone and helps to fix it when other muscles are acting from it. If the hyoid bone is held by the suprahyoid muscles, sternohyoid may act as an accessory muscle of inspiration [p. 367].

STERNOTHYROID AND THYROHYOID

Sternothyroid [FIGS. 5.21 and 5.24], which lies deep to sternohyoid, arises from the back of the manubrium sterni below the origin of sternohyoid, and also from the first costal cartilage. It is broader than sternohyoid, and instead of converging on its fellow as it rises, it swings laterally in front of the trachea and the thyroid gland to be inserted into the oblique line on the outer surface of the thyroid cartilage. It is deep to sternocleidomastoid and sternohyoid, and overlapped in its upper part by omohyoid.

Thyrohyoid is a short strap muscle which arises from the oblique line of the thyroid cartilage [FIG. 5.24], continuing the line of the fibres of sternothyroid. It runs upwards to be inserted into the lower border of the body and greater horn of the hyoid bone [FIG. 3.68].

Nerve supply

Sternothyroid is supplied from the ansa cervicalis: thyrohyoid is supplied from the ventral ramus of the first cervical nerve by fibres brought to the muscle through the hypoglossal nerve.

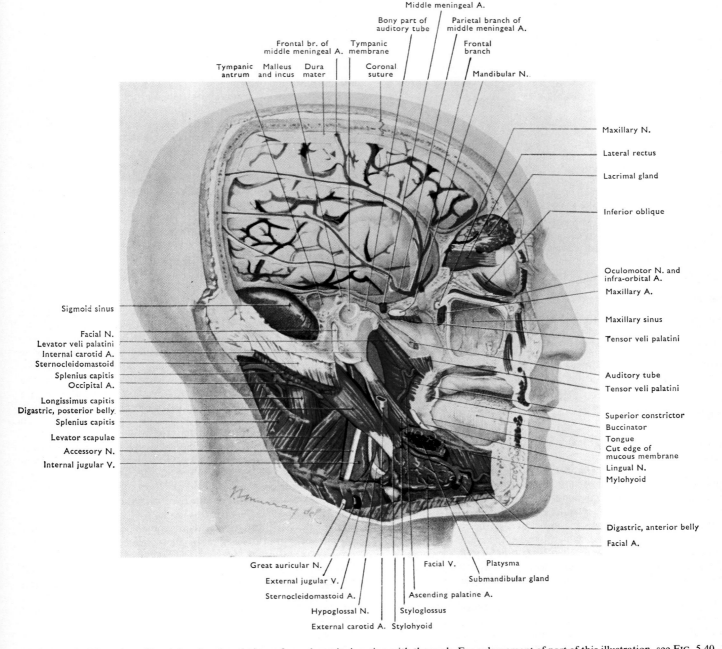

FIG. 5.25. Dissection of head showing the relations of muscles at its junction with the neck. For enlargement of part of this illustration, see FIG. 5.40.

Actions

Sternothyroid and thyrohyoid can act as a unit which has actions very similar to those of sternohyoid, except that the thyroid cartilage is pulled down as well as the hyoid bone. This unit can depress the mandible in conjunction with mylohyoid and digastric.

Acting by itself, thyrohyoid approximates the thyroid cartilage to the hyoid bone. This pulls the larynx up under the base of the tongue, and brings the arytenoid cartilages under cover of the epiglottis; it is therefore largely responsible for closing the laryngeal orifice and is of great importance in preventing food from entering the larynx in swallowing.

Acting on its own, sternothyroid pulls the thyroid cartilage away from the hyoid bone, so opening the laryngeal orifice. In forced inspiration the muscle contracts, perhaps to help in raising the sternum, but mainly to keep the laryngeal orifice fully open.

LEVATOR GLANDULAE THYROIDEAE

An occasional unpaired slip of muscle passes between the hyoid bone and the isthmus or the pyramidal lobe of the thyroid gland. It presumably helps to elevate the gland, but its frequent absence produces no disability, and the real elevators of the gland, because of its attachment by fascia to the thyroid cartilage, are the suprahyoid and the thyrohyoid muscles.

OMOHYOID

This muscle has two bellies; the inferior one arises from the upper border of the scapula and the superior transverse scapular ligament [FIG. 5.65] and passes forwards and slightly upwards across the scalene muscles to end in an intermediate tendon under cover of sternocleidomastoid [FIG. 5.21]. The tendon is tied down to the clavicle and first rib by a sling of the cervical fascia [p. 308], and from its anterior end the superior belly of the muscle runs upwards, crossing the carotid sheath at the level of the arch of the cricoid cartilage, to be inserted into the lower border of the body of the hyoid bone lateral to the sternohyoid [FIG. 3.68].

Nerve supply

Both bellies are supplied from the ansa cervicalis.

Actions

Like the other infrahyoid muscles, omohyoid pulls down the hyoid bone.

Regions of the neck

Sternocleidomastoid divides the side of the neck, for descriptive purposes, into two triangles. The posterior triangle [FIG. 5.21] is bounded by the middle third of the clavicle, the posterior border of sternocleidomastoid, and the anterior border of trapezius. Its apex is formed by the occipital bone, and in its floor lie the scalene muscles, levator scapulae, splenius capitis, and, usually, a portion of semispinalis capitis. Across the lower part of the triangle runs the inferior belly of omohyoid, which divides the triangle into two, a larger occipital triangle and a smaller supraclavicular triangle.

The anterior triangle, bounded by the midline of the neck, the lower border of the mandible, and the anterior border of sternocleidomastoid [FIG. 5.21], has its apex at the manubrium sterni, and is subdivided into four smaller triangles. The small submental triangle lies between the anterior bellies of the digastric muscles and the body of the hyoid bone, and the digastric triangle of each side lies between the mandible and the anterior and posterior bellies of

digastric. The muscular triangle is the part of the anterior triangle anterior to the superior belly of the omohyoid muscle, and the remainder of the anterior triangle, between the posterior belly of digastric above and the superior belly of omohyoid below, constitutes the carotid triangle.

Movements of the vertebral column

The muscles of the vertebral column are involved in the maintenance of the erect posture [p. 401]. The most important parts of the musculature in this respect are those which bridge the secondary curvatures of the vertebral column, for these have less intrinsic stability. In the erect 'stand-at-ease' position no single component of the postvertebral group is continuously active, but the constant small disturbances of posture produced by the normal swaying of the body result in the appropriate muscles contracting to restore balance. Movement of a limb, because it alters the centre of gravity of the body, results in greater activity in the vertebral muscles, and during walking the muscles of the back serve to steady the body above the supporting limb [p. 404]. They are also in action as fixators and synergists in every movement of the upper limb.

Because of the relatively recent acquisition of the upright posture in the course of evolution, the vertebral muscles have a task which imposes on them a considerable strain, and this in part explains the common occurrence of aches and pains referred to the musculature of the lower part of the back and to the neck—the two most mobile parts of the column and the ones in which muscular control is most necessary. Following injury, the muscles of the back readily pass into reflex contraction to prevent painful movements, and the difficulty experienced in the performance of a large variety of movements under such circumstances is indicative of the degree to which these muscles enter into all everday actions.

In flexion and extension of the vertebral column, each vertebra may be thought of as rocking backwards and forwards on the nucleus pulposus of the intervertebral disc. The spines which protrude backwards from the vertebrae therefore afford a considerable mechanical advantage to the muscles attached to them, and this factor, together with the great bulk of the postvertebral muscles as compared to the prevertebral muscles, indicates the importance of the maintenance of, and the ability to regain, the extended upright position. In active extension of the column, all the muscles posterior to the axis of movement can be involved, but once the normal erect position has been attained, further extension is largely controlled by the active relaxation of the abdominal muscles, the prevertebral muscles, and sternocleidomastoid. Similarly, the most important factor in flexion of the column from the upright position is active relaxation of the postvertebral group. If, however, flexion is performed against resistance, or against gravity, as in getting up from the supine position, the abdominal, prevertebral, and sternocleidomastoid muscles are called into action as prime movers. The abdominal muscles have a mechanical advantage even greater than that of the postvertebral muscles, since they operate on the ribs, at some considerable distance from the axis of movement. Their action in flexion of the spine involves fixation of the thorax by the muscles which elevate the ribs [p. 367].

During flexion of the trunk, erector spinae pays out excentrically until the position of full flexion has been reached. In this position, no muscular activity in the back can be detected, and the strain is taken entirely by the ligaments of the vertebral column (Floyd and Silver 1952; Pauly 1966). Much of the movement of bending forwards takes place at the hip joint, and a similar excentric contraction of the gluteus maximus and hamstring muscles occurs [p. 378].

Extending the trunk from the position of full flexion to the anatomical position is more complicated. Up to about the half-way stage of the movement, the lumbar spine is held compressed on to the sacrum by contraction of the abdominal and back muscles. It does not move very much during this phase, and the main straightening takes place at the hip joints, where the pelvis is rotated backwards on the heads of the femora. After this phase is completed, the abdominal muscles relax, and erector spinae now pulls the lumbar spine into its normal curvature, so straightening the back. Faults in this mechanism of regaining the erect posture often become apparent at the stage of transition from hip to spinal movement, when the maximum power of erector spinae is required.

Lateral bending and rotation of the vertebral column are intimately connected [p. 223], and are produced largely by muscles which lie lateral to the vertebral column and obliquely in relation to it, such as sternocleidomastoid in the neck, psoas major and quadratus lumborum in the lumbar region, and multifidus and rotatores throughout the length of the column. The following table indicates the main muscles which are anatomically in a position to be concerned in the various movements of the vertebral column, but it will be appreciated that there is an extremely complex interaction between them all, and it is not always possible to distinguish prime movers from synergists, or to pick out those whose primary function is that of an extensile ligament.

The table is not intended to be complete, and some muscles have been omitted for the sake of simplicity. Thus, many of the suprahyoid muscles (mylohyoid, stylohyoid, and digastric) may combine with the elevators of the mandible (to fix the jaw) and the infrahyoid group to form a unit capable of flexing the cervical part of the vertebral column.

All the muscles in the table have segmental innervation from the spinal nerves except for sternocleidomastoid and trapezius, whose motor innervation comes from the spinal part of the accessory nerve.

Muscles concerned in movements of the intervertebral joints

MOVEMENT	CERVICAL REGION	THORACIC AND/OR LUMBAR REGIONS
Flexion	Sternocleidomastoid Longus colli Longus capitis	Muscles of anterior abdominal wall [p. 354]
Extension	Splenius capitis Splenius cervicis Semispinalis capitis Semispinalis cervicis Iliocostalis cervicis Longissimus capitis Longissimus cervicis Trapezius [p. 311] Interspinales	Erector spinae as a whole Quadratus lumborum Trapezius [p. 311]
Rotation and Lateral Flexion	Sternocleidomastoid Scalene group Splenius capitis Splenius cervicis Longissimus capitis Longissimus cervicis Levator scapulae [p. 311] Longus colli Iliocostalis cervicis Multifidus Intertransversarii	Psoas major [p. 363] Quadratus lumborum [p. 363] Muscles of anterior abdominal wall [p. 354] Multifidus Iliocostalis lumborum Iliocostalis thoracis Intertransversarii Rotatores

Movements of the head on the neck

The muscles which lie behind the atlanto-occipital joint have a considerable postural responsibility, since the centre of gravity of the skull falls in front of this joint [p. 223]. It also follows that in the upright position flexion of the head on the neck is usually produced by the action of gravity, controlled by the relaxation of the extensor muscles, while extension of the skull is an active movement against gravity. It is therefore not surprising to find that the extensor muscles are much better developed than the flexors. Rather surprisingly, however, electrophysiological evidence indicates that the semispinalis capitis is not active as a postural muscle (Takebe, Vitti, and Basmajian 1974a).

As in the movements of the vertebral column, the precise role of the muscles acting at the atlanto-occipital and atlanto-axial joints is a matter for conjecture, and the table given is not to be interpreted too rigidly.

Again, some omissions have been made deliberately. The hyoid muscles may combine with the elevators of the mandible to form a unit capable of flexing the head on the neck, and some of the extensors are also capable of a degree of rotation and lateral flexion at the atlanto-axial joint (e.g. rectus capitis posterior major, semispinalis capitis, trapezius).

All the muscles in the table have a segmental innervation from the spinal nerves except for sternocleidomastoid and trapezius, whose motor innervation comes from the spinal part of the accessory nerve.

Muscles concerned in movements of the atlanto-occipital and atlanto-axial joints

MOVEMENT	
Flexion	Longus capitis Rectus capitis anterior Sternocleidomastoid (anterior fibres)
Extension	Semispinalis capitis Splenius capitis Rectus capitis posterior major Rectus capitis posterior minor Obliquus capitis superior Longissimus capitis Trapezius [p. 311] Sternocleidomastoid (posterior fibres)
Rotation and Lateral Flexion	Sternocleidomastoid Obliquus capitis inferior Obliquus capitis superior Rectus capitis lateralis Longissimus capitis Splenius capitis

MUSCLES OF THE SCALP AND FACE

These muscles [FIGS. 5.26 and 5.27] are derived from the mesoderm of the second pharyngeal arch, which in the course of evolution spread in the superficial fascia of the head and neck. They are all supplied by the facial nerve, the nerve of this arch.

EPICRANIUS (OCCIPITOFRONTALIS)

Epicranius has two pairs of bellies, frontal and occipital [FIG. 5.26], which are united by a strong aponeurosis, the galea aponeurotica, so named because it forms a sort of helmet for the skull. Each occipital belly arises from the outer part of the superior nuchal line of the occipital bone and the mastoid part of the temporal bone. The larger frontal bellies have no bony attachment; they extend forwards from the galea to the region of the eyebrow, where they become attached to the skin, interlacing with orbicularis oculi. Unlike the two occipital bellies, the two frontal bellies blend with each other in the midline, and some of the fibres pass down over the nose to become continuous with the procerus muscle [p. 289].

The galea aponeurotica is attached posteriorly, between the two occipital bellies, to the superior nuchal line. Anteriorly it sends a slip for some distance between the two frontal bellies, and laterally it becomes considerably thinner and fuses with the temporal fascia just above the zygomatic arch. It is very firmly bound to the skin by dense superficial fascia containing the blood vessels and nerves, but can move freely over the underlying periosteum of the skull since the areolar tissue between the two is loose.

Nerve supply

The frontal bellies are supplied by temporal branches, and the occipital bellies by posterior auricular branches of the facial nerve.

Actions

The occipital bellies pull the scalp backwards, and the frontal bellies can move it forwards. On the other hand, the frontal bellies can act from the aponeurosis on the eyebrows and the skin over the root of the nose, which they pull upwards in the expressions of surprise, fright, and horror, so producing horizontal wrinkling of the forehead.

EXTRINSIC MUSCLES OF THE AURICLE

Three small muscles are inserted into the cartilages of the external ear. Auricularis posterior arises from the mastoid process; auricularis superior, the largest of the three, comes from the galea aponeurotica and the temporal fascia, and auricularis anterior, the smallest, comes from the temporal fascia [FIG. 5.26].

Nerve supply

From the temporal and posterior auricular branches of the facial nerve.

Actions

The muscles are not normally used, but can be trained to move the external ear slightly; this power is greater in some people than in others.

ORBICULARIS OCULI

Orbicularis oculi is one of the two most complex members of the group of facial muscles and is functionally extremely important. It consists of three parts. The **orbital part** surrounds the orbit and spreads on to the forehead, temple, and cheek [FIG. 5.26]. Its fibres are reddish in colour, and its fasciculi are thicker than those of the other parts. They take origin from the nasal part of the frontal bone and the frontal process of the maxilla [FIG. 5.27], as well as from the medial palpebral ligament. The fibres arising above this ligament pass round the margin of the orbit in an ellipse and return to the bony attachment below the ligament; there is no lateral attachment to bone.

The **palpebral part** of the muscle lies in the eyelids; it arises from the medial palpebral ligament and the adjacent bone, and runs

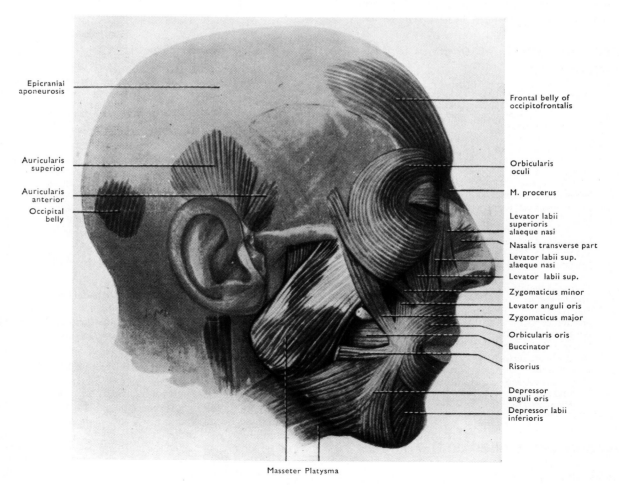

Epicranial aponeurosis

Auricularis superior

Auricularis anterior

Occipital belly

Frontal belly of occipitofrontalis

Orbicularis oculi

M. procerus

Levator labii superioris alaeque nasi

Nasalis transverse part

Levator labii sup. alaeque nasi

Levator labii sup.

Zygomaticus minor

Levator anguli oris

Zygomaticus major

Orbicularis oris

Buccinator

Risorius

Depressor anguli oris

Depressor labii inferioris

Masseter Platysma

FIG. 5.26. Muscles of face and scalp.

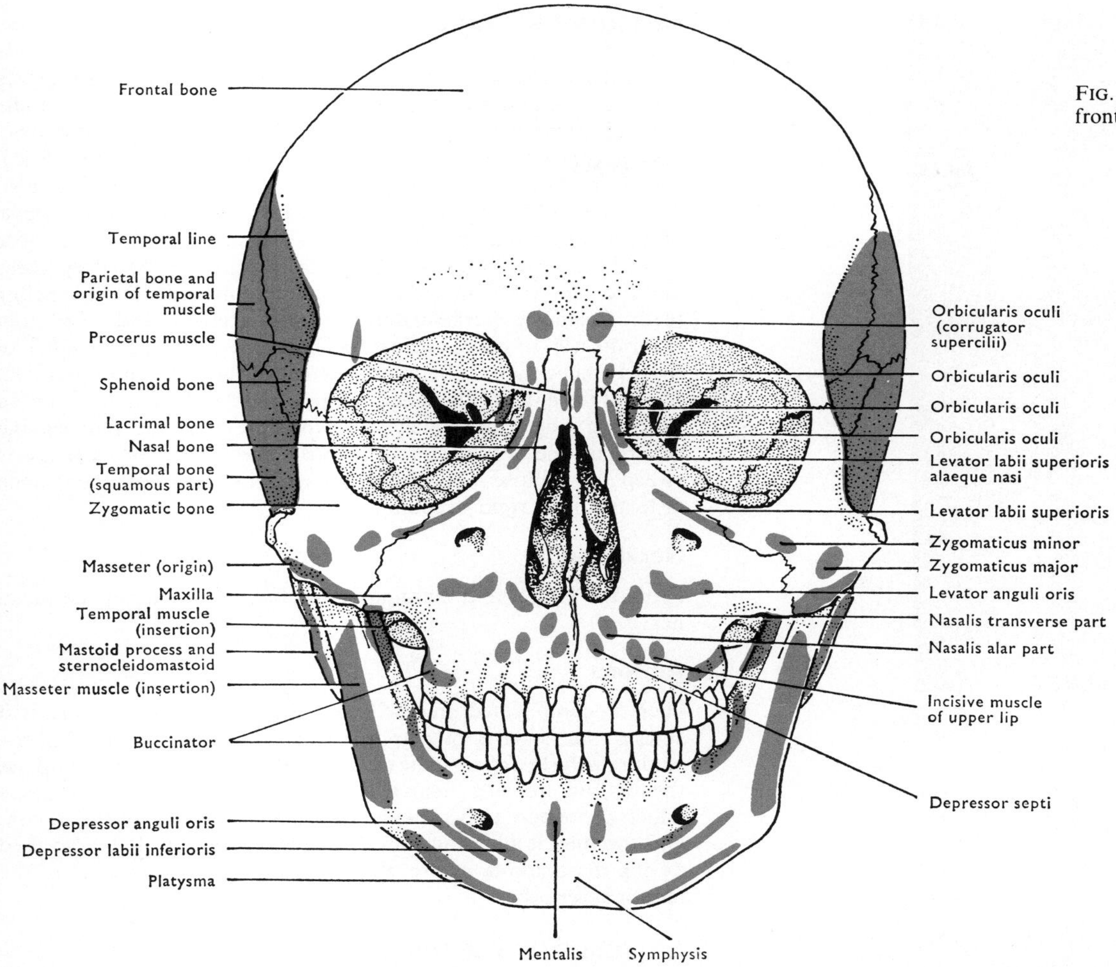
Frontal bone
Temporal line
Parietal bone and origin of temporal muscle
Procerus muscle
Sphenoid bone
Lacrimal bone
Nasal bone
Temporal bone (squamous part)
Zygomatic bone
Masseter (origin)
Maxilla
Temporal muscle (insertion)
Mastoid process and sternocleidomastoid
Masseter muscle (insertion)
Buccinator
Depressor anguli oris
Depressor labii inferioris
Platysma
Mentalis
Symphysis
Orbicularis oculi (corrugator supercilii)
Orbicularis oculi
Orbicularis oculi
Orbicularis oculi
Levator labii superioris alaeque nasi
Levator labii superioris
Zygomaticus minor
Zygomaticus major
Levator anguli oris
Nasalis transverse part
Nasalis alar part
Incisive muscle of upper lip
Depressor septi

FIG. 5.27. Muscle attachments to the front of the skull.

laterally in the lids, where the fibres eventually interlace to form the **lateral palpebral raphe**. A slip of fine fibres called the **ciliary bundle** runs along the free margin of each lid behind the eyelashes [FIG. 12.29].

The **lacrimal part** of orbicularis oculi [FIG. 5.49] comes from the fascia posterior to the lacrimal sac and from the crest of the lacrimal bone, and passes laterally to be inserted into the tarsal plates of both eyelids. These fibres are closely related to the lacrimal canaliculi.

Nerve supply

From the temporal and zygomatic branches of the facial nerve.

Actions

The muscle is an important protective mechanism surrounding the eye, and plays a vital part in the disposal of the tears. The orbital part, which can operate either together with the other parts or independently, draws the skin of the forehead and the cheek towards the medial angle of the orbit. By so doing it helps to close the eyelids firmly and 'screws up' the eye. It thus helps to prevent excess light entering the eye, and guards against injury to the eyeball. The contraction of this part of the muscle wrinkles the skin at the lateral margin of the orbit and these wrinkles become permanent and noticeable in old age, particularly in those whose occupation has constantly exposed them to the weather.

The palpebral part of orbicularis oculi closes the lids gently, and comes into action involuntarily in blinking. By drawing the eyelids medially the tears are wiped across the eyeball to the lacrimal puncta, which are held firmly to the conjunctival sac by the contraction of the muscle. Paralysis of orbicularis oculi causes the punctum of the lower lid to fall away from the eyeball, and tears overflow from the conjunctival sac on to the cheek.

Finally, the lacrimal part of the muscle pulls the eyelids medially and dilates the lacrimal sac.

CORRUGATOR SUPERCILII

This small muscle arises from the medial end of the superciliary arch, blending with orbicularis oculi, and passes upwards and laterally to be inserted into the skin of the eyebrow.

Nerve supply

From the facial nerve.

Actions

It draws the eyebrow medially and downwards, so producing vertical wrinkles in the forehead; this action is well seen in frowning and is used to protect the eyes from glare.

PROCERUS

The two procerus muscles form a continuous unit, which arises from the lower part of the nasal bones and the upper parts of the lateral nasal cartilages. The fibres insert into the skin over the lower part of the forehead, interlacing with the frontal belly of epicranius.

Nerve supply

From the facial nerve.

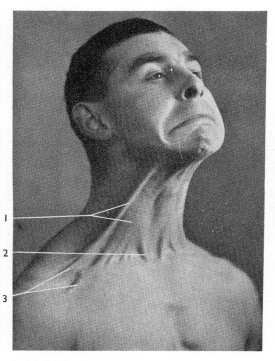

FIG. 5.28. The platysma in action.
(*By courtesy of Professor R. D. Lockhart.*)
1. Platysma.
2. Sternal head of sternocleido-mastoid.
3. Platysma.

Actions

Procerus pulls down the medial part of the eyebrow and so produces transverse wrinkles at the root of the nose.

NASALIS

Nasalis consists of a transverse and an alar portion. The transverse portion [FIGS. 5.26 and 5.27] runs from the upper end of the canine eminence of the maxilla to an aponeurosis on the cartilaginous part of the nose. The alar portion of the muscle [FIG. 5.27] comes from the maxilla above the lateral incisor tooth, and is inserted into the lateral part of the lower margin of the ala of the nose.

Nerve supply

From the facial nerve.

Actions

The alar part of the muscle draws the ala downwards and laterally, so widening the aperture of the nose: the transverse part has been thought to act as a compressor of the nose, but it is possible that both parts of the muscle are really dilators.

DEPRESSOR SEPTI

This muscle arises with the medial fibres of the alar part of nasalis, and is inserted into the mobile part of the nasal septum.

Nerve supply

From the facial nerve.

Actions

It draws the septum downwards, so narrowing the nostril.

ORBICULARIS ORIS

Orbicularis oris is a composite sphincter muscle which receives fasciculi from many other muscles. It encircles the mouth, its vertical extent being from the septum of the nose to a point midway between the chin and the free margin of the lower lip [FIG. 5.26]. The deepest fibres of the muscle are continuous with buccinator [p. 291], the middle fibres of which decussate posterolateral to the angle of the mouth so that the lower fibres of buccinator run into the upper lip and the upper fibres into the lower lip. Some of these deep fibres are attached to the maxilla and mandible near the lateral incisor teeth. The most superficial fibres of orbicularis oris are derived from other muscles of facial expression, and some of them decussate like those of buccinator. The interlacement of the deep and superficial fibres at the angle of the mouth forms a nodule which is easily felt in the living subject and which is known as the **modiolus**. Between the superficial and deep layers the intrinsic fibres of the orbicularis oris sweep in an ellipse round the mouth without being attached to bone or to any other rigid structure.

Nerve supply

From the buccal and marginal mandibular branches of the facial nerve.

Actions

Such a complicated muscle must inevitably have complicated actions. Orbicularis oris approximates the lips; its deep fibres compress the lips against the teeth, and its superficial fibres compress the lips and protrude them. Contraction of the various components which contribute to orbicularis oris alters the shape of the mouth, and the muscle plays an important part in speech. It is also used during mastication, to prevent food from being forced out of the mouth anteriorly.

MUSCLES OF THE UPPER LIP

Levator labii superioris alaeque nasi arises from the frontal process of the maxilla and runs down to be inserted into the ala of the nose and into the skin and muscle of the upper lip [FIGS. 5.26 and 5.27]. **Levator labii superioris** takes origin from the maxilla immediately superior to the infraorbital foramen, and is inserted into orbicularis oris in the upper lip.

    Zygomaticus minor is a small bundle which comes from the anterior surface of the zygomatic bone to be inserted into the skin and muscle of the upper lip. **Zygomaticus major**, somewhat larger, comes from the zygomatic portion of the zygomatic arch and passes to the angle of the mouth, where it is inserted partly into the skin and partly into the obicularis oris. **Levator anguli oris** comes from the maxilla inferior to the infraorbital foramen. It passes downwards and laterally to be inserted into orbicularis oris and the skin at the angle of the mouth; some of its fibres pass into orbicularis oris in the lower lip.

Nerve supply

All the muscles of the upper lip are supplied by branches of the facial nerve.

Actions

The two elevators of the upper lip raise and evert the lip; levator labii superioris alaeque nasi also acts as a dilator of the nostril by means of its attachment to the ala of the nose. Zygomaticus minor assists in raising the upper lip, and, together with levator labii superioris and levator anguli oris, is responsible for producing the nasolabial furrow, which passes from the side of the nose to the upper lip and, like all facial creases, becomes more marked in old age. Zygomaticus major raises the angle of the mouth and pulls it laterally, as in laughing, and in this it is assisted by levator anguli oris.

MUSCLES OF THE LOWER LIP

Depressor labii inferioris [FIGS. 5.26 and 5.27] arises from an oblique line on the outer surface of the mandible between the symphysis menti and the mental foramen. Its fibres pass up and medially to the skin of the lower lip and orbicularis oris, blending with those of its fellow of the opposite side. Depressor anguli oris takes origin just inferolateral to the depressor of the lower lip. Its fibres converge on the angle of the mouth to be inserted into the skin and orbicularis oris; some of them pass into the upper lip. Both these muscles are continuous with platysma and depressor anguli oris overlaps depressor labii inferioris.

Mentalis is a small muscle which arises from the mandible just below the incisor teeth and passes downwards to be inserted into the skin of the chin. It is the only muscle of the lips which usually has no connection with the orbicularis oris. Risorius is a thin muscle which passes from the fascia covering the parotid gland to the skin at the angle of the mouth: it is often completely fused with platysma.

Nerve supply

All the muscles of the lower lip are supplied by the marginal mandibular branch of the facial nerve.

Actions

Depressor labii inferioris and depressor anguli oris pull the mouth downwards and somewhat laterally. Mentalis pulls the skin of the chin upwards, and so causes a protrusion of the lower lip as in drinking or pouting. Risorius pulls the angle of the mouth laterally.

PLATYSMA

Platysma, like the muscles of the lips, lies in the superficial fascia, and it is very variably developed in different people. It arises from the skin and superficial fascia in the upper pectoral and anterior deltoid regions, crosses the clavicle and runs upwards over the front and side of the neck. The anterior fibres of the muscle decussate with those of the opposite side below the chin, while the intermediate fibres are inserted into the lower border of the body of the mandible. The posterior fibres cross the mandible superficially and sweep forwards to blend with depressor anguli oris and depressor labii inferioris, so becoming inserted into the modiolus and the upper lip [FIGS. 5.26 and 5.27]. The uppermost fibres are reinforced by risorius, with which they are continuous.

Nerve supply

From the cervical branch of the facial nerve.

Actions

Platysma may be seen in action in a runner finishing a race. In such circumstances it is conceivably being used as an accessory muscle of inspiration, helping to draw up the chest wall to the fixed mandible, though its utility in this respect can only be very slight. An alternative theory is that it prevents the soft tissues at the root of the neck from being sucked inwards by the violent inspiratory efforts; in this way it guards against compression of the great veins returning blood to the heart. The muscle wrinkles the skin of the neck obliquely, and can serve to depress the mandible against resistance; it also depresses the lower lip and the angle of the mouth. If the cervical branch of the facial nerve is cut, the resulting paralysis of the platysma causes the skin to fall away from the neck in slack folds.

BUCCINATOR

Buccinator [FIG. 5.35] forms the substance of the cheek, and lies in the same plane as the superior constrictor of the pharynx, being covered by an extension forwards of the buccopharyngeal fascia [p. 309]. It takes origin from the outer surfaces of both the maxilla and the mandible opposite the molar teeth [FIGS. 5.27 and 5.33], and from a fibrous line of union with the superior constrictor of the pharynx, the pterygomandibular raphe, which stretches from the pterygoid hamulus to the posterior end of the mylohyoid line.

The fibres of buccinator proceed forwards to blend with those of orbicularis oris almost at right angles. The intermediate fibres decussate at the modiolus, but the upper and lower ones pass into the upper and lower lips respectively. Posteriorly, buccinator is covered superficially by the masseter muscle [FIG. 5.26], from which it is separated by the buccal pad of fat. It is pierced by the duct of the parotid gland, and on its inner surface it is covered by the mucous membrane and buccal glands of the vestibule of the mouth.

Nerve supply

From buccal branches of the facial nerve.

Actions

Buccinator is named from the Latin *buccina* (a trumpet) because if the cheeks are filled with air it is buccinator which compresses the air through the lips as in whistling or blowing a trumpet. More important, however, is its use during mastication [p. 305], when it forces food escaping into the vestibule of the mouth back into the sphere of action of the teeth and the tongue.

If the lips are protruded by orbicularis oris, the buccinators cause the cheeks to 'cave in', producing the action of sucking. They can be used to pull orbicularis oris backwards against the teeth, or to retract the angles of the mouth to show the premolar and molar teeth.

Facial expression and facial paralysis

The individual actions of the muscles described are combined together to produce emotional expressions. Joy results from the contraction of one set of muscles and grief from the contraction of an opposing set. Determination is accompanied by a 'rigid' expression, and despair is expressed by a general relaxation of the facial musculature. The details of the muscular expression of individual emotions are too complex to discuss here, but a full account may be found in the works of Bell (1847), Darwin (1872), and Huber (1931). It is worth noting that different individuals may use different patterns of muscular activity to produce similar results (Isley and Basmajian 1973).

The part of the facial musculature surrounding the eye has the additional important task of protecting the eyeball and disposing of the tears [p. 289]. The muscles connected to the nose take part in contraction and dilation of the nostrils. The alae nasi do not move during quiet respiration, but when there is any obstruction to inspiration there is a visible movement of elevation and lateral expansion, due to the action of the dilators of the nose. If these muscles are weakened, as in terminal illnesses, the alae nasi are drawn inwards with each inspiration, giving to the nose a characteristic pointed appearance.

The functions of the facial muscles are perhaps best illustrated by what happens if they are paralysed. In the common form of peripheral paralysis of the facial nerve (Bell's palsy), the mouth is pulled to the opposite side by the unopposed action of the intact muscles. On the paralysed side, the lines of the face are smoothed out, due to the lack of tone in the muscles, and the face remains expressionless. The affected eye cannot be closed firmly, and the lower lid falls away from the eyeball under the pull of gravity. This results in the tears overflowing down the cheek because the lower lacrimal punctum is not functioning. Food and saliva cannot be retained by the affected cheek and angle of the mouth, and they escape through the lips on that side. Speech is usually only slightly affected, but the patient cannot whistle or show his teeth properly—the attempt to do so results in the mouth being pulled further to the healthy side. Nor can he wrinkle the affected side of his forehead on demand.

MUSCLES OF THE ORBIT

LEVATOR PALPEBRAE SUPERIORIS

This muscle arises from the roof of the orbit just in front of the optic canal. As it passes forwards it expands to form a triangular aponeurosis which covers over the superior rectus muscle [FIGS. 5.30 and 5.31]. The aponeurosis is loosely attached to the midpoints of the medial and lateral orbital margins, and between these two points passes into the upper lid. Here it divides into three; the most anterior and superficial part runs down into the eyelid to become attached to the front of the tarsus and also into the skin of the eyelid, blending with the orbicularis oculi. The middle part of the aponeurosis, which contains a layer of visceral muscle fibres, is attached to the upper margin of the tarsus, and the deepest of the three layers is attached to the superior fornix of the conjunctiva. The levator palpebrae superioris is unusual in containing both somatic and visceral muscle fibres; a similar condition is found in the middle third of the oesophagus [p. 438].

Nerve supply

The somatic portion of the muscle is supplied by the superior division of the oculomotor nerve. The visceral component is supplied by the sympathetic system.

Actions

The muscle elevates the upper eyelid, and its attachments ensure that the skin folds up and the conjunctiva is pulled out of the way as the lid is raised. The movement of blinking is normally reflex and involuntary, but the muscle is used under the control of the will when the eyes are raised to look at an object above the head. Levator

palpebrae superioris is the antagonist of the palpebral part of the orbicularis oculi, and paralysis of the smooth muscle in the levator results in the upper lid drooping down over the eyeball, a condition known as **ptosis**. Total paralysis closes the eye.

RECTUS MUSCLES

The four rectus muscles—superior, inferior, medial, and lateral—all take origin from a common tendinous ring [FIG. 11.12] which surrounds the opening of the optic canal and bridges the superior orbital fissure. Inside this ring lie the optic nerve and ophthalmic artery, emerging from the canal, and, passing through the medial part of the superior orbital fissure, the two divisions of the oculomotor nerve, the nasociliary nerve, and the abducent nerve.

Inferior rectus comes from the lowest part of the ring, medial rectus from its medial aspect, and superior rectus from its upper part. Lateral rectus is sometimes said to have two heads, but there is no separation between them, and the muscle arises from the lateral and inferior parts of the tendinous ring, as it crosses the fissure, and from the adjacent orbital surfaces of the greater and lesser wings of the sphenoid bone. All four muscles form flattened bands in the fat round the optic nerve and the eyeball [FIGS. 5.29–5.32] and end in short tendons which pierce the fascial sheath of the eyeball [p. 309] to be inserted into the sclera some 6 to 8 mm behind the margin of the cornea. The superior and inferior rectus muscles are inserted slightly medial to the vertical axis of the eyeball, but the lateral and medial rectus muscles are inserted exactly in its horizontal axis; all are attached in front of the equator of the eye.

Nerve supply

Superior rectus is supplied by the superior division of the oculomotor nerve, and the medial and inferior recti by its inferior division. Lateral rectus is supplied by the abducent nerve.

Actions

The medial and lateral rectus muscles turn the eyeball so that the cornea looks medially or laterally respectively. The movement is entirely in the horizontal plane. Superior and inferior recti turn the eyeball so as to make the cornea look upwards and downwards respectively, but because the pull of both muscles is on the medial side of the vertical axis of the eyeball, the cornea is also deviated medially [p. 294].

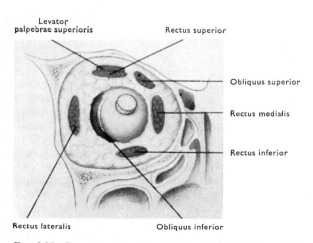

FIG. 5.29. Coronal section through left orbit behind the eyeball to show arrangement of muscles.

Labels: Levator palpebrae superioris; Rectus superior; Obliquus superior; Rectus medialis; Rectus inferior; Rectus lateralis; Obliquus inferior.

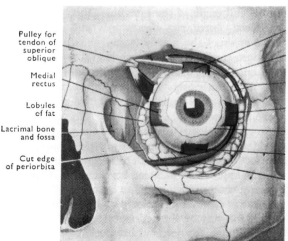

FIG. 5.30. Dissection of left orbit from the front.

Labels: Pulley for tendon of superior oblique; Medial rectus; Lobules of fat; Lacrimal bone and fossa; Cut edge of periorbita; Superior rectus; Cut edge of levator palpebrae superioris; Lacrimal gland; Lateral rectus; Cut edge of fascial sheath of eyeball; Inferior rectus; Inferior oblique.

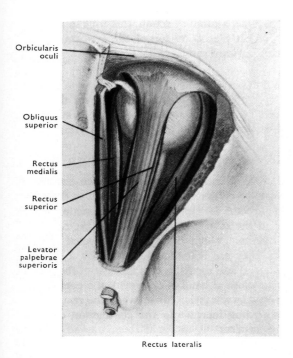

FIG. 5.31. Muscles of right orbit (from above).

SUPERIOR OBLIQUE

Superior oblique arises close to the margin of the optic canal above the origin of medial rectus. It is a fusiform muscle which passes forwards along the medial wall of the orbit superior to medial rectus. The narrow tendon of the muscle passes through a special fibrocartilaginous pulley called the **trochlea**, which is attached to the trochlear fossa [FIG. 3.75]. As it does so, it is enclosed in a synovial sheath. The trochlea changes the direction of pull of the superior oblique, the tendon of which now passes backwards and laterally between the tendon of the superior rectus and the eyeball, and becomes inserted into the sclera between superior rectus and lateral rectus, behind the equator of the eye [FIG. 5.32].

Nerve supply

From the trochlear nerve.

Actions

When the eye is looking straight forwards superior oblique turns it so that the cornea looks downwards and laterally. But when the cornea is already deviated medially, the line of pull of superior oblique is straightened out, so that in this position it acts as a pure depressor of the cornea. Conversely, when the cornea is already deviated laterally, the line of pull of superior oblique comes into a more transverse position in relation to the axis of the eye, and in this position it acts as an inward rotator; that is, it turns the eyeball so that an object placed on top of it would fall off towards the nose.

INFERIOR OBLIQUE

Unlike all the other orbital muscles, inferior oblique takes origin anteriorly from the floor of the orbit immediately lateral to the nasolacrimal canal. The muscle passes laterally and backwards inferior to inferior rectus [FIG. 5.30], and curves up, deep to lateral rectus, to be inserted between the superior and lateral rectus muscles behind the equator of the eyeball [FIG. 5.32].

Nerve supply

From the inferior division of the oculomotor nerve.

Actions

Inferior oblique turns the eye so that the cornea deviates upwards and laterally. But when the cornea is already deviated medially the line of pull of the muscle is such that it acts as a pure elevator, and when the cornea is already deviated laterally it has, like superior oblique, a rotatory action, this time in a direction such that an object placed on top of the eye would fall off towards the ear (outward rotation).

ORBITALIS

A scanty layer of visceral muscle bridges over the inferior orbital fissure and the infraorbital groove; it is supplied by the sympathetic system but its actions are obscure.

Movements of the eyes

The movements of both eyes are intimately co-ordinated by their nerve supply. When the gaze is directed to the right or left, the medial rectus of one eye acts in conjunction with the lateral rectus

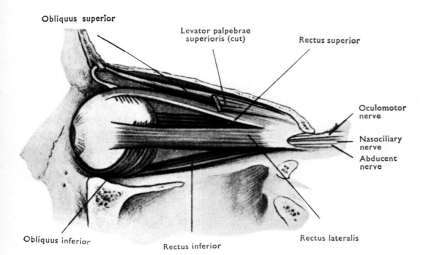

FIG. 5.32. Muscles of left orbit (from lateral aspect).

of the other; such movements are called **conjugate deviations**. When the eyes are directed towards a distant object, they are maintained parallel by a balance of power in the lateral and medial rectus muscles of the two eyes. When an object close at hand is examined, the eyes are made to **converge** on it by relaxation of the two lateral rectus muscles and a corresponding increase in the contraction of the two medial rectus muscles.

Movements in the vertical plane are more complicated. The cornea of each eye is turned straight downwards by contraction of inferior rectus and superior oblique, superior oblique being able to cancel out the unwanted medial deviation produced by inferior rectus. Movement of the cornea straight upwards is effected by simultaneous contraction of superior rectus and inferior oblique. In addition, the upper eyelid is raised by levator palpebrae superioris. If this is insufficient to clear the line of gaze, the skin of the eyebrows is raised by epicranius; still further elevation of the gaze is brought about by extension of the head on the neck.

When the cornea is deviated medially, its elevation by superior rectus becomes less efficient, since the line of action of the muscle now lies at a considerable angle to the optic axis, and it tends to produce inward rotation of the eyeball on this axis while increasing the medial deviation of the cornea. Accordingly, as the cornea moves medially, so the inferior oblique muscle, whose direction of pull now comes to lie more nearly in the optic axis, is found to be progressively more important in the movement of elevation of the cornea.

Exactly the converse happens if the cornea is deviated laterally. Superior rectus is now pulling more directly in the line of the axis of the eye and can effect a straight elevation of the cornea. Inferior oblique, on the other hand, is now pulling in a line which has a component at right angles to the axis of the eye, and its rotatory effect becomes more pronounced as its capacity to elevate the cornea weakens.

Similar considerations apply to the movement which depresses the cornea. When the cornea is medially deviated, depression in the vertical plane is largely due to the contraction of superior oblique, and inferior rectus becomes less important. Again, when the cornea is deviated laterally, it is inferior rectus which takes the major part in the movement.

Further complications are introduced by the contraction of two adjacent rectus muscles simultaneously. For example, lateral rectus can help to cancel out some of the unwanted medial deviation produced by the superior and inferior rectus muscles.

These factors are important when considering the origin of a squint. For example, if the superior oblique muscle is paralysed, the patient will still be able to turn the cornea downwards and laterally, by using his lateral and inferior rectus muscles in conjunction. But if the cornea is deviated medially and the patient is then asked to depress it, double vision at once appears, since in this position superior oblique is the only muscle which can satisfactorily depress the cornea.

MUSCLES OF MASTICATION

MASSETER

Masseter [FIG. 5.26], the most superficial of the muscles of mastication, is easily felt when the jaw is clenched. It is a flat muscle which stretches from its origin on the lower border and deep surface of the zygomatic arch downwards to the lateral surfaces of the coronoid process, ramus, and angle of the mandible [FIG. 5.33]. The superficial fibres run downwards and backwards to the angle of the mandible, and the deeper fibres run more vertically, being separated from the superficial ones by a cleft posteriorly, though they blend with them anteriorly. The deepest fasciculi of the muscle often partly unite with the subjacent fasciculi of the temporalis.

On its lateral surface the duct of the parotid gland runs forwards, and when the muscle is contracted the duct can be rolled under the finger as it turns inwards around the anterior border of masseter to pierce buccinator. The lateral surface of masseter is covered posteriorly by the parotid gland, further forward by the accessory parotid gland and zygomaticus major, and inferiorly by risorius and platysma. Deep to masseter lie the ramus and notch of the mandible and anteriorly buccinator is separated from it by the buccal pad of fat.

Nerve supply

From the anterior division of the mandibular nerve.

Actions

Masseter elevates the mandible, and its superficial fibres help to pull it forwards.

TEMPORALIS

This strong fan-shaped muscle arises from the temporal fossa and from the temporal fascia [p. 309] which covers it. The fasciculi in the anterior part of the muscle run more or less vertically, while those in the more posterior part are almost horizontal [FIG. 5.34].

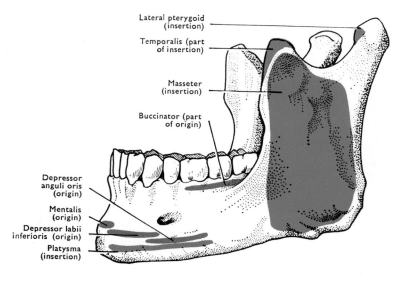

FIG. 5.33. Muscle attachments to superficial surface of mandible.

Lateral pterygoid (insertion)

Temporalis (part of insertion)

Masseter (insertion)

Buccinator (part of origin)

Depressor anguli oris (origin)

Mentalis (origin)

Depressor labii inferioris (origin)

Platysma (insertion)

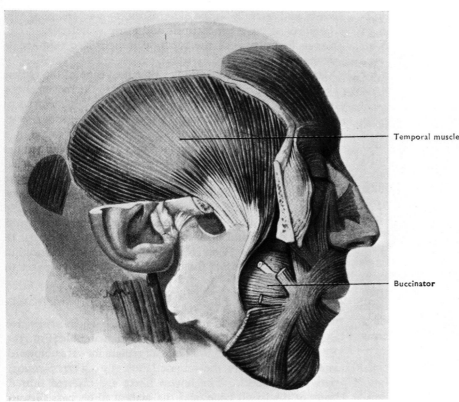

FIG. 5.34. Right temporal muscle. (The zygomatic arch and the masseter muscle have been removed.)

Temporal muscle

Buccinator

They all converge on a thick tendon which passes medial to the zygomatic arch and is inserted into the apex and the deep surface of the coronoid process and into the anterior border of the ramus of the mandible [FIGS. 5.23 and 5.33]. The fibres reach down almost as far as the last molar tooth, and some of them become continuous with buccinator. Others, in the more superficial fasciculi, fuse with masseter.

Nerve supply

By the deep temporal nerves from the anterior division of the mandibular nerve.

Actions

The vertical fibres elevate the mandible and close the mouth: they are in constant postural contraction to counteract the effect of gravity. The posterior fibres pull the mandible backwards after it has been protruded (MacDougall and Andrew 1953). This action is inportant since the head of the mandible passes forwards on to the articular eminence when the mouth is opened or the mandible protruded, and has to be retracted again when the mouth is closed [p. 226].

LATERAL PTERYGOID

Lateral pterygoid has two heads; the upper one arises from the inferior surface of the greater wing of the sphenoid bone, and the lower one from the lateral surface of the lateral pterygoid plate. From this double origin the fibres pass backwards and slightly laterally to be inserted into the front of the neck of the mandible, and into the capsule and the disc of the temporomandibular joint [FIGS. 5.23 and 5.35]. Between the two heads of the muscle pass the buccal nerve and the maxillary artery, and laterally it is covered by the ramus of the mandible, temporalis, and masseter.

Nerve supply

From the anterior division of the mandibular nerve.

Actions

The muscle pulls the head of the mandible, the disc, and the capsule of the temporomandibular joint forwards on to the articular tubercle [p. 116]. This movement occurs during: (1) protrusion of the jaw, and (2) opening of the mouth: in the latter case the head of the mandible also rotates on the articular disc. Protrusion occurs when the elevators of the jaw are in action at the same time as the lateral pterygoids, the lower incisors being carried forwards in front of the upper ones [p. 226]. When the elevators are not in action, the lateral pterygoids carry the head of the mandible forwards and rotate it on the disc so that the chin moves downwards and the mouth is opened. Simultaneous contraction of the lateral and medial pterygoid muscles of one side draws the head of the mandible forwards on that side, the depressing action of the lateral pterygoid being cancelled out by the elevating action of medial pterygoid. The result is to carry the chin towards the opposite side, so producing a grinding movement between the teeth. Alternate contraction of these muscles, first on one side and then on the other, is an important element of chewing.

Lateral pterygoid is the chief antagonist to the movement of retraction of the jaw (carried out by temporalis), and pays out to control it.

MEDIAL PTERYGOID

The main head of medial pterygoid arises from the medial surface of the lateral pterygoid plate and from the pyramidal process of the palatine bone. A much smaller head comes from the tuber of the maxilla. From these two origins, which embrace the inferior fibres of the lateral pterygoid [FIG. 5.35], the fasciculi of the muscle pass

Temporal muscle (reflected)

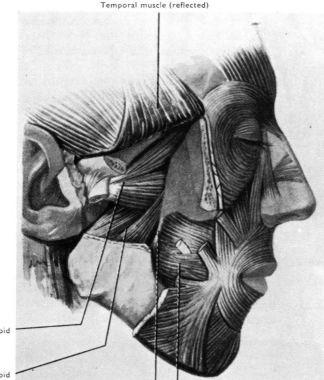

FIG. 5.35. Pterygoid muscles of the right side.

Lateral pterygoid

Medial pterygoid

Pterygomandibular raphe   Buccinator

downwards, backwards, and laterally to be inserted into a triangular rough impression on the medial surface of the mandible between the mylohyoid groove and the angle [FIG. 5.23].

Nerve supply

From the main trunk of the mandibular nerve.

Actions

Medial pterygoid elevates and protrudes the mandible. Acting with lateral pterygoid of the same side, medial pterygoid protrudes the mandible and swings the chin to the opposite side. Alternation of such movements is important in chewing.

MUSCLES OF THE TONGUE

The substance of the tongue consists mainly of somatic muscle. Some of the muscles which compose the tongue lie entirely within the organ, and are therefore known as intrinsic muscles; others pass into the tongue from origins outside it, and are called extrinsic muscles. The tongue is actively concerned in mastication, swallowing, and speaking. The posterior part of the tongue is attached to, and moves with, the hyoid bone, on which it rests. The anterior part lies on the mylohyoid and geniohyoid muscles, and therefore moves with the floor of the mouth.

INTRINSIC MUSCLES

There are four pairs of intrinsic muscles [FIG. 5.36], arranged on either side of a midline, vertical, septum of fibrous tissue, to which many of their fibres are attached (Abd-el-Malik 1939). The superior longitudinal muscle lies under the mucous membrane of the dorsum of the tongue, and extends from the back of the tongue, close to the epiglottis, to the tip and edges of the tongue, some of its fibres passing into the mucosa. It takes origin in part from the median

fibrous septum. The inferior longitudinal muscle lies in the lower part of the tongue between the extrinsic muscles genioglossus and hyoglossus, and blends with the fibres of another extrinsic muscle, styloglossus.

The transverse muscle of the tongue arises from the median septum and radiates upwards to the dorsum and laterally to the sides of the tongue, where it inserts into the submucous tissue and mucous membrane. The vertical intrinsic muscle arises from the mucous membrane of the dorsum and passes downwards and laterally to the sides of the tongue, mingling with the fibres of the other extrinsic and intrinsic muscles.

Nerve supply

All the intrinsic muscles are supplied by the hypoglossal nerve.

Actions

The intrinsic muscles alter the shape of the tongue, and are of great importance in speech and in the movements of mastication and swallowing [p. 305]. The longitudinal muscles pull the tip backwards towards the base, and, because the volume of the tongue does not change, this causes it to broaden from side to side and increases its vertical dimensions. Similarly, the transverse muscle narrows the tongue from side to side, so causing it to lengthen and become vertically thicker at the same time. The vertical muscle flattens, lengthens, and broadens the tongue.

Contraction of the superior longitudinal muscles in isolation will turn the tip and sides of the tongue upwards to render the dorsum concave. Contraction of the inferior longitudinal muscles pulls the tip of the tongue downwards and makes the dorsum convex.

GENIOGLOSSUS

Genioglossus is an extrinsic muscle which arises from the upper part of the mental spine of the mandible [FIG. 5.23]. From this small

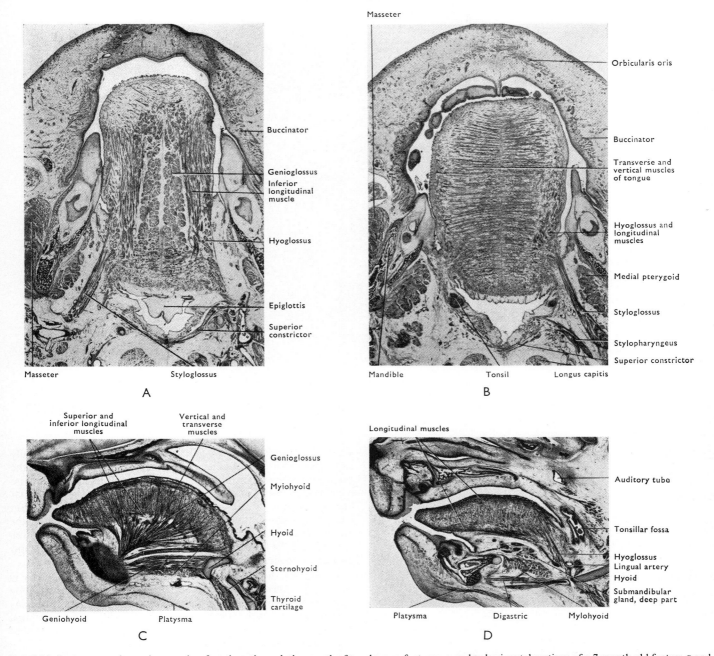

FIG. 5.36. Low-power photomicrographs of sections through the mouth of two human foetuses. A and B, horizontal sections of a 7-month-old foetus; C and D, sagittal sections of a 14 weeks-old-foetus.

origin, its fasciculi radiate forwards, upwards, and backwards into the tongue [FIGS. 5.24 and 5.36], forming a paramedian sheet. The lowest fibres are inserted into the body of the hyoid bone, and the highest fibres pass forwards into the tip of the tongue. The intermediate fibres permeate the substance of the tongue in its whole length. The two muscles are separated from each other by the median septum.

Nerve supply

From the hypoglossal nerve.

Actions

Like all the extrinsic muscles, genioglossus is concerned mainly in moving the tongue bodily about rather than in altering its shape. The anterior fibres of the muscle retract the tip of the tongue; the

posterior fibres protrude it. When the whole muscle contracts, the central part of the tongue is depressed. If the posterior fibres of genioglossus of one side act alone, the tip of the tongue is protruded and turned to the opposite side. Genioglossus is the only extrinsic muscle which can protrude the tongue, and its attachment to the mandible prevents the tongue from falling back and obstructing respiration. In an unconscious patient the airway can be kept clear by pulling the mandible forwards.

HYOGLOSSUS

This thin, quadrangular extrinsic muscle arises from the body and the greater horn of the hyoid bone, from which it passes upwards and forwards to be inserted into the side of the tongue, intermingling with the fibres of the inferior longitudinal muscle and styloglossus

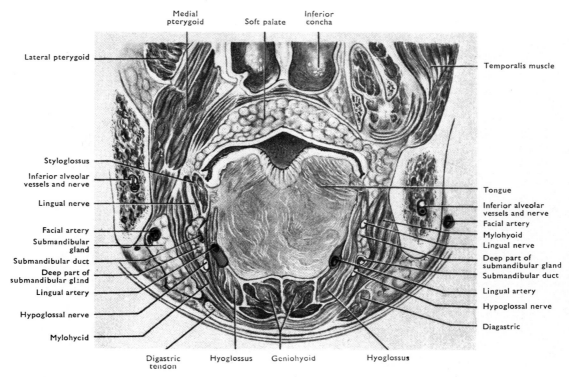

FIG. 5.37. Coronal section through the tongue and submandibular region behind the molar teeth.

[FIGS. 5.24, 5.36, and 5.37]. Sometimes a slip of hyoglossus arises from the lesser horn of the hyoid bone and is called the chondroglossus.

Nerve supply

From the hypoglossal nerve.

Actions

Hyoglossus pulls the side of the tongue downwards and retracts it. Acting with genioglossus, the muscle depresses the tongue; acting with the transverse intrinsic muscle it arches the tongue transversely.

STYLOGLOSSUS

This muscle arises mainly from the anterior border of the styloid process near its tip; some of the fibres come from the stylohyoid ligament. It passes forwards, medially, and downwards to be inserted into the side of the tongue, mingling with palatoglossus, hyoglossus, and the inferior longitudinal muscle [FIG. 5.24]. It lies superficial to stylopharyngeus, and also to the lower and anterior part of the superior constrictor.

Nerve supply

From the hypoglossal nerve.

Actions

Styloglossus retracts and elevates the tongue; acting with the vertical intrinsic muscle and genioglossus, it raises the sides of the tongue while leaving the centre depressed, so producing a transverse concavity. This movement is important in drinking [p. 306].

PALATOGLOSSUS

Palatoglossus, which also contributes fibres to the substance of the tongue, is considered along with the muscles of the soft palate [p. 302].

MUSCLES OF THE PHARYNX AND SOFT PALATE

The wall of the pharynx is a sandwich of muscle covered internally and externally with fascia. The outer layer of fascia is the buccopharyngeal fascia [p. 309], which extends forwards on to the buccinator muscle, and the inner layer is the pharyngobasilar fascia, which is thicker near the skull than elsewhere. The main bulk of the meat in the sandwich is provided by three constrictor muscles, and these are reinforced internally by a longitudinal layer consisting of the fibres of stylopharyngeus, salpingopharyngeus, and palatopharyngeus muscles.

In contrast, the soft palate is a sandwich with the fibrous tissue in the middle, surrounded by several muscles. The fibrous palatine aponeurosis is attached in front to the posterior border of the bony palate, and on each side to the pharyngobasilar fascia. At first horizontal, the soft palate hangs downwards as it is traced backwards, separating the pharynx from the mouth.

CONSTRICTOR MUSCLES

Superior constrictor takes origin from the lower half of the posterior border of the medial pterygoid plate, the pterygoid hamulus, the pterygomandibular raphe, the mylohyoid line of the mandible, and the musculature of the posterior part of the side of the tongue [FIG. 5.24]. The pterygomandibular raphe lies between it and buccinator, and is a fibrous line of junction between the two muscles [p. 291].

The fibres of superior constrictor curve backwards around the pharynx, to unite with their fellows in a median raphe in the posterior wall of the pharynx. The upper fibres curve upwards to reach the pharyngeal tubercle of the occipital bone [FIGS. 5.17 and 5.38], while the lowest fibres curve slightly downwards deep to middle constrictor.

On the inner surface of superior constrictor near its upper border there is a horizontal band of muscle fibres which elevates the mucous

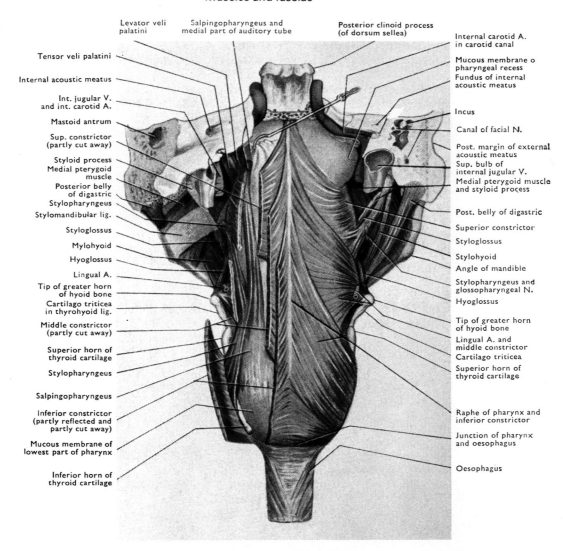

FIG. 5.38. Muscles of pharynx from behind. The constrictors of the left side have been cut to show the deeper muscles.

membrane to form a ridge. These fibres arise from the palatine aponeurosis and sweep backwards to surround the pharynx at the junction of its nasal part and its oral part: they are sometimes referred to as the **palatopharyngeal sphincter** (Whillis 1930), which is considerably enlarged in cases of complete cleft palate.

The curved upper border of superior constrictor is separated from the base of the skull by an interval occupied by the pharyngobasilar fascia, and through this interval pass the auditory tube and levator veli palatini muscle [FIGS. 5.40 and 5.41]. Posteriorly and laterally superior constrictor is overlapped inferiorly by middle constrictor. Between the lower border of superior constrictor, the upper border of middle constrictor, and the posterior border of hyoglossus is another muscular gap filled with fascia, through which pass stylopharyngeus, the glossopharyngeal nerve, and the stylohyoid ligament [FIG. 5.41].

The triangular **middle constrictor** [FIGS. 5.38 and 5.41] arises from the lower part of the stylohyoid ligament and from the greater and lesser horns of the hyoid bone. It curves backwards around the pharynx to be inserted into the median raphe of the pharynx, the upper (ascending) fibres overlapping superior constrictor and the lower (descending) fibres being overlapped posteriorly and laterally by inferior constrictor. Between the anterior part of the lower border of middle constrictor and the upper border of inferior constrictor

there is yet another muscular gap, triangular in shape, bounded by the thyrohyoid muscle in front and occupied by the posterior part of the thyrohyoid membrane and the lower part of the stylopharyngeus muscle [FIGS. 5.24 and 5.41]. Through the gap pass the superior laryngeal vessels and the internal branch of the superior laryngeal nerve.

Inferior constrictor [FIGS. 5.38 and 5.41] comes from the oblique line of the thyroid cartilage, from the side of the cricoid cartilage, and from the fascia on the lateral aspect of the cricothyroid muscle. Like those of the other constrictors, its upper fasciculi pass upwards and backwards to the median raphe of the pharynx, overlapping those of middle constrictor. The lowest fasciculi, which arise from the cricoid cartilage, run more horizontally, and blend inferiorly with the beginning of the circular muscle of the oesophagus. These fibres are sometimes classified as forming a separate **cricopharyngeus** muscle, and it has been claimed that this part of inferior constrictor has somewhat different functions from the remainder of the muscle (Batson 1955).

Nerve supply

All three constrictor muscles are supplied by the pharyngeal plexus: cricopharyngeus receives additional innervation from the recurrent laryngeal nerve, and sometimes also from the external laryngeal.

Actions

The constrictor muscles contract involuntarily during the process of swallowing. This contraction takes place sequentially from above downwards, and propels the bolus of food onwards into the oesophagus [see also p. 306]. Contraction of the palatopharyngeal sphincter helps to shut off the nasal part of the pharynx from the oral part during swallowing. The cricopharyngeus fibres of inferior constrictor are said to act as a sphincter preventing air from entering the oesophagus; they relax suddenly during swallowing.

STYLOPHARYNGEUS

Stylopharyngeus takes origin from the medial side of the root of the styloid process, and its fibres pass downwards, between the external and internal carotid arteries, to enter the wall of the pharynx between the superior and middle constrictor muscles [FIGS. 5.39 and 5.41]. Here the muscle is accompanied by the glossopharyngeal nerve winding round its lateral aspect. Stylopharyngeus then spreads out on the inner surface of middle constrictor to be inserted into the superior and posterior borders of the thyroid cartilage, the side of the epiglottis, and the musculature of the pharynx itself, becoming continuous with palatopharyngeus.

Nerve supply

From the glossopharyngeal nerve.

Actions

It elevates the larynx and helps to pull the wall of the pharynx up over the descending bolus of food in the process of swallowing [see also p. 306].

SALPINGOPHARYNGEUS

This small muscle [FIG. 5.38] comes from the lower part of the cartilage at the pharyngeal opening of the auditory tube. It passes downwards inside the constrictor muscles and blends with palatopharyngeus.

Nerve supply

From the pharyngeal plexus.

Actions

It raises the upper part of the lateral wall of the pharynx in swallowing, and at the same time pulls open the mouth of the auditory tube.

PALATOPHARYNGEUS

Palatopharyngeus [FIGS. 5.39 and 5.42] originates in the soft palate by two layers, between which lie levator veli palatini and the muscle of the uvula. The stronger of the two layers lies antero-inferiorly,

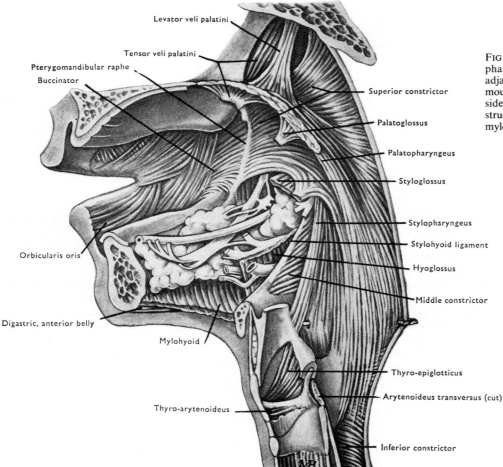

FIG. 5.39. Dissection of the constrictors of the pharynx and the associated structures which lie adjacent to the mucous membrane lining the mouth, pharynx, and larynx, from the medial side. The tongue has been removed to expose the structures which lie between it and the mylohyoid muscle.

between the levator veli palatini muscle and the tendon of tensor veli palatini, and it is attached to the palatine aponeurosis and the posterior border of the hard palate. The thin posterosuperior layer lies under the mucous membrane of the nasopharyngeal surface of the soft palate, joining with its fellow across the midline. The two layers blend at the lateral side of the soft palate and run downwards on the side of the pharynx, forming the substance of the palatopharyngeal arch. Later, the muscle is joined by the posterior part of salpingopharyngeus, and inserts partly on to the posterior border of the thyroid cartilage. Most of the fibres end by blending with the other muscles of the pharyngeal wall, and gaining insertion to the pharyngobasilar fascia; they decussate across the midline.

Nerve supply

From the pharyngeal plexus.

Actions

Palatopharyngeus pulls up the walls of the pharynx over the descending bolus during swallowing and also produces a constriction of the palatopharyngeal arch. This movement, assisted by the backward and upward movement of the tongue [p. 306] helps to shut off the pharynx from the mouth during swallowing. The pull of the vertical fibres of the muscle depresses the soft palate to separate the oral part of the pharynx from the mouth, so allowing breathing when the mouth is full: their insertion on the thyroid cartilage assists in raising the larynx.

TENSOR VELI PALATINI

This thin triangular muscle [Figs. 5.39, 5.41, and 5.42] has a linear origin on the posteromedial edge of the greater wing of the sphenoid bone from the scaphoid fossa to the spine of the sphenoid. It also arises from the adjacent cartilage of the auditory tube. Its fibres converge downwards on a tendon which changes direction abruptly

by hooking medially round the pterygoid hamulus; there is a small bursa here to facilitate its movement on the bone. The tendon then expands horizontally to form, with its fellow, the **palatine aponeurosis**, to which most of the other muscles of the soft palate are attached. The palatine aponeurosis is attached to the posterior border of the hard palate and to its under surface behind the palatine crest.

Nerve supply

From the mandibular nerve by the branch which supplies the medial pterygoid muscle.

Actions

The tensors tighten the soft palate and flatten its arch; this prevents the passage of a bolus of food from forcing the soft palate upwards into the nasal part of the pharynx. If one tensor is paralysed, the other pulls the palate over to its own side. The fibres attached to the auditory tubes may help to pull them open during swallowing.

LEVATOR VELI PALATINI

The elevator of the soft palate [Figs. 5.39, 5.40, and 5.41] arises from the rough area on the inferior surface of the temporal bone in front of the opening of the carotid canal. Other fibres come from the upper portion of the carotid sheath, and from the inferior aspect of the cartilaginous part of the auditory tube. The muscle descends medially above the upper border of superior constrictor to enter the soft palate between the two layers of palatopharyngeus. Its fibres are inserted into the palatine aponeurosis and blend with those of its fellow. It is separated from tensor veli palatini at its origin by the pharyngobasilar fascia.

Nerve supply

From the pharyngeal plexus.

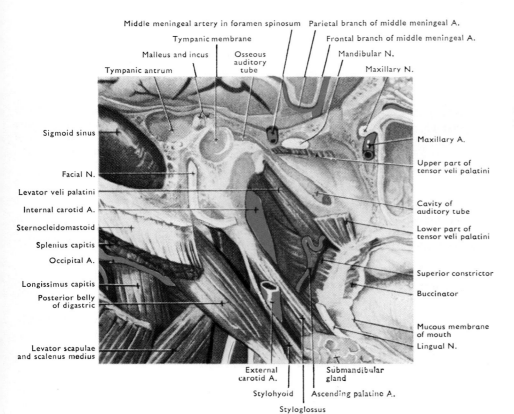

Middle meningeal artery in foramen spinosum — Parietal branch of middle meningeal A.

Tympanic membrane — Frontal branch of middle meningeal A.

Malleus and incus — Osseous auditory tube — Mandibular N.

Tympanic antrum — Maxillary N.

Sigmoid sinus

Facial N.

Levator veli palatini

Internal carotid A.

Sternocleidomastoid

Splenius capitis

Occipital A.

Longissimus capitis

Posterior belly of digastric

Levator scapulae and scalenus medius

Maxillary A.

Upper part of tensor veli palatini

Cavity of auditory tube

Lower part of tensor veli palatini

Superior constrictor

Buccinator

Mucous membrane of mouth

Lingual N.

External carotid A. — Submandibular gland

Stylohyoid — Ascending palatine A.

Styloglossus

FIG. 5.40. Region of auditory tube showing the levator and tensor veli palatini muscles. Enlargement of part of FIG. 5.25.

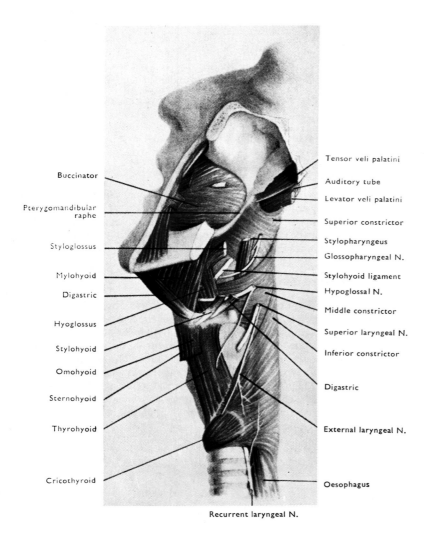

Buccinator

Pterygomandibular raphe

Styloglossus

Mylohyoid

Digastric

Hyoglossus

Stylohyoid

Omohyoid

Sternohyoid

Thyrohyoid

Cricothyroid

Tensor veli palatini

Auditory tube

Levator veli palatini

Superior constrictor

Stylopharyngeus

Glossopharyngeal N.

Stylohyoid ligament

Hypoglossal N.

Middle constrictor

Superior laryngeal N.

Inferior constrictor

Digastric

External laryngeal N.

Oesophagus

Recurrent laryngeal N.

FIG. 5.41. Lateral view of wall of pharynx.

Actions

The levators raise the soft palate and draw it backwards, and the fibres attached to the auditory tubes tend to open them during swallowing. The muscles co-operate with the palatopharyngeal sphincter [p. 299] to separate the nasal and oral parts of the pharynx.

PALATOGLOSSUS

This thin muscle [FIG. 5.42] arises from the under surface of the palatine aponeurosis, being continuous with its fellow of the opposite side, and then turns downwards, forming the substance of the palatoglossal arch, to be inserted into the dorsum and side of the tongue.

Nerve supply

From the pharyngeal plexus.

Actions

The palatoglossus muscles pull the palatoglossal arches towards the midline, and so narrow the isthmus of the fauces. In the movement of swallowing, this helps to close off the mouth cavity from the pharynx. They also assist in pulling upwards the posterior part of the tongue.

MUSCULUS UVULAE

The small muscle of the uvula [FIG. 5.42] is attached anteriorly to the posterior nasal spine, and is enveloped by the palatine aponeurosis, which splits to surround it. The muscle fibres travel backwards to end in the mucous membrane of the uvula.

Nerve supply

From the pharyngeal plexus.

Actions

The uvular muscle elevates the uvula in the movement of swallowing.

MUSCLES OF THE MIDDLE EAR

Two muscles operate on the auditory ossicles which cross the middle ear between the tympanic membrane and the fenestra vestibuli.

TENSOR TYMPANI

Originating from the upper surface of the cartilage of the auditory tube and the adjacent part of the greater wing of the sphenoid,

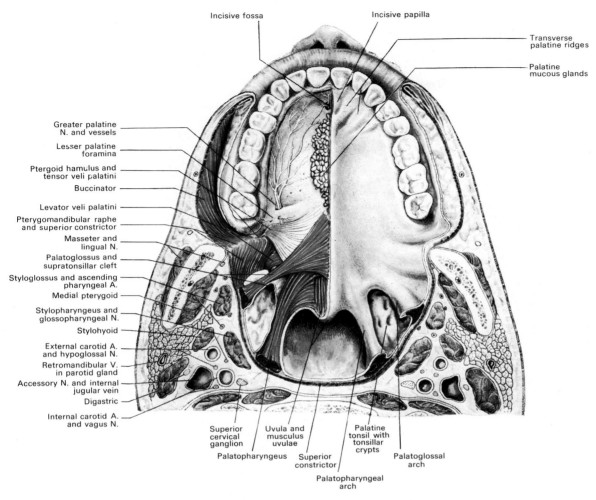

Incisive fossa

Incisive papilla

Transverse
palatine ridges

Palatine
mucous glands

Greater palatine
N. and vessels

Lesser palatine
foramina

Ptergoid hamulus and
tensor veli palatini

Buccinator

Levator veli palatini

Pterygomandibular raphe
and superior constrictor

Masseter and
lingual N.

Palatoglossus and
supratonsillar cleft

Styloglossus and ascending
pharyngeal A.

Medial pterygoid

Stylopharyngeus and
glossopharyngeal N.

Stylohyoid

External carotid A.
and hypoglossal N.

Retromandibular V.
in parotid gland

Accessory N. and internal
jugular vein

Digastric

Internal carotid A.
and vagus N.

Superior
cervical
ganglion

Uvula and
musculus
uvulae

Palatine
tonsil with
tonsillar
crypts

Palatoglossal
arch

Palatopharyngeus   Superior
constrictor

Palatopharyngeal
arch

FIG. 5.42. Horizontal section at the level of the oral fissure seen from
below. The right side has been dissected to show the muscles of the soft
palate.

tensor tympani runs backwards and laterally in a bony semicanal
superior to the bony part of the auditory tube; from this canal
additional fibres arise. The tendon, which emerges in the middle ear
above the auditory tube, turns laterally round the processus
cochleariformis and runs to be inserted into the handle of the
malleus, near its root.

Nerve supply

Like tensor veli palatini, tensor tympani is supplied by the branch
of the mandibular nerve to the medial pterygoid muscle.

Actions

The muscle draws the handle of the malleus towards the tympanic
cavity, so tensing the tympanic membrane.

STAPEDIUS

Stapedius is a tiny muscle which arises from the walls of a cavity in
the interior of the pyramidal eminence. The tendon emerges from
the apex of the eminence and is inserted directly into the posterior
surface of the neck of the stapes [FIG. 12.46].

Nerve supply

From the facial nerve.

Actions

Stapedius draws the stapes backwards, and tilts the anterior part of
the base of the bone towards the tympanic cavity, thus tightening
the anular ligament and reducing the amount of movement of the
footplate.

MUSCLES OF THE LARYNX

The larynx has a set of **extrinsic** muscles which move the apparatus
bodily about; these muscles have already been described
[pp. 283–6]. The **intrinsic** muscles of the larynx are those which alter
the length, tension, and distance apart of the vocal cords; in this
way they alter the size and shape of the space between the cords
(**rima glottidis**). Some of the intrinsic muscles also alter the size of
the inlet of the larynx. The intrinsic muscles are concerned in the
production of speech and in the regulation and protection of the air
inlet to the lungs.

With the exception of cricothyroid, the intrinsic muscles of the
larynx all lie between the cartilaginous framework and the mucous
membrane, and with the exception of the transverse arytenoid
muscle all the intrinsic muscles are paired. These muscles are also
mentioned in the section on the larynx [p. 504], where their actions
are further discussed, and the cartilages are described.

CRICOTHYROID

Cricothyroid [FIGS. 5.41 and 5.43] arises from the anterior and lateral aspects of the outer surface of the cricoid cartilage. The fibres radiate upwards and backwards to the lower border of the lamina and the inferior horn of the thyroid cartilage.

Nerve supply

From the external laryngeal branch of the superior laryngeal nerve.

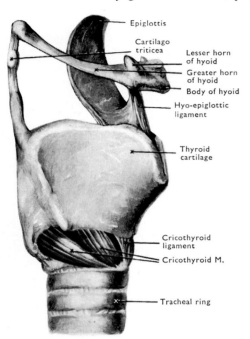

Epiglottis

Cartilago triticea

Lesser horn of hyoid

Greater horn of hyoid

Body of hyoid

Hyo-epiglottic ligament

Thyroid cartilage

Cricothyroid ligament

Cricothyroid M.

Tracheal ring

FIG. 5.43. Right cricothyroid muscle.

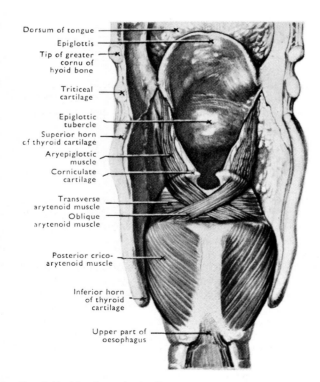

Dorsum of tongue

Epiglottis

Tip of greater cornu of hyoid bone

Triticeal cartilage

Epiglottic tubercle

Superior horn of thyroid cartilage

Aryepiglottic muscle

Corniculate cartilage

Transverse arytenoid muscle

Oblique arytenoid muscle

Posterior crico-arytenoid muscle

Inferior horn of thyroid cartilage

Upper part of oesophagus

FIG. 5.44. Muscles on back of larynx.

Actions

Cricothyroid is the chief tensor of the vocal cords. Contraction of the muscle pivots the thyroid and cricoid cartilages on each other, thus tilting back the upper border of the cricoid lamina. These movements separate the two attachments of the vocal ligaments (the vocal processes of the arytenoid cartilages and the angle of the thyroid cartilage), so elongating and tensing the ligaments.

POSTERIOR CRICO-ARYTENOID

This muscle arises from the back of the lamina of the cricoid cartilage [FIGS. 5.44 and 5.45] and its fibres run upwards and laterally to be inserted into the muscular process of the arytenoid cartilage. The highest fibres are virtually horizontal while the lowest ones are almost vertical.

Nerve supply

From the recurrent laryngeal branch of the vagus nerve.

Actions

It is the only abductor of the vocal folds, and it is interesting that the adductor muscles of the larynx are four times as heavy as the abductors (Bowden and Scheuer 1960). The more horizontal fibres rotate the arytenoid cartilages on their vertical axes, thus swinging the vocal processes outwards, and separating the posterior ends of the vocal folds. The more vertical fibres draw the arytenoids away from each other by pulling them downwards and laterally on the sloping upper border of the cricoid lamina. When the whole muscle is in action, abduction of the arytenoids by the vertical fibres is largely cancelled out by the tendency of the horizontal fibres to cause adduction: the net result is a pure rotation of the arytenoids around a vertical axis.

LATERAL CRICO-ARYTENOID

Arising from the upper border of the arch of the cricoid cartilage, the fibres of this muscle are inserted into the anterior aspect of the muscular process of the arytenoid cartilage [FIG. 5.45].

Nerve supply

From the recurrent laryngeal nerve.

Actions

It is one of the chief adductors of the vocal ligaments, since it rotates the arytenoid cartilage medially and approximates the vocal processes.

TRANSVERSE ARYTENOID

The only unpaired muscle of the larynx [FIG. 5.44] stretches between the posterior surfaces of the bodies and muscular processes of the two arytenoid cartilages.

Nerve supply

From the recurrent laryngeal nerve.

Actions

It pulls the arytenoid cartilages together and thus helps to close the posterior part of the opening of the glottis.

OBLIQUE ARYTENOID

Lying posterior to the transverse arytenoid, the oblique arytenoid crosses its colleague of the opposite side in a St. Andrew's cross

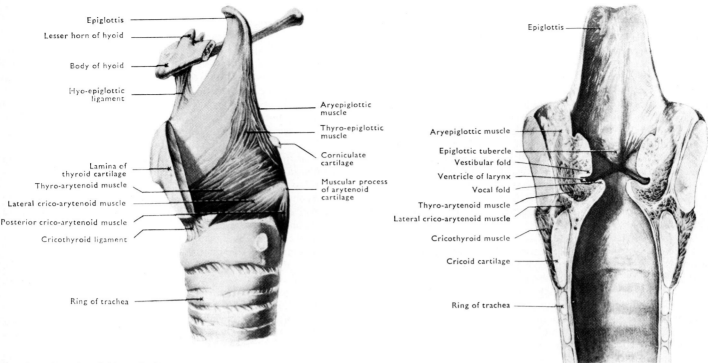

FIG. 5.45. Muscles of side-wall of larynx.

FIG. 5.46. Coronal section of larynx to show situation of muscles.

[FIG. 5.44], passing from the posterior aspect of the muscular process of one arytenoid cartilage to the apex of the cartilage of the opposite side. Some of the fibres run into the aryepiglottic fold, where they form the **aryepiglotticus** [FIGS. 5.44 and 5.45], and insert into the side of the epiglottis.

Nerve supply

From the recurrent laryngeal nerve.

Actions

The continuous fibres of the oblique arytenoid and the aryepiglottic muscle help to close the inlet of the larynx. They approximate the arytenoid cartilages, and pull them upwards towards the epiglottis, while at the same time they help to bend the epiglottis backwards over the laryngeal inlet.

THYRO-ARYTENOID

This flat muscle [FIG. 5.45] takes origin from the lower half of the internal surface of the thyroid cartilage at the angle, and also from the upper end of the cricothyroid ligament. Its fibres radiate upwards, backwards, and also horizontally. The lowest, horizontal fibres are attached to the lateral surface of the vocal process of the arytenoid cartilage, and the deepest members of this group, which are attached near the tip of the process, are called the **vocalis** muscle [FIG. 5.46]. This lies parallel with and just lateral to the vocal ligament, to which many of its fibres are attached.

The more superior fibres of thyro-arytenoid run into the aryepiglottic fold, and some of them reach the epiglottis, so forming **thyro-epiglotticus** [FIG. 5.45]. The rest merge with the entering fibres of the oblique arytenoid muscle to form aryepiglotticus.

Nerve supply

The whole composite mass is supplied by the recurrent laryngeal nerve.

Actions

The thyro-arytenoids pull the arytenoid cartilages forwards to the thyroid cartilage, so relaxing the vocal ligaments. They also rotate the arytenoid cartilages medially, approximating the vocal folds. The vocalis portion of the muscle is a fine adjustment on the tension of the vocal folds and is intimately concerned in vocalization.

Movements of mastication and swallowing

Food is bitten off by the incisor teeth, by means of the contraction of masseter, temporalis, and medial pterygoid. Once in the mouth, the food is then ground up by the premolar and molar teeth under the control of the muscles of mastication. Chewing movements are produced by the alternate contraction of the pterygoid muscles of first one and then the other side, and if the food escapes from between the teeth into the vestibule of the mouth it is returned by the contraction of buccinator. If it escapes medially into the cavity of the mouth it is triturated by the action of the tongue on the hard palate and pushed back between the teeth by the intrinsic and extrinsic muscles of the tongue. Mylohyoid helps to keep the tongue braced up towards the hard palate, and the orbicularis oris contracts to prevent food escaping through the lips.

Normally the downward movement of the mandible during mastication results from relaxation of the elevators, and contraction of lateral pterygoid, but more active depression of the mandible may be required if sticky food is being eaten, and this movement results from contraction of digastric, geniohyoid, and mylohyoid acting from a hyoid bone fixed by the infrahyoid muscles (sternohyoid, sternothyroid, omohyoid).

Once the food is judged by the sensory endorgans in the mouth and tongue to be of suitable consistency for swallowing, it is collected from the anterior portion of the mouth by the tip of the tongue, which is then squeezed up against the hard palate by the contraction of the superior longitudinal muscle, assisted by mylohyoid and styloglossus, the elevators of the mandible coming into play at the same time. This forces the food backwards towards the palatoglossal and palatopharyngeal arches, much as toothpaste is squeezed from a tube.

In this phase the hyoid bone, carrying with it the larynx, is raised by means of mylohyoid, stylohyoid, digastric, palatopharyngeus, and stylopharyngeus. Fractionally later, geniohyoid pulls the hyoid bone forwards. As the food passes through the palatoglossal and palatopharyngeal arches, palatoglossus and palatopharyngeus contract reflexly, bringing the pillars of the arches closer together. The dorsum of the tongue bulges backwards, under the influence of styloglossus, to occupy the remaining space between the arches, and in this way the oral part of the pharynx is shut off from the mouth cavity, so that food cannot be forced back into the mouth during subsequent stages of swallowing.

Now the soft palate, tensed by tensor veli palatini, is raised by levator veli palatini to block off the communication between the oral and the nasal parts of the pharynx, so preventing the subsequent contraction of the constrictor muscles from forcing food up into the nose. In this the palatal muscles are assisted by the muscular contraction of the palatopharyngeal sphincter, which pulls the pharynx forwards to meet the rising soft palate.

The elevation of the tongue and the hyoid bone brings the inlet of the larynx up under the cover of the epiglottis, and at the same time this is bent backwards and downwards by the backward pressure of the tongue and the contraction of the oblique arytenoid and the aryepiglottic muscles. It may come into contact with the posterior wall of the pharynx. The thyroid cartilage is pulled up to the level of the hyoid bone by thyrohyoid, so approximating the arytenoid cartilages to the bent epiglottis. The contraction of aryepiglotticus helps to narrow the inlet, and the whole arrangement is designed to prevent the entry of food into the larynx, for the food now slips over the oral surface of the epiglottis and the closed inlet of the larynx into the laryngeal part of the pharynx.

Respiration is now reflexly inhibited, and the constrictor muscles of the pharynx contract successively from above downwards, the constriction wave being preceded by a wave of relaxation. At the same time the pharynx is pulled upwards over the descending bolus of food by palatopharyngeus, stylopharyngeus, and salpingopharyngeus. The action of the constrictors is normally automatic, but they can be brought under voluntary control, so that a bolus of food can voluntarily be recovered from the lower part of the pharynx. Once it has come into the grip of the visceral muscle of the oesophagus, however, it is beyond recall except by vomiting.

Following the descent of the bolus past the aperture of the larynx, the hyoid bone and larynx are pulled down by the infrahyoid muscles, breathing can be resumed, and another mouthful can be dealt with.

The swallowing of fluid is essentially similar, except that in the initial stages the tongue is made to form a tube with the palate by the contraction of styloglossus and the transverse intrinsic muscle of the tongue. The fluid is then forced backwards in the same way as solid food, and cascades down the sides of the epiglottis, avoiding the entry to the larynx.

Gravity is of considerable assistance in causing the bolus of food to descend through the pharynx. However, the pharyngeal constrictors can still operate successfully even though the individual is standing on his head, and a glass of water can be drunk in this position. During the movement of the bolus through the pharynx, a considerable pressure is developed by its muscular wall, and the mucous membrane may be temporarily protruded through weak points in the pharyngeal wall where the fascial basis of the pharynx is inadequately guarded by musculature. In particular, it is possible for such bulging to take place above, below, or through the inferior

constrictor, and a 'pulsion diverticulum' may develop, in which food lodges during, and remains after, the descent of the bolus. Such diverticula may require surgical attention.

If the innervation of the soft palate or pharynx is damaged, the whole process of swallowing becomes difficult or impossible, and food may be returned through the nose after it is taken in by the mouth. In such cases feeding by means of a tube may be necessary. The movements of the soft palate are important not only in swallowing, but also in speech and in blowing or whistling. All these activities require an airtight seal between the nasal and the oral parts of the pharynx, since otherwise air would escape through the nose. For this reason paralysis of the soft palate results not only in difficulty in swallowing, but in a nasal tone of voice and an inability to develop any pressure on blowing through the mouth.

The complex mechanisms necessitated by the crossing over of the stream of air and the flow of food are quite liable to disturbance, and if food should succeed in penetrating the opening into the larynx, explosive coughing will be set up, and the whole delicate process may be temporarily disorganized, with food being forced up into the nose. The diversion of food or liquid to one or other side of the opening of the larynx forces it into the piriform recess [FIG. 7.19], and it is here that such objects as fish bones tend to lodge.

During the action of swallowing, salpingopharyngeus helps to open the orifice of the auditory tube, and in this it is probably assisted by levator and tensor veli palatini. This opening of the tube equalizes the pressure in the nasal part of the pharynx and the middle ear, and passengers in an aircraft can relieve the differential pressure on their eardrums during ascent or descent by swallowing.

The process of swallowing, like most common functions of the body, is extremely complicated, and the exact pattern of muscular activity varies with the individual and with the material being swallowed (Cunningham and Basmajian 1969); a useful review of the literature is that of Bosma (1957).

Muscles concerned in movements of the mandible

Movement	Muscle	Principal Nerve Supply
Elevation	Temporalis (anterior fibres)	Anterior division of mandibular
	Masseter	Anterior division of mandibular
	Medial pterygoid	Main trunk of mandibular
Depression	Lateral pterygoid	Anterior division of mandibular
	Digastric	Facial: mylohyoid branch of inferior alveolar
	Mylohyoid ⎫ acting in concert	Mylohyoid branch of inferior alveolar
	Geniohyoid ⎭	C. 1 through hypoglossal
	Infrahyoid group	Ansa cervicalis: thyrohyoid by C.1 through hypoglossal
Protraction	Lateral pterygoid	Anterior division of mandibular
	Medial pterygoid	Main trunk of mandibular
	Masseter (superficial fibres)	Anterior division of mandibular
Retraction	Temporalis (horizontal fibres)	Anterior division of mandibular
	Digastric	Facial: mylohyoid branch of inferior alveolar
Chewing	Medial pterygoid	Main trunk of mandibular
	Lateral pterygoid	Anterior division of mandibular
	Masseter	Anterior division of mandibular
	Temporalis	Anterior division of mandibular

Muscles concerned in movements of the larynx and hyoid bone

Movement	Muscle	Principal Nerve Supply
Elevation	Digastric	Facial: mylohyoid branch of inferior alveolar
	Stylohyoid	Facial
	Mylohyoid	Mylohyoid branch of inferior alveolar
	Geniohyoid	C. 1 through hypoglossal
	Thyrohyoid	C. 1 through hypoglossal
	Palatopharyngeus	Pharyngeal plexus
	Stylopharyngeus	Glossopharyngeal
	Inferior constrictor (possibly)	Pharyngeal plexus: recurrent laryngeal
Depression	Sternohyoid	Ansa cervicalis
	Sternothyroid	Ansa cervicalis
	Omohyoid	Ansa cervicalis
Protraction	Geniohyoid	C. 1 through hypoglossal
Retraction	Stylohyoid	Facial
	Inferior constrictor (possibly)	Pharyngeal plexus: recurrent laryngeal
	Middle constrictor (possibly)	Pharyngeal plexus

Fasciae of the vertebral column, neck, and head

SUPERFICIAL FASCIA

The superficial fascia of the back is thick and contains a considerable amount of fat held within a meshwork of fibres. Laterally, it is loosely connected to the skin, but in the midline, particularly in the upper part of the neck, it tacks the skin more tightly to the deep fascia.

The front and sides of the neck are covered by loose superficial fascia which allows free movement of the skin, containing the fibres of platysma, over the underlying structures. This muscle is attached to the skin, and if it is paralysed, the skin of the neck sags downwards [p. 291].

On the scalp, the superficial fascia forms a dense fibro-fatty layer, which is adherent both to the skin and to the galea aponeurotica, which replaces the deep fascia in the scalp [p. 287]. This arrangement means that the three superficial layers of the scalp (skin, superficial fascia, and galea aponeurotica) move as a unit over the periosteum of the skull, which is separated from the galea by a space filled by loose connective tissue. It may therefore be pulled away from the skull in a flap of skin following injuries or during surgical operations. The texture of the superficial fascia becomes somewhat more open as it is traced laterally, and it contains the superficial vessels and nerves of the scalp.

The superficial fascia of the face contains the muscles of facial expression as well as the regional nerves and blood vessels. Superficial to buccinator and between it and masseter lies the buccal pad of fat (Gaughran 1957), an accumulation which is large and prominent in the infant. Over most of the face the superficial fascia is fairly tightly attached to the skin, especially in the region of the nose and the external ear, where it is very scanty and ties the skin

down to the underlying cartilages. But in the eyelids it is lax and open, and a fascial 'space' exists in which fluid can easily accumulate following injuries to the region of the orbit.

DEEP FASCIA

The whole of the back is covered by a layer of deep fascia of variable thickness and strength. In the neck it is strong and dense, and it forms part of the investing layer of the cervical fascia. Lower down it is relatively thin. It covers over and ensheaths the superficial muscles which connect the upper limb to the trunk, and is attached to the spines of the thoracic and lumbar vertebrae, to the spine and the acromion of the scapula and to the iliac crests and the back of the sacrum. More laterally still, it becomes continuous with the deep fascia of the axilla, the thorax, and the abdomen, and blends with the investing deep fascia of the arm.

Deep to the superficial muscles of the back lies a very much stronger development of the deep fascia. This is well developed only in the sacral, lumbar, and lower thoracic regions, and is therefore termed the thoracolumbar fascia. It consists of three layers. The posterior layer, which is superficial to erector spinae, is attached medially to the thoracic, lumbar, and sacral spines and their supraspinous ligaments. It extends from the sacrum and the iliac crest upwards to be attached to the angles of the ribs lateral to the iliocostalis part of erector spinae. In the lower part of the back this layer forms a very strong membranous aponeurosis which gives attachment to the latissimus dorsi muscle.

The middle layer of the thoracolumbar fascia extends from the lower border of the twelfth rib and the lumbocostal ligament down to the iliac crest and iliolumbar ligament. It is attached medially to the tips of the transverse processes of the lumbar vertebrae and to the intertransverse ligaments, and laterally it joins the posterior layer at the lateral border of erector spinae [FIG. 5.47].

The anterior layer of the thoracolumbar fascia is the thinnest of the three. It lies in front of quadratus lumborum and is attached medially to the anterior aspect of the lumbar transverse processes. Laterally it fuses with the middle layer at the lateral border of the

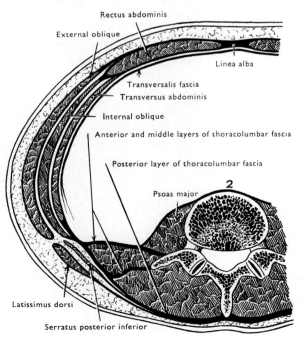

FIG. 5.47. Transverse section through the abdomen at level of second lumbar vertebra.

quadratus lumborum and becomes continuous with the posterior aponeurotic attachments of internal oblique and the transversus abdominis [FIG. 5.47]. Superiorly the anterior layer is attached to the twelfth rib and the first lumbar transverse process, between which it is thickened to form the lateral arcuate ligament. Inferiorly, it gains attachment to the iliolumbar ligament and the iliac crest.

CERVICAL FASCIA

In the neck, as elsewhere in the body, the muscles and viscera are surrounded by connective tissue packing material. In some places this is more dense than in others, and on this basis many fascial planes have been described which have at best a somewhat doubtful importance, and at worst a tenuous claim to reality. However, three main layers of cervical fascia can be recognized without much difficulty.

The investing layer surrounds the neck like a sleeve; it is less strong anteriorly, where it is covered by platysma, and posteriorly it splits to enclose the trapezius muscle. In the midline, the investing layer is attached to the external occipital protuberance, the ligamentum nuchae, and the spine of the seventh cervical vertebra. In front of trapezius it forms the roof of the posterior triangle [p. 286], and at the anterior margin of the triangle it splits again to invest sternocleidomastoid [FIG. 5.48]. Anterior to this muscle the investing layer passes across the structures in the anterior triangle of the neck [p. 286], becoming bound down to the body and the greater horn of the hyoid bone as it does so.

Superiorly, the investing layer is attached in a continuous line to the external occipital protuberance, the superior nuchal line of the occipital bone, and the mastoid process. Further forwards, it divides to form a sheath for the parotid gland, the superficial part of this sheath being attached to the zygomatic arch and disappearing anteriorly on the face. A portion of the deep part of the sheath forms the stylomandibular ligament, which ascends to the styloid process as a partition between the parotid and submandibular glands. Anterior to the parotid gland, the investing layer is thin and capable of being stretched; it is attached along the lower border of the mandible, but splits again to enclose the submandibular gland, the deep layer ascending to the mylohyoid line on the inner surface of the mandible.

Inferiorly, the investing layer is attached, along with the trapezius, to the spine and acromion of the scapula, and in front of this, in the roof of the posterior triangle, it splits inferiorly into two layers. The superficial one is attached to the upper surface of the clavicle, and the deeper one surrounds the inferior belly of the omohyoid, forming for this muscle a fascial sling which holds it approximately parallel to the clavicle. This deeper layer passes behind the clavicle, fusing with its periosteum, and is attached to the sheath of the subclavius muscle and the anterior end of the first rib.

In the anterior triangle, the inferior part of the investing layer splits again. The superficial layer so formed is attached to the anterior border of the manubrium sterni and the deeper layer to its posterior border, so enclosing between them a small compartment called the suprasternal space, which contains the interclavicular

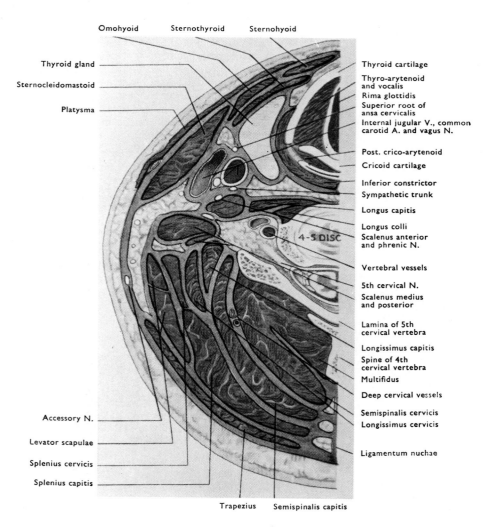

FIG. 5.48. Transverse section of neck between fourth and fifth cervical vertebrae.

ligament, the sternal heads of the sternocleidomastoid muscles and the jugular venous arch.

The deepest of the three layers of deep fascia in the neck is the **prevertebral fascia**, which extends from the basilar part of the occipital bone into the thorax, where it loses its identity by becoming continuous with the anterior longitudinal ligament. It lies in front of the prevertebral muscles [FIG. 5.48], and is connected to the buccopharyngeal fascia [q.v.] by a loose layer of areolar tissue; in this **retropharyngeal space** lie the retropharyngeal lymph nodes. Laterally, the fascia extends in front of the scalene muscles on to levator scapulae and splenius. Here it forms a floor for the posterior triangle of the neck, and lies in front of the cervical plexus. As the subclavian artery and the trunks of the brachial plexus emerge from behind the scalenus anterior, they take with them a prolongation of this fascia which forms the **axillary sheath**. Further laterally, deep to the trapezius, the prevertebral fascia becomes much thinner and loses its identity.

The third, and intermediate, layer of fascia is the **pretracheal fascia**, which is less definite than the other two, and forms a sheath for the thyroid gland. Anteriorly it is attached to the arch of the cricoid cartilage. Superiorly the pretracheal fascia is fixed to the oblique line of the thyroid cartilage, and fuses with the fascia on the inferior constrictor muscle. Inferiorly, the fascia passes downwards from the isthmus of the thyroid gland in front of the trachea into the thorax, where it blends with the fascial sheaths of the great vessels.

The **carotid sheath** is merely a condensation of rather more dense connective tissue around the commmon carotid artery, the internal jugular vein, and the vagus nerve. The sheath is thick in the region of the artery, and relatively thin elsewhere. Anteromedially it blends loosely with the pretracheal fascia, and anterolaterally with the portion of the investing layer on the deep surface of sternocleidomastoid [FIG. 5.48]. Posteriorly, the carotid sheath is connected to the prevertebral fascia by relatively loose tissue in which lies the sympathetic trunk.

These layers and sheaths of fascia to a certain extent determine the way infection will spread in the neck. Perhaps the most important of the compartments which they form is the **visceral compartment**, which contains the larynx and the trachea, the pharynx, and the oesophagus. This is bounded anteriorly by the pretracheal fascia, posteriorly by the prevertebral fascia, and laterally by the carotid sheath. It is open below, so that infection

entering the space from the larynx or oesophagus may travel downwards into the superior mediastinum of the thorax. On the other hand, fluid forming posterior to the prevertebral fascia will be directed laterally towards the posterior triangle, where it may burst the fascia and appear on the surface, or may be conducted down the axillary sheath to the axilla.

The **buccopharyngeal fascia** clothes the outer surface of the pharynx; it is continuous above with the fascia on the buccinator muscle and below with that on the surface of the oesophagus. Laterally it fuses with the carotid sheath. Between the mucosa of the pharynx and its muscular wall there is another fibrous layer which is given the name of the **pharyngobasilar fascia**. This is well developed superiorly, where it forms the lateral wall of the pharynx above the superior constrictor [p. 298], and gradually fades out as it is traced downwards, though it is still distinguishable at the level of the tonsil, for which it helps to form a fibrous bed. It is attached to the base of the skull, to the pterygomandibular raphe, and to the posterior end of the mylohyoid line of the mandible, and some of its fibres are attached to the cartilaginous part of the auditory tube. Anteriorly and inferiorly extensions of the fascia pass forwards on to the hyoid bone and the thyroid cartilage.

FASCIA OF HEAD

In the central portion of the **scalp**, the deep fascia is replaced by the galea aponeurotica [p. 287]. In the temple, a strong layer of **temporal fascia** passes from the superior temporal line to the zygomatic arch. It covers the temporalis muscle and extends to the posterior border of the frontal process of the zygomatic bone, but does not enter the face. The deep fascia of the face is almost non-existent, though a recognizable layer occurs over masseter.

In the orbit the deep fascia forms a sort of cup for the eyeball, the **vagina bulbi**, but is separated from it by the **episcleral space**, which allows the eyeball to move slightly within the cup. The vagina bulbi and the eyeball move together in the surrounding soft orbital fat.

The vagina bulbi adheres anteriorly to the conjunctiva and the sclera close to the corneal margin. The extra-ocular muscles pierce it and each carries backwards from it a sleeve of fascia which is attached to the wall of the orbit, so anchoring the fascial cup. The strongest of these attachments are the medial and lateral **check ligaments**, derived from the sheaths of the medial and lateral rectus

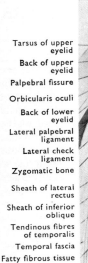
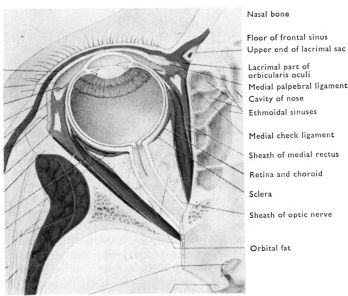

Tarsus of upper eyelid
Back of upper eyelid
Palpebral fissure
Orbicularis oculi
Back of lower eyelid
Lateral palpebral ligament
Lateral check ligament
Zygomatic bone
Sheath of lateral rectus
Sheath of inferior oblique
Tendinous fibres of temporalis
Temporal fascia
Fatty fibrous tissue

Optic nerve and canal   Anterior clinoid process

Nasal bone
Floor of frontal sinus
Upper end of lacrimal sac
Lacrimal part of orbicularis oculi
Medial palpebral ligament
Cavity of nose
Ethmoidal sinuses
Medial check ligament
Sheath of medial rectus
Retina and choroid
Sclera
Sheath of optic nerve
Orbital fat

FIG. 5.49. Horizontal section through left orbit to show arrangement of fascial sheath of eyeball and check ligaments.

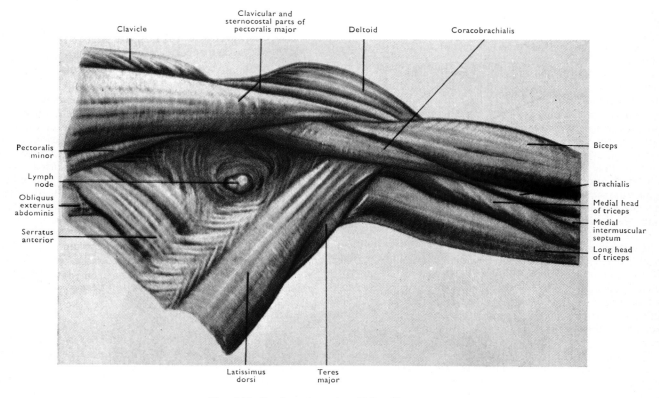

Clavicle · Clavicular and sternocostal parts of pectoralis major · Deltoid · Coracobrachialis

Pectoralis minor · Lymph node · Obliquus externus abdominis · Serratus anterior

Biceps · Brachialis · Medial head of triceps · Medial intermuscular septum · Long head of triceps

Latissimus dorsi · Teres major

FIG. 5.50. Fascia and muscles of left axilla.

muscles [FIG. 5.49]. The medial check ligament is attached to the lacrimal bone, to the fascia of the eyelids behind the lacrimal part of the orbicularis oculi, and to the medial fornix of the conjunctiva. The lateral check ligament is attached to the middle of the lateral margin of the orbit and to the lateral palpebral ligament and conjunctival fornix [FIG. 5.49]. The portion of the vagina bulbi which surrounds the inferior rectus muscle and is connected to the medial and lateral check ligaments is thickened to form a sling (the **suspensory ligament**) which helps to support the eyeball.

If the eye has to be removed, the vagina bulbi forms a socket for an artificial eye, and it also prevents the severed muscles from retracting very far. It is pierced by the vessels and nerves of the eyeball, and posteriorly it is continuous with the dural sheath of the optic nerve.

Muscles of the upper limb

In the course of evolution the hand has become a manipulating instrument of great precision, power, and delicacy, and the upper limb is a system of joints and levers designed to bring this instrument to bear at any desired point in space and to hold it there steadily and securely while it carries out its task. The muscles of the upper limb therefore exert their primary functions from proximal origins on distal insertions. However, the upper limb is still on occasion an instrument of locomotion, and by grasping some convenient immovable object, the body can be pulled towards the hand, as in climbing. In this case the muscles operate from a distal fixed point on a proximal movable one.

Since the upper limb is also used for carrying loads, and occasionally for supporting the weight of the body, it is necessary to

think of the way in which force is transmitted—usually by the tension of muscles—from the hand to the trunk and vice versa. Lastly, because the upper limb is itself heavy, it must be borne in mind that every movement of it has to be accompanied by postural contractions of the muscles of the trunk and lower limb to compensate for the shift in the centre of gravity of the body.

Many upper-limb muscles cross two or more joints, and can therefore operate on all of them; a complex system of synergists and fixators is thus provided to cancel out the unwanted movements, and the procedure of testing for suspected paralysis of a given muscle is often quite complicated.

MUSCLES ATTACHING THE UPPER LIMB TO THE TRUNK

Movements of the limb relative to the trunk take place at the sternoclavicular, acromioclavicular, and shoulder joints, and this system of three joints is largely responsible for the great mobility of the upper limb. The muscles which run from the trunk to the scapula operate exclusively on the shoulder girdle and have no action on the shoulder joint; these are trapezius, levator scapulae, the rhomboids, serratus anterior, pectoralis minor, and subclavius. Two, latissimus dorsi and pectoralis major, pass from the trunk to the humerus, so their primary action is on the shoulder joint, but they also produce movements of the shoulder girdle. Others—deltoid, supraspinatus, infraspinatus, teres minor, subscapularis, teres major, and coraco-brachialis—pass from the scapula to the humerus, and thus act exclusively on the shoulder joint.

Many of these muscles form the boundaries of a pyramidal space, the axilla, which lies between the arm and the upper part of the thorax [FIG. 5.50]. The apex of the axilla is at the medial side of the root of the coracoid process and the base looks downwards. The axilla is bounded anteriorly by pectoralis major and minor, posteriorly by subscapularis, teres major, and latissimus dorsi, medially by serratus anterior and the upper five ribs, and laterally by the humerus, coracobrachialis, and the short head of biceps brachii. Lining this space is the axillary fascia [p. 346].

TRAPEZIUS

Trapezius is a roughly triangular muscle, which forms a trapezium with its fellow. It takes origin from the medial third of the superior nuchal line of the occipital bone, the external occipital protuberance, the ligamentum nuchae, the spines of the seventh cervical and all the thoracic vertebrae, and the intervening supraspinous ligaments. It is inserted into the posterior border of the lateral third of the clavicle [FIGS. 5.51 and 5.58], the medial border of the acromion, the upper border of the crest of the spine of the scapula, and the tubercle on this crest.

The uppermost fibres descend to the clavicle and form the posterior border of the posterior triangle of the neck. The lowest fibres run up from the lower thoracic spines to the medial end and tubercle of the spine of the scapula, and form the medial boundary of the **triangle of auscultation**—a portion of the chest wall relatively thinly covered by muscle and free of the bony obstruction of the scapula. It is completed below by the upper border of latissimus dorsi, and laterally by the medial border of the scapula [FIG. 5.52].

There may be a small bursa deep to the lower and almost horizontal intermediate fibres of trapezius as they cross the medial part of the scapular spine. There is a small diamond-shaped tendinous area over the lower part of the neck and the upper part of the thorax [FIG. 5.52], and this corresponds to a hollow in this vicinity seen in the living subject [FIG. 5.53].

Nerve Supply

Trapezius receives its motor supply from the spinal portion of the accessory nerve, which enters it from the posterior triangle. Sensory fibres reach it from the ventral rami of the third and fourth cervical nerves through the cervical plexus.

Actions

The upper fibres pull the lateral part of the shoulder girdle up towards the skull (elevation) and help to prevent depression of the shoulder girdle when a weight is being carried on the shoulder or in the hand; the importance of this action varies in different subjects (Bearn 1961), and when the shoulder is lowered to the fully depressed position, activity in trapezius decreases dramatically and may stop, the strain presumably being taken by the capsule of the sternoclavicular joint (Bearn 1967). The lower fibres, on the other hand, depress the medial part of the scapula and lower the shoulder, particularly against resistance, as when using the hands to get up from a chair, or when the weight of the body is taken on the hands by leaning on a table. The intermediate fibres pull the scapula backwards towards the midline (retraction).

When the upper and lower fibres contract together they rotate the scapula laterally round a point approximately half way along the spine, the glenoid socket turning to face upwards and forwards. This rotatory action is of importance in the movement of raising the arm above the head [p. 321]. The spine of the scapula affords a good purchase for the muscle, and turns like a winged nut in response to the couple produced by the trapezius [FIG. 5.53].

Paralysis of trapezius allows the scapula to drift forwards on the chest wall, and to rotate so that its inferior angle swings medially (Lockhart 1930); the smooth curve of the upper border of the muscle from occiput to acromion process eventually becomes converted into an unsightly angulation. The usual test for integrity of trapezius is to ask the patient to shrug the shoulders against resistance, when

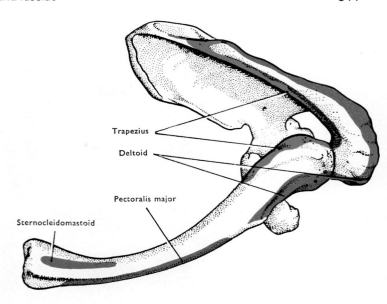

FIG. 5.51. Shoulder girdle from above showing muscle attachments.

the upper fibres can be seen and felt in action. Even though trapezius is paralysed, a full range of abduction of the upper limb is often possible, for the serratus anterior is an even more powerful rotator of the scapula than the trapezius.

LEVATOR SCAPULAE

The upper third of levator scapulae is concealed by sternocleidomastoid; the middle third forms part of the floor of the posterior triangle of the neck [FIG. 5.21] and the lower third is hidden by trapezius. It arises from the posterior tubercles of the transverse processes of the first three or four cervical vertebrae posterior to scalenus medius, and descends to the medial margin of the scapula between the superior angle and the spine [FIGS. 5.52 and 5.55].

Nerve supply

From the ventral rami of C. 3, 4, and 5, the latter via the dorsal scapular nerve.

Actions

Levator scapulae elevates the scapula and helps to fix the upper angle in scapular rotation [p. 321]. It also helps to support the weight of the limb and to transmit to the axial skeleton the forces produced by weights carried in the hand.

RHOMBOIDEUS MINOR

This small muscle comes from the lower part of the ligamentum nuchae, the spines of the seventh cervical and first thoracic vertebrae, and the supraspinous ligament between them. Its fibres pass downwards and laterally, parallel to those of levator scapulae, to the medial border of the scapula at the level of the spine [FIGS. 5.52 and 5.55].

Nerve supply

From the ventral ramus of C. 5, through the dorsal scapular nerve.

Actions

It is mainly a retractor but also an elevator of the scapula.

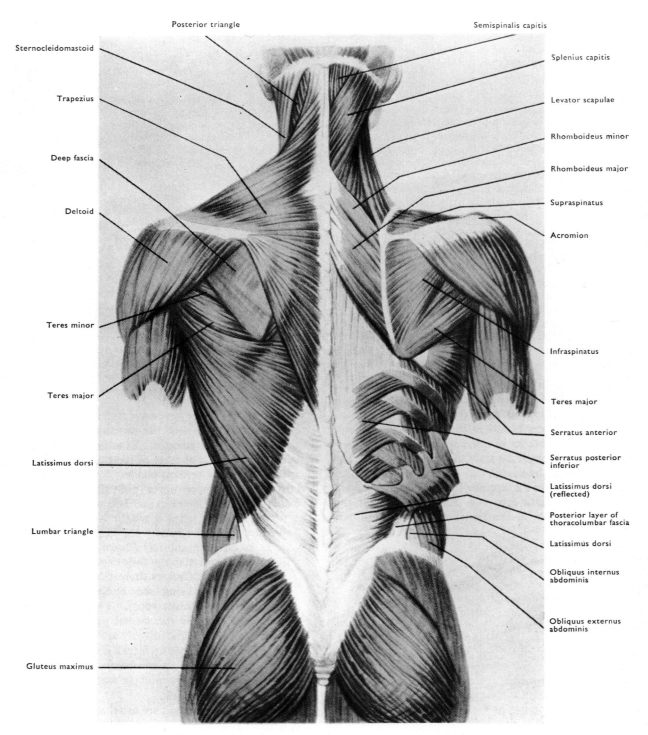

Posterior triangle

Sternocleidomastoid

Trapezius

Deep fascia

Deltoid

Teres minor

Teres major

Latissimus dorsi

Lumbar triangle

Gluteus maximus

Semispinalis capitis

Splenius capitis

Levator scapulae

Rhomboideus minor

Rhomboideus major

Supraspinatus

Acromion

Infraspinatus

Teres major

Serratus anterior

Serratus posterior inferior

Latissimus dorsi (reflected)

Posterior layer of thoracolumbar fascia

Latissimus dorsi

Obliquus internus abdominis

Obliquus externus abdominis

FIG. 5.52. Superficial muscles of the back, and vertebroscapular muscles.

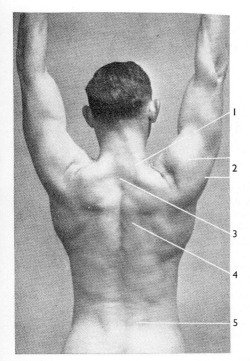

FIG. 5.53. Action of trapezius and deltoid muscles in elevation of the upper limb. (*By courtesy of Professor R. D. Lockhart.*)
1. Upper fibres of trapezius.
2. Deltoid.
3. Diamond-shaped tendinous area of trapezius muscles.
4. Lower fibres of trapezius.
5. Erector spinae.

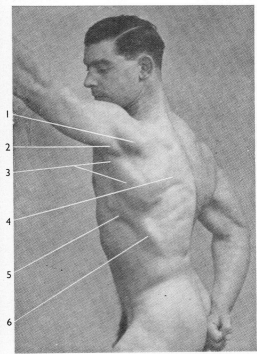

FIG. 5.54.. Action of latissimus dorsi as depressor and adductor of upper limb, subject pulling on a rope. (*By courtesy of Professor R.D. Lockhart.*)
1. Posterior part of deltoid.
2. Long head of triceps.
3. Teres major and latissimus dorsi.
4. Rhomboideus major.
5. Serratus anterior.
6. Latissimus dorsi.

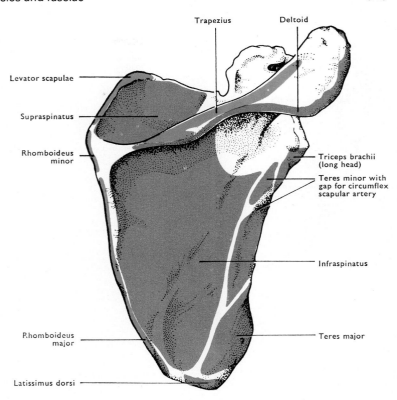

FIG. 5.55 Muscle attachments to dorsal surface of right scapula.

RHOMBOIDEUS MAJOR

Rhomboideus major is often continuous with rhomboideus minor, and arises from the spines and supraspinous ligaments of the thoracic vertebrae from the second to the fifth inclusive. It runs parallel to rhomboideus minor to be inserted into the medial border of the scapula between the root of the spine and the inferior angle. Its lower border appears from under cover of trapezius to form the floor of the triangle of auscultation.

Nerve supply

From the ventral ramus of C. 5, via the dorsal scapular nerve.

Actions

It is principally a retractor but also a medial rotator of the scapula. The rhomboid muscles are tested by asking the patient to brace his shoulders back against resistance, when the muscles may be felt and sometimes seen to contract. It is difficult to distinguish their activity from that of the overlying part of trapezius.

SERRATUS ANTERIOR

Serratus anterior is a large muscle which arises from the outer surfaces of the upper eight ribs and the fascia covering the spaces between them by a series of slips [FIG. 5.56]; the lower ones interdigitate with the origin of the external oblique muscle of the abdomen [FIG. 5.58]. The first slip runs almost straight backwards to the costal surface of the scapula at the superior angle. The next three go to the costal surface of the medial border of the scapula, and the last four pass obliquely upwards and backwards to the inferior angle of the scapula on its costal surface.

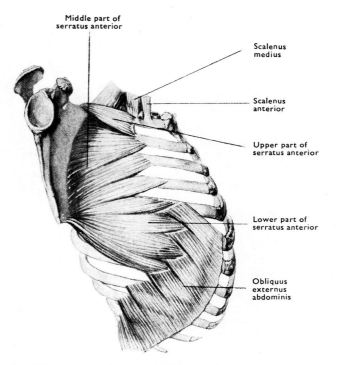

FIG. 5.56. Serratus anterior and origin of external oblique. The scapula is drawn away from the side of the chest.

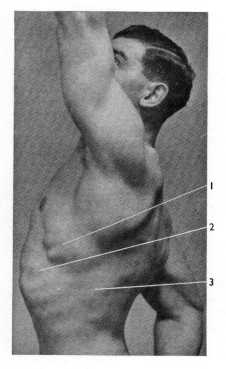

FIG. 5.57. Action of serratus anterior in elevation of upper limb. (*By courtesy of Professor R. D. Lockhart.*)
1. Serratus anterior.
2. External oblique.
3. Latissimus dorsi.

Serratus anterior forms the medial wall of the axilla, and is partly covered by the inferolateral part of the mammary gland. Its upper digitation lies behind the clavicle, and its lower border is crossed by the latissimus dorsi.

Nerve supply

From the ventral rami of C. 5, 6, and 7 via the long thoracic nerve, which enters the muscle on its superficial aspect. The highest digitations are supplied by the fifth cervical nerve and the lowest by the seventh.

Actions

In quadrupeds serratus anterior acts from its insertion on its origin and transfers the weight of the trunk to the scapula and thence to the forelimb. This action is also of some importance in man when the weight is taken on the hands, as in crawling on all fours. In the erect posture, the muscle thrusts the shoulder girdle forwards, and can push a stalled car, or deliver a punch in boxing. In all such movements of **protraction** serratus anterior pulls the scapula forwards on the curve of the chest and holds it closely in to the chest wall. Because the concentration of protracting power brought to bear at the lower angle of the scapula is greater than that achieved at the superior angle, the muscle operates with trapezius in rotating the scapula laterally so that the glenoid socket looks upwards and forwards [FIG. 5.57].

With the rhomboids and levator scapulae, serratus anterior steadies the scapula to form a stable platform from which other muscles can act. If serratus anterior is paralysed, the medial border of the scapula stands further away from the chest wall, especially when the body is supported by the arms or when pushing movements are carried out. This is called 'winging' of the scapula, and is an important test for detecting paralysis of the muscle. In the absence of the rotating action of serratus anterior the upper limb cannot be abducted from the side through more than about 90 degrees. Contrary to former belief, the serratus anterior does not appear to play any part in forced respiration (Catton and Gray 1951).

PECTORALIS MINOR

This flat triangular muscle lies in the anterior wall of the axilla posterior to pectoralis major, and concealed by it [FIGS. 5.50 and 5.58]. It arises from the outer surfaces of the third, fourth, and fifth ribs near their anterior ends and from the fascia of the corresponding intercostal spaces. There may be an additional origin from the second rib or the sixth rib; less commonly from both. The short, flat tendon passes into the medial border and upper surface of the coracoid process of the scapula. Rarely, the tendon continues over the coracoid process, separated from it by a bursa, to pierce the coracoacromial ligament and become continuous with the coraco-humeral ligament [FIG. 4.25].

Nerve supply

The medial pectoral nerve, a branch of the medial cord of the brachial plexus, pierces the muscle and supplies it. The lateral pectoral nerve, a branch of the lateral cord of the plexus, communicates with the medial pectoral nerve in the axilla, and this communication ensures that pectoralis minor, like pectoralis major, is supplied by both pectoral nerves. The segmental supply is probably from C. 6, 7, and 8.

Actions

The muscle draws the shoulder girdle forwards and downwards, and helps to transfer the weight of the trunk to the upper limb when leaning on the hands. It could also be used to pull the ribs up as an

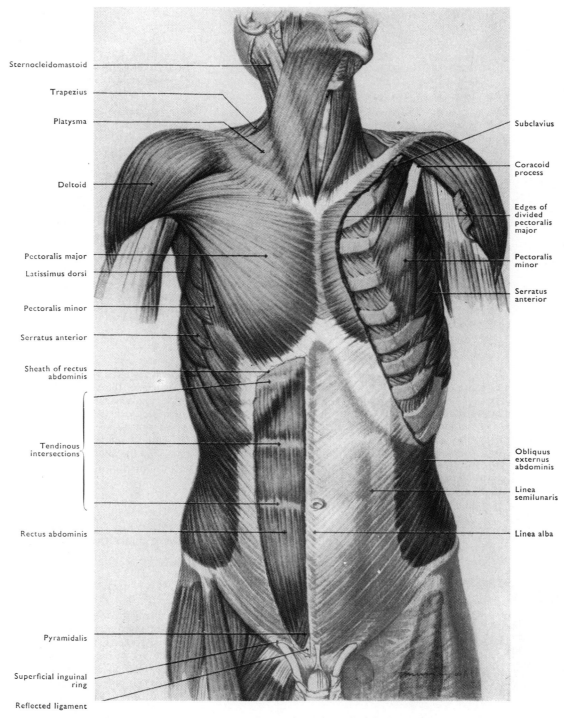

Sternocleidomastoid

Trapezius

Platysma

Deltoid

Pectoralis major

Latissimus dorsi

Pectoralis minor

Serratus anterior

Sheath of rectus abdominis

Tendinous intersections

Rectus abdominis

Pyramidalis

Superficial inguinal ring

Reflected ligament

Subclavius

Coracoid process

Edges of divided pectoralis major

Pectoralis minor

Serratus anterior

Obliquus externus abdominis

Linea semilunaris

Linea alba

FIG. 5.58. Muscles of anterior wall of the trunk.

accessory muscle of inspiration. It is virtually impossible to test for the integrity of pectoralis minor, since it is wholly under cover of pectoralis major, and is never paralysed alone.

SUBCLAVIUS

The tendinous origin of subclavius is from the junction of the first rib and the first costal cartilage in front of the costoclavicular ligament [FIG. 5.58]. The small fleshy belly is inserted into the floor of a groove on the inferior surface of the clavicle [FIG. 5.62]. The muscle lies wholly under cover of pectoralis major and the clavicle.

Nerve supply

The nerve to subclavius comes from the upper trunk of the brachial plexus (C. 5, 6), and descends in front of the subclavian artery to reach the muscle.

Actions

Paralysis of subclavius produces no demonstrable effect, but its position suggests that it is a 'shunt' muscle which may exert a slight protective action on the sternoclavicular joint during shoulder girdle movement.

FIG. 5.59. Folds of the axilla and the medial aspect of the upper arm, showing the position of the long and the medial heads of the triceps.

Note the concentration of the sternocleidomastoid and compare with FIGURES 5.22 and 5.112. (*By courtesy of Professor R. D. Lockhart.*)
1. Deltoid.
2. Infraclavicular fossa.
3. Biceps.
4. Medial head of triceps.
5. Long head of triceps.
6. Anterior and posterior folds of axilla.
7. Serratus anterior.

LATISSIMUS DORSI

Latissimus dorsi passes between the trunk and the humerus, and therefore acts on both the shoulder joint and the shoulder girdle. It arises from the posterior layer of the thoracolumbar fascia, which is attached to the spines of the lower thoracic and all the lumbar and sacral vertebrae, as well as to the intervening supraspinous ligaments. In addition, it takes origin from the posterior part of the outer lip of the iliac crest, from the outer surfaces of the lower three or four ribs [FIG. 5.52] and the inferior angle of the scapula, for which the muscle forms a kind of pocket.

From this widespread origin, latissimus dorsi narrows down to a thin flattened tendon 3–4 cm in width, which winds round and adheres to the lower border of teres major, and inserts into the floor of the intertubercular sulcus of the humerus [FIG. 5.67] anterior to the tendon of teres major and separated from it by a bursa. Because of this spiral turn, the anterior surface of the tendon of insertion is continuous with the posterior surface of the rest of the muscle. The superior border of latissimus dorsi forms the inferior border of the triangle of auscultation, and the lateral border of the muscle forms the medial border of the lumbar triangle [p. 355 and FIG. 5.52].

Nerve supply

From the thoracodorsal nerve (C. 6, 7, 8), which supplies it on its deep surface. The nerve comes from the posterior cord of the brachial plexus.

Actions

The muscle retracts the shoulder girdle (if the humerus is fixed to the scapula by the scapulohumeral muscles) and is a strong extensor of the flexed arm. It is also a powerful adductor of the humerus, and helps to rotate it medially. Acting from its insertion, latissimus dorsi is one of the chief climbing muscles, since it pulls the trunk up on the arms [FIG. 5.54], and it is also a powerful factor in rowing and in the downstroke in swimming. But the muscle also contracts during violent expiration, such as coughing or sneezing, apparently to compress the thorax and abdomen. The best test of the muscle is to ask the patient to cough while holding the muscle between finger and thumb as it reaches the posterior axillary fold. Alternatively, the muscle can be felt when the patient adducts the horizontally abducted arm against resistance.

Pectoralis minor and latissimus dorsi transfer the weight to the upper limb when the body is pushed upwards by the hands. In patients with paralysis of the lower limbs resulting from spinal injury the nerve supply of the muscle, coming from the cervical region, may remain intact, and latissimus dorsi can be used from its insertion to move the pelvis. Accordingly, efforts are made to strengthen the muscle, through exercise such as archery, in order to help these patients to achieve some independence of movement.

FIG. 5.60. Activity of the pectoral muscles as adductors against reciprocal resistance of hands. (*By courtesy of Professor R. D. Lockhart.*)
1. Infraclavicular fossa.
2. Deltoid, (a) anterior fibres, (b) middle fibres.
3. Pectoralis major, (a) clavicular part, (b) sternocostal part.

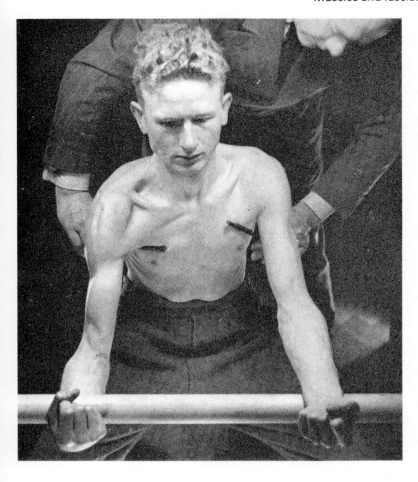

FIG. 5.61. Actions of pectoralis major. The subject's right arm is being raised and the left arm depressed against resistance. The clavicular part of the muscle is active on the right, inactive on the left; the reverse is the case with the sternocostal part. (From *Living anatomy* by R. D. Lockhart, 6th edn. Faber and Faber, London (1963).)

PECTORALIS MAJOR

Like latissimus dorsi, pectoralis major extends from the thoracic wall to the humerus, and therefore acts on both the shoulder girdle and shoulder joint; its major importance relates to the shoulder joint. The smaller clavicular head of the muscle arises from the front of the clavicle in its medial half or two thirds [FIGS. 5.51, 5.58, and 5.62]; the larger sternocostal portion arises from the front of the manubrium and the body of the sternum, and from the anterior aspects of the first six costal cartilages as well as from the anterior wall of the rectus sheath [FIGS. 5.58 and 5.63]. The two parts of the

muscle are usually continuous with each other, but may be separated by a groove.

Pectoralis major is inserted into the crest of the greater tubercle of the humerus. The clavicular portion of the muscle is attached in front of and distal to the sternocostal portion, the fibres of which twist on themselves, forming the rounded **anterior fold of the axilla**, so that the fibres which arise lowest on the chest wall pass behind the upper fibres to be inserted posterior to them. In this way the tendon comes to resemble a U in cross section. The posterior limb of the U is blended with the capsule of the shoulder joint, and the clavicular fibres blend with the insertion of the deltoid.

Superiorly pectoralis major is separated from deltoid by a depression known as the **infraclavicular fossa** [FIG. 5.60], in which lie the cephalic vein and the deltoid branches of the thoraco-acromial artery.

Nerve supply

The lateral pectoral nerve (C. 5, 6, 7), from the lateral cord of the brachial plexus, supplies the clavicular and the upper part of the sternocostal portion; the medial pectoral nerve (C. 8; T. 1), from the medial cord, supplies the lower part of the sternocostal portion.

Actions

Pectoralis major is a powerful adductor and medial rotator of the humerus. The clavicular portion helps to flex the humerus to the horizontal, while the sternocostal portion, acting against resistance, extends the flexed humerus [FIG. 5.61]. In such actions the two parts of the muscle act quite independently, but they combine in movements of pushing or throwing. Pectoralis major is one of the

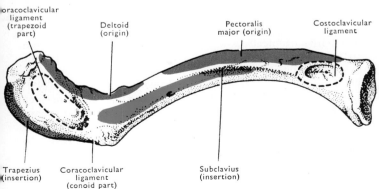

FIG. 5.62. Muscle attachments to inferior surface of right clavicle.

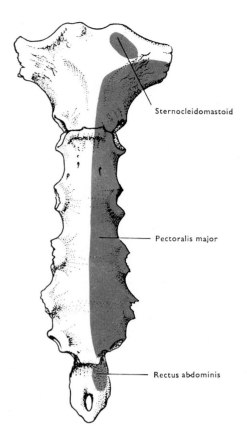

Sternocleidomastoid

Pectoralis major

Rectus abdominis

FIG. 5.63. Muscle attachments to front of sternum.

main climbing muscles, and pulls the body up to the fixed arm. This action is also used in respiratory difficulties; the upper limb is fixed by the hand grasping the bed rail, and pectoralis major becomes an accessory muscle of inspiration, pulling up the upper ribs.

Despite its numerous and important actions, absence of pectoralis major, which may rarely occur as a congenital anomaly, causes virtually no disability, though the deformity caused by the absence of the anterior axillary fold is striking. The integrity of the muscle is tested by adduction of the arm against resistance [FIG. 5.59].

MUSCLES OF THE SHOULDER JOINT

DELTOID

Deltoid [FIGS. 5.52 and 5.58] is a coarse, thick, multipennate muscle with an extensive origin from the anterior border of the lateral third of the clavicle [FIG. 5.51], the lateral border of the acromion, the lower lip of the crest of the spine of the scapula, and the fascia covering infraspinatus. This attachment is similar to that of trapezius, though deltoid is on the outside and the trapezius on the inside of the bony arch to which they are attached.

As its name implies, the muscle is triangular, and its apex is inserted into the deltoid tuberosity half way down the lateral surface of the body of the humerus [FIG. 5.67]. Its fibres blend with the anterior aspect of the tendon of pectoralis major.

Deltoid is composed of three functional parts, anterior, posterior, and intermediate. Only the intermediate portion, which abducts the shoulder, is multipennate, and this is probably because the

mechanical disadvantage under which it must act requires considerable strength.

Nerve supply

From the axillary nerve, a branch of the posterior cord of the brachial plexus containing fibres from the ventral rami of C. 5 and 6.

Actions

The intermediate portion is the chief abductor of the humerus, elevating the arm away from the side in the plane of the scapula, but it can only do so efficiently after the movement has been initiated by supraspinatus, since when the arm is by the side deltoid of necessity directs its pull upwards rather than outwards. Its tendency to produce upward displacement of the humerus is resisted by the muscles of the 'rotator cuff'. Deltoid also controls gravitational adduction of the upper limb [FIGS. 5.11 and 5.12].

The anterior portion of the muscle is a strong flexor and medial rotator of the humerus; the posterior portion is a strong extensor and lateral rotator, and can help to transfer to the shoulder girdle the strain imposed by heavy weights carried in the hand (Bearn 1961). In adduction, the posterior fibres are very active, probably to resist the medial rotation which latissimus dorsi and pectoralis major would otherwise produce (Scheving and Pauly 1959). To test deltoid, the patient's arm is abducted in the plane of the scapula to about 45 degrees, and he is asked to hold it there against resistance. The muscle is then easily seen and felt. It is interesting that when deltoid is paralysed it is sometimes possible for the upper limb to be abducted through a full normal range (Sunderland 1944b).

SUPRASPINATUS

The origin of supraspinatus is from the medial two-thirds of the supraspinous fossa of the scapula [FIG. 5.55], and from the fascia which covers the muscle. It is inserted into a facet on the superior aspect of the greater tubercle of the humerus and also into the capsule of the shoulder joint [FIG. 5.67]. Supraspinatus lies deep to trapezius, the acromion, and the coraco-acromial ligament.

Nerve supply

From the suprascapular nerve (C. 5 and 6), a branch of the upper trunk of the brachial plexus.

Actions

Supraspinatus initiates the process of abduction at the shoulder joint, and is more important in the early stages of this movement than it is later, when deltoid takes over. If it is paralysed, the patient may produce this initial abduction by leaning slightly to the side, so using gravity to replace the supraspinatus, or he may 'kick' the elbow out by a jerk of his hip. In all movements of the joint supraspinatus acts with infraspinatus, teres minor, and subscapularis as extensile ligaments [p. 271] holding the head of the humerus in contact with the shallow glenoid socket and preventing dislocation in any position of the joint. The whole muscle group is sometimes known as the 'rotator cuff' of the shoulder joint (Moseley 1951). Damage to supraspinatus tendon is a common form of derangement of the shoulder joint, and causes pain on abduction of the humerus: the tendon occasionally ruptures. To test the muscle, the patient attempts to abduct the arm from the side against resistance; the muscle belly can be felt above the spine of the scapula.

INFRASPINATUS

Infraspinatus comes from the infraspinous fossa of the scapula, except for the portion next the neck of the bone [FIG. 5.55], and from the thick fascia covering the muscle. The tendon inserts into the middle facet on the greater tubercle of the humerus and into the capsule of the shoulder joint. Between infraspinatus and the neck of the scapula there is a bursa which occasionally communicates with the shoulder joint.

The upper part of infraspinatus lies deep to the acromion, deltoid, and trapezius, but the lower part is superficial [FIG. 5.52].

Nerve supply
From the suprascapular nerve (C. 5 and 6), a branch of the upper trunk of the brachial plexus.

Actions
Infraspinatus is a lateral rotator of the humerus, but its main action is as a member of the 'rotator cuff' [p. 318]. To test infraspinatus,

the patient is asked to rotate the arm laterally against resistance, while keeping the elbow close to the side. This is conveniently done if the forearm is flexed to a right angle and used to push backwards against the examiner's hand. The muscle belly can be felt contracting during this movement.

TERES MINOR

Teres minor comes by fleshy fibres from the upper two-thirds of the lateral border of the dorsal surface of the scapula, alongside the lateral border of infraspinatus [FIG. 5.55]. It is inserted into the lowest of the three facets on the greater tubercle of the humerus and into the capsule of the shoulder joint. Some fibres extend downwards on to the body for a short distance. The long head of triceps separates

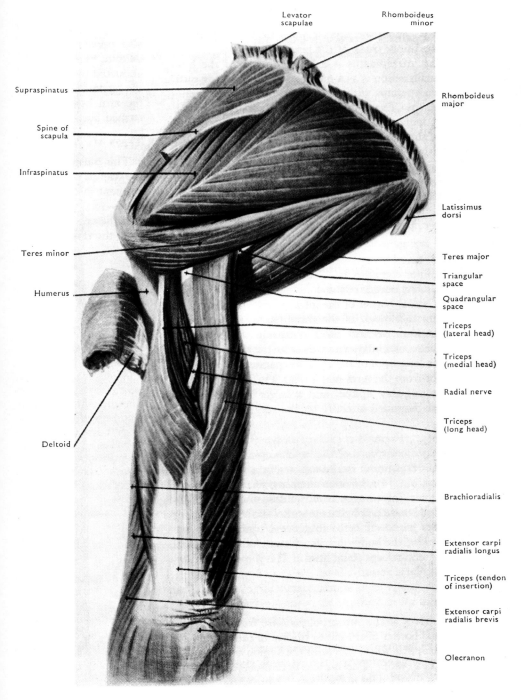

FIG. 5.64. Left scapular muscles and triceps.

Levator scapulae
Rhomboideus minor
Supraspinatus
Spine of scapula
Infraspinatus
Teres minor
Humerus
Deltoid
Rhomboideus major
Latissimus dorsi
Teres major
Triangular space
Quadrangular space
Triceps (lateral head)
Triceps (medial head)
Radial nerve
Triceps (long head)
Brachioradialis
Extensor carpi radialis longus
Triceps (tendon of insertion)
Extensor carpi radialis brevis
Olecranon

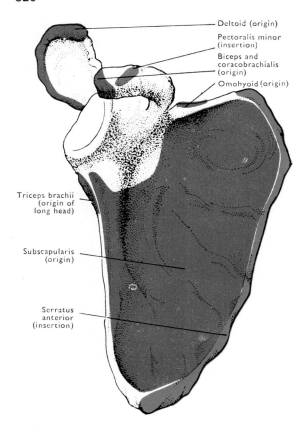

FIG. 5.65. Muscle attachments to costal surface of right scapula.

the tendon of teres minor from teres major [FIG. 5.64], so producing a very small **triangular space**, and a larger **quadrangular space** which is bounded laterally by the surgical neck of the humerus.

Nerve supply

From the axillary nerve (C. 5 and 6), a branch of the posterior cord of the brachial plexus.

Actions

Teres minor is a lateral rotator and an adductor of the humerus, but, like the other members of the rotator cuff [p. 318], its main importance is in holding the head of the humerus in contact with the glenoid socket during movements of the shoulder joint. It helps to prevent upward displacement of the head of the humerus by such muscles as deltoid, biceps brachii, and triceps (long head). It is impossible to test the teres minor satisfactorily.

SUBSCAPULARIS

The fleshy fibres of subscapularis arise from the groove along the lateral border of the costal surface of the scapula and from the subscapular fossa, except in the region of the angles of the bone [FIG. 5.65]. The muscle is reinforced by fibrous septa springing from ridges in the fossa. A broad thick tendon inserts into the lesser tubercle of the humerus, the capsule of the shoulder joint, and the front of the body of the humerus for 2–3 cm below the tubercle [FIG. 5.67].

Subscapularis forms the greater part of the posterior wall of the axilla, and it is separated from the neck of the scapula by the subscapular bursa, which nearly always communicates with the shoulder joint. It constitutes the upper boundary of the triangular and the quadrangular spaces as seen from the front [FIG. 5.66].

Nerve supply

From the upper and lower subscapular nerves (C. 5, 6, and 7), branches of the posterior cord of the brachial plexus.

Actions

The muscle is a medial rotator of the humerus but its chief function is as a member of the 'rotator cuff' [p. 318]. It also prevents the head from being pulled upwards by deltoid, biceps, and long head of triceps. If the humerus is abducted through 90 degrees, subscapularis, together with pectoralis major and the anterior part of deltoid, pulls the arm horizontally forwards, a movement which has been termed horizontal flexion (see below).

TERES MAJOR [FIG. 5.54]

This muscle comes from a raised oval area on the dorsum of the scapula near the inferior angle [FIG. 5.55], and also from the adjacent fascia. Its fibres are adherent to those of latissimus dorsi [FIG. 5.66] and it is inserted by a broad flat tendon into the crest of the lesser tubercle of the humerus, separated from the tendon of latissimus dorsi by a bursa.

Teres major, and the tendon of latissimus dorsi which sweeps round it, form the **posterior fold of the axilla** [p. 316]. The muscle also constitutes the lower boundary of the triangular and quadrangular spaces in the posterior wall of the axilla [FIG. 5.64]. Through the quadrangular space pass the axillary nerve and the posterior humeral circumflex vessels, and through the triangular space pass the circumflex scapular vessels [FIG. 5.71].

Nerve supply

From the lower subscapular nerve (C. 6 and 7), a branch of the posterior cord of the brachial plexus which also supplies the lower portion of the subscapularis.

Actions

It adducts and medially rotates the humerus, and can help to extend it from the flexed position. If the arm is adducted against resistance, teres major can be felt contracting in the posterior axillary fold.

Movements of the shoulder region

Movements occurring at the shoulder girdle and at the shoulder joint normally complement each other, and are almost impossible to dissociate under normal conditions. From the anatomical position, the scapula can be elevated or depressed, protracted or retracted round the chest wall, or rotated so that the inferior angle travels laterally or medially. All such movements involve the sternoclavicular and acromioclavicular joints, and the muscles concerned in them are shown in the first table below. The second table shows the muscles concerned in movements of the shoulder joint.

The movements of the shoulder joint are best defined in relation to the plane of the scapula rather than in relation to the three basic anatomical planes. Thus abduction is considered to take place away from the body in the plane of the scapula, the arm travelling upwards, outwards, and forwards. Adduction is the reverse of this, and flexion and extension take place at right angles to this plane. Thus, in flexion the arm passes not only forwards but medially, and in extension it travels laterally as well as backwards.

Two additional useful movements require special definition. The swing of the arm from a position of 90 degrees of abduction forwards to a position of 90 degrees flexion is called **horizontal flexion**, and the reverse movement is **horizontal extension**. These movements are used in such activities as drawing horizontal lines on a blackboard and dusting a shelf.

The way in which movement of the shoulder girdle combines with movement of the shoulder joint may be illustrated by considering abduction of the upper limb above the head. The muscles which rotate the scapula are active throughout this movement, and so are the muscles which abduct the humerus at the shoulder joint. About half the total range of movement occurs as a result of the glenoid socket being turned to face upwards, the other half is due to abduction of the humerus [FIGS. 4.26 and 4.28].

When abduction begins, the weight of the upper limb may cause a slight medial movement of the inferior angle of the scapula, but it then begins to move laterally under the influence of trapezius and serratus anterior. Levator scapulae steadies the upper angle of the scapula and prevents it moving laterally. Rotation of the scapula is necessarily accompanied by some rotation and elevation of the clavicle, under the influence of the upper fibres of trapezius, and the subclavius is probably also in action to hold the clavicle into the socket of the sternoclavicular joint [p. 315].

At the same time supraspinatus and deltoid abduct the humerus at the shoulder joint, while the rotator cuff of muscles, particularly supraspinatus, holds the head of the humerus into the glenoid socket of the scapula. The downward pull of subscapularis, infraspinatus, and teres minor prevents upward displacement of the head of the humerus due to the action of the strong deltoid muscle. As the movement proceeds, the long head of the triceps comes to lie inferior to the capsule of the joint, and acts as a support preventing downward dislocation of the head of the humerus. If abduction is undertaken in the coronal plane, the summated action of the muscles operating from the scapula and trunk on the humerus causes an

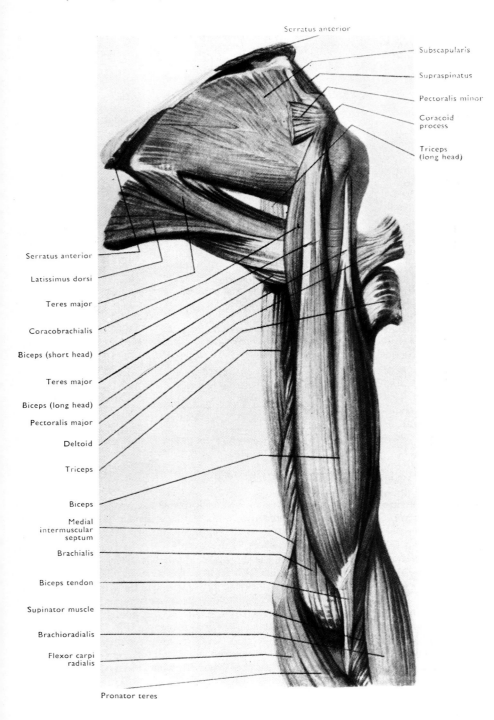

Serratus anterior

Subscapularis

Supraspinatus

Pectoralis minor

Coracoid process

Triceps (long head)

FIG. 5.66. Muscles of posterior wall of left axilla and front of arm.

Serratus anterior

Latissimus dorsi

Teres major

Coracobrachialis

Biceps (short head)

Teres major

Biceps (long head)

Pectoralis major

Deltoid

Triceps

Biceps

Medial intermuscular septum

Brachialis

Biceps tendon

Supinator muscle

Brachioradialis

Flexor carpi radialis

Pronator teres

Muscles concerned in movements of the shoulder girdle

Movement	Muscle	Principal Nerve Supply
Elevation	Trapezius (upper fibres)	Spinal part of accessory
	Levator scapulae	Ventral rami of C. 3, 4, 5
	Rhomboideus major	Ventral ramus of C. 5
	Rhomboideus minor	Ventral ramus of C. 5
	Sternocleidomastoid	Spinal part of accessory
Depression	Trapezius (lower fibres)	Spinal part of accessory
	Pectoralis minor	Medial and lateral pectoral
	Pectoralis major (sternocostal portion)	Medial and lateral pectoral
	Latissimus dorsi	Thoracodorsal
Protraction	Serratus anterior	Long thoracic
	Pectoralis minor	Medial and lateral pectoral
	Pectoralis major	Medial and lateral pectoral
Retraction	Trapezius (middle fibres)	Spinal part of accessory
	Rhomboideus major	Ventral ramus of C. 5
	Rhomboideus minor	Ventral ramus of C. 5
	Latissimus dorsi	Thoracodorsal
Lateral displacement of inferior angle of scapula	Serratus anterior	Long thoracic
	Trapezius (upper and lower fibres)	Spinal part of accessory
Medial displacement of inferior angle of scapula	Pectoralis minor	Medial and lateral pectoral
	Rhomboideus major	Ventral ramus of C. 5
	Rhomboideus minor	Ventral ramus of C. 5
	Latissimus dorsi	Thoracodorsal

Muscles concerned in movements of the shoulder joint

Movement	Muscle	Principal Nerve Supply
Flexion	Deltoid (anterior portion)	Axillary
	Pectoralis major (clavicular portion: sternocostal portion flexes the extended humerus as far as the position of rest)	Medial and lateral pectoral
	Coracobrachialis	Musculocutaneous
	Biceps brachii	Musculocutaneous
Extension	Deltoid (posterior portion)	Axillary
	Teres major (of flexed humerus)	Lower subscapular
	Latissimus dorsi (of flexed humerus)	Thoracodorsal
	Sternocostal portion of pectoralis major (of flexed humerus)	Medial and lateral pectoral
	Long head of triceps (to position of rest)	Radial
Abduction	Deltoid (middle portion)	Axillary
	Supraspinatus	Suprascapular
	Biceps brachii (long head)	Musculocutaneous
Adduction	Pectoralis major	Medial and lateral pectoral
	Latissimus dorsi	Thoracodorsal
	Teres major	Lower subscapular
	Triceps (long head)	Radial
	Coracobrachialis	Musculocutaneous
Lateral rotation	Deltoid (posterior portion)	Axillary
	Infraspinatus	Suprascapular
	Teres minor	Axillary
Medial rotation	Pectoralis major	Medial and lateral pectoral
	Teres major	Lower subscapular
	Latissimus dorsi	Thoracodorsal
	Deltoid (anterior portion)	Axillary
	Subscapularis	Upper and lower subscapular
Horizontal flexion	Deltoid (anterior portion)	Axillary
	Pectoralis major	Medial and lateral pectoral
	Subscapularis	Upper and lower subscapular
Horizontal extension	Deltoid (posterior portion)	Axillary
	Infraspinatus	Suprascapular

automatic lateral rotation of the humerus (Martin 1940). If this rotation is deliberately prevented, it is impossible to abduct the arm to an angle of more than about 90 degrees because of the obstruction of the greater tubercle of the humerus. If abduction takes place in the plane of the scapula no such rotation occurs.

Deltoid, pectoralis major, latissimus dorsi, and teres major are the chief prime movers of the shoulder joint, while the components of the 'rotator cuff'—supraspinatus, infraspinatus, teres minor, and subscapularis—have the primary role of extensile ligaments. Deltoid and supraspinatus help to transfer the strain produced by weights carried in the hand. All the muscles passing from the trunk or scapula to the humerus can act in the reverse direction, and this is well seen when a gymnast performs exercises on the parallel bars or on the rings. The two chief climbing muscles are pectoralis major and latissimus dorsi, and both are used in the downstroke in swimming.

The shoulder joint is involved in many movements occurring primarily at other joints. For example, in supination and pronation it is brought into play to enlarge the range of these movements when the elbow joint is extended. Paralysis of the lateral rotators of the shoulder means that when writing the patient has to pull the paper along, since without this shoulder movement he cannot write a continuous line. It is perhaps strange that the medial rotators are apparently so much more powerful than the lateral rotators. Infraspinatus, the posterior portion of deltoid, and teres minor (lateral rotators) correspond roughly to the subscapularis, anterior portion of deltoid, and teres major (medial rotators). This leaves a surplus of two extremely strong muscles (pectoralis major and latissimus dorsi) favouring medial rotation. Yet this movement does not appear to be, functionally, any more important than lateral rotation.

MUSCLES OF THE ARM

BICEPS BRACHII

This muscle operates on three joints. It has two tendinous heads at its upper end, and two tendinous insertions at its lower end. The short head is attached, in common with coracobrachialis, to the tip of the coracoid process of the scapula [FIGS. 5.65 and 5.66]. The long head arises from the supraglenoid tubercle of the scapula and from the labrum glenoidale of the shoulder joint. Its tendon runs through the shoulder joint, invested by a covering of the synovial membrane, and leaves the joint deep to the transverse humeral ligament. It then lies in the intertubercular sulcus of the humerus, covered over anteriorly by the tendon of pectoralis major [FIG. 5.66].

The two fleshy bellies derived from the separate heads unite just below the middle of the arm. At the elbow, the resultant fusiform mass gives rise to a strong tendon which is inserted into the posterior part of the tuberosity of the radius [FIGS. 5.73 and 5.75], from the remainder of which it is separated by a small bursa. The second

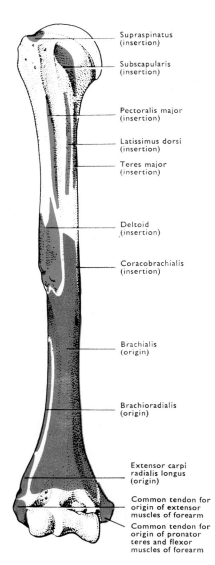

Supraspinatus
(insertion)

Subscapularis
(insertion)

Pectoralis major
(insertion)

Latissimus dorsi
(insertion)

Teres major
(insertion)

Deltoid
(insertion)

Coracobrachialis
(insertion)

Brachialis
(origin)

Brachioradialis
(origin)

Extensor carpi
radialis longus
(origin)

Common tendon for
origin of extensor
muscles of forearm

Common tendon for
origin of pronator
teres and flexor
muscles of forearm

FIG. 5.67. Muscle attachments to front of right humerus.

and the usual test of its power is to ask the patient to flex, against resistance, the supinated forearm, bent at right angles to the abducted arm. The muscle is then easily seen and felt.

Biceps can also flex the shoulder joint, and the long head is part of the important stabilizing mechanism for the joint. If deltoid is paralysed, the long head of the biceps can occasionally effect some useful abduction of the humerus.

CORACOBRACHIALIS [FIG. 5.74]

Coracobrachialis [FIGS. 5.50 and 5.66] arises from the tip of the coracoid process of the scapula in company with the short head of the biceps, and ends in a flat tendon inserted into the middle of the medial aspect of the body of the humerus [FIG. 5.67]. Some of the fibres may continue into the medial intermuscular septum of the arm.

Nerve supply

The nerve to coracobrachialis (C. 6, 7) may arise directly from the lateral cord of the brachial plexus, though it is usually a branch of the musculocutaneous nerve.

Actions

It is a flexor and adductor of the shoulder joint.

BRACHIALIS

The main flexor of the elbow lies posterior to the biceps brachii, and comes from the distal two thirds of the front of the body of the humerus and from the anterior aspects of both intermuscular septa [FIGS. 5.67 and 5.69]; its proximal end clasps the insertion of deltoid. The muscle inserts into the inferior part of the coronoid process and the tuberosity of the ulna [FIGS. 5.68 and 5.75] by a tendon which forms the floor of the cubital fossa [p. 325]. Some fibres of brachialis are inserted into the articular capsule of the elbow joint, and the muscle may be partly fused with brachioradialis [p. 331].

Nerve supply

From the musculocutaneous nerve (C. 5, 6). Another branch is received from the radial nerve as it runs along the lateral border of the brachialis. This branch is almost certainly sensory.

Actions

It flexes the elbow. The fibres attached to the capsule of the elbow joint pull the redundant folds of the articular capsule upwards during flexion of the joint, and prevent them from becoming trapped between the moving bones. During flexion of the elbow against resistance, brachialis can be felt contracting alongside the tendon of biceps and on either side of its belly.

TRICEPS BRACHII

Triceps is the only muscle on the back of the arm, and arises by three heads. The tendinous long head comes from the infraglenoid tubercle of the scapula and the labrum glenoidale, and its fleshy belly occupies the middle of the back of the arm [FIGS. 5.64 and 5.71]. The fleshy lateral and medial heads are attached to the posterior aspect of the humerus between the insertion of teres minor and the olecranon fossa [FIG. 5.70]. The lateral head arises above and lateral to the groove for the radial nerve, and the larger medial head arises inferior and medial to it. The lateral head therefore passes posterior to the groove to join the medial head. The medial head, which lies deep to the other two, has additional attachments

insertion is the **bicipital aponeurosis**, a strong membranous band which runs downwards across the cubital fossa [FIGS. 5.72 and 5.74], to join the deep fascia covering the origins of the flexor muscles of the forearm. The upper portion of this band can easily be felt in the living body as a crescentic border crossing in front of the brachial artery and median nerve.

Occasionally there may be a third head of the biceps; this arises at the insertion of the coracobrachialis and passes into the bicipital aponeurosis on the medial side of the main belly of the muscle.

Nerve supply

From the musculocutaneous nerve (C. 5, 6).

Actions

Biceps is an important flexor of the elbow joint, and a powerful supinator of the forearm. It has been described as the muscle which puts in the corkscrew and pulls out the cork. Under normal circumstances both actions occur together, and the unwanted one is cancelled out by synergists. Biceps exerts its maximum power in both supination and flexion when the elbow is flexed at right angles, to the posterior aspects of the lateral and medial intermuscular septa.

All three heads are inserted together by a broad flattened tendon into the posterior part of the proximal surface of the olecranon of the ulna and into the deep fascia of the forearm on each side of it. Some of the fibres of the medial head are attached to the posterior part of the capsule of the elbow joint, and a small bursa separates the tendon of triceps from the remainder of the capsule.

Triceps is superficial in almost its whole length [FIGS. 5.59 and 5.74], and the long head passes between teres major and teres minor in the posterior wall of the axilla.

Nerve supply

All three heads are supplied separately by branches of the radial nerve. The lateral head derives its supply from C. 6, 7, 8, and the long and medial heads from C. 7, 8. The medial head receives two branches, one of which has a long extramuscular course in company with the ulnar nerve before it enters the distal part of the muscle; the other enters its proximal part and continues through the substance of the muscle to end in the anconeus [FIG. 5.71].

Actions

It extends the elbow joint and pulls the posterior part of its capsule out of danger as it does so. The long head can adduct the humerus and extend it from the flexed position; it has a stabilizing action on the shoulder joint, and is an important extensile ligament on the under surface of its capsule during abduction. To test triceps, the arm is abducted to a right angle to bring the forearm parallel to the ground and so eliminate the effect of gravity on extension of the elbow. The flexed forearm is then extended against resistance.

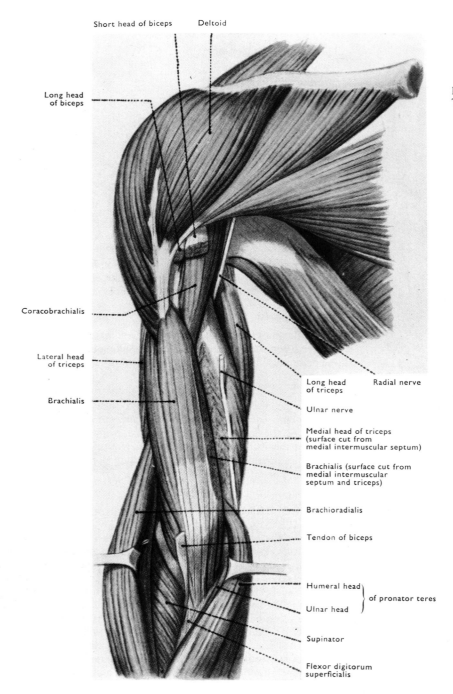

FIG. 5.68. Dissection of muscles of front of arm. The biceps has been removed to show the brachialis.

Short head of biceps   Deltoid

Long head of biceps

Coracobrachialis

Lateral head of triceps

Brachialis

Long head of triceps   Radial nerve

Ulnar nerve

Medial head of triceps (surface cut from medial intermuscular septum)

Brachialis (surface cut from medial intermuscular septum and triceps)

Brachioradialis

Tendon of biceps

Humeral head ⎫
             ⎬ of pronator teres
Ulnar head ⎭

Supinator

Flexor digitorum superficialis

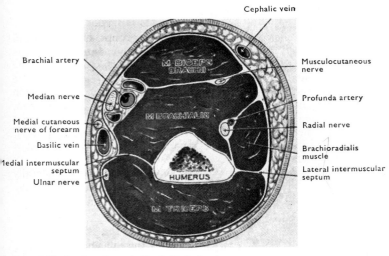

FIG. 5.69. Section through distal third of right arm.

Movements of the elbow

There is little difficulty in sorting out the muscles concerned in the movements of flexion and extension of the elbow joint, and they are indicated in the table below. The movements are considerably influenced by gravity. From the anatomical position the elbow can be flexed and then extended by the flexors alone, even though the triceps and anconeus are paralysed. Again, a patient with paralysed flexors can both flex and extend the elbow by the triceps alone if the upper limb is raised above the head.

The more important of the two movements is flexion, which depends on several groups of muscles. Some of the muscles are not normally used to flex the elbow, but can assist in the movement in case of difficulty, particularly pronator teres and flexor carpi radialis. Flexion has thus the safeguard of being mediated by three nerves—musculocutaneous, radial, and median, whereas extension depends on one nerve alone, the radial.

Flexion at the elbow is usually limited by the collision of the soft parts of the forearm with those of the arm: the greater the muscular development, the smaller the range of movement. A similar situation occurs in relation to the knee joint.

Muscles concerned in movements at the elbow joint

Movement	Muscle	Principal Nerve Supply
Flexion	Brachialis	Musculocutaneous
	Biceps brachii	Musculocutaneous
	Brachioradialis	Radial
	Extensor carpi radialis longus	Radial
	Pronator teres	Median
	Flexor carpi radialis	Median
Extension	Triceps brachii	Radial
	Anconeus	Radial

MUSCLES OF THE FOREARM

Muscles of the front of the forearm

There are three functional groups of muscles on the front of the forearm—the pronators of the forearm, the flexors of the wrist, and the long flexors of the fingers and thumb. They are arranged in three

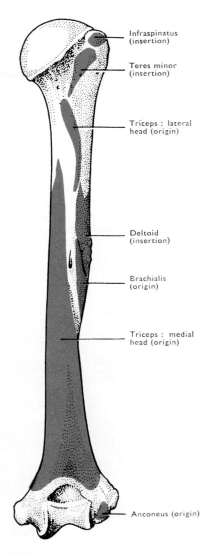

FIG. 5.70. Muscle attachments to back of right humerus.

layers. The **superficial layer** consists of four muscles which radiate from a common tendon (the common flexor origin) attached to the front of the medial epicondyle of the humerus: these are pronator teres, flexor carpi radialis, palmaris longus, and flexor carpi ulnaris [FIGS. 5.72 and 5.74]. Pronator teres forms the medial boundary of a depression in front of the elbow known as the **cubital fossa**; the lateral boundary of this fossa is the medial edge of the brachioradialis muscle [FIG. 5.72], and its superior boundary is an imaginary line drawn between the medial and lateral epicondyles of the humerus. The floor is formed by the tendon of the brachialis and the supinator, and the fossa contains the tendon of insertion of the biceps brachii, in addition to parts of the brachial, radial, and ulnar arteries and the radial and median nerves.

Deep to the four superficial muscles lies a **middle layer** consisting of flexor digitorum superficialis only. Deep to this again is the **third layer**, comprising flexor digitorum profundus, flexor pollicis longus, and pronator quadratus [FIG. 5.80].

PRONATOR TERES

The humeral head of pronator teres comes from the lower third of the medial supracondylar ridge of the humerus, the corresponding part of the medial intermuscular septum, and from the front of the medial epicondyle of the humerus and the adjacent fascia. The small ulnar head runs from the coronoid process of the ulna [FIG. 5.75] to join the humeral head on its deep surface [FIG. 5.73]. The median nerve passes into the forearm between the two heads.

The fasciculi of pronator teres pass downwards and laterally to a roughened oval impression on the middle of the lateral surface of the body of the radius deep to the brachioradialis [FIGS. 5.73 and 5.75].

Nerve supply

From the median nerve by a branch (C. 6, 7) which enters it above the elbow.

Actions

Pronation of the forearm is the chief action, but it also assists in flexing the elbow. To test the muscle the patient resists an attempt to supinate the forearm with the elbow flexed to a right angle. The muscle belly can be felt and occasionally seen contracting.

FLEXOR CARPI RADIALIS

The radial flexor of the wrist arises from the common flexor origin and from the fascia which covers it and separates it from its neighbours. The muscle runs downwards and obliquely across the forearm from the medial to the lateral side. About the middle of the forearm the fleshy belly is replaced by a flattened tendon which becomes cord-like as it approaches the wrist [FIG. 5.72]. The tendon, surrounded by a synovial sheath [p. 336], runs in a special compartment in the lateral border of the flexor retinaculum [p. 347], and plays in the groove on the trapezium. It inserts into the front of the bases of the second and third metacarpal bones.

At the wrist, the tendon lies between the radial vessels laterally and the median nerve medially, and is an important guide to the position of both.

Nerve supply

By a branch (C. 6, 7) from the median nerve, which usually enters it just below the elbow joint.

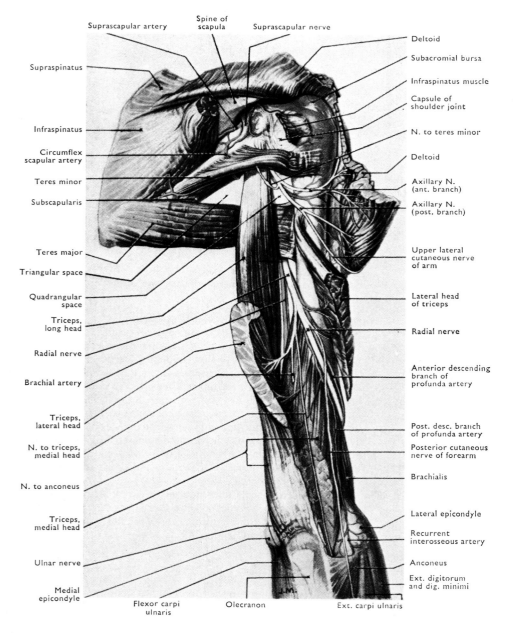

FIG. 5.71. Dissection of back of shoulder and arm.

Actions

It flexes and abducts the carpus and also helps to flex the elbow and to pronate the forearm. It is an important synergist during extension of the fingers [pp. 333, 337]. The muscle is tested with the dorsum of the forearm flat on a table: the patient then flexes the wrist against resistance while the examiner feels the tendon of the muscle.

PALMARIS LONGUS

This small elongated muscle is frequently absent. It arises from the common flexor origin, medial to flexor carpi radialis, and from the septa between it and its neighbours. Half way down the forearm the thin flat tendon appears, and at the wrist it adheres to the superficial surface of the flexor retinaculum and inserts into the apex of the palmar aponeurosis [FIG. 5.72]. A prolongation from this insertion may cover over the short muscles of the thumb. If the muscle is present, the median nerve lies posterior to its tendon at the wrist.

Nerve supply

By a branch (C. 8) from the median nerve, which enters it just distal to the elbow joint.

Actions

It is a flexor of the wrist, and, by its attachment to the palmar aponeurosis and thus to the digital fibrous sheaths [p. 348], it may

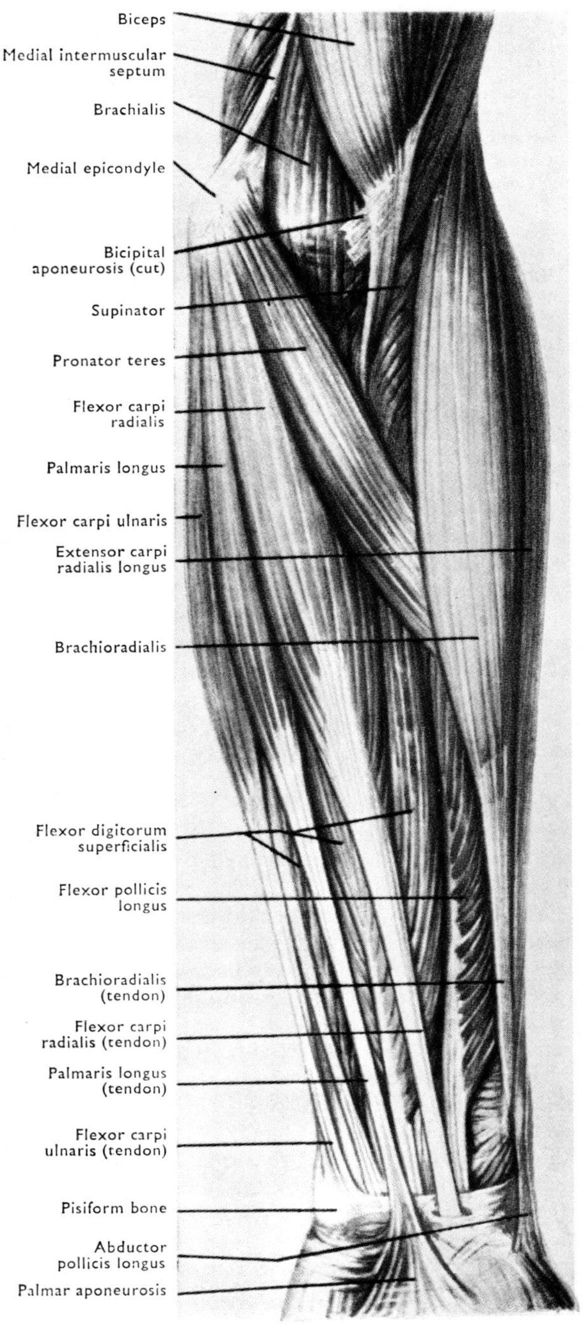

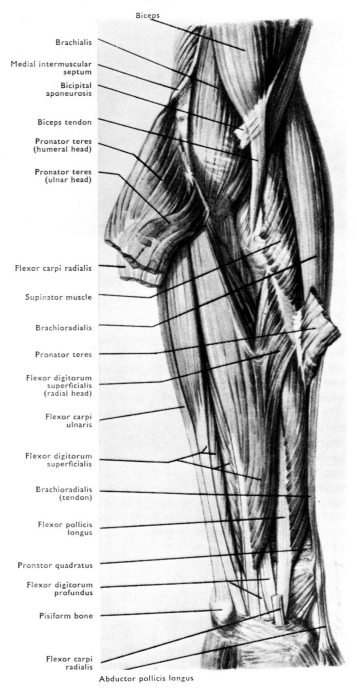

FIG. 5.72. Superficial muscles of front of left forearm.

FIG. 5.73. Deeper muscles of front of left forearm.

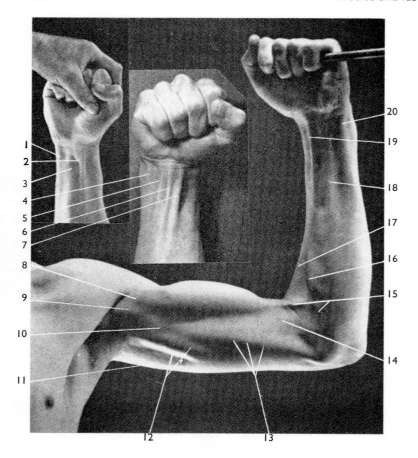

FIG. 5.74. Flexors of wrist and elbow in action. In the top left figure the median nerve, escaping from under flexor carpi radialis to lie upon flexor digitorum superficialis tendons, is seen well because palmaris longus is absent. (From *Living anatomy* by R. D. Lockhart, 6th edn. Faber and Faber, London (1963).)

1. Flexor carpi ulnaris.
2. Median nerve.
3. Flexor digitorum superficialis.
4. Flexor carpi ulnaris.
5. Flexor digitorum superficialis, ring finger tendon.
6. Palmaris longus.
7. Flexor carpi radialis.
8. Biceps, short head.
9. Coracobrachialis.
10. Neurovascular bundle.
11. Triceps, long head.
12. Triceps, medial head.
13. Intermuscular septum.
14. Brachialis.
15. Bicipital aponeurosis.
16. Pronator teres.
17. Brachioradialis.
18. Flexor digitorum superficialis.
19. Flexor carpi radialis.
20. Flexor carpi ulnaris.

have some slight action in flexing the metacarpophalangeal joints of the fingers. The muscle is tested like flexor carpi radialis.

FLEXOR CARPI ULNARIS

Like pronator teres [p. 325] flexor carpi ulnaris has two heads, and between them the ulnar nerve enters the forearm. The humeral head of the muscle is small, and arises from the common flexor origin and the adjacent fascia. The ulnar head takes aponeurotic origin from the medial border of the olecranon and the posterior border of the ulna in its upper two-thirds. The tendon, which appears approximately half way down the forearm, is attached to the pisiform bone, and so, through the **pisohamate** and **pisometacarpal ligaments**, to the hook of the hamate and the base of the fifth metacarpal [FIG. 5.90]. Some fibres may also be prolonged into abductor digiti minimi [p. 341]. The tendon is a guide to the ulnar nerve and vessels, which lie on its lateral side.

Nerve supply

From the ulnar nerve by several branches (C. 7, 8) which enter it at intervals in the forearm.

Actions

It flexes and adducts the carpus, and fixes the pisiform during abduction of the little finger in order to stabilize the origin of abductor digiti minimi [p. 341]. It is an important synergist during extension of the fingers and extension of the thumb. The muscle is tested like flexor carpi radialis.

FLEXOR DIGITORUM SUPERFICIALIS

Flexor digitorum superficialis [FIG. 5.73] has a long linear origin from the common flexor tendon, the adjacent fascia and septa, the

ulnar collateral ligament of the elbow joint, the medial border of the coronoid process of the ulna, a thin fibrous arch which crosses the median nerve and ulnar vessels, and the upper two-thirds of the anterior border of the radius [FIG. 5.75].

Four tendons arise from the muscle half way down the forearm: two, destined for the middle and ring fingers, lie side by side superficial to those for the index and little fingers. They maintain this relationship as they pass under the flexor retinaculum in the carpal tunnel, lying superficial to the tendons of flexor digitorum profundus. In the palm, the tendons diverge, and each one, in company with the corresponding tendon of flexor digitorum profundus, enters the fibrous flexor sheath [p. 348] of its own digit [FIG. 5.76]. At the level of the metacarpophalangeal joint, each tendon of flexor digitorum superficialis splits into two. The two halves pass posteriorly, with a spiral twist, round the sides of the tendon of flexor profundus, into the margins of the palmar surface of the middle phalanx. Some of the tendinous fibres decussate posterior to the tendon of the profundus, forming for it a tunnel which cannot be obliterated by tension.

The tendons of flexor superficialis are also attached to the bones of the fingers by **vincula tendinum** which convey to them the small blood vessels necessary for their nutrition (Brockis 1953). The **vincula brevia** are small triangular bands of connective tissue attached to the front of the interphalangeal joint and the distal part of the proximal phalanx. They occupy the interval between the tendon and the digit close to the insertion [FIGS. 5.77 and 5.78] and contain some collagen and elastic fibres. The **vincula longa** are variable, narrow bands extending from the back of the tendon to the proximal part of the palmar surface of the proximal phalanx.

In the most distal part of the forearm, the carpal tunnel, and the palm, the tendons of flexor digitorum superficialis and profundus are enclosed in a common synovial sheath [p. 336].

Nerve supply

By branches (C. 7, 8; T. 1) from the median nerve in the forearm.

Actions

The muscle is a flexor of the metacarpophalangeal and proximal interphalangeal joints of the fingers. Because it crosses the radiocarpal and midcarpal joints it can flex the wrist, and it can also help to flex the elbow. It is always difficult to be sure that this muscle is working if flexor digitorum profundus is active, since the profundus tendon can produce all the actions of the superficialis. The best method is to ask the patient to bend the proximal interphalangeal joint without bending the distal one.

FLEXOR DIGITORUM PROFUNDUS

The deep flexor of the fingers, with the long flexor of the thumb, is in the deepest layer of muscles on the front of the forearm [FIGS. 5.79 and 5.80]. It arises from the upper two-thirds of the medial and anterior surfaces of the ulna, the medial side of the olecranon process, the medial half of the anterior surface of the interosseous membrane in its middle third [FIG. 5.75], and from the aponeurosis attaching flexor carpi ulnaris to the posterior border of the ulna.

Halfway down the forearm, the portion of the muscle coming from the interosseous membrane forms a separate tendon destined for the index finger [FIG. 5.80]. Those for the remaining fingers usually separate at the flexor retinaculum, behind which they pass to the palm in the carpal tunnel, posterior to the tendons of flexor digitorum superficialis. Each tendon enters the fibrous sheath of its digit behind the tendon of flexor superficialis, and pierces this tendon opposite the proximal phalanx. It inserts into the anterior surface of the base of the distal phalanx. The tendons are provided with vincula; the **vincula brevia** are attached to the capsule of the distal interphalangeal joint, the **vincula longa** to the tendons of the flexor superficialis and their vincula brevia [FIGS. 5.77 and 5.78].

In the palm, the tendons of flexor digitorum profundus afford origins for the lumbrical muscles. Proximal to this, they are enveloped with the superficialis tendons in a common synovial sheath [p. 336].

Nerve supply

The medial part of the muscle, destined for the little and ring fingers, is supplied by branches (C. 8; T. 1) from the ulnar nerve, and the lateral part, destined for the index and middle fingers, is supplied by branches (C. 7, 8; T. 1) from the anterior interosseous nerve. Sometimes the ulnar nerve supplies the whole muscle, and more rarely the anterior interosseous nerve does so.

Actions

Flexor digitorum profundus is the only muscle which can flex the distal interphalangeal joints of the fingers. Although this is its primary action, it helps to flex all the joints across which it passes—the proximal interphalangeal, the metacarpophalangeal, the carpometacarpal, midcarpal, and radiocarpal joints. To test the muscle, the dorsum of the hand and fingers is laid flat on a table, the middle phalanx is pressed firmly against the table by the examiner's finger, and the patient tries to bend the distal phalanx.

FLEXOR POLLICIS LONGUS

Like flexor digitorum profundus, flexor pollicis longus lies deeply in the forearm, arising by fleshy fibres from the anterior surface of the body of the radius between the radial tuberosity superiorly and the upper border of pronator quadratus inferiorly. It also comes from the corresponding lateral portion of the front of the interosseous membrane, and often also by a small slip from the medial border of the coronoid process of the ulna [FIG. 5.75].

The muscle descends lateral to flexor digitorum profundus [FIG. 5.80]. A tendon leaves it just above the wrist and passes into the hand posterior to the flexor retinaculum, enveloped in a special synovial sheath [p. 336]. The tendon enters the fibrous digital sheath of the thumb and inserts into the anterior aspect of the base of the distal phalanx of the thumb [FIG. 5.85].

Nerve supply

By a branch (C. 8; T. 1) from the anterior interosseous nerve.

Actions

Flexor pollicis longus is the only flexor of the interphalangeal joint of the thumb. It also flexes the metacarpophalangeal and carpometacarpal joints of the thumb, and can assist in flexion of the wrist. (Because the thumb is placed at right angles to the fingers, flexion is a movement which takes place in the plane of the palm.) The muscle can be tested by holding the proximal phalanx of the thumb and asking the patient to flex the terminal phalanx.

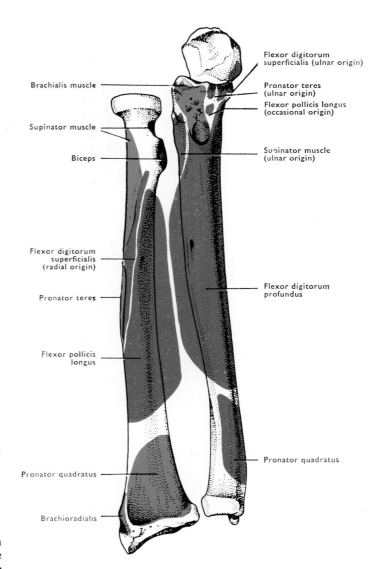

FIG. 5.75. Muscle attachments to front of right radius and ulna.

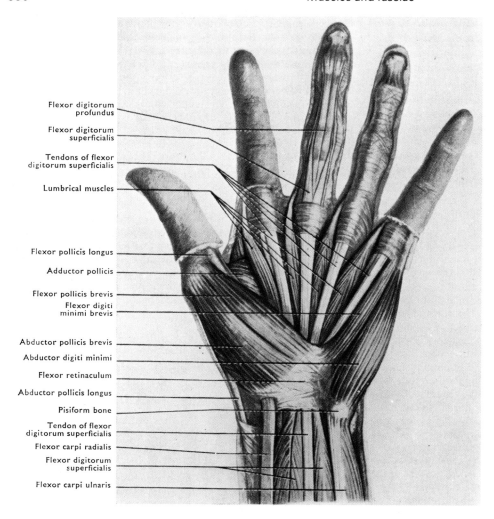

Flexor digitorum
profundus

Flexor digitorum
superficialis

Tendons of flexor
digitorum superficialis

Lumbrical muscles

Flexor pollicis longus

Adductor pollicis

Flexor pollicis brevis

Flexor digiti
minimi brevis

Abductor pollicis brevis

Abductor digiti minimi

Flexor retinaculum

Abductor pollicis longus

Pisiform bone

Tendon of flexor
digitorum superficialis

Flexor carpi radialis

Flexor digitorum
superficialis

Flexor carpi ulnaris

FIG. 5.76. Superficial muscles and tendons in palm of left hand.

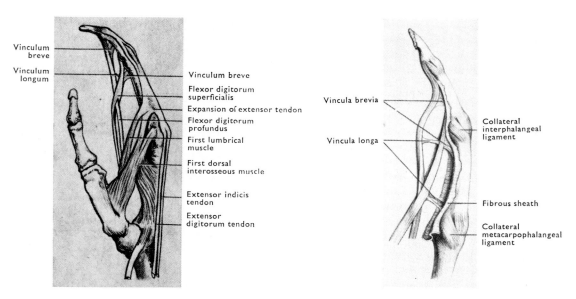

Vinculum
breve

Vinculum
longum

Vinculum breve

Flexor digitorum
superficialis

Expansion of extensor tendon

Flexor digitorum
profundus

First lumbrical
muscle

First dorsal
interosseous muscle

Extensor indicis
tendon

Extensor
digitorum tendon

FIG. 5.77. The tendons attached to the index finger.

Vincula brevia

Vincula longa

Collateral
interphalangeal
ligament

Fibrous sheath

Collateral
metacarpophalangeal
ligament

FIG. 5.78. Flexor tendons of finger with vincula tendinum.

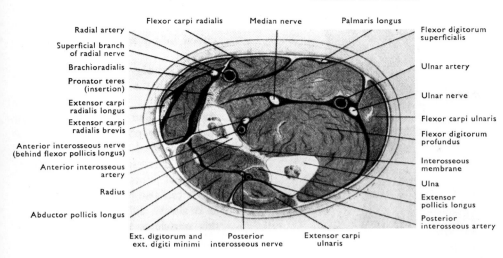

Radial artery
Superficial branch of radial nerve
Brachioradialis
Pronator teres (insertion)
Extensor carpi radialis longus
Extensor carpi radialis brevis
Anterior interosseous nerve (behind flexor pollicis longus)
Anterior interosseous artery
Radius
Abductor pollicis longus

Flexor carpi radialis   Median nerve   Palmaris longus

Flexor digitorum superficialis
Ulnar artery
Ulnar nerve
Flexor carpi ulnaris
Flexor digitorum profundus
Interosseous membrane
Ulna
Extensor pollicis longus
Posterior interosseous artery

Ext. digitorum and ext. digiti minimi   Posterior interosseous nerve   Extensor carpi ulnaris

Fig. 5.79. Distal surface of section through middle third of left forearm.

PRONATOR QUADRATUS

This fleshy quadrangular muscle lies deeply in the distal part of the forearm, passing from the distal quarter of the anterior surface of the body of the ulna to the distal quarter of the anterior surface of the body of the radius [Fig. 5.75].

Nerve supply

From the anterior interosseous nerve (C. 8; T. 1).

Actions

It is said to be a more powerful pronator of the forearm than pronator teres (Basmajian and Travill 1961), and it also holds the radius and ulna together, thus protecting the inferior radio-ulnar joint. The muscle is tested as for pronator teres, but cannot be seen or felt.

Muscles of the back of the forearm

The muscles of the back of the forearm are arranged in two groups, superficial and deep. In the **superficial group**, from the radial to the ulnar side, lie brachioradialis, extensor carpi radialis longus, extensor carpi radialis brevis, extensor digitorum, extensor digiti minimi, extensor carpi ulnaris, and anconeus [Fig. 5.81]. The **deep group** consists of five muscles, supinator, abductor pollicis longus, extensor pollicis brevis, extensor pollicis longus, and extensor indicis. This group is largely concealed by the superficial group until just above the wrist, where abductor pollicis longus and extensor pollicis brevis emerge to wind round the lateral side of the radius to the thumb [Fig. 5.81].

BRACHIORADIALIS

Brachioradialis [Fig. 5.72] arises from the proximal two-thirds of the front of the lateral supracondylar ridge of the humerus [Fig. 5.67] and the adjoining part of the lateral intermuscular septum. Brachioradialis forms the lateral boundary of the cubital fossa, and its fasciculi converge in the middle of the forearm into a narrow flat tendon which is inserted into the lateral side of the lower end of the radius [Fig. 5.75], just above the styloid process.

Nerve supply

From the radial nerve by a branch (C. 5, 6) which enters its medial aspect above the elbow.

Actions

Brachioradialis is a flexor of the elbow, and is regarded as a shunt muscle, in contrast to biceps and brachialis, which are spurt muscles [p. 271]. In normal circumstances it plays no part in either pronation or supination, but cases have been reported in which it has been able to effect some pronation from the fully supinated position when the other pronators were paralysed (Sunderland 1944b). When testing the muscle, the forearm is bent to a right angle in a position midway between pronation and supination, and then flexed against resistance. The muscle belly is readily seen and felt.

EXTENSOR CARPI RADIALIS LONGUS

This muscle, which is often blended with brachioradialis, arises from the anterior aspect of the distal third of the lateral supracondylar ridge of the humerus and the adjacent part of the lateral intermuscular septum [Figs. 5.67 and 5.81]. The muscle ends, about the middle of the forearm, in a flat tendon closely related to the lateral side of the tendon of extensor carpi radialis brevis (in lower animals the two muscles are fused). The two tendons descend side by side into the wrist, and are enveloped in a single synovial sheath under the extensor retinaculum [p. 347]. They groove the distal end of the radius and are crossed obliquely by the tendon of extensor pollicis longus [Fig. 5.88]. Eventually extensor carpi radialis longus inserts into the dorsal surface of the base of the second metacarpal bone on its radial side [Fig. 5.92]; there may be a small bursa under the tendon close to its insertion.

Nerve supply

From the radial nerve, by a branch (C. 6, 7) which enters the muscle above the elbow.

Actions

It extends and abducts the radiocarpal and midcarpal joints, and is an important synergist during flexion of the fingers [p. 337]. Because of its origin from the lateral supracondylar ridge, the muscle helps to flex the elbow joint. To test the muscle the patient attempts, with the fingers extended, to extend the wrist to the radial side against resistance. The muscle belly can be felt and sometimes seen.

EXTENSOR CARPI RADIALIS BREVIS

The short radial extensor [Fig. 5.81], which is often fused with its longer colleague at its origin, comes from the anterior aspect of the lateral epicondyle of the humerus by a tendon which it shares with the extensor digitorum, the extensor digiti minimi, and the extensor carpi ulnaris [FIG. 5.67]. It also arises from the fascia covering this **common extensor origin**, and from the intermuscular septa which penetrate between it and its neighbours.

The tendon of insertion, which appears approximately half way down the forearm, runs with that of extensor carpi radialis longus, and is inserted into the base of the dorsal surface of the third metacarpal bone with a small bursa between it and the styloid process of the metacarpal.

Nerve supply

By the deep branch of the radial nerve (C. 6, 7).

Actions

Extension and abduction of the radiocarpal and midcarpal joints: it helps the long extensor to prevent flexion of the wrist during flexion of the fingers. The muscle is tested in the same way as extensor carpi radialis longus.

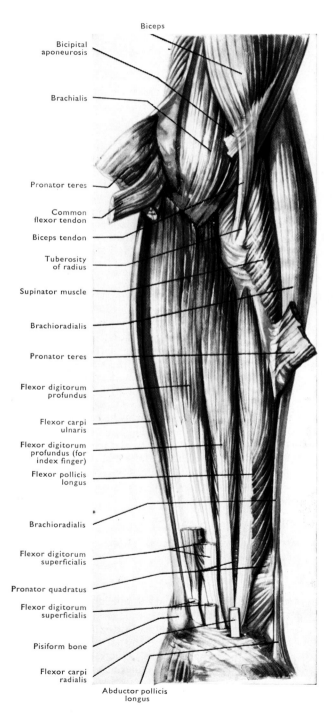

FIG. 5.80. Deepest muscles of front of left forearm.

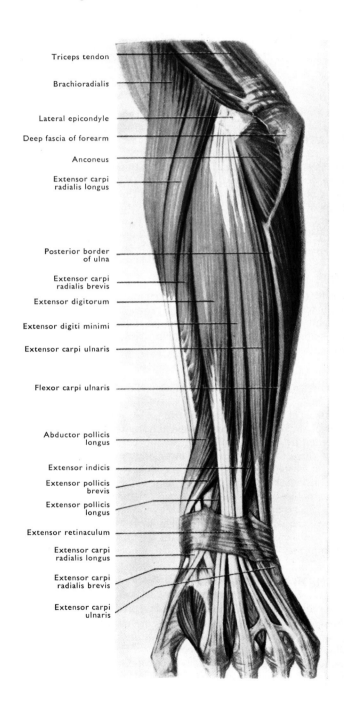

FIG. 5.81. Superficial muscles on dorsum of left forearm.

EXTENSOR DIGITORUM

Extensor digitorum arises from the common extensor origin on the humerus [FIGS. 5.67 and 5.81] from the fascia covering it, and from the intermuscular septa at its sides. Above the wrist the muscle ends in four tendons which pass deep to the extensor retinaculum [p. 347] in a compartment along with extensor indicis and surrounded by a common synovial sheath which ends about the middle of the dorsum of the hand [FIG. 5.88]. On the back of the hand the tendons of extensor digitorum are interconnected by obliquely placed fibrous bands. The arrangement of these bands is very variable, but the tendons for the little and ring fingers are usually quite closely attached until just proximal to the metacarpophalangeal joints, and that for the index finger is frequently not attached to the tendon for the middle finger [FIGS. 5.81 and 5.88].

As each individual tendon approaches the metacarpophalangeal joint it forms a triangular membranous expansion (the **extensor hood**) over the metacarpophalangeal joint and the dorsum of the proximal phalanx. This hood forms the dorsal part of the fibrous capsule of the joint, and its base, which lies proximally, extends forwards on each side of the head of the metacarpal to fuse with the deep transverse metacarpal ligament. Distally it narrows towards the dorsal surface of the proximal interphalangeal joint, and is reinforced at each side by receiving the insertions of the tendons of the interosseous and lumbrical muscles of that particular finger [FIGS. 5.77 and 5.82].

At the distal end of the proximal phalanx the hood splits into somewhat ill-defined central and collateral slips, which replace the dorsal ligament of the proximal interphalangeal joint. The central

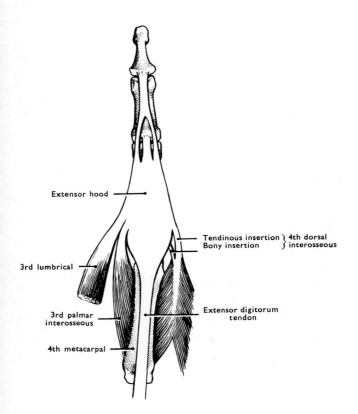

FIG. 5.82. The extensor hood of the right ring finger. The complicated system of local strengthening bands in the hood revealed by detailed dissection has been omitted for the sake of clarity. For the same reason the sides of the hood have been shown in the same plane as its central portion, though in fact they are more or less at right angles to it.

Labels on figure:
Extensor hood
3rd lumbrical
3rd palmar interosseous
4th metacarpal
Tendinous insertion } 4th dorsal
Bony insertion } interosseous
Extensor digitorum tendon

slip, directly continuous with the fibres of the extensor tendon itself, inserts into the dorsum of the base of the middle phalanx. The two collateral slips, which are continuous with the fibres of the lumbrical and interossei tendons, cross the dorsal surface of the proximal interphalangeal joint and unite on the dorsum of the middle phalanx. They are inserted into the base of the distal phalanx (Haines 1951).

Nerve supply

By branches (C. 7, 8) from the posterior interosseous nerve.

Actions

The muscle extends the metacarpophalangeal joints of the fingers. By hyperextending these joints (i.e. extending them beyond 180 degrees), it can abduct (diverge) the fingers away from an imaginary line drawn through the middle finger.

Extensor digitorum can also help to extend the proximal and distal interphalangeal joints. However, the main extensors of these joints, the tendons of which form the collateral slips of the extensor hood, are interossei and lumbricals, which also help to prevent extensor digitorum from hyperextending the metacarpophalangeal joint [p. 341]. If these muscles are paralysed, extensor digitorum will hyperextend this joint, and, having done so, cannot extend the distal or proximal interphalangeal joints until the hyperextension is undone.

Since extensor digitorum crosses the radiocarpal and midcarpal joints it will also extend these joints, and if it is desired to extend the fingers without extending the wrist, it is necessary to call into play the flexors of the radiocarpal and midcarpal joints. Although the muscle crosses the elbow joint anteriorly, its action as an elbow flexor is negligible. The tendon running to the index finger can deviate the finger radially when the first dorsal interosseous is paralysed (Duchenne 1867).

If an attempt is made by the examiner to flex the extended fingers at the metacarpophalangeal joints, the belly of the muscle can be felt contracting as the subject resists the movement.

EXTENSOR DIGITI MINIMI

The extensor of the little finger arises from the common extensor origin and from the fascia surrounding it. It runs distally between extensor digitorum and extensor carpi ulnaris [FIG. 5.81] to end in a tendon which traverses a special compartment of the extensor retinaculum in a groove between the radius and the ulna, surrounded by a synovial sheath [FIG. 5.88]. On the dorsum of the hand the tendon splits into two, and both parts are inserted into the extensor hood on the back of the proximal phalanx of the little finger [FIG. 5.83], the lateral one joining the tendon of extensor digitorum to the little finger.

Nerve supply

By branches (C. 7, 8) from the posterior interosseous nerve.

Actions

Extensor digiti minimi helps to extend the metacarpophalangeal joint of the little finger, and acts similarly to the corresponding tendon of extensor digitorum, except that it is said to be capable of deviating the little finger in an ulnar direction (Duchenne 1867; Jones 1919).

The presence of a separate extensor for the little finger does confer on it some independence of movement, but the tendons of extensor digitorum passing to the ring and little fingers are usually firmly adherent to each other, and pulling on the one always involves pulling on the other. Like extensor digitorum, extensor digiti minimi helps to extend the radiocarpal and midcarpal joints.

EXTENSOR CARPI ULNARIS

This muscle arises from the common extensor origin on the humerus, and from the adjacent fascia. It has a second aponeurotic head, partly shared with flexor carpi ulnaris, from the middle part of the posterior border of the ulna. The tendon emerges from the muscle belly nearer to the wrist than the other extensor tendons [FIG. 5.81] and runs in a groove on the back of the lower end of the ulna in a special compartment of the extensor retinaculum, surrounded by a short synovial sheath [FIG. 5.88]. It is inserted into the medial side of the base of the fifth metacarpal bone [FIG. 5.92].

Nerve supply

By branches (C. 7, 8) from the posterior interosseous nerve.

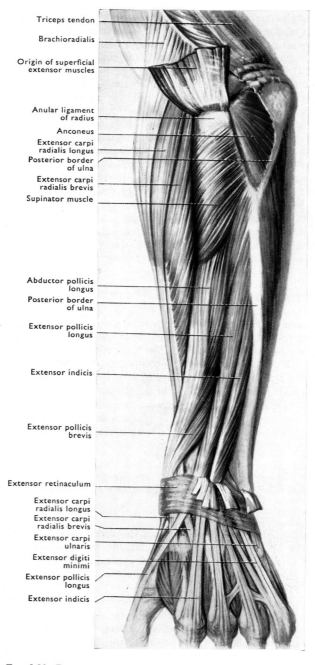

Triceps tendon
Brachioradialis
Origin of superficial extensor muscles
Anular ligament of radius
Anconeus
Extensor carpi radialis longus
Posterior border of ulna
Extensor carpi radialis brevis
Supinator muscle
Abductor pollicis longus
Posterior border of ulna
Extensor pollicis longus
Extensor indicis
Extensor pollicis brevis
Extensor retinaculum
Extensor carpi radialis longus
Extensor carpi radialis brevis
Extensor carpi ulnaris
Extensor digiti minimi
Extensor pollicis longus
Extensor indicis

FIG. 5.83. Deep muscles on dorsum of left forearm.

Actions

Extension and adduction of the carpus. It is an important synergist during flexion of the fingers [p. 339]. To test the muscle, the patient is asked to extend the wrist to the ulnar side against resistance: the tendon stands out and can be palpated by the examiner.

ANCONEUS

This small muscle arises from the back of the lateral epicondyle of the humerus [FIG. 5.81] and from the adjacent part of the capsule of the elbow joint. The insertion is to a triangular surface on the lateral side of the olecranon and to the posterior surface of the ulna as far down as an oblique ridge [FIG. 5.84], and also into the fascia which covers it.

Nerve supply

From the branch (C. 7, 8) of the radial nerve to the medial head of the triceps.

Actions

It helps to extend the elbow joint.

SUPINATOR

Supinator lies deeply, almost wholly concealed by the superficial muscles. It has two heads. The humeral one arises from the inferior aspect of the lateral epicondyle of the humerus, from the radial collateral ligament of the elbow joint, and from the anular ligament of the radius. The ulnar head comes from the supinator crest of the ulna [FIG. 5.75]. Between the two layers of muscle derived from these heads the deep branch of the radial nerve runs into the back of the forearm.

From its composite origin the muscle fibres run distally and laterally posterior to the upper third of the radius [FIG. 5.83]. Wrapping round this almost completely, they become inserted into the posterior, lateral, and anterior aspects of the radius as far forwards as the anterior margin, as far proximally as the neck, and as far distally as the insertion of pronator teres [FIGS. 5.75 and 5.84].

Nerve supply

From the deep branch of the radial nerve (C. 5, 6).

Actions

Supination of the forearm. Supinator is probably the main prime mover in this movement, biceps brachii being an auxiliary (Travill and Basmajian 1961). It is difficult to test supinator apart from biceps brachii, but the patient is asked to resist pronation of the forearm while the elbow is held straight.

ABDUCTOR POLLICIS LONGUS

The long abductor of the thumb arises from the dorsal surfaces of the ulna and radius distal to the origin and insertion of the supinator [FIG. 5.84], and also from the intervening portion of the interosseous membrane. Although deeply situated under the other extensor muscles at its origin, the muscle emerges as it passes towards the wrist to become superficial in the distal part of the forearm [FIGS. 5.81 and 5.83]. Its tendon, together with that of extensor pollicis brevis, crosses those of extensor carpi radialis brevis and extensor carpi radialis longus and the insertion of brachioradialis. It then passes through the most lateral compartment of the extensor retinaculum in a common synovial sheath with the tendon of extensor pollicis brevis [FIG. 5.87], and after crossing the radial artery, is inserted, sometimes by several slips, into the lateral side of the base of the first metacarpal [FIG. 5.90]. It usually gives off a slip which passes to abductor pollicis brevis and to the fascia over the thenar eminence, and sometimes another which goes to the trapezium (Baba 1954).

Nerve supply

From the posterior interosseous nerve (C. 7, 8).

Actions

It pulls the metacarpal bone of the thumb into a position midway between extension and abduction [p. 345], and the tendon stands out during this movement. It is also quite a strong assistant in resisted flexion and abduction of the wrist, and this action is important in 'trick' flexion of the wrist when the normal prime movers are paralysed.

EXTENSOR POLLICIS BREVIS

The short extensor of the thumb comes from an impression on the posterior surface of the radius, and from the adjacent part of the interosseous membrane [FIG. 5.84]. It lies distal to abductor pollicis longus, to which it is closely adherent, and the tendons of the two muscles pass down to the wrist and run deep to the extensor retinaculum in a common synovial sheath [FIG. 5.87]. The tendon of extensor pollicis brevis helps to replace the dorsal part of the capsule of the metacarpophalangeal joint of the thumb, and is inserted into the base of the dorsal surface of the proximal phalanx. Occasionally it proceeds further and may be inserted into the distal phalanx along with extensor pollicis longus.

Nerve supply

By a branch from the posterior interosseous nerve (C. 7, 8).

Actions

It extends the carpometacarpal and metacarpophalangeal joints of the thumb. Because of its position on the anterior aspect of the lateral side of the wrist, the muscle can be called into play to help in flexing and abducting the wrist against resistance. To test the muscle, the patient tries to resist flexion of the thumb at the metacarpophalangeal joint; the tendon of the muscle will then stand out.

EXTENSOR POLLICIS LONGUS

Extensor pollicis longus arises from the lateral part of the posterior surface of the ulna in its middle third, and also from the interosseous membrane somewhat more distally [FIG. 5.84]. Above the wrist a tendon arises and is enclosed in a synovial sheath as it grooves the posterior surface of the radius in a special compartment deep to the extensor retinaculum [FIG. 5.88]. At this point it changes direction by turning laterally round the medial side of the **dorsal tubercle of the radius** [FIG. 3.131], and passes obliquely across the back of the hand over the tendons of extensor carpi radialis brevis and longus and over the radial artery. Extensor pollicis longus tendon replaces the dorsal part of the capsule of the metacarpophalangeal joint of the thumb, and is here joined by slips from the tendons of abductor pollicis brevis on the lateral side and adductor pollicis on the medial side, forming a triangular expansion somewhat similar to the extensor hoods of the fingers (Napier 1952). This expansion may be joined by the tendon of extensor pollicis brevis. It is finally inserted into the dorsal surface of the base of the distal phalanx of the thumb.

As the tendon passes laterally across the back of the hand, it forms the posterior boundary of a triangular hollow known as the **anatomical snuff box**, which is best seen when the muscle is in

action. The other boundaries of this hollow are the distal end of the radius proximally, and the tendons of abductor pollicis longus and extensor pollicis brevis anteriorly. In the hollow lies the radial artery, which can be felt pulsating against the scaphoid bone.

Nerve supply

By a branch (C. 7, 8) from the posterior interosseous nerve.

Actions

Extensor pollicis longus extends all the joints of the thumb and can assist in extension and abduction of the wrist. In full abduction or extension of the thumb it can act as a 'trick' adductor of the thumb. The tendon stands out when an attempt is made to resist flexion at the interphalangeal joint of the thumb.

EXTENSOR INDICIS

Unlike extensor digiti minimi, extensor indicis lies deeply in the dorsum of the forearm. It arises from the lowest impression on the posterior surface of the ulna, and sometimes also from the interosseous membrane [FIG. 5.84]; the origin is just distal to that of

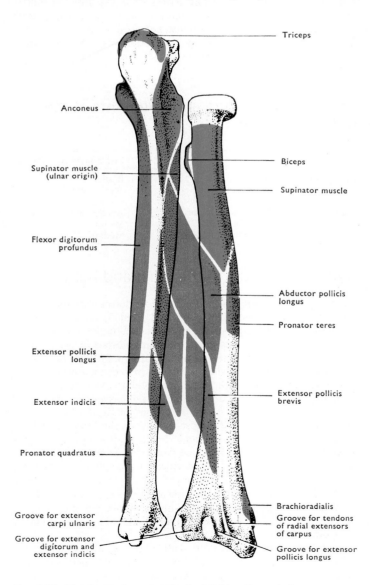

FIG. 5.84. Muscle attachments to back of right radius and ulna.

extensor pollicis longus. The tendon of extensor indicis passes through the extensor retinaculum in common with extensor digitorum and surrounded by the same synovial sheath [FIG. 5.88]. On the back of the hand the tendon lies on the ulnar side of the index finger tendon of extensor digitorum, and is inserted into the extensor hood on the back of the proximal phalanx of the index.

Nerve supply

By branches (C. 7, 8) from the posterior interosseous nerve.

Actions

Extensor indicis produces independent extension of the index finger. In all other respects the actions of extensor indicis are similar to those of extensor digitorum, except that it has been claimed as an ulnar deviator of the index.

Movements of the radio-ulnar joints

The two strong supinators are biceps brachii and supinator.

The chief pronators are pronator quadratus and pronator teres, but flexor carpi radialis, which passes obliquely from medial to lateral side on the front of the forearm, is capable of assisting. Both supination and pronation are most powerful when the elbow is flexed to a right angle, and in the majority of shoulder-elbow positions isometric pronation is a stronger movement than isometric supination (Darcus 1951).

Muscles concerned in movements at the radio-ulnar joints

Movement	Muscle	Principal nerve supply
Supination	Supinator	Radial
	Biceps brachii	Musculocutaneous
Pronation	Pronator quadratus	Anterior interosseous
	Pronator teres	Median
	Flexor carpi radialis	Median

Synovial sheaths in the wrist and hand

At the wrist and in the fingers, the flexor tendons are surrounded by synovial sheaths [FIGS. 5.85, 5.87, and 5.88]. Each sheath has a visceral layer which surrounds and is attached to the tendon, and a parietal layer, which lines the surrounding fibrous or osteofibrous compartment through which the tendon runs [p. 277]. In front of the wrist there are three such sheaths, all of which begin approximately 2 or 3 cm proximal to the flexor retinaculum. Two of them, the separate sheaths which surround flexor pollicis longus and flexor carpi radialis, terminate at the insertions of these muscles. The third, which is by far the largest, surrounds the tendons of both flexor digitorum profundus and flexor digitorum superficialis [FIG. 5.85]. These tendons may be thought of as being pushed into a closed bursa from the lateral side, so producing a somewhat complicated-looking arrangement [FIG. 5.86]. In about 50 per cent of people there is a communication between this common flexor synovial sheath and the synovial sheath of flexor pollicis longus. Distally, the common flexor sheath terminates about the middle of the palm.

Each fibrous digital sheath [p. 348] is lined by a synovial sheath which envelops the tendons of flexor digitorum superficialis and flexor digitorum profundus of that digit. These synovial sheaths extend from the insertion of the profundus tendon distally to the

neck of the metacarpal proximally, except that in about 70 per cent of cases the digital synovial sheath of the little finger communicates with the common flexor synovial sheath [FIG. 5.85]. The flexor tendons give the appearance of having invaginated their synovial sheath from behind, so that the vincula longa and brevia [FIG. 5.78], representing the remains of their original mesotendon and conveying their blood supply, reach the tendons posteriorly, and attach them to the bone.

On the dorsum of the wrist and hand, the synovial sheaths of the extensor tendons begin just superior to the extensor retinaculum. Those of the tendons going to the metacarpal bones usually extend to their insertions, and the others reach half way down the hand [FIGS. 5.87 and 5.88].

The digital synovial sheaths are often infected. Such an infection may, because of the confined space within which the swelling develops, obliterate the blood supply of the tendons and consequently kill them, so disorganizing the finger. An infection of the synovial sheath of the little finger is potentially more dangerous in this respect than those of the other fingers, since the infection may spread to involve the common flexor synovial sheath, so endangering the viability of all the flexor tendons to the fingers. Similarly, an infection of the synovial sheath of flexor pollicis longus may, in about 50 per cent of cases, affect not only the thumb, but also all the flexor tendons of the fingers. An infection developing in a digital synovial sheath may burst into one of the potential spaces of the palm [p. 349], whence it may spread up into the forearm posterior to the flexor retinaculum, again endangering by pressure the viability of the tendons and the other structures, such as the median nerve, which lie in the carpal tunnel [p. 347].

A chronic infection in the common flexor synovial sheath may produce a swelling in the palm and another extending up in front of the wrist for a short distance into the forearm, with a constriction between the two at the site of the flexor retinaculum. The fluid may be pressed from one to the other behind the retinaculum.

Muscles concerned in movements at the radiocarpal and midcarpal joints

Movement	Muscle	Principal nerve supply
Flexion	Flexor carpi radialis	Median
	Flexor carpi ulnaris	Ulnar
	Palmaris longus	Median
	Flexor digitorum profundus	Anterior interosseous: ulnar
	Flexor digitorum superficialis	Median
	Flexor pollicis longus	Anterior interosseous
	Abductor pollicis longus	Posterior interosseous
	Extensor pollicis brevis	Posterior interosseous
Extension	Extensor carpi radialis longus	Radial
	Extensor carpi radialis brevis	Radial
	Extensor carpi ulnaris	Posterior interosseous
	Extensor digitorum	Posterior interosseous
	Extensor pollicis longus	Posterior interosseous
	Extensor indicis	Posterior interosseous
	Extensor digiti minimi	Posterior interosseous
Abduction	Extensor carpi radialis longus	Radial
	Extensor carpi radialis brevis	Radial
	Flexor carpi radialis	Median
	Abductor pollicis longus	Posterior interosseous
	Extensor pollicis brevis	Posterior interosseous
	Extensor pollicis longus	Posterior interosseous
Adduction	Flexor carpi ulnaris	Ulnar
	Extensor carpi ulnaris	Posterior interosseous

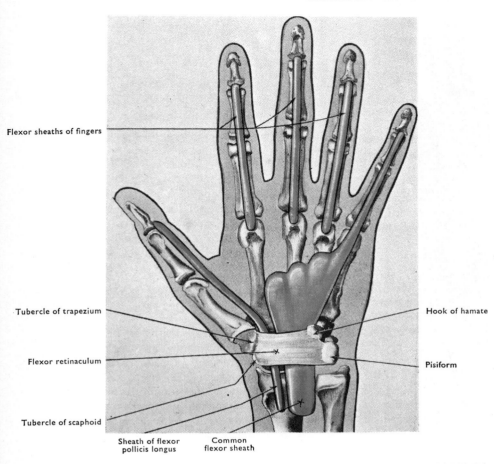

Flexor sheaths of fingers

Tubercle of trapezium

Flexor retinaculum

Tubercle of scaphoid

Sheath of flexor pollicis longus

Common flexor sheath

Hook of hamate

Pisiform

FIG. 5.85. Synovial sheaths of flexor tendons of digits.

Movements at the wrist

Movements in the region of the wrist take place at the radiocarpal and midcarpal joints. The radiocarpal joint is more concerned in extension and adduction at the wrist than the midcarpal, which is responsible for the greater part of flexion and abduction. The flexors of the carpus fix the wrist during extension of the fingers, and the extensors of the wrist fix it during flexion of the fingers. The object in both cases is to prevent the muscles operating on the fingers from losing power and efficiency through also moving the radiocarpal and midcarpal joints. In powerful movements of the fingers, the wrist is rigidly fixed by simultaneous contraction of the flexors and extensors of the carpus.

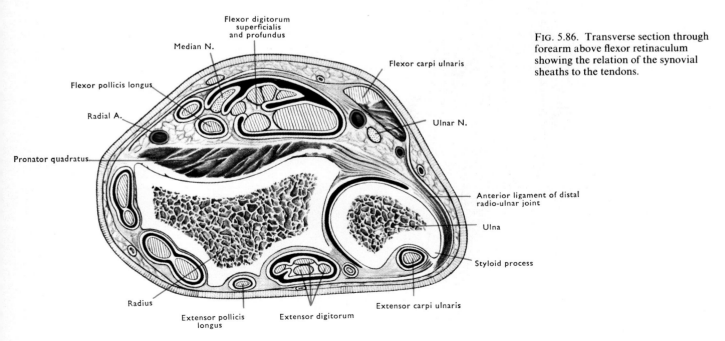

Flexor digitorum superficialis and profundus

Median N.

Flexor pollicis longus

Radial A.

Pronator quadratus

Radius

Extensor pollicis longus

Extensor digitorum

Flexor carpi ulnaris

Ulnar N.

Anterior ligament of distal radio-ulnar joint

Ulna

Styloid process

Extensor carpi ulnaris

FIG. 5.86. Transverse section through forearm above flexor retinaculum showing the relation of the synovial sheaths to the tendons.

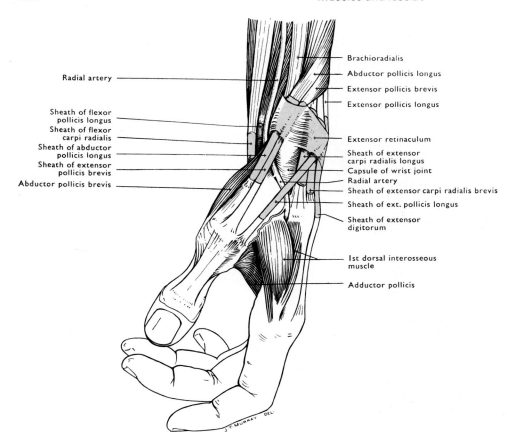

Radial artery

Sheath of flexor
pollicis longus
Sheath of flexor
carpi radialis
Sheath of abductor
pollicis longus
Sheath of extensor
pollicis brevis
Abductor pollicis brevis

Brachioradialis
Abductor pollicis longus
Extensor pollicis brevis
Extensor pollicis longus

Extensor retinaculum
Sheath of extensor
carpi radialis longus
Capsule of wrist joint
Radial artery
Sheath of extensor carpi radialis brevis
Sheath of ext. pollicis longus
Sheath of extensor
digitorum

1st dorsal interosseous
muscle

Adductor pollicis

FIG. 5.87. Dissection of lateral side of left wrist and hand, showing synovial sheaths of tendons.

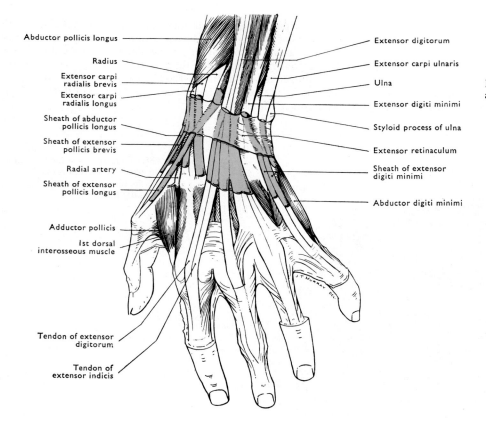

Abductor pollicis longus

Radius

Extensor carpi
radialis brevis
Extensor carpi
radialis longus
Sheath of abductor
pollicis longus
Sheath of extensor
pollicis brevis
Radial artery
Sheath of extensor
pollicis longus

Adductor pollicis
1st dorsal
interosseous muscle

Tendon of extensor
digitorum

Tendon of
extensor indicis

Extensor digitorum
Extensor carpi ulnaris
Ulna
Extensor digiti minimi
Styloid process of ulna
Extensor retinaculum
Sheath of extensor
digiti minimi

Abductor digiti minimi

FIG. 5.88. Dissection of back of forearm, wrist, and hand, showing synovial sheaths of tendons.

The importance of these synergic actions of the wrist muscles is easily seen when an attempt is made to grip an object with the long flexors of the fingers when the wrist is fully flexed. Because of the slackening of the flexor tendons caused by the wrist flexion, they cannot shorten sufficiently, and the power they can exert is greatly diminished. This is the basis of the fact that forcible flexion of the wrist can cause an assailant to drop a weapon. Conversely, extension of the wrist stretches the flexors of the fingers and increases the range of their finger movement.

A patient with paralysis of the radial nerve can often extend his wrist to some extent by flexing the fingers fully at the metacarpophalangeal joints; the tendons of extensor digitorum are tightened and pull the wrist into extension. This trick movement is inadequate to permit a satisfactory grip.

MUSCLES OF THE HAND

The 'intrinsic' muscles of the hand include the interossei in the intermetacarpal spaces and the lumbricals, which spring from the flexor digitorum profundus tendons in the middle of the palm and act on the four fingers. The muscles which operate solely on the thumb or on the little finger are grouped together to form the rounded swellings, on either side of the palm, known as the thenar and hypothenar eminences. The muscles of the hypothenar eminence—abductor digiti minimi, flexor digiti minimi brevis, and opponens digiti minimi—have their thenar counterparts in abductor pollicis brevis, flexor pollicis brevis, and opponens pollicis, but the thumb has an additional muscle, adductor pollicis, which lies deep in the palm. All the small muscles of the hand are supplied by the first thoracic segment of the spinal medulla (Highet 1943).

Abduction of the fingers is defined as movement, in the plane of the nails, away from an imaginary line drawn through the middle finger; adduction is movement towards this line. The middle finger can thus be abducted in two directions, either towards the little finger, or towards the index. Further, adduction of the index finger is a movement in the opposite direction to adduction of the ring finger and little finger. This apparently difficult convention in fact makes the actions and attachments of the interossei easier to remember.

LUMBRICALS

The four lumbricals, one for each finger, are small cylindrical muscles resembling the earthworm from which they are named. The lateral two each arise, just distal to the flexor retinaculum, by a single head from the lateral side of the tendons of flexor digitorum profundus running to the index and middle fingers. The medial two muscles arise by two heads from the adjacent sides of the profundus tendons between which they lie—that is, from those going to the middle, ring, and little fingers [FIG. 5.76]. Each lumbrical then passes in a plane anterior to the deep transverse metacarpal ligament and sweeps obliquely dorsally, on the lateral side of the metacarpophalangeal joint of the finger concerned, to join the lateral edge of the extensor hood at the side of the proximal phalanx [FIG. 5.77]. Some of its fibres can be traced to the central attachment of the hood to the dorsum of the base of the middle phalanx, but most of them run in the lateral part of the hood, over the dorsal surface of the proximal interphalangeal joint, and ultimately reach the dorsum of the base of the terminal phalanx [p. 333].

Nerve supply

This varies. The most usual arrangement is that the lateral two lumbricals are supplied by twigs (T.1) from the median nerve; the medial two being supplied by twigs (T.1) from the ulnar nerve. However, the number of lumbricals supplied by the ulnar nerve may be increased to four or decreased to one.

Actions

Each lumbrical is a flexor of the metacarpophalangeal joint of the corresponding finger and an extensor of both interphalangeal joints. Acting from its insertion on its origin, the lumbrical slackens off the tension in the flexor digitorum profundus tendon.

Theoretically, all the lumbricals could rotate the fingers on which they act at the metacarpophalangeal joint (Sunderland 1953). This action, however, is only well marked in the index finger. It brings the pad of the index finger to face medially, at an angle of about 30 degrees to the pad of the middle finger, and this movement is used in precision gripping [p. 345]. The first dorsal interosseous muscle is associated with the first lumbrical in producing this rotation, and is probably the major factor. Although they run on the lateral side of each digit, it is still not certain whether the lumbricals have any

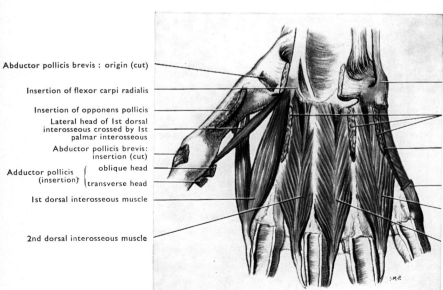

Abductor pollicis brevis : origin (cut)

Insertion of flexor carpi radialis

Insertion of opponens pollicis

Lateral head of 1st dorsal interosseous crossed by 1st palmar interosseous

Abductor pollicis brevis: insertion (cut)

Adductor pollicis { oblique head
(insertion)   { transverse head

1st dorsal interosseous muscle

2nd dorsal interosseous muscle

Insertion of flexor carpi ulnaris

Origins of 2nd, 3rd, and 4th palmar interosseous muscles

Insertion of opponens digiti minimi

Insertion of abductor digiti minimi

4th dorsal interosseous muscle

3rd dorsal interosseous muscle

FIG. 5.89. Dorsal interosseous muscles of right hand (seen from the palmar aspect).

action in radially deviating the fingers (Braithwaite, Channell, Moore, and Whillis 1948; Backhouse and Catton 1954; Matheson, Sinclair, and Skene 1970).

INTEROSSEI

Four palmar and four dorsal interossei occupy the spaces between the metacarpal bones. Each palmar interosseous comes from the metacarpal of the digit on which it acts; each dorsal interosseous comes from the adjacent surfaces of the metacarpals between which it lies: the dorsal muscles are about twice the size of the palmar ones (Eyler and Markee 1954), and are composite muscles evolved from the fusion of primitive dorsal abductors and short flexors of the fingers (Lewis 1965). The tendons of all the interossei, except the first palmar, pass to one or other side of the appropriate metacarpophalangeal joint and (except for the first pair) posterior to the deep transverse metacarpal ligament. They become inserted partly into the base of the proximal phalanx and partly into the extensor hood. The attachment to the hood is largely responsible for prolonging the hood to the base of the distal phalanx [p. 333].

The first palmar interosseous is a small slip of muscle which arises from the ulnar side of the base of the first metacarpal and passes between the first dorsal interosseous and the oblique head of adductor pollicis to insert into the ulnar side of the base of the proximal phalanx of the thumb in common with the tendon of adductor pollicis [FIG 5.89].

The second palmar interosseous arises from the ulnar side of the body of the second metacarpal and is inserted into the ulnar side of the extensor hood of the index; the third and fourth palmar interossei arise from the radial sides of the bodies of the fourth and fifth metacarpals respectively, and insert into the radial sides of the extensor hoods of the ring and little fingers [FIGS. 5.90 and 5.91]. (The middle finger has no palmar interosseous, and is deviated to

either side by the action of the two dorsal interossei which are attached to it.)

The arrangement of the dorsal interossei is shown in FIGURES 5.77 and 5.89, and their attachments in FIGURES 5.90 and 5.92. The first dorsal interosseous, occupying the space between the thumb and the index finger, is larger than the others, and has an interval between its two heads through which the radial artery passes into the palm: its fibres are mostly inserted into the radial side of the proximal phalanx of the index finger, and though some join the extensor hood, they are not prolonged to the distal phalanx (Braithwaite *et al.* 1948). The tendon of the second dorsal interosseous inserts into the radial side of the extensor hood and proximal phalanx of the middle finger; the third and fourth are inserted into the ulnar sides of the extensor hoods of the middle and ring fingers, and in a minority of instances into the proximal phalanges (Eyler and Markee 1954).

Nerve supply

All the interossei are supplied by branches (T.1) from the deep branch of the ulnar nerve.

Actions

The dorsal interossei abduct (diverge) the fingers, in the plane of the nails, away from an imaginary line drawn through the middle finger. The first dorsal interosseous can rotate the index finger at the metacarpophalangeal joint [p. 339]. This muscle may also assist adductor pollicis in the movement of adduction of the thumb.

The first palmar interosseous assists flexor pollicis brevis in flexion of the metacarpophalangeal joint of the thumb and adductor pollicis in adduction of the thumb; the other three palmar interossei adduct (converge) the index, ring, and little fingers towards the line of the middle finger. Such movements play an important part in all kinds of manipulation and grasping.

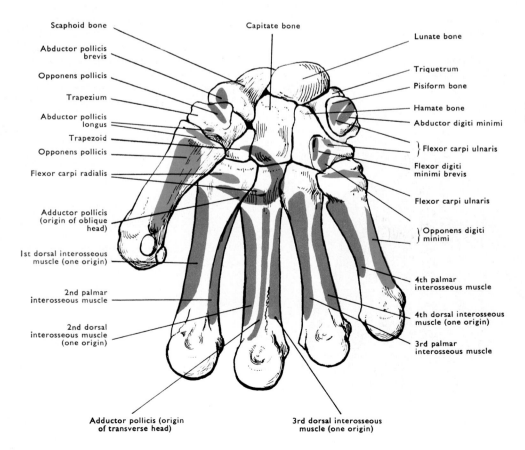

Scaphoid bone
Abductor pollicis brevis
Opponens pollicis
Trapezium
Abductor pollicis longus
Trapezoid
Opponens pollicis
Flexor carpi radialis
Adductor pollicis (origin of oblique head)
1st dorsal interosseous muscle (one origin)
2nd palmar interosseous muscle
2nd dorsal interosseous muscle (one origin)

Capitate bone

Lunate bone
Triquetrum
Pisiform bone
Hamate bone
Abductor digiti minimi
} Flexor carpi ulnaris
Flexor digiti minimi brevis
Flexor carpi ulnaris
} Opponens digiti minimi
4th palmar interosseous muscle
4th dorsal interosseous muscle (one origin)
3rd palmar interosseous muscle

Adductor pollicis (origin of transverse head)

3rd dorsal interosseous muscle (one origin)

FIG. 5.90. Muscle attachments to front of carpus and metacarpus.

Along with the lumbricals, interossei flex the metacarpophalangeal joints and extend the proximal and distal interphalangeal joints. They are responsible for the fine control over these movements, and this action is well seen in the upward stroke in writing; it is essential in all such activities as typing, sewing, and playing musical instruments. When interossei and lumbricals are paralysed, extensor digitorum pulls the metacarpophalangeal joints into a position of hyperextension [p. 333], and the interphalangeal joints become flexed, so producing one variety of 'claw hand'. In such a hand the individual, independent movements of the affected fingers, whether abduction or adduction, flexion or extension, are largely lost. For this reason, it becomes very difficult or impossible to do up buttons, to pick up small objects, or to perform any delicate controlled movement with the hand [p. 345].

The interossei and the lumbrical muscle acting on a given finger are tested by hyperextending and fixing the metacarpophalangeal joint and then asking the patient to extend the proximal interphalangeal joint against resistance. Individual interossei are also tested by the patient's ability to make the appropriate movement of abduction or adduction of the finger concerned, the hand and finger being placed palm down on the table.

ABDUCTOR DIGITI MINIMI

This is the most superficial of the muscles which form the hypothenar eminence, and is morphologically in series with the dorsal interossei (Lewis 1965). It arises from the pisiform bone, from the tendon of the flexor carpi ulnaris, and from the pisohamate and pisometacarpal ligaments [FIGS. 5.76 and 5.90]; the tendon is inserted into the medial side of the base of the proximal phalanx of the little finger and into the extensor hood of this finger.

Nerve supply

From the deep branch of the ulnar nerve (T. 1).

Actions

It abducts the little finger from the ring finger, and helps to flex its metacarpophalangeal joint (Forrest and Basmajian 1965). The attachment to the extensor hood of the little finger probably helps in

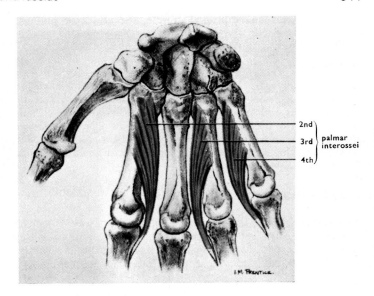

FIG. 5.91. The second, third, and fourth palmar interosseous muscles (right side). See FIG. 5.89 for first palmar interosseous muscle.

extending the interphalangeal joints of that digit. Abductor digiti minimi is a surprisingly powerful muscle, and comes into play when the fingers are spread to grasp a large object [p. 345]. To test it, the back of the hand and fingers is laid flat on the table and the patient abducts the little finger against resistance: the muscle belly can be felt and usually seen. (It is important to remember that a similar but less extensive movement can often be produced by transmitted tension from flexor carpi ulnaris even though abductor digiti minimi is paralysed.)

OPPONENS DIGITI MINIMI

The opponens [FIG. 5.93] lies deep to the abductor, and arises from the flexor retinaculum and the hook of the hamate bone. It is inserted into the medial half of the palmar surface of the fifth metacarpal in its distal two-thirds [FIG. 5.90].

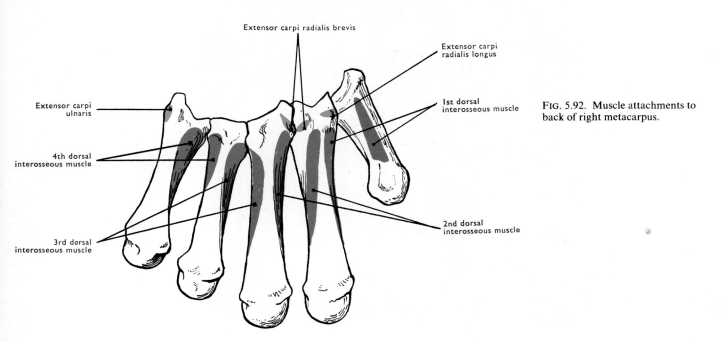

FIG. 5.92. Muscle attachments to back of right metacarpus.

Extensor carpi radialis brevis

Extensor carpi radialis longus

Extensor carpi ulnaris

1st dorsal interosseous muscle

4th dorsal interosseous muscle

3rd dorsal interosseous muscle

2nd dorsal interosseous muscle

Nerve supply

From the deep branch of the ulnar nerve (T. 1).

Actions

It pulls the metacarpal of the little finger forwards and rotates it laterally, so deepening the hollow of the hand. This is the basis of the test for the muscle, in which the patient is asked to make a cup of the palm while keeping the fingers extended. Contact between the pads of the thumb and the little finger during opposition [p. 345] depends on this muscle as well as upon the thenar musculature (Sunderland 1944a).

FLEXOR DIGITI MINIMI BREVIS

This rather variable muscle [FIG. 5.76] which may be absent or fused with one or other of its neighbours, arises by tendinous fibres from the front of the flexor retinaculum and the hook of the hamate bone [FIG. 5.90]. It is inserted, along with abductor digiti minimi, to

the medial side of the base of the proximal phalanx of the little finger; the tendon usually contains a sesamoid bone.

Nerve supply

From the deep branch of the ulnar nerve (T. 1).

Actions

It flexes the little finger at the carpometacarpal and metacarpophalangeal joints. During abduction of the little finger all three hypothenar muscles are active (Forrest and Basmajian 1965).

PALMARIS BREVIS

Palmaris brevis is a small subcutaneous muscle [FIG. 5.95] overlying the hypothenar eminence. It arises from the medial border of the palmar aponeurosis and the front of the flexor retinaculum, and is inserted into the skin of the medial border of the hand, covering the ulnar artery and nerve.

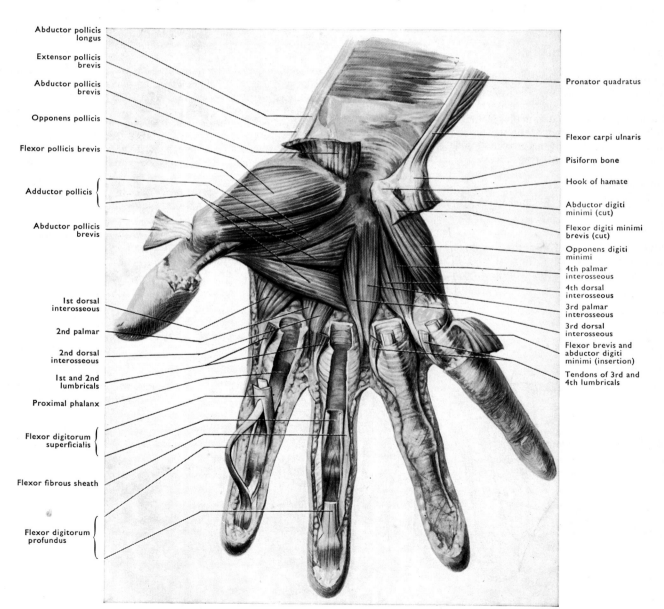

FIG. 5.93. Palmar muscles of right hand.

Nerve supply

From the superficial branch of the ulnar nerve (T. 1).

Actions

It wrinkles up the skin on the medial border of the hand, and deepens the hollow of the hand, assisting the opponens digiti minimi to obtain a firm palmar grip (Kirk 1924).

ABDUCTOR POLLICIS BREVIS

This muscle [FIGS. 5.76 and 5.93] is the most superficial of those which form the thenar eminence, and takes origin by fleshy fibres mainly from the front of the flexor retinaculum. Other fibres come from the tubercles of the scaphoid and trapezium, and there may be a contribution from the tendon of abductor pollicis longus, or from the insertion of palmaris longus. The muscle is inserted by a short tendon into the radial side of the proximal phalanx of the thumb, at its base; some fibres pass dorsally to reach the expansion of the tendon of extensor pollicis longus.

Nerve supply

The short abductor is usually supplied by a recurrent branch (T. 1) of the median nerve in the palm. In a minority of cases it may be supplied by the deep branch of the ulnar nerve.

Actions

It abducts the thumb at the carpometacarpal and metacarpophalangeal joints, causing it to travel anteriorly at right angles to the plane of the palm and to rotate medially as in typing or playing the piano. When the thumb is fully abducted, there is an angulation of some 30 degrees between the proximal phalanx and the metacarpal. Because of the direction of the muscle, abduction involves medial rotation of the metacarpal, and the abductor is used along with opponens pollicis in the initial stages of the movement of opposition of the thumb.

OPPONENS POLLICIS

Deep to the abductor pollicis brevis lies a fleshy mass imperfectly separated into two components. The proximal and lateral of these [FIG. 5.93] is opponens pollicis, and the distal and medial part of the mass is flexor pollicis brevis. The opposer of the thumb arises from the front of the flexor retinaculum and from the tubercle of the trapezium, and is inserted into the whole length of the radial half of the anterior surface of the first metacarpal [FIG. 5.90].

Nerve supply

Like abductor pollicis brevis, the opponens is usually supplied by the recurrent branch of the median nerve (T. 1): in a minority of cases it may be supplied by the ulnar nerve.

Actions

It pulls the metacarpal bone of the thumb into the position of opposition; that is, it rotates it at the carpometacarpal joint so that the pad of the thumb can be brought into contact with the pads of the fingers. This movement involves, first abduction away from the plane of the palm, secondly medial rotation at the carpometacarpal joint, and finally flexion and adduction of the abducted and medially rotated thumb. In the first two phases opponens is active, helped in abduction by abductor pollicis brevis, and in medial rotation by abductor pollicis brevis and flexor pollicis brevis. In the last phase the thumb, held in position by these muscles, is flexed to the little finger, or adducted towards the index finger by adductor pollicis [p. 343].

Opposition is tested by asking the patient to touch the palmar surface of the thumb to the palmar surface of the little finger, without bending the tip of the thumb.

Like that of an interosseous muscle, the tendon of abductor pollicis brevis lies in front of the metacarpophalangeal joint but inserts partly into the long extensor, and so the muscle can help in flexion of the metacarpophalangeal joint and in extension of the interphalangeal joint. This extension is often of some importance as a 'trick' movement when the extensors are paralysed.

To test the muscle, the patient's hand is placed with its dorsum flat on the table, the thumb nail being at right angles to the other nails. The patient is then asked to raise the thumb vertically at right angles to the table.

FLEXOR POLLICIS BREVIS

This small muscle is usually partly fused with opponens pollicis [FIG. 5.93], and arises by a superficial head from the distal border of the flexor retinaculum and the tubercle of the trapezium, and a deep head from the trapezoid and capitate bones (Day and Napier 1961). It is inserted into the radial side of the base of the proximal phalanx of the thumb just medial to the insertion of abductor pollicis brevis. The tendon of insertion contains a sesamoid bone.

Nerve supply

Flexor pollicis brevis is more variable in its nerve supply than the abductor and the opponens. Usually the recurrent branch of the median nerve (T. 1) is responsible but this is often replaced by the deep branch of the ulnar nerve (T. 1), and occasionally both nerves may supply the muscle.

Actions

It flexes the thumb at the carpometacarpal and metacarpophalangeal joints. It is also involved in the movement of opposition of the thumb, since flexion of the carpometacarpal joint automatically involves medial rotation. It is difficult to test this muscle apart from flexor pollicis longus and opponens pollicis.

ADDUCTOR POLLICIS

The oblique head of adductor pollicis comes from the sheath of the tendon of the flexor carpi radialis, the bases of the second, third, and fourth metacarpal bones, and the anterior aspects of the trapezoid and capitate bones [FIG. 5.90]. The transverse head of the muscle comes from the longitudinal ridge on the front of the body of the third metacarpal.

The fibres of the two heads, which close to their origin are separated by a gap which transmits the radial artery, converge on a tendon, containing a sesamoid bone, which inserts into the ulnar side of the base of the proximal phalanx of the thumb. Some fibres often pass deep to the tendon of flexor pollicis longus to be inserted along with flexor pollicis brevis into the radial side of the proximal phalanx, and others may run into the ulnar side of the extensor tendon.

Nerve supply

From the deep branch of the ulnar nerve (C. 8; T. I).

Actions

It adducts the thumb into the plane of the palm and pulls it towards the midline: it is active in the later stages of the movement of opposition. If the muscle is paralysed the patient will be unable to prevent the examiner withdrawing a piece of paper gripped between the thumb and the index finger, always provided the thumb nail is

Muscles concerned in movements of fingers

A. Metacarpophalangeal Joints

Movement	Muscle	Principal nerve supply
Flexion	Flexor digitorum profundus	Anterior interosseous: ulnar
	Flexor digitorum superficialis	Median
	Lumbricals	Median: ulnar
	Interossei	Ulnar
	Flexor digiti minimi	Ulnar
	Abductor digiti minimi	Ulnar
	Palmaris longus (through palmar aponeurosis)	Median
Extension	Extensor digitorum	Posterior interosseous
	Extensor indicis	Posterior interosseous
	Extensor digiti minimi	Posterior interosseous
Abduction and Adduction	Interossei	Ulnar
	Abductor digiti minimi	Ulnar
	Lumbricals may assist in radial deviation	Median: ulnar
	Extensor digitorum (abducts by hyperextending; tendon to index radially deviates)	Posterior interosseous
	Flexor digitorum profundus } Adduct by flexing	Anterior interosseous: ulnar
	Flexor digitorum superficialis	Median
Rotation	Lumbricals (specially first)	Median: ulnar
	Interossei (movement slight except index: only effective when phalanx is flexed)	Ulnar
	Opponens digiti minimi (rotates little finger at carpometacarpal joint)	Ulnar

B. Interphalangeal Joints

Movement	Muscle	Principal nerve supply
Flexion	Flexor digitorum profundus (both joints)	Anterior interosseous: ulnar
	Flexor digitorum superficialis (proximal joint only)	Median
Extension	Extensor digitorum	Posterior interosseous
	Extensor digiti minimi	Posterior interosseous
	Extensor indicis	Posterior interosseous
	Lumbricals	Median: ulnar
	Interossei	Ulnar

Muscles concerned in movements of the thumb

A. Carpometacarpal Joint

Movement	Muscle	Principal nerve supply
Flexion	Flexor pollicis brevis	Median or ulnar
	Flexor pollicis longus	Anterior interosseous
	Opponens pollicis	Median
Extension	Extensor pollicis brevis	Posterior interosseous
	Extensor pollicis longus	Posterior interosseous
	Abductor pollicis longus	Posterior interosseous
Abduction	Abductor pollicis longus	Posterior interosseous
	Abductor pollicis brevis	Median
Adduction	Adductor pollicis	Ulnar
	First dorsal interosseous	Ulnar
	Extensor pollicis longus } in full extension	Posterior interosseous
	Flexor pollicis longus } or abduction	Anterior interosseous
Opposition	Opponens pollicis	Median
	Abductor pollicis brevis	Median
	Flexor pollicis brevis	Median or ulnar
	Flexor pollicis longus	Anterior interosseous
	Adductor pollicis	Ulnar

B. Metacarpophalangeal Joint

Movement	Muscle	Principal nerve supply
Flexion	Flexor pollicis longus	Anterior interosseous
	Flexor pollicis brevis	Median or ulnar
	First palmar interosseous	Ulnar
	Abductor pollicis brevis	Median
Extension	Extensor pollicis longus	Posterior interosseous
	Extensor pollicis brevis	Posterior interosseous
Abduction	Abductor pollicis brevis	Median
Adduction	Adductor pollicis	Ulnar
	First palmar interosseous	Ulnar

C. Interphalangeal Joint

Movement	Muscle	Principal nerve supply
Flexion	Flexor pollicis longus	Anterior interosseous
Extension	Extensor pollicis longus	Posterior interosseous
	Abductor pollicis brevis	Median
	Adductor pollicis	Ulnar
	Extensor pollicis brevis (occasional insertion)	Posterior interosseous

kept at right angles to the other nails and the tip of the thumb is not allowed to bend (flexor pollicis longus). This test is not reliable unless care is taken to exclude the trick action of extensor pollicis longus in adducting the thumb [p. 335].

Movements of the fingers

The muscles primarily involved in movements of the fingers are shown in the table above. Because there is so little movement at the carpometacarpal joints of the three radial fingers, no special heading has been included for these joints. But the carpometacarpal joint of the little finger is more mobile, and can be flexed by flexor digiti minimi brevis and the long flexors, and rotated by opponens digiti minimi.

Because the knuckle is in each case formed by the proximal bone, the length of the dorsal surface of the finger increases when the finger is bent, and the extensor hoods are pulled distally. This in turn alters the line of pull of the lumbricals and interossei which are inserted into them, and at the same time drags these insertions distally. The power and range of contraction of lumbricals is reduced in this situation because they pull the flexor profundus tendons distally. Interossei are maximally efficient in abduction and adduction of the fingers when the metacarpophalangeal joints are flexed to an angle of 135 degrees, but they are unable to adduct or abduct when flexion is continued to 90 degrees (Matheson, Sinclair, and Skene 1970).

Movement of the distal part of the finger is further complicated by the existence of an **oblique band** of fibrous tissue which runs from the front of the proximal phalanx and the fibrous digital sheath to the side of the terminal portion of the extensor hood. When the distal interphalangeal joint is flexed, this band is pulled upon, and causes the proximal joint to become flexed also, quite apart from any muscular action. When the proximal joint is extended, the oblique bands cause the distal joint to be extended also.

Clinical evidence has made it clear that extension of the interphalangeal joints depends very largely on lumbricals and interossei, and that extensor digitorum, which is the undisputed extensor of the metacarpophalangeal joints, has much less effect on the interphalangeal joints. Exactly why this should be is not easy to explain.

The movements of the fingers in fact present one of the most difficult mechanical problems in anatomy, and the exact actions of the muscles involved are still open to question. For further information see Landsmeer (1949, 1955) and Stack (1962).

A patient with paralysis of the radial nerve can weakly extend his metacarpophalangeal joints by the trick action of fully flexing the wrist, so tightening the tendons of extensor digitorum. Similarly, a patient with a complete paralysis of the median and ulnar nerves can flex his fingers by hyperextending the wrist. A patient with an isolated paralysis of the median nerve may sometimes be able to flex the distal interphalangeal joint of the middle or middle and index fingers. This may be due either to the ulnar nerve supplying the whole of flexor digitorum profundus or to incomplete separation of its tendons from each other. In this case the trick action can only be produced if the distal interphalangeal joints of the ring and little fingers are free to move.

Movements of the thumb

Because the metacarpal of the thumb is set at right angles to those of the fingers, all the movements of the thumb take place in planes at right angles to the corresponding movements of the fingers. Thus, **flexion** and **extension** are in the plane of the palm: flexion brings the thumb medially across the palm and extension moves it laterally. In **abduction** the thumb stands away from the palm at right angles; **adduction** closes the thumb on the index finger. **Opposition**, the most important movement, is a composite movement involving first abduction of the thumb, then flexion and medial rotation, usually followed by adduction, resulting in the palmar surface of the thumb being brought into contact with the palmar surface of one of the other digits.

Because of the set of the articular surfaces of the carpometacarpal joint, flexion and abduction of the joint are associated with medial rotation, and extension and adduction with lateral rotation.

Abductor pollicis brevis produces abduction at both the carpometacarpal and metacarpophalangeal joints. Abductor pollicis longus, which operates only on the carpometacarpal joint, pulls the thumb into a position midway between abduction and extension (i.e. at an angle of about 45 degrees to the plane of the palm). It is accordingly involved in both abduction and extension of the thumb. In full extension extensor pollicis longus can, as a trick action, be used to adduct the thumb after the palmar ligament of the carpometacarpal joint has become tight (Sunderland 1944b). If this muscle is paralysed, the interphalangeal joint of the thumb can often be extended, as a trick movement, by abductor pollicis brevis through its insertion into the extensor tendon. Such extension is, of course, usually accompanied by abduction of the thumb.

In extension of the thumb flexor and extensor carpi ulnaris contract to cancel the abduction produced at the wrist by extensor pollicis longus and abductor pollicis longus.

Flexion of the distal phalanx of the thumb is often used as a test for the integrity of the median nerve, which supplies flexor pollicis longus. It is therefore important to bear in mind a trick movement

of flexion produced by strongly extending the wrist and thumb. This puts flexor pollicis longus on the stretch and so pulls the terminal phalanx into a position of flexion (Sunderland 1944b).

The use of the hand

However fine the movements produced in the hand, they must be controlled from a stable base, and, in order to fix the origins of the intrinsic musculature of the hand, the forearm muscles are brought into play. These in turn require fixation of the elbow by the muscles of the arm, and in turn this requires stabilization of the scapula and the shoulder girdle. Thus even the movements of writing and sewing involve shoulder muscles as well as those of the hand.

The **position of rest** of the hand is one in which the palm of the hand is hollowed, the fingers are flexed, and the thumb slightly opposed; the wrist is adducted, so that when the elbow is extended the axis of the middle finger is parallel to that of the humerus and not to that of the bones of the forearm. Flexion of the fingers increases progressively from the index to the little finger.

From this position the hand can be used in three different ways. In the first place the digits may be used individually. The thumb is fully separated functionally from the fingers, and its movements are not necessarily associated with movements of the other digits. The index finger is less completely autonomous, but it has its own extensor tendon, and the tendon of extensor digitorum which runs to the index is often not firmly fixed to that of the middle finger. The tendon of flexor digitorum profundus to the index is also usually separated higher in the forearm. These arrangements allow the finger to be extended separately, and this capacity is used in pointing movements. Again, it has a much greater range of abduction and active rotation (about 30 degrees) at the metacarpophalangeal joint than any other finger except the little finger; this relative freedom is of use in grasping.

When the interphalangeal joints are straight, individual movements of flexion at the metacarpophalangeal joints of the fingers are carried out largely by the small muscles—lumbricals and interossei (Long and Brown 1964). Such movements are of very considerable importance in typing, playing musical instruments, etc., though of course the power exerted is relatively small, and when much force is required the long flexors must be called upon. The little finger has a wider range of abduction at the metacarpophalangeal joint than any other finger; its metacarpal is more mobile, and can be opposed towards the thumb.

In the second type of hand movement, the fingers and thumb combine in a 'precision grip'. This kind of grip is used to handle an object which is small or breakable. The object is seized with the pads of the fingers, which spread around it, the pads adapting themselves to its surface. This entails a considerable amount of rotation, some active, some passive, at the metacarpophalangeal joints. This rotation is sufficient to bring the pad of the index finger at right angles to the pad of the little finger should it be necessary, with the pads of the other fingers occupying an intermediate position. The thumb lies in a position of opposition. The motive power for the precision grip comes largely from the small muscles of the hand—those of the thenar and hypothenar eminences, the lumbricals and the interossei, though flexor digitorum profundus is probably also in action, since the distal interphalangeal joints are usually slightly flexed. In consequence, the grip depends very largely on the integrity of the ulnar nerve, and is grossly impaired if that nerve is paralysed.

The third type of activity of the hand comes into play when force is required, as in manipulating tools, climbing, and so on. The long flexor muscles, in association with the extensors of the wrist, contract to produce a 'power grip' in which the fingers are brought towards the palm round the object gripped and the thumb is flexed at the carpometacarpal joint and extended at the interphalangeal joint to grip the shaft or handle of the object concerned. When maximum power is exerted, all the muscles round the wrist, including the wrist flexors, are set in contraction to stabilize it. If the object gripped is small enough, the tips of the fingers come into contact with the palm and in such a clenched fist they all reach the palm together, even though the lengths of the individual fingers are very different. The power grip, involving the long flexor muscles and their synergists, depends very greatly on the integrity of the median nerve, and suffers if this nerve is paralysed. The two types of grip are thus largely independent of each other, and can be separately abolished (Napier 1956; Bowden and Napier 1961).

A fourth type of grip, the 'hook grip' (Napier 1956), closely resembles the power grip, but the thumb plays little or no part. It is used for carrying suitcases, and in rock climbing, and depends on the long flexor muscles and their synergists.

The versatility of the human hand (Capener 1956) has in the past been considered evidence of its unspecialized character, in contrast to the foot, which has altered greatly in the course of evolution in response to weight bearing. The contrary view is well expressed by Bishop (1964), who considers that the human hand, 'far from being a primitive organ which has consistently avoided the dangers of specialization, on the contrary, has first adapted to the branch, then to the ground, and lastly specialized beyond all other Primates to the tool'.

Fasciae of the upper limb

SUPERFICIAL FASCIA

The superficial fascia of the shoulder region and the arm contains a variable amount of fat; secondary sexual fat is deposited in the back of the female arm, and after middle age large quantities may be found here. At the back of the elbow a subcutaneous bursa separates the skin from the olecranon of the ulna. An enlargement of this bursa in people who lean on their elbow a great deal is called 'student's elbow'.

In the forearm the superficial fascia is in no way remarkable, but in the hand it undergoes several specializations. On the back of the hand it is loose and thin and contains little fat. In the palm it embodies tough strong strands of connective tissue which enclose the fat in loculi and connect the skin firmly to the thickened deep fascia, the palmar aponeurosis [p. 348], especially at the flexure lines. The fixation of the skin by this means is very marked in the centre of the palm, but much less so in the thenar and hypothenar

eminences. Here, however, the superficial fascia forms thicker and less fibrous pads of tissue which play an important part in the gripping mechanism of the hand, for they are readily adaptable to the contours of objects which are grasped in it. Other such pads are found in the palm opposite the metacarpophalangeal joints, and here a band of transverse fibres, the **superficial transverse metacarpal ligament**, crosses the distal part of the palm in the superficial fascia of the webs of the fingers, being connected to the palmar surfaces of the fibrous flexor sheaths of the digits [p. 348].

A similar, and very important, pad occurs on the palmar surfaces of the distal phalanges of the digits. Here the skin is tacked down to the distal two-thirds of the phalanx by strong fibrous strands between which are loculi containing fat. In this way the chief tactile exploratory instruments of the body are provided with a resilient and strong backing. Through the specialized tissue of the digital pads runs the blood supply to the shaft of the distal phalanx, and should the pad become infected, the pressure generated within it by the confined pus is likely to compress the artery and cause death of this part of the bone. On the dorsum of the distal phalanx strength is provided by the nail, and there is no superficial fascia deep to this.

In the superficial fascia of the hypothenar eminence lie the fibres of palmaris brevis, which help to wrinkle the skin of the medial side of the hand and give it a better grip [p. 342].

DEEP FASCIA

In front of pectoralis major is a layer of deep fascia continuous with the fascia of the anterior abdominal wall; superiorly, it is attached to the front of the clavicle. Lateral to the pectoralis major it thickens to form the **axillary fascia** [FIG. 5.50], which constitutes the floor of the axilla. As it is traced laterally it becomes continuous with the deep fascia of the arm.

Posterior to pectoralis major in the anterior wall of the axilla lies the dense **clavipectoral fascia**. This membrane extends from the first costal cartilage medially to the coracoid process and the coracoclavicular ligament laterally. Superiorly, it splits around subclavius to become attached to the margins of its insertion to the under surface of the clavicle. Inferiorly it splits to form an investment for pectoralis minor. The deep surface of the clavipectoral fascia is connected with the sheath of the axillary vessels [FIG. 5.94].

At the inferolateral border of pectoralis minor there is a further extension of this stratum of deep fascia downwards to join the fascia of the axillary floor. This extension is continued laterally into the fascia covering biceps and coracobrachialis.

In the shoulder region, the deep fascia is very strong over infraspinatus and teres minor, being firmly attached to the medial and lateral borders of the scapula. Superiorly, it forms a sheath for deltoid, and is attached along the line of the clavicle, the acromion and the spine of the scapula.

In the arm, the deep fascia forms a less dense **investing layer** which encloses the muscles of the arm. The strength of this layer is enhanced by the fact that it receives fibres from pectoralis major, latissimus dorsi, and deltoid. At the elbow the fascia is attached to the medial and lateral epicondyles of the humerus and to the olecranon process, and becomes continuous with the deep fascia of the forearm. From the investing layer two intermuscular septa run to be attached to the supracondylar ridges of the humerus [FIG. 5.69]. The **medial intermuscular septum** is the stronger of the two; it lies between brachialis and the medial head of the triceps, and gives origin to both [FIG. 5.68]. It is usually reinforced by some fibres from coracobrachialis, and does not extend superiorly beyond the insertion of this muscle into the humerus.

The **lateral intermuscular septum** lies between brachialis and brachioradialis in front, and the medial and lateral heads of triceps

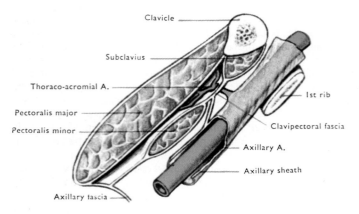

FIG. 5.94. Diagram of clavipectoral fascia in oblique section.

Clavicle

Subclavius

Thoraco-acromial A.

Pectoralis major

Pectoralis minor

1st rib

Clavipectoral fascia

Axillary A.

Axillary sheath

Axillary fascia

behind; it gives origin to fibres of all three muscles. This septum extends up to the insertion of deltoid and receives some fibres from it.

The deep fascia of the forearm is of varying density and strength in different parts of the region. Superiorly, it is strengthened by the attachment of the bicipital aponeurosis in front and the insertion of triceps behind. The flexor and extensor muscles of the forearm are covered by a strong layer of fascia from which they take part of their origin. For these reasons, the deep fascia in the upper part of the forearm next the elbow is very strong and dense. Lower down the forearm, as the muscle bellies give way to tendons and are no longer attached to it, the deep fascia becomes thinner, although the attachment to the posterior border of the ulna is thickened by the attachments of flexor and extensor carpi ulnaris and flexor digitorum profundus. At the wrist, it develops strong transverse retinacula within its substance; these bands prevent the flexor and extensor muscles of the wrist and digits from 'bowstringing' away from the bones during flexion or extension.

The **extensor retinaculum** [FIGS. 5.83, 5.87, and 5.88] is a thickening of the deep fascia on the back of the forearm and wrist; it runs more obliquely than transversely, and is continuous with the deep fascia of the dorsum of the forearm and hand. Laterally it is attached to the distal part of the anterior border of the radius, and medially to the distal end of the ulna, the carpus, and the ulnar collateral ligament of the wrist. From its deep surface septa run to the distal ends of the radius and ulna, so producing six separate compartments, each lined by a synovial sheath. From the lateral to the medial side these compartments transmit: (1) abductor pollicis longus and extensor pollicis brevis; (2) extensors carpi radialis longus and brevis; (3) extensor pollicis longus; (4) extensor digitorum and extensor indicis; (5) extensor digiti minimi; (6) extensor carpi ulnaris. Superficial to the retinaculum lie the dorsal veins of the hand and forearm, the superficial branch of the radial nerve, and the dorsal branch of the ulnar nerve.

On the dorsum of the hand, the deep fascia is arranged in two layers. The superficial layer covers the extensor tendons and is continuous with the distal border of the extensor retinaculum. Over the fingers, this layer fuses with the extensor hoods [p. 333], which replace the deep fascia on the dorsum of the digits. The deep layer of deep fascia is somewhat stronger, and lies over the interossei, blending with the superficial layer at the interdigital clefts.

The **flexor retinaculum** [FIGS. 5.85 and 5.96] is about 2.5 cm in length and in breadth. Laterally it is attached to the tubercle of the scaphoid and to both lips of the groove on the anterior surface of the trapezium, so completing an osteofascial compartment through which the flexor carpi radialis tendon and its synovial sheath pass into the hand. Medially, the retinaculum is attached to the hook of the hamate and to the pisiform bone.

The flexor retinaculum converts the hollow on the anterior aspect of the carpus into the **carpal tunnel**, through which run the flexor pollicis longus tendon, together with its synovial sheath, the flexor tendons of the fingers and their synovial sheath, and the median nerve. The anterior surface of the retinaculum gives origin to the muscles of the thenar and hypothenar eminences, and is crossed by the palmar branches of the median and ulnar nerves. The tendon of palmaris longus is partly inserted into this surface of the retinaculum.

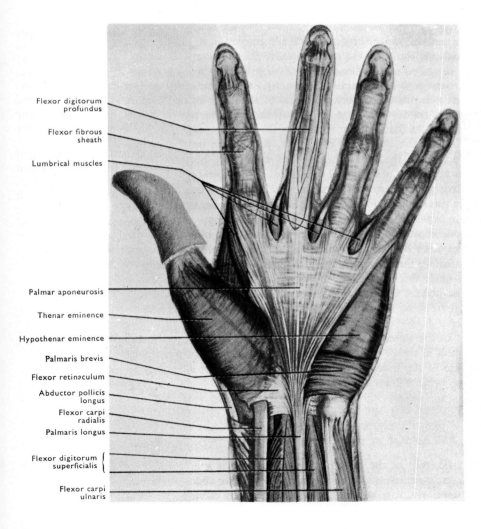

Flexor digitorum profundus
Flexor fibrous sheath
Lumbrical muscles
Palmar aponeurosis
Thenar eminence
Hypothenar eminence
Palmaris brevis
Flexor retinaculum
Abductor pollicis longus
Flexor carpi radialis
Palmaris longus
Flexor digitorum superficialis
Flexor carpi ulnaris

FIG. 5.95. Palmar aponeurosis.

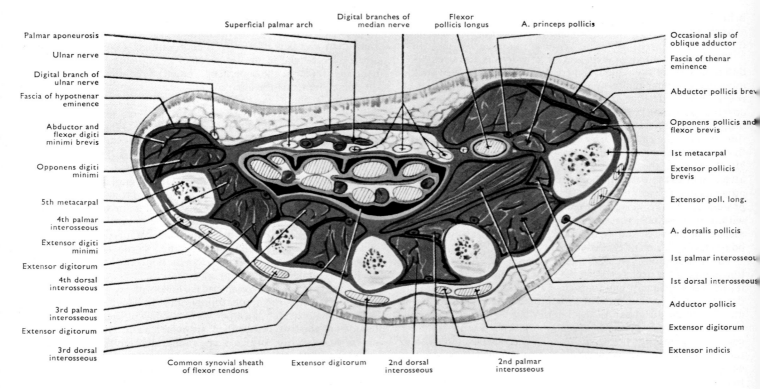

FIG. 5.96. Oblique cross-section through hand, showing fascial compartments.

The ulnar nerve and artery cross it on its medial side, often covered superficially by a thin layer of fascia which passes from the front of the pisiform bone to the fascia over the palmaris longus. Distally the retinaculum is loosely attached to the deep surface of the palmar aponeurosis.

In the palm of the hand, as on the dorsum, there are two layers of deep fascia. The deep layer covers the interosseous muscles and encloses the adductor muscle of the thumb [FIG. 5.96]. The superficial layer has a strong central portion known as the **palmar aponeurosis** [FIG. 5.95] and on each side it thins out to allow the maximum freedom of movement to the muscles of the thenar and hypothenar eminences. The palmar aponeurosis, which is continuous with the insertion of the palmaris longus [p. 327], is part of the apparatus designed to strengthen the hand for gripping, and to prevent the tendons of the flexor muscles from 'bowstringing' away from the joints they control. It is therefore dense and thick, consisting of longitudinal and transverse fibres which are firmly connected through the superficial fascia to the skin.

The palmar aponeurosis is triangular in shape, with its apex proximal and its base at the webs of the fingers. Four slips are prolonged from it into the fingers to become continuous with the fibrous digital sheaths of the flexor tendons. At their origin, these slips are connected together on their deep surfaces by transverse fibres which cross in front of the lumbrical muscles and the digital nerves and vessels. Each slip is also attached, by dorsally running fibres, to the capsule of the metacarpophalangeal joint and to the deep transverse metacarpal ligament.

The **fibrous digital sheaths** [FIG. 5.95] are attached to the raised lateral and medial edges of the palmar surfaces of the proximal and middle phalanges and to the palmar surface of the distal phalanx. Between these attachments the fibrous tissue composing each sheath arches from one side of the finger to the other, so completing an osteofibrous tunnel in which the flexor tendons run, surrounded by their synovial sheath. This tunnel is closed distally by the attachment of the sheath to the distal phalanx, but is open proximally, deep to the palmar aponeurosis. At the interphalangeal joints, the sheath is

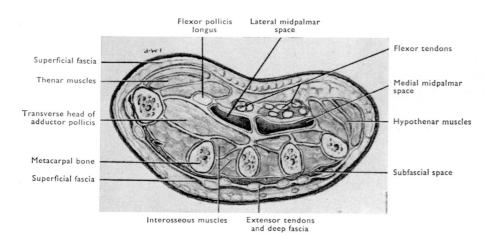

FIG. 5.97. Transverse section through hand to show the position of the (potential) midpalmar spaces.

attached to the palmar ligament, and is relatively thin and loose. On the phalanges, it is thick and strong and holds the tendons closely applied to the bones during flexion of the digit.

In the condition known as Dupuytren's contracture, the medial portion of the palmar aponeurosis and the fibrous digital sheaths of the ring and little fingers may become shortened, so pulling these fingers down towards the palm. In severe cases their tips may be pressed firmly into the palm and they become useless. The lateral two fibrous sheaths are less commonly affected.

From the medial and lateral borders of the aponeurosis septa pass backwards to fuse with the fascia on the interossei, so dividing the palm into three compartments. The most lateral of these, the thenar compartment, contains the short muscles of the thumb; the most medial, the hypothenar compartment, contains the short muscles of the little finger, and the intermediate compartment contains the long flexor tendons and the lumbrical muscles, surrounded by thin and relatively loose connective tissue. From the deep aspect of this connective tissue sheath a septum runs to the front of the body of the third metacarpal [FIG. 5.97]. Although this septum is relatively fragile, and often incomplete, it serves to divide the intermediate compartment into two potential spaces in which infected fluid may accumulate following injury to the hand. These are the lateral midpalmar and the medial midpalmar spaces, and they communicate proximally, behind the deep flexor tendons, with a similar potential space in front of the pronator quadratus. The spaces are bounded laterally and medially by the septa connecting the palmar

aponeurosis with the deep layer of deep fascia; posteriorly, by the interossei and adductor pollicis [FIG. 5.97]. Distally, prolongations of loose connective tissue round the tendons of the lumbrical muscles allow a spread of fluid into the web of the finger; this web space is freely continuous with the dorsal subaponeurotic space on the back of the hand, deep to the extensor tendons, and with the dorsal subcutaneous space, in which the fluid may then travel up the dorsum of the forearm in the loose superficial fascia (Bojsen-Møller and Schmidt 1974).

Muscles of the trunk

MUSCLES OF THE THORAX

The wall of the thorax is composed of ribs, between which lie the intercostal muscles, and to which are attached many other muscles. Some of these are considered with the upper limb, on which they act; others with the abdominal wall, in which their main bulk lies. The muscles dealt with in this section are all small and are all primarily concerned with movements of the ribs.

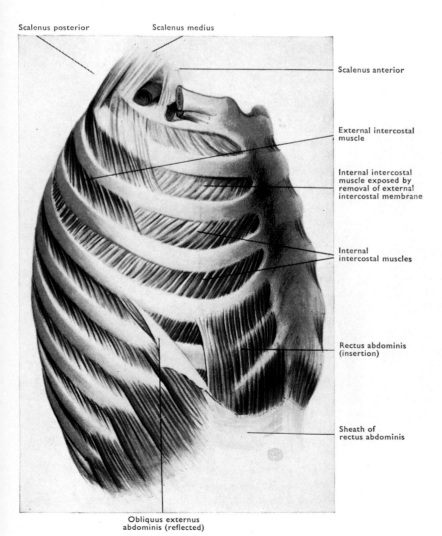

Scalenus posterior

Scalenus medius

Scalenus anterior

External intercostal muscle

Internal intercostal muscle exposed by removal of external intercostal membrane

Internal intercostal muscles

Rectus abdominis (insertion)

Sheath of rectus abdominis

Obliquus externus abdominis (reflected)

FIG. 5.98. Muscles of the right side of thoracic wall.

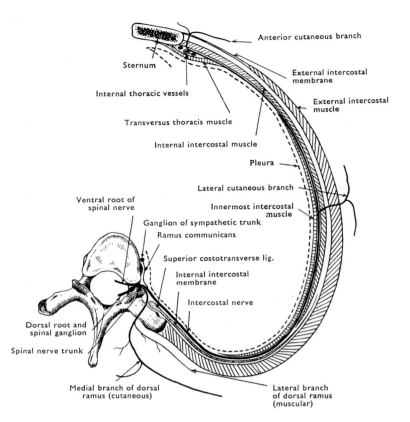

FIG. 5.99. Diagram of intercostal space to show relations of intercostal muscles, membranes, and nerve.

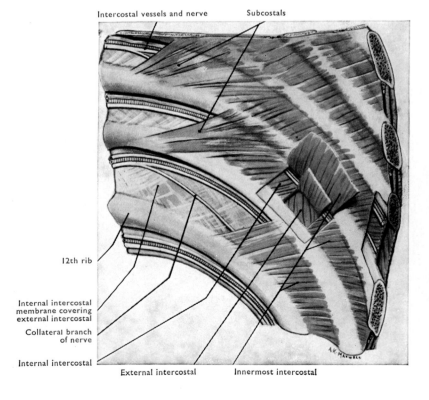

FIG. 5.100. Dissection of inner surface of chest wall to show intercostal muscles, vessels, and nerves.

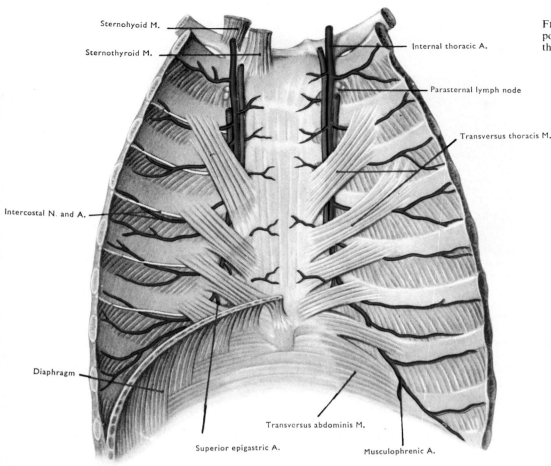

Sternohyoid M.

Sternothyroid M.

Internal thoracic A.

Parasternal lymph node

Transversus thoracis M.

Intercostal N. and A.

Diaphragm

Transversus abdominis M.

Superior epigastric A.

Musculophrenic A.

FIG. 5.101. A dissection of the posterior surface of the anterior thoracic wall.

INTERCOSTAL MUSCLES

Each **external intercostal** muscle is attached to the lower border of one rib and passes downwards and forwards to gain attachment to the upper border of the rib next below. Posteriorly it extends to the tubercles of the ribs, and anteriorly it thins out gradually to become continuous with the external intercostal membrane, a transition which takes place roughly where the rib becomes continuous with the costal cartilage. The membrane is a thin fibrous sheet which fills the remainder of the intercostal space, bounded by the adjacent costal cartilages and the sternal margin [FIG. 5.99].

In the lower intercostal spaces the external intercostal muscles may blend with the fibres of the external oblique muscle of the abdomen, which overlap them, but run in an exactly similar direction [FIG. 5.98]. The external intercostals and the external oblique may indeed be looked on as one continuous sheet of muscle, which in the region of the thorax has become stranded in between the ribs.

The **internal intercostal** muscle of each space is attached to the lower border of the costal cartilage and the costal groove of one rib and to the upper border of the rib below. The fibres lie deep to those of the external intercostal, and run downwards and backwards, crossing those of the external intercostal obliquely [FIGS. 5.98 and 5.100]. They run in the same direction as those of the internal oblique muscle of the abdomen, and in the lower two spaces the fibres of the two muscles are continuous [FIG. 5.107].

The internal intercostal is thicker anteriorly than posteriorly, and runs from the side of the sternum to the angle of the rib, where it is replaced by the internal intercostal membrane [FIG. 5.99]. This

extends back to the tubercle of the rib and becomes continuous with the superior costotransverse ligament.

The **innermost intercostals** [FIGS. 5.99 and 5.100] are variable layers separated off from the internal intercostals by the intercostal nerve and vessels. They lie deep to the internal intercostals, and their fibres run in the same direction. The **subcostal** muscles [FIG. 5.100] are irregular slips, varying considerably in number and size, which lie on the inner surface of the lower ribs near their angles, and pass over one or two intercostal spaces. They run in the same direction as the innermost intercostal muscles of these spaces, and may be continuous with them.

Nerve supply

All the intercostals and subcostals are supplied by the corresponding intercostal nerves.

Actions

The anatomical arrangement of the intercostals is simple, but their actions and functions are disputed. It is generally accepted that at least some of the external intercostals are active during inspiration, when they are believed to pull the rib below up towards the rib above [p. 367]. There is also electrophysiological evidence to suggest that they are active during the first phase of expiration, possibly to resist the elastic recoil of the lungs to maintain a uniform flow of air over the vocal cords in speaking or singing.

The actions of the deeper groups of intercostals are even more uncertain. It has been claimed that they act with the external intercostals to elevate the ribs in inspiration, but others hold that

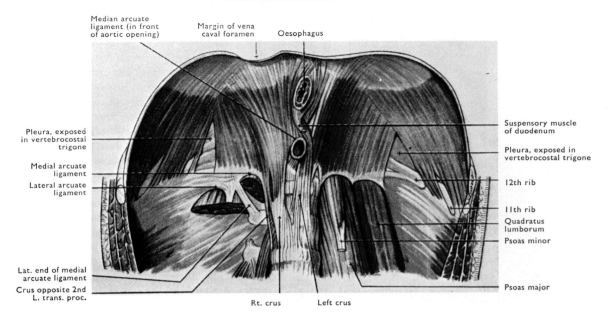

Median arcuate
ligament (in front
of aortic opening)
Margin of vena
caval foramen
Oesophagus

Pleura, exposed
in vertebrocostal
trigone

Medial arcuate
ligament

Lateral arcuate
ligament

Lat. end of medial
arcuate ligament

Crus opposite 2nd
L. trans. proc.

Rt. crus   Left crus

Suspensory muscle
of duodenum

Pleura, exposed in
vertebrocostal trigone

12th rib

11th rib

Quadratus
lumborum

Psoas minor

Psoas major

FIG. 5.102. Dissection showing posterior origin of the diaphragm.

they act from the rib below on the rib above to depress the thoracic cage during forced expiration. The reason for this discrepancy probably lies in the difficulty with which activity in one set of muscles can be distinguished from activity in the other. It is also fairly certain that the type and the amount of such activity may vary with depth of respiration (Campbell 1955b; Campbell, Agostini, and Davis 1970).

Both the external and the internal groups are apparently active during speech, and both groups are inactive in quiet expiration. It is also undisputed that both groups have a function in preventing the bulging or caving-in of the intercostal spaces during the variations in the intrathoracic pressure produced by respiratory movements. Contraction of the intercostals occurs during various movements of the trunk, and it has been suggested that this activity is perhaps more important than their actions in respiration (Jones and Pauly 1957).

TRANSVERSUS THORACIS

This muscle originates from the posterior aspect of the xiphoid process and the body of the sternum [FIG. 5.101] and is inserted into the inner aspect of the costal cartilages, from the second to the sixth inclusive. The lower fibres are more or less horizontal, and are parallel with the uppermost part of the transversus abdominis; the higher fibres pass obliquely upwards and laterally.

Nerve supply

From the corresponding intercostal nerves.

Actions

The transversus thoracis is thought to depress the costal cartilages articulating with the sternum, and so to contribute to expiration (Taylor 1960).

LEVATORES COSTARUM

Each levator costae [FIG. 5.14] is a small triangular muscle arising from the transverse processes of a vertebra from the 7th cervical to the 11th thoracic inclusive. It fans out as it descends to be inserted into the external surface of the rib below, lateral to the tubercle.

Nerve supply

From the ventral rami of the thoracic nerves.

Actions

Elevation of the ribs [p. 228].

SERRATUS POSTERIOR SUPERIOR

This thin, wide, unimportant muscle lies deep to the vertebroscapular muscles and superficial to splenius and erector spinae. It is attached to the lower part of the ligamentum nuchae and the spines of the seventh cervical and the upper three or four thoracic vertebrae, and runs downwards and laterally to be inserted into the second, third, fourth, and fifth ribs lateral to the angles.

Nerve supply

From the second to the fourth intercostal nerves.

Actions

Elevation of the upper ribs.

SERRATUS POSTERIOR INFERIOR

Another thin and unimportant muscle, lying deep to latissimus dorsi, and taking origin from the thoracolumbar fascia [FIG. 5.52]. The fibres pass almost horizontally to be inserted into the last four ribs.

Nerve supply

From the ninth to the eleventh intercostal nerves.

Actions

It may help to draw the lower ribs downwards and backwards.

THE DIAPHRAGM

The diaphragm is a dome-shaped partition separating the thoracic and abdominal cavities. It consists of a **central tendon**, shaped not unlike a boomerang, into which muscle fibres are inserted from all

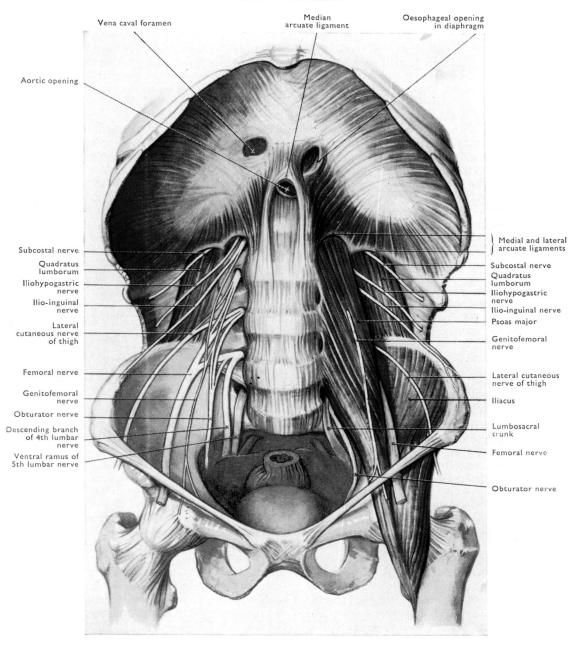

Vena caval foramen
Median arcuate ligament
Oesophageal opening in diaphragm
Aortic opening
Subcostal nerve
Quadratus lumborum
Iliohypogastric nerve
Ilio-inguinal nerve
Lateral cutaneous nerve of thigh
Femoral nerve
Genitofemoral nerve
Obturator nerve
Descending branch of 4th lumbar nerve
Ventral ramus of 5th lumbar nerve

Medial and lateral arcuate ligaments
Subcostal nerve
Quadratus lumborum
Iliohypogastric nerve
Ilio-inguinal nerve
Psoas major
Genitofemoral nerve
Lateral cutaneous nerve of thigh
Iliacus
Lumbosacral trunk
Femoral nerve
Obturator nerve

FIG. 5.103. Diaphragm and posterior wall of abdomen.

parts of the circumference of the inner aspect of the body wall. The fibres attached to the posterior part of the body wall extend much further inferiorly than do the anterior fibres, so that the concavity of the diaphragm looks forwards as well as downwards [FIGS. 5.102 and 5.103]. The posterior, concave border of the central tendon receives the fibres from the vertebral column, and the convex anterior border receives those from the thoracic margin. The tendon is dense and fibrous, particularly towards the midline, where its fibres interlace in a cruciform arrangement, and the pericardium is firmly attached to its upper surface.

The sternal portion of the diaphragm consists of two small slips which arise from the back of the xiphoid process and pass backwards to the central tendon. The costal portion arises from the deep surfaces of the lower six costal cartilages on each side, interdigitating with the slips of origin of the transversus abdominis, and is inserted into the anterolateral border of the central tendon.

The lumbar portion of the diaphragm arises in part by two crura from the anterolateral surfaces of the bodies of the first three lumbar vertebrae on the right side and the first two lumbar vertebrae on the left. The crura embrace the aorta as it enters the abdominal cavity [FIGS. 5.102 and 5.103], and are attached firmly to the margins of the vertebrae and the intervertebral discs. In between these attachments the upper lumbar arteries separate the fibres from the bodies of the vertebrae. The two crura are connected by a tendinous band, the median arcuate ligament, which arches in front of the aorta and gives origin to fibres of the right crus. This, which is usually the larger of the two, spreads out to form a thick triangular sheet of muscle directed upwards and somewhat to the left to be inserted into the middle part of the concave border of the central tendon on both sides of the median plane. The oesophagus passes through this part of the right crus [FIGS. 5.102 and 5.103], and is surrounded by its decussating fibres (Bowden and El-Ramli 1967). This arrangement may have a mild sphincteric action, though the importance of this is disputed. Close to the oesophageal opening the **suspensory muscle of the duodenum** is attached to the right crus (Low 1907). This somewhat unusual mixture of somatic and plain muscle, elastic and fibrous tissue descends to be attached to the ascending part of the duodenum, which it tethers in place.

The bulk of the fibres of the left crus pass to the left of the oesophageal opening, separated from it by the left margin of the right crus. Some fibres may pass to the right, behind the right crus [FIG. 5.102] and between the oesophageal and aortic openings.

The remainder of the lumbar portion of the diaphragm arises from the **medial** and **lateral arcuate ligaments**, which are thickenings of the fascia covering the anterior surface of psoas major and quadratus lumborum respectively [p. 372].

The sternocostal and lumbar portions of the diaphragm are distinct developmentally, and in 80 per cent of bodies they are separated by a hiatus in the muscular sheet called the **vertebrocostal trigone**. This gap [FIG. 5.102] lies above the twelfth rib, so that the upper pole of each kidney is separated from the pleura by loose areolar tissue only.

Several structures pass from the thorax to the abdomen or vice versa either through the diaphragm or between it and the body wall. The **aortic opening** lies behind the diaphragm at the level of the lower border of the twelfth thoracic vertebra slightly to the left of the midline. It transmits the aorta and the thoracic duct. Posterolateral to the thoracic duct and either accompanying it through the opening or passing behind the right crus is the azygos vein, and each crus is pierced by a greater and a lesser splanchnic nerve. The sympathetic trunk runs from the thorax to the abdomen behind the medial end of the medial arcuate ligament, and the subcostal nerve runs posterior to the lateral arcuate ligament. Between the sternal and costal origins of the diaphragm, the superior epigastric artery passes in front of the diaphragm to enter the sheath of the rectus abdominis. Similarly, the musculophrenic artery runs between the attachments of the diaphragm to the seventh and eighth costal cartilages.

The **oesophageal opening** in the diaphragm lies to the left of the midline in the muscular part of the right crus at the level of the tenth thoracic vertebra; it is elliptical in shape, and transmits the oesophagus, the trunks of the vagus nerves, and oesophageal branches of the left gastric vessels. The **opening for the inferior vena cava** lies in the central tendon at the level of the disc between the eighth and ninth thoracic vertebrae; its wall is adherent to the margins of the opening. It transmits branches of the right phrenic nerve as well as the inferior vena cava. The left phrenic nerve runs off the pericardium to pierce the muscular part of the diaphragm in front of the left limb of the central tendon.

The three major structures passing between chest and abdomen all have different relationships to the diaphragm and are thus liable to be affected differently by its contraction. The inferior vena cava, passing through the central tendon, is stretched and dilated, so helping the return of blood to the heart. The oesophagus is, if anything, constricted by its muscular surroundings, and the aorta is completely unaffected, since it lies behind the diaphragm.

Nerve supply

The phrenic nerves (ventral rami of C. 3, 4, 5) supply the diaphragm with motor and sensory innervation. Additional sensory fibres are supplied by branches from the lower six or seven intercostal nerves which run to the peripheral part of the muscle.

Actions

The diaphragm is the chief muscle of inspiration [p. 367]. Acting from the ribs and from its lumbar origins the radiating fibres of the diaphragm pull the central tendon downwards and tend to flatten the dome, thereby increasing the vertical diameter of the chest. The abdominal viscera are compressed against the anterior abdominal wall and pelvic diaphragm, and this eventually brings the tendinous dome to a halt. After this stage, the costal fibres act from the central tendon to pull upwards the lower ribs, thereby raising and everting the sternum and the costal margin. The thoracic capacity is thus increased transversely and anteroposteriorly as well as vertically [p. 228].

Normally the dome of the diaphragm bulges into the thorax higher on the right side than on the left, and on both sides the level of the diaphragm in relation to the ribs and vertebrae varies considerably according to the phase of respiration, the posture adopted, and the size and degree of distension of the abdominal viscera. It is highest when the body is supine, for in this position gravity causes the viscera to press on the under surface of the diaphragm, forcing it up into the chest. In the standing position the diaphragm falls, and it may fall even lower in the sitting position, presumably because of relaxation of the abdominal muscles. When the patient lies on his side the half of the diaphragm which lies uppermost sinks to a lower level still. In contrast the half of the diaphragm nearest the bed rises higher in the thorax because of the gravitational push of the abdominal viscera. As might be imagined, therefore, the excursion of the diaphragm varies very considerably according to circumstances [p. 367].

MUSCLES OF THE ANTERIOR ABDOMINAL WALL

The anterior abdominal wall, unsupported by bone, is required to form a strong but expansible support for the viscera. It must be able to stretch to accommodate a heavy meal, a distended urinary bladder, or a pregnant uterus, and yet to recoil in the expiratory phase of respiration, and to contract to expel the contents of the abdomen in defaecation, micturition, childbirth, or vomiting. In addition, it has to be capable of producing bending and twisting movements of the trunk. These demands are met by a three-layered arrangement of muscle corresponding to the three layers of muscle in the wall of the thorax, and having fibres running in the same directions. The outermost layer, formed by the external oblique muscle, has fibres which cross the middle layer, formed by the internal oblique, like a St. Andrew's cross, and the innermost layer, formed by the transversus abdominis, runs more or less horizontally, like the transversus thoracis. The mechanical strength afforded by this construction is increased by a thickening of the deepest layer of the superficial fascia [p. 371] over the lower part of the abdominal wall, which is subject to the greatest pressure from the pull of gravity on the abdominal contents.

On each side of the midline lies a muscle, rectus abdominis, which has no counterpart in the thorax. Each of these muscles runs vertically from the pubis to the rib cage and xiphoid process, and is enclosed in a strong fibrous sheath formed by contributions from the other muscles.

OBLIQUUS EXTERNUS ABDOMINIS

External oblique [FIG. 5.104] is the outermost of the three layers of muscle on the anterior and lateral aspects of the abdominal wall. It arises from the outer surfaces of the lower eight ribs by slips which interdigitate with serratus anterior and latissimus dorsi, and which can easily be seen in a muscular subject [FIG. 5.57]. The fibres may be continuous with those of the external intercostal muscles, and they radiate downwards and forwards towards the midline, the most posterior ones being almost vertical.

The posterior vertical fibres are commonly overlapped by latissimus dorsi, but occasionally there may be an interval between the two, known as the **lumbar triangle**. This triangle is bounded medially by latissimus dorsi, laterally by external oblique, and inferiorly by the iliac crest; it is a weak point in the abdominal wall [FIG. 5.52].

The lower and posterior part of the external oblique is inserted by fleshy fibres into the anterior two thirds of the outer lip of the iliac crest [FIG. 5.104], and the remainder gives rise to an aponeurosis which is broader below than above. Towards the midline the aponeurosis fuses with the underlying aponeurosis of internal oblique. Above the umbilicus this fusion extends for almost the whole width of rectus abdominis, and forms the anterior wall of its sheath. Below the umbilicus the area of fusion gradually narrows, and the two layers remain separate until they reach the midline. From the xiphoid process to the pubis the aponeurosis is inserted into the linea alba [p. 361].

The upper part of the aponeurosis of external oblique covers the insertion of rectus abdominis, and affords an origin for fibres of pectoralis major. The lower, free border of the aponeurosis, between the anterior superior iliac spine and the pubic tubercle, forms the **inguinal ligament**. This ligament lies anterior to the structures passing into the lower limb in front of the hip bone—the iliacus, the psoas major, and the pectineus muscles, the femoral nerve and vessels, and the femoral canal [p. 405]. The inguinal ligament is convex downwards because of the pull exerted on it by the fascia lata (the deep fascia of the thigh), which is attached along its length. It gives attachment in its whole length to the fascia transversalis and in its lateral part to the fascia iliaca [p. 372] and the fibres of internal oblique and transversus abdominis. Its medial part, which is rolled backwards on itself like a gutter, forms a floor for the inguinal canal [p. 362].

The medial end of the inguinal ligament has a triangular expansion backwards to the pecten pubis [FIG. 5.105]; this is the **lacunar ligament**. The free crescentic lateral margin of the lacunar ligament forms the medial boundary of the femoral ring, and it sends an extension laterally along the pecten which is termed the **pectineal ligament**. Still other fibres pass obliquely from the pecten across the linea alba [p. 361] to mingle with their opposite equivalents; this is the **reflected** part of the inguinal ligament, and is of comparatively little importance.

Just above the pubic tubercle there is a triangular cleft in the aponeurosis of external oblique; its base is at the pubic crest, and its apex is directed upwards and laterally. This **superficial inguinal**

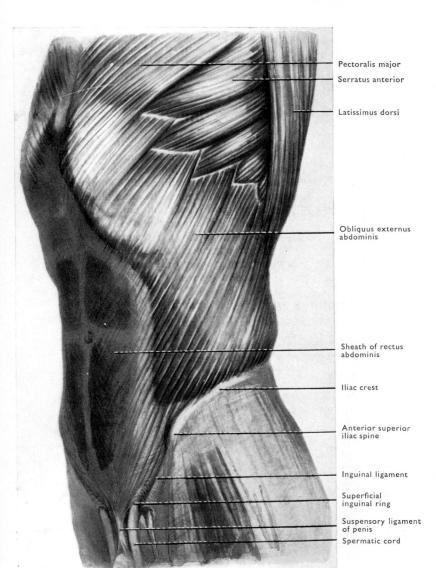

Pectoralis major

Serratus anterior

Latissimus dorsi

FIG. 5.104. Left obliquus externus abdominis.

Obliquus externus abdominis

Sheath of rectus abdominis

Iliac crest

Anterior superior iliac spine

Inguinal ligament

Superficial inguinal ring

Suspensory ligament of penis

Spermatic cord

ring transmits the spermatic cord in the male or the round ligament of the uterus in the female, and is bounded by lateral and medial **crura**. The lateral crus is formed by the fibres of the external oblique aponeurosis passing to the pubic tubercle. The medial crus is the part of the external oblique aponeurosis attached to the pubic crest and the symphysis pubis. Arching across between the two crura and strengthening the opening are a number of intercrural fibres

[FIG. 5.106], and the margins of the superficial ring are continuous with the **external spermatic fascia**. This thin evagination of the external oblique aponeurosis, which surrounds the spermatic cord (or round ligament) after it has passed through the abdominal wall, is the outermost of the three coverings of the cord. If a finger is placed so as to invaginate the skin of the scrotum upwards and laterally, the spermatic cord can be followed up to the superficial

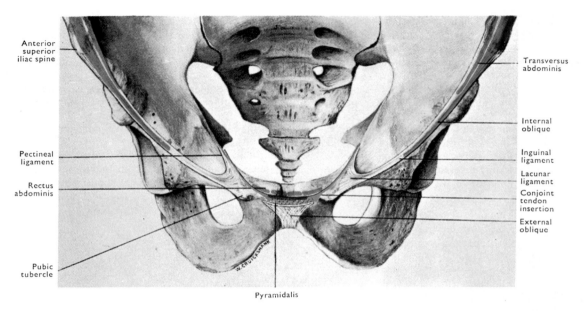

FIG. 5.105. Muscle attachments to inguinal ligament and pubis.

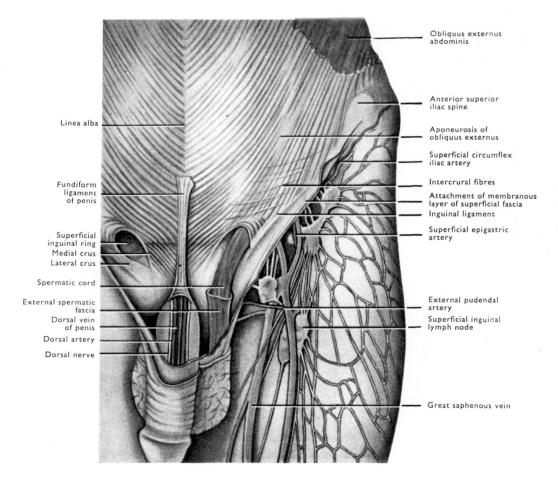

FIG. 5.106. Superficial anatomy of the groin.

inguinal ring. By pressing backwards, the finger can then detect the crura of the ring, and the size of the aperture can be estimated; normally it just admits the tip of the little finger.

Nerve supply

External oblique is supplied by the ventral rami of the lower six thoracic nerves.

Actions

Like the other muscles of the anterolateral part of the abdominal wall, external oblique helps to support the abdominal viscera against the pull of gravity. If the pelvis is fixed, the external oblique muscles forcibly depress the ribs and help to compress the chest, as for instance in the sudden violent expiratory efforts of coughing and sneezing. If both the ribs and pelvis are fixed, external oblique acting with its fellow of the opposite side can bend the trunk forwards and flex the lumbar vertebral column. If the muscle of one side acts on its own, the trunk is bent laterally to that side and rotated to the opposite side. If these movements are prevented by their antagonists, the external obliques help to compress the abdominal viscera, an action which is brought into play in the movement of expiration against resistance, since it forces the diaphragm upwards, so decreasing the volume of the thoracic cavity. The compressing action of the abdominal muscles is also used in defaecation, parturition, micturition, and vomiting, and the rise in intra-abdominal pressure forces together the walls of the inguinal canal. When heavy weights are lifted, compression of the abdominal contents by the abdominal muscles and the diaphragm helps to make the trunk rigid and protects the vertebral column. Finally, with the other abdominal muscles, external oblique steadies the pelvis during walking, so providing a stable origin for the muscles operating from the pelvis on the thigh. It can also flex the pelvis on the vertebral column during jumping or climbing.

OBLIQUUS INTERNUS ABDOMINIS

Internal oblique, the second of the three muscular layers of the abdominal wall, arises by fleshy fibres from the thoracolumbar fascia [FIG. 5.47], the anterior two thirds of the intermediate line of the iliac crest, and the lateral two-thirds of the inguinal ligament, and the related fascia on the iliopsoas muscle. The fibres fan out from this origin [FIG. 5.107], the highest and most posterior fibres

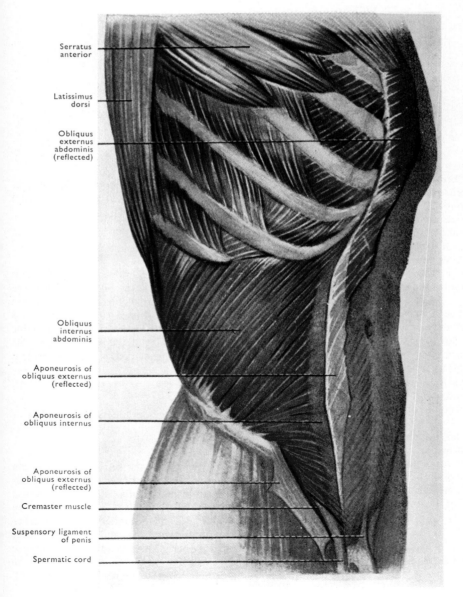

Serratus anterior

Latissimus dorsi

Obliquus externus abdominis (reflected)

Obliquus internus abdominis

Aponeurosis of obliquus externus (reflected)

Aponeurosis of obliquus internus

Aponeurosis of obliquus externus (reflected)

Cremaster muscle

Suspensory ligament of penis

Spermatic cord

FIG. 5.107. Right obliquus internus abdominis.

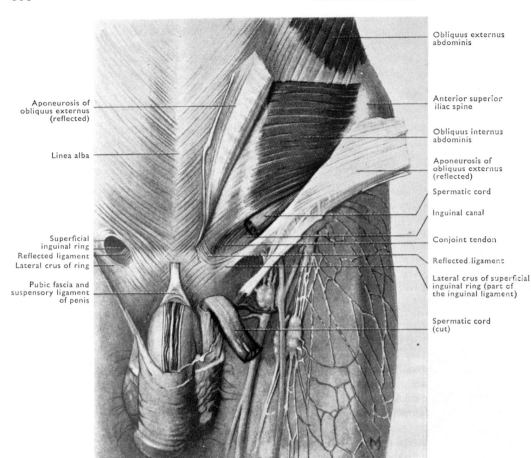

Aponeurosis of
obliquus externus
(reflected)

Linea alba

Superficial
inguinal ring
Reflected ligament
Lateral crus of ring

Pubic fascia and
suspensory ligament
of penis

Obliquus externus
abdominis

Anterior superior
iliac spine

Obliquus internus
abdominis

Aponeurosis of
obliquus externus
(reflected)

Spermatic cord

Inguinal canal

Conjoint tendon

Reflected ligament

Lateral crus of superficial
inguinal ring (part of
the inguinal ligament)

Spermatic cord
(cut)

FIG. 5.108. Left inguinal
canal. Structures seen on
reflecting obliquus
externus.

being inserted into the cartilages of the last three ribs in line with the internal intercostals. The more anterior and lower fibres give way to an aponeurosis along a line which extends downwards and medially from the tenth costal cartilage to the body of the pubis. The lower quarter of this aponeurosis, which represents the fibres arising from the inguinal ligament, passes in front of rectus abdominis, and unites with the lower part of the aponeurosis of transversus abdominis to form the conjoint tendon, which arches downwards to be attached to the pubic crest and the pecten pubis [FIG. 5.105]. Its fibres are continuous with those of the anterior wall of the sheath of rectus abdominis [p. 361].

The upper three-quarters of the internal oblique aponeurosis splits almost immediately into anterior and posterior layers which enclose rectus abdominis and at its medial margin are attached to the linea alba and the xiphoid process. The anterior layer blends with the aponeurosis of external oblique and extends up in front of rectus abdominis on to the chest wall. The posterior layer reaches up as far as the margins of the seventh, eighth, and ninth costal cartilages; above this level the posterior surface of rectus abdominis is in direct contact with the chest wall.

The lower fibres of internal oblique, as they arch across the spermatic cord to join the conjoint tendon, lie first in the anterior wall of the inguinal canal [FIG. 5.108], and later form its roof; medially, the conjoint tendon is the main component of the posterior wall of the canal. Some fibres pass off the inferomedial part of internal oblique and run along the spermatic cord to form the cremaster muscle.

Nerve supply

From the ventral rami of the lower six thoracic and the first lumbar nerves.

Actions

Similar to those of external oblique [p. 357], except that in rotation of the body the internal oblique of one side acts with the external oblique of the opposite side.

CREMASTER

This is a loose arrangement of muscle fibres forming festoons connected by the thin cremasteric fascia [p. 372]. The fibres are continuous with the inferior edge of internal oblique muscle and the adjacent part of the inguinal ligament [FIG. 5.109], and they loop round the spermatic cord and the testicle, returning to the pubic tubercle, which is the other fixed point of the muscle. The male cremaster is usually reasonably well developed, but in the female it is scanty or absent.

Nerve supply

From the genital branch of the genitofemoral nerve (L. 1 and 2).

Actions

The cremaster pulls the testis up from the scrotum towards the superficial inguinal ring. Voluntary control over the muscle is not usually possible, but in the cremasteric reflex the testis is raised when the medial side of the thigh is stroked. This reflex is very active in the infant and young child, and it is important to remember its existence when examining an infant in whom it is suspected that the testis has not descended properly, since the mere performance of the examination may cause the testis to disappear into the superficial inguinal ring or up on to the anterior abdominal wall,

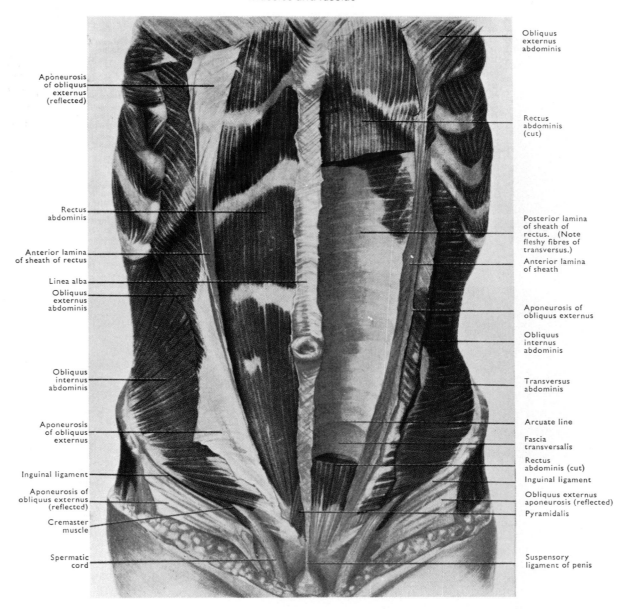

FIG. 5.109. Deep dissection of the abdominal wall. Rectus muscle and its sheath.

leaving that side of the scrotum empty. By the age of puberty this mobility is much reduced.

Together with dartos [p. 372], cremaster forms a mechanism for controlling the temperature of the testis. When the testis is allowed to hang down at a distance from the superficial inguinal ring, its temperature falls; when it is pulled closer to the ring, it becomes warmer. This is important, since the proper formation of spermatozoa requires a fairly constant temperature below that of the remainder of the body.

TRANSVERSUS ABDOMINIS

Transversus abdominis [FIGS. 5.109, 5.110, and 5.113], corresponding to the transversus thoracis, arises from the deep surfaces of the costal cartilages of the lower six ribs, interdigitating with the origin of the diaphragm, from the thoracolumbar fascia [FIG. 5.47], from the anterior two-thirds of the inner lip of the iliac crest, from the lateral third of the inguinal ligament, and from the fascia covering iliacus. As its name suggests, its fibres run horizontally round the

abdominal wall, and they end in an aponeurosis which fuses with the posterior layer of the aponeurosis of internal oblique. The upper three-quarters of the resultant fibrous sheet passes behind rectus abdominis to reach the linea alba [p. 361] and the xiphoid process, but the lower quarter is fused with the entire thickness of internal oblique aponeurosis and passes with it anterior to rectus abdominis, to the linea alba. This changeover takes place approximately half way between the umbilicus and the pubis, at the **arcuate line** [p. 361]. Inferior to this there is no aponeurotic posterior layer of the rectus sheath, and the posterior surface of rectus is in contact with the fascia lining transversus abdominis (**transversalis fascia**). The lowest fibres of the aponeurosis, together with those of internal oblique, turn downwards to form the **conjoint tendon** [FIGS. 5.110 and 5.113], and are inserted into the pubic crest, in front of the rectus abdominis, and into the pecten pubis. There they form the medial boundary of the femoral ring [p. 405], with the lacunar ligament.

The lower, free border of transversus abdominis is concave downwards between its attachment to the inguinal ligament and to

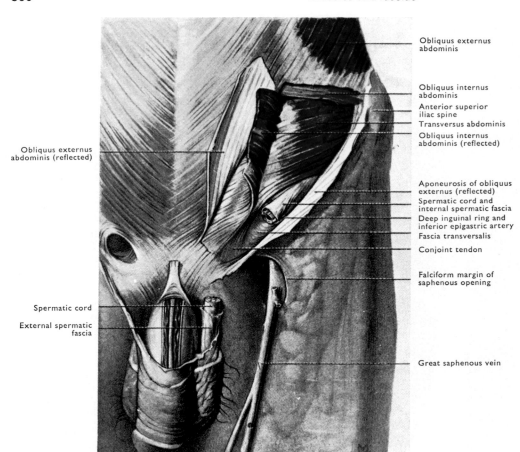

Obliquus externus
abdominis

Obliquus externus
abdominis (reflected)

Spermatic cord

External spermatic
fascia

Obliquus internus
abdominis
Anterior superior
iliac spine
Transversus abdominis
Obliquus internus
abdominis (reflected)

Aponeurosis of obliquus
externus (reflected)
Spermatic cord and
internal spermatic fascia
Deep inguinal ring and
inferior epigastric artery
Fascia transversalis
Conjoint tendon

Falciform margin of
saphenous opening

Great saphenous vein

FIG. 5.110. Dissection of inguinal
canal.

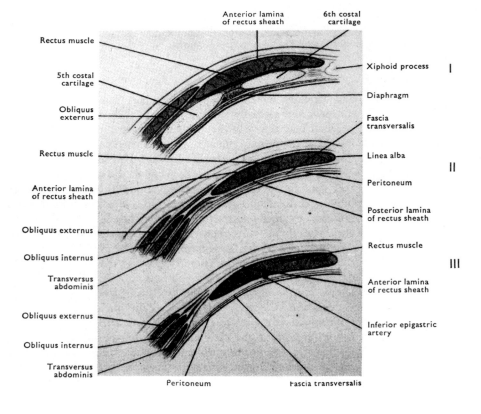

Anterior lamina
of rectus sheath

6th costal
cartilage

Rectus muscle

5th costal
cartilage

Obliquus
externus

Rectus muscle

Anterior lamina
of rectus sheath

Obliquus externus

Obliquus internus

Transversus
abdominis

Obliquus externus

Obliquus internus

Transversus
abdominis

Peritoneum

Fascia transversalis

Xiphoid process   I

Diaphragm

Fascia
transversalis

Linea alba

Peritoneum   II

Posterior lamina
of rectus sheath

Rectus muscle

III

Anterior lamina
of rectus sheath

Inferior epigastric
artery

FIG. 5.111. Sheath of rectus abdominis muscle.
   I. In the thoracic wall.
  II. In the upper three-quarters of the abdominal
      wall [cf. FIG. 5.109].
 III. In the lower fourth of the abdominal wall.
      (Below the umbilicus, the fusion of the
      aponeuroses of external and internal oblique
      muscles in front of the rectus is incomplete.)

the pecten. Between this border and the inguinal ligament is an elliptical interval in which the transversalis fascia comes into contact with the inferomedial part of internal oblique, the aponeurosis of external oblique, and the inguinal ligament. The spermatic cord evaginates the transversalis fascia in the lateral part of this interval at the **deep inguinal ring**, where it enters the inguinal canal, the inferior margin of transversus abdominis arching medially over the cord as it does so.

Nerve supply

From the ventral rami of the lower six thoracic and the first lumbar nerves.

Actions

These are similar to those of the external and internal oblique muscles [pp. 357 and 358], except that it has little flexing action on the vertebral column, does not depress the ribs, and has virtually no stabilizing effect on the pelvis. The transversus can pull the umbilicus to its own side.

RECTUS ABDOMINIS

This long strap muscle arises from the pubic crest and adjacent symphysis. Widening as it passes upwards, it is inserted into the anterior surfaces of the xiphoid process and of the fifth, sixth, and seventh costal cartilages. In the anterior part of its substance, but not extending through the entire depth of the muscle, are three or sometimes more transverse tendinous intersections, which are firmly adherent to the anterior wall of the rectus sheath. The lowest is at the level of the umbilicus, the highest at the xiphoid process, and the third about half way between the other two. The convex lateral border of the muscle shows as a groove on the skin of the lean living subject; this is the **linea semilunaris** [FIG. 5.112].

The muscle is enclosed in a **sheath** formed by the aponeuroses of the other abdominal muscles. At its lateral margin, the aponeurosis of internal oblique splits into two layers, one of which passes in front of the rectus and the other behind it. The anterior layer is joined by the aponeurosis of external oblique, and the posterior layer by the aponeurosis of transversus abdominis, as well as by some fibres of the muscle itself in the upper lateral part of the sheath [FIG. 5.109]. Above the costal margin, the posterior layer of the sheath is deficient, and rectus lies directly on the thoracic wall. Below a point about midway between the umbilicus and the symphysis pubis, the internal oblique aponeurosis does not split, and the posterior wall of the sheath is again deficient. Its lower limit is marked by a crescentic border, the **arcuate line** [FIG. 5.113]. Inferior to this, rectus is directly in contact with the fascia transversalis, and the anterior wall of the sheath in this region is formed by the aponeuroses of external oblique, internal oblique, and transversus abdominis, the inferior parts of the last two fusing to form the conjoint tendon [FIG. 5.113].

Within the sheath are found the pyramidalis muscle, the superior and inferior epigastric vessels, and the terminal portions of the lower five intercostal and the subcostal vessels and nerves.

Nerve supply

From the ventral rami of the lower six or seven thoracic nerves.

Actions

Rectus abdominis plays comparatively little part in compressing the abdominal contents (Floyd and Silver 1950), but it is a powerful flexor of the lumbar spine and depressor of the rib cage. During walking, it stabilizes the pelvis, so fixing the origins of the large muscles of the thigh which are concerned in propulsion. A similar fixation of the pelvis occurs when the body is supine and the feet are lifted off the bed; rectus abdominis contracts to prevent the pelvis being tilted by the weight of the lower limbs.

PYRAMIDALIS

This small muscle, which is not always present, lies on the front of the lower part of rectus abdominis, arising from the pubic crest and inserting into the linea alba [FIGS. 5.105 and 5.109].

Nerve supply

From the ventral ramus of the subcostal nerve.

Actions

It tenses the linea alba, but the purpose this serves is not known.

LINEA ALBA

The linea alba is a midline raphe extending from the xiphoid process to the symphysis pubis, and formed by the fusion of the aponeuroses of external oblique, internal oblique, and transversus abdominis. The raphe is relatively narrow below the umbilicus, but widens out to separate the rectus abdominis muscles as they diverge from each other above the umbilicus; here it forms a visible groove in the lean living subject [FIG. 5.112]. Approximately half way down the linea alba lies the **umbilicus**, a puckered scar through which, in the foetus, the umbilical vessels, the urachus, and the vitello-intestinal duct were transmitted. On its deep surface remnants of these structures are still attached. The ligamentum teres of the liver is derived from the left umbilical vein, the lateral umbilical ligaments from the

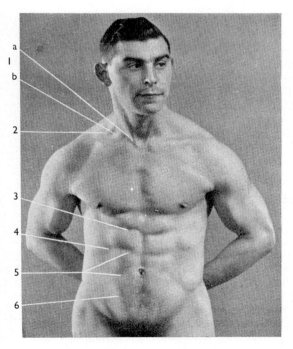

FIG. 5.112. Rectus abdominis in action. (*By courtesy of Professor R. D. Lockhart.*)
1. Sternocleidomastoid, (a) sternal head, (b) clavicular head.
2. Clavicular fibres of trapezius.
3. Tendinous intersection of rectus abdominis.
4. Linea semilunaris.
5. Tendinous intersections.
6. Linea semilunaris.

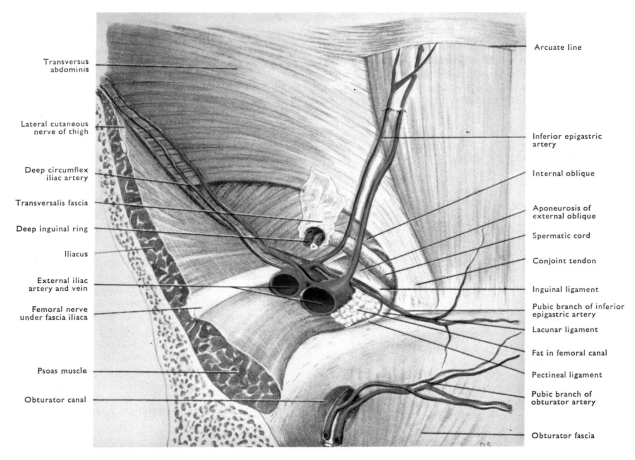

Transversus
abdominis

Lateral cutaneous
nerve of thigh

Deep circumflex
iliac artery

Transversalis fascia

Deep inguinal ring

Iliacus

External iliac
artery and vein

Femoral nerve
under fascia iliaca

Psoas muscle

Obturator canal

Arcuate line

Inferior epigastric
artery

Internal oblique

Aponeurosis of
external oblique

Spermatic cord

Conjoint tendon

Inguinal ligament

Pubic branch of inferior
epigastric artery

Lacunar ligament

Fat in femoral canal

Pectineal ligament

Pubic branch of
obturator artery

Obturator fascia

FIG. 5.113. Deep inguinal ring, femoral ring, and obturator canal seen from above and behind inside the abdomen. The anterior abdominal wall has been pulled forwards.

umbilical arteries, and the median umbilical ligament from the urachus. In addition, a fibrous band representing the remains of the vitello-intestinal duct may connect the umbilicus with the ileum or with the diverticulum ilei when this is present. If the urachus or the vitello-intestinal duct remains patent, urine or faeces may be discharged through the umbilicus of the newborn infant.

The linea alba permits no transverse stretching under normal circumstances, but stretches longitudinally sufficiently to allow considerable extension of the lumbar vertebral column.

INGUINAL CANAL

The inguinal canal [FIGS. 5.108, 5.110, and 5.113] is the oblique passage in the abdominal wall which transmits the spermatic cord in the male or the round ligament of the uterus in the female. In both sexes it also contains the ilio-inguinal nerve. The canal begins at the deep inguinal ring, which is an area of evagination in the fascia transversalis [p. 361] about 1.5 cm above the inguinal ligament midway between the anterior superior iliac spine and the pubic symphysis. From this point the canal travels for about 4 cm medially and slightly downwards to the superficial inguinal ring, which is an area of evagination of the external oblique aponeurosis immediately above the pubic tubercle and the medial end of the inguinal ligament.

The anterior wall of the canal is formed throughout its extent by the aponeurosis of external oblique, which is reinforced in its lateral one-third by the muscular fibres of internal oblique. The posterior wall of the canal is formed throughout its extent by the fascia transversalis, and in its medial one-third or so this is reinforced by

the conjoint tendon, which therefore lies deep to the superficial ring. It follows that the anterior wall is strongest opposite the deep inguinal ring, and the posterior wall strongest opposite the superficial inguinal ring. The weakest portion of the canal is the intermediate one-third. The floor of the canal is formed by the incurved inguinal ligament and the roof by the fibres of internal oblique and transversus arching medially to cross the spermatic cord obliquely from front to back.

The spermatic cord (or the round ligament of the uterus) enters the canal at the deep inguinal ring, carrying with it a covering of internal spermatic fascia from the fascia transversalis [p. 372]. The cord runs along the canal and acquires from the lower border of the internal oblique muscle a second covering, the cremasteric fascia, which contains the cremaster muscle. A third covering is added as the cord escapes through the superficial inguinal ring, for from the edges of this the external spermatic fascia is carried off as a continuation of the aponeurosis of external oblique.

The inguinal triangle is bounded inferiorly by the inguinal ligament, medially by the lateral border of the rectus abdominis, and laterally by the inferior epigastric artery, which runs upwards and medially behind the fascia transversalis from the external iliac artery to the rectus sheath, passing immediately medial to the deep inguinal ring. The triangle is divided into medial and lateral parts by the lateral umbilical ligament.

The inguinal canal is a weak point in the abdominal wall, and contraction of the abdominal muscles tends to force mobile abdominal contents along it. However, the same muscles, by their contraction, narrow the canal and tend to close the rings. Contraction of transversus causes the sling of transversalis fascia surrounding the medial margin of the internal ring to move laterally and upwards

(Blunt 1951), while the arched fibres of the internal oblique which form its lateral margin move medially and downwards, so closing the ring like the shutter of a camera (Patey 1949). The pull of external oblique causes the inguinal ligament to tighten, so supporting the lower edge of the external ring and somewhat narrowing its calibre. The fact that internal oblique is in more or less continuous contraction during the maintenance of the upright posture (Floyd and Silver 1950) may thus have a protective as well as a supportive significance. However, it has been suggested that the movement of rotation of the trunk, in which the external oblique of one side co-operates with the internal oblique of the other, may actually pull open the canal. Thus, when the trunk rotates to the left, as the patient pulls sideways on a rope, the linea alba becomes convex to the left (Parry 1966) and transversus and internal oblique fibres on the right side are presumably pulled away from the inguinal ligament.

MUSCLES OF THE POSTERIOR ABDOMINAL WALL

Behind the posterior margin of the anterior abdominal muscles, between the iliac crest and the last rib, the abdominal wall is formed by quadratus lumborum [FIG. 5.103]. Medial to this, clothing the fronts of the lumbar transverse processes and the sides of the vertebral bodies, lies the origin of psoas major, a long muscle which runs downwards to be joined by iliacus, a fleshy mass which lines the iliac fossa. The result is a composite unit, the iliopsoas, which leaves the abdomen behind the inguinal ligament to become the chief flexor of the hip joint. A fourth muscle, psoas minor, is sometimes present in front of the psoas major.

Together, these muscles form a soft padded bed for many important abdominal viscera, and their position lateral to the lumbar vertebrae allows them to act as lateral flexors of the vertebral column.

QUADRATUS LUMBORUM

Quadratus lumborum runs between the posterior part of the iliac crest, the iliolumbar ligament, and the transverse processes of the lower lumbar vertebrae to the medial part of the lower border of the last rib and the transverse processes of the upper lumbar vertebrae. It lies between the anterior and middle layers of the thoracolumbar fascia [FIG. 5.47], and in front of it lie the ascending or descending colon, the kidney, psoas major and minor, and the diaphragm. Behind, it is overlapped by erector spinae.

Nerve supply

From the ventral rami of the subcostal nerve and the upper three or four lumbar nerves.

Actions

Quadratus lumborum is a lateral flexor of the vertebral column, and fixes the twelfth rib during deep respiration. When both muscles act together they help to extend the lumbar part of the vertebral column, as well as giving it lateral stability.

PSOAS MINOR

This weak muscle is absent in about 40 per cent of bodies. It arises from the sides of the bodies of the twelfth thoracic and first lumbar vertebrae, and from the disc between them. It has a short fleshy belly [FIG. 5.102] and a long tendon, which runs down anterior to the psoas major to be inserted into the iliopubic eminence and the iliac fascia.

Nerve supply

From the ventral ramus of the first lumbar nerve.

Actions

It is a weak flexor of the lumbar spine.

PSOAS MAJOR

In contrast to psoas minor, psoas major is a long thick powerful muscle. It arises from the lateral margins of the bodies of the vertebrae from the lower border of the twelfth thoracic to the upper border of the fifth lumbar, from the intervertebral discs between them, from tendinous arches over the sides of the bodies of the upper four lumbar vertebrae and the lumbar vessels, and from the roots of the transverse processes of all the lumbar vertebrae [FIGS. 5.47 and 5.103].

The muscle narrows in a fusiform fashion to pass along the margin of the superior aperture of the lesser pelvis where it is joined by the fibres of iliacus. It enters the thigh posterior to the inguinal ligament, and ends in a tendon inserted into the lesser trochanter of the femur [FIGS. 5.124 and 5.135]. The tendon is separated from the pubis and the capsule of the hip joint by a large bursa, which may be continuous with the cavity of the hip joint.

Nerve supply

From the ventral rami of the upper three or four lumbar nerves.

Actions

In conjunction with iliacus, it is the chief flexor of the hip joint. Acting from its insertion, psoas major bends the trunk and pelvis forwards, as in sitting up from the supine position, and it is also a powerful lateral flexor of the lumbar spine. This action is used to maintain the balance of the trunk while sitting. Psoas major may come into play as a postural muscle preventing hyperextension of the hip joint (Basmajian 1958); but usually it is relaxed in the standing position (Joseph and Williams 1957; Keagy, Brumlik, and Bergan 1966).

ILIACUS

This large fleshy muscle, which lies along the lateral side of psoas major in the greater pelvis, arises mainly from the floor of the iliac fossa, but fibres also take origin from the ala of the sacrum and the anterior ligaments of the lumbosacral and sacroiliac joints. The fasciculi run downwards and medially behind the inguinal ligament and the majority of them are attached to the lateral side of the tendon of psoas major, so forming a composite muscle commonly known as the iliopsoas. Others run to the lesser trochanter of the femur and to a short linear insertion below this [FIG. 5.124]. A few of the most lateral fibres are inserted into the capsule of the hip joint.

Nerve supply

From branches of the femoral nerve (L. 2 and 3).

Actions

Iliacus acts with psoas major as a flexor of the hip joint.

Anatomy of hernia

The musculature of the wall of the abdominal cavity presents several areas of weakness, through which the more mobile abdominal viscera may occasionally be protruded, so forming a hernia. The most obvious region for such herniation is the inguinal canal, and a contributory factor to the frequency of hernia in this region is the fact that in the erect posture the pressure inside the lower part of the abdomen becomes considerably greater than that inside the upper part. Out of every 100 hernias through the anterior abdominal wall seen in practice, about 90 are in the inguinal region. Anything which raises the intra-abdominal pressure, such as heavy muscular effort, or severe and prolonged coughing, may precipitate the protrusion of some of the abdominal contents through the inguinal canal or other weak spots in the abdominal wall. The most important of these other weak spots are the femoral canal and the umbilicus.

Hernias are about six times as common in males as in females, and in males about 97 per cent of hernias occur in the inguinal region, 2 per cent in the femoral canal, and 1 per cent in the umbilical region. In females, on the other hand, only about 50 per cent of hernias occur in the inguinal region, 34 per cent in the femoral canal, and 16 per cent at the umbilicus.

Inguinal hernia

The reason for the preponderance of inguinal hernia in males is the large canal that houses the spermatic cord: the small canal in the female contains only the round ligament of the uterus. The testis in the course of its descent through the inguinal canal into the scrotum, which it reaches about the time of birth, is preceded by a process of peritoneum, the **processus vaginalis**, which is normally closed off after the descent by adherence of its walls, leaving only a slight dimple in the peritoneum at the site of the deep inguinal ring. But if the funnel-shaped processus vaginalis persists, some of the gut may descend within it along the spermatic cord, so acquiring the same coverings as the cord. Because such a protrusion traverses the inguinal canal, it is called an **oblique**, or indirect inguinal hernia, and it always enters the canal lateral to the inferior epigastric artery. Oblique hernias may also occur in the absence of a patent processus vaginalis.

A **direct** inguinal hernia, as its name suggests, bulges outwards in the region of the weak point in the abdominal wall at the inguinal triangle. Here, in the middle of the canal, both its anterior and its posterior walls are relatively weak, and the protrusion occurs medial to the inferior epigastric artery and usually lateral to the edge of the conjoint tendon. There is never a pre-formed tunnel of peritoneum along which the gut may protrude, and it pushes in front of it the peritoneum and the transversalis fascia. It may then either add the external oblique aponeurosis to its coverings or bulge through the superficial inguinal ring, carrying with it the external spermatic fascia.

Umbilical hernia

The umbilicus is a weak point in the linea alba, through which a hernia of the midgut into the umbilical cord always occurs in the early stages of embryonic development. Usually this hernia is 'reduced' (i.e. the gut returns within the abdominal cavity) by the eleventh week of foetal development, but occasionally it may still remain outside the abdominal wall at birth, forming a **congenital umbilical hernia**. Gross herniation is unusual, but a small knuckle of gut may be protruded in this way, and it normally retreats within the abdomen within a week or so after birth, though in some races it may persist until adolescence.

In the adult, umbilical hernia is most common in females after middle age, and this may be associated with the distension of the abdomen during pregnancy and with the violent efforts of the abdominal wall during childbirth, for the raised abdominal pressure so produced may further weaken the linea alba in this region. The strain is further increased by intra-abdominal deposition of fat.

Epigastric hernia

It is much less common for a hernia to occur elsewhere in the linea alba. Such protrusions are usually above the umbilicus, and are known as epigastric hernias.

Femoral hernia

This is discussed in relation to the anatomy of the femoral canal [p. 405].

Diaphragmatic hernia

This is most common in middle age, when the intra-abdominal pressure may rise because of the accumulation of fat in the abdominal cavity. This is associated with a weakening of the musculature, which may allow part of the stomach to be pushed from the abdomen into the thoracic cavity. Such a protrusion may rarely occur between the costal and vertebral origins of the diaphragm, above the last rib [FIG. 5.102], but this weak point in the diaphragmatic partition is normally closed by the kidney on the left side and by the kidney and the liver on the right.

Much more usual is herniation through or alongside the oesophageal opening in the diaphragm. Several varieties of this type of hernia are described. As might be expected, the protrusion occurs and symptoms tend to develop when the pressure in the upper part of the abdomen is raised, as for example when the patient lies down in bed or bends down as in gardening.

MUSCLES OF THE PELVIS

The adoption of the erect posture in the course of evolution necessitated the provision of a strong and resilient pelvic floor to maintain the pelvic viscera in position (Thompson 1899). The levator ani and coccygeus muscles, which are concerned in this function, together make a supporting hammock called the **pelvic diaphragm**.

LEVATOR ANI

The two levator ani muscles [FIGS. 5.114, 5.115, and 5.116] form a thin, gutter-like sheet between the pelvic cavity and the perineum. Anteriorly, the two muscles are separated by a narrow interval occupied in the male by the lower part of the prostate gland, and in the female by the urethra and vagina.

On each side levator ani takes origin anteriorly from the pelvic surface of the body of the pubis, and posteriorly from the pelvic surface of the ischial spine. Between these two bony attachments it arises from the fascia covering the inner surface of the obturator internus muscle [p. 373]. From this linear origin the fibres run backwards, medially, and downwards to be inserted into the central tendon of the perineum, the sidewall of the anal canal, and the fibromuscular **anococcygeal ligament** stretching from the anal canal to the coccyx. The muscle consists of two distinct portions, iliococcygeus and pubococcygeus; the latter is somewhat artificially divided for descriptive purposes into three imperfectly separated parts. These are pubococcygeus proper, puborectalis, and levator prostatae (pubovaginalis in the female).

Pubococcygeus is that part of levator ani arising in front of the obturator canal [FIG. 5.115]. Its most medial fibres pass back inferior to the capsule of the prostate gland (**levator prostatae**) or to be attached to the side of the vagina (**pubovaginalis**), and are inserted into the central tendon of the perineum [p. 368]. More laterally the fibres of **puborectalis** run backwards past the prostate or vagina and the upper part of the anal canal to form with its fellow a U-shaped loop which holds the anorectal junction forwards towards the symphysis pubis [FIGS. 5.115 and 5.116]. The lowest fibres of

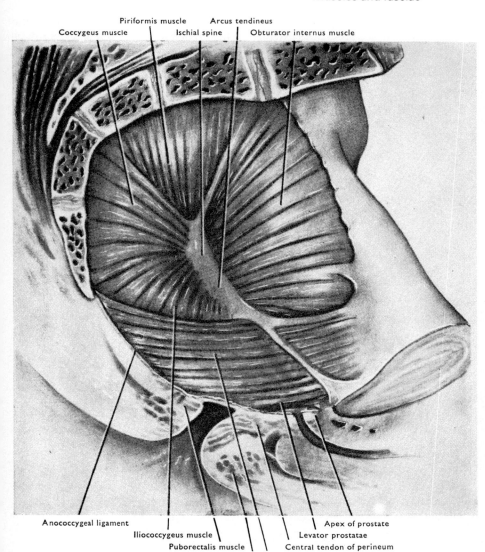

Coccygeus muscle   Piriformis muscle   Arcus tendineus
                    Ischial spine   Obturator internus muscle

FIG. 5.114. Left levator ani muscle.

Anococcygeal ligament   Apex of prostate
        Iliococcygeus muscle   Levator prostatae
            Puborectalis muscle   Central tendon of perineum
        External sphincter muscle   Pubococcygeus proper

insertion of puborectalis fuse with the involuntary longitudinal muscle coat of the rectum, which then gives way to a series of fibro-elastic slips which thread their way through the external sphincter of the anus and become attached to the skin round the anus.

More laterally still are the fibres of **pubococcygeus** proper, which run to be inserted into the anococcygeal ligament and into the sides and front of the coccyx.

Iliococcygeus is the most posterior part of the levator ani, and arises from the strong fascia on the internal surface of obturator internus [FIG. 5.122]. Its fibres run medially to be inserted into the side of the coccyx and the anococcygeal ligament; they are covered on their pelvic aspect by the backward-running fibres of pubococcygeus [FIGS. 5.114 and 5.115].

Nerve supply

Levator ani has a double supply, from the ventral rami of the third and fourth sacral nerves by branches which enter its pelvic surface, and by a branch from the perineal branch of the pudendal nerve, which enters the muscle on its perineal surface.

Actions

With coccygeus, levator ani plays an extremely important part in supporting the pelvic viscera and retaining them in position, especially in the female. It is constantly active, even during sleep (Porter 1962). Contraction of the pelvic diaphragm helps to increase intra-abdominal pressure, and, conversely, any increase in this pressure results in reflex contraction of the pelvic diaphragm to prevent the viscera being forced downwards. Puborectalis and pubovaginalis reinforce the sphincters of the anal canal and the vagina respectively. Levator prostatae is responsible for holding the prostate forwards, and can raise it as the pelvic diaphragm is contracted. The insertion of levator ani into the central tendon of the perineum draws it upwards and forwards in the later stages of defaecation and parturition, and part of the puborectalis muscle pulls the anal canal up over the descending faeces. When the rectum is full, puborectalis maintains the angle between the rectum and the anal canal, and may be thought of as an accessory sphincter preventing faeces from entering the anal canal.

COCCYGEUS

Coccygeus [FIGS. 5.114–5.116] lies posterior to and edge to edge with levator ani. It stretches as a triangular sheet of muscle and fibrous tissue from the inner surface of the ischial spine to the margin of the lower two pieces of the sacrum and the side of the coccyx. Its fibrous gluteal surface forms the **sacrospinous ligament**.

Nerve supply

From the ventral ramus of the fourth sacral nerve.

Actions

It assists levator ani in supporting the pelvic viscera and in maintaining intra-abdominal pressure. It also pulls forwards the coccyx after it has been pressed backwards during defaecation or parturition.

OBTURATOR INTERNUS

This muscle [FIGS. 5.114, 5.115, and 5.122] arises from the inner surface of the obturator membrane and from the surrounding bony margin of the obturator foramen except in the region of the obturator groove. An extension of this origin runs backwards and upwards on the pelvic surface of the ilium. The fasciculi converge on a tendon where the muscle turns round the lesser sciatic notch, deep to the sacrotuberous ligament. As it does so, the surface of the bone is covered with cartilage and grooved to accommodate the tendon, which is separated from the cartilage by a bursa. The tendon then passes over the back of the hip joint and is inserted into the medial surface of the greater trochanter of the femur above the trochanteric fossa [FIGS. 5.125 and 5.127]. It is closely associated with the two gemelli [p. 378], which virtually form another belly for the muscle.

The inner surface of the muscle is covered by the obturator fascia [p. 373] from which levator ani arises. The part of obturator internus above this origin lies in the lateral wall of the lesser pelvis, and the part below lies in the lateral wall of the ischiorectal fossa [p. 368].

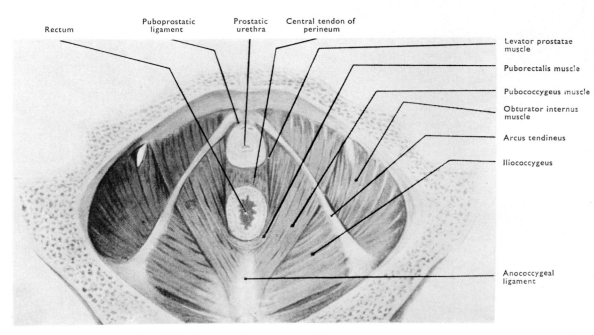

FIG. 5.115. Levator ani muscles viewed from above.

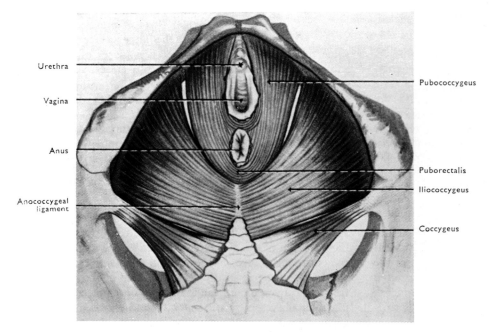

FIG. 5.116. Dissection of female pelvic diaphragm from below.

Nerve supply

From the nerve to obturator internus, a branch from the anterior aspect of the ventral rami of the fifth lumbar and the first two sacral nerves.

Actions

It is a lateral rotator of the thigh, but when the thigh is flexed, the alteration in its line of pull allows it to assist in abduction. Its main function is probably to hold the head of the femur in the acetabulum.

PIRIFORMIS

Piriformis [FIGS. 5.114, 5.125, and 5.128] comes from the anterior aspect of the second to the fourth pieces of the sacrum between and lateral to the pelvic sacral foramina. It leaves the pelvis through the greater sciatic foramen, where it has a further origin from the margin of the greater sciatic notch and the pelvic surface of the sacrotuberous ligament. The tendon is inserted into the upper border and medial side of the greater trochanter of the femur [FIG. 5.124] close to the insertion of obturator internus and the gemelli, with which it may be merged. If the common peroneal nerve leaves the pelvis separately it often pierces piriformis, the anterior surface of which is directly in contact with the sacral plexus.

Nerve supply

Branches from the ventral rami of the fifth lumbar and the first two sacral nerves.

Actions

It is a lateral rotator of the thigh when the hip joint is extended; when it is flexed, piriformis abducts the thigh. Its main function is probably to hold the head of the femur in the acetabulum.

Movements of respiration

The capacity of the thorax can be altered in a transverse, an anteroposterior, and a vertical direction. Alterations in the vertical diameter are due mainly to movements of the diaphragm, though a limited decrease or increase can be achieved by bending or straightening the thoracic part of the vertebral column; alterations in the other two dimensions are due to movements of the ribs.

A distinction must be made between quiet and forced respiration. In quiet inspiration the diaphragm is the main muscle concerned. The central tendon is pulled downwards, and the pressure in the thoracic cavity is thereby lowered at the same time as the pressure in the abdominal cavity is increased. When the resistance of the raised abdominal pressure rises to a certain level the descent of the diaphragm slows and stops, and the contraction of its costal fibres can then raise and evert the costal margin, thus increasing the transverse diameter of the lower part of the chest [p. 354]. The scalene muscles hold the first rib in position during these movements, and the upper external intercostals are in action, probably to prevent the intercostal spaces being sucked in by the changes in pressure within the chest. There is very little actual movement of the upper ribs.

In quiet expiration the main factor is the elastic recoil of the lungs and the chest wall. Activity in the diaphragm continues well into the expiratory phase, the muscle gradually relaxing to control the expiratory rate.

If there is any obstruction to breathing, other muscles are brought into play. In deep or forced inspiration, the first rib is raised by scalenus anterior and medius, though the movement is often slight (Jones, Beargie, and Pauly 1953). The intercostal muscles (certainly the external ones and possibly the internal ones also) pull the second rib up to the first, the third up to the second, and so on, in a cascade effect which reaches its greatest magnitude round about the seventh or eighth rib and then fades away as it continues downwards to the twelfth rib, which is now steadied by quadratus lumborum to provide a stable origin for increased efforts of the diaphragm. The levatores costarum help to raise the ribs, and it has indeed been claimed that these are the chief muscles of costal inspiration, but their mechanical advantage is poor.

In severe inspiratory difficulties the components of the erector spinae straighten out the thoracic curvature of the vertebral column, and so extend the depth of the thorax. At the same time accessory muscles of inspiration come into action. The sternum is pulled upwards by sternocleidomastoid acting from its insertion and by sternohyoid and sternothyrohyoid acting from a fixed hyoid bone; the clavicle is pulled up by trapezius and sternocleidomastoid, carrying with it the manubrium and the first rib. The scapula may be fixed by trapezius, rhomboids, and levator scapulae, so allowing pectoralis minor to act from its insertion to elevate the ribs; the humerus may be fixed by the patient grasping the head of the bed, so allowing pectoralis major to raise the sternum and ribs. Towards the end of maximal voluntary inspiration the abdominal muscles contract to limit the movement (Mills 1950); this contraction does not occur when breathing is increased by asphyxia (Campbell 1952).

In forced expiration, the normal reflex recoil of the anterior abdominal wall and the pelvic diaphragm becomes converted into stronger contraction, particularly of the oblique and transverse abdominal muscles; rectus abdominis is less concerned. The increased intra-abdominal pressure so produced forces the diaphragm up and decreases the vertical diameter of the chest cavity. At the same time, the abdominal muscles pull down the lower ribs, so decreasing the transverse and anteroposterior diameters of the chest [p. 357]. In sudden sharp expiratory movements, such as coughing, the abdominal muscles contract against the resistance of a closed glottis [p. 504], so building up a considerable expiratory pressure which is then suddenly released by relaxation of the laryngeal muscles. In such movements latissimus dorsi [p. 316] comes into sudden contraction, probably to help to compress the abdominal and thoracic contents, and the scalene muscles may contract, possibly to prevent downward displacement of the upper ribs (Campbell 1955a; Jones and Pauly 1957).

During quiet breathing the central tendon of the diaphragm moves up and down about 1 to 2 cm; in deep respiration it may achieve as much as 10 cm. Because of the varying level of the diaphragm in different postures [p. 354], and because of the pressure of the abdominal viscera on the under surface of the diaphragm when the body is supine, patients with inspiratory difficulties breathe more easily when propped up in bed.

In summary, the pattern of activity in quiet breathing is a balance between active muscular inspiration and passive elastic expiration; the expiratory muscles are involved only in forced expiration or in expiration against some resistance. In old age, as the elastic recoil of the lungs and body wall becomes progressively impaired, expiration is also impeded and the lungs tend to be incompletely emptied.

The muscles of respiration, although somatic, are not wholly under voluntary control. It is possible to vary the rate and depth of the breathing at will, and also to stop breathing for a short time, but there is very little voluntary control of the degree of contraction of the muscles engaged (Wade 1954).

MUSCLES OF THE PERINEUM

The perineum is a diamond-shaped area between the pubic arch in front and the coccyx behind. Laterally it is bounded by the inferior ramus of the pubis, the ramus of the ischium, the ischial tuberosity, and the sacrotuberous ligament. It is conveniently divided, for descriptive purposes, into a posterior **anal region** and an anterior **urogenital region**, by a transverse line just in front of the ischial tuberosities [FIGS. 5.117 and 5.118]. The urogenital part of the perineum is blocked in by the inferior fascia of the urogenital diaphragm [p. 373], and is divided by it into **superficial** and **deep perineal spaces**, each of which contains a group of muscles. The inferior fascia of the urogenital diaphragm is pierced in both sexes by the urethra, and in the female by the vagina. The arrangement of the muscles is therefore somewhat different in the two sexes.

Posteriorly, in the anal compartment, the anus is surrounded by sphincter ani externus, and on each side of this lies an **ischiorectal fossa** [FIGS. 5.117 and 5.118] bounded medially by levator ani, and laterally by obturator internus, the ischial tuberosity, the sacrotuberous ligament, and gluteus maximus [p. 375]. Inferiorly, the fossa is bounded by the skin of the perineum, and superiorly by the origin of levator ani from the fascia covering obturator internus. Posteriorly, the ischiorectal fossa extends backwards for a little distance above the sacrotuberous ligament and gluteus maximus, and anteriorly there is a variable extension of the fossa above the deep perineal space [FIG. 5.122], sometimes as far as the pubis. These extensions, like the rest of the fossa, are occupied by loose fatty tissue [p. 372], and this is sometimes infected from the lower part of the rectum or the anal canal.

Between the anal canal and the urogenital diaphragm lies the **central tendon of the perineum**. This fibromuscular node blends with the urogenital diaphragm and it contains both somatic and visceral muscle fibres. Most of the perineal muscles meet at this central point, which in the female is of great importance during childbirth, since it may be torn and damaged by the descending head of the child [p. 371].

SPHINCTER ANI EXTERNUS

This elliptical muscle surrounds and clasps the conical lower end of the puborectalis part of levator ani, and consists of three imperfectly separated layers. The subcutaneous lamina contains fibres which surround the anus, a few being attached to the central tendon of the perineum and to the anococcygeal ligament. The main bulk of the muscle is in the superficial lamina which has the only bony attachment. It runs from the coccyx and anococcygeal ligament forwards round the internal sphincter of the anus (a specialization of the circular layer of visceral muscle in the gut wall) to the central tendon of the perineum [FIGS. 5.117 and 5.118]. The deep layer of sphincter ani externus encircles the upper part of the internal sphincter. It blends anteriorly with the central tendon of the perineum; its deeper fibres fuse with those of puborectalis. The superficial and the subcutaneous parts of the muscle are penetrated by fibres derived from the longitudinal muscle of the rectum [p. 365], on their way to be attached to the skin.

Nerve supply

From the perineal branch of the fourth sacral nerve, and from the inferior rectal branch of the pudendal nerve (S. 2, 3).

Actions

The muscle is normally contracted (Floyd and Walls 1953), even in sleep, and with the internal sphincter it holds the anal canal and its orifice closed. It is under voluntary control, and can be made to shut the orifice firmly. It also tethers the central tendon of the perineum to the coccyx.

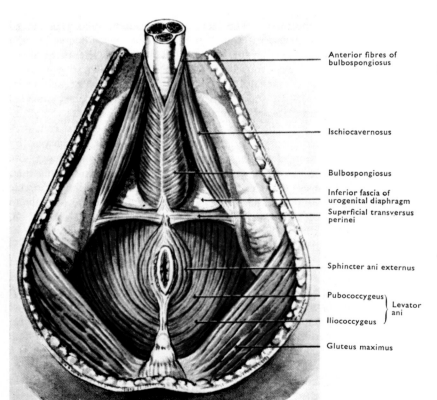

FIG. 5.117. Muscles of the male perineum.

Anterior fibres of bulbospongiosus

Ischiocavernosus

Bulbospongiosus

Inferior fascia of urogenital diaphragm
Superficial transversus perinei

Sphincter ani externus

Pubococcygeus ⎫
                ⎬ Levator ani
Iliococcygeus ⎭

Gluteus maximus

TRANSVERSUS PERINEI SUPERFICIALIS

This thin strand of muscle is often difficult to detect in older people. It arises from the medial side of the ischial tuberosity and passes medially to be inserted into the central tendon of the perineum [FIGS. 5.117 and 5.118]. Like the other muscles of the superficial perineal space it lies inferior to the inferior fascia of the urogenital diaphragm, to which it is attached. Some of its fibres are continuous with decussating fibres of sphincter ani externus.

Nerve supply

From the perineal branch of the pudendal nerve (S. 2, 3, 4).

Actions

It helps to steady the central tendon of the perineum.

BULBOSPONGIOSUS

The male bulbospongiosus muscles unite with each other in the midline by a fibrous raphe which covers inferiorly the bulb of the penis and the adjoining part of the corpus spongiosum [FIGS 5.117 and 5.122]. The muscles arise from this raphe and from the central tendon of the perineum, and curve upwards and forwards round the side of the bulb. The more posterior fibres are inserted into the lower surface of the ·inferior fascia of the urogenital diaphragm, the intermediate ones to the dorsal surface of the corpus spongiosum of the penis, and the most anterior ones sweep round the corpora cavernosa of the penis and are inserted into the deep fascia on the dorsum of the penis.

In the female, the bulbospongiosus muscles are separated by the openings of the vagina and the urethra [FIG. 5.118], and each muscle is a thin layer covering the bulb of the vestibule at the side of the lower part of the vagina. They arise from the central tendon of the perineum, and are inserted into the sides of the pubic arch and into the root and dorsum of the clitoris.

Nerve supply

From the perineal branch of the pudendal nerve (S. 2, 3, 4).

Actions

The male muscles contract at the end of the act of micturition to expel the last drops of urine. They also contract rhythmically during ejaculation, to help to propel the semen along the urethra. The middle fibres of the muscle assist in erection of the penis by compressing the erectile tissue of the bulb, and the anterior fibres compress the corpora cavernosa and the dorsal vein of the penis.

The female bulbospongiosus muscles provide a sphincter for the orifice of the vagina. The most anterior fibres can compress the dorsal vein of the clitoris and are concerned in the erection of the organ.

ISCHIOCAVERNOSUS

Ischiocavernosus comes from the medial side of the ischial tuberosity and passes forwards inferior to the crus of the penis or the clitoris to be inserted into the margin of the pubic arch on each side of the crus, and also into the corpus cavernosum of the penis or clitoris [FIGS. 5.117, 5.118, and 5.122].

Nerve supply

From the perineal branch of the pudendal nerve (S. 2, 3, 4).

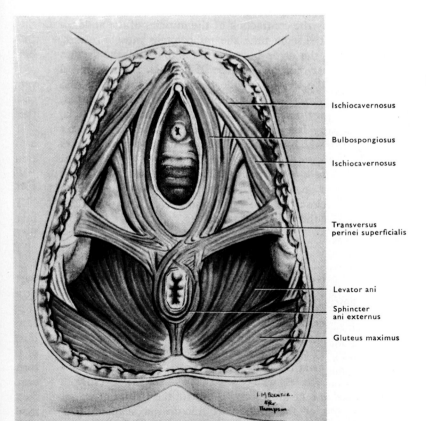

FIG. 5.118. Muscles of the female perineum.

Ischiocavernosus

Bulbospongiosus

Ischiocavernosus

Transversus
perinei superficialis

Levator ani

Sphincter
ani externus

Gluteus maximus

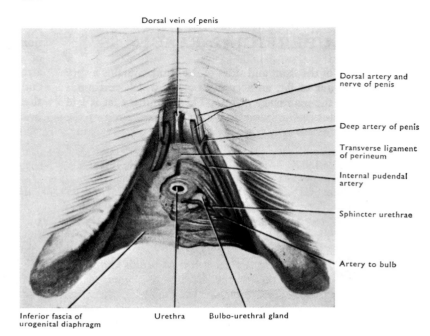

Dorsal vein of penis

Dorsal artery and
nerve of penis

Deep artery of penis

Transverse ligament
of perineum

Internal pudendal
artery

Sphincter urethrae

Artery to bulb

Inferior fascia of   Urethra   Bulbo-urethral gland
urogenital diaphragm

FIG. 5.119. Inferior fascia of the urogenital diaphragm
(perineal membrane) and sphincter urethrae.

Actions

By compressing the crus of the penis or clitoris the muscle contributes to the process of erection.

SPHINCTER URETHRAE

This muscle arises from the inferior pubic ramus, and from the walls of the deep perineal space [FIGS. 5.119 and 5.122] in which it lies. Its fibres run medially towards the urethra, and in the male are inserted into a median raphe, mostly posterior to the urethra. The fibres which surround the urethra form a muscular collar for the canal and have no bony attachments. In the female, sphincter urethrae encircles the lower end of the urethra as in the male, and the outer fibres decussate with each other between the urethra and the opening of the vagina; some of them are attached to the vaginal wall. Together with the deep transverse muscle of the perineum, with which it is often fused, sphincter urethrae forms the muscular basis of the urogenital diaphragm.

Nerve supply

From the perineal branch of the pudendal nerve (S. 2, 3, 4).

Actions

The muscle, which is under voluntary control, compresses the membranous portion of the urethra and prevents the onset of micturition when the bladder wall contracts. It is relaxed during micturition but can be brought into action to shut off the stream.

TRANSVERSUS PERINEI PROFUNDUS

This is another thin muscle which blends with the posterior fibres of sphincter urethrae and is inserted into the central tendon of the perineum, some of its fibres being inserted into the vaginal wall in the female. Its origin is from the medial surface of the ramus of the ischium [FIG. 5.119] and it runs parallel with, and superior to, the superficial transverse muscle.

Nerve supply

From the perineal branch of the pudendal nerve (S. 2, 3, 4).

Actions

It helps to steady the central tendon of the perineum.

Mechanism of expulsive movements

DEFAECATION

Like most commonplace actions, defaecation is in reality an extremely complicated process. Our knowledge of the mechanism is derived largely from radiographic examination, and detailed information regarding the function of the various muscles is not available. Nevertheless, it can be said that when the rectum has filled from the sigmoid colon—an entirely involuntary process—the abdominal pressure is increased by the contraction of the muscles of the abdominal wall. At the same time puborectalis relaxes, so decreasing the angle between the ampulla of the rectum and the upper portion of the anal canal. The general relaxation of levator ani allows the colon and the rectum to pass downwards and to become narrow and elongated. The involuntary muscle of the wall of the rectum contracts, the involuntary internal sphincter relaxes, and sphincter ani externus, which represents the voluntary control, allows the faeces to pass. At the same time fibres of levator ani draw up the walls of the anal canal over the descending faeces. Subsequent to evacuation, the puborectalis and the sphincter ani externus close the anal canal, and the internal sphincter contracts after the passage of the faeces.

MICTURITION

Even less is known about the movements concerned in micturition. Radiographic studies suggest that urine is held up at the level of the neck of the bladder, not at sphincter urethrae, and certainly sphincter urethrae can be destroyed without necessarily producing incontinence. The sphincter is under voluntary control, and can be used to cut off the stream either during or at the end of micturition. In the female the sphincter is much less well defined than in the male, yet there is no diminution of the ability to control the stream of urine.

One theory of the process of micturition suggests that relaxation of the levator ani allows the neck of the bladder to descend, and this initiates reflexes resulting in contraction of the bladder wall. This is supported by the finding that the constant resting activity in the muscles of the pelvic floor falls to a very low level during micturition (Porter 1962). Voluntary cessation of micturition can be produced by contraction of the levator ani, the stream being cut off at the neck of the bladder.

It is possible that in the male the bulbospongiosus helps to propel the urine along the urethra, and this muscle is certainly concerned in the expulsion of the last few drops of urine from the urethra.

The contraction of the bladder musculature is assisted by a rise in the intra-abdominal pressure due to the contraction of the muscles of the abdominal wall.

PARTURITION

The main propulsive force in childbirth is the massive involuntary muscle of the wall of the uterus, but this is assisted by strong contractions of the abdominal muscles. As the head of the baby descends through the birth canal, the muscles forming its walls are stretched and may become torn. This is particularly so in the case of the perineal muscles, which have little time to stretch gradually, as once the head has reached them the final stages of expulsion are usually rapid. As a result of this, the central tendon of the perineum is commonly torn in uncontrolled childbirth. It is one of the main objectives of the midwife to prevent this happening, since should it be torn, control of the perineal musculature is lost, and such catastrophes as prolapse of the pelvic organs may result. It is for this reason that before the actual birth an incision is often made through the posterolateral part of the wall of the vaginal orifice and the adjacent part of the perineum lateral to the central tendon. This operation of episiotomy enlarges the outlet in a planned direction, and prevents ragged tearing; the incision is sewn up after the birth and the perineum can be well reconstituted.

Fascia of the trunk

SUPERFICIAL FASCIA

The fascia of the front and sides of the trunk contains a very variable amount of fat, an accumulation of which surrounds and protects the mammary gland. This mammary fat is laid down at puberty as a secondary sexual characteristic of the adult female, in whom it extends from the side of the sternum to the mid-axillary line, and from the second to the sixth ribs in the vertical plane. The male breast is surrounded by a much smaller amount of fat.

In the superficial fascia of the anterior abdominal wall fat is commonly deposited in middle age, the favoured site in males being the upper part of the abdomen and in females the lower part. In really obese people this layer may reach several centimetres in thickness.

As the superficial fascia is traced down towards the thigh, it separates into two layers, between which lie the superficial vessels and nerves and the superficial inguinal lymph nodes. The superficial layer is continuous with the superficial fascia of the thigh without interruption.

The deeper layer, which is a thin membrane containing elastic fibres, is an elastic tunic which may well have a function in supporting the abdominal contents. In the lower part of the abdominal wall this membranous layer is loosely attached to the aponeurosis of external oblique, and more closely to the linea alba and to the symphysis pubis; there is very liitle deep fascia in this region, and the membranous layer of the superficial fascia may be looked on as a substitute. Traced downwards, it passes superficial to the inguinal ligament to become attached to the deep fascia [fascia lata, p. 405] of the thigh just distal to, and roughly parallel with, the inguinal ligament.

In the male, the membranous layer is prolonged on to the dorsum of the penis, where it forms a relatively dense band called the

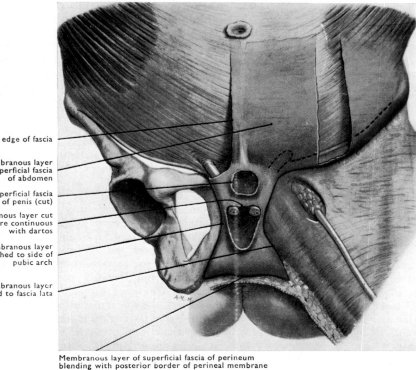

Cut edge of fascia

Membranous layer
of superficial fascia
of abdomen

Superficial fascia
of penis (cut)

Membranous layer cut
where continuous
with dartos

Membranous layer
attached to side of
pubic arch

Membranous layer
attached to fascia lata

Membranous layer of superficial fascia of perineum
blending with posterior border of perineal membrane

FIG. 5.120. Diagram showing continuity of membranous layer of the superficial fascia of abdominal wall and perineum.

fundiform ligament, and both layers of the superficial fascia become thinner as they pass downwards over the spermatic cord to become continuous with the superficial fascia of the scrotum. This, like that of the penis, contains very little fat. It is reddish in colour and in its substance lie the fibres of the involuntary **dartos** muscle, which assists in suspending the testis and corrugates the skin of the scrotum. This muscle is supplied by the genital branch of the genitofemoral nerve. The superficial fascia also helps to form the septum of the scrotum, which extends upwards between the two testes and incompletely separates them and their coverings.

In the female, the superficial fascia of this region contains a considerable quantity of fat, which is responsible for much of the bulk of the mons pubis and the labia majora.

In the anterior part of the perineum of both sexes the superficial fascia is again composed of two recognizable components, a superficial fatty layer, and a deep membranous layer. The fatty layer is continuous with the superficial fascia in the thigh, and also posteriorly with the fat of the ischiorectal fossae and buttocks. The membranous layer, on the other hand, is firmly attached laterally to the inferior ramus of the pubis and the ischial ramus [FIG. 5.120]. Posteriorly it fuses with the posterior border of the urogenital diaphragm and with the central tendon of the perineum. In the midline it is attached to the septum of the scrotum and to the median raphe of the bulbospongiosus muscle [FIG. 5.122].

The attachments of the membranous layer of the superficial fascia are of some importance in cases of rupture of the male urethra in the perineum, as may occur when a small boy falls astride a railing. The fascial attachments prevent urine which leaks out into the surrounding tissues from passing backwards into the ischiorectal fossae or sideways into the thigh. The escaping fluid is therefore directed forwards into the subcutaneous tissue of the scrotum. It then fills up the 'space' within the fascial sheath of the penis and mounts upwards round the spermatic cord to the anterior abdominal wall. Here it may pass freely upwards, but it cannot enter the thigh because of the attachment of the membranous layer of the superficial fascia to the fascia lata and to the body and inferior ramus of the pubis [FIGS. 5.120–5.122].

In each ischiorectal fossa the superficial layer of the superficial fascia forms a pad of fat, which separates the levator ani from the perineal part of the wall of the pelvis, lined by the obturator internus [p. 366]. These ischiorectal pads form a semifluid packing material which allows the anal canal to distend as the faeces pass along it during defaecation. Over the ischial tuberosities the fat is infiltrated by bands of fibrous tissue which adhere to the skin and to the bone.

The membranous layer of the superficial fascia in each ischiorectal fossa is thin and for the most part difficult to define; it lines the walls of the fossa, covering over the fatty pad superiorly. On the lateral wall of the fossa it is somewhat thickened to form the medial wall of the **pudendal canal**, which extends from the lesser sciatic foramen to the posterior margin of the urogenital diaphragm. The canal conveys the internal pudendal vessels and the pudendal nerve and its branches, and its lateral wall is formed by the deep fascia covering the inferior part of obturator internus [p. 373].

DEEP FASCIA

The deep fascia covering the anterior and lateral surfaces of the walls of the chest and abdomen presents no special features; it is thin and relatively elastic to meet the need for the thorax and abdomen to expand, and in places is virtually replaced by the aponeurosis of external oblique and by the membranous layer of the superficial fascia. Superiorly it is attached to the clavicle and to the side of the sternum; inferiorly, it is attached to the iliac crest. At the superficial inguinal ring [p. 355] the external oblique aponeurosis

and the deep fascia covering it are prolonged downwards and medially as the **external spermatic fascia**, the outermost of the three fascial coats which in the male invest the spermatic cord and the testis, and in the female surround the round ligament of the uterus. Deep to this outer layer lies the **cremasteric fascia**, a layer pulled away from internal oblique, and containing the fibres of the cremaster muscle; deep to this again is the **internal spermatic fascia**, a coating derived from the transversalis fascia (see below).

The cavities of the chest, abdomen, and pelvis all have a fascial lining, though the strength and importance of this fascia vary from place to place. Lining the chest wall, outside the pleura, lies the **endothoracic fascia**. This is a thin layer of loose areolar tissue, which is often only recognizable posteriorly. However, superior to the first rib, it becomes a more imposing fibrous layer which is attached to the inner border of the first rib, and runs from this to the anterior border of the transverse process of the seventh cervical vertebra. This layer is known as the **suprapleural membrane**, and is reinforced and made tense by some fibres from the scalene muscles under cover of which it lies (**scalenus minimus**).

The fascial lining of the abdominal and pelvic cavities is a continuous membranous bag lying outside the peritoneum; the different regions of the bag are, perhaps confusingly, given different names. It encloses the abdominal and pelvic cavities, the viscera, and the great vessels, but the nerves which emerge from the vertebral canal lie posterior to the bag, as do the lumbar and lumbosacral plexus which they form. The **transversalis fascia** is the name given to the portion of this fibrous bag which lines the deep surface of transversus abdominis; it is thick and strong inferiorly, but becomes thinner as it ascends to become continuous with the fascia on the under surface of the diaphragm. Posteriorly the transversalis fascia is continuous with the anterior layer of the thoracolumbar fascia on quadratus lumborum, and hence with the fascia on the anterior surface of psoas [FIG. 5.47]. Inferiorly it becomes continuous with the pelvic fascia, but along the line of the inguinal ligament and the iliac crest, to both of which it is firmly attached, it changes its name to become the **iliac fascia**. As the femoral vessels pass from their position inside the fascial bag of the abdomen into the upper part of the thigh, they drag out with them an investment of this fascia, the **femoral sheath** [FIGS. 5.163 and 5.164]. The anterior wall of this sheath is derived from the transversalis part of the abdominal fascia, and the posterior wall from the iliac part. Above the inguinal ligament, at the deep inguinal ring, the transversalis fascia is evaginated by the spermatic cord in the male and the round ligament of the uterus in the female. Its prolongation into the inguinal canal around these structures forms the **internal spermatic fascia**.

The fascia on the posterior wall of the abdomen is strengthened in two places. The **medial arcuate ligament** is a thickening of the fascia covering the psoas major and minor muscles, and stretches from the side of the body of the first lumbar vertebra to its transverse process. The **lateral arcuate ligament** is a similar band in the anterior layer of the thoracolumbar fascia covering quadratus lumborum, and runs from the transverse process of the first lumbar vertebra to the twelfth rib. Both arcuate ligaments serve to give origin to fibres of the posterior part of the diaphragm [FIG. 5.103].

Between the lining fascia of the abdominal wall and the peritoneum lies a layer of areolar **extraperitoneal tissue** which is relatively scanty anteriorly, but abundant posteriorly, where it contains a good deal of fat. This tissue surrounds and forms sheaths for the kidneys, the ureters, the suprarenal glands, the abdominal aorta, inferior vena cava, and their branches and tributaries. It is continuous with similar tissue in the posterior mediastinum of the thorax through the aortic opening in the diaphragm, and is also prolonged into the pelvis. In both abdomen and pelvis it is found between the layers of peritoneum forming the mesenteries which tether the viscera to the body wall. The extraperitoneal tissue on the anterior abdominal wall below the level of the umbilicus contains the inferior epigastric vessels, the lateral and median umbilical ligaments, and the upper part of the full bladder as it rises into the abdominal cavity. There is no extraperitoneal tissue on the under surface of the diaphragm.

The ramifications and detailed topography of the fascia in the pelvis are exceedingly complicated. However, in broad terms its arrangement is simple. A **parietal layer** covers the muscles which line the pelvis and form its floor. Some of the pelvic viscera are surrounded by a **visceral layer** which forms a sheath for them. Thirdly, specialized condensations of the fascia or extraperitoneal tissue serve to form 'ligaments' tethering the viscera to the pelvic wall.

The **parietal pelvic fascia** covers the pelvic surfaces of obturator internus, piriformis, coccygeus, sphincter urethrae, deep transverse perinei, and levator ani muscles; the part which lines obturator internus is thicker than the rest. This **obturator fascia** is continuous with the iliac fascia above, and is attached anteriorly to the back of the body of the pubis. Levator ani takes origin from a thickened part of it, the **arcus tendineus**, and the part of the obturator fascia above this linear origin is tough and strong, being reinforced by the aponeurotic origin of levator ani. The part below the origin, which

forms the lateral wall of the ischiorectal fossa, is thin and relatively weak.

The fascia lining the superior surface of levator ani is called the **superior fascia of the pelvic diaphragm**. The portion of this fascia which stretches between the symphysis pubis and the neck of the bladder and the urethra in the female or the prostate gland in the male is thickened to form the **pubovesical ligaments** or **puboprostatic ligaments** respectively. These bands lie on each side of the median plane, and between them the superior fascia of the pelvic diaphragm blends with the **superior fascia of the urogenital diaphragm**, which lines the superior surfaces of the deep transversus perinei and the sphincter urethrae muscles [FIGS. 5.121 and 5.122].

The **inferior fascia of the pelvic diaphragm** forms the medial wall of the ischiorectal fossa, lying on the under surface of levator ani [FIG. 5.122]. It is continuous with the fascia on the medial surface of obturator internus, and with the fascia on the inferior surface of sphincter ani externus. The fascia on the under surface of sphincter urethrae and the deep transversus perinei muscles is very dense and strong, and is known as the **inferior fascia of the urogenital diaphragm** (perineal membrane). This membrane stretches almost horizontally across the pubic arch, attached to the inferior ramus of the pubis and the ramus of the ischium: the crura

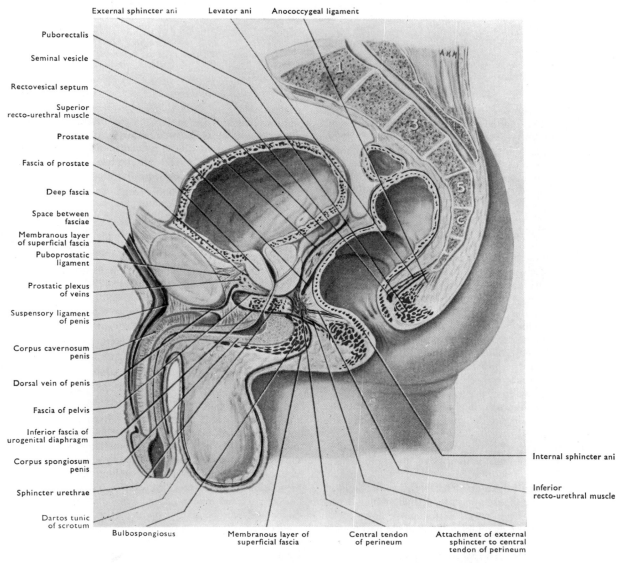

FIG. 5.121. Sagittal section of male pelvis showing fasciae in blue (semidiagrammatic).

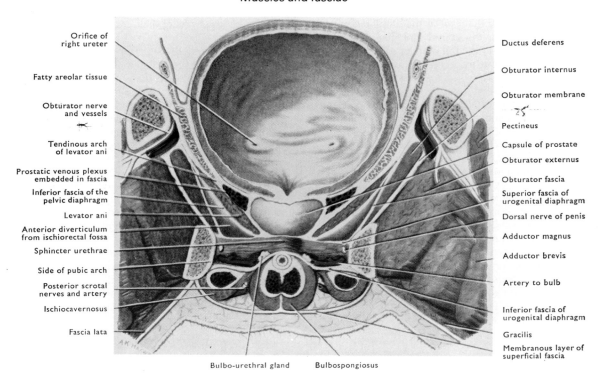

Orifice of
right ureter

Fatty areolar tissue

Obturator nerve
and vessels

Tendinous arch
of levator ani

Prostatic venous plexus
embedded in fascia

Inferior fascia of the
pelvic diaphragm

Levator ani

Anterior diverticulum
from ischiorectal fossa

Sphincter urethrae

Side of pubic arch

Posterior scrotal
nerves and artery

Ischiocavernosus

Fascia lata

Ductus deferens

Obturator internus

Obturator membrane

Pectineus

Capsule of prostate

Obturator externus

Obturator fascia

Superior fascia of
urogenital diaphragm

Dorsal nerve of penis

Adductor magnus

Adductor brevis

Artery to bulb

Inferior fascia of
urogenital diaphragm

Gracilis

Membranous layer of
superficial fascia

Bulbo-urethral gland   Bulbospongiosus

FIG. 5.122. Section of male pelvis in a coronal plane to show the fasciae of pelvis and perineum. The fascia of the prostate has been removed in front to show the prostatic capsule.

of the penis or clitoris are attached to its inferior surface. It is roughly triangular in shape, and the short anterior border at its apex is thickened to form the **transverse perineal ligament** [FIG. 5.119]. Between this and the symphysis pubis the dorsal vein of the penis or clitoris enters the pelvis. Posteriorly the base of the triangle is connected to the central tendon of the perineum and becomes continuous with both the superior fascia of the urogenital diaphragm and the membranous layer of the superficial fascia [FIG. 5.121], thus closing the deep and superficial perineal spaces posteriorly.

Around the prostate gland, the pelvic fascia becomes thickened and condensed to form the **fascial sheath of the prostate**, which surrounds the veins forming the prostatic venous plexus. This sheath is continuous below with the superior fascia of the urogenital diaphragm, and on each side with the superior and inferior fasciae of the pelvic diaphragm round the free medial border of levator ani [FIG. 5.122]. Anteriorly it is attached to the pelvic surface of the symphysis pubis by the puboprostatic ligaments, and superiorly it becomes continuous with the connective tissue surrounding the plexus of veins on the inferolateral surfaces of the bladder. Posteriorly the sheath of the prostate is continuous with the **rectovesical septum** (rectovaginal septum in the female) which extends from the central tendon of the perineum to the floor of the rectovesical pouch of peritoneum, separating the prostate, the seminal vesicles, and the ductus deferentia anteriorly from the ampulla of the rectum posteriorly [FIG. 5.121]. The fascial sheath of the prostate is of importance when the prostate has to be removed surgically, since if the sheath and the capsule of the gland are left intact much haemorrhage can be avoided.

In the female, the cervix of the uterus is attached to the front of the sacrum by a thickening of extraperitoneal tissue which forms the **uterosacral ligaments**. These contain a small amount of visceral muscle, and are enclosed in the rectouterine folds of the peritoneum [FIG. 8.27].

Elsewhere in the pelvis the extraperitoneal tissue contains the internal iliac vessels and their branches and tributaries, the visceral nerves and plexuses, the ureters, and the ductus deferentia. Around the lower part of the rectum it forms a thick sheath of fat which allows expansion of the rectum during defaecation. The same tissue forms a packing for the parts of the bladder that are not covered with peritoneum. In the female the broad ligament, a folded sheet of peritoneum which connects the uterus to the side wall of the pelvis [FIG. 8.53] contains a packing of extraperitoneal tissue called the **parametrium** [p. 570], which surrounds the various ligaments, vessels, and embryonic remnants which the broad ligament contains. Passing from each side of the cervix of the uterus to the wall of the pelvis within the base of the broad ligament is a thickening of the extraperitoneal tissue, the **lateral cervical ligament** [p. 570], around the uterine arteries passing to the cervix uteri. This strong band is probably of importance in holding the cervix of the uterus in place.

Muscles of the lower limb

Unlike the upper limb, the lower limb is concerned with support and locomotion, and although the 'origins' of the muscles are, by convention, located proximally, the fixed points from which they act are as often as not situated distally. In standing the feet are anchored to the ground by gravity and friction, and the tendency of the body to overbalance is corrected by muscles acting from the foot on the lower limb and from the limb on the trunk. In walking, the propulsive thrust derives from the limb in contact with the ground and the actions of a given muscle on its insertion are often less important than the actions on its origin.

Another difference between the lower and the upper limbs is that the components of the pelvic girdle are fused together so that virtually all the movements between the limb and the trunk take place at a single joint—the hip joint.

Thirdly, the rotation of the lower limb in the course of development results in its extensor and flexor surfaces coming to lie anteriorly and posteriorly respectively. This rotation, which is contrary to that of the upper limb, is reflected in the arrangement of the innervation. Thus the muscles on the front of the thigh and leg are supplied by nerves coming from the posterior aspect of the lumbosacral plexus, and those at the back of the thigh and leg and in the sole of the foot from the anterior aspect of the plexus.

As in the upper limb, many muscles cross over several joints and exert their action on each. It is most exceptional for one joint of the lower limb to be moved in isolation, and in standing and walking the joints and muscles form a co-ordinated mechanism which cannot be rigidly subdivided into individual functional groups.

MUSCLES OF THE BUTTOCK

The buttock is largely formed by the gluteus maximus muscle, which lies posterior to a number of smaller muscles. Some of these, like the rotator cuff of the shoulder joint, probably have as their main function the protection of the hip joint by their action as extensile ligaments, but others, such as the gluteus minimus and medius, are important prime movers. Two muscles, obturator internus and piriformis, whose tendons enter the buttock from the pelvis, have been dealt with elsewhere [pp. 366 and 367].

GLUTEUS MAXIMUS

Gluteus maximus is the heaviest and most coarsely fibred muscle in the body, and forms a soft pad covering the ischial tuberosity

[p. 273]. It originates from the area behind the posterior gluteal line of the ilium, from the posterior layer of the thoracolumbar fascia, from the dorsal surfaces of the sacrum and coccyx, from the posterior surface of the sacrotuberous ligament, and from the fascia which covers the muscle [FIGS. 5.123 and 5.126].

The deep fibres of the lower part of the muscle, comprising about a quarter of its bulk, are inserted into the gluteal tuberosity of the femur [FIG. 5.124]. The remaining fibres are inserted into the iliotibial tract of the fascia lata [p. 405]. This tract runs down the lateral side of the thigh, and blending with the capsule of the knee joint, becomes fixed to the lateral condyle of the tibia and to the head of the fibula [FIGS. 3.176, 5.134, 5.139, 5.140, and 5.142].

The lower border of gluteus maximus runs obliquely across the fold of the buttock [FIG. 5.123]. Three bursae separate the muscle from the ischial tuberosity, the lateral side of the greater trochanter of the femur, and the upper part of the origin of the vastus lateralis.

Nerve supply

From the inferior gluteal nerve (L.5; S.1, 2).

Actions

Acting from its origin it extends and rotates the thigh laterally at the hip joint; its lowest fibres can adduct the femur, and the upper fibres may help in abduction. Its action as an extensor is not much used in ordinary walking, but comes into play in running. The force which the muscle applies to the iliotibial tract is a powerful support on the lateral side of the knee joint (see tensor fasciae latae).

Acting from its insertion, the muscle extends the trunk on the lower limb. This action is of the greatest importance in rising from a seated or stooped position, or in climbing stairs. Gluteus maximus

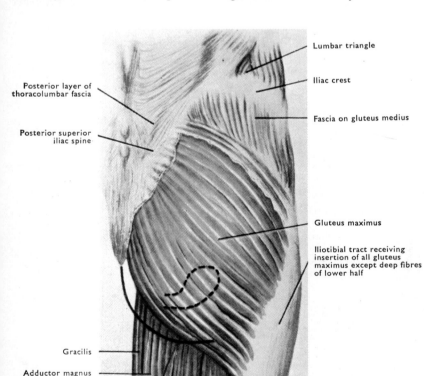

Posterior layer of thoracolumbar fascia

Posterior superior iliac spine

Lumbar triangle

Iliac crest

Fascia on gluteus medius

Gluteus maximus

Iliotibial tract receiving insertion of all gluteus maximus except deep fibres of lower half

Gracilis

Adductor magnus

Semimembranosus

Semitendinosus

Long head of biceps femoris

FIG. 5.123. Right gluteus maximus muscle. The broken line indicates the site of the ischial tuberosity. Note that the curved line indicating the site of the fold of the buttock does not correspond to the lower border of gluteus maximus.

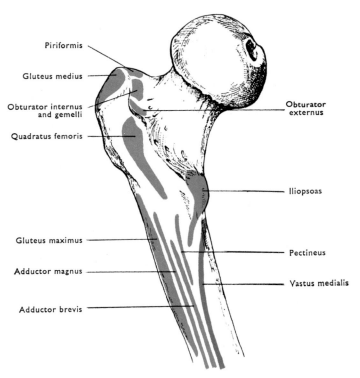

FIG. 5.124. Muscle attachments to back of upper part of left femur.

may be tested by making the patient lie prone and then asking him to lift the thigh off the couch; the gluteus maximus can be felt contracting.

TENSOR FASCIAE LATAE

In front of gluteus maximus, on the lateral side of the hip, lies tensor fasciae latae [FIG. 5.134], which also inserts into the iliotibial tract. The muscle takes origin from the anterior part of the outer lip of the iliac crest [FIG. 5.126], and joins the iliotibial tract just below the level of the greater trochanter [FIGS. 5.136 and 5.139]. The layer of fascia on its deep surface extends medially to fuse with the capsule of the hip joint.

Nerve supply

By a branch (L. 4, 5) from the superior gluteal nerve.

Actions

It helps to flex, abduct, and medially rotate the femur; it also straightens out the somewhat backward pull of gluteus maximus on the iliotibial tract. With gluteus maximus it affords support to the lateral side of the knee joint: the balanced pull of gluteus maximus on the posterior end of the iliac crest and tensor fasciae latae on the anterior end also stabilizes the pelvis on the femur in the anteroposterior plane.

GLUTEUS MEDIUS [FIG. 5.143]

Gluteus medius shows on the surface between gluteus maximus and tensor fasciae latae [FIG. 5.123], but most of its substance lies under

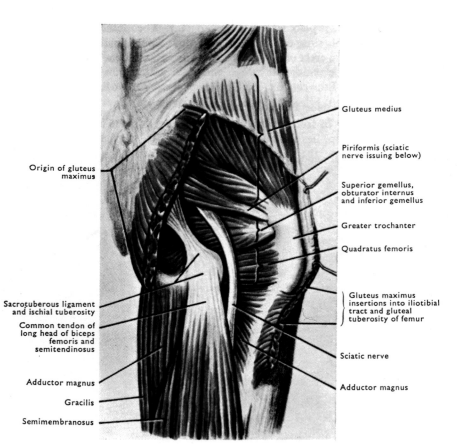

FIG. 5.125. Muscles of right gluteal region deep to gluteus maximus.

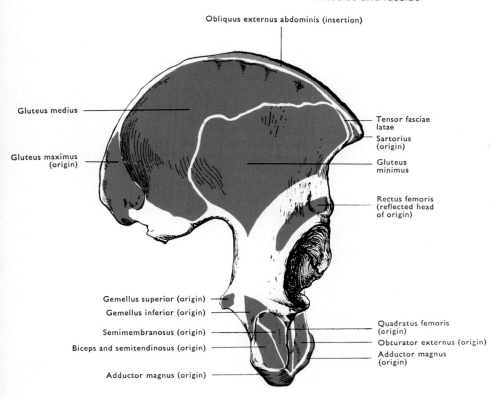

Obliquus externus abdominis (insertion)

Gluteus medius

Gluteus maximus (origin)

Tensor fasciae latae

Sartorius (origin)

Gluteus minimus

Rectus femoris (reflected head of origin)

Gemellus superior (origin)
Gemellus inferior (origin)
Semimembranosus (origin)
Biceps and semitendinosus (origin)
Adductor magnus (origin)

Quadratus femoris (origin)
Obturator externus (origin)
Adductor magnus (origin)

FIG. 5.126. Muscle attachments to outer surface of right hip bone.

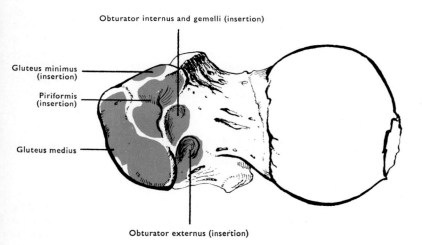

Obturator internus and gemelli (insertion)

Gluteus minimus (insertion)
Piriformis (insertion)

Gluteus medius

Obturator externus (insertion)

FIG. 5.127. Femur (looked at from above), showing muscle attachments to greater trochanter.

cover of gluteus maximus [FIG. 5.125]. It is shaped like a fan, and arises from the outer surface of the ilium between the posterior gluteal line superiorly, and the anterior gluteal line inferiorly [FIG. 5.126]. Many of its fibres are attached to the strong fascia covering its anterior portion. The muscle is inserted by tendon into an area running diagonally downwards and forwards across the lateral surface of the greater trochanter of the femur [FIG. 5.127]; it is separated from the trochanter by a bursa.

Nerve supply

By a branch (L. 4, 5; S. 1) from the superior gluteal nerve.

Actions

The muscle abducts the thigh in all positions of the lower limb. In walking it abducts the pelvis on the supporting thigh, and so pulls the centre of gravity of the body over the supporting limb [p. 404].

GLUTEUS MINIMUS

This muscle lies antero-inferior to and under cover of gluteus medius [FIG. 5.128]. Its origin is from the outer surface of the ilium between the anterior and inferior gluteal lines [FIG. 5.126]; its insertion is into the anterosuperior angle of the greater trochanter, extending on to its anterior surface [FIGS. 5.127 and 5.137], from which it is separated by a bursa.

Nerve supply

By a branch (L. 4, 5; S. 1) from the superior gluteal nerve.

Actions

Gluteus minimus is an abductor of the thigh on the pelvis or of the pelvis on the thigh. Its anterior fibres produce medial rotation of the thigh. Gluteus minimus acts with gluteus medius in walking, to pull the centre of gravity forwards and over the supporting limb. Paralysis of the two muscles (by section of the superior gluteal nerve) makes it impossible for the patient to support the pelvis when the opposite foot is raised from the ground. This he overcomes by flexing the trunk strongly to the paralysed side to bring the centre of gravity over the supporting limb—a movement which results in a characteristic waddling gait. The same effect is produced by bilateral congenital dislocation of the hip joints, a condition which prevents gluteus medius and minimus from functioning properly.

Gluteus medius and minimus and tensor fasciae latae are tested together. The patient lies on his back with the knee straight, and is asked to abduct the thigh against resistance. The bellies of all three muscles can then usually be felt. Alternatively, the patient lies face down with the knee bent to a right angle and is asked to push the foot laterally against resistance. This is a test of the capacity of the muscles to rotate the femur medially.

GEMELLI

These small 'twin' muscles are accessory to the obturator internus [p. 366] and give it additional origins from the margins of the lesser sciatic notch [FIGS. 5.125 and 5.126]. The superior gemellus arises from the gluteal surface of the ischial spine, and the inferior gemellus from the upper part of the ischial tuberosity. They insert into the tendon of the obturator internus, one above and the other below, and so obtain an attachment to the medial surface to the greater trochanter above the trochanteric fossa [FIG. 5.127].

Nerve supply

The superior gemellus is supplied by the nerve to obturator internus (L. 5; S. 1); the inferior gemellus by the nerve to quadratus femoris (L. 5; S. 1).

Actions

They help the obturator internus to rotate the thigh laterally when it is extended, and to abduct it when it is flexed. They most probably function as extensile ligaments [p. 271].

QUADRATUS FEMORIS

This is a flat quadrilateral muscle [FIGS. 5.125 and 5.132] passing from the lateral margin of the ischial tuberosity to be inserted into the intertrochanteric crest and quadrate tubercle of the femur [FIGS. 5.124 and 5.130]. It lies between the inferior gemellus above and the upper fibres of the adductor magnus below, and is often fused with either or both.

Nerve supply

By a special branch of the lumbosacral plexus (L. 4, 5; S. 1), which supplies both it and the inferior gemellus.

Actions

It laterally rotates the thigh on the trunk, or medially rotates the trunk on the thigh. Like most of the short muscles in relation to the hip joint, it probably has a more important function as an extensile ligament [p. 271].

MUSCLES OF THE THIGH

The muscles of the thigh fall into three main groups by virtue of their situation, actions, and nerve supply. On the back of the thigh (the flexor surface) lies the hamstring group, corresponding to the

flexors of the elbow in the upper limb. They are the semitendinosus, semimembranosus, and biceps femoris. The second group corresponds to coracobrachialis in the upper limb, but is much more extensive. It is known as the adductor group because the main action of its components is adduction of the thigh. It consists of the adductors magnus, brevis and longus, gracilis, obturator externus, and pectineus. Finally, on the front of the thigh lies the anterior group, corresponding to the triceps brachii of the upper limb and consisting of sartorius and the four components of the quadriceps femoris. The front of the thigh also contains the terminations of the main flexors of the hip, iliacus and psoas major, which have been described elsewhere [p. 363].

The boundaries of these groups are not rigidly exclusive; for example, part of adductor magnus [p. 380] is to be considered as a hamstring muscle in view both of its action and of its nerve supply, and pectineus [p. 383], though it is functionally an adductor, usually receives its innervation from the femoral nerve, which supplies the anterior group.

Hamstring group

SEMITENDINOSUS

Semitendinosus arises by a short tendon in common with the long head of biceps femoris from the medial impression (Martin 1968) on the ischial tuberosity [FIGS. 5.126 and 5.130]. The long round tendon of insertion begins approximately two-thirds of the way down the thigh [FIG. 5.128] and runs as a cord in a furrow made for it by semimembranosus. Over the medial side of the knee it spreads out to form a flattened aponeurosis which inserts into the upper part of the medial surface of the body of the tibia, behind the insertions of gracilis and sartorius [FIG. 5.129] and into the deep fascia of the leg. A complex of intercommunicating bursae (the bursa anserina) intervenes between these tendons and separates them from the bone.

Nerve supply

By two branches (L. 5; S. 1, 2) from the tibial part of the sciatic nerve.

Actions

Semitendinosus extends the hip joint, and is important in controlling flexion of the hip when the body is bent forwards from the erect position. It is a flexor of the knee joint and a medial rotator of the tibia on the femur (particularly when the knee is flexed), or, if the foot is on the ground, a lateral rotator of the pelvis and femur on the tibia. The muscle cannot exert its full range on both hip and knee joints simultaneously; if the knee is fully flexed, it is so shortened that it cannot contract further and act on the hip. Conversely, if it is initially stretched by fully flexing the hip it will not allow enough further stretching to permit full simultaneous extension of the knee unless it has been specially trained to do so [p. 272].

SEMIMEMBRANOSUS

The strong membranous tendon which gives the muscle its name arises from the lateral facet on the ischial tuberosity [FIG. 5.126], and continues downwards along the lateral margin of the muscle (Martin 1968). The fleshy belly medial to the tendon is deep to semitendinosus and the long head of biceps femoris. It runs downwards and medially with semitendinosus grooving its superficial surface [FIGS. 5.128 and 5.138], and ends at the back of the knee

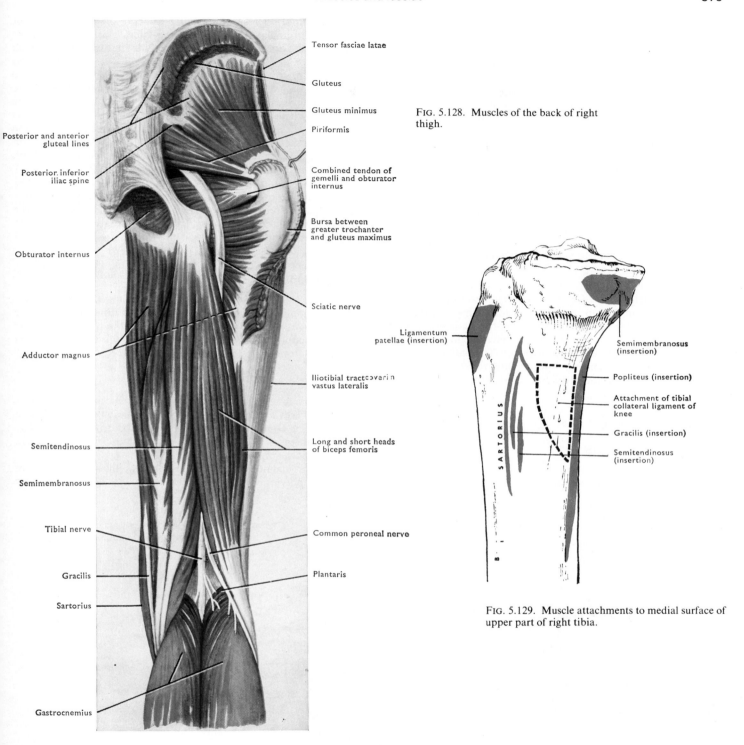

Tensor fasciae latae

Gluteus

Gluteus minimus

Piriformis

Posterior and anterior gluteal lines

Posterior, inferior iliac spine

Combined tendon of gemelli and obturator internus

Bursa between greater trochanter and gluteus maximus

Obturator internus

Adductor magnus

Sciatic nerve

Iliotibial tract covering vastus lateralis

Semitendinosus

Long and short heads of biceps femoris

Semimembranosus

Tibial nerve

Gracilis

Common peroneal nerve

Sartorius

Plantaris

Gastrocnemius

FIG. 5.128. Muscles of the back of right thigh.

Ligamentum patellae (insertion)

Semimembranosus (insertion)

Popliteus (insertion)

Attachment of tibial collateral ligament of knee

Gracilis (insertion)

Semitendinosus (insertion)

SARTORIUS

FIG. 5.129. Muscle attachments to medial surface of upper part of right tibia.

in a tendon which is inserted into a horizontal groove on the posteromedial surface of the medial condyle of the tibia. A bursa separates it from the medial head of gastrocnemius, and another from the tibia near its insertion [FIGS. 4.55 and 5.129].

From the tendon a fascial band runs to join the posterior border of the tibial collateral ligament of the knee joint; another, running downwards and laterally, forms the fascia covering the popliteus muscle [p. 406], and is attached to the soleal line of the tibia. Finally, a strong band extends upwards and laterally to the back of the lateral condyle of the femur, forming the **oblique popliteal ligament** of the knee joint [FIG. 4.50].

Nerve supply

By a branch (L. 5; S. 1, 2) from the tibial part of the sciatic nerve.

Actions

As for semitendinosus, except that its rotatory action on the knee joint only operates when the joint is flexed.

BICEPS FEMORIS

The long head of biceps arises with semitendinosus from the medial impression on the ischial tuberosity [FIGS. 5.125 and 5.126] and from the sacrotuberous ligament. The fleshy belly of the long head becomes tendinous in the distal third of the thigh and is then joined by the short head of the muscle [FIGS. 5.128 and 5.132]. The short head arises separately from the lateral lip of the linea aspera and the upper half of the lateral supracondylar line of the femur; some of its fibres come from the lateral intermuscular septum.

The rounded tendon of insertion can easily be felt [FIGS. 5.139 and 5.143] as it passes to its attachment to the head of the fibula; it is split in two by the fibular collateral ligament of the knee joint, which is separated from it by a bursa. Some of its fibres gain insertion to the lateral condyle of the tibia and to the deep fascia on the lateral side of the leg [FIG. 5.142].

Nerve supply

The long head of the muscle, like the other hamstring muscles, receives a branch from the tibial portion of the sciatic nerve (L. 5; S. 1, 2). The short head is separately supplied by a branch (L. 5; S. 1, 2) from the common peroneal part of the sciatic nerve.

Actions

The long head helps semitendinosus, semimembranosus, and gluteus maximus to extend the hip joint. Both heads of the muscle flex the knee, and, when this has been done, rotate the leg laterally on the thigh or, if the foot is fixed, the thigh and pelvis medially on the leg.

The three hamstrings are tested together. The patient lies on his face and attempts to bend the knee against resistance, when the tendons of the biceps and semitendinosus stand out and can be palpated [FIG. 5.140].

Adductor Group

ADDUCTOR MAGNUS

Adductor magnus [FIG. 5.131] is the largest of the adductor group of muscles; and is really a composite muscle, part hamstring, part

adductor. It arises from the femoral aspect of the conjoint rami of the ischium and pubis, and from the lateral part of the inferior surface of the ischial tuberosity [FIG. 5.130]. The upper fibres, which arise most anteriorly, run horizontally, parallel to and often fused with the fibres of quadratus femoris; the lowest fibres, which arise from the ischial tuberosity, pass vertically downwards, and the intervening ones radiate out obliquely [FIG. 5.132]. It inserts linearly along the whole length of the femur from the insertion of quadratus femoris to the adductor tubercle, including the linea aspera and the medial supracondylar line. The posterior fibres of the muscle reach the adductor tubercle, and some continue downwards to fuse with the tibial collateral ligament of the knee joint [FIG. 5.129].

The perforating branches of the profunda femoris artery run between the insertion of the muscle and the femur, deep to a series of tendinous arches which protect these vessels. A much larger opening, the hiatus tendineus, interrupts the attachment of the muscle to the supracondylar line; it allows the passage of the femoral vessels into the popliteal fossa.

Nerve supply

The vertical fibres belong morphologically to the hamstring group, and are supplied by a branch (L. 4) from the tibial portion of the sciatic nerve; like the rest of the adductors, the remainder of the muscle is supplied from the obturator nerve, in this case by a branch (L. 2, 3, 4) from its posterior division [FIG. 5.133].

Actions

The proximal portion of adductor magnus acts with adductors longus and brevis in adduction and medial rotation of the thigh, and in preventing lateral overbalancing at the hip during walking. The vertical fibres of the ischial part of the muscle are also weak extensors of the hip joint.

Adductor magnus is tested together with the whole group of adductor muscles. The patient lies on his back with the knee

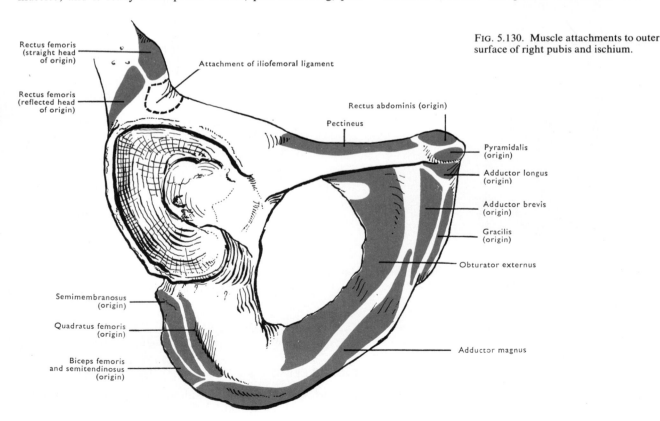

FIG. 5.130. Muscle attachments to outer surface of right pubis and ischium.

Rectus femoris (straight head of origin)

Attachment of iliofemoral ligament

Rectus femoris (reflected head of origin)

Rectus abdominis (origin)

Pectineus

Pyramidalis (origin)

Adductor longus (origin)

Adductor brevis (origin)

Gracilis (origin)

Obturator externus

Semimembranosus (origin)

Quadratus femoris (origin)

Biceps femoris and semitendinosus (origin)

Adductor magnus

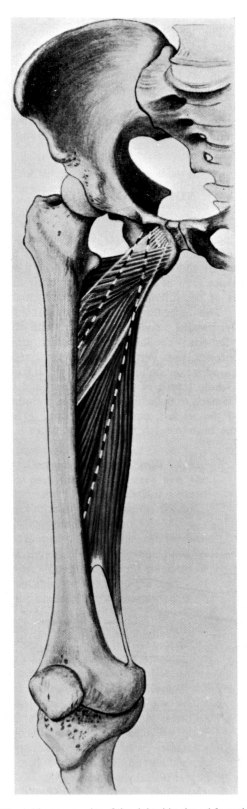

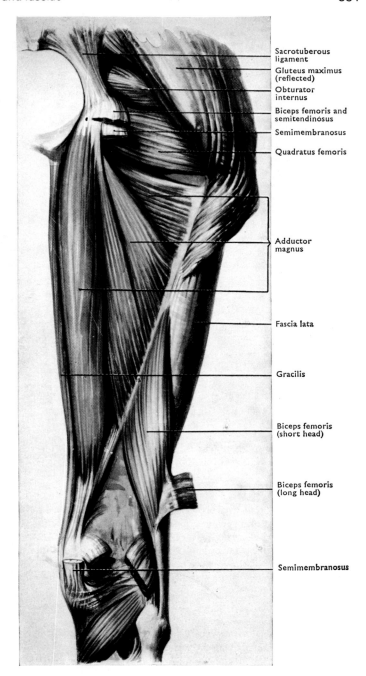

Sacrotuberous ligament
Gluteus maximus (reflected)
Obturator internus
Biceps femoris and semitendinosus
Semimembranosus
Quadratus femoris

Adductor magnus

Fascia lata

Gracilis

Biceps femoris (short head)

Biceps femoris (long head)

Semimembranosus

FIG. 5.132. Deep muscles of back of right thigh.

FIG. 5.131. Adductor muscles of the right side viewed from the front. Adductor longus, interrupted outline; adductor brevis, white shadow; adductor magnus, red.

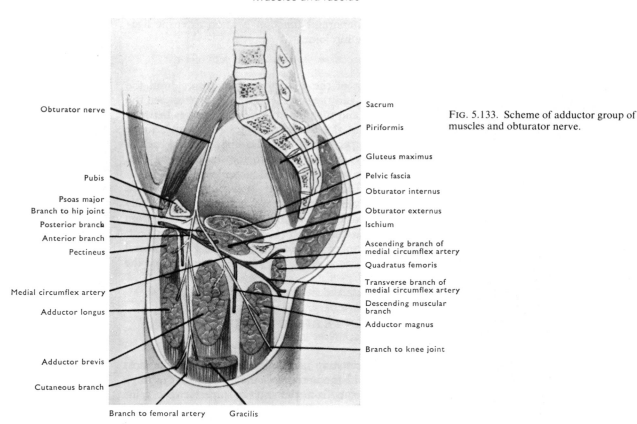

Obturator nerve

Pubis

Psoas major
Branch to hip joint
Posterior branch
Anterior branch
Pectineus

Medial circumflex artery

Adductor longus

Adductor brevis

Cutaneous branch

Branch to femoral artery   Gracilis

Sacrum

Piriformis

Gluteus maximus

Pelvic fascia

Obturator internus

Obturator externus

Ischium

Ascending branch of
medial circumflex artery

Quadratus femoris

Transverse branch of
medial circumflex artery

Descending muscular
branch

Adductor magnus

Branch to knee joint

FIG. 5.133. Scheme of adductor group of
muscles and obturator nerve.

extended, and then adducts the thigh against resistance while the examiner feels the bellies of the adductor muscles.

ADDUCTOR BREVIS

Adductor brevis, lying in front of adductor magnus, has a linear origin from the femoral surfaces of the body and inferior ramus of the pubis [FIGS. 5.130 and 5.131]. Its fibres pass downwards and laterally, and are inserted, behind pectineus and adductor longus, into the upper half of the linea aspera and the lower two-thirds of the line leading from it to the lesser trochanter of the femur [FIG. 5.124].

Nerve supply

From the anterior division (L. 2, 3, 4) of the obturator nerve [FIG. 5.133]: sometimes the posterior division also supplies a branch to it.

Actions

It adducts the femur and medially rotates it. It can also flex the extended femur and extend the flexed femur. Like the other adductors, it is an important factor in preventing lateral overbalancing on the supporting leg during walking [p. 403].

ADDUCTOR LONGUS

The triangular adductor longus [FIGS. 5.131 and 5.134] is the most anterior of the three adductor muscles, arising by a cylindrical tendon from the femoral surface of the body of the pubis in the angle between the crest and the symphysis [FIG. 5.130]. The fibres, like those of adductor brevis, pass downwards and laterally, and are inserted into the medial lip of the linea aspera between adductors brevis and magnus posteriorly and vastus medialis anteriorly. It is often difficult to separate its fibres from those of vastus medialis and adductor magnus [FIG. 5.138].

Nerve supply

From the anterior division (L. 2, 3, 4) of the obturator nerve [FIG. 5.133].

Actions

As for adductor brevis.

GRACILIS

Gracilis [FIGS. 5.138 and 5.146] the last large member of the adductor group, is a long flat muscle which tapers somewhat from above downwards. It takes origin from the femoral surface of the body of the pubis, and from its inferior ramus [FIG. 5.130]. It descends, immediately deep to the deep fascia, along the medial side of the thigh between sartorius anteriorly and semimembranosus posteriorly, ending in a tendon which expands to be inserted into the upper part of the medial surface of the body of the tibia between sartorius and semitendinosus [FIG. 5.129]. There is a bursa deep to the tendon, between it and semitendinosus, and another between it and sartorius. If they communicate, they are given the name of bursa anserina.

Nerve supply

By a branch (L. 2, 3) from the anterior division of the obturator nerve [FIG. 5.133].

Actions

It adducts the thigh, flexes the knee joint, and rotates the tibia medially on the femur or, if the foot is on the ground, rotates the pelvis and the femur laterally on the tibia. Like the other adductors, its chief importance is as a balancing muscle during walking [p. 403].

OBTURATOR EXTERNUS

Obturator externus lies deeply in the front of the thigh, taking origin from the margins of the pubis and ischium surrounding the obturator membrane and from the inferior surface of the membrane itself [FIG. 5.130]. The fasciculi of the muscle converge on the groove inferior to the acetabulum, ending in a tendon which passes obliquely across the back of the neck of the femur to be inserted into the trochanteric fossa [FIG. 5.124].

Nerve supply

From the posterior division (L. 3, 4) of the obturator nerve.

Actions

It is a lateral rotator and adductor of the thigh. Probably its most important function is as an extensile ligament [p. 271] of the hip joint.

PECTINEUS

Pectineus arises from the pecten of the pubis and the surface in front of it, between the iliopubic eminence and the pubic tubercle. It passes between psoas major and adductor longus [FIG. 5.134], to be inserted into the upper half of a line leading from the back of the lesser trochanter of the femur to the linea aspera [FIG. 5.124], anterior to the upper part of adductor brevis.

Nerve supply

This is variable. It is normally supplied by a branch (L. 2, 3) from the femoral nerve, but occasionally receives an additional branch from the obturator nerve or from the accessory obturator nerve (L. 3) when this is present. Sometimes the pectineus is divided into a medial and a lateral part which are innervated respectively by the obturator and the femoral nerves.

Actions

It adducts and flexes the femur. It has been claimed as a lateral rotator of the thigh, but electromyography indicates rather that it is in action during medial rotation (Takebe, Vitti, and Basmajian 1974b).

Anterior Group

SARTORIUS

This is the longest strap muscle in the body, and the most superficial muscle in the anterior compartment of the thigh [FIG. 5.134]. It extends from the anterior superior iliac spine to the upper part of the medial surface of the body of the tibia, in front of the insertions of gracilis and semitendinosus [FIG. 5.129]. A bursa separates it from the insertion of gracilis, and some fibres run from the tendon of insertion to the tibial collateral ligament of the knee joint and to the fascia of the leg.

The medial border of the upper third of sartorius forms the lateral boundary of the **femoral triangle**, a region bounded medially by the medial border of adductor longus and superiorly by the inguinal ligament. The floor of the triangle is formed laterally by iliopsoas and medially by pectineus, adductor longus, and, occasionally, a small part of adductor brevis [FIG. 5.134].

The middle third of sartorius, and the strong fascia underlying it, form the roof of the **adductor canal**, in which lie the femoral vessels, the saphenous nerve, and the nerve to vastus medialis. The canal is bounded anterolaterally by vastus medialis, and posteromedially by

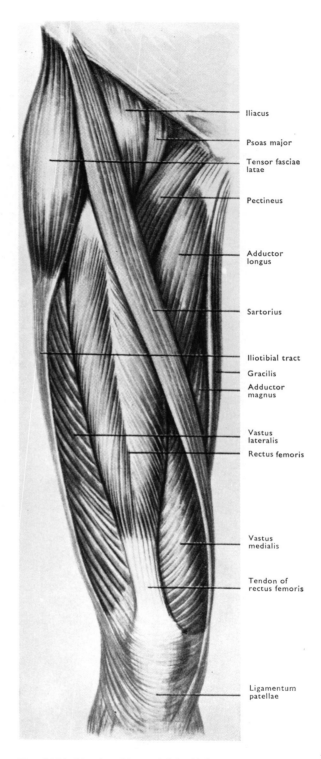

FIG. 5.134. Muscles of front of right thigh.

adductor longus and adductor magnus; it is triangular in cross-section [FIG. 5.138].

Nerve supply

It receives two branches (L. 2, 3) from the femoral nerve, conveyed to the muscle by its anterior cutaneous branches.

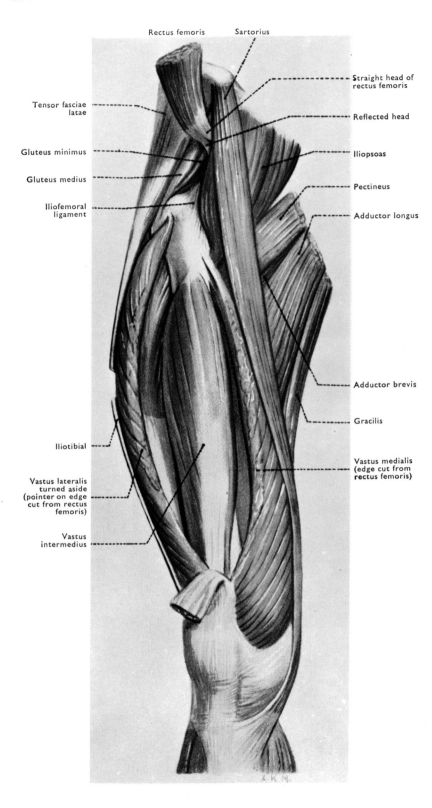

Rectus femoris   Sartorius

Straight head of
rectus femoris

Tensor fasciae
latae

Reflected head

Gluteus minimus

Iliopsoas

Gluteus medius

Pectineus

Iliofemoral
ligament

Adductor longus

Adductor brevis

Gracilis

Iliotibial

Vastus medialis
(edge cut from
rectus femoris)

Vastus lateralis
turned aside
(pointer on edge
cut from rectus
femoris)

Vastus
intermedius

FIG. 5.135. Dissection of muscles of front of right thigh. The rectus femoris has been removed and the vastus lateralis and vastus medialis pulled apart to show the vastus intermedius.

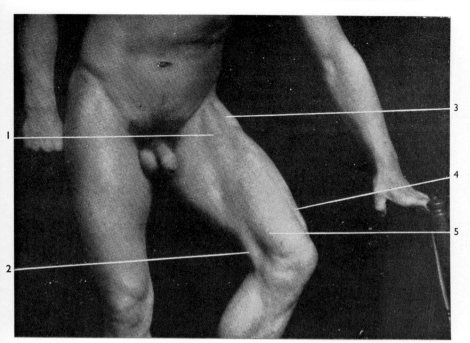

FIG. 5.136. Sartorius in action. (*By courtesy of Professor R. D. Lockhart.*)
1. Sartorius.
2. Semitendinosus applied to semimembranosus covered by sartorius and gracilis.
3. Tensor fasciae latae.
4. Vastus lateralis.
5. Vastus medialis.

Actions

It is a flexor of the hip and of the knee; it abducts the thigh slightly and rotates it laterally [FIG. 5.136] and it assists in medial rotation of the tibia on the femur. These actions may be summarized by saying that it places the heel on the knee of the opposite limb.

QUADRICEPS FEMORIS

Quadriceps femoris is a composite muscle which includes four distinct parts usually described as separate muscles—rectus femoris, vastus lateralis, vastus medialis, and vastus intermedius [FIGS. 5.134–5.136 and 5.139].

Rectus femoris is a spindle-shaped bipennate muscle which is described as having two heads of origin, although the two heads are continuous with each other. The straight head of the muscle comes from the anterior inferior iliac spine, and the reflected head from a rough groove on the ilium immediately above the acetabulum. The reflected head is the one which is in the line of pull of the muscle in four-footed animals, and the straight head is a human acquisition, developed with the upright posture. From this continuous origin, a single tendon descends to a fleshy belly on the front of vastus intermedius [FIG. 5.134]. From the lower end of this belly a tendon runs into the upper border of the patella; on its lateral and medial sides it receives fibres of insertion of vastus lateralis and medialis respectively. Because rectus femoris is capable of movement independent of vastus intermedius (it acts on the hip joint as well as on the knee) it has a smooth tendinous deep surface which can slip over the similar tendinous superficial surface of vastus intermedius [FIG. 5.135].

The patella is really a sesamoid bone in the tendons of rectus femoris and vastus intermedius, and the true insertion of the muscle is via the patellar ligament, a strong band which attaches the inferior angle and the margins of the patella to the tuberosity of the tibia [FIG. 5.129]. The patella is also anchored to the anterior borders of the tibial condyles by the lateral and medial patellar retinacula [p. 247].

Vastus lateralis has a linear origin from the capsule of the hip joint, the upper part of the intertrochanteric line, the lower border of the greater trochanter, the lateral margin of the gluteal tuberosity,

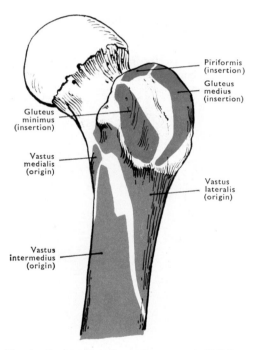

FIG. 5.137. Muscle attachments to front of upper part of left femur.

and the upper half of the linea aspera; it also comes from the fascia lata, and from the lateral intermuscular septum [FIGS. 3.157, 3.160, and 5.138]. The fibres run downwards and forwards, and give way to a broad tendon which is inserted into the tendon of rectus femoris, the lateral border of the patella, and the front of the lateral condyle of the tibia, blending with the iliotibial tract, and to a large extent replacing the capsule of the knee joint in this region [FIGS. 5.134 and 5.135]. These aponeurotic fibres blend with the deep fascia to form the lateral patellar retinaculum [p. 247]. The muscle exerts its main pull on the tibia through the patella ligament.

Vastus medialis is larger and heavier than vastus lateralis, and has a long origin from the lower half of the intertrochanteric line,

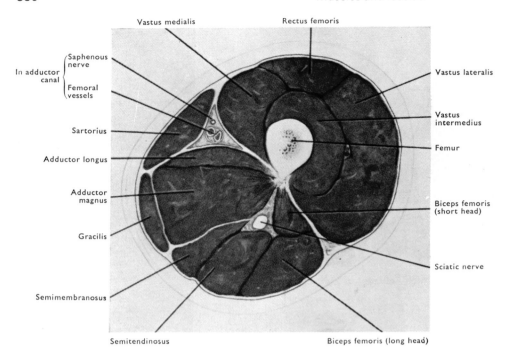

FIG. 5.138. Transverse section of thigh through the adductor canal.

the linea aspera, the upper two-thirds of the medial supracondylar line, the medial intermuscular septum, and the tendon of adductor magnus [FIGS. 3.157 and 3.160]. Corresponding to vastus lateralis on the lateral side, it is inserted into the tendon of rectus femoris, the upper and medial borders of the patella and the front of the medial

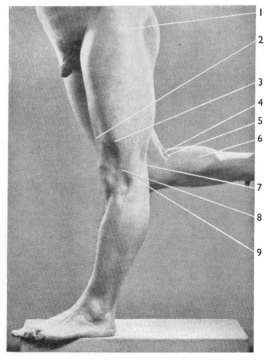

FIG. 5.139. Activity of muscles in extension and resisted flexion of leg. (*By courtesy of Professor R. D. Lockhart.*)

1. Tensor fasciae latae.
2. Vastus medialis.
3. Vastus lateralis.
4. Semimembranosus.
5. Semitendinosus.
6. Gastrocnemius in action as flexor of knee.
7. Biceps femoris.
8. Expansion from vastus lateralis.
9. Iliotibial tract.

condyle of the tibia, replacing the anteromedial part of the capsule of the knee joint, and becoming fused with the deep fascia to form the **medial patellar retinaculum** [p. 405]. Muscular fibres of vastus medialis extend to the patella, and the lowest ones are almost horizontal [FIGS. 5.134 and 5.135].

Vastus intermedius, the deepest portion of the quadriceps, arises by fleshy fibres from the upper two-thirds of the body of the femur on its anterior and lateral surfaces. It also comes from the lower half of the lateral lip of the linea aspera and the upper part of the lateral supracondylar line, as well as from the lateral intermuscular septum [FIGS. 3.157, 3.160, and 5.137]. The muscle has a membranous tendon on its anterior surface, which separates it from the deep surface of rectus femoris, and allows movement between the two [FIG. 5.135]. It is inserted into the deep surface of the tendons of rectus and the other vasti muscles. Vastus intermedius is difficult to separate from vastus lateralis in the middle of the thigh, and impossible to separate from vastus medialis lower down [FIG. 5.138]. Some of its deepest fibres in the lower third of the thigh, **articularis genus**, are inserted into the walls of the **suprapatellar bursa** (synovial pouch) which lies deep to vastus intermedius in the lowest part of the thigh.

Nerve supply

All the components of quadriceps femoris are supplied by branches (L. 2, 3, 4) of the femoral nerve. The branch to vastus medialis is specially large, and reaches it by way of the adductor canal.

Actions

Rectus femoris is a strong flexor of the hip joint, and it assists iliopsoas to flex the trunk on the thigh. Together with all the other components of the quadriceps, it extends the limb at the knee joint. The vasti, acting without rectus femoris, pay out to control the movements of sitting down. The quadriceps is used to straighten the knee when rising from the sitting position, during walking, and in climbing. Without it, the knee becomes grossly unstable. Because the pull of rectus femoris and vastus intermedius is directed in the line of the femur, at an angle to the line of the patellar ligament and to the shaft of the tibia, there is a tendency for the patella to dislocate laterally during extension of the knee. This tendency is resisted by

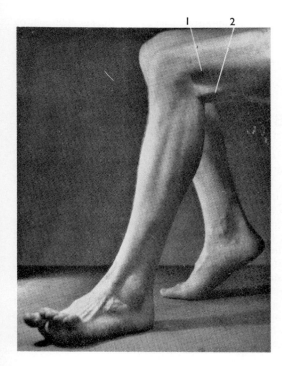

FIG. 5.140. Lateral rotation of leg upon thigh. Note the interval between the tendon of biceps femoris and the iliotibial tract covering vastus lateralis. (*By courtesy of Professor R. D. Lockhart.*)
1. Iliotibial tract.
2. Biceps femoris.

the pull of the horizontal fibres of vastus medialis, and the raised patellar surface of the lateral condyle. Quadriceps femoris is particularly vulnerable to disuse atrophy, and if the lower limb is enclosed in plaster, perhaps for a fracture, the muscle rapidly loses its bulk and power; for this reason it has to be kept exercised in such circumstances.

The quadriceps is tested with the patient lying on his back; he tries to extend the knee against resistance, and the muscle bellies can be seen and felt.

Movements at the hip joint

At the hip, as at all joints of the lower limb, the movements must always be looked at from two points of view. The muscles operating on the joint act from one end when they are moving the limb upon the trunk, and from the other when they are moving the trunk upon the limb during walking or running. It is also somewhat unrealistic to consider the hip joint in isolation, since (for example) flexion of the hip very seldom occurs without a corresponding flexion of the knee; indeed, if the knee is voluntarily kept extended, flexion of the hip will be limited because of the tension of the hamstrings [p. 378].

Secondly, the action of certain muscles alters as the hip is moved from its anatomical position of extension. For example, obturator internus is a lateral rotator of the thigh when the joint is extended, and an abductor when the joint is flexed.

At the hip the extensors are stronger than the flexors, and the adductors are much stronger than the abductors. As in any ball and socket joint, most muscles acting on the hip joint are active in several of the artificially defined movements which such joints can undertake. One of the main functions of the muscles around the hip joint is to stabilize the trunk on the limb when the other foot is raised from the ground [p. 403].

The table shows the muscles chiefly involved in these movements: iliopsoas, piriformis, and obturator internus are described elsewhere [pp. 363, 366–7].

Muscles concerned in movements of the hip joint

Movement	Muscle	Principal nerve supply
Flexion	Iliopsoas	Ventral rami of L. 1, 2, 3
	Rectus femoris	Femoral
	Tensor fasciae latae	Superior gluteal
	Sartorius	Femoral
	Adductor longus	Obturator
	Adductor brevis	Obturator
	Pectineus	Femoral
Extension	Gluteus maximus	Inferior gluteal
	Semimembranosus	Tibial part of sciatic
	Semitendinosus	Tibial part of sciatic
	Biceps femoris (long head)	Tibial part of sciatic
	Adductor magnus (ischial fibres)	Tibial part of sciatic
Abduction	Gluteus medius	Superior gluteal
	Gluteus minimus	Superior gluteal
	Tensor fasciae latae	Superior gluteal
	Obturator internus (in flexion)	Ventral rami of L. 5; S. 1, 2
	Piriformis (in flexion)	Ventral rami of L. 5; S. 1, 2
Adduction	Adductor magnus	Obturator
	Adductor longus	Obturator
	Adductor brevis	Obturator
	Pectineus	Femoral
	Gracilis	Obturator
	Gluteus maximus (lower fibres)	Inferior gluteal
	Quadratus femoris	Ventral rami of L. 4, 5; S. 1
Lateral rotation	Gluteus maximus	Inferior gluteal
	Obturator internus and gemelli	Ventral rami of L. 5; S. 1, 2
	Obturator externus	Obturator
	Quadratus femoris	Ventral rami of L. 4, 5; S. 1
	Piriformis	Ventral rami of L. 5; S. 1, 2
	Sartorius	Femoral
Medial rotation	Adductor magnus	Obturator
	Adductor longus	Obturator
	Adductor brevis	Obturator
	Iliopsoas (in initial stage of flexion)	Ventral rami of L. 1, 2, 3
	Tensor fasciae latae	Superior gluteal
	Gluteus minimus (anterior fibres)	Superior gluteal

Muscles concerned in movements of the knee joint

Movement	Muscle	Principal nerve supply
Flexion	Biceps femoris	Sciatic
	Semitendinosus	Tibial part of sciatic
	Semimembranosus	Tibial part of sciatic
	Gastrocnemius	Tibial
	Plantaris	Tibial
	Sartorius	Femoral
	Gracilis	Obturator
	Popliteus	Tibial
Extension	Quadriceps femoris	Femoral
Medial rotation of tibia on femur (equivalent to lateral rotation of femur on tibia)	Popliteus	Tibial
	Semitendinosus	Tibial part of sciatic
	Semimembranosus	Tibial part of sciatic
	Sartorius	Femoral
	Gracilis	Obturator
Lateral rotation of tibia on femur (equivalent to medial rotation of femur on tibia)	Biceps femoris	Sciatic

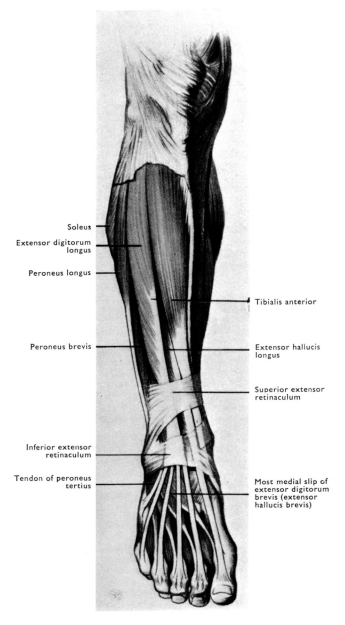

Soleus

Extensor digitorum
longus

Peroneus longus

Tibialis anterior

Peroneus brevis

Extensor hallucis
longus

Superior extensor
retinaculum

Inferior extensor
retinaculum

Tendon of peroneus
tertius

Most medial slip of
extensor digitorum
brevis (extensor
hallucis brevis)

FIG. 5.141. Muscles of front of leg and dorsum of foot of right side.

Movements at the knee joint

The table shows the muscles operating on the knee joint. Quadriceps femoris is the only extensor of the joint, whereas there are several flexors, only two of which do not operate on other joints—the short head of biceps, and popliteus [p. 394]. The hamstrings, which flex the leg at the knee, are also extensors of the hip. Because of this, many people are unable to touch their toes while the knee joints are kept extended. Tight hamstrings may also be a source of discomfort to a patient sitting up in bed, unless the knees are flexed by a pillow under them. It is possible to train the hamstrings to relax, as any contortionist can demonstrate.

Rotation of the tibia at the knee joint is a much stronger movement when the knee is flexed to a right angle, since in this position the tendons of the muscles concerned are placed at right angles to the tibia. The iliotibial tract, which represents the main insertion of gluteus maximus and the only insertion of tensor fasciae

latae, has an important supportive role as an extensile ligament on the lateral side of the knee joint, and can also maintain the joint braced back in full extension, so that standing, and even clumsy walking, may be possible even though quadriceps femoris is paralysed.

MUSCLES OF THE LEG

The leg contains three groups of muscles: the extensor (dorsiflexor) muscles anteriorly, the peroneal muscles on the lateral side, and the flexor (plantar flexor) muscles posteriorly. These groups are all contained within osteofascial compartments [FIG. 5.144].

The muscles in the anterior compartment are tibialis anterior, extensor hallucis longus, extensor digitorum longus, and peroneus tertius. In the peroneal compartment are peroneus longus and brevis, and in the posterior compartment are gastrocnemius, soleus, plantaris, popliteus, flexor digitorum longus, flexor hallucis longus, and tibialis posterior.

As in the upper limb, the muscles acting on the digits fall into two categories; long muscles with their bellies in the leg, and short muscles fully contained within the foot. The big and little toes, like the thumb and little fingers, have separate muscles of their own in the sole, and the other toes share a common muscular apparatus.

Muscles of the anterior compartment

TIBIALIS ANTERIOR

Tibialis anterior [FIGS. 5.141, 5.142, 5.144, and 5.147] arises from the lateral condyle and the upper two-thirds of the lateral surface of the body of the tibia, from the interosseous membrane, and from the deep fascia surrounding the muscle. The tendon, which becomes free in the lower third of the leg, passes downwards and medially over the front of the distal end of the tibia, posterior to both extensor retinacula [p. 406], and enclosed in a separate synovial sheath 5.158 and 5.160]. It inserts into the medial sides of the medial cuneiform and the base of the first metatarsal bone [FIG. 5.154], reaching the ventral surface to blend with the insertion of peroneus longus [p. 391].

Nerve supply

By a branch (L. 4, 5) from the deep peroneal nerve.

Actions

It dorsiflexes the ankle joint, and is used to pull the body forward on the foot of the supporting limb during walking, as well as to raise the fore part of the foot of the free limb as it swings forwards. If tibialis anterior is paralysed, the remaining dorsiflexors are unable to raise the toes sufficiently to prevent them dragging along the ground; this is the condition of 'foot-drop', and can be remedied either by flexing the knee more than is normally needed during walking, or by fitting a 'toe-raising' spring to the toe of the shoe.

Tibialis anterior is also a powerful invertor of the foot: that is, it turns the sole to face medially. This action, which occurs at the joints between the talus and the other tarsal bones, allows the foot to adapt itself to uneven ground. Acting from the foot, tibialis anterior pulls the leg medially, and so prevents lateral overbalancing during walking [p. 404]. The tendons of peroneus longus and tibialis anterior together form a sling under the foot which is an important support for the transverse arch of the foot.

The integrity of the muscle is tested by asking the patient to dorsiflex the ankle; the muscle belly and the tendon can be seen and felt if the muscle is working.

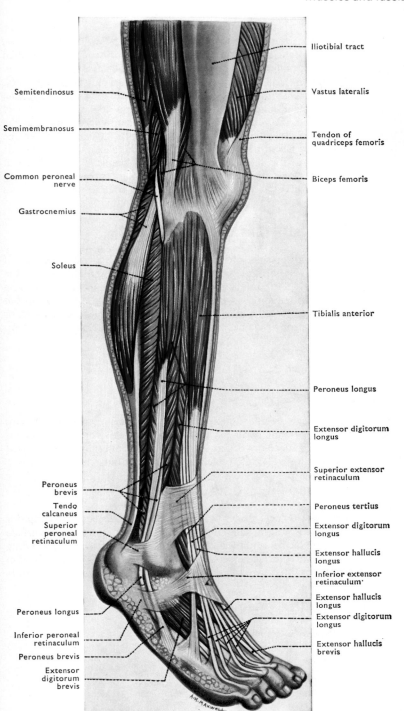

Semitendinosus

Semimembranosus

Common peroneal nerve

Gastrocnemius

Soleus

Peroneus brevis

Tendo calcaneus

Superior peroneal retinaculum

Peroneus longus

Inferior peroneal retinaculum

Peroneus brevis

Extensor digitorum brevis

Iliotibial tract

Vastus lateralis

Tendon of quadriceps femoris

Biceps femoris

Tibialis anterior

Peroneus longus

Extensor digitorum longus

Superior extensor retinaculum

Peroneus tertius

Extensor digitorum longus

Extensor hallucis longus

Inferior extensor retinaculum

Extensor hallucis longus

Extensor digitorum longus

Extensor hallucis brevis

FIG. 5.142. Muscles of knee, leg, and dorsum of foot seen from lateral side.

EXTENSOR DIGITORUM LONGUS

Extensor digitorum longus lies lateral to tibialis anterior [FIGS. 5.141 and 5.142] and is a pennate muscle with a tendon which appears on its medial side as it courses down the leg. Its origin is from the lateral side of the lateral condyle of the tibia, the upper two-thirds of the anterior surface of the body of the fibula, the deep fascia, and the upper part of the interosseous membrane.

The tendon passes deep to the superior extensor retinaculum without dividing, but under the inferior extensor retinaculum [FIG. 5.141] it gives rise to four tendons which run to the lateral four toes. In exactly the same manner as the corresponding tendons in the hand [p. 333] they form extensor hoods on the dorsum of the metatarsophalangeal joints and proximal phalanges. These hoods are joined by the tendons of the lumbricals and extensor digitorum brevis (2nd–4th toes), though not by the interossei (Jones 1944; Manter 1945). Each hood is then inserted by its central portion into the base of the middle phalanx, and by two collateral portions into the base of the distal phalanx, just as in the hand [p. 333].

A common synovial sheath surrounds the four tendons as they lie deep to the inferior extensor retinaculum [FIGS. 5.158 and 5.159].

Nerve supply

By a branch (L. 5; S. 1) from the deep peroneal nerve.

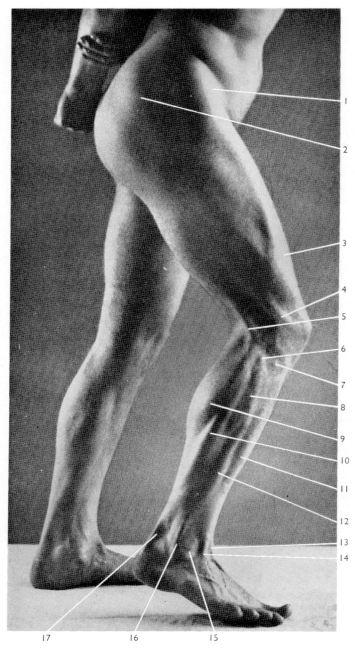

FIG. 5.143. The subject is 'setting' the muscles of the leg. Note the concavity of the popliteal fossa behind the flexed knee and the fullness of that region behind the straight knee. (From *Living anatomy* by R. D. Lockhart, 6th edn. Faber and Faber, London (1963).)

1. Tensor fasciae latae.
2. Gluteus medius.
3. Vastus lateralis.
4. Iliotibial tract.
5. Tendon of biceps femoris.
6. Head of fibula.
7. Prominence of side of condyle of tibia.
8. Peroneus longus.
9. Gastrocnemius.
10. Soleus.
11. Extensor digitorum longus.
12. Peroneus brevis.
13. Extensor hallucis longus.
14. Extensor digitorum longus.
15. Lateral malleolus.
16. Peroneus longus superficial to peroneus brevis.
17. Tendo calcaneus.

Actions

It extends the lateral four toes at the metatarsophalangeal joints, and assists with the extension of the interphalangeal joints, which are principally extended by the lumbricals. The arrangement is very similar to that in the fingers, but the interossei play no part [p. 400]. Like the extensor digitorum of the upper limb [p. 345] it is unable on its own to extend the interphalangeal joints. If the lumbricals are paralysed, the action of the muscle produces hyperextension of the metatarsophalangeal joint and the interphalangeal joints become passively flexed through tension in the long flexors. Because of its position at the ankle, extensor digitorum longus is a dorsiflexor of the ankle joint, and helps to pull the body forward on the foot during walking. To test the muscle, the patient tries to extend the toes against resistance; the tendons may be seen and felt.

PERONEUS TERTIUS

This small muscle [FIG. 5.142] is a partially separated portion of extensor digitorum longus, and arises in common with it from the lower part of the anterior surface of the fibula and from the adjoining fascia. Its tendon runs with those of extensor digitorum longus but is inserted into the dorsum of the fifth metatarsal near its base, or into the deep fascia nearby [FIG. 5.159]. The muscle is often absent.

Nerve supply

By a branch (L. 5; S. 1) from the deep peroneal nerve.

Actions

It assists in dorsiflexion of the ankle joint and in everting the foot. It may be that peroneus tertius is a muscle which evolution is rendering more important, for eversion (turning the sole so that it faces laterally) is a characteristically human movement.

EXTENSOR HALLUCIS LONGUS

Extensor hallucis longus lies deep to, and between, tibialis anterior and extensor digitorum longus, and arises from the anterior surface of the fibula in its middle half [FIG. 3.182] and from the corresponding portion of the interosseous membrane [FIG. 5.144]. The tendon arises on the medial side of the muscle belly, and appears

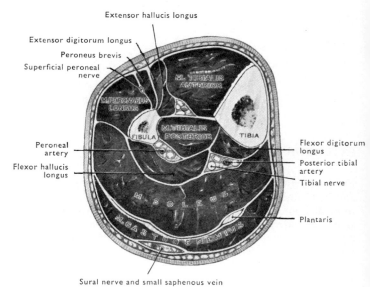

FIG. 5.144. Transverse section through middle of leg.

superficially in the lower part of the leg [FIG. 5.141]. It occupies the same compartment under the superior extensor retinaculum as extensor digitorum longus and peroneus tertius, but under the inferior extensor retinaculum it has a special compartment lined with a separate synovial sheath [FIG. 5.158]. The tendon does not form a fully developed extensor hood on the great toe, but runs to be inserted into the dorsal surface of the base of the distal phalanx, occasionally sending slips to the proximal phalanx and to the metatarsal bone of the great toe.

Nerve supply

By a branch (L. 5; S. 1) from the deep peroneal nerve.

Actions

It is a strong dorsiflexor of the ankle joint, and it extends all the joints of the big toe. Acting from its insertion, it is used to pull the body weight forwards on the supporting foot during walking.

The great toe has neither a lumbrical nor an interosseous muscle, and extension of the interphalangeal joint thus depends entirely on extensor hallucis longus. If extensor hallucis longus is paralysed, the contraction of flexor hallucis longus in the last phase of the propulsive movement in walking will therefore flex the joint and buckle up the toe. In testing the muscle the patient tries to extend the great toe against resistance while the examiner observes the tendon.

Muscles of the peroneal compartment

PERONEUS LONGUS

Peroneus longus [FIGS. 5.142, 5.144, 5.145, and 5.156], arises from the lateral condyle of the tibia, from the head and the upper two-thirds of the lateral surface of the body of the fibula, and from the fascia which surrounds the muscle [p. 406]. The tendon lies superficial to that of peroneus brevis, and is enclosed with it in a common synovial sheath which runs posterior to the lateral malleolus, deep to the superior peroneal retinaculum. A separate sheath prolonged from this envelops the peroneus longus tendon as it runs inferior to the peroneal trochlea on the calcaneus and enters the groove on the inferior surface of the cuboid bone [FIG. 5.159]. This groove is converted into a tunnel by a floor consisting of fibres derived from the long plantar ligament and the tibialis posterior tendon. The tunnel conveys the peroneus longus tendon obliquely across the sole of the foot to be inserted into the lateral side and plantar surface of the medial cuneiform bone and of the base of the first metatarsal [FIG. 5.154]. Here it is virtually continuous with the insertion of tibialis anterior [p. 388]. As it enters the sole of the foot the tendon develops a sesamoid fibrocartilage (occasionally a bone) where it plays over the smooth surface of the cuboid bone.

Nerve supply

By a branch (L. 5; S. 1) from the superficial peroneal nerve.

Actions

Peroneus longus everts the foot; that is, it turns the sole to face laterally [FIGS. 5.143 and 5.146]. During walking, the muscle helps to prevent overbalancing medially on the supporting foot. The course of the tendon obliquely forwards and medially across the sole of the foot means that it helps to maintain the transverse and lateral longitudinal arches of the foot [p. 403], and its position posterior to the fibula allows peroneus longus to assist in plantar flexing the ankle.

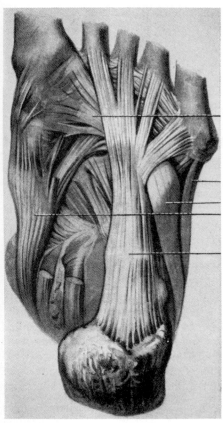

FIG. 5.145. Insertions of tibialis posterior and peroneus longus in sole of left foot.

Peroneus longus

Peroneus brevis

Peroneus longus

Tibialis posterior

Long plantar ligament

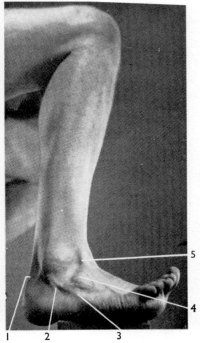

FIG. 5.146. Eversion with dorsiflexion of foot. The tendons of the peronei and extensor digitorum longus stand out actively. (From *Living anatomy* by R. D. Lockhart, 6th edn. Faber and Faber, London (1963).)
1. Tendo calcaneus.
2. Peroneus longus.
3. Peroneus brevis.
4. Extensor digitorum brevis.
5. Extensor digitorum longus.

PERONEUS BREVIS

Enclosed in the same osteofascial compartment as peroneus longus [FIG. 5.144], peroneus brevis arises from the lower two-thirds of the lateral surface of the fibula and from the intermuscular septa at its sides: its upper fibres lie anterior to the lower fibres of peroneus longus [FIG. 5.142]. In a synovial sheath shared with the tendon of peroneus longus, its tendon grooves the back of the lateral malleolus [FIG. 5.159]. Inferior to the superior peroneal retinaculum, it passes above the peroneal trochlea to be inserted into the dorsal surface of

the base of the fifth metatarsal bone [FIG. 5.159]. A slip from the muscle often joins the long extensor tendon of the little toe and is then known as the **peroneus digit minimi**. Other slips may run to the calcaneus, the cuboid, or the peroneus longus tendon.

Nerve supply

By a branch (L. 5; S. 1) from the superficial peroneal nerve.

Actions

It plantar flexes the ankle joint, everts the foot, and helps to prevent medial overbalancing of the body on the supporting foot during walking. Except that it has no action on the arches of the foot, its actions are identical with those of peroneus longus.

The peroneal muscles are tested by asking the patient to evert the foot against resistance: the tendons can be felt working behind the lower end of the fibula.

Muscles of the posterior compartment

GASTROCNEMIUS

Gastrocnemius [FIGS. 5.142, 5.144, and 5.147], with soleus and plantaris, makes up a composite muscle, sometimes called **triceps surae**, which forms the prominence of the calf, and in the upper part of the leg conceals the deeper muscles.

The two bellies of origin of gastrocnemius, together with plantaris, form the inferior boundaries of a diamond-shaped region at the back of the knee known as the **popliteal fossa**. Superiorly the fossa is bounded by the tendon of biceps laterally and by semimembra-

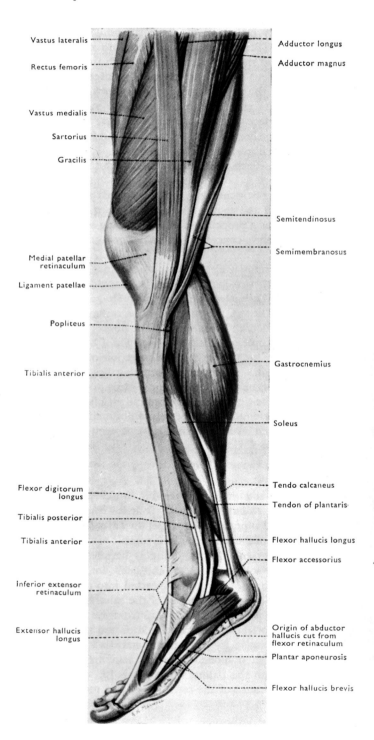

FIG. 5.147. Muscles of knee, leg, and foot seen from medial side.

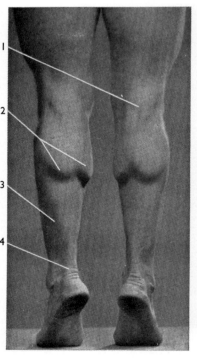

FIG. 5.148. Activity of the calf muscles in rising on the toes. (*By courtesy of Professor R. D. Lockhart.*)
1. Prominence in popliteal region during extension.
2. Medial and lateral bellies of gastrocnemius.
3. Soleus.
4. Tendo calcaneus.

nosus and semitendinosus on the medial side [FIG. 5.128]. The fossa only exists as such when the knee is flexed; when it is extended, the fleshy semimembranosus pushes laterally and bulges posteriorly to help to form a prominence at the back of the knee [FIG. 5.148].

The lateral head of gastrocnemius comes from the upper and posterior part of the lateral surface of the lateral condyle of the femur; the medial head comes from the popliteal surface of the femur above its medial condyle. Either head may contain a sesamoid bone (sometimes known as the **fabella**), and each head has an additional origin from the capsule of the knee joint; inferior to this it is separated from the capsule by a bursa. The bursa related to the medial head frequently communicates with the cavity of the knee joint, but the bursa under the lateral head seldom does so.

The two fleshy bellies arising from the tendons of origin remain separate [FIG. 5.148], and insert into the posterior surface of a broad membranous tendon which fuses with the tendon of soleus to make up the **tendo calcaneus** [FIGS. 5.147 and 5.149]. This, the strongest tendon in the body, runs downwards for about 15 cm to be inserted into the middle part of the posterior surface of the calcaneus. A bursa lies between the tendon and the bone above the insertion, and inferior to the insertion lies the fibro-fatty pad of the heel.

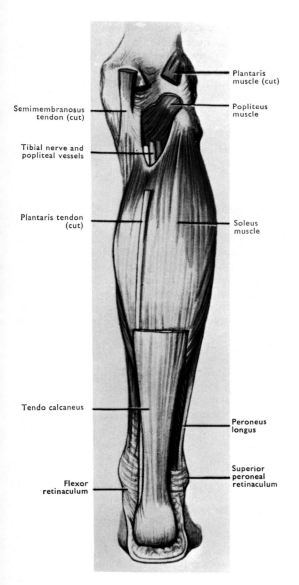

Semimembranosus tendon (cut)

Tibial nerve and popliteal vessels

Plantaris tendon (cut)

Tendo calcaneus

Flexor retinaculum

Plantaris muscle (cut)

Popliteus muscle

Soleus muscle

Peroneus longus

Superior peroneal retinaculum

FIG. 5.149. Right soleus muscles.

Nerve supply

Each head receives a branch (S. 1, 2) from the tibial nerve.

Actions

Gastrocnemius and soleus are the chief plantar flexors of the foot, and probably gastrocnemius plays more part as a propelling force in walking and running; soleus is more concerned in maintaining the upright posture [p. 394].

Gastrocnemius is also a flexor of the knee joint [FIG. 5.139] but, being a 'two-joint' muscle, is incapable of exerting its full range on both joints simultaneously [p. 272]. Thus, if the knee is fully flexed, gastrocnemius cannot completely plantar flex the foot, and vice versa. The fibres of gastrocnemius may become considerably shortened if high heels are constantly worn, for this brings the insertion of the muscle closer to its origin. Once such shortening has taken place, difficulties may occur when an attempt is made to walk without a high heel, for the tightness of the muscle may hamper dorsiflexion at the ankle joint. Conversely, full dorsiflexion at the ankle may stretch even a normal gastrocnemius enough to pull the knee into flexion.

PLANTARIS

This small, variable muscle has a fleshy belly only about 10 cm in length. It arises from the lowest part of the lateral supracondylar ridge, the adjacent part of the popliteal surface of the femur, and the capsule of the knee joint. The long slender tendon runs obliquely between gastrocnemius and soleus [FIG. 5.149] and later along the medial side of the tendo calcaneus to be inserted into it or into the medial side of the posterior surface of the calcaneus.

Nerve supply

By a branch (S. 1, 2) from the tibial nerve.

Actions

It is a feeble flexor of the knee and plantar flexor of the ankle joint. In digitigrade animals plantaris passes into the plantar aponeurosis in the same way as palmaris longus passes into the palmar aponeurosis [p. 327], but the functional significance of the retention of this muscle in man is unknown.

SOLEUS

This muscle receives its name from its shape, which resembles that of the fish [FIGS. 5.147 and 5.149]. It takes origin from the posterior surfaces of the head and the upper third of the body of the fibula, from the soleal line of the tibia and the middle third of its medial border, and from a tendinous arch between the tibia and fibula [FIG. 5.149]. The fibres run downwards to join the anterior surface of a membranous tendon which faces posteriorly, and glides on a similar tendon on the anterior surface of gastrocnemius. This arrangement facilitates independent movement of the two major components of triceps surae. Inferiorly, the two tendons of soleus and gastrocnemius fuse to form the tendo calcaneus, which is inserted into the middle part of the posterior surface of the calcaneus.

Nerve supply

By two branches (S. 1, 2) from the tibial nerve. The first arises in the popliteal fossa and runs to its superficial surface; the other arises in the back of the leg and enters its deep surface.

Actions

Soleus is a plantar flexor of the ankle joint. Because the centre of gravity of the body is so placed [p. 402] as to cause the body to tend to fall forwards at the ankle during standing, soleus is frequently in contraction to prevent this happening. It is probably mainly devoted to this postural duty, in contrast to gastrocnemius, which is a propulsive muscle during walking and running. Soleus stands out very clearly in the living body when the weight is supported on tiptoe, and gastrocnemius is also recruited to help.

Triceps surae is tested with the patient lying prone: he attempts to plantar flex the foot against resistance, and the gastrocnemius belly and tendon can be seen and felt.

POPLITEUS

Popliteus takes origin by tendon from a pit on the lateral surface of the lateral condyle of the femur just below the attachment of the

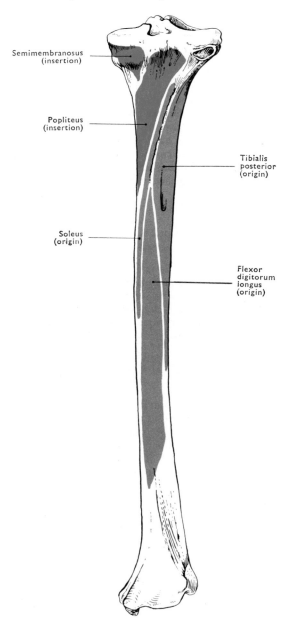

Semimembranosus (insertion)

Popliteus (insertion)

Tibialis posterior (origin)

Soleus (origin)

Flexor digitorum longus (origin)

FIG. 5.150. Muscle attachments to posterior surface of right tibia.

fibular collateral ligament of the knee. Other fibres are attached to the back of the lateral meniscus of the knee joint and to the fibula. The tendon, which lies inside the capsule of the joint at its origin, receives an investment from the synovial membrane, which runs along it for some distance. It passes downwards and medially, and emerges from under the arcuate popliteal ligament of the knee, from which the muscle takes an additional fleshy origin [FIG. 4.50]. The insertion is by fleshy fibres into a triangular surface on the back of the tibia superior to the soleal line [FIGS. 5.150 and 5.151] and into the strong fascia covering the muscle. In full flexion of the knee, popliteus tendon lies in a groove on the lateral femoral condyle [FIG. 3.161].

Nerve supply

By a branch (L. 5) from the tibial nerve winding round the inferolateral border of the muscle to enter its anterior surface.

Actions

It is a lateral rotator of the femur on the tibia when the foot is fixed on the ground, or a medial rotator of the tibia on the femur in the free limb. When the knee is fully extended and bearing weight the ligaments of the joint are 'wound up' tight by a movement of medial rotation of the femur on the tibia. In the first stage of flexion of the joint this rotation has to be undone, and popliteus has been said to have a special responsibility for this task. Its attachment to the lateral meniscus of the knee draws this meniscus out of the way during lateral rotation of the femur, so protecting it against being nipped between the moving bones; this is said to be a factor responsible for its relative freedom from injury.

Popliteus can assist in flexion of the knee joint, and is active during crouching (Barnett and Richardson 1953), though the explanation of this activity is not yet certain.

FLEXOR DIGITORUM LONGUS

Flexor digitorum longus lies in the intermediate layer of muscles on the back of the leg, concealed by soleus, and only emerging from under cover of it near the ankle. It has a fleshy origin from the medial part of the posterior surface of the body of the tibia inferior to the soleal line, and from the fascia surrounding it [FIGS. 5.144, 5.150, and 5.151]. The tendon crosses the lower end of the tibia lateral to that of tibialis posterior, and then passes deep to the flexor retinaculum in an individual synovial sheath [FIG. 5.160]. It runs on the medial margin of the sustentaculum tali [FIG. 5.152] and enters the sole superior to abductor hallucis [FIG. 5.147].

In the sole it travels forwards superior to flexor digitorum brevis, and crosses inferior to the tendon of flexor hallucis longus [FIG. 5.155] which separates it from the plantar calcaneonavicular ligament. As the tendons cross each other, a slip from flexor hallucis longus passes into the medial two of the four tendons into which flexor digitorum divides. In the middle of the sole these tendons receive the insertion of flexor accessorius [p. 398] which converts their oblique pull towards the medial side of the ankle into a straight one towards the heel. A little further forward the lumbricals [p. 398] have their origin.

The insertion of the tendons of flexor digitorum longus into the lateral four toes exactly parallels the insertion of flexor digitorum profundus in the hand [p. 329]. Each tendon enters the fibrous flexor sheath of the toe, perforates the tendon of flexor digitorum brevis, and is inserted into the inferior surface of the base of the distal phalanx. As in the hand, vincula longa and brevia pass to it from the sheath, conveying blood vessels to the tendon.

Nerve supply

By a branch (S. 1, 2) from the tibial nerve.

Actions

Its primary action is to flex all the joints of the lateral four toes. This is an important action during walking, for it grips the supporting foot firmly to the ground [p. 403]. The muscle also has a plantar flexing effect on the ankle joint, and helps to maintain the longitudinal arch of the foot [p. 403]. To test the muscle, the patient tries to flex the terminal joints of the lateral four toes against resistance: the muscle belly can be felt.

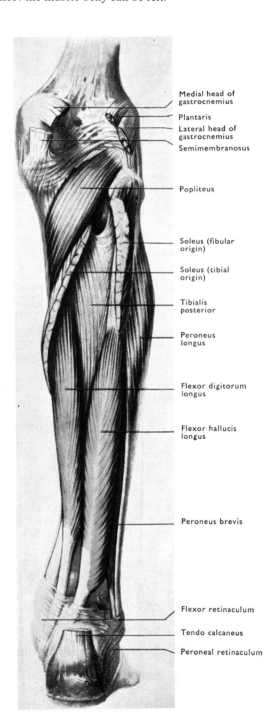

FIG. 5.151. Deep muscles on back of right leg.

FLEXOR HALLUCIS LONGUS

This powerful bipennate muscle [FIGS. 5.151 and 5.155] arises from the lower two-thirds of the posterior surface of the body of the fibula and from the adjacent fascia [FIG. 5.144]. Its tendon passes deep to the flexor retinaculum enclosed in an individual synovial sheath [p. 401], and grooves (1) the posterior surface of the lower end of the tibia, (2) the talus, and (3) the plantar surface of the sustentaculum tali [FIG. 5.152], to which it is strapped by a fibrous sheath lined by a synovial sheath. It runs forwards in the sole of the foot inferior to the plantar calcaneonavicular ligament and crosses the superior surface of the tendon of flexor digitorum longus, sending a slip to its medial two tendons. Eventually it runs into the fibrous digital sheath of the great toe [FIGS. 5.153 and 5.160] and is inserted into the plantar surface of the base of its distal phalanx.

Nerve supply

By a branch (S. 1, 2) from the tibial nerve.

Actions

The muscle flexes all the joints of the great toe, and is of great importance in the final propulsive thrust of the foot during walking [p. 404]. It is also a factor in maintaining the medial longitudinal arch of the foot [p. 402], and helps to plantar flex the ankle joint.

The test for integrity of flexor hallucis longus is similar to that for integrity of the flexor digitorum longus.

TIBIALIS POSTERIOR

Tibialis posterior, the deepest muscle on the back of the leg, takes origin from the upper and lateral part of the posterior surface of the tibia lateral to the soleal line and the origin of flexor digitorum longus, and from the posterior surface of the shaft of the fibula between the medial crest and the interosseous border [FIGS. 5.144, 5.150, and 5.151]. It also arises from the interosseous membrane, and from the fascial septum which covers it posteriorly [FIG. 5.144].

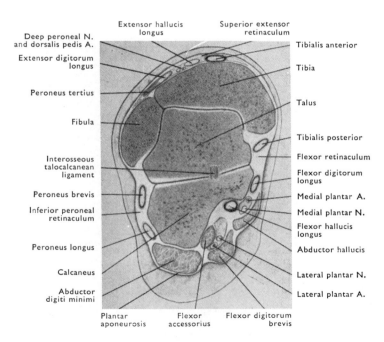

FIG. 5.152. Frontal section through left ankle joint, talus, and calcaneus.

The tendon of the muscle crosses anterior to flexor digitorum longus, grooves the back of the medial malleolus enclosed in a synovial sheath [FIGS. 5.152 and 5.160], and turns forwards between the flexor retinaculum and the deltoid ligament of the ankle joint. As it lies inferior to the plantar calcaneonavicular ligament, it usually contains a sesamoid fibrocartilage or bone, and it then spreads out to be inserted chiefly into the tuberosity of the navicular and the plantar surface of the medial cuneiform bone. A tendinous band passes backwards to the tip of the sustentaculum tali, and others pass anterolaterally to all the tarsal bones (except the talus) and to the bases of the middle three metatarsals [FIG. 5.154].

Nerve supply

By a branch (L. 4, 5) from the tibial nerve.

Actions

It is a plantar flexor of the ankle joint, and because of its attachments to the lateral tarsal bones, is also an invertor of the foot; that is, it turns the sole of the foot to face medially. Its many extensions help to maintain the arches of the foot [p. 403]. Acting from its insertion, tibialis posterior acts with tibialis anterior in preventing lateral overbalancing on the supporting foot. The peroneal muscles and the two tibial muscles are of considerable use when walking on rough or irregular ground, since they allow the foot to maintain a wide contact with the ground while the body remains upright.

To investigate tibialis posterior the patient, lying supine, tries to invert the plantar flexed foot against resistance. The tendon can then be seen and felt behind the medial malleolus.

Movements of the ankle joint

The muscles operating on the ankle joint, which permits only dorsiflexion and plantar flexion, are shown in the table. The plantar flexors are much more powerful than the dorsiflexors, partly because in the erect posture the body tends to fall forwards at the ankle, and partly because they have to provide the main propulsive thrust during walking. The main dorsiflexor of the joint is tibialis anterior, and the main plantar flexor is the composite triceps surae. However, several other muscles which cross the joint can also assist in these movements.

Muscles concerned in movements of the ankle joint

Movement	Muscle	Principal nerve supply
Dorsiflexion	Tibialis anterior	Deep peroneal
	Extensor digitorum longus	Deep peroneal
	Extensor hallucis longus	Deep peroneal
	Peroneus tertius	Deep peroneal
Plantar flexion	Gastrocnemius	Tibial
	Soleus	Tibial
	Plantaris	Tibial
	Tibialis posterior	Tibial
	Flexor hallucis longus	Tibial
	Flexor digitorum longus	Tibial
	Peroneus longus	Superficial peroneal
	Peroneus brevis	Superficial peroneal

Movements at the intertarsal joints

The main tarsal movements occur between the talus and the other tarsal bones at the talocalcaneonavicular and subtalar joints, but a limited amount of movement occurs at the other intertarsal joints. The muscles crossing these joints, whether they be ultimately

inserted into the digits or into the tarsal bones, are therefore capable of producing intertarsal movements, the most important of which are those of inversion and eversion. The table shows the muscles chiefly concerned. The flexors and extensors of the digits play a minor part, and the chief prime movers are the peronei and tibialis anterior and posterior.

The movements of inversion and eversion allow the foot to adapt itself to uneven surfaces during walking [p. 404], and the reversed action of the invertors and evertors is of great importance in preventing overbalancing during standing or walking [p. 403].

Muscles concerned in movements of the intertarsal joints

Movement	Muscle	Principal nerve supply
Inversion	Tibialis anterior	Deep peroneal
	Tibialis posterior	Tibial
Eversion	Peroneus longus	Superficial peroneal
	Peroneus brevis	Superficial peroneal
	Peroneus tertius	Deep peroneal
Other movements	Sliding movements summating to allow some dorsiflexion, plantar flexion, abduction and adduction, are produced by the muscles acting on the toes. Tibialis anterior, tibialis posterior, and peroneus longus are also concerned.	

MUSCLES OF THE FOOT

Like the hand, the foot contains a large number of important 'intrinsic' muscles on its plantar surface, but, unlike the hand, it also contains one such muscle on its dorsal surface, the extensor digitorum brevis. Whereas the chief importance of the intrinsic muscles in the hand lies in their control of the movements of individual digits, the intrinsic muscles of the sole are important mainly because of their massed action in the movements of walking; fine control of individual toes is of much less value than fine control of individual fingers. Indeed individuals usually have little ability in this direction, and testing the small muscles of the foot satisfactorily is virtually impossible. In contrast to the corresponding movements in the hand, abduction and adduction of the toes are considered to be carried out relative to an imaginary line drawn through the second (not the third) toe.

The muscles of the sole of the foot are arranged in four layers. The first, and most inferior, contains abductor hallucis, flexor digitorum brevis, and abductor digiti minimi. Above this, in the second layer, lie the lumbricals and flexor accessorius, together with the tendons of flexor hallucis longus and flexor digitorum longus. Superior to this again, in the third layer, lie flexor hallucis brevis, adductor hallucis, and flexor digiti minimi brevis. Finally, in the uppermost layer, lie the interossei and the tendons of tibialis posterior and peroneus longus.

ABDUCTOR HALLUCIS

The abductor of the great toe is a member of the most inferior layer of muscles in the sole, and arises from the medial side of the tuber of the calcaneus, the flexor retinaculum, the plantar aponeurosis, and the intermuscular septum between it and flexor digitorum brevis [FIG. 5.153]. The muscle lies superficially along the medial border of the sole, and its tendon inserts into the medial side of the base of the proximal phalanx of the great toe, along with the medial head of flexor hallucis brevis.

Nerve supply

By a branch (S. 1, 2) from the medial plantar nerve.

Actions

It abducts the great toe at the metatarsophalangeal joint and helps to flex it at this joint. Electromyography indicates that it plays no part in maintaining the medial side of the longitudinal arch of the foot, although it is anatomically extremely well placed to do so. However, it comes into action in the final phase of the propulsive thrust in walking [p. 404].

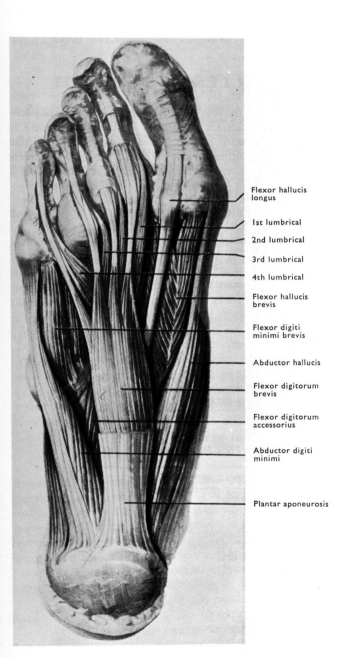

FIG. 5.153. Superficial muscles of sole of foot.

FLEXOR DIGITORUM BREVIS

The short flexor of the toes corresponds to flexor digitorum superficialis in the upper limb. It lies between abductor hallucis and abductor digiti minimi, superior to the tough central portion of the plantar aponeurosis, and inferior to flexor accessorius and the tendons of flexor digitorum longus [FIG. 5.153]. It arises from the medial process of the tuber of the calcaneus, the plantar aponeurosis, and the muscular septa on either side of it. The fleshy belly runs directly forwards in the middle of the sole, and separates into four tendons which are inserted into the middle phalanges of the lateral four toes after being perforated by the long flexor tendons, in exactly the same way as flexor digitorum superficialis inserts in the hand [p. 328].

Nerve supply

By a branch (S. 2, 3) from the medial plantar nerve.

Actions

It flexes all the joints of each of the lateral four toes except the distal interphalangeal joint. This action supplements that of flexor digitorum longus in walking [p. 403].

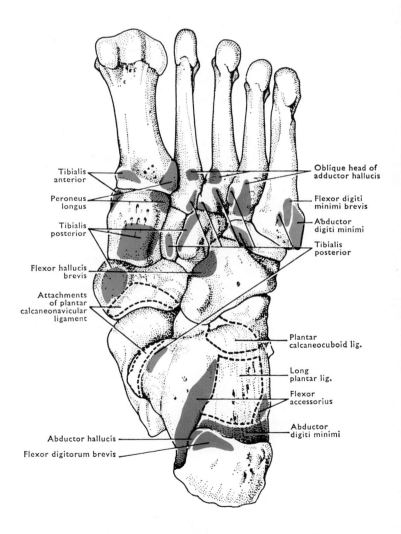

FIG. 5.154. Muscle attachments to left tarsus and metatarsus (plantar aspect).

ABDUCTOR DIGITI MINIMI

The abductor of the little toe is the most lateral member of the three muscles in the inferior layer of the sole. It arises by fleshy and tendinous fibres [FIG. 5.153] from both processes of the tuber of the calcaneus, from the fascia covering it, from the lateral part of the plantar aponeurosis, and from the intermuscular septum between it and flexor digitorum brevis. The tendon runs along the fifth metatarsal bone to be inserted with flexor digiti minimi brevis into the lateral side of the base of the proximal phalanx of the little toe.

Nerve supply

By a branch (S. 2, 3) from the lateral plantar nerve.

Actions

It abducts the little toe at the metatarsophalangeal joint, and helps to flex it at this joint.

FLEXOR ACCESSORIUS

This muscle, which has no counterpart in the hand, converts the oblique pull of flexor digitorum longus into a straight one. It lies immediately superior to flexor digitorum brevis, and arises by two heads, one on each side of the long plantar ligament, and attached both to it and to the medial and lateral surfaces of the calcaneus [FIG. 5.154]. The two heads converge to form a flattened band inserted into the tendons of flexor digitorum longus in the middle of the sole [FIG. 5.155].

Nerve supply

By a branch (S. 2, 3) from the lateral plantar nerve.

Actions

It helps the long flexor to flex all the joints of the lateral four toes, and alters its direction of pull so that the toes are flexed towards the heel and not towards the medial side of the ankle.

LUMBRICALS

Like flexor digitorum profundus in the hand, flexor digitorum longus has four lumbrical muscles arising from its component tendons [FIG. 5.155]. The first of these arises from the medial side of the tendon for the second toe, but all the others spring by two heads from the adjacent sides of two tendons. Each then passes forwards on to the tibial side of the corresponding toe, inferior to the deep transverse metatarsal ligament, and winds obliquely upwards and forwards to be inserted by tendinous fibres into the medial side of the corresponding extensor hood.

Nerve supply

The lateral three lumbricals are supplied by branches (S. 2, 3) from the lateral plantar nerve, and the first lumbrical is supplied by the medial plantar nerve (S. 1, 2).

Actions

The lumbricals of the foot, like those in the hand, flex the metatarsophalangeal joints and extend the interphalangeal joints of the lateral four toes. This action prevents the long flexors of the toes from 'clawing' the toes underneath themselves during the propulsive phase in walking [p. 403]. If they are paralysed, the extensors of the toes pull the metatarsophalangeal joints into hyperextension and the toes become 'clawed' at rest.

FLEXOR HALLUCIS BREVIS

Flexor hallucis brevis appears between flexor digitorum brevis and abductor hallucis on the sole of the foot [FIG. 5.153], but it belongs to a deeper layer of muscles and lies superior to the tendon of flexor hallucis longus. It arises from the medial part of the plantar surface of the cuboid bone [FIG. 5.154] posterior to the groove for peroneus longus, from the adjacent part of the lateral cuneiform bone, and from the tendon of tibialis posterior. The fleshy belly separates into two parts, one on each side of the flexor hallucis longus tendon. Each part gives rise to a tendon inserted into the corresponding side of the base of the proximal phalanx of the great toe; the medial one unites with the insertion of abductor hallucis, and the lateral one with the insertion of adductor hallucis. In each tendon a sesamoid bone is developed.

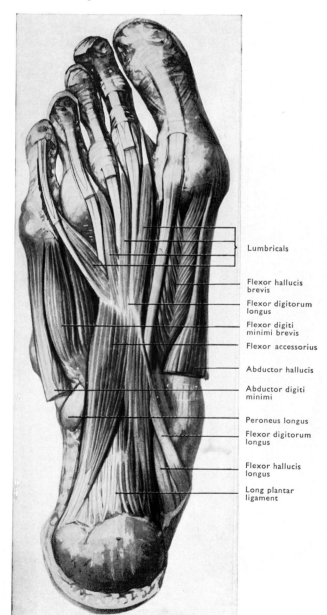

FIG. 5.155. Muscles of right foot (second layer).

Lumbricals

Flexor hallucis brevis

Flexor digitorum longus

Flexor digiti minimi brevis

Flexor accessorius

Abductor hallucis

Abductor digiti minimi

Peroneus longus

Flexor digitorum longus

Flexor hallucis longus

Long plantar ligament

Nerve supply

By a branch (S. 1, 2) from the medial plantar nerve.

Actions

It flexes the metatarsophalangeal joint of the toe. During walking the great toe, which is the most important part of the forefoot, pivots on the sesamoid bones in the tendons of the muscle [p. 403].

ADDUCTOR HALLUCIS

Like the adductor of the thumb, the adductor of the great toe has two heads. The oblique head arises from the sheath of the peroneus longus tendon, and from the plantar surfaces of the bases of the second, third, and fourth metatarsal bones [FIGS. 5.154 and 5.156]. The transverse head arises from the plantar surface of the lateral four metatarsophalangeal joints and from the deep transverse metatarsal ligament. The two heads unite with the lateral part of flexor hallucis brevis at their insertion into the lateral side of the base of the proximal phalanx of the great toe. The muscle lies inferior to the interosseous muscles, on the same plane as flexor hallucis brevis, and superior to the long flexor tendons and the lumbricals of the toes.

Nerve supply

By a branch (S. 2, 3) from the lateral plantar nerve.

Actions

It adducts the great toe towards the second toe (not a powerful movement), flexes the metatarsophalangeal joint of the great toe, and assists in maintaining the transverse arch of the foot.

FLEXOR DIGITI MINIMI BREVIS

The short flexor of the little toe [FIG. 5.156] arises from the sheath of the peroneus longus tendon and from the base of the fifth metatarsal. It runs along the fifth metatarsal bone to be inserted, in common with abductor digiti minimi, to the lateral side of the base of the proximal phalanx of the little toe.

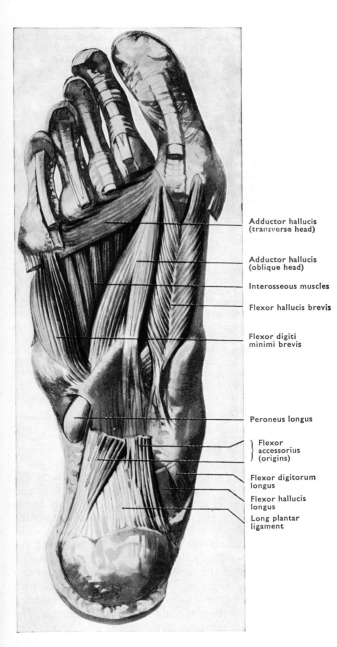

Adductor hallucis (transverse head)

Adductor hallucis (oblique head)

Interosseous muscles

Flexor hallucis brevis

Flexor digiti minimi brevis

Peroneus longus

} Flexor accessorius (origins)

Flexor digitorum longus

Flexor hallucis longus

Long plantar ligament

FIG. 5.156. Deep muscles of sole of foot.

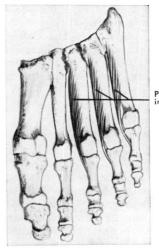

Plantar interossei

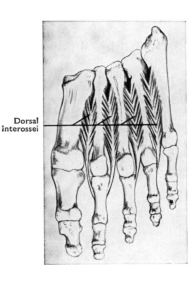

Dorsal interossei

FIG. 5.157. Interosseous muscles of right foot.

Nerve supply

By a branch (S. 2, 3) from the lateral plantar nerve.

Actions

It flexes the little toe at the metatarsophalangeal joint.

INTEROSSEI

In the most superior layer of muscles in the sole lie the interossei, which are essentially similar to those of the hand. There are four dorsal interossei but only three plantar ones. The dorsal ones diverge the toes away from an imaginary line drawn through the second (not the third) toe, and the plantar ones converge the toes on this line.

The four dorsal muscles, as in the hand, are larger than their plantar colleagues, and arise by two heads from the adjacent bones. Each runs forwards superior to the deep transverse metatarsal ligament to be inserted into the side of the proximal phalanx and into the capsule of the metatarsophalangeal joint, but not, as in the hand, into the extensor hood on the dorsum of the toe concerned (Jones 1944; Manter 1945). The first dorsal interosseus inserts into the medial side, and the second dorsal interosseus into the lateral side of the second toe. The third and fourth muscles are inserted into the lateral sides of the third and fourth toes.

The three plantar muscles lie in the lateral three interosseous spaces. Each arises by a single head from the medial sides of the third, fourth, and fifth metatarsal bones respectively. The tendons pass forwards superior to the deep transverse metatarsal ligament and are inserted in the same manner as those of the dorsal muscles into the medial sides of the proximal phalanges of the third, fourth, and fifth toes [FIG. 5.157].

Nerve supply

By branches (S. 2, 3) from the lateral plantar nerve.

Actions

Apart from the abduction and adduction on the line of the second toe, which is of small importance in the foot, the interossei have the function of flexing the metatarsophalangeal joints. They may also help to hold the metatarsals together to maintain the transverse arch of the foot. By opposing the action of the extensor muscles on the proximal phalanx, they allow the extensors to pull the leg forward on the fixed toes during walking.

EXTENSOR DIGITORUM BREVIS

This is the only muscle springing from the dorsum of the foot [FIGS. 5.141, 5.142, and 5.146]. It arises on the anterior part of the upper surface of the calcaneus and the deep surface of the fascia which covers the muscle. Four thin tendons run forwards and medially to be inserted into the medial four toes. The lateral three tendons join the hoods of extensor digitorum longus, and the most medial tendon, sometimes known as extensor hallucis brevis, is inserted separately into the base of the proximal phalanx of the great toe.

Nerve supply

By a branch (S. 1, 2) from the deep peroneal nerve.

Actions

It helps extensor hallucis longus and extensor digitorum longus to extend the metatarsophalangeal joints of the medial four toes. Like extensor digitorum longus, it is unable to extend the interphalangeal joints without the assistance of the lumbricals.

Movements of the toes

The table shows the muscles involved in movements of the toes. The situation is essentially similar to that obtaining in the hand, although in the hand the actions of the muscles are directed towards forming a grasping mechanism, and in the foot they are directed towards providing a stable support for the body in locomotion. This support has to be flexible and able to contribute towards the thrust of the foot while at the same time capable of taking the enormous strains involved in landing on the foot during running, jumping, and other athletic exercises.

Muscles concerned in movements of the toes

Metatarsophalangeal Joints

Movement	Muscle	Principal nerve supply
Flexion	Flexor hallucis longus	Tibial
	Flexor hallucis brevis	Medial plantar
	Flexor digitorum longus	Tibial
	Flexor accessorius	Lateral plantar
	Flexor digitorum brevis	Medial plantar
	Flexor digiti minimi brevis	Lateral plantar
	Lumbricals	Lateral and medial plantar
	Interossei	Lateral plantar
Extension	Extensor hallucis longus	Deep peroneal
	Extensor digitorum longus	Deep peroneal
	Extensor digitorum brevis	Deep peroneal
Abduction and adduction	Abductor hallucis	Medial plantar
	Adductor hallucis	Lateral plantar
	Interossei	Lateral plantar
	Abductor digiti minimi	Lateral plantar

Interphalangeal Joints

Movement	Muscle	Principal nerve supply
Flexion	Flexor hallucis longus	Tibial
	Flexor digitorum longus	Tibial
	Flexor accessorius	Lateral plantar
	Flexor digitorum brevis (proximal joint only)	Medial plantar
Extension	Extensor hallucis longus	Deep peroneal
	Extensor digitorum longus	Deep peroneal
	Extensor digitorum brevis (not in great toe)	Deep peroneal
	Lumbricals	Lateral and medial plantar

Synovial sheaths of the ankle and foot

The synovial sheaths of the tendons entering the foot [FIGS. 5.158-5.160] are not so important as those in the region of the wrist and hand. They are not so liable to injury or infection, and they do not communicate with the synovial sheaths in the fibrous sheaths of the flexor tendons in the toes.

In front of the ankle and on the dorsum of the foot there are three separate sheaths. One is for the tendon of tibialis anterior, beginning near the upper border of the superior extensor retinaculum and reaching almost to the insertion of the tendon. The second is for extensor hallucis longus; it begins between the two extensor retinacula and extends to the proximal phalanx of the great toe. The third is a sheath common to extensor digitorum longus and peroneus tertius. This begins about the lower border of the superior extensor retinaculum and ends about the middle of the dorsum of the foot.

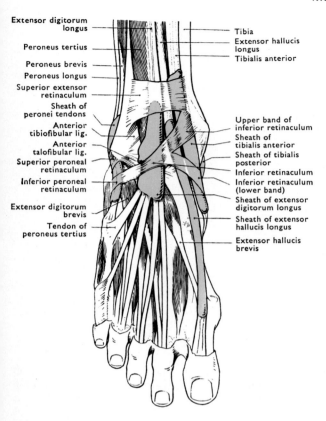

Extensor digitorum longus
Peroneus tertius
Peroneus brevis
Peroneus longus
Superior extensor retinaculum
Sheath of peronei tendons
Anterior tibiofibular lig.
Anterior talofibular lig.
Superior peroneal retinaculum
Inferior peroneal retinaculum
Extensor digitorum brevis
Tendon of peroneus tertius

Tibia
Extensor hallucis longus
Tibialis anterior
Upper band of inferior retinaculum
Sheath of tibialis anterior
Sheath of tibialis posterior
Inferior retinaculum
Inferior retinaculum (lower band)
Sheath of extensor digitorum longus
Sheath of extensor hallucis longus
Extensor hallucis brevis

FIG. 5.158. Synovial sheaths of dorsum of foot.

On the lateral side of the ankle and foot, peroneus longus and brevis share a common sheath which begins at the back of the lateral malleolus, and bifurcates as it approaches the peroneal trochlea, so that each tendon receives a separate prolongation from it. Both these continue round their respective tendons to their insertions.

On the medial side of the ankle and foot there are three synovial sheaths. That of tibialis posterior begins about 5 cm above the tip of the medial malleolus and runs to its insertion into the navicular

bone. The sheath of flexor digitorum longus begins at about the same level, and extends rather further forwards, reaching the attachment of flexor accessorius. The third is the sheath of flexor hallucis longus, which begins rather lower down than the other two, and extends to the point at which the slip is given to flexor digitorum longus [FIG. 5.160].

Within the fibrous flexor sheaths of the toes, the long flexor tendons, as in the fingers, are surrounded by short synovial sheaths extending proximally as far as the heads of the metatarsals.

Mechanism of standing

In the anatomical position the centre of gravity of the skull lies in front of the atlanto-occipital joint, so that the skull tends to fall forwards, and has to be restrained from doing so by postural contraction of the muscles at the back of the neck. Similarly the lumbar curvature of the vertebral column is maintained by postural contraction in the various portions of erector spinae [p. 278]; detectable activity is very variable, but is most marked in the lower thoracic region (Joseph and McColl 1961). Equilibrium in the erect posture is not very stable, and there are constant slight swaying movements of the trunk. These movements are immediately checked by contraction of the antagonistic muscles of the vertebral column [p. 286].

The centre of gravity of the whole body lies near the body of the second sacral vertebra. A perpendicular dropped from this point to the ground falls behind the axis of movement of the hip joint, and the weight of the body thus tends to cause the pelvis to roll backwards at the hip. This tendency is resisted by the disposition of the ligaments, which, especially the iliofemoral ligament [p. 245], are taut during standing and take most of the strain. Iliopsoas, the chief flexor of the hip joint, is apparently not called upon when standing at ease (Joseph and Williams 1957; Keagy et al. 1966).

The perpendicular line of gravity passes through the front of the knee joint, and this factor tends to produce full extension of the joint, but in the natural standing position the joint is not fully extended [p. 253], and, because of the apparent instability of the joint surfaces and the fact that the ligaments are not 'wound up' tight unless the joint is fully extended, one would expect the muscles of the thigh to be in continuous postural contraction to prevent overbalancing at this joint. However, no constant activity can be

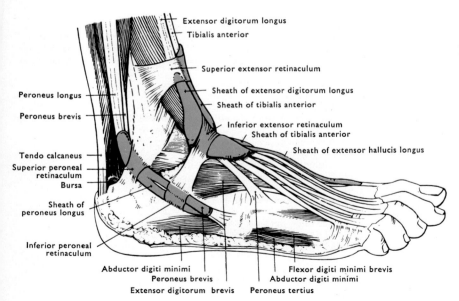

Extensor digitorum longus
Tibialis anterior
Superior extensor retinaculum
Sheath of extensor digitorum longus
Sheath of tibialis anterior
Inferior extensor retinaculum
Sheath of tibialis anterior
Sheath of extensor hallucis longus

Peroneus longus
Peroneus brevis
Tendo calcaneus
Superior peroneal retinaculum
Bursa
Sheath of peroneus longus
Inferior peroneal retinaculum

Abductor digiti minimi
Peroneus brevis
Extensor digitorum brevis
Flexor digiti minimi brevis
Abductor digiti minimi
Peroneus tertius

FIG. 5.159. Dissection showing synovial sheaths of tendons of foot.

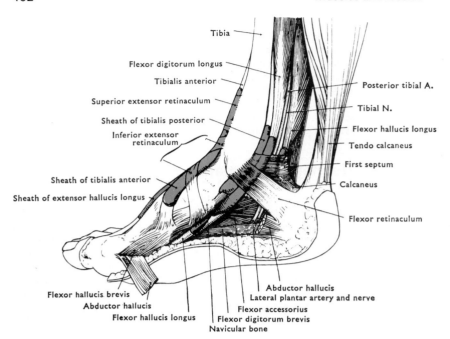

Tibia

Flexor digitorum longus

Tibialis anterior

Superior extensor retinaculum

Sheath of tibialis posterior

Inferior extensor
retinaculum

Sheath of tibialis anterior

Sheath of extensor hallucis longus

Posterior tibial A.

Tibial N.

Flexor hallucis longus

Tendo calcaneus

First septum

Calcaneus

Flexor retinaculum

Flexor hallucis brevis
Abductor hallucis
Flexor hallucis longus
Navicular bone

Flexor digitorum brevis
Flexor accessorius
Lateral plantar artery and nerve
Abductor hallucis

FIG. 5.160. Dissection of leg and foot showing synovial sheaths.

detected in the muscles of the thigh during standing (Joseph and Nightingale 1954).

At the ankle joint, the line of gravity passes well forward of the axis of movement, and there is thus a tendency to fall forwards. This tendency is resisted by muscular contraction of triceps surae, in particular of soleus, which comes into play as a postural muscle more often than any other during standing (Joseph and Nightingale 1952). Wearing high-heeled shoes, which tip the body still further forwards, increases the activity in soleus (Joseph and Nightingale 1956) and may produce a measurable increase in the circumference of the calf.

In the standing position the weight of the body is shared approximately equally by the heel pad and the pads underlying the heads of the metatarsals. About one third of the total weight taken

by the metatarsal pads falls on the pad under the ball of the great toe, and the remainder is shared between the other toes (Morton 1935). Between the posterior and the anterior weight-bearing areas lies the longituinal arch of the foot. The lateral side of the sole is usually in contact with the ground in the normal standing position [FIG. 5.161], but this contact may be incomplete or absent, and the region does not normally bear weight. Nevertheless, this lateral side of the longitudinal arch is much less elevated than the medial side, under which the vessels and nerves pass into the sole. The longitudinal arch is maintained particularly by the dense intermediate part of the plantar aponeurosis, by the ligaments underlying the intertarsal joints—especially by the plantar calcaneonavicular ligament—and by the long tendons going to the toes, which stretch like tie-beams along the sole of the foot. The action of flexor hallucis

CMS. INS.

FIG. 5.161. Comparative series of footprints.

1–4. Footprints of a man. (1) Walking print; (2) stationary, foot highly arched; (3) with the muscles relaxed as much as possible; (4) walking on the toes. Note the prominence of the bar between the ball and phalangeal pad of the big toes.

5. Footprint of male athlete.
6. Athletic girl.
7. Ballet dancer in attention attitude.
8. The same showing the difference between the arched and relaxed state of the foot, the increased area of contact indicated by the dotted area.
9. Negro male.
10. Australian aboriginal woman.
11. Chimpanzee, female, 3½ years of age, walking print.
12. Hamadryas baboon, female.
13. Rhesus macacus monkey. (Observe the narrow heel, 11, 12, and 13.)
14. Infant before walking, at 3 months old.
15. The same child after walking at 2 years 9 months old.
16. The same child at 3 years 7 months old.
17. The same child at 7½ years old.
18. This infant, after beginning to walk, was confined to bed through illness and lost the art. The footprint is taken from a series while the art of walking was being learned anew. The eminence in front of the heel on the medial side is caused by the impression of the area under the head of the navicular bone.

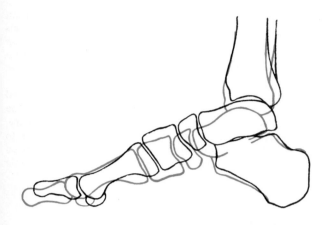

FIG. 5.162. Tracings of radiographs superimposed to show the position of the longitudinal arch of the foot during relaxation (red) and contraction (black) of the muscles. (See also FIG. 5.161 footprints 2 and 3.)

longus and flexor digitorum longus in supporting the medial side of the longitudinal arch is reinforced by the tendon of tibialis posterior as it hooks under the medial side of the ankle [FIG. 5.160]. The short muscles of the sole appear to be of little importance as supports of the longitudinal arch (Smith 1954; Basmajian 1961) though they become very active when standing on tip-toes.

The anterior part of the foot shows a transverse arch, which is particularly well marked between the shafts of the metatarsals. This arch is maintained by the ligaments and by the activity of tibialis anterior, adductor hallucis, the interossei, and the tendons of tibialis posterior and peroneus longus, which sling across the sole transversely (Hicks 1961). Activity in these muscles is minimal with the foot off the ground, but as soon as any movement takes place or pressure is applied they come into play to protect the nerves and vessels which run forwards in the interdigital clefts to supply the toes.

The standing posture consists of a cycle of practically static phases, each lasting about half a minute, alternating with brief phases of movement (Smith 1954; Thomas and Whitney 1959). During the static phases, all the muscles directly associated with the foot, except for the calf muscle, are inactive, and the support for the arches of the foot is presumably provided mainly by the ligaments.

In the phase of movement, alterations in the distribution of muscle contraction occur to prevent the tendency of the body to overbalance. The only permanently static part of the body is the skin of the sole of the foot in contact with the ground. As postural sway occurs the centre of gravity is pulled back into its original position relative to the feet by contraction of muscles which first of all balance the leg on the foot, then the thigh on the leg, and lastly the trunk on the thigh. For example, should the body begin to topple to the left side, tibialis anterior and tibialis posterior of the left lower limb, acting together, would pull the left leg medially on the foot, while the adductor group in the thigh, acting from their insertions, would pull the pelvis over again to the medial side. These movements might be assisted by contraction of the corresponding antagonist muscles in the right lower limb; peroneus longus, peroneus brevis, and peroneus tertius would pull the right leg laterally on the foot, and gluteus medius and minimus might come into play to tilt the pelvis over towards the right side again. At the same time the right erector spinae muscle would pull the centre of gravity towards the right.

Anteroposterior sway is dealt with in a similar fashion by the dorsiflexors and plantar flexors of the ankle, the extensors and flexors of the knee, and the extensors and flexors of the hip joint; the balancing action of gluteus maximus and tensor fasciae latae acting

on the posterior and anterior ends of the iliac crest is an important factor.

Standing is therefore not merely a matter of attaining a fixed position, and the erect posture necessarily involves a rotation of activity in the various muscles of the whole of the lower limb and trunk. Standing on one foot is a much more complicated achievement than standing with the weight equally distributed between the two, and standing on the toes of one foot is more difficult still; the principles, however, are exactly the same. It is worth noting that when standing on one limb the force which keeps the pelvis level on the supporting hip is not entirely the pull of the abductors of the hip; much is contributed by passive tension of the iliotibial tract of the fascia lata (Inman 1947).

Mechanism of walking and running

In walking, the centre of gravity of the body is thrust upwards and forwards by the propulsion of the lower limb muscles. As one foot advances, the upper limb of the opposite side swings forwards, thus helping to balance the trunk, and the foot of the opposite side remains on the ground until the advancing foot re-establishes contact. The trunk is thus always supported by at least one lower limb.

The rise in the centre of gravity with each step amounts to about 5 cm (Saunders, Inman, and Eberhart 1953), and is brought about by a combination of extension at the hip and knee with plantar flexion at the ankle, the heel being raised from the ground by the contraction of triceps surae. In this way an upward and forward impetus is achieved, which is now assisted by gravity, so that the body tends to fall forwards.

The hip and knee of the thrusting limb now flex, and the ankle dorsiflexes. At the same time the pelvis on that side is elevated by the contraction of the opposite hip abductors. These movements lift the thrusting foot clear of the ground and bring the limb forward, so altering the centre of gravity and increasing the tendency of the body to fall forwards. The hip and knee now extend again, so that the limb returns to the ground in the fully extended position, with the ankle partly dorsiflexed. The heel therefore comes down first, and the weight of the body is then transferred, first along the lateral side of the longitudinal arch of the foot to the lateral side of the metatarsal pad, and then across the transverse arch to the ball of the great toe, which takes most of the strain as the cycle begins again with another propulsive thrust of the extensors of hip and knee and the plantar flexors of the ankle. The great toe rocks on the two sesamoid bones in the substance of flexor hallucis brevis, which act as roller bearings. As the heel is raised, the metatarsophalangeal joint of the great toe extends, so that the last parts of the lower limb to retain contact with the ground are the ball and phalangeal pad of the great toe. This extension 'winds up' the plantar aponeurosis by pulling it forwards round the heads of the metatarsals, like a cable being wound on a windlass (Hicks 1954). This helps to support the medial side of the longitudinal arch, which takes the main strain and also transmits the pull of triceps surae to the fore part of the foot.

The long flexors of the toes have an important part to play in maintaining the frictional grip between the foot and the ground. Their pull tends to flex all the joints of the toes, but this action is resisted by the extensors of the interphalangeal joints, which hold them in a position of extension. If the extensors of these joints are paralysed, the resulting curling up of the toes causes a considerable loss in the efficiency of the propulsive thrust, as well as corns, sores, and ulcers caused by unspecialized areas of skin being called upon to bear weight and take friction. At the beginning of the thrust, the flexor muscles of the toes are fully extended, and, though exerting themselves powerfully with the toes applied to the ground, they do not in fact shorten [FIG. 4.73]. The inner toes exert the greatest pressure and the importance of the big toe in walking may be estimated from the fact that flexor hallucis longus exerts three times more force upon it than is sustained by any other toe [Fig. 5.161].

Abductor hallucis is active during the final phase of the thrust, and in some adults and in all children it is strong enough to cause abduction of the great toe at this stage. In those who wear unphysiological tight pointed shoes, and who walk 'toeing out', the prevention of this movement of abduction leads to a drift of the great toe laterally, a condition known as hallux valgus (Barnett 1962). This condition tends to be aggravated by the oblique pull of extensor hallucis brevis.

The short intrinsic muscles of the foot, considered as a whole, are most active during the final propulsive thrust (Basmajian 1961).

As the limb leaves the ground, its support for the trunk is suddenly withdrawn, so that the trunk tends to fall over away from the supporting limb. Actual falling is prevented by the contraction of the gluteus medius and minimus of the supporting limb, which abduct the trunk so that the centre of gravity is brought back over its supporting pillar. At the same time the erector spinae and the abdominal muscles on the side of the free limb contract, apparently to keep the trunk erect on the tilting pelvis. They are not entirely successful in this, so that there is a sideways lurch of about 5 cm towards the supporting limb with each step. If the gluteus medius and minimus are paralysed, this lurch becomes greatly exaggerated for the trunk is flexed towards the supporting limb to bring the centre of gravity over it, thus tilting the pelvis to raise the other limb from the ground. This gives the patient a characteristic waddling gait.

As the trunk moves forward, it not only becomes flexed and abducted at the supporting hip, but also rotates to the side of the supporting limb (this is equivalent to a medial rotation of the femur on the pelvis). This movement, which allows the free limb to pass directly forwards in the line of progression, is largely carried out by gluteus minimus. At the same time, and for the same reason, the thigh of the free limb is rotated laterally on the pelvis. Walking movements at the hips are thus sequences or combinations of abduction, flexion, extension, and rotation.

Rotation movements also occur at the knee. For example, in the final 30 degrees of extension of the knee of the free limb, there is an automatic lateral rotation of the tibia on the femur, and in the similar phase of extension when the foot is on the ground the femur rotates medially on the tibia. As the knee flexes again, these rotations are rapidly undone (Barnett 1953; Langa 1963).

In such a complicated situation it is not surprising that there should be a tendency to overbalance, and this is rectified by the lower limb muscles acting from their insertions on their origins, just as they do to prevent overbalancing during standing [p. 403]. The peronei and tibialis posterior and anterior have an additional value because of their ability to allow the greatest possible surface of the foot to remain in contact with the ground in spite of local irregularities in the surface. For example, a firm grip can be maintained and the body can be held upright even though the feet are traversing a steep incline.

In running, the movements are all exaggerated, the time of each movement is diminished, and their force and distance are increased. Both feet are off the ground for a time during each stride, and the whole foot does not strike the ground, but only the metatarsal pads and the toes. The trunk is inclined forward much more than in walking, so that the centre of gravity is well forward of the point of support. The basic principles are, however, similar to those obtaining during walking.

It is a curious and so far unexplained fact that in walking women take shorter steps in relation to the length of their legs than do men, and that this is associated with a lower expenditure of energy (Booyens and Keatinge 1957). Women increase their walking speed mainly by increasing the frequency of their stride.

Walking on high heels involves greater activity in soleus than walking on low heels. There is less powerful but more continuous activity in tibialis anterior, intermittent or continuous contraction of quadriceps femoris during the supporting phase, and contraction of gluteus medius in the free limb (Joseph 1968). These changes are less marked than might be expected.

Walking is one of the most complicated actions the general musculature of the body is called upon to perform, and the mechanisms concerned are readily disturbed by disease or injury, often producing an altered gait so characteristic as to lead to immediate diagnosis. The disturbances of normal gait are discussed by Steindler (1935, 1955), and the effects of flat foot on muscle action by Gray and Basmajian (1968).

Fasciae of the lower limb

SUPERFICIAL FASCIA

The superficial fascia of the lower limb is continuous with that of the perineum, abdominal wall, and back. The two layers of superficial fascia found in the lower part of the abdominal wall [p. 371] and in the perineum are also represented in the uppermost part of the anterior surface of the thigh, in the groin. The external fatty layer presents no special features in this region; the deep membranous layer crosses superficial to the inguinal ligament and fuses with the deep fascia of the thigh along a line which runs laterally from the region of the pubic tubercle, roughly parallel to the inguinal ligament, and about a finger's breadth inferior to it. It also has a linear attachment to the antero-inferior surface of the body of the pubis and to the margins of the inferior pubic and ischial rami. This attachment prevents the passage into the thigh of fluid collected in the perineum or deep to the superficial fascia of the abdominal wall [Fig. 5.120].

In the gluteal region the superficial fascia is very thick and fatty. In the vicinity of the ischial tuberosity, where it takes the weight of the sitting body, it has the characteristic structure found where pressure is borne, with dense strands of fibrous tissue enclosing between them loculi containing fat. The fat of the buttock contributes to its contour, and is responsible for the formation of the gluteal fold. On the outer side of the female thigh there is a deposit of secondary sexual fat [p. 275], and the whole of the thigh may be the site of considerable fat storage.

The superficial fascia of the leg and foot is not remarkable except in the sole. Here it is greatly thickened by characteristic loculated pads of fat, particularly on the weight-bearing areas of the heel, the balls and pads of the toes. These digital pads are exactly similar to those which occur in the palm and fingers [p. 346], and serve to protect the deeper structures against the pressure of the weight of the body. The heel pad underlying the calcaneus may be as much as 2 cm in thickness. In the webs of the toes, the superficial fascia contains some weak transversely running fibres which form the superficial transverse ligament of the sole.

On the dorsum of the foot, as on the dorsum of the hand, the superficial fascia is thin and composed of very loose connective tissue containing little fat.

DEEP FASCIA

The investing layer of the deep fascia of the lower limb is attached superiorly to the inguinal ligament, the crest of the ilium, the sacrotuberous ligament, the ischium, the pubic arch, and the anterior surface of the pubis and the pubic tubercle. *In the thigh*, where it is known as the fascia lata, there are considerable differences in its thickness. On the medial side of the thigh it is thin and almost transparent, but on the lateral side it is immensely tough and strong, particularly in a band extending from the tubercle of the iliac crest to the lateral condyle of the tibia. This is the iliotibial tract of the fascia lata, and into it are inserted three-quarters of gluteus maximus and the whole of tensor fasciae latae; it may be looked upon as a common aponeurosis of these muscles.

In the lower part of the thigh the investing layer sends two intermuscular septa to be attached to the femur. The lateral intermuscular septum extends from the iliotibial tract to the lateral supracondylar line and the linea aspera of the femur; it gives attachment to vastus lateralis and vastus intermedius anteriorly, and to the short head of the biceps posteriorly. The medial intermuscular septum, which runs to the medial supracondylar line and the linea aspera, is greatly strengthened by the tendon of insertion of adductor magnus, and separates the hamstrings from the adductors.

Higher up in the thigh the fascial septa become more indistinct, but there is a well-marked thickening of a layer of fascia deep to sartorius. This layer binds vastus medialis to adductor longus and adductor magnus, and forms a roof for the adductor canal [p. 383], which transmits the femoral vessels.

At the knee, the deep fascia is continuous with that of the leg, and is attached to the condyles of the tibia, the head of the fibula, and the patella. The patella is held to the tibial condyles by the medial and lateral patellar retinacula [p. 385], which are thickened bands of deep fascia, reinforced by the tendons of vastus medialis and vastus lateralis respectively. At the back of the knee, covering over the popliteal fossa [p. 392], the deep fascia is strengthened by transverse fibres.

In the upper part of the front of the thigh there is an oval gap in the fascia lata, through which the great saphenous vein and some of its tributaries pass on their way to join the femoral vein. The centre of this saphenous opening [FIG. 5.110] lies approximately three fingerbreadths inferior and lateral to the pubic tubercle. The opening itself is about 4 cm long and 1–2 cm wide, and it has a smooth sloping medial margin formed by the fascia covering the anterior surface of pectineus. In contrast, the upper, lateral, and

lower boundaries of the opening form a sharp crescentic edge. This falciform margin is joined to the medial boundary of the opening by the cribriform fascia, a thin layer of fibrous and fatty tissue which is perforated by the great saphenous vein, small arteries, and lymph vessels [FIG. 5.163].

Deep to the investing layer in the groin, an oval funnel-shaped tube, the femoral sheath, is prolonged downwards from the abdomen into the thigh around the femoral vessels [FIG. 5.164]. This sheath is formed anteriorly by a prolongation from the transversalis fascia and posteriorly from a prolongation of the fascia iliaca [p. 372]. The femoral sheath is divided into three compartments by septa which pass from the anterior to the posterior wall. In the lateral compartment is the femoral artery; in the intermediate one is the femoral vein, and in the medial one, which is called the femoral canal, there is little except some fat, a few lymph vessels, and possibly a lymph node. This canal, which is necessary to allow for the expansion of the femoral vein, is a weak spot in the wall of the abdomen, through which a femoral hernia may protrude. Laterally, the femoral sheath is closely applied to the walls of the femoral vessels, and terminates about 4 cm inferior to the inguinal ligament, deep to the saphenous opening, by becoming fused with their adventitial coats. The upper end of the femoral canal admits the tip of the little finger and is known as the femoral ring; its wall is firmly attached to the inguinal ligament, the lacunar ligament [FIG. 5.164], the pectineal ligament, and the fascia in front of pectineus. Inferiorly, the canal, which is about 2 cm long, narrows down to nothing because its medial wall slopes laterally to become fused with the wall of the femoral vein.

A hernia passing down the femoral canal will push in front of it a covering of peritoneum, and the contents of the femoral canal. It will then distend the wall of the canal and bulge outwards and forwards, pushing in front of it the relatively weak cribriform fascia. Finally it turns upwards over the inguinal ligament on to the anterior abdominal wall. As the hernia passes down through the narrow femoral ring it is in a position of considerable danger, since the sharp crescentic edge of the lacunar ligament may interfere with its blood supply.

Lateral to the femoral sheath, a thickening of the fascia passes from the inguinal ligament to the superior ramus of the pubis. This thickening, sometimes known as the iliopectineal ligament, divides

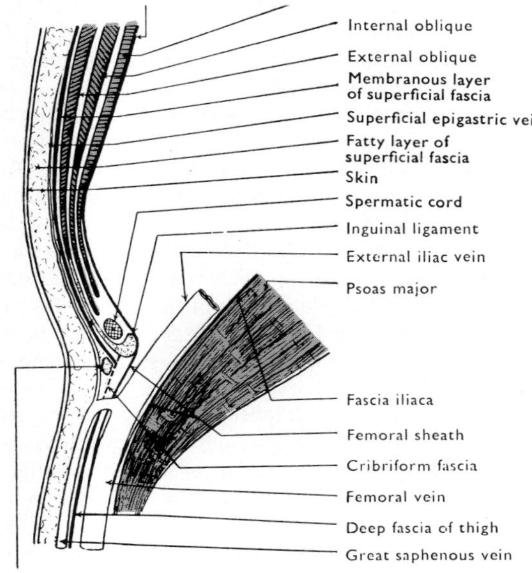

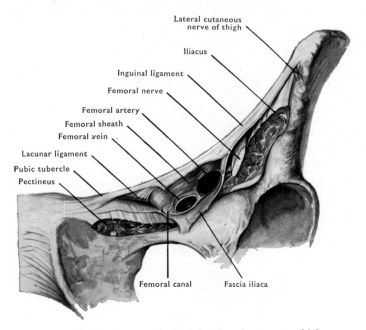

FIG. 5.163. Diagram of fasciae and muscles of inguinal and subinguinal regions in the line of the saphenous opening.

FIG. 5.164. Dissection to show femoral sheath and structures which pass between inguinal ligament and hip bone.

the space between the inguinal ligament and the pelvis into two. Medial to it lies the femoral sheath and its contents, and lateral to it psoas major and iliacus pass into the thigh along with the femoral nerve and the lateral cutaneous nerve of the thigh [FIG. 5.164]. Along the pathway afforded by the fascial sheath of psoas major, pus produced in the vertebrae to which the muscle is attached may track down into the thigh behind the fascia lata.

The deep fascia of the leg is continuous with the fascia lata and is strengthened in the region of the knee by expansions from the tendons of the sartorius, gracilis, semitendinosus, and biceps femoris. Anteriorly it is reinforced by the fibres of the vastus medialis and the vastus lateralis as they form the patellar retinacula [p. 247].

The investing layer extends round the leg, binding together the muscles, affording them origins, and blending with the periosteum on the subcutaneous surfaces of the tibia and fibula. Superiorly it is attached to the tuberosity of the tibia and the head of the fibula.

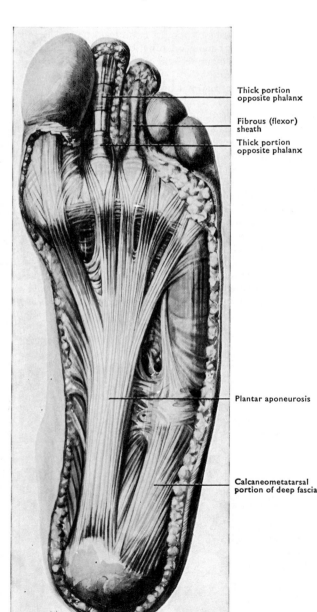

Thick portion opposite phalanx

Fibrous (flexor) sheath

Thick portion opposite phalanx

Plantar aponeurosis

Calcaneometatarsal portion of deep fascia

FIG. 5.165. Left plantar fascia.

Two intermuscular septa pass from its deep surface to the fibula—the anterior to the anterior border and the posterior to the posterior border [FIG. 5.144] thus separating the peroneal group of muscles from the extensor group in front and the flexor group behind.

Another septum extends across the back of the leg, forming a partition between the superficial and deep muscles; superiorly it is attached to the tibia and fibula deep to the origins of soleus. Deep to this again is a septum separating tibialis posterior from the flexors of the toes [FIG. 5.144]. Above soleus, a strong fascial covering is prolonged from the insertion of semimembranosus down over popliteus to the soleal line [FIG. 4.50].

At the ankle the deep fascia is strengthened by transverse fibres which help to form the thickened bands of the flexor, extensor, and peroneal retinacula. The **flexor retinaculum** [FIG. 5.160] runs between the medial malleolus and the medial process of the tuber of the calcaneus. From its deep surface fibrous septa pass to the back of the lower end of the tibia and to the capsule of the ankle joint, dividing the space into four tunnels. These tunnels transmit, from medial to lateral side, the tendon of tibialis posterior, the tendon of flexor digitorum longus, the posterior tibial vessels and tibial nerve, and the tendon of flexor hallucis longus; each of the tendons is enclosed in a separate synovial sheath. The upper border of the flexor retinaculum is continuous with the investing deep fascia of the leg, and also with the septum covering the deep muscles of the calf. Its lower border is continuous with the deep fascia of the sole, and abductor hallucis arises from its outer surface.

There are two **peroneal retinacula** [FIG. 5.159]; the superior one holds down the tendons of peroneus longus and peroneus brevis as they pass together in a single synovial sheath posterior to the lateral malleolus. The inferior peroneal retinaculum, which is attached by a septum to the peroneal trochlea, holds the same two tendons separately to the surface of the calcaneus, each having its own prolongation from the common synovial sheath.

There are also two **extensor retinacula** [FIGS. 5.158 and 5.159]. The superior one runs between the body of the tibia and the body of the fibula just above the ankle; it binds down the tendons of tibialis anterior and the extensor muscles. The inferior extensor retinaculum is a Y-shaped thickening of the deep fascia on the dorsum of the foot. The stem of the Y is attached to the anterior part of the upper surface of the calcaneus and to the floor of the sinus of the tarsus [FIG. 4.68]. The upper of the two limbs of the Y is attached to the medial malleolus, and the lower of the two blends with the deep fascia on the medial side of the foot [FIGS. 5.158 and 5.160]. Deep to the stem of the Y, peroneus tertius and extensor digitorum longus run distally in a common synovial sheath. The tendons of extensor hallucis longus and tibialis anterior usually run deep to the limbs of the Y, though they may pierce them; they are enclosed in separate synovial sheaths [p. 400]. Also running deep to the limbs of the Y are the deep peroneal nerve, the anterior tibial artery, and its continuation, the dorsalis pedis artery.

The deep fascia of the sole, like that of the palm, has a greatly thickened central portion, known as the **plantar aponeurosis** [FIG. 5.165]. There is also a dense thickening of the plantar fascia on the lateral side of the sole, attaching the calcaneus to the base of the fifth metatarsal and covering, as well as giving origin to, the abductor digiti minimi.

The plantar aponeurosis is attached to the medial process of the tuber of the calcaneus, and spreads out anteriorly into five slips which run forwards to become continuous with the digital fibrous flexor sheaths of the toes, just as in the hand. The slip for each toe is attached on each side to the deep transverse metatarsal ligament, and to the base of the proximal phalanx; other fibres pass to the skin of the web of the toes (Bojsen-Møller and Flagstad 1976).

From the margins of the plantar aponeurosis strong septa pass upwards to separate the flexor digitorum brevis from the abductor digiti minimi laterally and the abductor hallucis medially. These septa give origin to the muscles between which they lie and divide the sole of the foot into osteofascial compartments.

The plantar aponeurosis is a most important factor in the maintenance of the longitudinal arch of the foot. When the toes are extended by the extensors, or when standing on the toes, the proximal phalanx winds its slip of the aponeurosis round the head of the metatarsal, a windlass effect which tightens the plantar aponeurosis and raises the longitudinal arch (Hicks 1961).

References and further reading

General references and review articles are marked with an asterisk

ABD-EL-MALIK, S. (1939). Observations on the morphology of the human tongue. *J. Anat.* (*Lond.*) **73**, 201.

BABA, M. A. (1954). The accessory tendon of the abductor pollicis longus. *Anat. Rec.* **119**, 541.

BACKHOUSE, K. M. and CATTON, W. T. (1954). An experimental study of the functions of the lumbrical muscles in the human hand. *J. Anat.* (*Lond.*) **88**, 133.

BARNETT, C. H. (1953). Locking at the knee joint. *J. Anat.* (*Lond.*) **87**, 91.

—— (1962). The normal orientation of the human hallux and the effect of footwear. *J. Anat.* (*Lond.*) **96**, 489.

—— and RICHARDSON, A. T. (1953). The postural function of the popliteus muscle. *Ann. phys. Med.* **1**, 177.

BARRETT, B. (1962). The length and mode of termination of individual muscle fibres in the human sartorius and posterior femoral muscles. *Acta anat.* (*Basel*) **48**, 242.

BASMAJIAN, J. V. (1958). Electromyography of iliopsoas. *Anat. Rec.* **132**, 127.

—— (1959). 'Spurt' and 'shunt' muscles: an electromyographic confirmation. *J. Anat.* (*Lond.*) **93**, 551.

*—— (1961). Electromyography of postural muscles. In *Biomechanical studies of the musculoskeletal system* (ed. F. G. Evans). C. C. Thomas, Springfield, Illinois.

*—— (1974). *Muscles alive* (3rd edn). Williams and Wilkins, Baltimore.

—— and TRAVILL, A. (1961). Electromyography of the pronator muscles in the forearm. *Anat. Rec.* **139**, 45.

BATSON, O. V. (1955). The cricopharyngeus muscle. *Ann. Otol. Rhinol. Laryngol.* **64**, 47.

BEARN, J. G. (1961). An electromyographic study of the trapezius, deltoid, pectoralis major, biceps and triceps muscles, during static loading of the upper limb. *Anat. Rec.* **140**, 103.

—— (1967). Direct observations on the function of the capsule of the sternoclavicular joint in clavicular support. *J. Anat.* (*Lond.*) **101**, 159.

*BEEVOR, C. E. (1904). *The Croonian lectures on muscular movements*, London; (1903). *Adlard. Abstracts*, *Br. med. J.* **1**, 1357, 1417, 1480: **2**, 12; edited and printed (1951) for the guarantors of *Brain*, London.

*BELL, C. (1847). *The anatomy and philosophy of expression as connected with fine arts* (4th edn). London.

BISHOP, A. (1964). Use of the hand in lower Primates. In *Evolutionary and genetic biology of primates (*ed. J. Buettner-Janusch) (2 vols.). Academic Press, New York.

BLOMFIELD, L. B. (1945). Intramuscular vascular patterns in man. *Proc. roy. Soc. Med.* **38**, 617.

BLUNT, M. J. (1951). The posterior wall of the inguinal canal. *Br. J. Surg.* **39**, 230.

BOJSEN-MØLLER, F. and FLAGSTAD, K. E. (1976). Plantar aponeurosis and internal architecture of the ball of the foot. *J. Anat.* (*Lond.*) **121**, 599.

—— and SCHMIDT, L. (1974). The palmar aponeurosis and the central spaces of the hand. *J. Anat.* (*Lond.*) **117**, 55.

BOOYENS, J. and KEATINGE, W. R. (1957). The expenditure of energy by men and women walking. *J. Physiol.* (*Lond.*) **138**, 165.

*BOSMA, J. (1957). Deglutition: pharyngeal stage. *Physiol. Rev.* **37**, 275.

*BOURNE, G. H. (1960). *The structure and function of muscle*, Vols. 1, 2, 3. Academic Press, New York.

BOWDEN, R. E. M. and EL-RAMLI, H. A. (1967). The anatomy of the oesophageal hiatus. *Br. J. Surg.* **54**, 983.

—— and NAPIER, J. R. (1961). The assessment of hand function after peripheral nerve injuries. *J. Bone Jt Surg.* **43B**, 481.

—— and SCHEUER, J. L. (1960). Weights of abductor and adductor muscles of the human larynx. *J. Laryngol. Otol.* **74**, 971.

BRAITHWAITE, F., CHANNELL, G. D., MOORE, F. T., and WHILLIS, J. (1948). The applied anatomy of the lumbrical and interosseous muscles of the hand. *Guy's Hosp. Rep.* **97**, 185.

BRASH, J. C. (1955). *Neurovascular hila of limb muscles*. E. & S. Livingstone, Edinburgh.

BRIDGE, D. T. and ALLBROOK, D. (1970). Growth of striated muscle in an Australian marsupial (*Setonyx brachyurus*). *J. Anat.* (*Lond.*) **106**, 285.

BROCKIS, J. G. (1953). The blood supply of the flexor and extensor tendons of the fingers in man. *J. Bone Jt Surg.* **35B**, 131.

BUCHTHAL, F., ERMINIO, F., and ROSENFALCK, P. (1959). Motor unit territory in different human muscles. *Acta physiol. scand.* **45**, 72.

CAMPBELL, E. J. M. (1952). An electromyographic study of the role of the abdominal muscles in breathing. *J. Physiol.* (*Lond.*) **117**, 222.

—— (1955a). The role of the scalene and sternomastoid muscles in breathing in normal subjects: an electromyographic study. *J. Anat.* (*Lond.*) **89**, 378.

—— (1955b). An electromyographic examination of the role of the intercostal muscles in breathing in man. *J. Physiol.* (*Lond.*) **129**, 12.

—— AGOSTINI, E., and DAVIS, J. N. (1970). *The respiratory muscles; mechanisms and neural control* (2nd edn). Lloyd-Luke, London.

*CAPENER, N. (1956). The hand in surgery. *J. Bone Jt Surg.* **38B**, 128.

CATTON, W. T. and GRAY, J. E. (1951). Electromyographic study of the action of the serratus anterior muscle in respiration. *J. Anat.* (*Lond.*) **85**, 412P.

CHURCH, J. C. T. (1969). Satellite cells and myogenesis: a study in the fruit-bat web. *J. Anat.* (*Lond.*) **105**, 419.

*CLOSE, R. I. (1972). Dynamic properties of mammalian skeletal muscles. *Physiol. Rev.* **52**, 129.

CUNNINGHAM, D. P. and BASMAJIAN, J. V. (1969). Electromyography of genioglossus and geniohyoid muscles during deglutition. *Anat. Rec.* **165**, 401.

DARCUS, H. D. (1951). The maximum torques developed in pronation and supination of the right hand. *J. Anat.* (*Lond.*) **85**, 55.

*DARWIN, C. (1872). *The expression of the emotions in man and animals* (2nd edn; 1889) (ed. J. Darwin). London.

DAY, M. H. and NAPIER, J. R. (1961). The two heads of flexor pollicis brevis. *J. Anat.* (*Lond.*) **95**, 123.

*DUCHENNE, G. B. A. (1867). *Physiologie des mouvements démontrée à l'aide de l'expérimentation électrique et de l'observation clinique*. Paris. Translated and edited by E. B. Kaplan as *Physiology of motion*. Philadelphia (1949).

DUNN, N. (1920). The causes of success and failure in tendon transplantation. *J. orthop. Surg.* **2**, 554.

EDWARDS, D. A. W. (1946). The blood supply and lymphatic drainage of tendons. *J. Anat.* (*Lond.*) **80**, 147.

EYLER, D. L. and MARKEE, J. E. (1954). The anatomy and function of the intrinsic muscles of the fingers. *J. Bone Jt Surg.* **36A**, 1.

FEINSTEIN, B., LINDEGÅRD, B., NYMAN, E., and WOHLFART, G. (1955). Morphologic studies of motor units in normal human muscles. *Acta anat.* (*Basel*) **23**, 127.

FLOYD, W. F. and SILVER, P. H. S. (1950). Electromyographic study of patterns of activity of the anterior abdominal wall muscles. *J. Anat.* (*Lond.*) **84**, 132.

—— —— (1952). Electromyography of the erectores spinae muscles in flexion of the lumbar vertebrae. *J. Anat.* (*Lond.*) **86**, 484.

—— —— (1955). The function of the erectores spinae muscles in certain movements and postures in man. *J. Physiol.* (*Lond.*) **129**, 184.

—— and WALLS, E. W. (1953). Electromyography of the sphincter ani externus in man. *J. Physiol.* (*Lond.*) **122**, 599.

FORREST, W. J. and BASMAJIAN, J. V. (1965). Functions of human thenar and hypothenar muscles. *J. Bone Jt Surg.* **47A**, 1585.

GALLAUDET, B. B. (1931). *A description of the planes of fascia of the human body*. New York.

GAUGHRAN, G. R. L. (1957). Fasciae of the masticator space. *Anat. Rec.* **129**, 383.

GRAY, E. G. and BASMAJIAN, J. V. (1968). Electromyography and cinematogaphy of leg and foot ('normal' and flat) during walking. *Anat. Rec.* **161**, 1.

HAINES, R. W. (1934). On muscles of full and of short action. *J. Anat.* (*Lond.*) **69**, 141.

—— (1951). The extensor apparatus of the fingers. *J. Anat.* (*Lond.*) **85**, 251.

HICKS, J. H. (1954). The mechanics of the foot. II. The plantar aponeurosis and the arch. *J. Anat.* (*Lond.*) **88**, 25.

*—— (1961). The three weight-bearing mechanisms of the foot. In *Biomechanical studies of the musculo-skeletal system*. (ed. F. G. Evans) C. C. Thomas, Springfield, Illinois.

HIGHET, W. B. (1943). Innervation and function of the thenar muscles. *Lancet* i, 227.

HILL, A. V. (1956). The design of muscles. *Br. med. Bull.* **12**, 165.

*HUBER, E. (1931). *Evolution of facial musculature and facial expression.* Williams and Wilkins, Baltimore.

HUGH-JONES, P. (1947). The effect of limb position in seated subjects on their ability to utilize the maximum contractile force of the limb muscles. *J. Physiol. (Lond.)* **105**, 332.

INMAN, V. T. (1947). Functional aspects of the abductor muscles of the hip. *J. Bone Jt Surg.* **29**, 607.

ISLEY, C. L. and BASMAJIAN, J. V. (1973). Electromyography of the human cheeks and lips. *Anat. Rec.* **176**, 143.

JONES, D. S., BEARGIE, R. J., and PAULY, J. E. (1953). An electromyographic study of some muscles of costal respiration in man. *Anat. Rec.* **117**, 17.

—— and PAULY, J. E. (1957). Further electromyographic studies on muscles of costal respiration in man. *Anat. Rec.* **128**, 733.

JONES, F. W. (1919). Voluntary muscular movements in cases of nerve lesions. *J. Anat. (Lond.)* **54**, 41.

*—— (1941). *The principles of anatomy as seen in the hand* (2nd edn). Ballière, London.

—— (1944). *Structure and function as seen in the foot.* Baillière, London.

*JOSEPH, J. (1960). *Man's posture.* Springfield, Illinois.

—— (1968). The pattern of activity of some muscles in women walking on high heels. *Ann. phys. Med.* **9**, 295.

—— and MCCOLL, I. (1961). Electromyography of muscles of posture: posterior vertebral muscles in males. *J. Physiol. (Lond.)* **157**, 33.

—— and NIGHTINGALE, A. (1952). Electromyography of muscles of posture: leg muscles in males. *J. Physiol. (Lond.)* **117**, 484.

—— —— (1954). Electromyography of muscles of posture: thigh muscles in males. *J. Physiol. (Lond.)* **126**, 81.

—— —— (1956). Electromyography of muscles of posture: leg and thigh muscles in women, including the effects of high heels. *J. Physiol. (Lond.)* **132**, 465.

—— and WILLIAMS, P. L. (1957). Electromyography of certain hip muscles. *J. Anat. (Lond.)* **91**, 286.

KEAGY, R. D., BRUMLIK, J., and BERGAN, J. J. (1966). Direct electromyography of the psoas major muscle in man. *J. Bone Jt Surg.* **48A**, 1377.

KIRK, T. S. (1924). Some points in the mechanism of the human hand. *J. Anat. (Lond.)* **58**, 228.

LANDSMEER, J. M. F. (1949). The anatomy of the dorsal aponeurosis of the human finger and its functional significance. *Anat. Rec.* **104**, 31.

—— (1955). Anatomical and functional investigations on the articulations of the human fingers. *Acta anat. (Basel)* Suppl. **24**, 1.

LANGA, G. S. (1963). Experimental observations and interpretations on the relationship between the morphology and function of the human knee joint. *Acta anat. (Basel)* **55**, 16.

*LAST, R. J. (1954). The muscles of the mandible. *Proc. roy. Soc. Med.* **47**, 571.

*LE DOUBLE, A. F. (1897). *Traité des variations du système musculaire de l'homme.* Schleicher, Paris.

LEWIS, O. J. (1965). The evolution of the mm. interossei in the Primate hand. *Anat. Rec.* **153**, 275.

LOBLEY, G. E., WILSON, A. B., and BRUCE, A. S. (1977). An estimation of the fibre type composition of eleven skeletal muscles from New Zealand White rabbits between weaning and early maturity. *J. Anat. (Lond.)* **123**, 501.

LOCKHART, R. D. (1930). Movements of the normal shoulder joint and of a case with trapezius paralysis studied by radiogram and experiment in the living. *J. Anat. (Lond.)* **64**, 288.

—— (1963). *Living anatomy* (6th edn paperback, 1970). Faber and Faber, London.

—— and BRANDT, W. (1938). The length of striated muscle fibres. *J. Anat. (Lond.)* **72**, 470.

LONG, C. and BROWN, M. E. (1964). Electromyographic kinesiology of the hand: muscles moving the long finger. *J. Bone Jt Surg.* **46A**, 1683.

LOW, A. (1907). A note on the crura of the diaphragm and the muscle of Treitz. *J. Anat. Physiol.* **42**, 93.

MACCONAILL, M. A. (1946). Some anatomical factors affecting stabilising functions of muscles. *Irish J. med. Sci.* March, 160.

—— and BASMAJIAN, J. V. (1969). *Muscles and movements.* Williams and Wilkins, Baltimore.

MACDOUGALL, J. D. B. and ANDREW, B. L. (1953). An electromyographic study of the temporalis and masseter muscles. *J. Anat. (Lond.)* **87**, 37.

MANTER, J. T. (1945). Variations of the interosseous muscles of the human foot. *Anat. Rec.* **93**, 117.

MARTIN, B. F. (1968). The origins of the hamstring muscles. *J. Anat. (Lond.)* **102**, 345.

MARTIN, C. P. (1940). The movements of the shoulder joint, with special reference to rupture of the supraspinatus tendon. *Am. J. Anat.* **66**, 213.

MATHESON, A. B., SINCLAIR, D. C. and SKENE, W. G. (1970). The range and power of ulnar and radial deviation of the fingers. *J. Anat. (Lond.)* **107**, 439.

*MEDICAL RESEARCH COUNCIL (1951). *War Memorandum No. 7: Aids to the investigation of peripheral nerve injuries* (2nd edn). London.

MILLS, J. N. (1950). The nature of the limitation of maximal inspiratory and expiratory efforts. *J. Physiol. (Lond.)* **111**, 376.

MORRIS, J. M., BENNER, G., and LUCAS, D. B. (1962). an electromyographic study of the intrinsic muscles of the back in man. *J. Anat. (Lond.)* **96**, 509.

*MORTON, D. J. (1935). *The human foot.* New York.

MOSELEY, H. F. (1951). Ruptures of the rotator cuff. *Br. J. Surg.* **38**, 340.

NAPIER, J. R. (1952). The attachments and function of the abductor pollicis brevis. *J. Anat. (Lond.)* **86**, 335.

—— (1956). The prehensile movements of the human hand. *J. Bone Jt Surg.* **38B**, 902.

PARRY, E. (1966). The influence of rotation of the trunk on the anatomy of the inguinal canal. *Br. J. Surg.* **53**, 205.

PASSAVANT, G. (1869). Ueber die Verschliessung des Schlundes beim Sprechen. *Virchows Arch. path. Anat.* **46**, 1.

PATEY, D. H. (1949). Some observations on the functional anatomy of inguinal hernia and its bearing on the operative treatment. *Br. J. Surg.* **36**, 264.

PAULY, J. E. (1966). An electromyographic analysis of certain movements and exercises. 1. Some deep muscles of the back. *Anat. Rec.* **155**, 223.

PETERSEN, F. B. (1960). Muscle training by static, concentric, and eccentric contractions. *Acta physiol. scand.* **48**, 406.

PORTER, N. H. (1962). A physiological study of the pelvic floor in rectal prolapse. *Ann. roy. Coll. Surg. Engl.* **31**, 379.

PULLEN, A. H. (1977). The distribution and relative sizes of fibre types in the extensor digitorum longus and soleus muscles of the adult rat. *J. Anat. (Lond.)* **123**, 467.

ROWE, R. W. D. and GOLDSPINK G. (1969). Muscle fibre growth in five different muscles in both sexes of mice. *J. Anat. (Lond.)* **104**, 519.

*SALTER, N. (1955a). Methods of measurement of muscle and joint function. *J. Bone Jt Surg.* **37B**, 474.

—— (1955b). The effect on muscle strength of maximum isometric and isotonic contractions at different repetition rates. *J. Physiol. (Lond.)* **130**, 109.

*SAUNDERS, J. B. de C. M., INMAN, V. T., and EBERHART, H. D. (1953). The major determinants in normal and pathological gait. *J. Bone Jt Surg.* **35A**, 543.

SCHEVING, L. E. and PAULY, J. E. (1959). An electromyographic study of some muscles acting on the upper extremity of man. *Anat. Rec.* **135**, 239.

SCHWARZACHER, H. G. (1959). Über die Länge und Anordnung der Muskelfasern in menschlichen Skeletmuskeln. *Acta anat. (Basel)* **37**, 217.

*SINGER, E. (1935). *Fasciae of the human body, and their relations to the organs they envelop.* Baltimore.

SMITH, J. W. (1954). Muscular control of the arches of the foot in standing: an electromyographical assessment. *J. Anat. (Lond.)* **88**, 152.

STACEY, M. J. (1969). Free nerve endings in skeletal muscle of the cat. *J. Anat. (Lond.)* **105**, 231.

*STACK, A. J. (1962). Muscle function in the fingers. *J. Bone Jt Surg.* **44B**, 899.

STANIER, D. I. (1977). The function of muscles around a simple joint. *J. Anat. (Lond.)* **123**, 827.

*STEINDLER, A. (1935). *Mechanics of normal and pathological locomotion in Man.* Baillière, London.

*—— (1955). *Kinesiology of the human body.* C. C. Thomas, Springfield, Illinois.

STILWELL, D. L. (1957). The innervation of tendons and aponeuroses. *Am. J. Anat.* **100**, 289.

SUNDERLAND, S. (1944a). The significance of hypothenar elevation in movements of opposition of the thumb. *Aust. N.Z. J. Surg.* **13**, 155.

—— (1944b). Voluntary movements and the deceptive action of muscles in peripheral nerve lesions. *Aust. N.Z. J. Surg.* **13**, 160.

—— (1953). Rotation of the fingers by the lumbrical muscles. *Anat. Rec.* **116**, 167.

—— (1978). *Nerves and nerve injuries.* (2nd edn). E. & S. Livingstone, Edinburgh.

TAKEBE, K., VITTI, M., and BASMAJIAN, J. V. (1947a). The functions of semispinalis capitis and splenius capitis: an electromyographic study. *Anat. Rec.* **179**, 477.

———— ———— ———— (1974b). Electromyography of pectineus muscle. *Anat. Rec.* **180**, 281.

TAYLOR, A. (1960). The contribution of the intercostal muscles to the effort of respiration in man. *J. Physiol.* (*Lond.*) **151**, 390.

TAYLOR, K. and CALVEY, T. N. (1977). Histochemical characteristics and contractile properties of the spinotrapezius muscle in the rat and the mouse. *J. Anat.* (*Lond.*) **123**, 67.

*TESTUT, L. (1884). *Les anomalies musculaires chez l'homme expliquées par l'anatomie comparée*. Paris.

THOMAS, D. P. and WHITNEY, R. J. (1959). Postural movements during normal standing in man. *J. Anat.* (*Lond.*) **93**, 524.

*THOMPSON, P. (1899). *The myology of the pelvic floor*. London.

TRAVILL, A. and BASMAJIAN, J. V. (1961). Electromyography of the supinators of the forearm. *Anat. Rec.* **139**, 557.

WADE, O. L. (1954). Movements of the thoracic cage and diaphragm in respiration. *J. Physiol.* (*Lond.*) **124**, 193.

WHILLIS, J. (1930). A note on the muscles of the palate and the superior constrictor. *J. Anat.* (*Lond.*) **65**, 92.

WHILLIS, T. (1946). Movements of the tongue in swallowing. *J. Anat.* (*Lond.*) **80**, 115.

*WHITNALL, S. E. (1932). *The anatomy of the human orbit and accessory organs of vision* (2nd edn). Oxford University Press, London.

*WRIGHT, W. G. (1928). *Muscle Function*. New York.

*ZIMMERMAN, L. M. and ANSON, B. J. (1967). *Anatomy and surgery of hernia* (2nd edn). Williams and Wilkins, Baltimore.

6 The digestive system

F. R. JOHNSON

THE digestive system is responsible for the reception, digestion, and absorption of food, and for the retention and excretion of the unabsorbed residue. It consists of the alimentary canal and several associated glandular organs. The **alimentary canal**, measuring in Man some 9 metres in length, has the following successive parts— mouth, pharynx, oesophagus, stomach, small intestine, and large intestine. The associated **glandular organs** are the salivary glands, which have ducts opening into the mouth, and the liver and pancreas, which discharge their exocrine secretions through ducts opening into the duodenum—the first part of the small intestine. The general arrangement, position and size of these various parts are shown in FIGURE 6.1.

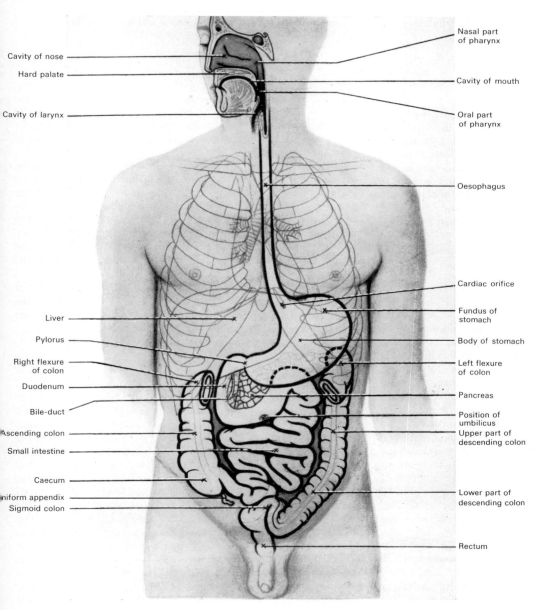

FIG. 6.1. Diagram of general arrangement of digestive system. The transverse colon is not represented in order that the duodenum and pancreas, which lie behind it, may be seen.

Cavity of nose

Hard palate

Cavity of larynx

Liver

Pylorus

Right flexure of colon

Duodenum

Bile-duct

Ascending colon

Small intestine

Caecum

niform appendix

Sigmoid colon

Nasal part of pharynx

Cavity of mouth

Oral part of pharynx

Oesophagus

Cardiac orifice

Fundus of stomach

Body of stomach

Left flexure of colon

Pancreas

Position of umbilicus

Upper part of descending colon

Lower part of descending colon

Rectum

The mouth

The characteristic feature of the mammalian mouth is the highly mobile, muscular lips and cheeks, correlated with the functions of mastication and, in Man, of articulated speech. In the mouth food is mixed with saliva from the salivary glands and, after mastication, is passed into the oral part of the pharynx by the process of swallowing or deglutition. There are two parts of the mouth, the vestibule, external to the teeth, and the mouth cavity proper, internal to the teeth. When the teeth are closed, these two parts communicate with one another by a narrow cleft on each side, behind the third molar teeth, and by variable spaces between adjacent teeth.

The mouth is lined by a mucous membrane, consisting of stratified squamous epithelium, and a layer of connective tissue, the lamina propria. In man the epithelium is non-keratinized, apart from the hard palate and the gingivae (gums), areas subjected to the friction of mastication (Landay and Schroeder 1979). Except in the gingivae, there is a layer of loose connective tissue, the submucosa, of varying quantity and composition. It contains the larger blood vessels, lymph vessels, and nerves, and it attaches the mucous membrane to the underlying structures (Scott and Symons 1978).

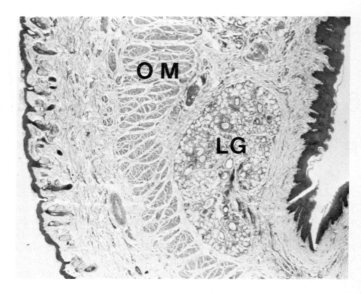

FIG. 6.2. Sagittal section through part of the lower lip. The skin and dermis are on the left and the thick mucosa and submucosa on the right. The orbicularis oris muscle (OM) has a mucous labial gland (LG) deep to it.

VESTIBULE

The vestibule is a narrow cleft enclosed externally by the cheeks and lips, and internally by the dental arches which consist of the alveolar processes of the maxillae and the mandible, the teeth, and the gingivae. It communicates with the exterior by the **oral fissure**. The anterior and infratemporal surfaces of the maxillae and the body of the mandible can be palpated by a finger placed in the vestibule, and the small papilla marking the opening of the parotid duct can be felt with the tip of the tongue on each cheek, opposite the second upper molar tooth.

LIPS

The lips surround the oral fissure. Laterally, they join to form the angle of the mouth which usually lies just in front of the first premolar tooth. The line of contact of the closed lips is at the cutting edge of the upper incisor teeth. On its external surface, the upper lip has a median vertical groove, the **philtrum**. This widens below to end in the labial tubercle, a projection on the free edge of the lip. From without inwards, each lip consists of: (i) skin with underlying fascia; (ii) voluntary muscle (mainly orbicularis oris); (iii) submucosa of loose connective tissue containing vessels, nerves, and small labial salivary glands which can be felt with the tip of the tongue on the oral surface; and (iv) mucous membrane [FIG. 6.2]. The mucous membrane of lips is extremely sensitive, having large numbers of nerve fibres, from the trigeminal nerve. An arterial ring, formed by the labial branches of the facial arteries, is situated in the submucosa. A midline fold of mucous membrane, the **frenulum**, connects the inner surface of each lip to the corresponding gum [FIG. 6.3].

The red transitional zone between the outer skin and inner mucous membrane is found only in Man. In it the stratum lucidum of the epidermis is relatively thick and the stratum corneum is thin. The resulting transparent epithelium, in association with the highly vascular underlying papillae, account for the reddish colour of this region. Hair follicles and sweat glands are absent in the transitional zone and sebaceous glands are sparse, except at the corners of the lips. The near absence of glandular structures results in the transitional zone being normally dry, except when moistened by the tongue, and this accounts for 'chapped' lips during periods of low humidity.

The upper lip develops as a result of the growth forwards of the maxillary process to meet and fuse with the medial and lateral nasal folds of the frontonasal process. Thus, the lateral parts of the primitive upper lip are derived from the maxillary processes and the central part from the lower end of the frontonasal process. The definitive lip becomes separated by development of the labiogingival groove [p. 419]. Failure of fusion of the maxillary process with the frontonasal process, on one or both sides, results in unilateral and bilateral hare-lip, respectively.

The lower lip arises in a similar manner by the median fusion of the mandibular processes and the subsequent formation of the labiogingival sulcus. Failure of fusion of the mandibular processes with resulting cleft lip and mandible is extremely rare.

Cheeks

The cheek forms the lateral wall of the vestibule and is continuous anteriorly with the lips. In structure it is similar to the lip, having a central framework of muscle with skin externally, and mucous membrane supported on a submucosa internally. The principal muscle of the cheek is the buccinator. It is connected posteriorly to the superior constrictor through the pterygomandibular raphe and it blends in front with the fibres of the orbicularis oris. Superficial to buccinator is the thin elastic buccopharyngeal fascia, and superficial to that the **buccal pad of fat**, which is particularly prominent in the infant, helping to produce the rounded contours of the cheek. The submucosa of the cheeks, like that of the lips, contains many elastic fibres, which allow stretching of the tissues when the mouth is opened or the cheeks are blown out, and prevent the mucous membrane from being caught between the teeth during biting and chewing. Movements of the cheeks and lips are necessary for normal mastication. They keep the food between the biting surfaces of the teeth and prevent it from accumulating in the vestibule—a feature of paralysis of their muscles which results from injury to their motor supply, the facial nerve.

The parotid duct traverses the cheek and, having crossed the masseter, turns inwards piercing the buccal pad of fat and the buccinator before passing forwards for a short distance deep to the mucous membrane to open into the vestibule on a small papilla on the oral surface of the cheek [FIG. 6.22]. The cheeks contain some small salivary glands, the **buccal glands**, which, like those of the lips are both mucous and serous in type, and are present in the submucosa. Four or five of the larger ones, known as **molar glands**, are situated near the entry of the parotid duct.

The **sensory nerves** of the upper lip, the cheeks, and the lower lip are the infraorbital, the buccal, and the mental branches of the trigeminal nerve, respectively. **Lymph vessels** from cheeks and lips drain into the submandibular lymph nodes [FIG. 13.159].

GINGIVAE (GUMS)

The gingiva is the firm mucous membrane covering the apical regions of the alveolar processes of the maxillae and mandible, and surrounding the necks of the teeth like a collar [FIGS. 6.10 and 6.25B]. The gingiva lacks a submucosa as such, and is firmly bound by collagenous fibres of its lamina propria to the periosteum of the alveolar processes, and around each tooth to: (i) the alveolar crests; (ii) the periodontal ligament; and (iii) the cementum.

The mucous membrane of the gingiva is pinkish-grey, and in the healthy state has a stippled appearance. Its stratified squamous epithelium is subjected to considerable mechanical trauma and is slightly keratinized. As the gingiva surrounds the necks of the teeth, the epithelium forms an attachment to the enamel, the **epithelial attachment**, near the junction of the enamel and cementum [FIG. 6.10]. This junction of soft oral tissues and hard dental tissues is of considerable clinical importance, as it tends to be a site of reduced resistance to mechanical and bacterial attack. During later life the gingiva commonly recedes, due to a shifting of the epithelial attachment to the dental cementum. In coloured races the gums may show varying degrees of pigmentation.

The **nerve supply** to the upper gums is derived from the greater palatine, nasopalatine, and the anterior, middle, and posterior superior alveolar nerves, all of which are branches of the maxillary nerve. The lower gum is innervated by the buccal, inferior alveolar, and lingual branches of the mandibular nerve.

MOUTH CAVITY PROPER

The mouth cavity proper is bounded above by the hard and soft palates, and laterally and in front by the upper and lower dental arches [FIG. 6.3]. Below, it is limited by the anterior part of the tongue and by the mucous membrane passing from the undersurface of the tongue to the internal surface of the body of the mandible [FIG. 6.4]. This forms a median fold passing to the anterior part of the tongue, the **frenulum of the tongue**. On each side of the frenulum, the sublingual gland raises the mucous membrane of the floor of the mouth to form the **sublingual fold**, at the anterior end of which the submandibular duct opens on the **sublingual papilla**, close to the frenulum [FIG. 6.4]. The minute ducts of the sublingual gland open on the surface of the sublingual fold. Posteriorly, the isthmus of the fauces (see below) connects the mouth with the oral part of the pharynx [FIG. 6.6].

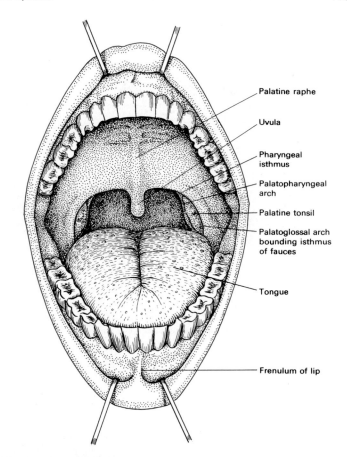

Palatine raphe
Uvula
Pharyngeal isthmus
Palatopharyngeal arch
Palatine tonsil
Palatoglossal arch bounding isthmus of fauces
Tongue
Frenulum of lip

FIG. 6.3.   Open mouth showing palate and tonsils.

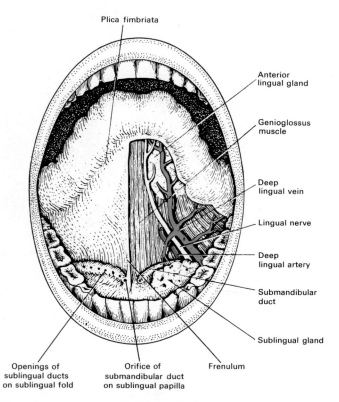

Plica fimbriata
Anterior lingual gland
Genioglossus muscle
Deep lingual vein
Lingual nerve
Deep lingual artery
Submandibular duct
Sublingual gland
Openings of sublingual ducts on sublingual fold
Orifice of submandibular duct on sublingual papilla
Frenulum

FIG. 6.4.   Open mouth with tongue raised and mucous membrane removed from right side to show deep lingual vessels.

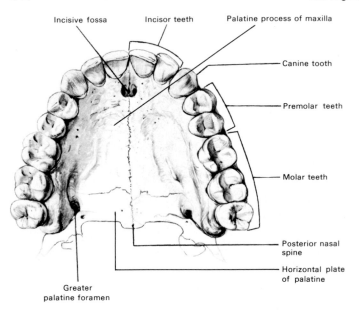

FIG. 6.5.   Hard palate and upper permanent teeth.

PALATE

The complete palate is characteristic of mammals and its development is associated with the ability to suck. Its anterior two-thirds, the hard palate, has a bony framework while its posterior one-third, the soft palate, has a fibromuscular basis [FIG. 6.3]. The palate projects backwards into the pharynx and separates the mouth from the nasal cavities and the isthmus of the fauces from the nasal part of the pharynx. It is arched both anteroposteriorly and transversely, the height of the arching varying considerably between individuals [FIG. 6.27].

HARD PALATE

The bony framework of the hard palate is formed by the palatine processes of the maxillae in front, and the horizontal plates of the palatine bones behind [FIG. 6.5]. Anteriorly and laterally it is continuous with the maxillary alveolar arch, and posteriorly it gives attachment to the soft palate. Immediately behind the central incisor teeth is the deep incisive fossa, into which the incisive canals open and transmit the terminal branches of the greater palatine artery and nasopalatine nerve between the nasal cavity and the palate [FIG. 3.57]. Posterolaterally, the greater and lesser palatine

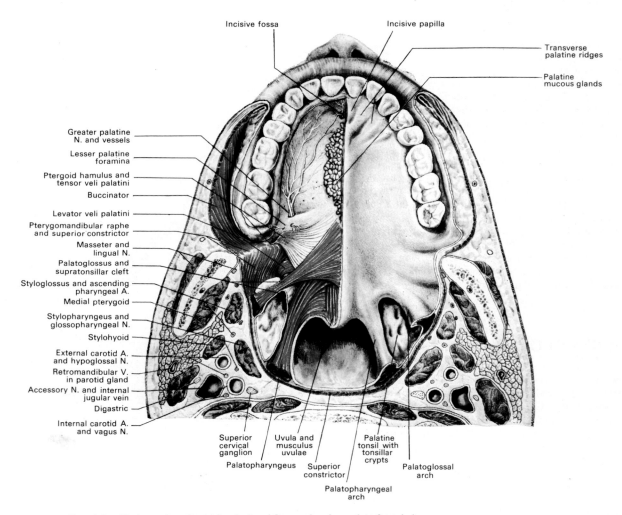

FIG. 6.6.   Horizontal section at level of oral fissure showing palate from below.
The right half of the palate has been dissected to show the muscles of the soft palate by the removal of the majority of the palatal glands.

foramina transmit the greater and lesser palatine nerves and vessels to the palate.

The mucous membrane covering the buccal surface of the palate is firmly attached to the periosteum of the underlying bone by collagenous fibres of the lamina propria. As the fibres pass through the submucosa to reach the bone, they subdivide it into compartments which anteriorly contain adipose tissue, and posteriorly contain palatine mucous salivary glands. These collections of fat and glandular tissue are thought to have a cushioning effect, so that the mucosa can bear the mechanial stresses imposed upon it during mastication without being damaged against the underlying bone. In the midline the mucous membrane forms a median ridge (raphe), deep to which the submucosa is absent and the lamina propria is directly attached to the periosteum. The raphe terminates anteriorly, in the incisive papilla overlying the incisive fossa; transverse ridges (rugae) of mucous membrane radiate from its anterior part [FIG. 6.6]. These are more prominent in lower mammals and in the child, and tend to disappear in old age. The stratified squamous epithelium of the mucosa of the hard palate is thick and keratinized.

SOFT PALATE

This is a fibromuscular curtain enabling the mouth to be cut off from the oral part of the pharynx, as during breathing with the mouth full, or separating the oral and nasal parts of the pharynx in a number of other activities (see below). It is attached anteriorly, to the hard palate but posteriorly it is free with a short, conical, midline process, the uvula, hanging down from its posterior border [FIGS. 6.3 and 6.6].

In the resting state its anterior part continues the curvature of the hard palate, while the posterior part turns downwards, following the curvature of the dorsum of the tongue. Laterally, it is continuous with the palatoglossal and palatopharyngeal arches, with which and the dorsum of the tongue, it forms the isthmus of the fauces. Superiorly, it forms the floor of the nasal part of the pharynx.

The soft palate is covered by mucous membrane on its upper and lower surfaces. The epithelium on the upper surface is a mixture of stratified squamous and pseudostratified ciliated columnar, whereas on the lower surface it is nonkeratinized stratified squamous in type. In addition to the palatal muscles and the associated palatine aponeurosis [p. 301], the soft palate also contains mucous salivary glands and lymph tissue; the glands lie deep to the oral mucosa of the anterior part, where they are continuous with those of the hard palate.

Movements of the soft palate

The muscles of the soft palate and the individual movements produced by them are described on page 300.

A major function of the soft palate is to close the pharyngeal isthmus and thereby to separate the cavities of the oral and nasal parts of the pharynx. This activity occurs in the acts of swallowing, coughing, blowing, and during the production of explosive consonants, such as 'p' and 'b'. It is produced by the levator veli palatini muscles, which pull the soft palate upwards and backwards, towards the posterior wall of the pharynx. At the same time, the upper fibres of the superior constrictor pull the posterior pharyngeal wall forwards as a ridge, Passavant's ridge, to meet the soft palate, and the isthmus of the fauces is narrowed by contraction of the palatopharyngeus muscle. During quiet respiration the palate is relaxed, as it is also during the production of nasal sounds, such as 'm' and 'n'. In saying 'ah', the isthmus of the fauces widens and the posterior wall of the pharynx comes more clearly into view. Paralysis of the musculature of the soft palate results in a nasal intonation to the voice and regurgitation of food into the nose during swallowing (Bosma 1957; Whillis 1946).

VESSELS, NERVES, AND LYMPHATICS OF THE PALATE

The main source of blood to the palate comes from the **greater** and **lesser palatine arteries**, with a contribution from the facial (ascending palatine) and ascending pharyngeal arteries.

The **sensory nerves** are branches of the maxillary nerves. The anterior part of the hard palate receives its sensory supply from the nasopalatine nerves, which reach the palate by traversing the incisive canal [FIG. 11.15]. The remainder of the hard palate and the soft palate receive their sensory supplies from the greater and lesser palatine nerves, which pass through the greater and lesser palatine canals, respectively. In the palatine nerves are parasympathetic nerve fibres which arise in the pterygopalatine ganglion and supply the glands of the palate. They also transmit postganglionic sympathetic fibres.

The **lymph vessels** of the palate run lateral to the tonsil and the palatoglossal arch, to drain into the upper deep cervical lymph nodes.

DEVELOPMENT OF THE PALATE AND MOUTH [see also pp. 54 and 496]

The palate develops from palatal outgrowths, the palatal processes, which arise from the maxillary processes. At first, these are vertically disposed, due to the presence of the developing tongue but later, as the tongue sinks, they become horizontal and fuse with each other and with the lower end of the nasal septum. In front, the palatal processes fuse with the primitive palate (posterior part of the frontonasal process), the site of fusion being indicated in the adult by the incisive fossa. Posteriorly, the fused palatal processes extend behind the nasal septum and mainly form the connective tissue of the soft palate. This is subsequently invaded by mesenchyme from the pharyngeal arches, which differentiates into the palatal musculature (Ferguson 1978; Waterman and Meller 1974).

Complete or partial failure of the palatal processes to meet and fuse with each other, with the nasal septum, and with the primitive palate results in defects varying from a cleft uvula to a bilateral cleft palate, in which the nasal septum hangs down in the midline, with a free lower border. The more severe forms of cleft palate are generally associated with bilateral hare-lip.

The mouth develops from the stomadeum [FIG. 2.59] and the cephalic end of the foregut, which are separated by the **buccopharyngeal membrane** up to the fourth intra-uterine week. The cheeks are formed by the fusion of the posterior portions of the maxillary and mandibular processes, and the tongue projects upwards from the floor of the mouth.

TEETH

In the early vertebrates, as in most present-day fish, the dentition tends to consist of simple conical teeth (haplodont teeth) of similar shape (homodont dentition), and with unlimited succession of teeth (polyphyodont dentitions). Mammals have heterodont dentitions, composed of specialized forms of teeth evolved in different parts of the mouth, with succession limited to two dentitions, deciduous (milk) and permanent (diphyodont dentitions). In heterodont dentitions the teeth are adapted for specialized functions, hence the incisors for cutting, the canines for piercing and the molars for grinding.

In man, there are five deciduous teeth in each half of each jaw (two incisors, one canine and two molars) and eight permanent teeth (two incisors, one canine, two premolars, and three molars).

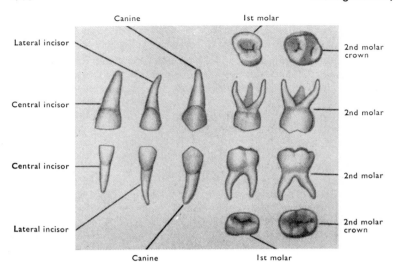

FIG. 6.7.   Deciduous teeth of left side.
The masticating surfaces of the two upper molars are shown above. In the second row the upper teeth are viewed from the outer or labial side. In the third row the lower teeth are shown from the same side; and below are the masticating surfaces of the two lower molars. In the specimen from which the first upper molar was drawn the two labial turbercles were not distinctly separated, as is often the case.

The permanent molars have no predecessors and erupt behind the deciduous molars, the successional teeth of the two deciduous molars being the two premolars.

The teeth of both dentitions on each side of the upper and lower jaws are indicated by the combined dental formula:

	RIGHT							
Deciduous				M_2	M_1	C	I_2	I_1
Permanent	M_3	M_2	M_1	P_2	P_1	C	I_2	I_1
Permanent	M_3	M_2	M_1	P_2	P_1	C	I_2	I_1
Deciduous				M_2	M_1	C	I_2	I_1
	Distal							Mesial
	LEFT							
Deciduous	I_1	I_2	C	M_1	M_2			
Permanent	I_1	I_2	C	P_1	P_2	M_1	M_2	M_3
Permanent	I_1	I_2	C	P_1	P_2	M_1	M_2	M_3
Deciduous	I_1	I_2	C	M_1	M_2			
	Distal							Mesial

I_1 = the central incisors, I_2 = the lateral incisors, C = the canines, P_1, P_2 = the premolars, M_1 = the first molars, M_2 = the second molars, and M_3 = the third molars.

The bulk of a tooth consists of **dentine**, a hard, mineralized substance enclosing the pulp cavity. The dentine of the anatomical **crown** of the tooth is covered by **enamel**, while that of the roots is covered by **cementum**. Where the enamel and cementum meet on the surface of the dentine, there is a slight constriction, the cervix or **neck of the tooth**. The strong, fibrous, periodontal ligament [p. 419] anchors each root of every tooth in a separate socket in the alveolar processes of the maxilla and mandible. The part of the tooth projecting above the gingiva is known as the **clinical crown** [FIG. 6.10].

The surfaces of the incisor and canine teeth are: labial (lip side), lingual (tongue side), incisal (cutting edge), mesial (nearest the midline of the dental arch), and distal (furthest from the midline of the arch). The molar and premolar teeth have the following surfaces—buccal (cheek side), lingual, occlusal (chewing surface), mesial, and distal.

CROWN MORPHOLOGY

The incisors, particularly the permanent ones when they first erupt, show three small tubercles on their incisal edges, which wear away with use. The margins of the lingual surfaces of the upper incisors

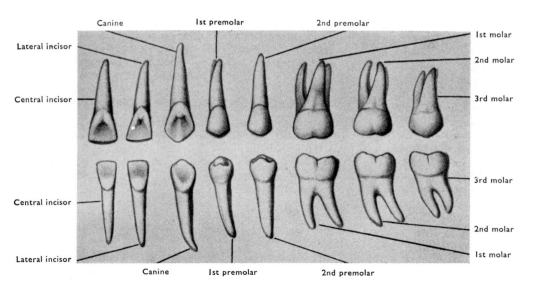

FIG. 6.8.   Permanent teeth of right side, lingual surfaces.
The cingulum is distinct on the upper incisors and both canines, the lingual tubercle on the upper lateral incisor and the upper canine.

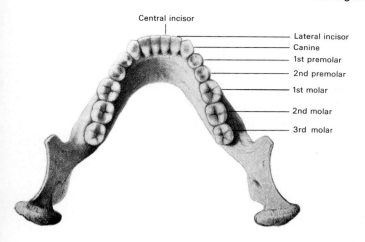

Central incisor
Lateral incisor
Canine
1st premolar
2nd premolar
1st molar
2nd molar
3rd molar

FIG. 6.9.   Mandible and lower permanent teeth.

are elevated so that the surface is slightly hollowed. The cervical part of this elevated margin is known as the **cingulum**.

In the **permanent dentition**, the occlusal surfaces of the upper premolars have two projections or **cusps**, buccal and lingual, with a dividing fissure between [FIG. 6.5]. Those of the lower premolars resemble the upper ones to some extent, but the lingual cusps are smaller, that of the first being merely an enlarged cingulum [FIG. 6.8]. The occlusal surfaces of the upper first permanent molars are quadrilateral with four cusps and there is an oblique ridge joining the distobuccal and mesiolingual cusps. The upper second and third molars conform to this same pattern but with a diminution, or even absence of the distolingual cusp, so that they tend to be tricuspid; a tendency which is more marked in the third molar. The lower first molar has typically five cusps, but quite commonly only four cusps may be present [FIG. 6.9]. The second molar is smaller and usually has four cusps, although a fifth is sometimes present. The third molar is usually smaller still, but it is not uncommon for it to be as large as the first and to have a variable number of small cusps.

In the **deciduous dentition** the occlusal surfaces of the first molars present a cusp form that does not resemble either that of the premolars that succeed them, or the permanent molars. They possess a distinct prominence or bulge on their buccal surfaces. The morphology of the crowns of the second deciduous molars closely resembles that of the first permanent molars [FIG. 6.7]. For further details students are referred to texts on dental anatomy (Scott and Symons 1978).

ROOT MORPHOLOGY

The roots of the incisors, canines, and most premolars are single, though the first upper premolar is often bifurcated to a variable extent and may even have separate buccal and lingual roots. The lower molars have two roots, a mesial and distal, but those of the third are frequently fused into one. Upper molars have three roots, two buccal which are flattened like those of lower molars, and a lingual which is conical. The three roots are divergent, the divergence being less in the second and third molars; the roots of the third molar usually show various degrees of fusion [FIG. 6.8].

The canines, have the longest roots and form a corner to the dental arches, particularly in the upper jaw where they produce a prominence on the anterior surface of the maxilla. The roots of the upper molars, especially those of the first, are close to the maxillary air sinus and often produce conical elevations on its floor.

SOCKETS

The labial walls of the sockets of the anterior teeth in both jaws are thin, but the lingual walls are thicker and in the upper jaw are supported by the palatal processes of the maxilla [FIG. 6.5]. The buccal roots of the upper first molars are buttressed by the root of the zygomatic process of the maxilla [FIG. 3.48]. The socket of the upper third molar may extend into the maxillary tuberosity, in which case extraction of the tooth may result in fracture of the tuberosity.

The buccal walls of the sockets of the lower second and third molars are somewhat buttressed by the oblique line or ridge of bone which passes from the anterior margin of the ramus to the body of the mandible [FIG. 3.63], whereas the lingual walls are thin and easily broken during extraction of these teeth. Fracture of the inner plate of the alveolar process in the region of the lower third molar, close to the lingual nerve and to the nearby attachments of the lower border of the superior constrictor and posterior border of the mylohyoid muscle, especially if associated with infection, may lead to serious complications.

OCCLUSION

The dental arches are said to be catenary in shape, that is they correspond to the curve assumed by a chain hanging from its two extremities. However, variation in the shape of the dental arch is common, particularly in contemporary people, in whom ideal shapes are rare. Teeth out of place and abnormal arch relationships result in malocclusion.

The lower arch is narrower than the upper, so that it occludes within the upper; consequently, the buccal and labial surfaces of the upper teeth are outside those of the lower and conceal them partially from view (centric occlusion). In the front of the mouth, the upper teeth overlap about one-third of the crowns of the lower ones.

All of the teeth of the upper arch incline slightly outwards, while those of the lower arch incline correspondingly inwards. The occlusal plane between the two arches is not flat but forms a complex curve with transverse and anteroposterior components.

The upper incisors are broader than the lower, so, on closure each upper tooth, with the exception of the third molar, meets part of the distal tooth as well as the corresponding tooth in the lower arch. The upper third molar is usually smaller than the lower, with the result that the upper and lower arches end opposite each other posteriorly.

When the jaws are at rest the arches are normally in a centric occlusal relationship but held slightly apart, though making momentary contact in speech and when saliva is swallowed.

DENTINE

Dentine is yellowish-white, avascular, and harder than bone. It consists of a highly mineralized (70 per cent by weight) organic matrix almost entirely composed of collagen fibres. The mineral, like that of bone, consists of a calcium phosphate closely related to hydroxyapatite, $Ca_{10}(PO_4)_6(OH)_2$, which contains traces of carbonates, magnesium, chloride, and fluoride (Miles 1967).

Dentine contains a system of evenly spaced tubules, **dentinal tubules**, about $1-2\,\mu m$ in diameter, but tapering as they extend from the pulp towards the dentine surface. The tubules pursue large S-shaped primary curves which on examination with the light microscope show small, spiral, secondary curvatures. Each tubule contains, at least for the greater part of its length, a protoplasmic process of an odontoblast (Holland 1976; Whittaker 1978) [p. 420].

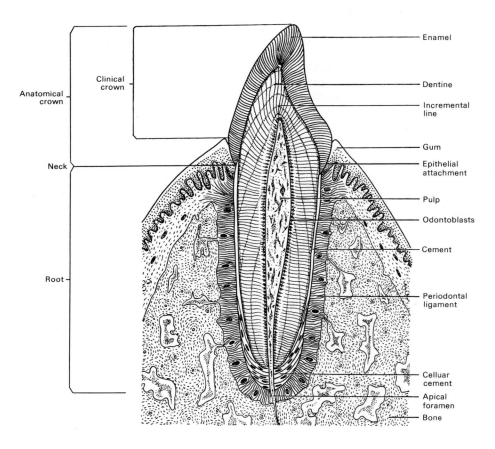

FIG. 6.10.   Vertical section through a tooth.

Except in disease states and other causes of interruption of dentine formation, the pulpal surface of dentine always shows a zone of **predentine**, which is not fully mineralized. This corresponds to the osteoid of bone surfaces where active osteogenesis is occurring.

Dentine is exquisitely sensitive to touch and temperature, especially at the dentine–enamel junction. Although nerve endings have been demonstrated adjacent to the odontoblasts of the pulp and in the proximal ends of the dentinal tubules, controversy still exists as to whether they reach the periphery of the dentine or whether some other sensory mechanism exists in which the odontoblast plays some kind of receptor role.

With advancing age and also in response to peripheral changes, such as tooth wear and dental caries, the dentinal tubules may undergo centripetal mineralization with consequent narrowing. This increases the degree of mineralization and results in a translucent and more impermeable dentine. Dentine contains incremental lines, which represent the forming front at various stages in the history of the tooth.

ENAMEL

Enamel, the hardest and most mineralized (97 per cent by weight) substance in the body, covers the crowns of the teeth. It is thickest over the occlusal surfaces and tapers sharply at the necks of the teeth where it meets the cementum covering the roots [FIG. 6.10]. It is translucent and, consequently, the colour of the crowns depends mainly on that of the underlying dentine.

Enamel is composed of a system of rods (prisms), approximately 4 μm in diameter, extending from the enamel–dentine junction to the surface of the tooth. The rods are arranged roughly at right angles to the junction, so that in the cervical region they lie almost horizontally, while in the crown they run either vertically or obliquely in relation to the long axis of the tooth. The mineral matter of enamel consists of hydroxyapatite crystals, similar in composition to those of dentine and bone. The crystals are tightly packed and in general are arranged with their long axes parallel to the long axes of the rods. The organic matter is rather unique in that its amino-acid composition differs from other major structural proteins, such as keratin (epidermal in origin) and collagen (mesodermal in origin). Though small in quantity by weight, the organic matter is distributed throughout the enamel and probably serves to cement the hydroxyapatite crystals together (Fearnhead 1967; Fearnhead and Stach 1971).

Ground sections of enamel show striae or incremental lines of Retzius. These follow the general contour of the crown of the tooth and illustrate the rhythmic formation of the enamel during development.

CEMENTUM

Cementum is a bone-like tissue that covers the roots of the teeth and by virtue of its collagen fibres (Sharpey fibres) provides a means of attachment of the periodontal ligament to the tooth. Since cementum can be added to by the cementoblasts, within the ligament, a mechanism exists for the repair and strengthening of its attachment throughout life, as for instance, in response to physiological tooth movements and to those brought about by orthodontic appliances.

Cementum has a chemical composition similar to that of bone. It has an irregularly lamellar structure and, although that over the greater part of the root is acellular, the thicker cementum over the apices of the roots and in forks between the roots is cellular. In these situations it is much more like bone, though never showing the regularity of structure of typical lamellar bone.

PERIODONTAL LIGAMENT

The periodontal ligament holds the root of the tooth in its bony sockets [FIG. 6.10]. It consists of fibrous connective tissue, which allows a limited amount of movement of the tooth during mastication. The ligament is formed of bundles of collagenous fibres, embedded at one end in the cementum and attached at the other to either the wall of the socket, the roots of adjacent teeth, or the lamina propria of the gingiva. Most of the fibres are obliquely orientated so that the roots are suspended within the alveolar bone, an arrangement resulting in the transfer of the forces of mastication to the walls of the socket and in the prevention of pressure on the vessels and nerves entering the apical foramen of the root of the tooth. In addition, transeptal fibres of the periodontal ligament unite the adjacent surfaces of the teeth above the alveolar septum.

Fibres attached to the gingiva are important in helping to maintain the close contact of the gingival epithelium and neck region of the tooth. There is continuous loss and replacement of the fibres as the ligaments adjust to functional or growth changes in the teeth and alveolar bone. Varying degrees of destruction of the fibres of the periodontal ligaments, with consequent loosening of the teeth, are common in later life. This is almost invariably associated with destruction of the epithelial attachment of the gingiva and a spread of infection along the roots of the teeth (Melcher and Bowen 1969). The nerves of the periodontal ligament provide important mechano-receptor information (Anderson, Hannar, and Mathews 1970).

PULP CAVITY AND PULP

The pulp cavity, enclosed by dentine, extends into the cusps and is prolonged into the roots as **root canals**, narrowing to the **apical foramen** [FIG. 6.10] which may have multiple openings. It is filled with tooth pulp, consisting of loose connective tissue, with collagen and reticular fibres. The surface of the pulp is covered with **odontoblasts**, which form a single layer of cells in young teeth, but become crowded and form a pseudostratified layer in the smaller adult pulp cavity. The odontoblasts send processes into the dentinal tubules. They are concerned with the nourishment and reaction of the dentine to injury; thus the dentine and pulp form a complex, similar to that of bone and its marrow.

Blood and lymph vessels and nerves enter the pulp via the apical foramina. The nerves at first lie in the central zone of the pulp and are myelinated, but spread out peripherally in the crown to form a plexus of unmyelinated nerves deep to and around the odontoblasts. Variable numbers of nerve fibres pass into the dentinal tubules accompanying the processes of the odontoblasts.

As dentine is formed slowly throughout life, and more rapidly in response to peripheral injury to the dentine, the pulp cavity and pulp become smaller as age advances.

DEVELOPMENT OF TEETH

The first sign of tooth development appears during the sixth week of intra-uterine life, when proliferative activity produces thickenings, the **primary epithelial bands**, in the epithelium on the opposing surfaces of the primitive jaws. Shortly after their formation, each band divides on its deep surface into an outer, vestibular lamina and an inner, dental lamina [FIG. 6.11].

The **vestibular lamina** grows into the underlying mesoderm and its central cells soon degenerate, converting it into a furrow, the **labiogingival groove**, which separates the lips and cheeks from the alveolar process and becomes the vestibule of the mouth. Within each **dental lamina**, localized cellular proliferations produce ten swellings, known as **tooth buds**, which are the first indication of the

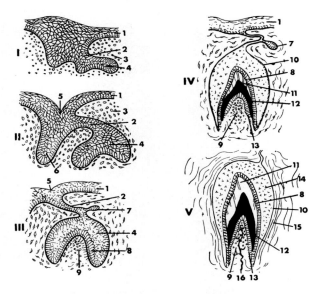

FIG. 6.11. Diagram to illustrate development of a tooth.
I. Shows the dental lamina (2), the surface epithelium (1), and the tooth bud (4).
II. Shows the further growth of the tooth bud and its invagination to form the enamel organ.
III. The enamel organ is more invaginated, and its inner layer of cells (8) becomes columnar. The dental lamina is thinner, but near its posterior or lingual edge there is an enlargement (7) which is the bud of a permanent tooth. The superficial cells of the dental papilla (9) are becoming columnar.
IV. The inner columnar cells of the enamel organ (ameloblasts, 8) have formed a cap of enamel (11) inside which the superficial cells of the papilla (odontoblasts, 13) have formed a layer of dentine (12).
V. Shows a still more advanced stage. The deposit of dentine is extending downwards and enclosing the papilla to form the future pulp, in which a vessel (16) is seen.

1. Epithelium.	9. Dental papilla.
2. Dental lamina.	10. Outer enamel epithelium.
3. Mesoderm.	11. Enamel.
4. Tooth bud.	12. Dentine.
5. Labiogingival groove.	13. Odontoblasts.
6. Labiogingival lamina.	14. Stellate reticulum.
7. Bud of permanent tooth.	15. Periodontal ligament.
8. Ameloblasts.	16. Blood vessel.

formation of the individual deciduous teeth. The factors initiating the localized proliferations are poorly understood, but there is accumulating evidence that a mesoderm of neural crest origin (ectomesoderm) is involved in this process.

As each tooth bud increases in size, it sinks more deeply into the substance of the jaw, although still maintaining its connection with the surface epithelium by cords of cells. The deep surface of the bud soon becomes invaginated and encloses a mass of proliferating mesoderm, the **dental papilla**. Since the epithelial bud rests on top of the dental papilla, this stage of development is often referred to as the cap stage. Later, the margins of the bud extend further to enclose the dental papilla, and the bud assumes a bell shape, hence the bell stage of development. When the epithelial portion of the developing tooth has reached this stage of differentiation, it is referred to as the **enamel organ**, indicating that it later produces enamel. The enamel organ and dental papilla, along with the surrounding condensed mesoderm (dental follicle), form the three component parts of a **tooth germ**.

During the bell stage of development, fluid accumulates between the central cells of the enamel organ, resulting in their separation

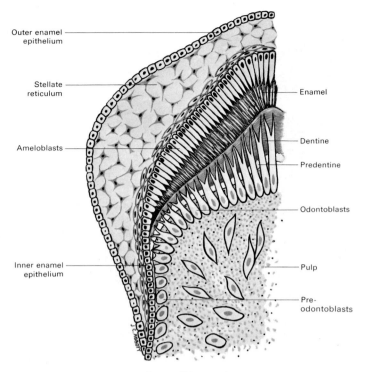

FIG. 6.12.   An enlargement of part of FIGURE 6.11v.

and the production of the **stellate reticulum**. The cells lining the hollow of the bell become cuboidal and form an orderly row, the **inner enamel epithelium**, which is continuous with the cuboidal cells on the outside, known as the **outer enamel epithelium**. As development of the enamel organ continues, the inner enamel epithelium induces the surface cells of the dental papilla to differentiate into **odontoblasts**, whose function is to produce the dentine of the tooth. These cells form a single layer abutting on the overlying inner enamel epithelium; the remaining cells of the papilla differentiate into the pulp [FIG. 6.12].

Dentine formation

Dentine formation is a rhythmic process which can be thought of as occurring in two stages; (1) the laying down of an organic matrix, the **predentine**, and (2) the mineralization of this matrix. In the growing tooth these two processes proceed at approximately the same rate, so that a uniform layer of predentine always lies internal to the mineralizing process. As the odontoblasts produce the predentine, it is deposited on the inner enamel epithelium, which, over the occlusal surface, establishes the cusp morphology of the tooth that is to be formed. With the mineralization of the predentine, this morphology becomes fixed. During the production of dentine the odontoblasts elongate, leaving behind processes which become enclosed by the dentine matrix, thus forming the dentinal tubules (Katchburian 1973; Weinstock and Leblond 1974).

Enamel formation

This commences as soon as a layer of mineralized dentine approximately 10 μm thick has formed. The first signs that it is about to commence are seen in the cells of the internal enamel epithelium, which become columnar in shape and acquire the organelles associated with the production of protein for export. The cells, now known as **ameloblasts**, become polarized in such a way

that their secretion, the enamel precursor, **amelogenin**, is deposited on the surface of the already formed dentine (Josephensen and Fejerskov 1977; Moe 1971).

At first enamel is deposited as an organic matrix, which later becomes mineralized with hydroxyapatite crystals. These crystals are deposited in a highly orientated fashion which gives rise to the rod structure of the enamel. When first deposited the enamel contains only about a third the mineral it will eventually contain. The rest is added by a process of secondary mineralization (maturation), during which not only does mineral seep into the enamel from the enamel organ, but at the same time most of the protein matrix is removed.

During enamel and dentine formation, the two layers of formative cells are gradually separated by the accumulation of their own secretions. Thus the oldest enamel and dentine are in apposition. As mineralization of the enamel becomes complete, the ameloblasts lose their columnar morphology and, together with the other cells of the enamel organ, form a layer, two to three cells thick, of polygonal cells, known as the **reduced enamel epithelium**. This, together with the last formed layer of enamel (primary cuticle) which is not mineralized, is known as **Nasmyth's membrane**. Nasmyth's membrane is present on the tooth surface at the time of eruption and is quickly worn away by attrition.

Root formation

At the cervical limit of the crown the stellate reticulum of the enamel organ is lost, but the inner and outer enamel epithelia continue to proliferate into the underlying mesoderm as a sheet of epithelium (epithelial root sheath of Hertwig), which has an inductive function on the marginal cells of the dental papilla inducing them to become odontoblasts. These produce the dentine of the root in a manner similar to the dentine of the crown. When its inductive function is complete the epithelial root-sheath degenerates, forming a network of cell rests (Malassez) between which mesodermal cells of the follicle approach the newly formed dentinal surface of the root. Here these cells differentiate into cementoblasts.

As root formation becomes complete the open end of the developing root becomes narrowed, persisting as a canal or canals through which vessels and nerves pass to the dental pulp.

Formation of cementum and the periodontal ligament

The cementum and periodontal ligament develop within the dental follicle. Cementum is deposited on the surface of the root dentine by **cementoblasts**, which resemble osteoblasts of bone. During its formation cementum incorporates collagen fibres and Sharpey fibres, which are continuous with the fibrous tissue of the rest of the follicle. This fibrous tissue differentiates further and changes its orientation to become the periodontal ligament.

Formation of the permanent teeth

In the advanced bell stage of tooth development a proliferation of epithelium on the lingual side of the deciduous enamel organ, close to its attachment to the dental lamina, gives rise to the permanent tooth bud. Subsequent development follows the same sequence as for the deciduous teeth. The three permanent molars have no deciduous predecessors, and the tooth buds develop successively from a subepithelial extension of the deep surface of the posterior end of the dental lamina.

The deciduous tooth germs at first lie in a continuous trough in the bone of the developing jaws, but gradually septa appear between the germs so that each lies in a separate socket or crypt. Successional tooth germs lie at first in the same crypts as their predecessors, but in due course they become surrounded by their own bony crypts [FIG. 6.13].

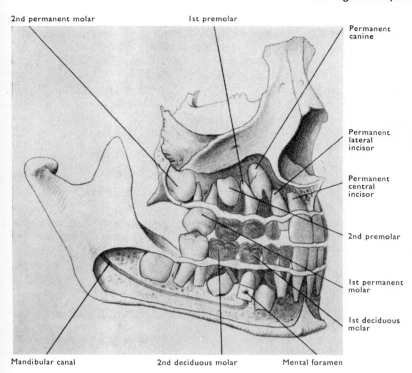

2nd permanent molar   1st premolar

Permanent canine

Permanent lateral incisor

Permanent central incisor

2nd premolar

1st permanent molar

1st deciduous molar

Mandibular canal   2nd deciduous molar   Mental foramen

(i)

FIG. 6.13.   Teeth of child over 7 years old. (Modified from Testut.)

By the removal of the bony outer wall of the alveoli, the roots of the teeth which have been erupted, and the permanent teeth which are still embedded in the mandible and maxilla, have been exposed. The deciduous teeth are shown dark, the permanent teeth light.

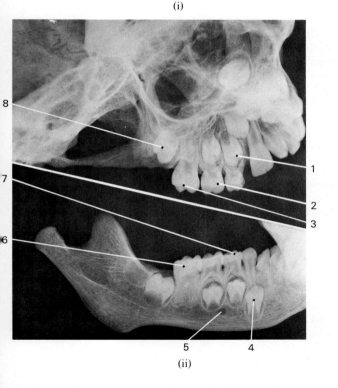

(ii)

FIG. 6.14.   (i) Oblique radiograph of maxilla at 7 years. (ii) Radiograph of half mandible at 7 years.
1. Crown of 1st premolar in its bony crypt.
2. 2nd deciduous molar.
3. 1st permanent molar.
4. Crown of permanent canine in its bony crypt.
5. Mental foramen.
6. 1st permanent molar.
7. 1st deciduous molar.
8. Crown of 2nd permanent molar in its bony crypt.

ERUPTION

Eruptive movements leading to the emergence of the tooth through the gingiva to reach occlusion with teeth in the opposite jaw begin as soon as the crowns of the deciduous teeth are complete. The pre-emergent movements are closely correlated with growth of the jaws and are by no means solely in the direction of the tooth axis; e.g. the occlusal surfaces of the upper permanent molars at first are directed backwards and outwards. However, with the backward extension of the maxilla, the molars gradually rotate downwards, so that shortly before emergence (clinical eruption) the occlusal surface faces downwards.

Growth in length of the roots occurs at the same time as eruption; bone resorption and bone remodelling must also be associated. However, none of these alone is the mechanism or force that causes the tooth to erupt. Teeth can, for instance, erupt long after their roots are complete in length. The force seems to be cellular activity of some kind and most theories of eruption attempt to explain it in terms of a force that pushes the tooth into position. (For a fuller discussion, see specialized texts.)

As the cusps of a tooth finally emerge through the gingiva, the reduced enamel epithelium fuses with the epithelium of the gingiva, which together with the primary cuticle forms the epithelial attachment.

As soon as cusps of opposing erupting teeth come into contact, they exert mutual guiding forces which help to establish the intercuspid relationship. With the first emergence of the primary dentition, opposing pressures of tongue, lips, and cheeks also help to guide the teeth into their arcades.

The roots of the primary teeth, according to a time sequence related to the stages of development of their successors, begin to undergo resorption which finally results in the tooth being shed. The successor then emerges to take its place.

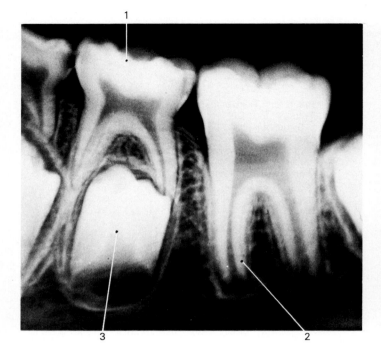

FIG. 6.15.   Intra-oral radiograph of mandibular molars at 7 years—mesial to the left.
1. 2nd deciduous molar. The roots are being absorbed.
2. Root of 1st permanent molar incompletely formed.
3. 2nd premolar in its bony crypt. Only the crown is heavily calcified.

Approximate times of eruption of deciduous and permanent teeth

Deciduous		Permanent	
I_1	6–8 months	I_1	6–8 years
I_2	8–10 months	I_2	7–9 years
C	16–20 months	C	9–12 years
M_1	12–16 months	P_1	10–12 years
M_2	20–30 months	P_2	10–12 years
		M_1	6–7 years
		M_2	11–13 years
		M_3	17–21 years

NERVE SUPPLY OF THE TEETH AND GUMS

The pulps of the lower molar and premolar teeth are supplied by branches of the inferior alveolar nerve, and those of the canine and incisors by the incisive branch of this same nerve. The gingiva and supporting tissues of the lower teeth are supplied on the lingual aspect by the lingual nerve, while on the buccal aspect they are supplied in the molar region by branches of the inferior alveolar nerve, together with branches of the buccal nerve. The supporting tissues anterior to the molars are supplied by branches of the mental nerve. Anaesthesia of the lower teeth, but not all the gum, can be obtained by injection of the inferior alveolar nerve as it enters the mandible.

The teeth of the upper jaw are innervated by a plexus of nerves, which lies within the substance of the maxilla and is formed from the anterior and posterior superior alveolar nerves and the inconstant middle superior alveolar nerve. The posterior superior alveolar nerves enter canals on the posterior and lateral walls of the maxilla, while the anterior and middle nerves, branches of infra-orbital nerve, descend in canals in the anterior wall of the maxillary

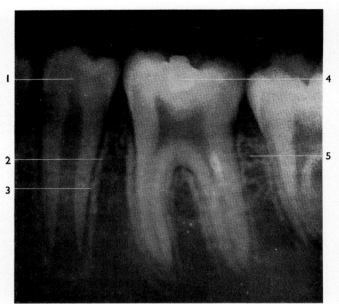

FIG. 6.16.   Intra-oral radiograph of mandibular premolar and molars at 14 years—mesial to left.
1. 2nd premolar, its root has not yet been completely formed.
2. Lamina dura (compact bone) of socket.
3. Periodontal space, for periodontal membrane.
4. Radiopaque filling in crown of 1st molar.
5. Bone trabeculae of interalveolar septum.

sinus. The gingiva and supporting tissues of the upper teeth are supplied on the lingual aspect, as far forward as the canines, by the greater palatine nerve, and by the nasopalatine nerve in the incisor region. The buccal aspects of the gingiva receives branches of the superior dental plexus and additionally, as a rule, in the molar region, branches from the buccal nerve.

Tongue

The tongue is a muscular organ involved in sucking, chewing, swallowing, and speech. It is covered by a mucous membrane which, like the lips and anterior part of the palate, is more sensitive to touch, pain, and temperature than the tips of the fingers, and in addition subserves the sensation of taste by means of its contained taste buds.

In shape the tongue is sometimes likened to an inverted boot with a root through which extrinsic muscles pass to attach it to the soft palate, the styloid process, the hyoid bone, and the mandible and a body consisting largely of muscle intermingled with lingual salivary glands, connective tissue, nerves, vessels, and lymph follicles. The body has a convex dorsum which comes into contact with the hard and soft palates when the mouth is closed, an inferior surface resting on the floor of the mouth and a highly mobile tip [FIG. 6.27].

The dorsum of the tongue is divided by an ill-defined V-shaped groove, the sulcus terminalis, into an anterior two-thirds, or oral part, facing upwards and a posterior one-third, or pharyngeal part, facing backwards into the pharynx. The apex of the sulcus terminalis points posteriorly and coincides with a pit, the foramen caecum, representing the site of origin of the thyroglossal duct [p. 597]. When traced laterally the limbs of the sulcus meet the lateral borders of the tongue at the attachment of the palatoglossal arch [FIG. 6.17].

The mucous membrane of the tongue is formed of a stratified squamous epithelium resting on a lamina propria of connective tissue. It varies in structure in different regions. Over the anterior

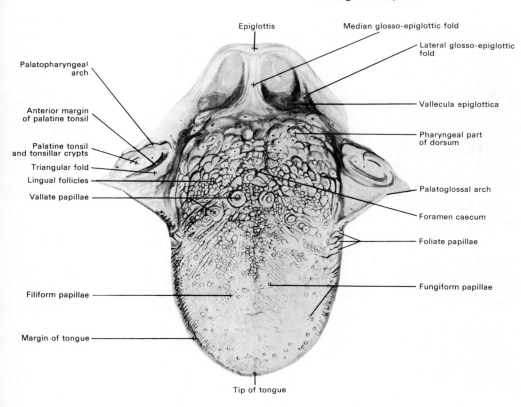

Epiglottis

Median glosso-epiglottic fold

Lateral glosso-epiglottic fold

Palatopharyngeal arch

Anterior margin of palatine tonsil

Palatine tonsil and tonsillar crypts

Triangular fold

Lingual follicles

Vallate papillae

Filiform papillae

Margin of tongue

Tip of tongue

Vallecula epiglottica

Pharyngeal part of dorsum

Palatoglossal arch

Foramen caecum

Foliate papillae

Fungiform papillae

FIG. 6.17.   Dorsum of tongue and the palatine tonsils.

two-thirds of the dorsum the epithelium is thick and keratinized. It is firmly attached through the lamina propria to the connective tissue surrounding the underlying muscle, and has a velvety or slightly rough appearance due to the presence of large numbers of epithelial papillae.

In Man three types of papillae are recognized, namely, filiform, fungiform, and vallate [FIG. 6.17]. Filiform papillae are conical in shape, measure 2–3 mm in length, and are by far the most numerous. Except near the tip of the tongue they tend to be arranged in rows parallel to the sulcus terminalis. Each papilla has a branching core of connective tissue which is intimately covered by the epithelium. On the apices of the papillae the surface cells of the epithelium are transformed into hard scales which give the oral part of the tongue its normal furred appearance and contribute to the slight roughness helpful in moving food within the mouth. Fungiform papillae are relatively few in number and are more concentrated at the sides and tip of the tongue. They are larger than the filiform papillae, are globular in shape, and in the living subject appear red due to their rich blood supply. Vallate papillae are the largest, measuring 1– 3 mm across, and varying from eight to twelve in number. They are situated immediately in front of the sulcus terminalis, often with a particularly prominent one anterior to the foramen caecum. Each papilla resembles an inverted truncated cone which projects slightly above the general level of the tongue and is surrounded by an epithelial-lined furrow or sulcus. Ducts of serous glands situated in the lamina propria and submucosa pour saliva into the bases of the sulci and thereby help to expel particles of food and debris. The epithelium on the sides of the furrow and especially that on the side of the papilla contains numerous taste buds.

At the sides of the tongue, opposite the molar teeth, the mucous membrane forms four or five vertical, red folds, the foliate papillae. These are rudimentary in man but they may be subjected to trauma by the teeth if they are enlarged in pathological processes.

On the posterior one-third the mucous membrane lacks papillae and has a relatively smooth surface. Less firmly attached than that of the anterior two-thirds, it bulges due to underlying collections of lymph follicles which together form the lingual tonsil. The epithelium invaginates the follicles to form lingual crypts, which may contain collections of lymphocytes that have migrated through the epithelium. Posteriorly, the mucous membrane is reflected onto the anterior surface of the epiglottis in a median and two lateral glossoepiglottic folds. The two depressions thus formed on each side of the median fold are known as the epiglottic valleculae [FIG. 6.17].

The undersurface of the tongue is covered by a smooth mucous membrane with a relatively thin epithelium. It is bound to the musculature of the tongue and in the midline it forms a fold, the frenulum, running from the tip of the tongue to the floor of the mouth. Lateral to the frenulum, on each side, there is a fringe of mucous membrane, the plica fimbriata, and medial to this the deep lingual vein can be seen as it courses backwards to unite with the sublingual vein [FIG. 6.4].

Taste buds have a wide distribution in the mouth and pharynx. They are most numerous on the sides of the vallate papillae and on the foliate papillae in front of the palatoglossal arch. They are also present on the fungiform papillae, the posterior third of the tongue, the inferior surface of the soft palate, the palatoglossal arch, the posterior wall of the oral part of the pharynx, and the posterior surface of the epiglottis. Their structure is described on page 868.

As already noted, the substance of the tongue consists largely of extrinsic and intrinsic muscles, arranged both parallel and at right angles to the long axis of the tongue. When the muscles are paralysed, or in the unconscious patient, the tongue may fall backwards into the pharynx and obstruct respiration. Paralysis of the muscles on one side results in the tongue being pulled to the opposite side in the mouth and protruded to the paralysed side. Fibrous connective tissue containing blood and lymph vessels and nerves surrounds the muscle bundles and gives attachment to the dense lamina propria of the mucous membrane. In the midline there is a fibrous septum which extends throughout the length of the

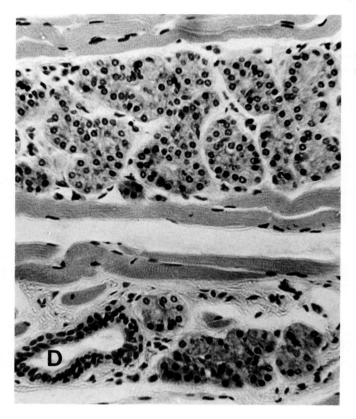

FIG. 6.18.   Serous salivary glands situated among striated muscle fibres of the tongue. A duct (D) is present in the lower region of the section.

organ. It fails to reach the dorsum of the tongue, whilst posteriorly it fans out and is attached to the hyoid bone. Numerous small salivary glands are also present within the tongue [FIG. 6.18]. They are most common in the pharyngeal region, although on the under surface near the tip there is a prominent group, the anterior lingual glands, on each side of the frenulum. These glands are mainly mucous and their ducts open on the plica fimbriata. The glands of the pharyngeal region of the tongue are mostly serous and many of their ducts open into the sulci surrounding the vallate papillae. Some of the ducts also open into the epithelial crypts of the lingual tonsil.

BLOOD VESSELS AND NERVES OF THE TONGUE

The main blood supply to the tongue is derived on each side from the lingual artery, a branch of the external carotid. In its course the artery gives off dorsal branches to the posterior part of the tongue. It then continues as the deep artery on the inferior surface close to the frenulum to the tip of the tongue, where it anastomoses with the artery of the opposite side. Apart from this anastomosis there is little communication between the arteries of the two sides. In addition to the dorsal branches of the lingual artery, the posterior part of the tongue receives blood from the tonsillar and ascending palatine branches of the facial artery. The venous blood is drained by the dorsal lingual veins which join the lingual veins accompanying the lingual artery, and by the deep lingual veins which course backwards from the tip of the tongue, deep to the mucous membrane on the inferior surface to unite with the sublingual vein and accompany the hypoglossal nerve.

The lymph vessels of the tongue are described on page 1002.

The sensory nerves of the tongue are as follows: (i) the lingual branch of the mandibular division of the trigeminal nerve, which is the nerve of common sensation to the anterior two-thirds; (ii) the chorda tympani, a branch of the facial nerve, which is the nerve of taste to the anterior two-thirds with the exception of the vallate papillae; (iii) the lingual branch of the glossopharyngeal nerve, which is the nerve of common sensation and of taste to the posterior one-third and to the vallate papillae; and (iv) the superior laryngeal nerve, which is the nerve of taste in the region of the valleculae [FIG. 6.19]. With the exception of the palatoglossus (innervated by the pharyngeal plexus) all the extrinsic and intrinsic muscles receive motor nerve fibres from the hypoglossal nerve.

DEVELOPMENT OF THE TONGUE

The mucous membrane, connective tissue, blood vessels, lymph tissue, and lingual salivary glands develop from the first, third, and fourth pharyngeal arches. The striated muscle is derived from myotomes of the occipital somites which in migrating forwards bring their segmental nerves with them, these later become grouped together to form the hypoglossal nerve.

As a result of the proliferation of the branchial mesoderm a number of elevations are formed in the floor of the primitive pharynx [FIG. 2.68]. The first to appear, the tuberculum impar, lies in the midline between, and slightly caudal to, the first arch. It is quickly followed by bilateral lingual swellings in the ventromedial ends of this arch and these rapidly increase in size and merge with each other and with the tuberculum impar. The resulting mass of tissue forms the anterior two-thirds of the tongue and since it is derived from the mandibular arch (first arch) its nerve supply comes from the mandibular nerve. Immediately caudal to the tuberculum impar the entoderm is invaginated to form the thyroglossal duct, the site of this invagination is indicated in the adult by the foramen caecum.

The second arch makes little or no contribution except that a branch of its nerve (the chorda tympani branch of the facial nerve) grows forwards to innervate most of the taste buds of the anterior two-thirds of the tongue.

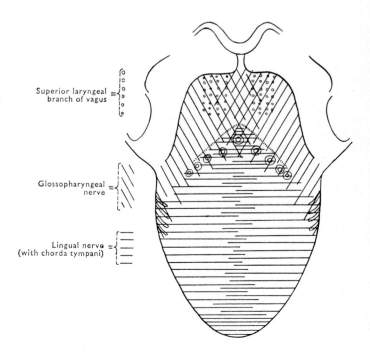

Superior laryngeal branch of vagus =

Glossopharyngeal nerve =

Lingual nerve (with chorda tympani) =

FIG. 6.19.   Areas of afferent nerve supply of tongue.

A further elevation known as the **hypobranchial eminence** gives rise to the posterior one-third of the tongue. It appears in the midline between the ventral ends of the third and fourth arches. The mesoderm of its cranial end grows forwards over the ends of the second arch to fuse with the lingual swellings and the tuberculum impar; the line of fusion corresponding to the sulcus terminalis. As the cranial end of the hypobranchial eminence, derived from the third arch, makes the major contribution to the posterior third, the nerve supply to this area comes from the glossopharyngeal nerve, the nerve of the third arch. It should be noted that whilst the line of fusion of mesoderm corresponds to the sulcus terminalis that of the epithelium occurs somewhat more anteriorly, hence the innervation of the vallate papillae by the glossopharyngeal nerve. The most posterior part, derived from the contribution of the fourth arch to the hypobranchial eminence, receives its nerve supply from the superior laryngeal branch of the vagus, the nerve of the fourth arch.

ANOMALIES

An excessively large tongue, **macroglossia**, due to hypertrophy of the lingual muscles is sometimes present in conditions such as cretinism and mongolism. Failure of fusion of the lateral lingual swellings gives rise to a **bifid tongue**, or when less severe, a **cleft tongue**. Abnormalities related to the development of the thyroid are dealt with on page 597.

Glands of the digestive system

The digestive system contains both **exocrine** and **endocrine glands**. Exocrine glands discharge their secretion through ducts on to the surfaces of the body, whereas endocrine glands, lacking a duct system, release their secretions into adjacent blood or lymph vessels. Both types of gland may be either unicellular, for example goblet cells and argentaffin cells, or multicellular, for example salivary glands and islets of Langherhans. Unicellular exocrine glands are incorporated in the lining epithelium of viscera. Multicellular exocrine glands are in continuity with the epithelium by means of ducts which convey and sometimes modify the secretion produced by the secretory cells of the gland.

Exocrine glands are classified according to the complexity of their ducts and the shape of their secretory units [FIG. 6.20]. Glands with unbranched ducts are called **simple glands**; those with branched ducts, **compound glands**. Secretory units may be either tubular or flask-like in shape, or a combination of both. Flask-like units are usually referred to as **alveoli** or **acini**.

The classification of exocrine glands can be further modified according to the type of secretion they produce. This may be either mucous or serous, or sometimes both, in which case the gland is said to be a **mixed gland**.

In mixed glands the individual alveoli may be composed solely of mucous or of serous cells, or they may contain both. Where both are present the serous cells form a cresentic cap, the **demilune**, on top of the mucous cells and their secretion reaches the alveolar lumen by canaliculi between the mucous cells.

Microscopically, the secretory cells of serous and mucous glands show considerable differences. Each **serous cell** contains a basally-situated, spherical or ovoid nucleus and, with conventional dyes such as haematoxylin and eosin, the cytoplasm shows basophilia due to the presence of ribonucleic acid. The apical half of the cell usually contains large numbers of secretory granules, which can individually be resolved with the light microscope. Ultrastructural study of these cells shows the cytoplasm to contain much rough

endoplasmic reticulum (RER) for the synthesis of secretory protein, and a well-developed supranuclear Golgi complex which packages the secretion in membrane bound granules. These are stored in the apical cytoplasm and their contents are discharged by reverse pinocytosis (exocytosis) (Barrowman 1975).

In the **mucous cell** the nucleus stains deeply with haematoxylin and eosin and, because of the presence of a large number of secretion granules, is usually flattened and applied to the basal plasma membrane. The granules of secretion, mucinogen droplets, tend to coalesce during specimen preparation and cannot easily be resolved by light microscopy. (They are best demonstrated by stains which show their polysaccharide content, e.g. the periodic acid–Schiff reaction.) The cytoplasm, containing much RER for the synthesis of the protein of mucin, is compressed to the periphery of the cell. A highly-developed Golgi complex is present and this not only packages the protein formed in the RER, but also adds a sugar moiety, thus converting the protein into a glycoprotein. In certain situations, such as the gastric pits, it is also responsible for the sulphation of the glycoprotein. As in serous cells, discharge of mucus is by reverse pinocytosis.

All multicellular glands are embedded in connective tissue which may be condensed to form a capsule. Except in the smallest glands connective tissue septa subdivide the gland into lobes and lobules. Septa dividing the gland into lobes are relatively thick and are known as interlobar septa; those dividing the lobes into lobules are extremely thin and are called interlobular septa. As well as providing mechanical support for the gland, the capsule and septa convey blood and lymph vessels, nerves, and ducts from the secretory tissue.

The **duct system** of compound glands may be extremely complex resembling a tree with its main trunk, the main excretory duct, and a succession of branches of diminishing size known as lobar, interlobular, intralobular, and intercalated ducts, the last being continuous with and draining the secretory units. In the large paired

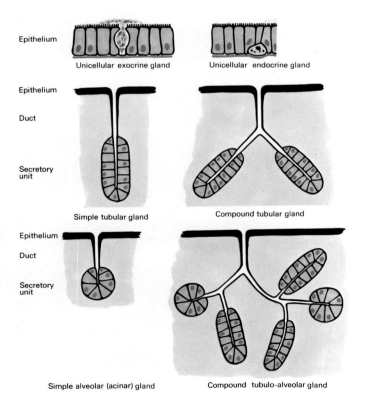

Epithelium — Unicellular exocrine gland — Unicellular endocrine gland

Epithelium

Duct

Secretory unit

Simple tubular gland — Compound tubular gland

Epithelium

Duct

Secretory unit

Simple alveolar (acinar) gland — Compound tubulo-alveolar gland

FIG. 6.20.   Various types of glands.

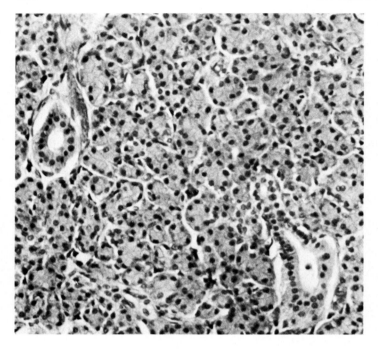

FIG. 6.21.   Photomicrograph of a section of the parotid gland showing serous alveoli and two ducts lined by cuboidal epithelium.

salivary glands an additional, striated duct is present between the intercalated and intralobular duct. The cells of the intercalated duct are low cuboidal and contain secretory granules similar to those of the alveolus being drained. In the striated duct the epithelium is composed of columnar cells in which the mitochondria are vertically orientated in compartments, produced by the invagination of the basal plasma membrane. These cells are largely responsible for the final adjustment of the composition of the saliva. This they do by absorbing sodium and water and secreting potassium. The lobar, interlobular, and intralobular ducts are lined with tall columnar epithelium.

The discharge of secretion from the salivary glands, sweat glands, and mammary glands is aided by myoepithelial cells. These are large branching cells with numerous processes that surround the alveoli and smaller ducts. They have many of the characteristics of smooth muscle fibres and are believed to be capable of expressing the secretion by contraction of their processes.

Salivary glands

The salivary glands secrete saliva which is a colourless, opalescent, hypotonic fluid. It contains water, mucopolysaccharides, immuno-globulins, and inorganic substances including sodium, calcium, phosphorous, potassium chloride, and traces of iron and iodine. It also contains the enzyme amylase, which initiates the breakdown of carbohydrate to maltose and a small amount of glucose. Apart from its seemingly insignificant role in digestion, saliva lubricates and cleanses the mouth and pharynx, and by moistening the food aids in swallowing and in the appreciation of taste.

There are three pairs of large salivary glands, namely the **parotid**, the **submandibular**, and the **sublingual**, and numerous small glands in the palate, tongue, lips, and cheeks, which open directly on to the surface of the overlying mucous membrane. The composition of the saliva produced by the various pairs of large glands differs and reflects mainly their relative proportion of mucous and serous cells. That from the parotid is watery and rich in amylase, whilst that from the sublingual is extremely viscous, containing much mucus and little amylase. The composition of the saliva from the submandibular is intermediate between these two extremes.

Approximately 750 ml of saliva are produced daily, of which 25 to 35 per cent is contributed by the parotids, 60 to 70 per cent by the

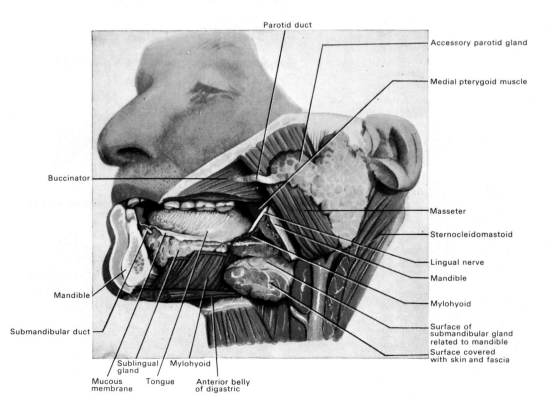

FIG. 6.22.   Salivary glands and their ducts.

submandibulars, and about 3 per cent by the sublinguals. The production of secretion by the large glands is intermittent and is under nervous control, whereas that from the small glands appears to be more or less continuous (Amsterdam, Ohad, and Schramm 1969).

PAROTID GLANDS

The parotid is the largest of the paired salivary glands. It is a compound gland and in the adult contains only serous acini [FIG. 6.21]; in the infant occasional mucous units may be found. It occupies an irregular wedge-shaped space between the ramus of the mandible anteriorly, the styloid process medially, the external acoustic meatus superiorly, and the mastoid process posteriorly. Inferiorly the space narrows by the approximation of the sternocleidomastoid muscle to the angle of the mandible [FIG. 6.22]. The gland is surrounded by a condensation of the investing layer of deep **fascia** of the neck which, being strong and firmly attached to surrounding bony points, resists swelling of the gland in pathological states. A condensation of this fascia, the **stylomandibular ligament,** separates the parotid from the submandibular gland [FIG. 4.21]. For descriptive purposes, the gland is described as having four surfaces, namely superficial or lateral, anteromedial, posteromedial, and superior, and three borders—anterior, medial, and posterior.

The **superficial surface** is nearly flat and extends almost to the zygomatic arch above and downwards to the angle of the mandible. Anteriorly it extends over the surface of the masseter muscle whilst posteriorly it encroaches on the anterior border of sternocleidomastoid [FIG. 6.22]. A few small lymph nodes are embedded in the surface of the gland, others are in its substance.

The concave **anteromedial surface** is in contact with the masseter, the ramus of the mandible, and the medial pterygoid [FIG. 6.6]. It usually sends an extension forwards between the ramus of the mandible and the medial pterygoid. The **posteromedial surface** is in contact with the mastoid process, sternocleidomastoid, the posterior belly of the digastric, the styloid process and its attached muscles, and, at a deeper level, the internal carotid artery and internal jugular vein along with the last four cranial nerves. The medial border between the posteromedial and anteromedial surfaces may project inwards in front of the styloid process to come close to the pharyngeal wall. The superior surface is relatively small and is applied to the floor and anterior wall of the external acoustic meatus as well as the posterior surface of the temporomandibular joint. It is this part of the gland especially which is affected by movements of the jaw and if the gland is inflamed gives rise to considerable pain (Cope and Williams 1973).

The **parotid duct** leaves the anterior border of the gland about its middle and runs horizontally forwards sometimes associated with accessory parotid tissue on the surface of the masseter [FIG. 6.22]. At the anterior edge of the masseter it turns medially and, having pierced the buccal pad of fat and the buccinator muscle, opens on a small papilla opposite the second upper molar tooth. Its course is indicated by the middle part of a line from the lower margin of the concha [FIG. 12.35] to the mid-point between the ala of the nose and the red margin of the upper lip. In its terminal part the duct runs deep to the mucous membrane of the cheek for about a quarter of an inch thus forming a valvular mechanism preventing the entrance of air or food during increased intraoral pressure. The walls of the duct are thick and are composed of fibrous tissue with some smooth muscle fibres. Towards its opening the columnar epithelium first becomes pseudostratified, then stratified, and eventually stratified squamous just before it joins with the epithelium of the mouth. The duct can be palpated on the surface of the contracting masseter and it and its ramifications may be examined radiologically after injecting a radiopaque dye through a canula placed in its orifice.

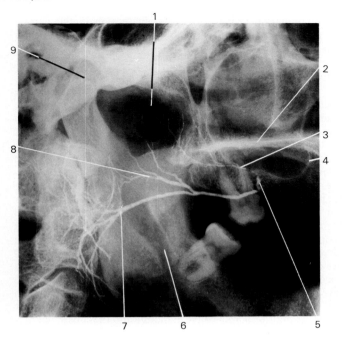

FIG. 6.23. Parotid sialogram produced by injection of X-ray opaque material into the oral opening of the duct.
1. Nasal part of pharynx, closed below by soft palate.
2. Hard palate.
3. Accessory duct.
4. Wall of maxillary sinus.
5. Opening of main duct into mouth.
6. Mandibular canal.
7. Main parotid duct branching into gland.
8. Accessory duct.
9. Temporomandibular joint.

Structures traversing the gland

The **facial nerve** enters the posteromedial surface and passes forwards superficial to the contained vessels. Within the gland it divides into its five terminal branches which radiate from the anterior border along with the parotid duct and the **transverse facial artery.** The external carotid artery also enters the posteromedial surface and after ascending to the level of the neck of the mandible divides into the maxillary and superficial temporal arteries [FIG. 13.43]. The **maxillary artery** emerges from the anteromedial surface and the **superficial temporal artery** leaves the superior corner of the gland along with its vein, the temporal branch of the facial nerve, and the **auriculotemporal nerve.** Between the facial nerve and the external carotid artery the **retromandibular vein** is formed from the union of the superficial temporal and maxillary veins. At the lower border of the gland it emerges as two branches, the anterior to join the facial vein and the posterior to join the posterior auricular vein.

Blood and lymph vessels and nerves

The parotid receives its blood directly from the external carotid artery and its branches and it drains into the retromandibular vein. Its lymph vessels pass to the parotid lymph nodes and thence to the superficial and deep cervical nodes. Secretomotor nerve fibres to the gland travel with the glossopharyngeal nerve and pass to the otic ganglion via the tympanic branch, the tympanic plexus, and the lesser petrosal nerve. They relay in the otic ganglion and the postganglionic fibres reach the gland via the auriculotemporal nerve. Sympathetic fibres travel to the gland with the external carotid artery. Pain felt in rapidly growing tumours and swellings due to acute inflammation probably arises from tension in the parotid fascia or pressure on the auriculotemporal and great auricular nerves.

SUBMANDIBULAR GLANDS

The submandibular gland is approximately half the size of the parotid. It is a mixed gland in which serous cells predominate. Both mucous and serous alveoli are present in addition to which serous cells also form demilunes on the surface of mucous alveoli [FIG. 6.24]. The gland has a large superficial part continuous, round the posterior border of the mylohyoid, with a small deep part.

The superficial part occupies the digastric triangle and is loosely surrounded by superficial and deep layers of the deep cervical fascia which pass respectively to the inferior border and the mylohyoid line of the mandible. It has three surfaces, a lateral, an inferior, and a medial. Its lateral surface comes into contact anteriorly with a shallow fossa on the medial surface of the body of the mandible and posteriorly with the insertion of the medial pterygoid. The inferior surface is covered by skin, superficial fascia with the platysma, the facial vein and the cervical branch of the facial nerve, and deep fascia. Three or four submandibular lymph nodes are present between the inferior border of the mandible and the superficial part

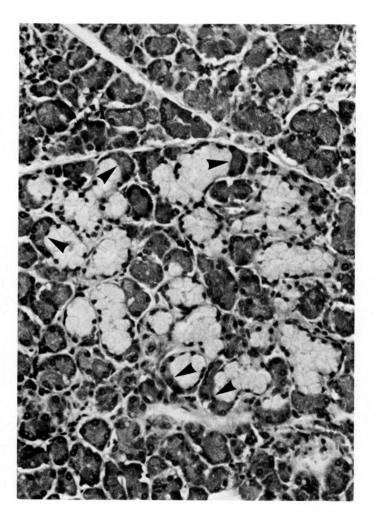

FIG. 6.24.   Section of the submandibular salivary glands showing serous alveoli above and below, and mucous alveoli with serous demilunes (arrow heads) in the centre. Connective tissue septa pass through the upper part of the section.

of the gland or sometimes embedded within the gland. The medial surface is in contact from before backwards with (i) mylohyoid, the mylohyoid nerve and vessels intervening, and (ii) hyoglossus, from which it is partly separated by the hypoglossal nerve and the deep lingual vein, digastric, and stylohyoid. Behind the hyoglossus the gland is separated from the pharynx by the stylohyoid ligament, middle constrictor, and the glossopharyngeal nerve. Posteriorly, the gland is separated from the parotid by the stylomandibular ligament. The facial artery courses upwards on the medial surface of the gland and then loops downwards between its lateral surface and the insertion of the medial pterygoid to reach the inferior border of the mandible.

The deep part of the gland extends anteriorly on the hyoglossus, superior to the mylohyoid, almost to the posterior end of the sublingual gland. The lingual nerve lies above it on the hyoglossus and inferiorly the hypoglossal nerve passes forwards to its distribution to the muscles of the tongue. Superiorly this part of the gland lies under the mucous membrane of the floor of the mouth, where it may be palpated.

The submandibular duct begins in the superficial part and winds round the posterior border of the mylohyoid in the gland substance. It passes forwards between the deep part of the gland and the hyoglossus with the lingual nerve above and the hypoglossal nerve below. Near the anterior border of the hyoglossus it is crossed superficially by the lingual nerve before passing on to the genioglossus deep to the sublingual gland. The duct opens on the floor of the mouth on the summit of the sublingual papilla at the side of the frenulum [FIG. 6.4]. It is 4 to 5 cm long and like the parotid duct its orifice is its narrowest part. Its wall is thin and elastic (Shackleford and Schneyer 1971).

Blood and lymph vessels and nerves

The gland receives its blood from branches of the facial and lingual arteries and is drained by the accompanying veins. Its lymphatics pass directly to the submandibular lymph nodes. Preganglionic parasympathetic fibres travel in the chorda tympani branch of the facial nerve and relay in the submandibular ganglion. From here the postganglionic fibres pass direct to the gland. Sympathetic fibres reach the gland in company with the facial artery.

SUBLINGUAL GLANDS

The sublingual gland is the smallest of the three major glands. It is a mixed gland in which mucous cells predominate [FIG. 6.26]. Few purely serous alveoli are present, the serous cells being arranged almost entirely as demilunes. The gland rests medially on the submandibular duct and lingual nerve which separate it from the underlying genioglossus [FIG. 6.25]. Above it raises the mucous membrane of the floor of the mouth to form the sublingual fold whilst laterally it is in contact with a shallow depression on the mandible above the mylohyoid line. Its anterior end meets the gland of the opposite side near the midline. It is drained by ten to twenty small ducts which pass from its upper border to open on the sublingual fold, occasionally some of these open into the submandibular duct [FIG. 6.4].

The gland receives its blood supply from the submental and sublingual arteries. Its nerve supply is from the same source as that of the submandibular gland.

RADIOLOGY

Plane films of the salivary glands are of value in demonstrating calculi (stones) in the duct system. These can be seen because of

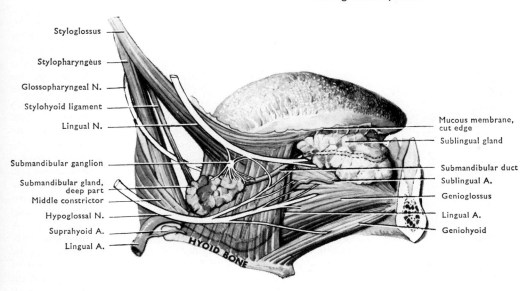

Styloglossus
Stylopharyngèus
Glossopharyngeal N.
Stylohyoid ligament
Lingual N.
Submandibular ganglion
Submandibular gland, deep part
Middle constrictor
Hypoglossal N.
Suprahyoid A.
Lingual A.
HYOID BONE

Mucous membrane, cut edge
Sublingual gland
Submandibular duct
Sublingual A.
Genioglossus
Lingual A.
Geniohyoid

FIG. 6.25A. Lingual and hypoglossal nerves in the submandibular region.

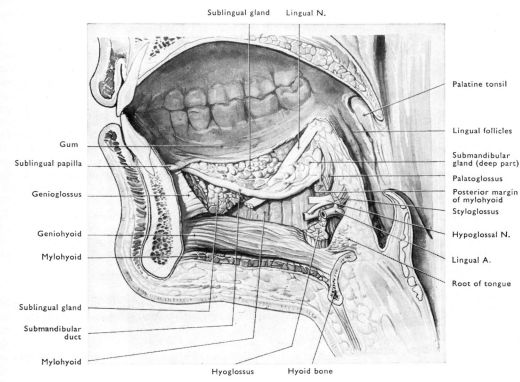

Sublingual gland   Lingual N.

Gum
Sublingual papilla
Genioglossus
Geniohyoid
Mylohyoid
Sublingual gland
Submandibular duct
Mylohyoid

Hyoglossus   Hyoid bone

Palatine tonsil
Lingual follicles
Submandibular gland (deep part)
Palatoglossus
Posterior margin of mylohyoid
Styloglossus
Hypoglossal N.
Lingual A.
Root of tongue

FIG. 6.25B. Side-wall and floor of mouth after removal of tongue.

their high radiopacity, and most are found in the submandibular gland or duct.

The parotid and submandibular main and intraglandular ducts can be shown by catheterization of the main ducts and the injection of oily or water soluble media (sialography) [FIG. 6.23].

The sublingual glands cannot be demonstrated radiographically because of their multiple ducts.

Pharynx

The pharynx is a fibromuscular tube, part of which forms a common pathway for food and air. It is conical in general shape with the base upwards, at the base of the skull and the apex downwards, at the level of the cricoid cartilage. In the adult it is about 13 cm long and narrows from 3.5 cm in width above to 1.5 cm below where it joins the oesophagus, 15 cm from the incisor teeth.

The pharynx lies in front of the upper six cervical vertebrae from which it is separated by the prevertebral muscles and fascia and the loose areolar tissue of the retropharyngeal space. Anteriorly, it lies behind the nose, mouth, and larynx with which it communicates and is correspondingly divided into nasal, oral, and laryngeal parts [FIG. 6.27]. Laterally, it is related on each side to the styloid process and its muscles, the carotid sheath and its contained vessels, the last four cranial nerves, and the superior pole of the thyroid gland [FIGS. 6.30 and 6.33].

The **wall of the pharynx** is formed of four layers. From within outwards these are: (1) a lining of mucous membrane, (2) a layer of

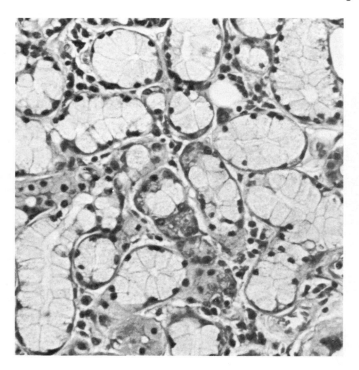

FIG. 6.26. Photomicrograph of the sublingual salivary gland consisting mainly of mucous alveoli.

fibrous connective tissue, called in its upper part the pharyngobasilar fascia, (3) a layer of striated muscle which is deficient anteriorly, and (4) an external covering of connective tissue, the buccopharyngeal fascia.

The mucous membrane is continuous anteriorly with that of the nose, mouth, and larynx and laterally with the lining of the auditory tube. The pharyngobasilar fascia is thick and strong above where the superior constrictor is deficient, but inferiorly, it thins and is not recognizable as a distinct layer. Through it the pharynx gains attachment above to the basilar portion of the occipital bone and the petrous part of the temporal bone, and anteriorly on each side to the posterior border of the medial pterygoid plate, the pterygomandibular raphe, the side of the tongue, the mylohyoid line of the mandible, the hyoid bone, and the thyroid and cricoid cartilages. Posteriorly, the fascia forms a strong fibrous raphe in the midline which is attached above to the pharyngeal tubercle of the occipital bone and gives attachment to the constrictor muscles.

The musculature of the pharynx consists of the superior, middle, and inferior constrictor muscles along with the fibres of the stylopharyngeus, salpingopharyngeus, and palatopharyngeus [p. 298]. Its investing layer of fascia, the buccopharyngeal fascia, is thick inferiorly but thins superiorly where it blends above the upper border of the superior constrictor with the pharyngobasilar fascia. Anteriorly it is continuous with the fascia on the buccinator. The pharyngeal plexus of veins, draining the pharynx and soft palate, and the pharyngeal nerve plexus, innervating the mucous membrane and muscles of the pharynx, are embedded in the buccopharyngeal fascia.

Nasal part of the pharynx

The nasal part of the pharynx, lined by ciliated pseudostratified columnar epithelium, may be described as having a roof, a floor, lateral, anterior, and posterior walls [FIG. 6.27]. Apart from the

floor, its walls are relatively fixed and its lumen is always patent. The anterior wall is somewhat of a misnomer for it consists only of the posterior openings of the nasal cavities (choanae) separated by the posterior edge of the nasal septum. (Hence a mirror placed in the nasal part of the pharynx and pointing forwards looks directly into the nasal cavities [FIG. 6.30].) The floor is formed anteriorly by the relatively mobile soft palate and behind by the pharyngeal isthmus through which the nasal and oral parts communicate. During swallowing the pharyngeal isthmus is closed by the elevation of the soft palate in conjunction with the contraction of the palatopharyngeal sphincter. On the lateral wall, at the level of the inferior concha, there is the opening of the auditory tube. It is bounded above and behind by a rounded prominence, the tubal elevation, caused by the projecting end of the cartilage of the tube. In the region of the tubal elevation there is a variable amount of lymph tissue, the tubal tonsil, which is continuous posteriorly with the pharyngeal tonsil. A vestigial fold of mucous membrane, the salpingopharyngeal fold, containing the corresponding muscle, passes downwards from the lower end of the elevation to disappear gradually on the side wall of the pharynx. Immediately below the tubal opening there is a slight bulge produced by the levator veli palatini muscle. Behind the tubal elevation the pharyngeal recess passes posterolaterally below the petrous portion of the temporal bone on each side.

The roof and posterior wall of the nasal part of the pharynx form a continuous sloping surface related to the sphenoid and occipital bones, and the anterior arch of the atlas. On the upper part of the posterior wall and the adjacent roof there is a prominence produced by an accumulation of lymph tissue which is known as the pharyngeal tonsil or adenoids [FIG. 6.27]. The lymph tissue increases in amount up to about the age of six years and then gradually atrophies. It causes the mucous membrane to be thrown into folds radiating forwards from a blind median recess known as the pharyngeal bursa. This bursa is associated with the attachment of the notochord to the entoderm of the embryonic pharynx. In the adult it may give rise to cysts which sometimes become infected. Enlargement of the pharyngeal tonsil is common in children and may partially or completely prevent nasal breathing, as well as obstructing the openings of the auditory tubes.

Oral part of the pharynx

The oral part of the pharynx extends from the soft palate above to the upper border of the epiglottis below. It opens anteriorly into the mouth through the isthmus of the fauces and below that its anterior wall is formed by the pharyngeal surface of the tongue, with the median and lateral glosso-epiglottic folds bounding the epiglottic valleculae. The posterior wall, visible through oropharyngeal isthmus, lies in front of the second and upper part of the third cervical vertebrae. Its lateral wall has two prominent diverging folds of mucous membrane, known as the palatoglossal and palatopharyngeal arches, which, along with the dorsum of the tongue, bound a triangular recess, the tonsillar fossa containing the palatine tonsil [FIGS. 6.29 and 6.33]. The mucous membrane of this part of the pharynx, like that of the laryngeal part, is subjected to mechanical stress by the passage of food and is lined by a stratified squamous epithelium.

PALATINE TONSILS

The palatine tonsils develop in the late months of foetal life and grow rapidly in the child up to six or seven years of age. In the adult they gradually atrophy and in old age they are almost completely absent. Along with the lingual and pharyngeal tonsils they form a

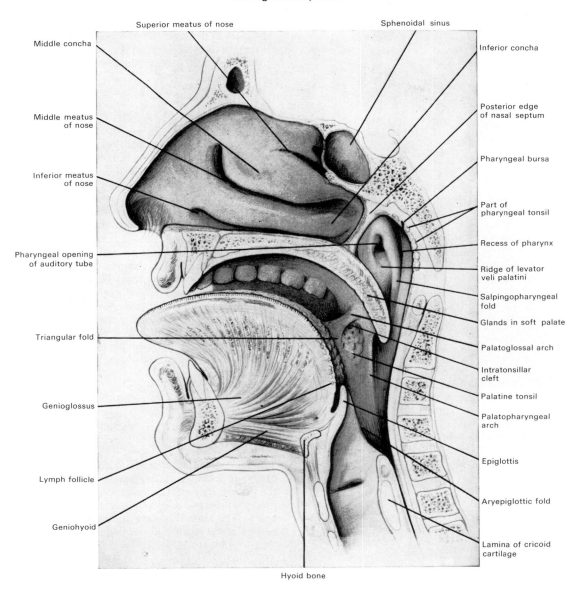

FIG. 6.27.   Sagittal section through mouth, tongue, larynx, pharynx, and nasal cavity.
    The section was slightly oblique, and the posterior edge of the nasal septum has been preserved. The specimen is viewed slightly from below; hence the apparently low position of the inferior concha.

ring of lymph tissue surrounding the entrances to the respiratory and digestive tubes. When fully formed they are almond-shaped with their long axis in the vertical plane and measure in the healthy state approximately 25 mm in height, 15 mm in width, and 10 mm in thickness. Each palatine tonsil has two surfaces, a medial which in the child and young person projects into the pharynx, and a lateral embedded in the wall of the pharynx [FIG. 6.33]. The extent of the projection of the medial surface of the tonsil is a poor guide to its total size for its lateral aspect may burrow upwards into the soft palate, forwards into the palatoglossal arch or downwards into the dorsum of the tongue. In the healthy state the medial surface does not project beyond the palatal arches but when both tonsils are enlarged in inflammatory states they may meet in the midline. On the medial surface, the epithelium forms fifteen to twenty **tonsillar crypts** which penetrate deeply into the substance of the tonsil. These may become enlarged and may contain infected debris. Superiorly there is a deep **intratonsillar cleft** which may reach a large size and extend into the soft palate [FIG. 6.29].

In the child a triangular fold of mucous membrane, the **plica triangularis**, extends from the palatoglossal arch to cover the antero-inferior part of the tonsil. It usually becomes incorporated into the tonsil and may not be recognizable in the adult. The lateral or deep surface of the tonsil is covered by a firmly adherent, thin layer of fibrous tissue called the **capsule of the tonsil**. This sends septa into the substance of the tonsil and is loosely attached by means of areolar tissue to the superior constrictor of the pharynx. Antero-inferiorly the capsule is also attached to the side of the tongue and receives fibres from the palatoglossus and palatopharyngeus muscles.

The tonsil receives its main blood supply from the **tonsillar branch of the facial artery**, which pierces the superior constrictor to enter its lower pole, accompanied by one or two veins. Its veins pierce the superior constrictor muscle and join the external palatine, pharyngeal, or facial veins. The **external palatine vein** is immediately lateral to the tonsil as it descends from the soft palate to pierce the superior constrictor and join the pharyngeal plexus, and is often the site of haemorrhage in tonsillectomy [FIG. 6.32]. The tonsil

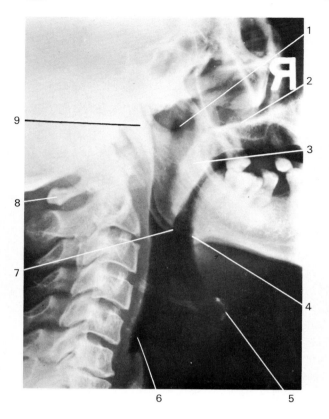

FIG. 6.28.   Lateral radiograph of pharynx.
1. Opening of auditory tube, nasal part of pharynx.
2. Hard palate.
3. Soft palate.
4. Dorsum of tongue.
5. Hyoid bone.
6. Laryngeal part of pharynx.
7. Oral part of pharynx.
8. Posterior arch of atlas.
9. Pharyngeal tonsil.

receives its sensory **nerve supply** from a plexus formed by branches of the lesser palatine and glossopharyngeal nerves. **Lymph vessels** drain into the deep cervical nodes, in particular the **jugulodigastric node**, immediately below the angle of the mandible.

Laryngeal part of the pharynx

The laryngeal part of the pharynx extends from the superior border of the epiglottis to the lower border of the cricoid cartilage, at which level it is continuous with the oesophagus. It narrows rapidly, especially in its lower part, where it becomes slit-like in conformity with the oesophagus, except during the passage of food, and apart from the vermiform appendix, forms the narrowest region of the alimentary canal. Its anterior wall is formed above by the posterior surface of the epiglottis and the entrance to the larynx, and below by the mucous membrane on the posterior surface of the arytenoid and cricoid cartilages. On each side of the laryngeal entrance there is a small recess called the **piriform recess**. It is bounded medially by the aryepiglottic fold and laterally by the medial surface of the thyroid cartilage and the thyrohyoid membrane. Branches of the internal laryngeal nerve lie external to the mucous membrane of the recess and are liable to be damaged in the clumsy removal of foreign bodies, which may lodge in this part of the pharynx. The posterior wall lies in front of the third, fourth, fifth, and sixth cervical vertebrae.

Blood and lymph vessels and nerves of the pharynx

The main **arterial supply** comes from branches of the ascending pharyngeal, the ascending palatine, the maxillary, the facial, and the lingual arteries. The **veins** form a plexus which drains into the pterygoid plexus above and into the internal jugular or facial veins below. The **lymph vessels** drain into the deep cervical nodes, in particular the jugulodigastric node, immediately below the angle of the mandible [FIG. 13.159].

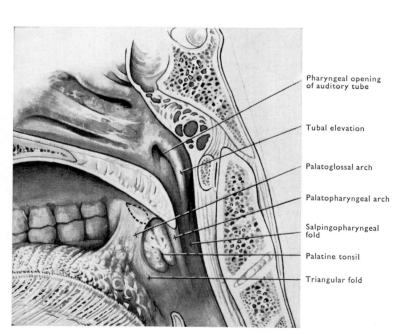

Pharyngeal opening of auditory tube

Tubal elevation

Palatoglossal arch

Palatopharyngeal arch

Salpingopharyngeal fold

Palatine tonsil

Triangular fold

FIG. 6.29.   Median section of pharynx to show the right palatine tonsil. The palatine tonsil is large and divided below by an oblique cleft. The orifice of the intratonsillar cleft is near its upper pole, and the extent of the cleft in this specimen is indicated by a dotted line.

The nerve supply to the pharynx is derived mainly from the pharyngeal plexus which is formed by the vagus, the glossopharyngeal and the sympathetic nerves. Some fibres in this plexus are derived from the cranial part of the accessory and travel to the pharynx in the recurrent laryngeal branches of the vagus. They supply all the muscles, with the exception of the stylopharyngeus which is innervated by the glossopharyngeal nerve, and the tensor veli palatini, supplied by the mandibular nerve. The sensory supply of the mucous membrane in the nasal part is provided mainly by the maxillary nerve, in the oral part by the glossopharyngeal and lesser palatine nerves, and in the laryngeal part by the internal and recurrent laryngeal branches of the vagus.

Swallowing

The movements of the soft palate and pharynx in swallowing are described in relation to the individual muscles involved in this activity [pp. 305–7].

RADIOLOGY

Swallowing

Fluoroscopic observation of the normal act of swallowing is commonly unrewarding because of the speed of the movement. The experienced observer can sometimes detect an abnormality although he may not be able to indicate its nature. In such cases specialized examinations such as cineradiography may be helpful.

DEVELOPMENT OF THE PHARYNX

The pharynx develops from the foregut, which up to the end of the fourth week is separated from the stomodaeum by the buccopharyngeal membrane. At first it is funnel shaped with a broad cranial end and a tapering caudal end. As described on page 55, proliferation of bars of mesoderm in the lateral walls of the primitive pharynx results in the formation of the branchial arches, separated internally by pharyngeal pouches and externally by corresponding pharyngeal grooves [FIG. 2.68]. The mesoderm of each pharyngeal arch gives rise to a cartilaginous bar, a muscle element, and an artery; a nerve grows into it from the developing brain.

The structures developed from the components of the pharyngeal arches are summarized in the Table overleaf.

The mesoderm and the overlying entoderm of the ventral ends of the branchial arches take part in the formation of the tongue, in addition to forming cartilaginous, muscular, and vascular elements. That of the first arch forms the lingual swellings, which fuse with the single centrally placed swelling, the tuberculum impar and gives rise to the epithelium and connective tissue of the anterior two

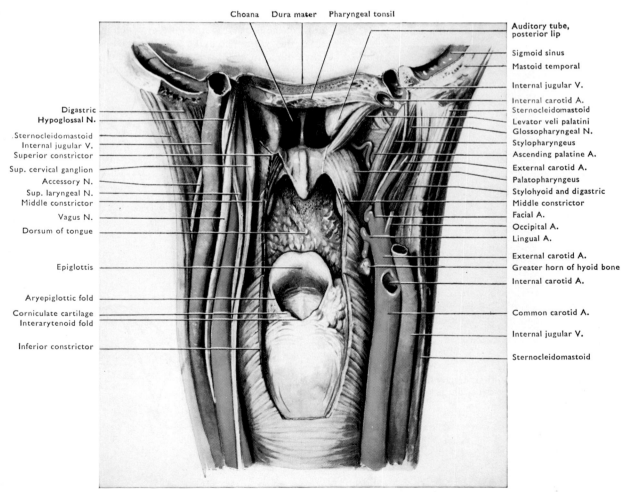

FIG. 6.30.   Interior of pharynx and the structures in relation to its sidewalls, viewed from behind.

Derivatives of branchial arch components

Arch	Muscle element	Nerve	Skeletal derivatives	Blood vessel
1st	Muscles of mastication, Ant. belly digastric, Mylohyoid, Tensor veli palatini, Tensor tympani	Mandibular (V)	Malleus Incus, Meckel's cartilage, Sphenomandibular and ant. malleolar ligaments	Part of maxillary artery
2nd	Muscles of facial expression, Post. belly digastric, Stapedius, Stylohyoid,	Facial (VII)	Upper part of body and lesser cornu of hyoid, Styloid process, Stapes, Stylohyoid ligament	Stapedial artery
3rd	Stylopharyngeus, Muscle of upper pharynx	Glossopharyngeal (IX)	Lower part of body and greater cornu of hyoid	Internal carotid artery
4th (5th) 6th	Pharyngeal and laryngeal muscles. (In man and most animals the 5th arch does not develop to any appreciable extent)	Superior and recurrent laryngeal branches of vagus (X)	Cartilages of larynx with the exception of the epiglottis	Left part of arch of aorta, Prox. part of right subclavian artery. No derivatives. Prox. part of left pulmonary artery and ductus arteriosus, Prox. part of right pulmonary artery

thirds of the tongue [p. 424]. The second arch makes no contribution to the tongue but the third and fourth enter a central swelling, the **hypobranchial eminence**, the anterior part of which forms the epithelium and connective tissues of the posterior one third of the tongue. The remaining arches are less well developed and end at the side of the laryngotracheal groove, they help in the formation of the larynx and lower part of the pharyngeal wall.

Each pharyngeal pouch develops a dorsal and ventral recess. However, the ventral recess of the first becomes obliterated by the developing tongue. The dorsal recesses of the first and second join to form the **tubotympanic recess**, which gives rise to the auditory tube, the tympanic cavity, and the mastoid antrum and air cells. The entoderm of the ventral recess of the second pouch grows as solid cords into the surrounding mesoderm. Later, the centres of the

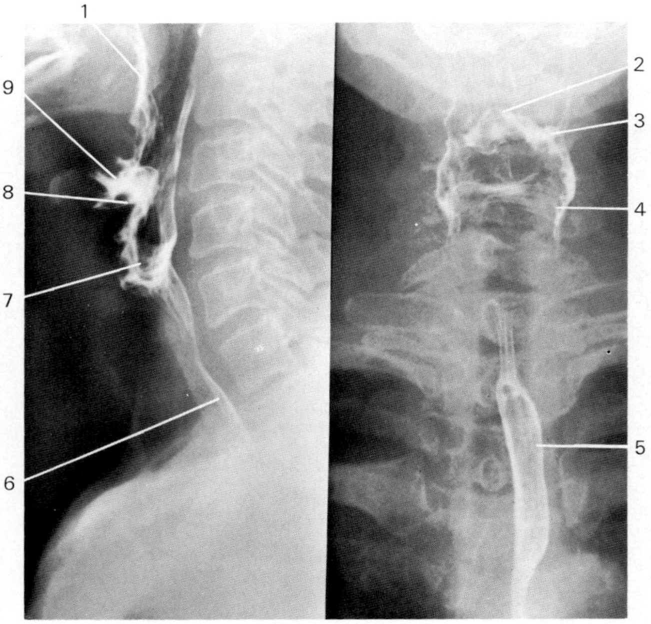

FIG. 6.31. Lateral (i) and anteroposterior (ii) radiographs of the pharynx. The mucous membrane has been coated with barium as a result of a barium swallow.
1. Dorsum of tongue.
2. Epiglottis.
3. Epiglottic vallecula.
4. Piriform recess.
5. Oesophagus.
6. Oesophagus.
7. Piriform recess.
8. Epiglottis.
9. Epiglottic vallecula.

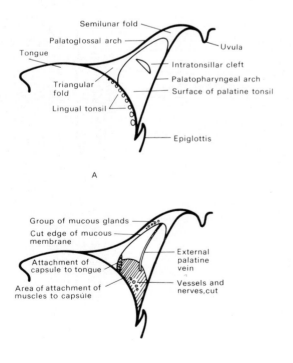

FIG. 6.32. Diagrams of surroundings of palatine tonsil (Browne 1928).
A, Tonsil *in situ*.
B, After its removal.

cords degenerate, giving rise to the **crypts** of the palatine tonsil, and the mesoderm differentiates into lymph follicles. The intratonsillar cleft is a remnant of the second pouch. The derivatives of the pharyngeal pouches are summarized below:

Pharyngeal pouch derivatives

	Recess	Derivative
1st	Dorsal (Ventral obliterated)	Auditory tube, tympanic cavity, mastoid antrum and air cells
2nd	Dorsal	
	Ventral	Palatine tonsil
3rd	Dorsal	Inferior parathyroids
	Ventral	Thymus
4th	Dorsal	Superior parathyroids
	Ventral	Thymus (inconstant) ultimobranchial body—parafollicular thyroid cells
5th	(This pouch is transitory in man. If it develops, it either disappears or becomes incorporated into the fourth pouch)	

The thyroid gland develops as a midline entodermal outgrowth, the **thyroglossal duct**. It is attached to the floor of the pharynx immediately behind the tuberculum impar at the site of the adult foramen caecum. Its development is further considered on page 597.

ANOMALIES

Apart from those associated with the pharyngeal derived endocrine glands, anomalies of the pharynx are relatively rare. Congenital sinuses and fistulae may arise in connection with the development of the pharyngeal pouches. The most common fistula extends from the tonsillar fossa to the anterior triangle of the neck and results from a persistence of parts of the second pouch and groove. In its course it passes between the internal and external carotid arteries.

STRUCTURAL PLAN OF THE ALIMENTARY CANAL

The wall of the alimentary canal has a similar structure throughout most of its length. It is formed of four main layers which, from within outwards, are known as the mucous membrane or mucosa, the submucosa, the muscularis externa, and the adventitia or serosa [FIG. 6.34].

The **mucous membrane** consists of; (1) an epithelium, which varies in its specialization from one region to another and rests on a basal lamina (basement membrane), (2) a supporting layer of loose connective tissue, the lamina propria, which contains blood and lymph vessels, nerve fibres, and accumulations of lymph tissue, and (3) a double layer of smooth muscle, the muscularis mucosae,

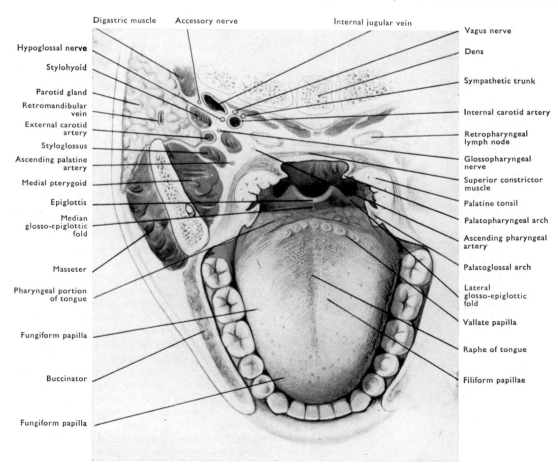

FIG. 6.33. Horizontal section through mouth and pharynx at the level of the palatine tonsils.
The prevertebral muscles and stylopharyngeus (which is shown immediately to the medial side of the external carotid artery) are not indicated by reference lines.

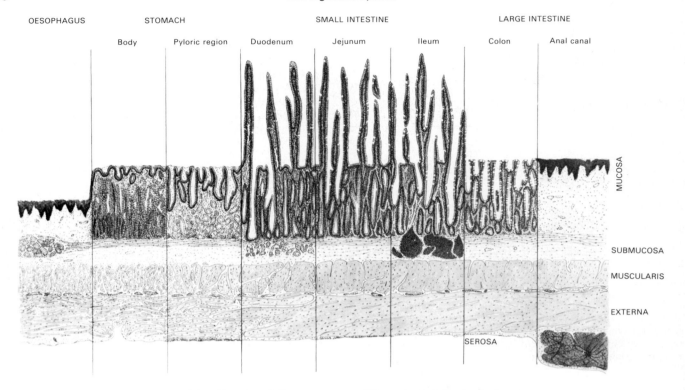

| OESOPHAGUS | STOMACH | | SMALL INTESTINE | | | LARGE INTESTINE | |
| Body | Pyloric region | Duodenum | Jejunum | Ileum | Colon | Anal canal |

MUCOSA

SUBMUCOSA

MUSCULARIS

EXTERNA

SEROSA

FIG. 6.34. Drawing of alimentary tract at different levels in longitudinal section.

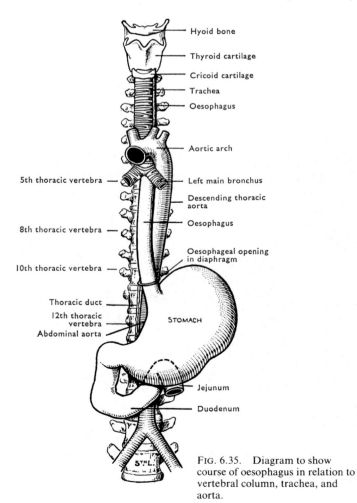

Hyoid bone

Thyroid cartilage

Cricoid cartilage

Trachea

Oesophagus

Aortic arch

5th thoracic vertebra — Left main bronchus

Descending thoracic aorta

8th thoracic vertebra — Oesophagus

10th thoracic vertebra — Oesophageal opening in diaphragm

Thoracic duct

12th thoracic vertebra

Abdominal aorta

STOMACH

Jejunum

Duodenum

5TH L.

FIG. 6.35. Diagram to show course of oesophagus in relation to vertebral column, trachea, and aorta.

which is capable of producing localized movements of the mucous membrane; it is absent in the mouth and pharynx.

The submucosa connects the mucosa to the underlying muscularis externa. It is formed of loose connective tissue, containing larger blood and lymph vessels and lymph nodules, and, especially in the upper part of the tract, considerable numbers of elastic fibres. Where the muscularis mucosae is absent, the connective tissue of the submucosa blends with that of the lamina propria.

The muscularis externa, along most of the length of the alimentary canal, consists of smooth muscle arranged as an outer longitudinal and an inner circular layer of fibres. Co-ordinated relaxation and contraction of the muscularis externa produces peristalsis, which is responsible for the movement of the contents of the canal from the pharynx to the large intestine. In certain localized zones the circular fibres are increased in amount to form sphincters which help to control the passage of the contents from one major region to the next.

The outermost coat of the alimentary canal is a layer of loose connective tissue which in some regions is directly continuous with that of adjacent organs and so is known as the adventitia. In other regions it is covered by mesothelium forming a moist serous membrane, the serosa or peritoneum, which lines the abdominal and pelvic cavities and covers the structures lying in them. It allows the viscera to slide over each other during movement. The serosa never completely surrounds a viscus, but passes from it as two apposed layers of peritoneum either to the body wall or to another viscus. Such double layers of peritoneum are known as mesenteries, omenta, or sometimes as ligaments, even though they do not have the strength or function commonly associated with that term. These frequently contain considerable accumulations of fat, and transmit vessels and nerves to and from the viscera to which they are attached. Some organs are applied to the posterior abdominal wall and held there by a layer of peritoneum which covers them anteriorly. Such organs have no mesentery and are said to be retroperitoneal. The general plan of the alimentary canal is depicted in FIGURE 6.34.

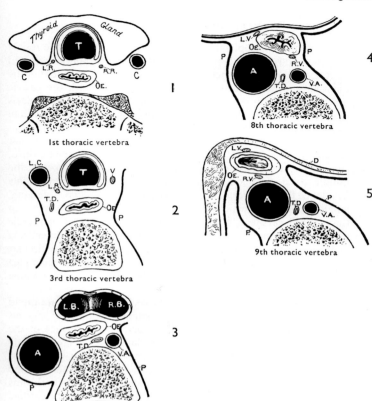

Ist thoracic vertebra

3rd thoracic vertebra

5th thoracic vertebra

8th thoracic vertebra

9th thoracic vertebra

FIG. 6.36.   Tracings from frozen sections to show relations of oesophagus at the levels of the first, third, fifth, eighth, and ninth thoracic vertebrae.

1. At level of the superior part of the first thoracic vertebra, showing the chief relations of the oesophagus in the neck and also its divergence to the left.
2. At the third thoracic vertebra, showing the thoracic duct lying on the left side of the oesophagus. V, right vagus nerve.
3. At level of the fifth thoracic vertebra. The left principal bronchus is seen in relation to the anterior surface of the oesophagus.
4. At level of the eighth thoracic vertebra, showing the oesophagus passing behind the pericardium.
5. At the ninth thoracic vertebra, showing the oesophagus inclining to the left just before piercing the diaphragm.

A, aorta; C, common carotid artery; D, diaphragm; L.B., left principal bronchus; L.C., left subclavian artery; L.R., left recurrent laryngeal nerve; L.V., left vagus; OE, oesophagus; P, pleura; Pc, pericardium; R.B., right principal bronchus; R.R., right recurrent laryngeal nerve; R.V., right vagus; T, trachea; T.D., thoracic duct; V.A., vena azygos.

The alimentary canal has a copious blood and lymph supply. As well as meeting nutritional and defence requirements, these are responsible for the transport of the absorbed products of digestion to the liver and other organs. They are also involved in local hormonal mechanisms for the regulation of the activity of the digestive tract. The wall has a rich nerve supply largely concerned with the control of its movements. It is derived from both the sympathetic and parasympathetic divisions of the autonomic nervous system and is distributed through the myenteric plexus (of Auerbach), situated between the outer and inner layers of the muscularis externa and the submucosal plexus (of Meissner) in the submucosa (Gunn 1968).

THE OESOPHAGUS

The oesophagus is a muscular tube approximately 25 cm long, extending from the pharynx to the stomach. In its course it follows the curvature of the vertebral column, and traverses the neck and thorax to the abdomen. It begins in the midline, at the level of the cricoid cartilage and the sixth cervical vertebra. As it descends, it deviates slightly to the left returning again to the midline at the level of the fourth thoracic vertebra. At the seventh thoracic vertebra it passes forwards and to the left to pierce the diaphragm approximately at the level of the tenth thoracic vertebra [FIGS. 6.35 and 6.36].

In the neck it is applied posteriorly to the prevertebral fascia and the longus colli, anteriorly to the trachea with the right and left recurrent laryngeal nerves intervening and laterally on each side to the lobes of the thyroid gland. In the thorax it passes through the superior and posterior mediastina. In this part of its course it is in contact with the following structures from above downwards. Anteriorly, (1) the trachea and left recurrent laryngeal nerve, (2) the left principal bronchus, (3) the pericardium posterior to the left atrium, and (4) the diaphragm. Posteriorly, (1) the vertebrae, longus cervicis, and prevertebral fascia, (2) the thoracic duct as it crosses to the left on the fourth thoracic vertebra, (3) the azygos vein and its connexions with the hemiazygos veins and the right posterior intercostal arteries, and (4) the descending thoracic aorta posterior to the diaphragm. To the left, (1) the left subclavian artery, the thoracic duct, and the mediastinal pleura, (2) the aortic arch, (3) the left vagus and descending thoracic aorta, and (4) mediastinal pleura. On the right, (1) mediastinal pleura, (2) the azygos vein as it arches forwards above the root of the right lung to the superior vena cava, and the right vagus, and (3) the mediastinal pleura.

The abdominal part of the oesophagus, measuring 1.5 cm in length, is covered with peritoneum on its anterior surface and on its left side. On entering the abdomen it turns sharply to the left and its right border becomes continuous with the lesser curvature of the stomach; its left border forms an acute angle, the cardiac notch, with the fundus. Anteriorly, it lies on the oesophageal impression on the left lobe of the liver [FIG. 6.102] and posteriorly on the left crus of the diaphragm. It carries on its anterior surface the anterior vagal trunk, formed largely by fibres from the left vagus, and on its posterior surface the posterior trunk, derived mainly from the right vagus.

The oesophagus is somewhat constricted; (1) at its beginning, 15 cm from the incisor teeth, (2) where it is crossed by the arch of the aorta, 22.5 cm from the incisor teeth, (3) where it is crossed by the left principal bronchus, 27 cm from the incisor teeth, and (4) where it pierces the diaphragm 40 cm from the incisor teeth. These constrictions are of clinical importance for they are sites which may present difficulties in the passage of surgical instruments [FIG. 6.37].

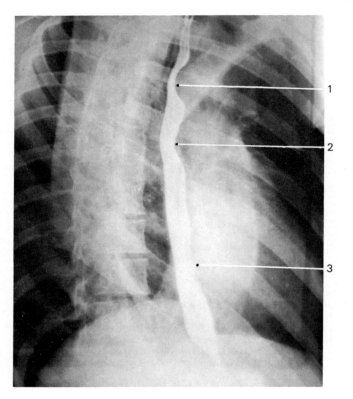

FIG. 6.37.   Right oblique radiograph of thorax with oesophagus filled with barium.
1. Impression by arch of aorta.
2. Impression by left principle bronchus.
3. Impression by left atrium of heart, below which is the diaphragm.

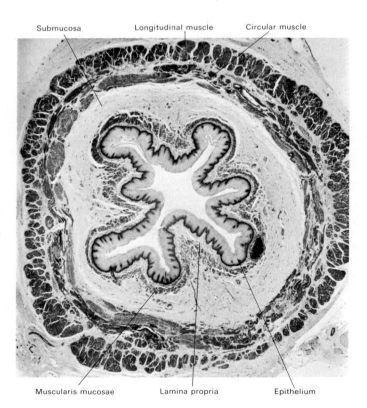

FIG. 6.38.   Photomicrograph of transverse section of oesophagus to show the structure of its wall.

STRUCTURE OF THE OESOPHAGUS

The wall of the oesophagus follows the basic pattern of the rest of the alimentary canal [FIG. 6.38]. The lining of non-keratinized stratified squamous epithelium is continuous above with that of the pharynx and at the gastro-oesophageal junction with the gastric mucosa at a sharply demarcated, crenated line. The thick muscularis mucosae and elastic fibres in the submucosa produce longitudinal folds of the mucosa in the undistended oesophagus.

The muscularis externa consists of outer longitudinal and inner circular muscle fibres. These are striated in the upper one-third, smooth in the lower one-third, and a mixture of both types in the middle third. The longitudinal layer of muscle is more highly developed than the circular. In its upper part it forms two, diverging, longitudinal bands which are attached to the posterior surface of the cricoid cartilage, thus exposing the circular layer of muscle posteriorly and producing a potentially weak area where diverticula of the oesophagus may occur [FIGS. 6.39 and 6.40]. The circular layer of muscle is continuous above with the cricopharyngeal part of the inferior constrictor and below with the circular and oblique muscular fibres of the stomach.

Two types of small, compound, tubulo-alveolar glands are found in the human oesophagus. Both are composed entirely of mucous secreting cells and have ducts which open on the surface of the lining epithelium. (1) The **oesophageal glands** proper are unevenly distributed and are present in the submucosa, their ducts pass through the muscularis mucosae. (2) The **oesophageal cardiac glands** are present in the lamina propria and are found at both ends of the oesophagus. Histologically these closely resemble the cardiac glands of the stomach.

BLOOD AND LYMPH VESSELS AND NERVES OF THE OESOPHAGUS

The **arteries** of the oesophagus are branches of the inferior thyroid and bronchial arteries, the thoracic aorta, and the left gastric and left inferior phrenic arteries.

The **veins** begin in a submucosal plexus and pass through the oesophageal wall to terminate in the inferior thyroid and vertebral veins above, the azygos and hemiazygos veins lower down, and the left gastric vein in the abdomen. The latter vein, a tributary of the portal vein, communicates with the azygos vein by means of anastomotic channels in the submucosa of the lower oesophagus, and so forms an important *connection between the portal and systemic circulations.* In obstruction of the portal circulation the anastomotic channels may become varicose and rupture into the oesophagus with serious loss of blood. The **lymph vessels** drain to cervical, posterior mediastinal, and left gastric lymph nodes.

The **nerves** to the oesophagus are derived from the vagus and the sympathetic. The upper part receives vagal fibres from the recurrent laryngeal nerves and sympathetic fibres accompanying the inferior thyroid arteries from the cervical sympathetic trunks. The thoracic part of the oesophagus receives its fibres from the oesophageal plexus formed by branches of the vagus nerves, the thoracic sympathetic trunks and the greater splanchnic nerves. The abdominal oesophagus is supplied by the anterior and posterior vagal trunks reconstituted from the oesophageal plexus, and by branches from the greater splanchnic nerves. Within the walls of the oesophagus, as elsewhere in the gut, there is a submucosal and a myenteric plexus containing mainly parasympathetic neurons.

GASTRO-OESOPHAGEAL JUNCTION

When empty, the oesophagus is collapsed. Fluid passes directly, with no hold-up, into the stomach, but with solid food there is a

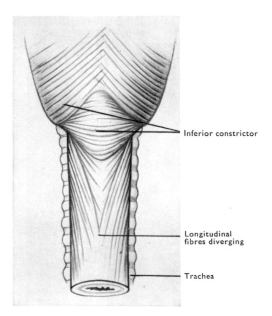

FIG. 6.39. Arrangement of muscular fibres on back of oesophagus and pharynx.

Traced upwards, the longitudinal muscle fibres of the oesophagus are seen to separate; passing round to the sides, they form two longitudinal bands which meet in front and are united to the cricoid cartilage, as shown in FIG. 6.40.

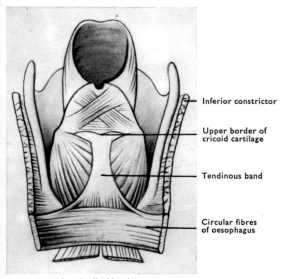

FIG. 6.40. Relation of muscular coat of oesophagus to cricoid cartilage and pharynx.

The pharynx and oesophagus have been slit up from behind and the mucous membranes removed to show the muscular fibres. The two longitudinal bands are seen passing by a common tendon to the upper border of the cricoid cartilage. The circular fibres are continuous with the inferior constrictor.

delay of about 2 seconds at the gastric end. This fact, along with the normal absence of reflux from the stomach, even when the body is inverted, implies some form of sphincteric mechanism at the gastro-oesophageal junction. However, unlike other regions of the alimentary canal, where there is control of the passage of the gut contents, there is no increase in the mass of the circular fibres at this junction. Numerous explanations have been put forward and these include; (1) the reinforcing of the circular fibres of the oesophagus and stomach by those of the right crus of the diaphragm, (2) the valve-like fold of stomach wall produced by the cardiac notch, and (3) a possible funnel-shaped process of gastric mucosa pulled into the orifice by the muscularis mucosae and capable of acting as a one-way valve (Code 1968; Di Dio and Anderson 1968).

DEVELOPMENT OF THE OESOPHAGUS

The oesophagus develops from the foregut immediately caudal to the laryngotracheal groove. At first it is relatively short but later with the descent of the heart and the development of the lungs it elongates and this is accompanied by a rapid proliferation of its epithelium that temporarily obliterates its lumen. The musculature of its cranial part is derived from the branchial arches, that of its caudal part from the surrounding splanchnic mesoderm.

ANOMALIES

The most common abnormality is a tracheo-oesophageal fistula which results from an incomplete division of the foregut into separate respiratory and digestive portions. There are several varieties of this condition which usually occur with some form of oesophageal atresia (closure). Atresia may also result from failure of recanalization of the obliterated lumen of the developing oesophagus.

RADIOLOGY

Radiographic examination of the oesophagus is carried out by a barium swallow under fluoroscopic control with cineradiographic filming of this rapid process when necessary.

The barium filled oesophagus is in contact with several structures in the mediastinum and provides a means of assessing changes in them. Four indentations are commonly seen in radiographs of the barium filled oesophagus taken in the oblique position but all are not always obvious. From above downwards these are:

(1) The **aortic arch** indents the left anterolateral wall, approximately at the level of the fourth thoracic vertebra. This tends to become more obvious in later life because the aorta widens as it loses its elasticity.

(2) Immediately inferior to the aortic arch is an anterior impression caused by the **left principal bronchus** pressed posteriorly by the beginning of the right pulmonary artery.

(3) A smooth rounded concavity caused by the posterior surface of the heart (**left atrium**) impinging on the oesophagus approximately at the level of the sixth to eighth thoracic vertebrae. On fluoroscopic screening this impression can be seen to move with respiration and cardiac pulsations [FIG. 6.37].

(4) The most caudal oesophageal impression is on the posterior wall of the oesophagus. It is formed by the oesophagus passing in front of the **aorta**.

Anomalous vessels, for example, a **right subclavian artery** arising from the descending aorta and passing behind the oesophagus, create an abnormal impression on the posterior wall of the barium-filled oesophagus above the level of the aortic arch.

The abdominal cavity

The abdominal cavity consists of a large upper part, the abdomen proper, and a smaller, lower part, the cavity of the lesser pelvis. The abdomen proper extends upwards into the inferior aperture of the thorax and downwards to the superior aperture of the lesser pelvis, where it is continuous with the cavity of the lesser pelvis. Within the confines of the thoracic cage, its walls are formed by the vault of the diaphragm, while below they are formed by the lumbar part of the vertebral column, the iliac bones with the iliopsoas muscles, and the musculature of the anterolateral abdominal wall. The cavity of the lesser pelvis is bounded by the rigid bony wall of the pelvis and the muscle attached to its inner surface; inferiorly, the muscular pelvic diaphragm separates it from the perineum.

For descriptive purposes the abdomen is conventionally divided into nine regions by two vertical and two horizontal planes [FIG. 6.41]. The vertical planes on each side pass through the midinguinal point, lying half way between the anterior superior iliac spine and the pubic symphysis. The horizontal planes are, (1) the transtubercular, which passes through the tubercles of the iliac crests at the level of the fifth lumbar vertebra, and, (2) the transpyloric, passing through the midpoint of a line joining the sternal, jugular notch and the pubic crest, at the lower border of the first lumbar vertebra. The regions delineated by these planes are named in FIGURE 6.41. It should be noted that the hypochondriac regions are largely under cover of the lower ribs and that during expiration they extend upwards in the midclavicular line on the right as far as the fifth rib and on the left to a slightly lower level. Apart from their value in the study of topographical anatomy, these regions are frequently referred to in clinical medicine when they are used to give some degree of exactness to the description of the localization of signs and symptoms in the abdomen. A further subcostal plane passes through the tip of the tenth costal cartilage on each side. Its level is described as the upper border of the third lumbar vertebra, but it often almost overlies the transtubercular plane, especially in the elderly.

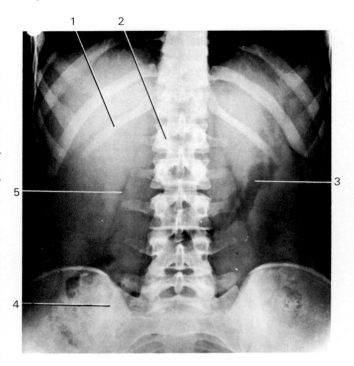

FIG. 6.42.   Anteroposterior radiograph of the abdomen. This shows how few abdominal organs are visible in such a 'straight' radiograph. Sometimes the kidneys are visible because of a thick layer of perirenal fat, and other organs may be outlined because of a content of gas, particularly the stomach and large intestine.
1. 12th rib.
2. 1st lumbar vertebra.
3. Gas in stomach.
4. Sacro-iliac joint.
5. Margin of psoas major.

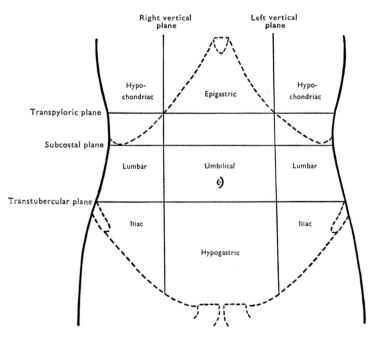

FIG. 6.41.   Planes of subdivision of the abdomen proper, with the names of the nine abdominal regions.

The umbilicus tends to be a somewhat unreliable landmark except in the young adult when it indicates the level of the disc between the third and fourth lumbar vertebrae. In conditions associated with decreased tone of the abdominal musculature, and in the child before the development of the pelvic region has caught up with the rest of the abdomen, it lies at a lower level. In transverse sections the cavity of the abdomen proper is kidney shaped, owing to the marked forward projection of the vertebral column [FIGS. 6.44 and 6.45]. This projection results in a deep paravertebral groove on each side.

THE CONTENTS OF THE ABDOMEN

The abdominal cavity contains most of the components of the digestive and genito-urinary systems, the suprarenal glands, and the spleen. In their general arrangement the digestive organs lie in front of the genito-urinary organs in the abdomen proper and behind them in the pelvis. The components of the digestive system within the abdominal cavity are the liver, the pancreas and the gastro-intestinal tract. The tract consists of the stomach, the small intestine formed by the duodenum, jejunum, and ileum, and the large intestine composed of the vermiform appendix, caecum, ascending, transverse, descending, and sigmoid colons, the rectum, and the anal canal [FIGS. 6.50 and 6.69].

The liver, divided into a larger right and a smaller left lobe, occupies the right hypochondriac and much of the epigastric and left hypochondriac regions of the abdomen. It is largely concealed by the thoracic wall, and it comes into contact with and moves with the diaphragm, to which it is attached by short but broad peritoneal folds (ligaments). Its inferior border extends across the epigastric region from the ninth right costal cartilage to the eighth left costal cartilage. Usually the fundus of the gall-bladder projects below the liver to come into contact with the abdominal wall at the level of the right ninth costal cartilage, where the lateral edge of the rectus abdominis crosses it.

A triangular peritonal fold, the **falciform ligament**, passes from the median deep surface of the anterior abdominal wall, above the umbilicus, to the anterior and upper surfaces of the liver, a little to the right of the midline. In its lower free edge, between its two layers of peritoneum there is a fibrous cord, the **ligamentum teres** of the liver, which passes to the visceral surface of the liver at its inferior margin. This represents the remains of the left umbilical vein which brought the blood back from the placenta to the foetus.

The **stomach** varies considerably in shape and position, depending among other things upon body form, posture, and degree of filling. It lies to the left, behind and below the liver and occupies parts of the left hypochondriac, epigastric, and umbilical regions [FIG. 6.50]. It has anterior and posterior surfaces and two borders known as the lesser and greater curvatures. The **lesser curvature** is concave to the right and superiorly. Passing from it and the first 2–3 cm of the duodenum, there is a double layer of peritoneum, the **lesser omentum**, which attaches the stomach to the postero-inferior or visceral surface of the liver. The **greater curvature** is convex to the left and inferiorly. From it extends another double sheet of peritoneum part of which hangs downwards as an apron-like fold, the **greater omentum**. This varies considerably in size and has a wide range of movement. In the healthy state it usually lies in front of the coils of the small intestine, but in pathological conditions it often migrates to the affected site and by adhering to it helps to isolate it from the peritoneal cavity.

The first part of the **small intestine**, the **duodenum**, is continuous proximally with the stomach at the pyloric orifice and distally with the jejunum at the duodenojejunal junction. It is some 25 cm in length and is relatively fixed, being largely retroperitoneal [p. 436] and applied to the structures on the posterior abdominal wall. The jejunum and ileum together measure approximately 6 m in the adult and terminate in the caecum at the ileocaecal valve in the right iliac fossa. While the coils of the small intestine are extremely mobile, their mesenteric attachment tends to keep the jejunum in the upper left part of the abdomen and the ileum in the lower right part and pelvis.

The **large intestine**, measuring 1.5 to 1.8 m, begins as the blind-ended **caecum** in the right iliac fossa, with the vermiform appendix attached to it. The **ascending colon**, passes upwards from the caecum through the right lumbar region to the under surface of the liver where it turns to the left, forming the **right colic flexure**, just below the transpyloric plane, and becomes the transverse colon. The **transverse colon** passes across the abdomen below or behind the stomach to meet the spleen in the left hypochondriac region. Here it turns sharply downwards at the **left colic flexure** and passes through the left lumbar and iliac regions as the **descending colon**. At the superior aperture of the lesser pelvis the descending colon loops upwards as the curved **sigmoid colon** which then descends to the third sacral vertebra. Here it is succeeded by the **rectum** which terminates in the anal canal.

The **spleen** lies in the left hypochondriac region in contact with the abdominal surface of the diaphragm and sheltered by the thoracic cage [FIG. 6.65]. It is in contact with the stomach and left kidney, and is attached to them by peritoneal ligaments. The **pancreas** rests on the posterior abdominal wall in the curvature formed by the duodenum and extends to the left making contact with the left kidney and spleen. The **kidneys**, each capped by a suprarenal gland, are both retroperitoneal. In the lesser pelvis, the urinary bladder is in front and the rectum behind. In the female they are separated by the vagina and uterus centrally, and by the broad ligaments containing the uterine tubes and the ovaries laterally. In the male, the seminal vesicles and deferent ducts on the posterior surface of the bladder form only a partial separation for these two viscera.

The peritoneum

The peritoneum is the serous membrane which lines the abdominal wall (*parietal peritoneum*) and is reflected to cover the contained viscera partly or almost completely (*visceral peritoneum*). The partly covered viscera simply raise the peritoneum from the abdominal wall, whereas the almost completely covered viscera are contained in folds of peritoneum which attach them to the abdominal wall or to other viscera.

Folds of peritoneum are given specific names signifying the viscera to which they are attached. A fold passing between the stomach and another abdominal viscus is usually referred to as an **omentum**, hence the **greater omentum** between the stomach and the transverse colon, and the **lesser omentum** between the stomach and the liver. A fold connecting the intestine and the abdominal wall is called a **mesentery**, hence the mesenteries of the small intestine, vermiform appendix, transverse colon, and sigmoid colon. Other folds of peritoneum are referred to as **ligaments**. These connect viscera which are not part of the intestine to each other, to the abdominal wall, or to the diaphragm; hence the ligaments of the liver, urinary bladder, and uterus. The gastrocolic, gastrophrenic and gastrosplenic ligaments are some of the exceptions in that they connect the stomach to the colon, diaphragm, and spleen respectively.

The space enclosed by the peritoneum is the **peritoneal cavity** [FIG. 6.43]. It is a closed sac except for the openings of the uterine tubes in the female, and is completely filled by the abdominal viscera. It is therefore only a potential cavity, for the viscera are everywhere in contact with each other and with the abdominal walls, and no space exists apart from that occupied by a small amount of serous exudate produced by the peritoneum. The peritoneum prevents adhesions developing between adjacent viscera or between viscera and the abdominal walls, and the serous exudate lubricates the contacting surfaces, so facilitating the constant movement of the viscera which occurs during respiration and during the activity of the organs themselves.

The peritoneal cavity consists of two compartments, the greater sac and a diverticulum from this, the omental bursa [FIGS. 6.43 and 6.44]. The **omental bursa** is said to be hour-glass shaped, consisting of a small superior recess and a larger inferior recess. It lies behind the stomach, the lesser omentum, and the caudate lobe of the liver, and it extends for a variable distance into the greater omentum. Its posterior wall is formed by peritoneum covering the diaphragm, the upper part of the left kidney, the left suprarenal gland, the pancreas and the anterior of the two posterior layers of the greater omentum. It is bounded on the left by the spleen, the gastrosplenic and lienorenal ligaments, and by the left margin of the greater omentum. On the right it is limited above by the reflection of the peritoneum from the posterior abdominal wall to the liver to the left of the inferior vena cava, and below by the greater omentum. Between the superior and inferior recesses there is a constriction produced by the superior and inferior **gastropancreatic folds**. These are formed by the left gastric and common hepatic arteries, arching forwards to the cardiac and pyloric ends of the stomach, respectively. The omental bursa communicates with the greater sac through the **epiploic foramen** situated behind the right or free border of the lesser omentum.

The disposition of the parietal and visceral layers of peritoneum and their continuity with each other can best be understood by tracing them in vertical and horizontal planes.

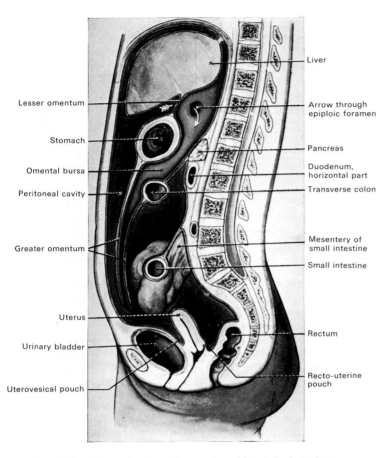

FIG. 6.43.   Diagrammatic median section of female body to show abdominopelvic cavity and peritoneum on vertical tracing. The greater sac of peritoneum is blue and the omental bursa is red; both are represented as much larger spaces than in nature.

DISPOSITION OF PERITONEUM IN THE MEDIAN PLANE

In the median vertical plane [FIG. 6.43] the two layers of peritoneum forming the **lesser omentum** leave the inferior surface of the liver and pass to the lesser curvature of the stomach. Here they separate to cover the anterior and posterior surfaces of the stomach, and at the greater curvature come together again and pass downwards as the two anterior layers of the **greater omentum**. These then turn backwards and ascend to the transverse colon as the posterior two layers of the greater omentum. Here they pass in front of the transverse colon and its mesentery (to which they adhere in the adult) and reach the anterior surface of the pancreas [FIG. 6.46]. Having reached the pancreas the two layers separate. The anterior passes upwards to the under surface of the diaphragm, behind the left lobe of the liver, as the posterior wall of the omental bursa. From here it is reflected forwards to the caudate lobe of the liver which it covers and from which it continues as the posterior of the two layers forming the lesser omentum. The posterior layer of peritoneum passes downwards and is immediately reflected as the upper layer of the **transverse mesocolon**. It surrounds the colon and ascends to the pancreas as the inferior layer of the transverse mesocolon. It then turns downwards on the anterior surface of the pancreas, covers the horizontal part of the duodenum, and descends until it is carried forwards by the superior mesenteric vessels as **the mesentery of the small intestine** which it covers and returns to the attachment. Thence the peritoneum passes downwards on the posterior abdominal wall and into the pelvis to cover the anterior surface and sides of the upper part of the rectum.

In the male the peritoneum passes from the rectum to the upper parts of the seminal vesicles on the posterior surface of the bladder and from there to the superior surface of the bladder. As it passes from the rectum to the seminal vesicles it dips to form the **rectovesical pouch**, the bottom of which is about 7.5 cm from the anus. In the female the peritoneum is reflected from the rectum to the posterior fornix of the vagina and uterus, thus forming the **recto-uterine pouch**, which reaches to within 5 cm of the anus. From the posterior fornix it passes to the uterus covering its intestinal (posterosuperior) and the upper part of its vesical (antero-inferior) surfaces. It is then reflected to the superior surface of the bladder as the floor of a shallow depression, the **uterovesical pouch**.

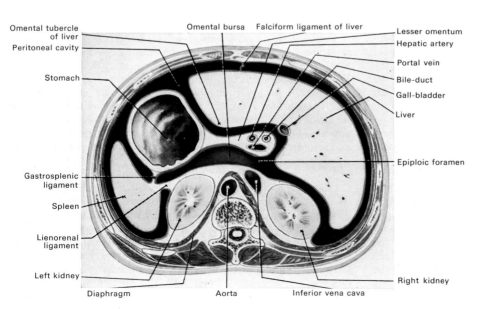

FIG. 6.44.   Transverse section of abdomen at level of the opening into the omental bursa.

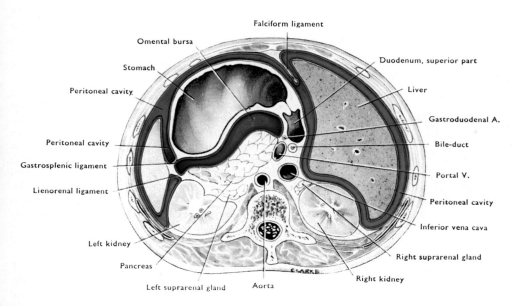

FALCIFORM LIGAMENT
Omental bursa
Stomach
Peritoneal cavity
Peritoneal cavity
Gastrosplenic ligament
Lienorenal ligament
Left kidney
Pancreas
Left suprarenal gland
Aorta

Duodenum, superior part
Liver
Gastroduodenal A.
Bile-duct
Portal V.
Peritoneal cavity
Inferior vena cava
Right suprarenal gland
Right kidney

FIG. 6.45. Transverse section of abdomen immediately below the opening into the omental bursa.

The peritoneum next ascends on the posterior surface of the anterior abdominal wall, on the lower part of which it is raised into five folds converging on the umbilicus. In the centre the **median umbilical fold**, containing the fibrous remnant of the urachus, stretches upwards from the apex of the bladder. Lateral to it, on each side, there is the **medial umbilical fold**, formed by the underlying obliterated umbilical artery, and more lateral still the **lateral umbilical fold**, produced by the inferior epigastric artery as it ascends to reach the sheath of rectus abdominis. On each side, just above the inguinal ligament, these folds separate three depressions. The most lateral, the **lateral inguinal fossa**, lies immediately lateral to the inferior epigastric artery and is situated deep to the deep inguinal ring. It indicates the site where the **processus vaginalis** passed through the abdominal wall during the descent of the testis [p. 556]. The intermediate depression, the **medial inguinal fossa**, lies between the medial and lateral umbilical folds and the most medial, the **supravesical fossa**, is situated between the median and medial umbilical folds. These three fossae may determine the site of exit of an inguinal hernia. Above the umbilicus the peritoneum of the anterior abdominal wall ascends to the under surface of the diaphragm, where it is reflected to the upper surface of the liver as the superior layer of the **coronary ligament** and the anterior layer of the **left triangular ligament**. It then covers the superior and anterior surfaces of the liver and passes round its sharp inferior border to the porta hepatis on the visceral surface, where it leaves the liver as the anterior layer of the lesser omentum.

DISPOSITION OF PERITONEUM IN HORIZONTAL PLANES

(1) *At the level of the epiploic foramen* [FIG. 6.46 BB]. It will be seen [FIG. 6.44] that the omental bursa and the greater sac communicate at the epiploic foramen. This is bounded anteriorly by the right, free border of the lesser omentum, inferiorly by the superior part of the duodenum, superiorly by the caudate process of the liver and posteriorly by the inferior vena cava. The anterior and posterior layers of peritoneum of the lesser omentum are continuous at the free border and enclose within them the hepatic artery, portal vein, and bile duct. When traced towards the left the two layers separate at the lesser curvature of the stomach and cover the anterior and posterior surfaces. At the upper part of the greater curvature they unite again to form the **gastrosplenic ligament**. This passes to the

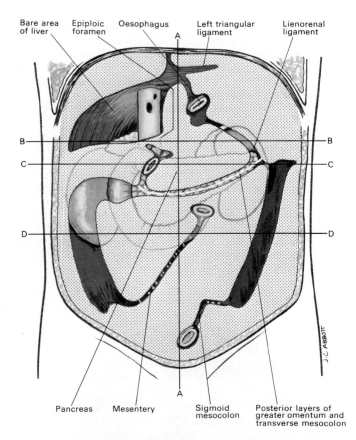

Bare area of liver
Epiploic foramen
Oesophagus
Left triangular ligament
Lienorenal ligament

Pancreas
Mesentery
Sigmoid mesocolon
Posterior layers of greater omentum and transverse mesocolon

FIG. 6.46. Peritoneal reflections from the posterior abdominal wall as seen after the removal of the liver, stomach, and small and large intestines. The lines AA, BB, CC, and DD correspond to the planes illustrated in FIGURES 6.43, 6.44, 6.45, and 6.48, respectively.

hilus of the spleen where the left layer proceeds to envelop the spleen and join with the right to form the **lienorenal ligament**. On the anterior surface of the upper part of the kidney the two layers separate. One layer passes to the right as the posterior wall of the omental bursa and at the epiploic foramen becomes continuous with the peritoneum of the greater sac. From the epiploic foramen it

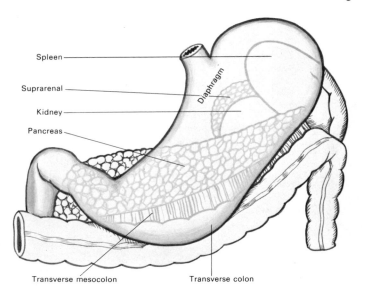

Spleen

Suprarenal

Kidney

Pancreas

Transverse mesocolon   Transverse colon

FIG. 6.47.   Structures lying posterior to the stomach.

passes to the right, over the anterior surface of the right kidney, to line the under surface of the diaphragm. Near the midline it is reflected to the anterior surface of the liver as the right layer of the falciform ligament of the liver. From here it passes to the right and completely surrounds the liver to become continuous with the left layer of the falciform ligament.

The left layer of the lienorenal ligament covers the left part of the anterior surface of the kidney and passes to the under surface of the diaphragm. It continues round the abdominal wall until it reaches the falciform ligament, the left layer of which it forms.

(2) *Immediately below the opening into the omental bursa* [FIG. 6.46CC]. At this level, which passes through the first lumbar vertebra, continuity between the peritoneum of the omental bursa and that of the greater sac is absent [FIG. 6.45]. The section passes through the inferior recess of the omental bursa, which lies behind the stomach and in front of the structures on the posterior abdominal wall that form the stomach bed [FIG. 6.47]. The right border of the inferior recess is formed by the reflection of the peritoneum from the head

of the pancreas to the posterior surface of the superior part of the duodenum. As it passes forwards to become continuous with the peritoneum on the posterior surface of the stomach and initial segment of the duodenum, it is in intimate contact with the gastroduodenal artery. On the left, the border consists of the gastrosplenic and lienorenal ligaments.

(3) *At the level of the fourth lumbar vertebra* [FIG. 6.46DD]. The peritoneum on the posterior abdominal wall is reflected as the mesentery of the small intestine. Its line of reflection, sometimes referred to as the root of the mesentery, passes downwards and to the right from the second lumbar vertebra to the right sacro-iliac joint [FIG. 6.46]. Laterally, the peritoneum covers the ascending and descending colons, forming as it does the medial and lateral paracolic sulci on each side, and then passes forwards to line the anterolateral abdominal wall [FIG. 6.48]. The greater omentum is present in front of the coils of small intestine, and at this level its anterior and posterior layers of peritoneum are usually fused, but in some individuals remain separate with the result that the omental bursa extends to a lower level.

A consideration of the various relections of peritoneum will indicate that the greater sac of the peritoneal cavity is partially divided into a number of compartments which may be of clinical importance, particularly in the spread of infected material within the abdomen. The projecting transverse mesocolon incompletely divides the greater sac into a supracolic and an infracolic space. The supracolic space is bounded above by the diaphragm and below by the colon and its transverse mesocolon. Above it is partly divided into the right and left subphrenic recesses by the peritoneal attachments of the liver to the diaphragm [p. 443]. The posterior part of the right subphrenic space forms a deep pouch between the liver in front and the right kidney and suprarenal behind, known as the hepatorenal recess or right subhepatic space [FIG. 6.49]. It is limited above by the inferior layer of the coronary ligament and the right triangular ligament, and below it leads into the right lateral paracolic sulcus. It and the pelvic cavity are the two most dependent parts of the peritoneal cavity in the supine position, and are the areas in which fluid collects in patients nursed in the recumbent position. The left subhepatic recess is, in fact, the omental bursa and is usually referred to by that name. The infracolic space is partially divided into right superior and left inferior parts by the attachment of the mesentery of the small intestine. The left inferior is in free communication with the pelvis.

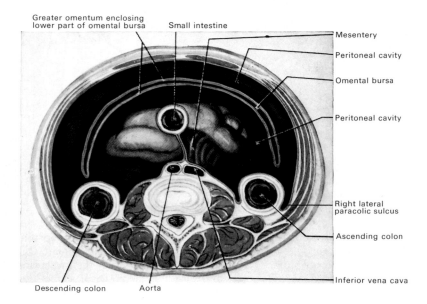

Greater omentum enclosing
lower part of omental bursa   Small intestine

Mesentery

Peritoneal cavity

Omental bursa

Peritoneal cavity

Right lateral
paracolic sulcus

Ascending colon

Inferior vena cava

Descending colon   Aorta

FIG. 6.48.   Transverse section of abdomen through fourth lumbar vertebra.

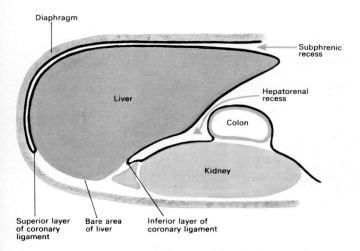

FIG. 6.49. Peritoneal recesses in the region of the liver, as seen in a sagittal section to the right of the midline.

THE STOMACH

The stomach is the dilated first part of the gastro-intestinal tract. In it food is mixed with the gastric secretion by the peristaltic activity of its muscular wall, and digestion is initiated. Food remains in the stomach for varying periods of time, depending on a number of factors, including its composition, and is then ejected through the pyloric orifice into the duodenum. The stomach produces digestive enzymes, hydrochloric acid, and mucus, as well as hormones involved in the control of its own muscular and secretory activity and a glycoprotein, the gastric intrinsic factor [p. 450].

The *shape and position* of the stomach vary considerably, tending to be short, high, and transversely oriented in broad individuals (steerhorn-shaped stomach), and elongated and vertically oriented in thin individuals (J-shaped stomach). Its position is also influenced by its degree of distention and that of other viscera, by posture, and by respiration. Situated in the upper left portion of the abdominal cavity the stomach partly occupies the left hypochondriac, epigastric, and umbilical regions [FIG. 6.50]. Its capacity in the adult is about 1.5 l, and in the newborn infant about 30 ml, but it is capable of considerable distension.

The abdominal oesophagus enters it through a vertical, oval opening, the cardiac orifice, which lies at the level of the eleventh thoracic vertebra, approximately 3 cm to the left of the midline and about 10 cm deep to the 7th left costal cartilage. The cardiac orifice is the most fixed region of the stomach, moving only with movements of the diaphragm. The pyloric orifice, through which the stomach communicates with the duodenum, lies, in the recumbent position, at or slightly to the right of the midline on the transpyloric plane, opposite the first lumbar vertebra. It is more mobile than the cardiac orifice, descending in the erect posture to the second or third lumbar vertebra and being displaced as much as 5 cm to the right when the stomach is full.

The stomach is described as consisting of a fundus, a body, and a pyloric region [FIG. 6.51]. The fundus is that part extending above the cardiac orifice and coming into contact with the under surface of the left cupola of the diaphragm at the level of the fifth intercostal space. Normally, in the erect position, it contains air, referred to as the 'gas bubble'. The body forms the major part of the stomach and extends from the fundus proximally to an imaginary line running downwards and to the left from the angular notch of the lesser curvature. The long axis of the body is directed downwards,

forwards, and to the right. An ill defined, small, but variable area of the body surrounding the entrance of the oesophagus is sometimes called the cardiac part; its true extent can only be recognized histologically. The pyloric region has a wide antrum narrowing to the 2.5 cm long canal, the terminal part of which is surrounded by a thick ring of muscle, the pyloric sphincter. This controls the passage of the stomach contents into the duodenum. The term pylorus is frequently used to indicate the pyloric canal with its surrounding sphincter. The long axis of the pyloric region runs upwards and to the right (Di Dio and Anderson 1968).

It is usual to describe the stomach as having anterior and posterior surfaces meeting at the greater and lesser curvatures. However, these become less distinct when the stomach is distended. The anterior surface is covered by peritoneum of the greater sac, and is in contact with the inferior surface of the left lobe of the liver, the diaphragm, and the anterior abdominal wall [FIG. 6.43]. The posterior surface, covered except for a small area to the left of the oesophageal entrance by peritoneum of the anterior wall of the omental bursa, is in contact with a number of viscera, which collectively are referred to as the 'stomach bed'. These include the diaphragm, the spleen, the upper part of the left kidney, the left suprarenal, the body and tail of the pancreas, the transverse mesocolon, and to a variable extent the transverse colon [FIGS. 6.47 and 6.53].

The lesser curvature forms the right border of the stomach and gives attachment to the lesser omentum. It is continuous at the cardiac orifice with the right border of the oesophagus and at the pyloric orifice with the upper border of the superior part of the duodenum. At the junction of the body and the pyloric antrum the curvature increases abruptly to form the angular notch, which is relatively permanent and is more marked in J-shaped stomachs. [FIG. 6.51]. As the two layers of the lesser omentum approach the lesser curvature they separate to enclose the right and left gastric arteries and veins, and pass to the anterior and posterior surfaces of the stomach.

The greater curvature forms an acute permanent angle, the cardiac notch, with the left border of the oesophagus. From here it extends over the fundus and then downwards and to the right, so forming the left border of the stomach. At its lowest point, which in the J-shaped stomach distended with food may be considerably below the umbilicus, it turns upwards and to the right to become continuous with the lower border of the superior part of the duodenum.

The peritoneum on the anterior and posterior surfaces of the stomach unites at the greater curvature to form the gastrosplenic ligament above [FIG. 6.45] and the anterior layers of the greater omentum below. When the inferior recess of the omental bursa is reduced in size, the greater omentum fuses with the transverse mesocolon to form what is known as the gastrocolic ligament [FIG. 6.46]. In the region of the oesophagus and adjacent fundus the two layers diverge and pass to the diaphragm as the gastrophrenic ligament, thus leaving a small area of the stomach devoid of peritoneum.

STRUCTURE OF THE STOMACH

The outermost layer of the stomach wall, the serosa or peritoneum, is firmly attached to the underlying muscularis externa, except at the curvatures where the connection is looser.

The muscularis externa differs from that of the rest of the alimentary tract in having, in addition to circular and longitudinal layers, an oblique layer placed internal to the other two [FIGS. 6.52 and 6.54]. The fibres of the longitudinal layer are thinly scattered over the anterior and posterior surfaces of the body of the stomach, but form thicker masses along the curvatures and a complete layer

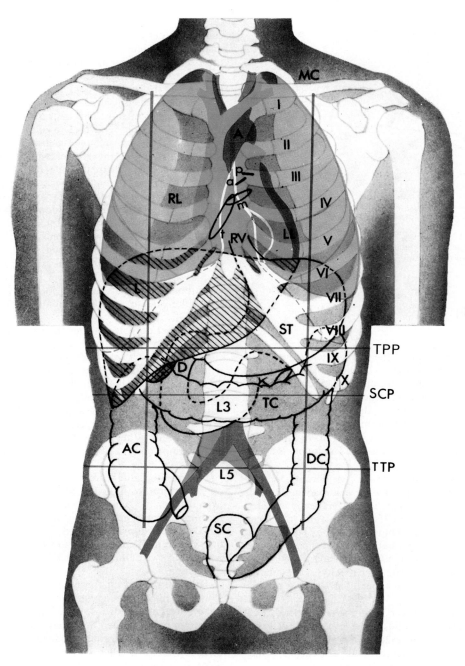

FIG. 6.50.   Anterior aspect of trunk, showing surface topography of viscera.

A.	Aorta.	PR.	Opposite line of pleural reflexion.
a.	Aortic orifice.	RL.	Right lung.
AC.	Ascending colon.	RV.	Right ventricle.
D.	Duodenum.	SC.	Sigmoid colon.
DC.	Descending colon.	SCP.	Subcostal plane.
L.	Liver.	ST.	Stomach.
LL.	Left lung.	t.	Tricuspid orifice.
m.	Mitral valve.	TC.	Transverse colon.
MC.	Midclavicular line.	TPP.	Transpylonic plane.
p.	Pulmonary orifice.	TTP.	Transtubercular plane.

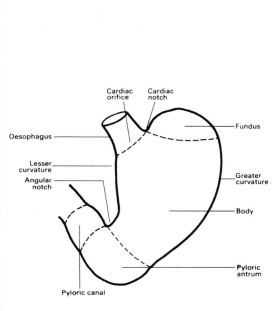

FIG. 6.51.   Diagram to show the different regions of the stomach.

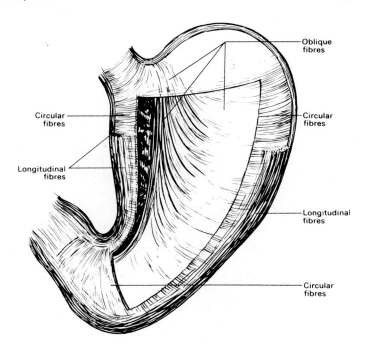

FIG. 6.52.   The arrangement of the muscle fibres of the stomach wall.

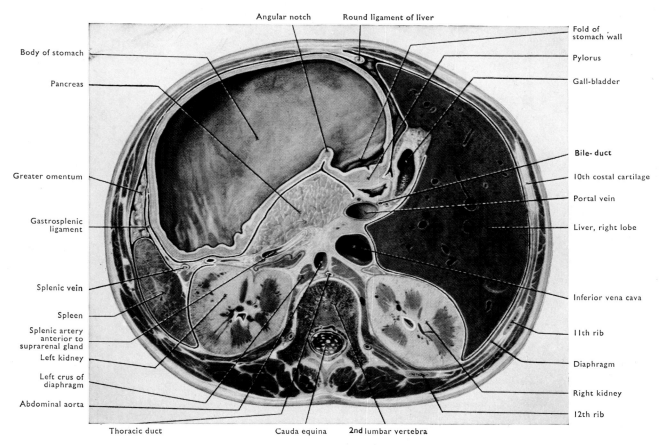

FIG. 6.53.   Transverse section of trunk at level of second lumbar vertebra.

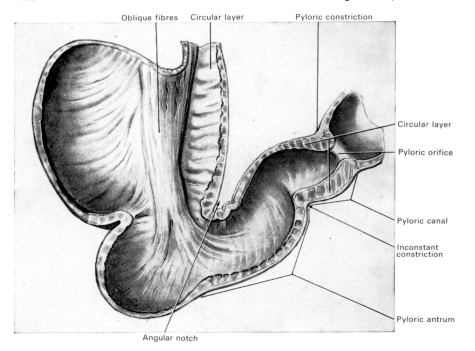

FIG. 6.54.   Muscular coat of stomach, seen from within after removal of mucous and submucous coats (Cunningham 1906). The anterior half of the stomach is shown, viewed from behind.

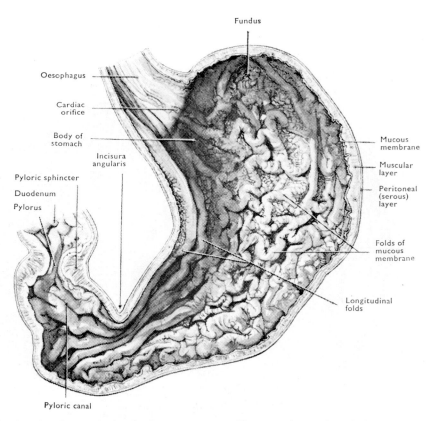

FIG. 6.55.   Interior of stomach, showing its mucous coat and its parts when moderately filled.

on the pyloric canal. Proximally, the longitudinal fibres are continuous with those of the oesophagus. Inferiorly, the majority turn inwards to reinforce the circular fibres of the pyloric sphincter and only some continue to the duodenum. The circular fibres form a continuous layer, except in the fundus, where they are sparse. In the pyloric canal they are markedly increased in number to form the pyloric sphincter. They are continuous with the circular fibres of the oesophagus but distally they are separated from those of the duodenum by a connective tissue septum. The oblique fibres are largely confined to the fundus and body of the stomach. They are best developed in the region of the cardiac orifice, from which they sweep downwards on the anterior and posterior surfaces of the stomach. The uppermost fibres turn laterally to blend with the circular fibres of the body, whilst the remainder pass downwards more or less parallel with the lesser curvature as far as the pyloric antrum. There is some radiological evidence that contraction of these vertical fibres produces a gastric canal, or gutter, which directs fluid entering the stomach into the pyloric antrum.

In character the submucosa is similar to the general plan outlined on page 436.

The mucosa is relatively thick, measuring up to 2 mm in the adult. In the living subject it is reddish-grey in colour and is thrown into numerous folds or rugae. These are arranged mainly longitudinally and, with the exception of those adjacent to the greater curvature, tend to disappear when the stomach is distended [FIG. 6.55].

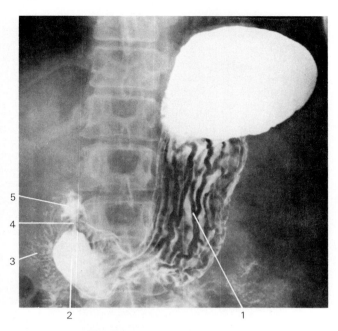

FIG. 6.57. Radiograph of barium meal taken in the supine position. The gas bubble lies in the body of the stomach anterior to the vertebral column. Here the small amount of barium outlines longitudinal mucosal folds.
1. Longitudinal mucosal folds of stomach.
2. Pyloric antrum.
3. Descending part of duodenum.
4. Pylorus.
5. Superior part of duodenum, contracted.

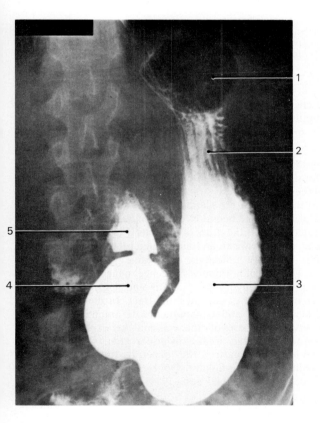

FIG. 6.56. Radiograph of barium meal. The barium is in the stomach and superior part of the duodenum. Taken in the erect posture. Note indentations caused by gastric contractions.
1. Gas bubble in fundus of stomach.
2. Mixture of gastric juice and barium which outlines the mucosal pattern of the stomach.
3. Body of stomach.
4. Pyloric antrum.
5. Superior part of duodenum.

The mucosa is largely composed of tubular glands surrounded by lamina propria which occasionally contains small lymph follicles. The glands open into the bottoms of gastric pits, or foveolae gastricae. The pits are lined by an epithelium which is continuous with that on the surface of the mucosa and consists entirely of a single layer of tall, columnar, mucous secreting cells [FIG. 6.60].

The glands of the mucosa are said to number approximately 15 million in man and they are consequently tightly packed, being separated by only small amounts of lamina propria and occasional smooth muscle fibres. They are perpendicular to the surface of the gastric mucosa and three to four glands open into each gastric pit. Histologically it is possible to distinguish three different types of gland, namely gastric or fundic, pyloric, and cardiac. These are localized to specific areas of the stomach, although there is some variation in the extent of these areas and in the degree of overlap in different stomachs.

Gastric or fundic glands are by far the most numerous and are found throughout the fundus and body of the stomach. Each gland extends from a slightly narrowed neck at the bottom of the gastric pit to the muscularis mucosae, where its base may be coiled and sometimes divided into two or three branches.

Four main cell types are readily recognized in the gastric glands, but this is undoubtedly an oversimplification of a cell population which as yet is only partly understood (Ito 1968).

Chief or zymogen cells are present mainly in the basal regions of the glands. They produce pepsinogen and the other enzymes of gastric secretion. They are cuboidal in shape and their basophilic cytoplasm contains much rough endoplasmic reticulum as in other protein synthesizing cells [FIGS. 6.61 and 6.62]. A prominent Golgi complex for the concentration and packaging of the synthesized protein is also present. In the apical cytoplasm the secretory products are stored in membrane-bound granules.

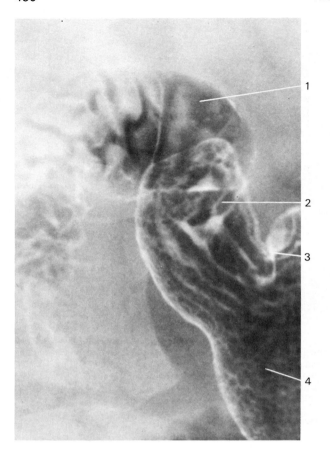

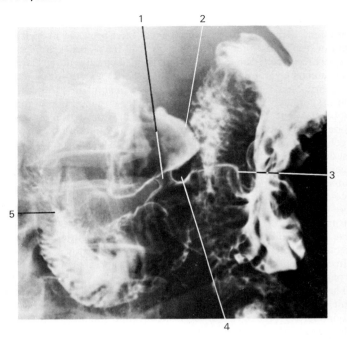

FIG. 6.59. Oblique radiograph of pyloric part of stomach and duodenum. The mucosal surface is outlined by a thin layer of barium. Taken in the supine position with gas in stomach and duodenum.
1. Pylorus.
2. Superior part of duodenum.
3. Pyloric antrum of stomach.
4. Pyloric sphincter.
5. Descending part of duodenum.

FIG. 6.58. Right oblique radiograph of pyloric antrum and proximal duodenum. Taken with a mixture of gas and barium (double contrast) to show the mucosal patterns.
1. Superior part of duodenum.
2. Pyloric antrum, longitudinal mucosal folds.
3. Peristaltic contraction.
4. Area gastricae in pyloric part of stomach.

Parietal or **oxyntic cells** occur throughout the gastric gland but are most common in the neck and adjacent regions. They are large, round, or pyramidal-shaped and they tend to be insinuated between the bases of other cells. They usually have centrally placed, single nuclei, although binucleate forms are sometimes present. When stained with haematoxylin and eosin, the cytoplasm appears intensively eosinophilic [FIG. 6.61].

Ultrastructural studies reveal the presence of extensive invaginations of the apical plasma membrane, somewhat erroneously called **intracellular canaliculi**. The surface area of these canaliculi is increased by the presence of microvilli [FIG. 6.63]. The cytoplasm contains numerous mitochondria with tightly packed cristae (indicative of the considerable energy requirements of these cells), smooth surfaced tubules of unknown function, lysosomes, and free ribosomes (Schofield, Ito, and Bolender 1979).

Parietal cells produce **hydrochloric acid** and **gastric intrinsic factor**, the latter known to be essential for the absorption of vitamin B_{12} from the ileum. Reduction of the total parietal cell population either at surgery or as a result of pathological conditions, such as atrophic gastritis, results in a decrease in the production of acid and may cause pernicious anaemia, due to vitamin B_{12} deficiency.

Neck mucous cells are comparatively few in number and are found interspersed with parietal cells in the necks of the glands [FIG. 6.63]. Apart from their position, they differ from surface mucous cells in that they are irregular in shape and their mucus has different

staining characteristics. These cells are relatively undifferentiated and contain considerable quantities of free ribosomes for the synthesis of structural proteins. They frequently show mitotic activity.

Argentaffin or **enterochromaffin cells**, so called because of their affinity for silver and chromium salts, are distributed throughout the length of the gastro-intestinal tract. In the stomach they are confined largely to the bases of the glands. They are characterized by the presence of membrane bound electron-opaque granules which are concentrated towards the bases of the cells [FIG. 6.62]. Immunocytochemical, histochemical, and ultrastructural studies suggest that argentaffin cells are unicellular endocrine glands, capable of producing a variety of hormones. Evidence so far strongly indicates that they produce 5-hydroxytryptamine and a glucagon-like substance throughout the gastro-intestinal tract. In the stomach they also produce **gastrin** and **histamine**. Most argentaffin cells are situated between the bases of other cells and do not reach the lumen of the gland. However, in the pyloric region, where they are considered to produce gastrin, they penetrate the full thickness of the epithelium (Baetens, Sukant, Dobbs, Unger, and Orci 1976; Solcia, Capella, Vassallo, and Buffa 1975).

Pyloric glands are confined to the pyloric antrum and pyloric canal. Here the gastric pits are long and the glands are relatively short. They branch more frequently and their basal parts are more extensively coiled than those in the body and fundus of the stomach. The pyloric glands contain only mucus-secreting and argentaffin cells. The mucus-secreting cells resemble the neck mucous cells of the gastric glands and also the submucosal glands of the duodenum (Johnson and McHinn 1970; Johnson and Young 1968).

Cardiac glands occupy an ill-defined narrow zone of mucosa adjacent to the cardiac orifice. They are less tightly packed than

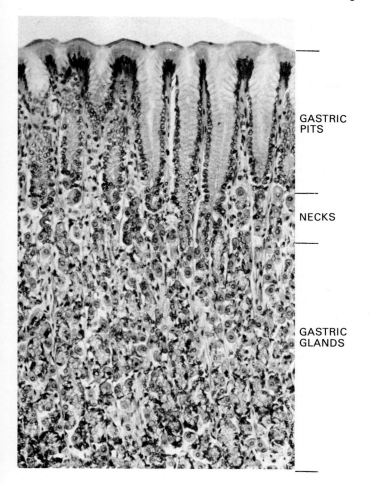

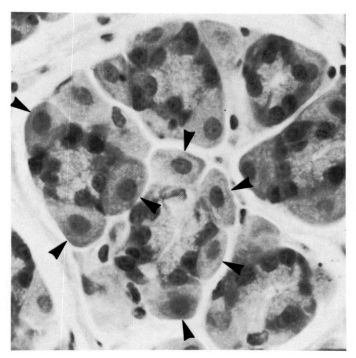

FIG. 6.61. High power photomicrograph of a transverse section of gastric glands illustrating the peripheral position of the parietal cells (arrows). The remaining cells in this section are all zymogen cells.

FIG. 6.60. Low power photomicrograph of the mucosa of the body of the stomach. The luminal epithelium and that lining the gastric pits is composed of surface mucous cells. The gastric glands are tightly packed.

glands elsewhere but in other respects resemble those of the pyloric region.

Epithelial cell renewal in the stomach

In common with other regions of the alimentary canal, the epithelium of the stomach undergoes frequent renewal. This is effected by mitotic activity in the bases of the gastric pits and in the neck mucous cells of the gastric glands. The cells formed in the gastric pits migrate upwards and replace those lost from the surface of the mucosa. Autoradiographic studies indicate that the surface epithelium of the stomach is replaced every three to four days. Mitotic activity in the neck mucous cells is mainly concerned with the replacement of the cells of the gastric glands. Since the gland is formed of different types of cells, it is obvious that a selective differentiation must also be involved so that a constant population of cells is maintained (Eastwood 1977; Chen and Withers 1975).

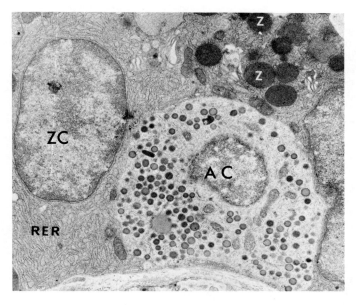

FIG. 6.62. Electron micrograph of gastric gland. An argentaffin cell (AC) is insinuated between the bases of two zymogen cells (ZC). The zymogen cells contain much rough endoplasmic reticulum (RER) and some zymogen granules (Z).

BLOOD AND LYMPH VESSELS AND NERVES OF THE STOMACH

The arteries of the stomach are derived entirely from the coeliac trunk [FIG. 13.68]. The **left gastric artery** passes to the cardiac region of the stomach in the superior gastropancreatic fold. It sends branches to the oesophagus, and then usually divides into anterior and posterior branches. These pass to the right in the lesser omentum and anastomose with similarly disposed branches from the **right gastric** branch of the proper hepatic artery.

Short gastric arteries are branches of the splenic artery which reach the fundus of the stomach by passing forwards in the gastrosplenic ligament. **Right and left gastro-epiploic arteries** pass in the greater omentum, parallel to and a short distance from the

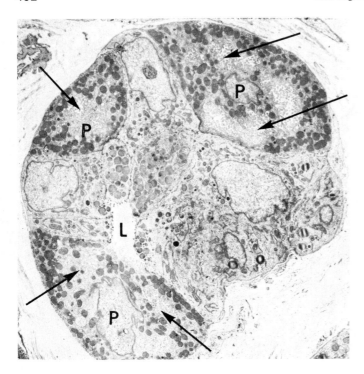

FIG. 6.63. Electron micrograph of a transverse section passing through the neck of a gastric gland. Three parietal cells (P) with intracellular canaliculi (arrows) are present. The remaining cells are neck mucous cells. One of the parietal cells opens directly into the lumen (L) of the gland.

greater curvature. They arise respectively from the gastroduodenal and splenic arteries, and usually form a continuous vessel. Branches of all these arteries run at first deep to the peritoneum on both surfaces of the stomach and then supply and pierce the muscularis externa to form a submucosal plexus which is distributed mainly to the mucosa (Piasecki 1975).

The named arteries are all accompanied by veins which eventually carry the blood back to the portal vein. The right and left gastric

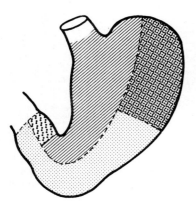

FIG. 6.64. Diagrammatic representation of lymph drainage areas of the stomach. For details see text.

veins pass directly to the portal vein and the right gastro-epiploic terminates in either the superior mesenteric or its middle colic tributary. The left gastro-epiploic and the short gastric veins join the splenic vein. Obstruction to the portal vein usually results in the opening-up of anastomotic channels joining the portal and systemic circulations. These exist between oesophageal branches of the left gastric and azygos veins.

The lymph vessels of the stomach wall form a dense anastomotic, submucosal plexus which drains by intermuscular vessels to a subserosal plexus on the surface of the stomach. This in turn is drained by larger lymph vessels which follow the main arteries of the greater and lesser curvatures.

There are four areas of lymph drainage of the stomach, each with its own group of regional lymph nodes [FIGS. 13.146 and 13.147].

The largest area, comprising the abdominal part of the oesophagus and the right two-thirds of the stomach from the fundus as far as the pyloric canal, drains into the left gastric nodes lying along the left gastric artery [FIG. 6.64]. From the upper part of the pyloric region drainage is to right gastric and hepatic nodes in the lesser omentum. The lower part of the pyloric region drains to right gastro-epiploic nodes in the greater omentum and to pyloric nodes on the head of the pancreas below the duodenum. The fundus and left portion of the body drain to the pancreaticosplenic nodes at the hilus of the spleen and along the course of the splenic artery. Efferents from all these regional lymph nodes pass to the coeliac group of pre-aortic nodes.

The stomach receives both sympathetic and parasympathetic nerves. The sympathetic nerve supply comes from the coeliac ganglia which give rise to postganglionic fibres that accompany the arteries to the stomach. The parasympathetic supply is derived from branches of the anterior and posterior vagal trunks which pass mainly to the corresponding surfaces of the stomach. Most of these branches reach the lesser curvature with the left gastric artery. However, the anterior vagal trunk sends a hepatic branch to the liver and a branch from this passes to supply the pyloric region of the stomach. A large coeliac branch runs on the left gastric artery from the vagal trunk (mainly the posterior) to the coeliac plexus and is distributed to the intestinal tract as far as the left colic flexure. The terminal vagal fibres to the stomach synapse in the myenteric and submucosal plexuses. Postganglionic fibres from these plexuses are distributed to the muscular tissue and the mucous membrane. The postganglionic sympathetic fibres also join the myenteric and submucosal plexuses before being distributed to the arteries, muscular tissue, and glands of the stomach wall. It is claimed that they are the main pathway for pain fibres from the stomach.

Stimulation of the vagus nerve increases peristalsis, and causes relaxation of the pyloric sphincter. It also results in increased secretion of acid and pepsinogen. Because of this, section of the vagus (vagotomy) is a surgical procedure sometimes adopted to reduce (acid) secretion in the treatment of gastric and duodenal ulcers. Stimulation of the sympathetic supply causes vasoconstriction and inhibition of peristalsis.

Anomalies of the stomach are relatively rare. The most common is congenital hypertrophic pyloric stenosis, in which the pyloric sphincter is greatly increased in thickness and the contents of the stomach have difficulty in passing into the duodenum. Symptoms do not usually arise until 4 to 6 weeks after birth, so there is doubt if this condition is truly congenital in nature. The stomach may also be congenitally herniated through the diaphragm, or be involved in oesophagal hiatal hernias.

RADIOLOGY

It should be remembered that the alimentary canal is a muscular tube which can contract. The contractions can produce appearances which may mimic pathology but which are transient.

STOMACH

Two distinct patterns of the gastric mucosa are demonstrable radiographically. The first is formed by the *folds* which the dissector

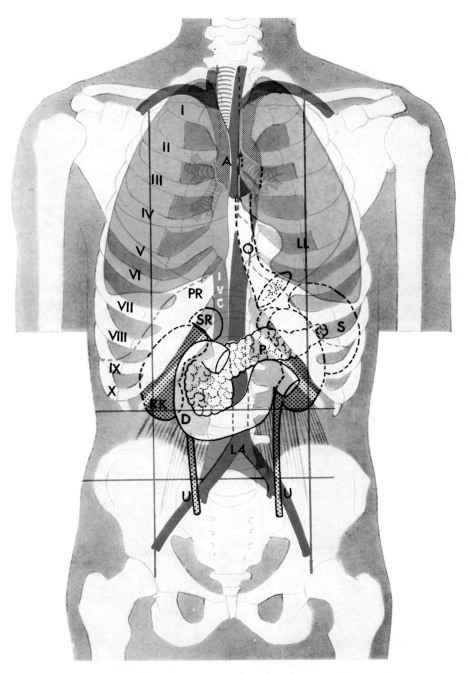

FIG. 6.65. Anterior aspect of trunk, showing surface topography of viscera.

A.	Arch of aorta.	PR.	Line of pleural reflexion.
D.	Duodenum.	RK.	Right kidney.
IVC.	Inferior vena cava.	S.	Spleen.
LL.	Left lung.	SR.	Suprarenal.
O.	Oesophagus.	U.	Ureter.
P.	Pancreas.		

can see in his specimen. The other is the *areae gastricae* pattern, which is seen when the gastric mucosal folds have been obliterated by distension with air or gas, and a small quantity of barium, previously ingested, is manipulated under fluoroscopic control so that it coats the mucosa thinly (double contrast examination) [FIG. 6.58]. This pattern appears as a series of fine lines outlining the areae gastricae. Study of this pattern can sometimes lead to earlier recognition of disease than is possible by other radiological means.

The shape of the filled stomach in general corresponds to the body habitus, but it is extremely variable, even in the same subject at different times and in different positions [FIGS. 6.56 and 6.57].

Certain areas which are often easily recognized on radiographs, such as the pylorus and pyloric antrum are not always so readily defined when a specimen is viewed from the serosal aspect.

THE SMALL INTESTINE

Food, mixed with gastric secretion and partially digested in the stomach to form chyme, passes in a carefully regulated manner into the small intestine. Here the chyme is mixed with pancreatic and intestinal enzymes and digestion continues to completion. The products of digestion, along with water, electrolytes, and minerals such as calcium and iron, are then absorbed by the intestinal mucosa and pass either into the portal circulation, or the lymph system. As well as digestive enzymes, the small intestine produces hormones, which are involved in regulating its peristaltic and secretory activity, and large quantities of mucus, thought to function primarily in protecting the epithelium from mechanical and other forms of damage.

Duodenum

The duodenum is the first and widest part of the small intestine, with a diameter of 4–5 cm. It is continuous proximally with the stomach at the pyloric orifice, and distally with the jejunum at the duodenojejunal junction. It is 25 cm long and is shaped like the letter C. [FIGS. 6.65 and 6.66]. For descriptive purposes it is divided into four parts, namely superior, descending, horizontal, and ascending. These measure approximately 5, 7.5, 10, and 2.5 cm in

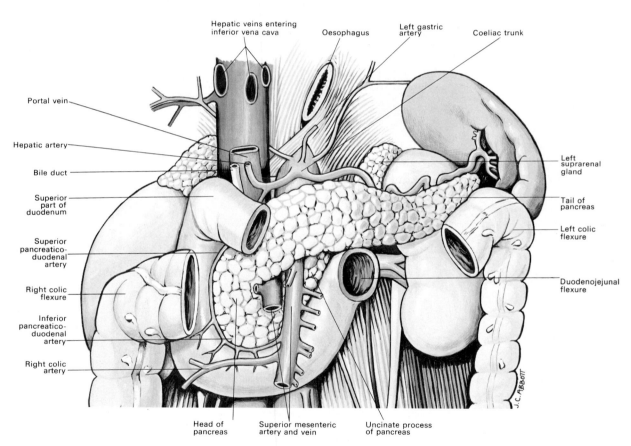

FIG. 6.66.   A dissection showing the duodenum and pancreas. The stomach, small intestine, and transverse colon have been removed.

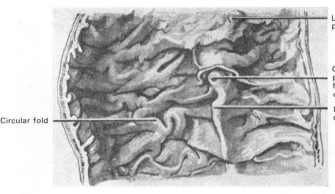

FIG. 6.67. Interior of descending part of duodenum (viewed from the front) showing duodenal papilla and opening of accessory pancreatic duct (lesser duodenal papilla).

length, respectively. The first 2 to 3 cm of the duodenum is enclosed in peritoneum and is free to move with the pyloric region of the stomach; the remainder, with the exception of the terminal 1 to 2 cm, is retroperitoneal and fixed to the structures on the posterior abdominal wall.

The **superior part** of the duodenum begins at or slightly to the right of the midline, and sometimes on the transpyloric plane (level of the first lumbar vertebra). It passes to the right, backwards, and slightly upwards in conformity with the posterior abdominal wall, and is therefore foreshortened in anteroposterior radiographs [FIG. 6.53]. When filled with barium the proximal region appears in radiographs as a triangular shaped shadow, the **duodenal cap** [FIG. 6.59]. The anterior surface is in contact with the quadrate lobe of the liver and the neck of the gall-bladder. Posteriorly, its proximal part is separated from the head of the pancreas by the omental bursa and its distal part touches the bile-duct and the gastroduodenal artery with, behind these, the portal vein and the inferior vena cava [FIG. 6.45]. Above, its first 2–3 cm gives attachment to the lesser omentum and forms the lower boundary of the epiploic foramen, while below the greater omentum is attached to its proximal end. The superior part turns downwards on the right side of the first lumbar vertebra to become continuous with the descending part.

The **descending part** of the duodenum passes downwards on the right side of the vertebral column to the level of the lower border of the third lumbar vertebra. It is crossed anteriorly by the commencement of the transverse colon, which is usually devoid of a mesentery in this region and consequently is directly in contact with the duodenum. Above the transverse colon the anterior surface is applied to the right lobe of the liver and the gall-bladder; below the transverse colon it is overlaid by coils of jejunum. Posteriorly, the descending part lies on the medial part of the anterior surface of the right kidney, its vessels, the right ureter, the right psoas muscle, and the inferior vena cava. Its lateral surface lies close inferiorly to the ascending colon and its medial to the head of the pancreas, which overlaps it anteriorly and posteriorly to a varying extent. The superior and inferior pancreaticoduodenal vessels run between the head of the pancreas and its medial surface. The bile and pancreatic ducts pierce the posteromedial wall obliquely just below the middle of the descending part. Here they join to form the **hepatopancreatic ampulla**, which opens through its distal end on the summit of the **greater duodenal papilla** [FIG. 6.67] 8–10 cm from the pylorus and 2–3 cm inferior to the opening of the accessory pancreatic duct on the **lesser duodenal papilla**.

The **horizontal part** passes transversely to the left with a slight inclination upwards. From right to left it crosses in front of the right ureter, the right psoas major, the right gonadal vessels, and the inferior vena cava; it terminates in front of the aorta. Its anterior surface is crossed by the superior mesenteric vessels and by the root of the mesentery of the small intestine.

The **ascending part** runs upwards, on or slightly to the left of the aorta to the level of the second lumbar vertebra, where it terminates by bending abruptly forwards and to the right as the **duodenojejunal flexure**. Posteriorly, are the left sympathetic trunk, the left psoas major, and the left gonadal vessels. Anteriorly, it gives attachment to the upper part of the root of the mesentery, while the left kidney lies laterally and the uncinate process of the pancreas medially. The region of the duodenojejunal flexure is fixed in position by the **suspensory muscle of the duodenum**. This fibromuscular band blends with the musculature of the flexure and passes upwards deep to the pancreas to gain attachment to the right crus of the diaphragm [FIG. 5.102] (Jit and Grewal 1977).

Peritoneal recesses

In the region of the duodenojejunal flexure there may be a number of peritoneal recesses or evaginations, which occasionally can be sites of intestinal herniation. The **superior** and **inferior duodenal recesses** occur most commonly and are formed by folds of peritoneum stretching from the ascending part of the duodenum to the posterior abdominal wall [FIG. 6.68]. The **paraduodenal recess** is relatively rare, being present in only 2 per cent of bodies. It lies to the left of the ascending part of the duodenum and is formed by a fold of peritoneum, the free margin of which usually encloses the inferior mesenteric vein, and a branch of the left colic artery.

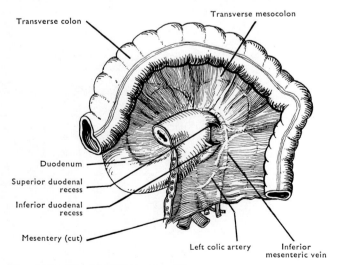

FIG. 6.68. Duodenal folds and recesses.
The transverse colon and mesocolon have been thrown up, and the mesentery has been turned to the right and cut. The paraduodenal recess is situated to the medial side of the inferior mesenteric vein, between it and the terminal part of the duodenum. It is not shown in the illustration.

RADIOLOGY

The first 2 cm or so of the small intestine constitutes the 'duodenal cap'. This part of the gut is easy to identify radiographically because of its characteristic shape which has been likened to the ace of spades or a triangle. It does not appear similar in the body, possibly because anteroposterior radiographs show the superior part of the duodenum in an oblique view. The pylorus often projects into the duodenum. The duodenal cap is the most frequent site of peptic ulceration, and is therefore important [FIGS. 6.56 and 6.59].

The *mucosal pattern* of the duodenum consists largely of transverse folds with the exception of a single longitudinal fold in the second part which extends downwards from the greater duodenal papilla. The horizontal or third part of the duodenum crosses the vertebral column to the left and then ascends to the duodenojejunal flexure.

Jejunum and ileum

The remainder of the small intestine, consisting of an upper three-fifths, the jejunum, and a lower two-fifths, the ileum, measures approximately 6 m in length in the living subject. It gradually decreases in diameter from approximately 3 cm at the duodenojejunal junction to 2 cm at the terminal ileum, where the walls are also thinner and less vascular. The small intestine is arranged in coils and loops occupying the infracolic part of the greater peritoneal sac and separated from the stomach, spleen, and liver by the transverse mesocolon and colon. Normally, the greater omentum hangs down in front of the coils of intestine and separates them from the anterior abdominal wall.

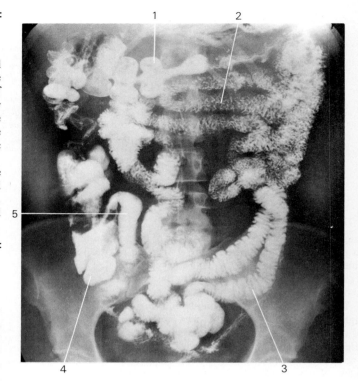

FIG. 6.70. Anteroposterior radiograph of abdomen with small amount of barium in small intestine, caecum, and proximal colon to show the mucosal pattern.
1. Haustra of transverse colon.
2. Jejunum.
3. Ileum.
4. Caecum.
5. Terminal ileum.

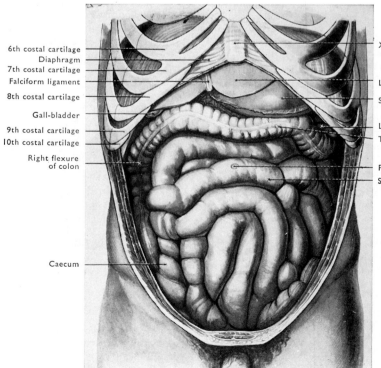

6th costal cartilage
Diaphragm
7th costal cartilage
Falciform ligament
8th costal cartilage
Gall-bladder
9th costal cartilage
10th costal cartilage
Right flexure of colon
Caecum

Xiphoid process
Liver
Stomach
Left flexure of colon
Transverse colon
Position of umbilicus
Small intestine

FIG. 6.69. Abdominal viscera, after removal of greater omentum. Note the high position of the stomach and transverse colon in this subject.

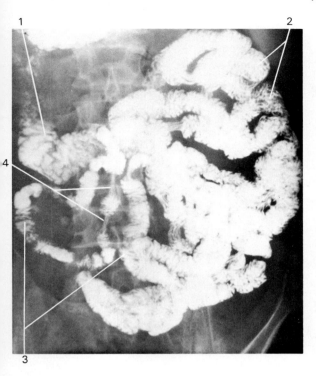

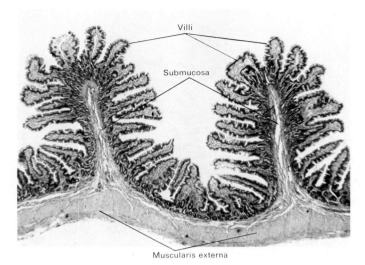

Villi

Submucosa

Muscularis externa

FIG. 6.72. Longitudinal section of small intestine passing through two circular folds.

FIG. 6.71. Anteroposterior radiograph of barium in small intestine. The amount of barium is small and so outlines mucosal folds. Note narrowings due to peristaltic waves which change direction of folds but do not obliterate them.
1. Duodenum distended prior to emptying. Coarse transverse folds.
2. Jejunum, fine feathery mucosal pattern.
3. Ileum, coarser transverse folds.
4. Narrowings due to peristattic waves.

In general, the coils of jejunum lie above and to the left, occupying mainly the umbilical and left lumbar regions, while those of the ileum lie below and to the right in the hypogastric and pelvic regions. The jejunum and ileum are attached to the posterior abdominal wall by the mesentery which allows the coils of intestine considerable freedom of movement. The root of the mesentery, measuring about 15 cm in length, runs obliquely from the left side of the second lumbar vertebra to the right iliac fossa, crossing the horizontal part of the duodenum, the aorta, the inferior vena cava, the right gonadal vessels, the right ureter, and the right psoas major muscle [FIG. 6.46]. The mesentery varies in breadth, being shorter at the fixed ends of the small intestine and longer, especially where it is attached to coils of ileum that may lie in the pelvis. It contains between its two layers of peritoneum the blood and lymph vessels, lymph nodes, and nerves of the intestine, along with a variable amount of fat [FIG. 6.79]. At the jejunal end of the intestine the fat may not reach the intestinal wall, with the result that there are transparent areas in the mesentery, whereas at the ileal end, it reaches and often overlaps the intestinal wall.

STRUCTURE OF THE SMALL INTESTINE

The wall of the small intestine has several structural specializations which greatly increase its luminal surface area and hence facilitate its absorptive and secretory activities. These specializations are confined to the mucosa and submucosa, and it is here that the small intestine differs significantly from the rest of the digestive tract. The serosa and muscularis externa have the general structure described for the alimentary canal on page 436 [FIG. 6.34].

(1) Circular folds are permanent structures involving the mucosa and submucosa [FIG. 6.72]. They appear first 2.5 to 5.0 cm from the pylorus and reach their fullest development in the distal half of the duodenum and upper part of the jejunum. Here they reach a height of 8 mm, and are responsible for the feather-like appearance seen on radiographs. In the distal small intestine they gradually decrease in size and frequency, and are absent from the terminal ileum. The folds are in general transversely orientated and usually do not extend more than two-thirds of the way round the wall [FIGS. 6.71 and 6.72].

(2) Intestinal villi. These finger-like projections are present throughout the entire length of the small intestine. Being less than half a millimetre in height, they are barely visible to the naked eye but en masse they give the mucous membrane its velvety appearance [FIG. 6.72]. Each villus consists of a central core of lamina propria covered by a layer of simple columnar epithelium. The lamina propria contains blood vessels, a central lacteal (lymph vessel), smooth muscle fibres, plasma cells, lymphocytes, macrophages, mast cells, eosinophilic leucocytes, and non-myelinated nerve fibres. The smooth muscle fibres belong to the muscularis mucosae and are peripherally disposed in the long axis of the villus. By their contraction they are thought to be responsible for the emptying of the central lacteal into the submucosal lymph plexus.

(3) Intestinal glands or crypts of Lieberkühn are simple tubular glands which invaginate the lamina propria almost as far as the muscularis mucosae. They are extremely numerous and open on the surface of the mucosa between the bases of the villi. Their lining epithelium is continuous with that covering the villus. In the duodenum the ducts of the submucosal glands open into the bases of the intestinal glands.

The epithelium of the mucosa rests on a basal lamina separating it from the lamina propria which forms the cores of the villi and is continuous below with that surrounding the intestinal glands [FIG. 6.73]. It contains five different types of cell, namely absorptive cells, goblet cells, Paneth cells, undifferentiated cells, and argentaffin or enterochromaffin cells. In addition, small lymphocytes are frequently present between the bases of the epithelial cells; they originate in the lamina propria and migrate in and out of the epithelium where they probably have an immunological role to play (Ferguson 1977).

Absorptive cells, sometimes also called principal cells or enterocytes, are present in the epithelium covering the villi and lining the upper regions of the intestinal glands [FIG. 6.73]. They are characterized by a well-developed striated border consisting of large

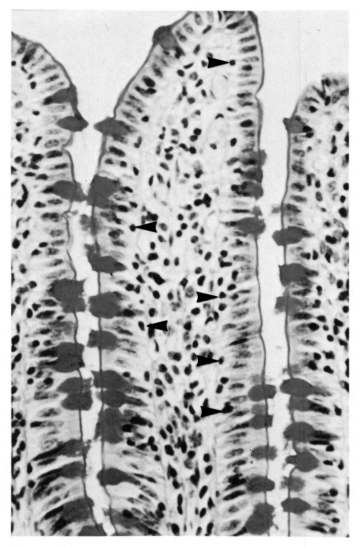

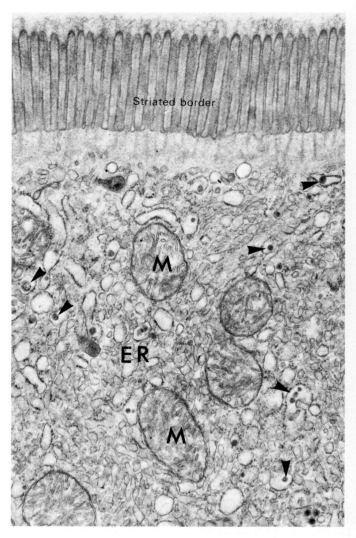

Striated border

FIG. 6.73.   Photomicrograph of section of intestinal villi stained by the periodic-acid–Schiff technique. The epithelium is composed of columnar absorptive cells and goblet cells filled with red-staining mucus. Lymphocytes (arrowheads) are present between the bases of the epithelial cells. The centres of villi are occupied by lamina propria.

FIG. 6.74.   Electron micrograph of absorptive cell. The striated border consists of tightly packed microvilli, and the apical cytoplasm contains mitochondria (M) and profiles of rough and smooth endoplasmic reticulum (ER). Chylomicrons (arrowheads) are present in some of the vacuoles of the ER.

numbers of tightly packed microvilli, which greatly increase the surface area and thereby aid digestion and absorption [FIG. 6.74]. Each microvillus is surrounded by plasma membrane containing proteins involved in transport mechanisms and enzymes which take part in the final stages of digestion of carbohydrates and proteins. In addition, the external surface of the plasma membrane has a firmly attached layer of glycoprotein which is thought, amongst other things, to provide a protective microenvironment for the surfaces of the cells [FIG. 6.75]. Each microvillus has a central core of longitudinally arranged actin filaments. They pass from the plasma membrane at the tip of the microvillus to a band of filamentous material, the terminal web, stretching across the apex of the cell [FIG. 6.76] (Crane 1968; Holmes 1971; Mooseker and Tilney 1975; Bretscher and Weber 1978).

The absorptive cells have well-developed junctional complexes on their lateral surfaces [FIG. 6.76]. These act: (1) as areas of attachment between adjacent cells; (2) as barriers hindering the movement of substances from the lumen of the intestine into the intercellular epithelial compartment; and (3) as sites of low electrical

resistance, where diffusion of ions and small molecules can occur between cells (Farquhar and Palade 1963).

Deep to the terminal web the cytoplasm contains profiles of smooth endoplasmic reticulum, which are in continuity with the more deeply placed and more extensive rough endoplasmic reticulum. Immediately above and at the sides of the basally situated nucleus, there is a well-developed Golgi complex and, throughout the cell, there are many randomly distributed mitochondria (Johnson 1975).

Although actively involved in the absorption of products of digestion and other essential substances, the absorptive cell shows structural changes in the adult only during the transport of lipid [FIG. 6.74]. In this process electron opaque chylomicrons, consisting of triglyceride, phospholipid, cholesterol, and β-lipoprotein, appear in the profiles of the smooth and adjoining rough endoplasmic reticulum. The chylomicrons subsequently travel in vacuoles, via the Golgi complex, to the lateral and basal plasma membranes, where they are discharged into the intercellular space and lamina propria. From here the chylomicrons pass into the central lacteal, eventually to be transported to the general circulation through the lymph vessels. It should be noted that lipid in the intestinal lumen is finally broken down to fatty acids and monoglycerides, which are not electron opaque and which are therefore not detected as they cross the plasma membrane into the apical cytoplasm. It is only when they are taken up by the smooth endoplasmic reticulum, re-esterified and combined with the other components of the chylomicrons that they become detectable with the electron microscope. The enzymes necessary for the re-esterification of lipid and the formation of chylomicrons are present within the membranes of the smooth endoplasmic reticulum (Cardell, Badenhausen, and Porter 1967; Friedman and Cardell 1977).

In the newborn of certain species, which receive gamma globulins from their mother's milk, the absorptive cells show marked structural changes indicative of the process of pinocytosis. By this mechanism it is possible to transfer the large immunoglobulin molecules across the intestinal epithelium without first subjecting them to digestion, and consequently to transfer immunity from the mother to the offspring by this route (Brambell 1970; Clarke and Hardy 1971).

Goblet cells owe their name to their characteristic shape. They occur singly and appear to be scattered at random amongst the absorptive cells on the sides of the villi and upper parts of the glands [FIG. 6.73]. Their shape is largely due to the accumulation of large numbers of membrane-bound, mucigen granules within the apical half of the cell. These are discharged from the cell by a process known as *exocytosis*, in which the membrane of the granule fuses with the apical plasma membrane and then an opening appears through which the mucus flows into the intestinal lumen without the cell loosing its integrity. The rough endoplasmic reticulum, involved in the synthesis of mucin, is compressed by the accumulated secretion to the peripheral and basal cytoplasm. Goblet cells have a well-developed Gogli complex, lying immediately above the basally situated nucleus. In addition to packaging the mucus, it modifies it by adding carbohydrate and sulphate moieties.

Argentaffin cells are now considered to be unicellular endocrine glands [p. 450]. Present throughout the gastro-intestinal tract, the biliary tract, and ducts of the pancreas, they are particularly common in the duodenum and the vermiform appendix. In the small intestine they are present on the villi and in the intestinal glands. They tend to be ovoid or triangular in shape, insinuated between the bases of other cells and the basement membrane, and frequently do not extend for the full thickness of the epithelium. They contain membrane bound granules of varying electron opacity, which are occasionally seen to be discharged through the basal plasma membrane into the lamina propria (Buchan, Polak, Solcia, Capella, Hudson, and Pearse 1978; Pearse, Polak, and Bloom 1977).

In addition to producing 5-hydroxytryptamine along the length of the small intestine, the argentaffin cells of the duodenum produce *gastrin*, which has a similar action to that formed in the stomach, and the hormones *secretin* and *cholecystokinin-pancreozymin* (CPZ).

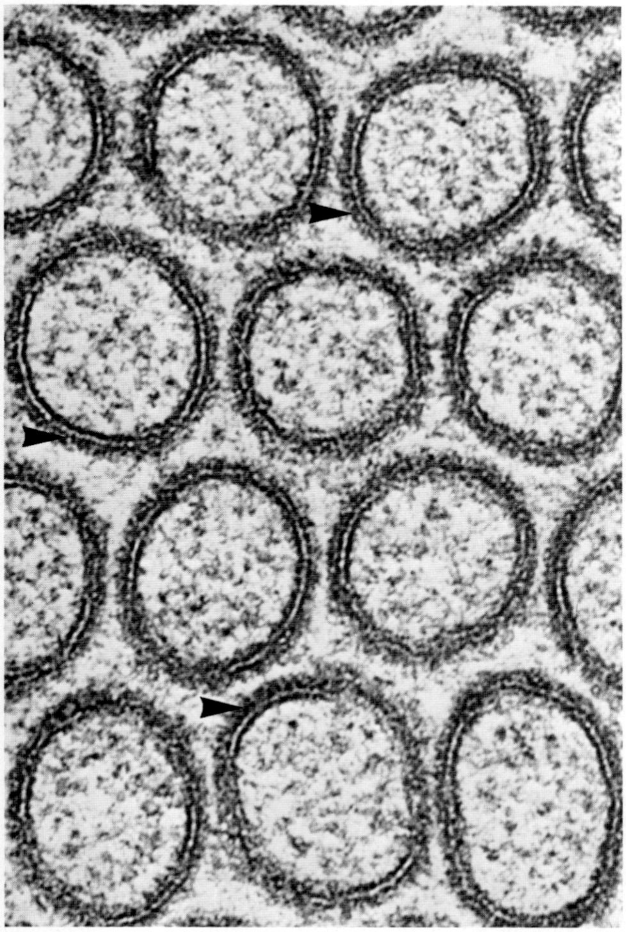

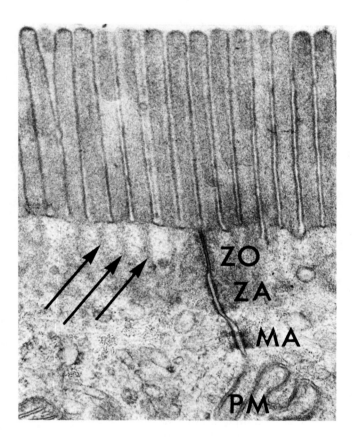

FIG. 6.75. Electron micrograph of transversely sectioned microvilli of an absorptive cell. Each microvillus is surrounded by plasma membrane with a fuzzy layer of glycoprotein on its external surface (arrow heads).

FIG. 6.76. Electron micrograph of adjacent apical regions of two absorptive cells. The plasma membranes (PM) of the two cells form a junctional complex consisting of three components, the zonula occludens (ZO), the zonula adhaerens (ZA), and the macula adhaerens (MA). Cores of the microvilli (arrows) enter the apical cytoplasm.

The latter two hormones are liberated when acid chyme passes into the duodenum. Secretin stimulates the pancreas to produce a watery secretion rich in bicarbonate but low in enzyme content, whereas cholecystokinin-pancreozymin stimulates both the production of an enzyme-rich secretion and contraction of the gall-bladder.

Undifferentiated cells. The name given to these cells is somewhat of a misnomer for, like neck mucous cells in the stomach, they are specialized for epithelial replacement. They are found in the lower halves of the intestinal glands and are cuboidal. They have a poorly developed striated border and an abundance of free ribosomes. Mitosis occurs frequently and they are the only cells in the intestinal epithelium capable of dividing. Daughter cells differentiate into either goblet or absorptive cells and migrate upwards from the glands along the sides of the villi to replace those constantly shed from the tips of the villi. Autoradiographic studies indicate that the cells of the epithelium on the villi are replaced every two to three days (Cheng and Leblond 1974; Clarke 1970).

Paneth cells line the bases of the intestinal glands [FIG. 6.77]. They are pyramidal in shape, have basally situated nuclei and contain large numbers of secretory granules. They have a well-developed rough endoplasmic reticulum and closely resemble the exocrine cells of the pancreas. A current view of their function is that they may play a role in regulating the intestinal flora by producing the enzyme, *lysozyme*, which is capable of breaking down (lysing) bacteria (Sandow 1979).

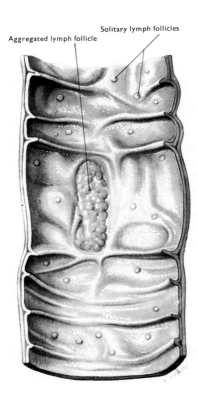

FIG. 6.78.   Internal surface of part of the distal ileum.

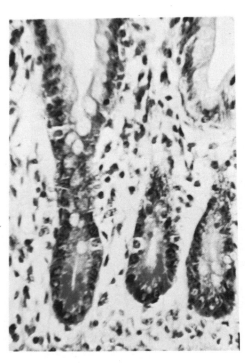

FIG. 6.77.   Bases of intestinal glands. The Paneth cells contain red-stained granules. Section stained by phloxine-tartrazine.

Lymph tissue

The wall of the small intestine contains large numbers of **solitary lymph follicles** measuring up to 3 or 4 mm in diameter. Small follicles are confined to the lamina propria of the mucous membrane, but larger ones extend through the muscularis mucosae into the submucosa and may cause the overlying epithelium to bulge into the intestinal lumen. Villi and crypts of Lieberkuhn are either rudimentary or absent over the surface of the follicles. The follicles

are more plentiful in the ileum and here they may be aggregated in considerable masses called **aggregated lymph follicles** or Peyer's Patches [FIG. 6.78]. These occur in the anti-mesenteric border of the intestine and normally number 30 to 40. They are usually oval in outline, with their long axis in the long axis of the intestine, and they may be up to 2.0 to 2.5 cm in length. The lymph tissue of the intestine, like that elsewhere, tends to undergo gradual involution after reaching its maximum development in childhood.

There is evidence that the solitary and aggregated lymph follicles of the intestine and vermiform appendix may provide the precursors of the intestinal plasma cells that produce the immunoglobulin A (IgA) and other antibodies. In this role the lymph follicles are behaving as the equivalent of the bursa of Fabricus, found in the intestine of birds, and known to be essential for the development of antibody-producing lymphocytes and plasma cells. IgA produced in the plasma cells of the lamina propria passes into the intestinal lumen by way of the epithelial cells. Here it combines with a secretory protein to form a complex known as secretory IgA, which may control the growth of bacteria as well as preventing the attachment of antigens to the epithelial surface (Fichtelius 1968).

REGIONAL SPECIALIZATIONS IN THE SMALL INTESTINE

In addition to the specializations already noted, there are a number of others of a functional and histological nature.

(1) The submucosa of the upper two-thirds of the duodenum contains branched tubuloalveolar glands, often known as Brunner's glands, which pour their secretion into the bases of the intestinal glands via ducts which pass through the muscularis mucosae. Their alkaline, mucous secretion is thought to protect the duodenal mucosa against the acid secretion of the stomach (Treasure 1978).

(2) The number of goblet cells increases along the length of the small intestine.

(3) There is histochemical and ultrastructural evidence that differentiation of the absorptive cell continues as it migrates along the side of the villus. This correlates with experiments which show that the cells on the upper part of the villus have greater absorptive powers than those at the base of the villus.

(4) While most products of digestion can be absorbed along the length of the intestine, glucose, iron, and folic acid are preferentially absorbed by the duodenum and upper part of the jejunum, and vitamin B_{12} and bile salts are absorbed only by the ileum.

BLOOD AND LYMPH VESSELS AND NERVES OF THE SMALL INTESTINE

The blood supply to the small intestine comes from two sources. That to the upper half of the duodenum is derived from the coeliac trunk, the artery of the foregut, and that to the lower half of the

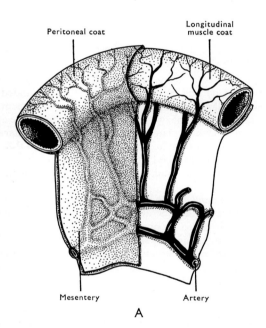

Peritoneal coat   Longitudinal muscle coat

Mesentery   Artery

A

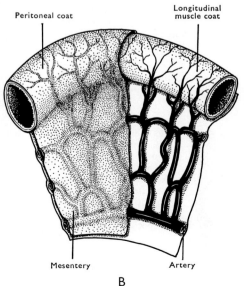

Peritoneal coat   Longitudinal muscle coat

Mesentery   Artery

B

FIG. 6.79.   Portion of A, jejunum, and B, ileum.

duodenum, the jejunum and the ileum is derived from the superior mesenteric artery, the artery of the midgut. Blood reaches the upper half of the duodenum in the **superior pancreaticoduodenal artery**, and in small branches of the hepatic and right gastric arteries. The lower half of the duodenum is supplied by the **inferior pancreaticoduodenal artery**, a branch of the superior mesenteric. This artery divides into anterior and posterior branches which pass to the right, between the head of the pancreas and the duodenum, to anastomose with the corresponding branches of the superior pancreaticoduodenal artery. The jejunum and ileum receive their blood supply from ten to sixteen branches of the **superior mesenteric artery**. Within the mesentery these vessels form anastomosing arterial arcades which increase in complexity towards the terminal ileum [FIG. 6.79]. From the convexity of the peripheral arcades, straight arteries run to the intestine and, having partially encircled it, perforate the muscular wall to supply it and terminate in the submucosal plexus. The straight arteries supplying the upper part of the intestine are longer and fewer than those supplying the lower part (Chevral and Guérand 1978).

The **veins** for the most part follow the course of the arteries; those from the upper part of the duodenum pass either directly to the portal vein or its two main tributaries, the splenic and superior mesenteric veins. Those from the remainder join the superior mesenteric vein. A small vein, the **prepyloric vein**, helps to drain the superior part of the duodenum. As it passes in front of the pylorus to join the right gastric or portal vein, it acts as a useful landmark for the pyloric orifice.

The **lymph vessels** of the small intestine begin as the central lacteals of the villi. These open into a submucosal plexus which is drained by large vessels that perforate the muscularis externa. From the duodenum the vessels run upwards to the hepatic nodes and downwards to the mesenteric nodes. Lymph vessels from the jejunum and ileum drain into a series of nodes lying along the arteries to the intestine, the efferents from which also pass to the superior mesenteric nodes.

The **nerves** of the small intestine are postganglionic-sympathetic fibres from the coeliac and superior mesenteric ganglia and preganglionic parasympathetic fibres from the vagus. These accompany the arteries to the intestine and pass to the **myenteric plexus** (Auerbach's plexus), between the circular and longitudinal muscle layers, and the **submucosal plexus** (Meissner's plexus), where the parasympathetic fibres synapse. In this dual innervation, which exists throughout the digestive tract, the parasympathetic system and the gastro-intestinal hormones play a major role in the control of digestive processes. Both divisions of the autonomic nervous system also transmit afferent fibres from the various coats of the intestine. Those in the vagus initiate visceral reflexes and carry sensations such as hunger, distension, and nausea, while those passing with the sympathetic system are mainly concerned with the conduction of pain impulses. The efferent parasympathetic fibres are secretomotor to the intestinal glands and in general augment peristalsis and inhibit sphincter activity. The sympathetic fibres tend to have an opposing action on the smooth muscle of the intestine and also produce vasoconstriction (Gunn 1968).

RADIOLOGY

The filling of the small intestine is largely governed by pyloric control of gastric emptying. The normal *mucosal pattern* of the jejunum consists of delicate, transverse, feathery folds [FIG. 6.70]. These gradually become less pronounced towards the ileum, the terminal part of which appears as a smooth tube extending into the caecum at the ileocaecal valve when filled with barium. When partially filled, mucosal folds are visible which usually run in a longitudinal direction [FIG. 6.83]. The radiological appearances of the **ileocaecal valve** depend on its position. In partial filling in profile, it has a prominent upper lip [FIG. 6.83] and a lower lip which is often scarcely visible. *En face*, the valve appears as a central rosette of contrast within a filling defect.

With gaseous distention, the mucosal pattern of the small intestine appears as a series of almost straight, transverse ridges. This is the characteristic appearance in obstruction of the small intestine.

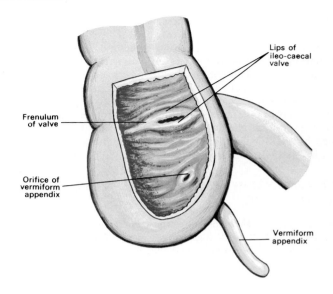

FIG. 6.81.   Interior of caecum showing the ileocaecal valve and orifice of the vermiform appendix.

THE LARGE INTESTINE

The large intestine consists of the caecum, to which the vermiform appendix is attached, the ascending, transverse, descending, and sigmoid colons, the rectum, and the anal canal [FIG. 6.84]. It varies in length from 1.5 to 1.8 m and, with the exception of the rectum, its calibre gradually decreases from the caecum to the anal canal. It receives its name from the fact that, on average, its diameter is greater than that of the small intestine. The proximal part of the large intestine absorbs water and electrolytes from the unabsorbed chyme passing into it through the ileocaecal orifice and the distal part stores the faeces thus formed until they are defaecated.

Apart from its calibre and position, the large intestine can be distinguished from the small by: (1) the thickening of the longitudinal muscle to form three bands, the **taeniae coli**, (2) the presence of sacculations or **haustrations** produced by the relative shortness of the taeniae compared with that of the gut, and (3) the **appendices epiploicae**, small peritoneal masses of protruding fat present everywhere except on the appendix, the caecum, and the rectum [FIGS. 6.66 and 6.84]. The taeniae coli are continuous with the longitudinal muscle of the appendix, and on the caecum, ascending, descending, and sigmoid colons are situated on their anterior, posterolateral, and posteromedial aspects [FIG. 6.82]. On the transverse colon they occupy anterior, superior, and inferior positions. They become less marked on the sigmoid colon, at the lower end of which they are reduced to two broad bands which continue down on the anterior and posterior surfaces of the rectum.

The Caecum

The caecum is a sacculated diverticulum of the large intestine, lying below the level of the ileocaecal junction [FIG. 6.80]. It is about 6 cm

long and 7.5 cm wide and is usually completely surrounded by peritoneum, which allows it considerable freedom of movement. In the foetus it is conical in shape, with the appendix forming the apex of the cone. After birth its lateral wall undergoes considerable growth, while the medial remains relatively stationary, so that the original apex with the attached appendix comes to lie on the posteromedial surface of the adult caecum (Fitzgerald, Nolan, and O'Neill 1971).

The caecum is situated in the right iliac fossa, lying obliquely above and parallel with the lateral half of the inguinal ligament [FIG. 6.84]. When distended it completely fills the fossa, but when empty coils of small intestine may slide in front of it. It rests on the psoas and iliacus muscles, the femoral and lateral femoral cutaneous nerves, and, anteriorly, it is in contact with either the greater omentum or directly with the anterior abdominal wall. Its posteromedial surface is adjacent to the terminal ileum as it rises up out of the pelvis to reach the ileocaecal junction at, or just to the right of the intersection of the right vertical and transtubercular planes. The base of the caecum is continuous above with the ascending colon, the junction between the two being frequently marked by a slight transverse groove, corresponding to the position of the frenula of the ileocaecal valve.

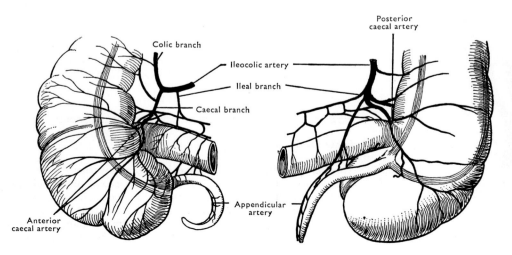

FIG. 6.80.   Blood supply of caecum and vermiform appendix. (Modified from Jonnesco 1895.)

The illustration on the left is from the front, the right from behind. In the latter the appendicular artery and the three taeniae coli springing from the root of the appendix should be specially noted.

The opening of the ileum into the ascending colon points forwards and to the right. It is oval or slit-like and is surrounded by the **ileocaecal valve** consisting of two folds which project into the lumen of the colon [FIG. 6.81]. At the medial and lateral ends of the orifice the folds join to form ridges, the **frenula of the valve**, which continue for a variable distance round the circumference of the gut. Each fold is formed by a duplication of the mucous membrane and circular fibres of the intestinal wall. The longitudinal fibres are only partly involved and most pass directly from the terminal ileum to the ascending colon. The mucous membrane of the opposing surfaces of the folds is similar to that of the small intestine, while that on the other surfaces resembles, and is continuous with, the mucous membrane of the large intestine. Radiological evidence indicates that the valve is extremely ineffective in preventing regurgitation from the caecum to the ileum. It is probable that its main function is to delay the passage of ileal contents into the large intestine. Relaxation of the circular muscle of the valve occurs as a result of the *gastro-ileal reflex*, which is initiated by the passage of food into the stomach.

The vermiform appendix

The vermiform appendix is a worm-like tube about 0.5 cm in diameter and varying from 2–15 cm in length, the average being about 9 cm. It is attached by its base to the posteromedial surface of the caecum where the taeniae coli converge to become continuous with its outer layer of muscle, 2 to 3 cm below the ileocaecal orifice [FIGS. 6.80 and 6.82]. The base or attachment of the appendix lies approximately at the junction of the lateral one-third and medial

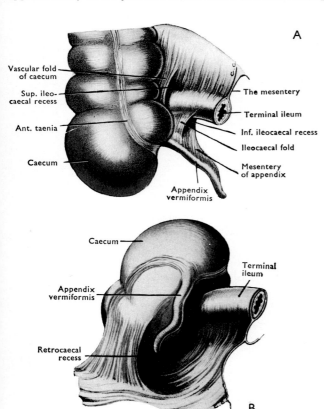

FIG. 6.82. Folds and recesses in the ileocaecal region.
A, The caecum viewed from the front; the mesentery of the vermiform appendix is distinct.
B, The caecum turned upwards to show a retrocaecal recess which lies behind it and behind the beginning of the ascending colon.

two-thirds of a line joining the anterior superior iliac spine and the umbilicus (McBurney's point). The appendix has a triangular-shaped mesentery, the **mesoappendix**, which attaches it to the undersurface of the mesentery of the terminal ileum. The **appendicular artery**, a terminal branch of the ileocolic, passes to the appendix behind the ileum in, or near, the free edge of the mesoappendix. Its terminal part usually lies on the wall of the appendix and may become thrombosed in inflammatory states of the appendix.

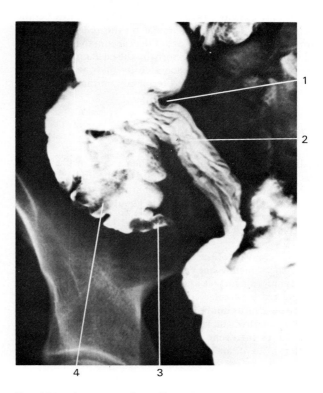

FIG. 6.83. Anteroposterior radiograph of small quantity of barium in terminal ileum.
1. Superior lip of ileocaecal valve.
2. Terminal ileum, longitudinal folds.
3. Vermiform appendix incompletely filled.
4. Caecum.

The appendix enjoys considerable freedom of movement, being limited only by the attachment of its base and its mesentery. According to Wakeley (1933) it lies behind the caecum and ascending colon in some 65 per cent of cases. Its next most common position (31 per cent of cases) is hanging over the brim of the lesser pelvis where, in the female, it may come into contact with the uterine tube and ovary. In the remaining 4 per cent it lies either below the caecum, or in front of, or behind the terminal ileum. Its variable position accounts for the different localizations of pain associated with the spread of inflammation from the appendix to the adjacent abdominal wall.

PERITONEAL FOLDS AND RECESSES

In the region of the caecum and appendix there are a number of moderately constant folds of peritoneum which bound recesses or fossae of the peritoneal cavity [FIG. 6.82].

The **superior ileocaecal recess** opens medially and downwards just above the terminal part of the ileum. It is bounded in front by the **vascular fold of the caecum**, which contains the anterior caecal vessels and stretches from the mesentery of the ileum to the medial surface of the upper part of the caecum and ascending colon. Posteriorly, the recess is bounded by the terminal ileum and its mesentery.

The **inferior ileocaecal recess** opens downwards and medially below the terminal ileum. Its anterior wall is formed by the **ileocaecal fold** extending from the lower border of the ileum to the caecum, and anterior surface of the mesentery of the appendix. Its posterior wall is formed by the mesoappendix.

The **retrocaecal recess**, as its name implies, lies behind the caecum but it may extend upwards for a few centimetres behind the ascending colon. It is limited above by the reflection of the peritoneum from the posterior surface of the caecum to the posterior abdominal wall. It is bounded anteriorly by the caecum, posteriorly by parietal peritoneum, and on each side by the **caecal folds** of peritoneum which run from the posterior abdominal wall to the caecum.

Ascending colon

The ascending colon is continuous with the caecum at the level of the ileocaecal junction and extends upwards to the undersurface of the right lobe of the liver, where it bends sharply forwards and to the left to form the **right colic flexure**. It is about 15 cm long and is somewhat narrower than the caecum but broader than the descending colon. It lies in the right lumbar region [FIGS. 6.50 and 6.84].

The lower 2.5 cm or so of the ascending colon may be completely surrounded by peritoneum, but apart from this, peritoneum usually covers only its anterior surface and sides. Posteriorly, the ascending colon is connected by areolar connective tissue to the iliacus, the iliac crest, the quadratus lumborum, the transversus abdominis and the inferolateral part of the right kidney. The iliohypogastric and ilioinguinal nerves intervene between it and the quadratus lumborum. Anterior to it are coils of small intestine, the greater

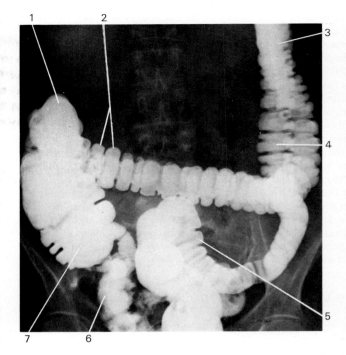

FIG. 6.85.   Colon and terminal ileum filled by barium enema.
1. Right colic flexure.
2. Haustra of transverse colon.
3. Left colic flexure.
4. Descending colon.
5. Sigmoid colon.
6. Terminal ileum.
7. Caecum.

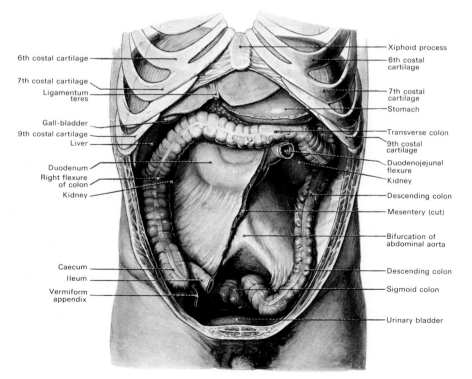

FIG. 6.84.   Abdominal viscera after removal of jejunum and ileum.

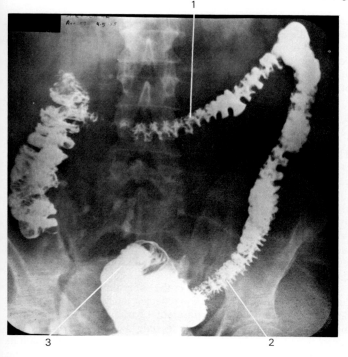

FIG. 6.86.   Anteroposterior radiograph showing colon contracted after a barium enema.
1. Transverse colon.
2. Lower end of descending colon.
3. Sigmoid colon.

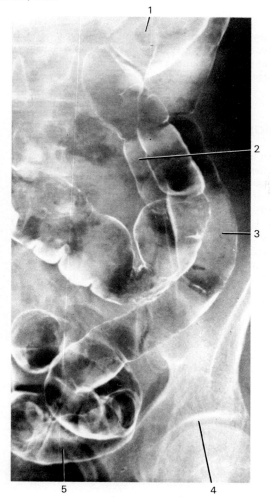

FIG. 6.87.   Distal colon. Double contrast (air and barium) enema. A thin layer of barium lines the mucosa so that overlapping loops of colon can be seen.
1. Gas bubble.
2. Transverse colon.
3. Descending colon.
4. Hip joint.
5. Sigmoid colon.

omentum, and the anterior abdominal wall. On the sides of the caecum and ascending colon there are the deep, peritoneal lined, **paracolic sulci** [FIG. 6.48]. As the colon bends forwards and medially at the right colic flexure to become continuous with the transverse colon, it rests on the anterior surface of the kidney and the descending part of the duodenum [FIG. 6.66].

Transverse colon

The transverse colon is about 50 cm long and extends across the abdomen from the right colic flexure, just below the transpyloric plane, to the left colic flexure which lies at a slightly higher level [FIG. 6.84]. Its extremities are deeply placed and relatively fixed, but the remainder suspended by the **transverse mesocolon**, has a considerable range of movement and usually descends to the level of the umbilicus or lower. Its position, however, is subject to variation from individual to individual and is influenced by factors such as posture and the degree of distension of other abdominal viscera. Except for the first few centimetres, the transverse colon is completely covered by peritoneum and its mesentery, the transverse mesocolon, is fused to the posterior surface of the greater omentum, which must, therefore, be turned up in order to expose it from in front [FIG. 6.43]. It is in contact superiorly with the liver, gall-bladder, stomach, and spleen; anteriorly with the greater omentum; posteriorly with the descending part of the duodenum, the head of the pancreas, the small intestine, and the left kidney.

At its left extremity the transverse colon bends sharply backwards and downwards, forming the **left colic flexure**, which is continuous with the descending colon [FIG. 6.66]. The left colic flexure lies behind the stomach and comes into contact with the spleen, the left kidney, and the tail of the pancreas. It is attached to the diaphragm by a peritoneal fold, the **phrenicocolic ligament**.

Descending colon

The descending colon, which is approximately 25 cm long, is almost twice the length of the ascending. It begins at the left colic flexure and descends through the left hypochondriac, lumbar, and iliac regions, terminating at the superior aperture of the lesser pelvis. Its course is almost vertical as far as the iliac crest, but then it deviates medially, close to the inguinal ligament, and joins the sigmoid colon at the superior aperture of the lesser pelvis, anterior to the external iliac vessels [FIGS. 6.50 and 6.84].

It is usually empty and contracted, and is covered by peritoneum only on its anterior surface and sides. The reflections of peritoneum from its sides to the abdominal wall create the medial and lateral left **paracolic sulci**. Anteriorly, it is overlaid by coils of small intestine and the greater omentum, but when distended its lower part comes into contact with the anterior abdominal wall through which it may be palpated [FIG. 6.84]. Posteriorly, it rests on the lower lateral region of the kidney, the quadratus lumborum, the iliacus, and the psoas major. In its course it crosses the subcostal, iliohypogastric, ilioinguinal, lateral femoral cutaneous, femoral, and genitofemoral nerves, and near its termination the gonadal vessels and the external iliac artery.

Sigmoid colon

The sigmoid colon has a greater variation in length than any other part of the large intestine, but averages about 40 cm. It is continuous above at the superior aperture of the lesser pelvis with the descending colon, and below at the level of the third sacral vertebra with the rectum [FIG. 6.89]. Between its fixed ends it is surrounded by peritoneum and is suspended by a mesentery, the **sigmoid mesocolon**, which carries its nerves and vessels. The mesocolon has an inverted V-shaped attachment with the apex near the point of division of the common iliac artery, the left or upper limb ascending on the medial side of the left psoas, and the right limb passing downwards to the third sacral vertebra at or near the midline. The left ureter passes into the pelvis behind the apex of the mesocolon, where there may be a small orifice leading into an intersigmoid peritoneal recess of variable size.

In the adult the sigmoid colon usually forms a transversely orientated loop occupying the lesser pelvis, but when the pelvic viscera are distended it passes upwards into the abdomen, and in the child, before the pelvis has reached its full development, it is an

abdominal viscus. On its left it is related to the structures on and forming the left lateral pelvic wall; posteriorly it comes into contact with pyriformis, the sacral plexus, the left ureter, and internal iliac vessels, inferiorly it rests on the bladder in the male and on the uterus and bladder in the female; on its right it is related to the coils of small intestine.

Rectum

The rectum is continuous above with the sigmoid colon at the level of the third sacral vertebra, and below with the anal canal as it passes through the pelvic diaphragm, 2.5 cm antero-inferior to the tip of the coccyx, at the level of the apex of the prostate gland. It is about 12 cm long and increases in diameter from above downwards, its lower part being dilated to form the **ampulla of the rectum**. It is devoid of a mesentery and in its course it follows the curvature of the sacrum and coccyx at first downwards and then forwards on the levatores ani, before bending sharply backwards to join the anal canal at the anorectal junction.

The anterior surface and sides of the upper third of the rectum are covered by peritoneum, which is reflected laterally to the posterior pelvic wall, so forming the **pararectal fossae** [FIG. 6.90]. In the middle third the peritoneum covers only the anterior surface and passes forwards as the lower boundary of the rectovesical pouch in the male and the recto-uterine pouch in the female [FIG. 6.43]. The lower third of the rectum has no peritoneal covering. The level of reflection of peritoneum in the male is about 7.5 cm from the anus

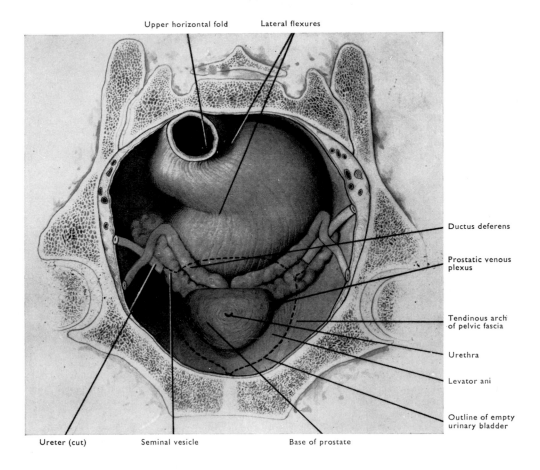

FIG. 6.88.   Distended rectum *in situ*.
   The bladder has been removed, but its form is shown by a dotted line. The rectum is very much distended and almost completely occupies the pararectal fossae.

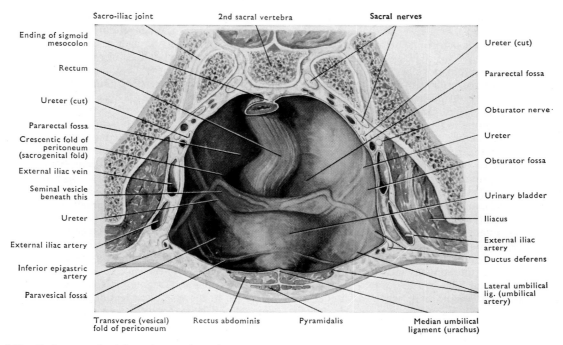

Sacro-iliac joint — 2nd sacral vertebra — Sacral nerves

Ending of sigmoid mesocolon

Rectum

Ureter (cut)

Pararectal fossa

Crescentic fold of peritoneum (sacrogenital fold)

External iliac vein

Seminal vesicle beneath this

Ureter

External iliac artery

Inferior epigastric artery

Paravesical fossa

Ureter (cut)

Pararectal fossa

Obturator nerve

Ureter

Obturator fossa

Urinary bladder

Iliacus

External iliac artery

Ductus deferens

Lateral umbilical lig. (umbilical artery)

Transverse (vesical) fold of peritoneum — Rectus abdominis — Pyramidalis — Median umbilical ligament (urachus)

FIG. 6.89.   Peritoneum of pelvic cavity seen from above.

The pelvis of a thin male subject, aged 60, was sawn across obliquely. Owing to the absence of fat the various pelvic organs were visible through the peritoneum, though not quite so distinctly as represented here. The urinary bladder and rectum are both empty and contracted; the paravesical and pararectal fossae, as a result, are very well marked.

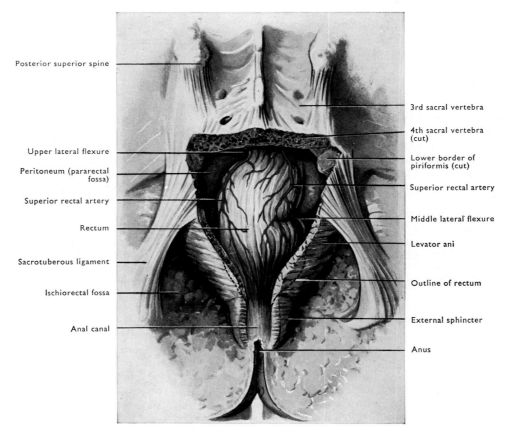

Posterior superior spine

Upper lateral flexure

Peritoneum (pararectal fossa)

Superior rectal artery

Rectum

Sacrotuberous ligament

Ischiorectal fossa

Anal canal

3rd sacral vertebra

4th sacral vertebra (cut)

Lower border of piriformis (cut)

Superior rectal artery

Middle lateral flexure

Levator ani

Outline of rectum

External sphincter

Anus

FIG. 6.90.   Rectum exposed by dissection from behind.

The sacrum has been sawn across the fourth sacral vertebra, and its inferior part removed with the coccyx. The posterior portions of the coccygei, levatores ani, and of the external sphincter have been cut away. The 'pinching in' of the lower end of the rectum by the medial edges the levatores ani, resulting in the formation of the flattened anal canal, is suggested in the illustration. The lateral flexures are also seen.

and so lies at the tip of the index finger during rectal examination, whereas in the female the level is 2–3 cm lower, with the result that the contents of the recto-uterine pouch can be palpated in this form of examination.

The rectum is a midline structure at its beginning and end; in between it follows a sinuous course resulting in three lateral **flexures**, which are accentuated when it is distended [FIG. 6.88]. The upper and lower flexures are convex to the right and the middle one, the largest, to the left. Corresponding to the lateral flexures, there are usually three internal sickle-shaped **transverse folds** (valves of Houston), consisting of mucous membrane, submucosa, and circular and some longitudinal fibres of the muscularis externa. The uppermost of these permanent folds lies near the beginning of the rectum and may project from either side or sometimes completely encircle it. The middle fold, situated on the right side just above the ampulla, is the most prominent, while the lowest, the least constant, is on the left near the anorectal junction. It is sometimes suggested that the rectal folds help to support the contents of the rectum. However, most workers believe that the faeces accumulate in the sigmoid colon and that the rectum normally fills only immediately prior to defaecation. The rectal folds can be seen by means of a proctoscope and the lower and middle ones, which can hinder the passage of instruments such as the sigmoidoscope, may also be felt on digital examination.

The rectum is in contact posteriorly in the median plane with the lower three sacral vertebrae, the coccyx, the anococcygeal ligament, the superior rectal vessels, and the ganglion impar. Posterolaterally, expecially when it is distended, it comes into contact on each side with the lower part of the sacral plexus and the sympathetic trunk lying on the piriformis, and with the coccygeus and levator ani muscles [FIG. 6.90]. Inferiorly, the levator ani muscles separate it from the ischiorectal fossae. The rectum is surrounded by a fascial sheath, which, at the front and sides, is loosely attached to the muscularis externa. Posteriorly, this fascia is condensed around the superior rectal vessels and attaches the rectum to the anterior surface of the sacrum. Similar condensations around the middle rectal vessels and nerves constitute, on each side, the lateral ligament of the rectum.˗

In front, the structures in contact with the rectum differ in the two sexes. In the male, the upper two thirds, covered by firmly adherent peritoneum, has coils of the small intestine or the sigmoid colon separating it from the upper parts of the base of the bladder and seminal vesicles. Below the reflection of the peritoneum forming the floor of the rectovesical pouch, it is applied to a small area of the base of the bladder, the seminal vesicles, the deferent ducts, and the prostate gland with only the **rectovesical fascia** intervening [FIGS. 6.88 and 8.18]. In the female, the recto-uterine pouch, with its contents of coils of small intestine or sigmoid colon, separates the upper two thirds from the posterior fornix of the vagina and the uterus [FIG. 8.27]. Below the pouch the **rectovaginal septum** intervenes between it and the lower part of the vagina.

RADIOLOGY

The **caecum** and **vermiform appendix** can be demonstrated by barium meal or barium enema [FIG. 6.83 and 6.85]; both are very variable in position. The appendix varies in length also. Frequently it is not visible either because it fails to fill with contrast medium, or because it lies behind the filled caecum.

Barium enema with complete filling of the colon is seldom practised because it shows gross features only, however, it gives some information about the structure and function of the colon. The **rectum**, viewed in the lateral position, lies close to the sacrum. The folded **sigmoid colon** is difficult to see in its entirety because of its overlapping parts. The **descending colon**, being relatively fixed, is seen without difficulty, though its lower part, in the left iliac fossa, may have a mesentery and, being mobile, become confused with the sigmoid colon. The mobile **transverse colon** tends to overlap the ascending and descending parts, especially at the flexures. The **ascending colon** and caecum are easily seen and there may be some reflux into the ileum.

The colon shows a typical haustral pattern [FIG. 6.85] produced by sacculation between contracting strips of circular muscle which tend to segment its contents, while the taeniae shorten the colon without compressing the sacculations between them. The large intestine does not show peristaltic contractions, but occasionally filling of the caecum produces a simultaneous contraction of the proximal colon extending to the middle of the transverse colon or beyond. More usually this process is less dramatic, but involves the mass contraction of a considerable length of the colon [FIG. 6.86]. A radiograph taken after evacuation of the enema [FIG. 6.86] shows evidence of shortening of the redundant segments, and sometimes demonstrates part of the *mucosal surface* thrown into a complex of short transverse and longitudinal folds. However, the mucosa is best inspected in the colon and rectum by the introduction of a small amount of barium followed by distention with air (double contrast) [FIG. 6.87]. This technique virtually obliterates any mucosal pattern so that detection even of small pathological changes is possible.

The small gas bubble in FIGURE 6.87 illustrates the fact that all variations in the radiograph are not alterations in the mucosa. Such a bubble is readily displaced by pressure on the abdomen or a change of posture, but other causes of such appearances (e.g. a small faecal mass adherent to the mucosa) may be more difficult to interpret.

Anal canal

The anal canal begins at the **anorectal junction** (where the rectum narrows abruptly and bends sharply downwards and backwards) and ends at the anus. It is about 4 cm long and when empty its side walls are compressed together to form an anteroposterior slit by the tonic contraction of the surrounding muscle. At the **anus** the moist, hairless mucosa of the canal changes to dry, hairy, perianal skin.

Posteriorly, it is in contact with the **anococcygeal ligament** which extends from it to the coccyx, while anteriorly the central perineal tendon lies between it and the membranous urethra and bulb of the penis in the male, and the lower part of the vagina in the female. On each side the external sphincter separates it from the fat in the ischiorectal fossa which allows the canal to distend during defaecation [FIG. 6.90].

The mucous membrane of the upper half is raised to form five to ten permanent vertical folds known as the **anal columns** [FIG. 6.92] each containing a terminal branch of the superior rectal artery and vein. At their lower ends the columns are joined by semilunar folds of epithelium, the **anal valves**, behind which are the shallow **anal sinuses**. The line produced by the upper borders of the anal valves and lower ends of the columns is known as the **pectinate line**. It approximates to the line of junction of the columnar epithelium of the upper part of the canal and the stratified squamous of the lower part.

In the living subject the first 15 mm or so of the mucosa below the pectinate line has a shiny bluish appearance. This region is called the transitional zone or **pecten**, and its stratified squamous epithelium is thin and lacks sweat glands; it overlies the internal rectal venous plexus. The pecten ends at a narrow wavy white line (of Hilton), which is indistinct in the living and corresponds to a

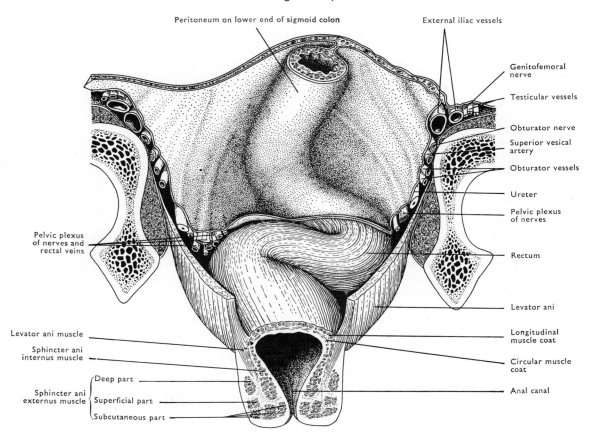

Peritoneum on lower end of sigmoid colon

External iliac vessels

Genitofemoral nerve

Testicular vessels

Obturator nerve

Superior vesical artery

Obturator vessels

Ureter

Pelvic plexus of nerves

Rectum

Pelvic plexus of nerves and rectal veins

Levator ani

Levator ani muscle

Longitudinal muscle coat

Sphincter ani internus muscle

Circular muscle coat

Sphincter ani externus muscle

Deep part

Superficial part

Subcutaneous part

Anal canal

FIG. 6.91.   Drawing of rectum and anal canal seen from the front to show peritoneum and relations to the pelvic wall.

palpable intersphincteric groove where the lower border of the internal sphincter and the subcutaneous part of the external sphincter meet. Between the intersphincteric groove and the anus, the epithelium is thicker and contains both sweat and sebaceous glands (Walls 1958).

The musculature of the anal canal forms two sphincters [FIGS. 6.91 and 6.92]. The internal sphincter, consisting of a thickening of the circular layer of the muscularis externa, surrounds the upper three-quarters of the canal and extends from the anorectal junction to the intersphincteric groove. The external sphincter, composed of voluntary muscle, surrounds the entire length of the canal. It is divisible into deep, superficial, and subcutaneous parts. The deep and subcutaneous encircle the gut without bony attachment; the superficial is attached behind to the coccyx and in front to the perineal body. The subcutaneous part lies between the lower border of the internal sphincter and the perianal skin.

The internal and external sphincters are separated by the longitudinal layer of the muscularis externa. Towards the anus this becomes progressively more fibro-elastic and eventually breaks into a number of strands. Some of these pass outwards between the superficial and subcutaneous parts of the external sphincter to become lost in the fat of the ischiorectal fossa or the central perineal tendon; others pass medially and gain attachment to the lining of the anal canal at the intersphincteric groove. The remainder end in the dermis, deep to the lining of the lower part of the canal and the perianal skin.

At the anorectal junction the puborectalis fibres of the levator ani muscles form a U-shaped sling by passing on either side of the rectum to unite posteriorly. Contraction of the puborectalis pulls the anorectal junction forwards, thus accentuating the angle of the junction and helping to prevent the passage of faeces into the anal

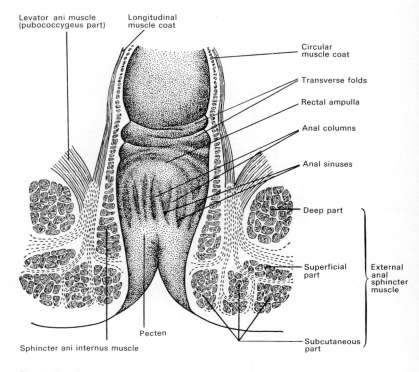

Levator ani muscle (pubococcygeus part)

Longitudinal muscle coat

Circular muscle coat

Transverse folds

Rectal ampulla

Anal columns

Anal sinuses

Deep part

Superficial part

Subcutaneous part

External anal sphincter muscle

Sphincter ani internus muscle

Pecten

FIG. 6.92.   Rectum and anal canal with the sphincter muscles. See text.

canal. As the fibres of the puborectalis pass round the gut they blend with the deep part of the external and with the internal sphincter, forming a muscular band, the **anorectal ring**, which can be felt on rectal examination. Integrity of the anorectal ring is essential for the maintenance of rectal continence (Dickinson 1978).

The desire to defaecate is initiated when the rectum becomes distended with faeces from the sigmoid colon. Should the opportunity not be available, defaecation can be delayed by the voluntary contraction of the external sphincter and the puborectalis. During defaecation all activity of the anal sphincters is inhibited and the mass action of the muscle of the large intestine, plus increased abdominal pressure, forces the faeces through the anal canal.

STRUCTURE OF THE LARGE INTESTINE

The large intestine differs from the small intestine in that its mucosa does not contain villi, and its thin outer longitudinal layer of muscle is thickened to form three strips, **taeniae coli**, throughout most of its length [FIGS. 6.82 and 6.84]. The mucosa, as far as the pectinate line, has a simple columnar epithelium which invaginates the lamina propria to form numerous tubular **intestinal glands** or crypts of Lieberkühn. Below the pectinate line it has a stratified squamous epithelium and lacks intestinal glands. The lamina propria contains scattered lymph follicles, and though more cellular, is in general similar to that of the small intestine.

The glands of the large intestine are longer than those of the small intestine and have a higher proportion of mucous secreting goblet cells in their epithelium [FIG. 6.93]. As well as goblet cells, the epithelium contains absorptive cells largely concerned with the absorption of water, argentaffin cells, and, in the bases of the glands, undifferentiated cells which divide to replace the epithelium every few days (Chang and Nadler 1975).

The structure of the **vermiform appendix** is essentially similar to that of the large intestine. However, the longitudinal layer of the muscularis externa forms a continuous, thick coat of muscle and the muscularis mucosae is poorly developed. Its most characteristic feature is a large number of **lymph follicles**, which frequently completely surround its lumen [FIG. 6.94]. With increasing age the lumen tends to become obliterated and, even in the healthy

FIG. 6.94.   Photomicrograph of transverse section of vermiform appendix showing numerous lymph follicles (L). ME, muscularis externa.

appendix, may contain masses of dead cells and acellular detritus. Villi are absent and the glands are relatively sparse and irregular in shape. Its lining epithelium and that of the glands contain absorptive and goblet cells, as well as numerous argentaffin cells (Gorgollon 1978).

BLOOD AND LYMPH VESSELS AND NERVES OF THE LARGE INTESTINE

The appendix, caecum, ascending colon and most of the transverse colon, being developed from the midgut, receive their blood supply from the ileocolic, right colic, and middle colic branches of the **superior mesenteric artery** [FIG. 13.70]. The appendicular artery, a branch of the ileocolic, passes behind the terminal ileum and runs to the tip of the appendix in the mesoappendix [FIG. 6.80] (Solanke 1968). The left colic flexure, descending and sigmoid colons, the rectum, and the anal canal as far as the level of the pectinate line, being developed from the hindgut, receive blood from the left colic, sigmoid, and superior rectal branches of the **inferior mesenteric artery** [FIG. 13.73]. The branches of the superior and inferior mesenteric arteries join to form a continuous arcade, the **marginal artery**, a short distance from the inner side of the colon. A series of straight arteries pass from it to the wall of the intestine.

The rectum and anal canal receive blood from three main sources. (1) The **superior rectal artery**, the terminal branch of the inferior mesenteric divides into two branches that pass down one on each side of the rectum. Branches from these pierce the muscularis externa and form a plexus in the submucosa which supplies the mucosa down to the level of the anal valves. (2) The **middle rectal** branches of the internal iliac artery reach the rectum in the lateral

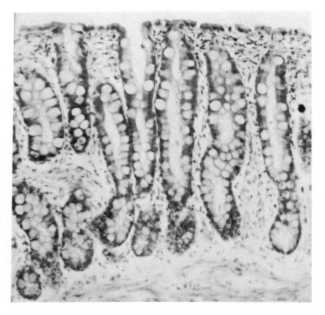

FIG. 6.93.   Photomicrograph of the mucosa of the colon. The lining of the intestinal glands contains numerous goblet cells.

rectal ligaments and are distributed mainly to the musculature of the rectum and anal canal. The extent and significance of their contribution to the blood supply of the rectum varies considerably. (3) The **inferior rectal** branches of the internal pudendal arteries supply the anal canal below the anal valves, the perianal skin and to a variable extent the muscle of the anal sphincters. Anastomoses occur between the three rectal arteries but these are more extensive between the superior and inferior vessels.

The **veins** of the large intestine follow the corresponding arteries. Those draining the colon up to the left colic flexure terminate in the superior mesenteric. Those draining the remainder, except for the lower part of the anal canal, terminate in the inferior mesenteric. In the rectum and anal canal the veins form a **submucosal plexus** deep to the mucous membrane and an external plexus outside the muscularis externa. The submucosal plexus above the pectinate line is drained by the **superior rectal veins**. Dilatation of the veins of this plexus causes them to bulge into the lumen of the rectum and anal canal forming internal haemorrhoids. The veins of the inferior part of the submucosal plexus join the **inferior rectal veins**— tributaries of the internal pudendal veins. Dilatation of these results in external haemorrhoids. The external plexus drains the muscularis externa and opens into the middle rectal veins—tributaries of the internal iliac veins. Since blood from the upper and lower parts of the submucosal plexus enters veins that are tributaries of the portal and systemic circulations, the plexus forms an anastomotic site between the two circulations.

Nerve supply

The preganglionic parasympathetic fibres to the caecum, ascending, and proximal two-thirds of the transverse colon are derived from the **vagus nerves**; those to the remainder of the large intestine, with the exception of the anal canal below the pectineal line, come from the **pelvic splanchnic nerves** (nervi erigentes). All of these fibres synapse in the submucosal and myenteric plexuses; the postganglionic fibres are motor to the smooth muscle of the intestinal wall. Postganglionic sympathetic fibres from **mesenteric** or **hypogastric plexuses** reach the gut directly or with its blood supply. They are mainly vasomotor in function. The internal anal sphincter receives both sympathetic and parasympathetic fibres. The sympathetic produce contraction, while the parasympathetic inhibit the muscle. The external sphincter is supplied by the inferior rectal branch of the **pudendal nerve** and by the perineal branch of the **fourth sacral nerve**.

Afferent nerve fibres subserving the sensation of distension travel with the parasympathetic nerves, while those concerned with pain are conveyed mainly with the sympathetic fibres. Sensory impulses from the anal canal below the pectinate line and from the perianal skin travel in the inferior rectal branches of the **pudendal nerves** and the **perineal branch of the fourth sacral nerve**.

The lymph vessels of the large intestine and of the rectum and anal canal are described on pages 994 and 989, respectively.

THE DEVELOPMENT OF THE ALIMENTARY CANAL

The formation of the head, tail, and lateral folds [p. 48] leads to the envelopment of the dorsal part of the yolk sac within the embryo and to the formation of the primitive gut. This consists of three parts: (1) A foregut, extending cranially from the yolk sac to the buccopharyngeal membrane, caudal and dorsal to the septum transversum and the developing heart; (2) a midgut, communicating with the yolk sac through a wide opening which subsequently

narrows to form the vitello-intestinal duct; and (3) a hindgut, running caudally from the yolk sac to the cloacal membrane [FIG. 2.63].

The epithelial lining of the alimentary canal together with its associated glandular structures are derived from the entoderm of the primitive gut; the muscular and connective tissue elements develop from the surrounding splanchnopleuric mesoderm [FIG. 2.54B]. Caudal to the septum transversum, the splanchnopleuric mesoderm also forms the dorsal mesentery which suspends the gut from the midline of the dorsal abdominal wall. Ventrally, the abdominal part of the foregut is in contact with the mesoderm of the septum transversum from which it draws out a ventral mesentery. Since this is the only part of the gut to have such a mesentery, it has a caudal free border.

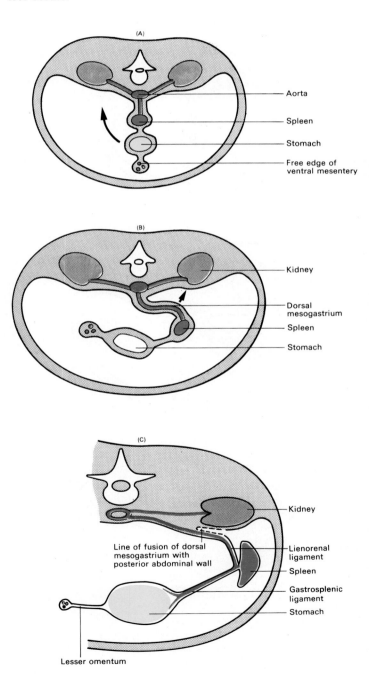

FIG. 6.95. Drawings illustrating the formation of the gastrosplenic and lienorenal ligaments.

The derivatives of the foregut are the pharynx [p. 433], respiratory system [p. 525], oesophagus [p. 437], stomach, and cranial half of the duodenum as well as the liver [p. 485] and pancreas [p. 487]. The primitive stomach appears first as a fusiform dilatation of the foregut but soon its dorsal wall grows more rapidly than its ventral and it develops the characteristic shape of the adult stomach. Its dorsal wall becomes the greater curvature and the attached dorsal mesentery (dorsal mesogastrium) eventually forms the gastrosplenic and lienorenal ligaments and the greater omentum. Its ventral wall becomes the lesser curvature and has connected to it the ventral mesentery (ventral mesogastrium) which subsequently gives rise to the lesser omentum. Later, the stomach rotates to the right with the result that the left surface becomes ventral, and the right, dorsal

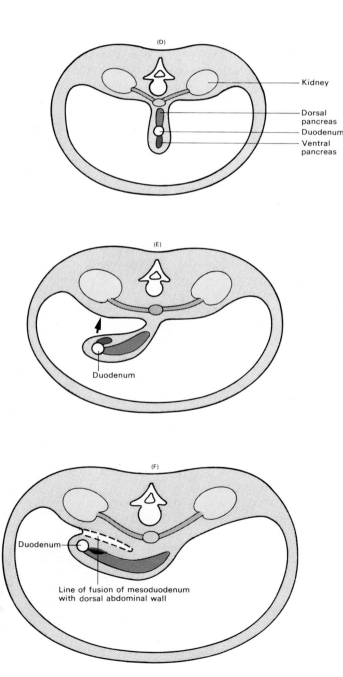

FIG. 6.96.   Drawings illustrating the rotation of the duodenum and the fusion of the mesoduodenum with the posterior abdominal wall.

[FIG. 6.95]. In so doing, the left and right vagus nerves are carried ventrally and dorsally and hence the definitive anterior and posterior surfaces of the stomach receive most of their nerve supply from the left and right vagus nerves respectively. The artery to the foregut is the coeliac whose branches supply the stomach, the proximal part of the duodenum, the liver, most of the pancreas, and the spleen which develops in the dorsal mesogastrium.

The derivatives of the midgut are the caudal half of the duodenum, the jejunum, the ileum, and the large intestine as far as the left colic flexure. Its blood supply comes from the superior mesenteric artery.

The cranial end of the midgut and caudal end of the foregut grow rapidly and form a loop which becomes the *duodenum*. Its dorsal mesentery (mesoduodenum) elongates opposite the apex of the loop but at its ends, and especially its caudal end, it remains short and anchors the duodenum to the dorsal abdominal wall. Later the duodenal loop falls to the right and its mesoduodenum containing the developing pancreas fuses with the peritoneum of the dorsal abdominal wall [FIG. 6.96].

Distal to the duodenum, the midgut rapidly elongates and because of a shortage of space within the abdominal cavity projects or herniates as a U-shaped loop, the midgut loop, into the extra-embryonic coelom in the base of the umbilical cord. The loop has cranial or proximal and caudal or distal limbs with the vitello-intestinal duct attached to its apex [FIG. 6.97]. The superior mesenteric artery, the artery of the midgut, extends into the elongated mesentery of the loop supplying both limbs and continuing along the vitello-intestinal duct as far as the yolk sac. Normally the duct becomes obliterated and transformed into a fibrous cord, finally disappearing together with the accompanying artery.

Within the umbilical cord, the cranial limb grows quickly forming coils of jejunum and ileum whilst the caudal limb undergoes little change, except for a bulbous outgrowth, the caecal diverticulum. As this proceeds the midgut loop rotates anticlockwise as seen from the front through 90 degrees (eventually 180 degrees) around the axis of the superior mesenteric artery [FIG. 6.98]. During the third month the midgut returns rapidly to the abdominal cavity, a process in which the relative decrease in size of the liver and kidneys, together with the increase in the size of the abdominal cavity may be important factors. The coils of intestine formed from the cranial limb return first and pass behind the superior mesenteric artery to occupy the lower and dorsal part of the abdominal cavity, to the right of the mesentery of the hindgut. This is followed by the withdrawal of the caudal limb and its derivatives which are forced ventrally above the coils of small intestine. In this withdrawal the caecum returns last, possibly because of its size, and comes to occupy the upper right quadrant of the abdomen close to the liver. As the caecum and the ascending colon take up this position they pass in front of the duodenal loop which is pushed to the right against the dorsal body wall. Finally, towards the end of intra-uterine life the caecum and appendix descend into the adult position in the right iliac fossa.

The caecum arises in the five-week-old embryo as a diverticulum of the antimesenteric border of the caudal limb of the midgut loop. In its growth the extremity of the diverticulum fails to keep pace with the proximal part so that the viscus as a whole becomes conical in shape, the wider proximal part forming the caecum and the narrower blind extremity the vermiform appendix. This conical form of caecum and appendix, known as the 'infantile caecum', is retained for some time after birth and may in 2 or 3 per cent of the population persist throughout life. Normally, however, the anti-mesenteric border of the caecum grows faster than the mesenteric, with the result that the appendicular attachment moves medially and posteriorly. The appendix remains relatively small (Fitzgerald *et al.* 1971).

The hindgut gives origin to the left flexure of the transverse colon, the descending and sigmoid colons, the rectum and upper part of the anal canal. It receives its blood supply from the inferior mesenteric artery.

At first the hindgut opens into the wide **cloaca** but later this is divided into ventral and dorsal portions by the **urorectal septum** which grows caudally between the allantois and the hindgut to fuse with the cloacal membrane. The ventral portion with the allantois forms the **urogenital sinus** and the dorsal portion forms the **rectum** and upper part of the **anal canal** [FIGS. 2.72 and 2.73]. With the division of the cloaca into two by the urorectal septum the cloacal membrane now consists of a ventral part, the **urogenital membrane** and a dorsal part, the **anal membrane**. The anal membrane separates the terminal part of the hindgut from a shallow ectoderm-lined depression, the **anal pit** which forms the lower part of the anal canal. During the seventh week of intra-uterine life, the anal membrane ruptures and thus the rectum comes to open to the exterior through the anal canal. At the site of attachment of the anal membrane the entodermal lining of the gut becomes continuous with the ectoderm of the anal pit. In the adult, the junction between the two epithelia corresponds roughly to the pectinate line at the bases of the anal valves. The origin of the anal canal from two different sources, namely hindgut and anal pit, is reflected not only in its lining epithelium but also in its arterial supply, venous and lymph drainages, and nerve supply.

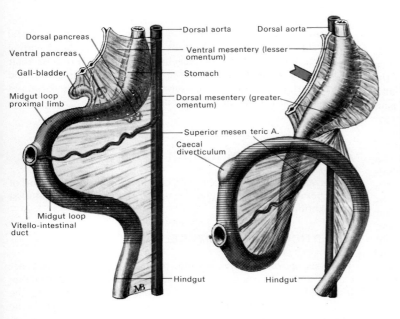

FIG. 6.97. A diagram to show two early stages in the development of the abdominal part of the gut tube and its mesenteries. Note the rotation of the midgut loop and the ballooning of the dorsal mesentery of the stomach to the left by the extension of the omental bursa (indicated by the arrow) into that mesentery. The proximal limb of the midgut has been spotted in blue. The distal limb is hatched in blue.

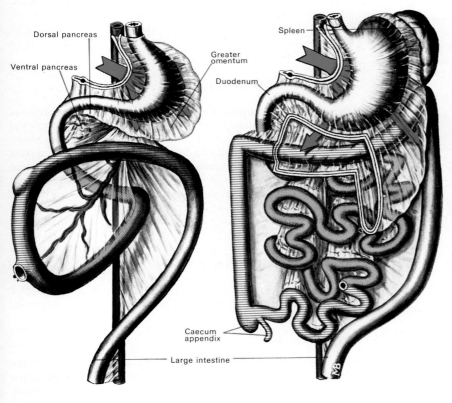

FIG. 6.98. A diagram of two later stages in the development of the abdominal part of the gut tube and its mesenteries. Note the changes in position of the various parts of the gut tube and the extension of the omental bursa with consequent formation of the greater omentum and the ligaments of the spleen from the dorsal mesentery of the stomach. The duodenum and the ascending and descending parts of the colon become fused to the posterior abdominal wall while the other parts of the gut tube retain their mesenteries.

PERITONEUM AND PERITONEAL CAVITY

The peritoneal cavity is formed from the intraembryonic coelom which in the somite embryo arises by the splitting of the lateral plate mesoderm into somatopleuric and splanchnopleuric mesoderm [FIG. 2.57]. Adjacent to the coelom, these give origin to the parietal and visceral layers of the peritoneum respectively. At first, the peritoneal cavity has a simple form having near equal right and left halves communicating with each other ventral to the developing gut. However, later, as a result of the growth, rotation and fixation of the gut, the development of the liver, and the formation of the omental bursa, the cavity increases in complexity, and the disposition of the peritoneum becomes less easy to follow.

OMENTAL BURSA

The omental bursa arises from the confluence of mesodermal spaces in the right side of the thick dorsal mesogastrium continuous with a pit (epiploic foramen) on this surface. The space (vestibule of the omental bursa) thus formed increases in size by extending caudally, dorsal to the stomach, and cranially, to the right of the oesophagus and behind the liver, to the developing lung, as the pneumato-enteric recess. The upper part of the pneumato-enteric recess is usually obliterated but its lower part becomes the superior recess of the omental bursa. The caudal extension develops into the inferior recess of the omental bursa and communicates with the vestibule and superior recess through a narrow neck between the gastropan-creatic folds, formed by the left gastric and the common hepatic arteries. Partly as a result of the formation of the omental bursa, the stomach rotates into a transverse position and the ventral mesogas-trium, part of which becomes the lesser omentum, forms part of the anterior wall of the superior recess and vestibule of the omental bursa. The free edge of the ventral mesogastrium becomes the right free margin of the lesser omentum and so forms the anterior boundary of the epiploic foramen.

The further development of the bursa is influenced by a number of events involving the dorsal mesogastrium.

1. The attachment of the cranial part of the dorsal mesogastrium shifts to the ventral surface of the left kidney, due to its fusion with the peritoneum on the dorsal abdominal wall [FIG. 6.95].
2. The spleen develops in the cranial part of the mesentery and divides it into a portion extending from the spleen to the kidney,

the lienorenal ligament, and a portion between the spleen and the greater curvature of the stomach, the gastrosplenic ligament [FIG. 6.95].
3. The inferior recess of the omental bursa extends into the expanding caudal part of the dorsal mesogastrium thus forming the greater omentum. This passes downwards in front of the transverse colon and the coils of the small intestine and separates them from the anterior abdominal wall [FIGS. 6.98 and 6.99].
4. The posterior layers of the greater omentum, enclosing the pancreas, fuse with the peritoneum of the dorsal wall so that the pancreas becomes fixed behind the omental bursa.
5. The posterior layers of the greater omentum fuse with the anterior surface of the transverse mesocolon to form the gastrocolic ligament [FIG. 6.99].

After birth the anterior and posterior layers of the greater omentum usually fuse inferiorly so that in the adult the omental bursa seldom extends below the transverse colon. Once the ascending and descending colons have reached their adult positions their mesenteries normally fuse with the posterior abdominal wall and they become retroperitoneal; in contrast, the transverse and sigmoid colons retain their mesenteries throughout life.

Abnormalities of the gut

Abnormalities of the gut occur with considerable frequency, most of them being due to diverticula, atresia, or abnormalities of rotation and fixation of the gut.

DIVERTICULUM ILEI (MECKEL'S)

This sausage-shaped diverticulum occurs in approximately 2 per cent of the population and is caused by a failure of the vitello-intestinal duct to atrophy and disappear. It is attached to the antimesenteric border of the ileum, about 1 metre proximal to the ileocaecal junction, and its wall is similar to that of the ileum, although it may contain patches of gastric mucosa or pancreatic tissue. It may be attached to the umbilicus by a fibrous cord which rarely contains a continuation of the cavity, thus producing a vitelline fistula.

ATRESIA AND STENOSIS

These abnormalities can occur throughout the gut but they are most common in the duodenum and ileum. They are usually attributed to a complete or partial failure of recanalization of the mucosa which obliterates the gut lumen during development. When this is complete the lumen of the gut is occluded by a diaphragm of epithelial tissue (atresia). Stenosis is simple narrowing. In the duodenum atresia occurs most frequently at the level of the major duodenal papilla, where constriction of the wall may be a contributory factor. Failure of normal recanalization can also give rise to duplication of different parts of the intestine.

ANOMALIES OF ROTATION

In these the viscera are usually normal, but their positions relative to each other and in the abdominal cavity show varying alterations of a permanent nature.

Situs inversus

In this condition, which may affect the thorax as well as the abdomen, the viscera are transposed in such a way that they form a mirror image of the normal arrangement with the heart, stomach,

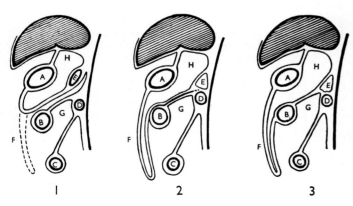

FIG. 6.99.   Diagrams to illustrate development of greater omentum. (After Hertwig 1892.)

(1) shows the beginning of the greater omentum and its independence of the transverse mesocolon; in (2) the two come into contact; and in (3) they have fused along the line of contact.

A, stomach; B, transverse colon; C, small intestine; D, duodenum; E, pancreas; F, greater omentum; G, placed in peritonal cavity; H, in omental bursa.

and spleen on the right side and the liver, gall-bladder, caecum, and appendix vermiformis on the left.

Reversed rotation

Here the intestines rotate in a clockwise, rather than an anticlockwise direction, with the result that the small intestine lies in front of the superior mesenteric artery and the colon lies behind. In the latter position the colon may be obstructed by pressure from the artery.

Non-rotation

In non-rotation the midgut loop returns to the abdomen without any essential change in its position. Consequently the small intestine lies cranially and to the right side of the abdomen and the colon lies caudally and to the left; both remain suspended by their original mesenteries.

Subhepatic caecum

The proximal part of the colon fails to elongate and the caecum and appendix remain immediately under the liver. More frequently the ascending colon may remain mobile from a failure of its mesentery to fuse with the peritoneum on the posterior abdominal wall. In such cases the caecum and appendix may not be confined to the right iliac fossa.

CONGENITAL UMBILICAL HERNIA

This condition arises when the midgut loop fails to return to the abdominal cavity. Its severity varies considerably from cases in which the hernial sac contains most of the intestine to those in which only a small part, often the caecum, is present.

ANOMALIES OF THE RECTUM AND ANAL CANAL

Anomalies of the rectum and anal canal are said to occur once in about 5000 births. The most common is an *imperforate anus* which in its simplest form arises from a persistence of the anal membrane with consequent failure of communication between the rectum and anal canal. Rarely this condition may be associated with a failure of development of the lower part of the rectum.

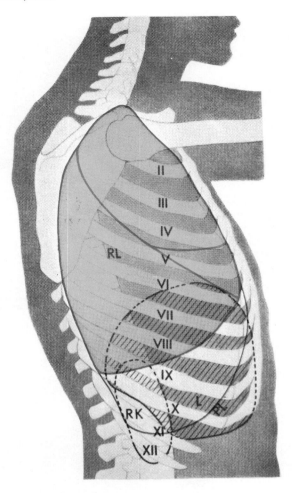

FIG. 6.100.   Right lateral aspect of trunk, showing surface topography of viscera.
L.   Liver.
PL. Line of pleural reflexion.
RK. Right kidney.
RL. Right lung.

THE LIVER

The liver is the largest glandular structure in the body, forming approximately $\frac{1}{40}$ of the weight of the adult and proportionately more in the child. Apart from acting as a storage organ, the liver is involved in the metabolism of carbohydrates, lipids, and proteins. It produces bile salts and bile pigments which it passes to the duodenum through its system of hepatic and bile ducts. In the foetus it is a site of haemopoiesis. In the living it is reddish-brown in colour, soft and pliable to touch, moulding itself to the surrounding viscera. It has a thin, connective tissue capsule (Glisson's capsule), which completely surrounds it and becomes continuous at the porta hepatis with connective tissue distributed throughout the liver substance with the branches of the portal vein, hepatic artery, and bile-duct.

The liver is roughly pyramidal in shape, with its base on the right and its apex directed to the left. It occupies the right hypochondrium, the upper part of the epigastrium, and extends to a limited extent into the left hypochondrium [FIG. 6.44 and 6.50]. Its position is influenced by the degree of distension of surrounding viscera, movements of the diaphragm, and posture. In the recumbent position and during quiet respiration, the upper surface of the liver can be represented by a horizontal line crossing the abdomen just below the xiphosternal joint. To the left, the line passes horizontally to the lower border of the fifth rib and ends in the fifth intercostal space, 8 to 9 cm from the midline, near the apex of the heart. To the right, it rises upwards, reaching the fifth rib in the midclavicular line [FIGS. 6.50 and 6.100].

The **lower margin** of the liver can be represented by an oblique line following the right costal margin as far as the tip of the ninth costal cartilage, then crossing to the tip of the left eighth costal cartilage, and ending in the fifth intercostal space at the left midclavicular line. In the recumbent position the only palpable part of the liver is normally that in the infrasternal angle. It will be noted that the liver is largely under cover of the ribs and costal cartilages, and in its upper part is overlapped by the lungs, visceral and parietal pleura, and the diaphragm [FIGS. 6.50 and 6.100].

The liver has anterior, superior, right, posterior, and postero-inferior surfaces. The postero-inferior surface is the **visceral surface**, and the other four together form a curved **diaphragmatic surface**. The borders are rounded and ill-defined, except the *inferior*, which is sharp and separates the postero-inferior surface from the anterior and right surfaces.

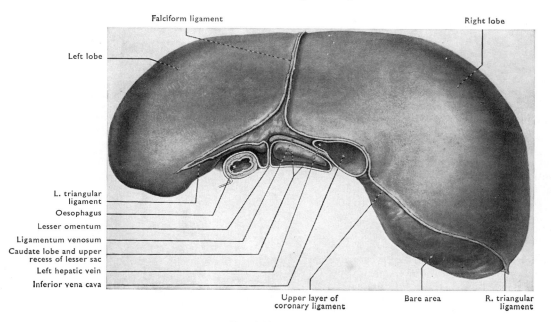

Falciform ligament   Right lobe

Left lobe

L. triangular
ligament
Oesophagus
Lesser omentum
Ligamentum venosum
Caudate lobe and upper
recess of lesser sac
Left hepatic vein
Inferior vena cava

Upper layer of   Bare area   R. triangular
coronary ligament   ligament

FIG. 6.101.   Liver viewed from above.

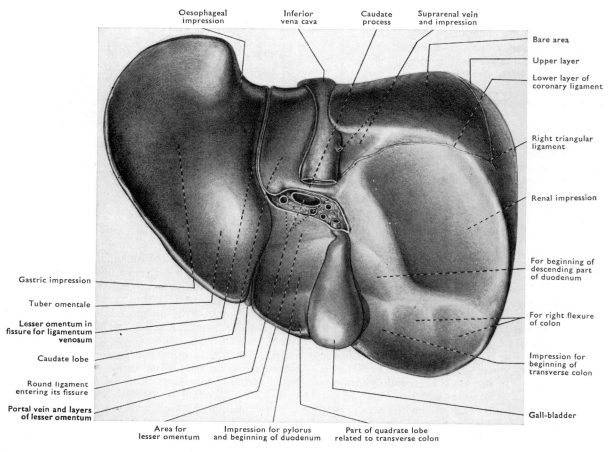

Oesophageal   Inferior   Caudate   Suprarenal vein
impression   vena cava   process   and impression

Bare area

Upper layer

Lower layer of
coronary ligament

Right triangular
ligament

Renal impression

For beginning of
descending part
of duodenum

For right flexure
of colon

Impression for
beginning of
transverse colon

Gall-bladder

Gastric impression

Tuber omentale

Lesser omentum in
fissure for ligamentum
venosum

Caudate lobe

Round ligament
entering its fissure

Portal vein and layers
of lesser omentum

Area for   Impression for pylorus   Part of quadrate lobe
lesser omentum   and beginning of duodenum   related to transverse colon

FIG. 6.102.   Lower (visceral) and posterior surfaces of liver.

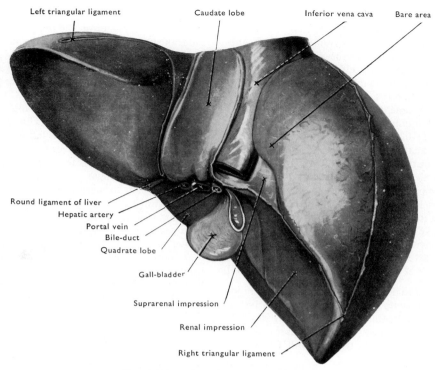

Left triangular ligament Caudate lobe Inferior vena cava Bare area

Round ligament of liver
Hepatic artery
Portal vein
Bile-duct
Quadrate lobe

Gall-bladder

Suprarenal impression

Renal impression

Right triangular ligament

FIG. 6.103. Liver viewed from behind.

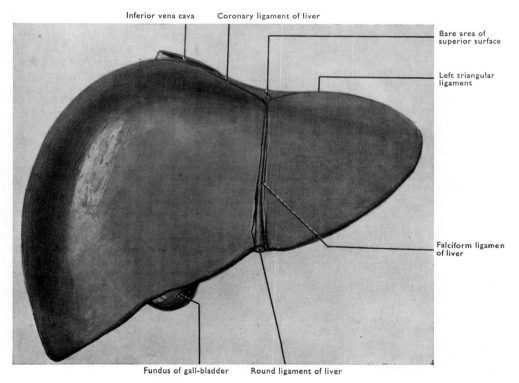

Inferior vena cava Coronary ligament of liver

Bare area of
superior surface

Left triangular
ligament

Falciform ligamen
of liver

Fundus of gall-bladder Round ligament of liver

FIG. 6.104. Liver viewed from the front.

The *anterior surface* is triangular in shape with the apex directed towards the left [FIG. 6.104]. Above the costal margins it is in contact with the diaphragm but in the epigastrium, where it escapes below the costal margins, it is applied to the anterior abdominal wall. The falciform ligament extends obliquely from the midline of the anterior abdominal wall and the diaphragm to the anterior and superior surfaces of the liver, slightly to the right of the midline. The attachment of the falciform ligament, along with the fissures for the ligamentum teres and ligamentum venosum on the postero-inferior and posterior surfaces [FIG. 6.102], divide the liver into right and left lobes. This division is an anatomical one and is not reflected in either its development or in the distribution of its vessels. In the adult the right lobe forms about three-quarters of the liver, but in the child the lobes are more equal in size. The two layers of peritoneum of the falciform ligament separate on reaching the liver and pass to invest the right and left lobes. Posterosuperiorly the layers of peritoneum diverge and leave between them a small area of liver devoid of peritoneum. In this region the left layer is continuous with the upper layer of the left triangular ligament and the right with the upper layer of the coronary ligament [FIG. 6.101].

The *right surface* is convex and lies in contact with the diaphragm, opposite the seventh to the eleventh ribs in the mid-axillary line. It can be divided into three parts, the diaphragm separating its upper part from the lung and pleural cavity, its middle part from the pleural cavity, and its lowest part from the thoracic wall [FIG. 6.100].

The *superior surface* is slightly convex to the right and to the left of the midline, in keeping with the shape of the cupulae of the diaphragm with which it is in contact. Between the convexities there is a slight depression, the cardiac depression, which is separated from the pericardium and heart by the central tendon of the diaphragm [FIG. 6.104].

The *postero-inferior* (visceral) *surface* points downwards, backwards, and to the left [FIG. 6.102]. Near its posterior border there is a centrally placed transverse fissure, the porta hepatis which serves for the passage, in order from before backwards, of the common hepatic duct, the hepatic artery accompanied by the hepatic plexus of sympathetic nerves, lymph vessels, and the portal vein. Attached to the margins of the porta hepatis is the right or free border of the lesser omentum in which the structures traversing the porta hepatis pass in front of the epiploic foramen. Between the right end of the porta hepatis and the inferior border of the liver there is a shallow depression, usually devoid of peritoneum, which is occupied by the gall-bladder. From the left end of the porta hepatis a deep groove, the fissure for the ligamentum teres, extends to the inferior border of the liver. It contains the ligamentum teres, the fibrous remains of the left umbilical vein, which runs in the free edge of the falciform ligament and in the fissure to join the umbilicus to the left branch of the portal vein. The ligamentum teres is accompanied by some small para-umbilical veins, which connect the veins of the anterior abdominal wall with the left branch of the portal vein. These form a site of anastomosis between the portal and systemic circulations and may be dilated in conditions causing increased portal venous pressure—portal hypertension. The quadrangular area of the *postero-inferior surface* bounded by the porta hepatis, the fossa for the gall-bladder, the fissure for the ligamentum teres and the inferior border of the liver, forms the quadrate lobe.

The postero-inferior surface of the liver, fixed and hardened *in situ*, retains the impressions of the viscera with which it comes in contact. The left lobe has a shallow gastric impression formed by the fundus and upper part of the body of the stomach, while to its right, near the superior end of the fissure for the ligamentum teres, there is a smooth elevation, the tuber omentale, where the inferior surface is related to the lesser omentum. The quadrate lobe comes into contact with the pylorus and the superior part of the duodenum. To the right of the gall-bladder the inferior surface has two well-defined impressions, a posterior, the renal, formed by the upper part of the anterior surface of the right kidney, and an anterior or colic, formed by the right colic flexure of the colon. Medial to the renal impression and posterolateral to the neck of the gall-bladder, there is a slight concavity for the beginning of the descending part of the duodenum.

The *posterior surface* of the liver [FIG. 6.103] points backwards and is moulded on the diaphragm anterior to the tenth and eleventh thoracic vertebrae [FIG. 6.101]. This surface is more extensive over the right lobe, and towards its left extremity it is little more than a sharp margin projecting over the fundus of the stomach. It comes into contact with the posterior part of the diaphragm and with the abdominal oesophagus.

Much of the posterior surface of the right lobe is devoid of peritoneum thus forming a 'bare area' of the liver. This is triangular in shape with the base bordered by the inferior vena cava and the apex pointing downwards and to the right. The bare area is limited above and below by the reflection of the peritoneum from the diaphragm to the liver as the superior and inferior layers of the coronary ligament and its apex corresponds to the right triangular ligament. It is connected to the undersurface of the diaphragm by some loose connective tissue and a number of small veins. Its lower medial part is slightly indented by the right suprarenal gland.

The inferior vena cava runs in a vertical groove which varies considerably in depth, sometimes being bridged by liver tissue and converted into a tunnel. The groove is devoid of peritoneum and is pierced by the hepatic veins passing to the inferior vena cava, its lower end is separated from the porta hepatis by the caudate process [FIG. 6.102].

Passing vertically on the posterior surface between the right and left lobes there is a deep cleft, the fissure for the ligamentum venosum. This is continuous inferiorly with the left end of the porta hepatis and with the fissure for the ligamentum teres [FIG. 6.102], and superiorly it passes to the right towards the groove for the inferior vena cava. The lesser omentum is attached to the bottom of this fissure and to the margins of the porta hepatis. The ligamentum venosum contained within the fissure is the fibrous remains of the ductus venosus which passes from the left branch of the portal vein to the left hepatic vein just before it enters the inferior vena cava. *In utero* the ductus venosus conveys much of the oxygenated blood from the umbilical vein direct to the inferior vena cava, thus short-circuiting the hepatic sinusoids. It closes shortly after birth.

The fissure for the ligamentum venosum and the groove for the inferior vena cava demarcate a part of the liver known as the caudate lobe. It extends below to the porta hepatis and its upper border is bounded by the ligamentum venosum turning to the right towards the inferior vena cava. The caudate lobe faces into and forms the anterior wall of the upper recess of the omental bursa. Its lower extremity is connected to the remainder of the right lobe by the caudate process, which lies in front of the inferior vena cava and forms the upper boundary of the epiploic foramen. Immediately to the left of the upper part of the fissure for the ligamentum venosum the posterior surface is grooved by the oesophagus.

RADIOLOGY

Plain radiographs of the abdomen rarely show the whole outline of the liver. The upper surface of the right lobe is outlined by the right dome of the diaphragm, the rest is rarely seen. The thin inferior border may sometimes be seen but gas and other material in the colon often obscure this.

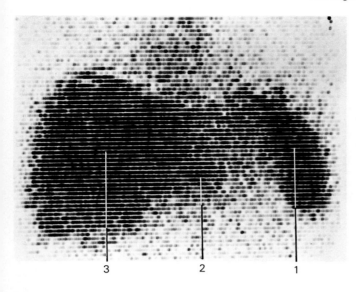

FIG. 6.105. Radioactive isotope scan of upper abdomen. The liver and other viscera take up the isotope and the intensity of X-ray emission from each organ is proportional to its thickness. Hence the liver is ill-defined at its thin edges and the pancreas is poorly outlined. The spleen is unusually well shown in this scan. Increased or diminished uptake by an organ usually indicates abnormality.
1. Spleen.
2. Pancreas.
3. Liver.

Ultrasound examination allows the size of the organ to be estimated and gives information about the larger vessels and ducts within it.

Radioisotope scanning shows an indefinite outline of the liver [FIG. 6.105] but its chief usefulness is in pathological conditions.

For the examination of the internal structure of the liver selective catheterization of the hepatic artery and the injection of contrast medium (hepatic angiography) will show its fine vessels in detail.

Injection of contrast medium into the spleen allows filling of the splenic vein and the intrahepatic branches of the portal vein (splenoportography) [FIG. 13.112].

The intrahepatic (and extrahepatic) bile ducts can be made visible by injection of contrast medium into the bile-duct system after puncture of the liver by a fine needle—percutaneous transhepatic cholangiography [FIG. 6.114].

BLOOD AND LYMPH VESSELS AND NERVES OF THE LIVER

Blood reaches the liver in the **portal vein** and **hepatic artery**. The portal vein, formed by the union of the splenic and superior mesenteric veins, carries approximately 60 per cent of the total blood which enters the liver. This fraction has already lost some of its oxygen and contains most of the absorbed products of digestion and the waste products of the spleen. The common hepatic branch of the coeliac artery carries fully oxygenated blood which is delivered at a higher pressure than that in the portal vein. After circulating through the liver the blood is drained by two or three **hepatic veins**, which join the inferior vena cava (Masselot and Leborgne 1978).

Traced into the liver, the portal vein, hepatic artery, and common hepatic duct each divide into a right and left branch at, or

immediately within, the porta hepatis. Thereafter, there is a regular sequence of branching, forming first segmental and eventually interlobular vessels. Collectively, these three interlobular vessels are known as portal triads; they are accompanied by lymph vessels and surrounded by connective tissue, the whole complex forming **portal canals** or **portal areas**. Three or four of such portal canals are present at the periphery of each hepatic lobule. Branches of the interlobular vein and artery open into the sinusoids of the liver lobules [see below] where the portal and arterial bloods mix. From the sinusoids the blood passes into the central vein and hence by way of sublobular veins to the hepatic veins and the inferior vena cava.

Study of the distribution within the liver of the right and left branches of the portal vein, hepatic artery, and common hepatic duct indicates that, from a functional point of view, the liver should be divided into right and left lobes of approximately equal size. This is because the quadrate lobe and most of the caudate lobe are supplied (or drained) by the left branches of the vessels and so belong to the functional left lobe of the liver. This division is emphasized by the fact that there is little or no anastomosis between the vessels supplying the functional right and left lobes (Gupta and Gupta 1976; Gupta, Gupta, and Arora 1977).

Functional segments supplied by main branches of the right and left lobar vessels are also recognized. Each functional lobe has two segments, those in the left being called the left lateral and the left medial, and those in the right the right anterior and right posterior. The left lateral segment corresponds to the left anatomical lobe and the left medial to the quadrate and most of the caudate lobe. The plane of separation between these segments corresponds to the fissures for the ligamentum teres and ligamentum venosum. The plane of separation between the segments of the right lobe is less sharply outlined and appears to run obliquely from the middle of the anterior surface of the right anatomical lobe to the groove for the inferior vena cava. With the exception of the hepatic veins, the arterial, portal, and biliary vessels do not cross to adjacent segments—a feature of considerable surgical importance (Braash 1958).

The **nerves** of the liver are derived from the **coeliac plexus** and filaments from the right and left **vagus** nerves and the **right phrenic** nerve. The branches from the coeliac plexus form the hepatic plexus which accompanies the hepatic artery and portal vein to be distributed with the divisions of these vessels.

The **lymph vessels** of the liver are described on page 995. The flow of lymph from the liver accounts for up to half that in the thoracic duct. It is rich in protein.

STRUCTURE OF THE LIVER

The liver consists of a large number of lobules, each being a polyhedral prism, which, on section, is usually hexagonal in outline. The lobule contains a **central vein**, an indirect tributary of the hepatic veins, and its periphery is indicated by the presence of portal canals (or areas) which are shared with contiguous lobules. In some species including the pig, a distinct layer of connective tissue, continuous with that of the portal canals, encapsulates each lobule and makes its identification easier [FIG. 6.106]. In man, however, no visible connective tissue surrounds the lobules so their outlines must be inferred by imaginary lines joining the portal canals surrounding each central vein. The units of liver as defined above are called either hepatic or 'classic' lobules.

The **hepatic lobule** consists largely of parenchymal or epithelial glandular cells (hepatocytes) which are arranged as interconnected plates or laminae. Each plate is usually only one cell thick and, when viewed edge-on, appears as a cord of cells [FIG. 6.107]. In a general manner the plates radiate from the centre towards the periphery of

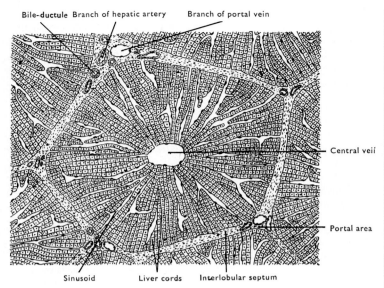

FIG. 6.106.   Diagram of the liver lobule (pig).

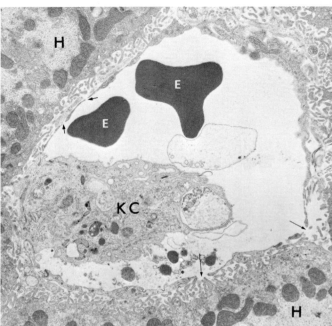

FIG. 6.108.   Electron micrograph of a liver sinusoid. The lumen contains two erythrocytes (E) and the remains of degenerating blood cells. A highly active phagocytic cell (KC) projects into the sinusoid. The lining endothelium is extremely attenuated and there are fenestrations in some areas (arrows). Microvilli project from the hepatocytes (H) into the space (space of Disse) between them and the endothelium.

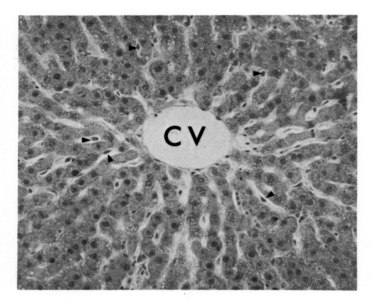

FIG. 6.107.   Section of liver showing central vein (CV), with converging plates of parenchymatous cells. Kupffer cells (arrow heads) are present with the sinusoids.

the lobule and are separated by spaces or lacunae which are occupied by the hepatic sinusoids and outside these by the perisinusoidal space of Disse. The lacunae open at the centre of the lobule into a space occupied by the central vein, but peripherally they fail to communicate freely with the portal areas, due to the presence of *limiting plates* of hepatocytes surrounding these areas (Elias 1949, 1955).

The **hepatic sinusoids** communicate through perforations in the plates and are lined by endothelial cells and by phagocytic cells, known as **Kupffer cells**. Normally the endothelial cells are more common than the Kupffer cells and, whilst they have much in common with endothelial cells elsewhere, they are characterized by the presence of numerous large fenestrations and by the absence of a supporting basal lamina. These fenestrations, along with spaces between adjacent endothelial cells, allow the plasma of the sinusoidal

blood free access to the perisinusoidal space separating the sinusoids from hepatocytes of the hepatic plates [FIG. 6.108]. The Kupffer cells have considerable phagocytic activity and often contain the remains of effete erythrocytes, and other cellular debris. In common with other phagocytic cells of the reticulo-endothelial system, they contain lysosomes which increase in number during phagocytosis. Stimulation of the reticulo-endothelial system causes the Kupffer cells to increase in number largely as a result of local mitotic activity, and also by augmentation from other sources. Blood reaches the hepatic sinusoids in branches of the interlobular portal vein and hepatic artery, after these have perforated the limiting plate surrounding the portal areas. It courses centrally in the lobule and is drained by the central vein (Wisse 1970).

The **perisinusoidal space**, with its contained plasma, is an important site of exchange between the sinusoidal blood and the hepatocyte. With the electron microscope the space can be seen to contain large numbers of microvilli which, by increasing the surface area of the hepatocyte, undoubtedly aid in the process of exchange [FIG. 6.108]. Whilst allowing free movement of plasma, the perforations in the walls of the sinusoids are not sufficiently large to allow the entry of cells into the perisinusoidal space. The space also contains a close meshwork of reticular fibres which supports the laminae of hepatocytes and may be responsible for keeping the sinusoids open. The origin of these fibres has been a matter of debate but recent evidence indicates that the perisinusoidal space contains occasional fibroblasts which could be their source. In certain pathological states the space may be enlarged and become detectable with the light microscope (Matter, Orci, and Rouiller 1969).

The **hepatocyte**, sometimes referred to as the parenchymatous liver cell, is polygonal in shape and measures approximately 20 μm

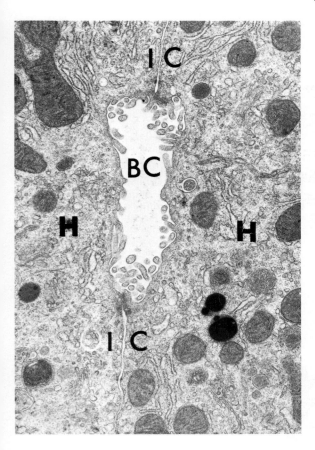

FIG. 6.109. Electron micrograph showing a bile canaliculus (BC) produced by the dilation of the intercellular space (IC) between two hepatocytes (H).

in width. Two of its surfaces possess microvilli and face into the perisinusoidal spaces of adjacent sinusoids. Its other surfaces are in contact with the surfaces of neighbouring hepatocytes, except near their centres, where the intercellular spaces are dilated to form bile canaliculi. Desmosomes and gap junctions are present on these surfaces [p. 458].

Hepatocytes contain large, round, centrally placed nuclei with finely dispersed chromatin and large nucleoli. Up to one quarter of the cells may be binucleate but mitotic activity is rarely seen in the healthy liver. The cytoplasm contains a wide spectrum of organelles including some hundreds of mitochondria, lysosomes, cisternae of rough endoplasmic reticulum, branching and anastomising tubules of smooth endoplasmic reticulum, and, near to the bile canaliculi, vesicles and vacuoles belonging to the Golgi apparatus. In addition, hepatocytes contain membrane-bound, spherical bodies known as **microbodies** or **peroxisomes** and, in the well-fed state, accumulations of glycogen and lipid (Babcock and Cardell 1974, 1975).

Bile that is formed by the hepatocyte passes into **bile canaliculi**. These vessels, measuring approximately 1 μm in diameter, are formed by a dilatation of the central zone of the intercellular space between contacting cells [FIG. 6.109]. The plasma membranes participating in their formation possess short microvilli and are joined at the sides of the canaliculi by tight junctions, similar in structure to the zonula occludens found in various lining epithelia. Thus the wall of the canaliculus consists of the plasma membranes of adjacent hepatocytes. The fusion of the plasma membranes at the sides of the canaliculi prevents bile from escaping into the perisinusoidal space and the constituents of the sinusoidal blood from passing into the bile without first entering the hepatocytes.

When selectively stained and viewed *en face*, the bile canaliculi of the liver laminae have the appearance of a network, the spaces of which are occupied by the hepatocytes. At the periphery of the lobule the canaliculi join to form intralobular ductules which perforate the limiting plate and open into the **interlobular ductules** of the portal canals. In this system of ducts the direction of flow of bile is opposite to that of the blood within the sinusoids (Motta and Fumagalli 1975; Richards 1975).

Lymph vessels are apparently absent from the liver lobule but begin as blind capillaries in the portal canals. It is thought that the voluminous lymph produced by the liver comes from the space of Disse and that it reaches the lymph capillaries in the portal canals by filtering through the limiting plate.

The above description of the liver is based on what has been termed the hepatic or 'classical' lobule [FIG. 6.106]. Its axis is the central vein and it is polygonal or hexagonal in shape. At its periphery there are usually four to six portal canals and the flow of blood from the periphery to the centre is thought to induce the radial pattern of the hepatic plates characteristic of the lobule. As the blood flows along the sinusoids there is a fall in its oxygen content and this is reflected in a gradient of metabolic activity within the lobule and in a variation in response of different areas of the lobule to toxic agents. The classical lobule emphasizes the endocrine function of the liver.

A more satisfactory way of looking at liver structure, in the eyes of some workers, is to place the emphasis on its exocrine role and to include in the lobule those parts that secrete bile into a common interlobular ductule. The 'portal' lobule thus defined is the triangular area surrounding the portal canal and bounded by lines joining the three adjacent central veins. It contains segments of these three hepatic lobules. In the portal lobule blood flows peripherally and bile centrally.

Recently, the emphasis on liver structure has swung to the concept of the '*liver acinus*' [FIG. 6.110]. This is thought of as an oval or diamond-shaped mass of tissue surrounding the terminal branches of the interlobular vessels between two adjacent hepatic lobules and whose periphery is indicated by the corresponding two central veins. Blood and bile flows are in the same directions as in the portal lobule. In some respects this unit resembles the portal lobule but it has decided advantages in permitting a more physiological interpretation of liver morphology. It has also proved useful in explaining certain aspects of liver pathology, especially those associated with bile-duct occlusion, and cirrhosis, as well as the pattern of liver regeneration (Rappaport 1958; Popper and Udenfriend 1970).

THE EXTRAHEPATIC BILIARY APPARATUS

The extrahepatic biliary apparatus is concerned with: (1) the conduction of bile from the liver, (2) the storage and modification of bile, and (3) the regulation of the discharge of bile into the duodenum. It consists of a right and a left **hepatic duct** which unite to form the **common hepatic duct**, and the **gall-bladder** with its **cystic duct**, which joins the common hepatic duct to form the **bile-duct**. Before opening into the duodenum the bile-duct is joined by the pancreatic duct.

Gall-bladder and cystic duct

The gall-bladder is shaped like a pear, about 8 cm long and 3 cm at its widest part and has a capacity between 30 and 50 ml. It is described as having a fundus, body, and neck [FIG. 6.102]. The **fundus**, or expanded end, projects below the inferior border of the liver and comes into contact anteriorly with the posterior surface of

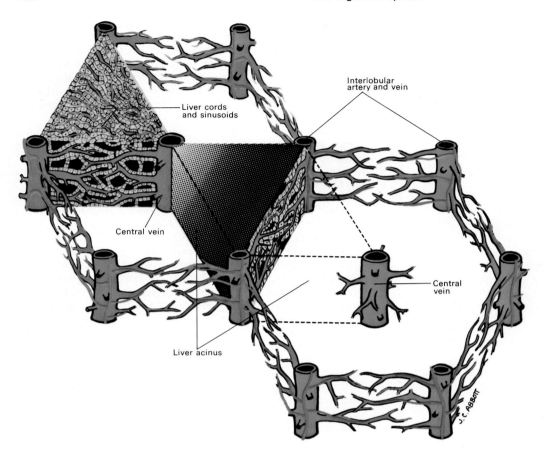

Interlobular
artery and vein

Liver cords
and sinusoids

Central vein

Central
vein

Liver acinus

FIG. 6.110. Diagrammatic representation of segments of adjacent hepatic lobules. The corners of the lobules are demarcated by interlobular veins and arteries (the interlobular bile ducts have been omitted for simplicity). The centre of each lobule is occupied by a central vein. Branches of adjacent interlobular vessels anastomose and further help to demarcate the lobules. One half of a liver acinus is shown in detail on the left, the shading shown in the opposite half acinus indicates the fall in oxygen tension towards the central vein.

the abdominal wall, near the tip of the ninth right costal cartilage. It is completely surrounded by peritoneum and posteriorly is in touch with the beginning of the transverse colon [FIG. 6.84]. The **body** lies in the shallow fossa on the inferior surface of the right lobe of the liver, to which it is attached by loose areolar tissue, some small veins, and lymph vessels. Occasionally it is partly, or completely embedded in the substance of the liver and rarely, it may be suspended by a short mesentery. It is directed upwards, posteriorly, and to the left. Its inferior surface is covered by peritoneum and is applied to the superior and descending parts of the duodenum and the transverse colon [FIG. 6.111A]. The **neck** is continuous with the body at the right end of the porta hepatis and forms an S-shaped curve, the last part of which is completed in the cystic duct. At its junction with the body, the neck sometimes has a small pouch (Hartman's pouch) projecting from its posteromedial wall towards the superior part of the duodenum. Although relatively common, its presence is now generally accepted to be due to pathological changes. The neck is attached to the liver by areolar tissue. It is constricted at its point of continuity with the cystic duct and its mucous membrane is thrown into a number of oblique shelf-like folds resembling a **spiral valve**. Bile is conveyed to and from the gall-bladder in the cystic duct.

The **cystic duct**, the narrowest extrahepatic biliary duct, is about 4 cm in length [FIG. 6.111A]. From the neck of the gall-bladder it passes downwards, backwards, and to the left and runs parallel to the common hepatic duct before joining it, usually 2–3 cm below the porta hepatis. It should be noted that the cystic duct is prone to considerable variation in its length and site of termination. It may be as short as 1 cm and join either the right hepatic duct or the junction of the two hepatic ducts, or it may be longer than usual and not join the common hepatic duct until it is near or even behind the descending part of the duodenum. The mucous membrane of the cystic duct has similar oblique folds to those in the neck of the gall-bladder, these are thought to have little influence on the flow of bile in the duct and possibly exist only to reinforce its wall.

Hepatic ducts

Bile is drained from the liver segments by five or six large ducts. They unite in groups at each end of the porta hepatis to form the right and left hepatic ducts. The caudate and quadrate lobes, being developmentally and functionally part of the left lobe of the liver, drain into the left duct. The hepatic ducts usually unite immediately outside the porta hepatis to form the **common hepatic duct** which descends in the free edge of the lesser omentum and is joined on its right side by the cystic duct from the gall-bladder [FIG. 6.111A]. Sometimes the union of the hepatic ducts is at a lower level in which case the common hepatic duct is reduced in length.

Bile-duct

The bile-duct, formed by the union of the cystic and common hepatic ducts, is about 5 mm in diameter and usually 7 to 8 cm in length. Its length, however, is inversely related to that of the cystic duct. Its first or upper part lies in the free border of the lesser omentum with the hepatic artery on its left and the portal vein posterior. On reaching the duodenum it passes behind the superior part, accompanied posteriorly by the portal vein and on its left by the gastroduodenal artery. From here it runs inferiorly and to the right to reach the posteromedial wall of the duodenum near the

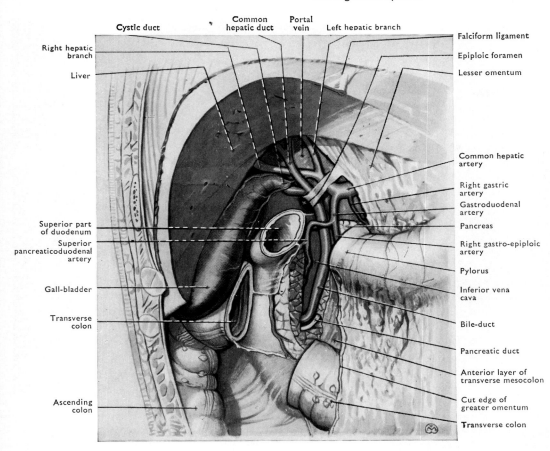

Fig. 6.111A. Dissection showing relations of gall-bladder and bile passages.

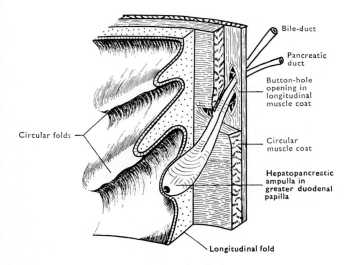

Fig. 6.111B. Diagrammatic section of wall of duodenum to show passage of bile and pancreatic ducts and their junction to form the hepatopancreatic ampulla.

middle of its descending part [FIG. 6.111A]. In this section of its course it grooves, or is sometimes embedded in, the posterior surface of the head of the pancreas, in front of the inferior vena cava. As the bile-duct approaches the descending part of the duodenum it comes into close contact with the pancreatic duct which it accompanies as they run obliquely through the duodenal wall for about 2 cm. Within the wall the two ducts unite to form the hepatopancreatic ampulla which opens through a constricted orifice on the summit of the greater duodenal papilla, about 10 cm from the pylorus [FIG. 6.111B]. Occasionally, however, the two ducts may remain separate and open independently on the duodenal papilla. At its termination the circular smooth muscle of the bile duct is increased in amount to form a sphincter. Similar, but much less constant, increases in smooth muscle are also associated with the termination of the pancreatic duct and with the hepatopancreatic ampulla. These combined sphincters are traditionally known as the sphincter of Oddi. There is now some evidence for the reciprocal functioning of the musculature of the gall-bladder and the sphincter whereby, with the emptying of the gall-bladder, there is inhibition of the sphincter. Contraction of the gall-bladder appears to be brought about by the hormone, *cholecystokinin*, which is liberated from the duodenum when the gastric contents come into contact with the intestinal mucosa. In Man there is little evidence for neural control of either the gall-bladder or the bile-duct sphincter (Boyden 1957).

Although the discharge of the bile into the duodenum is intermittent, being related to digestion, its production by the liver is a continuous process. In the absence of digestion the sphincter of Oddi remains in a state of tonic contraction and bile accumulates in the gall-bladder. Here it is concentrated by the selective absorption of water and certain ions and reduced to about one tenth of its original volume.

RADIOLOGY

The gall-bladder cannot be seen by X-rays unless its contents are abnormal or contrast medium has been administered to fill its

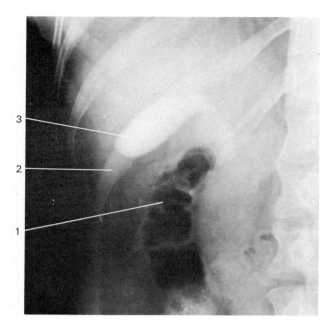

FIG. 6.112.   Cholecystography. Prone radiograph of the filled gall-bladder showing its relatively high position in this posture.
1. Gas in colon.
2. 11th rib.
3. Gall-bladder.

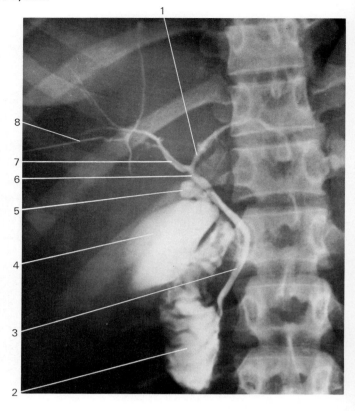

FIG. 6.114.   Transhepatic cholangiogram showing a normal biliary system. The contrast medium is introduced through a needle puncturing the liver.
1. Left hepatic duct.
2. Duodenum.
3. Bile-duct.
4. Gall-bladder.
5. Cystic duct.
6. Common hepatic duct.
7. Right hepatic duct.
8. Needle in liver.

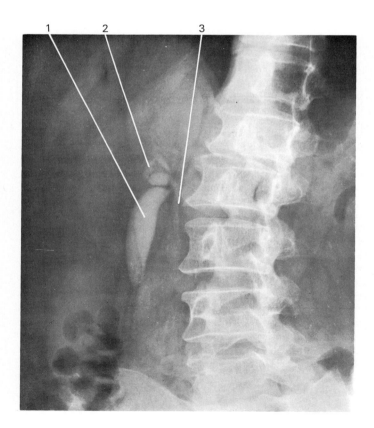

FIG. 6.113.   Cholecystography. Oblique radiograph of the gall-bladder contracting after a fatty meal. Note the low level of the gall-bladder in this case—its position is very variable as it moves with the liver.
1. Gall-bladder.
2. Cystic duct.
3. Bile-duct.

lumen. Rarely the gall-bladder wall may calcify or the lumen may contain sufficient calcium to show in a radiograph (lime bile).

An orally administered, iodine-containing, cholecystographic agent allows the gall-bladder to be visualized some hours after ingestion, provided that the agent is satisfactorily absorbed, then excreted by the liver, and adequately concentrated in the gall-bladder. This is a useful and safe method. After the administration of a fatty meal, contraction of the gall-bladder occurs and its opacified contents outline the cystic duct, the spiral valve, and frequently the bile-duct.

An intravenous preparation of contrast medium can be used to demonstrate the extrahepatic biliary ducts and, later in the course of the examination, the gall-bladder. This method is free of errors of interpretation that may result from failure of absorption of the oral contrast medium, but requires adequate liver function for success.

Should these methods fail and the visualization of the biliary duct system is thought to be essential, the following means can be used. The extrahepatic biliary ducts can be filled by passing a catheter through the duodenal papilla into the bile-duct from an endoscope in the duodenum. (The pancreatic duct can be filled in similar fashion.) This procedure is known as endoscopic retrograde cholangiopancreatography. Alternatively the ducts can be filled by percutaneous puncture of an intrahepatic duct and injection of contrast medium, as described in the section on the liver.

During operations on the gall-bladder, the biliary passages are usually filled with contrast medium by direct injection. This method is useful for the detection of residual stones in the bile or hepatic ducts, as these can occasionally escape the most careful probing or palpation.

BLOOD VESSELS AND NERVES OF THE GALL-BLADDER AND BILIARY DUCTS

The gall-bladder is supplied by the cystic artery which arises from the right branch of the hepatic artery. It normally passes behind the common hepatic and cystic ducts and divides into anterior and posterior branches supplying the respective surfaces of the gall-bladder. The cystic artery is subject to considerable variation in its origin and course. Commonly it arises from the proper hepatic artery and crosses in front of, or behind, the bile-duct or the common hepatic duct. The blood supply to the biliary ducts is derived from the hepatic, cystic and superior pancreaticoduodenal arteries. The veins from the gall-bladder and ducts pass either directly into the adjacent liver substance or form tributaries of the portal vein.

The nerve supply of the gall-bladder and biliary ducts consists of sympathetic and parasympathetic fibres passing in the hepatic plexus and being joined at the porta hepatis by branches from the anterior vagal trunk. The sympathetic fibres are vasoconstrictor and also transmit pain fibres, especially from the gall-bladder and bile-duct. Stimulation of the parasympathetic causes contraction of the gall-bladder and relaxation of the sphincter of Oddi but normally this activity is produced by the release of cholecystokinin from the duodenum. As well as autonomic fibres the gall-bladder receives fibres from the right phrenic nerve via the coeliac and hepatic plexuses. It is thought that these fibres account for the 'referred' shoulder pain sometimes associated with diseases of the gall-bladder.

STRUCTURE OF THE EXTRAHEPATIC BILIARY APPARATUS

The wall of the gall-bladder consists of a mucosa, a layer of muscle and a serosa. The mucosa is thrown into a large number of folds many of which disappear when the gall-bladder is distended. It is formed by an epithelium resting on a layer of loose connective tissue, the lamina propria, directly applied to the layer of smooth muscle. The epithelium consists of tall columnar cells, some of which secrete mucus. The cells carry short, somewhat irregular microvilli and are attached near their apices by junctional complexes including zonulae occludentes. During the absorption of water and inorganic ions the intercellular spaces increase in size especially at their bases. The muscular layer is made up of irregularly orientated bundles of smooth muscle fibres interdigitating with collagenous and elastic fibres. External to it there is a layer of loose connective tissue which on its anterior surface is continuous with the capsule of the liver and on its postero-inferior surface is covered by peritoneum, thus forming a serosa. In the region of the neck there may be a few simple mucus-secreting glands (Hayward 1968).

The walls of the extrahepatic ducts are similar to those of the gall-bladder in having a mucosa consisting of tall columnar epithelium resting on a lamina propria. However, they differ in that the muscular coat is absent in the hepatic ducts and first becomes apparent in the bile-duct, where it forms an incomplete layer. At the lower end of the duct the circular fibres are greatly increased in number and form part of the sphincter of Oddi. The bile-duct usually contains scattered clusters of mucus-secreting glands within its lamina propria (Boyden 1957; Riches 1972).

Development of the liver and biliary apparatus

The liver and biliary apparatus develop from an entodermal outgrowth, the hepatic diverticulum [FIG. 2.73] which arises from the ventral surface of the caudal end of the foregut. Its site of attachment later becomes the apex of the duodenal loop. The hepatic diverticulum grows into the septum transversum and divides into a larger cranial branch and a smaller caudal branch. From the cranial part anastomosing cords of cells are formed and these give rise to the hepatocytes and the lining of the intrahepatic biliary vessels. As the cords of cells infiltrate and occupy much of the septum transversum they associate with, and break up the anastomotic plexus between the vitelline veins, thus forming primitive hepatic sinusoids. Later this process extends laterally to involve the umbilical veins as they pass through the septum transversum. The Kupffer cells of the hepatic sinusoids and the haemopoietic cells are derived from mesodermal cells of the septum, these also form the connective tissue and capsule of the liver.

The caudal branch of the hepatic diverticulum, together with its undivided stem, forms the gall-bladder, cystic duct, and bile-duct. Originally the bile-duct is attached to the ventral aspect of the duodenum but, as a result of the rotation of the duodenum to the right and a differential growth of its walls, the point of attachment is carried to its adult position on the posteromedial wall [FIG. 6.116].

The liver grows rapidly during the early stages of its development and occupies much of the abdominal cavity. At first its two main lobes are of equal size but later, as its relative rate of growth decreases, its lobes are unequally affected resulting eventually in the establishment of the adult proportions.

Haemopoiesis commences shortly after the formation of the endodermal cords of cells and is partly responsible for the early rapid growth of the liver. During the last two months of foetal life it gradually subsides and at birth only isolated islands of blood-forming cells are present. Bile is first formed at about 4 months of intra-uterine life. It is responsible in the later months of pregnancy for the characteristic colour of the foetal intestinal contents (meconium).

As the liver and stomach develop they draw away from the septum transversum bringing with them mesoderm which forms the ventral mesentery. This extends caudally to the umbilicus as the septum does. Since this is the only ventral mesentery associated with the developing gut it has a caudal free border. Between the liver and the developing diaphragm the mesentery persists as the coronary and triangular ligaments. Caudal to this it forms the falciform ligament between the liver and the anterior abdominal wall, and has the left umbilical vein in its free border. The portion of the ventral mesentery between the liver and the ventral border of the stomach and first 2–3 cm of the duodenum forms the lesser omentum and contains the bile-duct in its free border.

THE PANCREAS

The pancreas is a soft, lobulated organ extending transversely and slightly upwards across the posterior abdominal wall from the duodenum on the right to the spleen on the left. It is 12 to 15 cm in length and lies in the epigastric and left hypochondriac regions, behind the omental bursa, at the level of the first and second lumbar vertebrae. It weighs approximately 90 g and is described as having a head, body, and tail.

The **head** lies within the concavity of the duodenum which it overlaps to a variable extent, and from which it is often separated on the right and inferiorly by the superior and inferior pancreaticoduodenal arteries. It is flattened from before backwards and rests posteriorly on the inferior vena cava, the right crus of the diaphragm, and the terminal parts of the renal veins. The bile-duct grooves the upper lateral part of the posterior surface or sometimes runs within the substance of the pancreas [FIG. 6.111A]. Anteriorly, the head is crossed by the transverse colon, attached only by some loose areolar tissue, and below this, it is separated from the coils of jejunum by peritoneum continuous with the inferior layer of the transverse mesocolon. The lower part of the head extends to the left behind the superior mesenteric vein and artery and in front of the aorta. It is known as the **uncinate process** [FIG. 6.66].

The slender first part of the **body** extends upwards and to the left from the upper part of the head. It is relatively short, lies behind the pylorus, and is deeply grooved posteriorly by the beginning of the portal vein.

The remainder of the **body** is triangular in section and has anterior, posterior, and inferior surfaces. The anterior surface is covered by the peritoneum of the posterior wall of the omental bursa and forms part of the stomach bed [FIG. 6.47]. There is often a prominence at the right extremity of this surface, the **tuber omentale**, which projects above the lesser curvature of the stomach and comes into contact with the lesser omentum. Posteriorly, the body crosses the front of the aorta covering the origins of the coeliac and superior mesenteric arteries and lies in contact with the left crus of the diaphragm, the left suprarenal and the left kidney and its vessels. The splenic vein runs from left to right behind the posterior surface of the body while the splenic artery follows a wavy course along the upper border of the body. The inferior surface is relatively narrow and is covered by peritoneum of the greater sac. From right to left it rests on the duodenojejunal flexure, some coils of jejunum, and the left colic flexure. The **tail** of the pancreas lies within the lienorenal ligament with the splenic vessels and usually comes into contact with the inferior part of the gastric surface of the spleen.

The transverse mesocolon, much of it fused with the greater omentum, is attached to the anterior border of the pancreas between the anterior and inferior surfaces [FIG. 6.46] over which its two layers of peritoneum pass.

The **main pancreatic duct** commences in the tail and runs to the right in the body of the gland, nearer its posterior than its anterior surface. Along its course it receives tributaries from the surrounding lobules and on reaching the head of the gland it turns downwards, backwards and to the right to meet the left side of the bile-duct. Together they penetrate the duodenal wall and in the submucosa join to form the **hepatopancreatic ampulla** (ampulla of Vater) which opens on the summit of the greater duodenal papilla. An **accessory duct** is frequently present beginning in the lower part of the head and running upwards in front of the main pancreatic duct to which it sends a communicating branch. It opens on the lesser duodenal papilla approximately 2 cm above the greater papilla.

BLOOD AND LYMPH VESSELS AND NERVES OF THE PANCREAS

The **arteries** of the pancreas are branches of the splenic artery and of the superior and inferior pancreaticoduodenal arteries. Its **veins** are tributaries of the splenic, portal, and superior mesenteric veins. The **lymph vessels** drain into the pancreaticosplenic nodes, pyloric nodes, and nodes associated with the pancreaticoduodenal arteries, and finally into the coeliac and superior mesenteric nodes associated with the corresponding arteries. The **nerves** are preganglionic vagal fibres and postganglionic sympathetic fibres which travel to the pancreas in the plexuses on the blood vessels on which are scattered

autonomic ganglion cells. It is thought that the sympathetic fibres are purely vasomotor and that the parasympathetic are secretomotor to both the acinar and islet cells. Precisely how the nerves help in regulating the pancreatic secretion is not yet known. There is little doubt that the regulation of both the exocrine and endocrine secretion is largely under hormonal control. The pancreas and the root of the mesentery contain a number of **Pacinian corpuscles** each innervated by a single, thickly myelinated nerve fibre. These are found in a number of other areas in the body and are thought to have a mechanoreceptor function.

RADIOLOGY

The normal pancreas is not visible in plain films of the abdomen or in the examinations commonly undertaken in the upper abdomen, though occasionally part of the pancreatic duct may fill during investigations of the biliary tract.

Other methods are specialized and require brief mention only.

Ultrasound is the simplest type of examination available. It can sometimes give valuable information, especially when part of the gland is enlarged.

Computerized axial tomography usually shows the pancreas adequately if several sections are used [ATLAS, FIGS. 19A–21A]. The limitation arises because abnormal masses have to be of a certain size before they can be identified with certainty.

The pancreatic duct may be visualized either by retrograde catheterization of the duct through an endoscope passed into the duodenum (see biliary ducts) or by injection at operation, exploration having failed to determine the site of the pathology. These methods are useful only if the disease is close to the ducts or their main tributaries.

Selective angiography of the coeliac and sometimes the superior mesenteric artery can often give excellent results. It requires a detailed knowledge of the blood supply of the pancreas and its many variations which is greater than that expected of the undergraduate student of anatomy.

STRUCTURE OF THE PANCREAS

The pancreas is the second largest glandular organ in the body. It resembles the liver in having exocrine and endocrine functions, but differs in that the hepatocyte of the liver performs both functions, whereas two different cell types are involved in the pancreas.

The volume of the exocrine secretion produced in the **acinar cells** in Man ranges from 800 to 1200 ml per day and contains at least nine digestive enzymes, as well as water, bicarbonate, and salts. In the duodenum it takes part in the breakdown of proteins, carbohydrates, and lipids and helps to neutralize the acid chyme from the stomach. The endocrine secretion is produced in isolated groups of cells, the **pancreatic islets** or the islets of Langerhans, which number approximately one million in Man and are more numerous in the tail of the pancreas than elsewhere. Their secretion contains two hormones, *insulin* which increases the permeability of certain plasma membranes to glucose and partly through this profoundly influences carbohydrate, fat, and protein metabolism, and *glucagon*, which to a large extent, has opposite actions to those of insulin.

The cells of the **exocrine pancreas** form spherical or tubular acini which are surrounded by loose connective tissue containing blood and lymph vessels and excretory ducts. Each acinus is formed by a single layer of pyramidal-shaped cells with their bases resting on a basal lamina and their apices bounding the central lumen. The acini

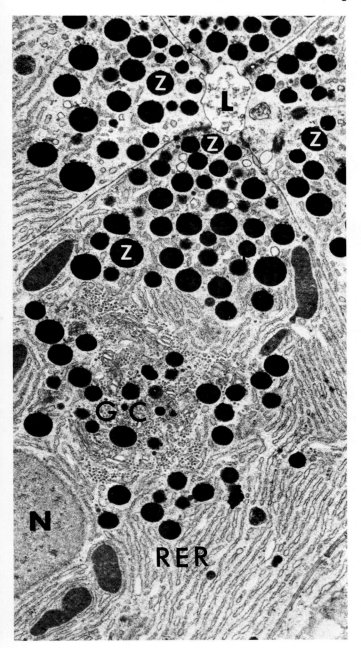

FIG. 6.115. Electron micrograph of the exocrine pancreas. The lumen (L) of the acinus is surrounded by the apices of four exocrine cells. The lower cell contains numerous zymogen granules (Z), rough endoplasmic reticulum (RER), a nucleus (N), and an extensive Golgi complex (GC).

are drained by intercalated (intralobular) ducts whose terminal cells protrude to a limited extent into the lumen of each acinus and are consequently known as centroacinar cells. These cells produce the water and electrolytes of the pancreatic secretion.

The base of the acinar cell stains intensely with basic dyes indicating the presence of large amounts of ribonucleoprotein and correlating with the marked accumulation of rough endoplasmic reticulum seen with the electron microscope [FIG. 6.115]. In the supranuclear area there is a well-developed Golgi apparatus, and above this the cytoplasm contains numerous refractile zymogen granules which stain with acid stains. These granules are electron dense and are each surrounded by membrane. They contain the digestive enzymes, usually in a precursor state and discharge them into the acinar lumen by reverse pinocytosis.

Numerous studies using ultrastructural, autoradiographic, and biochemical techniques have now shown that in the acinar cell the synthesis of protein for export takes place on the ribosomes of the rough endoplasmic reticulum. From here it passes into the cisternae of the reticulum and is later transferred to the Golgi apparatus, where it is condensed and packaged to form the mature zymogen granules of the apical cytoplasm. Evidence so far indicates that each acinar cell is capable of producing all the digestive enzymes simultaneously (Caro and Palade 1964; Jamieson and Palade 1968, 1971).

In contrast to the acinar cells, the centroacinar cells are characterized by their pale staining and the relative paucity of their organelles.

The intercalated ducts, draining the individual acini, are lined by low columnar cells similar to the centroacinar cells. They open into larger interlobular ducts, whose columnar epithelium increases in height as the ducts increase in diameter, and contains occasional goblet and argentaffin cells. The interlobular ducts drain into the major pancreatic ducts. These have an epithelium similar to that of the largest interlobular ducts and are surrounded by a moderately thick layer of connective tissue containing some elastic fibres.

The islets of Langerhans are round or oval in shape and can easily be recognized in histological sections due to their contrasting pale staining and circumscribed outline. Each is composed of irregular cords of cells and, like all endocrine glands, they are highly vascular with capillaries probably touching each cell. With special histological methods for demonstrating the secretory granules it is possible to distinguish three different types of cell, which also have characteristic ultrastructural appearances. These are called alpha, beta, and delta cells. They occur roughly in the ratio of 20:75:5 respectively. The alpha cells produce the hormone glucagon and tend to be situated at the periphery of the islet, whilst the beta cells produce insulin and are normally centrally placed. Although the delta cells can be distinguished morphologically there is still considerable doubt about their function. In the view of some workers they represent a stage in the secretory cycle of one of the other endocrine cells (Lazarow 1957; Like 1967).

DEVELOPMENT OF THE PANCREAS

The pancreas develops from a ventral and a dorsal entodermal outgrowth from the duodenum. The ventral arises close to, or in common with, the hepatic diverticulum, and the larger, dorsal outgrowth arises slightly cranial to the ventral extending into the mesoduodenum and dorsal mesogastrium [FIG. 6.116]. As a result of differential growth in the duodenal wall the point of attachment of the ventral outgrowth with the undivided stem of the hepatic diverticulum (the bile-duct) and the ventral mesentery are displaced round the right side of the duodenum to a dorsal position. As a result the two pancreatic rudiments come together and eventually fuse to form a single organ. In this process the ventral rudiment forms the uncinate process and adjacent region of the head, while the dorsal forms the anterior part of the head, the body, and the tail of the pancreas.

At the time of fusion of the two rudiments their ducts anastomose in such a way that most of the dorsal rudiment is drained by the duct of the ventral rudiment and the composite duct, thus formed, becomes the main pancreatic duct [FIG. 6.116]. Consequently the proximal or terminal part of the duct of the dorsal rudiment remains relatively small or may even disappear. It normally forms the accessory pancreatic duct and opens into the duodenum above the main duct. It may retain or lose its continuity with the main duct.

The parenchymatous cells differentiate from the entoderm of the pancreatic ducts. The exocrine cells form alveoli which retain their connection with the ducts, but by the third month of foetal life the endocrine cells separate and form isolated islets of Langerhans.

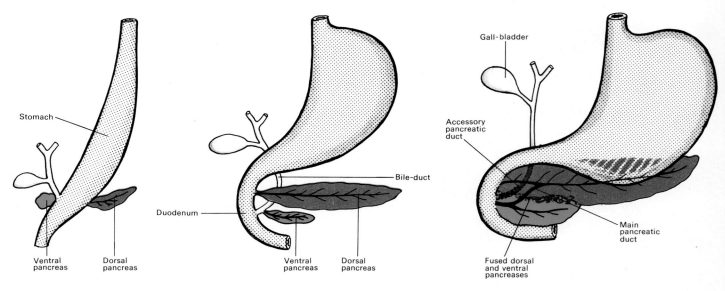

Fig. 6.116. Development of the pancreas.

Shortly afterwards, alpha and beta cells, containing their characteristic granules, can be recognized. The pancreas at first lies between the two layers of the dorsal mesentery but, with the rotation of the duodenum and stomach to the right and the fusion of much of the dorsal mesentery with the posterior abdominal wall, it finally becomes a retroperitoneal structure with the ventral pancreas dorsally placed (Laitio, Lev, and Orlic 1974; Like and Orci 1972).

ANOMALIES

The commonest anomalies consist of accessory pancreatic tissue in the walls of the stomach and small intestine, and of variations in the arrangement of the pancreatic ducts. Extremely rarely there may be an annular pancreas consisting of a band of pancreatic tissue surrounding the descending part of the duodenum. It is of little clinical significance except that in pathological states it may cause duodenal obstruction.

References

AMSTERDAM, A., OHAD, I., and SCHRAMM, M. (1969). Dynamic changes in the ultrastructure of the acinar cell of the rat parotid gland during the secretory cycle. *J. Cell. Biol.* **41**, 753.

ANDERSON, D. J., HANNAM, A. G., and MATHEWS, B. (1970). Sensory mechanisms in Mammalian teeth and their supporting structures. *Physiol. Rev.* **50**, 171.

BABCOCK, M. B. and CARDELL, R. R. (1974). Hepatic glycogen patterns in fasted and fed rats. *Am. J. Anat.* **140**, 299.

—— —— (1975). Fine structure of hepatocytes from fasted and fed rats. *Am. J. Anat.* **143**, 399.

BAETENS, D., SUKANT, B. C., DOBBS, R., UNGER, R., and ORCI, L. (1976). Identification of glucagon producing cells (A-cells) in dog gastric mucosa. *J. Cell. Biol.* **69**, 455.

BARROWMAN, J. (1975). The secretory cell, in *The cell in medical science* (eds. F. Beck and J. Lloyd) Vol. 3. Academic Press, London.

BOSMA, J. F. (1957). Deglutition: pharyngeal stage. *Physiol. Rev.* **37**, 275.

BOYDEN, E. A. (1957). The anatomy of the choledochoduodenal junction in man. *Surg. Gynaecol. Obstet.* **140**, 641.

BRAASH, J. N. (1958). Surgical anatomy of the liver and pancreas. *Surg. Clin. North Am.* **38**, 747.

BRAMBELL, F. W. R. (1970). *The transmission of passive immunity from mother to young.* North-Holland, Amsterdam.

BRETSCHER, E. R. and WEBER, K. (1978). Localization of actin and microfilament-associated proteins in the microvilli and terminal web of the intestinal brush border by immunofluoresence microscopy. *J. Cell Biol.* **79**, 839.

BUCHAN, A. J. M., POLAK, J. M., SOLCIA, E., CAPELLA, C., HUDSON, D., and PEARSE, A. G. E. (1978). Electron immunohistochemical evidence for the human intestinal I cell as the source of CCK. *Gut* **19**, 403.

CARDELL, R. R., BADENHAUSEN, S., and PORTER, K. R. (1967). Intestinal triglyceride absorption in the rat. An electron microscopical study. *J. Cell Biol.* **34**, 123.

CARO, L. G. and PALADE, G. E. (1964). Protein synthesis, storage and discharge in the pancreatic exocrine cell. An autoradiographic study. *J. Cell Biol.* **20**, 473.

CHANG, W. W. L. and NADLER, N. J. (1975). Renewal of the epithelium in the descending colon of the mouse. *Am. J. Anat.* **144**, 39.

CHEN, K. Y. and WITHERS, H. R. (1975). Proliferative capability of parietal and zymogen cells. *J. Anat.* (*Lond.*) **120**, 421.

CHENG, H. and LEBLOND, C. P. (1974). Origin, differentiation and renewal of the four main epithelial cell types in the mouse small intestine V. Unitarian theory of the origin of the four epithelial cell types. *Am. J. Anat.* **141**, 537.

CHEVREL, J. P. and GUÉRAND, J. P. (1978). Arteries of the terminal ileum. *Anat. Clin.* **1**, 95.

CLARKE, R. M. (1970). Mucosal architecture and epithelial cell production rate in the small intestine of the albino rat. *J. Anat.* (*Lond.*) **107**, 519.

—— and HARDY, R. N. (1971). Histological changes in the small intestine of the young pig and their relation to macromolecular uptake. *J. Anat.* (*Lond.*) **108**, 63.

CODE, C. F. (1968). Oesophagogastric junction. *Handbook of physiology.* American Physiological Society, Washington, DC.

COPE, G. H. and WILLIAMS, M. A. (1973). Exocrine secretion in the parotid gland: a stereological analysis at the electron microscopic level of the zymogen granule content before and after isoprenaline-induced degranulation. *J. Anat.* (*Lond.*) **116**, 269.

CRANE, R. K. (1968). A concept of the digestive absorptive surface of the small intestine. In *Handbook of physiology* (eds. C. F. Code and W. Heidel) Section 6, Vol. 5, p. 2535. American Physiological Society, Washington DC.

DICKINSON, V. A. (1978). Maintenance of anal continence: a review of pelvic floor physiology. *Gut* **19**, 1163.

DI DIO, L. J. A. and ANDERSON, M. C. (1968). *The 'sphincters' of the digestive system.* Williams and Wilkins, Baltimore.

EASTWOOD, G. L. (1977). Gastro-intestinal epithelial renewal. *Gastroenterol.* **72**, 962.

ELIAS, H. (1949a). A re-examination of the structure of the mammalian liver, 1. Parenchymal architecture. *Am. J. Anat.* **84**, 311.

—— (1949b). A re-examination of the structure of the mammalian liver, 2. The hepatic lobule and its relation to the vascular and biliary systems. *Am. J. Anat.* **85**, 379.

—— (1955). Liver morphology. *Biol. Rev.* **30**, 263.

FARQUHAR, M. G. and PALADE, G. E. (1963). Junctional complexes in various epithelia. *J. Cell Biol.* **17**, 375.

FEARNHEAD, R. W. (1967). In *Structural and chemical organization of teeth* (ed. A. E. W. Miles) Vol. II. Academic Press, London.

—— and STACK, M. V. (1971). *Tooth enamel,* Vol. II. Wright, Bristol.

FERGUSON, A. (1977). Intra-epithelial lymphocytes of the small intestine. *Gut* **18**, 921.

FERGUSON, M. W. J. (1978). Palatal shelf elevation in the Wistar rat fetus. *J. Anat. (Lond.)* **125**, 555.

FICHTELIUS, K. E. (1968). The gut epithelium—a first level lymphoid organ? *Exp. Cell Res.* **49**, 87.

FITZGERALD, M. J. T., NOLAN, J. P., and O'NEILL, M. N. (1971). The position of the human caecum in fetal life. *J. Anat. (Lond.)* **109**, 71.

FRIEDMAN, H. I. and CARDELL, R. R. (1977). Alterations in the endoplasmic reticulum and Golgi complex during fat absorption and after termination of this process: a morphological and morphometric study. *Anat. Rec.* **188**, 77.

GORGOLLON, P. (1978). The normal human appendix: a light and electron microscopic study. *J. Anat. (Lond.)* **126**, 87.

GUNN, M. (1968). Histological and histochemical observations on the myenteric and submucous plexuses of mammals. *J. Anat. (Lond.)* **102**, 223.

GUPTA, C. D. and GUPTA, S. C. (1976). Evaluation of intrahepatic arterial branching patterns in corrosion casts. *J. Anat. (Lond.)* **122**, 31.

GUPTA, S. C., GUPTA, C. D., and ARORA, A. K. (1977). Subsegmentation of the human liver. *J. Anat. (Lond.)* **124**, 413.

HAYWARD, A. F. (1968). The structure of the gall-bladder epithelium. *Int. Rev. Gen. Exp. Zool.* **3**, 205.

HOLLAND, G. R. (1976). An ultrastructural survey of cat dentinal tubules. *J. Anat. (Lond.)* **122**, 1.

HOLLINSHEAD, W. H. (1968). The head and neck. In *Anatomy for surgeons*, Vol. 1. Harper & Row, New York.

—— (1971). The thorax, abdomen and pelvis. In *Anatomy for surgeons*, Vol. 2. Harper & Row, New York.

HOLMES, R. (1971). Progress report. The intestinal brush border. *Gut* **12**, 668.

ITO, S. (1968). Anatomical structure of the gastric mucosa. In *Handbook of physiology* (eds. C. F. Code and M. E. Grossman) Section 6, Vol. 2. American Physiological Society, Washington DC.

JAMIESON, J. D. and PALADE, G. E. (1968). Intracellular transport of secretory proteins in the pancreatic exocrine cell. *J. Cell Biol.* **39**, 580.

—— —— (1971). Condensing vacuole conversion and zymogen granule discharge in pancreatic exocrine cells: metabolic studies. *J. Cell Biol.* **48**, 503.

JIT, I. and GREWAL, S. S. (1977). The suspensory muscle of the duodenum and its nerve supply. *J. Anat. (Lond.)* **123**, 397.

JOHNSON, F. R. (1975). The absorptive cell. In *The cell in medical science* (eds. F. Beck and J. Lloyd) Vol. 3. Academic Press, London.

—— and McMINN, R. M. H. (1970). Microscopic structure of pyloric epithelium of the cat. *J. Anat. (Lond.)* **107**, 67.

—— and YOUNG, B. A. (1968). Undifferentiated cells in gastric mucosa. *J. Anat. (Lond.)* **102**, 541.

JOSEPHENSEN, K. and FEJERSKOV, O. (1977). Ameloblast modulation in the maturation zone of the rat incisor enamel organ. A light and electron microscope study. *J. Anat. (Lond.)* **124**, 45.

KATCHBURIAN, E. (1973). Membrane-bound bodies as initiators of mineralization of dentine. *J. Anat. (Lond.)* **116**, 285.

LAITIO, M., LEV, R., and ORLIC, D. (1974). The developing human fetal pancreas: an ultrastructural and histochemical study with special reference to exocrine cells. *J. Anat. (Lond.)* **117**, 619.

LANDAY, M. A. and SCHROEDER, H. E. (1979). Differentiation in normal human buccal mucosa epithelium. *J. Anat. (Lond.)* **128**, 31.

LAZAROW, A. (1957). Cell types of the islets of Langerhans and the hormones they produce. *Diabetes* **6**, 222.

LIKE, A. A. (1967). The ultrastructure of the secretory cells of the islets of Langerhans in Man. *Lab. Invest.* **16**, 937.

—— and ORCI, L. (1972). The embryogenesis of the human pancreatic islets, a light microscopic and electron microscopic study. *Diabetes* **21** (Suppl. 2), 511.

MASSELOT, R. and LEBORGNE, J. (1978). Anatomical studies of the hepatic veins. *Anat. Clin.* **1**, 109.

MATTER, A. L., ORCI, C., and ROUILLER, A. (1969). A study on the permeability barrier between Disse's space and the bile canaliculus. *J. Ultrastruct. Res.* **11** (supp.).

MELCHER, A. H. and BOWEN, W. H. (1969). *The biology of the periodontium.* Academic Press, New York.

MILES, A. E. (1967). *Structural and chemical organization of teeth.* Academic Press, London.

MOE, H. (1971). Morphological changes in the infranuclear portion of the enamel-producing cells during their life cycle. *J. Anat. (Lond.)* **108**, 43.

MOOSEKER, M. S. and TILNEY, L. G. (1975). Organization of an active filament membrane complex. Filament polarity and membrane attachment in the microvilli of intestinal epithelial cells. *J. Cell Biol.* **67**, 725.

MOTTA, P. and FUMAGALLI, G. (1975). Structure of rat bile canaliculi as revealed by scanning electron microscopy. *Anat. Rec.* **182**, 499.

PEARSE, A. G. E., POLAK, J. M., and BLOOM, S. R. (1977). The newer gut hormones: cellular sources, physiology, pathology and clinical aspects. *Gastro-enterol.* **72**, 746.

PIASECKI, C. (1975). Observations on the submucous plexus and mucosal arteries of the dog's stomach, and the first part of the duodenum. *J. Anat. (Lond.)* **119**, 133.

POPPER, H. and UDENFRIEND, S. (1970). Hepatic fibrosis. Correlation of biochemical and morphological investigations. *Am. J. Med.* **49**, 407.

RAPPAPORT, A. M. (1958). The structural and functional unit in the human liver (liver acinus). *Anat. Rec.* **130**, 673.

RICHARDS, T. G. (1975). The liver cell in bile formation. *Anat. Rec.* **130**, 673.

RICHES, D. J. (1972). Ultrastructural observations on the common bile duct epithelium of the rat. *J. Anat. (Lond.)* **111**, 157.

SANDOW, M. J. (1979). The Paneth cell. *Gut* **20**, 420

SCHOFIELD, G. C., ITO, S., and BOLENDER, R. P. (1979). Changes in membrane surface areas in mouse parietal cells in relation to high levels of acid secretion. *J. Anat. (Lond.)* **128**, 669.

SCOTT, J. H. and SYMONS, N. B. B. (1978). *Introduction to dental anatomy*, 7th edn. E. & S. Livingstone, London.

SHACKLEFORD, J. M. and SCHNEYER, L. H. (1971). Ultrastructural aspects of the main excretory duct of rat submandibular gland. *Anat. Rec.* **169**, 679.

SOLANKE, T. F. (1968). The blood supply of the vermiform appendix in Nigerians. *J. Anat. (Lond.)* **102**, 353.

SOLCIA, E., CAPELLA, C., VASSALLO, G., and BUFFA, R. (1975). Endocrine cells of the gastric mucosa. *Int. Rev. Cytol.* **42**, 223.

TREASURE, T. (1978). The ducts of Brunner's glands. *J. Anat. (Lond.)* **127**, 299.

WAKELEY, C. P. C. (1933). The position of the vermiform appendix as ascertained by an analysis of 10,000 cases. *J. Anat. (Lond.)* **67**, 227.

WALLS, E. W. (1958). Observations on the microscopic anatomy of the human anal canal. *Br. J. Surg.* **45**, 504.

WATERMAN, R. E. and MELLER, S. M. (1974). Alterations in the epithelial surface of human palatal shelves prior to and during fusion: a scanning electron microscope study. *Anat. Rec.* **180**, 111.

WEINSTOCK, M. and LEBLOND, C. P. (1974). Synthesis, migration and release of precursor collagen by odontoblasts as visualized by autoradiograph after (^{3}H) proline administration. *J. Cell Biol.* **60**, 92.

WHILLIS, J. (1946). Movements of the tongue in swallowing. *J. Anat. (Lond.)* **80**, 115.

WHITTAKER, D. K. (1978). The enamel–dentine junction of human and *Macaca irus* teeth: a light and electron microscope study. *J. Anat. (Lond.)* **125**, 323.

WISSE, E. (1970). An electron microscopic study of the fenestrated endothelial lining of rat liver sinusoids. *J. Ultrastruct. Res.* **31**, 125.

7 The respiratory system

R. J. SCOTHORNE

The general arrangement of the respiratory system is shown in FIGURE 7.1. It consists of:

(i) the paired **nasal cavities**, right and left, which open in front to the air through the nostrils or anterior **nares**, and behind,

through the **choanae** (posterior nares) into the **nasal part of the pharynx**. This continues below into:

(ii) the **oral and laryngeal parts** of the **pharynx**, which are passages common to the respiratory and digestive systems;

(iii) the **larynx**;

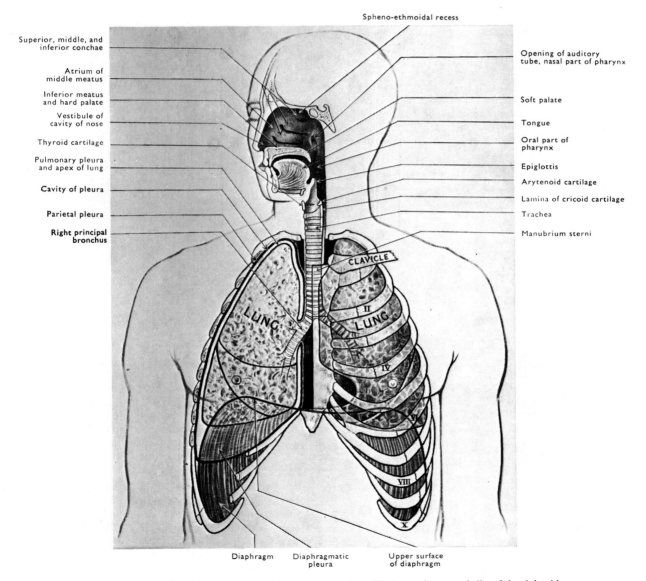

Spheno-ethmoidal recess

Superior, middle, and inferior conchae

Atrium of middle meatus

Inferior meatus and hard palate

Vestibule of cavity of nose

Thyroid cartilage

Pulmonary pleura and apex of lung

Cavity of pleura

Parietal pleura

Right principal bronchus

Opening of auditory tube, nasal part of pharynx

Soft palate

Tongue

Oral part of pharynx

Epiglottis

Arytenoid cartilage

Lamina of cricoid cartilage

Trachea

Manubrium sterni

CLAVICLE

LUNG

LUNG

Diaphragm   Diaphragmatic pleura   Upper surface of diaphragm

FIG. 7.1.   Diagram of general arrangement of respiratory system. The lung, pleura, and ribs of the right side are represented as cut in a coronal plane.

(iv) the trachea and extrapulmonary bronchi;
(v) the branching tree of intrapulmonary bronchi in the lungs. The terminal branches of the tree (bronchioles) lead to the air sacs or alveoli. Exchange of respiratory gases between air and blood occurs through the alveolar walls. The lungs also contain branching vascular trees, and are covered externally by a serous membrane, the pulmonary (visceral) layer of the pleura.

The respiratory tract may be divided into a respiratory portion, which bears alveoli, and is directly involved in respiratory exchange, and a conducting portion, consisting of passages for transport of air into and out of the respiratory portion. A continuous column of air extends from the nostrils, through the respiratory tract, to the alveoli. Here the layer of tissue intervening between air and blood (the blood–air barrier) measures about 1 μm or less in thickness. The potential hazards of this close proximity of external environment to circulating blood are guarded against by several structural specializations.

THE NOSE

External nose

The nasal cavities are, for the most part, deeply placed in the head, but project forwards on to the face within the external nose. The dorsum of this is the anterosuperior surface, which extends from the junction of the root with the forehead to the apex (tip). The skeleton of the upper part of the external nose is bony. It is formed by the paired nasal bones and by the frontal process of each maxilla. The skeleton of the lower part consists of hyaline cartilage in several

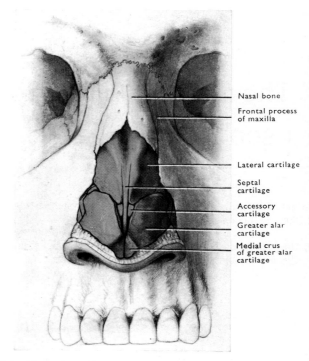

FIG. 7.3.   Front view of bony and cartilaginous skeleton of external nose.

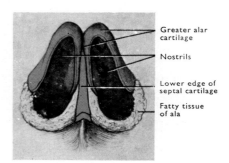

FIG. 7.4.   Cartilages of nose from below.

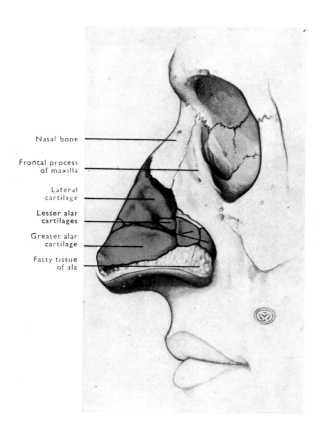

FIG. 7.2.   Profile view of bony and cartilaginous skeleton of external nose.

pieces which permit movement of the alae—the lateral boundaries of the nostrils. These cartilages are fully illustrated in FIGURES 7.2–7.5. They are attached to adjacent bones and to one another by fibrous tissue, and they do not extend to the free margins of the nostrils, which are formed by fibrofatty tissue. The skin over the bony skeleton of the external nose is mobile but it is firmly bound to the cartilaginous skeleton.

Vessels and nerves

The arteries are branches of the facial, ophthalmic, and maxillary arteries. The veins drain with the arteries and have important connections within the orbit with the ophthalmic veins, and through them, with the cavernous sinus. The principal lymph vessels accompany the facial vein and drain into submandibular lymph nodes, but a few vessels run laterally to the superficial parotid lymph nodes. The muscles of the nose belong to the mimetic group and are innervated by the facial nerve. The skin is innervated by the infra-orbital branch of the maxillary nerve, and by the infratrochlear and external nasal branches from the ophthalmic nerve.

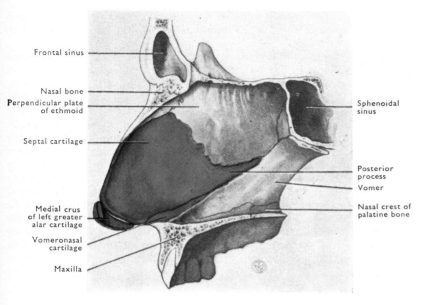

Frontal sinus

Nasal bone
Perpendicular plate
of ethmoid

Septal cartilage

Sphenoidal
sinus

Posterior
process
Vomer

Medial crus
of left greater
alar cartilage

Nasal crest of
palatine bone

Vomeronasal
cartilage

Maxilla

FIG. 7.5.   View of nasal septum from the left side.

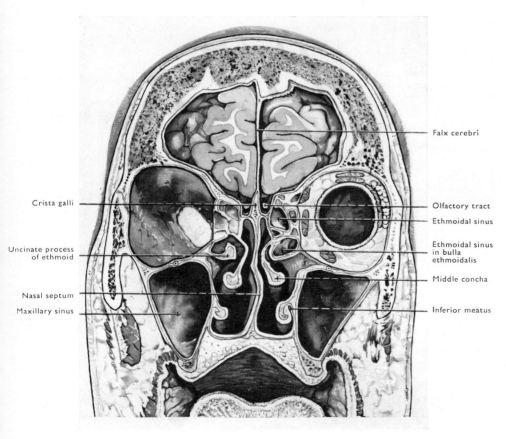

Falx cerebri

Crista galli

Olfactory tract

Ethmoidal sinus

Uncinate process
of ethmoid

Ethmoidal sinus
in bulla
ethmoidalis

Middle concha

Nasal septum

Maxillary sinus

Inferior meatus

FIG. 7.6.   Coronal section of head, viewed from the front. The contents of the right orbit have been removed.

The nasal cavities

A description of the bony boundaries of the nasal cavities and of the paranasal air sinuses which open into them, is given on pages 121–7.

The paired nasal cavities for the most part lie deeply in the head. Their position is shown in FIGURES 7.6 and 7.7. Each cavity has a medial wall, a floor, a roof and a lateral wall. The medial wall, between the two cavities, is the nasal septum whose major components are the perpendicular plate of the ethmoid, the vomer, and the septal cartilage, with minor contributions from those bones which abut upon the septum (nasal, frontal, sphenoid, maxilla, and palatine; FIG. 7.5). The septal cartilage frequently deviates from the midline. The roof consists of a sloping anterior part in the external nose formed by the nasal bone; a horizontal middle part, the cribriform plate of the ethmoid, which separates the nasal cavities from the meninges and brain; and an angled posterior part (at first vertical, then nearly horizontal), the body of the sphenoid. The floor is broader than the roof, and consists of the palatine process of the maxilla and the horizontal lamina of the palatine bone; together these form the hard palate, which completely separates the nasal cavities from the mouth, thus allowing the infant to breathe and suckle at the same time.

The lateral wall [FIGS. 7.6–7.8] separates the nasal cavity from the orbit and from the maxillary air sinus. Its most anterior part, just inside the nostril, is part of the skin-lined **vestibule**. The upper limit of the vestibule is marked, on the lateral wall, by a curved ridge (limen nasi) formed by the upper margin of the lower nasal cartilage. This is the region of smallest cross-sectional area of the nasal cavity; here, skin gives way to mucous membrane which covers the smooth lateral wall of the **atrium**. Behind this, three conchae project inwards from the lateral wall. Each **concha** consists of a delicate scroll of bone, covered by a thick mucoperiosteum. In coronal section [FIG. 7.6] it is seen that each concha is attached to the lateral wall by its upper edge and that it projects medially and downwards, overhanging like a canopy, a space or **meatus**, which it partially separates from the rest of the nasal cavity. The three conchae are named

inferior, middle, and superior; a fourth, small, supreme concha is not uncommon. Each meatus is named from the concha beneath which it lies.

High up in each nasal cavity, above the superior concha, is the **spheno-ethmoidal recess**, into which the sphenoidal air sinus opens through the anterior wall of the body of the sphenoid [FIG. 7.8]. Below the short, obliquely orientated, superior concha is the **superior meatus** which receives the openings of the posterior ethmoidal air sinuses. The **middle meatus** lies below the middle concha and is continuous in front with the atrium. When the middle concha is removed [FIG. 7.8], the meatus is seen to be much deeper in front than behind. Its lateral wall shows two elevations: the **ethmoidal bulla**, a balloon of bone containing middle ethmoidal air sinuses; and a curved ridge, formed by mucous membrane covering the uncinate process of the ethmoid, and separated from the bulla by the **semilunar hiatus**. This hiatus leads forwards into the ethmoidal infundibulum, a shallow trough of mucous membrane lateral to the uncinate process. Several air sinuses open into the middle meatus [FIG. 7.8]: the frontal and anterior ethmoidal sinuses

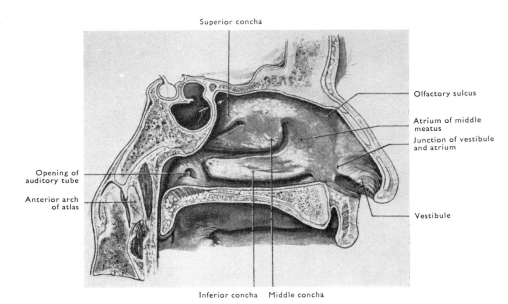

Superior concha

Olfactory sulcus

Atrium of middle meatus

Junction of vestibule and atrium

Opening of auditory tube

Anterior arch of atlas

Vestibule

Inferior concha   Middle concha

FIG. 7.7   Lateral wall of left nasal cavity.

The arrow passes from the sphenoidal sinus to the spheno-ethmoidal recess.

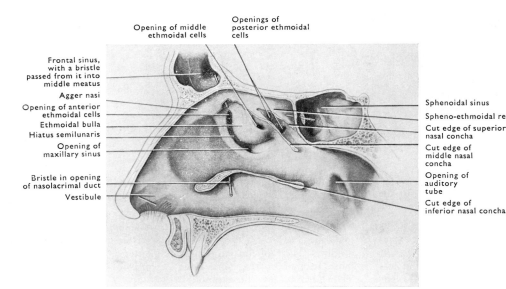

Opening of middle ethmoidal cells

Openings of posterior ethmoidal cells

Frontal sinus, with a bristle passed from it into middle meatus

Agger nasi

Opening of anterior ethmoidal cells

Ethmoidal bulla

Hiatus semilunaris

Opening of maxillary sinus

Bristle in opening of nasolacrimal duct

Vestibule

Sphenoidal sinus

Spheno-ethmoidal recess

Cut edge of superior nasal concha

Cut edge of middle nasal concha

Opening of auditory tube

Cut edge of inferior nasal concha

FIG. 7.8.   Lateral wall of right nasal cavity—the nasal conchae have been removed.

For the bones of the lateral wall, see FIGURE 3.57.

into, or close to, the upper end of the infundibulum; the middle ethmoidal sinuses on, or above, the surface of the bulla; and the maxillary sinus by a principal ostium close to or into the lower end of the semilunar hiatus or of the infundibulum, where it is hidden by the uncinate process.

The inferior meatus lies below the inferior concha and receives the nasolacrimal duct anteriorly.

The mucous membrane of the nasal cavities

The vestibule is lined with skin, bearing coarse hairs and sebaceous and sweat glands; the rest of the nasal cavity is lined by mucous membrane. This has the usual two layers—epithelium and tunica propria—and is firmly attached to the underlying bony or cartilaginous skeleton, forming respectively a mucoperiosteum or a mucoperichondrium. It is continuous with the mucous membrane of the paranasal sinuses, of the nasopharynx, and of the nasolacrimal duct. The mucous membrane of the olfactory region (roof, upper septum, superior concha), is described on page 867 [FIGS. 12.65 and 12.66]. That of the respiratory region is richly glandular and highly vascular, and is thickest over the septum and the conchae. The epithelium is pseudostratified columnar ciliated, and contains abundant goblet cells. Since this type of epithelium lines much of the conducting portion of the respiratory tract, it is often called respiratory epithelium. It is described more fully later [p. 520]. From the surface epithelial layer, numerous branched, tubulo-alveolar glands, with mixed mucous and serous secretory cells, extend deeply into the tunica propria of collagenous connective tissue which blends with the underlying periosteum or perichondrium. The surface of the epithelium is covered with a film of mucus; this floats on a less viscid serous film in which the cilia beat. Air is cleaned as it passes through the nose: particulate material sticks to the mucous film, while gaseous pollutants, such as sulphur dioxide, dissolve in it. The direction of ciliary beat is mainly backwards and the mucous film is moved at 5–10 μm/s towards the pharynx, where it is usually disposed of by swallowing. The mucous film, and the cilia which move it, together provide an important mucociliary defence mechanism.

The vascular arrangements in the nasal mucosa are important in protecting the lungs by warming and moistening inspired air. The sphenopalatine artery is the major supply: it arises from the maxillary artery and enters the region through the sphenopalatine foramen in its lateral wall. Its branches run forwards deep in the mucosa covering the septum (posterior septal nasal arteries), and the lateral wall (posterior lateral nasal arteries). The nasal walls are also supplied, above and anteriorly by the ethmoidal arteries; behind, by small branches of the descending palatine artery and, in the region of the nostrils, by branches of the facial and superior labial arteries. Branches of these major nasal arteries anastomose in the lower anterior part of the septum, a site commonly associated with nose bleed.

From the deep arterial plexus, arterial branches run towards the surface, to supply the glands and end in a subepithelial capillary plexus. Blood returns into rich venous plexuses lying superficial to the arteries, from which they also receive blood directly, via arteriovenous anastomoses. These are particularly abundant in the mucosa of the inferior and middle conchae which lie in the main respiratory air stream, and here the superficial veins are large and thin-walled (Dawes and Prichard 1953). Increased blood flow through these veins (which collectively constitute the 'swell body') leads to swelling of the nasal mucosa which may temporarily occlude the airway. There is some evidence of alternating phasic occlusion of the two nasal cavities. The periods of rest in each cavity are thought to guard against drying of the mucous membrane (Swift and Proctor 1977).

Venous blood from the nasal mucosa drains forwards into the facial vein, backwards into the sphenopalatine vein and thence into the pterygoid plexus, and upwards into the ethmoidal veins. The ethmoidal veins communicate with the ophthalmic and dural veins, and, directly or indirectly, with cerebral veins.

The anatomical design of the nasal cavities facilitates conditioning of the inspired air by providing for its maximal contact with the nasal mucosa. The flow of inspired air is laminar until it reaches the end of the vestibule, where there is an abrupt increase in cross-sectional area. Reduced air velocity and viscous retardation of air produce turbulent flow, principally through the lower part of the cavity, at the level of the middle meatus. Most of the airstream is within 1 mm of the mucosal surface, and it is this close proximity which allows effective warming, moistening, and cleansing of the air. Regardless of ambient temperature and humidity, inspired air is already at about body temperature, with a relative humidity of 75–80 per cent, by the time it reaches the nasal part of the pharynx.

Lymph vessels form a rich plexus in the mucous membrane. From the anterior part of the nasal cavity they run forwards and join with those of the skin of the nose to drain into the submandibular nodes. The rest of the nasal cavity, and the paranasal sinuses, drain into retropharyngeal and upper deep cervical lymph nodes.

Vomeronasal Organ [FIG. 7.12]

Within the mucous membrane on either side of the septum is a blind pouch, the vomeronasal organ. Each opens on to the surface through a minute orifice just above the incisive canals. It is vestigial in Man, but is well developed in many animals and lined with olfactory epithelium innervated by the olfactory nerve.

Development of the nose

SURFACE FEATURES

The development of the nose and nasal cavities is closely linked with that of the face, and changes in surface features are shown in FIGURE 7.9. Each nasal cavity is preceded in the fourth week by an olfactory placode, an ovoid area of thickened ectoderm on the ventrolateral aspect of the head. In the fifth week each placode sinks into the underlying loose mesenchyme to form a nasal pit, which opens on to the surface at the primitive nostril. The nostrils are at first widely separated and, around the margin of each, the mesenchyme grows more rapidly than elsewhere, forming surface elevations called lateral and medial nasal processes [FIG. 7.9A]. It must be emphasized that these 'processes', and the maxillary process referred to later, are simply elevations produced by localized centres of mesenchymal proliferation, and that they are covered by a continuous sheet of ectodermal epithelium. The lateral nasal process flanks the dorsolateral margin of the nostril and eventually forms the ala of the nose. The medial nasal processes are at first less conspicuous and are separated from one another by a broad area which is continuous in front with the future forehead and behind with the roof of the primitive mouth. The oral end of each medial nasal process is sometimes called the premaxillary process [FIG. 7.9A and B]. As the medial nasal processes grow, they become confluent and form a common mass between the two nostrils; from this will arise the anterior part of the nasal septum. The two premaxillary growth centres together provide the mesenchymal basis for the central part of the upper lip and for the premaxillary portion of the palate, called, in the embryo, the primary palate.

The maxillary processes also contribute to facial development. Each is a surface ridge formed by proliferation of the underlying

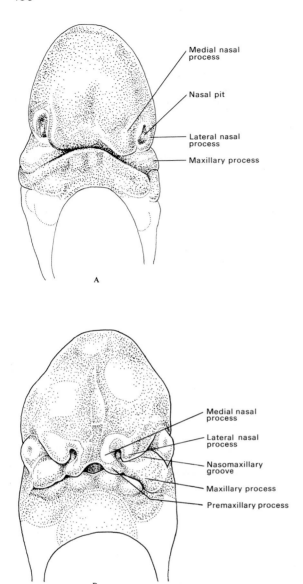

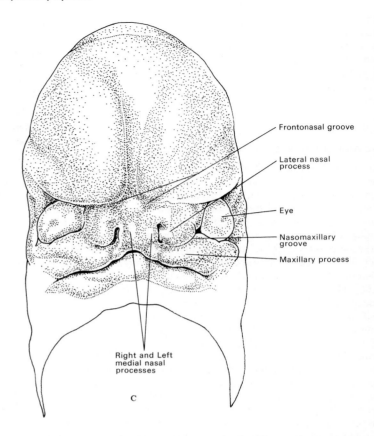

FIG. 7.9.   Three stages in facial development. In A, the olfactory pits are widely spaced, and each is flanked by lateral and medial nasal processes. (Stage 15; approx. 7–9 mm; approx. 33 days.) In B, the anterior nares lie closer together and the two medial nasal processes form a common midline elevation. On each side, the maxillary process meets the lateral nasal process at the shallow nasomaxillary groove, and the premaxillary part of the medial nasal process, a shallow groove intervening here also. (Stage 16; approx. 8–11 mm; approx. 37 days). In C, the lateral nasal process forms the ala of the nose, the medial nasal processes together form the anterior free edge of the nasal septum and the central part of the upper lip. The lateral parts of the upper lip are formed by the maxillary processes. (Stage 18; approx. 13–17 mm; approx. 44 days.) All × 12.5. (Redrawn after Streeter 1948.)

branchial mesenchyme. It is continuous behind with the dorsal end of the mandibular arch and extends forwards and ventrally below the eye. It meets the lateral nasal process along its whole length; between the two is a shallow furrow, the **nasomaxillary groove**, running between the inner angle of the eye and the oral end of the nostril. A lamina of epithelium sprouts from the deep surface of the groove, separates from it, and eventually forms the lining of the nasolacrimal duct.

At the eighth week the nose is like the snout of a pug dog: the nostrils face forwards, and the tip of the nose is separated from the bulging forehead only by a deep frontonasal groove [FIG. 7.9C]. This furrow is slowly obliterated by proliferation of underlying paraxial mesenchyme, which forms the basis of the dorsum of the nose. As the dorsum grows, the nostrils gradually assume a more human orientation, facing downwards rather than forwards. This process is not finally completed until adolescence.

THE NASAL CAVITIES

When these external changes first begin (fifth intra-uterine week) each nasal pit is a shallow depression which expands and deepens to form a **nasal sac** ending blindly behind. The epithelium of the oral wall of the sac proliferates to form a plate of cells—the **nasal fin**—which maintains continuity between the epithelium of the nasal sac and that of the roof of the primitive mouth [FIG. 7.10A and B]. The anterior end of the nasal fin meets the surface at a shallow groove which marks the boundary between the premaxillary and maxillary growth centres; it normally disappears and the two mesenchymes then become continuous and form the mesenchymal basis of the upper lip. The posterior part of the fin continues to grow backwards; cavities appear within it and coalesce, providing for the backwards expansion of the nasal sac, which is then separated from the primitive mouth only by a thin epithelial membrane, the **bucconasal membrane** [FIG. 7.10C]. This soon breaks down, to give a **primitive choana** which links the nasal and buccal cavities. At first the choanae are merely small holes, but they become elongated slits in the roof of the stomatodaeum as the nasal cavities extend backwards. Between the nasal cavities is the **nasal septum**, whose lower free edge is seen in the roof of the stomatodaeum, between the primitive choanae [FIG. 7.11].

From the medial surface of each maxillary process arises a **palatal process**, a thick sheet of mesenchyme covered by epithelium,

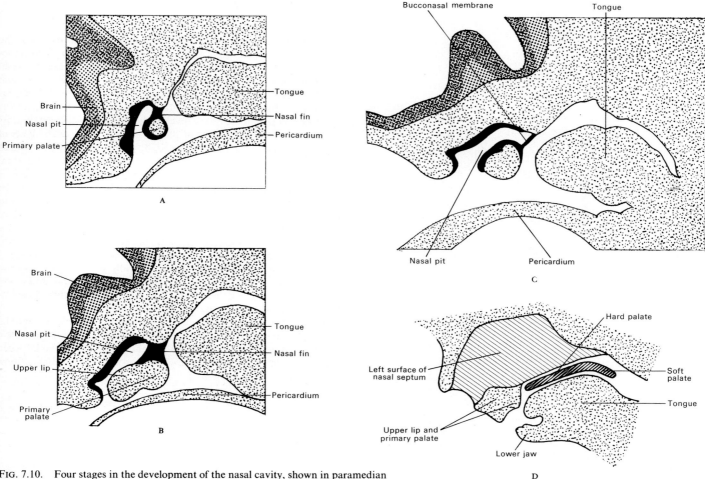

FIG. 7.10. Four stages in the development of the nasal cavity, shown in paramedian section. The epithelial lining of the pit is shown in A, B, and C in solid black. In A, premaxillary and maxillary mesenchymes have fused to form the primitive palate, posterosuperior to which is the nasal fin. In B, the nasal fin extends further posteriorly through proliferation of its cells. In C, the nasal pit enlarges by cavitation of the nasal fin, and is separated from the stomatodaeum only by the bucconasal membrane. D is a diagrammatic paramedian section through the left nasal cavity. The lower free edge of the nasal septum lies horizontally in the roof of the mouth. The anterior part of the palate will fuse with the free edge of the septum, its posterior part is dependent and forms the soft palate. (A, B, and C, after Streeter 1948; D after Frazer 1940.)

hanging downwards at first, with its free edge at the side of the developing tongue. In the seventh week of gestation, the palatal processes swing (or hinge) upwards into a horizontal position, and fuse with one another along their whole length, in the ninth week, to form the basis of the **secondary palate**. The larger, anterior part of this, the future hard palate, fuses with the lower, free edge of the nasal septum, which lies horizontally in the roof of the mouth [FIGS. 7.10D and 7.12]. The smaller, posterior part of the secondary palate, the future soft palate, does not fuse with the nasal septum but hangs freely downwards below it. The choanae lie in a horizontal plane on either side of the posterior part of the lower free edge of the septum. The more oblique plane of the definitive choanae is acquired as the septum increases in height and the face is carried forwards on the growing skull base—a process still continuing after birth.

The initial skeletal framework of the nose and nasal cavities is provided by the **nasal capsule**, which develops by chondrification within the mesenchyme surrounding the cavities. It gives rise, by endochondral ossification, to the ethmoid bone and to the inferior conchae, and part of it persists as the septal cartilage. Of the other bones which contribute to the nasal skeleton, the maxilla, palatines, nasals, and vomer develop in membrane, while the body of the

sphenoid develops within the cartilaginous base of the skull. The conchae develop as a series of ridges on the lateral wall of the cavity, and acquire a cartilaginous and, later, a bony basis. In the late foetus there are five or six conchae on each side; fusion of upper members of the series reduces this number to the definitive three or four.

The epithelial lining of the nasal cavity is of dual origin. The olfactory epithelium arises from the olfactory placode, which may be regarded as neurectoderm, since it gives rise to the olfactory receptors and to their supporting cells. The other epithelia of the nasal cavity arise from non-neural surface ectoderm, apparently secondarily incorporated within the nasal sac as it expands.

The paranasal sinuses develop as the result of invasion of neighbouring bones by outgrowths of nasal respiratory mucosa. The sites of their definitive openings indicate the positions of original outgrowth, but it should be noted that a sinus may transgress the boundaries of the bone which gives it its name. At birth, the ethmoidal, frontal, and maxillary sinuses are rudimentary diverticula; full development of the frontal sinuses occurs only after puberty, and of the maxillary sinuses after acquisition of the permanent dentition. Invasion of the sphenoid bone by the sphenoidal sinuses does not begin until the third year.

For pharynx see pages 429–35.

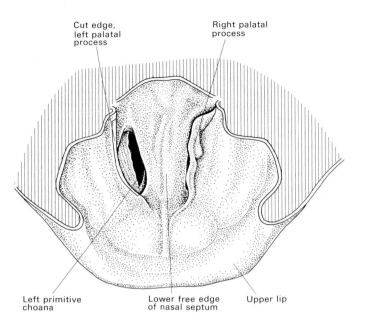

FIG. 7.11   Roof of stomatodaeum in a 28 mm CR embryo, seen from below and behind. The primitive posterior nares are elongated slits separated by the lower free edge of the nasal septum. The right palatal process hangs downwards, the left has been cut away to show the posterior narial opening more clearly. (After Frazer 1940.)

THE LARYNX

The larynx communicates above, through the pharynx, with the mouth and nasal cavities, and below with the trachea. It provides mechanisms for excluding from the lower respiratory tract material being swallowed, for regulating air flow to and from the lungs, and for the production of the voice. Its position is seen best in sections of the head. In sagittal section [FIG. 7.13] its upper end, the tip of the epiglottis, is seen to lie at the level of the third cervical vertebra, its lower end, the lower border of the cricoid cartilage, at the level of the sixth cervical vertebra. It lies slightly higher in children, and in infants its upper border is level with the second cervical vertebra. The larynx is raised relative to the vertebrae when the neck is extended and lowered when it is flexed and on deep inspiration. Its major excursion is during swallowing, when the upward movement of the thyroid cartilage can be readily felt and seen. The position of the larynx relative to adjacent structures is best seen in transverse section [FIG. 5.48]. Posteriorly, is the pharynx and the retropharyngeal space, which allows for vertical movement of the pharynx and larynx on the prevertebral fascia. Laterally is the thyroid gland, covered anterolaterally by the infrahyoid muscles, and bound to the larynx by pretracheal fascia so that they move together. Posterolaterally are the carotid sheaths and their contents.

The larynx has a framework of **cartilages**, linked by synovial joints and ligaments, and moved on each other by various **intrinsic muscles**. It is lined by a mucous membrane. In the plane between the mucous membrane and the cartilages is a discontinuous fibroelastic membrane (see below).

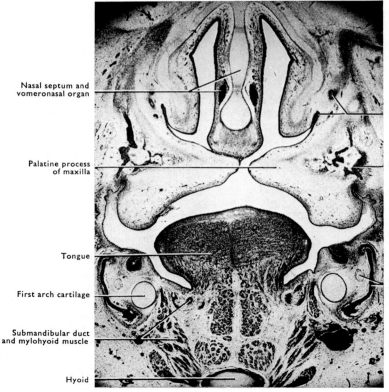

FIG. 7.12.   Section through nose and mouth of 25 mm human embryo to show position of vomeronasal organ and formation of palate. × 28.

Cartilages of the larynx

THYROID CARTILAGE

This is the largest cartilage of the larynx and forms most of its anterior and lateral walls. Its two laminae join one another in the midline anteriorly, at an angle of about 90 degrees in men and about 120 degrees in women, but diverge posteriorly. The superior thyroid notch and the laryngeal prominence (Adam's apple) can be seen and felt in the midline, particularly in men [FIG. 7.15]. From the posterior border of each lamina, a long superior horn projects upwards and a shorter inferior horn curves downwards and medially to articulate with the lateral surface of the cricoid cartilage. On the outer surface of the lamina is an oblique line [FIG. 7.16] to which are attached the sternothyroid and thyrohyoid muscles, the inferior constrictor of the pharynx and the pretracheal fascia. The inner surface is smooth and, in its upper posterior part, is lined by mucous membrane.

CRICOID CARTILAGE

This is smaller than the thyroid cartilage and lies mainly below it, clasped by its two inferior horns. It is signet ring shaped, with the narrower arch in front, where it is readily felt, and the broader quadrate lamina behind. The lower border of the arch is horizontal, the upper inclines steeply to join the horizontal upper border of the lamina. On each side, at the junction of arch and lamina, is an articular surface for the inferior horn of the thyroid cartilage [FIG. 7.18]. On the upper, outer angle of the lamina is a facet for articulation with the base of the arytenoid cartilage [FIGS. 7.17 and 7.18]. The lamina lies between the pharynx and the larynx [FIG. 7.13].

EPIGLOTTIS

A median section [FIG. 7.22] shows the epiglottis lying behind the thyroid cartilage, thyrohyoid membrane, hyoid bone, and the base

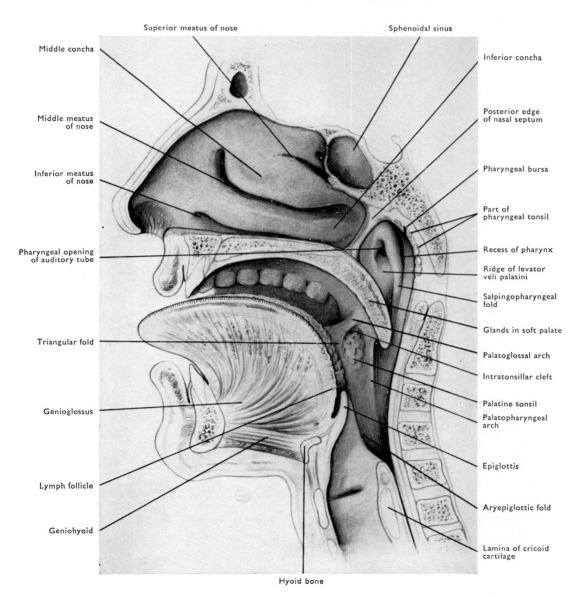

FIG. 7.13. Sagittal section through mouth, tongue, larynx, pharynx, and nasal cavity.
     The section is slightly oblique, and the posterior edge of the nasal septum is preserved. The specimen is viewed slightly from below, hence the apparently low position of the inferior concha.

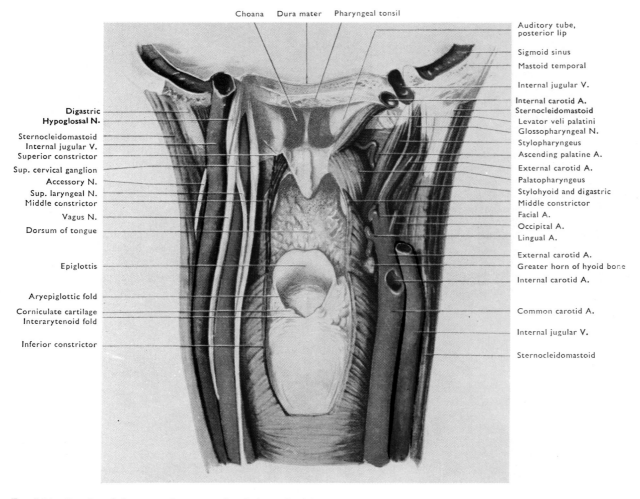

Choana   Dura mater   Pharyngeal tonsil

Auditory tube, posterior lip
Sigmoid sinus
Mastoid temporal
Internal jugular V.
Internal carotid A.
Sternocleidomastoid
Levator veli palatini
Glossopharyngeal N.
Stylopharyngeus
Ascending palatine A.
External carotid A.
Palatopharyngeus
Stylohyoid and digastric
Middle constrictor
Facial A.
Occipital A.
Lingual A.
External carotid A.
Greater horn of hyoid bone
Internal carotid A.
Common carotid A.
Internal jugular V.
Sternocleidomastoid

Digastric
Hypoglossal N.
Sternocleidomastoid
Internal jugular V.
Superior constrictor
Sup. cervical ganglion
Accessory N.
Sup. laryngeal N.
Middle constrictor
Vagus N.
Dorsum of tongue
Epiglottis
Aryepiglottic fold
Corniculate cartilage
Interarytenoid fold
Inferior constrictor

FIG. 7.14.   Interior of pharynx and structures in relation to its side-walls viewed from behind.

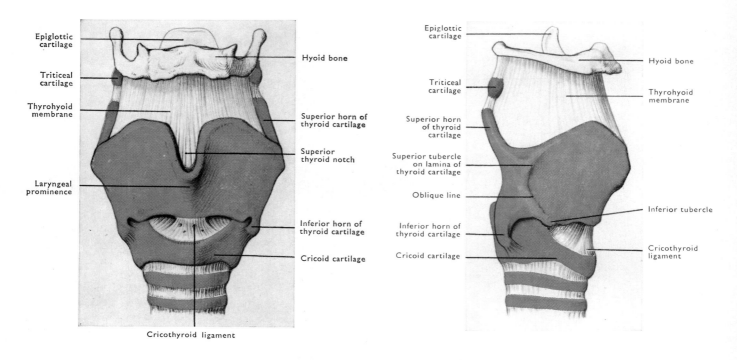

Epiglottic cartilage
Triticeal cartilage
Thyrohyoid membrane
Laryngeal prominence
Cricothyroid ligament

Hyoid bone
Superior horn of thyroid cartilage
Superior thyroid notch
Inferior horn of thyroid cartilage
Cricoid cartilage

Epiglottic cartilage
Triticeal cartilage
Superior horn of thyroid cartilage
Superior tubercle on lamina of thyroid cartilage
Oblique line
Inferior horn of thyroid cartilage
Cricoid cartilage

Hyoid bone
Thyrohyoid membrane
Inferior tubercle
Cricothyroid ligament

FIG. 7.15.   Anterior aspect of cartilages and ligaments of larynx.

FIG. 7.16.   Profile view of cartilages and ligaments of larynx.

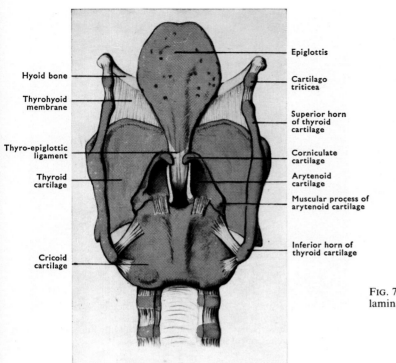

FIG. 7.17.   Posterior aspect of cartilages and ligaments of larynx.

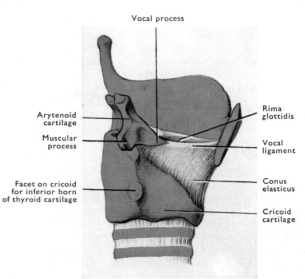

FIG. 7.18.   Dissection to show laryngeal fibro-elastic membrane. The right lamina of the thyroid cartilage has been removed.

of the tongue, and forming the anterior wall of the upper larynx. Its basis is a perforated leaf-shaped lamina of elastic fibrocartilage with a narrow stalk connected to the inner surface of the angle of the thyroid cartilage by the elastic **thyro-epiglottic ligament**. The epiglottic cartilage is also attached to the hyoid bone by an elastic **hyo-epiglottic ligament**. Mucous membrane covers the anterior surface of the upper broad part of the cartilage and is continuous in front with that lining the epiglottic valleculae and the pharyngeal surface of the tongue. The valleculae are bounded by a median and two lateral **glosso-epiglottic folds** of mucous membrane [FIG. 7.19] and may be a trap for the unwary when attempting to pass a tube into the larynx or the oesophagus. Mucous membrane also covers the posterior surface of the epiglottis, and is reflected from its lateral borders downwards and backwards to the arytenoid cartilages, forming the **aryepiglottic folds** [FIGS. 7.19 and 7.20]. These project backwards into the pharynx [FIG. 7.14] and separate the vestibule of the larynx from the piriform recesses of the pharynx, which extend forwards on each side, medial to the mucous membrane lining the internal surfaces of the thyroid cartilage laminae [FIG. 7.19]. In the midline on the posterior surface is a prominent bulge, the **tubercle of the epiglottis** [FIG. 7.20].

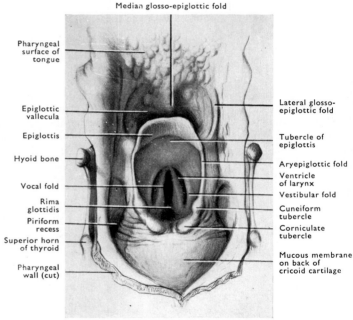

FIG. 7.19.   Laryngeal inlet exposed by removal of posterior wall of pharynx and viewed from above and behind.

ARYTENOID CARTILAGES

Each cartilage is a three sided pyramid, with a base which articulates with the cricoid lamina, and approximately triangular posterior, anterolateral, and medial surfaces. The **muscular process** [FIGS. 7.17 and 7.18] projects backwards and laterally from the base; attached to it are the major muscles which move the arytenoid cartilage. The **vocal process** projects forwards from the base; attached to it is the vocal ligament.

CORNICULATE AND CUNEIFORM CARTILAGES

These are small paired nodules in the lower part of the aryepiglottic folds [FIG. 7.19], which they reinforce.

Joints and ligaments of the larynx

The thyroid cartilage tilts backwards and forwards about a horizontal axis passing transversely through the two cricothyroid joints [FIGS. 7.16 and 7.18]. At the crico-arytenoid joints two types of movement occur: gliding movements of the arytenoid cartilages on the cricoid cartilage (laterally and medially) and rotatory movements, about a vertical axis, which result in their vocal processes swinging away from or towards one another. The thyrohyoid membrane [FIGS. 7.15 and 7.16] is a sheet of fibro-elastic tissue, attached below to the upper border of the thyroid cartilage, above to the internal surface of the upper border of the hyoid bone. It is separated from the posterior surface of the body of the hyoid by a bursa. It is thickened in the midline (median thyrohyoid ligament) and in its posterior margins (lateral thyrohyoid ligaments).

The intrinsic ligaments together constitute a discontinuous fibro-elastic membrane, which lies in the plane between the mucous membrane, internally, and many of the intrinsic muscles, externally. Its upper part is a thin sheet, the quadrangular membrane, which extends into the aryepiglottic fold above, and the vestibular fold below, where it is thickened to form the vestibular ligament. The membrane is attached to the side of the epiglottic cartilage and to the arytenoid cartilage.

The much thicker lower part of the fibro-elastic membrane is the conus elasticus, which is attached, below, to the upper border of the arch of the cricoid cartilage, and slopes superomedially to a free upper border. This lies deep to the lamina of the thyroid cartilage, and extends from the deep surface of the thyroid angle, close to the midline, to the vocal process of the arytenoid. This upper free border, the vocal ligament, consists of thick longitudinal bands of elastic tissue and forms the basis of the corresponding vocal fold [FIG 7.18].

The cavity of the larynx

A median section shows clearly the subdivisions of the laryngeal cavity [FIG. 7.22]. The inlet, from pharynx to larynx, lies almost vertically, and is bounded by the epiglottis above and the arytenoids below and behind, and by the aryepiglottic folds on each side [FIG.

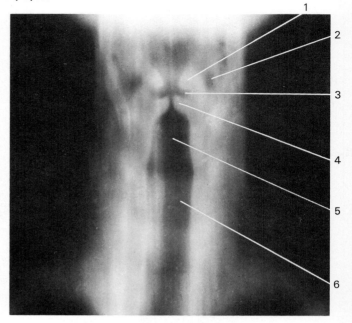

FIG. 7.21.   Anteroposterior tomogram of larynx. Only the structures lying in the plane of the optical section are sharp.
1. Vestibular fold.
2. Piriform recess.
3. Ventricle of larynx.
4. Vocal fold separated from its fellow by the rima glottidis.
5. Infraglottic part of larynx.
6. Trachea.

7.19]. The vestibule extends from the inlet to the vestibular folds (false vocal cords). These are thick, pink folds of mucous membrane, lying one on each side-wall of the larynx. Each contains a vestibular ligament (the lower part of the quadrangular membrane), fibrous tissue, and abundant glands. The vestibular ligament is attached to the thyroid angle in front and to the anterolateral surface of the arytenoid behind. The interval between the vestibular folds is the rima vestibuli.

The middle compartment

This lies between the vestibular folds above and the vocal folds below. Here the mucous membrane bulges outwards to form the ventricle of the larynx [FIG. 7.20], from whose anterior end a narrow diverticulum—the laryngeal saccule—extends upwards between the vestibular fold and the thyroid cartilage. The saccules are large in certain of the great apes, and may extend as air sacs into the tissues of the neck. In man they do not normally extend beyond the upper border of the thyroid cartilage.

THE VOCAL FOLDS

Each vocal fold, or true vocal cord, is a prominent fold of mucous membrane, with a pearly white, sharp edge attached in front to the angle of the thyroid cartilage and behind to the vocal process of the arytenoid cartilage [FIG. 7.22]. Each fold includes, in its edge, the vocal ligament (the thickened free upper border of the conus

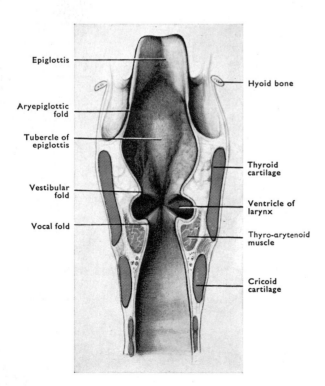

Epiglottis

Hyoid bone

Aryepiglottic fold

Tubercle of epiglottis

Thyroid cartilage

Vestibular fold

Ventricle of larynx

Vocal fold

Thyro-arytenoid muscle

Cricoid cartilage

FIG. 7.20.   Coronal section through larynx to show its compartments.

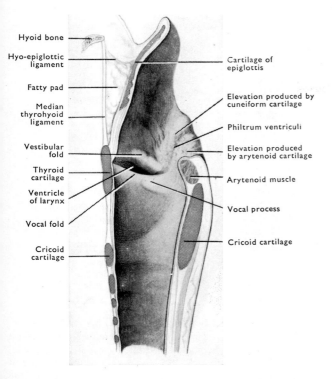

Hyoid bone
Hyo-epiglottic ligament
Fatty pad
Median thyrohyoid ligament
Vestibular fold
Thyroid cartilage
Ventricle of larynx
Vocal fold
Cricoid cartilage

Cartilage of epiglottis
Elevation produced by cuneiform cartilage
Philtrum ventriculi
Elevation produced by arytenoid cartilage
Arytenoid muscle
Vocal process
Cricoid cartilage

FIG. 7.22.   Median section through larynx to show side-walls of its right half.

elasticus) and abundant skeletal muscle elsewhere. The folds are triangular in cross-section and their thin, free edges project upwards and medially, and consist of epithelium firmly adherent to the underlying vocal ligament [FIG. 7.18]. The glottis, at the level of the vocal folds, is the narrowest part of the cavity of the larynx. The rima glottidis is described in two parts: its anterior three-fifths lies between the vocal folds and is therefore called intermembranous; its posterior two-fifths is flanked on each side by the vocal processes and bases of the arytenoid cartilages and is therefore intercartilaginous [FIG. 7.24].

Infraglottic compartment

This extends downwards from the vocal folds, gradually increasing in transverse diameter, to the junction of the larynx with the trachea at the lower border of the cricoid cartilage.

Mucous membrane of the larynx

Most of the epithelium is columnar ciliated like that of the trachea, but this is replaced by non keratinized stratified squamous epithelium in areas frequently in contact with other surfaces, e.g. the upper half of the posterior surface of the epiglottis, and the margins of the aryepiglottic and vocal folds. Small branched tubulo-alveolar seromucous glands arise from the epithelium, particularly of the aryepiglottic and vestibular folds, of the laryngeal ventricles and of the epiglottis. Glands are absent from the vocal folds. The tunica propria of most of the laryngeal mucosa is loose connective tissue, and rests on a loose submucosa. This can readily become swollen with tissue fluid (oedema of the larynx) and may obstruct the airway. The oedema does not involve the vocal folds, where the epithelium is firmly bound to the vocal ligament.

The mucous membrane of the larynx has a rich sensory innervation from laryngeal branches of the vagus nerves. These nerves provide the afferent limb for reflexes involved in coughing and swallowing, important to the protection of the airway.

RADIOLOGY

The upper air passages are often remarkably well demonstrated in a lateral soft tissue radiograph of the neck. This also shows many structures belonging to other systems. Appearances can vary from person to person and in the same individual at different times, according to the amount of air present [FIG. 7.23].

Structures in the respiratory system commonly seen are:

The epiglottis

Its tip is often surrounded by air. In front are the epiglottic valleculae, behind is the vestibule of the larynx. Most of the posterior surface of the epiglottis is seen as an oblique line against air in the vestibule. This line leads the eye to a horizontal, cigar-shaped translucency, the ventricle of the larynx. Below this there is a fairly wide translucent area mainly caused by the trachea, though its upper part is the infraglottic portion of the larynx, seen better in an anteroposterior tomogram [FIG. 7.21]. The cartilaginous rings of

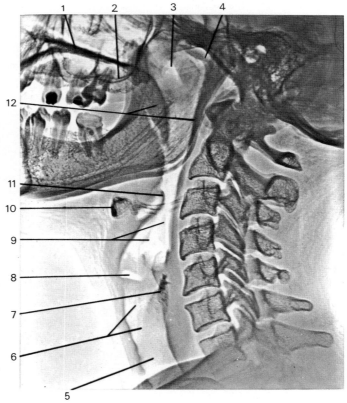

FIG. 7.23.   Lateral soft tissue radiograph (xerogram) of neck.
1. Hard palate.
2. Floor of maxillary sinus.
3. Opening of auditory tube.
4. Roof of nasal part of pharynx.
5. Trachea.
6. Infraglottic larynx.
7. Cricoid and arytenoid cartilages, ossified.
8. Ventricle of larynx.
9. Aryepiglottic fold and vestibule of larynx.
10. Hyoid bone.
11. Tip of epiglottis.
12. Soft palate, posterior margin of ramus of mandible, and posterior pharyngeal wall.

the trachea often cause its lumen to appear uneven anteriorly. Posterior to the infraglottic part of the larynx, in FIGURE 7.23, there is a dense vertical area—the partly ossified **cricoid cartilage** surmounted by ossification in the **arytenoid cartilage**. The cricoid and thyroid cartilages may be extensively ossified, the thyroid often obscuring the ventricle. It is sometimes possible to identify the arytenoid cartilages projecting upwards from the lamina of the cricoid cartilage, the mucous membrane covering them being continuous above with the aryepiglottic folds.

The body of the **hyoid bone** lies just below the mandible, and the greater horns extend backwards to the level of the anterior margin of the vertebral column, crossing the upper part of the epiglottis. The dorsum of the tongue is seen forming the anterior wall of the oral part of the pharynx. Inferiorly it can be traced to the epiglottic valleculae, and superiorly, in this radiograph, it is separated from the soft palate by a narrow air space. At the posterior border of the hard palate it is possible to see the superimposed **pterygoid laminae** of the sphenoid bone. Their anterior edges are well defined and give an approximate indication of the position of the choanae, posterior to the pterygomaxillary fissure.

Behind this is the nasal part of the pharynx. In FIGURE 7.23 a little air has been trapped in the orifice of the auditory tube. In many cases, though not in the one illustrated here, the **pharyngeal tonsil** appears as a density at the junction of the roof and posterior wall of the nasal part of the pharynx [FIG. 6.28].

The prevertebral soft tissue density is narrow superiorly, but abruptly widens about the level of the cricoid cartilage where the anterior pharyngeal and posterior laryngeal walls are applied to it.

Tomography is needed for examination of the larynx from the front. A series of sections at adjacent levels are taken during the phonation of 'ee'. By this device the vocal cords are approximated. A typical tomogram [FIG. 7.21] shows the **vocal folds, ventricles of larynx, vestibular folds, piriform recesses**, infraglottic portion of larynx, and the trachea.

Examination of the larynx by the introduction of contrast medium through a nasal catheter is sometimes useful. The aim is to coat the various structures with a thin layer of medium. When successful this gives a very good demonstration of the anatomy [FIG. 6.31].

Muscles of the larynx

These may be classified as *extrinsic*, i.e. those which move the larynx as a whole and *intrinsic*, i.e. those producing movement of laryngeal structures relative to one another. The extrinsic muscles are described on pages 283–6, the intrinsic on pages 303–5. These descriptions should be read before the following account of the actions of these muscle groups in the protection of the airway from entry of material being swallowed, in regulating the passage of air to and from the lungs, and in the production of the voice [FIGS. 7.25–7.28].

PROTECTION OF THE AIRWAY

The laryngeal inlet is closed during swallowing by several mechanisms. The larynx is elevated towards the hyoid bone, epiglottis, and tongue by the thyrohyoid muscle. This movement compresses the laryngeal aditus supero-inferiorly. The lateral portions of the thyro–arytenoid muscles draw the arytenoid cartilages towards the epiglottis. The transverse arytenoid muscle approximates the arytenoid cartilages. The inlet is narrowed in all directions by contraction of a weak **inlet sphincter** made up of the oblique arytenoid and aryepiglottic muscles.

Closure of the inlet may be accompanied by bending of the epiglottis downwards and backwards over the inlet. It should be

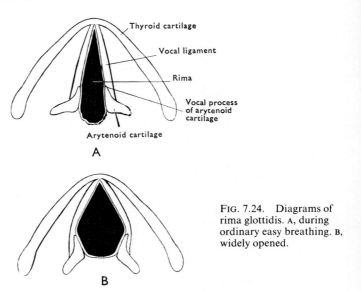

FIG. 7.24. Diagrams of rima glottidis. A, during ordinary easy breathing. B, widely opened.

noted however that this bending does not invariably occur; particularly when liquids are swallowed, the epiglottis may remain upright 'like a rock in a waterfall', with the swallowed material cascading past it in two streams. Moreover, even when bending of the epiglottis does occur, it is largely irrelevant to closure of the inlet: surgical amputation of the epiglottis above the tubercle does not impair the swallowing mechanism.

The cough reflex is the second line of defence. It involves a deep inspiration, followed by forcible expiration against the closed glottis: the vocal cords are tightly approximated (adducted), principally by the lateral crico-arytenoid muscle, which rotates the arytenoid cartilage medially and approximates the vocal processes. Closure of the rima glottidis is reinforced, in violent coughing, by closure of the rima vestibuli (approximation of the vestibular folds), possibly by contraction of the thyro-epiglottic and thyro-arytenoid muscles. The glottis is now opened and the explosive release of air into the upper respiratory tract is a cough, which may clear the offending object from the larynx. The reflex has its dangers: the inspiration which precedes the cough may carry the material through the glottis, or cause a larger object to become lodged in the glottis and obstruct the airway.

REGULATION OF ENTRY OF AIR

In quiet respiration, the vocal folds are partially separated [FIG. 7.24A]; during forced inspiration, they are further separated (*abducted*) [FIG. 7.29B] by contraction of the posterior crico-arytenoid muscles, which rotate the arytenoid cartilages laterally, swinging their vocal processes laterally. The posterior crico-arytenoids are the only abductors of the vocal folds: they are therefore vitally important since unopposed adduction of the vocal folds will obstruct the airway, and may produce asphyxia.

THE PRODUCTION OF VOICE

Voice production is a complex function, and involves not only the larynx but also the other components of the **vocal tract**: the mouth, with tongue, teeth and palate; the nasal cavities; the pharynx; and the lungs and respiratory muscles. **Voiced sound** is produced by expiration through a closed glottis; the edges of the adducted vocal folds are forced apart by the subglottic pressure, and drawn together again by the elastic recoil of stretched elastic fibres of the vocal ligaments. The vocal folds therefore vibrate, and allow puffs of air

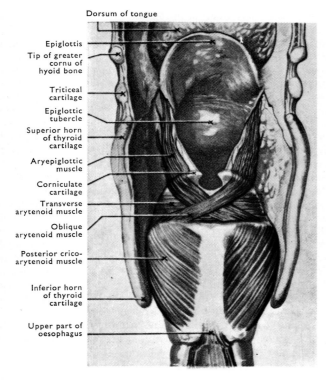

Dorsum of tongue

Epiglottis

Tip of greater
cornu of
hyoid bone

Triticeal
cartilage

Epiglottic
tubercle

Superior horn
of thyroid
cartilage

Aryepiglottic
muscle

Corniculate
cartilage

Transverse
arytenoid muscle

Oblique
arytenoid muscle

Posterior crico-
arytenoid muscle

Inferior horn
of thyroid
cartilage

Upper part of
oesophagus

FIG. 7.25.   Muscles on back of larynx.

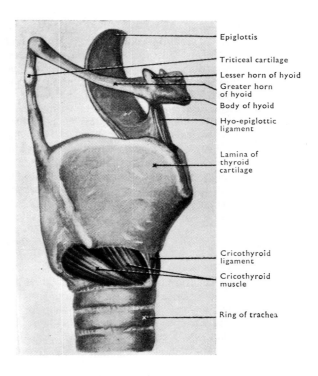

Epiglottis

Triticeal cartilage

Lesser horn of hyoid

Greater horn
of hyoid

Body of hyoid

Hyo-epiglottic
ligament

Lamina of
thyroid
cartilage

Cricothyroid
ligament

Cricothyroid
muscle

Ring of trachea

FIG. 7.27.   Right cricothyroid muscle.

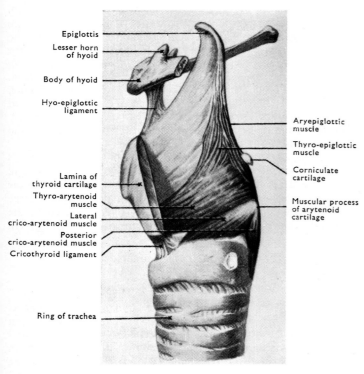

Epiglottis

Lesser horn
of hyoid

Body of hyoid

Hyo-epiglottic
ligament

Aryepiglottic
muscle

Thyro-epiglottic
muscle

Corniculate
cartilage

Muscular process
of arytenoid
cartilage

Lamina of
thyroid cartilage

Thyro-arytenoid
muscle

Lateral
crico-arytenoid muscle

Posterior
crico-arytenoid muscle

Cricothyroid ligament

Ring of trachea

FIG. 7.26.   Muscles on side-wall of larynx.

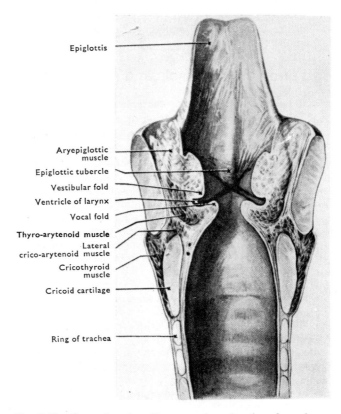

Epiglottis

Aryepiglottic
muscle

Epiglottic tubercle

Vestibular fold

Ventricle of larynx

Vocal fold

Thyro-arytenoid muscle

Lateral
crico-arytenoid muscle

Cricothyroid
muscle

Cricoid cartilage

Ring of trachea

FIG. 7.28.   Coronal section of larynx to show situation of muscles.

to pass intermittently through the glottis. The pitch of the sound thus produced is determined by the frequency of vibration of the vocal folds, and this in turn is determined by the mass, length, and tension of the vocal ligaments. The vocal folds are shorter and thinner in women and boys than in men, with obvious effect on the quality of voice. Within the limits set by the dimensions of the larynx, the pitch and volume of voiced sound may be varied by many factors: volume increases with increased approximation of the vocal folds and increased expiratory pressure; pitch may be varied by altering the tension, length, or mass of the vocal ligaments, principally through the nicely balanced activity of the cricothyroid and thyro-arytenoid muscles. Part of the thyro-arytenoid muscle is of special importance, and is specially designated as **vocalis muscle**. It lies close to the vocal ligament and many of its fibres are attached to the conus elasticus, just below the ligament. The vocalis probably produces local variations of tension within the vocal ligament; it may, for example, tense the posterior parts of the folds so holding them together and permitting air to be expelled only between their anterior parts.

Voiced sounds, produced essentially by vibration of the vocal folds, are modified by supraglottic parts of the vocal tract, which act as resonators. Some of these resonators are of variable dimensions: upward and downward movement of the larynx shortens or lengthens the resonating chamber provided by the laryngeal vestibule and the pharynx, and permits 'tuning' of the resonator. The nasal cavity and paranasal sinuses, on the other hand, resonate only to particular frequencies since their dimensions are fixed. The mouth provides the most subtly variable resonator of all; it is also the principal source of speech sounds, through movements of jaws, tongue, lips, and soft palate. The mouth is principally responsible for production of some sounds in normal speech. In whispering, the vocal folds are abducted but vibrate nevertheless, and the mouth adds its contribution.

Blood and lymph vessels and nerves of the larynx

Two pairs of arteries supply the larynx—the superior and inferior laryngeal arteries. The **superior laryngeal artery**, a branch of the superior thyroid, enters the larynx by penetrating the thyrohyoid membrane with the internal laryngeal branch of the superior laryngeal nerve, under cover of thyrohyoid muscle. The artery descends beneath the mucosa of the lateral wall and the floor of the piriform recess to supply the mucous membrane and muscles of the larynx. The **inferior laryngeal artery**, a branch of the inferior thyroid, enters the larynx beneath the lower border of the inferior constrictor muscle with the inferior laryngeal nerve. This artery supplies the mucous membrane and muscles of the lower part of the larynx, and anastomoses with the superior laryngeal artery. The superior and inferior **laryngeal veins** accompany the arteries to join the superior and inferior thyroid veins respectively.

The **lymph vessels** are numerous except over the vocal folds, and are divided into a superior and an inferior group. The vessels of the superior group accompany the superior laryngeal artery and join the upper deep cervical nodes. The efferents from the inferior group accompany the inferior laryngeal artery and join the lower deep cervical nodes, some of them going to the supraclavicular nodes.

The **nerves** of the larynx are the internal and external branches of the **superior laryngeal nerve** and the terminal branches of the **recurrent laryngeal nerve**. The external branch of the superior laryngeal nerve accompanies the superior thyroid artery and supplies the cricothyroid muscle. The internal branch of the superior laryngeal nerve pierces the thyrohyoid membrane, breaks into a number of branches in the lateral wall of the piriform recess, and is the sensory nerve supply of the mucous membrane of the epiglottis,

of the most posterior part of the tongue, of the larynx down to the vocal folds, and of the upper part of the laryngeal pharynx. The terminal part of the recurrent laryngeal nerve—the inferior laryngeal nerve—enters the larynx just behind the corresponding cricothyroid articulation close to the inferior thyroid artery. It divides into an anterior and a posterior branch, communicates with the internal laryngeal nerve, and supplies the remaining laryngeal muscles and the mucous membrane of the larynx below the vocal folds, and of the lower part of the laryngeal pharynx. Injury to the laryngeal nerves, especially the recurrent, may occur during operations on the thyroid gland, or to the left recurrent laryngeal nerve in aortic aneurysm. Injury to the superior laryngeal nerves is seldom bilateral but even unilateral paralysis of cricothyroid muscle may markedly alter the quality of the voice, converting a soprano into a husky contralto. Injury to both recurrent laryngeal nerves results in paralysis of the vocal folds with partial loss of voice, and some difficulty in swallowing (involvement of inferior pharyngeal constrictors and upper oesophagus). Paralysis of one recurrent laryngeal nerve is, however, partly compensated for by movements of the normal vocal fold, which may be abducted to widen the airway, or adducted across the midline for phonation. Nevertheless, the voice is hoarse and the cough abnormal in quality. Temporary paralysis may occur due to post-operative oedema affecting the nerves. Partial damage to the recurrent laryngeal nerve generally affects the abductor muscles, i.e. the posterior crico-arytenoids, more severely than the other muscles.

Examination of the larynx

The larynx can be examined in the living subject through the mouth using a laryngoscopic mirror. The vestibular and vocal folds are easily recognized and between them on each side the entrance to the laryngeal ventricle appears as a dark line on the side-wall of the larynx. The cavity seen in this way is much foreshortened. FIGURE 7.29A shows the shape of the rima glottidis during quiet respiration, and FIGURE 7.29B represents a condition of extreme abduction during forced inspiration, when the arytenoid cartilages are moved and rotated laterally so that the vocal processes are widely separated. During certain phases of phonation the rima glottidis is reduced to a linear chink.

There is considerable variation in the size of the larynx which is quite independent of stature. The male larynx is absolutely and relatively larger than the female larynx, more particularly in the anteroposterior diameter. In the new-born child the larynx is large compared with the calibre of the trachea. At adolescence the androgen-controlled general increase in growth rate (the 'growth spurt') also involves the larynx.

THE TRACHEA

The trachea is the continuation of the larynx. It is 4 cm long at birth; 4.5 cm between the first and second years; 5.5 cm between the sixth and eighth years; 7 cm between the fourteenth and sixteenth years; and from 9 to 15 cm in the adult. The maximum anteroposterior and transverse internal diameters of the trachea are: 5×6 mm at birth; 9×9 mm between the first and second years; 10×11 mm between the sixth and eighth years; 10×13 mm between the fourteenth and sixteenth years; and 16×14 mm in the adult. Hence, in general, the coronal diameter is greater than the sagittal until adult life. The external diameter is about 2 cm in the adult male and rather less in the female.

Until the third or fourth year the cross-sectional area of the trachea is equal to the sum of that of the two principal bronchi; but from then until the tenth year these bronchi together have slightly

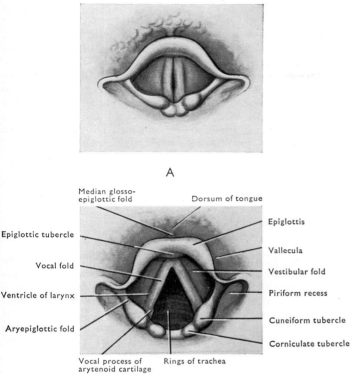

FIG. 7.29.   Cavity of larynx seen with laryngoscope. A, rima glottidis closed. B, rima glottidis widely opened.

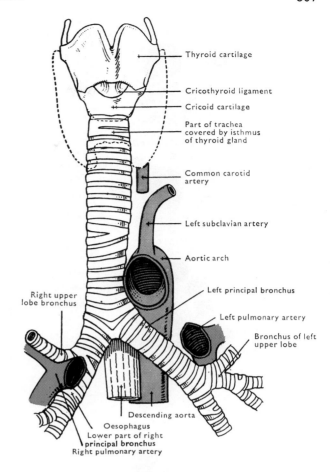

FIG. 7.30.   Trachea and bronchi.
The dotted area gives the outline of the thyroid gland.

the larger area. After puberty, the transverse sectional area of the two principal bronchi continues its relative increase and is 40 per cent greater than that of the trachea in the adult.

The trachea begins at the lower border of the cricoid cartilage at the level of the sixth cervical vertebra and ends at the upper border of the fifth thoracic vertebra where it bifurcates into a right and a left principal bronchus. Anteriorly this level corresponds to the sternal angle and the second costal cartilage. In the living subject during deep inspiration in the erect position, the level of the bifurcation descends to the sixth thoracic vertebra, in part due to a lengthening of the trachea itself by about 2 cm. Fifteen to twenty U-shaped rings of cartilage, incomplete behind, encircle the anterior and lateral parts of its circumference [FIGS. 7.30 and 7.54].

Throughout its length the trachea is in the midline; but it inclines slightly to the right at the bifurcation. Posteriorly it is separated from the vertebral column by the oesophagus. *In the neck* it is covered in front by the pretracheal part of the cervical deep fascia and by the sternohyoid and sternothyroid muscles. Lying on its anterior surface are the following: the isthmus of the thyroid gland at the level of the second, third, and fourth tracheal cartilages; below this the inferior thyroid veins, the tracheal lymph nodes, the thyroidea ima artery (if present), and occasionally accessory thymic glandular tissue extending upwards as far as the lower border of the thyroid cartilage.

Lateral to the trachea are the lobes of the thyroid gland and the great vessels of the neck in the carotid sheath, both under cover of the anterior part of the sternocleidomastoid muscle. In the groove between the trachea and the oesophagus, on each side, is a recurrent laryngeal nerve (the left as far down as the arch of the aorta, the right to a level slightly above the subclavian artery) intermingled with branches of the inferior thyroid artery immediately below the larynx. The trachea may be opened surgically above the level of the

isthmus of the thyroid gland—a high tracheostomy—or at the level of the sixth or seventh tracheal rings—a low tracheostomy—when, however, the inferior thyroid veins are encountered.

In the thorax, the jugular venous arch lies anterior to the trachea at the level of the jugular notch of the sternum, with the infrahyoid muscles and inferior thyroid veins intervening. The brachiocephalic trunk, the left common carotid artery, and, below and to the left, the arch of the aorta, all lie anterior to the trachea, with the left brachiocephalic vein crossing obliquely from left to right in front of the arteries. The remains of the thymus gland are directly behind the sternum, and the deep cardiac plexus of nerves lies on the trachea at its bifurcation behind the aortic arch [FIG. 7.30].

On the right side the trachea is in contact with the mediastinal pleura; between them are the right vagus which passes downwards and backwards to reach the root of the lung and the azygos vein which arches forwards to join the superior vena cava. On the left of the trachea are the arch of the aorta with the left recurrent laryngeal nerve looping round it, and posterolaterally the left subclavian artery. Contact with the aortic arch makes the trachea particularly liable to compression from dilatations of the aorta (aneurysms) or from abnormal vascular rings which may partially encircle the oesophagus and trachea [p. 899].

THE PRINCIPAL BRONCHI

The right and left principal bronchi differ in length, diameter, and the angle they make with the trachea [FIGS. 7.30, 7.51, and 7.52].

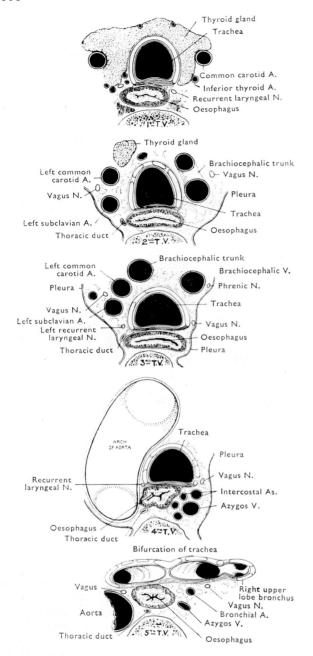

FIG. 7.31.   Transverse sections through the trachea and neighbouring structures at levels of upper five thoracic vertebrae.

The **left principal bronchus** is about 5 cm long from its origin to its first branch (the bronchus of the superior lobe), is narrower than the right bronchus, and leaves the trachea at an angle of 135 degrees. It lies at first immediately in front of the oesophagus at the level of the fifth thoracic vertebra [FIG. 7.31] and passes inferolaterally beneath the arch of the aorta and then in front of the descending aorta. The left pulmonary artery crosses in front of the bronchus to lie above it at the hilus of the lung, with no branch of the bronchus superior to the artery. Below the level of the artery, the upper left pulmonary vein lies in front of the bronchus which at the hilus, continues downwards immediately postero-inferior to the oblique fissure as the left inferior lobar bronchus. The left bronchus is situated behind and a little above the transverse sinus of the pericardium and the left atrium of the heart. The **right principal bronchus** is about 2.5 cm in

length from the tracheal bifurcation to its first branch which it gives off above the right pulmonary artery. It is wider than the left bronchus, forms an angle of 155 degrees with trachea, and passes behind the ascending aorta and then behind the superior vena cava where the terminal portion of the azygos vein arches above the bronchus. As it reaches the hilus of the lung, both the pulmonary artery and vein cross it anteriorly after it has given off its branch to the upper lobe. The stem branch then enters the lung medial to the oblique fissure. In spite of its shorter length, the right principal bronchus is easier to approach surgically than the left which is close to the aorta. The structure of the tracheobronchial tree is described later, with that of the lungs.

The thoracic cavity

The thoracic cavity is enclosed by the ribs and costal cartilages, the vertebral column, and the sternum. The superior aperture slopes downwards and forwards, and is bounded by the first thoracic vertebra, the manubrium of the sternum, the first ribs, and their costal cartilages. The inferior aperture slopes downwards and backwards, is closed by the diaphragm, and is bounded by the last thoracic vertebra, the twelfth ribs, the last six costal cartilages, and the xiphoid process.

The mediastinum

The main contents of the thoracic cavity are the heart and right and left lungs. Each lung is covered by a serous membrane, pulmonary pleura, and lies within a sac of parietal pleura. Between the two pleural sacs lie most of the remaining thoracic contents which together constitute the mediastinum. This is a thick septum which lies between the two pleural sacs and so is covered on its left and right surfaces by mediastinal pleura. The mediastinum of the embalmed cadaver appears immobile, but in life it may move from side to side, with changes in position of the recumbent body, or, abnormally, as the result of differences of pressure in the two pleural sacs. It may also lengthen or shorten. Thus the heart, the great vessels, the tracheal bifurcation, and the oesophagus all descend during inspiration or on movement from a recumbent to an upright position. They ascend in expiration and in abdominal distension. The trachea normally lies in the midline at the jugular notch of the sternum and so is a useful indicator of lateral deviation of the mediastinum. The mediastinum is continuous above with the space occupied by the midline visceral structures of the neck—oesophagus and trachea, pharynx and larynx [FIG. 7.31]. Infections may spread between neck and mediastinum, and the thyroid gland may enlarge downwards into the mediastinum. The mediastinum is closed below by the diaphragm, which separates it from the abdominal cavity, and extends from the sternum in front to the vertebral column behind, except where the pleural sacs are in contact behind the sternum [FIG. 7.37].

For descriptive purposes the mediastinum is divided into a superior and an inferior mediastinum by an anteroposterior plane through the disc between the fourth and fifth thoracic vertebrae and the manubriosternal joint. The inferior mediastinum is further subdivided into the anterior mediastinum in front of the pericardium, the middle mediastinum occupied by the heart and pericardium, and the posterior mediastinum behind the pericardium superiorly and the diaphragm inferiorly.

The **superior mediastinum** is separated from the lower part of the manubrium by the pleural sacs. Anteriorly, loose strands of fibrous

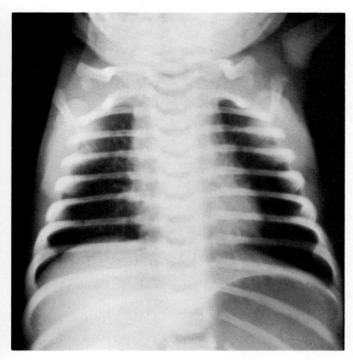

FIG. 7.32.   Anteroposterior radiograph of a new-born baby in inspiration.

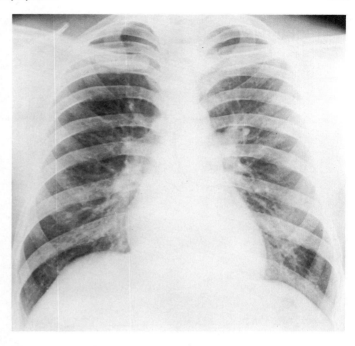

FIG. 7.34.   Postero-anterior radiograph of the chest of an adult in inspiration.

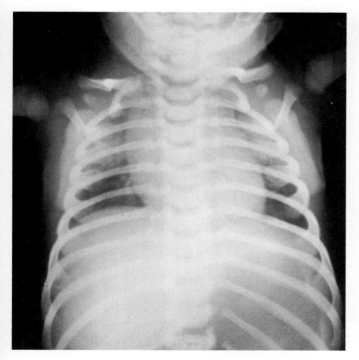

FIG. 7.33.   Radiograph of a new-born baby in expiration. Note the small movement of the ribs and the greater movement of the diaphragm. Even in expiration the ribs are very horizontal. Note also the widening of the mediastinum on expiration. Compare these Figures with FIGURES 7.34 and 7.35.

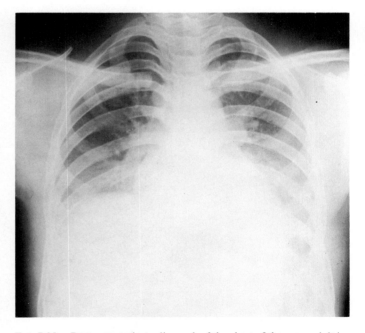

FIG. 7.35.   Postero-anterior radiograph of the chest of the same adult in expiration. Note that the lung bases become dense owing to the loss of air, that the heart shadow increases in transverse diameter as the diaphragm rises, and that the costodiaphragmatic recesses are obliterated.

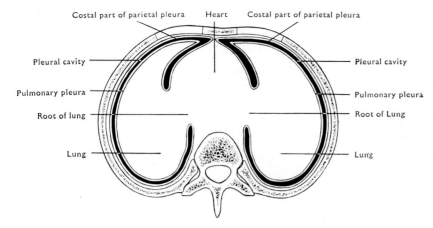

FIG. 7.36. Diagram showing arrangement of pleural sacs as seen in transverse section.

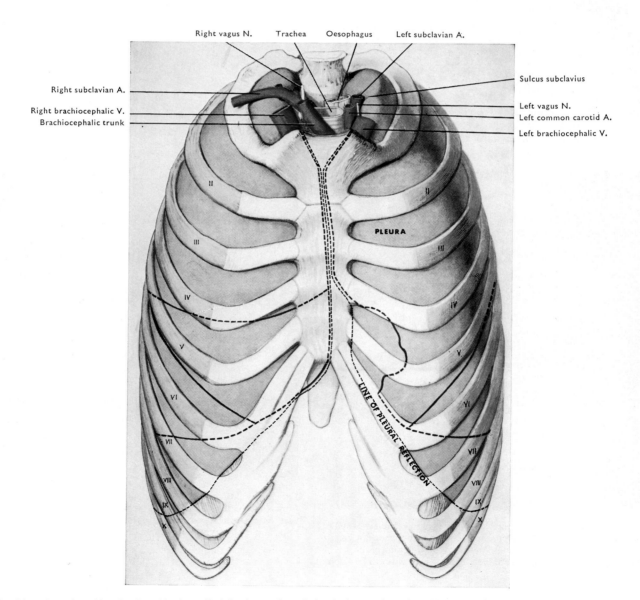

FIG. 7.37. Pleural sacs in subject hardened by formalin injection to show their relations to thoracic wall and superior aperture.

The anterior and diaphragmatic lines of pleural reflexion are exhibited by broken black lines, whilst the outlines of the lungs and their fissures are indicated by the blue lines. Blue, parietal pleura.

tissue, the **superior sternopericardial ligament**, connect the upper parts of the manubrium of the sternum and pericardium. The superior mediastinum contains several structures: the thymic remnants, together with a number of lymph nodes, the aortic arch and its three branches—the brachiocephalic trunk, the left common carotid, and the left subclavian arteries; the upper part of the superior vena cava, the arch of vena azygos, the brachiocephalic veins, and the left superior intercostal vein; the vagus, phrenic, cardiac, and left recurrent laryngeal nerves; the trachea, its bifurcation and principal bronchi, the oesophagus, and the thoracic duct [FIG. 7.31].

The **anterior mediastinum** is separated from the sternum by the pleural sacs above the level of the fourth costal cartilages. It contains a few lymph nodes, small branches of the internal thoracic (mammary) artery, and is also bridged by fibrous tissue—the **inferior sternopericardial ligament**—attaching the xiphoid process to the inferior extremity of the pericardium.

The **middle mediastinum** contains the heart and the intrapericardial parts of the great vessels—the ascending aorta, the pulmonary trunk, the lower part of the superior vena cava, the uppermost part of the inferior vena cava, and the pulmonary veins. It also contains the termination of the azygos vein, the phrenic nerves, and some lymph nodes.

In the **posterior mediastinum** are the descending thoracic aorta and its branches, the oesophagus, the azygos and hemiazygos veins, the vagus and splanchnic nerves, lymph nodes and the main lymph channel—the thoracic duct.

The pleura

Each lung is covered by **pulmonary pleura** and lies within a closed pleural sac which is covered by **parietal pleura**. On each side the pulmonary and parietal pleurae are continuous with one another at the root of the lung [FIG. 7.36], the parietal pleura lining the chest wall and covering the upper surface of the diaphragm and the lateral surface of the mediastinum. Each pleural cavity lies in a closed, separate sac. The cavity contains a thin layer of fluid between the

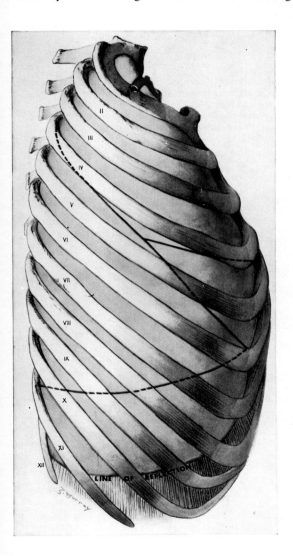

FIG. 7.38.   Lateral view of right pleural sac in subject hardened by formalin injection.
The blue lines indicate the outline of the right lung, and also the position of its fissures. Blue, parietal pleura.

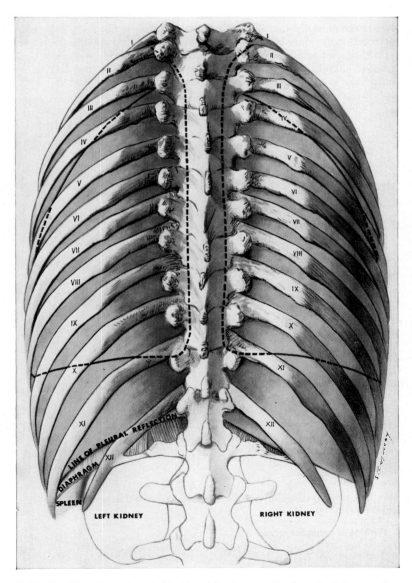

FIG. 7.39.   Pleural sacs from behind in subject hardened by formalin injection. The blue lines indicate the outlines and the fissures of the lungs. Blue, parietal pleura.

pulmonary and parietal layers of pleura, and this allows smooth
movement between the two layers during respiration. It also permits
an even fall of pressure all round the lung during inspiration,
irrespective of the part of the thoracic wall which is moving. Fibrous
adhesion may occur between the pulmonary and parietal layers
following inflammation, or they may be widely separated by fluid
exudates (hydrothorax) or by air (pneumothorax). Different parts of
the parietal pleura are identified by the surfaces on which they lie.
The costal pleura lines the inner surface of the chest wall; the
diaphragmatic pleura covers the upper surface of the diaphragm;
the mediastinal pleura covers the right or left face of the
mediastinum. The cervical pleura projects upwards into the neck.

Cervical pleura (cupula pleurae)

This projects upwards through the superior thoracic aperture into
the root of the neck, and forms the dome-like roof of the pleural
cavity [FIGS. 7.38 and 7.43]. Its profile may be marked on the surface
by a line, convex upwards, from the sternoclavicular joint to the
junction of the medial and middle thirds of the clavicle. Its highest
point is 2–3 cm above the medial one third of the clavicle, at the
level of the neck of the first rib and of the seventh cervical spine. It
is strengthened by the suprapleural membrane, a fibrous sheet
attached to the inner border of the first rib and continuous with the
endothoracic fascia which lines the thoracic cavity. The cupula is in
contact laterally with the scalenus anterior and medius muscles;
anteriorly, with the brachiocephalic and subclavian veins, and the
subclavian artery.

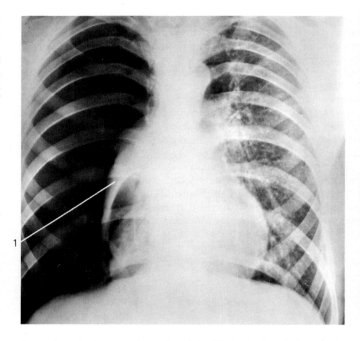

FIG. 7.41.   Postero-anterior radiograph of thorax in inspiration. Air has
entered the right pleural cavity (pneumothorax) causing the lung to
collapse on to its root and ligament. The image of the collapsed lung is
dense, but the ribs are unusually clear on the right because of the absence
of overlying lung and its contained blood vessels.
1. Shadow of collapsed lung.

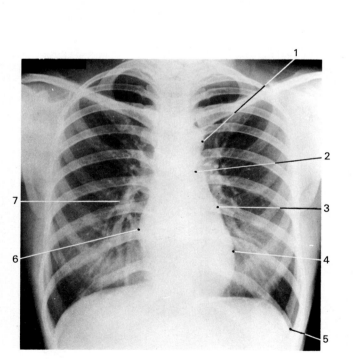

FIG. 7.40.   Postero-anterior radiograph of male thorax of average
physique in inspiration.
1. Arch of aorta.
2 Position of pulmonary trunk.
3. Left auricle.
4. Left ventricle.
5. Costodiaphragmatic recess.
6. Right atrium.
7. Image of blood vessels in lung.

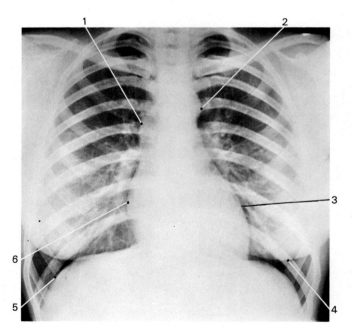

FIG. 7.42.   Postero-anterior radiograph of female chest in inspiration.
Note how the presence of the breasts makes the lower thorax relatively
opaque compared with the portion below the breast. Large breasts,
asymmetrically compressed against the film may make the base of one lung
seem much more dense than the other.
1. Superior vena cava.
2. Arch of aorta.
3. Left ventricle.
4. Margin of shadow of breast.
5. Right dome of diaphragm.
6. Right atrium.

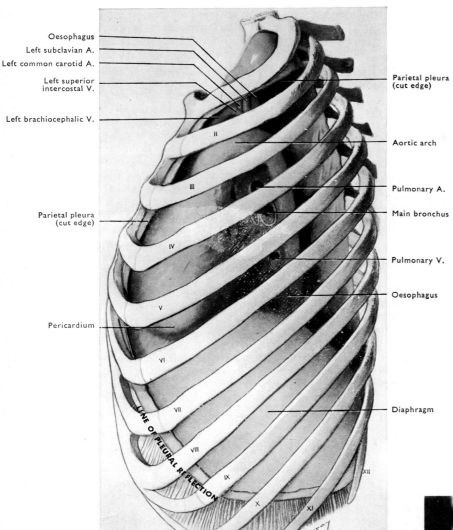

Oesophagus
Left subclavian A.
Left common carotid A.
Left superior
intercostal V.
Left brachiocephalic V.
Parietal pleura
(cut edge)
Pericardium
LINE OF PLEURAL REFLECTION

Parietal pleura
(cut edge)
Aortic arch
Pulmonary A.
Main bronchus
Pulmonary V.
Oesophagus
Diaphragm

II, III, IV, V, VI, VII, VIII, IX, X, XI, XII

FIG. 7.43.   Left pleural sac in subject hardened by formalin injection.
    The sac has been opened by the removal of the costal part of the parietal pleura; and the lung also has been removed to display the mediastinal pleura. Blue, pleura.

Costal pleura

This, the strongest part of the parietal pleura, extends round the inner surface of the thoracic wall, from the sternum in front to the sides of the vertebral bodies behind, where it is firmly attached. The line along which it leaves the chest wall to become continuous with the mediastinal and diaphragmatic portions of the parietal pleura is called the line of pleural reflexion. This line, projected on to the outer surface of the body, is the surface marking of the boundary of each pleural sac. Starting on each side behind the sternoclavicular joint, the lines converge, to meet in the midline above the sternal angle, and descend together to the level of the fourth costal cartilage. Here the right line continues downwards to the level of the xiphosternal joint, while that on the left deviates laterally, to descend along the left border of the sternum. Each line then runs downwards and laterally, reaching the eighth costochondral junction in the midclavicular line and the tenth rib in the midaxillary line [FIG. 7.38]. This is the lowest point reached by the line of pleural reflexion, and therefore by the pleural sac. Thereafter it passes backwards and slightly upwards, reaching the vertebral column at a level which varies, as seen radiologically in the living, between the upper border of the twelfth thoracic vertebra and the lower border

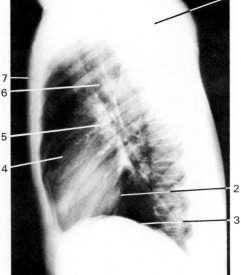

FIG. 7.44.   Left lateral radiograph of chest (male) taken with the arms raised. The shoulders obscure the posterosuperior parts of the lungs, but other parts are more clearly shown, e.g. the parts of the lungs posterior to diaphragm.

1. Opacity due to shoulders.
2. Left ventricle as seen in full inspiration.
3. Dome of diaphragm.
4. Sternocostal surface of heart.
5. Superimposed lung roots.
6. Trachea.
7. Manubriosternal joint

of the first lumbar. It is usually below the level of the inner end of the twelfth rib and therefore at risk in operations on the kidney [FIGS. 7.39 and 8.4].

Diaphragmatic pleura

This thin layer covers the thoracic surface of the diaphragm, except for that part which is in contact with the fibrous pericardium and posterior mediastinum, and for a narrow peripheral rim where the pleura does not reach the costal attachment of the diaphragm (except posteromedially). This means that potentially the thoracic cavity extends below the lowest part of the pleural sac. This margin of the pleural sac is called the costodiaphragmatic recess. In turn, this recess extends below the lower border of the lung, except in deep inspiration, and is tethered to the apposed surfaces of the diaphragm and the ribs by the phrenicopleural fascia, part of the endothoracic fascia.

Mediastinal pleura

On each side, this clothes the face of the mediastinum and is continuous in front and behind with the costal pleura, below with the diaphragmatic pleura, and round the root of the lung with the pulmonary pleura [FIG. 7.36]. The most anterior part of each pleural cavity, along the sternal line of pleural reflexion, is the costo-mediastinal recess, into which the anterior margin of the lung extends only on deep inspiration. Above the root of the lung the mediastinal pleura passes directly from the sternum to the vertebral column, and on the left side [FIG. 7.46] is in contact with the arch of the aorta, the left phrenic and vagus nerves, the left superior

intercostal vein, the left common carotid and subclavian arteries, the oesophagus, and the thoracic duct. On the right side [FIG. 7.45] it is in contact with the right brachiocephalic vein, the superior vena cava, the phrenic nerve, the brachiocephalic trunk, the trachea, the right vagus nerve, the oesophagus, and the arch of the azygos vein.

Below the root of the lung a fold of pleura—the pulmonary ligament—stretches from the mediastinal surface of the lung to the pericardium and ends below in a free border [FIGS. 7.49 and 7.50]. In front of the root of the lung the mediastinal pleura covers and adheres to the pericardium, except where the phrenic nerve and pericardiacophrenic artery intervene. Behind the root of the lung on the right side, the mediastinal pleura passes over the oesophagus, azygos vein, and vertebral bodies, while on the left side it lies on the descending thoracic aorta and vertebral bodies. Immediately above the diaphragm, the right mediastinal pleura is in contact with the inferior vena cava, and on both sides the mediastinal pleura covers the oesophagus and extends medially behind it, forming a recess. These retro-oesophageal recesses, right and left, meet one another, behind the oesophagus. Here the two pleural cavities are separated only by a thin partition consisting of two layers of pleura. Each recess contains lung, and is demonstrable by X-rays on inspiration.

THE LUNGS

There are two lungs, right and left. Each lies free within its pleural cavity, attached by its root and pulmonary ligament to the trachea, heart, and pericardium. The lungs are highly elastic, rose pink in the

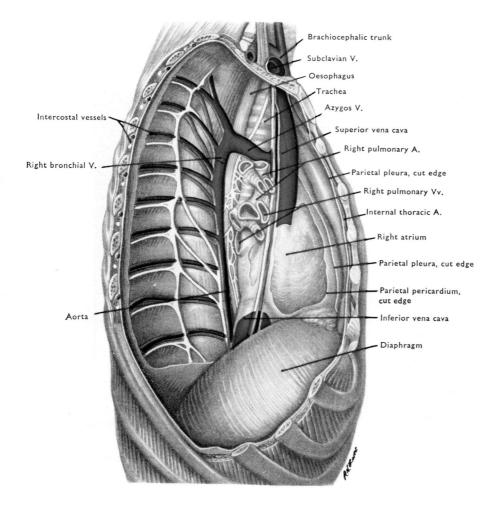

FIG. 7.45.   The right side of the mediastinum and thoracic vertebral column. The pleura has been removed together with part of the parietal pericardium.

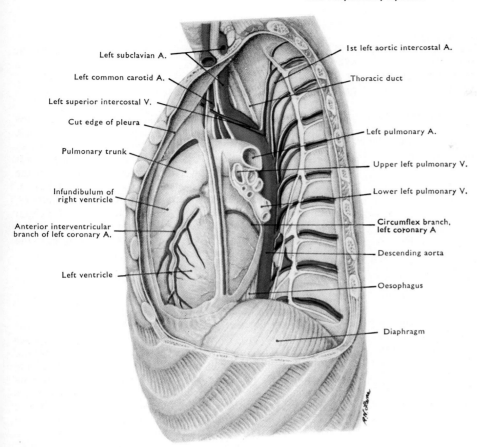

Left subclavian A.

Left common carotid A.

Left superior intercostal V.

Cut edge of pleura

Pulmonary trunk

Infundibulum of right ventricle

Anterior interventricular branch of left coronary A.

Left ventricle

1st left aortic intercostal A.

Thoracic duct

Left pulmonary A.

Upper left pulmonary V.

Lower left pulmonary V.

Circumflex branch, left coronary A

Descending aorta

Oesophagus

Diaphragm

FIG. 7.46A.   The left side of the mediastinum and thoracic vertebral column. The pleura and part of the parietal pericardium have been removed.

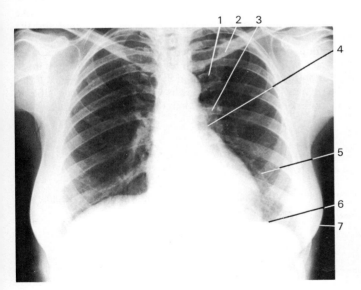

FIG. 7.46B.   Postero-anterior radiograph of female chest.
1. Partly ossified first costal cartilage.
2. Clavicle.
3. Partly ossified second costal cartilage.
4. Pulmonary trunk.
5. Vascular markings of lung.
6. Diaphragm.
7. Outline of breast.

infant but mottled in the adult, owing to the accumulation of inhaled particles in subpleural tissue. After the onset of breathing at birth, the air-filled lungs float in water and are soft, spongy and crepitant when handled. The future air passages of the foetal lung are filled with fluid, so it has the consistency of liver and sinks in water.

The lungs form about one thirty-seventh of the body weight in men and one forty-third in women and are absolutely heavier in men; on average, the right weighs 620 g, the left 560 g. In the fresh body, the air pressure within the lung is atmospheric, and greater than that in the pleural cavity; the lung is therefore expanded and

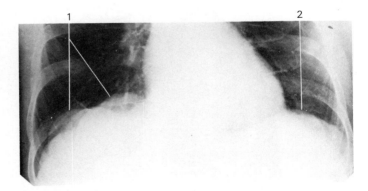

FIG. 7.47.   Postero-anterior radiograph to show irregularity of the right dome of the diaphragm—a normal feature.
1. Right dome of diaphragm.
2. Left dome of diaphragm.

stretched and, on deep inspiration, fills the pleural cavity. When the chest is opened, the pressures within and without the lung become equal and the fresh lung collapses, through the recoil of its stretched, elastic, connective tissue. The lungs of the embalmed cadaver are in the expiratory position, hard and stiff, and moulded to the form of the thoracic cavity. Each is conical, with an **apex**, a **base**, a **costal**, and a **medial surface**. The inferior border is sharp where it separates the base and costal surface, blunt and rounded between the base and medial surface. A sharp anterior border separates the costal and medial surfaces. The right lung is partially divided into three lobes by two fissures, the left into two lobes by one fissure [FIG. 7.48].

APEX

The apex of the lung projects upwards, through the superior aperture of the thorax, into the root of the neck. It fills the cupula pleurae, and its highest point is at the level of the body of the first thoracic vertebra and of the seventh cervical spine behind, and about 2.5 cm above the medial third of the clavicle in front. Posteriorly, it lies against the neck of the first rib, with the cervicothoracic sympathetic ganglion, the ventral ramus of the first thoracic nerve and the highest intercostal artery intervening. Laterally, are the scalenus anterior and medius muscles; anteriorly, the brachiocephalic and subclavian veins and above, the subclavian artery, as it arches laterally to cross the first rib. Medially, on the right side, are the brachiocephalic artery, the right brachiocephalic vein, the trachea and oesophagus, and on the left side, the left common carotid and subclavian arteries, the oesophagus and the thoracic duct [FIG. 7.37].

BASE

The base of the lung conforms to the thoracic surface of the diaphragm and is deeply concave. The *right* base is separated by the diaphragm from the right lobe of the liver, and the *left* base from the left lobe of the liver, the fundus of the stomach, the spleen and the left colic flexure. The inferior border of the lung, between the costal surface and the base, moves downwards and laterally in deep inspiration, to fill the costodiaphragmatic recess of the pleura [FIGS. 7.32–7.35].

COSTAL SURFACE

This conforms to the thoracic wall and may be grooved by the overlying ribs [FIG. 7.49].

MEDIAL SURFACE

The medial surface of each lung conforms to, and carries the impress of, the lateral surface of the mediastinum and of the thoracic vertebral column. These reciprocal mouldings are fully illustrated, for the right lung, in FIGURES 7.45 and 7.50; for the left in FIGURES 7.46A and 7.49. Centrally is the hilus, where the structures of the **root** enter or leave the lung. It is surrounded by a short tube of pleura, narrowing inferiorly to become the **pulmonary ligament**, and connecting the mediastinal and the pulmonary pleurae. Behind the hilus, the lung is applied to the vertebral column and to the structures of the posterior mediastinum; in front of the hilus to the structures of the middle and anterior mediastina.

ANTERIOR BORDER

The anterior part of each lung extends forwards as a thin wedge into the costomediastinal recess of the pleural cavity and ends at the sharp anterior border. The anterior border of the right lung follows the costomediastinal line of pleural reflexion, starting above behind the sternoclavicular joint and extending vertically near the midline to the xiphosternal joint. The anterior border of the left lung follows a similar course to the level of the fourth costal cartilage, where it is interrupted by the **cardiac notch**, and curves at first laterally to a point 3.5 cm from the border of the sternum and then medially, to reach the sixth costal cartilage 2–3 cm from the sternum in the adult. At the cardiac notch the pericardium and heart are separated from the chest wall, in expiration, only by the left costodiaphragmatic recess. This area is therefore less resonant on percussion, and is known as the *area of superficial cardiac dulness*. The tongue-like portion of the left upper lobe, between the cardiac notch and the oblique fissure, is the **lingula** [FIG. 7.48].

In childhood, the anterior borders of the two lungs are separated from one another in expiration by the large thymus, but on inspiration they slide forwards and approach one another anterior to the thymus. The thymic shadow is normally no longer visible in radiographs of children of more than three years.

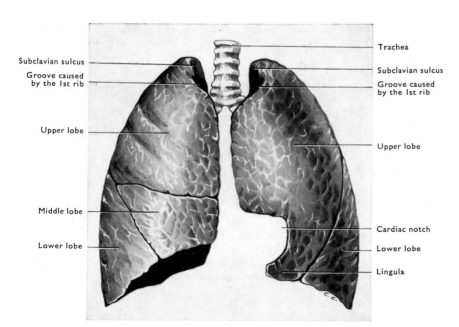

Subclavian sulcus
Groove caused by the 1st rib

Upper lobe

Middle lobe

Lower lobe

Trachea

Subclavian sulcus
Groove caused by the 1st rib

Upper lobe

Cardiac notch

Lower lobe

Lingula

FIG. 7.48.   Trachea, bronchi, and lungs of a child, hardened by formalin injection.

INFERIOR BORDER

The surface projection of the costodiaphragmatic part of the inferior border lies at the sixth costal cartilage in the midclavicular line, at the eighth rib in the midaxillary line and then runs horizontally to a point about 2 cm lateral to the tenth thoracic spine [FIGS. 7.38 and 7.39].

ROOTS OF THE LUNGS

The bronchi, the pulmonary arteries, veins, lymph vessels and nerves, and the bronchial vessels constitute the roots of the lungs, and towards the hilus they are enclosed within the pulmonary pleura where it becomes continuous with the parietal pleura. The roots of the lungs lie opposite the bodies of the fifth, sixth, and seventh thoracic vertebrae, with the phrenic nerve, the pericardiacophrenic artery and vein in front, and the vagus nerves behind. On the right, the superior vena cava lies in front of the root of the lung and the azygos vein lies behind and arches over its superior surface. On the left side the aorta arches over the lung root and the thoracic aorta descends behind it.

The relative positions of the main structures within the root of the lung are seen in FIGURES 7.49 and 7.50. The arrangement from before backwards on both sides is: the upper of the two pulmonary veins, the pulmonary artery, the principal bronchus, and the bronchial vessels. From above downwards on the right side there is the superior lobar bronchus and pulmonary artery, the stem of the middle and inferior lobar bronchi, and the lower pulmonary vein, while on the left side the pulmonary artery is superior, then the principal bronchus, then the lower pulmonary vein.

FISSURES AND LOBES

The lungs are incompletely divided into lobes by fissures, which are lined by extensions of pulmonary pleura and are of variable depth. Each lobe is supplied, in general, by its own bronchus and associated blood vessels, but these may pass from one lobe to another at the hilus, where the lobes are never completely separate. Since certain disease processes may be confined to a single lobe, knowledge of the position of lobes and fissures relative to the surface of the body is of some importance [FIGS. 7.37–7.39].

The oblique fissure may be indicated posteriorly by the position of the medial border of the scapula when the hand is on top of the head. In the *right lung* it generally begins somewhat lower than in the left lung. Its position is indicated approximately by a line beginning at the third to fifth rib in the paravertebral line, and following the curve of the sixth rib downwards and forwards across the midaxillary line. It meets the inferior border of the lung 7 to 8 cm from the midline behind the sixth costal cartilage.

The oblique fissure of the *left lung* is similar to that of the right lung. It has a more vertical course, usually crossing either the fifth or sixth rib at the midaxillary line, and running forwards parallel to the sixth interspace or the upper border of the seventh rib, ends deep to the sixth or seventh costal cartilage.

The horizontal fissure of the right lung commonly begins at the oblique fissure in the midaxillary line at the level of the sixth rib, and runs almost horizontally forwards to end about the chondrosternal junction of the fourth rib. This fissure is rarely complete.

The superior lobes lie above and anterior to the oblique fissures and include the apex and the entire anterior border in the left lung. The posterior surface of the right superior lobe is adjacent to the upper three to five ribs posteriorly, but anteriorly descends only as far as the fourth costal cartilage. With the more vertical position of the left oblique fissure the superior lobe of the left lung is in contact with the upper three to five ribs posteriorly, but anteriorly it descends to the sixth or seventh costal cartilage.

The right middle lobe is wedge-shaped, with its apex posteriorly deep to the sixth rib at the midaxillary line, and anteriorly it is in contact mainly with the fifth, sixth, and seventh costal cartilages. Each inferior lobe is in contact behind with the fifth to eleventh ribs and in front of the midaxillary line its highest point is at the sixth or seventh rib.

Bronchi and bronchopulmonary segments

The right and left principal bronchi divide into lobar (secondary) bronchi (three on the right and two on the left) each of which is distributed to a pulmonary lobe. Within the lobes the secondary bronchi in turn give off tertiary bronchi, and the region of lung supplied by each tertiary (segmental) bronchus constitutes a bronchopulmonary segment. A detailed knowledge of the segmental anatomy of the lungs is essential to the chest specialist; the undergraduate student should at least be familiar with the principles involved. Infections and malignant disease may be localized to a particular segment and it may be possible for the surgeon to remove that segment (segmental resection) rather than a lobe or the entire lung. Drainage of infected material (pus) from the bronchial tree under the influence of gravity (postural drainage) depends on correct orientation of the involved segmental bronchi. Blockage of a segmental bronchus may lead to collapse of the segment of lung which it supplies. It does not always have this effect because small communications between segments at bronchiolar or alveolar level may permit a limited amount of collateral ventilation (Hislop and Reid 1974). Moreover, the segments are not strictly independent bronchovascular units since a segment is not exclusively supplied by a single segmental bronchus, nor by its accompanying branch of the pulmonary artery: intersegmental connective tissue planes may be crossed by small bronchi and by small pulmonary arterial branches from neighbouring segments. Moreover tributaries of the pulmonary veins lie at the periphery of the segments or in an intersegmental position and therefore drain adjacent segments. Finally, although a standard pattern of segments is recognized, there is considerable individual variation in their disposition within the lobe, in the precise pattern of branching of the bronchi, and in their blood supply.

The short superior lobar bronchus to the upper lobe of the right lung is given off before or just as the right principal bronchus enters the hilus, and after about a centimetre divides into three segmental bronchi, an apical directed superiorly towards the apex of the lung, a posterior directed backwards and laterally, and an anterior passing downwards and forwards. These are distributed to the corresponding bronchopulmonary segments [FIG. 7.51].

The bronchus to the right middle lobe arises from the stem bronchus very close to the origin of the superior segmental bronchus of the inferior lobe. It divides into a medial and a lateral segmental bronchus distributed to the corresponding bronchopulmonary segments. The first branch to the right lower lobe is the superior segmental bronchus. Thereafter, the stem bronchus gives off medial and anterior basal segmental bronchi, and then divides into lateral and posterior basal segmental bronchi, each distributed to the bronchopulmonary segments of the same name.

The left principal bronchus divides, inferior to the left pulmonary artery, into superior and inferior lobar bronchi. The superior lobar bronchus divides into apicoposterior and anterior segmental bronchi. In addition this bronchus gives off a stem which bifurcates into superior and inferior lingular bronchi distributed to the lingula.

The left inferior lobar bronchus continues into the lower lobe where it has the five branches corresponding to those of the lower lobe of the right lung, though the anterior and medial basal may have a common stem.

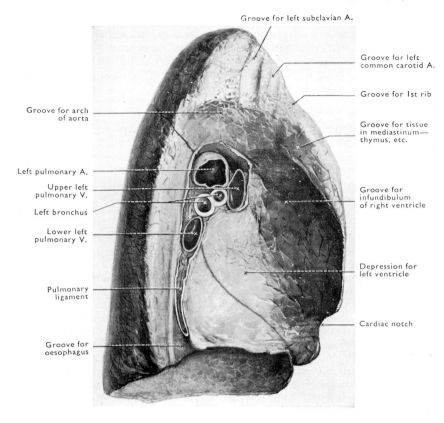

Groove for left subclavian A.

Groove for left
common carotid A.

Groove for 1st rib

Groove for tissue
in mediastinum—
thymus, etc.

Groove for arch
of aorta

Left pulmonary A.

Upper left
pulmonary V.

Left bronchus

Lower left
pulmonary V.

Groove for
infundibulum
of right ventricle

Depression for
left ventricle

Pulmonary
ligament

Cardiac notch

Groove for
oesophagus

FIG. 7.49.   Medial surface of left lung hardened *in situ.*

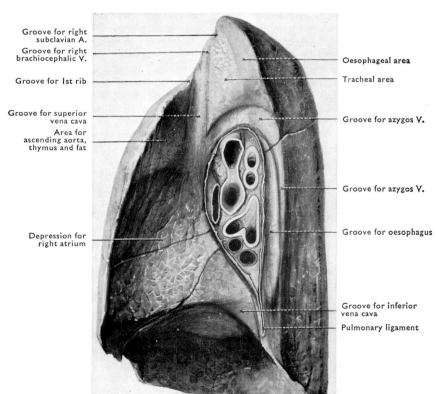

Groove for right
subclavian A.

Groove for right
brachiocephalic V.

Groove for 1st rib

Groove for superior
vena cava

Area for
ascending aorta,
thymus and fat

Depression for
right atrium

Oesophageal area

Tracheal area

Groove for azygos V.

Groove for azygos V.

Groove for oesophagus

Groove for inferior
vena cava

Pulmonary ligament

FIG. 7.50.   Medial surface of right lung hardened *in situ.*

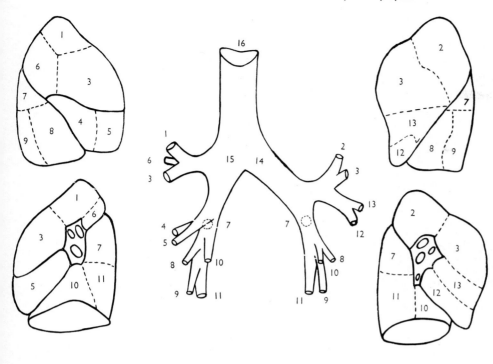

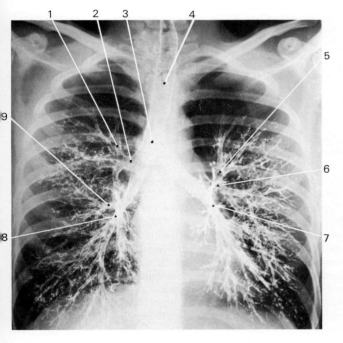

FIG. 7.51. The bronchopulmonary segments. (After Jackson and Huber 1943.)

The asymmetry of the main bronchial pattern is imposed on the bronchial tree by the asymmetry of the heart.

Each bronchial branch is designated by the name of the subdivision of the lung supplied by it.

1. Apical.
2. Apicoposterior.
3. Anterior.
4. Lateral.
5. Medial.
6. Posterior.
7. Superior.
8. Anterior basal.
9. Lateral basal.
10. Medial basal.
11. Posterior basal.
12. Inferior lingular.
13. Superior lingular.
14. Left principal bronchus.
15. Right principal bronchus.
16. Trachea.

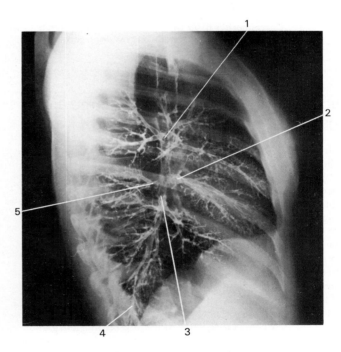

FIG. 7.52. Postero-anterior bronchogram. Compare with FIGURE 7.51. The bronchi are made visible by a thin layer of opaque material on their internal surfaces.
1. Apical bronchus of right upper lobe—a branch viewed end on.
2. Right superior lobe bronchus.
3. Tracheal bifurcation.
4. Trachea.
5. Apicoposterior bronchus.
6. Left superior lobe bronchus.
7. Left inferior lobe bronchus.
8. Right inferior lobe bronchus.
9. Right middle lobe bronchus.

FIG. 7.53. Lateral bronchogram of right lung.
1. Right upper lobe bronchus.
2. Right middle lobe bronchus.
3. Right lower lobe bronchus.
4. Branches of right posterior basal bronchus in costodiaphragmatic recess.
5. Apical segmental bronchus of lower lobe.

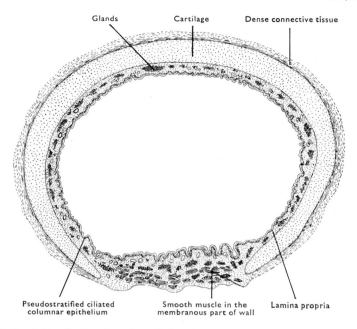

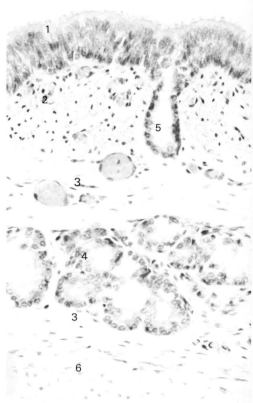

Fig. 7.54.   Section through wall of trachea.

Fig. 7.55.   Photomicrograph of part of wall of trachea, showing: (1) pseudostratified columnar ciliated epithelium; (2) tunica propria; (3) loose connective tissue (submucosa), containing (4) tubolo-alveolar glands whose duct (5) opens at the surface; (6) cartilage. (× 384.)

Structure of the tracheobronchial tree and lungs

The structure of the upper respiratory tract has already been described [p. 495]. The lower respiratory tract, distal to the larynx, consists of a **conducting portion**—a branching system of tubes collectively called the tracheobronchial tree—which carries air into and out of the lungs, and a **respiratory portion** with thin walls for exchange of respiratory gases. The **trachea** and the **extrapulmonary bronchi** (i.e. those parts of the tree outside the lung) are similar in structure [Figs. 7.54 and 7.55]. The **trachea** is lined by a **mucous membrane**, consisting of an epithelium resting upon a conspicuous basement membrane, and a tunica propria of collagenous and elastic fibre networks. The **elastic fibres** are more abundant in the outer part of the tunica propria, where they run mostly longitudinally, and constitute an inner fibro-elastic membrane. External to this is a submucosa of loose connective tissue, containing seromucous glands. The main component of the wall is a series of U-shaped rings of hyaline cartilage which stiffen the airway and prevent its collapse when the pressure is lowered internally, e.g. in forced inspiration. The gap in each ring faces posteriorly and abuts on the oesophagus; it is bridged by transverse and oblique bundles of smooth muscle fibres, which are attached to the inner side of the posterior end of each ring. The rings are joined to one another, to form a flexible and extensible tube, by an outer fibro-elastic membrane, which blends with the perichondrium of each ring. The trachea is covered externally by an adventitial layer of loose connective tissue.

The epithelial lining of the trachea and extrapulmonary bronchi is characteristic of most of the conducting portion of the tract and is often referred to as **respiratory epithelium**. It is pseudostratified columnar and ciliated. The ciliated cells are the most abundant and occur in groups interspersed with non-ciliated cells. Each bears about 200 cilia which are about 5 μm in length, and beat about 1000 times/minute. The cilia beat in waves towards the oral part of the pharynx, in a layer of serous fluid on which floats mucus, either as a continuous blanket or, according to some, as isolated droplets or flakes (Jeffrey and Reid 1977).

The mucus is secreted by **goblet cells**, and by the branched tubulo-alveolar submucosal glands. Other non-ciliated cells in the epithelium include: (1) **serous cells** which, together with the serous elements of submucosal glands, provide the watery fluid film in which the cilia beat; (2) **basal cells**, which give rise, through a common **intermediate cell**, to both ciliated and goblet cells; (3) **brush cells**, so-called because of their conspicuous brush border of microvilli, possibly indicating an absorptive function. Long-lived lymphocytes migrate slowly through the epithelium.

From the extrapulmonary bronchi onwards, there is progressive diminution in diameter and in thickness of the wall and a gradual modification in the relative contribution of each of the structural components. In the intrapulmonary bronchi [Fig. 7.57A] the cartilage is in the form of plates of irregular outline, like scattered pieces of a jig-saw puzzle. The muscle now forms a meshwork, of criss-crossing spirals, lying in a plane internal to the cartilage plates in a position in which it can vary the calibre of the bronchus. The muscle fibres are intermingled with abundant elastic connective tissue fibres, and the two elements collectively form a **myo-elastic net**. (In histological sections the epithelium and subjacent connective tissue are longitudinally folded by contraction and elastic recoil of this net on fixation of the lung.) The epithelium becomes thinner and eventually simple columnar in type; the tunica propria is also thinned, so that the epithelium rests on the inner elastic membrane.

FIG. 7.56. Photograph of a cast of a terminal bronchiole and its branches.
1. Alveolus.
2. Respiratory bronchiole.
3. Terminal bronchiole.

The most distal branches of the conducting portion of the tree, approximately the seventeenth generation of branching, are the terminal bronchioles [FIGS. 7.56 and 7.57B]. These are 1 mm or less in diameter, and lack a cartilaginous component; the muscular component is relatively thicker than in any other part of the tree. The mucous membrane has thinned further and consists of a simple columnar or cubical ciliated epithelium, without goblet cells, and a thin tunica propria, from which glands are now absent. Interspersed among the ciliated cells are non-ciliated bronchiolar cells, which are serous in type, but unusual in containing abundant smooth endoplasmic reticulum. They may be important in keeping the bronchioles clear of mucus secreted further up the airway.

Each terminal bronchiole leads into three or more generations of respiratory bronchioles [FIGS. 7.56 and 7.57C and D], which bear scattered alveoli as sac-like out-pouchings of their walls. From each respiratory bronchiole there arise several generations of alveolar ducts, short channels each ending in one or more alveolar sacs. The walls of alveolar ducts and alveolar sacs consist of little more than the mouths of the continuous series of alveoli which open from them [FIG. 7.57D].

A pulmonary alveolus is a spherical or polyhedral sac about 250 μm in diameter in the adult, with an open mouth connecting it to a part of the airway. Each alveolus usually abuts upon several other alveoli, with each of which it shares a common wall, an interalveolar septum. The alveolus is lined by an epithelium, continuous with that of the alveolar duct, but so thin that it can only be seen with the electron microscope [FIG. 7.58]. The alveolar epithelium has a thin basal lamina and is in close contact with a dense capillary plexus, which is supported by elastic and reticular fibres and forms the principal component of the interalveolar

septum. The *blood–air barrier* across which there occurs exchange of respiratory gases between air and blood has a mean thickness of 1.5 μm. At its thinnest it measures only about 0.6 μm and consists of alveolar epithelium, capillary endothelium and their respective basal laminae, fused together. The capillary endothelium is non-fenestrated, and tight junctions between the cells alternate with gaps 4 nm in width. By contrast, the alveolar epithelial cells are united by continuous tight junctions. This arrangement helps to reduce to a minimum the passage of fluid into the alveoli, although the principal safety factor here is the low pressure in pulmonary capillaries. Small quanta of protein may traverse the alveolar epithelium, in each direction, within pinocytotic vesicles.

Alveolar epithelial cells are of two types: Type I pneumonocytes are large squamous cells which provide a very thin smooth lining for most of the alveolar wall [FIG. 7.58]. Type II pneumonocytes by contrast are cubical and are involved in the synthesis and secretion of surface active agents (surfactant) which reduce the surface tension of fluid on the alveolar surface and prevent alveolar collapse on expiration [FIG. 7.59].

Adherent to the inner surface of the alveolar wall are alveolar macrophages [FIG. 7.60]. These are large mononuclear phagocytes, which ingest inhaled particles and bacteria. They are derived mainly from monocytes, which are produced in the bone marrow, circulate in the blood, migrate through the blood–air barrier, and transform into macrophages within the alveoli. Some remain here for long periods: others migrate, or are carried in the thin film of alveolar fluid to the bronchioles, where they are carried upwards through the tracheobronchial tree and larynx by the mucociliary mechanism to the pharynx, and are then swallowed. Many millions of macrophages follow this route each day, and are principally responsible for maintenance of the normal sterility of the lungs. Macrophages are also associated with the interstitial connective tissue of the lung, principally that around the bronchovascular tree; only small numbers lie in interalveolar septa (Brain, Godlewski, and Sorokin 1977).

A second major phagocytic defence mechanism in the lungs is provided by polymorphonuclear leucocytes, which leave the blood-stream to enter the interstitial tissue and the alveoli, particularly in suppurative conditions (Sorokin 1977).

RADIOLOGY

The commonest method of investigation of the lower respiratory tract is the postero-anterior radiograph of the chest. This examination is expected to provide information about the lungs, but parts are obscured by the heart, great vessels, diaphragm, and constituents of the chest wall. Sometimes additional methods are necessary in order to overcome these disadvantages, but a great deal of information can be obtained from this simple examination, provided that its limitations are remembered and the visible structures are carefully scrutinized.

For detailed study the student should have available a number of full-sized normal chest radiographs, though the illustrations in this book will serve.

There is considerable variation in the general appearance of chest radiographs—some short and almost square in shape, others long and narrow, the majority intermediate.

Soft tissues of chest wall and bones of thorax and upper limb

Little of the vertebral column can be seen but occasionally the margin of the manubrium sterni is visible at the sternoclavicular joints. The clavicles usually slope slightly down towards the sternum

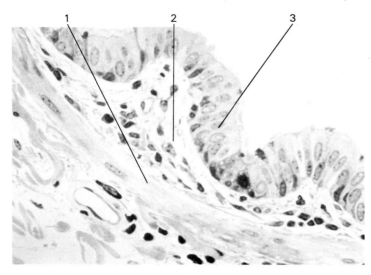

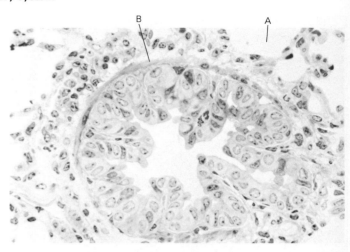

FIG. 7.57. A, Photomicrograph of part of wall of small bronchus, with (3) simple columnar epithelium, mostly of ciliated cells, with occasional mucous cells; (2) a thin, loosely cellular tunica propria, and (1) a thin layer of smooth muscle. (× 384.)

C, Photomicrograph of transverse section of respiratory bronchiole without alveolar outpouchings at this level. The mucosa consists of a simple cubical epithelium and a very thin propria. It is folded longitudinally by contraction of the thin, discontinuous layer of smooth muscle. Most of the epithelial cells are bronchiolar cells, interspersed with occasional ciliated cells. A, alveolus; B, smooth muscle. (× 384.)

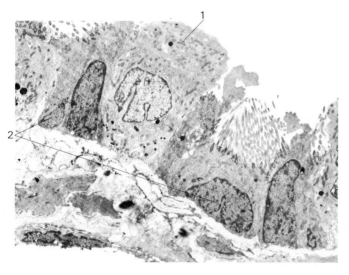

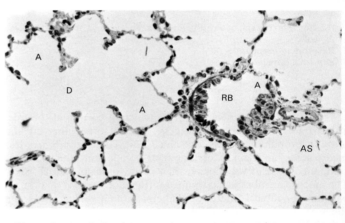

B, Electron micrograph of epithelium of terminal bronchiole, showing a bronchiolar cell (1) and several low columnar or cuboidal ciliated cells (2). The bronchiolar cell bulges characteristically above the general level of the epithelial surface. (× 7200.)

D, Photomicrograph showing: a respiratory bronchiole (RB), from which arises an alveolar sac (AS); an alveolar duct (D) into which open several alveoli (A). Adjacent alveoli abut on one another, forming interalveolar septa. (× 250.)

(steeper in films taken in recumbency). The film is taken with the shoulders rotated forwards to reduce the overlap of the scapulae. Irregular ossification in costal cartilages may simulate opacities in the lung. The muscles of the shoulder girdle often make the axillary area relatively opaque. In muscular individuals, the lower margin of the pectoralis major muscle can sometimes be seen as a line crossing the axilla and lateral part of the thorax, while the lung behind the muscle appears denser than elsewhere. The thickness of the chest wall can be assessed by observing the amount of tissue at the side; when considerable, the lung markings are intensified and less sharp [FIG. 7.42]. Similarly, the lower parts of the lungs are covered by the breasts which cause intensification of the lung density in proportion to their thickness. Thus asymmetry due, for example, to positioning, difference in size of the two breasts, or absence of one breast may cause apparent alterations in lung density; careful inspection usually reveals the cause [FIG. 7.42].

Diaphragm

The upper surface of each dome usually appears as a smooth arc meeting the chest wall at an acute angle, the costodiaphragmatic angle. The right dome is usually the higher. The central portion is not visible where it is continuous with the opacity of the heart. The upper surface is occasionally uneven and rarely quite large humps may be seen in the normal subject [FIG. 7.47]. Gas in the stomach or colon is often seen beneath the left dome in radiographs taken in the upright position.

The trachea

The trachea appears as a translucent band because of its air content. It may be traced upwards in the midline to the jugular notch of the sternum. The principal bronchi are seldom seen in a normal postero-anterior chest radiograph, but sometimes a circular translucency in

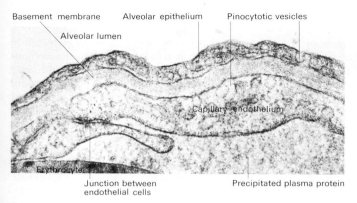

FIG. 7.58. Electron micrograph of the blood–air barrier at its thinnest. It consists of the attenuated cytoplasm of a Type I alveolar epithelial cell, and the endothelium of a pulmonary capillary. Between the two epithelia are their respective basal laminae, fused to form a single layer. (× 30 000.)

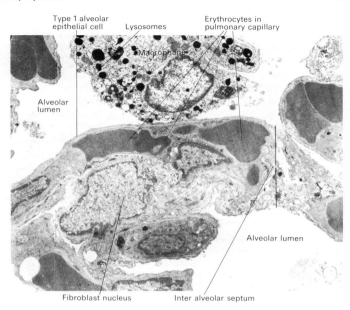

FIG. 7.60. Electron micrograph of an alveolar macrophage lying within an alveolus in contact with an interalveolar septum. Its cytoplasm is vacuolated, and contains many densely stained lysosomes. (× 9000.)

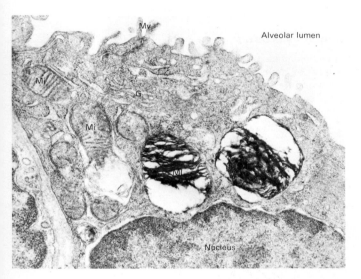

FIG. 7.59. Electron micrograph of part of a Type II alveolar epithelial cell, showing multilamellar bodies (ML), mitochondria (Mi), microvilli (Mv), and Golgi apparatus (G). (× 40 000.)

A **lung root** projects on each side, the left being slightly the higher. Although many structures make up these roots, the principal structures which produce the density are the **pulmonary vessels** [FIG. 13.18]. Their peripheral branches form the lung markings which extend outwards from the roots and become smaller as they are traced towards the chest wall. They are larger towards the base of each lung than at the apex because of the greater volume of lung to be supplied. Sometimes the main **pulmonary veins** can be seen entering the heart, but usually form indefinable parts of the lung roots. Radioisotope scanning is a convenient and safe method of detecting blockage of a pulmonary artery or branch by the absence of emission from the avascular area. Evidence of this is usually not apparent on a radiograph.

Examinations in the wards

In radiographs taken in the recumbent or semi-recumbent position with the film posteriorly, the diaphragm is relatively elevated, the image of the heart is enlarged because of its distance from the film, and there is a greater degree of filling of the veins at the root of the neck, with consequent broadening of the upper mediastinal opacity.

The chest in infancy

In healthy young babies the mediastinal prominences are not so readily distinguishable as in the adult. Sometimes the upper mediastinum appears abnormally wide, although this may not be significant. In the supine position, which sometimes has to be adopted, the brachiocephalic veins are distended and thus produce broadening. Occasionally the **thymus** may cause displacement of these veins or their tributaries into the lung translucency [p. 602] to a greater degree than shown in FIGURE 7.33.

The lateral film of the chest [FIG. 7.44]

The amount of lung below the highest part of the diaphragm should be noted. This is concealed by the upper abdominal contents in the postero-anterior radiograph. Detail in the upper posterior parts of the lungs is obscured by the density of the shoulders, the arms being raised in the taking of the film. The right and left lungs and lung

the lung, close to the mediastinum, indicates a large bronchus lying 'end-on' to the X-ray beam. Apart from the trachea, the structures forming the mediastinum cannot be differentiated from each other (because of their similar density) except where they project into the translucent lung areas from the right or left surfaces of the mediastinum. *On the left*, the **aortic arch** forms a well-defined protrusion behind the first costal cartilage. Sequentially below this (1) the **pulmonary trunk** forms a flatter protrusion often obscured by the density of the left lung root, (2) the **left auricle** may form a small prominence, and (3) the **left ventricle** produces the most obvious convexity. This meets the diaphragm below at an acute **cardiodiaphragmatic angle** (open to the left) which may be filled by a pericardial fat pad [FIG. 1.8] usually distinguishable from the cardiac apex by its shape and lesser density. *On the right*, the **superior vena cava** in sometimes seen as a vertical edge extending upwards from the convex right border of the heart (**right atrium**) below which the **inferior vena cava** may form a short, vertical part of this surface extending to the diaphragm.

roots are superimposed and are usually indistinct. The posterior surface of the heart, formed by the **left atrium**, with the left ventricle below in inspiration is usually identifiable, but the sternocostal surface (mostly right ventricle) may not be clearly seen. The trachea is usually visible. The denser area in front of it is formed by structures of the chest wall. The aortic arch can occasionally be seen as it crosses the trachea. Among the bones visible are the lower thoracic vertebrae, the scapulae, sternum, occasionally the manubriosternal joint, and the ribs nearer the film.

Fluoroscopic examination does not have the importance formerly accorded to it, although it has been simplified by the development of image intensification and television, but brief reference to the most striking features will be made. During systole there is an obvious contraction of the ventricles, followed almost immediately afterwards by a slight expansion of the aortic arch and pulmonary trunk. (This may account for the variable prominence of these structures in radiographs.) Movements of the **diaphragm** and chest wall can also be observed. The excursion of the **dome of the diaphragm** is variable but is often no more than 2–3 cm in deep respiration; after a period of breathing exercises this can often be increased.

The **bronchi** may be studied by coating their walls with an opaque substance [FIGS. 7.52 and 7.53]. The segmental division of the bronchial tree can be appreciated readily by this method.

Tomography can be of great value in pathological conditions because it shows deep structures clearly by removing confusing superimposed images.

MOVEMENTS OF THE LUNG IN RESPIRATION

On inspiration, the capacity of the thoracic cavity is increased by descent of the diaphragm and by movement of the ribs and sternum. This reduces the pressure within the pleural cavities and air is sucked into the lungs. The root of the lung moves downwards and forwards and the bronchi and bronchioles are elongated and dilated. Most of the increased volume of the lungs in inspiration seems to be due to distention of the alveolar ducts; the alveoli change little in size. Expiration depends principally on cessation of inspiratory effort and on the elastic recoil of stretched connective tissue within the lung. The geodetic arrangement of the myoelastic net in the walls of the bronchial tree is of particular functional importance. Active contraction of the muscle and passive recoil of the stretched elastic fibres serve both to shorten and to constrict the elongated, dilated bronchi and bronchioles. In forced expiration the muscles of the thoracic and abdominal walls are also involved.

BLOOD AND LYMPH VESSELS OF THE LUNG

Pulmonary arteries

Venous blood is conveyed to the lung by the pulmonary artery. At the hilus of each lung the pulmonary artery divides into branches corresponding to the divisions of the bronchi, so that there is an artery accompanying the bronchus of each bronchopulmonary segment and generally lying posterolateral to it [FIG. 7.61]. The right

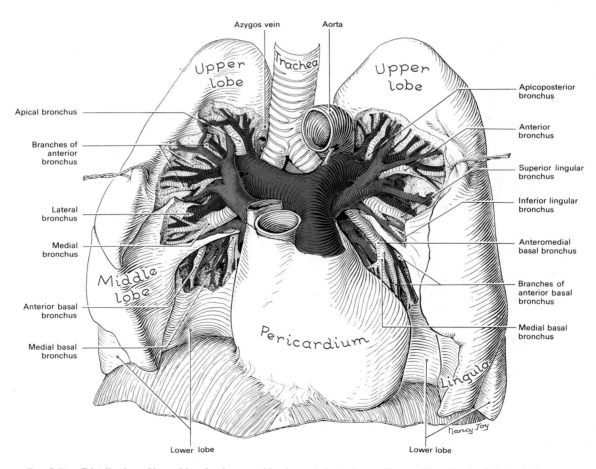

FIG. 7.61.   Distribution of bronchi and pulmonary blood vessels in the lungs. (From a dissection by Miss A. I. Scott.)

pulmonary artery gives off its first branch to the upper lobe at the hilus, and thereafter the stem artery lies just below the medial part of the oblique fissure and in this situation it gives off branches to the middle and lower lobes. The **left pulmonary artery** gives off its branches to the upper lobe as it lies in the hilus, medial to the oblique fissure. The branches of the principal artery to each segment generally accompany the bronchi as far as the respiratory bronchiole where they taper rapidly to about 50 μm in diameter before breaking up into the capillary networks in the interalveolar septa. Pulmonary capillaries have an average diameter of 7–9 μm, just large enough to allow erythrocytes to squeeze through. The spaces between the capillaries are often smaller than the diameter of the capillaries themselves.

Pulmonary veins

The oxygenated blood is carried away from the capillary plexuses by the pulmonary veins. At the periphery, the veins are intersegmental and drain adjacent bronchopulmonary segments. Towards the hilus, the veins fuse to form about ten principal vessels, one emerging from each bronchopulmonary segment and running with the segmental bronchus. The veins lie below and in front of the bronchi, while the arteries are generally above and behind. The **right superior pulmonary vein** is formed close to the hilus by the union of the veins from the upper and middle lobes, and the **right inferior pulmonary vein** drains the lower lobe. The **left inferior pulmonary vein** lies at the upper end of the pulmonary ligament. Some veins run superficially underneath the pulmonary pleura, particularly on the mediastinal surface or within the fissures. When the fissures are incomplete it is common to find a vein passing from one lobe to another.

Bronchial vessels

The walls of the air ducts within the lungs as far as the respiratory bronchioles are supplied by bronchial **arteries**, generally one for the right lung with a variable origin—from the aorta or posterior intercostal arteries—and two for the left lung, more constantly arising from the descending thoracic aorta. Branches of these arteries accompany the bronchioles, forming capillary plexuses in their walls, and communicate with branches of the pulmonary artery. Much of the blood carried by the bronchial arteries is returned by the pulmonary veins. However, the larger bronchial **veins** lie on the posterior aspect of the main bronchi and drain, on the right side, into the azygos and the superior intercostal veins, and on the left side, into the hemiazygos system.

Lymph vessels

The loose areolar tissue of the pulmonary pleura contains a rich **superficial plexus** of lymph vessels, whose polygonal meshes outline the bases of those pulmonary lobules which abut on the surface of the lung. The valves of these vessels are randomly orientated and there is free circulation of lymph in the plexus. Collecting trunks run within the pleura towards the hilus of the lung. The **deep (intrapulmonary) lymph vessels** begin as lymph capillary plexuses around the pulmonary arterioles, respiratory bronchioles, and pulmonary venules, and follow the bronchovascular tree and the pulmonary veins towards the hilus. Lymph capillaries are absent from the interalveolar septa, but alveoli adjacent to terminal airways and to pulmonary arterioles may abut on lymph vessels. There are valved connections between the superficial and deep lymph vessels; flow is normally from superficial to deep, but may be reversed if the deep lymph vessels are blocked, for example by cancer cells. The volume of lymph draining from the lung is normally very small compared with that from the liver or gastro-intestinal tract; it is greatly increased immediately after birth, when the fluid which fills the airways and alveoli is rapidly absorbed. It is also increased in pulmonary oedema (Leak 1977).

Superficial and deep collecting lymph vessels drain into the **bronchopulmonary lymph nodes** at the hilus. Lymph flows then to the **tracheobronchial nodes** on the principal bronchi and at the tracheal bifurcation and thence to right and left **bronchomediastinal trunks**, which frequently open independently at the junction of the internal jugular and subclavian veins of their own side. A few vessels from the lower lobe of each lung drain to the posterior mediastinal lymph nodes. Lymph tissue is scattered within the lung in association with the bronchovascular tree and a few pulmonary nodes are interposed along the deep lymph pathway.

INNERVATION OF THE LUNG

Nerve fibres reach the lungs through the pulmonary plexuses, which lie in front of but mainly behind the root of the lung. They include preganglionic parasympathetic (vagal) and post ganglionic sympathetic efferents and visceral afferents travelling with the vagus. Within the lung they form plexuses around the bronchial and arterial trees. Vagal efferents relay on postganglionic neurones within the bronchial plexuses. Increased secretion of submucosal glands is usually attributed to vagal activity, but recent evidence suggests sympathetic involvement also. Goblet cells are generally thought not to be innervated, but to secrete in response to direct stimulation. Vagal stimulation causes bronchoconstriction. Sympathetic fibres seem to be principally vasoconstrictor; an action on bronchial muscle is doubtful, although sympathomimetic drugs cause bronchodilatation. There are three principal groups of afferent nerve endings in the lung, all of vagal origin (Widdicombe 1977).

(i) 'Irritant receptors' are associated with small, myelinated fibres and are found within the airway epithelium, where they reach almost to the surface. They are most numerous in the trachea and larger bronchi, but reach as far distally as respiratory bronchioles. They are stimulated by intraluminal irritants, by large volume changes in the lung, and by local deformation of the epithelium. Stimulation of irritant receptors in the trachea and at major bronchial divisions, causes coughing; further distally it causes hyperpnoea and bronchoconstriction.

(ii) Receptors with non-myelinated fibres. These are found within interalveolar septa, and appear to be stimulated by irritant gases, by increased interstitial fluid, and by blocking of small arteries—microembolism.

(iii) Stretch receptors in the smooth muscle of the tracheobronchial tree are involved in reflex responses in full inspiration. Sensations from the lung itself, as from other viscera, are generally vague and ill-localized; strong stimulation of 'irritant receptors' causes unpleasant or even painful sensations.

Development of the respiratory tract

In embryos of 2–3 mm length, in the fourth week of development, a gutter-like outgrowth appears in the midline on the ventral wall of the foregut, at the caudal end of the primitive pharynx [FIG. 7.62]. This entodermal **laryngotracheal groove**, is the source of the epithelial lining of the entire respiratory tract distal to the level of the glottis. At first the groove opens along its whole length into the foregut, but the two become separated, except at the cephalic end, by the union of ridges of proliferating entodermal epithelium on each side of the wall of the foregut. These ridges are invaded by migrating mesenchyme which forms the tracheo-oesophageal septum in the coronal plane, and separates the laryngotracheal epithelial tube, ventrally, from the oesophagus dorsally. Separation begins caudally, but does not reach the cephalic end of the groove, which retains its continuity with the primitive pharynx through a slit-like opening. This corresponds to the future **glottis**, and lies between the paired sixth pharyngeal arches (Smith 1957).

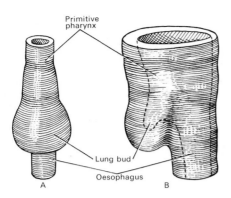

FIG. 7.62. Development of respiratory tube in 4.25 mm human embryo. (After Grosser 1912.) A, ventral view. B, lateral view.

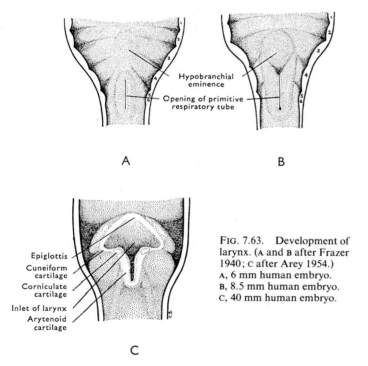

FIG. 7.63. Development of larynx. (A and B after Frazer 1940; C after Arey 1954.)
A, 6 mm human embryo.
B, 8.5 mm human embryo.
C, 40 mm human embryo.

THE LARYNX

The supraglottic larynx (i.e. that part above the vocal folds) arises secondarily as an upward extension into the pharynx, with contributions of pharyngeal mesoderm from three sources. (1) The hypobranchial eminence, a midline swelling in the floor of the pharynx behind the tongue and in front of the glottic opening, gives rise to the epiglottis. (2) The fourth pharyngeal arches provide the aryepiglottic folds. (3) The sixth pharyngeal arches form the arytenoid cartilages and the vocal folds [FIG. 7.63].

The infraglottic larynx develops around the epithelium of the cephalic end of the laryngotracheal groove, in direct continuity with the trachea. The thyroid cartilage arises from several initially separate centres of chondrification within the ventral part of the fourth pharyngeal arches. The cricoid cartilage arises within the condensed mesenchyme of the lower part of the sixth pair of arches; the arytenoids and the vocal folds develop in the upper part of these arches.

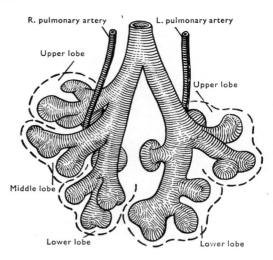

FIG. 7.64. Development of lobes of lungs in 14 mm human embryo. (After Arey 1954, after Ask.)

The cricothyroid muscles develop from the fourth pharyngeal arches. All the other intrinsic muscles of the larynx from the sixth pharyngeal arches. Their motor innervation is by branchial efferent fibres arising in the nucleus ambiguus, and distributed through the nerves of the corresponding pharyngeal arches: the external laryngeal nerve (fourth arch) and the inferior laryngeal nerve (sixth arch).

THE TRACHEO BRONCHIAL TREE AND LUNGS

The tracheal and oesophageal epithelial tubes are embedded in a thick median sheet of splanchnic mesoderm, flanked on each side by a pericardioperitoneal canal, or primitive pleural cavity. The lower end of the tracheal epithelial tube divides into right and left principal bronchi, which project laterally into the primitive pleural cavity, carrying with them a covering of mesenchyme. Each bronchial epithelial tube, together with its investing mesenchyme, will form a lung, and is called a lung bud. The epithelial tube grows and branches, under the inductive influence of the mesenchyme, and forms successively the epithelial lining of lobar bronchi (5th week) and segmental bronchi (6th week; FIG. 7.64). By the 16th week of gestation, the definitive pattern of branching of the conducting portion of the airway is fully established up to and including the future terminal bronchioles. The number of generations in different segments varies between 15 and 26, as in the adult. The branching tubular tree is lined proximally by a stratified columnar epithelium; distally it ends in terminal buds, blind-ending sacs lined by a simple cubical to columnar epithelium. The epithelial tree is embedded within a loose vascular mesenchyme and at this stage the lung bears a superficial resemblance to a branched tubulo-acinar gland. This is therefore called the pseudo glandular stage of development, which continues until about 16 weeks of gestation [FIG. 7.65A]. Between the sixteenth and twenty-fourth weeks of gestation, the future respiratory portion of the lung develops through further growth and branching of the terminal buds. Capillary blood vessels develop rapidly within the mesenchyme and the epithelial lining becomes progressively thinner, apparently where capillaries appear to press against it [FIG. 7.65B]. At twenty-four weeks of gestation, the end of this canalicular stage of development, the lung can provide for respiratory exchange if required, and the foetus is viable.

At twenty-four weeks, the lung enters the alveolar phase of its prenatal development. Growth and branching of the epithelial tree continues so that, at birth, the respiratory portion of the lung

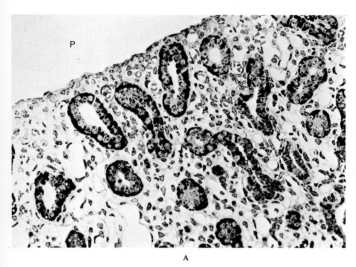

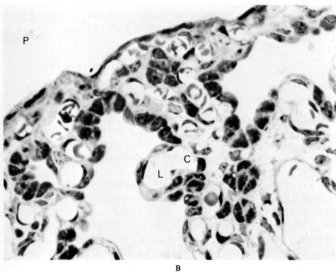

FIG. 7.65.   A, Foetal lung, subpleural region, pseudoglandular stage. The terminal buds of the airway, lined by simple cubical epithelium, are embedded in a loose cellular mesenchyme, which contains moderate numbers of primitive blood vessels. The basal ends of the epithelial cells contain abundant glycogen (black in photograph). B, Foetal lung, subpleural region, canalicular stage. Expansion of the airways is associated with thinning of their epithelial lining and increased vascularity of the mesenchyme. Thin-walled blood vessels, covered by thinning endodermal epithelium, bulge into the airways. C, capillary; L, lumen of terminal bud; P, pleural cavity.

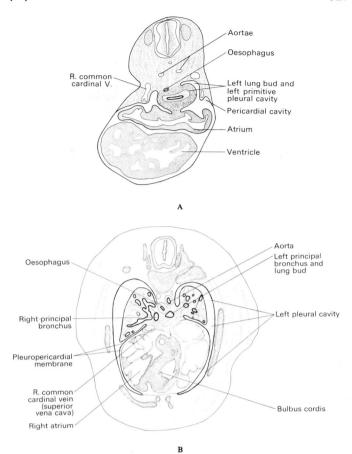

FIG. 7.66.   Two stages in development of pleural cavities, shown in transverse section. In A (6 mm human embryo), at the level of the tracheal bifurcation, the lung buds bulge into the narrow primitive pleural cavities, which open into the pericardial cavity. (× 26.) In B (22 mm human embryo), the pleural cavities have extended laterally and forwards round the heart and approach one another anteriorly. The lung buds do not fill the pleural cavities. (× 9.)

consists of several generations of tubular passages (the future alveolar ducts and alveolar sacs) which bear a series of shallow saucer-like evaginations, the primitive alveoli. The epithelium of the respiratory division of the tract is already thinned beyond the resolution of the optical microscope. The future airways are filled with fluid which is probably derived mainly from the lung itself, and whose hydrostatic pressure may play a part in the thinning of the epithelium before birth. Although respiratory movements can and do occur before birth, they do not move much fluid into or out of the lung because of the large frictional forces; nevertheless, they are an important rehearsal for post-natal life (Godfrey 1974). After birth the first breaths expand the primitive alveoli, further thinning the interalveolar septa, and reduction of the pulmonary vascular resistance results in increased pulmonary circulation. Lung fluid is rapidly removed, partly by squeezing of the chest during passage through the birth canal, but mainly through uptake by pulmonary

lymph vessels. Once the lungs are aerated, the alveoli are lined by a fluid film of surface active agent, surfactant. This is secreted in increasing quantity by Type II pneumonocytes from about 23 weeks of gestation; its presence is essential to survival because it lowers the surface tension at the alveolar surface, prevents alveolar collapse on expiration, and therefore greatly reduces the work of breathing. The quantity (and probably the quality) of surfactant present is an important factor in determining the viability of infants born prematurely (McDougall and Smith 1975).

After birth, existing alveoli enlarge and there is further branching of one or two generations of alveolar ducts and development of new alveoli from distal respiratory bronchioles or even from terminal bronchioles. The adult pattern of the respiratory portion is completed by about the 6th year; thereafter lung growth involves no new developments but merely interstitial growth of the airways and enlargement of the alveoli, to a final size double that in the newborn (Hislop and Reid 1974).

PLEURAL CAVITIES

In embryos of 4–5 mm, each pleural cavity is represented by a short, narrow pericardioperitoneal canal or primitive pleural cavity which connects the pericardial and peritoneal cavities over the dorsal edge of the septum transversum [FIG. 7.67]. The two canals open into the dorsal wall of the pericardial cavity, separated only by the free ventral edge of a thick median septum of condensed mesenchyme which contains the oesophagus and trachea and lies between the canals caudal to the opening. At this stage the canals are dorsal to the pericardial cavity and septum transversum [FIG. 7.66A]. On each side, the **common cardinal vein** lies in the mesenchyme of the lateral wall of the canal and there raises a low pulmonary ridge as it runs ventrally, caudally, and medially to the sinus venosus in the septum transversum.

Each primitive pleural cavity enlarges laterally to form a separate, definitive pleural cavity by a process of extension into the loose mesenchyme of the body wall, dorsal to the pulmonary ridge. The pericardial and peritoneal cavities are each separated from this extension by an intervening sheet of mesenchyme, the **pleuropericardial** and **pleuroperitoneal membranes** [Fig. 7.67], though still continuous through the primitive pleural cavity medial to the free edges of the two membranes. These edges form the lateral margins of the **pleuropericardial** and **pleuroperitoneal openings**, and meet ventrally in the septum transversum. The pleuropericardial membrane lies in the coronal plane and contains the common cardinal vein in its free margin. The pleuroperitoneal membrane lies horizontally in the same plane as the septum transversum.

As the pleural cavities expand, they extend forwards to surround the pericardial cavity and heart, separated from them by the pleuropericardial membrane. Anterior to the pericardium, the cavities approach one another, but do not come into contact everywhere, the **sternopericardial ligaments** developing from mesenchyme which persists between them [FIG. 7.66B].

The growing **lungs** follow the expanding pleural sacs; they are substantially smaller than the sacs throughout foetal life, and do not begin to fill them until they are expanded and stretched at birth with the onset of respiration.

The openings of the pleural cavities into the pericardial cavity are closed in embryos of about 12 mm (sixth week) by fusion of the free medial edges of the pleuropericardial membranes with one another and with the ventral edge of the mesenchymal septum which surrounds the oesophagus and the trachea. The **right common cardinal vein**, persisting as the lower part of the superior vena cava, is fused with the anterior surface of the right lung root which develops posterior to the free edge of the membrane [FIG. 7.66B]. The pleuroperitoneal openings are closed in embryos of approximately 16 mm (seven to eight weeks) by fusion of the medial, free edges of the pleuroperitoneal membranes with the mesenchymal septum which contains the oesophagus at the level of the septum transversum (Wells 1954).

ANOMALIES OF DEVELOPMENT

Cleft palate results from failure of fusion of the palatal processes with each other and with the free margin of the nasal septum [FIG. 7.12]. **Cleft (hare) lip**, which may occur on one or both sides, independently or in association with cleft palate, appears to result from defective growth of the maxillary process mesenchyme. The anterior part of the nasal fin, which is normally obliterated by growth pressure from the mesenchymes of the maxillary and medial nasal processes, persists and subsequently breaks down, leaving an epithelial-lined cleft.

In **choanal atresia**, the definitive posterior nares are small or absent. Early recognition of this defect is important because nasal breathing is obligatory in young infants during suckling.

Tracheo-oesophageal fistula is an abnormal communication between the trachea and oesophagus, presumably due to defective development of the tracheo-oesophageal septum [FIG. 7.62]. It is frequently associated with atresia of the oesophagus.

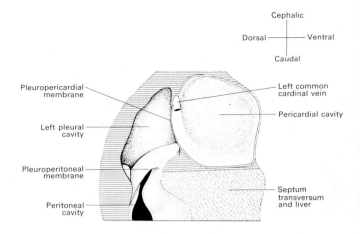

FIG. 7.67.   Diagrammatic three dimensional view of the left pleural cavity in an embryo of 16 mm CR. The pleuropericardial membrane lies coronally and is continuous below with the septum transversum. The pleuroperitoneal membrane lies nearly horizontally and is continuous anteriorly with the septum transversum. (After Frazer 1940.)

Lung

Variations in the lung are relatively common, but are usually of no functional importance. **Accessory fissures** may subdivide otherwise normal lobes. They are usually incomplete and shallow, and tend to correspond to divisions between bronchopulmonary segments, more commonly between the superior and basal segments of the lower lobes and between the lingular and superior lobes of the left lung. **Aberrant lobes** are associated with abnormal branches of the tracheobronchial tree. Thus a **tracheal lobe** is associated with a supernumerary bronchus arising from the trachea, or an **azygos lobe** with an unusual proximal branch of the right principal bronchus passing superior to the arch of the azygos vein, which is then enclosed in an accessory fissure in the medial surface of the right upper lobe. Lower **accessory lungs** have been described attached to the pulmonary ligament and without any bronchial connection. They may arise from an aberrant bronchus which has lost its connection with the tracheobronchial tree or, rarely, from a bronchus which arises ectopically in early development from the entodermal oesophagus. In either case, the accessory lung may be aerated by collateral ventilation from normal lung.

Agenesis—failure of development—of a lung presumably results from failure of initial branching of a lung bud.

Polycystic lung, sometimes associated with polycystic kidney, may reflect a general disturbance in the interaction of epithelial and mesenchymal components.

References

ALCORN, D., ADAMSON, T. M., LAMBERT, T. F., MALONEY, J. E., RITCHIE, B. C., and ROBINSON, P. M. (1977). Morphological effects of chronic tracheal ligation and drainage in the fetal lamb lung. *J. Anat. (Lond.)* **123**, 649.

ASKIN, F. B. and KUHN, C. (1971). The cellular origin of pulmonary surfactant. *Lab. Invest.* **25**, 260.

BODDY, K. and DAWES, G. S. (1975). Foetal breathing. *Br. Med. Bull.* **31**, 3.

BOYDEN, E. A. (1955). *Segmental anatomy of the lungs.* New York.

BRAIN, J. D., GODLEWSKI, J. L., and SOROKIN, S. P. (1977). Quantification, origin and fate of pulmonary macrophages. In *Respiratory defense mechanisms* (eds. J. D. Brain, D. F. Proctor, and L. M. Reid). Marcel Dekker, New York.

BROCK, R. C. (1954). *The anatomy of the bronchial tree*, 2nd edn. London.

DAWES, J. D. K. and PRICHARD, M. M. L. (1953). Studies of the vascular arrangements of the nose. *J. Anat. (Lond.)* **87**, 311.

FRAZER, J. E. (1940). *A manual of embryology. The development of the human body*, 2nd edn. London.

GODFREY, S. (1974). Growth and development of the respiratory system. Functional development. In *Scientific foundations of paediatrics* (eds. J. A. Davis and J. Dobbing). Heinemann, London.

GREENWOOD, M. F. and HOLLAND, P. (1972). Mammalian respiratory tract surface. *Lab Invest.* **27**, 296.

GROSSER, O. (1912). The development of the pharynx and of the organs of respiration. In *Kiebel and Mall's manual of human embryology*, Vol. 2 London.

HISLOP, A. and REID, L. M. (1974). Growth and development of the respiratory system. Structural development. In *Scientific foundations of paediatrics* (eds. J. A. Davis and J. Dobbing). Heinemann, London.

JACKSON, C. L. and HUBER, J. R. (1943). Correlated applied anatomy of the bronchial tree and lungs with a system of nomenclature. *Dis. Chest* **9**, 319.

JEFFREY, P. K. and REID, L. M. (1977). The respiratory mucous membrane. In *Respiratory defense mechanisms* (eds. J. D. Brain, D. F. Proctor, and L. M. Reid). Marcel Dekker, New York.

KUHN, C. and FINKE, E. H. (1972). The topography of the pulmonary alveolus. *J. Ultrastruc. Res.* **38**, 161.

LAUWERYNS, J. (1971). The blood and lymphatic micro-circulation of the lung. In *Pathology annual* (ed. S. C. Sommers) Vol. 6, p. 365. Appleton Century Crofts, New York.

LEAK, L. V. (1977). Pulmonary lymphatics. In *Respiratory defense mechanisms* (eds. J. D. Brain, D. F. Proctor, and L. M. Reid). Marcel Dekker, New York.

MCDOUGALL, J. and SMITH, J. F. (1975). The development of the human type II pneumocyte. *J. Path.* **115**, 245.

NAI-SAI WANG (1975). The preformed stomas connecting the pleural cavity and the lymphatics in the parietal pleura. *Ann. R. Resp. Dis.* **3**, 12.

O'RAHILLY, R. and BOYDEN, E. A. (1973). The timing and sequence of events in the development of the human respiratory system during the embryonic period proper. *Z. Anat. Entwickl.-Gesch.* **141**, 237.

RYAN, S. F., CIANNELLA, A., and DUMAIS C. (1969). The structure of the interalveolar septum of the human lung. *Anat. Rec.* **165**, 467.

SMITH, E. I. (1957). The early development of the trachea and oesophagus in relation to atresia of the oesophagus and tracheo-oesophageal fistula. *Contrib. Embryol. Carnegie Instn.* **36**, 41.

SOROKIN, S. P. (1977a). The respiratory system. In *Histology* (eds. L. Weiss and R. D. Greep) 4th edn. McGraw-Hill, New York.

—— (1977b). Phagocytes in the lungs. In *Respiratory defense mechanisms* (eds. J. D. Brain, D. F. Proctor, and L. M. Reid). Marcel Dekker, New York.

STREETER, G. L. (1945). Developmental horizons in human embryos: XII and XIII. *Contrib. Embryol. Carnegie Instn.* **31**, 27.

—— (1948). Developmental horizons in human embryos: XIV–XVII. *Contrib. Embryol. Carnegie Instn.* **32**, 133.

SWIFT, D. L. and PROCTOR, D. F. (1977). Access of air to the respiratory tract. In *Respiratory defense mechanisms* (eds. J. D. Brain, D. F. Proctor, and L. M. Reid). Marcel Dekker, New York.

VIDIC, B. (1971). The morphogenesis of the lateral nasal wall in the early prenatal life of man. *Am. J. Anat.* **130**, 121.

WARBRICK, J. G. (1960). The early development of the nasal cavity and upper lip in the human embryo. *J. Anat. (Lond.)* **94**, 351.

WEIBEL, E. R. (1963). *Morphometry of the human lung.* Academic Press, New York.

WELLS, L. J. (1954). Development of the human diaphragm and pleural sacs. *Contrib. Embryol. Carnegie Instn.* **35**, 107.

—— and BOYDEN, E. A. (1954). The development of the broncho-pulmonary segments in human embryos of horizons XVII to XIX. *Am. J. Anat.* **95**, 163.

WESSELLS. N. K. (1970). Mammalian lung development: interactions in formation and morphogenesis of tracheal buds. *J. exp. Zool.* **175**, 455.

WIDDICOMBE, J. G. (1977). Respiratory reflexes and defense. In *Respiratory defense mechanisms* (eds. J. D. Brain, D. F. Proctor, and L. M. Reid). Marcel Dekker, New York.

8 The urogenital system

R. G. HARRISON

THE urogenital system consists of: (1) the **urinary organs** proper, which are concerned with the formation, temporary storage, and discharge of the urine; and (2) the **reproductive or genital organs**, some of which arise from the same embryonic tissues as the urinary organs and retain an intimate relationship to them.

The urinary organs

The urinary organs are: the **kidneys**, where the urine is elaborated from the blood; the **ureters**, which convey the urine from the kidneys to the urinary bladder; the **urinary bladder**, where the urine is temporarily stored, and the **urethra** through which it is voided.

THE KIDNEYS

Form and size

The kidneys are a pair of bean-shaped organs, brownish-red when fresh, and of a glistening appearance due to the fact that each of them is covered with a thin but tough **fibrous capsule** which is close-fitting though easily stripped off. They are approximately symmetrical in size and shape. Each kidney is about 11 cm long, 5 cm wide, and about 3 cm thick; and it weighs about 130 g. It possesses **upper** and **lower extremities** or poles (of which the upper is generally the more bulky), **anterior** and **posterior surfaces**, and **medial** and **lateral borders**. The lateral border is uniformly convex from the upper to the lower pole; the medial border has a deep concavity which leads into a hollow that extends almost half-way through the kidney. This hollow is the **sinus of the kidney**, and the entrance into it is called the **hilus of the kidney**. The renal artery enters the kidney through the hilus; the renal vein and the **pelvis of the kidney** (the upper, expanded part of the ureter) leave it by the same route.

Position

Each kidney lies in the upper, deep part of a gutter alongside the projecting vertebral bodies which are covered laterally by the psoas major muscle. Inferiorly, these gutters diverge and become shallower as the psoas major muscles, increasing in bulk, pass laterally and forwards. Thus each kidney, lying against the sloping lateral surface of the psoas major, has its long axis parallel to the lateral surface of that muscle [FIG. 8.2], while its medial border faces forwards as well as medially and its anterior surface laterally as well as forwards [FIG. 8.1].

The two kidneys are not quite symmetrical in position; almost always the left is about 1.5 cm higher than the right and it is therefore slightly more medial, on account of the slope of the gutter in which it lies [FIG. 8.2]. The level of each kidney, however, varies slightly both with the respiratory movements and with posture, but

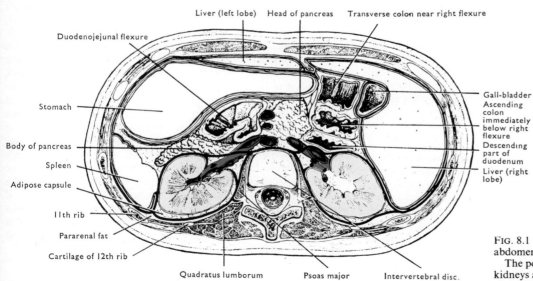

FIG. 8.1 Transverse section through the abdomen of a child.

The position and the general relations of the kidneys are well seen, and the arrangement of the renal fascia is indicated. The fascia is shown as a solid black line.

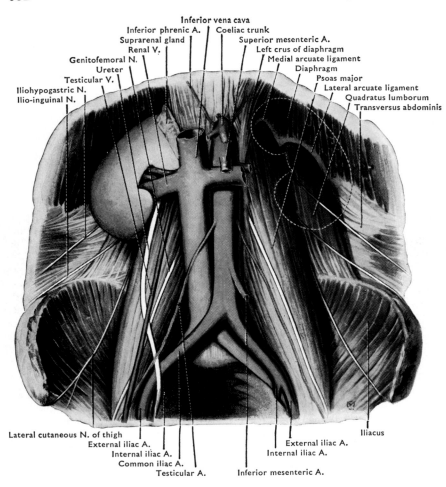

FIG. 8.2. Dissection to show the relation of the kidney and ureter to the posterior abdominal wall. The dotted outline indicates the position of the left kidney.

the range of this normal physiological movement does not exceed 2.5 cm. During quiet respiration in the recumbent attitude, the hilus of the left kidney is approximately in the transpyloric plane about 3.5 cm from the median plane of the body, or, anteriorly, about a finger's breadth medial to the tip of the ninth costal cartilage; the hilus of the right kidney is about 1.5 cm below this point. The lower extremity of the kidney is about 7.5 cm from the median plane—the left in the subcostal plane and the right 1.5 cm lower.

Seen from behind, the hilus of the left kidney is at the level of the spine of the first lumbar vertebra, or at the level of the line where the lateral border of the erector spinae muscle meets the twelfth rib, and the lower extremity is 5 cm or less above the highest point of the iliac crest [FIG. 8.4]. Corresponding points on the right kidney are about 1.5 cm lower.

Renal fascia and fat

Each kidney has an intimate relationship to the corresponding suprarenal gland and is enclosed with it in a common investment of weak fascia, known as the renal fascia, which is formed by the splitting of the transversalis fascia into anterior and posterior layers [FIG. 8.1]. Outside this investing fascia and between it and the kidney (surmounted by the suprarenal gland) there is a variable amount of fat called the renal fat. The fat outside the renal fascia is known as pararenal fat; it is most abundant behind the lower extremity of the kidney. The fat within the fascia is called the adipose capsule (perirenal fat) and is most abundant at the margins of the organ; it is traversed by weak fibrous strands which connect the renal fascia with the fibrous capsule of the kidney.

The two layers of the renal fascia fuse together a short distance above the suprarenal gland and become continuous with the fascia on the under surface of the diaphragm; lateral to the kidney also they fuse together and are continued as the transversalis fascia. Below the kidney the two layers remain separate, the anterior gradually fading into retroperitoneal areolar tissue and the posterior joining the fascia over the iliacus muscle. The posterior layer, passing over the quadratus lumborum, becomes continuous medially with the fascia of the psoas major at its posterior border. As it reaches the medial border of the kidney it gives off a lamina which turns laterally into the hilus of the kidney and is attached to the back of the pelvis of the kidney; the remainder continues over the psoas and gains an attachment to the bodies of the vertebrae just in front of the muscle. The anterior layer of the renal fascia extends across the aorta and inferior vena cava as a thin layer which loses its identity superiorly in the connective tissue around the coeliac trunk. From the anterior layer a distinct lamina also passes backwards medial to the kidney and blends with the anterior lamina of the posterior layer. This lamina is pierced by the renal vessels and the ureter (Martin 1942); its presence accounts for the limitation of perirenal effusions to one side of the body and for the statement (Mitchell 1939) that the anterior layer as well as the posterior is attached to the bodies of the vertebrae. See also Mitchell (1950), and FIGURES 8.1 and 9.4.

Structures posterior to the kidneys [FIGS. 8.1–8.5]

The medial part of the kidney overlaps the lateral side of the psoas muscle. Lateral to this, and forming the posterior relation of most of the kidney, lies the quadratus lumborum muscle covered by the anterior layer of the thoracolumbar fascia, and further laterally the aponeurosis of the transversus abdominis muscle, on to which the kidney extends for a short distance. Towards the upper end of the kidney, the posterior margin of the diaphragm (medial and lateral arcuate ligaments) crosses the psoas major and quadratus lumborum muscles, so that the superolateral third of the kidney lies in front of the diaphragm.

Emerging from behind the lower border of the lateral arcuate ligament and running downwards and laterally across the front of the quadratus lumborum muscle is the subcostal nerve accompanied by its vessels. A little lower down the iliohypogastric and, below it, the ilio-inguinal nerve emerge from the lateral side of the psoas major and run a parallel, inferolateral course. The subcostal and

iliohypogastric nerves pierce the aponeurosis of the transversus abdominis a short distance lateral to the quadratus lumborum, and the ilio-inguinal nerve pierces it a little further forwards [FIG. 8.2].

In this region, some bony structures lie posterior to the diaphragm and quadratus lumborum, and are therefore more distantly posterior to the kidneys. The twelfth rib runs downwards and laterally behind the diaphragm, and its lateral part passes for a variable distance into the abdominal wall just lateral to the upper end of quadratus lumborum. The transverse process of the first lumbar vertebra lies at the level of the junction of the medial and lateral arcuate ligaments, and those of the other vertebrae at correspondingly lower levels—their tips being behind the medial edge of the quadratus lumborum. If the kidneys have been hardened in situ shortly after death in a thin subject, their posterior surfaces are generally moulded

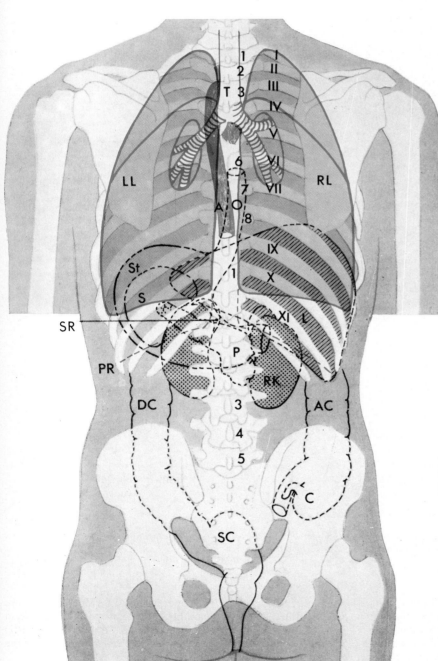

FIG. 8.3. Posterior aspect of trunk, showing surface topography of viscera. Roman numbers, ribs; arabic numbers, vertebrae.

A.   Aorta, descending.
AC. Ascending colon.
C.   Caecum.
DC. Descending colon.
L.   Liver.
LL. Left lung.
O.   Oesophagus.
P.   Pancreas.
PR. Line of pleural reflexion.
RK. Right kidney.
RL. Right lung.
S.   Spleen.
SC. Sigmoid colon.
SR. Suprarenal.
St. Stomach.
T.   Trachea.

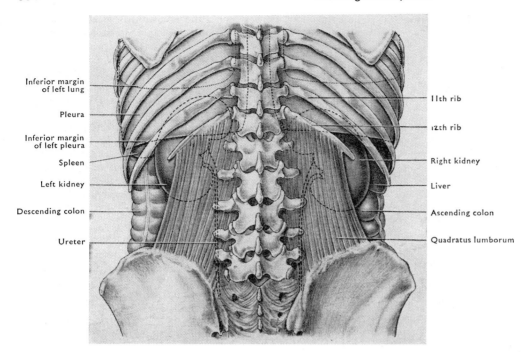

FIG. 8.4. Dissection from behind to show the relation of the pleural sacs to the kidneys.

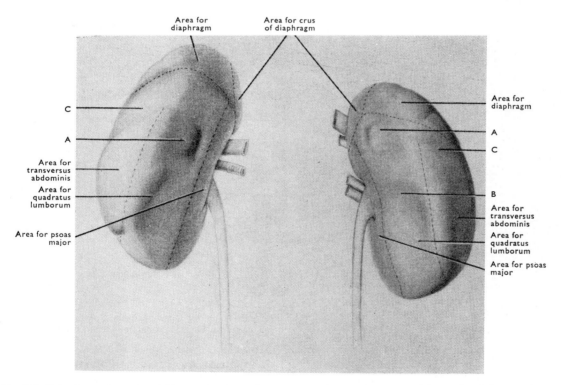

FIG. 8.5. Posterior surfaces of the kidneys. The dotted lines mark out the areas in contact with muscles of the posterior abdominal wall.
A and B, depressions corresponding to transverse processes of first and second lumbar vertebrae.
C, depression corresponding to the twelfth rib.

by the muscles and bony structures against which they lie [FIG. 8.5], but this is merely a post-mortem phenomenon. Moreover, the part of the twelfth rib which lies behind the diaphragm is separated from the diaphragm by the costodiaphragmatic recess of the pleura, which extends down as far as a line drawn from the middle of the twelfth rib to a point midway between the head of the twelfth rib and the transverse process of the first lumbar vertebra [FIG. 8.4].

The pararenal fat behind the upper part of the kidney is therefore only separated from the pleura by a very thin sheet of diaphragm which is often partly deficient (vertebrocostal triangle, FIG. 5.102) especially on the left side of the body. Thus the pararenal fat and the pleura may be in direct contact.

More remote posterior relations of the kidneys are the middle layer of the thoracolumbar fascia, the upper three lumbar arteries, the erector spinae muscle, and the posterior layer of the thoraco-lumbar fascia giving origin to the serratus posterior inferior and the latissimus dorsi muscles.

Structures anterior to the kidneys [FIGS. 8.6 and 8.7].

The anterior aspect and medial border of the upper extremity of each kidney is covered by the corresponding suprarenal gland. The other structures anterior to the two kidneys are quite different; and, as already pointed out, all other organs in front of the kidney are separated from it by the renal fascia, whereas the suprarenal gland is enclosed with the kidney inside this fascia.

RIGHT KIDNEY. Below the area covered by the suprarenal gland, about two-thirds of the anterior surface of the right kidney is in contact with the posterior surface of the right lobe of the liver. The uppermost part of this hepatic area may be related to the bare area of the liver, from which it is separated by some loose areolar tissue, but most of the hepatic area is covered by peritoneum which is reflected from the back of the liver on to the front of the kidney. Below the hepatic area, the right flexure of the colon is directly in contact near the lower extremity, and medial to both, a vertical strip, including the hilus, is covered by the descending part of the duodenum. The peritoneum on the hepatic area passes forwards to

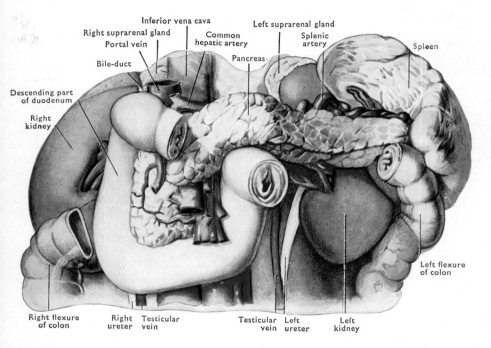

FIG. 8.6. Dissection to show relations of the kidneys.

The greater part of the stomach has been removed by an incision made close to the pylorus. The transverse colon has been taken away, and the small intestine has been cut across close to the duodenojejunal flexure. (From a model by Birmingham.)

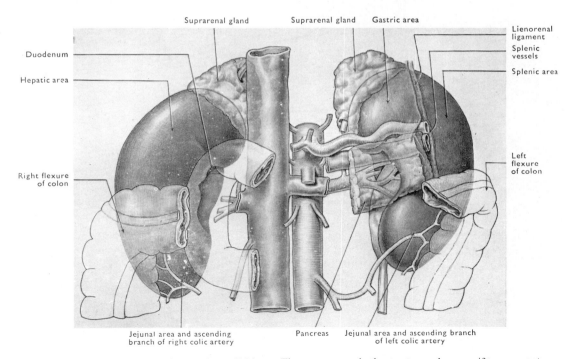

FIG. 8.7. Diagram of anterior relations of kidneys. The pancreas and other parts are shown as if transparent.

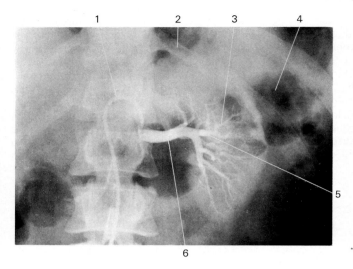

FIG. 8.8. Selective renal angiogram. The injection is made through a catheter passed into the abdominal aorta from the femoral artery. The segmental and interlobar arteries are well seen, but only parts of the arcuate system are visible.
1. Catheter in aorta.
2. 11th rib.
3. Interlobar artery.
4. Gas in left flexure of colon.
5. Segmental artery.
6. Renal artery.

cover the colon and the descending part of the duodenum. At the lower extremity of the kidney there is sometimes a small gap between the colic and duodenal areas. When present, this is covered with peritoneum that passes up from the posterior abdominal wall on to the front of the kidney and separates this area from coils of small intestine.

LEFT KIDNEY. Below and lateral to the area covered by the suprarenal gland there is a small triangular region covered with peritoneum of the omental bursa which separates it from the posterior surface of the stomach. Below this gastric area the body of the pancreas lies across the middle of the kidney. Lateral to the gastric area is the line of attachment of the lienorenal ligament, which runs approximately from the upper extremity of the kidney to the junction of its middle and lower thirds. The peritoneum which covers the gastric area passes downwards on to the front of the body of the pancreas, whence it continues as the anterior layer of the transverse mesocolon. Laterally it is reflected forwards as the medial layer of the lienorenal ligament. The body of the pancreas is separated from the kidney by loose areolar tissue in which the splenic vein runs medially, while the artery runs its tortuous course along the upper border of the pancreas. The area lateral to the line of attachment of the lienorenal ligament is separated from the spleen by the peritoneum which covers both and is continuous with the lateral layer of the lienorenal ligament. Below the pancreatic area, the lateral part is covered by the left flexure of the colon and the upper part of the descending colon. The medial part is covered by the peritoneum of the posterior abdominal wall. Above, this is reflected forwards and downwards as the posterior layer of the

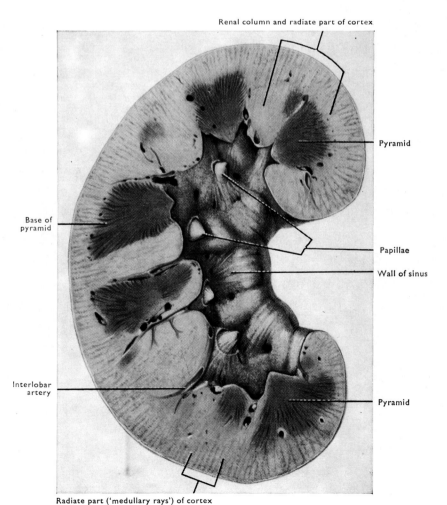

Renal column and radiate part of cortex

Base of pyramid

Interlobar artery

Radiate part ('medullary rays') of cortex

Pyramid

Papillae

Wall of sinus

Pyramid

FIG. 8.9. Longitudinal section through the kidney.
   The vessels and fat have been removed to show the wall of the sinus of the kidney. The points where the vessels enter the substance of the kidney are seen as holes in the wall of the sinus.

transverse mesocolon, and laterally it passes over the left flexure and the descending colon. This medial area is covered by coils of jejunum in front of the peritoneum and by the ascending branch of the left colic artery, which crosses it behind the peritoneum. The obliquity of the anterior surface of the kidney has already been pointed out; and, if this is borne in mind, it will be realized that though the spleen and left flexure of the colon are described as anterior relations of the left kidney, they are really more lateral than anterior.

Structures medial to the kidneys [FIGS. 8.1 and 8.2].

Medial to each kidney are the corresponding psoas major muscle and psoas minor (if present); and along the medial margin of the psoas lies the abdominal part of the sympathetic trunk. Antero-medially, each kidney is related to the corresponding suprarenal and renal vessels, ovarian or testicular vein, and upper end of the ureter. Further medially, is the inferior vena cava on the right side and the abdominal aorta on the left.

Sinus of the kidney [FIGS. 8.9 and 8.10]

The sinus is the hollow within the kidney, and its long axis corresponds to that of the kidney itself. Its thick walls are formed by the substance of the kidney, and, except where they are covered by the minor calyces [p. 542], they are lined with a continuation of the capsule of the kidney, which is reflected into the sinus over the lips of the hilus. The floor of the sinus is not flat but presents a series of small conical elevations, called **renal papillae**, which vary from six to fifteen in number. Radiating from each papilla there are several small ridges of kidney substance separated by depressed areas. The blood vessels and nerves enter and leave the kidney by piercing the

depressed areas. On the summit of each renal papilla there are many minute openings, which are the terminal apertures of the tubules, of which the kidney is mainly composed, and through which urine escapes into the minor calyces which are cupped over the papillae. The cavity of the sinus is filled with fat, embedded in which lie the major and minor calyces, part of the pelvis of the kidney, branches of the renal artery, and tributaries of the renal vein. The artery usually branches and the tributaries of the vein unite a short distance outside the hilus; and, within the sinus, the tributaries of the vein are most anterior, while further posteriorly are the calyces of the kidney, with branches of the artery both in front and behind.

Fixation of the kidneys

The kidneys are held in position partly by the renal vessels and the surrounding fat and fascia, but mainly by the tension of the muscles of the anterior abdominal wall transmitted through the other abdominal viscera and acting against the inclined plane formed by the posterior abdominal wall.

The kidney in section

Sections of the kidney [FIGS. 8.9 and 8.10] show that it is composed to a large extent of a number of conical masses—known as **renal pyramids**. The pyramids constitute, collectively, the **medulla** of the kidney; their bases are directed towards the surface and their apices project into the minor calyces as the renal papillae. The pyramids are more numerous than the papillae, two or three usually ending in each papilla in the middle part of the kidney, and sometimes as many as six or more in a single papilla near the extremities of the kidney. The bases of the pyramids are separated from the surface by

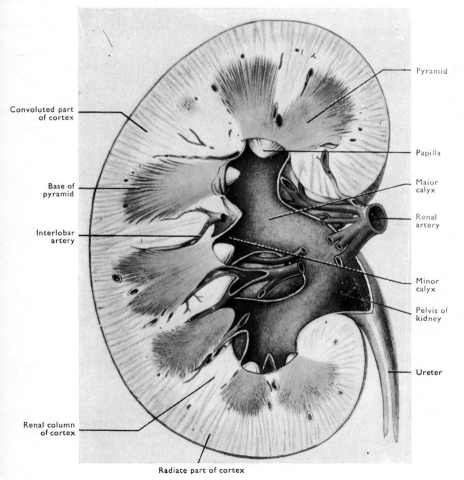

Convoluted part of cortex

Base of pyramid

Interlobar artery

Renal column of cortex

Radiate part of cortex

Pyramid

Papilla

Major calyx

Renal artery

Minor calyx

Pelvis of kidney

Ureter

FIG. 8.10. Longitudinal section of the kidney and its pelvis to show the relations of the calyces to the renal papillae.

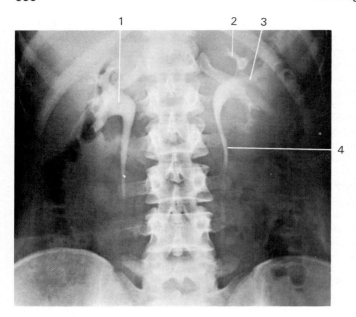

FIG. 8.11. Intravenous pyelogram a few minutes after the intravenous injection of contrast medium. The pelvis of the kidney and the upper end of the ureter are filled on each side.
1. Pelvis of kidney.
2. Minor calyx.
3. Major calyx.
4. Ureter.

a layer called the **cortex** of the kidney. The cortex not only covers the bases of the pyramids, but also sends prolongations, known as **renal columns**, between the pyramids, towards the sinus. In section, both parts have a striated appearance, but the cortex is less markedly striated and is usually darker in colour. The base of each pyramid appears to be composed of alternate dark and light streaks while the papillary part is often of a lighter colour, and is very faintly striated.

There are five main branches ('segmental' arteries) of the renal artery within the sinus. These divide into **interlobar arteries** which enter the renal tissue and are visible between the pyramids. Some of their main branches can be seen passing across the bases of the pyramids parallel to the surface of the kidney as the **arcuate arteries**, which do not anastomose with one another.

In the foetus and young child, and sometimes, though much less distinctly, in the adult, the surface of the kidney is marked by a number of grooves that divide it into polygonal areas [FIG. 9.1]. The areas represent the **lobes** or **reniculi** of which the kidney is originally composed, and each corresponds to one papilla with its pyramids and the surrounding cortex. There are fourteen such lobes, seven anterior and seven posterior (Löfgren 1949). In addition there is sometimes a longitudinal sulcus situated along the lateral margin of the kidney which separates the anterior from the posterior lobes.

An examination with an ordinary pocket lens shows that it is the lighter striae of the bases of the pyramids which are continuing into the cortex becoming progressively less distinct. The parts of the cortex which seem in this way to be continuations of the medulla are called medullary rays or the **radiate part** of the cortex. The part of the cortex which intervenes between them is darker in colour and forms what is known as the **convoluted part** or labyrinth. The appearance presented by the cortex in section varies according to the plane in which the section has been taken. If the section is parallel to the axis of a pyramid, the medullary rays will appear as isolated streaks directed from the base of the pyramid towards the surface of the kidney, and separated from one another by narrow strips of the convoluted part. In sections made at right angles to the

axis of a pyramid, the rays are circular in outline, and are surrounded by the convoluted part.

Kidney tubules

The glandular substance of the kidney is composed of a vast number of minute **renal tubules**, each of which has a complicated course [FIG. 8.12]. The wall of each tubule consists throughout of a basement membrane with an epithelial lining, but the lumen of the tubule and the character of the epithelium vary much in their different parts. Every tubule begins in a thin-walled spherical dilatation, known as a **glomerular capsule**, in which lies a complicated tuft of capillary blood vessels or **glomerulus** covered by a reflexion of the delicate wall of the capsule (visceral layer) [FIG. 8.13]. The capsules with their enclosed capillaries are called **renal corpuscles**, and they are all placed in the convoluted portion of the cortex, where they may be recognized as minute red points just visible to the unaided eye and best marked when the renal vessels are congested. The part of the tubule that leads from the capsule—the **proximal (first) convoluted tubule**—is very tortuous and lies within the convoluted part of the cortex. Passing from the convoluted

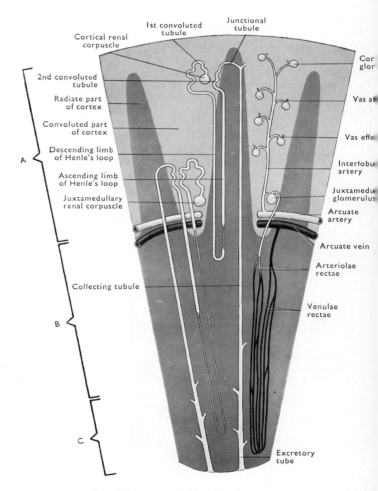

FIG. 8.12. Diagrammatic representation of the structures forming a kidney lobe. On the left of the Figure the course of a renal tubule connected with a cortical renal corpuscle, and one connected with a juxtamedullary renal corpuscle, are shown. Note the differences in position and variations in calibre in Henle's loop of cortical and juxtamedullary renal corpuscles. On the right of the Figure are displayed the vascular arrangements within the kidney. The vasa efferentia of the cortical renal corpuscles are shown cut across for purposes of clarity, but they actually break up into capillaries around the renal tubules.

A, cortex; B, basal portion; C, papillary portion of pyramid.

part, the tubule enters a medullary ray, in which its course becomes less complicated, and here it receives the name of the spiral tubule. Leaving the medullary ray, the tubule enters the basal portion of the pyramid, and, diminishing in diameter, it pursues a straight course towards the apex of the pyramid, forming the so-called descending

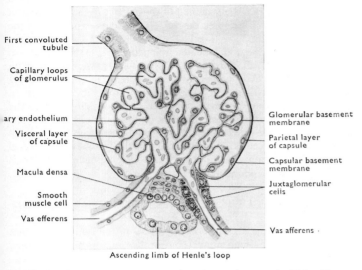

FIG. 8.13. Diagrammatic representation of a renal corpuscle. (After Ham 1957.)

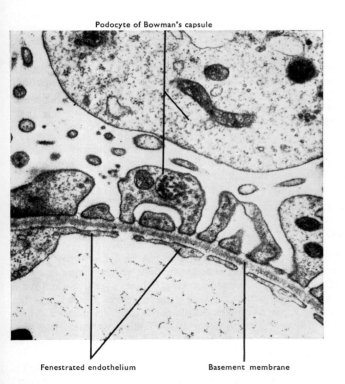

FIG. 8.14. An electron micrograph showing the filtration apparatus of the kidney. The section shows the structure separating the blood in the glomerular capillaries and the cavity of Bowman's capsule. It consists of a thick basement membrane partly covered on the vascular aspect by a layer of fenestrated endothelium, and on the capsular side by the interdigitating foot processes of the podocytes which form the internal epithelium of Bowman's capsule. As can be seen, there are spaces between the foot processes where they expand on to the basement membrane.

limb of Henle's loop. Within the apical portion of the pyramid the tubule suddenly bends upon itself, forming the loop of Henle, and, reversing its direction, passes back again through the base of the pyramid into a medullary ray of the cortex as the ascending limb of Henle's loop. Leaving the medullary ray again, the ascending limb enters the convoluted part of the cortex, where it returns to the glomerulus of its own renal tubule, and lies in between the afferent and efferent vessels of the glomerulus [FIG. 8.13]. The nuclei of the epithelial cells of the tubule are concentrated in one wall of the tubule at this point [FIG. 8.13] to form the macula densa. While still within the convoluted part, the tubule again becomes tortuous as the distal (second) convoluted tubule; this tubule finally ends in a short junctional tubule, which passes back into a medullary ray and joins a collecting tubule. Each collecting tubule receives numerous renal tubules, and pursues a straight course through a medullary ray of the cortex and the pyramid. Finally, several collecting tubules, uniting together, form a papillary duct, which opens on the summit of a renal papilla into a minor calyx by one of the minute foramina already described. In microscopic sections the various portions of the renal tubule may be distinguished by the position which they occupy and by the character of the lining epithelium. Each medullary ray with its surrounding glomeruli and proximal and distal convoluted tubules which empty into the collecting tubule forming the centre of the ray, constitutes a lobule of kidney tissue. It is demarcated on each side by the interlobular arteries, fine vessels which arise from the arcuate arteries and mostly pass centrifugally, although a few course centripetally to supply the juxtamedullary glomeruli [FIG. 8.12].

Supporting tissue of the kidney

The tubules and blood vessels are all united together by a very small amount of areolar tissue which completely surrounds each tubule and blood vessel and binds it to its neighbours. The areolar tissue forms a continuous network, the spaces in which faithfully reproduce the outlines and the arrangement of the renal tubules. The network of the stroma is continuous with the capsule of the kidney.

VESSELS AND NERVES OF THE KIDNEY

The renal arteries, one on each side, arise directly from the aorta. The left is usually slightly higher than the right and they are both large vessels relative to the size of the kidneys. Each artery lies behind the corresponding renal vein, and at or just outside the hilus of the kidney it divides typically into anterior and posterior divisions. The posterior division passes behind the pelvis of the kidney while the anterior division passes in front of it and divides into two to five branches. Within the sinus each branch divides further. The territories of distribution of the anterior and posterior divisions within the kidney meet at the junction of its anterior two-thirds and posterior one-third (Brödel's line) which does not correspond to the longitudinal sulcus (Sykes 1964a). Having entered the substance of the kidney, the larger arteries lie in the intervals between the pyramids and are called interlobar arteries. These vessels divide and form a series of incomplete arterial arches—the arcuate arteries—which pass across the bases of the pyramids. No anastomosis between the branches of the interlobar arteries takes place, each artery which pierces the wall of the kidney sinus being an end-artery. The arcuate arteries give off a number of vessels which pass through the convoluted part of the cortex towards the surface of the kidney; they are known as the interlobular arteries, and they lie at very regular intervals. From them a number of short branches arise, termed vasa afferentia, each of which proceeds to a renal corpuscle. There, the vas afferens breaks up into the capillaries of the glomerulus, which are contained within the invaginated glomerular capsule. Just before breaking up into the glomerular capillaries, the cells of the vas afferens change in character, and are termed **juxtaglomerular cells** [FIG. 8.13], which are reputed to secrete a substance called renin. The small vessel—**vas efferens**—which issues from the glomerulus, instead of running directly into a larger vein, breaks up, after the manner of an artery, into capillaries which supply the convoluted tubules of the cortex and the longitudinal tubules of the medullary rays. At the bases of the pyramids there are some 'juxtamedullary' glomeruli whose efferent vessels pass as fine arterioles into the pyramids where they break up into a capillary network around the tubules. These clusters of fine arterioles, known as the **arteriolae rectae**, are just visible to the naked eye, and they give the bases of the pyramids their striated appearance [FIGS. 8.9, 8.10, and 8.12]. The juxtamedullary glomeruli are larger than those in the convoluted part of the cortex and their afferent and efferent vessels are of equal calibre; they are apparently non-functional. In certain conditions the renal blood flow is said to be shunted through the vessels of these non-functional glomeruli and the arteriolae rectae into the renal vein without passing through the functional glomeruli of the cortex proper (Trueta *et al.* 1947). Franklin (1949) gives an account of the history of research on the renal circulation.

A few interlobar arteries continue their radial course, perforate the fibrous capsule, and supply the adipose capsule.

Veins corresponding to the interlobular arteries and arteriolae rectae collect the blood from the capillaries around the tubules and unite to form a series of complete arcuate veins which anastomose freely across the bases of the pyramids. From the venous arcades, vessels arise which traverse the intervals between the pyramids and reach the sinus of the kidney, where they unite to form the tributaries of the renal vein. Some small veins in the superficial part of the cortex (**stellate venules**) communicate through the fibrous capsule with minute veins in the renal fat. Typically three main tributaries issue from the sinus and unite to form the **renal vein**, which runs a direct course to end in the inferior vena cava.

Variations in the renal vessels are common. Supernumerary arteries and veins or supernumerary branches of the renal artery and vein are frequent. The abnormal vessels may enter or leave the kidney within the sinus or at some other point. Small accessory renal veins from the lower extremity of the kidney are especially common.

The **lymph vessels** of the kidney end in the aortic lymph nodes.

The **nerves** of the kidney are derived mainly from the coeliac plexus (Mitchell 1956) and accompany the branches of the artery, forming a renal plexus. The branches of this plexus form bundles of nerve fibres which travel along the branches of the renal artery into the kidney and also run between and supply the tubules.

From clinical evidence it would appear that the afferent nerve fibres which supply the kidney are connected with the tenth, eleventh, and twelfth thoracic nerves.

VARIATIONS

A marked difference in the size of the two kidneys is sometimes observed, a small kidney on one side of the body being usually compensated for by a large kidney on the opposite side. Congenital absence of one or other kidney is recorded. A few cases are on record in which an extra kidney was found on one or other side.

Traces of the superficial lobulation of the kidney, present in the foetus and young child, are often retained in the adult, representing remnants of the seven anterior and seven posterior lobes which, with their corresponding papillae and calyces, are present by the twenty-eighth week of development (Sykes 1964b), but are subsequently reduced in number by fusion.

Horseshoe-kidney is a not infrequent abnormality. In such cases the two kidneys are united at their lower ends across the median plane by a connecting bar of kidney substance. The amount of fusion between the two kidneys varies considerably; it is sometimes

very complete, while in other cases it is slight, the connection being composed chiefly of fibrous tissue. The hilus of each kidney looks forwards, and the ureters always pass in front of the connecting piece which lies across the vertebral column caudal to the inferior mesenteric artery. Fusion between the two kidneys may also occur at other points on their medial borders.

Very rarely the kidney appears to be almost entirely surrounded by peritoneum and to be attached to the abdominal wall by a kind of mesentery which encloses the vessels and nerves passing to the hilus. The condition is believed to be congenital.

One or both kidneys may be at a much lower level than usual—in the iliac fossa or even in the pelvic cavity. This is associated with an arrest in the normal change in position, relative to surrounding structures, which the kidney experiences during development. In such cases the kidney does not receive its blood supply from the usual source but from vessels which arise from the lower end of the aorta, or from an iliac artery, or the median sacral artery. Such abnormally situated kidneys do not usually possess the typical outline of the normal organ but vary much in shape, and the hilus is often malformed and misplaced.

METHODS OF CLINICAL ANATOMICAL EXAMINATION

A kidney of normal size and occupying its normal position cannot be palpated in the living subject, but, if the organ is much enlarged or abnormally mobile, it can usually be felt at the end of a deep inspiration. To search for a palpable kidney, place the patient on his back, pass one hand under him below the last rib, and with the other hand press the anterior abdominal wall opposite the fingers of the first hand. If the kidney is palpable it can be felt slipping between the opposed fingers when the patient takes a deep breath. The outline of the kidney cannot be defined by percussion as the organ is deeply placed and adjacent to other organs of similar consistency to itself.

In the angle between the lateral border of the erector spinae muscle and the lower border of the twelfth rib, the kidney lies fairly close to the posterior surface of the body, for there it is separated from the surface only by relatively thin structures—the quadratus lumborum, the layers of the lumbar fascia, the serratus posterior inferior, and the latissimus dorsi. Lesions, e.g. an abscess in the region of the kidney, can cause tenderness and swelling in this area. The close relationship of the psoas major muscle and the kidney explains why extension of the thigh may increase pain due to inflammation in the kidney region.

RADIOLOGY

The internal structure of the kidney can be demonstrated by the intravenous injection of a contrast medium which is excreted by the kidneys. This allows the renal cortex, pyramids, calyces, and renal pelvis to be seen within minutes of the injection. Retention of the medium in the renal tubules and glomeruli gives the kidney a denser appearance in the early stages of the examination (*nephrogram*). This nephrographic effect can be intensified, if required, by increasing the dose of medium.

Alternatively the renal pelvis and calyces can be filled through a ureteric catheter introduced through a cystoscope in the bladder. Because of the high quality of intravenous pyelograms now obtainable, this method is now less often required [FIG. 8.15].

The blood supply of the kidney can be demonstrated by selective angiography after the passage of a catheter through the femoral artery into the renal artery. Serial films show successive filling of the renal arteries, the interlobar arteries and sometimes the arcuate and interlobular arteries and veins. The later films of the series show a

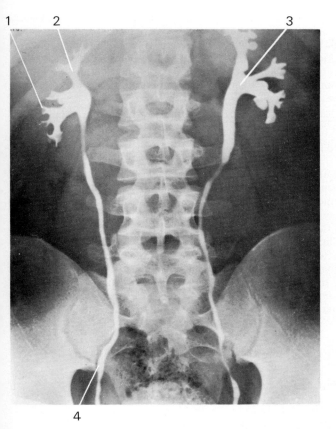

FIG. 8.15. Retrograde pyelogram. The renal pelves and ureters have been filled through catheters inserted into the lower ends of both ureters through a cystoscope.
1. Minor calyx.
2. Major calyx.
3. Pelvis of kidney.
4. Ureter.

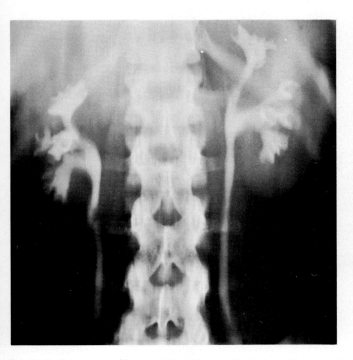

FIG. 8.16. Tomogram of intravenous pyelogram. Note the obliteration of overlying shadows, e.g. gas in the intestine.

dense nephrographic outline [FIG. 8.8]. This method shows defects in the vascular tree and also demonstrates other pathology by the altered vascular pattern.

Ultrasound examination of the kidney is a simple non-invasive method of demonstrating the renal outline and renal pelvis; though the picture is less clear than that obtainable by radiography, it may be more informative. For example the distinction between solid and cystic masses is often more reliable when this method is used.

Congenital anomalies of the kidney and renal pelvis include congenital absence, partial fusion (horseshoe kidney) and complete fusion, ectopic kidney, and double ureter. The renal artery is often double, the accessory artery usually arising from the aorta near the lower pole of the kidney ('polar artery') which it supplies. Such an artery to the right kidney passes anterior to the inferior vena cava and may be seen indenting it in a cavogram [p. 959].

THE URETERS

The ureters are the ducts that lead from the kidneys to the urinary bladder. Each ureter begins above in a thin-walled, funnel-shaped expansion, called the **pelvis of the kidney** which is placed partly inside and partly outside the sinus of the kidney. Towards the lower end of the kidney the part of the pelvis which lies outside the sinus diminishes in calibre, and forms a tube, the ureter, which conveys the urine to the bladder.

PELVIS OF THE KIDNEY

Within the sinus of the kidney the pelvis lies among the larger renal vessels [FIG. 8.10]. Its volume is about 8 ml. It is formed by the

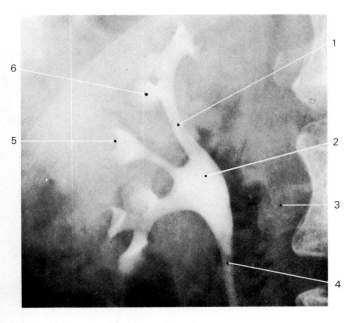

FIG. 8.17. Details of pelvis of kidney visualized by intravenous pyelography.
1. Major calyx.
2. Pelvis of kidney.
3. Transverse process of 2nd lumbar vertebra.
4. Ureter.
5. Minor calyx cupped by renal pyramid.
6. Minor calyx, oblique view.

junction of three (more rarely, two or four) thin-walled tubes—the major calyces—each of which has a number of branches. The branches, called minor calyces, are short, and increase in diameter as they approach the wall of the sinus, to which they are attached. Their wide, funnel-like ends enclose the renal papillae and receive the urine. The minor calyces vary in number from five to twenty, but there are usually eight or nine in the adult kidney (Sykes 1964b) one calyx sometimes surrounding two or even four papillae, the result of a greater degree of fusion of the calyces than the papillae during development. The portion of the pelvis that lies outside the kidney has anterior to it, in addition to the renal vessels, on the right side, the descending part of the duodenum, and on the left side, a part of the body of the pancreas and the peritoneum of the posterior abdominal wall [FIGS. 8.6 and 8.7] or, sometimes, the duodenojejunal flexure.

RADIOLOGY

The renal pelvis usually divides into two or three major calyces.

The minor calyces are variable in number and size. When viewed in profile they appear cupped, to receive the renal papillae, but at an angle they appear somewhat different [FIG. 8.17]. The mobility of the kidneys and renal pelves, as shown by comparing their levels in the erect and supine positions and in inspiration and expiration, amounts to a few centimetres, but occasionally it may be much more.

URETER

The ureter proper is a pale-coloured, thick-walled duct with a small lumen. When *in situ* it has a total length of about 25 cm, and lies behind, and closely adherent to the peritoneum throughout most of its course. The upper half of the ureter lies in the abdominal cavity, and the lower half in the lesser pelvis [FIGS. 8.2 and 8.18].

The abdominal portion of the ureter, about 12.5 cm in length, is directed downwards and slightly medially, and lies on the psoas major muscle, crossing it obliquely from the lateral to the medial side [FIG. 8.2]. Anteriorly, the abdominal portions of both ureters are crossed obliquely by the testicular or ovarian vessels, while posteriorly, the genitofemoral nerve passes inferolaterally in almost

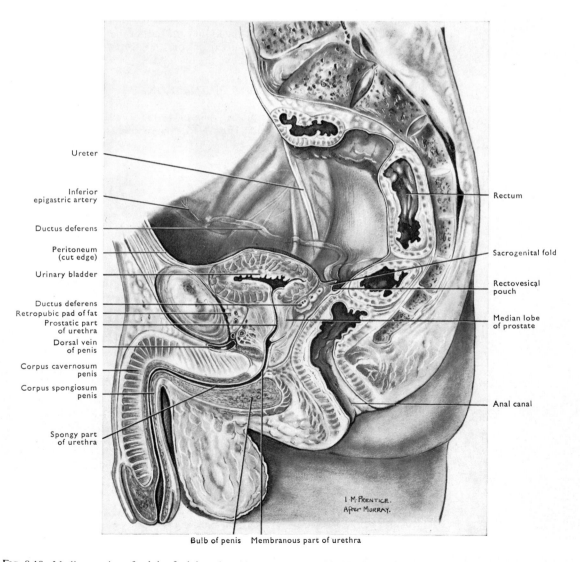

FIG. 8.18. Median section of pelvis of adult male subject. The urinary bladder is empty and firmly contracted [cf. FIG. 8.26].

the same line as the vessels. The right ureter has the descending part of the duodenum in front of its upper part, and below this is crossed anteriorly by the right colic and ileocolic vessels, and by the root of the mesentery shortly before the ureter enters the lesser pelvis. The left ureter is crossed anteriorly by the left colic vessels and by the root of the sigmoid mesocolon.

Crossing the common iliac or the external iliac artery, the ureter enters the lesser pelvis. Usually the left ureter crosses the common iliac artery, and the right ureter the external iliac; but that arrangement is by no means constant.

The **pelvic portion** of the ureter is about 12.5 cm in length; it passes backwards and downwards on the side-wall of the lesser pelvis, describing a curve which is convex backwards and laterally [FIG. 8.18], the most convex point being close to the inferior extremity of the greater sciatic notch at the level of the ischial spine [FIG. 13.75].

In its course within the pelvis, the ureter lies below and in front of the internal iliac artery, and crosses the medial side of the obturator nerve and vessels and of the umbilical artery.

IN THE MALE, about the level of the ischial spine, the ureter is crossed from before backwards by the ductus deferens, and from that point onwards is less intimately related to the peritoneum. It then bends forwards and slightly medially to reach the lateral angle of the bladder, where it is anterior to the upper extremity of the seminal vesicle [FIG. 8.24]. The ductus deferens having crossed the ureter also turns medially, and as it does so it lies on a plane posterior to the ureter. The lower end of the ureter is surrounded by a dense plexus of veins which brings the vesical plexus into communication with the internal iliac vein. The hypogastric nerve, which connects the superior hypogastric plexus with the inferior hypogastric plexus, divides to pass on each side of the ureter where it is crossed by the ductus deferens [FIGS. 8.18 and 11.85].

The right and left ureters reach the bladder 5 cm apart, and piercing the wall of the bladder very obliquely in an inferomedial direction, are embedded within its muscular tissue for nearly 2 cm. Finally, they open into the bladder by a pair of minute, slit-like apertures which are of a valvular nature and prevent a backward passage of fluid from the bladder. It is probable, however, that an exaggerated idea of the valvular nature of the openings of the ureters into the bladder is obtained by an examination of the parts in the dead subject. When the bladder is empty the openings of the ureters are placed about 2.5 cm apart, but when the bladder is distended they may be 5 cm apart, or more. As the ureter pierces the wall of the bladder the muscular fibres of the bladder and ureter remain quite distinct, and so the ureter, remaining a thick-walled tubular structure, appears to pass through a gap in the muscular wall of the bladder. The mucous membrane of the ureter becomes continuous with that of the bladder.

The lumen of the ureter proper is not uniform throughout; it is slightly constricted near the renal pelvis and where it crosses the iliac artery; and it is narrowest where it passes through the bladder wall.

IN THE FEMALE, the ureter descends in a peritoneal fold which forms the posterior boundary of the ovarian fossa [FIG. 8.52]. Near its termination it is accompanied by the uterine artery, and passing below the root of the broad ligament of the uterus and that artery, lies about 2 cm lateral to the cervix uteri, above the lateral fornix of the vagina. Finally it inclines medially to lie anterior to the lateral margin of the vagina [FIG. 8.56], but since the vagina deviates to one side, usually the left, more of one ureter than the other lies anterior to the vagina (Brash 1922).

Structure of the ureter

The wall of the ureter, which is thick and of a whitish colour, is composed of mucous, muscular, and fibrous coats. The **mucous coat** is lined with a transitional epithelium composed of many layers of cells, those nearest the lumen being of large size. When the canal is empty the mucous coat is thrown into numerous longitudinal folds, and so its lumen has a stellate outline in transverse section. The **muscular coat** consists of plain muscle fibres collected into bundles which are separated by a considerable amount of fibro-areolar tissue, and are arranged some longitudinally, some circularly. In the upper part of the ureter a relatively large amount of fibro-areolar tissue is present deep to and among the bundles of muscle fibres, which are arranged in three strata—an inner longitudinal, middle circular, and an outer longitudinal. In the middle part of the tube the same layers may be recognized, but the circularly disposed bundles of fibres are more numerous than higher up. This may be due to the fibres of the middle stratum, which are helical, being more tightly coiled. In the lower part of the ureter the fibro-areolar tissue is relatively scanty and the inner longitudinal fibres lie close to the mucous coat; in this region also the longitudinal folds of the mucous coat become fewer and less marked. A short distance above the bladder, a number of coarse bundles of longitudinally arranged muscle fibres are applied to the outer surface of the muscular coat. These form the so-called **sheath of the ureter**, and are continuous with the superficial part of the muscular wall of the bladder. In the portion of the ureter which traverses the wall of the bladder, nearly all the fibres of the muscular coat are disposed longitudinally. The muscle fibres lie immediately outside the epithelium, and end just where the mucous coats of the bladder and ureter become continuous. The outer **adventitial coat** of fibrous tissue varies in thickness at different levels. In its lower part it blends with the fibro-areolar tissue which lies among the muscle fibres of the sheath of the ureter just mentioned.

The mucous coat of the calyces and of the pelvis of the kidney has an epithelium like that of the ureter. Where each renal papilla projects into one of the calyces a deep, circular recess is formed between the wall of the calyx and the papilla. At the bottom of this recess the epithelium of the calyx becomes continuous with that covering the papilla, and in this recess the epithelium is poorly supported and liable to rupture on sudden changes of pressure in the calyces. At the apex of the papilla the epithelium joins that of the renal tubules. The muscular fibres in the wall of the calyces and of the pelvis are collected into loosely arranged bundles separated by wide intervals occupied by fibrous tissue. As in the ureter proper, the outermost and innermost fibres are longitudinal, and the middle ones are circular. The circular fibres alone form a distinct layer.

VESSELS AND NERVES OF THE URETER

The most constant arteries supplying the ureter are branches from the uterine artery in the floor of the lesser pelvis in the female. In the male, similar branches are derived from the inferior vesical artery. Branches from the renal arteries supply the upper end of the ureters. Usually intermediate arteries are derived either from the aorta, gonadal, common iliac, or internal iliac arteries. In the abdomen the arteries supplying the ureter approach it from the medial side; in the pelvis they approach it from the lateral side. In both abdomen and pelvis the supplying arteries are intimately connected with the peritoneum (Racker and Braithwaite 1951), and form a longitudinal plexus on the ureter.

The **nerves** of the ureter are derived from the renal, testicular or ovarian, and hypogastric plexuses. The afferent fibres reach the spinal medulla through the eleventh and twelfth thoracic and first lumbar nerves.

VARIATIONS [FIG. 8.19]

The ureter is sometimes duplicated in its upper portion. In rarer cases it is double throughout the greater part of its extent, or even in

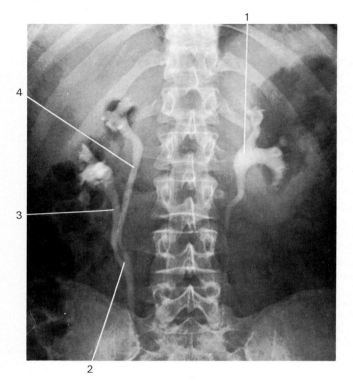

FIG. 8.19. Intravenous pyelogram showing the congenital abnormality of double ureter and renal pelvis. In this condition, the two separate ureters on one side each drain a separate renal pelvis. These ureters may unite anywhere throughout their abdominopelvic course, or may enter the bladder separately.
1. Pelvis of kidney.
2. Point of junction of ureters.
3. Lower ureter.
4. Upper ureter.

its whole length with two openings into the bladder. Asymmetry in such abnormalities is very common.

Variations in the form of the pelvis of the kidney are of frequent occurrence. Most usually the pelvis divides into three major calyces—upper, middle, and lower—though commonly only two are present. In some cases the two calyces spring directly from the ureter without the intervention of a pelvis, or a marked subdivision may lead to the formation of two pelves.

RADIOLOGY

The ureters can be seen during intravenous pyelography as contrast filled structures of slightly variable calibre.

Compression of the ureters in the lower abdomen [FIG. 8.11] is frequently used to improve filling of the intra-abdominal portion of the ureter and the renal pelvis.

When the pressure is released the lower ureters are filled and their pelvic course can readily be seen [FIG. 8.23].

In radiographs, the ureters in the abdomen usually overlie the lumbar transverse processes. In the pelvis, they take a curved course (wider in the female) first laterally and then medially to reach the trigone a short distance from the midline.

Variation of position and appearance is common in the upper urinary tract. For example the right renal pelvis rarely appears as a mirror image of the left and sometimes is very different.

THE URINARY BLADDER

Position

The urinary bladder is a hollow muscular organ situated below the peritoneum on the anterior part of the pelvic floor behind the pubic symphysis. The space in which it lies is three-sided. On each side, it is bounded above by the obturator internus and lower down by the levator ani muscles; and these two side-walls meet in front at the pubic symphysis. As the two levator ani muscles slope downwards, backwards, and medially they form the walls of a gully which is deeper behind than in front owing to the slope of its floor. This floor is defective in front because of the slight gap that exists between the two levator ani muscles, but the upper surface of these muscles is covered with a layer of fascia which bridges over the gap. Laterally where the fascia lies on the muscles it is sometimes known as the lateral puboprostatic (pubovesical in the female) ligaments; medially where it covers the gap between the two muscles the fascia is thickened and forms the **puboprostatic** (pubovesical in the female) **ligaments**. The gap is also partly closed at a lower level by the urogenital diaphragm. The space for the bladder is bounded behind in the male by the fascial tissue which stretches across the back of the bladder and is thickened and condensed as it passes from the bladder to be connected with the fascia of the levator ani on each side, the **rectovesical septum**.

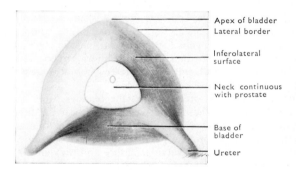

FIG. 8.20. Inferior aspect of empty male urinary bladder. From a subject in which the viscera had been hardened *in situ*.

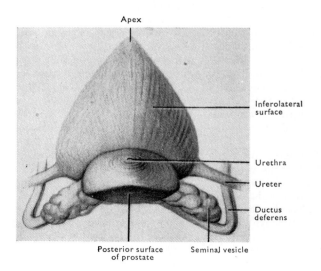

FIG. 8.21. Urinary bladder, prostate, and seminal vesicles, viewed from below.

Taken from a subject in which the viscera were hardened *in situ*. Same specimen as in FIG. 8.22A. The bladder contained a very small amount of fluid.

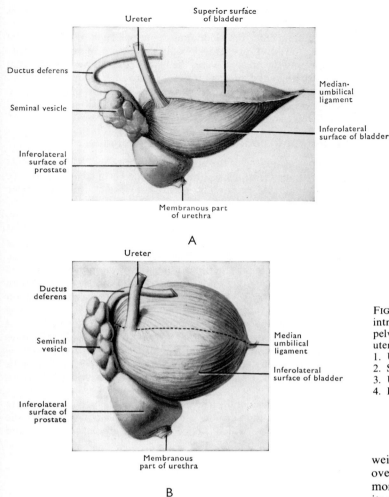

FIG. 8.22. Urinary bladder, prostate, and right seminal vesicle viewed from the right side.

Drawn from specimens in which the viscera were hardened *in situ*. In A the bladder contained a very small quantity of fluid; in B the quantity was a little greater. In A the peritoneum is shown covering the superior surface of the bladder, and its cut edge is seen where it is reflected along the lateral border. In B the level of the peritoneal reflexion is indicated by a dotted line.

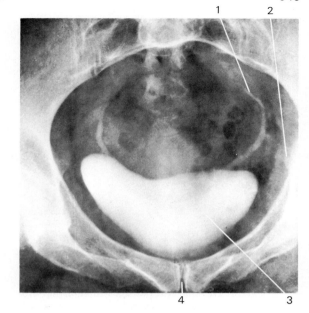

FIG. 8.23. Urinary bladder containing contrast medium following intravenous pyelography [FIG. 8.11]. Radiograph taken in axis of lesser pelvis. The impression on the upper surface of the bladder is caused by the uterus.
1. Ureter.
2. Spine of ischium.
3. Urinary bladder.
4. Pubic symphysis.

Shape and parts [FIGS. 8.20–8.22].

The empty adult bladder lies completely or almost completely within the above-mentioned space, and in dissecting-room fixed specimens it is usually moulded to a pyramidal shape by the walls of the space and by the pressure of overlying viscera. It is therefore usual to describe the empty adult bladder as possessing four surfaces: a **superior surface** looking straight upwards, a pair of **inferolateral surfaces** facing downwards, laterally, and forwards, and a **posterior surface** directed backwards and slightly downwards. The posterior surface is the **base**; the anterior extremity of the bladder, where the superior and the two inferolateral surfaces meet, is the **apex**; and the lowest part of the organ, where the base and the two inferolateral surfaces come together, is the **neck**. In life the bladder almost always contains some fluid so that it is more or less rounded, though its spherical shape may be slightly altered either by pressure of surrounding organs or by its attachment to surrounding structures. Hence, radiographs of the bladder in living subjects show it as a rounded organ, with its upper aspect frequently flattened by the

weight of overlying viscera, or even slightly depressed by a heavy overlying viscus, e.g. a gravid uterus. It would therefore probably be more correct to speak of the various **aspects** of the bladder than of its **surfaces**; for its true shape appears to be that of a sphere or a spheroid, except when temporarily modified by pressure of surrounding structures [FIG. 8.33].

Attachments

In both sexes the **neck** is the most firmly fixed part of the bladder. In the male it is held in place chiefly by the puboprostatic ligaments and by its firm connection with the prostate, with which it is structurally continuous. Each side of the neck gives attachment to a lateral puboprostatic ligament which is merely the fascia on the anterior part of the levator ani muscle. The **puboprostatic ligaments** are a pair of short, strong bands that lie side by side over the anterior part of the gap between the two levatores ani in front of the neck of the bladder and attach it firmly to the back of the pubic bones. These ligaments contain bundles of plain muscle fibres which are continuous with the muscular coat of the bladder and are named, therefore, the **pubovesical muscle**. The prostate lies in the posterior part of the gap between the two levator ani muscles. It is enclosed in a strong fibrous sheath which also is firmly connected with the puboprostatic ligaments anteriorly and with the superior fascia of the urogenital diaphragm inferiorly [p. 551]. In the female the neck of the bladder and upper part of the urethra are connected with the pubis and the levator ani by similar fascial bands which, in the absence of a prostate, are named **pubovesical** ligaments (Zacharin 1963); and these parts are also directly connected with the superior fascia of the urogenital diaphragm.

The **apex** of the bladder is continuous with the median umbilical ligament, which extends up on the posterior surface of the anterior abdominal wall to the umbilicus. This ligament is the fibrous

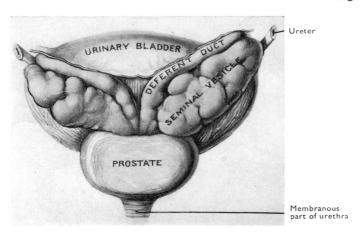

FIG. 8.24. Urinary bladder, seminal vesicles, and prostate, viewed from behind.

remnant of the urachus—the passage which, in the embryo, connects the developing bladder with the part of the allantoic diverticulum in the umbilical cord.

The **base** of the bladder is attached by means of the condensation of fascia which, on each side, blends with the fascia on the levator ani muscle. The whole condensation is a collection of fibrous tissue around the internal iliac artery and its branches, and it passes medially to be attached to the base of the bladder and seminal vesicles in the male, and the base of the bladder and the sides of the vagina and cervix uteri in the female.

In addition to these connections, the base is loosely attached by areolar tissue to the anterior wall of the vagina in the female, but

firmly to the seminal vesicles and ampullae of the ductus deferentes in the male.

The ureters enter the bladder at the upper and lateral aspects of the base, the points of entrance being about 5 cm apart in the empty bladder; and the urethra leaves the bladder at its lowest point—that is the neck—by the internal urethral orifice.

In dissecting-room specimens, once the median umbilical ligament has been cut, the whole anterior part of the bladder is very mobile but the base and neck remain fixed.

Structures in contact with the male bladder

The superior surface of the bladder [FIGS. 8.18 and 8.25] is covered with peritoneum, which separates it from coils of small intestine or from the sigmoid colon. The peritoneum passes from the upper surface of the empty bladder forwards on to the posterior surface of the anterior abdominal wall at the level of the pubic crest; laterally, it passes slightly downwards and then upwards on to the side-wall of the pelvis, thus forming a shallow **paravesical fossa** on each side of the bladder. Posteriorly it passes backwards for about 1.5 cm, forming a prominent shelf-like fold—the **sacrogenital fold**—and then turns abruptly downwards for about 2.5 cm to be reflected on to the front of the rectum, thus forming the **rectovesical pouch** [FIG. 8.25]. As the peritoneum turns downwards it may, in the median plane, cover a small portion of the base of the bladder but more laterally it is separated from the base by the ductus deferentes and the upper ends of the seminal vesicles. When the bladder is completely empty, the peritoneum on its upper aspect may be thrown into a transverse fold—the **transverse vesical fold** [FIG. 8.25].

The two inferolateral surfaces and the border between them are separated from the surrounding firm structures by a space—the **retropubic space**—which is filled with loose areolar tissue, fat, and

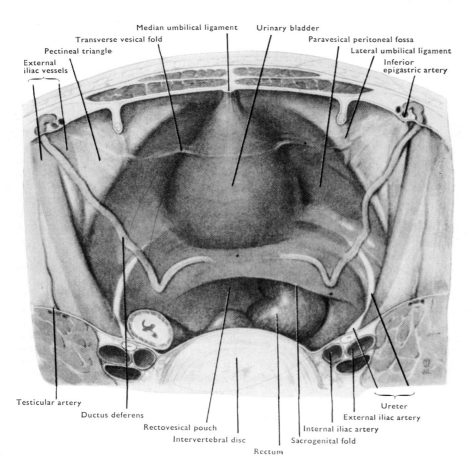

FIG. 8.25. Cavity of male pelvis seen from above and behind.

From a specimen in which the bladder was firmly contracted and contained only a small amount of fluid. The peritoneal pouch in front of the rectum is bounded on each side by the sacrogenital folds, which meet together in the median plane some distance behind the posterior border of the bladder.

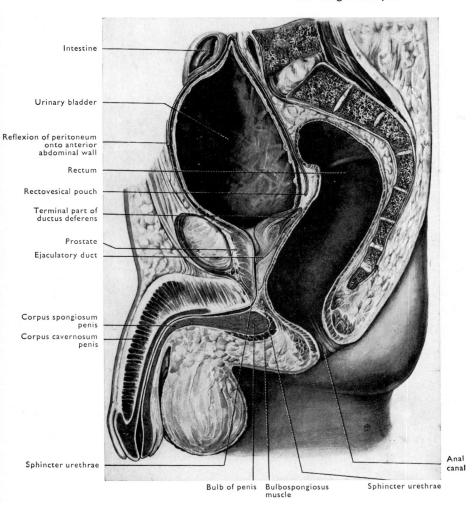

Intestine

Urinary bladder

Reflexion of peritoneum
onto anterior
abdominal wall

Rectum

Rectovesical pouch

Terminal part of
ductus deferens

Prostate

Ejaculatory duct

Corpus spongiosum
penis

Corpus cavernosum
penis

Sphincter urethrae

Bulb of penis   Bulbospongiosus
muscle

Sphincter urethrae

Anal
canal

FIG. 8.26. Median section of pelvis of adult male subject. The urinary bladder and rectum are both distended [cf. FIG. 8.18].

a plexus of veins [FIG. 8.18]. This mass of tissue acts as a packing to fill the gaps between the bladder and the neighbouring, rather rigid, pelvic walls, and is known as the **retropubic pad of fat**. The retropubic space is bounded antero-inferiorly by the back of the pubis, at each side by the levator ani muscle below and the obturator internus muscle above, inferiorly by the puboprostatic ligaments, and posteriorly by the fascial condensation which passes from the base of the bladder to the side-walls of the pelvis. Superiorly, it is bounded by the peritoneum which passes from the superior surface of the bladder to the anterior wall of the abdomen and the side-walls of the pelvis. As the bladder fills and expands, it lifts the peritoneum off the anterior abdominal wall and occupies the area between the lateral umbilical ligaments (obliterated umbilical arteries). Thus the retropubic space potentially extends upwards on to the back of the anterior abdominal wall and as far laterally as these ligaments. The obturator nerve and vessels extend into the posterior and upper part of the side-wall of the retropubic space even when the space is unexpanded, but as the bladder fills and the space enlarges the ductus deferens also may be included. The plexus of veins in the retropubic pad of fat is especially rich in the neighbourhood of the neck of the bladder.

The **base** of the bladder faces backwards and slightly downwards [FIGS. 8.18 and 8.20]. Its inferolateral parts are in contact with the seminal vesicles [FIG. 8.24], which diverge from each other at rather more than a right angle as they pass laterally and upwards from the back of the prostate. Medial to each seminal vesicle, the base is in contact with the ampulla of the ductus deferens of the same side.

Between these two ducts it is separated by some loose areolar tissue from the front of the rectum, but the uppermost part of that interval may be covered with peritoneum.

The **neck** is firmly attached to the prostate, and is structurally continuous with it, though a well-marked groove is present where the surfaces of the two organs meet.

Structures in contact with the female bladder [FIGS. 8.27 and 8.56].

The bladder occupies a slightly lower position than in the male. The superior surface is covered with peritoneum, which, as in the male, passes forwards on to the back of the anterior abdominal wall, and also laterally to the side-wall of the pelvis, forming, as it does so, a shallow **paravesical fossa**. Posteriorly, the peritoneum passes from the superior surface of the bladder upwards to be attached to the uterus at the junction of its isthmus and body, and then passes forwards on the antero-inferior surface of the body of the uterus. In the female, therefore, the body of the uterus overlies the bladder and the peritoneal recess between the two organs is known as the **uterovesical pouch**. This pouch is usually slit-like, but in some cases coils of small intestine pass into it and partly separate the bladder and uterus.

The base of the bladder in the female is only loosely attached by areolar tissue to the anterior wall of the vagina. The upper part of the base, just as it passes into the superior surface of the organ, is separated from the anterior surface of the supravaginal part of the

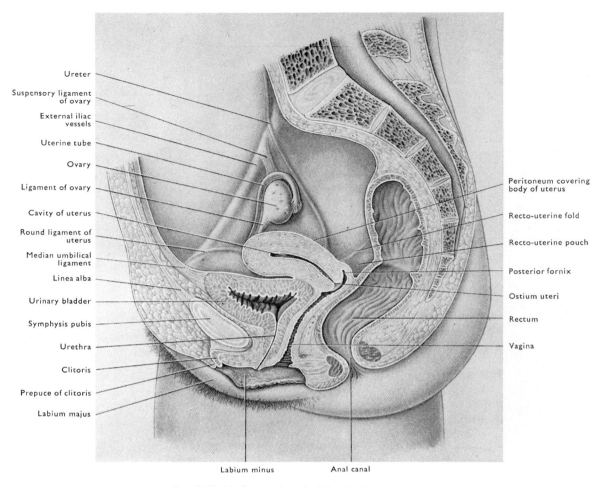

Ureter

Suspensory ligament of ovary

External iliac vessels

Uterine tube

Ovary

Ligament of ovary

Cavity of uterus

Round ligament of uterus

Median umbilical ligament

Linea alba

Urinary bladder

Symphysis pubis

Urethra

Clitoris

Prepuce of clitoris

Labium majus

Peritoneum covering body of uterus

Recto-uterine fold

Recto-uterine pouch

Posterior fornix

Ostium uteri

Rectum

Vagina

Labium minus   Anal canal

FIG. 8.27. Median section of pelvis of adult female.

cervix uteri by loose areolar tissue; and laterally on each side the uterine vessels are about 2.5 cm distant [FIG. 8.55].

The inferolateral surfaces and the apex have the same general relations in the female as in the male.

The neck, in the absence of the prostate, is continuous only with the urethra, and the pubovesical ligaments are connected with it through the considerable amount of smooth muscle, glands, and fibrous tissue that surrounds it.

Bladder in the new-born child and infant

At birth the empty bladder is spindle-shaped; its long axis extends from the apex to the internal urethral orifice, and is directed downwards and backwards [FIG. 8.28]. The lateral and posterior borders seen in the adult organ cannot be recognized at birth, nor is there any part of the bladder wall directed backwards and downwards, as in the base of the adult organ. In the foetus and young child the bladder occupies a much higher level than it does in the adult, and, even when empty, it extends upwards into the abdominal cavity. Its anterior surface is in contact with the back of the anterior abdominal wall. At birth the peritoneum forming the rectovesical pouch covers the whole of the posterior surface of the bladder, and reaches as low as the upper limit of the prostate. The internal urethral orifice is placed at a high level, and sinks gradually after birth [FIGS. 8.29 and 8.30]. In the new-born child the opening is on a level with the upper margin of the pubic symphysis and the

openings of the ureters are almost in the plane of the superior aperture of the pelvis. The obliterated umbilical arteries are more intimately related to the bladder in the child than in the adult, and lie close against its sides as they pass upwards on the back of the anterior abdominal wall towards the umbilicus [FIG. 8.39].

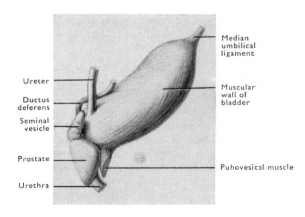

Median umbilical ligament

Ureter

Ductus deferens

Seminal vesicle

Prostate

Urethra

Muscular wall of bladder

Pubovesical muscle

FIG. 8.28. Urinary bladder of new-born male child, viewed from right side.

The drawing is from a specimen which had been hardened *in situ* [cf. FIG. 8.22].

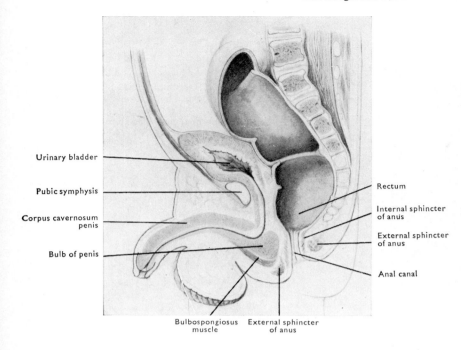

Urinary bladder

Pubic symphysis

Corpus cavernosum
penis

Bulb of penis

Rectum

Internal sphincter
of anus

External sphincter
of anus

Anal canal

Bulbospongiosus   External sphincter
muscle   of anus

FIG. 8.29. Median section of new-born male pelvis.

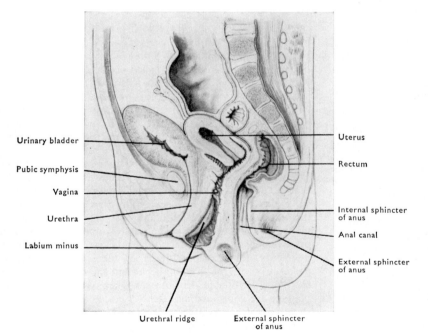

Urinary bladder

Pubic symphysis

Vagina

Urethra

Labium minus

Uterus

Rectum

Internal sphincter
of anus

Anal canal

External sphincter
of anus

Urethral ridge   External sphincter
of anus

FIG. 8.30. Median section of new-born female pelvis.

Interior of the bladder [FIGS. 8.27, 8.31, and 8.32]

The mucous membrane which lines the bladder is loosely connected to the muscular coat, and when the bladder is contracted it is thrown into a number of prominent folds, except in a triangular area of the base (**trigone of the bladder**) where the mucous membrane is firmly attached to the muscular coat and remains smooth. The apex of the triangle is at the internal urethral orifice, and the base is a line drawn between the openings of the ureters. Immediately above and behind the internal urethral orifice the bladder wall in the male sometimes bulges slightly into the cavity owing to the presence of the median lobe of the prostate, which lies outside the mucous coat in that position. When well marked, as it often is in old men, the bulging is termed the **uvula of the bladder**. Stretching between the openings of the ureters there is usually a smooth **interureteric fold**,

due to the presence of a transverse bundle of muscle fibres. It may be deficient near the median plane, and it is curved so as to be convex downwards and forwards. Lateral to the opening of each ureter there is a ridge, called the **ureteric fold**, produced by the terminal parts of the ureters as they traverse the bladder wall and lie outside the mucous coat of the bladder. In old people the region above and behind the trigone often bulges backwards and forms a shallow **retro-ureteric fossa**. A less distinct depression may sometimes be observed on each side of the trigone. Around the urethral orifice there are several minute, radially disposed folds which, disappearing into the urethra, become continuous with the longitudinal folds of the mucous membrane of the upper part of that canal. The ureter pierces the bladder wall very obliquely, and so the ureteral orifice has an elliptical outline. The lateral boundary of each opening is a thin, crescentic fold which, when the bladder is

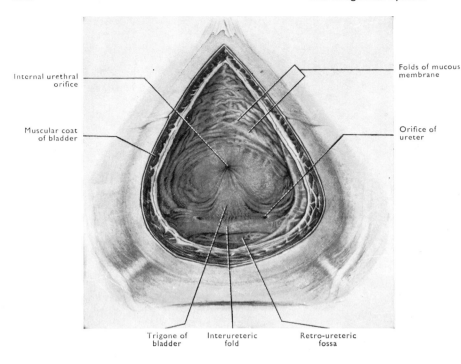

Internal urethral orifice

Muscular coat of bladder

Folds of mucous membrane

Orifice of ureter

Trigone of bladder   Interureteric fold   Retro-ureteric fossa

FIG. 8.31. Empty and contracted urinary bladder, opened by removal of its superior wall.

The peritoneum is seen spreading out from the lateral and posterior borders of the organ [cf. FIG. 8.25].

distended with urine or with fluid introduced in the process of cystoscopy, acts as a valve preventing entry to the ureter and back pressure on the kidney. In the empty bladder the internal urethral orifice and the openings of the two ureters lie at the angles of an approximately equilateral triangle the sides of which are about 2.5 cm in length. When the bladder is distended, the distance between the openings may be increased to 5 cm or more.

Position of the internal urethral orifice

During the various changes in shape and size which the bladder undergoes, the region of the internal urethral orifice remains almost fixed in position. In the male the internal urethral orifice is immediately above the prostate, and behind and below the level of the upper margin of the pubic symphysis, from which it is about 5–6.5 cm distant. It can be reached by a finger introduced into the bladder through the abdominal wall above the pubic symphysis. It is usually 1.5–2.5 cm above the level of a plane which passes through the lower margin of the symphysis and the lower end of the sacrum, but in some cases it is slightly lower. In the female the internal urethral orifice is normally at a lower level than in the male.

Capacity of the bladder

The distended bladder may contain 550 ml or more; but in most cases the organ is emptied when it contains 160–300 ml of fluid (Thompson 1919). When the bladder contains 450 ml it is in contact with the back of the anterior abdominal wall for a distance of about 7–8 cm above the pubic crest and the peritoneal reflexion from its upper surface on to the abdominal wall has been raised above this level.

Structure of the bladder

The wall of the bladder from without inwards is composed of a serous, a muscular, a submucous, and a mucous coat. The **serous coat** is the peritoneum, and covers only the upper surface of the bladder [FIGS. 8.18 and 8.27].

A considerable amount of loose fibro-areolar tissue surrounds the **muscular coat**, and, penetrating it, divides it into numerous coarse

bundles of muscle fibres. All the muscle fibres are of the plain variety, and the bundles which they form are arranged in three very imperfectly separated strata called external, middle, and internal, which together constitute the 'detrusor' muscle.

1. The **external stratum** is for the most part made up of fibres which are directed longitudinally. It is most marked near the median plane on the superior and inferolateral surfaces of the bladder. Further from the median plane, on the sides of the bladder, the fibres composing the external stratum run more obliquely, and frequently cross one another. In the male many of the fibres of the external stratum pass into the prostate in front of and behind the urethral opening, and in the female the corresponding fibres join the dense tissue which, in this sex, forms the upper part of the wall of the urethra. Other fibres of this stratum on each side of the median plane join the lower part of the pubic symphysis and constitute the **pubovesical muscle**, which follows the course of the puboprostatic ligaments. Lastly, some fibres of the external stratum blend posteriorly with the front of the rectum and receive the name of the **rectovesical muscle**. In both sexes many muscular fibres of the bladder are continued along the course of the ureter to form the sheath of the ureter already mentioned [p. 543].

2. The **middle stratum** is composed of irregular interlacing bundles of muscle fibres which for the most part run circularly, and form the greater part of the thickness of the muscular coat. At the internal urethral orifice, the bundles of fibres are finer and more densely arranged (**sphincter vesicae**), and surround the opening in a plane which is directed obliquely downwards and forwards. Inferiorly the fibres of the sphincter vesicae are continuous with the muscular tissue of the prostate in the male, and with the muscular wall of the urethra in the female. In other parts of the bladder the bundles of the middle stratum are coarser and separated by intervals filled with fibro-areolar tissue.

3. The **internal stratum** is composed of a thin layer of muscle fibres directed for the most part longitudinally.

The **submucous coat** is composed of areolar tissue which contains numerous fine elastic fibres.

The **mucous coat** is loosely attached by the submucous layer to the muscular coat except in the region of the trigone. There, the muscular fibres are firmly adherent to the mucous coat. The mucous coat of the bladder is continuous with that of the ureters and urethra. Its epithelial covering is of the variety known as transitional epithelium; it varies much in appearance as the organ expands and contracts, being essentially an elastic epithelium, impervious to urine.

VESSELS AND NERVES OF THE BLADDER

The arteries on each side are the **superior** and **inferior vesical arteries**. Braithwaite (1952) has given details of the distribution of these arteries to the different parts of the bladder. The largest veins are found above the prostate and in the region where the ureter reaches the bladder. They form a dense plexus which pours its blood into tributaries of the internal iliac vein, and communicates below with the prostatic venous plexus.

The **lymph vessels** from the bladder join the iliac groups of lymph nodes [FIG. 13.140].

The **nerve supply** of the bladder is derived on each side from the **vesical plexus**, the fibres of which come from two sources: (1) from the **upper lumbar nerves** through the hypogastric plexuses (sympathetic); and (2) from the **pelvic splanchnic nerves** (parasympathetic), which spring from the second and third, or the third and fourth, sacral ventral rami. These splanchnic nerves join the vesical plexus directly.

METHODS OF CLINICAL ANATOMICAL EXAMINATION

The normal empty bladder cannot be palpated, and its outlines cannot be determined by percussion, but, as the organ fills and rises into contact with the anterior abdominal wall, its upper boundary can be mapped out by percussion. The base of the female bladder can be palpated by a vaginal examination.

The interior of the bladder can be inspected by means of a cystoscope introduced through the urethra; and the outline of its cavity may be seen radiographically [FIG. 8.33].

RADIOLOGY

In plain anteroposterior radiographs the filled or partially filled bladder is often seen. It is more clearly outlined when filled with contrast medium in the course of intravenous pyelography. The bladder under these conditions has a triangular or rounded appearance and a smooth outline, but being partly behind the pubic symphysis is best seen when the X-ray beam is directed downwards into the pelvic cavity. This projects the lower margin of the bladder at or near the upper border of the pubic bones [FIG. 8.23].

The bladder may also be filled through a catheter passed along the urethra (cystogram), thus allowing more control over the amount introduced so that smaller abnormalities in the bladder may be seen.

Ultrasound examination of the bladder is also a valuable method, particularly in assessing the thickness of the bladder wall in pathological conditions.

THE URETHRA

The urethra is the channel which conveys the urine from the bladder to the exterior. In the male its proximal part, less than 2.5 cm in length, extends from the bladder into the prostate where it is joined by the ducts of the reproductive glands. A much longer distal portion serves as a common passage for the urine and the semen. The female urethra represents only the proximal part of the male canal. It is a short passage that leads from the bladder to the external urethral orifice—an aperture placed within the pudendal cleft immediately in front of the opening of the vagina.

The male urethra [FIG. 8.33]

The male urethra [FIGS. 8.18 and 8.32] is a channel, about 20 cm in length, which leads from the bladder to the external urethral orifice at the end of the glans penis. The canal not only serves for the passage of urine; it also affords an exit for the seminal products, which enter it by the ejaculatory ducts, and for the secretion of the prostatic and bulbo-urethral glands; in addition, numerous minute **urethral glands** pour their mucous secretion into the urethra.

As it passes from the internal urethral orifice to its external opening the urethra describes an ∽-shaped course. Within the lesser pelvis it has a nearly vertical course through the prostate. Turning more forwards, the urethra leaves the pelvis by piercing the urogenital diaphragm 2.5 cm below and behind the pubic symphysis, and enters the bulb of the penis. Throughout the rest of its course it lies in the erectile tissue of the corpus spongiosum and of the glans penis. The part of the urethra which is embedded in the prostate is called the prostatic part; the short part which pierces the urogenital diaphragm is called the membranous part; and the part surrounded by the corpus spongiosum is the spongy part. The spongy part is much the longest, and the membranous is the shortest (slightly less than 1.5 cm).

1. The **prostatic part** descends through the prostate from the base towards the apex, describing a slight curve which is concave forwards. It is about 2.5 cm in length, and is narrower above and below than in the middle portion, which is the widest part of the whole urethral canal. Except while fluid is passing, the canal is contracted, and the mucous membrane of the anterior and posterior walls is in contact and thrown into a series of longitudinal folds. When distended, the widest middle part of the canal may normally have a diameter of about 0.8 cm. The prostatic urethra is crescentic in transverse section [FIG. 8.51] because its posterior wall projects forwards forming a median ridge, the **urethral crest** [FIG. 8.32], with a groove on each side, the **prostatic sinus**. Numerous small ducts of the prostate gland open into the sinuses. A few ducts from the middle lobe open nearer the median plane, on the sides of the urethral crest. The urethral crest is highest half-way down, forming an eminence called the **seminal colliculus**, on which is a small, slit-like opening which leads backwards and upwards for about 0.5 cm into the substance of the prostate as a blind pouch. This **prostatic utricle** represents the fused caudal ends of the paramesonephric ducts, from which the uterus and vagina are developed. On each side of the mouth of the utricle there is the much smaller opening of the ejaculatory duct [p. 559]. Traced upwards, the urethral crest becomes indistinct, but often reaches as far as the uvula of the bladder. In the opposite direction the ridge again diminishes, but enters the membranous portion of the canal, where it divides into two before disappearing [FIG. 8.32]. The curvature and, to a lesser degree, the length of the prostatic urethra depend upon the amount of distension of the bladder and of the rectum [compare FIGS. 8.18 and 8.26].

2. The **membranous part** of the urethra, with the exception of the external orifice, is the least dilatable part. It runs downwards and forwards through the urogenital diaphragm. This diaphragm consists of two layers of fascia enclosing the sphincter urethrae and deep transverse perineal muscles. Its thin, upper layer is the **superior fascia of the urogenital diaphragm**. Its lower layer, the **inferior fascia of the urogenital diaphragm or perineal membrane**, is weak

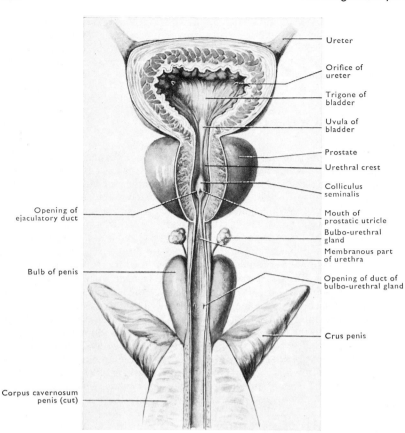

Ureter

Orifice of
ureter

Trigone of
bladder

Uvula of
bladder

Prostate

Urethral crest

Colliculus
seminalis

Mouth of
prostatic utricle

Bulbo-urethral
gland

Membranous part
of urethra

Opening of duct of
bulbo-urethral gland

Crus penis

Opening of
ejaculatory duct

Bulb of penis

Corpus cavernosum
penis (cut)

FIG. 8.32. Dissection showing trigone of bladder and floor of urethra in its prostatic, membranous, and proximal spongy parts.

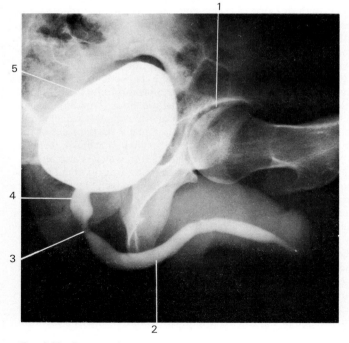

FIG. 8.33. Cysto-urethrogram.
1. Hip joint.
2. Spongy part of urethra.
3. Membranous part of urethra.
4. Prostatic part of urethra.
5. Urinary bladder.

in the female, but strong in the male for it forms a basis for attachment of the muscles of the penis. It is continuous with the superior fascia anterior and posterior to the muscle layer, and fuses with the central perineal tendon in the midline posteriorly. The membranous urethra begins just in front of the apex of the prostate, and immediately pierces the superior fascia of the diaphragm. Within the diaphragm it is surrounded by the sphincter urethrae with a bulbo-urethral gland and a deep artery of the penis lying about 0.5 cm laterally on each side. It then pierces the perineal membrane and ends about 0.5 cm below the fascia by entering the bulb.

In addition to the continuation of the urethral crest there are other longitudinal folds of mucous membrane seen when the canal is empty, and the lumen of the empty tube is therefore stellate in transverse section.

Although the membranous part of the urethra appears to be firmly held, it is not strongly attached to the surrounding fascia, but is so mobile and elastic that it can be drawn upwards for a distance of about 4 cm above the perineal membrane (Souttar 1947). This is of some importance in the operation of retropubic prostatectomy.

Immediately after the urethra has pierced the perineal membrane it turns forwards and sinks into the corpus spongiosum of the penis. Thus its inferior surface is overlapped by the erectile tissue of the bulb, but its superior wall remains uncovered for about 0.5 cm [FIG. 8.18]. Here the wall of the urethra is very thin, and the passage is more readily dilatable than in other parts. In this region the urethral wall may readily be torn if undue force is used in passing an instrument, particularly when an attempt is being made to pass it into the narrower, more fixed part of the membranous urethra.

3. The **spongy part** of the urethra is about 15 cm in length and is much the longest of the three divisions. It begins at the point where

the urethra passes into the substance of the corpus spongiosum, and it ends at the external urethral orifice. Its proximal or perineal portion has a fixed position and direction, while its distal part varies with the position of the penis. It traverses the corpus spongiosum, including the bulb and the glans, and is therefore surrounded by erectile tissue in the whole of its length. Directed at first forwards through the bulb, in the flaccid penis it turns downwards and forwards at the point where it comes to lie beneath the anterior part of the pubic symphysis [FIG. 8.18]. The bend in the canal corresponds approximately to the place of attachment of the suspensory ligament to the dorsum of the penis. When the penis is erect, the whole spongy urethra becomes more uniformly concave posterosuperiorly.

The urethra lies at first in the upper part of the erectile tissue, but as it passes forwards it sinks deeper and comes to occupy the middle part of the corpus spongiosum. In the glans, on the other hand, the erectile tissue lies mainly on the dorsal and lateral aspects of the urethra [FIG. 8.49]. Like the other parts of the urethra, the spongy portion is closed except during the passage of fluid, the closure being effected by the apposition of the dorsal and ventral walls [FIG. 8.48] except in the glans penis, where the side-walls of the canal are in contact. The spongy part of the urethra is wider in the bulb and glans than in the corpus spongiosum. In the glans the urethra expands to form a terminal dilated part named the **fossa navicularis** which opens on the surface by the slit-like **external urethral orifice**. This is the narrowest and least dilatable part of the whole urethra.

The **ducts of the bulbo-urethral glands** [p. 565] open by very minute apertures in the inferior wall of the proximal part of the spongy urethra [FIG. 8.32]. Before opening into the canal, they lie for some distance immediately outside its mucous membrane. A number of small pit-like recesses, called the **urethral lacunae**, also open into the spongy part of the urethra by openings which face towards the external urethral orifice.

A valve-like fold of the mucous membrane, the valve of the fossa navicularis, is sometimes found in the upper wall of the fossa. Its free edge is directed towards the external orifice, and it may engage the point of a fine instrument introduced into the urethra.

Structure

The mucous membrane of the urethra contains numerous elastic fibres and varies in thickness in different parts of the canal. The lining epithelium is stratified columnar in type in the membranous and spongy parts of the canal; in the prostatic part of the urethra it changes to transitional epithelium which is continuous through the internal urethral orifice with the transitional epithelium of the bladder. In the region of the fossa navicularis the lining cells become stratified squamous.

Numerous minute, mucous, urethral glands lie in the mucous coat and form flask-like depressions with very short ducts. They are most numerous in the membranous part and the anterior half of the spongy part, especially in the upper walls, but also in the floor and side walls.

There are also larger mucous glands deeply placed outside the mucous coat, which communicate with the urethra by long, slender, obliquely placed, branching ducts. The ducts of some of the glands open into the lacunae, but many lacunae have no connection with the urethral glands.

Frequently, two or more elongated ducts belonging to some of the larger glands open into the urethra quite close to its termination. They are termed **para-urethral ducts**, and may be traced backwards for some distance outside the mucous membrane of the roof of the urethra. Morphologically they do not correspond to the ducts in the female which have received the same name.

The muscular wall of the proximal part of the prostatic urethra consists of two layers of plain muscle fibres, longitudinal internally and circular externally. Both layers are continuous above with the muscle of the bladder. The oblique, circular layer decreases in amount at the level of the seminal colliculus, and both it and the longitudinal layer fade out as the urethra enters the corpus spongiosum. At a lower level, in front of the prostatic urethra, there is a band of striated muscular fibres which is continuous inferiorly with the inner, circular, voluntary sphincter urethrae.

The female urethra

The female urethra [FIG. 8.27] is a canal, about 4 cm in length, which follows a slightly curved direction downwards and forwards, behind and below the lower border of the pubic symphysis. Except during the passage of urine the canal is closed by the apposition of its anterior and posterior walls. The **external urethral orifice** is placed between the labia minora, immediately in front of the opening of the vagina, about 2.5 cm below and behind the clitoris [FIGS. 8.27 and 8.61]. The opening is slit-like, and is bounded by ill-defined lips. The mucous coat of the canal is raised into a number of low longitudinal folds, one of which, more distinct than the others and placed on the posterior wall of the passage, receives the name of **urethral crest**.

The upper part of the urethra is separated from the front of the vagina by a **vesicovaginal space** filled with loose connective tissue. Lower down, however, the urethra becomes very intimately connected with the vagina, so that, as it approaches the external urethral orifice, it appears to be embedded in the anterior vaginal wall. This is due to the fact that the lower part of the urethra is bound to the front of the vagina by the fusion of their fasciae into a single dense layer.

Structure

The muscular coat of the female urethra is continuous above with that of the bladder, and is composed of layers of longitudinally and obliquely disposed plain muscle fibres. In the lower part, some of the fibres do not completely surround the urethra but are attached to the anterior wall of the vagina. Within the muscular coat the wall of the urethra is very vascular, and the canal itself is lined with a pale mucous membrane. The epithelium of the canal, in its upper part, is of the transitional variety, like that of the bladder; it then becomes stratified columnar in type, and finally, stratified squamous near the external urethral orifice. Numerous minute **urethral glands** and pit-like **urethral lacunae** open into the urethra. One group of these glands on each side possesses a minute common duct, known as the **para-urethral duct**, which opens into the pudendal cleft by the side of the external urethral orifice. These glands represent the prostatic glands of the male. The vascular layer which lies between the muscular coat and the mucous membrane contains elastic fibres, and in appearance resembles erectile tissue. Striated muscle fibres are present on the outer surface of the muscular coat of the urethra. In the upper part of the canal these fibres form a complete ring-like sphincter, so that the urethra is encircled by both plain and striated fibres. Lower down, some of the striated fibres are attached to the anterior vaginal wall, and in the lower third, where they are specially developed, they pass backwards on the outer surface of the vagina to enclose that passage together with the urethra in a single loop of muscle tissue, to form a sphincter.

RADIOLOGY

Radiographic examination of the urethra is carried out after the introduction of contrast medium into the bladder (cystogram).

X-ray films are taken whilst the subject is micturating. In the male, the prostatic, membranous, and spongy parts of the urethra can be seen in oblique projection (micturating urethrogram) [FIG. 8.33].

In the female a similar type of examination is often requested in cases of incontinence. The female urethra is relatively short and does not show many details, although information about the state of the pelvic floor and of possible downward displacement of the bladder can be obtained by this method.

The genital organs

THE MALE GENITAL ORGANS

The male reproductive organs are: (1) the testes together with (2) their coverings and (3) their ducts and the seminal vesicles, (4) the prostate, (5) the bulbo-urethral glands, (6) the external genital organs, and (7) the male urethra.

The **testes** are the essential reproductive glands (gonads) of the male, and are a pair of nearly symmetrical, oval bodies situated in the scrotum. The duct of each testis is at first much twisted and convoluted to form a structure known as the **epididymis**, which is applied to the back of the testis. This duct emerges from the epididymis as the **ductus deferens**, which passes upwards towards the lower part of the anterior abdominal wall and pierces it obliquely to enter the abdominal cavity. There each ductus deferens is covered with peritoneum, and crosses the linea terminalis of the pelvis to run on the side-wall of the lesser pelvis towards the spine of the ischium. On reaching the sacrogenital fold of peritoneum, the ductus deferens turns abruptly towards the base of the bladder, where it is in contact with a branched tubular structure termed the **seminal vesicle**. Joined by the duct of the seminal vesicle, the ductus deferens forms a short canal, called the **ejaculatory duct**, which pierces the prostate and opens into the prostatic part of the **urethra**. The **prostate** and the **bulbo-urethral glands** are accessory organs connected with the male reproductive system. The ducts of the bulbo-urethral glands and those of the prostate, like the ejaculatory ducts, open into the urethra, which thus serves not only as a passage for urine, but also for the generative products. The external genital organs are the **penis** and **scrotum**.

The testes

The testes are a pair of oval, slightly flattened bodies of whitish colour, measuring an average 4 cm in length, 2.5 cm from before backwards, and rather less in thickness, but there is great variability in their size and weight (Harrison and de Boer 1977). Each testis lies in the scrotum, the left usually a little lower than the right. The long axis of the testis is nearly vertical, while the lateral and medial surfaces are somewhat flattened, and the anterior and posterior margins are rounded. The posterior margin is attached to the wall of the scrotum, while the remainder is covered by the visceral (serous) layer of the tunica vaginalis, and projects into the cavity of that tunic. The epididymis and the lowest part of the **spermatic cord** [p. 557] are attached to the posterior margin of the testis [FIGS. 8.35 and 8.37].

Epididymis

The epididymis is a comma-shaped structure that clasps the posterior margin of the testis and to some extent overlaps the

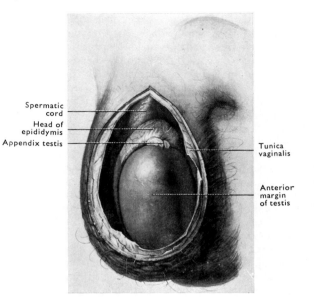

FIG. 8.34. Right testis and epididymis, exposed by removal of anterior wall of scrotum.

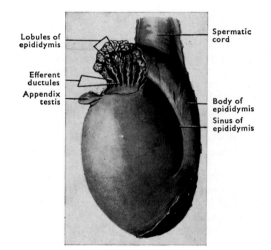

FIG. 8.35. Left testis and epididymis.
Part of the tunica vaginalis has been removed in order to show the efferent ducts and the lobules of the epididymis.

posterior part of its lateral surface [FIG. 8.34]. The upper, larger part is the **head** of the epididymis and overhangs the upper extremity of the testis, to which it is directly connected by several **efferent ductules**, by fibro-areolar tissue, and by the serous covering of the organ. The inferior and smaller part is the **tail of the epididymis**; it is attached by loose areolar tissue and by the serous covering to the lower extremity of the testis. The intervening **body** of the epididymis, is applied to the posterior part of the lateral surface of the testis but is separated from it by a slit-like recess of their serous covering termed the **sinus of the epididymis**.

The epididymis is composed of an irregularly twisted tube, called the **duct of the epididymis** [FIGS. 8.36 and 8.37].

A minute sessile or pedunculated body is often found attached to the head of the epididymis, and another to the adjacent part of the testis. They have a developmental interest. This **appendix of the testis** [FIG. 8.35] is a minute body which lies on the upper end of the testis and develops from the free end of the paramesonephric duct of the embryo; it is usually sessile. The **appendix of the epididymis** is attached to the head of the epididymis and is believed to represent a remnant of the cephalic end of the mesonephric duct. Other vestigial remains lying along the posterior margin of the testis, the **ductuli aberrantes** and the **paradidymis**, are believed to be derived from some of the mesonephric tubules.

Tunica vaginalis

The walls of the cavity within which the testis and epididymis are placed are lined with a serous membrane—the tunica vaginalis—which resembles the peritoneum, from which it was originally derived [FIG. 8.40]. The cavity tapers to a point or ends blindly as it is traced upwards. The tunica vaginalis is a sac doubled in on itself from behind by the testis and epididymis. It lines the scrotal chamber and it is reflected forwards from the posterior wall to cover

the testis, epididymis, and lower part of the spermatic cord immediately above the testis. The inner layer is known as the visceral layer and is closely applied to the enclosed organs; the outer or parietal layer lines the other coverings of the testis and is fairly strongly attached to them. Between the two layers there is a closed cavity that contains a little serous fluid. The visceral layer dips into the narrow interval between the body of the epididymis and the lateral surface of the testis to form the **sinus of the epididymis** [FIGS. 8.35 and 8.38]. The posterior margin of the testis is not in contact with the tunica vaginalis, being covered above by the head and below by the tail of the epididymis, and posteriorly by the spermatic cord from which blood vessels and nerves enter the testis.

Structure of the testis

Under cover of the tunica vaginalis the testis is invested by an external coat of dense, white, inelastic fibrous tissue called the **tunica albuginea**, from the deep surface of which a number of thin fibrous **septula** dip into the gland. These septula imperfectly divide the organ into a number of wedge-shaped **lobules** [FIG. 8.38]. All the septula end posteriorly in a mass of fibrous tissue called the **mediastinum testis**, which is directly continuous with the tunica albuginea and projects forwards into the testis. The mediastinum is traversed by an exceedingly complicated network of fine canals, the **rete testis**, into which the tubules of the testis open. The arteries, veins, and lymph vessels enter the posterior border of the testis, traverse the mediastinum, and spread out on the deep surface of the tunica albuginea to form the tunica vasculosa.

The mediastinum, the septula, and the tunica albuginea form a framework enclosing a number of imperfectly isolated spaces which are filled by a substance of a light brown colour called the **parenchyma testis**. The parenchyma is composed of large numbers of **convoluted seminiferous tubules** which look like fine threads to the unaided eye. Each tubule is a continuous loop with its convexity anterior. Its markedly convoluted limbs pass towards the mediastinum, and becoming less convoluted, unite with adjacent tubules to

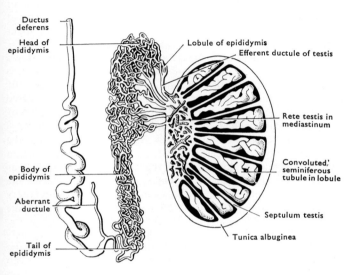

FIG. 8.36. Diagram to illustrate structure of testis and epididymis.

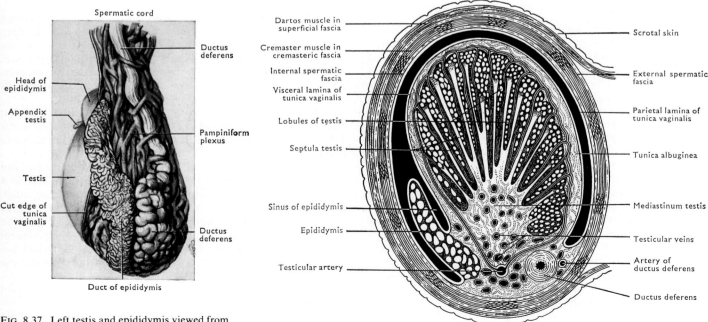

FIG. 8.37. Left testis and epididymis viewed from behind, showing duct of epididymis and first part of ductus deferens.

FIG. 8.38. Diagrammatic transverse section through the testis and scrotum.

form a smaller number of slender, straight seminiferous tubules, which open into the rete testis [FIG. 8.36]. Usually two to four tubules are found in each lobule of the gland where they are loosely held together by delicate areolar tissue. The total number of tubules in the testis has been estimated at more than 800.

Histological sections show that the walls of the seminiferous tubules are composed of a basement membrane and of an epithelial lining formed of several layers of cells. Outside the basement membrane is a layer of smooth muscle-like cells which cause contractions of the seminiferous tubules; the contractions are thought to promote the transport of spermatozoa and fluid within the tubules towards the rete testis (see Suvanto and Kormano 1970). In the adult, certain cells of the epithelium, called spermatogonia [p. 21], divide to form cells which mature into spermatozoa, and the histological appearance and spermatogenic activity of the tubules vary greatly with the age of the individual.

Structure of the epididymis

The spermatozoa formed in the convoluted seminiferous tubules are carried through the straight tubules into the rete testis, and leave the rete, to reach the duct of the epididymis, through fifteen to twenty minute tubules called the efferent ductules of the testis. These efferent ductules pierce the tunica albuginea and enter the head of the epididymis. Each ductule is at first straight, but soon becomes much convoluted, and forms a little conical mass called a lobule of the epididymis. Within the head of the epididymis the duct of each lobule opens into the single, much-convoluted duct which constitutes the chief bulk of the epididymis. This duct of the epididymis, which is about 600 cm in length, begins in the head of the epididymis, and ends, after an extraordinarily tortuous course, at the tail by becoming the ductus deferens [FIGS. 8.36 and 8.37].

The duct has a muscular coat composed of an inner stratum of circular fibres and an outer stratum of longitudinally directed fibres. This muscular coat and its peristaltic activity (Macmillan and Aukland 1960) are responsible for the passage of spermatozoa along the duct. The wall, at first thin, becomes much thicker as the canal approaches the ductus deferens.

VESSELS AND NERVES OF THE TESTIS

The testis is supplied by the testicular artery, a branch of the aorta. It is a slender vessel which, after a long course, reaches the posterior border of the testis, where it breaks up into branches. Harrison and Barclay (1948) found that there are commonly two main branches which pass forwards, one on each side of the organ, to ramify on the deep surface of the tunica albuginea in the tunica vasculosa. From this vascular plexus the small terminal arteries pass backwards into the substance of the testis along the septula and converge on the mediastinum. The artery of the ductus deferens and the cremasteric artery anastomose with the testicular artery (Harrison 1949).

The veins in the septula, and those of the tunica vasculosa, converge on the posterior border of the testis where they form a dense plexus, called the pampiniform plexus, which finally pours its blood through the testicular vein, on the right side, into the inferior vena cava, and, on the left side, into the left renal vein. The pampiniform plexus is also drained by the cremasteric vein which forms a venous plexus within and external to the cremasteric fascia (Harrison 1966).

The lymph vessels of the testis pass upwards in the spermatic cord and end in the lymph nodes at the sides of the aorta and inferior vena cava below the renal veins [FIG. 13.140].

The nerves for the testis and epididymis accompany the artery, and are derived through the aortic and renal plexuses from the tenth thoracic spinal nerve.

RADIOLOGY

Radiographic examination of the testes is inadvisable. When the contents of the scrotum are swollen, the differentiation of a collection of fluid in the tunica vaginalis from a swelling of the testis may be made by the use of ultrasound, if simpler examinations such as palpation and transillumination are inadequate.

DESCENT OF THE TESTIS

The peculiar course pursued by the ductus deferens in the adult, and the manner in which it is related to the anterior abdominal wall, are explained by the development of the testis. Until nearly the end of intra-uterine life the testes are in the abdominal cavity. The testis at first lies on the posterior abdominal wall at the level of the upper lumbar vertebrae. To its lower pole is attached a ridge of tissue called the gubernaculum testis (Hunter 1786) which extends down to the inguinal region and, passing through the abdominal wall, is attached to the skin. Both testis and gubernaculum lie behind the primitive peritoneum; the gubernaculum thus forms a ridge covered with peritoneum, and the testis—almost completely surrounded by peritoneum—is attached to the posterior abdominal wall by a peritoneal fold called the mesorchium. As the foetus grows and the gubernaculum becomes relatively shorter, it carries the peritoneum on its anterior surface down through the anterior abdominal wall as a blind tube, the processus vaginalis, which thus traverses the inguinal region and reaches the genital swelling (rudimentary scrotum) (Backhouse and Butler 1960). As the processus vaginalis descends, the testis is guided by the gubernaculum down the posterior abdominal wall and the back of the processus vaginalis (Wells 1943) into the scrotum. By the third month of foetal life the testis lies in the iliac fossa, and by the seventh it is near the deep inguinal ring [FIGS. 8.39 and 8.40]. Normally the communication between the peritoneal cavity and that part of the processus vaginalis surrounding the testis is obliterated, leaving the testis isolated in the lower part of the processus, which is known as the tunica vaginalis.

A small fibrous band may be found in the adult passing through the inguinal canal and joining the peritoneum superiorly at the deep inguinal ring. Sometimes the band is connected below with the tunica vaginalis, but more often it cannot be traced so far down. When present it represents the obliterated portion of the processus vaginalis, and is therefore known as the vestige of the processus vaginalis.

The processus vaginalis normally closes shortly after birth but occasionally persists into adult life as a channel freely open to the peritoneal cavity above. If it fails to close completely, it may give rise to one or more cysts within the coats of the spermatic cord.

It sometimes happens that the descent of the testis is arrested, and then it either remains on the posterior abdominal wall or in a persisting upper part of the processus vaginalis in the inguinal canal. The term cryptorchism is applied to such undescended testes, which may be incapable of normal spermatogenesis. On the other hand the testis may be ectopic, through irregular descent into the perineum or even into the upper part of the thigh. In its descent the testis takes with it its ducts and vessels and nerves, which together form the spermatic cord.

In some mammals, such as the elephant, the testes remain permanently within the abdominal cavity; in others, such as the hedgehog, the peritoneal pouches remain widely open throughout life, and the testes are periodically withdrawn into the abdomen. The descent of the mammalian testes ensures an intratesticular

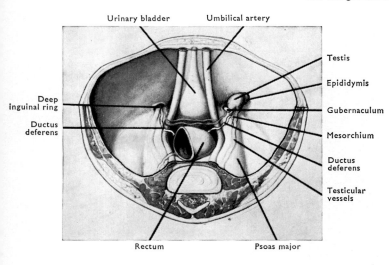

Urinary bladder   Umbilical artery

Deep inguinal ring

Ductus deferens

Rectum   Psoas major

Testis

Epididymis

Gubernaculum

Mesorchium

Ductus deferens

Testicular vessels

FIG. 8.39. View from above of cavity of pelvis and lower part of abdomen in male foetus about the 7th month.

On the left side, which represents a slightly more advanced condition than the right, the testis has entered the inguinal canal; on the right side the testis is still within the abdominal cavity.

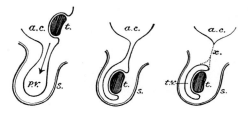

FIG. 8.40. Diagram to illustrate descent of the testis and origin of tunica vaginalis.
a.c., abdominal cavity; *p.v.*, processus vaginalis; *s.*, scrotum; *t.*, testis; *t.v.*, tunica vaginalis; *x.*, vestige of processus vaginalis.

temperature lower than that in the abdomen (Harrison and Weiner 1949; Harrison 1975). The testes of normal children may frequently be retracted into the inguinal canal by the action of the cremaster muscle, particularly in cold weather, or after stroking the skin on the medial surface of the thigh (cremasteric reflex).

The ductus deferens and spermatic cord

The ductus (*vas*) deferens is the direct continuation of the duct of the epididymis. It begins at the lower end of the epididymis and ends,

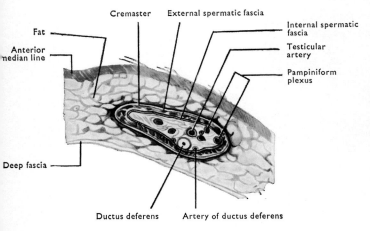

Cremaster   External spermatic fascia

Fat

Anterior median line

Internal spermatic fascia

Testicular artery

Pampiniform plexus

Deep fascia

Ductus deferens   Artery of ductus deferens

FIG. 8.41. Transverse section of spermatic cord immediately below superficial inguinal ring.

after a course of nearly 45 cm, by joining the duct of the seminal vesicle to form the ejaculatory duct, which opens into the prostatic part of the urethra. In parts of its course it is slightly convoluted, and the actual distance traversed is not more than 30 cm.

In the scrotum, the ductus deferens ascends along the back of the testis on the medial side of the epididymis, and it is here that it is most convoluted [FIG. 8.37]. At the upper extremity of the testis it meets the vessels and nerves of the testis and epididymis, and is bound together with them by areolar tissue in a loose bundle called the spermatic cord.

The **spermatic cord** is composed of: (1) the ductus deferens and its own artery and vein; (2) the testicular artery and the pampiniform plexus of veins; (3) the lymph vessels and nerves of the testis and epididymis; and (4) remnants of the processus vaginalis—the plexus of veins being by far the bulkiest of these constituents. The cord extends from the testis to the deep inguinal ring enclosed in three tubular sheaths or **coats**—the external spermatic fascia, the cremasteric muscle and fascia, and the internal spermatic fascia [FIG. 8.41]. When these coats reach the scrotum they expand to enclose the testis in the tunica vaginalis. The cremasteric coat carries with it its blood supply and nerve supply—the **cremasteric artery, vein**, and the **genital branch** of the **genitofemoral nerve**.

Between the scrotum and the superficial inguinal ring, the cord lies on the deep fascia of the muscles that spring from the pubis, and it is crossed by the superficial external pudendal vessels. In this part of the course the ductus is in the posterior part of the cord, and is easily distinguished from the other constituents by its hard, firm feel when the cord is gripped between the finger and thumb. When the cord enters the inguinal canal it lies on the lateral crus of the superficial ring immediately lateral to the pubic tubercle; and here it loses its external spermatic coat, which blends with the external oblique aponeurosis at the margins of the ring [FIG. 8.42].

During its passage through the inguinal canal, the ductus lies in the lower part of the cord on the upper surfaces of the lacunar and inguinal ligaments. The external oblique aponeurosis is anterior throughout and the lower fleshy fibres of the internal oblique are also in front in the lateral half of the canal. The conjoint tendon and the transversalis fascia are behind the medial part. Half-way along the canal the cord loses its cremasteric coat, since this is derived from the lower border of the internal oblique muscle, which arches over the cord to join transversus abdominis and form the conjoint tendon.

1.3 cm above the midinguinal point, at the lateral side of the inferior epigastric artery, the cord reaches the deep inguinal ring. Here, the cord loses its internal spermatic fascia which blends with

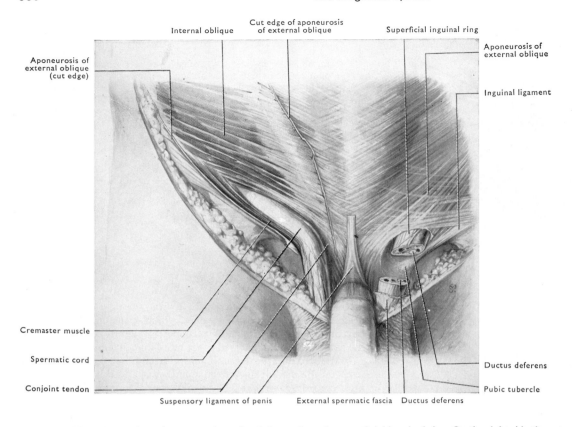

FIG. 8.42. Dissection to show the spermatic cord as it issues from the superficial inguinal ring. On the right side the external oblique muscle has been removed.

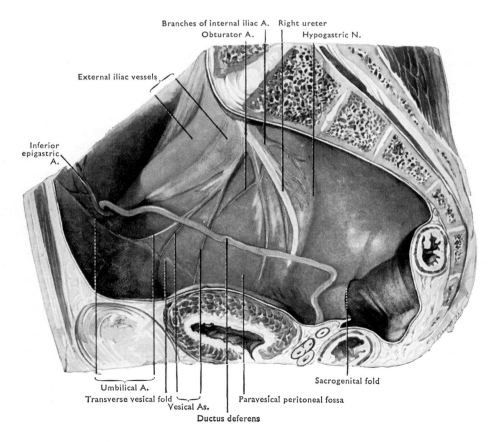

FIG. 8.43. Median section of adult male pelvis to show course of ductus deferens on side-wall of pelvic cavity.

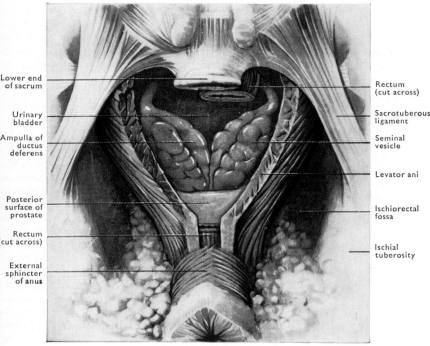

Lower end of sacrum
Urinary bladder
Ampulla of ductus deferens
Posterior surface of prostate
Rectum (cut across)
External sphincter of anus

Rectum (cut across)
Sacrotuberous ligament
Seminal vesicle
Levator ani
Ischiorectal fossa
Ischial tuberosity

FIG. 8.44. Dissection from behind to display the seminal vesicles, the ampullae of the ductus deferentes, and the prostate.

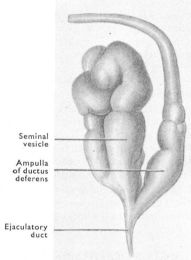

Seminal vesicle
Ampulla of ductus deferens
Ejaculatory duct

the transversalis fascia at the margins of the ring, and the ductus deferens parts company with the vessels and nerves.

The ductus deferens now curves medially across the external iliac artery behind the root of the inferior epigastric artery, and changing its direction, runs for a short distance backwards, medially, and upwards, immediately outside the peritoneum, to a point 4–5 cm from the pubic tubercle, where it crosses the linea terminalis and enters the lesser pelvis. In that part of its course the ductus usually lies at first in front of the external iliac vessels, and then in the floor of a small triangular fossa—the **pectineal triangle**—which lies between the external iliac vessels and the linea terminalis [FIG. 8.25]. On the side-wall of the pelvis the ductus continues backwards, and a little downwards and medially, towards the ischial spine, and lies immediately external to the peritoneum, through which it can usually be seen. In the pelvic part of its course the ductus crosses the medial side of: (1) the umbilical artery; (2) the obturator nerve and vessels; (3) the vesical vessels; and (4) the ureter [FIG. 8.43].

Crossing the ureter the ductus deferens hooks round it and passes downwards and medially. There, the ductus lies a short distance behind the terminal part of the ureter and immediately in front of the sacrogenital fold of peritoneum and upper end of the corresponding seminal vesicle [FIGS. 8.24 and 8.44]. Reaching the interval between the base of the bladder in front and the rectum behind, the two ductus deferentes occupy the angle between the right and left seminal vesicles [FIG. 8.44]. As they approach one another each ductus becomes slightly tortuous, sacculated, and dilated, and appears similar to the seminal vesicle. The dilated part of the ductus is termed its **ampulla**. Immediately above the base of the prostate the ductus deferens becomes once more a slender tube and is joined by the duct of the corresponding seminal vesicle to form the **ejaculatory duct** [FIG. 8.45].

In some cases the ductus deferens crosses the obliterated umbilical artery before it enters the cavity of the lesser pelvis; it normally does so in the foetus [FIG. 8.39].

Ejaculatory duct

This very slender tube is formed by the union of the ductus deferens with the duct of the corresponding seminal vesicle [FIG. 8.45]. It is

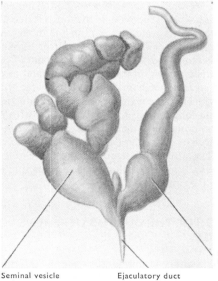

Ampulla of ductus deferens
Seminal vesicle
Ejaculatory duct

FIG. 8.45. Drawings of seminal vesicle and ampulla of ductus deferens taken from different subjects.

less than 2.5 cm in length, and lies very close to its fellow as it passes downwards and forwards through the prostate behind its median lobe. The ducts open by slit-like apertures into the prostatic part of the urethra, one on each side of the mouth of the **prostatic utricle**. They are well seen in sections through the upper part of the prostate [FIG. 8.51].

The seminal vesicles

The seminal vesicles are a pair of hollow sacculated structures placed on the base of the bladder in front of the rectum [FIGS. 8.22 and 8.44]. Each vesicle is about 5 cm in length, and has its long axis directed downwards, medially, and slightly forwards. The large, rounded, upper end of the vesicle, partly covered with peritoneum, lies at a considerable distance from the median plane, behind the lower end of the ureter, and is separated from the rectum by the peritoneum of the rectovesical pouch. Below the level of the peritoneal cavity the seminal vesicle and rectum are more intimately related, but are still separated by a partition of smooth muscle fibres and fascia which is thicker than that separating the seminal vesicles from the bladder. The vesicle tapers towards its lower end, which is placed not far from the median plane immediately above the prostate. In this position the vesicle becomes constricted to form a short duct which joins the lateral side of the corresponding ductus deferens at an acute angle. The common duct thus formed is the **ejaculatory duct**. The medial side of each vesicle is related to the ductus deferens, and the lateral side, when the bladder is empty, lies close to the levator ani. The seminal vesicles are more intimately related to the bladder than to the rectum.

The seminal vesicles develop as small, pouched outgrowths of the ductus deferentes. The dense tissue in which the seminal vesicles are embedded contains much plain muscle tissue which sweeps round in the side-wall of the rectovesical pouch and is attached inferiorly to the capsule of the prostate. The large veins draining the prostatic and vesical plexuses are closely related to the seminal vesicles.

STRUCTURE OF THE DUCTUS DEFERENS AND SEMINAL VESICLE

The ductus deferens is a thick-walled tube with a very small lumen. The hard cord-like sensation which it conveys to the touch is due to the thickness and denseness of its wall. The wall of the ductus is composed of three layers—an outer **adventitious coat** of fibrous tissue, a middle **muscular coat**, and an inner **mucous coat**. The thick coat of plain muscle is arranged in three layers, an outer and an inner of longitudinal fibres, and a middle layer—by far the thickest—of circular fibres. The mucous membrane of the ductus exhibits a number of slight longitudinal folds and is lined with a ciliated epithelium. The **ampulla** has a much thinner wall, and its mucous membrane is thrown into ridges so that it presents a honeycomb appearance. The wall of the seminal vesicle resembles that of the ampulla in being thin, and in having a mucous lining with uneven, honeycomb-like ridges.

VESSELS AND NERVES OF THE DUCTUS DEFERENS AND SEMINAL VESICLE

The ductus receives its main **artery to the ductus** from the inferior vesical artery. This artery supplies the ductus, the seminal vesicle, the lower ureter, and the bladder, and could be called the vesiculodeferential artery (Braithwaite 1952). The artery accompanies the ductus as far as the testis, where it ends by anastomosing with branches of the testicular artery. The **nerves** to the ductus seem to

come mainly from the sympathetic through the hypogastric plexuses, and each of its muscle fibres appears to have a separate innervation (Richardson 1962)—a unique arrangement in smooth muscle. The preganglionic sympathetic nerve fibres to the seminal vesicles emerge in the upper lumbar nerves, the parasympathetic in the pelvic splanchnics (S. 2, 3, 4), and both traverse the inferior hypogastric plexus.

The scrotum

The scrotum varies in appearance in different subjects, and in the same person at different times. As the result of cold or of exercise, the wall of the scrotum becomes contracted and firm, and its skin becomes wrinkled. At other times the wall may be relaxed and flaccid, the scrotum then assuming the appearance of a pendulous bag. The left side of the scrotum reaches to a lower level than the right, in correspondence with the lower level of the left testis. The skin of the scrotum is darker than the general skin of the body and is covered sparsely with hair. It is marked in the median plane by a ridge called the **raphe** of the scrotum, which is continued backwards towards the anus and forwards on the urethral surface of the penis. The difference in the appearance of the scrotum at different times is due to the degree of contraction or relaxation of a layer of plain muscular fibres, the **dartos**, situated in the superficial fascia. The layer of fascia containing the dartos is continuous superiorly with the superficial fascia of the penis, and with the deep layer of the superficial fascia of the abdomen, and is attached laterally to the sides of the pubic arch. The muscle fibres are arranged in a thick layer of interlacing bundles, and many of the deeper fibres are continued into the **septum** of the scrotum, which divides it into a pair of chambers, one for each testis. The wall of each chamber is formed by the fusion of the corresponding tunica vaginalis, internal spermatic fascia, cremasteric muscle and fascia, and external spermatic fascia, while the skin, the superficial fascia and the dartos muscle form coverings which are common to the whole scrotum, and they enclose both chambers. Immediately internal to the dartos is a layer of exceedingly loose and easily stretched areolar tissue; throughout it and the superficial fascia of the scrotum there is no fat.

The scrotum in the foetus has no cavity until the processus vaginalis enters it, but, like the labia majora which correspond to it in the female, it is composed entirely of vascular and fatty areolar tissue.

VESSELS AND NERVES OF THE SCROTUM

On each side there are **posterior scrotal branches** from the internal pudendal artery, which reach it from behind, and **anterior scrotal branches** from the external pudendal arteries, which supply its upper and anterior part.

The **nerves** of the scrotum are derived on each side from the posterior scrotal branches of the pudendal nerve, from the perineal branch of the posterior cutaneous nerve of the thigh, and from the ilio-inguinal nerve. The branches from the pudendal and posterior cutaneous nerves reach the scrotum from behind, while the ilio-inguinal supplies its upper and anterior part.

The penis

The penis [FIGS. 8.18 and 8.46] is composed chiefly of cavernous (erectile) tissue, and is traversed by the urethra. Engorgement of the cavernous tissue with blood produces considerable enlargement of the penis and its erection. The posterior surface of the flaccid penis is nearest the urethra and is called the **urethral surface**; the opposite

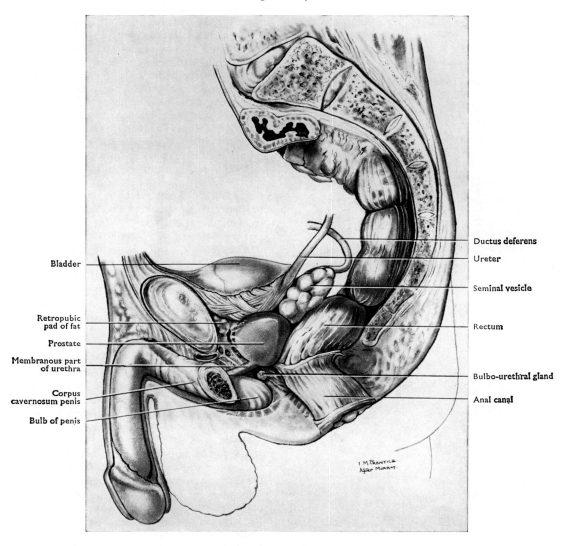

FIG. 8.46. Dissection of penis and male pelvic organs from left side.

and more extensive aspect is the **dorsum**. The erectile tissue is mainly disposed in three longitudinal columns, which in the body of the organ are closely united, while at the root of the penis they separate from one another and become attached to the perineal membrane and the sides of the pubic arch. Two of the columns of erectile tissue are placed alongside the median plane; they form the dorsum and sides of the penis, and are called the **corpora cavernosa** of the penis. The third column is situated in the median plane near the urethral surface, and is called the **corpus spongiosum** of the penis. The corpus spongiosum is the part of the penis traversed by the urethra. It is considerably smaller than the corpora cavernosa, which form the chief bulk of the organ [FIGS. 8.47 and 8.48].

In the **body** of the penis each corpus cavernosum presents a rounded surface, except where it is flattened by contact with its fellow. They are separated on the dorsal surface by a shallow groove, and on the urethral aspect by a deeper and wider furrow in which the corpus spongiosum lies [FIG. 8.47]. Towards the distal end of the penis the corpus spongiosum expands, and, spreading towards the dorsal surface, forms a cap—the **glans penis**—which covers the conical end of the united corpora cavernosa [FIG. 8.49]. The prominent margin of the glans, called the **corona**, projects backwards and laterally over the ends of the corpora cavernosa. The groove so formed between the corona and the ends of the corpora

cavernosa is the neck of the glans. The glans is traversed by the terminal part of the urethra, which ends near the summit of the glans in a vertical slit-like opening—the **external urethral orifice**. The skin of the body of the penis is thin, delicate, and freely movable, and, except near the pubis, is free from hairs; on the urethral aspect the skin is marked by a median raphe, continuous with the raphe of the scrotum. Traced towards the base of the glans, the skin forms a free fold—the **prepuce** or foreskin—which overlaps the glans to a variable extent. From the deep surface of the prepuce the skin is reflected on the neck of the glans, and is continued over the entire glans to the external urethral orifice [FIG. 8.18]. A small median fold—the **frenulum of the prepuce**—passes to the deep surface of the prepuce from a point immediately below the external urethral orifice. The skin of the glans is firmly attached to the underlying erectile tissue, and here, as well as on the deep surface of the prepuce, it has some resemblance to mucous membrane.

The secretion which tends to collect beneath the prepuce is known as the **smegma**, and has its source in the desquamated and broken-down epithelial cells derived from the surface of the glans and prepuce, and the secretion of preputial glands. A long, narrow prepuce, which cannot be retracted from the glans and may be adherent to it, may have to be removed by the operation of **circumcision**.

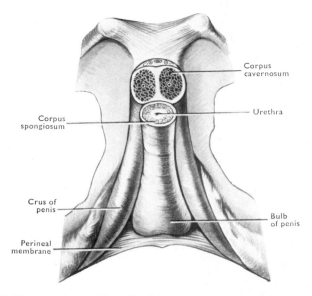

FIG. 8.47. Root of penis. The body of the penis is seen in section.

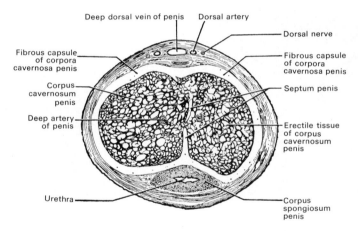

FIG. 8.48. Transverse section through anterior part of body of penis to show its structure.

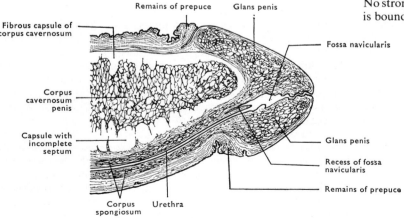

At the **root of the penis** its three corpora separate from one another [FIG. 8.47]. The corpora cavernosa, diverging sidewards, at first become slightly swollen, and then, gradually tapering, gain a firm, fibrous attachment to the periosteum on the medial surface of the sides of the pubic arch. These diverging parts of the corpora cavernosa are the **crura** of the penis and each is covered by the corresponding ischiocavernosus muscle. The corpus spongiosum, lying between the crura, is enlarged to form a globular mass, the **bulb** of the penis. The bulb is variable in size and is attached to the inferior surface of the perineal membrane. Postero-inferiorly the bulb has a median notch, and deep to this a poorly developed, median fibrous septum—both features which indicate its development by the fusion of symmetrical halves; the cutaneous fusion being marked by the overlying raphe. The urethra, having pierced the perineal membrane, enters the bulb obliquely a short distance in front of its posterior end [FIGS. 5.117 and 8.46]. The superficial surface of the bulb is covered by the bulbospongiosus muscles.

A triangular band of strong fibrous tissue, the **suspensory ligament of the penis**, is attached to the front of the pubic symphysis and fuses with the fascial sheath of the penis [FIG. 8.42]. In addition the penis is suspended by the **fundiform ligament** which extends downwards from the rectus sheath and linea alba to loop around the penis.

Structure of the penis

Each **corpus cavernosum** penis is enclosed in a dense, white fibrous capsule (the tunica albuginea) which fuses with that of the opposite side to form the median **septum** of the penis. The septum is very incomplete, especially near the end of the penis, where it is interrupted by a number of nearly parallel slits [FIGS. 8.48 and 8.49]. Through these slits the erectile tissue of the two corpora cavernosa is continuous.

The **tunica albuginea** contains some elastic fibres, and numerous fibrous **trabeculae** pass from its deep surface across the interior of the corpus cavernosum to form a fine sponge-like framework whose interstices, the cavernous venous spaces, communicate freely with one another and are filled with blood. These spaces lead directly into the veins of the penis, and, like the veins, have a lining of flat endothelial cells. The size of the penis varies with the amount of blood in the erectile tissue (see Harrison and de Boer 1977). The structure of the corpus spongiosum resembles that of the corpora cavernosa, but its tunica albuginea is much thinner and more elastic, and the spongework is finer [FIG. 8.48].

The **glans penis** is also composed of erectile tissue which communicates by a rich venous plexus with the corpus spongiosum. No strongly marked fibrous capsule is present, and the erectile tissue is bound to the thin, firmly adherent skin. The urethra in the glans

FIG. 8.49. Median section through terminal part of circumcised penis to show its structure.

penis is dilated and is compressed laterally into a slit-like passage, the **fossa navicularis**, surrounded by a mass of fibro-elastic tissue which forms a median septum within the glans. This septum is continued backwards to join the tunica albuginea of the conical end of the corpora cavernosa, and ventrally it gives attachment to the frenulum of the prepuce. It divides the erectile tissue of the glans imperfectly into right and left portions, which, however, are in free communication dorsally. From the septum, trabeculae pass out in all directions into the tissue of the glans.

A fascial sheath, containing numerous elastic fibres, forms a loose common envelope for the corpora cavernosa and the corpus spongiosum. It is termed the **deep fascia of the penis**, and reaches as far as the base of the glans, where it becomes fixed to the neck of the glans. In its proximal part the sheath gives insertions to many of the fibres of the bulbospongiosus and ischiocavernosus muscles.

Superficial to the deep fascia of the penis is the **superficial fascia of the penis** composed of a layer of extremely lax areolar tissue continuous with the dartos tunic of the scrotum, which lies underneath the delicate skin of the penis. This fascia contains no fat. Numerous sebaceous glands are present in the skin, especially on the urethral surface of the penis.

In some mammals, such as the walrus, dog, bear, and baboon, a bone called the **os penis** is developed in the septum between the corpora cavernosa penis.

VESSELS AND NERVES OF THE PENIS

The arteries are derived from the internal pudendal artery. The erectile tissue of the corpora cavernosa is supplied chiefly by the **deep arteries** of the penis, and that of the corpus spongiosum by the **arteries to the bulb**. Branches of the **dorsal arteries** of the penis pierce the tunica albuginea of the corpora cavernosa and furnish additional twigs to the erectile tissue. The glans receives its chief blood supply from branches of the dorsal arteries. The small branches of the arteries run in the trabeculae of the erectile tissue, and the capillaries into which they lead, open directly into the

cavernous spaces. In the finer trabeculae, the smaller branches are often tortuous—**helicine arteries**.

The **veins** with which the cavernous spaces communicate carry the blood either directly into the **prostatic plexus**, or into the **deep dorsal vein** and so to that plexus. The deep dorsal vein of the penis begins in tributaries from the glans and prepuce, ascends in the groove between the corpora cavernosa, and passes beneath the pubic arcuate ligament to join the prostatic plexus. On each side of it lies a dorsal artery, and, still further from the median plane, a dorsal nerve [FIGS. 8.48 and 8.50]. The **superficial dorsal vein** of the penis is either unpaired or double and runs on the deep fascia of the penis towards the symphysis pubis where it bends laterally to drain into the external pudendal vein.

The **lymph vessels** of the penis are arranged in a deep and superficial series, and they end in the medial groups of the **superficial inguinal lymph nodes**.

The **nerve supply** of the penis is derived from the pudendal (second, third, and fourth sacral nerves) and ilio-inguinal (L. 1) nerves, and from the pelvic autonomic plexuses. The branches of the pudendal are the **dorsal nerve of the penis**, and branches from the perineal nerves. They supply the cutaneous structures of the penis with the ilio-inguinal (anterior scrotal) nerve, while the filaments from the **hypogastric plexuses**, which reach the penis through the prostatic nerve plexus, end in the erectile tissue.

The prostate

The prostate is a partly glandular, partly muscular organ surrounding the beginning of the urethra in the male. It lies within the pelvis behind the pubic symphysis and is enclosed by a dense fascial sheath. Through the various connections of this sheath the prostate is firmly fixed within the pelvic cavity. The ejaculatory ducts traverse the upper, posterior part of the prostate in their course to join the urethra. The size of the prostate varies considerably, but usually its greatest transverse diameter is 4 cm, its anteroposterior diameter is 2 cm, and its vertical diameter 3 cm. Superficially, the prostate is separated from the bladder by deep, wide lateral grooves and by a narrow posterior groove [FIGS. 8.22 and 8.46].

The prostate has an apex which is directed downwards, a base superiorly, a posterior surface, an anterior surface, and a pair of inferolateral surfaces. The **base** adjoins the inferior aspect of the bladder, surrounding its urethral opening. The greater part of the base is structurally continuous with the bladder wall; only a narrow portion remains free on each side and forms the lower limit of the deep groove which marks the separation of the bladder and prostate [FIG. 8.22]. The **inferolateral surfaces** of the prostate are convex and prominent, and rest against the fascia covering the levator ani muscles. They also face slightly forwards, and meet in front in a rounded anterior surface. The **posterior surface** is flat and triangular, and faces backwards and very slightly downwards and is in contact with the rectum. Consequently it may be felt in the living subject by rectal examination. The **apex** points downwards and is in contact with the urogenital diaphragm. From the apex, the **anterior surface** passes upwards in the median plane behind the pubic symphysis and retropubic pad of fat.

The urethra enters the prostate at a point near the middle of its base, and leaves it at a point on its anterior surface immediately above the apex.

The **ejaculatory ducts** enter a slit immediately in front of the posterior border of the base, in the groove between the bladder and prostate, and run downwards and forwards to open into the prostatic portion of the urethra on each side of the mouth of the prostatic utricle.

The wedge-shaped portion of the prostate which separates these ducts from the urethra and urinary bladder is called the **median**

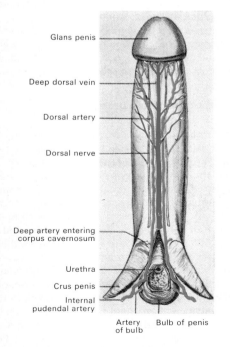

Glans penis

Deep dorsal vein

Dorsal artery

Dorsal nerve

Deep artery entering corpus cavernosum

Urethra

Crus penis

Internal pudendal artery

Artery of bulb   Bulb of penis

FIG. 8.50. Dorsal surface of penis showing the main blood vessels and nerves.

lobe [FIG. 8.18]. It projects upwards against the bladder, and is continuous with the bladder wall immediately behind the internal urethral orifice. When hypertrophied, as it often is in old men, the median lobe of the prostate may cause an elevation in the cavity of the bladder, the uvula of the bladder, which may obstruct the flow of urine. The remaining part of the prostate is described as a pair of large lateral lobes.

In front of the prostate there is a close venous network called the prostatic plexus, which receives the dorsal vein of the penis. This plexus is continuous round the sides of the prostate with the large, thin-walled veins which lie principally in the deep sulcus between the bladder and the prostate, and communicate with the vesical plexus. Most of the veins of the plexus lie embedded in the fascial sheath of the prostate [FIG. 8.51].

Fascial sheath of the prostate

This sheath is a dense fibrous portion of the pelvic fascia, and closely invests the prostate. Inferiorly, the sheath becomes continuous with the superior fascia of the urogenital diaphragm and, through it, gains attachment to the sides of the pubic arch. In front and at the sides, it is fused with the puboprostatic ligaments, by which it is connected with the pubic bones and the fascia on the levatores ani. Between the puboprostatic ligaments of the two sides there is a shallow depression, the floor of which is formed by a thin layer of fascia which connects the anterior part of the sheath of the prostate with the back of the pubic symphysis. The medial edges of the levator ani muscles are immediately below the puboprostatic

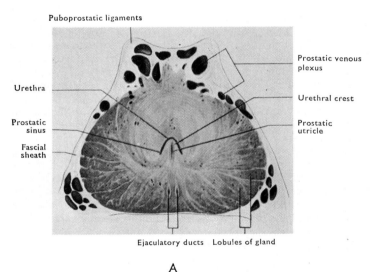

FIG. 8.51. Horizontal sections through prostate. Section A is at a higher level than B.

ligaments, and embrace the lower part of the prostate posteriorly to form the levator prostatae. The lower fibres of the pubovesical muscle [p. 550], which follow the course of the puboprostatic ligament, constitute the puboprostatic muscle.

Posteriorly the sheath is continuous above with the fascial layers which enclose the ampullae of the ductus deferentes and the seminal vesicles, and it is adherent to the peritoneum of the rectovesical pouch. In this position it is spoken of as the rectovesical septum.

Structure of the prostate

Inside the fascial sheath the superficial part of the prostate is largely composed of interlacing bundles of plain muscle fibres and fibrous tissue which form the capsule of the organ. The capsule is not sharply defined, for fibrous and muscular strands pass inwards from it, towards the posterior wall of the urethra, to become continuous with the mass of plain muscle which surrounds this canal. These radially arranged strands divide the prostate into a number of incompletely defined lobules, of which there are about fifty. The yellowish-coloured glandular tissue of the lobules is composed of minute, slightly branched tubules, the walls of which show numerous saccular dilatations. In the upper portion of the prostate the tubules are more convoluted, slightly dilated, and shorter than in the lower part. The glandular tubules lead into twenty or thirty minute prostatic ducts which open on the posterior wall of the urethra, for the most part into the prostatic sinuses [p. 551].

The bulk of the glandular tissue is situated at the sides of the urethra and behind it. In front of its upper part there is a mass of plain muscle which is continued upwards and backwards on its sides to form a part of the sphincter vesicae. Below the level of the opening of the ejaculatory ducts and prostatic utricle, striated muscular tissue occupies a position in front of the urethra and is continuous with the deep part of the sphincter urethrae. Behind the urethra is a layer of transversely arranged smooth muscle which is thickest above and in front of the ejaculatory ducts (Clegg 1957).

The muscular tissue of the prostate is to be regarded as the thickened muscular layer of the wall of the urethra broken up and invaded by the prostatic glands, which are developed from the lining layer of that canal during foetal life. The large amount of plain muscle in the ductus deferens, prostate, prostatic urethra, and seminal vesicles is concerned with ejaculation.

Small at birth, the prostate enlarges rapidly at puberty. After the fourth decade the glandular tissue may atrophy, and the organ becomes more fibrous and diminishes in size. In many individuals, however, the prostate progressively enlarges (Swyer 1944).

VESSELS AND NERVES OF THE PROSTATE

The arteries are constant branches of a trunk of variable origin. This trunk, the prostatovesical artery (Clegg 1955), divides into two terminal branches, the prostatic and inferior vesical arteries. It arises most usually from the common origin of the internal pudendal and inferior gluteal arteries from the internal iliac artery, but may take origin either from the umbilical artery or the vesiculodeferential artery. Within the prostate the tortuous intrinsic vessels form an outer (capsular) plexus, an intermediate zone of vessels, and a urethral plexus (Clegg 1956). The veins are wide and thin-walled; they form the prostatic plexus, which communicates with the vesical plexus and is drained into the internal iliac veins. The prostatic plexus communicates with the vertebral venous plexuses; a tumour in the prostate may therefore give rise to a secondary growth in the vertebral column. In old men the veins of the prostate usually become enlarged. The lymph vessels [FIG. 13.140] are associated with those of the seminal vesicle and the neck of the bladder. The nerves of the prostate are derived from the inferior hypogastric plexuses.

The bulbo-urethral glands

The bulbo-urethral glands are a pair of small bodies placed behind the membranous part of the urethra among the fibres of the sphincter urethrae. In old age they are often difficult to find without microscopical examination; in young adults they are each about the size of a pea and are of a yellowish-brown colour. Placed within the urogenital diaphragm, deep to the perineal membrane, the glands lie below the level of the apex of the prostate and above that of the bulb of the penis [FIGS. 8.32 and 8.46]. Each gland is made up of a number of closely applied lobules, and is of the compound racemose type. The ductules of the gland unite to form a single **duct** which pierces the perineal membrane and enters the bulb of the penis. After a course of about 2.5 cm, it ends by opening into the inferior surface of the spongy portion of the urethra by a minute aperture. The secreting alveoli are lined with columnar epithelium, whose secretion is mostly mucous in character.

The glands receive their arterial supply from the arteries to the bulb.

Methods of clinical anatomical examination

The testis, epididymis, and lower end of the spermatic cord can be palpated easily in the living subject; and the back of the prostate and lower ends of the seminal vesicles and ductus deferentes can be palpated by rectal examination.

THE FEMALE GENITAL ORGANS

The female genital organs are: (1) the ovaries; (2) the uterine tubes; (3) the uterus; (4) the vagina; (5) the external genital organs; (6) the greater vestibular glands; and (7) the mammary glands.

The reproductive glands (gonads) in the female [FIG. 8.27] are a pair of **ovaries** placed one on each side in the cavity of the lesser pelvis. In connection with each ovary there is an elongated tube—the **uterine tube**—which leads to the uterus and opens into its cavity. There is no direct continuity between the ovary and the uterine tube, such as exists between the other glands of the body and their ducts, but the ova, when shed from the ovary, pass into the peritoneal cavity and thence into the pelvic opening of the tube, whence they are transported to the uterine cavity. The **uterus** is a hollow muscular organ which occupies a nearly median position in the lesser pelvis. It is joined by the uterine tubes anterosuperiorly, and it communicates with the vagina postero-inferiorly. The ovum, having passed through the tube, reaches the cavity of the uterus, and undergoes development there if it has been fertilized. The **vagina** is the passage which leads from the uterus to the exterior, and has its external opening behind that of the urethra, within the **pudendal cleft** [FIG. 8.61]. In relation to the cleft there are a number of structures, included under the term **external genital organs**. They are the labia majora and the mons pubis, the labia minora, the clitoris, and the bulbs of the vestibule. The greater vestibular glands, placed one on each side of the lower part of the vagina, are accessory organs of the female reproductive system.

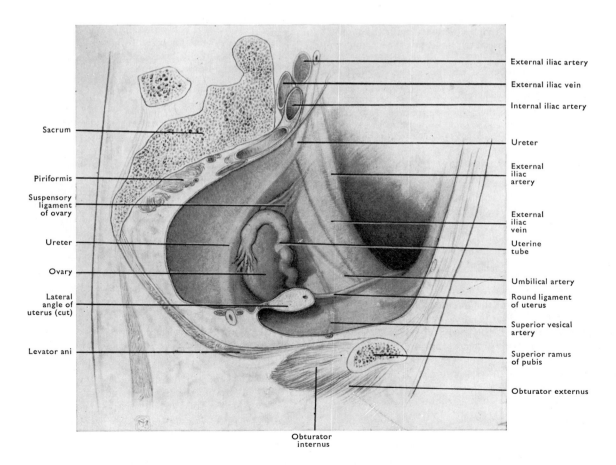

FIG. 8.52. Side-wall of female pelvis, showing position of the ovary and its relation to the uterine tube. The pelvis was cut parallel to the median plane, but at some distance from it.

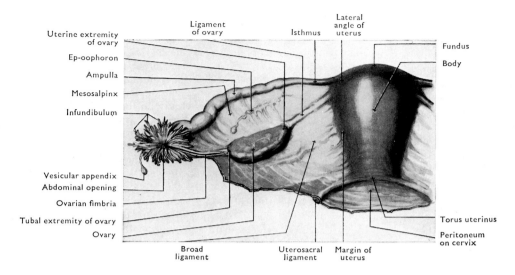

Fig. 8.53. Intestinal surface of uterus and broad ligament. The broad ligament has been spread out.

The ovaries

The ovaries are a pair of solid, flattened ovoids, 2.5–3.5 cm long, 1.5 cm broad, and 1 cm thick in the multiparous woman. In the adult the ovary is placed against the side-wall of the pelvic cavity, and is connected by a peritoneal fold with the broad ligament of the uterus [p. 570].

The ovary has two extremities—upper and lower. The upper, **tubal extremity**, is the larger and more rounded, and is most intimately connected with the uterine tube. The lower, **uterine extremity**, is more pointed and is connected with the uterus by a fibrous cord termed the **ligament of the ovary**. The surfaces of the ovary are medial and lateral, and the borders separating them are the anterior or **mesovarian border** and the posterior or **free border**. The free border is convex; the mesovarian border is straighter and narrower, and it is connected by two very short peritoneal layers, called the **mesovarium**, with the posterosuperior layer of the broad ligament of the uterus. The vessels and nerves enter the ovary at the mesovarian border, which is therefore termed its hilus.

Position and relations of the ovary [Figs. 8.27 and 8.52]

In the nulliparous woman the long axis of the ovary is vertical. Its lateral surface lies against the wall of the lesser pelvis, and its medial surface faces the pelvic cavity. The peritoneum of the pelvic wall, where the ovary lies against it, is depressed to form a shallow fossa between the ureter and the obliterated umbilical artery. Outside the peritoneum forming the floor of this fossa are the obturator nerve and vessels. The tubal extremity of the ovary lies a little below the level of the external iliac vessels, and its uterine extremity is immediately above the peritoneum of the pelvic floor. The mesovarian border is behind the line of the obliterated umbilical artery; the free border is immediately in front of the ureter [Fig. 8.52]. The medial surface of the ovary is largely covered by the uterine tube, which, passing upwards on it near its mesovarian border, arches over the tubal extremity and then turns downwards in relation to the free border and posterior part of the medial surface.

In women who have borne children the position of the ovary is variable, and its long axis may be horizontal.

Connections of the ovary

Anteriorly the ovary is connected with the broad ligament by a very short mesentery, or **mesovarium**, along the whole length of its mesovarian border. A small peritoneal fold passes upwards from the tubal end of the ovary to the peritoneum over the external iliac vessels and the psoas major muscle [Figs. 8.27 and 8.52]. It is the **suspensory ligament** and is continuous below with the mesovarium and, in front of it, with the broad ligament of the uterus. Between its two layers it contains the ovarian vessels and nerves as they pass down into the lesser pelvis to enter the mesovarium and thus reach the hilus of the ovary. The uterine extremity of the ovary is connected with the lateral border of the uterus by a round cord called the **ligament of the ovary**; this band lies in the free border of the medial part of the mesovarium, and is attached to the lateral angle of the uterus posterosuperior to the point of entrance of the uterine tube. It is composed of fibrous tissue containing some plain muscle fibres continuous with those of the uterus. The tubal end of the ovary is directly connected with the **ovarian fimbria**—one of the largest of the fimbriae of the uterine tube [Fig. 8.53].

Descent of the ovary

Like the testes, the ovaries at first lie in the abdominal cavity and only later descend to the lesser pelvis. At birth the ovary lies partly in the abdomen and partly in the lesser pelvis; soon, however, it takes up a position entirely within the lesser pelvis. As in the male, a **gubernaculum** is present in the early stages of development and is connected inferiorly with the tissues that become the labium majus; but in the female the gubernacula become attached to the paramesonephric ducts, from which the uterus and tubes are developed. The gubernaculum becomes the round ligament of the uterus in front and the ligament of the ovary behind. The ovary descends only as far as the lesser pelvis: it is a rare abnormality for the ovary, instead of entering the pelvis, to take a course similar to that of the testis, and pass through the inguinal canal into the labium majus.

Structure of the ovary

The ovary is composed of a fibro-areolar tissue, called the **stroma of the ovary**. Superficially the stroma is dense and consists of fibroblasts and reticular fibres; centrally it is looser in texture, richly supplied by blood vessels and nerves, and contains many elastic fibres. The surface of the ovary is covered with an epithelium composed of columnar or cubical cells, and is continuous with the flat-celled mesothelium of the peritoneum which covers the mesovarium near the hilus of the ovary. Shining through the epithelium of the fresh adult ovary (except in old age) a variable number of small vesicles can usually be seen—the **vesicular ovarian follicles**—in each of which is a developing egg-cell or oocyte. The number of follicles visible, and also the size which each follicle reaches before it ruptures and sheds it contents, are by no means constant. When a follicle ruptures and discharges the oocyte, its walls at first collapse, but later the cavity becomes filled with cellular tissue of a yellowish colour—a **corpus luteum**. This slowly degenerates unless pregnancy occurs, when it develops and becomes larger during the first 3 months. As it atrophies, the cells of the corpus luteum disappear, and the structure loses its yellow colour and becomes white as it is replaced by collagen—the **corpus albicans**. After a time the corpus albicans contracts to a mere scar. Owing to the periodic rupture of the ovarian follicles, the surface of the ovary, which is at first smooth and even, becomes dimpled and puckered in old age.

A histological section through the ovary, especially in the young child, shows that its superficial (cortical) part possesses large numbers of small **primary (primordial) ovarian follicles**, embedded in the fibro-areolar tissue of the stroma. Each primary follicle consists of a primary oocyte surrounded by a layer of investing follicular cells. It was believed that the primary follicles were formed anew at each menstrual cycle from the epithelium covering the ovary, which was therefore termed the germinal epithelium. But these primary oocytes are present from the time of birth, and develop from the primordial germ cells which migrate into the developing ovary from the wall of the yolk sac (Brambell 1956). After puberty, some of the primary follicles grow to form vesicular ovarian follicles during each menstrual cycle. One of these usually undergoes further maturation and ruptures at the time of ovulation to liberate a secondary oocyte in the process of undergoing the second maturation division [p. 18] while the others degenerate. As a result there is a progressive decline in the total number of oocytes in the ovary with increasing age (Zuckerman 1956, 1962). In the deeper part (medulla) of the ovary the blood vessels are most numerous, and some plain muscle fibres are found.

The ripe follicle contains a relatively large amount of fluid, and the surrounding stroma becomes differentiated to form a capsule or **theca folliculi**. The capsule is composed of an inner vascular coat, the **theca interna**, and an outer coat, the **theca externa**, composed of fibrous tissue.

Vessels and nerves of the ovary

The **ovarian arteries**, which correspond to the testicular arteries of the male, are a pair of long, slender vessels which spring from the front of the aorta below the level of origin of the renal vessels. Each artery descends to the suspensory ligament of the ovary, and runs through it to enter the ovary at its mesovarian border. Close to the ovary the ovarian artery anastomoses with branches of the **uterine artery**. The smaller arteries within the adult ovary are spiral (Reynolds 1948). The blood is returned by a series of communicating **veins**, which eventually form one vein draining into the inferior vena cava on the right side, and the renal vein on the left side.

The **lymph vessels** of the ovary follow the blood vessels and run with those from the uterine tube and upper part of the uterus to end in the lymph nodes beside the aorta and inferior vena cava [FIGS. 13.141 and 13.142].

The **nerves** of the ovary are derived chiefly from a plexus which accompanies the ovarian artery and is continuous above with the renal plexus. Other fibres are derived from the lower part of the aortic plexus and join the plexus on the ovarian artery. The afferent impulses from the ovary reach the central nervous system through the fibres of the dorsal root of the tenth thoracic nerve.

The uterine tubes

The uterine tubes (Fallopian tubes) are a pair of ducts which convey the oocytes, discharged from the follicles of the ovaries, to the cavity of the uterus [FIG. 8.57]. Each tube is about 10 cm in length, and opens at one end into the peritoneal cavity near the ovary, and at the other end by a smaller opening into the anterolateral part of the uterine cavity. The tube lies in the margin of a fold of peritoneum called the **mesosalpinx**, which is a part of the broad ligament of the uterus.

The **abdominal opening** of the tube is only about 2 mm in diameter when its walls are relaxed, and much narrower when the muscular coat of the tube is contracted. It is placed at the bottom of a funnel-like expansion of the ovarian extremity of the tube called the **infundibulum**, the margins of which are continued as a number of irregular processes, called **fimbriae**, many of which are branched or fringed. The surfaces of the fimbriae which look into the cavity of the infundibulum are covered with a mucous membrane continuous with that of the tube, while the outer surfaces are clothed with peritoneum. The mucous surfaces of the larger fimbriae present ridges and grooves which are continued into the folds and furrows of the mucous coat of the tube. One of the fimbriae, usually much larger than the rest, is connected either directly or indirectly with the tubal end of the ovary, and to it the name **ovarian fimbria** is applied. The part of the tube continuous with the infundibulum is called the **ampulla**. It is the widest and longest portion of the tube, and is usually slightly tortuous and of varying diameter, being slightly constricted in some places. The ampulla ends in the narrower, thicker-walled, and much shorter **isthmus** of the tube, which joins the lateral angle of the uterus. The **uterine part** of the tube is embedded in the uterine wall, which it traverses to reach the cavity of the uterus [FIG. 8.57]. The **uterine opening** is smaller than the pelvic opening, being about 1 mm in diameter. The lumen of the canal gradually increases in width as it is traced from the uterus towards the ovary.

Course of the uterine tube

From the uterus, the uterine tube is directed at first horizontally towards the uterine extremity of the ovary, and then covers most of the medial surface of the ovary [p. 566]. The fimbriated extremity of the uterine tube lies in the abdominal cavity until the ovary in its descent has entered the lesser pelvis.

Structure of the uterine tubes

The wall of each tube is composed of a number of concentric layers. First there is a **serous coat** of peritoneum, under which lies a thin **subserous coat** of loose areolar tissue containing many vessels and nerves. Internal to the subserous coat there is the **muscular coat** composed of two strata of plain muscle fibres—a more superficial, thin stratum of longitudinal fibres and a deeper, thicker layer of circular fibres. Inside the circular muscular coat is a **submucous layer**, and then the lining membrane or **mucous coat**. In the part of the tube near the uterus the muscular layer is thicker than towards the ovarian end, and in the isthmus it forms the chief part of the wall. The mucous membrane, on the contrary, is thickest towards the fimbriated extremity, and there it forms the chief part of the wall. The mucous membrane is thrown into numerous longitudinal folds, which in the ampulla are exceedingly complex, the larger folds bearing on their surface numerous smaller ones which project into the lumen of the tube and almost completely fill it. The mucous membrane is covered with an epithelium which is partly ciliated, the cilia of which tend to drive the fluid contents of the tube towards the uterus.

Vessels and nerves of the uterine tube

The **arteries** supplying the uterine tube are derived from the anastomosis between the uterine and ovarian arteries which course along the uterine tube in the mesosalpinx. The **veins** of the tube pour their blood partly into the **uterine** and partly into the **ovarian**

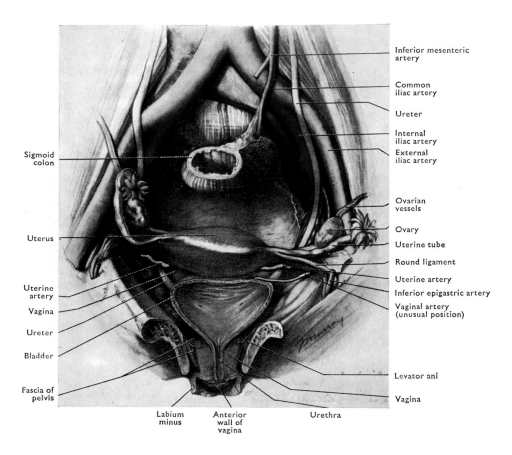

Sigmoid colon

Uterus

Uterine artery

Vagina

Ureter

Bladder

Fascia of pelvis

Labium minus

Anterior wall of vagina

Urethra

Inferior mesenteric artery

Common iliac artery

Ureter

Internal iliac artery

External iliac artery

Ovarian vessels

Ovary

Uterine tube

Round ligament

Uterine artery

Inferior epigastric artery

Vaginal artery (unusual position)

Levator ani

Vagina

FIG. 8.54. Dissection of pelvis of multiparous female showing some of the relations of the internal genital organs.

veins. The **lymph vessels** run with those of the ovary and the upper part of the uterus to the lymph nodes beside the aorta. The **nerves** are derived from the plexus that supplies the ovary, and also from the plexus in connection with the uterus. The afferent fibres are believed to run to the eleventh and twelfth thoracic and first lumbar nerves.

The uterus

The uterus or womb [FIGS. 8.27 and 8.54] is a hollow, thick-walled muscular organ that projects upwards and forwards above the bladder from the upper part of the vagina [FIGS. 8.55 and 8.57].

Shape and parts

The upper two-thirds of the uterus, known as the **body**, is pear-shaped with the blunt end anteriorly, and slightly flattened from above downwards. The lower third, called the **cervix**, is cylindrical and tapers slightly to its blunt lower end, which projects through the anterior vaginal wall into the upper part of the cavity of the vagina. The cervix is therefore divisible into a **vaginal** and a **supravaginal** part; the latter communicates with the body at the **isthmus** of the uterus which is indicated by a slight constriction.

In the centre of the vaginal extremity of the cervix is a small opening—the **ostium uteri**—by which the cavity of the uterus communicates with that of the vagina. This part of the cervix is slightly flattened from before backwards, and the ostium of a uterus from a woman who has not been pregnant has the appearance of a short transverse slit bounded above and below by thick **anterior** and **posterior lips**; in women who have borne children the slit is larger and has an irregular outline. The posterior lip is thicker and more

rounded than the anterior, and both lips are usually in contact with the posterior wall of the vagina.

The uterine tubes enter the organ at its lateral angles, where it attains its maximum breadth; and the rounded part above a line joining the points of entrance of the two tubes is the **fundus**. The uterus is described as possessing an antero-inferior or **vesical surface** and a posterosuperior or **intestinal surface**, separated by right and left borders. Both surfaces are convex, but the intestinal is much more rounded. The length of the body is about 5 cm; its maximum breadth also is 5 cm; and its thickness or depth is 2.5 cm. The cervix is 2.5 cm long and about 2.5 cm in diameter. During pregnancy the whole organ is greatly enlarged.

Position and attachments

The non-pregnant uterus lies completely within the cavity of the pelvis. It rarely lies exactly in the median plane, but is usually inclined slightly to one or other side, more frequently to the left. The cervix is fixed to, and structurally continuous with, the wall of the vagina where it pierces that organ. The cervix passes upwards and forwards from the vagina to become continuous with the isthmus and body which continue in a similar but more horizontal direction so that the body is flexed downwards on the cervix. In its usual position, therefore, the uterus is said to be **anteverted** and **anteflexed**. Although anteversion and anteflexion of the uterus are regarded as the normal condition, it should be realized that in many women the uterus is retroflexed or retroverted without producing any symptoms.

The peritoneum which covers the upper surface of the urinary bladder is reflected upwards near the posterior edge of that organ on to the uterus at the junction of the isthmus with the body and then passes upwards and forwards, covering the inferior or vesical surface

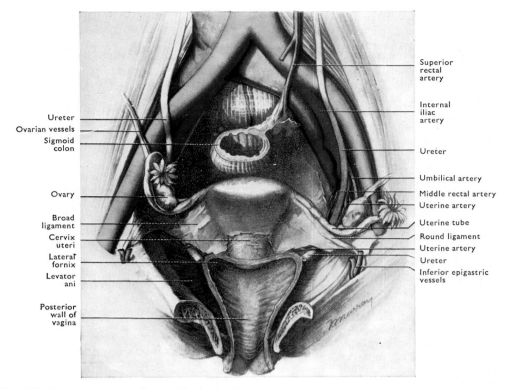

FIG. 8.55. Uterus and vagina displayed by further dissection of pelvis of multiparous female shown in FIG. 8.54.

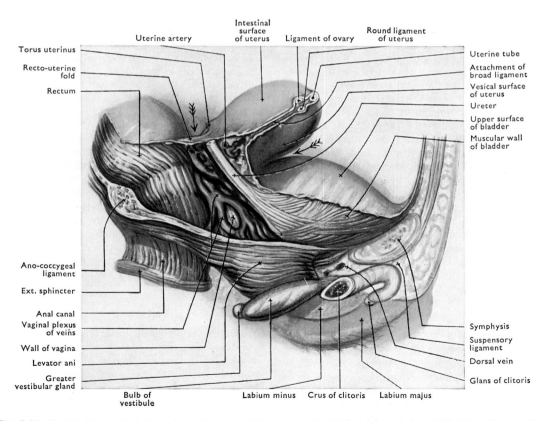

FIG. 8.56. Model of dissection of female pelvic organs. The arrows point to the vesico-uterine and recto-uterine pouches.

of the body. Thus, a rather deep **vesico-uterine pouch** is formed [FIG. 8.56]. From the vesical surface of the uterus the peritoneum passes over the fundus and then downwards and backwards, covering the intestinal surface of the body, the isthmus and supravaginal part of the cervix. It then passes on to the upper part of the vagina and is reflected from it upwards and backwards on the anterior wall of the rectum [FIGS. 8.27 and 8.60]. The peritoneal pouch between the back of the uterus and the front of the rectum is known as the **recto-uterine pouch**, the bottom of which is about 7.5 cm or rather less above the level of the anus. The layers of peritoneum which cover the vesical and intestinal surfaces of the uterus come together at its borders and extend laterally as broad peritoneal folds to the side-walls of the pelvis. These folds are known as the **broad ligaments of the uterus** [FIGS. 8.53 and 8.55], and they enclose various important structures associated with the uterus.

The two layers of each broad ligament are continuous with each other at a free edge which is directed forwards and upwards and surrounds the uterine tube. Elsewhere they enclose some loose areolar tissue and a little plain muscle—the **parametrium**. In the upper part of the broad ligament the parametrium is small in quantity and the two layers of peritoneum are close together, but towards its root the parametrium increases considerably in amount as the two layers diverge from each other. The anterior layer passes forwards into the floor of the paravesical fossa; the posterior layer forms a prominent, shelf-like **recto-uterine fold**, which corresponds to the sacrogenital fold in the male and curves backwards from the cervix uteri to the posterior wall of the pelvis a little lateral to the rectum. The two recto-uterine folds sometimes meet across the back of the cervix, forming a ridge known as the **torus uterinus** [FIG. 8.53].

Enclosed in the free edge of each broad ligament is the **uterine tube**; and just below the entrance of the tube into the uterus two other structures are attached to the uterus, the **round ligament of the uterus** antero-inferiorly and the **ligament of the ovary** posterosuperiorly [FIG. 8.56]. These three structures are all enclosed in the broad ligament, and while they are very close together at their attachments to the uterus, they diverge widely as they approach the pelvic wall. Thus the attachment of the broad ligament to the uterus is almost a straight line but its attachment to the pelvic wall is drawn out into a sort of tripod [FIGS. 8.54 and 8.55] the diverging legs of which are the round ligament of the uterus, the uterine tube shrouded in peritoneum, and the recto-uterine fold. The anterior part of the ligament is drawn forwards by the round ligament of the uterus, which is a narrow flat band of fibrous tissue containing some plain muscle at its uterine end. It passes anterolaterally and very slightly upwards from the upper part of the border of the uterus to the deep inguinal ring, crossing the obliterated umbilical artery and external iliac vessels. After hooking round the lateral side of the inferior epigastric artery and traversing the inguinal canal, it ends in the skin and subcutaneous tissue of the labium majus. In some cases a minute diverticulum of the peritoneal cavity accompanies the round ligament through the inguinal canal; it is the vestige of the **processus vaginalis**, which is formed in the foetus in both sexes.

The part of the broad ligament between the ligament of the ovary, the ovary, and the uterine tube is known as the **mesosalpinx**.

Laterally the mesosalpinx is continuous with the **suspensory ligament of the ovary**, which crosses the external iliac vessels about 2.5 cm below their upper end. The lateral part of the mesosalpinx is more free than the medial part, and this permits the lateral part of the uterine tube to curve down along the posterior border of the ovary and almost enclose that organ in a fold of the mesosalpinx, thus forming a peritoneal compartment known as the **bursa ovarica**. The mesosalpinx contains the epoophoron and the paroophoron (see below) and the anastomosis between the ovarian and uterine arteries.

The part of the broad ligament that covers the ligament of the ovary is really the medial continuation of the mesovarium [FIG. 8.53], and, as the mesovarium is continuous laterally with the suspensory ligament of the ovary also, the complete fold forms the base of the mesosalpinx. The **ligament of the ovary** is a round fibrous band, about 2.5 cm long [p. 566].

At the root of the broad ligament, where the two layers of peritoneum diverge to pass into the pelvic floor, a considerable amount of fibrous tissue and plain muscle occurs. This mass is the upper end of the fascial condensation which passes from the side of the vagina and the base of the urinary bladder to the fascia on the levator ani muscles. Its upper and more bulky part is attached medially to the side of the supravaginal part of the cervix and to the side of the upper third of the vagina, while laterally it fans out. Its posterior part passes backwards in the recto-uterine fold of peritoneum and gains an attachment to the front of the sacrum; this part is the **uterosacral ligament** and it contains nerves supplying the uterus from the inferior hypogastric plexuses [FIG. 11.85]. The anterior part, passing directly laterally to blend with the fascia on the levator ani muscle, is called the **lateral cervical ligament** (Mackenrodt 1895). The ureter and uterine artery lie in its upper part (Power 1944). It is a condensation of fascia around the internal iliac artery and its branches.

Epoophoron and paroophoron

The epoophoron and paroophoron are two vestigial structures sometimes found between the layers of the broad ligament.

The **epoophoron** [FIG. 8.53] lies in the mesosalpinx between the uterine tube and the ovary. In the adult it consists of a number of small rudimentary blind tubules lined with epithelium. One of the tubules—the **longitudinal duct of the epoophoron**—lies close to the uterine tube and runs nearly parallel with it. It is joined by a number of **transverse ductules** which enter it at right angles from the neighbourhood of the ovary. The longitudinal duct is a persistent portion of the mesonephric duct, and represents the duct of the epididymis; the transverse ductules are derived from the mesonephric tubules and represent the efferent ductules of the testis (and probably also the ductuli aberrantes).

One or more small pedunculated cystic structures, called **vesicular appendices**, are often seen near the infundibulum of the uterine tube. They are supposed to represent portions of the cephalic end of the mesonephric duct.

The **paroophoron** is a collection of very minute tubules that lie in the mesosalpinx nearer the uterus than the epoophoron. They represent the paradidymis in the male, and are derived from the part of the mesonephros which lies nearer the caudal end of the embryo. Though sometimes visible in the child at birth, the paroophoron in the adult can be recognized only by microscopic examination.

Relations of the uterus

Anteriorly the body of the uterus is separated from the bladder by the uterovesical pouch of peritoneum; the front of the supravaginal part of the cervix is separated from the bladder by loose areolar tissue. Posteriorly the body and supravaginal part of the cervix are separated by a layer of peritoneum from the sigmoid colon or coils of small intestine. On each side, there is the broad ligament of the uterus with its contained structures. Of particular importance is the close relationship of the **ureter** which is crossed superiorly by the uterine artery at the side of the cervix. The uterus rises and falls as the urinary bladder fills and empties. The vaginal part of the cervix pierces the anterior wall of the vagina and projects into its cavity, where it is covered with vaginal mucous membrane.

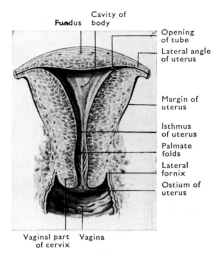

FIG. 8.57. Diagrammatic representation of uterine cavity. The tissue at the side of the cervix represents the lateral cervical ligament.

Cavity of the uterus

In comparison with the size of the organ, the cavity of the uterus is small, owing to the great thickness of the uterine wall. The **cavity of the body** of the uterus when viewed from the side, is a mere slit [FIG. 8.27]; from behind it is triangular in outline, the base in the fundus and the apex continuous with the canal of the cervix, with the sides of the triangle convex inwards. The lateral angles of the triangle mark the entrance of the two uterine tubes. The **canal of the cervix** extends from the isthmus to the **ostium**, where it opens into the vagina. The canal is narrower at the ends than in the middle, and is flattened anteroposteriorly. In the body of the uterus the walls of the cavity are smooth and even, but the mucous membrane of the nulliparous cervix forms a remarkable series of folds called the **palmate folds**. They consist of an anterior and a posterior longitudinal fold or ridge from which a large number of secondary folds or rugae branch off obliquely upwards and laterally [FIG. 8.57].

Isthmus of the uterus

The lowest part of the cavity of the body, about 1 cm in length, is constricted and is commonly known as the isthmus of the uterus [FIG. 8.57]. The separate identity of this part of the uterus has a functional basis, and the structure of its mucous coat is less complex than in the main part of the body (Frankl 1933). Not only is it much less affected by the periodic changes of the menstrual cycle, but in the last months of pregnancy it becomes more distinct and during labour is transformed into the 'lower uterine segment' of the obstetrician.

Supports of the uterus

The uterine supports are mainly the bony pelvis and the pelvic floor, for although the uterus itself does not rest directly on these structures, yet it lies on the urinary bladder and rectum which are supported by them. The strength of the pelvic floor is therefore essential to the support of the uterus, and the strength of the floor is dependent on an intact **central perineal tendon** [p. 368]. This fibromuscular mass is liable to injury and laceration during parturition; and it is of such importance in obstetrical practice that some obstetricians refer to it simply as 'the perineum'. The lateral cervical and uterosacral ligaments are simple condensations of the pelvic fascia passing to

the cervix (which is the part of the uterus most firmly held in position), but they are unable to resist sustained stress if the pelvic floor has given way (Harrison and de Boer 1977). The fundus of the uterus is mobile, and though the round ligaments play some part in maintaining the anteverted position, they do not prevent the uterus from enlarging into the abdomen in pregnancy.

Structure of the uterus

The uterine wall is composed of three chief layers—the serous, the muscular, and the mucous coats.

The **serous coat**, or perimetrium, is the peritoneal covering. Over the fundus and body it is very firmly adherent to the deeper layers, and cannot be peeled off without tearing either it or the underlying muscular tissue. Near the borders of the uterus the peritoneum is less firmly attached, and over the back of the cervix it may be stripped easily without injury to the underlying structures.

The **muscular coat**, or myometrium, is composed of interlacing bundles of plain fibres, and forms the chief part of the uterine wall. Inferiorly the muscular coat of the uterus becomes continuous with that of the vagina. The more superficial layer of the muscular coat sends prolongations into the recto-uterine folds, into the round ligaments of the uterus, and into the ovarian ligaments. Other fibres join the walls of the uterine tubes. The main branches of the blood vessels and nerves of the uterus lie among the muscle fibres. The internal layers of the muscular coat contain a considerable amount of fibrous tissue and some elastic fibres. At the isthmus the muscular fibres decrease and the fibrous tissue increases. The middle coat of the cervix contains mostly fibrous tissue with small amounts of muscular and elastic tissue; hence its greater firmness and rigidity.

The **mucous coat**, or endometrium, in the body of the uterus is smooth and soft and covered with columnar, partially ciliated epithelium. Simple tubular glands, also lined with a columnar epithelium, are present in the mucous membrane and extend to the muscle, as there is no submucosa. In the isthmus the glands are fewer, less deep, stain less intensely, and are less affected by the menstrual cycle. In the cervix the mucous coat is firmer and more fibrous than in the body, and its surface is ridged by the palmate folds. The mucous membrane of the cervix is covered with columnar epithelium showing occasional ciliated cells, and this changes to squamous epithelium just inside the ostium of the uterus. In addition to unbranched tubular glands, numerous slightly branched glands are found in the cervix uteri. Both kinds of glands are lined with the same epithelium as the cervical canal. In many cases small, clear retention cysts are to be seen in the cervical mucous membrane. They arise as a result of obstruction at the mouths of the glands.

Differences in the uterus at different ages

At birth the cervix uteri is relatively large, and its cavity is not distinctly marked off from that of the body by an isthmus. At that time also, the uterus is mainly an abdominal organ and the palmate folds extend throughout the whole of its length. The organ grows slowly until just before puberty, when its growth is rapid for a time. As it increases in size the mucous membrane of the body becomes smooth and the palmate folds are restricted to the cervix. In women who have borne children the cavity remains permanently a little wider and larger than in those who have never been pregnant. In old age the uterine wall becomes harder and has a paler colour than in the young subject.

Variations

In rare cases the uterus may be divided by a septum into two distinct cavities, or the upper aspect of one or both of its borders may be prolonged into straight or curved processes (uterine cornua) continuous with the uterine tubes. The latter abnormality recalls the appearance of the bicornuate uterus of some animals. Both the above conditions arise from an arrest in the fusion of the two separate tubes—the paramesonephric ducts—parts of which normally unite in the embryo to form the uterus.

Periodic changes in the uterine wall

At each menstrual period (approximately every 28 days) there is a shedding of the superficial layers (stratum compactum and stratum spongiosum) of the endometrium. In the early part of the menstrual cycle the mucous membrane gradually thickens; after ovulation it becomes more vascular and its glands tortuous. Soon the superficial parts of the mucous membrane disintegrate and haemorrhage takes place from the endometrial blood vessels. When menstruation is over the mucous membrane is rapidly regenerated from the basal layer (stratum basalis) of the endometrium which remains unchanged throughout the menstrual cycle.

The pregnant uterus

The pregnant uterus increases rapidly in size and weight, so that from being a pelvic organ 7.5 cm in length and about 40 g in weight, it rises into the abdomen and becomes by the eighth month of pregnancy about 20 cm in length and sometimes as much as 1 kg in weight. In shape the uterus is then oval or rounded, with a thick wall composed chiefly of greatly enlarged, smooth muscle fibres arranged in distinct layers. The fundus is very round and prominent. The round ligaments are longer, and the layers of the medial part of the broad ligaments become separated by the growth of the uterus between them. The blood vessels, especially the arteries, are very large and tortuous. The changes which occur in the mucous membrane of the pregnant uterus are described on pages 35–40. For changes in the pelvic floor during parturition, see Power (1946).

Involution of uterus

Immediately after delivery the uterus contracts, but it is still many times the size of the normal resting organ. It weighs about 1 kg and lies against the anterior abdominal wall almost as high as the umbilicus. The peritoneum on its surface is thrown into wrinkles, and owing to the slackness of its ligaments the whole organ is extremely mobile. The cervix is lengthened and is flaccid, with thin walls and a widely open ostium. From the second day after delivery onwards the uterus rapidly shrinks. By the end of the first week it has diminished in weight to about 500 g, the wrinkles in its peritoneal covering have disappeared, and the cervix has shortened and become firmer. The ligaments of the uterus shorten as the uterus shrinks, and the pelvic floor also regains tone, so that the organ is less mobile. By the end of the second week it weighs 350 g and by the end of the eighth week it should be reduced to its resting size and should weigh 40 g or less. This process of gradual reduction of the uterus to its normal size is known as involution.

VESSELS AND NERVES OF THE UTERUS

The uterus receives its arterial supply mainly from the **uterine arteries**, which are branches of the internal iliac arteries, but also from the **ovarian arteries**, derived from the aorta. The vessels from these two sources communicate freely with one another.

Numerous thin-walled **veins** form a plexus at the side of the cervix and pour their blood into the tributaries of the internal iliac vein.

The **lymph vessels** from the fundus of the uterus join those from the ovary and end in aortic lymph nodes. One or two run along the round ligament to the superficial inguinal lymph nodes. The lymph vessels from the cervix run to the sacral nodes and all the iliac groups; those from the body end chiefly in the external iliac nodes [FIG. 13.142].

The **nerves** of the uterus are derived chiefly from the **uterovaginal plexus** which is placed in the neighbourhood of the cervix uteri. Superiorly this plexus is continuous with the pelvic sympathetic plexuses, but it receives parasympathetic fibres also from the second and third or the third and fourth sacral nerves through the pelvic splanchnic nerves.

Clinical observations indicate that afferent impulses reach the central nervous system from the uterus through the posterior roots of the tenth, eleventh, and twelfth thoracic and the first lumbar nerves (Whitehouse and Featherstone 1923).

RADIOLOGY

Although the uterus can be seen as an ovoid opacity on the upper surface of the contrast filled bladder during the course of an intravenous pyelogram [FIG. 8.23], the nature of its lumen can be appreciated only after the performance of a hysterosalpingogram, i.e. the introduction of contrast medium by cannulation of the cervix uteri [FIGS. 8.58 and 8.59].

This shows the lumen of the body of the uterus with the thread-like uterine tubes extending laterally towards the ovarian fossae, where the tubes widen at the **ampullae**. If the tubes are patent, the

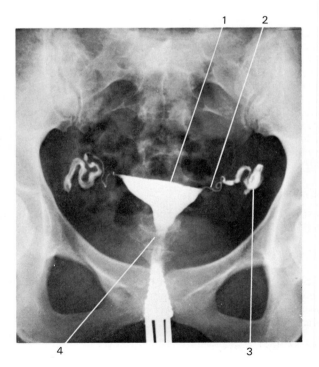

FIG. 8.58. Hysterosalpingogram in axis of lesser pelvis, produced by injecting water-soluble contrast medium through the cervix of the uterus. The body of the uterus appears foreshortened (cf. FIG. 8.59) due to its anteversion (normal). The narrow isthmus and wider ampullae of the uterine tubes are well shown. There is no escape of contrast medium from the abdominal openings of the tubes in this radiograph.
1. Body and fundus of uterus.
2. Isthmus of uterine tube.
3. Ampulla of uterine tube.
4. Cervix of uterus.

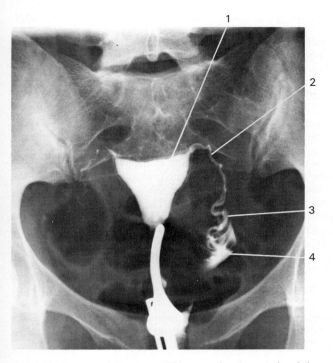

FIG. 8.59. Hysterosalpingogram. This uterus is retroverted and shows the shape of its lumen. Only the left uterine tube has filled completely, allowing some contrast medium to enter the peritoneal cavity.
1. Uterus.
2. Isthmus of uterine tube.
3. Ampulla of uterine tube.
4. Contrast medium in peritoneal cavity.

contrast medium escapes immediately, or after a short delay, into the peritoneal cavity, an important fact to establish in the investigation of sterility. The appearance of the uterus varies with the angle it makes with the horizontal plane. If much **anteverted** (bent forwards) the cavity of the body is seen 'end on' and appears sausage shaped and if **retroverted** a triangular outline is revealed [FIG. 8.59]. Most uteri occupy an intermediate position.

Ultrasound allows the uterus (and ovaries, if enlarged) to be seen; in early pregnancy this is the only permissible method.

The ovaries and uterus can be shown by the introduction of air or carbon dioxide into the peritoneal cavity. This is rarely used because of advances in ultrasonic technique.

The vagina

The vagina [FIGS. 8.27 and 8.55] is a passage of variable size, but usually about 9 cm in length, open at its lower end, and communicating above with the cavity of the uterus. The vagina is directed upwards and backwards, describing a slight curve which is convex forwards. When the urinary bladder is empty, the axis of the vagina makes with the axis of the uterus an angle of slightly more than 90 degrees, but this angle increases as the bladder fills and raises the fundus of the uterus. Its anterior and posterior walls are in contact except at its upper end, where the cervix is inserted; and that is its widest part also. In transverse section the lower part is usually an H-shaped cleft, the middle part of a simple transverse slit, while the lumen of the upper portion is more open. The cervix uteri enters the vagina through the upper portion of its anterior wall [FIG. 8.27]. As more of the posterior part of the cervix than of the anterior

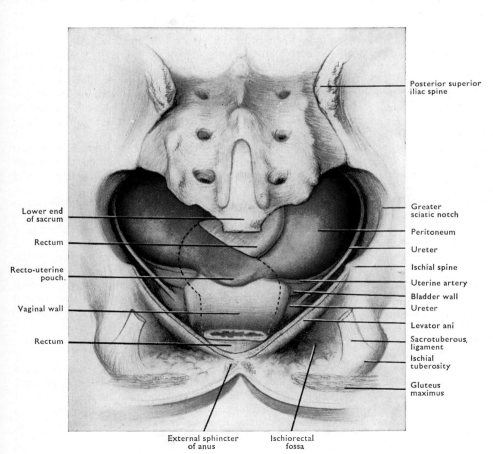

FIG. 8.60. Vagina, base of bladder, and recto-uterine pouch from behind.

The coccyx and sacrotuberous and sacrospinous ligaments have been removed. The levatores ani have been separated and drawn laterally. A considerable portion of the rectum has been removed, but its position is indicated by the dotted lines. The recto-uterine pouch is probably not quite so deep as usual. The triangular interval between the ureter and uterine artery was filled by a mass of fibromuscular tissue.

part projects into the vagina, the recess between the vaginal wall and the cervix is deeper behind than in front or laterally. The term **anterior fornix** is applied to the recess in front, **posterior fornix** to the recess behind, and **lateral fornix** to the recess on each side, though the four fornices are parts of a continuous recess surrounding the cervix. The **anterior wall** of the vagina is about 7.5 cm in length; the **posterior wall** is a little longer—about 9 cm. At its lower end the vagina opens into the pudendal cleft between the labia minora. The opening is partly closed in the virgin by a thin, crescentic or annular fold, called the **hymen**, torn fragments of which persist around the opening, as the **carunculae hymenales**, after the fold itself has been ruptured.

Relations of the vagina

The vagina is in contact **anteriorly** from above downwards with the cervix uteri, the base of the bladder and ends of the ureters, from which it is separated by areolar tissue, and the urethra, to which it is intimately connected. Its upper end usually deviates a little to one side, with consequent asymmetrical relation to the ureters [p. 543]. **Posteriorly** it is in contact from above downwards with the recto-uterine peritoneal pouch for 1.3 cm or more, then with the ampulla of the rectum, from which it is separated by a thin layer of very loose areolar tissue, and then with the central perineal tendon. **Laterally** on each side is the root of the broad ligament, and the ureter crossed by the uterine vessels runs just above the lateral fornix. Lower down is the levator ani muscle, a plexus of veins intervening; and lower still is the greater vestibular gland and the bulb of the vestibule covered by the bulbospongiosus muscle. As the perineal membrane approaches the median plane, it turns downwards on each side of the vagina between it and the bulb of the vestibule and fuses below with the root of the labium minus.

Structure of the vagina

The vaginal wall has a **muscular coat** composed of plain muscle fibres, most of which are longitudinally disposed. Towards the lower end there are circular bundles of striated muscle fibres, some of which are continuous with those which form a part of the urethral wall. The **mucous coat** is thick and is corrugated by a number of transverse ridges or elevations known as **vaginal rugae** covered with stratified squamous epithelium. In addition to the transverse rugae, longitudinal ridges are present on the anterior and posterior walls. These receive the name **columns of the rugae**, and, like the transverse rugae, they are seen best in young subjects and in the lower part of the vagina. The urethra lies in close relationship to the anterior column in its lower part, and hence that portion of the column is called the **urethral ridge**. Between the muscular and mucous coats there is a thin layer of vascular tissue which resembles erectile tissue; and in the mucous coat there are small nodules of lymph tissue.

The vaginal wall is surrounded by a layer of loose, vascular areolar tissue which contains numerous large communicating veins.

VESSELS AND NERVES OF THE VAGINA

The arteries on each side are the **vaginal artery**, the vaginal branch of the **uterine artery**, the vaginal branches of the **middle rectal artery**, and branches of the **internal pudendal artery**. The veins form a plexus around the vaginal wall, and drain their blood into the tributaries of the internal iliac veins.

The **lymph vessels** from the upper part of the vagina join the external and internal iliac nodes, and those from the lower part end in the superficial inguinal nodes [FIG. 13.141].

The **nerves** of the vagina are derived from the uterovaginal plexus and from the vesical plexus. Parasympathetic fibres are derived from the second to fourth sacral nerves.

METHODS OF CLINICAL ANATOMICAL EXAMINATION

The interior of the vagina and the vaginal part of the cervix can be inspected through a vaginal speculum. By a bimanual examination—two fingers of one hand in the vagina and the other hand on the anterior abdominal wall above the pubic crests—the fundus of the uterus can be felt when the fingers in the vagina press forwards and upwards. Normal uterine tubes cannot usually be felt, but if thickened or inflamed they can often be palpated in a bimanual examination. The ovaries may be palpated by placing the examining fingers in the appropriate lateral fornix.

The ostium of the uterus is usually at the level of the spines of the ischium, which can be felt from the vagina. In parous women the uterus may prolapse so that the ostium lies lower, and may even protrude through the vaginal orifice. The vaginal part of the cervix and the ostium can be felt by a finger in the rectum. The pregnant uterus can be palpated as soon as it has risen above the pelvic cavity and lies against the anterior abdominal wall.

The female external genital organs

The collective term **pudendum femininum** or **vulva** is applied to the female external genital organs, i.e. to the mons pubis, the labia majora, and the structures which lie between the labia [FIG. 8.61].

Labia majora

The labia majora represent the scrotum in the male, and they form the largest part of the female external genital organs. They are the boundaries of the **pudendal cleft**, into which the urethra and vagina open. Each labium is a prominent, rounded fold of skin, narrow behind where it approaches the anus, but increasing in size as it passes forwards and upwards to meet the other labium at the **anterior commissure** and end in a median elevation—the **mons pubis**. The mons pubis lies over the pubic symphysis, and, like the labia majora, it is composed chiefly of fatty and areolar tissue, and is covered with hair. The skin of the lateral, convex surface of each labium majus contains numerous sebaceous glands and corresponds to that of the scrotum in the male, but the medial surface is flatter and smooth and has a more delicate cutaneous covering. The posterior ends of the labia majora are connected across the middle line, in front of the anus, by the **posterior commissure**.

Usually the mons and the labia majora are the only visible parts of the external genital organs, since the labia are in contact with each other and completely conceal the structures within the pudendal cleft.

The round ligament of the uterus ends in the skin and fibrofatty tissue of the labium majus. The superficial layer of the subcutaneous tissue resembles that of the scrotum, but contains no muscular fibres.

The **nerve supply** corresponds with that of the scrotum—the anterior part of each labium being supplied by branches of the ilio-inguinal nerve, and the posterior part by branches from the pudendal nerve and by the perineal branch of the posterior cutaneous nerve of the thigh. The **blood vessels** are derived from the external pudendal branches of the femoral and from the labial branches of the internal pudendal vessels.

Labia minora

The labia minora are a pair of much smaller and narrower longitudinal folds of skin usually completely hidden in the cleft between the labia majora. Their posterior parts diminish in size and end by gradually joining the medial surfaces of the labia majora. In the young subject a slightly raised transverse fold, the **frenulum of the labia**, is usually seen connecting the posterior ends of the labia minora. Traced forwards, each labium minus divides into two

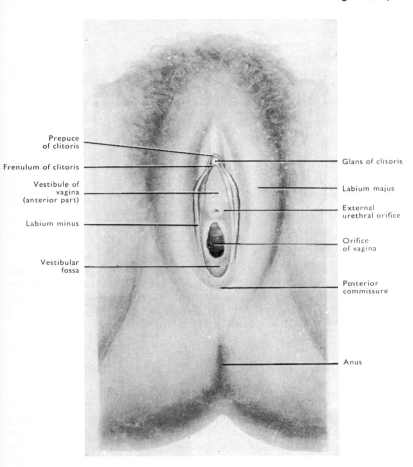

Prepuce of clitoris

Frenulum of clitoris

Vestibule of vagina (anterior part)

Labium minus

Vestibular fossa

Glans of clitoris

Labium majus

External urethral orifice

Orifice of vagina

Posterior commissure

Anus

FIG. 8.61. Female external genital organs.
The frenulum labiorum is seen stretching across behind the vestibular fossa.

portions—a lateral and a medial. The lateral portions of the two labia unite over the glans of the clitoris and form for it a fold or covering called the **prepuce of the clitoris**. The medial portions, uniting at an acute angle, join the ventral surface of the glans and form the **frenulum of the clitoris**. The skin of the labia minora resembles that on the medial surface of the labia majora, being smooth, moist, and pink in colour. The medial surfaces of the labia minora are in contact with each other, and their lateral surfaces with the labia majora. The labia minora and the frenulum are devoid of fat.

The openings of the urethra and vagina are in the median plane, between the labia minora, which must be separated to bring them into view.

Vestibule of the vagina

The vestibule is the name applied to the space between the labia minora. In its roof are the openings of the urethra, the vagina, and the ducts of the greater vestibular glands.

The **vestibular fossa** is the part of the vestibule between the vaginal orifice and the frenulum of the labia.

The **external urethral orifice** is immediately in front of that of the vagina, and is about 2.5 cm behind the glans clitoridis. The opening has the appearance of a vertical slit or of an inverted V-shaped cleft whose margins are prominent and in contact with each other. Sometimes on each side of the urethral orifice there may be seen the minute opening of the para-urethral duct [p. 553].

The **orifice of the vagina** varies in appearance with the condition of the **hymen**. When the hymen is intact the opening is small, and is seen only when that membrane is stretched. When the hymen has been ruptured the opening is much larger, and round its margins are

often seen small projections—**carunculae hymenales**—which are fragments of the hymen.

The **hymen** is a thin membranous fold which partially closes the lower end of the vagina, and is perforated usually in front of its middle point, though rarely it may be complete. The position of the opening gives the fold, when stretched, a crescentic appearance. The opening in the hymen is sometimes cleanly cut, sometimes fringed. The membrane is not stretched tightly across the lower end of the vagina, but is so ample that it projects downwards into the pudendal cleft. The opening is thus a median slit whose margins are normally in contact.

On each side of the vaginal opening, in the angle between the labium minus and the inferior surface of the carunculae hymenales, is the opening of the duct of the **greater vestibular gland** [see below]. It is usually just large enough to be visible to the unaided eye [FIG. 8.62].

Numerous minute mucous glands—the **lesser vestibular glands**—open into the vestibule between the urethral and vaginal orifices.

Clitoris

The clitoris is the morphological equivalent of the penis, and it has a body, a pair of crura, and a minute glans at the distal end of the body. Unlike the penis, the clitoris is not traversed by the urethra.

The **body** is composed for the most part of erectile tissue resembling that of the penis. It is about 2.5 cm in length, and is bent upon itself [FIG. 8.62]. It tapers towards its distal end, which is covered by the glans. It is enclosed in a dense fibrous fascia and is divided by an incomplete **septum** into a pair of cylindrical **corpora cavernosa**, which diverge from each other at the root of the clitoris

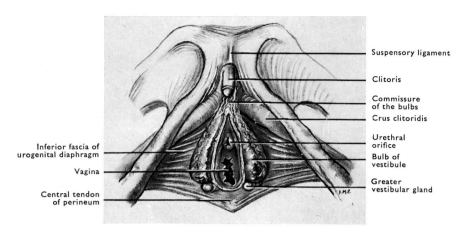

Suspensory ligament

Clitoris

Commissure
of the bulbs

Crus clitoridis

Inferior fascia of
urogenital diaphragm

Vagina

Central tendon
of perineum

Urethral
orifice

Bulb of
vestibule

Greater
vestibular gland

FIG. 8.62. Dissection of female perineum to show clitoris, bulb of the vestibule, and the greater vestibular glands.

to form the crura. A fibrous **suspensory ligament** passes from the body of the clitoris to the pubic symphysis [FIG. 8.62].

The **glans** of the clitoris is a small mass of erectile tissue fitted over the pointed end of the body, and, like the glans penis, it is covered with a very sensitive epithelium. The prepuce, which hoods over the glans, and the frenulum, which is attached to it inferiorly, are continuous with the labia minora [FIG. 8.61].

The **crura** of the clitoris diverge from the body posteriorly, and are attached to the sides of the pubic arch. Each is continuous with one of the corpora cavernosa, and has a firm fibrous fascia which is covered by the corresponding ischiocavernosus muscle. In structure the crura and body of the clitoris resemble the corpora cavernosa of the penis, while the glans more closely resembles the bulbs of the vestibule, with which it is connected by a slender band of erectile tissue [FIG. 8.62].

In the seal and some other animals a bone—the **os clitoridis**, which represents the os penis of the male—is developed in the septum of the clitoris.

The **arterial supply** of each crus is the deep artery of the clitoris, a branch of the internal pudendal. The glans is supplied by the dorsal arteries of the clitoris.

The **nerve supply** of the clitoris is derived partly from the inferior hypogastric plexuses and partly from the dorsal nerves of the clitoris, which are branches of the pudendal nerves.

Bulbs of the vestibule

These are a pair of masses of erectile tissue which correspond developmentally to the two halves of the bulb of the penis, but are almost completely separated from each other by the vagina and the urethra—being only slightly connected in front by a narrow median **commissure**. Each bulb is thick posteriorly, and more pointed in front, where it joins the commissure. It rests against the lateral wall of the vagina and the lower surface of the perineal membrane. Superficially it is covered by the bulbospongiosus muscle. The commissure lies in front of the opening of the urethra, and is connected by a slender band of erectile tissue with the glans clitoridis. The bulb is mostly composed of minute convoluted blood vessels which anastomose frequently with one another, and with the vessels of the commissure and the glans clitoridis and are held together by a small amount of areolar tissue.

The **blood supply** of the bulb is derived from a special branch of the internal pudendal artery.

Greater vestibular glands

The greater vestibular glands are placed one on each side of the lower part of the vagina, below the perineal membrane and concealed by the posterior parts of the bulbs of the vestibule. Each

is about the size and shape of a small bean; and it has a long slender duct which pierces the perineal membrane and opens into the vestibule in the angle between the labium minus and the hymen [FIGS. 8.56 and 8.62]. They are tubulo-alveolar mucous glands, similar to the bulbo-urethral glands [p. 565].

The development of the urogenital organs

General account

The urogenital system develops from intermediate mesoderm (intermediate cell mass), situated originally between the paraxial mesoderm medially and the lateral plate mesoderm, with the coelomic epithelium (splanchnopleuric intra-embryonic mesoderm) covering its surface.

In an embryo which is 2.5 mm in length, the intermediate mesoderm in the cervical region has differentiated into a mass of cells from which a duct is beginning to extend in a caudal direction. This mass of cells is the **pronephros**. It corresponds to the functional kidney of some chordates and forms a simple tubular system which connects the coelom with the pronephric duct in the embryos of some mammals; in the human embryo, however, it is probably functionless and soon becomes vestigial. The duct which grows caudally from it is the **pronephric duct**, and this is then utilized by the intermediate mesoderm in the thoracic and upper lumbar segments of the embryo which forms an embryologically functional kidney known as the **mesonephros**. The part of the pronephric duct utilized by this structure then becomes the **mesonephric (Wolffian) duct**. With the development of the mesonephros the pronephros atrophies.

MESONEPHRIC DUCT AND EMBRYONIC EXCRETORY ORGAN

The mesonephric duct lies in the lateral aspect of the mesonephros. At the end of the fourth week this duct in its growth reaches the cloaca. As soon as the cloaca is divided into dorsal and ventral parts by the urorectal septum of intermediate mesoderm, the mesonephric ducts end in the ventral division, thereby dividing it into two parts [FIG. 8.63B]. The cephalic part forms most of the urinary bladder, and all (in the female) or the cephalic part (in the male) of the urethra, while the caudal portion forms the urogenital sinus [FIGS. 8.63–8.69].

The **mesonephros** consists of a number of minute, transversely arranged tubules [FIG. 8.66], each of which opens laterally into the mesonephric duct, while its medial end is blind. The transverse

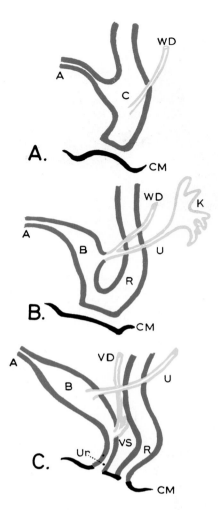

FIG. 8.63. Diagram to illustrate manner in which ureter, ductus deferens, and urinary bladder arise in the embryo.

The structures developed from the cloaca are indicated in red, those from the mesonephric duct in yellow, and the ectoderm in black.

The manner in which the rectum and bladder become separated and acquire openings to the exterior is shown in B and C.

A.	Allantois.	R.	Rectum.
B.	Bladder	Ur.	Urogenital canal.
C.	Cloaca.	U.	Ureter.
CM.	Ectoderm of cloacal	VD.	Ductus deferens.
	membrane.	VS.	Seminal vesicle.
K.	Pelvis of kidney.	WD.	Mesonephric duct.

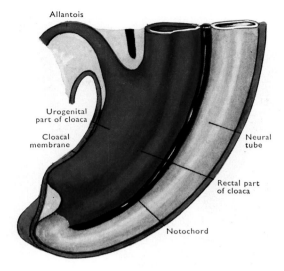

FIG. 8.64. Tail-end of 3-mm human embryo, before the time at which the mesonephric ducts reach the cloaca. (Keibel 1896.)

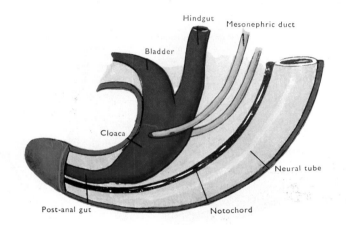

FIG. 8.65. Tail-end of 4.2-mm human embryo. The mesonephric ducts open into the urogenital (anterior) part of the cloaca. (Keibel 1896.)

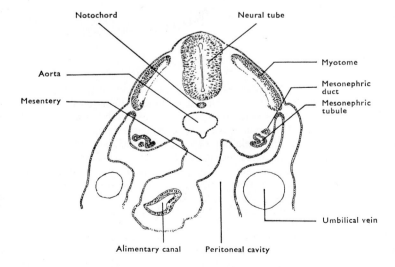

FIG. 8.66. Transverse section of 5-mm human embryo.

tubules, like the canal into which they open, are at first solid cellular structures, and only later acquire a lumen. Increasing rapidly in length and number, the tubules become twisted and tortuous, and the blind end of each dilates to form a capsule invaginated upon itself and containing a bunch of capillary blood vessels similar to the glomeruli of the adult kidney. It would appear that primitively, as in the pronephros, one tubule is developed in the portion of the intermediate cell mass (nephrotome) which corresponds to each mesodermal somite, but in higher vertebrates the mesonephros shows no sign of segmentation and there are several tubules opposite each somite, particularly at its caudal end where the tubules are very numerous. The tubules of the mesonephros arise in all segments from the sixth cervical to the third lumbar. The tubules in the headward part atrophy and disappear very early in development, even while others are being formed towards the caudal end of the embryo. When at its greatest development (fifth to eighth week) the

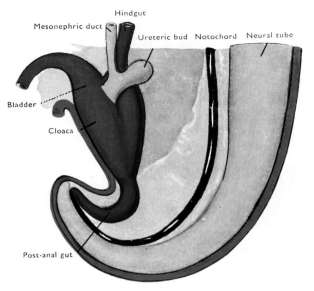

FIG. 8.67. Tail-end of 6.5-mm human embryo.
The cloaca is dividing into rectal and urogenital parts. The ureter is arising as a bud from the mesonephric duct. (Keibel 1896.)

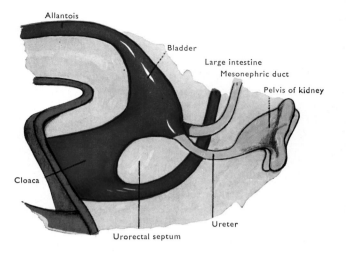

FIG. 8.68. Tail-end of 11.5-mm human embryo.
The cloaca is becoming separated into rectal and urogenital portions by the formation of the urorectal septum. The ureter has acquired a separate opening into the urinary bladder. (Keibel 1896.)

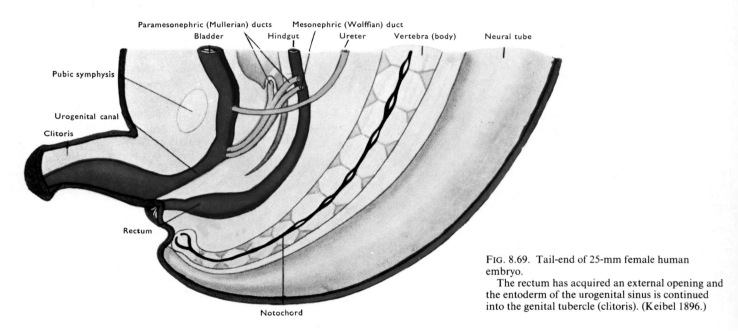

FIG. 8.69. Tail-end of 25-mm female human embryo.
The rectum has acquired an external opening and the entoderm of the urogenital sinus is continued into the genital tubercle (clitoris). (Keibel 1896.)

mesonephros is a relatively large structure composed of tubules which resemble, but are shorter than, those of the adult kidney. At that time it bulges into the dorsal part of the body cavity, and extends from the region of the liver to the caudal end of the cavity.

In fishes and amphibia the mesonephros persists as the excretory organ of the adult, but when the metanephros develops in reptiles, birds, and mammals the mesonephros atrophies; yet some of its tubules persist in the male as the **efferent ductules** of the testis, the **ductuli aberrantes**, and the **paradidymis**.

The mesonephric duct becomes the duct of the epididymis, the ductus deferens, and the ejaculatory duct of the adult male [see p. 581]. In the female, when the permanent kidney is formed, the mesonephros and its duct atrophy to a greater extent than in the male, and they are represented only by vestigial structures, the epoophoron and paroophoron in the broad ligament of the uterus [p. 570].

URETER AND PERMANENT KIDNEY

The permanent kidney or **metanephros** has a double origin. The renal tubules are developed from a part of the intermediate cell mass caudal to that from which the mesonephros is derived; the ureter, the pelvis, calyces, and the collecting tubules of the kidney are formed from the ureteric bud of the mesonephric duct.

The **ureteric bud** arises as a tubular diverticulum from the mesonephric duct close to the point where it joins the cloaca [FIGS. 8.67 and 8.68]. The diverticulum appears during the fifth week, and grows headwards, dorsal to the body cavity. The portion of the outgrowth nearest the mesonephric duct is slender and forms the adult ureter. The expanded distal part forms the pelvis of the kidney and gives rise to branches which become the calyces of the kidney. If the ureteric bud branches too early, a partial or complete double ureter may result. From the calyces numerous collecting tubules grow out and acquire connections with the renal tubules, which arise independently in the metanephric cap—a condensation of the intermediate mesoderm covering the expanded ureteric bud. The blind, proximal end of each renal tubule soon dilates to form a capsule invaginated on itself to enclose a tuft of capillary blood vessels. The renal corpuscles arising in this manner are found in the human kidney as early as the eighth week. The renal tubules join the collecting tubules at their distal ends, the site of the junctional tubule [p. 539] in the adult. If the tubules fail to join, the secreting tubules have no outlet and a congenital cystic kidney is the result; there are, however, other views on the origin of this condition (Willis 1958).

The metanephric cap, in which the renal tubules arise, lies at first on the medial side of the ureteric bud and later on its dorsal aspect. The ureteric bud and metanephric cap grow in a cephalic direction dorsal to the mesonephros. If this process fails to occur, the kidney remains in the pelvic region in the adult. The right and left metanephric caps are closely related to one another at their caudal ends and fusion between them in this situation results in the formation of a 'horseshoe' kidney [p. 540]. Variations in the origin of the renal arteries are due to the change in position of the kidney during development; the blood supply is at first from the internal iliac artery, but as the kidney moves headwards it is supplied successively by branches from the common iliac and finally from the aorta, and any of these may persist, the most common in a normally placed kidney being an accessory artery to the lower pole.

As the ureteric bud divides to form the calyces, the metanephric cap breaks up into numerous cell masses—one for each calyx, and later one for each of the collecting tubules which grow out from the calyces. The formation of renal tubules within the cell masses is continued until shortly after birth but their differentiation continues long after this. The kidney is at first distinctly lobulated; at birth, and sometimes even in the adult, distinct traces of its original division into lobules may be seen [FIG. 9.1].

URINARY BLADDER

As the ureter increases in length, the part of the mesonephric duct uniting it to the ventral division of the cloaca is absorbed into that structure. Thus the mesonephric duct and ureter acquire independent openings into that part of the cloaca; the ureter entering nearer the head of the embryo into the part which will become the urinary bladder [FIG. 8.63].

The main portion of the urinary bladder is formed from the cephalic part of the ventral division of the cloaca. Early in development this becomes flattened laterally and has the two mesonephric ducts opening on either side of its narrow dorsal aspect [FIG. 8.68]. The tailward part of the ventral division of the cloaca, caudal to the openings of the mesonephric ducts, becomes constricted to form the urogenital sinus. The caudal ends of the mesonephric ducts and ureters, now entering separately, are further absorbed into the expanding ventral part of the cloaca, so that their attachments move further apart—the mesonephric ducts caudally to the urogenital sinus, the ureters cranially to the bladder. In the adult, the mesonephric ducts, therefore, open close to the median plane in the prostatic part of the urethra. The urinary bladder has therefore a double origin. Its main portion is derived from the cloaca and is therefore entodermal; the remainder, which corresponds approximately to the trigone and the posterior wall of the urethra down to the entry of the mesonephric (i.e. the ejaculatory) ducts is mesodermal in origin for it arises from the parts of the mesonephric ducts and ureters absorbed into its walls. The extreme cephalic end of the ventral part of the cloaca tapers gradually, and is continuous with the allantoic diverticulum. This part of the cloaca

is the urachus; it loses its lumen and in the adult is represented by a fibrous cord—the median umbilical ligament.

The cavity of the urachus is sometimes not lost so early, and in rare cases it persists in the child or adult as a patent channel that extends from the apex of the bladder to the umbilicus. The entoderm of the cloaca forms the mucous coat of the bladder except in the trigone. The muscular coat is derived from mesoderm which surrounds the bladder during development and which also contributes to the formation of the infra-umbilical region of the body wall (Wyburn 1937; Glenister 1958).

Malformations

Simple malformations of the bladder are uncommon. The gross malformation called ectopia vesicae or extroversion of the bladder is probably due to failure of fusion of the bilateral contributions of mesoderm which form the muscular coat of the bladder and the infra-umbilical region of the body wall. The lower part of the anterior abdominal wall is therefore deficient (in complete extroversion the pubic symphysis is absent) and the bladder is represented by an exposed, irregular area of mucous membrane—including the trigone—continuous with the surrounding skin. The ureters open on the exposed surface, and, in the male, this malformation is commonly associated with epispadias [p. 580].

MALE URETHRA

The posterior wall of the first part of the male urethra, down to the openings of the ejaculatory ducts, has an origin similar to that of the trigone of the bladder. The remaining portion—inferior to the openings of the ejaculatory ducts of the adult—is derived from the urogenital sinus or caudal division of the ventral part of the cloaca. The urogenital sinus is separated from the rectal part of the cloaca by a mesodermal septum, the urorectal septum, formed by fusion of the caudal ends of the columns of intermediate mesoderm, which also divides the original cloacal membrane into urogenital and anal membranes. Both of these membranes rapidly break down. The urogenital sinus itself is early divided into a pelvic part which lies in the future lesser pelvis and forms the prostatic urethra inferior to the entry of the ejaculatory ducts and the membranous urethra, and a phallic part which occupies the region where the corpus spongiosum is developed. The opening of the urogenital sinus on the surface is flanked on each side by a genital fold and outside that by a genital swelling. Anteriorly lies the genital tubercle which, growing forwards, carries the anterior wall of the sinus on its inferior surface (the urethral plate) flanked by the anterior extremities of the genital folds [FIG. 8.74]. The genital tubercle is formed by bilateral proliferation of the mesoderm of the infra-umbilical part of the anterior abdominal wall which forms the corpora cavernosa, while the apex develops into the glans penis. The genital folds fuse together beneath the opening of the urogenital sinus, and this fusion extends forwards on the inferior surface of the penis, closing off the urethral plate to form the spongy part of the urethra, and forming the corpus spongiosum from its mesoderm (Paul and Kanagasuntheram 1956; Glenister 1958). This closure extends to the proximal margin of the glans penis where it is joined by a column of ectodermal cells which grows through the glans from its apex. The column canalizes to form the terminal part of the urethra (including the fossa navicularis) and unites proximally with the spongy part, the opening of which on the inferior surface of the penis is then closed. The genital swellings, enlarging to accommodate the testes, extend beneath the posterior part of the fused genital folds, and meeting in the midline, fuse to form the scrotum; the line of fusion being indicated by the raphe and the septum extending inwards from it.

MALFORMATIONS

Malformations of the male urethra are not uncommon. They are due mainly to failure of closure of the genital folds at the two critical points—the junction of the membranous and spongy parts, and at the neck of the glans. Such abnormal openings of the urethra on the urethral surface of the penis constitute the malformation known as **hypospadias**, which varies in extent; abnormal formation of the fossa navicularis with hypospadias at the neck of the glans is the simplest and commonest variety. Complete hypospadias, associated with cryptorchism and failure of formation of the scrotum, occurs in male pseudohermaphroditism.

The cause of **epispadias**, in which the urethra most commonly opens on the dorsal surface of the glans penis, is related to failure of fusion of the bilateral contributions of mesoderm which form the corpora cavernosa of the penis as well as the infra-umbilical region of the body wall; it is therefore often associated with extroversion of the bladder.

FEMALE URETHRA

In the female the urogenital sinus opens out on to the surface and is continued forwards as a sulcus between the genital folds. The genital folds do not unite, but form the labia minora of the adult, the genital tubercle forms the clitoris, and the genital swellings develop into the labia majora. Later, a shortening and spreading out of the lower portion of the urogenital sinus, to form the vestibule of the adult, is responsible for bringing the definitive opening of the urethra to the surface. The female urethra corresponds to the part of the male passage which lies above the opening of the prostatic utricle and ejaculatory ducts, since it is derived from the cephalic part of the ventral division of the cloaca caudal to the portion forming the bladder, and cranial to the opening of the mesonephric ducts.

GONADS

In the development of the gonads, male and female, the coelomic epithelium (splanchnopleuric intra-embryonic mesoderm) on the medial side of the mesonephros thickens to form a genital (gonadal) ridge. The **genital ridge** is soon found to have numerous **rete cells** embedded in its stroma of embryonic connective tissue. Some of these appear to originate, in both sexes, by a proliferation from the deep surface of the epithelium that covers the ridge.

The tissue which gives rise to the genital ridge occurs in all the body segments from the sixth thoracic to the second sacral, but the cephalic end of the ridge atrophies before the germinal epithelium can be recognized in the more caudal segments, and only about one-fourth to one-half of the ridge gives origin to the permanent gonad. As the gonad develops it becomes suspended by a mesentery (mesorchium in the case of the testis, mesovarium for the ovary) from the medial aspect of the mesonephros [FIG. 8.70].

IN THE MALE, as early as the thirty-third day, the rete cells embedded in the stroma of the developing testis have become arranged into a network of **rete cords**. The rete cords undergo direct transformation into the **convoluted** and **straight seminiferous tubules**, and the **rete testis**. At a very early stage the superficial part of the stroma of the developing testis becomes denser, and gives origin to the **tunica albuginea**. The tissue surrounding the rete cords becomes converted into the septula of the testis and the mediastinum. A lumen can first be recognized in the seminiferous tubules in the seventh month. The rete testis becomes connected secondarily with the efferent ductules of the testis, which are derived from those tubules of the mesonephros which lie opposite the definitive gonad; these are connected with the mesonephric duct which becomes the epididymis and ductus deferens.

IN THE FEMALE, numerous large **primordial oocytes** are found in the stroma of the developing ovary beneath the germinal epithelium as early as the thirty-third day. They are formed outside the developing ovary, from the entoderm of the yolk sac, and later

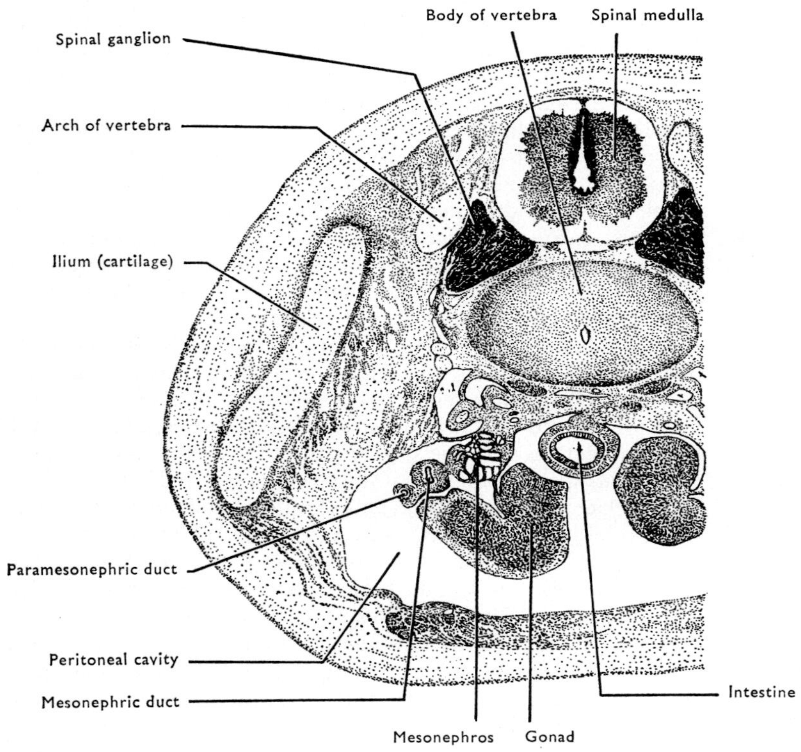

FIG. 8.70. Transverse section through lower part of trunk of human embryo of about 7 weeks. (Symington.)

migrate through the root of the mesentery to occupy their permanent position within the stroma of the ovary (Witschi 1948). The primordial spermatogenic cells, which become located in the rete cords, may have a similar origin. At first the primordial oocytes are isolated, but later—about the fifth week—they become surrounded by smaller cells which arise from the coelomic epithelium. Each primordial oocyte, surrounded by its cells, becomes a **primordial follicle**, the further development of which has already been described [p. 17]. The proliferation of cells from the coelomic epithelium goes on until birth, but after birth it is doubtful whether oocytes are produced in this manner in the human ovary. There is considerable evidence to suggest that the ovary possesses its full complement of oocytes at birth and that these mature in successive menstrual cycles during adult life [p. 567].

Descent of the testis and ovary

The descent of the testis into the scrotum and of the ovary into the pelvis have already been considered [pp. 556 and 566].

GENITAL DUCTS

The male genital ducts arise from the mesonephros and its duct; but those of the female arise from the paramesonephric ducts. The embryos of both sexes at first possess both mesonephric and paramesonephric ducts, which are arranged in a very definite manner. The mesonephric ducts, communicating directly with the tubules of the mesonephros, lie at first parallel to each other, and at a considerable distance apart. As they pass towards the caudal end of the embryo they approach each other, and each becomes enclosed in a fold of peritoneum called the **plica urogenitalis**. More caudally the ducts become closely approximated, and finally open into the dorsal aspect of the ventral division of the cloaca.

The **paramesonephric (Müllerian) ducts** are formed later by an evagination of the coelomic epithelium at the cephalic end of the mesonephros (Faulconer 1951), and therefore open freely into the body cavity. Each duct extends caudally on the lateral side of the corresponding mesonephric duct, and crossing ventral to it at the caudal extremity of the mesonephros, passes medially in the upper part of the urorectal septum to meet and fuse with its fellow. The median canal so formed descends in the septum to end blindly against the dorsal wall of the urogenital sinus between the points of entry of the two mesonephric ducts [FIGS. 8.69 and 8.71]. Here it forms a solid tubercle of cells which invaginates the posterior wall of the urogenital sinus (Müllerian tubercle).

Ducts in the male

By the third month the seminiferous tubules of the testis become connected with the mesonephric duct through a fusion of the tubules

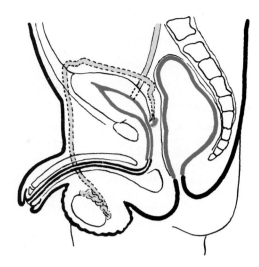

FIG. 8.72. Male urogenital passages.
Ureter, epididymis, and ductus deferens: yellow. Prostatic utricle: orange. Bladder, pelvic part of urethra and rectum: red. Penile part of urethra: black.

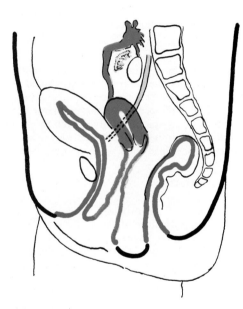

FIG. 8.73. Female urogenital passages.
Derivatives of paramesonephric duct: orange. Ureter: solid and dotted yellow. The epoophoron is indicated in yellow between the uterine tube and the ovary. Bladder, urethra, and rectum: red.

of the mesonephros adjacent to the gonad with the rete testis. The number of tubules that take part varies considerably, but each forms one of the efferent ductules found in the adult. The efferent ductules, becoming more convoluted where they join the mesonephric duct, form the **lobules of the epididymis**, while the **duct of the epididymis** is formed from the cephalic part of the mesonephric duct, and the **ductus deferens** from the more caudal portion. The ductuli aberrantes and the paradidymis are to be looked upon as persistent tubules of a more caudal portion of the mesonephros which have failed to become connected with the tubules of the testis, while the appendix of the epididymis represents the more cephalic remnants of the mesonephros and its duct.

The **seminal vesicles** are developed in the third month as evaginations from the mesonephric ducts near their caudal ends.

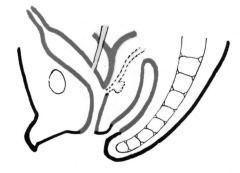

FIG. 8.71. Urogenital passages at the indifferent stage of development.
Ureter: solid yellow. Mesonephric duct: dotted yellow; the origin of the seminal vesicle is indicated. Paramesonephric ducts: orange. Rectum, bladder, and urogenital sinus: red.

The duct of the seminal vesicle therefore joins the ductus deferens to form the ejaculatory duct.

The paramesonephric ducts atrophy in the male embryo, but the appendices of the testes are vestigial remains of their cephalic portions and the prostatic utricle represents the fused caudal portions [FIG. 8.72].

Ducts in the female

The paramesonephric ducts in the female retain their openings into the body cavity, and their cephalic portions become the uterine tubes. Their fused caudal parts, which at first abut on the urogenital sinus, give rise to the uterus and at least the greater part of the vagina. The mechanism whereby the definitive female urethra comes to open into the vestibule, involving shortening and opening out of the urogenital sinus to form the major part of the vestibule [p. 580], is also responsible for the lower end of the fused paramesonephric ducts being carried caudally till it lies above the vestibule and therefore forms the eventual opening of the vagina into the vestibule, the remnants of the intervening tissue forming the hymen. The epithelium lining the vagina may be derived from entodermal cells of the pelvic part of the urogenital sinus (Bulmer 1959).

The mesonephric ducts and the mesonephros atrophy in the female, but traces of them are found as the epoophoron and paroophoron [FIG. 8.73]. In the foetus the mesonephric duct can be traced along the side of the uterus as far as the upper end of the vagina, where it is known as the duct of Gärtner.

Malformations

Variations of the uterus have been mentioned on page 571 as due to the arrest of the fusion of the paramesonephric ducts. Such failure of union may be so complete that the uterus and vagina appear to be double, or a dividing septum may persist in one or other of these organs or in both. The vagina may fail to reach the surface, or the simple malformation of 'imperforate hymen' may result from failure of the fused paramesonephric ducts to open into the vestibule.

ACCESSORY GLANDS

The glandular portion of the prostate arises as a series of solid outgrowths from the epithelium of the urogenital sinus during the third month. The outgrowths are simple at first, but become branched and finally acquire a lumen. They are arranged in three groups—an upper and lower dorsal, and a ventral group. The glands of the ventral group soon become reduced in number and often completely disappear; those of the upper dorsal group form the chief part of the prostate. The prostatic glands arise in both sexes; but in the female, where they are known as para-urethral glands, they are few in number and not densely packed as in the male. The muscular tissue of the prostate is derived from the mesoderm surrounding it during development.

The bulbo-urethral glands arise in the third month as outgrowths from the epithelium of the urogenital sinus. The greater vestibular glands arise in the same manner.

EXTERNAL GENITAL ORGANS [FIG. 8.74]

The external genital organs are developed in the region of the cloacal membrane, and the male and female are alike in the earlier stages. Part of this membrane at first extends on the ventral aspect of the body almost from the tail to the umbilical cord. At its cephalic end there is a tubercle known as the genital tubercle, and at its caudal end there is a coccygeal tubercle. The cloacal membrane becomes divided by the urorectal septum into an anal membrane, which lies immediately in front of the coccygeal tubercle, and a urogenital membrane situated between the urorectal septum and the genital tubercle. Both membranes rapidly break down: the anal orifice thus formed becomes surrounded completely by a circular ridge and therefore comes to lie at the base of an ectodermal pit, the proctodaeum, which forms the lower one-third of the anal canal. Anteriorly, through the urogenital membrane, the urogenital sinus comes to open on the surface by a slit-like aperture, the primitive urogenital opening, the margins of which are raised up to form the genital folds. A further pair of prominent folds, the genital swellings (labioscrotal folds), appear lateral to the genital folds.

IN THE FEMALE, the genital swellings give rise to the labia majora. The genital folds give origin to the labia minora, and the genital tubercle becomes the clitoris. On the clitoris at a very early date the glans is marked off by a surrounding sulcus.

IN THE MALE, the genital swellings grow backwards and, meeting behind the urogenital opening, fuse together to form the scrotum. The genital tubercle develops into the glans penis, and the genital folds fuse together to complete the ventral wall of the penis, and the penile urethra.

Hermaphroditism

True hermaphroditism—the presence of both testis and ovary—is exceedingly rare in the human subject; the production of an ovary and testis in the same individual, or the two constituents of a mixed ovotestis, is probably determined genetically.

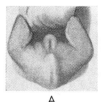

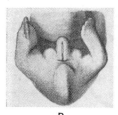

A

B

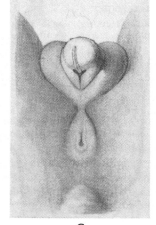

C

D

FIG. 8.74. External genital organs of human embryo.
A, embryo of 20 mm. B, slightly larger.
Indifferent stage. The genital folds and genital swellings are evident.
C, male embryo.
Formation of scrotum. The genital swellings have grown backwards and united behind the primitive urogenital opening to form the scrotal raphe. The genital folds are fusing below the base of the genital tubercle to form the body of the penis. The glans, formed from the genital tubercle, is very prominent.
D, male embryo at later stage.
Behind the glans penis the urethra opens in a diamond-shaped fossa at the proximal end of which the median raphe ends. The prepuce is forming behind the constriction which marks off the glans, and later grows forwards to cover it.

Pseudohermaphroditism

This is defined as the association of the gonads of one sex with external genital organs that resemble those of the other sex. Complete hypospadias in the male [p. 580] gives rise to the commonest variety; the external genital organs, through failure of the testes to descend and of the genital swellings to unite, may closely resemble those of the female. The opposite condition, with enlargement of the clitoris to simulate a penis and possible descent of the ovaries, is much less common. It is probable that these conditions, in which secondary sex characters also are affected, are due to disturbance of normal regulation of development by sex hormones. Such a disturbance is usually caused by abnormal activity of the suprarenal cortex.

The mammary glands

The mammary glands are accessory organs of the female reproductive system. Each gland is situated on the front of the thorax in the superficial fascia of the hemispherical elevation known as the mamma or **breast**. It usually extends from the second rib to the sixth rib and lies on the pectoralis major and, to a lesser extent, on the obliquus externus abdominis and the serratus anterior. A part of the gland extends towards the axilla partly under cover of the lateral border of the pectoralis major, and is known as the **axillary tail**. The **nipple** is situated near the summit of the breast, and tends to lie at the level of the fourth intercostal space in the midclavicular line, though this is very variable; the lactiferous ducts open on it by minute apertures, and it is surrounded by a coloured, circular area of skin called the **areola**. The skin of the nipple is thrown into numerous wrinkles, and on the areola it exhibits many minute, rounded projections due to the presence of underlying **areolar glands**. The colour of the nipple and areola varies with the complexion, but in young subjects it is usually a rosy-pink, and it changes to a deep brown during the second and third months of the first pregnancy. Also, during pregnancy, the areola increases in size and its glands become more marked. The nipple contains a considerable number of plain muscle fibres, and it becomes firmer and more prominent as a result of mechanical stimulation.

The size and appearance of the breasts vary much, not only in different races of mankind, but also in the same person under different conditions. In the young child they are small, and there is little difference between those of the male and female. Their growth is slow until the approach of puberty, and then the female mammary glands increase rapidly in size. At each pregnancy the breasts enlarge, and attain their greatest development during lactation. The size of the breasts depends partly on the amount of superficial fat and partly on the amount of glandular tissue present, though the amount of glandular tissue is very small in the non-lactating breast.

Structure of breast

The mamma is composed of a mass of glandular tissue traversed and supported by strands of fibrous tissue, and covered with skin and a thick layer of fat. The glandular tissue forms a conical mass, the apex of which corresponds to the position of the nipple, while its base is loosely connected to the deep fascia on which the gland lies. In section it is readily distinguished from the surrounding fat by its firmer consistence and by its pinkish-white colour. It is composed of from fifteen to twenty lobes divided into lobules, which make its superficial surface and edges very uneven—the inequalities of its surface being filled up by processes of the fatty tissue which covers the gland. The fatty covering is incomplete in the region of the areola, and here the lactiferous ducts pass into the nipple. The **lobes** radiate from the nipple, each lobe being quite distinct from the others and possessing its own duct; the **lobules** are bound together and supported by a considerable amount of fibro-areolar tissue, which forms the **stroma** of the gland [FIG. 8.75].

The alveoli of the gland and the secretory epithelium which lines them vary much under different conditions. At puberty the glandular tissue is composed chiefly of the ducts; at that time the alveoli are small and few in number. During lactation, when the gland is fully functional, the alveoli are enlarged, distended with fluid, and much more numerous. The epithelial cells are cubical and filled with fat globules. When the gland is not secreting, the alveoli become small and reduced in number, and the cells of the lining epithelium, which are then small, do not contain fat globules.

The **lactiferous ducts**, passing towards the nipple, enlarge to form small, spindle-shaped dilatations called **lactiferous sinuses**; then, becoming once more constricted, each duct passes, without communicating with its neighbours, to the summit of the nipple, where it opens on the surface.

In the male the various parts of the breast are represented in a rudimentary condition.

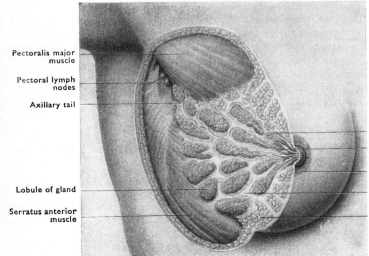

Pectoralis major muscle

Pectoral lymph nodes

Axillary tail

Lobule of gland

Serratus anterior muscle

Lactiferous duct

Lactiferous sinus

Areolar gland

Stroma

Subcutaneous fatty tissue

FIG. 8.75. Dissection of the right mammary gland.

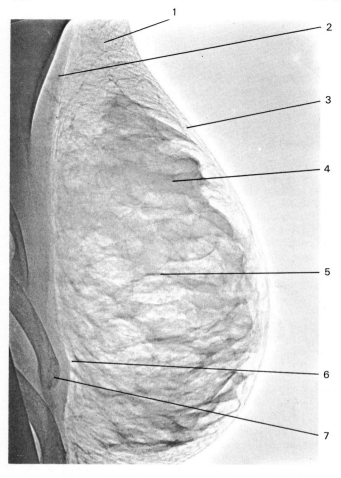

Fig. 8.76. A lateral mammoxerogram. Most of the glandular tissue has
been replaced by fat (clear spaces) or fibrous tissue (dark areas).
1. Superficial fascia.
2. Pectoralis major.
3. Skin and subcutaneous tissue.
4. Remnants of glandular tissue.
5. Fibrous tissue strand.
6. Retromammary fat.
7. Rib.

most part join the lymph nodes of the axilla. They take origin from an extensive **perilobular plexus**. The majority follow the lactiferous ducts and thus converge towards the nipple, and there join a plexus situated beneath the areola (subareolar plexus). From this plexus the main efferent vessels pass to the **pectoral axillary nodes**, but it is important to remember that free communication exists between all the subgroups of axillary nodes. In addition, vessels pass directly from the deep surface of the lateral part of the breast either through, above, or behind the pectoral muscles to the **apical axillary nodes**. A few vessels from the medial part of the breast follow the course of the perforating arteries to the **parasternal lymph nodes** situated along the course of the internal thoracic artery. A few, probably irregular, communications exist across the median plane with the lymph vessels of the opposite breast; and, under diseased conditions, communications may exist with lymph vessels in the upper part of the sheath of the rectus abdominis, and so with the hepatic nodes, via the falciform ligament, and parasternal nodes. The surgical importance of the facts regarding the lymphatic drainage of the breast cannot be exaggerated.

The **nerve supply** of the gland is derived from the **intercostal nerves** of the fourth, fifth, and sixth intercostal spaces. Along the course of these nerves sympathetic filaments reach the breast from the thoracic part of the sympathetic trunk. The nipple has a particularly rich nerve supply (Cathcart, Gairns, and Garven 1948).

RADIOGRAPHY

Radiographs of the breasts are obtained from above down (craniocaudal) and from the side (lateral). This type of examination is largely confined to patients who are approaching or are beyond the menopause, because then atrophy of glandular tissue begins to take place with fibrous tissue replacement and variable infiltration by fat. This fat renders the remaining glandular tissue visible and also aids the recognition of abnormalities [Fig. 8.76].

This is a valuable technique because it can often detect abnormalities that may not be apparent clinically.

In the younger person the glandular tissue is usually of homogeneous opacity so that little information is gained.

DEVELOPMENT OF THE MAMMARY GLANDS

The mammary glands are developed as ingrowths of the ectoderm into the underlying mesodermal tissue. In the human embryo a thickened raised area of the ectoderm can be recognized in the region of the future gland at the end of the fourth week. The thickened ectoderm becomes depressed into the underlying mesoderm, and thus the surface of the mammary area soon becomes flat, and finally sinks below the level of the surrounding epidermis. The

Variations

Asymmetry in the development of the breasts is very common—the left mamma being often larger than the right. Absence of one or both mammae is a very rare abnormality which may or may not be associated with absence of the nipples. When one nipple only is present it is usually the left. The presence of supernumerary glands or nipples is uncommon. The term **polymastia** has been applied to cases in which more than the normal number of mammae are present, and **polythelia** to those in which accessory nipples occur. Usually the accessory glands, or nipples, are present on the front of the thorax, and in most instances they occur below and a little to the medial side of the normal site. When the abnormal glands are found above the normal site they generally lie further from the median plane. Much more rarely, accessory glands have been found on the abdomen, in the axilla, or in some other situation, such as the medial aspect of the thigh. Asymmetry is very common in such abnormal structures. Examples of polymastia and polythelia occur in the male as well as in the female.

VESSELS AND NERVES OF THE BREAST

The mamma receives its arterial supply from the perforating branches of the **internal thoracic artery** and from the lateral mammary branches of the **lateral thoracic**. Additional supply is derived from some of the intercostal vessels. The veins from the gland pour their blood into the **axillary** and **internal thoracic veins**. Some small superficial veins from the breast join tributaries of the external jugular.

The **lymph vessels** of the breast are very numerous, and for the mesoderm in contact with the ingrowth of the ectoderm, is compressed, and its elements become arranged in concentric layers which, at a later stage, give rise to the stroma of the gland. The ingrowing mass of ectoderm cells soon becomes flask-shaped and then grows out into the surrounding mesoderm as a number of solid processes which represent the future ducts of the gland. These processes, by dividing and branching, give rise to the future lobes and lobules, and, much later, to the alveoli. The mammary area becomes gradually raised again in its central part to form the nipple. A lumen is formed in each part of the branching system of cellular processes only at birth, and the secretion of a fluid resembling milk may take place at this time as a result of maternal hormones which have passed across the placenta into the child's circulation. The lactiferous sinuses appear before birth as swellings of the developing ducts.

In those animals which possess a number of mammary glands—such as the cat and pig—the thickening of the ectoderm, which is the first indication of the development of these glands, takes the form of a pair of ridges that extend from the level of the fore limb towards the inguinal region. These mammary ridges converge caudally, and at their terminations lie not far from the median line. By the absorption of the intervening portions, the ridges become divided into a number of isolated areas in which the future glands arise. Similar linear thickenings of the ectoderm have been recognized in the human embryo also, and the usual positions assumed by the accessory glands when present suggest that polymastia and polythelia are caused by abnormal persistence of portions of the mammary ridges.

References

BACKHOUSE, K. M. and BUTLER, H. (1960). The gubernaculum testis of the pig (Sus scropha). J. Anat. (Lond.) 94, 107.

BRAASCH, W. F. and EMMETT, J. L. (1951). Clinical urography. Saunders, Philadelphia.

BRAITHWAITE, J. L. (1952). The arterial supply of the male urinary bladder. Br. J. Urol. 24, 64.

BRAMBELL, F. W. R. (1956). Ovarian changes. In Marshall's physiology of reproduction (ed. A. S. Parkes) Vol. I, Part I. London.

BRASH, J. C. (1922). The relation of the ureters to the vagina: with a note on the asymmetrical position of the uterus. Br. med. J. ii, 790.

BULMER, D. (1959). The epithelium of the urogenital sinus in female human foetuses. J. Anat. (Lond.) 93, 491.

CATHCART, E. P., GAIRNS, F. W., and GARVEN, H. S. D. (1948). The innervation of the human quiescent nipple, with notes on pigmentation, erection and hyperneury. Trans. roy. Soc. Edinb. 61, 699.

CLEGG, E. J. (1955). The arterial supply of the human prostate and seminal vesicles. J. Anat. (Lond.) 89, 209.

—— (1956). The vascular arrangements within the human prostate gland. Br. J. Urol. 28, 428.

—— (1957). The musculature of the human prostatic urethra. J. Anat. (Lond.) 91, 345.

FAULCONER, R. J. (1951). Observations on the origin of the Müllerian groove in human embryos. Contrib. Embryol. Carneg. Inst. 34, 159.

FRANKL, O. (1933). On the physiology and pathology of the isthmus uteri. J. Obstet. Gynaecol. Br. Emp. 40, 397.

FRANKLIN, K. J. (1949). The history of research upon the renal circulation. Proc. roy. Soc. Med. 42, 721.

GLENISTER, T. W. (1958). A correlation of the normal and abnormal development of the penile urethra and of the infra-umbilical abdominal wall. Brit. J. Urol. 30, 117.

HARRISON, R. G. (1949). The distribution of the vasal and cremasteric arteries to the testis and their functional importance. J. Anat. (Lond.) 83, 267.

—— (1966). The anatomy of varicocele. Proc. roy. Soc. Med. 59, 763.

—— (1975). Effect of temperature on the mammalian testis. In Handbook of physiology, Section 7, Vol. 5, p. 219. American Physiological Society, Washington.

—— and BARCLAY, A. E. (1948). The distribution of the testicular artery (internal spermatic artery) to the human testis, Br. J. Urol. 20, 57.

—— and DE BOER, C. H. (1977). Sex and infertility. Academic Press, London.

—— and WEINER, J. S. (1949). Vascular patterns of the mammalian testis and their functional significance. J. exp. Biol. 26, 304.

HUNTER, J. (1786). A description of the situation of the testis in the foetus, with its descent into the scrotum, Palmer's (1837) edition of Works of John Hunter, Vol. IV, p. 1. London.

LÖFGREN, F. (1949). Das topographische System der Malpighischen Pyramiden der Menschenniere. Lund.

MACKENRODT, A. (1895). Über die Ursachen der normalen und pathologischen Lagen des Uterus. Arch. Gynäk. 48, 393.

MACMILLAN, E. W. and AUKLAND, J. (1960). The transport of radiopaque medium through the initial segment of the rat epididymis. J. Reprod. Fert. 1, 139.

MARTIN, C. P. (1942). A note on the renal fascia. J. Anat. (Lond.) 77, 101.

MITCHELL, G. A. G. (1939). The spread of retroperitoneal effusions arising in the renal regions. Br. med. J. ii, 1134.

—— (1950). The renal fascia. Br. J. Surg. 37, 257.

—— (1956). Cardiovascular innervation. E. & S. Livingstone, Edinburgh.

PAUL, M. and KANAGASUNTHERAM, R. (1956). The congenital anomalies of the lower urinary tract. Br. J. Urol. 28, 118.

POWER, R. M. H. (1944). The exact anatomy and development of the ligaments attached to the cervix uteri. Surg. Gynecol. Obstet. 79, 390.

—— (1946). The pelvic floor in parturition. Surg. Gynecol. Obstet. 83, 296.

RACKER, D. C. and BRAITHWAITE, J. L. (1951). The blood supply to the lower end of the ureter and its relation to Wertheim's hysterectomy. J. Obstet. Gynaecol. Br. Emp. 53, 609.

REYNOLDS, S. R. M. (1948). Morphological determinants of the flow-characteristics between an artery and its branch, with special reference to the ovarian spiral artery in the rabbit. Acta anat. (Basel) 5, 1.

RICHARDSON, K. C. (1962). The fine structure of autonomic nerve endings in smooth muscle of the rat vas deferens. J. Anat. (Lond.) 96, 427.

SOUTTAR, H. S. (1947). On complete removal of the prostate: a preliminary communication. Br. med. J. i, 917.

SUVANTO, O. and KORMANO, M. (1970). The relation between in vitro contractions of the rat seminiferous tubules and the cyclic stage of the seminiferous epithelium. J. Reprod. Fert. 21, 227.

SWYER, G. I. M. (1944). Postnatal growth changes in the human prostate. J. Anat. (Lond.) 78, 130.

SYKES, D. (1964a). Some aspects of the blood supply of the human kidney. Symp. Zool. Soc. Lond. 11, 49.

—— (1964b). The morphology of renal lobulations and calyces, and their relationship to partial nephrectomy. Br. J. Surg. 51, 294.

THOMPSON, R. (1919). The capacity of, and the pressure of fluid in, the urinary bladder. J. Anat. (Lond.) 53, 241.

TRUETA, J., BARCLAY, A. E., DANIEL, P. M., FRANKLIN, K. J., and PRITCHARD, M. M. L. (1947). Studies of the renal circulation. Oxford.

WELLS, L. J. (1943). Descent of the testis: anatomical and hormonal considerations. Surgery 14, 436.

WHITEHOUSE, B. and FEATHERSTONE, H. (1923). Certain observations on the innervation of the uterus. Br. med. J. ii, 406.

WILLIS, R. A. (1958). The borderland of embryology and pathology. Butterworths, London.

WITSCHI, E. (1948). Migration of the germ cells of human embryos from the yolk sac to the primitive gonadal folds. Contrib. Embryol. Carneg. Instn. (No. 209) 32, 67.

WYBURN, G. M. (1937). The development of the infra-umbilical portion of the abdominal wall, with remarks on the aetiology of ectopia vesicae. J. Anat. (Lond.) 71, 201.

ZACHARIN, R. F. (1963). The suspensory mechanism of the female urethra. J. Anat. (Lond.) 97, 423.

ZUCKERMAN, S. (1956). The regenerative capacity of ovarian tissue. Ciba Colloquia on Ageing 2, 31.

—— (1962). The ovary. Academic Press, London.

9 The ductless glands

R. G. HARRISON

THE term ductless is applied to certain glands whose products reach the circulation without passing through any special channels or ducts. They differ widely in structure, function, and development, exhibiting a diversity similar to that of glands in general.

Glands

From the functional point of view glands may belong to one of two main categories:

1. **Exocrine**, in which the secretion leaves the gland by way of a duct (e.g. salivary glands);

2. **Endocrine** or ductless glands, in which the secretion is poured into the circulation without passing through any special channels or ducts (e.g. thyroid and suprarenal glands).

The **exocrine glands** may be further subdivided into: (i) **holocrine**, in which the cells composing the gland break down completely in order to liberate their secretion (e.g. sebaceous glands); (ii) **apocrine**, in which some of the cytoplasm at the free surface of the cells composing the gland is discharged with the secretion (e.g. axillary sweat glands); (iii) **merocrine**, which discharge their secretion without any fragmentation of the cytoplasm of their cells. Endocrine glands are merocrine in character. Exocrine glands may also be described as **simple** or **compound** depending on the complexity of their ducts, and **tubular** or **alveolar** according to the shape of their secretory units [FIG. 6.20].

Glands may be classified also with reference to their development. Thus, they may arise from one or other of the primary germ layers of the embryo—ectoderm, mesoderm, or entoderm, or from certain regions, e.g. the thyroid and parathyroid glands which are **pharyngeal derivatives**. In a similar manner the hypophysis cerebri and the pineal body have been associated in a cerebroglandular group.

ENDOCRINE GLANDS

These glands differ from exocrine glands in not retaining as a duct any connection with the epithelium from which they develop, but they come into intimate contact with the intraglandular blood vessels. Their cells are arranged in columns, vesicles, or clusters, and they have a more profuse blood supply, often with specialized vascular arrangements such as a fenestrated capillary endothelium and arteriovenous anastomoses.

A single gland may contain both exocrine and endocrine parts (e.g. pancreas), and the gonads (testis and ovary) have a dual function to produce germ cells, and secrete hormones, the testis manufacturing **androgenic hormones** and the ovary **oestrogens**.

Endocrine glands produce substances, commonly known as hormones, which control and modify the functional activity of the cells of many other tissues in specific ways. They are the basis of a great chemical system which co-ordinates the activities of the various tissues; and their physiological and clinical importance is such that special textbooks are devoted to them (Selye 1947; Pincus 1947–67; Pincus and Thimann 1948–64; Astwood 1968–76; Greep 1977–8). The study of the various aspects of these organs is known as endocrinology.

Histologically the endocrine glands exhibit various grades of specialization (Cowdry 1932, 1944) which probably illustrate the evolutionary steps by which the structural and functional differentiation of endocrine organs took place. The production of a hormone has, indeed, fundamental resemblances to an activity displayed by all living cells, for these all give off substances into the circulation which affect the activities of other cells. Hormones are distinguished by the specific nature of the effects they produce; in some instances they are recognized only by these effects. For example, the adrenocorticotrophic hormone of the hypophysis exerts a direct effect only on the cortex of the suprarenal gland, and can only be recognized by virtue of this effect.

The first stage in the differentiation of a tissue as an endocrine organ is illustrated by certain tissues (e.g. gastric and duodenal mucosa and placenta) which, though primarily adapted by their structure to some other function, also produce a hormone, or hormones, as a secondary function without any evident cytological differentiation of the cells concerned.

A second stage in differentiation is illustrated in organs such as the pancreas, in which there is a recognizable specialization of structure as well as of function. In this organ some of the cells remain epithelial in structure and act as an exocrine secretory organ, whilst other cells are budded off from the epithelium in the course of development to form dissociated clusters, the islets of Langerhans, which act purely as an endocrine organ.

A further stage of differentiation is seen in the thyroid and parathyroid glands, for in the development of these glands there comes about a complete separation from the pharyngeal epithelium of masses of cells destined solely for an endocrine function.

Lastly, two masses of cells, each of specialized structure and function, may come into such intimate relation in the course of development as to constitute a single organ from the topographical point of view. Examples are furnished by the stomodaeal and neural parts of the hypophysis, and by the cortex and medulla of the suprarenal gland. It has been suggested that such close approximation may perhaps allow important interactions between the two parts, thus influencing their respective functions.

In this section only the more specialized endocrine glands are described, namely those which belong to the third and fourth grades. The glomus caroticum, the corpus coccygeum, the pineal body, and the thymus are also included as a matter of convenience, although there is less evidence that these structures serve as endocrine glands.

The organs described in this section are taken in the following order:

1. The **Suprarenal Glands**, which are compound organs including chromaffin and cortical tissue.

(i) The **Chromaffin System**, of ectodermal origin, includes the medulla of the suprarenal gland and various 'paraganglionic' masses of chromaffin tissue.

(ii) The **Cortical System**, of mesodermal origin, is represented by the cortex of the suprarenal gland and by the occasional 'accessory cortical bodies'.

2. The **Organs of the Pharyngeal Pouches** (pharyngeal derivatives), of entodermal origin and developed from the walls of the pharynx, include the thyroid and parathyroid glands and the thymus.

3. The **Cerebroglandular Organs**, of ectodermal origin, consisting of the hypophysis cerebri and the pineal body.

SHAPE AND DEVELOPMENT OF GLANDS

In the living body, glands are soft, plastic structures, and their shape is continuously subject to modification by the pressure of adjacent structures.

The special secreting cells of an exocrine gland are usually developed from a tubular epithelial growth. This may remain simple [FIG. 6.20] or produce an elaborate system of branches such as is seen in compound racemose glands. The tubular structure with its branches eventually forms the gland duct and its tributaries. In some instances the tubular form is not immediately apparent while the secreting cells are developing, since elongated columns of cells are first formed which only later become canalized (e.g. in the development of the liver, page 485). The secreting cells of some of the endocrine glands likewise arise from the walls of tubular structures which later disappear—a mode of development which results from a progressive specialization of function (e.g. the organs of the pharyngeal pouches).

As glands grow, the glandular tissue accommodates itself between adjacent structures, and may become moulded to reflect their contours, e.g. the thyroid [FIG. 9.10]. A fixed gland from the cadaver thus frequently shows impressions made by adjacent structures; or it may even surround structures during development, e.g. the parotid gland.

The majority of glands are extremely vascular, especially endocrine glands. The glandular cells are thus provided with the raw materials of secretion by the blood stream which also carries away the secretions of endocrine glands.

The nerves which supply glands terminate in two ways: (1) they may end in the walls of blood vessels and influence the functions of the gland through their control over the blood supply (e.g. the thyroid and the suprarenal cortex); (2) they may end in direct relation with the actual secreting cells (e.g. the suprarenal medulla).

THE SUPRARENAL GLANDS

The paired suprarenal (adrenal) glands lie one on each side of the vertebral column in intimate relation with the superomedial aspects of each kidney [FIGS. 9.1, 9.2, and 9.4].

Each gland consists of a relatively thick layer of cortex enclosing a medulla, and is extremely vascular. Arterial blood enters the gland from a plexus on the surface of the cortex and passes through the cortical tissue to the medulla, whence it is drained usually by a single vein [FIG. 9.7].

The cortex and medulla are of different developmental origin and function; they are in effect two distinct endocrine organs. The **cortex** plays an indispensable role in the body. Interference with its function affects carbohydrate metabolism and the distribution of sodium, potassium, and water in the body. It secretes hormones which are closely related to the sex hormones. At least twenty-eight steroids have been isolated from suprarenal glands processed after death (Reichstein and Shoppee 1943) but not all of them may be secreted by the suprarenal cortex in life. Of the steroids known to be present in the cortex, the mineralocorticoids, **aldosterone** and **deoxycortone**, control the metabolism of sodium and potassium. The glucocorticoid, **cortisone**, affects carbohydrate metabolism by increasing the deposition of glycogen in the liver. Oestrogenic, progestational, and androgenic hormones are also secreted by the cortex. The injection of cortisone causes a reduction in the number of eosinophils in the blood and involution of lymph tissue,

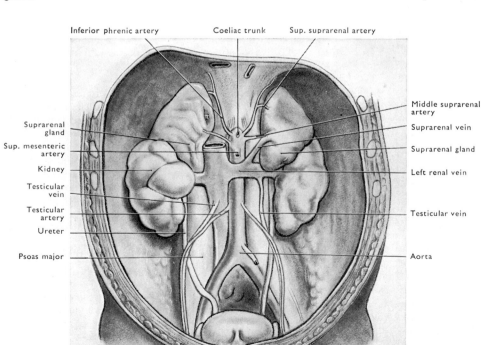

Inferior phrenic artery   Coeliac trunk   Sup. suprarenal artery

Middle suprarenal artery

Suprarenal vein

Suprarenal gland

Left renal vein

Testicular vein

Aorta

Suprarenal gland
Sup. mesenteric artery
Kidney
Testicular vein
Testicular artery
Ureter
Psoas major

FIG. 9.1 Posterior abdominal wall of full-term foetus, illustrating the relatively large size of the suprarenal glands and the lobulation of the kidneys.

particularly of the thymus. The cortex is also concerned in the reaction of the body to stress (Selye 1950), and stores vitamin C and lipoids. The medulla secretes two hormones, **adrenaline** and **noradrenaline**, which have dissimilar actions, and are secreted separately from the medulla under different conditions (von Euler 1955). Noradrenaline, also a regular constituent of postganglionic sympathetic nerves, causes general vasoconstriction and a rise in blood pressure. Adrenaline facilitates carbohydrate metabolism, and although it constricts the blood vessels in skin, it produces vasodilatation of the blood vessels of skeletal muscle. Both adrenaline and noradrenaline are liberated from the suprarenal gland when the splanchnic nerves of the cat are stimulated. Adrenaline may be a later product of evolution, for noradrenaline appears to be the hormone secreted in foetal and neonatal life and later replaced by adrenaline. Histochemical observations demonstrate that some cells of the suprarenal medulla contain much adrenaline and little noradrenaline, whilst others contain noradrenaline but little or no adrenaline.

The cortex is essential to life, unlike the medulla. The medulla does not display the functional changes of size which are seen in the cortex under various circumstances, e.g. during menstruation, after extirpation of the contralateral gland, or the influence of stress (Selye 1950), and under the influence of the **adrenocorticotrophic hormone (ACTH)** of the hypophysis.

In colour the cortex is yellow, owing to the contained lipoid substances. The medulla is seen in a section of the fresh, healthy gland as a dark streak within the yellow cortex; it becomes brown on treatment with a dilute solution of chromic acid or potassium dichromate (chromaffin reation).

The post-mortem size of the gland varies within wide limits, mainly on account of the great modifications produced by certain pathological conditions.

The average dimensions of the suprarenal glands are as follows: height, 3–5 cm; breadth, 2.5–3 cm; thickness, slightly under 1 cm; and the weight is 7–12 g. The medulla forms only about one-tenth of the whole gland. The glands are relatively much larger in the foetus than in the adult. Even at birth they are still relatively large [FIG. 9.1], being little smaller than in the adult.

As a rule the glands are of unequal size (the left is more frequently the larger), occasionally the difference is extreme, and rarely one gland is absent. Sometimes the two glands are fused (cf. horseshoe-kidney).

Frequently there are accessory glands. These develop in the neighbourhood of the main gland and usually remain there, but may become attached early in embryonic life to adjacent organs which subsequently change their position. As a result, they may be found not only beside the main gland but also in the broad ligament of the uterus, in the spermatic cord, or even attached to the epididymis. Like the main glands, true accessory suprarenals are compounded of cortex and medulla, and require to be distinguished from the purely chromaffin bodies and accessory cortical bodies which may be found in any of the positions in which accessory suprarenal glands occur.

Shape and position

The right gland is pyramidal; the left is semilunar and extends further down the medial side of the kidney than the right one, and both are moulded by adjacent structures. Each presents posterior, anterior, and renal surfaces, and the right has an anteromedial surface. The renal surface is moulded on the medial aspect of the upper pole of the kidney, while the posterior surface lies against the diaphragm. The anterior surface of the **left gland** [FIGS. 9.2 and 9.5] is situated behind the omental bursa, which separates it from the stomach; the inferomedial extremity is, however, sometimes separated from the omental bursa by the splenic artery and the body of the pancreas. The anterior surface of the **right gland** is applied to the liver, being in immediate contact with the bare area in its upper part but separated from it in its lower part by peritoneum. The anteromedial surface is in contact with the inferior vena cava [FIGS. 9.1 and 9.4]. The coeliac plexus and ganglia are situated between the two glands.

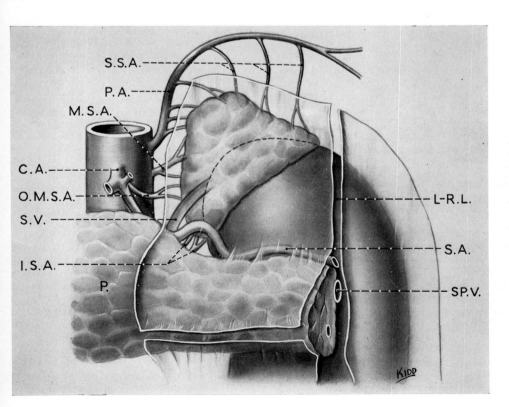

FIG. 9.2. Relations and blood supply of the left suprarenal gland.

C.A., coeliac trunk; I.S.A., inferior suprarenal arteries; L-R.L., lienorenal ligament; M.S.A., middle suprarenal artery; O.M.S.A. occasional middle suprarenal artery arising from coeliac trunk; P., pancreas; P.A., inferior phrenic artery; S.A., splenic artery; SP.V., splenic vein; S.S.A., superior suprarenal arteries; S.V., suprarenal vein.

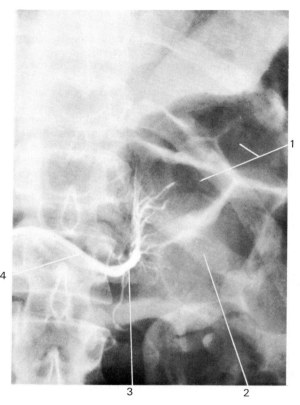

FIG. 9.3 Anteroposterior radiograph of selective injection of left suprarenal vein through catheter in left renal vein.
1. Gas in colon.
2. 12th rib
3. Catheter in left suprarenal vein.
4. Catheter in left renal vein.

The suprarenal gland is separated from the kidney by a small amount of the adipose capsule surrounding the kidney, and is enclosed with the kidney in the renal fascia [FIG. 9.4].

A cleft is found in each gland where the suprarenal vein leaves it. The cleft or hilus is situated in the right gland on the anteromedial surface, and near the lower end of the medial margin in the left gland.

The level of the gland varies in relation to the adjacent structures. Thus, the inferomedial part of the right gland is sometimes overlapped by the duodenum, and the area covered with peritoneum, is correspondingly reduced. On the left side the pancreas and splenic artery usually lie at a lower level than the gland. In the infant the spleen makes contact with the superolateral part of the left suprarenal and sometimes it may do so in the adult. The suprarenal is displaced with the kidney by respiratory movements of the diaphragm.

RADIOLOGY

Formerly these glands were outlined by the introduction of air retroperitoneally, usually in association with tomography. Recent work has shown that aortic angiograms involving the suprarenal arteries or selective catheterization of the suprarenal vein show these structures more clearly [FIG. 9.3].

Vessels and nerves

The blood supply of the suprarenal glands is very abundant. Each gland receives a variable number of separate **arteries** from the three sources—0–10 from the aorta, 0–27 from the inferior phrenic artery,

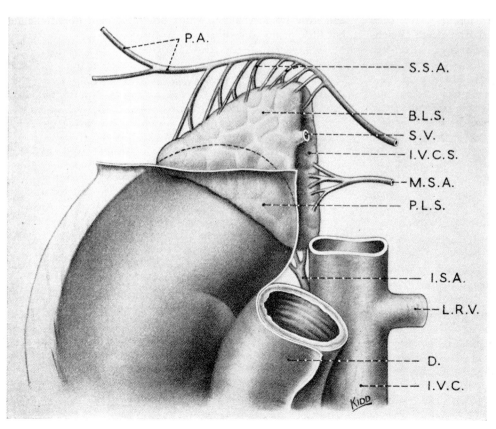

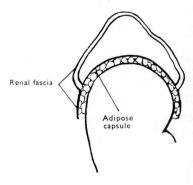

FIG. 9.4A. Relations and blood supply of the right suprarenal gland, seen from the anterior aspect.

B.L.S., part of anterior surface related to bare area of liver; D., descending part of duodenum; I.S.A., inferior suprarenal artery; I.V.C.S., surface related to inferior vena cava (I.V.C.); L.R.V., left renal vein; M.S.A., middle suprarenal artery; P.A., inferior phrenic artery; P.L.S., peritoneal covered portion of anterior surface related to liver; S.S.A., superior suprarenal arteries; S.V., suprarenal vein.

FIG. 9.4B. Schematic vertical section illustrating relation of the suprarenal gland to the kidney and the renal fascia.

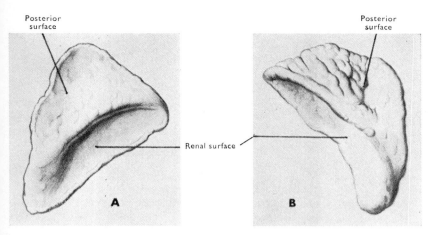

Posterior
surface

Posterior
surface

Renal surface

A

B

FIG. 9.5. Suprarenal glands isolated and viewed from behind
A, right
B, left.

and 10–30 from the renal artery [FIGS. 9.2 and 9.4]. Individual arteries are end-arteries to the zona fasciculata of the cortex (Harrison 1951). Each gland may thus receive as many as sixty arteries, but is drained by a single large central **vein** which emerges through the hilus; the right, after a very short course, joins the inferior vena cava, the left enters the left renal vein behind the body of the pancreas after receiving the inferior phrenic vein, and anastomosing with the first lumbar vein. The suprarenal veins have little or no circular muscle in their walls. The arrangement of the blood vessels within the gland is described with its structure. Numerous **lymph vessels** pass from the suprarenal glands to the aortic lymph nodes.

The **medulla** of the gland is more richly supplied with **nerves** than any other organ; they are derived from the sixth thoracic to the first lumbar ventral rami (Crowder 1957) and reach the gland through the **splanchnic nerves** and the **coeliac plexus**. The nerve filaments form a plexus of medullated and non-medullated nerve fibres in the fibrous capsule before they pass into the gland. Some of the fibres are postganglionic fibres from nerve cells scattered along the sympathetic nerves or in the gland itself (Swinyard 1937), but the majority are preganglionic fibres which end in synaptic relation with the secretory cells of the medulla (Hollinshead 1936), which can therefore be considered as postganglionic sympathetic neurons.

There is no evidence that cells in the **cortex** receive any nerve supply; nerve fibres may end in relation to the arteriae medullae.

Structure

The suprarenal gland consists of a highly vascular central mass of chromaffin tissue—the **medulla**—enclosed within a thick layer of **cortex**, which in turn is enveloped in a capsule of fibrous tissue.

The **cortex** of the human suprarenal [FIGS. 9.6 and 9.7] is so folded or convoluted as to increase its surface of contact with the medulla. From the deep aspect of the capsule, trabeculae of fibrous tissue pass inwards to support the glandular parenchyma. In the superficial part of the cortex the trabeculae interlace freely so as to enclose a series of small rounded clusters of cortical cells, the **zona glomerulosa**. In the intermediate region of the cortex, elongated cell columns, usually formed of two or three rows of cortical cells, lie at right angles to the surface, the **zona fasciculata**. In the deepest part of the cortex the cell columns are more irregularly arranged and form a reticulum—the **zona reticularis**. There is some evidence for functional as well as morphological independence of the cortical zones; thus the mineralocorticoids are formed in the zona glomerulosa, the glucocorticoids in the zona fasciculata, and sex hormones in the zona reticularis.

The cortical parenchyma consists of large polyhedral cells arranged in the interstices of the fibrous trabeculae. The cells

contain lipoid material, and are richer in cholesterol than any other tissue in the body. Ascorbic acid (vitamin C) is stored in the fasciculate and reticular zones. As the blood passes between the columns of the fasciculate zone it gains cortical hormones; the regulation of this process involves several mechanisms (Harrison and Hoey 1960).

The **medulla** [FIG. 9.7] is formed of a spongework of cell columns separated by anastomosing venous sinusoids. The cells are large and granular and exhibit the characteristic chromaffin reaction. In a fresh gland the medulla is of dark-red colour owing to the presence of blood in its sinusoidal spaces.

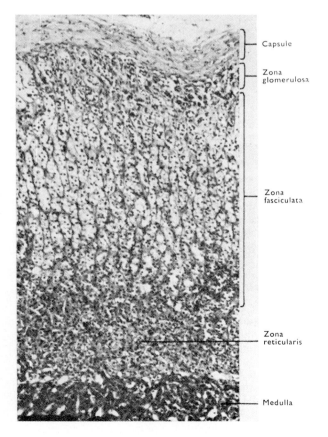

Capsule

Zona glomerulosa

Zona fasciculata

Zona reticularis

Medulla

FIG. 9.6 Histological section through a human suprarenal gland. × 36.

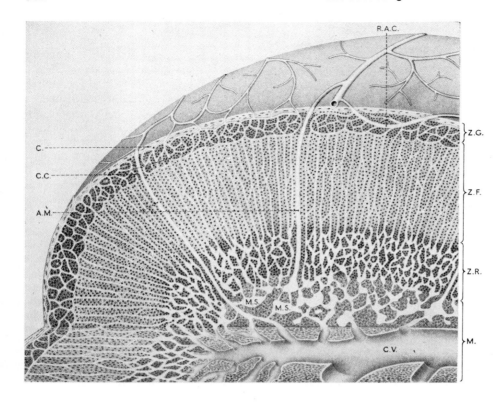

FIG. 9.7 Stereogram of the suprarenal gland, showing the medulla (M.) with its sinusoids (M.S.) and central vein (C.V.), and the cortex with its three zones, the zona glomerulosa (Z.G.), zona fasciculata (Z.F.) and zona reticularis (Z.R.), enclosed by the capsule (C.). Two arteriae medullae (A.M.), and a recurrent cortical artery (R.A.C.) are shown. The columns of cells are mostly separated by cortical capillary sinusoids (C.C.). (After Harrison 1963.)

Vascular pattern

From the main blood vessels, smaller vessels enter at numerous points in the fibrous capsule and run in the trabeculae, forming a close network around and between the cell masses and columns of the zona glomerulosa and zona fasciculata [FIG. 9.7]. In the zona reticularis the blood vessels open up to form a venous plexus which is continuous with the sinusoidal plexus in the medulla, and thus with the central efferent vein of the medulla which emerges at the hilus of the organ as the suprarenal vein. Some of the arterioles that enter the cortex pass directly through into the medulla, the **medullary arteries**. These act as channels to provide a blood supply to the medulla independent of that through the cortex. Constriction of these medullary arteries diverts blood into the capillaries of the cortex (Harrison and Hoey 1960). The **cortical arteries** pass into the cortex to varying depths, and there end in capillaries, or form a loop which recurves to the capsule. Blood which reaches the medulla may therefore do so directly or first pass through capillaries in the cortex. Endothelial cells which line these capillaries show phagocytic properties. The medulla is, therefore, the first tissue in the body to receive suprarenal cortical hormones or vitamin C from the cortical tissue.

Development of the suprarenal glands

The cortical system is a derivative of the coelomic epithelium (splanchnic mesoderm), which proliferates in the medial coelomic bay between the root of the mesentery and the mesonephros at the 7-mm stage, when numerous buds of cells develop from the deep surface of the mesothelium. These cells rapidly form a mass of cortical cells separate from the mesothelium. In Man the greater part of that tissue is ultimately included in the suprarenal cortex, but small masses may separate off to form either independent cortical bodies or portions of accessory suprarenals [p 589].

Meanwhile the sympathochromaffin primordium of the medulla, derived mainly from the neural crest, has appeared (5-mm stage). The cells of that tissue early make contact with the cortex

primordium and commence to invade it at the 14-mm stage. Not until the 19-cm stage does the immigrant chromaffin tissue reach the neighbourhood of the central vein and form a true medulla (Zuckerkandl 1912). Envelopment of the medulla by the cortex is incomplete for a long time; this is probably due as much to progressive cortical overgrowth as to chromaffin invasion.

The relatively bulky **foetal cortex** degenerates progressively during the last 10 weeks of intra-uterine life. At birth two parts of the cortex are distinguished, a still bulky foetal cortex and a thin, overlying, true cortex [FIG. 9.8]. The characteristic cortical tissue of the foetus disappears in the course of the first year and the volume of the entire cortex thus diminishes rapidly for a time. The final specialization of the cortex is not complete until much later. The definitive zones of the permanent cortex are probably formed from the thin rim of adult cortex surrounding the foetal cortex at birth (Barr 1954). In the prepubertal period there is accelerated growth of the cortex; and in middle age its involution begins.

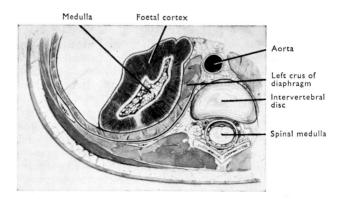

FIG. 9.8. Transverse section through suprarenal gland of a new-born child *in situ*.

The chromaffin system

This system comprises a large number of masses of tissue similar in development, structure, and histochemical reactions to the medulla of the suprarenal gland.

The tissue is called chromaffin on account of its supposed specific affinity for chromium salts, which give a brownish reaction with it. Chromaffin tissue, however, contains substances such as the catechol amines and indole amines which form coloured compounds following treatment with oxidants other than bichromate or chromic acid (Boyd 1960). The term chromaffin is therefore inappropriate, but still used. Although the reaction is given by adrenaline and noradrenaline, it is not specific for these substances. The masses of tissue which form the system all originate in intimate association with the sympathetic nervous system. Indeed, the characteristic chromaffin cells and the neurons of the sympathetic ganglia are derived mainly from a common mother cell, the cell of the neural crest, though cells of the neural tube may also contribute.

The chromaffin system includes the medulla of the suprarenal gland and the extra-suprarenal chromaffin bodies which may occur in any part of the sympathetic system, though they are most numerous in the abdomen. The majority of the extra-suprarenal chromaffin bodies are adjacent to the aorta, the **para-aortic (chromaffin) bodies**. Some cells in the wall of the intestine contain granules giving the chromaffin reaction and are therefore termed **enterochromaffin cells**.

The extra-suprarenal chromaffin bodies are encapsulated structures, but non-encapsulated collections of chromaffin cells may be associated with any of the sympathetic ganglia. The abdominal para-aortic chromaffin bodies are dispersed during prepubertal growth due to relative growth changes. Some collections of chromaffin cells also remain in the abdominal prevertebral sympathetic plexuses in the adult, the largest collections being found in the coeliac and superior mesenteric plexuses. This dispersion also affects the largest discrete para-aortic bodies which lie adjacent to the aorta in the region of the origin of the inferior mesenteric artery and are sometimes known as the organs of Zuckerkandl (Coupland 1952, 1954). In the new-born child they are paired, elongated structures, usually about a centimetre in length. They begin to disperse soon after birth, and by the period of puberty they have practically ceased to be visible to the naked eye, though vestiges may be recognized, at least histologically, to a much later period in life. Their regression is roughly paralleled by the maturation of the suprarenal medulla. In foetal life they contain from two to three times the quantity of noradrenaline present in the suprarenal medulla, but this decreases in postnatal life as the quantity of adrenaline and noradrenaline in the suprarenal medulla steadily increases.

Thus the function of the para-aortic bodies is taken over by the suprarenal medulla in postnatal life, and the character of the secretion changes, adrenaline rather than noradrenaline becoming the predominant hormone.

GLOMUS CAROTICUM

The carotid body (glomus caroticum) is situated close to the bifurcation of the common carotid artery [FIG. 9.9], adjacent to the carotid sinus. Frequently the glomus lies deep to the bifurcation; sometimes it is wedged in between the roots of the internal and external carotids, or it may be placed between them at a slightly higher level.

It is a small, neurovascular structure of a shape that varies with its position. When free from pressure from its surroundings it is oval; when situated between the internal and external carotids, it is wedge-shaped. On average its height is about 7 mm, its breadth 1.5–

5 mm. It sometimes comprises two or more separate nodules. Its colour is yellowish-grey to brownish-red.

Structure

The carotid body consists of groups of 'epithelioid' cells, some of which are argyrophil, surrounded by and interspersed with fibro-areolar tissue containing sinusoids. The cells and the sinusoids are richly innervated by sensory nerve endings. The carotid body was formerly regarded as a part of the chromaffin system, but its cells show only a slight chromaffin reaction (Boyd 1960). Its structure and the abundant sensory nerves which pass into the sinocarotid nerve are in keeping with its functions as a chemoreceptor, responsive to oxygen lack and carbon dioxide excess, although it is less sensitive to the latter.

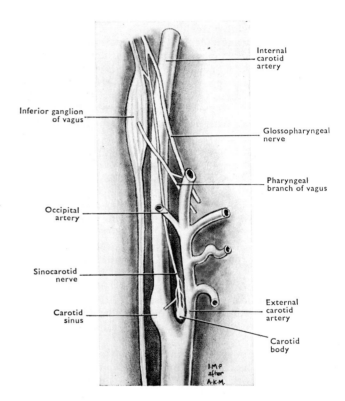

FIG. 9.9 Carotid body and the sinocarotid nerve with its principal connections.

Nerve supply

Numerous afferent nerves pass from the carotid sinus and carotid body to a sinocarotid plexus. The nerve to the carotid sinus [FIG. 9.9] passes upwards with the pharyngeal branch of the glossopharyngeal nerve and divides into two branches, one of which joins the inferior ganglion of the vagus, the other the glossopharyngeal nerve. Other nerve filaments from the plexus pursue an independent but similar course, and join the glossopharyngeal and vagus nerves and the cervical sympathetic trunk. The afferent fibres pass mainly to the glossopharyngeal nerve; but some of those from the **carotid sinus** join the vagus, and it is probable that some afferent fibres from the carotid body do so also.

Development

The carotid body is developed as a condensation of mesenchyme around the third aortic arch (Boyd 1937). Nerve fibres of the glossopharyngeal nerve enter this condensation, and neuroblasts derived from the ganglion of the nerve subsequently migrate along the fibres into the developing organ.

Similar structures are developed in relation to the fourth and sixth aortic arches and form smaller bodies which, on the left side, are adjacent to the aorta and the ductus arteriosus.

CORPUS COCCYGEUM

This is a small vascular body measuring 2–3 mm in diameter. It lies immediately anterior to the tip of the coccyx in intimate relation with a branch of the median sacral artery and with the ganglion impar of the sympathetic trunks. Usually it is associated with a group of smaller bodies of similar structure.

Structure

The corpus and its satellite bodies are enclosed in a fibrous capsule which ensheaths them individually. The characteristic feature of the corpus is the presence of masses of polyhedral 'epithelioid' cells, with large nuclei, surrounding the lumina of tortuous sinusoidal blood spaces. From these spaces capillary channels extend among the masses of closely packed cells. They resemble the cells of the carotid body and do not give the characteristic chromaffin reaction; hence the corpus cannot be included with the chromaffin system.

DEVELOPMENT OF THE CHROMAFFIN SYSTEM

All chromaffin tissue develops in intimate relation with the sympathetic nervous system. The cells of both appear to be derivatives of a common mother-tissue whose elements originate from cells which migrate at an early period from the neural crest and neural tube.

In the 16-mm embryo, sympathochromaffin tissue is met with in situations which more or less correspond to the distribution of the sympathetic nervous system. Differentiation of chromaffin cells from sympathetic neuroblasts begins about the 18-mm stage, but is not completed till late in the period of gestation, if then. The process is characterized by increase in the size of the chromaffin cells and a diminution of the intensity of their reaction to ordinary stains; the chromaffin reaction does not develop until later.

The aortic bodies develop as the chief masses of a paired discontinuous series associated with the prevertebral plexuses. Cranially this series is originally continuous with the primordium of the medulla of the suprarenal gland on each side. The aortic bodies are prominent structures in the 2-month embryo. The paraganglia associated with the ganglia of the sympathetic trunk appear a little later.

The cortical system

The single constant representative of this system in higher vertebrates is the mass of tissue that forms the cortex of the suprarenal gland. Masses of similar tissue, however, are not infrequently met with in various situations, forming accessory cortical bodies.

Accessory cortical bodies may be associated with chromaffin medullary tissue, as is the case in the suprarenal itself, but the majority, including the smaller bodies, are purely cortical. The justification for regarding the suprarenal cortex, with the occasional accessory cortical masses, as representing a distinct system, is based upon phylogenetic and ontogenetic considerations. These warrant the concept of a distribution of cortical tissue which was originally more extensive and distinct from the chromaffin system. In the higher vertebrates the cortical system would appear to have undergone concentration, thus forming mainly the cortex of the suprarenal gland. In the cartilaginous fishes, such as the dogfish, the chromaffin bodies are arranged segmentally, while the cortical tissue is represented by a pair of interrenal bodies between the caudal ends of the kidneys.

It is only in Tetrapoda (vertebrates possessing paired limbs) that cortical tissue and chromaffin tissue come into intimate topographical relation. In mammals the cortical tissue encloses the chromaffin tissue.

THE ORGANS OF THE PHARYNGEAL POUCHES

This group of organs is developed from epithelium derived from the embryonic pharynx, and they illustrate an advanced stage in the histological differentiation of endocrine glands [p. 587].

In an early phase of vertebrate evolution, entodermal epithelium lines the recesses of the pharyngeal cavity, and in one instance—the thyroid—it is sufficiently elaborated to justify the appellation of an exocrine gland (as still seen in ammocoetes, the larva of the lamprey, *Petromyzon*). Subsequently the epithelium of the recesses acquires specific functions and is isolated during embryonic development from the general pharyngeal epithelium to form separate extra-pharyngeal organs.

Early development of pharynx

In the section on embryology, a description is given of the processes which lead to the formation of the embryonic mouth and pharynx [p. 55]. The primitive pharynx is formed from the cephalic part of the foregut and is brought into continuity with the stomodaeum by the breaking down of the buccopharyngeal membrane [FIG. 2.63]. The position of that membrane is such that the greater part of the floor of the mouth is derived from the foregut and is therefore lined with entoderm.

The part of the foregut which forms the primitive pharynx is flattened dorsoventrally and has narrow side-walls that contain thick, vertical bars of mesoderm [FIG. 2.68]. The thickenings are termed pharyngeal arches, and they contain the arterial aortic arches [FIG. 2.78].

At a stage when the cephalic part of the foregut is still cut off from the exterior by the buccopharyngeal membrane, there is a minute diverticulum in its floor known as the median thyroid diverticulum [FIG. 2.72]. This lies immediately posterior to a swelling called the tuberculum impar [p. 59]—a mass of tissue which is eventually incorporated in the tongue, in which the original site of formation of the median thyroid diverticulum is marked permanently by the foramen caecum.

In each side-wall of the pharynx, between the arches, are five entodermal depressions, the pharyngeal pouches. Corresponding ectodermal pharyngeal grooves are found on the external aspect of the pharynx. Between each groove and pouch is a separating membrane.

Only the first four separating membranes become so thinned that entoderm and ectoderm make contact, and in the case of the fourth membrane this phase is of short duration. The fifth pharyngeal pouch is vestigial and becomes associated with the fourth pharyngeal pouch as the ultimobranchial (telobranchial) body [p. 58]. The fourth pouch and ultimobranchial body are together termed the caudal pharyngeal complex.

Owing to the greater prominence of the first and second pharyngeal arches externally, the other arches are more deeply set, so that a cervical sinus, and later a cervical vesicle [p. 55], are formed. The epithelial cells of the vesicle later acquire an intimate relation with the epithelium of the third pharyngeal pouch, forming with it the primordium of the thymus (Norris 1938). Portions of the cervical sinus may occasionally persist in the form of a slender tubular channel (branchial sinus) which opens to the exterior on the lower part of the anterior triangle of the neck, or a portion may be closed off within the neck to form a branchial cyst. If the separating membrane breaks down, a branchial sinus becomes continuous with the pharynx, and forms a branchial fistula.

Fate of pharyngeal pouches [FIG. 2.70]

The first and second pharyngeal pouches are concerned in the development of the tympanic cavity and the auditory tube [pp. 56–7].

The third and fourth pouches, in embryos 10 mm in length, have become small cavities connected to the pharynx only by narrow canals (the pharyngobranchial ducts) [FIG. 2.70]; shortly afterwards these slender connexions become solid and break, leaving small nests of cells (of entodermal origin) embedded in the pharyngeal wall. At the time of their isolation from the pharyngeal wall the third pouches each consist of: (1) a small, hollow, dorsal diverticulum, from which a parathyroid gland develops; and (2) a caudally directed cylinder of cells, the ventral side of which grows and is destined to form a lateral lobe of the thymus [p. 601].

The fourth pharyngeal pouch (with the ultimobranchial body) of each side forms a small, hollow mass at the time of its separation from the pharynx. This becomes differentiated into three parts—a dorsal portion from which a parathyroid gland is developed [p. 598], a ventral diverticulum which corresponds to the thymic part of the third pouch, and the ultimobranchial body.

The thyroid gland

The thyroid gland is situated in the lower part of the front of the neck, and is enclosed in a fascial compartment formed by the sheath of pretracheal fascia which fixes it firmly to the trachea and larynx by its attachment to the oblique line on the thyroid cartilage. The thyroid is an endocrine gland which produces a hormone of the greatest importance for the proper growth and function of most of the tissues in the body. This hormone is known as thyroxine; its metabolically active form is probably triiodothyronine. Characteristic of this organ is the storage of its secretion within small closed cavities—the follicles of the thyroid. It is an organ of very ancient history, yet one that exhibits remarkably little evolutionary change, for its histology and endocrine action are similar throughout the whole series of vertebrates. The gland is yellowish-red, soft, and extremely vascular. It varies in size with age, sex, and general nutrition, being relatively large in youth, in women, and in the well nourished. In women it increases temporarily with menstruation and pregnancy. Its average dimensions are: height 5 cm, breadth 6 cm, thickness of each lobe 1–2 cm; and its weight is 18–31 g (Nolan 1938). The maximum weight of the normal gland is attained in the young adult.

Usually the thyroid gland consists of a pair of conical lobes united across the median plane by a short band of gland tissue called the isthmus. But to many thyroid glands this description is inapplicable. In men and thin, elderly women the gland is not uncommonly horseshoe-shaped; in young, well-nourished women and during pregnancy its general contour is more rounded, deeply notched above to accommodate the larynx and deeply grooved behind for the trachea and oesophagus. Rarely, the gland is in two separate parts. Not infrequently, it is asymmetrical. In about 40 per cent of specimens, a process of gland tissue called the pyramidal lobe extends upwards from the upper border of the isthmus, in front of the cricoid and thyroid cartilages, towards the hyoid bone. This process is seldom median, lying more often on the left than on the right; in rare cases, it is double; less rarely, it is double below and single above; sometimes it is represented by a strip of fibrous tissue or a narrow muscle (levator glandulae thyroideae). The muscle may, however, be present independently of the pyramidal lobe [p. 286].

Small, oval, accessory thyroid glands are common in the region of the hyoid bone, and are occasionally met with in relation to the right and left lobes. They may occur also in the superior mediastinum of the thorax.

Position

The gland itself has an external layer of fibrous tissue—the fibrous capsule—which is loosely connected by areolar tissue to the fascial sheath; the gland with its capsule is thus readily taken out of its sheath of pretracheal fascia. It is under cover of the sternothyroid and sternohyoid muscles, and the superior bellies of the omohyoid muscles, and, more laterally, it is overlapped by the sternocleido-mastoids [FIG. 9.10]. The isthmus is, however, comparatively superficial between the margins of the sternothyroid muscles a short distance above the sternum. The isthmus usually lies on the front of the second, third, and fourth rings of the trachea; but occasionally it lies as high as the cricoid cartilage or as low as the fourth, fifth, and sixth rings of the trachea. The lobes extend down to the level of the sixth ring of the trachea or even lower, and up as far as the oblique line of the thyroid cartilage. Superficially they are covered by the above muscles with twigs from the ansa cervicalis, and by the anterior jugular vein. The upper part of each lobe is moulded medially on the thyroid and cricoid cartilages, and the cricothyroid and inferior constrictor muscles, the latter crossed by the external branch of the superior laryngeal nerve. The lower part is moulded to the sides of the trachea and oesophagus and thus comes into close relation with the recurrent laryngeal nerve and inferior laryngeal artery.

The posterior surface of each lobe is variable in width. It is in contact with the prevertebral fascia covering the longus cervicis muscle, and more laterally with the carotid sheath, which becomes displaced in a posterolateral direction by simple enlargements of the gland. The inferior end of the left lobe approaches or even makes contact with the thoracic duct. Closely applied to the posterior surface of each lobe, and within the fibrous capsule, is a pair of parathyroid glands; they may even be embedded within the substance of the thyroid.

Vessels and nerves

The extraordinarily rich arterial supply is effected through the superior and inferior thyroid arteries. Occasionally (10 per cent of cases) a fifth artery—the thyroidea ima, normally present in the embryo—persists in the adult as a branch of the brachiocephalic trunk. The pyramidal lobe, if well developed, receives a special branch from one of the superior thyroid arteries, usually the left. The arteries are remarkable for their large size and for the frequency of their anastomoses, which may be arterial or capillary in nature. An anastomosing trunk courses up over the back of each lobe within the fascial sheath and unites the inferior and superior thyroid arteries; it is a landmark for the identification of the parathyroid glands. The anterior branches of the superior thyroid arteries anastomose on the upper border of the isthmus. The anastomoses between the tracheal and oesophageal branches of the inferior thyroid artery and bronchial arteries are probably of importance in maintaining the arterial supply to thyroid tissue following partial thyroidectomy. Typically, three pairs of veins drain the gland. The upper two pairs—the superior and middle thyroid veins—join the internal jugular veins; the lower pair—the inferior thyroid veins—join the brachiocephalic veins. The veins take origin from the venous plexus on the surface of the gland or, in the case of the

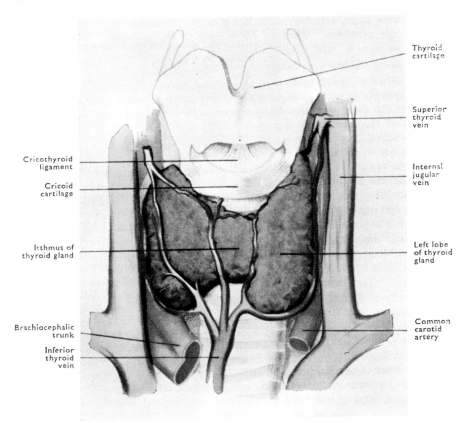

Thyroid
cartilage

Superior
thyroid
vein

Cricothyroid
ligament

Cricoid
cartilage

Internal
jugular
vein

Isthmus of
thyroid gland

Left lobe
of thyroid
gland

Brachiocephalic
trunk

Inferior
thyroid
vein

Common
carotid
artery

FIG. 9.10. Dissection of thyroid gland and of structures in immediate relation to it.

inferior, from a downward extension of the plexus in front of the trachea. When the gland is very large, accessory veins are present, sometimes in considerable numbers. Most of them pass to the internal jugular veins. A free, transverse, venous anastomosis is effected along the borders of the isthmus through superior and inferior communicating veins.

The lymph vessels pass directly from the subcapsular plexus to the deep cervical lymph nodes; a few descend in front of the trachea to the pretracheal lymph nodes, through which they are connected with sternal nodes behind the manubrium sterni.

The nerves are non-medullated postganglionic fibres which come from the superior and middle cervical sympathetic ganglia. They reach the gland by way of the cardiac and the superior and recurrent laryngeal nerves, and along the superior and inferior thyroid arteries. All of the nerve fibres are vasomotor, and influence thyroid secretion indirectly by their action on blood vessels, particularly the arteriovenous anastomoses present in the thyroid.

Structure

The gland consists of a mass of minute, rounded follicles of various sizes, commonly about 300 μm in diameter [FIGS. 9.11 and 9.12], groups of follicles being enclosed in a connective tissue capsule to form lobules. Each follicle consists of a layer of epithelial cells enclosing a variable quantity of colloid. This acts as a store of thyroid hormone in a resting follicle, and becomes depleted in an active follicle. Individual follicles within a thyroid gland have different appearances due to variations in quantity of contained colloid and the heights of the epithelial cells. In addition to the epithelial cells, 'light' or 'C' cells are to be found between the thin basement membrane of the follicle and the epithelial cells. They may be found either singly or in groups, and are usually larger than the epithelial cells. They are responsible for the secretion of the

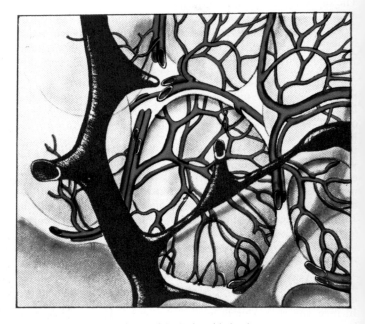

FIG. 9.11. Injected specimen of dog's thyroid gland.
Enlarged drawing demonstrating the relative size of the lymph-plexus and blood-capillary plexus and their relation to the individual follicles. Note that the lymphatic plexus (black) lies external and in less intimate contact with the individual follicles than the blood-capillary plexus, which is specific for each follicle. (Rienhoff 1931.)

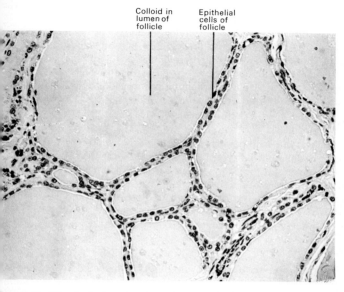

Colloid in lumen of follicle · Epithelial cells of follicle

FIG. 9.12. Histological section through the relatively inactive thyroid gland of a man aged 62. × 275.

hormone *thyrocalcitonin*, which inhibits the mobilization of calcium from bone and thereby depresses the amount of calcium in the blood. Microscopic study of living cells has shown that the epithelial cells produce cytoplasmic processes which extend into the lumen of the follicle, a process which is associated with the addition of colloid to the lumen (Williams 1941; de Robertis 1942). The presence of a fenestrated vascular endothelium in this and other ductless glands suggests that the control of secretion into the blood stream is a function of the gland cells and not of the endothelium. The follicles are embedded in a framework or stroma of fibrous tissue which is continuous externally with the capsule. Numerous lymph vessels, arteries, veins, and nerves course in the stroma. The lymph vessels appear to be mainly concerned in the drainage of the interfollicular tissue [FIG. 9.11]; the larger trunks pass to a dense plexus on the surface of the gland.

Development

The thyroid gland is developed from a median, ventral diverticulum of the floor of the pharynx [p. 58; FIG. 2.72]. It appears in embryos less than 2 mm in length (possessing about six mesodermal somites) as a small outpocketing of the pharyngeal floor which passes downwards between the ventral ends of the first and second pharyngeal arches. The diverticulum by rapid downgrowth soon forms a plate-like structure ventral to the trachea, connected with the pharyngeal floor by a narrow tube called the thyroglossal duct. This duct disappears normally in embryos of 6 weeks, but the site of its original opening into the pharynx is marked in the adult by the foramen caecum of the tongue [FIG. 6.17]. In the sixth week the fourth pharyngeal pouch, with its ventral diverticulum, is still connected by the fourth pharyngobranchial duct with the pharynx. At this stage the caudal pharyngeal complex [p. 594] on each side becomes closely applied to the lateral portion of the thyroid gland; the cells of the ultimobranchial body proliferate and develop into cells within the thyroid gland (Bejdl and Politzer 1953) which form the 'light' or 'C' cells. These cells therefore become associated with the thyroid follicles secondarily (Hirsch and Munson 1969). It has been stated that the pharyngeal region as a whole has the potentiality for developing thymus and parathyroid tissue, so that parathyroid may develop from thymus, and thymus from parathyroid or thyroid

tissue. Thymic tissue may develop from the ventral diverticulum of the fourth pouch, and there is also a theoretical possibility of thyroid tissue developing from this pouch (Kingsbury 1939).

Histological differentiation of the thyroid gland proceeds by the arrangement of the embryonic epithelial cells in two sheets, with disappearance of the original lumen of the diverticulum, and, in the 50-mm foetus, definite follicles are formed which later enlarge. Colloid material is present in the follicles, and the lining cells are apparently active, in the foetus of 11 weeks. Additional 'secondary' follicles are added throughout foetal life but the process slows down before birth and probably ceases about puberty. Postnatal growth of the gland is mainly by an increase in size of the follicles.

The development of the gland affords a ready explanation of its variations in the adult. Thus the pyramidal lobe, with partial or complete duplication, is due to the development of gland tissue from the inferior part of the thyroglossal duct, which may bifurcate at a higher or a lower level, with more or less complete fusion or separation of the masses thus formed. Accessory thyroid glands near the hyoid bone, or within the tongue, are the result of development of gland tissue from isolated remnants of the duct. Their occurrence behind the sternum above the arch of the aorta is probably due to a downward displacement of thyroid tissue along with the arteries (common carotid) of the third pharyngeal arch which, at their origin from the truncus arteriosus, have the median thyroid diverticulum closely attached at an early stage.

The occurrence in the adult of a duct that leads from the foramen caecum to, or towards, the hyoid bone (lingual duct) is due to a persistence of the upper part of the thyroglossal duct. Similarly, thyroglossal cysts are due to the persistence of short segments of the duct. The thyroglossal duct may be continuous with the pyramidal lobe [p. 595], and it usually passes in front of the hyoid bone since it is initially anterior to the second pharyngeal arch of the embryo. Any change in this relation is due to extension and growth of the primordium of the hyoid which may lead to the duct becoming embedded within the body of the hyoid or even lying posterior to it. Cyst-like structures, derived from the ultimobranchial body, may sometimes be found in the adult in close relation with the lower pole of the lateral lobe of the gland.

Methods of clinical anatomical examination

The lateral lobe of the thyroid gland is felt indistinctly as a soft mass at the anterior border of the sternocleidomastoid muscle on the side of the lamina of the cricoid cartilage and the adjacent part of the thyroid cartilage. The isthmus also can be felt as a soft swelling extending over a small part of the front of the trachea a short distance below the cricoid cartilage. When the larynx and trachea are pulled up in the movement of swallowing, the thyroid gland accompanies them. The gland is often large enough to produce an evident fullness of the neck; the sternocleidomastoid can then be felt to lie superficial to the postero-inferior part of the lateral lobe.

RADIOLOGY

The normal thyroid gland is not usually obvious in radiographs of the neck, but the isthmus is sometimes visible anterior to the trachea in lateral radiographs.

Abnormal thyroid glands and other pathology in this region may displace and distort the trachea, or push the internal jugular veins laterally into the translucent areas of the lung apices. Radioisotope scans may be used to demonstrate the presence or absence of active thyroid tissue.

The parathyroid glands

The parathyroid glands are two pairs of small glands closely applied to the back of the thyroid gland within its fibrous capsule [FIG. 9.13]. They are distinguished as the superior and inferior parathyroid glands; but with reference to their developmental origin [see below], the upper pair are known also as the parathyroids IV and the lower pair as the parathyroids III.

The parathyroid glands produce a hormone which mobilizes calcium from bone and therefore elevates the level of calcium in the blood. Extirpation of all four of the glands in a mammal causes death within a few days, but such is the margin of safety that removal of two of them produces no obvious effect.

The parathyroid glands are yellowish-brown, ovoid or lentiform structures, varying in size between extremes of 1 and 20 mm in their long (generally vertical) axis. Most commonly they are from 3 to 6 mm in length, from 2 to 4 mm in width, and from 0.5 to 2 mm in thickness, and their individual weights vary within wide limits. The total amount of parathyroid gland-tissue has been stated to show a mean variation from 0.10 to 0.14 g.

The normal number of parathyroids may be diminished to three, two, or even one. In some such cases, however, it is not easy to be certain that a minute ectopic gland may not have escaped notice. On the other hand, the number may be increased, probably as a result of division of the original primordia. Thus from five to eight—even in an extreme case as many as twelve—have been recorded. Six per cent of subjects have only three parathyroid glands, and in 6 per cent also there are five (Gilmour 1938).

The superior parathyroid gland is usually embedded in the capsule of the thyroid gland at the back of the corresponding lobe, about its middle. The inferior parathyroid gland is similarly embedded in the back of the inferior end of the lobe. As a rule the anastomosing arterial channel which connects the inferior and superior thyroid arteries passes near both parathyroids and furnishes the best guide to them; but the range of exceptional positions which the glands may occupy is wide.

The superior parathyroid may be found: (1) behind the pharynx or oesophagus; (2) in the fibrous tissue at the side of the larynx, above the level of the thyroid gland; (3) behind any part of the corresponding lobe of the thyroid gland or even embedded in the thyroid substance (internal parathyroid).

The inferior parathyroid may be found: (1) near the bifurcation of the common carotid artery; (2) behind any part of the corresponding lobe of the thyroid gland; (3) on the side of the trachea; or (4) in the thorax close to the thymus. This wide range of variation in position is explicable by the close association of the inferior parathyroid with the main thymus element in development.

Vessels and nerves

The artery to each parathyroid may spring from any branch of the inferior or superior thyroid arteries, but most commonly is a branch of the large anastomosing vessels between them. The lymph vessels drain with those of the thyroid gland. The nerve supply is abundant and comes from the plexus surrounding the thyroid vessels [see above]. The parathyroids produce their hormone after transplantation; the nerves are apparently vasomotor and not secretory.

Structure

The parathyroids are built up of interconnected trabeculae of epithelial cells with only a little vascular tissue between them. Sometimes the cellular arrangement is of a more compact character, or again, some of the cells may be arranged in follicle-like clumps enclosing a colloid material (devoid of iodine). The 'principal' cells, which contribute the majority, are large and clear; but, after the age of 8 or 9 years, some of the cells contain acidophil granules and are believed to be degenerating principal cells (Morgan 1936).

Development

The parathyroid glands develop from the dorsal diverticula of the third and fourth pharyngeal pouches [FIG. 2.70]. The first indication of their development is a proliferation and thickening of the epithelium on the cephalic and lateral aspects of the diverticula, which may be seen in a 10-mm embryo. Cords of cells grow out from the thickening, and fibrous tissue penetrates between the outgrowing cords, which soon lose their connection with the pharynx.

Parathyroid III—the inferior parathyroid of the adult—being developed from the same pouch as the thymus, is normally drawn caudal to parathyroid IV by the migrating thymus. As a rule it halts at the caudal end of the lobe of the thyroid gland, but may continue with the thymus into the thorax. On the other hand it may not descend at all. In the latter case it remains near the bifurcation of the common carotid artery, where it is liable to be mistaken for the carotid body. (For details of development, consult Weller 1933 and Boyd 1950.)

The thymus

The thymus is situated in the superior mediastinum between the sternum and the great vessels; and it usually extends down for a short distance into the anterior mediastinum in front of the pericardium. In the new-born child (in which it is relatively much larger than in the adolescent or adult) it commonly extends laterally between the thoracic wall and the anterior borders of the pleurae and lungs [FIG. 9.14].

The thymus is essentially an organ of the growth period of life. In the course of a few years after birth it shows a rapid diminution in the rate of its growth and hence of its relative size; but it continues to grow until puberty, and it is only after that period of life that it undergoes a gradual diminution in absolute size (Young and Turnbull 1931).

During its period of growth the thymus appears as a pinkish mass (yellower in later life), consisting of a pair of laterally compressed

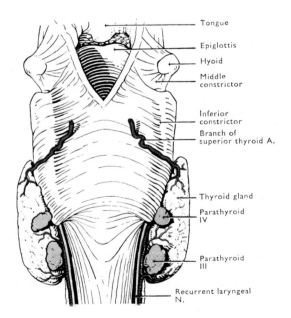

FIG. 9.13. Dissection showing thyroid and parathyroid glands of adult from behind.

Tongue

Epiglottis

Hyoid

Middle constrictor

Inferior constrictor

Branch of superior thyroid A.

Thyroid gland

Parathyroid IV

Parathyroid III

Recurrent laryngeal N.

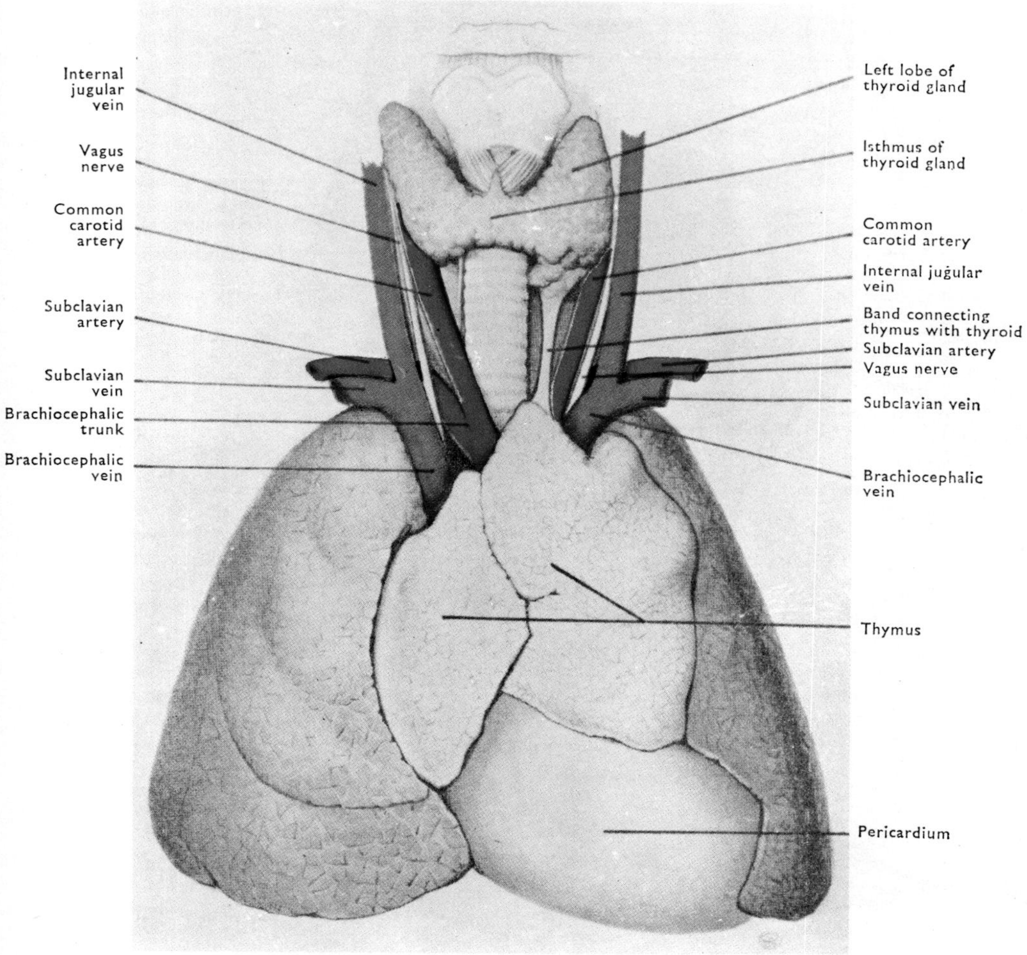
Internal jugular vein
Vagus nerve
Common carotid artery
Subclavian artery
Subclavian vein
Brachiocephalic trunk
Brachiocephalic vein
Left lobe of thyroid gland
Isthmus of thyroid gland
Common carotid artery
Internal jugular vein
Band connecting thymus with thyroid
Subclavian artery
Vagus nerve
Subclavian vein
Brachiocephalic vein
Thymus
Pericardium

FIG. 9.14. Thymus and thyroid gland in a full-term foetus hardened by formalin injection.

more or less pyramidal, asymmetrical lobes. The lobes are connected with each other, not by any bridge of glandular tissue but by areolar tissue. The surface of the organ, in the young, more actively glandular condition, is finely lobulated.

The thymus is soft in consistence, and the details of its shape are determined by its size and by the structures upon which it is moulded, namely, the pericardium and the great vessels of the superior mediastinum and the root of the neck. Its shape varies with its size and the age of the individual. In infants (with a short thorax) it is broad and squat; in adults (with a long thorax) it is drawn out into two irregular but more or less flattened bands [FIG. 9.15].

Posteriorly it is in relation from below upwards with the pericardium, ascending aorta, left brachiocephalic vein, the brachiocephalic trunk, the trachea, and inferior thyroid veins. Anteriorly it is related to the sternum and the lower ends of the sternothyroid muscles. In the young child the lateral margin is insinuated between the pleura and the upper costal cartilages, intercostal spaces, and internal thoracic artery.

There is great individual variation in its size at any given age. Thus, at birth it ranges in weight from 2 to 17 g with an average of about 13 g. At puberty the thymus is usually at the height of its development with an average weight of 37 g; but sometimes it has already undergone considerable retrogression. In the young adult the average weight is reduced to about 25 g, but the organ is occasionally still quite large.

In old age it commonly weighs only about 10 g. These age changes may be secondary to the hormonal changes occurring in other endocrine glands.

Variations of the shape as well as in the size of the thymus sometimes occur, attributable to a partial persistence of the slender stalks by which at first the developing thymus remains attached to the third pharyngeal pouches. Thus, the thymus may exhibit slender prolongations into the neck on each side, anterolateral to the trachea. These processes may be connected to the lower parathyroid glands by strands of fibrous tissue [FIG. 9.18]; or again, the whole of the cervical processes may be represented by fibrous strands (atrophic cervical thymus). An isolated portion of thymus tissue may persist in close relation with the lower parathyroid (accessory cervical thymus III).

Vessels and nerves

The arterial supply of the thymus is effected through inconstant branches, chiefly of the **internal thoracic arteries** and their branches. Its **veins** are irregular and mostly join the internal thoracic and left brachiocephalic veins. Its **lymph vessels** are abundant. Although they are not related to the organ in the same manner as in a lymph node, they have been found filled with lymphocytes presumably derived from the gland. The lymph vessels enter the **anterior mediastinal lymph nodes.**

Its **nerves** are derived from the vagus and the sympathetic. The branches of the **vagus** descend directly to the thymus from about the level of the thyroid cartilage; the **sympathetic** fibres run with the blood vessels. The fibrous capsule of the thymus receives small irregular branches from the phrenic nerves, but they do not supply the gland tissue.

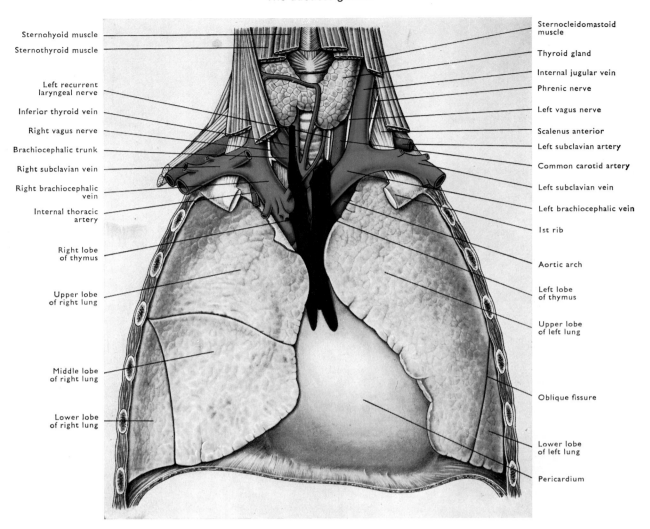

Sternohyoid muscle
Sternothyroid muscle
Left recurrent laryngeal nerve
Inferior thyroid vein
Right vagus nerve
Brachiocephalic trunk
Right subclavian vein
Right brachiocephalic vein
Internal thoracic artery
Right lobe of thymus
Upper lobe of right lung
Middle lobe of right lung
Lower lobe of right lung

Sternocleidomastoid muscle
Thyroid gland
Internal jugular vein
Phrenic nerve
Left vagus nerve
Scalenus anterior
Left subclavian artery
Common carotid artery
Left subclavian vein
Left brachiocephalic vein
1st rib
Aortic arch
Left lobe of thymus
Upper lobe of left lung
Oblique fissure
Lower lobe of left lung
Pericardium

FIG. 9.15. Dissection to show the thymus in an adult female.

Structure

The thymus is invested by a fibrous capsule which sends septa into its substance between its constituent lobules. The two 'lobes', though really independent paired glands, are also bound together more or less intimately by their capsular investments. If the young gland is dissected so as to liberate its lobules from the fibro-areolar tissue between them , it will be found to consist of an elongated series of lobules, each connected with a central parenchymatous cord along which they are arranged in irregular necklace fashion. The cord, mainly of medullary substance, represents the original thymic diverticulum of which the lobular masses are secondary derivatives.

The lobules of the thymus are further subdivided into secondary lobules which average a little over a millimetre in diameter. They are imperfectly separated from one another by delicate areolar tissue. These are the units of the thymic structure: each shows a cortex and a medulla. The cortex consists of closely packed lymphocytes embedded in a delicate fibrous reticulum. It contains no lymph vessels. The cortex completely surrounds the medulla, which is continuous with that of adjacent lobules and, through the lobular stem, with the central cord of the gland [FIG. 9.16].

In its intimate structure the medulla differs from the cortex by the presence of many reticular cells and fibres but only few lymphocytes. The medulla, further, contains the so-called corpuscles of Hassall,

which are concentrically arranged nests of epithelioid cells varying in size from about 25 to 75 μm in diameter. The central cells often show granular degeneration. They vary in number not only in different individuals but also at different periods of life. In a general way the structure of the thymus follicle resembles that of a lymph node, but there are neither lymph sinuses nor germinal centres. The vascular tissue elements and some associated areolar tissue are derived from invading mesenchyme during the course of development.

The thymus appears, then, to have some special relation with the cellular constituents of the vascular system. It has an endocrine function in that thymosin, a cell-free thymic extract, stimulates lymphopoiesis. Its presence is necessary during embryonic and early neonatal life for the maturation of lymph tissue and the development of cell-mediated (T-cell) immune responses. In the adult it is concerned with the production of lymphocytes which are immunologically uncommitted. The process of involution, by which there comes about a great reduction in the size of the gland after puberty, may be influenced by various endocrine glands, but is probably intrinsic to the thymus. A persistently enlarged thymus is associated with delayed sex development and obesity. For recent reviews on thymic function see Metcalf (1966), Miller and Osaba (1967), and Goldstein, Asanuma, and White (1970).

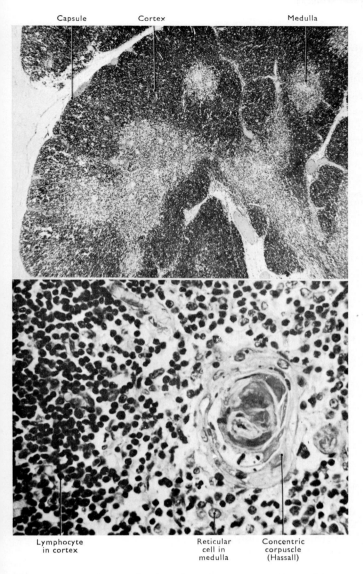

Capsule   Cortex   Medulla

Lymphocyte   Reticular   Concentric
in cortex   cell in   corpuscle
                            medulla   (Hassall)

FIG. 9.16. Photomicrographs of the thymus of a young adult man (above, × 40; below, × 750). (By courtesy of the late Professor A. R. Muir.)

Development

The paired thymus glands, or 'lobes', right and left, arise from the ventral diverticula of the third pharyngeal pouches and, in the pig, from ectoderm of the third pharyngeal cleft. The first indication of the thymus is seen in the 9-mm embryo, when the third pouch has developed a caudally directed cylindrical elongation which migrates tailwards along the third pharyngeal arch arteries to reach the pericardium at the 15-mm stage. As a result of the migration, the cephalic part, to which parathyroid III is attached, becomes drawn out and normally disappears [FIG. 2.70]. The relative time of the disappearance determines the permanent level of parathyroid III, for until it happens that gland is dragged in the wake of the thymus [p. 598]. Sometimes a small detached mass of thymic tissue may persist beside parathyroid III, and may differentiate to form an accessory cervical thymus III [FIG. 9.18].

The thymus soon becomes isolated from the parathyroid (22-mm embryo) and forms a closely packed, cellular mass. The mass of cells is penetrated shortly afterwards by mesenchyme and blood vessels; reticular fibres, apparently derived from the blood vessels, make their appearance between entodermal cells (Norris 1938). Thymocytes appear in the eighth week, probably derived from the invading mesenchyme. The differentiation of cortex and medulla is visible soon afterwards. The concentric corpuscles are derived from the cellular mass which developed from the entoderm of the third pharyngeal pouch.

The growth of the thymus and its involution after puberty have been described above.

Thymus IV vestiges [FIG. 9.18]

Although extremely rare in Man, accessory thymic derivatives from the fourth pharyngeal pouch are not uncommonly met with in other mammals. They are found close to parathyroids IV, and must not be confused with the accessory thymic lobules referred to above, which are to be regarded as detached portions of the cervical part of the main thymus.

Methods of clinical anatomical examination

In children the apex of the thymus extends above the level of the manubrium sterni; but it is only occasionally recognizable on palpation as a soft mass situated in front of the trachea slightly to the left of the median line and behind the infrahyoid muscles.

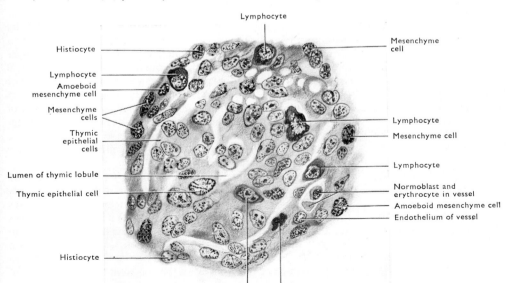

Lymphocyte

Histiocyte

Lymphocyte

Amoeboid
mesenchyme cell

Mesenchyme
cells

Thymic
epithelial
cells

Lumen of thymic lobule

Thymic epithelial cell

Histiocyte

Lymphocyte   Mesenchyme cell

Mesenchyme
cell

Lymphocyte

Mesenchyme cell

Lymphocyte

Normoblast and
erythrocyte in vessel

Amoeboid mesenchyme cell

Endothelium of vessel

FIG. 9.17. Section of embryonic thymus of rabbit. (Maximow and Bloom, *Textbook of histology.*)
A hollow lobule is surrounded by mesenchyme in which some histiocytes of the reticulo-endothelial system are seen. Lymphocytes are present both around and among the thymic epithelial cells.

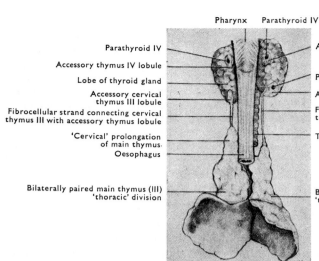

Parathyroid IV
Accessory thymus IV lobule
Lobe of thyroid gland
Accessory cervical thymus III lobule
Fibrocellular strand connecting cervical thymus III with accessory thymus lobule
'Cervical' prolongation of main thymus
Oesophagus
Bilaterally paired main thymus (III) 'thoracic' division

Pharynx   Parathyroid IV

Accessory thymus IV lobule
Parathyroid III
Accessory cervical thymus III lobule
Fibrocellular strand connecting cervical thymus III with accessory thymus lobule
Trachea
Bilaterally paired main thymus (III) 'thoracic' division

FIG. 9.18. Illustrating accessory thymus lobules described by Groschuff (1900).

RADIOLOGY

The thymus in the newborn is large and is sometimes visible as a sail-like projection from the upper mediastinal margin, mainly into the right lung. It may displace the right brachiocephalic vein into the translucent area of the right lung. In children and adults, the thymus is a thin structure, so there is little evidence of it in lateral radiographs.

THE HYPOPHYSIS

The hypophysis is a small gland of dual origin which forms a median basal appendage of the hypothalamus [FIGS. 9.19 and 9.20], and is known as the pituitary gland from an old notion that it secreted the nasal mucus.

Position

It occupies the hypophysial fossa [FIG. 3.55] between the dorsum sellae behind, and the tuberculum sellae in front. This cavity is lined with dura mater and roofed over by the diaphragma sellae. According to Wislocki (1937), no separate arachnoid or pial covering is formed around the hypophysis [FIG. 9.19] below the level of the diaphragma sellae. Through a small aperture in the diaphragma sellae, the infundibulum connects the hypophysis with the tuber cinereum of the floor of the third ventricle. Lateral to the hypophysis are the cavernous sinuses which are united by the intercavernous sinuses passing below, behind, and in front of the hypophysis. The internal carotid artery and abducent nerve in each cavernous sinus, and the oculomotor, trochlear, and ophthalmic nerves in its lateral wall are also close to the hypophysis. The diaphragma sellae intervenes between the hypophysis and the optic chiasma, which lies about 8 mm above the diaphragma. The interpeduncular cistern, with the contained circulus arteriosus, surrounds the infundibulum above the diaphragma sellae. The sphenoidal sinus is situated below and in front of the hypophysial fossa, separated merely by a thin wall of bone; if small, it is only antero-inferior. The close proximity of important structures to the hypophysis accounts for various symptoms associated with pathological enlargement of the gland (as in acromegaly). Enlargement of the anterior part of the hypophysis tends to thrust the forepart of the diaphragma sellae upwards and to cause visual defects through pressure on the optic chiasma [FIG. 9.20]. An enlarged hypophysis sometimes also gives rise to symptoms which are due to encroachment upon the roof of the sphenoidal sinus or to pressure upon the cavernous sinus and the abducent and oculomotor nerves. Enlargement may be recognized radiographically through the expansion of the hypophysial fossa produced by it.

The hypophysis is a rounded structure with the following average dimensions: 14 mm transversely, 9 mm anteroposteriorly, and 6 mm vertically. It usually weighs a little over half a gramme. It undergoes some enlargement during pregnancy.

The following parts are distinguished [FIGS. 9.19 and 9.20];

1. The anterior lobe, including: (i) the distal part; (ii) the infundibular part; (iii) the intermediate part.
2. The posterior lobe.

The anterior lobe forms the greater part of the entire gland; its distal part [FIG. 9.20] is separated by an intraglandular cleft or a series of cysts from the intermediate part, which is a very thin layer of tissue applied to the surface of the posterior lobe [FIG. 9.25]. The infundibular part is a thin layer of tissue which encircles the front of the infundibulum, and extends as far as the tuber cinereum. When the infundibulum is cut across in the removal of the brain, most of the infundibular part usually remains adherent to the tuber cinereum. The anterior lobe is slightly heavier in women and is five times as large as the posterior lobe, which contributes most of the remainder of the entire organ.

The two lobes of the hypophysis are of different developmental origin [FIG. 9.24]. The anterior lobe is derived from Rathke's pouch, a diverticulum of the stomodaeum [p. 52]; the posterior lobe arises from a diverticulum of the floor of the diencephalon. The two lobes are different in structure and function, but the line of demarcation between them becomes obscured by their very intimate union [FIG. 9.26]. The hypophysis occupies a position of pre-eminence in the endocrine system by virtue of the control it exerts over most of the other endocrine glands by means of the various hormones it produces. For this reason it has been termed the master gland.

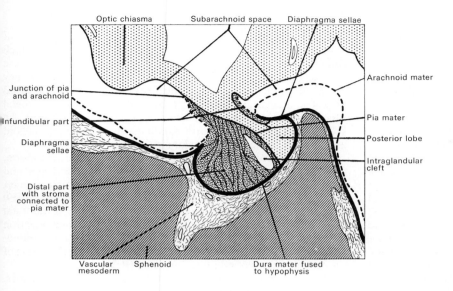

FIG. 9.19 Diagram of relation of meninges to hypophysis in 160-mm human foetus. (After Wislocki 1937.) The relative positions of the structures are not the same as in the adult.

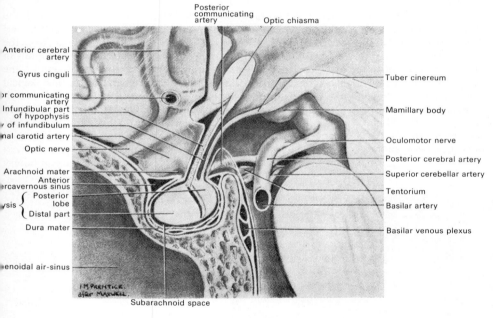

FIG. 9.20. Diagrammatic median section of hypophysis *in situ*, showing the parts of the organ and its relations to adjacent structures. (The existence of a subarachnoid space around the gland in the hypophysial fossa is doubtful [cf. FIG. 9.19].)

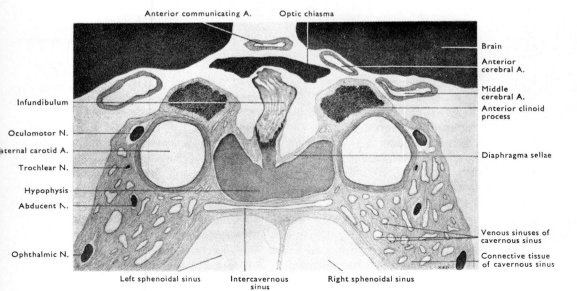

FIG. 9.21. Camera lucida drawing of a coronal section through the cavernous sinus showing the relations of the hypophysis.

Vessels and nerves

The anterior and posterior lobes of the gland have an abundant, but relatively independent, **arterial supply** (Stanfield 1960). The anterior lobe is supplied by the **superior hypophysial arteries**, branches of the internal carotid artery immediately after this vessel has pierced the dura mater; these vessels mainly enter the infundibulum and there break up into capillaries. From this first capillary bed **hypophysial portal vessels** run down to the sinusoids of the anterior lobe. By this means a large part of the blood supply of the anterior lobe passes first through capillaries in the hypothalamus before being collected by the portal vessels and being conveyed to the anterior lobe. Releasing hormones secreted by the hypothalamus [p. 662] are conveyed to the cells of the anterior lobe through these vessels. The posterior lobe is supplied by the **inferior hypophysial arteries**, which arise on each side from the internal carotid artery as it lies within the cavernous sinus. They divide into medial and lateral branches, and anastomose with the branch of the superior hypophysial artery which supplies the lower part of the infundibulum and gives branches direct to the distal part. Blood leaves the gland by means of veins which pass into the **cavernous** and **intercavernous sinuses. Lymph vessels** have not been demonstrated in the gland.

The **nerves** of the anterior lobe are a rich supply of nonmedullated fibres from the carotid plexus. While many of these are vasomotor in function, there are some nerve fibres which have terminals in intimate relation with the epithelial cells and may be secretory in function. That the nervous connections are not essential to all its endocrine functions is shown by the fact that grafts of the gland produce active secretions. The posterior lobe receives nerve fibres from the supra-optic and paraventricular hypothalamic nuclei by a well-marked tract which runs along the infundibulum. The fibres end by forming a dense network around the small cells of the posterior lobe. Functionally there is a close connection between the hypophysis and the hypothalamus.

Structure and function

The **anterior lobe** is composed in the adult mainly of columns of epithelial cells supported by two bands of areolar tissue which contain vessels of considerable size. Capillaries pass from these vessels into intimate relation with the epithelial cells, and they eventually drain into veins emerging from the periphery of the distal part. The tubular arrangement of the epithelial cells seen in foetal life disappears before birth. Vesicles of colloid material are to be seen at all ages from the fourth month of intra-uterine life; colloid also appears in the intraglandular cleft.

Four principal kinds of cells are recognized: **chromophobe, acidophil, basophil,** and **amphophil** (containing both acidophil and basophil types of granules), depending on the reaction of their cytoplasmic granules to acid Schiff–orange G stain (Sheehan 1959) [FIG. 9.22]. Acidophil cells comprise about one-third of the total number of cells, basophils only about one-tenth of the total. Electron microscopy, and techniques of immunohistochemistry and immunofluorescence have recently enabled more precise determination of the cellular origins of pituitary hormones (Baker 1974).

The anterior lobe is known to produce seven distinct hormones, several of which influence the activity of other endocrine glands. The growth hormone and prolactin appear to be derived from the acidophil cells; thyrotrophin, corticotrophin, melanotrophin, and the gonadotrophic hormones (which influence the gonads), from the basophils. The unequal distribution of basophils and acidophils in the anterior lobe (the latter being situated mainly laterally, the basophils centrally) assists the study of alterations in histological appearance of the hypophysis in disease and experimental investigations. Tumours consisting of the different types of cell produce different effects—basophil tumours cause obesity and various other changes, known collectively as Cushing's syndrome, acidophil tumours cause acromegaly. Various conditions have been recognized which appear to depend on hyposecretion, e.g. hypopituitary

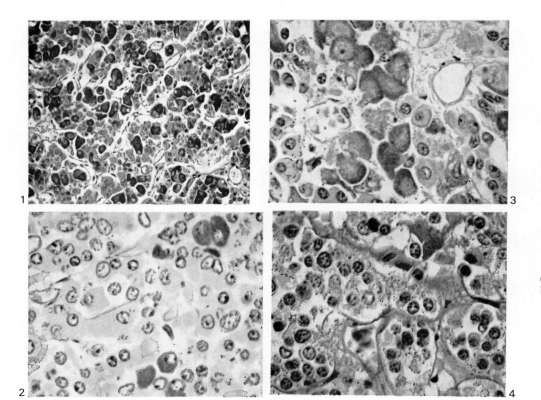

FIG. 9.22. Photomicrographs demonstrating the cell types to be found in the anterior lobe of the human hypophysis, coloured with acid Schiff–orange G.
1. Low power view showing eosinophils (yellow) and basophils (purple). × 170.
2–4. High power views showing (2) eosinophils (yellow) with five basophils) (3) basophils flanked by a few yellow eosinophils; (4) chromophobe cells. × 780.

(Preparations and photomicrographs 2–4 by courtesy of Professor H. L. Sheehan.)

dwarfism. A relation to periods of the oestrous cycle is indicated by the cyclic variations of cell types in guinea-pigs. The condition of diabetes insipidus is thought to be due to a lesion of the posterior lobe or of the hypothalamus or of both.

Absence of the acidophils is found to be related in mice to a defect of a certain gene; the effect on the gonads is then found to be present, but not the growth-stimulating effect.

The **intermediate part** appears to be the least specialized portion of the hypophysis; it grows less, and shows cysts which contain colloid. These cysts are occasionally lined by ciliated cells. Mostly chromophobes and amphophils are present, but there are also a few basophils and acidophils. The pars intermedia is a major source of melanotrophin. In the foetus, gland tubules are present and open into the cleft, which is a vestige of the lumen of the stomodaeal diverticulum (Rathke's pouch). These gland tubules subsequently extend into the posterior lobe [FIG. 9.26] and basophil cells also migrate from the intermediate part into the posterior lobe (Lewis and Lee 1927). The **infundibular part** consists mainly of amphophils and chromophobe cells.

The **posterior lobe**, sometimes called the nervous part, consists of a mass of neuroglia, and of small cells that resemble neuroglial cells. The invasion of this part by basophil cells from the intermediate part of the anterior lobe has been mentioned above. Small hyaline bodies, described by Herring (1908), are found among these cells [p. 606]. Potent hormones are present in the posterior lobe; the **pressor** hormone (**vasopressin**) raises the blood pressure, the **antidiuretic** hormone diminishes urinary excretion of water and increases the excretion of chlorides, and **oxytocin** causes uterine muscle to contract. There is evidence suggesting that the pressor and antidiuretic effects are manifestations of the action of only one hormone. It is not known whether the immigrant basophil cells have the same function as those of the anterior lobe. The posterior lobe hormones are produced by the supra-optic and other hypothalamic nuclei, and pass along the axons of the hypothalamo-hypophysial tracts (such as the supra-opticohypophysial, p. 662) into the posterior lobe, which serves as a storage-release centre rather than a gland of internal secretion (Scharrer and Scharrer 1954; Stutinsky 1967). Damage of the supra-opticohypophysial tract certainly causes **diabetes insipidus**, but this has always been considered previously as being due to release of the control over the secretions of the posterior lobe by the supra-optic nucleus.

Development

As already mentioned, the hypophysis originates from two entirely distinct rudiments. Both of these are hollow ectodermal diverticula, one neuro-ectodermal, derived from the floor of the primary forebrain, the other from the ectodermal lining of the stomodaeum. The stomodaeal diverticulum, known as Rathke's pouch, gives origin to the whole of the anterior lobe. The other diverticulum gives origin to the posterior lobe and the infundibulum.

The first indication of the appearance of the hypophysis is met with in embryos of 2–3 mm in the shape of an angular depression from the stomodaeum [p. 52] immediately in front of the dorsal attachment of the buccopharyngeal membrane. It deepens progressively, and in the 7-mm embryo it has become a deep and wide saccular diverticulum, compressed anteroposteriorly, and opening by an aperture of equal width into the primitive mouth cavity. The earliest indication of the appearance of the posterior lobe is a slight funnel-like depression in the floor of the forebrain vesicle [FIG. 2.72] in the 9 mm embryo. At the 10–12-mm stage the recess has become deeper and more sharply marked, and its anterior wall is in close apposition with the fundus of the saccular anterior lobe. After the 12-mm stage the portion of the stomodaeal diverticulum nearest the mouth rapidly narrows and elongates to form a slender tubular stalk. By the stage of the 20-mm embryo this stalk has become interrupted, separating the embryonic anterior lobe from its original connexion with the epithelium of the mouth. Remnants of the obliterated stalk persists, not only at that stage but up to much later periods of intra-uterine life, and may sometimes be detected in the postnatal period, particularly in the pharyngeal roof at the posterior margin of the nasal septum, which marks the site of origin of Rathke's pouch. Very rarely the base of the skull remains incomplete immediately around the stem of the hypophysis, so that a **craniopharyngeal canal** may be found later in the osseous cranial base and sometimes even contains vestiges of the epithelial tissue of the stalk.

At the end of the second month the anterior lobe is a broad compressed sac, deeply notched for the reception of the posterior lobe. On each side of that notch the fundus of the sac is prolonged backwards in the form of paired hollow cornual extensions of the anterior lobe at the sides of the neck of the posterior lobe (they are already conspicuous in that situation in transverse sections as early

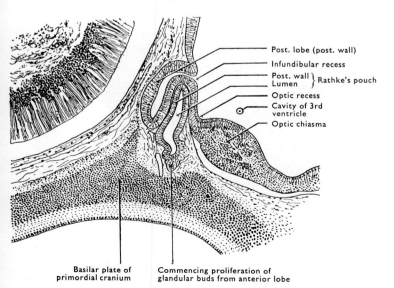

Post. lobe (post. wall)
Infundibular recess
Post. wall
Lumen } Rathke's pouch
Optic recess
Cavity of 3rd ventricle
Optic chiasma

Basilar plate of primordial cranium

Commencing proliferation of glandular buds from anterior lobe

FIG. 9.23. Paramedian sagittal section of hypophysial region of a human embryo (20 mm CR length).

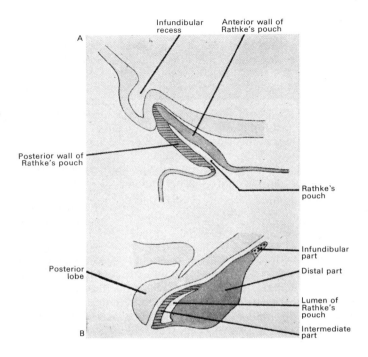

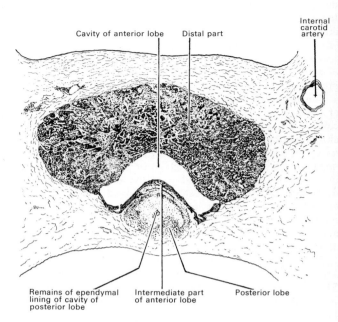

FIG. 9.25. Horizontal section of hypophysial region of a human foetus (71 mm CR length).

FIG. 9.24. Diagrammatic median section of hypophysis in early stages of development, showing A, approximation of the infundibular recess to Rathke's pouch, and B, the origin of the various parts of the organ. The posterior wall of Rathke's pouch develops into the intermediate part of the hypophysis, its anterior wall into the distal and infundibular parts.

as the 14-mm stage). At the 2-month period the original anterior surface of the sac has become secondarily cupped, its concavity now looking towards the floor of the brain in front of the infundibular region. The concavity is partially divided by a prominence in the middle into two fossae which are occupied by tissue continuous with the surrounding mesoderm. These fossae are presently invaded by the proliferating glandular cell cords of the anterior wall of the sac during the process of formation of the distal part of the organ. The bilateral pockets of areolar tissue, representing these mesodermal accumulations, were noted on page 604 in the course of the description of the distal part of the adult gland.

The recurved margin of the anterior lobe extends forwards and upwards towards the brain and eventually may reach the tuberal region of the floor of the diencephalon, where it spreads out to a greater or lesser extent as the infundibular part.

Up to about the 20-mm stage the wall of the anterior lobe preserves its simple epithelial character. Already, however, it has begun to show indications of a process of proliferative budding [FIG. 9.23] which intermingles the cell cords with highly vascular mesenchyme which forms the sinusoidal stroma of the fully developed distal part. In a foetus of the third month [FIG. 9.25] the epithelial posterior wall of the original Rathke's pouch is closely applied to the anterior surface of the posterior lobe and shows very little, if any, proliferative increase in thickness. It is this epithelial lamina which forms the intermediate part of the anterior lobe. In front of it the lumen of the anterior lobe of the hypophysis is still quite roomy. Later, it undergoes reduction and at least partial obliteration, the intraglandular cleft or clefts which represent it becoming occupied by globules and irregular masses of colloid. The lumen shown in FIGURE 9.25 does not, however, represent the whole of the earlier cavity of the anterior lobe, for portions of the sac become involved in the proliferative activity that produces the solid tissue of the anterior lobe.

During the course of development from the 20-mm stage onwards the cavity of the posterior lobe and infundibulum tends to be obliterated in the human hypophysis, and they become solid except in the root of the stalk which is occupied by the infundibular recess of the third ventricle.

In the later stages of antenatal growth there is a progressive invasion of the posterior lobe by groups of cells and by individual basophil cells that belong to the intermediate part of the anterior lobe, i.e. to the posterior wall of the original buccal diverticulum. These cells tend to undergo atrophy in the adult, but may persist throughout life. The basophil cells give rise to colloid and the hyaline bodies of Herring (1908) may represent the products of secretion of

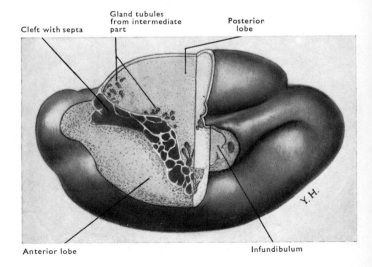

FIG. 9.26. Hypophysis of a child, aged 1 year, with a portion removed to show the stalk and glandular tubules projecting from intermediate part of anterior lobe into posterior lobe. (Lewis and Lee 1927.)

the supraoptic nucleus. Replacement of worn-out cells is apparently not a feature of the adult hypophysis, for cell division and degenerating cells are seldom seen. The intraglandular cleft, commonly filled with colloid in the child, becomes obliterated in many adults, though in others it persists as a space filled with colloid.

In the 20-mm embryo the two lobes of the hypophysis are surrounded by mesenchyme, except where the infundibulum remains connected to the brain. This mesenchyme eventually differentiates into a lamina of dura mater which is firmly united both to the hypophysis and to the hypophysial fossa [FIG. 9.19].

The pineal body

The pineal body is a small oval or cone-shaped structure, attached by a short, hollow stalk to the hinder end of the roof of the diencephalon, between the habenular and posterior commissures [FIGS. 9.27 and 10.70]. It is moderately firm in consistence, reddish-brown in colour, and measures 5–10 mm anteroposteriorly and 3–7 mm in its other diameters. It weighs about 0.2 g. Much discussion has centred around its endocrine function, and it contains several pharmacologically active indole hormones (Altschule 1975). One of these hormones has an inhibitory effect on the gonads. It commonly becomes calcified in middle age [see below].

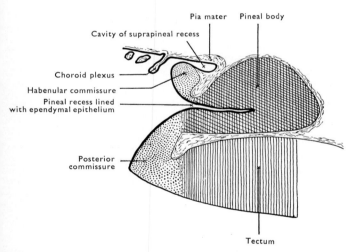

FIG. 9.27. Diagram of median sagittal section through pineal body and its immediate attachments

Position

It is covered with pia mater, and it lies in the sulcus between the superior colliculi. It is below the splenium of the corpus callosum, from which it is separated by the great cerebral vein. Its posterior end is free, and is a short distance in front of the vermis of the cerebellum and the free edge of the tentorium [FIG. 10.70].

The cavity of the stalk is the pineal recess of the third ventricle. On the dorsal lip of this recess is the habenular commissure. Immediately above the habenular commissure and the stalk of the pineal body is another and larger recess of the third ventricle called the suprapineal recess.

Structure

The pineal body consists mainly of a mass of distinctive cells (the pineal parenchyma) interspersed with neuroglial cells. This tissue becomes divided into lobules by the ingrowth of septa of fibrous tissue from the pia mater—a process which takes place mostly

during childhood. Vascularized fibrous tissue invades the organ during early intra-uterine life.

The characteristic parenchyma cells have long processes which end in bulbous extremities, some of which are on the surfaces of the lobules next to the intervening fibrous septa, and others within the lobule on the fibrous tissue which encloses the blood vessels. The body of each cell is rounded and has a large round nucleus and granular cytoplasm. They offer a marked contrast with nerve cells—which are absent from the pineal body except close to the habenular commissure. The granules of the cytoplasm have been claimed as being secretory in character, but they show no variation in number at different ages, nor does the pineal body display any differences in appearance under different physiological conditions. There is no evidence of histological change at puberty. The numerous, non-myelinated nerves which are found in the pineal body arise in the superior cervical sympathetic ganglion and appear to be concerned with the activity of its cells as well as being distributed to its vessels. Calcareous deposits usually make their appearance in the organ from adolescence onwards, and the pineal body thus becomes visible in radiographs.

Development

The pineal body develops as a pouch-like evagination of the hinder part of the roof of the diencephalon, separated from the midbrain by the posterior commissure only. Its epithelial wall becomes thickened, with restriction of the lumen to the stalk of attachment. Later, this epithelial mass is invaded by cellular sprouts from the ependyma around the cavity; these form the characteristic pineal parenchyma. Blood vessels also invade the organ. The development of the septa takes place actively in early postnatal life.

References

ALTSCHULE, M. D. (1975). *Frontiers of pineal physiology*. London.
ASTWOOD, E. B. (1968–76). *Recent Prog. Horm. Res.* **24–32**.
BAKER, B. L. (1974). Functional cytology of the hypophysial pars distalis and pars intermedia. In *Handbook of physiology*, Section 7, Vol. IV, p. 45. American Physiological Society, Washington.
BARR, H. S. (1954). Foetal cortex of the adrenal glands. *Lancet* **i**, 670.
BEJDL, W. and POLITZER, G. (1953). Über die Frühentwicklung des telobranchialen Körpers beim Menschen. *Z. Anat. Entwickl.-Gesch.* **117**, 136.
BOYD, J. D. (1937). The development of the human carotid body. *Contrib. Embryol. Carneg. Instn* (No. 152) **26**, 1.
—— (1950). Development of the thyroid and parathyroid glands and the thymus. *Ann. roy. Coll. Surg. Engl.* **7**, 455.
—— (1960). Origin, development and distribution of chromaffin cells. In Ciba Foundation Symposium on *Adrenergic mechanisms*, London.
COWDRY, E. V. (1932). *Special cytology. The form and functions of the cell in health and disease*, 2nd edn, Vol. 2. New York.
—— (1944). *A textbook of histology. Functional significance of cells and intercellular substances*, 3rd edn. London.
COUPLAND, R. E. (1952). The prenatal development of the abdominal para-aortic bodies in man. *J. Anat. (Lond.)* **86**, 357.
—— (1954). Post-natal fate of the abdominal para-aortic bodies in man. *J. Anat. (Lond.)* **88**, 455.
CROWDER, R. E. (1957). The development of the adrenal gland in man, with special reference to origin and ultimate location of cell types and evidence in favour of the 'cell migration' theory. *Contrib. Embryol. Carneg. Instn* **36**, 195.
DE ROBERTIS, E. (1942). Intracellular colloid in the initial stages of thyroid activation. *Anat. Rec.* **84**, 125.
EULER, U. S. VON (1955). Noradrenaline in hypotensive states and shock. *Lancet* **ii**, 151.
GILMOUR, J. R. (1938). The gross anatomy of the parathyroid glands. *J. Path. Bact.* **46**, 133.

GOLDSTEIN, A. L., ASANUMA, Y., and WHITE, A. (1970). The thymus as an endocrine gland: properties of thymosin, a new thymus hormone. *Recent Prog. Horm. Res.* **26**, 505.

GREEP, R. O. (1977–8). *Recent prog. Horm. Res.* **33–4**.

HARRISON, R. G. (1951). A comparative study of the vascularization of the adrenal gland in the rabbit, rat and cat. *J. Anat. (Lond.)* **85**, 12.

—— and HOEY, M. J. (1960). *The adrenal circulation.* Oxford.

HERRING, P. T. (1908). The histological appearances of the mammalian pituitary body. *Q. J. exp. Physiol.* **1**, 121.

HIRSCH, P. F. and MUNSON, P. L. (1969). Thyrocalcitonin. *Physiol. Rev.* **49**, 548.

HOLLINSHEAD, W. H. (1936). The innervation of the adrenal glands. *J. comp. Neurol.* **64**, 449.

KINGSBURY, B. F. (1939). The question of a lateral thyroid in mammals with special reference to man. *Am. J. Anat.* **65**, 333.

LEWIS, D. and LEE, F. C. (1927). On the glandular elements in the posterior lobe of the human hypophysis. *Bull. Johns Hopk. Hosp.* **41**, 241.

METCALF, D. (1966). *The thymus.* Berlin.

MILLER, J. F. A. P. and OSABA, D. (1967). Current concepts of the immunological function of the thymus. *Physiol. Rev.* **47**, 437.

MORGAN, J. R. E. (1936). The parathyroid glands. I. A study of the normal gland. *Arch. Path.* **21**, 10.

NOLAN, L. E. (1938). Variations in the size, weight and histologic structure of the thyroid gland. *Arch. Path.* **25**, 1.

NORRIS, E. H. (1937). The parathyroid glands and the lateral thyroid in Man: their morphogenesis, histogenesis, topographic anatomy and prenatal growth. *Contrib. Embryol. Carneg. Instn* (No. 159) **26**, 247.

—— (1938). The morphogenesis and histogenesis of the thymus gland in Man: in which the origin of the Hassall's corpuscles of the human thymus is discovered. *Contrib. Embryol. Carneg. Instn* (No. 166) **27**, 191.

PINCUS, G. (1947–67). *Recent Prog. Horm. Res.* **1–23**.

—— and THIMANN, K. V. (1948–64). *The hormones* **1–5**.

REICHSTEIN, T. and SHOPPEE, C. W. (1943). The hormones of the adrenal cortex. *Vitam. Horm.* **1**, 345.

SCHARRER, E. and SCHARRER, B. (1954). Hormones produced by neurosecretory cells. *Recent Prog. Horm. Res.* **10**, 183.

SELYE, H. (1947). *Text-book of endocrinology.* Montreal.

—— (1950). *Stress.* Montreal.

SHEEHAN, H. L. (1959). La glande pituitaire dans la cirrhose du foie (modifications histologiques). *C.r. Soc. franç. Gynéc.* **29**, 211.

STANFIELD, J. P. (1960). The blood supply of the human pituitary gland. *J. Anat. (Lond.)* **94**, 257.

STUTINSKY, F. (1967). *Neurosecretion.* Berlin.

SWINYARD, C. A. (1937). The innervation of the suprarenal glands. *Anat. Rec.* **68**, 417.

WELLER, G. L. (1933). Development of the thyroid, parathyroid and thymus glands in Man. *Contrib. Embryol. Carneg. Instn* (No. 141), **24**, 93.

WILLIAMS, R. G. (1941). Studies of vacuoles in the colloid of thyroid follicles in living mice. *Anat. Rec.* **79**, 263.

WISLOCKI, G. B. (1937). The meningeal relations of the hypophysis cerebri. I. The relations in adult mammals. *Anat. Rec.* **67**, 273.

WISLOCKI, G. B. and KING, L. S. (1936). The permeability of the hypophysis and hypothalamus to vital dyes, with a study of the hypophyseal vascular supply. *Am. J. Anat.* **58**, 421.

YOUNG, M. and TURNBULL, H. M. (1931). An analysis of the data collected by the status lymphaticus investigation committee. *J. Path. Bact.* **34**, 213.

ZUCKERKANDL, E. (1912). The development of the chromaffin organs and of the suprarenal glands. In Keibel and Mall's *Manual of human embryology*, Vol. II, Chap. 15. Philadelphia.

10 The central nervous system

G. J. ROMANES

INTRODUCTION

THE nervous system is concerned with controlling and integrating the activity of the various parts of the body, and provides a mechanism by which the individual can react to a changing external environment, while maintaining an internal environment delicately balanced within the narrow limits which are necessary for survival.

The mechanisms by which the nervous system controls the tissues of the body are at first sight manifold, but in essence consist of the release of substances which alter the activity of the target organs.

This is achieved:

1. By the release of minute quantities of such substances from the terminations of nerve cells on the tissues in the presence of a mechanism (e.g. enzyme) which rapidly inactivates the substance, so that the tissue response is localized and transitory. This is the mechanism by which muscle and some gland cells are stimulated and by which many nerve cells stimulate or inhibit other nerve cells at synapses [p. 614].

2. By the release of substances into the bloodstream, either from the nervous system itself, or from ductless glands directly under its control. These substances, known as hormones, are widely disseminated throughout the body and have more general effects which persist often for long periods. They may produce their effects directly on the tissues or through the intermediary of other glands which, in their turn, secrete varying concentrations of their hormones into the blood stream.

Nerve cells are therefore basically secretory in character, and though different substances are secreted by different nerve cells the mechanism is the same throughout.

The ability of an organism to react to change presupposes the presence of a mechanism which can detect such change and set in motion the appropriate response. This requires a sensory system which is able to record the wide range of changes imposed on the animal from without, and the most minute changes in its internal processes, whether these are alterations of its chemical or physical structure. Thus control of gases and ions in the blood is dependent on a method for measuring them, as indeed the control of movement is dependent on a mechanism for measuring the progress of the movement from moment to moment. Such a sensory or receptive system must be able to differentiate between an enormous number of different circumstances and be capable of activating a mechanism which can integrate all the information showered upon it by the sensory system. Only in this way can the nervous system produce a response which is appropriate to all the conditions prevailing at the time. Thus manual dexterity in Man is not simply the result of the development of a hand with a wide range of movement and a complex nervous mechanism to control it, but depends on the fact that the hand is also an important sensory organ. Loss of this sensation removes the ability to use the hand properly, even though the muscles, tendons, and joints which are essential to the production of the fine movements are in no way impaired. Similarly the full utilization of the hand in highly skilled activities requires that there should be adequate integration of the sensory information which records the position, speed, force, and direction of movements of the fingers with the visual or auditory impressions which these movements produce. Loss of part or all of this sensory information or failure to integrate it in the nervous system can disrupt the whole pattern of activity.

In every activity involving the nervous system, however complex or simple, there are therefore three elements involved—the sensory or receptive, the integrative, and the effector or motor. Of these three it is the integrative element which undergoes the greatest degree of development in higher animals to become the major part of the nervous system.

It is usual, for descriptive purposes, to divide the nervous system into central and peripheral parts. Nevertheless such a division is quite arbitrary, since the nerves which comprise the peripheral part, and the brain and spinal medulla which make up the central nervous system, are both essential parts of a functioning unit and cannot be clearly separated from each other on anatomical grounds.

A detailed knowledge of the structural organization of the nervous system is essential to an understanding of its function; yet the information which we possess is too scanty to allow of a full interpretation of the significance even of those structural details which are well known.

PHYLOGENY

The central nervous system in Man, and indeed in vertebrates in general, is so complex that a brief survey of the early stages of its evolutionary development may help to clarify the basis on which it is constructed.

In simple invertebrates, the nervous system consists of a network of interconnected, branching cells which lies between the external and internal epithelia and spreads thoughout the body. Such a network [FIG 10.1B] is in contact with sensory cells which form part of the epithelial surfaces, and is capable of receiving nervous impulses from them and transmitting these to activate the muscles. Such a diffuse nervous system is admirably suited to the protective closure of a sea-anemone and to the rhythmical activity by which a jelly-fish progresses, but it can also produce localized reactions such as the movement of a single tentacle. Indeed, in all animals including the protozoa, the response obtained is related to the nature, intensity, and duration of the sensory stimulus. Thus even in this simple nervous system the sensory cells are able to discriminate between stimuli, and some degree of interaction (integration) of different and coincident stimuli can take place.

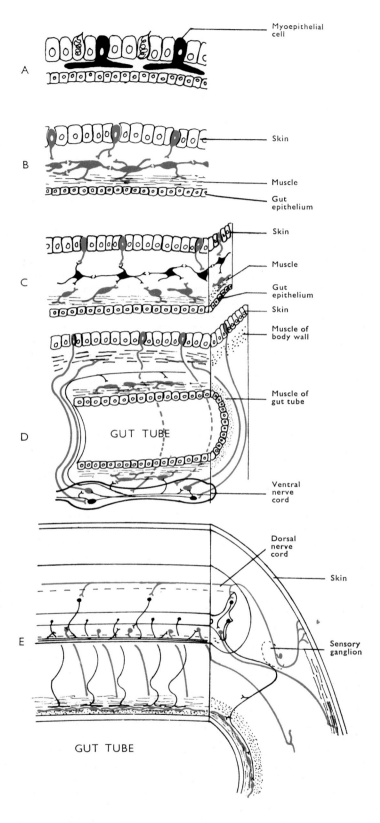

This diffuse type of nervous system is found in coelenterates. Its first significant modification appears in the flatworms. Here movement of the animal in one direction causes the development of a leading or head end which receives a disproportionate number of sensory stimuli. This is associated with an aggregation of nerve cells at the head end—the head ganglion—with the rest of the net streaming tailwards from it as a series of interconnected, longitudinal strands. The head ganglion is further increased in size by the development of eye spots and gravity measuring organs (statocysts) in close association with it. The preponderance of sensory input gives the head ganglion a dominant role in a net which is orientated longitudinally, and through which impulses continue to travel in all directions and interact with each other.

In all higher animals, the nerve net is concentrated into a single longitudinal cord extending tailwards from the head ganglion. This centralization of the nervous system permits more rapid interaction between nerve cells which are closely packed and dissociates the majority of the nerve cells from the sensory receptors and the effector organs (e.g. the muscles) imposing on them a longitudinal orientation in the cord so that they conduct predominantly headwards and tailwards. The dissociation of the central nervous system from the sensory receptors and the effector organs necessitates the development of cells with long processes to bridge the gap. In many invertebrates the sensory nerve cell bodies remain close to the receptor organs and grow a long process (axon) into the central nervous system [FIG. 10.1D]. In higher forms the cell bodies migrate towards the central nervous system and form compact groups (sensory ganglia) adjacent to it with each cell body sending a long peripheral process to the sensory receptor and a central process entering the central nervous system [FIG 10.1E]. The cell bodies of the motor cells which innervate the muscles of the body wall and limbs lie inside the central nervous system and send long peripheral processes to these muscles. Together with the processes of the sensory cells they form a large part of the nerves of the peripheral nervous system [see below].

The majority of the nerve cells in the central nervous system are not directly connected either with sensory receptors or with effector organs. They transmit impulses from the central processes of the sensory cells to the motor cells either directly or through a complex of other cells of the same type. They are, therefore, intercalated or connector nerve cells. They may send impulses to adjacent motor cells or transmit them longitudinally through the net of cells, which forms the cord and head ganglion, to motor cells at more headward or tailward levels or to interact with sensory impulses entering at the other levels. Some intercalated neurons, especially in the head ganglion, develop long processes which bypass the chains of short neurons to make direct contact with other intercalated neurons or motor cells at a distance. Thus a dual mechanism of long, rapidly conducting pathways is superimposed on the original, slowly conducting, multineuronal pathways of the net—an arrangement which permits more rapid association of nerve impulses arising in widely separated parts of the body.

The process of centralization and the replacement of a diffuse net of connections with long specific pathways is never complete. Thus, on the motor side, the cells of that part of the nervous system (autonomic part) which controls the involuntary organs of the body, e.g. the gut tube and heart, lie outwith the central nervous system and form nerve cell nets or plexuses on or near the organs which they innervate. These neurons are controlled by the central nervous system through the processes of intercalated neurons which pass from the central nervous system to the autonomic plexuses. On the sensory side, the olfactory cells retain a primitive position buried in the nasal mucous membrane even in Man. Within the central nervous system, long specific pathways exist side by side with complex multineuronal connections which are poorly understood because they are very difficult to analyse anatomically.

FIG. 10.1. A schematic diagram to indicate some of the features of the phylogeny of the nervous system.

A, hydra; B, coelenterate; C, flatworm; D, higher invertebrate, e.g. earthworm; E, vertebrate.

In A, the solid black cells are myoepithelial cells. In the remainder: blue, sensory cells; black, connector or internuncial cells; red, motor cells.

While it is true that the gross appearances and position of the vertebrate and invertebrate nervous systems are very different, nevertheless the basic structural arrangements are similar in both. The vertebrates show a great increase in the number of nerve cells at all parts of the nervous system compared with the invertebrates. This increase is mainly at the cephalic extremity which is influenced by the volume of sensory information entering it from the snout, eye, ear, and olfactory apparatus as well as from the heart, great vessels, and gills. In this way the cephalic part of the vertebrate nervous system becomes the largest correlating centre, or **brain**, whose activity is integrated with that of more caudal levels by the longitudinally directed intercalated cells. The processes of some of these intercalated cells carry information from the body to the brain while others give it a measure of control over local mechanisms in the caudal part of the nervous system.

Once such a pattern has been established, the increased sensory experience which arises from the assumption of a terrestrial existence is associated with further increases in the size and complexity of the brain so that more complex behavioural patterns are possible. Thus in Man the assumption of the erect posture and the freeing of the upper limb from its supporting functions, with the consequent development of the hand as a potent new sensory organ, may well have played a significant part in the growth of the human nervous system to its present complexity.

THE CELLULAR ELEMENTS OF THE NERVOUS SYSTEM

The central nervous system consists of nerve cells or **neurons**, and **neuroglia**. The neurons are the active elements of the nervous system: the neuroglia form a cellular connective tissue which is peculiar to the central nervous system and is intimately related to the nervous elements both anatomically and functionally (Hyden 1967). The neuroglia cover the nerve cells and isolate them from each other, from the surrounding tissues, and from the blood vessels which enter the nervous system. They also form the framework of the nervous system and carry out phagocytic activities within it.

Neurons and neuroglial cells are both characterized by having **cytoplasmic processes** in the form of fine filaments which extend from the **cell body** with its contained nucleus [FIGS. 10.3 and 10.13]. In neurons these processes are often long enough to allow them to propagate nervous impulses at great speed over relatively large distances.

The prodigious length of the processes (axons) of some neurons, which may exceed 1 metre, and their close association with neuroglial cells in the central nervous system and another type of sheath cell, **cells of Schwann**, in the peripheral nerves, led some investigators to assume that each axon was formed not by one cell but by many. Thus these investigators considered that the nervous system was a true syncytium whose cells were in continuity with each other without intervening cell membranes. In 1891 Waldeyer enunciated the **neuron theory**, which stated that the nervous system consisted of separate cells which, though they came into contact with each other, were never in continuity. This theory stimulated one of the most bitter controversies of neurohistology, which has continued almost to the present day despite the support it received from Cajal, the greatest of all neurohistologists, and the demonstration in 1907 by R. G. Harrison that nerve cells could be grown outside the body in clotted plasma. Their processes then extended into areas devoid of cells from which they could have been formed. The solution of this controversy has had to await the high resolution of the electron microscope. This has made it possible to show that neurons and neuroglia remain separate cells however close their

association, and that individual neurons are never in cytoplasmic continuity with each other though they come into contact at specific points (synapses).

Neurons (Nerve Cells)

Neurons, like other cells, consist of a nucleus and its surrounding cytoplasm. The **nucleus** appears almost empty in stained sections. It has a large, usually single, central nucleolus and close to it a small satellite present only in the cells of female animals, the 'sex chromatin' (Barr 1960). A distinct nuclear membrane separates the nucleus from the cytoplasm which contains well-defined bodies,

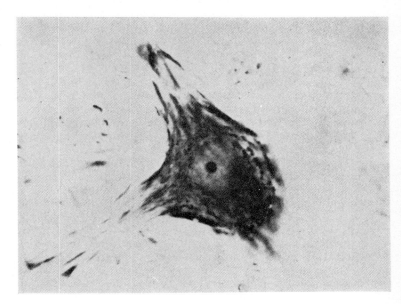

A

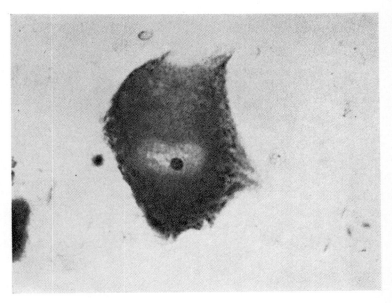

B

FIG. 10.2. Photomicrographs showing A, a normal motor nerve cell of the spinal medulla, and B, a motor nerve cell which has undergone chromatolysis two weeks after its axon has been cut.

Note that the granules of Nissl substance which occupy the cell body and extend into the dendrites in A, have disappeared or dispersed as fine particles in B, and that the latter cell is swollen.

stainable with basic aniline dyes, and known as Nissl substance after their discoverer. The Nissl substance is arranged in blocks which vary in size from cell to cell, but tend to have a similar appearance in cells of like function. Thus in the large motor neurons which supply striated muscle [FIG. 10.2A] the blocks tend to be of considerable size, while in the equally large sensory cells they are smaller and appear as a fine dust when stained for light microscopy. They may be virtually absent in some neurons or present only in the peripheral cytoplasm. Nissl substance [FIG. 1.2] consists of particles of ribose nucleic acid arranged in a manner similar to that in secretory cells and seems to be concerned with protein synthesis in the cell. It may disappear from cells which have been kept in a state of activity for a prolonged period, or subjected to injury [FIG. 10.2].

Another characteristic of nerve cells is the fibrillary structure of the cytoplasm which can be seen in cells stained with silver. Very fine fibres, neurofibrillae, appear to course through the cytoplasm and extend into the processes. Such fibrils have been described in living cells, and though fine tubules and filaments are visible under the electron microscope, it is not possible to be certain that they form continuous structures. Their function is unknown, but they have been implicated in the transport of substances within the cell, and show that there is orientation of the elements of the cytoplasm parallel to the long axis of the processes. The neurotubules are finer in axons than they are in dendrites.

The cytoplasmic processes of nerve cells are normally described as being of two types called axons and dendrites. In many cells it is possible to differentiate these from each other on histological grounds [FIGS. 10.3 and 10.5], though this is not invariably so. The terms indicate the direction in which these processes normally conduct nervous impulses. The dendrites are afferent or receptive and normally conduct towards the cell body, while the axon is efferent and conducts away from it.

The axon is usually a thin, single process of uniform calibre throughout its length. It differs from the rest of the cytoplasm in not containing Nissl substance.

The dendrites are much more variable in structure, being usually thicker and shorter than the axon. In cells with many dendrites [multipolar cells, FIG. 10.4] they are of irregular outline, branch frequently and gradually taper to their terminations, often forming a very complex arrangement in the vicinity of the cell body [FIG. 10.5]. Some cells, especially the sensory cells in the spinal and cranial

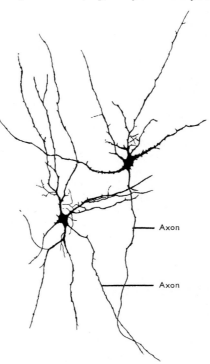

FIG. 10.4. Two multipolar nerve cells. (From a specimen prepared by the Golgi method.)

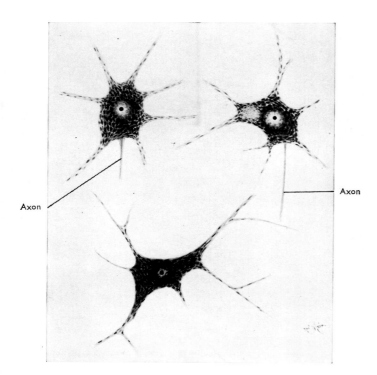

FIG. 10.3. Three nerve cells from anterior grey column of human spinal medulla, stained to show the Nissl substance. The dendrites have many more branches.

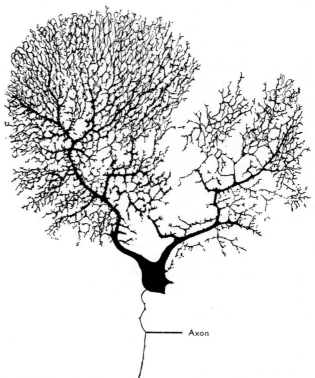

FIG. 10.5. Nerve cell from cerebellum (cell of Purkinje) showing the branching of the dendrites (Symington).

nerve ganglia, have only two processes which may either arise separately from the cell (vestibulocochlear nerve) making it **bipolar** in type, or as a single process which subsequently divides into two. These are **unipolar cells** which are normally derived from bipolar cells [FIG. 10.6]. In this case both processes have the same structure as an axon, though that which conducts towards the cell body is essentially a dendrite. The presence of many dendrites indicates that the cell receives impulses from several different sources, while single dendrites are found in those which are activated from a single source.

Every **axon** is enclosed in a **cellular sheath** and forms a **nerve fibre** with it. In peripheral nerves the sheath is composed of elongated cells each of which encloses a certain length of one or more axons, the complete sheath of any one axon being formed by a chain of these **Schwann cells**, arranged end to end along it. An almost identical arrangement exists within the central nervous system, but here the axons are ensheathed by neuroglial cells [p. 617].

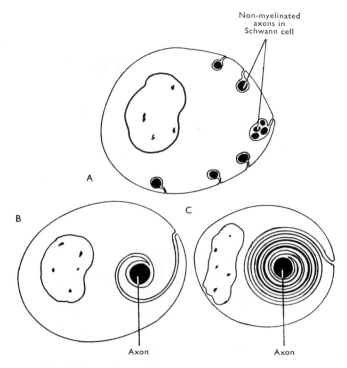

FIG. 10.7. Diagram to show the relation between peripheral axons and their covering Schwann cells.

A, non-myelinated fibres; B, developing myelin sheath; C, final stage of myelination.

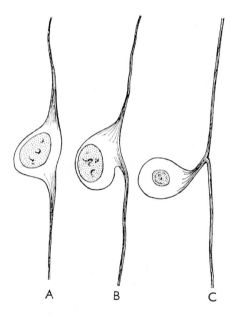

FIG. 10.6. Three stages in the development of a unipolar cell from a bipolar cell (spinal ganglion).

All nerve fibres fall into two groups on account of the arrangement of their sheaths, which are either simple or complex. Axons vary in diameter from 0.0001 mm to 0.01 mm. Those less than 0.001 mm in diameter are simply enveloped by the sheath cells. In this case there may be several such axons surrounded by each sheath cell either as one bundle (e.g. olfactory nerves) or more usually each lying separately [FIG. 10.7A]. Axons of larger diameter are enclosed singly within a chain of sheath cells each of which has its surface membrane wound spirally around its segment of the axon to form one part of a discontinuous **myelin sheath** [FIGS. 10.7B, C and 10.8]. The number of turns of the spiral which each sheath cell produces is directly related to the diameter of the axon it encloses. The myelin sheath gives the nerve fibres a white, glistening appearance, unlike the grey colour of the nerve cell body, the axon, and the sheath cells. Nerve fibres which possess a myelin sheath are said to be **myelinated**, as distinct from the finer nerve fibres which are **non-myelinated** and grey in colour.

The majority of peripheral nerves contain a mixture of myelinated and non-myelinated fibres and are consequently white. Some autonomic nerves are composed entirely of non-myelinated fibres

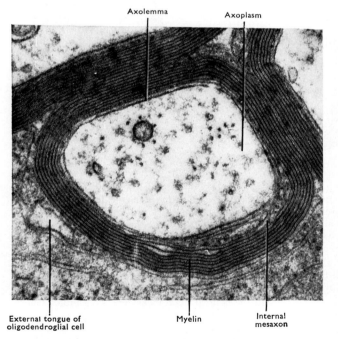

FIG. 10.8. An electron micrograph of a myelinated axon in the central nervous system. (By courtesy of Professor A. Peters.)

and are grey. In the central nervous system the same mixture of fibres forms the **white matter**, but there are also regions formed predominantly of nerve cell bodies and dendrites comprising the **grey matter**. This is pink in colour by transmitting light owing to the high concentration of cytochrome.

Nerve fibres vary in diameter. When myelinated, the thickness of the myelin sheath and the length of the axon enclosed by each sheath cell and its myelin segment are approximately in direct proportion to the diameter of the axon. Thus, as the diameter increases, the number of sheath cells per unit length of a myelinated nerve fibre decreases, hence the number of junctional regions (**nodes of Ranvier**) between adjacent myelin segments also decreases. In peripheral nerve fibres the portions of the axon between the myelin segments are covered by interdigitating processes of the adjacent Schwann cells, while in the central nervous system they are covered by neuroglial elements other than those concerned with the formation of the myelin [FIG. 10.9]. The amount of cytoplasm of the myelin-forming cell external to the myelin is much less in the central than in the peripheral nervous system.

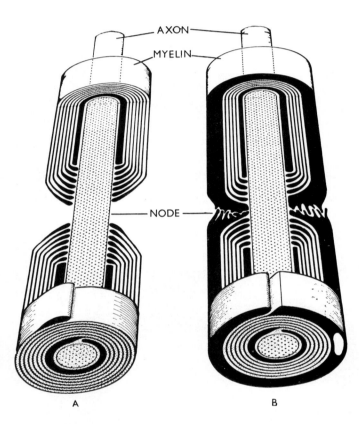

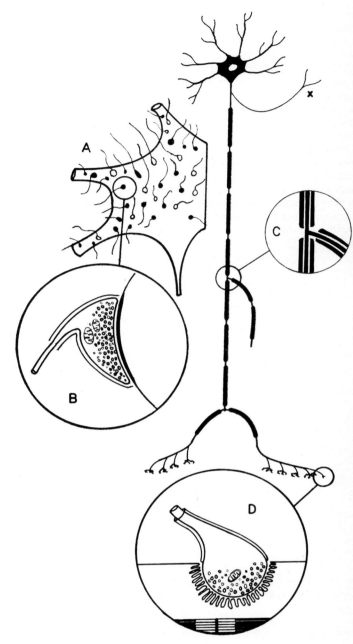

FIG. 10.9. Diagrams to show the relation between the sheath cell and axon in myelinated fibres at nodal regions.

A, in the central nervous system; B, in the peripheral nervous system. Solid black, sheath cell cytoplasm.

At the nodes, axons may give rise to branches or **collaterals** [FIG. 10.10]. It is this feature which allows a sensory nerve cell to activate many nerve cells in the central nervous system, or a motor cell to innervate 200 or more muscle fibres.

Axons and their collaterals terminate within the central nervous system by coming into contact with other nerve cells. Such **contacts**, or **synapses**, are found on the dendrites or body of the cell but only rarely on the axon of the receiving cell. The axon terminals (presynaptic part) either form expanded knobs (**boutons**) or touch the surface of the receiving cell (postsynaptic part) as they pass, or are coiled round the cell body, sometimes forming a basket-like structure. In every case the surface of the axon comes into contact with the receiving cell without any intervening sheath, though there may be dense material on the cytoplasmic surface of the postsynaptic

FIG. 10.10. Diagram to show the parts of a spinal motor neuron and its connections. x, recurrent collateral.

A, enlarged part of the cell body to show the terminations of other nerve fibres on it forming boutons.

B, drawing of electron micrograph of a bouton. Note the expanded axon termination containing mitochondria and vesicles, and the dense postsynaptic zone in the motor cell.

C, to indicate axon and myelin sheath arrangements at a branch.

D, diagram of motor ending on striated muscle. Note vesicles in the expanded axonal termination, the sheath of the axon and the complicated, folded, subneural apparatus which contains cholinesterase.

membrane and between the membranes. Each presynaptic terminal contains a large number of minute vesicles which are believed to enclose the substance responsible for transmitting the impulse across such chemical synapses [FIGS. 10.10 and 10.11]. These vesicles may be clear and spherical or flattened, or they may contain a dense central core. Each type is believed to release a different substance

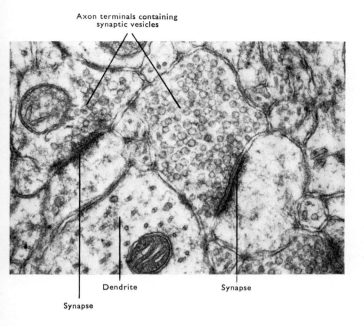

Axon terminals containing
synaptic vesicles

Dendrite   Synapse

Synapse

FIG. 10.11. An electron micrograph to show a synapse on each of two dendritic spines. Cerebral cortex. (By courtesy of Professor A. Peters.)

against the surface of the postsynaptic membrane of the cell when an impulse reaches the synapse, but each presynaptic ending tends to contain vesicles of one type only. Some of the substances (e.g. acetylcholine or certain amines) tend to stimulate the postsynaptic cell, others (e.g. glycine or gamma-aminobutyric acid) inhibit it. It is the presence of synapses on the dendrites and cell body which determines that these are the receptive parts of the neuron. An impulse generated in this region can spread over the surface of the neuron and be carried away by the axon. Thus directional conduction within the nervous system is established as a result of the anatomical organization of its cells, even though the individual cells are able to conduct in either direction. Similar vesicles are found in the terminal parts of axons which innervate muscle or gland cells and the same mechanism of chemical transmission is responsible for activating these structures [FIG. 10.10], e.g. at motor nerve endings on a muscle. In some synapses the presynaptic and postsynaptic membranes are in close apposition without vesicles or dense material. These are believed to be electrical synapses for electrotonic spread of impulses.

There is considerable evidence for the presence of a continuous flow of material in normal axons, particularly in development or regeneration. This movement is presumably responsible for carrying those substances which maintain every branch of the cell, and for bringing to the terminations the materials for chemical transmission of nerve impulses. Thus the nerve cell can be looked upon as a type of gland cell in which the duct is the axon, and whose secretions are released in minute quantities in contact with the tissues to be affected. A special example of this type of activity is found in some nerve cells, e.g. in the hypothalamus. In these cells, droplets akin to those found in gland cells are present in the cell bodies and extend down the axons to accumulate within the posterior lobe of the hypophysis. This forms a storehouse from which these secretory products (hormones) can be released into the bloodstream to produce their effects at a distance, thus involving sensitive tissues throughout the body. In keeping with the view that nerve cells are akin to gland cells is the large concentration of Nissl substance (with its contained ribose nucleic acid) which they have in their cytoplasm, indicating a considerable synthetic ability such as can be seen in the cells of the pancreas and other glands.

The significance of the centripetal flow is not clear. It may play a part in the formation of connections within the central nervous system during development, and seems to transmit virus particles from the periphery to the cell body. Both directions of flow are now used for tracing the origin, course, and distribution of axons by the introduction of appropriate substances into the cell bodies or the terminals of the axons.

DEGENERATION AND REGENERATION IN NERVE CELLS

In all cells every part of the cytoplasm is dependent for its survival on being connected, however indirectly, with the nucleus, in which certain essential substances are produced. Thus if the axon of a nerve cell is divided, the part separated from the cell body undergoes a process of degeneration which ends in complete fragmentation and loss. If the axon is myelinated, the myelin sheath also degenerates although the myelin-producing sheath cells survive, proliferate, and may once again form myelin sheaths for the new processes which grow out from the central end of the divided axon. The process of **degeneration**, first described by Waller in 1850, has allowed anatomists to follow nerve fibres through the complexities of the central nervous system, and even to demonstrate their minute terminations when they have been damaged by pathological processes or by surgical intervention in Man and in experimental animals. This has been made possible by the development of staining methods which specifically demonstrate the degenerating fibres, their myelin sheaths, or their terminals.

In addition to the degeneration of the distal part of a divided axon, certain changes may take place in the cell body from which it arises (**retrograde changes**). Thus the blocks of Nissl substance may fragment and disappear from the central cytoplasm, the cell body swells, and the nucleus on occasion is displaced to the periphery of the cell. In extreme cases there may be complete loss of Nissl substances and even cell death. This process, known as **chromatolysis** [FIG. 10.2B], occurs in many pathological conditions, but its main significance for anatomy is that it allows of the identification of the cell bodies from which divided axons have arisen. Unfortunately this reaction of the cell to injury is not invariable and, since many normal cells show a structure similar to that described above, the experimental results are liable to misinterpretation. The presence of chromatolysis in pathological conditions and the death of certain nerve cells following injury to their processes, has led to the assumption that chromatolysis invariably indicates an injured cell. However, the chromatolytic appearance of some normal nerve cells, and the fact that all nerve cells go through a phase of profound chromatolysis during their development, makes this unlikely. It has been shown, too, that motor cells of the spinal medulla which have undergone chromatolysis following injury to their axons do not succumb to the virus of poliomyelitis, as do their unchanged neighbours (Howe and Bodian 1942).

Theoretically the study of axonal degeneration and chromatolysis should give a clear picture of the origin, course, and termination of nerve fibres which have been divided for this purpose. The methods are, however, fraught with difficulties because the silver stains, which demonstrate degenerating axons and boutons, frequently stain normal structures, and the classical **Marchi** method for staining degenerating myelin not only produces many artefacts but also is incapable of showing the exact point of termination of the fibres, since all axons lose their myelin sheath some distance before ending. In addition, some nerve cells, notably in the forebrain of adult animals and almost every cell in young animals, react to injury by complete degeneration not only of the peripheral part of the axon but also of its central part and the cell body. In such cases the direction in which the fibres are passing cannot be determined by simple injury methods and this causes particular difficulties when lesions are placed in regions where fibres are passing in many different directions. The more recent methods which use axonal transport of labelled or stainable substances overcome many of the difficulties of the older methods.

In some situations, notably in the lateral geniculate body and the auditory and olfactory systems, nerve cells undergo extensive atrophy, with loss of cytoplasm and processes, when the nerve fibres leading to and synapsing with them are severed. This **transneuronal atrophy** may be much more common than has been supposed, and it cannot be assumed that because a nucleus or group of cells has decreased in size that the cells within it have sustained direct injury. The time taken for transneuronal atrophy to appear is much longer than the primary degeneration of severed axons, but in the human nervous system studies have often to be made on brains in which lesions have persisted for long periods of time. It is generally assumed that the degeneration of severed axons is complete in a few weeks at most, but evidence has been produced that it may take much longer than this (Van Crevel and Verhaart 1963).

When a peripheral nerve fibre is divided the central stump usually degenerates at least to the nearest node but subsequently begins to grow out a new process which frequently divides to give rise to a number of minute fibres which grow rapidly into the surrounding tissue. Some of these may eventually reach and re-innervate the structure which was supplied by the parent fibre, though this is unlikely unless the original lesion is close to the organ supplied or the outgrowing processes enter the peripheral part of the divided nerve and are led to their destination by the surviving Schwann cells and the connective tissue sheaths of the nerves. Thus in crushed nerves, where the axons are divided but the connective tissue sheaths may still be in continuity, the regeneration may well be very satisfactory. Where all the elements are divided the chance of recovery is much less, either because the outgrowing fibres are directed by the peripheral part of the nerve to an incorrect destination (e.g. motor nerve fibres entering a cutaneous branch) or because the growing nerve sprouts escape into the surrounding tissue, where they form a complex tangle or **neuroma** and may even turn back and grow towards the central nervous system, reaching and invading it. Thus regeneration in a divided nerve is likely to be less satisfactory when the peripheral part contains many different kinds of branches than when it is either purely cutaneous or muscular and the distribution is to muscles of similar function. When part of a peripheral nerve is removed, the chances of satisfactory regeneration are much reduced even if the ends can be brought together. This is because the plexiform nature of the bundles of nerve fibres which the nerve contains makes the proximal and distal cut surfaces very different from each other (Sunderland 1968).

Regeneration does not occur in the central nervous system, owing, it was believed, to the absence of Schwann cells. It now seems likely that the real reason is the rapid proliferation of neuroglial cells which blocks the outgrowth of axons. Certainly it is possible to grow cells of the central nervous system in tissue culture and some regeneration can be promoted in experimental animals which have been subjected to pyrogens. These raise the temperature of the animal and delay the proliferation of neuroglia.

TECHNIQUES IN THE STUDY OF THE NERVOUS SYSTEM

In spite of the difficulties described above, it is possible to obtain much information on the central nervous system by the application of a number of staining methods to the normal nervous system. A very large number of such methods exist but they fall into four main complementary groups:

1. The staining of **Nissl substance** and nuclei with basic aniline dyes. This give a clear picture of the cell bodies and the larger dendrites but no indication of axons or the finer ramifications of the dendrites.

2. **Normal myelin** may be stained with fat stains in frozen sections, e.g. with osmic acid which is reduced to a lower, black oxide of osmium by the unsaturated fats. It may also be stained by a number of methods which show the benzene insoluble fractions left behind after the preparation of sections in paraffin or celloidin. These methods show only the myelin sheath, not the axon, the cell body or its dendrites. **Degenerating myelin** is demonstrated by the Marchi method which depends on the fact that degenerating myelin will reduce osmic acid in the presence of a strong oxidizing agent, e.g. potassium dichromate, while normal myelin will not. This method has been extensively used to follow degenerating fibres, but is subject to artefact production, precludes the use of other staining methods on the same tissue, is valueless with non-myelinated fibres, and does not stain the terminal non-myelinated parts of myelinated fibres.

3. **Axons, dendrites, cell bodies, boutons, etc.**, may be stained with silver. These methods show most features of the nerve cells but are difficult to interpret because they show all the elements in the complex tangle of nerve cells and fibres which forms the central nervous system. Modifications of silver methods are used to stain degenerating axons, boutons, etc., and though they are often unreliable they allow a more detailed analysis of the central nervous system than is possible with the Marchi method.

4. The **Golgi method** is a particular silver technique which picks out and stains in their entirety a few nerve cells only. These show up as opaque structures on a clear background and demonstrate the immense complexity of the branching patterns of different nerve cells [FIG. 10.5]. Although this is a method of great importance, it is not possible to be certain which, if any, nerve cells will be stained and, like most other methods used on the nervous system, lacks a rationale which would allow of its use with experimental material.

5. The introduction into nerve cell bodies or the terminations of their axons of substances which are transmitted through the nerve cell by axonal flow (e.g. radioactive amino acids or horse-radish peroxidase) permits accurate mapping of the connections of the neurons in many parts of the nervous system of experimental animals. It is not applicable to Man where the more conventional methods have to be used following accidental or surgical interference.

6. The histochemical indentification of particular transmitter substances in nerve fibres or presynaptic endings may permit the identification of inhibitory and excitatory synapses.

These methods are mentioned here to make it clear that anatomical analysis of the nervous system is a task of great difficulty, and that those details which are available give only a sketchy indication of the complexity of this system. In fact there are very few parts of the nervous system in which the anatomy is sufficiently well known to make it possible to link structure and function beyond reasonable doubt.

THE NEUROGLIA

This is the supporting tissue of the central nervous system and consists of the cells which line the cavities of the brain and spinal medulla (the **ependyma**) and which pervade the central nervous system and separate it from the surrounding tissues (**external limiting membrane**) and from the blood vessels which enter it.

The **ependyma** is formed of columnar cells, some of which are ciliated. Processes of some of these cells pass through the central nervous system and reach the external surface at least in the early stages of development [FIG. 10.12]. This connection appears to be broken in the adult except where the ependyma comes close to the surface as in the anterior median fissure of the spinal medulla. Some ependymal cells retain processes which extend into the nervous system and may transfer substances into it from the fluid (cerebrospinal fluid) which fills the cavities of the brain (see tanycytes). Others secrete substances into the cerebrospinal fluid and the activity and structure of some of them seem to be altered by hormone levels in the blood. The role of the cerebrospinal fluid in transporting substances within the central nervous system is not yet understood. In each ventricle of the brain, one wall is formed only of ependyma which here separates the cavity from the pia mater [p. 733] investing the brain. In these regions tufts of vascular pia mater invaginate the ventricles carrying non-ciliated ependyma before them to form the **choroid plexuses**. These are secretory structures responsible for forming a large part of the cerebrospinal fluid which fills the cavities of the brain and bathes its external surface. Since the capillaries in the choroid plexuses are separated from the ventricular cavities by ependyma, this epithelium probably plays a part in the formation of the fluid.

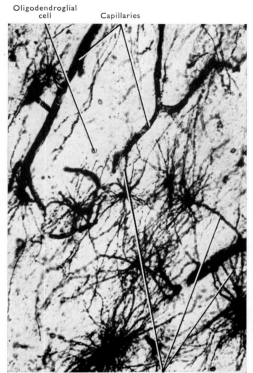

FIG. 10.13. Photomicrograph of fibrillary astrocytes. Astrocytes, capillaries, and a few oligodendroglial cells are stained.

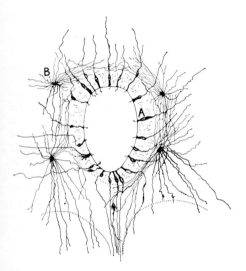

FIG. 10.12. Section through central canal of spinal medulla of human embryo, showing ependymal (A) and neuroglial (B) cells. (After v. Lenhossek.)
(Note that the dorsal (posterior) aspect is *below*.)

The **neuroglia proper** consists of astrocytes, oligodendroglia, and microglia. Of these the first two are derived from the ectoderm which forms the nervous system, while the microglia are mesodermal cells with active phagocytic properties. They become swollen and filled with the debris of degenerating nerve cells and myelin sheaths when this tissue is destroyed.

Astrocytes are found throughout the central nervous system and consist of a cell body from which processes radiate in every direction. In the grey matter the processes are relatively thick and covered with branches giving the appearance of a dense bush (**protoplasmic astrocytes**), while in the white matter the processes are thinner, longer, contain minute fibrillae, and branch much less frequently, tending to run parallel to the nerve fibres among which they lie (**fibrillary astrocytes**). Both these types have at least one larger process which extends to an adjacent blood vessel over which it expands as an end foot [FIG. 10.13]. Such end feet form a complete layer outside the endothelium of capillaries in the central nervous system. Thus the astrocytes may be responsible for transporting substances to the neurons, or, together with the endothelium, may be responsible for the 'blood–brain barrier' which prevents certain substances from entering nervous tissue which pass freely into other tissues from the circulation. Recent evidence suggests that the endothelium is the essential element in the barrier to certain substances (Reese and Karnovsky 1967) and electron microscopy indicates that the endothelial cells of the brain capillaries are tightly bound to each other in a manner not found in other tissues (Peters, Palay, and Webster 1976). On the surface of the central nervous system is a layer of neuroglia intimately adherent to the overlying pia mater and consisting of small astrocytes and the processes of other astrocytes lying in the subjacent nervous tissue. This external membrane is continued as a sleeve around the blood vessels which pierce the surface of the brain and becomes continuous with the layer of end feet around the intraneural blood vessels.

In spite of their complicated form, astrocytes are capable of rapid proliferation in regions of the nervous system which have been injured and they form the scar tissue in it.

The **oligodendroglia** [FIG. 10.14] are smaller cells with fewer processes which branch infrequently. They lie in rows among the nerve fibres and are responsible for the formation of myelin sheaths in the central nervous system (Bunge, Bunge, and Pappas 1962). They are therefore the counterparts of the Schwann cells of the peripheral nerves, but each oligodendroglial cell forms myelin segments on several different axons.

The **microglia** [FIG. 10.15], the smallest of the neuroglia, are found throughout the central nervous system often as slender, rod-like cells with a moderate number of branched processes. They are highly mobile and are easily demonstrated only during pathological processes when their macrophage activity is brought into play. It is not known if they have any function in the normal, uninjured nervous system.

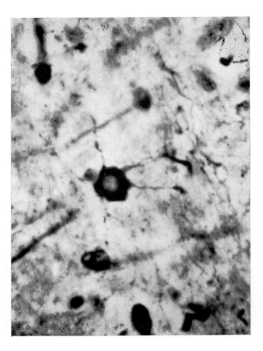

FIG. 10.14. Photomicrograph of an oligodendroglial cell.

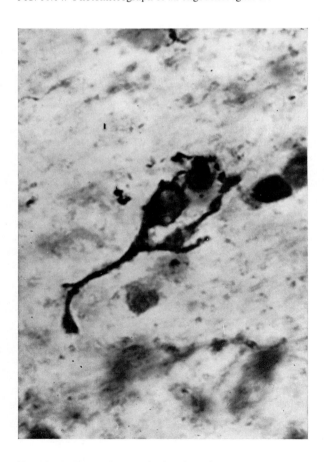

FIG. 10.15. Photomicrograph of a microglial cell.

The development and gross anatomy of the nervous system

The following section deals with the major features of the development of this system to clarify the general arrangement in the adult. Many of the details have been omitted for the sake of clarity and no attempt has been made to discuss the mechanisms responsible for the complicated morphogenetic processes which are involved, even in those regions where they are slightly understood.

The early stages of the formation of the neural tube and its differentiation into forebrain, midbrain, hindbrain, and spinal medulla have already been described [p. 50]. This structure is essentially a dorsal, longitudinal tube which shows no significant evidence of the segmentation that affects the surrounding mesoderm.

Lying along the dorsal aspect of the recently closed neural tube is a ridge of cells, the **neural crest**. These cells migrate from their original site and pass around each side of the neural tube towards the alimentary canal. They give rise to a large number of different structures (Horstadius 1950), including the sheath cells of the peripheral nerves and all the nerve cells whose cell bodies lie outside the central nervous system, with the exception of some which migrate from the neural tube. This migration of the neural crest cells [FIG. 10.16] occurs throughout the length of the neural tube and is responsible for forming all the collections of nerve cells which lie beside the hindbrain and spinal medulla and produce the **afferent (sensory) nerve fibres** in the cranial and spinal nerves. These and other peripheral clumps of nerve cells are termed **ganglia**, in contrast to similar aggregations within the central nervous system which are called **nuclei**. Other cells of the neural crest migrate further than those which form the sensory ganglia. They produce the efferent ganglia of the **autonomic nervous system** whose processes (**postganglionic nerve fibres**) grow out to innervate involuntary structures. These autonomic nerve cells are formed only from the parts of the neural crest related to the brain stem and to those regions of the spinal medulla which do not supply the neck and limbs. Hence they are only associated with some cranial and spinal nerves and these are the only nerves which subsequently transmit preganglionic nerve fibres from cells in the central nervous system to control the activity of the autonomic nerve cells. These nerves and the distribution of postganglionic nerve fibres from the corresponding autonomic nerve cells are:

1. The oculomotor, facial, glossopharyngeal, and vagus cranial nerves whose autonomic territories are the eye and the fore and mid-parts of the gut tube, together with the involuntary structures which develop from its epithelium (e.g. glands) and from its mesoderm (e.g. heart and great vessels). The second, third, and fourth sacral nerves which have a similar autonomic distribution to the hindgut and its derivatives. This **cranio-sacral part** of the autonomic nervous system with its limited distribution to the eye and to the gut tube and associated structures constitutes the **parasympathetic part**. It is defined by virtue of the emergence of its preganglionic nerve fibres through the above cranial and sacral nerves, and its ganglion cells lie in or near the organs which they innervate.

2. The thoracic and upper two or three lumbar spinal nerves. The ganglia of this **thoraco-lumbar (sympathetic) part** of the autonomic nervous system spread throughout the body and innervate all the involuntary structures in it, including those innervated by the parasympathetic part. Yet the preganglionic nerve fibres of this part only emerge through these spinal nerves—a feature which defines this part of the autonomic nervous system. The ganglia of this part tend to lie at a distance from the structures which they innervate, usually in association with the blood vessels to these organs. The postganglionic nerve fibres are distributed to the organs along these

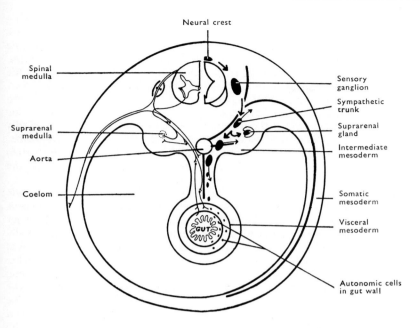

FIG. 10.16. Diagrammatic transverse section through the trunk of an embryo to show, on the right, the route of migration of neural crest cells (heavy arrows), the peripheral nerve cells derived from it (solid black bodies), the direction of growth of their postganglionic fibres (thin arrows), and their relation to the primary branches of the aorta (solid black lines).

On the left, the distribution of pre- and postganglionic fibres is indicated.

blood vessels and along any convenient nerves. Thus the sympathetic ganglia can be divided into three more or less separate groups, each associated with one of the primary branches of the dorsal aorta which supply the three primary divisions of the mesoderm—the mesoderm of the body wall; the intermediate or urogenital mesoderm; the mesoderm of the gut tube [FIG. 2.54B and 10.16]. However, these ganglia are formed from a mass of migrating cells and so are merely local concentrations within a scattering of cells rather than clearly separate elements.

The ganglia concerned with the supply of involuntary structures in the body wall form the **sympathetic trunk**. At first this is limited to the thoracolumbar region (T.1–L.2), but later extends headwards and tailwards throughout the length of the body. Thus the cells forming it are able to send processes along all the body wall arteries and into every spinal (**grey rami communicantes**) or cranial nerve for distribution through them. The remaining ganglia of the sympathetic part supply the viscera and are therefore the visceral or **splanchnic ganglia**.

The **neural tube** consists initially of a thick layer of ectodermal cells or ependyma and is divided into quadrants by the thin, median **roof** and **floor plates** and a groove on each lateral wall, the **sulcus limitans**, which separates an **alar lamina** dorsally from a **basal lamina** ventrally. The ependyma of the basal lamina is mainly concerned with the formation of the efferent or motor cells, while the alar lamina gives rise to the receptive cells. Shortly after the closure of the neural tube, the ependyma of the basal lamina proliferates and produces, on its anterolateral aspect, a mass of daughter cells which forms a second or mantle layer covering it. This **mantle layer**, which is subsequently formed from the alar lamina also, becomes the part of the neural tube composed mostly of cell bodies, the **grey matter**, as distinct from the nerve fibres which form on its external surface and later, developing myelin sheaths, become white in colour and constitute the **white matter**.

THE SPINAL MEDULLA (SPINAL CORD)

In that part of the neural tube which forms the spinal medulla, many of the cells in the mantle layer of the basal lamina begin to spin out

processes, or axons, which pierce the surface of the neural tube over a narrow longitudinal strip and pass towards the surrounding tissues [FIG. 10.17]. These **ventral rootlets** gather together in bundles and pass out between the developing vertebrae, thus forming the ventral roots of individual spinal nerves. The ventral rootlets, which combine to form the efferent fibres of a single spinal nerve, arise from a length of the spinal medulla which also gives attachment to the corresponding dorsal root fibres which form its sensory component [FIG.11.31]. This is the part of the spinal medulla supplying one segment of the body and is known as a **segment of the spinal medulla**, even though the nervous system shows no definite segmentation.

The **ventral root** fibres are of two distinct types and are formed separately from neurons in the dorsal and ventral parts of the basal lamina. Those from the ventral part pass out into the body wall or limb and supply striated muscles in these situations (**somatic motor**), while those from the dorsal part follow the general line of migration of the neural crest cells to end in one or more of the autonomic ganglia derived from the corresponding part of the neural crest. Thus they form a connection between the central nervous system and the peripherally situated autonomic motor ganglia, and constitute the **preganglionic** or **visceral efferent fibres**. The cells giving origin to these fibres are found only in those parts of the neural tube where the neural crest contributes significantly to the autonomic nervous system, and it is therefore only through the nerves in these regions that preganglionic nerve fibres emerge from the neural tube (see above). In the fully developed nervous system the preganglionic fibres are of smaller calibre (not more than 3 μm) than most of the somatic motor fibres but are also myelinated and therefore white in colour. In the thoracolumbar region they form the **white rami communicantes** of the sympathetic trunk.

In following the line of migration of the neural crest cells, some of the preganglionic fibres of the sympathetic part end in the sympathetic trunk, spreading headwards and tailwards within that trunk so as to reach its extremities. Other fibres pass through the trunk to the splanchnic ganglia, thus forming the **splanchnic nerves**. In the parasympathetic part there is nothing equivalent to the sympathetic trunk and no supply to the body wall, only the splanchnic type of preganglionic nerve bundles being present.

While the motor cells are developing, each of the cells in the sensory ganglia (spinal and cranial) grows out two processes in

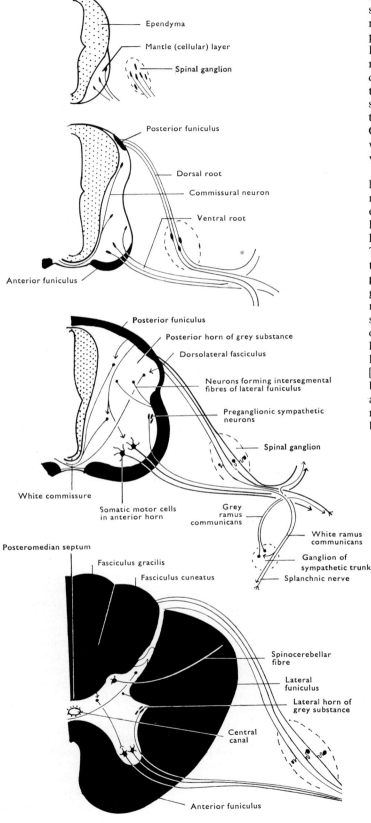

opposite directions. Each peripheral process is destined to reach a sensory ending either in the body wall, by passing with the somatic motor cells, or in relation to the viscera, by coursing with the preganglionic fibres through the ganglia of the autonomic system. Each central process is directed towards the alar lamina of the neural tube to form part of the **dorsal root of a spinal nerve** or part of a cranial nerve [FIG. 10.47]. All the sensory cells of the body, with the exception of the olfactory and visual systems and some of the sensory cells of the trigeminal nerve [p. 650], lie in ganglia of this type and all send nerve fibres into the alar lamina of the neural tube. Originally bipolar, these cells become unipolar in type [FIG. 10.6] with the outstanding exception of the ganglion cells of the vestibulocochlear nerve which retain their bipolar form.

The ependyma of the alar lamina proliferates to form a mantle layer of receptive cells, as the central processes of the sensory cells run on to its surface in the form of linear row of rootlets. Without entering the mantle layer these axons divide and run longitudinally headwards and tailwards forming, in aggregate, a third or marginal layer, composed of nerve fibres, on the dorsal surface of the tube. These fibres form a bundle on each side, extending from the midline to the point of entry of the dorsal rootlets, and constitute the **posterior funiculi** of the adult spinal medulla [FIG. 10.17]. The greater part of this mass consists of the thicker fibres in the dorsal roots. Many of them extend upwards throughout the length of the spinal medulla to enter the caudal part of the brain, the medulla oblongata, and are so arranged that those fibres entering over the lowest dorsal roots lie most medially while fibres from successively higher roots are laid down progressively in a more lateral position [FIG. 10.18]. In the midline, the fibres of the two sides are separated by a **posterior median septum** of glial tissue which meets the surface at the **posterior median sulcus**. In the cephalic half of the spinal medulla, the bundle on each side is further subdivided by a similar but incomplete **posterior intermediate septum** into a thin, medial,

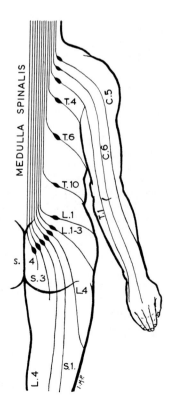

FIG. 10.17. Diagram to show the formation of the spinal medulla, its white (solid black) and grey (white) matter, and the general arrangement of the processes of the sensory, motor (somatic and visceral), and internuncial cells.

FIG. 10.18. Diagram to show the manner in which fibres of dorsal nerve roots enter and ascend in the posterior funiculus.

fasciculus gracilis carrying dorsal root fibres from the lower half of the body, and a thicker, lateral, wedge-shaped **fasciculus cuneatus**, containing fibres which enter the spinal medulla through the upper thoracic and cervical dorsal roots [FIG. 10.21].

At the lateral edge of each posterior funiculus, pierced by the entering dorsal rootlets, is a smaller bundle in which the thinner dorsal root fibres run a short course of a few segments only. This is the **dorsolateral fasciculus** which remains of approximately the same size throughout its length, since the fibres within it end as rapidly as others enter it. The remainder of the posterior funiculus steadily increases in size as it is followed headwards, owing to the continued addition of fibres to its lateral side.

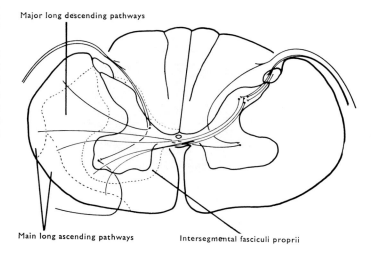

FIG. 10.20. Diagrammatic transverse section through the spinal medulla to show the position of the fasciculi proprii, and the general arrangement of the long ascending and descending pathways, in the anterior and lateral funiculi, formed by internuncial neurons.

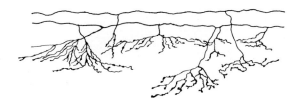

FIG. 10.19. Drawing, after Cajal, of a few of the collaterals of longitudinal fibres in the posterior funiculi of the spinal medulla.

These groups of longitudinal fibres subsequently give off large numbers of branches or collaterals into the mantle layer [FIG. 10.19] to make contact with its cells. The longitudinal course of the fibres allows each of them to make a variety of different connections throughout a variable length of the spinal medulla and even to send branches as far ventrally as the motor cells [FIG. 10.17].

The **receptive cells of the alar lamina**, together with similar cells of the basal lamina which take no part in the formation of the ventral root, therefore form a link between the incoming sensory fibres on the dorsal surface and the cells giving rise to the efferent fibres of the ventral roots. They are therefore known as **connector cells** or **interneurons** or **internuncial cells**, and the processes which they develop may link them either with other interneurons, or with efferent cells at the same level or, by extension along the spinal medulla, with similar cells at other levels of the nervous system, including the brain. The connections which these interneurons produce are by no means simple but usually consist of complex chains of cells, often involving several pathways through which different sensory stimuli entering over several dorsal roots may interact with each other and play their part in producing the final reaction. To some extent the complexity of the response to any stimulus is related to the number of interneurons involved and the complexity of their connections, and hence, in part, to the volume of active nervous tissue.

The **interneurons** [FIG. 10.17] in the alar and basal laminae send their processes to other levels mainly on the surface of the spinal medulla. Some pass to the same side, others cross to the opposite side forming the **white commissure** (crossing) ventral to the cavity of the neural tube. All split into ascending and descending branches which form a **marginal layer** of fibres covering the mantle layer (grey matter). Branches (collaterals) of these fibres re-enter the grey matter at various levels so that they form **intersegmental tracts** (**fasciculi proprii**) of the spinal medulla [FIG. 10.20].

At a later stage, further fibres are added to the surface of the intersegmental tracts. These are processes of interneurons in the brain which enter the spinal medulla—**long descending tracts**— and similar fibres which ascend from the spinal medulla to the brain—**long ascending tracts**.

Thus the lateral and anterior parts of the marginal layer (funiculi of white matter) of the spinal medulla, separated from each other

only by the emergent ventral rootlets, consist of a mixture of long ascending and long descending tracts overlying a narrow zone of intersegmental tracts, while the posterior funiculi, lying between the dorsal rootlets, consist of longitudinally-running dorsal root fibres with a few intersegmental fibres in their anterior parts [FIG. 10.20; for details of ascending tracts see pp. 691–5].

FIGURE 10.17 gives an indication of the processes involved in the formation of the spinal medulla as seen in transverse section. It shows how the cavity of this part of the neural tube becomes relatively smaller at each stage, until it is represented by the small **central canal** lying in the narrow strip of grey matter connecting the two sides. In the spinal medulla of the adult, the central canal is frequently missing, being represented by a scattered group of ependymal cells without any cavity. This figure also indicates the formation of the **horns of the grey matter** (anterior, lateral, and posterior) and the arrangement of the surrounding white matter. It does not show that virtually all the structures seen in transverse section extend throughout the length of the spinal medulla, and that they are therefore arranged in columns of grey matter and funiculi of white matter.

The **grey columns** and **funiculi** (white columns) of the spinal medulla are not of the same size throughout their length [FIG. 10.21]. The funiculi increase steadily in size on ascending the spinal medulla owing to the greater number of the fibres in each of the long ascending and descending tracts in its cephalic parts, while the amount of grey matter in each segment is proportional to the amount of tissue to be supplied by that segment. Thus, opposite the limbs, where the volume of muscle and the area of skin are greatly increased, the anterior and posterior horns of grey matter are correspondingly increased in size in comparison with similar structures in the thoracic and upper cervical regions. Similarly the lateral horn, containing the cells of origin of the preganglionic autonomic fibres, is formed only where such fibres arise, that is in the thoracolumbar and midsacral spinal regions.

The increase in size of the anterior and posterior horns of the grey matter, greatest in the lumbosacral region, is responsible for the **enlargements** of the spinal medulla opposite the origin of the great limb plexuses, a feature missing from the spinal medulla in cases of congenital absence of the limbs, in animals such as snakes, and in the lumbosacral region of whales. The cervical enlargement has

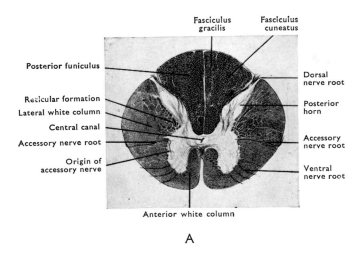

Fasciculus gracilis   Fasciculus cuneatus

Posterior funiculus

Reticular formation
Lateral white column
Central canal
Accessory nerve root
Origin of accessory nerve

Dorsal nerve root
Posterior horn
Accessory nerve root
Ventral nerve root

Anterior white column

A

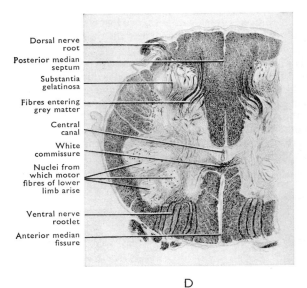

Dorsal nerve root
Posterior median septum
Substantia gelatinosa
Fibres entering grey matter
Central canal
White commissure
Nuclei from which motor fibres of lower limb arise
Ventral nerve rootlet
Anterior median fissure

D

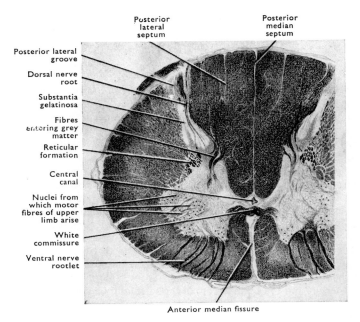

Posterior lateral septum   Posterior median septum

Posterior lateral groove
Dorsal nerve root
Substantia gelatinosa
Fibres entering grey matter
Reticular formation
Central canal
Nuclei from which motor fibres of upper limb arise
White commissure
Ventral nerve rootlet

Anterior median fissure

B

Dorsal nerve root
Posterior median septum
Substantia gelatinosa
Grey commissure
White commissure
Anterior median fissure

E

FIG. 10.21. Sections through spinal medulla, not to scale. (From specimens prepared by the Weigert–Pal method, by which the white matter is made dark and the grey matter bleached.)

A, transverse section of the upper part of the cervical region.
B, at the level of the fifth cervical nerve.
C, through the midthoracic region.
D, at the level of the fourth lumbar nerve.
E, at the level of the fourth sacral nerve.

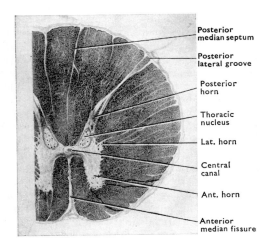

Posterior median septum
Posterior lateral groove
Posterior horn
Thoracic nucleus
Lat. horn
Central canal
Ant. horn
Anterior median fissure

C

greater overall dimensions than the lumbosacral, although the volume of the cervical grey matter is less; the ratio of grey to white matter being 1:2 in the sacral region and 1:5 in the cervical region. It is this preponderance of fibres in the cervical region which makes the superior limit of that enlargement less well defined, so that it is described as beginning at the first cervical segment though the highest limb nerve arises from the fifth.

The posterior horn varies markedly in shape in the different regions of the spinal medulla, but is capped throughout by a layer of cells, difficult to stain, which has a gelatinous appearance, the substantia gelatinosa. This is the most superficial of a series of layers of cells in the posterior horn. It lies immediately deep to the fibres of the dorsolateral fasciculus which send collaterals into it and through it to the body of the posterior horn. Here the impulses which they carry are subject to the influence of those from the fasciculi proprii, the long descending tracts, and the posterior funiculi whose collaterals enter the deeper parts of the posterior horn [FIG. 10.21] and so are able to modify reflexes generated by impulses from the dorsolateral fasciculus. Towards the junction of the posterior horn with the anterior and lateral horns, there are certain well-defined nuclei such as the **thoracic nucleus** which indents the posterior funiculi and is found between the first thoracic and second lumbar segments. It is composed of large cells which give rise to the posterior spinocerebellar tract. Between the anterior and posterior horns lies the **intermediate substance** (grey matter) which is divided into central and lateral parts. It contains many large internuncial neurons and its lateral portion lies against the medial edge of the lateral funiculus and extends into it to a greater or lesser degree, giving a mixture of nerve cells and fibres known as reticular formation.

The **reticular formation** is best developed in the upper cervical part of the spinal medulla where it gradually increases in amount and becomes continuous with the much more voluminous reticular formation of the medulla oblongata. The **lateral horn**, where present (T.1–L.2; S.2–S.4), is also an extension of the lateral intermediate substance, and contains the cells of origin of the preganglionic autonomic fibres [FIGS. 10.17 and 10.47].

The **anterior horn** is small in the thoracic and upper cervical regions but shows a marked lateral expansion in the enlargements. It is this lateral portion which contains the motor cells supplying the muscles of the limbs [FIG. 10.22]. At any level of the developing spinal medulla the ventromedial cells are at a more advanced stage of development than those situated dorsolaterally, while the

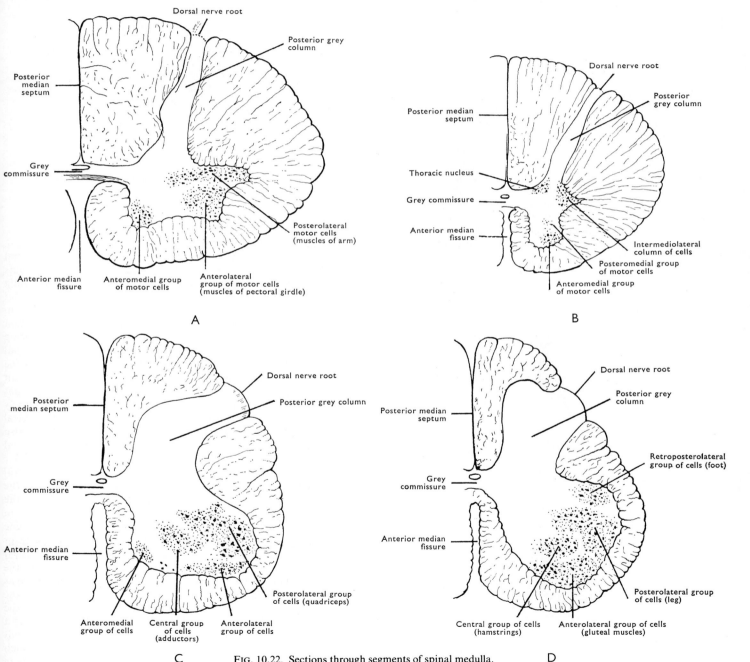

FIG. 10.22. Sections through segments of spinal medulla.
A, fifth cervical segment.
B, eighth thoracic segment.
C, third lumbar segment.
D, second sacral segment.
(Founded on Plates in Bruce's *Atlas of the spinal cord*.)

corresponding groups at more caudal regions are at a less advanced stage. Thus, of the motor cells which are arranged around the anterior horn, the anteromedial are the first to spin out their axons and these pass into the most proximal or trunk muscles. The remaining motor cells, which develop in sequence around the periphery of the anterior horn in the enlargement, supply successively more peripheral muscles (Romanes 1951; Sharrard 1955). Thus the girdle muscles are supplied by cells in a ventrolateral position while the muscles of the proximal (arm or thigh), intermediate (forearm or leg), and distal (hand or foot) segments of the limb are innervated by groups of cells which lie in sequence posterior and caudal to this. Each of these groups, except the last, is split into lateral and medial parts which supply respectively the extensor and flexor muscles of the segment. Thus the small muscles of the hand and foot are supplied by cells lying in the extreme dorsolateral part of the anterior horn in the first thoracic and second and third sacral segments respectively. Each group of motor cells, seen in transverse section, forms a column extending for a variable length of the anterior horn, and is formed by a mixture of large, multipolar motor cells and smaller cells some of which are the source of the fine motor fibres (γ-efferents). These accompany the large motor fibres to somatic muscles and are responsible for the motor supply to the small muscle fibres which form the core of the sensory **muscle spindles**. The presence of such a motor supply to the muscle spindles allows these sensory organs to be adjustable in length and tension. Since they are tension recorders, the activation of their muscle fibres increases their sensitivity, and heightens, or may initiate a discharge of, sensory impulses from them. This is one proven example of control by the central nervous system over the sensitivity of a sensory ending, though others exist [p. 701].

The spinal medulla extends throughout the length of the vertebral column up to the third month of intra-uterine life [FIG. 10.23], but

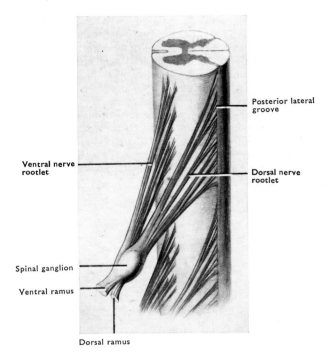

FIG. 10.24. Roots of the seventh thoracic nerve (semidiagrammatic).

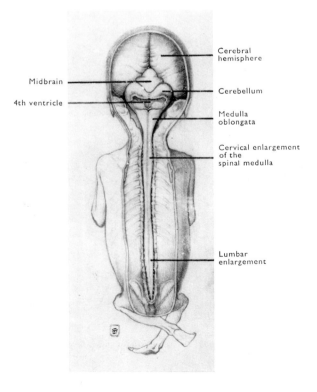

FIG. 10.23. Human foetus in third month of development, with brain and spinal medulla exposed from behind.

subsequently grows less rapidly than that structure, so that a disproportion in length between the two steadily develops. In normal development the junction of the brain and spinal medulla remains fixed at the foramen magnum of the skull, and the relative shortening of the spinal medulla causes it to rise in the vertebral canal so that its caudal extremity lies at the level of the third lumbar vertebra in the new-born, and in the adult has risen to the disc between the first and second lumbar vertebrae, the whole spinal medulla measuring from 43 to 45 cm in length. This relative shortening of the spinal medulla is not uniform throughout its length, but affects the lumbosacral segments most and thoracic segments least. As a result the lumbosacral roots are closely packed together at their attachment to the spinal medulla and the corresponding segments are extremely short; all the sacral segments taking up little more than the vertical extent of the first lumbar vertebra in the adult. The shortening also produces a considerable lengthening and obliquity of the spinal roots which is in proportion to the extent to which any particular segment has risen within the vertebral canal. In the second sacral and more caudal spinal nerves, the spinal ganglia leave the corresponding intervertebral foramina and come to lie within the sacral canal, so reducing the lengthening of these roots which, in common with all spinal roots, join to form the spinal nerves at the spinal ganglia [FIG. 10.24]. Thus the length of the roots varies from 8 mm in the second cervical nerve to 260 mm in the fifth sacral nerve, and the length of the spinal segments from 25 mm in the thoracic region, through 15 mm in the cervical region, to as little as 4 mm in the sacral region. As the spinal medulla rises within the vertebral canal, it leaves behind it a thin filament attached to the dorsal surface of the coccyx, the **filum terminale** [FIG. 10.25], and a leash of dorsal and ventral roots descending to their point of exit from the vertebral canal, the **cauda equina** [FIG. 10.26], which surrounds the filum terminale. The filum terminale is 18 cm in length and consists of the continuation of the immediate investment (the pia mater) of the spinal medulla [p. 728] from its tapering caudal extremity, the **conus medullaris**. This encloses a few nerve fibres and cells which may represent the remains of coccygeal nerves and, for a variable distance, the central canal which often expands into a small sac or **terminal ventricle** at the beginning of the filum. Since the spinal medulla rises relative to the vertebral column, the segments of the spinal medulla no longer lie opposite the corresponding vertebrae. The relative levels in the adult are indicated in FIGURE 10.29, but the position of the spinal medulla changes in flexion and extension of the vertebral column, the conus rising in flexion and descending in extension as the length of the anterior surface of the vertebral canal, composed of vertebral bodies and intervertebral discs, increases and decreases.

MYELOGRAPHY

Recently discovered water soluble media give much clearer delineation of the spinal subarachnoid space and its contents than

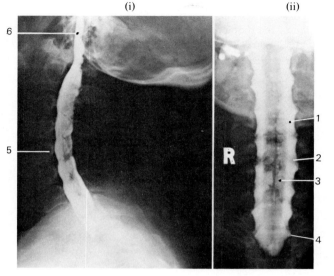

FIG. 10.27. Cervical myelograms. These are produced by injecting X-ray opaque material into the subarachnoid space by lumbar puncture. (i) lateral, (ii) anteroposterior.
1. Subarachnoid space.
2. Nerve roots.
3. Central shadow due to spinal medulla. No such shadow is visible below the 1st lumbar vertebra where the spinal medulla ends.
4. Extension of subarachnoid space along nerve roots.
5. 5th cervical vertebra.
6. Basal cistern.

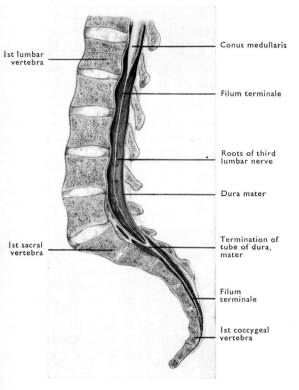

FIG. 10.25. Conus medullaris and the filum terminale.

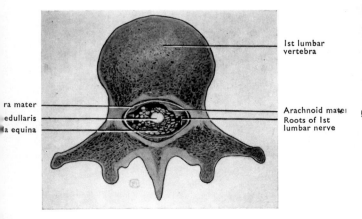

FIG. 10.26. Section through conus medullaris and cauda equina as they lie in the vertebral canal.

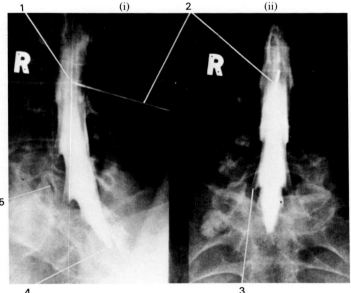

FIG. 10.28. Lumbar myelogram. A water soluble contrast medium introduced into the lumbar subarachnoid space by lumbar puncture, viewed in lateral oblique (i) and anteroposterior (ii) views.
1. Roots forming cauda equina show dark against light contrast medium.
2. Lumbar puncture needle.
3. 1st sacral nerve roots with subarachnoid extension surrounding them.
4. End of subarachnoid space. (S.1–2.)
5. Lumbosacral synovial articulation.

the oily media. The spinal nerve roots are shown as a translucent leash against the dense medium and the extensions of the subarachnoid space around the nerve roots appear as thorn-like projections from the column of medium [FIG. 10.28]. It is easier to detect deformities and displacements with these media.

At present, these substances are applicable to the lumbosacral region only, because at higher levels there is risk of toxic effects on the central nervous system. Therefore oily media are required for investigation of the thoracic and cervical regions [FIG. 10.27] and for the subarachnoid cisterns.

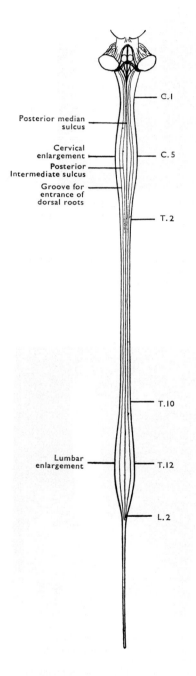

Posterior median sulcus

Cervical enlargement

Posterior Intermediate sulcus

Groove for entrance of dorsal roots

Lumbar enlargement

C.1

C.5

T.2

T.10

T.12

L.2

FIG. 10.29. Diagram of the spinal medulla as seen from behind. C.1, C.5, etc., indicate levels in relation to the vertebrae.

The blood supply of the spinal medulla
[FIG. 10.30]

The arteries of the spinal medulla reach it along the roots of the spinal nerves and are therefore divisible into ventral and dorsal root arteries which arise from the vertebral, intercostal, lumbar, and lateral sacral arteries in the different regions of the body. The **dorsal radicular arteries** supply the spinal ganglia and continue on the dorsal roots to the spinal medulla where they anastomose along the line of entry of the dorsal roots, the **posterior spinal arteries**. With the ventral radicular arteries they form an irregular network of vessels on the pial surface. Radial branches from this pierce the spinal medulla to supply the tips of the posterior horns and all but the deepest parts of the white matter. The **ventral radicular arteries** in the embryo not only assist in the formation of the surface anastomosis but also send a large branch ventrally to unite with similar branches from all the other ventral root arteries in a median, ventral, longitudinal vessel, the **anterior spinal artery**. Branches of this vessel, larger than those from the surface anastomosis, pierce the spinal medulla through the anterior median fissure and, passing alternately to right and left, supply the greater part of the grey matter and the deeper parts of the lateral and posterior funiculi. As development proceeds, many ventral root arteries cease to have a significant connection with the anterior spinal artery and persist as smaller vessels which end principally in the surface anastomosis. Thus a reduced number of ventral root arteries supply blood to the anterior spinal artery, and only those on the ventral roots of the first cervical nerves, which are enlarged to form the upper parts of the **vertebral arteries**, retain the embryonic arrangement. These give rise to the cranial part of the anterior spinal artery running caudally and its equivalent on the ventral surface of the pons, the **basilar artery**, passing headwards [FIG. 13.52]. This reduction of large feeding vessels to the anterior spinal artery proceeds until only three to ten persist, most commonly on the left side. Thus the greater part of the grey matter of the lumbosacral enlargement may be supplied by a single vessel entering on a ventral root between the ninth thoracic and second lumbar, supported to some extent through surface anastomoses with the posterior spinal arteries.

All the arteries which pierce the spinal medulla are end-arteries without anastomosis. This pattern of a surface anastomosis with

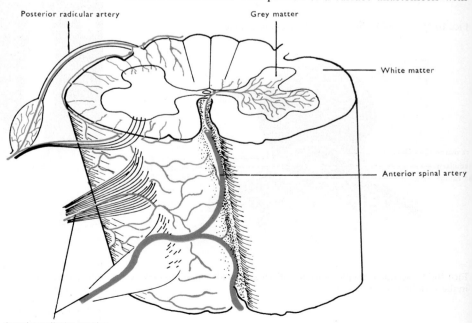

Posterior radicular artery

Grey matter

White matter

Anterior spinal artery

Anterior radicular arteries

FIG. 10.30. Semidiagrammatic drawing to show the blood supply of the spinal medulla.

multiple small end-arteries supplying the outer layers and larger end-arteries entering the ventral surface to supply the interior is characteristic of the entire central nervous system.

The veins of the spinal medulla drain mainly into anterior and posterior median channels which connect with the intersegmental veins at various levels along the ventral and dorsal roots. A smaller, more variable, lateral spinal vein is also present.

The brain (encephalon)

The part of the neural tube which is enclosed in the developing skull is the brain. Even before the closure of the neural tube is complete, the brain shows three swellings which are only indistinctly separable. These are the hindbrain, midbrain, and forebrain. The hindbrain is continuous with the spinal medulla at the foramen magnum and is separated from the midbrain cranially by a constricted zone or isthmus. At first a relatively large part of the brain, the midbrain becomes relatively smaller and narrower than the hindbrain or the forebrain. The midbrain and hindbrain together form the brain stem.

The cephalic extremity of the forebrain is closed by a thin membrane, the lamina terminalis, which represents the most rostral part of the neural tube until the cerebral hemispheres grow out from each side of the original forebrain and extending beyond it, hide it between them.

Basically the structure of the brain is the same as that of the spinal medulla but it is greatly modified, firstly by the development of folds or flexures and secondly by the formation of three additional masses of nervous tissue which overshadow the original neural tube and come to make up the major part of the brain. These are:

1. The two cerebral hemispheres (telencephalon) one on each side, which develop from the forebrain as hollow extensions of the full thickness of the neural tube, so that their cavities, the lateral ventricles, are continuous with the remainder of the cavity of the neural tube.

2. The cerebellum forms as a solid organ on the dorsal aspect of the hindbrain. It expands to fill the greater part of the posterior cranial fossa of the skull, and completely hides the hindbrain and midbrain from the dorsal aspect.

The cerebellum and the two cerebral hemispheres expand independently in the cranial cavity leaving thin but tough sheets of connective tissue between them as they grow. One of these, the falx cerebri, lies sagittally between the two hemispheres; the other, the almost horizontal tentorium cerebelli, separates both hemispheres from the cerebellum which lies behind and below them [p. 729].

The first or cephalic flexure [FIG. 10.31] appears at the midbrain and allows the forebrain to fold ventrally. This greatly elongates the dorsal surface of the midbrain while restricting its ventral surface. Although in later development this flexure is partly straightened, a marked angulation persists in the midbrain region. Thus the long axes of the fore- and hindbrains lie at an angle of approximately 140 degrees to each other and the difference in extent of the dorsal and ventral surfaces of the midbrain is maintained. In the second or pontine flexure [FIG. 10.32] the middle of the hindbrain bulges ventrally towards the forebrain, increasing the cephalic flexure and forming an acute ventral flexion between the hindbrain and the spinal medulla, the cervical flexure, where previously only the gentle curve due to the flexion of the embryo as a whole was present.

It should be appreciated that the three parts of the brain are developed from a single longitudinal tube and the line of demarcation between them is not sharp. Many structures pass longitudinally through the tube without interruption of any kind in the same way as the alar and basal laminae extend from the spinal medulla to the forebrain.

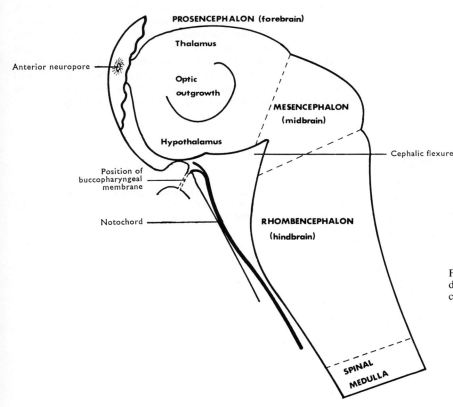

FIG. 10.31. Drawing to show an early phase in the development of the brain. The anterior neuropore is not yet closed.

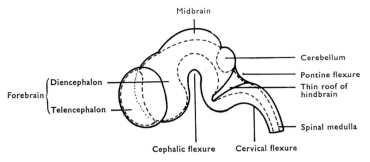

FIG. 10.32. Diagram to show the parts of the embryonic brain after the formation of the telencephalon. The optic outgrowth is not shown. Broken lines indicate the ventricular system; the dotted line, the roof of the diencephalon.

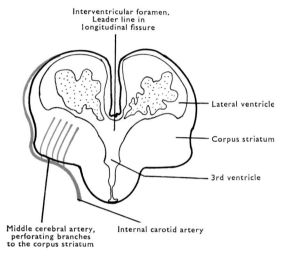

FIG. 10.33. Transverse section through the forebrain at the level of the interventricular foramen of the brain shown in FIG. 10.34.

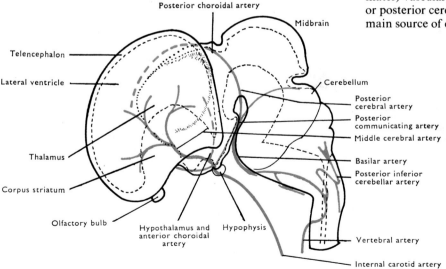

The blood supply of the brain [FIG. 13.52]

The blood vessels which supply the brain are limited in number by comparison with the multiple vessels of the spinal medulla. The general arrangement is [FIG. 10.34] of four main arteries supplying the longitudinal anastomosis continued from the spinal medulla. The two vertebral arteries join to form the basilar artery, which supplies the hindbrain, cerebellum, midbrain, and the posterior part of the forebrain. It anastomoses with the cephalic ends of the dorsal aortae, the future internal carotids, which enter the cranial cavity to supply the remainder of the forebrain. Here, as in the spinal medulla, the vessels are divisible into: (1) a surface anastomosis from which many small perforating end-arteries pass radially into the brain; (2) larger end-arteries which pierce the ventral aspect of the neural tube to supply its deeper parts. The absence of multiple radicular arteries in this region causes the surface anastomosis to be fed entirely from the ventral longitudinal vessels. Thus, as the brain increases in size, parts of the anastomosis develop into larger branches of the ventral vessels to supply the lateral and dorsal surfaces of the neural tube and the masses (cerebrum and cerebellum) which arise from them. These masses give their names to the arteries which, nevertheless, continue to supply the corresponding parts of the parent brain-tube. Thus the three pairs of **cerebellar arteries** (superior, anterior inferior, and posterior inferior) also supply the lower midbrain, pons, and medulla oblongata and anastomose with each other on the surface. Similarly, the three **cerebral arteries** which supply each hemisphere all give ventral perforating branches to the interior of the brain [FIG. 10.33] and together form a common anastomosis on the surface. Thus the **posterior cerebral artery** sends multiple ventral branches into the midbrain and caudal forebrain, and supplies the lateral and dorsal surfaces of the upper midbrain and the dorsal part (thalamus) of the original forebrain tube (**posterior choroidal artery**; FIG. 10.34) in addition to supplying the posterior part of the cerebral hemisphere.

In the early stages of development the forebrain and hindbrain have many features in common. In particular, the roof plate which forms along the line of closure of the neural tube consists of a single layer of ependymal cells which does not give rise to neural elements. This later forms the **choroid plexus** on each side when it is tucked into the cavity of the tube by the overlying connective tissue (pia mater) vascularized by the posterior inferior cerebellar (hindbrain) or posterior cerebral (forebrain) arteries. This choroid plexus is the main source of cerebrospinal fluid in the cavities of the brain.

FIG. 10.34. Drawing of reconstruction of early human foetal brain derived from serial sections. The arteries are shown diagrammatically.

THE HINDBRAIN

With the development of the pontine flexure the shape of the hindbrain is markedly altered, in much the same manner as a soft rubber tube is distorted when it is flexed after being cut along one surface [FIG. 10.35]. The neural tube is flattened at the line of flexure, and the thin roof plate drawn out into an extensive sheet stretching from side to side and even slightly everted at the lateral extremities of the flexure. Headwards and tailwards of the flexure, the amount of flattening progressively decreases until the margins of the alar laminae once again fuse with each other at the caudal margin of the midbrain and at the middle of the lower half of the hindbrain. In this fashion, the slit-cavity of the hindbrain is converted into a diamond-shaped or rhomboid **fourth ventricle**, and the hindbrain is known as the **rhombencephalon**. It is divisible at the line of the pontine flexure into two parts; the caudal is the **myelencephalon** or **medulla oblongata**, tapering towards the spinal medulla, the cranial is the **metencephalon** or **pons**. The cephalic and caudal extremities of the fourth ventricle are continuous, respectively, with the aqueduct of the cerebrum in the midbrain and the central canal of the lower medulla oblongata and spinal medulla. The lateral angles lie at the widest part of the ventricle and constitute the **lateral recesses of the fourth ventricle**. Here, and at its caudal extremity, the thin roof subsequently breaks down, forming the two **lateral apertures** and the single **median aperture** which constitute the only communications between the cavity of the nervous system and the surrounding tissues. Cerebrospinal fluid emerging from these apertures percolates the surrounding loose connective tissue to form the subarachnoid space by incompletely splitting the tissue into pia mater internally and arachnoid externally [p. 728]. The choroid plexus on each side forms an L-shaped invagination of the thin roof extending from the median aperture headwards to the line of the flexure. Here each turns laterally and passes into the lateral recess to reach the lateral aperture through which it may present on the surface of the brain. The evagination of the thin roof at the lateral recess forms a sleeve-like process extending ventrally around the side of the hindbrain with the terminal aperture facing ventrally and the choroid plexus bulging through it [FIG. 10.50]. With the development of the cerebellum on its cephalic margin, each lateral recess is restricted and forms a funnel-shaped extension of the fourth ventricle passing over the posterolateral aspect of the upper medulla oblongata and separating it from the cerebellum. In the adult the cephalic margin of the median aperture is drawn posteriorly on the cerebellum, the aperture thus forming a triangular opening which faces inferiorly between the cerebellum and the medulla oblongata [FIG. 10.38] so that air rising through the spinal subarachnoid space enters the fourth ventricle. On the ventricular aspect of this portion of the thin roof lies the choroid plexus which curves posteriorly on the anterior surface of the cerebellum.

The cerebellum

The cerebellum develops in the roof of the fourth ventricle as a result of proliferation of cells at the margins of the alar laminae of the metencephalon. The masses so formed extend into the thin roof and meet in the midline, leaving parts of the thin roof of the

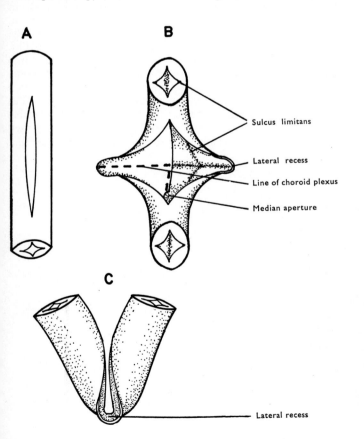

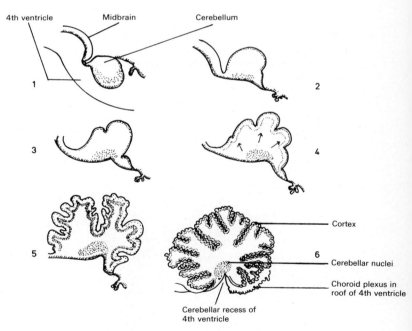

FIG. 10.36. Six stages in the development of the cerebellum as shown in median sagittal section. Note the formation of cells on the ventricular aspect, their later migration to form the cortex and the folding of the cerebellum to form the cerebellar recess of the fourth ventricle. The drawings are not to scale.

FIG. 10.35. Diagram to illustrate the probable mechanical effects of the pontine flexure in the formation of the fourth ventricle.

A, before folding. B, posterior view after folding, the thin roof has been removed on the right side. C, lateral view after folding with the thin roof removed.

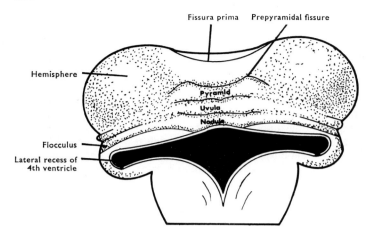

Fissura prima Prepyramidal fissure

Hemisphere

Pyramid
Uvula
Nodule

Flocculus

Lateral recess of
4th ventricle

FIG. 10.37. Dorsal view of cerebellum and medulla oblongata from human foetus of 4 months. The ependymal roof of the fourth ventricle has been removed. At this stage the cerebellum is little more than a simple band or plate which arches over the back of the superior part of the cavity of the hindbrain.

metencephalon cephalic and caudal to the fused cerebellar rudiments. These parts are later invaded by nerve fibres of the cerebellum and, becoming white in colour, are known respectively as the **superior** and **inferior medullary vela**. The inferior medullary velum is continuous with the thin roof of the myelencephalon which carries the choroid plexus.

At first there is a great proliferation of cells on the ventricular aspect of the cerebellar rudiment which causes it to bulge into the ventricle [FIG. 10.36]. Subsequently many of these cells, formed on its ependymal aspect, migrate to the posterior surface of the cerebellum and form a superficial layer (or cortex) of nerve cells which is never covered with white matter and hence remains grey in colour, the **cortex of the cerebellum**. Subsequently this aspect of the cerebellum enlarges rapidly, increasing its surface area partly by the development of transversely placed folds [FIGS. 10.36 and 10.37], but also by a general increase in size which eventually leads to the

cerebellum overlying the posterior surface of the whole extent of the hindbrain and midbrain. Despite this enormous increase in size, and the complex folding of its surface to form the many deep and subsidiary transverse fissures which, on section, give the cerebellum the appearance of a tree (arbor vitae) with many leaves (folia [FIGS. 10.36 and 10.38]) and confer on the cerebellum an enormous surface area, the extent of its attachment to the metencephalon is not increased. Thus the mass of the cerebellum as it expands posteriorly becomes folded on itself with its caudal part rolled round against the thin roof of the inferior part of the fourth ventricle, and its cranial part against the dorsal surface of the midbrain. This type of growth carries the fourth ventricle backwards towards the centre of the cerebellum in the tent-like **cerebellar recess** [FIG. 10.36] and deepens the fourth ventricle at this level. Superior to this lies the narrow, transverse attachment of the cerebellum to the roof of the fourth ventricle. The nerve cells which are produced on the ventricular aspect of the cerebellar rudiment, and which do not migrate to form the cortex, remain close to the ventricle and adjacent to the cerebellar recess. These cells develop into four groups on each side extending laterally from the midline and forming the nuclei of the cerebellum. The most lateral and largest of these is the **dentate nucleus**, so called because it is formed from a folded lamina of grey matter in the shape of a wrinkled bag [FIG. 10.39], which gives it a ridged or toothed appearance on surface view [FIG. 10.41]. It lies over the lateral extremity of the cerebellar recess. The dentate and the other nuclei, which are much less well defined and are named from lateral to medial the **emboliform, globose**, and **fastigial** [FIG. 10.40], give rise to the nerve fibres which leave the cerebellum for other regions of the brain.

The cerebellum receives and emits large numbers of nerve fibres which are gathered together in a considerable bundle at the lateral margin of the metencephalon on each side, and emerge from the anterior parts of the **horizontal fissure** of the cerebellum [FIG. 10.45]. Leaving the cerebellum, each mass of fibres divides into three separate bundles of which the most lateral (the **middle cerebellar peduncle** [FIG. 10.46]) runs anterosuperiorly to arch across the anterior surface of the pons and fuse with its fellow, thus forming the bulging **ventral part of the pons** which is responsible for the name given to this part of the hindbrain. The other two bundles diverge, the most medial or **superior peduncle** passes in a

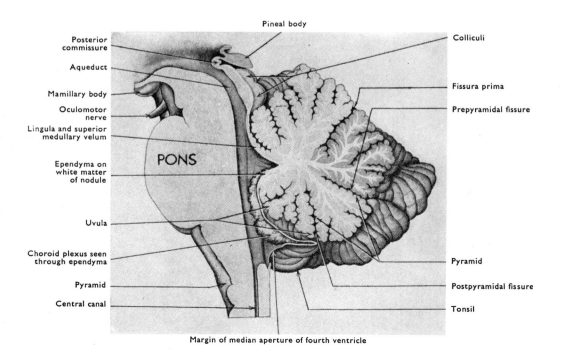

Posterior
commissure

Aqueduct

Mamillary body

Oculomotor
nerve

Lingula and superior
medullary velum

Ependyma on
white matter
of nodule

Uvula

Choroid plexus seen
through ependyma

Pyramid

Central canal

Pineal body

PONS

Colliculi

Fissura prima

Prepyramidal fissure

Pyramid

Postpyramidal fissure

Tonsil

Margin of median aperture of fourth ventricle

FIG. 10.38. Median section through brain stem and cerebellum, showing arbor vitae of vermis and the fourth ventricle.

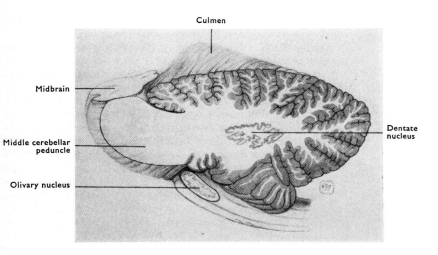

Culmen

Midbrain

Middle cerebellar
peduncle

Olivary nucleus

Dentate
nucleus

FIG. 10.39. Sagittal section through left hemisphere of cerebellum, showing the arbor vitae and dentate nucleus.

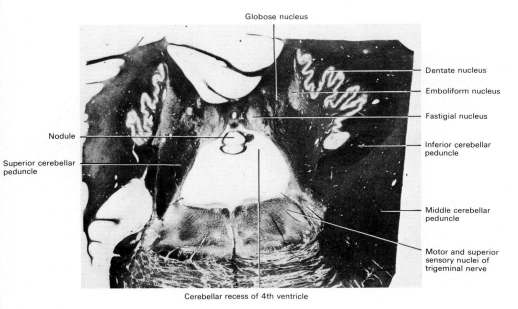

Globose nucleus

Dentate nucleus

Emboliform nucleus

Fastigial nucleus

Inferior cerebellar
peduncle

Nodule

Superior cerebellar
peduncle

Middle cerebellar
peduncle

Motor and superior
sensory nuclei of
trigeminal nerve

Cerebellar recess of 4th ventricle

FIG. 10.40. Oblique transverse section, parallel to the cerebellar recess of the fourth ventricle, to show the cerebellar nuclei.

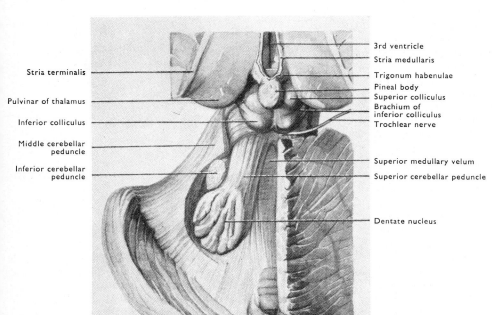

Stria terminalis

Pulvinar of thalamus

Inferior colliculus

Middle cerebellar
peduncle

Inferior cerebellar
peduncle

3rd ventricle
Stria medullaris
Trigonum habenulae
Pineal body
Superior colliculus
Brachium of
inferior colliculus
Trochlear nerve

Superior medullary velum
Superior cerebellar peduncle

Dentate nucleus

FIG. 10.41. Dissection from above to show dentate nucleus, cerebellar peduncles, and lateral lemniscus (between inferior colliculus and middle cerebellar peduncle).

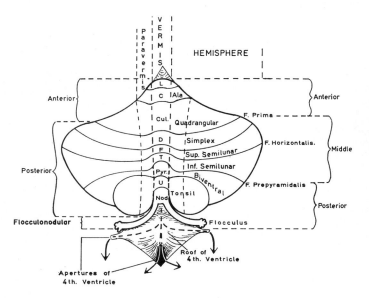

FIG. 10.42. Diagram of cerebellum. This has been unfolded so that the lingula (L) and the nodule (Nod) lie at opposite ends. The attachment of the superior and inferior medullary vela to these two parts respectively, and the latter to the flocculus and its peduncle is shown. The thin roof of the fourth ventricle and its apertures (arrows) are indicated cf. FIG. 10.38.

C, lobulus centralis; Cul., culmen; D, declive; F, folium; T, tuber; Pyr., pyramid; U, uvula.

and a caudal **nodule** with its lateral extension the **flocculus**, both continuous inferiorly with the inferior medullary velum.

The third phase in the phylogenetic development of the cerebellum is represented by a growth *pari passu* with that of the cerebral hemispheres and with the development of increasingly voluminous connections between them [see middle cerebellar peduncle, p. 710]. This third or 'cerebral' part divides the 'spinal' part into cephalic and caudal regions, and becomes the largest part of the cerebellum in Man. In other mammals where cerebral development is not so advanced, the lateral parts of the cerebellum are much smaller and the median portion stands out as a raised, sinuous ridge which, on account of the transverse fissures, has the appearance of a worm and is known as the **vermis**. The growth of the lateral lobes, or hemispheres, in Man makes the vermis much less obvious except in the caudal part. This is displaced to the inferior surface, and lying in a notch between the hemispheres, is known as the **inferior vermis** [tuber, pyramid, uvula, and nodule, FIG. 10.45]. The more cephalic part is continuous with the hemispheres without interruption and is known as the **superior vermis** [FIG. 10.44].

cephalic direction along the margin of the superior medullary velum and sinks into the substance of the midbrain [FIG. 10.41]. The **inferior peduncle**, lying between the other two, runs anteriorly then turns caudally, anterior to the lateral recess of the fourth ventricle, to pass along the dorsolateral aspect of the medulla oblongata and gradually disappear as its constituent fibres diverge into the medulla oblongata.

The size and intricacy of the fissural pattern of the human cerebellum, as well as the number of fanciful descriptive terms applied to its several parts, all serve to emphasize the complexity of this organ. In addition, its mode of development and consequent curved nature make it difficult to describe. For this reason it is common practice to unfold the cerebellum diagrammatically, so that the superior and inferior medullary vela lie at the extremities of the diagram, instead of close together in the roof of the fourth ventricle [FIG. 10.42].

Phylogenetically, three arbitrary stages may be recognized in the formation of the cerebellum and though these are not independent parts functionally, consideration of them does help to clarify the arrangement and connections of this organ. Initially [FIG. 10.43] the cerebellum is an extension of the vestibular nuclei of the hindbrain and is concerned with them in the muscular adjustments necessary for the control of balance and posture resulting from information conveyed to it from the internal ear, the principal organ for recording changes in position in aquatic vertebrates. Control of muscular activity requires information regarding the state of the muscles, and from an early phase the cerebellum receives this from the spinal medulla by ascending nervous pathways, the **spinocere-bellar tracts**, which also send fibres to the vestibular nuclei. With the increasing importance of muscle, joint, and other senses for recording position, expecially in terrestrial vertebrates, the spino-cerebellar system and that part of the cerebellum to which it passes becomes much enlarged. This 'spinal part' is formed in the centre of the original cerebellum [FIG.10.43B] splitting the 'vestibular part' into a cephalic **lingula**, attached to the superior medullary velum,

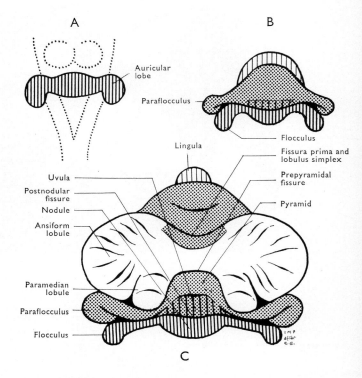

FIG. 10.43 Diagrams illustrating evolutionary development of cerebellum and the constitution of its three lobes.

A, hypothetical primitive cerebellum with vestibular connections only.

B, cerebellum of reptile showing addition of part that receives spinocerebellar tracts.

C, mammalian cerebellum showing addition of part (lobus medius) that receives cortical impulses by way of the pontine nuclei.

Vestibular cerebellum, vertical hatching; spinal cerebellum, stipple; neocerebellum (lobus medius), white.

FIGURE 10.42 shows the parts of the cerebellum belonging to each of the above divisions, the main fissures and the lobes. The general arrangement of these can readily be understood by reference to FIGURES 10.38, 10.44, and 10.45. FIGURE 10.42 shows two of the methods of division of the cerebellum into lobes, that on the left being less open to criticism.

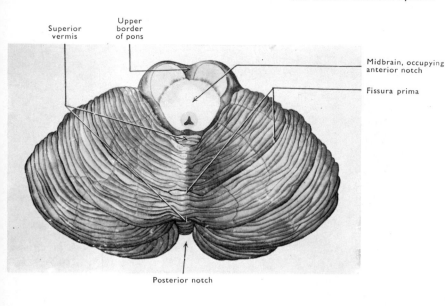

Superior vermis

Upper border of pons

Midbrain, occupying anterior notch

Fissura prima

FIG. 10.44. Superior surface of the cerebellum.

Posterior notch

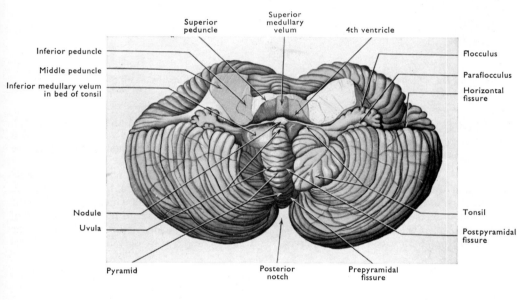

Superior peduncle

Superior medullary velum

4th ventricle

Inferior peduncle

Middle peduncle

Inferior medullary velum in bed of tonsil

Flocculus

Paraflocculus

Horizontal fissure

FIG. 10.45. Inferior surface of the cerebellum. The right tonsil has been removed so as to display more fully the inferior medullary velum.

Nodule

Uvula

Tonsil

Postpyramidal fissure

Pyramid

Posterior notch

Prepyramidal fissure

THE LATERAL WALLS AND FLOOR OF THE FOURTH VENTRICLE

The lateral walls consist mainly of the cerebellar peduncles. The inferior of these, together with the gracile and cuneate tubercles [FIG. 10.46], form the inferolateral wall, while the superolateral wall is composed of the superior cerebellar peduncle and the nervous tissue anterior to its inferior part, between the upper end of the inferior peduncle and the ventricle. This is mainly the superior vestibular nucleus [p. 707] and the juxtarestiform body—scattered nerve fibres passing to and from the cerebellum on the medial side of the inferior cerebellar peduncle (restiform body).

The floor of the fourth ventricle is rhomboid in shape. The lateral recess extends over the posterior surface of the inferior cerebellar peduncle at the widest part of the ventricle. In FIGURE 10.46 note the median sulcus and lateral to it in the medial eminence. The eminence represents the basal lamina, and is a ridge extending throughout the length of the ventricle on each side of the median sulcus and showing in its lower half a series of ridges and shallow grooves dividing it into:

1. The facial colliculus is just cephalic to the level of the lateral recess. Here the facial nerve passing to its exit partly encircles the abducent nucleus (VIth cranial nerve) which forms the colliculus [FIG. 10.59].

2. The hypoglossal triangle, covering the hypoglossal nucleus.

3. The vagal triangle, covering the dorsal nucleus of the vagus.

4. The area postrema, just cephalic to the opening of the central canal of the lower half of the medulla oblongata. This consists of a mass of capillaries with incompletely fenestrated endothelium and nerve cells which have synapse-like endings on the ependyma of the floor of the ventricle.

5. Foveae. Lateral to the medial eminence lies an ill-defined groove, the remains of the sulcus limitans, often represented only by the superior and inferior foveae.

6. Lateral to the sulcus limitans and filling out the floor in the lateral angle is the vestibular area. This represents the position of the groups of cells, vestibular nuclei, which receive sensory fibres from the vestibular part of the internal ear via the vestibular part of the vestibulocochlear nerve.

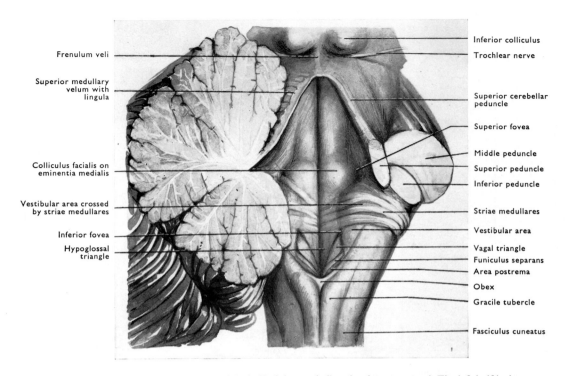

Frenulum veli

Superior medullary
velum with
lingula

Colliculus facialis on
eminentia medialis

Vestibular area crossed
by striae medullares

Inferior fovea

Hypoglossal
triangle

Inferior colliculus

Trochlear nerve

Superior cerebellar
peduncle

Superior fovea

Middle peduncle

Superior peduncle

Inferior peduncle

Striae medullares

Vestibular area

Vagal triangle

Funiculus separans

Area postrema

Obex

Gracile tubercle

Fasciculus cuneatus

FIG. 10.46. Floor of the fourth ventricle. The right half of the cerebellum has been removed. The left half is drawn over to the left so as to expose the floor of the ventricle fully.

The brain stem (medulla oblongata, pons, and midbrain) [FIGS. 10.54–10.58]

THE MEDULLA OBLONGATA

The pontine flexure produces the fourth ventricle and causes marked changes in the shape and general position of the parts of the neural tube. This is further altered by: (1) a considerable proliferation of the internuncial nerve cells, many of which are associated with the cerebellum; and (2) a change in the position of two of the major longitudinal pathways of the spinal medulla, the posterior funiculi, and the lateral corticospinal tract [FIG. 10.52F and G] both of which decussate to a ventromedial position in the medulla oblongata (see below).

The flattening of the neural tube in the pons and upper medulla oblongata causes the alar lamina to lie lateral to the basal lamina. Consequently the cephalic sensory ganglia which lie lateral to the hindbrain send their central processes on to the lateral surface of this part of the brain as the equivalent of the dorsal roots of the spinal nerves. Ventral root fibres grow outwards from the basal

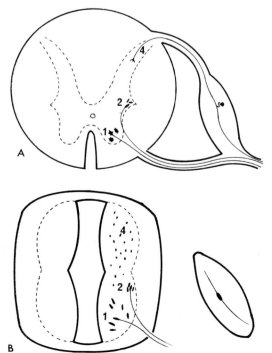

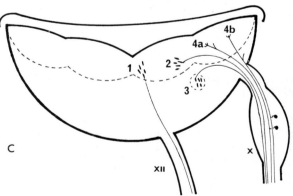

FIG. 10.47. Diagram to illustrate the difference in development between the spinal medulla and the hindbrain.

B, indicates the basic structure from which both A, thoracic spinal medulla, and C, medulla oblongata, arise.

1. Somatic motor supply to striated, somite muscle.
2. Preganglionic autonomic connector cells. ⎫
3. Supply to striated visceral, or branchial muscle. ⎬ Basal lamina
4. Sensory terminations: ⎭
   (a) Visceral sensory, tractus solitarius;
   (b) Somatic sensory, trigeminal system.

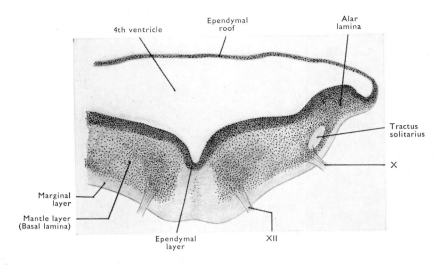

FIG. 10.48. Transverse section across the medulla oblongata of a human embryo. (From His, slightly modified.)

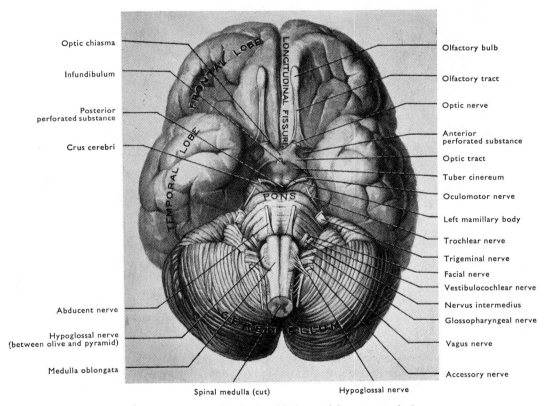

FIG. 10.49. Base of the brain with the cranial nerves attached.

lamina, as in the spinal medulla, but fail to join the corresponding dorsal roots [FIG. 10.47]. Thus, excluding the olfactory and optic nerves which are formed in an entirely different manner and belong to the forebrain, two distinct groups of nerves are attached to the brain stem and constitute the **cranial nerves**. These two groups are:

1. *Nerves attached to the ventral aspect of the brain stem.* These consist of motor fibres supplying the head somites, which form the extrinsic ocular muscles and the muscles of the tongue except palatoglossus. These nerves lie in a row in series with the spinal ventral roots and are the **oculomotor** (III), the **abducent** (VI), the **hypoglossal** (XII), and, belonging to the same group but arising

from the dorsal surface of the brain after decussation, the **trochlear** (IV) nerve [FIGS. 10.50 and 10.51].

2. *Nerves attached in a row to the lateral aspect of the hindbrain.* They are ganglionated and contain sensory nerve fibres and are the **trigeminal** (V), the **facial** (VII), the **vestibulocochlear** (VIII), the **glossopharyngeal** (IX), the **vagus** (X), and the **accessory**) (XI).

FIGURES 10.49, 10.50, and 10.51 show the points of attachment of the cranial nerves and demonstrate that the largest, the trigeminal, is the only nerve attached to the pons. Since the nerves are numbered more or less serially from above downwards, it follows that the oculomotor and trochlear are attached to the midbrain, and those

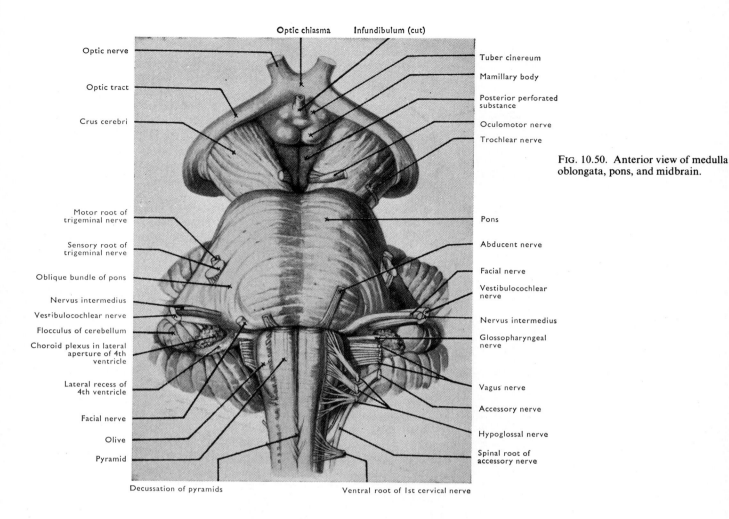

Optic chiasma   Infundibulum (cut)

Optic nerve

Optic tract

Crus cerebri

Motor root of
trigeminal nerve

Sensory root of
trigeminal nerve

Oblique bundle of pons

Nervus intermedius

Vestibulocochlear nerve

Flocculus of cerebellum

Choroid plexus in lateral
aperture of 4th
ventricle

Lateral recess of
4th ventricle

Facial nerve

Olive

Pyramid

Decussation of pyramids

Tuber cinereum

Mamillary body

Posterior perforated
substance

Oculomotor nerve

Trochlear nerve

Pons

Abducent nerve

Facial nerve

Vestibulocochlear
nerve

Nervus intermedius

Glossopharyngeal
nerve

Vagus nerve

Accessory nerve

Hypoglossal nerve

Spinal root of
accessory nerve

Ventral root of 1st cervical nerve

FIG. 10.50. Anterior view of medulla oblongata, pons, and midbrain.

with numbers greater than V to the medulla oblongata. The deep connections of these nerves are dealt with on pages 646–51.

In its caudal half, the medulla oblongata retains many of the features of the spinal medulla, including the presence of a **central canal** which opens out into the fourth ventricle about the middle of the medulla oblongata, having moved posteriorly in its lower half. This posterior movement of certain structures within the medulla oblongata is the result of the decussation of the corticospinal tracts and medial lemnisci [p. 695 and FIG. 10.52]. In the first cervical segment a round bundle of fibres (the lateral corticospinal tract) bulges medially into the intermediate grey matter from each lateral funiculus [FIG. 10.52], and swings into the position of the white commissure of the spinal medulla in the lowest part of the medulla oblongata [FIG. 10.52]. Here the bundles from the two sides interlace in alternate fascicles of considerable size and cross to the ventral surface of the medulla oblongata on which they can be followed upwards until they plunge into the inferior border of the pons. Because of the manner of crossing, the medullary bundles are at first narrow but rapidly increase in size as each fascicle enters from the commissure so that they form the narrow elongated **pyramids** which are a characteristic feature of the ventral aspect of the medulla oblongata. This massive **decussation of the pyramids** [FIGS. 10.50, 10.53, and 10.58] virtually fills the lower third of the anterior median fissure in the medulla oblongata and, cutting through the grey matter, separates the anterior horn of the first cervical segment of

the spinal medulla from its equivalent in the medulla oblongata, the hypoglossal nucleus. It also cuts off the spinal nucleus of the accessory nerve from the nucleus ambiguus [FIG.10.52F–H], both of which belong to a single elongated group of cells supplying motor fibres to the glossopharyngeal, vagus, and accessory nerves [FIG. 10.59].

On each side, just superior to the decussation of the pyramids, the fasciculi cuneatus and gracilis enlarge to form the corresponding tubercles by the appearance of nerve cells (the nuclei **cuneatus and gracilis**) among their fibres. The smaller, more protuberant, **gracile tubercle** extends to the obex [FIG. 10.46] while the **cuneate tubercle** passes upwards lateral to the most caudal part of the fourth ventricle. The fibres of the fasciculi cuneatus and gracilis end on the cells of the corresponding nuclei. These cells give rise to axons which sweep ventrally across the midline (**decussation of the lemnisci**; FIG. 10.52) and turn headwards as a vertical sheet, the **medial lemniscus**. This abuts on the median raphe, dorsal to the corresponding pyramid and ventral to the superior continuation of the fibres of the anterior funiculus of the spinal medulla [FIG. 10.58, mlf].

Traced superiorly, the decussations of the pyramids and lemnisci transfer a large number of fibres from the lateral and dorsal parts of the spinal medulla to a ventromedial position in the medulla oblongata [FIG. 10.52]. This displaces the central canal, hypoglossal nucleus and the fibres of the anterior funiculus dorsally, while the lateral funiculus is moved dorsolaterally towards the dorsolateral

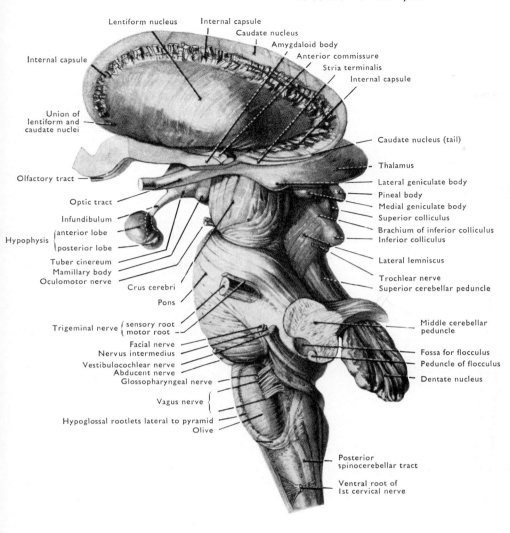

Lentiform nucleus
Internal capsule
Caudate nucleus
Amygdaloid body
Anterior commissure
Stria terminalis
Internal capsule
Internal capsule
Union of lentiform and caudate nuclei
Caudate nucleus (tail)
Olfactory tract
Thalamus
Lateral geniculate body
Optic tract
Pineal body
Infundibulum
Medial geniculate body
Hypophysis {anterior lobe / posterior lobe}
Superior colliculus
Brachium of inferior colliculus
Inferior colliculus
Tuber cinereum
Mamillary body
Oculomotor nerve
Lateral lemniscus
Crus cerebri
Trochlear nerve
Superior cerebellar peduncle
Pons
Trigeminal nerve {sensory root / motor root}
Middle cerebellar peduncle
Facial nerve
Nervus intermedius
Vestibulocochlear nerve
Abducent nerve
Glossopharyngeal nerve
Fossa for flocculus
Peduncle of flocculus
Dentate nucleus
Vagus nerve
Hypoglossal rootlets lateral to pyramid
Olive
Posterior spinocerebellar tract
Ventral root of 1st cervical nerve

FIG. 10.51. Left lateral aspect of brain after removal of the cerebral hemisphere (except corpus striatum) and the cerebellum (except dentate nucleus).

fasciculus [FIG. 10.52G] which is enlarged in the medulla oblongata by descending fibres of the trigeminal nerve and is known as the spinal tract of the trigeminal nerve. On the lateral aspect of the medulla oblongata this **spinal tract of the trigeminal nerve** forms a faint ridge with the roots of the cranial nerves attached to the lateral surface emerging through its ventral part. Immediately ventral to the spinal tract of the trigeminal nerve a shallow groove, the posterior lateral sulcus, separates it from an oval swelling, the **olive**, which lies in the upper two-thirds of the medulla oblongata. The olive is separated from the pyramid ventrally by the **anterior lateral sulcus** which transmits the rootlets of the hypoglossal nerve inferiorly and the abducent nerve superiorly [FIG. 10.50]. In section the olive is a crinkled bag of grey matter (olivary nucleus) which is open medially (hilus). Its efferent fibres emerge through the hilus, decussate, then sweep posterolaterally into the opposite inferior cerebellar peduncle on the posterolateral aspect of the medulla oblongata. A thin plaque of grey matter medial to the **olivary nucleus**, but separated from it by the emergent rootlets of the hypoglossal nerve, is the **medial accessory olivary nucleus**, and a similar piece of grey matter dorsal to the olivary nucleus forms the **dorsal accessory olivary nucleus** [FIG. 10.62].

The appearance of the olive lateral to the pyramid displaces the fibres from the lateral funiculus further posteriorly so that the most dorsal fibres, the **posterior spinocerebellar tract** [FIG. 10.51], move posteriorly over the surface of the spinal tract of the trigeminal

nerve [FIG. 10.143] into the inferior cerebellar peduncle [FIG. 10.57B]. Here the posterior spinocerebellar tract may be visible on the surface as an ill-defined ridge.

The change in position of the tracts, the appearance of the olive, and the outfolding which produces the fourth ventricle together make the grey matter, corresponding to that in the spinal medulla, into a horizontal layer in the floor of the ventricle which contains the nuclei of the cranial nerves [FIG. 10.47]. In addition to this, many fibres of the compact funiculi of the spinal medulla spread out in the region between the olivary nucleus and the cranial nerve nuclei. These are mixed with scattered nerve cells to form the reticular formation of the medulla oblongata.

THE RETICULAR FORMATION [FIGS. 10.57 and 10.58]

This apparently confused jumble of nerve cells and fibres is roughly divisible into lateral and medial parts, the former lying dorsolateral to the olive, the latter dorsomedial to it. The dispersal of the deeper fibres of the anterior and lateral funiculi of the spinal medulla into the reticular formation obscures many of the nuclei in the medulla oblongata and makes the differentiation of grey and white matter less obvious except in certain well-defined nuclei, e.g. the olivary nucleus. The arrangement of the nerve cells and fibres which constitute the reticular formation is such that it would permit

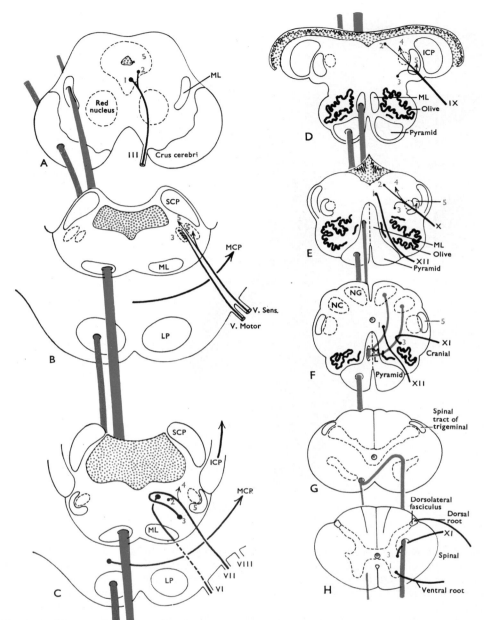

FIG. 10.52. Diagram showing the variations in shape of the parts of the brain stem including its cavities (blue dots), the change in position of the long ascending (F, blue, medial lemniscus) and descending (G, red, corticospinal) pathways in the lower medulla oblongata, and the positions of the cranial nerves and their nuclei. Cranial nerve nuclei indicated by the same red numeral are of the same functional type and form parts of one column cells, all lying in the same relative position: 1, somatic motor; 2 visceral motor (preganglionic); 3, branchial motor; 4, visceral sensory (tractus solitarius); 5, somatic sensory (parts of trigeminal).

ICP, inferior cerebellar peduncle; LP, longitudinal fibres of pons; MCP, middle cerebellar peduncle; ML, medial lemniscus; NC, nucleus cuneatus; NG, nucleus gracilis; SCP, superior cerebellar peduncle. Roman numerals indicate cranial nerves.

A, midbrain. B and C, mid and low pons. D, E, F, and G, medulla oblongata. H, upper cervical spinal medulla. The different drawings are not to scale.

multiple connections between the cells and ascending and descending impulses in the predominantly longitudinal bundles of fibres. It is certainly concerned with a considerable number of different functions [p. 707].

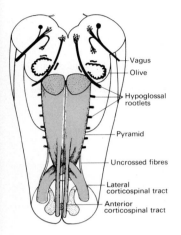

Vagus

Olive

Hypoglossal rootlets

Pyramid

Uncrossed fibres

Lateral corticospinal tract

Anterior corticospinal tract

FIG. 10.53. Diagram of the medulla oblongata and spinal medulla to show the decussation of the pyramids (corticospinal tracts).

THE PONS

The pons is the part of the hindbrain lying ventral to the upper half of the fourth ventricle. Externally if forms a rounded ridge which bulges ventrally separating the medulla oblongata below from the midbrain above [FIGS. 10.51 and 10.56]. Its anterior surface is grooved longitudinally by the basilar artery, and is continuous on each side of the hindbrain with a smooth, rounded middle cerebellar peduncle, which gradually sinks into the ventral extremity of the horizontal fissure of the cerebellum. The pons and middle cerebellar peduncles are directly continuous, but are arbitrarily separated by the trigeminal nerve which pierces them at their junction.

The pons consists of two parts:

1. The above ridge of tissue, or ventral part of the pons, appears to pass uninterruptedly from one cerebellar hemisphere to the other. It consists of groups of nerve cells (pontine nuclei) traversed and covered by the nerve fibres which arise from them. These fibres cross the midline horizontally and pass backwards to the cerebellum in the opposite middle cerebellar peduncle. The pontine nuclei on each side are further split up by a number of bundles of fibres which run vertically through the ventral part of the pons from the crus cerebri on the ventral surface of the midbrain. The fibres in these bundles end on the pontine nuclei except for a minority which pass through the pons and emerge as the corresponding pyramid on the medulla oblongata.

2. The dorsal part of the pons lies between the fourth ventricle and the ventral part. It unites the greater part of the medulla oblongata to the midbrain, and transmits pathways between them with the exception of those in the crura cerebri which traverse the ventral part of the pons. It also contains the superior part of the vesibular area, nuclei of the trigeminal, abducent, and facial cranial nerves, and the pontine reticular formation which is directly continuous below with that in the medulla oblongata and above with that in the midbrain.

THE MIDBRAIN [FIGS. 10.54 and 10.55]

The midbrain forms the narrow neck which connects the hindbrain (including the cerebellum) to the forebrain and passes through the aperture in the tentorium cerebelli [tentorial notch, p. 729]. It is the site of the cephalic flexure of the embryonic brain. Hence the dorsal aspect of the midbrain is the more extensive, and the forebrain lies anterior as well as superior to it.

In the midbrain the roof plate is thick and the proliferation of cells of the alar and basal laminae leads to a steady reduction in the relative size of its cavity so that it comes to form the narrow, tubular aqueduct of the cerebrum. This traverses the midbrain from end to end, and connecting the median cavity of the forebrain (third ventricle) to the fourth ventricle, forms a vital link through which the cerebrospinal fluid formed in the ventricles of the forebrain can reach the fourth ventricle and escape into the subarachnoid space [p. 728] through the apertures in its roof.

The aqueduct is surrounded by a thick cylindrical sheath of grey matter, the central grey substance which contains many nuclei including those of the oculomotor and trochlear nerves and the mesencephalic nucleus of the trigeminal nerve [pp. 646–51].

Tectum

This the part of the midbrain dorsal to the aqueduct. It is derived from the alar lamina and consists of the dorsal part of the central grey substance and four small swellings of grey matter which protrude from the dorsal surface, two on each side, the superior and inferior colliculi. The superior pair are broader and flatter than the inferior pair which rise abruptly above the superior medullary velum and overhang the emerging trochclear nerves and the superior cerebellar peduncles as they enter the midbrain [FIG. 10.51]. Here the lateral surface of each peduncle is covered by a ribbon of white matter (the lateral lemniscus) which carries auditory impulses to the inferior colliculus. These are transmitted from the colliculus by a similar but narrower ridge of white matter, the brachium of the inferior colliculus, which runs forwards and upwards over the lateral surface of the midbrain to enter a rounded swelling on the posterior edge of the upper end of the crus cerebri. This medial geniculate body is a projection from the inferior surface of the thalamus [FIGS. 10.51 and 10.54].

On the same surface of the thalamus above the medial geniculate body is a bundle of nerve fibres, often invisible on the surface, which sweeps backwards and medially to the dorsal surface of the upper midbrain reaching the superior colliculus and the pretectal region immediately anterior to it. This brachium of the superior colliculus is composed of nerve fibres from the optic tract carrying visual impulses to the midbrain.

Cerebral peduncle

The part of the midbrain ventral to the aqueduct is much larger than the tectum and each half, followed superiorly, expands laterally to become continuous with the corresponding half of the cerebrum. Each half of this part of the midbrain is known, therefore, as a cerebral peduncle and has a broad, thick strip of white matter, the crus cerebri, on its anterolateral surface. The crura cerebri originate in the cerebral hemispheres and converge as they pass inferiorly on the midbrain to enter the ventral part of the pons and split into longitudinal bundles. The posterior part of the interpeduncular fossa lies on the ventral aspect of the midbrain between the crura cerebri, and has the oculomotor nerves arising from its lateral walls

FIGS. 10.54–10.58. Photographs of transverse sections, posterior (left) and lateral (right) views of the human brain stem, to show the surface features and internal structure of its various regions. The posterior and lateral views show the position of the individual sections.
FIG. 10.54A has most of the surface features labelled, but in the subsequent figures only those surface features at the level of the section are lettered.
[See also FIG. 10.52.]

a.	Aqueduct [FIG. 10.54]; accessory cuneate nucleus [FIG. 10.58].
an.	Arcuate nucleus.
ap.	Area postrema.
asc.	Anterior spinocerebellar tract.
bic.	Brachium of inferior colliculus.
bsc.	Brachium of superior colliculus.
c.	Central canal [FIG. 10.58]; central nucleus of thalamus [FIG. 10.54].
cc.	Crus cerebri.
cgs.	Central grey substance.
ch.	Optic chiasma.
ci.	Internal capsule.
cn.	Cochlear nuclei.
dn.	Dentate nucleus.
do.	Dorsal accessory olivary nucleus.
fc.	Facial colliculus.
gt.	Gracile tubercle.
h.	Habenular triangle.
ic.	Inferior colliculus.
icc.	Intercollicular commissure.
icp.	Inferior cerebellar peduncle.
ipf.	Interpeduncular fossa.
ivn.	Inferior vestibular nucleus.
l.	Lateral nucleus of thalamus.
lfp.	Longitudinal fibres of pons.
lg.	Lateral geniculate body.
ll.	Lateral lemniscus.
m.	Mamillary body.
mcp.	Middle cerebellar peduncle.
me.	Medial eminence.
mg.	Medial geniculate body.
ml.	Medial lemniscus.
mlf.	Medial longitudinal fasciculus.
mo.	Medial accessory olivary nucleus.
nc.	Nucleus cuneatus.
ng.	Nucleus gracilis.
nt.	Dorsal nucleus of trapezoid body.
o.	Optic nerve [FIGS. 10.54 and 10.55]; obex [FIG. 10.57].
ol.	Olive.
ot.	Optic tract.
p.	Pineal [FIG. 10.54]; pons [FIGS. 10.55 and 10.56].

pc.	Posterior commissure.
pd.	Pyramidal decussation.
po.	Central tegmental fasciculus.
psc.	Posterior spinocerebellar tract.
pyr.	Pyramid.
r.	Reticular nucleus of thalamus.
rf.	Reticular formation.
rn.	Red nucleus.
rs.	Rubrospinal tract.
s.	Subthalamic nucleus.
sc.	Superior colliculus.
scp.	Superior celebellar peduncle.
sm.	Striae medullares of fourth ventricle.
smt.	Stria medullaris of thalamus.
smv.	Superior medullary velum.
sn.	Substantia nigra.
st.	Spinothalamic tract.
tb.	Decussating fibres of trapezoid body.
tfp.	Transverse fibres of pons.
th.	Hypoglossal triangle.
ts.	Tractus solitarius.
tsv.	Thalamostriate vein.
tv.	Vagal triangle.
va.	Vestibular area.
vm.	Vestibulomesencephalic tract.
vn.	Vestibular nuclei.
vpl.	Nucleus ventralis posterolateralis of thalamus.
vpm.	Nucleus ventralis posteromedialis of thalamus.
vs.	Superior vestibular nucleus.
z.	Zona incerta.
III.	Oculomotor nerve.
IV.	Trochlear nerve.
V.	Trigeminal nerve.
VI.	Abducent nerve.
VII.	Facial nerve.
VIIn.	Nucleus of facial nerve.
VIII.	Vestibulocochlear nerve.
IX.	Glossopharyngeal nerve.
X.	Vagus.
XIc.	Accessory nerve, cranial root.
XII.	Hypoglossal nerve.

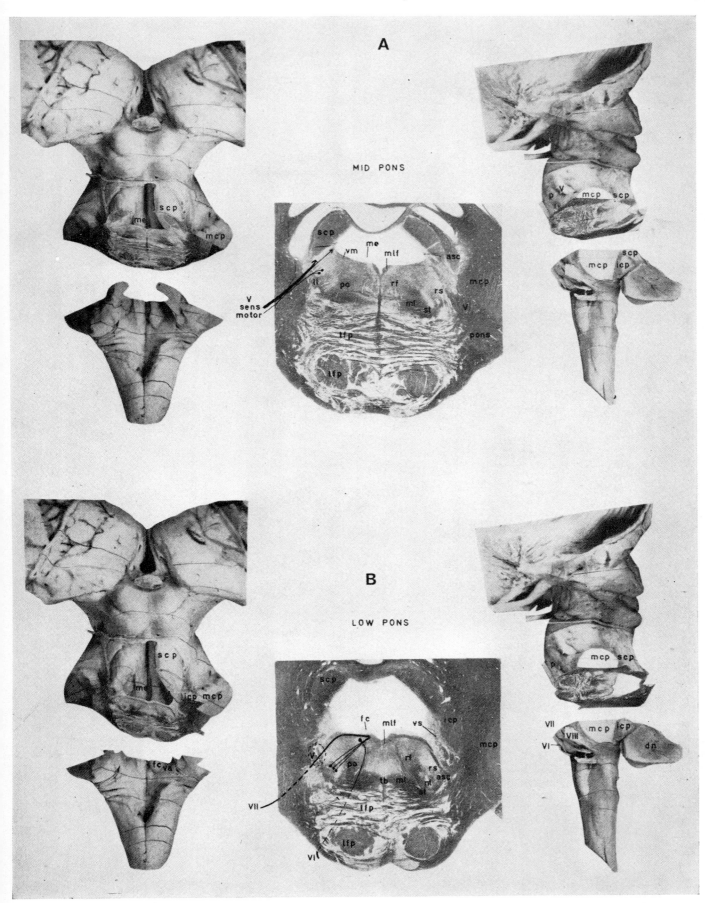

FIG. 10.56.

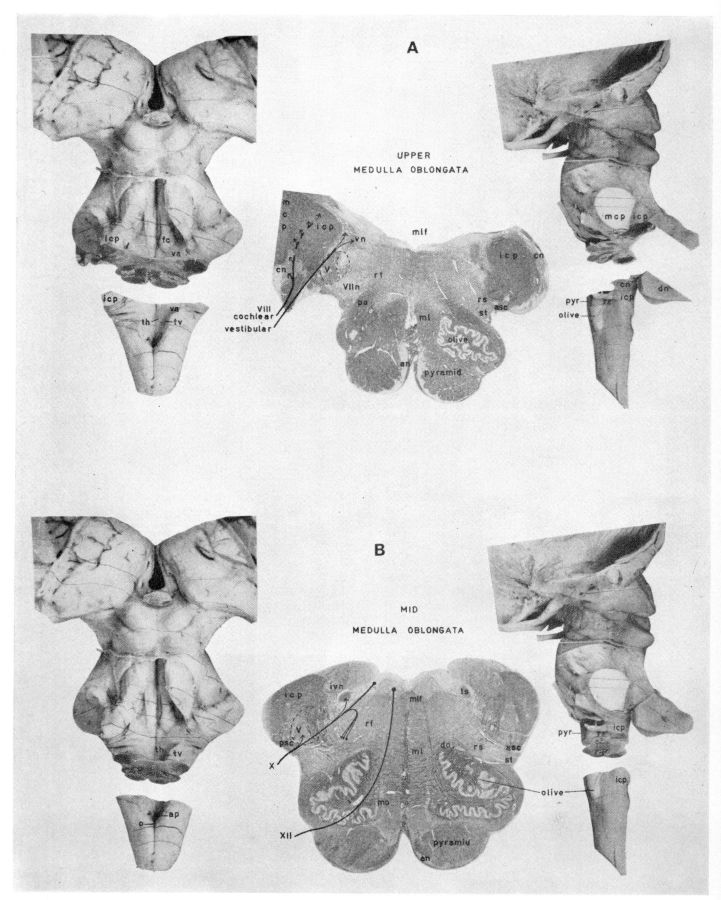

Fig. 10.57.

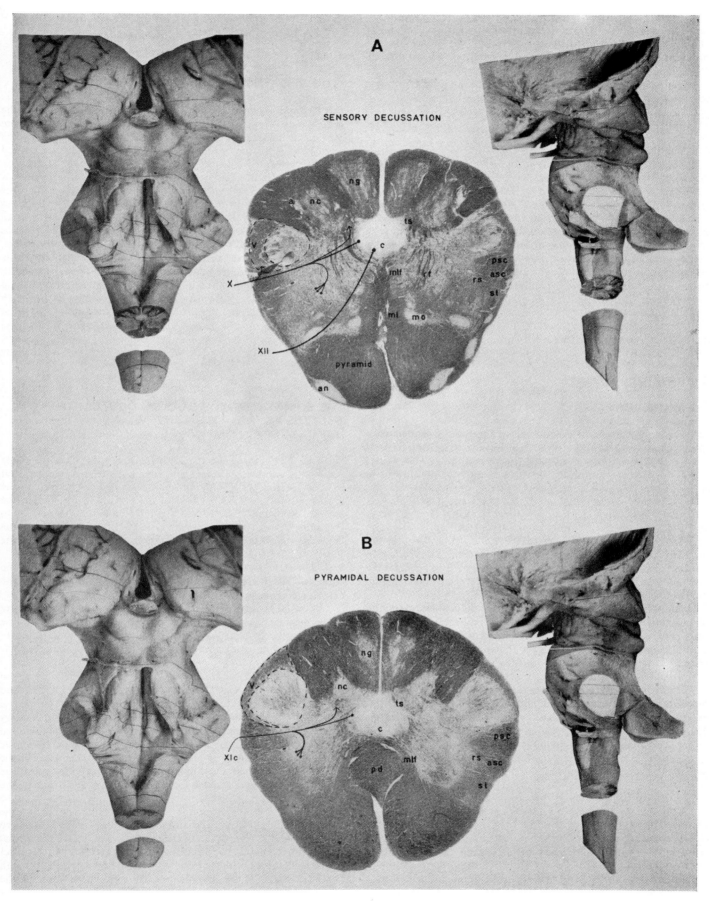

FIG. 10.58.

at the level of the superior colliculi [FIG. 10.54]. A leash of small branches from the posterior cerebral arteries pierce the median part of the fossa, and enter the ventral surface of the midbrain, thus producing the **posterior perforated substance**. These arteries supply the median part of the midbrain and the posterior part of the hypothalamus which lies immediately anterior to the midbrain and completes the roof of the interpeduncular fossa [FIG. 10.70].

Tegmentum

Dorsal to the crura cerebri, the remainder of the cerebral peduncles is composed of the tegmentum of the midbrain. It is separated from the crura on each side by a layer of grey matter composed for the most part of melanin pigmented cells, the **substantia nigra**. The tegmentum of the midbrain is essentially the upward continuation of the dorsal part of the pons, from which it is in no way separated except by an arbitrary line at the upper border of the pons. It contains the **reticular formation of the midbrain**, the crossing of the superior cerebellar peduncles in its lower part, and the large **red nuclei** in its upper part close to the midline [FIGS. 10.54 and 10.55]. Superiorly the tegmentum of the midbrain becomes continuous with the forebrain without any clear dividing line separating them; in fact the red nucleus and the substantia nigra extend upwards into the forebrain as do the long ascending tracts which have traversed the medulla, pons, and midbrain without interruption.

The cranial nerves

COMPONENTS

In their arrangement, the cranial nerves are akin to the primitive spinal nerves of the lamprey in which the dorsal and ventral roots do not join but run a separate course to the periphery. The cranial nerves of all higher forms are also reminiscent of these primitive spinal nerves in that the preganglionic or **visceral motor** (parasympathetic) nerve fibres, which supply the cephalic part of the gut tube, emerge through the laterally attached, ganglionated nerves and not through the ventral series, except in the case of the oculomotor (III) nerve. The laterally attached cranial nerves are therefore both motor and sensory in function, though equivalent in position to the dorsal roots of spinal nerves [FIG. 10.47], while the ventral series, including the trochlear, are almost exclusively **somatic motor**.

In addition, striated muscle is developed in the cephalic extremity of the gut tube, particularly in relation to the branchial arches [p. 55] which are supplied by the laterally attached cranial nerves. In these cranial nerves, therefore, as a special modification of the visceral efferent group, there are nerve fibres which pass directly to this striated muscle without any interposed ganglion cells, the **branchial motor** cells [p. 648]. They are identical with the nerve fibres (somatic motor) which supply the striated muscle of the body wall and limbs, except that they emerge in the laterally attached cranial nerves. Thus each laterally attached cranial nerve can contain two types of motor nerve fibres, in addition to the sensory fibres which, as in the spinal nerves, are divisible into those supplying the gut tube (**visceral sensory**), and those supplying the skin, muscles, etc. (**somatic sensory**).

The central connections of the cranial nerves, like their spinal counterparts, are arranged in such a manner that the motor nerve fibres originate in groups of cells (nuclei) lying opposite the point of emergence of the nerve to which they contribute, while the entering sensory fibres run longitudinally before ending on the receptive, internuncial cells or 'sensory nuclei' which may, therefore, lie at some distance from the point of entry of the fibres. *Likewise there is within the brain a separate nucleus of origin (motor) or of termination* *(sensory) for each of the different functional groups of nerve fibres which each nerve contains.*

Nuclei of the cranial nerves

In the spinal medulla the rootlets which gather together into the ventral roots of the nerves form a continuous series and arise from continuous columns of motor nerve cells which develop from the basal lamina. In the brain only the IXth, Xth, and XIth cranial nerves are formed from a continuous series of rootlets and it is therefore only in these nerves that the motor nuclei form continuous columns of cells. The remaining nerves, separated from each other by varying intervals, arise from separate groups of cells. Nevertheless each motor nucleus of the same functional group lies in the same relative position, as though it formed a part of a continuous longitudinal column or rod of cells. Thus the **somatic motor nuclei**, being derived from the ventral part of the basal lamina, always lie adjacent to the midline ventral to the central canal, the fourth ventricle, or the aqueduct; while the **visceral motor nuclei** take up a more dorsal and lateral position except in the region of the fourth ventricle where the lateral wall of the tube is folded outwards so that these nuclei lie directly lateral to the somatic motor nuclei. The **branchial motor nuclei** take up a more ventral position and, except in the case of the trigeminal nerve, send their fibres curving dorsally to join one of the laterally attached cranial nerves [FIG. 10.52, 3].

THE SOMATIC MOTOR NUCLEI—OCULOMOTOR (III), TROCHLEAR (IV), ABDUCENT (VI), AND HYPOGLOSSAL (XII) NERVES [FIGS. 10.52–10.62]

Of these the IVth [see also p. 635], VIth, and XIIth have the single function of supplying striated somatic muscle (somatic motor) and each arises from a single motor nucleus located respectively in the lower midbrain, the lowest part of the pons, and throughout the medulla oblongata [FIG. 10.59]. The IIIrd cranial nerve belongs to this same series, and the major portion of its nucleus lies in the upper midbrain in line with, and almost connected to, that of the IVth. It differs from the others in two respects:

1. The right and left IIIrd nerve nuclei are fused ventral to the aqueduct and interchange somatic motor fibres in the supply of the extrinsic ocular muscles [FIGS. 10.54 and 10.60].

2. The IIIrd cranial nerve also contains visceral motor fibres concerned with the supply of the intrinsic ocular muscles. These arise in an incompletely fused mass, the **accessory** or Edinger–Westphal (autonomic) **nucleus**, lying dorsal to the main nuclear complex.

The **emergent fibres** from these nuclei pass ventrally, except in the case of the trochlear nerve where the fibres pass laterally and wind spirally in a caudal direction around the margin of the central grey substance to reach the superior part of the superior medullary velum. Here the two nerves meet and decussate deep to the frenulum of the superior medullary velum [FIG. 10.46]. Each emerges on the opposite side dorsal to the superior cerebellar peduncle. As the fibres of the trochlear nerve wind round the central grey substance they pass close to the **mesencephalic nucleus** and tract of the trigeminal nerve and send some fibres into it. These may be proprioceptive fibres from the superior oblique muscle to the mesencephalic nucleus (Tarkhan 1934). The fibres of the IIIrd nerve pass through the tegmentum of the midbrain and pierce the medial part of the red nucleus as a series of separate bundles which collect in the medial part of the substantia nigra before emerging through the lateral wall of the interpeduncular fossa. The fibres of the abducent pass slightly caudally, as they run ventrally, and thus

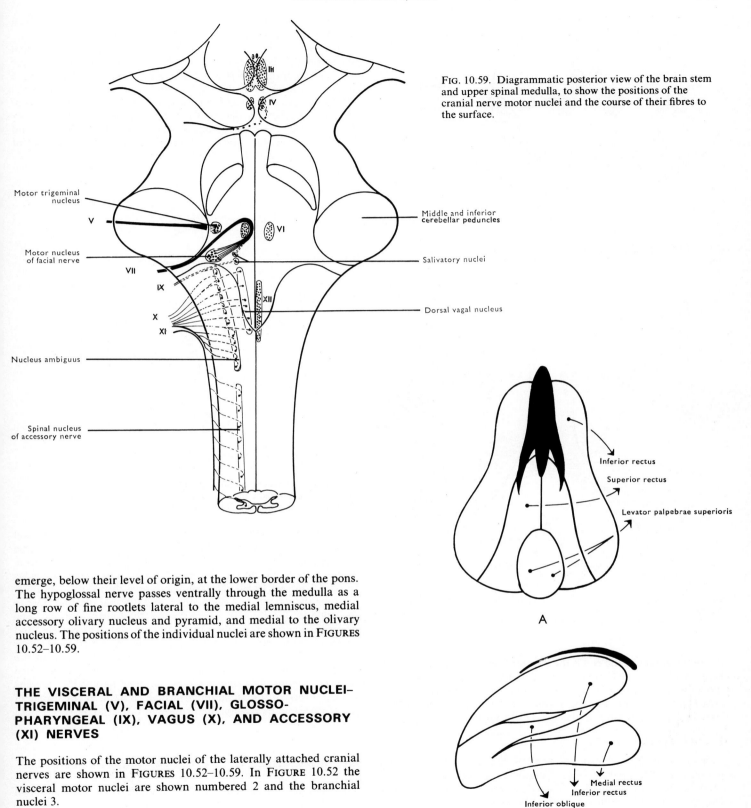

FIG. 10.59. Diagrammatic posterior view of the brain stem and upper spinal medulla, to show the positions of the cranial nerve motor nuclei and the course of their fibres to the surface.

FIG. 10.60. Diagrams of the oculomotor nucleus showing its subdivisions and their functional significance. After Warwick (1953).

A, dorsal view. B, lateral view.

Accessory (autonomic) nucleus, solid black.

emerge, below their level of origin, at the lower border of the pons. The hypoglossal nerve passes ventrally through the medulla as a long row of fine rootlets lateral to the medial lemniscus, medial accessory olivary nucleus and pyramid, and medial to the olivary nucleus. The positions of the individual nuclei are shown in FIGURES 10.52–10.59.

THE VISCERAL AND BRANCHIAL MOTOR NUCLEI–TRIGEMINAL (V), FACIAL (VII), GLOSSO-PHARYNGEAL (IX), VAGUS (X), AND ACCESSORY (XI) NERVES

The positions of the motor nuclei of the laterally attached cranial nerves are shown in FIGURES 10.52–10.59. In FIGURE 10.52 the visceral motor nuclei are shown numbered 2 and the branchial nuclei 3.

Visceral motor nuclei

Of the laterally attached cranial nerves only the VIIth, IXth, Xth, and cranial portion of the XIth contain visceral motor or preganglionic parasympathetic fibres, and of these the last three receive such fibres from a single nucleus lying lateral to the hypoglossal nucleus and known as the **dorsal nucleus of the vagus**.

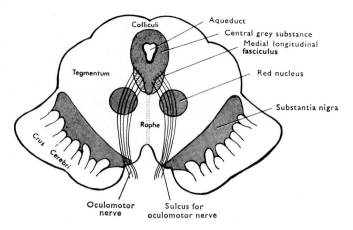

FIG. 10.61. Diagrammatic transverse section through upper part of midbrain.

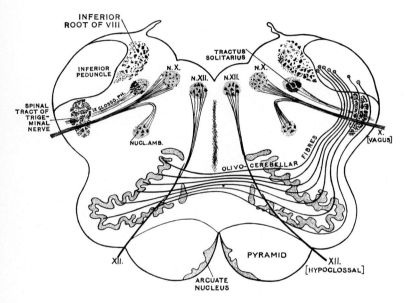

FIG. 10.62. Diagram of section of medulla oblongata to show the cranial nerve nuclei.

N.X., dorsal nucleus of vagus; N.XII, hypoglossal nucleus.

At its extreme cephalic end is the **inferior salivatory nucleus** which supplies fibres to the IXth nerve, predominantly concerned with the innervation of the parotid and posterior intralingual salivary glands. Just cephalic to this lies the **superior salivatory nucleus** which contributes similar fibres to the facial (VII) nerve.

Branchial motor nuclei

This system of nuclei contributes to all the laterally attached cranial nerves except the purely sensory vestibulocochlear nerve. At the cephalic end of this column, in the middle of the pons, lies the motor trigeminal nucleus; at the lower border of the pons the motor nucleus of the facial nerve, and caudal to this is a continuous column of cells, broken into two parts by the decussation of the pyramids. A cephalic part, the **nucleus ambiguus** [FIG. 10.62], sends a few fibres into the glossopharyngeal, but the majority into the vagus and cranial part of the accessory for the supply of the pharyngeal and laryngeal muscles. The caudal part, the **spinal nucleus of the accessory nerve**, extends from the decussation to the fifth cervical segment

and gives rise to the spinal roots of the accessory nerve, which supply the sternocleidomastoid and trapezius muscles.

The fibres arising in the nucleus ambiguus and in the spinal nucleus of the accessory nerve, which lies in the posterior part of the cervical anterior horn [FIG. 10.63], pass dorsally and then curve laterally. Fibres from the nucleus ambiguus join those from the dorsal nucleus of the vagus and emerge through the ventral part of the spinal tract of the trigeminal nerve. Fibres of the spinal accessory nerve emerge between the ligamentum denticulatum [p. 729] and the dorsal roots, turn cranially, and join to form the spinal part of the accessory nerve [FIG. 10.50].

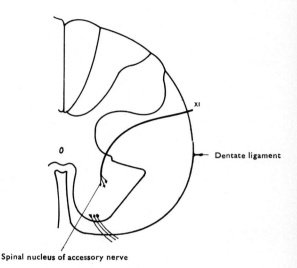

FIG. 10.63. Transverse section through the first cervical segment of the spinal medulla to show the nucleus of origin, and the mode of emergence of the spinal roots of the accessory nerve, XI.

The **motor nucleus of the facial nerve** [FIGS. 10.56, 10.59, and 10.64] lies at the level of the lower border of the pons medial to the emergent root of the facial nerve. From the nucleus fibres converge on the inferior part of the abducent nucleus in the facial colliculus [FIGS. 10.56 and 10.59]. The compact nerve then turns superiorly between the nucleus and the midline, and runs parallel and close to the opposite facial nerve in the floor of the fourth ventricle. The nerve then curves laterally, superior to the abducent nucleus, and turns ventrolaterally to emerge ventral to the vestibulocochlear nerve. The loop around the abducent nucleus is the **genu of the facial nerve**. At its point of emergence, the facial nerve is separated from the vestibulocochlear nerve by the small **nervus intermedius (sensory root of the facial nerve)** which contains at least some of the sensory and parasympathetic fibres of the facial nerve.

Sensory fibres of cranial nerves [FIG. 10.65]

These behave in a fashion very similar to those in the dorsal spinal roots with three major exceptions:

1. The majority of the **somatic sensory fibres** for the cranial region pass with the trigeminal nerve [FIG. 10.67]. For this reason it gives its name to the longitudinal bundles of somatic sensory fibres which course throughout the length of the brain stem from the upper midbrain to the upper cervical spinal medulla, even though similar fibres enter the same system in smaller numbers through the other laterally attached cranial nerves, just as the several dorsal roots of

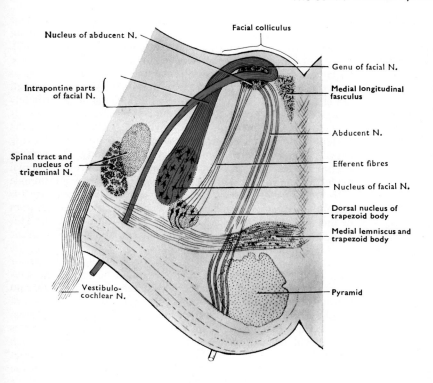

Nucleus of abducent N.

Facial colliculus

Genu of facial N.

Intrapontine parts of facial N.

Medial longitudinal fasiculus

Abducent N.

Efferent fibres

Spinal tract and nucleus of trigeminal N.

Nucleus of facial N.

Dorsal nucleus of trapezoid body

Medial lemniscus and trapezoid body

Vestibulo-cochlear N.

Pyramid

FIG. 10.64. Diagram of intrapontine course of facial nerve, seen from above.

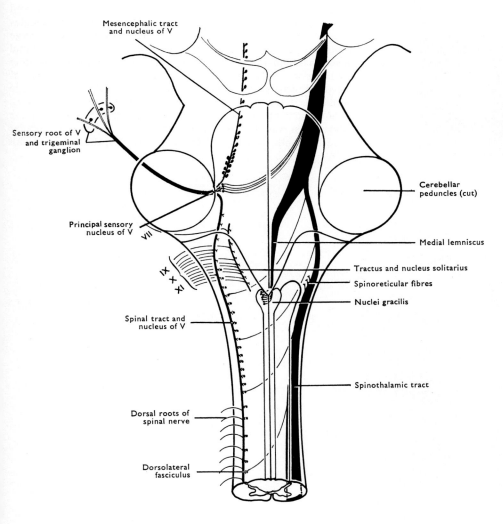

Mesencephalic tract and nucleus of V

Sensory root of V and trigeminal ganglion

Principal sensory nucleus of V

Cerebellar peduncles (cut)

Medial lemniscus

Tractus and nucleus solitarius

Spinoreticular fibres

Nuclei gracilis

Spinal tract and nucleus of V

Spinothalamic tract

Dorsal roots of spinal nerve

Dorsolateral fasiculus

FIG. 10.65. Diagrammatic posterior view of the brain stem and upper spinal medulla, to show the positions of the sensory nuclei of termination of the cranial nerves, except the eighth. The secondary ascending pathways are also shown.

the spinal medulla send their fibres into the same longitudinal system, the dorsolateral fasciculus.

2. The midbrain or **mesencephalic portion** of this system differs from all other sensory systems in that the primary sensory cell bodies lie within the midbrain and upper pons, forming a long **mesencephalic nucleus**, instead of lying in the trigeminal ganglion [FIG. 10.65].

3. The **visceral sensory fibres**, including those subserving taste, also form a well-defined longitudinal tract which, being surrounded by its nucleus of termination, forms an isolated bundle, the **tractus solitarius** [FIGS. 10.62 and 10.65] lateral to the dorsal nucleus of the vagus.

SOMATIC SENSORY FIBRES

The **sensory fibres of the trigeminal nerve** enter the brain stem immediately caudal to the motor root. Having pierced the superficial part of the pons at its junction with the middle cerebellar peduncle

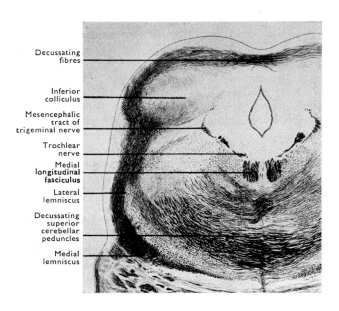

FIG. 10.66. Section through inferior colliculus and tegmentum below the level of nucleus of trochlear nerve.

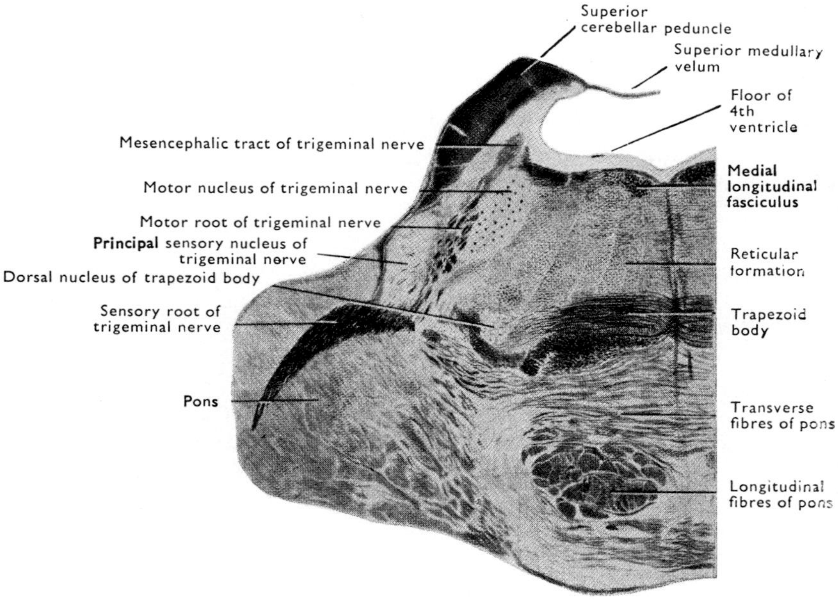

and entered the dorsal part of the pons, many of the fibres end immediately on a large (**superior**) **principal sensory nucleus** lying just lateral to the motor nucleus of the trigeminal nerve [FIG. 10.67]. The fibres which end here are believed to be concerned with tactile sensations. The majority of the remaining fibres turn caudally to form the **spinal tract of the trigeminal nerve**. This descends through the lateral part of the pons and medulla oblongata immediately ventral to the inferior cerebellar peduncle and nucleus cuneatus. It may form an ill-defined surface ridge which is obliterated where the posterior spinocerebellar tract crosses it [FIG. 10.57B]. Moving dorsally as the fourth ventricle closes inferiorly, it becomes continuous with the **dorsolateral fasciculus** in the cervical spinal medulla. Here the fibres of the spinal tract of the nerve intermingle with similar fibres from the cervical dorsal roots. The spinal tract is accompanied on its medial side by the **nucleus of the spinal tract of the trigeminal nerve** and sends collaterals on to its cells. This nucleus is continuous inferiorly with the substantia gelatinosa of the cervical spinal medulla.

The fibres of the spinal tract which descend below the inferior limit of the fourth ventricle (obex) seem to be concerned with relaying impulses which elsewhere are interpreted as painful and thermal sensations, since division of the spinal tract at the level of the obex abolishes these sensations in the ipsilateral trigeminal area. It is, therefore, functionally equivalent to the dorsolateral fasciculus [p. 694] with which it is continuous. The function of the fibres which terminate in the more cranial part of the nucleus of the spinal tract is not known, but there are differences between the cells in the cranial and caudal parts of the nucleus. The spinal tract of the trigeminal nerve is pierced by the facial, glossopharyngeal, and upper vagal rootlets in the same manner as the dorsolateral fasciculus is pierced by the dorsal spinal roots. The accessory rootlets emerge ventral to it [FIG. 10.63]. The remaining fibres of the trigeminal nerve pass cranially into the upper pons and midbrain constituting the **mesencephalic tract**. This lies near the anterolateral corner of the upper part of the fourth ventricle and ascends through the midbrain at the lateral edge of the central grey substance [FIG. 10.66]. This portion is believed to be concerned with proprioceptive information from the muscles of mastication, temporomandibular joint, the tooth sockets, and the facial muscles, the latter through

FIG. 10.67. Transverse section through pons at level of trigeminal nerve.

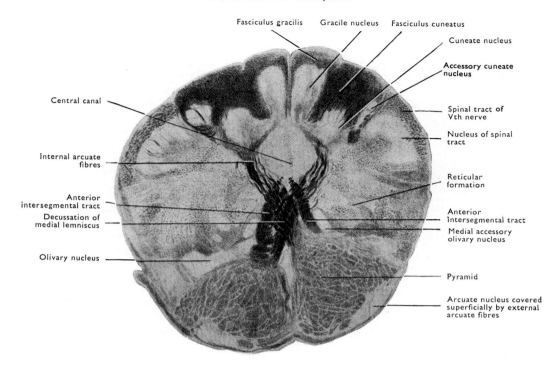

Fasciculus gracilis   Gracile nucleus   Fasciculus cuneatus

Cuneate nucleus

Accessory cuneate nucleus

Central canal

Internal arcuate fibres

Anterior intersegmental tract

Decussation of medial lemniscus

Olivary nucleus

Spinal tract of Vth nerve

Nucleus of spinal tract

Reticular formation

Anterior intersegmental tract

Medial accessory olivary nucleus

Pyramid

Arcuate nucleus covered superficially by external arcuate fibres

FIG. 10.68. Section through adult medulla oblongata immediately above the decussation of the pyramids (Weigert–Pal specimen).

communications with the facial nerve in the face, and by fibres entering with the facial nerve which have been described ascending through the pons to the mesencephalic nucleus. It may also receive proprioceptive fibres from the extrinsic ocular muscles [p. 646] but the physiological evidence indicates that such fibres terminate in the pons.

VISCERAL SENSORY FIBRES

These enter the brain stem through the facial, glossopharyngeal, vagus, and cranial parts of the accessory nerves [FIG. 10.65]. The complete picture of the termination of these fibres is not known, but the majority run caudally in the tractus solitarius and end in the surrounding nucleus of the tractus solitarius. The tractus solitarius extends longitudinally throughout the medulla oblongata in a position between the dorsal nucleus of the vagus and the nucleus of the spinal tract of the trigeminal nerve [FIGS. 10.57 and 10.58]. The cranial part receives taste fibres from the facial, glossopharyngeal, and vagus nerves. The caudal part receives other visceral sensory fibres from the vagus and cranial accessory (cardiovascular, respiratory, and alimentary).

The positioning of the nuclei of the vagus and hypoglossal nerves is shown in FIGURE 10.62. Apart from the special development of the nucleus ambiguus, solitary tract, and vestibulocochlear nuclei, the arrangement is identical with that in the grey matter of the thoracic spinal medulla, except that the grey matter lies horizontally in the upper medulla oblongata after the eversion of the lateral wall of the neural tube in the formation of the fourth ventricle [FIG. 10.47].

The table in the Peripheral Nervous System section gives the functions of the cranial nerves [p. 743]. From this and from an understanding of the general plan of the cranial nerve nuclei, the nuclei of origin and termination of fibres in the cranial nerves may be plotted with some accuracy.

NUCLEI OF VESTIBULOCOCHLEAR NERVE

The nuclei of termination of the vestibulocochlear nerve lie at the level of entry of the nerve and are divided into cochlear and vestibular nuclei. The cochlear fibres, lying in the inferior part of the nerve, end in a curved mass of cells on the lateral and posterior surfaces of the inferior cerebellar peduncle. The vestibular fibres pass anterior to the inferior cerebellar peduncle and are distributed to the vestibular nuclei [FIG. 10.57] which form the raised vestibular area. This roughly triangular mass extends from the mid-pons almost to the obex between the sulcus limitans and the lateral margin of the floor of the fourth ventricle [FIG. 10.46 and p. 707].

THE FOREBRAIN [FIGS. 10.71–10.75]

Initially the forebrain or prosencephalon consists simply of the anterior extremity of the neural tube. It has a deep, slit-like, median cavity closed in front by the thin lamina terminalis and continuous caudally with the aqueduct from which the sulcus limitans passes into its wall as the hypothalmic sulcus [FIG. 10.70]. In the early stages of its development the prosencephalon has many features in common with the hindbrain, including the presence of a thin roof plate which is subsequently invaginated on both sides by the overlying connective tissue to form choroid plexuses vascularized by the terminal branches of the basilar artery. These arteries, part of the surface circular anastomosis, pass dorsally at the junction of the midbrain and forebrain, supplying this region and continuing on to the dorsal surface of the median forebrain. They are similar in their arrangement to the posterior inferior cerebellar arteries of the hindbrain, and are later involved in the supply of the posterior part of the expanding cerebral hemispheres as the posterior cerebral arteries. The original supply to the choroid plexus and dorsal part of

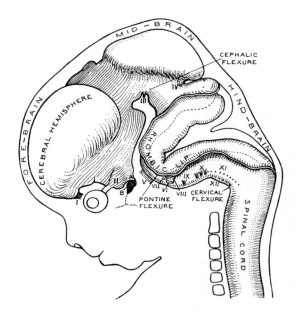

FIG. 10.69. Profile view of brain of human embryo of ten weeks. The cranial nerves are indicated by numerals. (After His.)

A, cerebral diverticulum (posterior lobe) of hypophysis cerebri; B, oral diverticulum, i.e. Rathke's pouch (anterior lobe) of hypophysis.

the median forebrain tube (thalamus) remains as the posterior choroidal branches of these posterior cerebral arteries [FIG. 10.87]. A posterior communicating artery connects each posterior cerebral artery near its origin to the corresponding internal carotid artery as it reaches the forebrain just superior to the buccopharyngeal membrane.

A series of outgrowths of the whole thickness of the forebrain wall forms six subsidiary structures; the two eyes, the paired telencephalic vesicles forming the greater part of the cerebral hemispheres, and two unpaired median structures, the hypophysis and the pineal body. The remaining, paramedian parts of the forebrain surround the third ventricle and form the diencephalon.

The Diencephalon

The first paired outgrowths from the forebrain are the optic vesicles [FIG. 10.31], followed shortly by the appearance of the telencephalic vesicles (cerebral hemispheres) from the upper anterior part of the forebrain [FIG. 10.88]. The optic outgrowths [FIG. 12.32], although hollow at first, rapidly become solid. The telencephalic vesicles remain hollow and therefore contain lateral extensions of the cavity of the forebrain—the lateral ventricles—which are continuous with each other through the part of the original median cavity immediately adjacent to the upper part of the lamina terminalis. This region of communication between the three cavities of the forebrain is known as the interventricular foramen. With the appearance of the lateral ventricles the median cavity is known as the third ventricle. Its walls (the diencephalon) form the thalamus superiorly and the hypothalamus inferiorly, separated on the ventricular aspect by the hypothalamic sulcus which extends forwards from the aqueduct to the interventricular foramen [FIGS. 10.70 and 10.77].

THE HYPOPHYSIS

The closure of the extreme anterior end of the neural tube is not as simple as elsewhere since this process, spreading headwards from

the cervical region, fails to reach the extreme anterior end of the neural plate before that part begins to close from the buccopharyngeal membrane caudally [FIG. 10.31]. The two processes of closure meet at the anterior neuropore in the region of the lamina terminalis; the roof plate between this and the buccophyaryngeal membrane is thus formed separately. In the roof plate adjacent to the buccopharyngeal membrane, the skin and neural ectoderm are in contact at the time of closure of the neural tube. This contact is maintained as proliferating mesoderm of the base of the skull lifts the neural tube away from the roof of the mouth, so that a finger-like pouch of ectoderm is carried upwards from the stomodaeal roof. This Rathke's pouch forms the hypophysis together with the part of the floor of the neural tube with which it is in contact [p. 605]. The hypophysis lies immediately behind the roots of the optic outgrowths (the optic chiasma [p. 703]) at the inferior end of the lamina terminalis. This chiasma is the meeting point of the stalks of the optic outgrowths which form the optic nerves [FIG. 10.69] when filled with nerve fibres from the eyes.

THE PINEAL BODY

This median diverticulum arises from the roof of the forebrain caudal to the thin roof of the third ventricle, at its junction with the midbrain. The cells of the ependymal diverticulum proliferate and form a small, reddish body in the shape of a flattened ovoid approximately 7–10 mm long and 5–7 mm wide, but very variable in size. The cells of which it is composed are secretory in type, and it contains many blood vessels but no duct system, thus suggesting that it produces a hormone or hormones. It is now known to contain several substances which are pharmacologically active, including noradrenaline, 5-hydroxytryptamine (serotonin), and melatonin. Melatonin produces marked contraction of melanophores when injected into frogs, and has many physiological effects in mammals (including inhibition of ovarian growth) produced mainly by direct action on the brain. The concentrations of serotonin and its derivative melatonin vary cyclically and reciprocally in the pineal body; serotonin increasing on exposure to light, melatonin in darkness. Surprisingly these effects are produced by the sympathetic supply to the pineal body and not by the optic tectum on which it lies. Thus the structure which is the pineal eye of the lower forms has an important role in the hormonal reactions to light though no longer directly exposed to it in mammals (Wurtman, Axelrod, and Kelly 1968). After middle age the pineal body is frequently calcified, and it may therefore be possible to demonstrate a shift of the brain to one side in a simple anteroposterior radiograph of the skull. The pineal body extends caudally to rest between the superior colliculi beneath the thick posterior end of the corpus callosum [p. 685] from which it is separated by an extension of the third ventricle, the suprapineal recess, and the junction of the internal cerebral veins to form the great cerebral vein [FIG. 10.70]. On either side of the base of the pineal, the posterior choroidal arteries pass forwards in company with the internal cerebral veins above the third ventricle and thalamus. A small pineal recess of the third ventricle extends into the base of the pineal between two commissures which cross close to its attachment—the posterior commissure below and the habenular commissure above. The habenular commissure is continuous with a bundle of fibres which passes forwards as a ridge, the habenula, on either side of the thin roof of the third ventricle to reach a triangular area on the posteriomedial aspect of the thalamus, the habenular triangle [FIG. 10.76]. The two habenulae, the habenular triangles and commissure, and the pineal body, constitute the epithalamus.

The posterior commissure, larger than the habenular commissure, belongs to the midbrain and allows passage of nerve fibres between the two sides of its cephalic extremity.

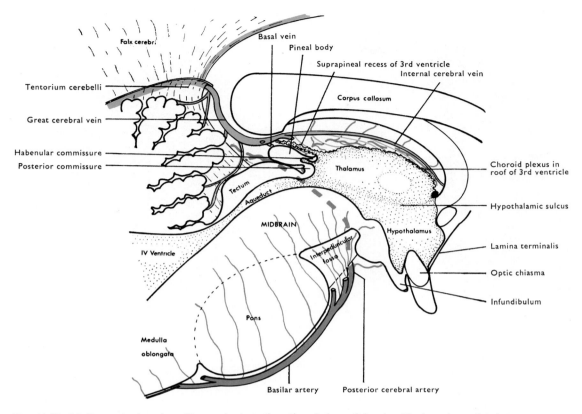

FIG. 10.70. Median sagittal section of human brain to show the relations of the pineal body and associated structures.

KEY TO FIGS. 10.71–10.75

a.	Adhesio interthalamica.
c.	Caudate nucleus.
cc2.	Genu of corpus callosum.
cc3.	Trunk of corpus callosum.
cc4.	Splenium of corpus callosum.
cplr.	Tuft of choroid plexus appearing through lateral aperture of 4th ventricle.
cr.	Corona radiata, visual radiation.
d.	Dentate nucleus of cerebellum.
f.	Fornix.
fl.	Flocculus.
gc.	Gyrus cinguli.
gr.	Gyrus rectus.
h.	Hippocampus.
ht.	Hypothalamus.
ic.	Internal capsule.
l.	Lateral nucleus of thalamus.
lv.	Lateral ventricle.
m.	Medial nucleus of thalamus.

mb.	Midbrain.
mo.	Medulla oblongata.
o.	Olfactory bulb.
ot.	Olfactory tract.
p.	Putamen of lentiform nucleus.
py.	Pyramid.
sfg.	Medial surface of superior frontal gyrus, medial frontal gyrus.
sp.	Septum pellucidum.
t.	Thalamus.
tp.	Pulvinar of thalamus.
v.	Ventral nucleus of thalamus.
II.	Optic nerve.
III.	Oculomotor nerve.
V.	Trigeminal nerve.
VII.	Facial nerve.
VIII.	Vestibulocochlear nerve.
IX.	Glossopharyngeal nerve.

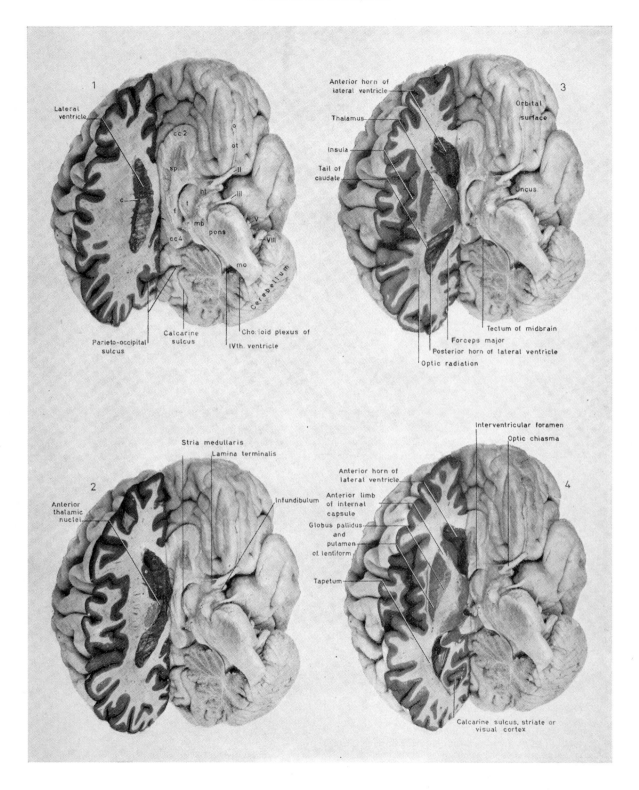

FIGS. 10.71 and 10.72. A series of photomontages showing stained horizontal sections of the brain in their correct relation to the hemispheres.
1. Through the central part of the lateral ventricle.
2. Through the superior part of thalamus.
3. Through the superior margin of the lentiform nucleus.
4. Through the splenium of the corpus callosum.

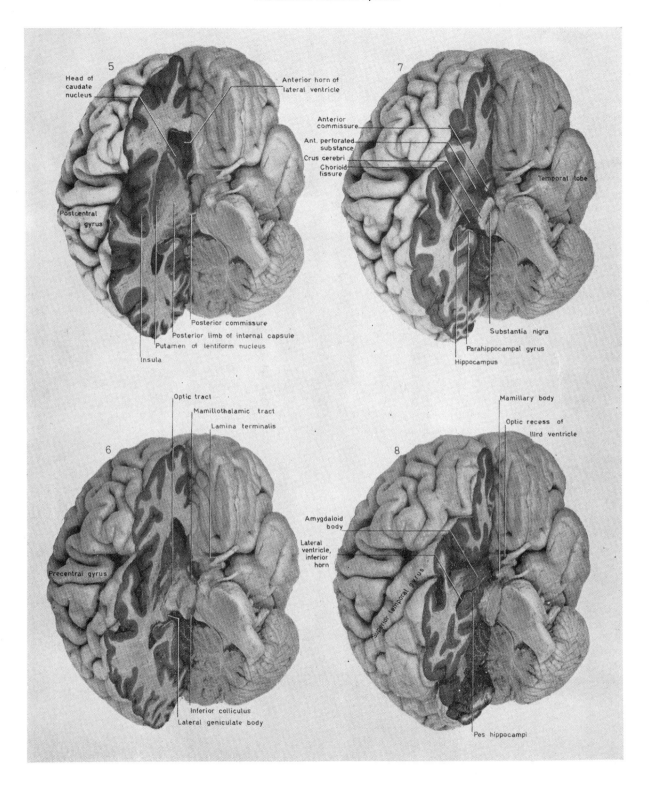

5. Through the posterior commissure.
6. Through the lateral geniculate body.

7. Through the lower midbrain.
8. Through the amygdaloid body.

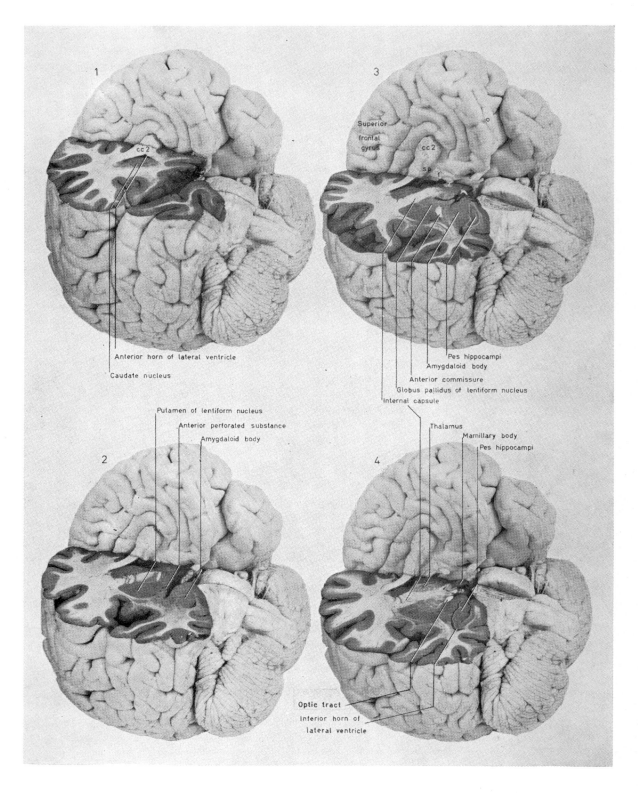

FIGS. 10.73–10.75. A series of photomontages showing stained coronal sections of the brain in their correct relation to the hemispheres.
1. Through the genu of the corpus callosum.   3. Through the tip of the inferior horn of the lateral ventricle.
2. Through the optic chiasma.   4. Through the mamillary body.

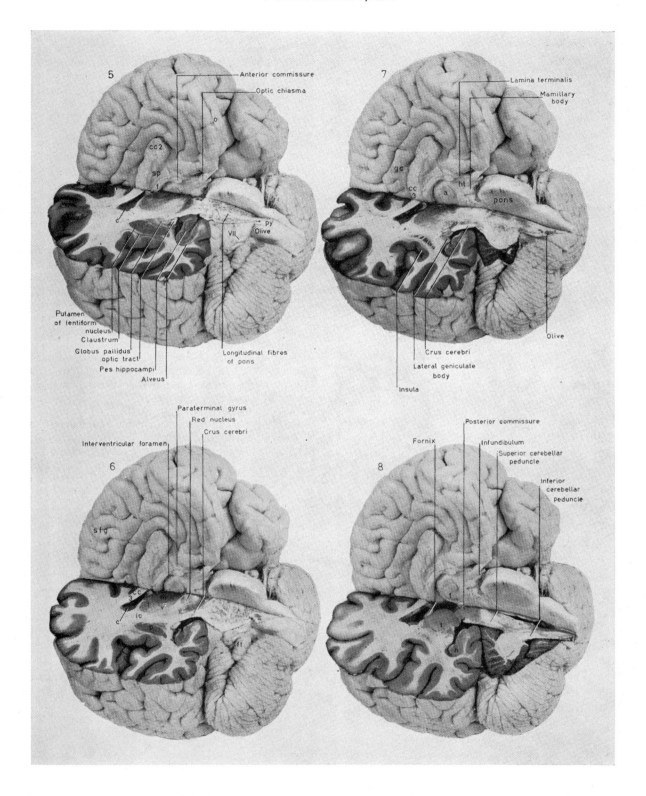

5. Showing the continuity of the corona radiata, the internal capsule, the crus cerebri, the longitudinal fibres of the pons, and the pyramid.
6. Through the anterior end of the red nucleus.
7. Through the posterior extremity of the lentiform nucleus.
8. Through the posterior commissure and the 4th ventricle.

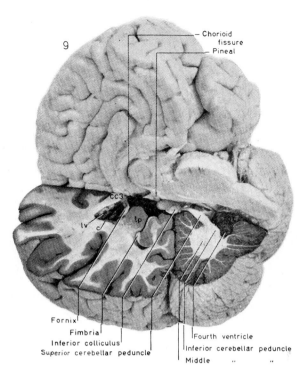

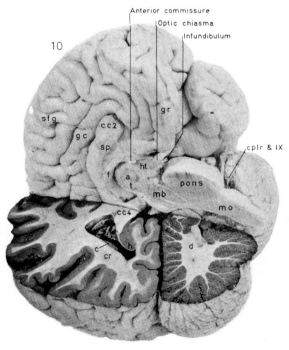

9. Through the pulvinar of the thalamus.
10. Through the junction of the inferior and posterior horns with the central part of the lateral ventricle.

THE THALAMUS

The thalamus grows rapidly, forming a considerable mass of cells which bulges into the third ventricle to meet and fuse with the opposite thalamus, partially dividing the ventricle into dorsal and ventral parts by this **interthalamic adhesion** [FIGS. 10.74, 10.75, and 10.77]. This growth is most marked at the posterior end so that the thalamus becomes triangular in shape (3 cm long by 1.5 cm wide in the adult) when seen from above [FIG. 10.76] with the apex at the interventricular foramen and the broad, rounded base (**pulvinar**) posteriorly, slightly overhanging the upper midbrain. The thalamus has therefore a medial surface covered by the ependyma of the third ventricle, and a dorsal surface, the lateral part of which is covered by the floor of the lateral ventricle [p. 685], while the medial part is covered with pia mater [FIG. 10.124]. The line of junction of these two areas is marked in the adult by a shallow groove which lodges the medial edge of the cerebral hemisphere, the **fornix** [FIGS. 10.76 and 10.80], and marks the line of reflection of the ependyma of the lateral ventricle from the dorsal surface of the thalamus on to the choroid plexus of the lateral ventricle. In the same fashion, the ependyma of the third ventricle covering the medial wall of the thalamus is reflected at the junction of the medial and dorsal surfaces on to the thin roof and choroid plexus of the third ventricle. These two reflections are known as the **taeniae thalami**, and are continuous at the interventricular foramen through which the choroid plexuses of the third and lateral ventricles unite. Along the line of attachment of the roof of the third ventricle to each thalamus lies a small ridge of white matter, the **stria medullaris of the thalamus**, which extends posteriorly to the habenular triangle from the region of the interventricular foramen [FIG. 10.77].

On the dorsal surface of the thalamus at its apex is a small rounded swelling, the **anterior tubercle**, formed by the **anterior nuclei**. Two small rounded **geniculate bodies** [FIGS. 10.54B and 10.78] protrude

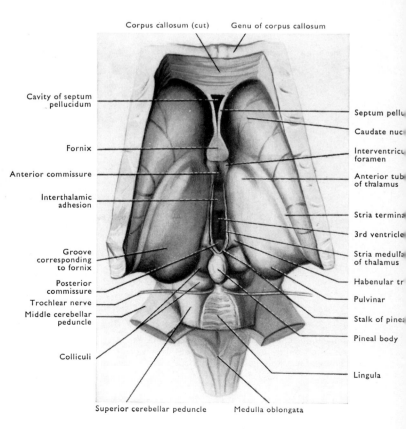

FIG. 10.76. Colliculi and neighbouring parts.

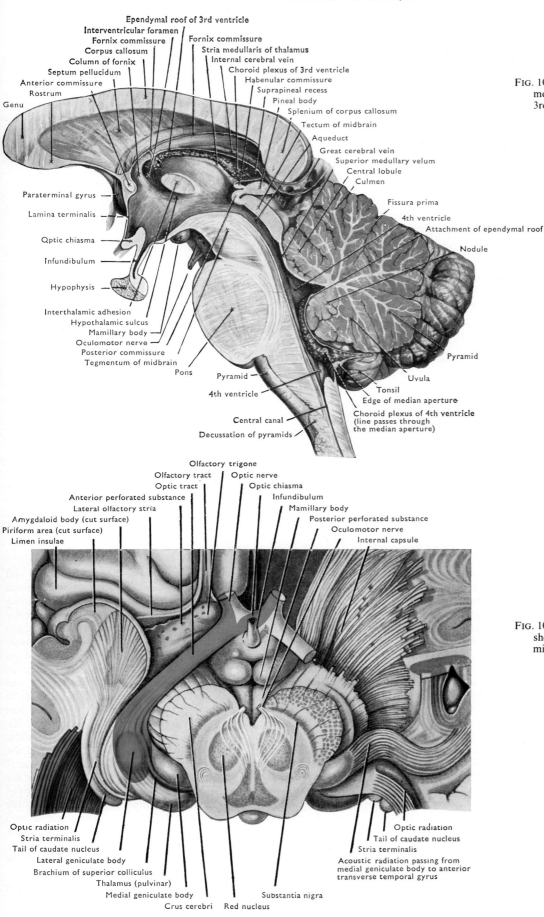

Ependymal roof of 3rd ventricle
Interventricular foramen
Fornix commissure
Corpus callosum
Column of fornix
Septum pellucidum
Anterior commissure
Rostrum
Genu

Fornix commissure
Stria medullaris of thalamus
Internal cerebral vein
Choroid plexus of 3rd ventricle
Habenular commissure
Suprapineal recess
Pineal body
Splenium of corpus callosum
Tectum of midbrain
Aqueduct
Great cerebral vein
Superior medullary velum
Central lobule
Culmen
Fissura prima
4th ventricle
Attachment of ependymal roof
Nodule

Paraterminal gyrus
Lamina terminalis
Optic chiasma
Infundibulum
Hypophysis

Interthalamic adhesion
Hypothalamic sulcus
Mamillary body
Oculomotor nerve
Posterior commissure
Tegmentum of midbrain
Pons
Pyramid
4th ventricle

Central canal
Decussation of pyramids

Pyramid
Uvula
Tonsil
Edge of median aperture
Choroid plexus of 4th ventricle
(line passes through
the median aperture)

FIG. 10.77. The parts of the brain in
median section. The side-walls of the
3rd and 4th ventricles are shown.

Olfactory trigone
Olfactory tract
Optic tract
Anterior perforated substance
Lateral olfactory stria
Amygdaloid body (cut surface)
Piriform area (cut surface)
Limen insulae

Optic nerve
Optic chiasma
Infundibulum
Mamillary body
Posterior perforated substance
Oculomotor nerve
Internal capsule

Optic radiation
Stria terminalis
Tail of caudate nucleus
Lateral geniculate body
Brachium of superior colliculus
Thalamus (pulvinar)
Medial geniculate body
Crus cerebri
Red nucleus
Substantia nigra

Optic radiation
Tail of caudate nucleus
Stria terminalis
Acoustic radiation passing from
medial geniculate body to anterior
transverse temporal gyrus

FIG. 10.78. Part of lower surface of forebrain,
showing olfactory and optic tracts. The
midbrain has been cut across.

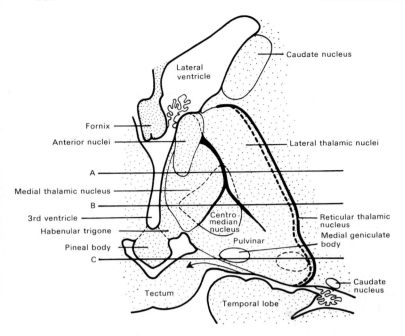

FIG. 10.79. Diagram to show the main subdivisions of the thalamus. A, B, and C are coronal sections indicated by lines A, B, and C on the horizontal section shown above.

In all figures the lateral medullary lamina is shown as a broken line, the medial medullary lamina by a solid line which splits to enclose the central nucleus postero-inferiorly and the anterior nuclei anterosuperiorly. In A, the medial medullary lamina is thickened by the presence of the mamillothalamic tract. The arrow indicates the brachium of the superior colliculus.

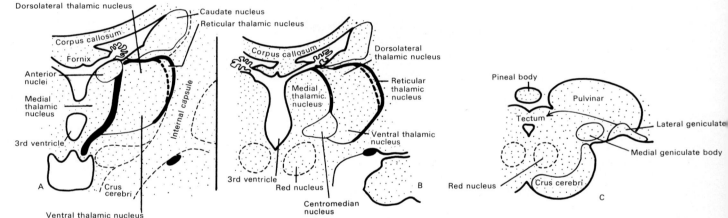

from its postero-inferior surface and comprise the **metathalamus**. Of the two geniculate bodies the **medial** is the more obvious from the surface. It lies wedged between the lateral surface of the midbrain, the dorsal margin of the crus cerebri, and the inferior surface of the pulvinar, and receives the brachium of the inferior colliculus [p. 639]. The **lateral geniculate body** forms a less well-defined swelling on the inferior surface of the pulvinar. This body is continuous anteriorly with the optic tract which enters it by sweeping posteriorly around the superior limit of the midbrain to reach the inferior surface of the thalamus. As it reaches the lateral geniculate body the optic tract appears to split into medial and lateral roots which seem to become continuous with the corresponding geniculate bodies, though both roots enter the lateral geniculate body [p. 704]. The optic tract also gives rise to a slender bundle of fibres which sweeps dorsomedially over the posterior surface of the thalamus, in a groove separating the pulvinar from the medial geniculate body, to reach the anterolateral part of the superior colliculus. This **brachium of the superior colliculus** conveys collaterals of nerve fibres in the optic tract to the most cephalic parts of the tectum of the midbrain [FIG. 10.54B and 10.78].

In addition to the nuclei, described above, which form well-defined protrusions from the surface of the thalamus, the main mass consists of cells divided more or less clearly into a series of subsidiary nuclei. This division is partially produced by two layers of white matter, the **medullary laminae**, which run approximately in a

sagittal plane and are best visualized in sections [FIGS. 10.79 and 10.80]. The medial medullary lamina separates the **medial nucleus** from the **lateral nuclear mass** which is less clearly divisible into a **dorsolateral nucleus**, whose caudal end forms the rounded posterior projection of the thalamus or **pulvinar**, and a **ventral nucleus** which can be subdivided into **arterolateral-ventral**, **intermediate-ventral**, and **posterior-ventral**. The last is divisible into lateral and medial parts, mainly on account of their different connections. A posterior group of nuclei is described between the posterior-ventral and medial geniculate nuclei. The medial medullary lamina splits to enclose the anterior nuclei anterosuperiorly, and the **centromedian nucleus** posteroventrally, and has in it some small groups of cells, the **intralaminar nuclei**. It is thickened anteriorly by the mamillo-thalamic tract [p. 680] running in it to the anterior nuclei. At the external margin of the thalamus is a thin curved sheet of cells lateral to the lateral and ventral nuclei and separated from them by the **lateral medullary lamina**. This rim is traversed by many groups of fibres entering and leaving the thalamus, and is known as the **reticular nucleus of the thalamus**. It lies against the posterior limb of the internal capsule [FIG. 10.80 and p. 671]. See pages 672–3 for the general arrangement of the thalamic connections.

Inferiorly the thalamus lies on the hypothalamus partially separated from it by a horizontal layer of nerve fibres, the thalamic fasciculus, through which certain ascending fibre bundles reach the ventral nucleus.

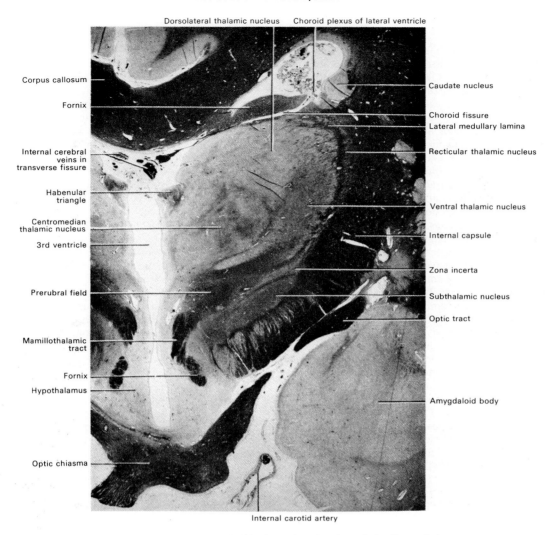

Dorsolateral thalamic nucleus   Choroid plexus of lateral ventricle

Corpus callosum

Fornix

Internal cerebral
veins in
transverse fissure

Habenular
triangle

Centromedian
thalamic nucleus

3rd ventricle

Prerubral field

Mamillothalamic
tract

Fornix

Hypothalamus

Optic chiasma

Caudate nucleus

Choroid fissure
Lateral medullary lamina

Recticular thalamic nucleus

Ventral thalamic nucleus

Internal capsule

Zona incerta

Subthalamic nucleus

Optic tract

Amygdaloid body

Internal carotid artery

FIG. 10.80. Low power photomicrograph of horizontal section through the diencephalon.

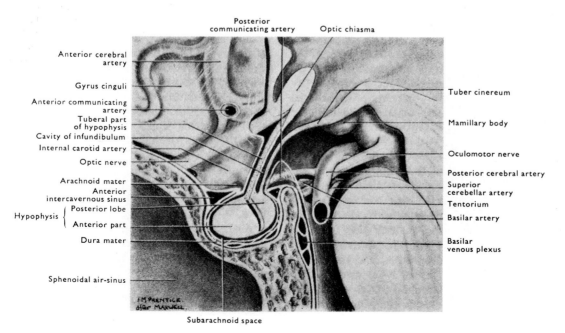

Posterior
communicating artery   Optic chiasma

Anterior cerebral
artery

Gyrus cinguli

Anterior communicating
artery
Tuberal part
of hypophysis
Cavity of infundibulum
Internal carotid artery

Optic nerve

Arachnoid mater
Anterior
intercavernous sinus
Hypophysis { Posterior lobe

Anterior part

Dura mater

Sphenoidal air-sinus

Subarachnoid space

Tuber cinereum

Mamillary body

Oculomotor nerve

Posterior cerebral artery
Superior
cerebellar artery
Tentorium

Basilar artery

Basilar
venous plexus

FIG. 10.81. Diagrammatic median section of hypophysis *in situ*, showing
the parts of the organ and its relations to adjacent structures. (The
existence of a subarachnoid space around the gland in the hypophysial
fossa is doubtful [cf. FIG. 9.19].)

THE HYPOTHALAMUS

This consists of two parts: (1) The inferomedial part, which forms the roof of the interpeduncular fossa and corresponds with the hypothalamus of the physiologist. (2) The lateral or subthalamic part, which is nowhere visible on the surface of the brain. Both parts are directly continuous caudally with the tegmentum of the midbrain.

The medial part forms the floor and lateral wall of the third ventricle inferior to the hypothalamic sulcus. It extends from the lamina terminalis and optic chiasma anteriorly to the tegmentum of the midbrain posteriorly. On the inferior aspect of the brain it consists of the optic chiasma and tracts anterolaterally, the tuber cinereum, the mamillary bodies, and the posterior perforated substance [FIG. 10.78]. It is limited posteriorly by the crura cerebi and the superior border of the pons, and forms the roof of the interpeduncular fossa. The **tuber cinereum** is a rounded swelling which projects inferiorly from the floor of the third ventricle immediately posterior to the optic chiasma. On the lateral aspect of the tuber cinereum are a series of small protuberances caused by groups of cells, the **tuberal nuclei**, lying immediately under the surface. Inferiorly the tuber cinereum tapers to a narrow stalk, the **infundibulum** [FIG. 10.78], by which it is continuous with the posterior (neural) lobe of the hypophysis [FIG. 10.81] and through which a considerable number of nerve fibres pass from the hypothalamus to the hypophysis. Posterior to this are two spherical masses of grey matter, the **mamillary bodies**, lying side by side, and still further posteriorly is a triangular area extending caudally on the inferior surface of the midbrain, recessed between the crura cerebri, and reaching the pons at its apex. This area, pierced by a large number of small branches from the posterior cerebral arteries, is the **posterior perforated substance**. These vessels supply the midbrain and posterior hypothalamus, and lateral to them emerge the oculomotor cranial nerves [FIG. 10.78].

THE OPTIC CHIASMA AND TRACTS

In the adult brain the optic chiasma or crossing lies at the inferior end of the lamina terminalis and receives the two optic nerves at its anterolateral margins where they are crossed superiorly by the anterior cerebral arteries. Lateral to the chiasma lie the internal

carotid arteries after they emerge from the cavernous sinus, and on a slightly superior level the anterior perforated substance. The optic tracts pass posterolaterally from the chiasma [FIG. 10.78] forming the anterolateral margins of the exposed part of the hypothalamus. Each tract is covered inferiorly by the medial margin of the temporal lobe of the cerebral hemisphere and by the posterior communicating artery. The anterior choroidal branch of each internal carotid artery follows the corresponding optic tract, and turns laterally into the antero-inferior extremity of the choroid plexus of the lateral ventricle in the tip of its inferior horn [p. 667]. Reaching the anterior margin of the crus cerebri, the optic tract crosses its lateral surface at the junction of the midbrain and forebrain to reach the lateral geniculate body on the postero-inferior surface of the thalamus.

INTERNAL STRUCTURE OF THE HYPOTHALAMUS

The ventromedial portion consists of a medial part containing several nuclei, separated by the fornix [p. 680] (which curves through the hypothalamus) from a lateral part consisting of longitudinally directed fibres running through scattered nerve cells. The fibres constitute the **medial forebrain bundle**, a complicated collection of fibres originating from the olfactory cells in the region of the anterior perforated substance and connecting them to the hypothalamus. It extends caudally to the midbrain, carrying with it other fibres which originate in the hypothalamus. In addition the hypothalamus may be divided from before backwards into supra-optic, tuberal, and mamillary regions, corresponding approximately to the optic chiasma, tuber cinereum, and mamillary bodies.

The nuclei of the hypothalamus [FIG. 10.82] are not easily separated from each other, except in the case of the **supra-optic**. This lies over the beginning of the optic tract, lateral to the pre-optic nucleus which extends superiorly from the chiasma close to the midline. It is also possible to outline the **paraventricular nucleus** because of its rich supply of capillaries. From this nucleus, and from the tuberal and supra-optic nuclei, nerve fibres pass to every part of the neurohypophysis, that is, to the median eminence (in contact with the pars tuberalis), infundibulum, and posterior lobe.

The **paraventricular** and **supra-optic nuclei** send nerve fibres to the posterior lobe of the hypophysis. These act as ducts for the cells of the nuclei, transmitting respectively oxytocin and antidiuretic hormone to the posterior lobe whence they are released into the blood-stream. The nuclei of the **tuber cinereum** send their axons to the infundibulum. Here they come into contact with blood sinusoids supplied by the superior hypophysial artery. These nerve fibres are believed to secrete 'releasing factors' into the sinusoids which drain through the portal vessels of the stalk and break up into capillary vessels in the anterior lobe where each releasing factor specifically stimulates the release of a particular hormone. This brings the anterior lobe of the hypophysis under the control of the hypothalamus without the presence of nerve fibres connecting them. In addition to this route of control by the secretion of hormones, the **mamillary** and **posterior hypothalamic nuclei** have efferent connections to the cerebral cortex respectively through the anterior and medial thalamic nuclei, and both send fibres to the midbrain tegmentum.

The medial hypothalamus is concerned with visceral activities and the control of the internal environment. Thus stimulation of the anterior part causes increased parasympathetic activity, while stimulation of the posterior part evokes increased sympathetic activity. The anterior part is also concerned with water balance (thirst and urinary excretion—antidiuretic hormone) and heat loss, while the posterior part is concerned with sleep mechanisms and heat generation. This brief statement is only to highlight the importance of this small area, it gives no indication of the complex interactions of its cells which are necessary for its functions or of the sensitivity of its cells to such varied stimuli as the concentrations of

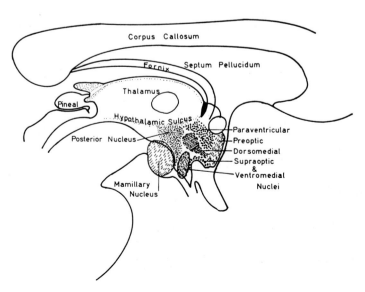

FIG. 10.82. Diagram to show the main hypothalamic nuclei and their position. After Le Gros Clark (1938).

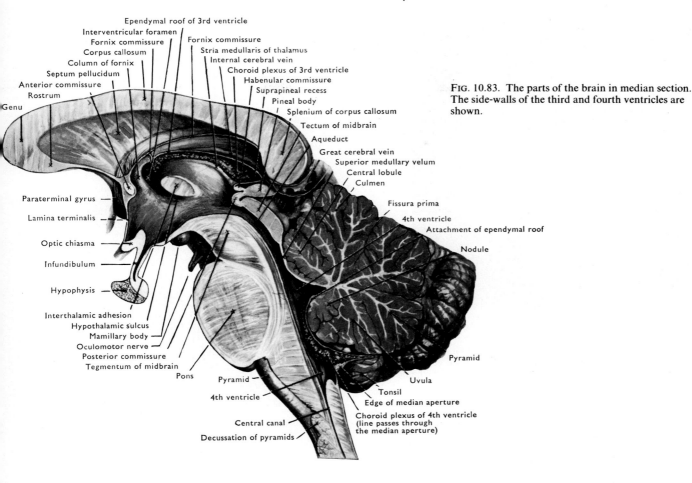

Ependymal roof of 3rd ventricle
Interventricular foramen
Fornix commissure
Corpus callosum
Column of fornix
Septum pellucidum
Anterior commissure
Rostrum
Genu
Fornix commissure
Stria medullaris of thalamus
Internal cerebral vein
Choroid plexus of 3rd ventricle
Habenular commissure
Suprapineal recess
Pineal body
Splenium of corpus callosum
Tectum of midbrain
Aqueduct
Great cerebral vein
Superior medullary velum
Central lobule
Culmen
Fissura prima
4th ventricle
Attachment of ependymal roof
Nodule
Paraterminal gyrus
Lamina terminalis
Optic chiasma
Infundibulum
Hypophysis
Interthalamic adhesion
Hypothalamic sulcus
Mamillary body
Oculomotor nerve
Posterior commissure
Tegmentum of midbrain
Pons
Pyramid
4th ventricle
Central canal
Decussation of pyramids
Pyramid
Uvula
Tonsil
Edge of median aperture
Choroid plexus of 4th ventricle
(line passes through
the median aperture)

FIG. 10.83. The parts of the brain in median section. The side-walls of the third and fourth ventricles are shown.

water, sugar, hormones, etc. in the blood through which it can monitor the efficacy of its activities or stimulate further change.

The lateral or subthalamic part of the hypothalamus also lies inferior to the thalamus separating it from the inferior part of the internal capsule [p. 671]. It consists of the **subthalamic nucleus** which is oval in section and lies against the internal capsule [FIG. 10.80], which separates it from the medial segment of the globus pallidus of the lentiform nucleus [p. 674]. Numerous fibres pass through the internal capsule or around its ventral margin between these two nuclei and constitute a part of the **ansa lenticularis** [p. 718]. Dorsal to the subthalamic nucleus lies the **zona incerta**, a nucleus which is separated from the thalamus by a layer of fibres, the **thalamic fasciculus**, consisting of ascending tracts passing to the thalamus. Medially this contains fibres from the capsule of the red nucleus [p. 712], which extends into the subthalamic region. Here fibres from the capsule of the red nucleus mingle with those of the thalamic fasciculus and with fibres from the ansa lenticularis which pass medially between the subthalamic nucleus and zona incerta, and enter the thalamus, the red nucleus, and the substantia nigra. Medially this fused mass of fibres abuts on the mamillothalmic tract [p. 680].

THE THIRD VENTRICLE [FIGS. 10.76 and 10.83]

This is the cephalic extremity of the original median cavity of the neural tube from which the two lateral ventricles grow out. It is a deep slit-like cavity, lined with ependyma and all but separates the two halves of the diencephalon. It is limited *anteriorly* by the thin **lamina terminalis**, stretching between the **anterior commissure** superiorly and the **optic chiasma** inferiorly [FIG. 10.83]. This lamina

lies at the anterior limit of the hypothalamus and not only joins the two hypothalami, but also the two cerebral hemispheres.

Inferiorly the floor of the third ventricle forms the roof of the interpeduncular fossa. Anteriorly it is superior to the optic chiasma (**optic recess**) then dips down into the infundibulum for a variable distance (**infundibular recess**) before passing posterosuperiorly towards the aqueduct, rapidly decreasing the vertical extent of the ventricle. Throughout most of its length, the **roof** of the third ventricle is composed of ependyma only, and is invaginated on both sides of the midline by minute tufts of choroid plexus. The thin roof begins anteriorly at the interventricular foramen, separated from the anterior commissure by the columns of the fornix [FIG. 10.83] descending into the hypothalamus. Here the choroid plexus is tucked into the foramen, each half passing laterally through it into the corresponding lateral ventricle. Posteriorly, the thin roof bulges backwards over the superior surface of the pineal body (**suprapineal recess**), and curves forwards to the habenular commissure [FIG. 10.83]. The thin roof ends here, but its ependyma is continuous with that on the ventricular surfaces of the habenular commissure, the **pineal recess** of the third ventricle, and the posterior commissure before becoming continuous with the lining of the aqueduct in the midbrain. The thin roof is attached on each side to the **stria medullaris** on the superomedial margin of the thalamus. Each lateral wall of the ventricle is formed by the thalamus and hypothalamus, separated by the hypothalamic sulcus. The right and left thalami are commonly, but not invariably, joined across the midline by the **interthalamic adhesion**. The anterosuperior part of the third ventricle communicates on each side with a lateral ventricle and forms the **interventricular foramen** where the three ventricles of the forebrain communicate with each other [FIG. 10.84].

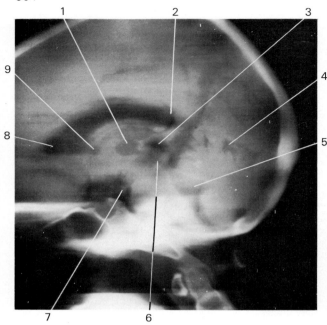

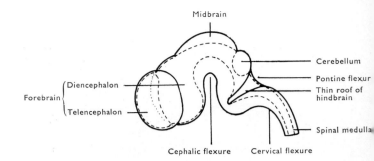

FIG. 10.85. Diagram to show the parts of the embryonic brain after formation of the telencephalon. The optic outgrowth is not shown. Broken lines indicate the ventricular system; the dotted line, the roof of the diencephalon.

FIG. 10.84. Tomogram of air encephalogram. There is air in the upper anterior part of the lateral ventricle, the upper part of third ventricle, and in the cerebral aqueduct and fourth ventricle. Air in the subarachnoid space is principally in the interpeduncular cistern, cisterna ambiens, subtentorial subarachnoid space, and cisterna magna.

1. 3rd ventricle.
2. Air surrounding splenium of corpus callosum. It also passes forwards above and below it.
3. Air between pineal body and colliculi with posterior and habenular commissures anterior to it. This air extends towards interpeduncular cistern in cisterna ambiens.
4. Infratentorial air.
5. 4th ventricle.
6. Aqueduct of cerebrum.
7. Oculomotor nerve and basilar artery in interpeduncular cistern.
8. Lateral ventricle.
9. Interventricular foramen.

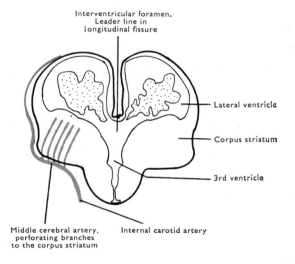

FIG. 10.86. Transverse section through the forebrain at the level of the interventricular foramen of the brain shown in FIG. 10.85.

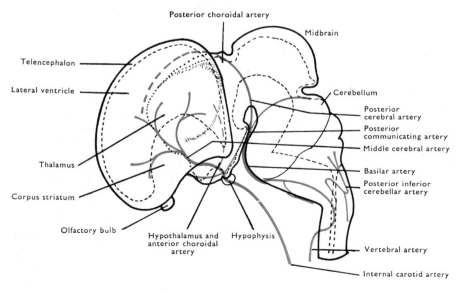

FIG. 10.87. Drawing of reconstruction of early human foetal brain derived from serial sections. The arteries are shown diagrammatically.

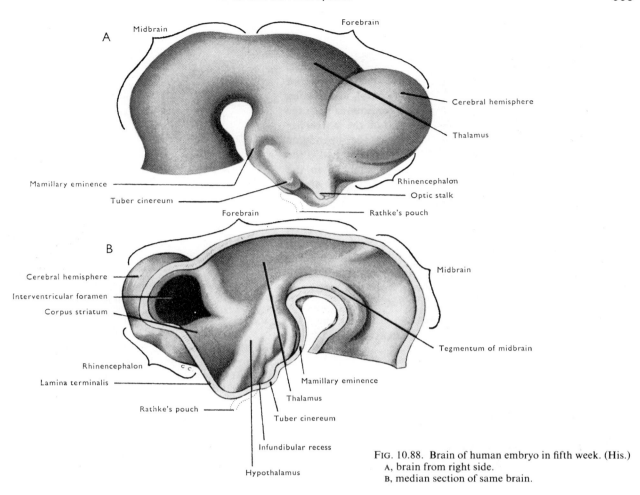

FIG. 10.88. Brain of human embryo in fifth week. (His.)
A, brain from right side.
B, median section of same brain.

The telencephalon [FIGS. 10.71–10.75]

The two telencephalic vesicles arise as outgrowths of the whole thickness of the walls of the forebrain at its anterosuperior part [FIG. 10.85]. These vesicles are thin-walled except inferiorly where the solid rudiment of the corpus striatum lies. Inferior to this is the cephalic extremity of the dorsal aorta (internal carotid artery) which runs on to the surface of the corpus striatum, and sends several minute striate arteries into it, forming the anterior perforated substance [FIG. 10.78] while the parent vessel runs on to the lateral surface of the vesicle [FIG. 10.86].

The growth of the telencephalon, or endbrain, consists of a great expansion of the thin upper part of the vesicle, which thereby overgrows the thick basal part, rolling around it, so that before long it has virtually encircled the corpus striatum and all but hidden it from view on the ventral aspect. FIGURE 10.89 shows how the posterior end of the vesicle gradually becomes inferior in position and forms the future temporal lobe, lying below and behind the anterior end of the vesicle which forms the frontal lobe. Because of this method of growth, each cerebral hemisphere and its contained ventricle become C-shaped with the open part of the C facing downwards and forwards. The middle cerebral artery, as it passes to reach the lateral surface of the telencephalic vesicle, is caught between the two limbs of the C in the stem of the lateral sulcus [FIG. 10.91] formed by the meeting of the frontal and temporal lobes. The artery coursing laterally in this sulcus gives off its striate branches [FIG. 10.107]. Further growth of the surface of the hemisphere in the sagittal plane leads to the protrusion of the intermediate part posteriorly [FIG. 10.89], to form the occipital lobe

containing an extension of the ventricle. In this fashion the general shape of the hemisphere is produced, and the contained ventricle comes to have anterior, posterior, and inferior horns, occupying the frontal, occipital, and temporal lobes respectively with a central part uniting them.

SULCI AND GYRI

Each telencephalic vesicle also expands transversely. Its lateral expansion causes the thin-walled part of the vesicle to bulge over the surface of the slowly expanding corpus striatum by forming folds which progressively hide it from view [FIGS. 10.90–10.95]. Since the corpus striatum is virtually surrounded, this overgrowth occurs in front, above, and below [FIGS. 10.89–10.95]. The covering folds, or opercula, gradually approach each other till they meet and completely hide from view the surface of the brain overlying the corpus striatum. This forms a buried island or insula with the main branches of the middle cerebral artery running on it [FIG. 10.92]. As a result the branches of the middle cerebral artery now appear between the lips of the opercula which meet but do not fuse, thereby forming a deep triradiate sulcus on the lateral aspect of the hemisphere between the four opercula: orbital, frontal, frontoparietal, and temporal. The longest (posterior) ramus of the sulcus extends towards the occipital lobe which forms no operculum because of its freedom to expand posteriorly. The anterior end of this ramus is continuous with the stem of the lateral sulcus, which contains the first part of the middle cerebral artery. At their roots the deep surfaces of the opercula become continuous with the surface of the insula and the groove at which this occurs (circular

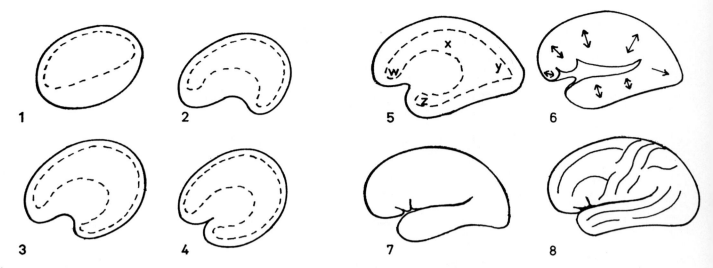

FIG. 10.89. Diagrams, not to scale, to show the growth of the cerebral hemisphere from the lateral view. The broken line indicates the position of the ventricle. Arrows indicate direction of cortical expansion which forms lateral sulcus.
w, anterior horn; x, central part; y, posterior horn; z, inferior horn of lateral ventricle.

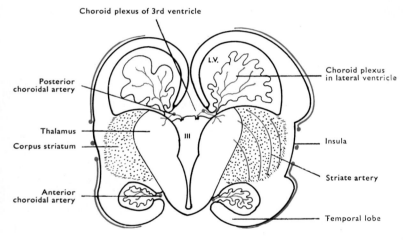

FIG. 10.90. Diagrammatic transverse section of developing cerebral hemisphere. The right side is further developed than the left, and the stipple indicates the corpus striatum.

sulcus) surrounds the insula, except at the lateral sulcus where a narrow tongue of insular cortex (the limen insulae) extends into its stem, towards the anterior perforated substance.

When the opercula meet, further expansion of the surface of the telencephalon can take place only by folding. The pattern of this folding is partly the result of local growth patterns and partly the result of continued transverse expansion. This leads to the formation of a complicated series of grooves or sulci, with intervening ridges or gyri. Two major groups of these are formed on the superolateral surface in addition to the lateral sulcus and many subsidiary sulci which vary greatly from individual to individual.

1. Those running parallel to the posterior ramus of the lateral sulcus in the temporal and frontal lobes, and a single similar sulcus in the parietal lobe, the intraparietal sulcus [FIGS. 10.89, 8 and 10.96].

2. Three coronal sulci are formed between the frontal and parietal lobes. These are the central, precentral, and postcentral sulci. The number of sulci formed in any region of the brain is a measure of the extent of the growth processes in that region, and these are never the same in any two brains. Thus, though the basic plan is similar in all brains [FIGS. 10.96–10.98], the details of the folding are so different that two or three major horizontal, frontal sulci may be formed, and there may be two, three, or four coronal sulci [FIG. 10.89, 8] lying transversely between the frontal and parietal lobes. Only if the normal pattern of three sulci is present and well defined can the central sulcus (between the frontal and parietal lobes) be determined readily.

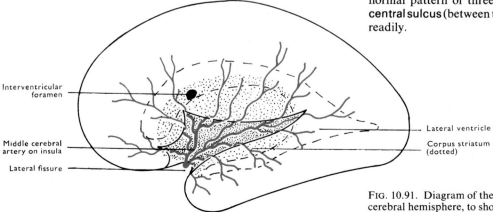

FIG. 10.91. Diagram of the lateral view of the developing cerebral hemisphere, to show the formation of the insula and the course of the middle cerebral artery. The positions of the lateral ventricle (broken line) and the corpus striatum (stipple) are shown.

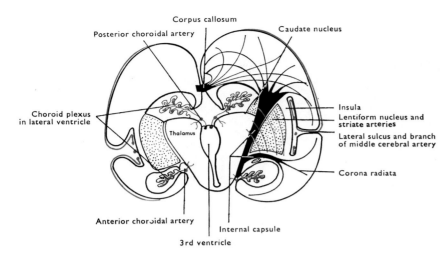

FIG. 10.92. Diagrammatic transverse section through the developing cerebral hemisphere at a later stage than that shown in FIG. 10.90. The right side is shown at a later stage of development than the left, and indicates the effects on the structure of the hemisphere of the development of the corpus callosum and internal capsule. Stipple, corpus striatum.

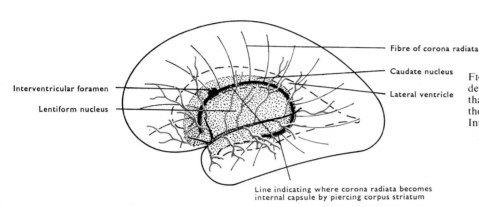

FIG. 10.93. Diagram of the lateral view of the developing cerebral hemisphere at a later stage than FIG. 10.91. The insula is entirely covered by the growth of the surrounding cerebral cortex. Internal structures shown as FIG. 10.91.

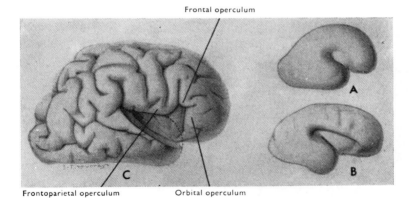

FIG. 10.94. Right hemispheres of human foetuses showing three stages in development of insula and insular opercula.

A, right cerebral hemisphere from a foetus in latter part of fourth month; B, right hemisphere in fifth month; C, right hemisphere in latter part of eighth month.

In C, the temporal operculum has been removed and a large part of the insula is thus exposed. The outline of the temporal operculum is indicated by a dotted line.

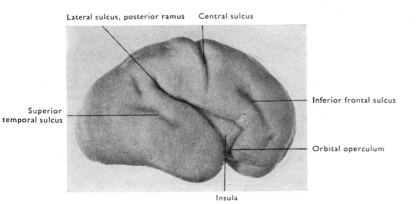

FIG. 10.95. Right cerebral hemisphere from early seventh-month foetus.

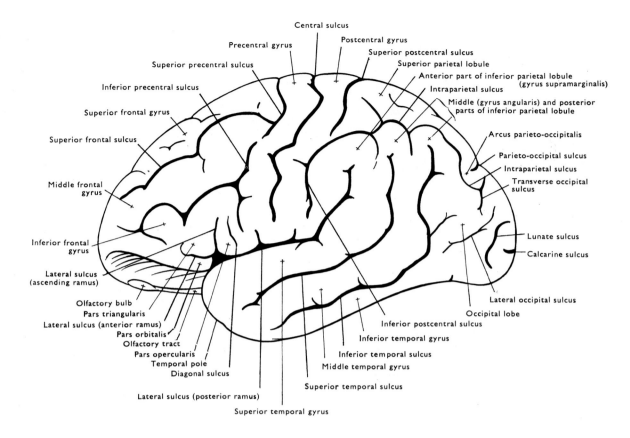

FIG. 10.96. Diagram of sulci and gyri on superolateral surface of hemisphere. The middle frontal sulcus, sometimes found between the superior and inferior frontal sulci, is not shown.

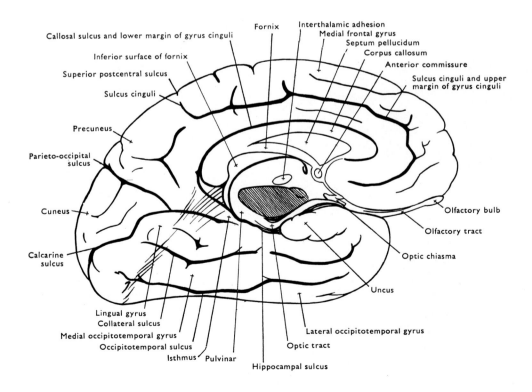

FIG. 10.97. Diagram of sulci and gyri on medial and tentorial surfaces of hemisphere.

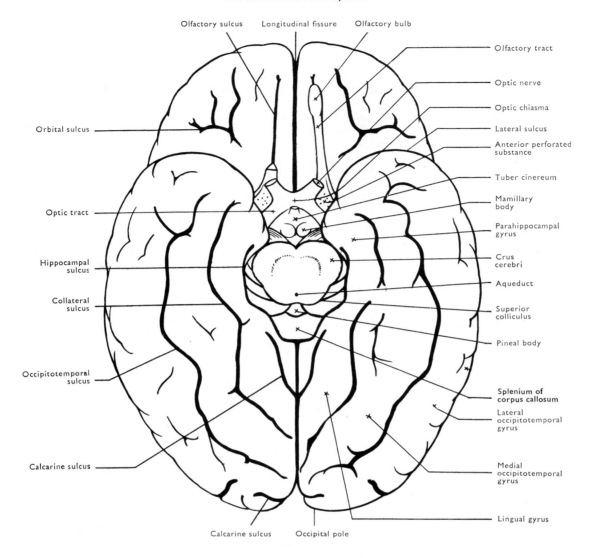

Olfactory sulcus   Longitudinal fissure   Olfactory bulb

Olfactory tract

Optic nerve

Optic chiasma

Lateral sulcus

Anterior perforated substance

Tuber cinereum

Mamillary body

Parahippocampal gyrus

Crus cerebri

Aqueduct

Superior colliculus

Pineal body

Splenium of corpus callosum

Lateral occipitotemporal gyrus

Medial occipitotemporal gyrus

Lingual gyrus

Orbital sulcus

Optic tract

Hippocampal sulcus

Collateral sulcus

Occipitotemporal sulcus

Calcarine sulcus

Calcarine sulcus   Occipital pole

FIG. 10.98. Diagram of sulci and gyri on inferior surface of the hemispheres.

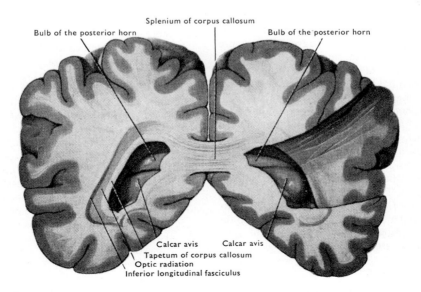

Splenium of corpus callosum

Bulb of the posterior horn   Bulb of the posterior horn

Calcar avis   Calcar avis
Tapetum of corpus callosum
Optic radiation
Inferior longitudinal fasciculus

FIG. 10.99. Coronal section through posterior horns of lateral ventricles, viewed from the front.

The pattern and names of the sulci and gyri on the medial aspect of the hemisphere are shown in FIGURE 10.97. In the occipital region two deep, complicated sulci with subsidiary sulci and gyri in their depths are formed, the parieto-occipital and the calcarine, which are separated by a wedge-shaped area of cerebral cortex, the cuneus. The parieto-occipital sulcus separates the parietal and occipital lobes on the medial aspect, cutting the superomedial margin, while the calcarine sulcus is of such depth that it bulges into the medial wall of the posterior horn of the lateral ventricle, forming the calcar avis [FIG. 10.99]. Further anteriorly [FIGS. 10.97 and 10.114] the cingulate sulcus runs parallel to the corpus callosum. In congenital absence of the corpus callosum, the medial aspect of the hemisphere is divided by a number of radially arranged sulci into a series of wedge-shaped gyri over the whole medial surface, suggesting that the gyrus and sulcus cinguli are formed as a result of the growth of the hemisphere against the corpus callosum.

LOBES OF CEREBRUM

Some of the sulci [FIG. 10.96] are used to divide the various lobes of the hemisphere from each other; thus the frontal and parietal lobes are separated by the central sulcus, the frontal and temporal lobes by the lateral sulcus, the parietal and occipital lobes by an imaginary line joining the superior part of the parieto-occipital sulcus to the pre-occipital notch. This line also separates the temporal and occipital lobes, and the extension of the lateral sulcus backwards to meet the line separates the parietal and temporal lobes. The pre-occipital notch is formed where the inferolateral margin crosses the petrous temporal bone. It is perhaps unfortunate that the central sulcus is the line of separation between the frontal and parietal lobes and also between two areas of the cerebral cortex ('motor' and 'sensory') which are significantly different in their structure and function. This tends to suggest that the division of the hemispheres into lobes has a functional significance whereas it is purely for descriptive purposes.

Initially the two telencephalic vesicles lie some distance apart. As they expand transversely, they continue to be held apart posteriorly and inferiorly by the brain stem and diencephalon between them, but above the diencephalon they overgrow the thalamus so that it appears to be inserted into the medial side of the telencephalon in the adult, with a free edge of the hemisphere encircling it. This expansion flattens the upper parts of the two hemispheres against each other or against the falx cerebri [p. 729] and allows the hemispheres to fuse above and in front of the third ventricle, below the edge of the falx. This fusion permits the growth of nerve fibres from one hemisphere to the other, thus forming the great commissure [p. 683] of the hemisphere, the corpus callosum [FIG. 10.92].

The convex superolateral surfaces of the hemispheres are fitted to the growing skull, and the occipital lobes extend backwards above the cerebellum which, bulging the tentorium cerebelli [p. 729] upwards, separates the inferior parts of these lobes. Thus each occipital lobe is a three-sided pyramid with a convex superolateral surface, a vertical medial surface applied to the falx cerebri, and a sloping inferomedial or tentorial surface [FIG. 10.99].

The frontal lobes lie adjacent to each other [FIG. 7.6] with their inferior surfaces on the floor of the anterior cranial fossa, superior to the nasal and orbital cavities. The temporal lobes fill the lateral parts of the middle cranial fossa, posterior to the orbits. They are widely separated from each other by the hypophysial fossa and the cavernous sinuses [FIG. 11.9] and above these by the interpeduncular fossa and the diencephalon [FIG. 10.110].

Choroid plexus of lateral ventricle

When the telencephalic vesicles first grow out from the original forebrain, they carry with them an extension of the thin roof of the third ventricle into their medial walls [FIG. 10.90]. As each vesicle expands backwards and downwards from the site of the original outgrowth, the interventricular foramen, this thin medial wall extends around the corpus striatum into the medial wall of the inferior horn of the ventricle, with one edge attached to the convexity of the corpus striatum, and the other to the medial margin of the hemisphere. When the hemisphere expands medially over the diencephalon, the thin medial wall is carried over the diencephalon and secondarily becomes attached to it. Thus the lateral part of the thalamus comes to lie in the floor of the central part of the lateral ventricle, and the hypothalamus in the roof of its inferior horn [p. 686; FIG. 10.90]. As in the third and fourth ventricles, the thin medial wall is invaginated into the ventricle by the surrounding vascular connective tissue to form the choroid plexus of the lateral ventricle. This is continuous with the choroid plexus of the third ventricle at the interventricular foramen [FIGS. 10.122 and 10.129], and passes from this point to the tip of the inferior horn of the lateral ventricle, forming a C-shaped curve following the line of the ventricle. It is invaginated between the medial edge of the hemisphere and the thalamus through a fissure known as the choroid fissure [FIG. 10.111].

Thus the cerebral hemisphere, its contained ventricle, and choroid plexus form C-shaped structures which overlap the diencephalon medially. Superiorly, each comes into contact with its fellow of the opposite side, but posteriorly they diverge around the lateral aspects of the brain stem. The formation of the insula displaces the corpus striatum from its position close to the superolateral surface, but it remains close to the inferior surface at the anterior perforated substance in the stem of the lateral sulcus.

THE WHITE MATTER OF THE CEREBRUM

At an early stage in the development of the hemisphere, cells derived from the ependymal lining of the vesicle migrate towards the surface and form the outer rind or cortex which eventually covers the entire surface. This process does not occur uniformly, but appears first in the margin of the hemisphere (limbic cortex) bounding the choroid fissure, somewhat later in the olfactory cortex of the temporal lobe (piriform area) and finally in the remainder of the cerebral vesicle; the polar region of the frontal and occipital lobes being the last parts to develop a cortex. Also it is the deepest layers of the cortex which are formed first. Subsequent migrations of cells from the ependyma pass between those which have already reached the cortex to take up a more superficial position.

With the appearance of the cerebral cortex and its gradual differentiation into a series of layers, the walls of the cerebral vesicle increase rapidly in thickness so that the ventricle and choroid plexus which originally filled the greater part of the vesicle become relatively reduced in size. This thickening is hastened by the outgrowth from the cortical cells of processes. Some of the axons remain within the cortex while others pass beneath it to form a layer of nerve fibres which separates the cortex from the ependyma of the ventricle and from the corpus striatum in those parts which overlie it. The axons of the cortical cells are very numerous and can be divided into three groups on the basis of their destination.

1. Processes which pass from one part of a cerebral hemisphere to another, the association fibres of the adult brain. These may be very short, looping between adjacent areas of the cerebral cortex; or of considerable length, even stretching throughout the length of the hemisphere [p. 676].

2. Processes which pass from one cerebral hemisphere to the corresponding parts of the other. These are commissural fibres which cross the midline. The majority of these fibres run in the corpus callosum [p. 683], though a smaller number pass by other

routes, especially the anterior commissure lying in the upper part of the lamina terminalis which stretches between the two hemispheres.

3. Processes which leave the cerebral cortex and pass to other parts of the central nervous system, the projection fibres.

Projection fibres

These fibres arise from all areas of the cerebral cortex and, passing deeply, take part in the formation of the white matter of the cerebral hemisphere. They converge radially on the corpus striatum as the corona radiata, which also contains fibres passing in the opposite direction, i.e. from the subjacent parts of the nervous system to the cerebral cortex. The fibres of the corona radiata reach the corpus striatum as a compact sheet which pierces it on a C-shaped curve [FIGS. 10.93 and 10.101], passing inferomedially through it to reach the lateral aspect of the thalamus, on which it lies medial to the inferior part of the corpus striatum. It thus splits the corpus striatum into an outer rim, the comma-shaped caudate nucleus, and a central part, the lens-shaped lentiform nucleus. The mass of fibres therefore covers the convex internal surface of the lentiform nucleus forming its internal capsule, which separates the anterior part and periphery of the lentiform nucleus from the caudate nucleus and the remainder of it from the thalamus [FIG. 10.110]. In horizontal sections, taken through the level of the interventricular foramen, the internal capsule presents two limbs sharply angled (the genu) on each other around the apex of the lentiform nucleus which lies opposite the interventricular foramen [FIG. 10.102]. The anterior limb lies between the caudate and lentiform nuclei and the posterior limb between the thalamus and the lentiform nucleus. The genu is only clearly visible in this plane of section and at this level, being absent both superiorly and inferiorly [FIGS. 10.71 and 10.72].

Inferiorly, each fan-shaped internal capsule, stripped of the fibres which join it to the diencephalon and corpus striatum, is concentrated into a thick bundle which emerges from the inferior surface of the forebrain between the hypothalamus medially and the medial edge of the hemisphere laterally [FIGS. 10.74 and 10.92]. Thus it separates this part of the diencephalon from the choroid fissure of the inferior horn of the lateral ventricle. Since the diencephalon and the midbrain are directly continuous, the internal capsule emerges from the forebrain, medial to the optic tract, directly on to the anterolateral surface of the midbrain, as the crus cerebri. Caudally the crus cerebri pierces the ventral part of the pons and splits into the longitudinal bundles within its substance. The majority of these fibres end in the pons, the remainder emerge from its caudal margin as the pyramid on the ventral surface of the medulla oblongata,

abutting on the pyramid of the opposite side. Followed caudally, the pyramids taper gradually in the lower medulla oblongata as fascicles from each interdigitate and cross the midline (pyramidal decussation) obliterating the anterior median fissure between them, and passing posterolaterally to enter the lateral funiculi of the spinal medulla. Here they form the lateral corticospinal tracts, which reach the caudal extremity of the spinal medulla. A variable proportion of fibres does not cross the midline in this region but continues down to the thoracic region, as the anterior corticospinal tract, in the anterior funiculus of the spinal medulla. A smaller number of fibres pass dorsally without decussating to enter the lateral corticospinal tract of the same side [FIG. 10.53].

Thus the corona radiata, the internal capsule, the crus cerebri, the longitudinal fibres in the ventral part of the pons, the pyramid, and the corticospinal tracts represent the parts of a single efferent pathway from the cerebral cortex. The fibres forming this system converge radially through the corona radiata into the internal capsule and are concentrated into the crus cerebri. Thus the relative positions of fibres arising in the different parts of the cerebral cortex within this system remain unchanged [FIG. 10.100]. The fibres from the frontal pole run almost horizontally backwards and have a long course through the anterior part of the internal capsule before reaching the anteromedial part of the crus cerebri. The fibres from the occipital and temporal lobes, which hook round the posterior and inferior parts of the caudate nucleus, have only a transitory contact with the lentiform nucleus and thalamus in the retrolentiform and sublentiform parts of the internal capsule respectively, and almost immediately enter the posterolateral part of the crus cerebri. The intermediate fibres occupy intermediate positions; those arising from the cerebral cortex in the region of the central sulcus occupy a central position [FIG. 10.103].

It should be appreciated that the corona radiata is a fountain, the fibres of which pass in both directions between the internal capsule and all parts of the cortex. FIGURES 10.92, 10.100, and 10.103 show the general arrangement and indicate that fibres arising from the precentral gyrus or passing to the postcentral gyrus do not form a cylindrical bundle in the internal capsule but are spread out in proportion to the obliquity of the central sulcus. The fibres which arise from, or end in the inferior extremities of these gyri tend to lie anterior to those of their superior extremities [FIG. 10.103]. Thalamocortical fibres, which form a considerable portion of the internal capsule, enter its medial side from the thalamus and are at first medial to those descending to regions caudal to the thalamus, though superiorly they become mixed.

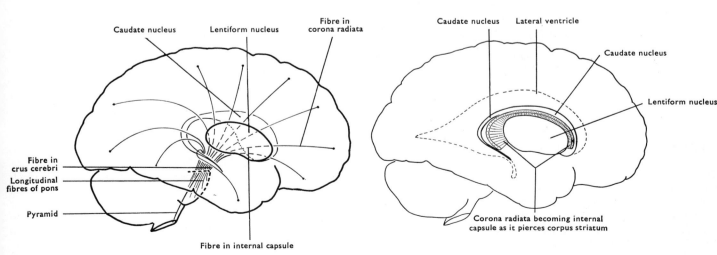

FIG. 10.100. Diagram to show arrangement of fibres in corona radiata, internal capsule, crus cerebri, pons, and pyramid.

FIG. 10.101. Diagram to show the position of the corona radiata piercing the corpus striatum as the internal capsule.

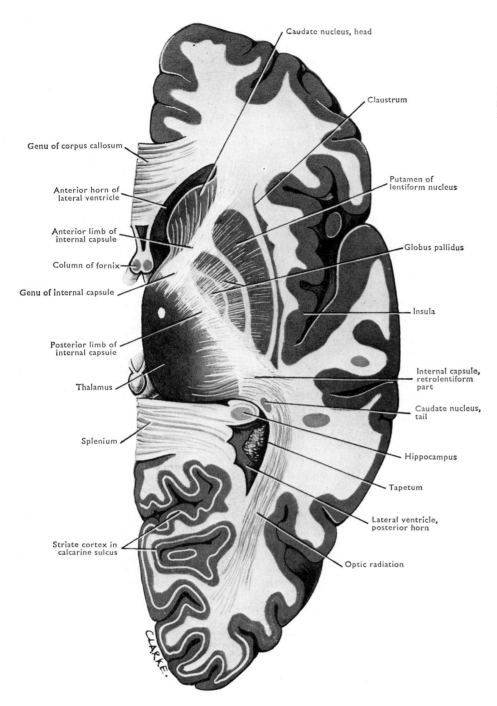

Caudate nucleus, head

Claustrum

FIG. 10.102. Horizontal section through the
right cerebral hemisphere at the level of the
interventricular foramen.

Genu of corpus callosum

Anterior horn of
lateral ventricle

Anterior limb of
internal capsule

Column of fornix

Genu of internal capsule

Putamen of
lentiform nucleus

Globus pallidus

Insula

Posterior limb of
internal capsule

Thalamus

Internal capsule,
retrolentiform
part

Caudate nucleus,
tail

Splenium

Hippocampus

Tapetum

Lateral ventricle,
posterior horn

Striate cortex in
calcarine sulcus

Optic radiation

CLARKE.

Thalamocortical connections

The fibres which enter the internal capsule from the thalamus on
their way to the cerebral cortex, and similar fibres passing in the
opposite direction (corticothalamic fibres) have the same radial
arrangement and follow the other fibres of the internal capsule in
their distribution. In horizontal sections through the thalamus [FIGS.
10.104 and 10.105] it can be seen that each part sends fibres to the
corresponding parts of the cerebral cortex. The medial, anterolateral-
and intermediate-ventral nuclei project to the frontal lobe; the
posterior-ventral to the parietal lobe; the pulvinar and lateral

geniculate body to the occipital lobe; the medial geniculate body to
the temporal lobe [see also p. 713]. In this fashion the most posterior
part of the thalamus projects to the part of the cerebral cortex which
was originally the posterior end of the cerebral vesicle—the temporal
lobe. Similarly in the coronal plane [FIG. 10.106] the thalamus
projects to the cortex in an orderly fashion; its ventromedial cells to
the cortex of the upper wall of the posterior ramus of the lateral
sulcus, its dorsomedial cells to the cortex on the medial surface of
the hemisphere, and its intermediate cells to the intermediate
regions. Thus the anterior nuclei project to the cingulate gyrus while
the medial part of the posterior-ventral nucleus (arcuate nucleus

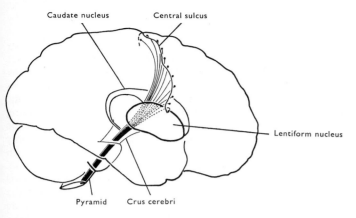

FIG. 10.103. Diagram to show the course of fibres arising in the precentral gyrus and passing through corona radiata, internal capsule, crus cerebri, pons, and pyramid.

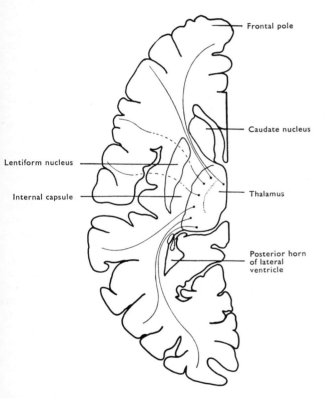

FIG. 10.104. Diagram of the arrangement of the thalamocortical radiations to the upper parts of the cerebral hemisphere. The broken lines indicate fibres arching superiorly out of the section over the lentiform nucleus.

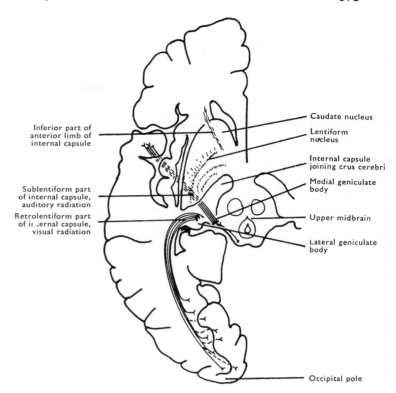

FIG. 10.105. Diagram of the thalamocortical radiations from the geniculate bodies. The broken lines indicate the fibres passing below the lentiform nucleus to the temporal lobe (auditory radiation).

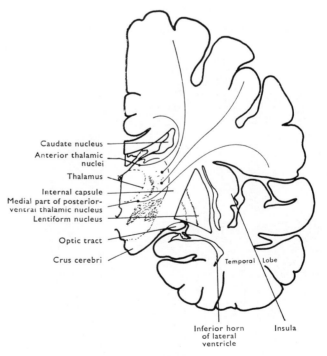

FIG. 10.106. Diagram to show the arrangement of thalamocortical radiations in a coronal section of the human brain.

[FIG. 10.80]) projects to the inferior part of the postcentral gyrus. In addition to these specific thalamocortical connections, which are exactly mirrored by corresponding corticothalamic connections, the intralaminar and **centromedian nuclei** project to the caudate and lentiform nuclei. They also receive fibres from the cerebral cortex and send diffuse efferent connections to it (Jones and Leavitt 1974)—findings which give anatomical confirmation of the long-standing physiological evidence that they are concerned with the ascending activating system by transmitting impulses from the brain stem reticular formation which modify the electroencephalo-gram and arouse the sleeping animal. For the reticular nucleus, see page 714.

THE CAUDATE AND LENTIFORM NUCLEI (CORPUS STRIATUM), THE AMYGDALOID BODY, AND THE CLAUSTRUM [FIGS. 10.71–10.75]

As a result of the growth of the internal capsule, the main part of the corpus striatum is divided into the caudate and the lentiform nuclei.

The caudate nucleus lies between the internal capsule and the lateral ventricle and follows the concavity of the ventricle from the anterior horn to the tip of the inferior horn. Anteriorly the head of the caudate nucleus is large and extends almost to the inferior surface of the brain, being separated from it only by a thin sheet of nerve fibres from the rostrum of the corpus callosum [p. 685] and the anterior perforated substance. Here it forms the lateral wall of the anterior horn of the ventricle and is fused with the anterior part of the putamen of the lentiform nucleus below the anterior limb of the

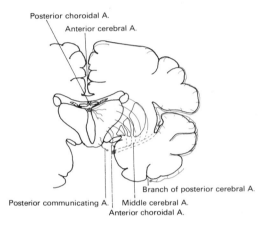

FIG. 10.107. Diagrammatic coronal section of the cerebrum to show the distribution of some of the arteries to this structure. The middle cerebral artery (broken red lines) and the origins of its striate branches are shown in a more anterior plane.

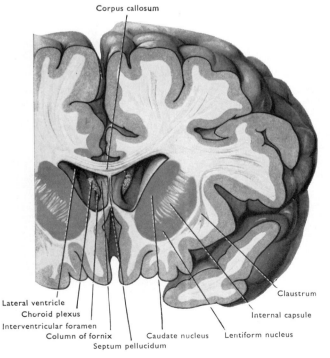

FIG. 10.108. Coronal section through the cerebral hemispheres. The section (viewed from the front) cuts the putamen of the lentiform nucleus in front of the globus pallidus.

internal capsule which separates them superiorly [FIG. 10.108]. The fused parts of the caudate and putamen are pierced horizontally by the anterior commissure [p. 682] and vertically by the striate branches of the middle cerebral artery and the striate veins. Many of these arteries course posterosuperiorly to supply more posterior parts of both nuclei, the internal capsule between them, and the anterolateral part of the thalamus [FIG. 10.107]. The caudate nucleus rapidly tapers from its head to a narrow body which forms the lateral wall of the central part of the lateral ventricle, and, becoming the tail, sweeps inferiorly, turning forwards into the roof of the inferior horn of the lateral ventricle as a very slender strip of grey matter. At the tip of the inferior horn it becomes continuous with the amygdaloid body which overlies this part of the ventricle. Throughout its length the caudate nucleus lies immediately external to the ventricular ependyma. A number of veins emerge from the hemisphere and run medially between the caudate nucleus and the ependyma to join the thalamostriate vein on the medial edge of the caudate nucleus [FIG. 10.122]. This vein courses towards the interventricular foramen with a bundle of fibres from the amygdaloid body—the stria terminalis. At the foramen, the vein turns posteriorly, and escaping from beneath the ventricular ependyma, forms the anterior extremity of the internal cerebral vein. The thalamostriate vein and the stria terminalis separate the body of the caudate nucleus from the superior and posterior surfaces of the thalamus, but in the roof of the inferior horn of the lateral ventricle the tail of the caudate with the vein and the stria are separated from the thalamus and the lentiform nucleus by the sublentiform part of the internal capsule and the fibres of the anterior commissure spreading posteriorly into the temporal lobe.

The amygdaloid body [FIGS. 10.73 (2 and 3) and 10.110] is an ovoid mass of nerve cells anterosuperior to the tip of the inferior horn of the lateral ventricle. It forms a solid mass of grey matter, continuous posteriorly with the tail of the caudate nucleus, superiorly with the lentiform nucleus, and medially with the cortex of the temporal lobe (piriform) which overlies it. It receives olfactory and other fibres and gives rise to the fibres of the stria terminalis. These pass with the caudate nucleus and thalamostriate vein, to end in the septal region [FIG. 10.120] and the anterior perforated substance. The amygdaloid body has several subdivisions.

The lentiform nucleus, approximately triangular in shape, underlies the insula and is separated from its cortex by two layers of nerve fibres—the external capsule and the white matter of the insula—between which is a thin sheet of grey matter, the claustrum [FIG. 10.110]. The smooth, rounded, lateral surface of the lentiform nucleus [FIG. 10.109] is easily separated from the external capsule except where it is pierced by striate vessels, thus suggesting that few fibres enter or leave this surface. The superomedial surface, angled to fit the genu of the internal capsule [FIG. 10.102] is firmly adherent to, and grooved by, the bundles of fibres of the internal capsule except antero-inferiorly where it fuses with the head of the caudate nucleus. Elsewhere strands of grey matter unite the outer margin of the lentiform nucleus to the caudate nucleus between the bundles of the internal capsule. The inferior surface is horizontal and is separated from the inferior horn of the lateral ventricle and the tail of the caudate nucleus by white matter composed of fibres passing to and from the temporal lobe of the hemisphere. This contains the sublentiform part of the internal capsule, fibres of the anterior commissure, and a part of the optic radiation [FIGS. 10.153 and 10.157], passing forwards over the inferior horn of the ventricle before sweeping backwards to the occipital lobe.

On section the lentiform nucleus is composed of three (or occasionally four) parts separated by two (or three) layers of white matter which extend from the internal capsule to its inferior surface [FIG. 10.110]. The lateral of these medullary laminae separates the larger, external, darker putamen from the paler, wedge-shaped, medial globus pallidus. The medial medullary lamina divides the

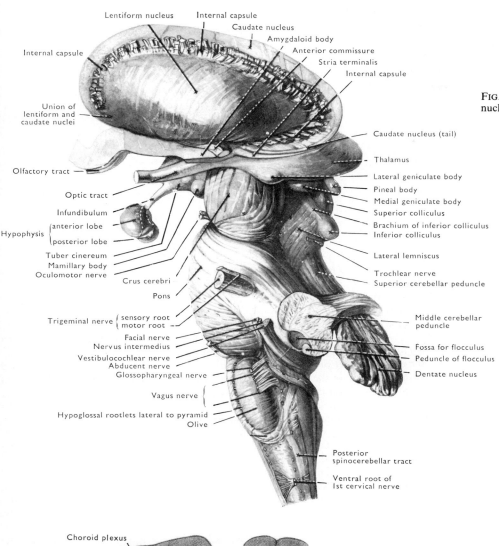

Lentiform nucleus
Internal capsule
Caudate nucleus
Amygdaloid body
Anterior commissure
Stria terminalis
Internal capsule
Internal capsule
Union of lentiform and caudate nuclei
Olfactory tract
Optic tract
Infundibulum
Hypophysis { anterior lobe / posterior lobe }
Tuber cinereum
Mamillary body
Oculomotor nerve
Crus cerebri
Pons
Trigeminal nerve { sensory root / motor root }
Facial nerve
Nervus intermedius
Vestibulocochlear nerve
Abducent nerve
Glossopharyngeal nerve
Vagus nerve
Hypoglossal rootlets lateral to pyramid
Olive

Caudate nucleus (tail)
Thalamus
Lateral geniculate body
Pineal body
Medial geniculate body
Superior colliculus
Brachium of inferior colliculus
Inferior colliculus
Lateral lemniscus
Trochlear nerve
Superior cerebellar peduncle
Middle cerebellar peduncle
Fossa for flocculus
Peduncle of flocculus
Dentate nucleus

Posterior spinocerebellar tract
Ventral root of 1st cervical nerve

FIG. 10.109. Dissection exposing the lentiform nucleus of the left hemisphere.

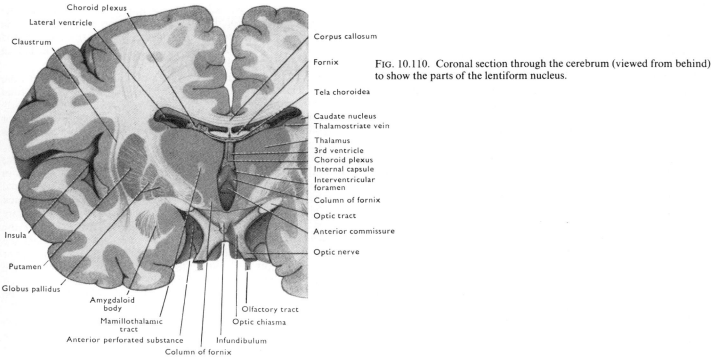

Choroid plexus
Lateral ventricle
Claustrum

Corpus callosum

Fornix

Tela choroidea

Caudate nucleus
Thalamostriate vein

Thalamus
3rd ventricle
Choroid plexus
Internal capsule
Interventricular foramen
Column of fornix
Optic tract
Anterior commissure
Optic nerve

Insula
Putamen
Globus pallidus
Amygdaloid body
Mamillothalamic tract
Anterior perforated substance
Column of fornix
Olfactory tract
Optic chiasma
Infundibulum

FIG. 10.110. Coronal section through the cerebrum (viewed from behind) to show the parts of the lentiform nucleus.

globus pallidus into internal and external segments. The globus pallidus is almost entirely in contact with the posterior limb of the internal capsule. The putamen, which has the same colour and structure as the caudate nucleus, is the portion of the lentiform which fuses and communicates with the caudate. If the lentiform nucleus is torn from the internal capsule, branches of the striate vessels can be seen passing from it through the internal capsule, and the medullary laminae are seen to be continuous with the internal capsule. Even when it is entirely separated from the internal capsule the lentiform nucleus remains attached to the brain by a bundle of fibres which emerges from the inferomedial angle of the globus pallidus and curves round the antero-inferior border of the internal capsule to enter the hypothalamic region. This is the **ansa lenticularis**, part of the main bundle of fibres leaving the lentiform nucleus [FIG. 10.169].

The **external capsule** is a thin sheet of white matter which covers the lateral surface of the lentiform nucleus and is continuous at the outer margin of that nucleus with the surrounding white matter of the hemisphere. This consists mainly of the intermingled fibres of the corpus callosum and corona radiata, but the rostrum of the corpus callosum beneath the anterior surface of the lentiform nucleus and the deep surface of the fasciculus uncinatus at the lateral sulcus. Fibres of the external capsule enter the claustrum and appear to form association pathways between the temporal lobe and the frontal and parietal lobes.

The **claustrum** is a thin sheet of grey matter of unknown function which lies between the external capsule and the white matter of the insula. It is smooth medially, scalloped on its lateral surface, and fades away at the superior margins of the insula, though inferiorly it is thickened and extends between the external capsule and those fibres of the fasciculus uncinatus [FIG. 10.113] which sweep posteriorly with the anterior commissure into the temporal and occipital lobes.

From the above description and the figures, it can be seen that the general structure of the telencephalon arises directly from its mode of development. In its final form the telencephalon consists of a centrally placed lentiform nucleus almost completely surrounded by a series of concentric, C-shaped structures; the internal capsule, the caudate nucleus, the lateral ventricle and its choroid plexus, and the cerebral cortex.

THE ASSOCIATION FIBRES OF THE CEREBRAL HEMISPHERE

These fibres arise in the cells of the cerebral cortex and pass from one part of the hemisphere to another. Some are very short, looping from one gyrus to the next (**arcuate fibres**) while others are longer and may stretch from end to end of the hemisphere. The longer association fibres are for the most part gathered together into certain bundles related to the corona radiata as follows:

1. On the sides of the corona radiata, covered by its fibres which arch over to the medial and lateral sides of the hemisphere are the cingulum and superior longitudinal (fasciculus) bundle [FIG. 10.112].

2. Along its inferior margin is the inferior longitudinal (fasciculus) bundle.

3. Beneath its anterior part is the uncinate fasciculus [FIG. 10.113].

The first group consists of two bundles, the cingulum on the medial side and the superior longitudinal bundle on the lateral side. Each of these is C-shaped and curves from the anterior part of the frontal lobe through the parietal lobe to turn downwards and forwards into the temporal lobe, thus following the general configuration of the hemisphere. Each receives fibres from and gives fibres to the cortex from its convex surface, while the concave surface is relatively free of such connections and is capable of being dissected from its immediate surroundings.

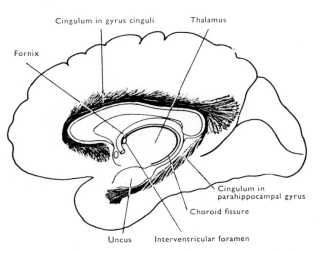

FIG. 10.111. Drawing of the medial aspect of the hemisphere to show the position of the cingulum (schematic) and the choroid fissure.

The **cingulum** [FIGS. 10.111 and 10.126] starts in the medial part of the frontal lobe below the rostrum of the corpus callosum [p. 685]. It sweeps over the superior surface of the corpus callosum enclosed in the gyrus cinguli [FIG. 10.112]. Hooking round the splenium of the corpus callosum, where a large number of fibres leave its convex surface, it runs antero-inferiorly in the medial margin of the temporal lobe (parahippocampal gyrus) to end, inferior to the uncus, by spreading into the tip of the temporal lobe. In this latter part of its course it has a rope-like appearance, the bundles of fibres running spirally.

The **superior longitudinal (fasciculus) bundle** [FIGS. 10.112 and 10.113] has a shape and arrangement very similar to the cingulum, but encircles the insula on the lateral aspect of the corona radiata. Beginning in the frontal lobe it runs posteriorly in the root of the frontoparietal operculum, and turns inferiorly, fanning out posteriorly into the occipital lobe and anteriorly into the temporal lobe. In the parietal part of the operculum it forms a compact bundle which is easily exposed by dissection, but in the frontal lobe it is traversed by many other fibres of the cerebral cortex, including fibres to and from the internal capsule and corpus callosum.

The **inferior longitudinal (fasciculus) bundle** runs between the temporal and occipital lobes along the inferior margin of the corona radiata. Its fibres blend with the corona radiata and extend anteriorly into the temporal lobe medial to the inferior part of the superior longitudinal bundle and inferior to the temporal part of the corona radiata.

The **uncinate fasciculus** hooks over the stem of the lateral sulcus from the frontal to the temporal and occipital lobes. It mingles in the frontal lobe with the anterior part of the superior longitudinal bundle. Posteriorly its fibres converge to form a narrow, thick bundle between the lateral sulcus and the anterior commissure where that commissure escapes from the lateral surface of the lentiform nucleus. The anterior fibres hook forwards beneath the stem of the lateral sulcus towards the temporal pole. The posterior fibres run almost horizontally towards the occipital lobe over the antero-inferior part of the claustrum. They mingle with the expanded extremity of the anterior commissure and spread laterally into the posterior temporal and occipital lobes [FIG. 10.113].

Each of these bundles consists of long association fibres which run a variable distance in them. The majority of the fibres in the cingulum and superior longitudinal bundle seem to run between the frontal lobe and the posterior parietal and occipital lobes.

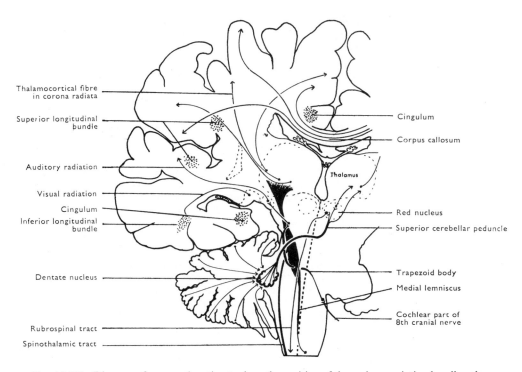

FIG. 10.112. Diagram of a coronal section to show the position of the main association bundles, the corona radiata, and the corpus callosum.

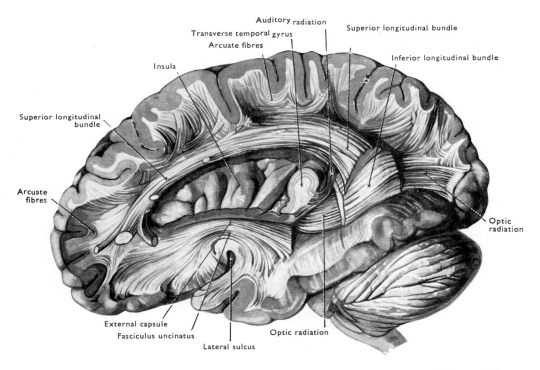

FIG. 10.113. Dissection of left cerebral hemisphere to display the optic and auditory radiations and the general direction of the principal association bundles.

THE MEDIAL MARGIN OF THE CEREBRAL CORTEX AND THE COMMISSURES

The extension of the telencephalic vesicle medially [p. 665] results in the diencephalon becoming invaginated into the medial aspect of the telencephalon. Thus the telencephalon develops a circular edge (or limbus) which surrounds the thalamus and hypothalamus and is attached to them only by the choroid plexus of the lateral ventricle invaginated through the choroid fissure between the limbus and the diencephalon. The cortex of the telencephalon adjacent to the limbus constitutes the 'limbic system'. Its continuity with the olfactory bulb through the olfactory striae at the anterior perforated substance (see below) is responsible for its association with olfaction and its older name of 'rhinencephalon' or 'smell-brain'—a view which is only partly true as evidenced by the large size of this part of the brain in anosmatic aquatic mammals and microsmatic Man.

The gross structure of this simple, ring-like limbus is distorted: (1) by the downgrowth of the temporal lobe which bends the part containing the lateral olfactory stria over the stem of the lateral sulcus and forms the hooked uncus from the part immediately posterior to this [FIG. 10.114]; (2) by the growth of the corpus callosum through its upper part which is thereby split into the septum pellucidum and the indusium griseum [FIG. 10.114]. The main claim for this limbus to be considered a separate 'limbic system' rests on the difference in the structure of its cortex (allocortex) from that of the rest of the hemisphere (isocortex) [FIG. 10.119] and on the presence of a large bundle of nerve fibres, the fimbria and fornix, which forms a connecting bundle between its parts and carries efferent fibres from it to the hypothalamus. Yet the gyrus cinguli and much of the parahippocampal gyrus are usually included in the limbic system though the structure of their cortex is akin to that of the rest of the hemisphere with which they have many connections. The parts which are histologically different from the remainder of the cortex [FIG. 10.114] are: the olfactory bulb, tract, and striae; the anterior perforated substance; the piriform area [FIG. 10.116]; the uncus; the hippocampus; the dentate gyrus; the gyrus fasciolaris; the indusium griseum; the septum pellucidum; the paraterminal gyrus; and part of the parahippocampal gyrus.

The **olfactory bulb** and tract are formed as an outgrowth from the ventral aspect of the developing frontal part of the cerebral vesicle and at first contain an extension of the lateral ventricle which is later obliterated. The olfactory bulb receives bundles of non-myelinated olfactory nerve fibres which grow through the cribriform lamina of the ethmoid bone from the sensory cells buried in the olfactory mucous membrane. These fibres make contact with the numerous dendrites of the large mitral cells of the bulb, forming little knots of interlacing processes, known as glomeruli, close to the inferior surface of the bulb [FIG. 10.115]. The mitral cells give rise to the fibres in the olfactory tract. It runs posteriorly on the inferior surface of the frontal lobe, wedged into the olfactory sulcus, and splits into medial and lateral striae [FIG. 10.121] which enclose between them the **olfactory trigone** [FIGS. 10.78 and 10.116]. The **olfactory tubercle** in the trigone is not usually visible in Man as a separate entity but is incorporated in the anterior perforated substance. It is large in mammals with a well-developed sense of smell. Many of the olfactory tract fibres end in the olfactory trigone, the remainder skirt the anterolateral aspect of the anterior perforated substance in the lateral olfactory stria.

In most mammals the lateral olfactory stria enters the pear-shaped **piriform area** on the inferior surface of the hemisphere. This is limited laterally by the curved rhinal sulcus [FIG. 10.116]. In the poorly developed olefactory system of Man, the anterior part of the

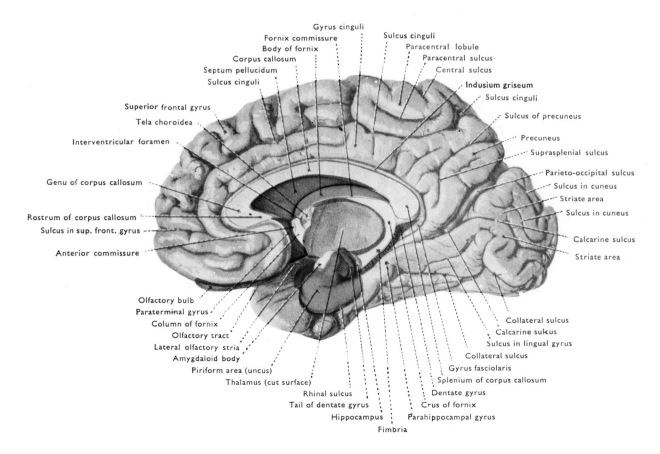

FIG. 10.114. Medial aspect of right cerebral hemisphere, with the 'rhinencephalon' coloured.

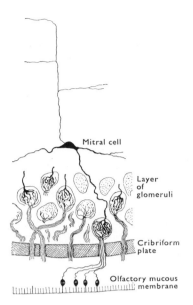

FIG. 10.115. Diagram of minute structure of olfactory bulb.

Mitral cell

Layer of glomeruli

Cribriform plate

Olfactory mucous membrane

the remainder of the temporal lobe by the rhinal sulcus [FIG. 10.114]. Fibres from the olfactory bulb extend into the anterior piriform area, but do not reach the posterior part.

The **uncus** is essentially a recurved portion of the anterior extremity of the parahippocampal gyrus and is continuous posteriorly with the inferior part of the limbic cortex of the hemisphere (the hippocampus and dentate gyrus) which curves under the hypothalamus and lentiform nucleus but is separated from them by the choroid fissure and by the crus cerebri and the sublentiform part of the internal capsule respectively.

FIGURE 10.117 shows the mode of development of the hippocampus and dentate gyrus on the margin of the hemisphere and the formation of the fimbria.

In the adult, the **dentate gyrus** begins as a narrow ribbon of grey matter on the posteromedial surface of the uncus [FIG. 10.114]. It runs inferiorly on the uncus (tail of the dentate gyrus), and then passes posteriorly, under cover of the fimbria in the hippocampal sulcus, as a crinkled strip of cortex lying above the superior margin of the parahippocampal gyrus (subiculum). Turning superiorly it reaches the splenium of the corpus callosum around which it curves as the **gyrus fasciolaris** to become continuous with a thin sheet of grey matter on the superior surface of the corpus callosum, the **indusium griseum**.

The **hippocampus** is a portion of the marginal cortex invaginated as a rounded ridge into the medial wall of the inferior horn of the lateral ventricle by the hippocampal sulcus. The hippocampus begins in the floor of the inferior horn of the lateral ventricle near its tip, where it forms the **pes hippocampi** [FIG. 10.130]. This extends laterally from the uncus as a ridged structure owing to the folded nature of the hippocampal cortex in this region. Turning posteriorly it runs parallel to the dentate gyrus to reach the splenium of the

piriform area corresponds to the lateral olfactory stria. It curves laterally on the antero-inferior part of the insula, the limen insulae, and then inferomedially behind the stem of the lateral sulcus to become continuous with the uncus [FIG. 10.114] and the anterior extremity of the parahippocampal gyrus which are separated from

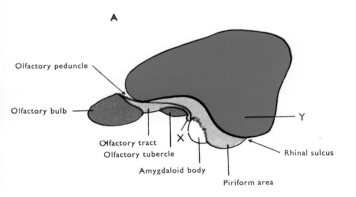

A

Olfactory peduncle

Olfactory bulb

Olfactory tract
Olfactory tubercle

Amygdaloid body

X

Y

Rhinal sulcus

Piriform area

FIG. 10.116. Relation of olfactory part of hemisphere to neopallium.

A, lateral aspect of left cerebral hemisphere; B, inferior aspect of right half of rabbit's brain; C, corresponding view of human foetal brain at fifth month.

In Man, the great growth of the neopallium (phylogenetically the most recent part of the cerebral cortex which receives afferent fibres from the thalamus rather than from the olfactory system) produces a well-defined temporal lobe (Y) and not only flexes the more ancient piriform area at X but also relegates it to the medial surface of the hemisphere.

Olfactory areas, red; neopallium, blue.

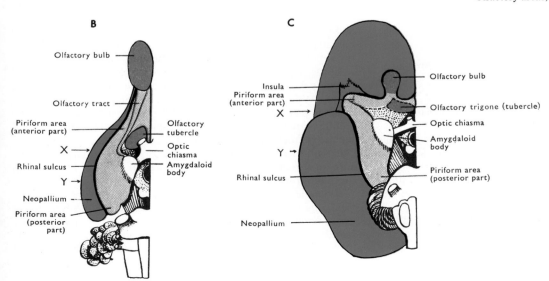

B

Olfactory bulb

Olfactory tract

Piriform area (anterior part)

X

Rhinal sulcus

Y

Neopallium

Piriform area (posterior part)

Olfactory tubercle

Optic chiasma

Amygdaloid body

C

Insula
Piriform area (anterior part)

X

Y

Rhinal sulcus

Neopallium

Olfactory bulb

Olfactory trigone (tubercle)

Optic chiasma

Amygdaloid body

Piriform area (posterior part)

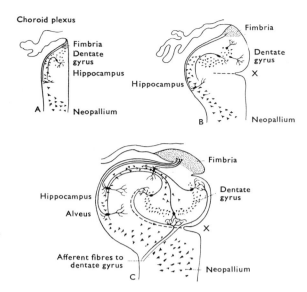

FIG. 10.117. Three stages in the development of the right hippocampal formation.

X Hippocampal sulcus.

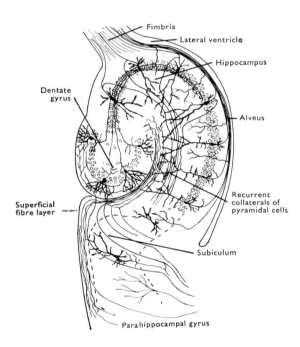

FIG. 10.118. Drawing of Golgi preparation of the left hippocampus, dentate gyrus, and subiculum. Redrawn from Cajal (1955).

corpus callosum where it also becomes continuous with the indusium griseum and probably also with the septum pellucidum.

The hippocampus and dentate gyrus [FIG. 10.118] differ from the remainder of the cerebral cortex in having a simpler structure, consisting essentially of three layers. The superficial layer contains few cells but a considerable number of tangentially running fibres which give this cortex its characteristic superficial white layer. In the dentate gyrus the middle layer consists of small granule cells which send their axons to the large pyramidal cells in the middle layer of the hippocampus. The deepest layer of both parts of this cortex is formed of scattered cells of irregular shape.

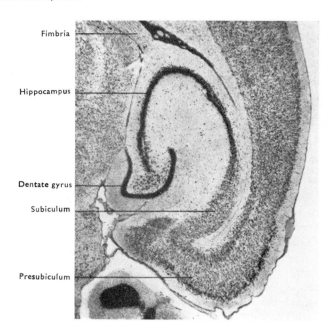

FIG. 10.119. Horizontal section of rat brain to show transition from hippocampus to neopallium. Cell bodies only are stained.

The fornix

The efferent fibres of the pyramidal cells of the hippocampus pass to its ventricular surface forming a thin sheet of white matter, the alveus. The fibres of the alveus pass into a bundle (the fimbria) on the dorsomedial aspect of the hippocampus immediately inferior to the choroid fissure. The fimbria sweeps posteriorly between the hippocampus and the choroid fissure, and turning superiorly as the crus of the fornix passes beneath the splenium of the corpus callosum [FIGS. 10.111, 10.114, and 10.130]. Thence it follows the choroid fissure over the dorsal surface of the thalamus fused medially with the opposite fornix, body of the fornix, as far as the interventricular foramen. Throughout this part of their course the fimbria and fornix form the rim of the marginal cortex and have the ependyma of the choroid plexus of the lateral ventricle attached to them, the taenia of the fornix. At the interventricular foramen most of the fornix turns inferiorly (column of the fornix) between that foramen and the anterior commissure to curve postero-inferiorly from the superior part of the lamina terminalis into the hypothalamus. Here it ends in the mamillary body, separated from the opposite column of the fornix by the third ventricle. The fimbria and fornix thus form one complete turn of a spiral, starting in the medial part of the temporal lobe and ending close to the midline in the hypothalamus, having encircled the thalamus [FIG. 10.121].

As the crura of the two fornices approach each other beneath the corpus callosum they exchange fibres—the commissure of the fornix. At the interventricular foramen a number of fibres leave the fornix to pass ventrally in front of the anterior commissure into the septal region [FIG. 10.120]. These extend to the anterior perforated substance from which other fibres pass upwards to enter the fornix. Other fibres which leave it in this region pass posteriorly to the anterior thalamic nuclei and the fornix gives off many fibres within the anterior hypothalamus.

A bundle, the mamillothalamic tract, almost equal in size to the fornix, passes laterally from the mammillary body, its fibres giving rise to branches, the mamillotegmental tract, which turn inferiorly into the midbrain. The mamillothalamic tract then turns superiorly in the medial medullary lamina of the thalamus to reach the parts of

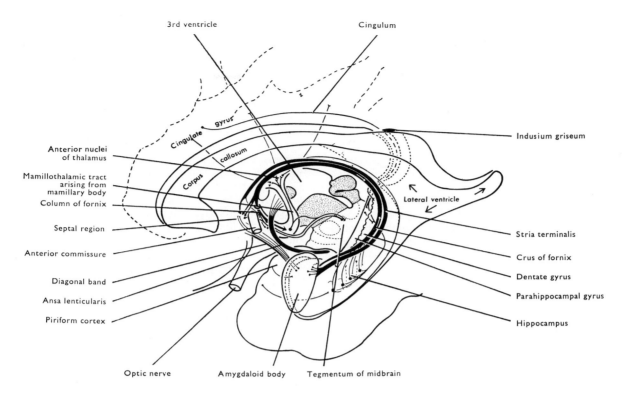

FIG. 10.120. Diagram to show the marginal structures of the hemisphere based on a dissection. The left thalamus and hypothalamus (except the mamillary body) have been removed exposing the third ventricle, interthalamic adhesion, and the upper part of the midbrain. The left half of the midbrain has been transected. Arrows in fornix indicate the passage of fibres in both directions within it.

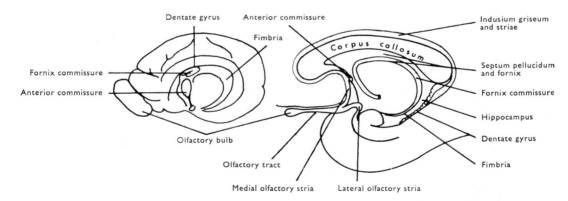

FIG. 10.121. Drawings of the medial surface of the hemisphere of a kangaroo (left) and of Man (right). The absence of a corpus callosum in the marsupials leaves the limbic ring intact.

the anterior nuclei of the thalamus. The anterior thalamic nuclei in their turn have connections with the cingulate gyrus through the internal capsule. Thus the hippocampus has connections with other parts of the limbic cortex, the hypothalamus, thalamus, midbrain, and cerebral cortex.

The hippocampus has long been considered an important part of the olfactory apparatus and, while this may be true in lower vertebrates, it seems certain that in Man the secondary olfactory fibres from the olfactory bulb do not reach the hippocampus, but end in the olfactory trigone and the anterior part of the piriform area. It seems certain that it is concerned with other functions and its massive hypothalamic and septal connections should not be forgotten. Physiological and clinical investigations seem to suggest that it may be concerned with emotion and recent memory (Scoville and Milner 1957). Certainly it has many connections with the remainder of the temporal lobe which has also been implicated in the latter function.

In marsupials, where there is no corpus callosum, the hippocampus and dentate gyrus encircle the thalamus and can be traced forwards above it and the commissure of the fornix to become almost continuous with the medial olfactory stria anterior to the lamina terminalis. In eutherian mammals, including Man, the corpus callosum [FIG. 10.121] grows through the dorsal part of the medial margin of the hemisphere, obliterating the hippocampus and dentate gyrus in this region, but leaving a portion of the marginal cortex (septum pellucidum) uniting the fornix to the inferior surface of the corpus callosum. The part of the marginal cortex remaining superior to the corpus callosum persists as a thin layer of grey matter, the indusium griseum, together with two longitudinal bundles of nerve fibres, the **striae longitudinales** [FIG. 10.126].

The **septum pellucidum** in the fully developed brain appears to be a single membrane which connects both fornices to the corpus callosum and separates the anterior parts of the two lateral ventricles. It is, however, formed from right and left parts, one from each hemisphere, and though these are usually adherent over a large part of their area, a cavity (**cavum septi pellucidi**) is frequently present anteriorly; this represents the part of the longitudinal fissure lying inferior to the corpus callosum [FIG. 10.122]. The septum pellucidum extends anteriorly in the concavity of the genu of the corpus callosum between the anterior part of the trunk and the rostrum. It is covered on its ventricular surface with ependyma and is separated from this by a number of veins coursing posteriorly from the genu of the corpus callosum and the white matter of the frontal lobe to join the internal cerebral veins.

The septum pellucidum is extensive anteriorly, but narrows posteriorly to the commissure of the fornices applied to the corpus callosum. The size of the septum is directly related to the size of the lateral ventricles and is usually enlarged in compensatory hydrocephalus, when the ventricle expands with the loss of neural tissue in the aged.

The **indusium griseum** probably represents a remnant of the hippocampus [FIG. 10.114]. In the depths of the callosal sulcus, lateral to the indusium griseum, is a small strip of cortex covered by a layer of white matter. This is continuous over the splenium of the corpus callosum with the medial part of the parahippocampal gyrus, the subiculum [FIG. 10.119], which also has a superficial fibrous layer. At the rostrum of the corpus callosum the indusium griseum and septum pellucidum become continuous with the subcallosal area and **paraterminal gyrus** which lie anterior to the lamina terminalis. It is into this area that the medial olfactory stria runs, thus completing a continuous ring of grey matter (limbic cortex) and nerve fibres forming the medial margin of the hemisphere and having the olfactory tract entering its antero-inferior part.

THE ANTERIOR COMMISSURE

This small, cylindrical bundle of variable diameter (2–4 mm) crosses the midline in the superior part of the lamina terminalis anterior to the columns of the fornices and the interventricular foramen [FIG. 10.83]. Traced laterally it pierces the head of the caudate nucleus and the anterior part of the putamen of the lentiform nucleus inferior to the anterior limb of the internal capsule and above the anterior perforated substance. In this part of its course it may be joined by fibres from the olfactory tract which turn medially and, crossing to the opposite side, act as a commissure for the olfactory bulbs. This **anterior part** of the commissure is the major part in macrosmatic mammals, but is small in Man where the majority of the fibres incline posteriorly (**posterior part**) while still in the putamen of the lentiform nucleus, and emerge from its lateral surface [FIG. 10.109] passing posterolaterally into the white matter of the temporal lobe in which its fibres fan out, some reaching as far posteriorly as the inferior part of the occipital lobe.

THE TRANSVERSE FISSURE AND CORPUS CALLOSUM [FIG. 10.114]

The corpus callosum, found only in eutherian mammals, is formed of fibres which join corresponding parts of the cerebral cortex of the two hemispheres. It crosses the midline where the two hemispheres fuse superior to the roof of the third ventricle. It thus overlies the roof of the third ventricle and cuts off the inferior portion of the **longitudinal fissure** (which separates the two hemispheres) to form the **transverse fissure** and cavity of the septum pellucidum, between it and the diencephalon. The transverse fissure contains the connective tissue (pia mater) lying immediately superior to the roof of the third ventricle, which contains the posterior choroidal arteries and the internal cerebral veins [FIG. 10.80]. The pia mater in the

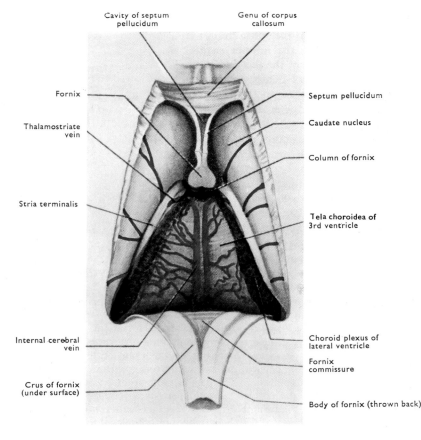

Cavity of septum pellucidum

Genu of corpus callosum

Fornix

Thalamostriate vein

Stria terminalis

Internal cerebral vein

Crus of fornix (under surface)

Septum pellucidum

Caudate nucleus

Column of fornix

Tela choroidea of 3rd ventricle

Choroid plexus of lateral ventricle

Fornix commissure

Body of fornix (thrown back)

FIG. 10.122. Dissection to show tela choroidea of third and lateral ventricles, and the parts near it.

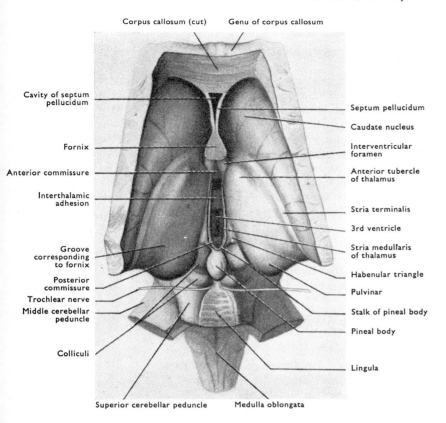

Corpus callosum (cut)   Genu of corpus callosum

Cavity of septum pellucidum

Fornix

Anterior commissure

Interthalamic adhesion

Groove corresponding to fornix

Posterior commissure

Trochlear nerve

Middle cerebellar peduncle

Colliculi

Superior cerebellar peduncle   Medulla oblongata

Septum pellucidum

Caudate nucleus

Interventricular foramen

Anterior tubercle of thalamus

Stria terminalis

3rd ventricle

Stria medullaris of thalamus

Habenular triangle

Pulvinar

Stalk of pineal body

Pineal body

Lingula

FIG. 10.123. Thalami and the parts of the brain around them.

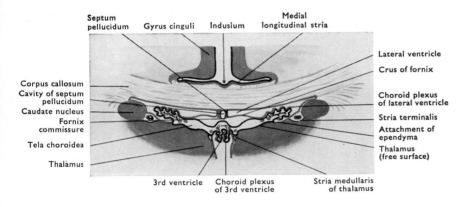

Septum pellucidum   Gyrus cinguli   Indusium   Medial longitudinal stria

Corpus callosum
Cavity of septum pellucidum
Caudate nucleus
Fornix commissure
Tela choroidea
Thalamus

Lateral ventricle

Crus of fornix

Choroid plexus of lateral ventricle

Stria terminalis

Attachment of ependyma

Thalamus (free surface)

3rd ventricle   Choroid plexus of 3rd ventricle   Stria medullaris of thalamus

FIG. 10.124. Diagram of coronal section through central parts of lateral ventricles.

transverse fissure invaginates the ependymal wall of the lateral ventricles through the choroid fissure at its lateral margins, and inferiorly, on either side of the midline, bulges into the third ventricle to form the vascular connective tissue of the third ventricular choroid plexuses; thus it is the **tela choroidea of the lateral and third ventricles.** The transverse fissure is closed above and anteriorly by the fusion of the two fornices; the enclosed sheet of pia mater forming the tela choroidea, narrows anteriorly as the lateral ventricles approach each other, and ends in an apex immediately posterior to the interventricular foramen [FIG. 10.122]. Here the **choroid plexus** of each lateral ventricle on the lateral margin of the tela choroidea becomes continuous, through the interventricular foramen, with the corresponding choroid plexus of the third ventricle on the inferior surface of the tela. As the lateral ventricles diverge posteriorly, the tela choroidea widens to cover the medial part of the superior surface of the thalamus [FIG. 10.124] and becomes continuous with the pia mater on the remainder of the brain by emerging between the splenium of the corpus callosum [p. 685] and the pineal body on the upper midbrain [FIG. 10.83]. In

this region the two internal cerebral veins unite to form the **great cerebral vein** which turns superiorly on the splenium of the corpus callosum and enters the junction of the free edges of the falx and tentorium to form the straight sinus with the inferior sagittal sinus. The straight sinus runs posteriorly in the midline between the layers of dura mater at the junction of the falx cerebri with the tentorium cerebelli [p. 729] and reaches the internal occipital protuberance.

The fornix and the cortex immediately adjacent to it (septum pellucidum), which together form the medial part of the floor and medial wall of the lateral ventricle in this region, are separated from the remainder of the medial surface of the hemisphere by the corpus callosum [FIG. 10.129].

Nerve fibres from most parts of the cerebral cortex converge radially on the developing **corpus callosum** in a manner similar to the fibres running to the internal capsule. However, they are at right angles to the fibres of the internal capsule, and run over the roof of the lateral ventricle [FIG. 10.125] instead of passing within its concavity [FIG. 10.101]. Thus the fibres from the superolateral surface of the hemisphere above the lateral sulcus pass through the

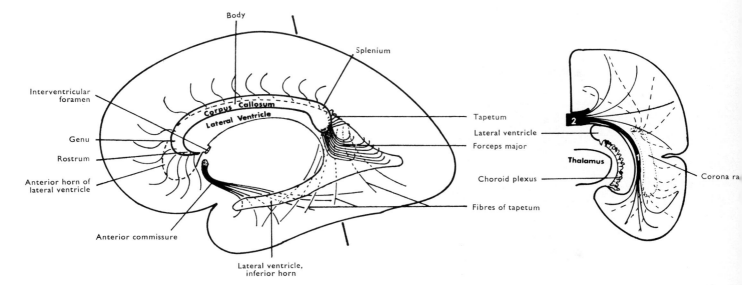

FIG. 10.125. Diagram to illustrate the course and position of the major commissural fibres of the hemisphere. In the right figure, which is a coronal section along the line in the left figure, (1) is the tapetum and (2) the splenium of the corpus callosum.

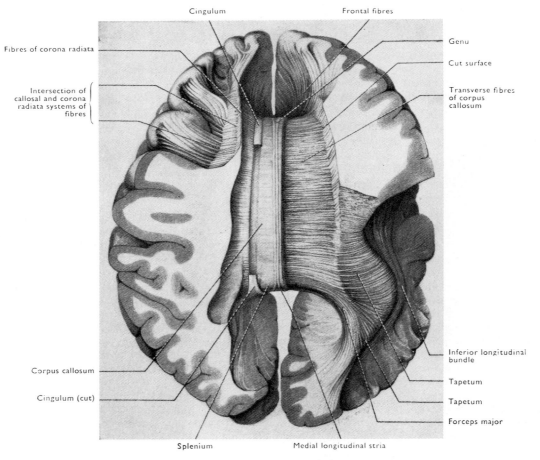

FIG. 10.126. Corpus callosum, exposed from above and right half dissected.

fibres of the corona radiata and converge on the corpus callosum by passing superior to the caudate nucleus and the lateral ventricle [FIG. 10.112]. The callosal fibres do not mix, but diverge again on the opposite side so that each passes to the part of the opposite cerebral cortex corresponding to that from which it arose [FIG. 10.125].

In view of this arrangement of the callosal fibres it might be expected that the corpus callosum would develop the same C-shaped structure as the other parts of the hemisphere. However, the brain stem holds the hemispheres apart postero-inferiorly and prevents the formation of any part of the corpus callosum inferior to the dorsal surface of the upper midbrain, so that it extends only from the superior part of the lamina terminalis anteriorly to the superior surface of the midbrain posteriorly. Hence callosal fibres from the temporal and inferior parts of the occipital lobes sweep superiorly to the posterior extremity of the corpus callosum, which is thereby increased in size to form the thick **splenium** [FIG. 10.125]. These ascending fibres form a compact layer, **the tapetum**, immediately lateral to the posterior and inferior horns of the lateral ventricle and medial to the corona radiata. Fibres from the superior part of the occipital lobe also enter the splenium by coursing anteriorly over the superior surface of the posterior horn of the lateral ventricle, while those from its medial aspect are divided by the deep calcarine sulcus, which bulges into the posterior horn of the lateral ventricle forming the **calcar avis** [FIG. 10.99]. Callosal fibres arising inferior to the calcarine sulcus loop beneath the posterior horn of the lateral ventricle to enter the tapetum. Those arising in the cuneus, superior to the calcarine sulcus, pass anteriorly around the deep parieto-occipital sulcus, and so bulge far laterally into the medial wall of the ventricle to form the **bulb of the posterior horn** [FIG. 10.129] immediately superior to the calcar avis [FIG. 10.99] before turning medially into the splenium of the corpus callosum. When these fibres are dissected on both sides they are seen to form the wide **forceps major** [FIG. 10.126]. Elsewhere on the medial sides of the hemispheres the corresponding fibres are not forced laterally by the presence of a deep sulcus and so form a smaller curve—the **forceps minor**. Fibres of the tapetum and forceps major virtually surround the posterior horn of the lateral ventricle.

Fibres from all surfaces of the frontal lobe enter the anterior part of the corpus callosum. They converge on the **rostrum, genu**, and anterior part of the body or **trunk** of the corpus callosum [FIG. 10.125] from the inferior, anterior, and posterior parts of the lobe respectively, approaching the midline on the corresponding surfaces of the anterior horn of the lateral ventricle. Thus this horn of the ventricle is surrounded by fibres of the corpus callosum except on its lateral side where the head of the caudate nucleus is lodged [FIGS. 10.108 and 10.125].

THE LATERAL VENTRICLE [FIGS. 10.127 and 10.128]

The general shape of this ventricle has already been described [p. 665], but its position can now be understood. The **anterior horns** of the two ventricles lie antero-inferior to the interventricular foramen and are separated from each other only by the **septum pellucidum** (and its cavity) which forms their medial walls. The lateral walls are formed by the large convex heads of the **caudate nuclei**. The remaining walls of the anterior horns are formed by fibres of the corpus callosum [FIGS. 10.102 and 10.125]. Posterior to the interventricular foramen the septum pellucidum continues to form the medial wall of the ventricle but rapidly decreases in height. The fornix, now on the inferior margin of the septum pellucidum, extends laterally into the floor of the **central part of the lateral ventricle** to the line of the choroid plexus, which is invaginated into the lateral ventricle between the fornix and the dorsal surface of the thalamus. The thalamus and the narrowing caudate nucleus, with

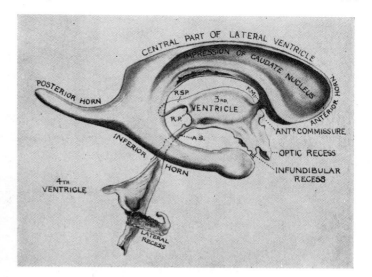

FIG. 10.127. Profile view of a cast of the ventricles of the brain. (From G. Retzius (1900). *Biol. Untersuch.* 9, 45.)
R.SP., suprapineal recess; A.S., aqueduct; R.P., pineal recess; F.M., interventricular foramen.

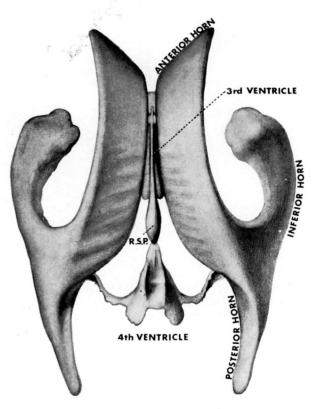

FIG. 10.128. Drawing of cast of ventricular system of the brain, seen from above. (After Retzius.)
R.SP., suprapineal recess.

the stria terminalis and thalamostriate vein between them, form the remainder of the floor of the ventricle [FIG. 10.129]. Laterally, the floor slopes superiorly to meet the roof, which is formed by the trunk of the corpus callosum and posteriorly by its splenium.

The two ventricles diverge posteriorly because of the intervening brain stem, and the central part of the ventricle turns inferiorly around the posterior end of the thalamus to become continuous with

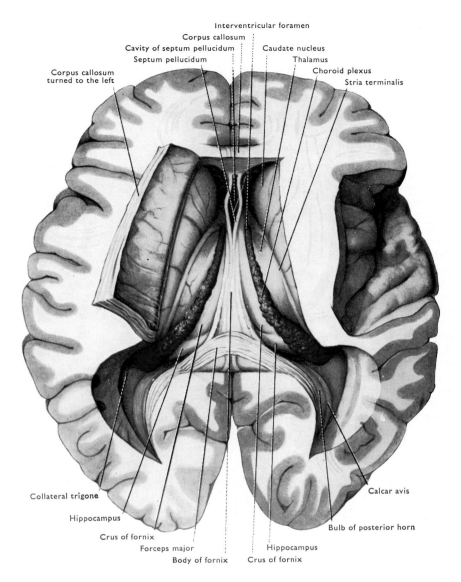

Corpus callosum
turned to the left

Corpus callosum
Cavity of septum pellucidum
Septum pellucidum

Interventricular foramen
Corpus callosum
Caudate nucleus
Thalamus
Choroid plexus
Stria terminalis

Collateral trigone
Hippocampus
Crus of fornix
Forceps major
Body of fornix

Crus of fornix
Hippocampus

Calcar avis
Bulb of posterior horn

FIG. 10.129. Dissection to show the fornix and lateral ventricles.

the inferior horn anteriorly, and the posterior horn posteriorly. The inferior horn [FIG. 10.130] with the choroid fissure in its medial wall curves anteriorly on the crus cerebri. The choroid fissure follows the line of the optic tract antero-inferiorly from the lateral geniculate body to end against the uncus; the cavity of the ventricle, continuing anteriorly for a further 1 cm, turns inferiorly over the pes hippocampi, posterior to the amygdaloid body. In the medial wall of the inferior horn is the rounded ridge of the hippocampus, covered by the alveus, with the fimbria on its superior margin forming the lower edge of the choroid fissure. The narrow floor is formed by fibres of the inferior longitudinal bundle, the corona radiata, and the tapetum passing to the medial aspect of the temporal lobe. Posteriorly, where the three parts of the ventricle meet, the floor widens into a triangular area slightly raised from below by the collateral sulcus, the collateral trigone. The roof of the inferior horn contains the stria terminalis and the tail of the caudate nucleus—a narrow ribbon of grey matter often broken in places, but usually continuous with the amygdaloid body anteriorly. Superior to both of these are the fibres of the auditory radiation and the most anterior fibres of the optic radiation in the sublentiform part of the internal capsule.

The posterior horn varies greatly in size, and is indented medially by the calcarine sulcus (calcar avis) and forceps major. It is virtually surrounded by fibres of the corpus callosum [p. 685].

THE CHOROID PLEXUSES OF THE THIRD AND THE LATERAL VENTRICLES

The choroid plexuses of the third ventricle, consist of a core of vascular pia mater covered on the ventricular aspect by ependyma. They are small fringes, one on each side of the midline, which increase slightly in size in the suprapineal recess of the ventricle. Anteriorly, each choroid plexus of the third ventricle turns laterally on the posterior wall of the interventricular foramen and becomes continuous with the choroid plexus of the corresponding lateral ventricle. In the lateral ventricle, the choroid plexus sweeps posterolaterally, attached to the dorsal surface of the thalamus and the lateral edge of the fornix [FIG. 10.129]. At first the choroid plexus is small, but it increases in size as it turns inferiorly on the posterior surface of the thalamus into the meeting of the central part of the ventricle with the posterior and inferior horns. Here it expands into the posterior horn and continues anteriorly in the

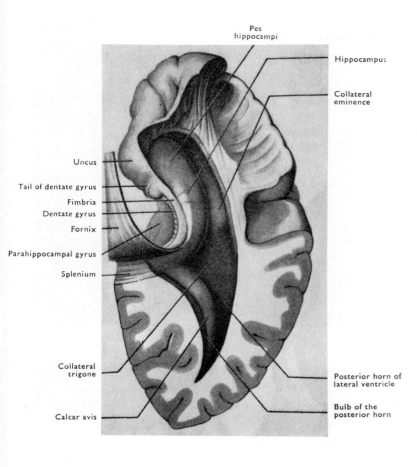

Pes
hippocampi

Hippocampus

Collateral
eminence

Uncus

Tail of dentate gyrus

Fimbria

Dentate gyrus

Fornix

Parahippocampal gyrus

Splenium

Collateral
trigone

Calcar avis

Posterior horn of
lateral ventricle

Bulb of the
posterior horn

FIG. 10.130. Dissection from above to show posterior and inferior horns of lateral ventricle.

inferior horn to the notch between the fimbria and the uncus as a thin ribbon. The **anterior choroidal artery** enters the plexus at this point and supplies it in the inferior horn. Further posteriorly, the choroid fissure sweeps over the lateral surface of the upper midbrain parallel to the posterior cerebral artery from which several branches pass laterally into it. In the central part of the ventricle, the choroid plexus is supplied by a branch or branches of the same artery passing forwards in the tela choroidea dorsal to the thalamus. All these are **posterior choroidal** branches of the posterior cerebral artery. A large **vein** drains along each plexus towards the interventricular foramen. At the foramen, it leaves the plexus, and joining the thalamostriate vein, forms the **internal cerebral vein**.

RADIOLOGY

The various components of the **central nervous system** cannot be differentiated on the plain radiograph.

By the introduction of air or other contrast media into the ventricular system or by opacification of the blood vessels, much valuable information may be obtained.

Computerized axial tomography has revolutionized and simplified the study of intracranial anatomy and pathology in the living. The method is complex, but its effects are easily understood. If horizontal sections are cut at various levels in the head and radiographs of these sections are made, the end result is akin to the findings in *computerized axial tomography* (CAT) [Atlas, FIGS. 1A–7A]. The grey and white matter can be distinguished and the ventricles studied. It

is sometimes possible to assess the condition of the subarachnoid space.

Air studies of the ventricular and subarachnoid space (*encephalography*) are made by the introduction of air, sometimes in quite small amounts, into the spinal subarachnoid space by lumbar puncture. By varying the position of the head in the sitting patient it is possible to direct the rising air mainly into the ventricles through the median aperture of the fourth ventricle, or mainly into the subarachnoid space as required.

Ventriculography requires the injection of air directly into the ventricles through a small opening in the skull. It is rarely performed except when an opaque medium is also introduced to outline the third ventricle, aqueduct, and fourth ventricle, which are not always adequately demonstrated in air studies, although tomography is sometimes helpful in these difficult areas [FIG. 10.84].

In the 'brow up' position [FIG. 10.132A] the anterosuperior regions of the **lateral ventricles** appear as two triangles separated by the **septum pellucidum**. The inferolateral sides are indented by the heads of the **caudate nuclei**. The anterior horns form fainter images extending downwards from these, giving a butterfly appearance.

The slit-like cavity of the **third ventricle** can be seen in the midline, below the septum pellucidum.

When the patient is turned to the 'brow down' position, the diverging central portions of the lateral ventricles are seen, though in FIGURE 10.131A the septum is still visible because of the large size of the ventricle. The **inferior horns** pass downwards and laterally from them. A darker area, representing the **posterior horn**, is sometimes seen superimposed on the inferior horn. Often the posterior horn is small (and occasionally absent) so that it does not show. The third ventricle, wider posteriorly, presents as a rounded translucency.

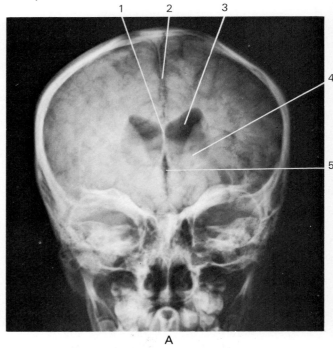

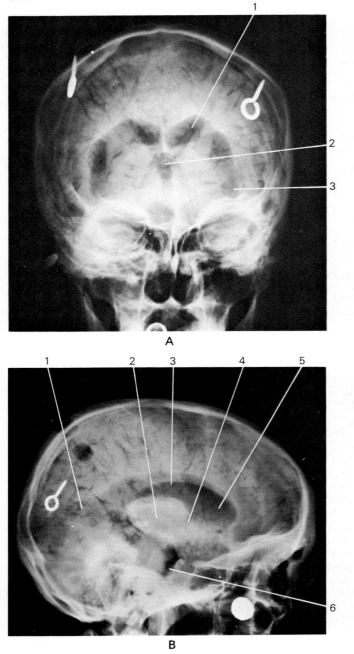

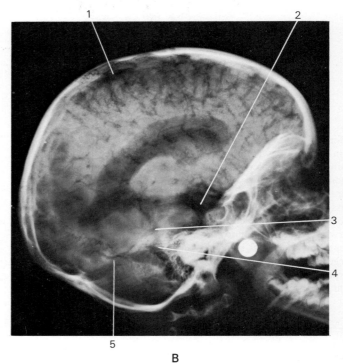

FIG. 10.131. Ventriculograms taken after the introduction of air into the ventricular cavity by a needle passed through a burr hole in the skull.

A, Brow down. 1. Central part of lateral ventricle.
                2. 3rd ventricle.
                3. Inferior horn of lateral ventricle.

     B. 1. Posterior horn of lateral ventricle.
                2. 3rd ventricle.
                3. Central part of lateral ventricle.
                4. Interventricular foramen.
                5. Anterior horn of lateral ventricle.
                6. Cisterna pontis.

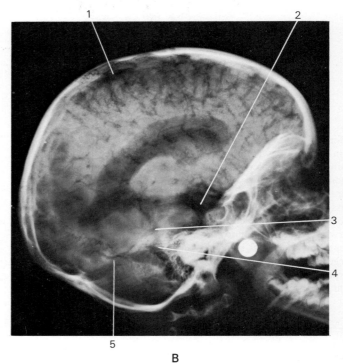

FIG. 10.132. Encephalograms taken after the introduction of air into the lumbar subarachnoid space. Some air has passed into the ventricular system through the median aperture of the fourth ventricle, the remainder outlines the subarachnoid space.

A, Brow up. 1. Septum pellucidum.
              2. Falx cerebri.
              3. Central part of lateral ventricle.
              4. Anterior horn of lateral ventricle.
              5. Third ventricle.

     B. 1. Air in subarachnoid space, mainly sulci.
              2. Interpeduncular cistern.
              3. Aqueduct of cerebrum.
              4. Fourth ventricle.
              5. Transverse sinus.

In the lateral view, taken with the patient lying on his side, the third ventricle is often poorly filled, though the interventricular foramen can often be identified. In the lateral encephalogram, the outline of the fourth ventricle and aqueduct can occasionally be seen, but the positive contrast ventriculogram usually shows these structures more clearly.

Some of the structures in the basal subarachnoid cisterns can be identified in encephalograms [FIG. 10.84].

For cerebral angiography, see blood vascular system.

Ultrasound has little anatomical relevance, but provides a simple and useful method of determining if there is any shift of the midline structures, e.g. falx cerebri. This can often help if there is doubt whether to proceed with further investigations.

Radioisotope Scanning is relatively simple to carry out and is valuable in certain pathological states, especially if computerized tomography is not available.

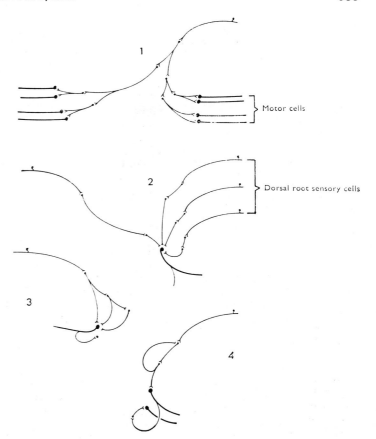

FIG. 10.133. Four diagrams to indicate some of the types of pathways which impulses, entering the spinal medulla over the dorsal roots, may follow through the internuncial neurons.
1. Divergent effect of internuncial neurons.
2. Convergent effect of internuncial neurons.
3. Arrangement of internuncial neurons capable of producing repetitive stimulation of a motor neuron from a single sensory impulse.
4. Shows recurrent collaterals of internuncial and motor neurons. The former can produce reverberating circuits, the latter are believed to stimulate internuncial neurons, Renshaw (1941) cells, which inhibit adjacent motor neurons.
These diagrams represent concepts derived from electrophysiological experiments, and bear no relation to the actual arrangement or appearance of these neurons under the microscope.

THE MICROSCOPIC ANATOMY OF THE CENTRAL NERVOUS SYSTEM

The previous section has shown the arrangement of the main masses of nerve cells and fibres which comprise the central nervous system and of the nerves which are attached to it. The gross anatomy gives little indication of the functions of this system beyond those which can be inferred from the peripheral connections of the various nerves to be described in the next section. Knowledge of the functional significance of the various parts of this system, based on their structure, can only be obtained with any certainty from a study of the connections of the groups of cells which comprise the central nervous system. However, no complete picture of the connections of the cells in the nervous system is available, and the anatomical approach to the study of its function cannot yet produce more than a very sketchy outline.

The divisions of the central nervous system which are based on gross anatomy, i.e. forebrain, midbrain, hindbrain, and spinal medulla, do not necessarily have any functional significance, but rather represent parts of a unit which is concerned with integrating the activity of every part of the body. Such a system requires a high degree of integration of its parts, and this is attained by nerve fibres which connect the various regions of the central nervous system reciprocally.

Basically the central nervous system consists of three types of nerve cells: sensory, motor, and connector cells or interneurons.

1. **Sensory neurons** have two processes one of which passes to a sensory ending. Virtually all of their cell bodies lie outside the central nervous system, either in the spinal ganglia of the dorsal roots, or in the corresponding ganglia of the cranial nerves adjacent to the brain, or in the retinae, or in the olfactory mucous membrane. The other process enters the central nervous system, runs longitudinally within it, and gives rise to a number of branches through which it can activate many interneurons.

2. **Motor neurons** are connected with effector organs such as muscle, glands, etc., and have their cell bodies either in the motor nuclei of the cranial nerves, or in the anterior grey column of the spinal medulla, or in the peripheral autonomic ganglia. With the exception of certain autonomic ganglion cells which produce inhibition when they are stimulated (e.g. the action of the vagus on the heart), these cells invariably produce a response of a motor nature in the muscles which they innervate.

3. **Interneurons** [p. 621] are not directly connected either with sensory endings or with effector organs, but are interposed between the sensory and motor cells. They compose almost all the grey matter of the central nervous system, including such massive structures as the cerebrum and cerebellum, the thalamus, the corpus striatum, and a multitude of smaller nuclei and more scattered groups of nerve cells. Interneurons are of every shape and size. They lie entirely within the central nervous system with the single exception of the preganglionic cells of the autonomic nervous system. These send their axons to the peripherally situated autonomic ganglia which contain the motor cells of that system.

The complex communications which the numerous interneurons have with each other, and the multiple routes through which impulses can reach them through the branching processes of the sensory cells, make it potentially possible for such impulses to traverse many different pathways through the interneurons before they reach and modify motor-cell activity. This also allows different sensory impulses to interact with each other in the interneuron complex (by the blocking (inhibition) of some connections and the activation (facilitation) of others) and permits them to be modified by previous events which have altered the pattern of interneuron activity. Thus the visible outcome (e.g. muscular contraction) of the activity (the reflex) may be looked on as a sort of algebraic sum of all the prevailing influences.

It is relatively simple to envisage this type of activity in a limited region of the central nervous system such as a spinal segment (or related group of segments) with its sensory input and its motor cells. However, sensory impulses enter almost every level of the central nervous system so that it would be possible to have a large number of local reflexes occurring at different levels and each interfering with the other—a problem which is overcome by the communications between interneurons at different levels which ensure that activity at one level can alter, or be altered by that at another. This requires complex connections which ascend and descend to link all levels of the central nervous system. Such interregional links are very variable in length, ranging from fibres joining adjacent segments of the spinal medulla (intersegmental fibres of the fasciculi proprii) or adjacent gyri of the cerebral cortex (short association or arcuate fibres), to nerve fibres which extend from the spinal medulla to the thalamus (spinothalamic tracts) or from the cerebral cortex even to the most caudal part of the spinal medulla (corticospinal tracts). Long tracts of the latter type may form short-distance connections by means of branches (collaterals), while neurons with short processes may be linked together in series to form long pathways. Such multineuronal pathways conduct slowly and tend to be less specific in their connections than those formed by the long processes of individual cells.

The formation of the major mass of interneurons (the brain) at the cranial end of the central nervous system [p. 611] is associated with the production of long ascending and descending connections linking it with every other level of the central nervous system. Some of these are extensions of multineuronal linkages, others consist of long nerve fibres each of which passes specifically from an interneuron, e.g. in the spinal medulla, to a particular part or parts of the brain and vice versa. Such long nerve fibres are often gathered in bundles known as **tracts**. These tracts (and the corresponding multineuronal pathways) are primarily concerned with integrating the activities of the parts of the central nervous system, and may produce alterations of motor activity. However, the presence of specific regions in the brain which alone are capable of evoking subjective awareness of a sensory stimulus—in Man the cerebral cortex and possibly the thalamus—means that some ascending tracts activate these regions though this may not be their sole function. They therefore become known as 'sensory tracts' because the most obvious outcome of their destruction is the loss of conscious appreciation of particular sensations. This tends to make the student assume that every ascending tract is a 'sensory tract'—a view which is very far from the truth. Similarly it is possible to evoke movements by stimulation of some parts of the brain from which long descending tracts arise. Such pathways (e.g. from the cerebral cortex to the spinal medulla—corticospinal tracts) are sometimes called 'motor tracts', but they have many other functions and, like the ascending tracts, are composed of the processes of interneurons and not motor cells. The assumption that all descending tracts are 'motor' is as erroneous as the assumption that all ascending tracts are 'sensory'. To avoid confusion, the terms 'motor' and 'sensory' should be limited to the cells so-named above.

The presence of the long ascending and descending pathways, which are relatively simple to investigate anatomically, tends to overshadow the multisynaptic pathways, which are not readily amenable to anatomical techniques. However, the multisynaptic pathways form an important part of the conducting routes even in the central nervous system of Man, and are not replaced by the longer and more direct tracts.

The remarkable variability in the structure of the nervous system, even within the mammals, makes it clear that experimental findings obtained in one animal cannot readily be transferred to Man, yet the great majority of the anatomical and physiological findings concerning the central nervous system have been obtained from mammals other than Man. It is therefore essential to view with caution some of the inferences drawn from animal sources which are inevitably included in the following account.

Studies in Man depend mainly on naturally occurring injuries and on the effects of surgical intervention for clinical purposes. It is extremely difficult to interpret the significance of functional changes arising from either of these types of injury. It should always be remembered that the consequences of damage to any part of the central nervous system indicate what the intact parts of that system are able to perform in the absence of the damaged part. They do not show the functions of the damaged part.

The sensory cells and long ascending pathways

DORSAL ROOTS

The dorsal root fibres of the spinal nerves and the fibres of the laterally attached cranial nerves have already been shown to form a series of rootlets entering the hindbrain and spinal medulla, and the distribution of the sensory fibres which they contain has been shown to follow the same general plan throughout [p. 650]. The **dorsal roots** of the spinal nerves form a linear series of rootlets attached to the posterior lateral sulcus of the spinal medulla. The fibres in these rootlets arise in the spinal ganglia from cells which vary markedly in size. The largest cells give rise to processes up to 10 μm or more in diameter, with a correspondingly thick myelin sheath, while the smallest cells give rise to non-myelinated fibres many of which are 0.1 μm or less in diameter, and between these there is a range of intermediate sizes. The dorsal roots of spinal nerves transmit all types of sensory fibres from the peripheral nerves with which they are connected, including visceral sensory fibres. The area of skin supplied by each dorsal root, a **dermatome**, overlaps the similar areas supplied by adjacent dorsal roots to such an extent that the division of a single dorsal root does not usually produce an area of absolute sensory loss, though the quality of the sensation aroused by stimulation of the partially denervated skin is altered, because of the smaller number of sensory endings remaining in connection with the central nervous system.

Efferent fibres have been described in the dorsal roots, and although these occur in birds and some lower vertebrates, there is no anatomical or physiological evidence for their presence in mammals. Vasodilatation which results from stimulation of the distal end of a cut dorsal root is attributed, without complete anatomical proof, to the distribution of cutaneous sensory fibres also to blood vessels: a feature which would explain the localized vasodilatation following cutaneous stimulation, even after recent complete denervation—the **axon reflex**. However, the presence of large numbers of extremely small fibres in the dorsal roots has not been satisfactorily explained.

As the dorsal rootlets curve over into the dorsal surface of the spinal medulla the coarser fibres pass to the **posterior funiculus** of *the same side* while the finer fibres enter a smaller bundle lying lateral to this, the **dorsolateral fasciculus** [FIG. 10.134]. Each of the fibres on entering the spinal medulla divides into an ascending and a descending branch from which numerous collaterals are given off into the grey matter, thus allowing each fibre to have several different connections within the spinal medulla. Some of these collaterals can activate neurons responsible for producing local spinal reflexes, or more distant spinal activity through the

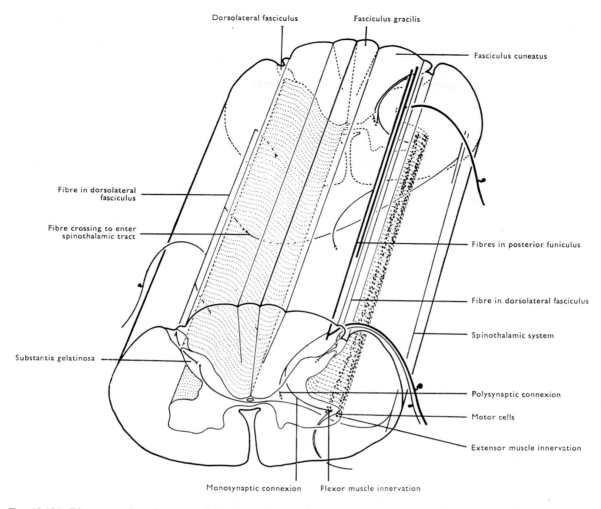

FIG. 10.134. Diagram to show the course of dorsal root fibres of spinal nerves entering the cervical spinal medulla, and some of their collateral connections.

intersegmental tracts; others are responsible for the transmission of impulses headwards to the brain.

THE POSTERIOR FUNICULI OF THE SPINAL MEDULLA

The coarser fibres of the dorsal root, on entering the posterior funiculus divide into long ascending processes, many of which reach headwards to the medulla oblongata, and shorter descending processes which may terminate almost at once or extend for a few segments. Both processes rapidly decrease in diameter.

The long ascending processes of fibres which enter the spinal medulla through successively higher dorsal roots lie lateral to those which enter at lower levels [FIG. 10.18]. Hence, in its upper parts, the posterior funiculus on each side is greatly enlarged and composed of fibres, the most medial of which are processes of spinal ganglion cells in the sacral region, while the most lateral fibres come from the cervical region, the fibres from intermediate regions lying between [FIG. 10.135]. There may be some reorganization of the fibres as they ascend. In the monkey, the simple spinal segmental arrangement is modified so that adjacent fibres are concerned more with parts of the body than with individual spinal nerves.

Throughout its length, the posterior funiculus on each side is separated from its neighbour by the posterior median septum, but above the midthoracic region of the spinal medulla, each posterior

funiculus is further subdivided by a posterior intermediate septum into a medial fasciculus gracilis and a lateral fasciculus cuneatus [FIG. 10.134]. The former contains fibres from the dorsal roots entering the caudal half of the spinal medulla, the latter those entering its cephalic half. In Man, unlike lower vertebrates, many of the long ascending branches reach the medulla oblongata, though probably fewer from the caudal than from the cranial half of the body, as evidenced by the small size of the fasciculus gracilis in the cervical region of the spinal medulla as compared with the fasciculus cuneatus. In most mammals and probably in Man, no muscle or joint sensory fibres from the lower half of the body reach the nucleus gracilis directly in the fasciculus gracilis, but they do so through collaterals of the posterior spinocerebellar tract (see below). In the medulla oblongata, the fibres of the fasciculi gracilis and cuneatus end on the cells of the nuclei gracilis and cuneatus which are visible on the dorsal surface of the medulla oblongata as small swellings, the gracile and cuneate tubercles [FIG. 10.46]. These are not simple relay nuclei but have complex synaptic arrangements which vary in the different parts of the nuclei concerned with cutaneous and deep sensations. Like the posterior horns of the spinal medulla, they receive fibres from the postcentral region of the cerebral cortex and from the reticular formation of the brain stem. They also contain interneurons whose processes do not leave the nuclei—all features which suggest considerable interaction of impulses, but do not prevent transfer of specific information

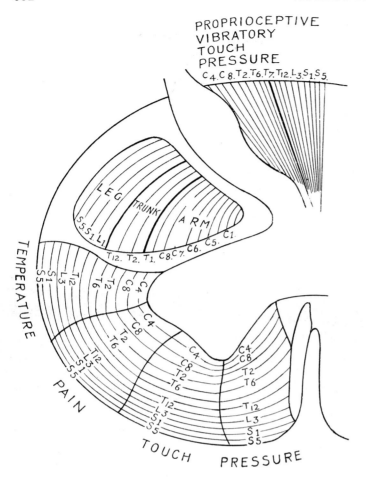

FIG. 10.135. Diagram to show: (1) the peripheral position in the anterior and lateral funiculi of fibres ascending from the lowest segments of the spinal medulla and the position progressively nearer the grey matter of those from higher segments; (2) the medial position in the posterior funiculi of fibres ascending from the lowest segments and the progressively lateral position of those from higher segments; (3) lamination of fibres within the lateral corticospinal tract. (After Foerster 1936.)

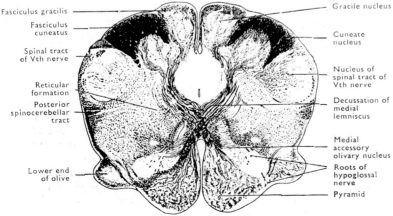

FIG. 10.136. Transverse section through foetal medulla oblongata immediately above the decussation of the pyramids (Weigert–Pal specimen).

onwards to the thalamus via a specific pathway—the medial lemniscus.

Primary sensory fibres in the posterior funiculi arise from cells in the spinal ganglia which innervate hair follicles, complex encapsulated endings of the skin, muscle spindles, tendon organs, and the endings found in the capsules of joints. They also contain second order fibres which arise in the posterior horn of the same side. Some of these are intersegmental fibres of the spinal medulla, but others ascend to the nucleus cuneatus or gracilis. Since these fibres arise from cells which may receive impulses from any of the fibres of the dorsal roots, they may transmit impulses other than those in the primary sensory fibres of the posterior funiculi.

The descending branches of fibres entering the posterior funiculi tend to be gathered together into two bundles: one, derived from fibres entering the fasciculus gracilis, lies adjacent to the postero-median septum, the **septomarginal fasciculus**; the other lies between the fasciculi gracilis and cuneatus and is the **fasciculus interfascicularis**, formed primarily by fibres entering the fasciculus cuneatus [FIG. 10.138]. Both bundles end as collaterals which enter the grey matter of the spinal medulla.

The finer fibres of the dorsal roots enter the **dorsolateral fasciculus** and again divide into ascending and descending branches which extend only for three or four segments in the fasciculus. Because of the short course of its fibres, the dorsolateral fasciculus does not increase in size as it is traced cranially. It is largest in the enlargements of the spinal medulla, where the number of fibres entering it is increased, and in its cephalic extension through the medulla oblongata and the lower half of the pons where it is known as the **spinal tract of the trigeminal nerve** [FIGS. 10.65 and 10.136], and receives a large number of sensory fibres which enter it principally from the trigeminal nerve.

Collaterals

Each primary sensory fibre running longitudinally in the spinal medulla gives off numerous collaterals into the spinal grey matter [FIG. 10.19]. Thus multiple routes are open to it for the production of local reflexes or transmission to other levels of the central nervous system.

Collaterals of fibres in the **dorsolateral fasciculus** enter the tip of the posterior horn and synapse with cells of the substantia gelatinosa. Collaterals of primary sensory fibres in the **posterior funiculi** pass deep into the grey matter of the spinal medulla, but some curve back into the superficial layers of the posterior horn, while others end in its deeper parts, and some reach as far ventrally as the motor cells [FIG. 10.139]. Thus they straddle the cells concerned with the transmission of impulses from the dorsolateral fasciculus and are in an anatomical position to play a dominant role in spinal reflexes, because the impulses which they transmit reach the spinal medulla and are widely distributed in it before those carried by the thinner, more slowly conducting fibres which enter the dorsolateral fasciculus and have a more restricted distribution. Impulses traversing the cells of the posterior horn are also subject to control by descending fibres which reach the posterior horn from the reticular formation and from the sensory (postcentral) cerebral cortex.

In other mammals (notably the cat) and possibly also Man (Truex *et al.* 1970) recurrent collaterals from fibres in the posterior funiculi which transmit impulses from hair follicles enter the posterior horn and activate cells which form the **spinocervical tract** (Brown and Gordon 1977). This pathway ascends in or immediately ventral to the dorsolateral fasciculus (Webster 1977) and enters the **lateral cervical nucleus**. This lies in the lateral funiculus of the upper cervical spinal medulla and lower medulla oblongata. It is not always present in Man (Truex *et al.* 1970) but in animals it transmits information from hair follicles through the medial lemniscus to the thalamus.

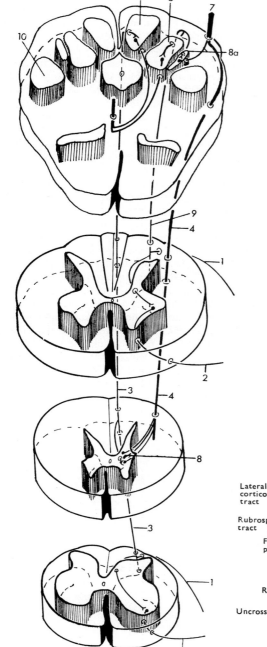

POSTERIOR SPINOCEREBELLAR TRACT

In the thoracic and upper lumbar regions of the spinal medulla, collaterals or entire fibres of the fasciculus gracilis enter a well-defined nucleus of large cells in the medial part of the base of the posterior horn—the thoracic nucleus [FIG. 10.137]. The large axons of these cells enter the lateral funiculus of the same side and form an ascending bundle immediately ventral to the dorsolateral fasciculus—the posterior spinocerebellar tract [FIG. 10.138]. This rapidly conducting tract carries proprioceptive, tactile, and pressure senses from the lower half of the body to the cerebellum, but only the muscle sense elements seem to arise in the thoracic nucleus. In the lower part of the medulla oblongata, it gives off collaterals to a nucleus at the superior pole of the nucleus gracilis. This nucleus is believed to send fibres into the medial lemniscus and to be responsible for transmission to the thalamus of that information from the lower half of the body which is missing in the upper part of the fasciculus gracilis (see above) and which has been believed not to reach consciousness (but see Matthews 1977). Corresponding information from the upper half of the body is relayed to the cerebellum by collaterals of the fasciculus cuneatus which enter the accessory cuneate nucleus on the lateral side of the nucleus cuneatus. This consists of cells similar to those in the thoracic nucleus. They give rise to the posterior external arcuate fibres, which ascend on the dorsolateral part of the medulla oblongata to join the posterior spinocerebellar tract and enter the cerebellum by the inferior cerebellar peduncle [FIG. 10.137].

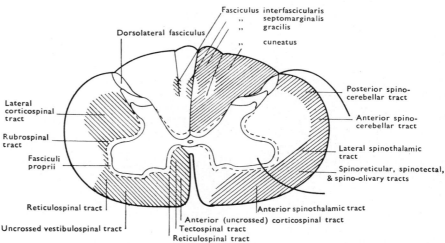

FIG. 10.138. Diagram to show the main ascending (right) and descending (left) pathways in the white matter of the spinal medulla. These have not been outlined individually because of the mixture of different pathways in the anterior and lateral funiculi.

FIG. 10.137. Diagram, not to scale, composed of transparent slices of the lower medulla oblongata, cervical, thoracic, and sacral spinal medulla, to show the collaterals of the fibres in the fasciculi gracilis and cuneatus.
1. Dorsal root fibres.
2. Ventral root fibres.
3. Fibre in fasciculus gracilis.
4. Posterior spinocerebellar tract.
5. Nucleus gracilis.
6. Nucleus cuneatus.
7. Inferior cerebellar peduncle receiving posterior spinocerebellar tract and posterior external aucuate fibres.
8. Thoracic nucleus.
8a. Accessory cuneate nucleus.
9. Fibres in fasciculus cuneatus.
10. Nucleus of spinal tract of trigeminal nerve.

Immediately ventral to the posterior spinocerebellar tract lies the anterior spinocerebellar tract [FIG. 10.138]. The exact origin and hence the significance of the fibres in this tract in Man is unknown. Some may arise from cells in the dorsolateral part of the anterior horn, but the majority come from the base of the posterior horn in the lumbosacral region in experimental mammals. They may be activated by collaterals from the posterior funiculi, or carry impulses from the motor apparatus of the spinal medulla as a feedback to the cerebellum.

The two spinocerebellar tracts form the surface of the greater part of the lateral funiculus. Their fibres convey impulses from sensory fibres in the peripheral nerves, but division of these pathways does *not* interfere with conscious appreciation of these sensations, though it interferes with cerebellar function [but see page 691, nucleus Z]. There may also be a rostral spinocerebellar tract. This arises in experimental mammals from the posterior horn mainly in the cervical region. It enters the same side of the cerebellum through the inferior and superior cerebellar peduncles, and corresponds to the anterior spinocerebellar tract from the lumbosacral region.

THE SPINOTHALAMIC SYSTEM

Fibres of the dorsolateral fasciculus send their collaterals into the subjacent tip of the posterior horn to the small cells of the substantia gelatinosa [layers I–III, FIG. 10.139]. These cells are concerned either with local spinal reflexes, or give rise to intersegmental and long ascending fibres in the fasciculi proprii particularly in the dorsolateral fasciculus [FIG. 10.138] and the adjacent part of the lateral funiculus. Such fibres include the spinocervical tract (see above). Some cells in the superficial layer of the substantia gelatinosa [layer I, FIG. 10.139] give rise to fibres which pass to the spinothalamic tracts of the opposite side (see below) and so form a specific pathway for transmission of painful and thermal impulses to the thalamus (Webster 1977). The majority of the dorsolateral fasciculus fibres act through chains of short neurons in the posterior horn. These are influenced by numerous other nerve fibres so that the structure exists for: (1) convergence of impulses from a number of dorsal root fibres either directly or through the intersegmental tracts, and (2) modification of the impulses (facilitation or inhibition) by intersegmental tracts, posterior funiculi fibres, reticulospinal fibres, and corticospinal fibres, all of which reach the deeper parts of the posterior horn. This arrangement also permits the transmission of impulses cranially and caudally through the intersegmental tracts.

The majority of the fibres in the spinothalamic system arise from the same layers in the body of the posterior horn [principally layer V; FIG.10.139] as the spinocervical tract (Webster 1977). The cells which give rise to those fibres are in a position to transmit impulses from a wide variety of sources. They therefore form a relatively non-specific pathway by comparison with the primary sensory fibres in the posterior funiculi or the fibres from layer I of the substantia gelatinosa. All the fibres entering the spinothalamic system pass obliquely headwards and *cross the midline* of the spinal medulla in the white commissure, ventral to the central canal. Here they are joined by fibres arising in the medial part of the anterior horn. Together they enter the anterolateral region of the white matter on the opposite side of the spinal medulla and ascend in this region forming the so-called spinothalamic tracts [FIG. 10.138]. Unlike the posterior spinocerebellar tract and the posterior funiculi which form well-defined pathways, the fibres forming the spinothalamic tracts are mixed with other axons including some descending fibres, and are widely scattered around the region of the emerging ventral rootlets which divide them into lateral and anterior parts. The exact extent of the fibres forming this system is not clear and it is certain that many of its fibres do not reach the thalamus directly since, on division of this tract in Man, degeneration studies seem to indicate many fewer fibres in the pontine and mesencephalic parts of this pathway than are present in the spinal medulla. This may be due to the large proportion of the fibres which end in the grey matter at lower levels, e.g. in the reticular formation of the brain stem. Certainly there is good evidence that the fibres ascending in this part of the white matter of the spinal medulla, or their collaterals, end at various levels in the spinal medulla and brain stem.

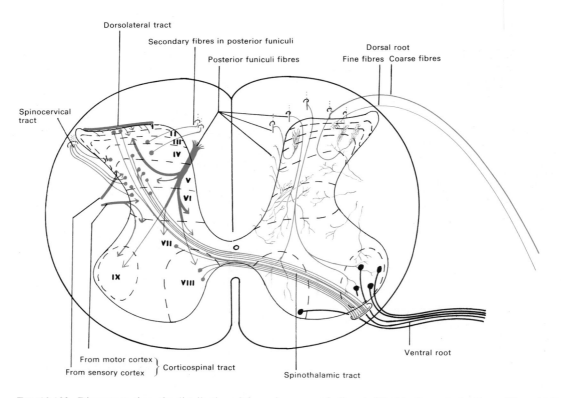

FIG. 10.139. Diagram to show the distribution of the main groups of afferents (blue) to the posterior horn of the spinal medulla and of efferents (red) from it. The intersegmental tracts are not shown. On the right, the courses of the afferents is indicated. On the left, only the general distribution is given. It shows the restricted distribution of fibres in the dorsolateral tract compared with that of the fibres in the posterior funiculi. (After Schibel and Schibel 1967; Webster 1977.) The broken lines and roman numerals indicate the layers (not to scale) described by Rexed (1954).

In cats, interruption of the fine fibres in the lateral parts of the dorsal roots at their entry to the spinal medulla, while leaving the larger medial fibres entering the posterior funiculi intact, abolishes reflexes produced by painful and thermal stimuli. In Man, destruction of the anterolateral region of the white matter of the spinal medulla interferes with similar sensations from the opposite side of the body below the level of injury. These two facts, combined with the fact that at least some of the pain spots and all of the heat receptors (Darian-Smith and Johnson 1977) in the skin are innervated by fibres of the smallest calibre, indicate that these fibres are responsible for the transmission of many of the impulses which evoke thermal and painful subjective sensations at a higher level. Experience with division of the spinothalamic system in Man seems to indicate that the lateral part is particularly concerned with the above sensory mechanisms and it is believed that the ventral part is concerned with tactile and proprioceptive sensations, since division or pathological degeneration of the posterior funiculi, though it interferes profoundly with the conscious appreciation of the latter sensations, does not totally abolish them (but see spinocervical tract).

Unilateral division of the spinothalamic system in Man does not always abolish painful sensations from the opposite side of the body, but has been reported to produce bilateral retrograde changes in the cells of the posterior horn, and to be associated with a raised pain threshold on the side of the division. These facts seem to suggest that this pathway is composed partly of uncrossed fibres, which could account for failure to obtain relief after unilateral surgical division of the spinothalamic system. This failure may be due either to the scattered nature of the system so that some fibres are spared, or to the presence of other pathways (possibly multisynaptic, or the second order fibres of the posterior columns, or the spinocervical tract (Brown and Gordon 1977), or to the passage of fibres through the sympathetic trunk to enter the spinal medulla above the level of injury (Nathan 1956). However, this system of fibres gives off many collaterals throughout its length, so that many other pathways are potentially possible.

Since most of the fibres of the spinothalamic system are crossed, and more fibres are added to it at each level of the spinal medulla and of the brain stem (mainly from the trigeminal system), the most superficial fibres in the spinal medulla are those derived from the sacral region, while the deepest are those which have just entered the spinothalamic tracts. In keeping with this is the experience of surgical division (chordotomy) of this system in Man under local anaesthesia, when it can be shown that, as the cut is deepened into the anterolateral region, the loss of pain and temperature sensation on the opposite side rises towards the level of the section. The sensory loss never reaches that level since the fibres cross obliquely to enter the anterolateral region. This obliquity of crossing is believed to be greater in tactile than in pain and temperature fibres. Chordotomy also indicates that the fibres in the spinothalamic tracts which evoke conscious appreciation of painful and thermal stimuli at higher levels tend to lie in the posterior part of this system, while fibres concerned with tactile and other sensations lie further anteriorly, mainly anterior to the ventral roots.

The division of a particular group of fibres in the main ascending pathways of the spinal medulla, either of the posterior funiculi or of the spinothalamic system, produces the same topographical distribution of sensory loss whatever the level of division. Thus superficial involvement of the spinothalamic system or injury to the midline fibres of the posterior funiculi, even in the cervical region, produces sensory loss limited to the lowest part of the body, ipsilateral in the latter case and contralateral in the former. However, the emergence of ventral rootlets through the spinothalamic tracts frequently leads to coincident injury of both—the resulting paralysis of muscles indicating the true level of the injury. A midline injury of fibres crossing to the spinothalamic system produces a bilateral belt of diminution of pain and temperature sensation corresponding to the longitudinal extent of the lesion, but sparing this sensation both above and below the level of the injury; such a condition is found in lesions in the region of the central canal, e.g. syringomyelia.

The ascending pathways of the spinal medulla together form a mass of fibres which virtually covers its entire surface except in the midline anteriorly [FIG. 10.138]. These pathways retain this arrangement throughout the length of the spinal medulla except in its uppermost segments where the lateral corticospinal tract comes to the surface and temporarily displaces the spinocerebellar tracts ventrally, though elsewhere it lies deep to them [FIG. 10.143].

The connections of the cells which give rise to nerve fibres in the main ascending tracts indicate the functional differences between them, but also show that each is potentially capable of transmitting many varieties of sensory information though with different degrees of refinement. This may account for the degree of recovery which usually follows damage to one of these pathways (Brown and Gordon 1977).

ASCENDING PATHWAYS IN THE MEDULLA OBLONGATA

Spinothalamic tracts and medial lemniscus

The redistribution of white matter which occurs between the spinal medulla and the medulla oblongata [p. 636; FIGS. 10.52, 10.57, 10.58, 10.140, and 10.141] displaces the spinothalamic tracts from their position around the exit of the ventral rootlets of the spinal nerves so that they lie close to the spinal tract and nucleus of the trigeminal nerve and the laterally attached cranial nerves [FIG. 10.57], while the medial part of the anterior funiculus of the spinal medulla is displaced dorsally with the central canal—the funiculus becoming continuous with the medial longitudinal bundle [FIG. 10.141]; the central canal opening into the fourth ventricle. Some of this displacement is due to the formation of the medial lemnisci.

Immediately superior to the pyramidal decussation, the posterior funiculi expand into the corresponding nuclei in which their fibres end. From these nuclei (gracilis and cuneatus on both sides) secondary fibres arise which sweep ventrally (internal arcuate fibres) to cross the midline, forming the decussation of the lemnisci [FIGS. 10.137 and 10.141]. Leaving the decussation, the fibres on each side form a flat vertical sheet or fillet, the medial lemniscus, situated between the pyramid ventrally and the medial longitudinal fasciculus dorsally [FIG. 10.141]. Only the median raphe separates the two medial lemnisci in the medulla oblongata.

Thus neurons in the nuclei cuneatus and gracilis which transmit impulses from the posterior funiculi to the thalamus send their axons to the opposite medial lemniscus, so that both the ascending systems (spinothalamic and posterior funiculi/medial lemniscus) are crossed in the medulla oblongata, but are still widely separated from each other. Both medial lemnisci lie between the emerging rootlets of the hypoglossal nerves [FIG. 10.141], while a spinothalamic system lies on each lateral surface of the medulla oblongata between the olive and the spinal tract of the trigeminal nerve. Thus either system may be damaged independently of the other (as in the spinal medulla), but both medial leminsci are likely to be involved in a single injury because of their proximity. Such injuries, often of vascular origin, usually involve adjacent structures. Thus the hypoglossal nerves and the pyramids may be involved with the medial lemnisci, while the VII, IX, X, and XIth nerves, the spinal tract of the trigeminal nerve and its nucleus, and the spinocerebellar tracts are damaged with the spinothalamic system (e.g. posterior inferior cerebellar artery thrombosis). Injury to the latter complex produces the bizarre loss of pain and temperature senses on the opposite side of the body (spinothalamic), and on the same side of the face (spinal root of the trigeminal nerve), combined with symptoms and signs referable to

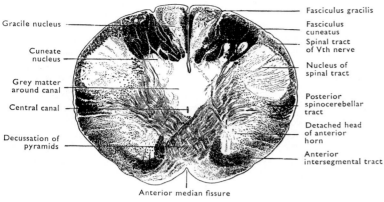

Gracile nucleus

Cuneate nucleus

Grey matter around canal

Central canal

Decussation of pyramids

Fasciculus gracilis

Fasciculus cuneatus

Spinal tract of Vth nerve

Nucleus of spinal tract

Posterior spinocerebellar tract

Detached head of anterior horn

Anterior intersegmental tract

Anterior median fissure

FIG. 10.140. Transverse section of lower end of medulla oblongata of full-term foetus. Treated by the Weigert–Pal method. The grey matter is bleached white, and the tracts of myelinated fibres are black.

Gracile nucleus

Fasciculus cuneatus

Cuneate nucleus

Tractus solitarius

Spinal tract of trigeminal nerve

Nucleus of spinal tract

Internal arcuate fibres

Roots of hypoglossal nerve

External arcuate fibres

Olivary nucleus

Medial accessory olivary nucleus

Pyramid

Central canal

Hypoglossal nucleus

Medial longitudinal bundle

Hypoglossal nerve

Raphe

Medial lemniscus

External arcuate fibres

FIG. 10.141. Transverse section of medulla oblongata at the level of the olive.

destruction of one or more of the above nerves, together with cerebellar signs consequent on the destruction of the spinocerebellar tracts. The opposite side of the face is not affected because the fibres crossing from the spinal nucleus of the trigeminal nerve ascend obliquely through the medial lemniscus to reach the spinothalamic system at a higher level (ventral trigeminothalamic tract).

Spinocerebellar pathways

Dorsal to the spinal tract and nucleus of the trigeminal nerve, opposite the inferior extremity of the fourth ventricle, lie the **cuneate** and **accessory cuneate nuclei**. Cranial to these nuclei is the inferior cerebellar peduncle, formed by the posterior external arcuate fibres from the accessory cuneate nucleus, joined by the **posterior spinocerebellar tract** [FIG. 10.142], numerous olivocerebellar fibres from the opposite olivary and accessory olivary nuclei,

and several fibres connecting the reticular formation and vestibular nuclei with the cerebellum. The posterior spinocerebellar tract joins the inferior cerebellar peduncle by passing dorsally across the spinal tract of the trigeminal nerve, where it can frequently be seen with the naked eye as an ill-defined oblique ridge [FIG. 10.109].

FIGURE 10.143 shows the course of the main ascending tracts in the brain stem (except the medial lemniscus which is not shown in the medulla oblongata). It indicates how each tract in turn inclines dorsally to enter a derivative of the alar lamina, the first being the most dorsal of the laterally situated tracts [FIG. 10.138], the posterior spinocerebellar tract.

The passage of the posterior spinocerebellar tract into the inferior cerebellar peduncle leaves the anterior spinocerebellar tract and the **spinothalamic system** ventral to the spinal tract of the trigeminal nerve. In this position, the latter system gives off many fibres into the lateral reticular formation and the olive, thus forming what are sometimes described as separate bundles, the **spinoreticular** and **spino-olivary tracts**. Above this level the spinothalamic system is

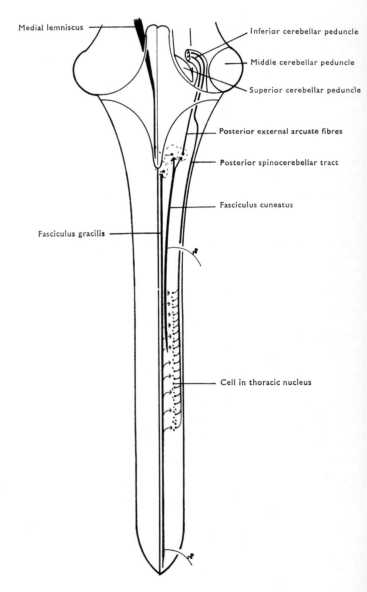

Medial lemniscus

Inferior cerebellar peduncle

Middle cerebellar peduncle

Superior cerebellar peduncle

Posterior external arcuate fibres

Posterior spinocerebellar tract

Fasciculus cuneatus

Fasciculus gracilis

Cell in thoracic nucleus

FIG. 10.142. A diagram of the origin and course of the spinocerebellar fibres entering by the inferior cerebellar peduncle.

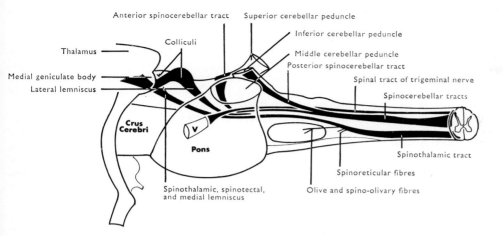

FIG. 10.143. Diagram of the course of the main ascending tracts in the brain stem. The tracts have been separated to show each group individually.

much reduced in size so that the number of direct fibres to the thalamus through this system is much smaller than might be expected from its size in the spinal medulla. This reduction in number of fibres (more obvious in experimental mammals than in Man) may suggest that some of the fibres transmit impulses from many sensory fibres (deep and superficial) as distinct from the more specific fibres in the spinothalamic tracts and the posterior funiculi. This is certainly possible owing to the convergence of impulses from many posterior root fibres in the deeper parts of the posterior horn. If it is the case, it would account for the poor localization of many sensations evoked by impulses transmitted in the spinothalamic tracts in individuals with injuries to the posterior funiculi or medial lemnisci. It may also explain the phenomenon of referred pain in which pain originating in the viscera may be referred erroneously to the skin innervated by the same spinal nerves. The impulses which enter the reticular formation are also transmitted to the thalamus by cells which send fibres on both sides of the brain stem mainly

through the central tegmental bundle [p. 643] to the posterior and intralaminar groups of thalamic nuclei. Presumably these could transmit an integrated assessment of unpleasant sensations from the entire body to both thalami—the basis of the ascending activating or arousal system by which the reticular formation awakens the sleeping animal [p. 673].

ASCENDING PATHWAYS IN THE PONS [FIGS. 10.55 and 10.56]

The medial lemniscus, spinothalamic, and anterior spinocerebellar tracts enter the dorsal part of the pons. Here the medial lemniscus twists through a right angle from its sagittal position to form a coronal sheet [FIGS. 10.65 and 10.144] with fibres from the nucleus gracilis in the lateral part in contact with the spinothalamic tract, while those from the nucleus cuneatus lie in the medial part and are

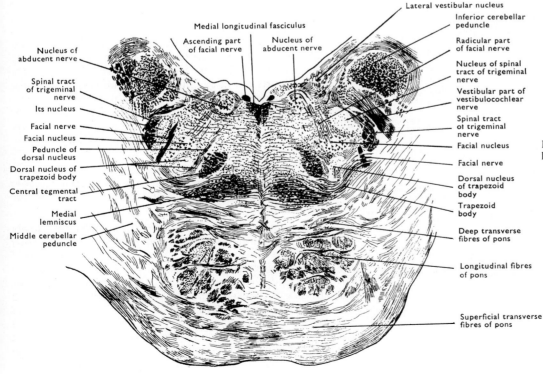

FIG. 10.144. Section through the lowest part of the pons.

at first close to the corresponding fibres in the opposite medial lemniscus. In this situation the medial lemniscus is pierced by the transversely running fibres of the **trapezoid body** [FIG. 10.149] passing laterally to form the **lateral lemniscus** (the auditory path) which runs headwards between the spinothalamic and anterior spinocerebellar tracts [FIG. 10.143]. Thus, lying ventrally in the dorsal part of the pons there is a horizontal sheet of fibres, the **lemniscus**, formed from medial to lateral by the medial lemniscus, spinothalamic tract (**spinal lemniscus**), lateral lemniscus, and the anterior spinocerebellar tract, extending from the midline to the spinal tract of the trigeminal nerve laterally. The two main ascending pathways from the spinal medulla to the thalamus thus come together for the first time in the pons and headwards of this point remain in close association. This entire sheet of fibres which can readily be dissected out in the fixed brain, slides laterally and posteriorly as it is traced headwards. The **anterior spinocerebellar tract** turning posteriorly [FIG. 10.143], cephalic to the entry of the trigeminal nerve, runs into the cerebellum across the lateral surface of the superior cerebellar peduncle and on to the superior medullary velum, where many of its fibres cross to enter the opposite half of the cerebellum.

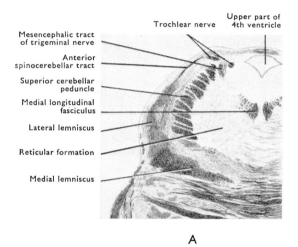

Mesencephalic tract
of trigeminal nerve

Anterior
spinocerebellar tract

Superior cerebellar
peduncle

Medial longitudinal
fasciculus

Lateral lemniscus

Reticular formation

Medial lemniscus

Trochlear nerve

Upper part of
4th ventricle

A

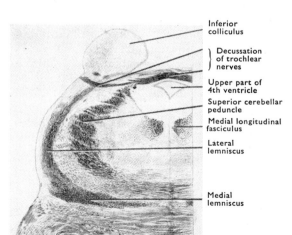

Inferior
colliculus

Decussation
of trochlear
nerves

Upper part of
4th ventricle

Superior cerebellar
peduncle

Medial longitudinal
fasciculus

Lateral
lemniscus

Medial
lemniscus

B

FIG. 10.145. Sections through dorsal part of pons close to the midbrain. A is at a slightly lower level than B.

ASCENDING PATHWAYS IN THE MIDBRAIN AND DIENCEPHALON

In the midbrain [FIGS. 10.54 and 10.55], the **lateral lemniscus**, now forming the lateral edge of the lemniscus, curves dorsally as a well-defined ridge which emerges from the upper border of the pons immediately dorsal to the crus cerebri and curves posterosuperiorly over the lateral surface of the superior cerebellar peduncle, to expand into the **inferior colliculus** [FIGS. 10.109 and 10.145]. Here the majority of its fibres end, though a few pass on into the **brachium of the inferior colliculus**. This is a bundle of fibres carrying auditory impulses. It originates principally in the inferior colliculus and forms a ridge passing anterosuperiorly on the lateral surface of the superior part of the midbrain towards the angle between the crus cerebri and the posterior end of the thalamus where it ends in a small round swelling, the **medial geniculate body** [FIGS. 10.143 and 10.147]—the thalamic relay station for auditory impulses.

In the lower part of the midbrain the **medial** and **spinal lemnisci** move further laterally and dorsally so that they come to lie deep to the lateral surface of the midbrain between the crus cerebri ventrally and the colliculi dorsally [FIGS. 10.146 and 10.147], covered by the brachium of the inferior colliculus. In this situation the spinal lemniscus gives off fibres dorsally into the superior colliculus (the **spinotectal tract**) and medially into the tegmentum of the midbrain. Both tracts then enter the posteroventral part of the thalamus where they fan out to enter and end in the medial and lateral parts of the **posterior-ventral nucleus of the thalamus** [FIGS. 10.54 and 10.80]. Because of the loss of fibres from the spinothalamic system to several levels of the brain stem the majority of the fibres ending in the posterior-ventral nucleus of the thalamus are derived from the medial lemniscus.

It has already been noted that the lemniscus receives crossed fibres from the spinal nucleus of the trigeminal nerve. The majority of fibres from the **principal sensory nucleus** of the trigeminal nerve cross into the medial portion of the medial lemniscus in the upper pons [FIG. 10.65], but a smaller number ascend ipsilaterally beside the periaqueductal grey matter of the midbrain—the **dorsal trigeminothalamic tract**. It is presumed that impulses reaching the mesencephalic nucleus are transmitted via the medial lemniscus. Together these fibres form the **trigeminal lemniscus** which passes into the medial part of the posterior-ventral nucleus of the thalamus, while fibres carrying impulses from the lower parts of the body terminate in the lateral parts of this nucleus and those from the intermediate parts end in the intervening regions. These terminations have been determined by anatomical means but more specifically by recording the site of arrival of impulses at the thalamus in various mammals when different parts of the body are stimulated. In the cat, Poggio and Mountcastle (1959), on physiological grounds, differentiate between:

1. The above terminations, which are arranged in strict relation to the parts of the opposite half of the body and receive impulses carried principally by the medial lemniscus, but also through the spinothalamic tract.

2. Endings in the posterior nuclei [p. 713] of the thalamus where the same topographical localization does not exist, and individual neurons may be activated by stimulating widely separated parts of the body, even on either side. The pathways responsible for this can have little to do with localizing the source of the stimulus by comparison with the detailed point to point information transmitted by the medial lemniscus. Because of the long delay in the arrival of these impulses at the thalamus, compared with those in the medial lemniscus, it seems that such impulses come from neurons in the brain stem (e.g. reticular formation) which transmit a more general picture of the sensory input to the body as a whole, having been activated by spinoreticular fibres (see above).

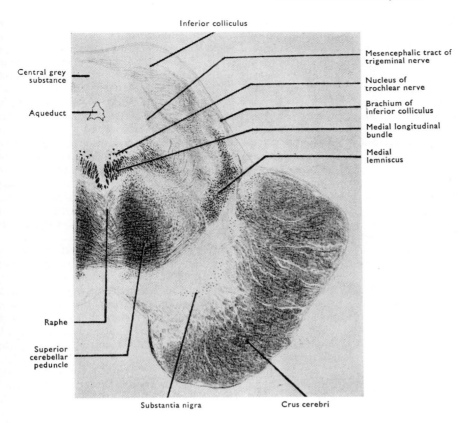

Inferior colliculus

Central grey substance

Aqueduct

Mesencephalic tract of trigeminal nerve

Nucleus of trochlear nerve

Brachium of inferior colliculus

Medial longitudinal bundle

Medial lemniscus

Raphe

Superior cerebellar peduncle

Substantia nigra

Crus cerebri

FIG. 10.146. Transverse section of midbrain at level of inferior colliculus.

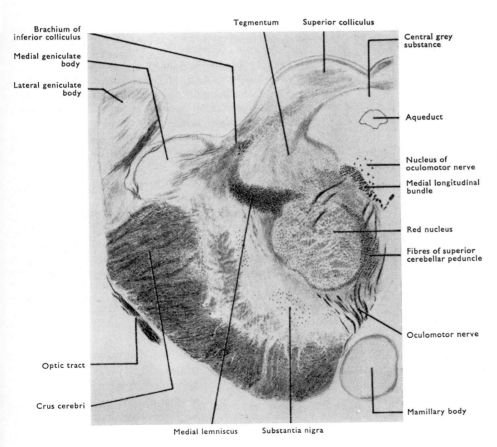

Brachium of inferior colliculus

Medial geniculate body

Lateral geniculate body

Tegmentum

Superior colliculus

Central grey substance

Aqueduct

Nucleus of oculomotor nerve

Medial longitudinal bundle

Red nucleus

Fibres of superior cerebellar peduncle

Oculomotor nerve

Optic tract

Crus cerebri

Medial lemniscus

Substantia nigra

Mamillary body

FIG. 10.147. Transverse section of midbrain at level of superior colliculus.

ASCENDING PATHWAYS TO THE CEREBRAL CORTEX

Thalamocortical fibres arise from almost every part of the thalamus including the posterior-ventral nucleus, which sends fibres through the posterior limb of the internal capsule to the posterior wall of the central sulcus, to the adjacent surface of the **postcentral gyrus**, and to the precentral gyrus. Within this area there is a marked difference between the arrangement of the cells in the cerebral cortex of the posterior wall of the central sulcus and the exposed part of the postcentral gyrus. Powell and Mountcastle (1959a, b) indicate that this structural difference corresponds to a functional difference, since the fibres ending on the posterior wall of the sulcus convey impulses from the superficial tissues, while the exposed part of the gyrus and the precentral gyrus receive impulses from the deep tissues.

Anatomical and physiological techniques demonstrate that the projections from the posterior-ventral nucleus of the thalamus to the postcentral gyrus follow a strictly defined pattern. Hence the pattern transmitted to the thalamus by the specific ascending pathways is passed on to the postcentral gyrus. Thus impulses from the trigeminal area and larynx reach the lowest part of the gyrus, followed superiorly by the cervical segments in reversed order, the thoracic segments, and the lumbosacral segments; impulses from

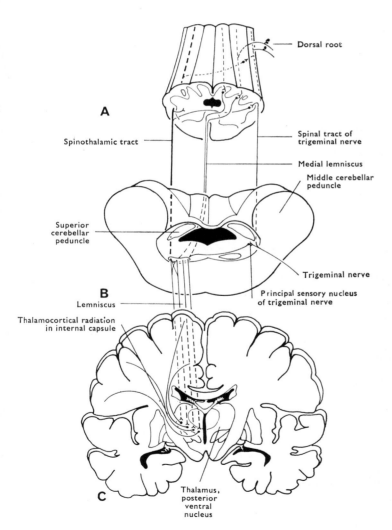

A, low medulla oblongata; **B**, pons; **C**, thalamocortical projections.

Dorsal root

Spinothalamic tract

Spinal tract of trigeminal nerve

Medial lemniscus

Middle cerebellar peduncle

Superior cerebellar peduncle

Trigeminal nerve

B

Lemniscus

Principal sensory nucleus of trigeminal nerve

Thalamocortical radiation in internal capsule

C

Thalamus, posterior ventral nucleus

FIG. 10.148. Diagram to show the course of the medial lemniscus and spinothalamic tracts through the brain, seen from above.

the perineum through the lowest sacral nerves arriving at the medial surface (paracentral lobule). Each horizontal level appears to be associated with one dermatome overlapped by adjacent dermatomes as it is at the periphery. In addition, the dorsal part of the dermatome is anterior while the ventral part is posterior. Physiological methods also demonstrate that the relative *area of the cerebral cortex concerned with each part of the body is not equivalent to the size of the part, but to its importance as a sensory organ* and hence to the number of nerve fibres innervating it. Thus in Man the areas which receive impulses from the face (particularly the lips), the hand and foot (particularly the thumb and great toe) are much larger than would be expected on the basis of their size alone, just as the snout takes up the major part of the cortical sensory area in the pig. The reversal of the cervical segmental representation brings the lower cervical segmental area into association with the trigeminal area; hence the two main human sensory organs, the hand and the oral region, are topographically associated in the postcentral cerebral cortex.

The postcentral cerebral cortex is the area concerned with general sensations from the body and is therefore known as the **somatic sensory area**. Other sensations are relayed to this area, e.g. taste and vestibular impulses to the inferior part of the gyrus. Visceral sensations, such as the sensations of fullness from bladder and rectum, known to be transmitted through the spinothalamic system, because of their loss when it is divided (Nathan and Smith 1951, 1953) are passed to levels in the gyrus corresponding to the appropriate dermatome.

Lesions of the postcentral gyrus do not produce significant alterations in the subjective appreciation of painful and thermal sensations, though they do interfere markedly with position, tactile, and vibration sensations from the opposite side of the body. For this reason, and because stimulation of the postcentral gyrus in the unanaesthetized patient produces numbness, or sensations of light touch, or tingling in the appropriate part of the body, but does not evoke the sensation of pain, it has been assumed that pain is appreciated in other parts of the brain, possibly in the second sensory area which receives afferents from the posterior group of thalamic nuclei. There is evidence of alteration of painful sensations following injuries to the postcentral gyrus. The failure to demonstrate it convincingly may well result from the presence of other sensory areas or a bilateral representation of these sensations.

The **second sensory area** has been demonstrated mainly by physiological means. It lies at the lowest extremity of the postcentral gyrus on the upper wall of the lateral sulcus, extending on to the insula. This area receives a reversed but bilateral topographical representation of the skin of all parts of the body—the corresponding parts of the two sides sending impulses to the same point. The significance of this and similar secondary auditory and visual areas associated with the corresponding primary areas is not certain, but the reversal of representation relative to that in the primary area seems to be common to all of them. The secondary somaesthetic area appears to be divisible into two. Each receives impulses from the skin of the entire body, and one also receives auditory and visual impulses. The first is activated from the ventroposterior thalamic nuclei, the second from the posterior thalamic group—probably the part concerned with pain appreciation.

The auditory system

The vestibulocochlear nerve enters the lateral aspect of the medulla oblongata at its junction with the pons. The cochlear fibres arch over the lateral and posterior surfaces of the inferior cerebellar peduncle to end in the cochlear nuclei which form a ridge of grey matter in this position, anterior to the lateral recess of the fourth ventricle.

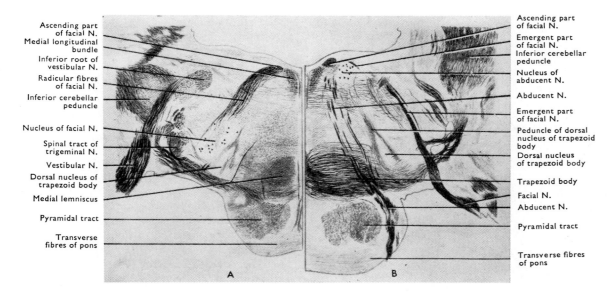

Ascending part of facial N.
Medial longitudinal bundle
Inferior root of vestibular N.
Radicular fibres of facial N.
Inferior cerebellar peduncle
Nucleus of facial N.
Spinal tract of trigeminal N.
Vestibular N.
Dorsal nucleus of trapezoid body
Medial lemniscus
Pyramidal tract
Transverse fibres of pons

Ascending part of facial N.
Emergent part of facial N.
Inferior cerebellar peduncle
Nucleus of abducent N.
Abducent N.
Emergent part of facial N.
Peduncle of dorsal nucleus of trapezoid body
Dorsal nucleus of trapezoid body
Trapezoid body
Facial N.
Abducent N.
Pyramidal tract
Transverse fibres of pons

FIG. 10.149. Two sections through the pons. A is taken at a level slightly inferior to B.

The cochlear nuclei [FIG. 10.57], on which the fibres from the cochlea end in a well-defined pattern, are usually divided into the dorsal and ventral parts according to their relation to the inferior cerebellar peduncle. Each gives rise to fibres which pass supero-medially, dorsal and ventral to the inferior cerebellar peduncle, to enter the dorsal part of the pons where they incline medially to form a mass of crossing fibres, the trapezoid body [FIGS. 10.149 and 10.150], in association with which are the nuclei of the trapezoid body on which some of the fibres end by forming complex synapses. The fibres which cross in the trapezoid body pass through the substance of the medial and spinal lemnisci to turn headwards between the latter and the anterior spinocerebellar tract as the lateral lemniscus [FIG. 10.151], the ascending auditory pathway [p. 698]. This lemniscus is recognizable in sections of the superior part of the pons by the presence of an elongated nucleus, the nucleus of the lateral lemniscus, within its substance [FIG. 10.56A].

The auditory pathway in the brain stem is characterized by the large number of cells associated with it throughout its course. Immediately anterior to the motor nucleus of the facial nerve at the lateral edge of the trapezoid body lies the dorsal nucleus of the trapezoid body, forming a slightly coiled mass of cells on which end fibres of the trapezoid body originating on both sides. This nucleus gives rise to fibres in the lateral lemniscus and to a bundle of fibres passing posteriorly towards the floor of the fourth ventricle. The latter bundle (peduncle of the dorsal nucleus of the trapezoid body) is said to make connections with other motor cranial nerve nuclei, e.g. the facial, through which blink reflexes in relation to loud sounds may be effected, and has been suggested as the source of the efferent cochlear fibres (Rasmussen 1953), which have been traced in the vestibulocochlear nerve to the cochlea. When these fibres are stimulated, they block the impulses in the cochlear nerve which would normally arise in response to noise (Galambos 1956). Groups

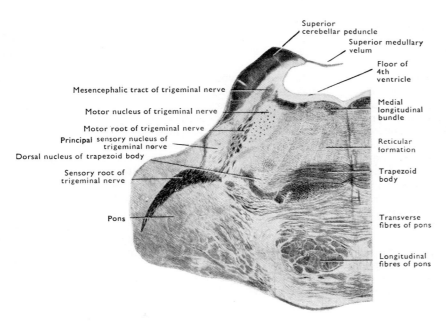

Superior cerebellar peduncle
Superior medullary velum
Floor of 4th ventricle
Mesencephalic tract of trigeminal nerve
Motor nucleus of trigeminal nerve
Motor root of trigeminal nerve
Principal sensory nucleus of trigeminal nerve
Dorsal nucleus of trapezoid body
Sensory root of trigeminal nerve
Pons
Medial longitudinal bundle
Reticular formation
Trapezoid body
Transverse fibres of pons
Longitudinal fibres of pons

FIG. 10.150. Section through the pons at level of entry of the trigeminal nerve.

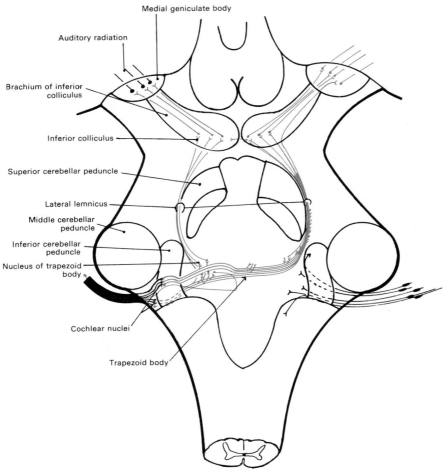

Medial geniculate body

Auditory radiation

Brachium of inferior colliculus

Inferior colliculus

Superior cerebellar peduncle

Lateral lemnicus

Middle cerebellar peduncle

Inferior cerebellar peduncle

Nucleus of trapezoid body

Cochlear nuclei

Trapezoid body

FIG. 10.151. Diagram of the main connections of the cochlear part of the vestibulocochlear nerve. First order fibres—black; second and fourth order—blue; third order—red.

of cells are also present among the fibres of the trapezoid body, the **ventral nucleus of the trapezoid body**. In their turn the fibres of the lateral lemniscus surround the nucleus of that name, and most of them ascending over the superior cerebellar peduncle [FIG. 10.109] end in the **inferior colliculus** [p. 698]. A few continue to the **medial geniculate body**, through the **brachium of the inferior colliculus** [FIGS. 10.55 and 10.109] which is mainly formed by fibres arising in the inferior colliculus.

The numerous nuclei on the course of the auditory pathway, cause the impulses that reach the medial geniculate body for transmission to the cerebral cortex (for conscious appreciation) to have passed through several synapses, as compared with the single synapse involved in some other ascending pathways to the thalamus, e.g. the medial lemniscus. In keeping with the important reflex function of the various nuclei in the auditory pathway is their extreme development in the bat, which uses auditory impulses, generated by the reflection of its high-frequency cries, to locate and catch insects in flight and to avoid obstacles. This echo-location requires very rapid reflex adjustments to the flight of the bat, and indicates the potential complexity of the connections of the auditory pathway for reflexes.

The fibres in each lateral lemniscus carry impulses from both right and left cochlear nuclei, the uncrossed reaching it through the nuclei of the trapezoid body. Thus each lateral lemniscus contains fibres carrying impulses from both sides—a mechanism which allows direct comparison of impulses arising in the two cochleae, and no doubt is a factor in sound localization. Naturally this is also

true of all parts of the auditory pathway cranial to the formation of the lateral lemniscus in the pons; hence unilateral lesions of this pathway or of the cerebral cortex to which it finally passes, the **superior temporal gyrus**, produce little or no loss of hearing. Unilateral deafness of nervous origin arises as a result of injury to the cochlear nerve, the cochlear nuclei, or the fibres passing from these nuclei to the trapezoid body, but not from unilateral injury superior to this.

The **medial geniculate body** [FIGS. 10.54 and 10.78] gives rise to fibres which leave its anterolateral aspect and passing laterally enter the sublentiform part of the internal capsule. In this they pass, superior to the inferior horn of the lateral ventricle, to mingle with fibres of the superior longitudinal bundle curving anteriorly into the temporal lobe and the posterior fibres of the fasciculus uncinatus and anterior commissure. The fibres curve superiorly to enter the superior temporal gyrus, and end in the transverse temporal gyrus on the superior and lateral surfaces of the superior temporal gyrus [FIG. 10.152]. Thus, much of the auditory receptive area is hidden from view in the intact brain, as indeed are the major parts of the somatic sensory and visual areas. Within this auditory area of the superior temporal gyrus, fibres carrying impulses from the various parts of the cochlea end in a definite topographical manner in animals, those from the apical parts, concerned with low frequencies, posterior to those from the basal parts; the arrangement in Man is unknown. A **second auditory area** has been described in animals. The localization in this is reversed, and its presence in Man is unconfirmed. If, as seems likely, this area exists in Man, it is

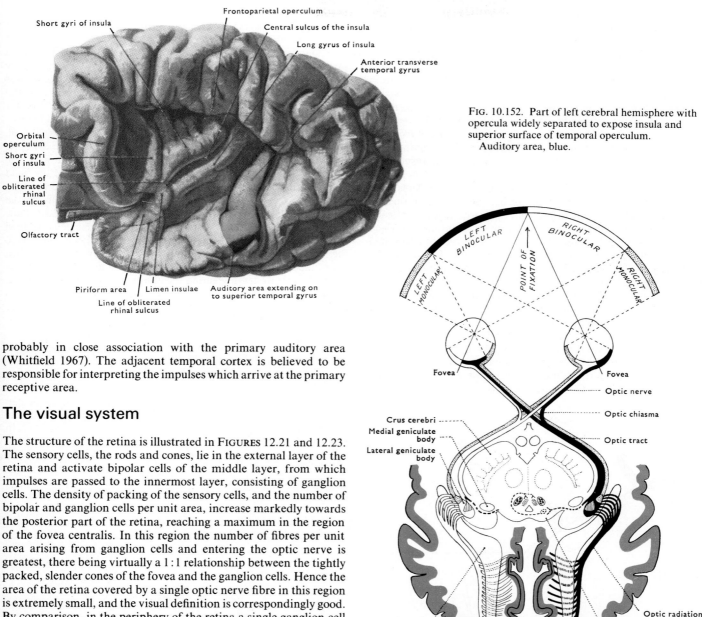

Short gyri of insula
Frontoparietal operculum
Central sulcus of the insula
Long gyrus of insula
Anterior transverse temporal gyrus
Orbital operculum
Short gyri of insula
Line of obliterated rhinal sulcus
Olfactory tract
Piriform area
Limen insulae
Auditory area extending on to superior temporal gyrus
Line of obliterated rhinal sulcus

FIG. 10.152. Part of left cerebral hemisphere with opercula widely separated to expose insula and superior surface of temporal operculum. Auditory area, blue.

LEFT BINOCULAR
RIGHT BINOCULAR
POINT OF FIXATION
LEFT MONOCULAR
RIGHT MONOCULAR
Fovea
Fovea
Optic nerve
Optic chiasma
Crus cerebri
Medial geniculate body
Lateral geniculate body
Optic tract
Lateral ventricle
Optic radiation
Superior colliculus
Striate area

FIG. 10.153. Composite diagram to show the projection of the visual field upon the retinae, and the relationship of retinae to lateral geniculate bodies and striate area of cerebral hemispheres.

The fibres from the maculae, some crossed and some uncrossed, are not shown separately in the optic nerves and optic tracts, but their place of delivery in the lateral geniculate body—the intermediate part of its posterior two-thirds—is shown by hatching. The remainder of the posterior two-thirds receives the fibres from non-macular regions of the binocular retinal field. The monocular retinal field is shown projecting into the anterior third of the lateral geniculate body. No attempt has been made in this Figure to localize the delivery within the striate area of fibres from these different regions of the lateral geniculate body.

(After I. Mackenzie (1934). *J. Path. Bact.* **39**, 113, and F. R. Winton and L. E. Bayliss (1948). *Human physiology*, 3rd edn. London.)

probably in close association with the primary auditory area (Whitfield 1967). The adjacent temporal cortex is believed to be responsible for interpreting the impulses which arrive at the primary receptive area.

The visual system

The structure of the retina is illustrated in FIGURES 12.21 and 12.23. The sensory cells, the rods and cones, lie in the external layer of the retina and activate bipolar cells of the middle layer, from which impulses are passed to the innermost layer, consisting of ganglion cells. The density of packing of the sensory cells, and the number of bipolar and ganglion cells per unit area, increase markedly towards the posterior part of the retina, reaching a maximum in the region of the fovea centralis. In this region the number of fibres per unit area arising from ganglion cells and entering the optic nerve is greatest, there being virtually a 1:1 relationship between the tightly packed, slender cones of the fovea and the ganglion cells. Hence the area of the retina covered by a single optic nerve fibre in this region is extremely small, and the visual definition is correspondingly good. By comparison, in the periphery of the retina a single ganglion cell receives impulses from a large number of rods spread over a not inconsiderable area, and the definition is correspondingly reduced, even though the sensitivity to light is increased.

The fibres of the ganglion cells converge radially on the optic nerve, with the exception of the fibres from the temporal parts of the retina which lie adjacent to a horizontal line passing through the fovea; these curve superiorly or inferiorly so as to pass round the fovea. Thus fibres from the foveal region pass directly medially and enter the optic nerve on its lateral side. In this region the axons of the ganglion cells develop a myelin sheath and course posteriorly in the optic nerve through the optic canal towards the **optic chiasma**. This is situated at the inferior end of the lamina terminalis, above the sella turcica and some distance posterior to the sulcus chiasmatis on the sphenoid bone [FIG. 10.183]. Within the optic nerve the fibres from the central region of the retina take up a central position, with the fibres from the more peripheral fields arranged around them more or less in the same topographical position as the area of the retina from which they originate.

At the chiasma the two optic nerves meet and an interchange of fibres takes place so that the **optic tracts** which arise from the posterolateral angles of the chiasma contain some fibres from both eyes. The arrangement of fibres within the chiasma is not simple, but in general the fibres from the lower parts of the retina enter the inferior part of the chiasma, and the upper fibres the superior part, while the fibres from the foveal region lie between. Fibres which originate in the temporal half of each retina, including those from the temporal half of the foveal region, do not cross in the chiasma but pass back into the optic tract of the same side. Fibres from the nasal halves of the retinae intermingle and cross to the opposite optic tract. In this manner each optic tract carries fibres from the parts of the two retinae which are stimulated by light coming from the opposite half of the field of vision, in fact from the corresponding parts of the two retinae [FIG. 10.153].

The **optic tracts** emerge from the chiasma with the fibres from the central regions of the corresponding halves of the two retinae in their intermediate part, the fibres from the lower quadrant of each retina lateral, and those from the superior quadrant medial. Each optic tract passes posterolaterally, medial to the anterior perforated substance and the internal carotid artery, and lateral to the hypothalamus. The anterior choroidal artery and the basal vein, which subsequently crosses the optic tract, lie immediately lateral to it in the groove which separates the tract from the uncus. Having passed lateral to the mamillary body the optic tract runs over the lateral side of the crus cerebri at its junction with the internal capsule, on to the postero-inferior part of the thalamus [FIG. 10.153]. Here the tract appears to split into two **roots** [FIG. 11.6]—the medial root passing towards the medial geniculate body, while the lateral is continuous with a poorly defined swelling, the lateral part of the lateral geniculate body. This apparent division is due to the **lateral**

geniculate body being concave inferiorly, so that the entire optic tract which expands into it seems to be divded by the furrow formed by this concavity. In addition to the fibres which end in the lateral geniculate body, a smaller bundle, which is not always visible on the surface of the brain, separates from the optic tract immediately anterior to the lateral geniculate body. It turns medially, and passes over the postero-inferior aspect of the thalamus, superior to the medial geniculate body, to enter the anterolateral part of the superior colliculus where it frequently forms an obvious ridge. This bundle, the **brachium of the superior colliculus** [FIGS. 10.54 and 10.154], consists of collaterals of the optic tract fibres which pass to the optic stratum of the superior colliculus and are concerned with the production of reflexes such as the involuntary turning of the head and eyes towards a sudden flash of light. Fibres from the complex laminated cortex of the superior colliculus sweep ventrally on the central grey substance which surrounds the cerebral aqueduct. They cross the midline ventral to the oculomotor nucleus (**dorsal tegmental decussation**) and turn caudally immediately anterior to the medial longitudinal fasciculus [p. 707]. Many of these fibres are distributed to the reticular formation of the brain stem (**tectoreticular tract**) the remainder descend in the anterior funiculus of the spinal medulla (**tectospinal tract**) and end in its grey matter. Fibres from the tectum also pass to the intermediate region of the cerebellar cortex, presumably through the superior cerebellar peduncle [FIGS. 10.41 and 10.176].

Fibres of the brachium of the superior colliculus also pass to the small pretectal area immediately anterior to the superior colliculus. These **pretectal nuclei** [FIG. 10.155] send fibres through the posterior commissure and the central grey substance of the midbrain to the visceral efferent cells (accessory nucleus) of both oculomotor nuclei. Preganglionic parasympathetic fibres from these nuclei run to the **ciliary ganglia** in the oculomotor nerves. Axons of the large cells in this ganglion innervate the circular muscle of the iris and cause constriction of the pupil, in this case as a result of light falling on the retina.

Other fibres (the accessory optic tract) from the optic tract or from the brachium of the superior colliculus pass ventrally around the crus cerebri to enter the interpeduncular region.

The fibres of the brachium of the superior colliculus are concerned with visual reflexes, while the fibres to the lateral geniculate body

Optic nerve

Optic tract

Optic radiation

Brachium of
superior colliculus

Pretectal region

Superior colliculus

Oculomotor
nucleus

Corticotectal fibre

Tectospinal and
tectoreticular
tracts

Occipital lobe of cerebrum

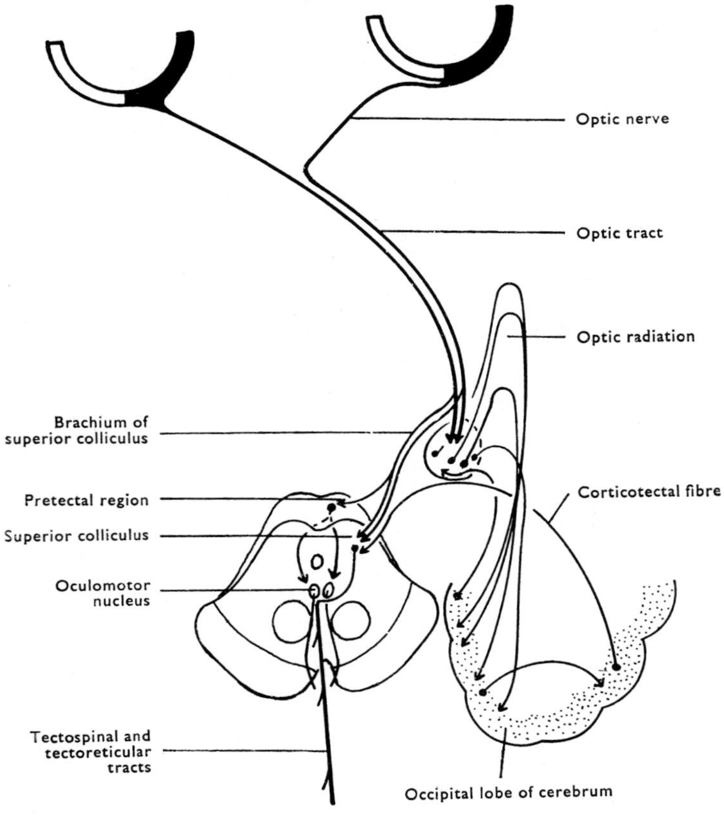

FIG. 10.154. Composite diagram of the origin and distribution of fibres in one optic tract. The midbrain has been turned so as to be seen from below.

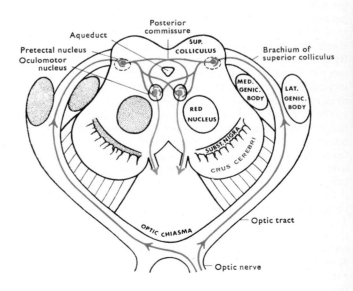

FIG. 10.155. Diagram of section through midbrain and optic chiasma showing the path of pupillary constrictor fibres. (After S. W. Ranson and H. W. Magoun (1933). *Arch. Neurol. Psychiat. (Chic.)* **30**, 1193.)

are concerned with transmission of retinal impulses to the cerebral cortex and their conscious appreciation.

The lateral geniculate body lies on the postero-inferior surface of the thalamus immediately medial to the choroid fissure in the descending central part of the lateral ventricle. In coronal section the body is a sinuous plate of grey matter, concave inferiorly in Man with the edges, especially the lateral, thinner and slightly everted. It consists essentially of six layers of cells separated by laminae of fibres consisting partly of fibres from the optic tract but mainly from the cells of the lateral geniculate body passing to the cerebral cortex. The six layers are most pronounced in the intermediate region of the posterior part, but tend to fuse together and become less distinct at the edges. They are numbered serially from the inferior surface. The functions of the various layers are probably different, firstly because layers 1 and 2 are composed of larger cells than the others and are not concerned with colour vision; and secondly because layers 1, 4, and 6 receive fibres from the nasal half of the contralateral retina, while layers 2, 3, and 5 receive fibres from the temporal half of the ipsilateral retina [FIG. 10.156]. These connections have been demonstrated by making restricted lesions in the retina and following the degenerating optic fibres to their termination. Such experiments have also shown that each spot on the retina is connected to the corresponding parts of all three layers. In Man each optic tract fibre may end predominantly on a single geniculate neuron, at least in the postero-intermediate part of the geniculate body which receives fibres from the central parts of the retina. The more peripheral parts of the retina send their fibres to the anteromedial (superior retina) and anterolateral (inferior retina) parts. The presence of three layers concerned with each spot on the retina, and the well-developed nature of these layers in the part of the geniculate body to which the fibres from the highly colour-sensitive central region of the retina pass, has led to the suggestion that each layer may be concerned with one primary colour. However, physiological experiments show that the different layers are concerned with other retinal functions (De Valois 1960).

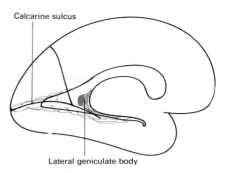

FIG. 10.157. A diagram of the course of nerve fibres in the optic radiation. Medial view of left cerebral hemisphere showing relation of radiation to lateral ventricle and calcarine sulcus.

The efferent fibres of the lateral geniculate body leave it as a compact bundle and run anteriorly, to pass within the concavity of the lateral ventricle [FIG. 10.157] and turn backwards towards the occipital lobe, separated from the ventricle by the tapetum. In general, the central fibres of this expanding sheet are those from the postero-intermediate part of the lateral geniculate body and are concerned with central vision, while the superior fibres carry impulses which originate in the upper peripheral parts of the retinae and the inferior fibres convey impulses from the lower peripheral parts. The lowest fibres run forwards into the temporal lobe, and turning posteriorly, sweep beneath the posterior horn of the ventricle and curve superiorly on its medial side to the inferior aspect of the calcarine sulcus. The superior fibres similarly turn over the superior aspect of the ventricle to the superior surface of the calcarine sulcus. The fibres at the margins of the radiation escape to the medial aspect of the occipital lobe anterior to the central fibres which pass towards the occipital pole, and may even extend with the sulcus to the cortex on the lateral surface of the pole. Thus the upper half of the radiation passes to the cerebral cortex on the superior surface of the calcarine sulcus while the lower half is distributed to its lower surface [FIG. 10.175, area 17]. The fibres carrying impulses from the central regions of the retina occupy a disproportionately large part of the cortex per unit area of the retina. Thus a large part of the polar cortex to which these fibres pass is taken up with macular vision which subtends only 1 degree of arc in the field of vision [see also p. 843].

The visual system is of great *clinical importance* because most defects are readily noticed by the patient. Also its intracranial extent is from the chiasma to the occipital pole, lying across the general direction of most of the long pathways in the brain stem, so that it is subject to damage throughout this extent. Injuries to the visual system produce loss of vision and of visual reflexes (e.g. pupillary constriction on exposure of the retina to light). The loss of vision is described in terms of the effect on the visual field of the patient. The defects are understood readily if the following points are remembered.

1. The pathways from the retina to the cerebral cortex and the visual cortex are essential for sight. The brachia of the superior colliculi which leave this pathway before it reaches the cerebral cortex are concerned with the production of some visual reflexes, but play no part in seeing. Hence destruction of the brachia (e.g. in the tectum of the midbrain) abolishes these reflexes but does not interfere with vision. Injuries to the visual pathway after the brachia have left it (e.g. optic radiation) interfere with vision but not with these reflexes, while both reflexes and vision are affected by injuries anterior to the origin of the brachia (e.g. optic nerves).

2. The eye is a simple lens system which inverts and reverses the image on the retina. Hence damage to any part of the retina (or to

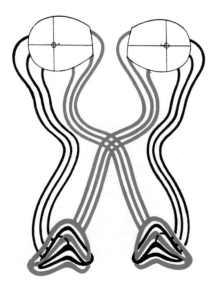

FIG. 10.156. Diagram to show relationship of crossed and uncrossed fibres from retinae to cell laminae of lateral geniculate bodies.

The macular fibres, some of which are crossed and some uncrossed, are not shown separately. They are delivered to the cells that lie between the medial and lateral edges of the body in its posterior two-thirds [cf. FIG. 10.153].

(From I. Mackenzie (1934). *J. Path. Bact.* **39**, 113, after Le Gros Clark.)

the nerve fibres which arise from it) of one eye produces a loss of vision in the diametrically opposite part of the field of vision of that eye.

3. Nerve fibres which form the visual pathway are so arranged that those respectively in the lower and upper parts of the optic nerves, chiasma, and radiations carry impulses from the lower and upper parts of the retinae. Hence, taking 2 into account, damage to the lower parts of these will cause upper visual field defects.

4. Only fibres from the nasal halves of the retinae cross in the optic chiasma. (i) Thus a midline injury to the chiasma causes loss of vision in the temporal fields of both eyes—bitemporal hemianopia. If the lower midline fibres are affected first, as by pressure from a pituitary tumour, the loss of vision will begin in the upper temporal parts of the field of vision (see 3 above) and vice versa. (ii) On each side, all parts of the visual pathway posterior to the chiasma (optic tract, lateral geniculate body, optic radiation, and visual cortex) are concerned with impulses from the corresponding parts of both retinae which survey the opposite half of the field of vision (ipsilateral temporal retina and contralateral nasal retina). Hence destruction of any of these parts produces blindness in the opposite half of the field of vision—homonymous hemianopia.

5. Nerve fibres from the corresponding parts of the two retinae (i.e. the parts concerned with the same small area of the field of vision) lie together in the optic tract, lateral geniculate body, optic radiation, and visual cortex. If small parts of these are damaged, a blind spot (scotoma) occurs in the opposite half of the field of vision involving the corresponding area in both eyes. This is usually obvious to the patient by comparison with a blind spot in one eye (e.g. a small retinal lesion) which may go unnoticed if it falls in the field of vision common to both eyes (binocular field) as is the case with the physiological blind spot caused by the absence of sensory elements in the area of the optic nerve head.

Of the reflex pathways, the connection with the pretectal region is important as an index of the integrity of the visual pathway. Since both pretectal regions, right and left, activate both third nerve nuclei, impulses passing through either optic tract and brachium produce an equal (consensual) contraction of both pupils. Both pretectal regions or the fibres leading to or from them would have to be destroyed to abolish the light reflex which would then be lost bilaterally, though there would be no interference with vision (see 1 above). Thus with simple methods of testing, such as exposure of one eye to a bright light, a bilateral response will always occur even if one optic tract or one brachium of the superior colliculus is damaged, or if there is a bilateral destruction of the optic radiation or visual cortex with total blindness, but no response in either eye will be elicited if the opitc nerve of the exposed eye is destroyed. In this case a response will be elicited in the pupil of both eyes if the other eye is exposed to light. Unilateral loss of pupillary response to light is probably only found in lesions involving the efferent pathways, i.e. the oculomotor nerve, the ciliary ganglion or its branches.

The reticular formation of the brain stem

[FIGS. 10.54–10.58]

It is difficult to define exactly what is meant by the above term which is used in various ways by anatomists and physiologists. It is an anatomical concept and is properly applied to those parts of the brain stem (midbrain, pons, and medulla oblongata) composed of scattered cells of various sizes intermingled with nerve fibres running mainly in a longitudinal direction. In the upper medulla oblongata it lies dorsal to the olive, extending from the region of the spinocerebellar tracts laterally to the medial lemniscus medially; further caudally it tapers to become continuous with the reticular

formation of the upper cervical spinal medulla which lies at the medial projection of the lateral funiculus. In the pons it forms a large portion of the dorsal part, between the lemniscus and the floor of the fourth ventricle; while in the midbrain it forms the central part of each half of the tegmentum, between the central grey substance and the substantia nigra, and from the red nucleus medially to the lateral aspect of the mesencephalon.

The term 'reticular formation' is not usually applied to well-defined nuclei such as the red nucleus, the olivary nucleus or the vestibular nuclei which are closely associated with it. It is a purely descriptive term and, to some extent, a cloak for ignorance, indicating, as it does, the inability to analyse this type of neural complex under the microscope. It is obvious, however, that the close intermingling of nerve cells and fibres in the reticular formation permits complex intercommunications which would allow its cells to be used in different combinations for several functions. In aggregate it represents a collection of internuncial cells, the internal arrangements of which are not readily susceptible to anatomical analysis. Though it may be thought of as an anatomical entity, it is functionally very closely associated with the well-defined nuclei in close topographical relation to it.

Superficially, the reticular formation forms a continuous mass of similar tissue throughout the brain stem, but an analysis of its cyto-architecture and fibre connections shows that its various parts are not as similar as would appear at first sight. Though the cells which lie in it are not collected into compact nuclei, certain groups can be recognized which have been shown to have different connections. In general, it is possible to divide the reticular formation into lateral and medial parts, and parts which send fibres to the cerebellum. The lateral parts receive most, but not all, of the afferent fibres, and the medial parts give rise to the ascending and descending pathways arising from the reticular formation.

Afferent fibres reach the reticular formation from a number of sources:

1. **Spinoreticular fibres** pass with the spinothalamic system. They are widely distributed in the reticular formation, but particularly to those medullary and pontine parts from which ascending fibres arise and pass ipsilaterally and contralaterally through the pons and midbrain, probably to the thalamus. Some spinoreticular fibres end on the lateral reticular nucleus of the medulla oblongata which seems to act as a relay station for impulses to the anterior lobe of the cerebellum [FIG. 10.160]. Other reticular afferents come from the nuclei cuneatus and gracilis.

2. **Cerebelloreticular fibres** arise principally in the fastigial nucleus and are distributed to the medulla oblongata through the inferior cerebellar peduncle. Other fibres in the superior cerebellar peduncle give off branches which descend from the lower midbrain into the pontine and medullary reticular formation, as the descending limb of the superior cerebellar peduncle [FIG. 10.164].

3. **Cerebroreticular fibres** arise principally from the region of the central sulcus, but to a lesser extent from other parts of the cerebral cortex. They descend with the fibres of the corticospinal system and end principally in the cranial parts of the pontine and medullary reticular formation (the regions which give rise to most of the reticulospinal tracts) and also in the reticular nuclei which project to the cerebellum.

4. Some fibres which end in the reticular formation arise in the vestibular nuclei, the superior colliculi, and the red nucleus; others from the subthalamus and from the globus pallidus [p. 718] of the lentiform nucleus end in its cephalic part.

Efferent fibres

Reticulocerebellar fibres pass from the pontine and medulla oblongata parts of the reticular formation to most parts of the cerebellar cortex. A large number of these come from the lateral reticular nucleus on which spinoreticular fibres end. These reticulocerebellar fibres may be responsible for the transmission of ascending impulses from the spinal medulla to the cerebellum in addition to transmitting information on reticular formation activity. According to Brodal (1957) long ascending fibres arise mainly from the caudal parts of the pontine and medullary reticular formation and ascend bilaterally in the central tegmental tract. Many of these fibres reach the intralaminar and posterior nuclear groups of the thalamus. Thence impulses are distributed to the cerebral cortex, the corpus striatum, and the hypothalamus. It has been claimed on physiological grounds with some anatomical confirmation, that the impulses passing in these ascending fibres are distributed throughout the cerebral cortex by a diffuse thalamic projection system, thought to originate in the intralaminar thalamic nuclei [FIG. 10.79] and possibly distributed through their connections with other thalamic nuclei (see page 673).

The reticulospinal tracts arise in the pons and medulla oblongata, particularly their superior parts. Bilateral pathways arise from both sides of the medulla oblongata. Each crossed pathway descends with the medial longitudinal fasciculus. The uncrossed pontine reticulospinal fibres enter the anterior funiculus of the spinal medulla. The uncrossed medullary pathway, or pathways, descend in the anterolateral region of the white matter as scattered fibres in the cat; some lie as far posteriorly as the rubrospinal tract. The reticulospinal fibres end principally in the region of the anterior horn of the grey matter of the spinal medulla, and many are believed to terminate on the cells which give rise to the motor fibres to muscle spindles (γ-efferents). Some enter the posterior horn where they are believed to be responsible for influencing the transmission of pain and possibly other impulses through the cells of the horn. Stimulation of some of the cells in the raphe of the medulla oblongata, or of the central grey substance of the midbrain which act through the raphe, produces a blockage of pain impulses and a consequent state of analgaesia. This action is inhibited by substances which counteract the action of morphine. It is believed that the substance released by these nerve fibres is akin to morphine and that it acts in the same way.

Thus the reticular formation is: (1) a link and even a controlling factor in ascending pathways to the cerebellum, thalamus, corpus striatum, and cerebral cortex; and (2) a station on the descending pathways for most of the major masses of grey matter in the brain. In those mammals which have a rudimentary corticospinal system, e.g. the sheep, the reticulospinal tracts form a major part of the connection between brain and spinal medulla, being reinforced by the rubrospinal [p. 712], tectospinal, vestibulospinal [FIG. 10.158], and olivospinal tracts.

In addition it is intimately concerned with the control of respiration and the cardiovascular system—a feature which is not unexpected in view of the fact that in lower forms (e.g. the fishes) the gills, heart, and great vessels lie opposite the hindbrain and are innervated by the nerves arising from it. The neural mechanisms developed for the control of these structures are retained in the brain stem of Man, even though gills are superseded by lungs and the heart is displaced caudally relative to the central nervous system.

The vestibular system

Nerve fibres of the vestibular part of the vestibulocochlear nerve are distributed to the vestibular nuclei. These lie under the raised, triangular vestibular area which fills out the lateral part of the floor of the fourth ventricle, medial to the lateral recess, and extends from the midpons almost to the obex. Other vestibular fibres pass directly to the cerebellum by coursing dorsally, medial to the inferior cerebellar peduncle.

THE VESTIBULAR NUCLEI [FIG. 10.158]

The four vestibular nuclei, lateral (Deiter's), medial, superior, and inferior, each receive fibres from different elements of the vestibular part of the vestibulocochlear nerve. These are distributed to those parts of the nuclei which do not give rise to efferent fibres. Other afferent fibres reach these nuclei from the posterior spinocerebellar tract as it passes in the inferior cerebellar peduncle, also from the cerebellum (mainly its fastigial nuclei and flocculonodular cortex), and from the reticular formation of the midbrain and medulla oblongata. Efferent fibres from the vestibular nuclei pass throughout the length of the midbrain, hindbrain, and spinal medulla and are distributed to both sides. They fall into three groups:

1. Direct connections with motor cells of the brain stem and spinal medulla. The lateral and superior nuclei give rise mainly to uncrossed fibres. The large cells of the lateral nucleus send their efferents inferomedially through the medulla oblongata to the anterolateral region of the white matter of the spinal medulla, whence they pass mainly into the anterior horn. Fibres from the superior nucleus pass superomedially under the floor of the fourth ventricle [FIG. 10.56], to reach the medial longitudinal fasciculus and run headwards in it through the midbrain to the level of the posterior commissure. Fibres are given off to the oculomotor, trochlear, and abducent nerve nuclei and to the tegmentum of the midbrain.

The medial and inferior nuclei give rise to fibres which pass into the medial longitudinal fasciculus of the same and of the opposite sides and extend longitudinally to the midbrain and spinal medulla.

The medial longitudinal fasciculus lies close to the midline and immediately beneath the floor of the fourth ventricle and the aqueduct in the midbrain. It is a cranial extension of the fasciculi proprii of the anterior funiculus of the spinal medulla, and like it forms a pathway for communications between the cranial nerve nuclei in the brain stem, and for transmission of fibres from the vestibular nuclei to the brain stem and spinal medulla.

Through their bilateral connections, the right and left vestibular nuclei can effect changes in the tone of the extrinsic ocular muscles and of the muscles of the neck, trunk, and limbs. Such arrangements allow for compensatory movements of the eyes and trunk in association with movements of the head. These include nystagmus—the oscillatory eye movements (rapid in one direction, slower in the other) which occur on rotation of the head and which persist after rotation so long as the endolymph is in motion in the appropriate semicircular ducts. There is evidence of a specific relationship between the parts of the vestibular apparatus of the internal ear, the vestibular nuclei, and the motor nuclei of the oculomotor, trochlear and abducent nerves. Thus, the major connections of the ampullae of the lateral semicircular ducts are with the contralateral abducent and the ipsilateral part of the third nerve nucleus supplying the medial rectus [FIG. 10.159]. Such connections would account for the slow component of the nystagmus, a co-ordinated movement of both eyes towards the opposite side. It should be appreciated that the semicircular ducts in the right and left vestibular apparatus are arranged to give complementary information since they form corresponding pairs lying in the same plane, e.g. right and left lateral, and right anterior with left posterior semicircular ducts, only one of each pair being maximally stimulated by rotation of the head in one direction parallel to a pair of canals, the other being activated by rotation in the opposite direction around the same axis.

2. The vestibular nuclei have reciprocal connections with the flocculonodular lobe and fastigial nuclei of the cerebellum, as well as many connections with the reticular formation of the brain stem. These connections seem to be concerned with the control of muscle tone necessary to maintain a particular posture in relation to gravitational forces acting on the maculae of the utricle and saccule. During movements they produce adjustments of muscle tone as a

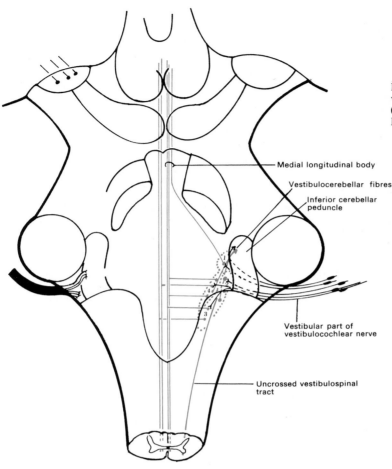

FIG. 10.158. Diagram of the main connections of the vestibular part of the vestibulocochlear nerve. (1) Superior, (2) lateral, (3) inferior, and (4) medial vestibular nuclei. Primary fibres—black; secondary fibres—red.

Medial longitudinal body
Vestibulocerebellar fibres
Inferior cerebellar peduncle
Vestibular part of vestibulocochlear nerve
Uncrossed vestibulospinal tract

result of afferent impulses mainly from the ampullae of the semicircular ducts. These help, for example, to maintain balance and to produce the rapid alterations in eye position which stabilize the visual image on the retina during movements of the head, e.g. in

walking and running. The anatomical pathways concerned in these reactions are obscure, but damage to this system makes all visual images blurred during head movements.

3. The pathway by which vestibular impulses are transmitted to the cerebral cortex is unknown. A group of fibres has been described which follows the auditory fibres, but vestibular impulses have been recorded in the lower part of the postcentral gyrus. It is, however, extremely difficult to analyse the efferent connections of the vestibular nuclei as a result of lesions placed in them, because of the coincident injury to auditory and other fibres passing through them.

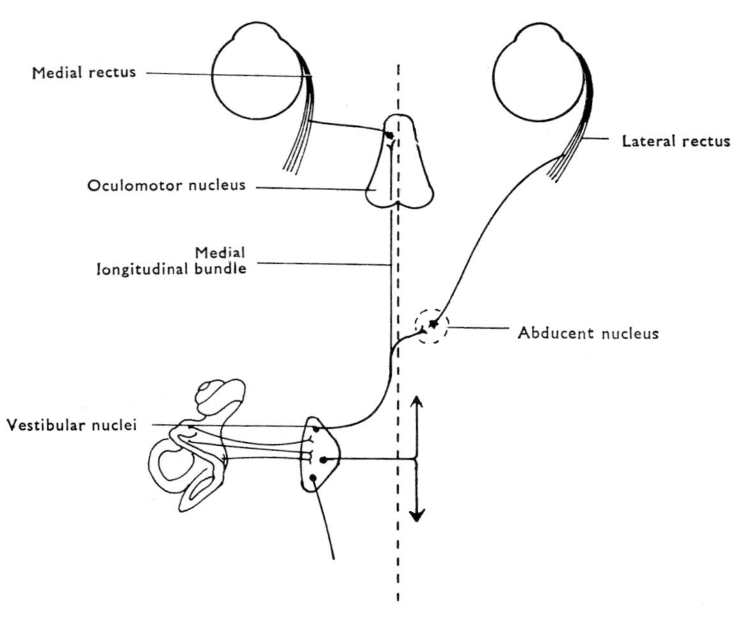

The cerebellum

The development and general arrangement of the cerebellum have already been considered [p. 629]. It has been noted that it consists of a highly convoluted cortex overlying a central mass of white matter in the depths of which are the nuclei of the cerebellum, closely related to the cerebellar recess of the fourth ventricle [FIGS. 10.36 and 10.40]. The cortex is divisible into a median part, or **vermis**, and the large lateral lobes, **hemispheres**, which are directly continuous with the vermis and of exactly the same histological structure. On the inferior surface, the vermis (pyramid, uvula, and nodule) is separated by a deep groove from the corresponding parts of the hemispheres, the biventral lobule, the tonsil, and the flocculus. On the superior surface no clear line of demarcation exists between vermis and hemispheres, the parts of which are indicated in FIGURES 10.42 and 10.44.

Within each cerebellar hemisphere, the largest nucleus is the laterally placed **dentate nucleus** formed from a crinkled layer of

Medial rectus
Oculomotor nucleus
Medial longitudinal bundle
Lateral rectus
Abducent nucleus
Vestibular nuclei

FIG. 10.159. Diagram of the vestibulo-ocular connections concerned in horizontal eye movements.

grey matter, arranged in the shape of an open bag with the mouth, the hilus, facing anterosuperiorly, and the superior cerebellar peduncle emerging from it. Medial to the dentate lie three smaller, more nearly spherical nuclei, the **emboliform**, **globose**, and **fastigial**, the last lying close to the midline [FIG. 10.40].

Afferent cerebellar connections

These are numerous, and are mostly distributed to the cerebellar cortex with collaterals to the deep nuclei. The great majority of the afferents form widely branched endings in the deepest (granular) layer of the cortex and hence are known as mossy fibres. Afferents from the olivary nuclei form the climbing fibres (see below).

1. Fibres pass to the fastigial nuclei, flocculonodular lobe, and lingua from the **vestibular nerve** and from the medial and inferior vestibular nuclei. These fibres run dorsally, medial to the inferior cerebellar peduncle, and turn across the roof of the fourth ventricle, between it and the beginning of the superior cerebellar peduncle, to reach the fastigial nuclei and the nodule. In the roof of the fourth ventricle, the fibres are in the medial part of the inferior medullary velum near its attachment to the **peduncles of flocculi** [FIG. 10.42] through which some pass to the flocculi [FIG. 10.45].

2. Fibres from the **spinal medulla** reach the cerebellum by way of the spinocerebellar tracts [FIG. 10.143] and the **posterior external arcuate fibres** [p. 693]. These arcuate fibres and the **posterior spinocerebellar tract** enter the cerebellum through the inferior cerebellar peduncle [FIG. 10.160], and are distributed to the central lobule, the culmen, the declive, the pyramid, and the uvula, i.e. to the anterior and posterior parts of the vermis. The **anterior** and part of the **rostral spinocerebellar tracts** enter the cerebellum by passing posteriorly over the lateral surface of the superior cerebellar peduncle on to the superior medullary velum. The anterior spinocerebellar tract passes through the white matter of the cerebellum to the same parts of the cerebellar cortex as the posterior spinocerebellar tract—the rostral tract ends more posteriorly in the anterior lobe. Anatomically there is only limited evidence of somatotopical localization in the distribution of these fibres to the cerebellar cortex. Nevertheless, physiological experiments in animals have shown that there is a definite localization of the various parts of the body at least in the anterior lobe of the cerebellum [FIG. 10.161]. Thus, when the lower limb is stimulated, potentials due to impulses arriving at the cerebellar cortex are found in the lobulus

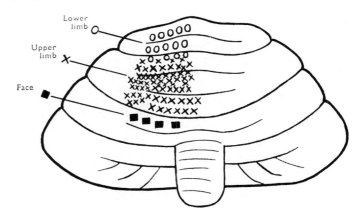

FIG. 10.161. Diagram of projection areas of afferent cerebellar impulses from limbs and face on the cerebellar cortex of the monkey. (E. D. Adrian (1943). *Brain*, **66**, 289.)

centralis and adjacent parts of its ala. When, however, the upper limb is stimulated, the culmen and medial parts of the quadrangular lobule are affected, and impulses from the head arrive at the anterior part of the declive and adjacent simple lobule. Fibres also enter these parts of the cerebellum from the lateral reticular nucleus [FIG. 10.160]. This receives many spinoreticular fibres, and those from different parts of the body end in separate parts of the lateral reticular nucleus (Brodal 1957). These parts project in a similar manner to the vermis of the cerebellum, an arrangement which could partly account for the discrepancy between the anatomical and physiological findings.

3. In addition to the lateral reticular nucleus, fibres from certain other nuclei of the **reticular formation** also project to the cerebellum

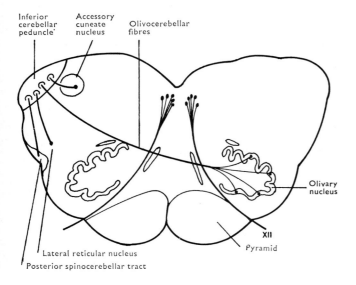

FIG. 10.160. Diagram of a transverse section of the medulla oblongata to show the fibres entering the inferior cerebellar peduncle.

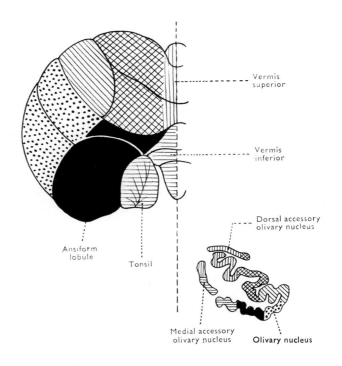

FIG. 10.162. Diagram to indicate the parts, respectively, of the right olivary nuclei in the medulla oblongata and of the left half of the cerebellum, which are linked together by olivocerebellar fibres. (After G. Holmes and T. G. Stewart (1908). *Brain* **31**, 125.)

through the inferior cerebellar peduncle. These nuclei have connections with the vermis and hemispheres of the cerebellum and receive spinoreticular and some cerebroreticular fibres.

4. A large number of afferent fibres reach the cerebellum (Brodal, Walberg, and Hoddevik 1975) from the opposite **olivary complex** in the medulla oblongata [FIG. 10.162]. These are distributed to all parts of the cerebellar cortex and are arranged in a definite point-to-point manner, each with a restricted area of distribution. The fibres arising in the medial accessory olive (which is situated medial to the hypoglossal rootlets) and the dorsal accessory olive pass to the vermis. These two nuclei form a much larger part of the olivary complex in lower mammals, e.g. the cat, than in Man, and are the site of termination of the spino-olivary fibres (Brodal 1957). The main mass of the olivary nucleus, which has increased in size *pari passu* with the increase in size of the cerebral and cerebellar hemispheres, sends fibres to the lateral parts of the cerebellum and receives fibres from headward regions of the central nervous system on the same side. According to Walberg (1954), these fibres arise in the frontal cerebral cortex, the globus pallidus, the red nucleus and the peri-aqueductal, central grey substance of the midbrain. Thus, the olivary nucleus links the cerebral cortex, the corpus striatum, and the tegmentum of the midbrain to the cerebellum; the fact that its fibres are crossed is in keeping with the crossed efferent cerebellar pathway to these regions.

5. The largest single mass of afferents reaches the cerebellum from the opposite **pontine nuclei** through the massive middle cerebellar peduncle. Fibres from the pontine nuclei pass to all parts of the cerebellar cortex except the flocculonodular lobe, and form a link between the cerebral and cerebellar cortices. Fibres enter the pons from all lobes of the cerebrum in the crus cerebri. A comparison of the size of the crus cerebri and the pyramid shows that most of the fibres in the crus end in the pons, since the only fibres which do not do so emerge in the pyramid.

On the ventral surface of the medulla oblongata, applied to the pyramids, are the **arcuate nuclei** [FIG. 10.177]. These give rise to fibres which pass dorsally in the raphe and laterally on the surface

of the medulla, over or around the olive. Those which enter the raphe pass posteriorly to the floor of the fourth ventricle and then turn laterally across it as the **striae medullares** [FIG. 10.46], to enter the cerebellum by the inferior cerebellar peduncle. The other fibres, passing on the surface of the medulla, **anterior external arcuate fibres**, also reach the cerebellum. Both sets of fibres represent aberrant pontocerebellar fibres, and the arcuate nuclei are caudal extensions of the pontine nuclei.

6. **Auditory** and **visual fibres** also reach the cerebellar cortex. These have been demonstrated by recording impulses in that cortex following exposures of the eye to flashes of light, or the ear to controlled noise. The intermediate parts, the declive, folium, and tuber, and the corresponding parts of the hemisphere receive these impulses (Snider 1950). The visual impulses from the superior colliculus probably enter through the superior cerebellar peduncle; the auditory impulses enter from the inferior colliculus by the same route, or through the inferior cerebellar peduncle from the cochlear nuclei (Fadiga and Pupilli 1964).

It will be clear from the foregoing brief account that:

1. The cerebellum is not simply concerned with proprioceptive information but with all types of afferent information.

2. So far as its afferent connections are concerned, each half of the cerebellum tends to be related to the same side of the spinal medulla and hindbrain, but to the opposite side of the cerebrum, the corpus striatum, and the midbrain.

3. The afferent fibres are not uniformly distributed throughout the cortex, but tend to be relayed to specific regions, e.g. the vestibular and spinal afferent fibres, but several different afferents project to the same areas of the cerebellum.

STRUCTURE AND CONNECTIONS OF THE CEREBELLAR CORTEX

The cerebellar cortex is of uniform structure throughout even though the connections of its different parts are not the same. The cortex consists of three layers: (1) an outer, **molecular layer**; (2) a middle

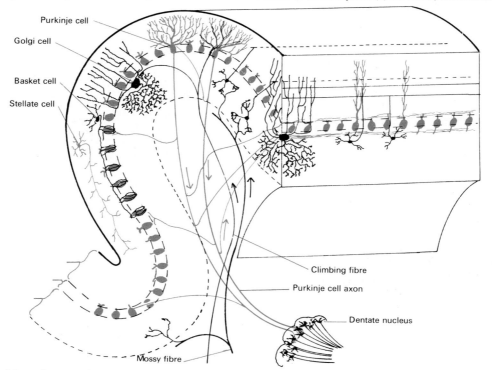

Purkinje cell
Golgi cell
Basket cell
Stellate cell
Climbing fibre
Purkinje cell axon
Dentate nucleus
Mossy fibre

FIG. 10.163. A diagram of the main types of nerve cells and their processes in the cerebellar cortex. A single folium is shown cut longitudinally (right) and transversely (left). The Golgi and stellate (basket) cells have inhibitory functions, while the mossy and climbing fibres are excitatory. Both mossy and climbing fibres send collaterals to the cerebellar nuclei. Purkinje and stellate cells are shown red.

layer consisting of the large cell bodies of the **Purkinje cells**; (3) an inner, **granular layer** [FIG. 10.163].

The granular layer receives endings of the **mossy fibres** each of which is widely branched and overlaps with the terminations of many others—a single mossy fibre even spreading to two adjacent folia. These form synapses with the dendrites of many small **granule cells** which send their axons into the molecular layer where they divide and run parallel to the surface in the long axis of the folia. Here they traverse and form facilitatory synapses (a) with the dendrites of many Purkinje cells which spread into the molecular layer in a plane at right angles to the long axis of the folia, (b) with the dendrites of stellate cells which lie in the molecular layer (see below), and (c) with the dendrites of large **Golgi cells** each of which has a cell body in the granular layer with dendrites passing vertically through the Purkinje and molecular layers to permeate a volume of cortex. The axon of each Golgi cell branches to pervade a volume of the granular layer deep to the distribution of its dendrites. These axons form inhibitory synapses with the dendrites of the granule cells, often in complexes (glomeruli) with facilitatory mossy fibre synapses.

Most of the outer and inner (basket) **stellate cells** of the molecular layer send their axons parallel to the surface and at right angles to the long axis of the folia. These give off radial branches which (a) run with the dendrites of the Purkinje and Golgi cells (outer stellate cells), or (b) pass to surround the Purkinje cell bodies and reach the beginning of their basal axons (inner stellate or basket cells). The synapses of both types are believed to be inhibitory.

Climbing fibres enter the cerebellum through the inferior cerebellar peduncle from the olivary nuclei (Szentagothai and Rajkovits 1959). Each passes to a very restricted area of the cerebellar cortex where it follows and branches with the dendrites

of a Purkinje cell, forming facilitatory synapses with them. It also sends branches to adjacent basket and Golgi cell dendrites.

The total outflow of the cerebellar cortex is formed by the axons of the Purkinje cells. These pass radially inwards towards the white matter giving off collaterals which turn back and run in the Purkinje cell layer to form inhibitory synapses with the processes of adjacent basket and Golgi cells. The axons then pass through the white matter to the cerebellar nuclei except those from the flocculonodular lobe and spinal part of the vermis which leave the cerebellum in the inferior cerebellar peduncle and pass directly to the vestibular nuclei (Walberg 1972). Those that pass to the cerebellar nuclei are believed to form inhibitory synapses with their cells—facilitatory synapses to these cells coming from collaterals of mossy and climbing fibres on their way to the cerebellar cortex.

The cerebellar cortex appears to be in a position to integrate many different incoming impulses, and to produce its effects by controlling the activity of the cerebellar (and vestibular) nuclei (for details see Eccles, Ito, and Szentagothai 1967).

Efferent cerebellar connections

The Purkinje cell axons (corticonuclear fibres) converge radially on the nuclei of the cerebellum, the vermis projecting to the fastigial nuclei, while the hemispheres send fibres principally to the dentate nuclei [FIG. 10.164]. That these fibres are arranged radially is indicated by the fact that the cephalic part of the fastigial nucleus receives fibres from the anterior part of the superior vermis, while its caudal part receives fibres from the inferior vermis. The same radial arrangement holds for the dentate nucleus, and the emboliform and globose (intermediate) nuclei receive fibres from the inter-mediate, **paravermal region** of the cerebellar cortex [FIG. 10.42].

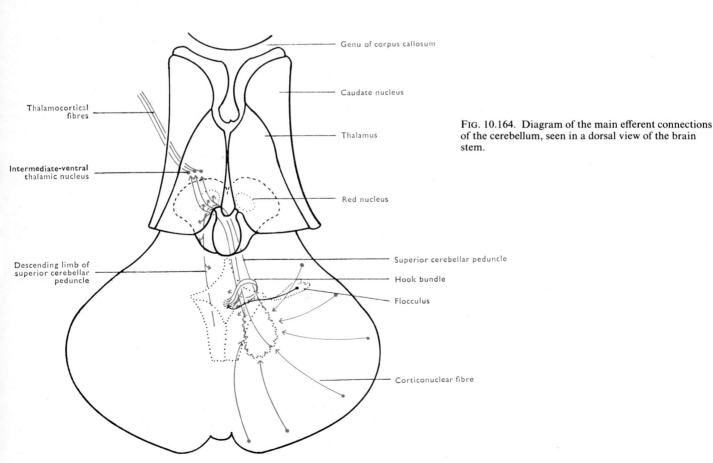

Thalamocortical fibres

Intermediate-ventral thalamic nucleus

Descending limb of superior cerebellar peduncle

Genu of corpus callosum

Caudate nucleus

Thalamus

Red nucleus

Superior cerebellar peduncle

Hook bundle

Flocculus

Corticonuclear fibre

FIG. 10.164. Diagram of the main efferent connections of the cerebellum, seen in a dorsal view of the brain stem.

The fastigial nuclei give rise to fibres to the vestibular and reticular nuclei of the medulla oblongata. These pass either directly through the ipsilateral inferior cerebellar peduncle, or by hooking over the contralateral superior cerebellar peduncle as the **hook bundle**.

The other cerebellar nuclei send efferent fibres into the **superior cerebellar peduncles** which pass anterosuperiorly into the midbrain beneath the inferior colliculi. In the lower midbrain these bundles sweep anteromedially and, intermingling in the midline [FIG. 10.55], cross to the opposite side ventral to the aqueduct, medial longitudinal bundle and trochlear nucleus. Having crossed, they ascend into the tegmentum of the upper midbrain where their fibres enter and surround the **red nuclei** [FIG. 10.147]. As the superior cerebellar peduncles are crossing, each gives rise to a descending bundle of fibres, the **descending limb of the superior cerebellar peduncle**, which passes to the pontine reticular formation.

The fibres of the superior cerebellar peduncle send collaterals to the red nucleus (mainly from the emboliform and globose nuclei) as they pass to the intermediate ventral and intralaminar nuclei of the thalamus. The cells of the intermediate ventral nucleus of the thalamus send their axons to the precentral region of the frontal cerebral cortex, while the intralaminar nuclei connect with other diencephalic structures and with the corpus striatum.

The **red nucleus** consists of two groups of cells, a small, caudal, magnocellular part which gives rise to the **rubrospinal** and **rubroreticular tracts**, and a large parvocellular part which gives rise to ascending fibres. In mammals other than primates, the magnocellular part of the red nucleus is of considerable size and gives rise to fibres which immediately cross the midline in the ventral part of the tegmentum of the midbrain, the **ventral tegmental decussation**. These fibres turn caudally through the dorsal parts of the pons and medulla oblongata to reach the lateral funiculus of the spinal medulla whence they pass to interneurons of its intermediate grey matter. This **rubrospinal tract** gives collaterals to the lateral reticular nucleus of the medulla oblongata—(the **rubroreticular tract**, which may be the largest part of this descending system in Man) and fibres leave its cephalic part to reach the emboliform and globose nuclei (**rubrocerebellar fibres**). The red nucleus is also connected to the olivary nucleus by fibres descending in the central tegmental bundle (**rubro-olivary fibres**). Both the red and olivary nuclei receive large numbers of fibres from the cerebral cortex in the region of the central sulcus. The parvocellular part of the red nucleus gives rise to fibres which are distributed to the thalamus, the midbrain reticular formation, and the cerebellum.

The cerebellum thus has extensive afferent and efferent fibres by which it is connected reciprocally with the vestibular nuclei, the reticular nuclei of the brain stem, the red nucleus, the thalamus, and through it with the cerebral cortex and corpus striatum. Functionally it has nothing to do with the conscious appreciation of sensation and has no direct mechanism for acting on the motor cells of the spinal medulla. All its activity must therefore be mediated through its action on the vestibular, reticular, and red nuclei, as well as on the cerebral cortex and corpus striatum. It is well known that the vestibulospinal, reticulospinal, and rubrospinal systems have marked effects on muscle tone in the trunk and limbs and are, presumably, the efferent pathways responsible for many of the righting reflexes known to be mediated in the brain stem. At least some of these pathways (e.g. reticulospinal) seem to act on muscle tone through the γ-efferent fibres, causing contraction or relaxation of muscle spindles and a consequent increase or decrease in the muscular contraction mediated by monosynaptic reflexes. There is evidence that the cerebellum plays a large part in controlling this activity, and it seems to be an essential element in the mechanism concerned with the correct distribution of tone in the muscles for the proper execution of voluntary and other activities. Cerebellar lesions

in Man are usually associated with a marked decrease in muscle tone (*atonia*), and stimulation of the anterior lobe in animals decreases the decerebrate rigidity resulting from transection of the midbrain. Sprague and Chambers (1959) produce evidence, in the cat, of a functional difference between the vermis, the paravermal, and the lateral zones of the cortex, which project respectively by the fastigial, intermediate (globose and emboliform), and lateral (dentate) nuclei. Thus the vermis and the fastigial nuclei play a part in regulating bilaterally 'tone, posture, equilibrium, and locomotion of the whole body'. The paravermal and the lateral regions have their effects principally on the ipsilateral limbs, the activity of the two sides apparently being integrated in the brain stem.

Cerebellar lesions are characterized by the transitory nature of the signs which they produce, and most of these signs are potentiated by a simultaneous lesion of the sensorimotor cerebral cortex, or made to reappear by such a lesion if already 'compensated' for. The *tremor* during voluntary movement, which results from cerebellar cortical lesions, is dependent on the presence of the sensorimotor cerebral cortex, since destruction of this region abolishes it. It seems likely, in view of the massive connections betweeen the cerebellum and cerebrum, both through the pons and the olive, and the fact that both send fibres to the same brain stem nuclei, that their interaction is necessary for the smooth execution of normal movements which, in its turn, is dependent on the correct balance of tone in the opposing muscles. Cerebellar injury decomposes this balance giving rise to jerky uncoordinated movements (*ataxia*) which frequently overshoot their aim. Speech is similarly affected, being often disjointed, slurred, or explosive.

The connections of the flocculonodular lobe with the vestibular apparatus suggest a degree of control by the cerebellum of this system. This is confirmed in injuries of the flocculonodular lobe, when disturbances of balance are common without other cerebellar signs. Damage to the cerebellum in Man frequently produces nystagmus, an oscillating movement of the eyes most usually in the horizontal plane but possible in any direction. If horizontal, it is made most obvious by looking to one side. Nystagmus is common following disturbances of the internal ear or lesions of the vestibular nuclei, and it may be that cerebellar nystagmus, which is similar in type, may represent a disturbance of the interaction of these two parts. On the other hand it seems likely that the cerebellum plays the same part in eye movement that it plays in other movements (Whitteridge 1960). If this is the case, the tremor of voluntary movement and of cerebellar nystagmus may have the same origin.

In general, the cerebellum is concerned in controlling, through the brain stem mechanisms, the adjustments in muscle tone which are necessary for posture and movement control (static and phasic tone). Each half of the cerebellum is predominantly concerned with the same side of the body and receives the sensory information necessary to carry out these adjustments through rapidly conducting, spinocerebellar pathways. In carrying out these actions, the cerebellum is intimately integrated with the other parts of the brain, which seem to be able to compensate for injuries to the cerebellum.

The diencephalon

The diencephalon consists of the **thalamus** dorsally and the **hypothalamus** ventrally, the latter being subdivided into a medial portion, the hypothalamus of the physiologist, and a lateral portion, the **subthalamus**. These parts have already been described [pp. 661–2], including their positions, major nuclear groups, and connections. The arrangement of the thalamic nuclei has been given in general [p. 658], and in more detail with each of the ascending pathways [pp. 698–705; FIGS. 10.79 and 10.80].

THE THALAMUS

The thalamus is the part of the diencephalon which forms the great relay station for pathways leading to the telencephalon. Virtually all ascending fibres relay in this structure, and from its cells fibres arise which pass to the cerebral cortex and to the corpus striatum and hypothalamus; the only exception to this rule being a group of ascending fibres which passes directly to the cerebral cortex through the pyramid, crus cerebri, and internal capsule. These form a very small proportion of the afferent fibres to the cerebral cortex and their origin and destination are uncertain, though it seems that they arise in the cervical spinal medulla in Man (Nathan and Smith 1955b), and possibly throughout the spinal medulla and from the nuclei cuneatus and gracilis in the cat (Brodal and Walberg 1952), the contribution from the cervical spinal medulla being larger than that from more caudal levels.

Each thalamic nucleus which projects to the cerebral cortex does so to a restricted area from which it receives corticothalamic fibres [FIG. 10.165]. Thus, the cerebral cortex can have direct effects on the thalamus, and it is possible to envisage these effects facilitating or inhibiting the passage of impulses through certain parts of the thalamus, or even being responsible for setting up continuous thalamocortical and corticothalamic activity.

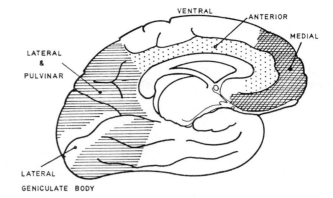

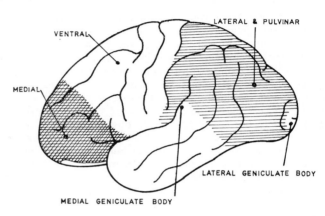

FIG. 10.165. Inferomedial and lateral views of the cerebral hemisphere, showing the projection areas of the thalamic nuclei.

Thalamic nuclei

1. NUCLEI TRANSMITTING ASCENDING PATHWAYS TO THE CEREBRAL CORTEX

(i) The **medial** (auditory) and **lateral** (visual) **geniculate bodies** which project to the superior temporal and calcarine (striate) cortex respectively [pp. 702 and 705].

(ii) The **posterior-ventral nucleus** relays the fibres of the spinal, medial, and trigeminal lemnisci to the postcentral gyrus of the parietal lobe, and is presumably also concerned with gustatory and vestibular sensations which find their way to the postcentral region [pp. 700 and 708]. It also projects to part of the second sensory area.

(iii) The **intermediate-ventral nucleus** which receives fibres from the superior cerebellar peduncle and red nucleus, and projects to the posterior part of the frontal lobe, including the precentral region.

2. NUCLEI WHICH PROJECT TO THE CEREBRAL CORTEX BUT DO NOT FORM THE SPECIFIC TERMINATIONS OF ASCENDING PATHWAYS

(i) The **medial nucleus** which has connections with the hypothalamus, other thalamic nuclei, and the anterior part of the frontal lobe of the cerebrum through the anterior limb of the internal capsule. These thalamofrontal and frontothalamic fibres have been divided in the operation of prefrontal leucotomy and may be responsible, at least in part, for the changes in emotional reactivity which this operation causes. Possibly because they form a route of communication between the frontal lobes and the hypothalamus, this system may be responsible for the visceral responses so often connected with emotional states, e.g. sweating, blushing, vomiting, increased pulse rate, etc.

(ii) The **dorsolateral nucleus** receives from other thalamic nuclei and projects to the parietal lobe (except for the post-central region to which the posterior-ventral nucleus sends fibres) and probably also to the temporal lobe. The posterior extension of the lateral nucleus, the **pulvinar**, has similar connections with the posterior

parietal, occipital, and temporal lobes and seems to be concerned with the integration of auditory and visual impulses.

(iii) The **anterior nuclei** receive fibres from the fornix which leave it in the region of the anterior commissure [FIG. 10.120], and from the hypothalamus in the **mamillothalamic tract**. The various parts of the anterior nuclei project to the gyrus cinguli. The significance of this system remains obscure. It forms a part of the limbic system. The afferent fibres to the mamillary body come mostly through the fornix from the hippocampus [p. 679]; the gyrus cinguli sends efferent fibres to the parahippocampal gyrus (entorhinal area), with which it is directly continuous, and the parahippocampal gyrus projects on to the hippocampus. Thus, a complex series of intercommunications is formed between the parts of the marginal cortex of the cerebral hemisphere. It should be remembered that in marsupials [FIG. 10.121], where there is no corpus callosum and the hippocampus and dentate gyrus extend to the anterior commissure, the cerebral cortex corresponding to the gyrus cinguli bears the same relation to the hippocampus as the parahippocampal gyrus does in Man, both lying immediately adjacent to it, separated only by the narrow strip of cortex known as the subiculum. There are therefore intimate connections between the parts of this marginal system [p. 678], which is connected to the hypothalamus by way of the fimbria and fornix, and with the brain stem, both indirectly through the mamillary body, and directly through the fornix.

(iv) The **posterior group** transmit impulses to part of the second sensory area. These may be concerned with painful and thermal impulses some of which reach these nuclei from the reticular formation. Some of the cells in these nuclei are activated by stimuli applied to any part of the surface of the body on either side. This degree of convergence appears to arise as a result of impulses reaching the nuclei bilaterally from the reticular formation or from the intralaminar nuclei which also receive bilateral impulses through extralemniscal ascending systems.

3. THALAMIC NUCLEI WITH SUBCORTICAL CONNECTIONS

(i) The anterolateral ventral nucleus receives fibres from the globus pallidus of the lentiform nucleus and sends efferent fibres to the corpus striatum as well as the cortex.

(ii) The habenular nucleus lies in the habenular triangle at the root of the pineal body, anterosuperior to the tectum of the midbrain and medial to the posterior part of the thalamus. Fibres enter the habenular triangle through the medullary stria of the thalamus which begins anteriorly in the anterior perforated substance (olfactory trigone) and amygdaloid body, and is reinforced by fibres from the septal region which join it near the interventricular foramen. From this point the bundle passes posteriorly to enter the habenular triangle, forming a ridge between the medial and superior surfaces of the thalamus with the thin roof of the third ventricle attached to it. Fibres from the habenular nucleus cross to the opposite side in the habenular commissure, which lies in the root of the pineal body above the pineal recess of the third ventricle. A large bundle of fibres, the fasciculus retroflexus, passes antero-inferiorly from each habenular triangle to the interpeduncular nucleus on the ventral surface of the midbrain, traversing the midbrain–diencephalic junction and the medial part of the red nucleus. The functions of this system are unknown but the interpeduncular nucleus probably forms part of the efferent pathway for this system, since it has connections with the tegmentum of the lower midbrain.

(iii) The centromedian nucleus is the largest nucleus in the medial medullary lamina. It lies between the medial part of the posterior ventral nucleus and the medial nucleus [FIG. 10.80]. In common with other intralaminar nuclei, the centromedian nucleus reacts to injury of the corpus striatum and parts of the limbic system (Cowan and Powell 1955) but not to injury of other parts of the cerebral cortex as thalamic nuclei with cortical connections do. This suggests that the centromedian nucleus has mainly subcortical connections. The main efferent fibres pass to the caudate nucleus, the putamen, and the anterolateral ventral nucleus of the thalamus. All the afferent connections are not known, but impulses are received from the long ascending systems (e.g. collaterals of the spinothalamic tracts—Getz 1952) often bilaterally, from the cerebral cortex (precentral gyrus), from the corpus striatum through the ansa lenticularis, and from the reticular formation of the brain stem (reticulothalamic fibres) (see also page 673).

THE RETICULAR NUCLEUS OF THE THALAMUS

This nucleus forms a thin shell external to the lateral medullary lamina and covers virtually the whole lateral surface of the thalamus.

It receives fibres from every part of the cerebral cortex (Carman, Cowan, and Powell 1964b) and projects mainly to other thalamic nuclei and to the midbrain reticular formation (Schibel and Schibel 1966). Since it forms a capsule for the thalamus, it is in a position to sample the thalamocortical and corticothalamic impulses traversing it. Thus it may form a feed-back mechanism to the brain stem reticular formation, but does not seem to be part of the direct pathway for the so-called diffuse thalamic projection through which the reticular formation of the brain stem is known to alter the activity of the entire cerebral cortex. The anatomical pathways for this diffuse system are uncertain, but the intralaminar and anterior-ventral nuclei of the thalamus have been considered to be important elements (see also page 673).

Hypothalamus

THE LATERAL PART—THE SUBTHALAMUS

This is essentially the upward continuation of the tegmentum of the midbrain. Laterally it consists of two nuclei, the subthalamic and the nucleus of the zona incerta, separated from each other by the ansa lenticularis [p. 718], with the zona incerta separated from the thalamus by the thalamic fasciculus [FIG. 10.80]. Medially it is composed of the subthalamic tegmental area, or prerubral field, in which the fibres from the capsule of the red nucleus, the ansa lenticularis, the trigeminothalamic tract, and medial lemniscus intermingle.

The subthalamic nucleus is pierced by many fibres of the ansa lenticularis, some of which end on its cells, and it also receives fibres from the cerebral cortex. Its efferents pass into the globus pallidus.

The nucleus of the zona incerta receives fibres from the ansa lenticularis and the cerebral cortex.

The fibres of the ansa lenticularis pass mainly to the thalamus. A small proportion descends to the midbrain reticular formation. The thalamic connections allow impulses originating in the lentiform nucleus to pass via the ansa lenticularis and thalamus to the cerebral cortex and, through the anterior-ventral and centromedian nuclei of the thalamus, back to the corpus striatum after interaction with other impulses in these nuclei.

The subthalamus is developed from the basal lamina and is intimately connected with the mechanisms concerned with muscle tone. Its exact function is unknown but thrombosis in the subthalamic nucleus is associated with the development of spontaneous, writhing or sometimes violently jerky movements of the opposite upper limb, hemiballismus.

THE MEDIAL PART OF THE HYPOTHALAMUS

The general arrangement and relations of this region have already been described [pp. 661–2].

Afferent fibres

These enter the hypothalamus from a large number of sources and the following account is by no means exhaustive:

1. **Connections with the cerebral cortex**. Direct connections with the cerebral cortex come from the orbital part of the frontal lobe and from the hippocampus through the fornix [p. 680]. The column of the fornix descends to the anterior commissure and splits around it. Most of the fibres pass posterior to the commissure and curve inferiorly to the mamillary body, sending many fibres to the anterior hypothalamus and anterior thalamic nuclei on the way. The precommissural part of the fornix enters the septal region. In Man, this is the grey matter ventral to but continuous with the anterior part of the septum pellucidum. It is united inferiorly with the anterior hypothalamus through the preoptic region, antero-superior to the optic chiasma. The septal region also receives the stria terminalis from the amygdaloid body. Several areas of the cerebral cortex adjacent to the principal sensory regions [p. 722] send fibres to the amygdaloid body, the cingulate gyrus, and the parahippocampal gyrus. Through these they are connected to the hypothalamus either directly (amygdaloid body) or through the hippocampus.

2. Afferent fibres enter the hypothalamus from the medial thalamic nucleus by running as periventricular fibres in the wall of the third ventricle. These would seem to form a pathway by which impulses from the frontal lobe of the cerebral cortex can reach the hypothalamus, and also a route by which fibres ascending in the brain stem to the thalamus can activate the hypothalamus, since collaterals of these fibres are said to pass to the medial nucleus.

3. The **medial forebrain bundle** arises in the region of the anterior perforated substance, septum, and preoptic region, and passes posteriorly through the lateral part of the medial hypothalamus conveying olfactory and septal impulses. This pathway may also transmit impulses which originate in the frontal lobe and relay in the septal region.

4. Fibres pass into the hypothalamus from the **amygdaloid body** by way of the stria terminalis and directly by fibres passing medially below the lentiform nucleus.

Efferent pathways

These fall into two groups:
  (i) Connections with the hypophysis
    (a) Neural connections with the posterior lobe.
    (b) Vascular connections with the anterior lobe.
  (ii) Connections with other parts of the brain.

Neural connections with the posterior lobe of the hypophysis arise in the supra-optic and paraventricular nuclei [FIG. 10.82], and descend in the stalk to the posterior lobe. The cells of the nuclei contain a substance which is passed along their axons and accumulates in the posterior lobe. The glial cells of the posterior lobe are not the source of its hormones, for though they remain intact after interruption of the stalk of the hypophysis, the posterior lobe loses its pharmacological activity, and the neurosecretory material collects proximal to the interruption. Interruption of the stalk causes excessive secretion of dilute urine due to the lack of antidiuretic hormone (vasopressin), and dehydration (Bargmann 1966) or substances which cause antidiuresis by acting through the hypophysis (Bodian 1951) cause a loss of neurosecretory substance from the hypophysis. Thus the supra-optic and paraventricular cells are secretory in type and pass hormones or their precursors along their axons to the posterior lobe for storage—a feature in keeping with the known activity of nerve cells. It appears that the antidiuretic hormone is produced by the supraoptic nuclei, while the paraventricular nuclei produce oxytocin (Olivecrona 1954; Heller 1966) a substance which causes uterine contractions.

Vascular connections with the anterior lobe of the hypophysis are present in the absence of neural connections. The anterior lobe of the hypophysis receives most of its blood supply indirectly from the hypophysial arteries through sinusoidal capillary glomeruli in the median eminence (a small median protuberance immediately posterior to the root of the hypophysial stalk), infundibulum, and stalk of the hypophysis. Axons from the **tuberal nuclei** of the hypothalamus end in contact with these glomeruli, and are believed to secrete releasing factors which are carried to the anterior lobe through the venules from the glomeruli. These unite into a series of **portal vessels** which descend on the stalk and become continuous with sinusoids in the anterior lobe. Here the factors control the release of the anterior lobe hormones. In addition, some specialized ependymal cells—tanycytes (Brawer 1972)—may absorb substances from the cerebrospinal fluid of the third ventricle and pass it to the sinusoids or to the cells of the hypothalamus.

Connections with other parts of the brain

(i) From the **mamillary body** a large bundle of fibres passes at first laterally, then turns dorsally in the medial medullary lamina of the thalamus to the anterior thalamic nuclei, the **mamillothalamic tract**. From the anterior nuclei, fibres pass to the gryus cinguli, particularly its anterior part. In its lateral course the mamillothalamic tract fibres bifurcate and a bundle of **mamillotegmental fibres** passes caudally into the tegmentum of the midbrain.

(ii) By way of the periventricular fibres the hypothalamus is connected with the **medial nucleus of the thalamus** and through this nucleus with the frontal lobe of the cerebral cortex, particularly the polar and orbital regions.

(iii) Fibres of the periventricular system pass from the posterior nucleus [FIG. 10.82] to the midbrain and run through the central grey substance as the **dorsal longitudinal fasciculus**, connecting with cells in the peri-aqueductal grey matter, other than those that the mamillotegmental fibres make contact with in the tegmentum.

(iv) The **medial forebrain bundle**, in the lateral part of the medial hypothalamus, is continued into the midbrain mainly by fibres which arise in the posterior hypothalamus. These form a descending pathway from this part of the brain produced mainly by cells with short axons.

(v) The posterior hypothalamic region is said to be able to activate the 'diffuse thalamic projection system', and can thus play a part in modifying the activity of much of the cerebral cortex.

Thus the hypothalamus receives fibres from the ascending systems, has communications with other parts of the brain, and gives rise to fibres which pass caudally to act through the mechanisms of the brain stem: a pattern common to every part of the nervous system.

Physiological stimulation of the hypothalamus can elicit many autonomic activities, e.g. changes in heart rate and blood pressure, hyperglycaemia, bladder contraction, etc., and for this reason the medial hypothalamus has been considered the part of the forebrain concerned with visceral activity. It is possible, however, to produce similar effects from other parts of the forebrain, and the multiple connections which the hypothalamus has with other regions of the brain, highlights the fact that it is an integral part of the brain mechanism working in association with the other parts. The function of the nervous system is to integrate the activity of all the organs of the body, including the viscera whose functions are inextricably bound up with somatic activity. Thus increased heart and respiratory rates, vasodilatation, vasoconstriction, etc., are the natural accompaniments of physical exercise.

The nuclei of the telencephalon [FIGS. 10.71– 10.75]

These consist of:
  The claustrum.
  The amygdaloid body.
  The corpus striatum—the caudate and lentiform nuclei.

THE CLAUSTRUM

This is a thin sheet of grey matter between the external capsule and the white matter of the insula. Its connections and function are relatively unknown, but it receives fibres from the cerebral cortex through the external capsule (Carman, Cowan, and Powell 1964a) in a manner similar to the corpus striatum.

THE AMYGDALOID BODY

This lies near the pole of the temporal lobe, over the anterior extremity of the inferior horn of the lateral ventricle. Medially it is continuous with the cortex of the temporal lobe where it abuts on the anterior perforated substance [FIG. 10.80], but laterally it lies deep to the cortex. It is continuous superiorly with the inferior surface of the lentiform nucleus, with the anterior commissure passing between them [FIG. 10.73, 3] and posteriorly with the tail of the caudate nucleus. The amygdaloid body is divisible into a considerable number of subnuclei which seem to have different connections and functions.

The connections and functions of the amygdaloid body are very incompletely known, and it is not possible to make any clear estimate of its function from those connections which are known. The afferent fibres which have been determined come mainly from the olfactory system. Direct fibres from the olfactory bulbs reach the dorsomedial portion through the lateral olfactory stria. Other afferent fibres have been described from the corpus striatum, the intralaminar nuclei of the thalamus, the olfactory trigone, the anterior piriform cortex, and from various regions of the cerebral cortex, including the temporal lobe. Electrophysiological studies indicate that ascending impulses from the spinal medulla reach the amygdaloid body, probably through multisynaptic pathways of the reticular formation.

The efferent fibres form the stria terminalis which arises in the amygdaloid body and, running on the concavity of the lateral ventricle medial to the caudate nucleus, is distributed to the septal region and the anterior hypothalamus. Other fibres pass in the diagonal band to the olfactory trigone, septum, and paraterminal and anterior cingulate gyri. The diagonal band [FIG. 10.120] originates in the amygdaloid body and, passing anteromedially beneath the lentiform nucleus crosses the posteromedial part of the anterior perforated substance, immediately above the optic tract. It turns superiorly on the medial surface of the hemisphere just anteriorly to the lamina terminalis in association with the medial olfactory stria.

Thus both routes connect the amygdaloid body with similar regions, bringing it into relation with parts of the brain (e.g. hypothalamus and medial forebrain bundle) through which descending pathways may run to the brain stem, or ascending pathways pass to the thalamus and cerebral cortex. Fibres also pass directly to the piriform lobe; thus the amygdaloid body may have connections with the hippocampus through this lobe.

From the above connections it might be assumed that the amygdaloid body was primarily a part of the olfactory system, but it is well developed in the anosmatic Cetacea (Breathnach and

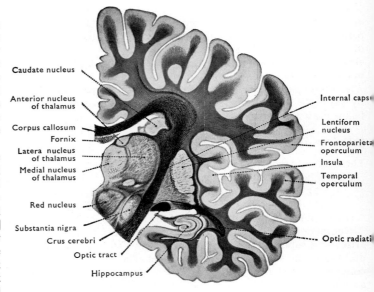

FIG. 10.167. Coronal section through cerebrum stained to show nerve fibres.

Goldby 1954) and in this respect is similar to the hippocampus and dentate gyrus. Bilateral lesions of the amygdaloid in monkeys produce disturbances of emotional behaviour, increased motor activity associated with an apparent inability to ignore any visual stimulus, altered sexual behaviour, etc. Similar effects occur in Man in cases of temporal lobe, or psychomotor, epilepsy. There is loss of memory of recent events, confusion, illusions related to one or other sense, sometimes marked alteration of personality and automatic actions, often of a complex type, during the attack. Many of the changes seen during the attacks can be reproduced at operation by stimulation of the amygdaloid.

THE CORPUS STRIATUM [FIGS. 10.166 and 10.167]

The corpus striatum consists of the caudate and lentiform nuclei. The lentiform nucleus having an external segment, the putamen, darker in colour than its inner part, the globus pallidus [FIG. 10.110], which is separated from the putamen by a thin layer of white matter, the lateral medullary lamina, and divided into internal and external segments by the medial medullary lamina. The putamen of the lentiform nucleus and the caudate nucleus have the same structure (being composed of small and medium-sized cells) and communicate with each other by strands of grey matter which pass between the bundles of the internal capsule separating them. The globus pallidus on the other hand contains many myelinated fibres and a considerable number of large cells. For this reason the caudate and putamen are sometimes combined under the term striatum, while the globus pallidus is called the pallidum.

It seems likely that the division of the caudate from the putamen of the lentiform nucleus by the internal capsule is quite fortuitous, and that the use of the term striatum is advisable since they really form a single unit.

The general disposition of the parts of the corpus striatum has already been described [p. 674], the following section deals with the nervous connections of its parts [FIG. 10.168].

Connections of the corpus striatum

Afferent fibres have already been described passing to the striatum from the intralaminar nuclei of the thalamus, especially from the centromedian nucleus. The fibres passing to this nucleus come

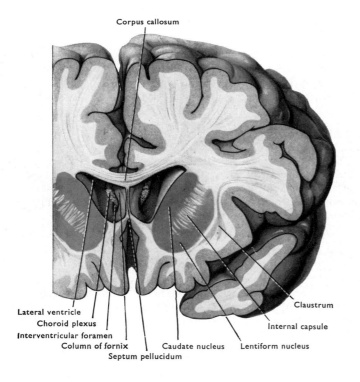

FIG. 10.166. Coronal section through the cerebral hemisphere. The section (viewed from the front) cuts the anterior horns of the lateral ventricles, through which the central part of the ventricles, the columns of the fornix, and the openings of the interventricular foramen can be seen.

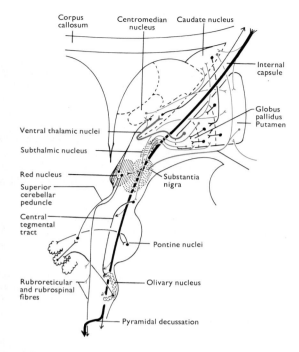

FIG. 10.168. Diagram to show some of the principal connections of the corpus striatum.

that the fibres which enter the striatum from the internal capsule are mostly non-myelinated, and possibly collaterals of fibres passing to other destinations. Thus, the striatum probably samples all the impulses leaving the cerebral cortex. Though the details of corticostriate fibres are not known in Man, it is likely that they converge radially on the striatum. Thus the head of the caudate would receive fibres from the medial, orbital, and anterior part of the superolateral surface of the frontal lobe; the body and tail of the caudate would receive similar fibres from the rest of the medial surface of the hemisphere, and the putamen from the remainder of the hemisphere (Carman, Cowan, and Powell 1963). Many crossed fibres to the caudate nucleus run in the subcallosal fasciculus, a longitudinal bundle of fibres in the angle between the corpus callosum and the caudate nucleus.

Efferent fibres from the striatum pass to both segments of the globus pallidus and to the substantia nigra.

The substantia nigra forms a slightly curved plate of cells between the crus cerebri and the tegmentum of the midbrain. It extends from the caudal midbrain into the subthalamus. To the naked eye the substantia nigra is dark grey in colour, because many of its cells contain melanin. The substantia nigra sends efferent fibres to the thalamus (anterolateral ventral nucleus), and to the striatum and pallidum. Dopamine which is present in the striatum seems to be derived from the substantia nigra, and is transmitted thence by efferent fibres to the putamen and caudate nuclei. These fibres have been seen by fluorescence microscopy because of their dopamine content, and the dopamine in the striatum disappears when the substantia nigra is destroyed. The melanin contained in the cells of the substantia nigra may well be explained by their activity in producing dopa, an intermediary between tyrosine and melanin.

Afferent fibres to the pallidum arise mainly from the caudate nucleus and putamen, the fibres from the caudate passing through the internal capsule to reach it. Other afferent fibres reach the pallidum from the subthalamic nucleus, the substantia nigra, and from the intralaminar nuclei of the thalamus. The pallidum forms the major efferent pathway for the striatum, apart from its connections with the substantia nigra. The medial segment of the pallidum receives fibres from the lateral segment.

from the globus pallidus, the midbrain reticular formation, the superior cerebellar peduncle (Jung and Hassler 1960), and possibly as collaterals of ascending sensory pathways. The striatum thus receives information from the spinal medulla, the reticular formation, and the cerebellum. See also substantia nigra.

Other afferent fibres reach the striatum from the substantia nigra and the cerebral cortex. Cortical afferents have long been postulated and denied because of the absence of Marchi degeneration in the striatum following injury to the cerebral cortex. It is clear, however,

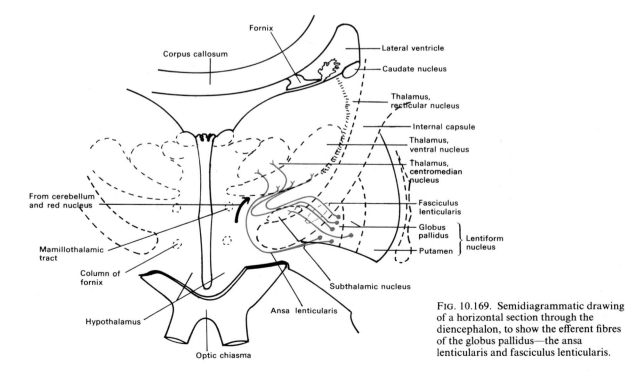

FIG. 10.169. Semidiagrammatic drawing of a horizontal section through the diencephalon, to show the efferent fibres of the globus pallidus—the ansa lenticularis and fasciculus lenticularis.

The efferent fibres of the pallidum form the ansa lenticularis. This is a mass of fibres which leaves the inferior and medial surfaces of the pallidum. Those leaving the inferior surface tend to hook around the ventral margin of the internal capsule, running parallel with the optic tract [FIGS. 10.120 and 10.169], and turn dorsally into the subthalamus, while those leaving the medial surface of the pallidum pierce the internal capsule to reach the subthalamus. The latter part is sometimes called the fascicularis lenticularis. Most of these fibres are destined for the thalamus. They form a well-defined bundle on the ventral aspect of the zona incerta [FIG. 10.80] and either pass medially towards the prerubral field, or turn dorsally to enter the thalamus, where they end in the anterolateral and intermediate ventral nuclei. These fibres come principally from the medial segment of the globus pallidus which also sends a small bundle to the caudal mesencephalic tegmentum. Fibres from the lateral segment pass mainly to the subthalamic nucleus and to the subthalamic tegmentum. Through this connection the corpus striatum can affect the reticular formation of the brain stem and through it, or through the olivary nucleus, the opposite cerebellar hemisphere. By its thalmic connections the corpus striatum affects the activity of the cerebral cortex.

The connections of the corpus striatum and cerebellum have certain features in common. Both have extensive reciprocal connections with the cerebral cortex, and both project to the brain stem. Neither has a direct pathway to the spinal medulla. The functions of the corpus striatum are not clear, but it seems certain that it acts mainly with the cerebral cortex, but also has some direct effect on brain stem mechanisms through the cranial part of the reticular formation—a feature which it has in common with the cerebellum and the cerebral cortex.

Clinically, lesions of the striatal mechanism are associated with alterations in muscle tone which tend to be the reverse of those caused by cerebellar lesions. There is a great increase in muscle tone with rigidity of the limbs and the same inco-ordination of antagonist muscles, so that the patient has to work against the unreleased tension of the antagonists. There is reduced activity with loss of associated movement (e.g. the swinging of the arms in walking, the smile of pleasure, the grimace of pain), and a tremor, particularly involving the limbs and neck, which is present at rest but tends to disappear during voluntary movement. To this extent it is the reverse of the cerebellar tremor, but it should be appreciated that the signs of disease of the striatal system have much in common with the signs of cerebellar disease, particularly when the midbrain–diencephalic region is involved since this ground is common to both.

The tremor of striatal disease seems to be primarily the result of damage to the substantia nigra, and it, like that of cerebellar disease, is abolished by damage to the efferent pathway of the motor cortex suggesting that an imbalance of the interaction of these systems is involved. It has been shown in Man that surgically produced lesions of the pallidum, the ansa lenticularis, or the anterior-ventral nucleus of the thalamus markedly reduce or even abolish the tremor and rigidity in striatal disease (Parkinson's disease). This is not understood, but it seems to suggest that the clinical condition represents an imbalance of the controlling mechanisms and that this is improved by lesions involving another element of the system normally in balance. Any explanation must remain speculative at this time, but it is highly probably that, in a delicately balanced system represented by the interacting masses of grey matter (corpus striatum, thalamus, cerebral cortex, cerebellum, etc.) which comprise the brain, lesions in different situations may produce similar objective effects. *It is always tempting to ascribe particular functions to particular parts of the central nervous system, but it must be remembered that the results of a lesion do not indicate the function of the injured part, only what the remaining nervous tissue is capable of performing.*

The cerebral cortex

The cerebral cortex is the convoluted layer of grey matter which covers the entire surface of the cerebral hemispheres and is believed to be the site of conscious appreciation of sensations, the origin of 'voluntary' activity, and many other functions, at least a proportion of which do not seem to be immediately under voluntary control. The cerebral cortex has a considerable area, approximately $2–2.5 \times 10^3$ cm^2, of which only about a third is visible on the surface, the remainder forming the walls of the sulci. This area contains a prodigious number of cells which has been estimated at approximately 5×10^9. This figure is only a very rough estimate, but it gives some indication of the possible number of neuronal complexes which can exist in this structure.

The cerebral cortex varies considerably in thickness in its different parts, being thickest (>4 mm) in the superior part of the precentral gyrus, and thinnest (<2 mm) in the regions receiving the ascending sensory pathways. The posterior part of the frontal lobe, the inferior parietal lobule, and the temporal pole also have a relatively thick cortex (>3.5 mm), while in the occipital and frontal poles the cortex is thin (<2.5 mm). The above measurements give only an approximate measure of thickness, since the cortex on the crown of any gyrus is approximately twice as thick as that in the depths of adjacent sulci.

Unlike the cerebellar cortex, the cerebral cortex varies considerably in structure from place to place, but it is possible to recognize two basically different types:

1. The isocortex which covers the greater part of the surface of the cerebral hemispheres and consists of nerve cells arranged in six layers. These layers lie parallel to the surface of the cortex and abut on each other, but are capable of differentiation because of the different types and densities of cells which each contains.

2. The allocortex forms a very small part of the surface of the hemisphere, limited to the limbic system, is very poorly laminated and has always less than six layers. The allocortex forms the margin of the hemisphere adjacent to the choroid fissure, and is continuous with the isocortex through transitional zones. Thus the allocortex of the dentate gyrus and hippocampus is continuous with the isocortex of the parahippocampal gyrus, through the adjacent subiculum and presubiculum [FIG. 10.119]. The allocortex is found in the above structures, in the gyrus fasciolaris with which the hippocampus and dentate gyrus are continuous, the indusium griseum (in the depths of the callosal sulcus), the paraterminal gyrus, the lateral and medial olfactory gyri which accompany the corresponding striae, the anterior perforated substance, and the uncus. Characteristic of the allocortex is the considerable amount of white matter on its external surface and this is most marked in the subiculum, through which fibres pass into the hippocampus [FIGS. 10.72, 10.118, and 10.130]. In the superficial layer of the uncus, the subiculum, and the presubiculum the cells are arranged in nests or islands often formed of alternate groups of small and medium sized cells. The general arrangement of cells and fibres in the dentate gyrus and hippocampus is shown in FIGURE 10.118, while the transition from these to the isocortex in the rat is shown in FIGURE 10.119.

THE ISOCORTEX

This cortex typically consists of six layers of cells which are most readily visible under the microscope when the cell bodies only are stained, as with a basic aniline dye. There are, however, several features readily visible to the naked eye when sections are made through a fresh brain. Even without magnification it is possible under these circumstances to recognize that the cortex is laminated because of a narrow white layer passing through certain parts of it

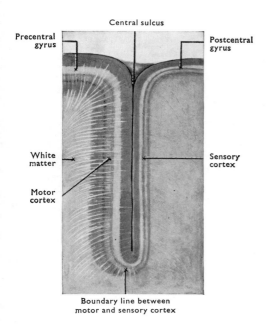

FIG. 10.170. Section across superior part of central sulcus in fresh brain.

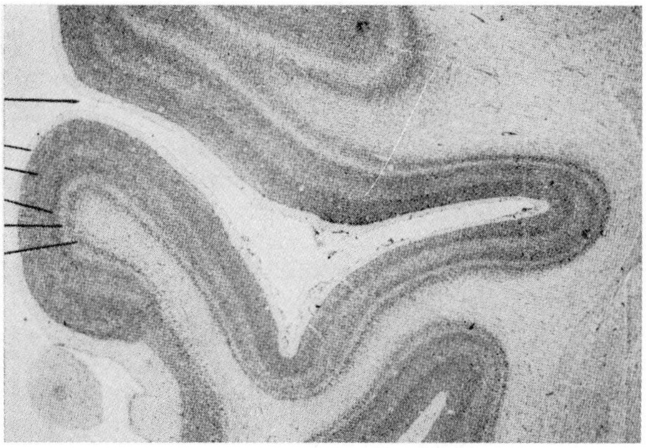

FIG. 10.171. Low power photomicrograph of section of visual (striate) area of occipital cortex. × 6.
The visual stria is situated in the lamina granularis interna, i.e. superficial to the deep stratum of that lamina indicated in the Figure. The junction with the surrounding (peristriate) cortex is clearly seen opposite **X**.

[FIG. 10.170]. This is the **outer band of Baillarger**, which separates an outer greyish layer, surmounted by a thin whitish lamina on the surface, from a deeper yellowish-red stratum extending to the underlying white matter. Occasionally it is possible to see a second thin layer of white matter, the **inner band of Baillarger**, passing through the inner part of the grey matter. The outer band of Baillarger is most obvious in the regions of the cortex which receive ascending sensory fibres, i.e. the postcentral, superior temporal, and occipital regions. It is most marked in the visual area (which receives fibres from the lateral geniculate body) where the band was first noticed, and where it is called the **visual stria**, giving to this part of the cerebral cortex the name **striate area**.

The histological structure of the isocortex

The following description is applicable to a large part of the cerebral cortex, but there are local differences in structure, and the architecture of the cells differs to a greater or lesser degree from that described below. The study of this cellular architecture (cyto-architectonics) has led some authors to describe a very large number of different areas, but minor variations, possibly of no great significance, are to be expected in a system of layers formed by scattered cells. Also, it is not possible to compare one area with another unless the sections used are exactly at right angles to the surface of the cortex: slight obliquity of a section increases the apparent thickness of the laminae and alters their appearance. In the convoluted cortex of Man and many other mammals, this is extremely difficult to achieve, and great care has to be exercised before differences of a minor nature can be accepted as significant. In some regions there is a sharp transition from one type of cortex to another, e.g. at the margins of the striate area and in the depths of the central sulcus [FIGS. 10.170 and 10.171]. Elsewhere the transitions are gradual (e.g. between the granular and agranular frontal cortex), and several intermediate types of cortex can be described. It remains to be seen whether some of these transitional areas are functionally different from the areas on either side of them before too much significance is attached to these structural differences. There is agreement, however, concerning the position

and extent of the main structural areas, and many of these have been shown to correspond with specific functional areas.

In a 'typical' part of the cortex the following laminae can be defined [FIGS. 10.172 and 10.173]:

1. The **lamina zonalis** is a thin, surface layer composed of non-myelinated and finely myelinated nerve fibres which form a closely meshed feltwork. They are derived from ascending axons of subjacent cortical cells, terminal branches of the apical dendrites of pyramidal cells, and processes of horizontally placed spindle-shaped cells, a few of which lie in this layer. The zonal lamina is probably of considerable importance in the diffusion of impulses in all directions over the surface of the cortex.

2. The **lamina granularis externa** is a thin and usually ill-defined layer of small cells. It is composed of the same morphological types of cells as those in layer 3.

3. The **lamina pyramidalis** contains many small, stellate cells, but is characterized by the presence of pyramidal-shaped cells, which become progressively larger in its deeper parts. Each pyramidal cell gives rise to an apical dendrite which ascends towards the surface, usually reaching the lamina zonalis, and to basal dendrites which spread out from the cell body parallel to the surface. The axon emerges from the base of the cell and descends to the white matter.

4. The **lamina granularis interna** is composed of numerous small stellate cells, each with several short, branching, dendritic processes, spreading in all directions. They also have a short axon which terminates in the vicinity of the cell. In addition there are minute pyramidal cells akin to those in layer 2.

5. The **lamina ganglionaris** contains cells of many types, but receives its name because of the very large pyramidal cells which are present in it in some areas. Throughout most of the cortex, the cells in layer 5 are smaller than those in the deeper part of layer 3.

6. The **lamina multiformis** lies adjacent to the white matter and is composed of cells of many types, though the majority are spindle-shaped or pyramidal. It also contains cells with axons which run through the more superficial layers towards the surface of the cortex. These are known as inverted pyramids and a few of them are to be found in the other layers of the cortex.

The presence of cellular laminae and of the bands of Baillarger suggests that the cortex has a laminar organization. In most parts of the cerebral cortex, however, bundles of nerve fibres can be seen running towards the surface from the white matter. The presence of

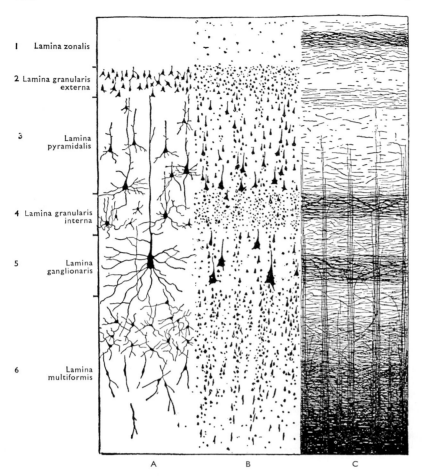

I	Lamina zonalis	
2	Lamina granularis externa	
3	Lamina pyramidalis	
4	Lamina granularis interna	
5	Lamina ganglionaris	
6	Lamina multiformis	

A   B   C

FIG. 10.172. Diagram of arrangement of cells and of myelinated nerve fibres in the cerebral cortex (cf. FIG. 10.173). (From C. J. Herrick (1938). *An introduction to neurology*, 5th edn. Philadelphia.)

A, cells shown by Golgi method; B, cells shown by Nissl method; C, nerve fibres shown by Weigert method.

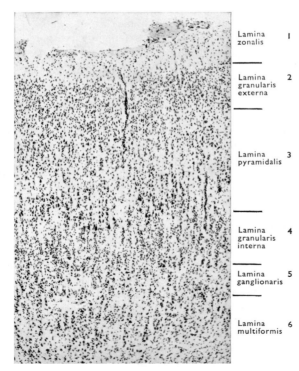

Lamina zonalis	I
Lamina granularis externa	2
Lamina pyramidalis	3
Lamina granularis interna	4
Lamina ganglionaris	5
Lamina multiformis	6

FIG. 10.173. Photomicrograph of section of peristriate area of occipital cortex, showing the laminar differentiation. The differentiation between the lamina ganglionaris and the lamina multiformis is much less distinct than it is in some other cortical areas. ×45.

such bundles at right angles to the cellular lamination, although to some extent to be expected, tends to suggest that the cortex may be organized in vertical columns—a view for which there is much physiological evidence. The bundles of fibres, horizontal or vertical, give very little indication of the complexity of the connections between the cells of the cortex and the fibres which enter it from many sources. When nerve fibres are stained in addition to cell bodies, the cellular lamination is obscured, and many connections between the cells of different layers have been established. This is clear from the fact that the apical dendrites of the pyramidal cells pass through all the more superficial layers and are covered with synapses throughout their length, so that they must be subjected to activity going on in these layers. The two types of organization are by no means exclusive and may well coexist as an integrating mechanism. Whatever the significance of the horizontal lamination of the cortex may be, it is heightened by the fact that the different layers tend to have different connections. Thus the lamina granularis interna (layer 4) is predominantly receptive in function. Not only is it particularly well developed in the sensory projection areas and absent from the cortical areas which are mainly efferent in function (e.g. precentral gyrus), but afferent fibres have been found, by direct histological examination, to terminate to a great extent among its cells. It is their presence in this layer which accounts for the well-developed outer layer of Baillarger in the sensory projection areas. Layers 5 and 6 (infragranular layers) are concerned primarily with the production of commissural and projection fibres, and layer 3 seems to receive a larger proportion of association fibres than the other layers. There is evidence then of some functional differences between the layers, though it must be understood that fibres of all

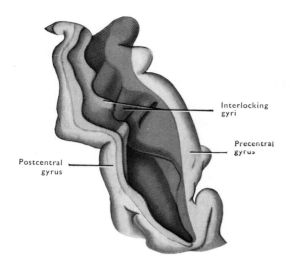

FIG. 10.174. Right central sulcus fully opened up, to exhibit interlocking and deep transitional gyri.
Motor cortex, red; sensory cortex, blue.

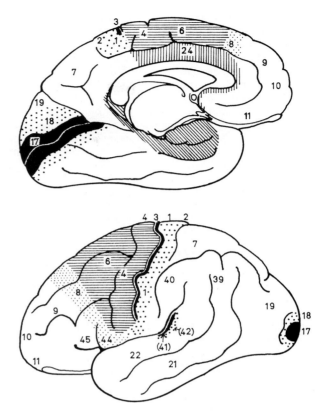

FIG. 10.175. Inferomedial and lateral views of the cerebral hemisphere, showing the distribution of the major types of cerebral cortex, based on histological structure. The numbers indicate some of those given by Brodmann, though the areas to which they correspond are not given.
▤ Agranular frontal cortex: ■ Koniocortex of the primary receptive areas, of 3 and 41 only a narrow strip is visible; ▦ frontal dysgranular cortex; ▦ parakoniocortex; ▥ cingulate agranular cortex; ▨ allocortex.

types may pass to, or arise from, every layer of the cortex, and the functional differences between layers are by no means clear cut.

Whatever the significance of the lamination, there is no doubt that its pattern changes and that such changes seem to be associated with different functions. Thus, for example, the 'sensory' areas of the cortex all have a very similar structure, and this is very different from the pattern in the 'motor area' of the precentral gyrus. Many areas of the cortex differ from each other in some subtle way, but because the significance of most of these differences is in doubt, and most of them have not been related to function [see, however, p. 700], only the major variations from the 'typical' (eulaminate) cortex described above will be considered. For further details see Bailey and Bonin (1951).

Variations in the structure of the cerebral cortex

1. The agranular cortex [horizontal hatching, FIG. 10.175], shows a virtual loss of layer 4 as a separate entity with an increase both in thickness of the cortex and in the number and size of the pyramidal cells, together with poorly differentiated laminae. It is found in the posterior part of the frontal lobe. In layer 5 the agranular frontal cortex contains very large pyramidal cells which may be up to 120 × 25 μm in size. The largest of these, giant cells of Betz, are found in the posterosuperior part of this area, but they decrease in size in its more anterior and inferior parts, there being a complete gradation between the giant cells and the large pyramidal cells in the anterior part of the agranular cortex.

Some giant pyramidal cells are also found in the fifth layer of the cortex on the posterior wall of the central sulcus.

The large pyramidal cells give rise to a small proportion (approximately 3 per cent, Lassek 1954) of the 1 200 000 fibres in the pyramid of the medulla (corticospinal tract), the remainder arising from other cells in the agranular frontal cortex (van Crevel 1958) and in the cortex of the postcentral region

A further variety of cortex, which is virtually agranular, with extremely ill-defined lamination, is found in the anterior part of the cingulate gyrus [vertical hatching, FIG. 10.175].

2. Koniocortex [solid black, FIG. 10.175]. At the other extreme there is a general reduction in the thickness of the cortex and of the size of its contained cells, with a considerable increase in thickness of layer 4. When the uniformly small cell bodies in this type of cortex are stained, the appearance is that of fine dust under the low

power of the microscope. Hence it is called dust cortex or koniocortex. This type of cortex is found in areas to which thalamocortical fibres pass which are concerned with specific sensations, i.e. the posterior wall of the central sulcus, the visual area, and the auditory area, but not the remainder of the cortex which receives thalamic efferents. In the visual area [FIG. 10.171] the fourth layer is subdivided by the presence in it of the visual stria (outer band of Baillarger). Also there are some large cells (cells of Meynert) in the infragranular layers. These are believed to send efferent fibres to the tectum of the midbrain and to be concerned with direct control over the visual tectal reflexes and, possibly, with some types of conjugate movements of the eyes. In the walls of the central sulcus the two extreme cortical modifications, argranular and koniocortex, lie adjacent to each other, and it is possible in a section to recognize with the naked eye the increase in thickness and the loss of the outer layer of Baillarger, in passing from the postcentral to the precentral cortex [FIGS. 10.170 and 10.174].

The eulaminate structure is found throughout the greater part of the remaining cerebral cortex, though it shows certain recognizable variations, e.g. the presence of particularly large cells in layers 3 and 5 of the inferior frontal gyrus. There are transitional zones between the eulaminate cortex and the agranular and koniocortex areas. In the former case there is a broad strip of so-called dysgranular cortex at the anterior margin of the agranular area; in this the laminae are poorly differentiated and the granular layer is often difficult to identify. Similar transitional zones (parakoniocortex) surround the koniocortices [large dots, FIG. 10.175]. The situations of the major types of cortex are shown in FIGURE 10.175. The above description is a much simplified version of that usually given and follows the general subdivision described by Bailey and Bonin (1951). While it is almost certain that further subdivision will become necessary as more information concerning the structure and functions of the various areas become available, it should be recognized that the map produced by Brodmann for the human brain has never been substantiated by histological evidence. As Bailey and Bonin (1951) and Walshe (1961) point out, it is surprising that this map has been so widely accepted in the absence of the necessary proof that it represents a true histological picture of the human cerebral cortex.

The connections of the cerebral cortex

The majority of the afferent connections have already been described with the ascending sensory tracts and the thalamus [pp. 700–15, FIG. 10.165]. Virtually all regions of the cortex receive fibres from the thalamus, except the anterior part of the temporal lobe, and each part of the cerebral cortex sends corticothalamic fibres back to the corresponding thalamic nucleus. In summary:

1. The frontal lobe receives fibres from the medial nucleus (granular cortex) and from the anterolateral and intermediate ventral nuclei (agranular area). The latter transmit impulses from the pallidum, cerebellum, and substantia nigra, while the connections of the medial nucleus are principally with the hypothalamus.

2. The koniocortex of the postcentral gyrus (primary sensory area) and the cortex adjacent to it including the secondary sensory area [p. 700] receive fibres from the posteromedial and posterolateral parts of the ventral nucleus. These relay impulses from the medial and trigeminal lemnisci, from the spinothalamic system, as well as gustatory impulses which seem to pass with the trigeminal lemniscus, and possibly vestibular impulses. In addition the secondary sensory area receives fibres from the posterior group of nuclei of the thalamus.

3. The remainder of the parietal lobe receives fibres from the lateral nucleus and pulvinar.

4. The visual area surrounding the calcarine sulcus receives fibres from the lateral geniculate body and projects to areas 18 and 19 [FIG. 10.175] which may represent the secondary and tertiary visual areas of other mammals where the representation is reversed [p. 700].

5. The remainder of the occipital lobe, and at least part of the temporal lobe receive fibres from the pulvinar.

6. The superior temporal gyrus (the primary auditory area and possibly a secondary auditory area adjacent to it) receives fibres from the medial geniculate body.

7. On the medial surface of the hemisphere the anterior nuclear complex projects to the cingulate gyrus connecting the hypothalamus (mamillary body) to this limbic cortex.

8. The exact distribution of the reticular thalamic nucleus is not known but it probably projects to the thalamus, midbrain, and other nuclei caudal to it. Other afferent fibres to the cerebral cortex are the commissural and association fibres.

The efferent fibres of the cerebral cortex can be divided into three separate groups:

1. Association fibres passing to other parts of the same hemisphere.

2. Commissural fibres passing to the opposite hemisphere.

3. Projection fibres passing to other parts of the central nervous system.

ASSOCIATION FIBRES

The main bundles (fasciculi) of long association fibres have already been described (p. 676). Detailed information regarding the organization of this system is lacking but in general it may be said that the primary receptive areas of the cerebral cortex, visual, auditory, and somaesthetic, send short association (arcuate) fibres into the adjacent cortex. From these areas, or from those adjacent to them, longer association fibres arise which pass to more distant regions of the cortex through one or other of the long association bundles. Thus the visual cortex (area 17, Brodmann) projects to the adjacent area 18; area 18 to area 19 external to it [FIG. 10.175], and the latter has long association fibres which pass as far afield as the agranular frontal cortex. There seems to be some difference in the functions of these three areas; thus stimulation of the primary visual area (17) produces flashes of light in the corresponding part of the opposite half of the field of vision [p. 706], while stimulation of the surrounding areas (which may correspond to the secondary and tertiary areas of the cat) seem to give rise to more organized visual images. In cats with all the commissures of the forebrain, including the optic chiasma, divided in the midline (split brain preparations) and trained to discriminate various objects with one eye (i.e. with the hemisphere of the same side; Sperry 1958), subsequent ablations of various parts of the trained hemisphere result in decreased ability to make the discriminations even when the ablations involve the temporal and frontal lobes. Initial ablation of the temporal lobe followed later by ablation of the frontal lobe leads to a deficit equal to the simultaneous ablation of both. This suggests that the complete hemisphere and not just the occipital lobe is necessary for full visual discrimination, though this does not seem to be true for sensations reaching the somaesthetic area. When the ablation leaves only the visual area intact, learned discriminations are lost, though training can still be effective but at a much lower level of performance. It is not clear what part the association fibres play in this mechanism, but they are probably responsible for implicating the entire cortex in particular functions. It is clear that the cerebral cortex, like other regions of the central nervous system (e.g. the cerebellum) has a marked ability to compensate for the loss of neural tissue, because similar unilateral cortical ablations in animals do not interfere with shape discrimination provided the opposite hemisphere can be involved even by any small part of the corpus callosum remaining intact.

If the gross anatomy is any index of the connections, then the association bundles of the superolateral surface connect the regions of that surface with each other through the superior longitudinal bundle, the lateral and posterosuperior fibres of the fasciculus uncinatus, and the external capsule [FIG. 10.113]. On the inferior surface the fibres of the inferior longitudinal bundle (temporal-occipital) and the medial fibres of the fasciculus uncinatus (frontotemporal) connect the various parts of this surface, while on the medial aspect the fibres of the cingulum connect the medial areas with the hemisphere [FIG. 10.111]. If the topographical position of the primary receptive areas is significant, it is clear that the somaesthetic area can have immediate connections with the 'motor' area (precentral gyrus) and such connections are known to exist. On the other hand, the three sensory areas are arrayed around the inferior parietal lobule, and it seems not impossible that such a region might be concerned with the association of the various incoming sensory impulses. The inferior parietal lobule lies inferior to the intraparietal sulcus and has the upturned end of the posterior ramus of the lateral sulcus and those of the superior and inferior temporal sulci entering it. In this fashion it is divided into three parts, each lying around the end of one sulcus; of these the most anterior is the supramarginal gyrus and the middle, the angular gyrus [FIG. 10.96].

Damage to the inferior parietal lobule seems to interfere with the formation of the body image which is presumably the result of interaction of the various senses, and when the non-dominant hemisphere (usually the right in right-handed individuals) [p. 727] is injured the individual seems to be unable to recognize the presence of objects in the opposite half of the field of vision, being quite content to draw only one-half of objects (e.g. a bicycle wheel as a half-circle) and may not recognize his hand, on the side opposite to the lesion, as belonging to him.

COMMISSURAL FIBRES

Commissural fibres pass between the corresponding parts of the two hemispheres by way of the corpus callosum [p. 683] and the anterior commissure [p. 682]. The **anterior commissure** connects the temporal lobes, particularly their anterior and inferior parts, the olfactory bulbs, and the anterior perforated substance. The **corpus callosum** transmits fibres from all lobes of the cerebrum, including the posterior and superior parts of the temporal lobes. It contains relatively few fibres from the anterior parts of the frontal lobes, and none either from the striate cortex [area 17, FIG. 10.175] or from the parts of the sensory areas which receive impulses from the distal parts of the limbs (Meyers 1965; Jones and Powell 1968). The auditory area also has relatively few callosal fibres.

From the size of the corpus callosum in the human brain it is reasonable to assume that this is a structure of great importance to the function of the brain, yet division of the corpus callosum or its congenital absence does not seem to be associated with commensurate signs or symptoms. It might be expected that there would be some loss of co-ordination of the two sides, but such defects are surprisingly slight and require careful testing to demonstrate any abnormality even in known cases—congenital absence of the corpus callosum usually being an incidental finding during air encephalography for other reasons. Experience with split brain preparations (see above) demonstrates that the commissures play a part in the transfer of learned activities from one hemisphere to the other in addition to co-ordinating their functions.

The **fornix commissure** lies where the crura of the fornices approach each other beneath the splenium of the corpus callosum. It forms a commissure between the two hippocampi.

PROJECTION FIBRES

Many of the projection fibres have been mentioned already in their course from the cerebral cortex through the subjacent white matter into the internal capsule, crus cerebri, pyramid and corticospinal tracts. As indicated above, these can be divided into two main groups, and even though this division is artificial it is made here to stress the significance of the pathways by which the various parts of the brain are co-ordinated.

1. Projection fibres which pass to other grey masses of the central nervous system, presumably to interrelate their functions with those of the cerebral cortex. The cerebral cortex receives reciprocal connections from these areas through the thalamus. Such pathways are the corticostriate, corticothalamic, corticopontocerebellar, corticorubral, corticoreticular, and cortico-olivary.

2. Projection fibres which pass to the brain stem and spinal medulla and have their effects on the internuncial and motor cells of those levels, the corticonuclear and corticospinal tracts.

(i) The largest single group of projection fibres about which there is evidence in Man is the **corticopontocerebellar system**. Fibres enter this system from all the lobes of the brain, though not necessarily from all parts of each lobe. They pass through the internal capsule and the corresponding crus cerebri (where they form at least two thirds of the total fibre complement) into the ventral part of the pons. Here they synapse with the cells of the pontine nuclei which send their axons principally to the opposite cerebellar hemisphere.

There is very little evidence concerning the **corticostriate** fibres in Man, though it seems likely that they are arranged on the same principle as those in laboratory animals (Carman, Cowan, and Powell 1963) in which they converge radially on the striatum from the cerebral cortex. They may be separate fibres or collaterals of other fibres passing through the internal capsule. There is evidence that some at least are collaterals, and though their origin is not known it may well be that they are branches of the corticopontine, corticothalamic, or even corticospinal fibres. Although the origin of the corticostriate fibres is not known in Man, the position of the striatum relative to the internal capsule makes it not unlikely that the striatum receives fibres from the greater part of the cerebral cortex, though the large size of the head of the caudate nucleus and the anterior part of the putamen of the lentiform nucleus makes it likely that the majority of the fibres enter the striatum from the cortex of the frontal lobe.

Cortical projections to brain stem nuclei

Fibres have been described passing from the cerebral cortex to the subthalamic nucleus, the red nucleus, the pontine nuclei, the olivary nucleus, and to the reticular nuclei of the pons and medulla oblongata. The corticoreticular fibres end principally on those reticular nuclei which give rise to the reticulospinal tracts (Brodal 1957). It is not clear from which parts of the cerebral cortex these fibres arise, though it is likely that most of them originate in the posterior part of the frontal lobe. These connections form an important part of the efferent cortical pathway through which it can affect the brain stem mechanisms responsible for muscle tone. It is interesting that these projections correspond with those already described for the cerebellum and possibly the corpus striatum. It is probable that the impulses from these masses of grey matter can interact in the brain stem nuclei, in addition to the direct pathways which exist for interaction between the masses of grey matter themselves. FIGURE 10.176 summarizes the distribution of the ascending 'sensory' systems (other than the olfactory) to the major masses of grey matter in the brain. A similar figure could show that these masses are reciprocally interconnected and that each of them has the common feature of projecting to the reticular formation of the brain stem. The positions of the major descending pathways are indicated in FIGURES 10.138 and 10.177.

Descending fibres from the cerebral cortex also end in relation to many primary **sensory nuclei** in the brain stem and spinal medulla (Brodal, Szabo, and Torvik 1956; Walberg 1957a; Kuypers and Lawrence 1967). Following injury to the cerebral cortex in the region of the central sulcus (principally the postcentral gyrus; Powell 1977), degenerating nerve terminals have been described in the sensory nuclei of the trigeminal nerve, the nuclei cuneatus and gracilis, the nucleus of the tractus solitarius, and the posterior horns of the spinal medulla. Like the corticothalamic fibres these may be concerned with facilitating or inhibiting the passage of impulses through the sensory nuclei, thus affecting the information reaching the cerebral cortex, and the passage of impulses through the local internuncial neurons. This could form a mechanism whereby the cortex could facilitate or inhibit brain stem or spinal reflexes almost at their source, thus preventing inappropriate activity during voluntary movement.

(ii) The **corticospinal** and **corticonuclear** tracts [FIG. 10.178] enter the **internal capsule** and lie in the anterior part of its posterior limb, with the fibres from the lower part of the precentral gyrus

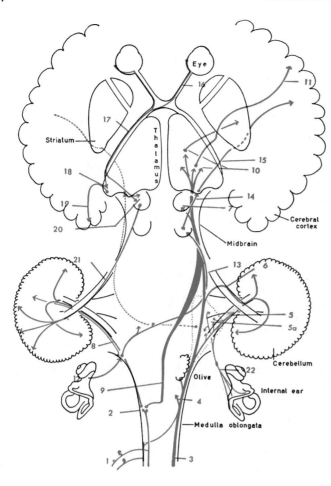

FIG. 10.176. Diagram of the 'sensory' pathways (other than olfactory) in the brain stem and cerebrum, seen from the dorsal aspect. The cerebellum has been split sagittally and each half moved to the side.

(1) Posterior root fibres (a) passing to nucleus cuneatus (2) in the fasciculus cuneatus, and (b) relaying in the posterior horn from which a fibre passes into the anterolateral ascending system (spinothalamic and spinocerebellar) of the spinal medulla (3). This gives off:

(4) Spinoreticular fibres.

(5) Posterior spinocerebellar tract fibres, some of which enter the vestibular nuclei (5a, broken outline) as spinovestibular fibres. Fibres from the vestibular nerve (22) and nuclei enter the cerebellum.

(6) Anterior spinocerebellar tract.

(7) Spinotectal fibres to superior colliculus.

(2) Nucleus cuneatus gives rise to posterior external arcuate fibres (8) and to the medial lemniscus (9). The medial lemniscus joins the spinothalamic fibres and is distributed to the thalamus (10), from which thalamic radiations pass to many regions including the cerebral cortex (11) and the striatum.

The cochlear nerve (12) ends on the cochlear nuclei, from which fibres pass to the cerebellum, and through the trapezoid body to the lateral lemniscus (13). The fibres in the lateral lemniscus reach the thalamus (medial geniculate body) via the brachium of the inferior colliculus (14) after relay in the inferior colliculus. The medial geniculate body gives rise to the auditory radiation (15) passing beneath the lentiform nucleus (in the sub-lentiform part of the internal capsule) to the superior temporal gyrus.

Fibres in the optic nerves (16) enter the optic tract (17) from the corresponding halves of the two retinae. The optic tract is distributed to the lateral geniculate body (18) and to the superior colliculus (20). The former sends fibres through the optic radiation (19) to the occipital lobe, the latter to the cerebellum (21) and brain stem.

anterior to those from its upper part. As they descend into the **crus cerebri**, the bundle turns almost through a right angle so that the anterior fibres take up a medial position in the central part of the crus with the remainder in a more lateral position. Entering the **ventral part of the pons** these fibres form part of the longitudinal bundles and gather together again at the lower border of the pons to emerge on to the ventral aspect of the medulla oblongata as the

pyramid. Throughout this part of their course, bundles are given off which cross the midline into the region of the motor nuclei of the cranial nerves (corticonuclear fibres). The available experimental evidence shows that most of these do not end on the motor cells, but pass to the internuncial neurons adjacent to them (Walberg 1957b; Kuypers 1958). Thus the fibres which pass to activate the oculomotor nucleus end in the tegmental reticular formation of the midbrain, in

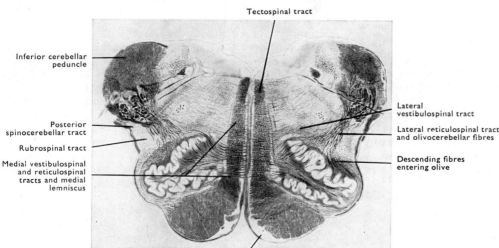

FIG. 10.177. To show the approximate areas occupied by the various long descending tracts in the medulla oblongata.

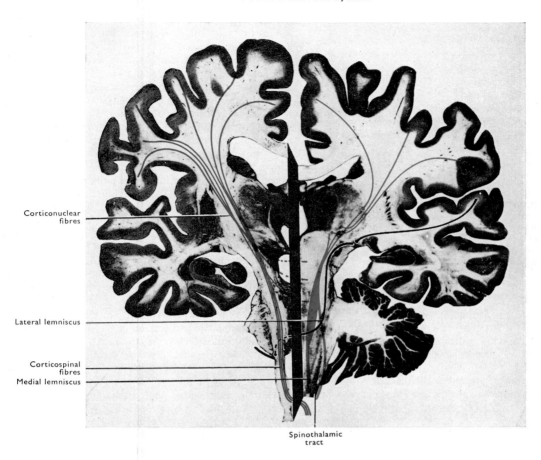

Corticonuclear
fibres

Lateral lemniscus

Corticospinal
fibres
Medial lemniscus

Spinothalamic
tract

FIG. 10.178. Coronal sections through two levels of one brain. Left half shows corticonuclear and corticospinal fibres and is anterior to right half showing main ascending 'sensory' systems.

common with fibres from the superior colliculus (tectoreticular tract) and probably other systems acting on the oculomotor nuclei. The corticonuclear fibres concerned with eye movements are activated from the frontal eye fields which produce conjugate movements of the eyes when stimulated. Similar movements can be produced by stimulation of the visual cortex, through fibres which traverse the visual radiation and brachium of the superior colliculus (Powell 1976) to act through the superior colliculi on the oculomotor, trochlear, and abducent nuclei and on other structures.

In the lower part of the medulla oblongata most of the fibres of both pyramids pass in large bundles into the white commissure and, interdigitating, cross to the opposite side, entering the lateral funiculi of the spinal medulla, to form the lateral corticospinal tracts. This pyramidal decussation is rarely complete and may be asymmetrical. A variable number of fibres do not cross in the decussation. Some of these enter the lateral corticospinal tract (Fulton and Sheehan 1935), while others descend in the anterior funiculus of the spinal medulla. In most cases this uncrossed, anterior corticospinal tract does not reach further caudally than the midthoracic region, most of its fibres crossing in the white commissure of the spinal medulla to enter the grey matter of the opposite side. Cases have been described where there was only a small pyramidal decussation and most of the pyramid passed into the anterior funiculus (Marie 1895). Whatever the arrangement at the pyramidal ducussation, the majority of the fibres in this system end on the opposite side of the brain stem and spinal medulla. Thus the cerebral cortex produces its effects predominantly on the opposite side of the body, and lesions of the cerebral cortex similarly produce contralateral disturbances.

The lateral corticospinal tract at first bulges medially into the grey matter of the first cervical spinal segment then sinks laterally into the lateral funiculus. In the second and third cervical segments it reaches the surface of the spinal medulla and displaces the spinocerebellar tracts ventrally [FIG. 10.143] (Smith 1957). Caudal to this it is more medially placed in the posterior part of the lateral funiculus, and gives off fibres to every level of the spinal medulla as it passes caudally. These fibres enter the intermediate part of the grey matter of the spinal medulla to end on the internuncial cells in the lateral part of the base of the posterior horn and the lateral intermediate substance [FIG. 10.179]. In most experimental animals, only a few of the fibres end directly on the motor cells of the anterior horn (Hoff and Hoff 1934), but these are in the minority. Thus the corticospinal system produces its effects principally through the same internuncial cells that are in use by the local spinal reflexes. It could, therefore, be said to activate or inhibit the spinal reflex mechanisms rather than the motor cells. Such connections persist in higher mammals, but in the primates an increasing number of corticospinal fibres end directly on the motor cells of the spinal medulla, particularly on those with axons passing to the distal muscles of the limbs (Kuypers 1960). Though the extent of these in Man is not known, it is likely that they are more numerous than in the apes: a feature in keeping with the greater involvement of the muscles of the peripheral parts of the limbs in the paralysis following a cerebral injury (e.g. a stroke).

The origin of the corticonuclear and corticospinal tracts (Walshe 1961) has been a source of controversy ever since Holmes and May (1909) demonstrated retrograde changes in the giant cells of Betz following injury to this descending system. It is natural that these

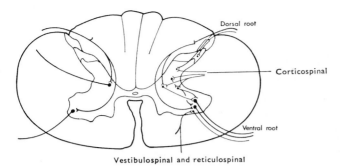

FIG. 10.179. Diagram to show the sites of termination of some of the major descending pathways in the spinal medulla. The corticospinal fibres to the motor cells are not shown.

Note that the corticospinal system ends predominantly in the posterior horn and intermediate substance. The single fibre shown representing the vestibulospinal and reticulospinal tracts, is indicated ending on a small motor cell, which is intended to indicate the source of a γ-efferent fibre.

cells should have been considered for many years to be the source of this system, since they had a structure and size akin to the motor cells of the spinal medulla, and lay in the excitable 'motor' cortex of the precentral gyrus. However, certain facts made it clear that there must be other cells of origin:

(a) Giant cells of Betz are not present in the inferior part of the precentral gyrus, though stimulation of this area causes movements of the head, neck, and upper limb which are abolished by section of the pyramids.

(b) The number of large diameter fibres (30 000) in the million which each pyramid contains is similar to the number of Betz cells in one hemisphere (Lassek 1954). It is assumed, therefore, that they are the origin of the large fibres and that the remainder arise from smaller cells. Many of these fibres arise in the precentral gyrus and in the frontal agranular cortex anterior to it (van Crevel 1958; van Crevel and Verhaart 1963), but there are some Betz cells in the postcentral gyrus which send descending fibres in the corticospinal system to sensory nuclei in the brain stem and spinal medulla (Powell 1977). In addition, a small number of fibres have been described as descending in the pyramid from other parts of the cerebral cortex, some even crossing in the corpus callosum before doing so (Walberg and Brodal 1953).

Ascending fibres have been described passing to the cerebral cortex through the lateral corticospinal tract, the pyramid, crus cerebri, and internal capsule, without relay in the thalamus (Brodal and Walberg 1952, in the cat; Nathan and Smith 1955b, in Man). These fibres arise in the contralateral brachial and lumbosacral regions, and the ipsilateral brachial region of the spinal medulla, and are distributed widely in the cerebral cortex, to agranular frontal cortex, and more posterior parts [FIG. 10.175]. The presence of these fibres has been confirmed by physiological methods in the cat (Brodal and Kaada 1953), and it has been shown that they transmit all varieties of ascending impulses from the spinal medulla.

Thus the pyramid is not a simple efferent pathway from the posterior part of the frontal lobe. The significance of the different groups of fibres which it contains has still to be assessed, and the presence of all of them in Man requires confirmation.

Threshold stimulation of the 'motor' cortex in Man has shown that there is a well-defined relationship between the parts of this cortex, and the site of movements produced. The arrangement is similar to that in the 'sensory', postcentral gyrus. Thus, the inferior part of the precentral gyrus is concerned with the face, tongue, chewing, swallowing, and laryngeal movements, while the neck, upper limb, trunk, lower limb, and perineal areas follow in sequence,

reaching the medially placed paracentral lobule superiorly. This has led to the belief that there is a mosaic of points in the precentral gyrus, each connected with a particular group of motor cells in the brain stem or spinal medulla. The anatomical evidence does not support such a simple view. (1) The majority of fibres which pass from the cerebral cortex to the brain stem and spinal medulla end on interneurons rather than motor cells. (2) Sherrington (1889) showed that localized lesions of the motor cortex led to degeneration of nerve fibres to all levels of the spinal medulla, though predominantly to a particular area—a finding which may be explained by the widespread involvement of muscles in any movement (e.g. stabilization of the lower limbs and trunk in arm movements). It may also suggest that the cerebral cortex is more concerned with movement complexes rather than the individual muscles, though this may simply be a matter of degree, e.g. the isolated contraction of orbicularis oculi in winking. None of this invalidates the clinical observation that injuries to the forebrain which involve the motor cortex and/or the corticospinal system produce a loss of voluntary movement (paralysis) in that part of the opposite side of the body corresponding to the damaged area ('upper motor neuron lesion'). In the paralysed area there is usually an increase of muscle tone (spastic paralysis) and exaggerated myotatic reflexes (e.g. knee and ankle jerks), but only the limited muscle wasting of disuse. On the other hand, injuries to the motor cells innervating particular muscles ('lower motor neuron lesions') produce a flaccid paralysis of muscles on the same side with rapid severe wasting, loss of all types of reflexes involving the paralysed muscles, and sometimes irregular contractions of bundles of fibres in the muscles (fasciculation). It should be appreciated that apparent paralysis may arise from sensory loss in the same manner as a reflex is abolished by the destruction of the afferent fibres. Thus a monkey with all the dorsal roots of the spinal nerves to one upper limb divided does not use the limb even though the corticospinal system and the spinal medulla are intact. The significance of localization in the precentral gyrus is discussed critically by Walshe (1948) and Clark (1948).

Stimulation of the cerebral cortex has uncovered the presence of two regions, adjacent to, but outside the classical motor area, from which motor responses can be elicited in a point-to-point fashion on threshold stimulation. The supplementary 'motor' area lies on the medial aspect of the cortex, antero-inferior to the leg region of the 'motor' area, while the second 'motor' area lies anterior to the second sensory area [p. 700] extending towards the insula from the motor area. It is difficult to estimate the significance of these and other 'motor' areas, though injury of the supplementary motor area in the monkey causes hypertonia (Travis 1955). The term 'motor' must be used with caution in this situation since the production of movement by stimulation of a particular cortical area does not make that area 'motor', any more than the dorsal (sensory) roots of the spinal nerves are 'motor' because stimulation of them causes movements. The cells of the cortex are interneurons, so the results of their stimulation depend on the nature and timing of the stimulus and on the activity of the cortex existing at the time of stimulation. In keeping with this is the fact that subliminal stimulation of the 'motor' cortex, in unanaesthetized monkeys, produces relaxation when applied to the same point from which a stronger stimulus would produce motor effects (Clark and Ward 1948).

Damage to the corticospinal system in Man commonly occurs as a result of injuries to the internal capsule, when a number of different pathways are involved, often including the sensory radiations. Such a lesion produces contralateral paralysis of voluntary movement which involves the peripheral parts of the limbs more than their proximal parts, and has little or no effect on the trunk. Voluntary movements normally produced by the activation of certain cranial nerves are affected, e.g. the hypoglossal, producing tongue paralysis on the opposite side; the nucleus ambiguus producing contralateral laryngeal and pharyngeal paralysis; part of the facial nucleus, resulting in paralysis of the contralateral facial muscles from the orbicularis oculi inferiorly, but not of the occipitofrontalis. The paralysis of voluntary movement affecting some muscle groups more than others, has been ascribed to a bilateral supply of cortical fibres to some motor nuclei, and not to others. The corticospinal system, however, has direct connections mainly with the motor cells innervating the distal muscles of the limbs, while other descending tracts (reticulospinal, vestibulospinal, tectospinal) end principally on those innervating the trunk and proximal limb muscles. The sparing of the latter in injury to the corticospinal system may be a measure of the extent to which their control is by these other systems. Nevertheless, the ending of the corticospinal system also on interneurons would allow it to affect motor cells on both sides of the brain stem and spinal medulla—a feature which may well be necessary for those movements in which the corresponding muscle groups of the opposite side are also involved. Thus movements of the trunk invariably use the muscles of both sides, while the movements of one hand may be entirely independent of the other. The absence of any commissural fibres between the parts of the 'sensory' cerebral cortex which receive impulses from the distal parts of the limbs (Jones and Powell 1968) may have a bearing on this. However, the same absence of commissural fibres is seen in the visual cortex which receives impulses from the highly co-ordinated eyes. It is interesting that animals with little or no independent movement of the distal parts of the limbs tend to have a very restricted corticospinal system which, for example, does not extend beyond the upper cervical region in the sheep.

In addition to the paralysis, lesions of the corticospinal system are associated with loss of the abdominal reflexes, and an alteration of the plantar reflexes. The former consist of contraction of the underlying abdominal muscles when the abdominal skin is lightly scratched. In the latter, the flexion of all the toes which normally follows scratching the outer side of the sole of the foot is reversed— the big toe being extended and the other toes flexed and spread, the **Babinski response**. It is possible that the ascending spinocortical fibres in the pyramid form the sensory pathway for these reflexes.

Fibres passing from the cerebral cortex to other parts of the brain have already been described, e.g. to corpus striatum, tectum of the midbrain, red nucleus, cerebellum, and reticular formation, through which the cerebral cortex can have indirect effects on the motor cells of the brain stem, and spinal medulla. These pathways form part of an extensive system, known as the **extrapyramidal system**, because the pathways involved do not pass through the pyramid in the medulla oblongata. It is concerned with complex movements of a more automatic nature, such as the maintenance of balance during movements, the correct distribution of muscle tone necessary for normal movements, etc. The long descending pathways of the extrapyramidal system are the reticulospinal, the vestibulospinal, the rubrospinal, and the tectospinal tracts. These are almost certainly reinforced by many other paths consisting of chains of short neurons, such as those which connect the hypothalamus and subthalamus with the reticular formation. The extrapyramidal system is an extremely complex apparatus and the method by which its numerous parts interact is not clear. It is characterized by multiple feedback systems linking its several parts. The major parts of this system have been described, but it should be realized that the separation of the pyramidal and extrapyramidal systems is completely artificial, and that the two systems are complementary parts of the normal nervous system.

CEREBRAL DOMINANCE

The two cerebral hemispheres have been treated as though they were simple mirror images of each other. This is true of their general

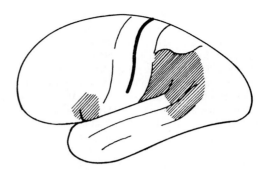

FIG. 10.180. Diagram of the primary speech areas in the dominant hemisphere. Redrawn from Rasmussen and Milner (1975).

arrangement and of their connections so far as these are known, but they are no more symmetrical than the two halves of the face. More significant than this gross asymmetry is the functional inequality of the two hemispheres in relation to speech and in the usual preference (or dominance) for the use of the right hand (controlled by the left hemisphere) over the left (controlled by the right hemisphere) in most individuals.

In 1861 Broca stated that injuries to the posterior part of the *left* inferior frontal gyrus (opercular part or Broca's area [FIG. 10.96] anterior to the head and larynx parts of the precentral gyrus) led to an interference with the execution of speech (motor aphasia) which did not occur after injuries to the same area in the right hemisphere. This is different from the interference produced by damage to the adjacent motor cortex on either side. The 'dominance' of the left hemisphere in speech and handedness in most individuals has been amply confirmed and extended in the case of speech to include parietal and temporal areas of the cerebral cortex [FIG. 10.180; Rasmussen and Milner 1975]. The speech dominance of the left hemisphere can be demonstrated by injection of sodium amytal into the left internal carotid artery in the conscious patient. This grossly interferes with speech, whereas there is no such defect when the opposite artery is injected in the majority of individuals. This and other methods have shown that both hemispheres may be partly involved in speech in some individuals (as they may in hand movements of the ambidextrous) and that transfer of handedness to the right hemisphere (left handed individuals) is not necessarily associated with right cerebral dominance in relation to speech (Zangwill 1960). It appears that left cerebral dominance is an acquired characteristic, for speech defects arising from damage to the left hemisphere in children under the age of 8 years usually recover. There is then evidence of transfer of speech mechanisms to the right hemisphere, though in older individuals only limited, if any, transfer occurs, and recovery is partial or absent.

Some experts believe that the term 'cerebral dominance' is inappropriate. They prefer 'cerebral specialization' because they believe that the two hemispheres contribute different elements in the activities of the cerebral cortex—the two halves being integrated by commissural fibres principally in the corpus callosum. Thus the dominant hemisphere is looked on as 'executive', while the non-dominant is considered the 'conceptual' part of the mechanism. This is almost certainly too simple an interpretation to be generally applicable, but injuries to the inferior partietal lobule of the non-dominant hemisphere interfere with concepts such as those of the body image and of surrounding space, while injuries to the dominant hemisphere interfere with the expression of concepts (e.g. by speech; see also Sperry 1974).

As yet too little is known of the cerebral cortex to do more than indicate some of the areas involved in particular processes, e.g. speech areas. However, a useful if simplistic attitude to cerebral function in relation to speech arises from noting that it may be a response to the spoken word (auditory input), the written word or lip reading (visual input), or to general sensory input (e.g. reading Braille, or the sensation of speaking such as that obtained by a child who, having learnt to speak entirely from an auditory input, begins to read by doing so out loud, then progresses to a soundless mouthing of the words (general sensory input), and eventually abandons these inputs for a purely visual one). It is then possible to relate failure of the speech mechanism (aphasia) to: (1) a breakdown of the input mechanism, e.g. deafness; (2) a failure to understand the input; (3) an inability to formulate a proper response; (4) an incapacity to activate the mechanism which performs the response; (5) paralysis of the muscles of the larynx etc. In (1) there is no response to the appropriate stimulus; in (2) there is an inappropriate but correctly performed response; in (3) a meaningless jumble of words; in (4) a series of sounds which may bear little relation to speech; in (5) the individual is mute if the paralysis is total.

A speech response may be triggered by one of many inputs. Hence it may be reasonable to expect these inputs to be integrated in a part of the cortex surrounded by the receptive areas—the occipital lobe, the superior temporal gyrus, and the postcentral gyrus—and receiving association fibres directly or indirectly from them. Such an area is the inferior parietal lobule and the adjacent temporal lobe [FIG. 10.180]. Clearly the response generated by such an integrating area to any of the inputs may not be speech but another motor activity (e.g. the production of a written reply) which would involve a different output pathway. Damage to the integrating apparatus might therefore be expected to involve other responses, and writing is often disturbed in a manner similar to speech. In injuries involving the input apparatus, the individual may fail to respond to one input (e.g. the word of command, yet still understand and react to another (e.g. the written command). In involvement of the executive apparatus, the individual may fail to make a response by one mechanism (e.g. speech) while able to do so in another way, e.g. by writing. Such oversimplifications of the complexities of cerebral cortical activities serve no other purpose than that of indicating a general concept.

The meninges

The meninges are three layers of fibrous tissue which enclose the central nervous system. The outer layer is the thick, fibrous **dura mater**, or pachymeninx. It is separated from the middle layer—the thin, transparent **arachnoid**—by a capillary interval, the **subdural space**. The apposed surfaces of the arachnoid and dura mater are covered with layers of flattened meosthelial cells which permit these meninges to slide freely on each other. Thus the subdural space forms a large bursa surrounding the central nervous system. The **pia mater** is the inner meningeal layer. It immediately surrounds the central nervous system and follows its every contour. The pia mater is separated from the arachnoid by the **subarachnoid space**.

The arachnoid and pia mater (leptomeninges) are believed to develop from the neural crest (Horstadius 1950) but are certainly formed from a single mass of loose tissue. Cerebrospinal fluid escaping from the apertures in the fourth ventricle percolates through this tissue forming and filling the subarachnoid space by splitting the tissue into an outer layer (arachnoid) applied to the dura mater and an inner layer (pia mater) applied to the central nervous system, while leaving a variable number of strands (trabeculae) uniting them across the subarachnoid space. The depth of the subarachnoid space varies with the closeness of fit of the dura mater to the brain. Where they are widely separated, the difference is taken up by the subarachnoid space which then forms deep spaces or **cisterns**. Trabeculae are then few in number, though where the

space is shallow the number of trabeculae is much greater, e.g. over the surfaces of the cerebral hemispheres.

Every structure which enters or leaves the brain or spinal medulla must traverse the subarachnoid and subdural spaces and pierce all three meninges. Where they do so, the continuity of the subdural space is interrupted and the arachnoid is bound to the dura by the structure traversing both. This interferes with the sliding of the brain, together with the pia mater and arachnoid, on the dura in violent movements of the head. Where the thin-walled veins of the central nervous system cross the subdural space they are likely to be torn by sudden movements of the brain, leading to a slow venous leakage of blood into the subdural space which produces gradually increasing compression of the brain—*subdural haemorrhage*. This should be compared with rupture of the arteries on the pial surface of the brain (*subarachnoid haemorrhage*) or within the brain (*cerebral haemorrhage*) and with *extradural haemorrhage* due to injuries involving the 'meningeal' arteries of the skull. All these are sudden dramatic events because of the rapid leakage of blood from a high pressure system.

The blood vessels of the central nervous system run on the external surface of the pia mater, usually anastomosing freely with each other. End-artery branches of these vessels pierce the brain or spinal medulla and carry a sleeve of pia mater and an extension of the subarachnoid space (**perivascular space**) for a short distance into the nervous tissue.

The meninges, as protective coverings of the central nervous system, have two entirely different mechanical problems to overcome:

1. The protection of the relatively firm spinal medulla within the moving vertebral column, by slinging it in a fluid-filled tube, protected externally by fat and extensive venous plexuses.

2. The support and protection of the voluminous, soft brain within the skull is achieved by the formation of rigid compartments within the skull and a special fluid-filled 'sponge' immediately surrounding the brain.

THE SPINAL MENINGES

The spinal dura mater forms a long tubular sheath extending from the margin of the foramen magnum to the upper part of the sacrum. It is not of uniform width throughout, being widest in the cervical region and narrowest in the thoracic region where the sac is closest to the spinal medulla. Throughout the length of the vertebral canal it is separated from the periosteum lining the bony elements of the vertebral canal by a variable amount of soft fat and a considerable number of venous spaces, the **internal vertebral venous plexus**. The fat and complex venous plexuses are found mostly over its dorsal surface forming a buffer between it and the moving laminae, and the large extradural veins form valveless channels on either side extending the length of the vertebral canal. These communicate with the intersegmental veins through the intervertebral foramina, and with the basivertebral veins anterior to the posterior longitudinal ligament. Principally in the thoracic region, the dural sac is lightly attached to the posterior longitudinal ligament by fine collagenous strands which hold the sac close to the posterior surface of the vertebral bodies and thus to the axes of vertebral movements.

Throughout its length the spinal dura mater is lined by arachnoid while the pia mater forms a thick, tough membrane on the surface of the spinal medulla. It is thickened into a glistening ribbon, the **linea splendens**, which runs anterior to the anterior median fissure and the anterior spinal artery. The pia mater also forms a linear flange projecting from each side of the spinal medulla between the ventral and dorsal roots of the spinal nerves. The thickened free edge of each of these flanges forms a series of sharp, lateral projections, each attached through the arachnoid to the dura mater

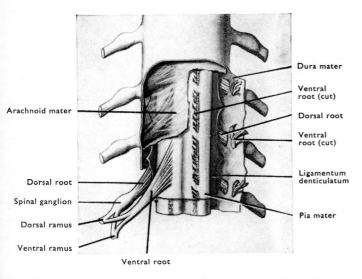

FIG. 10.181. Membranes of spinal medulla and mode of origin of spinal nerves. The lowest spinal nerve on the left is shown semidiagrammatically dissected, to indicate the participation of both roots in the formation of each ramus.

between the points where two adjacent spinal nerves pierce the dura mater. Hence each flange has a scalloped or saw-toothed margin, and is known as a ligamentum denticulatum [FIG. 10.181]. These ligaments suspend the spinal medulla within the spinal subarachnoid space at some twenty to twenty-two points on each side, and permit a considerable range of movement of the dura mater without a corresponding movement of the spinal medulla. The spinal subarachnoid space is relatively wide with only a few delicate trabeculae uniting the pia and arachnoid on the posterior surface.

The spinal medulla ends immediately below the first lumbar vertebra, but the dural and arachnoid sheaths extend inferiorly enclosing the cauda equina and extending the subarachnoid space to the second part of the sacrum. Here they close down on the filum terminale (the thread-like continuation of the spinal medulla and its covering of pia mater) and ensheathing it, pass to an attachment on the dorsal surface of the coccyx, thus holding down the caudal part of the dural sheath [FIGS. 10.25 and 10.26].

At its cranial end, the dural sheath is more firmly attached to the second and third cervical vertebrae and at the foramen magnum becomes continuous with the cranial dura mater which is fused with the periosteum lining the skull. Here the ligamentum denticulatum continues into the cranial cavity just anterior to the spinal accessory nerve and is attached to the dura posterior to the hypoglossal canal.

THE CRANIAL MENINGES

The cranial dura mater

This is closely applied to the interior of the skull without any intervening fat, and is fused with the periosteum except where the venous sinuses of the dura mater intervene and where the dura is folded inwards between the parts of the brain. The combined periosteum and dura mater, sometimes erroneously called the outer and inner layers of the dura mater, are less firmly adherent over the cranial vault than they are at the base of the brain, where both pass through the various foramina, the one to become continuous with the periosteum on the external surface of the skull, the other with the epineurium of the nerves. The venous sinuses of the dura mater are endothelial tubes lying in the same layer as the internal vertebral venous plexus (between the dura and the periosteum) and

are continuous with it through the foramen magnum. Like the internal vertebral venous plexus, the venous sinuses of the dura mater communicate:

1. With veins outside the skull:
   (i) Through the various foramina for the cranial nerves.
   (ii) Through emissary foramina.
2. With the diploic veins in the cranial bones.

The method of growth of the brain, whereby its three main masses, two cerebral hemispheres and the cerebellum, expand separately into the surrounding tissues, results in the formation of three separate pockets in that tissue, pockets which communicate with each other in the plane of the original brain tube. In this fashion an incomplete septum of tissue, the falx cerebri, is left between the two hemispheres and a similar septum, the tentorium cerebelli, lies between the cerebellum below and the occipital lobes of the hemispheres above. These two septa meet in a λ-shaped arrangement in the midline posteriorly where the two occipital lobes and the cerebellum come closest together. A similar but much smaller fold (falx cerebelli) extends anteriorly from the internal occipital crest between the cerebellar hemispheres; another, the diaphragma sellae, roofs over the hypophysis in the sella turcica. Each of these septa is in the form of a fold of dura mater with the arachnoid applied to its cerebral or cerebellar surface. *In situ*, the falx cerebri and tentorium cerebelli form rigid structures which act as baffles within the cranial cavity, helping to prevent movements of the brain within the skull.

The falx cerebri [FIGS. 10.182 and 10.183] is a median, sickle-shaped structure attached to the cerebral surface of the cranial vault from the crista galli of the ethmoid to the internal occipital protuberance. It passes into the longitudinal fissure between the hemispheres, increasing in depth from before backwards so that posteriorly its free margin approaches the corpus callosum (splenium) and joins the anterior extremity (apex) of the tentorium cerebelli [FIG. 10.183]. The superior sagittal sinus lies in the fixed border enclosed between the diverging surfaces of the falx and the periosteum, the lateral lacunae of the sinus separating the dura and periosteum further laterally. The inferior sagittal sinus lies between the layers of the free border of the falx, and is joined by the great cerebral vein where that border meets the tentorium. The straight sinus, so formed, passes postero-inferiorly in the line of junction of the tentorium and the falx to the internal occipital protuberance. The anterior part of the falx, and the dura mater of the superior sagittal sinus and its lacunae are perforated or cribriform. The cribriform parts of the dura mater allow the superior cerebral veins and protrusions of the arachnoid (arachnoid villi and granulations; p. 731) to pass through it and come into association with the endothelial lining of the sinus and its lacunae.

The tentorium cerebelli [FIGS. 10.182 and 10.183] is a large, crescentic, transversely placed fold of dura mater which roofs the posterior cranial fossa and is shaped like a tent, the apex of which lies at its meeting with the free edge of the falx cerebri, postero-superior to the splenium of the corpus callosum. It is accurately fitted to the superior surface of the cerebellum and every part of it slopes inferiorly from its apex towards its fixed margin. The fixed margin passes posterolaterally from the posterior clinoid process of the sphenoid bone, along the superior margin of the petrous temporal bone and the superior petrosal sinus. Passing over the mastoid part of the temporal bone to the postero-inferior angle of the parietal bone, it straddles the broad groove formed by the transverse sinus, and is pierced from above by inferior anastomotic and other inferior cerebral veins passing to the transverse sinus. The fixed margin then sweeps medially along the transverse sinus to the internal occipital protuberance where it meets the fixed margin of the falx cerebri. The free margin forms the anterosuperior border of the tentorium and clasps the midbrain, thus forming the tentorial

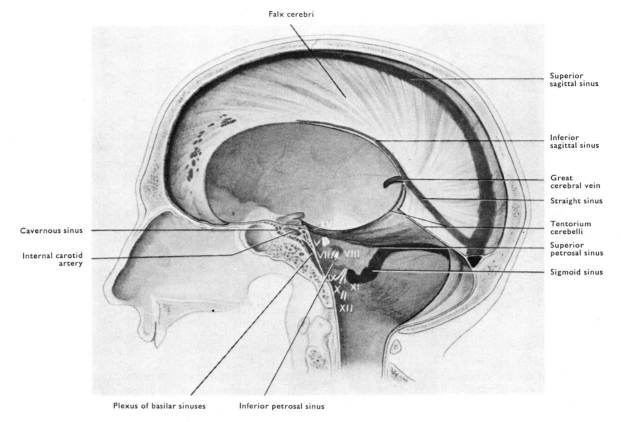

FIG. 10.182. Sagittal section of skull, a little to the left of median plane, to show arrangement of dura mater. Cranial nerves are indicated by numerals.

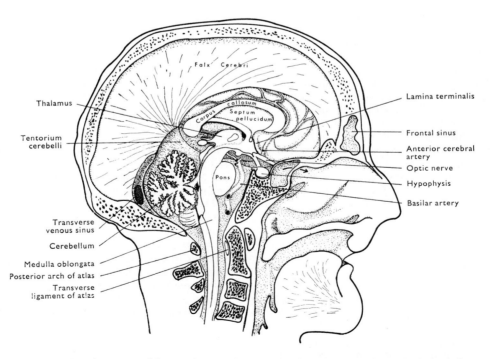

FIG. 10.183. Drawing prepared from a photograph of a section (supplied by Professor R. Walmsley) to show the relation between the skull, meninges, and brain. The brain stem is slightly displaced posteriorly, the pons normally lies closer to the clivus.

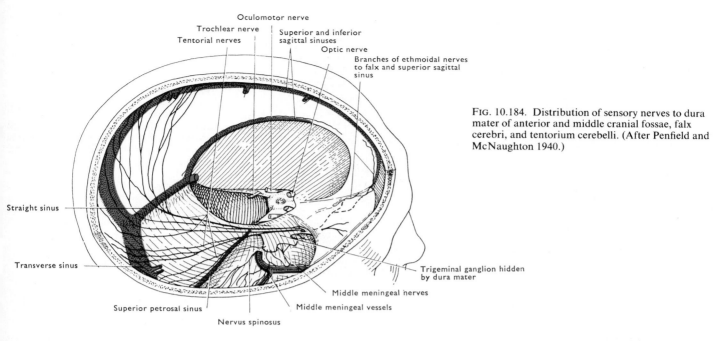

FIG. 10.184. Distribution of sensory nerves to dura mater of anterior and middle cranial fossae, falx cerebri, and tentorium cerebelli. (After Penfield and McNaughton 1940.)

notch. The free magin is attached antero-inferiorly to the anterior clinoid processes from which it passes posteriorly as a slightly raised ridge, lateral to the oculomotor nerve, to cross the fixed margin of the tentorium lateral to the posterior clinoid process. Above the anterior part of the free edge lies the medial side of the temporal lobe, and further posteriorly the posterior cerebral artery, while below this margin lies the trochlear nerve and the superior cerebellar artery. This relation of the temporal lobe to the tentorium results in the uncus being displaced downwards medial to it when there is any space-occupying lesion in the supratentorial compartment of the skull.

The innervation of the dura mater

The cranial dura mater is innervated by branches of the trigeminal, vagus, and by sympathetic nerves on the meningeal arteries which supply the dura mater and skull but play no part in the vascular supply to the brain. The anterior cranial fossa and the anterior part of the falx cerebri are supplied by the anterior (and posterior) ethmoidal nerves, passing from the cribriform plates of the ethmoid. The middle cranial fossa is supplied by branches of the maxillary and mandibular divisions of the trigeminal nerves, one of which (the nervus spinosus) is a recurrent branch of the mandibular nerve entering by the foramen spinosum. The tentorium and the posterior part of the falx cerebri are supplied by recurrent tentorial branches of the ophthalmic divisions of the trigeminal nerves, which run backwards near the free margin of the tentorium [FIG. 10.184]. The posterior cranial fossa receives a recurrent meningeal branch of the vagus which enters the skull through the jugular foramen.

The spinal dura mater is supplied by the recurrent meningeal branches of the spinal nerves.

The cranial arachnoid mater

This differs from the spinal arachnoid in two ways:

1. On the cerebral and cerebellar hemispheres it is connected by a dense meshwork of processes to the pia mater. The interstices of this mesh are filled with cerebrospinal fluid so that it forms a sort of fluid-filled sponge which may act as a cushion helping to prevent injury to the brain when the head is moved suddenly.

2. Arachnoid villi and granulations [FIGS. 10.185 and 10.186]. Along the line of the superior sagittal sinus, where the venous pressure is at its lowest, finger-like prolongations of the arachnoid pass through the fenestrations in the dura and project into the sinus and its lateral lacunae. These arachnoid villi are present and very small in the new-born but enlarge progressively throughout life until in elderly individuals they may be so large as to indent the overlying bone (foveolae granulares). The villi and granulations are covered on their vascular surfaces by the endothelium of the sinus which becomes continuous with capillary-like tubules leading through these structures to the subarachnoid space. In the villi and granulations the tubules are surrounded by the subarachnoid meshwork which becomes distended with cerebrospinal fluid when the pressure of that fluid is greater than the pressure in the sinus. The strands of the meshwork hold open the tubules which discharge cerebrospinal fluid directly into the sinus or its lacunae. When the venous pressure exceeds that of the cerebrospinal fluid, the villi or granulations collapse, the tubules are closed, and blood cannot pass into the subarachnoid space (Jayatilaka 1965). Thus these structures are valvular openings from the subarachnoid space into the venous system, and they prevent increases in intracranial pressure. This arrangement explains the great increase in intracranial pressure which follows clotting (thrombosis) of the blood in the superior sagittal sinus.

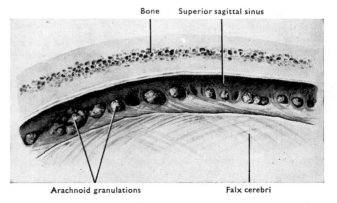

FIG. 10.185. Median section through cranial vault in parietal region (enlarged). Displays a portion of the superior sagittal sinus and the arachnoid granulations protruding into it.

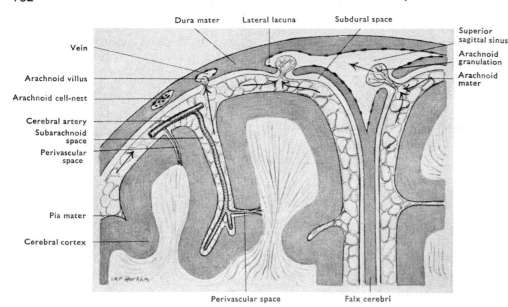

FIG. 10.186. Diagram to show relation of meninges to brain and of subarachnoid space to arteries, neuronal elements, and venous channels of dura mater. The necks of the villi are fused with the dura mater, and there is no continuity between the venous sinus and the subdural space.

The cranial subarachnoid space

Because the brain and skull show marked differences in shape, there are several situations where the subarachnoid space is deep, forming a series of subarachnoid cisterns within the skull. These are found particularly on the base of the brain, and in them the trabeculae joining arachnoid to pia mater are very much reduced so that the arachnoid can easily be lifted away from the pia.

The largest of the cisterns is the **cerebellomedullary** which lies betweeen the cerebellum and medulla oblongata, behind the medulla oblongata. It can readily be entered by a needle introduced immediately below the posterior margin of the foramen magnum and passed upwards (**cisternal puncture**). Deep in its uppermost part lies the median aperture of the fourth ventricle. It is continuous inferiorly with the spinal subarachnoid space, and around the sides of the medulla oblongata with a shallower but extensive cistern surrounding the pons and medulla oblongata. This pontine cistern is continuous anteriorly on the base of the brain with a deep **interpeduncular cistern**, formed by the arachnoid mater uniting the temporal lobes immediately above the diaphragma sellae. The interpeduncular cistern contains the arterial circle of the base of the brain, and extensions of this cistern follow the main branches of this circle:

1. With the posterior cerebral arteries around the superior part of the midbrain at the margin of the tentorium, cisterna ambiens.

2. With the middle cerebral artery into the lateral sulcus and its branches, the **cistern of the lateral fossa**.

3. With the anterior cerebral arteries, surrounding the superior and inferior surfaces of the optic chiasma, the **cisterna chiasmatis**. Within the lateral sulcus the middle cerebral artery is so closely surrounded by brain tissue that, it is believed, the pulsations of the artery drive the cerebrospinal fluid laterally from the interpeduncular fossa over the lateral surface of the hemisphere. Possibly the other major arteries have similar effects in causing directional flow of cerebrospinal fluid in the subarachnoid space.

The formation and circulation of cerebrospinal fluid

The bulk of the cerebrospinal fluid is formed in the choroid plexus of the lateral ventricles and in lesser amounts in the third and fourth ventricles. It is probably also produced by ependyma throughout the ventricular system and some may even be derived from the capillaries on the surface of the brain and spinal medulla. It is a clear, protein-free fluid which contains two or three lymphocytes per mm^3 and is in osmotic equilibrium with blood plasma, principally because of a higher concentration of chlorides. The normal course of the fluid is from the lateral ventricles through the interventricular foramen to the third ventricle, thence through the midbrain in the **cerebral aqueduct** to the fourth ventricle. Here the

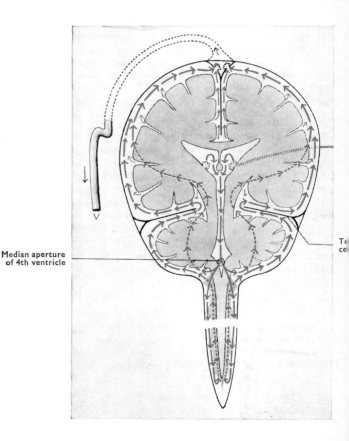

FIG. 10.187. Diagram to show the main points relating to the production, course, and absorption of the cerebrospinal fluid. (After N. M. Dott (1928). *Edinb. med. J.* **35**, 182.)

cerebrospinal fluid escapes through the median and the two lateral apertures [p. 629] into the subarachnoid space, entering the cerebello-medullary and pontine cisterns respectively. From this site the fluid may pass downwards around the spinal medulla, but most of it flows slowly through the tentorial notch and spreads over the cerebral hemispheres, partly aided by the active pulsatisons of the cerebral arteries, to reach the arachnoid granulations where it can pass into the venous system [FIG. 10.187]. Absorption has been shown to occur into the perineural lymph vessels and the veins on the spinal nerves. In particular, cerebrospinal fluid passes into:

1. The orbital lymph vessels, by passing along the subarachnoid sheath which surrounds the optic nerve to the lamina cribrosa of the sclera.

2. The lymph vessels of the nose from the region of the olfactory bulb, also along the vestibulocochlear, facial, and spinal nerves, along which some is also passed into the veins. At one time it was thought that every branch of the arteries in the subarachnoid space carried a sleeve of the space with it to its terminal ramifications in the brain, and that such spaces communicated with perineuronal spaces in the brain. This is now known not to be the case, since no such spaces are visible under the electron microscope and the perivascular spaces only accompany the larger vessels for a short distance as they pierce the nervous system.

The pia mater of the brain

In most situations this is a much more delicate membrane than its spinal counterpart, though over the medulla oblongata it is thick and firmly adherent. Elsewhere it forms the intimate covering of the brain, dipping into every sulcus. In certain places it lies in apposition with the ependymal lining of the brain, i.e. in the roofs of the fourth and third ventricles and in the medial walls of the lateral ventricles. Here it is known as the **tela choroidea**, and forms the vascular connective tissue of the choroid plexuses, being invaginated into the ventricle with a covering of ependyma in these situations [pp. 628 and 683].

The relation of the meninges to the nerves [FIG. 10.189]

In the spinal nerves the dura mater is prolonged separately over the ventral and dorsal roots to fuse with the epineurium at the distal part of the spinal ganglion. The arachnoid also forms a separate sheath for both roots, the subarachnoid space extending to the proximal part of the spinal ganglion. It is in this region that reabsorption of cerebrospinal fluid is believed to occur. Exactly the same arrangement is present in the cranial nerves, the dura fusing with the distal part of the ganglia and the arachnoid with their proximal parts. The arrangement around the trigeminal ganglion is the same as that of the spinal ganglia, only in this case the dura mater is folded over it, forming the floor of the middle cranial fossa, so that the cavum trigeminale has two layers of dura mater superior to it, but is otherwise the same as the space surrounding any spinal ganglion. The pia mater is continuous with the perineurium.

Craniocerebral topography

The relationship between the brain and the internal cranial base is readily visualized. In the midline [FIG. 10.183] the medulla and pons lie on the clivus with the upper border of the pons reaching almost to the dorsum sellae and its anterior surface separated from the bone by the basilar venous plexus [FIG. 13.100], the meninges, and the basilar artery in the subarachnoid space. The optic chiasma does

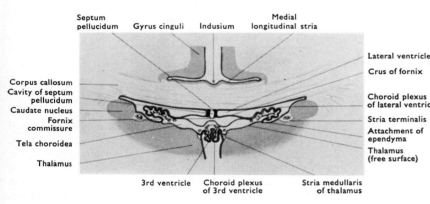

FIG. 10.188. Diagram of coronal section through tela choroidea of third ventricle.

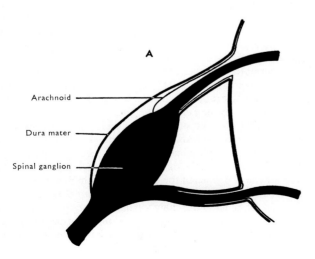

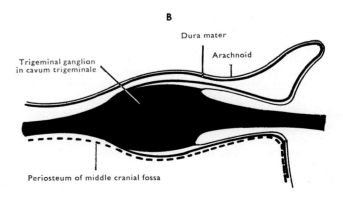

FIG. 10.189. Diagram to indicate the meningeal relation of A, a spinal nerve, and B, the trigeminal ganglion in the cavum trigeminale.

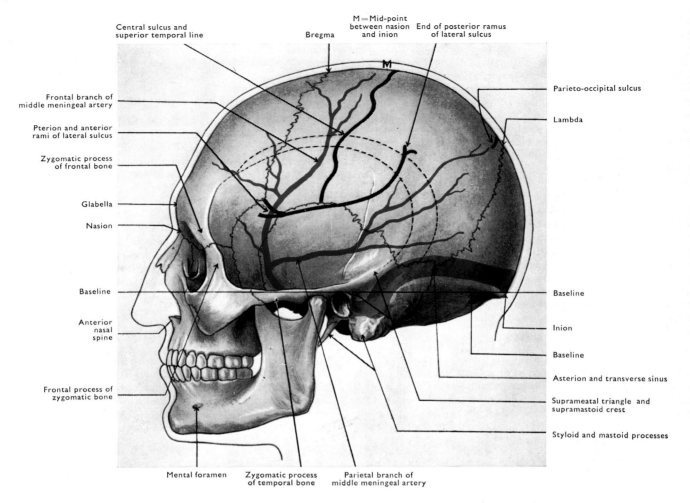

FIG. 10.190. Craniocerebral topography: landmarks of skull; chief sulci of cerebrum; transverse and sigmoid venous sinuses; middle meningeal artery.

not lie in the sulcus chiasmatis of the sphenoid bone but considerably posterior to this, almost vertically above the hypophysis. From the chiasma the optic nerves pass forwards and laterally, above the hypophysis in the sella turcica and below the corresponding anterior cerebral artery, to the optic canal immediately lateral to the sphenoidal air sinus. In the anterior cranial fossa lies the crista galli—the attachment of the falx cerebri to the ethmoid bone. Immediately on either side of this, the olfactory bulb lies on the cribriform lamina of the ethmoid which, together with the dura mater and nasal epithelium, separate it from the nasal cavity. Further laterally the orbital surface of the frontal lobe lies on the dura covering the orbital part of the frontal bone and the lesser wing of the sphenoid. The orbital part of the frontal bone separates the brain from the ethmoidal air sinuses medially and the orbit laterally. It is a thin plate of bone which is closely fitted to the surface of the frontal lobe and is marked with ridges and impresssions corresponding to the sucli and gyri. Posteriorly it is continuous with the lesser wing of the sphenoid, the sharp concave posterior edge of which is inserted into the lateral sulcus and overlies the extreme anterior part of the temporal lobe. The floor of the anterior cranial fossa is very thin throughout, except for the occasional extension of the frontal air sinus posteriorly between the inner and outer tables of the orbital part of the frontal bone. Because it is thin, the floor of the anterior cranial fossa is not uncommonly fractured in injuries to the skull. If the adherent dura is torn, blood and cerebrospinal fluid then may escape into the nose or orbit, or a communication may be established between the cranial cavity and the ethmoidal or frontal air sinuses.

The inferior surfaces of the temporal and occipital lobes, which cannot be separated from each other on this surface, rest on the greater wing of the sphenoid [FIG. 3.53] (separated from it by the middle meningeal artery and veins curving anterolaterally towards the pterion), the inferomedial part of the squamous temporal bone and the petrous temporal bone. On the latter it is separated from the middle ear by the thin tegmen tympani which is cartilaginous in early childhood. Posterior to the temporal bone the inferior surface is separated from the cerebellum by the tentorium cerebelli.

The superolateral surface of the brain [FIGS. 10.190 and 10.191] is related to the medial wall of the temporal fossa and the cranial vault and, over a large part of this surface, to the branches of the middle meningeal artery in the periosteum [FIG. 10.192]. The relationship of individual parts of the superolateral surface to this region of the skull is remarkably variable in different skull types, and exact positioning of any given sulcus from the surface is not possible. The approximate position of the main sulci and gyri is important, so that an estimate may be given of the part of the brain likely to be injured in direct blows producing depressed fractures, or to assist in localization of a subdural haemorrhage if signs of local brain damage are present. Much more detailed information is required in modern neurosurgery when it may be necessary to injure particular parts of the thalamus or corpus striatum by the introduction of a fine

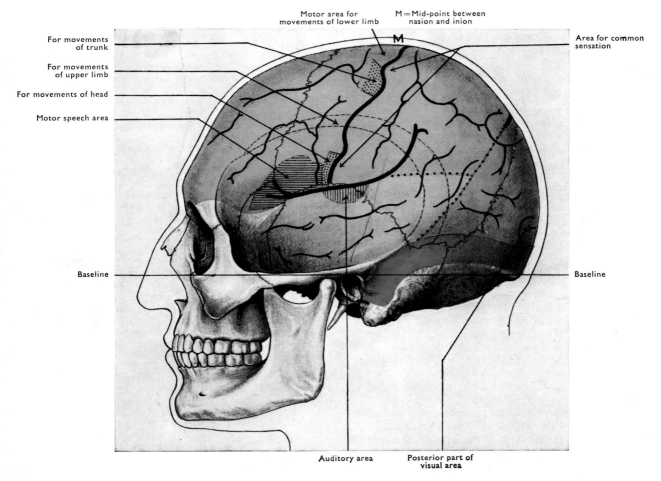

FIG. 10.191. Craniocerebral topography: lobes of cerebrum; motor and sensory areas. Frontal and occipital lobes, red; parietal lobe, blue; temporal lobe, pale blue.

electrode through a trephine hole in the skull. For this purpose, the surface configuration of the brain does not give sufficiently accurate information, and it is necessary to identify the position of deep structures in the brain by outlining the third and lateral ventricles radiographically after filling them with air [FIG. 10.192]. Then midline structures, e.g. the anterior and posterior commissures, may be visualized and used as a baseline from which to estimate the position of other structures. The relation of the lateral ventricle to the brain is shown in FIGURE 10.193.

On the surface of the head a baseline may be drawn from the lower margin of the orbit backwards through the upper margin of the external acoustic meatus. This line cuts the occiput close to the external occipital protuberance and has the cerebrum lying entirely above it, while that part of the cerebellum below its horizontal fissure lies in the posterior cranial fossa inferior to the posterior third of the line.

A point, two finger-breadths above the zygomatic arch and one finger-breadth posterior to the frontozygomatic suture (which can be felt as a notch) marks the inferior extremity of the coronal suture and the pterion. From this point the coronal suture slopes posterosuperiorly to the bregma, which may be felt as a slight depression and lies immediately anterior to the point at which a vertical line through the external acoustic meatus cuts the median plane. The central sulcus lies approximately parallel to the coronal suture, approaching it slightly in its inferior part, and at its superior

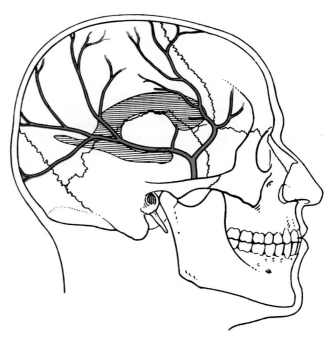

FIG. 10.192. Topography of normal lateral ventricle and middle meningeal artery. (By permission of the editor of *The British Journal of Surgery*.) [See also FIGS. 10.84 and 10.132.]

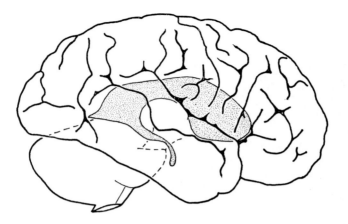

FIG. 10.193. Drawing to show the position of the lateral ventricle in the brain. Prepared from photographs taken during dissection, this ventricle is slightly enlarged and has a very small posterior horn.

extremity is approximately 1 cm posterior to the midpoint [M, FIG. 10.191] of a line from nasion to inion. The parieto-occipital sulcus is marked by a line extending 2 cm transversely from the midline at the lambda. The lambda can either be felt, or taken as 8–10 cm posterior to the superior end of the central sulcus, or 6–7 cm above the inion. A point 3.5 cm behind and 1 cm above the frontozygomatic suture marks the anterior end of the posterior ramus of the lateral sulcus. From this point the posterior ramus may be marked out as a posterior extension of the line joining it to the frontozygomatic suture and curving superiorly to the parietal eminence at its posterior end. The anterior ramus of the lateral sulcus passes forward from the point approximately to the lower end of the coronal suture, while the ascending ramus passes superiorly at right angles to the posterior ramus for 2 cm. 1 cm above a point 2.5 cm posterior to the origin of the posterior ramus from the lateral sulcus, lies the inferior end of the central sulcus.

The frontal branch of the middle meningeal artery, often in the form of several parallel vessels, lies between the central sulcus and the coronal suture.

The lateral ventricle [FIGS. 10.131 and 10.132] lies within the area of the temporal fossa except for its posterior horn. The interventricular foramen lies on the level of the posterior ramus of the lateral sulcus, 1.5 cm behind its anterior extremity. The anterior horn reaches the inferior part of the coronal suture, while the central part extends horizontally backwards, turns inferiorly with its dorsal surface parallel to the inferior temporal line, and divides into inferior and posterior horns. The inferior horn passes forwards in the line of the middle temporal gyrus to end about 3 cm posterior to the temporal pole, just posterior to the surface marking of the hypophysis and postero-inferior to the optic chiasma.

General references

BAILEY, P. and BONIN, G. (1951). *The isocortex of man.* Urbana.

BRODAL, A. (1957). *The reticular formation of the brain stem: anatomical aspects and functional correlations.* Edinburgh.

—— (1959). *The cranial nerves.* Oxford.

—— (1969). *Neurological anatomy in relation to clinical medicine,* 2nd edn. New York.

—— WALBERG, F., and POMPEIANO, O. (1961). *The vestibular nuclei and their connections. Anatomy and functional correlations.* Edinburgh.

BUCY, P. C. (1949). *The precentral motor cortex.* Chicago.

CAJAL, S. RAMÓN Y (1955). *Histologie du système nerveux de l'homme et les vertébrés.* Madrid.

CLARK, W. E. LE GROS (1942). The visual centres of the brain and their connexions. *Physiol. Rev.* **22**, 205.

—— *The tissues of the body; an introduction to the study of anatomy,* 4th edn., Chapter 13, p. 326. Oxford University Press.

DOW, R. S. (1942). The evolution and anatomy of the cerebellum. *Biol. Rev.* **17**, 179.

ECCLES, J. C., ITO, M., and SZENTAGOTHAI, J. (1967). *The cerebellum as a neuronal machine.* Berlin.

ECONOMO, C. VON (1929). *The cyctoarchitectonics of the human cerebral cortex,* trans. S. Parker. London.

FULTON, J. F. (1949). *Functional localization in the frontal lobes and cerebellum.* Oxford.

—— ARING, C. D., and WORTIS, S. B. (1948). The frontal lobes. *Res. Publ. Ass. nerv. ment. Dis.* **27**.

GLEES, P. (1955). *Neuroglia, morphology and function.* Oxford.

GORDON, G. (1977). Somatic and visceral mechanisms. *Br. Med. Bull.* **33**.

HAYMAKER, W., ANDERSON, E., and NAUTA, W. J. (1969). *The hypothalamus.* Springfield.

HERRICK, C. J. (1938). *An introduction to neurology,* 5th edn. Philadelphia.

HORSTADIUS, S. (1950). *The neural crest, its properties and derivatives in the light of experimental research.* London.

IGGO, A. (1973). *Handbook of sensory physiology.* Berlin.

JANSEN, J. and BRODAL, A. (1954). *Aspects of cerebellar anatomy.* Oslo.

JASPER, H. H., PROCTOR, L. D., KNIGHTON, R. S., NOSHAY, W. C., and COSTELLO, R. T. (1958). *Reticular formation of the brain.* London.

KAPPERS, C. U. A., HUBER, G. C., and CROSBY, E. C. (1936). *The comparative anatomy of the nervous system of vertebrates.* New York.

LASSEK, A. M. (1954). *The pyramidal tract.* Springfield.

OLSZEWSKI, J. and BAXTER, D. (1954). *Cytoarchitecture of the human brain stem.* New York.

PENFIELD, W. and JASPER, H. H. (1954). *Epilepsy and the functional anatomy of the human brain.* Boston.

—— and RASMUSSEN, T. (1950). *The cerebral cortex of man.* New York.

PETERS, A., PALAY, S. L., and WEBSTER, H. DE F. (1976). *The fine structure of the nervous system.* Philadelphia.

PURPURA, D. P. and YAHR, M. D. (1966). *The thalamus.* New York.

PUTNAM, T. J. (1942). The diseases of the basal ganglia. *Res. Publ. Ass. nerv. ment. Dis.* **21**.

RILEY, H. A. (1943). *An atlas of the basal ganglia, brain stem and spinal cord.* Baltimore.

SCHMITT, F. O. and WORDEN, F. G. (1974). *The neurosciences, third study programme. Hemispheric specialization and interaction.* Cambridge, Mass.

WALKER, A. E. (1938). *The primate thalamus.* Chicago.

WALSHE, F. M. R. (1948). *Critical studies in neurology.* Edinburgh.

WHITFIELD, I. C. (1967). *The auditory pathway.* London.

WOOLLAM, D. H. M. AND MILLEN, J. W. (1954). Perivascular spaces of the mammalian central nervous system. *Biol. Rev.* **29**, 251.

WURTMAN, R. J., AXELROD, J., and KELLY, D. E. (1968). *The pineal.* New York.

YOUNG, J. Z. (1942). The functional repair of nervous tissue. *Physiol. Rev.* **22**, 318.

ZULCH, K. J., CREUTZFELDT, O., and GALBRAITH, G. C. (1975). *Cerebral localization.* Berlin.

Specific references

BARGMANN, W. (1966). Neurosecretion. *Int. Rec. Cytol.* **19**, 183.

BARR, M. L. (1960). Sexual dimorphism in interphase nuclei. *Am. J. hum. Genet.* **12**, 118.

BODIAN, D. (1951). Nerve endings, neurosecretory substance and lobular organisation or the neurohypophysis. *Bull. Johns Hopk. Hosp.* **89**, 354.

BRAWER, J. R.. (1972). The fine structure of the ependymal tanycytes at the level of the arcuate nucleus. *J. comp. Neurol.* **143**, 411.

BREATHNACH, A. S. and GOLDBY, F. (1954). The amygdaloid nuclei, hippocampus and other parts of the rhinencephalon in the porpoise. *J. Anat. (Lond.)* **88**, 267.

BROCA, P. (1861). Perte de la parole. Ramollissement chronique et destruction partielle du lobe anterieure gauche du cerveau. *Bull. Soc. Anthropol.* **2**, 235.

BRODAL, A. and KAADA, B. R. (1953). Exteroceptive and proprioceptive ascending impulses in the pyramidal tract of cat. *J. Neurophysiol.* **16**, 567.

—— and WALBERG, F. (1952). Ascending fibres in the pyramidal tract of cat. *Arch. Neurol. Psychiat. (Chic.)* **68**, 1.

—— SZABO, T., and TORVIK, A. (1956). Corticofugal fibres to sensory trigeminal nuclei and nucleus of solitary tract. *J. comp. Neurol.* **106**, 527.

—— WALBERG, F., and HODDEVIK, G. H. (1975). The olivocerebellar projection in the cat studied with the method of retrograde axonal transport of horse-radish peroxidase. *J. comp. Neurol.* **164**, 449.

BROWN, A. G. and GORDON, G. (1977). Subcortical mechanisms concerned in somatic sensation. *Br. Med. Bull.* **33**, 121.

BUNGE, M. B., BUNGE, R. P., and PAPPAS, G. D. (1962). Electron microscopic demonstration of connections between glia and myelin sheaths in the developing mammalian central nervous system. *J. Cell. Biol.* **12**, 448.

CARMAN, J. B., COWAN, W. M., and POWELL, T. P. S. (1963). The organization of cortico-striate connexions in the rabbit. *Brain* **86**, 525.

—— —— —— (1964a). The cortical projection upon the claustrum. *J. Neurol. Neurosurg. Psychiat.* **27**, 46.

—— —— —— (1964b). Cortical connexions of the thalamic reticular nucleus. *J. Anat. (Lond.)* **98**, 587.

CLARK, G. (1948). The mode of representation in the motor cortex. *Brain* **71**, 320.

—— and WARD, J. W. (1948). Responses elicited from the cortex of monkeys by electrical stimulation through fixed electrodes. *Brain* **71**, 332.

CLARK, W. E. LE GROS (1952). A note on cortical cytoarchitectonics. *Brain* **75**, 96.

—— (1957). Enquiries into the anatomical basis of olfactory discrimination. *Proc. R. Soc. B* **146**, 299.

COWAN, W. M. and POWELL, T. P. S. (1955). The projection of the midline and intralaminar nuclei of the thalamus of the rabbit. *J. Neurol. Neurosurg. Psychiat.* **18**, 266.

CREVEL, H. VAN (1958). *The rate of secondary degeneration in the central nervous system.* Leiden.

—— and VERHAART, W. J. C. (1963). The rate of secondary degeneration in the central nervous system. 1. The pyramidal tract of the cat. *J. Anat. (Lond.)* **97**, 429.

DARIAN-SMITH, I. and JOHNSON, K. O. (1977). Temperature sense in the primate. *Br. Med. Bull.* **33**, 143.

DE VALOIS, R. L. (1960). Colour vision mechanisms in the monkey. *J. gen. Physiol.* **43**, 115.

FADIGA, E. and PUPILLI, G. C. (1964). Teleceptive components of the cerebellar function. *Physiol. Rev.* **44**, 432.

FULTON, J. F. and SHEEHAN, D. H. (1935). The uncrossed lateral pyramidal tract in higher primates. *J. Anat. (Lond.)* **69**, 181.

GALAMBOS, R. (1956). Suppression of auditory nerve activity by stimulation of efferent fibres to the cochlea. *J. Neurophysiol.* **19**, 424.

GETZ, B. (1952). The terminations of spino-thalamic fibres in the cat as studied by the method of terminal degeneration. *Acta anat. (Basel)* **16**, 271.

HARDING, B. N. and POWELL, T. P. S. (1977). An electron microscopic study of the centre-median and ventrolateral nuclei of the thalamus in the monkey. *Phil. Trans. R. Soc.* **279**, 357.

HARRISON, R. G. (1907). Observations on the living developing nerve fibre. *Anat. Rec.* **1**, 116.

HELLER, H. (1966). The hormone content of the vertebrate hypothalamo-neurohypophysial system. *Br. Med. Bull.* **22**, 227.

HOFF, E. C. and HOFF, H. E. (1934). Spinal terminations of the projection fibres from the motor cortex of primates. *Brain* **57**, 454.

HOLMES, G. and MAY, W. P. (1909). On the exact origin of the pyramidal tracts in Man and other mammals. *Brain* **32**, 1.

HOWE, H. A. and BODIAN, D. (1942). *Neural mechanisms in poliomyelitis.* The Commonwealth Fund, New York.

HUGHES, J. and KOSTERLITZ, H. W. (1977). Opioid peptides. *Br. Med. Bull.* **33**, 157.

HYDEN H. (1967). Dynamic aspects of the neuron-glia relationship, a study with microchemical methods. In *The neuron* (ed. H. Hyden). Amsterdam.

JAYATILAKA, A. D. P. (1965). An electron microscopic study of sheep arachnoid granulations. *J. Anat. (Lond.)* **99**, 635.

JONES, E. G. and LEAVITT, R. Y. (1974). Retrograde axonal transport and the demonstration of non-specific projections to the cerebral cortex from thalamic intralaminar nuclei in rat, cat and monkey. *J. comp. Neurol.* **154**, 349.

—— and POWELL, T. P. S. (1968). The commissural connexions of the somatic sensory cortex in the cat. *J. Anat. (Lond.)* **103**, 433.

JUNG, R. and HASSLER, R. (1960). The extrapyramidal motor system. In *Handbook of physiology*, Section 1, Vol. 2, p. 863. American Physiological Society, Washington.

KUYPERS, H. G. J. M. (1958). Corticobulbar connexions to the pons and lower brainstem in Man. *Brain* **81**, 364.

—— (1960). Central cortical projections to motor and somatic sensory groups. *Brain* **83**, 161.

—— and LAWRENCE, D. G. (1967). Cortical projections to the red nucleus and the brainstem in the rhesus monkey. *Brain Res.* **4**, 151.

McLARDY, T. (1950). Thalamic projection to the frontal cortex in Man. *J. Neurol. Neurosurg. Psychiat.* **13**, 198.

MARIE, P. (1895). *Lectures on diseases of the spinal cord*, trans. Lubbock, for New Sydenham Soc., London.

MATTHEWS, P. B. C. (1977). Muscle afferents and kinaesthesia. *Br. Med. Bull.* **33**, 137.

MYERS, R. E. (1965). In *Functions of the corpus callosum.* London.

NATHAN, P. W. (1956). Awareness of bladder filling with divided sensory tract. *J. Neurol. Neurosurg. Psychiat.* **19**, 101.

—— and SMITH, M. C. (1951). The centripetal pathway from the bladder and urethra within the spinal cord. *J. Neurol. Neurosurg. Psychiat.* **14**, 262.

—— —— (1953). Spinal pathways subserving defaecation and sensation from the lower bowel. *J. Neurol. Neurosurg. Psychiat.* **16**, 245.

—— —— (1955a). Long descending tracts in Man. *Brain* **78**, 248.

—— —— (1955b). Spino-cortical fibres in Man. *J. Neurol. Neurosurg. Psychiat.* **18**, 181.

—— —— (1959). Fasciculi proprii of the spinal cord in Man. *Brain* **82**, 610.

OLIVECRONA, H. (1954). Relation of the paraventricular nucleus to the pituitary gland. *Nature (Lond.)* **173**, 1001.

PAINTAL, A. S. (1977). Thoracic receptors connected with sensation. *Br. Med. Bull.* **33**, 169.

PENFIELD, W. and FAULK, M. E. (1955). The insula. Further observations on its function. *Brain* **78**, 445.

POGGIO, G. F. and MOUNTCASTLE, V. B. (1959). A study of the functional contributions of the lemniscal and spinothalamic systems to somatic sensibility. *Bull. Johns Hopk. Hosp.* **105**, 201.

POWELL, T. P. S. (1958). The organization and connexions of the hippocampal and intralaminar systems. *Rec. Progr. Psychiat.* **3**, 54.

—— (1976). Bilateral corticotectal projection from the visual cortex in the cat. *Nature (Lond.)* **260**, 526.

—— (1977). The somatic sensory cortex. *Br. Med. Bull.* **33**, 129.

—— and MOUNTCASTLE, V. B. (1959a). The cytoarchitecture of the postcentral gyrus of the monkey, *Macaca mulatta. Bull. Johns Hopk. Hosp.* **105**, 108.

—— —— (1959b). Some aspects of the functional organisation of the cortex of the postcentral gyrus of the monkey: a correlation of findings obtained in a single unit analysis with cytoarchitecture. *Bull. Johns Hopk. Hosp.* **105**, 133.

RASMUSSEN, G. L (1953). Further observations on the efferent cochlear bundle. *J. comp. Neurol.* **99**, 61.

RASMUSSEN, T. and MILNER, B. (1975). Clinical and surgical studies of the speech areas in Man. In *Cerebral localization* (eds. K. J. Zulch, O. Creutzfeldt, and G. C. Galbraith). Berlin.

REESE, T. S. and KARNOVSKY, M. J. (1967). Fine structural localization of a blood–brain barrier to exogenous peroxidase. *J. Cell. Biol.* **34**, 207.

RENSHAW, B. (1941). Influence of the discharge of motoneurons upon excitation of neighbouring motoneurons. *J. Neurophysiol.* **4**, 167.

REXED, B. (1954). A cytoarchitectonic atlas of the spinal cord in the cat. *J. comp. Neurol.* **100**, 297.

ROMANES, G. J. (1951). The motor cell columns of the lumbosacral spinal cord of the cat. *J. comp. Neurol.* **94**, 313.

SCHIBEL, M. A. and SCHIBEL, A. B. (1966). The organization of the nucleus reticularis thalami: a Golgi study. *Brain Res.* **1**, 43.

SCOVILLE, W. B. and MILNER, B. (1957). Loss of recent memory after bilateral hippocampal lesions. *J. Neurol. Neurosurg. Psychiat.* **20**, 11.

SHARRARD, W. J. W. (1955). The distribution of the permanent paralysis in the lower limb in poliomyelitis. A clinical and pathological study. *J. Bone Jt Surg.* **37-B**, 540.

SHERRINGTON, C. S. (1889). On the nerve tracts degenerating secondarily to lesions of the cortex cerebri. *J. Physiol. (Lond.)* **10**, 429.

SMITH, M. C. (1957). Observations on the topography of the lateral column of the human cervical spinal cord. *Brain* **80**, 263.

SNIDER, R. S. (1950). Recent contributions to the anatomy and phisiology of the cerebellum. *Arch. Neurol. Psychiat. (Chic.)* **64**, 196.

SPALDING, J. M. K. (1952). Wounds of the visual pathway. Part I. The visual radiation. *J. Neurol. Neurosurg. Psychiat.* **15**, 99.

SPERRY, R. W. (1958). Physiological plasticity and brain circuit theory. In *Biological and biochemical bases of behaviour* (eds. H. F. Harlow and C. N. Woolsey), p. 401.

—— (1974). Lateral specialization in the surgically separated hemispheres. In *The neurosciences, third study programme*. Cambridge, Mass.

SPRAGUE, J. M. and CHAMBERS, W. W. (1959). An analysis of cerebellar function in the cat, as revealed by its partial and complete destruction, and its interaction with the cerebral cortex. *Arch. Ital. Biol.* **97**, 68.

SUNDERLAND, S. (1968) *Nerves and nerve injuries*. Livingstone, Edinburgh.

SZENTAGOTHAI, J. and RAJKOVITS, K. (1959). Über den Ursprung der Kletterfasern des Kleinhirns. *Z. Anat. Entwickl.-Gesch.* **121**, 130.

TARKHAN, A. A. (1934). The innervation of the extrinsic ocular muscles. *J. Anat. (Lond.)* **68**, 293.

TRAVIS, A. M. (1955). Neurological deficiencies following supplementary motor area lesions in *Macaca mulatta. Brain* **78**, 174.

TRUEX, R. C., TAYLOR, M. J., SYMTH, M. Q., and GILDENBERG, P. L. (1970) The lateral cervical nucleus of cat, dog and Man. *J. comp. Neurol.* **139**, 93.

WALBERG, F. (1954). Descending connections to the inferior olive. In *Aspects of cerebellar anatomy* (eds. J. Jansen and A. Brodal). Oslo.

—— (1957a). Corticofugal fibres to the nuclei of the dorsal columns. *Brain* **80**, 273.

—— (1957b). Do the motor nuclei of the cranial nerves receive corticofugal fibres? *Brain* **80**, 597.

—— (1972). Cerebellovestibular relations: anatomy. *Prog. Brain Res.* **37**, 361.

—— and BRODAL, A. (1953). Pyramidal tract fibres from temporal and occipital lobes. An experimental study in the cat. *Brain* **76**, 491.

WALSHE, F. M. R. (1961) The problem of the origin of the pyramidal tract. In *Scientific aspects of neurology* (ed. H. Garland). Edinburgh.

WARWICK, R. (1953). Representation of the extra-ocular muscles in the oculomotor nuclei of the monkey. *J. comp. Neurol.* **98**, 449.

WEBSTER, K. E. (1977). Somaesthetic pathways. *Br. Med. Bull.* **33**, 113.

WHITTERIDGE, D. (1960). Central control of eye movements. In *Handbook of physiology*, Section 1, Vol. 2, P. 1089. American Physiological Society, Washington.

ZANGWILL, O. L. (1960). *Cerebral dominance in relation to psychological function*. Edinburgh.

11 The peripheral nervous system

G. J. ROMANES

THE peripheral nervous system is that part of the nervous system which lies external to the brain and spinal medulla. It consists of nerve fibres [p. 613] and their connective tissue sheaths which extend to all parts of the body. These nerve fibres are the processes of nerve cells the majority of which lie outside the brain and spinal medulla, though a proportion (somatic motor and preganglionic autonomic cells) lie inside these organs and send their processes into the peripheral nervous system.

Peripheral nerves consist of bundles of nerve fibres, both myelinated and non-myelinated, enclosed in three layers of connective tissue. The outermost layer, epineurium, is formed of collagen (and fibrocytes) which gives much of the strength to the peripheral nerve, and binds together the various bundles of nerve fibres which comprise it. The middle layer, perineurium, encloses each separate bundle of nerve fibres. It consists of a variable number of layers of squamous, epithelial cells which form continuous sheets separated by minute amounts of collagen (Gamble and Eames 1964; Shanthaveerappa and Bourne 1966). The innermost layer, endoneurium, is a delicate connective tissue which separates the individual nerve fibres of the nerve, inside the perineurium. Traced towards the central nervous system, the epineurium becomes continuous with the dura mater [p. 733] and the perineurium with the pia-arachnoid; also the number of layers of the perineurium increases as does the size of the individual nerve fibre bundles. Traced towards the periphery, the number of layers of the perineurium decreases, and it becomes continuous with the capsules of the various nerve endings (Shanthaveerappa and Bourne 1966). The perineurium thus isolates the peripheral nerve fibres from the surrounding tissues, and seems to form a layer which is impervious to many substances.

The peripheral nervous system forms the pathways through which the central nervous system receives sensory information from the periphery and produces changes in the activity of muscles, glands, etc. On topographical grounds it is broadly divisible into three parts.

1. **The spinal nerves** are attached to the spinal medulla by two separate rows of rootlets, dorsal and ventral. These rootlets unite in groups to form respectively the dorsal and ventral roots of the spinal nerves; one dorsal and one ventral root combine to form a spinal nerve. Each spinal nerve emerges through a separate foramen to play its part in the innervation of the trunk and limbs.

2. **The cranial nerves** are attached to the brain by a single root or by a single row of rootlets, and they leave the skull through its various foramina to be distributed predominantly to the head region.

Some nerves in both of the above groups also contain nerve fibres (preganglionic) which pass from cells in the central nervous system to control the activity of the nerve cells in 3.

3. **The autonomic nervous system** is divided into sympathetic and parasympathetic parts, and is concerned in the main with the non-volitional activities of the body. Both parts consist of groups of nerve cells and their processes which lie entirely outside the central nervous system. They are often arranged in complicated, loose meshworks (plexuses) unlike the compact bundles of nerve fibres which constitute the cranial and spinal nerves.

The **sympathetic part** receives preganglionic nerve fibres from the central nervous system through the thoracic and upper lumbar spinal nerves; it innervates involuntary structures throughout the body.

The **parasympathetic part** receives preganglionic nerve fibres from the central nervous system through certain cranial and sacral spinal nerves; it innervates only the viscera and some structures in the orbit.

The peripheral nervous system is also divisible on functional grounds into efferent (motor) and afferent (sensory) parts.

THE EFFERENT PART

(a) The nerve **supply to striated, voluntary muscles** which are developed from the somites is by nerve cells (**somatic motor**) which are situated within the central nervous system and send their thick myelinated axons to these muscles through certain cranial nerves and the ventral roots of all spinal nerves. These fibres are accompanied by smaller motor fibres which also arise from cells in the central nervous system but supply the muscle spindles [p. 836] in the voluntary muscles. The latter fibres form one example of efferent fibres which control the sensitivity of peripheral sensory organs. It is not known if this type of arrangement is common to all sensory endings, but efferent fibres, which may have the same function, are found in the optic and vestibulocochlear nerves which are commonly thought to be purely afferent.

(b) The **supply to involuntary (non-striated or smooth and cardiac) muscles and glands (visceral motor)** is by cells of the autonomic nervous system which lie entirely outside the central nervous system either in discrete clumps (**ganglia**) or scattered throughout the plexuses. These cells (autonomic ganglion cells) send their thin, usually non-myelinated axons (**postganglionic nerve fibres**) to the viscera through the plexuses, and to similar structures in the territories of the spinal and cranial nerves by running either in these nerves or on the arteries of the region. The activity of these autonomic nerve cells is controlled by thinly myelinated, **preganglionic nerve fibres** which pass to them from nerve cells in the central nervous system via certain cranial and spinal nerves.

THE AFFERENT PART

This part consists of the peripheral (dendrites) and central (axons) processes of sensory nerve cells. With three exceptions [p. 620], the cell bodies of these cells lie close to the central nervous system in the

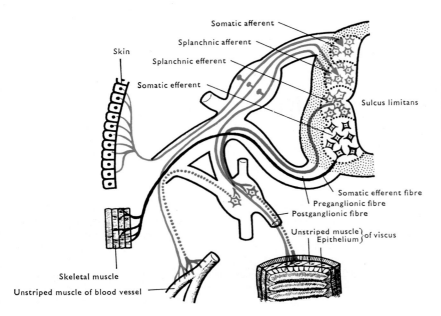

FIG. 11.1. Diagram of components of a spinal nerve and their relation to nuclei of origin or termination in section of embryonic spinal medulla. Compare with components of cranial nerves in FIG. 11.2 and also with FIG. 11.82.

ganglia of the cranial nerves and of the dorsal roots of the spinal nerves (**spinal ganglia**).

The diameters of these cells and their processes are very variable, and the smaller, slowly conducting processes ($<2\,\mu m$ in diameter) have no myelin sheath. The peripheral processes of the cells are distributed to the sensory endings of the body through the spinal and cranial nerves to which the sensory ganglia are attached. Those that pass to visceral structures course with the preganglionic and postganglionic fibres of the autonomic nervous system but have no direct functional connection with them. The central processes of all the sensory cells pass directly into the central nervous system; hence reactions to sensory stimuli are the result of the activity of the central nervous system and are not due to peripheral connections between the sensory and motor nerve cells.

Thus several different types of fibres are to be found in peripheral nerves. **Cutaneous nerves** are not only sensory to the skin and subcutaneous tissues but contain postganglionic sympathetic fibres to blood vessels, sweat glands, and the muscles of the hairs (arrectores pilorum). **Nerves to muscles** contain sensory fibres to muscles, tendons, bones, and joints, as well as motor fibres to the muscles and some sympathetic fibres to their blood vessels. The **autonomic nerve plexuses** transmit sensory fibres in addition to autonomic nerve fibres. *Thus almost every nerve, whatever its destination, consists of a mixture of efferent and afferent fibres.*

The artifical subdivisions given above and the separate sections dealing with the central and peripheral nervous systems are convenient for the purposes of study and description, but it is important to realize that the peripheral nervous system is an extension of the central nervous system, that neither system can function in the absence of the other, and that both are necessary for integration of the activities of the body. It is incorrect, therefore, to think of either system or their parts as though they acted independently.

The cranial nerves

There are twelve pairs of symmetrically arranged cranial nerves attached to the brain, and all are distributed in the head and neck with the exception of the tenth pair (the vagus nerves) which also supply structures in the thorax and abdomen. Each cranial nerve is attached to a specific region on the surface of the brain (superficial origin) and though some of the nerves consist almost entirely of one type of nerve fibre at this point, the majority contain a mixture of (1) **afferent** (incoming or sensory) **fibres** which enter the brain and end on various groups of nerve cells (nuclei) therein, and (2) **efferent** (outgoing or motor) **fibres** which arise from one or more nuclei in the brain. These afferent and efferent nuclei constitute respectively the deep terminations and origins of the nerves and are dealt with in the account of the brain. In this section the course of each cranial nerve will be described from the superficial origin to the peripheral distribution irrespective of the type or types of nerve fibres which it contains.

The table on page 743 shows the various types of nerve fibres in the cranial nerves, their distribution and function. It indicates that certain cranial nerves contain **branchial efferent nerve fibres** in addition to the types already described. These fibres, found only in some cranial nerves, are structurally identical with the somatic motor fibres of other cranial and spinal nerves which supply the striated muscles developed from the somites of the embryo. Unlike these, they supply the striated muscles of voluntary type which lie in the anterior extremity of the gut tube, and are used for the rapid movements of swallowing (pharynx and oesophagus), speaking (larynx), mastication, expression, etc. Not all these muscles are immediately under voluntary control.

At their **superficial attachments** the cranial nerves are arranged in two rows. (1) A ventral row consisting of the first, second, third, sixth, and twelfth which lie in line with the ventral roots of the spinal nerves. (2) A lateral row comprising the fifth and seventh to eleventh inclusive. These lie approximately in line with the dorsal roots of the spinal nerves and, like them, have sensory ganglia. The fourth (trochlear) nerve arises from the dorsal surface of the brain and does not lie in either row [FIG. 11.4].

The first two cranial nerves (olfactory and optic) differ from the others. The **olfactory nerves** consist of bundles of fine, non-myelinated fibres which arise in sensory cells situated in the olfactory epithelium. They enter the olfactory bulb through the cribriform plate of the ethmoid. The **optic nerves**, and the retinae from which their nerve fibres arise, are formed as an outgrowth of the entire thickness of the embryonic brain tube. Hence the nerves correspond to a brain tract rather than a peripheral nerve which is formed by

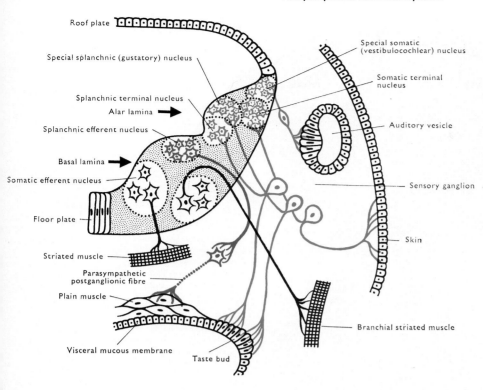

FIG. 11.2. Diagram of components of cranial nerves and their relation to nuclei of origin or termination in transverse section of embryonic hindbrain. Compare with components of spinal nerve in FIG. 11.1 and also with FIG. 10.47.

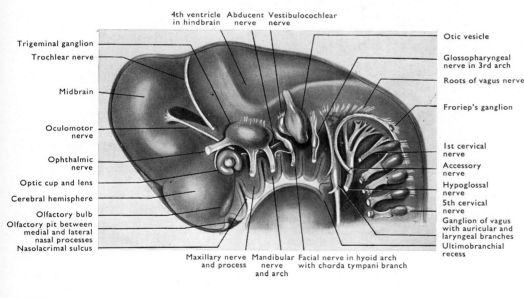

FIG. 11.3. Developmental arrangement of the cranial nerves. (After Streeter 1908.)

the outgrowth of nerve fibres into the surrounding tissues from nuclei within the brain or from ganglia external to it.

The remaining members of the ventral row and the fourth nerve resemble the ventral roots of the spinal nerves in that they supply the muscles derived from the head somites—the extrinsic ocular muscles and the muscles of the tongue. In addition they carry from these muscles a few sensory fibres, though the majority enter the brain through the lateral row of cranial nerves. The third nerve also transmits preganglionic fibres to the parasympathetic cells (ciliary ganglion) innervating certain structures in the eyeball.

The cranial nerves of the lateral row contain afferent fibres and efferent fibres of two types—branchial (V, VII, IX, X, and XI) and preganglionic parasympathetic (VII, IX, and X). However, the eighth or vestibulocochlear nerve is mainly afferent, and the spinal part of the accessory nerve (XI) is mainly branchial efferent.

OLFACTORY NERVE

The olfactory nerve consists of fine, non-myelinated fibres which arise as the central processes (axons) of slender bipolar nerve cells lying between the columnar epithelial cells of the olfactory mucous membrane on the upper third of the nasal septum and most of the superior concha [FIGS. 11.14 and 11.15]. The peripheral processes of these nerve cells [FIG. 12.66] are short and each forms a swelling between the apices of the epithelial (supporting) cells. Six to eight long, atypical cilia—the olfactory hairs—project from the swelling into the mucus covering the free surface of the epithelial cells (Reese 1965).

The olfactory nerve fibres gather into approximately twenty filaments which pass through the cribriform plate of the ethmoid

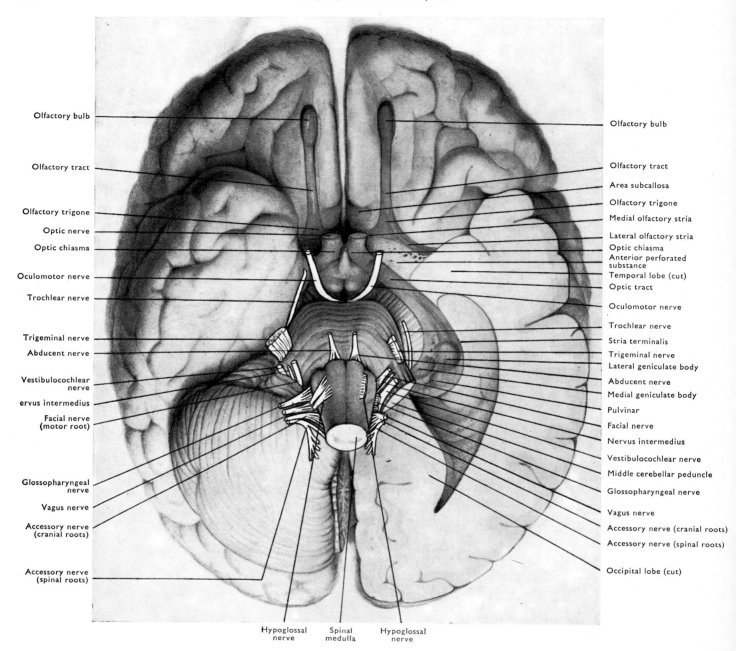

Olfactory bulb
Olfactory tract
Olfactory trigone
Optic nerve
Optic chiasma
Oculomotor nerve
Trochlear nerve
Trigeminal nerve
Abducent nerve
Vestibulocochlear nerve
ervus intermedius
Facial nerve (motor root)
Glossopharyngeal nerve
Vagus nerve
Accessory nerve (cranial roots)
Accessory nerve (spinal roots)

Olfactory bulb
Olfactory tract
Area subcallosa
Olfactory trigone
Medial olfactory stria
Lateral olfactory stria
Optic chiasma
Anterior perforated substance
Temporal lobe (cut)
Optic tract
Oculomotor nerve
Trochlear nerve
Stria terminalis
Trigeminal nerve
Lateral geniculate body
Abducent nerve
Medial geniculate body
Pulvinar
Facial nerve
Nervus intermedius
Vestibulocochlear nerve
Middle cerebellar peduncle
Glossopharyngeal nerve
Vagus nerve
Accessory nerve (cranial roots)
Accessory nerve (spinal roots)
Occipital lobe (cut)

Hypoglossal nerve   Spinal medulla   Hypoglossal nerve

and the meninges to enter the olfactory bulb. Here they form complicated synapses with the processes of the mitral cells of the bulb which send their axons to form the olfactory tract [pp. 678–9].

NERVUS TERMINALIS

Lying medial and parallel to each olfactory tract is a fine, plexiform bundle of nerve fibres interspersed with nerve cells. These terminal nerves are attached superficially to the anterior perforated substance and pass into the nose with the olfactory nerves [FIG. 11.5]. Their function is unknown, but they may be partly autonomic.

FIG. 11.5. Diagram of part of lower surface of adult human brain to show position of nervi terminales and their relations to olfactory tracts. (After Brookover 1914.)
    The olfactory bulb and part of the tract have been removed on the left side, and the right optic nerve and the greater part of the optic chiasma have been cut away to expose the lamina terminalis.

FIG. 11.4. Inferior surface of brain to show superficial attachments of cranial nerves. The lower portion of the left temporal and occipital lobes and the left half of the cerebellum have been removed.

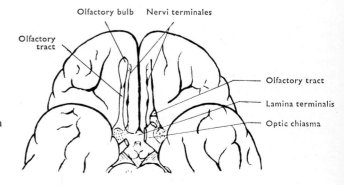

Olfactory bulb   Nervi terminales
Olfactory tract
Olfactory tract
Lamina terminalis
Optic chiasma

Component and functional analysis of cranial nerves

No.	Name	Components	Distribution	Functions
I.	Olfactory	Afferent	Olfactory mucous membrane	Smell
II.	Optic	Afferent	Retina	Vision
		Efferent	To retina	
III.	Oculomotor	Efferent:		
		i. Somatic	Sup., med., and inf. recti and inf. oblique muscles of eyeball; levator palp. sup. muscle	Movements of eyeball and elevation of upper eyelid
		ii. Parasympathetic preganglionic	Ciliary ganglion; ciliary and sphincter pupillae muscles of eyeball	Focusing; constriction of pupil
		Afferent?	The above muscles	Proprioceptive (? via ophthalmic part of trigeminal)
IV.	Trochlear	Efferent somatic	Sup. oblique muscle of eyeball	Depression of the medially rotated cornea.
		Afferent?	The above muscle	Proprioceptive (? via ophthalmic part of trigeminal)
V.	Trigeminal	Efferent branchial	Muscles of mastication	Mastication and swallowing
			Digastric (ant. belly) and mylohyoid muscles	
			Tensor veli palatini	Tenses anterior part of soft palate
			Tensor tympani	Reduces movements of tympanic membrane and malleus
		Afferent	Skin of face and anterior scalp	
			Mucous membrane of mouth including teeth, gums, and tongue (ant. $\frac{2}{3}$)	
			Mucous membrane of nasal cavities and paranasal sinuses	General sensibility
			Meninges (dura mater) and skull	
			Sensory to muscles of mastication and facial expression, and of tongue and eyeball	Proprioceptive
VI.	Abducent	Efferent somatic	Lateral rectus muscle of eyeball	Turns cornea laterally
		Afferent?	The above muscle	Proprioceptive (? via ophthalmic part of trigeminal)
VII.	Facial	Efferent:		
		i. Branchial	Muscles of facial expression and scalp	Facial expression, closure of eye, movements of lips
			Digastric (post. belly), stylohyoid and stapedius muscles	Elevation of hyoid bone in swallowing
				Control of stapes
		ii. Parasympathetic preganglionic	Pterygopalatine ganglion; glands of nasal cavity, of hard and soft palates, and lacrimal gland	
			Submandibular ganglion; submandibular and sublingual glands and glands of tongue (ant. $\frac{2}{3}$)	Secretomotor
		Afferent	Mucous membrane of tongue (ant. $\frac{2}{3}$) (excluding vallate papillae)	Taste
VIII.	Vestibulocochlear	Afferent	From duct of cochlea	Hearing
			From maculae of utricle and saccule	Equilibration (measurement of position of head relative to gravity and of changes in its angular velocity)
			From ampullae of semicircular ducts	
		Efferent	To organ of Corti and vestibular apparatus	Controls input
IX.	Glossopharyngeal	Efferent:		
		i. Branchial	Stylopharyngeus muscle	Pharyngeal movements and elevation of larynx
			Superior and middle constrictor	
		ii. Parasympathetic preganglionic	Otic ganglion	Secretomotor to parotid gland, and glands in posterior $\frac{1}{3}$ of tongue
			Lingual ganglia	
		Afferent	Mucous membrane of pharynx and tongue (post $\frac{1}{3}$) including vallate papillae, and tonsillar region	General sensibility and taste
			Mucous membrane of auditory tube, tympanic cavity and antrum, and mastoid air cells	General sensibility
			Carotid body and carotid sinus	Vasosensory (chemoreceptors and pressure receptors)

Component and functional analysis of cranial nerves (*cont.*)

No.	Name	Components	Distribution	Functions
X. (XI)	Vagus (and Cranial Root of Accessory)	Efferent: i. Branchial	Levator veli palatini muscle	Lifts soft palate
			Palatoglossus	Lowers soft palate
			Pharyngeal muscles	Swallowing and other pharyngeal movements
			Laryngeal muscles	Control of larynx in respiration and phonation
		ii. Parasympathetic preganglionic	Muscle of oesophagus, stomach, intestine (as far as transverse colon), and gall-bladder	Movements of these viscera
			Pancreas and gastric glands	Secretomotor
			Heart (nodal tissue and cardiac muscle)	Cardic depressor
			Lungs (plain muscle and glands of bronchi and bronchioles)	Bronchoconstriction
		Afferent	Mucous membrane of alimentary and respiratory passages from epiglottis to transverse colon	General and visceral sensibility, and some taste
			Part of auricle and ext. acoustic meatus	General sensibility
			Lungs	Respiratory reflexes
			Carotid body and sinus	
			Glomus aorticum and walls of aorta and great veins	Vasosensory
XI.	Accessory (Spinal Root)	Efferent branchial	Trapezius and sternocleidomastoid muscles	Movements of head and shoulder
XII.	Hypoglossal	Efferent somatic	Extrinsic and intrinsic muscles of tongue, except palatoglossus	Movements of tongue

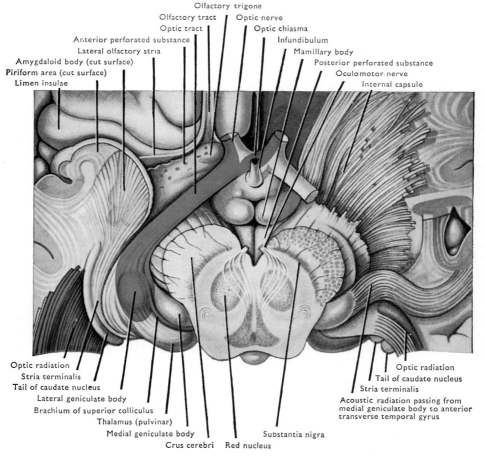

FIG. 11.6. Part of lower surface of forebrain, showing olfactory and optic tracts. The midbrain has been cut across.

OPTIC NERVE

The fibres of the optic nerve are the axons of the ganglion cells on the internal surface of the retina. The axons course over this surface and converge on the circular optic disc which lies 3 mm medial to the sagittal meridian of the eyeball and slightly above the horizontal meridian (Traquair 1949). At the optic disc each fibre acquires a myelin sheath; a feature which accounts for the white colour of the optic disc and occasionally for white streaks which extend outwards from the disc when some fibres myelinate before they reach it.

The fibres at the optic disc gather into bundles which pierce the sclera and transform it into a mesh (lamina cribosa). They then combine to form the compact optic nerve which is immediately enclosed in the three layers of the meninges and the spaces which separate them [p. 728]. It passes backwards and slightly medially to enter the optic canal within the tendinous ring from which the four rectus muscles of the eye arise. Approximately 1 cm posterior to the eyeball, the central artery and vein of the retina pierce the inferior surface of the optic nerve and its coverings and run in its substance to the optic disc as the only direct vascular supply to the retina.

In the optic canal, immediately lateral to the sphenoidal sinus, the nerve lies with the ophthalmic artery, and here may receive the central artery of the retina. It emerges from the canal into the middle cranial fossa, and ends by joining its fellow in the optic chiasma.

The optic chiasma lies in the anterior part of the floor of the third ventricle of the brain, (immediately anterior to the tuber cinereum and infundibulum [FIG. 9.20], a short distance superior to the diaphragma sellae and hypophysis [FIG. 9.21], and between the two internal carotid arteries. Thus pathological changes either in the carotid arteries, or in the hypophysis, or in the floor of the third ventricle may affect the chiasma.

In the chiasma the fibres from the nasal halves of the two retinae decussate, and joining the uncrossed fibres from the temporal half of the opposite retina, form an optic tract at each posterolateral angle of the chiasma. Each optic tract then sweeps posterolaterally around the hypothalamus and crus cerebri [FIG. 11.6] to the lateral geniculate body. Here the fibres concerned with vision end, but many of them give off branches concerned with visual reflexes. The latter continue (brachium of the superior colliculus) between the lateral and medial geniculate bodies to reach the superior colliculus and pretectal nucleus on the dorsal aspect of the midbrain [p. 704 and FIG. 10.155].

The effect which the course of these fibres has on the symptoms following injuries to the different parts of the optic pathways is dealt with on page 705.

There is no evidence that fibres cross in the chiasma to unite the two retinae (Woollard and Harpman 1939; Magoun and Ranson 1942; Wolff 1953) though extensive injuries to one eye may be followed by degenerative changes in the other (sympathetic ophthalmitis). In many animals and probably also in Man there are considerable numbers of **efferent fibres** to the retinae running in the optic tracts and nerves, but their function is unknown (Cowan and Powell 1963). Branching fibres in the optic chiasma have been described, but they seem to be too few to account for a significant bilateral pathway through both optic tracts from any part of the retina.

OCULOMOTOR NERVE

This is predominantly a **somatic motor nerve** which supplies the striated muscle in levator palpebrae superioris and all the extrinsic ocular muscles, except the lateral rectus and the superior oblique. It also carries **preganglionic parasympathetic fibres** to the ciliary ganglion which supplies the sphincter of the pupil and the ciliary muscle within the eyeball. Some **afferent nerve fibres** from the extrinsic ocular muscles may pass to the brain through the oculomotor nerve [p. 741], but the majority are probably transmitted through the trigeminal nerve via communications with the oculomotor nerve in the wall of the cavernous sinus. **Sympathetic postganglionic fibres** also join the nerve from the plexus on the internal carotid artery in the same region, and are distributed through the nerve, e.g. to the smooth muscle of levator palpebrae superioris.

SUPERFICIAL ORIGIN AND COURSE

The oculomotor nerve arises from the ventral surface of the midbrain by several closely packed rootlets attached to the medial sulcus of the crus cerebri, immediately lateral to the posterior perforated substance and just superior to the pons [FIGS. 10.50 and 11.4. Deep origin, p. 646]. The nerve passes horizontally forwards between the superior cerebellar and posterior cerebral arteries, then lateral to the posterior communicating artery in the interpeduncular cistern. At

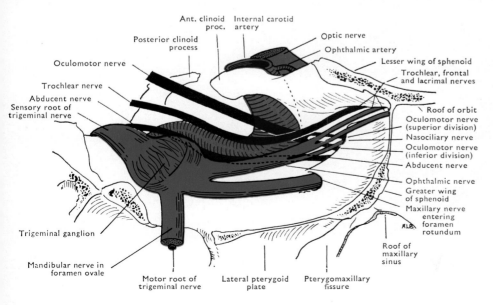

FIG. 11.7. Relations of structures in cavernous sinus and superior orbital fissure, viewed from the right side.

The section of the skull passes sagittally through the foramen ovale.

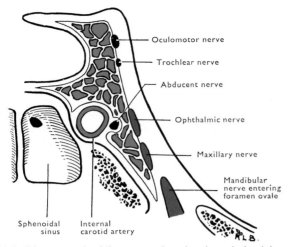

FIG. 11.8. Diagrammatic oblique coronal section through the right cavernous sinus, viewed from behind.

a point between the free and attached margins of the tentorium cerebelli the nerve pierces the small triangular area of dura mater which forms the roof of the **cavernous sinus**. It then runs forwards in the lateral wall of the cavernous sinus, superior to the trochlear nerve, and dividing into **superior** and **inferior branches**, enters the orbit through the superior orbital fissure within the tendinous ring from which the rectus muscles arise [FIG. 11.12]. The **preganglionic parasympathetic fibres** which enter the inferior branch lie in the superior part of the nerve at its superficial origin but are irregular in position after the nerve pierces the dura mater (Sunderland and Hughes 1946).

BRANCHES

The **superior branch** supplies the superior rectus and the levator palpebrae superioris.

The **inferior branch** passes forwards, and giving branches to the medial and inferior recti and to the ciliary ganglion (preganglionic parasympathetic fibres) continues to the inferior oblique muscle.

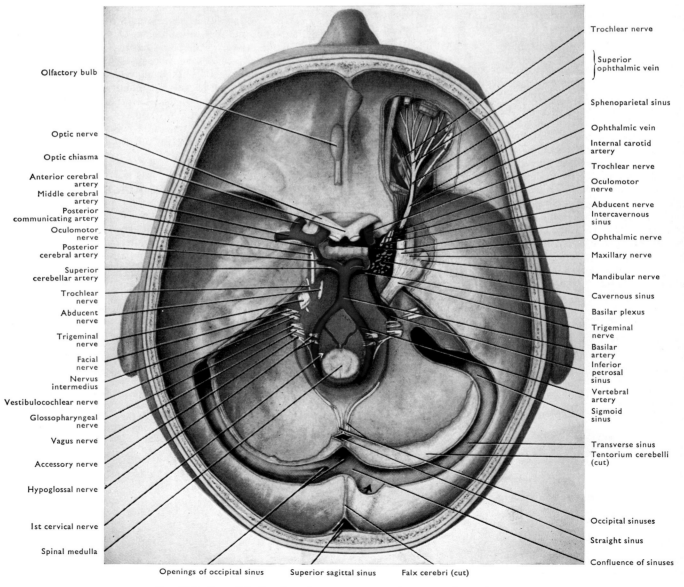

FIG. 11.9. Base of skull to show dura mater, venous sinuses, arteries, and nerves.

CILIARY GANGLION

This small reddish body about 1 mm in diameter lies in the posterior part of the orbit between the lateral rectus muscle and the optic nerve, anterior to the ophthalmic artery. It is one of the cranial parasympathetic ganglia, but does not contain all the ganglion cells of the orbit as some scattered cells are also found, particularly superior to the eyeball [p. 812]. The ganglion receives (1) **preganglionic parasympathetic fibres** from the nerve to the inferior oblique muscle (**oculomotor root**) and these are the only fibres which have functional, synaptic connections with the ganglion cells. (2) Some **postganglionic sympathetic fibres** from the **carotid plexus** pass through the ganglion on their way to supply the dilator pupillae and the vessels in the external coats of the eyeball. They may either reach it as branches of the nasociliary nerve or directly from the extension of the plexus on the ophthalmic artery (sympathetic branch to ciliary ganglion). (3) The **communicating branch from the nasociliary nerve** consists of sensory fibres of the trigeminal nerve which pass through the ganglion to the external coats of the eyeball.

The ciliary ganglion sends twelve to fifteen **short ciliary nerves** to pierce the sclera above and below the optic nerve. These nerves contain the fibres from the sympathetic branch to the ganglion and the communicating branch from the nasociliary nerve which have passed through the ganglion unchanged. In addition they carry the myelinated **postganglionic processes** of the large cells of the ciliary ganglion which pass to supply the sphincter pupillae muscle of the iris and the ciliary muscle. These fibres, together with the preganglionic fibres in the oculomotor nerve comprise the efferent pathway through which constriction of the pupil is achieved (e.g. in the light reflex) and changes in the focal length of the lens are produced, e.g. in focusing on a near object as in the convergence-accommodation reflex. Damage to the oculomotor nerve paralyses most of the extrinsic ocular muscles and levator palpebrae superioris, and disconnects the cells of the ciliary ganglion from the central nervous system. The unopposed action of the lateral rectus and superior oblique muscles turns the cornea laterally and downwards, yet the patient does not complain of diplopia (double vision) because the eye is shut owing to the paralysis of the levator. The dilated pupil on the injured side is not constricted either when the eyes are exposed to light, or in attempts at the convergence-accommodation reflex.

TROCHLEAR NERVE

SUPERFICIAL ORIGIN AND COURSE

This somatic motor nerve supplies the superior oblique muscle of the eyeball. The slender nerve arises from the dorsal surface of the brain immediately below the inferior colliculus and lateral to the frenulum veli [FIG. 10.76]. For deep origin, see page 646. It runs a long intracranial course fowards on the lateral aspect of the midbrain parallel to and immediately below the free edge of the tentorium, between the posterior cerebral and superior cerebellar arteries. It pierces the inferior surface of the tentorium close to the free edge, superior to the trigeminal nerve. It continues forwards to the superior orbital fissure in the lateral wall of the cavernous sinus [FIG. 11.8] between the oculomotor and ophthalmic nerves. Here it receives **sympathetic filaments** from the plexus on the internal carotid artery, and it may receive **sensory fibres** from the ophthalmic nerve. In the orbit it lies lateral to the fibrous ring from which the rectus muscles arise and turns medially between levator palpebrae superioris and the orbital periosteum near the posterior extremity of the orbit [FIG. 11.12]. It enters the orbital surface of the superior

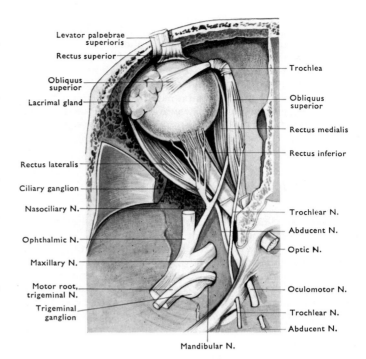

FIG. 11.10A. Dissection of the orbit and middle cranial fossa. The trigeminal nerve and ganglion have been turned laterally.

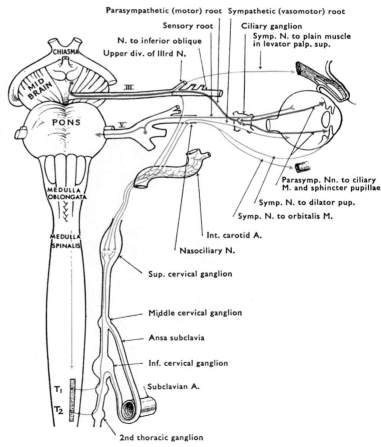

FIG. 11.10B. Diagram of ciliary ganglion and connections of autonomic nervous system with the eye.

Sympathetic, red; parasympathetic, green; sensory, blue.

oblique muscle, the only structure this nerve supplies. This nerve differs from the other somatic motor nerves (III, VI, XII) in that its fibres emerge dorsally, having decussated within the brain substance.

Damage to the trochlear nerve causes paralysis of the superior oblique muscle of the eye with consequent inability to turn the cornea inferiorly when it is already turned medially by the medial rectus. Thus in right superior oblique paralysis the patient complains of double vision when looking downwards and to the left. Failure of the right eye to follow the left results in the image of an object failing to fall on the corresponding parts of the two retinae, thereby producing double vision.

TRIGEMINAL NERVE

This is the major sensory nerve of the head but it is also motor to the muscles of mastication.

It is **sensory** to the face (except over the angle of the jaw), the temple, the anterior scalp as far as the vertex, and parts of the auricle and external acoustic meatus. Also to the contents of the orbit (except the retina of the eye), the walls of the nasal cavity and paranasal air sinuses, the mouth (except the posterior one-third of the tongue), the temporomandibular joint, and to portions of the nasal part of the pharynx, auditory tube, and the cranial dura mater and periosteum.

The trigeminal nerve divides into three nerves (ophthalmic, maxillary, and mandibular) each of which has a distribution which corresponds approximately to the structures that develop from one of the three processes (frontonasal, maxillary, and mandibular) which combine to produce the face in the embryo [p 52]. Thus the divisions supply three clearly defined cutaneous areas [FIG. 11.16] and the structures which lie deep to them.

Pain arising in the deep structures (e.g. teeth or air sinuses) may be referred incorrectly to other parts of the distribution and felt there. This may be because impulses carried by different fibres have a common pathway to the cerebral cortex in the central nervous system.

The **motor** fibres which emerge with the trigeminal nerve from the brain pass entirely into its mandibular division. They supply the muscles of the first pharyngeal (branchial) arch of the embryo, i.e. the muscles that move the mandible, the anterior belly of digastric, mylohyoid, tensor veli palatini, and tensor tympani.

Some of the peripheral branches of all three divisions of the trigeminal nerve are joined by preganglionic or postganglionic fibres of the parasympathetic nervous system and by postganglionic fibres of the sympathetic nervous system. These autonomic motor nerve fibres course with the fibres of the trigeminal nerve to supply salivary, sweat (Wilson 1936), and other glands, in the areas to which that nerve is sensory, but none of them emerges from the brain with that nerve.

The sympathetic fibres arise either in the cells of the superior cervical ganglion or in the cells of the internal carotid plexus. The preganglionic parasympathetic fibres leave the brain in the oculomotor, facial, and glossopharyngeal nerves, and passing to the ciliary, pterygopalatine, otic, and submandibular ganglia, end on their cells. The postganglionic parasympathetic nerve fibres arise from these cells and pass with the sensory (trigeminal) and sympathetic fibres which often traverse these ganglia but have no functional connexion with them.

SUPERFICIAL ORIGIN

The trigeminal nerve arises from the lateral surface of the pons in the posterior cranial fossa [for deep connections, see pp. 648 and

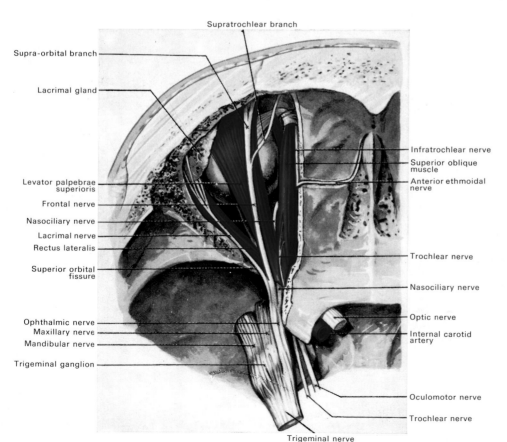

Supratrochlear branch

Supra-orbital branch

Lacrimal gland

Levator palpebrae superioris

Frontal nerve

Nasociliary nerve

Lacrimal nerve

Rectus lateralis

Superior orbital fissure

Ophthalmic nerve

Maxillary nerve

Mandibular nerve

Trigeminal ganglion

Trigeminal nerve

Infratrochlear nerve

Superior oblique muscle

Anterior ethmoidal nerve

Trochlear nerve

Nasociliary nerve

Optic nerve

Internal carotid artery

Oculomotor nerve

Trochlear nerve

FIG. 11.11. Nerves of orbit from above.

650]. It consists of a large, inferior, sensory root and a small, superior, motor root which emerge together. The roots twist through 180 degrees (Davies and Haven 1933) as they pass forwards to run beneath the dural floor of the middle cranial fossa [FIG. 10.189] carrying with them a loose sleeve of dura and arachnoid mater which surrounds the cavum trigeminale in which they lie. The dural sleeve begins where the roots leave the posterior cranial fossa grooving the superior border of the petrous temporal bone [FIG. 3.53] inferior to the superior petrosal sinus in the attached edge of the tentorium cerebelli [FIG. 13.100]. Anteriorly the dural sheath enlarges to enclose and fuse with the trigeminal ganglion into which the sensory root expands.

The trigeminal ganglion is a large, flattened, semilunar, sensory ganglion which extends obliquely from the lateral wall of the cavernous sinus to a shallow fossa on the anterior surface of the petrous temporal bone near its apex. Three large nerves (ophthalmic, maxillary, and mandibular [FIG. 11.7]) arise from the convex, distal margin of the ganglion, while the motor root passes deep to its inferolateral part to join the mandibular nerve.

Ophthalmic nerve

The ophthalmic nerve passes forwards to the orbit in the lateral wall of the cavernous sinus inferior to the trochlear nerve [FIG. 11.8]. Here it receives postganglionic sympathetic filaments from the plexus on the internal carotid artery for transmission through its branches. It also gives off a small, recurrent branch to the dura mater of the tentorium and posterior parts of the falx cerebri [FIG. 10.184]. Fine filaments which pass between the ophthalmic nerve and the oculomotor, trochlear, and abducent nerves may be purely temporary communications (Sunderland and Hughes 1946), but could transfer sensory fibres from the trigeminal nerve for distribution to the extrinsic ocular muscles. The ophthalmic nerve divides into lacrimal, frontal, and nasociliary branches [FIG. 11.7] which enter the orbit through the superior orbital fissure.

LACRIMAL NERVE

This nerve traverses the lateral angle of the superior orbital fissure and passes above and parallel to the lateral rectus muscle to the superolateral part of the orbital margin [FIG. 11.11]. It receives parasympathetic postganglionic nerve fibres for the lacrimal gland through a branch from the zygomatic nerve [FIG. 11.17] and divides into branches to (1) the lacrimal gland; (2) the lateral part of the upper eyelid.

FRONTAL NERVE

This branch passes forwards above the levator palpebrae superioris and divides at a variable point into a larger supraorbital and a smaller supratrochlear branch [FIGS. 11.11 and 11.13].

The supra-orbital nerve passes directly forwards through the supra-orbital groove or foramen, sends twigs to the frontal sinus and upper eyelid, and turns upwards to supply the forehead and scalp as far as the vertex.

The supratrochlear nerve passes anteromedially above the tendon of the superior oblique muscle. It turns superiorly on the medial part of the supra-orbital margin and supplies the medial parts of the upper eyelid and forehead.

NASOCILIARY NERVE

This nerve passes with the two divisions of the oculomotor nerve and the abducent nerve through the superior orbital fissure within the tendinous ring from which the rectus muscles arise [FIG. 11.12], to lie within the cone formed by the extrinsic ocular muscles. The nasociliary nerve then curves anteromedially, superior to the ophthalmic artery and the optic nerve, to leave the cone between the superior oblique and medial rectus muscles, close to the anterior ethmoidal foramen. Here it gives off the small infratrochlear nerve, and continues as the anterior ethmoidal nerve through that foramen between the frontal and ethmoid bones. It supplies the ethmoidal sinuses and enters the anterior cranial fossa on the cribriform plate of the ethmoid, supplying the adjacent dura mater. The anterior ethmoidal nerve then descends into the roof of the nasal cavity through the nasal slit [FIG. 3.91] and supplies the anterosuperior parts of the medial (septal) and lateral walls of that cavity by medial and lateral nasal branches [FIGS. 11.14 and 11.15]. The lateral branch sends an external nasal branch between the nasal bone and the upper nasal cartilage to supply the lower part of the dorsum and side of the external nose [FIG. 11.17].

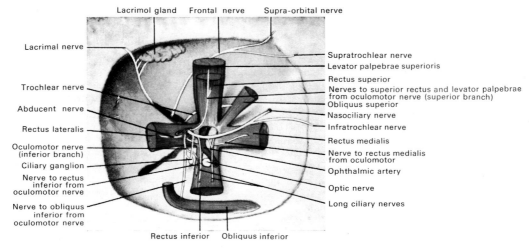

FIG. 11.12. Schematic representation of nerves which traverse the cavity of the orbit.
Note that the lacrimal, frontal, and trochlear nerves, after entering the orbit through the superior orbital fissure, pass outside the common tendinous ring from which the rectus muscles arise; whilst the oculomotor, abducent, and nasociliary nerves, together with the optic nerve and ophthalmic artery, enter the orbit within the ring, and therefore lie at first within the cone formed by the ocular muscles as they diverge forwards from it.

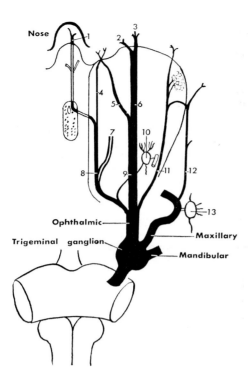

FIG. 11.13. Scheme to show the distribution of the ophthalmic nerve.

9, frontal nerve, giving off supratochlear branch (5) and proceeding as supra orbital (6), which may divide into medial (2) and lateral (3) branches before leaving the orbit.

8, nasociliary nerve; gives off afferent root to ciliary ganglion (10), two long ciliary nerves (7) to eyeball, and, just before leaving the orbit through the anterior ethmoidal foramen, the infratochlear nerve (4); the rest of its course up to its termination on the nose as the external nasal nerve (1) illustrates the description given in the text.

11, lacrimal nerve, communicates with zygomatic nerve (12). 13, pterygopalatine ganglion.

The **infratrochlear nerve** passes forwards inferior to the pulley (trochlea) of the superior oblique muscle to supply skin and fasciae of the most medial parts of the eyelids and adjacent region of the external nose [FIG. 11.17]. It *communicates* with the supratrochlear nerve in the orbit or on the face.

Within the cone of the extrinsic ocular muscles, the nasociliary nerve sends sensory and sympathetic postganglionic fibres to the coats of the eyeball and their vessels, including the dilator pupillae muscle of the iris. (1) The communicating branch to the ciliary ganglion arises while the nerve lies lateral to the optic nerve; its fibres pass directly through the ganglion into the **short ciliary nerves** [p. 747]. (2) The **long ciliary nerves** arise superior to the optic nerve and pass along it to pierce the sclera beside that nerve [FIG. 12.20].

The nasociliary nerve occasionally sends a minute posterior ethmoidal nerve through the foramen of the same name to supply the posterior ethmoidal and sphenoidal air sinuses.

Maxillary nerve

The maxillary nerve passes forwards in the lower part of the lateral wall of the cavernous sinus [FIG. 11.8], then through the foramen rotundum [FIG. 3.87] into the upper part of the pterygopalatine fossa. Here it turns anterolaterally in the pterygomaxillary fissure

and passing along the inferior orbital fissure, curves anteromedially through that fissure to enter and traverse the infra-orbital groove (infra-orbital nerve), canal, and foramen and emerge on the face [FIG. 11.18].

BRANCHES

1. In the cranial cavity it gives a minute **meningeal** branch to the dura mater of the middle cranial fossa.

2. As the nerve enters the pterygopalatine fossa, two large **pterygopalatine nerves** pass inferomedially to the pterygopalatine ganglion [p. 758]. These branches consist of sensory nerve fibres which, together with postganglionic sympathetic nerve fibres from the **deep petrosal nerve** [p. 818], traverse the ganglion and form most of the nerve fibres in its so-called branches, without being functionally related to its cells. The cells of the ganglion contribute postganglionic parasympathetic fibres to these branches which are, therefore, sensory, secretomotor, and vasomotor to the territories they supply.

The **branches of the pterygopalatine ganglion** are:

(i) The **pharyngeal branch** passes posteriorly through the palatovaginal canal to supply the roof of the pharynx, the sphenoidal air sinus, and the pharyngeal opening of the auditory tube.

(ii) Three **palatine nerves** descend vertically through the pterygopalatine fossa and the palatine canals.

(a) The **greater palatine nerve** emerges from the greater palatine foramen [FIGS. 6.5 and 6.6] and passes forwards near the lateral margin of the hard palate. It breaks into branches which supply the structures of the hard palate and adjacent gum as far as the incisor teeth, where it communicates with branches of the nasopalatine nerves. In the palatine canal the main nerve gives off small **posterior inferior nasal branches** which pierce the vertical plate of the palatine bone and supply the mucous membrane of the postero-inferior part of the nasal cavity including the inferior concha [FIG. 11.15]. Other small branches pierce the maxilla and supply the posterior wall of the **maxillary sinus**.

(b) Two **lesser palatine nerves** pierce the palatine bone behind the greater palatine foramen [FIG. 6.6]. The more medial supplies the anteromedial part of the soft palate, the more lateral innervates the posterolateral part, the adjacent part of the tonsil, and the posterior part of the gum.

(iii) The **nasopalatine nerve** passes medially through the sphenopalatine foramen into the roof of the posterior part of the nasal cavity. It supplies this part of the roof and runs antero-inferiorly over the nasal septum (vomer) giving branches to the posterior part of that septum and floor of the nasal cavity. It then traverses the incisive canal with its fellow, communicates with the branches of the greater palatine nerve, and supplies the anterior part of the hard palate and gum behind the incisor teeth [FIG. 11.14].

(iv) Small **posterior superior nasal branches** pass through the sphenopalatine foramen to supply the posterosuperior part of the lateral wall of the nasal cavity including the superior and middle conchae.

(v) Small **orbital branches** pass superiorly to supply the orbital periosteum, and may contain secretomotor nerve fibres for the lacrimal gland.

3. The **posterior superior alveolar nerves** descend through the pterygomaxillary fissure to the posterior surface of the maxilla [FIG. 11.18]. One branch passes to the gum and adjacent part of the cheek, while the other two enter canaliculi in the bone which course round the maxillary sinus between the zygomatic process and the alveolar magin. They supply the molar teeth and mingle (**dental plexus**) in the canaliculi with fibres of the **anterior** (and sometimes middle) **superior alveolar nerve** both of them supplying the first molar tooth (Westwater 1960), the wall of the maxillary sinus, the gum, and the alveolar periosteum.

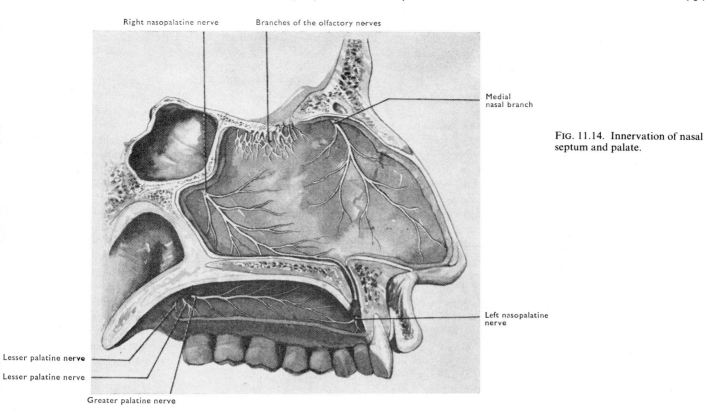

Right nasopalatine nerve   Branches of the olfactory nerves

Medial
nasal branch

FIG. 11.14. Innervation of nasal
septum and palate.

Left nasopalatine
nerve

Lesser palatine nerve

Lesser palatine nerve

Greater palatine nerve

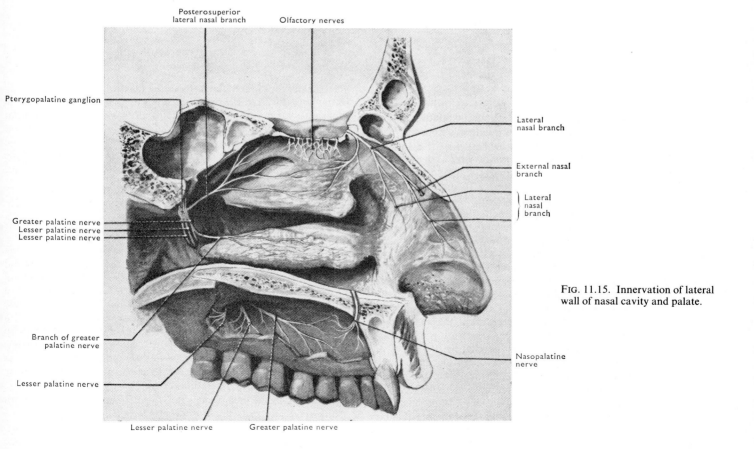

Posterosuperior
lateral nasal branch   Olfactory nerves

Pterygopalatine ganglion

Lateral
nasal branch

External nasal
branch

Lateral
nasal
branch

Greater palatine nerve
Lesser palatine nerve
Lesser palatine nerve

Branch of greater
palatine nerve

Lesser palatine nerve

Nasopalatine
nerve

FIG. 11.15. Innervation of lateral
wall of nasal cavity and palate.

Lesser palatine nerve   Greater palatine nerve

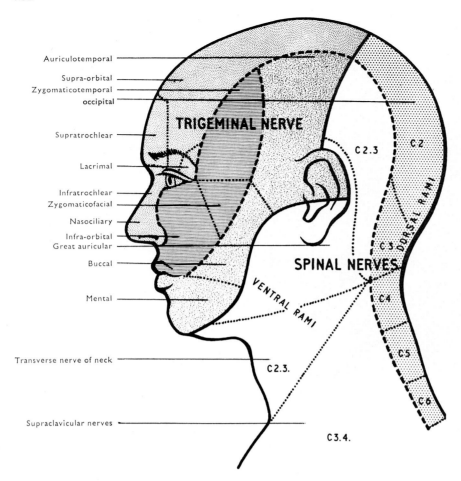

FIG. 11.16. Distribution of cutaneous nerves to head and neck. The ophthalmic, maxillary, and mandibular divisions of the trigeminal nerve are indicated by different shading.

4. The small **zygomatic nerve** arises in the pterygomaxillary fissure, passes through the inferior orbital fissure, and runs on the inferolateral wall of the orbit to enter the zygomatic bone and divide into: (1) a **zygomaticofacial branch** which emerges from the facial aspect of the zygomatic bone, and is sensory to the structures overlying it; (ii) a **zygomaticotemporal branch** which emerges from the temporal surface of the bone and pierces the temporal fascia to supply the anterior part of the temple [FIGS. 11.16 and 11.23]. In the orbit the zygomatic nerve **communicates with the lacrimal nerve** [FIG. 11.17] and may transmit postganglionic parasympathetic nerve fibres to the lacrimal gland from the pterygopalatine ganglion.

5. The **infra-orbital nerve** is the continuation of the maxillary through the infra-orbital groove and canal to the face.

The **anterior superior alveolar branch** arises from the infra-orbital nerve in the canal. It enters a sinuous canaliculus (Jones 1939) in the maxilla through which it passes forwards in the roof of the maxillary sinus, then medially in its anterior wall below the infra-orbital foramen to reach the lateral wall of the nasal cavity near the anterior end of the inferior concha. Here it turns inferiorly and then medially [FIG. 11.18] to enter the palatine process of the maxilla and emerge at the side of the nasal septum close to the anterior nasal spine. In the palate and anterior part of the maxilla the nerve gives **branches** to the incisor, canine, premolar, and first molar teeth, to the adjacent gum and periosteum, to the maxillary sinus, and to the lateral wall, floor, and septum of the nasal cavity anteriorly. The dental branches form a **plexus** with those of the posterior superior alveolar nerves [FIG. 11.18].

In 30 per cent of cases a **middle superior alveolar branch** arises from the infra-orbital nerve posterior to the anterior superior

alveolar nerve. It passes laterally in the maxilla and turns inferiorly in the lateral wall of the maxillary sinus to join the branches of the other alveolar nerves in the canaliculi. *In the face*, the infra-orbital nerve divides into a series of radiating *branches*: (i) **inferior palpebral**, to the lower eyelids; (ii) **external nasal**, to the side of the nose; (iii) **superior labial**, to the upper lip and anterior cheek. These branches form an infra-orbital plexus with the zygomatic branches of the facial nerve [FIG. 11.23].

Mandibular nerve

The mandibular nerve consists of a large sensory part from the trigeminal ganglion and the entire motor root of the trigeminal nerve. The two parts emerge through the foramen ovale combining into a short, thick nerve situated between the lateral pterygoid and tensor veli palatini muscles, immediately anterior to the middle meningeal artery. Two small branches arise from the nerve: (1) a small **meningeal branch** passes through the foramen spinosum with the middle meningeal artery to supply the periosteum, dura mater, and mastoid air cells; (2) the **nerve to the medial pterygoid** passes medially (through or beside the otic ganglion [p. 761]) to the tensor tympani and tensor veli palatini muscles.

The main nerve then divides into a small, mainly motor, anterior trunk and a large, mainly sensory, posterior trunk.

THE ANTERIOR TRUNK

This passes antero-inferiorly medial to the lateral pterygoid muscle.

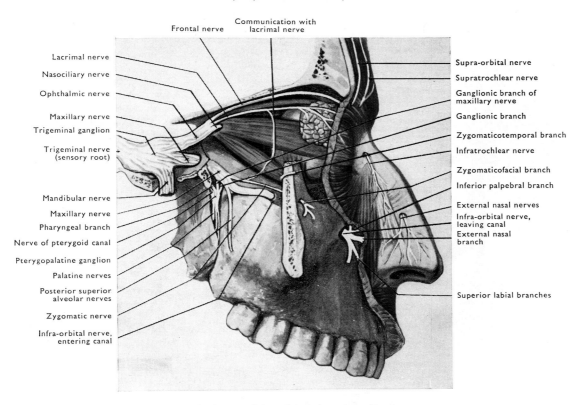

FIG. 11.17. Course of the ophthalmic and maxillary nerves.

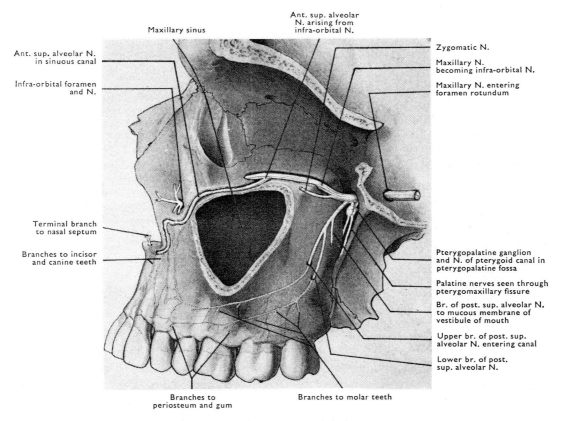

FIG. 11.18. Course and branches of maxillary nerve.

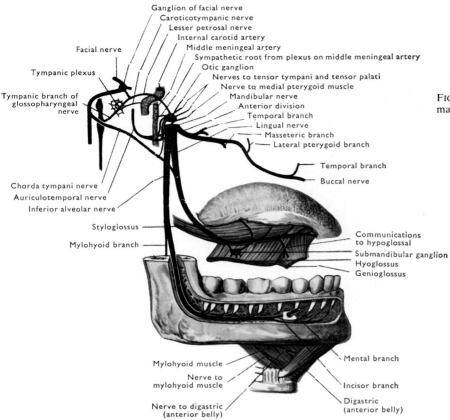

Ganglion of facial nerve
Caroticotympanic nerve
Lesser petrosal nerve
Internal carotid artery
Middle meningeal artery
Sympathetic root from plexus on middle meningeal artery
Otic ganglion
Nerves to tensor tympani and tensor palati
Nerve to medial pterygoid muscle
Mandibular nerve
Anterior division
Temporal branch
Lingual nerve
Masseteric branch
Lateral pterygoid branch
Temporal branch
Buccal nerve
Facial nerve
Tympanic plexus
Tympanic branch of glossopharyngeal nerve
Chorda tympani nerve
Auriculotemporal nerve
Inferior alveolar nerve
Styloglossus
Mylohyoid branch
Communications to hypoglossal
Submandibular ganglion
Hyoglossus
Genioglossus
Mylohyoid muscle
Nerve to mylohyoid muscle
Nerve to digastric (anterior belly)
Mental branch
Incisor branch
Digastric (anterior belly)

FIG. 11.19. Scheme of distribution and connections of mandibular nerve.

Branches

1. The **nerve to lateral pterygoid** passes into the deep surface of that muscle.

2. The **masseteric nerve** passes laterally between the lateral pterygoid muscle and the base of the skull, immediately anterior to the capsule of the temporomandibular joint. The nerve sends a filament to the joint and passes through the mandibular notch, posterior to the tendon of temporalis, into the deep surface of masseter.

3. Two or three **deep temporal nerves** run laterally between the lateral pterygoid muscle and the skull, and turn upwards into the deep surface of temporalis. The anterior nerve often passes with the buccal nerve.

4. The **buccal nerve**, the only purely sensory branch of the trunk, passes between the two heads of the lateral pterygoid muscle and descends towards the cheek. It grooves or pierces the lowest anterior fibres of the tendon of temporalis close to the posterior wall of the vestibule of the mouth. The branches of the nerve are sensory to the entire thickness of the cheek and to the gums. They may partly supply the premolar and the first molar teeth of the lower jaw. The motor fibres to the buccinator muscle reach it via the buccal branch of the facial nerve which communicates with the buccal nerve in the cheek.

THE POSTERIOR TRUNK

This trunk gives off the auriculotemporal nerve and then divides into the inferior alveolar and lingual nerves while still medial to the lateral pterygoid muscle.

Branches

1. The **auriculotemporal nerve** arises as two purely sensory roots which unite after embracing the middle meningeal artery. Each of these roots receives a small bundle of postganglionic parasympathetic fibres from the otic ganglion (secretomotor to the parotid gland, p. 814) and postganglionic sympathetic fibres which traverse the ganglion from the plexus on the middle meningeal artery. The nerve passes posterolaterally between the lateral pterygoid muscle and the spine of the sphenoid, then curves laterally on the posterior surface of the neck of the mandible. Here it sends branches to the temporomandibular joint and to the parotid gland (sensory and secretomotor). It then runs through the upper part of the parotid gland or its fascia, and sends branches to the tympanic membrane, the external acoustic meatus, and the lateral surface of the upper half of the auricle. It crosses the zygomatic process of the temporal bone, and supplies the skin and fascia of the temple and scalp almost to the vertex [FIG. 11.16].

2. The **lingual nerve** is the anterior and smaller of the descending branches of the posterior trunk. It is *entirely sensory*, but receives taste and preganglionic parasympathetic nerve fibres which join it with the **chorda tympani** branch of the facial nerve [FIG. 11.19] deep to the lateral pterygoid muscle. The lingual nerve curves anteriorly between the ramus of the mandible and the medial pterygoid muscle, and passes below the inferior margin of the superior constrictor muscle to lie between the mucous membrane of the mouth and the body of the mandible, immediately inferior to the last molar tooth, above the posterior end of the mylohyoid line [FIGS. 5.23 and 11.21]. Continuing antero-inferiorly between the lateral surface of the hyoglossus and the mylohyoid muscle, and

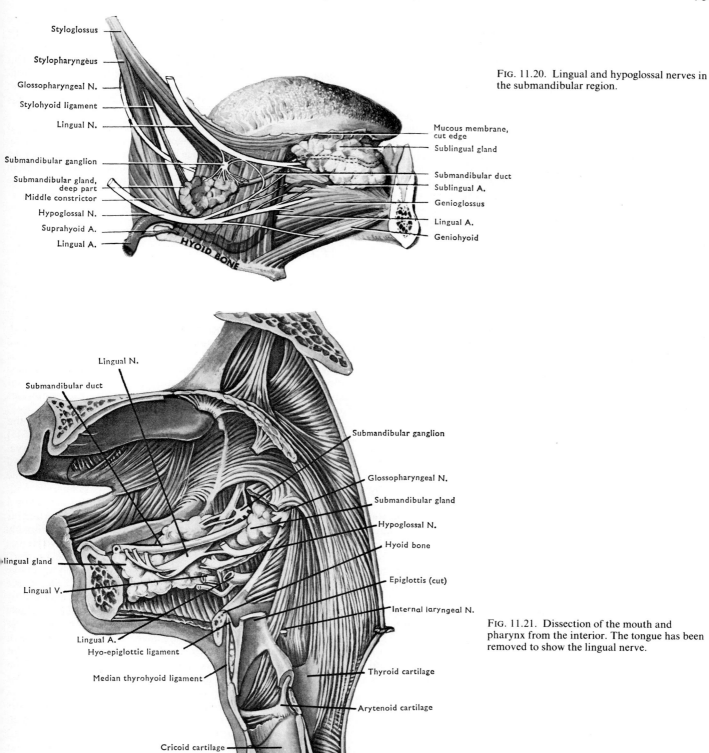

Styloglossus
Stylopharyngèus
Glossopharyngeal N.
Stylohyoid ligament
Lingual N.
Submandibular ganglion
Submandibular gland, deep part
Middle constrictor
Hypoglossal N.
Suprahyoid A.
Lingual A.
HYOID BONE

Mucous membrane, cut edge
Sublingual gland
Submandibular duct
Sublingual A.
Genioglossus
Lingual A.
Geniohyoid

FIG. 11.20. Lingual and hypoglossal nerves in the submandibular region.

Lingual N.
Submandibular duct
lingual gland
Lingual V.
Lingual A.
Hyo-epiglottic ligament
Median thyrohyoid ligament
Cricoid cartilage

Submandibular ganglion
Glossopharyngeal N.
Submandibular gland
Hypoglossal N.
Hyoid bone
Epiglottis (cut)
Internal laryngeal N.
Thyroid cartilage
Arytenoid cartilage

FIG. 11.21. Dissection of the mouth and pharynx from the interior. The tongue has been removed to show the lingual nerve.

separating the sublingual gland anteriorly from the deep part of the submandibular gland postero-inferiorly [FIG. 11.21], the nerve curves below the alveolingual sulcus and the submandibular duct to turn superiorly on the lateral surface of genioglossus and supply the anterior two-thirds of the tongue.

The lingual nerve gives off a series of **branches** on the lateral surface of hyoglossus.

(i) To the submandibular ganglion. This conveys sensory trigeminal fibres and preganglionic parasympathetic fibres from the chorda tympani.

(ii) To the floor and lateral wall of the alveololingual sulcus, the lingual aspect of the lower gum, and the premolar and first molar teeth. These branches may pass through the submandibular ganglion and receive postganglionic parasympathetic fibres from it.

(iii) A communicating branch with the hypoglossal nerve through which sensory fibres may be distributed with that nerve [FIG. 11.20].

The lingual nerve receives branches from the submandibular ganglion. These contain postganglionic parasympathetic fibres which pass to the glands in the territory of the lingual nerve.

3. The **inferior alveolar nerve** contains both motor and sensory fibres. It descends to the mandibular foramen posterior to the lingual nerve from which it receives a communication. At the foramen it lies between the sphenomandibular ligament and the bone, and gives rise to its only motor branch, the **mylohyoid nerve.** This pierces the ligament and descends with the corresponding vessels in the mylohyoid groove [FIG. 3.64] to pass between the inferior surface of the mylohyoid muscle and the submandibular gland. It supplies the mylohyoid muscle and the anterior belly of the digastric muscle.

The inferior alveolar nerve enters the mandibular canal with the inferior alveolar vessels. Here it gives branches which form a fine **inferior dental plexus** from which the mandibular teeth, alveolar bone, and gums are supplied. The branches that supply the molar, premolar, and canine teeth arise before the mental branch leaves the nerve, while the branches that form the plexus supplying the incisor (and canine) teeth arise distal to this (Starkie and Stewart 1931). The premolar and first molar teeth are also partly supplied by the lingual and buccal nerves (Stewart and Wilson 1928).

The **mental nerve** arises from the inferior alveolar nerve in the mandibular canal. It passes through the mental foramen, communicates with the marginal mandibular branch of the facial nerve [FIG 11.23] and immediately divides into numerous branches deep to the facial muscles. It supplies the skin of the lower lip and chin and the mucous membrane and glands of the alveololabial sulcus and adjacent gum.

Injury to the trigeminal nerve mainly produces sensory loss in the territory of the nerve, or of the part involved. If the mandibular nerve is destroyed, the muscles of mastication are paralysed. Thus the jaw cannot be protruded or retracted symmetrically; on protrusion the chin is swung towards the paralysed side (lateral pterygoid paralysis), on retraction towards the uninjured side (paralysis of temporalis).

ABDUCENT NERVE

The sixth cranial or abducent nerve supplies only the lateral rectus muscle of the eyeball. It emerges from the brain in the pontomedullary sulcus [FIG. 11.4] immediately superior to the pyramid [FIG. 10.50. For deep origin, see p. 646] and curves superiorly, anterior to the pons, and usually posterior to the anterior inferior cerebellar artery. At the apex of the petrous temporal bone it turns abruptly forwards through the arachnoid and dura mater of the posterior cranial fossa to enter or pass above, the inferior petrosal sinus and below the petroclinoid ligament [FIG. 13.47]. The nerve then enters

the cavernous sinus and runs forwards immediately lateral to the internal carotid artery, receiving a **sympathetic** filament from the internal carotid plexus. The nerve leaves the sinus anteriorly, communicates with the ophthalmic nerve, and enters the orbit through the superior orbital fissure within the tendinous ring from which the rectus muscles arise, lying inferior to the oculomotor and nasociliary nerves. It enters the medial surface of the lateral rectus muscle [FIG. 11.12].

The position of the nerve in the cavernous sinus makes it susceptible to pathological changes in the sinus, while its vertical course in the subarachnoid space results in increased tension on the nerve whenever the brain stem is displaced inferiorly. Both conditions lead to weakness or paralysis of the lateral rectus muscle with a consequent internal squint.

FACIAL NERVE

The seventh cranial or facial nerve supplies the structures derived from the second pharyngeal (branchial) arch of the embryo. It is predominantly an **efferent** nerve (1) to the muscles of facial expression, also to the posterior belly of digastric, the stylohyoid, and the stapedius muscles, and (2) to many of the glands of the head by preganglionic parasympathetic nerve fibres that pass to the cells of the pterygopalatine and submandibular ganglia. It also contains a few **afferent** fibres which originate in the cells of its genicular ganglion and are predominantly concerned with taste sensations from the anterior two-thirds of the tongue and the palate.

The facial nerve arises by two roots from the lateral part of the pontomedullary sulcus [FIG. 10.5l. For deep connections, see p. 648] immediately anterior to the vestibulocochlear (eighth cranial) nerve. The **roots** are the large, motor, facial nerve anterior to the small **nervus intermedius** which transmits sensory and preganglionic parasympathetic nerve fibres [FIG. 11.4]. They pass laterally with the eighth nerve into the internal acoustic meatus, surrounded by a sheath of the meninges. Here the branches of the nervus intermedius join the seventh and eighth nerves, though all its fibres probably enter the seventh nerve distally [FIG. 11.25]. The facial nerve pierces the meninges at the lateral end of the internal acoustic meatus [FIG. 12.51] and continues laterally in the bony facial canal lying above and between the cochlea and vestibule [FIG. 3.54 and 12.48]. At the hiatus for the greater petrosal nerve, the facial nerve is enlarged by the sensory **genicular ganglion** and gives off (1) the **greater petrosal nerve,** (2) a **branch to the tympanic plexus,** and (3) a branch to the sympathetic plexus on the middle meningeal artery. The facial nerve now turns abruptly backwards (**geniculum of the facial nerve**) in the bone of the upper part of the medial (labyrinthine) wall of the middle ear cavity, superior to the fenestra vestibuli and then inferior to the prominence caused by the lateral semicircular canal in the aditus to the mastoid antrum [FIG. 12.40]. Medial to the aditus, the nerve turns vertically downwards in the bony septum which separates the middle ear from the mastoid antrum and air cells, and gives off first the **nerve to stapedius**, then the **chorda tympani**, and finally a **communicating branch to the auricular branch of the vagus.** The last arises immediately before the nerve emerges from the stylomastoid foramen [FIG. 11.22].

The facial nerve emerges from the stylomastoid foramen under cover of the mastoid process; here it is more liable to injury in the child because that process is incompletely developed. The nerve passes anterolaterally between the styloid process and the posterior belly of digastric [FIG. 13.46], and gives off: (1) one or two descending **branches** which supply the posterior belly of the **digastric** and the **stylohyoid** muscle; and (2) the **posterior auricular nerve.** The facial nerve then enters the posteromedial surface of the

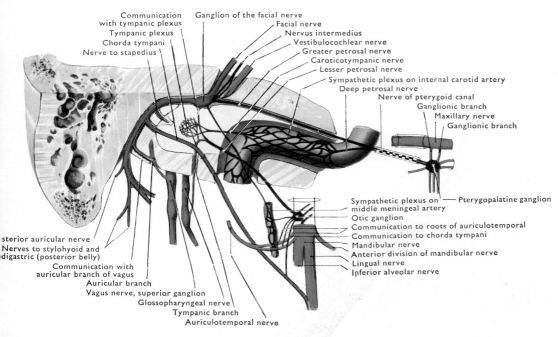

Fig. 11.22. Connections of facial nerve in the temporal bone.

parotid gland, and crossing superficial to the external carotid artery and the retromandibular vein, breaks into a number of branches which emerge separately from the gland [Fig. 11.23] and pass to supply the muscles of facial expression on their deep surfaces.

BRANCHES

1. *From the geniculum of the facial nerve;*

i. The **greater petrosal nerve** (sensory, taste to the palate, and preganglionic parasympathetic to pterygopalatine ganglion) passes forwards in its canal and then in the dural floor of the middle cranial fossa into the upper part of the foramen lacerum. Here it is joined by a branch (**deep petrosal nerve**) from the internal carotid plexus (sympathetic) and traverses the pterygoid canal to the pterygopalatine ganglion as the **nerve of the pterygoid canal** [Fig. 3.87].

ii. A small branch passes through the temporal bone to join the **tympanic plexus** of the glossopharyngeal nerve (q.v.).

iii. A minute, inconstant branch to the sympathetic plexus on the middle meningeal artery.

2. *In the descending part of the canal;*

i. The small **stapedial nerve** passes forwards to supply the stapedius muscle. If the facial nerve is destroyed proximal to this branch, paralysis of stapedius causes sounds to be heard louder (hyperacousia) in the affected ear in addition to the effects of destruction of the facial nerve distal to this branch.

ii. The **chorda tympani** passes forwards through its canaliculus into the middle ear, and crossing the medial aspect of the fibrous tympanic membrane and handle of the malleus, leaves it through the petrotympanic fissure [Fig. 3.81] medial to the temporomandibular joint. It then passes antero-inferiorly in the infratemporal fossa (medial to the spine of the sphenoid and the lateral pterygoid muscle), receives a filament from the otic ganglion, and joins the lingual nerve [Fig. 11.22]. The sensory (taste) fibres are distributed with that nerve to the anterior two-thirds of the tongue (excluding the vallate papillae). The preganglionic parasympathetic fibres pass to the cells of the submandibular ganglion [see below] in a branch from the lingual nerve.

iii. Just superior to the stylomastoid foramen, a fine branch passes to the **auricular branch of the vagus**. It presumably transmits sensory fibres to the external acoustic meatus.

3. *In the neck;*

i. A descending **digastric branch** (or branches) passes to supply the posterior belly of digrastic and the stylohoid muscle.

ii. The **posterior auricular nerve** bends backwards lateral to the mastoid process with the posterior auricular artery. It supplies the intrinsic and posterior auricular muscles, and sends an **occipital branch** on the superior nuchal line to the occipital belly of occipitofrontalis.

4. *In the parotid gland;*

i. **Temporal branches** pass anterosuperiorly over the zygomatic arch to supply the facial muscles above the level of the palpebral fissure. A small branch passes posterosuperiorly to the anterior and superior auricular muscles.

ii. The **zygomatic branches** supply the facial muscles between the levels of the palpebral and oral fissures. The smaller, upper branches cross the zygomatic arch towards the orbicularis oculi and may be indistinguishable from the temporal branches. The larger, lower branches pass with the parotid duct over masseter, and run medially to form the infra-orbital plexus with the **infra-orbital nerve**.

iii. The **buccal branches** run towards the angle of the mouth, and forming a plexus with the **buccal nerve**, supply buccinator and the other muscles in this region.

iv. The **marginal mandibular branch** passes along the inferior border of the mandible, often looping down into the neck. It ends after forming a plexus with the mental nerve by supplying the muscles in the lower lip and the chin.

v. The **cervical branch** emerges from the lower end of the parotid gland and passes antero-inferiorly, below the angle of the mandible, to supply platysma from its deep surface.

These branches are the only motor supply to the facial muscles, but the exact distribution of each cannot be given because they vary in size from individual to individual and communicate with each other on the face. Though these branches may transmit some

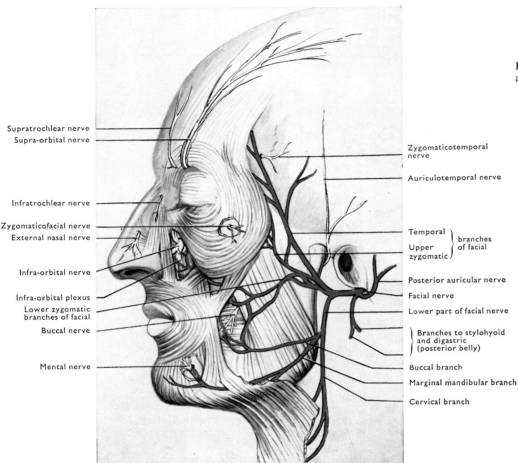

Supratrochlear nerve
Supra-orbital nerve

Infratrochlear nerve

Zygomaticofacial nerve
External nasal nerve

Infra-orbital nerve

Infra-orbital plexus
Lower zygomatic
branches of facial
Buccal nerve

Mental nerve

Zygomaticotemporal
nerve

Auriculotemporal nerve

Temporal   ⎫ branches
            ⎬ of facial
Upper   ⎭
zygomatic

Posterior auricular nerve
Facial nerve
Lower part of facial nerve

⎫ Branches to stylohyoid
⎬ and digastric
⎭ (posterior belly)

Buccal branch
Marginal mandibular branch
Cervical branch

FIG. 11.23. Distribution of trigeminal and facial nerves on the face.

proprioceptive fibres from the facial muscles, it is probable that the majority enter the brain through the trigeminal nerve via the communications on the face.

Injury to the facial nerve in the parotid gland causes paralysis of these muscles, and, amongst other things, inability to close the eye (orbicularis oculi) or mouth (orbicularis oris), or to use the lips or cheeks (buccinator) properly in eating or speaking. Thus the conjunctiva is not kept moist by blinking, while the drooping lower eyelid fills with lacrimal fluid which cannot reach the lacrimal puncta and consequently overflows on to the face. Speech is blurred, fluid and food collect in the vestibule of the mouth and escape between the lips. The mouth is pulled towards the normal side by the unopposed action of that half of the orbicularis oris. If the nerve is damaged in the bony canal there may also be loss of taste on the anterior two-thirds of the tongue (chorda tympani) and hyperacousia on the affected side (stapedius). Damage in the internal acoustic meatus is usually associated with injury to the eighth nerve and thus deafness, not hyperacousia, results.

PTERYGOPALATINE GANGLION [FIG. 11.15]

This parasympathetic ganglion is a small reddish body lying in the upper part of the pterygopalatine fossa [FIG. 11.18] close to the sphenopalatine foramen. The ganglion receives **preganglionic parasympathetic nerve fibres** from the facial nerve via the **nerve of the pterygoid canal** [FIG. 11.24]. These synapse with the cells of the ganglion which send postganglionic nerve fibres into its branches

and into the maxillary nerve for distribution through it. Thus **postganglionic parasympathetic nerve fibres** reach gland cells in the nasal mucosa, the paranasal sinuses, the palate, the roof of the pharynx, the auditory tube, the lacrimal gland, and the upper lip. Branches of the ganglion also transmit (1) **sensory fibres** which enter the brain through the maxillary (general sensation) and facial (taste from the palate, especially in children) nerves and have their cell bodies in the trigeminal and genicular ganglia, and (2) **postganglionic sympathetic nerve fibres** from the nerve of the pterygoid canal. Both these groups of fibres traverse the ganglion but have no functional connection with its cells.

SUBMANDIBULAR GANGLION

This parasympathetic ganglion lies on the hyoglossus muscle between the lingual nerve and the submandibular duct [FIG. 11.20]. Its connections are similar to those of the pterygopalatine ganglion. The **preganglionic parasympathetic fibres** come from the chorda tympani via the lingual nerve, while **sensory fibres** from the lingual nerve and **postganglionic sympathetic fibres** from the plexus on the facial artery traverse it. **Postganglionic parasympathetic fibres** from the cells of the ganglion pass with the sensory and sympathetic fibres to the submandibular and sublingual glands, to the lingual gum, and to the alveololingual sulcus, while others pass directly into the lingual nerve for distribution to the glands in the tongue. Possibly such fibres also pass through the communication which the lingual nerve has with the inferior alveolar nerve and thus pass to supply glands in the lower lip, etc.

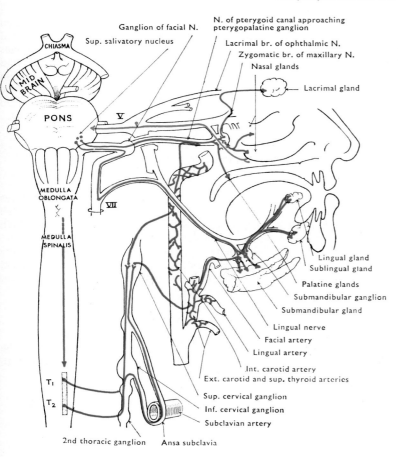

N. of pterygoid canal approaching
pterygopalatine ganglion

Ganglion of facial N.

Sup. salivatory nucleus

Lacrimal br. of ophthalmic N.

Zygomatic br. of maxillary N.

Nasal glands

Lacrimal gland

CHIASMA

MID BRAIN

PONS

V

MEDULLA OBLONGATA

VII

MEDULLA SPINALIS

Lingual gland
Sublingual gland
Palatine glands
Submandibular ganglion
Submandibular gland
Lingual nerve
Facial artery
Lingual artery
Int. carotid artery
Ext. carotid and sup. thyroid arteries

T₁

T₂

Sup. cervical ganglion
Inf. cervical ganglion
Subclavian artery

2nd thoracic ganglion   Ansa subclavia

FIG. 11.24. Diagram of connections of pterygopalatine and submandibular ganglia with distribution of parasympathetic components of seventh cranial nerve.
Sympathetic, red; parasympathetic, green; sensory, blue.

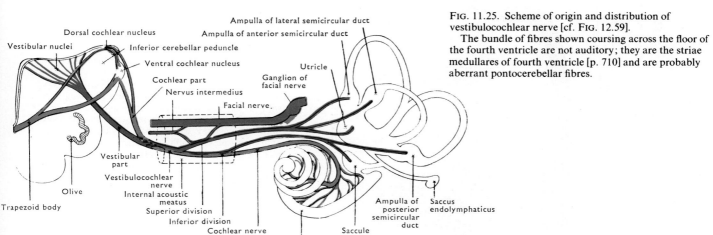

Dorsal cochlear nucleus
Vestibular nuclei
Inferior cerebellar peduncle
Ventral cochlear nucleus
Cochlear part
Nervus intermedius
Facial nerve

Ampulla of lateral semicircular duct
Ampulla of anterior semicircular duct
Utricle
Ganglion of facial nerve

Vestibular part
Vestibulocochlear nerve
Internal acoustic meatus
Superior division
Inferior division
Cochlear nerve

Olive

Trapezoid body

Ampulla of posterior semicircular duct
Saccus endolymphaticus
Saccule
Cochlear duct

FIG. 11.25. Scheme of origin and distribution of vestibulocochlear nerve [cf. FIG. 12.59].
The bundle of fibres shown coursing across the floor of the fourth ventricle are not auditory; they are the striae medullares of fourth ventricle [p. 710] and are probably aberrant pontocerebellar fibres.

VESTIBULOCOCHLEAR NERVE

The eighth cranial or vestibulocochlear nerve is sensory and supplies the organs of hearing (cochlea) and equilibration (utricle, saccule, and semicircular ducts) within the petrous temporal bone. It also contains efferent fibres to the cochlea from the dorsal nucleus of the trapezoid body [p. 701]. The nerve consists of a **superior** (vestibular) **root** and an **inferior** (cochlear) **root** which arise together from the pontomedullary sulcus of the brain stem immediately posterior to the facial nerve [FIGS. 10.51 and 11.25. For deep connections, see

p. 651]. The two roots unite to form the vestibulocochlear nerve and pass into the internal acoustic meatus inferior to the facial nerve and nervus intermedius with which they have temporary communications. Near the fundus of the meatus, two or more swellings appear on the eighth nerve. These consist of the bipolar cells of the **vestibular ganglion** the peripheral processes of which innervate the organs of equilibration, while the central processes form the **vestibular part of the nerve**. Distal to the ganglion, the vestibular part divides into superior and inferior parts, the inferior part lying with the cochlear part.

VESTIBULAR PART

The four branches of the superior vestibular part pierce the superior vestibular area in the fundus of the meatus [FIG. 12.51] and supply: (1) the macula of the utricle (utricular nerve), (2) and (3) supply the ampullae of the anterior and lateral semicircular ducts (anterior and lateral ampullary nerves), and (4) a small branch to the macula of the saccule.

The branches of the inferior vestibular part supply: (1) the macula of the saccule through the inferior vestibular area of the meatal fundus [FIG. 12.51]; and (2) the ampulla of the posterior semicircular duct through the foramen singulare (posterior ampullary nerve).

COCHLEAR PART

The cochlear part of the eighth nerve passes through the tractus spiralis foraminosus [FIG. 12.51] as a series of fine bundles which enter the modiolus [FIG. 12.50] and turning outwards towards the osseous spiral lamina expand into the spiral ganglion which fills the spiral canal medial to the base of the lamina [FIG. 12.55]. The spiral ganglion, the counterpart of the vestibular ganglion on the vestibular fibres, contains bipolar cell bodies, the peripheral processes of which innervate the hair cells of the cochlea.

The vestibulocochlear nerve may be injured in a number of pathological conditions, and the cells of its ganglia are particularly susceptible to certain drugs, e.g. streptomycin and salicylates.

Tumours may form from the sheath cells of this nerve and first irritate and then destroy it. The irritation causes noises in the head (cochlear part) and giddiness which is observed as a rapid nystagmus (oscillating movement) of the eyes, and a tendency to stagger and fall (vestibular part). Unilateral total deafness and loss of reaction from the vestibular part of the internal ear supervene when the nerve is destroyed. The facial nerve (q.v.) is also likely to be involved by such a tumour, especially if it is in the internal acoustic meatus.

GLOSSOPHARYNGEAL NERVE

The ninth cranial or glossopharyngeal nerve supplies the structures that develop from the third pharyngeal (branchial) arch of the embryo. It is a mixed nerve with a large afferent element which supplies part of the pharynx, the posterior one-third of the tongue, and the carotid body and sinus. The small efferent part supplies the stylopharyngeus and middle constrictor muscles (branchial motor) and also sends preganglionic parasympathetic nerve fibres predominantly to the otic ganglion.

SUPERFICIAL ORIGIN AND COURSE

The glossopharyngeal nerve arises by a few fine rootlets from the lateral surface of the upper medulla oblongata between the seventh

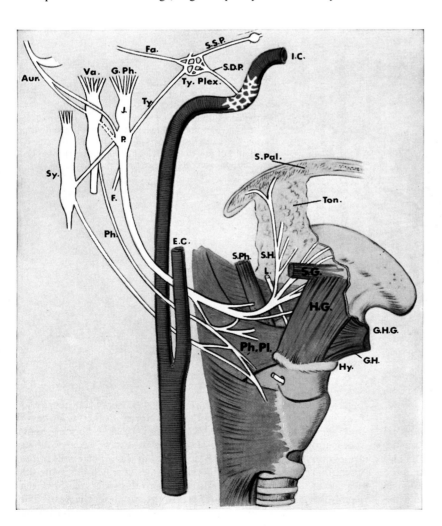

FIG. 11.26. Scheme of connections and distribution of glossopharyngeal nerve.

G.Ph., glossopharyngeal nerve; J., superior, and P., inferior ganglia; Ty., tympanic nerve; Ty. Plex., tympanic plexus; Fa., root from ganglion of facial nerve; S.S.P., lesser petrosal nerve to the otic ganglion; S.D.P., caroticotympanic nerve; I.C., internal carotid artery; Va., vagus nerve; Aur., auricular branch of vagus; Sv., superior cervical sympathetic ganglion; F., communicating branch to facial nerve; Ph., pharyngeal branch of vagus; E.C., external carotid artery; Ph.P., pharyngeal plexus; S.Ph., stylopharyngeus muscle; S.H.L., stylohyoid ligament; H.G., hyglossus, S.G., styloglossus; S.Pal., soft palate; G.H.G., genioglossus; G.H., geniohyoid; Hy., hyoid bone; Ton., palatine tonsil.

and tenth nerves [FIG. 10.50. For deep connections, see pp. 648 and 651]. The rootlets unite as they enter the jugular foramen between the sigmoid and inferior petrosal sinuses, close to the vagus and accessory nerves, but in a separate dural sheath [FIG. 11.29]. Within the foramen there are usually two enlargements of the nerve, the **superior** and **inferior (sensory) ganglia**. The nerve descends into the neck between the internal carotid artery and the internal jugular vein and, passing between the internal and external carotid arteries, curves behind the stylopharyngeus to reach its lateral aspect. It then runs forwards deep to the stylohyoid ligament and hyoglossus muscle, and enters the base of the tongue, close to the lower pole of the palatine tonsil, by passing below the inferior border of the superior constrictor muscle with the stylopharyngeus.

BRANCHES

1. The **tympanic nerve** arises from the inferior ganglion and superiorly in the bone between the jugular foramen and the carotid canal into the middle ear. Here it forms the **tympanic plexus** [FIG. 11.22] on the medial wall of the middle ear with sympathetic fibres from the internal carotid plexus (**caroticotympanic nerves**) and a twig from the genicular ganglion of the facial nerve. Branches from the plexus supply the walls of the middle ear, the mastoid air cells, and the auditory tube (**tubal branch**). Some of the glossopharyngeal (parasympathetic) and facial fibres unite to form the **lesser petrosal nerve**. This passes into the dural floor of the middle cranial fossa, and piercing the base of the skull close to the foramen ovale, enters the **otic ganglion**.

The **inferior ganglion** also receives sympathetic fibres from the superior cervical ganglion and has branches joining it to the auricular branch and superior ganglion of the vagus and to the facial nerve.

2. In the neck: (i) A **branch to stylopharyngeus** which also sends fibres through that muscle to supply pharyngeal mucous membrane (**tonsillar branches**); (ii) **pharyngeal branches** to the middle constrictor and the mucous membrane of the pharynx, which either pass directly through the superior constrictor or unite with pharyngeal plexus [FIG. 11.26]: (iii) the **carotid sinus branch** runs downwards anterior to the internal carotid artery [FIG. 9.9], communicates with the vagus and sympathetic (Sheehan, Mulholland, and Shafiroff 1941) then divides in the angle of bifurcation of the common carotid artery to supply the carotid body and carotid sinus. It carries impulses from the baroreceptors in the carotid sinus, and from chemoreceptors in the carotid body.

3. The **lingual branches** supply the mucous membrane of the posterior one-third of the tongue (including the vallate papillae) with taste and general sensory fibres. They also contain preganglionic parasympathetic fibres which may relay in small ganglia scattered in the substance of the tongue to supply the posterior lingual glands (Fitzgerald and Alexander 1969). **Tonsillar branches** supply the mucous membrane covering the tonsil, the adjacent part of the soft palate, and the palatine arches.

OTIC GANGLION

This small, parasympathetic ganglion lies immediately inferior to the foramen ovale, between the mandibular nerve and the tensor veli palatini muscle. **Preganglionic parasympathetic fibres** reach it from the glossopharyngeal nerve (and possibly also from the facial nerve) via the lesser petrosal nerve (q.v.). As in the other parasympathetic ganglia, these are the only fibres which synapse with the cells of the ganglion; the postganglionic sympathetic fibres from the plexus on the middle meningeal artery and the sensory and

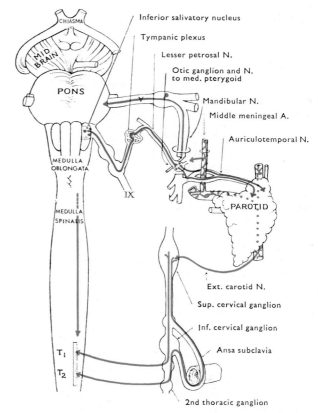

FIG. 11.27. Diagram of connection of otic ganglion with distribution of parasympathetic components of ninth cranial nerve.
    Sympathetic, red; parasympathetic, green; sensory, blue.

motor fibres (supply to the medial pterygoid, tensor veli palatini, and tensor tympani muscles) from the mandibular nerve merely pass through the ganglion.

The **postganglionic parasympathetic fibres** from the ganglion pass with the sympathetic fibres mainly into the auriculotemporal nerve to supply the parotid gland. Other small twigs joint the chorda tympani, the nerve of the pterygoid canal and branches of the mandibular nerve.

VAGUS NERVE

The tenth cranial or vagus nerve is mixed. It supplies (i) the structures developed from the fourth and sixth pharyngeal (branchial) arches of the embryo, and (ii) the foregut and midgut and the structures developed from the entoderm or mesoderm of these parts of the gut tube. It, therefore, carries **afferent nerve fibres** from the pharynx, oesophagus, stomach, small and part of the large intestine, larynx, trachea, lungs, heart and great vessels, pancreas and liver, though some of the sensory fibres from most of these organs enter the central nervous system by running through the sympathetic system into the spinal nerves. The **efferent fibres** of the vagus are: (1) preganglionic parasympathetic fibres distributed to the parasympathetic ganglia that innervate smooth or cardiac muscle and gland cells in the above organs; (2) fibres (branchial) direct to the striated muscles of the larynx and pharynx, some of which emerge from the brain in the cranial part of the accessory nerve.

SUPERFICIAL ORIGIN AND COURSE

The vagus arises from the lateral surface of the medulla oblongata by numerous rootlets in series with the glossopharyngeal above and the accessory nerve below [for deep connections, see pp. 647 and 651]. The rootlets unite and pass through the jugular foramen immediately posterior to the glossopharyngeal nerve, but in a separate dural sheath shared with the accessory nerve, a feature which makes the accessory nerve essentially a part of the vagus nerve. The vagus has a small, **superior ganglion** in the jugular foramen and a larger, **inferior ganglion** immediately below the foramen. Both ganglia are fusiform in shape and consist of afferent nerve cells.

In the neck, the vagus descends vertically in the carotid sheath, posterior to and between the internal jugular vein and first the internal and then the common carotid artery. It enters the superior mediastinum of the thorax posterior to the formation of the brachiocephalic vein: on the right side, after crossing anterior to the first part of the subclavian artery; on the left side, in the interval between the common carotid and subclavian arteries.

In the mediastinum both nerves pass postero-inferiorly to reach the posterior surface of the corresponding lung root, the right nerve on the trachea; the left at first between the two arteries and then on the left side of the arch of the aorta, posterior to the phrenic nerve. At the root of the lung each nerve gives branches to the **pulmonary plexuses** [q.v. FIG. 11.28] and appears inferior to the lung root as two or more branches. These pass inferomedially to unite with the corresponding branches of the opposite vagus and form a **plexus** surrounding the oesophagus which receives one or more branches from each greater splanchnic nerve (sympathetic). Near the oesophageal opening in the diaphragm, the plexus reforms into **anterior** and **posterior vagal trunks** each of which passes as one or more bundles on to the corresponding surface of the stomach (**gastric branches**). Each trunk contains fibres from both vagus nerves and from the sympathetic.

Communications and branches

SUPERIOR GANGLION

In the jugular foramen this ganglion communicates with the superior cervical sympathetic ganglion, the accessory nerve, and (sometimes) the inferior glossopharyngeal ganglion. It gives off two branches.

1. A meningeal branch passes posteriorly to supply the dura mater of the posterior cranial fossa.

2. The auricular branch receives a twig from the tympanic branch of the glossopharyngeal nerve, and entering the mastoid canaliculus on the lateral wall of the jugular fossa, is joined by a branch from the facial nerve. This complex emerges from the tympanomastoid fissure and supplies the inferior part of the tympanic membrane and the floor of the external acoustic meatus. The importance of this branch lies in the fact that irritation of the skin in this region, e.g. a foreign body in the external acoustic meatus of a child, may mimic irritation of other parts of the sensory distribution of the vagus and give rise to coughing or vomiting.

INFERIOR GANGLION

This ganglion communicates with the superior cervical sympathetic ganglion, the hypoglossal nerve, and the loop between the first and second cervical ventral rami. The **accessory nerve** is closely applied to the ganglion and sends its cranial fibres to it for distribution through the vagus to the muscles of the larynx. It gives off two branches.

1. The **pharyngeal branch** (often double, and said to contain fibres from the accessory nerve) passes inferomedially between the internal and external carotid arteries to form the pharyngeal plexus on the middle constrictor muscle with branches from the glossopharyngeal nerve and the superior cervical sympathetic ganglion [FIG. 11.26]. This plexus supplies most of the muscles of the pharynx and soft palate (except stylopharyngeus and tensor veli palatini) and sends a lingual branch to join the hypoglossal nerve in the anterior triangle of the neck.

2. The **superior laryngeal nerve** passes inferomedially, deep to both carotid arteries, towards the thyroid cartilage. It receives twigs from the sympathetic and pharyngeal plexus, and is said to send a branch to the internal carotid artery. It has two branches. (i) A larger, sensory, **internal branch** which passes anteromedially, deep to the thyrohyoid muscle, with the corresponding branch of the superior thyroid artery, and piercing the thyrohyoid membrane, divides into three branches immediately external to the mucous lining of the pharynx. It supplies the mucous membrane of the pharynx and larynx from the level of the vocal folds to the posterior part of the dorsum of the tongue (including the epiglottis), and communicates with a branch (**inferior laryngeal**) of the recurrent laryngeal nerve deep to the lamina of the thyroid cartilage. (ii) The external branch descends medial to the carotid sheath, on the inferior constrictor muscle of the pharynx. It supplies branches to that muscle, and ends in the cricothyroid muscle.

BRANCHES OF THE VAGUS IN THE NECK

1. The **superior** and **inferior cervical cardiac branches** arise at variable points and descend into the thorax. *On the right side*, they pass posterior to the subclavian artery on the side of the trachea to enter the deep cardiac plexus [p. 822]. *On the left*, the superior branch passes between the arch of the aorta and the trachea to the deep cardiac plexus; the inferior branch descends with the vagus and the superior cervical cardiac nerve (sympathetic) lateral to the arch of the aorta, to end in the superficial cardiac plexus [p. 822].

2. The **right recurrent laryngeal nerve** arises from the vagus anterior to the first part of the subclavian artery, hooks under it, and ascends superomedially, passing posterior to the common carotid artery and either posterior or anterior to the inferior thyroid artery. It then lies in the groove between the oesophagus and trachea, medial to the right lobe of the thyroid gland, and passes deep to the lower border of the inferior constrictor muscle as the **inferior laryngeal nerve**.

(i) **Cardiac branches** arise from the nerve on the subclavian artery. They descend beside the trachea to the deep cardiac plexus.

(ii) Communicating branches to the **cervicothoracic ganglion** arise from the nerve posterior to the subclavian artery.

(iii) Branches supply the **trachea**, **oesophagus**, and the **inferior constrictor** muscle of the pharynx.

(iv) The **inferior laryngeal nerve** supplies the intrinsic muscles of the larynx, except cricothyroid, and the mucous membrane of the larynx and pharynx inferior to the vocal folds. It communicates with the internal branch of the superior laryngeal nerve medial to the lamina of the thyroid cartilage.

BRANCHES OF THE VAGUS IN THE THORAX

1. The **left recurrent laryngeal nerve** arises from the vagus on aortic arch. It hooks round the inferior surface of the arch, posterior to the ligamentum arteriosum, and passes upwards through the superior mediastinum and lower part of the neck in the groove between the oesophagus and trachea, supplying both. Its branches are the same as those of the right nerve, but larger **cardiac branches** arise inferior to the aortic arch and pass to the deep cardiac plexus.

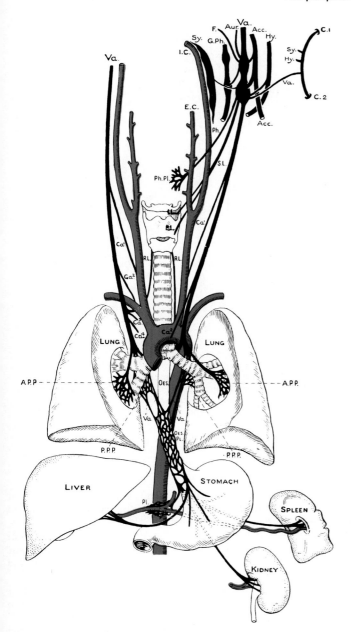

FIG. 11.28. Scheme of connections and distribution of vagus nerves.

Acc., accessory nerve; A.P.P., pulmonary plexus; Aur., auricular branch; C.1, C.2, loop between first and second cervical ventral rami; Ca.¹, Ca.², upper and lower cervical cardiac branches; Ca.³, cardiac branches of recurrent laryngeal; Ca.⁴, thoracic cardiac branch (right vagus); E.C., external carotid artery; E.L., external laryngeal branch; F., meningeal branch; G.Ph., glossopharyngeal nerve; Hy., hypoglossal nerve; Hy., cervical communication to hypoglossal nerve; I.C., internal carotid artery; I.L., internal laryngeal branch; Oes., oesophagus; Oes.Pl., oesophageal plexus; Ph., pharyngeal branch; Ph.Pl., pharyngeal plexus; Pl., coeliac plexus; P.P.P., pulmonary plexus; R.L., recurrent laryngeal nerve; S.L., superior laryngeal nerve; Sy., superior cervical ganglion of sympathetic; Sy., grey ramus communicans; Va., right and left vagi; Va., cervical communication with inferior ganglion of vagus.

2. The **right vagus** on the trachea sends thoracic cardiac branches to the deep cardiac plexus [p. 822], and **tracheal** and **oesophageal branches** medially to these structures.

3. As each vagus approaches the lung root it sends **bronchial** and **pulmonary branches** to form a complex **pulmonary plexus** [p. 823] mainly on the posterior surface of the lung root, though small branches pass to the anterior surface from the upper border of the root. The bronchial branches of the two sides communicate, and the plexus behind the lung root is joined by fine branches from the second, third, and fourth thoracic ganglia of the sympathetic trunk. The pulmonary plexus extends into the lung surrounding the bronchi and vessels and passing to the pulmonary pleura.

4. Each vagus emerges from behind the corresponding lung root as two or more trunks which pass inferomedially to break up and communicate on the oesophagus as the **oesophageal plexus**. This plexus receives filaments from the greater splanchnic nerve and its ganglion (sympathetic). It supplies the oesophagus, sends small filaments to the aorta (sensory), and reconstitutes into the anterior and posterior **vagal trunks** towards the inferior end of the thoracic oesophagus.

The cardiac, pulmonary, and oesophageal **plexuses** are predominantly parasympathetic. The fibres that reach them from the vagus are preganglionic in nature. They end on the ganglionic cells (parasympathetic) which are scattered throughout these plexuses and send their axons to the heart, lungs, and oesophagus. The vagal branches also contain many sensory fibres whose cell bodies lie in the superior and inferior ganglia of the vagus. These fibres convey impulses from the heart (but not pain), great vessels (including the glomus aorticum), and lung, and are of importance in vascular and respiratory reflexes.

BRANCHES OF THE VAGUS IN THE ABDOMEN

The anterior and posterior **vagal trunks** pass with the oesophagus through the diaphragm into the abdomen. Both trunks supply the corresponding surfaces of the stomach through several anterior and posterior **gastric branches**. In addition they send fibres along the left gastric artery to the coeliac and renal plexuses of the autonomic nervous system [p. 823] whence they are distributed along the visceral branches of the aorta to the abdominal structures. These include the gut tube as far as the transverse colon, caudal to which the parasympathetic innervation comes from the sacral region. The anterior gastric branches also send filaments between the layers of the lesser omentum to the liver and biliary system (**hepatic branches**), and to the pylorus and superior part of the duodenum.

All the **efferent vagal fibres** that enter the abdomen are preganglionic in type and they end on scattered ganglionic neurons, many of which lie in the intramural plexuses of the walls of the gut tube (submucous and myenteric plexuses) and in other viscera. These neurons innervate the gland cells and smooth muscle of these viscera with postganglionic parasympathetic nerve fibres. The sympathetic fibres which also innervate these viscera for the most part have their ganglia either at the origin from the aorta of the arteries that supply the viscera or scattered along these vessels. Thus the efferent nerve fibres entering a particular viscus consist mainly of postganglionic sympathetic and preganglionic parasympathetic fibres.

The *asymmetry* of certain parts of the vagus nerves is due to the developmental asymmetry of the pharyngeal (branchial) arch arteries between which the pharyngeal and laryngeal branches of the vagus nerves pass to innervate the cranial part of the foregut tube. On both sides the **recurrent laryngeal nerve** passes caudal to the dorsal part of the corresponding sixth arch artery, and thus forms a recurrent loop around the artery as it slides caudally with the heart relative to the foregut tube. *On the right*, the dorsal part of the sixth arch artery disappears allowing the nerve to loop round the fourth arch (subclavian) artery at a higher level. *On the left*, the dorsal part of the sixth arch artery persists (ligamentum arteriosum) and the nerve continues to pass caudal to it [FIG. 13.117]. The more caudal position of the left recurrent laryngeal nerve also explains the absence of thoracic cardiac branches of the left vagus, for these are incorporated in the branches of the recurrent nerve. The positions of the anterior and posterior **vagal trunks** and their gastric branches is explained by the rotation of the stomach which causes its left side to become anterior. The mixing of the fibres of both vagus nerves in the **oesophageal plexus** means that the anterior and posterior vagal trunks do not correspond to the left and right vagus nerves, but are a mixture of both.

Destruction of the vagus near its emergence from the brain produces defects which are visible mainly because of the paralysis of the pharyngeal and laryngeal muscles on the affected side. Thus there is difficulty in swallowing and in speech but relatively little to show for the loss of the preganglionic innervation of the heart, lungs, and abdominal viscera, though there may be a decrease in gastric and intestinal secretions. Vagotomy has been used to reduce the acid secretion from the stomach in cases of duodenal ulceration, but the bilateral distribution of the vagus in the abdomen probably reduces the effect of unilateral division of the nerve and makes drugs which block vagal impulses much more effective. Involvement of the laryngeal nerves in any pathological process may irritate the sensory fibres from the larynx and lead to protracted coughing of an ineffective, brassy type because there is partial or complete paralysis of the laryngeal muscles which prevents effective closure of the glottis.

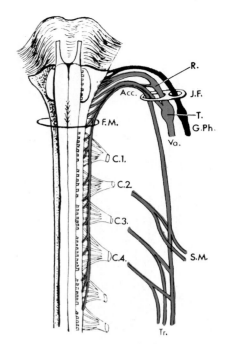

FIG. 11.29. Scheme of the origin, connections, and distribution of the accessory nerve.

Acc., accessory nerve; C.1–4, first four cervical nerves (dorsal roots); Va., vagus nerve; R., superior ganglion; T., inferior ganglion; G.Ph., glossopharyngeal nerve; S.M., nerves to sternocleidomastoid; Tr., nerves to trapezius; F.M., foramen magnum; J.F., jugular foramen.

ACCESSORY NERVE

The eleventh cranial or accessory nerve is essentially the caudal part of the vagus nerve. It therefore supplies the most caudal pharyngeal (branchial) arch, the muscle of which becomes the muscles of the larynx and caudal pharynx, and the sternocleidomastoid and trapezius muscles; the first two groups are supplied by the cranial roots, the last two muscles by the spinal roots.

The **cranial roots** arise from the lateral surface of the medulla oblongata caudal to and in series with the vagal rootlets [FIG. 10.49. For deep origin, see p. 648]. These rootlets join the spinal part and pass into the jugular foramen with it and the vagus. Here the cranial roots send a twig to the superior ganglion of the vagus, then join that nerve at the inferior ganglion, forming the **internal branch** of the accessory nerve. They are distributed through the laryngeal (and possibly pharyngeal) branches of the vagus (q.v.).

The **spinal roots** arise from the lateral aspect of the spinal medulla dorsal to the ligamentum denticulatum [p. 729] as far caudally as the fifth or sixth cervical segment [FIG. 11.29. For deep origin, see p. 648]. Turning cranially, these roots unite successively to form a trunk which ascends through the foramen magnum in the subarachnoid space, receives the cranial roots, and passes through the jugular foramen in the same dural sheath as the vagus [FIG. 11.9]. *In the foramen*, the cranial part joins the vagus, while the spinal part (**external branch**) emerging between the internal carotid artery and the internal jugular vein, passes postero-inferiorly either superficial or deep to the internal jugular vein, and crosses the transverse process of the atlas deep to the posterior belly of the digastric muscle. It receives a sensory **communicating branch** from the second cervical ventral ramus, and then supplies the sternocleidomastoid muscle and enters its deep surface. It emerges from the posterior border of the sternocleidomastoid at or below the junction of the upper and middle thirds, and usually has the lesser occipital nerve

turning round its inferior aspect to ascend along the posterior border of the muscle. The accessory nerve then runs postero-inferiorly in the investing fascia of the **posterior triangle**, receives **communicating branches** from the third and fourth cervical ventral rami (sensory fibres), and passes on to the deep surface of the trapezius muscle supplying it and again receiving communications from the same cervical nerves.

The nerve has a number of deep cervical lymph nodes in contact with it adjacent to the sternocleidomastoid muscle and is prone to damage in operations which require mobilization of the muscle.

When the external branch of the accessory nerve is destroyed, the sternocleidomastoid and trapezius muscles are paralysed, for the communicating branches from the cervical plexus are sensory in nature. The shoulder droops and its elevation is weak. The unopposed pull of the opposite sternocleidomastoid and trapezius draws the occiput towards the shoulder of that side, turning the face upwards and towards the paralysed side (wryneck).

HYPOGLOSSAL NERVE

The twelfth cranial or hypoglossal nerve is a pure motor nerve to the extrinsic and intrinsic muscles of the tongue, except palatoglossus. It arises by a number of rootlets from the ventral aspect of the medulla oblongata between the pyramid and the olive in line with abducent and oculomotor nerves and the ventral root of the first cervical nerve [FIG. 10.50. For deep origin, see p. 646]. The rootlets pierce the dura mater separately in two bundles which unite in the hypoglossal canal or outside the skull. Here the nerve lies posteromedial to the nerves emerging from the jugular foramen. It

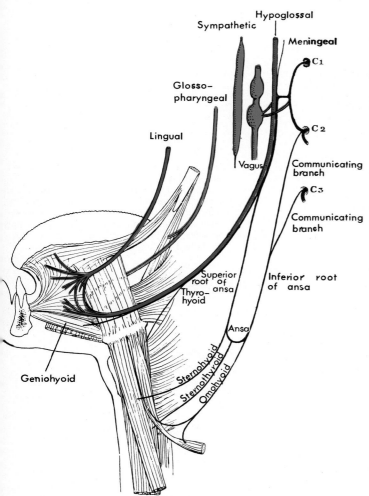

FIG. 11.30. Muscles of hyoid bone and styloid process, and extrinsic muscles of the tongue, showing connections and distribution of hypoglossal nerve. Most of the names of the muscles indicate the nerves to them.

2. The **meningeal branch** arises close to the skull, and passes to the dura mater and periosteum of the posterior cranial fossa.

3. The **superior root of the ansa cervicalis** arises where the nerve crosses the internal carotid artery. It descends anterior to the carotid sheath in the anterior triangle, and is joined, below the middle of the neck, either superficial or deep to the internal jugular vein, by the **inferior root** which consists of fibres from the ventral rami of the second and third cervical nerves. The loop which this junction forms is the **ansa cervicalis**. It sends branches to sternohyoid, sternothyroid, and both bellies of omohyoid.

4. The **nerve to the thyrohyoid muscle** arises in the carotid triangle and descends on the middle constrictor muscle, posterior to the greater cornu of the hyoid bone [FIG. 11.30].

If the hypoglossal nerve is destroyed, the intrinsic and extrinsic muscles of the corresponding half of the tongue are paralysed, as are the thyrohyoid and geniohyoid muscles unless the injury is superior to the communication from the first cervical nerve. The paralysis of tongue muscles causes difficulty with the lingual part of speech and swallowing, and the protruded tongue is thrust towards the paralysed side by the unopposed action of the muscles on the normal side. The paralysis of the thyrohyoid muscle interferes with laryngeal movements in swallowing.

The spinal nerves

Thirty-one pairs of spinal nerves arise from the spinal medulla and leave the vertebral canal through the intervertebral foramina. Eight of these are cervical, the first emerging between the occiput and the first cervical vertebra (**suboccipital nerve**), and the eighth between the seventh cervical and first thoracic vertebra. There are twelve thoracic, five lumbar, five sacral, and one coccygeal nerve, and each of these emerges below the corresponding vertebra. The coccygeal nerve is occasionally absent, but there may be one or two additional filaments caudal to it which remain within the sacral canal.

The size of a spinal nerve is directly proportional to the volume of tissue which it supplies. Thus the limb nerves (fifth cervical to first thoracic, and second lumbar to third sacral) are the largest, and of these the nerves innervating the lower limbs are the larger. The coccygeal nerve is the smallest, and the upper cervical nerves decrease in size from below upwards.

Origin

Each spinal nerve is attached to the spinal medulla by a **dorsal** (sensory or afferent) and a **ventral** (motor or efferent) root, except the first cervical which may have no dorsal root. Every root consists of a number of rootlets, each of which contains many nerve fibres surrounded by a layer of pia mater [p. 728]. The **dorsal rootlets** form a continuous linear series attached to the posterior lateral sulcus of the spinal medulla [FIG. 10.24]. The **ventral rootlets** arise irregularly from a strip on the anterolateral surface of the spinal medulla, and are separated from the dorsal rootlets by the ligamentum denticulatum [FIG. 10.181, p. 729], and above the sixth cervical segment also by the spinal part of the accessory nerve.

The dorsal and ventral rootlets of a spinal nerve separately converge on the arachnoid and dura opposite the corresponding intervertebral foramen to form a dorsal and a ventral root. The roots pass separately through these membranes and receive a sleeve from each of them [FIGS. 10.181 and 10.189]. The dorsal root, which is larger and contains more nerve fibres than the corresponding ventral

receives a bundle of fibres from the loop uniting the ventral rami of the first and second cervical nerves and from the superior cervical sympathetic ganglion. The hypoglossal nerve then passes inferolaterally and winds closely round the posterior surface of the inferior ganglion of the vagus (with which it communicates) to reach its lateral side deep to the posterior belly of digastric and the occipital artery. It then passes inferior to the digastric, and curving forwards, crosses the lateral surfaces of the internal carotid artery, the occipital and external carotid arteries near their point of separation, and the loop of the lingual artery on the middle constrictor muscle. It then runs on the hyoglossus muscle, and passing superior to the free posterior border of the mylohyoid muscle, lies between it and hyoglossus. Here it communicates with the lingual nerve and divides into its lingual branches which pass forwards between mylohyoid and genioglossus.

BRANCHES

1. The branches of the hypoglossal nerve proper are the **lingual** branches to the hyoglossus, styloglossus, genioglossus, and the intrinsic muscles of the tongue. The branches to **geniohyoid** and to **thyrohyoid** are formed by fibres which enter and run with the hypoglossal nerve from the loop between the first and second cervical ventral rami [FIG. 11.30].

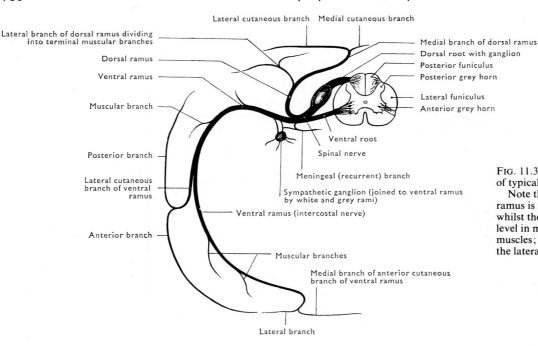

Lateral cutaneous branch   Medial cutaneous branch

Lateral branch of dorsal ramus dividing
into terminal muscular branches

Dorsal ramus

Ventral ramus

Muscular branch

Posterior branch

Lateral cutaneous
branch of ventral
ramus

Anterior branch

Medial branch of dorsal ramus
Dorsal root with ganglion
Posterior funiculus
Posterior grey horn

Lateral funiculus
Anterior grey horn

Ventral root
Spinal nerve

Meningeal (recurrent) branch

Sympathetic ganglion (joined to ventral ramus
by white and grey rami)

Ventral ramus (intercostal nerve)

Muscular branches

Medial branch of anterior cutaneous
branch of ventral ramus

Lateral branch

FIG. 11.31. Diagram of origin and distribution of typical spinal nerve.

Note that the medial branch of the dorsal ramus is represented as distributed to skin, whilst the lateral branch terminates at a deeper level in muscle. Both branches, however, supply muscles; and in the lower half of the body it is the lateral branch that supplies the skin.

root, immediately expands into the spinal ganglion in the inter-vertebral foramen.

The **spinal ganglion** consists of many spherical, unipolar [FIG. 10.6] **sensory nerve cells**, each of which gives rise to a short process that divides into a central branch which enters a dorsal rootlet, and a peripheral branch passing into the spinal nerve. Scattered ganglion cells may be found on the dorsal rootlets and even on the ventral roots (which have no ganglion) particularly in the lumbosacral region. The function of those on the ventral roots is unknown though they have the structure of sensory cells (Windle 1931). The ventral root grooves the anterior surface of the distal part of the ganglion, and beyond this *the two roots are fused into the spinal nerve*. The arachnoid sleeve of the dorsal root fuses with the proximal part of the ganglion, while the dural sleeves of the roots become continuous with the epineurium of the nerve. There are *no* synapses in sensory ganglia, so they are unaffected by synapse blocking drugs; cf. autonomic ganglia.

The spinal medulla is shorter than the vertebral canal, and ends between the first and second lumbar vertebrae. Thus the segment of the spinal medulla from which each pair of roots arises does not lie opposite the corresponding vertebra [FIGS. 10.29 and 11.32] so that the rootlets run with increasing obliquity from above downwards to reach the appropriate intervertebral foramen. Those roots which emerge from the vertebral column below the termination of the spinal medulla, together run longitudinally in the vertebral canal with the filum terminale [p. 624] and thus form the **cauda equina** [FIG. 10.26] enclosed with the cerebrospinal fluid of the subarachnoid space in the dura and arachnoid. The arachnoid and dura which surround the spinal medulla and cauda equina end at the level of the second sacral vertebra. Thus the nerves caudal to the first sacral pierce these meninges before reaching the appropriate intervertebral foramina. They also have their ganglia within the sacral part of the vertebral canal and not in the intervertebral foramina. The ganglion of the coccygeal nerve may be very small or consist of scattered cells on the dorsal root.

THE VENTRAL ROOTS

These consist of coarse and fine myelinated nerve fibres. The **coarse fibres** are motor to the striated muscles of the body-wall and limbs.

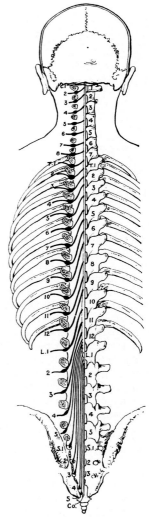

FIG. 11.32. Diagram to show level of origin of spinal nerve roots from spinal medulla relative to point of emergence from the vertebral column. The nerves are shown as thick black lines on the left side.

The fine fibres are of two types: (1) Motor fibres distributed to the intrafusal muscle fibres of the **muscle spindles**; (2) **Preganglionic autonomic nerve fibres.** (a) The **sympathetic** preganglionic fibres emerge from the spinal medulla with the first thoracic to second or third lumbar nerves inclusive, and pass from the ventral ramus of each of these nerves as a **white ramus communicans** to the corresponding ganglion of the sympathetic trunk. (b) The preganglionic **parasympathetic** fibres emerge with the second and third, or third and fourth, sacral nerves, and leave their ventral rami to form the **pelvic splanchnic nerves** (q.v.).

THE DORSAL ROOTS

These consist of nerve fibres of all diameters, both myelinated and non-myelinated. It seems certain that the nerve fibres in the dorsal roots are all processes of cells in the spinal ganglia. There is no conclusive evidence for efferent fibres within these roots, despite the fact that stimulation of the distal end of a divided dorsal root produces vasodilatation in its territory of distribution (for discussion, see Mitchell 1953).

In the **intervertebral foramen**, the spinal nerve and ganglion in their fibrous sheath are surrounded by veins which unite the **internal vertebral venous plexus** in the vertebral canal with the veins external to the vertebral column at every vertebral level. These structures in the intervertebral foramen lie posterior to the posterolateral part of the intervertebral disc and vertebral body, and anterior to the joint between the vertebral articular processes [FIG. 3.21]. They are liable to compression from a herniated nucleus pulposus [p. 221] or from the bony walls of the intervertebral foramen if the adjacent vertebral bodies are approximated, after rupture of an intervertebral disc. Pathological changes in the intervertebral joints may also involve the spinal nerves or their dorsal rami (see below) at this site. Neither the first nor the second cervical nerves are related to an intervertebral disc, but lie posterior to the superior and inferior parts of the lateral mass of the atlas. They may be involved in pathological changes in the joints which these form with the skull and axis. The sacral nerves pass between the fused sacral vertebra and are not liable to injury from the above causes.

RAMI OF THE SPINAL NERVES

As the spinal nerve emerges from the intervertebral foramen it gives off a recurrent **meningeal branch** and divides into dorsal and ventral rami, both of which contain motor and sensory fibres. The meningeal branch receives a branch from the sympathetic trunk, and re-enters the vertebral canal to supply the meninges and blood vessels therein.

The **dorsal rami** pass posteriorly, immediately lateral to the joint between the corresponding vertebral articular processes (except in the case of the first and second cervical nerves) and are distributed to the axial muscles, fascia, ligaments, and skin of the back. They play no part in the supply of limb muscles even though their cutaneous branches may pierce some of them, e.g. the rhomboids and trapezius [FIG. 5.52].

The **ventral rami** are larger than the dorsal rami. They supply the lateral and anterior parts of the body, the limbs, and the perineum, and they tend to unite with each other to form complicated plexuses particularly for the innervation of the limbs.

DISTRIBUTION OF SPINAL NERVES

The distribution of the spinal nerves is, like their origin, essentially segmental, though the pattern is modified and almost obliterated in some regions. This is because (1) most of the muscles are formed by the fusion of a number of segmental myotomes (e.g. the long muscles

of the back) and consequently are innervated by more than one spinal nerve; (2) the area of skin supplied by each spinal nerve (known as a **dermatome**) overlaps the areas of adjacent nerves. This overlap is sufficient to prevent total cutaneous sensory loss when a single spinal nerve is destroyed, though there is altered sensation in its area of distribution. When three adjacent spinal nerves are destroyed there is a total cutaneous sensory loss corresponding to the area supplied by the intermediate nerve (Head 1893; Foerster 1933; Keegan and Garrett 1948). Some indication of the cutaneous distribution of nerves may be obtained from the vesicular eruption of the skin which occurs in the virus infection of shingles (herpes zoster) which affects part or all of the cutaneous distribution of one spinal ganglion.

The outgrowth of the limbs, which can be looked upon as parts of the body-wall, carries certain dermatomes entirely into the limb so that they are missing from the expected sequence in the trunk. Yet the sequence remains unaltered if the limbs and trunk are looked on as a whole [FIGS. 11.33, 11.35, 11.48, 11.63, and 11.74].

In the description of nerves, the terms 'muscular' and 'cutaneous' are applied to the branches passing to muscles and skin. It is important to realize that **nerves to muscles** contain sensory nerve fibres (from muscle, tendon, joints, bone, and intermuscular fascial planes) and some postganglionic sympathetic fibres (to blood vessels) as well as motor fibres to the muscle. Also that **cutaneous nerves** invariably supply subcutaneous tissue and sometimes joints, ligaments, and bones (e.g. the digital nerves) and that they always contain postganglionic sympathetic fibres to blood vessels and sweat glands. Thus an area of skin which is anaesthetic due to a peripheral nerve injury does not sweat, and so is dry and has a high electrical resistance.

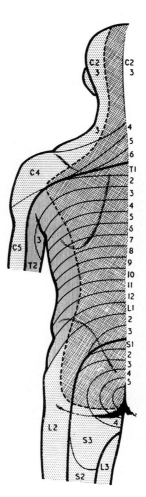

FIG. 11.33. Distribution of cutaneous nerves on the back of the trunk to illustrate the dermatomes.

DORSAL RAMI OF SPINAL NERVES

These branches supply the erector spinae and associated muscles, and the skin fasciae, ligaments, bones, and joints of the posterior part of the trunk and neck. Their cutaneous area also extends on to the back of the scalp and shoulder, and the gluteal region [FIG. 11.33].

The majority of the dorsal rami divide into medial and lateral branches [FIG. 11.31]. Both of these supply the erector spinae but only one reaches the skin—the medial in the upper half of the trunk, the lateral in the lower half. In the upper half the cutaneous branches pass posteromedially through the muscle to enter the skin close to the spines of the vertebrae, while in the lower half they extend inferolaterally and enter the skin further from the median plane and at a considerable distance below their level of origin.

CERVICAL DORSAL RAMI

First cervical (suboccipital) dorsal ramus

The dorsal root of this nerve may be small or absent. The dorsal ramus is larger than the ventral, does not divide into medial and lateral branches, and has no direct cutaneous branch.

The dorsal ramus arises from the nerve on the posterior arch of the atlas, inferior to the vertebral artery [FIG. 11.34]. It passes posteriorly into the suboccipital triangle deep to semispinalis capitis, gives muscular branches to the muscles of the triangle (obliquus capitis superior and inferior, rectus capitis posterior major and minor) and to semispinalis capitis, and communicates with the second cervical dorsal ramus over the posterior surface of obliquus capitis inferior. This communication carries sensory fibres to the muscles supplied by the first cervical dorsal ramus when the first cervical nerve has no dorsal root.

Second cervical dorsal ramus

This is the largest of the dorsal rami. It passes posteriorly between obliquus capitis inferior and the semispinalis cervicis muscles, gives communicating branches to the first and third cervical dorsal rami, and sends branches to semispinalis and splenius capitis, semispinalis cervicis, and multifidus. It pierces semispinalis capitis close to the midline, and running superolaterally between it and trapezius, pierces the tendinous origin of trapezius and the deep fascia approximately 2.5 cm lateral to the external occipital protuberance. It then runs with the occipital artery (as the **greater occipital nerve**) to supply the posterior part of the scalp as far as the vertex. It communicates in the scalp with the great auricular, lesser occipital, posterior auricular, supra-orbital, and third occipital nerves.

Third cervical dorsal ramus

This dorsal ramus is much smaller than the second cervical. It communicates with the second and fourth dorsal rami, and divides into medial and lateral branches. The lateral branch enters erector spinae; the medial branch passes posteromedially, pierces semispinalis capitis and trapezius close to the median plane as the **third occipital nerve**. It supplies the skin and fasciae of the upper part of the back of the neck and a small adjacent area of the scalp.

The **dorsal rami of the fourth, fifth and sixth cervical nerves** are still smaller. Deep to semispinalis capitis, they divide into medial and lateral branches to erector spinae, and the medial branches reach the skin close to the median plane. The sixth is the smallest, and the fifth and sixth may have no cutaneous branches.

Seventh and eighth cervical dorsal rami

These are the smallest of the cervical dorsal rami. They may have no cutaneous branches, though the eighth commonly has.

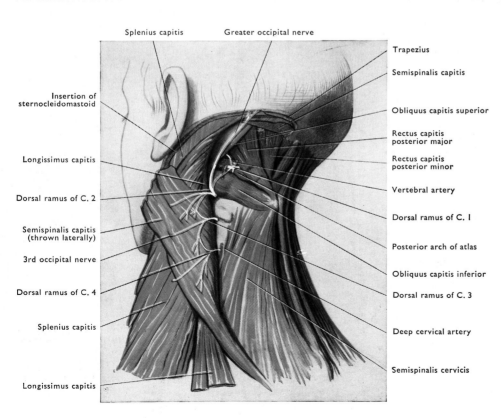

FIG. 11.34. Dorsal rami of upper cervical nerves.

THORACIC DORSAL RAMI

All the thoracic dorsal rami divide into medial and lateral branches. In the upper six or seven the medial branch is mainly cutaneous and innervates the scapular region. The medial emergence of these branches prevents them from interfering with medial movements of the scapula. The **first** is small, the **second** is very large and reaches the acromion. The rest diminish in size from above downwards and become progressively more oblique. In the lower five or six, the lateral branches curve downwards around or through the lateral part of erector spinae and pierce latissimus dorsi, emerging further laterally than the upper dorsal rami. They supply the skin over the back of the lower chest and loin, the **eleventh** and **twelfth** passing over the iliac crest into the gluteal skin. Because of the descent of the lower dorsal rami, the area of skin supplied lies below the point of emergence of the nerve from the vertebral column. This descent is commensurate with that of the corresponding ventral rami in the intercostal spaces, and thus the lower border of the dermatomes remains nearly horizontal despite the oblique course of the nerves [FIG. 11.33].

LUMBAR DORSAL RAMI

The **upper three** lumbar dorsal rami divide into medial and lateral branches which innervate the erector spinae. The lateral branches are mainly cutaneous. They pass inferolaterally lateral to or through erector spinae, and piercing the lumbar fascia immediately superior to the iliac crest, a short distance lateral to the posterior superior iliac spine, supply a strip of skin in the gluteal region. This strip extends from the posterior median line above the iliac crest to a point distal and posterior to the greater trochanter of the femur [FIG. 11.33]. Almost all the dorsal rami communicate in erector spinae, and the third lumbar nerve which may not send a direct branch to the skin, can transmit fibres to it through the second.

The dorsal rami of the **fourth** and **fifth** lumbar nerves usually supply only muscular branches to the erector spinae in the same manner as the lower cervical dorsal rami.

SACRAL DORSAL RAMI

The sacral dorsal rami emerge through the dorsal sacral foramina. The first three supply medial branches to multifidus, and lateral branches which pierce the sacrotuberous ligament and gluteus maximus to supply skin and fascia over the dorsal surface of the sacrum and the medial part of the gluteal region.

The dorsal rami of the fourth and fifth unite and are joined by the minute dorsal ramus of the coccygeal nerve. The combined nerve pierces the sacrotuberous ligament and supplies skin and fascia in the neighbourhood of the coccyx.

GENERAL ARRANGEMENT OF THE DORSAL RAMI

The **muscular distribution** is confined to erector spinae, splenius, and suboccipital muscles.

The **cutaneous distribution** for the greater part covers the area of the erectores spinae, but on the head and at the roots of the limbs the cutaneous branches extend beyond the regions supplied by their motor fibres as though drawn out by the growth of these regions. Thus the second and third cervical extend on to the scalp; the upper thoracic run laterally over the scapular region, and the upper lumbar and sacral dorsal rami extend their supply into the gluteal region.

At the level of each limb, where the ventral rami are particularly large, a variable number of the dorsal rami fail to innervate the skin (usually the fifth, sixth, and seventh cervical and the fourth and fifth lumbar); thus the regular sequence of dermatomes is interrupted dorsally at the level of the vertebra prominens and of the posterior superior iliac spines. Each of these interruptions may be indicated by a hypothetical **dorsal axial line** [FIG. 11.33]. This line can be continued into the limb where a similar interruption exists between the areas supplied by the ventral rami [p. 806]. Injuries to the nerves whose dorsal rami do not reach the skin will produce no alteration of cutaneous sensation in the areas of the dorsal rami, hence their integrity cannot be assured by testing cutaneous sensation in these areas.

VENTRAL RAMI OF SPINAL NERVES

The ventral rami are distributed to the lateral and anterior aspects of the body, including the limbs. Each separates from the dorsal ramus on emerging from the intervertebral foramen, and consists of the same mixture of dorsal and ventral root fibres except that the sympathetic preganglionic fibres in the thoracic and upper three lumbar nerves and the preganglionic parasympathetic fibres in the second to fourth sacral nerves pass only into the ventral rami. These fibres leave the ventral rami almost at once, the sympathetic passing to the sympathetic trunk as **white rami communicantes**, while the parasympathetic pass to the autonomic plexuses in the pelvis as **pelvic splanchnic nerves**.

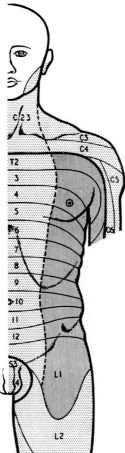

FIG. 11.35. Distribution of cutaneous nerves on the front of the trunk to illustrate the dermatomes.

Every ventral ramus is joined near its origin by a delicate bundle of non-myelinated, postganglionic fibres (grey ramus communicans) from the sympathetic trunk. These fibres are distributed to blood vessels, sweat glands, and arrectores pilorum muscles in the territory of both rami of the nerve they enter.

In the thoracic region the ventral rami are distributed in a regular segmental manner, but elsewhere they combine to form plexuses and consequently their arrangement is more complex.

A typical thoracic ventral ramus, e.g. the fifth intercostal nerve [FIG. 11.31], is first connected to the sympathetic trunk by a grey and a white ramus communicans, and sweeps forwards in the upper part of an intercostal space, frequently sending a collateral branch along the lower part of the same space. This branch either rejoins the main trunk or becomes cutaneous anteriorly. The ventral ramus supplies the muscles of the intercostal space and sends a lateral and an anterior cutaneous branch through the overlying muscles [FIG. 11.31]. Together these supply a strip of skin which extends posteriorly from the anterior median line to become continuous with the area supplied by the dorsal ramus of the same spinal nerve [FIGS. 11.33 and 11.35]. The entire strip of skin so innervated between the anterior and posterior median lines, constitutes a **dermatome**.

Cervical ventral rami

The ventral rami of the cervical nerves, together with parts of the first and second thoracic nerves, are distributed to the head, neck, and upper limbs. The first four cervical nerves innervate the neck and part of the head through the **cervical plexus**; the last four cervical and a large part of the first thoracic nerve supply each upper limb through the **brachial plexus**. The second thoracic ventral ramus may contribute to this plexus, and usually helps to innervate the arm through its lateral cutaneous branch.

The **first cervical ventral ramus** emerges from the vertebral canal inferomedial to the vertebral artery on the posterior arch of the atlas. It curves round the posterior and lateral surfaces of the lateral mass of the atlas, and emerges between rectus capitis anterior and rectus capitis lateralis. Here it descends anterior to the transverse process of the atlas to join the ascending branch of the second cervical ventral ramus and send a large part of its fibres to join the hypoglossal nerve (see below).

The **second cervical ventral ramus** emerges posterior to the **superior articular process** of the axis. It passes forwards, lateral to the vertebral artery and divides into: (1) an ascending branch which unites with the first cervical ventral ramus and contributes some fibres to the hypoglossal nerve; (2) a descending part which unites with the third cervical ventral ramus.

The remaining cervical ventral rami emerge though the intervertebral foramina, and pass between the anterior and posterior intertransverse muscles.

The cervical plexus

This plexus is formed by an irregular series of loop communications between the ventral rami of the first four cervical nerves, each of which is joined by one or more **grey rami communicantes** from the superior cervical sympathetic ganglion. The loop communications lie close to the vertebral column between the prevertebral muscles and those arising from the posterior tubercles of the transverse processes. The branches of the plexus arise from the loops posteromedial to the internal jugular vein, under cover of the sternocleidomastoid.

Cutaneous branches

(C. 2, 3)	(C. 3, 4)
Ascending to head	Descending to side of neck, shoulder and front of chest
Lesser occipital	
Great auricular	Medial, intermediate, and lateral supraclavicular
To front of neck	
Transverse nerve of neck	

Muscular branches

Lateral	Medial
Sternocleidomastoid (C. 2) sensory	Prevertebral muscles (C. 1, 2, 3, 4)
Trapezius (C. 3, 4) sensory	Geniohyoid, thyrohyoid (C. 1)
Levator scapulae (C. 3, 4)	Infrahyoid muscles (C. 1, 2, 3; through ansa cervicalis)
Scalenus medius and posterior (C. 3, 4)	Diaphragm (C. 3, 4, 5; phrenic nerve)

Communicating branches

Lateral	Medial
With accessory nerve (C. 2, 3, 4)	Vagus and hypoglossal nerves (C. 1, or C. 1, 2)
	Sympathetic (grey rami to C. 1, 2, 3, 4).

CUTANEOUS BRANCHES

These nerves appear in the posterior triangle a little above the midpoint of the posterior border of sternocleidomastoid [FIG. 11.38].

The **lesser occipital nerve** (C. 2, 3) is variable in size and is sometimes double. It passes posteriorly deep to sternocleidomastoid, and hooking below the accessory nerve at the posterior border of the muscle, ascends along that border to pierce the deep fascia at the apex of the posterior triangle. Here it divides into branches which supply the skin and fascia: (1) on the upper lateral part of the neck; (2) on the adjacent surfaces of the auricle and mastoid process; (3) on the adjoining part of the scalp. When double, it is usually only the smaller branch which is in contact with the accessory nerve.

The **great auricular nerve** (C. 2, 3, or rarely C. 3 alone) is usually the largest cutaneous branch of the plexus, but its size and territory varies reciprocally with the lesser occipital. It winds round the posterior border of sternocleidomastoid [FIG. 11.38] and passes anterosuperiorly across the superficial surface of that muscle, deep to platysma, towards the inferior part of the auricle. On the muscle it divides into a number of branches. (1) **Posterior branches** ascend over the mastoid process, communicate with the lesser occipital and posterior auricular nerves, and supply skin and fascia in this region. (2) **Intermediate branches** pass to supply the lower part of the auricle on both surfaces. (3) **Anterior branches** pass through the substance of the parotid gland and over the angle of the mandible, to supply skin and fascia over the postero-inferior part of the face [FIG 11.37]. It communicates with the facial nerve in the parotid gland and may send branches almost to the zygomatic bone.

The **transverse nerve of the neck** (C. 2, 3) winds horizontally round the posterior and lateral aspects of sternocleidomastoid, deep to platysma and the external jugular vein. It divides into **superior** and **inferior branches** near the anterior margin of sternocleidomastoid and supplies skin and fascia over the anterior triangle of the neck. The superior branches meet the area supplied by the trigeminal

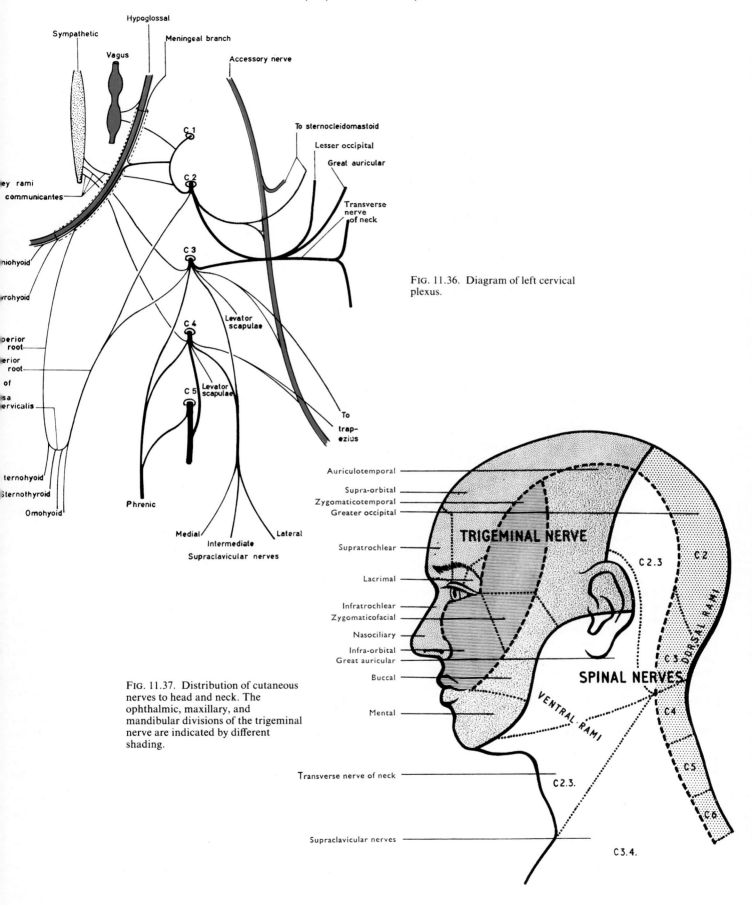

FIG. 11.36. Diagram of left cervical plexus.

FIG. 11.37. Distribution of cutaneous nerves to head and neck. The ophthalmic, maxillary, and mandibular divisions of the trigeminal nerve are indicated by different shading.

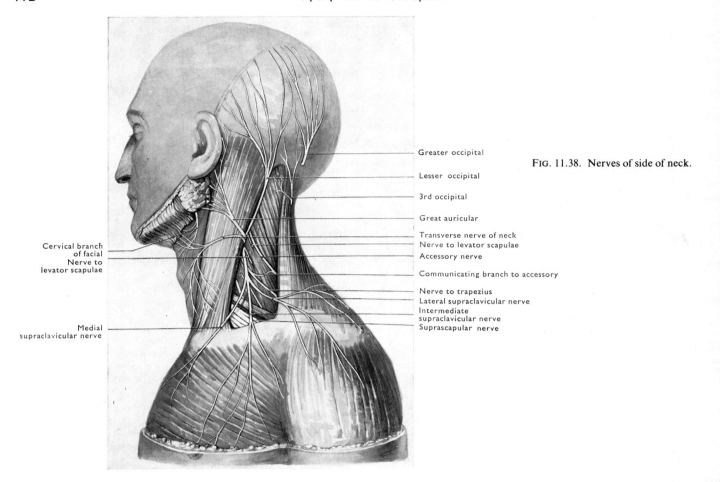

Greater occipital

Lesser occipital

3rd occipital

Great auricular

Transverse nerve of neck
Nerve to levator scapulae
Accessory nerve

Communicating branch to accessory

Nerve to trapezius
Lateral supraclavicular nerve
Intermediate
supraclavicular nerve
Suprascapular nerve

Cervical branch
of facial
Nerve to
levator scapulae

Medial
supraclavicular nerve

FIG. 11.38. Nerves of side of neck.

nerve at the inferior border of the body of the mandible, and communicate with the cervical branch of the facial nerve.

The **supraclavicular nerves** (C. 3, 4) appear at the posterior border of sternocleidomastoid as a large trunk which descends through the lower part of the posterior triangle and divides into **medial, intermediate,** and **lateral supraclavicular nerves.** These nerves pierce the deep fascia above the clavicle, supply the skin and fascia of the inferior part of the side of the neck, and descend superficial to the corresponding thirds of the clavicle, deep to platysma, to supply skin and fascia to the level of the sternal angle. The medial nerve also supplies the **sternoclavicular joint,** and the lateral nerve the **acromioclavicular joint.** Branches of the intermediate and lateral nerves may groove or pierce the clavicle, and some of the branches of the lateral nerve usually pass deep to trapezius and pierce it to reach the skin.

MUSCULAR BRANCHES

These arise with the other branches of the cervical plexus deep to sternocleidomastoid, and pass either laterally into the posterior triangle or medially into the anterior triangle.

Lateral branches

1. From the second cervical ventral ramus a sensory branch enters the deep surface of **sternocleidomastoid** having communicated with the accessory nerve deep to that muscle.

2. From the third and fourth cervical ventral rami. (i) Branches cross the posterior triangle to enter the deep surface of **trapezius.** They supply sensory fibres to that muscle and communicate with the

accessory nerve in the posterior triangle and deep to the trapezius. (ii) Separate branches supply the **levator scapulae** muscle by entering its lateral surface in the posterior triangle. (iii) Branches supply **scalenus medius** and **posterior.**

Medial branches

1. From the loop uniting the first and **second cervical ventral rami** anterior to the transverse process of the atlas. (i) A branch joins the **hypoglossal nerve** as it emerges from the skull [FIGS. 11.30 and 11.36]. A few sensory fibres of this branch pass superiorly in the hypoglossal nerve to supply the skull and dura mater of the posterior cranial fossa (meningeal branch). Most of the fibres descend in the hypoglossal nerve and form three branches of that nerve which probably do not contain any nerve fibres of the hypoglossal nerve proper. (a) The nerve to **thyrohyoid** [p. 765]. (b) The nerve to **geniohyoid** [p. 765]. (c) The **superior root of the ansa cervicalis** arises from the hypoglossal nerve anterior to the internal carotid artery. It descends in front of the internal and common carotid arteries, and joins the inferior root of ansa cervicalis (see below) to form this loop in the carotid sheath. The **ansa** sends branches to the **sternohyoid, sternothyroid,** and **omohyoid muscles.** The superior root of the ansa may occasionally arise from the vagus. In this case the fibres from the first two cervical ventral rami pass through the communication to the vagus and not to the hypoglossal.

(ii) A small branch supplies the rectus capitis lateralis and anterior, and the longus capitis muscles.

2. From the **second** and **third cervical ventral rami.** Each of these nerves sends a slender branch downwards along the internal jugular vein to unite on its anterior surface (**inferior root of the ansa**

cervicalis). This forms the ansa cervicalis with the superior root (see above) and supplies the infrahyoid muscles except thyrohyoid. Thus medial branches of the first three cervical ventral rami communicate to supply a paramedian strip of muscle from the chin to the sternum, and they also communicate with the hypoglossal nerve which supplies the tongue muscles immediately superior to this strip.

3. From the **second**, **third**, and **fourth cervical ventral rami**, small branches supply the intertransverse, longus colli, and longus capitis muscles.

4. From the **third**, **fourth**, and **fifth cervical ventral rami**. The **phrenic nerve** [FIGS. 11.36 and 11.39] arises mainly from the fourth, but receives branches from the third (either directly or through the nerve to sternohyoid) and fifth (either directly or from the nerve to subclavius as the **accessory phrenic nerve**). The phrenic nerve descends on scalenus anterior posterolateral to the internal jugular vein, and passing anterior to the cervical pleura and the second part of the subclavian artery, behind the subclavian vein (Qvist 1977) deviates medially in front of the internal thoracic artery (occasionally posterior to it on the left).

The nerve descends through the thorax to the diaphragm, separated from the pleural cavity only by the mediastinal pleura. In the *superior mediastinum*, the left nerve lies between the common carotid and subclavian arteries and crosses the aortic arch anterior to the vagus nerve; the right nerve lies on the right brachiocephalic vein and the superior vena cava, and is not in contact with the vagus nerve. Both nerves pass down the *middle mediastinum* between the pleura and pericardium, anterior to the lung root, the right reaching the diaphragm with the inferior vena cava. Most of the phrenic nerve fibres pierce the diaphragm and supply it from the inferior surface, but some pass over its pleural surface.

Branches of the phrenic nerve

1. Muscular to the diaphragm. 2. Sensory to mediastinal and diaphragmatic pleura and to pericardium (**pericardial branch**). 3. Sensory to diaphragmatic peritoneum and probably to liver, gallbladder, and inferior vena cava by **phrenico-abdominal branches**.

The **accessory phrenic nerve** arises from the fifth, or fifth and sixth cervical ventral rami, and passes inferomedially, deep to sternocleidomastoid to join the main nerve in the lower part of the neck or in the thorax. It commonly arises from the nerve to the subclavius muscle, but may be absent. In addition to this nerve, the phrenic may receive nerve fibres from the ansa cervicalis and from the cervical sympathetic trunk. In the abdomen it communicates with the coeliac plexus.

Communicating branches of the cervical plexus

In addition to the fibres which the plexus sends to the **accessory** and **hypoglossal** nerves, the loop between the first and second cervical ventral rami sends a branch to the **vagus**, and **grey rami communicantes** pass from the superior cervical ganglion of the sympathetic trunk to the first four cervical ventral rami.

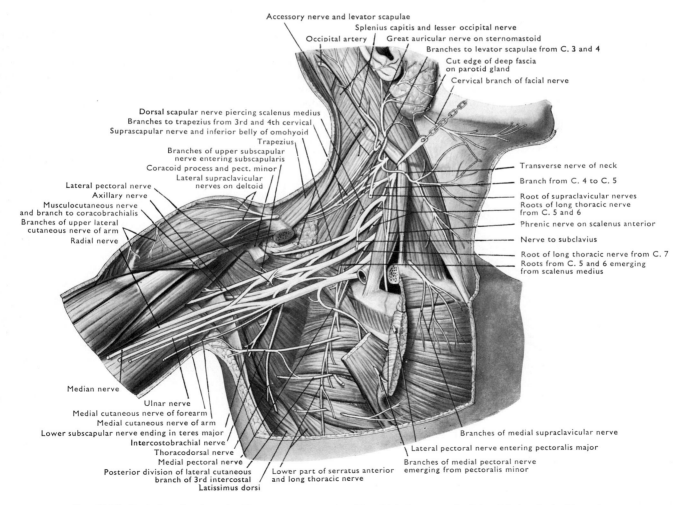

FIG. 11.39. Posterior triangle and axilla, showing cervical and brachial plexuses and origin of their principal branches.

SUMMARY OF DISTRIBUTION OF CERVICAL PLEXUS
[FIG. 11.36]

The cutaneous distribution, in which the first cervical nerve plays no part, extends from the side of the head (lesser occipital nerve, C. 2, 3), the lower part and posterior surface of the auricle and the skin over the parotid gland and angle of the jaw (great auricular nerve, C. 2, 3), and the anterior triangle of the neck (transverse nerve of the neck, C. 2, 3), to the front of the chest and to the shoulder down to the level of the sternal angle (supraclavicular nerves, C. 3, 4).

The muscular distribution is to the prevertebral and intertransverse muscles, to the other muscles attached to the transverse processes of the vertebra as far posteriorly as the levator scapulae, behind which the muscles are supplied by the dorsal rami. The cervical plexus also supplies the anterior strip of muscle from the chin to the sternum consisting of geniohyoid, sternohyoid, omohyoid, thyrohyoid, and sternothyroid, and the diaphragm with the assistance of the fifth cervical nerve. It also sends sensory fibres to sternocleidomastoid and trapezius.

THE BRACHIAL PLEXUS [FIGS. 11.39–11.41]

The brachial plexus extends from the neck to the axilla. It is formed by the ventral rami of the fifth, sixth, seventh, and eighth cervical nerves with the greater (ascending) part of that of the first thoracic nerve. Frequently a slender branch from the fourth cervical ventral ramus passes to the plexus, and the second thoracic ventral ramus not uncommonly sends a communication to the first thoracic ventral ramus. The lateral cutaneous branch of the second thoracic (intercostobrachial nerve) almost invariably innervates some skin on the medial side of the arm and the floor of the axilla.

POSITION OF PLEXUS

The ventral rami which form the plexus enter the lower part of the posterior triangle of the neck in series with the ventral rami of the cervical plexus [FIG. 11.39]. The second part of the subclavian artery lies immediately anterior to the lower two rami. The upper three rami intermingle and pass inferolaterally towards the axilla and the subclavian artery. They are enclosed with the artery and the lower two rami in an extension of the prevertebral fascia. In the neck, the plexus is deep to platysma, the supraclavicular nerves, the inferior belly of omohyoid, and the transverse cervical artery. It then passes deep to the clavicle and the suprascapular vessels, to enter the axilla, and surround the second part of the axillary artery.

CONSTITUTION OF BRACHIAL PLEXUS

The plexus is subject to some variation, but four stages in the redistribution of its contained nerve fibres can always be identified.

1. The ventral rami, the roots of the plexus, lie between scalenus anterior and medius.

2. As they enter the posterior triangle, the upper two (C. 5, 6) and the lower two (C. 8; T. 1) roots of the plexus unite to form the **upper** and **lower trunks** respectively, while the middle (C. 7) continues as the **middle trunk**. The lower trunk may groove the superior surface of the first rib posterior to the subclavian artery, and the root from the first thoracic ventral ramus is always in contact with it.

3. Each trunk divides [FIG. 11.40] into ventral and dorsal divisions which are destined to supply the anterior (flexor) and posterior (extensor) parts of the limb respectively. Occasionally the ventral rami divide before or during the formation of the trunks, it can then be seen that the dorsal division of the eighth is small and that of the first thoracic may be minute or absent, while the others divide equally.

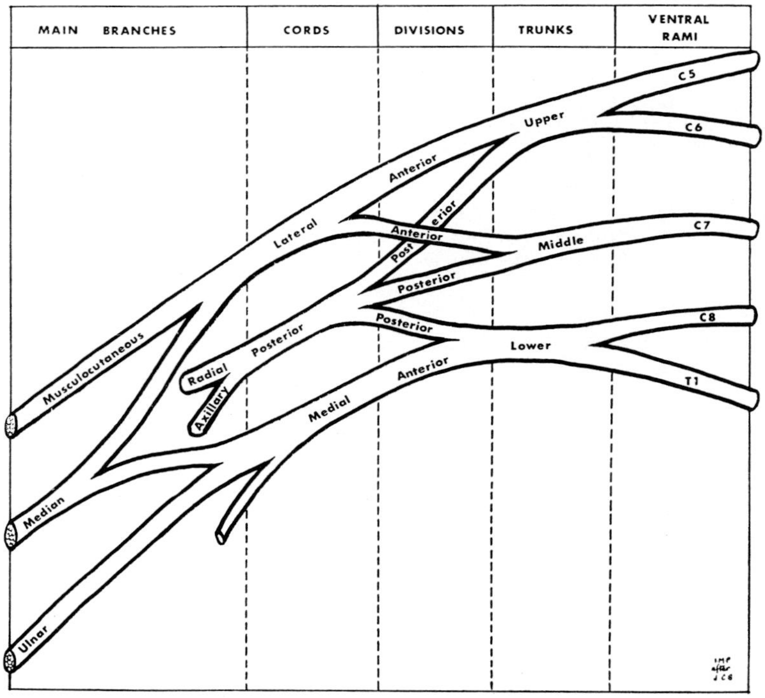

FIG. 11.40. Scheme to show stages of formation of brachial plexus.

4. The **cords (fasciculi)** of the plexus are formed in the axilla. The dorsal divisions unite to form the **posterior cord** (C. 5–8 $\pm$ T. 1). The ventral divisions of the upper and middle trunks unite to form the **lateral cord** (C. 5–7), while the ventral division of the lower trunk continues as the **medial cord** (C. 8 and T. 1). The various cords bear the relation suggested by their names to the second part of the axillary artery, the medial cord passing behind the artery to reach its medial aspect. Each cord ends by dividing into two main branches at the beginning of the third part of the artery.

Nerves arising from a cord do not necessarily contain fibres from all the ventral rami which send fibres into that cord; e.g. the musculo-cutaneous and axillary nerves, though arising from the lateral and posterior cords respectively, contain only fibres of the fifth and sixth cervical nerves. Each part of the brachial plexus, including its branches, shows an internal plexiform arrangement which prevents the dissection of individual fibres back through the plexus to determine their segmental origin.

Communications from sympathetic

The fifth and sixth cervical ventral rami receive grey rami communicantes from the middle cervical ganglion, while two or more grey rami pass from the inferior cervical or cervicothoracic ganglion to the seventh and eighth cervical ventral rami. These grey rami either pierce the prevertebral muscles or pass posterior to the medial border of the scalenus anterior muscle to reach the ventral rami. The first thoracic ventral ramus receives its grey ramus from the cervicothoracic ganglion. These sympathetic fibres to the brachial plexus are supplemented by branches from the sympathetic plexus on the vertebral artery to the fourth to sixth cervical ventral rami, while other sympathetic fibres enter the limb on the subclavian artery.

Variations

The plexus as a whole may arise from a slightly higher or lower level relative to the vertebral column than the usual description: (1) in approximately two-thirds, of all plexuses, a part of the fourth cervical ventral ramus joins the plexus; (2) in approximately one-third, there is no contribution from the fourth but all the fifth joins the plexus; (3) in less than 10 per cent of cases only part of the fifth joins the plexus (Kerr 1918); (4) in more than one-third of cases there is an intrathoracic communication between the second and first thoracic ventral rami. In type (1) the plexus is said to be **prefixed**; in types (3) and (4) it is **postfixed**. The presence of a

cervical rib is independent of the level of origin of the plexus, though plexus nerves which arise below a cervical rib loop over it to enter the plexus.

The pattern of the plexus may vary: (1) trunk division, or cord formation may be absent in one or other part of the plexus, though the constitution of the ultimate terminal branches is unchanged; (2) the lateral or medial cords may receive fibres from ventral rami below or above the normal levels respectively (see also, Herringham 1886; Harris 1904, 1939; Jones 1910).

Branches of the brachial plexus

ARISING FROM SUPRACLAVICULAR PART

These are nerves that arise from the plexus superior to the clavicle.

From the anterior surface

1. Muscular twigs to the scalenus anterior and longus colli muscles arise from the lower four cervical ventral rami as they emerge from the intervertebral foramina.
2. A branch to the **phrenic nerve** arises from the fifth cervical at the lateral border of scalenus anterior [see also accessory phrenic nerve, p. 773].
3. The **nerve to subclavius** arises from the upper trunk (C. (4), 5, 6). It descends in the posterior triangle of the neck anterior to the third part of the subclavian artery, and often communicates with the phrenic nerve [p. 773].

From the posterior surface

1. Small branches to the scalenus medius and posterior arise from the lower four cervical ventral rami as they emerge from the intervertebral foramina.
2. The **dorsal scapular nerve** arises from the fifth cervical ventral ramus. It pierces the scalenus medius, and descending in the posterior triangle of the neck, passes deep to the levator scapulae and to the rhomboid muscles close to the medial border of the scapula. It supplies these muscles and occasionally pierces the levator scapulae.
3. The **long thoracic nerve** arises from the fifth, sixth, and seventh cervical ventral rami close to the intervertebral foramina. The upper two roots unite after piercing scalenus medius and, descending on it, are joined inferiorly by the lowest root which

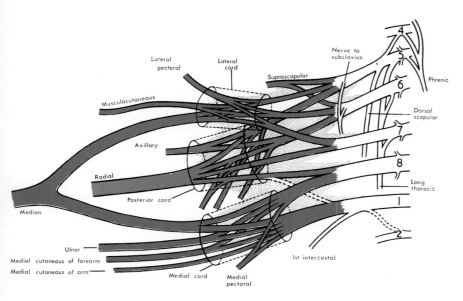

FIG. 11.41. Diagram of brachial plexus, showing the origins of its branches. Ventral divisions, red; dorsal divisions, blue. Broken lines indicate variable connections.

passes anterior to that muscle. The nerve descends posterior to the trunks of the brachial plexus, and enters the axilla between the axillary artery and the first digitation of serratus anterior on the first rib. It runs on the axillary surface of serratus anterior and supplies it.

4. The **suprascapular nerve** arises from the upper trunk of the plexus (C. (4), 5, 6). It lies superior to the trunks of the brachial plexus, and passes inferolaterally parallel to them to pass through the scapular notch on the superior border of the scapula deep to trapezius. It supplies and passes deep to supraspinatus, sends a slender filament to the shoulder joint, and ends in infraspinatus by traversing the spinoglenoid notch with the suprascapular artery.

ARISING FROM INFRACLAVICULAR PART

These branches arise from the cords of the plexus in the axilla. The anterior branches from the lateral and medial cords supply the front of the chest and limb; the posterior branches from the posterior cord supply the shoulder and back of the limb.

Branches of cords of brachial plexus

LATERAL CORD
    Lateral pectoral (C. 5, 6, 7)
    Musculocutaneous (C. 5, 6, 7)   } Main branches
    Lateral root of median (C. 5, 6, 7)

MEDIAL CORD
    Medial pectoral (C. 8; T. 1)
    Medial cutaneous of arm (T. 1)
    Medial cutaneous of forearm (C. 8; T. 1)
    Ulnar (C. 8; T. 1)
    Medial root of median (C. 8; T. 1)   } Main branches

POSTERIOR CORD
    Upper subscapular (C. (4), 5, 6, (7))
    Thoracodorsal (C. (6), 7, 8)
    Lower subscapular (C. 5, 6)
    Axillary (c. 5, 6)
    Radial (C. (5), 6, 7, 8; (T. 1))   } Main branches

BRANCHES OF THE MEDIAL AND LATERAL CORDS OF THE PLEXUS

Pectoral nerves

The **lateral pectoral nerve** arises from the ventral division of the upper and middle trunks (C. 5, 6, 7) just before they unite to form the lateral cord. It passes anteromedially, and giving a branch to the medial pectoral nerve, pierces the clavipectoral fascia and supplies pectoralis major.

The **medial pectoral nerve** arises from the medial cord (C. 8; T. 1), passes anteriorly between the axillary artery and vein, and receiving the communication from the lateral pectoral nerve anterior to the artery, supplies and pierces pectoralis minor to end in pectoralis major. Part of the nerve may pass below the inferior border of pectoralis minor.

Musculocutaneous nerve

This nerve arises from the lateral cord of the plexus (C. 5, 6, and C. 4 in more than 50 per cent of cases) and contains the nerve to the coracobrachialis muscle (C. 7, or 6 and 7) [FIG. 11.41]. The nerve descends between the axillary artery and coracobrachialis, sends a

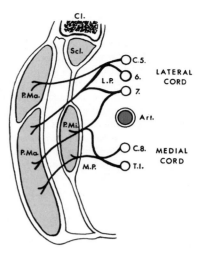

FIG. 11.42. Diagram of the origin and distribution of the nerves to the pectoral muscles.
    L.P., lateral pectoral nerve; M.P., medial pectoral nerve; C.5, 6, 7, C.8, T.1, nerves of the brachial plexus; Art., axillary artery; Cl., clavicle; Scl., subclavius muscle; P.Mi., pectoralis minor, joined to subclavius by clavipectoral fascia; P.Ma., pectoralis major.

branch to that muscle and then pierces it to continue distally between biceps and brachialis, often communicating with the median nerve. At the level of the elbow, it pierces the deep fascia between biceps and brachioradialis as the lateral cutaneous nerve of the forearm.

Muscular branches pass to coracobrachialis and to biceps and brachialis.

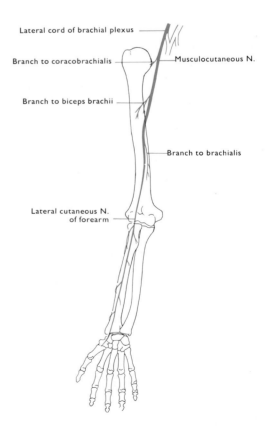

FIG. 11.43. The course and distribution of the musculocutaneous nerve.

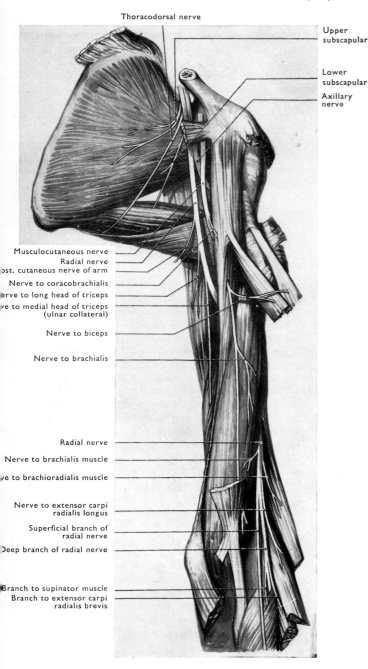

Thoracodorsal nerve

Upper subscapular

Lower subscapular

Axillary nerve

Musculocutaneous nerve
Radial nerve
ost. cutaneous nerve of arm
Nerve to coracobrachialis
erve to long head of triceps
ve to medial head of triceps
(ulnar collateral)
Nerve to biceps

Nerve to brachialis

Radial nerve
Nerve to brachialis muscle
ve to brachioradialis muscle

Nerve to extensor carpi
radialis longus
Superficial branch of
radial nerve
Deep branch of radial nerve

Branch to supinator muscle
Branch to extensor carpi
radialis brevis

FIG. 11.44. Deeper nerves of arm.
The nerve to the brachialis usually enters the medial side of the muscle as in FIG. 11.46.

The musculocutaneous nerve also sends twigs to the marrow and to the periosteum of the distal anterior surface of the humerus, and to the brachial artery.

The lateral cutaneous nerve of the forearm

The anterior branch descends along the front of the lateral aspect of the forearm. It supplies the skin and fascia of the lateral half of this surface, and extends to the ball of the thumb. It supplies branches to the radial artery and communicates with the superficial branch of the radial nerve proximal to the wrist. The posterior branch passes backwards to supply a variable area of skin and fascia over the

posterior surface of the forearm, wrist, and sometimes the first metacarpal bone. It communicates with the lower lateral cutaneous nerve of the arm and the superficial branch of the radial nerve.

Median nerve

This nerve is formed by the union of a lateral and a medial root respectively from the lateral (C. 5, 6, 7) and medial (C. 8 and T. 1) cords; the medial root passing anterior to the third part of the axillary artery. The median nerve descends lateral to the brachial artery, crosses to its medial side (usually anterior to the artery), and passes deep to the bicipital aponeurosis and the median cubital vein at the elbow [FIGS. 11.46 and 11.50]. It enters the forearm between the two heads of the pronator teres muscle, and runs on the deep surface of flexor digitorum superficialis within its fascial sheath. Near the wrist it becomes superficial between the tendons of flexor digitorum superficialis and flexor carpi radialis, deep to palmaris longus tendon. It enters the palm by passing deep to the flexor retinaculum anterior to the long flexor tendons. At the distal margin of the flexor retinaculum it separates into its six terminal branches deep to the palmar aponeurosis [FIG. 11.51]. In the forearm, the median nerve is accompanied by the small median artery.

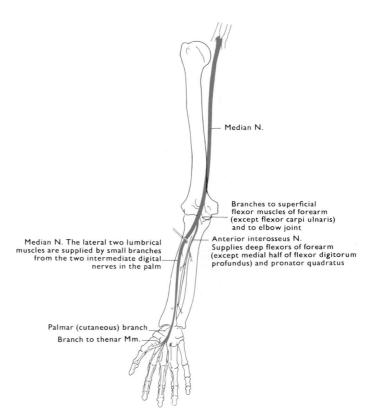

Median N.

Branches to superficial
flexor muscles of forearm
(except flexor carpi ulnaris)
and to elbow joint
Anterior interosseus N.
Supplies deep flexors of forearm
(except medial half of flexor digitorum
profundus) and pronator quadratus

Median N. The lateral two lumbrical
muscles are supplied by small branches
from the two intermediate digital
nerves in the palm

Palmar (cutaneous) branch
Branch to thenar Mm.

FIG. 11.45. Diagram of the course and distribution of the median nerve. The digital branches are not labelled so as to avoid confusion.

BRANCHES

There are no branches in the arm, though the nerve commonly communicates with the musculocutaneous nerve.

In the forearm

1. Muscular branches pass to pronator teres, flexor carpi radialis, palmaris longus, and flexor digitorum superficialis in the cubital

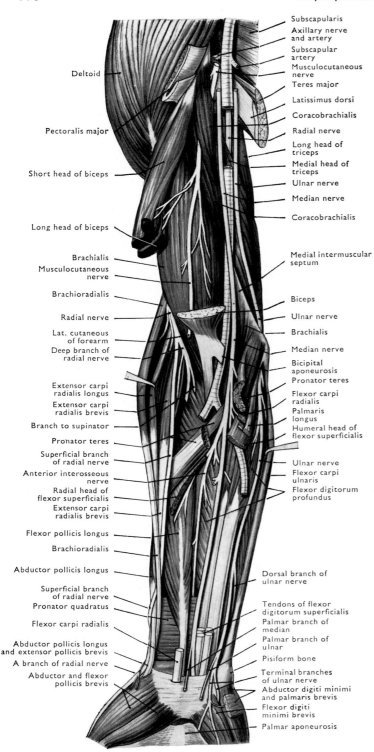

Deltoid

Pectoralis major

Short head of biceps

Long head of biceps

Brachialis
Musculocutaneous
nerve

Brachioradialis

Radial nerve

Lat. cutaneous
of forearm
Deep branch of
radial nerve

Extensor carpi
radialis longus
Extensor carpi
radialis brevis

Branch to supinator

Pronator teres

Superficial branch
of radial nerve
Anterior interosseous
nerve
Radial head of
flexor superficialis
Extensor carpi
radialis brevis

Flexor pollicis longus

Brachioradialis

Abductor pollicis longus

Superficial branch
of radial nerve
Pronator quadratus

Flexor carpi radialis

Abductor pollicis longus
and extensor pollicis brevis
A branch of radial nerve

Abductor and flexor
pollicis brevis

Subscapularis
Axillary nerve
and artery
Subscapular
artery
Musculocutaneous
nerve
Teres major
Latissimus dorsi
Coracobrachialis
Radial nerve
Long head of
triceps
Medial head of
triceps
Ulnar nerve
Median nerve
Coracobrachialis

Medial intermuscular
septum

Biceps
Ulnar nerve
Brachialis
Median nerve
Bicipital
aponeurosis
Pronator teres
Flexor carpi
radialis
Palmaris
longus
Humeral head of
flexor superficialis

Ulnar nerve
Flexor carpi
ulnaris
Flexor digitorum
profundus

Dorsal branch of
ulnar nerve

Tendons of flexor
digitorum superficialis
Palmar branch of
median
Palmar branch of
ulnar
Pisiform bone
Terminal branches
of ulnar nerve
Abductor digiti minimi
and palmaris brevis
Flexor digiti
minimi brevis
Palmar aponeurosis

FIG. 11.46. Deeper nerves of front of arm and forearm.

fossa. Branches are usually present here to the upper fibres of both flexor pollicis longus and flexor digitorum profundus. The nerve to pronator teres is the first to be given off and often arises independently. The index belly of flexor digitorum superficialis often receives a separate branch in the middle of the forearm.

2. The **anterior interosseous nerve** arises from the posterior surface of the median nerve in the cubital fossa. It descends on the anterior surface of the interosseous membrane with the anterior interosseous artery, supplying the flexor pollicis longus and the lateral part of the flexor digitorum profundus. It then passes posterior to the pronator quadratus, and supplies that muscle and the radiocarpal and intercarpal joints. The nerve also sends medullary and periosteal twigs to the radius and ulna and to the interosseous membrane.

3. **Articular branches** to the anterior surface of the elbow joint arise from the muscular branches in the cubital fossa.

4. The **palmar cutaneous branch** arises in the distal one-third of the forearm, pierces the deep fascia, and enters the palm of the hand superficial to the flexor retinaculum. It supplies a small area of the skin and fascia of the palm and adjacent thenar eminence, and communicates with the corresponding branch of the ulnar nerve. It may be absent.

In the hand

1. The **main muscular branch** arises near the distal border of the flexor retinaculum and curves laterally and proximally into the thenar muscles. It supplies branches to abductor pollicis brevis, opponens pollicis, and flexor pollicis brevis.

2. The **palmar digital branches** arise in a variable manner near the distal border of the flexor retinaculum. Frequently there is one **proper palmar digital nerve** to the radial side of the index finger (which also supplies the first lumbrical muscle), and three **common palmar digital nerves** each of which divides into two **proper palmar digital nerves**: (i) to both sides of the thumb; (ii) to the adjacent sides of the index and middle fingers (which supplies the second lumbrical muscle); (iii) to the adjacent sides of the middle and ring fingers. The last of these receives a communication from the common palmar digital branch of the ulnar nerve [FIG. 11.51].

The digital nerves lie deep to the palmar aponeurosis and the superficial palmar arch, but superficial to the flexor tendons in the palm. The nerves to the thumb and the radial side of the index finger emerge from under the lateral edge of the palmar aponeurosis; the other two pass with the corresponding lumbrical muscles, between the slips of the aponeurosis to the fingers. All the proper palmar digital nerves send branches to the dorsal digital nerves. These supply the skin on the dorsal surfaces of the distal phalanx of the thumb and the distal two phalanges of the lateral two and a half fingers. They also send **articular branches** to the interphalangeal and metacarpophalangeal joints, and vascular branches to the adjacent arteries. The nerves are beset with numerous lamellated corpuscles [FIG. 12.8] some of which are very large.

Communications

The median nerve frequently communicates with the musculocutaneous nerve in the arm, and sometimes with the ulnar nerve posterior to the flexor muscles of the forearm. The latter communication may transfer to the ulnar nerve fibres destined for the thenar muscles (Hovelacque 1927). It also communicates with the ulnar nerve in the palm of the hand.

Destruction of the median nerve above the elbow produces a cutaneous sensory loss on the palm of the hand, the palmar surfaces of the thumb, index, and middle fingers, and on the dorsal surfaces of the distal phalanx of the thumb, and the distal two phalanges of the index and middle fingers. There may also be some sensory loss on the radial surface of the ring finger, though the area of absolute sensory loss is always smaller than the anatomical distribution of the nerve because of the overlap from adjacent nerves (ulnar and radial). Pronation is severely affected (pronator teres, pronator quadratus, and flexor carpi radialis). Flexion of the wrist is associated with ulnar deviation (action of flexor carpi ulnaris). Flexion (flexor digitorum superficialis and lateral half of flexor digitorum profundus) is absent in the index and middle fingers and is weakened in the ring and little fingers. Flexion of the metacarpophalangeal joints with extension of the interphalangeal joints of index and middle fingers is weak (lumbricals) though the movement is maintained by the interossei. All movements of the thumb except adduction, extension, and abduction by abductor pollicis longus are lost; thus opposition and flexion are impossible.

Destruction of the median nerve at the wrist or pressure on it in the carpal tunnel (carpal tunnel syndrome) has similar sensory effects though the action of the flexors and pronators in the forearm is not affected. Opposition, true abduction, and flexion of the thumb at the metacarpophalangeal joint are severely affected (thenar muscles) as is lumbrical action on the index and middle fingers.

Medial cutaneous nerve of the arm

This nerve arises from the medial cord of the brachial plexus, principally from the first thoracic nerve [FIG. 11.41]. It descends between the axillary artery and vein, passes anterior or posterior, or even through the vein, to pierce the deep fascia and supply the skin and fascia on the medial side of the proximal half of the arm. The nerve usually communicates and varies inversely in size with the intercostobrachial nerve which may entirely replace it, together with the posterior cutaneous branch of the radial nerve to the arm.

Medial cutaneous nerve of the forearm

This nerve arises from the medial cord of the brachial plexus (C. 8 and T.1) and passes through the axilla and proximal half of the arm medial to the main artery. At the middle of the arm it pierces the deep fascia with the basilic vein, and accompanies it to the anterior surface of the elbow. Here it divides into anterior and ulnar branches.

BRANCHES

As it pierces the deep fascia the nerve gives a small branch to supply the skin on the anteromedial surface of the distal half of the arm. The terminal branches pass superficial or deep to the median cubital vein to supply the medial surface of the forearm [FIG. 11.47]. The anterior branch supplies the medial part of the front of the forearm as far as the wrist. The smaller ulnar (posterior) branch passes postero-inferiorly over the common flexor origin to supply skin and fascia on the proximal three-quarters of the medial part of the back of the forearm. Both branches may communicate with the ulnar nerve in the distal part of the forearm.

Ulnar nerve

The ulnar nerve arises from the medial cord of the brachial plexus (C. 8 and T. 1), and receives a communication (C. 7) either from the lateral cord of the plexus or from the lateral root of the median nerve in more than half the cases. The nerve descends between the axillary artery and vein, posterior to the medial cutaneous nerve of the forearm, and then lies anterior to triceps on the medial side of the brachial artery. In the distal half of the arm it passes backwards with an ulnar collateral artery, through the medial intermuscular septum, and continues between this structure and the medial head of triceps to enter the forearm between the medial epicondyle of the humerus and the olecranon, deep to the fibrous arch between the origins of flexor carpi ulnaris from these two bones. It descends,

deep to flexor carpi ulnaris, first on the ulnar collateral ligament and then on flexor digitorum profundus, and joins the ulnar artery above the middle of the forearm. In the distal half of the forearm it becomes comparatively superficial, sends branches to the ulnar artery, and pierces the deep fascia with it just proximal to the flexor retinaculum, lateral to flexor carpi ulnaris. It passes on the anterior surface of the flexor retinaculum immediately lateral to the pisiform bone, and divides into superficial and deep branches under cover of the palmaris brevis muscle.

BRANCHES

These arise in the forearm and hand.

An articular branch passes to the elbow joint as the nerve lies posterior to the medial epicondyle of the humerus.

Muscular branches in the forearm pass to flexor carpi ulnaris and the medial half of flexor digitorum profundus immediately distal to the elbow. Flexor carpi ulnaris receives two branches, one each to its olecranon and epicondylar heads, and one or two additional branches at a lower level.

Cutaneous branches in the forearm

1. The palmar cutaneous branch is variable in size and position. It pierces the deep fascia in the distal third of the forearm and supplies the skin of the medial part of the palm. It supplies the ulnar artery, and often communicates with the anterior branch of the medial cutaneous nerve of the forearm and the palmar cutaneous branch of the median nerve.

2. The dorsal branch arises in the distal half of the forearm. It passes postero-inferiorly between the ulna and flexor carpi ulnaris to become cutaneous on the medial side of the distal quarter of the forearm. At the medial side of the wrist it crosses the head of the ulna on to the triquetrum, against which it may be palpated, and gives branches to the skin and fascia of the dorsal surfaces of the wrist and hand. These branches may communicate with branches of the posterior cutaneous nerve of the forearm and with the corresponding branches from the superficial branch of the radial nerve. The areas supplied by these three nerves on the dorsum of the wrist and hand is very variable, though commonly the radial and ulnar nerves are the sole supply to the dorsum of the hand and share it almost equally. The dorsal branch of the ulnar nerve passes on to the base of the fifth metacarpal and divides into three dorsal digital nerves. One passes along the ulnar side of the dorsum of the hand and little finger as far as the nail bed; another divides at the cleft between the little and ring fingers and supplies their adjacent sides to the same extent; a third dorsal digital nerve supplies the adjacent sides of the middle and ring fingers as far as the proximal interphalangeal joints, the more distal parts being supplied by the palmar digital branches of the median nerve. This last branch communicates with the most medial dorsal digital branch of the radial nerve which may supply part of the cleft between the ring and middle fingers.

Branches in the wrist and hand

1. The nerve sends a small branch to palmaris brevis and then divides into superficial and deep branches.

2. The superficial branch lies deep to the palmaris brevis. It divides into two. (a) Medially, a proper palmar digital nerve which passes distally on the flexor digiti minimi brevis and the medial aspect of the little finger, supplying its palmar surface. (b) Laterally, a common palmar digital nerve which passes between the palmar aponeurosis and the long flexor tendons to the little finger. It sends a communicating branch to the medial common palmar digital nerve from the median, and reaches the loose tissue in the web

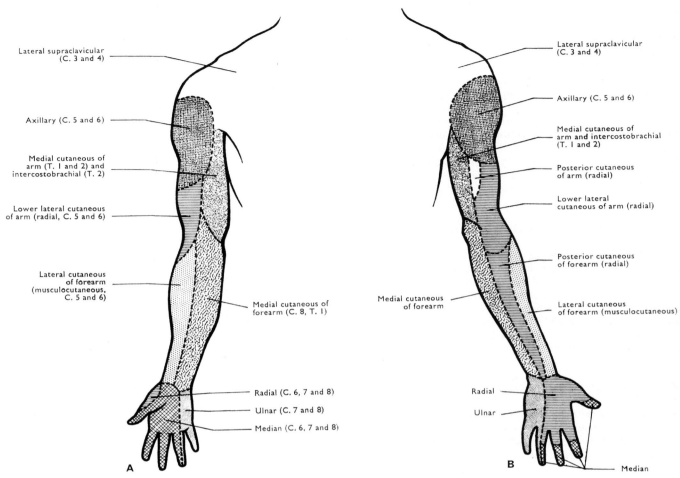

FIG. 11.47. Distribution of cutaneous nerves.
A, on the front; B, on the back of upper limb.

between the slips of the palmar aponeurosis to the ring and little fingers. Here, anterior to the fourth lumbrical muscle, it divides into two **proper palmar digital nerves** which pass to the adjacent sides, palmar surfaces, nail beds, and metacarpophalangeal and interphalangeal joints of these fingers [FIG. 11.51].

The ulnar nerve thus supplies the skin and fascia of the medial one-third of the palm and the palmar surfaces of one and a half fingers. On the dorsal surface it supplies skin of half of the hand and wrist, of the same one and a half fingers, and frequently of the cleft between the ring and middle fingers. Thus it commonly supplies all the skin on the dorsal surface of the proximal phalanx of the ring finger. This common arrangement (Stopford 1918) is subject to considerable variation. Rarely, the radial nerve may supply almost all the skin on the dorsum usually innervated by the ulnar nerve, or the ulnar may supply the greater part of the dorsum of the hand and the dorsal surfaces of the fingers not supplied by the median nerve. In some cases the posterior cutaneous nerve of the forearm may supply much of the dorsum of the hand. In all cases it is probable that these variations represent an abnormal distribution of fibres through the peripheral nerves rather than an abnormal innervation.

3. The **deep branch** arises from the nerve on the flexor retinaculum lateral to the pisiform bone [FIG. 11.51]. It passes posteriorly between the abductor and short flexor of the little finger, supplying them, and supplying and piercing the opponens digiti minimi near its origin from the flexor retinaculum, turns laterally over the distal surface of the hook of the hamate bone. It then runs laterally between the proximal parts of the metacarpal bones and

the long flexor tendons, to pass between the two heads of adductor pollicis with the deep palmar arch, and end in the first dorsal interosseous muscle. In the palm the deep branch supplies the interossei, the third and fourth lumbrical muscles, the adductor pollicis, and the deep vessels of the palm. The first dorsal interosseous muscle may receive a partial or complete nerve supply from the median nerve (Sunderland 1946). Very rarely the ulnar nerve may supply the thenar muscles, presumably by fibres which fail to pass to the median nerve through its medial root, or which enter the ulnar through an abnormal communication from the median in the forearm.

Destruction of the ulnar nerve at or above the elbow produces a sensory alteration in the medial part of the hand, and in the little and parts of the ring and middle (dorsal) fingers. The absolute sensory loss is usually less extensive than the anatomical distribution, indicating an overlap from adjacent nerves. Flexion of the wrist occurs with radial deviation (paralysis of flexor carpi ulnaris); the distal phalanges of the little and ring fingers cannot be flexed (medial half of flexor digitorum profundus); abduction and adduction of the fingers is missing (the interossei), though some adduction can still occur on flexion of the fingers (long flexor tendons). Adduction of the thumb is absent though a spurious adduction can be produced by flexor pollicis longus. Other movements of the thumb are normal. Paralysis of the third and fourth lumbricals results in extension of the metacarpophalangeal joints of the ring and little fingers (occasionally the middle finger also) with flexion of the interphalangeal joints (claw hand).

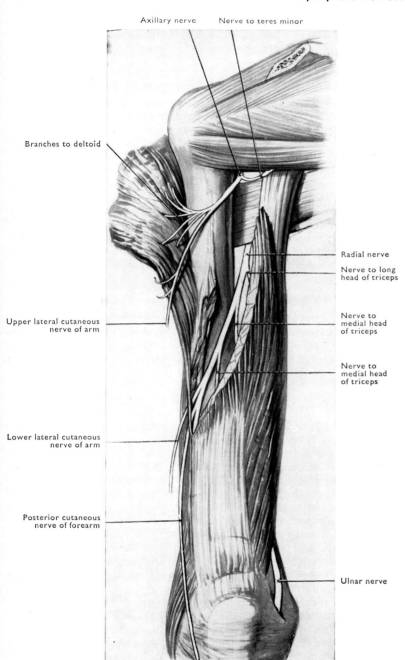

Axillary nerve

Nerve to teres minor

Branches to deltoid

Upper lateral cutaneous nerve of arm

Lower lateral cutaneous nerve of arm

Posterior cutaneous nerve of forearm

Radial nerve

Nerve to long head of triceps

Nerve to medial head of triceps

Nerve to medial head of triceps

Ulnar nerve

FIG. 11.53. Axillary and radial nerves.

Axillary nerve

This terminal branch (C. 5, 6) of the posterior cord descends posterior to the axillary artery to the lower border of subscapularis. Here it passes posteriorly through the quadrangular space [FIGS. 5.71 and 11.53] with the posterior humeral circumflex vessels, gives one or two branches to the shoulder joint, and divides into anterior and posterior branches.

The **anterior branch** winds round the surgical neck of the humerus, deep to deltoid. It sends branches into deltoid, a few of which descend through the muscle to supply skin over its lower part.

The **posterior branch** sends a nerve with a small swelling on it into the inferolateral surface of teres minor, and after giving a few branches to the posterior part of deltoid, descends round its posterior border as the **upper lateral cutaneous nerve of the arm**. This nerve supplies the skin and fascia that cover the lower part of deltoid and the lateral head of triceps as far as the middle of the arm [FIGS. 11.47 and 11.53].

Radial nerve

The radial nerve is the direct continuation of the posterior cord of the brachial plexus. It arises from the fifth to eighth cervical and the first thoracic ventral rami. The contributions from the first and last of these are small and inconstant [FIG. 11.41].

In the axilla it lies posterior to the axillary artery on the subscapularis, latissimus dorsi, and teres major, and descends into the arm between the brachial artery and the long head of triceps, medial to the humerus. It then inclines inferolaterally with the profunda brachii artery, and passing between the long head and the upper part of the origin of the medial head of triceps, spirals round the posterior surface of the humerus in the groove for the radial nerve, deep to the lateral head of triceps. Reaching the distal third of the arm at the lateral border of the humerus, the radial nerve pierces the lateral intermuscular septum and descends to the front of the lateral epicondyle, lying deeply between brachialis and brachioradialis. Here it divides into its terminal branches; deep and superficial.

BRANCHES

A. Medial to the humerus

1. The posterior cutaneous nerve of the arm arises in the axilla and pierces the deep fascia on the medial side of the arm near the posterior axillary fold. It supplies skin and fascia on the posterior surface of the proximal third of the arm [FIG. 11.47].

2. Four branches pass to the triceps in the following order: (i) to the long head, arising from the nerve in the axilla; (ii) to the distal part of the medial head, arising near the lower border of teres major it accompanies the ulnar nerve to the middle of the arm (ulnar collateral nerve); (iii) to the lateral head, arising just distal to the last-mentioned branch; (iv) the main nerve to the medial head, arising at the entry to the groove on the humerus for the radial

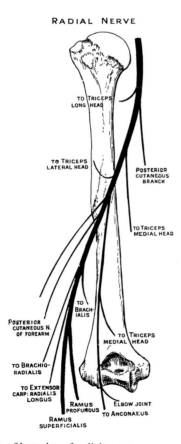

RADIAL NERVE

TO TRICEPS LONG HEAD

TO TRICEPS LATERAL HEAD

POSTERIOR CUTANEOUS BRANCH

TO TRICEPS MEDIAL HEAD

TO BRACHIALIS

POSTERIOR CUTANEOUS N. OF FOREARM

TO TRICEPS MEDIAL HEAD

TO BRACHIORADIALIS

TO EXTENSOR CARP. RADIALIS LONGUS

RAMUS PROFUNDUS

ELBOW JOINT

TO ANCONAEUS

RAMUS SUPERFICIALIS

FIG. 11.54. Diagram of branches of radial nerve.

nerve. This nerve supplies the medial head, and descending through the muscle, passes posterior to the lateral epicondyle of the humerus and ends by supplying the anconeus muscle.

B. Posterior to the humerus

1. The lower lateral cutaneous nerve of the arm [FIGS. 11.47 and 11.53] arises before the radial nerve pierces the lateral intermuscular septum. It emerges a little inferior to deltoid, and supplies the skin and fascia on the posterior surface of the distal third of the arm and a small area on the back of the forearm.

2. The posterior cutaneous nerve of the forearm arises with (or immediately distal to) the previous nerve, and pierces the deep fascia 2 cm lower down. It descends into the back of the forearm, and supplies a variable area of skin and fascia between the areas innervated by the musculocutaneous nerve and the ulnar branch of the medial cutaneous nerve of the forearm. It may reach the level of the wrist or extend on to the dorsum of the hand, supplementing the area supplied by the radial nerve there [p. 780].

C. Lateral to the humerus

As it lies between brachialis and brachioradialis, the radial nerve gives one or two branches to brachioradialis, a branch to extensor carpi radialis longus (occasionally to extensor carpi radialis brevis also), and an inconstant branch to brachialis. The latter appears to be entirely sensory to that muscle.

D. Terminal branches

1. The superficial branch is the direct continuation of the radial nerve along the lateral part of the anterior surface of the forearm and is entirely sensory. It lies successively on supinator, pronator teres, flexor digitorum superficialis, and flexor pollicis longus, deep to brachioradialis; with the radial artery on its medial side. It gives a branch to the artery, and passing obliquely backwards in the distal third of the forearm deep to the tendon of brachioradialis, pierces the deep fascia to supply skin and fascia of the dorsum of the wrist, the lateral side of the dorsum of the hand, and the dorsum of the thumb and lateral one (or two) and a half fingers [FIG. 11.47]. Of the four or five small dorsal digital nerves, two pass on the dorsum of the thumb as far as the interphalangeal joint, supplying also that joint, the metacarpophalangeal joint, and skin on the lateral part of the thenar eminence. One passes to the lateral side of the index finger as far as the middle phalanx. The remaining one or two divide at the clefts between the second and third and the third and fourth fingers. They innervate the adjacent sides of these fingers over the proximal phalanges. The area supplied by the superficial branch of the radial nerve is subject to wide variation [p. 780], but it usually does not supply skin medial to the second metacarpal bone (Stopford 1918).

2. The deep branch is entirely muscular and articular in its distribution. It passes postero-inferiorly, and giving branches to extensor carpi radialis brevis and supinator, enters the latter muscle [FIG. 11.55] and curves round the lateral and posterior surfaces of the radius either in the muscle or between it and the radius (Davies and Laird 1948). It emerges from supinator as the posterior interosseous nerve between the superficial extensor muscles and the lowest fibres of supinator. Here it gives recurrent branches which enter extensor digitorum, extensor digiti minimi, and extensor carpi ulnaris close to their origins, and occasionally to anconeus. It then passes with the posterior interosseous artery superficial to the abductor pollicis longus supplying it. It supplies branches to extensor pollicis longus and brevis and extensor indicis, and ends in a small gangliform enlargement on the back of the carpus from which the intercarpal joints are supplied.

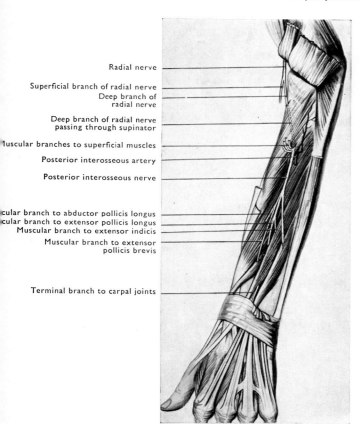

Radial nerve

Superficial branch of radial nerve
Deep branch of radial nerve

Deep branch of radial nerve passing through supinator

Muscular branches to superficial muscles

Posterior interosseous artery

Posterior interosseous nerve

Muscular branch to abductor pollicis longus
Muscular branch to extensor pollicis longus
Muscular branch to extensor indicis
Muscular branch to extensor pollicis brevis

Terminal branch to carpal joints

FIG. 11.55. Posterior interosseous nerve.

RADIAL NERVE INJURY

Damage to the radial nerve above the origin of the branches to triceps leads to paralysis of triceps, brachioradialis, supinator, and the extensors of the wrist, thumb, and fingers. Extension of the interphalangeal joints can still be produced by the lumbricals and interossei. The injury also produces impairment of sensation in the territories of distribution of the lower lateral cutaneous nerve of the arm, the posterior cutaneous nerves of the arm and forearm, and the superficial branch of the radial nerve [p. 784]. When the nerve is injured in the groove for the radial nerve on the humerus, the triceps is only partly paralysed and the posterior cutaneous nerve of the arm is unaffected, but the other effects are similar.

The outstanding defect in all injuries to the radial nerve is the inability to extend, or even straighten the wrist, which is held flexed by the flexor muscles (wrist drop). This position of the wrist renders the flexors incapable of acting on the fingers to produce a fist, so that objects cannot be firmly held in the hand (active insufficiency p. 272). The paralysis of supinator and brachioradialis is not significant because biceps and brachialis are adequate to produce, respectively, supination and flexion at the elbow. If, however, triceps is paralysed, supination cannot occur without flexion at the elbow.

MOTOR DISTRIBUTION

In the following tables are shown the positions in the limb of the muscles innervated by the various nerves, the muscles, and the effects of paralysis of these muscles following injury to the nerves or their branches. Where a nerve innervates muscles in more than one segment of the limb (shoulder, arm, forearm, hand) the effects of injury to the nerve will depend on the level of injury. Thus when the median nerve is destroyed at the wrist, the muscles supplied by it in the forearm are not paralysed though those in the hand are. Since forearm and hand muscles may act together on the same joints, the degree of paralysis of some movements may be increased by the more proximal destruction in addition to the paralysis of different movements. An attempt has been made to show this by linking the muscles which act together with brackets. In those cases where a number of muscles assist in an action and some are innervated by nerves other than those which are destroyed, these muscles which are still active are shown in brackets after the statement of the effect of the muscle paralysis.

Abbreviations

Fingers are simply shown by their name—e.g. 'index' for 'index finger' C.M. = carpometacarpal joint. M.P. = metacarpophalangeal joint. I.P. = interphalangeal joints. P.I.P. = proximal interphalangeal joint(s). D.I.P = distal interphalangeal joint(s). * = supplied by anterior or posterior interosseous branch.

Median nerve

SITUATION OF INNERVATED MUSCLES	MUSCLES	EFFECTS OF PARALYSIS BY NERVE DESTRUCTION
Shoulder	Nil	
Arm	Nil	
Forearm	Flexor carpi radialis Pronator teres Pronator quadratus* Flexor pollicis longus*	Weakened wrist flexion with ulnar deviation (flexor carpi ulnaris) *Pronation lost* Flexion of thumb lost at I.P. and weakened at C.M. & M.P. if only anterior interosseous nerve destroyed (flexor pollicis brevis, opponens pollicis). If total median nerve is destroyed above elbow, then thumb flexion is lost in all joints

SITUATION OF INNERVATED MUSCLES	MUSCLES	EFFECTS OF PARALYSIS BY NERVE DESTRUCTION
	Flexor digitorum profundus, lateral half ⎫ Flexor digitorum superficialis ⎬	D.I.P. flexion, lost, index and middle P.I.P. flexion lost, index and middle P.I.P. flexion weak, ring and little (flexor digitorum profundus) M.P. flexion weak all fingers: index and middle (interossei only); ring (flexor digitorum profundus, lumbrical, interossei); little (flexor digitorum profundus, flexor digiti minimi, interosseous)
Hand	Lumbricals, lateral two	I.P. extension weakened, index and middle (ext. digitorum and indicis, interossei)
	Abductor pollicis brevis ⎫ Flexor pollicis brevis ⎬ Opponens pollicis ⎭	C.M. & M.P. of thumb, weakened flexion (flexor pollicis longus) and abduction (abductor pollicis longus) *Opposition of thumb lost*

Ulnar nerve

Shoulder	Nil	
Arm	Nil	
Forearm	Flexor carpi ulnaris	Weakened wrist flexion with radial deviation (flexor carpi radialis)
	Flexor digitorum profundus	*D.I.P. loss of flexion* and M.P. & P.I.P. weakness of flexion in ring and little (flx. digitorum superficialis)
Hand	Abductor digiti minimi	M.P. abduction lost in little finger
	Flexor digiti minimi brevis	M.P. flexion weakened (flx. digitorum superficialis [and profundus if ulnar nerve destroyed distal to branches to that muscle])
	Opponens digiti minimi	C.M. opposition of little finger lost
	All interossei ⎫ Lumbricals, medial two ⎬	*M.P. abduction and adduction lost all fingers** I.P. extension weakened in all fingers (extensors digitorum, indicis, digiti minimi, and lumbricals of index and middle). No I.P. extension of ring and little if M.P. fully extended. These fingers then fixed in I.P. flexion—'claw'
	Adductor pollicis	Adduction of thumb lost, but long flexor and extensor of thumb acting together can mimic action of adductor

*Loss of abduction of little finger depends on the simultaneous paralysis of abductor digiti minimi, though extensor digiti minimi can mimic this action. Fingers are adducted in flexion and abducted in extension. This is due to the 'set' of the joints and is independent of the interossei. Hence interossei are tested with the fingers straight to overcome this and to avoid the restriction of the movement by tightening of the collateral ligaments in flexion.

Musculocutaneous nerve

Shoulder	Coracobrachialis ⎫ Biceps brachii ⎬ short head	Weakened shoulder flexion (deltoid; pectoralis major, clavicular part)
	Long head	Some instability of shoulder joint in abduction (supraspinatus, subscapularis, deltoid)
Arm	Biceps brachii ⎫	Weakened supination (supinator, brachioradialis)
	⎬	Severe weakening of elbow flexion (brachioradialis, ext. carpi. radialis longus, pronator teres, and flex. carpi radialis)
	Brachialis ⎭	

Axillary nerve

SITUATION OF INNERVATED MUSCLES	MUSCLES	EFFECTS OF PARALYSIS BY NERVE DESTRUCTION
Shoulder	Teres minor Deltoid	Severe weakening of abduction (supraspinatus), extension (if not flexed), and lateral rotation of humerus (infraspinatus)

Subscapular nerves and thoracodorsal nerve

Shoulder	Subscapularis Teres major Latissimus dorsi	Instability of shoulder joint with tendency to anterior dislocation Weakened medial rotation of humerus (pectoralis major, deltoid clavicular part) Weakened adduction (pectoralis major). The paralysis of latissimus dorsi shows up in an inability to pull the body upwards with the upper limb

Radial nerve

Shoulder	Triceps, long head	Minor effect on shoulder stability in abduction. Downward dislocation in this position more likely
Arm Forearm	Triceps Supinator Brachioradialis Extensor carpi radialis longus and brevis	*Loss of elbow extension* Weakening of supination of prone hand (biceps brachii) Weakening of elbow flexion in midprone position (brachialis, biceps brachii, pronator teres)
	Extensor carpi radialis longus and brevis	Markedly weakened radial deviation of wrist (flexor carpi radialis) *Loss of wrist extension—'wrist drop'*
	Extensor carpi ulnaris* Extensor digitorum*	Weakened ulnar deviation of wrist (flexor carpi ulnaris) *Loss of extension at M.P., all fingers,* with weakened I.P. extension (interossei, lumbricals)
	Extensor indicis* Extensor digiti minimi*	Loss of independent index extension Loss of independent little finger extension
	Extensor pollicis longus* Extensor pollicis brevis* Abductor pollicis longus*	Loss of C.M., M.P., & I.P. extension of thumb (abd. pollicis brevis may act at I.P. as an extensor) Weakness of thumb abduction (abductor pollicis brevis)

* Posterior interosseous branch.

Suprascapular nerve

Shoulder	Supraspinatus	Difficulty with initiating abduction of humerus (deltoid [teres minor and subscapularis can assist deltoid by holding down humeral head])
	Infraspinatus	Slight weakness of lateral rotation of humerus (post. fibres deltoid, teres minor)

Long thoracic nerve

Shoulder girdle	Serratus anterior	Weakened protraction of scapula (pectoralis major and minor) Weakened lateral rotation of scapula (trapezius) and hence weakened and restricted abduction of upper limb on trunk Scapula not held against ribs—'winged scapula'

THORACIC VENTRAL RAMI

Twelve thoracic ventral rami emerge inferior to the corresponding vertebra and rib; all of these are intercostal in position (intercostal nerves) except the twelfth which lies below the twelfth rib and is called the subcostal nerve. The first, second, third, and twelfth thoracic ventral rami show variations in their course and distribution from the typical arrangement of the remainder.

FIRST THORACIC VENTRAL RAMUS

This, the largest of the series, enters the first intercostal space below the neck of the first rib. It divides into (1) a smaller, inferior, intercostal nerve, and (2) a larger, ascending branch which passes superolaterally to the brachial plexus across the internal border of the first rib. It joins the eighth cervical ventral ramus between the scalenus medius muscle and the subclavian artery. The intercostal nerve runs forwards on the pleural surface of the first rib, and supplying the intercostal muscles of the first space enters that space near the costal cartilage. It then passes anterior to the internal thoracic artery and usually turns forwards to end in the superficial fascia as a small anterior branch which does not usually supply skin. A lateral branch is frequently absent; if present it communicates with the intercostobrachial nerve or, rarely, with the medial cutaneous nerve of the arm.

Communications

In addition to the above, the first thoracic ventral ramus is connected to the cervicothoracic ganglion (or first thoracic ganglion if separate) by grey and white rami communicantes. The first thoracic ventral ramus also receives a branch from the second thoracic ventral ramus. This branch varies in size and distribution, and may pass into either or both of the branches of the first thoracic ventral ramus; thus postganglionic sympathetic fibres from the second thoracic ganglion in addition to fibres of the second thoracic ventral ramus may reach the brachial plexus by this route.

SECOND THORACIC VENTRAL RAMUS

This ventral ramus runs forwards between the second rib and the parietal pleura and passes forwards between the innermost and internal intercostal muscles. Near the midaxillary line it gives off its large lateral cutaneous branch (intercostobrachial nerve), and then continues towards the sternum either as a single trunk or with a small collateral branch below [p. 770]. Anterior to the internal thoracic artery, it passes forwards (anterior cutaneous branch [FIG. 11.56]) through the internal intercostal muscle, the external intercostal membrane, and pectoralis major to supply skin and fascia in the region of the anterior part of the second interspace and the adjacent part of the sternum.

Branches

1. It supplies the muscles of the second intercostal space.
2. The intercostobrachial nerve pierces the intercostal and serratus anterior muscles, and passes postero-inferiorly in the axilla to pierce the deep fascia near the posterior axillary fold. It supplies the skin and fascia of the floor of the axilla and the posteromedial surface of the arm as far as the elbow [FIG 11.47]. The nerve varies in size inversely with the medial cutaneous nerve of the arm with which it communicates either before or after piercing the deep fascia. It also communicates with the posterior branch of the lateral cutaneous branch of the third intercostal nerve in the floor of the axilla, and communicates with the lateral cutaneous branch of the first intercostal nerve when that branch is present.
3. A branch to the first thoracic ventral ramus ascends anterior to the neck of the second rib.
4. A grey and a white ramus communicans connect it with the second thoracic ganglion of the sympathetic trunk.
5. Anterior cutaneous branch, see above.

THIRD THORACIC VENTRAL RAMUS

This nerve differs from the typical thoracic ventral ramus in that the posterior part of its lateral cutaneous branch communicates with the

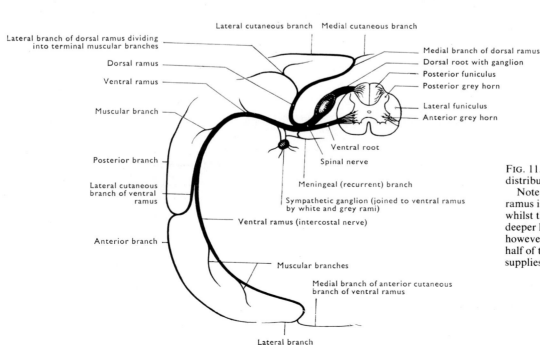

Lateral branch of dorsal ramus dividing into terminal muscular branches
Dorsal ramus
Ventral ramus
Muscular branch
Posterior branch
Lateral cutaneous branch of ventral ramus
Anterior branch

Lateral cutaneous branch   Medial cutaneous branch
Medial branch of dorsal ramus
Dorsal root with ganglion
Posterior funiculus
Posterior grey horn
Lateral funiculus
Anterior grey horn
Ventral root
Spinal nerve
Meningeal (recurrent) branch
Sympathetic ganglion (joined to ventral ramus by white and grey rami)
Ventral ramus (intercostal nerve)
Muscular branches
Medial branch of anterior cutaneous branch of ventral ramus
Lateral branch

FIG. 11.56. Diagram of origin and distribution of typical spinal nerve.
Note that the medial branch of the dorsal ramus is represented as distributed to skin, whilst the lateral branch terminates at a deeper level in muscle. Both branches, however, supply muscles; and in the lower half of the body it is the lateral branch that supplies the skin.

intercostobrachial nerve and supplies skin and fascia on the medial side of the root of the limb and the floor of the axilla.

FOURTH, FIFTH, AND SIXTH THORACIC VENTRAL RAMI

These ventral rami have a typical course and distribution as indeed have the second and third apart from the peculiarities mentioned above [FIG. 11.56].

At first lying in the posterior wall of the thorax between the ribs, each enters the costal groove of the upper rib, and passes forwards between the innermost and internal intercostal muscles inferior to the corresponding vessels. Anterior to the innermost intercostal muscle, each nerve is immediately external to the pleura, but then passes anterior to the transversus thoracis muscle and the internal thoracic artery. Close to the sternum, it pierces the internal intercostal muscle, the external intercostal membrane, and pectoralis major as the anterior cutaneous branch.

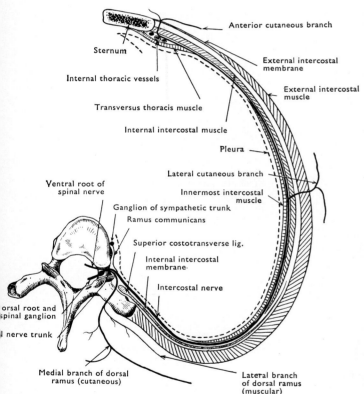

FIG. 11.57. Diagram to show relations of typical intercostal nerve to the intercostal muscles and membranes.

Branches

Each of these nerves supplies a segment of the thoracic wall from the skin to the parietal pleura, excluding the limb muscles which lie in it. There are direct muscular branches and, usually, a collateral branch [p. 770] to the intrinsic thoracic muscles. The skin and fascia are supplied by the anterior and lateral cutaneous branches. The lateral pierces the intercostal and serratus anterior muscles and divides into anterior and posterior branches to supply an area which is continuous with the areas innervated respectively by the anterior

cutaneous branch and by the dorsal ramus of the same spinal nerve. This whole area constitutes a dermatome which does not necessarily overlie the muscle strip (myotome) supplied by the nerve [FIGS. 11.33 and 11.35]. The second intercostal nerve reaches the skin mainly inferior to the sternal angle, the sixth at the base of the xiphoid process, while the fourth innervates the nipple.

Each ventral ramus receives a grey ramus communicans from the corresponding ganglion of the sympathetic trunk and sends a white ramus to it.

SEVENTH TO ELEVENTH THORACIC VENTRAL RAMI

These nerves differ from the preceding nerves in that their anterior parts run in the abdominal wall and supply its skin and fascia, muscles, and peritoneum. When these nerves reach the anterior end of their intercostal spaces they pass deep to the upturned part of the costal cartilage between the attachments to it of transversus abdominis and the diaphragm, and run antero-inferiorly in the abdominal wall between the transversus abdominis and the internal oblique muscles. Each nerve then pierces the rectus sheath, gives a small branch to the lateral part of the rectus abdominis, and sends a large branch deep to that muscle. The larger branch supplies the medial part of the muscle, and piercing it and the anterior wall of its sheath, forms an anterior cutaneous branch.

Branches

Muscular branches pass to the intercostal muscles and to the external and internal oblique, transverse, and rectus muscles of the abdomen. These branches may arise from the cutaneous branches. Branches are also given to the diaphragm, but these are sensory in nature.

The lateral and anterior cutaneous branches are similar to those of the other intercostal nerves. The lateral cutaneous branches emerge where the external oblique interdigitates with the serratus anterior or latissimus dorsi muscles. They descend further than the lateral branches of the higher nerves, the tenth extending as far inferiorly as the iliac crest. The lateral branch of the eleventh crosses the iliac crest into the gluteal region. The anterior cutaneous branches are small. The seventh to ninth supply the skin between the xiphoid process and the umbilicus, the tenth the umbilicus, and the eleventh immediately below the umbilicus.

Communications with the sympathetic trunk are by grey and white rami communicantes [pp. 814 and 817].

SUBCOSTAL NERVE

This ventral ramus of the twelfth thoracic nerve emerges below the twelfth rib, posterior to the upper part of the psoas muscle and the lowest part of the pleura. It passes inferolaterally into the abdomen posterior to the lateral arcuate ligament (edge of diaphragm), and lies on quadratus lumborum posterior to the kidney. It then pierces the transversus abdominis to pass between it and the internal oblique muscle in the abdominal wall. It passes anterior to rectus abdominis, and its anterior cutaneous branch pierces the rectus sheath midway between the umbilicus and the pubis.

Branches

Muscular branches pass to quadratus lumborum, to the four muscles of the abdominal wall, and to pyramidalis.

A large lateral cutaneous branch descends through the muscles of abdominal wall to become superficial immediately above the iliac crest, 5 cm behind the anterior superior iliac spine. It descends over the iliac crest to supply skin and fascia in the gluteal region down to the level of the greater trochanter of the femur [FIG. 11.61]. The **anterior cutaneous branch** supplies skin and fascia midway between the umbilicus and the pubis.

The subcostal nerve may receive a communicating branch from the eleventh intercostal near its origin, more frequently it sends a fine branch through psoas to the first lumbar ventral ramus. It may communicate with the iliohypogastric nerve in the abdominal wall. It communicates with the sympathetic trunk by grey and white rami communicantes [pp. 769 and 770].

LUMBAR, SACRAL, AND COCCYGEAL PLEXUSES

These plexuses are formed by the ventral rami of the lumbar (5), sacral (5), and coccygeal (1) nerves. The lower limb is supplied by the lumbar and sacral plexuses; the upper lumbar nerves also supply part of the trunk, and the lower sacral and coccygeal nerves also supply the perineum.

LUMBAR PLEXUS

This plexus is formed by the first four lumbar ventral rami in the substance of the psoas muscle and supplies that muscle, the quadratus lumborum, and the lowest parts of the muscles of the abdominal wall as well as part of the limb.

SACRAL PLEXUS

This plexus is formed by the ventral rami of the fourth and fifth lumbar and the upper four sacral nerves. The fourth lumbar splits to pass into both the lumbar and the sacral plexuses (nervus furcalis); the fourth sacral divides to enter the sacral and coccygeal plexuses. The sacral plexus lies on the dorsal wall of the pelvis and supplies the remainder of the lower limb and perineum.

COCCYGEAL PLEXUS

This minute plexus is formed from the ventral rami of the fourth and fifth sacral and coccygeal nerves. It lies in the dorsal wall of the

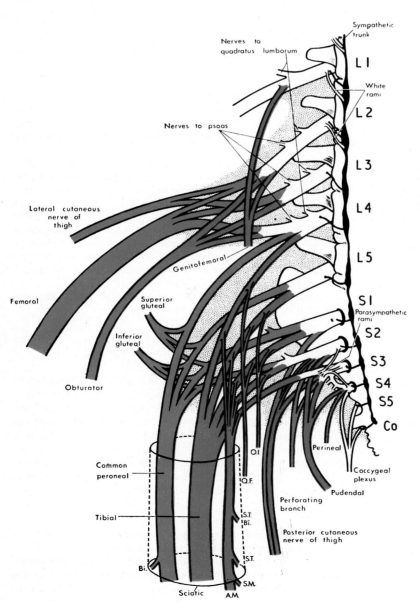

FIG. 11.58. Nerves of the lumbar, sacral, and coccygeal plexuses. Ventral divisions, red; dorsal, blue.
A.M. Adductor magnus.
O.I. Obturator internus.
Q.F. Quadratus femoris.
S.M. Semimembranosus.
Bi. Biceps.
S.T. Semitendinosus.

pelvis on coccygeus, and it supplies that muscle, the levator ani, and the skin over the coccyx.

COMMUNICATIONS WITH AUTONOMIC SYSTEM

Each of the nerves receives a grey ramus communicans from the sympathetic trunk. Because of the anterior position of the sympathetic trunk in the lumbar region, these rami are long and pass posteriorly between the body of the corresponding vertebra and the psoas muscle. One ganglion of the sympathetic trunk may send grey rami to two adjacent ventral rami, or one ventral ramus may receive two rami from adjacent ganglia.

The first two or three lumbar ventral rami each send a white ramus communicans to the sympathetic trunk. These either accompany the corresponding grey ramus communicans or are incorporated with it. The remaining ventral rami do not give rise to white rami communicantes, but the second and third, or third and fourth, sacral ventral rami send fine branches (pelvic splanchnic nerves) anteromedially to join the hypogastric plexuses of the autonomic system. These consist of preganglionic parasympathetic fibres which have no connection with the sympathetic trunk.

VARIATION IN ORIGIN OF LUMBAR AND SACRAL PLEXUSES

The lumbar plexus may begin as high as the eleventh thoracic ventral ramus or as low as the first lumbar (Eisler 1891). Similarly the lowest component to join the sciatic may be from the second, third, or fourth sacral ventral ramus. The table shows some of the extreme cases.

	PREFIXED VARIETY	NORMAL	POSTFIXED VARIETY
Nervus furcalis	L. 3, 4	L. 4	L. 5
Obturator	L. 1, 2, 3	L. 2, 3, 4	L. 2, 3, 4, 5
Femoral	T. 12, L. 1, 2	L. 2, 3, 4	L. 2, 3, 4, 5
Tibial	L. 3, 4, 5, S. 1, 2	L. 4, 5, S. 1, 2, 3	L. 5, S. 1, 2, 3, 4
Common peroneal	L. 3, 4, 5, S. 1	L. 4, 5, S. 1, 2	L. 5, S. 1, 2, 3

The commonest variation is towards the postfixed variety, and some evidence exists that vertebral anomalies (e.g. sacralization of the fifth lumbar vertebra) are often associated.

THE LUMBAR PLEXUS

This plexus is formed by the ventral rami of the first three lumbar spinal nerves and part of the fourth, with a small branch from the subcostal nerve in half the cases.

POSITION AND CONSTITUTION [Figs. 11.58 and 11.59]

The nerves emerge from the intervertebral foramina and their ventral rami communicate with the sympathetic trunk (see above). They then divide and recombine in the substance of the psoas muscle anterior to the plane of the transverse processes of the lumbar vertebrae.

The first two lumbar ventral rami divide into superior and inferior branches. The superior branch of the first, with or without a branch from the subcostal nerve, divides into the iliohypogastric and ilio-inguinal nerves. The inferior branch unites with the superior

branch of the second to form the genitofemoral nerve. The inferior branch of the second, the third, and the upper part of the fourth all divide into smaller ventral and larger dorsal divisions. The ventral divisions combine to form the obturator nerve (sometimes without the branch from the second nerve). The dorsal divisions unite to form the femoral nerve and the lateral cutaneous nerve of the thigh which arises from the second and third only. Irregular muscular branches pass to the psoas and quadratus lumborum muscles.

NERVES ARISING FROM THE LUMBAR PLEXUS
[FIG. 11.59]

1. Muscular branches to the quadratus lumborum (T. 12; L. 1–4) and psoas (L. (1), 2, 3, (4))
2. Iliohypogastric (L. 1, (T. 12))
3. Ilio-inguinal (L. 1, (T. 12))
4. Genitofemoral (L. 1, 2)
5. Lateral cutaneous of thigh (L. 2, 3)
6. Obturator (L. 2, 3, 4)
7. Femoral (L. 2, 3, 4)

DIRECT MUSCULAR BRANCHES

The nerves to quadratus lumborum and psoas arise separately from the ventral rami, though those to psoas are often associated at their origin with the nerve to iliacus from the femoral nerve. Psoas minor, when present, is supplied by the first or the second lumbar ventral ramus.

Iliohypogastric nerve

This nerve and the ilio-inguinal are in series with the subcostal nerve. They have a similar course and distribution.

The iliohypogastric nerve arises from the first lumbar ventral ramus with a small contribution from the twelfth thoracic. It passes inferolaterally through the substance of psoas major and over the anterior surface of quadratus lumborum posterior to the kidney. It then pierces transversus abdominis, and running forwards in the abdominal wall, superior to the iliac crest, between that muscle and the internal oblique, pierces the internal oblique 2–3 cm medial to the anterior superior iliac spine [FIG. 11.61]. It then passes medially, deep to the aponeurosis of the external oblique, and becomes cutaneous by piercing that aponeurosis superior to the superficial inguinal ring.

BRANCHES

1. To the muscles of the abdominal wall except rectus abdominis.
2. A small lateral cutaneous branch is usually present. It pierces obliquus internus and externus just superior to the iliac crest and posterior to the corresponding branch of the subcostal nerve. It supplies skin and fascia in the upper lateral part of the gluteal region adjacent to the area supplied by the dorsal ramus of the first lumbar nerve.
3. The anterior cutaneous branch supplies the skin and fascia of the anterior abdominal wall immediately superior to the pubis.

Ilio-inguinal nerve

This nerve arises from the first lumbar ventral ramus with the iliohypogastric nerve, and may also contain some fibres from the twelfth thoracic ventral ramus. It may be combined with the iliohypogastric nerve for a variable distance, but when they separate it follows the same course on a lower level, usually communicating with the iliohypogastric nerve. It pierces the internal oblique muscle

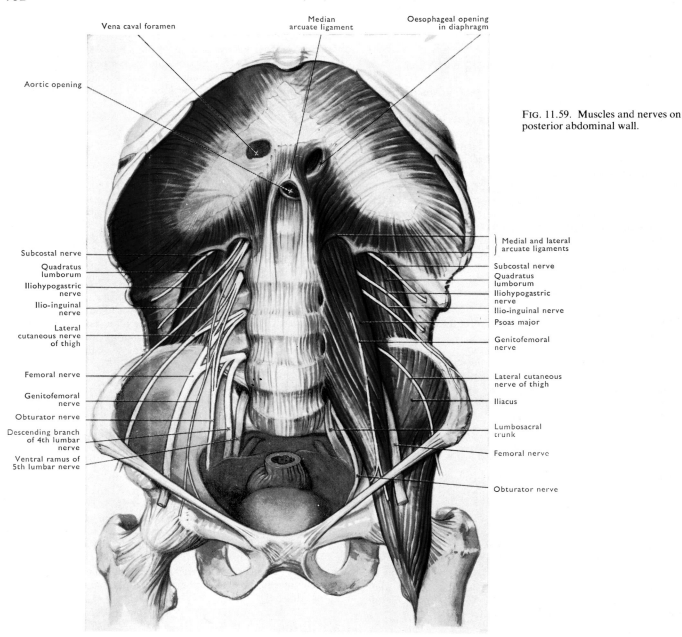

Vena caval foramen
Median arcuate ligament
Oesophageal opening in diaphragm

Aortic opening

Fig. 11.59. Muscles and nerves on posterior abdominal wall.

Subcostal nerve
Quadratus lumborum
Iliohypogastric nerve
Ilio-inguinal nerve
Lateral cutaneous nerve of thigh
Femoral nerve
Genitofemoral nerve
Obturator nerve
Descending branch of 4th lumbar nerve
Ventral ramus of 5th lumbar nerve

Medial and lateral arcuate ligaments
Subcostal nerve
Quadratus lumborum
Iliohypogastric nerve
Ilio-inguinal nerve
Psoas major
Genitofemoral nerve
Lateral cutaneous nerve of thigh
Iliacus
Lumbosacral trunk
Femoral nerve
Obturator nerve

further forward and lower down than the iliohypogastric [Fig. 11.61] and passing deep to the aponeurosis of the external oblique muscle close to the inguinal ligament, enters the inguinal canal, and reaches the skin through the superficial inguinal ring and the external spermatic fascia [Fig. 11.61].

BRANCHES

1. It supplies the muscles of the abdominal wall between which it passes.

2. Its terminal branches [Fig. 11.62] innervate the skin and fascia over (i) the pubic symphysis, (ii) the superomedial part of the femoral triangle, (iii) the anterior surface of the scrotum and root and dorsum of the penis (anterior scrotal nerves), or the mons pubis and labium majus in the female (anterior labial nerves). The last area abuts on that supplied by branches of the pudendal nerve. There is no lateral cutaneous branch.

Genitofemoral nerve

This nerve is formed in the substance of psoas major by the union of two roots which pass forwards from the first and second lumbar ventral rami. The nerve emerges through the anterior surface of psoas major medial to psoas minor, pierces the psoas fascia, and descending on that fascia posterior to the ureter, divides into two branches as it approaches the inguinal ligament lateral to the external iliac artery [Fig. 11.59].

BRANCHES

1. The small **genital branch** passes anterior to the external iliac artery through the deep inguinal ring into the inguinal canal. It gives branches to the external iliac artery, the cremaster muscle, the testicular autonomic plexus, and emerging with the spermatic cord (or the round ligament) supplies the skin and fascia of the scrotum

(or labium majus) and the adjacent part of the thigh. It communicates with the ilio-inguinal nerve in the inguinal canal.

2. The **femoral branch** enters the thigh posterior to the inguinal ligament, lateral to the femoral artery to which it gives a small branch. It then passes through the saphenous opening of the fascia lata and supplies an area of skin and fascia over the femoral triangle lateral to that innervated by the ilio-inguinal nerve [FIG. 11.62].

Lateral cutaneous nerve of the thigh

This cutaneous nerve arises from the posterior surface of the second and third lumbar ventral rami. It emerges from the lateral border of psoas major anterior to the iliac crest, and passing between iliacus and the iliac fascia, enters the thigh posterior to the lateral end of the inguinal ligament, medial to the anterior superior iliac spine. It then pierces (or passes superficial or deep to) the proximal part of sartorius [FIG. 11.61] and descends deep to the fascia lata giving off a few small branches to the skin. It pierces the fascia lata 10 cm inferior to the anterior superior iliac spine, and divides into anterior and posterior branches. The anterior branch supplies skin and fascia of the anterolateral surface of the thigh to the knee, where it may communicate with the patellar plexus. The smaller posterior branch supplies the skin and fascia on the lateral surface between the greater trochanter and distal third of the thigh [FIG. 11.62].

Obturator nerve

This nerve supplies the obturator externus and the adductor muscles of the thigh, gives articular branches to the hip and knee joints, and occasionally has a cutaneous branch. It arises by a branch from the ventral divisions of each of the second, third, and fourth lumbar ventral rami [FIG. 11.59]. The contribution from the second lumbar is occasionally small or absent. These branches unite in the substance of psoas and descend vertically through its posterior part to emerge from its medial border on the lateral part of the sacrum. It then crosses the sacro-iliac joint, enters the lesser pelvis, and descends on obturator internus a short distance posterior to the linea terminalis, to enter the obturator groove and divide into anterior and posterior branches, anterior to the obturator internus [FIG. 11.60]. In the lesser pelvis, the nerve is lateral to the internal iliac vessels and ureter, and is joined by the obturator vessels lateral to the **ovary** or the ductus deferens.

BRANCHES

1. The **anterior branch** descends into the thigh anterior to obturator externus and adductor brevis, posterior to pectineus and adductor longus [FIG. 11.61]. As it enters the thigh, it gives a lateral branch to the hip joint through the acetabular notch, and may send a communicating branch to the femoral nerve. As it lies between the muscles, it sends branches into adductor longus, gracilis, adductor brevis (usually), and pectineus (occasionally), and it divides into terminal cutaneous and vascular branches.

The **cutaneous branch** is frequently absent. When present, it passes between gracilis and adductor longus near the middle of the thigh, and supplies the skin and fascia of the distal two-thirds of its medial side. In the distal third of the thigh the nerve assists in the formation of a **plexus** with branches of the saphenous nerve and an anterior cutaneous nerve of the thigh, deep to sartorius. The **vascular branch** reaches the femoral artery in the adductor canal by passing along the medial border of adductor longus.

2. The **posterior branch** of the obturator nerve supplies and pierces obturator externus, and descends between adductor brevis

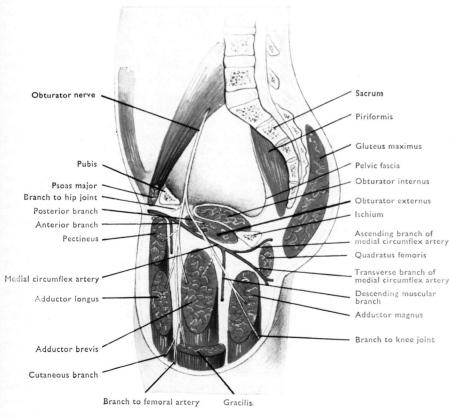

FIG. 11.60. Scheme of course and distribution of obturator nerve.

Obturator nerve
Sacrum
Piriformis
Gluteus maximus
Pubis
Psoas major
Branch to hip joint
Posterior branch
Anterior branch
Pectineus
Pelvic fascia
Obturator internus
Obturator externus
Ischium
Ascending branch of medial circumflex artery
Quadratus femoris
Transverse branch of medial circumflex artery
Descending muscular branch
Adductor magnus
Branch to knee joint
Medial circumflex artery
Adductor longus
Adductor brevis
Cutaneous branch
Branch to femoral artery
Gracilis.

and magnus. It runs obliquely through adductor magnus supplying its upper (adductor) part, and entering the popliteal fossa posterior to the popliteal artery, ends by piercing the oblique popliteal ligament to supply the posterior part of the knee joint, including the cruciate ligaments.

The accessory obturator nerve is only occasionally present (29 per cent, Eisler 1891; 8 per cent, Sunderland 1968). It arises from the third, or third and fourth lumbar ventral rami between the obturator and femoral nerves. It runs with the obturator nerve till that nerve emerges from psoas, and descending between the linea terminalis and the external iliac vessels, enters the thigh between the pubic bone and the femoral vessels.

The nerve commonly replaces the femoral branch to pectineus, sends a branch to the hip joint, and communicates with the anterior branch of the obturator nerve. It may only form the nerve to pectineus, or may make a significant contribution to the adductor muscles through the obturator nerve.

Femoral nerve

This nerve supplies the muscles of the front of the thigh, gives articular nerves to the hip and knee joints, and has an extensive cutaneous distribution to the anteromedial surface of the limb as far distally as the medial side of the foot.

It arises from the dorsal divisions of the second, third, and fourth lumbar ventral rami behind the obturator nerve in the substance of psoas major. It passes inferolaterally through the muscle and lies in the groove between psoas and iliacus immediately below the iliac crest [FIG. 11.59]. It descends in this groove, posterior to the fascia iliaca, and enters the thigh posterior to the inguinal ligament and lateral to the femoral sheath. In the femoral triangle it ends by dividing into a number of branches [FIG. 11.61].

BRANCHES

In the abdomen a lateral branch passes into iliacus, and a descending branch passes to the femoral artery, which receives further branches in the thigh.

In the femoral triangle

Muscular branches

1. The nerve to **pectineus**, frequently double, arises close to the inguinal ligament, and passing inferomedially posterior to the femoral vessels, enters the lateral border of the muscle. It may give off a small branch to the anterior branch of the obturator nerve. (See accessory obturator nerve.)

2. The nerves to **sartorius** arise from the anterior cutaneous branches of the femoral nerve. They consist of a lateral group of short branches and a medial group of longer branches which enter the muscle about its middle.

3. The nerves to the **quadriceps**. The vastus lateralis and rectus femoris are supplied on their deep surfaces by separate branches which accompany branches of the lateral femoral circumflex artery. The vastus intermedius is supplied by a branch which runs on its superficial surface, pierces it, and supplies the articularis genus muscle. The vastus medialis is supplied by a proximal branch which enters the superior part of the muscle and sends branches to the vastus intermedius. A distal branch to the same muscle descends on the lateral side of the femoral artery with the saphenous nerve and enters the medial surface of the muscle either from the adductor

canal or after piercing its fascial roof. The nerve sends a small branch through the nutrient canal of the femur, and to the articularis genus muscle.

Articular branches

A branch to the hip joint arises from the nerve to rectus femoris, and passes with branches from the lateral circumflex femoral artery. Four branches pass to the knee joint. One arises from each of the nerves to the three vastus muscles which send descending branches anterior to the femur. The fourth branch may arise from the saphenous nerve medial to the knee.

Cutaneous branches [FIGS. 11.61 and 11.62]

1. The anterior cutaneous nerves of the thigh arise as lateral and medial branches in the proximal part of the femoral triangle. Two lateral branches descend vertically on the anterior surface of the thigh, and become cutaneous by piercing the fascia lata over the proximal third of sartorius. They supply muscular branches to sartorius and the more lateral nerve may pierce the muscle. They supply the anterior surface of the thigh, communicate with the femoral branch of the genitofemoral nerve, and take part in the formation of the patellar plexus, anterior to the patella. In the femoral triangle, lateral to the femoral vessels, the medial branch gives off two or more twigs to the skin of the proximal part of the thigh in the region of the saphenous vein. It then runs inferomedially, superficial to the femoral vessels at the apex of the femoral triangle, and divides into anterior and posterior branches. The anterior branch passes downwards, superficial to and supplying sartorius, to end medial to the patella in the patellar plexus [FIG. 11.61]. It supplies the skin and fascia of the distal half of the medial side of the thigh. The posterior branch passes parallel to the medial edge of the sartorius muscle, and communicates with the saphenous and obturator nerves in the middle of the thigh to form a subsartorial plexus. It pierces the deep fascia on the medial side of the distal third of the thigh, and sends branches over the medial side of the knee to form part of the patellar plexus. The size of the medial branch varies inversely with that of the cutaneous branch of the obturator nerve and the infrapatellar branch of the saphenous nerve.

2. The saphenous nerve descends through the femoral triangle into the adductor canal on the lateral side of the femoral vessels. Here it gives a branch to a subsartorial plexus formed by the obturator and medial anterior cutaneous nerves, and passing obliquely across the anterior surface of the femoral artery, lies anterior to adductor magnus with the saphenous branch of the descending genicular artery. Here it gives off an infrapatellar branch which becomes superficial by piercing sartorius, then passes antero-inferiorly below the patella to supply skin and fascia over the front and medial side of the knee and proximal part of the leg. It communicates with the patellar plexus.

The saphenous nerve becomes superficial in the leg by passing between the sartorius and gracilis muscles, and may give an articular branch to the medial side of the knee joint. It then descends with the great saphenous vein on the medial side of the leg, and passing anterior to the medial malleolus, ends by supplying skin and fascia on the proximal half of the medial side and adjacent dorsum of the foot. Branches of the saphenous nerve supply skin and fascia on the front and medial side of the leg, some passing over the anterior border of the tibia.

The patellar plexus consists of fine intercommunicating branches of the nerves which supply the skin of the front of the knee, i.e. the infrapatellar branch of the saphenous nerve, the anterior cutaneous branches of the femoral nerve, and sometimes the lateral cutaneous nerve of the thigh [FIG. 11.61].

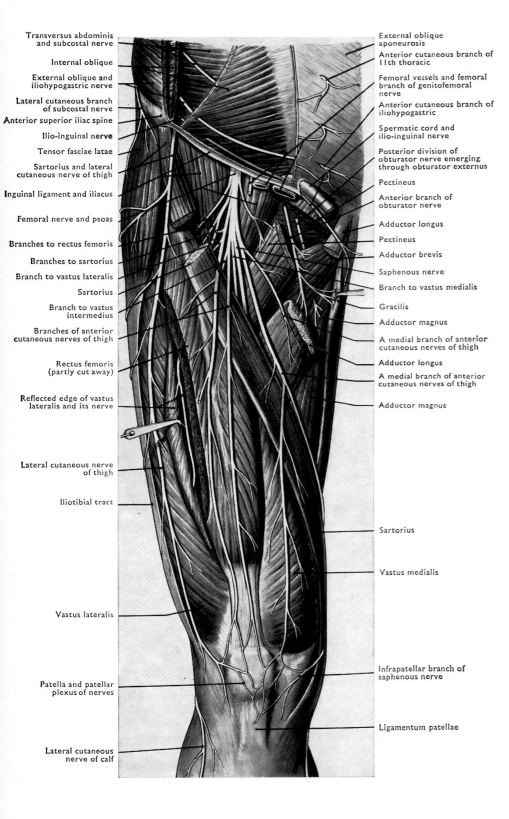

Transversus abdominis and subcostal nerve

Internal oblique

External oblique and iliohypogastric nerve

Lateral cutaneous branch of subcostal nerve

Anterior superior iliac spine

Ilio-inguinal nerve

Tensor fasciae latae

Sartorius and lateral cutaneous nerve of thigh

Inguinal ligament and iliacus

Femoral nerve and psoas

Branches to rectus femoris

Branches to sartorius

Branch to vastus lateralis

Sartorius

Branch to vastus intermedius

Branches of anterior cutaneous nerves of thigh

Rectus femoris (partly cut away)

Reflected edge of vastus lateralis and its nerve

Lateral cutaneous nerve of thigh

Iliotibial tract

Vastus lateralis

Patella and patellar plexus of nerves

Lateral cutaneous nerve of calf

External oblique aponeurosis

Anterior cutaneous branch of 11th thoracic

Femoral vessels and femoral branch of genitofemoral nerve

Anterior cutaneous branch of iliohypogastric

Spermatic cord and ilio-inguinal nerve

Posterior division of obturator nerve emerging through obturator externus

Pectineus

Anterior branch of obturator nerve

Adductor longus

Pectineus

Adductor brevis

Saphenous nerve

Branch to vastus medialis

Gracilis

Adductor magnus

A medial branch of anterior cutaneous nerves of thigh

Adductor longus

A medial branch of anterior cutaneous nerves of thigh

Adductor magnus

Sartorius

Vastus medialis

Infrapatellar branch of saphenous nerve

Ligamentum patellae

FIG. 11.61. Nerves of front of thigh.

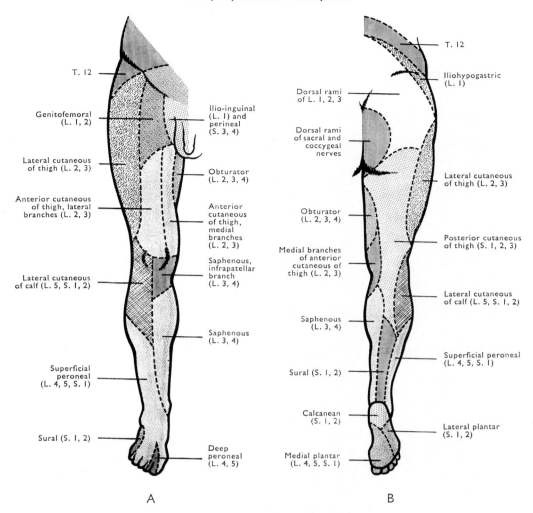

FIG. 11.62. Distribution of cutaneous nerves.
A, on the front; B, on the back of lower limb.

THE SACRAL PLEXUS [FIG. 11.58]

This plexus is formed from the ventral rami of the fourth lumbar to the fourth sacral spinal nerves [p. 790].

POSITION

The plexus lies on the dorsal wall of the pelvis, mainly between the piriformis muscle and its fascia. Anterior to that fascia lie the internal iliac vessels and the ureter, with the sigmoid colon on the left and coils of ileum on the right, further anteriorly.

The nerves entering the plexus converge to form a broad triangular band [FIG. 11.65] which passes through the inferior part of the greater sciatic foramen, below piriformis, and enters the gluteal region as the **sciatic nerve** [FIG. 11.68]. Numerous small branches arise from the anterior and posterior surfaces of the triangular band, and from its lower margin the pudendal nerve passes to the perineum.

The sciatic nerve divides into **tibial** and **common peroneal nerves** at a variable point in the thigh. These nerves are often distinct at their origin (in which case the common peroneal nerve may pierce the piriformis muscle), and are always separable to the

plexus when the sheath of the sciatic nerve is split. If this is done, the origin of these nerves from ventral (tibial nerve) and dorsal divisions (common peroneal nerve) can be confirmed.

FORMATION

The descending part of the fourth lumbar ventral ramus emerges from the medial border of the psoas muscle, medial to the obturator nerve, and posterior to the common iliac vessels. The nerve descends into the lesser pelvis, and joins the ventral ramus of the fifth lumbar nerve which has reached the pelvic surface of the sacro-iliac joint by passing over the lateral part of the sacrum. The combined nerves constitute the **lumbosacral trunk**. The first, second, and third sacral ventral rami pass laterally on piriformis from the pelvic sacral foramina. The fourth contributes to the sacral and coccygeal plexuses.

These nerves (of which the fifth lumbar and first sacral are the largest) converge on the lower part of the greater sciatic foramen, and are divisible into dorsal and ventral divisions. The dorsal divisions (L. 4, 5; S. 1, 2) form the common peroneal nerve, while the ventral divisions (L. 4, 5; S. 1, 2, 3) unite to form the tibial nerve [FIG. 11.58].

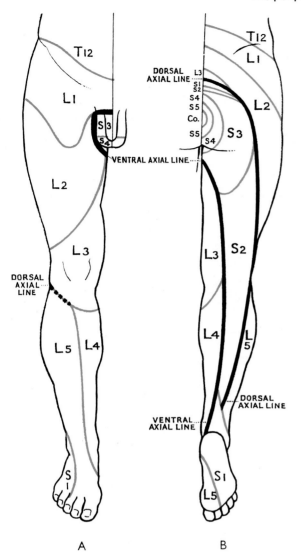

FIG. 11.63. Dermatomes of lower limb, showing the segmental cutaneous distribution of spinal nerves (T.12; L.1–5; S.1–5; Co.) on A, the front and B, the back of the limb and lower part of trunk. (After Head 1893; Foerster 1933.)

See page 767 for definition of 'dermatome' and page 806 for more particular reference to the segmental innervation of the lower limb.

BRANCHES OF THE SACRAL PLEXUS

ANTERIOR BRANCHES

Nerve to quadratus femoris and inferior gemellus (L. 4, 5; S. 1)
Nerve to obturator internus and superior gemellus (L. 5; S. 1, 2)
Pelvic splanchnic nerves (S. (2), 3, (4))
Posterior femoral cutaneous nerve (S. 2, 3)
Pudendal nerve (S. (1), 2, 3, 4, (5))

POSTERIOR BRANCHES

Muscular twigs to piriformis (S. 1, 2) and to coccygeus and levator ani (S. 3, 4)
Superior gluteal nerve (L. 4, 5; S. 1)
Inferior gluteal nerve (L. 5; S. 1, 2)
Posterior femoral cutaneous nerve (S. 1, 2)
Perforating cutaneous nerve (S. 2, 3)
Perineal branch of S. 4

SCIATIC NERVE (L. 4, 5; S. 1, 2, 3)

NERVE TO QUADRATUS FEMORIS

This nerve arises from the ventral divisions (L. 4, 5; S. 1), descends through the lower part of the greater sciatic foramen, and passing on to the posterior surface of the ischium anterior to the sciatic nerve, obturator internus, and the gemelli, sends a fine branch to the adjacent hip joint. It supplies gemellus inferior and ends in the anterior surface of quadratus femoris.

NERVE TO OBTURATOR INTERNUS [FIG. 11.65]

This nerve arises from the ventral divisions (L. 5; S. 1, 2) and descends into the gluteal region between the sciatic nerve (laterally) and the internal pudendal vessels. It sends a branch to the superior gemellus, and turns forwards below the ischial spine to enter the ischiorectal fossa on the medial surface of obturator internus.

PELVIC SPLANCHNIC NERVES

These consist of preganglionic parasympathetic nerve fibres. They arise as two slender filaments from the second and third, or third and fourth (or rarely fourth and fifth) sacral ventral rami, and pass forwards to join the autonomic plexuses of the pelvis, through which they supply the urogenital organs and the lower part of the large intestine [see p. 814].

PUDENDAL NERVE

This principal nerve of the perineum arises in the pelvis usually by three roots (S. (1), 2, 3, 4) [FIG. 11.58]. It enters the gluteal region through the inferomedial part of the greater sciatic foramen, to lie on the sacrospinous ligament medial to the internal pudendal artery. It passes through the lesser sciatic foramen into the perineum with the artery, entering the pudendal canal on the obturator fascia of the lateral wall of the ischiorectal fossa. Here it gives off the inferior rectal nerve, and shortly afterwards divides into the perineal nerve and the dorsal nerve of the penis or clitoris.

Inferior rectal nerve

This nerve usually arises as the pudendal nerve enters the pudendal canal, though it may arise separately from the third and fourth sacral ventral rami. It crosses the ischiorectal fossa inferomedially with the inferior rectal vessels, supplies levator ani, and reaching the anus, innervates the skin and fascia around it and the sphincter ani externus. The cutaneous branches communicate with the perineal branches of the (1) posterior femoral cutaneous, (2) perineal, and (3) fourth sacral nerves.

Perineal nerve

This nerve passes anteromedially and divides into posterior scrotal or posterior labial nerves and deep and superficial branches.

The posterior scrotal or posterior labial nerves pass superficially to supply the skin and fascia of the perineum and the scrotum or the labium majus by medial and lateral branches. The lateral may send branches to the skin of the medial aspect of the thigh, and it communicates with the inferior rectal and posterior femoral cutaneous nerves. The superficial and deep branches pass into the corresponding perineal spaces to supply the muscles and the erectile tissue and mucous membrane of the bulb and corpus spongiosum penis, extending as far as the glans penis.

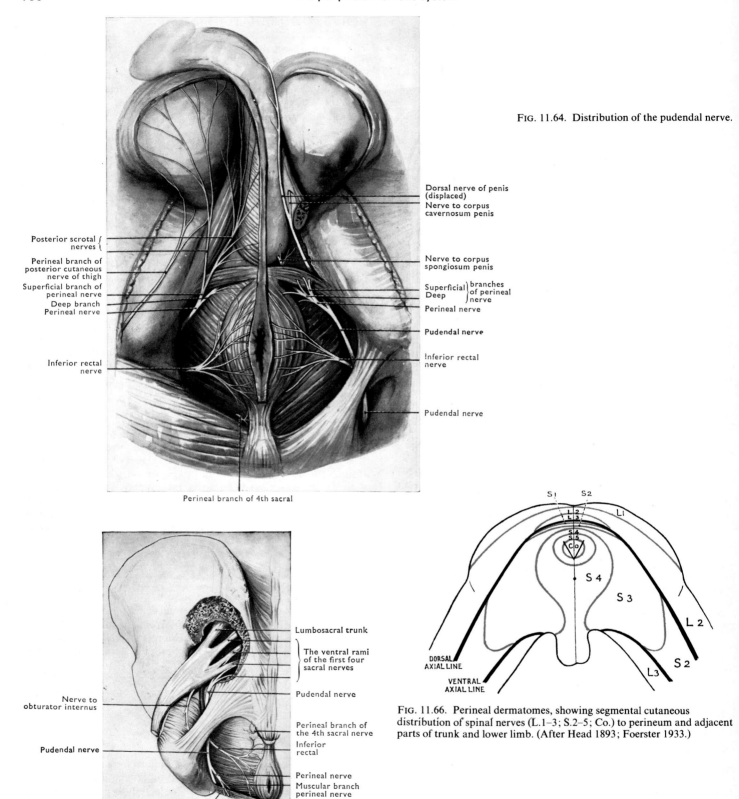

FIG. 11.64. Distribution of the pudendal nerve.

Dorsal nerve of penis (displaced)
Nerve to corpus cavernosum penis

Posterior scrotal nerves
Perineal branch of posterior cutaneous nerve of thigh
Superficial branch of perineal nerve
Deep branch
Perineal nerve

Nerve to corpus spongiosum penis

Superficial } branches
Deep } of perineal nerve
Perineal nerve

Pudendal nerve

Inferior rectal nerve

Inferior rectal nerve

Pudendal nerve

Perineal branch of 4th sacral

Lumbosacral trunk
The ventral rami of the first four sacral nerves
Pudendal nerve

Nerve to obturator internus

Perineal branch of the 4th sacral nerve
Inferior rectal

Pudendal nerve

Perineal nerve
Muscular branch perineal nerve

Posterior scrotal branches of perineal nerve

FIG. 11.65. Origin and course of the pudendal nerve.

FIG. 11.66. Perineal dermatomes, showing segmental cutaneous distribution of spinal nerves (L.1–3; S.2–5; Co.) to perineum and adjacent parts of trunk and lower limb. (After Head 1893; Foerster 1933.)

DORSAL AXIAL LINE
VENTRAL AXIAL LINE

S 1   S 2
L 1
L 2
L 3
S 4
S 5
Co
S 4
S 3
L 2
L 3
S 2

Dorsal nerve of the penis or clitoris

This nerve passes forwards through the deep perineal space with the internal pudendal artery. It lies close to the medial surface of the inferior pubic ramus, superior to the perineal membrane and the crus of the penis or clitoris, and inferior to the sphincter urethrae muscle. It sends a twig through the perineal membrane to the crus and corpus cavernosum of the penis or clitoris; then piercing the perineal membrane near its anterior border, passes on to the dorsal surface of the penis or clitoris, lateral to the dorsal artery. It is distributed to the distal two-thirds of the penis, sending branches round the sides to reach the inferior surface of that organ. The nerve is much smaller in the female than in the male.

NERVE TO PIRIFORMIS

This nerve arises from the dorsal surface of the first and second, or second sacral ventral rami, and enters the pelvic surface of the muscle, often as two separate nerves.

NERVES TO COCCYGEUS AND LEVATOR ANI

In addition to branches from the inferior rectal and the perineal nerves, nerves arise from a loop between the third and fourth sacral ventral rami, and descend to these muscles.

The perineal branch of the fourth sacral nerve arises from the lower part of the same loop. It supplies and pierces coccygeus, and entering the posterior angle of the ischiorectal fossa, passes anteriorly supplying levator ani, the posterior part of the external anal sphincter, and the overlying skin and fascia.

PERFORATING CUTANEOUS NERVE

This nerve arises (S. 2, 3) close to the lower roots of the posterior femoral cutaneous nerve [FIG. 11.58] and may be incorporated in the gluteal branches of that nerve or in the pudendal nerve. The nerve descends, pierces the sacrotuberous ligament, and becomes cutaneous close to the coccyx by turning round the inferior border or piercing the lower fibres of gluteus maximus. It supplies the skin and fascia of the lower part of the buttock and the medial part of the gluteal fold.

SUPERIOR GLUTEAL NERVE

This nerve arises from the dorsal surface of the plexus (L. 4, 5; S. 1) and passes posterolaterally into the gluteal region with the superior gluteal vessels, superior to piriformis. It passes forwards as an upper and a lower branch between gluteus medius and minimus. The upper branch supplies gluteus medius. The lower branch crosses the middle of gluteus minimus, sends branches to that muscle and to gluteus medius, and ends in the deep surface of tensor fasciae latae [FIG. 11.68]. It runs with the inferior branch of the deep division of the superior gluteal artery.

INFERIOR GLUTEAL NERVE

It arises from the dorsal surface of the plexus (L. 5; S. 1, 2) and running medial to the posterior femoral cutaneous nerve, enters the gluteal region at the inferior border of piriformis, superficial to the sciatic nerve. It divides into a number of branches which pass posteriorly into the deep surface of gluteus maximus.

POSTERIOR FEMORAL CUTANEOUS NERVE [FIG. 11.58]

This nerve arises from the upper three sacral ventral rami, or from the second and third. It is distributed to the skin and fascia of the posterior surface of the thigh and upper part of the leg, the perineum,

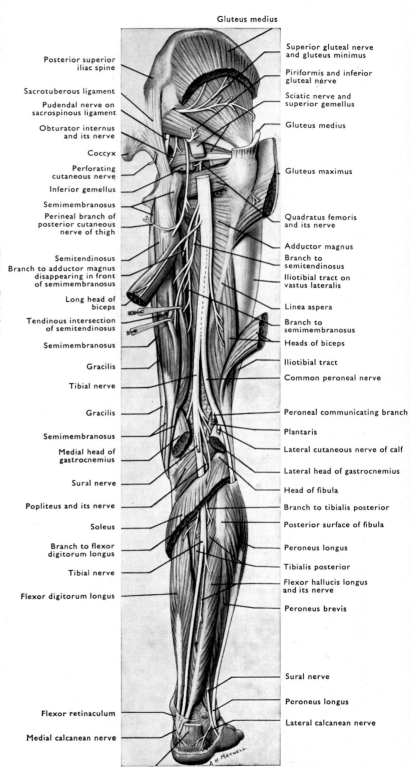

FIG. 11.67. Nerves of back of lower limb.

and the inferior half of the buttock. At its origin the upper roots of the nerve are associated with the origin of the inferior gluteal nerve, and the lowest root with the pudendal nerve. It enters the gluteal region on the posterior surface of the sciatic nerve with the inferior gluteal vessels and nerve. It descends to the inferior border of gluteus maximus, near which it gives off its **gluteal** and **perineal** branches, then passes down the back of the thigh deep to the fascia lata covering the hamstring muscles. Over the lower part of the popliteal fossa it pierces the deep fascia as two or more cutaneous branches which supply the skin and fascia over the upper half of the back of the leg, though they may reach the ankle, or stop at the lower border of the popliteal fossa [FIG. 11.62].

Branches

1. The **perineal branch** curves medially over the hamstring muscles [FIG. 11.68] and the pubic arch, and sends branches to the upper medial part of the thigh and to the perineum. Its branches [FIG. 11.64] pass forwards to supply the skin and fascia of the scrotum (labium majus) and root of the penis (clitoris), communicating with branches of the ilio-inguinal nerve. Other branches pass medially to the skin and fascia anterior to the anus, communicating with the inferior rectal and perineal branches of the pudendal nerve.

2. The **gluteal branches** arise deep to the inferior border of gluteus maximus, and pierce the deep fascia along that border. They supply the skin and fascia of the inferior half of the buttock, the lateral branches overlapping the areas supplied by the lateral femoral cutaneous nerve and the dorsal rami of the upper three lumbar nerves. The medial branches may pierce the sacrotuberous ligament and replace the perforating cutaneous nerve, or supply the area adjacent to it.

3. The **branches to the thigh** pierce the fascia lata at intervals, and supply the skin and fascia of its posterior surface. For the branches to the leg, see above.

When the two parts of the sciatic nerve are separated by piriformis, the posterior femoral cutaneous nerve is also in two parts. (i) A dorsal part (S. 1, 2), arising with the common peroneal nerve in association with the lower roots of the inferior gluteal nerve, supplies the buttock and lateral part of the back of the thigh. (iii) A ventral part (S. 2, 3), arising with the tibial nerve in association with the perforating cutaneous and pudendal nerves, supplies the perineum and medial part of the cutaneous area in the limb.

SCIATIC NERVE

This is the thickest nerve in the body. It consists of a medially placed tibial nerve and its branches to the hamstring muscles, and a laterally placed common peroneal nerve and its branch to the short head of biceps, bound together in a common sheath. It is formed by nerve fibres from the ventral rami of the fourth lumbar to third sacral spinal nerves, and emerges from the pelvis into the gluteal region through the greater sciatic foramen. At first, anterior to piriformis, it runs inferolaterally on the nerve to quadratus femoris and the ischium with the inferior gluteal and posterior femoral cutaneous nerves [FIG. 11.68]. Inferior to piriformis, it lies deep to gluteus maximus, and descends more vertically posterior to the tendon of obturator internus and the gemelli. It then passes on to quadratus femoris in the hollow between the ischial tuberosity and the greater trochanter of the femur, and enters the thigh by reaching the inferior border of gluteus maximus on the posterior surface of adductor magnus. Here the nerves to the hamstrings separate from the sciatic nerve, which is crossed posteriorly, from medial to lateral, by the long head of biceps femoris.

The sciatic nerve usually ends at the proximal angle of the popliteal fossa by dividing into the tibial and common peroneal nerves, but this division may occur at any higher level [FIGS. 11.67 and 11.68].

BRANCHES

The **nerve to the hamstring muscles** (L. 4, 5; S. 1, 2, 3) lies in the medial part of the sciatic nerve in the gluteal region, and leaves it near the lower border of quadratus femoris. It is distributed to the hamstring muscles (except the short head of biceps) by a series of branches, which are usually given off separately by the sciatic nerve [FIG. 11.68]. The nerve to semitendinosus is double; the branch to the part superior to the tendinous intersection may be associated with the nerve to the long head of biceps. The nerve to the ischial part of adductor magnus arises in common with the nerve to semimembranosus, and descends anterior to the lateral margin of that muscle supplying both.

The **nerve to the short head of biceps** (L. 5; S. 1, 2) is formed at the plexus in common with the inferior gluteal nerve, and leaves the lateral side of the common peroneal nerve above the middle of the thigh, usually with the branches of that nerve to the knee joint. It may have an independent course in the thigh, or be associated with the inferior gluteal nerve in the buttock.

An **articular branch** (L. 4, 5; S. 1) to the lateral and anterior parts of the knee joint usually arises with the nerve to the short head of biceps. It descends through the superior part of the popliteal fossa anterior to biceps femoris, and divides into proximal and distal branches, which pass with the superior and inferior lateral genicular arteries respectively.

Common peroneal nerve

This nerve supplies the skin and fascia of the anterolateral surface of the leg and the dorsum of the foot, the muscles of the anterior and peroneal compartments of the leg, extensor digitorum brevis, and the knee, ankle, and foot joints.

The nerve arises from the dorsal divisions of the sacral plexus (L. 4, 5; S. 1, 2) and is laterally placed in the sciatic nerve. From the bifurcation of that nerve it passes inferolaterally in the proximal and lateral part of the popliteal fossa, under cover of the medial edge of the biceps femoris and its tendon, to reach the back of the head of the fibula. It ends by dividing into the deep and superficial peroneal nerves where it winds round the lateral surface of the neck of the fibula in the substance of peroneus longus, approximately 2 cm distal to the apex of the head of the fibula. It is palpable posterior to the head of the fibula and posterolateral to its neck [FIGS. 11.67, 11.69, and 11.71].

BRANCHES

In the thigh. The nerve to the short head of biceps, and the articular branch to the knee joint are described above.

In the popliteal fossa. 1. The **lateral cutaneous nerve of the calf** often arises in common with the peroneal communicating nerve, pierces the deep fascia over the lateral head of gastrocnemius, and is distributed to the skin and fascia in the proximal two-thirds of the posterolateral surface of the leg. Its size and distribution varies inversely with that of the posterior femoral cutaneous and sural nerves [FIG. 11.62].

2. The **peroneal communicating branch** passes between the deep fascia and the lateral head of gastrocnemius to join the sural nerve in the middle third of the leg. Sometimes this union does not

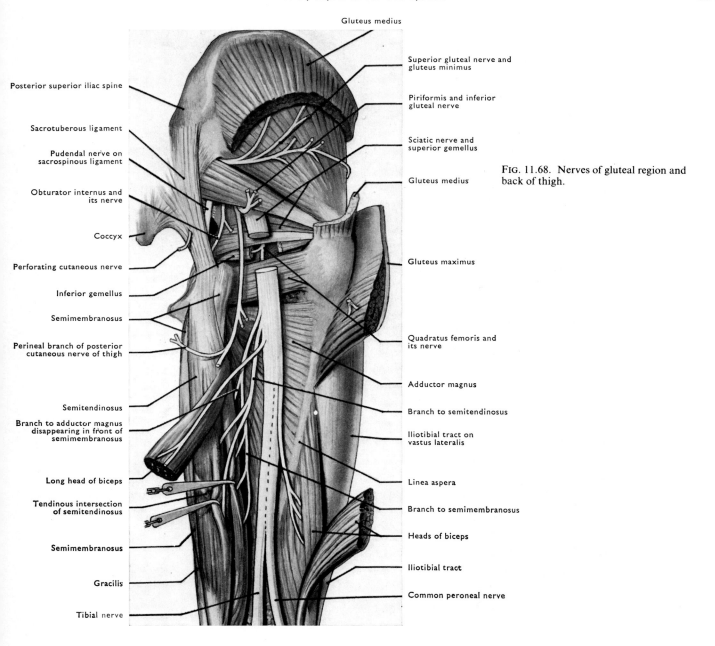

Gluteus medius

Posterior superior iliac spine

Sacrotuberous ligament

Pudendal nerve on
sacrospinous ligament

Obturator internus and
its nerve

Coccyx

Perforating cutaneous nerve

Inferior gemellus

Semimembranosus

Perineal branch of posterior
cutaneous nerve of thigh

Semitendinosus

Branch to adductor magnus
disappearing in front of
semimembranosus

Long head of biceps

Tendinous intersection
of semitendinosus

Semimembranosus

Gracilis

Tibial nerve

Superior gluteal nerve and
gluteus minimus

Piriformis and inferior
gluteal nerve

Sciatic nerve and
superior gemellus

Gluteus medius

Gluteus maximus

Quadratus femoris and
its nerve

Adductor magnus

Branch to semitendinosus

Iliotibial tract on
vastus lateralis

Linea aspera

Branch to semimembranosus

Heads of biceps

Iliotibial tract

Common peroneal nerve

FIG. 11.68. Nerves of gluteal region and
back of thigh.

take place; then the peroneal communicating nerve may supply skin and fascia on the lateral side of the leg, or pass to the area usually supplied by the sural nerve.

3. A small **recurrent branch** arises immediately above the termination of the common peroneal nerve. It passes forwards through the origin of peroneus longus and extensor digitorum longus, and divides, inferior to the lateral condyle of the tibia, into branches to the proximal part of the tibialis anterior muscle and the superior tibiofibular and knee joints [FIG. 11.69].

Deep peroneal nerve [FIG. 11.69]

This nerve begins between the neck of the fibula and peroneus longus. It runs inferomedially on the fibula, deep to extensor digitorum longus, to join the anterior tibial vessels on the anterior surface of the interosseous membrane, lateral to tibialis anterior. It then descends on the interosseous membrane medial to, and overlapped by extensor hallucis longus, and passes on to the distal part of the tibia deep to the muscle and the superior extensor retinaculum.

In the leg it supplies extensor digitorum longus, tibialis anterior, extensor hallucis longus, and peroneus tertius, and sends a filament to the anterior surface of the ankle joint.

At the ankle joint, the nerve is crossed obliquely by the tendon of extensor hallucis longus, deep to the inferior extensor retinaculum. It then becomes superficial between the tendons of extensor digitorum longus and extensor hallucis longus on the proximal part of the dorsum of the foot. Here it sends a lateral **muscular branch** obliquely between the tarsus and the extensor digitorum brevis. This ends in a gangliform enlargement from which branches pass to extensor digitorum brevis and to the tarsal, tarsometatarsal, and metatarsophalangeal joints. The lateral articular branches may only reach the tarsometatarsal joints, while the medial may give sensory branches also to the second and third dorsal interosseous muscles.

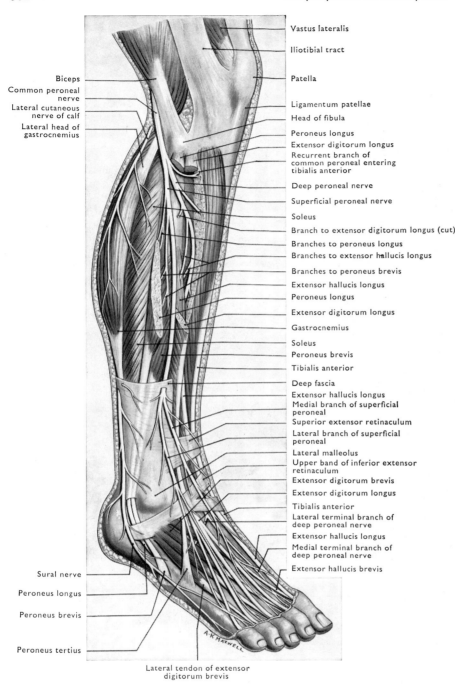

Biceps
Common peroneal nerve
Lateral cutaneous nerve of calf
Lateral head of gastrocnemius

Sural nerve
Peroneus longus
Peroneus brevis
Peroneus tertius

Lateral tendon of extensor digitorum brevis

Vastus lateralis
Iliotibial tract
Patella
Ligamentum patellae
Head of fibula
Peroneus longus
Extensor digitorum longus
Recurrent branch of common peroneal entering tibialis anterior
Deep peroneal nerve
Superficial peroneal nerve
Soleus
Branch to extensor digitorum longus (cut)
Branches to peroneus longus
Branches to extensor hallucis longus
Branches to peroneus brevis
Extensor hallucis longus
Peroneus longus
Extensor digitorum longus
Gastrocnemius
Soleus
Peroneus brevis
Tibialis anterior
Deep fascia
Extensor hallucis longus
Medial branch of superficial peroneal
Superior extensor retinaculum
Lateral branch of superficial peroneal
Lateral malleolus
Upper band of inferior extensor retinaculum
Extensor digitorum brevis
Extensor digitorum longus
Tibialis anterior
Lateral terminal branch of deep peroneal nerve
Extensor hallucis longus
Medial terminal branch of deep peroneal nerve
Extensor hallucis brevis

FIG. 11.69. Nerves of front and lateral side of leg and dorsum of foot.

The nerve continues along the lateral side of the dorsalis pedis artery and sends twigs to the medial tarsal and tarsometatarsal joints. Proximal to the first interdigital cleft, it divides into two **dorsal digital nerves**, one to the lateral side of the hallux, the other to the medial side of the second toe. Each of these nerves communicates with the terminal branches of the superficial peroneal nerve, supplies the metatarsophalangeal and interphalangeal joints, and the skin as far as the distal interphalangeal joint. A sensory branch enters the first dorsal interosseous muscle.

Superficial peroneal nerve

This nerve descends anterior to the fibula, in the intermuscular septum between the peronei and extensor digitorum longus, to

supply peroneus longus and brevis. As it enters the lower third of the leg, the nerve divides into **medial and intermediate dorsal cutaneous nerves** either before or after piercing the deep fascia anterior to it [FIG. 11.69]. These nerves descend in the subcutaneous tissue to the dorsum of the foot, giving branches to the skin and fascia of the distal third of the anterior surface of the leg and to the dorsum of the foot.

The **medial dorsal cutaneous nerve** divides into three branches. 1. The most medial branch supplies the skin and fascia of the medial side of the dorsum of the foot and of the hallux [FIG. 11.70] and communicates with the saphenous nerve. 2. The intermediate branch passes towards the cleft between the hallux and the second toe, and communicates with the branches of the deep peroneal

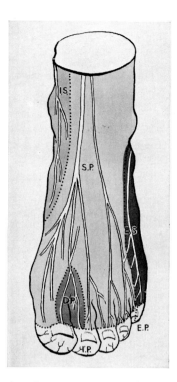

FIG. 11.70. Distribution of cutaneous nerves on dorsum of the foot.
  I.S., saphenous nerve; S.P., superficial peroneal nerve; D.P., deep peroneal nerve; E.S., sural nerve. The extremities of the toes are supplied by the medial and lateral plantar nerves (I.P., E.P.).

nerve to that cleft. 3. The lateral branch passes to the cleft between the second and third toes, and divides into dorsal digital nerves to their adjacent sides [FIG. 11.70].

The **intermediate dorsal cutaneous nerve** divides into two branches on the dorsum of the foot. They pass to the lateral two interdigital clefts to form dorsal digital nerves to the adjacent sides of the toes. The lateral branch communicates with the **lateral dorsal cutaneous nerve** from the sural nerve [FIG. 11.70].

The area supplied by the intermediate dorsal cutaneous nerve is variable. It may supply the adjacent sides of the second and third toes, or the lateral dorsal cutaneous nerve may extend into its territory and supply as many as two and a half toes (Kosinski 1926).

The dorsal digital nerves are much smaller than the plantar digital nerves which supply all the skin and fascia over the distal phalanges.

Tibial nerve

This nerve (L. 4, 5; S. 1, 2, 3) arises from the ventral surface of the plexus. It forms the medial part of the sciatic nerve, and continues the line of that nerve from its bifurcation, through the popliteal fossa to lie deep in the posterior part of the leg. In the popliteal fossa it is at first concealed by the semimembranosus, but then comes to lie immediately deep to the popliteal fascia, and passes obliquely across the posterior surface of the popliteal vessels to lie medial to them on popliteus, deep to gastrocnemius and plantaris [FIG. 11.71]. It continues deep to the tendinous arch of soleus, and lies in the intermuscular septum superficial to tibialis posterior, medial to the posterior tibial vessels. The nerve descends in this plane, between flexor digitorum longus and flexor hallucis longus, and crossing the posterior surface of the ankle joint between the tendons of these muscles, curves antero-inferiorly into the foot with them. Postero-

inferior to the medial malleolus it divides into lateral and medial plantar nerves [FIG. 11.73] and is covered superficially only by the flexor retinaculum and skin.

BRANCHES

1. While it is incorporated in the sciatic nerve it gives off its **nerves to the hamstrings** and an occasional **nerve to the hip joint**, see above.

2. In the popliteal fossa. (i) One slender **articular branch** pierces the oblique popliteal ligament, others pass with the inferior and sometimes superior medial genicular vessels to the knee joint.

(ii) **Muscular branches** enter the two heads of gastrocnemius and plantaris at the borders of the popliteal fossa [FIG. 11.71], and the nerve to soleus enters the superficial surface of that muscle. The nerve to popliteus descends over the posterior surface of that muscle, turns round its inferior border, and enters its anterior surface. At the inferior border of popliteus this nerve gives branches (a) to tibialis posterior, (b) to the whole length of the interosseous membrane, (c) to the tibiofibular joints, and (d) a medullary branch to the tibia.

(iii) The cutaneous branch is the **sural nerve**. It arises in the lower part of the popliteal fossa, and descending between the heads of gastrocnemius, pierces the deep fascia in the middle third of the posterior surface of the leg. It is then usually joined by the peroneal communicating branch of the common peroneal nerve, and passes on to the tendo calcaneus with the small saphenous vein, giving cutaneous branches to the back and lateral side of the lower third of the leg. At the ankle it lies posterior to the tendons of the peroneal muscles, and giving branches to the ankle and heel (**lateral calcanean branches**), passes forwards as the **lateral dorsal cutaneous nerve** of the foot. It supplies **articular branches** to the ankle and tarsal joints, and communicates with the intermediate cutaneous branch of the superficial peroneal nerve [FIG. 11.70], either reinforcing or replacing it. The peroneal communicating branch of the common peroneal nerve is distributed within the territory of the sural nerve if they do not unite.

3. Branches in the Leg. (i) **Muscular branches** (often by a common trunk) to the deep surface of soleus and to tibialis posterior, the latter sending branches to the posterior tibial artery. Separate branches pass to flexor digitorum longus and flexor hallucis longus; the nerve to flexor hallucis longus accompanying the peroneal artery and supplying it.

(ii) Cutaneous, **medial calcanean branches** pierce the flexor retinaculum and supply the skin and fascia of the heel and posterior part of the sole of the foot [FIG. 11.73]. (iii) A long, slender branch arises with the nerves to the deep muscles. It runs with the peroneal vessels as far as the ankle, gives a medullary branch to the fibula which also supplies its periosteum. One or two articular branches to the ankle arise from the tibial nerve at the ankle.

4. Terminal branches.

Medial plantar nerve

This nerve, homologous with the median nerve in the hand, is rather larger than the lateral plantar [FIGS. 11.72 and 11.73]. It passes forwards with the medial plantar artery under cover of the flexor retinaculum and abductor hallucis, to the interval between that muscle and the flexor digitorum brevis in the sole of the foot.

Branches

(i) **Muscular branches** to abductor hallucis and flexor digitorum brevis. (ii) Small **cutaneous branches** to the skin of the medial part of the sole of the foot pierce the plantar fascia between the above muscles. (iii) **Articular branches** to tarsal and tarsometatarsal joints. (iv) **Vascular nerves** to the adjacent arteries from the nerve and its branches. (v) One proper and three common plantar digital nerves.

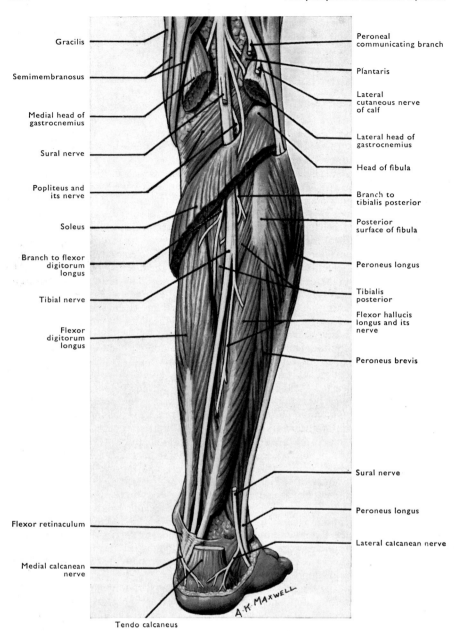

Gracilis

Semimembranosus

Medial head of gastrocnemius

Sural nerve

Popliteus and its nerve

Soleus

Branch to flexor digitorum longus

Tibial nerve

Flexor digitorum longus

Flexor retinaculum

Medial calcanean nerve

Tendo calcaneus

Peroneal communicating branch

Plantaris

Lateral cutaneous nerve of calf

Lateral head of gastrocnemius

Head of fibula

Branch to tibialis posterior

Posterior surface of fibula

Peroneus longus

Tibialis posterior

Flexor hallucis longus and its nerve

Peroneus brevis

Sural nerve

Peroneus longus

Lateral calcanean nerve

FIG. 11.71. Nerves of popliteal fossa and back of leg.

The **proper plantar digital nerve** [FIG. 11.73] separates from the nerve before the others, and passing forwards, pierces the deep fascia posterior to the ball of the big toe. It supplies a muscular branch to flexor hallucis brevis, and cutaneous branches to the medial side of the foot, the ball and medial aspect of the great toe [FIG. 11.70].

The three **common plantar digital nerves** arise together. The first (medial) supplies the first lumbrical muscle, and becoming superficial between the slips of the plantar aponeurosis, just proximal to the first interdigital cleft, gives a proper plantar digital nerve to the adjacent sides of the first and the second toes. The second and third common plantar digital nerves have no muscular branches, but otherwise behave as the first at the second and third interdigital clefts [FIG. 11.72]. The most lateral proper plantar digital nerve communicates with the most medial proper digital branch of the lateral plantar nerve.

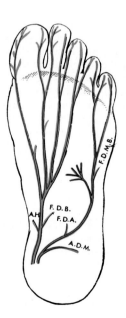

FIG. 11.72. Scheme of distribution of the plantar nerves.
   On the medial side, medial plantar nerve and its cutaneous and muscular branches; F.D.B., flexor digitorum brevis; A.H., abductor hallucis. On the lateral side, lateral plantar nerve and its cutaneous and muscular branches; F.D.A., flexor digitorum accessorius; A.D.M., abductor digiti minimi; F.D.M.B., flexor digiti minimi brevis.

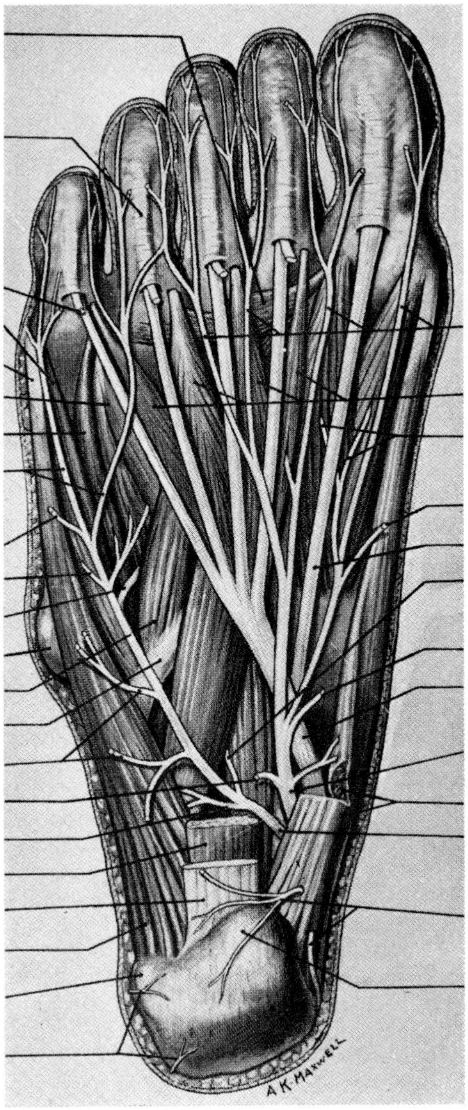
Transverse head of adductor hallucis
Fibrous flexor sheath
Tendons of short flexor (cut) and long flexor
Flexor digiti minimi
Abductor digiti minimi
4th dorsal interosseous
3rd plantar interosseous
Common digital branches of superficial branch of lateral plantar nerve
A branch to skin of sole
Branch to flexor digiti minimi
Deep branch of lateral plantar nerve
Base of 5th metatarsal bone
Oblique head of adductor hallucis
Sheath of peroneus longus
Branches to skin of sole
Branch to flexor digitorum brevis (cut)
Branch to abductor digiti minimi
Flexor digitorum brevis
Plantar aponeurosis
Abductor digiti minimi
Lateral process of tuber calcanei
Branches of lateral calcanean nerve
Proper and common digital branches of medial plantar nerve
Lumbricals
Branches to the two parts of flexor hallucis brevis
A branch to skin of sole
Flexor hallucis longus
Branch to flexor accessorius between its two heads
A branch to skin of sole
Flexor digitorum longus
Medial plantar nerve
Abductor hallucis
Lateral plantar nerve crossing flexor hallucis longus
Branches of medial calcanean nerve
Medial process of tuber calcanei
A. K. Maxwell

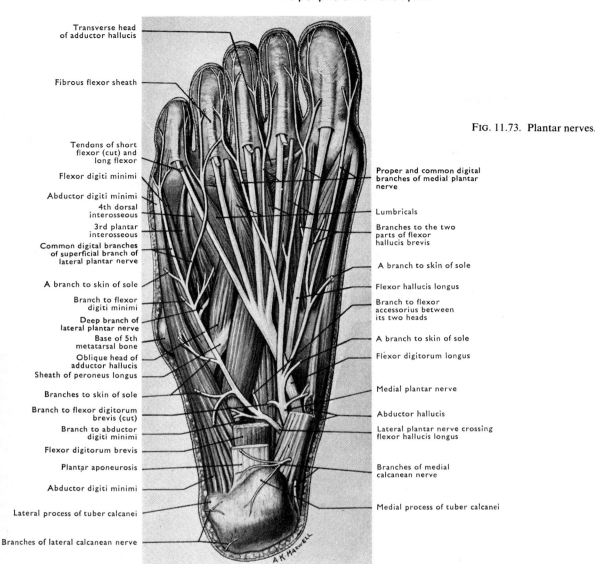

FIG. 11.73. Plantar nerves.

Lateral plantar nerve

This nerve, homologous with the ulnar nerve in the hand, passes anterolaterally towards the base of the fifth metatarsal between flexor digitorum brevis and flexor accessorius, on the medial side of the lateral plantar artery. It gives branches to flexor accessorius and abductor digiti minimi, and divides into superficial and deep branches medial to the base of the fifth metatarsal [FIG. 11.73].

Branches

(i) **Muscular**, see above. (ii) A line of small **cutaneous branches** pierces the plantar fascia between flexor digitorum brevis and abductor digiti minimi to supply the lateral part of the sole. (iii) Vascular branches to adjacent arteries.

The **superficial branch** passes forwards between flexor digitorum brevis and abductor digiti minimi, gives cutaneous branches to the sole. It divides into a **common** (medial) and a **proper** (lateral) **plantar digital nerve** which supplies flexor digiti minimi brevis and sometimes the fourth dorsal and third plantar interosseous muscles. It becomes superficial posterior to the ball of the little toe, supplies the skin on that area and on the plantar surface, lateral side, and distal part of the dorsum of the little toe. The **common plantar digital nerve** divides proximal to the fourth interdigital cleft into **proper plantar digital nerves** to the adjacent sides of the fourth and fifth toes. The more medial of these nerves communicates with the adjacent branch of the medial plantar nerve.

Each **proper plantar digital nerve** supplies the skin and fascia on the plantar surface, tip, and distal part of the dorsal surface of the toe. They also supply the metatarsophalangeal and interphalangeal joints.

The **deep branch** of the lateral plantar nerve runs medially with the plantar arch on the inferior surface of the metatarsal bases, superior to flexor digitorum accessorius and the oblique head of adductor hallucis. It supplies the tarsal and tarsometatarsal joints, the interossei (except any supplied by the superficial branch), the adductor hallucis, and the lateral three lumbricals. The muscular branches enter the superior surfaces of the muscles (except the interossei), that to the second lumbrical passing anteriorly above the transverse head of adductor hallucis.

THE COCCYGEAL PLEXUS

The fourth sacral ventral ramus descends to join the fifth sacral, which passes caudally beside the coccyx and is joined by the coccygeal nerve to form a plexiform cord, the coccygeal plexus. Fine twigs arise from the plexus, and some of these enter and supply coccygeus and the adjacent part of levator ani, while others pierce the coccygeus and the overlying ligaments (sacrospinous and sacrotuberous) to supply skin adjacent to the coccyx and posterior to the anus.

The sacral and coccygeal ventral rami receive grey rami communicantes from the sacral part of the sympathetic trunk which lies medial to their point of emergence from the vertebral column.

General arrangement of the perineal nerves

The perineum is supplied by the ventral rami of the spinal nerves which emerge from the vertebral column caudal to the lumbosacral plexus. It represents, therefore, the ventral part of the trunk which lies caudal to the lower limb and tapers to the coccyx. It is equivalent to that ventral part of the trunk caudal to the upper limb which is supplied by the ventral rami of the thoracic nerves inferior to the brachial plexus. Just as these nerves have cutaneous distributions serially arranged from the second thoracic caudally, so the perineal nerves are serially arranged so that the highest (S. 2, 3) innervates skin near the groin (adjacent to the area supplied by the first lumbar nerve) while the lowest (coccygeal) supplies skin in the region of the coccyx [FIG. 11.66]. Because of the overlap of all adjacent dermatomes [p. 767], no area of skin in the perineum is supplied by a single spinal nerve, but by two adjacent nerves at least. Between the areas of skin supplied by the ilio-inguinal nerve (L. 1) and the dorsal nerve of the penis (S. (2), 3) is the beginning of the ventral axial line (q.v.) of the lower limb [FIG. 11.63] and the beginning of the dorsal axial line lies between the areas of skin supplied by the dorsal rami of the third lumbar and first sacral nerves [FIG. 11.63, p. 769].

DISTRIBUTION OF SPINAL NERVES TO THE SKIN AND MUSCLES OF THE LIMBS

The evidence of dissection, experiment, and clinical observation indicate that virtually every peripheral nerve to the limbs contains nerve fibres derived from more than one ventral ramus, and that the cutaneous distribution of individual peripheral nerves is such that the area of skin supplied overlaps the areas of adjacent cutaneous nerves. The ventral rami of the spinal nerves are, nevertheless, distributed through these peripheral nerves in a standard manner such that certain general rules can be laid down regarding their cutaneous and muscular distribution (Herringham 1886). All ventral rami entering a limb plexus eventually divide into dorsal and ventral divisions which supply respectively the sheets of muscle and the overlying skin of the dorsal (extensor) and ventral (flexor) surfaces of the limbs prior to their rotation into the adult position [p. 50].

Innervation of the skin of the limbs [FIGS. 11.48 and 11.63]

The growing embryonic limb carries the overlying skin outwards from the trunk so that the dermatome supplied by the middle ventral

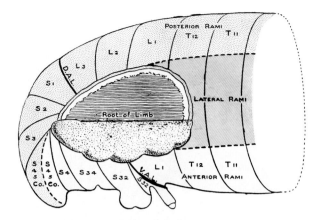

FIG. 11.74. Scheme of innervation of hinder portion of trunk and perineum. The scheme shows the interruption of the segmental arrangement of the nerves due to the formation of the limb. V.A.L., and D.A.L., ventral and dorsal axial lines. [See also FIGS. 11.63 and 11.66.]

ramus of the plexus overlies the distal part of the limb. The dermatomes cephalic and caudal to this are drawn out to a lesser extent so that they are arranged along the cranial (preaxial, thumb or big toe) and caudal (postaxial, little finger or little toe) borders of the limb respectively. Consequently a number of dermatomes are removed from the trunk so that non-adjacent nerves (C. 4 and T. 2, or L. 1 and S. 3) supply adjacent areas of skin on it [FIGS. 11.35 and 11.74]. The dermatomes missing from the trunk lie serially along the borders of the limbs [FIGS. 11.48 and 11.63] so that here also adjacent areas of skin on the cranial and caudal borders of the limb are supplied by non-adjacent nerves, such areas being separated by the imaginary dorsal and ventral axial lines. At the root of the limb, the axial lines separate areas of skin which are supplied by the ventral rami from the superior and inferior nerves of the plexus, and it is only at the periphery of the limbs that the dermatomes lie in numerical sequence from preaxial to postaxial borders.

Thus of two spots on the skin of the limb: (1) that in the preaxial part will be supplied by a higher ventral ramus than that in the postaxial part; (2) that more proximally placed on the preaxial border, or more distally placed on the postaxial border will also be supplied by a higher ventral ramus.

AXIAL LINES

In the upper limb the dorsal axial line extends from the vertebra prominens down the back of the limb to the elbow. The ventral axial line extends from the manubriosternal joint along the ventral surface of the arm and forearm [FIG. 11.48].

In the lower limb the dorsal axial line extends across the posterior superior iliac spine, the gluteal region, and the back of the thigh to the head of the fibula and the lateral side of the leg. The ventral axial line extends from the penis along the medial side of the thigh and knee and the posteromedial aspect of the leg to the heel [FIG. 11.63].

When the developing limbs rotate into the adult position, the flexor surfaces of the upper limbs and the extensor surfaces of the lower limbs become anterior, and the nerves of these respective surfaces form the preponderant cutaneous supply for each limb. Thus nerves arising from dorsal divisions of the ventral rami in the lower limb and from ventral divisions of the ventral rami in the upper limb supply a disproportionately large part of the skin of the limb [FIGS. 11.47 and 11.62].

Innervation of the muscles of the limbs

1. The limb muscles are entirely supplied by the ventral rami, and each muscle is supplied by more than one ventral ramus.

2. The dorsal (extensor) and ventral (flexor) strata of muscles are always supplied by the dorsal and ventral divisions of the ventral rami respectively. The ventral rami that supply the flexor muscles are more numerous, the extra ventral rami being inferior. Thus in the upper limb dorsal divisions supplying muscles arise from C. 5, 6, 7, 8, and ventral divisions from C. 5, 6, 7, 8; T. 1. In the lower limb the corresponding figures are L. 2, 3, 4, 5; S. 1, 2 (dorsal) and L. 2, 3, 4, 5; S. 1, 2, 3 (ventral).

3. The muscles of the preaxial part of the limb are supplied by nerves higher than those of the postaxial part of the limb.

4. The most distal muscles of the limb are supplied by the lowest nerve or nerves of the plexus; small muscles of the hand by the first thoracic nerve, and those of the foot by the second and third sacral nerves.

MUSCLES WITH A DOUBLE NERVE SUPPLY

Virtually all muscles of the limbs receive nerve fibres from more than one ventral ramus; in some cases (e.g. pectoralis major) from all the ventral rami of the plexus. In such cases, no mixing of fibres from dorsal or ventral divisions occurs. It is assumed in these cases that the muscles are formed from the fusion of muscle primordia from several segments. In a very few cases (e.g. biceps femoris and sometimes pectineus), the dual nerve supply is from different divisions of the ventral rami and this is thought to represent fusions between muscle rudiments of the flexor and extensor sheets. The terms 'flexor' and 'extensor' used in connection with the muscle sheets in the ventral and dorsal parts of the developing limb do not apply to the final functions of the muscles. Thus though most of the muscles of the dorsal sheet become extensors of joints, many may act as flexors and vice versa; e.g. brachioradialis is a flexor of the elbow, and rectus femoris and sartorius are flexors of the hip, though all these muscles belong to the 'extensor' muscle sheet and are supplied by dorsal divisions of the ventral rami.

The segmental innervation of the muscles of the upper limb

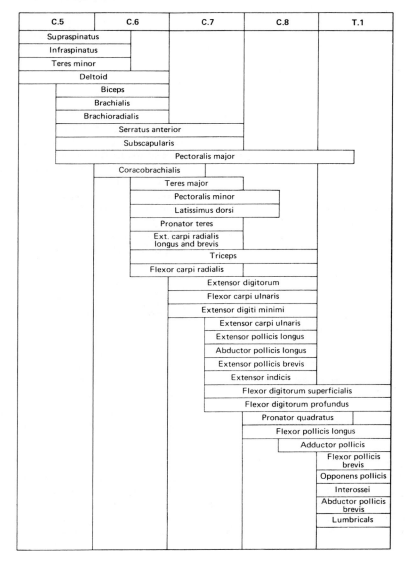

The segmental innervation of the muscles of the lower limb

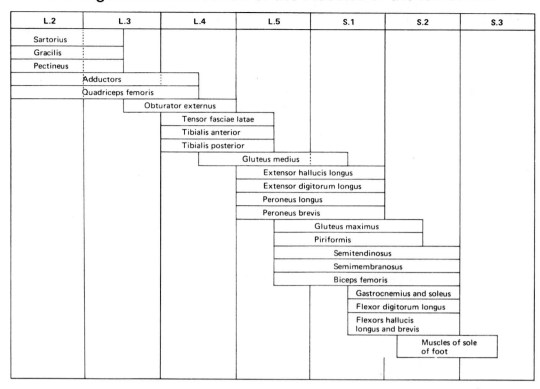

INTRANEURAL PLEXUSES

Within peripheral nerves, the bundles of nerve fibres are constantly changing by branching and recombining to form different fasciculi surrounded by perineurium. The significance of these intraneural plexuses is obscure (Sunderland 1968) since the amount of change is greater than necessary to rearrange the individual nerve fibres into the final peripheral branches. The variations which occur in the peripheral distribution of some nerves, and the communications between various nerves after they have left a common stem (e.g. the median and musculocutaneous nerves) show that the presence of these intraneural plexuses is not always associated with the standard distribution of individual fibres into specific branches. Whatever the reasons for these variable communications within nerves, their presence complicates the process of regeneration in a divided nerve where the outgrowths from the central stump have to reach the correct peripheral branches if regeneration is to be successful.

Blood supply of peripheral nerves

All peripheral nerves are supplied with blood vessels which are essential to their proper function. If the blood supply is interrupted, failure of conduction results; the large diameter fibres being more sensitive to anoxia than the small diameter fibres.

Every peripheral nerve receives a number of small arterial branches from adjacent vessels along its length. These are irregular in their arrangement, and they divide into longitudinal branches on the epineurium which anastomose with similar branches of adjacent nutrient vessels. The epineurial vessels send branches on to the perineurium of the fascicles in which they form a rich, longitudinal plexus of capillaries which passes along the whole length of the nerve in unbroken continuity, reinforced at intervals by the various nutrient arteries reaching the epineurium in such a manner that there is no one special nutrient artery (Durward 1948).

Special arteries associated with certain nerves, e.g. the sciatic and median, do not form a preferential supply, but are embryonic remnants. The arteries to nerves are accompanied by postganglionic sympathetic nerve fibres which supply their muscle coats (Adams 1942, 1943; Sunderland 1945 a, b, c).

Veins drain into adjacent vessels, but do not unite to form longitudinal channels of any size.

DEVELOPMENT OF SPINAL NERVES AND GANGLIA [FIG. 11.75]

The peripheral nervous system is derived from two sources, the cells of the neural crest [FIG. 11.75 and p. 618] and those of the neural tube, both of which are derived from the original neural plate.

The neural crest forms a considerable number of different varieties of cells (Horstadius 1950) and all of these are migratory in character. The crest extends throughout the length of the neural tube. Cells migrate outwards from it to form spongioblasts, neuroblasts, and dendritic cells of the skin among others. Spongioblasts (Schwann cells) are the cellular sheaths of the nerve fibres (Harrison 1924). Neuroblasts form the spinal and cranial nerve ganglia, and some migrate further from the central nervous system to form the autonomic ganglia. They are sometimes known as sympathoblasts, although they form both sympathetic and parasympathetic ganglia though from different regions of the neural crest. Dendritic cells are the pigment-forming cells of the skin. A further group is the phaeochromoblasts (see below).

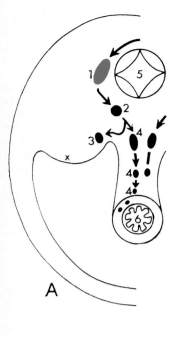

1. Spinal ganglion,
2. Ganglion of sympathetic trunk.
3. Sympathetic cells innervating kidneys and gonads (x) or forming suprarenal medulla.
4. Sympathetic ganglia of alimentary canal, and structures developed from it.
5. Spinal medulla.
6. Alimentary canal.
7. Dorsal ramus,
8. Ventral ramus.
9.
10. } Lateral and anterior cutaneous branches.
11.
12. } Dorsal and ventral divisions of ventral ramus to plexus.
13. Intercostal nerve.
14. Cutaneous branch of dorsal ramus.
15. Thoracic splanchnic nerve.
16. Anterior wall of trunk.
17. Cervical splanchnic nerve.
18. Ventral root.
19. Dorsal root.
20. Grey ramus communicans.
21. White ramus communicans.

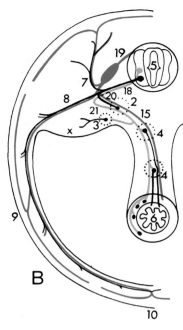

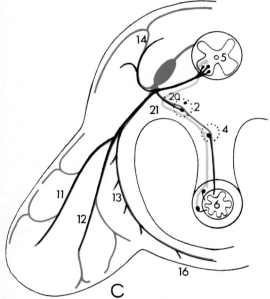

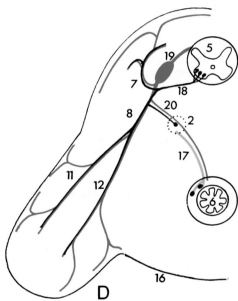

FIG. 11.75. Diagrams of the development and arrangement of certain spinal nerves. A, Arrangement of nerve cells derived from the neural crest. B, The elements of a typical spinal nerve. C, The elements of a spinal nerve adjacent to a limb plexus, e.g. 1st thoracic. D, The elements of a spinal nerve in a limb plexus. Black, somatic and sympathetic (postganglionic) motor cells. Red, preganglionic sympathetic cells. Blue, sensory cells.

The spinal ganglia appear as laterally directed, segmental swellings of the neural crest which extend ventrally to lie lateral to the neural tube. The neuroblasts which form these ganglia each send a process (axon) centrally to enter the dorsolateral aspect of the neural tube. Together they form the dorsal roots of the spinal nerves. The peripheral processes (dendrites) of these cells immediately join the corresponding ventral root to form a spinal nerve. By the continued outgrowth of their peripheral processes the spinal ganglion cells make contact with the various sensory areas of the body, and thus link them to the spinal medulla. The original bipolar cells of the spinal ganglia become unipolar by excentric growth of the cells [FIG. 10.6].

The sympathoblasts migrate ventrally from the spinal ganglia to form separate groups of sympathetic nerve cells, which are primarily associated with the three basic groups of arteries arising from the aorta: to the body wall mesoderm, to the intermediate mesoderm, and to the visceral mesoderm [FIG. 10.16]. By the outgrowth of axons, each of these groups of sympathetic neurons innervates the vessel adjacent to which it lies and the tissues supplied by that vessel [see also p. 618 and p. 825].

The phaeochromoblasts are intimately associated with the sympathoblasts in their migration, but form isolated clumps of cells which stain dark brown with chromium salts, chromaffin tissue. This forms the medulla of each suprarenal gland, and the para-aortic bodies which are temporary structures especially well developed in the foetus anterior to the abdominal aorta.

The ventral roots of the spinal nerves arise from neuroblasts situated in the mantle layer of the basal lamina [p. 619] of the neural tube. These neuroblasts send their axons out in small bundles from the neural tube to unite segmentally into the ventral roots, which join the dorsal roots at the spinal ganglia between the myotome and the neural tube. The ventral roots contain two types of nerve fibres. (1) Somatic motor nerve fibres are distributed to the striated muscles of the body wall and limbs, and are present in all ventral roots. (2) Preganglionic autonomic nerve fibres which grow out into the ventral rami of the first thoracic to third lumbar spinal nerves only. They leave the rami and grow along the line of migration of the sympathoblasts to reach and make synaptic connections with these cells and with the phaeochromoblasts. In the first part of this course they pass from the ventral rami to the sympathetic trunk, thus forming the white rami communicantes. Some of these fibres run in the sympathetic trunk, others pass through it to the visceral ganglia of the sympathetic system as splanchnic nerves. In the third, and fourth or second sacral spinal nerves, equivalent fibres leave the ventral rami and pass directly to the corresponding visceral ganglia (parasympathetic) as pelvic splanchnic nerves [p. 814]. They have no connection with the sympathetic trunk.

The body wall (somatic) part of each spinal nerve is formed by the lateral growth of its nerve fibres, which pass from the spinal ganglia and ventral roots into the corresponding dorsal and ventral rami. The dorsal ramus enters the myotome, supplies it, and branching, continues through it to the overlying skin. The ventral ramus grows ventrolaterally, anterior to the myotome, gives off its smaller visceral branch (white ramus communicans), and continues into the body wall giving rise to branches to muscles, skin, etc. [FIG. 11.75].

FORMATION OF THE LIMB PLEXUSES

The growing nerves in the regions of the limbs behave in the same manner, but the ventral rami give no visceral branches to the autonomic nervous system. They pass directly into the limb bud ventral to the myotomes, and divide into dorsal and ventral divisions which unite with the corresponding branches of adjacent ventral rami to form the nerves of the dorsal and ventral aspects of the limb bud. The dorsal and ventral divisions of the ventral rami that enter the limb plexuses may correspond to the lateral and anterior cutaneous branches of the typical thoracic ventral ramus.

Every spinal nerve distributes postganglionic sympathetic fibres which enter these nerves through grey rami communicantes from the sympathetic trunk.

The autonomic nervous system

The autonomic nervous system may be defined as that part of the nervous system which innervates glands, smooth muscle, cardiac muscle, and muscle which is histologically of the 'voluntary' type in the oesophagus. It has sometimes been called the 'involuntary' nervous system because the structures which it supplies do not appear to be immediately under voluntary control. Yet this is not a satisfactory name since there are smooth muscles which, though innervated by the autonomic nervous system, are obviously used as part of 'voluntary' acts, e.g. the urinary bladder in micturition, the ciliary muscle in focusing the lens of the eye, etc. Also many muscles which are histologically of the 'voluntary' type cannot usually be contracted at will, e.g. the pharyngeal constrictors, and any voluntary muscle may be used in 'involuntary' acts, e.g. shivering. In fact the functional arrangements which are dictated by the central nervous system are such that the structures innervated by the autonomic nervous system play their part in every kind of activity whether this activity is apparently initiated voluntarily or not. Thus changes occur in blood pressure and in the calibre of blood vessels so that extra blood is carried to active tissues, e.g. the voluntary muscles in muscular exercise, and in states of fear.

The only satisfactory anatomical definition of the autonomic nervous system, which allows it to be differentiated from the remainder of the nervous system, is that of Langley (1921). This regards it as a peripheral efferent system consisting of groups of nerve cells (ganglia) situated outside the central nervous system and interposed between it and the organs which are supplied by the non-myelinated axons (postganglionic nerve fibres) of the cells in these ganglia. It is in no sense functionally separate from the central nervous system, but receives myelinated axons (preganglionic nerve fibres) from cells within that system, and so forms one of the routes by which the central nervous system controls the tissues of the body. The significant difference between this efferent route (visceral efferent) and that which supplies the muscles of the body wall and limbs (somatic efferent) is that the terminal cells of the visceral efferent route lie outside the central nervous system, while those of the somatic efferent route lie inside the central nervous system.

This definition excludes all parts of the sensory system from the autonomic system. Yet sensory nerve fibres in large numbers run in the same fibrous sheaths as the nerve fibres of the autonomic nervous system, and are distributed to the same territories. However, they make no direct connections with the cells of the autonomic nervous system, but discharge directly into the central nervous system through the dorsal roots of the spinal nerves, or through the cranial nerves, and have their cell bodies in the sensory ganglia of these nerves. They are, therefore, known as visceral afferent, or visceral sensory fibres. Since the reactions their impulses produce through the central nervous system may be somatic or visceral or both combined, they cannot be considered as specifically autonomic. The information which the visceral afferent fibres transmit is, however, different from that transmitted by the somatic afferent fibres. Most of it does not reach a conscious level (though a sense of fullness and occasionally pain may be appreciated), but it is of profound importance for the maintenance of the internal environment, especially the information from pressure and chemical receptors related to the blood vessels. (Paintal 1977).

The autonomic nervous system is distributed throughout the body. Many of its cells are arranged in ganglia which can readily be dissected, but in other situations, particularly in association with the gut tube and blood vessels, the cells are scattered or in minute groups in a tangle of preganglionic, postganglionic, and sensory nerve fibres. Such arrangements constitute the autonomic plexuses. The ganglion cells greatly exceed in number the preganglionic fibres which pass to synapse with them. Thus each preganglionic fibre tends to synapse with many ganglion cells, hence widespread effects are likely to arise from stimulation of a few preganglionic nerve fibres. There must, however, be a topographical organization within the autonomic nervous system which allows for localized responses in particular functions.

The autonomic nervous system is divisible into sympathetic and parasympathetic parts because the preganglionic fibres to the

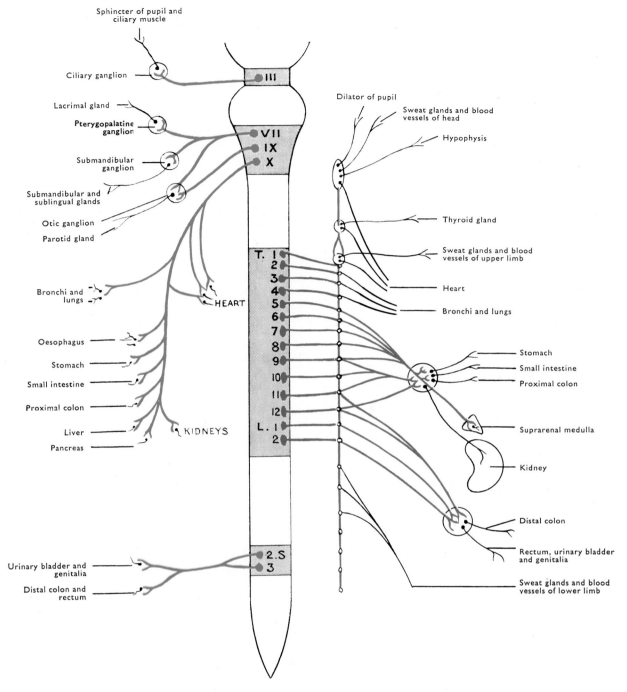

FIG. 11.76. Scheme of general arrangement of autonomic nervous system.
     Preganglionic fibres of the parasympathetic system, green, and of the sympathetic system, red. Postganglionic fibres of both systems, black. Afferent fibres not included.

sympathetic part emerge in the thoracic and upper two or three lumbar spinal nerves (**thoracolumbar outflow**), while those of the parasympathetic part emerge with certain cranial and sacral nerves (**craniosacral outflow**).

   The two parts can also be differentiated, for the sympathetic is distributed to all parts of the body, more especially to the blood vessels; while the parasympathetic has a distribution limited to the alimentary tract and structures developed in or from its walls, to the respiratory and urogenital systems, and to the eye, all of which are therefore innervated by both sympathetic and parasympathetic parts. *The parasympathetic plays no part in the innervation of the body wall or limbs.* It has been claimed that there are parasympathetic fibres in the dorsal roots of the spinal nerves (Kure 1931) because stimulation of the distal end of a cut dorsal root produces vasodilatation in its territory of distribution. There is, however, no evidence of efferent fibres in the dorsal roots, and the vasodilatation is believed to be due to retrograde stimulation of the sensory fibres to blood vessels.

Where sympathetic and parasympathetic parts supply the same structure, their effects are frequently opposite. Thus the heart is accelerated by the sympathetic, and slowed or arrested by the parasympathetic; the pupil is dilated by the sympathetic, and constricted by the parasympathetic; the parasympathetic increases peristalsis in the alimentary canal, while the sympathetic reduces it and causes contraction of the sphincters. This apparent antagonism is not real; the two parts co-operate to produce a balance within which it is possible to obtain the multiple gradations necessary for normal functioning in a wide variety of circumstances.

In the somatic tissues, the sympathetic part controls the calibre of blood vessels, causes sweat glands to secrete, and the arrectores pilorum to contract. In a general way the sympathetic part can be looked upon as preparing the body for reaction to some crisis, while the parasympathetic part is concerned with conservation and the building up of energy.

Chemical transmission of nerve impulses has mainly been studied in the autonomic nervous system because of the ease of access to the synapses between preganglionic axons and ganglion cells (Elliott 1905; Loewi 1935; Dale 1934, 1938). Acetylcholine has long been known to play an important part in the transmission at the above synapses, at some synapses in the central nervous system, and at the junctions both of postganglionic parasympathetic and of somatic motor axons with their effector organs. On the other hand noradrenaline is the chemical principally concerned with transmission of impulses at most of the endings of the sympathetic postganglionic fibres, though in the case of sweat glands the transmitter substance is acetylcholine. Cholinergic fibres cannot, therefore, be equated with postganglionic parasympathetic fibres, nor adrenergic fibres with postganglionic sympathetic fibres, though this basic chemical difference exists between most of them.

In general, the ganglion cells of the parasympathetic part differ from those of the sympathetic part in that they tend to lie closer to the organs which they supply. Also, the sympathetic ganglia are commonly associated with blood vessels, and their postganglionic fibres are frequently transmitted along blood vessels [FIG. 10.16]—features which are less common with the parasympathetic cells and fibres.

Not all cells in the autonomic ganglia give rise to postganglionic fibres. Some seem to act as interneurons, though their significance is not clear.

THE PARASYMPATHETIC PART

The preganglionic fibres of this part emerge from two widely separated regions of the central nervous system. The cranial fibres leave the brain through the oculomotor, facial, glossopharyngeal, and vagus nerves, while the preganglionic fibres of the sacral parasympathetic leave the spinal medulla in the third, and fourth or second sacral spinal nerves.

The preganglionic fibres of the cranial and sacral parasympathetic are relatively long and pass directly to parasympathetic ganglia. They have no connections with the sympathetic trunk or ganglia, and though they may be intermingled with sympathetic postganglionic fibres as they approach their termination, they are merely running together without any direct connections between them. In the case of the oculomotor, facial, and glossopharyngeal nerves, the preganglionic parasympathetic fibres end on discrete ganglia which have a circumscribed distribution. Hence the effects of stimulation of these fibres are localized, unlike the widespread effects of preganglionic sympathetic stimulation. On the other hand preganglionic parasympathetic fibres in the vagus have a very widespread distribution in the head, neck, thorax, and abdomen, extending as

far inferiorly as the left flexure of the colon where their territory of distribution meets that of the preganglionic fibres of the sacral outflow. The sacral parasympathetic preganglionic fibres are distributed to ganglionic neurons on or near the terminal parts of the alimentary and urogenital systems.

Cranial parasympathetic

OCULOMOTOR NERVE

The preganglionic parasympathetic nerve fibres in this nerve probably arise in the accessory nucleus of the oculomotor nerve in the midbrain [p. 646]. They pass in the oculomotor nerve [p. 745], and leaving the branch to the inferior oblique muscle in the orbit, run as the parasympathetic, oculomotor root to the ciliary ganglion [FIG. 11.77]. In the ganglion, they synapse with the large cells, from which myelinated postganglionic fibres enter the short ciliary nerves. These nerves pierce the sclera, and running forwards between it and the choroid, supply the ciliary muscle and the circular muscle of the iris (sphincter pupillae). Thus they decrease the focal length of the lens and the diameter of the pupil. When looking from a distant to a near object (convergence-accommodation) the eyes are converged on it (medial rectus muscles), the lens focused, and the pupil contracted; all these are activities produced by nerve fibres in the oculomotor nerve. The pupilloconstrictor fibres are also activated by light falling on the retina [p. 706]. Other ganglion cells have been

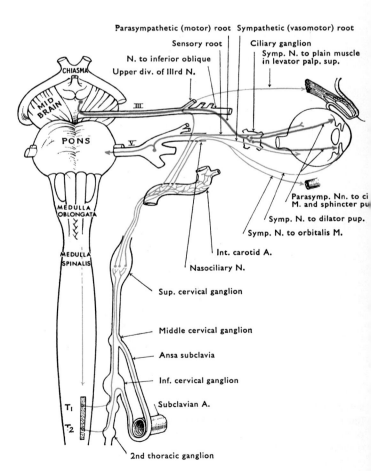

FIG. 11.77. Diagram of ciliary ganglion and connections of autonomic nervous system with the eye.

Sympathetic, red; parasympathetic, green; sensory, blue

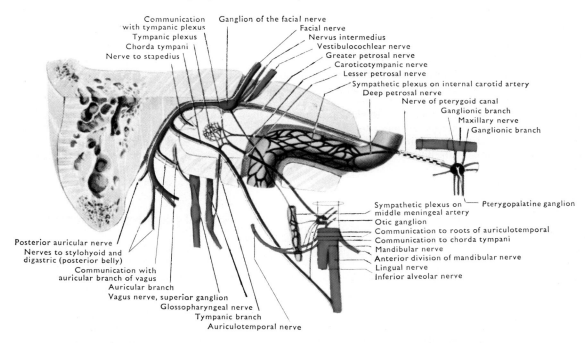

FIG. 11.78. Branches of facial and glossopharyngeal nerves in the temporal bone, and connections of pterygopalatine and otic ganglia.

described in the orbit (Nathan and Turner 1942), but their function, if different from those of the ciliary ganglion, is not known.

FACIAL NERVE

The preganglionic fibres in this nerve arise in the superior salivatory nucleus [p. 648] at the pontomedullary junction. They emerge in the nervus intermedius, and joining the facial nerve in the internal acoustic meatus, leave it in the greater petrosal and chorda tympani branches [FIG. 11.78].

1. The **greater petrosal nerve** [p. 757] is joined by the deep petrosal nerve (sympathetic postganglionic fibres) and together they form the nerve of the pterygoid canal to the **pterygopalatine ganglion**. Here the parasympathetic fibres synapse on the cells of the ganglion whose **postganglionic fibres** are distributed with the branches of the maxillary nerve to the palate, to the walls of the cavity and air sinuses of the nose, and to the lacrimal gland [FIG. 11.79], to increase the secretory activity of glands in these regions.

2. The **chorda tympani** [p. 757] joins the lingual nerve in the infratemporal fossa, the **preganglionic fibres** leave that nerve on the hyoglossus to pass to the **submandibular ganglion** and synapse with its cells. The **postganglionic fibres** from the cells of the ganglion pass: (i) directly to the submandibular [FIG. 11.20] and sublingual glands; (ii) into the lingual nerve for distribution to the glands in the anterior part of the tongue and to the sublingual gland [FIG. 11.79]. Some preganglionic fibres may pass directly to scattered ganglion cells adjacent to these glands (Fitzgerald and Alexander 1969). Stimulation of the chorda tympani causes increased secretion and vasodilatation in the glands it supplies, though their vasomotor supply is through sympathetic fibres from the plexuses on the arteries that supply them (Kuntz and Richins 1946).

Preganglionic fibres of the facial nerve are also described as passing to the **tympanic plexus**, and thence through the lesser petrosal nerve to the otic ganglion. If this is their course, then the parasympathetic fibres in the facial nerve may assist those of the glossopharyngeal nerve in supplying the parotid gland.

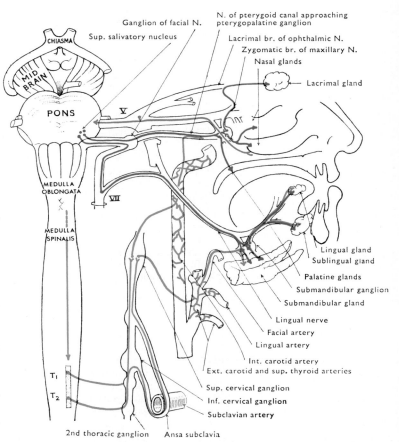

FIG. 11.79. Diagram of connections of pterygopalatine and submandibular ganglia with distribution of parasympathetic components of seventh cranial nerve.

Sympathetic, red; parasympathetic, green; sensory, blue.

GLOSSOPHARYNGEAL NERVE

The preganglionic parasympathetic nerve fibres pass into this nerve from the inferior salivatory nucleus [p. 648]. They are distributed through the **tympanic branch** to the tympanic plexus, and thence via the lesser petrosal nerve to the **otic ganglion** [FIG. 11.80], and directly to scattered ganglion cells in the posterior third of the tongue. The cells of the otic ganglion send their **postganglionic fibres** through the auriculotemporal nerve to the parotid gland, while those in the posterior third of the tongue supply the numerous glands in that part of the organ. Both increase the secretory activity of the gland cells.

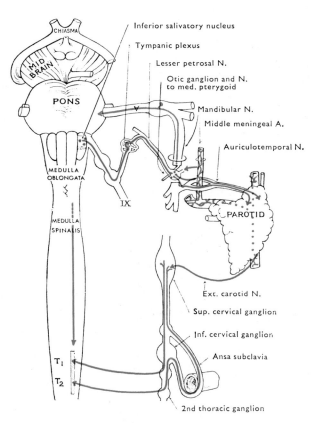

FIG. 11.80. Diagram of connections of otic ganglion with distribution of parasympathetic components of ninth cranial nerve.
   Sympathetic, red; parasympathetic, green; sensory blue.

VAGUS NERVE

This nerve contains the largest group of preganglionic parasympathetic fibres in any nerve, together with numerous visceral afferent fibres. The preganglionic fibres arise from the cells of the dorsal vagal nucleus [p. 647]. The ganglion cells to which the preganglionic fibres pass are all either scattered close to the organs they supply, or included in the autonomic plexuses to which the sympathetic fibres also run. The sympathetic fibres remain functionally separate though impossible to disentangle by disection. All branches of the vagus, except possibly the meningeal and auricular, contain preganglionic parasympathetic fibres. Through these they are distributed to: (1) scattered ganglion cells associated with the glands in the pharynx and larynx, and the glands and smooth muscle in the trachea and bronchi; (2) the cardiac, pulmonary, and

oesophageal plexuses; (3) the stomach and liver via the vagal trunks and gastric nerves; and (4) from the vagal trunks, fibres pass to the great abdominal plexuses, associated with the coeliac and superior mesenteric arteries and their branches [pp. 763 and 824]. Through these they are distributed to the myenteric and submucous plexuses of the gut tube as far as the left flexure of the colon, and to the liver and pancreas (Schofield 1956).

In general these fibres are concerned with: (1) decreasing the rate of cardiac contraction; (2) bronchoconstriction; (3) increasing peristalsis of the gut tube and relaxing the sphincters; and (4) causing increased secretion by the glands in the whole territory of the vagus nerve. Section of the vagus nerves decreases, for example, the gastric section of hydrochloric acid.

Sacral parasympathetic

The sacral parasympathetic nerves (**pelvic splanchnic nerves**) arise as delicate branches from the third and fourth sacral ventral rami, less commonly also from either the second or the fifth (Sheehan 1941). Their cells of origin lie in the sacral part of the spinal medulla at the level of the first lumbar vertebra, and thus their nerve fibres course through the cauda equina to reach the sacral foramina. The two (or three) pelvic splanchnic nerves consist of fine myelinated, preganglionic axons which pass directly forwards to the inferior hypogastric (pelvic) plexuses [p. 824]. Here they intermingle with sympathetic fibres, and synapse with the parasympathetic ganglionic neurons either in these plexuses, or in the walls of the large intestine from the left colic flexure caudally, or in the pelvic parts of the urogenital system. Thus they replace the vagus from the region of the left colic flexure distally, and co-operate with the sympathetic in the control of the above organs.

THE SYMPATHETIC PART

PREGANGLIONIC FIBRES

The preganglionic fibres of this part arise from cells which lie in the lateral grey column of the spinal medulla between the two limb plexuses, and emerge only in those spinal nerves which arise from this region, that is, only in the first thoracic to second or third lumbar spinal nerves (**thoracolumbar outflow**). The preganglionic fibres are slender and thinly myelinated, and traverse the ventral roots of the spinal nerves to enter the ventral rami. They leave the latter almost immediately to run to the adjacent part of the sympathetic trunk as the **white rami communicantes**. Each sympathetic trunk consists of a chain of more or less segmentally arranged ganglia (groups of sympathetic nerve cells) which extends along the whole length of the vertebral column. The ganglia of each trunk are linked by a bundle, or bundles, of nerve fibres (**interganglionic rami**). These contain longitudinally directed branches of the preganglionic fibres which enter the trunk through the white rami communicantes. In this way the ganglia of the sympathetic trunk which lie cranial to the first thoracic nerve and caudal to the second or third lumbar nerve receive preganglionic fibres, even though these fibres only emerge from the central nervous system through the restricted thoracolumbar outflow.

Each **white ramus communicans** usually enters the sympathetic trunk at a ganglion, not infrequently at a level inferior to the ventral ramus from which it emerges. Here branches of the individual preganglionic fibres pass to the ganglion cells and run longitudinally in the trunk to end by synapsing with ganglion cells at higher and/or lower levels. In the upper thoracic region the longitudinal branches are predominantly ascending, in the lower thoracic and upper

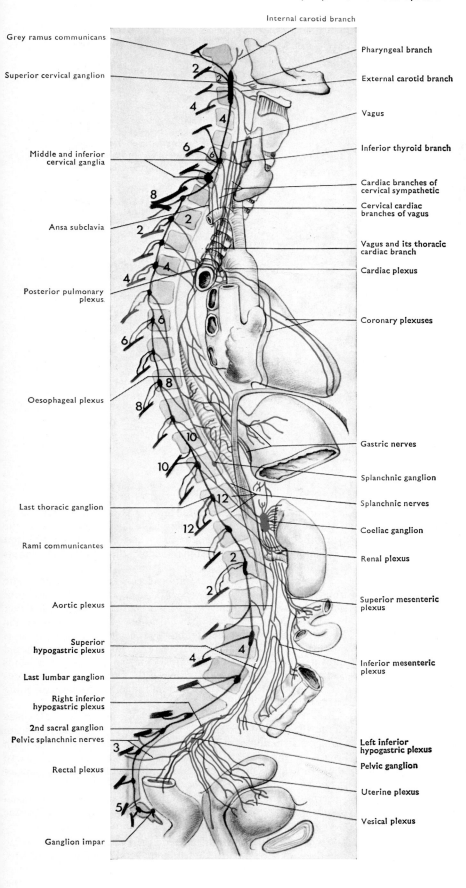

Internal carotid branch

Grey ramus communicans

Superior cervical ganglion

Middle and inferior
cervical ganglia

Ansa subclavia

Posterior pulmonary
plexus

Oesophageal plexus

Last thoracic ganglion

Rami communicantes

Aortic plexus

Superior
hypogastric plexus

Last lumbar ganglion

Right inferior
hypogastric plexus

2nd sacral ganglion
Pelvic splanchnic nerves

Rectal plexus

Ganglion impar

Pharyngeal branch

External carotid branch

Vagus

Inferior thyroid branch

Cardiac branches of
cervical sympathetic
Cervical cardiac
branches of vagus

Vagus and its thoracic
cardiac branch
Cardiac plexus

Coronary plexuses

Gastric nerves

Splanchnic ganglion

Splanchnic nerves

Coeliac ganglion

Renal plexus

Superior mesenteric
plexus

Inferior mesenteric
plexus

Left inferior
hypogastric plexus
Pelvic ganglion

Uterine plexus

Vesical plexus

FIG. 11.81. Topographical plan of main parts of
autonomic nervous system.

The spinal nerves, sympathetic trunk, and
rami communicantes, black; sympathetic
branches of distribution and plexuses, red;
parasympathetic nerves (vagus and sacral),
green. The vertebrae and spinal nerves are
numbered.

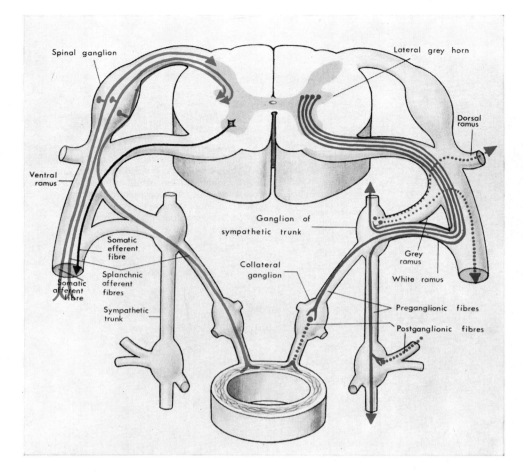

FIG. 11.82. Scheme showing relation of sympathetic system to spinal nerves and spinal medulla.

Sympathetic efferent fibres (preganglionic and postganglionic, red), are shown on the right side; a somatic efferent fibre (black) and somatic and visceral afferent fibres (blue) on the left. For relative position of white a grey rami, see page 817. [Cf. FIGS. 10.47 and 11.75.]

lumbar region they are predominantly descending. Other preganglionic fibres, sometimes accompanied by postganglionic fibres arising in the cells of the sympathetic trunk, pierce that trunk either directly or after running longitudinally in it for a variable distance, and pass anteromedially as visceral (splanchnic) nerves to reach the sympathetic ganglia (visceral or prevertebral) which do not lie in the sympathetic trunk.

The arrangement of the branches of the preganglionic fibres and of the ganglion cells (see below) is such that each preganglionic fibre can make synapses with a large number of ganglion cells at various levels in the sympathetic part of the autonomic nervous system.

GANGLIA

The cells of the sympathetic ganglia are all efferent, and there is no adequate evidence of afferent neurons in these ganglia. The nerve cells of the sympathetic ganglia, each surrounded by a cellular capsule, mostly have many dendrites (multipolar) though a few are bipolar and some are pseudo-unipolar. Some of the dendrites ramify within the capsule: others pierce it and branching in the ganglion make contact with the capsules or extracapsular dendrites of its cells, and are exposed to the preganglionic fibres entering the ganglion. The preganglionic fibres also pierce the capsules of the ganglion cells and make contact with the intracapsular dendrites. Thus a few preganglionic fibres entering a ganglion can synapse with a large number of the cells in the ganglion.

There is evidence (Matthews and Raisman 1969; Lever 1976) of interneurons in the ganglia, but it appears that most of the cells give rise to axons (postganglionic) which pass to effector organs. There

is no evidence that the visceral sensory fibres which traverse all parts of the sympathetic system have any synapses with the sympathetic cells of the ganglia or plexuses, but merely transmit their impulses direct to the spinal medulla through the dorsal roots of the spinal nerves.

There are twenty-one or twenty-two ganglia of the sympathetic trunk; three cervical, ten or eleven thoracic, four lumbar, and four sacral. At their caudal ends the two sympathetic trunks unite either by communicating strands or in a median, **ganglion impar** on the front of the coccyx. At their cranial ends each sympathetic trunk is prolonged into the cranial cavity as a plexus around the internal carotid artery. This plexus contains some ganglion cells, and sends postganglionic fibres into a number of the cranial nerves. It unites with its fellow in a small ganglion on the anterior communicating artery (Mitchell 1953).

Prevertebral ganglia are found in the large prevertebral **plexuses** of the thorax and abdomen, and their postganglionic fibres are mainly distributed along branches of these plexuses on the main arteries to the visceral structures. The plexuses consist of nerve fibres in which ganglion cells are scattered. Only a few groups of these are sufficiently large to be dissected and named.

Terminal ganglia are tiny groups of nerve cells in the distal ramifications of the prevertebral plexuses buried in the walls of the viscera. They belong mainly to the parasympathetic part.

While it is possible to differentiate these three types of ganglia, only those of the sympathetic trunk are clearly defined, the others are larger aggregations of ganglion cells within a gangliated plexus which extends from the splanchnic nerves to the terminations of the plexuses in the viscera. Even the ganglia of the sympathetic trunk have small **intermediate ganglia** scattered on the interganglionic rami, and two or more of the ganglia of the trunk may be fused. Every sympathetic ganglion cell, wherever situated, receives synapses from preganglionic fibres which emerge from the central nervous system in the thoracolumbar outflow.

POSTGANGLIONIC FIBRES

The fibres which arise from the ganglia of the **sympathetic trunk** are distributed predominantly to structures in the body wall, both in somatic nerves and as plexuses on the arteries supplying these structures. From the ganglia of the **visceral plexuses** they are distributed to the viscera through plexuses on the arteries which supply these structures. From the **terminal ganglia** they are distributed within the organs themselves.

Postganglionic fibres pass from the cephalic part of the sympathetic trunk (superior cervical ganglion) and its continuation, the internal carotid plexus, into most of the cranial nerves, and from all parts of the sympathetic trunk into the ventral rami of the corresponding spinal nerves (grey rami communicantes). In both cases these fibres are distributed through every branch of these nerves to blood vessels (**vasomotor fibres**), sweat (**sudomotor fibres**) and other glands, and arrectores pilorum (**pilomotor fibres**).

1. Grey rami communicantes

These bundles of postganglionic fibres pass from the cells of the sympathetic trunk into the ventral ramus of every spinal nerve. They arise directly from the ganglia or from the interganglionic rami, and usually pass transversely to enter the ventral ramus proximal to the point where the white ramus, if present (T. 1–L. 3), leaves it (Pick and Sheehan 1946; Botar 1932). Where present, the white ramus may be fused with the corresponding grey ramus, and both also transmit visceral sensory fibres. The grey ramus contains non-myelinated and a few finely myelinated postganglionic fibres. On entering the ventral ramus, some of the fibres pass into the dorsal ramus, but the distribution of a grey ramus is limited to the territory of a single spinal nerve, unlike the widespread distribution of the fibres of a white ramus communicans within the sympathetic trunk and through its visceral branches.

2. Vascular branches

The majority of these pass to the somatic arteries through the grey rami and spinal nerves, but direct branches from the sympathetic trunk pass to arteries (e.g. subclavian) which lie in close association with it. Most of the visceral postganglionic fibres pass with and supply the visceral arteries.

VISCERAL BRANCHES OF THE SYMPATHETIC TRUNK (SPLANCHNIC NERVES)

Such branches consist mainly of preganglionic fibres which have entered the sympathetic trunk in the white rami communicantes, and either pass directly through, or run longitudinally in the trunk for a variable distance before leaving it in the visceral branches. These branches arise from every level of the sympathetic trunk, and all course caudally towards the viscus they supply because of the caudal displacement of the viscera relative to the vertebral column and sympathetic trunks during development. In addition to the preganglionic fibres, some postganglionic fibres in the visceral branches arise from cells in the trunk, from ganglion cells scattered along the visceral branches, and in the prevertebral plexuses to which they are passing. The preganglionic fibres synapse with these scattered ganglion cells and with the macroscopically visible ganglia of the prevertebral plexuses. The postganglionic fibres of these cells

are distributed to the viscera principally as perivascular plexuses of nerves.

Visceral sensory fibres traverse all the visceral branches from sensory endings at the periphery. They enter the spinal nerves either through the grey or white rami communicantes and pass into the spinal medulla in the dorsal root, having their cell bodies in the spinal ganglion. These sensory fibres have no synapses with the ganglion cells of the autonomic nervous system, but their presence in its branches means that all contain a mixture of myelinated (preganglionic and sensory) and non-myelinated (postganglionic and sensory) fibres. It is impossible, therefore, to identify particular fibres by virtue of the presence or absence of a myelin sheath, and this complexity is further increased by the admixture of parasympathetic fibres in the peripheral parts of the distribution.

The sympathetic trunk

CERVICAL PART

This part consists of three (occasionally two or rarely four) cervical sympathetic ganglia, united by long, slender, interganglionic rami. It has no white rami communicantes, but it gives grey rami communicantes to all the cervical ventral rami. This part of the trunk lies on the prevertebral muscles, posterior to the carotid sheath. It is continued superiorly into the cranium as the internal carotid nerve which ascends with the internal carotid artery and breaks up into the internal carotid plexus on it. Inferiorly, the cervical part is continuous with the thoracic part anterior to the neck of the first rib, and here the trunk turns backwards to follow the vertebral curvature.

The **superior cervical ganglion** is a fusiform structure 2.5 cm or more in length. It lies on the fascia covering the longus capitis muscle posterior to the internal carotid artery, internal jugular vein, and the last four cranial nerves. It extends from a point 2 cm inferior to the entrance of the carotid canal, to the level of the angle of the mandible, but its lower limit varies with the level of bifurcation of the common carotid artery. It represents the fused upper four cervical segmental ganglia, and hence gives grey rami to the first four cervical nerves.

The **middle cervical ganglion** is the smallest and most variable of the cervical ganglia. It usually lies on the inferior thyroid artery as it passes between the common carotid artery and the sixth cervical transverse process. It may lie further inferiorly on the vertebral artery (Axford 1928) or even be partly or completely fused with the inferior cervical ganglion. The ganglion represents the fifth and sixth cervical ganglia and sends grey rami to the corresponding nerves.

The **cervicothoracic (stellate) ganglion** is large and irregular in shape. It is formed by the fusion of the inferior cervical ganglion (and sometimes the middle) with the first thoracic ganglion (and sometimes the second). The inferior cervical ganglion may form a separate element, in which case it represents the fused lower two cervical ganglia, and sends grey rami communicantes to the seventh and eighth cervical nerves. The cervicothoracic ganglion lies at the junction of the cervical and thoracic parts of the sympathetic trunk. It is anterior to the transverse process of the seventh cervical vertebra and the neck of the first rib, and posterior to the vertebral artery and veins, and the dome of the pleura, close to the termination of the costocervical trunk [FIG. 11.83]. The superior part of the ganglion may lie anterior to the vertebral artery (ganglion vertebrale). It is then often fused with the middle cervical ganglion, and united with the remainder by strands encircling the artery. The middle cervical and cervicothoracic ganglia are united by an interganglionic ramus (commonly multiple) and by a connection which loops around the subclavian artery (**ansa subclavia**) [FIG.

11.83]. When the two ganglia are fused, the ansa unites the upper and lower parts of the common ganglionic mass.

THORACIC PART

The thoracic part of the sympathetic trunk descends posterior to the pleura on the intercostal vessels. Superiorly it lies anterior to the necks of the ribs, but as the thoracic vertebrae widen inferiorly, the trunk passes on to the costovertebral joints (midthoracic region) and then on to the sides of the vertebral bodies (lower thoracic region), to become continuous with the lumbar part by passing behind the medial arcuate ligament.

There are usually ten or eleven ganglia united by thick interganglionic rami which are often duplicated.

The thoracic part receives most of the preganglionic fibres through the white ramus communicans which each thoracic ventral ramus sends to it. Of the fibres in these rami a large number ascend to provide the preganglionic supply to the cervical part of the sympathetic trunk, while others descend to aid those from the first and second lumbar nerves to form the preganglionic supply for the lower lumbar and sacral parts of the sympathetic trunk. Clinical evidence suggests that the preganglionic fibres which are concerned with particular peripheral structures emerge at constant levels from the spinal medulla. Thus (1) preganglionic fibres for the head and neck leave the spinal medulla in the upper two thoracic nerves, and passing in the white rami to the sympathetic trunk, ascend in it to the superior cervical sympathetic ganglion. (2) Preganglionic fibres concerned with the heart emerge with the upper four thoracic nerves, and ascend to all three cervical ganglia, whence they pass in cervical cardiac nerves to the cardiac plexus [p. 819]. Direct cardiac branches from the upper thoracic part of the trunk also pass to the cardiac plexus [p. 821]. Visceral sensory fibres from the heart pass in the latter branches and in the middle and inferior cervical cardiac nerves, to enter the spinal medulla in the first to fifth thoracic spinal nerves. (3) Preganglionic fibres for the upper limbs emerge in the third to seventh thoracic nerves, and pass in the trunk to the middle cervical and cervicothoracic ganglia before they synapse. (4) The preganglionic outflow for abdominal structures is in the lower seven thoracic spinal nerves via the splanchnic nerves. (5) Preganglionic fibres for the lower limbs emerge in the tenth thoracic to second or third lumbar spinal nerves, and synapse in the second lumbar to third sacral ganglia. (6) Preganglionic fibres for the pelvic viscera emerge in the upper lumbar nerves.

LUMBAR PART

This part of the trunk descends on the anterior surfaces of the lumbar vertebrae [FIG. 11.85], medial to the psoas muscle. The right trunk is overlapped by the inferior vena cava, the left by para-aortic, lumbar lymph nodes. The trunks pass posterior to the suprarenal, renal, gonadal, and common iliac vessels, and anterior to the lumbar vessels, but may be crossed anteriorly by some of the lumbar veins, thus complicating the removal of this part of the trunk in the operation of 'lumbar sympathectomy'. There are usually four lumbar ganglia, but a greater or lesser number is common (Pick and Sheehan 1946; Cowley and Yeager 1949).

The anterior position of the lumbar part of the trunk separates it from the emerging ventral rami of the lumbar nerves and increases the length of the rami communicantes. These pass with the lumbar vessels medial to psoas on the lumbar vertebral bodies. The upper two (or three) lumbar ventral rami send white rami communicantes to the trunk, and ganglion cells of the trunk send grey rami communicantes to all the lumbar ventral rami. The white and grey rami are frequently fused in this region, and commonly carry small,

intermediate ganglia on them (Wrete 1935); a feature which is commonest in this and the cervical regions. Such ganglia may be responsible for continued sympathetic activity in the territory of the lumbar nerves after lumbar sympathectomy (Pick and Sheehan 1946; Boyd and Monro 1949; Monro 1959).

PELVIC PART

The lumbar part of the sympathetic trunk becomes continuous with the pelvic part posterior to the common iliac vessels, and extends inferiorly and a little medially on the pelvic surface of the sacrum and coccyx. It lies medial to the first three pelvic sacral foramina, but crosses the fourth on the fourth sacral ventral ramus. Below this the two trunks end either by irregular communications on the terminal pieces of the sacrum and the coccyx, or in a small **ganglion impar** on the front of the coccyx, or independently. There are usually four small sacral ganglia, the main branches of which are the grey rami to the sacral and coccygeal nerves.

Branches of the sympathetic trunk

CERVICAL PART

FROM THE SUPERIOR CERVICAL GANGLION. **Branches accompanying Arteries.** (1) From the apex of the ganglion arise one or more **internal carotid nerves** which form the **internal carotid plexus** around the internal carotid artery in the carotid canal. The nerves consist of preganglionic and postganglionic fibres, and the plexus contains some small ganglia on which the preganglionic fibres end. The **postganglionic fibres** are distributed: (i) as **plexuses** on the branches of the internal carotid artery to brain, orbit, forehead, and hypophysis; (ii) to the third, fourth, fifth, and sixth **cranial nerves** at variable points, commonly between the cavernous sinus and orbit; (iii) from the internal carotid plexus in the upper part of the foramen lacerum, the **deep petrosal nerve** passes to join the greater petrosal branch of the facial nerve and form the nerve of the pterygoid canal, which runs to the pterygopalatine ganglion; (iv) the **caroticotympanic nerves** enter the middle ear cavity and run with fibres of the tympanic branch of the glossopharyngeal nerve; (v) the **sympathetic root of the ciliary ganglion** arises either from the nasociliary nerve or from the plexus on the ophthalmic artery. These postganglionic fibres pass straight through the ciliary ganglion to enter the eye in the short ciliary nerves. They supply the vessels of the choroid and iris, and the dilator of the pupil.

Thus the internal carotid plexus has a widespread distribution in the cranial cavity, orbit, forehead, nose, and palate (deep petrosal nerve); its interruption not only interferes with vasomotor control and sweating (forehead) in that area, but results in constriction of the pupil (paralysis of the dilator), drooping (ptosis) of the upper eyelid (paralysis of the smooth muscle in levator palpebrae superioris), and some sinking in of the eye (enophthalmos), said to be due to paralysis of smooth muscle in the orbit. The same effect on the eye (Horner's syndrome) can be produced by damage to the preganglionic fibres in any part of the cervical sympathetic trunk, but in addition there will be effects involving the other branches of the superior, middle, and cervicothoracic ganglia depending on the level of injury.

(2) Branches (postganglionic) also pass to the external carotid artery from the superior ganglion to form the **external carotid plexus**. This supplies blood vessels and glands (sweat and salivary) in the territory of that artery. Many of its fibres pass with branches of the trigeminal nerve near their termination, e.g. sudomotor fibres to the face (Wilson 1936).

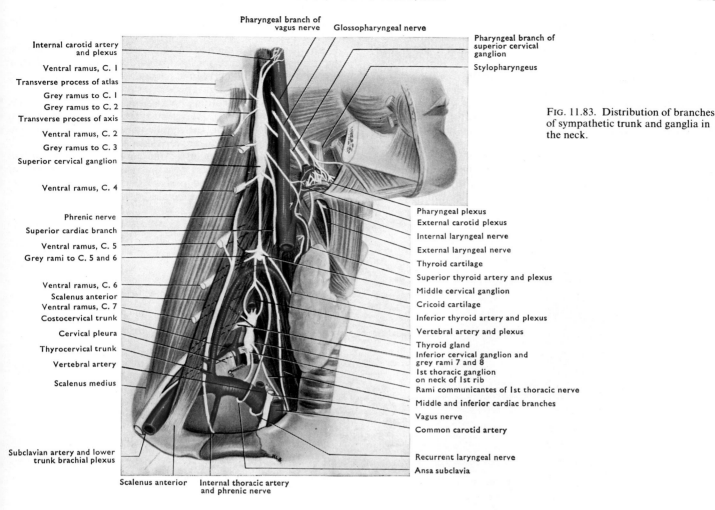

Pharyngeal branch of vagus nerve   Glossopharyngeal nerve

Internal carotid artery and plexus
Ventral ramus, C. 1
Transverse process of atlas
Grey ramus to C. 1
Grey ramus to C. 2
Transverse process of axis
Ventral ramus, C. 2
Grey ramus to C. 3
Superior cervical ganglion
Ventral ramus, C. 4
Phrenic nerve
Superior cardiac branch
Ventral ramus, C. 5
Grey rami to C. 5 and 6
Ventral ramus, C. 6
Scalenus anterior
Ventral ramus, C. 7
Costocervical trunk
Cervical pleura
Thyrocervical trunk
Vertebral artery
Scalenus medius
Subclavian artery and lower trunk brachial plexus

Scalenus anterior   Internal thoracic artery and phrenic nerve

Pharyngeal branch of superior cervical ganglion
Stylopharyngeus

Pharyngeal plexus
External carotid plexus
Internal laryngeal nerve
External laryngeal nerve
Thyroid cartilage
Superior thyroid artery and plexus
Middle cervical ganglion
Cricoid cartilage
Inferior thyroid artery and plexus
Vertebral artery and plexus
Thyroid gland
Inferior cervical ganglion and grey rami 7 and 8
1st thoracic ganglion on neck of 1st rib
Rami communicantes of 1st thoracic nerve
Middle and inferior cardiac branches
Vagus nerve
Common carotid artery
Recurrent laryngeal nerve
Ansa subclavia

FIG. 11.83. Distribution of branches of sympathetic trunk and ganglia in the neck.

Branches accompanying nerves

In addition to the grey rami communicantes to the upper four cervical ventral rami and the branches of the internal and external carotid plexuses to cranial nerves, the superior cervical ganglion gives a jugular branch to the glossopharyngeal and vagus nerves, and sends other postganglionic fibres to the hypoglossal and accessory nerves.

Visceral branches

The laryngopharyngeal branch passes anteromedially, deep to the carotid sheath to join the pharyngeal plexus of the glossopharyngeal and vagus nerves, and innervate glands and blood vessels in the pharynx and larynx. For the superior cervical cardiac nerve see below.

FROM THE MIDDLE CERVICAL GANGLION. Grey rami communicantes to the fifth and sixth cervical ventral rami.

Branches accompanying arteries

Medial branches of the ganglion form a plexus around the inferior thyroid artery. This supplies that artery and the gland, gives branches to the recurrent and external laryngeal nerves, and communicates with the superior cervical cardiac nerve. The ansa subclavia, which joins the ganglion to the cervicothoracic ganglion, gives branches to the subclavian artery (subclavian plexus), but these are probably derived mainly from the cerviothoracic ganglion.

Visceral branch

The middle cervical cardiac nerve; see below.

FROM THE CERVICOTHORACIC GANGLION. Grey rami communicantes to the last two cervical and the first thoracic ventral rami.

Branches accompanying arteries

(1) Several large branches (vertebral nerves) from the superior part of the ganglion (and from the ganglion vertebrale) pass with the vertebral artery (vertebral plexus). This plexus supplies the intracranial and extracranial branches of the vertebral artery, and gives postganglionic fibres to some of the cervical spinal nerves (usually 4th–6th). (2) Postganglionic fibres in the ansa subclavia pass on to the proximal part of the subclavian artery (subclavian plexus). This is reinforced in the axilla and further distally in the limb by postganglionic fibres which reach the arteries through adjacent peripheral nerves.

Visceral branch

The inferior cervical cardiac nerve; see below.

CARDIAC NERVES

There are six cervical cardiac (splanchnic) nerves, one arising from each of the three cervical sympathetic gangiia (or the interganglionic rami) on both sides. All of these nerves end in the deep part of the cardiac plexus except the left superior which enters the superficial part of the plexus after passing lateral to the arch of the aorta.

The superior cervical cardiac nerve descends from the superior ganglion in the posterior wall of the carotid sheath, and usually passes anterior to the inferior thyroid artery. *On the right*, in common with the middle nerve of that side, it lies either anterior or posterior to the subclavian artery, and descends behind the brachiocephalic trunk to reach the posterior surface of the arch of the aorta, anterior to the tracheal bifurcation. In the neck, both have connections with the external and recurrent laryngeal nerves, the vagus (superior cardiac branches), and the plexus on the inferior thyroid artery. *On the left*, the superior nerve descends with the common carotid artery,

and usually passing lateral to the arch of the aorta, enters the superficial part of the cardiac plexus inferior to the arch.

The middle cervical cardiac nerve passes inferomedially. On the right, it runs with the superior cervical cardiac nerve; on the left, it unites with the inferior nerve, and passes medial to the aortic arch, close to the recurrent laryngeal nerve.

The inferior cervical cardiac nerve passes antero-inferiorly; on the right it is posterior to the subclavian and brachiocephalic arteries; on the left it joins the middle nerve. Both communicate with the corresponding recurrent laryngeal nerve.

With the possible exception of the superior cervical cardiac nerves, all the cardiac nerves contain visceral sensory fibres in addition to preganglionic and postganglionic sympathetic fibres. For the arrangement of the cardiac plexus, see page 822.

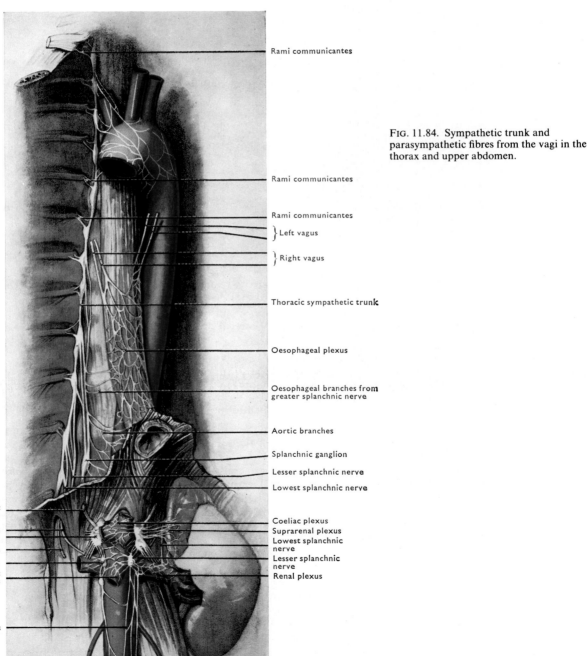

FIG. 11.84. Sympathetic trunk and parasympathetic fibres from the vagi in the thorax and upper abdomen.

Rami communicantes

Rami communicantes

Rami communicantes

Left vagus

Right vagus

Thoracic sympathetic trunk

Oesophageal plexus

Oesophageal branches from greater splanchnic nerve

Aortic branches

Splanchnic ganglion

Lesser splanchnic nerve

Lowest splanchnic nerve

Coeliac plexus

Suprarenal plexus

Lowest splanchnic nerve

Lesser splanchnic nerve

Renal plexus

Greater splanchnic nerve

Lesser splanchnic nerve
Coeliac ganglion

Lowest splanchnic nerve

Aorticorenal ganglion

Superior mesenteric plexus

Aortic plexus

THORACIC PART

Grey rami communicantes pass to every thoracic ventral ramus. Not infrequently a single ventral ramus may receive a grey ramus from more than one ganglion, and those that pass to the first (and second) thoracic ventral rami send postganglionic fibres into the brachial plexus. These are important in surgical operations designed to remove the sympathetic supply from the upper limb.

Branches accompanying blood vessels

Fine branches pass anteriorly from the upper five thoracic ganglia to form a plexus around the aorta [FIG. 11.84] and pulmonary trunk. These consist of postganglionic and visceral sensory fibres.
Visceral branches. 1. Several fine, **pulmonary** and **cardiac** branches pass anteromedially from the second to fourth thoracic ganglia to

join the corresponding plexuses. Those passing to the heart carry both efferent and afferent (pain) fibres (Mitchell 1956).
2. **The splanchnic nerves.** These arise from the fifth and subsequent thoracic ganglia. They pass antero-inferiorly on the thoracic vertebral bodies and consist of a mixture of preganglionic, postganglionic, and visceral sensory fibres, and scattered ganglion cells. The preganglionic fibres end partly on these cells, but principally on the ganglion cells in the prevertebral plexuses of the abdomen to which the splanchnic nerves are passing.

(i) The **greater splanchnic nerve** arises by branches from the fifth to ninth thoracic ganglia of the sympathetic trunk [FIGS. 11.81 and 11.84]. These unite as the nerve descends to pierce the crus of the diaphragm and enter the **coeliac ganglion** in the abdomen. It frequently carries a small, **splanchnic ganglion**, approximately at the level of the eleventh thoracic ganglion, and the nerve sends branches (postganglionic) to join the **oesophageal** and **aortic plexuses** in the thorax [FIG. 11.84].

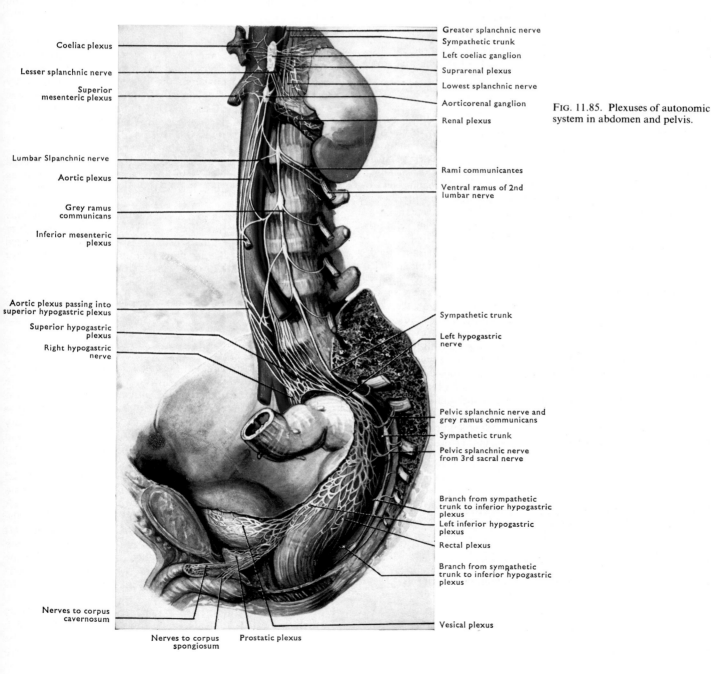

FIG. 11.85. Plexuses of autonomic system in abdomen and pelvis.

Coeliac plexus
Lesser splanchnic nerve
Superior mesenteric plexus
Lumbar Slpanchnic nerve
Aortic plexus
Grey ramus communicans
Inferior mesenteric plexus
Aortic plexus passing into superior hypogastric plexus
Superior hypogastric plexus
Right hypogastric nerve
Nerves to corpus cavernosum
Nerves to corpus spongiosum
Prostatic plexus

Greater splanchnic nerve
Sympathetic trunk
Left coeliac ganglion
Suprarenal plexus
Lowest splanchnic nerve
Aorticorenal ganglion
Renal plexus
Rami communicantes
Ventral ramus of 2nd lumbar nerve
Sympathetic trunk
Left hypogastric nerve
Pelvic splanchnic nerve and grey ramus communicans
Sympathetic trunk
Pelvic splanchnic nerve from 3rd sacral nerve
Branch from sympathetic trunk to inferior hypogastric plexus
Left inferior hypogastric plexus
Rectal plexus
Branch from sympathetic trunk to inferior hypogastric plexus
Vesical plexus

(ii) The lesser splanchnic nerve arises in the region of the ninth and tenth ganglia, and descends between the trunk and the greater splanchnic nerve. It pierces the diaphragm close to the greater splanchnic nerve, and enters the coeliac ganglion.

(iii) The lowest splanchnic nerve, if present, arises from the last thoracic ganglion, or from the lesser splanchnic nerve. It pierces the diaphragm lateral to the lesser splanchnic nerve, and enters the renal plexus.

LUMBAR PART

Grey rami communicantes pass to every lumbar ventral ramus. These long rami pass backwards medial to psoas, and may divide to pass to adjacent ventral rami, or one ventral ramus may receive more than one grey ramus from different levels of the trunk. They also transmit visceral sensory fibres to the ventral rami.

Branches accompanying blood vessels

Small branches pass on the lumbar arteries to the lumbar part of the abdominal aortic plexus. This is continuous with the thoracic aortic plexus, and is continued as plexuses around the common iliac and median sacral arteries and their branches.

Visceral branches

These lumbar splanchnic nerves arise from all the lumbar ganglia of the sympathetic trunk [FIG. 11.85]. They run into the abdominal aortic plexus and its continuation, the superior hypogastric plexus. Those from the upper two ganglia frequently unite on the aorta near the origin of the inferior mesenteric artery and send branches on to that artery, (inferior mesenteric plexus). They descend, anterior to the common iliac vessels, and are joined by the lumbar splanchnic nerve from the third ganglion, which may pass between the common iliac artery and vein. That from the fourth ganglion, passes deep to both vessels. The four nerves of both sides form a loose-meshed superior hypogastric plexus in the interval between the right and left common iliac vessels, anterior to the fifth lumbar vertebra.

The superior hypogastric plexus (presacral nerve) descends into the pelvis dividing into right and left hypogastric nerves, through which it is distributed to the inferior hypogastric (pelvic) plexuses. Thus the lumbar splanchnic nerves are predominantly distributed to the pelvic viscera, the only exceptions being their fibres which pass with the abdominal branches of the inferior mesenteric artery, and any filaments which they supply to the abdominal aortic plexus. Structurally the lumbar splanchnic nerves are similar to the cervical and thoracic splanchnic nerves and form a continuous series with them. Branches from the coeliac and renal ganglia, which receive their preganglionic fibres from the thoracic splanchnic nerves, pass to join the lumbar splanchnic nerves on the aorta [FIG. 11.86].

PELVIC PART

Grey rami communicantes pass irregularly from the pelvic part of the trunk or its ganglia to the ventral rami of the sacral nerves and to the coccygeal nerve.

Branches accompanying blood vessels

Small branches pass to the median and lateral sacral arteries.

Visceral branches

Two or three small sacral splanchnic nerves pass from the second and third ganglia to the inferior hypogastric plexus. Direct branches to the rectum and ureter have been described.

THE SYMPATHETIC PLEXUSES

The sympathetic plexuses are all of the same type but they differ from each other in their distribution, size, and complexity. From an early stage in its development [FIG. 10.16] the sympathetic nervous system is related to the arteries of the body, and every part of the body receives postganglionic sympathetic fibres through perivascular plexuses surrounding the arteries. In their simplest form (e.g. in the body wall and limbs) these plexuses consist of postganglionic fibres which pass on to the arteries (e.g. external carotid, vertebral, subclavian, intercostal, lumbar, and iliac) from the sympathetic trunk and extend along their branches. These are supplemented by postganglionic fibres from adjacent branches of the cranial or spinal nerves which also receive postganglionic fibres from the sympathetic trunk and distribute them either directly to the tissues or through the perivascular plexuses. The perivascular plexuses of the body wall and limbs also contain afferent fibres, but rarely any sympathetic ganglion cells (but see internal carotid plexus) and no parasympathetic fibres.

The prevertebral plexuses are concerned with the innervation of the thoracic, abdominal, and pelvic viscera. They are also closely associated with the blood vessels supplying these viscera, but they consist of a mixture of preganglionic, postganglionic, and sensory nerve fibres with many ganglion cells scattered among them, some of which are aggregated into large, readily visible ganglia. In addition, they contain parasympathetic fibres and cells, and it is through these plexuses that both parts of the autonomic nervous system are distributed to the viscera.

A number of separate plexuses are described, but they form what is virtually a continuous mass lying on the aorta and extending along its visceral branches from the cardiac plexus at the tracheal bifurcation to the hypogastric plexuses which extend beyond the aortic bifurcation into the pelvis. The parts of this mass are only differentiated by the territories to which their postganglionic fibres pass. Indeed the most peripheral part of this system, the gangliated plexuses in the wall of the gut tube, is continuous from the upper part of the oesophagus to the anal canal.

Thoracic plexuses

CARDIAC PLEXUS

The various cardiac branches of the vagus and sympathetic enter a large, gangliated cardiac plexus, which lies on the anterior surface of the lower part of the trachea and its bifurcation, behind the great vessels emerging from the base of the heart. Inferolaterally it is continuous along the principal bronchi with the pulmonary plexuses, and it extends antero-inferiorly around the aorta and pulmonary trunk to the coronary vessels of the heart. To the left, it extends into the concavity of the arch of the aorta, anterior to the ligamentum arteriosum. This part, frequently called the superficial cardiac plexus, receives the left superior cervical cardiac nerve (sympathetic) and the inferior cervical cardiac branch of the left vagus (parasympathetic). It is principally distributed along the pulmonary trunk to the right coronary plexus, and to the left pulmonary plexus. The deeper part, between the tracheal bifurcation and the aortic arch, receives all the other cardiac branches of the vagus and sympathetic of both sides, and is distributed to the heart partly along the aorta and pulmonary trunk, and partly along the superior vena cava (sinu-atrial node, etc.) and the pulmonary veins.

Early in development, the arterial and venous ends of the heart are widely separated, and each has its own mesocardium through which autonomic nerve fibres and cells reach these two ends of the heart separately. As the heart folds, the venous and arterial ends are approximated, thus bringing together both parts of the autonomic supply into a single mass (the cardiac plexus). These two parts cannot be differentiated in the adult heart, though the atrial part of the plexus passes predominantly with the great veins, the ventricular part with the great arteries. The folding of the heart [FIG. 13.113] and the asymmetrical development of the great veins [FIG. 13.126] and hence of the sinus venosus into which they discharge, leads to the left horn of that organ (with its part of the nodal tissue of the heart—atrioventricular node) becoming the coronary sinus, while the right horn (with its nodal tissue, sinu-atrial node) forms the right atrium adjacent to the superior vena cava. Thus the corresponding parts of the nodal tissue lie adjacent to the entry of these two veins, and while the atrioventricular node is supplied with nerves predominantly from the left side, the sinu-atrial node is supplied predominantly from the right.

The cardiac plexus, like the remainder of the prevertebral plexuses, is a mixture of sympathetic, parasympathetic, and sensory fibres. The ganglion cells in the plexus are predominantly parasympathetic (inhibitor) as are other ganglion cells which are found in the subepicardial tissues of the atria and ventricles and in the myocardium, principally with the conducting tissue. These ganglion cells are mixed in type (multipolar, bipolar, and unipolar) and it has been suggested that there may be sensory cells among them (Davies, Francis, and King 1952) though a similar variability of cells is found in the ganglia of the sympathetic trunk.

The coronary plexuses follow the right and left coronary arteries and are extensions of the cardiac plexus. Both sympathetic and parasympathetic fibres and ganglion cells are present; the sympathetic to dilate the coronary vessels, while the parasympathetic may constrict them. Sensory fibres are present with these and with the nerve fibres which innervate the subendocardial tissue (Stotler and McMahon 1947), though those which pass to the nodal tissue and follow the conducting tissue may be purely efferent.

PULMONARY PLEXUS

This plexus surrounds the root of each lung but is principally posterior to it. The plexus consists mainly of parasympathetic fibres derived from the vagus nerve [p. 763], but it also receives branches from the second, third, and fourth thoracic ganglia of the sympathetic trunk, which pass mainly to the posterior part. On each side, the plexus is continuous with the corresponding part of the cardiac plexus, and communicates with its fellow. The pulmonary plexuses are distributed to the lung (e.g. smooth muscle of the bronchial tree), its blood vessels, and the pulmonary pleura (Larsell 1922). Here sympathetic fibres dilate the bronchi, while parasympathetic (vagal) fibres constrict them.

OESOPHAGEAL PLEXUS

This plexus is formed principally by the vagus nerves [p. 763] and extends inferiorly from the pulmonary plexus. It is mainly a parasympathetic plexus, but it receives sympathetic fibres from the greater splanchnic nerves.

Abdominopelvic plexuses

This great mass of autonomic plexus lies principally in association with the abdominal aorta and the internal iliac arteries. Several parts corresponding to the visceral branches of these arteries are recognized, but they are continuous with each other, with the

splanchnic nerves, the oesophageal plexus in the thorax, and with the extensions of this complex along the visceral branches of the arteries. Major ganglia tend to be associated with the aorta and the internal iliac arteries, but smaller groups and scattered cells are found throughout the system.

In general the abdominal plexuses receive the splanchnic nerves from the thoracic part of the sympathetic trunk and vagal branches which enter the abdomen with the oesophagus from the oesophageal plexus. The hypogastric (pelvic) plexuses receive fibres from the sympathetic trunk through the lumbar splanchnic nerves, and parasympathetic fibres through the pelvic splanchnic nerves. The sacral splanchnic nerves are small and unimportant, but the pelvic splanchnic nerves (parasympathetic) also extend into the abdomen to innervate the descending and sigmoid parts of the colon.

THE COELIAC, SUPERIOR MESENTERIC, AND RENAL PLEXUSES

This complex is the largest of the prevertebral plexuses. It lies on the aorta at the level of the twelfth thoracic and first lumbar vertebrae, in close association with the origins of the coeliac, superior mesenteric, and renal arteries from the aorta. Minute ganglia are scattered throughout this plexus and its extensions along the aortic branches, but certain large ganglia are also present.

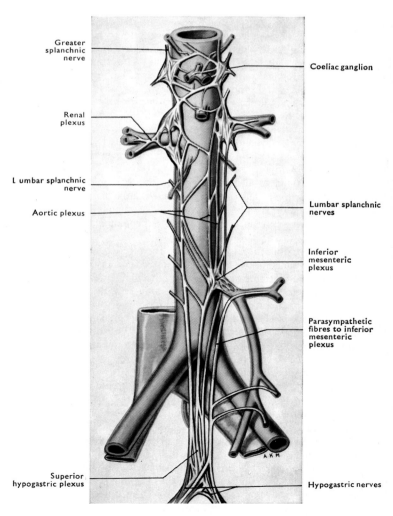

FIG. 11.86. The aortic and superior hypogastric plexuses.

The large coeliac ganglia flank the aorta at the level of origin of the coeliac artery. Each lies on the corresponding crus of the diaphragm; the right overlapped by the inferior vena cava, while the left lies posterior to the omental bursa superior to the pancreas.

The aorticorenal ganglia lie at the level of the renal arteries and may be partly posterior to them. They are often continuous with the coeliac ganglia, and are always linked by many nerve strands.

The superior mesenteric ganglia lie on the aorta around the origin of the superior mesenteric artery. They may form a single, lobulated mass, or a number of smaller ganglia.

All these ganglia are united by numerous nerve strands many of which pass from side to side across the aorta.

Nerves to the plexus

The greater splanchnic nerve pierces the crus of the diaphragm and enters the superior pole of the coeliac ganglion. Many of its fibres end here, but others pass through the ganglion into the branches of the plexus and to the medulla of the suprarenal gland. The lesser splanchnic nerve pierces the diaphragm lateral to the greater, and is distributed to all the named ganglia of the complex. The lowest splanchnic nerve, if present, joins the aorticorenal ganglion or the adjacent plexus.

The vagal trunks enter the abdomen with the oesophagus. They send branches along the left gastric artery to the coeliac plexus for distribution through the parts of that plexus, including the territory of the superior mesenteric artery.

Branches of the plexus

These consist of the same mixture of preganglionic parasympathetic, preganglionic and postganglionic sympathetic, and sensory nerve fibres intermingled with ganglion cells. The suprarenal plexus consists of preganglionic sympathetic fibres. The majority of these branches follow the branches of the aorta and receive their names from them. They are concerned with the control of blood vessels, glands, and the smooth muscle of the viscera to which the blood vessels are distributed; and they transmit afferent fibres from them. The arrangement of these plexuses is sufficiently irregular to make a detailed gross description valueless, but the general arrangement is indicated below.

1. Extensions of the plexus along the coeliac artery and its branches form: the left gastric plexus to stomach and oesophagus; the hepatic plexus to liver, gall-bladder, stomach, duodenum, and pancreas; the splenic plexus supplies the pancreas, spleen, and stomach.

2. The phrenic plexus passes with the inferior phrenic artery to the diaphragm, suprarenal plexus, inferior vena cava, and oesophagus. Some of its fibres unite with the phrenic nerve on the inferior surface of the diaphragm, and there may be a phrenic ganglion where it joins that nerve on the right side.

3. The suprarenal plexus arises from the coeliac plexus and from direct branches of the splanchnic nerves. The fibres are mostly preganglionic to the cells of the suprarenal medulla, which are developmentally equivalent to ganglionic neurons of the sympathetic system, and like them, secrete biologically active amines, mostly adrenaline (Elliott 1913; Young 1939; MacFarland and Davenport 1941).

4. The renal plexus is derived from the aorticorenal ganglion, the lesser and lowest splanchnic nerves, and the aortic plexus (Mitchell 1935a). It extends along the renal arteries to the hilus of the kidney, communicates with the suprarenal plexus along the inferior suprarenal artery, and forms the upper part of the ureteric plexus, which contains few if any ganglion cells (Notley 1969), but has both cholinergic and adrenergic fibres in it. The latter plexus extends along the ureter, and receives further fibres along the various arteries to the ureter.

5. The testicular or ovarian plexuses are extensions from the beginning of the renal plexus. They accompany the corresponding arteries and supply respectively the spermatic cord, epididymis, and testis, or ovary, broad ligament, and uterine tube. The ovarian plexus communicates with the uterine plexus from the inferior hypogastric plexus.

6. The superior mesenteric plexus is an extension along the superior mesenteric artery and its branches. It carries sympathetic and parasympathetic (vagal) fibres to the small intestine, pancreas, caecum, appendix vermiformis, and the ascending and transverse parts of the colon. For the most part the parasympathetic fibres synapse with the intramural plexuses (myenteric and submucous) in the gut tube.

Sympathetic strands extend inferiorly from the renal and superior mesenteric plexuses, along the aorta (abdominal aortic and intermesenteric plexuses). This part is joined by the upper two lumbar splanchnic nerves [FIG. 11.85] and gives off the inferior mesenteric plexus along the artery of that name. It continues anterior to the aorta and its bifurcation on to the anterior surface of the left common iliac vein and fifth lumbar vertebra as the superior hypogastric plexus.

The inferior mesenteric plexus is composed of sympathetic strands from the intermesenteric plexus in which are fibres from the upper lumbar splanchnic nerves. It receives parasympathetic fibres from the pelvic splanchnic nerves, though the route taken by these fibres is not known with certainty. It is claimed (Telford and Stopford 1934) that they ascend through the hypogastric plexuses [FIG. 11.86], but they may ascend from the pelvis along the branches of the inferior mesenteric artery. Ganglia are not usually visible on the inferior mesenteric artery; its plexus, mixed with scattered ganglion cells, is distributed to the descending and sigmoid parts of the colon, and the upper part of the rectum (superior rectal plexus).

SUPERIOR HYPOGASTRIC PLEXUS

This plexus is formed by the continuation of the abdominal aortic and intermesenteric plexuses to which are added the lower two lumbar splanchnic nerves in the interval between the two common iliac arteries. It consists of the same elements as the other parts of the autonomic plexuses, but the nerve fibres are predominantly sympathetic in origin, and join their parasympathetic counterparts in the lesser pelvis. It consists of a number of strands of gangliated plexus, and unites the abdominal and pelvic (inferior hypogastric plexuses) parts of the autonomic prevertebral plexuses. The superior hypogastric plexus is sometimes called the presacral nerve, and is occasionally surgically divided (presacral neurectomy) to relieve pain from pelvic viscera. The success of this procedure is evidence in favour of the passage of such sensory fibres by this route.

INFERIOR HYPOGASTRIC (PELVIC) PLEXUSES

In the lesser pelvis the superior hypogastric plexus splits right and left into a loose aggregation of nerve bundles, the hypogastric nerves. Each joins the corresponding inferior hypogastric plexus in the areolar tissue between the lateral wall of the pelvis and the rectum and adjacent parts of the bladder and seminal vesicle (male), or the rectum and vagina (female), and also in the sacrogenital fold.

The inferior hypogastric plexus receives fibres from: (1) the superior hypogastric plexus (sympathetic and sensory fibres); (2) the pelvic splanchnic nerves (sacral parasympathetic and sensory) from the third and fourth or second sacral ventral rami; (3) the small sacral splanchnic nerves from the second and third sacral ganglia of the sympathetic trunk.

Some of the parasympathetic fibres may run a recurrent course through the inferior and superior hypogastric plexuses to take part in the formation of the inferior mesenteric plexus (Telford and Stopford 1934; Mitchell 1935b).

Branches

As with other plexuses, the branches of the inferior hypogastric plexus pass with branches of the internal iliac artery to the pelvic viscera.

1. The **rectal plexus** is completed by plexuses along the middle and inferior rectal arteries (middle and inferior rectal plexuses) which mingle with the superior rectal plexus from the inferior mesenteric plexus. They supply rectum and anal canal and meet in the rectal wall.

2. The **vesical plexus** arises as a number of nerves from the anterior part of the inferior hypogastric plexus. They pass medially to surround the terminal part of the ureter and run on the walls of the bladder, the seminal vesicle, and of the ductus deferens, reaching as far as the epididymis. In the female, they pass medially in the base of the broad ligament, and giving branches to the uterus, pass forwards inferolateral to the cervix, below the uterine artery. Thus they reach the upper vagina and the base of the bladder through the vesicovaginal septum (Ballantyne and Smith 1968).

3. The **prostatic plexus** lies inferior to, and communicates with, the vesical plexus. It supplies the prostate, the structures that pierce it, and the seminal vesicle and membranous urethra. Sensory fibres from the pudendal nerve may enter this plexus.

4. The **uterine plexus** passes into the broad ligament with the uterine artery. It supplies all parts of the uterus and communicates with the ovarian plexus in supplying the uterine tube.

5. The **vaginal plexus** receives fibres which come mainly from the pelvic splanchnic nerves (parasympathetic). It supplies vagina and urethra, and together with the uterine plexus corresponds with the prostatic plexus in the male.

6. The **cavernous nerves.** Extensions of the prostatic plexus pass to the erectile tissue of the penis, and similar extensions of the vaginal plexus pass to the clitoris. Autonomic nerve fibres also reach the external genitalia through the pudendal nerve and along the internal pudendal artery.

INNERVATION OF THE URINARY BLADDER AND RECTUM

1. The **pudendal nerve** (S. (1), 2, 3, 4) supplies the voluntary muscle of the external anal and urethral sphincters which give voluntary control over these orifices. It gives sensory fibres to the mucosa of the anal canal and also to the urethra, probably including the prostatic part in the male which is sensitive to distension by urine entering it from the bladder.

2. **Sympathetic fibres** reach the rectum both by the inferior mesenteric and the inferior hypogastric plexuses, and the urinary bladder through the inferior hypogastric plexus. The preganglionic fibres emerge from the spinal medulla in the upper lumbar ventral rami.

3. **Parasympathetic fibres** reach the rectum and urinary bladder from the inferior hypogastric plexus. The preganglionic fibres enter this plexus by the pelvic splanchnic nerves.

Afferent fibres in the parasympathetic pathways seem to be associated with the normal sensations of bladder and rectal distension, those running with the sympathetic seem to convey pain from the pelvic organs, e.g. on over-distension.

The efferent parasympathetic fibres seem to be concerned with the contraction of urinary bladder and rectal muscle associated with micturition and defaecation, arising as a result of afferent impulses entering with the pelvic splanchnic nerves. The sympathetic efferents cause contraction of the internal anal and vesical sphincters, with relaxation of the muscular walls of these organs.

They play an important part in ejaculation for they cause contraction of the ductus deferentes, seminal vesicles, prostatic musculature, and internal vesical sphincter. The latter prevents the entry of urine into the prostatic urethra and the reflux of seminal fluid into the urinary bladder.

Innervation of blood vessels

All arteries, though not all veins (Lever, Spriggs, and Graham 1968) are accompanied by nerve fibres which form a plexus of varying density on them. These are postganglionic sympathetic and sensory fibres. Occasionally parasympathetic fibres appear to supply blood vessels; e.g. stimulation of the chorda tympani produces vasodilatation in salivary glands and the pelvic splanchnics in the erectile tissue of the external genitalia. For the most part the nervous control of vascular diameter depends on the sympathetic alone. Vasodilatation in the skin can be produced by stimulation of the distal end of cut dorsal roots, but it is not known to play any part in vasomotor control in normal circumstances.

The most extensive perivascular plexuses are the prevertebral plexuses, but only the minority of the fibres in these are concerned with control of the blood vessels. In the body wall and limbs, the perivascular plexuses are most extensive in the peripheral part of the limbs and exposed parts of the skin, particularly where there are **arteriovenous anastomoses** [FIG. 13.30] concerned with temperature control of such skin.

The vessels of the body wall and limbs receive a proximal plexus direct from the sympathetic trunk. This plexus is reinforced peripherally by branches from the nerves in the territory of each artery. These branches contain: (1) postganglionic sympathetic fibres which have entered the peripheral nerves through the grey rami communicantes; and (2) sensory nerve fibres.

Sensory fibres to blood vessels are widely distributed and are frequently myelinated. They do not usually give rise to conscious sensations, though pain may be appreciated from arteries and veins. The function of these nerve fibres is improperly understood except in certain regions, e.g. the pressure receptors (baroreceptors) in the aorta, carotid sinus, and great veins (glossopharyngeal and vagus nerves), and the chemoreceptors of the carotid body (glossopharyngeal nerve). The carotid body contains innervated nests of epithelioid cells in the walls of blood vessels. These cells seem to be sensitive to oxygen-lack and possibly also to excess of carbon dioxide in the blood. Similar structures found in relation to the heart and great vessels are innervated by the vagus nerve, and may also have a chemoreceptor function.

DEVELOPMENT OF THE AUTONOMIC NERVOUS SYSTEM

The cells of the autonomic ganglia and plexuses (sympathoblasts) migrate to their final position (1) from the neural crest and (2) from the neural tube (Raven 1937). The neural crest extends over the length of the neural tube and is concerned throughout with the formation of the sensory ganglia of the spinal and cranial nerves, while the formation of autonomic ganglion cells is limited mainly to the thoracolumbar (sympathetic) and craniosacral (parasympathetic) regions (Kuntz 1953).

In the thoracolumbar region the sympathoblasts which arise from the neural crest, separate from the ventral border of the primitive spinal ganglia and meet the cells which migrate outwards from the neural tube along the ventral roots. This mixture of cells, in which it is not possible to distinguish those of different origin, leaves the spinal nerves and migrates towards the aorta. Each group proliferates and fuses with adjacent groups to form an irregular, longitudinal rod of cells associated with the intersegmental body wall branches of the aorta. This primitive, thoracolumbar sympathetic trunk extends cranially and caudally by migration of cells from its two extremities, thus forming the definitive **sympathetic trunk**. With the growth of the embryo and the development of longitudinal fibres within the trunk, the rod breaks up into a number of separate ganglia united by interganglionic rami. This method of development accounts for the variable arrangement of the ganglia and of the grey rami communicantes which grow out from these cells to enter the ventral rami of all the spinal nerves.

Some cells continue to migrate ventromedially beyond the position of the sympathetic trunk [FIG. 11.75]. These come into relation first with the lateral, visceral branches of the aorta (suprarenal, renal, and gonadal) and then with the ventral, visceral branches of the aorta (coeliac, superior mesenteric, inferior mesenteric), and those of the internal iliac arteries. In association with all of these they form groups of cells of various sizes (ganglia), while other cells migrate along the arteries forming scattered cells and minor ganglia, even reaching the organs supplied by the arteries.

In this manner there is formed a scattering of cells between the ventral rami of the spinal nerves and the visceral organs, and within this are a certain number of larger ganglia, with smaller ganglia between them; e.g. the intermediate ganglia on rami communicantes, sympathetic trunk, and splanchnic nerves. The axon of each sympathoblast grows out along the line of the artery with which it is primarily associated, and is thus distributed to that artery and the other structures innervated by the sympathetic system within the territory of the vessel [FIG. 11.75].

Subsequently nerve cells in the grey matter of the spinal medulla (the future lateral grey horn) send processes through the ventral roots into the ventral rami (T. 1–L.3). These processes (preganglionic axons) leave the ventral rami, and follow all the migrating groups of sympathetic neuroblasts, thus reaching and synapsing with all of them. Thus some of the preganglionic axons end in the sympathetic trunk by synapsing with its cells at the level of entry, or at higher or lower levels by extending longitudinally within the trunk. Other preganglionic fibres pass straight through the trunk, or run longitudinally in it for some distance before leaving it to follow the further line of migration of sympathetic cells. These form the **splanchnic nerves**. These axons, predominantly preganglionic, synapse with the scattered and larger groups of ganglion cells along their route (coeliac, aorticorenal, etc.) and pass through or end in the prevertebral plexuses, where they are intermingled with the ganglion cells and their postganglionic axons. Thus the number of preganglionic axons steadily decreases from the sympathetic trunk to the peripheral distribution of the prevertebral plexuses, while the number of postganglionic axons steadily increases, only to diminish as they end on the organs which they supply.

Processes of the **spinal ganglion cells** grow outwards with the preganglionic fibres, and following the same routes, are distributed to the visceral organs which are innervated by the postganglionic fibres. They do not appear to develop any synaptic connections with the autonomic ganglion cells.

In the head region the autonomic (**parasympathetic**) cells are derived in a similar manner from the neural crest and neural tube. These cells migrate to their final destination along the pathway of one or other of the cranial nerves, including those through which the preganglionic fibres later pass to synapse with these cells. The ganglia formed by these cells lie close to the organs which they eventually innervate, and many of the cells are not aggregated into ganglia but lie in plexuses, often buried in the organs which they supply.

Many of the cells of the cranial outflow migrate over considerable distances and are responsible for the formation of most of the neurons in the **vagal plexuses** (cardiac, pulmonary, oesophageal,

and the submucous and myenteric plexuses of the abdomen as far as the left colic flexure). Within the head, some of the parasympathetic ganglion cells arise in association with the trigeminal ganglion, others migrate along the oculomotor, facial or glossopharyngeal nerves, which later transmit the preganglionic fibres to these ganglion cells.

GANGLION	ORIGIN OF GANGLION CELLS	ORIGIN AND COURSE OF PREGANGLIONIC FIBRES
Ciliary	From trigeminal via ophthalmic nerve; from oculomotor nerve	From accessory oculomotor nucleus via oculomotor nerve
Pterygopalatine	From trigeminal via maxillary nerve; from facial via greater petrosal nerve	From superior salivatory nucleus via facial nerve (greater petrosal nerve)
Submandibular	From trigeminal via mandibular nerve; from facial via chorda tympani	From superior salivatory nucleus via facial nerve (chorda tympani)
Otic	From trigeminal via mandibular nerve; from glossopharyngeal via lesser petrosal nerve	From inferior salivatory nucleus via glossopharyngeal (lesser petrosal nerve)

The origin of the sacral parasympathetic cells is not certain. It seems probable that they arise, in part at least, from the sacral parts of the neural crest and neural tube. They are mixed almost immediately with the thoracolumbar cells forming the inferior hypogastric plexus, and their identity is lost.

References

ADAMS, W. E. (1942). The blood supply of nerves, I. Historical review. *J. Anat.* (*Lond.*) **76**, 323.
—— (1943). The blood supply of nerves, II. The effects of exclusion of its regional sources of supply on the sciatic nerve of the rabbit. *J. Anat.* (*Lond.*) **77**, 243.
AXFORD, M. (1928). Some observations on the cervical sympathetic in Man. *J. Anat.* (*Lond.*) **62**, 301.
BALLANTYNE, B. and SMITH, P. H. (1968). Surgical anatomy of the extrinsic innervation of the human female urinary bladder. *J. Anat.* (*Lond.*) **103**, 199.
BOTAR, J. (1932). Etudes sur les rapports des rameaux communicants thoraco-lumbaires avec les nerfs viscéraux chez l'homme et chez l'animal. *Ann. Anat. path. norm. méd.-chir.* **9**, 88.
BOYD, J. D. and MONRO, P. A. G. (1949). Partial retention of autonomic function after paravertebral sympathectomy. *Lancet* **ii**, 892.
CLARK, W. E. Le Gros and WARWICK, R. T. T. (1946). The pattern of olfactory innervation. *J. Neurol. Neurosurg. Psychiat.* **9**, 101.
COWAN, W. M. and POWELL, T. P. S. (1963). Centrifugal fibres in the avian visual system. *Proc. R. Soc. B* **158**, 232.
COWLEY, R. A. and YEAGER, G. H. (1949). Anatomic observations on the lumbar sympathetic system. *Surgery* **25**, 880.
DALE, H. (1934). Pharmacology and nerve-endings (Dixon Memorial Lecture). *Proc. R. Soc. Med.* **28**, 319.
—— (1938). Chemical agents transmitting nervous excitation. *Irish J. Med. Sci.* **150**, 245.
DAVIES, F. and LAIRD, M. (1948). The supinator muscle and the deep radial (posterior interosseous) nerve. *Anat. Rec.* **101**, 243.
—— FRANCIS, E. T. B. and KING, T. S. (1952). Neurological studies of the cardiac ventricles of mammals. *J. Anat.* (*Lond.*) **86**, 130.
DAVIES, L. and HAVEN, H. A. (1933). Surgical anatomy of the sensory root of the trigeminal nerve. *Arch. Neurol. Psychiat.* (*Chic.*) **29**, I.
DURWARD, A. (1948). The blood supply of nerves. *Postgrad. med. J.* **24**, 11.
EISLER, P. (1891). Der Plexus lumbosacralis des Menschen. *Anat. Anz.* **6**, 274.
ELLIOTT, T. R. (1905). The action of adrenalin. *J. Physiol.* (*Lond.*) **32**, 401.
—— (1913). The innervation of the adrenal glands. *J. Physiol.* (*Lond.*) **46**, 285.
FITZGERALD, M. T. J. and ALEXANDER, R. W. (1969). The intramuscular ganglia of the cat's tongue. *J. Anat.* (*Lond.*) **105**, 27.

FOERSTER, O. (1933). The dermatomes in Man. *Brain* **56**, 1.

GAMBLE, H. J. and EAMES, R. A. (1964). An electron microscope study of the connective tissues of human peripheral nerve. *J. Anat. (Lond.)* **98**, 655.

HARRIS, W. (1904). The true form of the brachial plexus and its motor distribution. *J. Anat. Physiol.* **38**, 399.

—— (1939). *The morphology of the brachial plexus*. Oxford University Press, London.

HARRISON, R. G. (1924). Neuroblast versus sheath cell in the development of peripheral nerves. *J. comp. Neurol.* **37**, 123.

HEAD, H. (1893). On disturbances of sensation with especial reference to the pain of visceral disease. *Brain* **16**, I.

HERRINGHAM, W. P. (1886). The minute anatomy of the brachial plexus. *Proc. R. Soc. B* **41**, 423.

HORSTADIUS, S. (1950). *The neural crest*. Oxford University Press, London.

HOVELACQUE, A. (1927). *Anatomie des nerfs craniens et rachidiens et du système grand sympathique chez l'homme*. Paris.

JONES, F. W. (1910). On the relation of the limb plexuses to the ribs and vertebral column. *J. Anat. Physiol.* **44**, 377.

—— (1939). The anterior superior alveolar nerve and vessels. *J. Anat. (Lond.)* **73**, 583.

KEEGAN, J. J. and GARRETT, F. D. (1948). The segmental distribution of the cutaneous nerves in the limbs of Man. *Anat. Rec.* **102**, 409.

KERR, A. T. (1918). The brachial plexus of nerves in Man. The variations in its formation and branches. *Am. J. Anat.* **23**, 285.

KOSINSKI, C. (1926). The course, mutual relations and distribution of the cutaneous nerves of the metazonal region of leg and foot. *J. Anat. (Lond.)* **60**, 274.

KUNTZ, A. (1953). *The autonomic nervous system*, 4th edn. Philadelphia.

—— and RICHINS, C. A. (1946). Components and distribution of the nerves of the parotid and submandibular glands. *J. comp. Neurol.* **85**, 21.

KURE, K. (1931). *Über den Spinal-parasympathikus*. Basel.

LANGLEY, J. N. (1921). *The autonomic nervous system*. Heffer, Cambridge.

LARSELL, O. (1922). The ganglia, plexuses, and nerve-terminations of the mammalian lung and pleura pulmonalis. *J. comp. Neurol.* **35**, 97.

LEVER, J. D. (1976). Studies on sympathetic ganglia. *J. Anat. (Lond.)* **121**, 430.

—— SPRIGGS, T. L. B., and GRAHAM, J. D. P. (1968). A formol-fluorescence, fine-structure and autoradiographic study of the adrenergic innervation of the vascular tree in the intact and sympathectomised pancreas of the cat. *J. Anat. (Lond.)* **103**, 15.

LOEWI, O. (1935). Problems connected with the principle of humoral transmission of nervous impulses (Ferrier Lecture). *Proc. R. Soc. B* **118**, 299.

MACFARLAND, W. E. and DAVENPORT, H. A. (1941). Adrenal innervation. *J. comp. Neurol.* **75**, 219.

MAGOUN, H. W. and RANSON, M. (1942). The supraoptic decussations in the cat and monkey. *J. comp. Neurol.* **76**, 435.

MATTHEWS, M. R. and RAISMAN, G. (1969). The ultrastructure and somatic efferent synapses of small granule-containing cells in the superior cervical ganglion. *J. Anat. (Lond.)* **105**, 255.

MITCHELL, G. A. G. (1935a). The innervation of the kidney, ureter, testicle, and epididymis. *J. Anat. (Lond.)* **70**, 10.

—— (1935b). The innervation of the distal colon. *Edinb. med. J.* **42**, 11.

—— (1953). *Anatomy of the autonomic nervous system*. E. & S. Livingstone, Edinburgh.

—— (1956). *Cardiovascular innervation*. E. & S. Livingstone, Edinburgh.

MONRO, P. A. G. (1959). *Sympathectomy: an anatomical and physiological study*, Oxford University Press.

NATHAN, P. W. and TURNER, J. W. A. (1942). The efferent pathway for pupillary contraction. *Brain* **65**, 343.

NOTLEY, R. G. (1969). The innervation of the upper ureter in man and in the rat: an ultrastructural study. Ref. to ureteric plexus. *J. Anat. (Lond.)* **105**, 393.

NYBERG-HANSEN, R. (1956). Anatomical demonstration of gamma motoneurons in the cat's spinal cord. *Exp. Neurol.* **13**, 71.

PAINTAL A. S. (1977). Thoracic receptors connected with sensation. *Br. Med. Bull.* **33**, 169.

PICK, J. and SHEEHAN, D. (1946). Sympathetic rami in Man. *J. Anat. (Lond.)* **80**, 12.

QVIST, G. (1977). The course and relations of the left phrenic nerve in the neck. *J. Anat. Lond.* **124**, 803.

RAVEN, C. P. (1937). Experiments on the origin of the sheath cells and sympathetic neuroblasts in amphibia. *J. comp. Neurol.* **67**, 221.

REESE, T. S. (1965). Olfactory cilia in the frog. *J. Cell Biol.* **25**, 209.

SCHOFIELD, G. (1956). Observations on the distribution of extrinsic nerve fibres in the gut of the rat. *J. Anat. (Lond.)* **90**, 592.

SHANTHAVEERAPPA, T. R. and BOURNE, G. H. (1966). Perineurial epithelium: a new concept of its role in the integrity of the peripheral nervous system. *Science* **154**, 1464.

SHEEHAN, D. (1941). Spinal autonomic outflows in Man and monkey. *J. comp. Neurol.* **75**, 341.

—— MULHOLLAND, J. H., and SHAFIROFF, B. (1941). Surgical anatomy of the carotid sinus nerve. *Anat. Rec.* **80**, 431.

STARKIE, C. and STEWART, D. (1931). The intra-mandibular course of the inferior dental nerve. *J. Anat. (Lond.)* **65**, 319.

STEWART, D. and WILSON, S. L. (1928). Regional anaesthesia and innervation of teeth. *Lancet* **ii**, 809.

STOPFORD, J. S. B. (1918). The variation in the distribution of the cutaneous nerves of the hand and digits. *J. Anat. (Lond.)* **53**, 14.

—— (1930). *Sensation and the sensory pathway*. Longmans, London.

STOTLER, W. A. and MCMAHON, R. A. (1947). The innervation and structure of the conducting system of the human heart. *J. comp. Neurol.* **87**, 57.

SUNDERLAND, S. (1945a). Blood supply of the nerves of the upper limb in Man. *Arch. Neurol. Psychiat. (Chic.)* **53**, 91.

—— (1945b). Blood supply of peripheral nerves. *Arch. Neurol. Psychiat. (Chic.)* **54**, 280.

—— (1945c). Blood supply of the sciatic nerve and its popliteal divisions in Man. *Arch. Neurol. Psychiat. (Chic.)* **54**, 283.

—— (1946). The innervation of the first dorsal interosseous muscle of the hand. *Anat. Rec.* **95**, 7.

—— (1968). *Nerves and nerve injuries*. E. & S. Livingstone, Edinburgh.

—— and HUGHES, E. S. R. (1946). The pupilloconstrictor pathway and the nerves to the ocular muscles in Man. *Brain* **69**, 301.

TELFORD, E. D. and STOPFORD, J. S. B. (1934). The autonomic nerve supply of the distal colon. *Br. med. J.* **1**, 572.

TRAQUAIR, H. M. (1949). *An introduction to clinical perimetry*, 6th edn. London.

WESTWATER, L. A. (1960). The innervation of the pulps of the teeth. *Br. dent. J.* **109**, 407.

WILSON, W. C. (1936). Observations relating to the innervation of the sweat glands of the face. *Clin. Sci.* **2**, 273.

WINDLE, W. F. (1931). Neurons of the sensory type in the ventral roots of Man and of other mammals. *Arch. Neurol. Psychiat. (Chic.)* **26**, 791.

WOLFF, E. (1953). The so-called medial root of the optic tract is essentially a visual commissure. *Brain* **76**, 455.

WOOLLARD, H. H. and HARPMAN, A. (1939). The cortical projection of the medial geniculate body. *J. Neurol. Psychiat.* **2**, 35.

WRETE, M. (1935). Die Entwicklung der intermediären Ganglien beim Menschen. *Morph. Jb.* **75**, 229.

YOUNG, J. Z. (1939). Partial degeneration of the nerve supply of the adrenal. A study in autonomic innervation. *J. Anat. (Lond.)* **73**, 540.

12 The skin and the sensory organs

G. J. ROMANES

THE sense organs consist of the apparatus by which the individual is made aware of changes in the internal and external environments. These changes are monitored by special receptive tissues which may be highly developed, localized organs such as the eye and the internal ear, or may consist of structures of varying complexity scattered over the surface and in the deeper parts of the body. All of them receive the terminal parts of sensory nerve fibres through which impulses are transmitted to the central nervous system when appropriate stimuli are applied to the end-organ. The skin, because of its function as the largest single sensory organ, is included in this section.

The functions of many of the receptive end-organs is not known, but they are divided into the organs of the **special senses**, sight, hearing, equilibration, smell, and taste, and those concerned with **general sensations**, e.g. touch, pressure, pain, temperature, position

(joint sense), and others arising in muscles and tendons some of which do not appear to evoke conscious sensation.

Others which fail to reach consciousness in normal circumstances arise in the numerous sensory endings in the viscera. These include pressure receptors in arteries and veins, chemoreceptors in the carotid body and the brain itself, and others concerned with the continuous monitoring of visceral activity.

The skin

The skin is the largest single organ in the body. Apart from its important sensory function, it is a waterproof layer concerned with

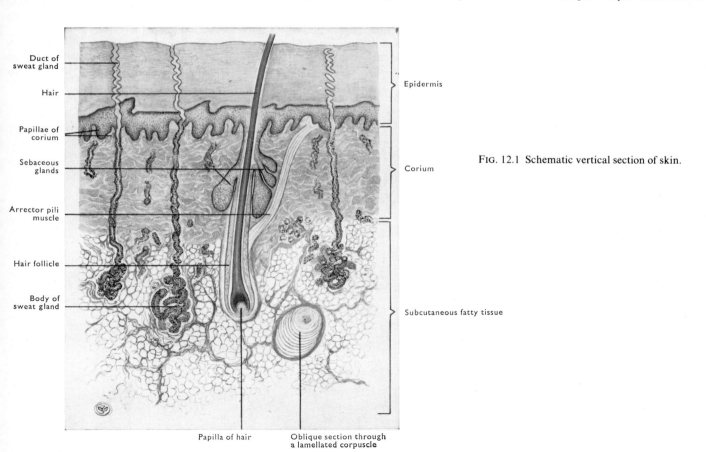

Duct of sweat gland

Hair

Papillae of corium

Sebaceous glands

Arrector pili muscle

Hair follicle

Body of sweat gland

Epidermis

Corium

Subcutaneous fatty tissue

FIG. 12.1 Schematic vertical section of skin.

Papilla of hair

Oblique section through a lamellated corpuscle

the prevention of fluid loss; a temperature regulating organ, both because of the variability of its vascular supply and the presence of sweat glands; an excretory organ for certain crystalloids in the sweat; and a secretory organ by means of its sebaceous glands whose fatty secretion not only helps to maintain the waterproofing but also may be acted on by sunlight to produce vitamin D. The skin can react to continued friction by increasing the thickness of its superficial layers, to sunlight and heat by increased pigmentation, and to wounds by increased growth and repair. It contains hairs over most of the surface, though these are more concerned with sensation than heat control in Man, and it produces the nails. Despite its waterproof nature, certain substances can be absorbed through the skin, particularly those which are fat soluble, and the amounts absorbed may be sufficient to cause acute poisoning particularly if a large area of skin is involved.

The skin is very **elastic**, though decreasingly so with age. Unless firmly adherent to the underlying tissues, it stretches readily, but most freely in one direction because of the tendency of the **collagen** in its deeper layers to run predominantly at right angles to that direction, i.e. parallel to the minute intercommunicating grooves which are present on most of the skin surface. When pierced by a cylindrical spike, it splits along the major direction of the collagen thus giving rise to a linear wound (Jones 1941). The colour of the skin depends on the presence of **pigment** (melanin) particularly in the deeper layers of the epidermis, and especially in certain regions, e.g. exposed skin, skin of the external genitalia, perianal region, axillae, and mammary areolae. There are also carotenoids in the horny layer, and in the fat of the skin and subcutaneous tissues. Blood, containing either reduced or oxyhaemoglobin depending on the state of the circulation, also colours the skin. In darker skinned races, the melanin is distributed throughout the layers of the epidermis.

The surface of the skin is pierced by the orifices of the sweat glands and hair follicles. Sebaceous glands discharge into the superficial parts of the hair follicles, though some sebaceous glands are found in the absence of hairs in the mammary areolae, the coronal sulcus of the penis, the labia minora, and the lips. On the palmar surface of the hand and the plantar surface of the foot are numerous, permanent, parallel **epidermal ridges** the arrangement of which is characteristic for each individual, is even different in detail in identical twins (Penrose 1968), and is responsible for the finger prints.

In most parts of the body the skin is freely mobile on the subjacent tissues, especially where there are cutaneous muscles or subcutaneous bursae. It is bound firmly to the deep fascia on the palms, soles, and lateral aspects of the fingers, at the skin creases over joints, and on the subcutaneous parts of the tibia.

Structure [Figs. 12.1 and 12.2]

The skin consists of a superficial, cellular layer of ectodermal origin, the **epidermis**, and a deeper layer derived from the mesoderm, the **dermis** or corium.

EPIDERMIS

This layer of stratified squamous epithelium varies in thickness from 0.3 mm to 1 mm or more. It is thickest on the palms and soles and thinnest on the eyelids and penis. In thick skin it may be divided into a number of layers from within outwards; stratum basale, stratum spinosum, stratum granulosum, stratum lucidum, and the stratum corneum.

The **stratum basale** consists of a single layer of columnar cells. Their cytoplasm contains tonofibrils which pass towards the basement membrane uniting them to the dermis. It is in this layer and the stratum spinosum that new cells are produced to replace the cells constantly being worn away from the surface—a process which is stimulated by the removal of the superficial layers.

The **stratum spinosum** consists of several layers of irregularly shaped cells which tend to become more flattened as the stratum granulosum is approached. These cells are readily separated by

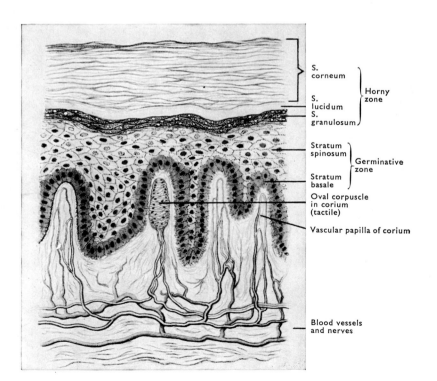

S. corneum
S. lucidum
S. granulosum
} Horny zone

Stratum spinosum
Stratum basale
} Germinative zone

Oval corpuscle in corium (tactile)

Vascular papilla of corium

Blood vessels and nerves

FIG. 12.2 Vertical section of epidermis and papillae of corium.

shrinkage, except where they are attached to each other by adhesion plaques or desmosomes; hence they have a spiny appearance in many histological preparations.

In the **stratum granulosum** the cells are further flattened, the cytoplasm contains many keratohyalin granules which are believed to be implicated in the formation of keratin, and the cells are in the process of dying.

Superficial to the stratum granulosum is the **stratum lucidum**, a transparent layer in which boundaries of the elongated flattened cells can only be seen with the electron microscope. Nuclei and intracellular organelles of the deeper layers have disappeared.

The **stratum corneum** consists of many layers of flattened, keratinized squames which are continually shed from the surface. This is the layer which is most responsible for the variations in thickness of the different regions of the skin.

In the deepest layers of the epidermis are **dendritic cells**, the branching processes of which pass superficially between the adjacent epithelial cells. These are the epidermal **melanocytes** which are responsible for the production of melanin and its transfer to the other cells of the epidermis. Clear Langerhans' cells with branching processes are also present in the stratum spinosum. They do not contain pigment but are thought either to be macrophages or possibly displaced dendritic cells.

The deep surface of the epidermis always interdigitates with irregularities of the surface of the dermis [FIG. 12.2]. The depth of the interdigitations varies with the degree of shearing force to which the skin is subjected, being deepest in the palms and soles. Here there are usually subsidiary epidermal ridges between the main extensions into the dermis, and it is through these smaller ridges that the ducts of the sweat glands pass to open on the surface epidermal ridges.

DERMIS

The basis of this is a felt of collagen and elastic fibres. The superficial, or **papillary layer**, interdigitates with the epidermis and is clearly demarcated from it. The deeper, **reticular layer** fades into the subcutaneous tissue without a clear line of demarcation. The dermis contains blood and lymph vessels, nerves, sebaceous glands, the superficial parts of the sweat glands and of the hair follicles, and the muscles of the hair follicles—the arrectores pilorum. Most of the cutaneous nerve endings lie in the dermis, especially its papillary layer, though some nerve fibres pass between the cells of the epidermis.

Vessels and nerves

From the arterial plexus in the subcutaneous tissue, branches pass into the reticular layer of the dermis to form a subsidiary plexus from which capillary loops enter the papillary layer. Veins and lymph vessels begin in the papillae of the dermis, pass to plexuses in the reticular layer, and then join the corresponding subcutaneous vessels. Most of the lymph vessels in the dermis are valveless. Within the dermis, especially in the exposed parts of the body, there are many **arteriovenous anastomoses** [FIG. 13.30]. These permit an increased flow of blood direct from the arterioles to the venules, thus increasing the heat supply to the superficial tissues (Lewis 1927; Popoff 1934) and promoting heat loss.

The skin is heavily innervated by **sensory nerve fibres** but it also receives many **postganglionic sympathetic fibres** for the supply of the blood vessels, the sweat glands, the arrectores pilorum, and the dartos muscle of the scrotum. The fibres that supply the sweat glands are cholingeric and do not release noradrenaline as do the fibres ending on the blood vessels.

Appendages of the skin

These are nails, hairs, sebaceous and sweat glands, and all are derived from the epidermis.

NAILS [FIGS. 12.3 and 12.4]

These consist of tightly packed, cornified cells which form a translucent plate, similar to but harder than the stratum lucidum. They are a special modification of the superficial epidermal layers, particularly of the stratum lucidum, but without a stratum granulosum. The visible part of the nail is its **body**; the hidden part, or **root**, from which growth of the nail principally occurs, is deeply buried in an invagination of the epidermis. The body and root both lie on the **nail bed** [FIG. 12.4], the epidermal surface of which is heavily cornified to form the nail (Clark and Buxton 1938).

The fold of skin which overlies the root of the nail is the **eponychium**, and its free, cornified margin forms the cuticle. A cornified layer under the free margin of the nail is the **hyponychium**. The downturned sides of the nail are partly buried by the **nail walls**. The epithelium of the nail bed consists only of stratum basale and spinosum, and is thrown into longitudinal folds corresponding to the longitudinal striations on the nail surface. The proximal part of this epithelium proliferates to form the nail over a region extending distally as far as the **lunule**, the white semilunar region which may extend beyond the eponychium.

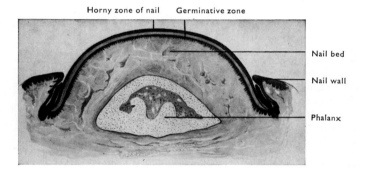

FIG. 12.3. Transverse section of nail.

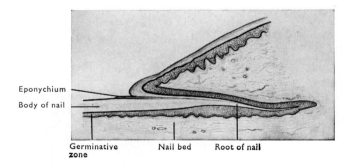

FIG. 12.4. Longitudinal section through root of nail.

HAIRS

These are present over the entire surface of the body except on the palms and soles, the sides of the fingers, feet and toes, the distal phalanges, the knuckles, the red parts of the lips, the glans and prepuce of the penis, the areolae and nipples, and the skin between

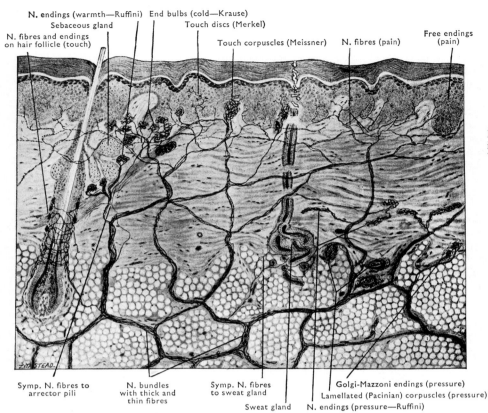

N. endings (warmth—Ruffini)   End bulbs (cold—Krause)
        Sebaceous gland   Touch discs (Merkel)
N. fibres and endings   Touch corpuscles (Meissner)   N. fibres (pain)   Free endings
on hair follicle (touch)   (pain)

FIG. 12.5. Composite diagram showing the innervation of the human skin. (Woollard, Weddell, and Harpman 1940.)

Symp. N. fibres to   N. bundles   Symp. N. fibres   Golgi-Mazzoni endings (pressure)
    arrector pili   with thick and   to sweat gland   Lamellated (Pacinian) corpuscles (pressure)
                           thin fibres   N. endings (pressure—Ruffini)
                                             Sweat gland

the labia majora. Most of the hairs are very fine (vellus hairs) so that the skin may appear hairless, but their numbers may be greater in some regions than in hairy mammals. There is a marked sexual differentiation in the distribution of coarse hairs, particularly on the face and trunk (Dupertuis, Atkinson, and Elftman 1945), and in its loss from the scalp. All hairs, except the eyelashes, emerge obliquely from the surface of the skin; the hairs in any one site point in the same direction forming a hair tract (Kidd 1903; Jones 1941; Clark 1971).

The root of the hair is ensheathed in a sleeve of epidermis called the **hair follicle** [FIG. 12.1] which extends into the subcutaneous tissue. The **shaft** is the free portion of the hair. Throughout most of its length the hair consists of the keratinized remains of cells. The outermost layer (**cuticle**) is formed of overlapping squames with their free edges pointing towards the tip of the hair. Internal to this is the **cortex** which forms the greater part of the hair. It consists of tightly packed hard keratin which contains pigment and a few small air spaces.

The **medulla** forms the core of the hair. It consists of soft keratin, may be broken into discontinuous pieces, and only forms a significant part in coarse hairs, though it is said to be present in all but the finest hairs (Montagna 1956). It also contains larger air

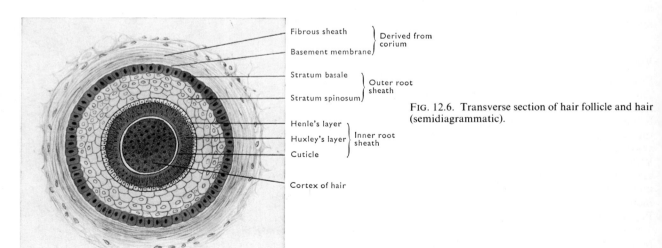

Fibrous sheath } Derived from
Basement membrane } corium

Stratum basale } Outer root
Stratum spinosum } sheath

Henle's layer } Inner root
Huxley's layer } sheath
Cuticle

Cortex of hair

FIG. 12.6. Transverse section of hair follicle and hair (semidiagrammatic).

spaces, fat globules and pigment, even in apparently white hairs.

In the growing hair, the deepest part of the hair follicle expands to form an onion-shaped cap (bulb of the hair) which almost encloses some loose, vascular connective tissue, the papilla [FIG. 12.1]. The cells of the follicle which surmount the papilla proliferate to form all three layers of the hair; the most central cells, which form the medulla, keratinize less rapidly than those forming the cortex, and hence retain their nuclei for a greater distance up the hair.

In the resting hair follicle, the bulb and papilla shrink and the deepest part of the follicle becomes irregular in shape.

The hair follicle extends obliquely through the skin into the superficial part of the subcutaneous tissue, and is spirally arranged in curly hair. It has a fibrous covering of outer longitudinal and inner circular layers. This is separated from the epidermal part of the follicle by a translucent layer which is continuous with the basement membrane of the epidermis. The epidermal part consists of inner and outer root sheaths, and expands laterally to form the sebaceous glands immediately below the papillary layer of the dermis, usually in the angle between the follicle and arrector pili muscle. Deep to the sebaceous glands only the two deeper layers of the epidermis can be identified, though all layers are visible superficial to them.

The outer root sheath is the thickest part of the follicle. It is continuous above with the stratum basale and spinosum of the epidermis, but is thinner in the depths of the follicle. The inner root sheath, which disappears in the region of the sebaceous gland consists of: (1) an inner thin cuticle composed of flattened, overlapping cells whose free edges face inwards and interlock with the squames of the hair cuticle, thus holding the hair in the follicle; (2) one or two layers of polyhedral, nucleated cells (Huxley's layer); (3) an outer (Henle's) single layer of nucleated, cubical cells.

ARRECTORES PILORUM MUSCLES [FIG. 12.5]

Bundles of smooth muscle fibres are attached to the fibrous sheath of the hair follicle deep to the sebaceous glands, and pass to the papillary layer of the dermis on the side towards which the exposed part of the hair slopes. These smooth muscle fibres are supplied by sympathetic postganglionic fibres in the cutaneous nerves. When they contract they erect the hairs, and pulling them slightly towards the surface of the skin, elevate the skin around the mouths of the hair follicles to produce the so-called gooseflesh. They also compress the sebaceous gland.

SEBACEOUS GLANDS

These glands are formed as diverticula of the superficial part of every hair follicle, but they also occur in the labia minora, the corona of the glans penis and the adjacent prepuce, and the mammary areolae in the absence of hairs. Those in the mammary areolae enlarge considerably in pregnancy. Each of the glands (1–4) associated with a hair follicle expands into a cluster of flask-shaped alveoli which lie in the angle between the shaft of the hair and the arrector pili muscle. The alveoli are filled with tightly packed cells containing oil droplets, and these disintegrate (holocrine secretion) to discharge their contents (sebum) into the hair follicle, or directly on to the surface where they are not associated with hairs. Disintegrating cells are replaced by proliferation of the neck cells of the alveoli, the daughter cells descending into the alveoli and gradually developing into mature sebaceous cells. The secretion waterproofs the hairs and skin, and helps to account for the relative ease of absorption of fat-soluble substances through the skin. Secretion of these glands is not thought to be affected by sympathectomy, but the skin becomes dry and scaly.

SWEAT GLANDS

These glands are found throughout the skin, but tend to be more numerous on its exposed parts, especially on the palms and soles where the ducts open on the summits of its epidermal ridges. Each gland consists of a long tube (duct) which may extend into the subcutaneous tissue, where it is coiled to form the secretory body of the gland, 0.1–0.5 mm in diameter [FIG. 12.1]. The duct takes a spiral course through the epidermis to open through a funnel-shaped sweat pore. Sweat is produced by the cells of the body of the gland without any loss of cytoplasm (eccrine secretion), except in the case of some large, apocrine sweat glands (1–4 mm in diameter) which tend to open into hair follicles in the axilla, scrotum, prepuce, labia minora, nipple, and perianal regions. In these glands the secretory cells have bulbous processes projecting into the lumen; these may be lost into the sweat (apocrine secretion).

The body of the gland is a tightly coiled tube with a single layer of pale-staining pyramidal, secretory cells which are partly separated from a thick basement membrane by flattened, branched cells which contain myofilaments similar to those in smooth muscle cells. These myo-epithelial cells are believed to assist with the expulsion of the secretion, but they are not found in the narrower duct which has two layers of darkly staining cells with many fine filaments in the internal cytoplasmic layer. Sweat glands are supplied by postganglionic, cholinergic, sympathetic nerve fibres. No sweating occurs after sympathetic denervation.

The ciliary glands which open at the margins of the eyelids are modified, uncoiled sweat glands, as also are the branched, apocrine ceruminous glands in the external acoustic meatus. The ducts of the latter glands open into hair follicles, and their cells contain a yellowish pigment which colours the wax secretion.

Development of the skin and its appendages

The dermis is developed from the mesoderm, the epidermis, nails, hair, sweat, sebaceous, and mammary glands from the ectoderm. By the first month of intra-uterine life the mesoderm forms a subectodermal layer which differentiates into two layers (dermis and subcutaneous tissue), by the third month. Dermal papillae appear in the fourth month.

The epidermis is a single layer of cells at first, but it differentiates into deep, cubical and superficial, flattened layers of cells, by the second intra-uterine month. Three layers are visible in the third month, the deeper polyhedral cells giving rise to the stratum basale and spinosum, while the flattened, superficial layer is known as the epitrichium because the growing hairs lie deep to it and push it off to form part of the greasy covering of the child at birth, the vernix caseosa. Formation of daughter cells by the deeper layers leads eventually to the production of the definitive epidermis of the skin (Montagna 1956).

Nails

These structures begin to develop in the third foetal month as a thickening of the epidermis on the dorsal surfaces of the terminal phalanges. This thickening is overgrown proximally and on each side by the surrounding epidermis to form the nail fold and walls. The nail thickening keratinizes without the appearance of a stratum granulosum to form the hard keratin of the nail, which grows in length by the proliferation of the cells of the stratum basale and spinosum of the nail bed under the nail fold and lunule (germinal matrix). Distal to this, the corresponding layers of the nail bed (sterile matrix) make no contribution to the nail. Nail growth is most rapid in the hand (3 mm per month) especially in well nourished boys, and particularly in the third digit (Clark and Buxton 1938; Gilchrist and Buxton 1939).

Hair

Solid downgrowths of the stratum basale of the epidermis begin to pass obliquely into the dermis in the third to fourth month of pregnancy. The deep extremity of each downgrowth expands to form the bulb, producing the germinal matrix over the dermal papilla. Growth of the cells of the matrix forms the hair and its inner root sheath; the outer root sheath is derived from the original downgrowth. Condensation of the surrounding mesoderm forms the fibrous sheath and vascular plexus of the follicle. The growing hairs appear at the surface about the fifth month, and pushing off the epitrichium, form the first crop of hairs, the lanugo. These very delicate hairs are fully formed by the seventh month, but are shed at birth or shortly afterwards; the eyelashes and scalp hairs, being the last to be shed, are replaced by stronger hairs.

The shedding and renewal of hairs continues throughout life, and periods of growth and quiescence alternate (Trotter 1924). In quiescence, proliferation of the germinal matrix ceases, the bulb and papilla shrink, the base of the hair becomes club-shaped, and eventually the deepest part of the follicle degenerates. Despite the absence of growth from below, the club hair is shed, and subsequently the outer root sheath cells proliferate to produce downgrowth of the follicle, reformation of the bulb and papilla, and regrowth of the hair.

Hair growth rates differ considerably in different parts of the body, but an average figure is probably 1.5–2 mm per week. This figure is greatly exceeded in areas which are shaved.

Sebaceous glands

Solid outgrowths of the outer root sheath cells of the hair follicles appear in the fifth foetal month. These enlarge distally forming a number of lobules. The sebaceous secretion plays a major part in the formation of vernix caseosa.

Sweat glands

These solid, yellowish downgrowths of the stratum basale of the epidermis pass through the dermis into the subcutaneous tissue. Here they become coiled to form the secretory part. They appear first on the palms and soles in the fourth month of pregnancy, but much later in the hairy regions. The ducts open on the surface during the seventh month.

Organs of general sensation

This group of sensory nerve endings is widely distributed throughout the body and is concerned with sensations such as pain, heat, cold, pressure, touch, and the afferent impulses which arise in muscles, tendons, and joints. These nerve endings may either be surrounded by capsules or form fine meshes of bare axons in the tissues (Cauna 1959).

Although many different varieties of encapsulated nerve endings have been described, most of them do not fall into clearly defined histological groups; attempts to correlate particular microscopic appearances with sensitivity to certain stimuli have not been outstandingly successful. Indeed the structure of nerve endings may alter with use, and some have been shown to increase in size and complexity with age while decreasing in number (Cauna 1964).

Encapsulated endings are found in many tissues, usually arranged in small groups. In the skin they are commonest in the hairless regions, and their relative absence in certain other regions is not associated with the inability to appreciate certain stimuli, though

the threshold may be higher. Thus it appears that highly differentiated sensory endings are not essential for the appreciation of most types of sensory stimuli, and though certain histological types of endings may be sensitive to particular stimuli, it does not follow that they are always concerned with producing the same kind of subjective sensation. Thus the known sensitivity of Pacinian corpuscles to deformation (Sinclair 1967) means that they could record any kind of tissue deformation, from pressure applied to the skin, to distension of a blood vessel under pressure.

Again, the majority of encapsulated sensory nerve endings are supplied by several nerve fibres, often both myelinated and non-myelinated. Some of these may be efferent, but this feature does not suggest a high degree of specificity between the nerve ending and the central nervous system which interprets its impulses. Indeed there is good evidence (Darian-Smith and Johnson 1977) that some endings respond to more than one type of stimulus, though others (e.g. temperature receptors) are highly specific. Very little is known about the chemical specificity of nerve endings; thus endings which appear similar under the microscope, may be chemically different and have a different stimulus specificity.

Under normal circumstances, stimuli applied to the body give rise to nerve impulses in a wide variety of nerve fibres from both superficial and deep tissues, and normal subjective sensation is not the result of stimulation of a few highly specific nerve endings, but rather results from the analysis of a shower of impulses delivered to the central nervous system in a particular pattern. Many problems remain to be solved before it is possible to interpret the significance of the structure of peripheral nerve endings in the genesis of subjective sensations (for review, see Sinclair 1967).

Free nerve endings [FIG. 12.7]

Such networks are found throughout the body, in skin, mucous membranes, and deep tissues. They form the only type of innervation in the cornea, teeth, tympanic membrane, and the walls of blood vessels, and they cannot be distinguished readily from sympathetic fibres which accompany them. They contain both myelinated and non-myelinated nerve fibres and tend to form coarser, deep plexuses, the meshes of which become finer as they approach the epidermis of the skin. While most of the nerve fibres in the skin end in the dermis, they also enter the epidermis and pass between its basal cells. Individual nerve fibres may form both deep and superficial endings, and though their over-all distribution is not known, it is clear that many fibres supply each area of skin.

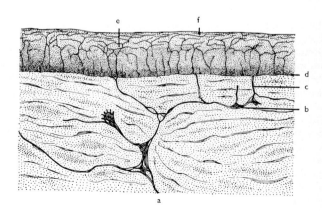

FIG. 12.7. Transverse section of cornea stained with gold chloride. a, b, plexus in substantia propria of cornea; c, branch passing to subepithelial plexus, d; e, intraepithelial plexus; f, termination of fibrils.

HAIRS

All hair follicles are innervated [FIG. 12.5], and these probably form the main tactile organs in hairy skin. Each hair receives branches from several nerve fibres, and these form an inner longitudinal and outer circular arrangement around the follicle. The manner of ending is not known, but there is evidence that single nerve fibres may innervate more than one hundred hairs (Weddell and Pallie 1955).

Special end-organs

LAMELLATED (PACINIAN) CORPUSCLES [FIG. 12.8B]

These corpuscles consist of many concentric layers of flattened epithelial-type cells which are continuous with the perineurium (Shanthaveerappa and Bourne 1963, 1966), but are separated from each other by layers of collagen. A single, thickly myelinated, nerve fibre, enters and extends in the central core, unlike the multiple innervation of most sensory corpuscles. Lamellated corpuscles vary greatly in size (1–4 mm or more), are often oval in shape, but may be lobulated or flattened. They have a very wide distribution, in the dermis (especially of hairless skin and more especially in the fingers), in fascial planes, in joint capsules, close to blood vessels, in the pleura and extraperitoneal tissues, in the external genitalia, in the mammary gland, and in many other situations. These corpuscles are sensitive to deformation, and may, therefore, record a wide variety of different changes depending on the tissue in which they are placed (Cauna and Mannan 1958; Sinclair 1967).

TACTILE (MEISSNER) CORPUSCLES [FIG. 12.8C]

These oval, circular, or lobulated corpuscles lie in small groups in the dermal papillae of the palmar surface of the fingers, the palm, and the sole of the foot. They consist of a stack of flattened cells enclosed in a capsule continuous with the perineurium of the nerves which enter it. Several medium-sized nerve fibres enter each corpuscle, and each of these fibres supplies several corpuscles. The nerve fibres lose their myelin sheath as they enter the corpuscle, and

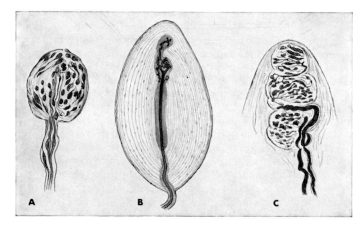

FIG. 12.8. Four types of special end-organs.
  A, bulbous corpuscle.
  B, lamellated corpuscle.
  C, oval corpuscle.

branching, pass between the cells of the corpuscle to form a complex mesh of fine fibres which do not unite. These corpuscles decrease in number, but increase in size and complexity with age. They are believed to be sensitive to touch, but absolute proof of this is lacking, though it is clear that several corpuscles are concerned with activating a single nerve fibre, and each corpuscle has many routes along which it can send impulses to the central nervous system.

TACTILE DISCS (MERKEL) [FIG. 12.9]

These disc-like expansions of branches of heavily myelinated fibres are usually found apparently surrounded by epithelial cells in the basal layers of the epidermis with the same distribution as the tactile corpuscles. They are also found in other regions, especially in other mammals, but though their functions are unknown, their close relation to the epidermis and the fact that each unit has an appearance like one element of a tactile corpuscle, suggests that their name may be appropriate.

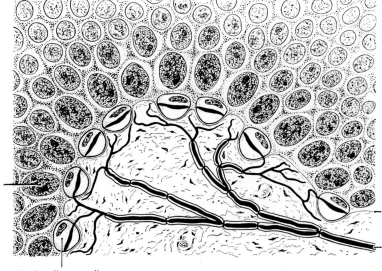

Nucleus of stratum germinativum

Tactile cell surrounding tactile disc

Basement membrane of epidermis

Myelinated nerve fibre

FIG. 12.9. Semidiagrammatic representation of a cutaneous tactile organ found in hairy skin and supplied by a single thick nerve fibre ending in a series of tactile discs.

BULBOUS CORPUSCLES (KRAUSE) [FIG. 12.8A]

These are small tufts formed by the non-myelinated branches of myelinated nerve fibres looped and coiled together in a cylindrical or spherical bundle. There is no capsule but often a condensation of the surrounding connective tissue, and individual fibres may leave the corpuscle and pass to take part with others in the formation of further bulbous corpuscles. They are found principally in mucocutaneous regions, e.g. lip, gum, external genitalia, perianal region, etc. Similar types of nerve endings are also found in other regions of the skin, but these are of such variable structure that it is often impossible to determine to which group of sensory endings they belong (Weddell and Miller 1962). It seems likely that the morphology of these and of the other special cutaneous endings is so variable that attempts to divide them into distinct groups is unrealistic, and merely leads to the assumption that such distinct morphological types have separate functions; an assumption which is not justifiable on the evidence (Quilliam 1966).

Sensory endings in muscle and tendon

Nerve endings of various kinds are found in intermuscular fascial planes, and in the capsules of joints especially near their attachments to the periosteum. The main sensory endings in muscle and tendon are the neuromuscular and neurotendinous spindles.

NEUROMUSCULAR SPINDLES

These complicated sensory organs are present in most of the somatic, striated muscles of the body [FIG. 12.10], though simpler varieties are present in the extrinsic ocular muscles (Cooper and Daniel 1949) [FIG. 12.11]. The fusiform spindles consist of a group (3–10) of thin intrafusal muscle fibres lying parallel to the ordinary muscle fibres, and enclosed together in a connective tissue sheath from which they are separated in their central region by a fluid-filled space. Within the space, each small muscle fibre contains a number of nuclei which are either arranged in a mass (nuclear bag) or in a row (nuclear chain) filling the muscle fibres (Boyd 1962). Two types of myelinated sensory fibres enter the spindle: (1) heavily myelinated fibres pass to the central region and form spiral (anulospiral) endings coiled around the nuclear region of the muscle fibres: (2) medium sized myelinated fibres pass to regions further from the centre and break up into a complicated series of branches (flower-spray endings) on the contractile part of the intrafusal muscle fibres. The anulospiral endings are sensitive to stretch, and it is

FIG. 12.11. Sensory nerve ending enveloping a fibre of an ocular muscle. (After Dogiel.)

probable that the flower-spray endings are sensitive to increased tension. However, the exact sensitivity of the parts of the spindle, if different, is not known, though they react to increasing length or tension.

In addition to the sensory fibres, fine motor fibres (γ-efferent fibres) innervate the intrafusal muscle fibres near their ends. These motor fibres arise independently from small cells in the anterior horn of the spinal medulla (Nyberg-Hansen 1965), and are used to increase the tension or decrease the length of the intrafusal muscle fibres. Thus they can adjust the length of the spindles to that of the surrounding muscle fibres so that the spindles are available to record stretching of the muscle in any position, and can increase the spindle tension so that it generates nerve impulses even when the muscle is not being stretched.

NEUROTENDINOUS SPINDLES

There are many types of endings on tendons, including simple branched nerve fibres, while complex and sometimes encapsulated endings are also present especially near the myotendinous junction. The latter have a resemblance to neuromuscular spindles, but are arranged in series with the muscle fibres. Presumably they record increasing tension in the tendon whether this is due to stretch or muscle contraction. Many of the tendon organs are arranged in the form of flower-spray endings, but their structure is variable.

The afferent impulses from muscle and tendon spindles were thought not to evoke conscious sensation; but it seems likely that they supplement information from joints in this function.

Development of nerve endings

Most sensory endings develop during foetal life (Jalowy 1939). Thus they are recognizable at birth, and though some of them continue to develop throughout postnatal life (Cauna 1964), there is a general decrease in the number of superficial sensory endings with age.

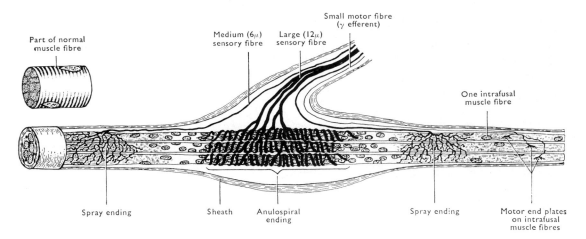

FIG. 12.10. Semidiagrammatic representation of the central region of a muscle spindle.

Organs of the special senses

ORGAN OF VISION—THE EYEBALL

The eyeball is composed of segments of two unequal spheres; the transparent, anterior, **corneal segment** has a smaller radius of curvature than the opaque, white, posterior, **scleral segment** [FIG. 12.12], and projects forwards from it. The two parts join at a slight, external groove, the **sulcus sclerae**. The central points of the anterior and posterior surfaces of the eyeball are known as the anterior and posterior **poles**; the line joining the poles is the **optic axis**; an imaginary line encircling the eyeball, midway between the poles, is the **equator**. These are purely descriptive terms and have little significance since the posterior pole cannot be determined in life. The **visual axis** is of more importance, and joins the centre of the cornea to the fovea centralis of the retina [FIG 12.12]. It represents the course taken through the eye by a ray of light from the centre point of vision. When looking at a distant object, the visual axes of the two eyes are parallel, though the optic axes are slightly, and the optic nerves markedly convergent posteriorly.

The eyeball measures approximately 2.5 cm in all directions, and has a volume just over 8 cm^3. The eyeball completes the greater part of its growth in the antenatal period. Hence relative to body size it is much larger in infants and children, and slightly larger in females than in males (Schultz 1940).

Fascial sheath of the eyeball

This is a thin fibrous sheath which surrounds the sclera, forms a socket for the eyeball, and separates it from the other contents of the orbit. The sheath fuses with the sclera both immediately posterior to the corneoscleral junction, and where the optic nerve, ciliary nerves, and posterior ciliary arteries pierce it. Elsewhere the sheath is very loosely united to the sclera by a few minute trabeculae of connective tissue which cross this potential, **episcleral (interva-**

ginal) space. The sheath is pierced by the venae vorticosae close to the equator [FIG. 12.20], and sends tubular expansions outwards over each of the extrinsic ocular muscles as they pass through it to the sclera [FIG. 5.49]. These **sheaths** become continuous with the epimysium of the muscles (in the case of the superior oblique with its trochlea), but each has an external extension (**check ligament**) close to its origin from the fascial sheath of the eyeball. The check ligaments of the medial and lateral rectus muscles are well developed. They form triangular masses of fibrous tissue which pass to the periosteum of the orbit. That of the superior rectus fuses with the fascial sheath of the levator palpebrae superioris. Muscle fibres of the rectus may pass into that sheath and thereby reach the tarsus, so that the muscle can elevate both the cornea and the upper eyelid. The check ligament of the inferior rectus fuses with the sheath of the inferior oblique and with the suspensory ligament (see below) to which some of the muscle fibres of the inferior rectus pass. Through the ligament it is attached to the inferior tarsus, so that it may depress the cornea and the lower lid together. The check ligament of the lateral rectus is attached to a tubercle on the orbital surface of the zygomatic bone and to the lateral palpebral ligament, while that of the medial rectus is attached to the lacrimal bone posterior to its crest. Between these two check ligaments, the fascial sheath inferior to the eyeball is thicker than elsewhere, is pierced by the inferior rectus and inferior oblique muscles, and forms the **suspensory ligament** (Lockwood 1885) which thus passes like a hammock under the eyeball from the medial to the lateral orbital walls.

The presence of the episcleral space between the eyeball and its fascial sheath allows some movement of the eyeball within the sheath, but in most movements of the eye, sheath and eyeball move together, the check ligaments limiting the range of movement.

At the margins of the orbit, the orbital periosteum (**periorbita**) is continued into the eyelids as the **palpebral fascia** which extends into the tarsal plates. This sheet, split by the palpebral fissure, constitutes the **orbital septum**. The part in the inferior eyelid receives an extension from the sheaths of the inferior rectus and inferior oblique muscles, while the sheath of the superior rectus, fused with that of levator palpebrae superioris, is inserted into the superior conjunctival fornix deep to the orbital septum.

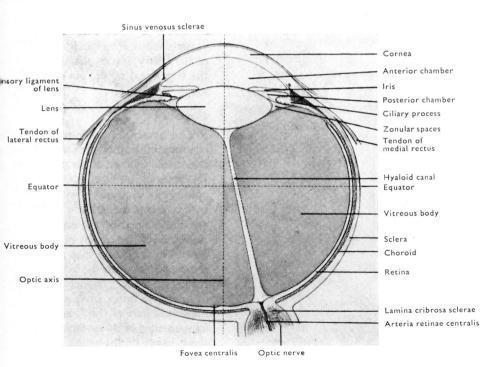

FIG. 12.12. Diagram of horizontal section through left eyeball and optic nerve. × 4.

COATS OF THE EYEBALL

There are three concentric coats forming the wall of the eyeball. (1) The outer **fibrous coat**, consisting of the sclera and cornea. (2) A middle, **vascular, pigmented coat** (or uvea), the choroid, ciliary body and iris. (3) An inner, **nervous coat**, the retina, which is continued forwards in a modified form on the ciliary body and iris [FIG. 12.17]. These coats surround and partly subdivide the contents of the eyeball, which consist of the **vitreous body** between the lens and the retina, and a space between the lens and the cornea which contains the **aqueous humour** and is partly subdivided by the iris into the **anterior and posterior chambers**.

Fibrous coat

SCLERA

This is the firm, opaque, white coat of the eye. It has a bluish tinge in young children due to the pigment of the choroid shining through, and is yellowish in old age. It is thickest posteriorly (1 mm), thinnest at the equator (0.5 mm), and intermediate anteriorly (0.6 mm). The optic nerve pierces the sclera as a considerable number of separate bundles slightly more than 3 mm medial to the fovea centralis, and just above the horizontal meridian (Traquair 1949). This part of the sclera is therefore known as the **cribriform part**. Around it the dural sheath of the optic nerve fuses with the sclera. External to this there are fifteen to twenty small apertures in the sclera for the **short ciliary nerves** and **short posterior ciliary arteries**, while a **long posterior ciliary artery** and a **long ciliary nerve** pierce it a little further distant on each side [FIG. 12.20]. The corresponding veins are the **venae vorticosae**, two superior and two inferior, which pierce the sclera at equidistant points near the equator. There are openings for the **anterior ciliary arteries** and **veins** near the sulcus sclerae [FIG. 12.20].

The deep surface of the sclera is attached to the choroid by a delicate mesh of pigmented areolar tissue, the **suprachoroid lamina**. The ciliary arteries and nerves run in this layer which is often looked on as a space (**perichoroidal space**) because of the delicate nature of its areolar tissue. When choroid and sclera are separated, some of the lamina remains adherent to the sclera, and is known as the **lamina fusca** of the sclera.

At the corneoscleral junction, the deepest part of the sclera projects inwards as a shelf (the **scleral spur**) posterior to a canal (**sinus venosus sclerae**) which encircles the eye in the junction [FIG. 12.13]. The scleral spur is continuous with the deepest part of the cornea which forms the fenestrated **pectinate ligament of the iridocorneal angle**, medial to the sinus venosus sclerae. The sinus, *which does not contain blood*, receives fluid from the aqueous humour through the fenestrations of the pectinate ligament (**spaces of the iridocorneal angle**), and communicates with the anterior ciliary veins.

Structure

The sclera consists of bundles of white fibrous tissue arranged in interlacing equatorial and meridional layers. There are some fine elastic fibres and many fibrous tissue cells, with some pigment cells between the bundles of the inner layers.

Vessels and nerves

The **arteries** are from the anterior and short posterior ciliary arteries; the **veins** enter the venae vorticosae and anterior ciliary veins. The **nerves** are branches of the ciliary nerves, mainly consisting of afferent fibres from the trigeminal nerve, but some are sympathetic fibres to the blood vessels.

CORNEA

This avascular, transparent, anterior part of the eye forms one sixth of the total eyeball surface, and is slightly thicker than the sclera, 0.9 mm at the centre and 1.2 mm at the periphery. Its anterior surface is formed by stratified squamous epithelium continuous with the **conjunctiva**; its posterior surface forms the anterior wall of the anterior chamber of the eye and is lined by the **endothelium** of that chamber. The curvature of the cornea differs in different individuals, and is greater in the young than in the aged; it is slightly greater in the vertical than in the horizontal plane, and diminishes from the centre to the periphery. As there is a considerable change in refractive index from air to cornea, this junction forms an important part of the focusing mechanism of the eye. Thus irregularities of its surface interfere with the ability to form a sharp image on the retina, and inequalities of curvature in the vertical or horizontal planes are associated with an inability to focus either vertical or horizontal lines sharply on the retina, a condition known as *astigmatism*.

The cornea fits inside the anterior margin of the sclera, the junction being oblique. The obliquity is not uniform around the entire junction, so that the cornea appears slightly elliptical from the front, the horizontal diameter being the greatest. The iris is attached immediately posterior to the internal junction of the cornea and sclera, and the angle formed here between the iris and the cornea is the **iridocorneal angle** [FIG. 12.13].

Structure [FIG. 12.14]

1. The **anterior epithelium of the cornea** consists of six to eight layers of cells continuous with the epithelium of the conjunctiva at the corneal margins. The deepest layer consists of columnar cells, the intermediate layer of polygonal cells with interdigitating processes attached by desmosomes, the superficial layer of nucleated squames. This is a highly sensitive layer [FIG. 12.7] which regenerates rapidly after injury.

2. The **anterior limiting layer**, approximately 8 μm thick, is a part of the substantia propria of the cornea which appears different

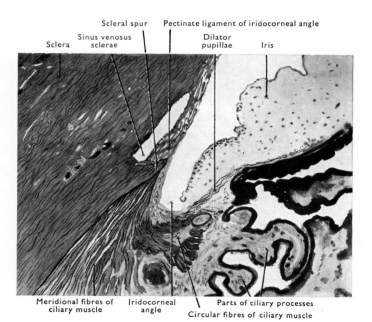

Scleral spur   Pectinate ligament of iridocorneal angle

Sinus venosus sclerae   Dilator pupillae

Sclera   Iris

Meridional fibres of ciliary muscle   Iridocorneal angle   Parts of ciliary processes

Circular fibres of ciliary muscle

FIG. 12.13. Section of iridocorneal angle. (Thomson 1912.)

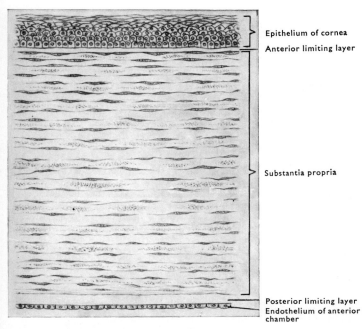

FIG. 12.14. Diagrammatic section of cornea.

Epithelium of cornea
Anterior limiting layer
Substantia propria
Posterior limiting layer
Endothelium of anterior chamber

forming a subepithelial plexus, send fine filaments into the surface layers of the anterior epithelium [FIG. 12.7].

Vascular coat

THE CHOROID

This vascular layer contains stellate pigment cells and phagocytes in a fibro-elastic mesh. It is loosely attached to the sclera by the suprachoroid lamina, is missing where the optic nerve pierces it, and is firmly bound to the sclera where the ciliary arteries enter it. The choroid extends forwards to the ora serrata of the retina [FIG. 12.17], where it becomes continuous with the ciliary body, and through that with the iris.

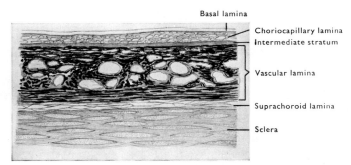

Basal lamina
Choriocapillary lamina
Intermediate stratum
Vascular lamina
Suprachoroid lamina
Sclera

FIG. 12.15. Tranverse section of choroid and inner part of sclera.

because of the irregular arrangement of its collagen fibres, and the absence of elastic fibres.

3. The substantia propria of the cornea consists of bundles of collagen fibres arranged in many layers in each of which the fibres run in different directions. The layers are woven together by the passage of fibres from one to another. There are also fine elastic meshes, fibrous tissue cells, and a polysaccharide containing ground substance which may be responsible for the transparency of the tissue. It is avascular, but lymphocytes can be found between the layers, and appear in large numbers in infections. In persons over middle age there may be a deposition of fat between the layers at the periphery, particularly inferiorly, and this forms an opaque, greyish ring, the *arcus senilis*.

4. The posterior limiting layer appears to be a thick, transparent basement membrane, probably formed from the endothelium of the anterior chamber. It separates easily from the remainder of the cornea, and contains an atypical collagen mesh. At the periphery of the cornea (limbus), this layer becomes continuous with the pectinate ligament whose fenestrations are lined by the endothelium of the anterior chamber and lead from that chamber to the sinus venosus sclerae. The deepest part of the pectinate ligament continues round the iridocorneal angle into the iris in close association with the myoepithelial fibres of dilator pupillae (Thomson 1912).

5. The endothelium of the anterior chamber consists of a single layer of large squamous cells which lines the entire anterior chamber, including the posterior surface of the cornea.

At the margin of the cornea, the anterior limiting layer is replaced by a layer of loose connective tissue containing many capillaries. These do not extend into the cornea, but are the blood vessels of the conjunctiva from which the cornea is partly nourished by diffusion.

Vessels and nerves

The cornea does not contain blood vessels. Spaces may be seen in it in microscopic preparations, and these have been thought to represent pathways for the passage of fluid, but they are probably artefacts.

The nerves are derived from the ciliary nerves which form a plexus round the limbus. Fibres enter the substantia propria, and

Structure

1. The suprachoroid lamina transmits the ciliary nerves, and long posterior ciliary arteries. It consists of the above structures.

2. The vascular layer of the choroid consists of an outer zone which contains large blood vessels (branches of the short posterior ciliary arteries) and superficial to these the whorled tributaries of the venae vorticosae [FIG. 12.20]. These are connected through an intermediate stratum of fibro-elastic tissue [FIG. 12.15] to the thin layer of wide bore capillaries which lie in the choriocapillary lamina and are separated from the pigmented cells of the retina by the basal lamina only.

The basal lamina is a thin, glassy membrane containing collagen and fine elastic fibres in its outer part. The inner, homogeneous layer is the basement membrane of the pigment cells of the retina.

Tapetum

In many mammals, but not in Man, the choroid contains a brilliant, iridescent layer by which light passing through the retina is reflected back into it, thus increasing its sensitivity in low intensity illumination. This may be a thin fibrous layer (horse, sheep) or several layers of flat, iridescent cells (seal) in the intermediate stratum.

CILIARY BODY

This body is the anterior extension of the choroid to the iris, and consists of the ciliary ring, the ciliary processes, and the ciliary muscle [FIG. 12.18]. It is lined by the ciliary part of the retina which consists of a double layer of columnar epithelium (the outer of which is pigmented) representing the two layers of the optic cup of the embryo. External to this, both ciliary ring and processes have a similar structure to the choroid, except that the suprachoroid lamina

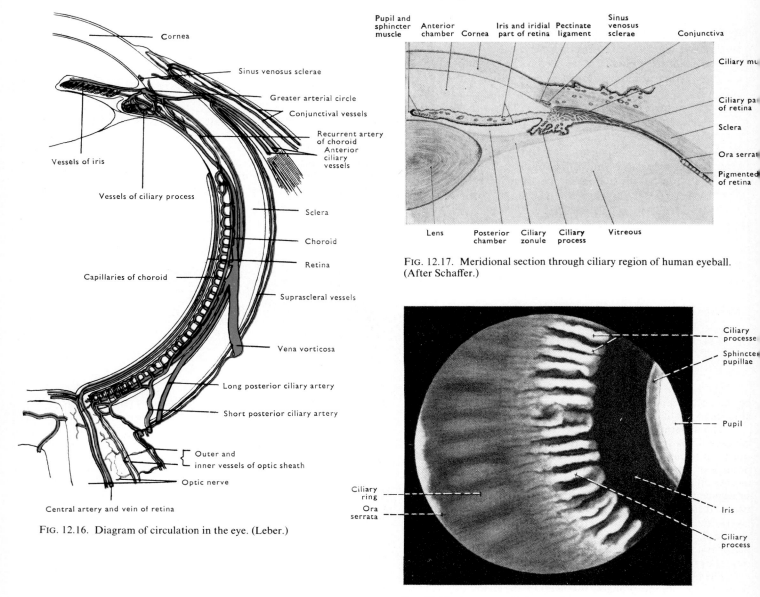

FIG. 12.16. Diagram of circulation in the eye. (Leber.)

FIG. 12.17. Meridional section through ciliary region of human eyeball. (After Schaffer.)

FIG. 12.18. A segment of the anterior part of eyeball seen from within.

is poorly developed, and the choriocapillary lamina is absent, in keeping with the absence of a functional retinal layer.

The ciliary ring is a slightly ridged zone, about 4 mm wide, which extends from the ora serrata of the retina [FIG. 12.18] to the external extremities of the ciliary processes.

The ciliary processes are about seventy pale, radially arranged folds which begin abruptly at the base of the iris (with their free ends facing the periphery of the lens) and extend outwards to merge into the ciliary ring. Each fold is approximately 2 mm long, and its surface is thrown into a number of subsidiary folds covered by the ciliary part of the retina. Close to the lens, the ciliary processes give attachment to some of the fibrils of the suspensory ligament of the lens. These fibrils appear to arise from the cells of the ciliary part of the retina, and passing towards the lens, fuse into thicker fibres which adhere to the ciliary processes on their way.

The ciliary muscle lies in the angle between the sclera and the ciliary processes and ring, external to the attached margin of the iris [FIG. 12. 17]. This smooth muscle consists of meridional fibres which radiate backwards from the scleral spur to the ciliary processes and ring. In addition, a small bundle of circular muscle fibres lies between the iridocorneal angle and the ciliary processes, medial to the meridional fibres.

When the ciliary muscle contracts, it pulls the ciliary ring and processes antero-internally. This relaxes the fibres of the suspensory ligament of the lens, and allows the elasticity of that organ to increase its convexity, thereby decreasing its focal length. The degree of development of the parts of the ciliary muscle varies; the smaller bundle of circular fibres is absent or rudimentary in myopic (short-sighted) eyes, but is usually well developed in hypermetropic eyes.

IRIS

The iris is the coloured part of the eye. It is a contractile, bilaminar diaphragm which extends inwards in front of the lens and consists anteriorly of a layer of vascular connective tissue (**stroma**) containing some circular and radial smooth muscle fibres and usually some pigment cells, and posteriorly of a double layer of pigmented epithelial cells, the **iridial part of the retina**. These layers are continuous respectively: (1) with the ciliary body and the pectinate ligament; and (2) with the ciliary part of the retina. Slightly to the nasal side of its centre, the iris is perforated by an almost circular aperture, **the pupil**. The size of the pupil is altered by contraction of the circular (**sphincter pupillae**) or radial (**dilator pupillae**) muscle fibres of the iris, so altering the amount of light entering the eye. The sphincter pupillae is said to be derived from the iridial part of the retina. It is a thin ring of smooth muscle in the pupillary margin of the iris which is innervated by **postganglionic parasympathetic fibres** from the ciliary ganglion, and tends to contract in unison with the ciliary muscle (which has the same nerve supply) in the **accommodation-convergence reaction**. The dilator pupillae is formed from the heavily pigmented cells of the anterior layer of the iridial part of the retina which develop myo-epithelial elements. It is supplied by **postganglionic sympathetic fibres** from the internal carotid plexus.

The iris tends to be slightly thicker in its intermediate part than at either margin, and its internal part lies on the anterior surface of the lens with only a narrow space between them through which the anterior and posterior chambers of the eye communicate [FIG. 12.17].

The anterior surface of the iris is said to be covered with the same endothelium of the anterior chamber which covers the posterior surface of the cornea. This surface is highly irregular with a number of jagged folds and crypts, and when the iris is fully dilated tends to bunch against the pectinate ligament, and may interfere with the drainage of aqueous humour through that ligament into the sinus venosus sclerae.

Vessels and nerves

The **arteries** of the ciliary body and iris are long posterior ciliary and anterior ciliary branches of the ophthalmic artery. The two former pierce the sclera medial and lateral to the optic nerve, and running forwards deep to the sclera, anastomose at the outer margin of the iris with four to six anterior ciliary arteries to form the **greater arterial circle** of the iris [FIG. 12.19]. Branches from this circle pass radially inwards towards the pupil in the stroma of the iris. They anastomose close to the pupillary margin to form the **lesser arterial circle** of the iris [FIG. 12.19]. Before birth the pupil is covered by a thin, vascular, **pupillary membrane** which extends inwards from the anterior surface of the iris and is initially supplied from the hyaloid artery, but later from the vessels of the iris. About the seventh month of pregnancy the central vessels begin to atrophy. This process extends outwards to the lesser arterial circle, the pupillary membrane breaks down and is lost. Fragments of the membrane (or even a complete membrane) may be present at birth and may persist into adult life. The **veins** of the iris pass radially outwards and communicate with the veins of the ciliary processes and the sinus venosus sclerae [p. 838].

The **nerves** of the choroid and iris are derived from the **long and short ciliary nerves** [FIG. 12.20]. The former (2–3) arise from the nasociliary nerve, the latter (8–14) from the ciliary ganglion. Both groups contain sensory (trigeminal) and sympathetic fibres, only the short ciliary nerves contain postganglionic parasympathetic fibres and occasional ganglion cells (Wolff 1948). The nerves pierce the sclera around the optic nerve, and pass forwards deep to the sclera to form a plexus on the ciliary muscle. Fibres from this are distributed to the ciliary body, iris, and cornea. Preganglionic parasympathetic fibres reach the ciliary ganglion through the **oculomotor nerve**.

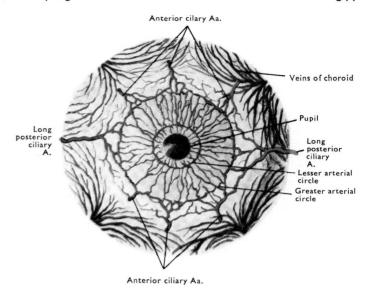

FIG. 12.19. Blood vessels of the iris and anterior part of the choroid.

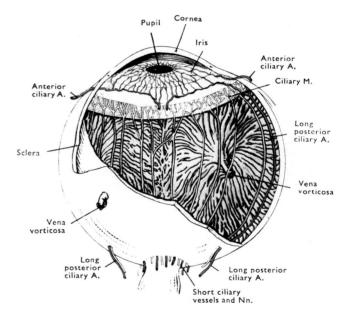

FIG. 12.20. Dissection to show the vascular coat and ciliary nerves and vessels.

Inner coat

RETINA

The retina is a double layer of epithelium derived from the optic cup of the embryo. It lines the internal surfaces of the choroid (**optic part**), the ciliary body and processes (**ciliary part**), and the iris (**iridial part**). Posteriorly its inner layer is greatly thickened to form the **cerebral layer**. This contains the light-sensitive elements of the retina, and gives origin to the optic nerve fibres. The cerebral layer extends peripherally from the optic nerve and is gradually reduced in thickness from 0.4 mm adjacent to that nerve, to 0.1 mm at the **ora serrata**—its irregular edge at the posterior border of the ciliary ring. [FIG. 12.17]. Here the cerebral layer loses its sensory elements, contains a number of spaces, and is then suddenly reduced to a single stratum of non-pigmented epithelial cells which passes over the ciliary body with the outer, pigmented epithelial layer of the retina. As these two layers pass on to the posterior surface of the iris, the inner layer also becomes pigmented.

The outer or **pigmented layer of the retina** is a single layer of heavily pigmented cells which have thick, internal processes which interdigitate with the photosensitive cells (rods and cones) of the cerebral layer. Pigment accumulates in the processes with increasing intensity of illumination of the retina. The pigment layer is sometimes considered to be part of the choroid to which its cells are more closely attached by a thick basement membrane than to the cerebral layer of the retina, from which it is sometimes separated in life (detachment of the retina). The pigment layer tends to absorb any light passing through the retina, and thus prevents it from being reflected back into the cerebral layer at various angles; a feature which would reduce the resolving power of that layer.

The cerebral layer has a small, oval, yellowish spot in the visual axis. This is the **macula**, and its central part is excavated to form the **fovea centralis**. About 3 mm to the nasal side and slightly above the macula (Fison 1920; Polyak 1957) there is the temporal edge of a circular white disc surrounded by a narrow ring of pigment. This **disc of the optic nerve** is formed by nerve fibres passing out of the eye and all obtaining their myelin sheaths at the disc. It has a slightly raised margin and a central excavation, is approximately 1.5 mm in diameter, and is non-pigmented and insensitive to light (**blind spot**) since both retinal layers are absent; the two layers fusing around the periphery of the disc at the pigment ring.

The cerebral layer is transparent during life, but in an animal kept in the dark, a purple tinge develops due to the accumulation of visual purple or rhodopsin in the rod cells. The purple is absent from the macula and in a narrow zone, 3–4 mm wide, close to the ora serrata.

Structure [FIGS. 12.21–12.23]

The **cerebral layer** of the retina consists of a number of layers of neural elements enclosed in the processes of slender neuroglial cells which extend radially between the external and internal limiting membranes [FIG. 12.22], forming these two 'membranes' by their extremities.

The nervous elements consist of three basic layers of cells with intermediate zones in which lie the synapses between the processes of these cells, which conduct impulses from the rods and cones to the nerve fibres in the optic nerve.

1a. The **layer of rods and cones** (stratum neuro-epitheliale). This is the outermost, light sensitive layer of cells which interdigitates with the processes of the cells of the pigment layer. Each cell consists of a body containing the nucleus with peripheral and central processes. The body lies internal to the external limiting membrane, while the peripheral process extends external to it. Under the electron microscope the peripheral process of each **rod cell** is seen to consist of an **outer segment**—a slender cylinder containing a stack of flattened membrane sacs joined to the thicker, mitochondria-filled, inner segment by a narrow, cilium-like neck. The **inner segment** is continued as a slender fibre (dendrite) into the **outer nuclear layer** [FIG. 12.21] to the **body of the rod cell**, whence an axon passes inwards to synapse with elements of the ganglionic layer of the retina. The outer segment of the **cone** processes are conical in shape, particularly towards the periphery of the retina where they are short and fat. Towards the fovea they increase in height and become thinner and more cylindrical; the fovea consisting solely of these tall, thin cones tightly packed together. The **outer segment** does not contain rhodopsin (hence the fovea is never purple in colour, but yellow), and is united to an inner segment similar to that in the rod cell except that it is directly fused to the cell body and its contained nucleus. The **body** therefore lies in the outer

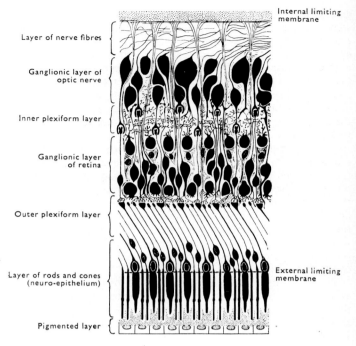

Layer of nerve fibres

Ganglionic layer of optic nerve

Inner plexiform layer

Ganglionic layer of retina

Outer plexiform layer

Layer of rods and cones (neuro-epithelium)

Pigmented layer

Internal limiting membrane

External limiting membrane

FIG. 12.21. Schematic reconstruction of human retina. (After Polyak 1941.)

nuclear layer immediately internal to the external limiting membrane, and forms a single row of nuclei with those of the other cone cells, while the rod cell nuclei lie further internally and are more scattered [FIG. 12.22]. The axon of each cone cell passes through the entire thickness of the outer nuclear layer to synapse with a cell or cells of the ganglionic layer of the retina.

In the **periphery** of the retina the cones are few and squat. They increase gradually in number and density towards the fovea centralis, becoming taller and thinner, while the relative number of rods decreases, though they are more densely packed. Over the retina as a whole there are approximately twenty times as many rods as cones. The rods have a lower threshold to light than the cones, but are not sensitive to colour. Hence the monochrome appearance of objects seen in low intensity illumination and the inability to see objects with the rod-free fovea in these conditions.

1b. The **outer nuclear layer** consists of the cell bodies of the rod and cone cells (the latter externally), and the processes of both.

2. The axons of the rod and cone cells enter the **outer plexiform** (synaptic) **layer**, and here they form synapses with the dendrites of the intermediate layer of cells; the bipolar cells of the **ganglionic layer of the retina**. The rod axons end in small spherules, the cone axons expand in the plane of the retina, and both synapse with a variety of different bipolar cells, some of which have contact with both rod and cone terminals by branching dendrites, while a few make contact with a single cone process or a group of rod spherules only. In addition there are **horizontal cells** which link various parts of this layer, and the processes of some bipolar cells which are believed to conduct peripherally.

3. The great majority of the **bipolar cells** conduct impulses through their axons to the inner plexiform layer. Here they synapse with the ganglion cells, from which the axons in the optic nerve arise. Cell bodies of the bipolar, horizontal, and amacrine cells form the **ganglionic layer of the retina** [FIG. 12.22]. The number of bipolar cells gradually increases from the periphery of the retina towards the fovea centralis; the number of those with a simple cone connection increasing most noticeably. The total increase is greater than the increase in number of light-sensitive elements, hence the

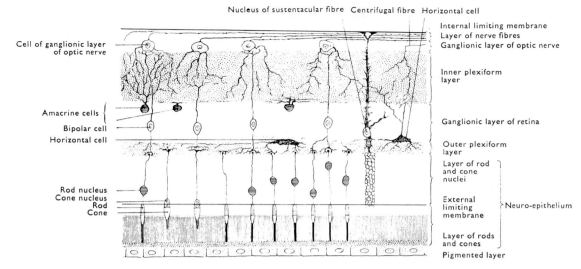

Nucleus of sustentacular fibre   Centrifugal fibre   Horizontal cell

Cell of ganglionic layer
of optic nerve

Internal limiting membrane
Layer of nerve fibres
Ganglionic layer of optic nerve

Inner plexiform
layer

Amacrine cells

Ganglionic layer of retina

Bipolar cell
Horizontal cell

Outer plexiform
layer

Layer of rod
and cone
nuclei

Rod nucleus
Cone nucleus
Rod
Cone

External
limiting
membrane

Neuro-epithelium

Layer of rods
and cones

Pigmented layer

FIG. 12.22. Plan of the retinal neurones. (After Cajal.)

size of the bipolar cell unit (the number of light-sensitive cells discharging on a single bipolar cell) decreases proportionately.

Horizontal cells are of several types. The small variety send dendrites to cone axons, and an axon which extends horizontally through the outer plexiform layer to end on other cone processes at some distance. The larger variety seem to have the same type of connection predominantly with the endings of rod cells. Similar cells are found in the inner plexiform layer. Here they unite the dendrites of the ganglion cells and the axons of the bipolar cells. Such cells could allow for considerable integration of impulses within adjacent parts of the retina.

Amacrine cells appear to have no axon. They are found adjacent to the inner plexiform layer in which they send their dendrites either in a single stratum or in a diffuse manner. Their function is unknown.

4. The **inner plexiform layer** consists of the branched, expanded axonal endings of the bipolar cells on the extensive dendritic trees

and on the cell bodies of the ganglion cells of the innermost cellular layer of the retina, and on the processes of horizontal and amacrine cells of the ganglionic layer of the retina. The bipolar cells which make a simple connection with a cone axon peripherally have a similar relation to a single ganglion cell centrally, their axons each being enclosed in the expanded dendrite of the ganglion cell [FIG. 12.23], while giving branches also to other ganglion, horizontal and amacrine cells.

5. The cells of the **ganglionic layer of the optic nerve** each give rise to one axon which passes over the internal surface of the retina to reach the optic disc and enter the optic nerve. These cells vary in size but are larger than the bipolar cells. At the periphery of the retina they are very sparse and widely separated, so that the area served by each is considerable. Further posteriorly they increase in number to form a continuous layer of cells, and near the fovea centralis they form several layers. Thus the average size of the retinal unit (the area stimulating a single ganglion cell) decreases considerably towards the fovea, and the number of single cone units increases markedly to reach its maximum in the fovea centralis which consists almost exclusively of such units.

Structure of the macula and fovea centralis

In this region the light-sensitive elements are purely cones, but their outer segments are greatly elongated, thin and tightly packed. At the fovea centralis the inner layers of the retina are swept aside so that the cones are virtually on the internal surface, and the light reaching them does not have to pass through the internal layers as in other parts of the retina. Hence the image thrown on the retina by the refracting tissues of the eye is least distorted when it reaches the light sensitive elements in this part. This feature combined with the tightly packed single cone elements gives this part of the retina the highest resolving power of any part, and it corresponds to the sharp area in the field of vision which covers only one degree of arc. At the fovea the axons of the cones pass obliquely outwards to the bipolar and ganglion cells which are heaped up around it. The large number of single cone units in the fovea results in a disproportionately large number of optic nerve fibres arising from this region and streaming away from its perimeter to the optic disc without crossing the fovea. Nerve fibres approaching the optic disc from the temporal parts of the retina sweep superior and inferior to the fovea, so that the otherwise radial arrangement of these fibres is distorted in this region.

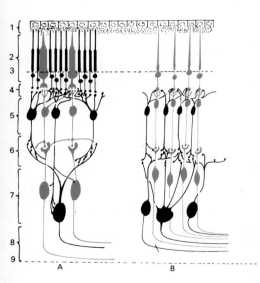

FIG. 12.23. Diagram of the arrangement of cells in the peripheral (A) and central (B) parts of the retina. Only the rods, cones, bipolar, and ganglion cells are shown. The cones and specific cone connections are shown in red. In B the tightly packed cones have been separated for diagrammatic purposes. (After Polyak.)

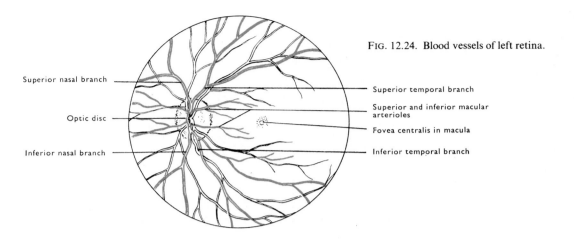

FIG. 12.24. Blood vessels of left retina.

Superior nasal branch

Optic disc

Inferior nasal branch

Superior temporal branch

Superior and inferior macular arterioles

Fovea centralis in macula

Inferior temporal branch

Vessels [FIG. 12.24]

The direct arterial supply to the retina is the **central artery of the retina**; a branch of the ophthalmic artery. At a variable point posterior to the eyeball it pierces the sheaths of the optic nerve to lie centrally in it. Passing distally in this position it appears in the centre of the optic disc and divides into superior and inferior branches. Each of these gives off a medial arteriole to the nasal retina and a lateral branch (**superior** and **inferior macular arterioles**) and then divides into **temporal** and **nasal retinal arterioles** which supply the corresponding quadrants of the retina. The superior and inferior temporal arterioles arch laterally above and below the macula sending small branches towards it. But neither these nor the macular arterioles reach the fovea centralis which is devoid of blood vessels. These arteries, and the corresponding veins which lie close to them, are in the nerve fibre layer of the retina, the veins external to the arteries. The veins consist of a layer of endothelium surrounded by delicate connective tissue, and are therefore easily distended. They unite to form the **central vein of the retina**. This lies with the central artery and either enters the superior ophthalmic vein or runs directly to the cavernous sinus.

The outer layers of the retina receive a supply from the choroid by diffusion. However, the inner layers are entirely supplied by the central vessels. Thus blockage of these vessels interrupts the pathway from the retina to the brain and causes blindness. Detachment of the retina from the choroid may interfere with the nutrition of the rods and cones, but it also makes the detached part of the retina useless because an image cannot be focused on it.

Refracting media

Light on its way to the retina passes through a number of structures with different refractive indices. These are cornea, aqueous humour, lens, and vitreous body [FIG. 12.12]. The change in refractive index from air to cornea is the greatest. Thus minor imperfections in the curvature of the cornea alter the course of light into the lens in such a manner that the lens may be unable to focus that light on the retina. It also explains the efficacy of contact lenses.

AQUEOUS HUMOUR AND CHAMBERS OF THE EYE

The aqueous humour is a watery fluid which contains 1.4 per cent sodium chloride and has a refractive index of 1.336. It fills the space between the cornea and the lens, and bathes both surfaces of the iris which incompletely divides that space into the anterior and posterior chambers of the eye. At the lateral margins of these chambers the fluid extends into the spaces of the iridocorneal angle and the zonular spaces [FIG. 12.26] respectively.

LENS

The lens is a biconvex, transparent body [FIG. 12.12] which lies between the iris and the vitreous body. The central points of its anterior and posterior surfaces are the anterior and posterior **poles**, and the line joining these is the **axis**; the peripheral circumference is the **equator**. The axis measures approximately 4 mm, the transverse diameter is 9–10 mm. The anterior surface lies immediately behind the posterior chamber. The posterior surface of the lens is lodged in the **hyaloid fossa** of the vitreous body. The anterior surface is less convex than the posterior surface. The shape of the lens is maintained by the constant tension of the ciliary zonule (q.v.). Contraction of the ciliary muscle relaxes this tension and allows the elastic lens to increase its convexity, thus shortening its

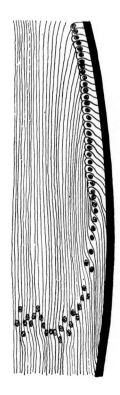

FIG. 12.25. Diagrammatic section through equator of lens, showing the gradual transition of the epithelium into lens fibres. (After Babuchin.)

focal length and bringing near objects into sharp focus on the retina. In the foetus the lens is soft, pink, and elastic, but with age it becomes increasingly harder, yellowish, and fixed in the flattened shape. Thus focusing on near objects becomes progressively more difficult; a condition known as presbyopia. The lens may also become opaque in the aged; a condition known as cataract.

Structure

The lens is formed from an ectodermal vesicle the posterior cells of which elongate greatly to fill the cavity. Thus they form the **lens fibres** which are arranged more or less concentrically around the axis of the lens, the outer fibres near the equator becoming continuous with the layer of cuboid cells **(epithelium of the lens)** which covers the anterior surfaces of the lens fibres [FIG. 12.25] and is derived from the anterior wall of the embryonic lens vesicle. New lens fibres are formed throughout life from the epithelium at the equator, but they fail to reach the poles, and so meet at lines on the surfaces of the lens. These **radii of the lens** become more complex with continuing growth and age [FIG. 12.27], the older lens fibres at the centre being harder than those at the periphery. The lens is enclosed in a thin (15 μm), homogeneous, transparent capsule of highly refractile material to which the fibres of the ciliary zonule are attached.

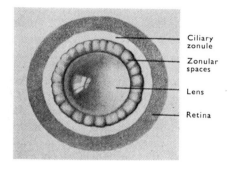

FIG. 12.26. The zonular spaces distended (viewed from the front).

Labels: Ciliary zonule; Zonular spaces; Lens; Retina

VITREOUS BODY

This transparent, jelly-like substance contains hyaluronic acid and a mesh of fine fibres. It fills the space between the lens and the retina **(vitreous chamber)**, and is especially adherent over the ora serrata and ciliary body. Here its fibres are more obvious and pass towards the equator of the lens with fibres which arise from the cells of the ciliary part of the retina. Together these form a ridged, fibrous **(zonular fibres)** sheet **(ciliary zonule** [FIG. 12.17]) which is intimately adherent to the surfaces of the ciliary processes but not to the grooves between them. The aqueous humour of the posterior chamber passes into the grooves anterior to the zonule. As the zonule approaches the equator of the lens, it splits into two layers separated by the **zonular spaces**. Coarse fibres pass to the anterior surface of the lens capsule while the fine fibres pass to its posterior surface. The thicker, anterior fibres are sometimes known as the suspensory ligament of the lens, and together with the posterior fibres of the zonule maintain tension on the lens. The vitreous body is traversed by the sinuous **hyaloid canal** [FIG. 12.12] from the middle of the optic disc to the posterior surface of the lens. This contains the remains of the **hyaloid artery**, which vascularized the lens in the earlier stages of its development, and an aqueous fluid which is of sufficiently different refractive index from the vitreous body to be seen in the living eye with a slit lamp microscope.

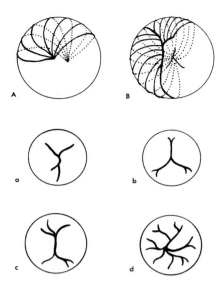

FIG. 12.27. Diagram showing formation of radii of lens. (After Ida Mann 1928.)

A, the simplest arrangement of lens fibres all of the same length and all extending from pole to pole, with resulting absence of radii. B, the manner in which lens radii develop when all the fibres are not of the same length and do not reach from pole to pole. a, b, c, d, varieties of radii.

Eyebrows

Each eyebrow is a fold of skin which is thickened by the presence of fatty areolar tissue between it and the periosteum of the skull. Above the margin of the fold is a zone covered with short, stout hairs, which extends below the supra-orbital ridges of the skull medially. Between the skin and fatty areolar tissue are the interlacing fibres of the frontalis and orbicularis oculi muscles.

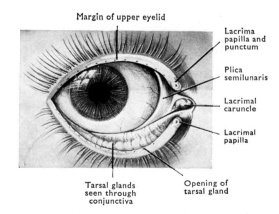

FIG. 12.28. Eye with lids slightly everted to show structures at the medial angle.

Labels: Margin of upper eyelid; Lacrima papilla and punctum; Plica semilunaris; Lacrimal caruncle; Lacrimal papilla; Tarsal glands seen through conjunctiva; Opening of tarsal gland

Eyelids

The eyelids are two thin, surface folds which are strengthened by the presence in each of a firm plate of condensed fibrous tissue, **the tarsus**. Each is covered by thin skin on its external surface and by conjunctiva on its deep surface. In the fat-free subcutaneous tissue

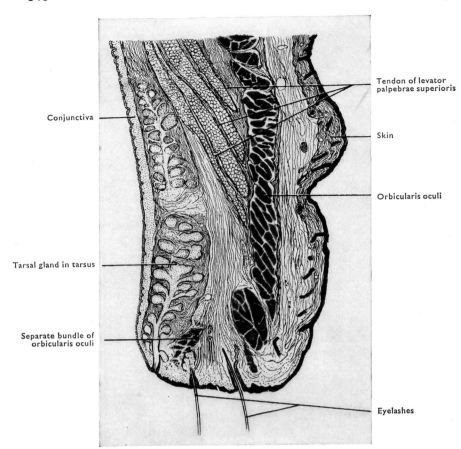

Conjunctiva

Tarsal gland in tarsus

Separate bundle of
orbicularis oculi

Tendon of levator
palpebrae superioris

Skin

Orbicularis oculi

Eyelashes

FIG. 12.29. Diagrammatic sagittal section
through upper eyelid.

both eyelids contain the palpebral fibres of the orbicularis oculi
muscle which approximate the flat, free margins (limbi) of the lids.
The eyelashes (cilia) with the ciliary glands [p. 833] opening into
their follicles arise from the anterior edge of the limbi, while the
ducts of 20–30 modified sebaceous tarsal glands (which lie on the
conjunctiva) open along the posterior edges [FIG. 12.29]. Between
the free margins is the palpebral fissure which forms a transverse
slit opposite the lower margin of the cornea when the eye is gently
closed; when open, the edge of the lower eyelid lies immediately
below the cornea, while the upper covers its superior quarter. The
eyelids meet (commissures of the eyelids) at the lateral and medial
angles of the eye. The eyelashes and tarsal glands cease some 6 mm
from the medial angle where there is a small posteriorly directed
elevation (lacrimal papilla) on the posterior edge of the free margin
of each lid. A minute opening (lacrimal punctum) on the apex of the
papilla marks the beginning of the lacrimal canaliculus through
which lacrimal fluid is transported to the lacrimal sac. Medial to the
papilla the margins of the lids are rounded and enclose a triangular
area (lacus lacrimalis) within which is a small raised island of
modified skin, the lacrimal caruncle. This is surmounted by tiny
hairs with minute sebaceous glands. Lateral to the caruncle, the
conjunctiva passing towards the sclera forms a vertical fold, the
plica semilunaris. This corresponds to the nictitating membrane of
lower vertebrates in which it may contain cartilage and smooth
muscle fibres.

The medial angle of the eye (medial canthus) is partly hidden in
some races, notably Mongolians, by a fold of skin which extends
from the area between the eyebrow and the upper eyelid to the side
of the nose. This epicanthic fold gives an oblique appearance to the
palpebral fissure, and may be well formed in so-called mongoloid
idiots (Down's syndrome).

THE TARSUS

The superior tarsus is larger and firmer than the inferior tarsus and
has a straight lower margin and a thin, curved upper edge into
which is inserted the deep lamella of levator palpebrae superioris.
The inferior tarsus is a strip about half the height of the superior
tarsus [FIG. 12.30], and both have the tarsal glands embedded in
their deep surface. Laterally and medially the tarsi unite in the
lateral and medial palpebral ligaments which are attached respectively to the orbital surfaces of the zygomatic bone and of the frontal
process of the maxilla anterior to the lacrimal fossa. The periosteum
of the orbital margin is continued as a fibrous sheet (orbital septum)
into each eyelid. In the upper lid it fuses with the superficial layer of
levator palpebrae superioris, anterior to the tarsus, but in the lower
lid it is thin and fuses with the inferior tarsus.

Among the follicles of the eyelashes are two or three rows of
modified sweat glands (ciliary glands) with relatively straight ducts
which open into the follicles. Immediately posterior to these is a
small, separate slip of orbicularis oculi in the lid margin, together
with some smooth muscle fibres (Whitnall 1932).

CONJUNCTIVA

This highly sensitive membrane covers the deep surfaces of the
eyelids and is reflected from them on to the anterior surface of the
eyeball at the superior and inferior conjunctival fornices. It is
continuous with the skin at the margins of the eyelids, and is closely
adherent to the deep surfaces of the tarsi and tarsal glands. Here it
is highly vascular and thicker than the transparent part which covers

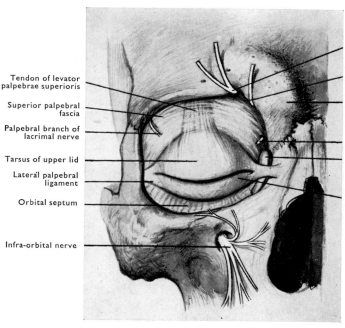

Tendon of levator palpebrae superioris

Superior palpebral fascia

Palpebral branch of lacrimal nerve

Tarsus of upper lid

Lateral palpebral ligament

Orbital septum

Infra-orbital nerve

Supra-orbital nerve

Supratrochlear nerve

Superciliary arch

Infratrochlear nerve

Lacrimal sac

Tarsus of lower lid

FIG. 12.30. Dissection of right eyelid. The orbicularis oculi has been removed.

the sclera and cornea. It is separated from the sclera by the fascial sheath of the eyeball, but is firmly adherent to the cornea where it forms the stratified squamous **anterior corneal epithelium**. Elsewhere its epithelium is stratified columnar, contains goblet cells, and varies in thickness.

VESSELS AND NERVES

A **medial palpebral artery** enters each eyelid from the ophthalmic artery by piercing the orbital septum superior and inferior to the medial palpebral ligament. Each forms a palpebral arch near the free margin of the eyelid, the superior reinforced from the supra-orbital and supratrochlear arteries, the inferior from the facial artery.

Conjunctival **veins** drain into the muscular tributaries of the ophthalmic veins, and pretarsal, palpebral veins enter the facial and superficial temporal veins. **Lymph** plexuses are present on both surfaces of the tarsus; some drain medially to join the lymph vessels that accompany the facial vein to the submandibular lymph nodes, but the main drainage is laterally to the superficial **parotid lymph nodes** (Burch 1939). The **sensory nerve** supply is from the trigeminal nerve via the supratrochlear and supra-orbital to the upper lid and the infra-orbital to the lower lid, while infratrochlear and lacrimal supply both at the corresponding angles [FIG. 12.30]. The **motor** supply to orbicularis oculi is from the facial nerve, that to levator palpebrae superioris is from the oculomotor (striated muscle fibres) and the sympathetic (smooth muscle fibres in that muscle and in the eyelids).

Lacrimal apparatus

LACRIMAL GLAND

The greater, **orbital part** of the lacrimal gland [FIG. 12.31] lies in the lacrimal fossa on the medial surface of the zygomatic process of the

frontal bone, extending down almost to the lateral angle of the eye lateral to the aponeurosis of levator palpebrae superioris. Here it becomes continuous with the **palpebral part** which arches upwards and medially, deep to the aponeurosis, in the root of the upper eyelid. It may extend downwards beyond the lateral angle of the eye. Three to nine **excretory ductules** open into the superolateral part of the superior conjunctival fornix. Numerous, small, **accessory lacrimal glands** open close to both conjunctival fornices (especially the superior) and there are mucous goblet cells in the conjunctival epithelium.

Structure

The lacrimal gland is a tubulo-alveolar, serous gland which moistens the conjunctival sac with a watery fluid containing lysozyme which destroys bacteria. It receives sympathetic, parasympathetic (facial nerve [FIG. 11.79]) and sensory (lacrimal nerve) **nerves**, and branches from the **lacrimal artery**. Its venous drainage is to the **ophthalmic vein**. The lacrimal fluid helps to remove particulate material and flows towards the medial angle of the eye, assisted by contraction of the orbicularis oculi.

LACRIMAL CANALICULI

Just lateral to the lacrimal caruncle the fluid enters the **puncta lacrimalia** [FIG. 12.31] to reach the lacrimal canaliculi. Each canaliculus (approximately 10 mm long) passes upwards or downwards into the corresponding eyelid, and then turning medially is distended to form an **ampulla** before coursing respectively superior or inferior to the medial palpebral ligament, to open into the lacrimal sac a little above its mid-point, close to its fellow. The canaliculi are lined with stratified squamous epithelium surrounded by connective tissue and the lacrimal part of the orbicularis oculi muscle. If the lower eyelid droops away from the eyeball either as a result of paralysis of orbicularis oculi or because of scarring contracture (e.g. after burns), the lacrimal fluid collects in the dependent eyelid and spills over its free margin without reaching the level of the inferior punctum.

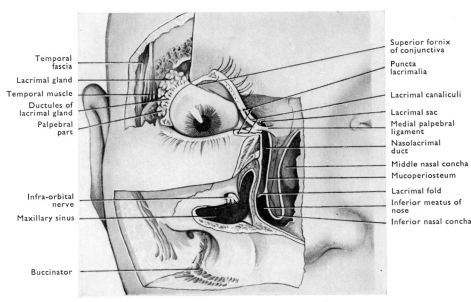

Temporal
fascia

Lacrimal gland

Temporal muscle

Ductules of
lacrimal gland

Palpebral
part

Infra-orbital
nerve

Maxillary sinus

Buccinator

Superior fornix
of conjunctiva

Puncta
lacrimalia

Lacrimal canaliculi

Lacrimal sac

Medial palpebral
ligament

Nasolacrimal
duct

Middle nasal concha

Mucoperiosteum

Lacrimal fold

Inferior meatus of
nose

Inferior nasal concha

FIG. 12.31. Dissection to show the lacrimal
apparatus (semidiagrammatic).

LACRIMAL SAC

The lacrimal sac [FIGS. 12.30 and 12.31] is the blind, upper part of
the nasolacrimal duct through which lacrimal fluid is conveyed to
the nasal cavity. The sac lies in a fossa formed by the lacrimal and
maxillary bones, and is covered laterally by a sheet of fascia which
joins the anterior and posterior edges of the fossa and gives origin to
the lacrimal part of orbicularis oculi. The medial palpebral ligament
and palpebral part of orbicularis oculi lie anterior to the upper half
of the sac.

NASOLACRIMAL DUCT

The nasolacrimal duct [FIG. 12.31] is approximately 18 mm long
and 3–4 mm wide. It passes downwards with a slight inclination
backwards and laterally through a canal formed by the maxilla,
lacrimal bone, and inferior concha, to open into the anterior part of
the inferior meatus of the nose, approximately 30 mm behind the
nostril. The opening, which is variable in size and position, is
guarded by a flap (lacrimal fold) of mucous membrane which
prevents the entry of air into the duct when the nose is blown [FIG.
12.31]. The lining of the sac and duct is stratified columnar
epithelium which is folded in the duct.

Vessels and nerves

The arteries of the lacrimal ducts are derived from the palpebral,
facial, and infra-orbital arteries. The veins form a plexus which
drains to face, orbit, and nose. The canaliculi and sac are supplied
by the infratrochlear nerve, the nasolacrimal duct receives fibres
from the anterior superior alveolar nerve.

RADIOLOGY

The naso-lacrimal duct and lacrimal sac can be demonstrated by
catheterization of the puncta lacrimalia and the injection of a small
quantity of oily contrast medium (dacrocystogram). The method is
particularly valuable in detecting obstructions and injury of the

nasolacrimal duct. The downward course of the duct from the
lacrimal sac to its opening in the inferior meatus of the nose is best
seen in the postero-anterior view.

Development of the eye

The retina and optic nerve are formed by an outgrowth of the full
thickness of the antero-inferior part of the diencephalon [p. 652].
The distal part is distended to form the optic vesicle; the proximal

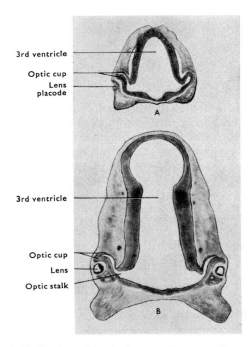

3rd ventricle

Optic cup

Lens
placode

3rd ventricle

Optic cup

Lens

Optic stalk

FIG. 12.32. Sections of developing eye of human embryo. × 20.
A, 5 mm; B, 10 mm.

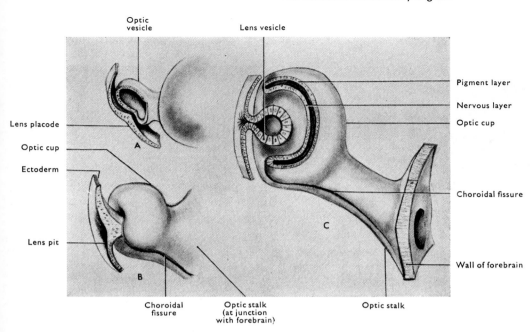

FIG. 12.33. Models of developing human eye. × 100. (From Arey's *Developmental anatomy*, after Ida Mann 1928.)

The lens has been sectioned, and the optic vesicle has been partly cut away in A and C.

A, 4.5 mm embryo; B, 5.5 mm; C, 7.5 mm.

part, elongated and narrow, forms the optic stalk, the rudiment of the optic nerve. The ectoderm overlying the optic vesicle thickens, invaginates, and finally separates from the surrounding ectoderm as the lens vesicle [FIG. 12.33]. As the lens vesicle approaches the optic vesicle, the inferolateral wall of the vesicle is invaginated, virtually obliterating the cavity of the vesicle and turning it into a two-layered optic cup with a deep choroidal fissure [FIG. 12.33] in its inferolateral wall. The fissure extends for some distance into the optic stalk, and the surrounding mesoderm enters the cup through it carrying the rudiments of the central vessels of the retina and assisting in the formation of the vitreous body. At first the artery is continued forwards as a number of branches which form a plexus on the surface of the lens, but by the fifth to sixth month only one remains (the hyaloid artery), and this, together with the remains of the lens plexus, atrophies during the last intra-uterine month, leaving only the hyaloid canal to mark its continuity with the central artery of the retina.

The edges of the choroidal fissure grow together and fuse, thus completing the optic cup and enclosing the future central artery and vein of the retina in the cup and stalk. The distal margin of the cup grows over the superficial surface of the lens to form the epithelium of the iris (iridial part of the retina), while posterior to this the two layers of the cup form the ciliary part of the retina, and further posteriorly the cerebral and pigment layers of the retina.

The vitreous body appears to be formed partly from the inner layer of the cup by fine protoplasmic fibrils which project from its cells. Some of these later become condensed to form the ciliary zonule and the walls of the hyaloid canal, though others remain throughout the vitreous with interfibrillar material probably derived from the mesoderm.

The lens [FIG. 12.25] is formed by the ectoderm in response to the presence of the optic vesicle. At first a somewhat flattened, spherical vesicle, the cells of its deeper wall elongate into and fill the cavity of the vesicle to form the lens fibres, while the superficial cells remain a simple cuboidal layer, the anterior epithelium of the lens. The lens fibres at the centre of the lens are the earliest formed and the longest, while growth in diameter is achieved as new lens fibres are continually formed at the equator by transformation of the cells of the anterior epithelium [FIG. 12.25]. During the greater part of

intra-uterine development the lens is surrounded by a vascular plexus fed by the hyaloid and iridial arteries. The part of this vascular coat superficial to the lens forms the pupillary membrane which normally disappears before birth [p. 841].

The walls of the hollow optic stalk are thickened by the ingrowth of nerve fibres from the cerebral layer of the retina, so that the original cavity is obliterated and the optic nerve formed.

The cerebral layer of the retina consists initially of a neuro-epithelial (ependymal) layer applied to the outer, pigment layer. Cells from the neuro-epithelial layer migrate towards the internal aspect of the optic cup (original external surface of optic vesicle) to form a second (inner) neuroblastic layer. This gives rise to the ganglion cells of the optic nerve, amacrine cells, and glial cells, while the outer layer gives rise to the rods and cones and the bipolar and horizontal cells of the ganglionic layer of the retina, parts of which are therefore derived from both of the primitive neuroblastic layers (Mann 1928).

Mesoderm surrounding the optic cup condenses to form the sclera, cornea, and choroid, while anterior to the lens a cleft (anterior chamber of the eye) appears in it between the cornea and the layer which forms the mesodermal stroma of the iris and the anterior vascular coat of the lens. The deep surface of the iris is the double layer of epithelium derived from the anterior part of the optic cup which also gives rise to the radial myoepithelial elements of the iris.

The eyelids arise as surface, ectoderm-covered folds above and below the cornea. These folds meet anterior to the cornea in the third intra-uterine month, and only separate shortly before birth; though they remain fused until after birth in many mammals. The ectoderm forms the epithelium of the conjunctiva, the anterior epithelium of the cornea, hair follicles of the eyelashes, and the tarsal, ciliary, and lacrimal glands.

The lacrimal sac and nasolacrimal duct are formed as a solid rod of ectodermal cells between the fusing maxillary and frontonasal processes [FIG. 12.60], which are separated superolaterally by the eye. The central cells of the rod disintegrate to form a tube, the lower end of which remains closed until shortly before birth. From the upper end, two small rods of cells which subsequently canalize, grow out to form the lacrimal canaliculi. The lacrimal caruncle is said to be a separated portion of the lower eyelid (Mann 1928).

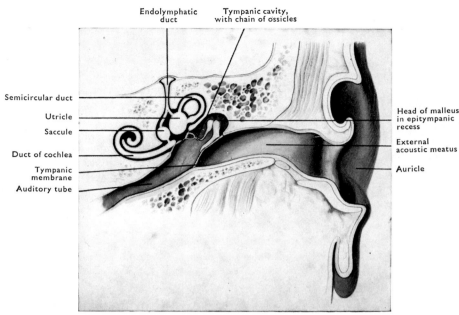

FIG. 12.34. Diagrammatic section of organ of hearing.

The organs of hearing and equilibration

The ear [FIG. 12.34] consists of three parts. (1) The **external ear** is a sound collecting apparatus. It consists of the auricle, continuous with a tube (external acoustic meatus) which conveys air vibrations to the tympanic membrane. This separates the meatus from the middle ear cavity. (2) The **middle ear (tympanic cavity)** is an extension of the nasal part of the pharynx. Vibrations of the tympanic membrane are transmitted across the middle ear cavity by a chain of three ossicles. The movements of the first and last of these are controlled by the actions of minute muscles, thus preventing extreme movements being transmitted to the delicate internal ear which is buried in the petrous temporal bone, medial to the middle ear. (3) The **internal ear** consists of a number of parts, but it is divisible functionally into (i) the **cochlea** concerned with hearing, and (ii) the **vestibular apparatus** concerned with assessing (a) the position of the head relative to gravity, and (b) movements of the head. The sensory endings in both these parts are supplied by the eighth cranial or vestibulocochlear nerve.

THE EXTERNAL EAR

Auricle

The auricle projects from the side of the head at a variable angle (approximately 30 degrees). Its skeleton is a folded plate of elastic fibrocartilage, and its lateral surface [FIG. 12.35] shows two irregular, concentric, C-shaped groves above and behind the opening of the external acoustic meatus. This lies at the antero-inferior end of the deeper, inner groove (the **concha**), partly overlapped from in front by the **tragus**. The two grooves are separated by the **anthelix**, which frequently splits above to enclose the **triangular fossa**, while the outer groove (**scaphoid fossa**) is limited by the rolled, outer margin of the auricle, the **helix**. The helix begins (**crus of the helix**) in the

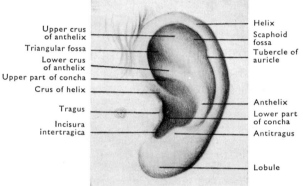

FIG. 12.35. Lateral surface of left auricle.

concavity of the concha above the external acoustic meatus. It ends postero-inferiorly by entering the lobule of the ear close to the anthelix which runs along the superior margin of the lobule and ends in a small protrusion. This **antitragus** usually points laterally, and partly overlapping the postero-inferior part of the concha, is separated from the tragus by the **incisura intertragica**.

The medial surface of the auricle shows a series of elevations which correspond to the depressions on the lateral surface and carry the same names, e.g. eminence of the concha.

The shape of the auricle is extremely variable. In the new-born its length is approximately one-third of that in the adult, and it increases in size and thickness in old age.

STRUCTURE OF THE AURICLE

The skin of the auricle is thin, smooth, and adherent to the perichondrium on the lateral aspect, but mobile on the medial surface. Fine hairs are found over the greater part of the auricle, and coarse hairs (tragi) are present on the tragus, antitragus, and in the incisura intertragica in men. **Sebaceous glands** are present on both surfaces, particularly in the concha and triangular fossa, while sweat glands are virtually limited to the medial surface.

The elastic fibrocartilage is absent from the lobule which is composed of fat and fibro-areolar tissue. The cartilage has the same

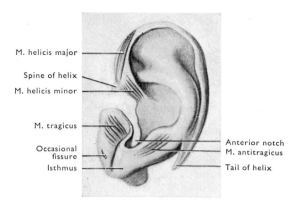

FIG. 12.36. Lateral surface of cartilage of left auricle, with muscles.

M. helicis major
Spine of helix
M. helicis minor
M. tragicus
Occasional fissure
Isthmus
Anterior notch
M. antitragicus
Tail of helix

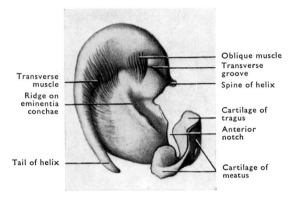

FIG. 12.37. Medial surface of cartilage of left auricle, with muscles.

Transverse muscle
Ridge on eminentia conchae
Tail of helix
Oblique muscle
Transverse groove
Spine of helix
Cartilage of tragus
Anterior notch
Cartilage of meatus

Lymph vessels pass: (1) forwards to the parotid lymph nodes, especially to a node immediately in front of the tragus; (2) downwards to lymph nodes along the external and internal jugular veins; (3) backwards to retro-auricular nodes on the mastoid process. The **nerve supply** to all the muscles is from the facial nerve. The sensory supply is from: (1) the **great auricular nerve** to most of the medial surface and the lower part of the lateral surface; (2) the **auriculotemporal nerve** to the tragus and above, principally on the lateral surface; and (3) to the upper part of the medial surface from the **lesser occipital nerve**. The latter is very variable in its extent and may supply most of the medial surface.

External acoustic meatus

This passage leads from the lower part of the concha to the tympanic membrane. In the adult it is 35 mm long from the tragus, and 24 mm from the medial part of the concha; of this the lateral cartilaginous

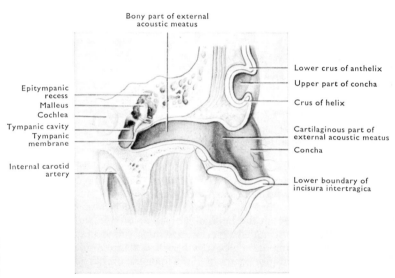

Bony part of external acoustic meatus
Epitympanic recess
Malleus
Cochlea
Tympanic cavity
Tympanic membrane
Internal carotid artery
Lower crus of anthelix
Upper part of concha
Crus of helix
Cartilaginous part of external acoustic meatus
Concha
Lower boundary of incisura intertragica

FIG. 12.38. Coronal section of right ear; anterior half of section, viewed from behind.

contours as the auricle and is continuous with that of the external acoustic meatus by a narrow strip (8–9 mm wide) medial to the deepest part of the incisura intertragica [FIG. 12.37], where the skin of auricle and meatus also become continuous. The cartilage of the helix ends anteriorly in the spine [FIGS. 12.36 and 12.37] and postero-inferiorly in the tail.

Ligaments

An **anterior ligament** extends from the spine of the helix and tragus to the zygomatic process, and a **posterior ligament** from the eminence of the concha to the mastoid part of the temporal bone. Small ligaments unite the various parts of the cartilage.

Muscles

The extrinsic muscles which attach the auricle to the skull and scalp are described elsewhere [p. 288]. In addition there are six rudimentary intrinsic muscles, four on the lateral surface and two on the medial surface [FIGS. 12.36 and 12.37].

Vessels and nerves

The **arteries** to the lateral surface come principally from the superficial temporal artery; those to the medial surface from the posterior auricular artery. The latter sends three or four branches which also give twigs to the lateral surface through the cartilage and round the margin of the helix. The **veins** drain to the corresponding veins, but some communicate with the mastoid emissary vein.

part is about 8 mm long, and the medial bony part about 16 mm but is relatively much shorter in the newborn where the bony part has not formed. The tympanic membrane lies obliquely across the end of the meatus, thus the anterior and inferior walls are longer than the posterior and superior walls [FIGS. 12.38 and 12.39]. The meatus passes medially: the lateral part also inclines forwards and upwards; the intermediate part backwards; and the medial part (the longest) forwards and slightly downwards. Throughout most of its course it is elliptical in transverse section with the greatest diameter directed downwards and backwards, but is more nearly circular close to the tympanic membrane. It is widest at the concha, becomes slightly narrower towards the medial end of the cartilaginous part, and is again constricted to its narrowest (isthmus) near the medial end, 20 mm from the concha. The bony part of the meatus lies between the mastoid process and the head of the mandible, which also abuts slightly on to the cartilaginous part, but is separated from it by a slip of the parotid gland. When the head of the mandible moves forwards, e.g. on opening the mouth, the cartilaginous part is widened.

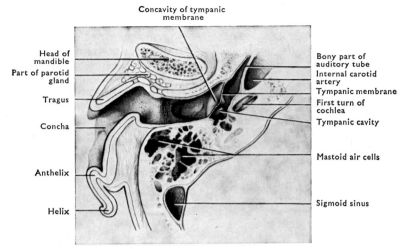

FIG. 12.39. Horizontal section through right ear; upper half of section, viewed from below.

STRUCTURE

The **cartilage of the meatus** is a curved plate which forms only the anterior and inferior walls [FIG. 12.37], the remainder being fibrous tissue. Medially the cartilage and fibrous tissue are fused to the bony part, laterally the cartilage is continuous with that of the tragus. The cartilage of the anterior wall contains two fissures filled with fibrous tissue.

The bony part is described on page 110. In the new-born it consists of the incomplete **tympanic ring** [FIG. 3.85] which is closed superiorly by part of the squamous temporal bone. The margin of the fibrous layer of the tympanic membrane is attached to the tympanic groove on the internal surface of the tympanic ring. A little inferior to its anterosuperior extremity, the medial surface of the tympanic ring has a groove which passes antero-inferiorly and transmits the **anterior ligament of the malleus**, the **chorda tympani nerve**, and the **anterior tympanic artery**. The tympanic ring is joined to the cartilage of the meatus by a fibrous tympanic plate (Symington 1885) which ossifies from the ring to form the bony tympanic part of the adult meatus.

In the **new-born** the lumen of the meatus is small. A lateral funnel-shaped part is continuous with the slit-like medial part which lies between the fibrous tympanic plate inferiorly and the tympanic membrane superiorly. The **tympanic membrane** is close to the surface, is more obliquely set than in the adult, and lies more nearly parallel to the inferior aspect of the skull.

The skin of the external acoustic meatus covers the tympanic membrane and is continuous with the skin of the auricle laterally. It is thick in the cartilaginous part where it carries fine hairs and sebaceous glands, the latter continuing along the posterosuperior wall of the bony part. Numerous enlarged 'sweat' glands are found in the cartilaginous part. These secrete wax or cerumen and are known as **ceruminous glands** [p. 833].

Vessels and nerves

The **arteries** are branches of the superficial temporal (anterior auricular), posterior auricular, and maxillary (deep auricular) arteries; the last also partly supplies the tympanic membrane. The **veins** join the maxillary, external jugular, and pterygoid veins. Lymph drainage is in common with the auricle. Sensory **nerves** are supplied by the auriculotemporal, vagus (auricular branch), and facial nerves [p. 757].

THE MIDDLE EAR

Tympanic cavity

The tympanic cavity is an air-filled space between the tympanic membrane and the internal ear [FIGS. 12.34 and 12.38]. The cavity is lined with **mucous membrane** the epithelium of which varies between simple squamous and ciliated columnar in its various parts. It is continuous with that of the nasal part of the pharynx through the auditory tube. The mucosa also covers: (1) the chain of ossicles (malleus, incus, and stapes) which extends across the cavity from the tympanic membrane to the wall of the inner ear; and (2) the ligaments and tendons passing to the ossicles from the walls of the cavity [FIG. 12.47]. With the exception of a few goblet cells, no glands are found in the tympanic cavity.

The tympanic cavity is little more than a vertical fissure most of which lies medial and parallel to the tympanic membrane [FIG. 12.39]. Its upper part (**epitympanic recess**) extends above the membrane and contains the greater part of the incus and the superior part of the malleus. The vertical and longitudinal measurements are each approximately 15 mm, while its width is 6 mm in the epitympanic recess, 4 mm near its floor, and 1.5 or 2 mm in its central part where the lateral and medial walls bulge into the cavity [FIG. 12.38].

Posteriorly the epitympanic recess is continuous through an opening (the **aditus**) with air-filled spaces in the mastoid process [FIGS. 12.40 and 12.42].

WALLS OF THE TYMPANIC CAVITY

The **roof** or **tegmental wall** is formed by a thin plate of bone (tegmen tympani) which separates it from the dura mater of the floor of the middle cranial fossa. Posteriorly, it covers the mastoid antrum, and anteromedially the semicanal for tensor tympani. It may contain air cells, or be deficient in part, and is formed of cartilage in the child. Anterolaterally it curves downwards to appear in the **tympanosquamous fissure** on the base of the skull.

The **floor** or **jugular wall** is narrower than the roof. It consists of a plate of bone which posteriorly separates the tympanic cavity from the jugular fossa, and anteriorly slopes upwards as the posterior wall of the carotid canal [FIG. 12.39]. The **canaliculus for the tympanic nerve** (branch of the glossopharyngeal nerve) pierces the floor close to the medial wall.

The **medial** or **labyrinthine wall** [FIG. 12.40] is the lateral wall of the internal ear. The bone covering the lateral surface of the base of the cochlea bulges into the middle ear cavity (**the promontory**) and is grooved by the tympanic plexus of nerves. There are two gaps in the medial wall. (1) Posterosuperior to the promontory is the **fenestra vestibuli** (3×1.5 mm) filled by the base of the stapes which is attached to the margin of the fenestra by the narrow elastic, **anular ligament** and has the vestibule of the bony labyrinth medial to it. (2) The **fenestra cochleae** lying postero-inferior to, and almost covered by the promontory, is filled by the **secondary tympanic membrane**—a layer of fibrous tissue covered laterally by the mucosa of the middle ear and medially by the epithelium of the cochlear part of the labyrinth. In the epitympanic recess, above the fenestra vestibuli, is an ill-defined, horizontal ridge produced by the bone covering the **canal for the facial nerve**. This prominence of the canal for the facial nerve passes back into the medial wall of the opening (aditus) to the mastoid antrum, and there lies inferior to the prominence of the **lateral semicircular canal**. Superior to the anterior part of the fenestra vestibuli lies the **processus cochleariformis**—the posterior end of a thin shelf of bone which forms the floor of the semicanal for tensor tympani, and here bends laterally

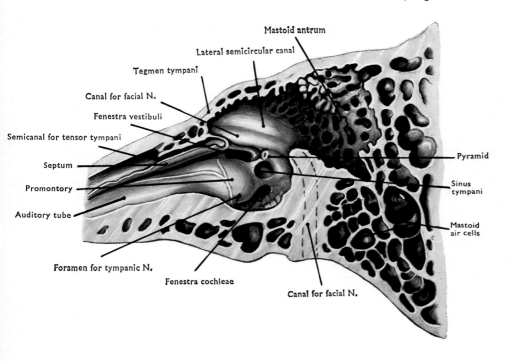

Mastoid antrum

Lateral semicircular canal

Tegmen tympani

Canal for facial N.

Fenestra vestibuli

Semicanal for tensor tympani

Septum

Promontory

Auditory tube

Foramen for tympanic N.

Fenestra cochleae

Canal for facial N.

Pyramid

Sinus tympani

Mastoid air cells

FIG. 12.40. The labyrinthine wall of the tympanic cavity.

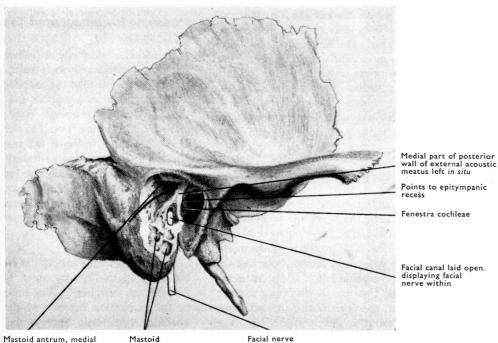

Medial part of posterior wall of external acoustic meatus left *in situ*

Points to epitympanic recess

Fenestra cochleae

Facial canal laid open. displaying facial nerve within

Mastoid antrum, medial wall of which is related to lateral semicircular canal

Mastoid air cells

Facial nerve

FIG. 12.41. Preparation displaying position and relations of right mastoid antrum.

The greater part of the posterior wall of the external meatus has been removed, leaving only a bridge of bone at its medial end; under this a bristle passes from the mastoid antrum to the tympanic cavity.

to form a pulley for the tendon of that muscle passing to the malleus. Immediately posterior to the promontory and between the two fenestrae is a small depression in the medial wall, the sinus tympani. It marks the position of the ampulla of the posterior semicircular canal.

The posterior or mastoid wall [FIGS. 12.40 and 12.46] is pierced superiorly by the round or triangular opening (aditus) into the mastoid antrum. Immediately below this is the fossa for the incus which receives the short crus and posterior ligament of that bone. Further inferiorly lies a small, conical projection (the pyramid) through the apex of which the tendon of stapedius passes anteriorly

to the stapes. The facial nerve descends vertically within the bony posterior wall [FIG. 11.22] and sends a fine branch arching upwards and forwards to supply stapedius. This branch may arise in the bony canal for the facial nerve or immediately after that nerve emerges from the base of the skull. Lateral to the pyramid the small posterior canaliculus for the chorda tympani opens close to the posterior edge of the tympanic membrane [FIG. 12.42].

The lateral or membranous wall [FIG. 12.42] is almost entirely formed by the tympanic membrane. This is inserted into the tympanic groove except superiorly where the groove is missing and the flaccid part of the membrane is attached to the margin of the

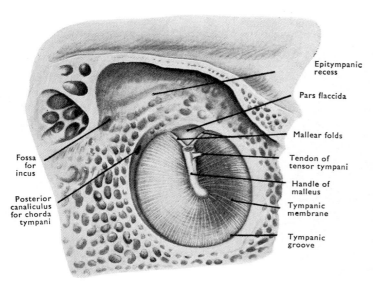

FIG. 12.42. View of left tympanic membrane and epitympanic recess from medial side.

tympanic notch. Superior to this is the lateral wall of the epitympanic recess formed by the squamous temporal bone. Immediately anterior to the upper part of the tympanic membrane, the tegmen tympani turns downwards in the tympanosquamous fissure. Here the anterior ligament of the malleus, the chorda tympani in the **anterior canaliculus**, and the anterior tympanic artery pass through the petrotympanic fissure, inferior to the tegmen tympani.

The carotid wall

Anteromedially the bony walls of the tympanic cavity converge on the openings of two canals which continue in this direction lateral to the carotid canal, to open on the base of the skull in the angle between the greater wing of the sphenoid and the petrous part of the temporal bone. The upper, smaller, **semicanal for tensor tympani** (2 mm in diameter) lies immediately below the tegmen tympani, and above a thin plate of bone which extends posteriorly to the processus cochleariformis. This separates it from the lower, larger **semicanal for** (i.e. bony part of) **the auditory tube** which expands as it enters the middle ear cavity. The floor of the tympanic cavity sloping upwards to the auditory tube is a thin plate of bone immediately above the ascending part of the carotid canal. It is sometimes deficient, and is always perforated by the **caroticotympanic nerves** passing to the tympanic plexus from the internal carotid plexus (sympathetic).

Tympanic membrane

This oblique, almost circular membrane lies at an angle of 55 degrees to the anterior and inferior walls of the external acoustic meatus [FIGS. 12.38 and 12.39]. Its thickened margin, the fibrocartilaginous ring, is inserted into the tympanic groove, but leaves the bone to pass to the lateral process of the malleus from the ends of the tympanic notch as the **anterior and posterior mallear folds** [FIG. 12.42]. Superior to these is the thin, **flaccid part of the tympanic membrane**, while the remainder (**tense part**) is tightly stretched. The handle of the malleus is firmly attached to the tympanic membrane, and its lower end lies opposite the point of maximum indrawing of the membrane, the **umbo** of the tympanic membrane.

The **chorda tympani** crosses the deep surfaces of the fibrous layer of the tympanic membrane and the neck of the malleus, raising a fold of the tympanic mucosa [FIG. 12.46].

STRUCTURE

The basis of the tympanic membrane is a fibrous sheet consisting of connective tissue bundles radiating outwards from the handle of the malleus (stratum radiatum) with an increasing number of circumferential fibres (stratum circulare) towards the periphery. Both groups stop at the fibrocartilaginous ring, and hence are absent from the pars flaccida.

The fibrous sheet is covered on its external surface by skin of the external acoustic meatus (cutaneous layer) and on its internal surface by the mucous membrane of the tympanic cavity (mucosal layer). This mucosa is thickest on the superior part of the membrane and on the bone immediately surrounding it.

OTOSCOPIC EXAMINATION OF TYMPANIC MEMBRANE [FIG. 12.43]

In life the tympanic membrane is pearl-grey in colour, sometimes with a yellowish tinge, and the posterior segment is more transparent than the anterior. The lateral process of the malleus appears as a whitish spot projecting into the meatus close to the anterosuperior edge of the membrane with the mallear folds passing from it to the ends of the tympanic notch. Below this the **handle of the malleus** forms a ridge extending down to a rounded extremity at the umbo. Posterior to the handle of the malleus, the long crus of the incus may be seen through the membrane descending parallel to the handle of the malleus as far as its mid-point. From the umbo a bright area, the cone of light, radiates antero-inferiorly.

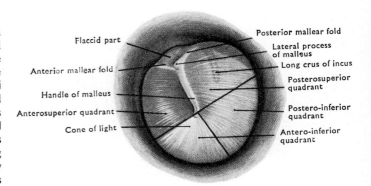

FIG. 12.43. Left tympanic membranes seen from lateral side. The four arbitrary quadrants are indicated by solid lines and by the handle of the malleus.

VESSELS AND NERVES

Separate **arteries** supply the two surfaces of the membrane, but they anastomose by small branches which pass mainly through the periphery. The inner surface is supplied mainly by the **auricular branch of the occipital artery** and the **anterior tympanic artery**, the outer surface by the **deep auricular artery**. The last two arise from the maxillary artery. The **veins** of the outer surface drain to the external jugular vein, those from the inner surface drain with the rest of the tympanic cavity [p. 857]. **Lymph vessels** drain in a comparable manner to the veins, those from both sides of the membrane communicating freely. The **auriculo-temporal nerve,** the auricular branch of the **vagus**, and the facial nerve supply the lateral aspect, while the medial aspect is supplied via the **tympanic plexus** by the facial (and glossopharyngeal) nerve. When a foreign body is lodged in the meatus, or it is subject to inflammation, the supply by the vagus accounts for the tendency to develop an intractable cough or even vomiting.

Mastoid antrum and mastoid cells [FIGS. 3.46 and 12.40]

The mastoid antrum is an air-filled space in the base of the mastoid process. It is continuous anteriorly with the epitympanic recess through the aditus. The antrum is approximately 12–15 mm in length, 8–10 mm high, and 6–8 mm wide. The **roof** is the tegmen tympani, the **floor** the mastoid process, the **medial wall** the petrous temporal bone, and the **lateral wall** the squamous temporal bone posterosuperior to the meatus. The latter wall is only 1–2 mm thick at birth, but increases to 10 mm by the ninth year, and 15 mm in the adult.

The **mastoid cells** form as air-filled extensions of the antrum as the mastoid process grows during childhood. They are very variable in size and extent. They may extend to the tip of the mastoid process and upwards into the squamous part, passing into the roof of the external acoustic meatus, and even medially into the jugular process of the occipital bone. When the cells are very extensive they may only be separated by a thin lamina of bone from the posterior cranial fossa and the sigmoid sinus. This lamina may be perforated, allowing infection to spread readily from the mastoid air cells to the sinus. In other cases the mastoid cells are small and fill only a fraction of the mastoid process. The antrum and cells are lined by an extension of the mucous membrane of the tympanic cavity consisting of flattened cells firmly bound by connective tissue to the bone.

Auditory tube

The auditory tube connects the tympanic cavity to the nasal part of the pharynx. By allowing air to enter or leave the cavity it balances the pressure on both sides of the tympanic membrane and so allows it to vibrate freely. If the tube is blocked, the air in the tympanic cavity is absorbed into the mucosal vessels with loss of pressure and increasing concavity of the tympanic membrane. If associated with infection, there may be a collection of fluid under increasing pressure in the tympanic cavity with bulging of the tympanic membrane into the meatus. In both cases there is increasing deafness due to the failure of the tympanic membrane to vibrate freely.

The tube is approximately 3.5 cm long. The cartilaginous, anteromedial two thirds passes backwards, laterally (45 degrees to the sagittal plane) and upwards (40 degrees to the horizontal) to join the more horizontal bony third at an angle of 160 degrees. The **cartilage** is a triangular plate with its base in the pharyngeal wall and its apex at the bony part. It is folded downwards to cover the medial and superolateral surfaces of the tube, the remaining surface being supported by fibrous tissue uniting the edges of the plate. The base of the cartilage produces a hook-like ridge (**tubal elevation** [FIG. 6.29]) above and posteromedial to the opening of the tube into the nasal part of the pharynx. A fold of mucous membrane (**salpingopharyngeal fold**, covering the salpingopharyngeus muscle) descends in the pharyngeal wall from the posterior edge of the elevation. The size of tube and cartilage gradually decreases towards the bony part, and its folded edge is firmly fixed to the base of the skull in the groove between the greater wing of the sphenoid and the petrous temporal bone. At its pharyngeal orifice, the tube is attached to the posterior edge of the medial pterygoid lamina and then lies medial to the tensor veli palatini muscle which partly arises from it and also separates it from the mandibular nerve and the middle meningeal artery. The levator veli palatini lies between the tube and the pharyngeal mucosa behind the tubal elevation. In swallowing and yawning, the tube is opened by the simultaneous contraction of tensor veli palatini and salpingopharyngeus which are attached to opposite sides of the cartilage plate. The **bony part** of the tube lies between the petrous and tympanic parts of the temporal bone, below the semicanal for tensor tympani muscle, lateral to the carotid canal. It increases in diameter as it approaches the tympanic cavity.

Mucous membrane

In the bony part this is thin, tightly bound to the periosteum, and lined with a single layer of columnar ciliated cells. In the

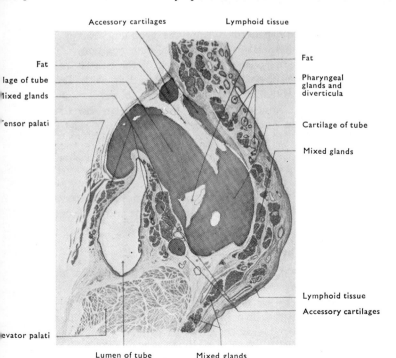

Accessory cartilages   Lymphoid tissue

Fat

Fat

...lage of tube

Pharyngeal glands and diverticula

...ixed glands

Cartilage of tube

...ensor palati

Mixed glands

Lymphoid tissue

Accessory cartilages

...evator palati

Lumen of tube   Mixed glands

FIG. 12.44. Transverse section through auditory tube near its pharyngeal end.

cartilaginous part it is thick, thrown into folds, and lined with pseudostratified columnar ciliated epithelium containing goblet cells. There are also some mucus-secreting tubulo-alveolar **glands** near the pharyngeal end, and scattered lymphocytes which may be aggregated to form a small **tubal tonsil** in the same region.

Vessels and nerves

The ascending pharyngeal, middle meningeal, and artery of the pterygoid canal supply the tube. Its **veins** drain into the pterygoid and pharyngeal plexuses. Sensory **nerves** are derived from the tympanic plexus and the pharyngeal branch of the pterygopalatine ganglion.

In the foetus the pharyngeal opening of the tube is below the level of the hard palate, at birth level with it, at the fourth year 3–4 mm above it, and in the adult 10 mm above it. In the child the orifice of the tube is small, but the tube is relatively wider, shorter, and more horizontal than that in the adult.

Auditory ossicles

The three ossicles, malleus, incus, and stapes—form a jointed arch between the tympanic membrane and the fenestra vestibuli.

THE MALLEUS [Fig. 12.45]

This is the largest of the three (8–9 mm long) with a head, a neck, a handle, and anterior and lateral processes.

The **handle** is fused to the tympanic membrane by its periosteum and a thin layer of cartilage. It begins at the umbo, where its inferior end is curved anterolaterally, and ascends with a forward inclination (125–150 degrees to the horizontal) to become continuous with the conical **lateral process** immediately inferior to the pars flaccida. This process has the mallear folds attached to it, is the upper limit of the attachment of the malleus to the tympanic membrane, and protrudes the membrane laterally. Above this the narrow **neck** and rounded **head** rise vertically into the epitympanic recess at a slight angle to the handle. On the medial surface of the upper part of the handle is a slight **tubercle** for the attachment of the tendon of tensor tympani. From the anterior surface of the neck, on the level of the pars flaccida, the slender **anterior process** passes towards the petrotympanic fissure, with the **anterior ligament** continuing from it through that fissure. The rounded **head** has an oval facet for articulation with the incus on the posteromedial surface below the summit. The facet is divided by an oblique ridge into a larger, upper, posterior surface, and a smaller, lower, medial surface. The inferior margin of the facet is prominent and forms a marked spur where it meets the oblique ridge. Immediately below the spur is the point of attachment of the **lateral ligament** of the malleus.

THE INCUS [Fig. 12.45]

This is named from its supposed likeness to an anvil on which the hammer (malleus) strikes. The anterior surface of the body of the incus has a saddle-shaped articulation for the head of the malleus, and a recess for its spur. The pyramidal **short crus** [Fig. 12.46] extends horizontally backwards from the body to the **fossa for the incus**, to which its cartilage tip is attached by the **posterior ligament of the incus**. The **long crus** extends vertically downwards from the body (posteromedial and parallel to the handle of the malleus) and in the same sagittal plane but approximately at right angles to the short crus. Inferiorly the long crus bends medially to end in the **lenticular process**, a small knob of bone which articulates with the stapes.

THE STAPES [Fig. 12.45]

This stirrup-shaped bone has a small, laterally directed head which articulates with the lenticular process of the incus, and medial to this a slightly constricted neck gives rise to the two crura which pass to the anterior and posterior ends of the base. The anterior crus is shorter and more nearly in line with the neck than the posterior crus which arches backwards before sweeping medially. In the fresh state the two crura are united by a membrane. The base is oval or kidney-shaped and fits into the fenestra vestibuli.

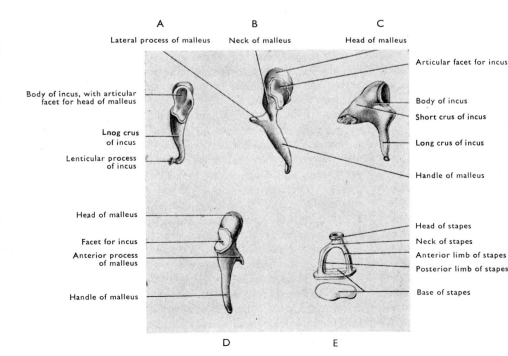

FIG. 12.45. Auditory ossicles of left ear (enlarged about three times).

A, incus, seen from the front; B, malleus, seen from behind; C, incus, and D, malleus, seen from the medial side; E, stapes, seen from above and from the medial side.

JOINTS OF THE AUDITORY OSSICLES

The incudomallear and incudostapedial joints are both synovial; the former is saddle-shaped, the latter of the ball-and-socket type, and both have considerable amounts of elastic tissue in their capsules.

Ligaments of the auditory ossicles

The malleus has anterior, superior, and lateral ligaments [FIG. 12.46]. The anterior ligament consists of (1) a part which passes forwards from the anterior process to run through the petrotympanic fissure to the spine of the sphenoid. Thence it is continued as the sphenomandibular ligament—the whole representing parts of the first pharyngeal arch (Meckel's) cartilage. It may contain some muscle fibres (Soemmerring 1806; Walls 1946). (2) The other part of the ligament passes to the anterior end of the tympanic notch. The superior ligament passes upwards from the head to the roof of the epitympanic recess. The lateral ligament is short and fan-shaped. Its fibres pass from the malleus to the posterior edge of the tympanic notch. With the anterior ligament it forms the axis around which the malleus moves (Helmholtz 1868).

The incus is held in position by the capsule of the incudomallear joint and its posterior ligament on which it can swing medially and laterally with the head of the malleus. The so-called superior ligament is a mucosal fold.

The base of the stapes is covered with hyaline cartilage on its rim and medial surface. That on the rim is attached to the margin of the fenestra vestibuli by the elastic anular ligament of the stapes.

MUSCLES OF THE AUDITORY OSSICLES [FIG. 12.46]

The tensor tympani arises from the upper surface of the cartilage of the auditory tube, the adjacent greater wing of the sphenoid, and the walls of its semicanal. In the semicanal it runs posterolaterally to end in a tendon which turns laterally around the processus cochleariformis, and crosses the tympanic cavity to the anteromedial surface of the handle of the malleus near its upper end. When the muscle contracts it pulls the handle of the malleus medially, tenses the tympanic membrane, and thus reduces the amplitude of its oscillations. Nerve supply: mandibular nerve.

The stapedius arises from the walls of a canal which leads postero-inferiorly from the pyramid. Its tendon passes forwards through the apex of the pyramid to the neck of the stapes. On contraction it pulls the neck of the stapes posteriorly, tilts the base in the fenestra vestibuli tightening the anular ligament and reducing the oscillatory range. Nerve supply: the facial nerve.

MOVEMENTS OF THE AUDITORY OSSICLES

The handle of the malleus moves with the tympanic membrane. When this moves inwards, the head of the malleus moves outwards around the axis of its anterior and lateral ligaments. This movement rotates the incus around its short crus, swinging the long crus inwards, and pressing the stapes into the fenestra vestibuli. This movement causes a wave of compression to pass through the perilymph of the internal ear, the displacement of this fluid by the stapes being allowed for by a corresponding bulging of the secondary tympanic membrane. The spur on the head of the malleus locks the incudomallear joint when the tympanic membrane is forcibly pushed inwards. In the opposite movement, excessive movement of the malleus is prevented by sliding of the head of the malleus on the incus, and thus the stapes is not torn from the fenestra vestibuli.

VESSELS AND NERVES OF THE TYMPANIC CAVITY

The arteries are: (1) the anterior tympanic, which enters through the petrotympanic fissure; (2) the posterior tympanic from the stylomastoid which enters by that foramen; (3) inferior tympanic [p. 903]; (4) superior tympanic which accompanies the lesser petrosal nerve. The veins drain to the pterygoid plexus, to the superior petrosal and sigmoid sinuses, and through the tegmen tympani to the dura mater. A plexus of lymph vessels in the mucous membrane drains to parotid and retropharyngeal lymph nodes. The sensory supply to the mucous membrane is through the tympanic plexus [p. 761] and from fibres in the chorda tympani nerve in the lateral wall [FIG. 12.46 and p. 757].

In foetal life much of the tympanic cavity is filled with loose mesenchymatous tissue, the epithelial-lined space being relatively small. By birth the epithelial cavity has its definitive arrangement but is filled with fluid which is absorbed shortly after birth. The

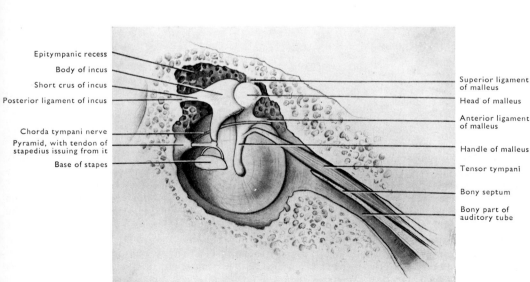

Epitympanic recess
Body of incus
Short crus of incus
Posterior ligament of incus
Chorda tympani nerve
Pyramid, with tendon of stapedius issuing from it
Base of stapes

Superior ligament of malleus
Head of malleus
Anterior ligament of malleus
Handle of malleus
Tensor tympani
Bony septum
Bony part of auditory tube

FIG. 12.46. Left tympanic membrane and chain of auditory ossicles (seen from medial side).

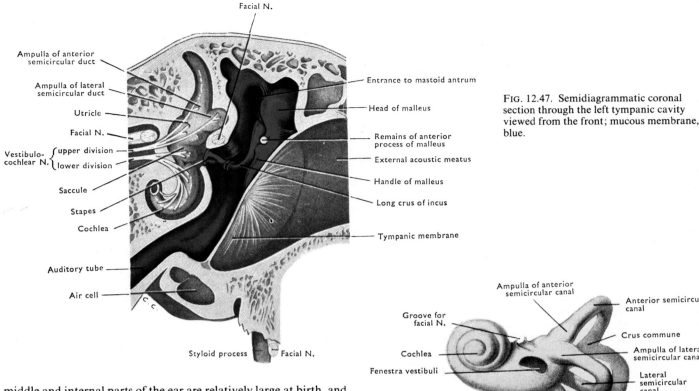

FIG. 12.47. Semidiagrammatic coronal section through the left tympanic cavity viewed from the front; mucous membrane, blue.

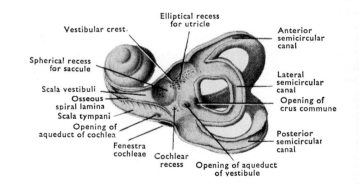

FIG. 12.48. Lateral view of left bony labyrinth.

FIG. 12.49. Lateral view of interior of left bony labyrinth.

middle and internal parts of the ear are relatively large at birth, and the amount of postnatal growth is small.

THE INTERNAL EAR

The internal ear lies buried in the petrous temporal bone. It consists of a complex of membranous tubes and spaces which are continuous with each other and are filled with endolymph—the membranous labyrinth. This fits very loosely in a similarly shaped bony cavity (the bony labyrinth). The space between the membrane and bone is filled with fluid (perilymph) and loose connective tissue. The parts of the internal ear from anterior to posterior are: (1) the spiral duct of the cochlea in the bony cochlea; (2) the saccule and utricle in the bony vestibule, the first continuous with the duct of the cochlea; (3) three semicircular ducts which arise from the utricle and are enclosed in bony semicircular canals.

The bony labyrinth

VESTIBULE

This ovoid middle part of the bony labyrinth communicates anteriorly with the bony cochlea and posteriorly with the semicircular canals [FIG. 12.49] by five circular openings. Its long axis (6 mm) lies anteroposteriorly, vertically it measures 4–5 mm, and transversely 3 mm. Its lateral wall, pierced by the fenestra vestibuli [FIG. 12.48], is part of the medial wall of the tympanic cavity. Its medial wall corresponds to the fundus of the internal acoustic meatus through which the various peripheral branches of the vestibulocochlear nerve pass to supply the sensory endings in the membranous labyrinth. On the medial wall of the vestibule anteriorly is a circular depression (spherical recess) pierced by 12–15 tiny foramina (macula cribrosa media) for the nerves to the saccule which lies in the depression [FIG. 12.49]. Above and behind the spherical recess is a curved ridge, the vestibular crest, which is

triangular in shape above (pyramid of the vestibule) but inferiorly splits to enclose a small depression (cochlear recess) pierced by nerves to the vestibular extremity of the cochlear duct. In the roof of the vestibule, posterosuperior to the vestibular crest, is the elliptical recess containing the utricle. The inferomedial part of the recess and the adjacent pyramid are pierced by 25–30 nerves to the utricle and the ampullae of the anterior and lateral simicircular ducts. Posterior to the cochlear recess is a furrow which deepens to a canal posteriorly. This aqueduct of the vestibule pierces the petrous temporal bone and opens through a slit-like fissure on its posterior surface midway between the internal acoustic meatus and the groove for the sigmoid sinus. It is 8–10 mm long, and transmits the endolymphatic duct and a small vein. The cochlea communicates with the vestibule through an elliptical opening which leads anteriorly from the spherical recess into the scala vestibuli of the cochlea [FIG. 12.49]. This opening is bounded inferolaterally by a thin plate of bone (the **osseous spiral lamina**) which begins immediately anterior to the cochlear recess, passes forwards into the cochlea, and partly separates the scala vestibuli from the scala tympani inferolateral to it.

SEMICIRCULAR CANALS

The three semicircular canals (anterior, posterior, and lateral) lie posterosuperior to the vestibule [FIGS. 12.48 and 12.49] into which they open. Each forms approximately two-thirds of a circle, and has one dilated end, the **ampulla**. The canals are oval in transverse section measuring 1–1.5 mm in diameter, while the ampullae are approximately 2 mm across. The anterior and posterior canals together open into the vestibule by three apertures, having one common stem, **crus commune**, while the lateral canal has two separate openings.

The **anterior semicircular canal** is 15–20 mm long, and lies in a vertical plane across the long axis of the petrous temporal bone deep to the arcuate eminence [FIGS. 3.54 and 3.82], with its convexity upwards. The ampulla lies at the anterolateral end, and opens into the vestibule immediately superior to the opening of the ampulla of the lateral canal. The posteromedial end joins the superior end of the posterior semicircular canal to form the **crus commune** (4 mm long) which opens into the superomedial part of the vestibule. The **posterior semicircular canal**, 18–20 mm long, lies in a vertical plane parallel to the long axis of the petrous temporal bone, with its convexity directed posteriorly. The ampulla is on the inferior extremity and opens into the posteroinferior part of the vestibule close to six to eight small foramina (**macula cribrosa inferior**) for the transmission of nerves to that ampulla. The **lateral semicircular canal**, 12–15 mm long, is nearly horizontal, but slopes upwards and forwards at approximately 30 degrees to the horizontal when the head is erect. Its convexity faces posterolaterally, and the ampulla at the anterolateral end opens into the vestibule immediately above the fenestra vestibuli. This end of the canal produces an elevation of the medial wall of the aditus to the mastoid antrum above that for the facial nerve [FIG. 12.47]. The semicircular canals of the two internal ears are arranged in pairs. The anterior canal of one side is parallel to the posterior canal of the other, and the two lateral canals are in the same plane.

COCHLEA

The bony cochlea surrounds and forms the central pillar (**modiolus**) for a spiral tube of $2\frac{1}{2}$–$2\frac{3}{4}$ turns arranged in the form of a short cone. The **base** of the cochlea lies at the fundus of the internal acoustic

meatus; the apex or **cupula** points anterolaterally and lies close to the semicanal for tensor tympani. Thus the axis of the cochlea is at right angles to the long axis of the petrous temporal bone. The diameter of the base is approximately 9 mm; from the base to apex is 5 mm. The tube is 32 mm long and 2 mm in diameter at its widest in the basal turn, the first part of which bulges towards the tympanic cavity forming the **promontory**.

The **modiolus** is about 3 mm long, conical in shape, and has a thin shelf of bone, the **osseus spiral lamina**, projecting from it into the spiral canal like the thread of a screw. This partly divides the canal into scala vestibuli and scala tympani [FIG. 12.50]; a division which is completed by the **basilar lamina**, which membrane passes from the free edge of the lamina to the outer wall of the canal. The osseous spiral lamina begins in the floor of the vestibule superomedial to the fenestra cochleae and scala tympani; it is separated from a smaller ridge of bone on the outer wall of the basal turn of the spiral canal (the **secondary spiral lamina**) by the slit which is closed by the basilar lamina [FIG. 12.50]. At first at right angles to the modiolus, the osseous spiral lamina becomes progressively angled towards the apex, and ends there in a sickle-shaped margin—the **hamulus** of the spiral lamina. The hamulus forms part of the boundary of an aperture, the **helicotrema**, through which the scalae vestibuli and tympani are continuous with each other in the fresh state.

The base of the modiolus forms the cochlear area of the fundus of the internal acoustic meatus [FIG. 12.51]. It is pierced by a large number of foramina arranged in a spiral form, **tractus spiralis foraminosus**. Each foramen transmits fibres of the cochlear part of the vestibulocochlear nerve longitudinally through the modiolus [FIG. 12.50] together with a branch of the labyrinthine artery. Where each foramen approaches the base of the corresponding part of the osseous spiral lamina it turns outwards to enter a small **spiral canal** of the modiolus which contains the cell bodies (**spiral ganglion**) of the sensory neurons of the cochlear part. These cells send peripheral processes outwards through a number of minute canals in the osseous spiral lamina to reach the basilar lamina.

The **scala tympani** begins at the fenestra cochleae inferolateral to the smaller **scala vestibuli** which begins in the vestibule. They take a spiral course around each other into the cochlea so that the scala vestibuli rapidly becomes lateral to the scala tympani (i.e. nearer the apex of the cochlea), and retains this position to the helicotrema, becoming larger than the scala tympani in the apical coils. Near the beginning of the scala tympani, a small orifice on its medial wall leads into a narrow canal, **aqueduct of the cochlea**, which passes inferomedially through the temporal bone to open on the upper border of the jugular fossa close to the posterior surface of that bone. This extension of the scala tympani has been said to communicate with the subarachnoid space.

The lateral end (fundus) of the **internal acoustic meatus** abuts on the medial wall of the vestibule and the base of the modiolus. It is divided into upper and lower areas by a **transverse crest** [FIG.

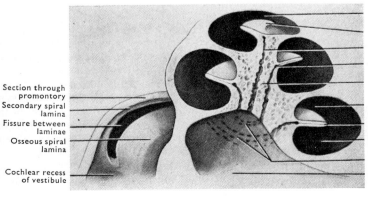

Section through promontory
Secondary spiral lamina
Fissure between laminae
Osseous spiral lamina

Cochlear recess of vestibule

Cupula
Hamulus
Central canal
Spiral canal
Modiolus
Scala vestibuli
Osseous spiral lamina
Scala tympani
Tractus spiralis foraminosus
Internal acoustic meatus

FIG. 12.50. Section of bony cochlea.

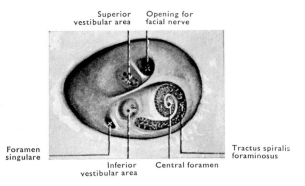

FIG. 12.51. Fundus of left internal acoustic meatus divided into upper and lower areas by transverse crest.

12.51]. The large single foramen for the facial nerve pierces the anterior part of the upper area, anterior to the **superior vestibular area** which transmits nerve fibres and minute arteries to the utricle and the ampullae of the anterior and lateral semicircular ducts (q.v.). The cochlear area lies in the anterior part of the lower area, while the posterior part (**inferior vestibular area**) transmits nerves to the saccule. Behind this, the **foramen singulare** carries vestibular nerve fibres to the ampulla of the posterior semicircular duct.

Membranous labyrinth

The membranous labyrinth [FIGS. 12.52 and 12.59] is an ectodermal cavity loosely fitted within the bony labyrinth and filled with endolymph. The spaces between it and the bony walls are filled with loose connective tissue and **perilymph**—a fluid akin to cerebrospinal fluid but containing more protein and much less chloride, and different in composition from **endolymph** which has a high potassium and low sodium content. The membranous labyrinth in the vestibule consists of two communicating sacs, the utricle and saccule. The semicircular ducts arise from the utricle and traverse the semicircular canals, while the saccule gives rise to a blind

cochlear duct which extends through the spiral cochlea sandwiched between the scalae vestibuli and tympani.

UTRICLE AND SACCULE

The **utricle** is an elongated sac in the posterosuperior part of the vestibule. The anterosuperior part of the utricle lies in the elliptical recess and receives the ampullae of the anterior and lateral semicircular ducts [FIG. 12.52], while the central part receives the crus commune and the non-ampullated end of the lateral semicircular duct. The inferomedial part receives the ampulla of the posterior semicircular duct. The **macula of the utricle** is a pale, oval (3 mm × 2.3 mm) thickening of the floor and adjacent anterior wall of the anterior part of the utricle. This sensory area is innervated by utricular nerve fibres of the vestibular part of the vestibulocochlear nerve which pierce the pyramid of the vestibule.

The **saccule** is an oval vesicle (3 mm × 2 mm), smaller than the utricle. It lies in the antero-inferior part of the vestibule, and has a thickening of its anterior wall, **macula of the saccule** (approximately 1.5 mm in diameter) innervated by the saccular fibres of the vestibulocochlear nerve, and similar but at right angles to the macula of the utricle. The posterosuperior part of the saccule is fused with the utricle (Dickie 1920), and the endolymphatic duct arises from the saccule just inferior to this. A small **utriculosaccular duct** passes from the utricle to the **endolymphatic duct**, which then traverses the aqueduct of the vestibule and ends as a blind dilatation, the **endolymphatic sac**, external to the dura mater of the posterior cranial fossa on the posterior surface of the petrous temporal bone.

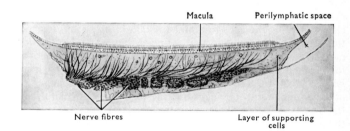

FIG. 12.53. Vertical section of wall of anterior part of utricle with the macula and bundles of nerve fibres.

Structure

The walls of the utricle and saccule consist of epithelium and a thin layer of connective tissue which is continuous through the connective tissue of the perilymph space with the endosteum of the bony vestibule. The lining epithelium is separated from the fibrous layer by a basement membrane, and is simple squamous in type except at the maculae where the epithelium (**neuroepithelium**) is of a columnar type and the connective tissue is greatly thickened and contains many nerve fibres. The epithelium of the maculae [FIG. 12.53] consists of supporting and **sensory cells**. Hair-like processes project from the sensory cells into a jelly-like mass of polysaccharide–protein complex. This **membrana statoconiorum** contains numerous, minute, crystalline bodies (the otoliths or **statoconia**) consisting of a calcium carbonate–protein complex. The free surfaces of the flask-shaped sensory cells have a mass of long microvilli projecting from them. These narrow markedly towards their bases, and increase progressively in length towards the side of the cell where a cilium is present. The deep parts of these cells contact the ends of the nerve fibres which enter the neuroepithelium. One type of

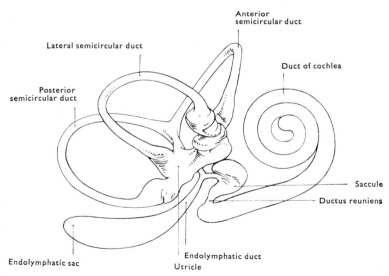

FIG. 12.52. Diagrammatic representation of model of right membranous labyrinth. (From a model by J. K. Milne Dickie.)

sensory cell is virtually enclosed in a cup-shaped nerve ending, while a second type appears to have a ribbon-like sensory ending together with another type of ending which has all the appearances of being efferent (Wersall 1956). The significance of these two types of sensory cell, one apparently capable of being modified by an efferent fibre, or fibres, is not known, but it is believed that the action of gravity on the statoconia is an adequate stimulus to the hair cells. The maculae of the utricle and saccule being at right angles to each other are differently affected by the pull of gravity whatever the position of the head relative to that force, and they may also be stimulated by inertial movements of the statoconia on acceleration or deceleration of the head.

SEMICIRCULAR DUCTS

The semicircular ducts are oval in transverse section, and are attached to one wall of the much larger semicircular canal in which they lie [FIG. 12.54]. In the bony ampulla, each duct is dilated and almost fills the space available: the wall of the **membranous ampulla** is folded transversely from one side into its lumen to form a ridge or crest (**crista ampullaris** [FIG. 12.58]). On the sides and apex of the crista the epithelial lining is thickened to form an innervated structure identical with that of the maculae but for the absence of statoconia; the processes of the hair cells also being embedded in a jelly-like mass, the **cupula** (Hillman 1974). The cupula surmounts the crest and may completely block the flow of endolymph through the ampulla. When the head is rotated in the plane of any pair of semicircular ducts, the inertia of the endolymph alters the pressures on the surfaces of the cupulae of the two ampullae so that they swing like doors carrying the microvilli of the hair cells with them—the two cupulae moving in opposite directions relative to the openings of their ampullae into the utricle. Movement of a cupula in one direction increases the rate of firing of its sensory cells, while movement in the opposite direction decreases it. Hence any rotation causes an asymmetrical flow of impulses in the vestibular nerve fibres. This is reversed either by rotation in the opposite direction or simply by stopping a rotation after the endolymph is moving with the semicircular ducts—the genesis of the subjective sensation of reversed rotation when a rotatory movement is suddenly stopped.

Nerve impulses generated in the hair cells enter the brain through the vestibular part of the vestibulocochlear nerve, and these evoke reflex movements of the eye, trunk, and limbs, and the conscious sensation of movement of the head.

DUCT OF THE COCHLEA

This spiral, blind, tube [FIG. 12.55] lies in the bony cochlea between the external part of the osseous spiral lamina and the outer wall of the cochlea. It is compressed between the scala vestibuli (which is separated from it by the delicate **spiral membrane—vestibular wall**), and the scala tympani which is separated from it by the basilar lamina and osseous spiral lamina (**tympanic wall**). These two scalae unite around the blind end of the cochlear duct (**caecum cupulare**) at the helicotrema. The basal part of the cochlear duct lies in the cochlear recess of the vestibule, and is connected to the lower part of the saccule by the short, conical **ductus reuniens** [FIG. 12.52]. The **basilar lamina** extends from the free edge of the osseous spiral lamina to the **basilar crest** on the outer wall of the cochlea [FIG. 12.56] where the endosteum is thickened to form the **spiral ligament** of the cochlea. The **spiral membrane** lies between the apical surface of the osseous spiral lamina and the outer wall of the cochlea; the duct of the cochlea enclosed between it and the basilar lamina being triangular in transverse section. The spiral membrane consists of little more than a layer of flattened mesothelium of the scala vestibuli applied to the somewhat thicker epithelial cells of the cochlear duct which bear short microvilli projecting into the cochlear duct.

On the outer wall of the cochlear duct, the epithelium attached to the endosteum of the bony cochlea, is thickened, stratified, and contains capillaries between its mitochondria rich cells. This **stria vascularis** is probably concerned with the secretion or maintenance of the endolymph. It bulges into the cochlear duct to form the **spiral prominence**, which contains a larger vessel (the **vas prominens**) close to the attachment of the basilar lamina, from which it is separated by the **external spiral sulcus** [FIG. 12.56].

The **epithelium of the cochlear duct** is modified to form the complex spiral organ on the internal part of the basilar lamina, while internal to this the endosteum on the apical surface of the osseous spiral lamina is considerably thickened (**limbus of the spiral lamina**) and grooved externally (**internal spiral sulcus** [FIG. 12.56]). The **tympanic** and **vestibular lips** of this groove pass respectively to the basilar lamina and the tectorial membrane. The dendrites of the spiral ganglion cells perforate the tympanic lip as they pass outwards through the osseous spiral lamina on their way to the spiral organ.

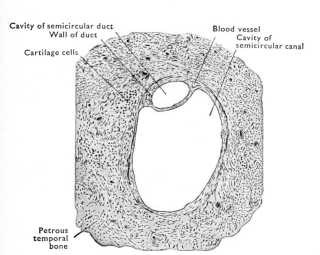

FIG. 12.54. Section of a semicircular canal and duct.

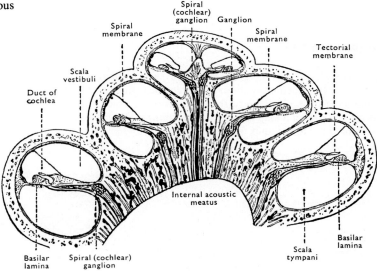

FIG. 12.55. Axial section of human cochlea. (Schaefer's *Essentials of histology*.)

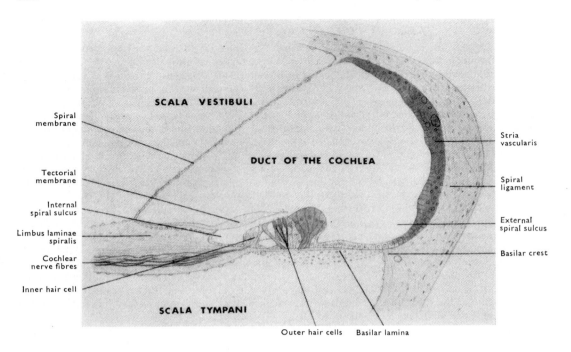

SCALA VESTIBULI

DUCT OF THE COCHLEA

SCALA TYMPANI

Spiral
membrane

Tectorial
membrane

Internal
spiral sulcus

Limbus laminae
spiralis

Cochlear
nerve fibres

Inner hair cell

Stria
vascularis

Spiral
ligament

External
spiral sulcus

Basilar crest

Outer hair cells   Basilar lamina

FIG. 12.56. Section across duct
of cochlea. (Retzius 1881.)

The surface of the limbus of the spiral lamina is markedly ridged towards the vestibular lip where the ridges project like a row of teeth (**acoustic teeth**). The ridges and grooves between them are covered by the epithelium of the cochlear duct, and these cells form a thin cuticle on their free surface, the **tectorial membrane**. This membrane thickens markedly as it projects outwards from the free edge of the vestibular lip to lie over the spiral organ [FIG. 12.57].

The **basilar lamina** lies between the tympanic lip of the limbus and the basilar crest of the spiral ligament, and carries the spiral organ on its thin internal part, while its outer part is thicker and striated. On the tympanic surface is a layer of vascular connective tissue with one larger vessel, the **spiral vessel**, coursing along the basilar lamina under the spiral organ. The width of the basilar lamina increases from the base (0.2 mm) of the cochlea to the apex (0.35 mm), and its overall length is 32 mm (Hardy 1938; Keen 1940).

SPIRAL ORGAN

This complex sensory organ consists of several elements formed from the thickened epithelium of the cochlear duct on the basilar lamina [FIG. 12.57]. Each of these elements seen on radial section forms one part of a long row of cells spiralling on the basilar lamina from the base to the apex of the cochlea.

1. Towards the internal edge of the spiral organ, two rows (internal and external) of tall pillar cells (rods of Corti) surround the innermost of three spaces in the spiral organ, the triangular **internal tunnel** (of Corti). This communicates between the external pillar cells with the paracuniculus (space of Nuel) immediately lateral to them. These supporting, pillar cells are filled with microtubules and have their bases widely spread on the basilar lamina, but lean towards each other so that they fuse at the free surface between the inner and outer hair and phalangeal cell groups. At the fusion of the pillar cells, the head of the inner cell splits into medial and lateral beak-like (coracoid) processes, the latter overlapping the laterally curved, finger-like (phalangeal) process of the corresponding lateral pillar cell, which separates the paracuniculus from the cochlear duct. The outer pillar cells lie more obliquely, are longer, and less numerous than the inner pillar cells, and groups of nerve fibres from the spiral ganglion pass between the intermediate portions of both rows of pillar cells, crossing the inner tunnel between them.

Hair cells and phalangeal (Deiters') cells

A smaller group of these cell complexes lies internal to the pillar cells, while a larger group lies external to them. The phalangeal cells have their bases on the basilar lamina and extend towards the free surface (**reticular membrane**) of the spiral organ. Approximately half way they suddenly narrow, thus forming a recess in which the base of a hair cell is situated; the narrowed part of the phalangeal cell passes between the surrounding hair cells to reach the reticular

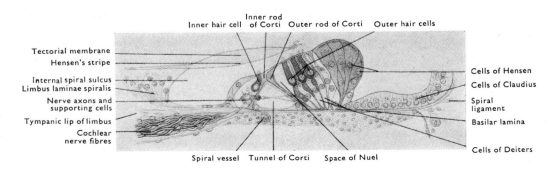

Inner rod
of Corti

Inner hair cell   Outer rod of Corti   Outer hair cells

Tectorial membrane

Hensen's stripe

Internal spiral sulcus
Limbus laminae spiralis

Nerve axons and
supporting cells

Tympanic lip of limbus

Cochlear
nerve fibres

Cells of Hensen

Cells of Claudius

Spiral
ligament

Basilar lamina

Cells of Deiters

Spiral vessel   Tunnel of Corti   Space of Nuel

FIG. 12.57. Transverse section
of spiral organ from central coil
of duct of cochlea. (Retzius
1881.)

membrane. Each phalangeal cell contains a bundle of fibrils, some of which pass from the basilar lamina to the base of the hair cell, while the remainder pass through the narrow process of the cell to expand on the reticular membrane and separate the free surfaces of adjacent hair cells. The inner hair cell–phalangeal cell complex (approximately 3500 hair cells) forms a single row along the length of the basilar lamina. The outer complex (approximately 12 000 hair cells) forms four rows in the apical region of the cochlea and three rows in the basal region, the inner and outer complexes being separated by the expanded laminae of the heads of the pillar cells. The hair cells are cylindrical in shape and their free ends protrude between the expanded ends of the phalangeal cells forming the reticular membrane. Forty or more microvilli project from the free ends and are arranged as in the hair cells of the vestibular apparatus. In life they lie against or may even be embedded in the tectorial membrane. Nerve fibres reach the hair cells by piercing the tympanic lip below the inner hair cells, and supplying them, pass outwards between the pillar cells and across the inner tunnel and the paracuniculus to reach the outer hair cells.

External to the outer hair cells is the **outer tunnel**, a space which communicates with the paracuniculus and is limited laterally by a mound of external limiting cells (of Hensen) which are larger and more numerous than the internal limiting cells internal to the inner hair cells [FIG. 12.57]. The external limiting cells are attached to the basilar lamina by thin processes, but expand towards the free surface to form a prominent ridge. On the basilar membrane external to the limiting cells is a layer (or layers) of large cuboidal supporting cells (of Claudius).

The **reticular membrane** consists of the heads of the outer pillar cells, the processes of the external phalangeal cells, and the surfaces of the external limiting cells. The free ends of the hair cells project through it in rows.

The tectorial membrane

This shelf of jelly-like material containing a mixture of delicate fibrils projects outwards from the vestibular lip, and thickening over the internal part of the spiral organ, lies on the processes of the hair cells. Externally it rapidly thins, and ends in an irregular margin close to the external limiting cells. Internally it lies on the limbus between the vestibular lip and the attachment of the spiral membrane.

Sound waves in the air cause movements of the tympanic membrane and ear ossicles so that similar waves are transmitted to the perilymph at the fenestra vestibuli, and so to the basilar membrane. It was long thought that this was a tuned membrane, the short basal parts of which would respond to high frequencies while the long apical parts would respond to low frequencies. However, the entire basilar membrane responds to low frequency vibrations; moreover the basilar membrane is lax and not taut. It has been shown that the basal part responds with nerve impulses to high and low frequencies, while the apical part responds to low frequencies only. Thus it is not clear exactly how individual frequencies are

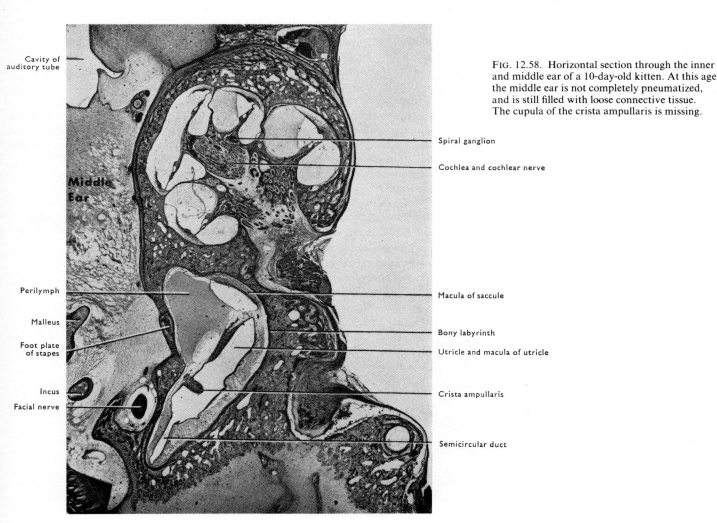

Cavity of auditory tube

Middle Ear

Perilymph

Malleus

Foot plate of stapes

Incus

Facial nerve

Spiral ganglion

Cochlea and cochlear nerve

Macula of saccule

Bony labyrinth

Utricle and macula of utricle

Crista ampullaris

Semicircular duct

FIG. 12.58. Horizontal section through the inner and middle ear of a 10-day-old kitten. At this age the middle ear is not completely pneumatized, and is still filled with loose connective tissue. The cupula of the crista ampullaris is missing.

recognized by the cochlea and transmitted to the brain. There are efferent fibres in the vestibulocochlear nerve which pass to the cochlea, and these fibres can reduce the nerve impulses arising from stimulation of the cochlea by sound (Rasmussen 1953; Galambos 1956).

Vestibulocochlear nerve [FIG. 12.59]

This nerve divides in the internal acoustic meatus into cochlear and vestibular parts.

The **cochlear part** consists of nerve fibres which are the central processes of cells in the spiral ganglion in the modiolus of the cochlea. From the ganglion they pierce the modiolus in bundles, those from the apical coil of the ganglion in the central foramen, the more basal parts sending fibres through the foramina of the tractus spiralis foraminosus and the cochlear recess. The peripheral processes of the ganglion cells pierce the osseous spiral lamina and the tympanic lip of the limbus to reach the inner hair cells or to pass between the pillar cells and across the inner tunnel of the spiral organ to the outer hair cells. Other coarser fibres accompany the former to the internal edge of the spiral organ, and then turn to take a spiral course on the internal part of the basilar membrane.

The **vestibular part** consists of the processes of the bipolar cells of the vestibular ganglion in the internal acoustic meatus. The ganglion is usually in two (or three) parts. The superior ganglion sends fibres through the superior vestibular area to the macula of the utricle, the ampullae of the anterior and lateral semicircular ducts, and the anterior part of the macula of the saccule. The inferior part sends fibres to the macula of the saccule, and through the foramen singulare to the ampulla of the posterior semicircular duct. A branch from the vestibular to the cochlear part of the nerve appears to contain some of the efferent cochlear fibres. A branch has been described passing to the macula of the saccule from the basal part of the spiral ganglion (Hardy 1934), but there is no evidence that fibres from that macula join the cochlear part of the nerve (Stein and Carpenter 1967).

Vessels of internal ear

The **arteries** are the labyrinthine, a branch of the basilar or anterior inferior cerebellar artery, and the stylomastoid, a branch of the posterior auricular or occipital artery. The former passes with the vestibulocochlear nerve and is distributed with its branches. The veins drain through the various apertures at the lateral extremity of the internal acoustic meatus, and unite to form a labyrinthine vein which passes to the inferior petrosal sinus or the sigmoid sinus. Small veins traverse the aqueducts of the vestibule and cochlea to enter the superior and inferior petrosal sinuses respectively.

DEVELOPMENT OF THE EAR

EXTERNAL EAR

The development of the external and middle ears is intimately connected with the pharyngeal arches and their intervening grooves and pouches [pp. 55–8]. The **auricle** is developed from the dorsal ends of the first and second pharyngeal arches each of which forms three temporary tubercles by proliferation of its mesoderm [FIG. 12.60]. These surround the upper part of the first ectodermal groove which forms the external acoustic meatus. The part played by each arch in the formation of the auricle is not clear, but it is likely that the first (mandibular) only forms the tragus, while the second forms the remainder of the auricle (Jones and I-Chuan 1934). The ectoderm of the dorsal part of the first groove initially is in contact with the entoderm of the first pouch, though later separated from it by the ingrowth of a considerable thickness of mesoderm. Subsequent proliferation of the ectoderm to form a solid **meatal plate** carries it medially towards the entoderm of the tympanic cavity, thinning the intervening mesoderm to form the fibrous part of the tympanic membrane and the cartilage of the handle of the malleus. The central cells of the meatal plate then break down, extending the meatus medially to the remaining layer of ectoderm which forms the cutaneous layer of the tympanic membrane. Extension of the entodermal lining of the tympanic cavity to cover the superomedial surface of the tympanic membrane completes that structure.

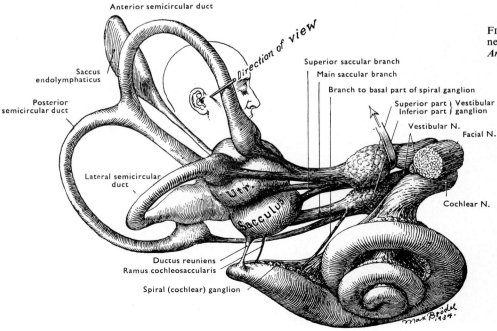

FIG. 12.59. Distribution of vestibulocochlear nerve to the membranous labyrinth. (M. Hardy, *Anatomical Record*, 1934.)

Anterior semicircular duct

Direction of view

Saccus endolymphaticus

Posterior semicircular duct

Lateral semicircular duct

Superior saccular branch
Main saccular branch
Branch to basal part of spiral ganglion
Superior part | Vestibular
Inferior part | ganglion
Vestibular N.
Facial N.
Cochlear N.

Utr.
Sacculus

Ductus reuniens
Ramus cochleosaccularis

Spiral (cochlear) ganglion

Max Brödel 1934.

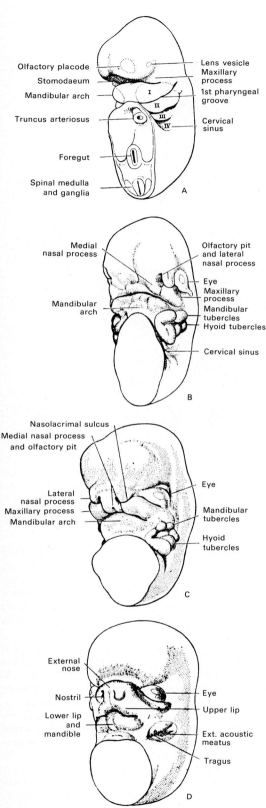

FIG. 12.60. Ventrolateral view of head in series of human embryos, showing development of face and external ear. Note the lateral and dorsal migration of the auricle coincident with the formation of the lower lip and mandible.

A, 6 mm embryo (×11); B, 12 mm (×7.5); C, 14 mm (×7.5); D, 18 mm (×6).

I, II, III, IV, pharyngeal arches.

MIDDLE EAR AND AUDITORY TUBE

This cavity is formed from the dorsal wings of the first and second pharyngeal pouches. The caudal growth of the second pharyngeal arch, combined with a forward growth of the third arch medial to it, draws the dorsal wings of the first and second pharyngeal pouches posterolaterally from the pharyngeal wall (lateral to the third arch). This compounds them into a single cavity with the second arch and its contained facial nerve in the posterolateral wall, while the first and third arches (with the corresponding nerves) respectively on the anterolateral and posteromedial walls. This complex extension of the pharyngeal cavity, the **tubotympanic recess**, forms the auditory tube and the middle ear cavity, and its subsequent extension in the sixth month of pregnancy produces the **mastoid antrum** and after birth the **mastoid cells**. At this stage the tubotympanic recess is very oblique with its future lateral and medial walls lying inferolateral and superomedial respectively, with the rudiment of the membranous labyrinth, the otic vesicle, on the superomedial wall. The later dorsal extension of the part forming the middle ear cavity brings the labyrinth into the medial wall and makes that wall more vertical. The dorsal ends of the cartilages of the first and second pharyngeal arches articulate superior to the tubotympanic recess. These cartilages form the **malleus, incus,** and **stapes**. At least part of the malleus is formed from the first arch cartilage (Lightoller 1939), and the stapes (perforated by the transitory artery of the second arch) is formed from the second, but the origin of the incus at least is in doubt (Hanson, Anson, and Strickland 1962). Subsequently the mesoderm surrounding these ear ossicles is absorbed as the tympanic cavity extends dorsolaterally to surround and cover them with the mucous lining of the middle ear. The ear ossicles begin to ossify in the third month of pregnancy; the malleus from one centre for the head and handle with another for the anterior process; the incus from a single centre in the upper part of the long crus; the stapes from a single centre in its base.

The tensor tympani and stapedius muscles are developed in the dorsal ends of the first and second pharyngeal arches respectively, and are supplied by the corresponding fifth and seventh cranial nerves.

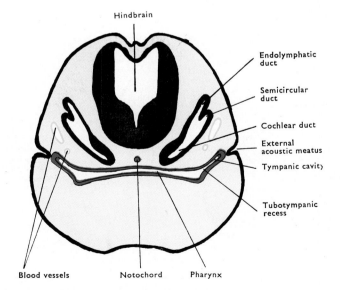

FIG. 12.61. Diagram of transverse section through the head of an embryo, showing the rudiments of the three parts of the ear and their relation to the tubotympanic recess and the first pharyngeal groove.

INTERNAL EAR

The epithelium of the membranous labyrinth is derived from the ectoderm by a thickening, **auditory placode**, which appears at the side of the hindbrain dorsal to the second pharyngeal arch. This placode sinks into the underlying mesoderm to form an auditory pit, which finally separates from the surface ectoderm as the **otic vesicle**. As it sinks into the mesoderm beside the hindbrain at the origin of the seventh and eighth cranial nerves, it elongates vertically and gives off from its medial side a dorsally directed diverticulum, the **endolymphatic duct** [FIG. 12.62]. Between the fifth and sixth weeks of embryonic life the ventral part of the vesicle extends anteromedially as a blind process, the beginning of the **cochlear duct**. This slowly increases in length, and coiling on itself, forms all the turns of the cochlear duct by the twelfth week. From the posterosuperior part of the vesicle three disc-like evaginations arise lateral to the endolymphatic duct. The central parts of each disc coalesce and disappear leaving only the peripheral **semicircular ducts**. Of these the first to develop is the anterior, and the last the lateral, though all three are formed ·by the beginning of the second

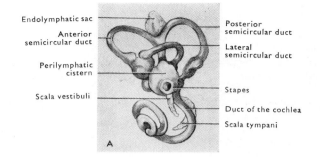

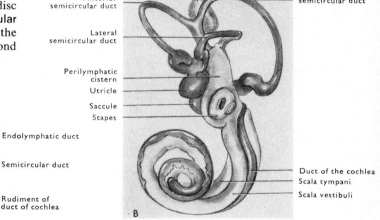

FIG. 12.64. Membranous labyrinths of human embryos (Streeter 1918.)

A, 50 mm; B, 85 mm.

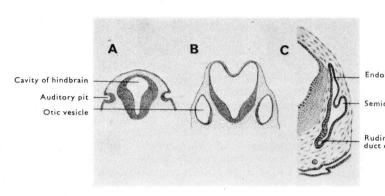

FIG. 12.62. Sections through hindbrain of foetal rabbits to illustrate development of membranous labyrinth.

A, B, and C, successive phases in the development of the auditory pit and the otic vesicle.

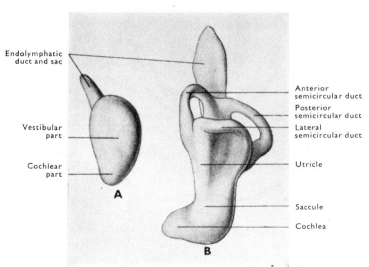

FIG. 12.63. The left membranous labyrinth seen from the lateral side.

A, at about the 4th week; B, at about the 5th week. (W.His, Jr.)

month. At the attachment of the endolymphatic duct the wall of the vesicle is constricted, thus separating the utricle posterosuperiorly from the saccule antero-inferiorly. This constriction extends into the base of the endolymphatic duct dividing it into a Y-shaped tube connecting the utricle and saccule. A further constriction occurs between the saccule and the cochlear duct to form the **ductus reuniens**.

The **epithelial lining**, at first columnar, becomes cubical except at the terminations of the vestibulocochlear nerve fibres. Here it forms the complex, columnar epithelium of the maculae and cristae, and of the spiral organ. Later much of the epithelium becomes more flattened, e.g. on the spiral membrane, in the semicircular ducts, and in the non-macular areas of the utricle and saccule. Two ridges appear on the basal surface of the cochlear duct. The inner, formed by connective tissue, is the limbus laminae spiralis, while the outer is epithelial and forms the spiral organ by differentiation of its cells (Bast, Anson, and Gardner 1947).

The connective tissue surrounding the otic vesicle differentiates into an inner, loose reticular layer, a middle precartilaginous layer, and an outer zone of cartilage, the otic (auditory) capsule [p. 135]. With the growth of the membranous labyrinth, the surrounding cartilage is eroded, and the loose reticular layer is progressively absorbed to produce the perilymphatic spaces surrounding the membranous labyrinth. Initially these are formed as a number of separate spaces which subsequently coalesce and expand at the expense of the precartilage and cartilaginous zones from which the loose connective tissue of the perilymphatic spaces is formed. The perilymphatic space first appears as a cistern around the utricle and saccule [FIG. 12.64], and from this the **scala vestibuli** extends along the cochlear duct to meet the **scala tympani** at the helicotrema about the sixteenth intra-uterine week. The scala tympani develops independently and slightly earlier than the scala vestibuli. The perilymphatic spaces in the semicircular canals contain more connective tissue than the scalae of the cochlea, but all communicate with the space surrounding the utricle and saccule.

Olfactory organ

The general anatomy of the nasal cavity and paranasal sinuses is dealt with in the respiratory system. This section is concerned only with the olfactory apparatus.

The olfactory epithelium lies in the posterosuperior part of the nasal cavity. This brownish-yellow mucous membrane covers the superior conchae and the upper 1 cm of the sides of the nasal septum, the total area being about 10 cm² but the area decreases greatly with age (Naessen 1970). The epithelial covering is tall pseudostratified columnar with three types of cells. (1) The cell bodies and dendrites of the bipolar **olfactory sensory cells.** These cells have their round nuclei in the middle of the thickness of the epithelium [FIG. 12.65] and conspicuous neurofibrils in their cytoplasm. The distal process (dendrite) passes superficially between the **supporting cells** (oval nuclei), and expanding into a rounded olfactory vesicle among the long microvilli on their surface, gives rise to 6–8 extremely long, non-motile cilia which contain a considerable amount of lipid, and run parallel to the surface of the epithelium buried in the surface mucus. (2) The supporting cells contain axial filaments and pigment granules and have a number of long microvilli projecting from their surface. They are packed around the projecting dendrites and are fused to them at the surface. (3) A thin layer of basal cells surrounds the axons of the sensory cells, which leave the epithelium surrounded by sheath cells, and form bundles that pierce the cribriform plate of the ethmoid to enter the olfactory bulb.

The olfactory mucous membrane contains many serous, tubuloalveolar **glands** whose secretion contains many enzymes. It appears that the secretions of these glands wash away substances dissolved in the surface mucus and make it sensitive to new substances passing into solution on the surface of the mucous membrane.

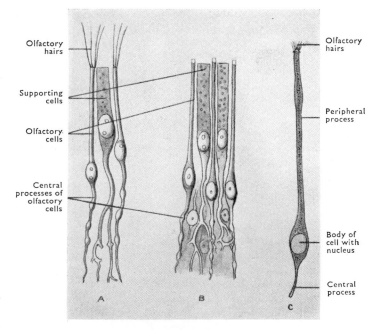

FIG. 12.66. Olfactory and supporting cells.
A, frog; B, human; C, human olfactory cell.

The organ of taste

The peripheral organs which are believed to be mainly responsible for the sensation of taste are oval or flask-shaped collections of cells, **taste buds,** in the epithelium of the oral and pharyngeal cavities. They are most numerous on the tongue, especially on the opposed walls of the vallate papillae [FIG. 12.67], and on the foliate papillae anterior to the palatoglossal arch. They are also present: (1) on the fungiform papillae of the back and sides of the tongue, though less numerous towards the tip; (2) on the anterior part of the soft palate; (3) on the posterior surface of the epiglottis (especially its lateral parts); (4) on the superior surfaces of the arytenoid cartilages; and (5) on the posterior wall of the oral part of the pharynx. The number of taste buds is maximal in children, but decreases rapidly with age

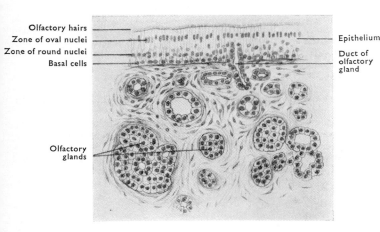

FIG. 12.65. Section through olfactory mucous membrane.

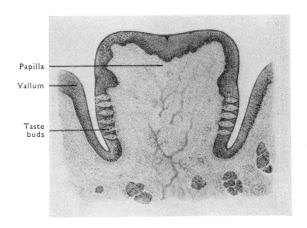

FIG. 12.67. Section through vallate papilla of human tongue.

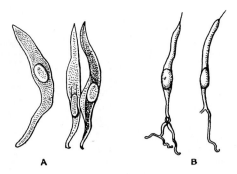

FIG. 12.68. Isolated cells from taste bud of rabbit (Engelmann.)
A, supporting cells; B, gustatory cells.

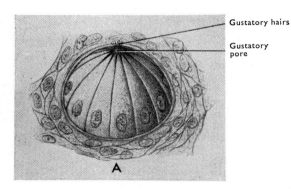

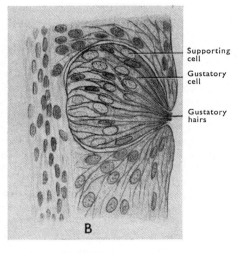

FIG. 12.69. Taste buds from folia linguae of rabbit.
A, three-quarter surface view; B, vertical section.

(Arey, Tremaine, and Monzingo 1935). Many salivary glands in the tongue open into the sulci of the vallate papillae. Their secretions first dissolve and subsequently wash away the substances which stimulate the taste buds.

STRUCTURE [FIG. 12.69]

The taste buds are pale-staining structures buried in the darker epithelium. They taper from expanded bases to end in a few fine, hair-like processes which project through a minute **gustatory pore** in the superficial surface of the epithelium. Most of the cells of the taste buds [FIG. 12.68] are elongated, slightly curved spindles (supporting cells) which enclose a group (5–15) of centrally placed gustatory cells. Each of these slender, rod-like cells has a nucleus near its middle, a peripheral process which ends in a gustatory hair at the gustatory pore, and a central process which branches at the base of the taste bud.

INNERVATION

Nerve fibres approaching taste buds in the epithelium branch frequently to form a subepithelial plexus from which some branches pass into the bases of the taste buds, while others closely surround them or pass into the epithelium between them. On the anterior two-thirds of the tongue they are innervated by fibres from the chorda tympani branch of the facial nerve (through the lingual nerve), while those on the posterior third, including the vallate and foliate papillae are supplied by the glossopharyngeal nerve. On the palate they are again supplied by the facial nerve through its greater petrosal branch which sends fibres into the palatine nerves, while the vagus (internal branch of the superior laryngeal nerve) supplies those on the epiglottis and arytenoid cartilages. Thus the nerves which supply taste buds also carry parasympathetic fibres to the associated glands.

It is not known if the taste buds are the sole sources of appreciation of the four taste sensations (salt, bitter, sweet, acid) for it seems that such sensations may be appreciated from areas where taste buds are not present. Whether the nerve fibres which end in the epithelium around the taste buds are also concerned with taste sensation is not known, but subjective testing is remarkably unreliable as the sensation of 'taste' is usually a combination of taste, texture (ordinary tactile sensation), and smell. Thus injuries to the trigeminal nerve, which is not thought to transmit any fibres concerned with the innervation of taste buds, may cause a subjective alteration in the sensation of taste, and interference with the sensation of smell markedly alters the subjective sense of taste.

References

AREY, L. B., TREMAINE, M. J., and MONZINGO, F. L. (1935). The numerical and topographical relations of taste buds to human circumvallate papillae throughout the life span. *Anat. Rec.* **64**, 9.

BAST, T. H., ANSON, B. J., and GARDNER, W. D. (1947). The developmental course of the human auditory vesicle. *Anat. Rec.* **99**, 55.

BOYD, I. A. (1962). The structure and innervation of the nuclear bag muscle fibre system and the nuclear chain muscle fibre system in mammalian muscle spindles. *Phil. Trans. B* **245**, 81.

BURCH, G. E. (1939). Superficial lymphatics of the human eyelids observed by injection *in vivo. Anat. Rec.* **73**, 443.

CAUNA, N. (1959) The mode of termination of the sensory nerves and its significance. *J. comp. Neurol.* **113**, 169.

—— (1964). In *Advances in biology of the skin*, Vol. VI, Aging. Pergamon, New York.

—— and MANNAN, G. (1958). The structure of human digital Pacinian corpuscles (*Corpuscula lamellosa*) and its functional significance. *J. Anat.* (*Lond.*) **92**, 1.

CLARK, W. E. LE GROS (1971). *The tissues of the body*, 6th edn. Oxford University Press.

—— and BUXTON, L. H. D. (1938). Studies in nail growth. *Br. J. Derm.* **50**, 221.

COOPER, S. and DANIEL, P. M. (1949). Muscle spindles in human extrinsic eye muscles. *Brain* **72**, 1.

DARIAN-SMITH, I. and JOHNSON, K. O. (1977). Temperature sense in the primate. *Br. Med. Bull.* **33**, 143.

DICKIE, J. K. M. (1920). Note on the anatomy of the membranous labyrinth. *J. Laryng.* **35**, 76.

DUPERTUIS, C. W., ATKINSON, W. B., and ELFTMAN, H. (1945). Sex differences in pubic hair distribution. *Hum. Biol.* **17**, 137.

FISON, J. (1920). The relative positions of the optic disc and macula latea to the posterior pole of the eye. *J. Anat. Physiol.* **54**, 184.

GALAMBOS, R. (1956). Suppression of auditory nerve activity by stimulation of efferent fibres to the cochlea. *J. Neurophysiol.* **19**, 424.

GILCHRIST, M. L. and BUXTON, L. H. D. (1939). The relation of finger-nail growth to nutritional status. *J. Anat. (Lond.)* **73**, 575.

HANSON, J. R., ANSON, B. J., and STRICKLAND, E. M. (1962). Branchial sources of the auditory ossicles in Man. *Arch. Otolaryng.* **76**, 200.

HARDY, M. (1934). Observations on the innervation of the macula sacculi in Man. *Anat. Rec.* **59**, 403.

—— (1938). The length of the organ of Corti in Man. *Am. J. Anat.* **62**, 291.

HELMHOLTZ, H. VON (1868). *The mechanism of the ossicles of the ear and membrana tympani*, trans. Buck, A. H. and Smith, N., 1873. New York.

HILLMAN, D. E. (1974). Cupular structure and its receptor relationship. *Brain Behav. Evol.* **10**, 52.

JALOWY, B. (1939). Über die Entwicklung der Nervenendigungen in der Haut des Menschen. *Z. ges. Anat. I. Z. Anat. Entwickl.-Gesch.* **109**, 344.

JONES, F. W. (1941). Tension lines, cleavage lines and hair tracts in Man. *J. Anat. (Lond.)* **75**, 248.

—— and I-CHUAN, W. (1934). The development of the external ear. *J. Anat. (Lond.)* **68**, 525.

KEEN, J. A. (1940). A note on the length of the basilar membrane in Man and in various mammals. *J. Anat. (Lond.)* **74**, 524.

KIDD, W. (1903). *The direction of hair in animals and man*. London.

LEWIS, T. (1927). *The blood vessels of the human skin and their responses*. London.

LIGHTOLLER, G. H. S. (1939). On the comparative anatomy of the mandibular and hyoid arches and their musculature. *Trans. zool. Soc. Lond.* **24**, 349.

LOCKWOOD, C. B. (1885). The anatomy of the muscles, ligaments and fasciae of the orbit, including an account of the capsule of Tenon, the check ligaments of the recti, and the suspensory ligament of the eye. *J. Anat. Physiol.* **20**, 1.

MANN, IDA C. (1928). *The development of the human eye*. Cambridge University Press.

MONTAGNA, W. (1956). *The structure and function of skin*. New York.

NAESSEN, R. (1971). An enquiry on the morphological characteristics and possible changes with age in the olfactory region of man. *Acta Otolaryng.* **71**, 49.

NYBERG-HANSEN, R. (1965). Anatomical demonstration of gamma motoneurons in the cat's spinal cord. *Exp. Neurol.* **13**, 71.

PENROSE (1968). Medical significance of finger-prints and related phenomena. *Br. med. J.* **2**, 321.

POLYAK, S. (1957). *The vertebrate visual system*. Chicago.

POPOFF, N. W. (1934). The digital vascular system. *Arch. Path.* **18**, 295.

QUILLIAM, T. A. (1966). Unit design and array patterns in receptor organs. In *Touch, heat and pain* (eds. A. V. S. de REUCK and J. KNIGHT). Ciba Symposium, London.

RASMUSSEN, G. L. (1953). Further observations on the efferent cochlear bundle. *J. comp. Neurol.* **99**, 61.

RETZIUS, G. (1881–4). *Das Gehörorgan der Wirbelthiere, morphologisch-histologische Studien*. Stockholm.

SCHULTZ, A. H. (1940). The size of the orbit and of the eye in primates. *Am. J. phys. Anthrop.* **26**, 389.

SHANTHAVEERAPPA, T. R. and BOURNE, G. H. (1963). New observations on the structure of the Pacinian corpuscle, and its relation to the perineurial epithelium of peripheral nerves. *Am. J. Anat.* **112**, 97.

—— —— (1966). Perineurial epithelium: a new concept of its role in the integrity of the nervous system. *Science* **154**, 1464.

SINCLAIR, D. (1967). *Cutaneous sensation*. Oxford University Press, London.

SOEMMERRING, S. T. (1806). *Icones Organi Auditus Humani*, Table 2, Fig. 10. Frankfurt.

STEIN, B. M. and CARPENTER, M. B. (1967). Central projections of portions of the vestibular ganglia innervating specific parts of the labyrinth in the rhesus monkey. *Am. J. Anat.* **120**, 281.

SYMINGTON, J. (1885). The external auditory meatus in the child. *J. Anat. Physiol.* **19**, 280.

THOMSON, A. (1912). *Atlas of the eye*. Oxford.

TRAQUAIR, H. M. (1949). *An introduction to clinical perimetry*, 6th edn. London.

TROTTER, M. (1924). The life cycles of hair in selected regions of the body. *Am. J. phys. Anthrop.* **7**, 427.

WALLS, E. W. (1946). The laxator tympani muscle. *J. Anat. (Lond.)* **80**, 210.

WEDDELL, G. and MILLER, S. (1962). Cutaneous sensibility. *Ann. Rev. Physiol.* **24**, 199.

—— and PALLIE, W. (1955). Studies on the innervation of skin. The number, size and distribution of hairs, hair follicles and orifices from which hairs emerge in the rabbit ear. *J. Anat. (Lond.)* **89**, 175.

WERSALL, T. (1956). Studies on the structure and innervation of the sensory epithelium of the cristae ampullares in the guinea pig. A light and electron microscopic investigation. *Acta Otolaryng. (Stockh.)* Suppl., **126**, 1.

WHITNALL, S. E. (1932). *The anatomy of the human orbit and accessory organs of vision*, 2nd edn. Oxford University Press, London.

WOLFF, E. (1948). *The anatomy of the eye and orbit*, 3rd edn. H. K. Lewis, London.

13 The blood vascular and lymphatic systems

E. W. WALLS

The blood vascular system

'THE movement of the blood is constantly in a circle, and is brought about by the beat of the heart.' When in 1628 William Harvey reached this conclusion the proof he offered was complete, and clearly demonstrated that the same blood which leaves the heart through **arteries** returns to it along **veins**. His reasoning had been masterly, and had forged a chain of evidence only one link of which awaited visual confirmation. This came when the introduction of the microscope enabled Malpighi (1661) to observe **capillaries** in the lung of the frog; tiny vessels which Harvey knew must exist between the smallest arteries, now called **arterioles**, and the smallest veins or **venules**.

The heart is a rhythmically contractile muscular pump divided by septa into two thin-walled receiving chambers, or **atria**, and two thicker-walled ejecting chambers or **ventricles**. During **diastole**, the interval between two contractions, blood carried by afferent vessels called veins enters the atria and continues on into the ventricles; and then by atrial contraction (**systole**) the onward flow is completed and the ventricles distended. Ventricular systole then follows and the blood is projected into efferent vessels called arteries.

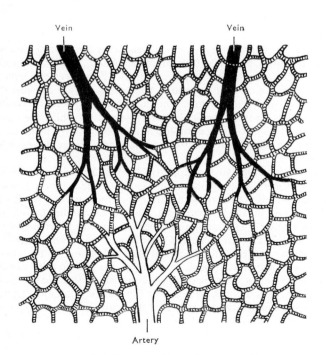

FIG. 13.1. Diagram of a capillary network.

Two great arteries leave the heart, the **aorta** and **pulmonary trunk**, and from them are given off branches which themselves divide, and so ever smaller and more numerous vessels are formed. As noted above, the arterioles end in plexuses of vessels known as capillaries [FIG. 13.1].

Capillary networks are found in all tissues except cartilage, epidermis, the cornea, and most epithelia. They are drained by venules which unite to form progressively larger veins the largest of which return the blood to the atria.

As blood flows along the capillaries, some of its fluid and other substances pass through their thin walls. In this way the interstitial fluid, which is the medium of metabolic exchange, is replenished. Means of draining this fluid away are provided in two ways. Part re-enters the blood capillaries towards their venous ends, and so is returned to the heart; and part is collected by quite another set of vessels, the **lymph capillaries**, which form a meshwork of blindly ending tubes of fine calibre. **Lymph vessels** drain this meshwork and eventually discharge into the great veins at the root of the neck. Thus, all the fluid which irrigates the tissues is eventually returned to the heart.

The food and water required by the tissues is taken by mouth and then absorbed in suitable form by the blood capillaries and lymph capillaries in the wall of the abdominal part of the alimentary canal. Elimination of fluid and of waste metabolic products is shared by the skin, lungs, and kidneys.

A constant supply of **oxygen** is indispensable for the continued activity, and indeed survival, of the tissues; equally necessary is the efficient discharge of **carbon dioxide**. In the higher vertebrates these needs are met by the lungs, through which all blood first passes before returning to the heart for distribution throughout the rest of the body. Such a system requires a double pump, one to receive venous blood from the body (right atrium) and pump it to the lungs (right ventricle) while another receives blood from the lungs (left atrium) and pumps it to the body (left ventricle). This completely separate, double system is only achieved in mammals and birds, and is quite different from the arrangement in fishes where the heart pumps blood through the gill capillaries to the body—a mechanism which precludes the circulation of blood to the rest of the body at a high pressure because of the capillary resistance in the gills. Thus in the fish all blood passes through two sets of capillaries before returning to the heart, while it only passes through one set in the higher forms, except in the portal circulations (see below). The arteries, capillaries, and veins of the lungs form the **pulmonary circulation**, and those of the body generally, the **systemic circulation** [FIG. 13.2].

A special arrangement exists for the return of blood from the abdominal part of the alimentary canal, the spleen and the pancreas. From the capillaries of these parts the blood flows into veins which unite to form the **portal vein**. This large vein passes to the liver where it breaks up into increasingly small branches, the smallest of

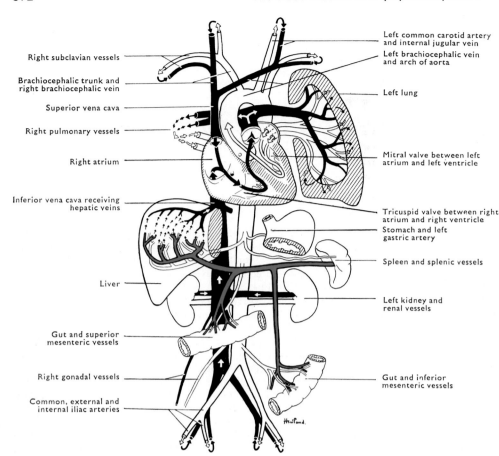

Right subclavian vessels

Brachiocephalic trunk and
right brachiocephalic vein

Superior vena cava

Right pulmonary vessels

Right atrium

Inferior vena cava receiving
hepatic veins

Liver

Gut and superior
mesenteric vessels

Right gonadal vessels

Common, external and
internal iliac arteries

Left common carotid artery
and internal jugular vein

Left brachiocephalic vein
and arch of aorta

Left lung

Mitral valve between left
atrium and left ventricle

Tricuspid valve between right
atrium and right ventricle

Stomach and left
gastric artery

Spleen and splenic vessels

Left kidney and
renal vessels

Gut and inferior
mesenteric vessels

FIG. 13.2. General plan of the
circulation.

The division of the pulmonary trunk is
indicated by a large white bifurcating
arrow, and only the left pulmonary veins
are shown entering the left atrium; the
liver is shown receiving the portal vein
and the hepatic artery; the three large
arteries which spring from the abdominal
aorta to supply the alimentary canal, i.e.
the coeliac trunk, and the superior and
inferior mesenteric arteries are shown
arising in succession from the front of the
aorta; the gonadal vessels are indicated
only on the right side.

which are capillary-like vessels called **sinusoids**. Accordingly when this blood leaves the liver by the hepatic veins it has passed through two sets of capillaries before returning to the heart. The term **portal circulation** is usually applied to this special arrangement which also occurs in the hypophysis.

There is, in addition, the system of lymph vessels already mentioned. Lymph nodes are associated with the lymph vessels, and together they constitute the lymphatic system [p. 981].

TISSUES OF THE VASCULAR SYSTEM

Of the tissues which compose the vascular system only endothelium is present throughout. The other constituents of the system are areolar tissue, white fibrous tissue, reticular tissue, elastic tissue, plain muscle, and cardiac muscle.

The **heart** consists mainly of cardiac muscle, the special features of which are considered on page 885.

Arteries

The arteries are all lined with endothelium, but the other tissues are present in different amounts according to functional needs. Although the diminution in arterial size from the aorta to the periphery is gradual, and the change in the structure of the wall usually

corresponding gradual, it is customary to recognize three orders of vessel, large arteries, medium arteries, and small arteries or arterioles. In all three categories the tissues are arranged in three coats, the **tunica intima**, **tunica media**, and **tunica adventitia**, from within outwards.

When the heart contracts, blood is suddenly forced into the aorta and its large branches (the brachiocephalic, common carotid, subclavian, renal, and common iliac). These vessels have a predominance of **elastic tissue** in their walls and are considerably distended with each systole. This distension protects the smaller vessels beyond from the sudden strain of systole, whilst the elastic recoil of the walls during diastole continues to drive the blood on. In this way the blood flow is converted to a steady stream in the smaller vessels.

The need to regulate the supply of blood to the various tissues and organs, according to their state of activity, is met by the **visceral muscle** which predominates in arteries of medium and small size. This muscle, in the tunica media, consists of spindle-shaped cells whose degree of contraction is under involuntary nervous control. Thus, should a given area need more blood, the muscle cells of the arteries concerned can be relaxed and the lumina of the vessels increased; should less blood be required, the muscle cells contract and the lumina are correspondingly diminished. The two processes, acting in different areas at the same time, allow of a more effective distribution of the limited output of the heart without causing a fall in the pressure of the circulating blood.

It is convenient to describe first the structure of the wall of medium-sized arteries, and thereafter to indicate the distinguishing features of the walls of large and small arteries.

MEDIUM-SIZED ARTERIES

Known also as distributing, or muscular, arteries these vessels all possess a tunica media which is predominantly muscular. Commencing as branches or continuations of large elastic vessels, they range down to something less than half a millimetre in diameter.

Tunica intima

The constituent parts of this **inner coat** are the endothelial lining, a thin subendothelial layer of fine areolar tissue, and the **internal elastic lamina**. This last consists of elastic fibres fused together in the form of a fenestrated membrane which, owing to the contracted state of the usual type of preparation, appears wavy in transverse section [FIG. 13.4].

LARGE ARTERIES

Known also as elastic or conducting arteries, these vessels all possess a very thick **tunica media** which consists of fifty or more concentrically arranged elastic laminae between which are collagen fibres, fine elastic fibres, and fibroblasts, all embedded in an amorphous ground substance. Small visceral muscle cells lie on the surfaces of the laminae [FIG. 13.3]. The **tunica intima** is noticeably thicker than in medium-sized arteries. It contains more elastic fibres, and so the boundary between the subendothelial tissue and the internal elastic lamina is ill defined, and the tunica intima is not sharply delimited from the tunica media. The **tunica adventitia** is relatively thin.

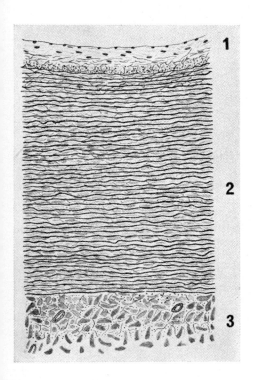

FIG. 13.3. Transverse section of wall of large artery.
1. Tunica intima.
2. Tunica media.
3. Tunica adventitia.

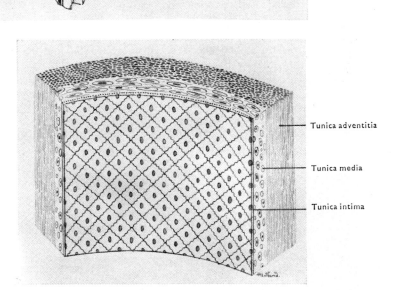

Tunica media

This **middle coat** consists mainly of visceral muscle fibres bound together in more or less spiral strata by reticular fibres, a few collagen fibres, and some delicate elastic fibres.

Tunica adventitia

This thick **outer coat**, although thinner than the tunica media, is a strong layer formed mainly of collagen fibres. It blends gradually with the surrounding fibro-areolar tissue which forms a sheath for the vessel. Longitudinal or tangential bundles of elastic fibres form the **external elastic lamina** at the outer border of the tunica media.

FIG. 13.4. Medium-sized artery and vein as seen in section.
The distinction between the tunica media and tunica adventitia of the vein is in fact less clear than shown, the diagram being designed to emphasize the relative poverty of muscular tissue in the media of the venous wall.

SMALL ARTERIES

Known also as **arterioles**, the smallest have a diameter only slightly greater than that of the capillaries. There is no universal agreement on the size of the largest arterioles, figures given ranging from 0.1 to 0.5 mm. The **tunica intima** consists simply of endothelium with a thin internal elastic lamina in all but the smallest. The **tunica media** consists of circularly disposed visceral muscle fibres arranged in layers which become fewer as the arterioles become smaller; ultimately only a single layer of small scattered fibres remains. The **tunica adventitia** in the larger arterioles may equal the tunica media in thickness, but in the smallest it becomes very thin and the external elastic lamina disappears.

Capillaries

Capillaries are on average 8–10 μm in diameter and about 0.75 mm in length. Their walls consist essentially of **endothelium**, i.e. pavement cells with sinuous, tapering edges which overlap and adhere to those of adjacent cells, and oval central nuclei [FIG. 13.5]. With the light microscope the cells appear cemented to one another along their margins by intercellular substance which stains with silver nitrate, but electron microscopy reveals no evidence of such material (Florey, Poole, and Meek 1959). Investing the capillary endothelium there is a layer of fine reticular fibres, with collagen fibres in places. Several types of cells occur in a pericapillary position, but there is not complete agreement as to their nature. Some are probably undifferentiated mesenchymal cells, while others seem to be fibroblasts and fixed macrophages. However, cells of a different character, with long branching processes, have been observed closely applied to capillary endothelium, and to them a contractile function has been ascribed. Rouget (1879) described such cells in the frog, and in Man somewhat similar cells are present in certain sites; but whether these cells, sometimes called **pericytes**, are responsible for capillary contractility is not yet agreed. Physiological evidence strongly suggests that capillaries possess inherent powers of contraction and dilatation, and it may be that swelling of their endothelial cells obliterates the vessel lumen which

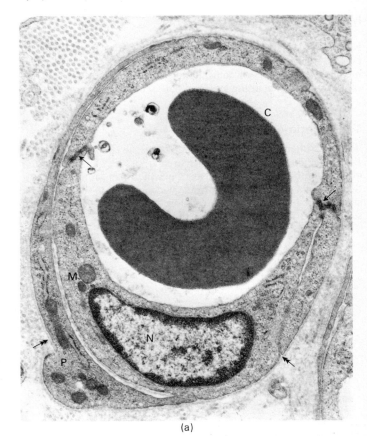

(a)

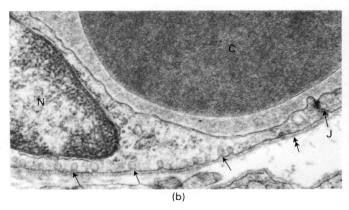

(b)

FIG. 13.7. Electron micrographs of transverse sections of rat capillaries (a) in the skin × 16 500, and (b) in the pancreas × 31 800.

(a) A complete transverse section is shown in which portions of three endothelial cells surrounded by a basal lamina (double arrow) form the capillary wall. The endothelial cells contain several mitochondria (M) and micropinocytotic vesicles; a nucleus (N) is seen in one of the cells. The cell membranes are closely apposed at the cell junctions (arrows) although they are probably of the 'gap' junction variety. The lumen is only slightly wider than the red blood corpuscle (C). A perivascular cell or pericyte (P) encompasses part of the outer aspect of the capillary wall; this cell is also covered by the basal lamina and may have a contractile function.

(b) Part of a capillary wall showing parts of two endothelial cells and the junction (J) between them. The blood plasma appears as a moderately electron dense flocculent material surrounding the red blood corpuscle (C). A nucleus (N) is seen in one of the endothelial cells and numerous flask-shaped micropinocytotic vesicles (arrows) line the cell margin. A distinct basal lamina (double arrow) covers the endothelium. (By courtesy of Dr R. A. Stockwell.)

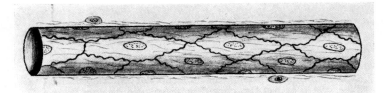

FIG. 13.5. Capillary blood vessels in surface view.

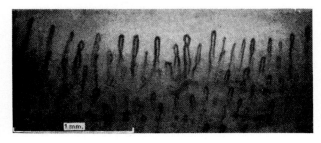

FIG. 13.6. Capillary loops of the nail fold photographed in life.

is restored when the cells become flattened again. The whole question is a difficult one and Clark and Clark (1943) should be consulted.

Capillaries are arranged in networks [FIG. 13.1] the meshes of which show great variations in size and form in different tissues, and also with functional activity. In the skin of the nail fold the capillaries form characteristic loops which may be examined with the microscope in the living [FIG. 13.6].

The general structure of capillaries as seen by the electron microscope is shown in FIGURE 13.7. Some capillaries, however, have special features, and those of the renal glomeruli are shown in FIGURE 8.14.

In some organs (e.g. liver, suprarenal glands, bone marrow, *inter alia*) the finest blood channels are not true capillaries but **sinusoids** (Minot 1900). These are formed by the growth of columns of cells into blood spaces which are thus broken up into narrow channels of somewhat irregular calibre. Wider than capillaries, they differ also in having phagocytic cells of the reticulo-endothelial system present amongst their endothelial cells.

In erectile tissue also there are no true capillaries. Small arteries open directly into cavernous spaces, lined with endothelium, from which blood is drained by small veins.

Veins

In veins the blood flows more slowly than in arteries and at a considerably lower pressure. Accordingly, their lumina are larger and their walls thinner than those of their accompanying arteries; and, being thin-walled, veins collapse when empty.

The walls of veins resemble those of arteries, but in general they are poorer in muscular and elastic tissue and richer in white fibrous tissue.

The structure of the three coats varies in different veins and it is not always possible to distinguish the boundaries between them [FIG. 13.4].

Tunica intima

This consists of the endothelial lining, a layer of subendothelial areolar tissue, less abundant than in arteries and often absent, and an outer layer of longitudinally arranged elastic fibres which seldom gives the appearance of a fenestrated membrane.

A feature of the inner coat in most veins is the presence of **valves** [FIG. 13.8] formed by folds of endothelium strengthened by a little fibrous and elastic tissue, and so disposed that they prevent flow of blood towards the periphery, and help to support a column of blood in veins in which there is an upward flow. The valve cusps are semilunar in shape, and usually are arranged in pairs. On the cardiac side of each valve the venous wall is, as a rule, dilated into a shallow pouch so that when the veins are distended they appear nodulated.

Competent valves do not occur in the large veins of the trunk or in those of the adult portal system [p. 962]. They are more numerous in the deep veins than in the superficial veins of the limbs, and in the veins of children than in those of adults. (See also Franklin 1937.)

The presence of valves in the superficial veins of the forearm is readily demonstrated by Harvey's simple, classical experiment. The venous return being obstructed by pressure on the upper arm, a finger is placed on one of the distended veins and held firmly in position while the vein is emptied of its blood as far as the next valve by stroking upwards along it. The column of blood will be sustained by that valve while the part of the vein below it remains empty. If the pressure of the finger is now removed, the vein fills from below, so that the flow of blood towards the heart is also demonstrated.

Tunica media

This is much thinner than the corresponding coat in an artery, and contains less muscle and more collagen fibres. In some veins, e.g. the majority of the veins of the brain, muscular tissue is absent.

Tunica adventitia

This consists of white fibrous and elastic tissue. In medium-sized veins it is thicker than the tunica media, and in large veins considerably so. Longitudinally disposed muscle fibres in the outer coat of some of the large veins doubtless allow of change of length.

Vascular and nervous supply of arteries and veins

The walls of large and medium-sized arteries and veins are supplied by small vessels, called **vasa vasorum**, which are distributed to the outer and middle coats. They may arise from the vessels they supply or from their branches, or from adjacent arteries, and after a short course enter the walls of the vessels in which they end in a capillary plexus. Draining the plexus are small veins which run to adjacent veins of larger size.

Lymph vessels, which drain into the perivascular lymph plexuses, are present in the outer coat of all the larger arteries and veins.

Arteries are well supplied with **nerves** which form a plexus in the outer coat. The non-medullated fibres, of sympathetic origin, are efferent and supply the muscular tissue of the middle coat. The myelinated fibres, which are considered to be afferent, end mainly in the outer coat but some may reach to the subendothelial tissue. A few lamellated (Pacinian) corpuscles occur in the outer coat of large vessels. The veins have a similar, but less rich, nerve supply. Sites of specially rich sensory innervation which are concerned with the reflex adjustment of heart rate and blood pressure are the carotid sinus [p. 901], the arch of the aorta, and the roots of the caval and pulmonary veins [p. 825].

The heart

The heart is a hollow, muscular organ, enclosed in a fibroserous sac called the pericardium, situated in the middle mediastinum. The cavity of the fully developed heart is separated into right and left halves by an obliquely placed longitudinal septum, and each half consists of a posterior, receiving chamber—the **atrium**—and an anterior, ejecting chamber—the **ventricle**. Externally the coronary sulcus indicates the separation of the atria from the ventricles.

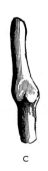

A   B   C

FIG. 13.8. Diagrams showing the valves of veins (W. Sharpey).
A, vein laid open, showing the folds of the inner coat forming a valve.
B, longitudinal section of a vein through a valve.
C, distended vein, showing the swellings opposite a valve. (Schäfer's *Essentials of histology*.)

Internally each atrium opens into the corresponding ventricle by a wide **atrioventricular orifice** provided with a valve which allows the passage of blood from atrium to ventricle, but effectually prevents its return.

In shape the heart is bluntly conical and somewhat flattened from front to back; a base (posterior surface), an apex, and three surfaces (sternocostal, diaphragmatic, and left or pulmonary) are described [FIG. 13.9]. The axis of the heart runs forwards, to the left and slightly downwards, from base to apex.

The **coronary sulcus** runs obliquely round the heart and separates a posterosuperior, atrial portion from an antero-inferior ventricular portion. Anterosuperiorly the sulcus is interrupted by the roots of the aorta and pulmonary trunk. On the base of the heart a faint vertical groove indicates the separation into right and left atria. Anterior and posterior interventricular grooves, which meet at the lower border of the sternocostal surface to the right of the apex, mark the division into right and left ventricles.

The **base of the heart** is formed by the atria—almost entirely the left atrium—and is directed mainly backwards. It is separated from the middle four thoracic vertebrae by the descending thoracic aorta, the oesophagus and the vagi, the thoracic duct, the azygos and hemiazygos veins [FIG. 13.17], and the pericardial cavity.

The base is more or less flat and somewhat quadrilateral in outline. It extends from the postero-inferior part of the coronary sulcus, in which the principal vein of the heart, the coronary sinus, lies [FIG. 13.10], to the pulmonary arteries above.

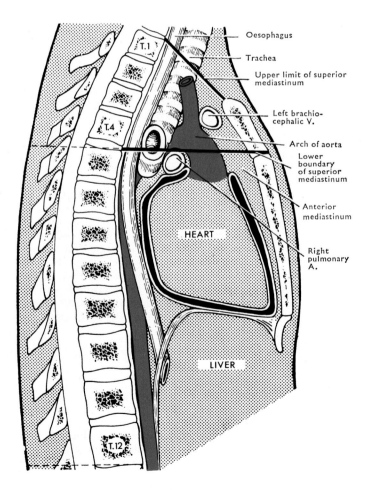

FIG. 13.9. Diagram of a median section to show the general distribution of the structures in the mediastinum.

Six large veins enter the base of the heart. The four pulmonary veins pierce the left atrium, the left pair near the left border of the base, and the right pair just to the left of the interatrial sulcus. The superior and inferior venae cavae enter the right atrium, the superior at the right upper angle of the base, and the inferior at its right lower angle. The part of the surface between the right and left pulmonary veins forms the anterior boundary of a cul-de-sac of the pericardial cavity called the oblique sinus [FIG. 13.17].

The **apex of the heart**, bluntly rounded, is formed by the left ventricle and is directed forwards, downwards, and to the left. It lies under cover of the left lung and pleura, behind the fifth left intercostal space within the midclavicular line (7.5–9 cm from the anterior median line in the average adult). The apex-beat may be seen and felt in that situation, where also the stethoscope is placed to hear the sound of the mitral valve. Age and bodily build influence the position of the apex beat; in children it is usually as high as the fourth interspace, while in adults with long narrow chests it may be as low as the sixth.

The **sternocostal surface** faces forwards, with a slight inclination upwards and to the left. It lies behind the body of the sternum, the medial ends of the third to the sixth right costal cartilages, and the greater parts of the corresponding cartilages of the left side. Separating it from these structures, and from the transversus thoracis muscles is the fibrous pericardium, which is itself overlaid by the lungs and pleurae save for a small area where it is in direct contact with the chest wall [p. 890]. The sternocostal surface is formed mainly by the right ventricle but partly by the other chambers. The ventricular portion of this surface, which is divided by the **anterior interventricular groove** into a smaller left and a larger right part, lies below and to the left of the anterosuperior part of the coronary sulcus [FIG. 13.11]. The atrial part of the surface is formed by the right atrium and by the auricles of both atria which embrace the roots of the ascending aorta and pulmonary trunk. These great vessels ascend in front of the left atrium so that only its auricle contributes to the sternocostal surface. If the ascending aorta and pulmonary trunk are removed, the left atrium is exposed on the sternocostal surface, and its upper margin (which is the **upper margin** of the heart) lies immediately inferior to the bifurcation of the pulmonary trunk and its right branch. The superior vena cava enters the heart at the right end of this margin.

The **right margin**, gently convex to the right, is formed by the right atrium. It is sometimes marked by a shallow groove, the **sulcus terminalis**, which runs from the front of the superior vena cava to the right side of the inferior vena cava.

The **lower margin** is where the sternocostal surface meets the diaphragmatic surface. It is sharp and, in the cadaver, usually concave corresponding with the curvature of the anterior part of the diaphragm. It is formed mainly by the right ventricle, but near the apex by the left ventricle. The anterior and posterior interventricular grooves meet on this border in a slight depression, called the **apical incisure** or notch, a little to the right of the apex. The lower margin lies in the angle between the diaphragm and the anterior wall of the thorax, and is commonly described as passing behind the xiphosternal joint from the lower end of the right border to the apex [but see p. 889]. To the left, the sternocostal surface becomes continuous with the left or pulmonary surface. The transition is gradual, and cannot be said to be marked by a left margin.

The **left surface**, convex from above downwards and from before backwards, is formed by the left ventricle and only to a small extent by the left atrium and its auricle. It is crossed at its widest part (i.e. above) by the left part of the coronary sulcus, and narrows towards the apex. It faces to the left and upwards and is separated from the left lung and pleura by the pericardium and phrenic nerve.

The **diaphragmatic surface** is formed by the ventricles and rests upon the diaphragm, chiefly on the central tendon. It is divided into a smaller right portion, and a larger left, by the **posterior**

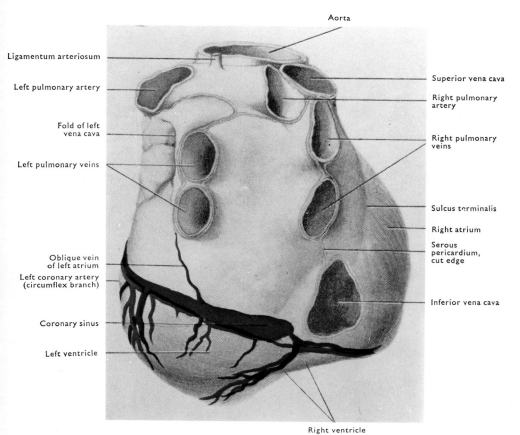

Aorta

Ligamentum arteriosum

Left pulmonary artery

Fold of left vena cava

Left pulmonary veins

Oblique vein of left atrium

Left coronary artery (circumflex branch)

Coronary sinus

Left ventricle

Superior vena cava

Right pulmonary artery

Right pulmonary veins

Sulcus terminalis

Right atrium

Serous pericardium, cut edge

Inferior vena cava

Right ventricle

FIG. 13.10. Base and diaphragmatic surface of heart.
The openings of the great vessels and the line of reflexion of the serous pericardium are shown.

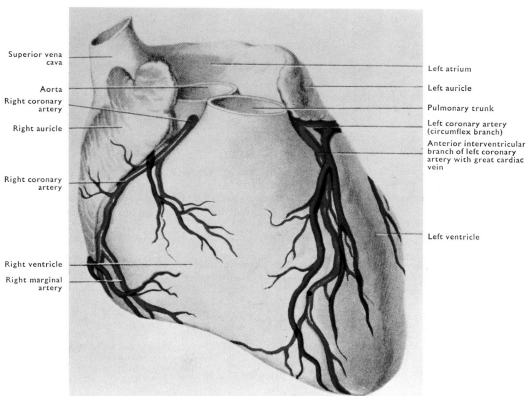

Superior vena cava

Aorta

Right coronary artery

Right auricle

Right coronary artery

Right ventricle

Right marginal artery

Left atrium

Left auricle

Pulmonary trunk

Left coronary artery (circumflex branch)

Anterior interventricular branch of left coronary artery with great cardiac vein

Left ventricle

FIG. 13.11. Sternocostal surface of heart.

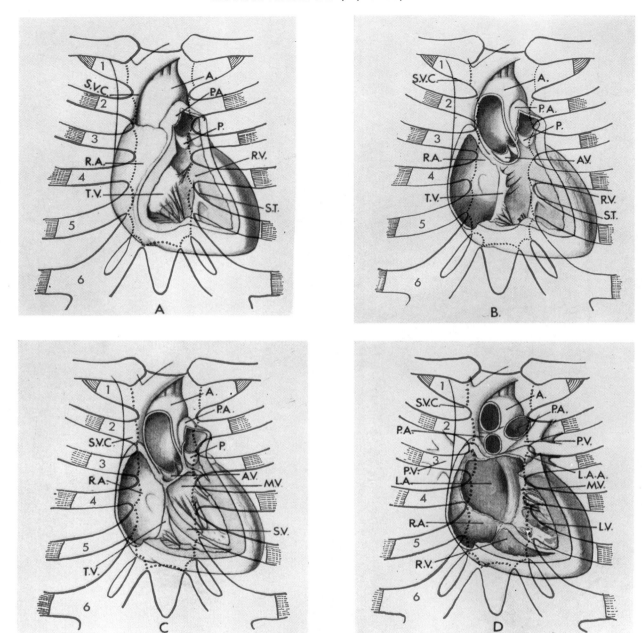

FIG. 13.12. Relations of cavities and valves of heart to anterior wall of thorax. (From photographs of a formalin-hardened subject, with the heart dissected *in situ*.)

In A, the anterior wall of the right ventricle has been removed and the pulmonary trunk opened.

In B, the anterior walls of the ascending aorta and of the right atrium have been removed; also the anterior cusp of the right atrioventricular (tricuspid) valve.

In C the greater part of the interventricular septum has been removed, and the anterior cusp of the left atrioventricular (mitral) valve exposed.

In D the ascending aorta. anterior cusp of left atrioventricular (mitral) valve, pulmonary trunk, and interatrial septum have been removed; the cavities of the left atrium and left ventricle are exposed.

A.	Aortic arch.		P.V.	Pulmonary vein.
A.V.	Aortic valve.		R.A.	Right atrium.
L.A.	Left atrium.		R.V.	Right ventricle.
L.A.A.	left auricle		S.T.	Septomarginal trabecula.
L.V.	left ventricle		S.V.	Interventricular septum.
M.V.	left atrioventricular (mitral) valve.		S.V.C.	Superior vena cava.
P.A.	Pulmonary trunk and arteries.		T.V.	Right atrioventricular (tricuspid) valve.
P.	Pulmonary valve.			

interventricular groove which is directed forwards and to the left. The postero-inferior part of the coronary sulcus separates it from the base.

The surface anatomy of the heart is described on page 889.

THE CHAMBERS OF THE HEART

Atria

The atrial portion of the heart is cuboidal in shape, but is bent round the lateral and posterior surfaces of the ascending aorta and pulmonary trunk. Its cavity is divided by the interatrial septum which runs from the anterior wall backwards and to the right.

The long axis of each atrium is vertical, and each possesses an ear-shaped prolongation, known as the auricle, which projects forwards from its anterior and upper angle. The margins of the auricles are crenated.

THE RIGHT ATRIUM [FIG. 13.19A]

The upper part of this chamber is in contact with the right side of the ascending aorta and its auricle overlaps the anterior surface of the root of that vessel. Its right side forms the right border of the heart, and is separated by the pericardium from the right pleura and lung (the right phrenic nerve and pericardiacophrenic vessels intervening) as is the part of the atrium which contributes to the sternocostal surface of the heart. Posteriorly and on the left the interatrial septum separates it from the left atrium. Joining the right atrium posteriorly there are the superior vena cava above and the inferior vena cava below; and a little above its middle, it is crossed behind by the lower right pulmonary vein [FIG. 13.17].

Interior of the right atrium

The walls are lined with a glistening membrane called endocardium, and are smooth except anteriorly and in the auricle, where muscular bundles, the musculi pectinati, form a series of small, oblique columns [FIG. 13.13]. They spring from a network of ridges in the auricle and run downwards and backwards to end behind in a crest, the crista terminalis. The crista terminalis corresponds in position with the sulcus terminalis, and indicates the junction of the right part of the primitive sinus venosus—the smooth-walled portion of the right atrium—with the atrium proper [p. 967].

In the posterior part of the cavity there is, above, the orifice of the superior vena cava, and below, that of the inferior vena cava—the inferior on a more posterior level. The former has no valve, but the latter is bounded anteriorly by the vestigial valve of the inferior vena cava. Immediately in front and to the left of that valve, between it and the atrioventricular orifice, is the opening of the coronary sinus, guarded by the unicuspid valve of the coronary sinus. The right atrioventricular orifice is guarded on the ventricular side by a tricuspid valve, and so is known also as the tricuspid orifice. It is situated in the antero-inferior part of the left wall of the atrium, and admits the tips of three fingers. Scattered over the walls are the orifices of the venae cordis minimae, tiny veins which return some of the blood from the substance of the heart. Two or three small anterior cardiac veins also open into the right atrium.

In the septal wall is an oval depression called the fossa ovalis. Its floor, which is formed from the septum primum of the embryonic heart, is bounded above and in front by a raised margin, the limbus fossae ovalis. The limbus represents the free edge of the septum secundum which faces the orifice of the inferior vena cava and partly overlaps the septum primum in the foetus [p. 966]. Before birth the two atria are in communication through the foramen ovale—the name given to the gap which passes between the thick, incomplete, septum secundum and the septum primum, and traverses the aperture in the septum primum [FIG. 13.14]. The pressure changes consequent upon birth bring the septa into

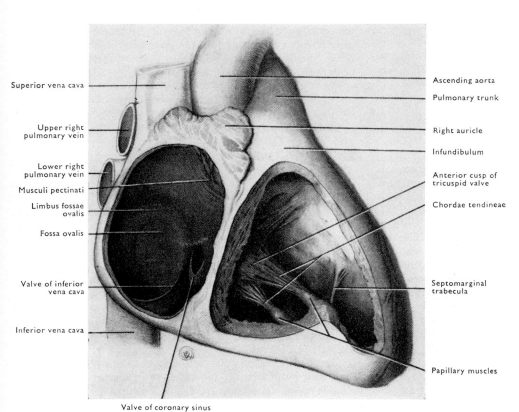

Superior vena cava

Upper right pulmonary vein

Lower right pulmonary vein

Musculi pectinati

Limbus fossae ovalis

Fossa ovalis

Valve of inferior vena cava

Inferior vena cava

Valve of coronary sinus

Ascending aorta

Pulmonary trunk

Right auricle

Infundibulum

Anterior cusp of tricuspid valve

Chordae tendineae

Septomarginal trabecula

Papillary muscles

FIG. 13.13. Cavities of right atrium and right ventricle of heart.

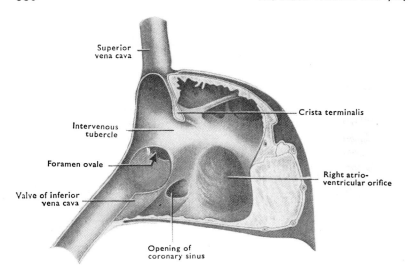

FIG. 13.14. Sagittal section of heart of full-term foetus.

apposition and fusion, but even in the adult, part of the foramen ovale sometimes persists at the upper part of the fossa ovalis as a small slit. On the posterior atrial wall a slight bulge may be seen, between the openings of the venae cavae, just above the upper part of the fossa ovalis. It is called the **intervenous tubercle**, and it has been supposed to help in directing the blood from the superior vena cava to the tricuspid orifice during foetal life [p. 977].

The **valve of the inferior vena cava** is a thin, sometimes fenestrated, sickle-shaped fold of endocardium and subendocardial tissue which is attached by its convex border to the anterior margin of the caval orifice. Its free concave margin is continuous with the anterior part of the limbus fossae ovalis to the left, and with the crista terminalis to the right. In the foetus the valve directs the greater part of the blood from the inferior vena cava through the foramen ovale into the left atrium; a measure of its functional importance is that, unlike the adult valve, it shows little variation. Variations in the form of the adult valve are described by Powell and Mullaney (1960).

The **valve of the coronary sinus** is a semilunar fold of endocardium at the right margin of the orifice of the sinus. It is very variable in size and form, and the role commonly ascribed to it of preventing reflux of blood into the sinus during atrial contraction is seldom possible.

THE LEFT ATRIUM [FIG. 13.19D]

This chamber is irregularly cuboidal in shape with its long axis running vertically. It forms most of the base of the heart and, posteriorly, is separated by the pericardium and its oblique sinus from the descending thoracic aorta and the oesophagus [FIG. 13.18]. The superior margin lies postero-inferior to the right pulmonary artery, and centrally is below the tracheal bifurcation. Its concave sternocostal surface is hidden by the ascending aorta and the pulmonary trunk. The interatrial septum, which faces forwards and to the right, forms its right side. Its left side forms a very small portion of the left surface of the heart, and its long, narrow left auricle overlaps the coronary sulcus and the left side of the root of the pulmonary trunk. Four pulmonary veins, two on each side, enter the upper, posterior part, but it is not rare to find the two veins of one or both sides united before their entry into the atrium.

Interior of the left atrium

The walls are smooth except in the auricle where musculi pectinati are present. The orifices of the pulmonary veins are devoid of valves.

Those on the left lie behind a muscular ridge that projects into the atrium and so cannot be seen from the front. On the septal wall there may be observed an oval area which corresponds to the fossa ovalis of the right atrium; it is more easily seen when this wall is held to the light. Foramina for venae cordis minimae are scattered over the atrial wall, and below and in front there is the **left atrioventricular** or **mitral orifice** which opens forwards, and slightly downwards and to the left. It admits the tips of two fingers, and is guarded on the ventricular side by a valve formed of two cusps arranged like a bishop's mitre, known as the **left atrioventricular** or **mitral valve** [FIG. 13.21].

Ventricles

The ventricular part of the heart forms a slightly flattened cone. Its base is directed backwards and upwards, and is pierced by the four orifices shown in FIGURE 13.15. In this region of junction there is a dense fibrous and fibrocartilaginous tissue, sometimes called the heart skeleton [p. 886].

The **interventricular septum** is placed obliquely, with one surface facing forwards and to the right and the other backwards and to the left. Its margins are indicated by the anterior and posterior interventricular grooves.

THE RIGHT VENTRICLE [FIG. 13.19A]

This chamber forms the greater part of the sternocostal surface of the heart and of its lower border, but the smaller part of the diaphragmatic surface. Above and to its right lies the right atrium, and leading out of the upper part of the ventricle is the pulmonary trunk. The left or septal wall bulges into its interior.

Interior of the right ventricle

Superiorly, between the pulmonary and atrioventricular orifices there is a thick arch of muscle called the supraventricular crest [FIG. 13.16]. This crest separates the lower or inflow part of the cavity from the anterosuperior part or **infundibulum**—so called because of its funnel shape—which carries the outflowing blood to the pulmonary trunk.

The **right atrioventricular** (or **tricuspid**) **orifice** is guarded by the **right atrioventricular** (or **tricuspid**) **valve** whose three cusps, of roughly triangular shape, are named **posterior** (inferior), **septal**, and **anterior** (superior). Small accessory cusps are frequently present

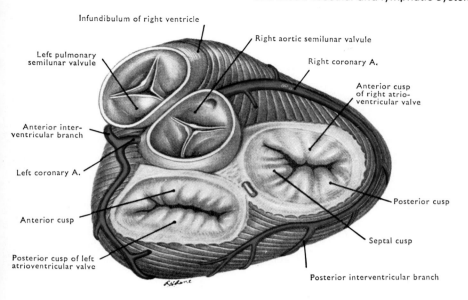

Infundibulum of right ventricle

Left pulmonary semilunar valvule

Anterior inter-ventricular branch

Left coronary A.

Anterior cusp

Posterior cusp of left atrioventricular valve

Right aortic semilunar valvule

Right coronary A.

Anterior cusp of right atrio-ventricular valve

Posterior cusp

Septal cusp

Posterior interventricular branch

FIG. 13.15. Base of the ventricular part of the heart with the atria and great vessels removed.

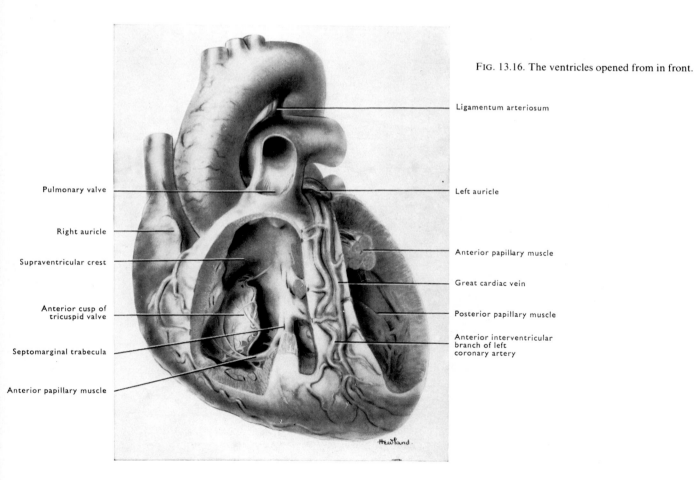

FIG. 13.16. The ventricles opened from in front.

Pulmonary valve

Right auricle

Supraventricular crest

Anterior cusp of tricuspid valve

Septomarginal trabecula

Anterior papillary muscle

Ligamentum arteriosum

Left auricle

Anterior papillary muscle

Great cardiac vein

Posterior papillary muscle

Anterior interventricular branch of left coronary artery

in the angles between the named cusps, and like them consist of a fold of endocardium strengthened by fibrous tissue. The bases of the cusps are attached to a fibrous ring at the atrioventricular orifice where they may be continuous with one another or separated by one or more of the accessory cusps; the cusps project into the ventricle. The cusp margins are irregularly notched and thinner than the central portions. The atrial surfaces over which the entering blood flows, are smooth, whereas the ventricular surfaces are roughened

and give insertion to the **chordae tendineae**. These last are fine tendinous cords which pass to the margins and ventricular surfaces of the cusps from conical muscular projections of the ventricle wall called **papillary muscles** [FIG. 13.13].

The **pulmonary orifice** is in front and to the left of the tricuspid orifice. It is guarded by a **pulmonary valve** composed of three semilunar, pocket-like valvules, named, from their positions, anterior, right, and left. The convex outer border of each valvule is

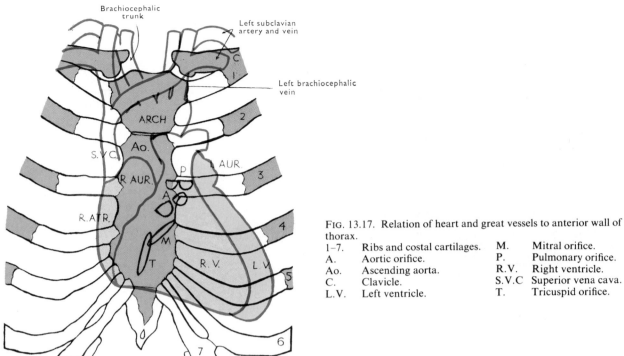

FIG. 13.17. Relation of heart and great vessels to anterior wall of thorax.

1–7.	Ribs and costal cartilages.	M.	Mitral orifice.
A.	Aortic orifice.	P.	Pulmonary orifice.
Ao.	Ascending aorta.	R.V.	Right ventricle.
C.	Clavicle.	S.V.C	Superior vena cava.
L.V.	Left ventricle.	T.	Tricuspid orifice.

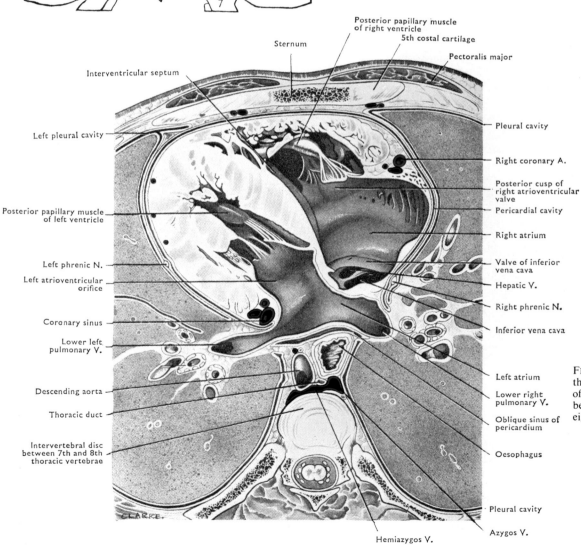

FIG. 13.18. Horizontal section through the thorax at the level of the intervertebral disc between the seventh and eighth thoracic vertebrae.

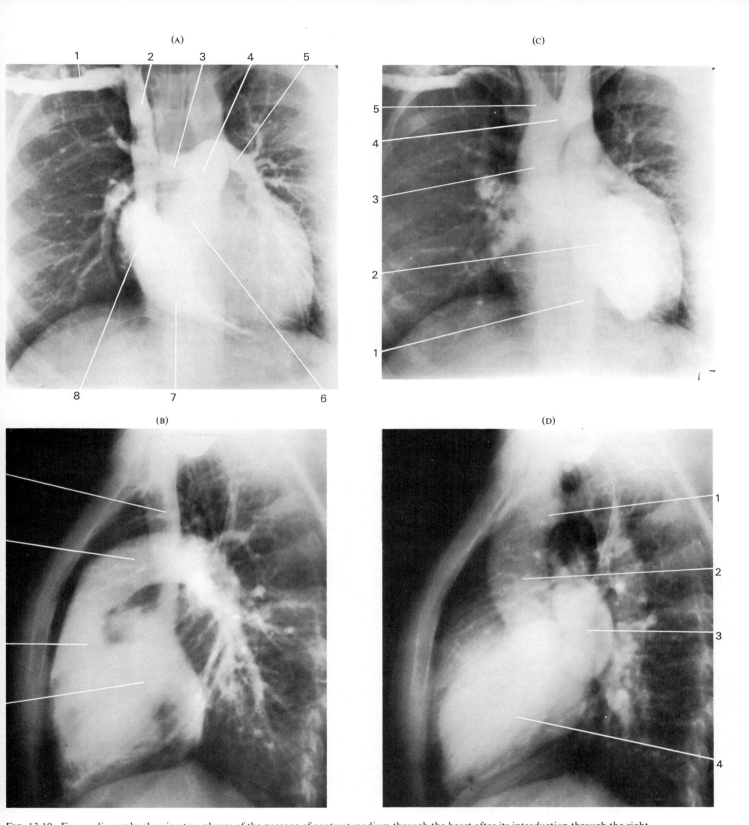

Fig. 13.19. Four radiographs showing two phases of the passage of contrast medium through the heart after its introduction through the right basilic vein. The method is being replaced by cardiac catheterization.

A and B. Anteroposterior and lateral views with right heart outlined.   C and D. Anteroposterior and lateral views with left heart outlined.

A.
1. Subclavian vein.
2. Right brachiocephalic vein.
3. Right pulmonary artery.
4. Pulmonary trunk.
5. Left pulmonary artery.
6. Infundibulum of right ventricle.
7. Right ventricle.
8. Right atrium.

B.
1. Right atrioventricular orifice.
2. Infundibulum of right ventricle.
3. Pulmonary trunk.
4. Superior vena cava.

C.
1. Descending aorta.
2. Left ventricle.
3. Ascending aorta.
4. Arch of aorta.
5. Brachiocephalic trunk.

D.
1. Arch of aorta.
2, Ascending aorta.
3. Left atrium.
4. Left ventricle.

attached to the root of the pulmonary trunk. The free inner border is thickened at its middle to form the **nodule** on each side of which there is a small, thin, crescentic area called the **lunule**. Each valvule of the valve is clad with endothelium on its arterial surface and endocardium on its ventricular surface, between which is a stratum of fibrous tissue. When the valve closes, the nodules are closely opposed, the lunules are pressed together, and both nodules and lunules project upwards into the lumen of the pulmonary trunk.

The wall of the ventricle is lined with endocardium. In the infundibulum, which is formed from part of the embryonic bulbus cordis [p. 966], the wall is smooth, but elsewhere it is ridged by many inwardly projecting muscular columns called the **trabeculae carneae**. Some are simply ridges, while others are rounded bundles attached at each end to the wall of the ventricle but free in the middle. Many of this second group cross the cavity especially towards its apical part where they form a meshwork. One member of this class runs from the septum to the sternocostal wall, where it joins the base of the anterior papillary muscle [FIG. 13.16]. It is called the **septomarginal trabecula** and transmits the right limb of the atrioventricular bundle to the anterior wall [p. 887]. The papillary muscles form a third type of trabecula. They are attached by their bases to the ventricular wall; from their apices numerous chordae tendineae run to their insertion on the tricuspid valve cusps.

The **papillary muscles** of the right ventricle are: (1) a large **anterior** muscle from which the chordae pass to the anterior and posterior cusps of the valve; (2) a **posterior** muscle, often represented by two or more parts, from which chordae pass to the posterior and septal cusps; and (3) a variable group of small **septal** papillary muscles whose chordae run to the anterior and septal cusps. When the ventricle contracts, the papillary muscles do so also, and their distribution ensures that the cusp margins are held together and that the cusps themselves are not driven into the atrium.

The wall of the right ventricle is one-third the thickness of the left—in harmony with their respective tasks.

THE LEFT VENTRICLE [FIG. 13.19c]

This ventricle is conical, and its apex forms the apex of the heart. At its base, the cavity communicates postero-inferiorly with the left atrium through the mitral valve, and with the aorta anterosuperiorly through the aortic valve. Inferiorly its surface is flattened and forms two-thirds of the diaphragmatic surface of the heart, but it is convex in front and on the left where it forms nearly one-third of the ventricular part of the sternocostal surface and almost all of the left surface. The right or septal wall is concave towards the cavity [FIG. 13.20]. The muscular wall is thickest around the widest part of the cavity, which is about one-quarter of its length from the base, and thinnest at the apex. But the upper, posterior part of the septum, which is membranous, is the thinnest region of the whole wall.

Interior of the left ventricle

Whereas in the right ventricle the 'inflow' and 'outflow' orifices—tricuspid and pulmonary—are 2.5 cm apart, the corresponding orifices in the left ventricle—mitral and aortic—are only separated by the anterior cusp of the mitral valve [FIGS. 13.15 and 13.21]. The part of the cavity which leads into the aorta is called the aortic vestibule; its tough fibrous wall is non-contractile.

The **left atrioventricular** (or mitral) **orifice** is oval, with its long axis running downwards, backwards, and to the right. It is guarded by the **left atrioventricular** or **mitral valve**, a bicuspid valve with triangular unequal cusps. The smaller **posterior cusp** is postero-inferior and to the left of the orifice; the larger **anterior cusp** is anterosuperior and to the right, between the mitral and aortic orifices. The bases of the cusps are attached to the fibrous ring round the mitral orifice where they are either continuous with one another or separated by accessory cusps. According to Rusted, Scheifley, and Edwards (1952) accessory cusps are seldom seen; these authors point out, however, that there is always junctional tissue between the two cusps of the valve over a depth of from 0.5 to 1 cm. The cusps are smooth on their atrial surfaces and their structure is the same as that of the tricuspid valve cusps; and like them their margins receive the insertions of chordae tendineae. The chordae do not extend far beyond the margin of the anterior cusp on its ventricular surface. This cusp is therefore smooth on both its surfaces over which blood flows into and out of the ventricle. Consult also Brock (1952, 1955).

The **aortic orifice** is circular and lies between the infundibulum and the right atrioventricular orifice, immediately above and in

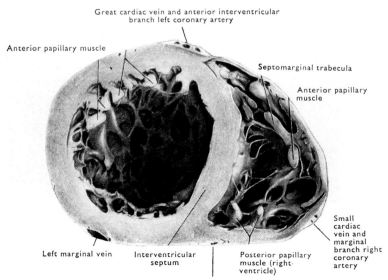

FIG. 13.20. Transverse section of ventricles of heart.

Great cardiac vein and anterior interventricular branch left coronary artery

Anterior papillary muscle

Septomarginal trabecula

Anterior papillary muscle

Left marginal vein

Interventricular septum

Posterior papillary muscle (right-ventricle)

Middle cardiac vein and posterior interventricular branch right coronary artery

Small cardiac vein and marginal branch right coronary artery

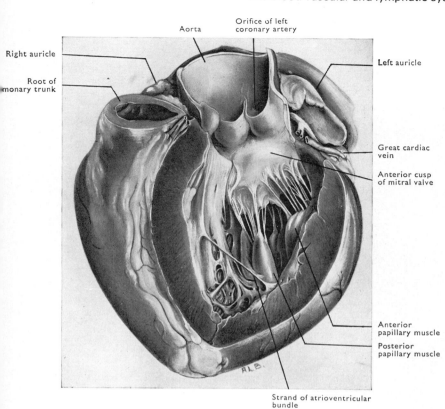

Right auricle

Root of
monary trunk

Aorta

Orifice of left
coronary artery

Left auricle

Great cardiac
vein

Anterior cusp
of mitral valve

Anterior
papillary muscle

Posterior
papillary muscle

Strand of atrioventricular
bundle

Fig. 13.21. Cavity of left ventricle of heart.
The greater part of the pulmonary trunk has been removed, and the aorta opened to show the aortic valves.

front of the mitral orifice from which it is separated by the anterior cusp of the mitral valve. It is guarded by the **aortic valve**, formed of three semilunar valvules, named, from their position, right, left, and posterior. The structure and attachments of the valvules of the aortic valve are similar to those of the valvules of the pulmonary valve [p. 881].

The walls of the left ventricle are lined with endocardium. Many fine **trabeculae carneae** form a meshwork which is especially rich at the apex and on the inferior wall, but the septum and upper part of the sternocostal wall are relatively smooth. There are two **papillary muscles**, an **anterior** arising from the sternocostal wall and a **posterior** from the inferior wall, and each is connected by chordae tendineae with both cusps of the mitral valve. Fine strands, which may be formed entirely of branches of the left limb of the atrioventricular bundle, are often found crossing the cavity from the septum to the papillary muscles [FIG. 13.21].

The **interventricular septum** extends from the right of the apex of the heart to the interval that separates the pulmonary and tricuspid orifices from the aortic and mitral orifices. Its borders correspond with the anterior and posterior interventricular grooves. It is so placed that one surface looks forwards and to the right and bulges into the right ventricle, whilst the other looks backwards and to the left and is concave towards the left ventricle. Almost the whole of the septum is thick and muscular, but its upper posterior part is fibrous and not more than 1 mm in thickness; it is called the **membranous part** of the interventricular septum and shows as a small oval transparency when held to the light. From the left side it is seen to lie just below the angle between the right and posterior valvules of the aortic valve; its right side is divided by the upper part of the attached border of the septal cusp of the tricuspid valve into anterior and posterior parts. The anterior part separates the right ventricle from the aortic vestibule, whereas the posterior part separates the lower part of the right atrium from the aortic vestibule.

THE STRUCTURE OF THE HEART

The heart consists mainly of muscle, called the myocardium, which is enclosed between the epicardium, or visceral layer of the pericardium, and the endocardium.

The **epicardium** consists of white fibrous and elastic tissue which is covered with flat polygonal mesothelial cells on its free surface.

The **endocardium** is continuous with the inner coats of the vessels. Much thinner than the epicardium, it consists of a lining of endothelial cells, a layer of fine areolar tissue, and an external elastic layer usually in the form of a fenestrated membrane.

The fibres of the **myocardium** differ from those of voluntary striated muscle in several ways: they are less distinctly striated; they are made up of short, cylindrical segments joined end to end by transverse discs; the nuclei lie centrally, one (seldom two) in each segment or cell; the sarcolemma is finer than in voluntary muscle fibres; finally, the cells of heart muscle give off branches which unite with branches of neighbouring cells and convert the whole into a network [FIG. 5.1].

Immediately subjacent to the subendocardial tissue are found the muscle **fibres of Purkinje**. First observed by Purkinje (1845) these fibres are part of the conducting system of the heart and in many animals are of considerably larger diameter than ordinary myocardial fibres; the central portion of each consists of granular protoplasm in which are embedded two or more nuclei, and the peripheral portion is transversely striated. Purkinje fibres of much smaller calibre also occur (Kugler and Parkin 1956). In the sheep and ox heart, typical large Purkinje fibres are readily observed, but in the human heart they are much smaller.

The network of cardiac muscle fibres forms sheets and strands which have a more or less characteristic and definite arrangement in different parts of the heart.

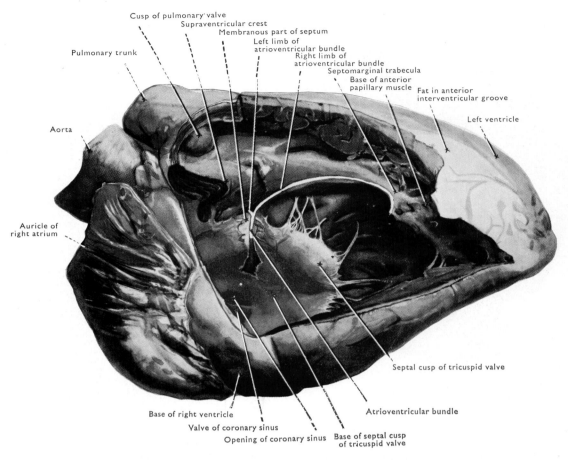

FIG. 13.22. Dissection of right ventricle of heart showing course and division of atrioventricular bundle.

In the atria the muscular fasciculi fall naturally into two groups: (1) superficial fibres common to both atria; (2) deep fibres peculiar to each atrium.

The superficial fibres run transversely across the atria and a few of them dip into the atrial septum.

The deep fibres are: (1) looped fibres; the ends of these are attached to the fibrous rings around the atrioventricular orifices, and the fibres pass anteroposteriorly over the atria; (2) anular fibres which surround (i) the extremities of the large vessels which open into the atria, (ii) the auricles, and (iii) the fossa ovalis. Cardiac muscle fibres extend for variable distances along the great veins entering the atria.

In the ventricles the muscular fasciculi form more or less definite V-shaped loops which begin and end at the fibrous rings at the bases of the ventricles. These loops embrace the cavities of either one or both ventricles, one limb of each loop lying on the outer surface of the heart and the other in the interior. The superficial fibres on the sternocostal surface pass towards the left, those on the inferior surface towards the right. At the apex all are coiled into a whorl or vortex through which they pass into the interior of the ventricular walls and run towards the base, some in the septum and others to end in the papillary muscles. The arrangement is exceedingly complex and for further details reference should be made to Mall (1911) and Thomas (1957).

The heart also possesses a fibrous skeleton which strengthens its orifices. It consists of:

1. Fibrous rings which surround the atrioventricular orifices.

2. Short tubular fibrous zones at the roots of the aorta and pulmonary trunk to which are attached, proximally, some of the ventricular muscle fibres, and distally, the arterial walls.

3. A mass of tough fibrous tissue, in which there may be some cartilage, which lies between the aortic root in front and the two fibrous rings: it is called the right fibrous trigone.

4. The left fibrous trigone which is a smaller mass joining the left side of the aortic root to the front of the left atrioventricular ring [FIG. 13.15].

5. The membranous part of the septum which is continuous above and behind with the right fibrous trigone.

The valves of the heart consist of a thin layer of fibrous tissue, continuous with the fibrous rings that surround the orifices and covered by a reduplication of the lining membrane. Atrial muscle fibres extend into the base of the atrioventricular valves, but the associated capillary vessels pass into the valves for a distance of not more than 3 mm in health (Dow and Harper 1932). Otherwise, these valves, like the pulmonary and aortic, are avascular.

Conducting system of heart

From the account, given above, of the fibrous skeleton of the heart, it would appear that the atrial and ventricular muscle fibres are everywhere effectively separated. This, however, is not the case, for there is connecting them a bundle of muscle fibres, called the atrioventricular bundle, which traverses the right fibrous trigone.

The bundle forms part of the conducting system of the heart which is composed of muscle fibres specially differentiated for their task of initiating the sequence of events in the cardiac cycle, of controlling its regularity, and of transmitting the impulses from atria to ventricles.

The parts of which the conducting system is composed are these:

The sinu-atrial node

This structure, first described by Keith and Flack (1907), is commonly called the pacemaker of the heart. It is situated in the wall of the right atrium in the upper part of the sulcus terminalis from which it extends over the front of the opening of the superior vena cava. The nodal fibres, of smaller calibre than the neighbouring atrial fibres with which they are continuous, are arranged in a meshwork intermingled with which there is a considerable quantity of connective tissue.

The atrioventricular node

First described by Tawara (1906), this structure lies in the atrial septum just above the opening of the coronary sinus. Its fibres form a dense meshwork which establishes continuity with the surrounding atrial muscle fibres. There is no special connection between the two nodes, the impulse spreading from the pacemaker through the atrial musculature.

The atrioventricular bundle

Known also as the bundle of His (1893), this slender strand springs from the atrioventricular node and runs forwards through the right fibrous trigone to reach the posterior border of the membranous part of the interventricular septum. Continuing forwards under cover of the septal cusp of the tricuspid valve, it runs along the lower border of the membranous septum before dividing into right and left limbs which straddle the upper border of the muscular septum [FIGS. 13.22 and 13.23]. The bundle is about 1 cm long and 2 mm in diameter.

The right limb continues on the right side of the muscular part of the septum, where it is covered by the endocardium and a thin layer of myocardium which, however, may be absent. From the septum the right limb passes in the septomarginal trabecula to the base of the anterior papillary muscle.

The left limb of the bundle descends immediately beneath the endocardium on the left side of the muscular interventricular septum, as a broad thin band which usually divides into two branches about half-way down the septum. These soon divide into lesser branches which pass in the trabeculae carneae to the papillary muscles. Strands of the left limb or its branches sometimes cross the cavity of the ventricle to reach the papillary muscles, forming as they do the so-called 'false tendons' [FIG. 13.21].

In many animals the bundle and its divisions are composed of large Purkinje fibres, but in the human heart such fibres do not appear until some way down the limbs; indeed, the fibres which compose the bundle and the upper parts of its limbs in Man are no larger than those of the ordinary ventricular myocardium.

In the heart of an ungulate the bundle and its branches are readily dissected, but in the human heart they are less easily displayed (Walls 1945).

The terminal subendocardial network (Purkinje) is continuous with the branches of the bundle at the bases of the papillary muscles from which it spreads out beneath the endocardium of the greater part of both ventricles. The subendocardial network and the course of the bundle and its branches may be shown by injecting suitable coloured liquid into the sheath which surrounds the whole system. This method gives good results on the ox or sheep heart [FIG. 13.23], but it is unsuccessful in the human heart because of the extreme thinness of the sheath.

When the antrioventricular bundle is severed, experimentally or by disease, the ventricles contract independently of the atria, and with a much slower rhythm. This condition is known as 'heart block'. However, since the bundle contains nerves, it has been argued that the impulse for cardiac contraction might well be carried

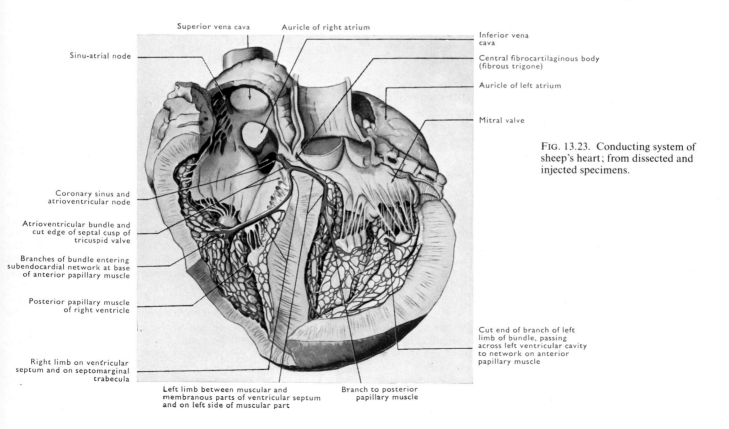

FIG. 13.23. Conducting system of sheep's heart; from dissected and injected specimens.

Labels on figure:

Superior vena cava
Auricle of right atrium
Inferior vena cava
Sinu-atrial node
Central fibrocartilaginous body (fibrous trigone)
Auricle of left atrium
Mitral valve
Coronary sinus and atrioventricular node
Atrioventricular bundle and cut edge of septal cusp of tricuspid valve
Branches of bundle entering subendocardial network at base of anterior papillary muscle
Posterior papillary muscle of right ventricle
Right limb on ventricular septum and on septomarginal trabecula
Cut end of branch of left limb of bundle, passing across left ventricular cavity to network on anterior papillary muscle
Left limb between muscular and membranous parts of ventricular septum and on left side of muscular part
Branch to posterior papillary muscle

out by these nerves rather than by the bundle itself. Evidence that conduction throughout the bundle and its limbs is indeed myogenic has been advanced by Davies, Francis, Wood, and Johnson (1956).

The size of the heart

The heart in health is roughly the size of the individual's closed fist. It is about 12.5 cm long, 8.5 cm broad, and its greatest depth from sternocostal to diaphragmatic surfaces is 6 cm.

Weight

At birth the heart weighs 20–25 g, in the adult male the average is 310 g and in the adult female 255 g.

The increase in weight with age is rapid up to the seventh year, then more gradual to the age of puberty when a second acceleration sets in. The gradual increase in adult heart weight commonly observed with increasing age is probably more to be linked with increasing body weight than advancing years (Gray and Mahan 1943).

The several parts of the heart also vary in their relation to one another at different periods of life. This is most marked in the ventricles which, at one stage of foetal development, are of equal bulk. After birth the left ventricle grows rapidly, while the right, after an initial reduction in absolute weight, grows more slowly. After the age of 6 months the proportion of the right ventricle to the left remains unaltered throughout infancy, adolescence, and normal adult life (Keen 1955).

Capacity

Although after death the cavity of the right ventricle appears larger than that of the left, during life their capacities are probably the same—90–120 ml. The atria are slightly less capacious.

The blood supply of the heart

The walls of the heart are supplied by the coronary arteries [p. 896], the capillaries forming a close-meshed network around the muscular fibres.

In most cases the sinu-atrial and atrioventricular nodes, and the bundle, are supplied by the right coronary artery. The limbs of the bundle derive their supply from both coronary arteries.

The veins of the heart are described on page 942.

Lymph vessels of heart

These are described on page 997.

Nerves of heart

The heart receives its nerves from the superficial and deep cardiac plexuses [p. 822]. These are formed by branches from the vagi and sympathetic trunks, all of which—with one exception—contain both efferent and afferent fibres; the exception is the superior cervical cardiac branch of the sympathetic which, on both sides, consists solely of efferent (postganglionic) fibres.

The cardiac plexuses send small branches to the heart along the great arteries and veins, and continue as the right and left coronary plexuses which innervate these arteries and their territories of supply.

The vagal efferent fibres (parasympathetic) exert restraint on the heart rate. They emerge from the brain stem in the roots of the vagus and accessory nerves, and run as preganglionic fibres to the

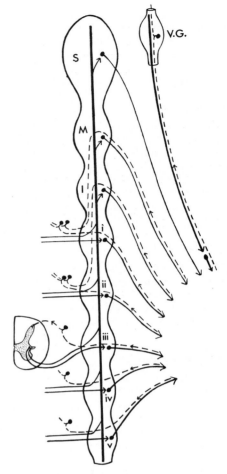

FIG. 13.24. The nerve supply of the heart (diagrammatic).
Afferent fibres are represented by interrupted lines. The superior, middle, and inferior cervical ganglia of the sympathetic trunk are marked S, M, and I, the first five thoracic ganglia i–v. The third thoracic segment of the spinal medulla is shown in section. V.G., inferior vagal ganglion.

ganglia present in the cardiac plexuses, and in the heart itself [see below] where they synapse.

The vagal afferent fibres have their cell bodies in the inferior vagal ganglion; they carry impulses from the heart, **arch of the aorta**, and the roots of the great veins, which bring about reflex adjustment of the heart rate.

The sympathetic (preganglionic) fibres emerge from the spinal medulla in the upper four or five thoracic nerves and pass in the white rami to the corresponding ganglia of the sympathetic trunk [FIG. 13.24]. Some of the fibres synapse in these ganglia, from which postganglionic fibres then proceed as small cardiac branches to the cardiac plexuses; most of the preganglionic fibres, however, ascend in the sympathetic trunk to relay in the cervical sympathetic ganglia, from which post-ganglionic branches run to the cardiac plexuses [p. 822].

Afferent fibres, whose cell bodies are in the spinal ganglia of the upper four or five thoracic spinal nerves, travel in all the sympathetic cardiac nerves except the superior cervical. They subserve pain sensibility, and the severe pain of angina pectoris may be greatly relieved by surgical removal of the related sympathetic ganglia.

Ganglion cells are widely distributed in the atria, including the regions close to the sinu-atrial and atrioventricular nodes, but appear to be absent from the ventricles (King and Coakley 1958).

Reports on the types of nerve ending found in the heart are contradictory, and Davies, Francis, and King (1952) should be consulted. These authors find no endings resembling the motor endings or muscle spindles of skeletal muscle. However, they find non-medullated fibres ending in spiral formations around individual muscle fibres; medullated fibres are present in the epicardium, endocardium, and myocardium, but all lose their myelin sheaths some distance from their terminations.

The surface anatomy of the heart

Several factors affect the position of the heart in health. The first is general body build, or habitus, with which the level of the diaphragm is commonly associated. In thick-set subjects the diaphragm is frequently at a higher level than in those of slender build, and in consequence the heart lies more horizontally in the first-mentioned group and more vertically in the second. Bodily build apart, there are factors which modify the position of the heart in any individual. Thus the beating of the heart itself, respiratory movements, and the position of the body all play their part; moreover, as the barrel-shaped, high-set thorax of infancy gradually assumes adult proportions, the heart comes to lie less horizontally, while undergoing a relative reduction in size.

It is therefore not possible to give a surface marking for the heart which is of general application.

In the *adult cadaver* the outline of the heart is shown in FIGURES 13.12, 13.17, and 13.25C, and may be drawn on the surface by connecting the following points: (1) at the lower border of the second left costal cartilage 3.5 cm from the median plane; (2) at the upper border of the third right cartilage 2.5 cm from the median plane; (3) on the sixth right cartilage 2.5 cm from the median plane; (4) over the fifth left intercostal space about 9 cm from the median plane, medial to the midclavicular line. The upper border is straight, and corresponds approximately to the upper level of the atria; the right border is gently convex to the right; the lower border is almost horizontal, and passes behind the xiphosternal joint; the left border is convex to the left and upwards.

The orifices of the valves may be represented on the surface by lines corresponding to those shown in FIGURE 13.17. Those for the pulmonary and aortic valves should be 2.5 cm in length, that for the mitral valve 3 cm, and for the tricuspid valve 4 cm.

THE POSITION OF THE HEART DURING LIFE

While traditional methods of examination such as inspection, palpation, and percussion unquestionably yield useful information, it cannot be disputed that radiology has provided the most exact knowledge of the position of the heart in the living subject. Using radiographs corrected for dispersion of X-rays, Mainland and Gordon (1941) investigated heart position in young adult males, standing and in the midphase of respiration. Their findings of the average relationships of the heart outline to the anterior thoracic wall are shown in FIGURE 13.25. Certain points call for special comment.

The lower border of the heart is just over 5 cm below the level of the xiphosternal joint, so that more than a third of the vertical cardiac height lies below the joint. The upper border of the heart lies a little above the middle of the body of the sternum, and slopes slightly downwards from left to right.

When lying supine the upper border is about a rib's width higher, and rather less than a third of the vertical cardiac height lies below the xiphosternal joint.

The range of variation between individuals is considerable.

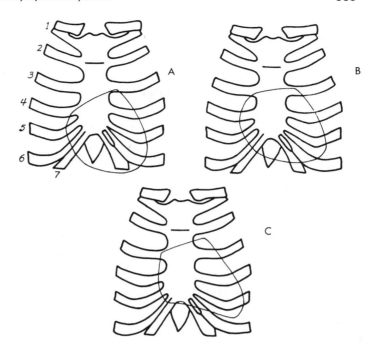

FIG. 13.25. Average relationships of heart outline to anterior thoracic wall. The upper two diagrams (A, living erect; B, living recumbent) are derived from radiographs corrected for dispersion of X-rays. Diagram C represents a common textbook statement. (After Mainland (1946) with kind permission.)

Heart sounds

The sounds produced by the individual valves are heard best, not directly over their anatomical situation, but over the area where the cavity in which the valve lies is nearest the surface. Hence the mitral sound is best heard over the apex (mitral area) which, though most often in the fifth left intercostal space about 9 cm from the midline, ranges from the fourth space to the sixth costal cartilage and from 6.5 to 10 cm from the midline; the tricuspid area is over the lower part of the body of the sternum or the xiphoid process; the aortic area is close to the sternum over the second right costal cartilage or second interspace; and the pulmonary area is also near the sternum over the third left costal cartilage or second left interspace.

RADIOLOGY

The conventional postero-anterior radiograph of the chest shows the silhouette of the heart only, but it does emphasize how it can vary in size and shape with the position of the diaphragm and with the intrathoracic pressure [p. 367], elongating and narrowing as the diaphragm falls, and subsequently increasing in volume with the influx of blood due to the lowered intrathoracic pressure.

From above downwards, the right border of the cardiovascular silhouette in the mediastinum is formed by the superior vena cava (unless obscured by the sternum) the right atrium, and occasionally the inferior vena cava. On the left side, the aortic arch is easily recognized, then the less prominent pulmonary trunk which is often difficult to identify, then an occasional small bulge produced by the left auricle. The main, convex part of the border is caused by the left ventricle which includes the cardiac apex. The inferior margin (formed mainly by the ventricles) merges with the opacity of the diaphragm and liver and cannot be identified.

The cavities of the heart can be seen only after the introduction of contrast medium by passage of a catheter along the basilic or femoral vein to the right atrium, thus filling the right ventricle and pulmonary vessels. The left atrium and left ventricle may be seen at a later stage of this examination. [FIGS. 13.19A–D]. To visualize the left ventricle and aorta a catheter is passed from the femoral artery along the aorta to reach the left ventricle through the aortic valve.

Rapid filming by cine radiography allows recording of the filling and emptying of the chambers of the heart, and their anatomical limits to be recognized [FIGS. 13.19A–D].

THE PERICARDIUM

The pericardium is the fibroserous sac which surrounds the heart in the middle mediastinum. It consists of a strong, conical, fibrous sac called the fibrous pericardium, within which is the serous pericardium arranged as shown schematically in FIGURE 13.26.

The **fibrous pericardium**. Inferiorly it is attached to the central tendon and to the adjacent muscular substance of the diaphragm, and is pierced by the inferior vena cava. Above, and behind, it becomes continuous with the adventitia of the great vessels which pierce it on their way to and from the heart. Thus it sheaths the aorta, the two branches of the pulmonary trunk, the superior vena cava, and the four pulmonary veins.

Its anterior surface forms the posterior boundary of the anterior mediastinum, and it is attached to the sternum, above and below, by the **sternopericardial ligaments**. In most of its extent it is separated from the thoracic wall by the anterior parts of the lungs and pleural sacs, but it is in contact with the left half of the lower portion of the body of the sternum and often with the medial ends of the left fourth, fifth and sixth costal cartilages and the left transversus thoracis muscle. The extent of its contact with the thymus decreases with age [FIGS. 9.14 and 9.15].

Its posterior surface forms the anterior boundary of the upper part of the posterior mediastinum, and each lateral surface is in contact with the mediastinal pleura, the phrenic nerve and its accompanying vessels intervening [FIG. 13.18].

The inner surface of the fibrous sac is closely adherent to the parietal part of the serous pericardium.

The **serous pericardium** is a transparent membrane which forms a closed sac containing a little fluid, and essentially forms a large

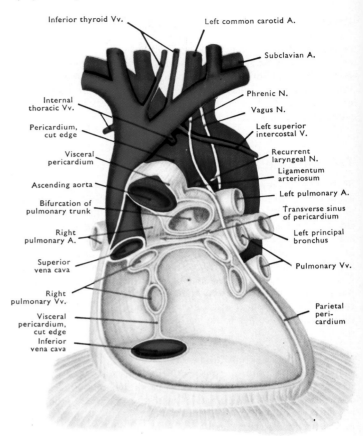

FIG. 13.27. The pericardium and great vessels after removal of the heart. The arrow lies in the transverse sinus of the pericardium, and the posterior wall of the oblique sinus lies between the right and left pulmonary veins.

FIG. 13.26. Diagram of the pericardial sac. The serous pericardium (interrupted line) covers the heart and lines the fibrous pericardium. P.C., pericardial cavity—a mere slit in life; D, diaphragm.

bursa which facilitates cardiac movement. It is invested by the fibrous pericardium and is itself invaginated by the heart in the manner of a finger pushed into a balloon. It is therefore described as having two parts—the parietal, which lines the fibrous pericardium, and the visceral (or epicardium) which partially ensheathes the heart and great vessels; but the two parts are continuous where the great vessels pierce the fibrous pericardium. By reference to FIGURE 13.27 it will be seen that there are two such sites of continuity: (1) at the arterial end of the heart where the aorta and pulmonary trunk are enclosed in one tube of the visceral layer; and (2) at the venous end where the two caval and four pulmonary veins are enclosed in a second tube. Whereas the investment of the visceral layer around the great arteries is almost cylindrical, that round the six veins is shaped like an inverted U with unequal limbs.

When the pericardial sac is opened from the front it is possible to pass a finger behind the aorta and pulmonary trunk, and in front of the atria, through the **transverse sinus** of the pericardium. This passage appears early in development as a result of the breakdown of the dorsal mesocardium which suspends the simple tubular heart within the sac of the serous pericardium: this breakdown affords greater freedom for growth and movement, and in the fully developed heart the sinus serves as a bursa, containing pericardial fluid, between the pulsating arteries in front and the contracting atria behind.

The **oblique sinus** of the pericardium [FIG. 13.27] forms a cul-de-sac bounded by the serous pericardium surrounding the pulmonary and caval veins. It owes its shape to the changes which attend the absorption of the sinus venosus and the pulmonary veins into the atria. The oblique sinus lies behind the left atrium, and in front of the oesophagus and the descending thoracic aorta.

A small fold of the serous pericardium, called the **fold of the left vena cava**, passes from the left pulmonary artery to the upper left pulmonary vein, behind the left end of the transverse sinus [FIG. 13.10]. It encloses a fibrous strand, called the ligament of the left vena cava, a remnant of the left superior vena cava which atrophies early in foetal life. From the lower end of the ligament the oblique vein of the left atrium descends to the coronary sinus. Together they represent the left common cardinal vein [p. 974].

Structure

The fibrous pericardium is a dense unyielding fibrous membrane. The serous pericardium is covered on its inner surface by a layer of flat mesothelial cells which rests upon a basis of mixed white and elastic fibres in which run numerous blood vessels, lymph vessels, and nerves.

THE PULMONARY CIRCULATION

The pulmonary trunk

This vessel arises from the infundibulum of the right ventricle at the level of the valvules of the pulmonary valve, i.e. posterior to the sternal end of the third left costal cartilage. Immediately above the valvules it is slightly dilated into three shallow **sinuses**. It passes upwards and backwards from the front of the ascending aorta to its left side, and after a course of just over 5 cm ends, within the concavity of the aortic arch, by dividing into right and left pulmonary arteries at the level of the sternal end of the second left costal cartilage.

Within the fibrous pericardium the pulmonary trunk is enveloped with the ascending aorta in a sheath of the visceral layer of the serous pericardium [FIG. 13.27]. It lies behind the sternal end of the second left intercostal space, from which it is separated by the anterior margins of the left lung and pleural sac, and by the pericardium. Posterior to it are first the ascending aorta, and then the left atrium. [FIGS. 13.11 and 13.15]. To its right are the ascending aorta and the auricle of the right atrium, and to its left the auricle of the left atrium. Between its bifurcation and the aortic arch, is the superficial part of the cardiac plexus and nerves. Both coronary arteries are close to its root—the right artery on its right, the left in contact at first behind and then on its left.

The **right pulmonary artery**, longer and wider than the left, commences in front of the oesophagus antero-inferior to the tracheal bifurcation [FIG. 13.29]; it passes to the hilus of the lung behind the ascending aorta, superior vena cava, and upper right pulmonary vein, and in front of the right bronchus immediately below its superior lobar branch.

The **left pulmonary artery** passes laterally and backwards in front of the left principal bronchus and descending aorta [FIG. 13.28].

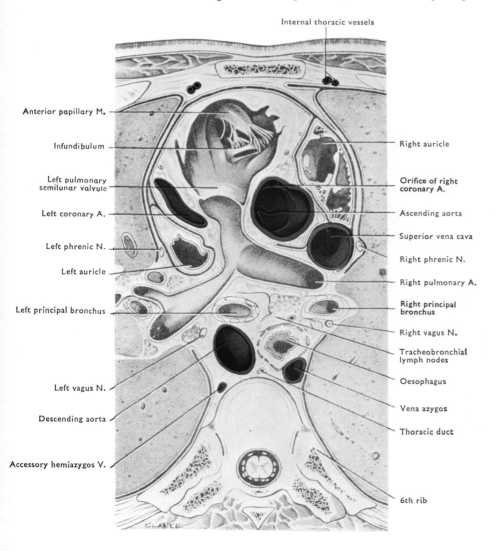

Anterior papillary M.

Infundibulum

Left pulmonary semilunar valvule

Left coronary A.

Left phrenic N.

Left auricle

Left principal bronchus

Left vagus N.

Descending aorta

Accessory hemiazygos V.

Internal thoracic vessels

Right auricle

Orifice of right coronary A.

Ascending aorta

Superior vena cava

Right phrenic N.

Right pulmonary A.

Right principal bronchus

Right vagus N.

Tracheobronchial lymph nodes

Oesophagus

Vena azygos

Thoracic duct

6th rib

CLARKE.

FIG. 13.28. Horizontal section between the fifth and sixth thoracic vertebrae.

Above are, the aortic arch—to which it is connected by the ligamentum arteriosum [p. 977]—and the left recurrent laryngeal nerve. Before entering the lung it is crossed anteriorly by the upper left pulmonary vein.

The pulmonary veins

The tributaries of the pulmonary veins arise in the capillary plexuses of the alveoli, and by the union of the smaller veins. Eventually one vein, formed by tributaries which correspond more or less to the bronchopulmonary segments (though many are intersegmental), passes from each lobe into the root of the lung. Thus are formed five pulmonary veins, but those from the upper and middle lobes of the right lung usually unite on entering the lung root, and so only four pulmonary veins open into the left atrium [FIGS. 13.10, 13.18, and 13.27]. Neither the main stems nor their tributaries possess valves.

In the root of each lung the upper pulmonary vein lies below and in front of the pulmonary artery [FIGS. 7.49 and 7.50]. The lower pulmonary vein is below and posterior to the upper vein.

On the right side the upper pulmonary vein passes behind the superior vena cava, and the lower behind the right atrium [FIG. 13.91]; both end in the upper and posterior part of the left atrium close to the atrial septum. On the left side both veins cross in front of the descending aorta [FIG. 13.33], to the upper, posterior part of the left atrium near its left border.

All four veins pierce the fibrous layer of the pericardium, and then receive partial coverings of the serous layer.

Fusion of the left veins into a single trunk before entry is not uncommon; on the other hand, there may be separate openings for the veins from all three lobes of the right lung.

The intrapulmonary course of the pulmonary vessels

In general, the branches of the pulmonary arteries accompany the bronchi, divide with them, and lie on their posterosuperior aspect. The veins behave differently: peripherally, they run apart from the arteries, but as larger veins are formed they join the arteries and bronchi and run with them to the lung root. They lie on the antero-inferior aspects of the corresponding bronchi. See also page 525.

Modern thoracic surgery demands a detailed knowledge of intrapulmonary vascular anatomy which is, in fact, variable (Boyden 1955).

RADIOLOGY

The pulmonary trunk is best demonstrated by injection of contrast medium into the outflow tract of the right ventricle through a catheter inserted into an arm vein. The narrow infundibulum of the right ventricle, which lies below the pulmonary valve, and the pulmonary trunk and arteries are well seen [FIG. 13.19A and B]. The branches of the pulmonary arteries in the lungs can be visualized by films taken in rapid succession as the contrast medium passes through the lungs.

The pulmonary veins drain into the left atrium, which can be seen on the posterior aspect of the heart of later films of the series in lateral projections [FIG. 13.19D]. The atrium appears as an oval opacity seen through the cardiac silhouette on anteroposterior views before ventricular filling.

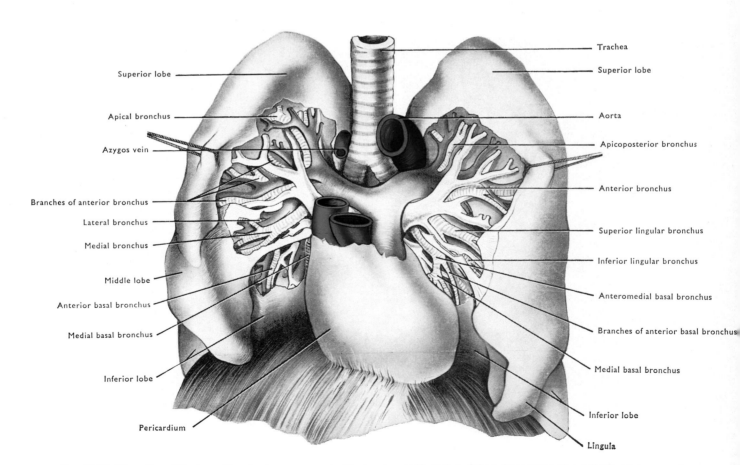

FIG. 13.29. Dissection of the bronchi and pulmonary vessels. The heart and bifurcation of the pulmonary trunk are displaced downwards.

The systemic circulation

ARTERIES

The systemic arteries all stem from the aorta, and by branching and rebranching form what is sometimes spoken of as the 'arterial tree'.

The situation of the larger arteries in the trunk and limbs affords them protection from injury. Thus the main artery of each limb runs distally on the flexor aspect; accordingly, as the limb is bent, the vessel is increasingly safeguarded. Further, its position lessens the tension exerted upon it during movements of the limb joints. Most main arteries pursue a more or less straight or evenly curved course, but some, for functional reasons, are very tortuous; for example, the facial artery which has to adapt itself to the movements of the mandible and of the facial muscles, and the uterine artery—probably in relation to necessary enlargement during pregnancy. The splenic artery, which runs behind the stomach, is another vessel whose tortuosity is linked with mechanical considerations, but such a direct association of tortuosity and adaption to changing position is not always apparent.

The cross-sectional area of the 'arterial tree' increases steadily towards the periphery, and the combined sectional area of all the arterioles greatly exceeds that of the aorta—a fact of profound haemodynamic importance. The manner in which arteries divide varies. The usual arrangement is illustrated by the arteries of the limbs, which give off successive branches yet retain their individuality as main vessels. Some arteries—the aorta is the most notable—end by dividing into two more or less equal branches; in some cases several arteries arise together from a short parent branch, e.g. the coeliac trunk.

Functional distribution

Some arteries, such as the pulmonary and renal, clearly have a functional distribution, but Hilton (1863) considered the distribution of other arteries, not confined to single organs, to have a general functional significance. For example, he named the maxillary artery the 'masticatory artery' because of its distribution to the bones, teeth, and muscles of mastication. The subclavian, apart from its continuation to the upper limb, he considered on an analysis of the distribution of its branches to be essentially a 'respiratory artery', with the vertebral artery giving supply to the respiratory nerve centres, the internal thoracic to the diaphragm and phrenic nerve, etc. Hilton's views are attractive, but it has to be recognized that structures which are linked functionally do tend to be in natural proximity.

Anastomoses

The great majority of arteries communicate with others through branches that unite to form anastomoses. In some regions anastomosis occurs directly between arteries of considerable size, so that there is—potentially—an even distribution of blood to the parts supplied, e.g. the formation of the 'arterial arcades' in the supply of the intestines. [FIG. 13.70]. At the base of the brain the circulus arteriosus is a remarkable example of such an anastomosis [p. 911]. But the most widespread anastomoses take place between the smaller arteries, and as a rule arterioles form extensive networks from which the capillaries arise. Around the limb joints the anastomoses between medium-sized arteries are of clinical importance as the means by which the blood supply is maintained distally by the opening up of a **collateral circulation** should the main artery be blocked by ligature or disease. Quiring (1949) provides a survey of the chief anastomoses, both arterial and venous, in the body.

Arteriovenous anastomoses

In certain regions the small arteries and veins are connected directly, as well as through the capillary bed. Such direct connections are called arteriovenous anastomoses and they consist of thick-walled, heavily innervated, contractile vessels. When they are fully closed all the blood reaching the site passes through the capillaries, but as they become open they will by-pass the capillary circulation partially or completely. The best-known are the complex anastomoses which form the basis of the 'glomera' found in the skin of the palmar surface of the digits [FIG. 13.30], but they have also been found in the skin of the lips, ears, and nose, and in the nasal mucous membrane [p. 495]. All these sites are subjected to temperature changes, and there is evidence that arteriovenous anastomoses are concerned *inter alia* with the regulation of body temperature. Their occurrence in the mucous membrane of the stomach and intestine is related to the intermittent activity of these parts, periods of digestion and absorption being accompanied by circulation through the capillary bed (when the anastomoses are closed), and periods of fasting by circulation through the anastomoses which are then open. For a general review of the subject Clark (1938) should be consulted; the part played by the anastomoses in temperature regulation is discussed by Prichard and Daniel (1956).

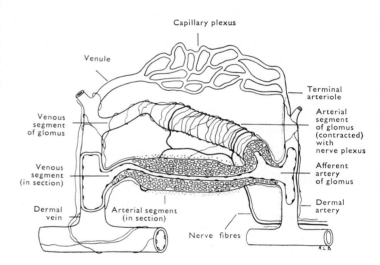

FIG. 13.30. Arteriovenous anastomosis in the skin of the hand. (After Masson 1937.)

End arteries

Small arteries which do not anastomose with others and therefore (apart from capillary communications) constitute the sole source of blood to the areas they supply are called end arteries; they are of pathological importance since occlusion of such an artery causes death of the tissues it supplies. End arteries are found in the kidney [p. 539] and spleen [p.1007]; and in the case of the blood supply of the spinal medulla and brain, although communications do exist, the smaller vessels probably act as 'functional end arteries' owing to the difficulty of establishing an effective collateral circulation. One very important end artery is the central artery of the retina [p. 907], since its occlusion causes blindness.

RADIOLOGY

Injection of contrast medium into the left ventricle via a catheter passed from the femoral artery through the aortic valve shows the left ventricle, the aortic sinuses at the root of the aorta, the aortic valves, and the arch of the aorta with the vessels arising from it.

With ageing the aorta loses its elasticity and consequently elongates and becomes tortuous.

The origins of the coronary arteries from the right and left sinuses respectively can be seen [FIG. 13.51]. Selective injection of the coronary arteries by catheterization allows the detail of each coronary artery to be shown with minimal overlap [FIGS. 13.36–13.38].

The course of the major vessels arising from the aortic arch is best seen by selective injection of each vessel, although a bolus injection of contrast material into the arch shows the origins of the main vessels (*arch aortography*; [FIG. 13.51]).

The branches of the arteries supplying the upper and lower extremeties are demonstrated by contrast injection of the subclavian and iliac arteries respectively (usually through the femoral artery, but in the upper limb direct injection is sometimes carried out) [FIG. 13.76].

The unpaired visceral branches of the abdominal aorta are shown by selective injection of the coeliac, superior [FIG. 13.71], or inferior mesenteric arteries. This demonstrates the distribution of each artery and also the extent of anastomoses between them, because forcible injection of one of the vessels, even in a normal subject, fills the adjacent vessels through anastomotic channels by temporarily reversing the flow in the adjoining vessel.

The circulation in certain paired organs such as the kidney may be studied in a similar manner [FIG. 8.8]. The blood supply of the suprarenal glands can sometimes be investigated by arteriography or venography [FIG. 9.3].

A general view of the abdominal aorta and its branches is also used [FIG. 13.35] though the filling of multiple branches tends to obscure the distal parts of all of them.

THE AORTA

The aorta is the main arterial trunk of the systemic circulation, and for descriptive purposes it is divided into the **ascending aorta**, the **arch of the aorta**, and the **descending aorta**, the last being further divided into **thoracic** and **abdominal** parts.

THE ASCENDING AORTA

The ascending aorta, about 5 cm long and 2.5 cm in diameter, springs from the base of the left ventricle, behind the left half of the sternum, at the level of the third intercostal space. Passing upwards, forwards, and to the right, it becomes the arch of the aorta behind the sternal angle and part of the second right costal cartilage. Its root shows three bulgings, called the **aortic sinuses**, each opposite a valvule of the aortic valve; and frequently the right side of the ascending aorta and the beginning of the arch show a slight dilation which is a common site of aneurysm (an aneurysm is a pathological dilation of an artery, of limited extent).

The ascending aorta is enclosed in the fibrous pericardium, and is enveloped, together with the pulmonary trunk, in a tube of serous pericardium. It lies at first behind the infundibulum and pulmonary trunk, but its upper part is nearer the sternum from which it is separated by the pericardium, the anterior margins of the right lung and pleura, and the remains of the thymus. Posteriorly it rests successively on the transverse sinus of the pericardium and left atrium, the right pulmonary artery, and the right bronchus. On its

left side are the left atrium below, and the pulmonary trunk above. On its right side are first the right atrium whose auricle overlaps it, and higher up the superior vena cava which it overlaps to a varying degree.

Branches

The right coronary artery springs from the right sinus of the aorta, and the left coronary artery from the left sinus [FIG. 13.15 and p. 896].

THE ARCH OF THE AORTA

The arch of the aorta lies in the superior mediastinum opposite the lower half of the manubrium sterni. It first runs upwards, backwards, and to the left; then backwards towards the left side of the body of the fourth thoracic vertebra; finally, turning downwards, it becomes continuous with the descending aorta at the lower border of that vertebra [FIG. 13.31]. The arch forms a marked upwards convexity [FIG. 13.33], and a slight concavity to the right [FIG. 13.31]. It lies more in a sagittal than in a coronal plane.

Anterior to the arch is the remains of the thymus gland, overlapped by the lungs and pleurae. On its left the following structures lie between it and the pleura: (1) four nerves, arranged from before backwards in this order: the left phrenic, the inferior cervical cardiac branch of the left vagus, the superior cervical cardiac branch of the left sympathetic, and the trunk of the left vagus; (2) the left superior intercostal vein which crosses it obliquely, superficial to the left vagus nerve and deep to the left phrenic nerve. Posteriorly and on the right of the arch [FIGS. 13.31 and 13.32] there are, from before backwards, the deep cardiac plexus, the trachea, the left recurrent laryngeal nerve, the left border of the oesophagus,

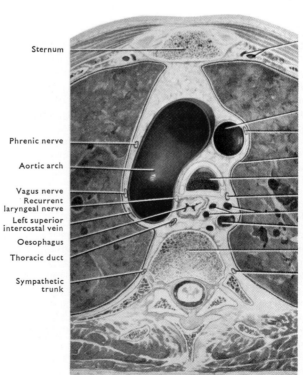

FIG. 13.31. Transverse section through superior mediastinum at level of fourth thoracic vertebra.

the thoracic duct, and the vertebral column. Above are its three large branches—the brachiocephalic trunk, the left common carotid and left subclavian arteries; and crossing anterior to them is the left brachiocephalic vein. Below the arch lie the root of the left lung, the bifurcation of the pulmonary trunk, and the ligamentum arteriosum [p. 977] with the left recurrent laryngeal nerve curving medially behind it from the left vagus, to ascend medial to the arch. The superficial cardiac plexus lies in front and to the right of the ligament.

Branches

The branches of the arch of the aorta are described on page 897.

Surface anatomy

The ascending aorta, the arch of the aorta, and the pulmonary trunk may be indicated on the surface as shown in FIGURE 13.17.

Variations

As a rule, the summit of the arch of the aorta reaches half-way up the manubrium, but it may reach the level of the upper border, or only as high as the sternal angle.

THE DESCENDING AORTA

Thoracic aorta

This part of the vessel lies in the posterior mediastinum [FIGS. 13.28 and 13.33]; it extends to the aortic opening in the diaphragm, where,

opposite the middle of the lower border of the twelfth thoracic vertebra, it become continuous with the abdominal aorta.

Posteriorly it lies on the vertebral column and on the hemiazygos and accessory hemiazygos veins as they pass to join the azygos vein. Anteriorly, from above downwards, are the root of the left lung, the pericardium over the left atrium, the oesophagus (which with its plexus of nerves crosses it from right to left) and the diaphragm. Along the whole length of the right side of the thoracic aorta are the thoracic duct and the azygos vein, and in its lower part the right lung and pleura; while the left lung and pleura are on its left side.

Branches

The branches of the thoracic aorta are described on page 922.

Surface anatomy

The thoracic aorta may be indicated by a band, 2.5 cm wide, centred above on the second left chondrosternal junction and below on the midline one finger's-breadth above the transpyloric plane. For variations see page 923.

The abdominal aorta [FIG. 13.34]

This vessel extends to the body of the fourth lumbar vertebra, where, to the left of the median plane, it bifurcates into the two common iliac arteries [FIG. 13.35]. The point of division is on a level with the highest points of the iliac crests.

Posteriorly it lies on the anterior longitudinal ligament covering the twelfth thoracic to fourth lumbar vertebrae, with the third and fourth left lumbar veins intervening. Anterior to it lies the aortic plexus of nerves [p. 823] and its extensions, and from above

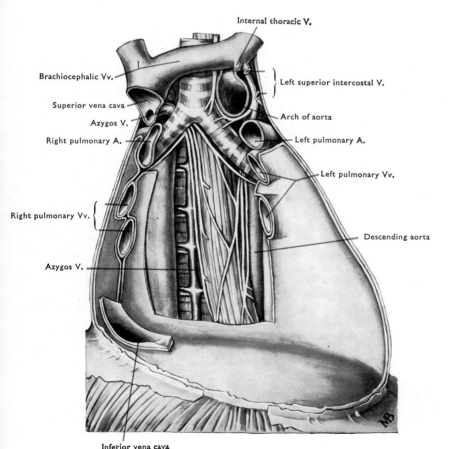

FIG. 13.32. Dissection of the upper part of the posterior mediastinum after removal of the heart and posterior wall of the pericardium.

Internal thoracic V.

Brachiocephalic Vv.

Left superior intercostal V.

Superior vena cava

Arch of aorta

Azygos V.

Right pulmonary A.

Left pulmonary A.

Left pulmonary Vv.

Right pulmonary Vv.

Descending aorta

Azygos V.

Inferior vena cava

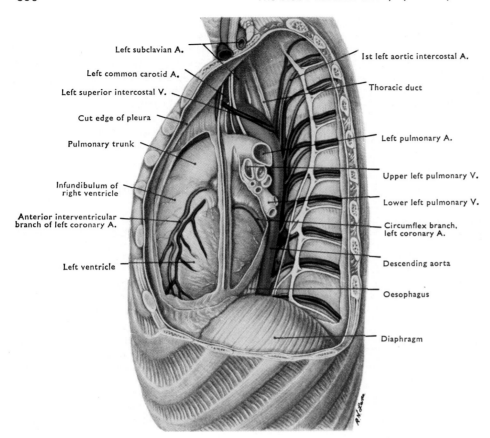

Left subclavian A.

Left common carotid A.

Left superior intercostal V.

Cut edge of pleura

Pulmonary trunk

Infundibulum of
right ventricle

Anterior interventricular
branch of left coronary A.

Left ventricle

1st left aortic intercostal A.

Thoracic duct

Left pulmonary A.

Upper left pulmonary V.

Lower left pulmonary V.

Circumflex branch,
left coronary A.

Descending aorta

Oesophagus

Diaphragm

FIG. 13.33. Left side of the mediastinum and thoracic vertebral column.

downwards, the omental bursa (which lies between it and the lower part of the caudate lobe of the liver and the lesser omentum), the body of the pancreas and the splenic vein, the left renal vein, the horizontal part of the duodenum, the root of the mesentery, the posterior parietal peritoneum, and coils of small intestine. It lies at first between the crura of the diaphragm. The thoracic duct and cisterna chyli, and the vena azygos lie between it and the right crus which separates it from the inferior vena cava. Below, it is in direct contact with the inferior vena cava. Lumbar lymph nodes lie in front and on both sides of it.

Branches

The branches of the abdominal aorta are described on page 923.

Surface anatomy

The abdominal aorta may be represented by a band, about 2 cm wide, centred above on the midline one finger's-breadth above the transpyloric plane, and below on a point 1.5 cm below and to the left of the umbilicus.

BRANCHES OF THE ASCENDING AORTA

Coronary arteries

There are two coronary arteries, right and left; they are distributed to the heart [FIGS. 13.10, 13.11, and 13.15] and great arteries.

The **right coronary artery** arises from the right aortic sinus, and passes forwards, between the pulmonary trunk and the right auricle, to the coronary sulcus. It follows the sulcus to the inferior extremity

of the right border of the heart giving branches to the right ventricle and atrium. Here it gives off a larger **right marginal branch** which runs along the lower border of the heart. Then, turning backwards and to the left, it continues in the coronary sulcus to the posterior interventricular groove where it divides into two branches. The larger is called the **posterior interventricular branch**, and it runs in the interventricular groove towards the apex of the heart, supplying branches to the ventricles and the interventricular septum. The smaller terminal branch continues in the coronary sulcus towards the termination of the circumflex branch of the left coronary artery, and ends in small branches to the left ventricle and atrium. The right coronary artery gives branches to the right atrium. The right coronary artery is accompanied by branches from the cardiac plexus and by lymph vessels; and in the second part of its course by the small cardiac vein and then the coronary sinus.

The **left coronary artery** arises from the left aortic sinus. It curves to the left on the posterior surface of the pulmonary trunk into the coronary sulcus. Here it divides into a larger **anterior interventricular branch**, and a smaller **circumflex branch**. The latter runs in the left part of the coronary sulcus, giving branches to the left atrium, the left (pulmonary) surface of the heart (**left marginal artery**), and the base of the left ventricle inferiorly. The anterior interventricular branch is an important vessel which passes down the anterior interventricular groove, and turns on to the posterior interventricular groove close to the apex. It supplies both ventricles and the interventricular septum, and is accompanied by cardiac nerves, lymph vessels, and the great cardiac vein [FIG. 13.11].

Variations

Rarely the coronary arteries arise by a single stem, or both may spring from the same aortic sinus. Such variability is not remarkable, as the arteries are essentially enlarged vasa vasorum.

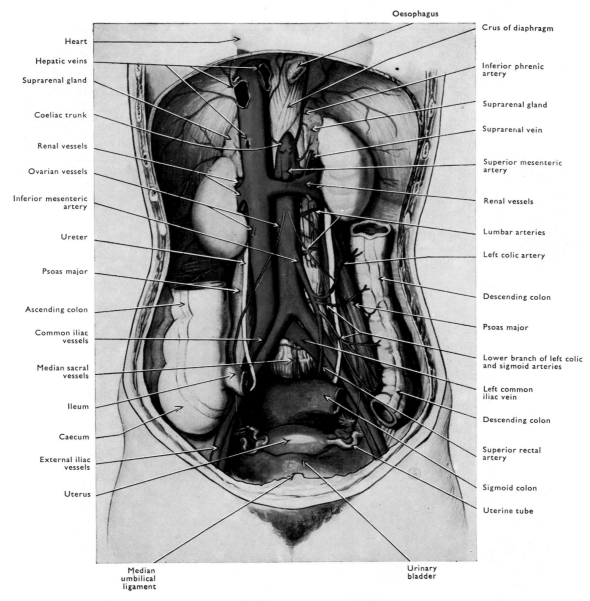

Heart — Hepatic veins — Suprarenal gland — Coeliac trunk — Renal vessels — Ovarian vessels — Inferior mesenteric artery — Ureter — Psoas major — Ascending colon — Common iliac vessels — Median sacral vessels — Ileum — Caecum — External iliac vessels — Uterus

Oesophagus — Crus of diaphragm — Inferior phrenic artery — Suprarenal gland — Suprarenal vein — Superior mesenteric artery — Renal vessels — Lumbar arteries — Left colic artery — Descending colon — Psoas major — Lower branch of left colic and sigmoid arteries — Left common iliac vein — Descending colon — Superior rectal artery — Sigmoid colon — Uterine tube

Median umbilical ligament — Urinary bladder

FIG. 13.34. Abdominal aorta and its branches.

Anastomoses

Obstruction of a coronary artery by the formation of blood clot in its lumen (thrombosis) is a common and very serious disease. Accordingly the extent to which the branches of the coronary arteries anastomose with one another has received much attention. Many anastomoses exist between arterioles of up to 200 μm in diameter, but they probably carry little blood in health. Gradual onset of obstruction to the coronary flow allows the collateral circulation provided by these small vessels to enlarge; but a sudden blockage is not supported by prior adaptation, and so the extent of cardiac damage is greater. Coronary arteriography [FIG. 13.36–13.38] is now a valuable clinical procedure.

Extracardiac anastomoses

Small twigs from the coronary arteries pass along the great vessels and form their vasa vasorum. They also pass to the pericardium where they anastomose, *inter alia*, with branches of the internal

thoracic artery and of its pericardiacophrenic and musculophrenic branches.

For more detailed information on the anatomy of the coronary arteries Fulton (1965) should be consulted.

BRANCHES OF THE ARCH OF THE AORTA

The branches which arise from the arch of the aorta supply the head and neck, the upper limbs, and part of the body wall.

They are three in number—the brachiocephalic trunk and the left common carotid and left subclavian arteries.

Variations

Six branches have been recorded—right and left subclavian, common carotid, and vertebral; less rarely, the left vertebral artery

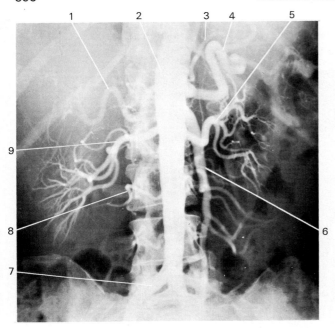

FIG. 13.35. Anteroposterior angiogram to show abdominal aorta and its branches.
1. Hepatic artery.
2. Aorta.
3. Left gastric artery.
4. Splenic artery.
5. Left renal artery.
6. Superior mesenteric artery.
7. Common iliac artery.
8. Lumbar artery.
9. Right renal artery.

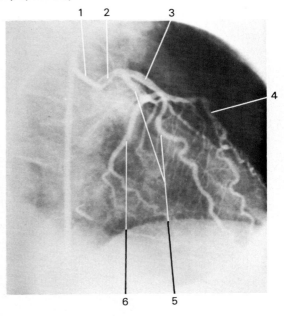

FIG. 13.37. Selective left coronary angiogram, right oblique view.
1. Catheter in aorta.
2. Left coronary artery.
3. Circumflex branch of L. coronary artery.
4. Marginal branch of circumflex.
5. Anterior interventricular branch.
6. Circumflex branch in coronary sulcus.

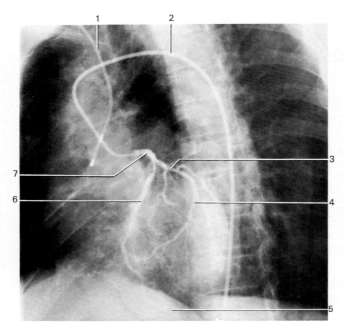

FIG. 13.36. Selective left coronary angiogram, left oblique view.
1. Catheter in superior vena cava.
2. Catheter in aorta.
3. Circumflex branch of left coronary artery.
4. Circumflex branch of left coronary artery.
5. Apex of heart.
6. Anterior interventricular branch of left coronary artery.
7. Left coronary artery.

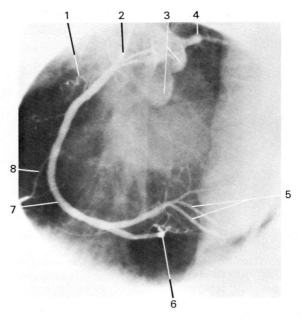

FIG. 13.38. Selective right coronary angiogram, left oblique view. Some of the medium has leaked into the aorta and entered the left coronary artery.
1. Auricular branch.
2. Catheter.
3. Aortic valvules.
4. Left coronary artery.
5. Branches to left ventricle.
6. Posterior interventricular artery viewed end on.
7. Right coronary artery.
8. Marginal branch.

forms an additional branch between the left common carotid and the left subclavian.

An artery called the thyroidea ima occasionally springs from the arch or from the brachiocephalic trunk.

An uncommon, but not unimportant, variation is that in which the right subclavian springs from the descending aorta and passes upwards and to the right behind the oesophagus.

Brachiocephalic trunk

Course and surface anatomy

The brachiocephalic trunk [FIG. 13.27] arises from the arch of the aorta behind the middle of the manubrium sterni. It is 3.5 to 5 cm long, and runs upwards, backwards, and to the right from the superior mediastinum into the root of the neck. It ends at the level of the upper part of the right sternoclavicular joint by dividing into the right subclavian and common carotid arteries. Its surface marking may be gauged by reference to FIGURE 13.17.

Anteriorly the left brachiocephalic vein crosses between it and the thymus; at a higher level, the sternothyroid muscle separates it from the sternohyoid and the sternoclavicular joint. Posteriorly it lies on the trachea and then on the right pleura. On its right side are the right brachiocephalic vein and the upper part of the superior vena cava. On its left side is the origin of the left common carotid artery, and at a higher level the trachea is in contact with it.

Branches

As well as its two terminal branches the brachiocephalic trunk may give off the thyroidea ima which runs on the front of the trachea to the thyroid gland.

ARTERIES OF THE HEAD AND NECK

The vessels to the head and neck are derived chiefly from the carotid arteries, but others arise from the main arterial stems of the upper limbs. The most important of those, the vertebral arteries, will be described with the carotid system since they supply structures in the neck and much of the brain.

Common carotid arteries

The right common carotid artery arises at the bifurcation of the brachiocephalic trunk, the left common carotid from the arch of the aorta; but each terminates at the level of the upper border of the thyroid cartilage (the lower border of the third cervical vertebra) by dividing into internal and external carotid arteries.

Left common carotid artery

This vessel arises immediately to the left and slightly behind the origin of the brachiocephalic trunk. Its thoracic portion, 2.5–3.5 cm in length, runs upwards and laterally to the left sternoclavicular joint.

Surface anatomy and course [FIG. 13.39].

A line from a point just below and to the left of the centre of the manubrium sterni to the sternoclavicular joint represents the thoracic portion of the left common carotid artery. The cervical portion of the left artery, and all of the right, is indicated by a line drawn from the appropriate sternoclavicular joint to a point 1 cm behind the superior horn of the thyroid cartilage.

In the thorax, the vessel is in contact, from below upwards, with the trachea, the left recurrent laryngeal nerve, and the left margin of the oesophagus [FIG. 13.57]. Anteriorly the left brachiocephalic vein runs across the artery, and cardiac branches from the left vagus and sympathetic descend in front of it. Those structures, together with the thymus and the anterior margins of the left lung and pleura, separate the artery from the manubrium sterni, and from the origins of sternohyoid and sternothyroid. Medially there are the brachiocephalic trunk below, and the trachea above; laterally is the left pleura; and posteriorly, the left phrenic and vagus nerves and the left subclavian artery.

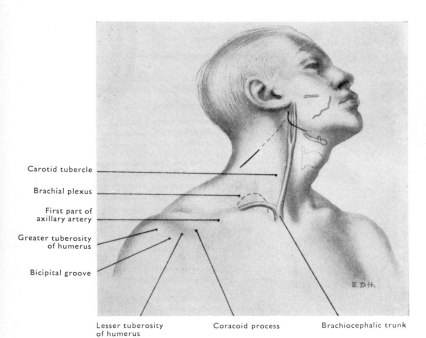

Carotid tubercle

Brachial plexus

First part of axillary artery

Greater tuberosity of humerus

Bicipital groove

Lesser tuberosity of humerus   Coracoid process   Brachiocephalic trunk

FIG. 13.39. Side of neck.
Of the structures shown, only the brachiocephalic trunk is labelled. The spinal portion of the accessory nerve is shown as an interrupted line between the transverse process of the atlas and the posterior border of the sternocleidomastoid, and as a continuous line between that muscle and the trapezius. The hypoglossal nerve is shown, above the hyoid bone, crossing the internal carotid and then the external carotid arteries. On the face the parotid duct is indicated above and the facial artery below. The apex of the lung is represented as an interrupted line arching above the clavicle.

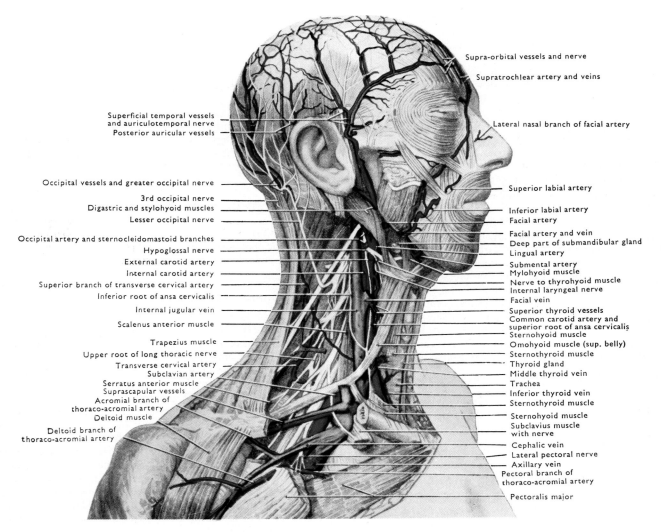

Supra-orbital vessels and nerve

Supratrochlear artery and veins

Superficial temporal vessels
and auriculotemporal nerve
Posterior auricular vessels

Lateral nasal branch of facial artery

Occipital vessels and greater occipital nerve

Superior labial artery

3rd occipital nerve
Digastric and stylohyoid muscles
Lesser occipital nerve

Inferior labial artery
Facial artery

Occipital artery and sternocleidomastoid branches
Hypoglossal nerve
External carotid artery
Internal carotid artery
Superior branch of transverse cervical artery
Inferior root of ansa cervicalis

Facial artery and vein
Deep part of submandibular gland
Lingual artery
Submental artery
Mylohyoid muscle
Nerve to thyrohyoid muscle
Internal laryngeal nerve
Facial vein

Internal jugular vein

Superior thyroid vessels
Common carotid artery and
superior root of ansa cervicalis
Sternohyoid muscle

Scalenus anterior muscle

Omohyoid muscle (sup. belly)

Trapezius muscle
Upper root of long thoracic nerve
Transverse cervical artery
Subclavian artery
Serratus anterior muscle
Suprascapular vessels
Acromial branch of
thoraco-acromial artery
Deltoid muscle

Sternothyroid muscle
Thyroid gland
Middle thyroid vein
Trachea
Inferior thyroid vein
Sternothyroid muscle

Sternohyoid muscle
Subclavius muscle
with nerve
Cephalic vein
Lateral pectoral nerve
Axillary vein
Pectoral branch of
thoraco-acromial artery

Deltoid branch of
thoraco-acromial artery

Pectoralis major

FIG. 13.40. Dissection of head and neck showing the carotid arteries.

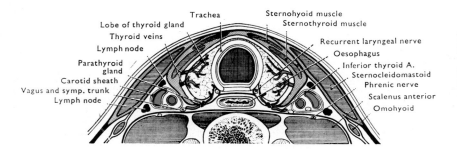

Trachea
Lobe of thyroid gland
Thyroid veins
Lymph node
Parathyroid
gland
Carotid sheath
Vagus and symp. trunk
Lymph node

Sternohyoid muscle
Sternothyroid muscle
Recurrent laryngeal nerve
Oesophagus
Inferior thyroid A.
Sternocleidomastoid
Phrenic nerve
Scalenus anterior
Omohyoid

FIG. 13.41. Diagram of relation of cervical fascia to
thyroid gland.
Deep cervical fascia (sternocleidomastoid layer),
blue; sheath of thyroid gland (pretracheal fascia), pale
red, and capsule of thyroid gland, deep to it, white.
(Modified from De Quervain 1924.)

In the neck, it runs upwards at first deep to the sternocleidomastoid, and then in the anterior triangle of the neck, enclosed with the internal jugular vein (lateral) and the vagus nerve (posterolateral), in the carotid sheath of cervical fascia [FIG. 13.41].

Posteriorly the sympathetic trunk intervenes between the artery and the prevertebral muscles and fascia. At the level of the seventh cervical vertebra the vertebral artery and the thoracic duct pass behind it, as does the inferior thyroid artery at the level of the cricoid cartilage.

The anterior tubercle of the sixth cervical transverse process is prominent and, since the common carotid artery can be compressed against it, is known as the **carotid tubercle**. It corresponds in level with the cricoid cartilage.

The superior root of the ansa cervicalis lies on the carotid sheath [FIG. 13.40], and opposite the sixth cervical vertebra the omohyoid muscle crosses superficial to the artery. Above that muscle it is overlapped by the anterior border of sternocleidomastoid and by deep cervical lymph nodes; and it is crossed by the superior thyroid vein. Below the omohyoid it is covered by the sternothyroid, sternohyoid, and sternocleidomastoid muscles, and may be overlapped by the thyroid gland; it is also crossed by the middle thyroid vein. Just above the sternum the anterior jugular vein is separated from it by sternohyoid and sternothyroid.

The trachea and oesophagus, with the recurrent laryngeal nerve between them, and the larynx and pharynx, are medial to the artery; and the carotid body lies just superior to its termination.

The internal jugular vein occupies the lateral part of the carotid sheath.

Branches

These are the external and internal carotid arteries and some minute twigs to the carotid body.

Right common carotid artery

The right common carotid artery corresponds with the cervical portion of the left common carotid, except that there is no thoracic duct on the right and the right lymph duct seldom exists as such [p. 984]; the left recurrent laryngeal nerve is posterior to the mediastinal part of the left artery, and medial to its cervical part, whereas the right nerve passes posterior to the lower part of the corresponding artery to reach its medial side. The oesophagus has a less intimate relation with the right artery than with the left.

The terminal portion of each common carotid artery and the root of its internal carotid branch are dilated to form the carotid sinus. This sinus is part of the mechanism that regulates blood pressure; its walls are specially innervated by the glossopharyngeal and vagus nerves, which also supply the carotid body [FIG. 9.9].

Variations

To what has been said earlier [p. 897], it is only necessary to add that the common carotid arteries may divide at a higher or lower level than usual—more often higher.

External carotid artery

Course and surface anatomy

The external carotid artery [FIGS. 13.40 and 13.45] extends from the upper border of the thyroid cartilage to the back of the neck of the mandible, where it ends by dividing into the superficial temporal and maxillary arteries.

It begins in the carotid triangle, and is at first anterior and medial to the internal carotid artery; it inclines backwards as it ascends, and thus becomes lateral to the internal carotid. Its course is indicated by a line drawn from a point just behind the tip of the greater horn of the hyoid bone to the lobule of the ear. Medially. At first the inferior constrictor muscle is close, but at a higher level the structures which intervene between it and the internal carotid—the stylopharyngeus muscle, the styloid process (or the styloglossus muscle), the glossopharyngeal nerve, the pharyngeal branch of the vagus and a portion of the parotid gland—separate it from the wall of the pharynx.
Superficially. In the carotid triangle it is overlapped by the anterior border of sternocleidomastoid, and is crossed by the superior thyroid (sometimes), lingual, and facial veins, and by the hypoglossal nerve. At the level of the angle of the mandible it passes under cover of the posterior belly of the digastric and stylohyoid muscles, which separate it from the inferior part of the parotid gland. It then passes laterally into the gland, within which it lies deep to the retromandibular vein and branches of the facial nerve.

Branches

Eight branches arise from the external carotid artery; three—the superior thyroid, the lingual, and the facial—from its anterior surface; two from its posterior surface, namely, the occipital and the posterior auricular; one from its medial side, namely, the ascending pharyngeal. Its terminal branches are the superficial temporal and maxillary arteries.

Branches of the external carotid artery

THE SUPERIOR THYROID ARTERY [FIGS. 13.40 and 13.45]

This vessel arises just below the tip of the greater horn of the hyoid bone, and runs downwards and forwards in contact medially with the inferior constrictor and the external laryngeal branch of the superior laryngeal nerve to the upper pole of the corresponding lobe of the thyroid gland. Its upper part is deep to the anterior border of sternocleidomastoid, and its lower part to the infrahyoid muscles.

Branches in the carotid triangle

The small infrahyoid branch runs along the lower border of the hyoid bone, deep to thyrohyoid.

The superior largyngeal artery pierces the thyrohyoid membrane and enters the lateral wall of the piriform recess, in company with the internal laryngeal nerve.

The sternocleidomastoid branch passes across the common carotid artery to the deep surface of the muscle.

Branches in the muscular triangle

The cricothyroid branch anastomoses in front of the cricothyroid ligament with its fellow of the opposite side.

The terminal glandular branches are anterior and posterior. (1) The anterior branch descends along the anterior border of the gland to the upper border of the isthmus where it anastomoses with its fellow. (2) The posterior branch is distributed to the deep surface of the lobe. (3) A lateral branch is inconstant. The branches anastomose with one another and with branches from the inferior thyroid artery.

THE LINGUAL ARTERY [FIGS. 11.20, 13.40, and 13.45]

This vessel arises opposite the tip of the greater horn of the hyoid bone, and ends beneath the tip of the tongue.

The first part of the artery is deep to platysma, and rests medially on the middle constrictor. It forms an upward loop which is crossed by the hypoglossal nerve, and then passes forwards on the middle constrictor, deep to hyoglossus which separates it from the hypoglossal nerve and its vena comitans, and from the lower part of the submandibular gland. It then ascends between genioglossus and the anterior border of hyoglossus, and having given off the sublingual artery, ends as the deep artery of the tongue. The deep artery is covered, on its lower surface, by the mucous membrane, and may be felt pulsating when the tongue is held between finger and thumb. At the tip of the tongue it anastomoses with its fellow of the opposite side [FIG. 6.4].

Branches

The suprahyoid branch is very small. It runs along the upper border of the hyoid bone.

The dorsal branches of the tongue—usually two—arise deep to hyoglossus, and ascend to the dorsum of the tongue. They supply the posterior part of the tongue, and send branches to the palatine tonsil.

The sublingual artery runs between mylohyoid and genioglossus to supply the sublingual gland.

THE FACIAL ARTERY [FIGS. 13.40 and 13.42]

This vessel arises immediately above the lingual and ends at the medial angle of the eye after a tortuous course. It first ascends on the lateral surface of the pharynx, and comes to lie deep to the posterior belly of digastric and stylohyoid, separated from the palatine tonsil only by the superior constrictor muscle. Just above stylohyoid it enters a groove in the submandibular gland, and runs downwards between the lateral surface of the gland and the medial pterygoid muscle to the lower border of the mandible. Turning round this border of the bone it enters the face at the anterior border of masseter where its pulsations may be felt. Thence it passes sinuously towards the angle of the mouth, giving off its labial branches 1.0 to 1.5 cm from the angle, and ascends more vertically to the medial angle of the eye (angular artery).

The facial vein is posterior to the artery in the face, and runs a straighter course.

Branches in the neck

The ascending palatine artery [FIG. 13.50] ascends between styloglossus and stylopharyngeus and turns downwards over the upper border of the superior constrictor, accompanying levator veli palatini [FIG. 13.46]. It supplies the pharynx, soft palate, tonsil, and auditory tube.

The tonsillar branch pierces the superior constrictor, and ends in the palatine tonsil.

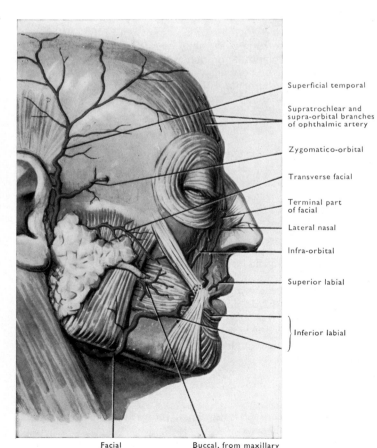

Superficial temporal

Supratrochlear and supra-orbital branches of ophthalmic artery

Zygomatico-orbital

Transverse facial

Terminal part of facial

Lateral nasal

Infra-orbital

Superior labial

Inferior labial

Facial   Buccal, from maxillary

FIG. 13.42. Arteries of the face.

The glandular branches pass into the submandibular gland.

The submental artery runs forwards on mylohyoid, medial to the submandibular gland, to the chin.

Branches in the face

The inferior labial artery runs medially, under cover of the muscles of the lower lip, and anastomoses with the mental artery, and with its fellow of the opposite side.

The superior labial artery runs medially, between orbicularis oris and the mucous membrane. It supplies the upper lip, and, by a septal branch, the lower and anterior part of the nasal septum.

The lateral nasal branch ramifies on the side of the nose [FIG. 13.42].

The labial arteries are readily palpable when the lips are held between finger and thumb.

The facial artery may arise from a stem common to it and the lingual artery—the linguofacial trunk.

THE OCCIPITAL ARTERY [FIGS. 13.40 and 13.45]

This vessel arises opposite the facial artery, and ends near the medial end of the superior nuchal line of the occipital bone, by dividing into medial and lateral branches. It runs deep and parallel to the lower fibres of the posterior belly of digastric, and at the base of the skull, deep to the attachment of that muscle, enters a groove on the temporal bone from which it passes on to the superior oblique muscle. It enters the superficial fascia at the junction of the medial and intermediate thirds of the superior nuchal line.

In its course it crosses the carotid sheath, the hypoglossal nerve (which hooks round it), and the accessory nerve. Near its termination it is crossed by the greater occipital nerve, and it passes either through trapezius or between trapezius and sternocleidomastoid.

Branches

Muscular branches include two sternocleidomastoid branches. One is looped downwards across the hypoglossal nerve; the other accompanies the accessory nerve into the muscle.

The descending branch is given off on the surface of superior oblique; it runs down among the muscles of the back of the neck, anastomosing with the deep cervical artery.

The meningeal branches enter the skull through the hypoglossal canal and the jugular foramen.

The auricular branch aids in the supply of the auricle.

The mastoid branch passes through the mastoid foramen; it supplies the mastoid air cells on its way to the dura mater.

The terminal occipital branches, medial and lateral, ramify tortuously in the scalp, where they anastomose with the posterior auricular and superficial temporal arteries.

THE POSTERIOR AURICULAR ARTERY [FIGS. 13.45 and 13.50]

From its origin just above the posterior belly of digastric it runs to the interval between the mastoid process and the auricle where it divides.

Branches

There are three named branches.

The stylomastoid artery enters the stylomastoid foramen and accompanies the facial nerve. It supplies the tympanic cavity and mastoid antrum, the vestibule and semicircular canals, the mastoid air cells (mastoid branches), and the stapedius muscle through the posterior tympanic artery, which also forms a vascular circle around the tympanic membrane with the anterior tympanic branch of the maxillary artery.

The **auricular branch** supplies the scalp and both surfaces of the auricle.

The **occipital branch** supplies the occipital belly of the occipito-frontalis and the adjacent skin.

THE ASCENDING PHARYNGEAL ARTERY [FIG. 13.50]

This long, slender vessel, usually the first or second branch of the external carotid, ascends on the pharynx, medial to the internal carotid artery.

Branches

Small **pharyngeal branches** supply the pharynx, the palatine tonsil, and the auditory tube.

Small **meningeal** branches enter the cranium by the hypoglossal canal, the jugular foramen (**posterior meningeal artery**), and the foramen lacerum.

The **inferior tympanic artery** accompanies the tympanic branch of the glossopharyngeal nerve.

THE SUPERFICIAL TEMPORAL ARTERY [FIG. 13.42]

This terminal branch of the external carotid begins between the parotid gland and the neck of the mandible, and ends in the scalp, from 2.5 to 5 cm above the zygomatic arch, by dividing into frontal and parietal branches. As it crosses the posterior root of the zygoma it may easily be compressed. The auriculotemporal nerve runs behind it, and the superficial temporal vein is usually superficial to it.

Branches

Small **parotid branches** supply the gland and the temporomandibular joint.

The **transverse facial artery** runs across masseter, below the zygomatic arch and above the parotid duct.

Small **anterior auricular branches** supply the auricle and the external acoustic meatus.

The **middle temporal artery** crosses the zygomatic arch, pierces the temporal fascia, and ascends in the temporal fossa grooving the skull wall.

The **zygomatico-orbital artery** runs forwards above the zygomatic arch.

The **frontal branch** runs sinuously towards the frontal tuber. It is frequently visible through the skin in the living person.

The **parietal branch**, less tortuous, runs towards the parietal tuber.

Wounds of the scalp bleed freely owing to the fixation of its vessels' walls to the dense subcutaneous tissue. Moreover, so freely do the arteries of the two sides anastomose, it is not possible to control haemorrhage from one side by ligature of the corresponding external carotid artery. The power of healing possessed by the scalp is remarkable—an index of its very rich vascularity.

THE MAXILLARY ARTERY

This, the larger terminal branch of the external carotid, begins between the parotid gland and the back of the neck of the mandible and ends in the pterygopalatine fossa [FIGS. 13.43, 13.45, and 13.46]. It is described in three parts. The **first part** extends forwards with the maxillary vein, between the sphenomandibular ligament and the neck of the mandible, below the auriculotemporal nerve, to the

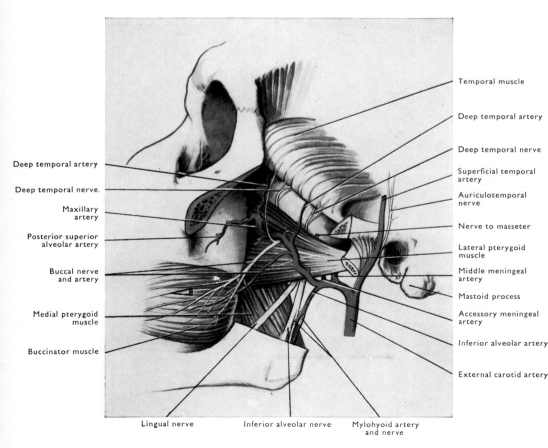

Deep temporal artery
Deep temporal nerve.
Maxillary artery
Posterior superior alveolar artery
Buccal nerve and artery
Medial pterygoid muscle
Buccinator muscle

Temporal muscle
Deep temporal artery
Deep temporal nerve
Superficial temporal artery
Auriculotemporal nerve
Nerve to masseter
Lateral pterygoid muscle
Middle meningeal artery
Mastoid process
Accessory meningeal artery
Inferior alveolar artery
External carotid artery

Lingual nerve   Inferior alveolar nerve   Mylohyoid artery and nerve

FIG. 13.43. Dissection showing termination of external carotid artery and first and second parts of maxillary artery.

middle of the lower border of the lateral pterygoid muscle. The **second part** is in the infratemporal fossa, and lies superficial (or deep) to the lateral pterygoid muscle. The **third part** passes between the heads of lateral pterygoid, and through the pterygomaxillary fissure into the pterygopalatine fossa.

Branches from the first part

The **deep auricular artery** pierces the wall of the external acoustic meatus and supplies its lining and the tympanic membrane.

The **anterior tympanic artery** traverses the petrotympanic fissure and enters the middle ear.

The **middle meningeal artery** [FIGS. 13.44 and 13.47], the largest of the meningeal arteries, is the principal artery to the vault of the skull. It ascends through the auriculotemporal nerve, and enters the skull by the foramen spinosum. In the middle cranial fossa it passes for a short distance forwards and laterally and divides into frontal and parietal branches at a point 1–2 cm above the centre of the zygomatic arch.

The middle meningeal artery sends a small **petrosal branch** through the hiatus for the greater petrosal nerve, and a small **superior tympanic artery** through either the canal for the tensor tympani or the petrosquamous suture. It also sends an **anastomotic branch** to the lacrimal artery through the lateral part of the superior orbital fissure, and may form that artery.

The **frontal branch**, the larger and more important of the two, passes upwards to the antero-inferior angle of the parietal bone,

where it is often enclosed in a bony canal. It continues upwards and backwards, parallel to and a variable distance (0.5–1.5 cm) behind the coronal suture, almost to the vertex of the skull, often as two branches. Here it lies in the periosteum lining the skull and is separated by the corresponding veins from the bone and by the meninges from the precentral gyrus [FIG. 13.44]. It sends branches forwards and backwards.

The **parietal branch** passes backwards on the squamous part of the temporal bone to reach the lower border of the parietal bone where it divides into its terminal branches.

Because of its liability to rupture from blows on the side of the head, with resultant haemorrhage between the skull and the dura mater, the surface marking of the middle meningeal artery [FIG. 13.44] is of surgical importance.

An **accessory meningeal branch** may arise from the first part of the maxillary or from the middle meningeal artery. It enters the skull through the foramen ovale.

The **inferior alveolar artery** passes downwards to enter the mandibular foramen with the inferior alveolar nerve. It descends in the mandibular canal and, after giving off the mental artery, is continued in the bone to the median plane. The **mylohyoid branch**, given off immediately above the foramen, descends with the mylohyoid nerve, in the mylohyoid groove to the superficial surface of the muscle. In the mandibular canal branches are given off to the teeth. The **mental artery** passes through the mental foramen, supplies the tissues of the chin, and anastomoses with the inferior labial artery.

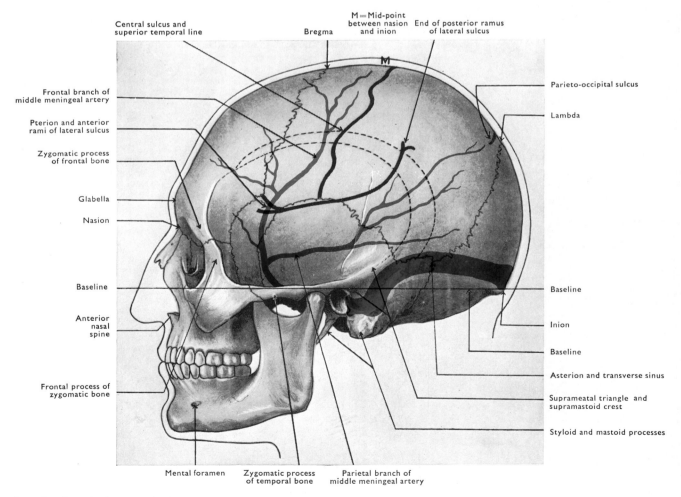

FIG. 13.44. Craniocerebral topography: landmarks of skull; chief sulci of cerebrum; transverse and sigmoid venous sinuses; middle meningeal artery. M = midpoint between nasion and inion.

Branches from the second part

Branches pass to the muscles of mastication. The masseteric artery passes through the mandibular notch to the deep surface of the muscle. The deep temporal arteries, anterior and posterior, ascend between the temporalis muscle and the skull. Small pterygoid branches supply the pterygoid muscles. The buccal artery [FIG. 13.43] supplies the buccinator muscle, and the skin and mucous membrane of the cheek.

Branches from the third part

One or more posterior superior alveolar arteries descend on the posterior surface of the maxilla to supply the molar and premolar teeth and the lining of the maxillary sinus.

The infra-orbital artery enters the orbit through the inferior orbital fissure and runs forwards, in the infra-orbital groove and canal, to the infra-orbital foramen, through which it emerges on the face. In the infra-orbital canal it gives twigs to the canine and incisor teeth (anterior superior alveolar arteries) and to the walls of the maxillary sinus. In the face it sends branches to the lower eyelid, upper lip, cheek, and side of nose.

The descending palatine artery descends through the greater palatine canal giving off several lesser palatine arteries which pass through the lesser palatine canals to the soft palate. Continuing as the greater palatine artery it passes through the greater palatine foramen and runs forwards on the bony palate, medial to the alveolar process. Its terminal portion ascends through the incisive canal to the nasal septum.

The artery of the pterygoid canal runs backwards through the canal with the corresponding nerve and supplies branches to the pharynx, to the auditory tube, and soft palate.

The pharyngeal branch runs through the palatovaginal canal to the roof of the pharynx.

The sphenopalatine artery enters the nose through the sphenopalatine foramen, and divides into the posterior lateral nasal and posterior septal nasal arteries. The latter accompany the nasopalatine nerve across the roof of the cavity and downwards and forwards in the groove on the vomer.

Internal carotid artery

The internal carotid artery [FIGS. 13.46, 13.47, and 13.50] springs from the common carotid opposite the upper border of the thyroid cartilage, and ascends to the base of the skull where it enters the carotid canal. From the canal it passes into the cavernous sinus above which it divides into the anterior and middle cerebral arteries [FIG. 13.52].

It is convenient to describe the vessel in three parts.

In the Neck. Posterior to the artery are longus capitis, the prevertebral fascia, and the sympathetic trunk, which separate it from the cervical transverse processes. The superior laryngeal nerve runs obliquely behind the vessel. Medial to the internal carotid is the external carotid artery for a short distance below, and above this is the wall of the pharynx with the ascending pharyngeal artery intervening throughout, and the superior laryngeal nerve dividing into its two branches inferiorly. Immediately lateral to the internal carotid artery are the internal jugular vein and, more posteriorly, the vagus nerve; but the vein becomes progressively more posterior until, just below the skull, it lies behind the artery, separated from it by the last four cranial nerves. Superficially, the artery is overlapped by sternocleidomastoid, and deep to this muscle is

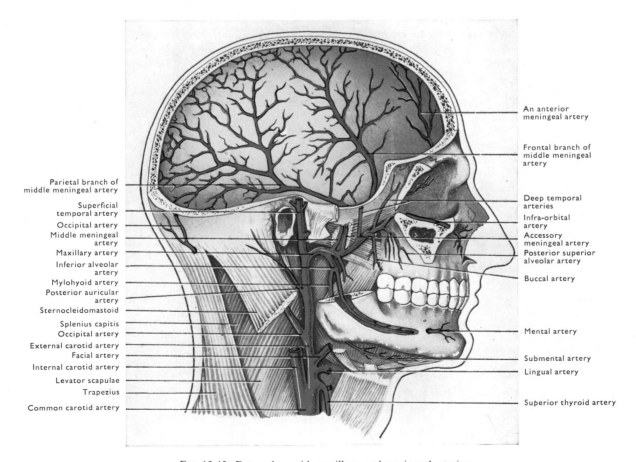

FIG. 13.45. External carotid, maxillary, and meningeal arteries.

crossed by the lingual and facial veins, and by the hypoglossal nerve from which the superior root of the ansa cervicalis descends upon it. The posterior belly of digastric, the occipital and posterior auricular arteries, and stylohyoid separate it from the parotid gland, and above this level it is separated from the external carotid artery by the deeper part of the gland and by the styloid process, styloglossus, stylopharyngeus, the glossopharyngeal nerve, and the pharyngeal branch of the vagus.

In the Carotid Canal. The artery ascends into the canal with the internal carotid plexus, and bending anteromedially is antero-inferior to the cochlea and the middle ear cavity, medial and then superior to the auditory tube, and then inferior to the trigeminal ganglion. Leaving the canal it comes to lie above the fibrocartilage which fills the lower part of the foramen lacerum during life. It then turns upwards into the cavernous sinus.

In the Cranial Cavity. The artery bends sharply forwards in the cavernous sinus [FIGS. 11.7 and 11.8] with the abducent nerve close to its inferolateral aspect. The oculomotor, trochlear, the ophthalmic division of the trigeminal, and, usually, the maxillary division, lie in the lateral wall of the sinus. Near the superior orbital fissure, it turns back on itself, and passing posteromedially, pierces the roof of the cavernous sinus inferior to the optic nerve. It then passes between the optic and oculomotor nerves to end below the anterior perforated substance by dividing into the anterior and middle cerebral arteries.

The carotid sinus [p. 901] includes the root of the internal carotid artery, and may be limited to it.

Variations

The internal carotid artery occasionally springs from the arch of the aorta. It may be tortuous in the neck and so come close to the tonsil.

Branches of internal carotid artery

No branches are given off in the neck.

Very small caroticotympanic branches perforate the posterior wall of the carotid canal and enter the tympanum.

In the cranium twigs are given to the trigeminal ganglion.

Branches to the hypophysis are constant vessels which have been fully described by Stanfield (1960); see page 604.

THE OPHTHALMIC ARTERY [FIGS. 13.47 and 13.50]

This vessel springs from the internal carotid immediately after it has pierced the dura and arachnoid. It runs through the optic canal below the optic nerve and within its arachnoid and dural sheaths. Piercing the sheaths, it runs forwards for a short distance on the lateral side of the nerve; it then crosses, between the optic nerve and

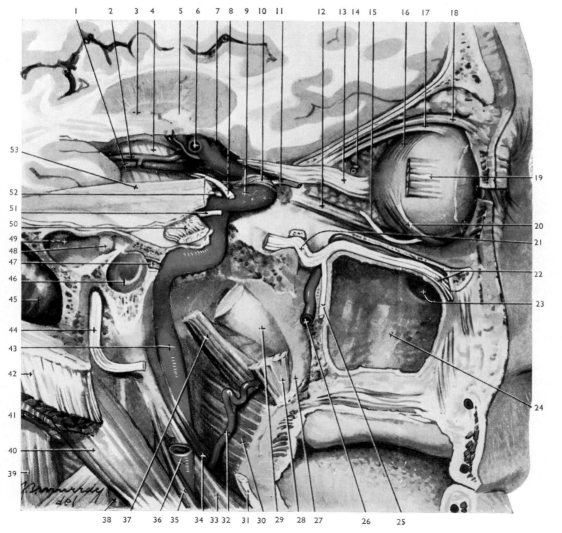

1. Posterior cerebral A.
2. Basal vein.
3. Lentiform nucleus.
4. Cerebral peduncle.
5. Anterior commissure.
6. Middle cerebral A.
7. Anterior cerebral A.
8. Oculomotor nerve.
9. Internal carotid A.
10. Interclinoid ligament.
11. Ophthalmic A.
12. Inferior rectus M.
13. Optic nerve.
14. Nasociliary nerve (cut).
15. Nerve to inf. oblique M.
16. Tendon of sup. oblique M.
17. Superior rectus M.
18. Levator palpebrae M.
19. Lateral rectus M.
20. Inferior oblique M.
21. Maxillary nerve.
22. Infra-orbital nerve and A.
23. Ostium of maxillary sinus.
24. Maxillary sinus.
25. Posterior superior alveolar N.
26. Maxillary artery.
27. Medial pterygoid lamina.
28. Tensor veli palatini M.
29. Auditory tube.
30. Superior constrictor M.
31. Lingual nerve.
32. Ascending palatine A.
33. Styloglossus M.
34. Stylopharyngeus M.
35. Stylohyoid M.
36. External carotid A.
37. Levator veli palatini M.
38. Internal jugular vein.
39. Longissimus capitis M.
40. Posterior belly of digastric M.
41. Occipital artery.
42. Sternocleidomastoid M.
43. Internal carotid A.
44. Facial nerve.
45. Sigmoid sinus.
46. Tympanic membrane.
47. Bony auditory tube.
48. Head of malleus.
49. Mastoid antrum.
50. Trigeminal ganglion.
51. Abducent nerve.
52. Trochlear nerve.
53. Tentorium cerebelli.

FIG. 13.46. Dissection showing course and relations of upper part of internal carotid artery.

the superior rectus, to the medial wall of the orbit, where it turns forwards to end by dividing into the supratrochlear and dorsal nasal arteries.

Branches

The central artery of the retina arises in, or close to, the optic canal and runs within the dural sheath of the optic nerve. It pierces the nerve about half-way along its intra-orbital course, and runs in the centre of the nerve with its companion vein to the retina [p. 844]. It is an 'end artery' and the most important branch of the ophthalmic.

The short posterior ciliary arteries, usually six to eight in number, run forwards beside the optic nerve; they divide into numerous branches which pierce the sclera around the nerve and end in the choroid coat. The long posterior ciliary arteries, two in number, pierce the sclera medial and lateral to the optic nerve. They then pass forward in the horizontal meridian, between sclera and choroid, to the periphery of the iris. Here their branches anastomose with anterior ciliary arteries to form the greater arterial circle at the periphery of the iris, from which secondary branches run to a lesser arterial circle near the pupillary border of the iris.

The lacrimal artery runs to the upper lateral corner of the orbit. It gives branches to the lacrimal gland and ends in the lateral palpebral arteries which supply the conjunctiva and the eyelids.

Small meningeal branches, of which one (the anterior meningeal artery) arises from the lacrimal artery, pass backwards through the superior orbital fissure and anastomose with the middle meningeal artery.

The muscular branches supply orbital muscles, and they also give off anterior ciliary arteries.

The anterior ciliary arteries pierce the sclera behind the corneoscleral junction, and join the greater arterial circle of the iris. [For the circulation in the eye and its accessory organs, see pages 839, 841, and 844.]

The supra-orbital artery runs forwards, between levator palpebrae superioris and the periosteum, to the supra-orbital notch or foramen. It supplies the skin of the forehead and anterior part of the scalp.

The anterior and posterior ethmoidal arteries. The posterior, much the smaller, traverses the posterior ethmoidal foramen and supplies the posterior ethmoidal cells. The anterior ethmoidal artery passes through the anterior ethmoidal foramen, crosses the cribriform plate to the nasal slit, and descends in a groove on the inner surface of the nasal bone. It passes between the lateral cartilage and the lower border of the bone to reach the tip of the nose. It supplies the anterior and middle ethmoidal cells, the meninges, the frontal sinus, the anterosuperior part of the nasal mucoperiosteum, and the skin of the nose.

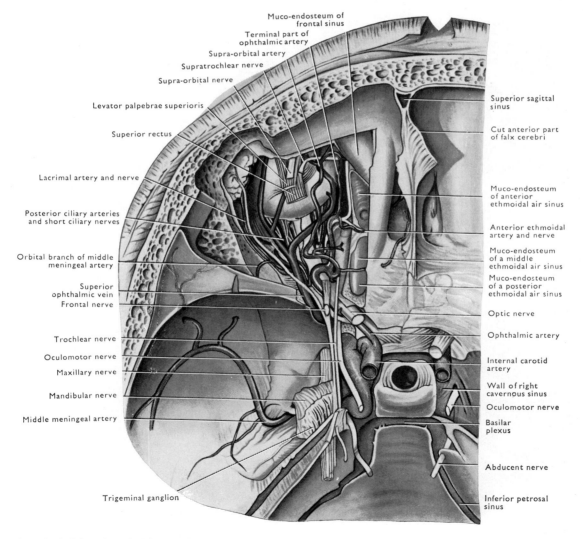

FIG. 13.47. Dissection of orbit and middle cranial fossa.
On the right side the trochlear nerve has been removed, and in the left orbit portions of the structures above the ophthalmic artery have been taken away.

Medial palpebral arteries, upper and lower, are distributed to the eyelids. With the lateral palpebral arteries they form the **superior** and **inferior palpebral arches**. Together with the anterior ciliary arteries, they send branches to the conjunctiva.

The **dorsal nasal artery** ends on the side of the nose.

The **supratrochlear artery** pierces the palpebral fascia and ascends in the superficial fascia of the forehead and scalp.

THE POSTERIOR COMMUNICATING ARTERY

This branch of the internal carotid forms part of the circulus arteriosus cerebri [p. 911]. It joins the posterior cerebral artery.

THE ANTERIOR CHOROIDAL ARTERY

This small branch passes backwards and laterally, and ends in the choroid plexus of the inferior horn of the lateral ventricle. It supplies the optic tract, the lateral geniculate body (Abbie 1938), the central

part of the crus cerebri, the greater part of the posterior limb and sublentiform parts of the internal capsule, and the adjacent parts of the globus pallidus and thalamus (Abbie 1937). For a full account of the blood supply of the visual pathway, see Abbie (1938).

THE ANTERIOR CEREBRAL ARTERY [FIG. 13.54]

This vessel passes forwards and medially above the optic nerve, and anterosuperior to the optic chiasma where it is joined to its fellow by the **anterior communicating artery**. It then ascends anterior to the lamina terminalis, and continues either on the external surface of the corpus callosum or in the sulcus cinguli. It ends by passing superiorly, anterior to the parieto-occipital sulcus.

Branches

Branches of all the cerebral arteries form two distinct groups which do not communicate with one another: (1) **central**, (2) **cortical**. The central vessels pierce the brain at once, but the cortical branches ramify and anastomose on the pia mater [p. 628].

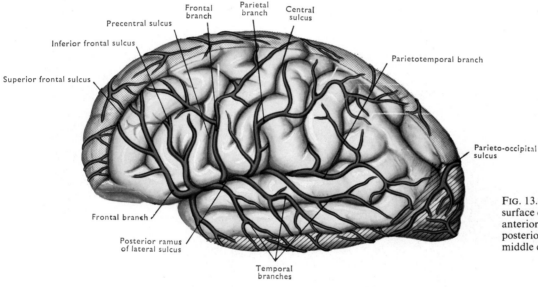

FIG. 13.48. Arteries of the superolateral surface of the cerebral hemisphere. Stipple, anterior cerebral artery; crosshatching, posterior cerebral artery; no markings, middle cerebral artery.

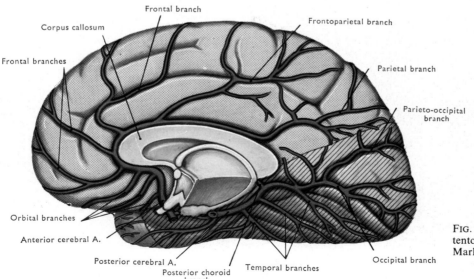

FIG. 13.49. Arteries of the medial and tentorial surfaces of the cerebral hemisphere. Markings as in Figure 13.48.

THE BRANCHES OF THE ANTERIOR CEREBRAL ARTERY

Central branches

These slender arteries pierce the lamina terminalis and rostrum of the corpus callosum, and supply the anterior parts of the caudate nucleus, lentiform nucleus, and internal capsule. Small branches from the region of the anterior communicating enter the optic chiasma and nerve. A larger branch arising lateral to the anterior communicating artery, runs back along the main vessel and enters the medial part of the anterior perforated substance, to supply the lower, anterior part of the corpus striatum and hypothalamus.

Cortical branches

The named cortical branches are orbital, frontal, and parietal, and their area of distribution is shown in FIGURES 13.48 and 13.49. It should be noted that this includes the superomedial parts of the motor and sensory areas of the cortex.

THE MIDDLE CEREBRAL ARTERY [FIGS. 13.53 and 13.54]

This, the larger terminal branch of the internal carotid artery, is in more direct continuation with it. It passes in the lateral sulcus to the surface of the insula, where it divides into numerous cortical branches.

Central branches pierce the anterior perforated substance. They may be divided into medial and lateral striate branches. The former traverse and supply the globus pallidus and internal capsule, reaching the lateral part of the thalamus. The latter pass over and through the putamen of the lentiform nucleus, reach the internal capsule further superiorly, and supply the upper part of the head and the body of the caudate nucleus. These vessels are commonly involved in cerebral thrombosis or haemorrhage.

The cortical branches are named orbital, frontal, parietal, and temporal, and their distribution is shown in FIGURES 13.48 and 13.52.

Vertebral artery

The vertebral artery [FIGS. 13.50 and 13.52] is the first branch of the subclavian artery, and is divisible into four parts.

The first part runs to the foramen in the sixth cervical transverse process. Posterior to it are the ventral rami of the seventh and eighth cervical nerves, and also, somewhat medial to its origin, the inferior cervical sympathetic ganglion. Anterior to it are the vertebral and internal jugular veins, and, on the left side, the terminal part of the thoracic duct.

The second part runs upwards through the foramina in the upper six cervical transverse processes. On passing through the transverse process of the axis, it turns laterally and then upwards to the atlas. It is accompanied by branches of the inferior cervical sympathetic ganglion, and by a plexus of veins.

The third part winds backwards and medially on the lateral mass of the atlas with the first cervical ventral ramus, and enters the suboccipital triangle in the groove on the upper surface of the posterior arch of the atlas. It then passes anterior to the edge of the posterior atlanto-occipital membrane and enters the vertebral canal.

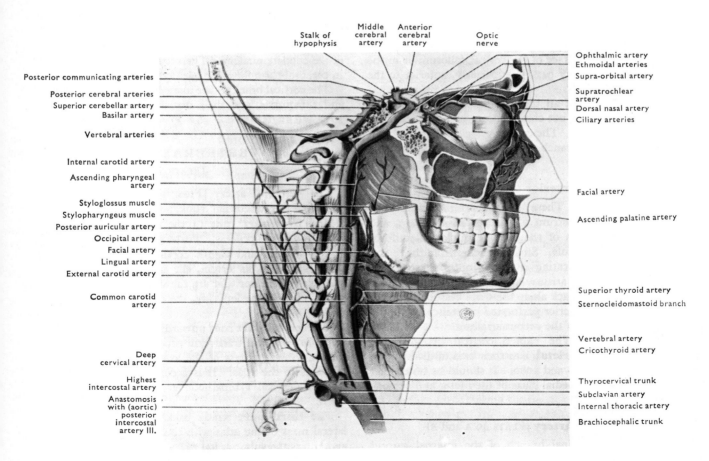

FIG. 13.50. Carotid, subclavian, and vertebral arteries and their main branches.

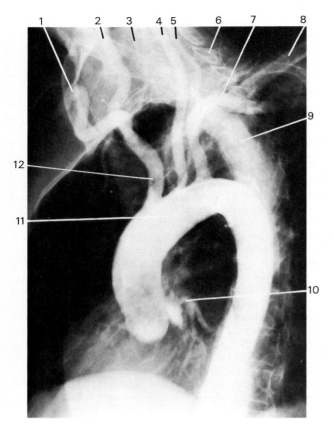

FIG. 13.51. Arch aortogram. Left oblique view of the aortic arch and its branches filled with contrast medium through a catheter passed along the aorta from the femoral artery.
  1. Right subclavian artery.
  2. Right common carotid artery.
  3. Right vertebral artery.
  4. Left common carotid artery.
  5. Left vertebral artery.
  6. Cervical lamina.
  7. Left subclavian artery.
  8. Left clavicle.
  9. Vertebral column.
  10. Left coronary artery.
  11. Arch of aorta.
  12. Brachiocephalic trunk.

The **fourth part** pierces the spinal dura and arachnoid and runs upwards and forwards on the side of the medulla oblongata, passing among the rootlets of the hypoglossal nerve. Reaching the lower border of the pons, it unites with its fellow to form the basilar artery [FIG. 13.52].

BRANCHES

Muscular branches from the vertebral artery supply the neighbouring muscles.

Spinal branches pass through the intervertebral foramina and may reinforce the anterior and posterior spinal arteries.

Small **meningeal** branches ascend into the posterior fossa of the skull.

The **posterior spinal artery** usually arises from the posterior inferior cerebellar branch of the vertebral artery, and seldom from the vertebral itself. It descends along the line of attachment of the dorsal nerve rootlets, being reinforced by spinal twigs of the vertebral, ascending cervical, posterior intercostal, lumbar, and lateral sacral arteries. Caudally the two vessels, which are much smaller than the anterior spinal artery, curve ventrally on the conus medullaris to join it. For their distribution see below.

The **anterior spinal artery** [FIG. 13.52] runs downwards and medially, and unites with its fellow in front of the decussation of the pyramids to form a single median vessel. A variable number (2–10) of radicular arteries (Romanes 1965) pass along the ventral roots and reinforce it.

The anterior spinal artery supplies all the grey matter of the spinal medulla, save part of the posterior columns which are supplied by the posterior spinal arteries, and the deepest parts of the lateral and posterior columns of white matter. It anastomoses on the pia mater with the posterior spinal arteries. The central branches enter the anterior median fissure, and though they do not anastomose with adjacent central vessels, their capillary territories overlap. The white matter receives most of its supply from the surface anastomotic vessels between the anterior and posterior spinal arteries. (See also Woollam and Millen 1955, 1958.)

The anterior spinal arteries also supply the paramedian parts of the medulla oblongata, including the hypoglossal nucleus.

The **posterior inferior cerebellar artery** pursues a tortuous course on the side of the medulla oblongata, and, entering the vallecula of the cerebellum, divides into lateral and medial terminal branches. The trunk of the artery gives branches to the posterolateral part of the medulla oblongata and to the choroid plexus of the fourth ventricle. For its detailed distribution consult Bury and Stopford (1913), and Shellshear (1922).

The clinical picture of thrombosis of the posterior inferior cerebellar artery is very typical. The following are a few of its features with, in parentheses, the structure involved: loss of pain and temperature sensation on the same side of the face (spinal tract of the fifth nerve), and on the opposite side of the body (lateral spinothalamic tract); muscular inco-ordination of the limbs on the same side (afferent fibres entering inferior cerebellar peduncle); difficulty in swallowing (nucleus ambiguus).

BASILAR ARTERY

The basilar artery is formed by the junction of the two vertebral arteries at the lower border of the anteromedian surface of the pons. It ends at the upper border of the pons by bifurcating into the two posterior cerebral arteries.

Branches

Small branches supply the pons.

The **labyrinthine arteries** arise either from the basilar or from the anterior inferior cerebellar artery of the same side. Each enters the internal acoustic meatus, and is distributed to the internal ear.

The **anterior inferior cerebellar arteries**, one on each side, pass backwards to the anterior surfaces of the cerebellar hemispheres.

The **superior cerebellar arteries** arise near the termination of the basilar. Each passes backwards on the lateral surface of the inferior part of the midbrain to the upper surface of the cerebellum.

The **posterior cerebral arteries** [FIGS. 13.46 and 13.52] are the terminal branches of the basilar. Each runs round the cerebral peduncle, parallel to the superior cerebellar artery from which it is separated by the oculomotor and trochlear nerves. Reaching the calcarine sulcus, it divides into occipital and parieto-occipital branches. It is connected with the internal carotid by the posterior communicating artery.

The branches of the posterior cerebral artery are of three kinds: (1) central; (2) choroidal; and (3) cortical.

1. A set of small **posteromedial central branches** pierces the posterior perforated substance; with similar branches of the

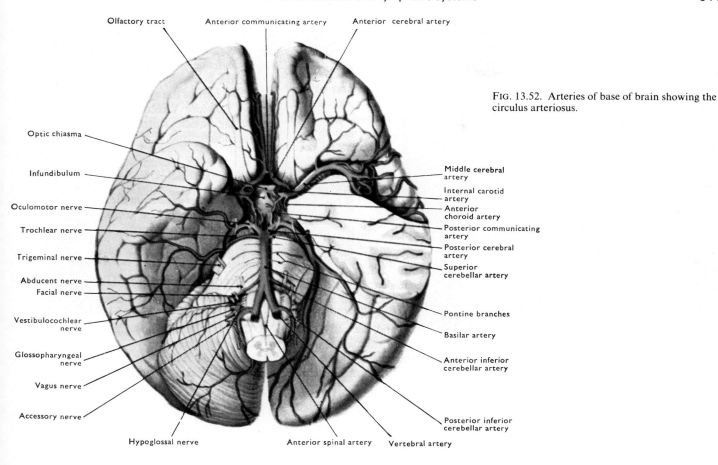

Olfactory tract
Anterior communicating artery
Anterior cerebral artery

Optic chiasma
Infundibulum
Oculomotor nerve
Trochlear nerve
Trigeminal nerve
Abducent nerve
Facial nerve
Vestibulocochlear nerve
Glossopharyngeal nerve
Vagus nerve
Accessory nerve

Hypoglossal nerve
Anterior spinal artery
Vertebral artery

Middle cerebral artery
Internal carotid artery
Anterior choroid artery
Posterior communicating artery
Posterior cerebral artery
Superior cerebellar artery
Pontine branches
Basilar artery
Anterior inferior cerebellar artery
Posterior inferior cerebellar artery

FIG. 13.52. Arteries of base of brain showing the circulus arteriosus.

posterior communicating artery they supply the midbrain and part of the floor and walls of the third ventricle.

A set of small **posterolateral central branches** supply the colliculi, the medial geniculate bodies, the pineal body, the cerebral peduncle, and the posterior part of the thalamus.

2. Small **posterior choroidal branches** enter the choroid fissure, and others run forwards in the tela choroidea of the third ventricle. They supply the choroid plexuses of the lateral and third ventricles.

3. The named cortical branches are **temporal, occipital,** and **parieto-occipital**; their positions and the areas which they supply are shown in FIGURES 13.49 and 13.52. The occipital branches are especially associated with the supply of the visual area of the cortex of the brain.

Circulus arteriosus [FIG. 13.52].

The cerebral arteries of opposite sides are intimately connected together at the base of the brain by anastomosing channels, an arrangement which provides for the continuation of a regular blood supply if one or more of the main trunks should be obstructed.

These vessels form the so-called **circulus arteriosus,** described by Willis (1664) and long known as the 'circle of Willis'. It is situated at the base of the brain in the interpeduncular cistern and is irregularly polygonal in outline. It is formed by the termination of the basilar and the two posterior cerebral arteries, the posterior communicating arteries and the internal carotids, the anterior cerebral arteries, and the anterior communicating artery. A comprehensive account of the circle is that by Alpers, Berry, and Paddison (1959). It is subject to many variations, the commonest being an enlargement of one (or both) posterior communicating artery so that a greater proportion (or all) of the blood in the posterior cerebral artery is derived from the internal carotid artery.

CEREBRAL ANGIOGRAPHY

Carotid angiography may be carried out by direct puncture of the common carotid artery in the neck, or by passage of a special catheter from the femoral artery through the iliac arteries and aorta to the appropriate artery. Multiple serial films are taken after injection of the contrast medium. Vessels of both cerebral hemispheres may be filled from one arterial puncture if the contralateral carotid artery is occluded.

FIGURE 13.54 is an unusual vertebral angiogram. It shows a large **posterior communicating artery** which has allowed filling of the middle and anterior cerebral arteries and their branches from the basilar artery, the internal carotid artery being occluded. It also illustrates variations in the course of the branches of the **anterior cerebral artery**—one branch lies on the upper surface of the corpus callosum, another runs parallel to it in the sulcus cinguli. These vessels are often more tortuous than shown here.

Arch aortography is often necessary to assess cerebral vascular supply because disturbances of the central nervous system may result from pathology in the aortic arch or in the branches as they arise from it [FIG. 13.51].

Vertebral angiography is achieved by a similar technique. It is not uncommon for both vertebral arteries to be filled by retrograde flow from the basilar into the uninjected artery. The anteroposterior radiograph is taken with the X-ray tube tilted slightly towards the feet so that the posterior cerebral arteries lie nearly parallel to the film and the basilar artery is greatly foreshortened [FIG. 13.56].

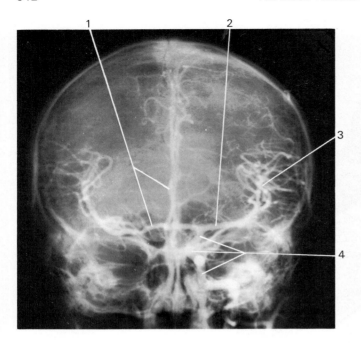

FIG. 13.53. Anterposterior cerebral angiograms produced by injection of the left internal carotid artery. The contralateral anterior and middle cerebral arteries have been filled through the anterior communicating artery because the right internal carotid artery is occluded.
1. Anterior cerebral artery.
2. Middle cerebral artery.
3. Branches of middle cerebral artery on insula.
4. Internal carotid artery.

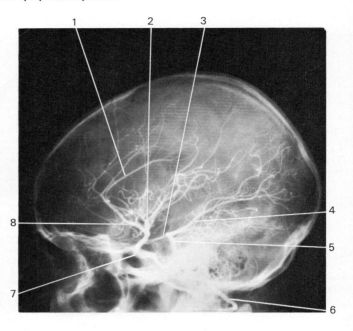

FIG. 13.54. Lateral cerebral angiogram with filling of vertebral and carotid arteries.
1. Anterior cerebral artery on corpus callosum.
2. Middle cerebral artery.
3. Posterior communicating artery (large).
4. Posterior cerebral artery.
5. Basilar artery.
6. Vertebral artery.
7. Internal carotid artery (carotid siphon).
8. Anterior cerebral artery.

ARTERIES OF THE UPPER LIMB

The main arterial stem of the upper limb passes through the root of the neck, the axilla, and the upper arm to the forearm, where, opposite the neck of the radius, it divides into the radial and ulnar arteries. In the root of the neck it is named the subclavian artery, in the axilla the axillary artery, and thereafter the brachial artery.

Subclavian arteries

The right subclavian artery [FIGS. 13.27, 13.40, and 13.50] arises from the brachiocephalic trunk behind the right sternoclavicular joint, and lies wholly in the root of the neck.

The left artery arises from the aortic arch in the superior mediastinum. In the root of the neck each artery arches over the cervical pleura behind the scalenus anterior muscle, and is described in three parts which lie respectively medial to, behind, and lateral to the muscle. The extent to which the artery rises above the clavicle varies from 1.2 to 2.5 cm. The first parts of the subclavian arteries differ both in extent and position, but the second and third parts are similar on the two sides.

FIRST PART OF THE RIGHT SUBCLAVIAN ARTERY
[FIG. 11.83]

Anteriorly the vessel is crossed by the right vagus, by cardiac branches of the vagus and sympathetic, and by the internal jugular and vertebral veins; the ansa subclavia curves beneath it. More superficially lie sternohyoid and sternothyroid, the anterior jugular vein, sternocleidomastoid, the medial supraclavicular nerves, and platysma. Posteriorly and inferiorly it is separated from the pleura over the apex of the lung by the suprapleural membrane and by the recurrent laryngeal nerve which winds round its inferior and posterior surfaces.

FIRST PART OF THE LEFT SUBCLAVIAN ARTERY
[FIG. 13.33]

1. The thoracic position of the artery is shown in FIGURE 13.57. All that need be added is that, medially, from below up, it is in contact with the trachea, the left recurrent laryngeal nerve, the oesophagus, and the thoracic duct.

2. In its cervical course the structures in contact with the first part differ from those of the corresponding part of the right artery in that the thoracic duct crosses it anteriorly, and the left recurrent laryngeal nerve does not curve below it.

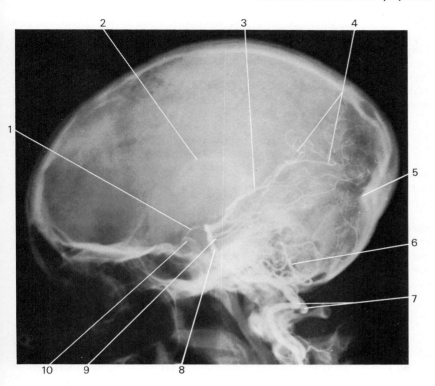

FIG. 13.55. Vertebral angiogram, lateral view. One vertebral artery has been filled by injection with reflux into the other from the basilar artery.
1. Posterior communicating artery.
2. Posterior choroidal arteries.
3. Posterior cerebral artery.
4. Branches in occipital lobe.
5. Internal occipital protuberance.
6. Posterior inferior cerebellar arteries.
7. Vertebral arteries on posterior arch of atlas.
8. Basilar artery.
9. Superior cerebellar arteries.
10. Ossification in tentorium bridging the sella turcica.

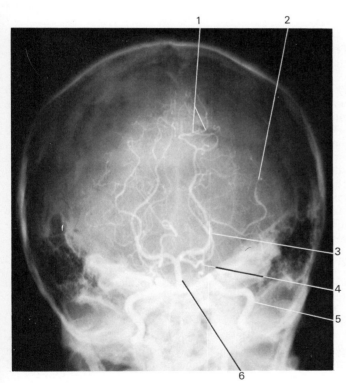

FIG. 13.56. Vertebral angiogram, anteroposterior view almost parallel to the clivus. Hence basilar artery is markedly foreshortened and posterior cerebral and superior cerebellar arteries overlap.
1. Posterior cerebral branches in occipital lobe.
2. Branch of posterior cerebral artery to tentorial surface of hemisphere.
3. Posterior cerebral artery on surface of midbrain.
4. Superior cerebellar artery.
5. Vertebral artery at foramen transversarium of atlas.
6. Basilar artery.

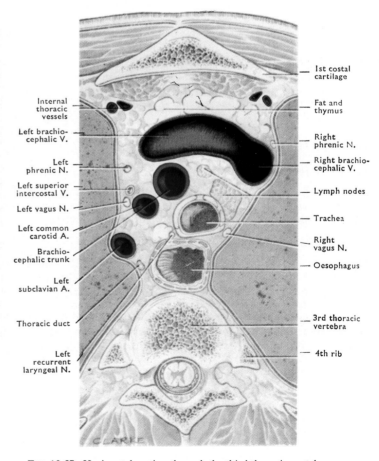

FIG. 13.57. Horizontal section through the third thoracic vertebra.

SECOND PART OF THE SUBCLAVIAN ARTERY

Postero-inferiorly it rests on the pleural sac and suprapleural membrane. Anteriorly it is covered by scalenus anterior, behind sternocleidomastoid. Scalenus anterior separates it from the subclavian vein, which lies at a slightly lower level, from the transverse cervical and suprascapular arteries, from the anterior jugular vein, and from the phrenic nerve.

THIRD PART OF THE SUBCLAVIAN ARTERY

This portion extends from the lateral border of scalenus anterior to the outer border of the first rib, lying partly in the posterior triangle of the neck and partly behind the clavicle, subclavius, and the suprascapular vessels.

It rests upon the upper surface of the first rib. Immediately posterior to it the lower trunk of the brachial plexus separates it from scalenus medius. Anterior to it, and at a slightly lower level, lies the subclavian vein. The external jugular vein crosses this portion of the artery, as does the nerve to subclavius.

SURFACE ANATOMY [FIG. 13.39]

To map out the course of the subclavian artery in the neck, a line is drawn, convex side upwards, from the upper border of the sternoclavicular joint to the middle of the clavicle, the highest part of the convexity to reach about 2 cm above the clavicle. The third part of the artery can be compressed against the first rib by pressure directed downwards and backwards, just behind the clavicle, a little lateral to the posterior border of sternocleidomastoid.

Branches of subclavian artery

Four branches arise from the subclavian artery—the vertebral artery, the thyrocervical trunk, the internal thoracic artery, and the costocervical trunk.

THE VERTEBRAL ARTERY [p. 909]

The thyrocervical trunk [FIGS. 13.50 and 13.93]

This short vessel arises close to the medial border of scalenus anterior. It ends by dividing into three branches: (1) the inferior thyroid; (2) the transverse cervical; and (3) the suprascapular.

The inferior thyroid artery [FIG. 11.83].

This sinuous artery first ascends behind the carotid sheath. It then turns medially, opposite the cricoid cartilage, to the middle of the posterior border of the thyroid gland, and then descends to the lower end of the lobe, where it terminates in ascending and inferior branches. The recurrent laryngeal nerve passes either in front of or behind the vessel, a variable relationship of great surgical importance (Bowden 1955); and, on the left side, the thoracic duct passes in front of it near its origin.

Branches

The ascending cervical artery ascends medial to the phrenic nerve, and sends spinal branches through the intervertebral foramina.

Small pharyngeal branches supply the pharynx, and twigs pass to the oesophagus and trachea.

An inferior laryngeal artery accompanies the recurrent laryngeal nerve to the larynx.

Inferior and ascending terminal glandular branches, which anastomose with branches of the superior thyroid artery [p. 901],

supply the posterior and lower parts of the thyroid gland, and give small branches to the parathyroid glands.

The transverse cervical artery [FIGS. 13.40 and 13.134].

This vessel crosses the scalenus anterior, and runs through the posterior triangle lying on the brachial plexus, scalenus medius, and levator scapulae. At the anterior border of levator scapulae it divides into a superficial branch which supplies trapezius, and a deep branch [FIG. 13.58] which descends close to the medial border of the scapula. The deep branch frequently arises from the third part of the subclavian artery, and is then called the descending scapular artery.

The suprascapular artery

This vessel crosses scalenus anterior, and continuing laterally behind the clavicle, anterior to the brachial plexus, reaches the scapular notch and passes over the superior transverse scapular ligament. It then descends through the supraspinous fossa deep to supraspinatus, and enters the infraspinous fossa lateral to the spine of the scapula [FIG. 13.58].

Branches

The suprascapular artery gives an acromial branch which joins the acromial rete [p. 917], and nutrient arteries to the clavicle and scapula.

It sends a branch to the subscapular fossa, and this, together with those given off in the supraspinous and infraspinous fossae, anastomose with branches of the subscapular and transverse cervical arteries.

THE INTERNAL THORACIC ARTERY [FIGS. 5.101, 13.50, and 13.57]

This vessel arises immediately below the thyrocervical trunk, and ends behind the sternal end of the sixth intercostal space by dividing into the musculophrenic and superior epigastric arteries.

It lies at first on the pleura and then on transversus thoracis, and runs behind the brachiocephalic vein, the sternal end of the clavicle, the first six costal cartilages and the intervening spaces. On both sides the phrenic nerve crosses in front of it, from lateral to medial. The artery is accompanied by lymph vessels and nodes, and by two veins which unite behind the third cartilage to form a single vessel that ends in the brachiocephalic vein. Below the first cartilage the surface marking of the artery is a vertical line drawn 1 cm from the border of the sternum.

Branches

The pericardiacophrenic artery accompanies the phrenic nerve to the diaphragm.

Mediastinal branches supply the back of the sternum, the thymus, and the pericardium.

The anterior intercostal arteries, two in each of the upper six spaces, pass laterally, first between transversus thoracis and the internal intercostal muscles, and then between these muscles and the intercostales intimi; they end by anastomosing with the posterior intercostal arteries.

The perforating branches, one in each of the upper six intercostal spaces, pass forwards to the skin. Those in the second, third, and fourth spaces give off mammary branches which are of special importance in the female.

The musculophrenic artery runs downwards and laterally on the thoracic surface of the diaphragm, but it pierces that muscle and ends on its abdominal surface. It gives branches to the diaphragm, and two anterior intercostal branches to each of the seventh, eighth, and ninth intercostal spaces.

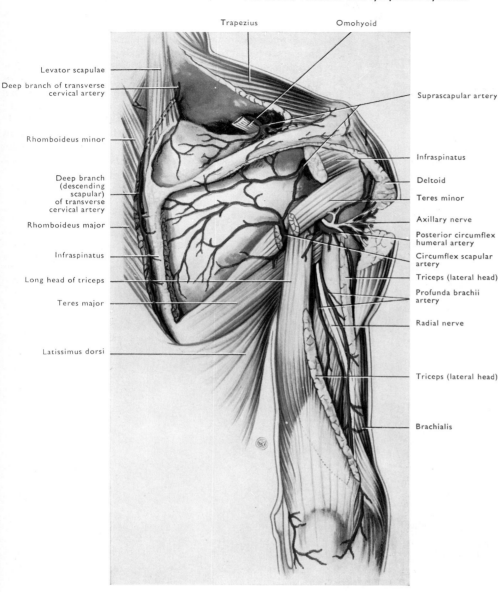

FIG. 13.58. Anastomosing arteries on dorsum of scapula and posterior humeral circumflex and profunda brachii arteries.

The **superior epigastric artery** leaves the thorax between the sternal and costal origins of the diaphragm, and enters the sheath of rectus abdominis. It ends by anastomosing with branches of the inferior epigastric artery. It gives branches to rectus, to the other muscles of the abdominal wall, to the diaphragm, and to the skin.

THE COSTOCERVICAL TRUNK [FIG. 13.50]

This vessel springs from the back of the second part of the subclavian artery on the right side and from the first part on the left side. It runs over the pleura to the neck of the first rib, where it divides into the deep cervical and highest intercostal arteries.

The **deep cervical artery** runs backwards between the seventh cervical transverse process and the neck of the first rib. It then ascends on semispinalis cervicis, and ends by anastomosing with the descending branch of the occipital artery.

The **highest intercostal artery** descends in front of the neck of the first rib. It gives off the posterior intercostal artery of the first space, then, crossing in front of the neck of the second rib, becomes the posterior intercostal artery of the second space. These intercostal

arteries are distributed in the same manner as the other posterior intercostal arteries [p. 922].

Axillary artery

The axillary artery extends from the outer border of the first rib to the lower border of teres major, where it becomes the brachial artery [FIG. 13.60].

For descriptive purposes the artery is divided into three parts: the first part lies above pectoralis minor, the second behind, and the third part below it.

The first part

The first part of the artery is enclosed, together with the vein and the cords of the brachial plexus, in a prolongation of the prevertebral fascia known as the **axillary sheath**. Posteromedially the sheath lies on the superior slip of serratus anterior, the first intercostal space, and the nerve to serratus anterior. Within the sheath, the medial cord of the brachial plexus lies behind the artery. Anteriorly it is covered by the clavipectoral fascia, which separates it from the cephalic vein and the clavicular part of pectoralis major. Laterally and above are the lateral and posterior cords of the brachial plexus. Medially and below is the axillary vein.

The second part

Posteriorly are the posterior cord of the plexus and subscapularis. Anteriorly are pectoralis minor and pectoralis major. Laterally lies the lateral cord of the brachial plexus. Medially the medial cord of the plexus lies close to the artery, between it and the axillary vein.

The third part

Posteriorly the artery rests upon subscapularis, latissimus dorsi, and teres major, separated from subscapularis by the axillary and radial nerves, and from the lower two muscles by the radial nerve alone. Anteriorly it is crossed by the medial root of the median nerve. Its upper half lies under cover of the lower part of pectoralis major, but its lower part is covered by skin and fasciae only. Laterally lie the median and musculocutaneous nerves and coracobrachialis. Medially is the axillary vein. The two vessels are, however, separated by the medial cutaneous nerve of the forearm in front and the ulnar nerve behind. The brachial veins end in the axillary vein at the lower border of subscapularis.

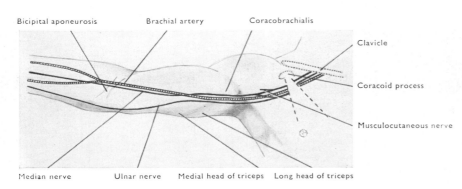

Bicipital aponeurosis　　Brachial artery　　Coracobrachialis

Clavicle

Coracoid process

Musculocutaneous nerve

Median nerve　　Ulnar nerve　　Medial head of triceps　　Long head of triceps

FIG. 13.59. Axilla and medial side of arm and elbow. The pectoralis minor is indicated by interrupted lines.

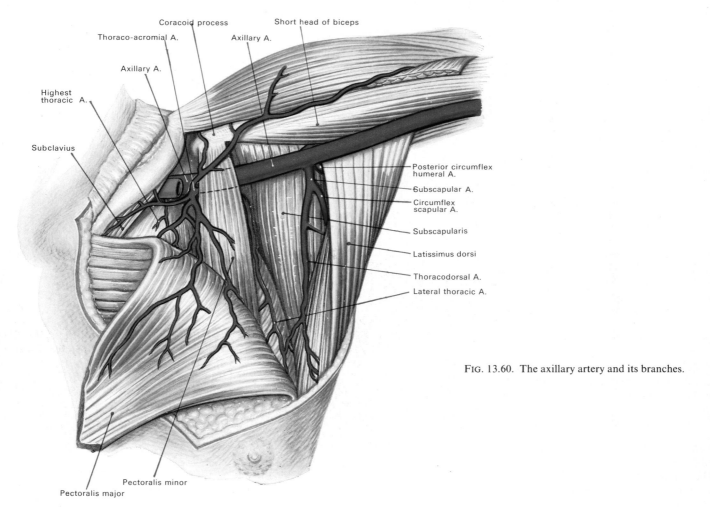

Coracoid process　　Short head of biceps

Thoraco-acromial A.　　Axillary A.

Axillary A.

Highest thoracic A.

Subclavius

Posterior circumflex humeral A.

Subscapular A.

Circumflex scapular A.

Subscapularis

Latissimus dorsi

Thoracodorsal A.

Lateral thoracic A.

Pectoralis minor

Pectoralis major

FIG. 13.60. The axillary artery and its branches.

Variations

The distal part of the artery may be crossed superficially by a muscular or tendinous 'axillary arch' passing between latissimus dorsi and pectoralis major; and sometimes the axillary artery divides into the radial and ulnar arteries.

Surface anatomy

With the arm abducted to a right angle, a line drawn from the middle of the clavicle to the medial border of the prominence of coracobrachialis indicates the position and direction of the artery which, in this position of the limb, is almost straight. With the arm hanging by the side the axillary artery describes a curve with the concavity directed downwards and medially, and the vein lies along its medial side.

Branches of axillary artery

HIGHEST THORACIC ARTERY

The highest thoracic artery arises from the first part of the axillary and runs across the first intercostal space.

THORACO-ACROMIAL ARTERY

The thoraco-acromial artery [FIG. 13.60] arises from the second part of the axillary artery. It is a short, wide trunk which pierces the clavipectoral fascia, and ends, deep to pectoralis major, by dividing into four terminal branches. The acromial branch runs to the acromion where it joins the acromial network or rete. The clavicular branch runs to the sternoclavicular joint. The deltoid branch descends alongside the cephalic vein, as far as the insertion of deltoid. The pectoral branch descends between and supplies the two pectoral muscles.

LATERAL THORACIC ARTERY

The lateral thoracic artery arises from the second part of the axillary, and descends along the lateral border of pectoralis minor. It supplies the adjacent muscles, and sends lateral mammary branches to the mammary gland.

SUBSCAPULAR ARTERY

This, the largest branch of the axillary, arises from the third part of the artery, and runs along the lower border of subscapularis. It gives off the circumflex scapular artery, and then, as the thoracordosal artery accompanied by the thoracodorsal nerve, continues to the wall of the thorax. A large branch enters latissimus dorsi with that nerve at a well-defined neurovascular hilus [p. 267].

The circumflex scapular artery passes backwards into the triangular space [p. 320]. Turning round the lateral border of the scapula, it enters the infraspinous fossa, where it anastomoses with the deep branch of the transverse cervical artery and the suprascapular artery. While in the triangular space it sends a branch into the subscapular fossa.

POSTERIOR CIRCUMFLEX HUMERAL ARTERY

This vessel arises from the third part of the axillary artery and passes backwards, with the axillary nerve, through the quadrangular space [FIG. 13.58]. It turns round the surgical neck of the humerus, and ends in branches which supply deltoid and anastomose with the anterior circumflex humeral artery.

It supplies the shoulder joint, and sends a branch to the acromial rete. A descending branch anastomoses with the ascending branch of the profunda brachii artery.

ANTERIOR CIRCUMFLEX HUMERAL ARTERY

This branch of the third part of the axillary passes deep to coracobrachialis and the two heads of the biceps, around the front of the surgical neck of the humerus, and ends by anastomosing with the posterior circumflex humeral. It sends a branch along the tendon of the long head of biceps to the shoulder joint.

Brachial artery

The brachial artery begins at the lower border of teres major, and ends, in the cubital fossa, anteromedial to the neck of the radius, by dividing into the radial and ulnar arteries [FIG. 13.62]. It lies at first on the medial side of the humerus and then in front of it.

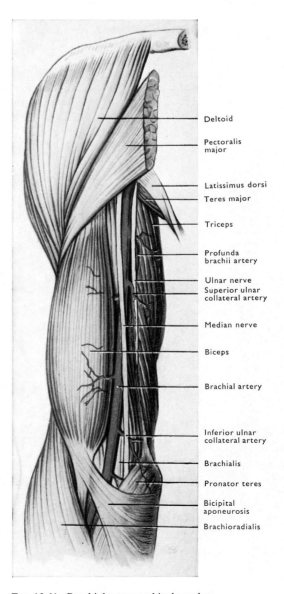

Deltoid

Pectoralis major

Latissimus dorsi

Teres major

Triceps

Profunda brachii artery

Ulnar nerve
Superior ulnar collateral artery

Median nerve

Biceps

Brachial artery

Inferior ulnar collateral artery

Brachialis

Pronator teres

Bicipital aponeurosis

Brachioradialis

FIG. 13.61. Brachial artery and its branches.

Posteriorly it lies, successively, on the long head of triceps with the radial nerve and the profunda vessels intervening; the medial head of the triceps; the insertion of coracobrachialis; and the brachialis. Anteriorly it is overlapped by the medial border of the biceps, and it is crossed, at the middle of the arm, by the median nerve. In the cubital fossa the bicipital aponeurosis separates it from the median cubital vein. Laterally it is in contact, proximally, with the median nerve, and, distally, with biceps. Medially it is in contact, proximally, with the basilic vein, the medial cutaneous nerve of the forearm, and the ulnar nerve, and, distally, with the median nerve. Two brachial veins accompany the artery, and freely communicate across it.

Variations

The brachial artery sometimes divides at a higher level than usual. In such cases the ulnar artery may cross superficial to the flexor muscles, and may even be subcutaneous; and the radial artery may descend in the superficial fascia of the forearm. In performing venesection at the elbow such variations have to be borne in mind.

Sometimes the brachial artery accompanies the median nerve behind a supracondylar process [p. 162], or ligament; and it may pass in front of the median nerve instead of behind it.

Surface anatomy

To mark the position of the brachial artery, the line of the axillary artery [p. 917] should be continued to a point 1 cm beyond the centre of the bend of the elbow [FIG. 13.59].

Branches of brachial artery

1. The **profunda brachii artery** runs with the radial nerve, and divides at the back of the humerus into two descending branches: (1) the **radial collateral artery** accompanies the radial nerve to the elbow, and in its course sends a branch to the back of the lateral epicondyle [FIG. 13.62]; (2) the **median collateral artery** descends in the substance of the medial head of triceps. One of the descending branches gives off a **nutrient branch** (occasional) to the humerus; and a **deltoid branch** which anastomoses with the descending branch of the posterior circumflex humeral artery.

2. **Muscular branches** are given to the adjacent muscles.

3. A small **nutrient branch** enters the nutrient canal on the anteromedial surface of the humerus.

4. The **superior ulnar collateral artery** runs with the ulnar nerve to the back of the medial epicondyle of the humerus, where it ends by anastomosing with the posterior ulnar recurrent and inferior ulnar collateral arteries.

5. The **inferior ulnar collateral artery** arises about 5 cm above the elbow. It takes part in the anastomoses in front and behind the medial epicondyle.

Radial artery

The radial artery [FIGS. 13.63–13.66] begins in the cubital fossa anteromedial to the neck of the radius, and ends by completing the deep palmar arch.

It is described in three parts. The **first part** runs down the forearm to the apex of the styloid process of the radius; the **second part** curves round the lateral side of the wrist, to the proximal end of the first interosseous space; the **third part** passes through that space to the palm.

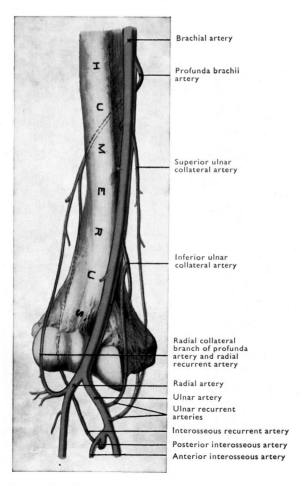

Brachial artery

Profunda brachii artery

Superior ulnar collateral artery

Inferior ulnar collateral artery

Radial collateral branch of profunda artery and radial recurrent artery

Radial artery

Ulnar artery

Ulnar recurrent arteries

Interosseous recurrent artery

Posterior interosseous artery

Anterior interosseous artery

FIG. 13.62. Diagram of anastomosis around elbow joint.

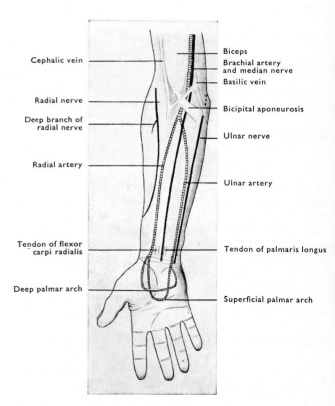

Cephalic vein

Radial nerve

Deep branch of radial nerve

Radial artery

Tendon of flexor carpi radialis

Deep palmar arch

Biceps

Brachial artery and median nerve

Basilic vein

Bicipital aponeurosis

Ulnar nerve

Ulnar artery

Tendon of palmaris longus

Superficial palmar arch

FIG. 13.63. Front of elbow, forearm, and hand.

First part

It is superficial immediately deep to the deep fascia. Posteriorly, it rests on the tendon of biceps, then on the supinator (fat usually intervening), the insertion of pronator teres, the radial origin of flexor digitorum superficialis, flexor pollicis longus, pronator quadratus, and finally on the lower end of the radius. To its radial side are brachioradialis which overlaps it proximally, and the superficial branch of the radial nerve which lies near its middle third. To its ulnar side are pronator teres proximally, and flexor carpi radialis, distally. Two veins accompany the artery throughout.

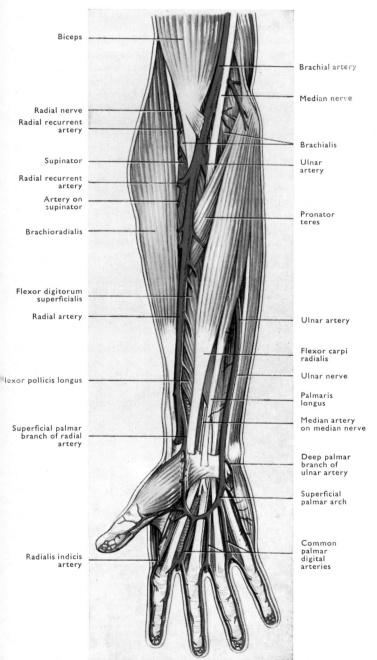

FIG. 13.64. Superficial dissection of forearm and hand, showing radial and ulnar arteries and superficial palmar arch with its branches.

Branches of first part

The radial recurrent artery arises in the cubital fossa. It runs between the superficial and deep branches of the radial nerve and then turns upwards to anastomose with the radial collateral artery.

Muscular branches supply the muscles on the radial side of the forearm.

The superficial palmar branch [FIG. 13.64] arises just above the wrist. It usually pierces the muscles of the thenar eminence, and ends either in their substance or by completing the superficial palmar arch.

The palmar carpal branch passes deep to the flexor tendons, and with the palmar carpal branch of the ulnar artery forms the palmar carpal arch.

Second part

It curves over the lateral surface of the wrist on the radial collateral carpal ligament and the scaphoid and the trapezium, deep to the tendons of abductor pollicis longus, extensor pollicis brevis, and then extensor pollicis longus. Superficial to these are the commencement of the cephalic vein and some filaments of the superficial branch of the radial nerve.

Branches of second part [FIG. 13.66].

The dorsal carpal branch, together with the corresponding branch of the ulnar artery, forms the dorsal carpal rete in which the anterior interosseous artery ends. From the rete three dorsal metacarpal arteries run distally: each divides into two dorsal digital arteries at the webs between the fingers.

The dorsal digital arteries of the thumb and of the radial side of the index finger take independent origin from the radial artery.

Each dorsal metacarpal artery is connected with the deep palmar arch by a proximal perforating branch, and with a common palmar digital branch from the superficial palmar arch by a distal perforating branch.

Third part

The radial artery passes into the palm between the heads of the first dorsal interosseous muscle, and then between the two heads of adductor pollicis to complete the deep palmar arch [FIG. 13.65].

Branches of third part

The princeps pollicis artery runs distally under cover of the long flexor tendon. It divides into the two proper palmar digital arteries of the thumb.

The radialis indicis artery descends between the first dorsal interosseous muscle and adductor pollicis, and along the radial side of the index finger.

Variations

The radial artery may pass to the back of the wrist superficial to the extensor tendons. [See also p. 918.]

Surface anatomy

To mark the course of the radial artery draw a line from the bifurcation of the brachial to the tubercle of the scaphoid; above the tubercle and below the tip of the styloid process the artery winds dorsally [FIG. 13.66]. Its pulsations can readily be felt at the wrist, lateral to the tendon of flexor carpi radialis.

Ulnar artery

The ulnar artery [FIGS. 13.64 and 13.65] begins in the cubital fossa, anteromedial to the neck of the radius, and ends at the radial side of

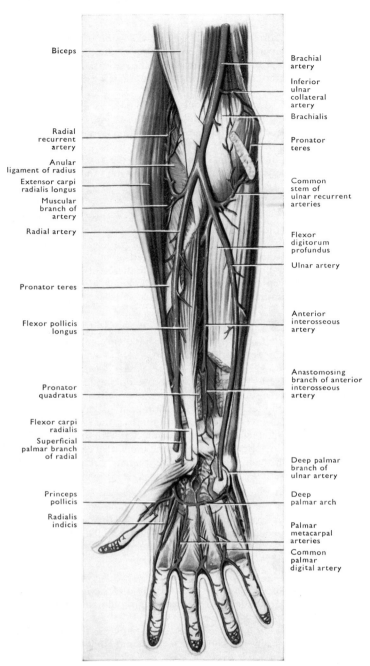

FIG. 13.65. Deep dissection of forearm and hand, showing radial and ulnar arteries and deep palmar arch with its branches.

Labels on figure:
Biceps
Radial recurrent artery
Anular ligament of radius
Extensor carpi radialis longus
Muscular branch of artery
Radial artery
Pronator teres
Flexor pollicis longus
Pronator quadratus
Flexor carpi radialis
Superficial palmar branch of radial
Princeps pollicis
Radialis indicis

Brachial artery
Inferior ulnar collateral artery
Brachialis
Pronator teres
Common stem of ulnar recurrent arteries
Flexor digitorum profundus
Ulnar artery
Anterior interosseous artery
Anastomosing branch of anterior interosseous artery
Deep palmar branch of ulnar artery
Deep palmar arch
Palmar metacarpal arteries
Common palmar digital artery

digitorum superficialis lies to its radial side, and the ulnar nerve is on its ulnar side. Two veins accompany the artery.

Branches.

The anterior ulnar recurrent artery and the posterior ulnar recurrent artery arise in the cubital fossa and ascend to the front and back of the medial epicondyle respectively. Both take part in the anastomosis round the joint.

The common interosseous artery arises in the distal part of the cubital fossa. It passes backwards, and divides into anterior and posterior interosseous branches.

The anterior interosseous artery descends on the anterior surface of the membrane to the proximal border of pronator quadratus. Here it pierces the membrane and continues downwards to join the dorsal carpal rete. It gives off muscular branches, nutrient arteries to the radius and ulna, and a communicating branch to the palmar carpal arch. The median artery, a slender branch from the proximal part of the anterior interosseous, runs with the median nerve to the palm.

The posterior interosseous artery passes backwards above the proximal border of the interosseous membrane; thereafter it descends between, and supplies, the superficial and deep muscles on the back of the forearm. It ends by anastomosing with the anterior interosseous artery and so with the dorsal carpal rete. It gives off an interosseous recurrent artery which runs to the back of the lateral epicondyle [FIG. 13.62].

The palmar carpal branch of the ulnar artery completes the palmar carpal arch.

The dorsal carpal branch passes to the dorsal carpal rete.

The deep palmar branch of the ulnar artery runs posteriorly between abductor and flexor digiti minimi, and, turning laterally, deep to opponens and the flexor tendons, joins the radial artery to complete the deep palmar arch.

Variations

The ulnar artery may arise higher than usual, and then commonly passes superficial to the muscles which spring from the medial epicondyle [p. 918]; but even when it begins at the usual level it may pass superficial to these muscles.

The median artery may be quite large, and may end either as digital arteries or by joining the superficial palmar arch.

Surface anatomy

The upper third of the ulnar artery takes a curved course towards the medial part of the front of the forearm; its lower two-thirds correspond to the lower two-thirds of a line drawn from the front of the medial epicondyle to the radial border of the pisiform [FIG. 13.63].

the pisiform bone, by dividing into its deep palmar branch and the superficial palmar arch.

It passes deeply on brachialis, and obliquely across flexor digitorum profundus, to meet the ulnar nerve at the middle of the medial side of the forearm, and descend with it on to the flexor retinaculum. Anteriorly, it is crossed by pronator teres, the median nerve (separated from it by the ulnar head of pronator teres), flexor carpi radialis, palmaris longus, and flexor digitorum superficialis. In the middle third of the forearm it is overlapped by flexor carpi ulnaris, and on the flexor retinaculum it is covered by a superficial layer from that structure [p. 348]. In its distal two thirds flexor

Arterial arches of hand

The superficial palmar arch [FIG. 13.64].

This is the main terminal branch of the ulnar artery. It begins on the flexor retinaculum immediately distal and radial to the pisiform, and descends superficial to the hook of the hamate deep to palmaris brevis. It then turns laterally and is completed by the superficial branch of the radial, or by the radialis indicis or the princeps pollicis artery. It is accompanied by two veins and lies just beneath the palmar aponeurosis. The arch is often irregular, and indeed may be absent.

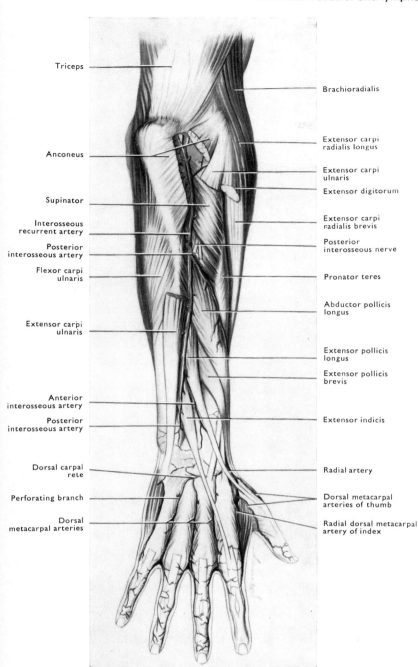

Triceps

Brachioradialis

Anconeus

Extensor carpi
radialis longus

Supinator

Extensor carpi
ulnaris

Interosseous
recurrent artery

Extensor digitorum

Posterior
interosseous artery

Extensor carpi
radialis brevis

Flexor carpi
ulnaris

Posterior
interosseous nerve

Extensor carpi
ulnaris

Pronator teres

Abductor pollicis
longus

Anterior
interosseous artery

Extensor pollicis
longus

Posterior
interosseous artery

Extensor pollicis
brevis

Extensor indicis

Dorsal carpal
rete

Radial artery

Perforating branch

Dorsal metacarpal
arteries of thumb

Dorsal
metacarpal arteries

Radial dorsal metacarpal
artery of index

FIG. 13.66. Posterior interosseous artery and second part of radial artery, with their branches.

Four arteries arise from the convexity of the arch, the **proper palmar digital artery** of the ulnar border of the little finger and three **common palmar digital arteries**. Each of the latter divides into two proper palmar digital branches which supply the contiguous sides of the related fingers. On the sides of the fingers they lie behind the corresponding digital nerves, and give off dorsal branches which anastomose with the dorsal digital arteries.

Each common palmar digital artery is joined, just before it divides, by a palmar metacarpal artery from the deep palmar arch and a distal perforating branch from a dorsal metacarpal artery. The digital artery to the ulnar side of the little finger is joined by a branch which arises either from the medial palmar metacarpal artery or from the deep palmar arch.

The deep palmar arch [FIG. 13.65].

This is formed by the radial artery and its anastomosis with the deep palmar branch of the ulnar. It lies deep to the long flexor tendons and their synovial sheaths on the proximal parts of the metacarpals.

From the arch three **palmar metacarpal arteries** pass distally to join the common palmar digital arteries.

Three small **perforating branches** join the dorsal metacarpal arteries, and small **recurrent branches** run to the palmar carpal arch.

Surface anatomy

The most distal part of the superficial palmar arch reaches to a point level with the distal border of the fully extended thumb. The deep

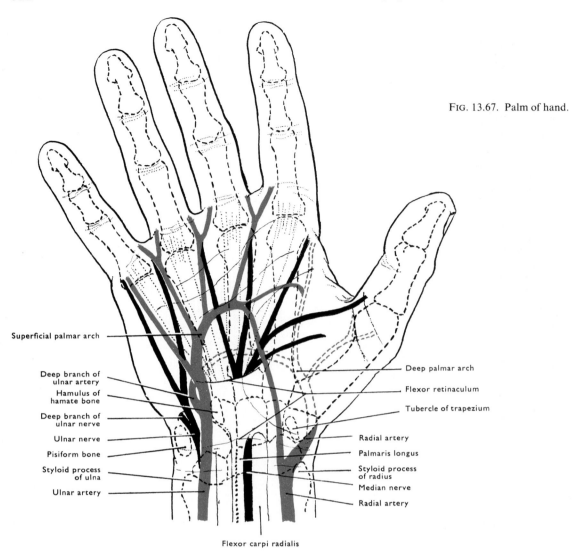

FIG. 13.67. Palm of hand.

Superficial palmar arch

Deep branch of
ulnar artery

Hamulus of
hamate bone

Deep branch of
ulnar nerve

Ulnar nerve

Pisiform bone

Styloid process
of ulna

Ulnar artery

Deep palmar arch

Flexor retinaculum

Tubercle of trapezium

Radial artery

Palmaris longus

Styloid process
of radius

Median nerve

Radial artery

Flexor carpi radialis

arch lies almost transversely about 1–2 cm proximal to the superficial arch [FIG. 13.67].

BRANCHES OF THE THORACIC AORTA

The branches of the thoracic aorta are distributed chiefly to the thoracic walls and viscera. They also help to supply the spinal medulla, the vertebral column, and the upper part of the abdominal wall.

Visceral branches of the thoracic aorta

The upper left bronchial artery arises from the front of the aorta opposite the fifth thoracic vertebra, and the lower left bronchial artery at a slightly lower level. Both vessels run to the back of the principal bronchus, and, dividing similarly, follow its ramifications [p. 525]. They supply the walls of the bronchial tubes, the connective tissue of the lungs, and the visceral pleura.

As a rule there is only one right bronchial artery. It arises from the third right posterior intercostal artery, and in its course and distribution corresponds to the bronchial arteries of the left side. It

may arise from the upper left bronchial artery or, more rarely, from the aorta.

The anastomoses between the branches of the bronchial and pulmonary arteries enlarge when one or other system is blocked (Tobin 1952).

Variations

Amongst the anomalous sites of origin of the bronchial arteries are the subclavian, internal thoracic, and inferior thyroid arteries. Consult also Nakamura (1924).

The oesophageal branches are small twigs that spring from the front of the aorta.

The pericardial branches are small, as are the mediastinal branches which pass to the lymph nodes in the posterior mediastinum.

Parietal branches of descending thoracic aorta

Posterior intercostal arteries

Nine pairs of posterior intercostal arteries (III–XI) arise from the back of the aorta. The arteries of opposite sides correspond, but, since the upper part of the descending thoracic aorta lies on the left of the vertebral column, most of the right posterior intercostal arteries cross the front of the column, behind the oesophagus, the thoracic duct, and the vena azygos. As each artery runs round the side of the vertebral column, to an intercostal space, it passes behind the pleura and the sympathetic trunk. The lower arteries are crossed by the splanchnic nerves also, and those on the left side by the hemiazygos or accessory hemiazygos vein.

Each artery passes laterally, and ascends to the upper border of the space to which it belongs; at the angle of the upper rib it enters the **costal groove** and continues forwards between the innermost intercostal and the internal intercostal muscles [p. 351]. In the groove the artery lies between the corresponding vein above and the intercostal nerve below, and it ends by anastomosing with an anterior intercostal artery. The lower two posterior intercostal arteries, on each side, continue into the abdominal wall.

Several branches arise from each posterior intercostal artery.

The **dorsal branch** passes backwards, between the necks of the ribs, and then between the transverse processes, to the erector spinae; here it divides into medial and lateral branches which supply the muscles and skin of the back. A **spinal branch** passes through the corresponding intervertebral foramen; twigs run medially, on the roots of the spinal nerve, and may reinforce the spinal arteries. Others anastomose with adjacent spinal branches and supply the vertebrae [Fig. 13.120].

A **collateral branch** arises near the angle of the rib and runs along the lower border of the intercostal space.

Several **nutrient branches** supply the rib above, and **muscular branches** arise from the main trunk and its collateral branch.

A **lateral cutaneous branch** accompanies the lateral cutaneous branch of the intercostal nerve. Those of the third, fourth, and fifth spaces give a **mammary branch** to the mammary gland.

The position of the intercostal arteries determines that in aspirating the chest the needle should pass immediately above a rib and should not be introduced medial to the rib angle.

Subcostal arteries

Each of these vessels, which has the same branches as the intercostal arteries, runs with the corresponding subcostal nerve along the lower border of a twelfth rib. Passing behind the lateral arcuate ligament, it crosses in front of quadratus lumborum, posterior to the kidney, and piercing the aponeurosis of origin of transversus abdominis, runs between that muscle and internal oblique.

Superior phrenic arteries

These small vessels arise from the lower part of the thoracic aorta and ramify on the surface of the diaphragm.

The **vas aberrans** is a remnant of the right dorsal aorta of the embryo. When present it arises near the upper left bronchial artery, and passes upwards and to the right behind the oesophagus. It may be enlarged and form the first part of the right subclavian artery.

Variations

Coarctation of the aorta, i.e. partial or complete obliteration of the lumen, occurs most commonly just beyond the origin of the left subclavian artery and proximal to the attachment of the ligamentum arteriosum. This part of the aorta is constricted in the foetus and is known as the aortic isthmus; it is succeeded by a fusiform dilation called the aortic spindle. Both these features persist, but much less markedly, in the adult. Coarctation leads in time to a well-marked collateral circulation. For example, the posterior intercostal arteries (which then transmit blood from the internal thoracic artery to the aorta) may become so enlarged and tortuous as to cause notching of the lower borders of the ribs; this is detectable on radiological examination.

BRANCHES OF THE ABDOMINAL AORTA

The branches of the abdominal portion of the aorta may be classified conveniently as follows:

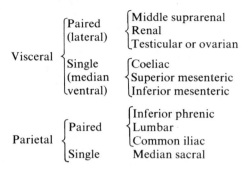

PAIRED VISCERAL BRANCHES OF THE ABDOMINAL AORTA

There are three sets of **suprarenal arteries** [p. 590]—the superior, middle, and inferior. Of these only the middle arise directly from the aorta; the superior spring from the inferior phrenic arteries, and the inferior from the renal arteries.

The **middle suprarenal arteries** arise close to the origin of the superior mesenteric artery. They run, one or more on each side, upon the crura of the diaphragm just above the renal arteries, to the suprarenal glands.

RENAL ARTERIES

These large arteries [Fig. 8.7] arise about 1.5 cm below the origin of the superior mesenteric artery and opposite the upper border of the second lumbar vertebra.

Both are large, and the right is longer, and frequently slightly lower than the left. Each runs to the hilus of the corresponding kidney, passing anterior to the crus of the diaphragm and the psoas muscle. The left artery lies posterior to the body of the pancreas; the right vessel passes behind the inferior vena cava, the head of the pancreas, and the descending part of the duodenum. The renal vein lies anterior to the artery and, on the left side, at a slightly lower level.

Approaching the hilus of the kidney, each artery divides into three or four segmental branches of which one passes behind the pelvis of the kidney and the rest in front of it. From these branches five segments of the kidney are supplied, and it is of surgical importance that anastomosis between segments does not occur (Graves 1956).

Each renal artery gives off small **ureteric branches** to the upper part of the ureter, and one or more **inferior suprarenal arteries**, which pass upwards to the lower part of the suprarenal gland.

Variations

Accessory renal arteries, more common on the left side, may be derived from the aorta, or from the common iliac or internal iliac arteries. They usually enter towards one pole of the kidney, and are 'end arteries' to a portion of the kidney substance (Graves 1956). As the kidney ascends during development it is not surprising that it may retain one or more of its primitive vessels of supply. For intrarenal distribution see page 539.

TESTICULAR OR OVARIAN ARTERIES

The testicular or ovarian arteries are long, slender vessels which spring from the front of the aorta a short distance below the renal arteries.

Each **testicular artery** runs on psoas major to the deep inguinal ring, and then accompanies the ductus deferens through the inguinal canal to the testis. Posteriorly, each artery passes in front of the genitofemoral nerve, the ureter, and the lower part of the external iliac artery. The right artery also crosses the inferior vena cava. Anterior to the right artery are the horizontal part of the duodenum, the right colic and ileocolic vessels, and the root of the mesentery with the distal parts of the superior mesenteric vessels. Crossing anterior to the left artery are the ascending part of the duodenum, the left colic and sigmoid vessels, and the terminal part of the descending colon.

In its lower abdominal course each testicular artery is accompanied by two veins which issue from the pampiniform plexus and enter the abdomen through the deep inguinal ring; at a higher level they fuse into a single stem which is usually lateral to the artery. Beyond the superficial inguinal ring the artery lies anterolateral to the ductus deferens and in close association with the anterior group of testicular veins. At the posterior border of the testis it breaks up into branches, some of which pass to the testis and others to the epididymis. Further details are given on page 556.

Each testicular artery gives off **ureteric branches** to the abdominal part of the ureter.

The course and the relations of each **ovarian artery**, as far as the external iliac artery, are the same as those of the corresponding testicular artery; it then enters the lesser pelvis by crossing the external iliac vessels, and passes between the layers of the suspensory ligament of the ovary to reach the broad ligament of the uterus. In the broad ligament it runs below the uterine tube, and turns backwards into the mesovarium, where it breaks up into branches which enter the ovary at its hilus. In the broad ligament each ovarian artery is accompanied by the pampiniform plexus of ovarian veins. In its lower abdominal course it is accompanied by two veins which issue from the plexus at the brim of the lesser pelvis and unite at a higher level into a single trunk [FIG. 13.74].

In the abdomen it gives branches to the ureter and, in the pelvis, branches which supply the uterine tube, the round ligament of the uterus, and anastomose with the uterine artery.

SINGLE VISCERAL BRANCHES OF THE ABDOMINAL AORTA

The coeliac trunk

This short, wide vessel arises immediately below the aortic orifice of the diaphragm and between its crura. It runs forwards for about 1 cm, and ends by dividing into three branches—the left gastric, the common hepatic, and the splenic [FIGS. 13.68, 13.69, and 13.74]. Developmentally it is the artery of the abdominal portion of the foregut.

The trunk lies behind the omental bursa below the caudate lobe of the liver and at the upper border of the pancreas above the splenic vein. It has a coeliac ganglion on each side and is surrounded by the coeliac plexus of nerves.

THE LEFT GASTRIC ARTERY

This is the smallest branch of the coeliac trunk. It runs upwards and to the left, behind the omental bursa, and turning forwards in the superior gastropancreatic fold, reaches the lesser curvature of the stomach close to the oesophagus. It then turns sharply downwards and to the right, and runs within the lesser omentum to anastomose with the right gastric branch of the proper hepatic artery.

The left gastric artery sends branches to both surfaces of the stomach, and also gives off **oesophageal branches** which anastomose with oesophageal branches of the thoracic aorta.

THE SPLENIC ARTERY [FIG. 13.68]

This is the largest branch of the coeliac trunk. It runs a tortuous course along the upper border of the body and tail of the pancreas,

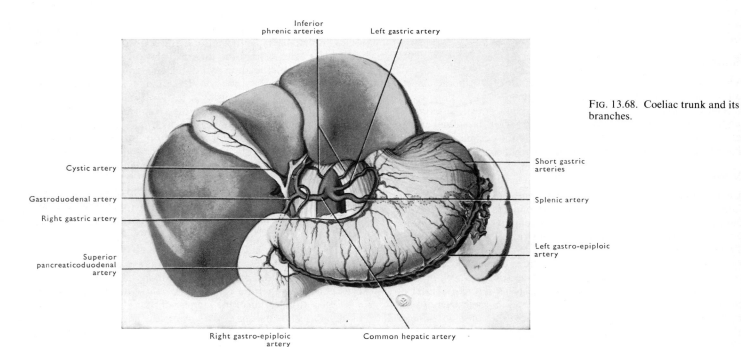

Cystic artery

Gastroduodenal artery

Right gastric artery

Superior pancreaticoduodenal artery

Inferior phrenic arteries

Left gastric artery

Short gastric arteries

Splenic artery

Left gastro-epiploic artery

Right gastro-epiploic artery

Common hepatic artery

FIG. 13.68. Coeliac trunk and its branches.

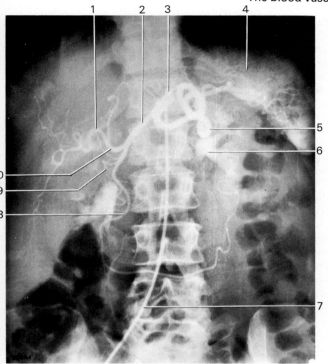

FIG. 13.69. Selective angiogram of coeliac trunk catheterized through the aorta, anteroposterior view.
  1. Right branch of proper hepatic artery.
  2. Common hepatic artery.
  3. Coeliac trunk.
  4. Spleen.
  5. Splenic vein.
  6. Pelvis of left kidney visualized by excreted material of test dose.
  7. Catheter in common iliac artery.
  8. Right gastro-epiploic artery.
  9. Superior pancreaticoduodenal artery.
  10. Proper hepatic artery.

behind the omental bursa, and superior to the splenic vein. It lies in front of the left suprarenal gland and left kidney, and passes forwards in the lienorenal ligament, with the splenic vein, dividing into five to eight terminal **splenic branches** which enter the hilus of the spleen. For their detailed distribution see page 1006.

Branches

Numerous small **pancreatic branches** are given off along its length.

Four or five **short gastric arteries** pass between the layers of the gastrosplenic ligament to the upper part of the greater curvature and adjacent surfaces of the stomach.

The **left gastro-epiploic artery** passes downwards and forwards in the gastrosplenic ligament, to run parallel to and a short distance from the greater curvature of the stomach, between the anterior two layers of the greater omentum. It anastomoses with the right gastro-epiploic artery, gives numerous branches to both surfaces of the stomach, and long, slender **omental (epiploic) branches** to the greater omentum.

THE COMMON HEPATIC ARTERY [Fig. 13.68]

This vessel runs forwards and to the right, below the epiploic foramen, to the upper border of the superior part of the duodenum.

Here it gives off the gastroduodenal artery and enters the lesser omentum as the **proper hepatic artery** which ascends near the free margin of the omentum, in front of the foramen. In the omentum it is in front of the portal vein and to the left of the bile-duct; and near the porta hepatis it divides into right and left terminal branches. It is accompanied by the hepatic plexus of nerves, and by lymph vessels.

Branches

The **gastroduodenal artery** arises at the upper border of the superior part of the duodenum, descends behind it, and ends opposite its lower border. At first it runs with the bile-duct on its right side, both being anterior to the portal vein; then for a short distance it is separated from the vein and duct by the pancreas. It ends by dividing into the right gastro-epiploic and superior pancreatico-duodenal arteries. The **right gastro-epiploic artery** passes to the left between the layers of the greater omentum to anastomose with the left gastro-epiploic artery. The **superior pancreaticoduodenal artery** passes to the right, and divides into anterior and posterior branches which descend between the duodenum and the head of the pancreas to anastomose with branches of the inferior pancreatico-duodenal artery. They supply the head of the pancreas, the descending part of the duodenum, and the bile-duct.

Branches of Proper hepatic artery

The **right gastric artery** is small and arises as the hepatic enters the lesser omentum. It runs within the omentum to the pylorus, and then along the lesser curvature. It supplies the stomach, anastomoses with the left gastric, and gives a branch to the superior part of the duodenum.

The **right branch** passes behind the common hepatic duct and the cystic duct, giving off the cystic artery, and reaching the right end of the porta hepatis, divides into two or more branches which enter the liver and accompany the branches of the portal vein and the right hepatic duct. The **cystic artery** runs along the cystic duct to the gall-bladder, where it divides into anterior and posterior branches; the anterior passes between the gall-bladder and the visceral surface of the liver, to both of which it gives offsets; the posterior branch is distributed on the posterior surface of the gall-bladder, and lies between it and the peritoneum. The **left branch** runs to the left end of the porta hepatis, gives branches to the caudate and quadrate lobes, crosses the fissure for the ligamentum venosum, and breaks up into branches which enter the left lobe of the liver.

VARIATIONS

The coeliac trunk may be absent, its branches arising separately from the aorta; the left gastric artery may give off the left branch of the hepatic, or an accessory hepatic artery. The accessory hepatic or the entire common hepatic artery may arise from the superior mesenteric artery. For further details Michels (1949) should be consulted.

The superior mesenteric artery

This, the artery of the embryonic midgut, arises about 1 cm below the coeliac trunk and opposite the first lumbar vertebra [Figs. 13.70, 13.73, and 13.74] behind the pancreas and splenic vein.

It passes downwards and forwards, anterior to the left renal vein, the uncinate process of the head of the pancreas, and the horizontal part of the duodenum. Here it enters the root of the mesentery, behind the transverse colon or its mesentery, and curves downwards to the right iliac fossa behind the coils of small intestine; crossing the inferior vena cava, the right ureter, and the right psoas major muscle. The superior mesenteric vein runs along its right side. It ends, much reduced in size, by anastomosing with the ileocolic artery.

BRANCHES

The branches of the superior mesenteric artery supply part of the duodenum and pancreas, the rest of the small intestine, and the large intestine approximately to the left colic flexure.

The **inferior pancreaticoduodenal artery** arises either from the superior mesenteric or from the first jejunal branch. It runs to the right, and ends in two branches, anterior and posterior, which anastomose with the corresponding branches of the superior pancreaticoduodenal artery.

The branches to the small intestine, from ten to sixteen in number, are separable into two groups—jejunal and ileal. They enter the mesentery, the branches of adjacent vessels uniting to form arcades from which secondary branches arise and may also unite in a similar manner. The upper jejunal branches form only one or two tiers of arcades, but the ileal branches form four or five in the longer, lower part of the mesentery. The branches from the terminal arcades pass to alternate sides of the wall of the intestine, and anastomose in it, but these intramural anastomoses are insufficient to maintain a segment of the jejunum divorced from the direct supply of an arcade (Barlow 1952).

The remaining branches of the superior mesenteric artery supply the caecum, appendix, ascending colon, and much of the transverse colon. The rest of the colon is supplied by branches of the inferior mesenteric artery [see below], and the major branches of both arteries to the colon unite to form a continuous vessel (**marginal artery**) which runs along the length of the colon and gives branches to it [FIGS. 13.70 and 13.73].

The **ileocolic artery** may arise with the right colic, or separately lower down. It passes downwards and to the right, and ends by dividing into an ascending branch which unites with the lower branch of the right colic, and a descending branch which supplies part of the ascending colon, the caecum, the appendix, and the terminal ileum, and ends by uniting with the continuation of the superior mesenteric trunk. The **anterior caecal branch** crosses the ileocaecal junction in the fold of peritoneum called the vascular fold of the caecum; the **posterior caecal branch** crosses posterior to the junction. The **appendicular artery** passes behind the terminal part of the ileum into the meso-appendix [FIG. 13.72].

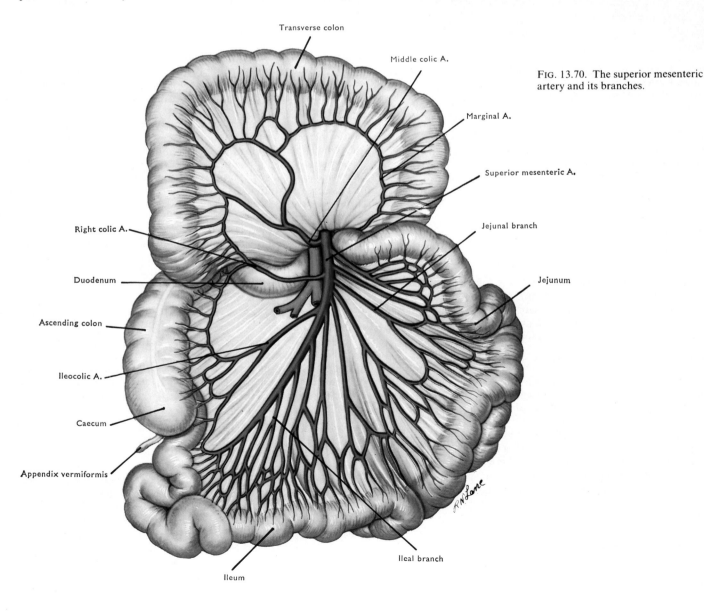

FIG. 13.70. The superior mesenteric artery and its branches.

The **right colic artery** arises either from the superior mesenteric as it crosses the duodenum [FIGS. 13.70 and 13.73] or below the duodenum from a common trunk with the ileocolic artery. It runs in front of the right ureter and gonadal vessels, towards the ascending colon, near which it divides into ascending and descending branches. The ascending unites with the middle colic artery, the descending

with the upper branch of the ileocolic, forming a part of the marginal artery from which branches pass to the ascending colon and the beginning of the transverse colon.

The **middle colic artery** arises from the superior mesenteric as it escapes from behind the pancreas. It enters the transverse mesocolon, and ends by dividing into two branches which unite with the right and left colic arteries, continuing the marginal artery from which branches are distributed to the transverse colon.

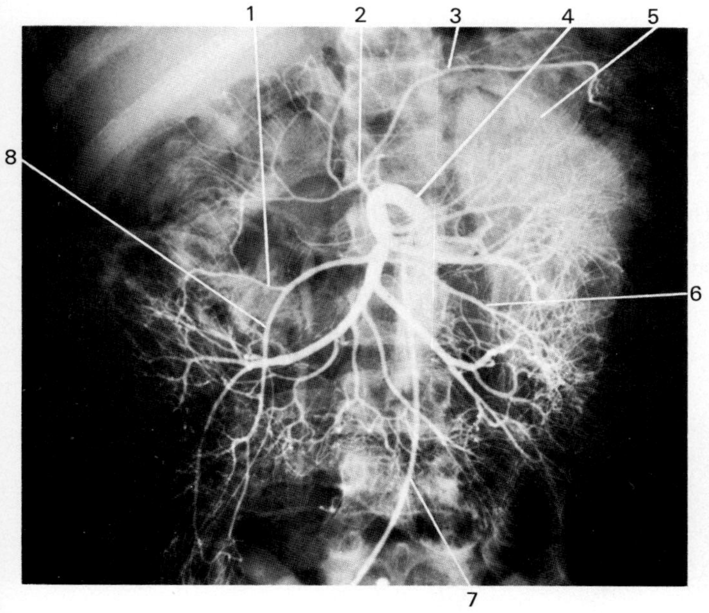

The inferior mesenteric artery [FIGS. 13.73 and 13.74]

This, the hindgut artery of the embryo, arises about 3–4 cm above the aortic bifurcation. It passes downwards, at first in front of the aorta, and then on its left side, and at the upper border of the left common iliac artery it becomes the superior rectal artery. Its vein, close to its left side below, separates from it above.

BRANCHES

The **left colic artery** arises from the inferior mesenteric near its origin, and, crossing anterior to the ureter and the gonadal vessels, divides into ascending and descending branches. The **ascending branch** runs across the lower part of the left kidney, and divides into two branches; one enters the transverse mesocolon to join the left branch of the middle colic artery, and the other joins the upper division of the descending branch. The **descending branch** also

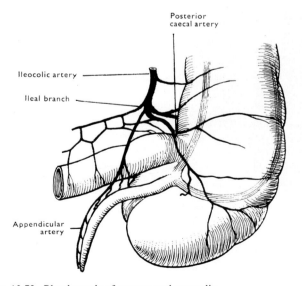

FIG. 13.71. Selective superior mesentric angiogram. The superior mesenteric artery is injected directly with contrast medium from a catheter in the aorta introduced through the femoral artery. Some contrast medium has entered the aorta and passed into the left kidney outlining it. The superior mesenteric artery is somewhat higher than normal.

1. Right colic artery.
2. Middle colic artery.
3. Marginal artery.
4. Superior mesenteric artery.
5. Left kidney.
6. Jejunal branch.
7. Catheter in aorta.
8. Ileocolic artery.

FIG. 13.72. Blood supply of caecum and appendix.
The caecum from behind. The artery of the vermiform appendix, and the three taeniae coli springing from its root, should be specially noted. (Modified from Jonnesco 1895.)

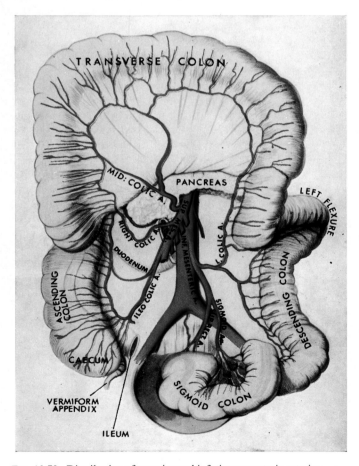

FIG. 13.73. Distribution of superior and inferior mesenteric arteries. There are usually two or more sigmoid arteries.

divides into two branches; one unites with the lower division of the ascending branch and the other with a branch of the first sigmoid artery.

The **sigmoid arteries** are usually three in number. The upper pass downwards and to the left in front of psoas major, the ureter, and the upper part of iliacus, the lowest entering the left limb of the sigmoid mesocolon. They end in branches which anastomose with those of the descending branch of the left colic above and of the superior rectal below, thus continuing the marginal artery from which branches pass to the colon.

The **superior rectal artery** is the direct continuation of the inferior mesenteric. It crosses the front of the left common iliac artery, enters the right limb of the sigmoid mesocolon, descends as far as the third piece of the sacrum, and divides into two branches which run on the sides of the rectum. Half-way down the rectum each terminal branch divides into two or more branches which pass through the muscular coats into the submucous tissue. From these, numerous small vessels pass vertically downwards, anastomosing

with one another, and with offsets from the middle rectal and inferior rectal arteries [p. 470].

The anastomosis between the lowest sigmoid artery and the superior rectal artery has long been surgically suspect; however, Griffiths (1956) states that this anastomosis is always adequate, but that the truly critical point is at the left colic flexure.

PARIETAL BRANCHES OF ABDOMINAL AORTA

THE INFERIOR PHRENIC ARTERIES [FIG. 13.74]

These small arteries, right and left, arise immediately below the diaphragm to which they are distributed. Each gives several **superior suprarenal arteries** to the gland of its own side [p. 590].

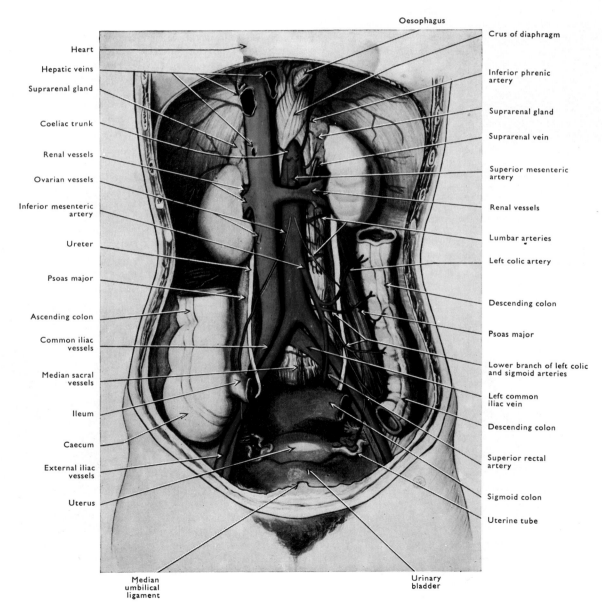

FIG. 13.74. Abdominal aorta and its branches.

THE LUMBAR ARTERIES

These vessels are in series with the posterior intercostal arteries, and their distribution is similar. The upper four pairs arise from the back of the aorta, the fifth pair from the median sacral artery, and all pass round the bodies of the lumbar vertebrae to reach the lateral part of the abdominal wall.

Each artery, except the fifth, passes backwards deep to one of the fibrous arches from which psoas major arises. All run behind the muscle and the lumbar plexus, then behind quadratus lumborum, and finally pierce the aponeurosis of origin of transversus abdominis to gain the interval between the transversus and internal oblique muscles. On the right side the arteries pass behind the inferior vena cava, but the upper two are separated from it by the right crus of the diaphragm.

The lumbar arteries anastomose with one another, with the lower posterior intercostal and the subcostal arteries, and with branches of the superior and inferior epigastric and of the deep circumflex iliac and iliolumbar arteries.

Branches

Each lumbar artery gives off muscular branches, and a dorsal branch which is distributed like that of a posterior intercostal artery [p. 923].

THE MEDIAN SACRAL ARTERY [FIG. 13.74]

This small vessel, originally the dorsal aorta in the sacral region, arises from the back of the aorta about 1 cm above its bifurcation, and ends in the coccygeal body. Its companion veins unite to form a single median sacral vein which joins the left common iliac.

It may give off on each side the **arteria lumbalis ima** (fifth lumbar artery), which is distributed like an ordinary lumbar artery. Small branches of the median sacral artery also pass to the rectum.

The common iliac arteries

The common iliac arteries [FIGS. 13.74 and 13.76] are the terminal branches of the abdominal aorta. They begin opposite the middle of the body of the fourth lumbar vertebra, a little to the left of the median plane, and passing inferolaterally, end anterior to the sacro-iliac joint at the level of the lumbosacral disc, by dividing into external and internal iliac branches.

Both arteries are covered by peritoneum in front and medially, and overlaid by small intestine. The superior hypogastric plexus [FIG. 11.86] lies anterior to their proximal ends, and the left artery is crossed by the superior rectal vessels. On each side, near its termination, the artery is often crossed by the ureter. The last two lumbar vertebral bodies and intervening disc, the sympathetic trunk and the medial part of psoas major lie behind each artery, but are separated from the right artery by the beginning of the inferior vena cava. Deep to psoas are the obturator nerve, the iliolumbar artery, and the lumbosacral trunk. Lateral to each artery are the ureter, gonadal vessels, and the genitofemoral nerve. The changing relation of each common iliac vein to its companion artery can be followed in FIGURE 13.74.

Branches

The external and internal iliac arteries are the only branches.

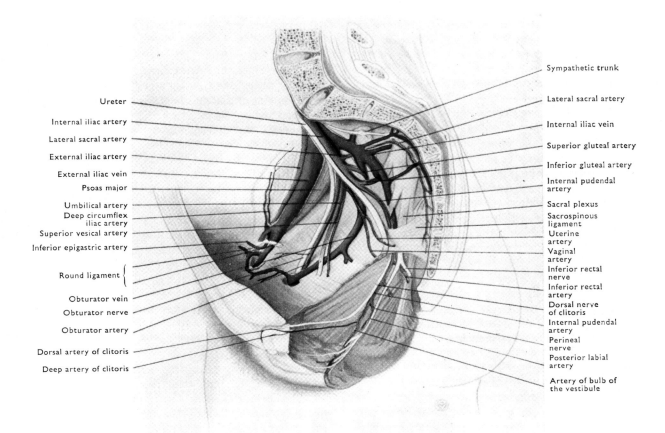

FIG. 13.75. Internal iliac artery and its branches in the female.

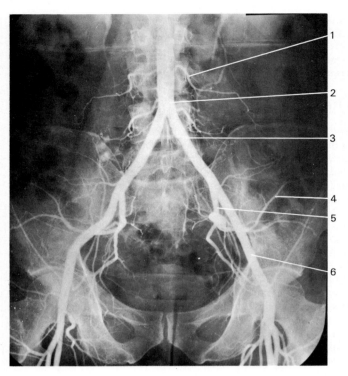

FIG. 13.76. Anteroposterior aortogram to show aortic bifurcation and main branches.

1. Lumbar artery.
2. Aorta.
3. Left common iliac artery.
4. Superior gluteal artery.
5. Internal iliac artery.
6. External iliac artery.

Surface anatomy

The common and external iliac arteries may be indicated by a line drawn with a slight lateral convexity, from a point opposite the bifurcation of the aorta [p. 896] to the midinguinal point. The upper third of the line corresponds to the common iliac artery, the lower two-thirds to the external iliac.

The internal iliac artery [FIGS. 13.75 and 13.79]

In the foetus this vessel is the direct continuation of the common iliac trunk to the umbilicus, and is prolonged through the umbilical cord as the umbilical artery [p. 932].

When the umbilical cord is severed at birth, the part of the internal iliac trunk between the pelvis and the umbilicus atrophies to a fibrous cord known as the lateral umbilical ligament. However, the proximal part of the artery remains patent to convey blood to its superior vesical branch. The permanent internal artery is short, and appears to end in an anterior and posterior division; the anterior is the original continuation of the vessel, and the posterior is simply a common stem of origin for some of the branches.

The artery arises from the common iliac anteromedial to the sacro-iliac joint at the level of the lumbosacral disc, and descends into the lesser pelvis, to end in its two divisions near the upper border of the greater sciatic notch.

Position

Anteromedially, each internal iliac artery is covered by peritoneum, beneath which the ureter descends along the anterior surface of the artery. The sigmoid colon usually overlaps the left vessel, and

the terminal ileum the right. In the female, the ovary and the infundibulum of the uterine tube usually lie close in front of it, separated by the peritoneum and the ureter [FIG. 13.79]. Posterior to it are the internal iliac vein and the commencement of the common iliac vein; behind these are the lumbosacral trunk and the sacro-iliac joint. Laterally, the external iliac vein separates it from psoas major, and at a lower level the obturator nerve lies between the artery and the side-wall of the pelvis.

Branches

The posterior division of the internal iliac artery gives rise to parietal branches, the anterior division to parietal and visceral branches.

BRANCHES OF THE POSTERIOR DIVISION OF THE INTERNAL ILIAC ARTERY

The iliolumbar artery

This vessel runs to the iliac fossa, passing over the lateral part of the sacrum between the lumbosacral trunk and the obturator nerve, and divides into an iliac and a lumbar branch. The iliac branch runs on the abdominal surface of iliacus giving offsets to it, and a large nutrient branch to the ilium. The lumbar branch ascends behind psoas major, anastomoses with, or forms the fifth lumbar artery, and ends in quadratus lumborum. Its spinal branch enters the intervertebral foramen between the fifth lumbar vertebra and the sacrum.

The lateral sacral arteries

Superior and inferior on each side, they descend in front of the sacral ventral rami, giving off spinal branches which pass through the pelvic sacral foramina.

The superior gluteal artery [FIGS. 13.75 and 13.78]

This large vessel, the continuation of the posterior division of the internal iliac artery, passes backwards between the lumbosacral trunk and the first sacral ventral ramus. Leaving the pelvis through the greater sciatic foramen, above the piriformis muscle, it divides under cover of gluteus maximus into superficial and deep branches.

The superficial branch supplies gluteus maximus and the skin over that muscle's origin.

The deep branch lies under gluteus medius and very soon divides into superior and inferior branches. The superior branch follows the upper border of gluteus minimus and ends under cover of tensor fasciae latae, by anastomosing with the ascending branch of the lateral circumflex femoral artery. It also anastomoses with the deep circumflex iliac artery, and gives branches to the adjacent muscles. The inferior branch runs with the superior gluteal nerve across gluteus minimus towards the greater trochanter. It supplies the gluteal muscles, and anastomoses with the ascending branches of the lateral and medial circumflex femoral arteries, and with the inferior gluteal artery.

Surface anatomy

The superior gluteal artery reaches the gluteal region opposite the junction of the upper and middle thirds of a line drawn from the posterior superior iliac spine to the upper border of the greater trochanter.

PARIETAL BRANCHES OF THE ANTERIOR DIVISION OF THE INTERNAL ILIAC ARTERY

The obturator artery [FIGS. 5.113 and 13.75].

This vessel runs along the side-wall of the lesser pelvis to the obturator canal through which it leaves the pelvis. Immediately on entering the thigh it divides into anterior and posterior terminal branches which surround the obturator foramen deep to obturator externus. In the pelvis the obturator nerve lies above the artery which is above the vein; the ureter, behind, and the ductus deferens, in front, lie between the artery and the peritoneum. In the female, the ovary and broad ligament are medial.

Branches

Near the obturator canal, the obturator artery gives off a pubic branch which anastomoses with the pubic branch of the inferior epigastric. This frequently passes medial to the femoral ring on the lacunar ligament, medial to the site of a femoral hernia. It becomes especially important when this branch of the inferior epigastric artery enlarges to form the obturator artery [p. 933]. The anterior and posterior terminal branches supply the adjacent muscles, and the posterior branch also supplies the hip joint and the ligament of the head of the femur by a branch which passes through the acetabular notch. According to Harty (1953) the head and neck of the femur derive their only significant blood supply from vessels on the neck of the femur.

The internal pudendal artery [FIGS. 13.75 and 13.77]

This vessel descends anterior to the sacral plexus to the lower part of the greater sciatic foramen, and passes between piriformis and coccygeus to enter the gluteal region. It is accompanied by two veins. From the gluteal region, where it lies on the ischial spine between the pudendal nerve medially and the nerve to obturator internus laterally, it passes through the lesser sciatic foramen and enters the perineum.

In the perineum it lies at first in the lateral wall of the ischiorectal fossa, where it is enclosed in the pudendal canal in the obturator fascia. This canal, which begins 3–4 cm above the lower margin of the ischial tuberosity, contains also the pudendal nerve, which lies above the artery, and the perineal nerve, which lies below it. From the ischiorectal fossa, the internal pudendal artery enters the deep perineal space, and runs forwards close to the inferior ramus of the pubis. About 1 cm posterior to the pubic symphysis it pierces the perineal membrane and immediately divides into its terminal branches—the deep and the dorsal arteries of the penis (or clitoris).

Branches

In the ischiorectal fossa the inferior rectal artery pierces the wall of the pudendal canal, and divides into two or three branches which pass across the fossa to levator ani and the anal canal. There they anastomose with the middle and superior rectal arteries.

Two posterior scrotal or posterior labial branches arise in the anterior part of the ischiorectal fossa, and pierce the posterior border of the perineal fascia to enter the superficial perineal space in which they pass forwards to the scrotum or labium majus. They supply the muscles and the covering skin of the superficial perineal space.

The perineal artery usually arises from one of the scrotal or labial branches, and runs to the central tendon of the perineum.

The artery of the bulb of the penis arises in the deep perineal space. It runs along the posterior border of sphincter urethrae, pierces the perineal membrane to enter the substance of the bulb, and passes onwards in the corpus spongiosum to the glans. It supplies sphincter urethrae, the bulbo-urethral gland, the corpus spongiosum penis, and the spongy part of the urethra. In the female, the corresponding artery supplies the bulb of the vestibule and the erectile tissue of the lower part of the vagina.

The deep artery of the penis (or clitoris) immediately enters the crus, and runs forwards in the corpus cavernosum of the penis (or clitoris) which it supplies.

The dorsal artery of the penis (or clitoris) ascends between the crus and the bone, passes forwards between the layers of the suspensory ligament of the penis (or clitoris), and runs along the

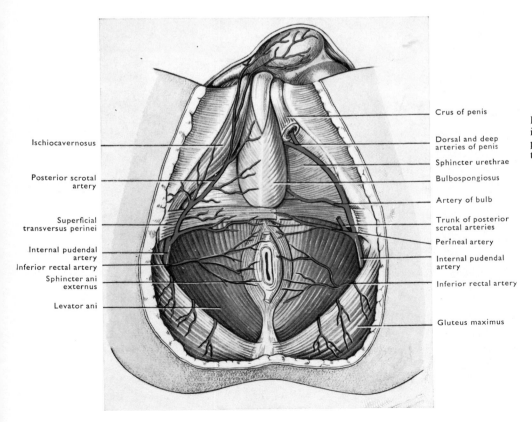

FIG. 13.77. Perineal distribution of internal pudendal artery in the male. The perineal membrane has been removed on the right.

dorsal surface of the organ, with the dorsal nerve immediately lateral to it, and separated from its fellow by the deep dorsal vein in the median plane. It supplies the superficial tissues on the dorsum of the penis, sends branches into the corpus cavernosum, and terminates in the glans and prepuce.

The inferior gluteal artery [FIGS. 13.75 and 13.78]

This vessel runs backwards and laterally between the first and second, or second and third, sacral ventral rami, and traverses the greater sciatic foramen to enter the gluteal region just below piriformis. It then descends along the posteromedial side of the sciatic nerve to reach the proximal part of the thigh.

Branches

Muscular branches pass to gluteus maximus, and one or two coccygeal branches pierce the sacrotuberous ligament to reach the skin over the coccyx. An anastomosing branch joins with branches of the medial and lateral femoral circumflex, and the first perforating

arteries, in the so-called 'cruciate anastomosis' of the thigh. The companion artery of the sciatic nerve is a long slender branch which runs on the surface of the sciatic nerve or in its substance.

Surface anatomy

The inferior gluteal artery enters the gluteal region just above the midpoint of a line joining the posterior superior iliac spine to the ischial tuberosity.

VISCERAL BRANCHES OF THE ANTERIOR DIVISION OF THE INTERNAL ILIAC ARTERY

Umbilical artery

The lumen of this vessel persists, though much diminished, as far as the origin of the superior vesical artery. Thereafter the umbilical artery is reduced to a fibrous cord which runs along the side of the bladder and then, as the lateral umbilical ligament, on the deep surface of the anterior abdominal wall to the umbilicus. In the pelvis

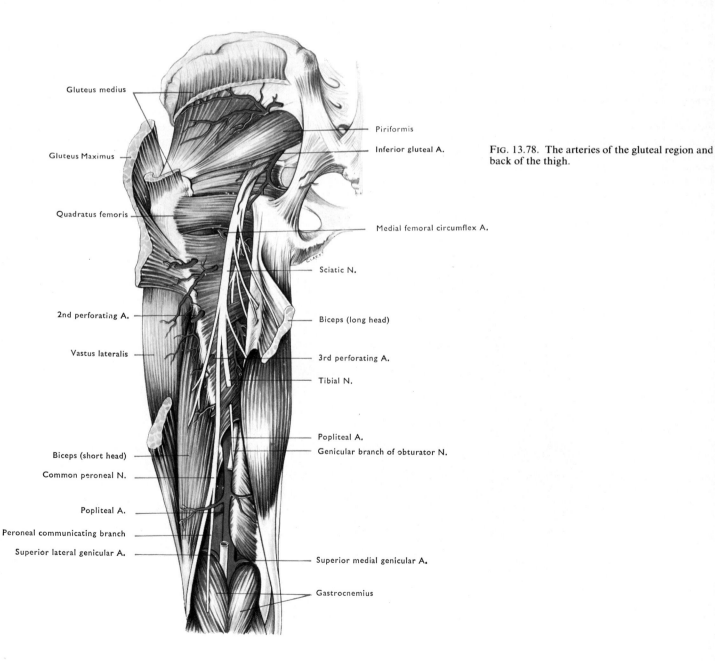

Gluteus medius

Gluteus Maximus

Quadratus femoris

2nd perforating A.

Vastus lateralis

Biceps (short head)

Common peroneal N.

Popliteal A.

Peroneal communicating branch

Superior lateral genicular A.

Piriformis

Inferior gluteal A.

Medial femoral circumflex A.

Sciatic N.

Biceps (long head)

3rd perforating A.

Tibial N.

Popliteal A.

Genicular branch of obturator N.

Superior medial genicular A.

Gastrocnemius

FIG. 13.78. The arteries of the gluteal region and back of the thigh.

it is crossed by the ductus deferens or by the round ligament of the uterus.

The **superior vesical artery** arises from the umbilical artery, and passes to the upper part of the urinary bladder.

Inferior vesical artery

This runs on levator ani to supply the lower part of the bladder, the seminal vesicle, the ductus deferens, the lower part of the ureter, and the prostate.

Artery of the ductus deferens

This may arise from the superior or inferior vesical artery. It is a slender vessel which accompanies the ductus deferens to the testis.

Middle rectal artery

This runs medially to the rectum, where it anastomoses with its fellow, and with the other rectal arteries.

Vaginal artery

This usually replaces the inferior vesical artery. It runs to the side of the vagina. The branches of opposite sides anastomose and form anterior and posterior longitudinal vessels—the so-called azygos arteries which anastomose above with the vaginal branches of the uterine arteries.

Uterine artery

This runs medially and slightly forwards on levator ani to the lower border of the broad ligament, and passes medially between its layers. About 3 cm from the midline it crosses above and in front of the ureter [FIGS. 8.55 and 13.75], and passes above the lateral fornix of the vagina to the side of the cervix. It then ascends in a tortuous manner alongside the uterus, and, turning laterally below the uterine tube, ends by anastomosing with the ovarian artery. The uterine artery sends many branches to the uterus, and aids in the supply of the vagina, uterine tube, and ovary.

VARIATIONS

The inferior gluteal artery may, as in the foetus, constitute the main artery of the lower limb, and become continuous with the popliteal artery; probably the normal companion artery of the sciatic nerve represents this original vessel [FIG. 13.78].

The obturator artery may arise from the inferior epigastric artery [p. 935]. The course of such an **accessory obturator artery** is of surgical importance. It descends on the medial side of the external iliac vein, and usually on the lateral side of the femoral ring; but in 30 per cent of cases, and more often in males, it descends on the medial side of the ring and may be injured in the operation for the relief of a strangulated femoral hernia [FIG. 5.113].

For further details of variations in the branches of the internal iliac artery Braithwaite (1952) should be consulted.

External iliac artery

The external iliac artery [FIGS. 13.74 and 13.79] extends from a point opposite the sacro-iliac joint, at the level of the lumbosacral disc, to the midinguinal point behind the inguinal ligament, midway between the anterior superior iliac spine and the pubic symphysis, where it becomes the femoral artery.

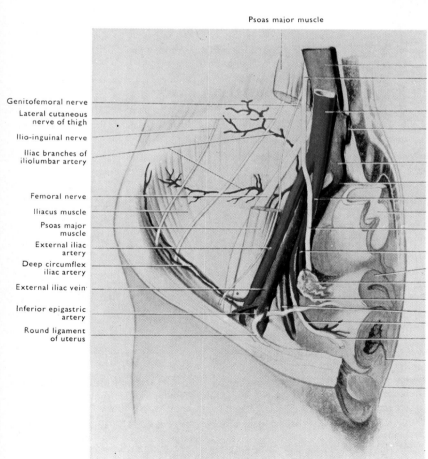

Psoas major muscle

Genitofemoral nerve
Lateral cutaneous nerve of thigh
Ilio-inguinal nerve
Iliac branches of iliolumbar artery
Femoral nerve
Iliacus muscle
Psoas major muscle
External iliac artery
Deep circumflex iliac artery
External iliac vein
Inferior epigastric artery
Round ligament of uterus

Inferior vena cava
Ureter
Right common iliac artery
Left common iliac vein
Right common iliac vein
Internal iliac vein
Internal iliac artery
Sigmoid colon
Ureter
Uterine artery
Uterus
Ovary
Uterine tube
Obturator artery
Superior vesical artery
Lateral umbilical ligament on urinary bladder
Urethra
Pubic symphysis

FIG. 13.79. Iliac arteries and veins in the female.

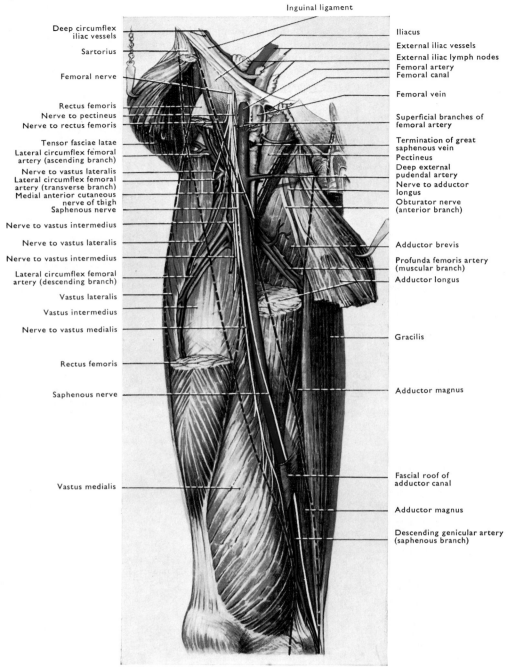

Inguinal ligament

Deep circumflex iliac vessels

Sartorius

Femoral nerve

Rectus femoris
Nerve to pectineus
Nerve to rectus femoris

Tensor fasciae latae
Lateral circumflex femoral artery (ascending branch)
Nerve to vastus lateralis
Lateral circumflex femoral artery (transverse branch)
Medial anterior cutaneous nerve of thigh
Saphenous nerve

Nerve to vastus intermedius

Nerve to vastus lateralis

Nerve to vastus intermedius

Lateral circumflex femoral artery (descending branch)

Vastus lateralis

Vastus intermedius

Nerve to vastus medialis

Rectus femoris

Saphenous nerve

Vastus medialis

Iliacus
External iliac vessels
External iliac lymph nodes
Femoral artery
Femoral canal

Femoral vein

Superficial branches of femoral artery

Termination of great saphenous vein
Pectineus
Deep external pudendal artery
Nerve to adductor longus
Obturator nerve (anterior branch)

Adductor brevis

Profunda femoris artery (muscular branch)
Adductor longus

Gracilis

Adductor magnus

Fascial roof of adductor canal

Adductor magnus

Descending genicular artery (saphenous branch)

FIG. 13.80. Femoral artery and its branches.
The outlines of the sartorius, the upper part of the rectus femoris, and the adductor longus are indicated by broken black lines.

It runs along the linea terminalis of the pelvis, and is enclosed, with its accompanying vein, in a thin fascial sheath.

Anteriorly, peritoneum separates it on the right side from the terminal ileum, and sometimes from the vermiform appendix, and on the left side from the small intestine, while distally is the descending colon. Several branches of the lower sigmoid arteries also cross it on the left side. The artery is crossed at its origin by the ureter and, 2.5 cm lower down, by the ovarian vessels. Near its lower end the testicular vessels and the genital branch of the genitofemoral nerve lie on the artery, and all are crossed by the deep circumflex iliac vein. That part of the artery is crossed also by the ductus deferens or the round ligament of the uterus. External iliac lymph nodes lie in front and at the sides of the artery. Posteriorly, it lies on the fascia iliaca, at first medial and then anterior to psoas

major. To its lateral side is the genitofemoral nerve, and to its medial side peritoneum. Below, the external iliac vein is medial to the artery, but, above, the vein comes to lie partly behind the artery.

BRANCHES

Two large branches spring from the external iliac artery: (1) the inferior epigastric; and (2) the deep circumflex iliac.

The inferior epigastric artery [FIGS. 5.113, 13.75, and 13.79].

This arises immediately above the inguinal ligament and ascends towards the umbilicus from the medial side of the deep inguinal ring

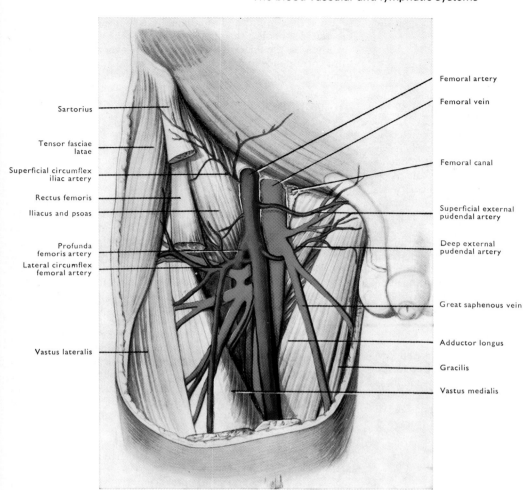

Sartorius

Tensor fasciae
latae

Superficial circumflex
iliac artery

Rectus femoris

Iliacus and psoas

Profunda
femoris artery
Lateral circumflex
femoral artery

Vastus lateralis

Femoral artery

Femoral vein

Femoral canal

Superficial external
pudendal artery

Deep external
pudendal artery

Great saphenous vein

Adductor longus

Gracilis

Vastus medialis

FIG. 13.81. Femoral vessels in the
femoral triangle.

to the deep surface of rectus abdominis. In its course it raises a fold
of peritoneum called the lateral umbilical fold—not to be confused
with the medial umbilical fold which is raised by the lateral
umbilical ligament. It then pierces the transversalis fascia, and
passes anterior to the arcuate line into the rectus sheath. It penetrates
the muscle, and its branches anastomose with those of the superior
epigastric artery and lower posterior intercostal arteries. At the
deep inguinal ring, the ductus deferens (or the round ligament of the
uterus) curves round the lateral and anterior surfaces of the artery.

Branches

Muscular and cutaneous branches supply the anterior abdominal
wall. The cremasteric artery in the male (artery of the round
ligament of the uterus in the female) is small. It accompanies the
spermatic cord, supplying its coverings; in the female it runs with
the round ligament. The pubic branch anastomoses with the pubic
branch of the obturator artery [p. 933].

The deep circumflex iliac artery [FIGS. 13.79 and 13.80]

This arises immediately above the inguinal ligament, and runs to
the anterior superior iliac spine. A little beyond the spine it pierces
transversus abdominis, and is continued between the transversus
and the internal oblique.

Branches

Of its muscular branches one is frequently large and is known as the
ascending branch. It pierces transversus muscle a short distance in

front of the anterior superior spine, and ascends vertically on
internal oblique.

ARTERIES OF THE LOWER LIMB

The main artery of the lower limb is the continuation of the external
iliac artery. It descends through the thigh to the lower end of the
popliteal fossa where it divides into the anterior and posterior tibial
arteries. In the proximal two-thirds of the thigh it is the femoral
artery; in the popliteal fossa it is the popliteal artery.

Femoral artery [FIG. 13.76]

The femoral artery [FIGS. 13.80 and 13.81] begins behind the
inguinal ligament, and becomes continuous with the popliteal artery
at the opening in adductor magnus.

 Its proximal half lies in the femoral triangle and is comparatively
superficial; at the apex of the triangle it passes deep to the sartorius,
enters the adductor canal, and is then more deeply placed.

 At their entry into the femoral triangle both the artery and its vein
are enclosed for a distance of 3 cm in a funnel-shaped fascial sheath
formed of the fascia transversalis in front and the fascia iliaca
behind. It is called the femoral sheath and is divided by septa into
three compartments. The lateral compartment is occupied by the
femoral artery and the femoral branch of the genitofemoral nerve;

the intermediate compartment contains the femoral vein; and the medial compartment is the **femoral canal** [p. 405].

IN THE FEMORAL TRIANGLE

Anteriorly, the femoral artery is covered by skin and fasciae, and by superficial inguinal lymph nodes. The femoral sheath and the cribriform fascia cover its proximal part, and the fascia lata its distal part; and near the apex of the triangle the artery is crossed by an anterior cutaneous nerve. Posteriorly and proximodistally it lies on the femoral sheath and psoas major, pectineus, and adductor longus. The nerve to pectineus passes between the artery and psoas major; the femoral vein and the profunda artery intervene between it and pectineus, and the femoral vein separates it from adductor longus also.

The femoral vein lies behind the artery in the lower part of the triangle, but passes to its medial side above. On the lateral side of the artery, proximally, are the femoral branch of the genitofemoral nerve, the femoral sheath, and the femoral nerve; more distally the saphenous nerve and the nerve to vastus medialis are continued on its lateral side.

IN THE ADDUCTOR CANAL

Adductors longus and magnus are posterior to the artery, but are separated from it by the femoral vein, which is posterior to the artery proximally, and posterolateral distally. Vastus medialis is anterolateral. The fascial roof of the canal, the subsartorial plexus of nerves, and sartorius are anteromedial. The saphenous nerve enters the canal lateral to the artery, and then crosses in front of it to become medial.

SURFACE ANATOMY

To map out the course of the femoral artery the thigh should be slightly flexed, abducted, and rotated laterally [FIG. 13.82]; a line is drawn from the **midinguinal point**, i.e. midway between the anterior

superior iliac spine and the symphysis pubis, to the adductor tubercle. Rather less than the upper third of the line corresponds to the artery in the femoral triangle and rather more than its middle third corresponds to the artery as it lies in the adductor canal. The femoral pulse is easily felt in the upper part of the vessel.

BRANCHES OF THE FEMORAL ARTERY

(a) The **superficial circumflex iliac artery** arises just below the inguinal ligament. It enters the superficial fascia lateral to the saphenous opening, and runs as far as the anterior superior iliac spine. It supplies lymph nodes and the skin of the groin.

(b) The **superficial epigastric artery** pierces the cribriform fascia, and passes upwards towards the umbilicus supplying the skin.

(c) The **superficial external pudendal artery** also pierces the cribriform fascia, and runs towards the pubic tubercle. It supplies the skin of the lower abdomen and pubis.

(d) The **deep external pudendal artery** runs medially, deep to the great saphenous vein, and either anterior or posterior to adductor longus. It pierces the deep fascia, and ends in the scrotum or the labium majus.

(e) The **profunda femoris artery** [FIG. 13.81], the largest branch of the femoral artery, arises from its lateral side usually about 4 cm distal to the inguinal ligament. It curves behind the femoral artery, and leaving the femoral triangle between pectineus and adductor longus, descends posterior to adductor longus, close to the medial aspect of the femur, anterior to adductor brevis and magnus. In the distal third of the thigh it pierces the latter muscle as the fourth perforating artery. Initially the profunda and femoral veins separate it from the femoral artery, while adductor longus separates them inferiorly.

Branches from the profunda supply the adductor muscles; some of them pass through adductor magnus to the hamstrings and vastus lateralis.

1. The **lateral circumflex femoral artery** [FIGS. 13.80 and 13.81] springs from the profunda near its origin. It runs laterally among the branches of the femoral nerve, then, passing posterior to sartorius and rectus femoris, it ends by dividing into three branches. The

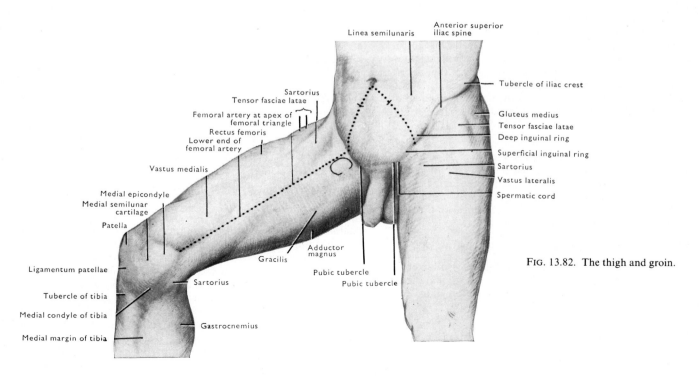

FIG. 13.82. The thigh and groin.

ascending branch runs upwards, deep to tensor fasciae latae, along the trochanteric line. It then passes between gluteus medius and gluteus minimus, and anastomoses with branches of the superior gluteal artery. It supplies the neighbouring muscles, and gives a branch to the hip joint. The transverse branch passes into vastus lateralis, winds round the femur and joins the cruciate anastomosis [p. 932]. The descending branch runs along the anterior border of vastus lateralis, accompanied by the nerve to that muscle. It gives branches to quadriceps femoris, and takes part in the anastomosis round the knee.

2. The medial circumflex femoral artery arises from the profunda at the same level as the lateral circumflex, and runs backwards, out of the femoral triangle, between psoas major and pectineus. It continues between obturator externus superiorly and adductor

brevis inferiorly, to the upper border of adductor magnus, where it divides into a transverse and an ascending branch. Before dividing it sends a branch through the acetabular notch to the hip joint. The ascending branch passes between obturator externus and quadratus femoris to the trochanteric fossa of the femur, where it anastomoses with branches of both gluteal arteries. The transverse branch passes backwards between quadratus femoris and adductor magnus. It takes part in the cruciate anastomosis [p. 932].

3. The perforating arteries [FIG. 13.83], including the terminal branch of the profunda, are usually four in number. They curve backwards and laterally, lying between the femur and tendinous arches which interrupt the insertion of adductor magnus to the linea aspera, and communicate freely with each other. The first perforating artery pierces the insertions of adductor brevis and

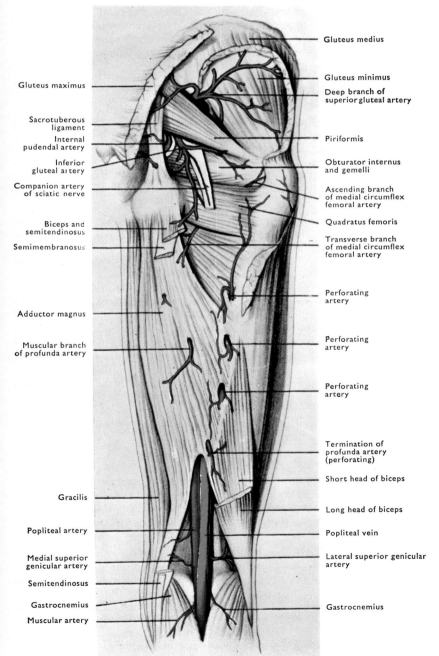

FIG. 13.83. Arteries of gluteal region and posterior surface of thigh and knee.

magnus, and usually that of gluteus maximus also. It takes part in the cruciate anastomosis. The second perforating artery pierces adductors brevis and magnus, and then passes between gluteus maximus and the short head of biceps femoris into vastus lateralis. The third and fourth perforating arteries pass through adductor magnus and the short head of biceps femoris into vastus lateralis.

A nutrient branch to the femur is given off from either the second or third perforating artery.

There is a chain of *anastomoses*, extending from the gluteal arteries to the muscular branches of the popliteal artery, by which a collateral circulation can be provided in cases of occlusion of the femoral artery [FIG. 13.83].

(f) The descending genicular artery arises low in the adductor canal, and divides into a superficial and a deep branch. The superficial or saphenous branch continues through the canal and appears superficially between gracilis and sartorius. It gives twigs to the skin of the proximal part of the leg. The deep branch runs in the substance of vastus medialis, and gives off articular branches which take part in the anastomosis around the knee.

VARIATIONS

In rare cases the femoral artery is small and ends in the profunda and circumflex branches, and the inferior gluteal artery forms the principal vessel of the lower limb. Commonly one or both circumflex arteries may arise from the femoral.

Popliteal artery

The popliteal artery is the direct continuation of the femoral. It begins at the medial and proximal side of the popliteal fossa, and ends by dividing into the anterior and posterior tibial arteries at the distal border of the popliteus muscle, on a level with the distal part of the tuberosity of the tibia.

Anteriorly, from above downwards, it rests on fat covering the popliteal surface of the femur, the back of the knee joint, and the fascia covering popliteus. Posteriorly, it is overlapped proximally by the lateral border of semimembranosus; it is crossed, about its middle, by the popliteal vein and the tibial nerve, the vein intervening between the artery and the nerve. Distally, it is overlapped by the adjacent borders of the heads of gastrocnemius, and is crossed by the nerves to soleus and popliteus and by the plantaris muscle. Laterally, it is in contact with the popliteal vein and the tibial nerve proximally, then with the lateral condyle of the femur and plantaris, and, distally, with the lateral head of gastrocnemius. Medially, it is in contact with semimembranosus proximally, then with the medial condyle of the femur, and distally with the tibial nerve, the popliteal vein, and the medial head of gastrocnemius. Popliteal lymph nodes are arranged irregularly around the artery.

SURFACE ANATOMY

To map out the popliteal artery a line is drawn from a point just medial to the upper angle of the fossa to a point midway between the condyles of the femur, and thence along the middle of the fossa to the level of the lower part of the tuberosity of the tibia. The pulsations of the vessel may be felt when the limb is flexed.

BRANCHES

Branches are given off to the hamstring muscles and to the muscles of the calf, those to the latter being called the sural arteries. Each head of the gastrocnemius receives a large branch which enters at a definite neurovascular hilus; these branches are virtually end arteries.

There are five genicular arteries [FIGS. 13.83 and 13.84].

The lateral superior genicular artery passes anterior to the biceps tendon, into vastus lateralis.

The medial superior genicular artery passes anterior to the tendon of adductor magnus, into vastus medialis.

The lateral inferior genicular artery runs across popliteus, and passes deep to the fibular collateral ligament.

The medial inferior genicular artery passes along the proximal border of popliteus, and then turns forwards between the tibia and the tibial collateral ligament.

The middle genicular artery pierces the oblique popliteal ligament of the knee joint, and supplies the cruciate ligaments and the synovial membrane.

The anastomosis round the knee joint is rich and is formed by the following vessels: the medial and lateral genicular, the descending genicular, the descending branch of the lateral circumflex femoral, the anterior tibial recurrent, and the circumflex fibular [see below].

Of the cutaneous branches, one, the superficial sural artery, accompanies the small saphenous vein.

VARIATIONS

The popliteal artery may, very rarely, form the direct continuation of the inferior gluteal artery; and it sometimes divides at a more proximal or more distal level than usual.

Posterior tibial artery

The posterior tibial artery, the larger terminal branch of the popliteal, begins at the distal border of popliteus and ends, midway between the tip of the medial malleolus and the most prominent part of the heel, by dividing into the medial and the lateral plantar arteries [FIGS. 13.85 and 13.86]. Two veins accompany it throughout.

Anteriorly, it is in contact, proximodistally, with tibialis posterior, flexor digitorum longus, and the posterior surfaces of the tibia and the ankle joint. Posteriorly, it is crossed by the tibial nerve about 2.5 cm distal to its origin, and elsewhere it is in contact with the fascia over the deep layer of muscles. The proximal half of the artery is also covered by soleus and gastrocnemius, but its distal half only by skin and fasciae. At its termination it lies deep to the flexor retinaculum. The tibial nerve lies at first on the medial side of the vessel, but soon crosses to its lateral side. The artery is separated from the medial malleolus by the tendons of tibialis posterior and flexor digitorum longus, whilst the tendon of flexor hallucis longus lies lateral to it.

SURFACE ANATOMY

The position of the posterior tibial artery corresponds with a line drawn from the lower angle of the popliteal fossa, at the level of the neck of the fibula, to a point midway between the medial malleolus and the most prominent part of the heel.

BRANCHES

The posterior tibial gives branches to the adjacent muscles and to the skin of the medial part of the leg.

A circumflex fibular branch passes over the neck of the fibula to the anastomosis round the knee. It may arise from the anterior tibial.

The nutrient artery to the tibia, the largest of the nutrient arteries to long bones, springs from the proximal part of the posterior tibial and enters the nutrient foramen on the posterior surface of the bone.

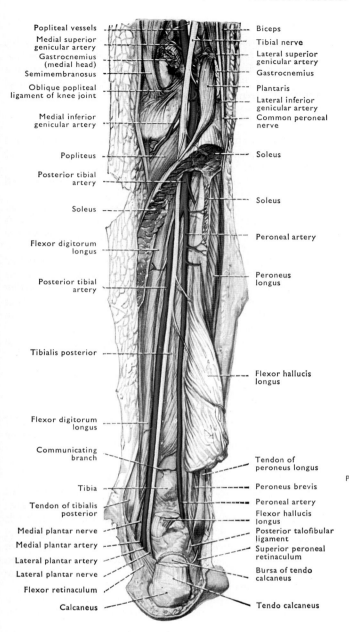

Popliteal vessels
Medial superior genicular artery
Gastrocnemius (medial head)
Semimembranosus
Oblique popliteal ligament of knee joint
Medial inferior genicular artery
Popliteus
Posterior tibial artery
Soleus
Flexor digitorum longus
Posterior tibial artery
Tibialis posterior
Flexor digitorum longus
Communicating branch
Tibia
Tendon of tibialis posterior
Medial plantar nerve
Medial plantar artery
Lateral plantar artery
Lateral plantar nerve
Flexor retinaculum
Calcaneus

Biceps
Tibial nerve
Lateral superior genicular artery
Gastrocnemius
Plantaris
Lateral inferior genicular artery
Common peroneal nerve
Soleus
Soleus
Peroneal artery
Peroneus longus
Flexor hallucis longus
Tendon of peroneus longus
Peroneus brevis
Peroneal artery
Flexor hallucis longus
Posterior talofibular ligament
Superior peroneal retinaculum
Bursa of tendo calcaneus
Tendo calcaneus

FIG. 13.84. Popliteal and posterior tibial arteries and their branches. [See also FIG. 5.151.]

The **communicating branch** unites the posterior tibial to the peroneal artery about 2.5 cm above the distal tibiofibular joint.

Calcanean branches pierce the flexor retinaculum and supply the tissues of the heel. Over the tuber calcanei they form a rete with branches of the peroneal artery,

A **malleolar branch** joins the network on the medial malleolus.

THE PERONEAL ARTERY [FIG. 13.84]

This is the largest branch of the posterior tibial [FIG. 13.86]. Arising about 2.5 cm below the distal border of popliteus, it descends along the medial crest of the fibula between tibialis posterior in front and flexor hallucis longus behind. About 2.5 cm proximal to the ankle joint it gives off a perforating branch, and then passes behind the

lateral malleolus to join the rete calcaneum. Its two companion veins anastomose around it.

Branches

The peroneal artery gives off muscular branches, the nutrient artery to the fibula, and a communicating branch which joins that of the posterior tibial.

The **perforating branch** passes through the interosseous membrane to the dorsum of the foot, where it anastomoses with branches of the anterior tibial and dorsalis pedis arteries.

A **malleolar branch** and **calcanean branches** take part in the anastomosis on the lateral side of the ankle and around the heel.

Variations

The perforating branch of the peroneal is almost invariably large when the anterior tibial artery is small; in some cases it replaces the whole of the dorsalis pedis continuation of the latter vessel.

Plantar arteries

The medial and lateral plantar arteries arise, under cover of the flexor retinaculum, midway between the tip of the medial malleolus and the most prominent part of the medial side of the heel [FIGS. 13.84–13.86].

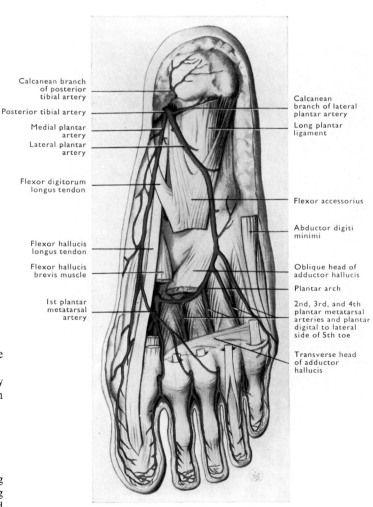

Calcanean branch of posterior tibial artery
Posterior tibial artery
Medial plantar artery
Lateral plantar artery
Flexor digitorum longus tendon
Flexor hallucis longus tendon
Flexor hallucis brevis muscle
1st plantar metatarsal artery

Calcanean branch of lateral plantar artery
Long plantar ligament
Flexor accessorius
Abductor digiti minimi
Oblique head of adductor hallucis
Plantar arch
2nd, 3rd, and 4th plantar metatarsal arteries and plantar digital to lateral side of 5th toe
Transverse head of adductor hallucis

FIG. 13.85. Plantar arteries and their branches.

MEDIAL PLANTAR ARTERY

The medial plantar artery, the smaller of the two, passes forwards, medial to the medial plantar nerve, in the interval between abductor hallucis and flexor digitorum brevis; it ends by uniting with the branch of the first plantar metatarsal artery to the medial side of the big toe. It gives off a superficial branch which ramifies on abductor hallucis, and branches to the adjacent muscles, articulations, and skin; it also gives three digital branches which join the medial plantar metatarsal arteries.

Surface anatomy

To mark the position of the medial plantar artery a line is drawn from the point of bifurcation of the posterior tibial artery to the medial side of the great toe. Just distal to the midpoint of this line the artery breaks up into its digital branches.

LATERAL PLANTAR ARTERY

The lateral plantar artery runs obliquely across the sole on the lateral side of the lateral plantar nerve, first between flexor digitorum brevis superficially and flexor accessorius deeply, and then, in the interval between flexor digitorum brevis and abductor digiti minimi. At the base of the fifth metatarsal bone it arches medially across the foot, on the metatarsals, to the lateral side of the first metatarsal bone, forming the plantar arch, which is completed by its junction with the dorsalis pedis.

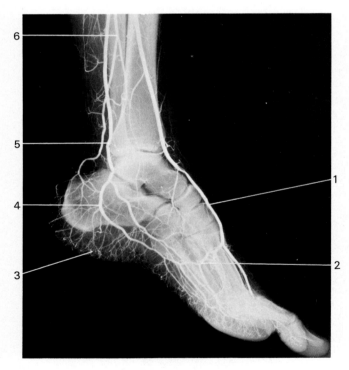

FIG. 13.86. Arteries of the ankle and foot injected after death.
1. Dorsalis pedis artery.
2. Plantar arch.
3. Arterial plexus of sole.
4. Lateral plantar artery.
5. Posterior tibial artery.
6. Peroneal artery.

Surface anatomy

To mark the position of the lateral plantar artery a line is drawn from the point of bifurcation of the posterior tibial artery towards the fourth toe. At the level of the tuberosity of the fifth metatarsal it is continued across the sole, with a distal convexity, to the proximal end of the first intermetatarsal space.

Branches

The lateral plantar artery gives off: a calcanean branch to the skin and subcutaneous tissue of the heel; muscular branches; and cutaneous branches to the lateral side of the foot.

The plantar arch gives off four plantar metatarsal arteries, and twigs to the tarsal joints and muscles.

The first plantar metatarsal artery [p. 942] arises where the plantar arch is joined by the deep plantar branch of the dorsalis pedis artery. The second, third, and fourth plantar metatarsal arteries run on the plantar surfaces of the interossei, and each gives off a perforating branch to the corresponding dorsal metatarsal artery before continuing as a common plantar digital artery. At the interdigital clefts, each of these vessels divides into two proper plantar digital arteries for the adjacent sides of the toes. The proper plantar digital artery to the lateral border of the little toe arises from the lateral plantar artery where it becomes the plantar arch.

Perforating branches also arise from the common plantar digital arteries just before they divide.

Anterior tibial artery

The anterior tibial artery begins opposite the distal border of popliteus, and ends in front of the ankle, where it is continued as the dorsalis pedis artery [FIG. 13.87].

Course

The artery passes forwards through an aperture in the interosseous membrane, and descends on the anterior surface of the membrane, the distal part of the tibia, and the front of the ankle joint. It lies lateral to tibialis anterior, and is first medial to extensor digitorum longus and then extensor hallucis longus, the tendon of which crosses anterior to it in its distal third. It then continues between that tendon and the medial tendon of extensor digitorum longus. Two veins accompany the artery.

The deep peroneal nerve approaches the vessel from the lateral side and comes to lie in front of its middle third; more distally the nerve again becomes lateral.

BRANCHES

The circumflex fibular branch may arise from the posterior tibial artery [p. 938].

The posterior tibial recurrent branch, small and inconstant, runs to the back of the knee joint. It arises behind the interosseous membrane.

The anterior tibial recurrent artery arises in front of the interosseous membrane. It runs upwards to the anastomosis round the knee joint [p. 938].

Muscular branches pass to the adjacent muscles, and cutaneous branches supply the skin of the front of the leg.

The medial anterior malleolar artery ramifies over the medial malleolus.

The lateral anterior malleolar artery anastomoses with the perforating branch of the peroneal artery and with the lateral tarsal artery.

On the surface of each malleolus the anastomosing vessels form a well-marked malleolar rete.

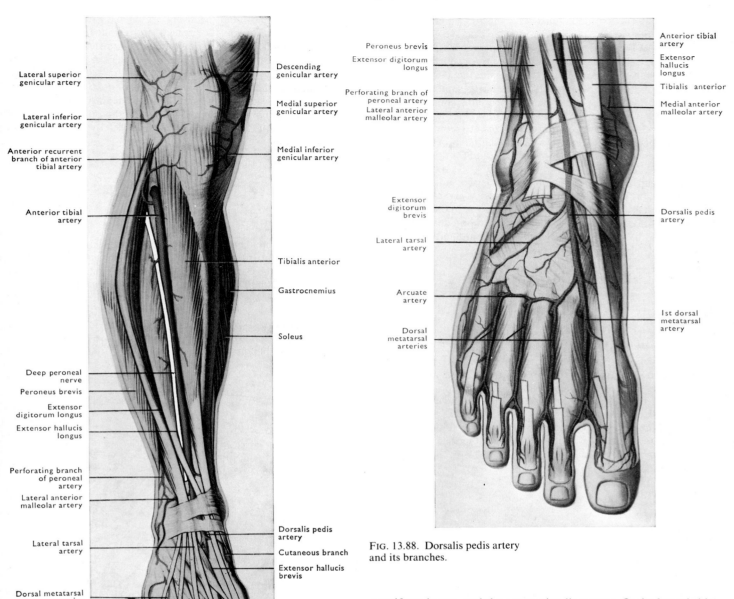

Lateral superior genicular artery

Lateral inferior genicular artery

Anterior recurrent branch of anterior tibial artery

Anterior tibial artery

Deep peroneal nerve

Peroneus brevis

Extensor digitorum longus

Extensor hallucis longus

Perforating branch of peroneal artery

Lateral anterior malleolar artery

Lateral tarsal artery

Dorsal metatarsal arteries

Descending genicular artery

Medial superior genicular artery

Medial inferior genicular artery

Tibialis anterior

Gastrocnemius

Soleus

Dorsalis pedis artery

Cutaneous branch

Extensor hallucis brevis

FIG. 13.87. Anterior tibial artery and its branches.

Peroneus brevis

Extensor digitorum longus

Perforating branch of peroneal artery

Lateral anterior malleolar artery

Extensor digitorum brevis

Lateral tarsal artery

Arcuate artery

Dorsal metatarsal arteries

Anterior tibial artery

Extensor hallucis longus

Tibialis anterior

Medial anterior malleolar artery

Dorsalis pedis artery

1st dorsal metatarsal artery

FIG. 13.88. Dorsalis pedis artery and its branches.

DORSALIS PEDIS ARTERY [FIG. 13.88]

This vessel, the direct continuation of the anterior tibial, begins on the front of the ankle joint and runs to the proximal end of the first intermetatarsal space. Here it gives off the arcuate artery and continues as the first dorsal metatarsal artery which sends its deep plantar branch between the two heads of the first dorsal interosseous muscle to unite with the plantar arch [FIG. 13.86]. Two veins accompany it throughout.

It is covered by skin and fasciae, including the inferior extensor retinaculum, and is crossed by extensor hallucis brevis. It rests upon the ankle joint, the head of the talus, the navicular and intermediate cuneiform bones, and the connecting ligaments. On its lateral side is the medial terminal branch of the deep peroneal nerve separating it from extensor digitorum brevis and the most medial tendon of extendor digitorum longus. On its medial side is the tendon of extensor hallucis longus.

Surface anatomy

The line of the dorsalis pedis artery is from a point opposite the ankle joint, midway between the malleoli, to the proximal end of the first intermetatarsal space. Because of its superficial position and its close contact with the tarsal bones (against which it can be felt pulsating) the artery is frequently divided in wounds of this region.

Branches

Cutaneous branches pass to the skin on the dorsum and medial side of the foot.

The lateral tarsal artery runs laterally, deep to extensor digitorum brevis.

The medial tarsal arteries are small branches which run to the medial border of the foot.

The arcuate artery runs laterally, on the bases of the metatarsals, deep to the long and short extensor tendons. It gives off three dorsal metatarsal arteries, second, third, and fourth, which run to the clefts of the toes, where each divides into two **dorsal digital arteries.** The lateral side of the little toe receives a branch from the most lateral artery. Each dorsal metatarsal artery gives off a **deep plantar branch** through the proximal part of the intermetatarsal space to the corresponding plantar metatarsal artery, and an anterior perforating branch to the end of the corresponding common plantar digital artery.

The **first dorsal metatarsal artery** runs forward from the dorsalis pedis, and ends by dividing into dorsal digital branches for the adjacent sides of the big toe and second toe, and for the medial side of the big toe.

As the deep plantar branch of the first space unites with the plantar arch, the **first plantar metatarsal artery** is given off. It passes forwards, and giving off the **plantar digital artery** to the medial side of the first (big) toe, becomes the **first common plantar digital artery.** This divides into proper **plantar digital arteries** to the adjacent sides of the first and second toes.

Veins

Veins commence in networks of capillaries; but whereas those of the **pulmonary circulation,** and those of the **systemic circulation,** progressively unite to form large trunks which open into the heart, the veins of the **portal system** both begin and end in capillaries [p. 962].

The four pulmonary veins return oxygenated blood from the lungs, and open into the left atrium. They are described on page 892.

The veins of the systemic circulation, and of the portal system, will now be considered.

SYSTEMIC VEINS

The systemic veins return blood to the right atrium of the heart through the **superior vena cava,** the **inferior vena cava,** and the **coronary sinus.** The coronary sinus drains blood only from the walls of the heart, while the venae cavae drain the rest of the body except the lungs.

The veins in general are much more numerous than the arteries and of greater capacity; also they anastomose more freely and in many regions form **venous plexuses** composed of networks of thin-walled, intercommunicating vessels, e.g. the pterygoid plexus [p. 948]. The presence of **valves** in the veins [p. 875] is of importance in the maintenance of the circulation particularly in the extremities.

GENERAL ARRANGEMENT

The veins of the head and neck, the body wall, and limbs form two groups: superficial and deep.

The **superficial veins** lie in the superficial fascia; they begin in the capillaries of the skin and subcutaneous tissues, and are very numerous. They freely anastomose with one another, and they also communicate with the deep veins, in which they end after piercing the deep fascia. They usually do not accompany superficial arteries.

The **deep veins** accompany arteries, except in the cranium. The large arteries have only one companion vein, but with the medium-sized and small arteries there are usually two (**venae comitantes**) which anastomose freely with each other by short transverse

channels which surround the artery—a mechanism which helps to warm blood flowing towards the heart from chilled extremities.

Visceral veins usually accompany the arteries which supply viscera in the head, neck, thorax, and abdomen. As a rule one vein runs with each visceral artery, and, with the exception of those which form the portal system, they end in the deep systemic veins.

The **venous sinuses** which are found inside the cranium [p. 950] differ from ordinary veins in that they have no muscular coat and are simply endothelial channels usually between the dura mater and the endocranium.

CORONARY SINUS AND VEINS OF THE HEART

The coronary sinus [FIGS. 13.10 and 13.89] is a short, but relatively wide, venous trunk which receives the majority of the veins of the heart. It lies in the posterior part of the coronary sulcus, between the left atrium and the left ventricle, and it is covered superficially by some of the muscular fibres of the atrium. Its opening into the right atrium has been described [p. 879].

The apertures of its tributaries are not provided with valves, except those of the great and small cardiac veins, and their valves are often incompetent.

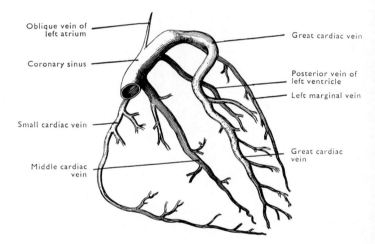

FIG. 13.89. Diagram of veins on surface of heart.

TRIBUTARIES

The **great cardiac vein** [FIG. 13.11] begins at the apex of the heart. It ascends in the anterior interventricular groove to the coronary sulcus, then passing round the left surface of the heart, ends in the left extremity of the coronary sinus. It receives the **left marginal vein** which ascends along the left (pulmonary) surface of the heart.

The **oblique vein of the left atrium** [FIG. 13.10] is a slender vessel which descends on the posterior wall of the left atrium, and ends in the coronary sinus. It is a remnant of the left common cardinal vein, as is the **fold of the left vena cava** [p. 891] with which it is continuous above.

The **posterior vein of the left ventricle** lies to the left of the posterior interventricular sulcus. It usually ends in the coronary sinus, but it may end in the great cardiac vein.

The **middle cardiac vein** begins at the apex of the heart, and runs in the posterior interventricular sulcus to end in the coronary sinus near its right extremity.

The small cardiac vein begins near the apex, and follows the inferior margin of the sternocostal surface to reach the antero-inferior part of the coronary sulcus. It then turns to the left in the posterior part of the sulcus, and ends in the right extremity of the coronary sinus.

VEINS OF THE HEART WHICH DO NOT END IN THE CORONARY SINUS

The anterior cardiac veins are small vessels which ascend on the anterior wall of the right ventricle. Having crossed the coronary sulcus, they end directly in the right atrium [FIG. 13.11]. The venae cordis minimae are small veins which begin in the walls of the heart and end directly in its cavities, principally in the atria. Direct observation of the interior of the left atrium during life has given proof of their functional importance (Butterworth 1954).

SUPERIOR VENA CAVA

The superior vena cava [FIGS. 13.27, 13.90, and 13.92] returns the blood from the upper half of the body. It is formed behind the lower border of the first right costal cartilage by the union of the two brachiocephalic veins, and it opens into the right atrium at the level of the third right cartilage. Its lower half is within the fibrous pericardium, and is covered in front and on the sides by serous pericardium.

Anteriorly, the margins of the right lung and pleura separate it from the chest wall and from the internal thoracic vessels. Posteriorly it is related to the right margin of the trachea, and then to the root of the right lung: and halfway along its length the vena azygos opens into it from behind. The ascending aorta and the brachiocephalic trunk lie to its left side (the aorta partly overlapping it), and on its

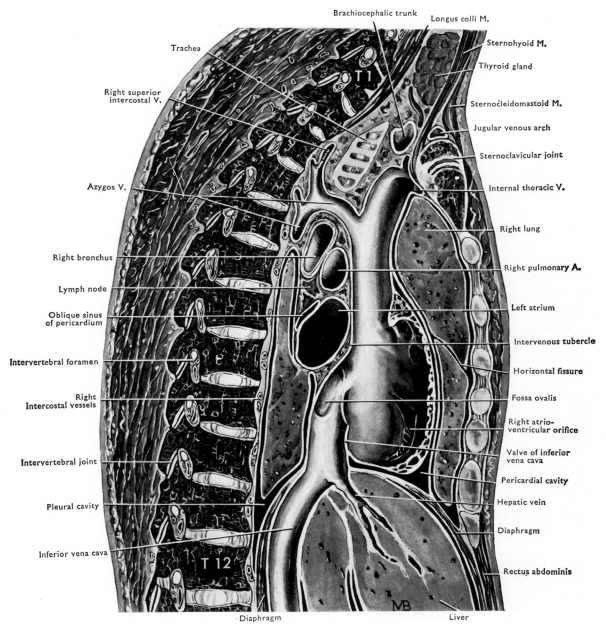

FIG. 13.90. Sagittal section through the superior and inferior venae cavae.

right side the phrenic nerve and the pericardiacophrenic vessels lie between it and the pleura.

Surface anatomy

The superior vena cava lies immediately to the right of the sternal margin, opposite the first and second interspaces and the second and third costal cartilages.

Tributaries

In addition to the brachiocephalic veins, the superior vena cava receives the vena azygos, and several small pericardial and mediastinal veins.

Variations

Persistence of both common cardinal veins results in the presence of two superior venae cavae, the transverse anastomosis which usually forms the left brachiocephalic vein being small or absent. In such cases the left brachiocephalic vein descends across the aortic arch, and being joined by the left superior intercostal vein, becomes the left superior vena cava which traverses the enlarged oblique vein of the left atrium and the coronary sinus to the postero-inferior part of the right atrium. More usually only the upper part of the left vein persists, particularly when, as an anomaly, it drains the corresponding pulmonary veins upwards to the left brachiocephalic vein. The right pulmonary veins may also drain into the corresponding vena cava.

The azygos vein and its tributaries

The vena azygos usually springs from the back of the inferior vena cava at the level of the renal veins; but it may begin as the continuation of the right subcostal vein, or from the junction of that vein and the right ascending lumbar vein. It enters the thorax through the aortic opening of the diaphragm, and having ascended through the posterior mediastinum, arches forwards above the root of the right lung, and enters the back of the superior vena cava at the level of the second costal cartilage [FIG. 13.91].

Posteriorly, it rests upon the lower eight thoracic vertebrae, and on the right posterior intercostal arteries. To its left side are the thoracic duct and the descending aorta, and, as it arches forwards above the root of the right lung, the oesophagus, trachea, and right vagus. To its right are the right lung and pleura and the greater splanchnic nerve. The diaphragm, oesophagus, right pleura and lung, and right lung root are anterior from below upwards in the posterior mediastinum.

TRIBUTARIES

1. The **accessory hemiazygos vein** begins as the continuation of the fourth left posterior intercostal vein. It descends alongside the descending aorta as far as the seventh thoracic vertebra, where, turning sharply to the right, it crosses behind the aorta to end in the azygos vein. It receives the left bronchial veins, and the left posterior

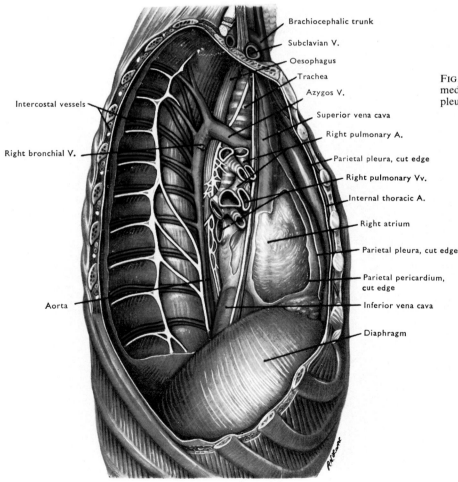

FIG. 13.91. The right side of the mediastinum and vertebral column with pleura and part of pericardium removed.

Brachiocephalic trunk
Subclavian V.
Oesophagus
Trachea
Azygos V.
Superior vena cava
Right pulmonary A.
Parietal pleura, cut edge
Right pulmonary Vv.
Internal thoracic A.
Right atrium
Parietal pleura, cut edge
Parietal pericardium, cut edge
Inferior vena cava
Diaphragm

Intercostal vessels
Right bronchial V.
Aorta

intercostal veins of the fourth, fifth, sixth, and seventh spaces. It communicates above with the left superior intercostal vein.

2. The hemiazygos vein usually springs from the back of the left renal vein and enters the thorax by piercing the left crus of the diaphragm; but it may begin as the continuation of the left subcostal vein or from the junction of that vein and the left ascending lumbar vein. It runs upwards to the eighth thoracic vertebra, where it turns to the right behind the aorta to end in the vena azygos. Its tributaries are the left ascending lumbar vein, the left subcostal vein, and the left posterior intercostal veins of the lower four spaces.

3. The bronchial veins are usually two on each side; those of the right side open into the azygos vein, and those of the left into the accessory hemiazygos vein, or into the left superior intercostal vein. Many veins from the smaller bronchi open into the pulmonary veins.

4. The ascending lumbar vein [p. 956] is a longitudinal channel that connects the lateral sacral, iliolumbar, and lumbar veins. It ascends posterior to psoas major, and ends in the azygos vein on the right side and the hemiazygos vein on the left side.

5. The posterior intercostal veins are eleven on each side. Each lies in the uppermost part of its intercostal space, and receives a lateral cutaneous tributary, and a dorsal tributary which passes forwards, between the transverse processes, from the vertebral plexuses.

The highest posterior intercostal vein ascends in front of the neck of the first rib, and arches over the pleura to end in either the brachiocephalic or the vertebral vein. The second and third (and sometimes the fourth) unite to form the superior intercostal vein. The right superior intercostal vein joins the azygos vein; the left crosses the aortic arch, superficial to the left vagus and deep to the left phrenic nerve, and joins the left brachiocephalic vein. The remaining posterior intercostal veins and the subcostal vein of the right side end in the azygos vein; those of the left side end in the hemiazygos and accessory hemiazygos veins.

The azygos vein also receives small twigs from the oesophagus, pericardium, and diaphragm.

VARIATIONS

The arch of the azygos vein is sometimes enclosed in a fold of pleura and sunk deeply into the right lung, cutting off an accessory lobe— the lobe of the azygos vein [p. 528] medial to it. Left posterior intercostal veins may drain individually into the azygos vein, making the pattern of the hemiazygos veins very variable.

For further information on the azygos and hemiazygos veins consult Gladstone (1929).

Brachiocephalic veins

The brachiocephalic veins [FIGS. 13.27 and 13.57] are two in number, right and left. Each is formed behind the medial end of the clavicle by the union of the internal jugular and subclavian veins; they terminate, at the lower border of the right first costal cartilage, by forming the superior vena cava. They do not possess valves.

THE RIGHT BRACHIOCEPHALIC VEIN

This vessel is just over 2.5 cm in length and runs almost vertically on the right side of the brachiocephalic trunk. Above, it rests on the pleura, and is anterior to the right phrenic nerve and internal thoracic artery; but lower down these structures become lateral relations, the phrenic nerve lying along its right side between it and

the pleura. The right vagus nerve passes behind the upper part of the vein to reach the side of the trachea, and so becomes posteromedial.

Its tributaries are the right vertebral and internal thoracic veins, the highest right intercostal vein, and sometimes the right inferior thyroid vein or a common trunk of the two veins. The right lymph duct (or separate lymph trunks) also opens into it.

THE LEFT BRACHIOCEPHALIC VEIN

This vessel, about 6.5 cm long, runs to the right and downwards behind the upper half of the manubrium sterni, and immediately above the arch of the aorta whose branches it crosses superficially. In the root of the neck it lies in front of the left pleura, the internal thoracic artery, and the phrenic and vagus nerves. Behind the manubrium it is covered by the thymus, and is separated from the trachea by the left common carotid artery and brachiocephalic trunk [FIG. 9.15].

Its tributaries are the thoracic duct (which opens into it at the angle of junction of the internal jugular and subclavian veins), the vertebral, internal thoracic, inferior thyroid, and superior intercostal veins of its own side, the highest left intercostal vein, and some pericardial and thymic veins. Sometimes the right inferior thyroid vein joins it.

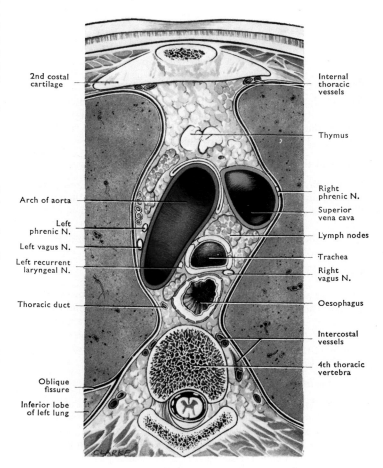

FIG. 13.92. A horizontal section at the level of the fourth thoracic vertebra.

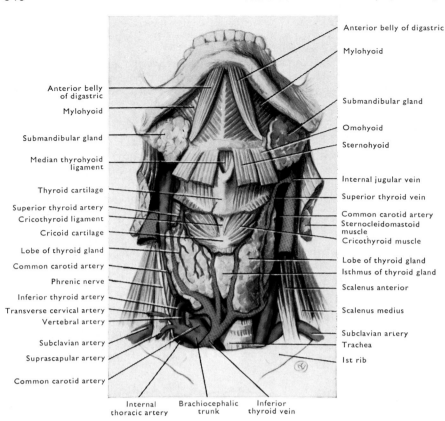

Anterior belly of digastric
Mylohyoid
Submandibular gland
Omohyoid
Sternohyoid
Internal jugular vein
Superior thyroid vein
Common carotid artery
Sternocleidomastoid muscle
Cricothyroid muscle
Lobe of thyroid gland
Isthmus of thyroid gland
Scalenus anterior
Scalenus medius
Subclavian artery
Trachea
Ist rib

Anterior belly of digastric
Mylohyoid
Submandibular gland
Median thyrohyoid ligament
Thyroid cartilage
Superior thyroid artery
Cricothyroid ligament
Cricoid cartilage
Lobe of thyroid gland
Common carotid artery
Phrenic nerve
Inferior thyroid artery
Transverse cervical artery
Vertebral artery
Subclavian artery
Suprascapular artery
Common carotid artery

Internal thoracic artery
Brachiocephalic trunk
Inferior thyroid vein

FIG. 13.93. Dissection of front of neck. The lower portions of the sternocleidomastoid muscles and right common carotid artery have been removed to show the deeper parts.

Surface anatomy

The left brachiocephalic vein lies behind the left sternoclavicular joint and the upper part of the manubrium sterni; the right vein lies behind the right sternoclavicular joint and the first right costal cartilage. The veins are about 1.5 cm wide.

Internal thoracic veins

Each internal thoracic artery is accompanied by two veins which unite behind the third costal cartilage; the single vessel then runs on the medial side of the artery to the brachiocephalic vein of the same side.

The tributaries of the internal thoracic veins correspond with the branches of the internal thoracic arteries [p. 914].

Vertebral veins

These vessels correspond only to the extracranial parts of the vertebral arteries. Each consists of a plexus of veins which surrounds the artery in the foramina of the cervical transverse processes. Inferiorly, the plexus forms a single trunk which emerges from the foramen in the transverse process of the sixth cervical vertebra, and descends to the brachiocephalic vein.

Tributaries

The vertebral vein receives tributaries from the external and internal vertebral plexuses, and also communicates with the suboccipital venous plexus. The anterior vertebral vein, which lies alongside the ascending cervical artery, ends in the vertebral vein immediately below the foramen in the sixth cervical transverse process. The deep cervical vein begins in the suboccipital venous plexus and descends beside the deep cervical artery. At the root of the neck it passes forwards between the seventh cervical transverse process and the neck of the first rib to join the vertebral vein.

The venous plexus around the vertebral artery may end below in

two terminal trunks. The second vessel, the accessory vertebral vein, passes through the foramen in the transverse process of the seventh cervical vertebra (which otherwise transmits a small tributary), and joins the brachiocephalic vein.

Inferior thyroid veins [FIG. 13.93]

These two vessels, formed by tributaries from the isthmus and lobes of the gland, descend in front of the trachea. The right vein ends either in the right brachiocephalic vein or in the junction of the two brachiocephalic veins, and the left in the left brachiocephalic vein. The two veins may unite to form a single trunk, which ends usually in the left brachiocephalic vein. The inferior thyroid veins are connected by anastomosing channels, and sometimes form a plexus in front of the trachea.

Veins of the head and neck

INTERNAL JUGULAR VEIN [FIGS. 13.40, 13.93, 13.98, and 13.159]

The internal jugular vein returns the blood from the brain, face, and much of the neck. Each begins in the jugular foramen as the continuation of the sigmoid sinus, and ends behind the medial end of the clavicle by uniting with the subclavian to form the brachiocephalic vein. Its commencement is dilated and forms the superior bulb. A second dilatation, called the inferior bulb, marks its lower end; it is bounded above by a valve of two cusps.

Surface anatomy

The position of the vein corresponds to a line drawn from the medial end of the clavicle to the interval between the ramus of the mandible and the mastoid process.

Course

Throughout the vein is enclosed within the carotid sheath, so that its course corresponds closely with that of its companion artery, and only certain points need be mentioned.

Immediately below the skull the internal carotid artery and the last four cranial nerves are anteromedial to the vein; but thereafter, it is in contact, medially, first with the internal and then with the common carotid artery, the vagus nerve lying posteromedially, between it and the large arteries.

Sternocleidomastoid overlaps the vein above, and covers it below; numerous deep cervical lymph nodes lie along its course, mostly on its superficial surface.

Just below the transverse process of the atlas it is crossed, on its lateral side, by the accessory nerve; about its middle it is crossed by the inferior root of the ansa cervicalis; and near its lower end by the anterior jugular vein which runs between it and the infrahyoid muscles.

Posterior to the vein are the cervical transverse processes and their attached muscles, the phrenic nerve as it descends on scalenus anterior, and the first part of the subclavian artery; and, on the left side, the terminal part of the thoracic duct.

Jugulogram

Needle puncture of the internal jugular vein with compression of its lower part results in reflux of the contrast medium upwards into the skull, thus filling the internal jugular vein, the intracranial venous sinuses and their tributaries. Such a procedure is undertaken in pathological conditions which may affect these venous sinuses.

Tributaries

The inferior petrosal sinus joins it at its commencement. Two or three pharyngeal veins drain the pharyngeal venous plexus. The common terminal trunk of the lingual veins returns part of the blood from the tongue, and may be joined by the vein which accompanies the hypoglossal nerve and usually ends in the facial vein [p. 424]. The superior thyroid vein accompanies the corresponding artery; it receives the superior laryngeal vein. The middle thyroid vein passes backwards across the common carotid artery.

The facial vein [FIG. 13.94] begins at the medial angle of the eye by the union of the supra-orbital and supratrochlear veins. It runs behind the facial artery, and with a much straighter course. After crossing the body of the mandible at the anterior border of masseter, it leaves the facial artery and passes superficial to the submandibular gland. A little below the angle of the jaw it is joined by the anterior division of the retromandibular vein [p. 948], and continues across the external and internal carotid arteries to end in the internal jugular vein.

The facial vein receives tributaries corresponding with all the branches of the facial artery, except the ascending palatine and the tonsillar which have no accompanying veins. It communicates with the cavernous sinus [p. 953] through the superior ophthalmic vein, and also through the deep facial vein which passes backwards from it, deep to the ramus of the mandible, to the pterygoid plexus. Thrombosis of the facial vein—consequent upon sepsis—may thus spread to the cavernous sinus.

SUBCLAVIAN VEIN

The subclavian vein is the continuation of the main vein of the upper limb, i.e. the axillary vein.

It commences at the outer border of the first rib, and ends behind the medial end of the clavicle by joining the internal jugular vein to form the brachiocephalic vein. Its surface marking may be gauged by reference to FIGURE 13.17.

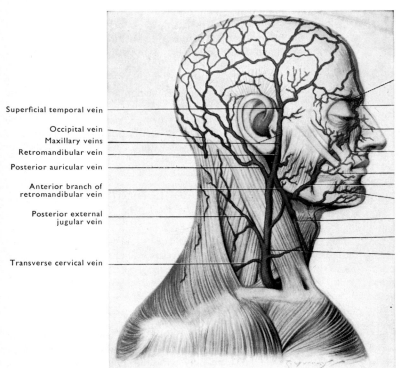

Superficial temporal vein
Occipital vein
Maxillary veins
Retromandibular vein
Posterior auricular vein
Anterior branch of retromandibular vein
Posterior external jugular vein
Transverse cervical vein

Supra-orbital vein
Facial vein
External nasal vein
Superior labial vein
Inferior labial vein
Facial vein
Inferior labial vein
Anastomosis between facial and anterior jugular veins
Anterior jugular vein
External jugular vein

FIG. 13.94. Superficial veins of head and neck.

Each subclavian vein possesses a bicuspid valve immediately lateral to the junction with the external jugular vein.

The subclavian vein lies behind subclavius and the medial part of the clavicle. It rests on the first rib, below and in front of the third part of the subclavian artery, and then on scalenus anterior which separates it from the second part of the artery. As soon as it reaches the medial border of the muscle it unites with the internal jugular vein.

Tributaries

The subclavian vein usually has only one named tributary—the external jugular vein.

EXTERNAL JUGULAR VEIN

The external jugular vein [FIG. 13.94] is formed by the union of the posterior auricular vein with the posterior division of the retromandibular vein. It descends vertically, deep to platysma, across sternocleidomastoid on which it lies in front of and parallel with the great auricular nerve. Reaching the subclavian triangle it pierces the deep fascia and ends in the subclavian vein.

It usually has two valves, one at its termination, and a second 4 cm higher.

Surface anatomy

The external jugular vein—usually visible through the skin, or easily made so by compressing it just above the clavicle, or by raising the intrathoracic pressure—runs in a line from the angle of the jaw to the junction of the medial and middle thirds of the clavicle.

Tributaries

The posterior auricular vein and the posterior external jugular vein drain the areas shown in FIGURE 13.94.

The transverse cervical and suprascapular veins accompany the corresponding arteries.

The anterior jugular vein is formed in the submental region, and descends, in the superficial fascia, at a variable distance from the median plane. Inferiorly, it enters the suprasternal space [p. 308] where it is connected with its fellow of the opposite side by a transverse channel, the jugular venous arch. It then turns laterally, deep to the sternocleidomastoid muscle, and superficial to the sternohyoid and sternothyroid muscles, to end in the external jugular vein.

The external jugular vein sometimes receives the cephalic vein or a communication from it.

Variations

The external jugular vein may be absent or smaller than usual, and if so the anterior or internal jugular vein is enlarged.

VEINS OF THE SCALP

The veins which drain the superficial parts of the scalp are the supratrochlear, the supra-orbital, the superficial temporal, the posterior auricular, and the occipital. The deeper part of the scalp, in the region of the temporal fossa, drains into the deep temporal veins, which pass medial to the temporalis muscle into the pterygoid plexus.

The supratrochlear and supra-orbital veins unite to form the facial vein [p. 947]. A branch of the supra-orbital vein passes through the supra-orbital notch (where it receives the frontal diploic vein—page 949) and joins the superior ophthalmic vein.

The superficial temporal vein [FIGS. 13.40 and 13.94] descends to the upper border of the zygomatic arch, immediately anterior to the auricle. Here it receives the middle temporal vein and crosses the arch to enter the parotid gland. Within the gland it unites with the transverse facial and maxillary veins to form the retromandibular vein.

The occipital vein [FIGS. 13.40 and 13.94] pierces trapezius, and passes into the suboccipital plexus which is drained by the vertebral and deep cervical veins. Occasionally it continues along the occipital artery to end in the internal jugular vein.

VEINS OF THE ORBIT, NOSE, AND INFRATEMPORAL REGION

The veins of these regions are closely associated, since the ophthalmic veins, which drain the orbit, are connected with the pterygoid plexus.

Veins of orbit

With the exception of the supra-orbital and supratrochlear, the veins of the orbit correspond with the branches of the ophthalmic artery. They converge to form two main trunks—a superior ophthalmic vein and an inferior ophthalmic vein—which pass to the cavernous sinus through the superior orbital fissure. The central vein of the retina joins the superior ophthalmic or runs direct to the sinus.

Veins of the nose

The majority of the internal nasal veins join to form the sphenopalatine vein which passes via the sphenopalatine foramen and pterygopalatine fossa to the pterygoid plexus; some end in the ethmoidal tributaries of the superior ophthalmic vein, and others in the septal tributaries of the superior labial and in the external nasal veins.

Pterygoid plexus and maxillary vein

The pterygoid plexus of veins lies in the infratemporal fossa on the lateral surface of medial pterygoid and around lateral pterygoid. Its tributaries correspond with many of the branches of the maxillary artery. It communicates with the cavernous sinus by emissary veins through the foramen ovale; with the inferior ophthalmic vein through the inferior orbital fissure; and with the facial vein by the deep facial branch. Posteriorly it forms and drains into the maxillary vein.

The maxillary vein accompanies the first part of the maxillary artery, and unites with the superficial temporal vein to form the retromandibular vein.

The retromandibular vein, so formed, descends through the parotid gland within which it bifurcates. The anterior division joins the facial vein, and the posterior division forms one of the vessels of origin of the external jugular vein.

Venous sinuses and veins of the cranium and contents

The venous channels met with in the cranial walls and cavity are: the diploic veins, the meningeal veins, the cranial venous sinuses, and the veins of the brain.

DIPLOIC AND MENINGEAL VEINS

Diploic veins are anastomosing spaces, lined with endothelium, in the marrow cavity, i.e. diploë, of the flat bones of the skull [FIG. 13.95] and may be seen especially in radiographs of older individuals. There are usually at least four different vessels on each side (Jefferson and Stewart 1928).

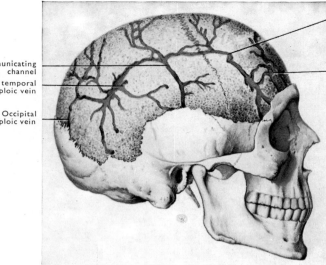

FIG. 13.95. Veins of the diploë.

The **frontal diploic vein** passes through a small aperture in the upper margin of the supra-orbital notch and ends in the supra-orbital vein.

The **anterior temporal diploic vein** ends either in the spheno-parietal sinus or in the anterior deep temporal vein [FIG. 3.58].

The **posterior temporal diploic vein** ends either in the transverse sinus, or, more commonly, in the mastoid emissary vein.

The **occipital diploic vein** is usually the largest of the series, and ends either in the occipital emissary vein or in the transverse sinus.

Small unnamed diploic veins pierce the inner table of the skull and enter the venous sinuses and meningeal veins.

Meningeal veins

These vessels arise from the diploë and from plexuses in the fused dura mater and periosteum. Some terminate in cranial venous sinuses, while others run with the meningeal arteries and end in extracranial veins. Those which accompany the middle meningeal artery separate that vessel from the bone, and are readily torn when the bone is fractured; they leave the skull through the foramen ovale or foramen spinosum and end in the pterygoid plexus.

Veins of the brain

The veins of the brain are thin-walled, valveless vessels which open into the cranial venous sinuses.

VEINS OF THE CEREBRUM

The cerebral veins are arranged in two groups—deep and superficial. The deep veins issue from the interior of the brain. The superficial veins lie upon its surface in the pia mater and the subarachnoid space. The terminal trunks of both sets pierce the arachnoid and the dura mater, and open into the venous sinuses.

The deep cerebral veins

These are the choroid veins, the thalamostriate veins, the internal cerebral veins, the great cerebral vein, and the striate veins.

Each **choroid vein** drains the choroid plexus of a lateral ventricle. It begins in the inferior horn and curves round with the plexus to the interventricular foramen; it there receives efferents from the choroid plexus of the third ventricle, and unites with the thalamostriate vein to form the internal cerebral vein.

The **thalamostriate vein**, on each side, is formed by tributaries from the corpus striatum, the cerebral white matter, and thalamus. It runs with the stria terminalis, immediately deep to the ependyma of the lateral ventricle, in the groove between the thalamus and caudate nucleus. It ends by joining the choroid vein.

Each **internal cerebral vein** runs backwards between the layers of the tela choroidea of the third ventricle. Beneath the splenium of the corpus callosum it unites with its fellow of the opposite side to form the great cerebral vein.

The **great cerebral vein** (Galen) curves upwards round the splenium and ends in the anterior extremity of the straight sinus. It receives the basal, midbrain, and superior cerebellar veins of each side [see below].

The superficial cerebral veins.

These vessels lie on the surface of the cerebrum, and are divisible into two sets—the superior and the inferior [FIG. 13.96].

The **superior cerebral veins**, from six to twelve in number on each side, lie on the upper and lateral aspect of the cerebral hemispheres. They converge on and enter the superior sagittal sinus, and those that encounter the lacunae laterales [p. 952] pass beneath them to reach the sinus.

The **inferior cerebral veins** lie on the tentorial and inferolateral aspects of the hemispheres, and they end in the cavernous, the superior petrosal, and the transverse sinuses.

The **superficial middle cerebral vein** runs superficially along the posterior ramus of the lateral sulcus and the sulcus itself to the cavernous sinus; frequently it is united either by a **superior anastomotic vein** with the superior sagittal sinus, or by an **inferior anastomotic vein** with the transverse sinus [FIG. 13.96].

The **basal vein** is formed at the anterior perforated substance by the union of: (1) the **anterior cerebral vein** which accompanies the corresponding artery; (2) the **deep middle cerebral vein** which lies deeply in the lateral sulcus and drains the insula; and (3) the **striate veins** which emerge through the anterior perforated substance. So formed, the basal vein passes backwards round the cerebral peduncle and ends in the great cerebral vein.

Veins of midbrain

These end either in the great cerebral vein or in the basal veins.

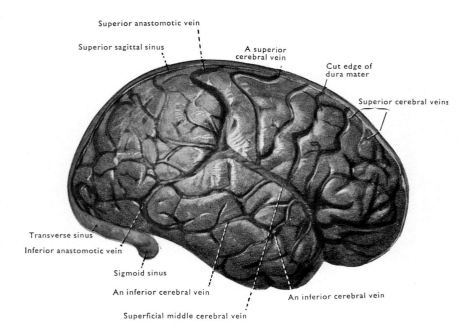

FIG. 13.96. Veins of superolateral surface of right cerebral hemisphere seen through arachnoid mater.

Cerebellar veins

Those veins form two sets, **superior** and **inferior**. The former end in the great cerebral vein, and in the transverse, straight, and superior petrosal sinuses; and the latter in the inferior petrosal, transverse, and occipital sinuses.

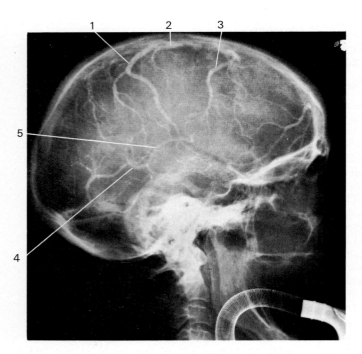

FIG. 13.97. Lateral view of cerebral angiogram showing the contrast medium in the veins after passing through the arteries.
1. Superior anastomotic vein.
2. Superior sagittal sinus.
3. Superior cerebral vein.
4. Great cerebral vein.
5. Internal cerebral vein.

Veins of pons

The veins of the pons pass to a ventral plexus which drains superiorly to the basal veins, and inferiorly to the cerebellar veins or petrosal sinuses.

Veins of medulla oblongata

The veins of the medulla end in a superficial plexus which is drained by an anterior and a posterior median vein and by radicular veins. The median veins are continuous with the corresponding veins of the spinal medulla; above, the anterior vein communicates with the plexus on the pons, and the posterior divides into two branches which join the inferior petrosal sinuses or the basilar plexus. The radicular veins run with the roots of the last four cranial nerves, and end in the inferior petrosal and occipital sinuses or in the upper part of the internal jugular vein.

Cerebral venography

The cerebral veins are often shown adequately in the later films of a series of angiograms [FIG. 13.97]. It is sometimes necessary to fill these veins through the internal jugular vein by retrograde injection (jugulogram; p. 947).

Venous sinuses of the dura mater

The cranial venous sinuses are spaces between the dura mater and the periosteum lining the skull (endocranium), and they are lined with an endothelium continuous with that of the veins. They receive the veins of the brain, communicate with the meningeal and diploic veins and with veins external to the cranium by emissary veins, and end directly or indirectly in the internal jugular vein. Some of the sinuses are unpaired, others are paired.

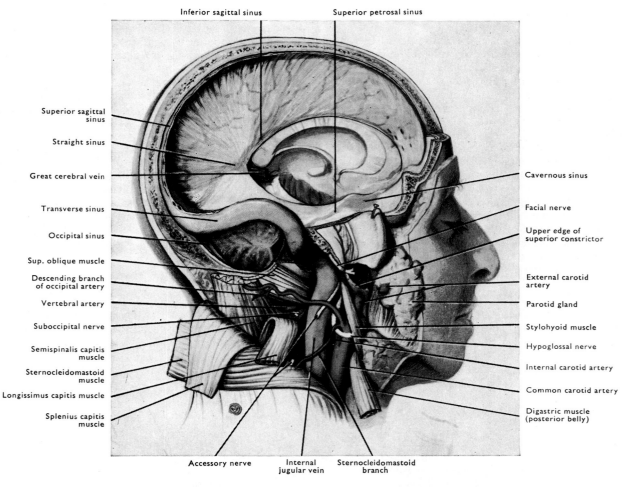

Inferior sagittal sinus   Superior petrosal sinus

Superior sagittal sinus

Straight sinus

Great cerebral vein

Transverse sinus

Occipital sinus

Sup. oblique muscle

Descending branch of occipital artery

Vertebral artery

Suboccipital nerve

Semispinalis capitis muscle

Sternocleidomastoid muscle

Longissimus capitis muscle

Splenius capitis muscle

Cavernous sinus

Facial nerve

Upper edge of superior constrictor

External carotid artery

Parotid gland

Stylohyoid muscle

Hypoglossal nerve

Internal carotid artery

Common carotid artery

Digastric muscle (posterior belly)

Accessory nerve   Internal jugular vein   Sternocleidomastoid branch

FIG. 13.98. Dissection of head and neck, showing venous sinuses of dura mater and superior part of internal jugular vein.

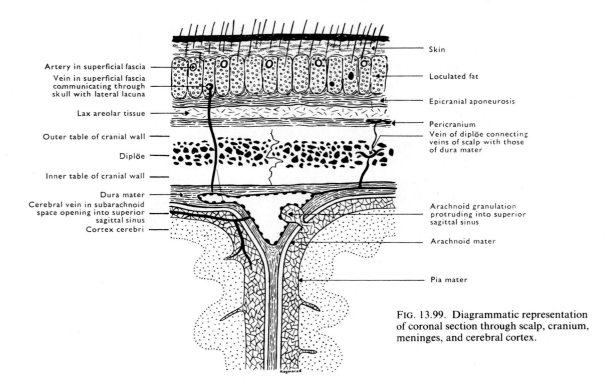

Artery in superficial fascia

Vein in superficial fascia communicating through skull with lateral lacuna

Lax areolar tissue

Outer table of cranial wall

Diplöe

Inner table of cranial wall

Dura mater

Cerebral vein in subarachnoid space opening into superior sagittal sinus

Cortex cerebri

Skin

Loculated fat

Epicranial aponeurosis

Pericranium

Vein of diplöe connecting veins of scalp with those of dura mater

Arachnoid granulation protruding into superior sagittal sinus

Arachnoid mater

Pia mater

FIG. 13.99. Diagrammatic representation of coronal section through scalp, cranium, meninges, and cerebral cortex.

UNPAIRED SINUSES

The **superior sagittal sinus** begins at the crista galli, where it may communicate, through the foramen caecum, with the veins of the frontal sinus, and sometimes with those of the nasal cavity. It arches backwards in the fixed margin of the falx cerebri, grooving the cranial vault [FIG. 13.98], and ends at the internal occipital protuberance by becoming the right transverse sinus. Occasionally it ends in the left transverse sinus, and sometimes in both. Its termination may show a dilatation, called the **confluence of the sinuses**, which is then connected by an anastomosing channel with a similar dilatation which marks the junction of the straight sinus with the transverse sinus of the opposite side. The superior cerebral veins open into the superior sagittal sinus, and it communicates on each side by slit-like openings with a series of spaces between the dura and the periosteum, the *lacunae laterales*, into which arachnoid granulations also project [p. 731; FIG. 10.185]. The parietal emissary veins [p. 953] link it with the veins on the exterior of the cranium. Its cavity is triangular in transverse section, and is crossed by several fibrous strands.

Surface anatomy

The superior sagittal sinus occupies the median plane of the vertex from the glabella to the inion. Narrow in front, it broadens to a width of 1 cm behind.

The small **inferior sagittal sinus** runs in the posterior two-thirds of the free edge of the falx cerebri. It receives tributaries from the falx and from the medial surface of each hemisphere, and ends behind by joining with the great cerebral vein to form the straight sinus.

The **straight sinus** runs downwards and backwards in the line of union of the falx cerebri with the tentorium cerebelli, and usually becomes continuous with the left transverse sinus. It may end in the right transverse sinus, and then the superior sagittal sinus ends in the left, or they all unite.

Two **intercavernous sinuses**, an anterior and a posterior, lie in the corresponding margins of the diaphragma sellae; others pass across the floor of the hypophysial fossa [FIG. 9.20].

The **basilar plexus** lies on the clivus of the skull. It connects the cavernous and the inferior petrosal sinuses of the two sides, and communicates with the internal vertebral venous plexus through the foramen magnum.

PAIRED SINUSES

Each **transverse sinus**, beginning at the internal occipital protuberance, passes laterally in the attached border of the tentorium cerebelli, in a groove which runs from the occipital bone to the posterior inferior angle of the parietal bone; it then leaves the

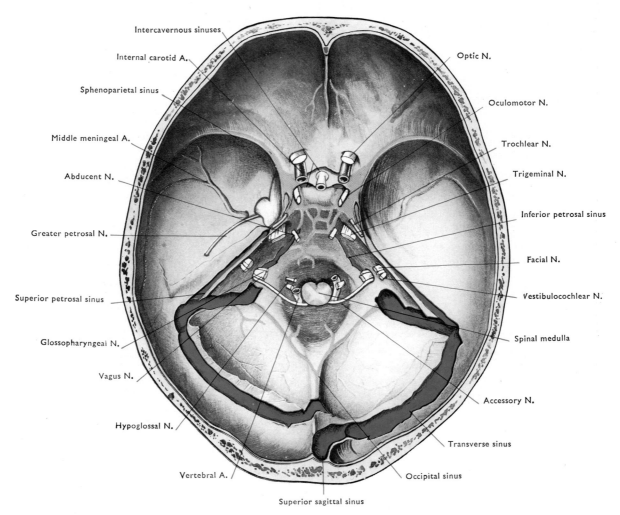

FIG. 13.100. The venous sinuses of the dura mater.

tentorium and becomes the sigmoid sinus. Its tributaries are some inferior cerebral and cerebellar veins, the posterior temporal diploic vein, the inferior anastomotic vein, and, as it becomes the sigmoid sinus, the superior petrosal sinus.

The sigmoid sinus first runs downwards and medially, then turning forwards enters the jugular foramen to become continuous with the internal jugular vein [FIG. 13.98]. It gives rise to emissary veins which pass through the mastoid foramen and the condylar canal.

The transverse sinus may be marked by a line, slightly convex upwards, drawn from just above the inion to the upper part of the back of the root of the auricle. The sigmoid sinus is indicated by a continuation of this line along the back of the root of the auricle to the level of the lower margin of the meatus, and then forwards to meet that part of the margin, opposite the jugular foramen [FIG. 13.44].

The occipital sinuses lie in the attached border of the falx cerebelli—in which they may be fused—and in the dura along the posterolateral boundaries of the foramen magnum. Each links the transverse and sigmoid sinuses of its own side, and both communicate with the posterior internal vertebral plexus through the foramen magnum.

The cavernous sinuses [FIG. 9.21] lie at the sides of the body of the sphenoid bone. Each sinus begins at the superior orbital fissure, and ends at the apex of the petrous temporal by dividing into the superior and inferior petrosal sinuses. Its cavity is so divided by fibrous strands as to give the appearance of cavernous tissue. In its lateral wall are embedded the oculomotor, the trochlear, the ophthalmic, and maxillary nerves, and passing through the sinus are the internal carotid artery with its sympathetic plexus, and the abducent nerve. The ophthalmic, and inferior cerebral veins (including the superficial middle cerebral vein) and the sphenoparietal sinus enter the cavernous sinus. It communicates with its fellow through the intercavernous sinuses; with the pterygoid plexus by emissary veins [see below]; with the internal jugular vein by the inferior petrosal sinus; and through the superior ophthalmic vein with the facial vein.

The sphenoparietal sinuses run along the lesser wings of the sphenoid bone close to their posterior borders. Each ends in the anterior part of the cavernous sinus.

Each superior petrosal sinus runs in the attached margin of the tentorium cerebelli to the junction of the transverse and sigmoid sinuses. It lies on the superior margin of the petrous temporal bone, and receives inferior cerebral, superior cerebellar, and small tympanic veins.

Each inferior petrosal sinus begins at the posterior end of the cavernous sinus; it runs backwards, laterally, and downwards, in the groove between the temporal and occipital bones, to the anterior compartment of the jugular foramen, through which it passes, to end in the internal jugular vein. Its tributaries include cerebellar veins, pontine veins, and the labyrinthine vein.

EMISSARY VEINS

Emissary veins connect the venous sinuses with the veins on the exterior of the skull. They are of clinical importance, for they may serve as pathways for the spread of infection from outside the skull to within.

The parietal emissary veins, one on each side, pass through the parietal foramina and connect the superior sagittal sinus with the occipital veins. A mastoid emissary vein connects the sigmoid sinus, through the mastoid foramen, with the occipital or the posterior auricular vein. A condylar emissary vein passes through the condylar canal when that canal is present, and connects the sigmoid sinus with the suboccipital plexus of veins.

The cavernous sinus is connected to the pterygoid and pharyngeal plexuses by veins which traverse the foramen ovale, the foramen lacerum, and the sphenoidal emissary foramen when it is present.

VEINS OF THE VERTEBRAL COLUMN

The veins of the vertebral column form freely communicating plexuses along its length, both inside and outside the vertebral canal. Accordingly, internal and external plexuses are described, in each of which anterior and posterior plexuses are recognized.

Internal vertebral plexuses

These form a continuous network between the dura mater and the walls of the vertebral canal.

In the anterior internal vertebral plexus there are two longitudinal channels which communicate with each other by transverse channels anterior to the posterior longitudinal ligament. Two less obvious longitudinal channels can sometimes be distinguished as the posterior internal vertebral plexus on the internal surfaces of the vertebral arches and the ligamenta flava. Both of these plexuses communicate with the corresponding venous sinuses of the dura mater through the foramen magnum.

External vertebral plexuses

The anterior external vertebral plexus lies on the anterior surfaces of the vertebral bodies; the posterior external vertebral plexus lies

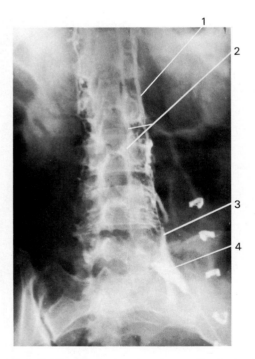

FIG. 13.101. Vertebral venous plexuses demonstrated by contrast medium introduced into the left femoral vein. The left common iliac vein is blocked by a blood clot so blood takes a retrograde course through the iliolumbar vein into the vertebral venous plexuses, ascending through them to the azygos system.
1. Ascending lumbar vein.
2. Internal vertebral venous plexus.
3. Iliolumbar vein.
4. Left common iliac vein.

on the posterolateral surfaces of the vertebrae, and around the spines and transverse processes.

The basivertebral veins are wide, thin-walled channels that arise within the haemopoietic tissue of the vertebral bodies. In front they communicate with the anterior external venous plexus, but their major drainage is through a large foramen in the back of the vertebral body into the transverse anastomoses of the anterior internal vertebral plexus.

Intervertebral veins

The intervertebral veins, which convey blood from the internal vertebral plexuses, pass through the intervertebral foramina and open into the vertebral veins, the dorsal branches of the posterior intercostal, lumbar, and the lateral sacral veins.

The veins of the vertebral plexuses possess few if any valves and are in widespread connection with veins of many parts of the body. The work of Batson (1957) strongly indicates that, *inter alia*, the mode of spread of cancer from such organs as the prostate and breast may well be explained by the nature of the vertebral plexuses and their connections. These connections also provide a collateral circulation should either vena cava become obstructed [see also p. 728].

VERTEBRAL VENOGRAPHY

Filling of the internal vertebral venous plexus through intervertebral veins occurs with forcible injection of the common iliac veins indicating that this system of veins is valveless and transmits blood in either direction. Where there is obstruction of the inferior vena cava, the whole of the venous return from the lower limbs and pelvis may ascend by this system, which normally drains the vertebral column and its contents and communicates with the external vertebral venous plexus through each intervertebral foramen, thus linking tributaries of the superior and inferior venae cavae [FIG. 13.101].

This system of veins is a route through which certain malignant tumour cells may be disseminated, particularly to the vertebral column.

Veins of the spinal medulla

The veins of the spinal medulla end on the pia mater in a plexus in which there are six longitudinal channels—one anteromedian,

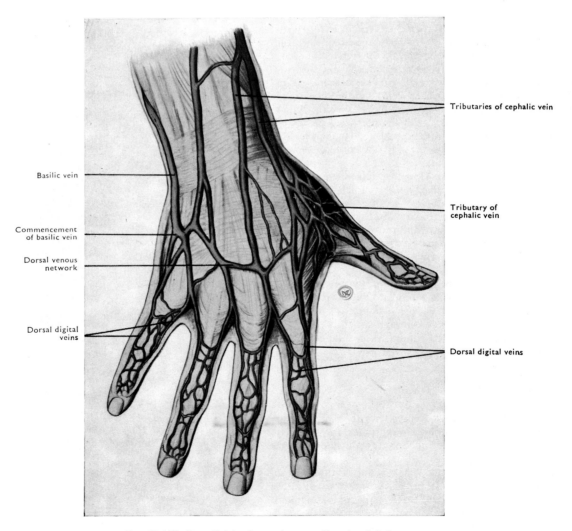

FIG. 13.102. Superficial veins on dorsum of hand and digits.

along the anterior median fissure; two **anterolateral**, immediately behind the ventral nerve roots; two **posterolateral**, immediately behind the dorsal nerve roots; and one **posteromedian**, behind the posterior median sulcus. Radicular efferent vessels pass along the nerve roots to communicate with the internal vertebral venous plexus. For further details Woollam and Millen (1958) should be consulted.

Veins of upper limb

The veins of the upper limb are divisible into two sets, **superficial** and **deep**, both sets of which open eventually into a common trunk known as the **axillary vein**.

DEEP VEINS OF UPPER LIMB

The deep veins, with the exception of the axillary vein, are arranged in pairs which accompany the arteries and are similarly named. They possess valves, and lie one on each side of their companion artery across which they are usually united by numerous transverse anastomoses.

AXILLARY VEIN

The **axillary vein** [FIGS. 13.40 and 13.153] begins, as the continuation of the basilic vein, at the lower border of teres major, and ends at the outer border of the first rib by becoming the subclavian vein.

It runs on the medial side of the axillary artery, overlapping it slightly. Between it and the artery are the medial cord of the brachial plexus, the medial cutaneous nerve of the forearm, and the ulnar nerve; but the medial cutaneous nerve of the arm, having crossed either superficial or deep to the vein runs on its medial side. The lateral group of axillary nodes lie close alongside.

Tributaries

Its tributaries correspond with the branches of the axillary artery, except the thoraco-acromial which joins the cephalic vein; it also receives the brachial veins at the lower border of subscapularis, and the cephalic vein above the upper border of pectoralis minor.

SUPERFICIAL VEINS OF UPPER LIMB

Veins of digits and hand

The palmar digital veins [FIG. 13.103] drain to a fine plexus in the superficial fascia of the palm, and also, by means of **intercapitular veins** which pass between the heads of the metacarpal bones, to the dorsal digital veins.

The **dorsal digital veins**, two in each digit, anastomose freely and unite to form a number of **dorsal metacarpal veins** which end in a **dorsal venous network** of variable shape [FIG. 13.102].

Veins of forearm and upper arm

The veins of the forearm emerge from the dorsal venous network and from the palmar venous plexus. They vary considerably, but as a rule there are two main longitudinal channels—the **cephalic vein** on the radial side and the **basilic vein** on the ulnar side. Sometimes there is a **median vein** on the front of the forearm.

The **cephalic vein** begins in the dorsal venous network, receives the dorsal veins of the thumb, and ascends in the anterolateral part of the forearm to the elbow. It then runs along the lateral border of biceps to the interval between deltoid and pectoralis major, along which it ascends to the infraclavicular fossa. Turning medially, it pierces the clavipectoral fascia and ends in the axillary vein.

A number of tributaries join it in the forearm; at the elbow it is connected with the basilic vein by the **median cubital vein**; and in the infraclavicular fossa it is joined by tributaries which correspond with the branches of the thoraco-acromial artery.

The **median cubital vein** slants upwards and medially superficial to the bicipital aponeurosis. It is connected to the deep veins of the forearm, and receives one or more superficial veins from the front of the forearm.

When the median cubital vein is large, the upper part of the cephalic vein is correspondingly small.

The **basilic vein** begins in the dorsal venous network of the hand. It ascends along the ulnar border of the back of the forearm for

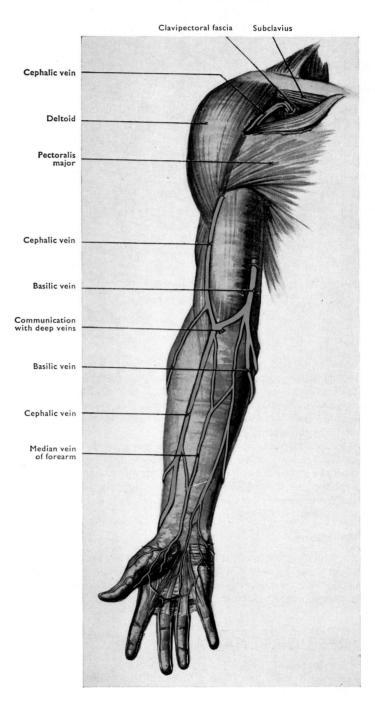

FIG. 13.103. Superficial veins on flexor aspect of right upper limb. The median cubital vein connects the cephalic and basilic veins.

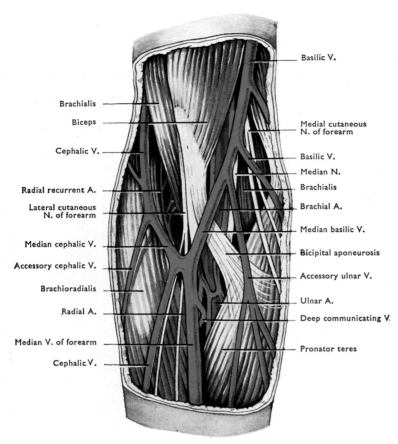

FIG. 13.104. Superficial veins at the elbow.
The median vein is large.

about two-thirds of its extent, and then inclines forwards to the medial bicipital groove. At the middle of the arm it pierces the deep fascia, and ascends on the medial side of the brachial artery to the axilla where it becomes the axillary vein.

It is joined by tributaries from the forearm, and in front of the elbow by the median cubital vein.

The median vein of the forearm, when present, begins in the palmar venous plexus and runs to the elbow. It may join either the median cubital vein or the basilic, or it may divide into the median cephalic and the median basilic veins which end in the cephalic and basilic veins respectively. The median basilic vein then replaces the median cubital vein [FIG. 13.104].

Variations

The superficial veins of the forearm are extremely variable, and any of them may be absent. The cephalic vein may end in the external jugular vein, or it may be connected with it by a vessel which passes over or through the clavicle.

INFERIOR VENA CAVA

The inferior vena cava [FIGS. 13.105 and 13.106] returns the blood from the lower limbs, and from the walls and viscera of the abdominopelvic cavity—most of the visceral return joining it after passing through the portal system.

It begins on the body of the fifth lumbar vertebra, immediately to the right of the median plane, and ends in the right atrium very shortly after passing through the diaphragm.

It first ascends on the vertebral column, and on the anterior edge of the right psoas major, and is in contact with the right side of the aorta; higher up, however, it becomes separated from the aorta by the right crus on which it continues forwards and upwards to the caval opening in the central tendon of the diaphragm at the level of the eighth thoracic vertebra. It covers the right lumbar sympathetic trunk, and the right renal and middle suprarenal arteries cross behind it. Anterior to it are the right common iliac artery, the mesentery and the superior mesenteric vessels, the right gonadal artery, and the horizontal part of the duodenum. It then passes behind the head of the pancreas, the bile-duct, the portal vein, the superior part of the duodenum, and the epiploic foramen. Finally it occupies a deep groove on the back of the liver, between the caudate and right lobes. To its right side there are the right ureter and kidney, and the right suprarenal gland which also lies partly behind it.

Obstruction of the inferior vena cava enlarges the collateral circulation whereby blood may pass to the heart via the superior vena cava. Thus the epigastric, circumflex iliac, lateral thoracic, thoraco-epigastric [p. 961], and internal thoracic veins all become enlarged and may become exceedingly prominent; and a deep by-pass is provided by the azygos and hemiazygos veins and their tributaries, principally the vertebral plexuses.

Variations and abnormalities

Occasionally the inferior vena cava is continuous with a much enlarged vena azygos and all the inferior caval blood is then carried to the superior vena cava. In these cases the hepatic veins open into the right atrium by a channel which represents the upper end of the inferior vena cava.

Such a variation—and there are many more, both of the inferior vena cava and of its tributaries—find an explanation in the complex mode of development of the inferior vena caval system [p. 975].

Tributaries

In addition to the common iliac veins, the inferior vena cava receives the third and fourth lumbar veins of both sides, the right gonadal vein, the right and left renal veins, the right suprarenal vein, the right inferior phrenic vein, and the hepatic veins.

There are usually five lumbar veins on each side, one with each aortic lumbar artery and one with the lumbar branch of the median sacral artery. Their mode of termination varies. The upper two may end in the ascending lumbar vein, or in the azygos or the hemiazygos vein; usually the third and fourth end in the inferior vena cava; the fifth ends in the iliolumbar vein. And they are all united by a longitudinal anastomosing vessel, the ascending lumbar vein, that lies deep to psoas major.

Each ascending lumbar vein begins in the lateral sacral vein of the same side, anastomoses with the iliolumbar vein, connects the lumbar veins together, and anastomoses with the inferior vena cava and the renal vein. The right ascending lumbar vein ends in the azygos vein and the left in the hemiazygos vein.

The testicular veins, on each side, issue from the testis and epididymis and form the pampiniform plexus in the spermatic cord. It consists of from eight to ten veins, most of which lie in front of the ductus deferens, and ends near the deep inguinal ring in two main trunks which ascend with the corresponding testicular artery; the two veins soon unite to form a single vein which on the right side opens into the inferior vena cava, and on the left side into the left renal vein.

The much greater frequency of varicocele, that is, a varicose condition of the pampiniform plexus, on the left side, is said to be

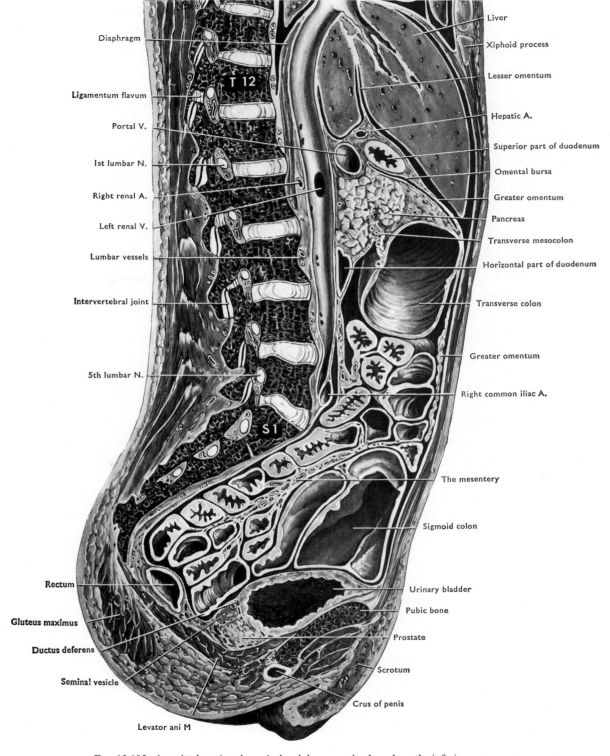

Diaphragm

Ligamentum flavum

Portal V.

1st lumbar N.

Right renal A.

Left renal V.

Lumbar vessels

Intervertebral joint

5th lumbar N.

Rectum

Gluteus maximus

Ductus deferens

Seminal vesicle

Levator ani M

T 12

S 1

Liver

Xiphoid process

Lesser omentum

Hepatic A.

Superior part of duodenum

Omental bursa

Greater omentum

Pancreas

Transverse mesocolon

Horizontal part of duodenum

Transverse colon

Greater omentum

Right common iliac A.

The mesentery

Sigmoid colon

Urinary bladder

Pubic bone

Prostate

Scrotum

Crus of penis

FIG. 13.105. A sagittal section through the abdomen and pelvis along the inferior vena cava.

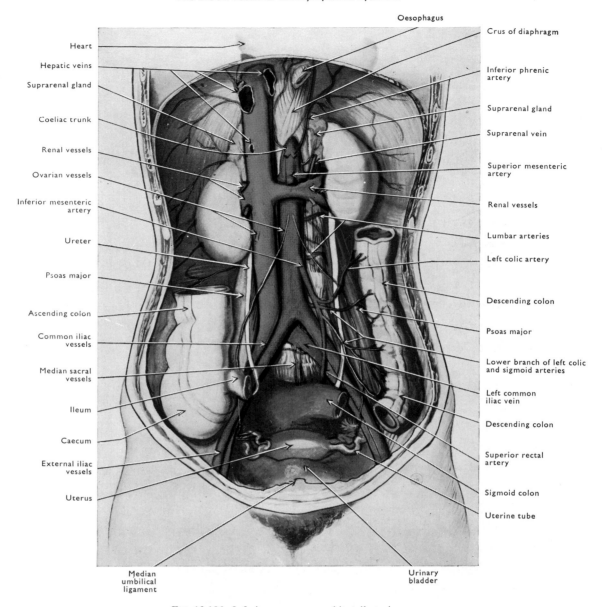

FIG. 13.106. Inferior vena cava and its tributaries.

associated with the fact that the left testicular vein joins the left renal at right angles; moreover, it is overlaid by the descending colon.

The **ovarian veins**, on each side, issue from the hilus of the ovary. Between the layers of the broad ligament they form the **pampiniform plexus**, from which two veins accompany the ovarian artery. They soon unite to form a single vein which ends, on the right side, in the inferior vena cava, and on the left side in the left renal vein.

The **renal veins** are formed by the union of five or six tributaries which issue from the hilus of each kidney.

The **right renal vein**, about 2.5 cm long, lies behind the descending part of the duodenum, and ends in the right side of the inferior vena cava.

The **left renal vein**, about 7.5 cm long, crosses in front of the aorta immediately behind the superior mesenteric artery. It runs behind the body of the pancreas and above the horizontal part of the duodenum to end in the left side of the inferior vena cava. The left gonadal vein and the left suprarenal vein open into it.

A single **suprarenal vein** issues from the hilus of each gland; the right vein is very short and ends in the back of the inferior vena cava; the left descends to join the left renal vein but may cross directly to the inferior vena cava.

The **inferior phrenic veins** correspond to the inferior phrenic arteries; the right vein ends in the inferior vena cava, the left usually in the left suprarenal vein.

The **hepatic veins** [FIG. 13.106] open into the inferior vena cava immediately below the diaphragm as it lies in the back of the liver. They have no extrahepatic course. They form two groups, upper and lower. The upper group may consist of only two veins, a right and a left; more frequently there is also a middle vein which issues from the caudate lobe. All three are large trunks. The veins of the lower group vary in number from six to twenty and are smaller; they return blood from the right and caudate lobes.

Detailed accounts of the hepatic veins are given by Elias and Petty (1952), and Gibson (1959) deals with their sphincter mechanisms.

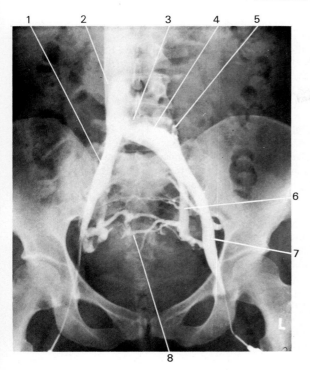

FIG. 13.107. Cavogram. An anteroposterior radiograph of pelvis and lower abdomen immediately after the simultaneous injection of contrast medium into both femoral veins. The filling defect in the left common iliac vein where it joins the inferior vena cava is caused by compression from the right common iliac artery.

1. Right common iliac vein.
2. Inferior vena cava.
3. Defect.
4. Left common iliac vein.
5. Iliolumbar vein.
6. Internal iliac vein.
7. External iliac vein.
8. Pelvic venous plexus.

Common iliac veins

The common iliac veins [FIGS. 13.79 and 13.106], right and left, are formed by the union of the external and internal iliac veins. Each begins at the brim of the lesser pelvis immediately behind the upper part of the internal iliac artery, and both veins pass superomedially to the front of the body of the fifth lumbar vertebra where they unite to form the inferior vena cava to the right of the midline.

The right common iliac vein is much shorter than the left; it lies anterior to the obturator nerve and the iliolumbar artery, and posterior to the right common iliac artery.

The left common iliac vein runs first on the medial side of its own artery, and then behind the right common iliac artery where it ends. Near its origin it is crossed by the sigmoid mesocolon and the superior rectal vessels.

Tributaries

Each common iliac vein receives the corresponding external iliac, internal iliac, and iliolumbar veins. The left common iliac vein receives, in addition, the median sacral vein, the single vessel formed by the fusion of the companion veins of the median sacral artery.

The iliolumbar vein receives the fifth lumbar vein, and ends in the back of the common iliac vein.

Cavograms

Simultaneous injection of contrast medium into the basilic veins of both arms results in opacification of both brachiocephalic veins and the superior vena cava as far as the right atrium where the contrast medium becomes diluted by the blood returning from the inferior vena cava. The inferior vena cava can likewise be filled by catheters inserted percutaneously into both common iliac veins through the femoral and external iliac veins [FIG. 13.107].

INTERNAL ILIAC VEIN [FIG. 13.75]

This short trunk is formed by the union of tributaries which correspond to all the branches of the internal iliac artery except the iliolumbar and umbilical arteries.

It begins at the upper border of the greater sciatic notch and ends at the margin of the superior aperture of the lesser pelvis by uniting with the external iliac to form the common iliac vein. It lies immediately posterior to the internal iliac artery, and is crossed laterally by the obturator nerve. The left vein is in contact medially with the sigmoid colon, the right with the lower part of the ileum.

Tributaries

These form extrapelvic and intrapelvic groups.

The extrapelvic tributaries are all parietal; they are:

1. The superior gluteal veins and the inferior gluteal veins. These large vessels accompany the corresponding arteries through the greater sciatic foramen. The inferior gluteal veins usually, and the superior gluteal veins frequently, unite to form a single vein before ending in the internal iliac.

2. The obturator vein enters the lesser pelvis through the obturator canal, and runs along the side-wall of the pelvis immediately below the obturator artery.

3. The internal pudendal veins accompany the internal pudendal artery, and usually unite on the ischial spine to form a single vessel that joins the internal iliac vein. They arise in the deep perineal space by the union of the deep vein of the penis, or clitoris, with the vein from the bulb, and at their origin are in communication with the prostatic or vesical venous plexus. They receive tributaries corresponding to the branches of the artery with the exception of the deep dorsal vein of the penis or clitoris.

The inferior rectal veins begin in the substance of the external anal sphincter and in the walls of the anal canal; they anastomose through the submucous rectal plexus with the middle and superior rectal veins, and so connect the portal and systemic veins.

The intrapelvic tributaries of the internal iliac vein are the parietal lateral sacral veins, and the visceral efferent vessels from the plexuses around the pelvic viscera.

1. The lateral sacral veins accompany the lateral sacral arteries and end in the corresponding internal iliac vein. They unite with the internal vertebral venous plexuses and with the visceral plexuses.

2. The pelvic venous plexuses form dense networks of thin-walled veins associated with the rectum, bladder, prostate, uterus, and vagina; they communicate freely with one another.

The rectal plexuses consist of an internal plexus in the submucosa of the rectum and anal canal, and an external plexus on the outer surface of their muscular coats; free communication exists between the two. The rectal plexuses are drained by the superior, middle, and inferior rectal veins; the superior rectal vein joins the portal system, whereas the middle and inferior are tributaries of the systemic veins. The main drainage of the internal plexus is upwards by a series of large vessels which ascend in the anal columns, and then, after a further upward course in the submucosa, pierce the muscle layers and unite to form the superior rectal vein. Varicosity of the internal rectal plexus results in haemorrhoids or piles.

3. The **middle rectal veins** are inconstant. When present they begin in the submucosa of the rectum, where they communicate with the superior and inferior rectal veins in the rectal plexus; they pass through the muscular coat, and fuse to form two middle rectal veins, right and left, which end in the corresponding internal iliac vein.

The **prostatic plexus** is situated mainly within the fascial sheath of the prostate [FIG. 8.51]. In front it receives the deep dorsal vein of the penis, and behind it communicates with the vesical plexus. Efferent vessels pass to the internal iliac and internal pudendal veins, but it drains mainly through the vesical plexus.

The **vesical plexus** lies on the outer surface of the muscular coat of the lower part of the bladder, and communicates either with the prostatic plexus or with the vaginal plexus; it receives the dorsal vein of the clitoris.

4. Several **vesical veins** pass from it on each side to the internal iliac vein.

The **uterine plexuses** lie within the broad ligament, and they communicate with the ovarian and vaginal plexuses.

5. The **uterine veins**, usually two on each side, issue from the uterine plexuses and end in the internal iliac vein.

The **vaginal plexuses** lie at the sides of the vagina. They communicate with the uterine, vesical, and rectal plexuses.

6. A **vaginal vein** issues from the vaginal plexus on each side, and ends in the internal iliac vein.

Dorsal veins of the penis

There are two dorsal veins of the penis.

The **superficial dorsal vein** runs immediately beneath the skin, to the pubic symphysis, where it divides into right and left branches which end in the superficial external pudendal veins.

The **deep dorsal vein** lies beneath the deep fascia. It begins in the sulcus behind the glans, and runs backwards in the mid-dorsal line in the sulcus between the corpora cavernosa. At the root of the penis it passes below the arcuate pubic ligament, and ends by dividing into two branches which join the prostatic plexus [FIG. 8.46].

The **dorsal vein of the clitoris** has a similar course to that of the deep dorsal vein of the penis. It ends in the vesical plexus.

EXTERNAL ILIAC VEIN [FIGS. 13.75, 13.79, and 13.106]

This vessel is the continuation of the femoral vein, and so begins immediately behind the inguinal ligament. It ascends along the margin of the superior aperture of the lesser pelvis, and, opposite the sacro-iliac joint, it joins the internal iliac vein to form the common iliac vein. The right vein, at first medial to its artery, becomes posterior to it above; the left vein is medial to its artery throughout. In its whole course, on each side, the vein lies anterior to the obturator nerve. Its tributaries, the deep circumflex iliac and inferior epigastric veins, open into it close to its commencement. In addition, it frequently receives the pubic vein which links the obturator and external iliac veins, and may form the main termination of the obturator vein.

Veins of lower limb

The veins of the lower limb, like those of the upper limb, are arranged in two groups, **superficial** and **deep**; but there are important differences in detail, for in the lower limb each main artery has two companion veins only as far as the knee, where a single trunk, the **popliteal vein**, is formed. It is continued through the thigh as the **femoral vein**.

Minor differences should not obscure certain basic similarities in the superficial venous drainage of the upper and lower limbs, for the basilic and cephalic veins have their counterparts, known respectively as the small and great saphenous veins.

DEEP VEINS OF LOWER LIMB

Two veins accompany each of the arteries of the limb, except the popliteal, femoral, and profunda femoris which have single veins similarly named. They possess numerous valves.

The **popliteal vein** [FIG. 13.83] is formed, at the distal border of popliteus, by the union of the anterior and posterior tibial veins. As it ascends through the popliteal fossa it crosses from medial to lateral behind the popliteal artery—to which it is bound by a dense fascial sheath. It then passes through the opening in adductor magnus and becomes the femoral vein.

Its tributaries correspond with the branches of the popliteal artery, and it receives also the small saphenous vein.

The **femoral vein** [FIG. 13.108] begins at the opening in adductor magnus. It ascends through the adductor canal and the femoral triangle, and ends a little medial to the midinguinal point by becoming the external iliac vein.

In the adductor canal it lies at first posterolateral to the femoral artery and then posterior to it. As it enters the femoral triangle it is still behind the artery, but in the proximal part of the triangle it is on the medial side of the femoral artery. About 3 cm below the inguinal ligament it enters the middle compartment of the femoral sheath.

Its **tributaries** are: (1) the profunda femoris vein; (2) the great saphenous vein; (3) the venae comitantes of the descending genicular, lateral, and medial circumflex femoral, and the deep external pudendal arteries. (4) several small tributaries from muscles.

The **profunda femoris vein** lies in front of its artery and separates it from adductor longus and the femoral vein. It receives tributaries corresponding to the branches of its artery, but only occasionally the circumflex femoral veins.

SUPERFICIAL VEINS OF LOWER LIMB

The superficial veins of the lower limb terminate in two trunks, one of which, the **small saphenous vein**, passes from the foot to the back of the knee; the other, the **great saphenous vein**, extends from the foot to the groin. These veins often become varicose (i.e. dilated and tortuous).

The superficial veins of the sole form a fine plexus (**plantar rete**), immediately under the skin, from which efferents pass medially to the great saphenous vein, laterally to the small saphenous vein, and anteriorly to the **plantar venous arch**. The arch, which lies at the root of the toes, also receives small **plantar digital veins**, and it communicates between the heads of the metatarsals with the veins on the dorsum.

The superficial veins on the dorsal aspect of each toe form two **dorsal digital veins**. Those of the adjacent borders of the interdigital clefts unite to form four **dorsal metatarsal veins** which end in the dorsal venous arch of the foot.

The **dorsal venous arch** lies opposite the anterior parts of the shafts of the metatarsal bones. It unites medially with the medial dorsal digital vein of the big toe to form the great saphenous vein, and laterally with the lateral dorsal digital vein of the little toe to form the small saphenous vein. The arch receives the dorsal metatarsal veins and efferents from the plantar venous arch; and numerous tributaries from the wide-meshed **dorsal venous rete** are continuous with it posteriorly.

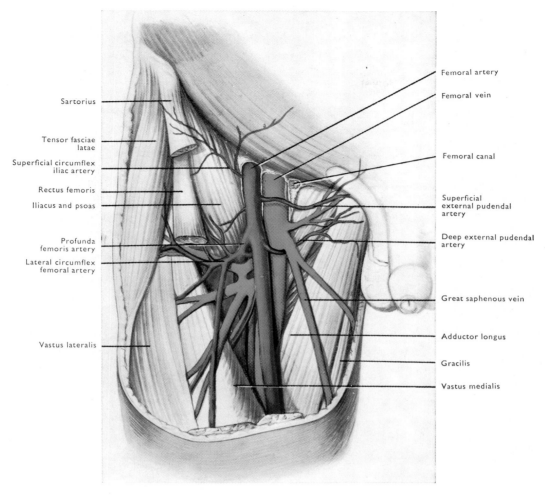

FIG. 13.108. Femoral vessels in the femoral triangle.

Labels on figure:

Sartorius

Tensor fasciae latae

Superficial circumflex iliac artery

Rectus femoris

Iliacus and psoas

Profunda femoris artery

Lateral circumflex femoral artery

Vastus lateralis

Femoral artery

Femoral vein

Femoral canal

Superficial external pudendal artery

Deep external pudendal artery

Great saphenous vein

Adductor longus

Gracilis

Vastus medialis

Great saphenous vein

The great saphenous vein [FIG. 13.109], formed as stated above, passes in front of the medial malleolus, ascends obliquely across the medial surface of the distal third of the shaft of the tibia, and reaching the knee, lies medial to the posterior part of the medial condyle of the femur. Continuing upwards, forwards, and laterally in the thigh, it perforates the cribriform fascia and the femoral sheath and ends in the femoral vein. In the foot and leg it is accompanied by the saphenous nerve. It has from eight to twenty bicuspid valves, but these become incompetent if the vein is varicose.

Tributaries

It communicates through the deep fascia with the deep, intermuscular veins at several points, particularly near the ankle and knee joints. Tributaries join it from the medial part of the sole and from the dorsal venous network, from the medial and posterior parts of the heel, the front of the leg, and the calf; and it anastomoses freely with the small saphenous vein. Cockett and Jones (1953) point out that the subcutaneous veins of the lower half of the leg on the inner side—the area prone to ulceration—normally drain not into the great saphenous vein but directly into the posterior tibial veins through three easily definable perforating veins. In the thigh it receives numerous unnamed tributaries, and also the **superficial circumflex iliac, superficial epigastric,** and **superficial external**

pudendal veins which join it just before it perforates the cribriform fascia [FIG. 13.109].

The tributaries of the superficial epigastric vein communicate above with those of the lateral thoracic veins which join the axillary vein; e.g. the **thoraco-epigastric vein** on the anterolateral surface of the trunk. These communications unite vessels which pass to the inferior vena cava and the superior vena cava respectively; and they may enlarge if there is any obstruction in either of the venae cavae [p. 956].

Small saphenous vein

The small saphenous vein [FIG. 13.110] passes with the sural nerve, along the lateral side of the foot and below the lateral malleolus. It then runs behind the lateral malleolus, and along the lateral border of the tendo calcaneus, to the middle of the calf, whence it reaches the inferior angle of the popliteal fossa, pierces the deep fascia and ends in the popliteal vein. It communicates with the great saphenous vein, and through the deep fascia with the deep veins. It contains from six to twelve bicuspid valves.

Tributaries

It receives tributaries from the lateral side of the foot, the lateral side and back of the heel, and the back of the leg. Just before it pierces the popliteal fascia it frequently gives off a branch which on the medial side of the thigh unites with a vein of some size to form

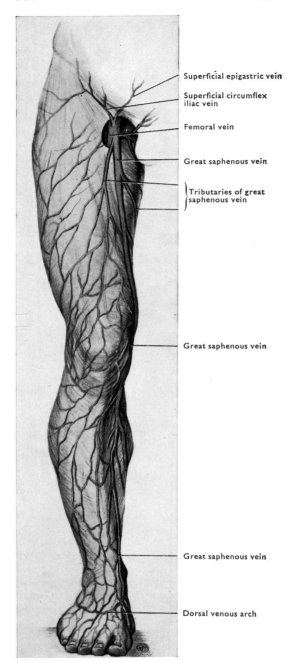

Superficial epigastric **vein**

Superficial circumflex **iliac** vein

Femoral vein

Great saphenous **vein**

Tributaries of **great** saphenous vein

Great saphenous vein

Great saphenous vein

Dorsal venous arch

FIG. 13.109. Great saphenous vein and its tributaries.

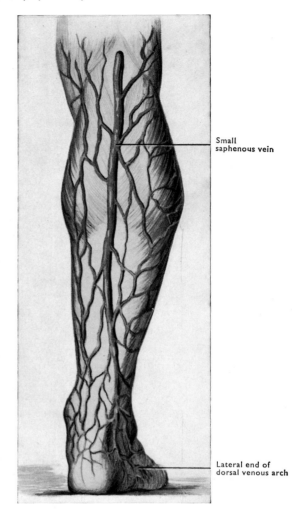

Small saphenous **vein**

Lateral end of dorsal venous **arch**

FIG. 13.110. Small saphenous vein and its tributaries.

the **accessory saphenous vein**, a channel through which a communication is established between the great and small saphenous veins.

THE PORTAL SYSTEM OF VEINS

The veins which, with their tributaries, form the portal system are the portal, the superior and inferior mesenteric, and the splenic. They convey blood to the liver: (1) from the whole of the abdominal and pelvic parts of the alimentary canal, except the terminal part of the anal canal; (2) from the pancreas; and (3) from the spleen. All the larger vessels of this system are devoid of valves in the adult.

Unlike other veins, the portal vein ends by breaking up into branches which ultimately terminate in the sinusoidal capillaries in the substance of the liver. From these capillaries, which also receive the blood conveyed to the liver by the hepatic artery, the hepatic veins arise and open into the inferior vena cava. Thus the portal blood ultimately reaches the general systemic circulation.

The portal vein conveys products of digestion from the intestine, and of red blood cell destruction from the spleen to the liver. In addition, the whole portal system also acts as a reservoir of blood for the needs of the general circulation. It may be demonstrated radiographically by means of a splenoportogram [FIG. 13.112].

The portal vein

This wide channel, about 7.5 cm long, is formed by the union of the superior mesenteric and the splenic veins behind and to the left of the head of the pancreas. It ascends in front of the inferior vena cava and behind the pancreas and the superior part of the duodenum, until, just above the duodenum, it passes forwards and enters the leser omentum. Continuing upwards, it lies behind the bile-duct and proper hepatic artery, and in front of the epiploic foramen. At the right end of the porta hepatis it ends by dividing into a short and wide right branch and a longer and narrower left branch.

The **right branch** receives the cystic vein, and then enters the right lobe of the liver in which it breaks up into numerous branches which end as interlobular veins; these are connected to the tributaries of the hepatic veins by the sinusoids [p. 480].

The **left branch** runs from right to left, in the porta hepatis, giving off branches to the caudate and quadrate lobes; it crosses the fissure for the round ligament of the liver, and ends in the left lobe of the liver in the same manner as the right branch. As it crosses the fissure it is joined in front by the **round ligament of the liver** and some small accompanying veins, and behind by the **ligamentum venosum**. The round ligament of the liver is a fibrous cord which passes from the umbilicus to the left branch of the portal vein; it is the remains of the umbilical vein of the foetus. The small para-umbilical veins which accompany the round ligament connect the left branch of the portal vein with the superficial veins round the umbilicus. The ligamentum venosum connects the left branch of the portal vein

with the upper part of the inferior vena cava. It is the remains of a foetal blood vessel—the **ductus venosus**—through which blood, carried from the placenta by the umbilical vein, is passed to the inferior vena cava without going through the liver sinusoids. The portal vein is accompanied by numerous lymph vessels, and it is surrounded, in the lesser omentum, by filaments of the hepatic plexus of nerves.

For further details on the distribution of the portal vein within the liver Elias and Popper (1955) should be consulted.

Tributaries

The **left gastric vein** runs with its artery as far as the coeliac trunk, and then with the common hepatic artery to the superior part of the duodenum where it joins the portal vein.

The **right gastric vein** runs with its artery along the lesser curvature, and ends in the portal vein as that vessel enters the lesser

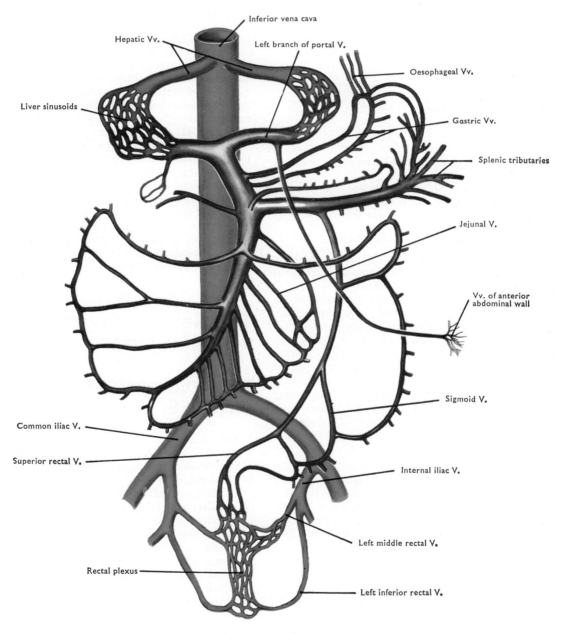

FIG. 13.111. Diagram of the portal venous system (grey) to show its anastomoses with the systemic venous system (blue).

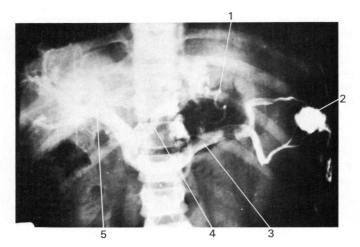

FIG. 13.112. Splenoportogram, contrast medium injected into the spleen enters the tributaries of the splenic vein, thus outlining the splenic and portal veins and some of their tributaries.
1. Left gastric vein, filled because of its communication with systemic veins.
2. Medium in splenic pulp.
3. Splenic vein.
4. Left gastric vein.
5. Branches of portal vein in porta hepatis.

omentum. It receives the prepyloric vein which marks the position of the pylorus [p. 445].

The cystic vein ascends along the cystic duct and ends as a rule in the right branch of the portal vein. Some small veins pass directly from the gall-bladder into the substance of the liver.

The superior mesenteric vein

This large vessel ascends along the right side of the superior mesenteric artery, and has a similar course. Behind the pancreas it unites with the splenic vein to form the portal vein.

Its tributaries are the veins that accompany the branches of the superior mesenteric artery, and also the right gastro-epiploic vein and pancreaticoduodenal veins, which enter it near its termination.

The right gastro-epiploic vein runs from left to right in the greater omentum, near the greater curvature of the stomach. Near the pylorus it turns downwards and backwards, and ends in the superior mesenteric vein.

The pancreaticoduodenal veins run with their corresponding arteries; the lower vein ends either in the superior mesenteric or in the right gastro-epiploic, and the upper vein either in the superior mesenteric or in the portal vein.

The splenic vein

This vessel is formed by the union of five or six tributaries which issue from the hilus of the spleen. It passes in the lienorenal ligament [p. 443] to the kidney, turns to the right and runs behind the body of the pancreas, below the splenic artery. It is separated from the aorta by the superior mesenteric artery and the left renal vein, and ends by joining the superior mesenteric vein to form the portal vein.

Its tributaries correspond to the branches of the splenic artery, and in addition it receives the inferior mesenteric vein.

The inferior mesenteric vein is continuous with the superior rectal vein on the left common iliac artery, and it receives the sigmoid veins from the sigmoid colon and lower portion of the descending colon, and the left colic vein from the upper part of the descending colon and the left flexure. It begins on the middle of the left common iliac artery, and ascends to the left of its own artery and

to the right of the ureter. It lies on the left psoas major, and crosses superficial to the left gonadal vessels. Near its upper end it passes behind, or to the left of the duodenojejunal flexure, and then, crossing in front of the left renal vein, it ends in the splenic vein behind the pancreas. Occasionally it ends in the junction of the splenic and superior mesenteric veins. Its relation to the duodenal peritoneal recesses has been mentioned on page 455.

The superior rectal vein is formed by tributaries of the rectal venous plexuses, through which it communicates with the middle and inferior rectal veins. It ascends with the superior rectal artery between the layers of the sigmoid mesocolon, and in front of the left common iliac artery it becomes the inferior mesenteric vein.

Communications between portal and systemic veins

The most notable of these venous anastomoses occur at the lower end of the oesophagus (between the left gastric vein and oesophageal veins which join the azygos system) and in the rectal plexus of veins (between the superior rectal vein and the middle and inferior rectal veins which pass to the internal iliac) [FIG. 13.111]. In both situations the communicating veins may enlarge; in the anal region such varicose enlargements of the submucous vessels are known as haemorrhoids or piles and in the oesophagus are called oesophageal varices. But whereas oesophageal varices are usually due to portal back pressure, piles seldom are.

The communication with superficial veins around the umbilicus by para-umbilical veins which pass along the round ligament of the liver has been mentioned already. Obstruction to the passage of blood through the liver may be indicated by the appearance of enlarged veins radiating from the umbilicus—the so-called caput medusae. Enlarged veins may also appear in the falciform ligament or on the 'bare area' of the liver; they communicate with the veins of the diaphragm and the internal thoracic veins. Other communications occur between intestinal tributaries of the portal vein and the renal or lumbar veins.

The development of the blood vascular system

THE DEVELOPMENT OF THE HEART AND ARTERIES

A general account of the development of the primitive vascular system has been given on pages 62–65; before considering the further development of the heart and blood vessels certain points will be amplified.

The first evidence of the heart is found between the primitive pericardial cavity dorsally and the yolk sac ventrally as a network of cells from which two endothelial tubes are formed. Later, the tubes fuse into a single heart tube, which, following the formation of the head fold, comes to lie dorsal to the pericardial cavity. Meanwhile, the pericardial wall next to the primitive heart tube becomes thickened to form the myo-epicardial mantle, and, for a time, it is separated from the endothelial heart tube by a loose mesh of endocardial tissue. Later, as the heart tube sinks into the pericardial cavity, the myo-epicardial mantle forms the myocardium and epicardium.

A series of dilatations now appear on the surface of the median tubular heart, and the following parts can be recognized from the

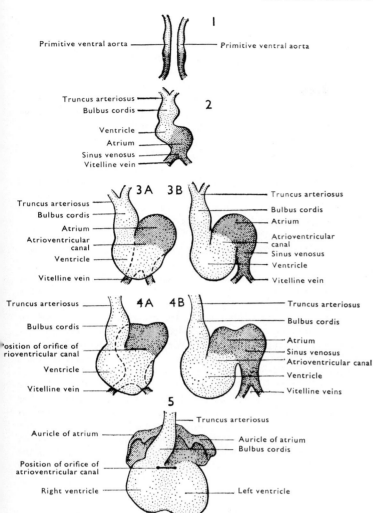

FIG. 13.113. Development of the heart.

Diagrams showing the changes in form and relation of parts at different stages. Modified from His's models. 3B and 4B are side views; the other figures represent the heart as seen from the front.

caudal end forwards: the sinus venosus in the septum transversum; the atrium, the narrow atrioventricular canal, the ventricle, and the bulbus cordis in the pericardial cavity; and the truncus arteriosus which lies beyond the pericardium.

It has been seen [pp. 63–64] that the truncus is linked through the short ventral aortae and first pair of aortic arches to the dorsal aortae. In the human embryo the ventral aortae soon fuse to form a dilated vessel, called the **aortic sac**, which is in continuity with the truncus arteriosus. When, in succession, the remaining five pairs of aortic arches are formed behind the first pair, they pass between the aortic sac and the dorsal aortae.

After its formation the heart elongates more rapidly than the pericardium and so assumes the form of a loop, the apex of which is formed by the ventricle [FIG. 13.113]. The interval between the limbs of the loop is occupied by the mesocardium—a fold of the lining pericardium.

As the heart folds longitudinally it is also bent in the transverse plane, and its parts take up the positions shown in FIGURE 13.113, 3A and 4A. Later, as the ventricle enlarges ventrally and tailwards, the right segment of the sinus venosus is absorbed into the atrium,

and the atrium, thus reinforced, expands round the sides of the bulbus cordis [FIG. 13.113].

Subsequently the bulbus cordis is absorbed, partly into the ventricle and partly into the truncus arteriosus. Thereafter the aortic sac and truncus divide into the pulmonary trunk and the ascending part of the aorta, the gap left by the earlier disappearance of the mesocardium now forming the transverse sinus of the pericardium between these great vessels ventrally and the atrium dorsally.

While these changes are taking place, the cavities of the atrium, atrioventricular canal, and ventricle are all being divided into right and left halves by means of septa, which it is convenient to consider separately although they are formed simultaneously and fuse with one another in parts of their extent.

DIVISION OF THE ATRIOVENTRICULAR CANAL

After the longitudinal folding of the primitive cardiac tube has occurred, the atrioventricular canal runs dorsoventrally from the atrium to the ventricle. Two thickenings, called endocardial cushions, gradually project into the canal from its cephalic and caudal walls, and by the fusion of their intermediate parts a septum of the atrioventricular canal is formed, dividing the canal into the right and left atrioventricular orifices [FIG. 13.114].

Later, by excavation of the ventricular aspect of the cushions the atrioventricular valve cusps are formed. Since the cushions are continuous with the loose meshwork of ventricular muscle, this surface of the cusps remains attached to the papillary muscles by chordae tendineae.

DIVISION OF THE ATRIUM

The primitive atrium is divided into right and left atria by the formation and later fusion of two septa—the septum primum and the septum secundum, which appear in that order—and by the fusion of the septum primum with the septum of the atrioventricular canal.

The **septum primum** is a thin sheet which grows ventrally from the dorsal and cephalic walls of the atrium. As it grows, the area of communication between the right and left parts of the atrium is gradually reduced to an aperture, called the **ostium primum** [FIG. 13.114]. It is obliterated when the septum primum fuses with the septum of the atrioventricular canal, but, before that occurs, an

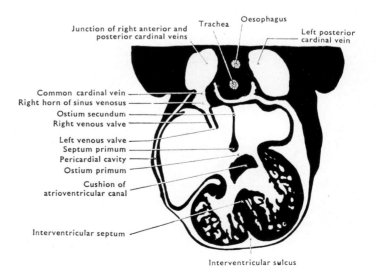

FIG. 13.114. Section of heart of human embryo, showing development of septa. (Edinburgh University collection.)

aperture called the ostium secundum appears in the dorsal part of the septum primum [Fig. 13.114].

The septum secundum appears immediately to the right of the septum primum, and is a thick, solid, incomplete septum, with its free edge facing towards the inferior vena cava.

The growth of the septum secundum ceases when its free margin overlaps the ostium secundum, and then the septum primum acts as a flap-valve against the septum secundum. This permits blood to pass through the foramen ovale—the name given to the interatrial communication after the formation of the septum secundum—from the right to the left atrium, but prevents its return [Fig. 13.14]. After birth, pulmonary venous blood returning to the left atrium raises the pressure in that chamber, the septum primum is thus held against the septum secundum and subsequently fuses with it, and so the orifice is permanently closed. As was pointed out in the description of the adult right atrium [p. 879], the floor of the fossa ovalis is formed from the septum primum, while the limbus fossae ovalis represents the free edge of the septum secundum.

DIVISION OF THE TRUNCUS ARTERIOSUS AND THE CRANIAL PART OF THE BULBUS CORDIS

While the bulbus cordis is being absorbed—partly into the truncus arteriosus, and partly into the ventricular portion of the heart—spiral ridges, similar in structure to the endocardial cushions, develop in it and in the truncus arteriosus and aortic sac. There are two such ridges, and they arise on opposite walls; later, two very short accessory ridges appear between them in the ventricular end of the part of the bulbus which will be absorbed into the truncus.

Within the aortic sac, truncus, and upper bulbus, the main ridges fuse to form a spiral septum which is so disposed that when the vessel splits longitudinally, the pulmonary trunk separating from the aorta, lies on a plane posterior to the aorta above, then on its left side, and is anterior to it below. At the level of the aortic and

pulmonary valves excavations of the main and accessory ridges produce the valvules of the corresponding valves.

Thus far the lower (ventricular) part of the bulbus and the ventricle are still undivided, and, clearly, the process of separation must be such as to bring the right ventricle into communication with the pulmonary trunk (and so with the sixth aortic arterial arches which become parts of the pulmonary arteries—see later), and the left ventricle into communication with the aorta (and so with the remaining aortic arches which help to form the main arterial trunks of the head, neck, and upper limbs) [Fig. 13.116].

Division of the lower part of the bulbus is effected by the continuous descent of the spiral septum, which separates a pulmonary channel (infundibulum) anterior and to the right, from an aortic channel, posterior and to the left. Until this septum is united to the developing interventricular septum, a communication between the two ventricles persists.

DIVISION OF THE VENTRICLE

The interventricular septum begins as a fold of the ventricle wall, immediately to the right of its most ventral point, and its position is indicated on the surface by a sulcus [Fig. 13.114] which sometimes remains to produce a bifid apex of the heart. At first the septum is formed mainly as a result of the expansion of the right and left ventricles on either side of it. It is semilunar in shape with anterior and posterior extremities, or horns. The posterior horn joins the septum of the atrioventricular canal to the right of the attachment of the septum primum superiorly, while the anterior horn joins the anterior part of the lower margin of the bulbar septum in front. The gap which still exists above the free edge of the interventricular septum (i.e. the interventricular foramen) is inferior to the posteriorly placed aortic part of the bulbus, and is now closed by tissue which grows forwards from the antero-inferior part of the septum of the atrioventricular canal [Fig. 13.115]. This not only

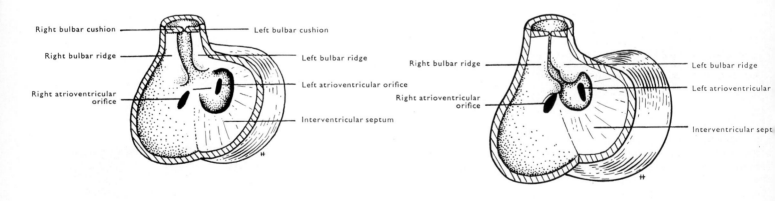

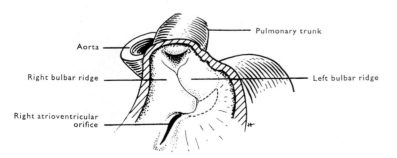

Fig. 13.115. Diagrams to show how the lower bulbar region is divided and the interventricular septum completed. (After Frazer.)

In the right Figure the contribution to the septum from the endocardial cushions is shown by an interrupted line.

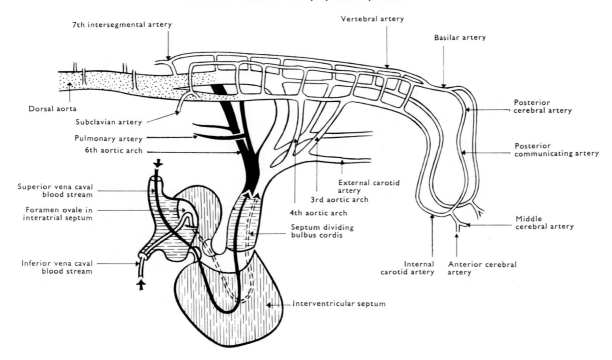

FIG. 13.116. Development of heart and main arteries.
    Diagram of the heart, showing the formation of its septa, and of the cephalic portion of the arterial system. Seen from right side.

unites the lower end of the bulbar septum to the interventricular septum, but also forms the right wall of the aortic part of the bulbus (aortic vestibule). The fully formed interventricular septum therefore consists of two parts—a ventral and larger muscular part formed from the ventricular wall, and a dorsal and smaller fibrous part, the **pars membranacea**, formed from extensions of the atrioventricular cushions, which form only connective tissue.

FATE OF THE SINUS VENOSUS

The sinus venosus was formed, in the septum transversum, by the fusion of the terminal parts of the common vitello-umbilical venous trunks. It consists of a single chamber possessing a right and left horn and a middle section.

Each horn receives a common cardinal vein, a vitelline vein, and an umbilical vein. Later, as the longitudinal folding of the cardiac tube occurs, the sinus venosus emerges from the septum transversum and appears in the caudal part of the pericardium. There it lies dorsal to the atrium into which its right part opens through an orifice guarded by right and left **venous valves** which fuse with each other at their cephalic and caudal ends.

As development proceeds, the right part of the sinus venosus is absorbed into the right atrium, the right common cardinal vein becoming the superior vena cava. With atrophy of the left common cardinal and right umbilical veins, and changes in the vitelline and left umbilical veins [p. 973], the left horn remains as the coronary sinus of the fully developed heart. The left venous valve disappears; the lower portion of the right valve persists in part to form the valve of the inferior vena cava and the valve of the coronary sinus, and its upper portion contributes to the formation of the crista terminalis.

The account given here of the development of the heart is a general outline of a very complex process. For further details the following should be consulted: Davis (1927), Odgers (1938), and Tandler (1912, 1913).

CONDUCTING TISSUE OF THE HEART

In mammalian embryos the heart begins to beat some time before the conducting tissue appears. At first the heart rate is slow, and no special means of propagating the impulse to cardiac contraction is required. Later, however, the heart rate increases considerably, and the conducting tissue provides the mechanism both for the rapid transmission of the impulse throughout the heart, and for the delay in transmission from the atria to the ventricles whose muscular continuity is broken by the fibrous ring.

The **atrioventricular node** and bundle can be recognized when the embryo is about 10 mm in length—by the sixth week or so. Opinions differ as to whether the node or the bundle appears first, for there are those (Shaner 1929; Walls 1947) who consider that the bundle grows from the node situated on the posterior wall of the atrioventricular canal, and others (Field 1951; Muir 1954) who are of the view that the bundle is the first structure to differentiate and that the node appears somewhat later.

The **sinu-atrial node** is the last part of the system to appear. As befits its role of pacemaker it is, from an early period, closely associated with nervous elements, but the special histological features which characterize the adult node do not appear until after birth (Duckworth 1952).

Gardner and O'Rahilly (1976) give a full account of the nerve supply and conducting system of the human heart at the end of the embryonic period proper, and Wenink (1976) has an interesting theory with regard to the origin of the conducting tissue.

FATE OF THE AORTIC ARCHES AND UNFUSED PARTS OF THE DORSAL AORTAE

Little need be added to the account already given [p. 64] of the formation of the aortic arches. The first arch of each side connects

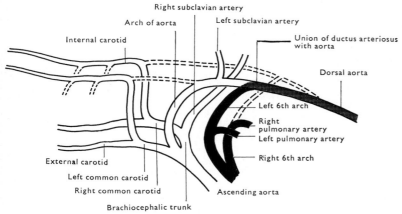

FIG. 13.117. Scheme of part of arterial system of foetus seen from left side.
Parts of the first and second arches, the ductus caroticus on both sides, the dorsal part of the right sixth arch, and the dorsal aorta caudal to the right fourth and fifth arches have atrophied. The position of the fifth arch is not indicated.

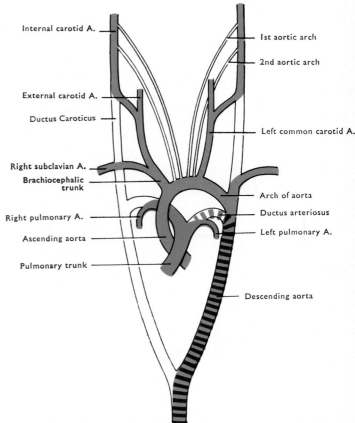

FIG. 13.118. Scheme of part of the foetal arterial system. Persisting vessels are shown in solid colour.

the corresponding dorsal aorta with the primitive ventral aorta which soon fuses with its partner to form the aortic sac. Additional pairs of arches then appear in this order, the second, the third, the fourth, and the sixth—a fifth pair appears briefly but rapidly disappears—and connect the aortic sac to the dorsal aortae. Not all the arches are present at the same time, for the first two pairs have disappeared before the fifth and sixth have formed.

By reference to FIGURES 13.117 and 13.118 the fate of the aortic arches and of the unfused parts of the dorsal aortae may be seen. Stated briefly, the first and second arch arteries disappear in great part. Their ventral roots contribute to the external carotid artery, which, however, is formed mainly as a new outgrowth from the third arch. In addition, the first arch artery probably contributes to the maxillary artery, and the second certainly forms the **stapedial artery**—a transient vessel which traverses the primordium of the stapes, and so helps to separate its crura.

The third aortic arches form the common carotid arteries and the proximal parts of the internal carotids; the distal parts of the latter vessels are formed by the extension headwards of the dorsal aortae. On the left side, the fourth aortic arch artery becomes the part of the arch of the aorta between the left common carotid and the left subclavian arteries, the rest of the arch of the aorta being formed from the aortic sac and from the left of two upward extensions of the sac, called horns, which result from lengthening of the neck. The right horn of the sac extends upwards as the brachiocephalic trunk and joins with the right third and fourth arch arteries, the latter of which becomes the root of the right subclavian artery. From the middle of each sixth arch artery a pulmonary artery arises and thereafter the dorsal part of the right sixth artery disappears. On the left side, however, this part persists as the **ductus arteriosus** through which blood passes from the pulmonary trunk into the aortic arch until birth. The ductus becomes obliterated after birth and is converted into the ligamentum arteriosum. The ventral part of the right sixth arch artery becomes the extrapulmonary portion of the right pulmonary artery, but on the left side this part is absorbed into the pulmonary trunk.

The fates of the dorsal aortae are indicated in FIGURE 13.118.

For further details consult Congdon (1922) and Padget (1948).

Congenital malformation of the heart and great vessels

Although such malformations have long been of academic interest, their recognition during life is now—thanks to advances in surgery—of great practical importance. Only the commoner types of anomaly will be mentioned here, and for further information Abbott (1936) should be consulted.

Many of the conditions listed are due to arrest or failure of development. On the other hand, parts which should normally regress sometimes persist. Again, a process of rotation may take place in a reverse direction.

Such defects may occur either singly or in combination, for the altered haemodynamics which result from certain types of malformation will be themselves causative of compensatory defects elsewhere.

The system of classification which follows is broadly topographical, but it should be evident which developmental error is responsible for the defects mentioned.

1. MALPOSITION
   (a) *Dextrocardia*. The heart is transposed and is simply a mirror image of the normal. It is usually associated with transposition of the other viscera (situs inversus). Life expectancy is not affected. This anomaly is very rare.

(b) *Ectopia cordis.* The heart projects outside the body wall. As yet surgical treatment is of no avail and the infant survives only a matter of weeks. Very rare.

2. ATRIAL SEPTAL DEFECTS

(a) *Patent foramen ovale.* In such cases, which are common, the opening is usually quite small and of no clinical importance. In a very few cases a wider opening is present, when, as a rule, it is associated with and probably due to other anomalies such as pulmonary stenosis (narrowing) or transposition of the aorta and pulmonary trunk.

(b) *Persistence of the ostium primum.* This results in a fairly large opening in the lower part of the interatrial septum.

(c) *Persistence of the primitive ostium secundum.* This results in a fairly large opening in the upper part of the septum.

(d) *Absence of the interatrial septum.* This results either in a three-chambered heart if there are two ventricles (cor trioculare biventriculare), or in a two-chambered heart if there is no interventricular septum (cor biloculare).

3. VENTRICULAR SEPTAL DEFECTS

(a) *Absence, or near absence, of the septum.* If two atria are present, the condition is called cor triloculare biatriatum. See also above.

(b) *Localized defects.* These occur just below the aortic valve and are most often associated with other abnormalities such as occur in the tetralogy of Fallot. In this condition the four essentials of the tetralogy are a high ventricular septal defect; apparent dextroposition of the aorta which overrides the defect, due to the absence of the membranous septum; pulmonary stenosis; and consequent hypertrophy of the right ventricle.

4. VALVULAR ATRESIA (IMPERFORATION) AND STENOSIS

(a) *Pulmonary.* Most often pulmonary valve atresia or stenosis is associated with septal defects. Its contribution to Fallot's tetralogy has been noted. The cause of pulmonary atresia and stenosis is failure of the bulbus cordis to develop normally.

(b) *Aortic.* Congenital stenosis of this valve is rare; atresia allows survival for only a few months after birth.

(c) *Atrioventricular.* Stenosis and atresia of these valves are rare congenital anomalies, usually associated with compensatory septal defects.

5. ANOMALIES OF THE AORTA

(a) *Coarctation.* This has been mentioned on page 923.

(b) *Transposition of the aorta and pulmonary trunk, with the aorta arising from the right ventricle and the pulmonary trunk from the left.* The condition is due to a reversal of the normal direction of rotation of the spiral, bulbar septum.

(c) *Double Aortic Arch.* This is due to the persistence of right and left aortic arches and of the associated parts of the dorsal aortae.

(d) *Right Aortic Arch.* This is due to the persistence of the fourth right aortic arch, and the associated portion of the right dorsal aorta, with disappearance of the corresponding vessels on the left.

(e) *Communications between Aorta and Pulmonary Artery.*
(i) Complete, i.e. absence of the spiral, bulbar septum with a resultant common arterial trunk.
(ii) Partial, due to a localized defect in the septum.

6. PATENT DUCTUS ARTERIOSUS

This may be compensatory to some other defect, but in many cases it seems to result from some arrest of development still to be explained. Its presence results in an abnormally high pulmonary arterial pressure and a consequent hypertrophy of the right ventricular wall which causes the heart to have a globular shape—a condition common to all abnormal communications between the right and left ventricles or the great arteries.

Descending aorta

The greater part of the descending aorta is formed by the progressive fusion of the primitive dorsal aortae down to the twenty-third body segment—the level of the fourth lumbar vertebra—where the common iliac arteries arise. Later the small terminal portions of the dorsal aortae fuse to form the median sacral artery.

BRANCHES OF THE DORSAL AORTAE

The branches of the primitive dorsal aortae are lateral splanchnic, ventral splanchnic, and dorsolateral or somatic intersegmental.

The **lateral splanchnic arteries** are distributed segmentally to organs developed in or near the intermediate cell masses. They become the suprarenal, renal, and gonadal arteries.

The **ventral splanchnic arteries** pass to the primitive alimentary canal and its diverticula, and, in the early stages, to the walls of the yolk sac. As the yolk sac atrophies, the branches to it disappear, and simultaneously the corresponding vessels of opposite sides fuse to form unpaired stem trunks from which the three great vessels of the abdominal part of the alimentary canal are derived, namely, the coeliac, the superior mesenteric, and the inferior mesenteric arteries. But all three vessels eventually arise from the abdominal part of the aorta at levels far below those at which they first took origin.

The superior mesenteric artery retains longest its connexion with the extra-embryonic part of the yolk sac, and occasionally a fibrous strand, representing the original channel, extends from one of its ileal branches to the umbilicus.

The **somatic intersegmental arteries** early form a series of paired vessels from the cervical to the sacral regions. However, only in the thoracic and lumbar regions are their original characters retained. The vessels pass backwards, and divide into dorsal and ventral branches which accompany the corresponding rami of the spinal nerves [FIGS. 13.119–13.121].

The **ventral branches**, together with the stems, form the main trunks of the intercostal and lumbar arteries in the adult. They are connected together by precostal and ventral anastomoses [see below] and each gives off a branch which runs with the lateral cutaneous branch of a spinal nerve.

The **dorsal branches** form the posterior branches of the intercostal and lumbar arteries of the adult; they are connected by postcostal and by post-transverse anastomoses [see below]. Each dorsal branch gives off a spinal offset which divides into a dorsal, a ventral, and a neural branch which are linked by longitudinal anastomoses as shown in FIGURES 13.119–13.121.

The precostal anastomoses persist in the thoracic region as the highest intercostal arteries; in the lumbar region they disappear. The ventral somatic anastomoses persist as the internal thoracic and superior and inferior epigastric arteries.

The postcostal and post-transverse anastomoses disappear in the thoracic and lumbar regions.

The anastomoses within the vertebral canal persist, the pre- and postneural anastomoses aiding in the formation of the anterior and posterior spinal arteries respectively.

In the cervical region, the first six pairs of somatic intersegmental arteries lose their connections with the dorsal aortae. The seventh pair, however, persist in their entirety; and from them are formed part of the right subclavian and the whole of the left subclavian artery. The continuation of the subclavian artery, beyond the inner margin of the first rib, is the persistent and enlarged lateral offset of the ventral branch of the seventh somatic intersegmental artery, which is continued into the upper limb. The thyrocervical trunk is

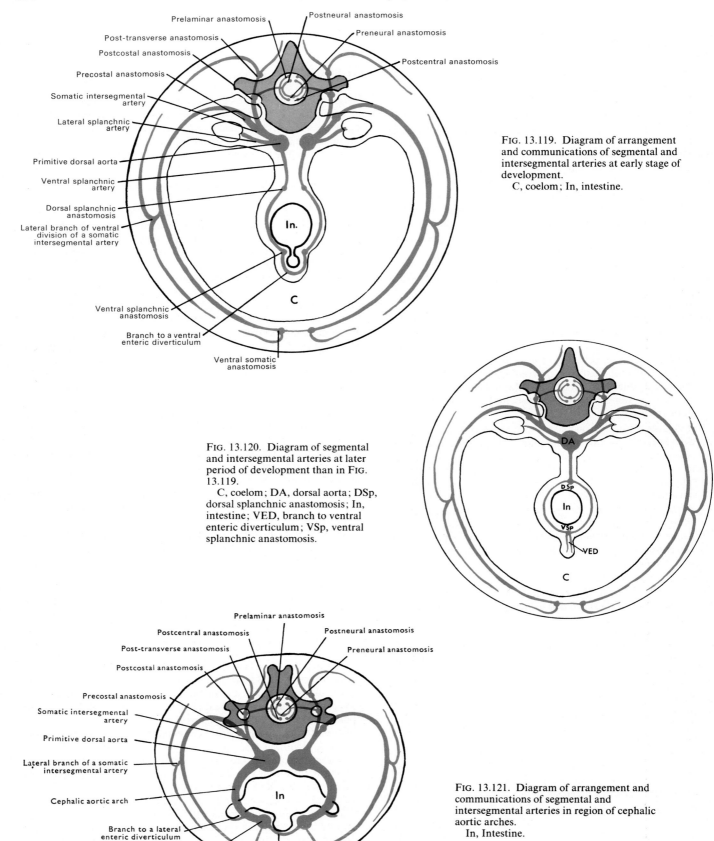

Prelaminar anastomosis
Postneural anastomosis
Post-transverse anastomosis
Preneural anastomosis
Postcostal anastomosis
Precostal anastomosis
Postcentral anastomosis
Somatic intersegmental artery
Lateral splanchnic artery
Primitive dorsal aorta
Ventral splanchnic artery
Dorsal splanchnic anastomosis
Lateral branch of ventral division of a somatic intersegmental artery
In.
C
Ventral splanchnic anastomosis
Branch to a ventral enteric diverticulum
Ventral somatic anastomosis

FIG. 13.119. Diagram of arrangement and communications of segmental and intersegmental arteries at early stage of development.
C, coelom; In, intestine.

FIG. 13.120. Diagram of segmental and intersegmental arteries at later period of development than in FIG. 13.119.
C, coelom; DA, dorsal aorta; DSp, dorsal splanchnic anastomosis; In, intestine; VED, branch to ventral enteric diverticulum; VSp, ventral splanchnic anastomosis.

DA
DSp
In
VSp
VED
C

Prelaminar anastomosis
Postcentral anastomosis
Postneural anastomosis
Post-transverse anastomosis
Preneural anastomosis
Postcostal anastomosis
Precostal anastomosis
Somatic intersegmental artery
Primitive dorsal aorta
Lateral branch of a somatic intersegmental artery
Cephalic aortic arch
In
Branch to a lateral enteric diverticulum
Primitive ventral aorta
Ventral somatic anastomosis
Branch to a ventral enteric diverticulum

FIG. 13.121. Diagram of arrangement and communications of segmental and intersegmental arteries in region of cephalic aortic arches.
In, Intestine.

a persistent precostal anastomosis, as is the ascending cervical artery. The vertebral artery is, morphologically, somewhat complex. The first part represents the dorsal branch of the seventh somatic intersegmental artery; the second part consists of persistent postcostal anastomoses; the part on the arch of the atlas is the spinal branch of the first somatic intersegmental artery and its neural continuation; whilst, finally, the part of the vertebral artery in the cranial cavity appears to represent a prolongation of the preneural anastomoses, which further upwards are probably represented by the basilar artery.

The deep cervical artery is a remnant of the post-transverse longitudinal anastomoses.

The origin of the seventh somatic intersegmental artery is, at first, caudal to the sixth aortic arch, but with the elongation of the neck and the descent of the heart, it comes to lie opposite the dorsal end of the fourth aortic arch.

Umbilical and iliac arteries

The umbilical arteries first arise, cranial to their final origin, from the ventral wall of the primitive dorsal aorta at the fourth lumbar segment. As each umbilical artery passes to the body stalk it lies medial to the pronephric duct; this, however, is temporary, for a new vessel arises on each side, from the lateral wall of the aorta. This new vessel passes lateral to the mesonephric duct, and then unites with the primitive umbilical artery. Thereafter the ventral

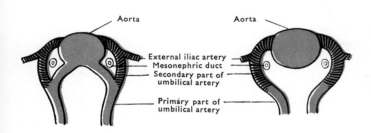

FIG. 13.122. Diagrams showing formation of secondary part of primitive umbilical artery.

origin of the umbilical artery disappears [FIG. 13.122]. From the newly formed vessel there arises the inferior gluteal artery, the first-formed main artery of the lower limb. Later, a second branch becomes the external iliac and the femoral arteries of the adult. The portion of the umbilical stem dorsal to the external iliac becomes the common iliac artery, and the ventral part becomes the internal iliac artery. But the part of the original umbilical artery that runs to the umbilicus, and on to the placenta, is still called the umbilical artery. After birth its intra-abdominal portion atrophies and becomes the lateral umbilical ligament; but a portion in the pelvis remains patent, and gives origin to the superior vesical artery.

Arteries of the upper limb

At first a number of branches pass from the dorsal aortae to the upper limb buds where they end in vascular plexuses.

Later, the only connection with the aorta, on each side, is the ventral branch of the seventh intersegmental artery; and a single axial artery, divisible into subclavian, axillary, brachial, and interosseous segments, serves the limb. Secondary vessels then appear, the first of which, the median artery, is destined to be superseded by the radial and ulnar arteries. The sequence of events can be followed very clearly in FIGURE 13.123.

Arteries of the lower limb

A primary axial artery springs from the secondary umbilical artery, and leaves the pelvis through the greater sciatic foramen. It descends between the hamstring and adductor muscles, passes through the popliteal fossa in front of the popliteus muscle, and descends along the interosseous membrane to the foot, where it ends in a capillary plexus [FIG. 13.124].

The main remnants of the primitive axial artery in the adult are—the inferior gluteal artery and its sciatic branch (the companion artery of the sciatic nerve), the proximal part of the popliteal artery, and the distal part of the peroneal artery.

A new branch, the external iliac artery, arises from the umbilical artery above the origin of the axial artery, and enters the front of the thigh. It becomes the femoral artery and is joined by a recurrent

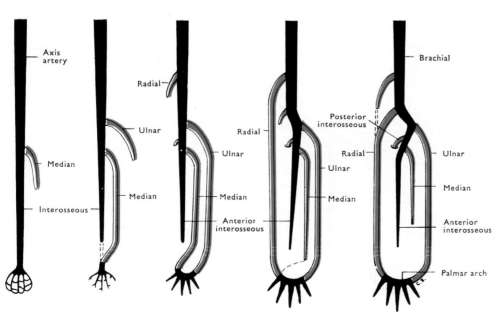

FIG. 13.123. Diagrams of stages in development of arteries of upper limb. (After Arey 1946.)

The original axial artery is black and the later-formed arteries red.

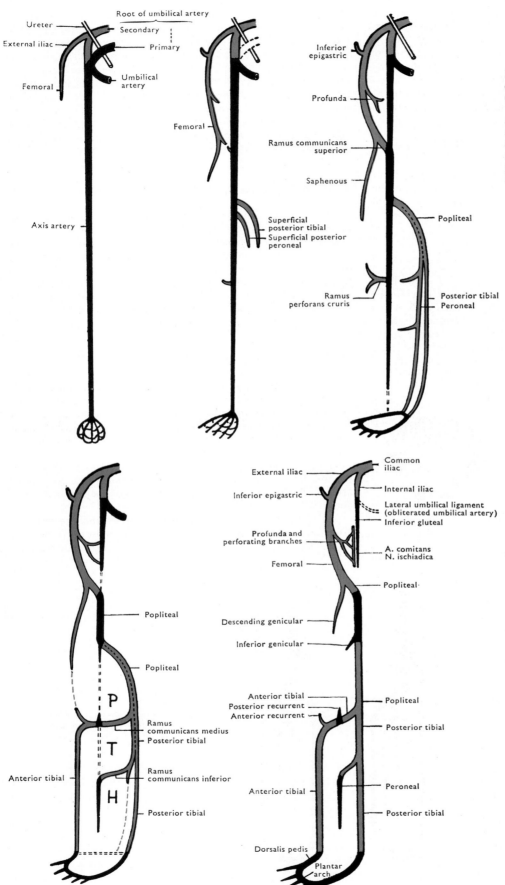

FIG. 13.124. Diagrams of stages in development of arteries of lower limb. (After Senior (1919) and Arey (1946).)

The original umbilical artery and the axial artery are black, the later-formed arteries red. The development of the arteries of the foot is schematic and details are not shown. P, T, and H: position of popliteus, tibialis posterior, and flexor hallucis longus muscles.

branch of the axial artery (ramus communicans superior) which pierces the adductor magnus so that the continuity between femoral and popliteal arteries is established.

The distal part of the **popliteal artery** and the proximal part of the **posterior tibial artery** are produced by the fusion of two branches of the axial artery. The **anterior tibial artery** arises from the 'perforating' branch of the axial artery and is joined to the popliteal by the **ramus communicans medius**. The **peroneal artery** obtains its connexion with the posterior tibial through the **ramus communicans inferior**. For further details Senior (1919a, b, 1920) should be consulted.

THE DEVELOPMENT OF VEINS

The veins which appear during embryonic life may be classified in two groups: (1) those which are partly extra-embryonic and partly intra-embryonic; (2) those which are entirely intra-embryonic.

VEINS PARTLY INTRA-EMBRYONIC AND PARTLY EXTRA-EMBRYONIC

The earliest veins to appear are the paired vitelline veins, which return blood from the yolk sac, and the paired umbilical veins, which return purified blood from the placenta [p. 35]. Their further history may now be considered.

Vitelline and umbilical veins, portal vein, hepatic veins, and cardiac end of the inferior vena cava

After the entodermal vesicle is separated into the embryonic alimentary canal, the yolk sac, and the vitello-intestinal duct, each vitelline vein passes along the duct to the umbilical orifice, where it passes at once into the septum transversum; there it unites with the cardiac end of the corresponding umbilical vein to form a common vitello-umbilical trunk, which terminates in the sinus venosus [FIGS. 2.75 and 2.76]. Subsequently, as the sinus venosus passes into the pericardial cavity, the common trunk is absorbed into it, and each vein thus acquires its own opening into the sinus [FIG. 2.77].

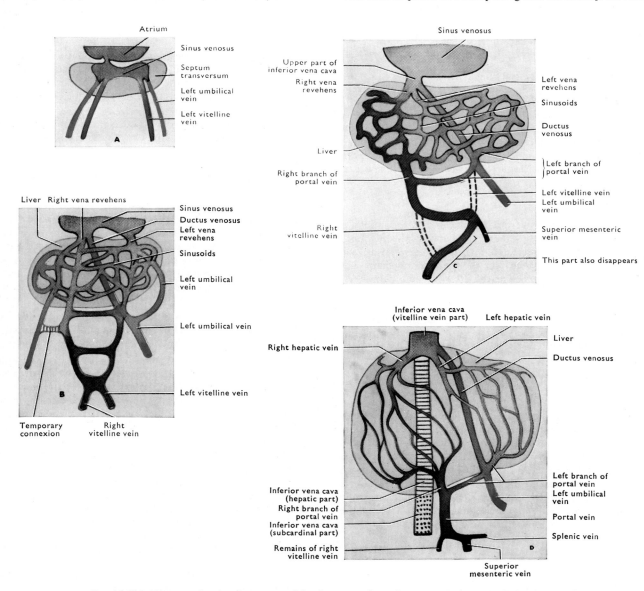

FIG. 13.125. Diagram showing four stages of development of portal system and parts of inferior vena cava.

The liver is developed in the septum transversum from the branchings of a diverticulum of the duodenum. As a result, the portions of the vitelline and umbilical veins which lie in the septum are expanded as the liver grows, and at the same time broken up into many anastomosing channels called sinusoids. Strands of liver cells multiply and invade the sinusoidal blood spaces until they are, in the main, reduced to the size of capillaries, and the liver becomes a comparatively solid organ.

As the duodenum is delimited, the vitelline veins pass along its sides and are connected around it by transverse anastomoses, two of which lie ventral and one dorsal to the duodenum. Passing through the septum transversum each vitelline vein is broken up into: a caudal part, the vena advehens, which enters the liver; an area of sinusoidal channels in the liver substance; and a cephalic part, the vena revehens, which passes from the liver to the heart. After a time the left vena revehens no longer opens directly into the sinus venosus, but into the cephalic end of the right vena revehens [FIG. 13.125]. This part of the right vena revehens becomes the terminal part of the inferior vena cava, and thus the left vena revehens forms the left hepatic vein; the right hepatic vein is formed from the intrahepatic part of the right vena revehens. Therefore all the blood passing to the liver by the vitelline veins eventually reaches the heart by the inferior vena cava, which also receives the ductus venosus [see below].

Caudal to the duodenum the vitelline veins are replaced by the superior mesenteric vein which opens into the left vitelline vein where, a little later, the splenic vein also enters. The final result is the formation of the portal vein from parts of both vitelline veins and of their transverse anastomoses [FIG. 13.125]. The right branch of the portal vein is the right vena advehens. The left branch of the portal vein is formed from the left vena advehens, and the more cephalic of the two ventral anastomoses between the vitelline veins.

Umbilical veins

A single umbilical vein passes from the placenta to the umbilical orifice, where it divides into the left and right umbilical veins. Each of these becomes directly connected with the sinus venosus, and later with sinusoidal spaces of the liver and the vena advehens. The right umbilical vein undergoes early atrophy and completely disappears. The left umbilical vein persists until birth. After the disappearance of the right umbilical vein, the left conveys blood from the placenta to the liver, where part of the placental blood passes into the left vena advehens and so through the sinusoids and left vena revehens to the inferior vena cava, and part passes into the ductus venosus.

Ductus venosus

The ductus venosus is developed as the left umbilical vein becomes united to the left vena advehens. It is formed from the sinusoidal spaces of the liver, and connects the commencement of the left vena advehens with the cephalic part of the right vena revehens. It forms the channel by which the greater part of the blood from the placenta is passed to the heart through the upper end of the inferior vena cava without passing through the liver sinusoids. After birth, the left umbilical vein becomes the ligamentum teres of the liver and the ductus venosus is converted into the ligamentum venosum [p. 963]. For further details consult Dickson (1957).

INTRA-EMBRYONIC VEINS

In the subsequent account the embryo and foetus are considered as being in the quadrupedal position. Therefore the terms anterior and posterior are equivalent to cephalic and caudal, and to superior and inferior in the erect posture; and the prefixes sub- and supra- are equivalent to ventral and dorsal, and to anterior and posterior in the erect posture.

Anterior cardinal veins

These extend, one on each side, from the region of the eye to the sinus venosus, and each is separable into cephalic, nuchal, and thoracic parts. The cephalic part, or primary head vein, runs medial to the trigeminal ganglion, but lateral to the otic vesicle and the seventh to the eleventh cranial nerves. Its anterior portion becomes the cavernous sinus, and its posterior portion disappears [see below].

The anterior tributaries of the primary head vein become the ophthalmic vein, whereas its dorsal tributaries form three plexuses, anterior, middle, and posterior, associated respectively with the forebrain and midbrain, with the cerebellar region, and with the region of the medulla oblongata [FIG. 13.126A]. A stem vessel drains from each plexus to the primary head vein; later, when an anastomosis forms between the middle and posterior stems, the part of the primary head vein lateral to the otic vesicle disappears [FIG. 13.126B].

The main parts of the plexuses are carried away from the brain with the membrane that will form the dura mater, and they later establish connections with the veins which appear on the surfaces of the cerebral hemispheres as these develop from the forebrain.

When the cerebral hemispheres increase in size, the dural tissue is compressed between them and between the hemispheres above and the midbrain and hindbrain below, in the form of folds. At the same time the anterior and middle plexuses (now conjoined) are carried towards those of the opposite side, fuse with them, and so form the superior and inferior sagittal sinuses and the straight sinus [FIG. 13.126C].

Meantime, the growth of the hemispheres forces the upper part of the middle stem tributary backwards and downwards until it becomes the transverse sinus, and the anastomosis above the otic region and the posterior stem tributary are converted into the sigmoid sinus. The anterior portion of the primary head vein which lies medial to the trigeminal ganglion becomes the cavernous sinus, and the lower part of the middle stem tributary is converted into the superior petrosal sinus.

The inferior petrosal sinus appears to be an independently formed anastomosis [FIG. 13.126C]. For a full account of the development of the cranial venous sinuses Streeter (1915) should be consulted.

The history of the nuchal and thoracic parts of the anterior cardinal veins is different on the two sides. On both sides the anterior cardinal vein is joined, at the root of the neck, by the chief vein of the upper limb [FIG. 13.126F]. The part cephalic to the junction becomes the internal jugular vein. Caudal to the junction a transverse anastomosis forms between the two anterior cardinal veins. The part between the anastomosis and the junction with the limb vein on the right side, becomes the right brachiocephalic vein; on the left side it becomes part of the left brachiocephalic vein, the remainder of which is formed by the transverse anastomosis. The right anterior cardinal, caudal to the transverse anastomosis, is joined by the right posterior cardinal vein. The part cephalic to that junction becomes the extrapericardial part of the superior vena cava. The part caudal to the junction is the right common cardinal vein and it becomes the intrapericardial part of the superior vena cava. On the left side, the anterior and common cardinal veins virtually disappear caudal to the transverse anastomosis, losing the connection with the left horn of the sinus venosus. The caudal part, the common cardinal vein, is represented by the oblique vein of the left atrium and the fold of the left vena cava, whilst the cephalic part becomes part of the left superior intercostal vein.

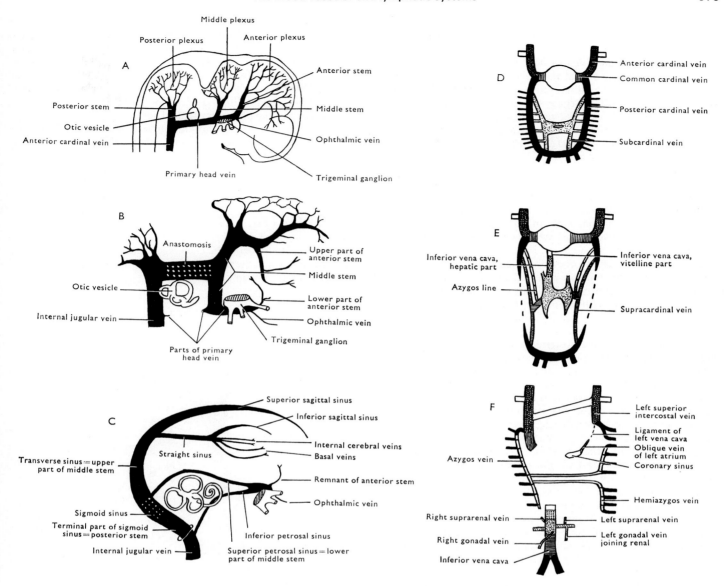

FIG. 13.126. Diagrams of the development of the head veins and venous sinuses (A, B, C), and of the veins of the trunk (D, E, F).

The **external jugular vein** is a new formation which receives for a time the cephalic vein of the upper limb.

Anomalies

The left superior vena cava may persist *in toto*, and then it drains into the coronary sinus (left horn of the sinus venosus). However, its upper part alone most commonly persists, particularly when the left pulmonary veins (or vein) enter it. In this case it drains upwards into the left brachiocephalic vein, and is joined by the left superior intercostal vein. Such partial or complete anomalous pulmonary venous drainage (which may occur from either or both lungs into the corresponding superior vena cava) leads to mixing of oxygenated and deoxygenated blood. Also, because of the greater volume of blood reaching the right atrium, there is a higher pressure in the right than the left atrium, and thus the foramen ovale remains open after birth, allowing deoxygenated venous blood, principally from the inferior vena cava, to enter the left atrium and pass to the systemic circulation.

DEVELOPMENT OF THE THORACO-ABDOMINAL VEINS CAUDAL TO THE HEART

The veins caudal to the heart form a complicated plexus from which three main longitudinal systems develop on the posterior thoracic and abdominal walls [FIG. 13.126D–F]. The definitive veins of the adult arise from parts of these systems, which because of their free anastomosis can take over each other's functions as development proceeds. The three systems are: (1) The **posterior cardinal veins** which lie lateral to the intermediate mesoderm and primarily drain the intersegmental veins of the body wall. (2) The **subcardinal veins** which lie ventromedial to the intermediate mesoderm and drain its derivatives—the kidneys, gonads, and suprarenals. The two subcardinal veins, right and left, anastomose ventral to the aorta. (3) The **supracardinal system** of veins which surrounds the sympathetic trunk and forms medial (azygos) and lateral (thoracolumbar) longitudinal channels. The azygos anastomoses with its fellow across the midline posterior to the aorta, and with the subcardinal vein of the same side lateral to the aorta. At their cranial and caudal ends the subcardinal and supracardinal systems anastomose with the posterior cardinal vein of the same side, and also have intermediate communications with it.

The posterior cardinal veins extend from the pelvis to unite with the anterior cardinal veins at the level of the septum transversum and form the common cardinal veins. As the supracardinal systems of veins take over the intersegmental veins of the trunk, the greater part of both posterior cardinal veins disappears; their most cranial parts (which still drain the other systems) remain as the upper parts of the azygos vein (on the right) and the left superior intercostal vein (on the left), while their caudal parts remain as the internal iliac veins.

The common iliac veins, which now drain the pelvic veins and the developing veins of the lower limbs, represent a communication between the posterior cardinal veins and the medial supracardinal veins; the left common iliac in the midline representing a communication between the two medial supracardinal veins, posterior to the aorta.

The subcardinal veins originally extend throughout the thorax and abdomen. As the thoracic part of the mesonephros disappears, the subcardinal veins become limited to the abdomen, where they not only anastomose with the medial supracardinal veins, but the right subcardinal also develops a communication ventrally with the hepatic venous system, thus forming an alternative route to the heart for the veins of the posterior abdominal wall. This rapidly enlarges to become the prerenal part of the inferior vena cava which drains the right subcardinal vein, the left subcardinal vein through an enlarged communication between these veins (the part of the left renal vein anterior to the aorta), and a large part of the supracardinal system of veins caudal to the renal veins. Probably as a result of the asymmetrical development of the prerenal part of the inferior vena cava, the right medial supracardinal vein and its communication with the right subcardinal enlarge to form the caudal, or postrenal part of the inferior vena cava, while its communications with the shrinking left medial supracardinal become the parts of the lumbar veins posterior to the aorta and the terminal part of the left common iliac vein.

As the mesonephros disappears and the more restricted metanephros reaches its definitive position, the subcardinal veins are limited to this level. On the left, the subcardinal vein forms a short length of the renal vein into which both gonadal and suprarenal veins drain, while on the right it forms a short segment of the inferior vena cava which receives the same tributaries from the right side. The lateral parts of the left renal vein and all of the right renal vein are formed by a communication from the posteriorly placed metanephros through the lateral and medial parts of the supracardinal system to the subcardinal. Hence the cranial parts of the supracardinal system, the hemiazygos and azygos veins, begin inferiorly in the left renal vein and the postrenal part of the inferior vena cava respectively, both of which are formed from the supracardinal veins.

Thus the inferior vena cava consists from below upwards of parts of: (1) the right medial supracardinal vein; (2) the right subcardinal vein; (3) the subcardinal-hepatic anastomosis; (4) hepatic sinusoids (vitelline veins); and (5) the right vitelline vein [Fig. 13.126F].

For further information on the development of the veins of the trunk, Gladstone (1929) and Franklin (1948) should be consulted.

Veins of limbs

In each limb the earliest veins are a superficial distal arch and a postaxial trunk; later, digital veins connect with the distal arch, and a pre-axial trunk is formed. In the upper limb the arch and its tributaries remain as the dorsal venous network and the digital veins, and the postaxial vein becomes the basilic, axillary, and subclavian veins. The pre-axial vein of the upper limb becomes the cephalic vein, which originally terminated in the external jugular vein, above the clavicle; its union with the axillary vein is secondary.

The distal arch and its tributaries in the lower limb persist as the dorsal venous arch of the foot and the digital veins. The postaxial vein becomes the small saphenous vein, which originally continued proximally as the popliteal and inferior gluteal veins to the internal iliac portion of the posterior cardinal vein.

The pre-axial vein of the lower limb becomes the great saphenous vein, which is continued as the proximal part of the femoral and the external iliac veins.

The companion veins of the arteries in the limbs are secondarily developed vessels.

THE FOETAL CIRCULATION

During foetal life metabolic exchange occurs between the maternal and foetal blood in the placenta; accordingly, the umbilical arteries and the umbilical vein persist till birth.

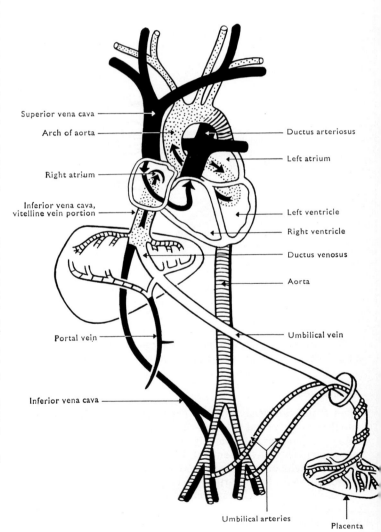

Superior vena cava

Arch of aorta

Right atrium

Inferior vena cava, vitelline vein portion

Portal vein

Inferior vena cava

Umbilical arteries

Ductus arteriosus

Left atrium

Left ventricle

Right ventricle

Ductus venosus

Aorta

Umbilical vein

Placenta

FIG. 13.127. Diagram of foetal circulation.
The inferior vena caval blood is represented as passing wholly through the foramen ovale into the left atrium, but recent work [see p. 977] has confirmed the older view that part of it is directed by the limbus fossae ovalis through the right atrioventricular orifice into the right ventricle.

The foetal lungs do not function as respiratory organs, and are bypassed by the foramen ovale and the ductus arteriosus, so that most of the blood may be transmitted to the aorta without passing through the lungs.

As a consequence, the course of the circulation is different before and after birth.

In the foetus, blood laden with food and oxygen is carried from the placenta by the umbilical vein, and transmitted to the inferior vena cava, partly directly through the ductus venosus, and partly indirectly through the liver by branches which are ultimately taken over by the left branch of the portal vein [FIG. 13.125]. In the liver and in the inferior vena cava it joins venous blood returning from lower levels, and so becomes mixed blood. The mixed stream then passes in the inferior vena cava into the right atrium, and the greater part of it traverses the foramen ovale into the left atrium, whence it passes to the left ventricle, which ejects it into the ascending aorta [FIG. 13.127].

However, the whole stream is not directed by the valve of the inferior vena cava through the foramen ovale, for it so impinges on the limbus fossae ovalis that a small part of it (probably less than one quarter) passes forwards and is carried into the right ventricle with the blood from the superior vena cava.

On the other hand, the blood returning by the superior vena cava to the right atrium passes wholly into the right ventricle and so to the pulmonary trunk. Only a small quantity of this stream enters the pulmonary arteries; the greater part of it passes through the ductus arteriosus—which is in direct line with the pulmonary trunk [FIG. 13.128]—into the descending aorta, beyond the origin of the left subclavian artery, where it mixes with the blood from the left ventricle which has not entered the branches of the arch of the aorta.

The part played in the human heart by the 'intervenous tubercle' [p. 880] in directing the stream from the superior vena cava through the right atrioventricular orifice has been exaggerated. The factors that prevent complete mixture of the blood entering the foetal right atrium through the venae cavae are the different directions of the two streams (the inferior vena cava lies on a plane posterior to the superior vena cava) so that the inferior caval blood impinges directly on the fossa ovalis. Doubts about the separation of the two streams of blood have been set at rest by the researches of Barclay and his colleagues on the foetal circulation as demonstrated by X-ray cinematography. Their method of injecting radiopaque media into the blood stream of foetal sheep has given a clear picture of the routes taken by the blood through the living heart; see Barclay, Barcroft, Barron, and Franklin (1939), and Barclay, Franklin, and Prichard (1944).

In the foetus, therefore, pure oxygenated blood is found only in the left umbilical vein and the ductus venosus; elsewhere it becomes mixed with venous blood. This mixed stream is delivered, via the ascending aorta to the head and neck and the upper limbs; the remainder passing into the descending aorta is joined, beyond the origin of the left subclavian artery, by the predominantly venous stream which has passed from the superior vena cava through the right side of the heart, the pulmonary trunk, and the ductus arteriosus. The latter blood, containing relatively less oxygen and food material and more waste products than the blood which is distributed from the arch of the aorta, passes to the walls of the thorax, to the abdomen and pelvis, to the lower limbs, and through the umbilical branches of the internal iliac arteries to the placenta.

Transition from foetal to adult circulation

The changes that occur in the vascular system at birth depend on changes in the lungs with the onset of breathing and on the cessation of the placental circulation.

At birth the ductus arteriosus begins to contract, and the **pulmonary arteries** dilate so that the pulmonary vascular resistance and pressure fall. Thus, for a while, blood passes through the ductus arteriosus from the aorta to the pulmonary artery. The consequent increase in blood flow through the lungs and the low pulmonary resistance raise the pressure in the left atrium. It exceeds that in the right during atrial systole because of the lower compliance of the left ventricle. Thus the foramen ovale ceases to function as a passage for the blood in the inferior vena cava, and the opposing surfaces of the valve-like aperture come together and begin to unite. Eventually the foramen is structurally as well as functionally closed in most cases, though the presence of an inadequate valve (failure of the septum secundum to grow sufficiently, or persistence of an ostium primum [p. 966], will now result in blood flowing from the left atrium to the right.

The functional closure of the **ductus arteriosus** occurs within a day or two, apparently by sphincter-like action of its heavily innervated but cholinesterase-free muscular wall. Thereafter, the ductus is gradually transformed into the **ligamentum arteriosum**.

The cessation of the placental circulation, completed by ligature of the umbilical cord, leads to the obliteration of the **umbilical arteries** and **vein** which are transformed into the lateral umbilical ligaments [p. 930] and the round ligament of the liver respectively. The closure of the **ductus venosus**, like that of the ductus arteriosus, seems in the first instance to be due to active sphincteric contraction, and it is gradually reduced to a fibrous cord—the ligamentum venosum.

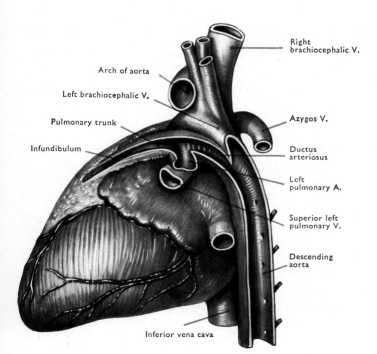

Arch of aorta

Left brachiocephalic V.

Pulmonary trunk

Infundibulum

Right brachiocephalic V.

Azygos V.

Ductus arteriosus

Left pulmonary A.

Superior left pulmonary V.

Descending aorta

Inferior vena cava

FIG. 13.128. The heart and great vessels of a foetus in which the pulmonary trunk, ductus arteriosus, and descending aorta have been opened.

References

ABBIE, A. A. (1937). The anatomy of capsular vascular disease. *Med. J. Aust.* **ii**, 564.

—— (1938). The blood supply of the visual pathways. *Med. J. Aust.* **ii,** 199.

ABBOTT, M. E. (1936). *Atlas of congenital cardiac disease.* American Heart Association, New York.

ALPERS, B. J., BERRY, R. G., and PADDISON, R. M. (1959). Anatomical studies of the circle of Willis in normal brain. *Arch. Neurol. Psychiat. (Chic.)* **81,** 409.

BARCLAY, A. E., BARCROFT, J., BARRON, D. H., and FRANKLIN, K. J. (1939). A radiographic demonstration of the circulation through the heart in the adult and in the foetus, and the identification of the ductus arteriosus. *Br. J. Radiol.* **12,** 505; *Am. J. Roentgenol.* (1942) **47,** 678.

—— FRANKLIN, K. J., and PRICHARD, M. M. L. (1944). *The foetal circulation.* Oxford.

BARLOW, T. E. (1952). Vascular patterns in the alimentary canal, in *Visceral circulation.* Ciba Foundation Symposium, **21,** London.

BATSON, O. V. (1957). The vertebral venous system. *Am. J. Roentgenol.* **78,** 195.

BOWDEN, R. E. M. (1955). The surgical anatomy of the recurrent laryngeal nerve. *Br. J. Surg.* **43,** 153.

BOYDEN, E. A. (1955). *Segmental anatomy of the lungs.* New York.

BRAITHWAITE, J. L. (1952). Variation in origin of the parietal branches of the internal iliac artery. *J. Anat. (Lond.)* **86,** 423.

BROCK, R. C. (1952). The surgical and pathological anatomy of the mitral valve. *Br. Heart J.* **14,** 489.

—— (1955). Control mechanisms in the outflow tract of the right ventricle in health and disease. *Guy's Hosp. Rep.* **104,** 356.

BURY, J. S. and STOPFORD, J. S. B. (1913). On a case of occlusion of the posterior inferior cerebellar artery. *Med. Chron.* **58,** 200.

BUTTERWORTH, R. F. (1954). The venous drainage of the left atrium. *J. Anat. (Lond.)* **88,** 131.

CLARK, E. R. (1938). Arterio-venous anastomoses. *Physiol. Rev.* **18,** 229.

—— and CLARK, E. L. (1943). Calibre changes in minute blood vessels observed in the living mammal. *Am. J. Anat.* **73,** 215.

COCKETT, F. B. and JONES, D. E. E. (1953). The ankle blow-out syndrome. A new approach to the varicose ulcer problem. *Lancet* **i,** 17.

CONGDON, E. D. (1922). Transformation of the aortic-arch system during the development of the human embryo. *Contrib. Embryol. Carneg. Instn.* **14,** 47.

DAVIES, F., FRANCIS, E. T. B., and KING, T. S. (1952). Neurological studies of the cardiac ventricles of mammals. *J. Anat. (Lond.)* **86,** 130.

—— WOOD, D. R., and JOHNSON, E. A. (1956). The atrioventricular pathway for conduction of the impulse for cardiac contraction in the dog. *Trans. R. Soc. Edinb.* **63,** 71.

DAVIS, C. L. (1927). Development of the human heart from its first appearance to the stage found in embryos of twenty paired somites. *Contrib. Embryol. Carneg. Instn.* **19,** 245.

DICKSON, A. D. (1957). The development of the ductus venosus in man and the goat. *J. Anat. (Lond.)* **91,** 358.

DOW, D. R. and HARPER, W. F. (1932). The vascularity of the valves of the human heart. *J. Anat. (Lond.)* **66,** 610.

DUCKWORTH, J. W. A. (1952). *The development of the sinu-atrial and atrioventricular nodes of the human heart.* MD Thesis, University of Edinburgh.

ELIAS, H. and PETTY, D. (1952). The gross anatomy of the blood vessels and ducts within the human liver. *Am. J. Anat.* **90,** 59.

—— and POPPER, H. (1955). Venous distribution in livers. *Arch. Path. Lab. Med.* **59,** 332.

FIELD, E. J. (1951). The development of the conducting system in the heart of the sheep. *Br. Heart J.* **13,** 129.

FLOREY, H. W., POOLE, J. C. F., and MEEK, G. A. (1959). Endothelial cells and cement lines. *J. Path. Bact.* **77,** 625.

FRANKLIN, K. J. (1937). *A monograph on veins.* Springfield, Ill.

—— (1948). *Cardiovascular studies.* Blackwell, Oxford.

FULTON, W. F. M. (1965). *The coronary arteries.* C. C. Thomas, Springfield, Ill.

GARDNER, E. and O'RAHILLY, R. (1976). The nerve supply and conducting system of the human heart at the end of the embryonic period proper. *J. Anat. (Lond.)* **121,** 571.

GIBSON, J. B. (1959). The hepatic veins in man and their sphincteric mechanisms. *J. Anat. (Lond.)* **93,** 368.

GLADSTONE, R. J. (1929). Development of the inferior vena cava in the light of recent research, with special reference to certain abnormalities, and current descriptions of the ascending lumbar and azygos veins. *J. Anat. (Lond.)* **64,** 70.

GRAVES, F. T. (1956). The anatomy of the intrarenal arteries in health and disease. *Br. J. Surg.* **43,** 605.

GRAY, H. and MAHAN, E. (1943). Prediction of heart weight in man. *Am. J. phys. Anthrop.* **1,** 271.

GRIFFITHS, J. D. (1956). Surgical anatomy of the blood supply of the distal colon. *Ann. R. Coll. Surg. Engl.* **19,** 241.

HARTY, M. (1953). Blood supply of the femoral head. *Br. med. J.* **2,** 1236.

HARVEY, W. (1628). *Exercitatio anatomica de motu cordis et sanguinis in animalibus.* Gulielmi Fitzeri, Frankfurt.

HILTON, J. (1863). *Lectures on rest and pain* (6th edn. (1950), eds. E. E. Philipp and E. W. Walls). London.

HIS, W. (1893). Die Thätigkeit des embryonalen Herzens. *Arb. med. Klin. Lpz.* (Cited by Mall, F. P., in *Am. J. Anat.* **13,** 278.)

JEFFERSON, G. and STEWART, D. (1928). On the veins of the diploë. *Br. J. Surg.* **16,** 70.

KEEN, E. N. (1955). The postnatal development of the human cardiac ventricles. *J. Anat. (Lond.)* **89,** 484.

KEITH, A. and FLACK, M. (1907). The form and nature of the muscular connexion between the primary divisions of the vertebrate heart. *J. Anat. (Lond.)* **41,** 172.

KING, T. S. and COAKLEY, J. B. (1958). The intrinsic nerve cells of the cardiac atria of mammals and man. *J. Anat. (Lond.)* **92,** 353.

KUGLER, J. H. and PARKIN, J. B. (1956). Continuity of Purkinje fibres with cardiac muscle. *Anat. Rec.* **126,** 335.

MAINLAND, D. and GORDON, E. J. (1941). The position of organs determined from thoracic radiographs of young adult males, with a study of the cardiac apex beat. *Am. J. Anat.* **68,** 457.

MALL, F. P. (1911). On the muscular architecture of the ventricles of the human heart. *Am. J. Anat.* **11,** 211.

MALPIGHI, M. (1661). *De pulmonibus.* Bononiae (Quoted by Cole, F. J. (1944). In *A history of comparative anatomy,* p. 180. Macmillan, London.)

MASSON, P. (1937). *Les glomus neuro-vasculaires.* Paris.

MICHELS, N. A. (1949). Variation in the blood supply of the supramesocolonic organs. *J. Int. Coll. Surg.* **12,** 625.

MINOT, C. S. (1900). On a hitherto unrecognized form of blood circulation without capillaries in the organs of the vertebrata. *Proc. Boston Soc. Nat. Hist.* **29,** 185.

MUIR, A. R. (1954). The development of the ventricular part of the conducting tissue in the heart of the sheep. *J. Anat. (Lond.)* **88,** 381.

NAKAMURA, N. (1924). Zur Anatomie der Bronchialarterien. *Anat. Anz.* **58,** 508.

ODGERS, P. N. B. (1938). The development of the pars membranacea septi in the human heart. *J. Anat. (Lond.)* **72,** 247.

PADGET, D. H. (1948). The development of the cranial arteries in the human embryo. *Contrib. Embryol. Carneg. Instn.* **32,** 205.

POWELL, E. D. U. and MULLANEY, J. M. (1960). The Chiari network and the valve of the inferior vena cava. *Br. Heart J.* **22,** 579.

PRICHARD, M. M. L. and DANIEL, P. M. (1956). Arteriovenous anastomoses in the human external ear. *J. Anat. (Lond.)* **90,** 309.

PURKINJE, J. E. VON (1845). Mikroskopisch-neurologische Beobachtungen. *Arch. Anat. Physiol.* **12,** 281.

QUIRING, D. P. (1949). *Collateral circulation (anatomical aspects).* London.

ROMANES, G. J. (1965). The arterial blood supply of the human spinal cord. *Paraplegia* **2,** 199.

ROUGET, C. (1879). Sur la contractilité des capillaires sanguins. *C. R. Acad. Sci. (Paris)* **88,** 916.

RUSTED, I. E., SCHIEFLEY, C. H., and EDWARDS, J. E. (1952). Studies of the mitral valve. I. Anatomic features of the normal mitral valve and associated structures. *Circulation* **6,** 825.

SENIOR, H. D. (1919a). The development of the arteries of the human lower extremity. *Am. J. Anat.* **25,** 55.

—— (1919b). An interpretation of the recorded arterial anomalies of the human leg and foot. *J. Anat. (Lond.)* **53,** 130.

—— (1920). The development of the human femoral artery, a correction. *Anat. Rec.* **17,** 271.

SHANER, R. F. (1929). The development of the atrio-ventricular node, bundle of His, and sino-atrial node in the calf: with a description of a third embryonic node-like structure. *Anat. Rec.* **44,** 85.

SHELLSHEAR, J. L. (1922). Blood supply of the dentate nucleus of the cerebellum. *Lancet* **i,** 1046.

STANFIELD, J. P. (1960). The blood supply of the human pituitary gland. *J. Anat. (Lond.)* **94,** 257.

STREETER, G. L. (1915). The development of the venous sinuses of the dura mater in the human embryo. *Am. J. Anat.* **18,** 145.

TANDLER, J. (1912). The development of the heart, in *Manual of human embryology* (ed. F. Keibel and F. P. Mall) Vol. II, p. 534. Philadelphia.

—— (1913). Anatomie des Herzens, *Bardeleben's Handbuch der Anatomie des Menschen,* Bd. III, Abth. I. Jena.

TAWARA, S. (1906). *Das Reizleitungssystem des Saugetierherzens.* Jena.

THOMAS, C. E. (1957). The muscular architecture of the ventricles of hog and dog hearts. *Am. J. Anat.* **101,** 17.

TOBIN, C. E. (1952). The bronchial arteries and their connections with other vessels in the human lung. *Surg. Gynec. Obstet.* **95**, 741.

WALLS, E. W. (1945). Dissection of the atrio-ventricular bundle in the human heart. *J. Anat. (Lond.)* **79**, 45.

—— (1947). The development of the specialized conducting tissue of the human heart. *J. Anat. (Lond.)* **81**, 93.

WALMSLEY, T. (1929). The heart, in *Quain's elements of anatomy*, (11th. edn.). Vol. IV, pt. III. London.

WENINK, A. C. G. (1976). Development of the human cardiac conducting system. *J. Anat. (Lond.)* **121**, 617.

WILLIS, T. (1664). *Cerebri anatome: cui accessit nervorum descriptio et usus.* London.

WOOLLAM, D. H. M. and MILLEN, J. W. (1955). The arterial supply of the spinal cord and its significance. *J. Neurol. Neurosurg. Psychiat.* **18**, 97.

—— —— (1958). The anatomical background to vascular disease of the spinal cord. *Proc. R. Soc. Med.* **51**, 540.

The lymphatic system

The role of the lymphatic system in absorption is mentioned on page 871. The fluid absorbed by its vessels is called lymph, and is colourless, except that from the intestine (chyle), which appears milky during digestion because of its fatty content. Not all parts of the body have lymph vessels, for they are absent from avascular tissues and from the central nervous system. In muscle they are present in the epimysium and perimysium but do not occur in the endomysium; and although present in the periosteum they have yet to be demonstrated in bone or bone marrow. In the spleen and thymus they are found only in the capsule and thickest trabeculae.

In their course lymph vessels are interrupted by lymph nodes which serve in part as filters (see below), and in part as sources of lymphocytes. Most of the lymph passes through at least one lymph node—generally more—before reaching the blood stream, but in a proportion of the animals studied by Engeset (1959a) lymph travelled from the testis to the thoracic duct without passing through any nodes.

The lymph capillaries from which lymph vessels arise are abundant in the skin, mucous membranes, glands, serous membranes, and synovial membranes and are essentially subepithelial in position.

Superficial lymph vessels lie in the skin and subcutaneous tissues. In the limbs, they join the deep vessels at constant sites.

On each side of the body the cutaneous lymph vessels converge from three large areas upon three groups of lymph nodes [FIGS. 13.130, 13.131, and 13.160]: (1) from the skin of the lower limb, perineum, external genital organs, and the trunk below the level of the umbilicus—to the superficial inguinal lymph nodes in the groin; (2) from the skin of the upper limb and the trunk above the umbilicus to the level of the clavicle in front and halfway up the back of the neck behind—to the lymph nodes in the axilla; (3) from the scalp, face, and the rest of the neck—to the cervical nodes.

Deep lymph vessels drain the lymph from parts deep to the deep fascia, and they tend to accompany the blood vessels of the region.

In the limbs and trunk they are relatively scanty and arise primarily in the synovial membranes of the joints.

Lymph nodes also are divided into superficial and deep groups. The former lie in the superficial fascia and are comparatively few; they are associated more particularly with the superficial lymph vessels of the limbs and the trunk. The deep nodes of the limbs also are comparatively few, but those of the head, neck, and trunk are numerous.

General plan of the lymphatic system

From the capillary plexuses, collecting vessels run towards lymph nodes. On reaching a node the vessels, called **afferent vessels**, penetrate the capsule at numerous points, and the lymph percolates slowly through a meshwork of lymph sinuses [p. 984] whose walls take up from the lymph particulate material, such as bacteria which have entered the tissues or inhaled carbon particles, and are an important part of the body's defence. **Efferent vessels** from the nodes either run with afferent vessels into another node of the same group or pass on to a different group. The efferent vessels from the most proximal group of each chain of nodes unite to form **lymph trunks** which are named: (1) lumbar; (2) intestinal; (3) bronchomediastinal; (4) subclavian; (5) jugular. Each trunk drains a definite territory of the body, and most empty into great terminal vessels—the thoracic duct and right lymph duct [p. 984]—which open into the great veins in the root of the neck [FIG. 13.129], forming the only

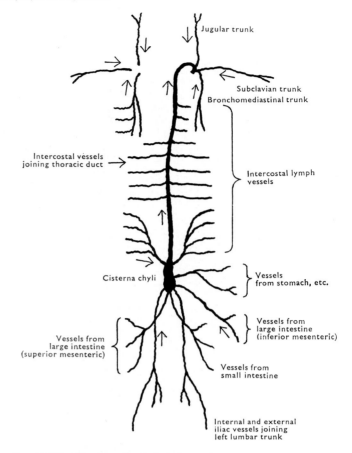

FIG. 13.129. Diagram of main lymph vessels.

lymphaticovenous communications, apart from the occasional direct entry into the great veins of the subclavian, jugular, or bronchomediastinal trunks (Engeset 1959b).

For a more detailed account of the lymphatic system than that which follows, Rouvière (1932) should be consulted. The physiological and clinical significance of the lymphatic system is fully discussed by Yoffey and Courtice (1970).

LYMPH VESSELS

The **capillaries** of origin are closed, and have no communication with blood vessels or with tissue spaces. They are wider than blood capillaries, more irregular in calibre, and have a similar structure except for the absence of pericytes. They are able to absorb particulate matter and larger molecules (e.g. protein) than can pass through blood capillary walls. The capillary plexuses have no valves, and are drained by vessels of the venule type; they form wider-meshed networks from which the collecting vessels arise.

The **collecting vessels** are small, and are distinguished by their beaded appearance due to valves placed at close and regular intervals which prevent backflow to the periphery. Unlike veins, they run in streams of individual vessels rather than by confluence into larger trunks [FIG. 13.154]. On reaching a node each afferent vessel usually breaks up into smaller branches which penetrate the capsule. The efferent vessels are only slightly larger than the afferents and have the same structure. Owing to the valves, the collecting lymph vessels can be injected only from the periphery (Jamieson and Dobson 1910).

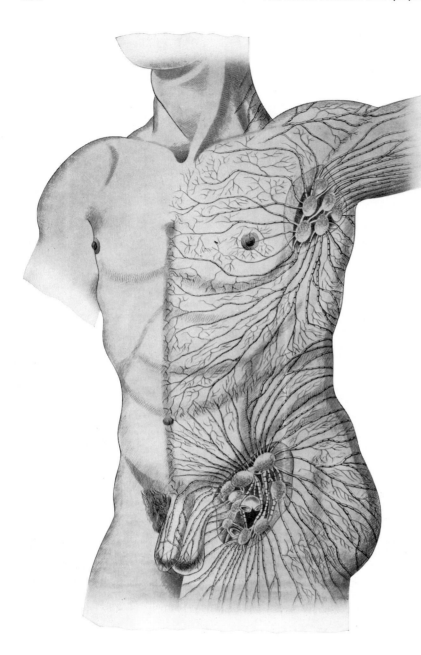

FIG. 13.130. Cutaneous lymph vessels of
anterior surface of trunk. (After Sappey
1874.)
   Note the 'lymphshed' between the vessels
that pass to the axillary and to the inguinal
lymph nodes.

In recent years techniques have been developed whereby lymph
vessels can be rendered visible in patients by the injection of a
harmless dye or X-ray opaque material. Information obtained in
this way is of value in assessing cases where there is faulty lymph
flow (Kinmonth 1954).

Structure.

The walls of the largest lymph vessels resemble those of veins. The
inner coat is a layer of endothelium covered by elastic fibres or a
fenestrated elastic membrane. The **middle coat** is formed of plain
muscle fibres, intermingled with elastic fibres. The **outer coat**
consists of fibrous tissue amongst which are plain muscle fibres.
   The bicuspid **valves** are semilunar folds of the inner coat similar
to those of veins. They are numerous in the collecting vessels and are
present also near the entrances of the great lymph channels into the
venous system.

LYMPH NODES

These are nodules which vary in size and shape; they are usually
ovoid or reniform, and from 1 to 25 mm in length. The larger nodes
show a depressed area, the hilus, through which the blood vessels
pass and the efferent lymph vessel emerges. Their colour is usually
greyish pink, but those of the lungs are commonly blackened by
carbon particles, and those of the liver and spleen are often
brownish. Lymph nodes tend to arrest for a while the central spread
of cancerous tumours—one reason why exact knowledge of the
lymphatic system is essential for surgical practice.

Structure of lymph nodes

Lymph nodes consist essentially of: (1) masses of lymph tissue,
supported by (2) a fibrous framework, which includes an external
capsule and internal trabeculae, separated from the lymph tissue by

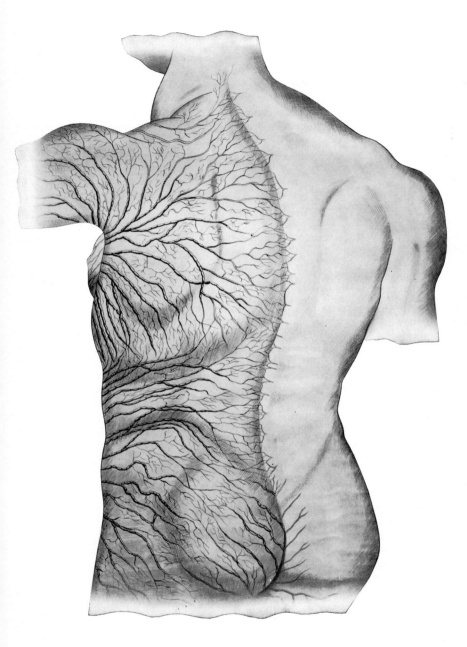

FIG. 13.131. Cutaneous lymph vessels of posterior surface of trunk. (After Sappey 1874.)

Note the 'lymphsheds' between the vessels that pass to the axillary and to the inguinal lymph nodes, and on the buttock between vessels that pass to the inguinal nodes by lateral and medial routes.

(3) spaces called lymph sinuses. Each node is separable into cortex and medulla. The **cortex** lies immediately internal to the capsule, except at the hilus, where it is absent. The **medulla** forms the internal part of the node and reaches the capsule at the hilus.

The **capsule** is formed mainly of white fibrous tissue and from its deep surface **trabeculae** pass towards the medulla. Within this collagenous framework there is a fine meshwork, composed of reticular fibres and their associated cells, which pervades the node and supports the lymph tissue [FIG. 13.132].

The lymph tissue in the medulla forms branching and anastomosing cords, but in the cortex it is in the form of rounded **lymph follicles** which consist of masses of densely packed lymphocytes supported by reticular tissue. The centre of each follicle consists of somewhat larger cells, called lymphoblasts, which by their division form lymphocytes; the name **germinal centre** is given to this part of the follicle.

The fact that the germinal centres do not develop in an animal reared in a germ-free environment, and appear only after the animal has been exposed to an infection or other form of antigen, is related to the role of the lymph nodes in immunity. In a section of a lymph node examined with the light microscope all the lymphocytes present look identical. Nevertheless, it is now certain that there are two types of lymphocytes, distinguishable both structurally—as seen with the electron microscope—and functionally—as shown by their response to an antigen. The two types are known as T- and B-lymphocytes, and both are derived from stem cells in the yolk sac or bone marrow, the marrow being the sole source of stem cells in late foetal and postnatal life. The letter T indicates that the cells spent some time in the thymus developing immune competence before re-entering the bloodstream to seed the lymph tissues of the body. The letter B indicates that the cells spent some time in an unknown site developing immune competence before seeding the lymph tissues.

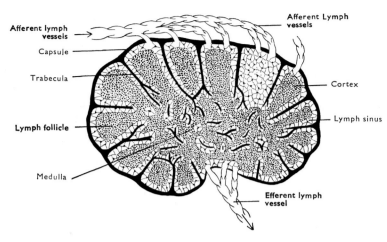

FIG. 13.132. Diagram of section of lymph node.

In birds the site is known, and is a lymph organ situated near the cloaca. It is called the bursa of Fabricius, and as with thymus the initial letter was chosen for the cell type. In mammals the analogue of the bursa awaits identification, although there are some who believe that the solitary lymph follicles [see below] have a claim to be so recognized.

Beneath the capsule, alongside the trabeculae, and in the medulla, the reticular tissue is more loosely arranged and forms lymph sinuses which afford passage to the lymph. The sinuses are known as subcapsular (or marginal), trabecular, and medullary, and their walls are formed by the reticular cells which clothe the reticular fibres. As noted earlier some of these cells are phagocytic.

The above account of lymph sinuses is based on light microscopy, and is at variance with the findings made by Nopajaroosri, Luk, and Simon (1971) using the electron microscope. These workers found the sinuses to have an endothelial lining—incomplete on one side— and to contain many free cells, mostly lymphocytes and macrophages. The arrangement clearly allows these cells easy access to the sinuses, and like the older account supports the concept of filtration taking place in the sinuses.

The lymphocytes produced in the germinal centres of the follicles are added to the lymph flowing through the node, and eventually enter the blood through the thoracic duct. Gowans and Knight (1964) have shown there is a continuous recirculation of lymphocytes from the blood back into the lymph. For an account of modern views on the role of lymphocytes in the immunological responses of the body, Roitt, Greaves, Torrigiani, Brostoff, and Playfair (1969) and Loor and Roelants (1977) should be consulted.

Solitary lymph follicles

Tiny collections of lymph tissue are found in the mucous coat of the alimentary canal. In the pharynx there are great aggregations of such units which form tonsils; in the ileum many of the follicles coalesce to form elongated raised patches (aggregated lymph follicles); and in the vermiform appendix they form masses which constitute the most striking feature of its structure.

LYMPHANGIOGRAPHY

Before lymph vessels are cannulated for injection of contrast medium they are first made visible by subcutaneous injection of a blue dye.

Very slow injection of oily contrast medium into these vessels on the dorsal aspects of both feet simultaneously allows the lymph vessels of the lower limb to be filled (lymphangiogram; FIG. 13.144).

The oily contrast medium aggregates in the lymph nodes and 12 and 24 hour films show filling reaching to the para-aortic lymph nodes as far as the cisterna chyli [FIGS. 13.144 and 13.145]. The thoracic duct is seldom filled because this sometimes results in overfilling and a rapid escape of contrast material into the systemic veins, which may ultimately reach the lungs, with the risk of causing blockage of the vessels there.

Terminal lymph vessels

The terminal lymph vessels are the thoracic duct and the right lymph duct (or the vessels that represent it).

THORACIC DUCT

The thoracic duct begins in the **cisterna chyli**, a dilatation some 6 mm wide and 6 cm long, which lies between the aorta and vena azygos on the bodies of the first two lumbar vertebrae [FIG. 13.133]. The duct passes from the cisterna through the aortic opening of the diaphragm, and enters the posterior mediastinum in which it lies on the vertebral column in, or just to the right of, the median plane. At the level of the fifth thoracic vertebra it crosses behind the oesophagus, and continues along its left side through the superior mediastinum. At the root of the neck [FIG. 13.134] the duct arches laterally over the apex of the pleura, and between the left carotid sheath in front and the vertebral vessels behind. It then descends across the first part of the left subclavian artery, and enters the angle of junction of the left internal jugular and subclavian veins.

The thoracic duct is about 2 mm in diameter, and has many paired valves—one near its termination.

Tributaries.

The cisterna chyli commonly receives: (1) the intestinal trunk [p. 992]; (2) a pair of lumbar trunks [p. 986]; and (3) a pair of descending intercostal lymph trunks [p. 995].

In the posterior mediastinum the thoracic duct receives efferents from some intercostal and posterior mediastinal nodes; and in the superior mediastinum the efferents of upper intercostal nodes of both sides open into it.

Immediately before its termination it may receive: (1) the efferents from nodes of the left upper limb, which frequently unite to form a **subclavian trunk**; and (2) the **left jugular trunk**, which conveys lymph from the left side of the head and neck; but those vessels may end separately in the brachiocephalic or the subclavian vein. The **left bronchomediastinal trunk** may join the thoracic duct in the neck, but it usually opens independently into the left brachiocephalic vein. It collects lymph from the left lung, the left side of the deeper parts of the anterior thoracic and upper abdominal walls, and from the anterior part of the diaphragm and the left side of the heart.

RIGHT LYMPH DUCT [FIG. 13.135]

This duct rarely exists as such, since the three vessels which may unite to form it usually open separately into the right internal jugular, subclavian, and brachiocephalic veins. These vessels are: (1) the **right jugular trunk**; (2) the **right subclavian trunk**; and (3) the **right bronchomediastinal trunk**. The jugular and subclavian trunks convey lymph from the right side of the head and neck and

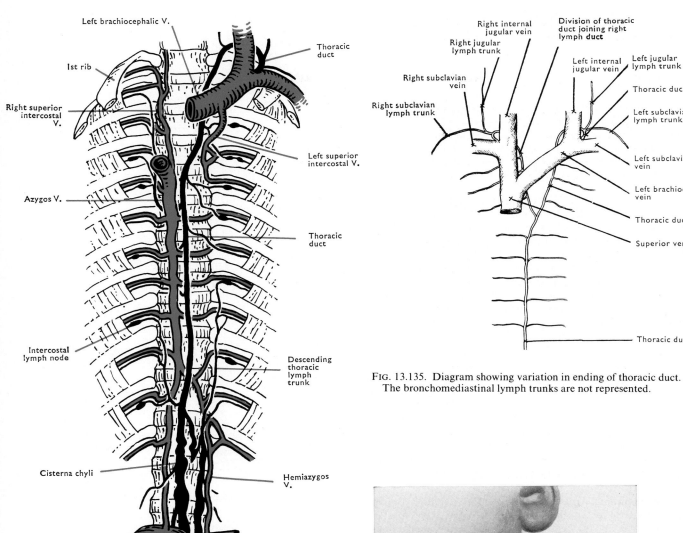

FIG. 13.133. The thoracic duct and cisterna chyli.

FIG. 13.135. Diagram showing variation in ending of thoracic duct. The bronchomediastinal lymph trunks are not represented.

FIG. 13.134. Deep dissection of root of neck on left side to show dome of pleura and relations of thoracic duct.

Parts of the sternocleidomastoid and the sternothyroid have been removed.

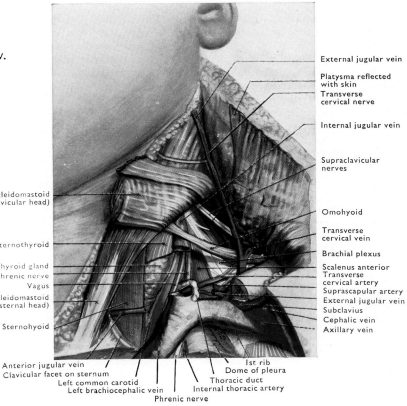

from the right upper limb respectively; the right bronchomediastinal trunk collects lymph from an area corresponding to that drained by the left vessel, but including also the upper part of the right lobe of the liver.

VARIATIONS

The cisterna chyli may be replaced by a plexus of vessels from which the thoracic duct takes origin by several roots.

The thoracic duct sometimes ends in a number of branches which enter the great veins separately; or it may divide in the upper part of the thorax into two stems, of which the right joins the right jugular trunk to form one variety of the right lymph duct [FIG. 13.135].

The arrangement and mode of termination of the three lymph trunks—subclavian, jugular, bronchomediastinal—are very variable, but the commonest arrangements are as follows.

1. On the left side the jugular trunk ends in the thoracic duct, and the other two trunks may do so also; but the subclavian trunk often ends in the subclavian vein, and the bronchomediastinal trunk usually ends in the brachiocephalic vein.

2. On the right side the trunks usually have separate entrances into the subclavian, internal jugular, and brachiocephalic veins; but it is not uncommon for the subclavian and jugular trunks to unite to form the right lymph duct.

Lumbar lymph trunks

These para-aortic lymph trunks receive the lymph from the lower limbs, the abdominal wall below the level of the umbilicus, the urogenital system, and the portion of the digestive tract supplied by the interior mesenteric artery [FIG. 13.142].

SUPERFICIAL LYMPH NODES OF LOWER LIMB

The superficial inguinal lymph nodes [FIGS. 13.130 and 13.136] lie in the subcutaneous fat and are normally palpable. They form a proximal set (five or six) below and parallel with the inguinal ligament and a distal set (four or five) associated with the great saphenous vein. Their afferents are derived from all the skin below the level of the umbilicus, except the area drained by the popliteal nodes; and from the lining membranes of the anal canal and, usually, of the penile urethra, or of the vulva and lower end of the vagina. They receive also a few vessels from the uterus which run along the round ligament through the inguinal canal. The efferents pass through the cribriform fascia; some enter deep inguinal nodes, but most pass through the femoral canal to end in external iliac nodes.

SUPERFICIAL LYMPH VESSELS OF THE LOWER LIMB AND LOWER TRUNK

These vessels are the afferents of superficial inguinal nodes upon which they converge in a great whorl [FIGS. 13.130, 13.131, 13.136, and 13.138].

Certain points must be noted: (1) lymph vessels from the lateral part of the plantar surface of the foot, from its lateral border, and from the heel, accompany the small saphenous vein and end in the popliteal nodes; (2) the lymph vessels from the glans penis end mainly in deep inguinal nodes, but a few run with the deep dorsal vein into the pelvis; (3) *the testis does not send lymph vessels to the inguinal nodes.*

The arrangement of the superficial lymph vessels of the toes and the foot is similar to that of the fingers and hand [p. 998]. From

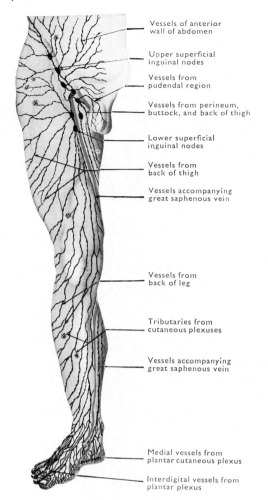

FIG. 13.136. Superficial lymph vessels of anterior surface of lower limb.

plexuses on the plantar surface, vessels pass to a dorsal network from which most of the collecting vessels ascend parallel to the great saphenous vein and end in the distal group of superficial inguinal nodes.

DEEP LYMPH NODES AND LYMPH VESSELS OF THE LOWER LIMB

The deep nodes are few and small. The anterior tibial lymph node is a tiny nodule on the upper part of the interosseous membrane.

The popliteal lymph nodes (six or seven) are small nodules in the popliteal fat alongside the great blood vessels. They receive the anterior and posterior tibial lymph vessels and others which issue from the knee joint. Lymph vessels running with the small saphenous vein end in a node which lies on the upper end of that vein under the deep fascia. The efferents of the popliteal nodes pass with the femoral vessels to the deep inguinal nodes.

The deep inguinal lymph nodes [FIGS. 13.140, 13.142, and 13.143] lie on the medial side of the femoral vein. They are small and few. They receive some of the efferents from the superficial inguinal nodes, and all the deep vessels in the region of distribution of the femoral artery. Their efferents pass to external iliac nodes.

The deep lymph vessels of the lower limb follow the main blood vessels. Some pass directly to deep inguinal nodes, others traverse

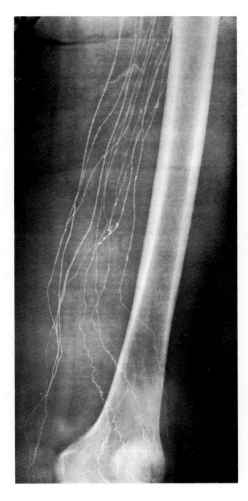

FIG. 13.137. Lymphangiogram to show some of the main superficial lymph channels of the thigh.

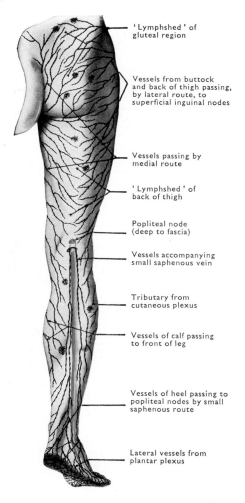

FIG. 13.138. Superficial lymph vessels of posterior surface of lower limb.

popliteal nodes. The deep lymph vessels of the gluteal region and perineum accompany branches of the internal iliac artery into the pelvis and will be noted later.

LYMPH VESSELS OF THE ANTERIOR WALL OF THE ABDOMEN

The *superficial lymph vessels* of the upper part of the anterior wall of the abdomen go mainly to the **pectoral** group of axillary nodes; but some pierce the wall of the thorax and end in **parasternal nodes** [p. 996]. Those of the lower part of the abdominal wall end in superficial inguinal nodes [FIG. 13.130].

The *deep lymph vessels* of the upper part of the anterior abdominal wall accompany the superior epigastric blood vessels and end in parasternal nodes. Those of the lower part accompany the inferior epigastric and deep circumflex iliac vessels, and end in external iliac nodes.

LYMPH VESSELS OF THE EXTERNAL GENITAL ORGANS

Lymph vessels of the scrotum or of the vulva pass to the medial, proximal superficial inguinal nodes. The superficial lymph vessels of

the penis go to the medial, proximal superficial inguinal group, but the deep vessels, including those of the penile urethra, end either in the medial, proximal superficial inguinal group or in the deep inguinal nodes. A few vessels follow the deep dorsal vein and join internal iliac nodes. The lymph vessels of the clitoris end like those of the penis.

Lymph nodes of the pelvis

The lymph nodes of the pelvis may be grouped with the blood vessels [FIGS. 13.139–13.142].

The **external iliac lymph nodes** (from eight to ten in number) lie along the external iliac vessels. They receive the efferents of the inguinal nodes, the deep lymph vessels of the iliac fossa and of the lower part of the anterior abdominal wall, and a number of afferents **direct** from the pelvic viscera (q.v.). Their **efferents** pass to the common iliac nodes.

The **internal iliac lymph nodes** lie along the trunk and branches of the internal iliac vessels. They receive **afferents** from all the pelvic viscera except the ovary and adjacent part of the uterine tube, from the deep structures of the perineum and the gluteal region, including most of the hip joint (the anterior part of the capsule

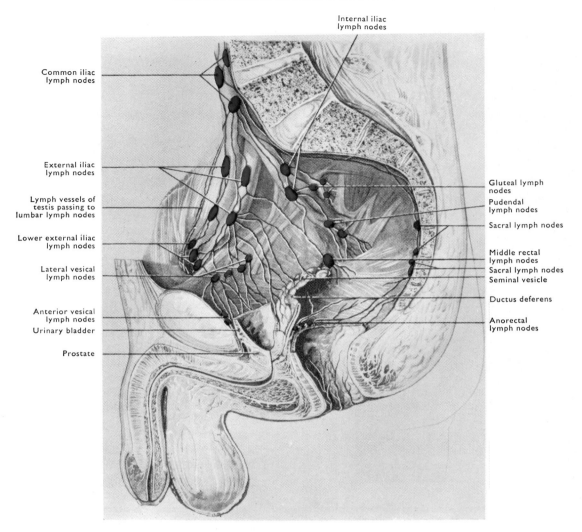

FIG. 13.139.   Lymph nodes of the pelvis.

drains to the deep inguinal nodes). Their **efferents** pass to the common iliac nodes. Only one subgroup of internal iliac nodes need be named—the **sacral lymph nodes,** which lie in the hollow of the sacrum.

The **common iliac lymph nodes** (from four to six in number) extend along the common iliac vessels to the aortic bifurcation where they form the median common iliac nodes. Their **efferents** pass to the lumbar nodes.

The nodes of the lesser pelvis thus form two major subdivisions: those of the **cavity**—the internal iliac nodes, and the external and common iliac nodes.

Both receive direct afferents from the pelvic viscera, and in addition the external and common iliac nodes receive the efferent vessels from the nodes of the cavity. *Nevertheless, vessels from the upper part of the rectum, from the upper part of the uterus, and from the uterine tube and ovary pass direct to the lumbar nodes in the abdomen with the inferior mesenteric and ovarian vessels.*

Lymph vessels of the pelvic viscera

The **lymph vessels of the urinary bladder** pass mainly to external iliac lymph nodes, but some from the base end in internal iliac

nodes. Those from the neck of the bladder run to the sacral and median common iliac nodes.

The **lymph vessels of the prostate** mainly end in internal iliac nodes; some pass to sacral nodes, and others follow the ductus deferens to external iliac nodes.

Lymph vessels of the urethra

In the male, the vessels of the spongy part of the urethra pass to superficial and deep inguinal nodes. Those of the membranous and prostatic parts of the urethra mainly end in internal iliac nodes.

The **lymph vessels of the female urethra** correspond with those of the membranous and prostatic parts of the male urethra and end mainly in internal iliac nodes, though its inferior extremity drains with the skin of the vulva.

The lymph vessels of the **ductus deferens** and of the **seminal vesicle** pass to external and internal iliac nodes.

Lymph vessels of the vagina

Below the hymen, the vessels pass with those of the vulva to superficial inguinal nodes. From the middle part they accompany the vaginal artery to internal iliac nodes; and from the upper part they accompany those of the cervix of the uterus to external and internal iliac nodes [FIGS. 13.141 and 13.142].

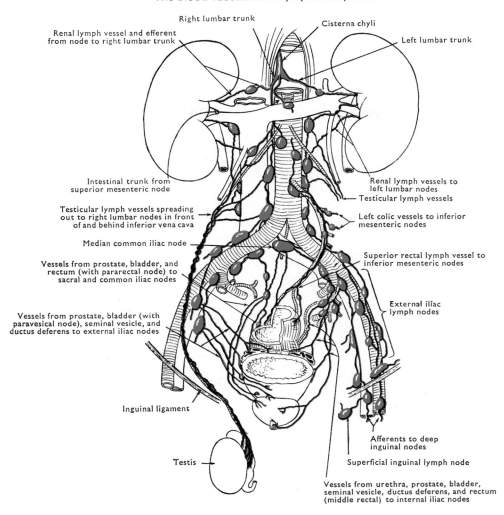

Right lumbar trunk

Cisterna chyli

Renal lymph vessel and efferent
from node to right lumbar trunk

Left lumbar trunk

Intestinal trunk from
superior mesenteric node

Renal lymph vessels to
left lumbar nodes

Testicular lymph vessels

Testicular lymph vessels spreading
out to right lumbar nodes in front
of and behind inferior vena cava

Left colic vessels to inferior
mesenteric nodes

Median common iliac node

Superior rectal lymph vessel to
inferior mesenteric nodes

Vessels from prostate, bladder, and
rectum (with pararectal node) to
sacral and common iliac nodes

External iliac
lymph nodes

Vessels from prostate, bladder (with
paravesical node), seminal vesicle, and
ductus deferens to external iliac nodes

Inguinal ligament

Afferents to deep
inguinal nodes

Superficial inguinal lymph node

Testis

Vessels from urethra, prostate, bladder,
seminal vesicle, ductus deferens, and rectum
(middle rectal) to internal iliac nodes

FIG. 13.140. Diagram of lymph vessels and lymph nodes of male pelvis and abdomen.

Lymph vessels of the uterus

The capillary plexus in the mucous coat communicates with a subserous plexus from which the collecting vessels pass to widely separated groups of lymph nodes [FIGS. 13.141 and 13.142].

From the cervix, vessels pass to both external and internal iliac nodes, and also to sacral and median common iliac nodes. From the lower part of the body most vessels go to external iliac nodes. From the upper part of the body and the fundus, the main outflow is with the vessels of the uterine tube and ovary to lumbar lymph nodes. In addition, a few vessels from the fundus and the body run along the round ligament of the uterus to superficial inguinal lymph nodes.

The lymph vessels of the ovary pass to lumbar lymph nodes from the bifurcation of the aorta to the level of the renal veins [FIG. 13.142].

Lymph vessels of the rectum and anal canal

Most of the vessels end in pelvic nodes, but those from the anal canal below the pectinate line pass to superficial inguinal nodes, and those from the upper part of the rectum pass upwards with the superior rectal vessels to the inferior mesenteric nodes.

From the greater part of the anal canal the lymph vessels run either across the ischiorectal fossa, or upwards with those of the lower part of the rectum along the middle rectal blood vessels, to

internal iliac nodes. Other vessels from the rectum end in sacral and median common iliac nodes. Vessels from the anal canal and lower rectum may be interrupted in small anorectal lymph nodes; and others from the middle and upper parts of the rectum pass through pararectal nodes which lie on the wall of the rectum.

Lumbar and inferior mesenteric lymph nodes

The lumbar lymph nodes [FIGS. 13.140 and 13.143] are continuous below with the common iliac nodes and the right and left lumbar lymph trunks emerge from their upper members. Their relation to the aorta allows of a subdivision into right and left lateral, pre-aortic, and retro-aortic groups, but they are all freely interconnected; moreover, they are not easily distinguished from the nodes around the stems of origin of the inferior and superior mesenteric arteries, and of the coeliac trunk.

The territory drained by the lumbar nodes corresponds to the distribution of the paired branches of the aorta, plus that of the inferior mesenteric artery; for although the nodes associated with the unpaired visceral branches of the abdominal aorta can be regarded as offshoots of the lumbar nodes, the inferior mesenteric group drains into the left lumbar lymph trunk while the coeliac and

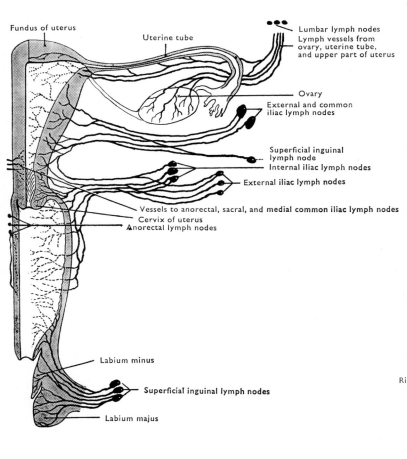

FIG. 13.141. Diagram of lymph vessels of female genital organs.

FIG. 13.142. Diagram of lymph vessels and lymph nodes of female pelvis and abdomen.

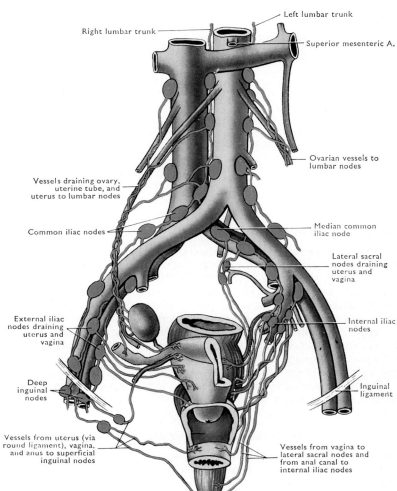

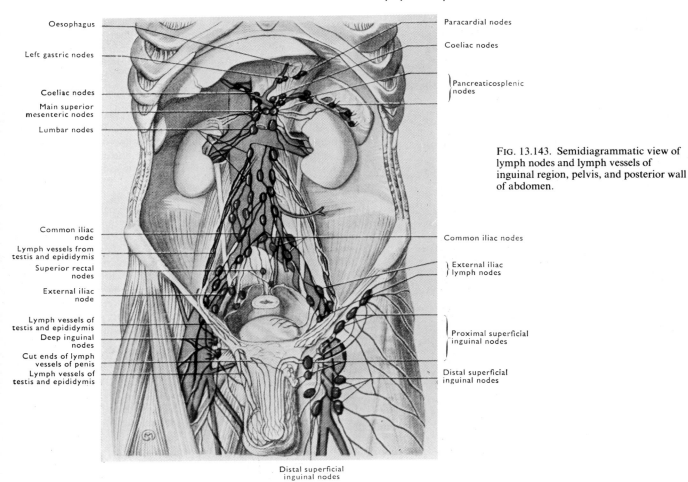

FIG. 13.143. Semidiagrammatic view of lymph nodes and lymph vessels of inguinal region, pelvis, and posterior wall of abdomen.

Oesophagus — Paracardial nodes
Left gastric nodes — Coeliac nodes
Coeliac nodes — Pancreaticosplenic nodes
Main superior mesenteric nodes
Lumbar nodes

Common iliac node — Common iliac nodes
Lymph vessels from testis and epididymis — External iliac lymph nodes
Superior rectal nodes
External iliac node — Proximal superficial inguinal nodes
Lymph vessels of testis and epididymis
Deep inguinal nodes — Distal superficial inguinal nodes
Cut ends of lymph vessels of penis
Lymph vessels of testis and epididymis

Distal superficial inguinal nodes

superior mesenteric groups reach the cisterna chyli separately through the intestinal trunk.

The lumbar nodes thus receive the following lymph vessels:

1. The efferents of the common iliac nodes.
2. The efferents of the inferior mesenteric nodes [see below].
3. Lymph vessels of the **testis and epididymis**. Some six to eight vessels ascend in the spermatic cord, and then follow the testicular vessels upwards. They end in lumbar nodes, from the bifurcation of the aorta to the level of the renal veins [FIG. 13.140]. The corresponding vessels from the **ovary**, the **uterine tube**, and upper part of the uterus have the same termination.
4. Lymph vessels of the **ureter**. From the part of the ureter in the pelvis vessels pass to the internal and external iliac nodes; from the middle part of the ureter vessels pass to lumbar nodes; and those from its upper part accompany the renal lymph vessels.
5. Lymph vessels of the **kidney**. These emerge from the hilus, and pass medially to lumbar nodes.
6. Lymph vessels of the **suprarenal gland**. These end mainly in upper lumbar nodes, but some pass through the diaphragm to posterior mediastinal nodes.
7. Lymph vessels of the lumbar synovial joints.
8. Lymph vessels of the **peritoneum**. The lymph vessels of the visceral peritoneum run with those of the organs it covers. Those from the parietal peritoneum tend to converge on the neighbouring blood vessels. In the lesser pelvis they pass to the nearest internal or external iliac nodes. On the anterior abdominal wall and in the iliac fossa they accompany the epigastric and the deep circumflex iliac blood vessels. On the posterior wall of the abdomen, in its lower part, they run to

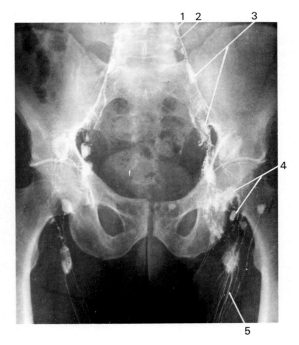

FIG. 13.144. Lymphangiogram of inguinal, external iliac, and lumbar regions, following the introduction of opaque medium into lymph vessels of the lower limb.
1. Lumbar lymph node.
2. Lumbar lymph trunk.
3. Common and external iliac nodes.
4. Inguinal nodes.
5. Lymph vessels of thigh.

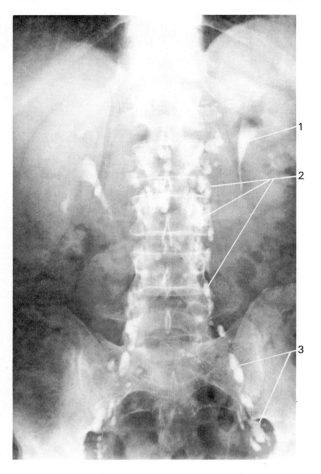

FIG. 13.145. Lymphangiogram taken a day later than FIG. 13.144. The contrast medium now fills many lumbar lymph nodes. An intravenous pyelogram at the same time outlines the right kidney with its surrounding fat.
1. Pelvis of kidney.
2. Lumbar lymph nodes.
3. Common and external iliac nodes.

nodes on the stems of the superior and inferior mesenteric arteries; laterally they accompany the lumbar blood vessels to lateral lumbar nodes; higher up they join the vessels from the kidneys. The lymph plexus on the abdominal surface of the diaphragm drains, partly, to the uppermost lumbar nodes; but it also communicates freely with the rich subpleural plexus on the upper surface of the muscle.

The **inferior mesenteric lymph nodes** lie along the stem of the artery of that name. They receive lymph from nodes scattered along the branches of the artery (including **left colic nodes**) as far as the wall of the intestine. These outlying nodes of the colon are arranged similarly to those along the colic branches of the superior mesenteric artery.

The **lymph nodes of the colon** [FIG. 13.148] form four groups. The **epicolic nodes** are tiny nodules in the appendices epiploicae and the wall of the gut. The **paracolic nodes** lie along the medial borders of the ascending and descending colon, and along the mesenteric borders of the transverse and sigmoid colon. The **intermediate colic nodes** lie along the branches of the colic arteries. The **main colic nodes** are situated on the stems of the colic arteries. In addition to the rectal lymph vessels that run with the superior rectal artery, the main inferior mesenteric nodes receive lymph from the descending and sigmoid colon; their **efferents** pass to the lumbar lymph nodes which drain mainly into the left lumbar lymph trunk.

Intestinal lymph trunk

The intestinal lymph trunk enters the cisterna chyli and is formed by the union of the efferents of the proximal members of the groups of lymph nodes situated on the coeliac and superior mesenteric arteries and their branches.

Coeliac and superior mesenteric lymph nodes

The coeliac lymph nodes surround the stem of the coeliac trunk, and drain the abdominal part of the foregut; but some lymph from the liver, duodenum, and pancreas passes to other nodes.

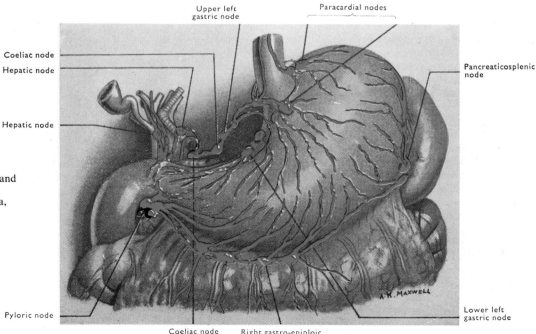

FIG. 13.146. Lymph vessels and lymph nodes of stomach. (Jamieson and Dobson 1907a, redrawn.)

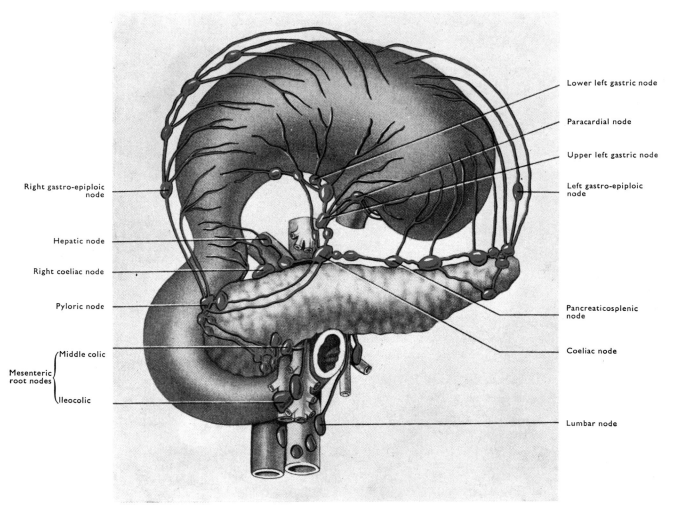

FIG. 13.147. Lymph vessels and lymph nodes of posterior surface of stomach. The stomach has been turned upwards to show the pancreaticosplenic and other groups of nodes. (Jamieson and Dobson 1907a, redrawn.)

In association with the branches of the coeliac trunk there are the following groups of lymph nodes: left gastric, right gastric, hepatic, pancreaticosplenic, and pyloric.

The **left gastric lymph nodes** form three subgroups. The **lower** nodes lie on the lesser curvature to the left of the angular notch. Around the cardiac end of the stomach **paracardial** nodes are disposed like a string of beads; the **upper** nodes lie on the trunk of the left gastric artery. Finally the chain merges with the coeliac group. The left gastric nodes drain the abdominal part of the oesophagus, and the right two-thirds of the surfaces of the stomach from the fundus as far as the pyloric canal.

The **right gastric lymph nodes** lie in the lesser omentum along the right gastric artery; their efferents pass to the hepatic group.

The **hepatic lymph nodes** lie along the proper hepatic artery as far as the porta hepatis; one of the group is the **cystic lymph node** which lies in the curve of the neck of the gall-bladder. Some nodes are arranged along the bile-duct and may be called **biliary lymph nodes**. The hepatic nodes receive afferents from the liver and gall-bladder, stomach, duodenum and pancreas; their efferents pass to the coeliac group.

The **pancreaticosplenic lymph nodes** lie along the splenic artery. The nodes in the hilus of the spleen receive lymph vessels from the capsule of that organ, and also vessels which accompany the left gastro-epiploic artery from the stomach; small left gastro-epiploic nodes may be found in the gastrosplenic ligament. The efferents from the pancreaticosplenic nodes pass to the coeliac group.

The **pyloric lymph nodes** lie on the head of the pancreas below the superior part of the duodenum, and receive efferent vessels from nodes between the layers of the greater omentum—the **right gastro-epiploic lymph nodes**. These nodes together drain the adjacent parts of the stomach and duodenum. The pyloric nodes send efferents to superior mesenteric, coeliac, and hepatic nodes [FIG. 13.147].

The efferent vessels of the coeliac and superior mesenteric nodes combine to form the (single or multiple) **intestinal trunk**.

SUPERIOR MESENTERIC LYMPH NODES

1. *On the stem of the artery* these are large and numerous. They receive the efferents of all the subsidiary groups, some efferents from pyloric nodes, direct afferents from the head of the pancreas and the duodenum, and some efferents from the upper left colic group [FIG. 13.148].

2. Subsidiary groups related to the branches of the artery.

(i) The **lymph nodes of the mesentery** (100–200 in number) are scattered throughout the mesentery: they drain the jejunum and the ileum (except its lower end).

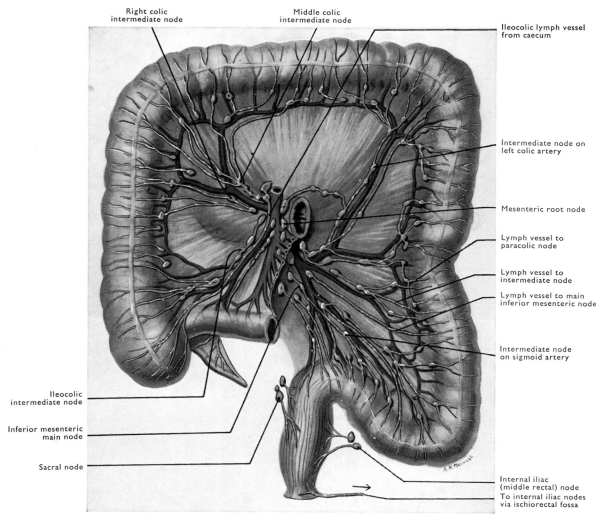

FIG. 13.148. Lymph vessels and lymph nodes of large intestine. (Jamieson and Dobson 1909, redrawn.)

(ii) The epicolic and paracolic members of the **ileocolic lymph nodes** form a cluster in the ileocolic angle, below which a few nodes descend with the caecal vessels in front of and behind the ileocaecal junction—**caecal lymph nodes**. One **appendicular node** may be found in the mesentery of the appendix. The intermediate group lies half-way up the stem of the ileocolic artery, and the main group lies on the origin of that artery. The ileocolic nodes drain the lower end of the ileum, the appendix, caecum, and lower end of the ascending colon.

(iii) The **right colic lymph nodes** lie along the right colic artery and receive the lymph from the ascending colon.

(iv) The **middle colic lymph nodes** accompany the branches and stem of the middle colic artery. They drain the right flexure and the right two thirds of the transverse colon.

Lymph vessels of the digestive system in the abdomen

The lymph vessels of the abdominal part of the alimentary canal arise from a capillary plexus under the lining epithelium and around the gastric and intestinal glands. In the small intestine they also have an important origin in the lacteals [p. 457]. The lymph vessels perforate the muscular coats, and are joined by tributaries from the subserous network; they then follow the blood vessels.

Lymph vessels of the large intestine

The blood supply of the large intestine is a guide to its lymph drainage. In general, the lymph vessels accompany the arteries, and, having traversed or bypassed intervening nodes, eventually end in the nodes round the stems of the inferior and superior mesenteric arteries. However, the lymph drainage from the **left third of the transverse colon** and the **left flexure** must be specially mentioned because the stream divides at the intermediate nodes [FIG. 13.148] to follow two paths—one leading to the inferior mesenteric nodes, the other to the superior mesenteric nodes. The course of the lymph vessels from the caecum and appendix is shown in FIGURES 13.148 and 13.149. Those from the appendix unite in its mesentery to form four or five trunks which follow the appendicular blood vessels behind the ileum and end in nodes along the whole length of the ileocolic chain.

Lymph vessels of the small intestine

From the ileum and jejunum the vessels run to the nodes of the mesentery. The lower half of the **duodenum** drains to the superior

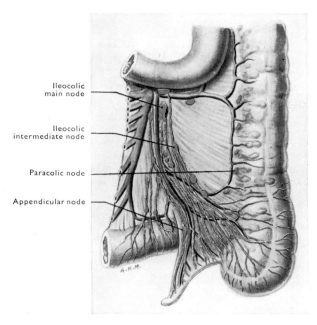

FIG. 13.149. Lymph vessels and lymph nodes of caecum and appendix, posterior aspect. (Jamieson and Dobson 1907b, redrawn.)

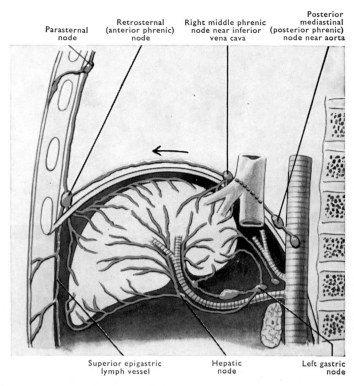

FIG. 13.150. Diagrammatic sagittal section showing lymph drainage of liver.

mesenteric nodes, the upper half to the pyloric nodes in front and to the biliary nodes behind.

Lymph vessels of the stomach

These run in three streams:
    1. From the part of the fundus and body lying to the right of a line (nearer the greater than the lesser curvature) from the fundus as far as the pyloric canal: to the left gastric nodes.
    2. From the area to the left of the line as far as the angular notch: to the pancreaticosplenic nodes.
    3. From the pyloric canal: downwards to the right gastro-epiploic and pyloric nodes, and upwards to the right gastric nodes, biliary chain, and coeliac nodes [FIGS. 13.146 and 13.147].

Lymph vessels of the pancreas

The pancreas drains to pancreaticosplenic nodes above, to superior mesenteric nodes below, and to lumbar nodes behind.

Lymph vessels of the liver

Superficial and deep collecting vessels pass from the capillaries of the liver lobules. The superficial vessels run beneath the peritoneum, and many traverse the falciform ligament and follow the superior epigastric blood vessels to the lower parasternal nodes [p. 996]. Others run round the lower edge of the liver towards the porta hepatis, where they pick up the vessels of the visceral surface and the gall-bladder, and a deep stream of vessels from the interior of the liver. Most of these vessels end in hepatic nodes, but some in left gastric nodes. From the posterior surface of the liver a few vessels run to coeliac nodes. Deep vessels also emerge with the hepatic veins and pass with the inferior vena cava to middle diaphragmatic nodes in the thorax [see below]. There are therefore two main lymph streams from the liver—one via the coeliac nodes to the cisterna chyli, the other via the thorax to the mediastinal trunks of both sides [FIG. 13.150]. Even so, figures for the hepatic contribution to thoracic duct lymph flow indicate that it contributes nearly half of the total (Yoffey and Courtice 1970).

Lymph vessels of the spleen

The capsule drains to nodes of the pancreaticosplenic group which lie near its hilus. The pulp drains via the splenic vein and not by lymph vessels.

Lymph drainage of the thorax

The chief pathway of the lymph vessels of the thorax is the bronchomediastinal lymph trunk [p. 984]. In addition, there are two descending intercostal lymph trunks [see below], and also direct tributaries of the thoracic duct from intercostal and posterior mediastinal lymph nodes.

    The drainage of the superficial tissues of the thoracic wall is mainly to axillary lymph nodes [FIGS. 13.130 and 13.131, and p. 997]. The deeper tissues drain mainly to intercostal, parasternal, and phrenic lymph nodes.

LYMPH NODES OF THE THORAX

The intercostal lymph nodes are one or two small nodules in the vertebral end of each space. They receive afferents from the parietal pleura and deep tissues of the posterior thoracic wall. The efferents of the upper nodes run to the thoracic duct. Those of the lower four or five spaces join on each side to form a descending intercostal trunk which passes through the aortic opening to the cisterna chyli.

    The posterior mediastinal lymph nodes lie around the lower part of the oesophagus. They receive afferents from the oesophagus, pericardium, and diaphragm; their efferents pass to the thoracic duct and to the descending intercostal lymph trunks.

    The phrenic lymph nodes [FIGS. 13.150 and 13.151] form: (1) a posterior group, better considered as the lowest members of the

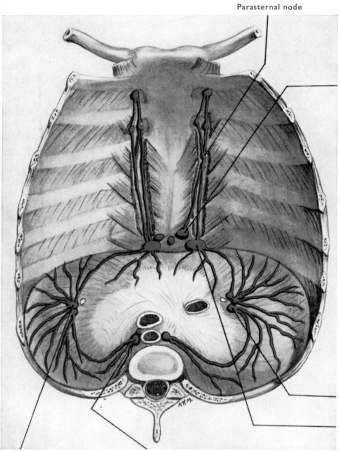

Parasternal node

Lower parasternal
(retrosternal) node

Middle phrenic node near right
phrenic nerve and inferior vena cava

Retrosternal (anterior
phrenic) node

Middle phrenic node
near left phrenic nerve

Posterior mediastinal (posterior phrenic)
node near aorta and oesophagus

FIG. 13.151. Parasternal and phrenic lymph
nodes, posterior aspect, and lymph drainage
of thoracic surface of diaphragm.

posterior mediastinal group; (2) an anterior group, better considered
as the lowest members of the parasternal chain; and (3) a middle
group on each half of the diaphragm which receives afferents from
the diaphragm and from the liver, and sends efferents to the
parasternal and posterior mediastinal nodes.

The **parasternal lymph nodes** [FIG. 13.151] lie along the internal
thoracic artery, in the intercostal spaces and behind the xiphoid
process, The lowest nodes receive vessels from the pericardium, the
diaphragm, the liver, and the upper anterior abdominal wall. Their
efferents pass to those alongside the sternum which receive the
lymph vessels running with the intercostal and perforating branches
of the internal thoracic artery. The intercostal afferents drain the
parietal pleura, and with the perforating arteries there travel lymph
vessels from the **mammary gland**. The importance of the parasternal
nodes in the surgery of malignant disease of the breast has been
stressed by Handley and Thackray (1954).

The **efferents** of the parasternal nodes unite into a single vessel
which joins with the efferents of the tracheobronchial and anterior
mediastinal nodes to form the bronchomediastinal lymph trunk.

In association with the pulmonary system and the heart there are
the following groups of nodes [FIG. 13.152].

1. **Bronchopulmonary lymph nodes** in the hilus of the lung.
Small **pulmonary nodes** occur on the larger bronchi within the
lung.

2. **Inferior tracheobronchial lymph nodes** in the angle of
bifurcation of the trachea.

3. **Superior tracheobronchial lymph nodes** continuous with the
upper bronchopulmonary nodes on each side.

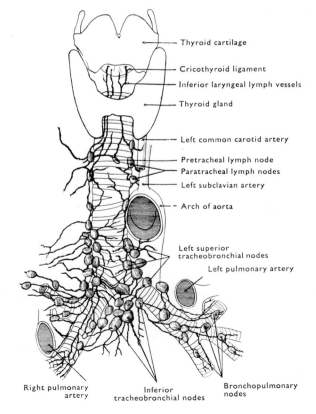

Thyroid cartilage

Cricothyroid ligament

Inferior laryngeal lymph vessels

Thyroid gland

Left common carotid artery

Pretracheal lymph node

Paratracheal lymph nodes

Left subclavian artery

Arch of aorta

Left superior
tracheobronchial nodes

Left pulmonary artery

Right pulmonary
artery

Inferior
tracheobronchial nodes

Bronchopulmonary
nodes

FIG. 13.152. Lymph nodes in relation to trachea and main bronchi.

These three groups receive the lymph vessels of the lung, visceral pleura, bronchi, and lower end of the trachea; and vessels from the left side of the heart end in the superior tracheobronchial group.

4. **Tracheal lymph nodes** extend along each side of the trachea into the neck. They drain the trachea and part of the oesophagus.

5. **Anterior mediastinal lymph nodes** lie in the vicinity of the brachiocephalic veins and the aortic arch. They receive lymph vessels from thymus, the pericardium, and the right side of the heart.

The ultimate efferents of all these nodes unite into a single vessel on each side which usually joins the single efferent of the parasternal nodes to form the bronchomediastinal lymph trunk [p. 984]. After a short course upwards, the trunk usually ends in the corresponding brachiocephalic vein. The right trunk may ascend deep to that vein, and passing behind the internal jugular vein, join the right lymph duct, while the left may join the thoracic duct.

LYMPH VESSELS OF THE DIAPHRAGM AND THORAX

Lymph vessels of the diaphragm [Fig. 13.151]

An extensive lymph plexus is present on each surface of the diaphragm, and the two plexuses are united by numerous vessels that pierce the diaphragm.

Lymph vessels from both surfaces of the diaphragm run to the parasternal, middle phrenic, and posterior mediastinal nodes: only a few vessels from the abdominal surface run to lumbar nodes.

Lymph vessels of the parietal pleura

The vessels from the named parts of the parietal pleura drain to the following nodes: from the costal pleura—mainly to the intercostal and parasternal nodes, but some vessels from the pleura lining the upper spaces, except the first, pass to the axillary nodes. The drainage of the costal pleura of the first space and the cervical pleura is to the lower deep cervical nodes, the mediastinal pleura to the tracheobronchial and mediastinal nodes, and the diaphragmatic pleura to the nodes draining the corresponding parts of the diaphragm.

LYMPH VESSELS OF THE LUNGS AND PULMONARY PLEURA

A plexus of vessels on the bronchial tree reaches the smallest bronchioles; but the alveoli have no lymph vessels. Beneath the pulmonary pleura there is also a plexus of vessels which extends over the lung surface to the hilus. Between the two plexuses there are communicating vessels which run in the interlobular septa, and in which lymph may run in either direction. All reach the hilus and enter pulmonary, bronchopulmonary, and tracheobronchial nodes.

Lymph vessels of the bronchi and trachea

The bronchi and the intrathoracic part of the trachea drain into the bronchopulmonary, tracheobronchial, and tracheal nodes.

Lymph vessels of the pericardium

The pericardium drains into the nearest nodes—anterior and posterior mediastinal, and the parasternal.

Lymph vessels of the heart

From a plexus under the endocardium efferents pierce the myocardium to join others from the richer plexus under the epicardium. The drainage of the atria is not well known, but seems to be chiefly from the upper parts of the atria to tracheobronchial nodes. The collecting vessels from the ventricles pass into the coronary sulcus and follow the coronary arteries towards their origin. Those with the right coronary artery usually form a single trunk which ascends in front of the ascending aorta to an anterior mediastinal node; those with the left coronary artery also usually end in a single vessel which passes behind the pulmonary trunk to a node of the right superior tracheo-bronchial group (Shore 1929; Patek 1939).

Lymph vessels of the oesophagus

The abdominal part of the oesophagus drains to left gastric nodes, the lower thoracic part to posterior mediastinal nodes, and the upper thoracic and cervical parts to tracheal nodes.

Lymph vessels of the thymus

From the interlobular connective tissue vessels pass to anterior mediastinal and parasternal nodes of both sides.

Subclavian lymph trunks

The subclavian lymph trunk [p. 984] drains the upper limb, a cutaneous area of the trunk from the umbilical plane to the clavicle in front and half-way up the neck behind, and most of the mammary gland.

The nodes in the territory of the subclavian trunk comprise a few superficial nodes and a greater number of deep nodes including the important axillary group.

Superficial lymph nodes of upper limb

One or two cubital nodes lie medial to the basilic vein a little above the medial epicondyle of the humerus: they receive afferents from the medial fingers and from the ulnar side of the hand and forearm. Their efferents run with the basilic vein and join the deep lymph vessels.

The infraclavicular lymph nodes, one or two in number, lie in the infraclavicular fossa on the cephalic vein and receive vessels from the skin over deltoid, and from the upper part of the mammary gland: their efferents perforate the clavipectoral fascia to join the apical axillary nodes. Outlying members occur in two situations.

A deltopectoral lymph node may be found in the groove between the deltoid and the pectoralis major, and small interpectoral lymph nodes are sometimes present between the pectoral muscles [Fig. 13.153].

Deep lymph nodes of the upper limb

Small nodules may occur along the radial, ulnar, and interosseous arteries, or as deep cubital nodes at the bifurcation of the brachial artery. Outlying axillary nodes may occur as brachial nodes along the upper part of the brachial artery. The efferents of all these nodes pass to the lateral axillary group.

The axillary lymph nodes are widely distributed in the axillary fat, and may be divided into five groups; four of the groups lie inferior to pectoralis minor, and one (the apical group) above.

The lateral axillary nodes lie alongside the axillary vein, and from them the pectoral and the subscapular nodes radiate along the lateral thoracic and subscapular arteries respectively. The central group is composed of members of the former three groups which lie superior to the fascia of the axillary floor. The afferents of these four groups are: (1) the deep vessels of the whole limb; (2) the cutaneous vessels of the limb and trunk above the umbilical plane; and (3) vessels of the mammary gland. Though all the groups receive vessels from all these areas, the main drainage of the limb is to the lateral nodes, of the mammary gland and the anterior thoracic wall to the pectoral nodes, and of the scapular region and the back of the chest to the subscapular nodes. The apical axillary nodes receive the

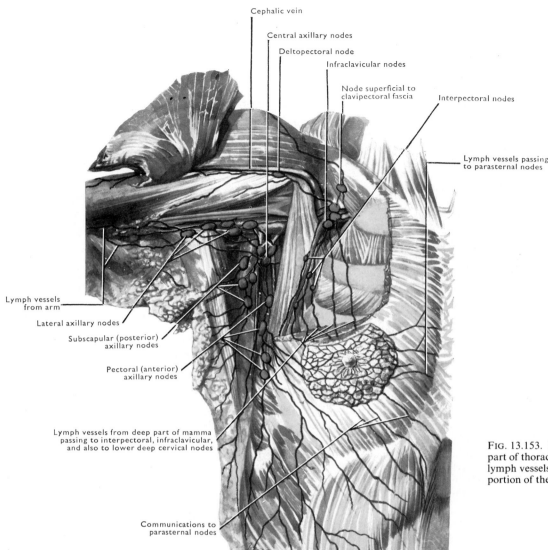

Cephalic vein

Central axillary nodes

Deltopectoral node

Infraclavicular nodes

Node superficial to
clavipectoral fascia

Interpectoral nodes

Lymph vessels passing
to parasternal nodes

Lymph vessels
from arm

Lateral axillary nodes

Subscapular (posterior)
axillary nodes

Pectoral (anterior)
axillary nodes

Lymph vessels from deep part of mamma
passing to interpectoral, infraclavicular,
and also to lower deep cervical nodes

Communications to
parasternal nodes

FIG. 13.153. Dissection of axilla and anterior part of thoracic wall, showing lymph nodes and lymph vessels. (Semidiagrammatic, with only a portion of the mamma represented.)

efferents of the lower groups and of the infraclavicular nodes. They are also in continuity with the lower deep cervical nodes [p. 1001], but their efferents go mainly to form the subclavian lymph trunk.

SUPERFICIAL LYMPH VESSELS OF THE UPPER LIMB AND TRUNK

The superficial lymph vessels of the upper limb run upwards on the front of the forearm with an inclination towards the axilla, and are reinforced by vessels that slope upwards from the back of the limb round its margins [FIGS. 13.154 and 13.155]. Some vessels are interrupted by superficial cubital nodes, and some pass through the deep fascia with the basilic vein to join the deep stream; but nearly all run to the axilla before they pierce the deep fascia. On the back of the elbow and of the arm there is a 'lymphshed' from which vessels pass round the lateral and medial margins to the front. A few vessels from the skin over the upper part of the cephalic vein pass to infraclavicular nodes, and may be interrupted in deltopectoral nodes.

The cutaneous plexuses are finest and most dense on the palmar surfaces of the fingers and hand. The efferents from the palmar plexus of each finger pass to the dorsum where they form dorsal digital vessels which run to the dorsum of the hand [FIG. 13.156].

The efferents from the plexus on the palm of the hand run in various directions. Those passing laterally and medially join the efferents of the thumb and little finger respectively; a few pass superiorly along the median vein of the forearm; and those that pass inferiorly turn backwards at the interdigital clefts to join the vessels on the dorsum of the hand [FIGS. 13.154 and 13.155].

The superficial lymph vessels of the trunk form a great whorl as they converge on the lower axillary nodes [FIGS. 13.130 and 13.131]. The plexus of origin is continuous over the median line, and many vessels draining to one side begin on the other. Several vessels penetrate deeply, and thus reach intercostal and parasternal nodes. A few ascend over the clavicle to reach the lower deep cervical nodes.

LYMPH VESSELS OF THE MAMMARY GLAND

The mammary gland receives its blood supply from three main sources—the axillary artery, the internal thoracic artery, and the lateral cutaneous branches of the posterior intercostals—and the

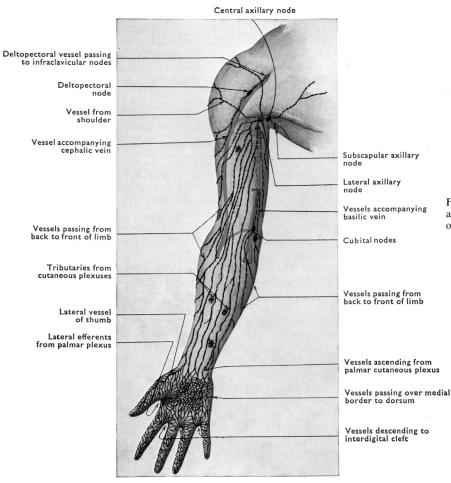

Central axillary node

Deltopectoral vessel passing to infraclavicular nodes

Deltopectoral node

Vessel from shoulder

Vessel accompanying cephalic vein

Vessels passing from back to front of limb

Tributaries from cutaneous plexuses

Lateral vessel of thumb

Lateral efferents from palmar plexus

Subscapular axillary node

Lateral axillary node

Vessels accompanying basilic vein

Cubital nodes

Vessels passing from back to front of limb

Vessels ascending from palmar cutaneous plexus

Vessels passing over medial border to dorsum

Vessels descending to interdigital cleft

FIG. 13.154. Superficial lymph vessels and lymph nodes of anterior surface of upper limb.

lymph drainage is approximately in proportion. Thus, most of the lymph from the breast drains to axillary nodes, a considerable proportion to parasternal nodes, and an inconstant, small amount to the intercostal nodes (Turner-Warwick 1959).

The lymph vessels of the main axillary pathway arise in the lobules of the mammary gland and most of them drain directly to pectoral nodes. They run within the substance of the gland and pass through the axillary fascia within the axillary tail of the mammary gland; at no time do they run in the deep fascia.

Drainage to parasternal nodes may proceed from both the medial and lateral halves of the gland, the lymph vessels running with the perforating branches of the internal thoracic artery. In a small proportion of cases some lymph vessels run with the lateral

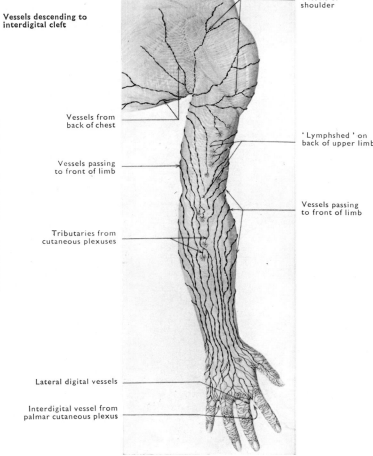

Vessels of shoulder

Vessels from back of chest

Vessels passing to front of limb

Tributaries from cutaneous plexuses

'Lymphshed' on back of upper limb

Vessels passing to front of limb

Lateral digital vessels

Interdigital vessel from palmar cutaneous plexus

FIG. 13.155. Superficial lymph vessels of posterior surface of upper limb.

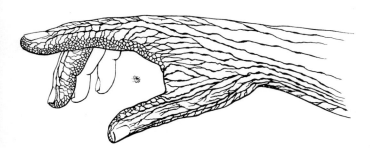

FIG. 13.156. Superficial lymph vessels of thumb and forefinger.

cutaneous branches of the posterior intercostal arteries to the intercostal nodes.

Because the various groups of axillary lymph nodes are interconnected, lymph vessels from the mammary gland may pass directly to any group. From the upper part of the mammary gland lymph vessels run to apical axillary nodes, either directly or via the infraclavicular group; some of these are interrupted in small interpectoral nodes.

The cutaneous lymph vessels of the breast are in continuity with neighbouring vessels across the midline, over the clavicles, and on the abdominal wall; and they anastomose with the deeper vessels of the mammary gland, especially in the nipple region. They drain to axillary, infraclavicular, and parasternal nodes, and a few from the upper part of the breast pass to lower deep cervical nodes.

DEEP LYMPH VESSELS OF THE UPPER LIMB

These accompany the deeper blood vessels. Some pass to deep cubital or brachial nodes, but most go directly to the lateral group of axillary nodes.

Jugular lymph trunks

The area of drainage of the jugular lymph trunk [p. 984] comprises the head and neck, except the skin of the lower part of the back of the neck. The trunk is formed by the efferents of deep cervical lymph nodes.

LYMPH NODES OF THE HEAD AND NECK

These nodes form one continuous group with numerous offshoots, but it is convenient to describe a main chain of nodes along the great vessels of the neck, and subsidiary groups along the course of the branches of the external carotid artery and on the inferior thyroid artery.

The nodes thus arranged are:

1. Deep cervical nodes (jugular chain).

2. Anterior cervical nodes: (i) infrahyoid; (ii) prelaryngeal; (iii) tracheal—pretracheal and paratracheal.

3. Nodes associated with the scalp and face, forming a 'collar-chain' between the head and the neck: (i) occipital; (ii) retro-auricular; (iii) parotid and superficial cervical; (iv) submandibular, with facial nodes; (v) submental.

1. **Deep cervical lymph nodes**. This chain extends along the carotid sheath from the base of the skull to the root of the neck. Most of the nodes lie under cover of sternocleidomastoid, but some appear in the anterior triangle of the neck, and quite a number extend into the posterior triangle, particularly in its lower part [FIGS. 13.157 and 13.159]. It is convenient to divide the continuous chain into superior and inferior parts, above and below the omohyoid muscle.

The highest members of the **superior deep cervical nodes** lie behind the lateral border of the nasopharynx, and are known as **retropharyngeal nodes**. In the anterior triangle, one of the nodes (jugulodigastric) is frequently palpable below the angle of the jaw,

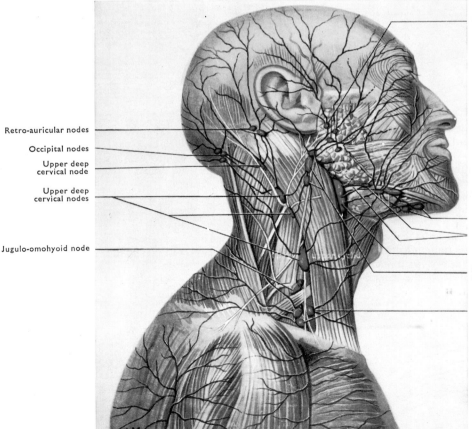

Retro-auricular nodes

Occipital nodes

Upper deep cervical node

Upper deep cervical nodes

Jugulo-omohyoid node

Superficial parotid nodes

FIG. 13.157. Lymph nodes of head and neck as seen with sternocleidomastoid in position.

The occipital, retro-auricular, and superficial parotid nodes are inserted in accordance with descriptions. The other nodes were present in one or other of the two bodies from which the drawing was made. [See also FIG. 13.159.]

Submental nodes

Facial vein and submandibular nodes

Upper deep cervical node

Superficial cervical nodes

Lower deep cervical nodes

between the digastric muscle and the internal jugular vein. It is especially associated with the lymph drainage of the tongue and palatine tonsil. Others of the superior group escape from beneath the sternocleidomastoid to appear in the posterior triangle along the accessory nerve. The jugulo-omohyoid node lies just above the tendon of the omohyoid, deep to the posterior border of sternocleidomastoid on the internal jugular vein. It is said to receive lymph vessels direct from the tongue.

The parts which drain, directly or indirectly, to the upper deep cervical nodes will be mentioned below. The efferents from these nodes pass to the lower deep cervical nodes and to the jugular lymph trunk.

The inferior deep cervical nodes [FIGS. 13.158 and 13.159] lie on the internal jugular vein deep to the heads of sternocleidomastoid, and also in front of the brachial plexus and the third part of the subclavian artery. The last nodes are in communication with the apical axillary nodes, and receive vessels from the skin of the breast. The efferents of all the inferior nodes pass to the jugular trunk.

2. Anterior cervical lymph nodes. These are small groups all of which send their efferents to the deep cervical nodes.

(a) With the superior thyroid artery: (i) infrahyoid nodes on the thyrohyoid membrane; (ii) prelaryngeal nodes on the cricothyroid ligament.

(b) With the inferior thyroid artery: (iiia) pretracheal nodes at the lower end of the thyroid gland; (iiib) paratracheal nodes in the groove between trachea and oesophagus.

They receive lymph from the larynx, trachea, cervical oesophagus, and thyroid gland.

3. The occipital lymph nodes (i) lie on the occipital vessels as they pierce trapezius; their afferents come from the scalp.

(ii) The retro-auricular lymph nodes (two or three) lie on the mastoid process, and the largest is frequently palpable. Their afferents drain the scalp and the back of the auricle.

(iii) The parotid lymph nodes lie on the superficial surface of the salivary gland both superficial and deep to its fascial sheath, and also in its substance. The superficial nodes receive afferents from the front of the auricle and from the scalp, forehead, eyebrow, eyelids, and cheek; and their efferents pass to the deep parotid nodes and to the upper deep cervical nodes. The deep parotid lymph nodes receive afferents from the external acoustic meatus, the auditory tube and the middle ear, the soft palate, the posterior part of the nasal cavity, and the deeper portions of the cheek. Their efferents pass to upper deep cervical nodes.

The superficial cervical lymph nodes [FIG. 13.157] are merely superficial parotid nodes scattered along the external jugular vein.

(iv) Four to six submandibular lymph nodes lie between the jaw and the submandibular salivary gland; they receive the vessels of the face below the eye, and many from the tongue. Their efferents run to deep cervical nodes, notably the jugulodigastric and jugulo-omohyoid.

Minute nodes may be found on the course of the vessels of the face [FIG. 13.160].

(v) The submental lymph nodes form one median group on the mylohyoid muscles; their afferents are derived from the floor of the mouth, mandible, and possibly also the tip of the tongue and the lower lip and chin; their efferents run to the deep cervical chain as far down as the jugulo-omohyoid node.

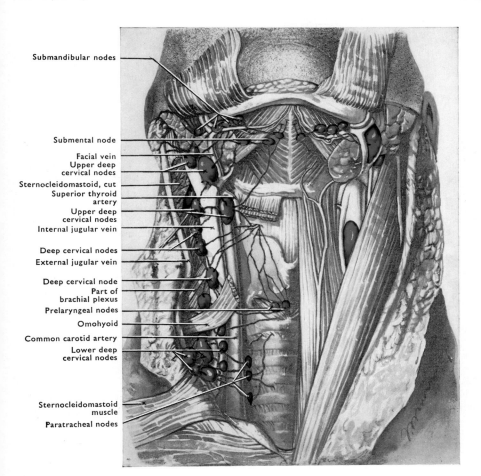

FIG. 13.158. Lymph nodes of the neck, anterior aspect. Infrahyoid nodes and pretracheal nodes were not present.

Submandibular nodes

Submental node
Facial vein
Upper deep cervical nodes
Sternocleidomastoid, cut
Superior thyroid artery
Upper deep cervical nodes
Internal jugular vein
Deep cervical nodes
External jugular vein
Deep cervical node
Part of brachial plexus
Prelaryngeal nodes
Omohyoid
Common carotid artery
Lower deep cervical nodes
Sternocleidomastoid muscle
Paratracheal nodes

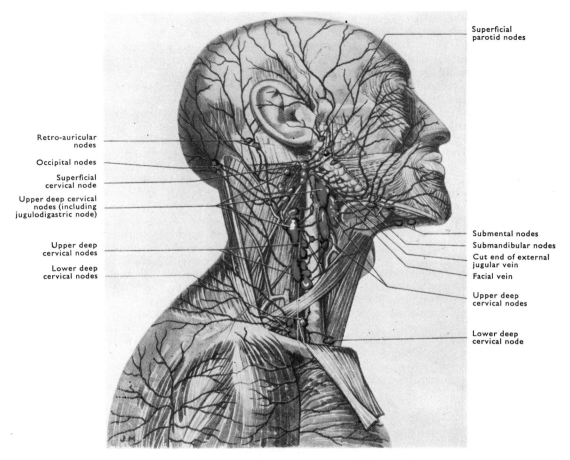

FIG. 13.159. Lymph nodes of head and neck as seen after removal of sternocleidomastoid.
The occipital, retro-auricular, and superficial parotid nodes are inserted in accordance with descriptions. The other nodes were present in one of both of the bodies from which the Figure was made. [See also FIG. 13.157.]

LYMPH VESSELS OF THE HEAD AND NECK

The lymph vessels of the scalp and forehead, and of the various parts of the face, pass to nodes of the 'collar-chain' [FIG. 13.160]; but many vessels proceed directly to deep cervical nodes [FIG. 13.159]. The superficial vessels of the neck perforate the deep fascia at numerous points to enter the deep cervical nodes; those from the lower part of the back of the neck pass with the vessels of the scapular region to axillary nodes.

The lymph vessels of the eyelids form two groups.

1. The medial vessels drain a small part of the skin only of the upper eyelid, and about half the skin and conjunctiva of the lower eyelid; they pass with the facial vessels to the submandibular lymph nodes.

2. The lateral vessels drain most of the skin of the upper eyelid and all its conjunctiva, and about half the skin and conjunctiva of the lower eyelid. They pass to parotid and superficial cervical nodes [see also p. 847, and Burch 1939].

The lymph vessels of the lacrimal gland run to parotid lymph nodes but from the lacrimal sac mainly to submandibular nodes.

The lymph vessels from the auricle and the external acoustic meatus end in parotid and retro-auricular nodes.

The lymph vessels of the lateral wall of the middle ear join those of the tympanic membrane and external acoustic meatus and end in parotid nodes. The vessels of the medial wall and auditory tube end in retropharyngeal, deep parotid, and upper deep cervical nodes.

The lymph vessels from the nose and from the lining of the anterior part of the nasal cavity end in submandibular nodes; those from the posterior part pass backwards through the pharyngeal wall to upper deep cervical (including the retropharyngeal) nodes.

The lymph vessels of the paranasal sinuses follow those from the walls of the nasal cavities. Channels along the course of the olfactory nerves allow the passage of cerebrospinal fluid to the lymph vessels of the nasal cavity, but such channels are common to many nerves.

The lymph drainage of the skin and mucous membrane of the median part of the lower lip may be to submental nodes; but, with that exception, the lymph vessels of the skin and mucous membrane of both lips end mainly in submandibular nodes—as do those of the cheeks.

The vessels of the teeth, and of the outer parts of both gums, pass to submandibular nodes. Some of the vessels from the inner part of the mandibular gum also end in these nodes; but others run to submental nodes from the incisor region, and from the more posterior part to upper deep cervical nodes which also receive the vessels from the inner part of the maxillary gum and from the palate.

The lymph vessels of the tongue pass from a rich capillary plexus in the mucous membrane. They are arranged in three groups.

1. The anterior vessels drain the anterior two-thirds of the dorsum and margins and pass to submandibular and upper deep cervical nodes [FIG. 13.161], though some vessels from the tip may pass to submental nodes or even to the jugulo-omohyoid node [FIGS. 13.161 and 13.162].

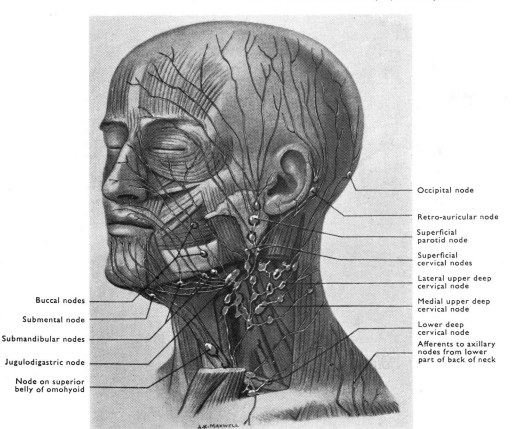

FIG. 13.160. Superficial lymph vessels of scalp and face, and superficial ('collar-chain') lymph nodes between head and neck.

Occipital node

Retro-auricular node

Superficial parotid node

Superficial cervical nodes

Lateral upper deep cervical node

Medial upper deep cervical node

Lower deep cervical node

Afferents to axillary nodes from lower part of back of neck

Buccal nodes

Submental node

Submandibular nodes

Jugulodigastric node

Node on superior belly of omohyoid

A.K. MAXWELL

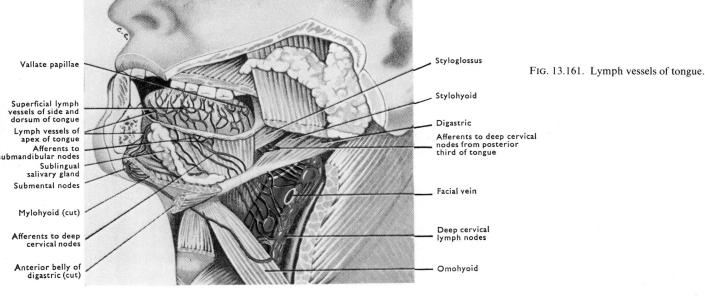

FIG. 13.161. Lymph vessels of tongue.

Vallate papillae

Superficial lymph vessels of side and dorsum of tongue

Lymph vessels of apex of tongue

Afferents to submandibular nodes

Sublingual salivary gland

Submental nodes

Mylohyoid (cut)

Afferents to deep cervical nodes

Anterior belly of digastric (cut)

Styloglossus

Stylohyoid

Digastric

Afferents to deep cervical nodes from posterior third of tongue

Facial vein

Deep cervical lymph nodes

Omohyoid

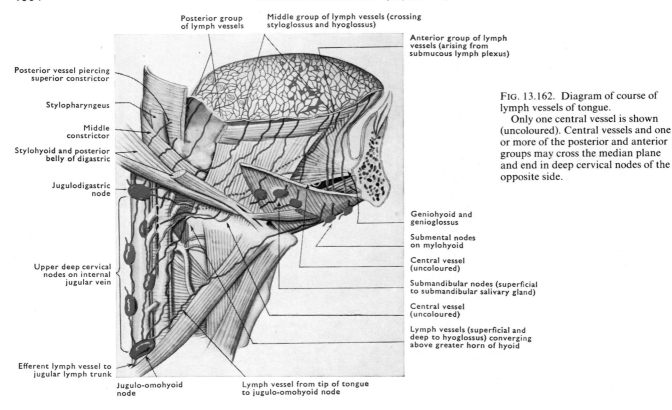

FIG. 13.162. Diagram of course of lymph vessels of tongue.

Only one central vessel is shown (uncoloured). Central vessels and one or more of the posterior and anterior groups may cross the median plane and end in deep cervical nodes of the opposite side.

2. The **posterior vessels** drain the posterior third of the dorsum, and also end in the deep cervical chain—principally in the jugulodigastric node.

3. The **central vessels** arise in the dorsal plexus near the median plane. They descend between the genioglossi, and with vessels from the deep parts of the tongue pass to upper deep cervical nodes below the digastric muscle [FIG. 13.164].

It is important to note that any upper deep cervical nodes below the digastric may receive vessels direct from the tongue; and, in general, the further forward the origin of the vessel the lower down is the node in which it ends. Note also that some anterior, posterior, and central vessels cross over to nodes of the opposite side [FIG. 13.164].

The lymph vessels from the mucous membrane of the anterior part of the floor of the mouth drain to submental nodes, while more posteriorly they accompany the vessels of the tongue.

The lymph vessels of the **parotid gland** end in parotid lymph nodes; those of the **submandibular** and **sublingual glands** end in submandibular and upper deep cervical nodes.

From the **pharyngeal walls**, lymph vessels pass to retropharyngeal and deep cervical nodes at the corresponding level. The nasal part of the **pharynx** and **pharyngeal tonsil** drain particularly to posterior upper, deep cervical nodes.

Lymph vessels of the **palatine tonsil** and the adjacent parts of the palatoglossal and palatopharyngeal arches end in upper deep cervical nodes, many in the jugulodigastric node.

From the **piriform fossae** and the posterior surface of the larynx, lymph vessels pierce the thyrohyoid membrane, and end in upper deep cervical nodes.

The lymph plexus of the **larynx** is divisible into upper and lower parts which are separated, laterally and anteriorly, by the vocal folds. The lymph vessels of the upper part pierce the thyrohyoid membrane and end in upper deep cervical nodes. From the lower part of the larynx some vessels pierce the cricothyroid ligament and end in prelaryngeal, pretracheal, and lower deep cervical nodes.

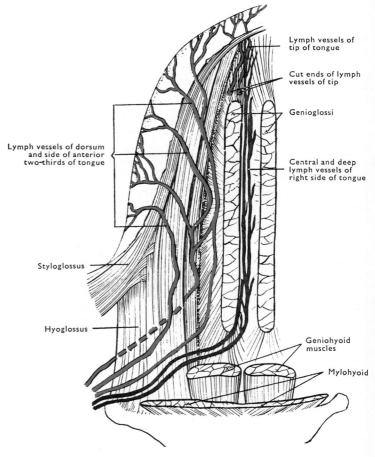

FIG. 13.163. Diagram of lymph vessels of anterior two-thirds of tongue, inferior aspect. (After Poirier and Cunéo 1902, modified.)

Anterior vessels, red; middle vessels, blue; central vessels, black.

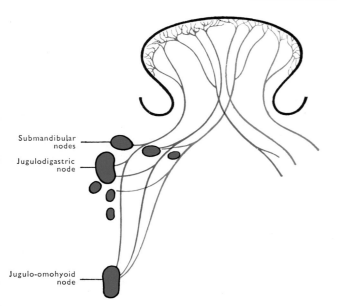

Submandibular nodes

Jugulodigastric node

Jugulo-omohyoid node

FIG. 13.164. Diagram of course of central lymph vessels of tongue to show crossing of median plane. (Jamieson and Dobson 1910.)

Others pierce the cricotracheal membrane and end in paratracheal and lower deep cervical nodes [FIG. 13.158].

From much of the **thyroid gland** the lymph vessels pass directly to deep cervical nodes up to the level of the superior thyroid artery. Those from the lower part of the isthmus and each lobe end in pretracheal and paratracheal nodes.

The lymph vessels of the cervical part of the **trachea** end in prelaryngeal, paratracheal, and inferior deep cervical nodes.

THE DEVELOPMENT OF THE LYMPHATIC SYSTEM

The first evidence of the lymphatic system is the appearance of structures known as **lymph sacs** which are closely related to the veins in certain situations. The first to appear is the jugular lymph sac, and six are present in the 3-cm embryo—an **anterior** or **jugular lymph sac** on each side at the union of the subclavian and anterior cardinal veins, a **posterior lymph sac** on each side at the union of the iliac and posterior cardinal veins, a **retroperitoneal lymph sac** at the root of the primitive mesentery, and another on the posterior wall of the abdomen is identified with the **cisterna chyli**.

With the exception of the anterior part of the sac from which the cisterna chyli is developed all become broken up into anastomosing channels and are later converted into groups of lymph nodes; and the jugular sacs acquire secondary connections with the veins.

According to Sabin (1916), the lymph sacs are developed as outgrowths of the endothelium of the veins near by, and from the sacs all the lymph vessels of the body sprout in a radiating manner. In this view, the lymph sacs lose their primary connections with the veins, and only later do the jugular sacs acquire connections again.

According to Huntington (1911) and McClure (1915), on the other hand, all lymph vessels are originally formed as clefts in the mesenchyme exactly as blood vessels are developed [p. 62]. These clefts acquire an endothelial lining and unite into plexuses of capillary vessels. In this view, the lymph sacs are formed by the running together of parts of the capillary plexuses along the primitive

venous trunks, the main lymph trunks by the development of channels in these plexuses, and all peripheral lymph vessels by the local fusion of mesenchymal spaces. Kampmeier (1960) strongly supports this concept.

Lymph nodes

Lymph nodes are formed from aggregations of cells in the mesenchymal strands surrounded by plexuses of lymph vessels. Around each nodule of lymph tissue so formed, the vessels are transformed into the peripheral portion of the lymph sinus. The lymph tissue is then broken up into cords by anastomising lymph channels which grow into it from the sinus.

In the early stages of their development, lymph nodes have a more extensive system of blood capillaries than of lymph vessels, though in postnatal life the latter predominate. If the blood vessels predominate in the later stages, the node has a reddish colour and is known as a haemal lymph node. Some such nodes are found in the neck, thorax, and retroperitoneal regions in Man, but are much more common in the sheep and the rat.

Thoracic duct

A rich plexus of lymph vessels is formed in the thorax around the aorta; it communicates below with the lymph sac that forms the cisterna chyli and above with both jugular lymph sacs. From the plexus two longitudinal vessels, with numerous transverse anastomoses, are developed, but as a rule the continuity of each vessel is broken at the level of the fifth thoracic vertebra, where one of the transverse anastomoses enlarges. The lower part of the left vessel and the upper part of the right disappear. From the primitive condition many varieties of the thoracic duct might be developed, and the majority of the possible variations have been found in adult bodies (Davis 1915).

THE SPLEEN

The spleen (lien) [FIG. 13.165] is a soft, freely movable organ of purplish colour, situated far back in the upper left part of the abdomen, behind the stomach. Its size varies greatly during life, depending on the amount of its contained blood, and this factor also affects its weight at death. As a rule the adult spleen is about 12.5 cm in length, 7.5 cm in breadth, and 3.5 cm in thickness; and its weight is about 150 g with a range of 50–250 g.

The **position** of the spleen varies with the excursions of the diaphragm, with the state of adjacent viscera, and with posture. It lies obliquely with its long axis parallel with the posterior parts of the ninth, tenth, and eleventh ribs, from which it is separated by the diaphragm and pleural cavity. Its posterior end is about 4 cm from the median plane at the level of the tenth thoracic spine, and its anterior end lies just behind the midaxillary line. The normal spleen is not palpable through the abdominal wall, but its **surface projection** can be assessed either from the above data, or by percussion which reveals an area of impaired resonance, the splenic dullness; this reaches almost to the midaxillary line where it gives way to the area of stomach resonance.

The **shape** of the spleen is influenced largely by the stomach and left colic flexure. When the stomach is distended the spleen resembles a segment of an orange, while distension of the colon causes it to become of irregular tetrahedral form. The spleen is described as having two **ends**, anterior and posterior; two **borders**, superior and inferior; and two principal **surfaces**, diaphragmatic and visceral.

The **diaphragmatic surface** is smooth, convex, and lies against the diaphragm which intervenes between it and the pleura and

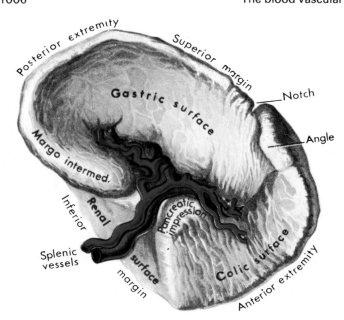

FIG. 13.165. The visceral surface of the spleen.

lower, posterior border of the lung. It is separated from the visceral surface by the superior and inferior borders. The superior border is notched near the anterior end, and when the spleen is sufficiently enlarged to be palpable, the notches can be felt easily.

The visceral surface [FIG. 13.165] is wedged between the stomach, left kidney, and left colic flexure, and is often touched by the tail of the pancreas. The gastric surface lies against the upper part of the posterior surface of the stomach; it is separated by a raised, blunt margin from the renal surface which is in contact with the upper lateral part of the left kidney. In the young child the spleen touches the left suprarenal gland also. The colic surface rests on the left colic flexure and the phrenicocolic ligament.

The linear hilus is on the lower part of the gastric surface, and here the splenic vessels enter and leave. The peritoneum which surrounds the spleen is firmly adherent to its capsule, and is continuous with the peritoneal folds known as the lienorenal ligament and the gastrosplenic ligament. Originally part of the dorsal mesogastrium, both of these are attached to the hilus, and connect it to the left kidney and stomach respectively, forming the left wall of the omental bursa in the adult. The tail of the pancreas lies in the lienorenal ligament, and may touch the spleen at the lower part of the hilus.

Small rounded accessory spleens are often present, usually on the gastrosplenic ligament.

RADIOGRAPHIC EXAMINATION OF SPLEEN [FIG. 13.112]

The normal spleen may be visible radiographically if there are large amounts of gas in both the stomach and the left colic flexure. The spleen, like the liver, may also be made semi-opaque to X-rays by the injection of an X-ray opaque medium into the splenic artery.

VESSELS AND NERVES

The splenic artery runs with the vein in the lienorenal ligament, and at the hilus breaks up into six or more branches which enter

independently. The splenic vein, which passes to the portal vein, is formed by several tributaries which emerge at the hilus. The lymph vessels, which arise only from the capsule and trabeculae, pass to pancreaticosplenic lymph nodes.

The nerves are vasomotor in function, and come through the coeliac plexus from the greater splanchnic nerve.

When the spleen is distended it acts as a reservoir from which blood can be returned to the general circulation in harmony with the body's needs; but the absence of smooth muscle in the capsule of the human spleen precludes active splenic contraction such as occurs in the dog.

STRUCTURE

Beneath its peritoneal coat the spleen is invested by a fibroelastic capsule. Trabeculae, of similar structure, pass into the organ both from the hilus (carrying the larger blood vessels with them) and from the capsule [FIG. 13.166]. The branches of the trabeculae ultimately become continuous with the reticular meshwork which pervades the main substance or pulp of the spleen. Associated with the reticular fibres are reticular cells of the reticulo-endothelial system [see below].

The pulp is composed of white pulp and red pulp. The white pulp consists of ovoid masses of lymph tissue called Malpighian corpuscles, or lymph follicles, within which may be seen germinal centres. The red pulp forms the greater part of the splenic substance. It consists of the reticular meshwork and venous sinuses between which are splenic cords of cells. In the cords are reticular fibres associated with reticular cells (which are to some extent phagocytic), numerous macrophages (which are strongly phagocytic), red and white blood corpuscles, and plasma cells.

VASCULAR PATTERN

From the trabeculae arteries pass into the red pulp and, almost immediately, each is invested by a lymph follicle (white pulp) through which it runs eccentrically. Having given capillaries to the follicle it re-enters the red pulp and divides into several parallel 'penicillate' vessels of fine calibre. These soon acquire thickened walls and have been termed sheathed arteries; the sheaths, which are poorly developed in Man, are composed of reticular cells and are known as ellipsoids or Schweigger–Seidel sheaths.

The sheathed arteries end in arterial capillaries, but beyond this point the question of whether the splenic circulation is 'open' or 'closed' has long remained unresolved. By 'open' is meant that the capillaries open into the pulp whence the extravasated blood finds its way into the venous sinuses, which, in this view, must have openings in their walls. According to the 'closed' view the blood passing through the spleen is, throughout, confined within endo-thelial-lined channels.

At the present time the view is gaining ground that it is of little functional significance whether the circulation is 'open' or 'closed': and for these reasons. First, the presence of slits between the endothelial cells of the sinus wall and of fenestrations in the basement membrane allows the passage of cells and plasma to and fro between the sinus lumen and the cords of the pulp. Second, the varying state of the spleen during life could, it is believed, favour an 'open' circulation when the organ is distended and a 'closed' circulation when it is collapsed. Interpretation of splenic structure as seen by the light or electron microscope is difficult, and made more so by the differences between the human spleen and those of laboratory animals. Chen and Weiss (1972) describe the electron microscopy of the red pulp of the human spleen, and Ham (1974) discusses the splenic circulation very fully.

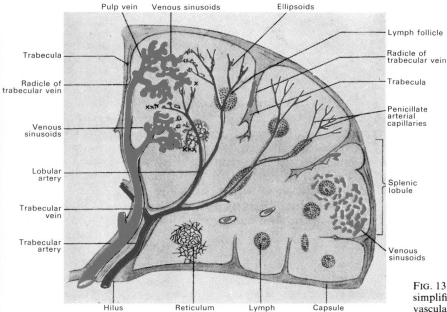

FIG. 13.166. Schematic representation of splenic structure, much simplified by the suppression of ramifications of the trabeculae and vascular tree. (After Stöhr, *Lehrbuch der Histologie*. Fischer.)

Note. Both the pulp reticulum and the system of venous sinusoids supported by it must be understood as pervading the entire extent of the area figured except the trabeculae.

DEVELOPMENT

The spleen is formed by the fusion of several condensations of mesoderm which appear in the dorsal mesogastrium about the sixth week (10-mm stage). Failure of fusion of one or more of the condensations accounts for the presence of accessory spleens [p. 1006]; moreover, the adult spleen still retains evidence of its multiple origin in possessing compartments each with a separate vascular supply from its own artery and vein (Braithwaite and Adams 1957). During intrauterine life the spleen is actively engaged in haemopoiesis, but by birth this function has been taken over entirely by the bone marrow; nevertheless, should an increased demand for blood formation arise after birth, the spleen may resume its former role.

THE RETICULO-ENDOTHELIAL OR MACROPHAGE SYSTEM

The cells of this system are widely distributed in the body, and serve as a protective and scavenging mechanism. They are actively phagocytic, and all have the property of taking up vital dyes, such as trypan blue, when they are injected into the body, and storing them as coarse granules in their cytoplasm. It is, indeed, this quality which distinguishes the cells of the system, rather than special morphological characters. Should circumstances demand, some are capable of detaching themselves from their usual sites and of acquiring amoeboid movement. They are engaged in the destruction of red blood corpuscles and platelets, in bile pigment production, in the metabolism of iron, fats, and proteins, and in clearing the body of bacteria, protozoa, and non-living particles.

The facts just stated have long been known, but recent work has brought recognition of the interdependence of lymph and reticulo-endothelial cells in immunological responses. Indeed it is suggested that in this respect they form an integral system which should be designated the lymphoreticular system (Symmers 1978).

The cells of the reticulo-endothelial system occur: (1) in the blood where they are represented by the monocytes; (2) in the reticular tissue of the spleen and lymph nodes; (3) in connective tissue where they are present as histiocytes or 'resting wandering cells'; (4) in the central nervous system where they are represented by the microglial cells; (5) lining the sinuses of lymph nodes, and lining the blood sinusoids in the liver, spleen, bone marrow, suprarenal cortex, and the anterior lobe of the hypophysis. In the case of the liver it is quite certain that between the endothelial cells, and forming part of the lining, there are true phagocytic cells—the so-called **stellate endothelial cells** of von Kupffer. But it seems established that the endothelium of lymph sinuses and of the blood sinusoids of bone marrow, the spleen, the suprarenal cortex, and of the anterior lobe of the hypophysis does not possess phagocytic powers. That phagocytosis of particulate matter traversing these channels occurs is not in dispute, but the cells responsible are macrophages which lie just outside the endothelial cells between which they push pseudopodia to reach the lumen of the vessel.

References

BRAITHWAITE, J. L. and ADAMS, D. J. (1957). The venous drainage of the rat spleen. *J. Anat. (Lond.)* **91**, 352.

BURCH, G. E. (1939). Superficial lymphatics of human eyelids observed by injection *in vivo*. *Anat. Rec.* **73**, 443.

CHEN, L.-T. and WEISS, L. (1972). Electron microscopy of the red pulp of human spleen. *Am. J. Anat.* **134**, 425.

DAVIS, H. K. (1915). A statistical study of the thoracic duct in man. *Am. J. Anat.* **17**, 211.

ENGESET, A. (1959a). The route of peripheral lymph to the blood stream. An X-ray study of the barrier theory. *J. Anat. (Lond.)* **93**, 96.

—— (1959b). Lymphatico-venous communications in the albino rat. *J. Anat. (Lond.)* **93**, 380.

GOWANS, J. L. and KNIGHT, E. J. (1964). The route of recirculation of lymphocytes in the rat. *Proc. R. Soc. B.* **159**, 257.

HAM, A. W. (1974). *Histology*. Lippincott, Philadelphia.

HANDLEY, R. S. and THACKRAY, A. C. (1954). Invasion of internal mammary lymph nodes in carcinoma of breast. *B. med. J.* **1**, 61.

HUNTINGTON, G. S. (1911). The anatomy and development of the systemic lymphatic vessels in the domestic cat. *Am. Anat. Mem.* **1**. Wistar Institute, Philadelphia.

JAMIESON, J. K. and DOBSON, J. F. (1907a). The lymphatic system of the stomach. *Lancet* **i**, 1061.

—— —— (1907b). The lymphatic system of the caecum and appendix. *Lancet* **i**, 1137.

—— —— (1909). The lymphatics of the colon. *Proc. R. Soc. Med.* **2**, 149.

—— —— (1910). On the injection of lymphatics by Prussian blue. *J. Anat. Physiol.* **45**, 7.

KAMPMEIER, O. F. (1960). The development of the jugular lymph sacs in the light of vestigial, provisional and definitive phases of morphogenesis. *Am. J. Anat.* **107**, 153.

KINMONTH, J. B. (1954). Lymphangiography in clinical surgery and particularly in treatment of lymphoedema. *Ann. R. Coll. Surg. Engl.* **15**, 300.

LOOR, F. and ROELANTS, G. E. (1977). *B and T cells in immune recognition.* London.

McCLURE, C. F. W. (1915). The development of the lymphatic system in the light of the more recent investigations in the field of vasculogenesis. *Anat. Rec.* **9**, 563.

NOPAJAROOSRI, C., LUK, S. C., and SIMON, G. T. (1971). Ultrastructure of the normal lymph node. *Am. J. Path.* **65**, 1.

PATEK, P. R. (1939). The morphology of the lymphatics of the mammalian heart. *Am. J. Anat.* **64**, 203.

POIRIER, P. and CUNÉO, B. (1902). Les lymphatiques, in Poirier and Charpy's *Traité d'anatomie humaine*, t. 11, fasc. 4. Paris. Engl. edn. trans. and ed. C. H. Leaf (1903). London.

ROITT, I. M., GREAVES, M. F., TORRIGIANI, G., BROSTOFF, J., and PLAYFAIR, J. H. L. (1969). The cellular basis of immunological responses. *Lancet* **ii**, 367.

ROUVIÈRE, H. (1932). *Anatomie des lymphatiques de l'homme.* Paris.

SABIN, F. R. (1916). The origin and development of the lymphatic system. *Johns Hopkins Hosp. Rep.* **17**, 347.

SAPPEY, P. C. (1874). *Traité d'anatomie, physiologie et pathologie des vaisseaux lymphatiques considérés chez l'homme et les vertébrés.* Paris.

SHORE, L. R. (1929). The lymphatic drainage of the human heart. *J. Anat. (Lond.)* **63**, 291.

SYMMERS, W. St.C. (1978). *Systemic pathology*, Vol. 2. Edinburgh.

TURNER-WARWICK, R. T. (1959). The lymphatics of the breast. *Br. J. Surg.* **46**, 574.

YOFFEY, J. M. and COURTICE, F. C. (1970). *Lymphatics, lymph and the lymphomyeloid complex.* E. Arnold, London.

Atlas

Introduction

The illustrations are photographs of transverse sections of a male body. Their inclusion in the Textbook arises out of the increasing use of Computerized Axial Tomography and ultrasound scans in many clinical situations, and the consequent need for the student to be able to visualize the body as it appears in such sections.

The sections of the head and neck are cut obliquely by flexing the head on the flexed neck to bring it into the plane most frequently used in computerized axial tomography (CAT) of this region. However, the degree of flexion is greater than that most frequently used in CAT scans, but is sufficiently close to be useful and demonstrates many anatomical features not readily seen in other planes. Some of the illustrations have corresponding CAT scans placed beside them for comparison, but it must be realized that the number of planes which are likely to be used will increase for specific purposes, and no atlas can hope to illustrate all of these. However, it is relatively easy to extrapolate from a number of transverse sections to an oblique section passing through them.

The illustrations are not at equidistant intervals, but several close or even adjacent sections are shown in regions where there is rapid anatomical change and where the CAT scans are of particular value. In all the illustrations the bones have been outlined with a thin, white line so that they may be differentiated more readily from the surrounding tissues. Similarly, some organs have also been outlined (e.g. the ureter) so as to make their positions more readily obvious, and the larger arteries have been coloured red. Because most of the smaller veins are filled with blood and appear black in the photographs, the larger veins are also shown in black rather then in the conventional blue—red being used for the pulmonary as well as the systemic arteries.

In keeping with the CAT scans, all the sections are viewed from below. Hence the right side of the body appears on the left of each illustration.

The head scans are from an individual with similar cerebral degeneration to that used for the sections. These scans are through the same levels as the sections, but are slightly less oblique, being at a higher level posteriorly.

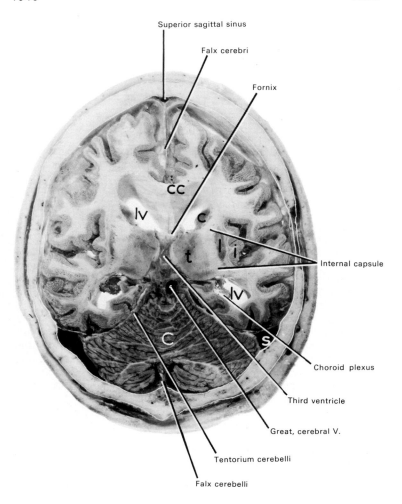

Superior sagittal sinus
Falx cerebri
Fornix
CC
lv
c
t
l
i
lv
s
Internal capsule
Choroid plexus
Third ventricle
Great, cerebral V.
Tentorium cerebelli
Falx cerebelli
C

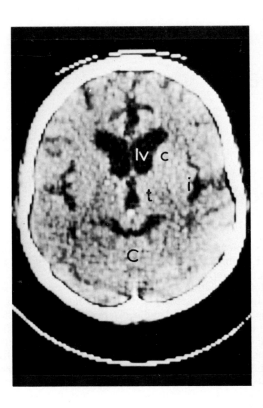

lv c
t
i
C

FIG. 1. Section through the upper parts of the cerebellum and thalami. All the sections through the brain show markedly distended ventricles due to the loss of neural tissue. This is also seen in the thin cerebral cortex and the wide cerebral sulci.

C, cerebellum; c, caudate nucleus; cc, corpus callosum; i, insula; l, lentiform nucleus, lv, lateral ventricle; s, transverse sinus; t, thalamus.

FIG. 1A. Section through the anterior horns of the lateral ventricles and the internal occipital protuberance. There is some pineal calcification posterior to the third ventricle. Keys as for FIG. 1.

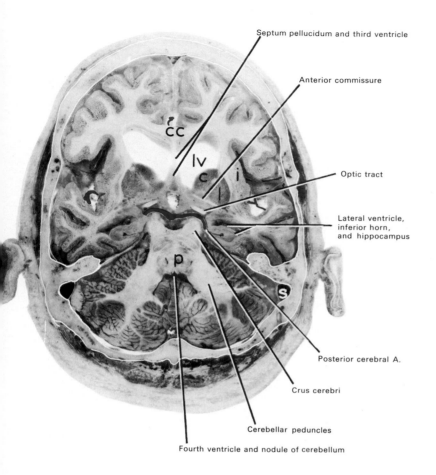

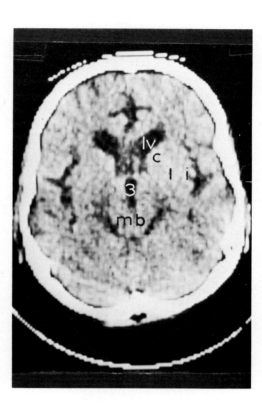

FIG. 2. Section through the bifurcation of the basilar artery into the two posterior cerebral arteries. Parts of the superior cerebellar, middle cerebral (on the insula), and anterior cerebral arteries are seen.

cc, corpus callosum; c, caudate nucleus; i, insula; l, lentiform nucleus; lv, anterior horn of lateral ventricle; p, pons; s, sigmoid sinus.

FIG. 2A. Section through upper midbrain and anterior horns of lateral ventricles. mb, midbrain; 3, third ventricle. Others as FIG. 2.

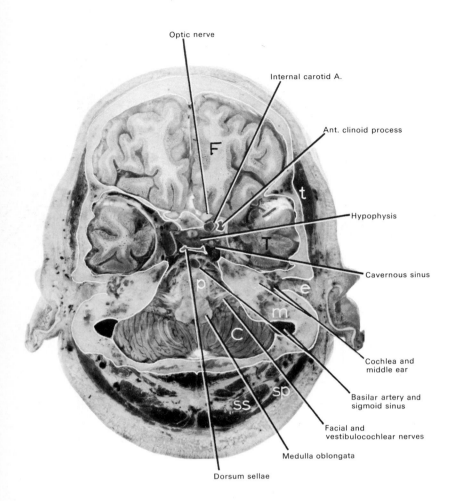

Optic nerve

Internal carotid A.

Ant. clinoid process

Hypophysis

Cavernous sinus

Cochlea and
middle ear

Basilar artery and
sigmoid sinus

Facial and
vestibulocochlear nerves

Medulla oblongata

Dorsum sellae

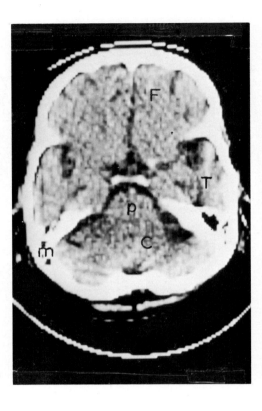

Fig. 3. Section through the pontomedullary junction, the external acoustic meatuses, and the clinoid processes. On the left side of the illustration (right side of the body) the internal carotid artery has been cut longitudinally in the cavernous sinus which has the sphenoparietal sinus entering it anterolaterally. The beginning of the ophthalmic artery is visible on the right of the illustration.

C, cerebellum; e, external acoustic meatus; F, frontal lobe; m, mastoid air cells; p, pons; sp, splenius; ss, semispinalis capitis; T, temporal lobe; t, temporalis.

Fig. 3A. Section through pons and orbital parts of frontal lobes. The petrous temporal bones are cut at a higher level than in the corresponding section (3). The dorsum sellae and posterior clinoid processes are visible in both, but the internal and middle ears are not present in the scan (3A) though the mastoid antrum is shown on the right of the illustration. Keys as Fig. 3.

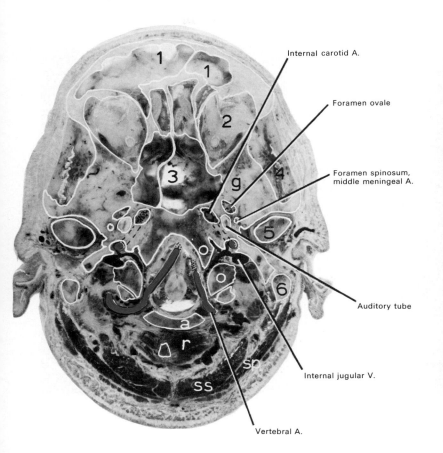

Internal carotid A.

Foramen ovale

Foramen spinosum,
middle meningeal A.

Auditory tube

Internal jugular V.

Vertebral A.

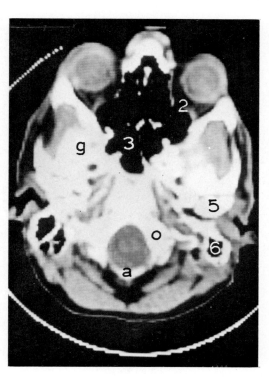

FIG. 4. Section through the posterior arch of the atlas, the heads of the mandible, and the floors of the middle cranial fossae. It is parallel to the third and fourth parts of the vertebral arteries. The internal carotid arteries are shown entering and within the carotid canals, the broken red lines indicating the course in the bone. The funnel-like fronto-nasal duct of each frontal sinus is shown in longitudinal section.

1, frontal sinus; 2, orbit; 3, sphenoidal sinus; 4, temporalis in temporal fossa; 5, head of mandible; 6, mastoid process.

a, posterior arch of atlas; g, greater wing of sphenoid; o, occipital bone; r, rectus capitis posterior major; sp, splenius; ss, semispinalis capitis.

FIG. 4A. Section through the floor of the middle cranial fossa. Keys as in FIG. 4.

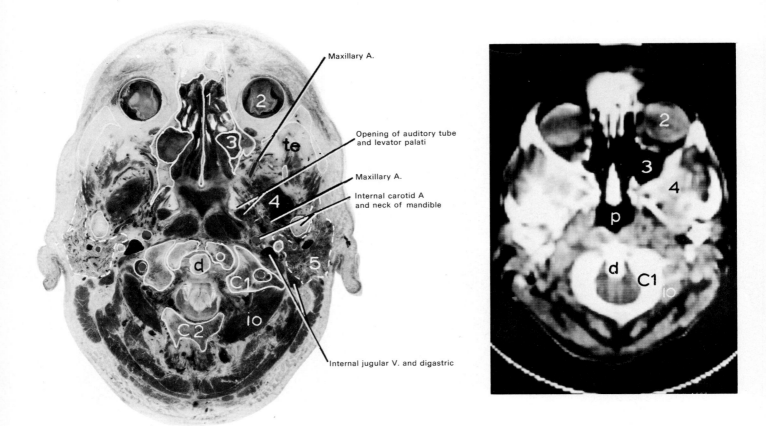

FIG. 5. Section through the spine of the axis and the posterior extremities of the maxillary sinuses. The styloid processes (outlined in white) are seen on the deep surfaces of the parotid glands which are outlined by a broken white line. The pterygoid venous plexus surrounds the lateral pterygoid muscle, and the alar ligaments can be seen passing from the dens to the occipital condyles.

1, nasal cavity; 2, eyeball; 3, maxillary sinus; 4, lateral pterygoid muscle; 5, parotid gland; C1, lateral mass of atlas; C2, spine of axis; d, dens of axis; io, inferior oblique muscle; o, occipital condyle; te, temporalis muscle.

FIG. 5A. Section through lower part of floor of middle cranial fossa and right temporomandibular joint (left side of illustration). Keys as in FIG. 5.

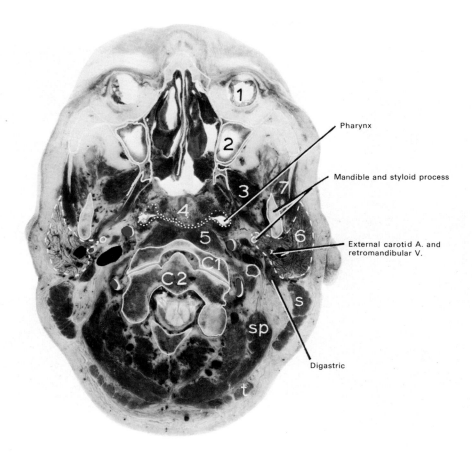

Fig. 6. Section through the anterior arch of the atlas and the body of the axis. The parotid glands are outlined by a broken white line, and the pharynx by a dotted white line.

1, eyeball; 2, maxillary sinus; 3, lateral pterygoid muscle; 4, soft palate; 5, longus capitis muscle; 6, parotid gland; 7, masseter muscle; C1, lateral mass of atlas; C2, body of the axis; s, sternocleidomastoid; sp, splenius; t, trapezius.

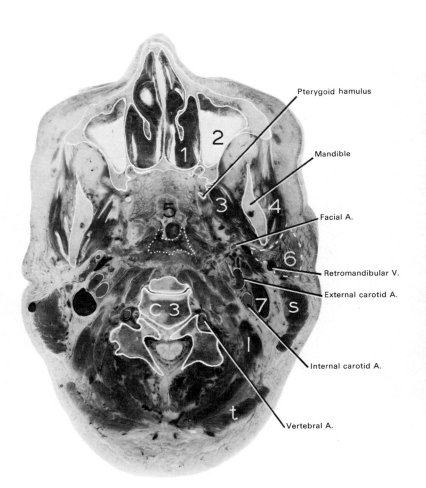

Pterygoid hamulus

Mandible

Facial A.

Retromandibular V.

External carotid A.

Internal carotid A.

Vertebral A.

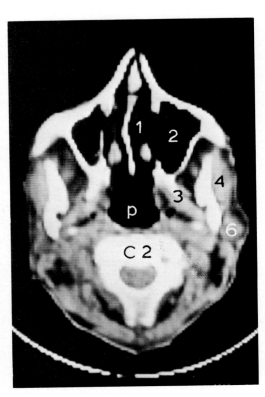

FIG. 7. Section through the soft palate and the pharyngeal isthmus (outlined with dotted white line). Part of the lingual artery is cut close to the external carotid artery on the left side of the body (right side of the illustration).

1, nasal cavity; 2, maxillary sinus; 3, medial pterygoid muscle; 4, masseter muscle; 5, soft palate; 6, parotid gland; 7, internal jugular vein; 1, levator scapulae; s, sternocleidomastoid; t, trapezius.

FIG. 7A. Section through the maxillary sinuses and the second cervical vertebra. The inferior concha is visible in the nasal cavity. p, nasal part of pharynx. Other keys as in FIG. 7.

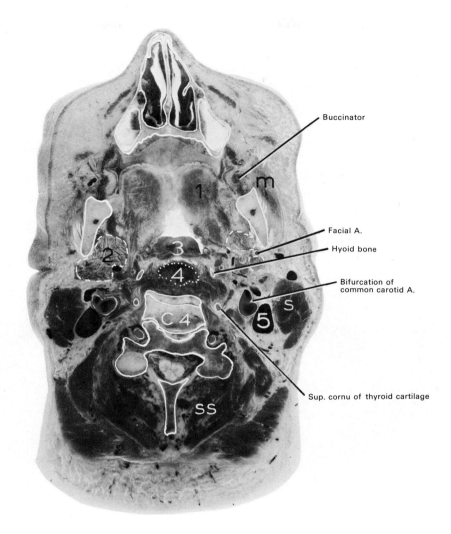

Buccinator

Facial A.

Hyoid bone

Bifurcation of
common carotid A.

Sup. cornu of thyroid cartilage

FIG. 8. Section through the body of the fourth cervical vertebra (C4). On the left side of the illustration (right side of the body) the lingual artery is sectioned transversely deep to the hyoglossus and close to the hyoid bone. The submandibular gland (outlined by a broken white line) is best seen on the left side of the illustration. On the right, its deep and superficial parts are separated by the facial artery.

1, tongue; 2, submandibular gland; 3, epiglottis; 4, pharynx; 5, internal jugular vein; m, masseter; s, sternocleidomastoid; ss, semispinalis.

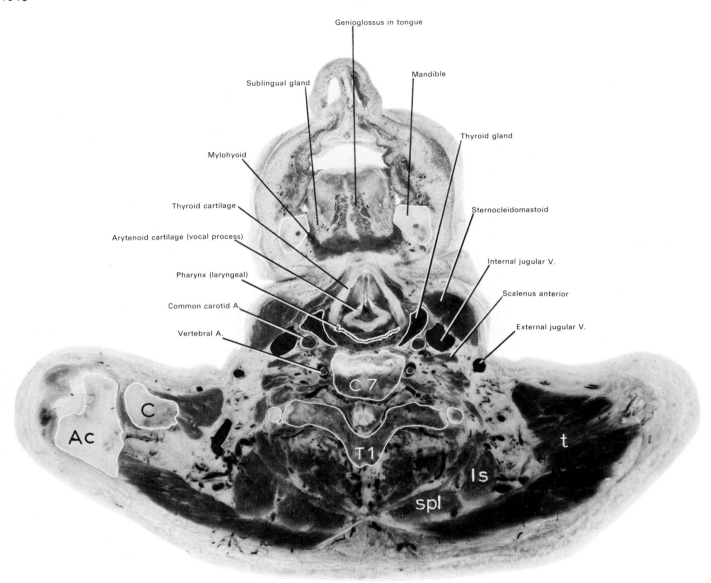

Genioglossus in tongue

Mandible

Sublingual gland

Thyroid gland

Mylohyoid

Thyroid cartilage

Sternocleidomastoid

Arytenoid cartilage (vocal process)

Internal jugular V.

Pharynx (laryngeal)

Scalenus anterior

Common carotid A.

External jugular V.

Vertebral A.

C 7

Ac

C

T1

ls

spl

t

FIG. 9. Section through the body of the seventh cervical vertebra (C7). The apparent low level of the larynx is due to the flexion of the neck. The lumen of the pharynx (laryngeal part) and the lobes of the thyroid gland are outlined in white. Branches of the facial artery are visible in the cheeks superficial to buccinator.

Ac, acromion; C, clavicle; ls, levator scapulae; spl, splenius; t, trapezius; T1, laminae and transverse processes of first thoracic vertebra.

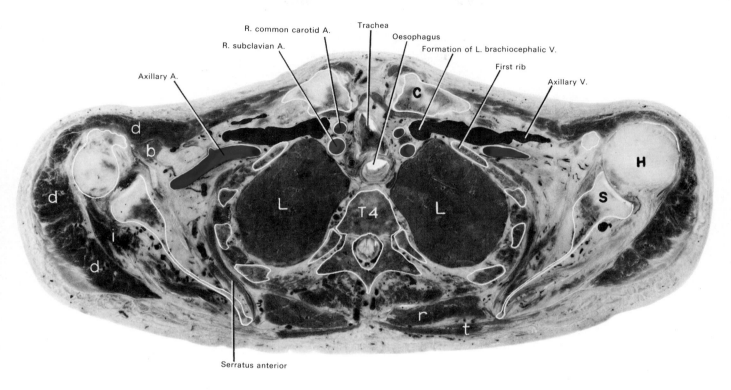

FIG. 10. Section through the fourth thoracic vertebral body (T4) and the sternoclavicular joints. The junction of the subclavian and internal jugular veins is present on both sides.

b, short head of biceps and coracobrachialis; C, clavicle; d, deltoid; H, head of humerus; L, lung; i, teres minor; r, rhomboid major; S, scapula; t, trapezius.

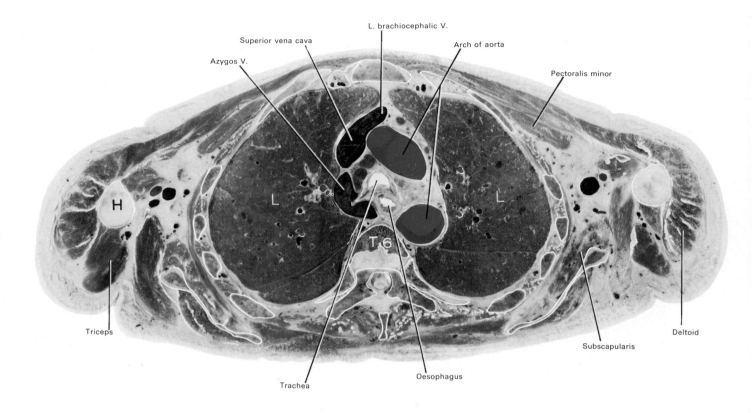

Fɪɢ. 11. Section through the lower parts of the sixth thoracic vertebra (T6) and
manubrium sterni. The oblique fissure is visible in both lungs, and tracheal lymph
nodes are present on both sides of the trachea. The origin of the brachiocephalic
trunk from the aortic arch is immediately posterior to the left brachiocephalic
vein. In this and the other sections, the plane is not horizontal, but passes
downwards as well as backwards.

H, humerus; L, lung.

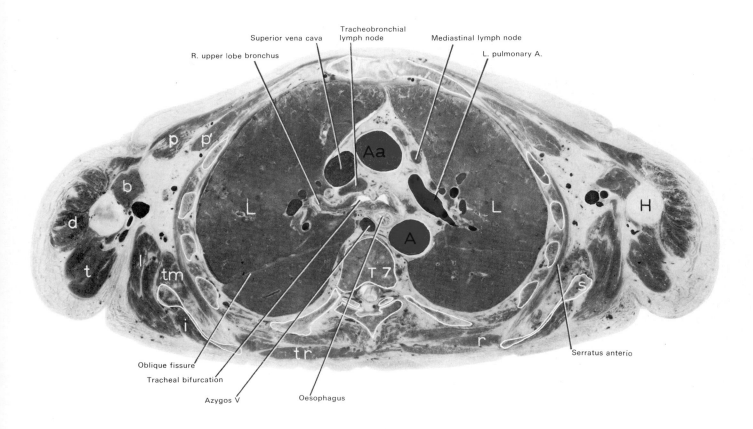

FIG. 12. Section through the seventh thoracic vertebra (T7) and the manubrio-sternal synchondrosis.

A, descending aorta; Aa, ascending aorta; b, biceps and coracobrachialis; d, deltoid; H, humerus; i, infraspinatus; L, lung; l, latissimus dorsi; p, pectoralis major; p', pectoralis minor; r, rhomboid major; s, scapula; t, triceps; tm, teres major; tr, trapezius.

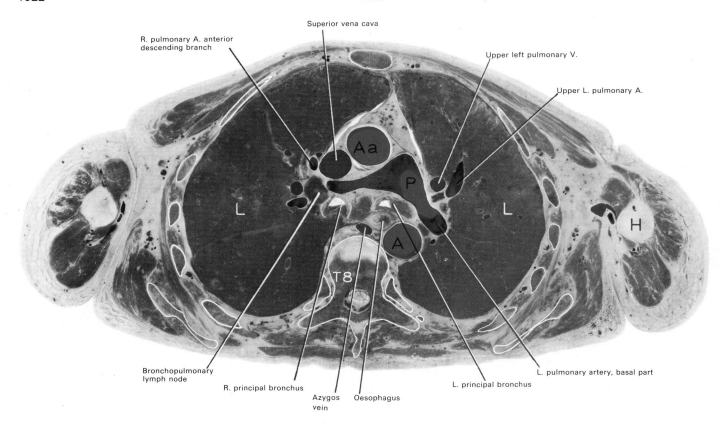

R. pulmonary A. anterior
descending branch

Superior vena cava

Upper left pulmonary V.

Upper L. pulmonary A.

Aa

P

A

T8

L

L

H

Bronchopulmonary
lymph node

R. principal bronchus

Azygos
vein

Oesophagus

L. principal bronchus

L. pulmonary artery, basal part

FIG. 13. Section through the eighth thoracic vertebra (T8) and the bifurcation of
the pulmonary trunk. A tracheobronchial lymph node lies between the two
principal bronchi.

A, descending aorta; Aa, ascending aorta; H, humerus; L, lung; P, bifurcation
of pulmonary trunk.

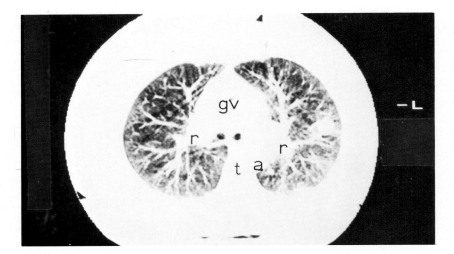

gv

r

r

t

a

—L

FIG. 13A. A CAT scan at the level of FIG. 13. The contrast has been adjusted so
that only the air in the lungs and the principal bronchi show dark. This makes the
pulmonary vessels visible as white lines in the dark lung areas. Note the vessels
running forwards in the upper lobes of both lungs, their line is visible in FIG. 13.

a, descending aorta; gv, great vessels; r, lung roots; t, thoracic vertebral column.

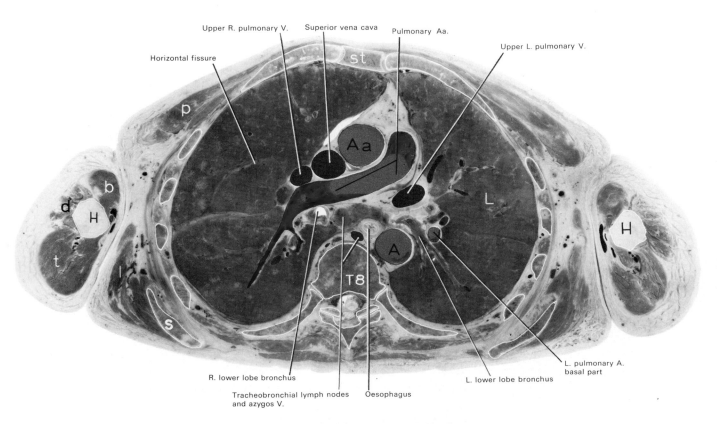

FIG. 14. Section through the body of the eighth thoracic vertebra (T8) and the third sternocostal articulation. The entire length of the right pulmonary artery and the beginning of its branch to the lower lobe of the right lung are shown.

A, descending aorta; Aa, ascending aorta; b, biceps and coracobrachialis; d, deltoid; H, humerus; L, lung; l latissimus dorsi; p, pectoralis major; s, scapula; st, sternum; t, triceps.

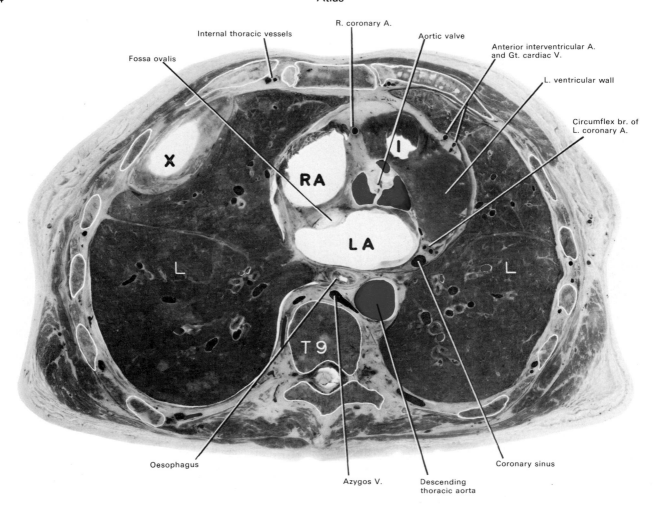

Fossa ovalis

Internal thoracic vessels

R. coronary A.

Aortic valve

Anterior interventricular A.
and Gt. cardiac V.

L. ventricular wall

Circumflex br. of
L. coronary A.

X

RA

I

LA

L

L

T 9

Oesophagus

Azygos V.

Descending
thoracic aorta

Coronary sinus

FIG. 15. Section through the body of the ninth thoracic vertebra (T9) and the
fourth sternocostal joint. The section shows the relative positions of the chambers
of the heart, particularly the atria and the outflow parts of the right and left
ventricles. X marks the position of a sterile abscess closed off from the pleural
cavity. It compresses the upper lobe of the right lung. The circumflex branch of the
left coronary artery is branching to give a left marginal branch.

I, infundibulum of right ventricle; L, lung; LA, left atrium; RA, right atrium.

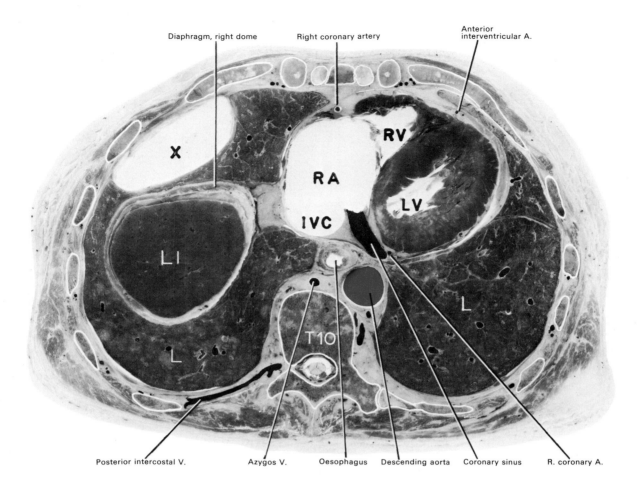

FIG. 16. Section through the tenth thoracic vertebral body (T10) and the seventh
sternocostal joint. It also shows the entry of the inferior vena cava (IVC) and the
coronary sinus into the right atrium, the right dome of the diaphragm and the liver
applied to it. For X see legend to FIG. 15.

IVC, inferior vena cava opening into right atrium; L, lung; LI, liver; LV, left
ventricle; RA, right atrium; RV, right ventricle; X, closed abscess.

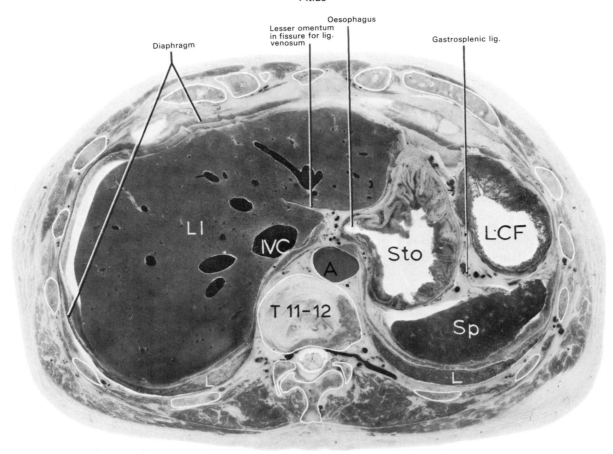

FIG. 17. Section through the intervertebral disc between the eleventh and twelfth thoracic vertebrae (T11–12) and the xiphoid process. The left lobe of the liver is unusually thick, and pushes the stomach to the left and forwards. The left colic flexure is also higher than usual so that it lies anterior to the greater part of the vertical extent of the spleen. Note the lung in the costodiaphragmatic recess of the pleural cavity.

A, descending aorta; IVC, inferior vena cava; L, lung; LCF, left colic flexure; LI, liver; Sp, spleen; Sto, stomach.

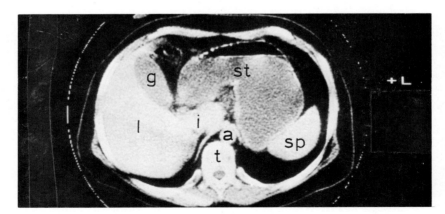

FIG. 17A. A CAT scan at approximately the same level as FIG. 17. However, the level is lower anteriorly than in that figure, and the liver is unusually small and the gall-bladder unusually high. The crura of the diaphragm are on each side of the aorta—see FIG. 19.

a, descending aorta; g, gall-bladder; i, inferior vena cava; l, liver; sp, spleen; st, stomach—the dark area anteriorly is the gas bubble, the individual is supine; t, eleventh thoracic vertebra.

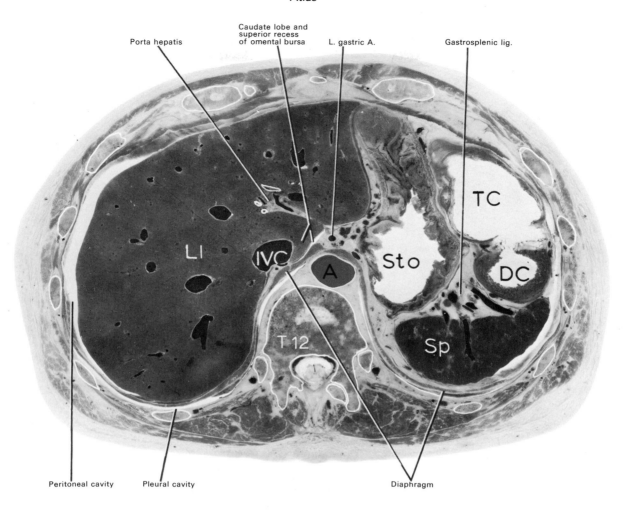

Porta hepatis — Caudate lobe and superior recess of omental bursa — L. gastric A. — Gastrosplenic lig.

TC

Sto

DC

LI

IVC

A

Sp

T 12

Peritoneal cavity — Pleural cavity — Diaphragm

FIG. 18. Section through the body of the twelfth thoracic vertebra (T12) and the tip of the xiphoid process. Tributaries of the hepatic ducts are outlined in white. The lesser omentum passes into the fissure for the ligamentum venosum in the liver between the left and caudate lobes of the liver.

A, descending aorta; DC, descending colon; IVC, inferior vena cava; Li, liver; Sp, spleen; Sto, stomach; TC, transverse colon.

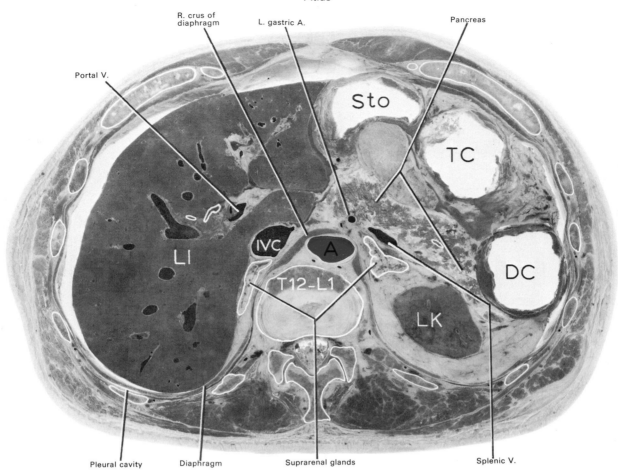

Portal V.   R. crus of diaphragm   L. gastric A.   Pancreas

Sto

TC

IVC   A

DC

LI

T12–L1

LK

Pleural cavity   Diaphragm   Suprarenal glands   Splenic V.

FIG. 19. Section through the intervertebral disc between the twelfth thoracic and first lumbar vertebrae (T12–L1). The suprarenal glands, the pancreatic duct, and the tributaries of the hepatic ducts are outlined in white. The common stem of the inferior phrenic arteries lies immediately posterior and to the right of the left gastric artery. The left crus of the diaphragm lies on the left posterior surface of the aorta.

A, descending aorta; DC, descending colon; IVC, inferior vena cava; LI, liver; LK, left kidney; Sto, stomach; TC, transverse colon.

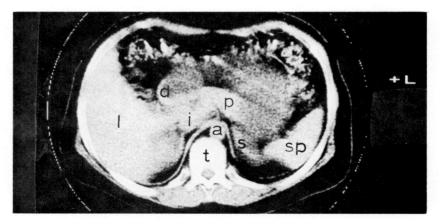

FIG. 19A. A CAT scan through the twelfth thoracic vertebra. The broken white masses in the anterior part of the abdomen represents faecal material in the transverse colon. The posterior parts of the diaphragm are easily seen. The liver is small.

a, descending aorta; d, duodenum; i, inferior vena cava; l, liver; p, pancreas; s, left suprarenal; sp, spleen; t, twelfth thoracic vertebra. The origin of part of the coeliac trunk is also visible anterior to the aorta.

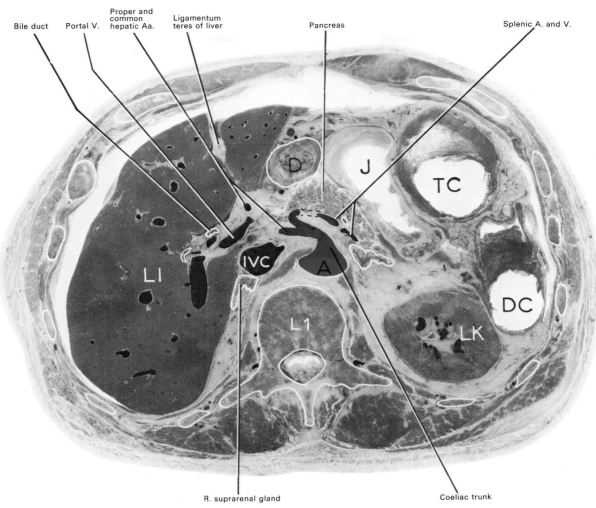

FIG. 20. Section through the body of the first lumbar vertebra (L1). Outlined in white are the suprarenal glands, the pancreatic duct, the duodenum, the common hepatic duct and one of its tributaries. The epiploic foramen lies between the portal vein and the inferior vena cava. The kidneys are both somewhat lower than usual, the right especially so, and both suprarenal glands are separated from their respective kidneys by a considerable amount of perirenal fat.

A, descending aorta; D, duodenum; DC, descending colon; IVC, inferior vena cava; J, jejunum; LI, liver; LK, left kidney; TC, transverse colon.

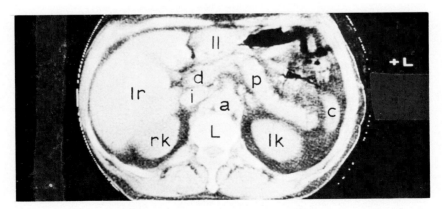

FIG. 20A. A CAT scan through the first lumbar vertebra. This is from another individual than that shown in FIGS. 17A and 19A. The liver is of more normal size. The fissure for the ligamentum teres of the liver lies between the right and left lobes of the liver. The neck of the gall-bladder projects from the visceral surface of the liver, and is separated from the fissure for the ligamentum teres by the quadrate lobe. The pancreas is more horizontal than usual.

a, aorta, with the superior mesenteric artery arising from its anterior surface; c, the left colic flexure, or the caudal extremity of the spleen; d, duodenum; i, inferior vena cava; L, first lumbar vertebral body; lk; left kidney; ll, left lobe of liver; lr, right lobe of liver; p, pancreas; rk, right kidney. The crura of the diaphragm lie on each side of the aorta.

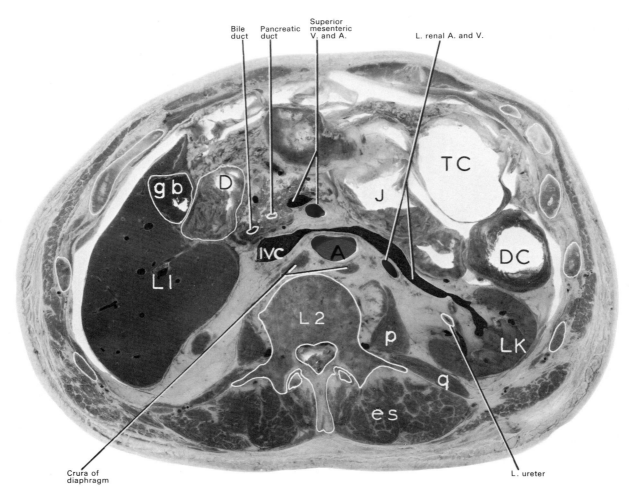

FIG. 21. Section through the second lumbar vertebral body (L2). Outlined in white are the left ureter, the pancreatic and bile-ducts, the duodenum, and the gall-bladder. The two ducts lie in the head of the pancreas to the right of the superior mesenteric vessels. The entire length of the left renal vein is seen. A large superior pancreaticoduodenal artery lies on the pancreas beside the duo-denum. The crura of the diaphragm are clearly seen, and also show in FIG. 21A.

A, descending aorta; D, duodenum; DC, descending colon; es, erector spinae; gb, gall-bladder; IVC, inferior vena cava; J, jejunum; LI, liver; LK, left kidney; p, psoas; q, quadratus lumborum; TC, transverse colon.

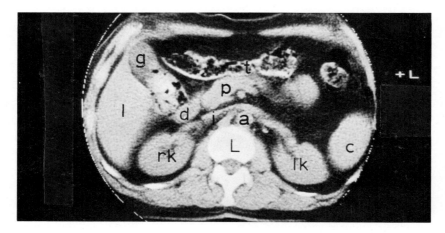

FIG. 21A. A CAT scan through the lower border of the first lumbar vertebra. This is from a different body than the other scans. The renal vessels, gall-bladder, and transverse colon are well shown. The superior mesenteric vessels lie between the aorta and the pancreas. The fundus of the gall-bladder lies at the costal margin against the upturned end of the ninth costal cartilage.

a, aorta, with right crus of the diaphragm between it and the inferior vena cava. c, descending colon filled with faecal matter; d, duodenum; g, gall-bladder; i, inferior vena cava; L, first lumbar vertebral body; l, liver; lk, left kidney; rk, right kidney; t, transverse colon.

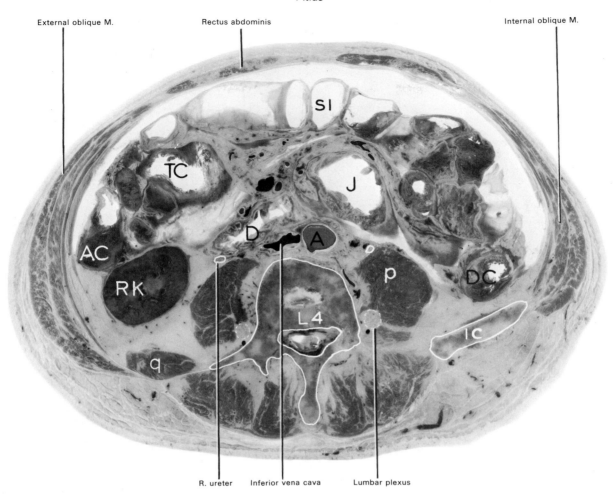

External oblique M.   Rectus abdominis   Internal oblique M.

R. ureter   Inferior vena cava   Lumbar plexus

FIG. 22. Section through the body of the fourth lumbar vertebra (L4). Outlined in white are the ureters and parts of the lumbar plexuses (broken line). The inferior mesenteric vessels lie close to the left ureter. Branches of the superior mesenteric vessels are present in the mesentery running to several parts of the small intestine which have been sectioned.

A, descending aorta; AC, ascending colon; D, duodenum; DC, descending colon; ic, iliac crest; J, jejunum; p, psoas; q, quadratus lumborum; RK, right kidney; SI, small intestine; TC, transverse colon.

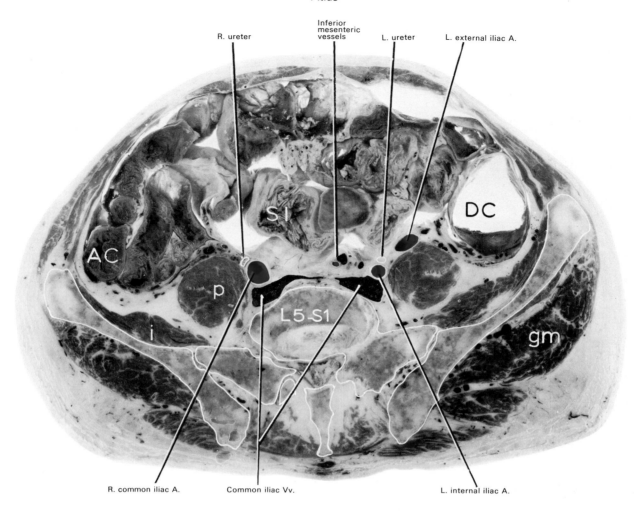

FIG. 23. Section through the lumbosacral intervertebral disc (L5–S1) just below
the formation of the inferior vena cava by the union of the two common iliac
veins. Outlined in white are the ureters, the left on the internal iliac artery, the
right on the common iliac artery. This difference indicates the slight obliquity of
the sections, the right side of the body (left side of the illustration) being cut at a
slightly higher level than the left.

AC, ascending colon; DC, descending colon; i, iliacus; gm, gluteus maximus;
p, psoas; SI, small intestine.

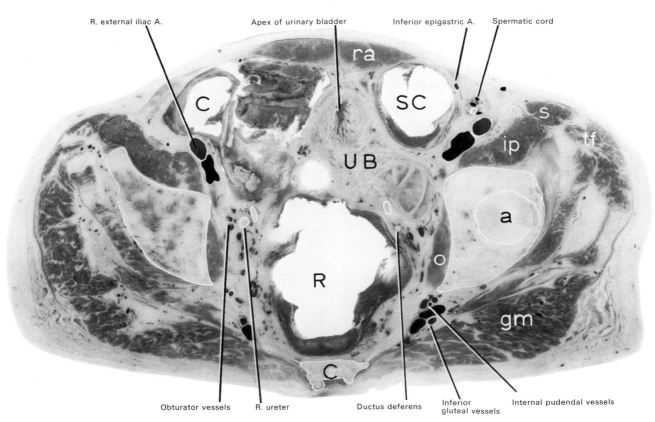

R. external iliac A.   Apex of urinary bladder   Inferior epigastric A.   Spermatic cord

Obturator vessels   R. ureter   Ductus deferens   Inferior   Internal pudendal vessels
                                                                              gluteal vessels

FIG. 24. Section through the coccyx, the upper part of the left hip joint, and the superior surface of the urinary bladder (left side and apex). Outlined in white is each ureter and ductus deferens. The right ductus deferens (left side of illustration) lies immediately anteromedial to the right ureter, the left is in the spermatic cord. The left inferior gluteal and internal pudendal vessels lie immediately posterior to the sacrospinous ligament, the spine of the ischium lying below the level of the section.

a, head of femur in acetabulum; C, caecum; C, coccyx; gm, gluteus maximus; ip, iliopsoas; o, obturator internus; R, rectum (distended); ra, rectus abdominis; s, sartorius; tf, tensor fasciae latae; UB, urinary bladder; SC, sigmoid colon.

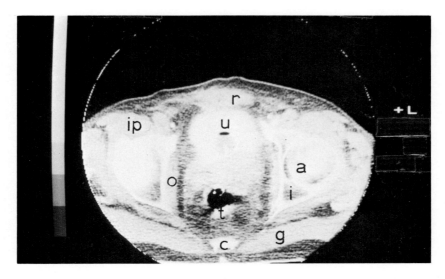

FIG. 24A. A CAT scan through the coccyx passing just above the pubic symphysis.
a, head of femur in acetabulum; c, coccyx; g, gluteus maximus; i, ischium; ip, iliopsoas; o, obturator internus; r, rectus abdominis; t, rectum; u, urinary bladder.

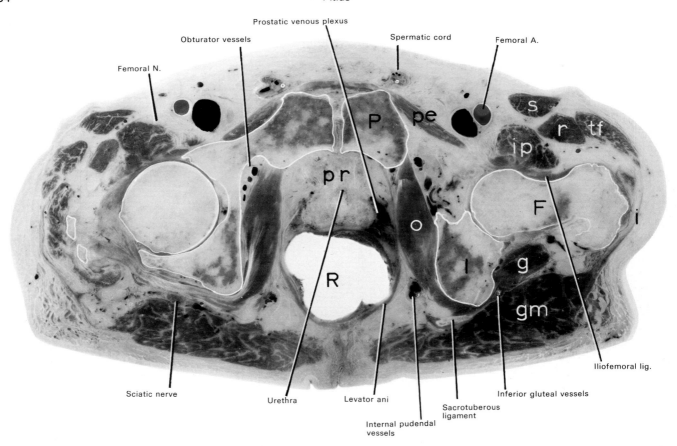

FIG. 25. Section through the pubic symphysis and the two hip joints. The left hip joint appears incomplete because the section passes through the acetabular notch. The entire length of the obturator internus is shown on the right side of the body, its tendon lying posterior to that of obturator externus as it approaches its insertion. The ductus deferens is outlined in white in each spermatic cord.

g, gemellus inferior; gm, gluteus maximus; ip, iliopsoas; o, obturator internus; P, body of pubic bone; pe, pectineus; pr, prostate gland; r, rectus femoris; R, rectum (distended); s, sartorius; tf, tensor fasciae latae.

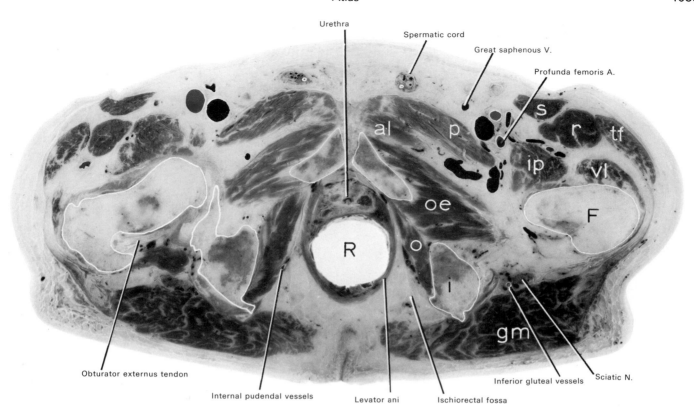

FIG. 26. Section through the trochanters of the femora, the greater on the right side of the body (left of illustration), the lesser on the left. Each ductus deferens is outlined in white.

al, adductor longus; F, femur; gm, gluteus maximus; ip, iliopsoas; o, obturator internus; oe, obturator externus; p, pectineus; r, rectus femoris; R, rectum; s, sartorius; tf, tensor fasciae latae; vl, vastus lateralis.

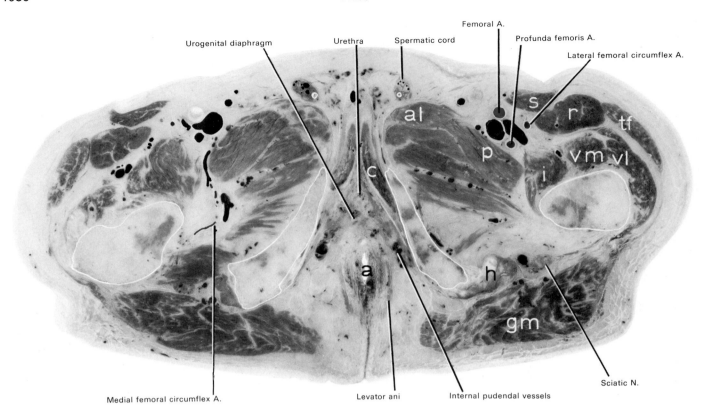

FIG. 27. Section through the inferior pubic and ischial rami, mainly in the plane of the urogenital diaphragm. Each ductus deferens is outlined in white.

a, anal canal; al, adductor longus; c, crus of penis; gm, gluteus maximus; h, hamstring muscles; i, iliacus; p, pectineus; r, rectus femoris; s, sartorius; tf, tensor fasciae latae; vl, vastus lateralis; vm, vastus medialis.

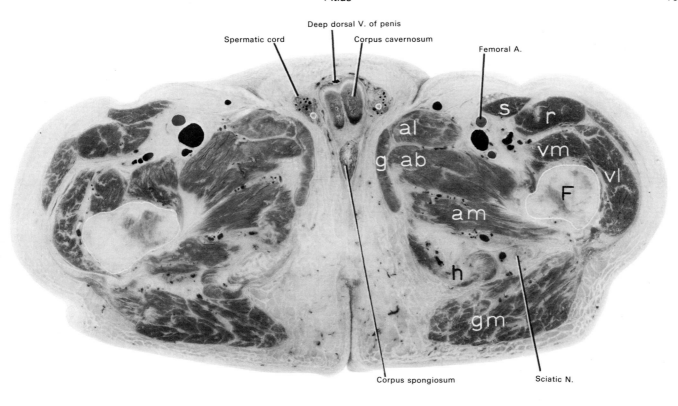

FIG. 28. Section through the proximal parts of the thighs and the root of the penis.
ab, adductor brevis; al, adductor longus; am, adductor magnus; F, femur; g, gracilis; gm, gluteus maximus; h, hamstring muscles; r, rectus femoris; s, sartorius; vl, vastus lateralis; vm, vastus medialis.

Index

Page numbers appearing in **bold** type indicate major descriptions

DATE DUE

DATE DUE			
FEB 2 7 2004			
GAYLORD			PRINTED IN U.S.A.